Chemical Process Equipment

BUTTERWORTHS SERIES IN CHEMICAL ENGINEERING

SERIES TITLES

Chemical Process Equipment *Stanley M. Walas*
Constitutive Equations for Polymer Melts and Solutions *Ronald G. Larson*
Gas Separation by Adsorption Processes *Ralph T. Yang*
Heterogeneous Reactor Design *Hong H. Lee*
Molecular Thermodynamics of Nonideal Fluids *Lloyd L. Lee*
Phase Equilibria in Chemical Engineering *Stanley M. Walas*
Transport Processes in Chemically Reacting Flow Systems *Daniel E. Rosner*
Viscous Flows: The Practical Use of Theory *Stuart Winston Churchill*

RELATED TITLES

Catalyst Supports and Supported Catalysts *Alvin B. Stiles*
Enlargement and Compaction of Particulate Solids *Nayland Stanley-Wood*
Fundamentals of Fluidized Beds *John G. Yates*
Liquid and Liquid Mixtures *J.S. Rowlinson and F.L. Swinton*
Mixing in the Process Industries *N. Harnby, M.F. Edwards, and A.W. Nienow*
Shell Process Control Workshop *David M. Prett and Manfred Morari*
Solid Liquid Separation *Ladislav Svarovsky*
Supercritical Fluid Extraction *Mark A. McHugh and Val J. Krukonis*

Chemical Process Equipment

Selection and Design

Stanley M. Walas

Department of Chemical and Petroleum Engineering
University of Kansas

Butterworths
Boston London Singapore Sydney Toronto Wellington

To the memory of my parents,
Stanislaus and Apolonia,
and to my wife, Suzy Belle

Library of Congress Cataloging-in-Publication Data
Walas, Stanley M.
Chemical process equipment.
(Butterworths series in chemical engineering)
Includes bibliographical references and index.
1. Chemical engineering—Apparatus and supplies.
I. Title. II. Series.
TP157.W334 1988 660.2′83 87-26795
ISBN 0-409-90131-8

British Library Cataloguing in Publication Data
Walas, Stanley M.
Chemical process equipment.—(Butterworths series in chemical engineering).
1. Chemical engineering—Apparatus and supplies
I. Title
660.2′8 TP157
ISBN 0-409-90131-8

Butterworth Publishers
80 Montvale Avenue
Stoneham, MA 02180

10 9 8 7 6 5 4 3 2 1

Printed in the United States of America

Contents

List of Examples

Preface

This book is intended as a guide to the selection or design of the principal kinds of chemical process equipment by engineers in school and industry. The level of treatment assumes an elementary knowledge of unit operations and transport phenomena. Access to the many design and reference books listed in Chapter 1 is desirable. For coherence, brief reviews of pertinent theory are provided. Emphasis is placed on shortcuts, rules of thumb, and data for design by analogy, often as primary design processes but also for quick evaluations of detailed work.

All answers to process design questions cannot be put into a book. Even at this late date in the development of the chemical industry, it is common to hear authorities on most kinds of equipment say that their equipment can be properly fitted to a particular task only on the basis of some direct laboratory and pilot plant work. Nevertheless, much guidance and reassurance are obtainable from general experience and specific examples of successful applications, which this book attempts to provide. Much of the information is supplied in numerous tables and figures, which often deserve careful study quite apart from the text.

The general background of process design, flowsheets, and process control is reviewed in the introductory chapters. The major kinds of operations and equipment are treated in individual chapters. Information about peripheral and less widely employed equipment in chemical plants is concentrated in Chapter 19 with references to key works of as much practical value as possible. Because decisions often must be based on economic grounds, Chapter 20, on costs of equipment, rounds out the book. Appendixes provide examples of equipment rating forms and manufacturers' questionnaires.

Chemical process equipment is of two kinds: custom designed and built, or proprietary "off the shelf." For example, the sizes and performance of custom equipment such as distillation towers, drums, and heat exchangers are derived by the process engineer on the basis of established principles and data, although some mechanical details remain in accordance with safe practice codes and individual fabrication practices.

Much proprietary equipment (such as filters, mixers, conveyors, and so on) has been developed largely without benefit of much theory and is fitted to job requirements also without benefit of much theory. From the point of view of the process engineer, such equipment is predesigned and fabricated and made available by manufacturers in limited numbers of types, sizes, and capacities. The process design of proprietary equipment, as considered in this book, establishes its required performance and is a process of selection from the manufacturers' offerings, often with their recommendations or on the basis of individual experience. Complete information is provided in manufacturers' catalogs. Several classified lists of manufacturers of chemical process equipment are readily accessible, so no listings are given here.

Because more than one kind of equipment often is suitable for particular applications and may be available from several manufacturers, comparisons of equipment and typical applications are cited liberally. Some features of industrial equipment are largely arbitrary and may be standardized for convenience in particular industries or individual plants. Such aspects of equipment design are noted when feasible.

Shortcut methods of design provide solutions to problems in a short time and at small expense. They must be used when data are limited or when the greater expense of a thorough method is not justifiable. In particular cases they may be employed to obtain information such as:

1. an order of magnitude check of the reasonableness of a result found by another lengthier and presumably accurate computation or computer run,
2. a quick check to find if existing equipment possibly can be adapted to a new situation,
3. a comparison of alternate processes,
4. a basis for a rough cost estimate of a process.

Shortcut methods occupy a prominent place in such a broad survey and limited space as this book. References to sources of more accurate design procedures are cited when available.

Another approach to engineering work is with rules of thumb, which are statements of equipment performance that may obviate all need for further calculations. Typical examples, for instance, are that optimum reflux ratio is 20% greater than minimum, that a suitable cold oil velocity in a fired heater is 6 ft/sec, or that the efficiency of a mixer-settler extraction stage is 70%. The trust that can be placed in a rule of thumb depends on the authority of the propounder, the risk associated with its possible inaccuracy, and the economic balance between the cost of a more accurate evaluation and suitable safety factor placed on the approximation. All experienced engineers have acquired such knowledge. When applied with discrimination, rules of thumb are a valuable asset to the process design and operating engineer, and are scattered throughout this book.

Design by analogy, which is based on knowledge of what has been found to work in similar areas, even though not necessarily optimally, is another valuable technique. Accordingly, specific applications often are described in this book, and many examples of specific equipment sizes and performance are cited.

For much of my insight into chemical process design, I am indebted to many years' association and friendship with the late Charles W. Nofsinger who was a prime practitioner by analogy, rule of thumb, and basic principles. Like Dr. Dolittle of Puddleby-on-the-Marsh, "he was a proper doctor and knew a whole lot."

RULES OF THUMB: SUMMARY

Although experienced engineers know where to find information and how to make accurate computations, they also keep a minimum body of information in mind on the ready, made up largely of shortcuts and rules of thumb. The present compilation may fit into such a minimum body of information, as a boost to the memory or extension in some instances into less often encountered areas. It is derived from the material in this book and is, in a sense, a digest of the book.

An Engineering Rule of Thumb is an outright statement regarding suitable sizes or performance of equipment that obviates all need for extended calculations. Because any brief statements are subject to varying degrees of qualification, they are most safely applied by engineers who are substantially familiar with the topics. Nevertheless, such rules should be of value for approximate design and cost estimation, and should provide even the inexperienced engineer with perspective and a foundation whereby the reasonableness of detailed and computer-aided results can be appraised quickly, particularly on short notice such as in conference.

Everyday activities also are governed to a large extent by rules of thumb. They serve us when we wish to take a course of action but are not in a position to find the best course of action. Of interest along this line is an amusing and often useful list of some 900 such digests of everyday experience that has been compiled by Parker (*Rules of Thumb,* Houghton Mifflin, Boston, 1983).

Much more can be stated in adequate summary fashion about some topics than about others, which accounts in part for the spottiness of the present coverage, but the spottiness also is due to ignorance and oversights on the part of the author. Accordingly, every engineer undoubtedly will supplement or modify this material in his own way.

COMPRESSORS AND VACUUM PUMPS

1. *Fans* are used to raise the pressure about 3% (12 in. water), *blowers* raise to less than 40 psig, and *compressors* to higher pressures, although the blower range commonly is included in the compressor range.
2. Vacuum pumps: reciprocating piston type decrease the pressure to 1 Torr; rotary piston down to 0.001 Torr, two-lobe rotary down to 0.0001 Torr; steam jet ejectors, one stage down to 100 Torr, three stage down to 1 Torr, five stage down to 0.05 Torr.
3. A three-stage ejector needs 100 lb steam/lb air to maintain a pressure of 1 Torr.
4. In-leakage of air to evacuated equipment depends on the absolute pressure, Torr, and the volume of the equipment, V cuft, according to $w = kV^{2/3}$ lb/hr, with $k = 0.2$ when P is more than 90 Torr, 0.08 between 3 and 20 Torr, and 0.025 at less than 1 Torr.
5. Theoretical adiabatic horsepower (THP) $= [(\text{SCFM})T_1/8130a][(P_2/P_1)^a - 1]$, where T_1 is inlet temperature in °F + 460 and $a = (k-1)/k$, $k = C_p/C_v$.
6. Outlet temperature $T_2 = T_1(P_2/P_1)^a$.
7. To compress air from 100°F, $k = 1.4$, compression ratio = 3, theoretical power required = 62 HP/million cuft/day, outlet temperature 306°F.
8. Exit temperature should not exceed 350–400°F; for diatomic gases ($C_p/C_v = 1.4$) this corresponds to a compression ratio of about 4.
9. Compression ratio should be about the same in each stage of a multistage unit, ratio $= (P_n/P_1)^{1/n}$, with n stages.
10. Efficiencies of reciprocating compressors: 65% at compression ratio of 1.5, 75% at 2.0, and 80–85% at 3–6.
11. Efficiencies of large centrifugal compressors, 6000–100,000 ACFM at suction, are 76–78%.
12. Rotary compressors have efficiencies of 70%, except liquid liner type which have 50%.

CONVEYORS FOR PARTICULATE SOLIDS

1. *Screw conveyors* are suited to transport of even sticky and abrasive solids up inclines of 20° or so. They are limited to distances of 150 ft or so because of shaft torque strength. A 12 in. dia conveyor can handle 1000–3000 cuft/hr, at speeds ranging from 40 to 60 rpm.
2. *Belt conveyors* are for high capacity and long distances (a mile or more, but only several hundred feet in a plant), up inclines of 30° maximum. A 24 in. wide belt can carry 3000 cuft/hr at a speed of 100 ft/min, but speeds up to 600 ft/min are suited to some materials. Power consumption is relatively low.
3. *Bucket elevators* are suited to vertical transport of sticky and abrasive materials. With buckets 20 × 20 in. capacity can reach 1000 cuft/hr at a speed of 100 ft/min, but speeds to 300 ft/min are used.
4. *Drag-type conveyors* (Redler) are suited to short distances in any direction and are completely enclosed. Units range in size from 3 in. square to 19 in. square and may travel from 30 ft/min (fly ash) to 250 ft/min (grains). Power requirements are high.
5. *Pneumatic conveyors* are for high capacity, short distance (400 ft) transport simultaneously from several sources to several destinations. Either vacuum or low pressure (6–12 psig) is employed with a range of air velocities from 35 to 120 ft/sec depending on the material and pressure, air requirements from 1 to 7 cuft/cuft of solid transferred.

COOLING TOWERS

1. Water in contact with air under adiabatic conditions eventually cools to the wet bulb temperature.
2. In commercial units, 90% of saturation of the air is feasible.
3. Relative cooling tower size is sensitive to the difference between the exit and wet bulb temperatures:

ΔT (°F)	5	15	25
Relative volume	2.4	1.0	0.55

4. Tower fill is of a highly open structure so as to minimize pressure drop, which is in standard practice a maximum of 2 in. of water.
5. Water circulation rate is 1–4 gpm/sqft and air rates are 1300–1800 lb/(hr)(sqft) or 300–400 ft/min.
6. Chimney-assisted natural draft towers are of hyperboloidal shapes because they have greater strength for a given thickness; a tower 250 ft high has concrete walls 5–6 in. thick. The enlarged cross section at the top aids in dispersion of exit humid air into the atmosphere.
7. Countercurrent induced draft towers are the most common in process industries. They are able to cool water within 2°F of the wet bulb.
8. Evaporation losses are 1% of the circulation for every 10°F of cooling range. Windage or drift losses of mechanical draft towers

are 0.1–0.3%. Blowdown of 2.5–3.0% of the circulation is necessary to prevent excessive salt buildup.

CRYSTALLIZATION FROM SOLUTION

1. Complete recovery of dissolved solids is obtainable by evaporation, but only to the eutectic composition by chilling. Recovery by melt crystallization also is limited by the eutectic composition.
2. Growth rates and ultimate sizes of crystals are controlled by limiting the extent of supersaturation at any time.
3. The ratio $S = C/C_{sat}$ of prevailing concentration to saturation concentration is kept near the range of 1.02–1.05.
4. In crystallization by chilling, the temperature of the solution is kept at most 1–2°F below the saturation temperature at the prevailing concentration.
5. Growth rates of crystals under satisfactory conditions are in the range of 0.1–0.8 mm/hr. The growth rates are approximately the same in all directions.
6. Growth rates are influenced greatly by the presence of impurities and of certain specific additives that vary from case to case.

DISINTEGRATION

1. Percentages of material greater than 50% of the maximum size are about 50% from rolls, 15% from tumbling mills, and 5% from closed circuit ball mills.
2. Closed circuit grinding employs external size classification and return of oversize for regrinding. The rules of pneumatic conveying are applied to design of air classifiers. Closed circuit is most common with ball and roller mills.
3. Jaw crushers take lumps of several feet in diameter down to 4 in. Stroke rates are 100–300/min. The average feed is subjected to 8–10 strokes before it becomes small enough to escape. Gyratory crushers are suited to slabby feeds and make a more rounded product.
4. Roll crushers are made either smooth or with teeth. A 24 in. toothed roll can accept lumps 14 in. dia. Smooth rolls effect reduction ratios up to about 4. Speeds are 50–900 rpm. Capacity is about 25% of the maximum corresponding to a continuous ribbon of material passing through the rolls.
5. Hammer mills beat the material until it is small enough to pass through the screen at the bottom of the casing. Reduction ratios of 40 are feasible. Large units operate at 900 rpm, smaller ones up to 16,000 rpm. For fibrous materials the screen is provided with cutting edges.
6. Rod mills are capable of taking feed as large as 50 mm and reducing it to 300 mesh, but normally the product range is 8–65 mesh. Rods are 25–150 mm dia. Ratio of rod length to mill diameter is about 1.5. About 45% of the mill volume is occupied by rods. Rotation is at 50–65% of critical.
7. Ball mills are better suited than rod mills to fine grinding. The charge is of equal weights of 1.5, 2, and 3 in. balls for the finest grinding. Volume occupied by the balls is 50% of the mill volume. Rotation speed is 70–80% of critical. Ball mills have a length to diameter ratio in the range 1–1.5. Tube mills have a ratio of 4–5 and are capable of very fine grinding. Pebble mills have ceramic grinding elements, used when contamination with metal is to be avoided.
8. Roller mills employ cylindrical or tapered surfaces that roll along flatter surfaces and crush nipped particles. Products of 20–200 mesh are made.

DISTILLATION AND GAS ABSORPTION

1. Distillation usually is the most economical method of separating liquids, superior to extraction, adsorption, crystallization, or others.
2. For ideal mixtures, relative volatility is the ratio of vapor pressures $\alpha_{12} = P_2/P_1$.
3. Tower operating pressure is determined most often by the temperature of the available condensing medium, 100–120°F if cooling water; or by the maximum allowable reboiler temperature, 150 psig steam, 366°F.
4. Sequencing of columns for separating multicomponent mixtures: (a) perform the easiest separation first, that is, the one least demanding of trays and reflux, and leave the most difficult to the last; (b) when neither relative volatility nor feed concentration vary widely, remove the components one by one as overhead products; (c) when the adjacent ordered components in the feed vary widely in relative volatility, sequence the splits in the order of decreasing volatility; (d) when the concentrations in the feed vary widely but the relative volatilities do not, remove the components in the order of decreasing concentration in the feed.
5. Economically optimum reflux ratio is about 1.2 times the minimum reflux ratio R_m.
6. The economically optimum number of trays is near twice the minimum value N_m.
7. The minimum number of trays is found with the Fenske–Underwood equation

$$N_m = \log\{[x/(1-x)]_{\text{ovhd}}/[x/(1-x)]_{\text{btms}}\}/\log \alpha.$$

8. Minimum reflux for binary or pseudobinary mixtures is given by the following when separation is esentially complete ($x_D \simeq 1$) and D/F is the ratio of overhead product and feed rates:

$$R_m D/F = 1/(\alpha - 1), \quad \text{when feed is at the bubblepoint,}$$
$$(R_m + 1)D/F = \alpha/(\alpha - 1), \quad \text{when feed is at the dewpoint.}$$

9. A safety factor of 10% of the number of trays calculated by the best means is advisable.
10. Reflux pumps are made at least 25% oversize.
11. For reasons of accessibility, tray spacings are made 20–24 in.
12. Peak efficiency of trays is at values of the vapor factor $F_s = u\sqrt{\rho_v}$ in the range 1.0–1.2 (ft/sec) $\sqrt{\text{lb/cuft}}$. This range of F_s establishes the diameter of the tower. Roughly, linear velocities are 2 ft/sec at moderate pressures and 6 ft/sec in vacuum.
13. The optimum value of the Kremser–Brown absorption factor $A = K(V/L)$ is in the range 1.25–2.0.
14. Pressure drop per tray is of the order of 3 in. of water or 0.1 psi.
15. Tray efficiencies for distillation of light hydrocarbons and aqueous solutions are 60–90%; for gas absorption and stripping, 10–20%.
16. Sieve trays have holes 0.25–0.50 in. dia, hole area being 10% of the active cross section.
17. Valve trays have holes 1.5 in. dia each provided with a liftable cap, 12–14 caps/sqft of active cross section. Valve trays usually are cheaper than sieve trays.
18. Bubblecap trays are used only when a liquid level must be maintained at low turndown ratio; they can be designed for lower pressure drop than either sieve or valve trays.
19. Weir heights are 2 in., weir lengths about 75% of tray diameter, liquid rate a maximum of about 8 gpm/in. of weir; multipass arrangements are used at high liquid rates.

20. Packings of random and structured character are suited especially to towers under 3 ft dia and where low pressure drop is desirable. With proper initial distribution and periodic redistribution, volumetric efficiencies can be made greater than those of tray towers. Packed internals are used as replacements for achieving greater throughput or separation in existing tower shells.
21. For gas rates of 500 cfm, use 1 in. packing; for gas rates of 2000 cfm or more, use 2 in.
22. The ratio of diameters of tower and packing should be at least 15.
23. Because of deformability, plastic packing is limited to a 10–15 ft depth unsupported, metal to 20–25 ft.
24. Liquid redistributors are needed every 5–10 tower diameters with pall rings but at least every 20 ft. The number of liquid streams should be 3–5/sqft in towers larger than 3 ft dia (some experts say 9–12/sqft), and more numerous in smaller towers.
25. Height equivalent to a theoretical plate (HETP) for vapor–liquid contacting is 1.3–1.8 ft for 1 in. pall rings, 2.5–3.0 ft for 2 in. pall rings.
26. Packed towers should operate near 70% of the flooding rate given by the correlation of Sherwood, Lobo, et al.
27. Reflux drums usually are horizontal, with a liquid holdup of 5 min half full. A takeoff pot for a second liquid phase, such as water in hydrocarbon systems, is sized for a linear velocity of that phase of 0.5 ft/sec, minimum diameter of 16 in.
28. For towers about 3 ft dia, add 4 ft at the top for vapor disengagement and 6 ft at the bottom for liquid level and reboiler return.
29. Limit the tower height to about 175 ft max because of wind load and foundation considerations. An additional criterion is that L/D be less than 30.

DRIVERS AND POWER RECOVERY EQUIPMENT

1. Efficiency is greater for larger machines. Motors are 85–95%; steam turbines are 42–78%; gas engines and turbines are 28–38%.
2. For under 100 HP, electric motors are used almost exclusively. They are made for up to 20,000 HP.
3. Induction motors are most popular. Synchronous motors are made for speeds as low as 150 rpm and are thus suited for example for low speed reciprocating compressors, but are not made smaller than 50 HP. A variety of enclosures is available, from weather-proof to explosion-proof.
4. Steam turbines are competitive above 100 HP. They are speed controllable. Frequently they are employed as spares in case of power failure.
5. Combustion engines and turbines are restricted to mobile and remote locations.
6. Gas expanders for power recovery may be justified at capacities of several hundred HP; otherwise any needed pressure reduction in process is effected with throttling valves.

DRYING OF SOLIDS

1. Drying times range from a few seconds in spray dryers to 1 hr or less in rotary dryers and up to several hours or even several days in tunnel shelf or belt dryers.
2. Continuous tray and belt dryers for granular material of natural size or pelleted to 3–15 mm have drying times in the range of 10–200 min.
3. Rotary cylindrical dryers operate with superficial air velocities of 5–10 ft/sec, sometimes up to 35 ft/sec when the material is coarse. Residence times are 5–90 min. Holdup of solid is 7–8%. An 85% free cross section is taken for design purposes. In countercurrent flow, the exit gas is 10–20°C above the solid; in parallel flow, the temperature of the exit solid is 100°C. Rotation speeds of about 4 rpm are used, but the product of rpm and diameter in feet is typically between 15 and 25.
4. Drum dryers for pastes and slurries operate with contact times of 3–12 sec, produce flakes 1–3 mm thick with evaporation rates of 15–30 kg/m^2 hr. Diameters are 1.5–5.0 ft; the rotation rate is 2-10 rpm. The greatest evaporative capacity is of the order of 3000 lb/hr in commercial units.
5. Pneumatic conveying dryers normally take particles 1–3 mm dia but up to 10 mm when the moisture is mostly on the surface. Air velocities are 10–30 m/sec. Single pass residence times are 0.5–3.0 sec but with normal recycling the average residence time is brought up to 60 sec. Units in use range from 0.2 m dia by 1 m high to 0.3 m dia by 38 m long. Air requirement is several SCFM/lb of dry product/hr.
6. Fluidized bed dryers work best on particles of a few tenths of a mm dia, but up to 4 mm dia have been processed. Gas velocities of twice the minimum fluidization velocity are a safe prescription. In continuous operation, drying times of 1–2 min are enough, but batch drying of some pharmaceutical products employs drying times of 2–3 hr.
7. Spray dryers: Surface moisture is removed in about 5 sec, and most drying is completed in less than 60 sec. Parallel flow of air and stock is most common. Atomizing nozzles have openings 0.012–0.15 in. and operate at pressures of 300–4000 psi. Atomizing spray wheels rotate at speeds to 20,000 rpm with peripheral speeds of 250–600 ft/sec. With nozzles, the length to diameter ratio of the dryer is 4–5; with spray wheels, the ratio is 0.5–1.0. For the final design, the experts say, pilot tests in a unit of 2 m dia should be made.

EVAPORATORS

1. Long tube vertical evaporators with either natural or forced circulation are most popular. Tubes are 19–63 mm dia and 12–30 ft long.
2. In forced circulation, linear velocities in the tubes are 15–20 ft/sec.
3. Elevation of boiling point by dissolved solids results in differences of 3–10°F between solution and saturated vapor.
4. When the boiling point rise is appreciable, the economic number of effects in series with forward feed is 4–6.
5. When the boiling point rise is small, minimum cost is obtained with 8–10 effects in series.
6. In backward feed the more concentrated solution is heated with the highest temperature steam so that heating surface is lessened, but the solution must be pumped between stages.
7. The steam economy of an N-stage battery is approximately $0.8N$ lb evaporation/lb of outside steam.
8. Interstage steam pressures can be boosted with steam jet compressors of 20–30% efficiency or with mechanical compressors of 70–75% efficiency.

EXTRACTION, LIQUID–LIQUID

1. The dispersed phase should be the one that has the higher volumetric rate except in equipment subject to backmixing where it should be the one with the smaller volumetric rate. It should be the phase that wets the material of construction less well. Since the holdup of continuous phase usually is greater, that phase should be made up of the less expensive or less hazardous material.

2. There are no known commercial applications of reflux to extraction processes, although the theory is favorable (Treybal).
3. Mixer–settler arrangements are limited to at most five stages. Mixing is accomplished with rotating impellers or circulating pumps. Settlers are designed on the assumption that droplet sizes are about 150 μm dia. In open vessels, residence times of 30–60 min or superficial velocities of 0.5–1.5 ft/min are provided in settlers. Extraction stage efficiencies commonly are taken as 80%.
4. Spray towers even 20–40 ft high cannot be depended on to function as more than a single stage.
5. Packed towers are employed when 5–10 stages suffice. Pall rings of 1–1.5 in. size are best. Dispersed phase loadings should not exceed 25 gal/(min) (sqft). HETS of 5–10 ft may be realizable. The dispersed phase must be redistributed every 5–7 ft. Packed towers are not satisfactory when the surface tension is more than 10 dyn/cm.
6. Sieve tray towers have holes of only 3–8 mm dia. Velocities through the holes are kept below 0.8 ft/sec to avoid formation of small drops. Redispersion of either phase at each tray can be designed for. Tray spacings are 6–24 in. Tray efficiencies are in the range of 20–30%.
7. Pulsed packed and sieve tray towers may operate at frequencies of 90 cycles/min and amplitudes of 6–25 mm. In large diameter towers, HETS of about 1 m has been observed. Surface tensions as high as 30–40 dyn/cm have no adverse effect.
8. Reciprocating tray towers can have holes 9/16 in. dia, 50–60% open area, stroke length 0.75 in., 100–150 strokes/min, plate spacing normally 2 in. but in the range 1–6 in. In a 30 in. dia tower, HETS is 20–25 in. and throughput is 2000 gal/(hr)(sqft). Power requirements are much less than of pulsed towers.
9. Rotating disk contactors or other rotary agitated towers realize HETS in the range 0.1–0.5 m. The especially efficient Kuhni with perforated disks of 40% free cross section has HETS 0.2 m and a capacity of 50 m^3/m^2 hr.

FILTRATION

1. Processes are classified by their rate of cake buildup in a laboratory vacuum leaf filter: rapid, 0.1–10.0 cm/sec; medium, 0.1–10.0 cm/min; slow, 0.1–10.0 cm/hr.
2. Continuous filtration should not be attempted if 1/8 in. cake thickness cannot be formed in less than 5 min.
3. Rapid filtering is accomplished with belts, top feed drums, or pusher-type centrifuges.
4. Medium rate filtering is accomplished with vacuum drums or disks or peeler-type centrifuges.
5. Slow filtering slurries are handled in pressure filters or sedimenting centrifuges.
6. Clarification with negligible cake buildup is accomplished with cartridges, precoat drums, or sand filters.
7. Laboratory tests are advisable when the filtering surface is expected to be more than a few square meters, when cake washing is critical, when cake drying may be a problem, or when precoating may be needed.
8. For finely ground ores and minerals, rotary drum filtration rates may be 1500 lb/(day)(sqft), at 20 rev/hr and 18–25 in. Hg vacuum.
9. Coarse solids and crystals may be filtered at rates of 6000 lb/(day)(sqft) at 20 rev/hr, 2–6 in. Hg vacuum.

FLUIDIZATION OF PARTICLES WITH GASES

1. Properties of particles that are conducive to smooth fluidization include: rounded or smooth shape, enough toughness to resist attrition, sizes in the range 50–500 μm dia, a spectrum of sizes with ratio of largest to smallest in the range of 10–25.
2. Cracking catalysts are members of a broad class characterized by diameters of 30–150 μm, density of 1.5 g/mL or so, appreciable expansion of the bed before fluidization sets in, minimum bubbling velocity greater than minimum fluidizing velocity, and rapid disengagement of bubbles.
3. The other extreme of smoothly fluidizing particles is typified by coarse sand and glass beads both of which have been the subject of much laboratory investigation. Their sizes are in the range 150–500 μm, densities 1.5–4.0 g/mL, small bed expansion, about the same magnitudes of minimum bubbling and minimum fluidizing velocities, and also have rapidly disengaging bubbles.
4. Cohesive particles and large particles of 1 mm or more do not fluidize well and usually are processed in other ways.
5. Rough correlations have been made of minimum fluidization velocity, minimum bubbling velocity, bed expansion, bed level fluctuation, and disengaging height. Experts recommend, however, that any real design be based on pilot plant work.
6. Practical operations are conducted at two or more multiples of the minimum fluidizing velocity. In reactors, the entrained material is recovered with cyclones and returned to process. In dryers, the fine particles dry most quickly so the entrained material need not be recycled.

HEAT EXCHANGERS

1. Take true countercurrent flow in a shell-and-tube exchanger as a basis.
2. Standard tubes are 3/4 in. OD, 1 in. triangular spacing, 16 ft long; a shell 1 ft dia accommodates 100 sqft; 2 ft dia, 400 sqft, 3 ft dia, 1100 sqft.
3. Tube side is for corrosive, fouling, scaling, and high pressure fluids.
4. Shell side is for viscous and condensing fluids.
5. Pressure drops are 1.5 psi for boiling and 3–9 psi for other services.
6. Minimum temperature approach is 20°F with normal coolants, 10°F or less with refrigerants.
7. Water inlet temperature is 90°F, maximum outlet 120°F.
8. Heat transfer coefficients for estimating purposes, Btu/(hr)(sqft)(°F): water to liquid, 150; condensers, 150; liquid to liquid, 50; liquid to gas, 5; gas to gas, 5; reboiler, 200. Max flux in reboilers, 10,000 Btu/(hr)(sqft).
9. Double-pipe exchanger is competitive at duties requiring 100–200 sqft.
10. Compact (plate and fin) exchangers have 350 sqft/cuft, and about 4 times the heat transfer per cuft of shell-and-tube units.
11. Plate and frame exchangers are suited to high sanitation services, and are 25–50% cheaper in stainless construction than shell-and-tube units.
12. Air coolers: Tubes are 0.75–1.00 in. OD, total finned surface 15–20 sqft/sqft bare surface, $U = 80$–100 Btu/(hr)(sqft bare surface)(°F), fan power input 2–5 HP/(MBtu/hr), approach 50°F or more.
13. Fired heaters: radiant rate, 12,000 Btu/(hr)(sqft); convection rate, 4000; cold oil tube velocity, 6 ft/sec; approx equal transfers of heat in the two sections; thermal efficiency 70–75%; flue gas temperature 250–350°F above feed inlet; stack gas temperature 650–950°F.

INSULATION

1. Up to 650°F, 85% magnesia is most used.
2. Up to 1600–1900°F, a mixture of asbestos and diatomaceous earth is used.

3. Ceramic refractories at higher temperatures.
4. Cyrogenic equipment (−200°F) employs insulants with fine pores in which air is trapped.
5. Optimum thickness varies with temperature: 0.5 in. at 200°F, 1.0 in. at 400°F, 1.25 in. at 600°F.
6. Under windy conditions (7.5 miles/hr), 10–20% greater thickness of insulation is justified.

MIXING AND AGITATION

1. Mild agitation is obtained by circulating the liquid with an impeller at superficial velocities of 0.1–0.2 ft/sec, and intense agitation at 0.7–1.0 ft/sec.
2. Intensities of agitation with impellers in baffled tanks are measured by power input, HP/1000 gal, and impeller tip speeds:

Operation	HP/1000 gal	Tip speed (ft/min)
Blending	0.2–0.5	
Homogeneous reaction	0.5–1.5	7.5–10
Reaction with heat transfer	1.5–5.0	10–15
Liquid–liquid mixtures	5	15–20
Liquid–gas mixtures	5–10	15–20
Slurries	10	

3. Proportions of a stirred tank relative to the diameter D: liquid level $= D$; turbine impeller diameter $= D/3$; impeller level above bottom $= D/3$; impeller blade width $= D/15$; four vertical baffles with width $= D/10$.
4. Propellers are made a maximum of 18 in., turbine impellers to 9 ft.
5. Gas bubbles sparged at the bottom of the vessel will result in mild agitation at a superficial gas velocity of 1 ft/min, severe agitation at 4 ft/min.
6. Suspension of solids with a settling velocity of 0.03 ft/sec is accomplished with either turbine or propeller impellers, but when the settling velocity is above 0.15 ft/sec intense agitation with a propeller is needed.
7. Power to drive a mixture of a gas and a liquid can be 25–50% less than the power to drive the liquid alone.
8. In-line blenders are adequate when a second or two contact time is sufficient, with power inputs of 0.1–0.2 HP/gal.

PARTICLE SIZE ENLARGEMENT

1. The chief methods of particle size enlargement are: compression into a mold, extrusion through a die followed by cutting or breaking to size, globulation of molten material followed by solidification, agglomeration under tumbling or otherwise agitated conditions with or without binding agents.
2. Rotating drum granulators have length to diameter ratios of 2–3, speeds of 10–20 rpm, pitch as much as 10°. Size is controlled by speed, residence time, and amount of binder; 2–5 mm dia is common.
3. Rotary disk granulators produce a more nearly uniform product than drum granulators. Fertilizer is made 1.5–3.5 mm; iron ore 10–25 mm dia.
4. Roll compacting and briquetting is done with rolls ranging from 130 mm dia by 50 mm wide to 910 mm dia by 550 mm wide. Extrudates are made 1–10 mm thick and are broken down to size for any needed processing such as feed to tabletting machines or to dryers.
5. Tablets are made in rotary compression machines that convert powders and granules into uniform sizes. Usual maximum diameter is about 1.5 in., but special sizes up to 4 in. dia are possible. Machines operate at 100 rpm or so and make up to 10,000 tablets/min.
6. Extruders make pellets by forcing powders, pastes, and melts through a die followed by cutting. An 8 in. screw has a capacity of 2000 lb/hr of molten plastic and is able to extrude tubing at 150–300 ft/min and to cut it into sizes as small as washers at 8000/min. Ring pellet extrusion mills have hole diameters of 1.6–32 mm. Production rates cover a range of 30–200 lb/(hr)(HP).
7. Prilling towers convert molten materials into droplets and allow them to solidify in contact with an air stream. Towers as high as 60 m are used. Economically the process becomes competitive with other granulation processes when a capacity of 200–400 tons/day is reached. Ammonium nitrate prills, for example, are 1.6–3.5 mm dia in the 5–95% range.
8. Fluidized bed granulation is conducted in shallow beds 12–24 in. deep at air velocities of 0.1–2.5 m/s or 3–10 times the minimum fluidizing velocity, with evaporation rates of 0.005–1.0 kg/m^2 sec. One product has a size range 0.7–2.4 mm dia.

PIPING

1. Line velocities and pressure drops, with line diameter D in inches: liquid pump discharge, $(5 + D/3)$ ft/sec, 2.0 psi/100 ft; liquid pump suction, $(1.3 + D/6)$ ft/sec, 0.4 psi/100 ft; steam or gas, $20D$ ft/sec, 0.5 psi/100 ft.
2. Control valves require at least 10 psi drop for good control.
3. Globe valves are used for gases, for control and wherever tight shutoff is required. Gate valves are for most other services.
4. Screwed fittings are used only on sizes 1.5 in. and smaller, flanges or welding otherwise.
5. Flanges and fittings are rated for 150, 300, 600, 900, 1500, or 2500 psig.
6. Pipe schedule number $= 1000P/S$, approximately, where P is the internal pressure psig and S is the allowable working stress (about 10,000 psi for A120 carbon steel at 500°F). Schedule 40 is most common.

PUMPS

1. Power for pumping liquids: HP = (gpm)(psi difference)/(1714)(fractional efficiency).
2. Normal pump suction head (NPSH) of a pump must be in excess of a certain number, depending on the kind of pumps and the conditions, if damage is to be avoided. NPSH = (pressure at the eye of the impeller − vapor pressure)/(density). Common range is 4–20 ft.
3. Specific speed $N_s = (\text{rpm})(\text{gpm})^{0.5}/(\text{head in ft})^{0.75}$. Pump may be damaged if certain limits of N_s are exceeded, and efficiency is best in some ranges.
4. Centrifugal pumps: Single stage for 15–5000 gpm, 500 ft max head; multistage for 20–11,000 gpm, 5500 ft max head. Efficiency 45% at 100 gpm, 70% at 500 gpm, 80% at 10,000 gpm.
5. Axial pumps for 20–100,000 gpm, 40 ft head, 65–85% efficiency.
6. Rotary pumps for 1–5000 gpm, 50,000 ft head, 50–80% efficiency.
7. Reciprocating pumps for 10–10,000 gpm, 1,000,000 ft head max. Efficiency 70% at 10 HP, 85% at 50 HP, 90% at 500 HP.

REACTORS

1. The rate of reaction in every instance must be established in the laboratory, and the residence time or space velocity and product distribution eventually must be found in a pilot plant.
2. Dimensions of catalyst particles are 0.1 mm in fluidized beds, 1 mm in slurry beds, and 2–5 mm in fixed beds.
3. The optimum proportions of stirred tank reactors are with liquid level equal to the tank diameter, but at high pressures slimmer proportions are economical.

4. Power input to a homogeneous reaction stirred tank is 0.5–1.5 HP/1000 gal, but three times this amount when heat is to be transferred.
5. Ideal CSTR (continuous stirred tank reactor) behavior is approached when the mean residence time is 5–10 times the length of time needed to achieve homogeneity, which is accomplished with 500–2000 revolutions of a properly designed stirrer.
6. Batch reactions are conducted in stirred tanks for small daily production rates or when the reaction times are long or when some condition such as feed rate or temperature must be programmed in some way.
7. Relatively slow reactions of liquids and slurries are conducted in continuous stirred tanks. A battery of four or five in series is most economical.
8. Tubular flow reactors are suited to high production rates at short residence times (sec or min) and when substantial heat transfer is needed. Embedded tubes or shell-and-tube construction then are used.
9. In granular catalyst packed reactors, the residence time distribution often is no better than that of a five-stage CSTR battery.
10. For conversions under about 95% of equilibrium, the performance of a five-stage CSTR battery approaches plug flow.

REFRIGERATION

1. A ton of refrigeration is the removal of 12,000 Btu/hr of heat.
2. At various temperature levels: 0–50°F, chilled brine and glycol solutions; −50–40°F, ammonia, freons, butane; −150–−50°F, ethane or propane.
3. Compression refrigeration with 100°F condenser requires these HP/ton at various temperature levels: 1.24 at 20°F; 1.75 at 0°F; 3.1 at −40°F; 5.2 at −80°F.
4. Below −80°F, cascades of two or three refrigerants are used.
5. In single stage compression, the compression ratio is limited to about 4.
6. In multistage compression, economy is improved with interstage flashing and recycling, so-called economizer operation.
7. Absorption refrigeration (ammonia to −30°F, lithium bromide to +45°F) is economical when waste steam is available at 12 psig or so.

SIZE SEPARATION OF PARTICLES

1. Grizzlies that are constructed of parallel bars at appropriate spacings are used to remove products larger than 5 cm dia.
2. Revolving cylindrical screens rotate at 15–20 rpm and below the critical velocity; they are suitable for wet or dry screening in the range of 10–60 mm.
3. Flat screens are vibrated or shaken or impacted with bouncing balls. Inclined screens vibrate at 600–7000 strokes/min and are used for down to 38 μm although capacity drops off sharply below 200 μm. Reciprocating screens operate in the range 30–1000 strokes/min and handle sizes down to 0.25 mm at the higher speeds.
4. Rotary sifters operate at 500–600 rpm and are suited to a range of 12 mm to 50 μm.
5. Air classification is preferred for fine sizes because screens of 150 mesh and finer are fragile and slow.
6. Wet classifiers mostly are used to make two product size ranges, oversize and undersize, with a break commonly in the range between 28 and 200 mesh. A rake classifier operates at about 9 strokes/min when making separation at 200 mesh, and 32 strokes/min at 28 mesh. Solids content is not critical, and that of the overflow may be 2–20% or more.
7. Hydrocyclones handle up to 600 cuft/min and can remove particles in the range of 300–5 μm from dilute suspensions. In one case, a 20 in. dia unit had a capacity of 1000 gpm with a pressure drop of 5 psi and a cutoff between 50 and 150 μm.

UTILITIES: COMMON SPECIFICATIONS

1. Steam: 15–30 psig, 250–275°F; 150 psig, 366°F; 400 psig, 448°F; 600 psig, 488°F or with 100–150°F superheat.
2. Cooling water: Supply at 80–90°F from cooling tower, return at 115–125°F; return seawater at 110°F, return tempered water or steam condensate above 125°F.
3. Cooling air supply at 85–95°F; temperature approach to process, 40°F.
4. Compressed air at 45, 150, 300, or 450 psig levels.
5. Instrument air at 45 psig, 0°F dewpoint.
6. Fuels: gas of 1000 Btu/SCF at 5–10 psig, or up to 25 psig for some types of burners; liquid at 6 million Btu/barrel.
7. Heat transfer fluids: petroleum oils below 600°F, Dowtherms below 750°F, fused salts below 1100°F, direct fire or electricity above 450°F.
8. Electricity: 1–100 Hp, 220–550 V; 200–2500 Hp, 2300–4000 V.

VESSELS (DRUMS)

1. Drums are relatively small vessels to provide surge capacity or separation of entrained phases.
2. Liquid drums usually are horizontal.
3. Gas/liquid separators are vertical.
4. Optimum length/diameter = 3, but a range of 2.5–5.0 is common.
5. Holdup time is 5 min half full for reflux drums, 5–10 min for a product feeding another tower.
6. In drums feeding a furnace, 30 min half full is allowed.
7. Knockout drums ahead of compressors should hold no less than 10 times the liquid volume passing through per minute.
8. Liquid/liquid separators are designed for settling velocity of 2–3 in./min.
9. Gas velocity in gas/liquid separators, $V = k\sqrt{\rho_L/\rho_V - 1}$ ft/sec, with $k = 0.35$ with mesh deentrainer, $k = 0.1$ without mesh deentrainer.
10. Entrainment removal of 99% is attained with mesh pads of 4–12 in. thicknesses; 6 in. thickness is popular.
11. For vertical pads, the value of the coefficient in Step 9 is reduced by a factor of 2/3.
12. Good performance can be expected at velocities of 30–100% of those calculated with the given k; 75% is popular.
13. Disengaging spaces of 6–18 in. ahead of the pad and 12 in. above the pad are suitable.
14. Cyclone separators can be designed for 95% collection of 5 μm particles, but usually only droplets greater than 50 μm need be removed.

VESSELS (PRESSURE)

1. Design temperature between −20°F and 650°F is 50°F above operating temperature; higher safety margins are used outside the given temperature range.
2. The design pressure is 10% or 10–25 psi over the maximum operating pressure, whichever is greater. The maximum operating pressure, in turn, is taken as 25 psi above the normal operation.
3. Design pressures of vessels operating at 0–10 psig and 600–1000°F are 40 psig.

4. For vacuum operation, design pressures are 15 psig and full vacuum.
5. Minimum wall thicknesses for rigidity: 0.25 in. for 42 in. dia and under, 0.32 in. for 42–60 in. dia, and 0.38 in. for over 60 in. dia.
6. Corrosion allowance 0.35 in. for known corrosive conditions, 0.15 in. for non-corrosive streams, and 0.06 in. for steam drums and air receivers.
7. Allowable working stresses are one-fourth of the ultimate strength of the material.
8. Maximum allowable stress depends sharply on temperature.

Temperature (°F)	−20–650	750	850	1000
Low alloy steel SA203 (psi)	18,750	15,650	9550	2500
Type 302 stainless (psi)	18,750	18,750	15,900	6250

VESSELS (STORAGE TANKS)

1. For less than 1000 gal, use vertical tanks on legs.
2. Between 1000 and 10,000 gal, use horizontal tanks on concrete supports.
3. Beyond 10,000 gal, use vertical tanks on concrete foundations.
4. Liquids subject to breathing losses may be stored in tanks with floating or expansion roofs for conservation.
5. Freeboard is 15% below 500 gal and 10% above 500 gal capacity.
6. Thirty days capacity often is specified for raw materials and products, but depends on connecting transportation equipment schedules.
7. Capacities of storage tanks are at least 1.5 times the size of connecting transportation equipment; for instance, 7500 gal tank trucks, 34,500 gal tank cars, and virtually unlimited barge and tanker capacities.

1

INTRODUCTION

Although this book is devoted to the selection and design of individual equipment, some mention should be made of integration of a number of units into a process. Each piece of equipment interacts with several others in a plant, and the range of its required performance is dependent on the others in terms of material and energy balances and rate processes. This chapter will discuss general background material relating to complete process design, and Chapter 2 will treat briefly the basic topic of flowsheets.

1.1. PROCESS DESIGN

Process design establishes the sequence of chemical and physical operations; operating conditions; the duties, major specifications, and materials of construction (where critical) of all process equipment (as distinguished from utilities and building auxiliaries); the general arrangement of equipment needed to ensure proper functioning of the plant; line sizes; and principal instrumentation. The process design is summarized by a process flowsheet, a material and energy balance, and a set of individual equipment specifications. Varying degrees of thoroughness of a process design may be required for different purposes. Sometimes only a preliminary design and cost estimate are needed to evaluate the advisability of further research on a new process or a proposed plant expansion or detailed design work; or a preliminary design may be needed to establish the approximate funding for a complete design and construction. A particularly valuable function of preliminary design is that it may reveal lack of certain data needed for final design. Data of costs of individual equipment are supplied in this book, but the complete economics of process design is beyond its scope.

1.2. EQUIPMENT

Two main categories of process equipment are proprietary and custom-designed. Proprietary equipment is designed by the manufacturer to meet performance specifications made by the user; these specifications may be regarded as the process design of the equipment. This category includes equipment with moving parts such as pumps, compressors, and drivers as well as cooling towers, dryers, filters, mixers, agitators, piping equipment, and valves, and even the structural aspects of heat exchangers, furnaces, and other equipment. Custom design is needed for many aspects of chemical reactors, most vessels, multistage separators such as fractionators, and other special equipment not amenable to complete standardization.

Only those characteristics of equipment are specified by process design that are significant from the process point of view. On a pump, for instance, process design will specify the operating conditions, capacity, pressure differential, NPSH, materials of construction in contact with process liquid, and a few other items, but not such details as the wall thickness of the casing or the type of stuffing box or the nozzle sizes and the foundation dimensions—although most of these omitted items eventually must be known before a plant is ready for construction. Standard specification forms are available for most proprietary kinds of equipment and for summarizing the details of all kinds of equipment. By providing suitable check lists, they simplify the work by ensuring that all needed data have been provided. A collection of such forms is in Appendix B.

Proprietary equipment is provided "off the shelf" in limited sizes and capacities. Special sizes that would fit particular applications more closely often are more expensive than a larger standard size that incidentally may provide a worthwhile safety factor. Even largely custom-designed equipment, such as vessels, is subject to standardization such as discrete ranges of head diameters, pressure ratings of nozzles, sizes of manways, and kinds of trays and packings. Many codes and standards are established by government agencies, insurance companies, and organizations sponsored by engineering societies. Some standardizations within individual plants are arbitrary choices from comparable methods, made to simplify construction, maintenance, and repair: for example, restriction to instrumentation of a particular manufacturer or to a limited number of sizes of heat exchanger tubing or a particular method of installing liquid level gage glasses. All such restrictions must be borne in mind by the process designer.

VENDORS' QUESTIONNAIRES

A manufacturer's or vendor's inquiry form is a questionnaire whose completion will give him the information on which to base a specific recommendation of equipment and a price. General information about the process in which the proposed equipment is expected to function, amounts and appropriate properties of the streams involved, and the required performance are basic. The nature of additional information varies from case to case; for instance, being different for filters than for pneumatic conveyors. Individual suppliers have specific inquiry forms. A representative selection is in Appendix C.

SPECIFICATION FORMS

When completed, a specification form is a record of the salient features of the equipment, the conditions under which it is to operate, and its guaranteed performance. Usually it is the basis for a firm price quotation. Some of these forms are made up by organizations such as TEMA or API, but all large engineering contractors and many large operating companies have other forms for their own needs. A selection of specification forms is in Appendix B.

1.3. CATEGORIES OF ENGINEERING PRACTICE

Although the design of a chemical process plant is initiated by chemical engineers, its complete design and construction requires the inputs of other specialists: mechanical, structural, electrical, and instrumentation engineers; vessel and piping designers; and purchasing agents who know what may be available at attractive prices. On large projects all these activities are correlated by a job engineer or project manager; on individual items of equipment or small projects, the process engineer naturally assumes this function. A key activity is the writing of specifications for soliciting bids and ultimately purchasing equipment. Specifications must be written so explicitly that the bidders are held to a uniform standard and a clear-cut choice can be made on the basis of their offerings alone.

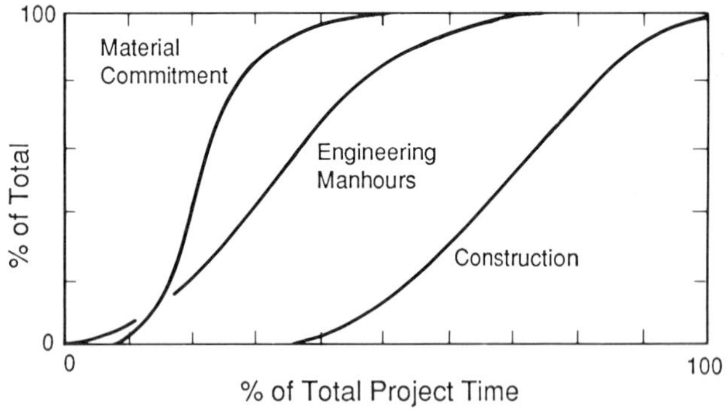

Figure 1.1. Progress of material commitment, engineering manhours, and construction [*Matozzi,* Oil Gas. J. p. **304,** (*23 March 1953*)].

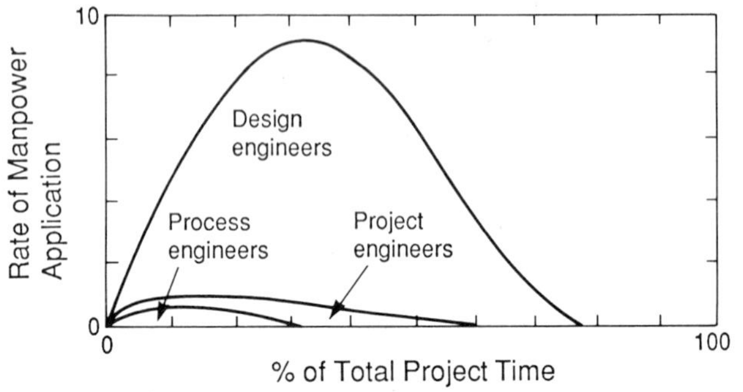

Figure 1.2. Rate of application of engineering manhours of various categories. The area between the curves represents accumulated manhours for each speciality up to a given % completion of the project [*Miller,* Chem. Eng., p. **188,** (*July 1956*)].

For a typical project, Figure 1.1 shows the distributions of engineering, material commitment, and construction efforts. Of the engineering effort, the process engineering is a small part. Figure 1.2 shows that it starts immediately and finishes early. In terms of money, the cost of engineering ranges from 5 to 15% or so of the total plant cost; the lower value for large plants that are largely patterned after earlier ones, and the higher for small plants or those based on new technology or unusual codes and specifications.

1.4. SOURCES OF INFORMATION FOR PROCESS DESIGN

A selection of books relating to process design methods and data is listed in the references at the end of this chapter. Items that are especially desirable in a personal library or readily accessible are identified. Specialized references are given throughout the book in connection with specific topics.

The extensive chemical literature is served by the bibliographic items cited in References, Section 1.2, Part B. The book by Rasmussen and Fredenslund (1980) is addressed to chemical engineers and cites some literature not included in some of the other bibliographies, as well as information about proprietary data banks. The book by Leesley (References, Section 1.1, Part B) has much information about proprietary data banks and design methods. In its current and earlier editions, the book by Peters and Timmerhaus has many useful bibliographies on classified topics.

For information about chemical manufacturing processes, the main encyclopedic references are Kirk-Othmer (1978–1984), McKetta and Cunningham (1976–date) and Ullmann (1972–1983) (References, Section 1.2, Part B). The last of these is in German, but an English version was started in 1984 and three volumes per year are planned; this beautifully organized reference should be most welcome.

The most comprehensive compilation of physical property data is that of Landolt-Börnstein (1950–date) (References, Section 1.2, Part C). Although most of the material is in German, recent volumes have detailed tables of contents in English and some volumes are largely in English. Another large compilation, somewhat venerable but still valuable, is the International Critical Tables (1926–1933). Data and methods of estimating properties of hydrocarbons and their mixtures are in the API Data Book (1971–date) (References, Section 1.2, Part C). More general treatments of estimation of physical properties are listed in References, Section 1.1, Part C. There are many compilations of special data such as solubilities, vapor pressures, phase equilibria, transport and thermal properties, and so on. A few of them are listed in References, Section 1.2, Part D, and references to many others are in the References, Section 1.2, Part B.

Information about equipment sizes and configurations, and sometimes performance, of equipment is best found in manufacturers' catalogs. Items 1 and 2 of References, Section 1.1, Part D, contain some advertisements with illustrations, but perhaps their principal value is in the listings of manufacturers by the kind of equipment. Thomas Register covers all manufacturers and so is less convenient at least for an initial search. The other three items of this group of books have illustrations and descriptions of all kinds of chemical process equipment. Although these books are old, one is surprised to note how many equipment designs have survived.

1.5. CODES, STANDARDS, AND RECOMMENDED PRACTICES

A large body of rules has been developed over the years to ensure the safe and economical design, fabrication and testing of equipment, structures, and materials. Codification of these rules has been done by associations organized for just such purposes, by professional societies, trade groups, insurance underwriting companies, and government agencies. Engineering contractors and large manufacturing companies usually maintain individual sets of standards so as to maintain continuity of design and to simplify maintenance of plant. Table 1.1 is a representative table of contents of the mechanical standards of a large oil company.

Typical of the many thousands of items that are standardized in the field of engineering are limitations on the sizes and wall thicknesses of piping, specifications of the compositions of alloys, stipulation of the safety factors applied to strengths of construction materials, testing procedures for many kinds of materials, and so on.

Although the safe design practices recommended by professional and trade associations have no legal standing where they have not actually been incorporated in a body of law, many of them have the respect and confidence of the engineering profession as a whole and have been accepted by insurance underwriters so they are widely observed. Even when they are only voluntary, standards constitute a digest of experience that represents a minimum requirement of good practice.

Two publications by Burklin (References, Section 1.1, Part B) are devoted to standards of importance to the chemical industry. Listed are about 50 organizations and 60 topics with which they are concerned. National Bureau of Standards Publication 329 contains about 25,000 titles of U.S. standards. The NBS-SIS service maintains a reference collection of 200,000 items accessible by letter or phone. Information about foreign standards is obtainable through the American National Standards Institute (ANSI).

A listing of codes and standards bearing directly on process

TABLE 1.1. Internal Engineering Standards of a Large Petroleum Refinery[a]

1	Appropriations and mechanical orders (10)
2	Buildings—architectural (15)
3	Buildings—mechanical (10)
4	Capacities and weights (25)
5	Contracts (10)
6	Cooling towers (10)
7	Correspondence (5)
8	Designation and numbering rules for equipment and facilities (10)
9	Drainage (25)
10	Electrical (10)
11	Excavating, grading, and paving (10)
12	Fire fighting (10)
13	Furnaces and boilers (10)
14	General instructions (20)
15	Handling equipment (5)
16	Heat exchangers (10)
17	Instruments and controls (45)
18	Insulation (10)
19	Machinery (35)
20	Material procurement and disposition (20)
21	Material selection (5)
22	Miscellaneous process equipment (25)
23	Personnel protective equipment (5)
24	Piping (150)
25	Piping supports (25)
26	Plant layout (20)
27	Pressure vessels (25)
28	Protective coatings (10)
29	Roads and railroads (25)
30	Storage vessels (45)
31	Structural (35)
32	Symbols and drafting practice (15)
33	Welding (10)

[a] Figures in parentheses identify the numbers of distinct standards.

TABLE 1.2. Codes and Standards of Direct Bearing on Chemical Process Design (a Selection)

A. American Institute of Chemical Engineers, 345 E. 47th St., New York, NY 10017
- **1.** Standard testing procedures; 21 have been published, for example on centrifuges, filters, mixers, firer heaters

B. American Petroleum Institute, 2001 L St. NW, Washington, DC 20037
- **2.** Recommended practices for refinery inspections
- **3.** Guide for inspection of refinery equipment
- **4.** Manual on disposal of refinery wastes
- **5.** Recommended practice for design and construction of large, low pressure storage tanks
- **6.** Recommended practice for design and construction of pressure relieving devices
- **7.** Recommended practices for safety and fire protection

C. American Society of Mechanical Engineers, 345 W. 47th St., New York, NY 10017
- **8.** ASME Boiler and Pressure Vessel Code. Sec. VIII, Unfired Pressure Vessels
- **9.** Code for pressure piping
- **10.** Scheme for identification of piping systems

D. American Society for Testing Materials, 1916 Race St., Philadelphia, PA 19103
- **11.** ASTM Standards, 66 volumes in 16 sections, annual, with about 30% revision each year

E. American National Standards Institute (ANSI), 1430 Broadway, New York, NY 10018
- **12.** Abbreviations, letter symbols, graphical symbols, drawing and drafting room practice

TABLE 1.2—(*continued*)

F. Chemical Manufacturers' Association, 2501 M St. NW, Washington, DC 20037
- **13.** Manual of standard and recommended practices for containers, tank cars, pollution of air and water
- **14.** Chemical safety data sheets of individual chemicals

G. Cooling Tower Institute, 19627 Highway 45 N, Spring, TX 77388
- **15.** Acceptance test procedure for water cooling towers of mechanical draft industrial type

H. Hydraulic Institute, 712 Lakewood Center N, 14600 Detroit Ave., Cleveland, OH 44107
- **16.** Standards for centrifugal, reciprocating, and rotary pumps
- **17.** Pipe friction manual

I. Instrument Society of America (ISA), 67 Alexander Dr., Research Triangle Park, NC 27709
- **18.** Instrumentation flow plan symbols
- **19.** Specification forms for instruments
- **20.** Dynamic response testing of process control instrumentation

J. Tubular Exchangers Manufacturers' Association, 25 N Broadway, Tarrytown, NY 10591
- **21.** TEMA standards

K. International Standards Organization (ISO), 1430 Broadway, New York, NY 10018
- **22.** Many standards

TABLE 1.3. Codes and Standards Supplementary to Process Design (a Selection)

A. American Concrete Institute, 22400 W. 7 Mile Rd., Detroit, MI 48219
- **1.** Reinforced concrete design handbook
- **2.** Manual of standard practice for detailing reinforced concrete structures

B. American Institute of Steel Construction, 400 N. Michigan Ave., Chicago, IL 60611
- **3.** Manual of steel construction
- **4.** Standard practice for steel buildings and bridges

C. American Iron and Steel Institute, 1000 16th St. NW, Washington, DC 20036
- **5.** AISI standard steel compositions

D. American Society of Heating, Refrigerating and Air Conditioning Engineers (ASHRE), 1791 Tullie Circle NE, Atlanta, GA 30329
- **6.** Refrigerating data book

E. Institute of Electrical and Electronics Engineers, 345 E. 47th St., New York, NY 10017
- **7.** Many standards

F. National Bureau of Standards, Washington, DC
- **8.** American standard building code
- **9.** National electrical code

G. National Electrical Manufacturers Association, 2101 L St. NW, Washington, DC 20037
- **10.** NEMA standards

design is in Table 1.2, and of supplementary codes and standards in Table 1.3.

1.6. MATERIAL AND ENERGY BALANCES

Material and energy balances are based on a conservation law which is stated generally in the form

$$\text{input} + \text{source} = \text{output} + \text{sink} + \text{accumulation}.$$

The individual terms can be plural and can be rates as well as absolute quantities. Balances of particular entities are made around a bounded region called a system. Input and output quantities of an entity cross the boundaries. A source is an increase in the amount of the entity that occurs without a crossing of the boundary; for example, an increase in the sensible enthalpy or in the amount of a substance as a consequence of chemical reaction. Analogously, sinks are decreases without a boundary crossing, as the disappearance of water from a fluid stream by adsorption onto a solid phase within the boundary.

Accumulations are time rates of change of the amount of the entities within the boundary. For example, in the absence of sources and sinks, an accumulation occurs when the input and output rates are different. In the steady state, the accumulation is zero.

Although the principle of balancing is simple, its application requires knowledge of the performance of all the kinds of equipment comprising the system and of the phase relations and physical properties of all mixtures that participate in the process. As a consequence of trying to cover a variety of equipment and processes, the books devoted to the subject of material and energy balances always run to several hundred pages. Throughout this book, material and energy balances are utilized in connection with the design of individual kinds of equipment and some processes. Cases involving individual pieces of equipment usually are relatively easy to balance, for example, the overall balance of a distillation column in Section 13.4.1 and of nonisothermal reactors of Tables 17.4–17.7. When a process is maintained isothermal, only a material balance is needed to describe the process, unless it is also required to know the net heat transfer for maintaining a constant temperature.

In most plant design situations of practical interest, however, the several pieces of equipment interact with each other, the output of one unit being the input to another that in turn may recycle part of its output to the inputter. Common examples are an absorber–stripper combination in which the performance of the absorber depends on the quality of the absorbent being returned from the stripper, or a catalytic cracker–catalyst regenerator system whose two parts interact closely.

Because the performance of a particular piece of equipment depends on its input, recycling of streams in a process introduces temporarily unknown, intermediate streams whose amounts, compositions, and properties must be found by calculation. For a plant with dozens or hundreds of streams the resulting mathematical problem is formidable and has led to the development of many computer algorithms for its solution, some of them making quite rough approximations, others more nearly exact. Usually the problem is solved more easily if the performance of the equipment is specified in advance and its size is found after the balances are completed. If the equipment is existing or must be limited in size, the balancing process will require simultaneous evaluation of its performance and consequently is a much more involved operation, but one which can be handled by computer when necessary.

The literature of this subject naturally is extensive. An early book (for this subject), Nagiev's *Theory of Recycle Processes in Chemical Engineering* (Macmillan, New York, 1964, Russian edition, 1958) treats many practical cases by reducing them to systems of linear algebraic equations that are readily solvable. The book by Westerberg et al., *Process Flowsheeting* (Cambridge Univ. Press, Cambridge, 1977) describes some aspects of the subject and has an extensive bibliography. Benedek in *Steady State Flowsheeting of Chemical Plants* (Elsevier, New York, 1980) provides a detailed description of one simulation system. Leesley in *Computer-Aided Process Design* (Gulf, Houston, 1982) describes the capabilities of some commercially available flowsheet simulation programs. Some of these incorporate economic balance with material and energy balances. A program MASSBAL in BASIC language is in the book of Sinnott et al., *Design,* Vol. 6 (Pergamon, New York, 1983); it can handle up to 20 components and 50 units when their several outputs are specified to be in fixed proportions.

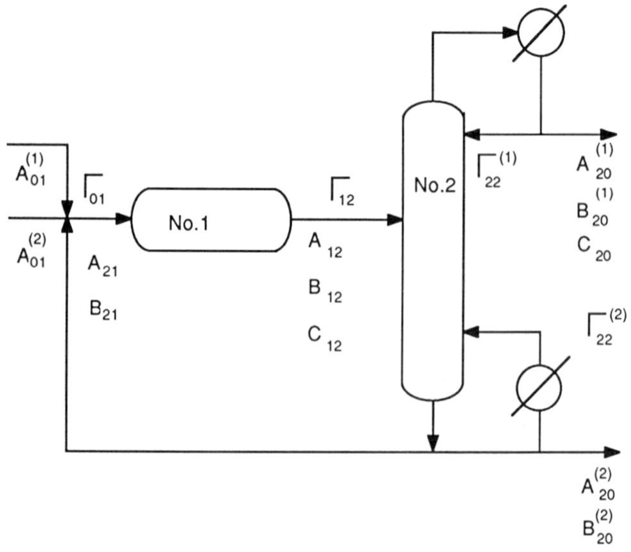

Figure 1.3. Notation of flow quantities in a reactor (1) and distillation column (2). $A_{ij}^{(k)}$ designates the amount of component A in stream k proceeding from unit i to unit j. Subscripts 0 designates a source or sink beyond the boundary limits. Γ designates a total flow quantity.

A key factor in the effective formulation of material and energy balances is a proper notation for equipment and streams. Figure 1.3, representing a reactor and a separator, utilizes a simple type. When the pieces of equipment are numbered i and j, the notation $A_{ij}^{(k)}$ signifies the flow rate of substance A in stream k proceeding from unit i to unit j. The total stream is designated $\Gamma_{ij}^{(k)}$. Subscript t designates a total stream and subscript 0 designates sources or sinks outside the system. Example 1.1 adopts this notation for balancing a reactor–separator process in which the performances are specified in advance.

Since this book is concerned primarily with one kind of equipment at a time, all that need be done here is to call attention to the existence of the abundant literature on these topics of recycle calculations and flowsheet simulation.

1.7. ECONOMIC BALANCE

Engineering enterprises always are subject to monetary considerations, and a balance is sought between fixed and operating costs. In the simplest terms, fixed costs consist of depreciation of the investment plus interest on the working capital. Operating costs include labor, raw materials, utilities, maintenance, and overheads which consists in turn of administrative, sales and research costs. Usually as the capital cost of a process unit goes up, the operating cost goes down. For example, an increase in control instrumentation and automation at a higher cost is accompanied by a reduction in operating labor cost. Somewhere in the summation of these factors there is a minimum which should be the design point in the absence of any contrary intangibles such as building for the future or unusual local conditions.

Costs of many individual pieces of equipment are summarized in Chapter 20, but analysis of the costs of complete processes is beyond the scope of this book. References may be made, however, to several collections of economic analyses of chemical engineering interest that have been published:

1. AIChE Student Contest Problems (annual) (AIChE, New York).

EXAMPLE 1.1
Material Balance of a Chlorination Process with Recycle
A plant for the chlorination has the flowsheet shown. From pilot plant work, with a chlorine/benzene charge weight ratio of 0.82, the composition of the reactor effluent is

A. C_6H_6	0.247
B. Cl_2	0.100
C. C_6H_5Cl	0.3174
D. $C_6H_4Cl_2$	0.1559
E. HCl	0.1797

Separator no. 2 returns 80% of the unreacted chlorine to the reactor and separator no. 3 returns 90% of the benzene. Both recycle streams are pure. Fresh chlorine is charged at such a rate that the weight ratio of chlorine to benzene in the total charge remains 0.82. The amounts of other streams are found by material balances and are shown in parentheses on the sketch per 100 lbs of fresh benzene to the system.

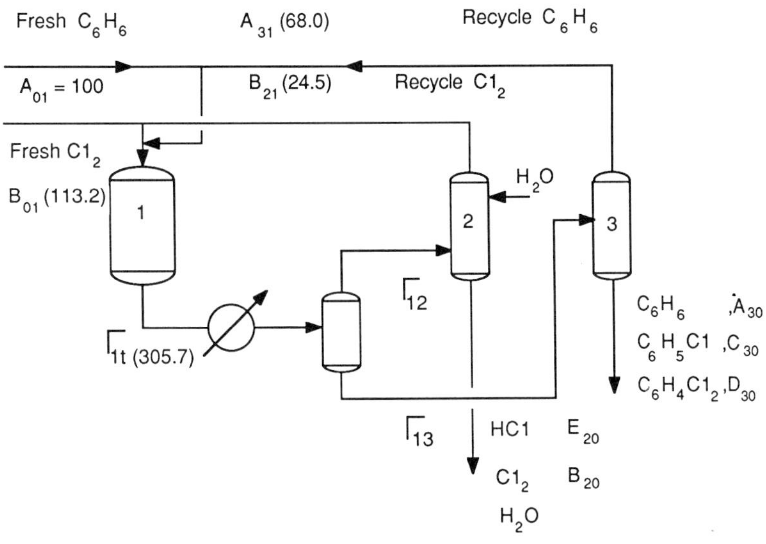

2. Bodman, *Industrial Practice of Chemical Process Engineering* (MIT Press, Cambridge, MA, 1968).
3. Rase, *Chemical Reactor Design for Process Plants, Vol. II, Case Studies* (Wiley, New York, 1977).
4. Washington University, St. Louis, *Case Studies in Chemical Engineering Design* (22 cases to 1984).

Somewhat broader in scope are:

5. Wei et al., *The Structure of the Chemical Processing Industries* (McGraw-Hill, New York, 1979).
6. Skinner et al., *Manufacturing Policy in the Oil Industry* (Irwin, Homewood, IL., 1970).
7. Skinner et al., *Manufacturing Policy in the Plastics Industry* (Irwin, Homewood, Il., 1968).

Many briefer studies of individual equipment appear in some books, of which a selection is as follows:

- Happel and Jordan, *Chemical Process Economics* (Dekker, New York, 1975):
  1. Absorption of ethanol from a gas containing CO_2 (p. 403).
  2. A reactor-separator for simultaneous chemical reactions (p. 419).
  3. Distillation of a binary mixture (p. 385).
  4. A heat exchanger and cooler system (p. 370).
  5. Piping of water (p. 353).
  6. Rotary dryer (p. 414).
- Jelen et al., *Cost and Optimization Engineering* (McGraw-Hill, New York, 1983):
  7. Drill bit life and replacement policy (p. 223).
  8. Homogeneous flow reactor (p. 229).
  9. Batch reaction with negligible downtime (p. 236).
- Peters and Timmerhaus, *Plant Design and Economics for Chemical Engineers* (McGraw-Hill, New York, 1980):
  10. Shell and tube cooling of air with water (p. 688).
- Rudd and Watson, *Strategy of Process Engineering* (Wiley, New York, 1968):
  11. Optimization of a three stage refrigeration system (p. 172).
- Sherwood, *A Course in Process Design* (MIT Press, Cambridge, MA, 1963):
  12. Gas transmission line (p. 84).
  13. Fresh water from sea water by evaporation (p. 138).
- Ulrich, *A Guide to Chemical Engineering Process Design and Economics* (Wiley, New York, 1984):
  14. Multiple effect evaporator for Kraft liquor (p. 347).
- Walas, *Reaction Kinetics for Chemical Engineers* (McGraw-Hill, New York, 1959):
  15. Optimum number of vessels in a CSTR battery (p. 98).

Since capital, labor, and energy costs have not escalated equally over the years since these studies were made, their conclusions are subject to reinterpretation, but the patterns of study that were used should be informative.

Because of the rapid escalation of energy costs in recent years, closer appraisals of energy utilizations by complete processes are being made, from the standpoints of both the conservation laws and the second law of thermodynamics. In the latter cases attention is focused on changes in entropy and in the related availability function, $\Delta B = \Delta H - T_0 \Delta S$, with emphasis on work as the best possible transformation of energy. In this way a second law analysis of a process will reveal where the greatest generation of entropy occurs and where possibly the most improvement can be made by appropriate changes of process or equipment. Such an analysis of a cryogenic process for air separation was made by Benedict and Gyftopolous [in Gaggioli (Ed.), *Thermodynamic Second Law Analysis,* ACS Symposium Series No. 122, American Chemical Society, Washington, DC, 1980]; they found a pressure drop at which the combination of exchanger and compressor was most economical.

A low second law efficiency is not always realistically improvable. Thus Weber and Meissner (*Thermodynamics for Chemical Engineers,* John Wiley, New York, 1957) found a 6% efficiency for the separation of ethanol and water by distillation which is not substantially improvable by redesign of the distillation process. Perhaps this suggests that more efficient methods than distillation should be sought for the separation of volatile mixtures, but none has been found at competitive cost.

Details of the thermodynamic basis of availability analysis are dealt with by Moran (*Availability Analysis,* Prentice-Hall, Englewood Cliffs, NJ, 1982). He applies the method to a cooling tower, heat pump, a cryogenic process, coal gasification, and particularly to the efficient use of fuels.

An interesting conclusion reached by Linnhoff [in Seider and Mah (Eds.), *Foundations of Computer-Aided Process Design,* AIChE, New York, 1981] is that "chemical processes which are properly designed for energy versus capital cost tend to operate at approximately 60% efficiency." A major aspect of his analysis is recognition of practical constraints and inevitable losses. These may include material of construction limits, plant layout, operability, the need for simplicity such as limits on the number of compressor stages or refrigeration levels, and above all the recognition that, for low grade heat, heat recovery is preferable to work recovery, the latter being justifiable only in huge installations. Unfortunately, the edge is taken off the dramatic 60% conclusion by Linnhoff's admission that efficiency cannot be easily defined for some complexes of interrelated equipment. For example, is it economical to recover 60% of the propane or 60% of the ethane from a natural gas?

1.8. SAFETY FACTORS

In all of the factors that influence the performance of equipment and plant there are elements of uncertainty and the possibility of error, including inaccuracy of physical data, basic correlations of behavior such as pipe friction or tray efficiency or gas–liquid distribution, necessary approximations of design methods and calculations, not entirely known behavior of materials of construction, uncertainty of future market demands, and changes in operating performance with time. The solvency of the project, the safety of the operators and the public, and the reputation and career of the design engineer are at stake. Accordingly, the experienced engineer will apply safety factors throughout the design of a plant. Just how much of a factor should be applied in a particular case cannot be stated in general terms because circumstances vary widely. The inadequate performance of a particular piece of equipment may be compensated for by the superior performance of associated equipment, as insufficient trays in a fractionator may be compensated for by increases in reflux and reboiling, if that equipment can take the extra load.

With regard to specific types of equipment, the safety factor practices of some 250 engineers were ascertained by a questionnaire and summarized in Table 1.4; additional figures are given by Peters and Timmerhaus (References, Section 1.1, Part B, pp. 35–37). Relatively inexpensive equipment that can conceivably serve as a bottleneck, such as pumps, always is liberally sized; perhaps as much as 50% extra for a reflux pump. In an expanding industry it is a matter of policy to deliberately oversize certain major equipment that cannot be supplemented readily or modified suitably for increased capacity; these are safety factors to account for future trends.

Safety factors should not be used to mask inadequate or careless design work. The design should be the best that can be made in the time economically justifiable, and the safety factors should be estimated from a careful consideration of all factors entering into the design and the possible future deviations from the design conditions.

Sometimes it is possible to evaluate the range of validity of measurements and correlations of physical properties, phase equilibrium behavior, mass and heat transfer efficiencies and similar factors, as well as the fluctuations in temperature, pressure, flow, etc., associated with practical control systems. Then the effects of such data on the uncertainty of sizing equipment can be estimated. For example, the mass of a distillation column that is related directly to its cost depends on at least these factors:

1. The vapor–liquid equilibrium data.
2. The method of calculating the reflux and number of trays.
3. The tray efficiency.
4. Allowable vapor rate and consequently the tower diameter at a given tray spacing and estimated operating surface tension and fluid densities.
5. Corrosion allowances.

Also such factors as allowable tensile strengths, weld efficiencies, and possible inaccuracies of formulas used to calculate shell and head thicknesses may be pertinent.

When a quantity is a function of several variables,

$$y = y(x_1, x_2, \cdots),$$

its differential is

$$dy = \frac{\partial y}{\partial x_1} dx_1 + \frac{\partial y}{\partial x_2} dx_2 + \cdots.$$

Some relations of importance in chemical engineering have the form

$$y = (x_1)^a (x_2)^b \cdots,$$

whose differential is rearrangable to

$$\frac{dy}{y} = a\frac{dx_1}{x_1} + b\frac{dx_2}{x_2} + \cdots,$$

that is, the relative uncertainty or error in the function is related linearly to the fractional uncertainties of the independent variables. For example, take the case of a steam-heated thermosyphon reboiler on a distillation column for which the heat transfer equation is

$$q = UA\Delta T.$$

The problem is to find how the heat transfer rate can vary when the other quantities change. U is an experimental value that is known

TABLE 1.4. Safety Factors in Equipment Design: Results of a Questionnaire

Equipment	Design Variable	Range of Safety Factor (%)
Compressors, reciprocating	piston displacement	11–21
Conveyors, screw	diameter	8–21
Hammer mills	power input	15–21[a]
Filters, plate-and-frame	area	11–21[a]
Filters, rotary	area	14–20[a]
Heat exchangers, shell and tube for liquids	area	11–18
Pumps, centrifugal	impeller diameter	7–14
Separators, cyclone	diameter	7–11
Towers, packed	diameter	11–18
Towers, tray	diameter	10–16
Water cooling towers	volume	12–20

[a] Based on pilot plant tests.
[Michelle, Beattie, and Goodgame, *Chem. Eng. Prog.* **50,** 332 (1954)].

only to a certain accuracy. ΔT may be uncertain because of possible fluctuations in regulated steam and tower pressures. A, the effective area, may be uncertain because the submergence is affected by the liquid level controller at the bottom of the column. Accordingly,

$$\frac{dq}{q} = \frac{dU}{U} + \frac{dA}{A} + \frac{d(\Delta T)}{\Delta T},$$

that is, the fractional uncertainty of q is the sum of the fractional uncertainties of the quantities on which it is dependent. In practical cases, of course, some uncertainties may be positive and others negative, so that they may cancel out in part; but the only safe viewpoint is to take the sum of the absolute values. Some further discussion of such cases is by Sherwood and Reed, in *Applied Mathematics in Chemical Engineering* (McGraw-Hill, New York, 1939).

It is not often that proper estimates can be made of uncertainties of all the parameters that influence the performance or required size of particular equipment, but sometimes one particular parameter is dominant. All experimental data scatter to some extent, for example, heat transfer coefficients; and various correlations of particular phenomena disagree, for example, equations of state of liquids and gases. The sensitivity of equipment sizing to uncertainties in such data has been the subject of some published information, of which a review article is by Zudkevich [*Encycl. Chem. Proc. Des.* **14,** 431–483 (1982)]; some of his cases are:

1. Sizing of isopentane/pentane and propylene/propane splitters.
2. Effect of volumetric properties on sizing of an ethylene compressor.
3. Effect of liquid density on metering of LNG.
4. Effect of vaporization equilibrium ratios, K, and enthalpies on cryogenic separations.
5. Effects of VLE and enthalpy data on design of plants for coal-derived liquids.

Examination of such studies may lead to the conclusion that some of the safety factors of Table 1.4 may be optimistic. But long experience in certain areas does suggest to what extent various uncertainties do cancel out, and overall uncertainties often do fall in the range of 10–20% as stated there. Still, in major cases the uncertainty analysis should be made whenever possible.

1.9. SAFETY OF PLANT AND ENVIRONMENT

The safe practices described in the previous section are primarily for assurance that the equipment have adequate performance over anticipated ranges of operating conditions. In addition, the design of equipment and plant must minimize potential harm to personnel and the public in case of accidents, of which the main causes are

a. human failure,
b. failure of equipment or control instruments,
c. failure of supply of utilities or key process streams,
d. environmental events (wind, water, and so on).

A more nearly complete list of potential hazards is in Table 1.5, and a checklist referring particularly to chemical reactions is in Table 1.6.

Examples of common safe practices are pressure relief valves, vent systems, flare stacks, snuffing steam and fire water, escape hatches in explosive areas, dikes around tanks storing hazardous materials, turbine drives as spares for electrical motors in case of power failure, and others. Safety considerations are paramount in the layout of the plant, particularly isolation of especially hazardous operations and accessibility for corrective action when necessary.

Continual monitoring of equipment and plant is standard practice in chemical process plants. Equipment deteriorates and operating conditions may change. Repairs sometimes are made with "improvements" whose ultimate effects on the operation may not be taken into account. During start-up and shut-down, stream compositions and operating conditions are much different from those under normal operation, and their possible effect on safety must be taken into account. Sample checklists of safety questions for these periods are in Table 1.7.

Because of the importance of safety and its complexity, safety engineering is a speciality in itself. In chemical processing plants of any significant size, loss prevention reviews are held periodically by groups that always include a representative of the safety department. Other personnel, as needed by the particular situation, are from manufacturing, maintenance, technical service, and possibly research, engineering, and medical groups. The review considers any changes made since the last review in equipment, repairs, feedstocks and products, and operating conditions.

Detailed safety checklists appear in books by Fawcett and Wood (Chap. 32, Bibliography 1.1, Part E) and Wells (pp. 239–257, Bibliography 1.1, Part E). These books and the large one by Lees (Bibliography 1.1, Part E) also provide entry into the vast literature of chemical process plant safety. Lees has particularly complete bibliographies. A standard reference on the properties of dangerous materials is the book by Sax (1984) (References, Section 1.1, Part E). The handbook by Lund (1971) (References, Section 1.1, Part E) on industrial pollution control also may be consulted.

TABLE 1.5. Some Potential Hazards

Energy Source
- Process chemicals, fuels, nuclear reactors, generators, batteries
- Source of ignition, radio frequency energy sources, activators, radiation sources
- Rotating machinery, prime movers, pulverisers, grinders, conveyors, belts, cranes
- Pressure containers, moving objects, falling objects

Release of Material
- Spillage, leakage, vented material
- Exposure effects, toxicity, burns, bruises, biological effects
- Flammability, reactivity, explosiveness, corrosivity and fire-promoting properties of chemicals
- Wetted surfaces, reduced visibility, falls, noise, damage
- Dust formation, mist formation, spray

Fire hazard
- Fire, fire spread, fireballs, radiation
- Explosion, secondary explosion, domino effects
- Noise, smoke, toxic fumes, exposure effects
- Collapse, falling objects, fragmentation

Process state
- High/low/changing temperature and pressure
- Stress concentrations, stress reversals, vibration, noise
- Structural damage or failure, falling objects, collapse
- Electrical shock and thermal effects, inadvertent activation, power source failure
- Radiation, internal fire, overheated vessel
- Failure of equipment/utility supply/flame/instrument/component
- Start-up and shutdown condition
- Maintenance, construction and inspection condition

Environmental effects
- Effect of plant on surroundings, drainage, pollution, transport, wind and light change, source of ignition/vibration/noise/radio interference/fire spread/explosion
- Effect of surroundings on plant (as above)
- Climate, sun, wind, rain, snow, ice, grit, contaminants, humidity, ambient conditions
- Acts of God, earthquake, arson, flood, typhoon, *force majeure*
- Site layout factors, groups of people, transport features, space limitations, geology, geography

Processes
- Processes subject to explosive reaction or detonation
- Processes which react energetically with water or common contaminants
- Processes subject to spontaneous polymerisation or heating
- Processes which are exothermic
- Processes containing flammables and operated at high pressure or high temperature or both
- Processes containing flammables and operated under refrigeration
- Processes in which intrinsically unstable compounds are present
- Processes operating in or near the explosive range of materials
- Processes involving highly toxic materials
- Processes subject to a dust or mist explosion hazard
- Processes with a large inventory of stored pressure energy

Operations

The vaporisation and diffusion of flammable or toxic liquids or gases
- The dusting and dispersion of combustible or toxic solids
- The spraying, misting or fogging of flammable combustible materials or strong oxidising agents and their mixing
- The separation of hazardous chemicals from inerts or diluents
- The temperature and pressure increase of unstable liquids

(Wells, *Safety in Process Plant Design,* George Godwin, London, 1980).

TABLE 1.6. Safety Checklist of Questions About Chemical Reactions

1. Define potentially hazardous reactions. How are they isolated? Prevented? (See Chaps. 4, 5, and 16)
2. Define process variables which could, or do, approach limiting conditions for hazard. What safeguards are provided against such variables?
3. What unwanted hazardous reactions can be developed through unlikely flow or process conditions or through contamination?
4. What combustible mixtures can occur within equipment?
5. What precautions are taken for processes operating near or within the flammable limits? (Reference: S&PP Design Guide No. 8.) (See Chap. 19)
6. What are process margins of safety for all reactants and intermediates in the process?
7. List known reaction rate data on the normal and possible abnormal reactions
8. How much heat must be removed for normal, or abnormally possible, exothermic reactions? (see Chaps. 7, 17, and 18)
9. How thoroughly is the chemistry of the process including desired and undesired reactions known? (See NFPA 491 M, *Manual of Hazardous Chemical Reactions*)
10. What provision is made for rapid disposal of reactants if required by emergency?
11. What provisions are made for handling impending runaways and for short-stopping an existing runaway?
12. Discuss the hazardous reactions which could develop as a result of mechanical equipment (pump, agitator, etc.) failure
13. Describe the hazardous process conditions that can result from gradual or sudden blockage in equipment including lines
14. Review provisions for blockage removal or prevention
15. What raw materials or process materials or process conditions can be adversely affected by extreme weather conditions? Protect against such conditions
16. Describe the process changes including plant operation that have been made since the previous process safety review

(Fawcett and Wood, *Safety and Accident Prevention in Chemical Operations,* Wiley, New York, 1982, pp. 725–726. Chapter references refer to this book.)

TABLE 1.7. Safety Checklist of Questions About Start-up and Shut-down

Start-up Mode (§4.1)

D1 Can the start-up of plant be expedited safely? Check the following:
- (a) Abnormal concentrations, phases, temperatures, pressures, levels, flows, densities
- (b) Abnormal quantities of raw materials, intermediates and utilities (supply, handling and availability)
- (c) Abnormal quantities and types of effluents and emissions (§1.6.10)
- (d) Different states of catalyst, regeneration, activation
- (e) Instruments out of range, not in service or de-activated, incorrect readings, spurious trips
- (f) Manual control, wrong routeing, sequencing errors, poor identification of valves and lines in occasional use, lock-outs, human error, improper start-up of equipment (particularly prime movers)
- (g) Isolation, purging
- (h) Removal of air, undesired process material, chemicals used for cleaning, inerts, water, oils, construction debris and ingress of same
- (i) Recycle or disposal of off-specification process materials
- (j) Means for ensuring construction/maintenance completed
- (k) Any plant item failure on initial demand and during operation in this mode
- (l) Lighting of flames, introduction of material, limitation of heating rate

TABLE 1.7—(*continued*)

(m) Different modes of the start-up of plant:
- Initial start-up of plant
- Start-up of plant section when rest of plant down
- Start-up of plant section when other plant on-stream
- Start-up of plant after maintenance
- Preparation of plant for its start-up on demand

Shut-down Mode (§§4.1, 4.2)

D2 Are the limits of operating parameters, outside which remedial action must be taken, known and measured? (C1 above)

D3 To what extent should plant be shut down for any deviation beyond the operating limits? Does this require the installation of alarm and/or trip? Should the plant be partitioned differently? How is plant restarted? (§9.6)

D4 In an emergency, can the plant pressure and/or the inventory of process materials be reduced effectively, correctly, safely? What is the fire resistance of plant (§§9.5, 9.6)

D5 Can the plant be shut down safely? Check the following:
(a) See the relevant features mentioned under start-up mode
(b) Fail-danger faults of protective equipment
(c) Ingress of air, other process materials, nitrogen, steam, water, lube oil (§4.3.5)
(d) Disposal or inactivation of residues, regeneration of catalyst, decoking, concentration of reactants, drainage, venting
(e) Chemical, catalyst, or packing replacement, blockage removal, delivery of materials prior to start-up of plant
(f) Different modes of shutdown of plant:
- Normal shutdown of plant
- Partial shutdown of plant
- Placing of plant on hot standby
- Emergency shutdown of plant

(Wells, *Safety in Process Plant Design,* George Godwin, London, 1980, pp. 243–244. Paragraph references refer to this book.)

1.10. STEAM AND POWER SUPPLY

For smaller plants or for supplementary purposes, steam and power can be supplied by package plants which are shippable and ready to hook up to the process. Units with capacities in a range of sizes up to about 350,000 lb/hr of steam are on the market, and are obtainable on a rental/purchase basis for emergency needs.

Modern steam plants are quite elaborate structures that can recover 80% or more of the heat of combustion of the fuel. The simplified sketch of Example 1.2 identifies several zones of heat transfer in the equipment. Residual heat in the flue gas is recovered as preheat of the water in an economizer and in an air preheater. The combustion chamber is lined with tubes along the floor and walls to keep the refractory cool and usually to recover more than half the heat of combustion. The tabulations of this example are of the distribution of heat transfer surfaces and the amount of heat transfer in each zone.

More realistic sketches of the cross section of a steam generator are in Figure 1.4. Part (a) of this figure illustrates the process of natural circulation of water between an upper steam drum and a lower drum provided for the accumulation and eventual blowdown of sediment. In some installations, pumped circulation of the water is advantageous.

Both process steam and supplemental power are recoverable from high pressure steam which is readily generated. Example 1.3 is of such a case. The high pressure steam is charged to a turbine-generator set, process steam is extracted at the desired process pressure at an intermediate point in the turbine, and the rest of the steam expands further and is condensed.

In plants such as oil refineries that have many streams at high temperatures or high pressures, their energy can be utilized to generate steam or to recover power. The two cases of Example 1.4

EXAMPLE 1.2
Data of a Steam Generator for Making 250,000 lb/hr at 450 psia and 650°F from Water Entering at 220°F

Fuel oil of 18,500 Btu/lb is fired with 13% excess air at 80°F. Flue gas leaves at 410°F. A simplified cross section of the boiler is shown. Heat and material balances are summarized. Tube selections and arrangements for the five heat transfer zones also are summarized. The term A_g is the total internal cross section of the tubes in parallel. (*Steam: Its Generation and Use,* 14.2, Babcock and Wilcox, Barberton, OH, 1972). (a) Cross section of the generator:

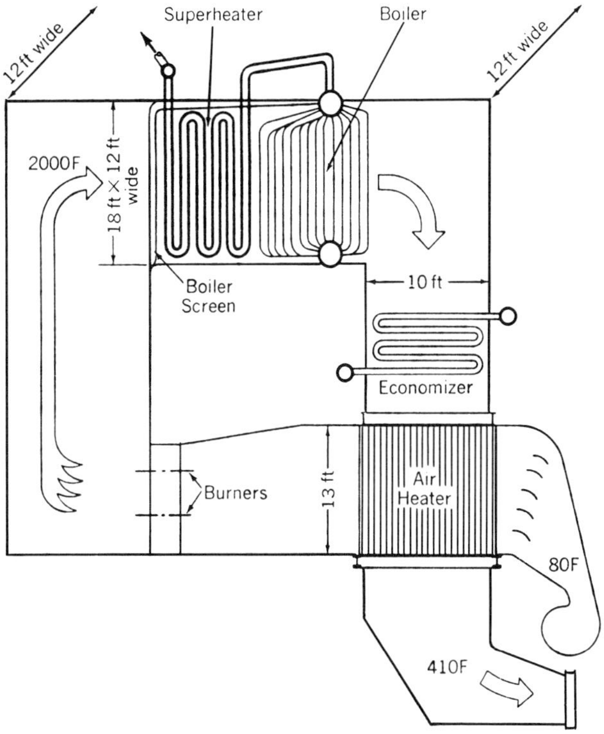

(b) Heat balance:

Fuel input	**335.5 MBtu/hr**
To furnace tubes	162.0
To boiler tubes	68.5
To screen tubes	8.1
To superheater	31.3
To economizer	15.5
Total to water and steam	285.4 Mbtu/hr
In air heater	18.0 MBtu/hr

(c) Tube quantity, size, and grouping:

Screen
- 2 rows of $2\frac{1}{2}$-in. OD tubes, approx 18 ft long
- Rows in line and spaced on 6-in. centers
- 23 tubes per row spaced on 6-in. centers
- $S = 542$ sqft
- $A_g = 129$ sqft

Example 1.2—(*continued*)

Superheater
- 12 rows of $2\frac{1}{2}$-in. OD tubes (0.165-in. thick), 17.44 ft long
- Rows in line and spaced on $3\frac{1}{4}$-in. centers
- 23 tubes per row spaced on 6-in. centers
- $S = 3150$ sqft
- $A_g = 133$ sqft

Boiler
- 25 rows of $2\frac{1}{2}$-in. OD tubes, approx 18 ft long
- Rows in line and spaced on $3\frac{1}{4}$-in. centers
- 35 tubes per row spaced on 4-in. centers
- $S = 10{,}300$ sqft
- $A_g = 85.0$ sqft

Economizer
- 10 rows of 2-in. OD tubes (0.148-in. thick), approx 10 ft long
- Rows in line and spaced on 3-in. centers
- 47 tubes per row spaced on 3-in. centers
- $S = 2460$ sqft
- $A_g = 42$ sqft

Air heater
- 53 rows of 2-in. OD tubes (0.083-in. thick), approx 13 ft long
- Rows in line and spaced on $2\frac{1}{2}$-in. centers
- 41 tubes per row spaced on $3\frac{1}{2}$-in. centers
- $S = 14{,}800$ sqft
- A_g (total internal cross section area of 2173 tubes) $= 39.3$ sqft
- A_a (clear area between tubes for crossflow of air) $= 70$ sqft
- Air temperature entering air heater = 80°F

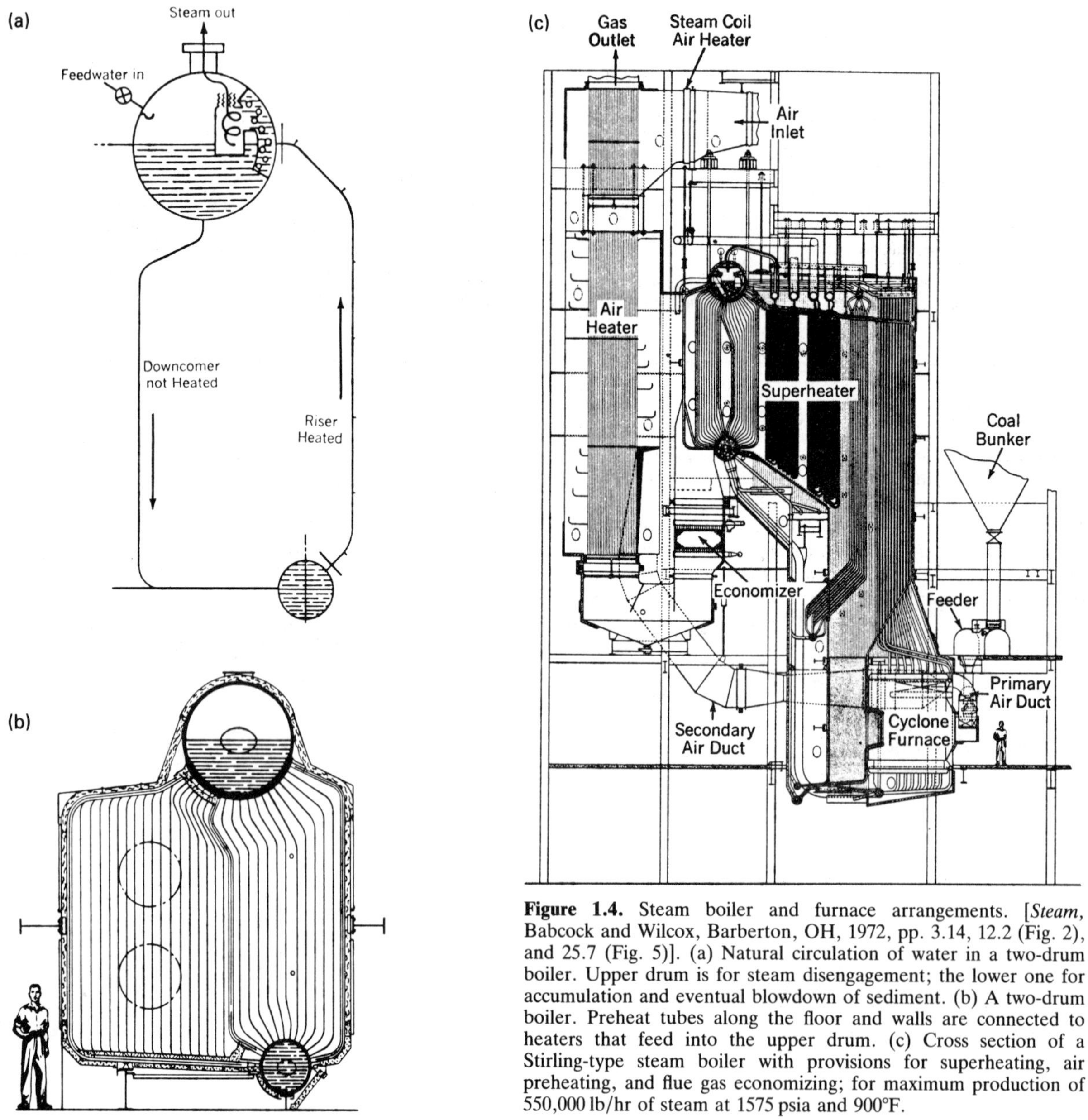

Figure 1.4. Steam boiler and furnace arrangements. [*Steam*, Babcock and Wilcox, Barberton, OH, 1972, pp. 3.14, 12.2 (Fig. 2), and 25.7 (Fig. 5)]. (a) Natural circulation of water in a two-drum boiler. Upper drum is for steam disengagement; the lower one for accumulation and eventual blowdown of sediment. (b) A two-drum boiler. Preheat tubes along the floor and walls are connected to heaters that feed into the upper drum. (c) Cross section of a Stirling-type steam boiler with provisions for superheating, air preheating, and flue gas economizing; for maximum production of 550,000 lb/hr of steam at 1575 psia and 900°F.

EXAMPLE 1.3
Steam Plant Cycle for Generation of Power and Low Pressure Process Steam

The flow diagram is for the production of 5000 kW gross and 20,000 lb/hr of saturated process steam at 20 psia. The feed and hot well pumps make the net power production 4700 kW. Conditions at key points are indicated on the enthalpy–entropy diagram. The process steam is extracted from the turbine at an intermediate point, while the rest of the stream expands to 1 in. Hg and is condensed (example is corrected from *Chemical Engineers Handbook,* 5th ed., 9.48, McGraw-Hill, New York, 1973).

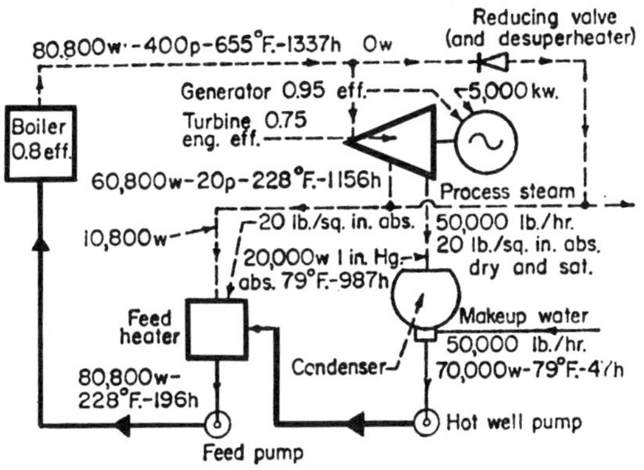

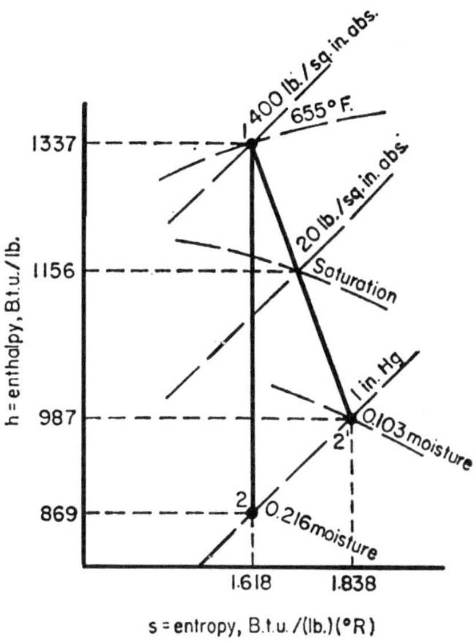

EXAMPLE 1.4
Pickup of Waste Heat by Generating and Superheating Steam in a Petroleum Refinery

The two examples are generation of steam with heat from a sidestream of a fractionator in a 9000 Bbl/day fluid cracking plant, and superheating steam with heat from flue gases of a furnace whose main function is to supply heat to crude topping and vacuum service in a 20,000 Bbl/day plant. (a) Recovery of heat from a sidestream of a fractionator in a 9000 Bbl/day fluid catalytic cracker by generating steam, $Q = 15{,}950{,}000$ Btu/hr. (b) Heat recovery by superheating steam with flue gases of a 20,000 Bbl/day crude topping and vacuum furnace.

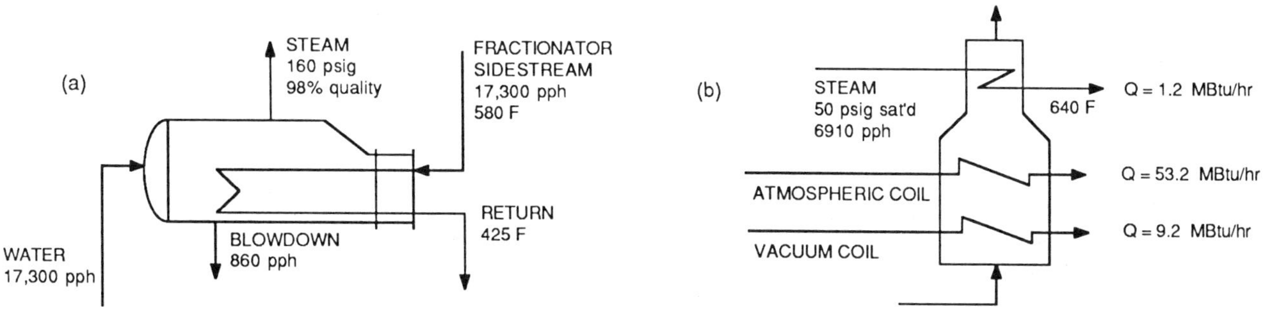

are of steam generation in a kettle reboiler with heat from a fractionator sidestream and of steam superheating in the convection tubes of a furnace that provides heat to fractionators.

Recovery of power from the thermal energy of a high temperature stream is the subject of Example 1.5. A closed circuit of propane is the indirect means whereby the power is recovered with an expansion turbine. Recovery of power from a high pressure gas is a fairly common operation. A classic example of power recovery from a high pressure liquid is in a plant for the absorption of CO_2 by water at a pressure of about 4000 psig. After the absorption, the CO_2 is released and power is recovered by releasing the rich liquor through a turbine.

Example 1.5
Recovery of Power from a Hot Gas Stream
A closed circuit of propane is employed for indirect recovery of power from the thermal energy of the hot pyrolyzate of an ethylene plant. The propane is evaporated at 500 psig, and then expanded to 100°F and 190 psig in a turbine where the power is recovered. Then the propane is condensed and pumped back to the evaporator to complete the cycle. Since expansion turbines are expensive machines even in small sizes, the process is not economical on the scale of this example, but may be on a much larger scale.

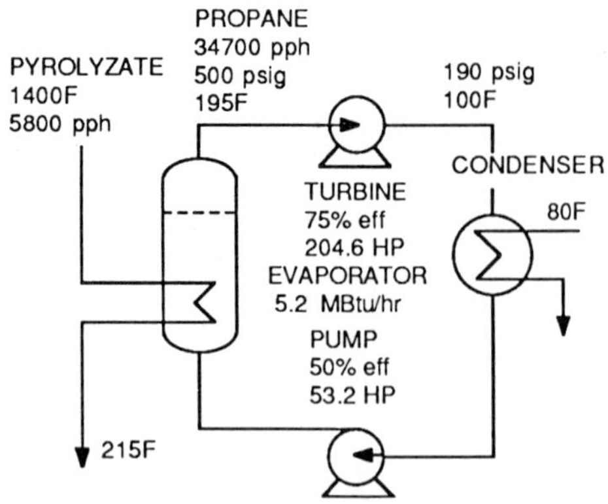

1.11. DESIGN BASIS

Before a chemical process design can be properly embarked on, a certain body of information must be agreed upon by all concerned persons, in addition to the obvious what is to be made and what it is to be made from. Distinctions may be drawn between plant expansions and wholly independent ones, so-called grassroots types. The needed data can be classified into specific design data and basic design data, for which separate check lists will be described. Specific design data include:

1. Required products: their compositions, amounts, purities, toxicities, temperatures, pressures, and monetary values.
2. Available raw materials: their compositions, amounts, toxicities, temperatures, pressures, monetary values, and all pertinent physical properties unless they are standard and can be established from correlations. This information about properties applies also to products of item 1.
3. Daily and seasonal variations of any data of items 1 and 2 and subsequent items of these lists.
4. All available laboratory and pilot plant data on reaction and phase equilibrium behaviors, catalyst degradation, and life and corrosion of equipment.
5. Any available existing plant data of similar processes.
6. Local restrictions on means of disposal of wastes.

Basic engineering data include:

7. Characteristics and values of gaseous and liquid fuels that are to be used.
8. Characteristics of raw makeup and cooling tower waters, temperatures, maximum allowable temperature, flow rates available, and unit costs.
9. Steam and condensate: mean pressures and temperatures and their fluctuations at each level, amount available, extent of recovery of condensate, and unit costs.
10. Electrical power: Voltages allowed for instruments, lighting and various driver sizes, transformer capacities, need for emergency generator, unit costs.
11. Compressed air: capacities and pressures of plant and instrument air, instrument air dryer.
12. Plant site elevation.
13. Soil bearing value, frost depth, ground water depth, piling requirements, available soil test data.
14. Climatic data. Winter and summer temperature extrema, cooling tower drybulb temperature, air cooler design temperature, strength and direction of prevailing winds, rain and snowfall maxima in 1 hr and in 12 hr, earthquake provision.
15. Blowdown and flare: What may or may not be vented to the atmosphere or to ponds or to natural waters, nature of required liquid, and vapor relief systems.
16. Drainage and sewers: rainwater, oil, sanitary.
17. Buildings: process, pump, control instruments, special equipment.
18. Paving types required in different areas.
19. Pipe racks: elevations, grouping, coding.
20. Battery limit pressures and temperatures of individual feed stocks and products.
21. Codes: those governing pressure vessels, other equipment, buildings, electrical, safety, sanitation, and others.
22. Miscellaneous: includes heater stacks, winterizing, insulation, steam or electrical tracing of lines, heat exchanger tubing size standardization, instrument locations.

A convenient tabular questionnaire is in Table 1.8. For anything not specified, for instance, sparing of equipment, engineering standards of the designer or constructor will be used. A proper design basis at the very beginning of a project is essential to getting a project completed and on stream expeditiously.

UTILITIES

These provide motive power and heating and cooling of process streams, and include electricity, steam, fuels, and various fluids whose changes in sensible and latent heats provide the necessary energy transfers. In every plant, the conditions of the utilities are maintained at only a few specific levels, for instance, steam at certain pressures, cooling water over certain temperature ranges, and electricity at certain voltages. At some stages of some design work, the specifications of the utilities may not have been established. Then, suitable data may be selected from the commonly used values itemized in Table 1.9.

1.12. LABORATORY AND PILOT PLANT WORK

The need for knowledge of basic physical properties as a factor in equipment selection or design requires no stressing. Beyond this, the state-of-the-art of design of many kinds of equipment and

TABLE 1.8. Typical Design Basis Questionnaire

1.101 Plant Location ______

1.102 Plant Capacity, lb or tons/yr ______

1.103 Operating Factor or Yearly Operating Hours
(For most modern chemical plants, this figure is generally 8,000 hours per year).

1.104 Provisions for Expansion ______

______

1.105 Raw Material Feed (Typical of the analyses required for a liquid)

- Assay, wt per cent min ______
- Impurities, wt per cent max ______
- Characteristic specifications
- Specific gravity ______
- Distillation range °F ______
- Initial boiling point °F ______
- Dry end point °F ______
- Viscosity, centipoises ______
- Color APHA ______
- Heat stability color ______
- Reaction rate with established reagent ______
- Acid number ______
- Freezing point or set point °F ______
- Corrosion test ______
- End-use test ______

For a solid material chemical assay, level of impurities and its physical characteristics, such as specific density, bulk density, particle size distribution and the like are included. This physical shape information is required to assure that adequate processing and material handling operations will be provided.

1.1051 Source

Supply conditions at process	Max	Min	Normal
plant battery limits	______	______	______
Storage capacity (volume or day's inventory)	______	______	______

Required delivery conditions at battery limits

- Pressure ______
- Temperature ______
- Method of transfer ______

1.106 Product Specifications

Here again specifications would be similar to that of the raw material in equivalent or sometimes greater detail as often trace impurities affect the marketability of the final product.

Storage requirements (volume or days of inventory) ______

Type of product storage ______

For solid products, type of container or method of shipment and loading facilities should be outlined. ______

1.107 Miscellaneous Chemicals and Catalyst Supply

In this section the operating group should outline how various miscellaneous chemicals and catalysts are to be stored and handled for consumption within the plant.

1.108 Atmospheric Conditions

Barometric pressure range ______

Temperature

- Design dry bulb temperature (°F) ______
- % of summer season, this temperature is exceeded. ______
- Design wet bulb temperature ______

% of summer season, this temperature is exceeded. ______

Minimum design dry bulb temperature winter condition (°F) ______

Level of applicable pollutants that could affect the process. Examples of these are sulfur compounds, dust and solids, chlorides and salt water mist when the plant is at a coastal location. ______

2.100 Utilities

2.101 Electricity

- Characteristics of primary supply ______
- Voltage, phases, cycles ______
- Preferred voltage for motors
- Over 200 hp ______
- Under 200 hp ______
- Value, ¢/kWh ______

(If available and if desired, detailed electricity pricing schedule can be included for base load and incremental additional consumption.)

2.102 Supply Water ______

Cleanliness ______

Corrosiveness ______

Solids content analysis ______

______

Other details ______

______

Pressure (at grade)	Maximum	Minimum
Supply	______	______
Return	______	______

2.103 Cooling Water

Well, river, sea, cooling tower, other. ______

Quality ______

Value ______

TABLE 1.8—*(continued)*

Use for heat exchanger design			
Fouling properties	___		
Design fouling factor	___		
Preferred tube material	___		
2.104 Steam	**Max**	**Normal**	**Min**
High pressure, psig	___	___	___
Temperature, °F	___	___	___
Moisture, %	___	___	___
Value per thousand lb	___	___	___
Medium pressure, psig	___	___	___
Temperature, °F	___	___	___
Moisture, %	___	___	___
Value per thousand lb	___	___	___
Low pressure, psig	___	___	___
Temperature, °F	___	___	___
Moisture, %	___	___	___
Value per thousand lb	___	___	___
2.105 Steam Condensate			
Disposition	___		
Required pressure at battery limits	___		
Value per thousand lb or gal	___		
2.106 Boiler Feed Water			
Quality			
Hardness, ppm	___		
Silica content	___		
Hardness			
Total solids, ppm	___		
Other details	___		
Chemical additives	___		
	Max		**Min**
Supply pressure	___		___
Temperature, °F	___		___
Value per thousand gal	___		___
2.107 Process Water			
(If the quality of the process water is different from the make-up water or boiler feed water, separate information should be provided.)			
Quality	**Max**		**Min**
Supply pressure, psig	___		___
Temperature, °F	___		___
Value per thousand gal	___		___

2.108 Inert Gas	**Max**	**Min**
Pressure, psig	___	___
Dew point, °F	___	___
Composition		
Per cent CO_2	___	
Per cent oxygen	___	
Per cent CO	___	
Other trace impurities	___	
Quantity available	___	
Value per thousand cu ft	___	
2.109 Plant Air		
Supply Source		
Offsite battery limits (OSBL)	___	
Portable compressor	___	
Process air system	___	
Special compressor	___	
Supply pressure, psig		
2.110 Instrument Air		
Supply source (OSBL)	___	
Special compressor	___	
Supply pressure, psig	___	
Dew point, °F	___	
Oil, dirt and moisture removal requirements	___	

In general a value of plant and instrument air is usually not given as the yearly over-all cost is insignificant in relation to the other utilities required.

3.101 Waste Disposal Requirements

In general, there are three types of waste to be considered: liquid, solid and gaseous. The destination and disposal of each of these effluents is usually different. Typical items are as follows:

Destination of liquid effluents	
Cooling water blowdown	___
Chemical sewer	___
Storm sewer	___
Method of chemical treating for liquid effluents	___
Preferred materials of construction for	
Cooling water blowdown	___
Chemical sewer	___
Storm sewer	___
Facilities for chemical treating for liquid effluents	___
Facilities for treatment of gaseous effluents	___
Solids disposal	___

(Landau, *The Chemical Plant*, Reinhold, New York, 1966).

TABLE 1.9. Typical Utility Characteristics

Steam		
Pressure (psig)	Saturation (°F)	Superheat (°F)
15–30	250–275	
150	366	
400	448	
600	488	100–150

Heat Transfer Fluids	
°F	Fluid
Below 600	petroleum oils
Below 750	Dowtherm and others
Below 1100	fused salts
Above 450	direct firing and electrical heating

Refrigerants	
°F	Fluid
40–80	chilled water
0–50	chilled brine and glycol solutions
−50–40	ammonia, freons, butane
−150–−50	ethane or propane
−350–−150	methane, air, nitrogen
−400–−300	hydrogen
Below −400	helium

Cooling Water
Supply at 80–90°F
Return at 115°F, with 125°F maximum
Return at 110°F (salt water)
Return above 125°F (tempered water or steam condensate)

Cooling Air
Supply at 85–95°F
Temperature approach to process, 40°F
Power input, 20 HP/1000 sqft of bare surface

Fuel
Gas: 5–10 psig, up to 25 psig for some types of burners, pipeline gas at 1000 Btu/SCF
Liquid: at 6 million Btu/barrel

Compressed Air
Pressure levels of 45, 150, 300, 450 psig

Instrument Air
45 psig, 0°F dewpoint

Electricity	
Driver HP	Voltage
1–100	220, 440, 550
75–250	440
200–2500	2300, 4000
Above 2500	4000, 13,200

processes often demands more or less extensive pilot plant effort. This point is stressed by specialists and manufacturers of equipment who are asked to provide performance guaranties. For instance, answers to equipment suppliers' questionnaires like those of Appendix C may require the potential purchaser to have performed certain tests. Some of the more obvious areas definitely requiring test work are filtration, sedimentation, spray, or fluidized bed or any other kind of solids drying, extrusion pelleting, pneumatic and slurry conveying, adsorption, and others. Even in such thoroughly researched areas as vapor–liquid and liquid–liquid separations, rates, equilibria, and efficiencies may need to be tested, particularly of complex mixtures. A great deal can be found out, for instance, by a batch distillation of a complex mixture.

In some areas, suppliers make available small scale equipment that can be used to explore suitable ranges of operating conditions, or they may do the work themselves with benefit of their extensive experience. One engineer in the extrusion pelleting field claims that merely feeling the stuff between his fingers enables him to properly specify equipment because of his experience of 25 years with extrusion.

Suitable test procedures often are supplied with "canned" pilot plants. In general, pilot plant experimentation is a profession in itself, and the more sophistication brought to bear on it the more efficiently can the work be done. In some areas the basic relations are known so well that experimentation suffices to evaluate a few parameters in a mathematical model. This is not the book to treat the subject of experimentation, but the literature is extensive. These books may be helpful to start:

1. R.E. Johnstone and M.W. Thring, *Pilot Plants, Models and Scale-up Methods in Chemical Engineering,* McGraw-Hill, New York, 1957.
2. D.G. Jordan, *Chemical Pilot Plant Practice,* Wiley-Interscience, New York, 1955.
3. V. Kafarov, *Cybernetic Methods in Chemistry and Chemical Engineering,* Mir Publishers, Moscow, 1976.
4. E.B. Wilson, *An Introduction to Scientific Research,* McGraw-Hill, New York, 1952.

REFERENCES

1.1. Process Design

A. Books Essential to a Private Library

1. Ludwig, *Applied Process Design for Chemical and Petroleum Plants,* Gulf, Houston 1977–1983, 3 vols.
2. *Marks Standard Handbook for Mechanical Engineers,* 9th ed., McGraw-Hill, New York, 1987.
3. Perry, Green, and Maloney, *Perry's Chemical Engineers Handbook,* McGraw-Hill, New York, 1984; earlier editions have not been obsolesced entirely.
4. Sinnott, Coulson, and Richardsons, *Chemical Engineering, Vol. 6, Design,* Pergamon, New York, 1983.

B. Other Books

1. Aerstin and Street, *Applied Chemical Process Design,* Plenum, New York, 1978.
2. Baasel, *Preliminary Chemical Engineering Plant Design,* Elsevier, New York, 1976.

3. Backhurst and Harker, *Process Plant Design,* Elsevier, New York, 1973.
4. Benedek (Ed.), *Steady State Flowsheeting of Chemical Plants,* Elsevier, New York, 1980.
5. Bodman, *The Industrial Practice of Chemical Process Engineering,* MIT Press, Cambridge, MA, 1968.
6. Branan, *Process Engineers Pocket Book,* Gulf, Houston, 1976, 1983, 2 vols.
7. Burklin, *The Process Plant Designers Pocket Handbook of Codes and Standards,* Gulf, Houston, 1979; also, Design codes standards and recommended practices, *Encycl. Chem. Process. Des.* **14,** 416–431, Dekker, New York, 1982.
8. Cremer and Watkins, *Chemical Engineering Practice,* Butterworths, London, 1956–1965, 12 vols.
9. Crowe et al., *Chemical Plant Simulation,* Prentice-Hall, Englewood Cliffs, NJ, 1971.
10. F.L. Evans, *Equipment Design Handbook for Refineries and Chemical Plants,* Gulf, Houston, 1979, 2 vols.
11. Franks, *Modelling and Simulation in Chemical Engineering,* Wiley, New York, 1972.
12. Institut Française du Petrole, *Manual of Economic Analysis of Chemical Processes,* McGraw-Hill, New York, 1981.
13. Kafarov, *Cybernetic Methods in Chemistry and Chemical Engineering,* Mir Publishers, Moscow, 1976.
14. Landau (Ed.), *The Chemical Plant,* Reinhold, New York, 1966.
15. Leesley (Ed.), *Computer-Aided Process Plant Design,* Gulf, Houston, 1982.
16. Lieberman, *Process Design for Reliable Operations,* Gulf, Houston, 1983.
17. Noel, *Petroleum Refinery Manual,* Reinhold, New York, 1959.
18. Peters and Timmerhaus, *Plant Design and Economics for Chemical Engineers,* McGraw-Hill, New York, 1980.
19. Rase and Barrow, *Project Engineering of Process Plants,* Wiley, New York, 1957.
20. Resnick, *Process Analysis and Design for Chemical Engineers,* McGraw-Hill, New York, 1981.
21. Rudd and Watson, *Strategy of Process Engineering,* Wiley, New York, 1968.
22. Schweitzer (Ed.), *Handbook of Separation Processes for Chemical Engineers,* McGraw-Hill, New York, 1979.
23. Sherwood, *A Course in Process Design,* MIT Press, Cambridge, MA, 1963.
24. Ulrich, *A Guide to Chemical Engineering Process Design and Economics,* Wiley, New York, 1984.
25. Valle-Riestra, *Project Evaluation in the Chemical Process Industries,* McGraw-Hill, New York, 1983.
26. Vilbrandt and Dryden, *Chemical Engineering Plant Design,* McGraw-Hill, New York, 1959.
27. Wells, *Process Engineering with Economic Objective,* Leonard Hill, London, 1973.

C. Estimation of Properties

1. *AIChE Manual for Predicting Chemical Process Design Data,* AIChE, New York, 1984–date.
2. Bretsznajder, *Prediction of Transport and Other Physical Properties of Fluids,* Pergamon, New York, 1971; larger Polish edition, Warsaw, 1962.
3. Lyman, Reehl, and Rosenblatt, *Handbook of Chemical Property Estimation Methods: Environmental Behavior of Organic Compounds,* McGraw-Hill, New York, 1982.
4. Reid, Prausnitz, and Poling, *The Properties of Gases and Liquids,* McGraw-Hill, New York, 1987.
5. Sterbacek, Biskup, and Tausk, *Calculation of Properties Using Corresponding States Methods,* Elsevier, New York, 1979.
6. S.M. Walas, *Phase Equilibria in Chemical Engineering,* Butterworths, Stoneham, MA, 1984.

D. Equipment

1. *Chemical Engineering Catalog,* Penton/Reinhold, New York, annual.
2. *Chemical Engineering Equipment Buyers' Guide,* McGraw-Hill, New York, annual.
3. Kieser, *Handbuch der chemisch-technischen Apparate,* Spamer-Springer, Berlin, 1934–1939.
4. Mead, *The Encyclopedia of Chemical Process Equipment,* Reinhold, New York, 1964.
5. Riegel, *Chemical Process Machinery,* Reinhold, New York, 1953.
6. *Thomas Register of American Manufacturers,* Thomas, Springfield IL, annual.

E. Safety Aspects

1. Fawcett and Wood (Eds.), *Safety and Accident Prevention in Chemical Operations,* Wiley, New York, 1982.
2. Lees, *Loss Prevention in the Process Industries,* Butterworths, London, 1980, 2 vols.
3. Lieberman, *Troubleshooting Refinery Processes,* PennWell, Tulsa, 1981.
4. Lund, *Industrial Pollution Control Handbook,* McGraw-Hill, New York, 1971.
5. Rosaler and Rice, *Standard Handbook of Plant Engineering,* McGraw-Hill, New York, 1983.
6. Sax, *Dangerous Properties of Industrial Materials,* Van Nostrand/Reinhold, New York, 1982.
7. Wells, *Safety in Process Plant Design,* George Godwin, Wiley, New York, 1980.

1.2. Process Equipment

A. Encyclopedias

1. Considine, *Chemical and Process Technology Encyclopedia,* McGraw-Hill, New York, 1974.
2. *Kirk-Othmer Concise Encyclopedia of Chemical Technology,* Wiley, New York, 1985.
3. *Kirk-Othmer Encyclopedia of Chemical Technology,* Wiley, New York, 1978–1984, 26 vols.
4. *McGraw-Hill Encyclopedia of Science and Technology,* 5th ed., McGraw-Hill, New York, 1982.
5. McKetta and Cunningham (Eds.), *Encyclopedia of Chemical Processing and Design,* Dekker, New York, 1976–date.
6. Ullmann, *Encyclopedia of Chemical Technology,* Verlag Chemie, Weinheim, FRG, German edition 1972–1983; English edition 1984–1994(?).

B. Bibliographies

1. Fratzcher, Picht, and Bittrich, The acquisition, collection and tabulation of substance data on fluid systems for calculations in chemical engineering, *Int. Chem. Eng.* **20**(1), 19–28 (1980).
2. Maizell, *How to Find Chemical Information,* Wiley, New York, 1978.
3. Mellon, *Chemical Publications: Their Nature and Use,* McGraw-Hill, New York, 1982.
4. Rasmussen and Fredenslund, *Data Banks for Chemical Engineers,* Kemiigeniorgruppen, Lyngby, Denmark, 1980.

C. General Data Collections

1. American Petroleum Institute, *Technical Data Book—Petroleum Refining,* API, Washington, DC, 1971–date.
2. Bolz and N. Tuve, *Handbook of Tables for Applied Engineering Science,* CRC Press, Washington, DC, 1972.
3. *CRC Handbook of Chemistry and Physics,* CRC Press, Washington, DC, annual.
4. Gallant, *Physical Properties of Hydrocarbons,* Gulf, Houston, 1968, 2 vols.
5. *International Critical Tables,* McGraw-Hill, New York, 1926–1933.
6. Landolt-Börnstein, *Numerical Data and Functional Relationships in Science and Technology,* Springer, New York, 1950–date.
7. *Lange's Handbook of Chemistry,* 13th ed., McGraw-Hill, New York, 1984.
8. Maxwell, *Data Book on Hydrocarbons,* Van Nostrand, New York, 1950.
9. Melnik and Melnikov, *Technology of Inorganic Compounds,* Israel Program for Scientific Translations, Jerusalem, 1970.
10. National Gas Processors Association, *Engineering Data Book,* Tulsa, 1987.
11. *Perry's Chemical Engineers Handbook,* McGraw-Hill, New York, 1984.
12. *Physico-Chemical Properties for Chemical Engineering,* Maruzen Co., Tokyo, 1977–date.

13. Raznjevic, *Handbook of Thermodynamics Tables and Charts (SI Units)*, Hemisphere, New York, 1976.
14. Vargaftik, *Handbook of Physical Properties of Liquids and Gases*, Hemisphere, New York, 1983.
15. Yaws et al., *Physical and Thermodynamic Properties*, McGraw-Hill, New York, 1976.

D. Special Data Collections

1. Gmehling et al., *Vapor–Liquid Equilibrium Data Collection*, DECHEMA, Frankfurt/Main, FRG, 1977–date.
2. Hirata, Ohe, and Nagahama, *Computer-Aided Data Book of Vapor–Liquid Equilibria*, Elsevier, New York, 1976.
3. Keenan et al., *Steam Tables*, Wiley, New York, English Units, 1969, SI Units, 1978.
4. Kehiaian, *Selected Data on Mixtures, International Data Series A: Thermodynamic Properties of Non-reacting Binary Systems of Organic Substances*, Texas A & M Thermodynamics Research Center, College Station, TX, 1977–date.
5. Kogan, Fridman, and Kafarov, *Equilibria between Liquid and Vapor* (in Russian), Moscow, 1966.
6. Larkin, *Selected Data on Mixtures, International Data Series B, Thermodynamic Properties of Organic Aqueous Systems*, Engineering Science Data Unit Ltd, London, 1978–date.
7. Ogorodnikov, Lesteva, and Kogan, *Handbook of Azeotropic Mixtures* (in Russian), Moscow, 1971; data of 21,069 systems.
8. Ohe, *Computer-Aided Data Book of Vapor Pressure*, Data Publishing Co., Tokyo, 1976.
9. Sorensen and Arlt, *Liquid–Liquid Equilibrium Data Collection*, DECHEMA, Frankfurt/Main, FRG, 1979–1980, 3 vols.
10. Starling, *Fluid Thermodynamic Properties for Light Petroleum Systems*, Gulf, Houston, 1973.
11. Stephen, Stephen and Silcock, *Solubilities of Inorganic and Organic Compounds*, Pergamon, New York, 1979, 7 vols.
12. Stull, Westrum, and Sinke, *The Chemical Thermodynamics of Organic Compounds*, Wiley, New York, 1969.
13. Wagman et al., *The NBS Tables of Chemical Thermodynamic Properties: Selected Values for Inorganic and C_1 and C_2 Organic Substances in SI Units*, American Chemical Society, Washington, DC, 1982.

2

Flowsheets

A plant design is made up of words, numbers, and pictures. An engineer thinks naturally in terms of the sketches and drawings which are his "pictures." Thus, to solve a material balance problem, he will start with a block to represent the equipment and then will show entering and leaving streams with their amounts and properties. Or ask him to describe a process and he will begin to sketch the equipment, show how it is interconnected, and what the flows and operating conditions are.

Such sketches develop into flow sheets, which are more elaborate diagrammatic representations of the equipment, the sequence of operations, and the expected performance of a proposed plant or the actual performance of an already operating one. For clarity and to meet the needs of the various persons engaged in design, cost estimating, purchasing, fabrication, operation, maintenance, and management, several different kinds of flowsheets are necessary. Four of the main kinds will be described and illustrated.

2.1. BLOCK FLOWSHEETS

At an early stage or to provide an overview of a complex process or plant, a drawing is made with rectangular blocks to represent individual processes or groups of operations, together with quantities and other pertinent properties of key streams between the blocks and into and from the process as a whole. Such block flowsheets are made at the beginning of a process design for orientation purposes or later as a summary of the material balance of the process. For example, the coal carbonization process of Figure 2.1 starts with 100,000 lb/hr of coal and some process air, involves six main process units, and makes the indicated quantities of ten different products. When it is of particular interest, amounts of utilities also may be shown; in this example the use of steam is indicated at one point. The block diagram of Figure 2.2 was prepared in connection with a study of the modification of an existing petroleum refinery. The three feed stocks are separated into more than 20 products. Another representative petroleum refinery block diagram, in Figure 13.20, identifies the various streams but not their amounts or conditions.

2.2. PROCESS FLOWSHEETS

Process flowsheets embody the material and energy balances between and the sizing of the major equipment of the plant. They include all vessels such as reactors, separators, and drums; special processing equipment, heat exchangers, pumps, and so on. Numerical data include flow quantities, compositions, pressures, temperatures, and so on. Inclusion of major instrumentation that is essential to process control and to complete understanding of the flowsheet without reference to other information is required particularly during the early stages of a job, since the process flowsheet is drawn first and is for some time the only diagram representing the process. As the design develops and a mechanical flowsheet gets underway, instrumentation may be taken off the process diagram to reduce the clutter. A checklist of the information that usually is included on a process flowsheet is given in Table 2.1.

Working flowsheets are necessarily elaborate and difficult to represent on the page of a book. Figure 2.3 originally was 30 in. wide. In this process, ammonia is made from available hydrogen supplemented by hydrogen from the air oxidation of natural gas in a two-stage reactor F-3 and V-5. A large part of the plant is devoted to purification of the feed gases of carbon dioxide and unconverted methane before they enter the converter CV-1. Both commercial and refrigeration grade ammonia are made in this plant. Compositions of 13 key streams are summarized in the tabulation. Characteristics of the streams such as temperature, pressure, enthalpy, volumetric flow rates, etc., sometimes are conveniently included in the tabulation. In the interest of clarity, however, in some instances it may be preferable to have a separate sheet for a voluminous material balance and related stream information.

A process flowsheet of the dealkylation of toluene to benzene is in Figure 2.4; the material and enthalpy flows and temperature and pressures are tabulated conveniently, and basic instrumentation is represented.

2.3. MECHANICAL (P&I) FLOWSHEETS

Mechanical flowsheets also are called piping and instrument (P&I) diagrams to emphasize two of their major characteristics. They do not show operating conditions or compositions or flow quantities, but they do show all major as well as minor equipment more realistically than on the process flowsheet. Included are sizes and specification classes of all pipe lines, all valves, and all instruments. In fact, every mechanical aspect of the plant regarding the process equipment and their interconnections is represented except for supporting structures and foundations. The equipment is shown in greater detail than on the PFS, notably with regard to external piping connections, internal details, and resemblance to the actual appearance.

The mechanical flowsheet of the reaction section of a toluene dealkylation unit in Figure 2.5 shows all instrumentation, including indicators and transmitters. The clutter on the diagram is minimized by tabulating the design and operating conditions of the major equipment below the diagram.

The P&I diagram of Figure 2.6 represents a gas treating plant that consists of an amine absorber and a regenerator and their immediate auxiliaries. Internals of the towers are shown with exact locations of inlet and outlet connections. The amount of instrumentation for such a comparatively simple process may be surprising. On a completely finished diagram, every line will carry a code designation identifying the size, the kind of fluid handled, the pressure rating, and material specification. Complete information about each line—its length, size, elevation, pressure drop, fittings, etc.—is recorded in a separate line summary. On Figure 2.5, which is of an early stage of construction, only the sizes of the lines are shown. Although instrumentation symbols are fairly well standardized, they are often tabulated on the P&I diagram as in this example.

2.4. UTILITY FLOWSHEETS

These are P&I diagrams for individual utilities such as steam, steam condensate, cooling water, heat transfer media in general,

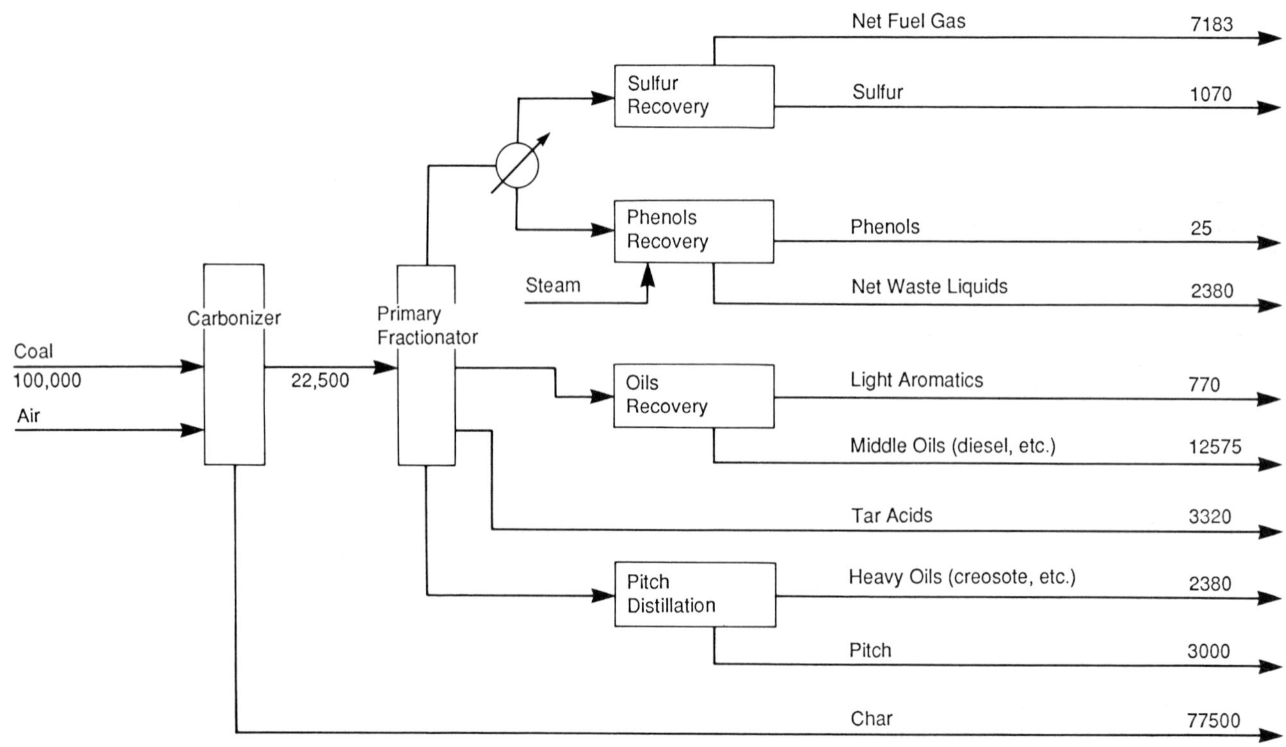

Figure 2.1. Coal carbonization block flowsheet. Quantities are in lb/hr.

compressed air, fuel, refrigerants, and inert blanketing gases, and how they are piped up to the process equipment. Connections for utility streams are shown on the mechanical flowsheet, and their conditions and flow quantities usually appear on the process flowsheet.

Since every detail of a plant design must be recorded on paper, many other kinds of drawings also are required: for example, electrical flow, piping isometrics, instrument lines, plans and elevations, and individual equipment drawings in all detail. Models and three-dimensional representations by computers also are now standard practice in many design offices.

2.5. DRAWING OF FLOWSHEETS

Flowsheets are intended to represent and explain processes. To make them easy to understand, they are constructed with a consistent set of symbols for equipment, piping, and operating conditions. At present there is no generally accepted industrywide body of drafting standards, although every large engineering office does have its internal standards. Some information appears in ANSI and British Standards publications, particularly of piping symbols. Much of this information is provided in the book by Austin (1979) along with symbols gleaned from the literature and some engineering firms. Useful compilations appear in some books on process design, for instance, those of Sinnott (1983) and Ulrich (1984). The many flowsheets that appear in periodicals such as *Chemical Engineering* or *Hydrocarbon Processing* employ fairly consistent sets of symbols that may be worth imitating.

Equipment symbols are a compromise between a schematic representation of the equipment and simplicity and ease of drawing. A selection for the more common kinds of equipment appears in Table 2.2. Less common equipment or any with especially intricate configuration often is represented simply by a circle or rectangle. Since a symbol does not usually speak entirely for itself but also carries a name and a letter-number identification, the flowsheet can be made clear even with the roughest of equipment symbols. The

TABLE 2.1. Checklist of Data Normally Included on a Process Flowsheet

1. Process lines, but including only those bypasses essential to an understanding of the process
2. All process equipment. Spares are indicated by letter symbols or notes
3. Major instrumentation essential to process control and to understanding of the flowsheet
4. Valves essential to an understanding of the flowsheet
5. Design basis, including stream factor
6. Temperatures, pressures, flow quantities
7. Weight and/or mol balance, showing compositions, amounts, and other properties of the principal streams
8. Utilities requirements summary
9. Data included for particular equipment
   a. Compressors: SCFM (60°F, 14.7 psia); ΔP psi; HHP; number of stages; details of stages if important
   b. Drives: type; connected HP; utilities such as kW, lb steam/hr, or Btu/hr
   c. Drums and tanks: ID or OD, seam to seam length, important internals
   d. Exchangers: Sqft, kBtu/hr, temperatures, and flow quantities in and out; shell side and tube side indicated
   e. Furnaces: kBtu/hr, temperatures in and out, fuel
   f. Pumps: GPM (60°F), ΔP psi, HHP, type, drive
   g. Towers: Number and type of plates or height and type of packing; identification of all plates at which streams enter or leave; ID or OD; seam to seam length; skirt height
   h. Other equipment: Sufficient data for identification of duty and size

TABLE 2.2. Flowsheet Equipment Symbols

Fluid Handling	Heat Transfer

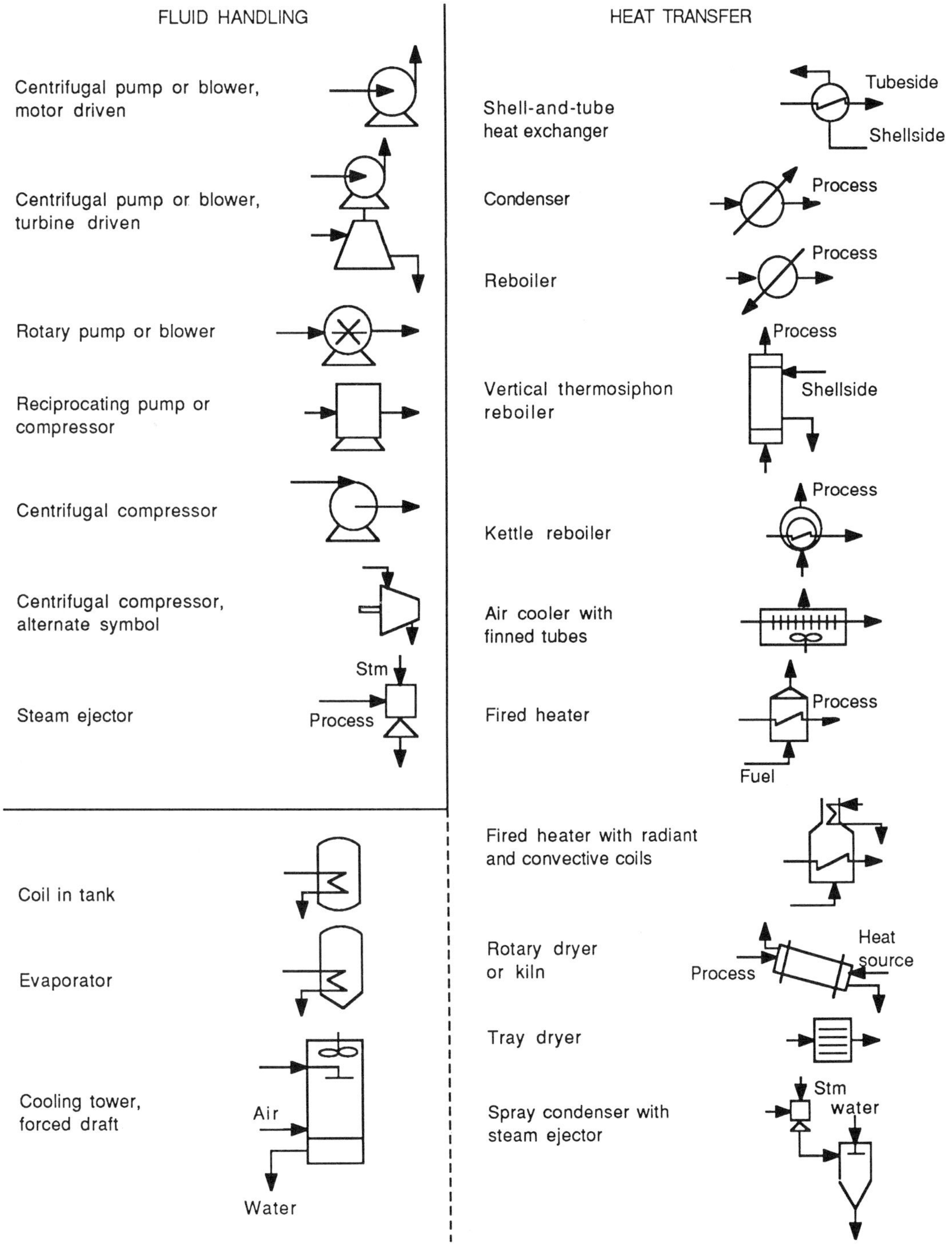
FLUID HANDLING
Centrifugal pump or blower, motor driven
Centrifugal pump or blower, turbine driven
Rotary pump or blower
Reciprocating pump or compressor
Centrifugal compressor
Centrifugal compressor, alternate symbol
Steam ejector
Stm
Process
Coil in tank
Evaporator
Cooling tower, forced draft
Air
Water
HEAT TRANSFER
Shell-and-tube heat exchanger
Tubeside
Shellside
Condenser
Process
Reboiler
Process
Vertical thermosiphon reboiler
Process
Shellside
Kettle reboiler
Process
Air cooler with finned tubes
Fired heater
Process
Fuel
Fired heater with radiant and convective coils
Rotary dryer or kiln
Process
Heat source
Tray dryer
Spray condenser with steam ejector
Stm
water

TABLE 2.2—(*continued*)

Mass Transfer	Vessels
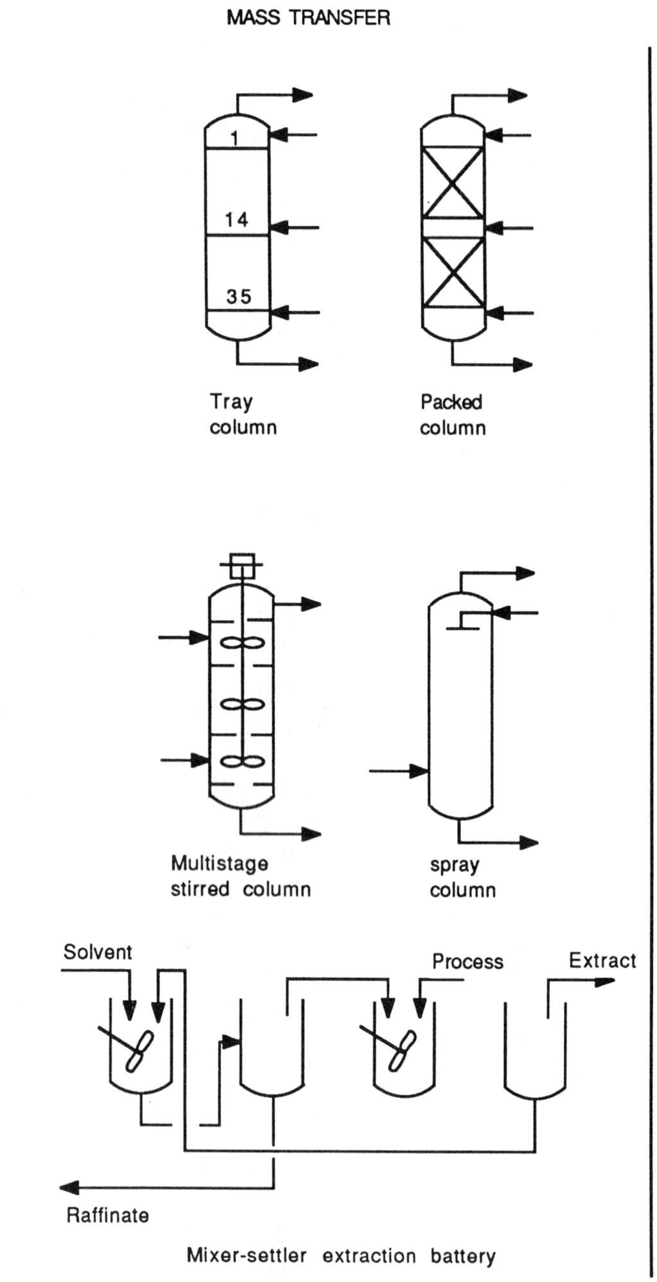	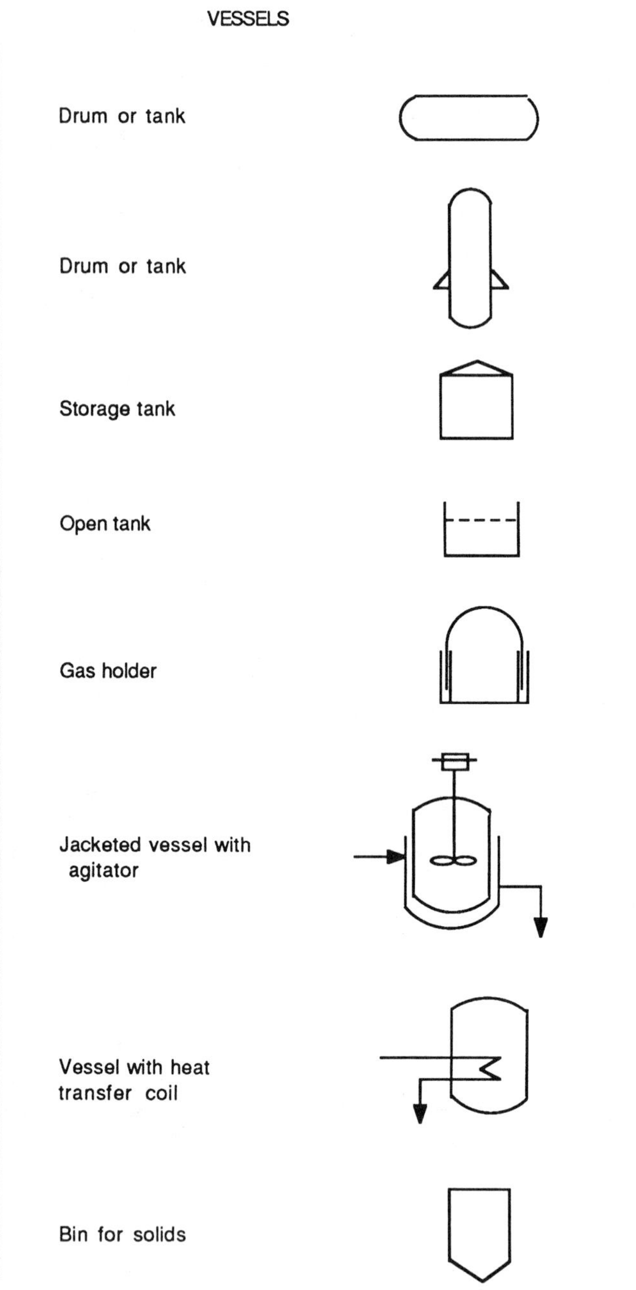

letter-number designation consists of a letter or combination to designate the class of the equipment and a number to distinguish it from others of the same class, as two heat exchangers by E-112 and E-215. Table 2.4 is a typical set of letter designations.

Operating conditions such as flow rate, temperature, pressure, enthalpy, heat transfer rate, and also stream numbers are identified with symbols called flags, of which Table 2.3 is a commonly used set. Particular units are identified on each flowsheet, as in Figure 2.3.

Letter designations and symbols for instrumentation have been

TABLE 2.2—(*continued*)

Conveyors and Feeders	Separators

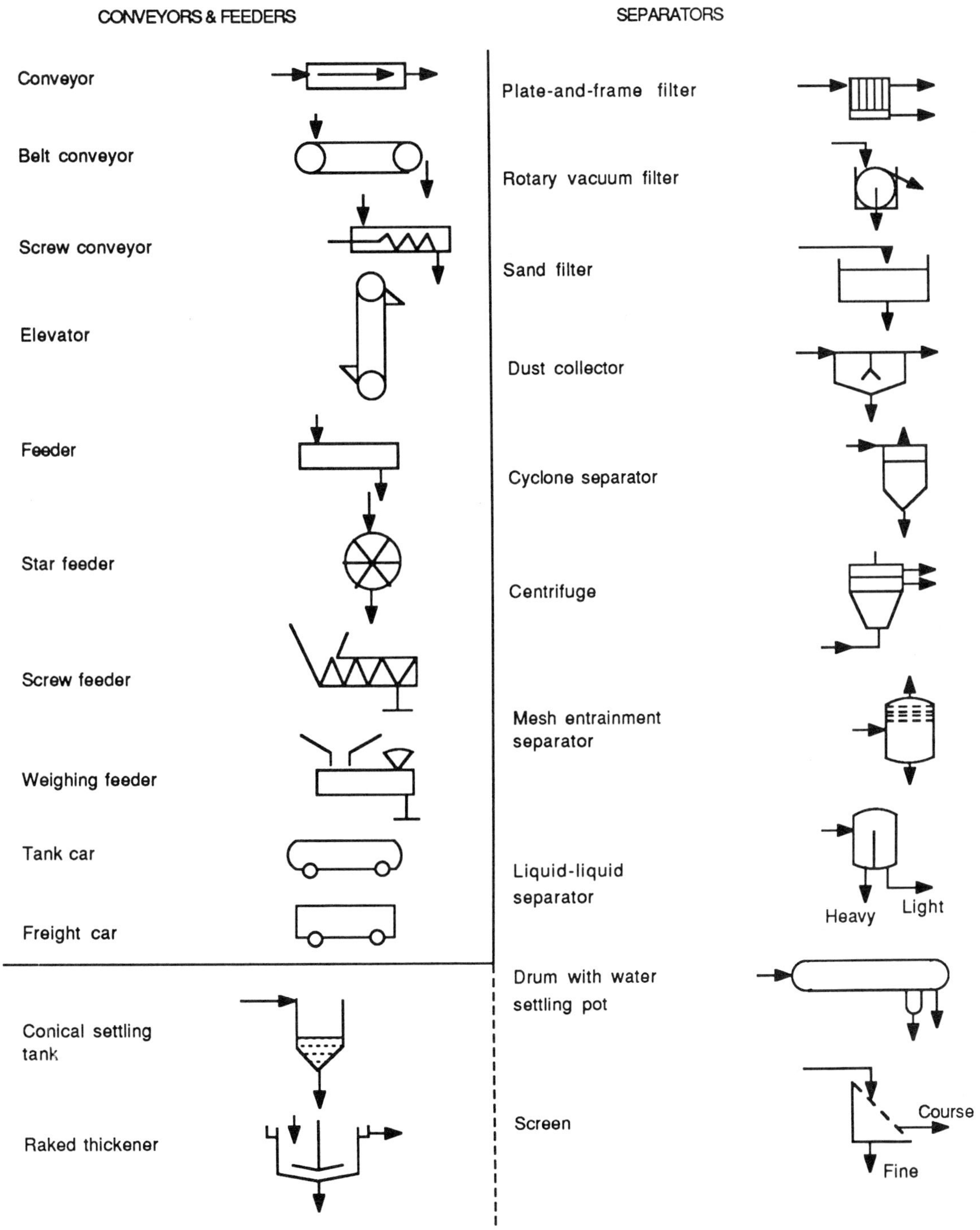

thoroughly standardized by the Instrument Society of America (ISA). An abbreviated set that may be adequate for the usual flowsketch appears on Figure 3.4. The P&I diagram of Figure 2.6 affords many examples.

For clarity and for esthetic reasons, equipment should be represented with some indication of their relative sizes. True scale is not feasible because, for example, a flowsheet may need to depict both a tower 150 ft high and a drum 2 ft in diameter. Logarithmic

TABLE 2.2—(*continued*)

Mixing and Comminution	Drivers

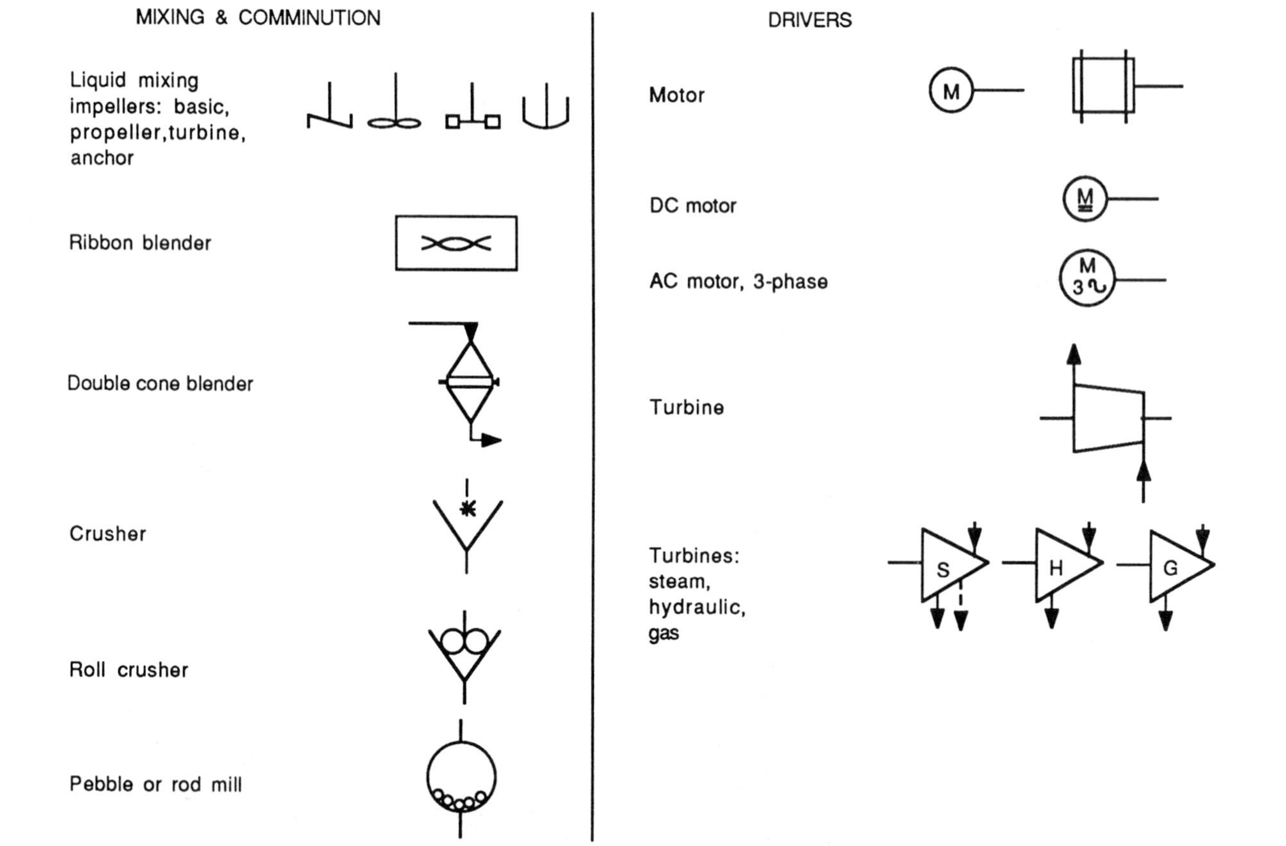

scaling sometimes gives a pleasing effect; for example, if the 150 ft tower is drawn 6 in. high and the 2 ft drum 0.5 in., other sizes can be read off a straight line on log–log paper.

A good draftsman will arrange his flowsheet as artistically as possible, consistent with clarity, logic, and economy of space on the drawing. A fundamental rule is that there be no large gaps. Flow is predominantly from left to right. On a process flowsheet, distillation towers, furnaces, reactors, and large vertical vessels often are arranged at one level, condenser and accumulator drums on another level, reboilers on still another level, and pumps more or less on one level but sometimes near the equipment they serve in order to minimize excessive crossing of lines. Streams enter the flowsheet from the left edge and leave at the right edge. Stream numbers are assigned to key process lines. Stream compositions and other desired properties are gathered into a table that may be on a separate sheet if it is especially elaborate. A listing of flags with the units is desirable on the flowsheet.

Rather less freedom is allowed in the construction of mechanical flowsheets. The relative elevations and sizes of equipment are preserved as much as possible, but all pumps usually are shown at the same level near the bottom of the drawing. Tabulations of instrumentation symbols or of control valve sizes or of relief valve sizes also often appear on P&I diagrams. Engineering offices have elaborate checklists of information that should be included on the flowsheet, but such information is beyond the scope here.

Appendix 2.1 provides the reader with material for the construction of flowsheets with the symbols of this chapter and possibly with some reference to Chapter 3.

TABLE 2.3. Flowsheet Flags of Operating Conditions in Typical Units

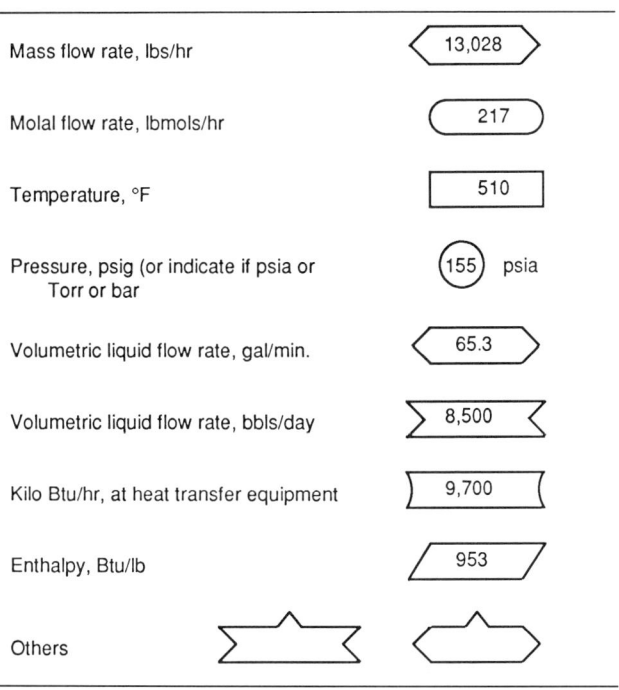

Mass flow rate, lbs/hr	13,028
Molal flow rate, lbmols/hr	217
Temperature, °F	510
Pressure, psig (or indicate if psia or Torr or bar	155 psia
Volumetric liquid flow rate, gal/min.	65.3
Volumetric liquid flow rate, bbls/day	8,500
Kilo Btu/hr, at heat transfer equipment	9,700
Enthalpy, Btu/lb	953
Others	

TABLE 2.4. Letter Designations of Equipment

Equipment	Letters	Equipment	Letters
Agitator	M	Grinder	SR
Air filter	FG	Heat exchanger	E
Bin	TT	Homogenizer	M
Blender	M	Kettle	R
Blower	JB	Kiln (rotary)	DD
Centrifuge	FF	Materials handling equipment	G
Classifying equipment	S	Miscellaneous[a]	L
Colloid mill	SR	Mixer	M
Compressor	JC	Motor	PM
Condenser	E	Oven	B
Conveyor	C	Packaging machinery	L
Cooling tower	TE	Precipitator (dust or mist)	FG
Crusher	SR	Prime mover	PM
Crystallizer	K	Pulverizer	SR
Cyclone separator (gas)	FG	Pump (liquid)	J
Cyclone separator (liquid)	F	Reboiler	E
Decanter	FL	Reactor	R
Disperser	M	Refrigeration system	G
Drum	D	Rotameter	RM
Dryer (thermal)	DE	Screen	S
Dust collector	FG	Separator (entrainment)	FG
Elevator	C	Shaker	M
Electrostatic separator	FG	Spray disk	SR
Engine	PM	Spray nozzle	SR
Evaporator	FE	Tank	TT
Fan	JJ	Thickener	F
Feeder	C	Tower	T
Filter (liquid)	P	Vacuum equipment	VE
Furnace	B	Weigh scale	L

[a] Note: The letter L is used for unclassified equipment when only a few items are of this type; otherwise, individual letter designations are assigned.

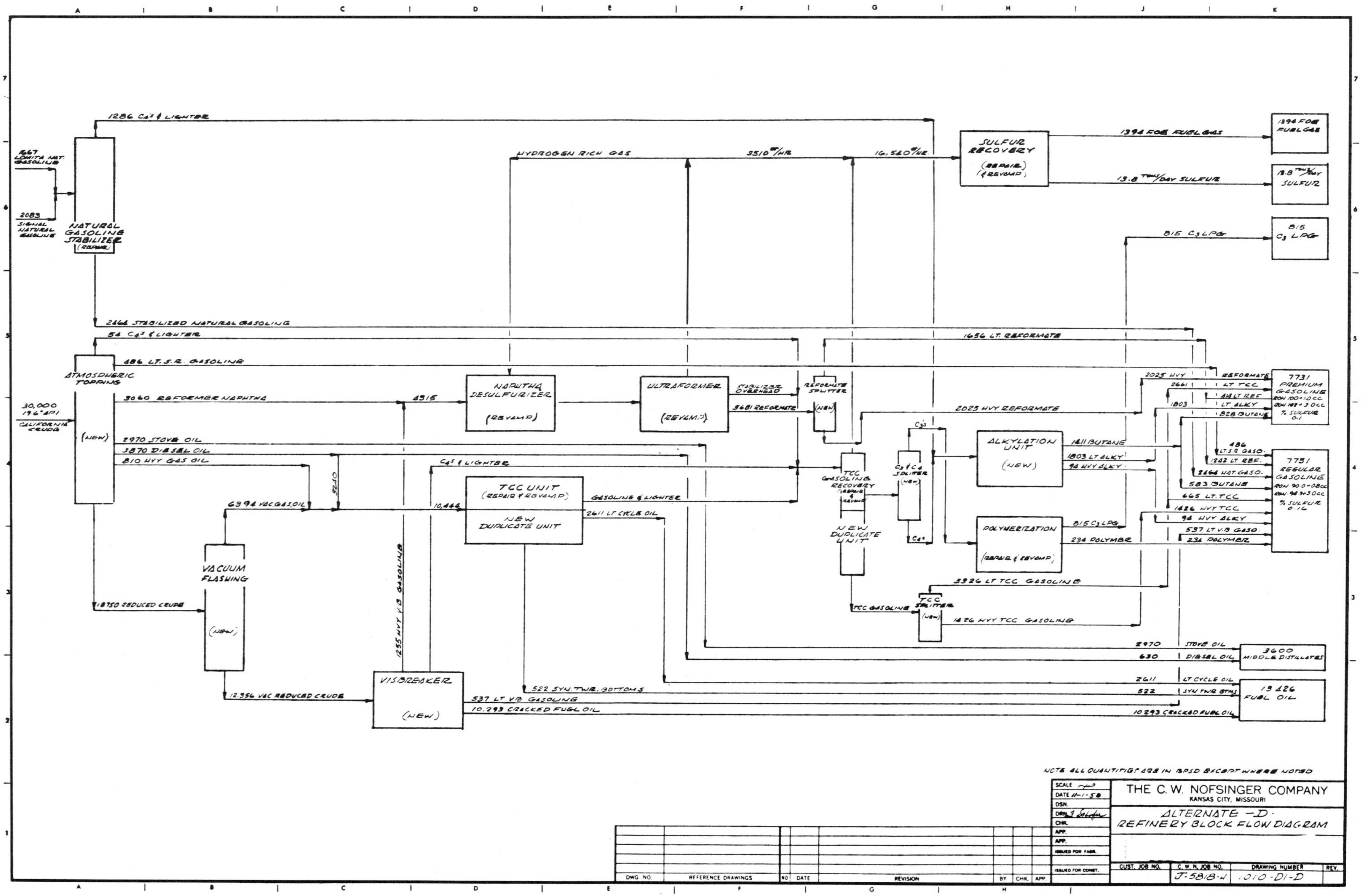

Figure 2.2. Block flowsheet of the revamp of a 30,000 Bbl/day refinery with supplementary light stocks (*The C. W. Nofsinger Co.*).

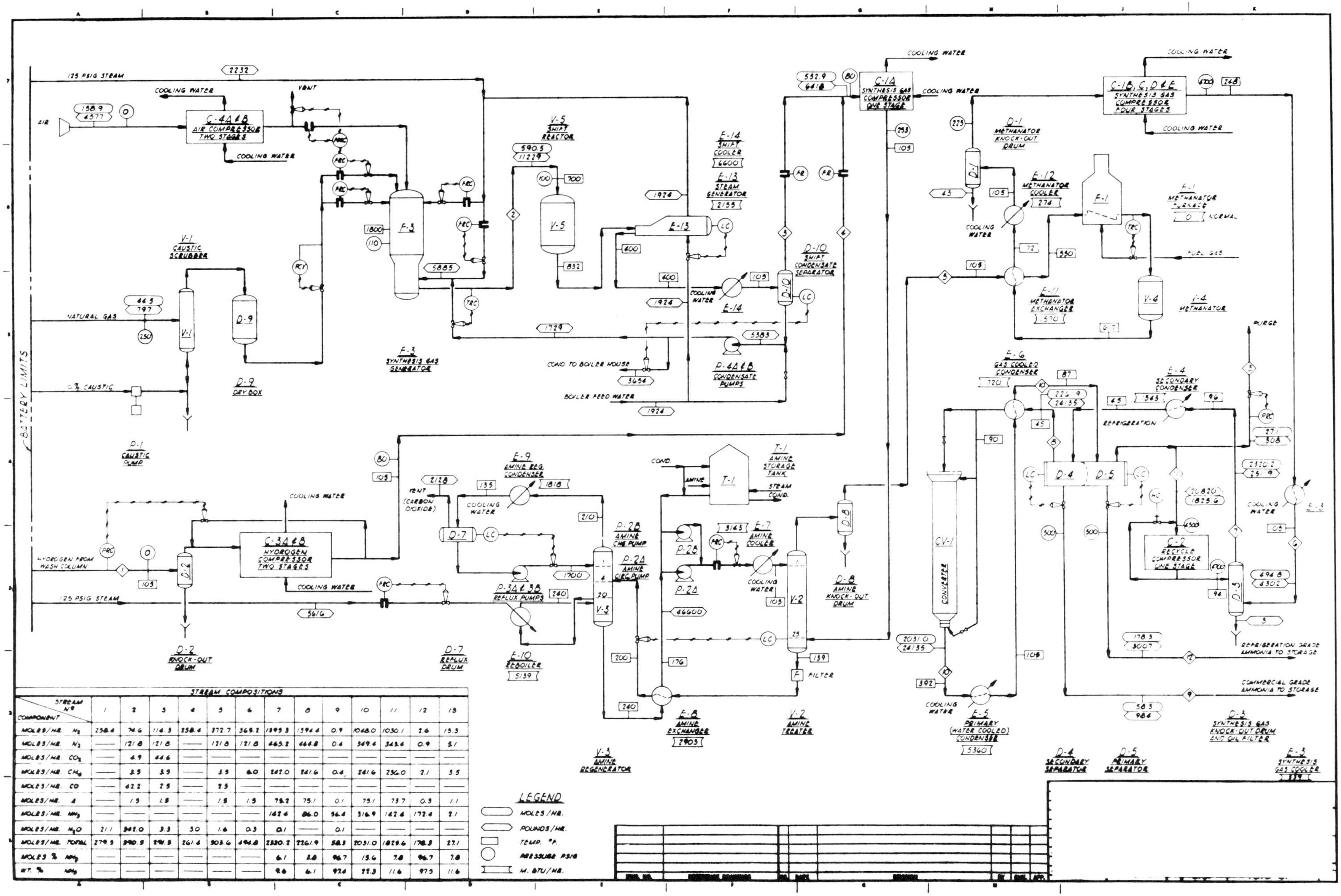

STREAM Nº / COMPONENT	1	2	3	4	5	6	7	8	9	10	11	12	13
MOLES/HR. H_2	258.4	74.6	114.3	258.4	372.7	369.2	1395.3	1394.4	0.9	1048.0	1030.1	2.6	15.3
MOLES/HR. N_2	—	121.8	121.8	—	121.8	121.8	465.2	464.8	0.4	349.4	343.4	0.9	5.1
MOLES/HR. CO_2	—	4.9	44.6	—	—	—	—	—	—	—	—	—	—
MOLES/HR. CH_4	—	3.5	3.5	—	3.5	6.0	247.0	241.6	0.4	241.6	236.0	2.1	3.5
MOLES/HR. CO	—	62.2	2.5	—	2.5	—	—	—	—	—	—	—	—
MOLES/HR. A	—	1.5	1.5	—	1.5	1.5	78.2	75.1	0.1	75.1	73.7	0.3	1.1
MOLES/HR. NH_3	—	—	—	—	—	—	162.4	86.0	56.4	316.9	162.4	172.4	2.1
MOLES/HR. H_2O	21.1	542.0	3.3	3.0	1.6	0.3	0.1	—	0.1	—	—	—	—
MOLES/HR. TOTAL	279.5	810.5	291.5	261.4	503.6	494.8	2350.2	2261.9	58.3	2031.0	1845.6	178.3	27.1
MOLES % NH_3	—	—	—	—	—	—	6.1	3.8	96.7	15.6	7.8	96.7	7.8
WT. % NH_3	—	—	—	—	—	—	9.6	6.1	97.4	22.3	11.6	97.5	11.6

Figure 2.3. Process flowsheet of a plant making 47 tons/day of ammonia from available hydrogen and hydrogen made from natural gas (*The C. W. Nofsinger Co.*).

Figure 2.4. Process flowsheet of the manufacture of benzene by dealkylation of toluene (*Wells,* Safety in Process Design, *George Godwin, London, 1980*).

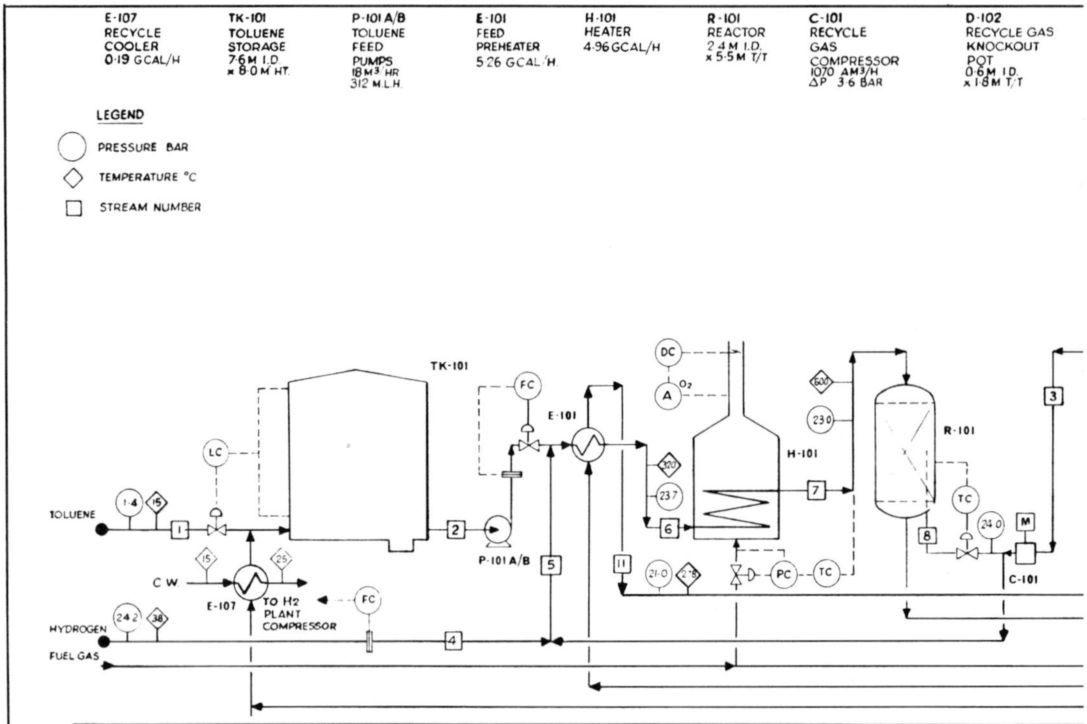

MATERIAL BALANCE

STREAM NUMBER		1	2	3	4	5	6	7	8	9	10
HYDROGEN	KMOL/H	0·0	0·0	470·80	285·00	730·80	730·80	730·80	25·00	647·30	647·30
METHANE	KMOL/H	0·0	0·0	329·40	15·00	32690	326·90	326·90	17·50	452·90	452·90
BENZENE	KMOL/H	0·0	0·40	6·10	0·00	5·80	6·20	6·20	0·30	115·00	115·00
TOLUENE	KMOL/H	108·70	143·90	0·43	0·00	0·41	144·31	144·31	0·02	35·83	35·83
TOTAL MOLAR FLOW	KMOL/H	108·70	144·30	806·73	300·00	1063·91	1208·21	1208·21	42·82	1251·03	1251·03
TOTAL MASS FLOW	KG/H	10,000	13,270	6727	810	7132	20452	20452	355	20 807	20 807
TEMPERATURE	°C	15	18	38	38	38	320	600	58	649	531
PRESSURE	BAR	1·4	1·1	20·4	24·0	24·0	23·7	23·0	24·0	22·0	21·2
TOTAL HEAT FLOW	GCALS/H	1·33	1·77	-3·36	0·40	-2·57	4·46	9·42	-0·17	9·13	7·24

Figure 2.5. Engineering (P&I) flowsheet of the reaction section of plant for dealkylation of benzene (*Wells,* Safety in Process Design, *George Godwin, London, 1980*).

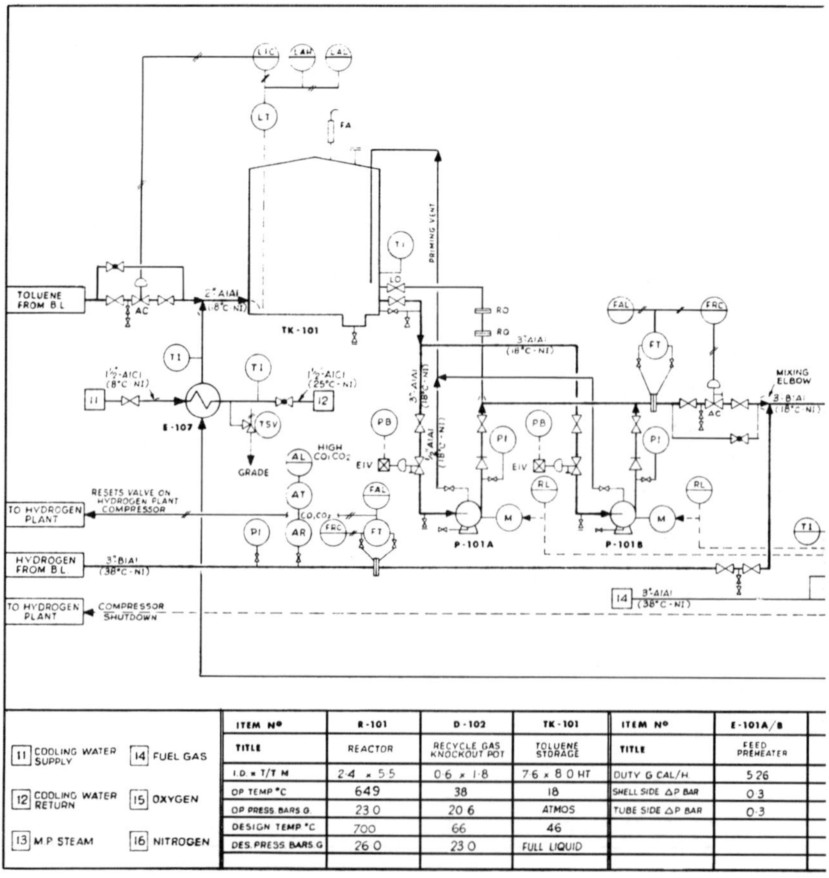

11	COOLING WATER SUPPLY	14	FUEL GAS
12	COOLING WATER RETURN	15	OXYGEN
13	M.P. STEAM	16	NITROGEN

ITEM Nº	R-101	D-102	TK-101	ITEM Nº	E-101A/B
TITLE	REACTOR	RECYCLE GAS KNOCKOUT POT	TOLUENE STORAGE	TITLE	FEED PREHEATER
I.D. × T/T M	2·4 × 5·5	0·6 × 1·8	7·6 × 8·0 HT	DUTY G CAL/H	5·26
OP TEMP °C	649	38	18	SHELL SIDE ΔP BAR	0·3
OP PRESS. BARS G.	23·0	20·6	ATMOS	TUBE SIDE ΔP BAR	0·3
DESIGN TEMP °C	700	66	46		
DES. PRESS. BARS G	26·0	23·0	FULL LIQUID		

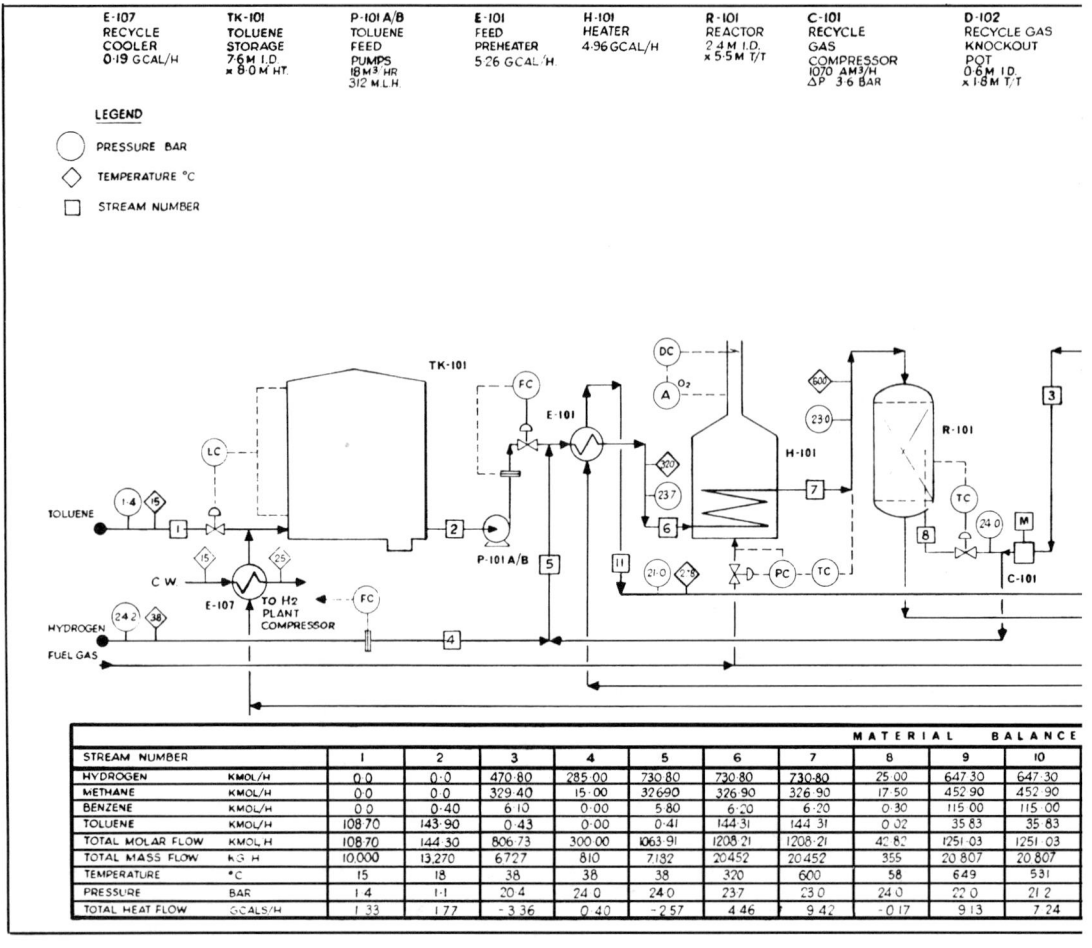

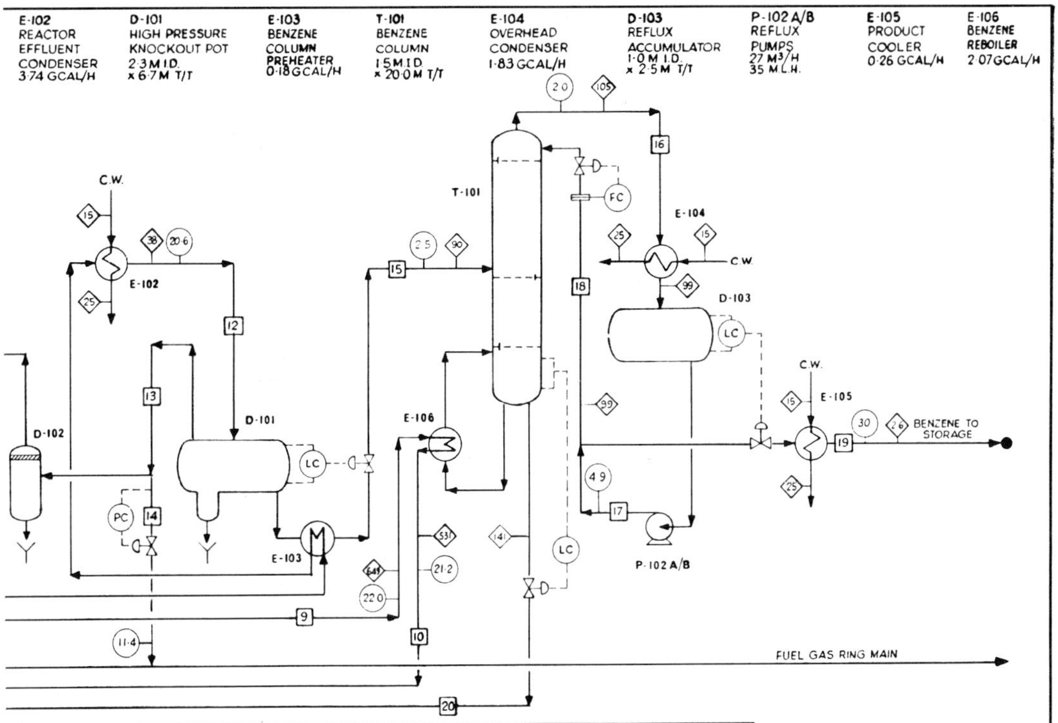

11	12	13	14	15	16	17	18	19	20
647·30	647·30	647·30	176·50	0·00	0·00	0·00	0·00	0·00	0·00
452·90	452·90	452·90	123·50	0·00	0·00	0·00	0·00	0·00	0·00
115·00	115·00	8·40	2·30	106·60	245·70	245·70	139·50	106·20	0·40
35·83	35·83	0·59	0·16	35·24	0·09	0·09	0·05	0·04	35·20
1251·03	1251·03	1109·19	302·46	141·84	245·79	245·79	139·55	106·24	35·60
20,807	20,807	9,250	2523	11,557	19,173	19,173	10,886	8,287	3,270
278	38	38	38	90	105	99	99	30	141
21·0	20·6	20·6	20·6	2·5	2·0	2·9	2·9	2·6	2·3
1·98	-1·94	-4·62	-1·26	2·86	7·18	5·39	3·06	2·06	0·63

XYZ ENGINEERING LTD.

TITLE: PROCESS FLOW DIAGRAM
BENZENE PLANT

Drawn By F.T.T. | Date | DRG No A-1001 | REV 6

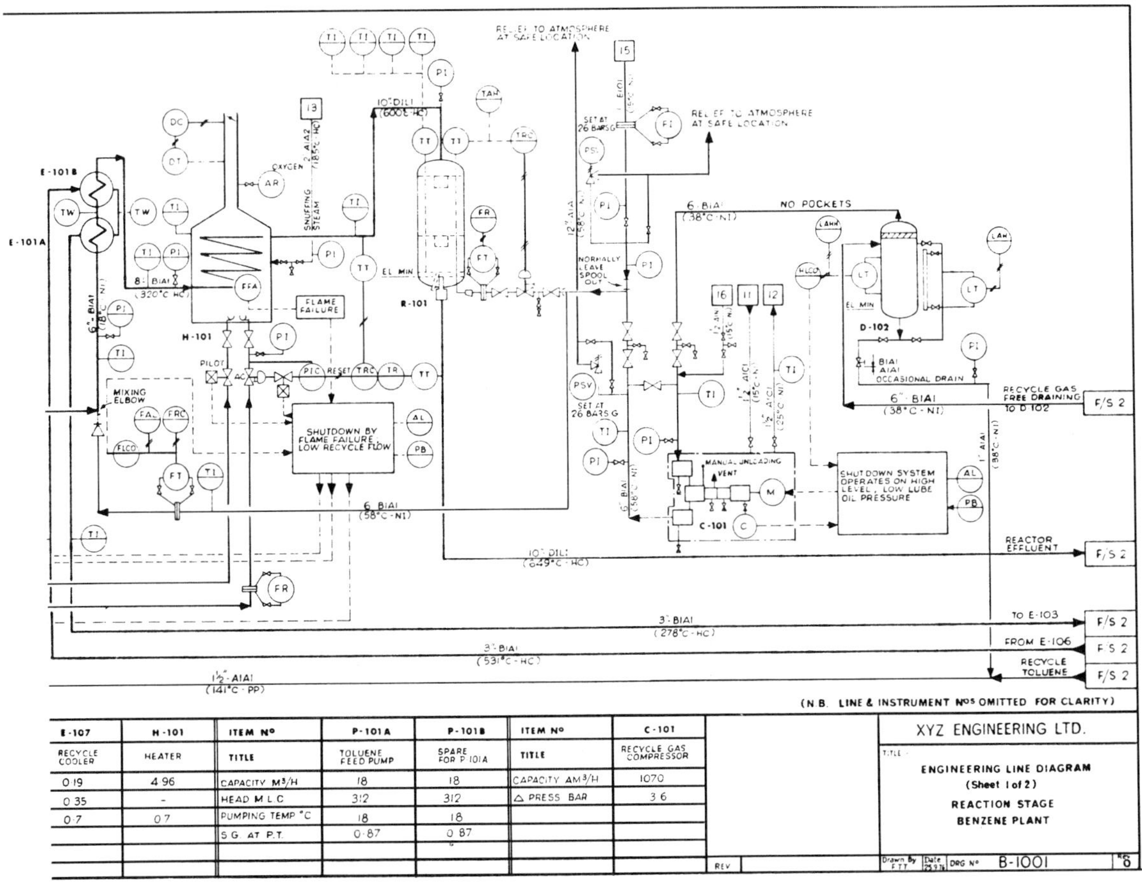

(N.B. LINE & INSTRUMENT Nos OMITTED FOR CLARITY)

E-107	H-101	ITEM No	P-101A	P-101B	ITEM No	C-101
RECYCLE COOLER	HEATER	TITLE	TOLUENE FEED PUMP	SPARE FOR P-101A	TITLE	RECYCLE GAS COMPRESSOR
0·19	4·96	CAPACITY M³/H	18	18	CAPACITY AM³/H	1070
0·35	-	HEAD M L.C.	31·2	31·2	△ PRESS BAR	3·6
0·7	0·7	PUMPING TEMP °C	18	18		
		S.G. AT P.T.	0·87	0·87		

XYZ ENGINEERING LTD.

TITLE: ENGINEERING LINE DIAGRAM
(Sheet 1 of 2)
REACTION STAGE
BENZENE PLANT

REV | Drawn By F.T.T. | Date 25.9.76 | DRG No B-1001 | REV 6

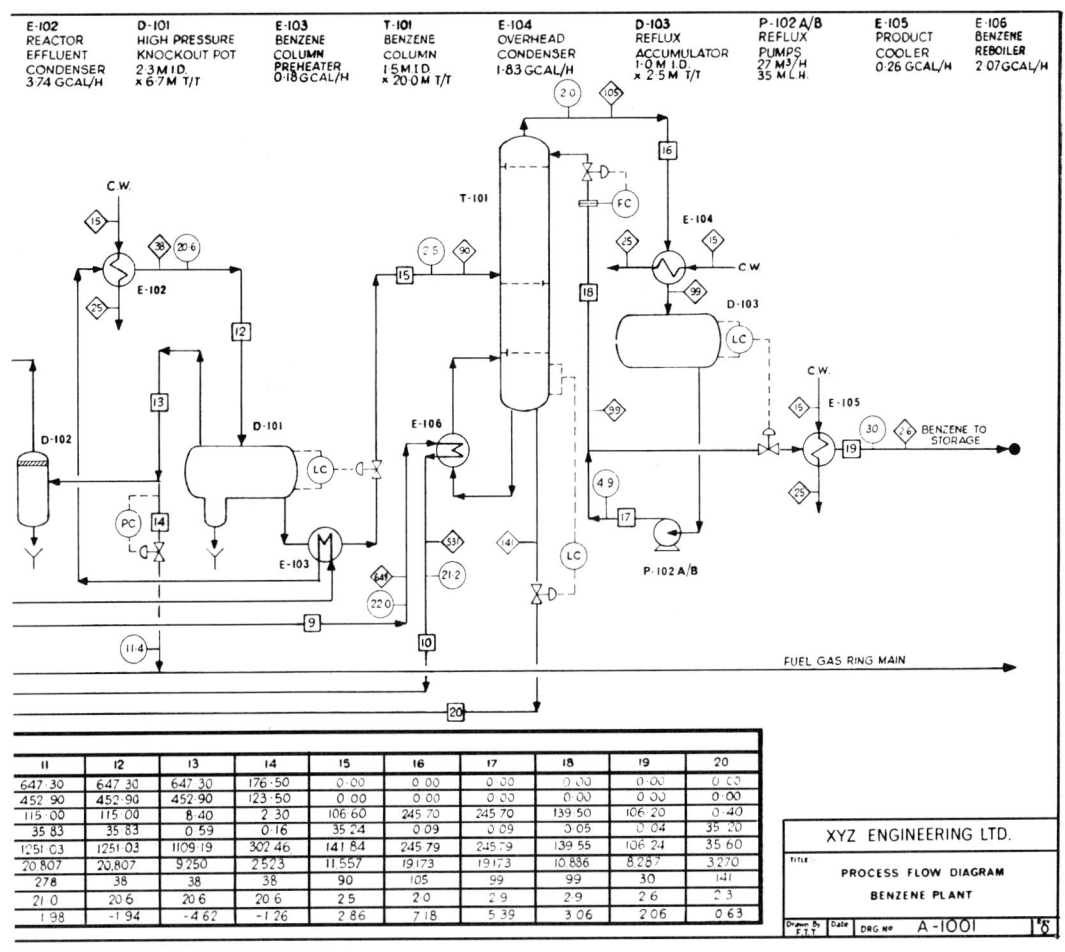

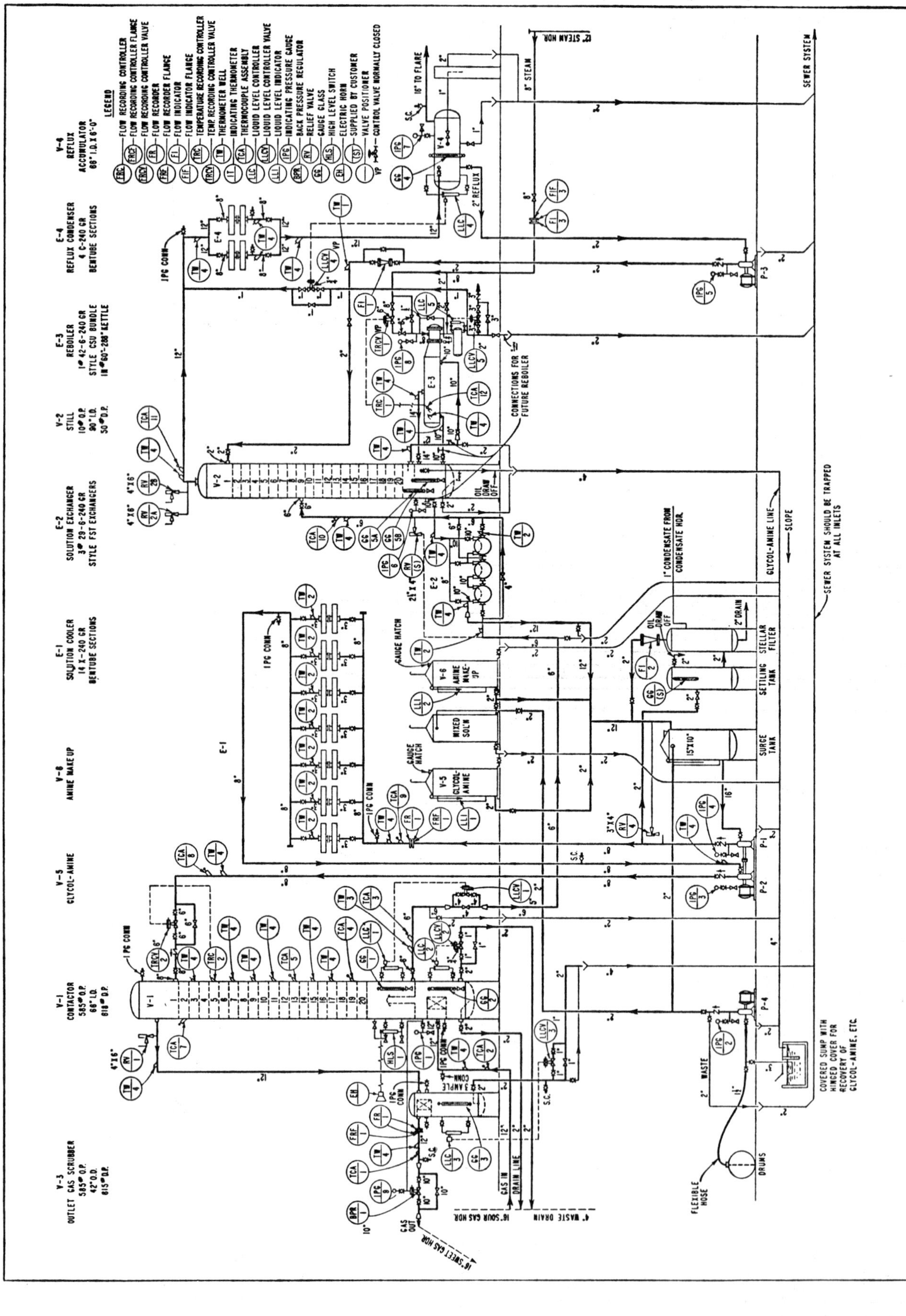

Figure 2.6. Engineering flowsheet of a gas treating plant. Note the tabulation of instrumentation flags at upper right (*Fluor Engineers, by way of Rase and Barrow,* Project Engineering of Process Plants, *Wiley, New York, 1957*).

REFERENCES

1. D.G. Austin, *Chemical Engineering Drawing Symbols,* George Godwin, London, 1979.
2. Graphical Symbols for Piping System and Plant, British Standard 1553: Part 1: 1977.
3. Graphical Symbols for Process Flow Diagrams, ASA Y32.11.1961, American Society of Mechanical Engineers, New York.
4. E.E. Ludwig, *Applied Process Design for Chemical and Petrochemical Plants,* Gulf, Houston, 1977, Vol. 1.
5. H.F. Rase and M.H. Barrow, *Project Engineering of Process Plants,* Wiley, New York, 1957.
6. R.K. Sinnott, Coulson, and Richardson, *Chemical Engineering, vol. 6, Design,* Pergamon, New York, 1983.
7. G.D. Ulrich, *A Guide to Chemical Engineering Process Design and Economics,* Wiley, New York, 1984.
8. R. Weaver, *Process Piping Design,* Gulf, Houston, 1973, 2 vols.

Appendix 2.1

Descriptions of Example Process Flowsheets

These examples ask for the construction of flowsheets from the given process descriptions. Necessary auxiliaries such as drums and pumps are to be included even when they are not mentioned. Essential control instrumentation also is to be provided. Chapter 3 has examples. The processes are as follows:

1. visbreaker operation,
2. cracking of gas oil,
3. olefin production from naptha and gas oil,
4. propylene oxide synthesis,
5. phenol by the chlorobenzene process,
6. manufacture of butadiene sulfone,
7. detergent manufacture,
8. natural gas absorption,
9. tall oil distillation,
10. recovery of isoprene,
11. vacuum distillation,
12. air separation.

1. VISBREAKER OPERATION

Visbreaking is a mild thermal pyrolysis of heavy petroleum fractions whose object is to reduce fuel production in a refinery and to make some gasoline.

The oil of 7.2 API and 700°F is supplied from beyond the battery limits to a surge drum F-1. From there it is pumped with J-1A&B to parallel furnaces B-1A&B from which it comes out at 890°F and 200 psig. Each of the split streams enters at the bottom of its own evaporator T-1A&B that has five trays. Overheads from the evaporators combine and enter at the bottom of a 30-tray fractionator T-2. A portion of the bottoms from the fractionator is fed to the top trays of T-1A&B; the remainder goes through exchanger E-5 and is pumped with J-2A&B back to the furnaces B-1A&B. The bottoms of the evaporators are pumped with J-4A&B through exchangers E-5, E-3A (on crude), and E-3B (on cooling water) before proceeding to storage as the fuel product.

A side stream is withdrawn at the tenth tray from the top of T-2 and proceeds to steam stripper T-3 equipped with five trays. Steam is fed below the bottom tray. The combined steam and oil vapors return to T-2 at the eighth tray. Stripper bottoms are pumped with J-6 through E-2A (on crude) and E-2B (on cooling water) and to storage as "heavy gasoline."

Overhead of the fractionator T-2 is partially condensed in E-1A (on crude) and E-1B (on cooling water). A gas product is withdrawn overhead of the reflux drum which operates at 15 psig. The "light gasoline" is pumped with J-5 to storage and as reflux.

Oil feed is 122,480 pph, gas is 3370, light gasoline is 5470, heavy gasoline is 9940, and fuel oil is 103,700 pph.

Include suitable control equipment for the main fractionator T-2.

2. CRACKING OF GAS OIL

A gas oil cracking plant consists of two cracking furnaces, a soaker, a main fractionator, and auxiliary strippers, exchangers, pumps, and drums. The main fractionator (150 psig) consists of four zones, the bottom zone being no. 1.

A light vacuum gas oil (LVGO) is charged to the top plate of zone 3, removed from the bottom tray of this zone and pumped to furnace no. 1 that operates at 1000 psig and 1000°F. A heavy vacuum gas oil (HVGO) is charged to the top plate of zone 2, removed at the bottom tray and charged to furnace no. 2 that operates at 500 psig and 925°F.

Effluents from both furnaces are combined and enter the soaker; this is a large vertical drum designed to provide additional residence time for conversion under adiabatic conditions. Effluent at 500 psig and 915°F enters the bottom zone of the main fractionator.

Bottoms from zone 1 goes to a stripping column (5 psig). Overhead from that tower is condensed, returned partly as reflux and partly to zone 3 after being cooled in the first condenser of the stripping column. This condensing train consists of the preheater for the stream being returned to the main fractionator and an air cooler. The cracked residuum from the bottom of the stripper is cooled to 170°F in a steam generator and an air cooler in series. Live steam is introduced below the bottom tray for stripping.

All of the oil from the bottom of zone 3 (at 700°F), other than the portion that serves as feed to furnace no. 1, is withdrawn through a cooler (500°F) and pumped partly to the top tray of zone 2 and partly as spray quench to zone 1. Some of the bottoms of zone 1 likewise is pumped through a filter and an exchanger and to the same spray nozzle.

Part of the liquid from the bottom tray of zone 4 (at 590°F) is pumped to a hydrogenation unit beyond the battery limits. Some light material is returned at 400°F from the hydrogenation unit to the middle of zone 4, together with some steam.

Overhead from the top of the column (zone 4) goes to a partial condenser at 400°F. Part of the condensate is returned to the top tray as reflux; the rest of it is product naphtha and proceeds beyond the battery limits. The uncondensed gas also goes beyond the battery limits. Condensed water is sewered.

3. OLEFIN PRODUCTION

A gaseous product rich in ethylene and propylene is made by pyrolysis of crude oil fractions according to the following description. Construct a flowsheet for the process. Use standard symbols for equipment and operating conditions. Space the symbols and proportion them in such a way that the sketch will have a pleasing appearance.

Crude oil is pumped from storage through a steam heated exchanger and into an electric desalter. Dilute caustic is injected into the line just before the desalting drum. The aqueous phase collects at the bottom of this vessel and is drained away to the sewer. The oil leaves the desalter at 190°F, and goes through heat exchanger E-2 and into a furnace coil. From the furnace, which it leaves at 600°F, the oil proceeds to a distillation tower.

After serving to preheat the feed in exchanger E-2, the bottoms proceeds to storage; no bottoms pump is necessary because the tower operates with 65 psig at the top. A gas oil is taken off as a sidestream some distance above the feed plate, and naphtha is taken off overhead. Part of the overhead is returned as reflux to the tower, and the remainder proceeds to a cracking furnace. The gas oil also is charged to the same cracking furnace but into a separate coil. Superheated steam at 800°F is injected into both cracking coils at their inlets.

Effluents from the naphtha and gas oil cracking coils are at 1300°F and 1200°F, respectively. They are combined in the line just before discharge into a quench tower that operates at 5 psig and 235°F at the top. Water is sprayed into the top of this tower. The bottoms is pumped to storage. The overhead is cooled in a water exchanger and proceeds to a separating drum. Condensed water and an aromatic oil separate out there. The water is sewered whereas the oil is sent to another part of the plant for further treating.

The uncondensed gas from the separator is compressed to 300 psig in a reciprocating unit of three stages and then cooled to 100°F. Condensed water and more aromatic distillate separate out. Then the gas is dried in a system of two desiccant-filled vessels that are used alternately for drying and regeneration.

Subsequently the gas is precooled in exchanger E-6 and charged to a low temperature fractionator. This tower has a reboiler and a top refluxing system. At the top the conditions are 280 psig and −75°F. Freon refrigerant at −90°F is used in the condenser. The bottoms is recycled to the pyrolysis coil. The uncondensed vapor leaving the reflux accumulator constitutes the product of this plant. It is used to precool the feed to the fractionator in E-6 and then leaves this part of the plant for further purification.

4. PROPYLENE OXIDE SYNTHESIS

Draw a process flowsheet for the manufacture of propylene oxide according to the following description.

Propylene oxide in the amount of 5000 tons/yr will be made by the chlorohydrin process. The basic feed material is a hydrocarbon mixture containing 90% propylene and the balance propane which does not react. This material is diluted with spent gas from the process to provide a net feed to chlorination which contains 40 mol % propylene. Chlorine gas contains 3% each of air and carbon dioxide as contaminants.

Chlorination is accomplished in a packed tower in which the hydrocarbon steam is contacted with a saturated aqueous solution of chlorine. The chlorine solution is made in another packed tower. Because of the limited solubility of chlorine, chlorohydrin solution from the chlorinator is recirculated through the solution tower at a rate high enough to supplement the fresh water needed for the process. Solubility of chlorine in the chlorohydrin solution is approximately the same as in fresh water.

Concentration of the effluent from the chlorinator is 8 lb organics/100 lb of water. The organics have the composition

Propylene chlorohydrin	75 mol %
Propylene dichloride	19
Propionaldehyde	6

Operating pressure of the chlorinator is 30 psig, and the temperature is 125°F. Water and the fresh gas stream are at 80°F. Heat of reaction is 2000 Btu/lb chlorine reacted. Percentage conversion of total propylene fed to the chlorinator is 95% (including the recycled material).

Overhead from the chlorinator is scrubbed to remove excess chlorine in two vessels in succession which employ water and 5% caustic solution, respectively. The water from the first scrubber is used in the chlorine solution tower. The caustic is recirculated in order to provide adequate wetting of the packing in the caustic scrubber; fresh material is charged in at the same rate as spent material is purged. Following the second scrubber, propylene dichloride is recovered from the gas by chilling it. The spent gas is recycled to the chlorinator in the required amount, and the excess is flared.

Chlorohydrin solution is pumped from the chlorinator to the saponifier. It is mixed in the feed line with a 10% lime slurry and preheated by injection of live 25 psig steam to a temperature of 200°F. Stripping steam is injected at the bottom of the saponifier, which has six perforated trays without downcomers. Propylene oxide and other organic materials go overhead; the bottoms contain unreacted lime, water, and some other reaction products, all of which can be dumped. Operating pressure is substantially atmospheric. Bubblepoint of the overhead is 60°F.

Separation of the oxide and the organic byproducts is accomplished by distillation in two towers. Feed from the saponifier contains oxide, aldehyde, dichloride, and water. In the first tower, oxide and aldehyde go overhead together with only small amounts of the other substances; the dichloride and water go to the bottom and also contain small amounts of contaminants. Two phases will form in the lower section of this tower; this is taken off as a partial side stream and separated into a dichloride phase which is sent to storage and a water phase which is sent to the saponifier as recycle near the top of that vessel. The bottoms are a waste product. Tower pressure is 20 psig. Live steam provides heat at the bottom of this column.

Overhead from the first fractionator is condensed and charged to the second tower. There substantially pure propylene oxide is taken overhead. The bottoms is dumped. Tower pressure is 15 psig, and the overhead bubblepoint is 100°F. Reactions are

$$Cl_2 + H_2O \rightarrow ClOH + HCl$$
$$C_3H_6 + Cl_2 + H_2O \rightarrow C_3H_6ClOH + HCl$$
$$C_3H_6 + Cl_2 \rightarrow C_3H_6Cl_2$$
$$C_3H_6ClOH \rightarrow C_2H_5CHO + HCl$$
$$2C_3H_6ClOH + Ca(OH)_2 \rightarrow 2C_3H_6O + CaCl_2 + 2H_2O$$

Show all necessary major equipment, pumps, compressors, refrigerant lines. Show the major instrumentation required to make this process continuous and automatic.

5. PHENOL BY THE CHLOROBENZENE PROCESS

A portion of a plant for the manufacture of phenol from monochlorbenzene and NaOH is in accordance with the following description.

a. Construct a flowsheet of the process, with operating conditions and the two control instruments mentioned.
b. Prepare a material balance showing the compositions of the process streams in the portion of the plant before the brine decanter V-103. The amount of phenol in this stream is 2000 lb/hr. Excess caustic (5%) is fed to the emulsifier.

Process description: The principal reactions in the plant are

$$C_6H_5Cl + 2NaOH \rightarrow C_6H_5ONa + NaCl + H_2O$$
$$C_6H_5ONa + HCl \rightarrow C_6H_5OH + NaCl$$
$$2C_6H_5OH \rightarrow (C_6H_5)_2O + H_2O$$

From storage, monochlorbenzene and 10% caustic are pumped together with diphenyl ether from decanter V-102 into emulsifier V-101 which is provided with intense agitation. The effluent from that vessel is pumped with a high pressure steam driven reciprocating pump P-103 at 4000 psig through a feed–effluent exchanger E-101 and through the tube side of a direct fired heater R-101. Here the stream is heated to 700°F and reaction 1 occurs.

From the reactor, the effluent is cooled in E-101, cooled further to 110°F in water cooler E-102, and then enters diphenyl ether decanter V-102. The lighter DPE phase is returned with pump P-104 to the emulsifier. The other phase is pumped with P-105 to another stirred vessel R-102 called a Springer to which 5% HCl also is pumped, with P-106; here reaction 2 occurs.

The mixture of two liquid phases is cooled in water cooler E-103 and then separated in brine decanter V-103. From that vessel the lighter phenol phase proceeds (P-108) to a basket type evaporator D-101 that is heated with steam. Overhead vapor from the evaporator proceeds beyond the battery limit for further purification. Evaporator bottoms proceeds to waste disposal. The aqueous phase from decanter V-103 is pumped with P-109 through a feed–bottoms exchanger E-104 to the top tray of the brine tower D-102. The overhead is condensed in E-105, collected in accumulator V-104 and pumped beyond the battery limits for recovery of the phenol. Tower D-102 is provided with a steam heated reboiler E-106. Bottom product is a weak brine that is pumped with P-110 through the feed–bottoms exchanger and beyond the battery limits for recovery of the salt.

Two important control instruments are to be shown on the flowsheet. These are a back pressure controller in the reactor effluent line beyond exchanger E-101 and a pH controller on the feed line of the 5% HCl that is fed to springer R-102. The pH instrument maintains proper conditions in the springer.

Note: There is a tendency to byproduct diphenyl ether formation in reactor R-101. However, a recycle of 100 pph of DPE in the feed to the reactor prevents any further formation of this substance.

6. MANUFACTURE OF BUTADIENE SULFONE

A plant is to manufacture butadiene sulfone at the rate of 1250 lb/hr from liquid sulfur dioxide and butadiene to be recovered from a crude C_4 mixture as starting materials. Construct a flowsheet for the process according to the following description.

The crude C_4 mixture is charged to a 70 tray extractive distillation column T-1 that employs acetonitrile as solvent. Trays are numbered from the bottom. Feed enters on tray 20, solvent enters on tray 60, and reflux is returned to the top tray. Net overhead product goes beyond the battery limits. Butadiene dissolved in acetonitrile leaves at the bottom. This stream is pumped to a 25-tray solvent recovery column T-2 which it enters on tray 20. Butadiene is recovered overhead as liquid and proceeds to the BDS reactor. Acetonitrile is the bottom product which is cooled to 100°F and returned to T-1. Both columns have the usual condensing and reboiling provisions.

Butadiene from the recovery plant, liquid sulfur dioxide from storage, and a recycle stream (also liquified) are pumped through a preheater to a high temperature reactor R-1 which is of shell-and-tube construction with cooling water on the shell side. Operating conditions are 100°C and 300 psig. The combined feed contains equimolal proportions of the reactants, and 80% conversion is attained in this vessel. The effluent is cooled to 70°C, then enters a low temperature reactor R-2 (maintained at 70°C and 50 psig with cooling water) where the conversion becomes 92%. The effluent is flashed at 70°C and atmospheric pressure in D-1. Vapor product is compressed, condensed and recycled to the reactor R-1. The liquid is pumped to a storage tank where 24 hr holdup at 70°C is provided to ensure chemical equilibrium between sulfur dioxide, butadiene, and butadiene sulfone. Cooling water is available at 32°C.

7. DETERGENT MANUFACTURE

The process of making synthetic detergents consists of several operations that will be described consecutively.

ALKYLATION

Toluene and olefinic stock from storage are pumped (at 80°F) separately through individual driers and filters into the alkylation reactor. The streams combine just before they enter the reactor. The reactor is batch operated 4 hr/cycle; it is equipped with a single impeller agitator and a feed hopper for solid aluminum chloride which is charged manually from small drums. The alkylation mixture is pumped during the course of the reaction through an external heat exchanger (entering at −10°F and leaving at −15°F) which is cooled with ammonia refrigerant (at −25°F) from an absorption refrigeration system (this may be represented by a block on the FS); the exchanger is of the kettle type. HCl gas is injected into the recirculating stream just beyond the exit from the heat exchanger; it is supplied from a cylinder mounted in a weigh scale. The aluminum chloride forms an alkylation complex with the toluene. When the reaction is complete, this complex is pumped away from the reactor into a storage tank with a complex transfer pump. To a certain extent, this complex is reused; it is injected with its pump into the reactor recirculation line before the suction to the recirculation pump. There is a steam heater in the complex line, between the reactor and the complex pump.

The reaction mixture is pumped away from the reactor with an alkymer transfer pump, through a steam heater and an orifice mixer into the alkymer wash and surge tank. Dilute caustic solution is recirculated from the a.w.s. tank through the orifice mixer. Makeup of caustic is from a dilute caustic storage tank. Spent caustic is intermittently drained off to the sewer. The a.w.s. tank has an internal weir. The caustic solution settles and is removed at the left of the weir; the alkymer overflows the weir and is stored in the right-hand portion of the tank until amount sufficient for charging the still has accumulated.

DISTILLATION

Separation of the reactor product is effected in a ten-plate batch distillation column equipped with a water-cooled condenser and a Dowtherm-heated (650°F, 53 psig) still. During a portion of the distillation cycle, operation is under vacuum, which is produced by a two-stage steam jet ejector equipped with barometric condensers. The Dowtherm heating system may be represented by a block. Product receiver drums are supplied individually for a slop cut, for toluene, light alkymer, heart alkymer, and a heavy alkymer distillate. Tar is drained from the still at the end of the operation through a water cooler into a bottoms receiver drum which is supplied with a steam coil. From this receiver, the tar is loaded at intervals into 50 gal drums, which are trucked away. In addition to the drums which serve to receive the distillation products during the operation of the column, storage tanks are provided for all except the slop cut which is returned to the still by means of the still feed pump; this pump transfers the mixture from the alkymer wash and surge tank into the still. The recycle toluene is not stored with the fresh toluene but has its own storage tank. The heavy alkymer distillate tank connects to the olefinic stock feed pump and is recycled to the reactor.

SULFONATION

Heart alkymer from storage and 100% sulfuric acid from the sulfuric acid system (which can be represented by a block) are pumped by the reactor feed pump through the sulfonation reactor. The feed pump is a positive displacement proportioning device with a single driver but with separate heads for the two fluids. The reactor is operated continuously; it has a single shell with three stages which are partially separated from each other with horizontal doughnut shaped plates. Each zone is agitated with its individual impeller; all three impellers are mounted on a single shaft. On leaving the reactor, the sulfonation mixture goes by gravity through a water cooler (leaving at 130°F) into a centrifuge. Spent acid from the centrifuge goes to storage (in the sulfuric acid system block); the sulfonic acids go to a small surge drum or can bypass this drum and go directly to a large surge tank which is equipped with an agitator and a steam jacket. From the surge drum, the material is sent by an extraction feed pump through a water cooler, then a "flomix," then another water cooler, then another "flomix" (leaving at 150°F), and then through a centrifuge and into the sulfonic acid surge tank. Fresh water is also fed to each of the "flomixers." Wash acid is rejected by the centrifuge and is sent to the sulfuric acid system. The "flomix" is a small vertical vessel which has two compartments and an agitator with a separate impeller for each compartment.

NEUTRALIZATION

Neutralization of the sulfonic acid and building up with sodium sulfate and tetrasodium pyrophosphate (TSPP) is accomplished in two batch reactors (5 hr cycle) operated alternately. The sodium sulfate is pumped in solution with its transfer pump from the sodium sulfate system (which can be represented by a block). The TSPP is supplied as a solid and is fed by means of a Redler conveyor which discharges into a weigh hopper running on a track above the two reactors. Each reactor is agitated with a propeller and a turbine blade in a single shaft.

Sodium hydroxide of 50% and 1% concentrations is used for neutralization. The 50% solution discharges by gravity into the reactor; the 1% solution is injected gradually into the suction side of the reactor slurry circulating pump. As the caustic is added to the reactor, the contents are recirculated through a water-cooled external heat exchanger (exit at 160°F), which is common to both reactors. When the reaction is completed in one vessel, the product is fed gradually by means of a slurry transfer pump to two double drum dryers which are steam-heated and are supplied with individual vapor hoods. The dry material is carried away from the dryers on a belt conveyor and is taken to a flaker equipped with an air classifier. The fines are returned to the trough between the dryer drums. From the classifier, the material is taken with another belt conveyor to four storage bins. These storage bins in turn discharge onto a belt feeder which discharges into drums which are weighed automatically on a live portion of a roller conveyor. The roller conveyor takes the drums to storage and shipping.

Notes: All water cooled exchangers operate with water in at 75°F and out at 100°F. All pumps are centrifugal except the complex transfer, and the sulfonation reactor feed, which are both piston type; the neutralization reactor recirculation pump and the transfer pumps are gear pumps.

Show all storage tanks mentioned in the text.

8. NATURAL GAS ABSORPTION

A gas mixture has the composition by volume:

Component	N_2	CH_4	C_2H_6	C_3H_8
Mol fraction	0.05	0.65	0.20	0.10

It is fed to an absorber where 75% of the propane is recovered. The total amount absorbed is 50 mol/hr. The absorber has four theoretical plates and operates at 135 psig and 100°F. All of the absorbed material is recovered in a steam stripper that has a large number of plates and operates at 25 psig and 230°F.

Water is condensed out of the stripped gas at 100°F. After compression to 50 psig, that gas is combined with a recycle stream. The mixture is diluted with an equal volume of steam and charged to a reactor where pyrolysis of the propane occurs at a temperature of 1300°F. For present purposes the reaction may be assumed to be simply $C_3H_8 \rightarrow C_2H_4 + CH_4$ with a specific rate $k = 0.28/\text{sec}$. Conversion of propane is 60%. Pressure drop in the reactor is 20 psi.

Reactor effluent is cooled to remove the steam, compressed to 285 psig, passed through an activated alumina drying system to remove further amounts of water, and then fed to the first fractionator. In that vessel, 95% of the unconverted propane is recovered as a bottoms product. This stream also contains 3% ethane as an impurity. It is throttled to 50 psig and recycled to the reactor. In two subsequent towers, ethylene is separated from light and heavy impurities. Those separations may be taken as complete.

Construct a flow diagram of this plant. Show such auxiliary equipment as drums, heat exchangers, pumps, and compressors. Show operating conditions and flow quantities where calculable with the given data.

9. TALL OIL DISTILLATION

Tall oil is a byproduct obtained from the manufacture of paper pulp from pine trees. It is separated by vacuum distillation (50 mm Hg) in the presence of steam into four primary products. In the order of decreasing volatility these are unsaponifiables (US), fatty acid (FA), rosin acids (RA), and pitch (P). Heat exchangers and reboilers are heated with Dowtherm condensing vapors. Some coolers operate with water and others generate steam. Live steam is charged to the inlet of every reboiler along with the process material. Trays are numbered from the bottom of each tower.

Tall oil is pumped from storage through a preheater onto tray 10 of the pitch stripper T-1. Liquid is withdrawn from tray 7 and pumped through a reboiler where partial vaporization occurs in the presence of steam. The bottom 6 trays are smaller in diameter and serve as stripping trays. Steam is fed below tray 1. Pitch is pumped from the bottom through steam generator and to storage. Overhead vapors are condensed in two units E-1 and E-2. From the accumulator, condensate is pumped partly as reflux to tray 15 and partly through condenser E-1 where it is preheated on its way as feed to the next tower T-2. Steam is not condensed in E-2. It flows from the accumulator to a barometric condenser that is connected to a steam jet ejector.

Feed enters T-2 at tray 5. There is a pump-through reboiler. Another pump withdraws material from the bottom and sends it to tower T-3. Liquid is pumped from tray 18 through a cooler and returned in part to the top tray 20 for temperature and reflux control. A portion of this pumparound is withdrawn after cooling as unsaps product. Steam leaves the top of the tower and is condensed in the barometric.

Tray 5 of T-3 is the feed position. This tower has two reboilers. One of them is a pumparound from the bottom, and the other is gravity feed from the bottom tray. Another pump withdraws material from the bottom, and then sends it through a steam generator and to storage as rosin acid product. A slop cut is withdrawn from tray 20 and pumped through a cooler to storage. Fatty acid product is pumped from tray 40 through a cooler to storage. Another stream is pumped around from tray 48 to the top tray 50 through a cooler. A portion of the cooled pumparound is sent to storage as another unsaps product. A portion of the overhead steam proceeds to the barometric condenser. The rest of it is boosted in pressure with high pressure steam in a jet compressor. The boosted steam is fed to the inlets of the two reboilers associated with T-3 and also directly into the column below the bottom tray.

The vapors leaving the primary barometric condenser proceed to a steam ejector that is followed by another barometric. Pressures at the tops of the towers are maintained at 50 mm Hg absolute. Pressure drop is 2 mm Hg per tray. Bottom temperatures of the three towers are 450, 500, and 540°F, respectively. Tower overhead temperatures are 200°F. Pitch and rosin go to storage at 350°F and the other products at 125°F. The steam generated in the pitch and rosin coolers is at 20 psig. Process steam is at 150 psig.

10. RECOVERY OF ISOPRENE

Draw carefully a flowsheet for the recovery of isoprene from a mixture of C_5 hydrocarbons by extractive distillation with aqueous acetonitrile according to the following description.

A hydrocarbon stream containing 60 mol % isoprene is charged at the rate of 10,000 pph to the main fractionator D-1 at tray 40 from the top. The solvent is acetonitrile with 10 wt % water; it is charged at the rate of 70,000 pph on tray 11 of D-1. This column has a total of 70 trays, operates at 10 psig and 100°F at the top and about 220°F at the bottom. It has the usual provisions for reboiling and top reflux.

The extract is pumped from the bottom of D-1 to a stripper D-2 with 35 trays. The stripped solvent is cooled with water and returned to D-1. An isoprene–acetonitrile azeotrope goes overhead, condenses, and is partly returned as top tray reflux. The net overhead proceeds to an extract wash column D-3 with 20 trays where the solvent is recovered by countercurrent washing with water. The overhead from D-3 is the finished product isoprene. The bottoms is combined with the bottoms from the raffinate wash column D-4 (20 trays) and sent to the solvent recovery column D-5 with 15 trays.

Overhead from D-1 is called the raffinate. It is washed countercurrently with water in D-4 for the recovery of the solvent, and then proceeds beyond the battery limits for further conversion to isoprene. Both wash columns operate at substantially atmospheric pressure and 100°F. The product streams are delivered to the battery limits at 100 psig.

Solvent recovery column D-5 is operated at 50 mm Hg absolute, so as to avoid the formation of an azeotrope overhead. The required overhead condensing temperature of about 55°F is provided with a propane compression refrigeration system; suction condition is 40°F and 80 psig, and discharge condition is 200 psig. Vacuum is maintained on the reflux accumulator with a two-stage steam ejector, with a surface interstage condenser and a direct water spray after-condenser. The stripped bottoms of D-5 is cooled to 100°F and returned to the wash columns. Some water makeup is necessary because of leakages and losses to process streams. The solvent recovered overhead in D-5 is returned to the main column D-1. Solvent makeup of about 20 pph is needed because of losses in the system.

Steam is adequate for all reboiling needs in this plant.

11. VACUUM DISTILLATION

This plant is for the distillation of a heavy petroleum oil. The principal equipment is a vacuum tower with 12 trays. The top tray is numbered 1. Trays 1, 2, 10, 11, and 12 are one-half the diameter of the other trays. The tower operates at 50 mm Hg.

Oil is charged with pump J-1 through an exchanger E-1, through a fired heater from which it proceeds at 800°F onto tray 10 of the tower. Live steam is fed below the bottom tray.

Bottoms product is removed with pump J-3 through a steam generator and a water cooled exchanger E-3 beyond the battery limits. A side stream is taken off tray 6, pumped with J-2 through E-1, and returned onto tray 3 of the tower. Another stream is removed from tray 2 with pump J-4 and cooled in water exchanger E-2; part of this stream is returned to tray 1, and the rest of it leaves the plant as product gas oil.

Uncondensed vapors are removed at the top of the column with a one-stage steam jet ejector equipped with a barometric condenser.

Show the principal controls required to make this plant operate automatically.

12. AIR SEPARATION

Make a flowsheet of an air purification and separation plant that operates according to the following description.

Atmospheric air at the rate of 6.1 million SCFD is compressed to 160 psig in a two-stage compressor JJ-1 that is provided with an intercooler and a knockout drum. Then it proceeds to a packed tower T-1 where it is scrubbed with recirculating caustic soda solution. Overhead from T-1 is cooled to 14°F in a refrigerated exchanger. After removal of the condensate, this stream proceeds to a dryer system that consists principally of two vessels F-1 and F-2 packed with solid desiccant.

After being precooled with product oxygen in exchanger E-1 and with product nitrogen in E-2, the air serves as the heating medium in reboiler E-3 of column T-2. Its pressure then is reduced to 100 psig, and it is fed to the middle of column T-2. Bottoms of T-2 is fed to the middle of column T-3. This stream contains 40% oxygen.

Columns T-3 and T-4 operate at 15 and 30 psig, respectively. Column T-3 is located above T-4. Elevations and pressure differentials are maintained in such a way that no liquid pumps are needed in the distillation section of the plant.

Part of the overhead from T-2 (containing 96% nitrogen) is condensed in E-4 which is the reboiler for column T-3, and the remainder is condensed in E-5 which is the reboiler for T-4. Part of the condensate from E-4 is returned as reflux to T-2 and the rest of the condensates from E-4 and E-5 serve as top reflux to T-3. Overhead from T-3 contains 99.5% nitrogen. After precooling the feed in E-2, this nitrogen proceeds to the battery limits.

Bottoms of T-3 proceeds to the top of stripper T-4. Vapor overhead from T-4 is recycled to the middle of T-3. The bottoms product (containing 99.5% oxygen) is sent partly to liquid storage and the remainder to precooler E-1 where it is vaporized. Then it is compressed to 150 psig in a two-stage compressor JJ-2 and sent to the battery limits. Compressor JJ-2 has inter- and aftercoolers and knockout drums for condensate.

3

PROCESS CONTROL

All processes are subject to disturbances that tend to change operating conditions, compositions, and physical properties of the streams. In order to minimize the ill effects that could result from such disturbances, chemical plants are implemented with substantial amounts of instrumentation and automatic control equipment. In critical cases and in especially large plants, moreover, the instrumentation is computer monitored for convenience, safety, and optimization.

For example, a typical billion lb/yr ethylene plant may have 600 control loops with control valves and 400 interacting loops with a cost of about $6 million. (Skrokov, 1980, pp. 13, 49; see Sec. 3.1); the computer implementation of this control system will cost another $3 million. Figure 3.1 shows the control system of an ethylene fractionator which has 12 input signals to the computer and four outgoing reset signals to flow controllers.

In order for a process to be controllable by machine, it must represented by a mathematical model. Ideally, each element of a dynamic process, for example, a reflux drum or an individual tray of a fractionator, is represented by differential equations based on material and energy balances, transfer rates, stage efficiencies, phase equilibrium relations, etc., as well as the parameters of sensing devices, control valves, and control instruments. The process as a whole then is equivalent to a system of ordinary and partial differential equations involving certain independent and dependent variables. When the values of the independent variables are specified or measured, corresponding values of the others are found by computation, and the information is transmitted to the control instruments. For example, if the temperature, composition, and flow rate of the feed to a fractionator are perturbed, the computer will determine the other flows and the heat balance required to maintain constant overhead purity. Economic factors also can be incorporated in process models; then the computer can be made to optimize the operation continually.

For control purposes, somewhat simplified mathematical models usually are adequate. In distillation, for instance, the Underwood–Fenske–Gilliland model with constant relative volatilities and a simplified enthalpy balance may be preferred to a full-fledged tray-by-tray calculation every time there is a perturbation. In control situations, the demand for speed of response may not be realizable with an overly elaborate mathematical system. Moreover, in practice not all disturbances are measurable, and the process characteristics are not known exactly. Accordingly feedforward control is supplemented in most instances with feedback. In a well-designed system (Shinskey, 1984, p. 186) typically 90% of the corrective action is provided by feed forward and 10% by feedback with the result that the integrated error is reduced by a factor of 10.

A major feature of many modern control systems is composition control which has become possible with the development of fast and accurate on-line analyzers. Figure 3.2 shows that 10 analyzers are used for control of ethylene composition in this plant within the purities shown. High speed on-line gas chromatographs have analysis times of 30–120 sec and are capable of measuring several components simultaneously with a sensitivity in the parts/million range. Mass spectrometers are faster, more stable, and easier to maintain but are not sensitive in the ppm range. Any one instrument can be hooked up to a half-dozen or so sample ports, but, of course, at the expense of time lag for controller response. Infrared and NMR spectrometers also are feasible for on-line analysis. Less costly but also less specific analyzers are available for measuring physical properties such as refractive index and others that have been calibrated against mixture composition or product purity.

The development of a mathematical model, even a simplified one that is feasible for control purposes, takes a major effort and is well beyond the scope of the brief treatment of process control that can be attempted here. What will be given is examples of control loops for the common kinds of equipment and operations. Primarily these are feedback arrangements, but, as mentioned earlier, feedback devices usually are necessary supplements in primarily feedforward situations.

When processes are subject only to slow and small perturbations, conventional feedback PID controllers usually are adequate with set points and instrument characteristics fine-tuned in the field. As an example, two modes of control of a heat exchange process are shown in Figure 3.8 where the objective is to maintain constant outlet temperature by exchanging process heat with a heat transfer medium. Part (a) has a feedback controller which goes into action when a deviation from the preset temperature occurs and attempts to restore the set point. Inevitably some oscillation of the outlet temperature will be generated that will persist for some time and may never die down if perturbations of the inlet condition occur often enough. In the operation of the feedforward control of part (b), the flow rate and temperature of the process input are continually signalled to a computer which then finds the flow rate of heat transfer medium required to maintain constant process outlet temperature and adjusts the flow control valve appropriately. Temperature oscillation amplitude and duration will be much less in this mode.

3.1. FEEDBACK CONTROL

In feedback control, after an offset of the controlled variable from a preset value has been generated, the controller acts to eliminate or reduce the offset. Usually there is produced an oscillation in the value of the controlled variable whose amplitude, period, damping and permanent offset depend on the nature of the system and the mode of action of the controller. The usual controllers provide one, two, or three of these modes of corrective action:

1. Proportional, in which the corrective action is proportional to the error signal.
2. Integral, in which the corrective action at time t is proportional to the integral of the error up to that time.

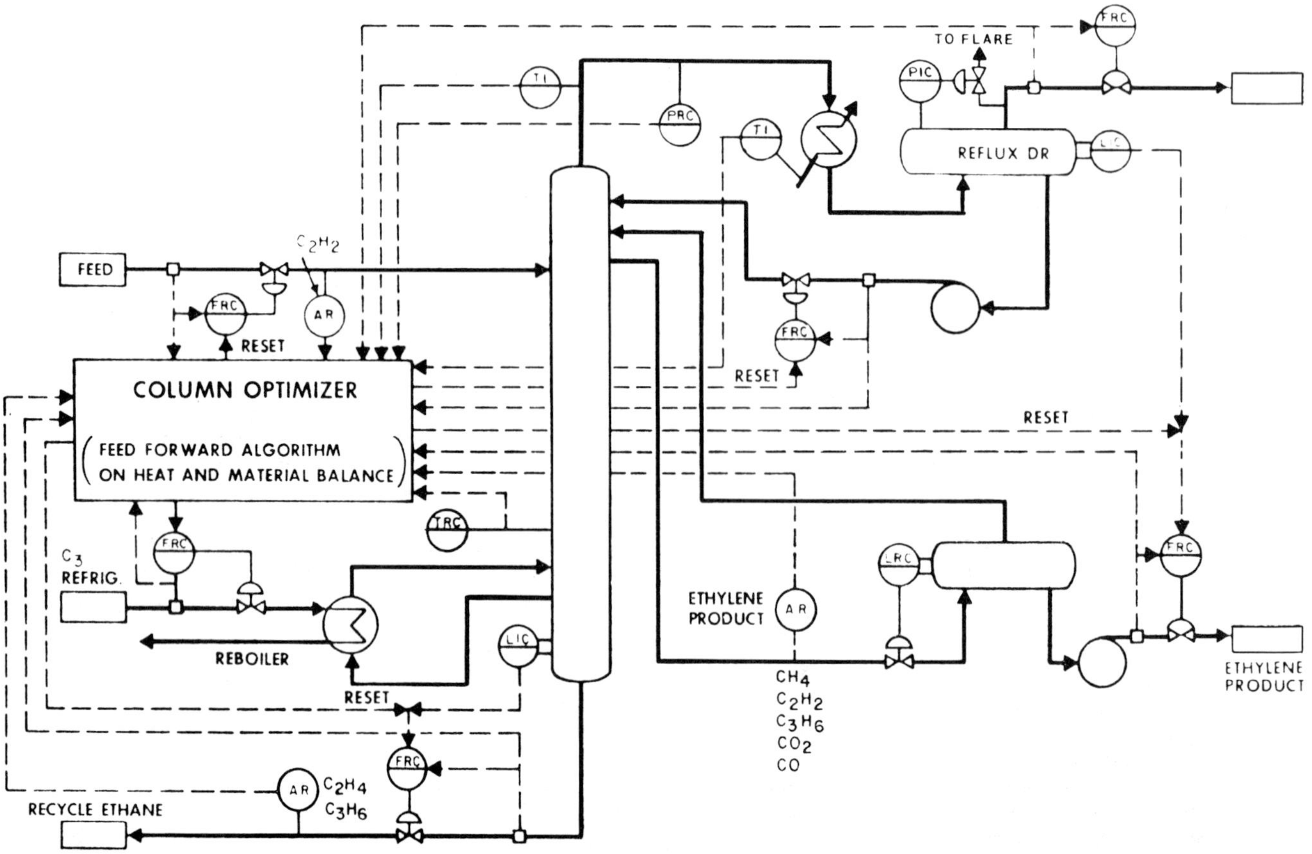

Figure 3.1. Optimized control of an ethylene tower (*Skrokov (Ed.),* Mini- and Microcomputer Control in Industrial Processes, *Van Nostrand/Reinhold, New York, 1980*).

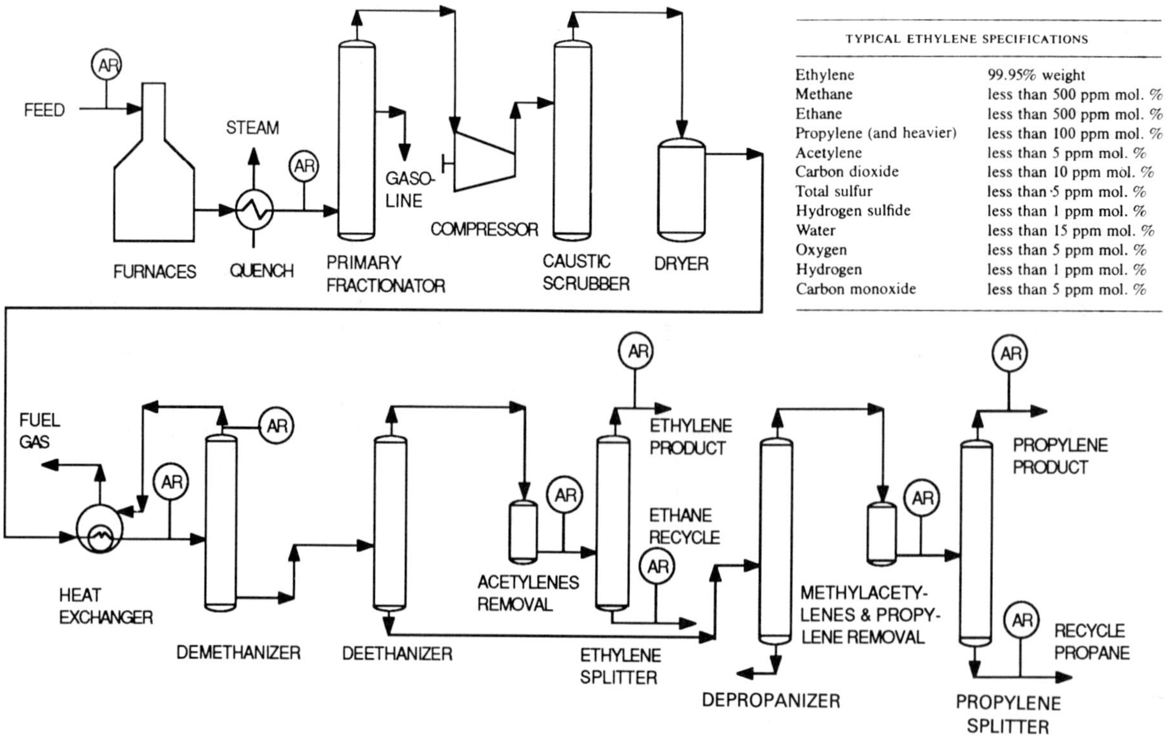

TYPICAL ETHYLENE SPECIFICATIONS

Ethylene	99.95% weight
Methane	less than 500 ppm mol. %
Ethane	less than 500 ppm mol. %
Propylene (and heavier)	less than 100 ppm mol. %
Acetylene	less than 5 ppm mol. %
Carbon dioxide	less than 10 ppm mol. %
Total sulfur	less than 5 ppm mol. %
Hydrogen sulfide	less than 1 ppm mol. %
Water	less than 15 ppm mol. %
Oxygen	less than 5 ppm mol. %
Hydrogen	less than 1 ppm mol. %
Carbon monoxide	less than 5 ppm mol. %

Figure 3.2. Flowsketch of an olefins plant and specifications of the ethylene product. AR designates a composition analyzer and controller (*after Skrokov (Ed.),* Mini- and Microcomputer Control in Industrial Processes, *Van Nostrand/Reinhold, New York, 1980*).

3. Derivative, in which the corrective action is proportional to the rate at which the error is being generated.

The relation between the change in output $m - m_0$ and input e signals accordingly is represented by

$$m - m_0 = K_p\left(e + \frac{1}{K_i}\int_0^t e\,dt + K_d\frac{de}{dt}\right).$$

Just how these modes of action are achieved in relatively inexpensive pneumatic or electrical devices is explained in books on control instruments, for example, that of Considine (*Process Instruments and Controls Handbook*, Sec. 17, 1974). The low prices and considerable flexibility of PID controllers make them the dominant types in use, and have discouraged the development of possibly superior types, particularly as one-shot deals which would be the usual case in process plants. Any desired mode of action can be simulated by a computer, but at a price.

A capsule summary of the merits of the three kinds of corrective action can be made. The proportional action is rapid but has a permanent offset that increases as the action speeds up. The addition of integral action reduces or entirely eliminates the offset but has a more sluggish response. The further addition of derivative action speeds up the correction. The action of a three-mode PID controller can be made rapid and without offset. These effects are illustrated in Figure 3.3 for a process subjected to a unit step upset, in this case a change in the pressure of the control air. The ordinate is the ratio of the displacements of the response and upset from the set point.

The reason for a permanent offset with a proportional controller can be explained with an example. Suppose the temperature of a reactor is being controlled with a pneumatic system. At the set point, say the valve is 50% open and the flow rate of cooling water is fixed accordingly. Suppose the heat load is doubled suddenly because of an increase in the reactor contents. At steady state the valve will remain 50% open so that the water flow rate also will remain as before. Because of the greater rate of heat evolution, however, the temperature will rise to a higher but still steady value. On the other hand, the corrective action of an integral controller depends on displacement of the temperature from the original set point, so that this mode of control will restore the original temperature.

The constants K_p, K_i, and K_d are settings of the instrument. When the controller is hooked up to the process, the settings appropriate to a desired quality of control depend on the inertia (capacitance) and various response times of the system, and they can be determined by field tests. The method of Ziegler and Nichols used in Example 3.1 is based on step response of a damped system and provides at least approximate values of instrument settings which can be further fine-tuned in the field.

The kinds of controllers suitable for the common variables may be stated briefly:

Variable	Controller
Flow and liquid pressure	PI
Gas pressure	P
Liquid level	P or PI
Temperature	PID
Composition	P, PI, PID

Derivative control is sensitive to noise that is made up of random higher frequency perturbations, such as splashing and turbulence generated by inflow in the case of liquid level control in a vessel, so that it is not satisfactory in such situations. The variety of composition controllers arises because of the variety of composition analyzers or detectors.

Many corrective actions ultimately adjust a flow rate, for instance, temperature control by adjusting the flow of a heat transfer medium or pressure by regulating the flow of an effluent stream. A control unit thus consists of a detector, for example, a thermocouple, a transmitter, the control instrument itself, and a control valve. The natures, sensitivities, response speeds, and locations of these devices, together with the inertia or capacity of the process equipment, comprise the body of what is to be taken into account when designing the control system. In the following pages will be described only general characteristics of the major kinds of control systems that are being used in process plants. Details and criteria for choice between possible alternates must be sought elsewhere. The practical aspects of this subject are treated, for example, in the References at the end of this chapter.

Curve	Mode of Control	Prop Sensitivity, K_p	Integral Time, K_i	Deriv Time, K_d	Period of Cycle, sec	Damping $e^{-t/T}$ T, sec	Max Error units	Offset units
1	Proportional derivative	16	—	0.9	32	15	0.18	0.06
2	Prop int deriv	10	22	2.0	44	20	0.21	0
3	Proportional	8	—	—	45	20	0.29	0.11
4	Proportional integral	4	3	—	66	30	0.37	0
5	Integral	—	35	—	210	100	0.69	0

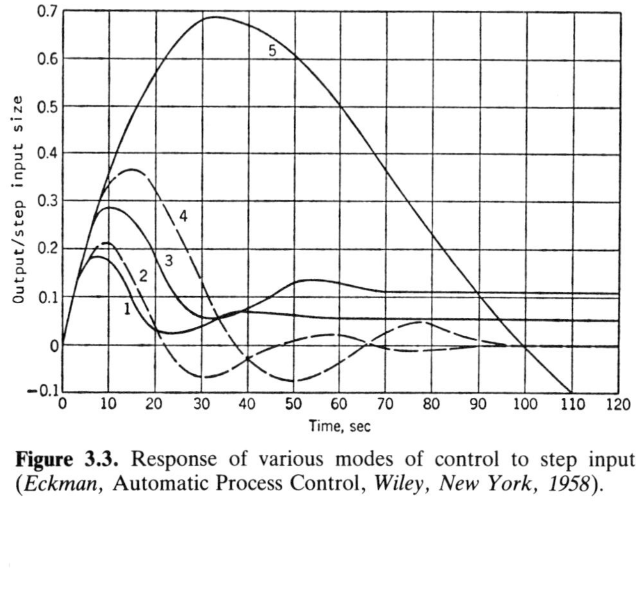

Figure 3.3. Response of various modes of control to step input (*Eckman,* Automatic Process Control, *Wiley, New York, 1958*).

SYMBOLS

On working flowsheets the detectors, transmitters, and controllers are identified individually by appropriate letters and serial numbers in circles. Control valves are identified by the letters CV- followed by a serial number. When the intent is to show only in general the kind of control system, no special symbol is used for detectors, but simply a point of contact of the signal line with the equipment or process line. Transmitters are devices that convert the measured variable into air pressure for pneumatic controllers or units appropriate for electrical controllers. Temperature, for instance, may be detected with thermocouples or electrical resistance or height of a liquid column or radiant flux, etc., but the controller can accept only pneumatic or electrical signals depending on its type. When the nature of the transmitter is clear, it may be represented by an encircled cross or left out entirely. For clarity, the flowsheet can include only the most essential information. In an actual design

EXAMPLE 3.1
Constants of PID Controllers from Response Curves to a Step Input
The method of Ziegler and Nichols [*Trans ASME,* (Dec. 1941)] will be used. The example is that of Tyner and May (*Process Engineering Control,* Ronald, New York, 1967). The response to a change of 2 psi on the diaphragm of the control valve is shown. The full range of control pressure is from 3 to 15 psi, a difference of 12 psi, and the range of temperature is from 100 to 200°F, a difference of 100°F. Evaluate the % displacement of pressure as

$$\Delta m = 100(2/12) = 16.7\%.$$

From the curve, the slope at the inflection point is

$$R = 17.5/100(7.8 - 2.4) = 3.24\%/\text{min},$$

and the apparent time delay is the intercept on the abscissa,

$$L = 2.40 \text{ min}.$$

The values of the constants for the several kinds of controllers are
Proportional: $100/K_p = \%\ \text{PB} = 100RL/\Delta m = 100(3.24)(2.4)/16.7 = 46.6\%$.
Proportional-integral: $\%\ \text{PB} = 110RL/\Delta m = 51.2\%$

$$K_i = L/0.3 = 8 \text{ min}$$

Proportional-integral-derivative:

$$\%\ \text{PB} = 83RL/\Delta m = 38.6\%,$$
$$K_i = 2L = 4.8 \text{ min},$$
$$K_d = 0.5L = 1.2 \text{ min}.$$

These are approximate instrument settings, and may need to be adjusted in process. PB is proportional band.

A recent improvement of the Ziegler–Nichols method due to Yuwana and Seborg [*AIChE J.* **28,** 434 (1982)] is calculator programmed by Jutan and Rodriguez [*Chem. Eng.* **91**(18), 69–73 (Sep. 3, 1984)].

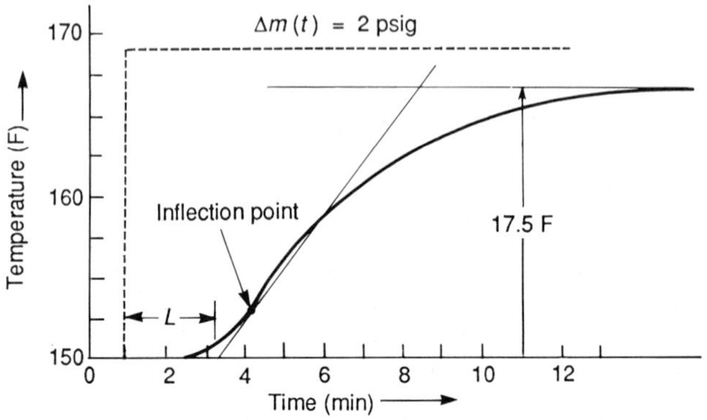

case, details of detectors and transmitters as well as all other elements of a control system are summarized on instrument specification forms. The simplified coding used in this chapter is summarized on Figure 3.4.

CASCADE (RESET) CONTROL

Some control situations require interacting controllers. On Figure 3.19(d), for instance, a composition controller regulates the setpoint of the temperature controller of a reactor and on Figure 3.15(g) the set point of the reflux flow rate is regulated by composition or temperature control. Composite systems made up of regions that respond with varying degrees of speed or sluggishness are advantageously equipped with cascade control. In the reactor of Figure 3.19(b), the temperature TT-1 of the vessel contents responds only slowly to changes in flow rate of the heat transfer medium, but the temperature TT-2 of the HTM leaving the cooling coil is comparatively sensitive to the flow rate. Accordingly, controller TC-2 is allowed to adjust the setpoint of the primary controller TC-1 with an overall improvement in control of the reactor temperature. The controller being reset is identified on flowsheets.

3.2. INDIVIDUAL PROCESS VARIABLES

The variables that need to be controlled in chemical processing are temperature, pressure, liquid level, flow rate, flow ratio, composition, and certain physical properties whose magnitudes may be influenced by some of the other variables, for instance, viscosity, vapor pressure, refractive index, etc. When the temperature and pressure are fixed, such properties are measures of composition which may be known exactly upon calibration. Examples of control of individual variables are shown in the rest of this chapter with the various equipment (say pumps or compressors) and processes (say distillation or refrigeration) and on the earlier flowsketches of this and the preceding chapters, but some general statements also can be made here. Most control actions ultimately depend on regulation of a flow rate with a valve.

TEMPERATURE

Temperature is regulated by heat exchange with a heat transfer medium (HTM). The flow rate of the HTM may be adjusted, or the condensing pressure of steam or other vapor, or the amount of heat transfer surface exposed to condensing vapor may be regulated by flooding with condensate, which always has a much lower heat transfer coefficient than that of condensing vapor. In a reacting system of appropriate vapor pressure, a boiling temperature at some desired value can be maintained by refluxing at the proper controlled pressure. Although examples of temperature control appear throughout this chapter, the main emphasis is in the section on heat exchangers.

PRESSURE

Pressure is controlled by regulating the flow of effluent from the vessel. The effluent may be the process stream itself or a non-condensable gas that is generated by the system or supplied for blanketing purposes. The system also may be made to float on the pressure of the blanketing gas supply. Control of the rate of condensation of the effluent by allowing the heat transfer surface to flood partially is a common method of regulating pressure in fractionation systems. Throttling a main effluent vapor line usually is not done because of the expense of large control valves. Figure 3.5 shows vacuum production and control with steam jet ejectors.

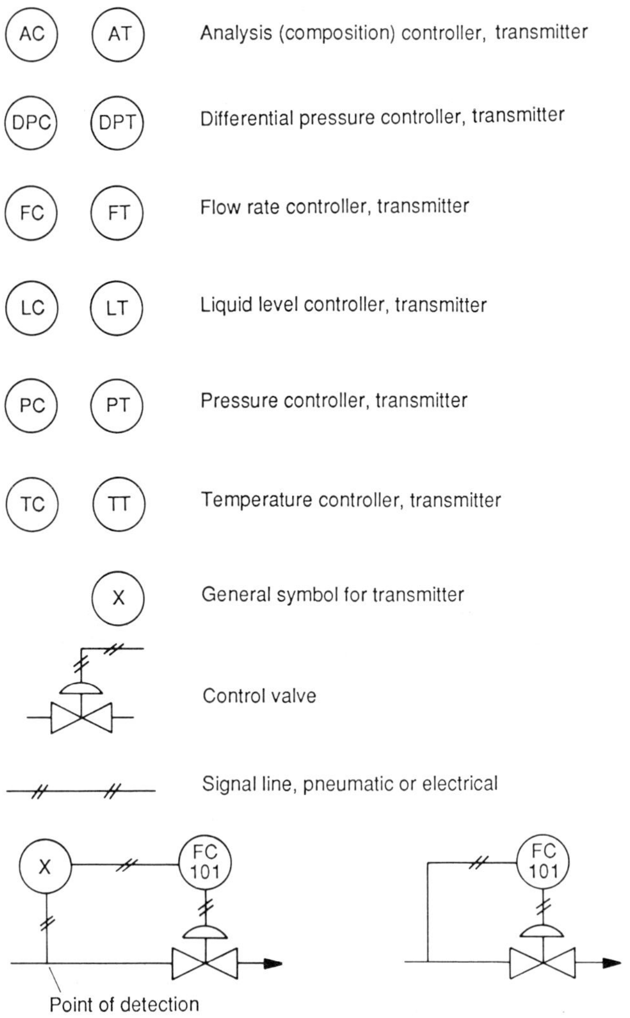

Figure 3.4. Symbols for control elements to be used on flowsheets. Instrument Society of America (ISA) publication no. S 51.5 is devoted to process instrumentation terminology.

LEVEL OF LIQUID

Level of liquid in a vessel often is maintained by permanent or adjustable built-in weirs for the effluent, notably on the trays of fractionators, extractors, etc., and in reactors and drums. Any desired adjustments of weir height, however, can be made only on shutdown. Control of the flow rate of effluent (sometimes of the input) is the most common other method of level control. Liquid levels often are disturbed by splashing or flow turbulence, so that rather sluggish controllers are used for this service. Conceivably, a level could be controlled by forcing effluent through an opening of fixed size with a controlled pressure, but there do not appear to be many such applications. Continual control of the weight of a vessel and its contents is another control method that is not used often. Figure 3.6 is devoted to level control.

FLOW RATE

A rate of flow is commonly measured by differential pressure across an orifice, but many other devices also are used on occasion. Simultaneous measurements of temperature and pressure allow the flow measurement to be known in mass units. Direct mass flow meters also are available. The flow measurement is transmitted to a controller which then adjusts the opening of a control valve so as to maintain the desired condition.

FLOW OF SOLIDS

Except for continuous weighing, control of the flow of solids is less precise than that of fluids. Several devices used for control of feed rates are shown schematically in Figure 3.7. They all employ variable speed drives and are individually calibrated to relate speed and flow rate. Ordinarily these devices are in effect manually set, but if the solid material is being fed to a reactor, some property of the mixture could be used for feed back control. The continuous belt weigher is capable ordinarily of ±1% accuracy and even ±0.1% when necessary. For processes such as neutralizations with lime, addition of the solid to process in slurry form is acceptable. The slurry is prepared as a batch of definite concentration and charged with a pump under flow control, often with a diaphragm pump whose stroke can be put under feedback control. For some applications it is adequate or necessary to feed weighed amounts of solids to a process on a timed basis.

FLOW RATIO

Flow ratio control is essential in processes such as fuel–air mixing, blending, and reactor feed systems. In a two-stream process, for example, each stream will have its own controller, but the signal from the primary controller will go to a ratio control device which adjusts the set point of the other controller. Figure 3.17(a) is an example. Construction of the ratioing device may be an adjustable mechanical linkage or may be entirely pneumatic or electronic. In other two-stream operations, the flow rate of the secondary stream may be controlled by some property of the combined stream, temperature in the case of fuel–air systems or composition or some physical property indicative of the proportions of the two streams.

COMPOSITION

The most common detectors of specific substances are gas chromatographs and mass spectrometers, which have been mentioned earlier in this chapter in connection with feedforward control. Also mentioned have been physical properties that have been calibrated against mixture compositions. Devices that are specific for individual substances also are sometimes available, for example pH, oxygen, and combustion products. Impregnated reactive tapes have been made as specific detectors for many substances and are useful particularly for low concentrations. Composition controllers act by adjusting some other condition of the system: for instance, the residence time in converters by adjusting the flow rate, or the temperature by adjusting the flow of HTM, or the pressure of gaseous reactants, or the circulation rate of regenerable catalysts, and so on. The taking of representative samples is an aspect of on-line analysis that slows down the responsiveness of such control. The application of continuously measuring in-line analyzers is highly desirable. Some physical properties can be measured this way, and also concentrations of hydrogen and many other ions with suitable electrodes. Composition controllers are shown for the processes of Figures 3.1 and 3.2.

3.3. EQUIPMENT CONTROL

Examples are presented of some usual control methods for the more widely occurring equipment in chemical processing plants. Other methods often are possible and may be preferable because of

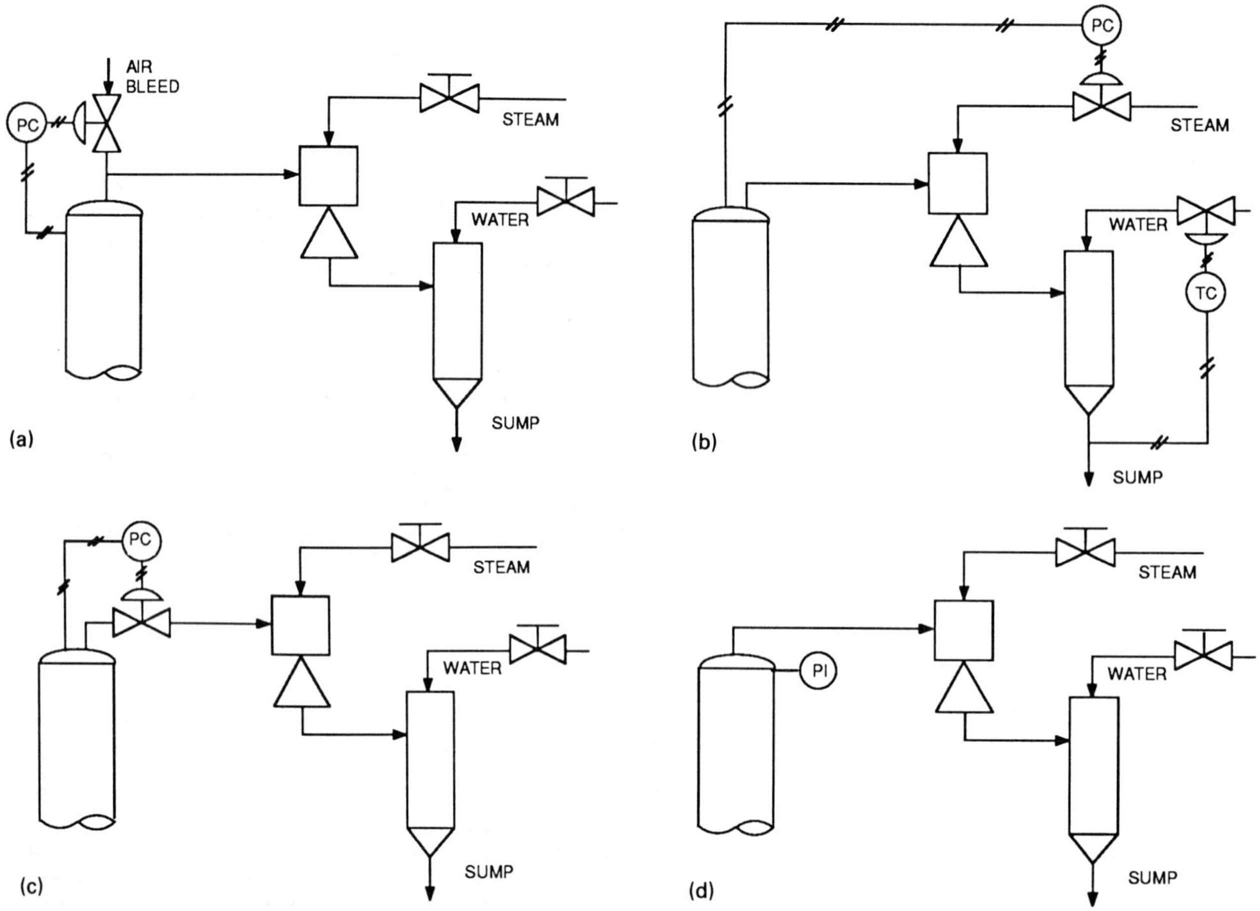

Figure 3.5. Vacuum control with steam jet ejectors and with mechanical vacuum pumps. (a) Air bleed on PC. The steam and water rates are hand set. The air bleed can be made as small as desired. This can be used only if air is not harmful to the process. Air bleed also can be used with mechanical vacuum pumps. (b) Both the steam and water supplies are on automatic control. This achieves the minimum cost of utilities, but the valves and controls are relatively expensive. (c) Throttling of process gas flow. The valve is larger and more expensive even than the vapor valve of case (a). Butterfly valves are suitable. This method also is suitable with mechanical vacuum pumps. (d) No direct pressure control. Settings of manual control valves for the utilities with guidance from pressure indicator PI. Commonly used where the greatest vacuum attainable with the existing equipment is desired.

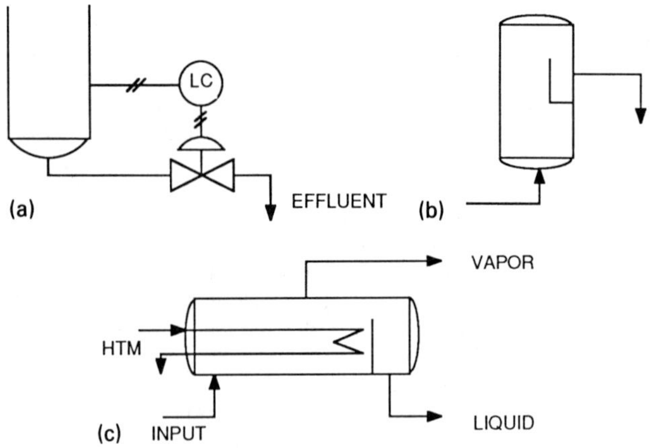

Figure 3.6. Some modes of control of liquid level. (a) Level control by regulation of the effluent flow rate. This mode is externally adjustable. (b) Level control with built in overflow weir. The weir may be adjustable, but usually only during shutdown of the equipment. (c) Overflow weir in a horizontal kettle reboiler. The weir setting usually is permanent.

greater sensitivity or lower cost. Also it should be noted that the choice of controls for particular equipment may depend on the kind of equipment it is associated with. Only a few examples are shown of feedforward control, which should always be considered when superior control is needed, the higher cost is justified, and the process simulation is known. Another relatively expensive method is composition control, which has not been emphasized here except for reactors and fractionators, but its possible utility always should be borne in mind. Only primary controllers are shown. The complete instrumentation of a plant also includes detectors and transmitters as well as indicators of various operating conditions. Such indications may be input to a computer for the record or for control, or serve as guides for manual control by operators who have not been entirely obsolesced.

HEAT TRANSFER EQUIPMENT

Four classes of this kind of equipment are considered: heat exchangers without phase change, steam heaters, condensers, and vaporizers or reboilers. These are grouped together with descriptions in Figures 3.8–3.11. Where applicable, comments are made about the utility of the particular method. In these heat

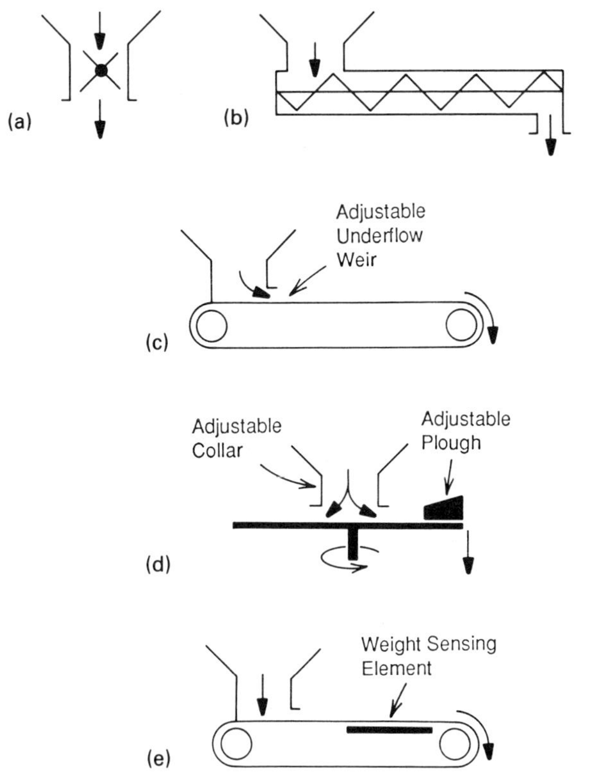

Figure 3.7. Solids feeders with variable speed drives. (a) Rotary vane (star) feeder with variable speed drive. (b) Horizontal screw feeder. (c) Belt feeder taking material from a bin with an adjustable underflow weir. (d) Rotary plate feeder: Rate of discharge is controlled by the rotation speed, height of the collar, and the position of the plow. (e) Continuously weighing feeder with variable speed belt conveyor.

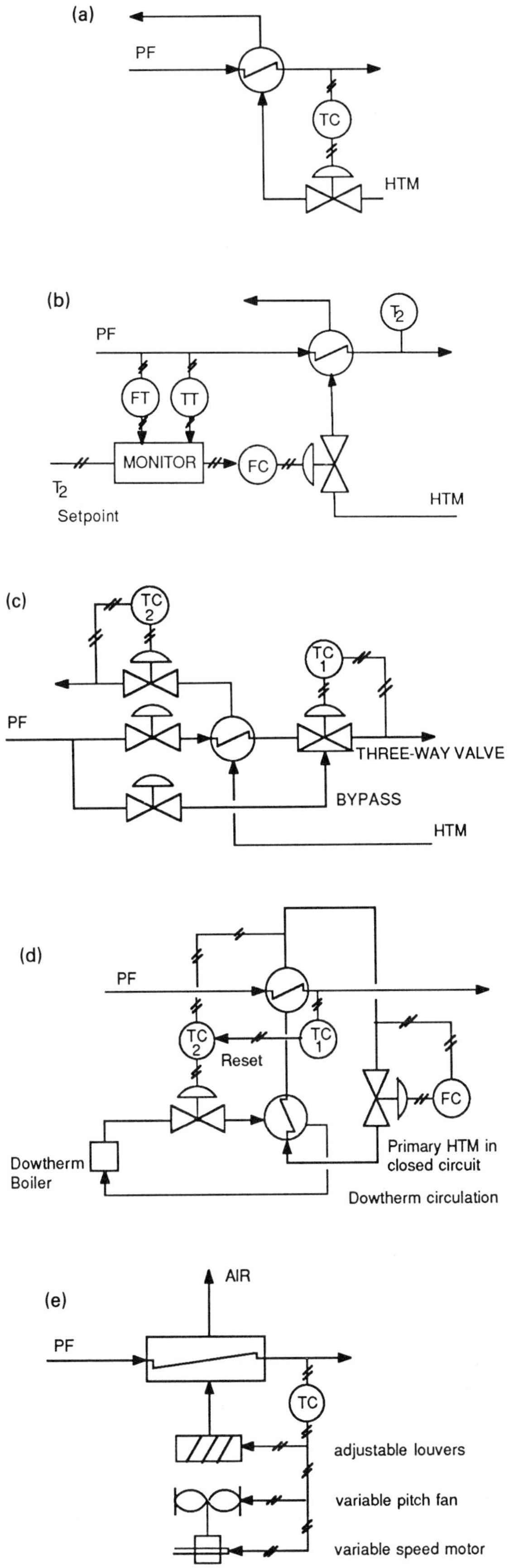

Figure 3.8. Heat exchangers without phase change. PF = process fluid, HTM = heat transfer medium. (a) Feedback control of PF outlet temperature. Flow rate of HTM is adjusted as the PF outlet temperature is perturbed. The valve may be in either the input or output line. (b) Feedforward control. PF outlet setpoint T-2 and perturbations of PF input flow and temperature are fed to the monitor which adjusts the flow rate of the HTM to maintain constant PF outlet temperature T2. (c) Exchanger with bypass of process fluid with a three-way valve. The purpose of TC-2 is to conserve on that fluid or to limit its temperature. When the inherent leakage of the three-way valve is objectionable, the more expensive two two-way valves in the positions shown are operated off TC-1. (d) A two-fluid heat transfer system. The PF is heated with the HTM which is a closed circuit heated by Dowtherm or combustion gases. The Dowtherm is on flow control acting off TC-2 which is on the HTM circuit and is reset by TC-1 on the PF outlet. The HTM also is on flow control. Smoother control is achievable this way than with direct heat transfer from very high temperature Dowtherm or combustion gases. (e) Air cooler. Air flow rate is controllable with adjustable louvers or variable pitch fan or variable speed motors. The latter two methods achieve some saving of power compared with the louver design. Multispeed motors are also used for change between day and night and between winter and summer. The switching can be made automatically off the air temperature.

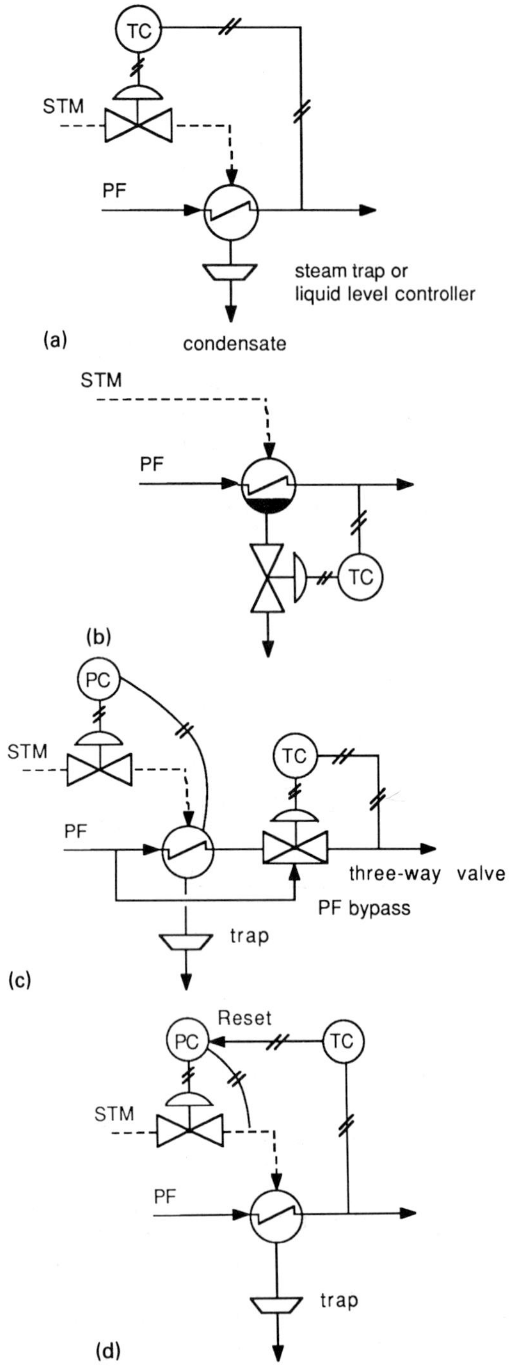

Figure 3.9. Steam heaters. (a) Flow of steam is controlled off the PF outlet temperature, and condensate is removed with a steam trap or under liquid level control. Subject to difficulties when condensation pressure is below atmospheric. (b) Temperature control on the condensate removal has the effect of varying the amount of flooding of the heat transfer surface and hence the rate of condensation. Because the flow of condensate through the valve is relatively slow, this mode of control is sluggish compared with (a). However, the liquid valve is cheaper than the vapor one. (c) Bypass of process fluid around the exchanger. The condensing pressure is maintained above atmospheric so that the trap can discharge freely. (d) Cascade control. The steam pressure responds quickly to upsets in steam supply conditions. The more sluggish PF temperature is used to adjust the pressure so as to maintain the proper rate of heat transfer.

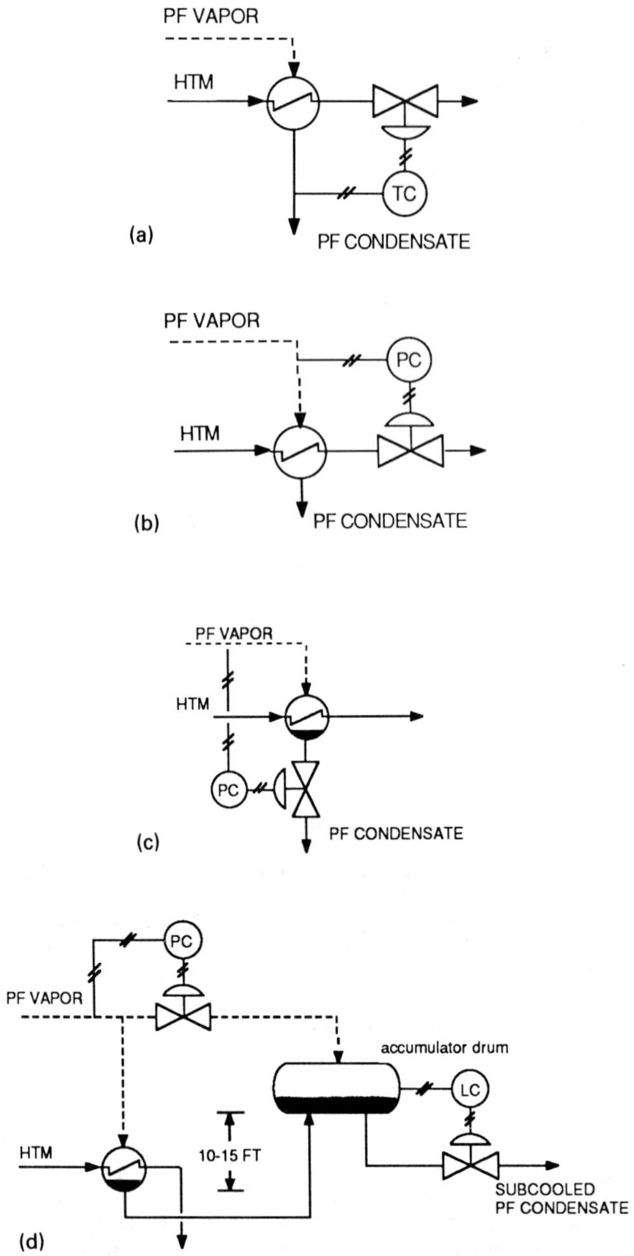

Figure 3.10. Condensers. (a) Condenser on temperature control of the PF condensate. Throttling of the flow of the HTM may make it too hot. (b) Condenser on pressure control of the HTM flow. Throttling of the flow of the HTM may make it too hot. (c) Flow rate of condensate controlled by pressure of PF vapor. If the pressure rises, the condensate flow rate increases and the amount of unflooded surface increases, thereby increasing the rate of condensation and lowering the pressure to the correct value. (d) Condenser with vapor bypass to the accumulator drum. The condenser and drum become partially flooded with subcooled condensate. When the pressure falls, the vapor valve opens, and the vapor flows directly to the drum and heats up the liquid there. The resulting increase in vapor pressure forces some of the liquid back into the condenser so that the rate of condensation is decreased and the pressure consequently is restored to the preset value. With sufficient subcooling, a difference of 10–15 ft in levels of drum and condenser is sufficient for good control by this method.

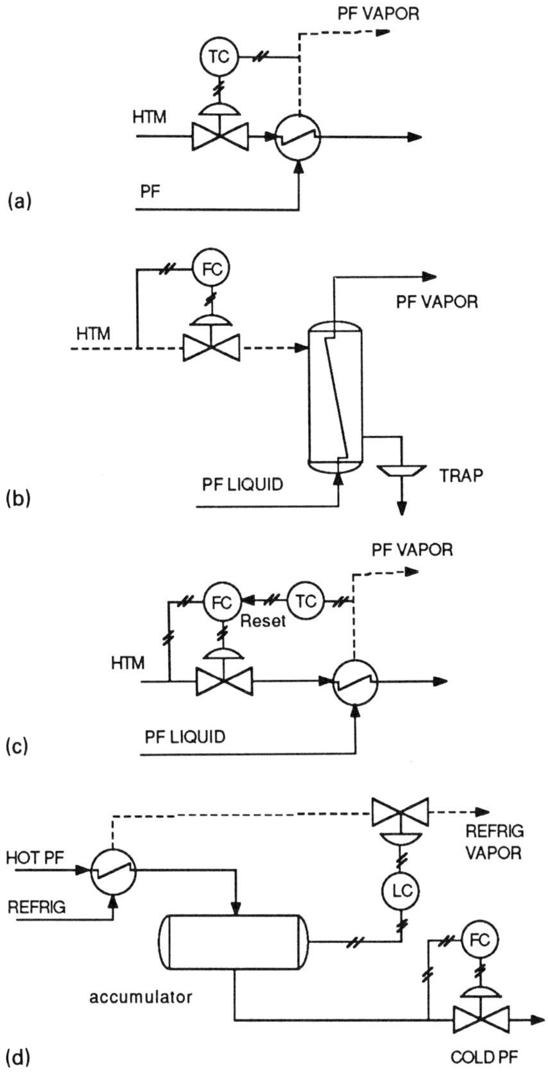

Figure 3.11. Vaporizers (reboilers). (a) Vaporizer with flow-rate of HTM controlled by temperature of the PF vapor. HTM may be liquid or vapor to start. (b) Thermosiphon reboiler. A constant rate of heat input is assured by flow control of the HTM which may be either liquid or vapor to start. (c) Cascade control of vaporizer. The flow control on the HTM supply responds rapidly to changes in the heat supply system. The more sluggish TC on the PF vapor resets the FC if need be to maintain temperature. (d) Vaporization of refrigerant and cooling of process fluid. Flow rate of the PF is the primary control. The flow rate of refrigerant vapor is controlled by the level in the drum to ensure constant condensation when the incoming PF is in vapor form.

transfer processes the object is to control the final temperature of the process fluid (PF) or the pressure of its source or to ensure a constant rate of heat input. This is accomplished primarily by regulation of the flow of the heat transfer medium (HTM). Regulation of the temperature of the HTM usually is less convenient, although it is done indirectly in steam heaters by throttling of the supply which has the effect of simultaneously changing the condensing pressure and temperature of the steam side.

DISTILLATION EQUIPMENT

As a minimum, a distillation assembly consists of a tower, reboiler, condenser, and overhead accumulator. The bottom of the tower serves as accumulator for the bottoms product. The assembly must be controlled as a whole. Almost invariably, the pressure at either the top or bottom is maintained constant; at the top at such a value that the necessary reflux can be condensed with the available coolant; at the bottom in order to keep the boiling temperature low enough to prevent product degradation or low enough for the available HTM, and definitely well below the critical pressure of the bottom composition. There still remain a relatively large number of variables so that care must be taken to avoid overspecifying the number and kinds of controls. For instance, it is not possible to control the flow rates of the feed and the top and bottom products under perturbed conditions without upsetting holdup in the system.

Two flowsketches are shown on Figures 3.1 and 3.12 of controls on an ethylene fractionator. On Figure 3.1, which is part of the complete process of Figure 3.2, a feedforward control system with a multiplicity of composition analyzers is used to ensure the high degree of purity that is needed for this product. The simpler diagram, Figure 3.12, is more nearly typical of two-product fractionators, the only uncommon variation being the use of a feed–overhead effluent heat exchanger to recover some refrigeration.

Crude oil fractionators are an example of a more elaborate system. They make several products as side streams and usually have some pumparound reflux in addition to top reflux which serve to optimize the diameter of the tower. Figure 3.13 is of such a tower operating under vacuum in order to keep the temperature below cracking conditions. The side streams, particularly those drawn off atmospheric towers, often are steam stripped in external towers hooked up to the main tower in order to remove lighter components. These strippers each have four or five trays, operate

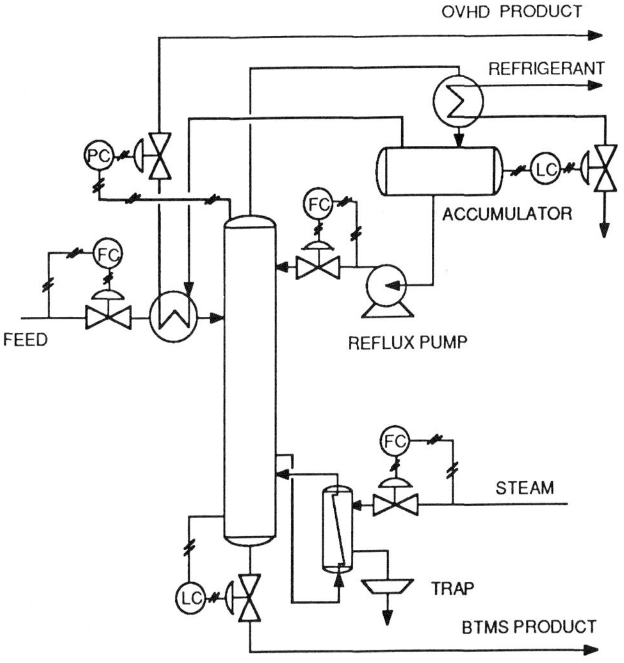

Figure 3.12. Fractionator for separating ethylene and ethane with a refrigerated condenser. FC on feed, reflux, and steam supply. LC on bottom product and refrigerant vapor. Pressure control PC on overhead vapor product.

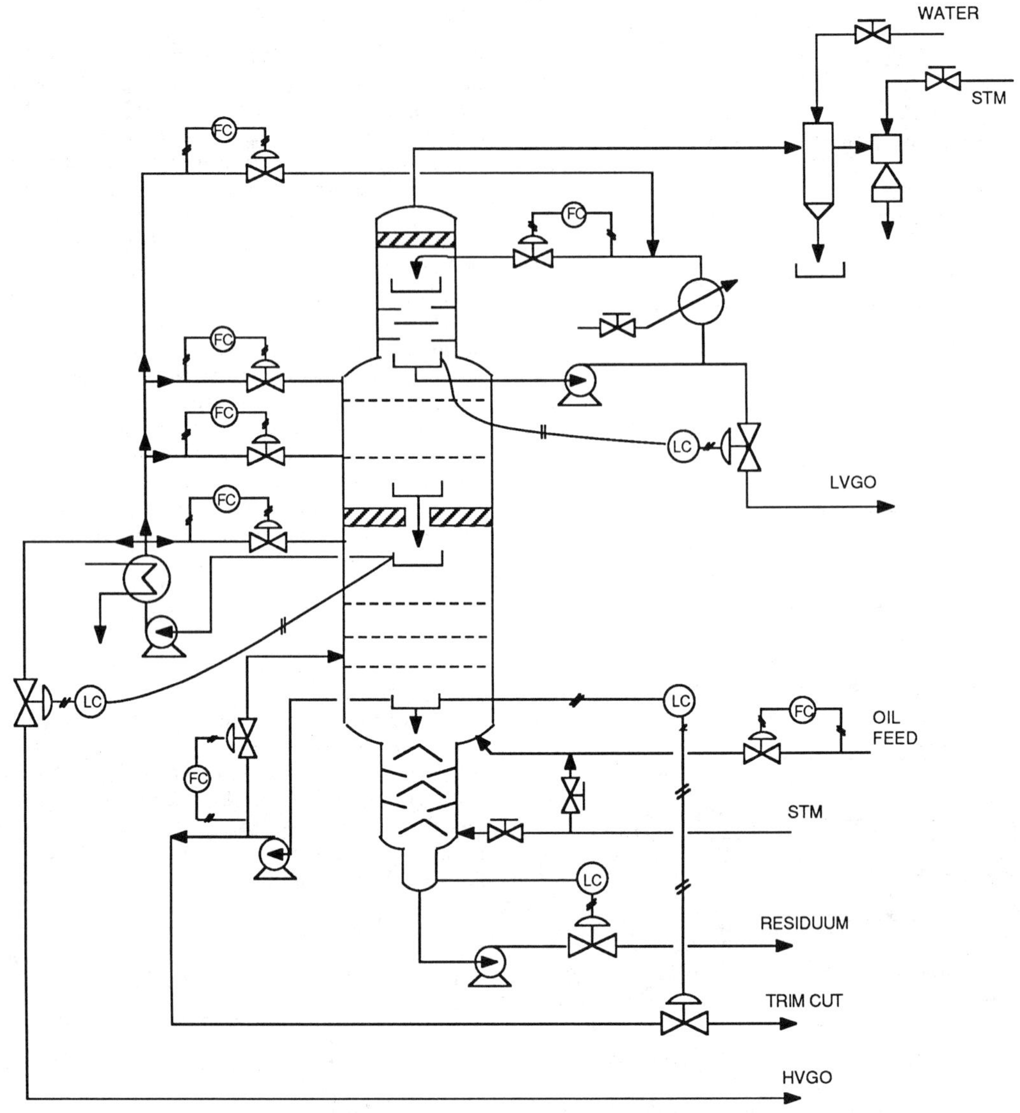

Figure 3.13. Crude oil vacuum tower. Pumparound reflux is provided at three lower positions as well as at the top, with the object of optimizing the diameter of the tower. Cooling of the side streams is part of the heat recovery system of the entire crude oil distillation plant. The cooling water and the steam for stripping and to the vacuum ejector are on hand control.

off level control on the main tower, and return their vapors to the main tower.

A variety of control schemes are shown separately in Figures 3.14 and 3.15 for the lower and upper sections of fractionators. To some extent, these sections are controllable independently but not entirely so because the flows of mass and heat are interrelated by the conservation laws. In many of the schemes shown, the top reflux rate and the flow of HTM to the reboiler are on flow controls. These quantities are not arbitrary, of course, but are found by calculation from material and energy balances. Moreover, neither the data nor the calculation method are entirely exact, so that some adjustments of these flow rates must be made in the field until the best possible performance is obtained from the equipment. In modern large or especially sensitive operations, the fine tuning is done by computer.

For the lower section of the fractionator, the cases of Figure 3.14 show the heat input to be regulated in these five different ways:

1. On flow control of the heat transfer medium (HTM),
2. On temperature control of the vapor leaving the reboiler or at some point in the tower,
3. On differential pressure between key points in the tower,
4. On liquid level in the bottom section,
5. On control of composition or some physical property of the bottom product.

Although only one of these methods can be shown clearly on a particular sketch, others often are usable in combination with the other controls that are necessary for completeness. In some cases the HTM shown is condensing vapor and in other cases it is hot oil, but the particular flowsketches are not necessarily restricted to one or the other HTM. The sketches are shown with and without pumps

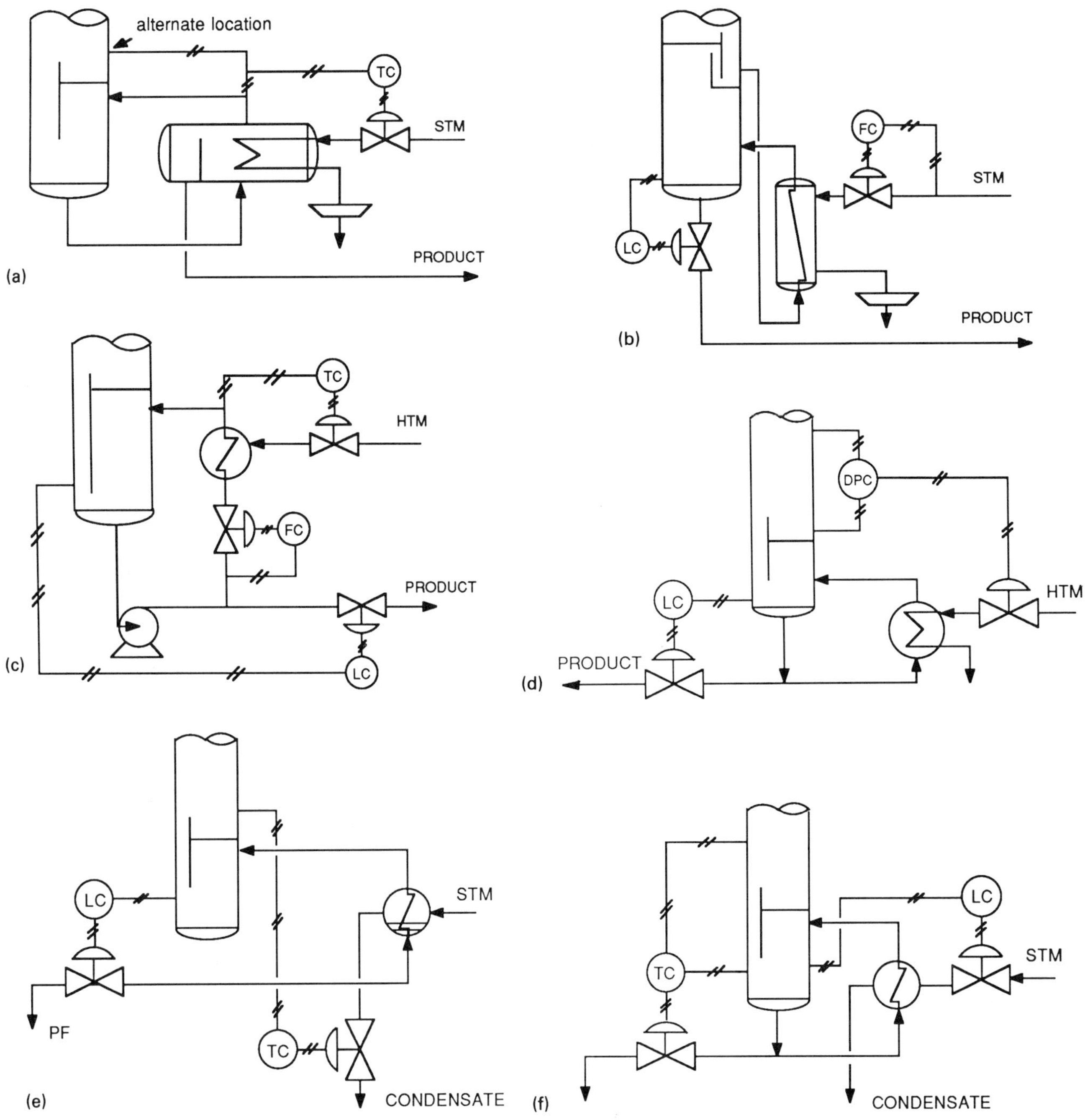

Figure 3.14. The lower ends of fractionators. (a) Kettle reboiler. The heat source may be on TC of either of the two locations shown or on flow control, or on difference of pressure between key locations in the tower. Because of the built-in weir, no LC is needed. Less head room is needed than with the thermosiphon reboiler. (b) Thermosiphon reboiler. Compared with the kettle, the heat transfer coefficient is greater, the shorter residence time may prevent overheating of thermally sensitive materials, surface fouling will be less, and the smaller holdup of hot liquid is a safety precaution. (c) Forced circulation reboiler. High rate of heat transfer and a short residence time which is desirable with thermally sensitive materials are achieved. (d) Rate of supply of heat transfer medium is controlled by the difference in pressure between two key locations in the tower. (e) With the control valve in the condensate line, the rate of heat transfer is controlled by the amount of unflooded heat transfer surface present at any time. (f) Withdrawal on TC ensures that the product has the correct boiling point and presumably the correct composition. The LC on the steam supply ensures that the specified heat input is being maintained. (g) Cascade control: The set point of the FC on the steam supply is adjusted by the TC to ensure constant temperature in the column. (h) Steam flow rate is controlled to ensure specified composition of the PF effluent. The composition may be measured directly or indirectly by measurement of some physical property such as vapor pressure. (i) The three-way valve in the hot oil heating supply prevents buildup of excessive pressure in case the flow to the reboiler is throttled substantially. (j) The three-way valve of case (i) is replaced by a two-way valve and a differential pressure controller. This method is more expensive but avoids use of the possibly troublesome three-way valve.

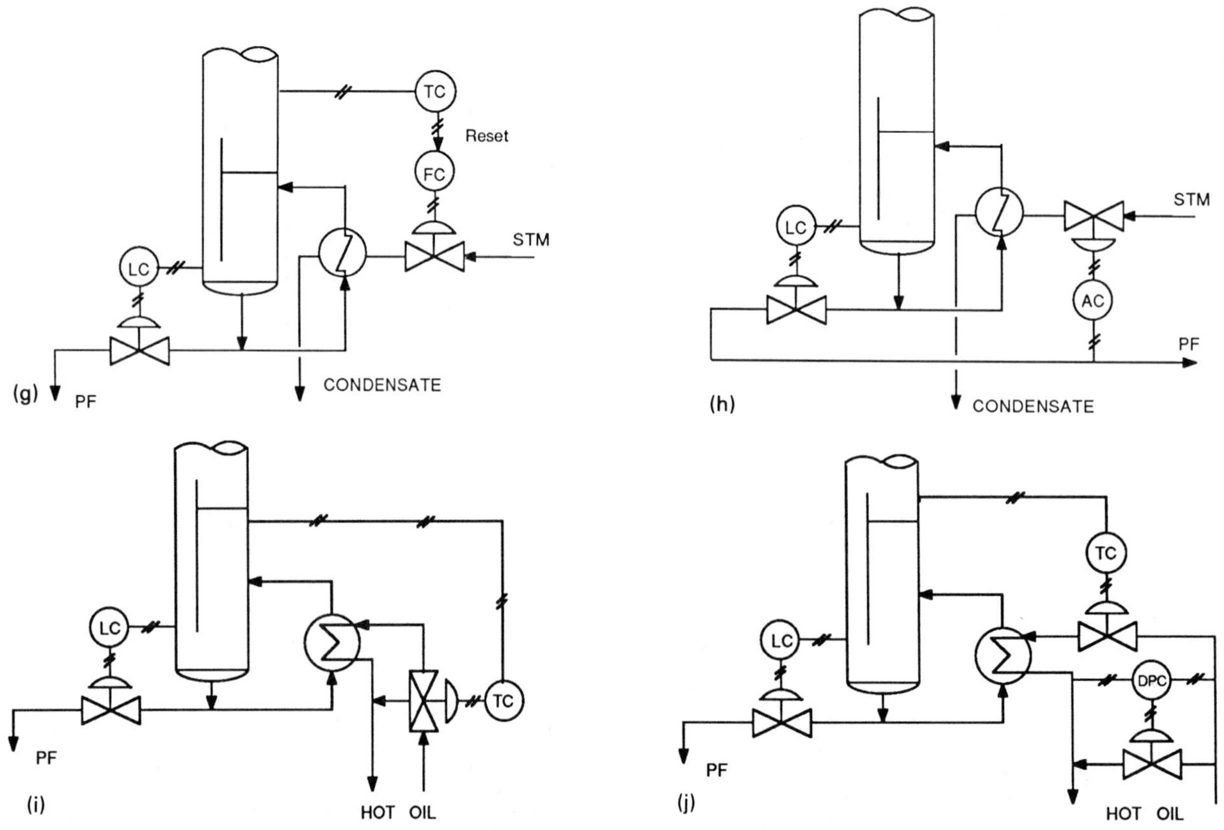

Figure 3.14—*(continued)*

for withdrawal of bottom product. When the tower pressure is sufficient for transfer of the product to the following equipment, a pump is not needed.

Upper section control methods are shown on Figure 3.15. They all incorporate control of the pressure on the tower, either by throttling some vapor flow rate or by controlling a rate of condensation. In the latter case this can be done by regulating the flow or temperature of the HTM or by regulating the amount of heat transfer surface exposed to contact with condensing vapor.

Flow control of reflux is most common. It is desirable in at least these situations:

1. When the temperature on a possible control tray is insensitive to the composition, which is particularly the case when high purity overhead is being made,
2. When the expense of composition control is not justifiable,
3. When noncondensables are present,
4. With tall and wide columns that have large holdup and consequently large lags in interchange of heat and mass between phases,
5. When the process coupling of the top and bottom temperature controllers makes their individual adjustments difficult,
6. When the critical product is at the bottom.

In all these cases the reflux rate is simply set at a safe value, enough to nullify the effects of any possible perturbations in operation. There rarely is any harm in obtaining greater purity than actually is necessary. The cases that are not on direct control of reflux flow rate are: (g) is on cascade temperature (or composition) and flow control, (h) is on differential temperature control, and (i) is on temperature control of the HTM flow rate.

LIQUID–LIQUID EXTRACTION TOWERS

The internals of extraction towers can be packing, sieve trays, empty with spray feeds or rotating disks. The same kinds of controls are suitable in all cases, and consist basically of level and flow controls. Figure 3.16 shows some variations of such arrangements. If the solvent is lighter than the material being extracted, the two inputs indicated are of course interchanged. Both inputs are on flow control. The light phase is removed from the tower on LC or at the top or on level maintained with an internal weir. The bottom stream is removed on interfacial level control (ILC). A common type of this kind of control employs a hollow float that is weighted to have a density intermediate between those of the two phases. As indicated by Figures 3.16(a) and 3.16(d), the interface can be maintained in either the upper or lower sections of the tower. Some extractions are performed with two solvents that are fed separately to the tower, ordinarily on separate flow controls that may be, however, linked by flow ratio control. The relative elevations of feed and solvents input nozzles depend on the nature of the extraction process.

Controls other than those of flow and level also may be needed in some cases, of which examples are on Figure 3.17. The scheme of part (a) maintains the flow rate of solvent in constant ratio with the main feed stream, whatever the reasons for variation in flow rate of the latter stream. When there are fluctuations in the composition of the feed, it may be essential to adjust the flow rate of the solvent to maintain constancy of some property of one or the other of the effluent streams. Figure 3.17(b) shows reset of the solvent flow rate by the composition of the raffinate. The temperature of an extraction process ordinarily is controlled by regulating the temperatures of the feed streams. Figure 3.17(c) shows the temperature of one of the streams to be controlled by TC-2 acting on the flow rate of the HTM, with reset by the temperature of a control point in the tower acting through TC-1. When the effluents are unusually sensitive to variation of input conditions, it may be inadvisable to wait for feedback from an upset of output performance, but to institute feedforward control instead. In this kind of system, the input conditions are noted, and calculations are made and implemented by on-line computer of other changes that are needed in order to maintain satisfactory operation.

Mixer–settler assemblies for extraction purposes often are preferable to differential contact towers in order to obtain very high extraction yields or to handle large flow rates or when phase

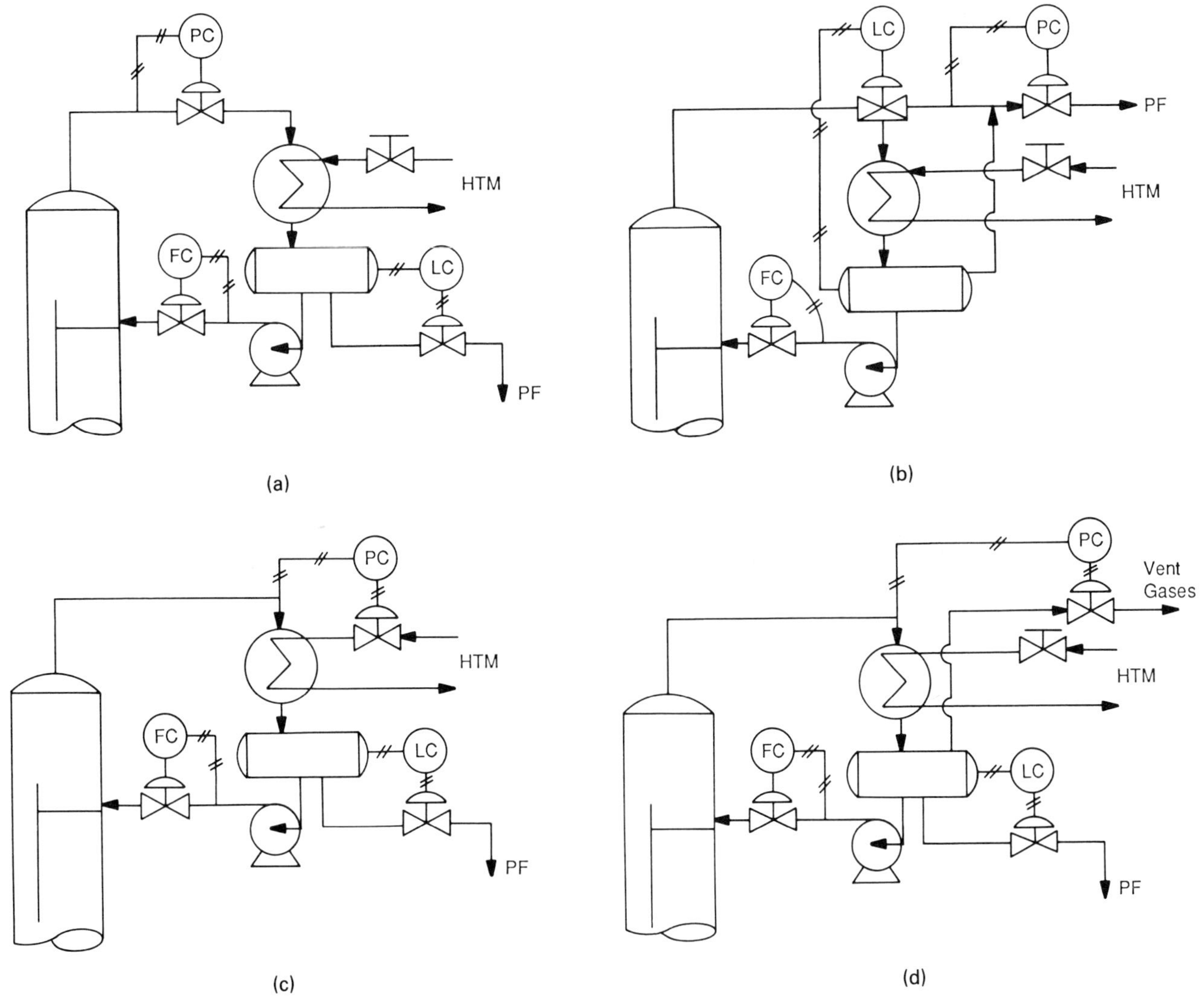

Figure 3.15. Control modes for the upper sections of fractionators. (a) Pressure control by throttling of the overhead vapor flow. The drawbacks of this method are the cost of the large control valve and the fact that the reflux pump operates with a variable suction head. The flow of HTM is hand set. (b) Applicable when the overhead product is taken off as vapor and only the reflux portion need be condensed. Two two-way valves can replace the single three-way valve. The flow of HTM is hand set. (c) Flow rate of the HTM is regulated to keep the pressure constant. One precaution is to make sure that the HTM, for example water, does not overheat and cause scaling. The HTM flow control valve is small compared with the vapor valve of case (a). (d) Pressure control is maintained by throttling uncondensed vapors. Clearly only systems with uncondensables can be handled this way. The flow of the HTM is manually set. (e) Bypass of vapor to the drum on PC: The bypassed vapor heats up the liquid there, thereby causing the pressure to rise. When the bypass is closed, the pressure falls. Sufficient heat transfer surface is provided to subcool the condensate. (f) Vapor bypass between the condenser and the accumulator, with the condenser near ground level for the ease of maintenance: When the pressure in the tower falls, the bypass valve opens, and the subcooled liquid in the drum heats up and is forced by its vapor pressure back into the condenser. Because of the smaller surface now exposed to the vapor, the rate of condensation is decreased and consequently the tower pressure increases to the preset value. With normal subcooling, obtained with some excess surface, a difference of 10–15 ft in levels of drum and condenser is sufficient for good control. (g) Cascade control: The same system as case (a), but with addition of a TC (or composition controller) that resets the reflux flow rate. (h) Reflux rate on a differential temperature controller. Ensures constant internal reflux rate even when the performance of the condenser fluctuates. (i) Reflux is provided by a separate partial condenser on TC. It may be mounted on top of the column as shown or inside the column or installed with its own accumulator and reflux pump in the usual way. The overhead product is handled by an after condenser which can be operated with refrigerant if required to handle low boiling components.

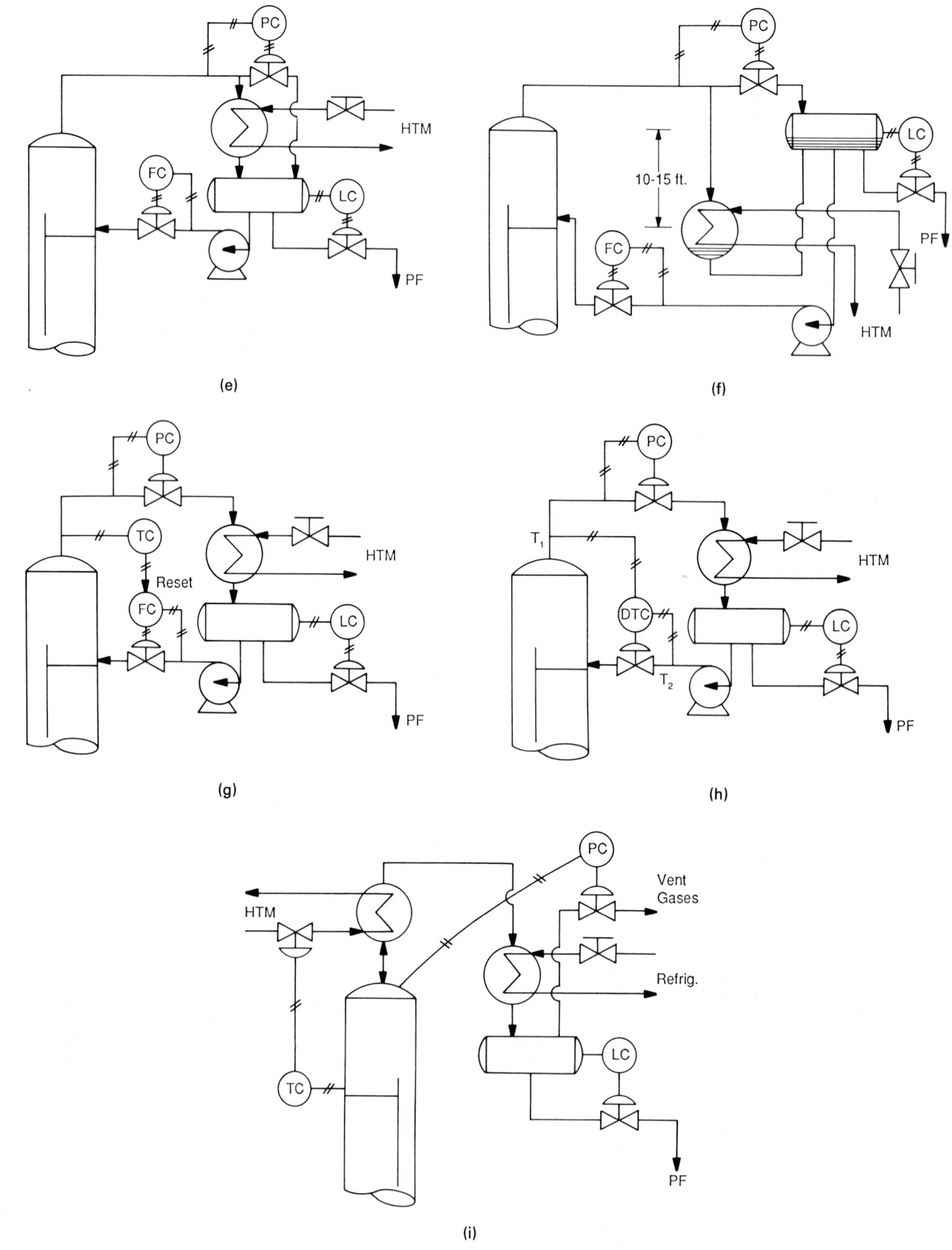

Figure 3.15—*(continued)*

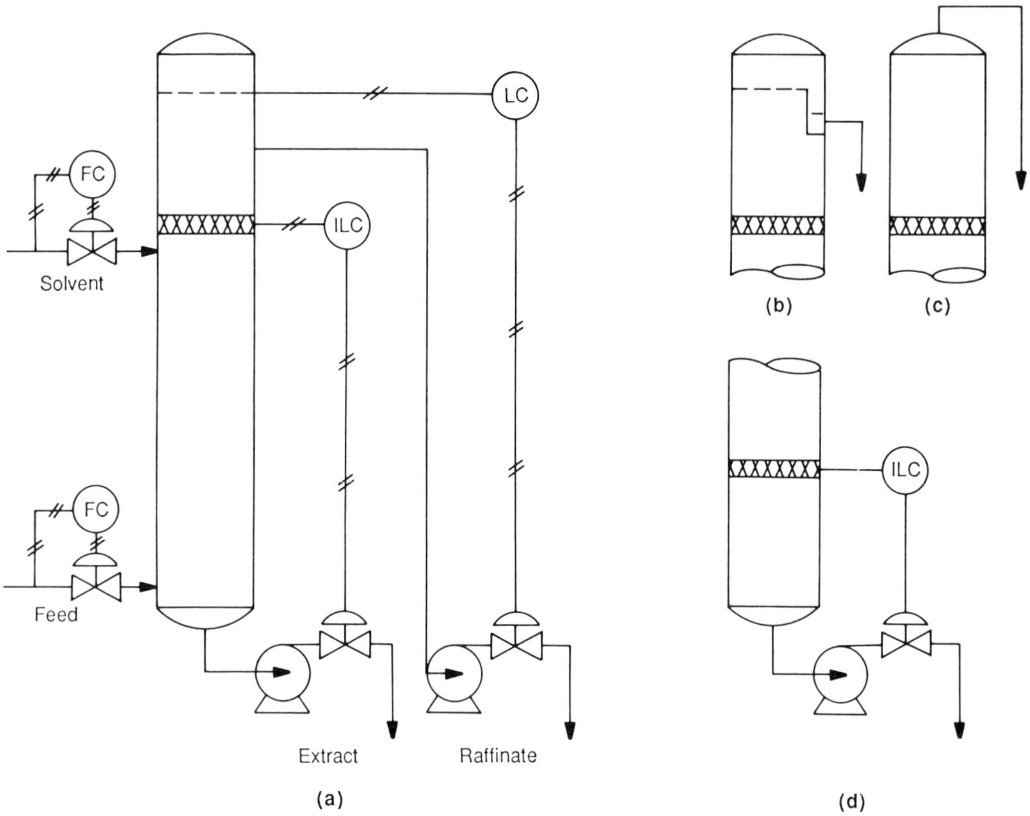

Figure 3.16. Extraction tower control. (a) Operation with heavy solvent, interface in the upper section, top liquid level on LC. (b) Same as part (a) but with overflow weir for the light phase. (c) Same as part (a) but with completely full tower and light phase out at the top. (d) Operation with interface on ILC in the lower section, removal of the light phase from the upper section by any of the methods of (a), (b), or (c).

separation is slow and much time is needed. Often, also, relatively simple equipment is adequate for small capacities and easy separations. Several designs of varying degrees of sophistication are available commercially, some of which are described by Lo, Baird, Hanson (*Handbook of Solvent Extraction,* Wiley, New York, 1983). The basic concept, however, is illustrated on Figure 3.18. The solvent and feed are thoroughly mixed in one chamber and overflow into another, partitioned chamber where separation into light and heavy phases occurs by gravity. Ordinarily the settling chamber is much the larger. The heavy phase is removed on interfacial level control and the light one on level control. The takeoffs also can be controlled with internal weirs or manually.

Several centrifugal contactors of proprietary nature are on the market. Their controls are invariably built in.

CHEMICAL REACTORS

The progress of a given reaction depends on the temperature, pressure, flow rates, and residence times. Usually these variables are controlled directly, but since the major feature of a chemical reaction is composition change, the analysis of composition and the resetting of the other variables by its means is an often used means of control. The possible occurrence of multiple steady states and the onset of instabilities also are factors in deciding on the nature and precision of a control system.

Because of the sensitivity of reaction rates to temperature, control of that variable often dominates the design of a reactor so that it becomes rather a heat exchanger in which a reaction occurs almost incidentally. Accordingly, besides the examples of reactor controls of this section, those of heat exchangers in that section may be consulted profitably. Heat transfer and holding time may be provided in separate equipment, but the complete assembly is properly regarded as a reactor. An extreme example, perhaps, is the two-stage heater-reactor system of Figure 3.19(f); three or more such stages are used for endothermic catalytic reforming of naphthas, and similar arrangements exist with intercoolers for exothermic processes.

Although the bulk of chemical manufacture is done on a continuous basis, there are sectors of the industry in which batch reactors are essential, notably for fermentations and polymerizations. Such plants may employ as many as 100 batch reactors. The basic processing steps include the charging of several streams, bringing up to reaction temperature, the reaction proper, maintenance of reaction temperature, discharge of the product, and preparation for the next batch. Moreover, the quality of the product depends on the accuracy of the timing and the closeness of the control.

Small installations are operated adequately and economically by human control, but the opening and closing of many valves and the setting of conditions at precise times clearly call for computer control of multiple batch installations. Computers actually have taken over in modern synthetic rubber and other polymerization industries. Interested readers will find a description, complete with

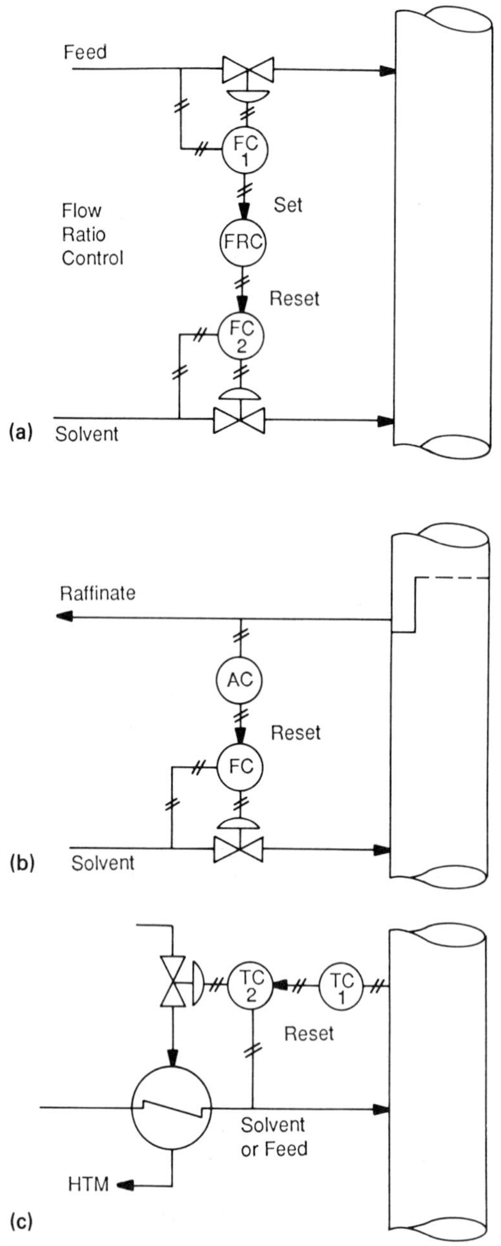

Figure 3.17. Some other controls on extraction towers. (a) Solvent flow rate maintained in constant ratio with the feed rate. (b) Solvent flow rate reset by controlled composition of raffinate. (c) Temperature of solvent or feed reset by the temperature at a control point in the tower.

logic diagrams for normal and emergency operations, of the tasks involved in generating a computer system for a group of batch reactors in the book of Liptak (1973, pp. 536–565). Control of discontinuous processes in general is treated in the book of Skrokov (1980, pp. 128–163).

In the present discussion, emphasis will be placed on the control of continuous reactors, concentrating on the several examples of Figure 3.19 in the order of the letter designations of individual figures used there.

(a) Stirred tanks are used either as batch or continuous flow reactors. Heat transfer may be provided with an external heat exchanger, as shown on this figure, or through internal surface or a jacket. Alternate modes of control may be used with the controls shown: (i) When the HTM is on temperature control, the pumparound will be on flow control; (ii) when the pumparound is on temperature control, the HTM will be on flow control; (iii) for continuous overflow of product, the control point for temperature may be on that line or in the vessel; (iv) for batch operation, the control point for temperature clearly must be in the vessel. Although level control is shown to be maintained with an internal weir, the product can be taken off with the pump on level control.

(b) This shows either direct or cascade control of the temperature of a reactor with internal heat transfer surface and an internal weir. The sluggishly responding temperature of the vessel is used to reset the temperature controller of the HTM. For direct control, the TC-2 is omitted and the control point can be on the HTM outlet or the product line or in the vessel.

(c) Quite a uniform temperature can be maintained in a reactor if the contents are boiling. The sketch shows temperature maintenance by refluxing evolved vapors. A drum is shown from which uncondensed gases are drawn off on pressure control, but the construction of the condenser may permit these gases to be drawn off directly, thus eliminating need for the drum. The HTM of the condenser is on TC which resets the PC if necessary in order to maintain the correct boiling temperature in the reactor. Other modes of pressure control are shown with the fractionator sketches of Figure 3.15 and on Figure 3.5 dealing with vacuum control.

(d) Flow reactors without mechanical agitation are of many configurations, tanks or tubes, empty or containing fixed beds of particles or moving particles. When the thermal effects of reaction are substantial, multiple small tubes in parallel are used to provide adequate heat transfer surface. The sketch shows a single tube provided with a jacket for heat transfer. Feed to the reactor is on flow control, the effluent on pressure control, and the flow of the HTM on temperature control of the effluent with the possibility of reset by the composition of the effluent.

(e) Heat transfer to high temperature reactions, above 300°C or so, may be accomplished by direct contact with combustion gases. The reaction tubes are in the combustion zone but safely away from contact with the flame. The control mode is essentially similar to that for case (d), except that fuel–air mixture takes the place of the HTM. The supply of fuel is on either temperature or composition control off the effluent stream, and the air is maintained in constant ratio with the fuel with the flow ratio controller FRC.

(f) High temperature endothermic processes may need several reaction vessels with intermediate heat input. For example, the inlet temperature to each stage of a catalytic reformer is about 975°F and the temperature drop ranges from about 100°F in the first stage to about 15°F in the last one. In the two-stage assembly of this figure, the input is on FC, the outlet of the last reactor on PC, and the fuel supply to each furnace is on TC of its effluent, with the air supply on flow ratio control, as shown for example (e).

(g) Very effective heat transfer is accomplished by mixing of streams at different temperatures. The cumene process shown here employs injection of cold reacting mixture and cold inert propane and water to prevent temperature escalation; by this scheme, the inlet and outlet temperature are made essentially the same, about 500°F. Although not shown here, the main feed is, as usual for reactors, on FC and the outlet on PC. The

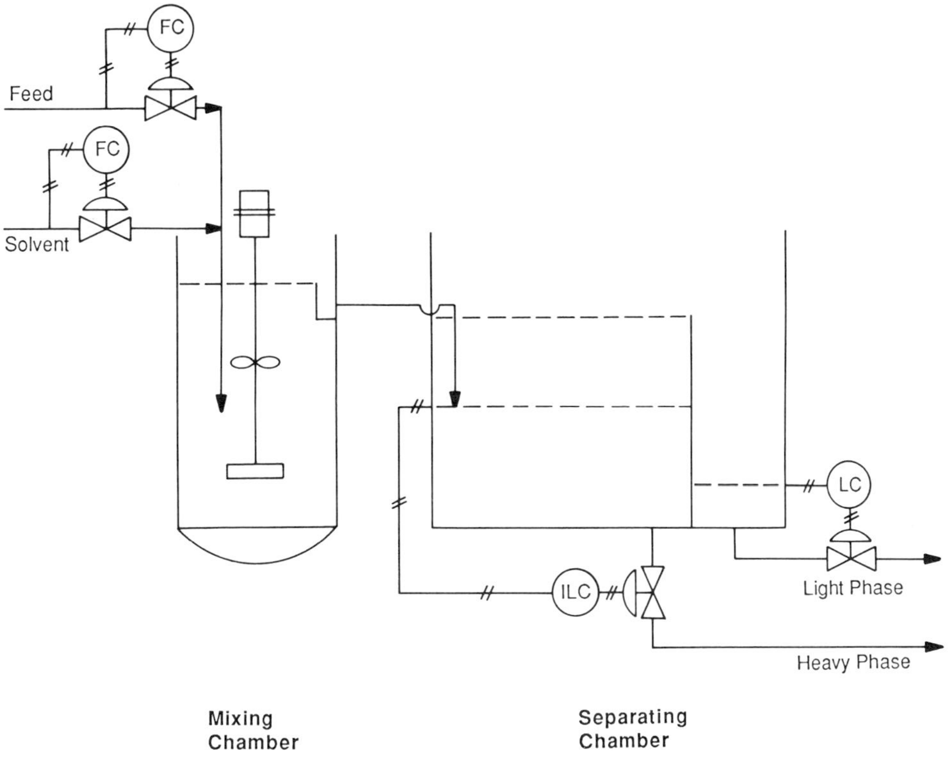

Figure 3.18. Functioning and controls of a mixer–settler assembly for liquid–liquid extraction.

sidestreams are regulated with hand-set valves by experienced operators in this particular plant, but they could be put on automatic control if necessary. Other processes that employ injection of cold process gas at intermediate points are some cases of ammonia synthesis and sulfur dioxide oxidation.

(h) In catalytic cracking of petroleum fractions, an influential side reaction is the formation of carbon which deposits on the catalyst and deactivates it. Unacceptable deactivation occurs in about 10 min, so that in practice continuous reactivation of a portion of the catalyst in process must be performed. As shown on this sketch, spent catalyst is transferred from the reactor to the regenerator on level control, and returns after regeneration under TC off the reactor temperature. Level in the regenerator is maintained with an overflow standpipe. Smooth transfer of catalyst between vessels is assisted by the differential pressure control DPC, but in some plants transfer is improved by injection of steam at high velocity into the lines as shown on this sketch for the input of charge to the reactor. Feed to the system as a whole is on flow control. Process effluent from the reactor is on pressure control, and of the regenerator gases on the DPC. Fuel to regeneration air preheater is on TC off the preheat air and the combustion air is on flow ratio control as in part (e).

LIQUID PUMPS

Process pumps are three types: centrifugal, rotary positive displacement, and reciprocating. The outputs of all of them are controllable by regulation of the speed of the driver.

Controllability of centrifugal pumps depends on their pressure-flow characteristics, of which Figure 3.20 has two examples. With the upper curve, two flow rates are possible above a head of about 65 ft so that the flow is not reliably controllable above this pressure. The pump with the lower curve is stable at all pressures within its range. Throttling of the discharge is the usual control method for smaller centrifugals, variable speed drives for larger ones. Suction throttling may induce flashing and vapor binding of the pump. Figures 3.21(a) and (b) are examples.

Rotary pumps deliver a nearly constant flow at a given speed, regardless of the pressure. Bypass control is the usual method, with speed control in larger sizes. Reciprocating pumps also may be controlled on bypass if a pulsation damper is provided in the circuit to smooth out pressure fluctuations; Figure 3.21(c) shows this mode.

Reciprocating positive displacement pumps may have adjustment of the length or frequency of the stroke as another control feature. These may be solenoid or pneumatic devices that can be operated off a flow controller, as shown on Figure 3.21(d).

SOLIDS FEEDERS

Several of the more common methods of controlling the rate of supply of granular, free-flowing solids are represented in Figure 3.7.

COMPRESSORS

Three main classes of gas compressors are centrifugal and axial, rotary continuous positive displacement, and reciprocating positive

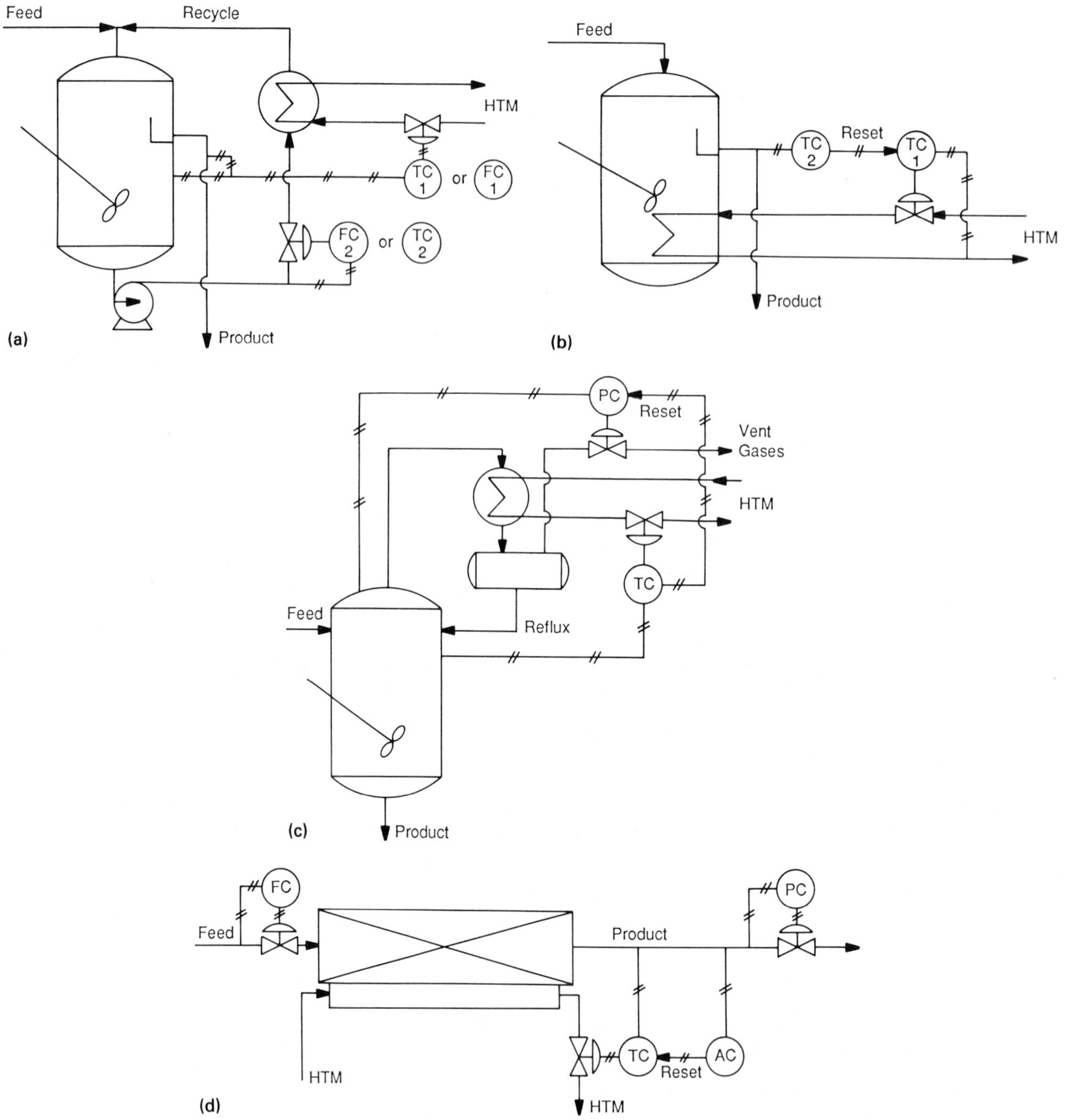

Figure 3.19. Chemical reactor control examples. (a) Temperature control of a stirred tank reactor with pumparound through an external heat exchanger, operable either in batch or continuously: Some alternate control modes are discussed in the text. Cascade control as in (b) can be implemented with external heat transfer surface. (b) Either cascade or direct control of temperature: For direct control, controller TC-2 is omitted, and the control point can be taken on the effluent line or in the vessel or on the HTM effluent line. A similar scheme is feasible with an external heat exchanger. (c) Reactor temperature control by regulation of the boiling pressure: The HTM is on TC off the reactor and resets the PC on the vent gases when necessary to maintain the correct boiling temperature. Although shown for batch operation, the method is entirely feasible for continuous flow. (d) Basic controls on a flow reactor: Feed on flow control, effluent on pressure control, and heat transfer medium flow rate on process effluent temperature or reset by its composition. (e) A fired heater as a tubular flow reactor: Feed is on FC, the product is on PC, the fuel is on TC or AC off the product, and the air is on flow ratio control. (f) A two-stage fired heater-reactor assembly: Details of the fuel-air supply control are in (e). (g) Control of the temperature of the exothermic synthesis of cumene by splitting the feed and by injection of cold propane and water into several zones. The water also serves to maintain activity of the phosphoric acid catalyst. (h) The main controls of a fluidized bed reactor–regenerator: Flow of spent catalyst is on level control, and that of regenerated catalyst is on TC off the reactor; these flows are assisted by maintenance of a differential pressure between the vessels. Details of the fuel-air control for the preheater are in (e).

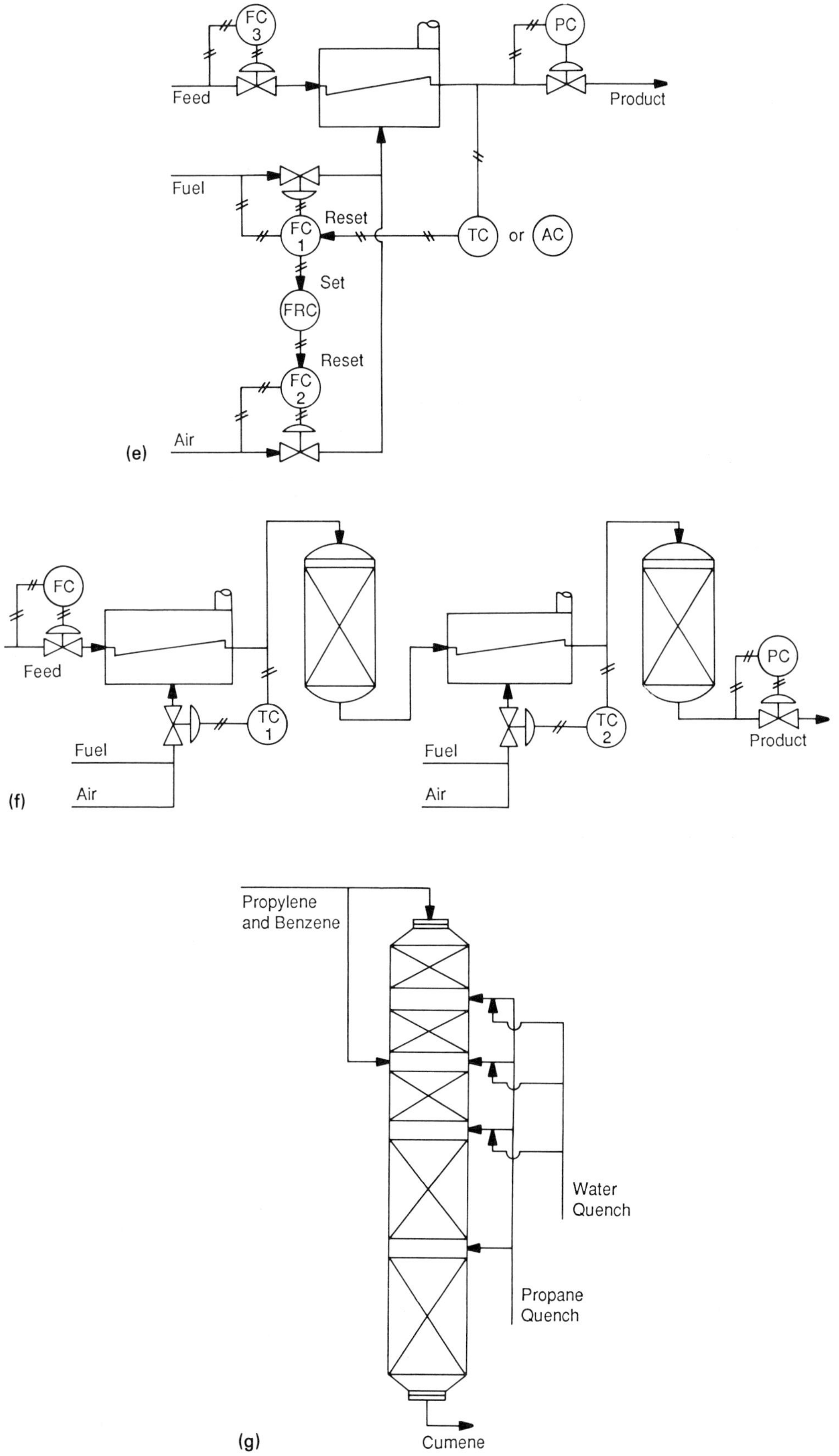

Figure 3.19—(*continued*)

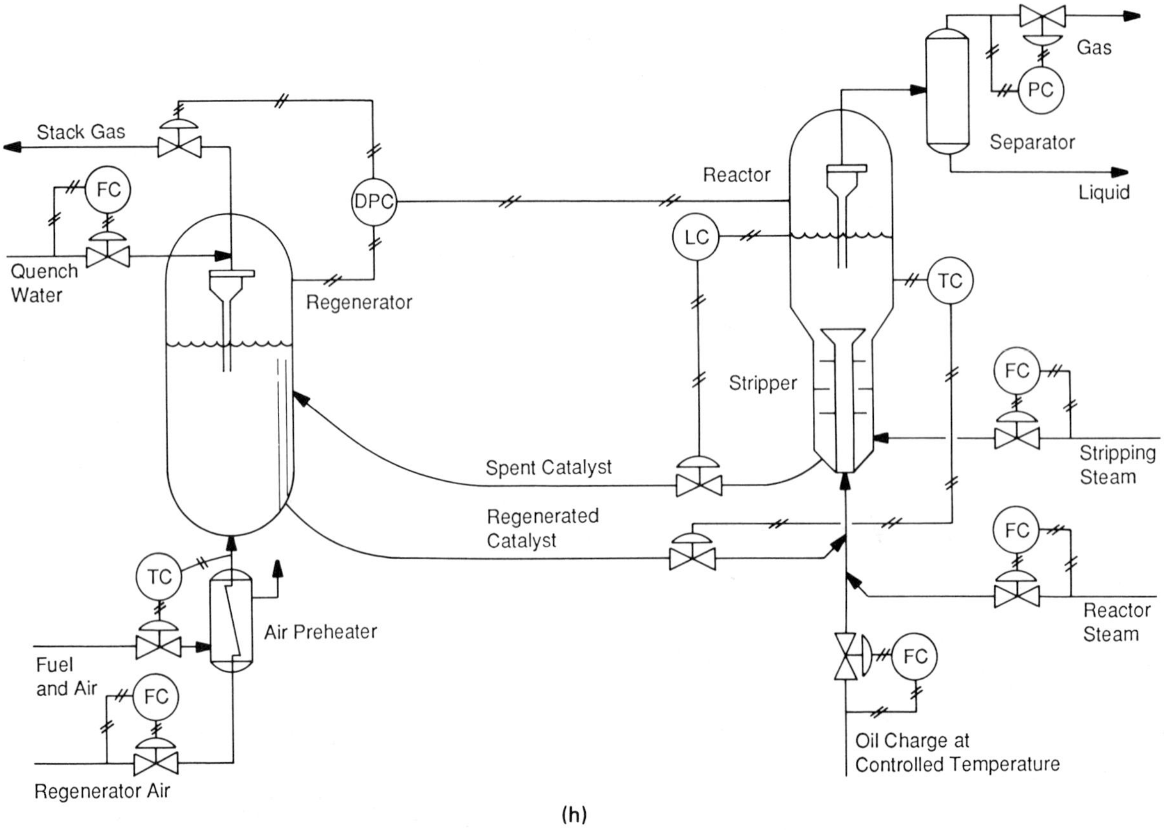

Figure 3.19—(*continued*)

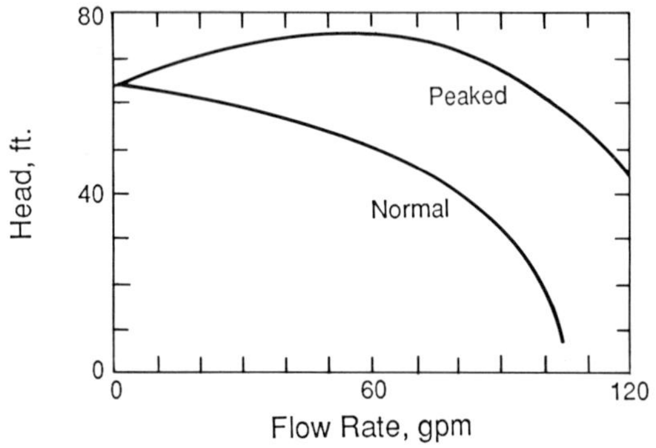

Figure 3.20. Characteristics curves of two centrifugal pumps.

displacement. The usual or feasible modes of control of pressure and flow may be tabulated:

Control Mode	Centrifugal and Axial	Rotary PD	Reciprocating PD
Suction throttling	x		
Discharge throttling	x		
Bypass	x	x	x
Speed	x	x	x
Guide vanes	x		
Suction valves			x
Cylinder clearance			x

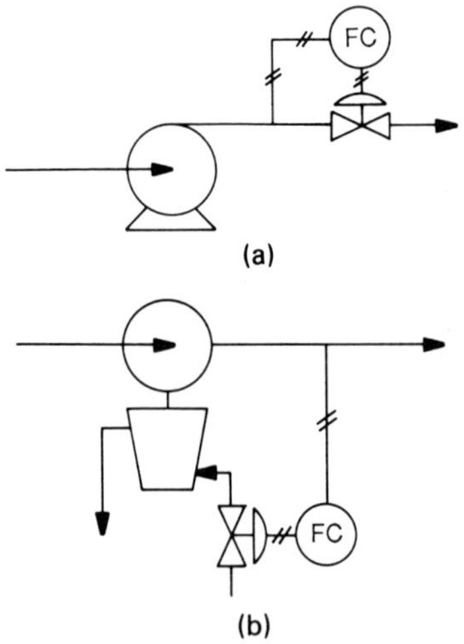

Figure 3.21. Control of centrifugal, rotary, and reciprocating pumps. (a) Throttling of the discharge of a centrifugal pump. (b) Control of the flow rate of any kind of pump by regulation of the speed of the driver. Although a turbine is shown, engine drive or speed control with gears, magnetic clutch, or hydraulic coupling may be feasible. (c) On the left, bypass control of rotary positive displacement pump; on the right, the reciprocating pump circuit has a pulsation dampener to smooth out pressure fluctuations. (d) Adjustment of the length or frequency of the stroke of a constant speed reciprocating pump with a servomechanism which is a feedback method whose action is control of mechanical position.

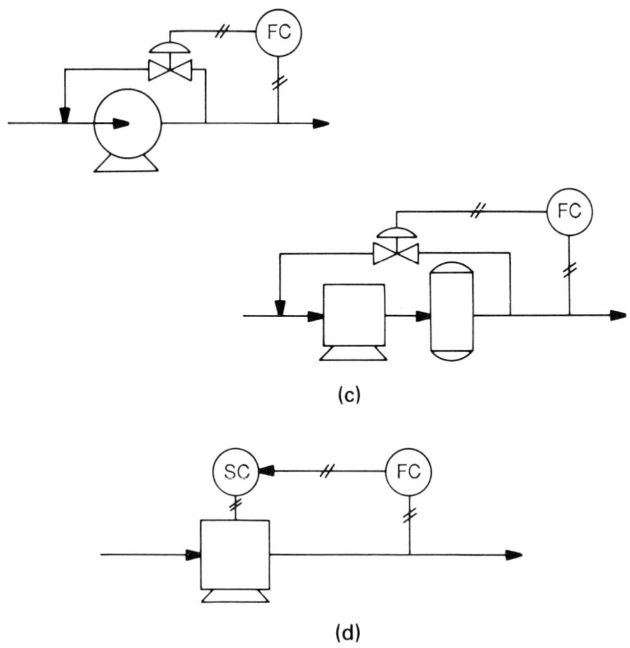

Figure 3.21—*(continued)*

Throttling of the suction of centrifugal and axial compressors wastes less power than throttling the discharge. Even less power is wasted by adjustment of built-in inlet guide vanes with a servomechanism which is a feedback control system in which the controlled variable is mechanical position. Speed control is a particularly effective control mode, applicable to large units that can utilize turbine or internal combustion drives; control is by throttling of the supply of motive fluids, steam or fuel.

Characteristic curves—pressure against flowrate—of centrifugal and axial compressors usually have a peak. Figure 3.22 is an example. In order to avoid surging, the flow through the compressor must be maintained above the magnitude at the peak in pressure. Figure 3.23(c) shows an automatic bypass for surge protection which opens when the principal flow falls to the critical minimum; recycle brings the total flow above the critical.

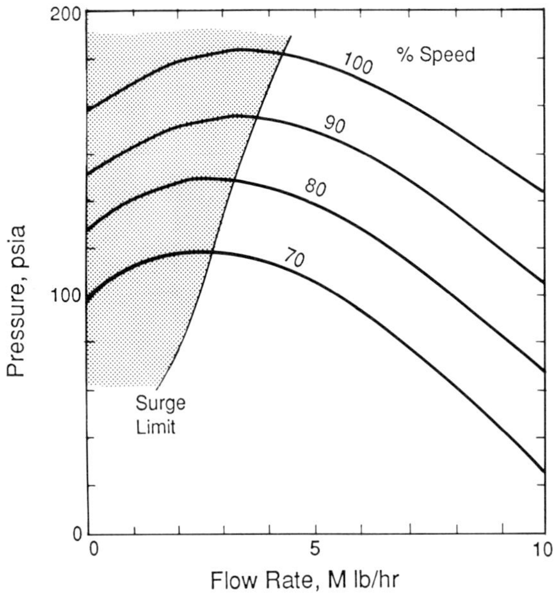

Figure 3.22. Characteristic curves of a centrifugal compressor at different speeds, showing surge limits.

Smaller rotary positive displacement compressors are controlled with external bypass. Such equipment usually has a built-in relief valve that opens at a pressure short of damaging the equipment, but the external bypass still is necessary for smooth control. Large units may be equipped with turbine or gas engine drives which are speed adjustable. Variable speed gear boxes or belt drives are not satisfactory. Variable speed dc motors also are not useful as compressor drives. Magnetic clutches and hydraulic couplings are used.

Reciprocating compressors may be controlled in the same way as rotary units. The normal turndown with gasoline or diesel engines is 50% of maximum in order that torque remains within

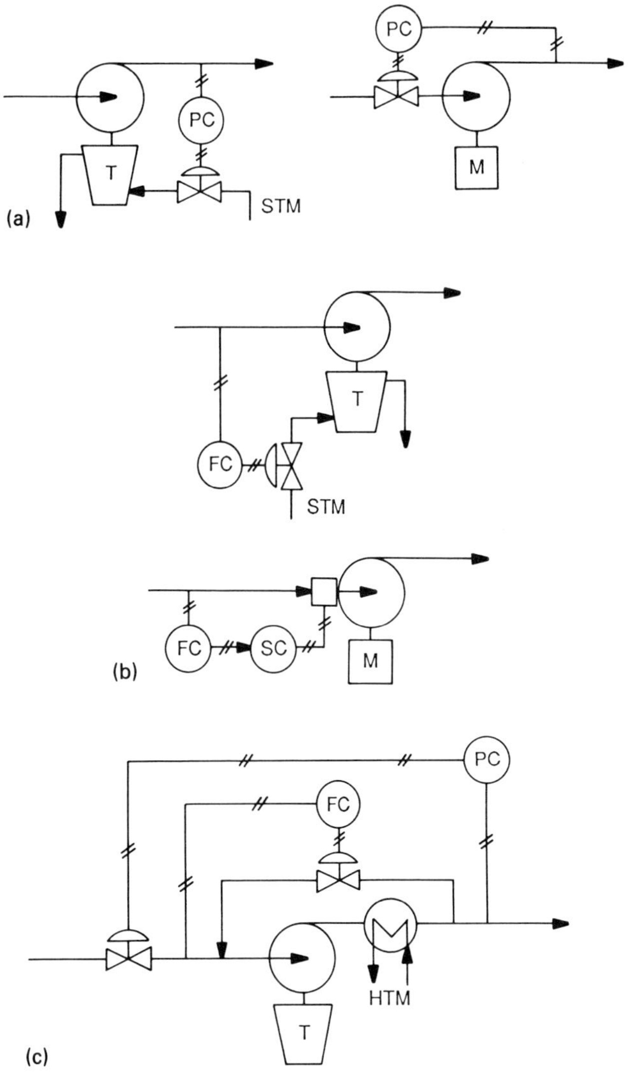

Figure 3.23. Control of centrifugal compressors with turbine or motor drives. (a) Pressure control with turbine or motor drives. (b) Flow control with turbine or motor drives. SC is a servomechanism that adjusts the guide vanes in the suction of the compressor. (c) Surge and pressure control with either turbine or motor drive. The bypass valve opens only when the flow reaches the minimum calculated for surge protection.

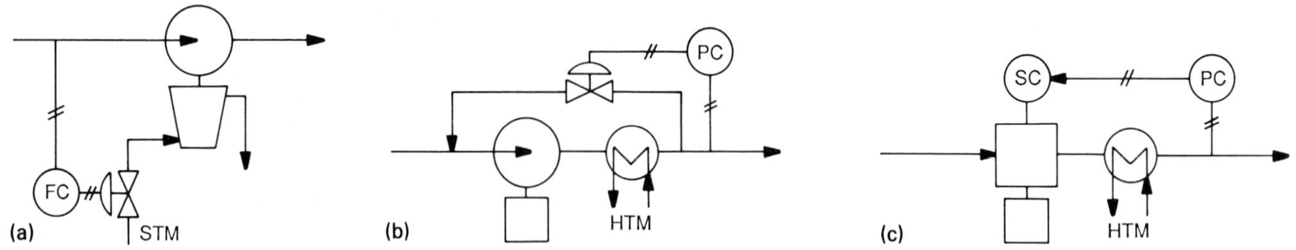

Figure 3.24. Control of positive displacement compressors, rotary and reciprocating. (a) Flow control with variable speed drives. (b) Pressure control with bypass to the suction of the compressor. (c) Reciprocating compressor. SC is a servomechanism that opens some of suction valves during discharge, thus permitting stepwise internal bypass. The clearance unloader is controllable similarly. These built-in devices may be supplemented with external bypass to smooth out pressure fluctuations.

acceptable limits. Two other aids are available to control of reciprocating units.

1. Valve unloading, a process whereby some of the suction valves remain open during discharge. Solenoid or pneumatic unloaders can be operated from the output of a control instrument. The stepwise controlled flow rate may need to be supplemented with controlled external bypass to smooth out pressure fluctuations.
2. Clearance unloaders are small pockets into which the gas is forced on the compression stroke and expands into the cylinder on the return stroke, thus preventing compression of additional gas.

Figure 3.24 shows control schemes for rotary and reciprocating compressors. Vacuum pumps are compressors operating between a low suction pressure and a fixed discharge pressure, usually atmospheric. Mechanical pumps are used for small capacities, steam jet ejectors for larger ones. Ejectors also are used as thermocompressors to boost the pressure of low pressure steam to an intermediate value. Control of suction pressure with either mechanical or jet pumps is by either air bleed [Fig. 3.5(a)] or suction line throttling [Fig. 3.5(c)]; air bleed is the more economical process. Up to five jets in series are used to produce high vacua. The steam from each stage is condensed by direct contact with water in barometric condensers or in surface condensers; condensation of steam from the final stage is not essential to performance but only to avoid atmospheric pollution. In a single stage ejector, motive steam flow cannot be reduced below critical flow in the diffuser, and water to the barometric condenser must not be throttled below 30–50% of the maximum if proper contacting is to be maintained. Control by throttling of steam and water supply, as on Figure 3.5(b), is subject to these limitations.

REFERENCES

1. *Chemical Engineering Magazine, Practical Process Instrumentation and Control,* McGraw-Hill, New York, 1980.
2. D.M. Considine, *Process Instruments and Controls Handbook,* McGraw-Hill, New York, 1985.
3. B. Liptak, *Instrumentation in the Process Industries,* Chilton, New York, 1973.
4. F.G. Shinskey, *Process Control Systems,* McGraw-Hill, New York, 1979.
5. F.G. Shinskey, *Distillation Control,* McGraw-Hill, New York, 1984.
6. M.R. Skrokov (Ed.), Mini- and Microcomputer Control in Industrial Processes, Van Nostrand Reinhold, New York, 1980.

4

DRIVERS FOR MOVING EQUIPMENT

Powered chemical processing equipment includes pumps, compressors, agitators and mixers, crushers and grinders, and conveyors. Drivers are electric motors, steam or gas turbines, and internal combustion engines. For loads under 150 HP or so electric motors are almost invariably the choice. Several criteria are applicable. For example, when a pump and a spare are provided, for flexibility one of them may be driven by motor and the other by turbine. Centrifugal and axial blowers and compressors are advantageously driven by turbines because the high operating speeds of 4000–10,000 rpm are readily attainable whereas electric motors must operate through a speed increasing gear at extra expense. When fuel is relatively cheap or accessible, as in the field, gas turbines and internal combustion engines are preferred drivers. Turbines, internal combustion engines, and direct current motors are capable of continuous speed adjustment over a wide range. Energy efficiencies vary widely with the size and type of driver as shown in this table.

	Efficiency (%)			
Driver	10 kW	100 kW	1000 kW	10,000 kW
Gas turbine and internal combustion engine		28	34	38
Steam turbine		42	63	78
Motor	85	92	96	97

Since the unit energy costs are correspondingly different, the economics of the several drive modes often are more nearly comparable.

4.1. MOTORS

Although each has several subclasses, the three main classes of motors are induction, synchronous, and direct current. Higher voltages are more efficient, but only in the larger sizes is the housing ample enough to accomodate the extra insulation that is necessary. The voltages commonly used are

Horsepower	Voltage
1–100	220, 440, 550
75–250	440
200–2500	2300, 4000
Above 2500	4000, 13,200

Direct current voltages are 115, 230, and 600.

The torque-speed characteristic of the motor must be matched against that of the equipment, for instance, a pump. As the pump comes up to speed, the torque exerted by the driver always should remain 5% or so above that demanded by the pump.

The main characteristics of the three types of motors that bear on their process applicability are summarized following.

INDUCTION

Induction motors are the most frequent in use because of their simple and rugged construction, and simple installation and control. They are constant speed devices available as 3600 (two-pole), 1800, 1200, and 900 rpm (eight-pole). Two speed models with special windings with 2:1 speed ratios are sometimes used with agitators, centrifugal pumps and compressors and fans for air coolers and cooling towers. Capacities up to 20,000 HP are made. With speed increasing gears, the basic 1800 rpm model is the economical choice as drive for centrifugal compressors at high speeds.

SYNCHRONOUS

Synchronous motors are made in speeds from 1800 (two-pole) to 150 rpm (48-pole). They operate at constant speed without slip, an important characteristic in some applications. Their efficiencies are 1–2.5% higher than that of induction motors, the higher value at the lower speeds. They are the obvious choice to drive large low speed reciprocating compressors requiring speeds below 600 rpm. They are not suitable when severe fluctuations in torque are encountered. Direct current excitation must be provided, and the costs of control equipment are higher than for the induction types. Consequently, synchronous motors are not used under 50 HP or so.

DIRECT CURRENT

Direct current motors are used for continuous operation at constant load when fine speed adjustment and high starting torque are needed. A wide range of speed control is possible. They have some process applications with centrifugal and plunger pumps, conveyors, hoists, etc.

Enclosures. In chemical plants and refineries, motors may need to be resistant to the weather or to corrosive and hazardous locations. The kind of housing that must be provided in particular situations is laid out in detail in the National Electrical Code, Article 500. Some of the classes of protection recognized there are in this table of differential costs.

Type	% Cost above Drip Proof	Protection Against
Drip proof		Dripping liquids and falling particles
Weather protected, I and II	10–50	Rain, dirt, snow
Totally enclosed fan cooled, TEFC, below 250 HP	25–100	Explosive and nonexplosive atmospheres
Totally enclosed, water cooled, above 500 HP	25–100	Same as TEFC
Explosion proof, below 3000 HP	110–140	Flammable and volatile liquids

TABLE 4.1. Selection of Motors for Process Equipment

Application	Motor Type[a] A.C.	D.C.
Agitator	1a, 1b, 2b	5a
Ball mill	1c, 2b, 3a	5b
Blower	1a, 1b, 2b, 3a, 4	5a
Compressor	1a, 1b, 1c, 3a, 4	5b, 7
Conveyor	1a, 1c, 2b, 3a	5b, 7
Crusher	1a, 1c, 1d	5a, 5b
Dough mixer	1a, 1b, 1c, 2b	5a, 5b
Fan, centrifugal and propeller	1a, 1b, 2c, 3a, 4	5a, 7
Hammer mill	1c	5a
Hoist	1d, 2a, 3b	6
Pulverizer	1c	5b
Pump, centrifugal	1a, 1b, 2b, 3a, 4	5b
Pump, positive displacement	1c, 2b, 3a	5b
Rock crusher	3a	5b, 6

[a] Code:
1. Squirrel-cage, constant speed
   a. normal torque, normal starting current
   b. normal torque, low starting current
   c. high torque, low starting current
   d. high torque, high slip
2. Squirrel-cage, multispeed
   a. constant horsepower
   b. constant torque
   c. variable torque
3. Wound rotor
   a. general purpose
   b. crane and hoist
4. Synchronous
5. Direct current, constant speed
   a. shunt wound
   b. compound wound
6. Direct current, variable speed series wound
7. Direct current, adjustable speed

(After Allis-Chalmers Mfg. Co., Motor and Generator Reference Book, Colorado Springs, CO).

Standard NEMA ratings for induction motors:

General purpose: $\frac{1}{2}$, $\frac{3}{4}$, 1, $1\frac{1}{2}$, 2, 3, 5, $7\frac{1}{2}$, 10, 15, 20, 25, 30, 40, 50, 60, 75, 100, 125, 150, 200, 250, 300, 350, 400, 450, 500.

Large motors: 250, 300, 350, 400, 450, 500, 600, 700, 800, 900, 1000, 1250, 1500, 1750, 2000, 2250, 2500, 3000, 3500, 4000, 4500, 5000 and up to 30,000.

TABLE 4.2. Checklist for Selection of Motors

Motor Data

General

Type of motor (cage, wound-rotor, synchronous, or dc)........
Quantity........ Hp........ Rpm......... Phase.........
Cycles........ Voltage........
Time rating (continuous, short-time, intermittent)............
Overload (if any) ......% for ...... Service factor ......%
Ambient temperature..........C Temperature rise..........C
Class of insulation: Armature... Field... Rotor of w-r motor...
Horizontal or vertical............ Plugging duty............
Full- or reduced-voltage or part-winding starting (ac)........
If reduced voltage—by autotransformer or reactor..........
Locked-rotor starting current limitations.....................
Special characteristics....................................

Induction Motors

Locked-rotor torque.........% Breakdown torque.........%
or for general-purpose cage motor: NEMA Design (A, B, C, D)
....

Synchronous Motors

Power factor...... Torques: Locked-rotor.....% Pull-in.....%
Pull-out......% Excitation......volts dc. Type of exciter......
If m-g exciter set, what are motor characteristics?............
Motor field rheostat........ Motor field discharge resistor.......

Direct-current Motors

Shunt, stabilized shunt, compound, or series wound.............
Speed range.......... Non-reversing or reversing.............
Continuous or tapered-rated...............................

Mechanical Features

Protection or enclosure.............. Stator shift.............

Mechanical Features (cont.)

Number of bearings............. Type of bearings............
Shaft extension: Flanged....... Standard or special length......
Press on half-coupling........ Terminal box.................
NEMA C or D flange........ Round-frame or with feet........
Vertical: External thrust load.....lbs. Type of thrust bearing.....
Base ring type................ Sole plates.................
Accessories..

Load Data

Type of load...
If compressor drive, give NEMA application number.............
Direct-connected, geared, chain, V-belt, or flat-belt drive..........
Wk^2 (inertia) for high inertia drives....................$lb\text{-}ft^2$
Starting with full load, or unloaded.........................
If unloaded, by what means?..............................
For variable-speed or multi-speed drives, is load variable torque, constant torque, or constant horsepower?......................
Operating conditions.......................................

(By permission, *Allis Chalmers Motor and Generator Reference Book*, Bul. 51R7933, and E.S. Lincoln (Ed.), *Electrical Reference Book*, Electrical Modernization Bureau, Colorado Springs, CO.

Clearly the cost increments beyond the basic drip-proof motor enclosures are severe, and may need to be balanced in large sizes against the cost of isolating the equipment in pressurized buildings away from the hazardous locations.

Applications. The kinds of motors that are being used successfully with particular kinds of chemical process equipment are identified in Table 4.1. As many as five kinds of AC motors are shown in some instances. The choice may be influenced by economic considerations or local experience or personal preference. In this area, the process engineer is well advised to enlist help from electrical experts. A checklist of basic data that a supplier of a motor must know is in Table 4.2. The kind of enclosure may be specified on the last line, operating conditions.

4.2. STEAM TURBINES AND GAS EXPANDERS

Turbines utilize the expansion of steam or a gas to deliver power to a rotating shaft. Salient features of such equipment are

1. high speed rotation,
2. adjustable speed operation,
3. nonsparking and consequently nonhazardous operation,
4. simple controls,
5. low first cost and maintenance, and
6. flexibility with regard to inlet and outlet pressures.

Single stage units are most commonly used as drivers, but above 500 HP or so multistage units become preferable. Inlet steam pressures may be any value up to the critical and with several

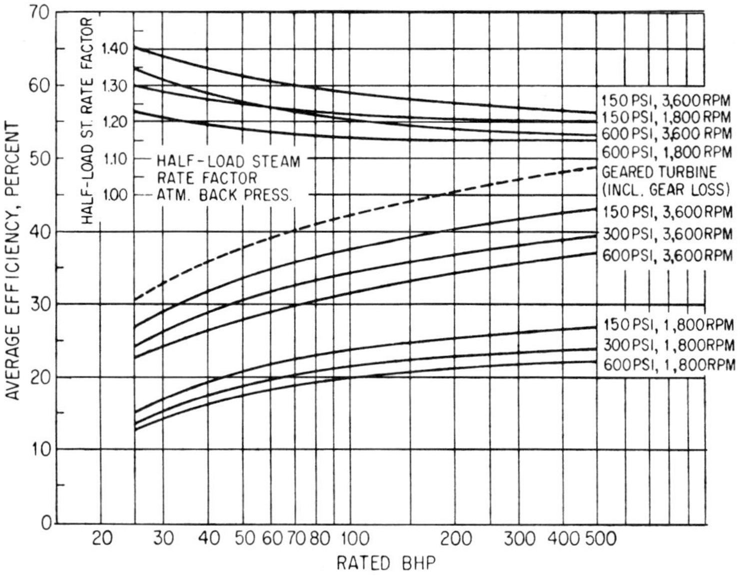

Average efficiency of single-stage turbines (noncondensing, dry, and saturated steam).

(a)

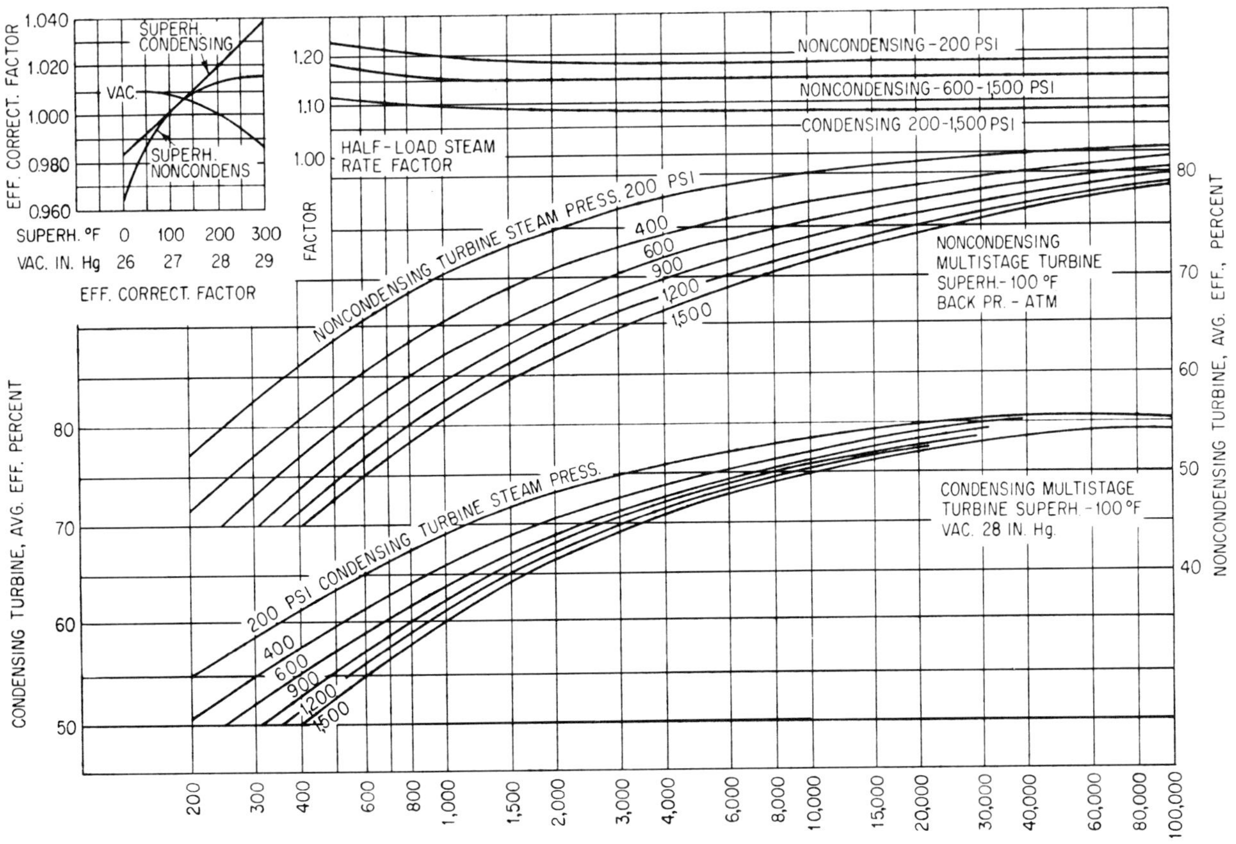

Average efficiency of multistage turbines (gear loss not included).

(b)

Figure 4.1. Efficiencies of (a) single-stage and (b) multistage turbines (*Gartmann,* De Laval Engineering Handbook, *McGraw-Hill, New York, 1970, pp. 5.8–5.9, Figs. 5.2 and 5.3*).

hundred degrees of superheat. In larger sizes turbines may be convenient sources of low pressure exhaust steam in the plant. From multistage units, steam may be bled at several reduced pressures. When the expansion is to subatmospheric conditions, the operation is called condensing because the exhaust steam must be condensed before removal from the equipment. Although the efficiency of condensing turbines is less, there is an overall reduction of energy consumption because of the wider expansion range.

Several parameters affect the efficiency of steam turbines, as shown partially on Figure 4.1. Closer examination will need to take into account specific mechanical details which usually are left to the manufacturer. Geared turbines [the dashed line of Fig. 4.1(b)] have higher efficiencies, even with reduction gear losses, because they operate with especially high bucket speeds. For example, for a service of 500 HP with 300 psig steam, a geared turbine has an efficiency of 49.5% and one with a direct drive at 1800 rpm has an efficiency of 24%.

The flow rate of steam per unit of power produced is represented by

$$\dot{m} = -\frac{2545}{\eta(H_2 - H_1)}\ \text{lb/HP hr}$$

$$= -\frac{3412}{\eta(H_2 - H_1)}\ \text{lb/kWh}$$

with the enthalpies in Btu/lb. The efficiency is η, off Figure 4.1, for example. The enthalpy change is that of an isentropic process. It may be calculated with the aid of the steam tables or a Mollier diagram for steam. For convenience, however, special tables have been derived which give the theoretical steam rates for typical combinations of inlet and outlet conditions. Table 4.3 is an abbreviated version.

Example 4.1 illustrates this kind of calculation and compares the result with that obtained by taking the steam to behave as an ideal gas. For nonideal gases with known PVT equations of state and low pressure heat capacities, the method of calculation is the same as for compressors which is described in that section of the book.

On a Mollier diagram like that with Example 4.1, it is clear that expansion to a low pressure may lead to partial condensation if insufficient preheat is supplied to the inlet steam. The final condition after application of the efficiency correction is the pertinent one, even though the isentropic point may be in the two-phase region. Condensation on the blades is harmful to them and must be avoided. Similarly, when carbon dioxide is expanded, possible formation of solid must be guarded against.

When gases other than steam are employed as motive fluids, the equipment is called a gas expander. The name gas turbine usually is restricted to equipment that recovers power from hot

TABLE 4.3. Theoretical Steam Rates for Typical Steam Conditions (lb/kWh)[a]

	Initial pressure, lb/in² gage															
	150	250	400	600	600	850	850	900	900	1,200	1,250	1,250	1,450	1,450	1,800	2,400
	Initial temp, °F															
	365.9	500	650	750	825	825	900	825	900	825	900	950	825	950	1000	1000
	Initial superheat, °F															
	0	94.0	201.9	261.2	336.2	297.8	372.8	291.1	366.1	256.3	326.1	376.1	232.0	357.0	377.9	337.0
Exhaust	Initial enthalpy, Btu/lb															
pressure	1,195.5	1,261.8	1,334.9	1,379.6	1,421.4	1,410.6	1,453.5	1,408.4	1,451.6	1,394.7	1,438.4	1,468.1	1,382.7	1,461.2	1,480.1	1,460.4
inHg abs																
2.0	10.52	9.070	7.831	7.083	6.761	6.580	6.282	6.555	6.256	6.451	6.133	5.944	6.408	5.900	5.668	5.633
2.5	10.88	9.343	8.037	7.251	6.916	6.723	6.415	6.696	6.388	6.584	6.256	6.061	6.536	6.014	5.773	5.733
3.0	11.20	9.582	8.217	7.396	7.052	6.847	6.530	6.819	6.502	6.699	6.362	6.162	6.648	6.112	5.862	5.819
4.0	11.76	9.996	8.524	7.644	7.282	7.058	6.726	7.026	6.694	6.894	6.541	6.332	6.835	6.277	6.013	5.963
lb/in² gage																
5	21.69	16.57	13.01	11.05	10.42	9.838	9.288	9.755	9.209	9.397	8.820	8.491	9.218	8.351	7.874	7.713
10	23.97	17.90	13.83	11.64	10.95	10.30	9.705	10.202	9.617	9.797	9.180	8.830	9.593	8.673	8.158	7.975
20	28.63	20.44	15.33	12.68	11.90	11.10	10.43	10.982	10.327	10.490	9.801	9.415	10.240	9.227	8.642	8.421
30	33.69	22.95	16.73	13.63	12.75	11.80	11.08	11.67	10.952	11.095	10.341	9.922	10.801	9.704	9.057	8.799
40	39.39	25.52	18.08	14.51	13.54	12.46	11.66	12.304	11.52	11.646	10.831	10.380	11.309	10.134	9.427	9.136
50	46.00	28.21	19.42	15.36	14.30	13.07	12.22	12.90	12.06	12.16	11.284	10.804	11.779	10.531	9.767	9.442
60	53.90	31.07	20.76	16.18	15.05	13.66	12.74	13.47	12.57	12.64	11.71	11.20	12.22	10.90	10.08	9.727
75	69.4	35.77	22.81	17.40	16.16	14.50	13.51	14.28	13.30	13.34	12.32	11.77	12.85	11.43	10.53	10.12
80	75.9	37.47	23.51	17.80	16.54	14.78	13.77	14.55	13.55	13.56	12.52	11.95	13.05	11.60	10.67	10.25
100		45.21	26.46	19.43	18.05	15.86	14.77	15.59	14.50	14.42	13.27	12.65	13.83	12.24	11.21	10.73
125		57.88	30.59	21.56	20.03	17.22	16.04	16.87	15.70	15.46	14.17	13.51	14.76	13.01	11.84	11.28
150		76.5	35.40	23.83	22.14	18.61	17.33	18.18	16.91	16.47	15.06	14.35	15.65	13.75	12.44	11.80
160		86.8	37.57	24.79	23.03	19.17	17.85	18.71	17.41	16.88	15.41	14.69	16.00	14.05	12.68	12.00
175			41.16	26.29	24.43	20.04	18.66	19.52	18.16	17.48	15.94	15.20	16.52	14.49	13.03	12.29
200			48.24	29.00	26.95	21.53	20.05	20.91	19.45	18.48	16.84	16.05	17.39	15.23	13.62	12.77
250			69.1	35.40	32.89	24.78	23.08	23.90	22.24	20.57	18.68	17.81	19.11	16.73	14.78	13.69
300				43.72	40.62	28.50	26.53	27.27	25.37	22.79	20.62	19.66	20.89	18.28	15.95	14.59
400				72.2	67.0	38.05	35.43	35.71	33.22	27.82	24.99	23.82	24.74	21.64	18.39	16.41
425				84.2	78.3	41.08	38.26	38.33	35.65	29.24	26.21	24.98	25.78	22.55	19.03	16.87
600						78.5	73.1	68.11	63.4	42.10	37.03	35.30	34.50	30.16	24.06	20.29

[a] From Theoretical Steam Rate Tables—Compatible with the 1967 ASME Steam Tables, ASME, 1969.

EXAMPLE 4.1
Steam Requirement of a Turbine Operation
Steam is fed to a turbine at 614.7 psia and 825°F and is discharged at 64.7 psia. (a) Find the theoretical steam rate, lb/kWh, by using the steam tables. (b) If the isentropic efficiency is 70%, find the outlet temperature. (c) Find the theoretical steam rate if the behavior is ideal, with $C_p/C_v = 1.33$.

(a) The expansion is isentropic. The initial and terminal conditions are identified in the following table and on the graph. The data are read off a large Mollier diagram (Keenan et al., *Steam Tables,* Wiley, New York, 1969).

Point	*P*	T^0F	*H*	*S*
1	614.7	825	1421.4	1.642
2	64.7	315	1183.0	1.642
3	64.7	445	1254.5	1.730

$$\Delta H_s = H_2 - H_1 = -238.4 \text{ Btu/lb}$$

Theoretical steam rate = 3412/238.4 = 14.31 lb/kWh. This value is checked exactly with the data of Table 4.3.

$$\text{(b) } H_3 - H_1 = 0.7(H_2 - H_1) = -166.9 \text{ Btu/lb}$$
$$H_3 = 1421.4 - 166.9 = 1254.5 \text{ Btu/lb}$$

The corresponding values of T_3 and S_3 are read off the Mollier diagram, as tabulated.

(c) The isentropic relation for ideal gases is

$$\Delta H = \frac{k}{k-1} RT_1[(P_2/P_1)^{(k-1)/k} - 1]$$
$$= \frac{1.987(1285)}{0.25}[(64.7/614.7)^{0.25} - 1]$$
$$= -4396 \text{ Btu/lbmol}, -244 \text{ Btu/lb}.$$

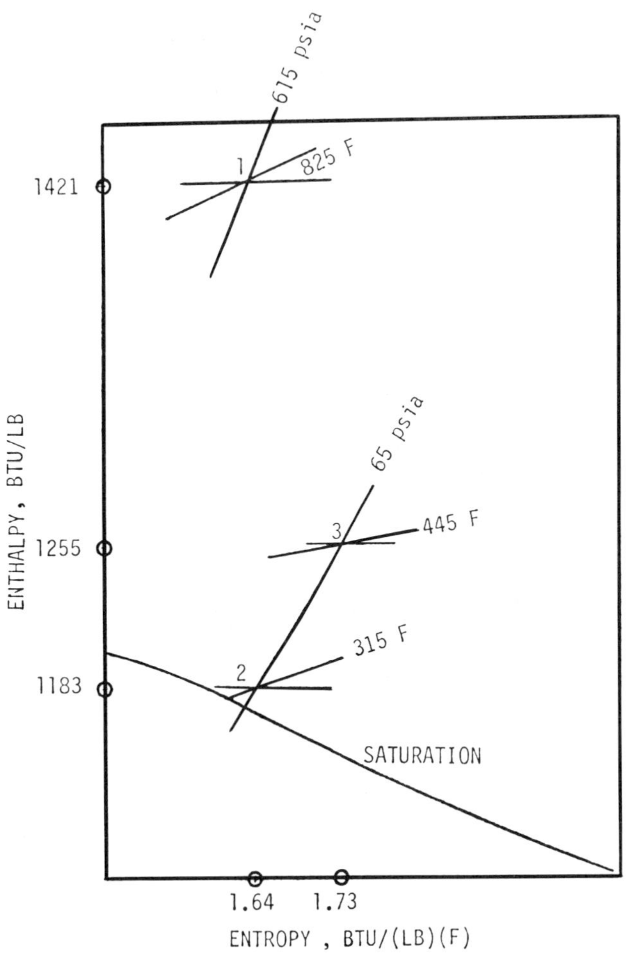

combustion gases. The name turboexpander is applied to machines whose objective is to reduce the energy content (and temperature) of the stream, as for cryogenic purposes.

Gas expanders are used to recover energy from high pressure process gas streams in a plant when the lower pressure is adequate for further processing. Power calculations are made in the same way as those for compressors. Usually several hundred horsepower must be involved for economic justification of an expander. In smaller plants, pressures are simply let down with throttling valves (Joule–Thomson) without attempt at recovery of energy.

The specification sheet of Table 4.4 has room for the process conditions and some of the many mechanical details of steam turbines.

4.3. COMBUSTION GAS TURBINES AND ENGINES

When a low cost fuel is available, internal combustion drivers surpass all others in compactness and low cost of installation and operation. For example, gas compression on a large scale has long been done with integral engine compressors. Reciprocating engines also are widely used with centrifugal compressors in low pressure applications, but speed increasing gears are needed to up the 300–600 rpm of the engines to the 3000–10,000 rpm or so of the compressor.

Process applications of combustion gas turbines are chiefly to driving pumps and compressors, particularly on gas and oil transmission lines where the low thermal efficiency is counterbalanced by the convenience and economy of having the fuel on hand. Offshore drilling rigs also employ gas turbines. Any hot process gas at elevated pressure is a candidate for work recovery in a turbine. Offgases of catalytic cracker regenerators, commonly at 45 psig and as high as 1250°F, are often charged to turbines for partial recovery of their energy contents. Plants for the manufacture of nitric acid by oxidation of ammonia at pressures of 100 psig or so utilize expanders on the offgases from the absorption towers, and the recovered energy is used to compress the process air to the reactors.

Combustion gas turbine processes are diagrammed on Figure 4.2 and in Example 4.2. In the basic process, a mixture of air and fuel (or air alone) is compressed to 5–10 atm, and then ignited and burned and finally expanded through a turbine from which power is recovered. The process follows essentially a Brayton cycle which is shown in Figure 4.2 in idealized forms on TS and PV diagrams. The ideal process consists of an isentropic compression, then heating at constant pressure followed by an isentropic expansion and finally cooling at the starting pressure. In practice, efficiencies of the individual steps are high:

Compressor isentropic efficiency, 85%
Expander isentropic efficiency, 85–90%
Combustion efficiency, 98%

TABLE 4.4. Data Sheet for General Purpose Steam Turbines, Sheet 1 of 2[a]

GENERAL-PURPOSE STEAM TURBINE DATA SHEET CUSTOMARY UNITS

CONTRACT NO.	
ITEM NO.	
REV. NO.	DATE
BY	REVIEWED
SHEET 1	OF 2
P.O. NO.	

Applicable to:	○ Proposal	○ Purchase	○ As – Built
For		Unit	
Site		No. Required	
Service		Driven Equipment	
Manufacturer		Model	Serial No.

NOTE: ○ Indicates Information Completed By Purchaser □ By Manufacturer

○ OPERATING CONDITIONS

Operating Point	Power, BHP	Speed, RPM
Normal		
Rated		
Other		

□ PERFORMANCE

Operating Point/ Steam Condition	No. Hand Valves Open (3.4.1.4)	Steam Rate, Lbs/HP - Hr.
Normal/Normal		
Rated/Normal		
Rated/Min. Inlet, Max. Exhaust		
Indicate Guarantee Point By *		

○ STEAM CONDITIONS

	MAX.	NORMAL	MIN.
Inlet Press, PSIG			
Inlet Temp, °F			
Exhaust Press (PSIG) (In. Hg)			

Unusual Conditions (2.12.2.6)			
Duty	○ Continuous	○ Standby	○ Auto Start
Eval. Steam Cost, $/1000 Lbs			
Payout Period, Years		Hrs/Yr	

TURBINE DATA

□ Minimum Allowable Speed, RPM		
□ Maximum Continuous Speed, RPM		
□ Trip Speed, RPM		
□ First Critical Speed, RPM		
○ Turbine Construction Safe For Runaway Speed (2.11.1)		
□ Exh. Temp. °F	Normal	No Load
□ Potential Max. Power, BHP		
□ Max. Nozzle Steam Flow, Lbs/Hr		
□ Max. Allowable Speed, RPM		
○ Rotation Facing Gov. End	○ CCW	○ CW
○ Driven Equipment Thrust, Lbs. (2.9.3)		Up
(Vertical Turbine)		Down
○ Mount Turbine on Baseplate Furnished by Driven Equipment Vendor		
○ Water Piping Furn. by	○ Vendor	○ Others
○ Oil Piping Furn. by	○ Vendor	○ Others

○ SITE AND UTILITY DATA

Location (2.1.13)	○ Indoor	○ Heated	○ Unheated
	○ Outdoor	○ Roof	○ Without Roof
Ambient Temp. °F	Min.	Max.	
Unusual Conditions	○ Dust	○ Salt Atmosphere	
○ Winterization Required			
Elect. Area (2.1.12)	Class	Group	Div
	○ Non-hazardous		
Control Power	V	Ph.	Hz
Aux. Motors	V	Ph.	Hz
Cooling Water:	Press, PSIG	ΔP, PSI	
	Flow, GPM	ΔT, °F	
Allowable Sound Level (2.1.11)		dBA @	ft.

□ CONSTRUCTION

Turbine Type	○ Horiz.	○ Vertical	
No. Stages		Wheel Dia., In.	
Rotor:	□ Built Up	□ Solid	
Blading	□ 2 Row	□ 3 Row	□ Re Entry
Casing Split	□ Axial	□ Radial	
Casing Support	□ Centerline	□ Foot	
	□ NEMA "P" Base		
Trip Valve	□ Integral	□ Separate	
Interstage Seals	□ Labyrinth	□ Carbon	
End Seals	□ Carbon Ring, No/Box		
	□ Labyrinth		
Type Radial Bearings			
Type Thrust Bearing (2.9.2)			
Thrust Collar (2.9.8)	□ Replaceable	□ Integral	□ None
Lube Oil Viscosity (2.10.2)	SUS @ 100°F	SUS @ 210°F	
Lubrication	○ Ring Oiled	○ Pressure	
	○ Purge Oil Mist	○ Pure Oil Mist	
○ Shaft Areas Suitable For Observing By Non-Contacting Type Vibration Probes (2.6.2.2)			

CASING DESIGN	INLET	EXHAUST
Max. Allow. Press, PSIG		
Max. Allow. Temp, °F		
Hydro Test Pressure, PSIG		

MATERIALS

High Pressure Casing			
Exhaust Casing			
Nozzles			
Blading			
Wheels			
Shaft			
Under Packing	Application Method		
Gov. Valve Trim			
Inlet Strainer	Mesh Size		
Governor Type	○ Mech.	○ Hydr.	○ Oil Relay
NEMA Class	Adj. Speed Range	+%	–%
Speed Changer	○ Manual	○ Pneum.	○ Elect.
Mfr.	Model		
Turbine Weight, Lbs			

[a] Also available in SI units (API Standard 611, January 1982). (Reprinted courtesy of the American Petroleum Institute.)

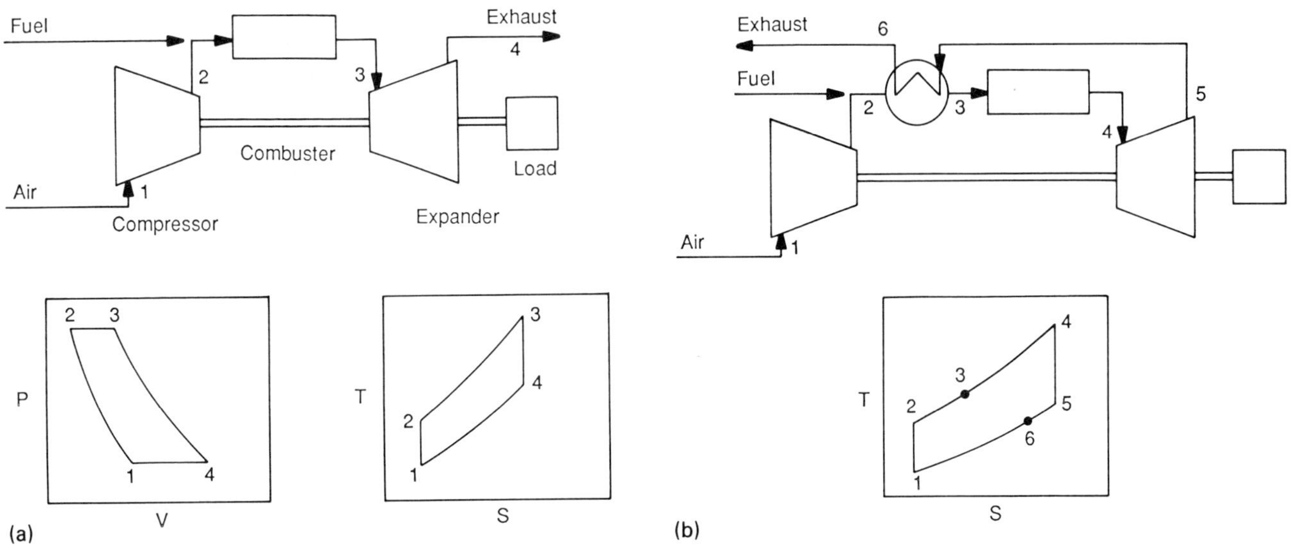

Figure 4.2. Combustion gas turbine arrangements and their thermodynamic diagrams. (a) Basic unit with PV and TS diagrams. (b) Unit with an air preheater and TS diagram.

EXAMPLE 4.2
Performance of a Combustion Gas Turbine

Atmospheric air at 80°F (305K) is compressed to 5 atm, combined with fuel at the rate of 1 kg/s, then expanded to 1 atm in a power turbine. Metallurgical considerations limit the temperature to 1700°F (1200K). The heat capacities of air and combustion products are

$$C_p = 0.95 + 0.00021T\ (\text{K})\ \text{kJ/kg},$$

the heat of combustion is 42,000 kJ/kg, the furnace efficiency is 0.975, the isentropic efficiency of the compressor is 0.84, and that of the expander is 0.89. Find

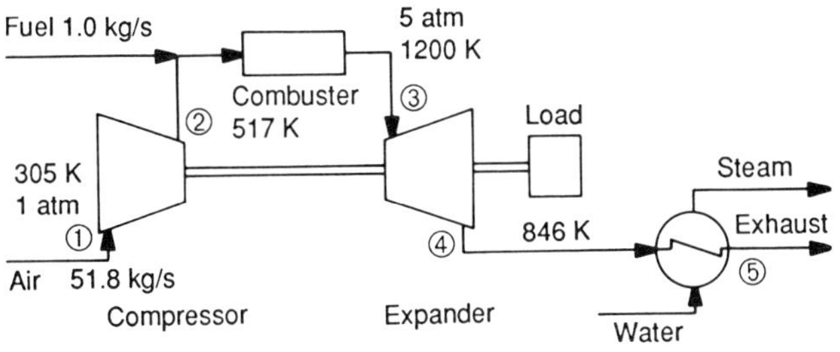

a. the required air rate,
b. the power loads of the compressor and expander, and
c. the overall efficiency as a function of the temperature of the exhaust leaving a steam generator.

Point	P	T_s	T
1	1		305
2	5	483	517
3	5		1200
4	1	802	846
5	1		400

Compression:

$$k = 1.4, \quad k/(k-1) = 3.5,$$

$$T_{2s} = T_1(P_2/P_1)^{1/3.5} = 305(5)^{1/3.5} = 483\text{K},$$

$$T_2 = 305 + \frac{483 - 305}{0.84} = 517\ \text{K}.$$

Combustion:

m'_a = flow rate of air, kg/kg fuel

$$0.975(42000) = \int_{305}^{1200} C_p\,dT + m'_a \int_{517}^{1200} C_p\,dT$$
$$= 991682 + 771985\,m'_a$$

$$m'_a = 51.8$$

Expansion:

$$k = 1.33, \quad k/(k-1) = 4.0$$

$$T_{4s} = T_3(P_4/P_1)^{0.25} = 1200(0.2)^{0.25} = 802°\text{K}$$

$$T_4 = 1200 - 0.89(1200 - 802) = 846°\text{K}$$

Power calculations:

$$\text{Compressor: } w'_c = -m'_a\Delta H = -51.8\int_{305}^{517} C_p\,dT$$
$$= -51.8(216.98) = -11.240\ \text{kJ/s}$$

$$\text{Expander: } w'_e = -52.8\int_{1200}^{517} C_p\,dT = 52.8(412.35) = 21{,}772\ \text{kJ/s}$$

$$\text{Steam generator: } Q' = 52.8\int_{T}^{846} C_p\,dT$$

$$\eta_t = \text{overall efficiency} = \frac{21772 - 11380 + Q'}{42000}$$

The tabulation shows efficiency with three different values of the exhaust temperature.

T	Q'	η_t
846	0	0.247
600	14311	0.588
500	19937	0.722

Other inefficiencies are due to pressure drops of 2–5%, loss of 1–3% of the enthalpy in the expander, and 1% or so loss of the air for cooling the turbine blades. The greatest loss of energy is due to the necessarily high temperature of the exhaust gas from the turbine, so that the overall efficiency becomes of the order of 20% or so. Some improvements are effected with air preheating as on Figure 4.2(b) and with waste heat steam generators as in Example 4.2. In many instances, however, boilers on 1000°F waste gas are economically marginal. Efficiencies are improved at higher pressure and temperature but at greater equipment cost.

Inlet temperature to the expander is controlled by the amount of excess air. The air/fuel ratio to make 1700°F is in the range of 50 lb/lb. Metallurgical considerations usually limit the temperature to this value. Special materials are available for temperatures up to 2200°F but may be too expensive for process applications.

REFERENCES

1. M.P. Boyce, *Gas Turbine Engineering Handbook,* Gulf, Houston, 1982.
2. F.L. Evans, *Equipment Design Handbook for Refineries and Chemical Plants,* Gulf, Houston, 1979, vol. 1.
3. H. Gartmann, *De Laval Engineering Handbook,* McGraw-Hill, New York, 1970.
4. R.T.C. Harman, *Gas Turbine Engineering,* Macmillan, New York, 1981.
5. E.E. Ludwig, *Applied Process Design for Chemical and Process Plants,* Gulf, Houston, 1983, vol. 3.
6. *Marks' Standard Handbook for Mechanical Engineers,* McGraw-Hill, New York, 1987.

5

TRANSFER OF SOLIDS

In contrast to fluids which are transferred almost exclusively through pipelines with pumps or blowers, a greater variety of equipment is employed for moving solids to and from storage and between process equipment. Most commonly, solids are carried on or pushed along by some kind of conveyor. Solids in granular form also are transported in pipelines as slurries in inert liquids or as suspensions in air or other gases.

5.1. SLURRY TRANSPORT

In short process lines slurries are readily handled by centrifugal pumps with large clearances. When there is a distribution of sizes, the fine particles effectively form a homogeneous mixture of high density in which the settling velocities of larger particles are less than in clear liquid. Turbulence in the line also helps to keep particles in suspension. It is essential, however, to avoid dead spaces in which solids could accumulate and also to make provisions for periodic cleaning of the line. A coal–oil slurry used as fuel and acid waste neutralization with lime slurry are two examples of process applications.

Many of the studies of slurry transfer have been made in connection with long distance movement of coal, limestone, ores, and others. A few dozen such installations have been made, in length from several miles to several hundred miles.

Coal–water slurry transport has been most thoroughly investigated and implemented. One of the earliest lines was 108 miles long, 10 in. dia, 50–60 wt % solids up to 14 mesh, at velocities of 4.5–5.25 ft/sec, with positive displacement pumps at 30-mile intervals. The longest line in the United States is 273 miles, 18 in. dia and handles 4.8–6.0 million tons/yr of coal; it is described in detail by Jacques and Montfort (1977). Other slurry pipeline literature is by Wasp, Thompson, and Snoek (1971), Bain and Bonnington (1970), Ewing (1978), and Zandi (1971).

Principally, investigations have been conducted of suitable linear velocities and power requirements. Slurries of 40–50 vol % solids can be handled satisfactorily, with particle sizes less than 24–48 mesh or so (0.7–0.3 mm). At low line velocities, particles settle out and impede the flow of the slurry, and at high velocities the frictional drag likewise increases. An intermediate condition exists at which the pressure drop per unit distance is a minimum. The velocity at this condition is called a critical velocity of which one correlation is

$$u_c^2 = 34.6 C_v D u_t \sqrt{g(s-1)/d}, \qquad \text{consistent units,} \tag{5.1}$$

where

u_c = critical flow velocity,
u_t = terminal settling velocity of the particle, given by Figure 5.1,
C_v = volume fraction of solids,
D = pipe diameter,
d = particle diameter,
s = ratio of densities of solid and liquid,
g = acceleration of gravity, 32.2 ft/sec^2, or consistent units.

The numerical coefficient is due to Hayden and Stelson (1971).

Another criterion for selection of a flow rate is based on considerations of the extent of sedimentation of particles of various sizes under flow conditions. This relation is developed by Wasp, Aude, Seiter, and Thompson (1971),

$$\frac{C}{C_0} = \exp(-2.55 u_t / k u \sqrt{f}), \tag{5.2}$$

where

C = concentration of a particular size at a level 92% of the vertical diameter,
C_0 = concentration at the center of the pipe, assumed to be the same as the average in the pipe,
f = Fanning friction factor for pipe flow

$$= 0.25 \frac{\Delta P}{\rho} \frac{L}{D} \frac{u^2}{2g_c} \tag{5.3}$$

At high Reynolds numbers, for example, Blasius' equation is

$$f = 0.0791 / N_{Re}^{0.25}, \qquad N_{Re} \geq 10^5 \tag{5.4}$$

k in Eq. (5.2) is a constant whose value is given in this paper as 0.35, but the value 0.85 is shown in a computer output in a paper by Wasp, Thompson, and Snoek (1971, Fig. 9). With the latter value, Eq. (5.2) becomes

$$C/C_0 = \exp(-3.00 u_t / u \sqrt{f}). \tag{5.5}$$

The latter paper also states that satisfactory flow conditions prevail when $C/C_0 \geq 0.7$ for the largest particle size. On this basis, the minimum line velocity becomes

$$u = \frac{3u_t}{\sqrt{f}\ln(C_0/C)} = 8.41 u_t / \sqrt{f}, \tag{5.6}$$

where u_t is the settling velocity of the largest particle present.

As Example 5.1 shows, the velocities predicted by Eqs. (5.1) and (5.6) do not agree closely. Possibly an argument in favor of Eq. (5.6) is that it is proposed by the organization that designed the successful 18 in., 273 mi Black Mesa coal slurry line.

Pressure drop in flow of aqueous suspensions sometimes has been approximated by multiplying the pressure drop of clear liquid at the same velocity by the specific gravity of the slurry. This is not borne out by experiment, however, and the multiplier has been correlated by other relations of which Eq. (5.7) is typical:

$$\Delta P_s / \Delta P_L = 1 + 69 C_v \left[\frac{gD(s-1)}{u^2 \sqrt{C_D}}\right]^{1.3}. \tag{5.7}$$

This equation is a modification by Hayden and Stelson (1971) of a series of earlier ones. The meanings of the symbols are

C_v = volume fraction occupied by the solids in the slurry,
d = particle diameter,
D = pipe diameter,
s = ratio of specific gravities of solid and liquid.

Example 5.1
Conditions of a Coal Slurry Pipeline

Data of a pulverized coal slurry are

$C_v = 0.4,$
$D = 0.333$ ft,
$f = 0.0045$ (Blasius' eq. at $N_{re} = 10^5$),
$s = 1.5.$

Mesh size	24	48	100	Mixture
d(mm)	0.707	0.297	0.125	0.321
Weight fraction	0.1	0.8	0.1	1
u_t (ft/sec)	0.164	0.050	0.010	0.0574

The terminal velocities are read off Figure 5.1, and the values of the mixture are weight averages.

The following results are found with the indicated equations:

Item	Eq.	24	48	100	Mixture
u_c	5.1	7.94	5.45	3.02	
u	5.6	20.6	6.27	1.25	
$\sqrt{c_D}$	5.8	1.36	2.89	9.38	3.39
$\Delta P_s/\Delta P_L$	5.11				1.539
$\Delta P_s/\Delta P_L$	5.13				1.296

Eq. (5.1): $$u_c^2 = 34.6(0.4)(0.333)\sqrt{32.2(0.5)}\frac{u_t}{\sqrt{d_{mm}/304.8}} = 323\frac{u_t}{\sqrt{d_{mm}}},$$

Eq. (5.6): $$u = \frac{8.41u_t}{\sqrt{0.0045}} = 125u_t,$$

Eq. (5.8): $$c_D = \frac{4}{3}\frac{32.2(1.5-1)}{u_t^2}\frac{d_{mm}}{304.8} = \frac{0.0704d_{mm}}{u_t^2},$$

Eq. (5.11): $$\frac{\Delta P_s}{\Delta P_L} = 1 + \frac{0.69}{0.4^{0.3}}\left[\frac{1}{0.0574}\sqrt{\frac{32.2(0.5)0.321}{304.8(3.39)^2}}\right]^{1.3} = 1.5391,$$

Eq. (5.13): $$\frac{\Delta P_s}{\Delta P_L} = 1 + 0.272(0.4)\left[\frac{0.0045(0.333)32.2(0.5)}{(0.0574)^2(3.39)}\right]^{1.3} = 1.296.$$

With coal of sp gr = 1.5, a slurry of 40 vol % has a sp gr = 1.2. Accordingly the rule, $\Delta P_s/\Delta P_L$ = sp gr, is not confirmed accurately by these results.

The drag coefficient is

$$C_D = 1.333gd(s-1)/u_t^2. \tag{5.8}$$

For mixtures, a number of rules has been proposed for evaluating the drag coefficient, of which a weighted average seems to be favored,

$$\sqrt{C_D} = \sum w_i\sqrt{C_{Di}}, \tag{5.9}$$

where the w_i are the weight fractions of particles with diameters d_i.

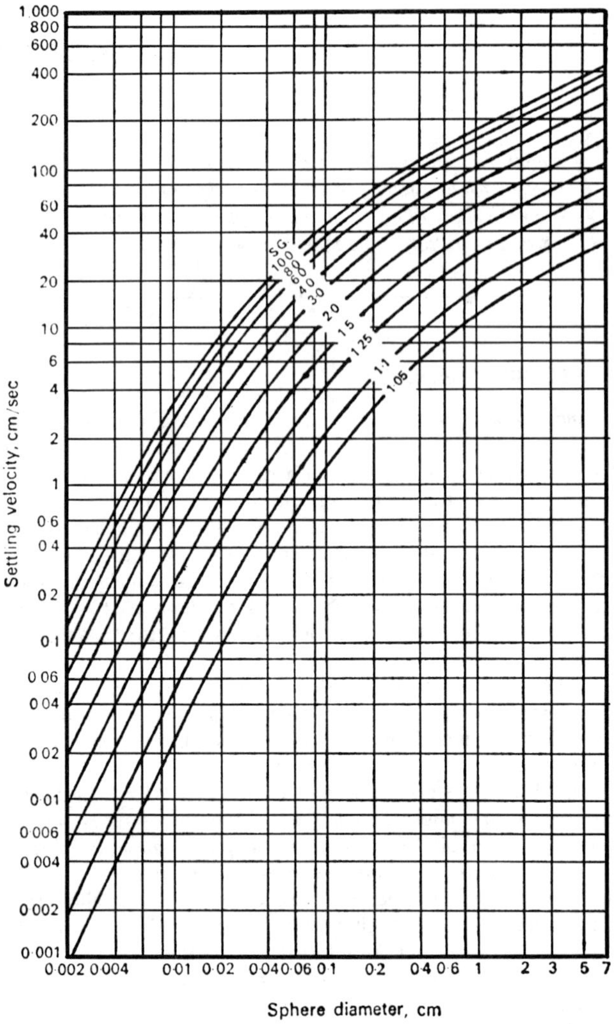

Figure 5.1. Settling velocities of spheres as a function of the ratio ot densities of the two phases. Stokes law applies at diameters below approximately 0.01 cm (*based on a chart of Lapple et al.*, Chemical Engineering Handbook, *McGraw-Hill, New York, 1984, p. 5.67*).

For particles of one size, Eqs. (5.7) and (5.8) combine to

$$\Delta P_s/\Delta P_L = 1 + 100C_v[(u_tD/u^2)\sqrt{g(s-1)/d}]^{1.3}, \quad \text{consistent units.} \tag{5.10}$$

The pressure drop relation at the critical velocity given by Eq. (5.1) is found by substitution into Eq. (5.7) with the result

$$\Delta P_s/\Delta P_L = 1 + \frac{0.69}{C_v^{0.3}}[(1/u_t)\sqrt{gd(s-1)/C_D}]^{1.3}. \tag{5.11}$$

With Eq. (5.10) the result is

$$\Delta P_s/\Delta P_L = 1 + 1/C_v^{0.3}. \tag{5.12}$$

With the velocity from Eq. (5.6), Eq. (5.7) becomes

$$\Delta P_s/\Delta P_L = 1 + 0.272C_v[fgD(s-1)/u_t^2\sqrt{C_D}]^{1.3} \tag{5.13}$$

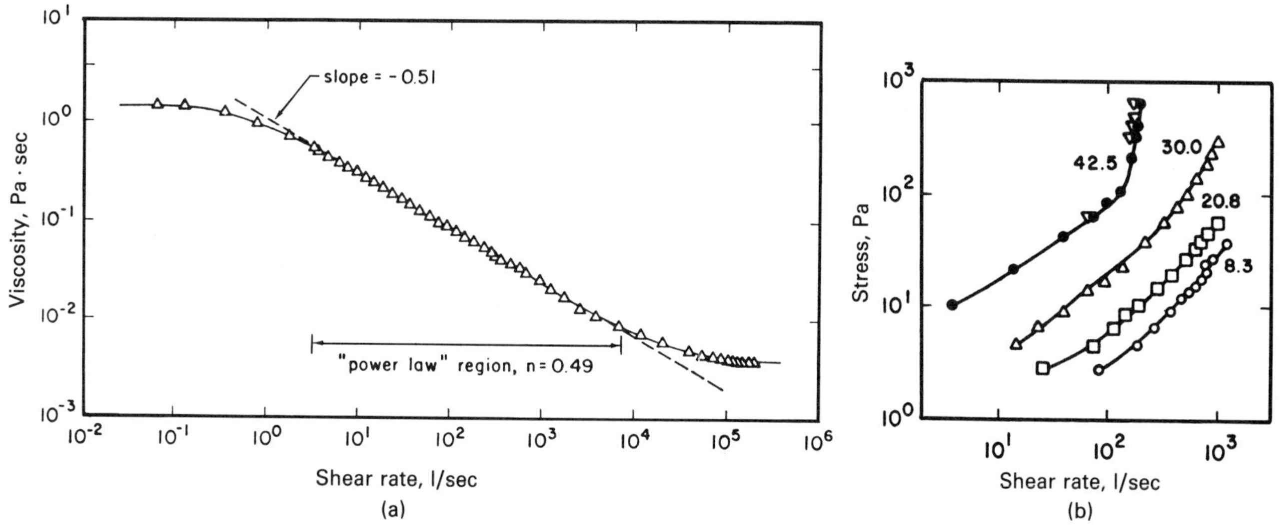

Figure 5.2. Non-Newtonian behavior of suspensions: (a) viscosity as a function of shear rate, 0.4 wt % polyacrylamide in water at room temperature; (b) shear stress as a function of shear rate for suspensions of TiO_2 at the indicated vol % in a 47.1 wt % sucrose solution whose viscosity is 0.017 Pa sec (*Denn,* Process Fluid Mechanics, *Prentice-Hall, Englewood Cliffs, NJ, 1980*).

and, for one-sized particles,

$$\Delta P_s/\Delta P_L = 1 + 0.394C_v[(fD/u_t)\sqrt{g(s-1)/d}]^{1.3}. \quad (5.14)$$

These several pressure drop relations hardly appear consistent, and the numerical results of Example 5.1 based on them are only roughly in agreement.

From statements in the literature, it appears that existing slurry lines were designed on the basis of some direct pilot plant studies.

Nonsettling slurries are formed with fine particles or plastics or fibers. Although their essentially homogeneous nature would appear to make their flow behavior simpler than that of settling slurries, they often possess non-Newtonian characteristics which complicate their flow patterns. In Newtonian flow, the shear stress is proportional to the shear strain,

$$\text{stress} = \mu(\text{strain}),$$

but in other cases the relation between these two quantities is more complex. Several classes of non-Newtonian behavior are recognized for suspensions. Pseudoplastic or power-law behavior is represented by

$$\text{stress} = k(\text{strain})^n, \qquad n < 1,$$

where k is called the consistency index. Plastic or Bingham behavior is represented by

$$\text{stress} = k_1 + \eta(\text{strain}),$$

where η is called the plastic viscosity. Data for some suspensions are given on Figure 5.2.

The constants of such equations must be found experimentally over a range of conditions for each particular case, and related to the friction factor with which pressure drops and power requirements can be evaluated. The topic of nonsettling slurries is treated by Bain and Bonnington (1970) and Clift (1980). Friction factors of power-law systems are treated by Dodge and Metzner (1959) and of fiber suspensions by Bobkowitz and Gauvin (1967).

5.2. PNEUMATIC CONVEYING

Granular solids of free-flowing natures may be conveyed through ducts in any direction with high velocity air streams. In the normal plant, such lines may be several hundred feet long, but dusty materials such as fly ash and cement have been moved over a mile in this way. Materials that are being air-veyed include chemicals, plastic pellets, grains, and powders of all kinds. The transfer of catalysts between regenerator and reactor under fluidized conditions is a common operation. Stoess (1983) has a list of recommendations for about 150 different materials, of which Table 5.1 is a selection. Basic equipment arrangements are represented in Figure 5.3.

The performance of pneumatic conveyors is sensitive to several characteristics of the solids, of which the most pertinent ones are

1. bulk density, as poured and as aerated,
2. true density,
3. coefficient of sliding friction (= tangent of the angle of repose),
4. particle size distribution,
5. particle roughness and shape,
6. moisture content and hygroscopicity, and
7. characteristics such as friability, abrasiveness, flammability, etc.

Sulfur, for example, builds up an electrostatic charge and may introduce explosive risks.

In comparison with mechanical conveyors, pneumatic types must be designed with greater care. They demand more power input per unit weight transferred, but their cost may be less for complicated paths, when exposure to the atmosphere is undesirable and when operator safety is a problem. Although in the final analysis the design and operation of pneumatic conveyors demands the attention of experienced engineers, a design for orientation purposes can be made by the inexpert on the basis of general knowledge and rules of thumb that appear in the literature. An article by Solt (1980) is devoted entirely to preventive trouble-shooting.

Some basic design features are the avoidance of sharp bends, a minimum of line fittings, provision for cleanout, and possibly electrical grounding. In many cases equipment suppliers may wish to do pilot plant work before making final recommendations. Figure

TABLE 5.1. Flow Rates and Power Requirements of Vacuum and Low Pressure Pneumatic Conveying Systems[a]

	Vacuum System (8–9 psia)											Low Pressure System (6–12 psig)							
		Conveying Distance											Conveying Distance						
	Wt per	100 ft		150 ft		250 ft		400 ft		Velocity	Pressure	100 ft		250 ft		400 ft		Velocity
Material	cu ft	Sat.	hp/T	Sat.	hp/T	Sat.	hp/T	Sat.	hp/T	ft/sec	Factor	Sat.	hp/T	Sat.	hp/T	Sat.	hp/T	(ft/sec)
Alum	50	3.6	4.5	3.9	5.0	4.3	5.7	4.7	6.3	110	4.0	1.6	2.7	2.0	3.4	2.2	3.8	65
Alumina	60	2.4	4.0	2.8	4.7	3.4	5.7	4.0	6.4	105	5.0	1.1	2.4	1.6	3.4	1.9	3.9	60
Carbonate, calcium	25–30	3.1	4.2	3.6	5.0	3.9	5.5	4.2	6.0	110	3.5	1.4	2.5	1.8	3.3	2.0	3.6	65
Cellulose acetate	22	3.2	4.7	3.5	5.1	3.8	5.7	4.1	6.0	100	3.0	1.4	2.8	1.7	3.4	1.9	3.6	55
Clay, air floated	30	3.3	4.5	3.5	5.0	3.9	5.5	4.2	6.0	105	4.0	1.5	2.7	1.8	3.3	1.9	3.6	50
Clay, water washed	40–50	3.5	5.0	3.8	5.6	4.2	6.5	4.5	7.2	115	4.5	1.6	3.0	1.9	3.9	2.1	4.4	60
Clay, spray dried	60	3.4	4.7	3.6	5.2	4.0	6.2	4.4	7.1	110	4.3	1.5	2.8	1.8	3.7	2.0	4.3	55
Coffee beans	42	1.2	2.0	1.6	3.0	2.1	3.5	2.4	4.2	75	5.0	0.6	1.2	0.9	2.1	1.1	2.5	45
Corn, shelled	45	1.9	2.5	2.1	2.9	2.4	3.6	2.8	4.3	105	5.0	0.9	1.5	1.1	2.2	1.3	2.6	55
Flour, wheat	40	1.5	3.0	1.7	3.3	2.0	3.7	2.5	4.4	90	2.5	0.7	1.8	0.9	2.2	1.1	2.7	35
Grits, corn	33	1.7	2.5	2.2	3.0	2.9	4.0	3.5	4.8	100	3.5	0.8	1.5	1.3	2.4	1.6	2.9	70
Lime, pebble	56	2.8	3.8	3.0	4.0	3.4	4.7	3.9	5.4	105	5.0	1.3	2.3	1.6	2.8	1.8	3.3	70
Lime, hydrated	30	2.1	3.3	2.4	3.9	2.8	4.7	3.4	6.0	90	4.0	0.6	1.8	0.8	2.2	0.9	2.6	40
Malt	28	1.8	2.5	2.0	2.8	2.3	3.4	2.8	4.2	100	5.0	0.8	1.5	1.1	2.0	1.3	2.5	55
Oats	25	2.3	3.0	2.6	3.5	3.0	4.4	3.4	5.2	100	5.0	1.0	1.8	1.4	2.6	1.6	3.1	55
Phosphate, trisodium	65	3.1	4.2	3.6	5.0	3.9	5.5	4.2	6.0	110	4.5	1.4	2.5	1.8	3.3	1.9	3.6	75
Polyethylene pellets	30	1.2	2.0	1.6	3.0	2.1	3.5	2.4	4.2	80	5.0	0.55	1.2	0.9	2.1	1.1	2.5	70
Rubber pellets	40	2.9	4.2	3.5	5.0	4.0	6.0	4.5	7.2	110								
Salt cake	90	4.0	6.5	4.2	6.8	4.6	7.5	5.0	8.5	120	5.0	2.9	3.9	3.5	4.5	4.0	5.1	83
Soda ash, light	35	3.1	4.2	3.6	5.0	3.9	5.5	4.2	6.0	110	5.0	1.4	2.5	1.8	3.3	1.9	3.6	65
Soft feeds	20–40	3.0	4.2	3.4	4.5	3.7	5.0	4.2	5.5	110	3.8	1.3	2.5	1.7	3.1	1.9	3.7	70
Starch, pulverized	40	1.7	3.0	2.0	3.4	2.6	4.0	3.4	5.0	90	3.0	0.8	1.7	1.1	2.4	1.5	3.0	55
Sugar, granulated	50	3.0	3.7	3.2	4.0	3.4	5.2	3.9	6.0	110	5.0	1.4	2.2	1.6	3.1	1.7	3.6	60
Wheat	48	1.9	2.5	2.1	2.9	2.4	3.6	2.8	4.3	105	5.0	0.9	1.5	1.1	2.1	1.3	2.6	55
Wood flour	12–20	2.5	3.5	2.8	4.0	3.4	4.9	4.4	6.5	100								

[a] HP/ton = (pressure factor)(hp/T)(sat.). The units of sat. are standard cuft of air/lb of solid transferred), and those of hp/T are horsepower/(tons/hr of solid transferred).
(Stoess, 1983).

5.4 shows a typical pilot plant arrangement. A preliminary design procedure is given by Raymus (1984). Many details of design and operation are given in books by Stoess (1983) and Kraus (1980) and in articles by Gerchow (1980), and Perkins and Wood (1974). Some of that information will be restated here. Pressure drop and power requirements can be figured largely on the basis of general knowledge.

EQUIPMENT

The basic equipment consists of a solids feeding device, the transfer line proper, a receiver, a solid–air separator, and either a blower at the inlet or a vacuum pump at the receiver. Four common kinds of arrangements are shown on Figure 5.3. Vacuum systems are favored for shorter distances and when conveying from several sources to one destination. Appropriate switching valves make it possible to service several sources and destinations with either a vacuum or pressure system. Normally the vacuum system is favored for single destinations and the pressure for several destinations or over long distances. Figure 5.3(b) shows a rotary valve feeder and Figure 5.3(c) a Venturi feeder which has a particularly gentle action suitable for friable materials. Figure 5.3(d) utilizes a fan to suck the solids from a source and to deliver them under positive pressure. Friable materials also may be handled effectively by the equipment of Figure 5.5 in which alternate pulses of granular material and air are transported.

Typical auxiliary equipment is shown on Figure 5.6. The most used blower in pneumatic conveying is the rotary positive displacement type; they can achieve vacua 6–8 psi below atmospheric or positive pressures up to 15 psig at efficiencies of about 65%. Axial positive displacement blowers also are used, as well as centrifugals for large capacities. Rotary feeders of many proprietary designs are available; Stoess (1983) and Kraus (1980) illustrate several types. Receivers may be equipped with fabric filters to prevent escape of fine particles; a dacron fabric suitable for up to 275°F is popular. Cyclone receivers are used primarily for entirely nondusting services or ahead of a filter. A two-stage design is shown in Figure 5.6(d). Typical dimensions are cited by Stoess (1983), for example:

line diameter (in.)	3	5	8
primary diameter (ft)	3.5	4.5	6.75
secondary diameter (ft)	2.75	3.5	5.0

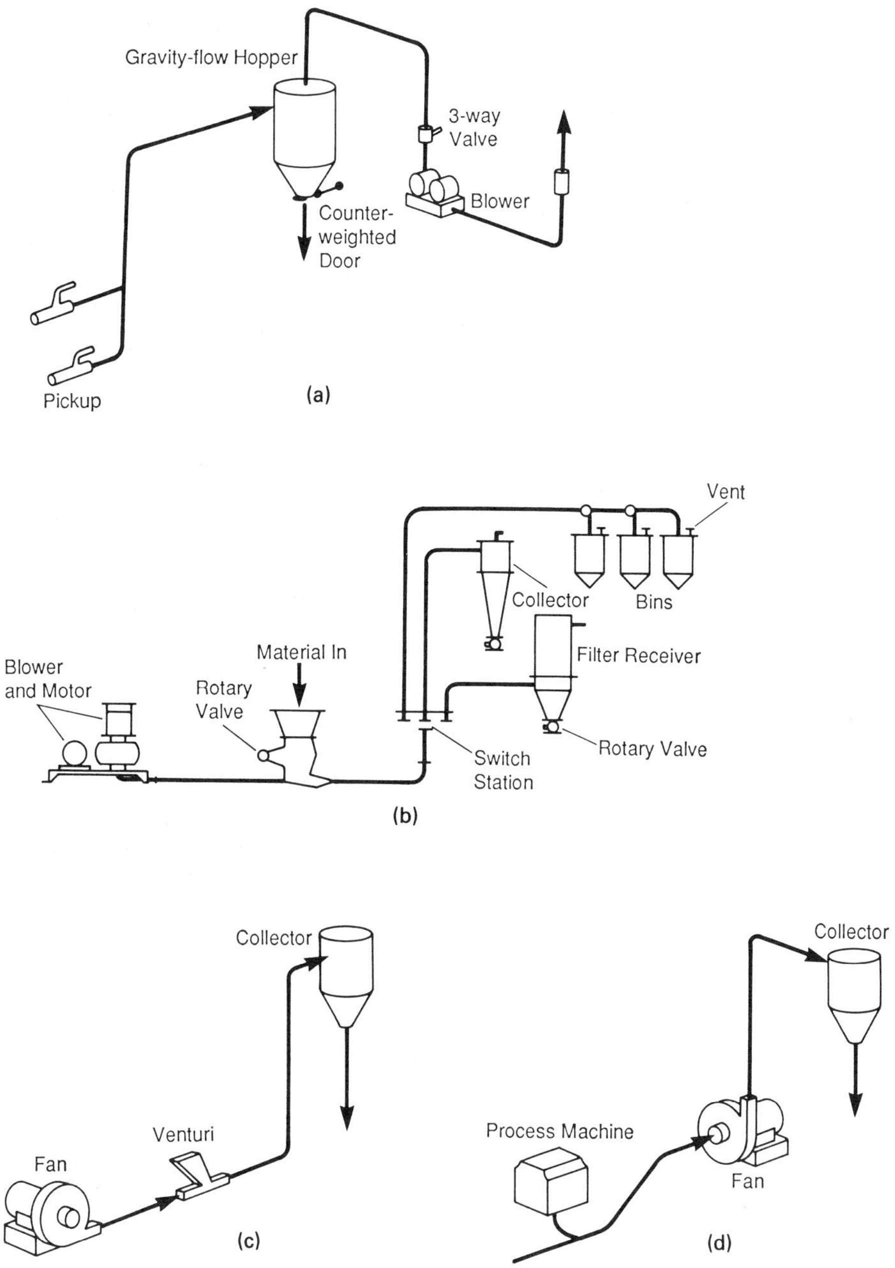

Figure 5.3. Basic equipment arrangements of pneumatic conveying systems. (a) Vacuum system with several sources and one destination, multiple pickup; (b) pressure system with rotary valve feeder, one source and several destinations, multiple discharge; (c) pressure system with Venturi feed for friable materials; (d) pull–push system in which the fan both picks up the solids and delivers them [*after F. J. Gerchow,* Chem. Eng. (*17 Feb. 1975, p. 88*)].

Piping usually is standard steel, Schedule 40 for 3–7 in. IPS and Schedule 30 for 8–12 in. IPS. In order to minimize pressure loss and abrasion, bends are made long radius, usually with radii equal to 12 times the nominal pipe size, with a maximum of 8 ft. Special reinforcing may be needed for abrasive conditions.

OPERATING CONDITIONS

Vacuum systems usually operate with at most a 6 psi differential; at lower pressures the carrying power suffers. With rotary air lock feeders, positive pressure systems are limited to about 12 psig. Other feeding arrangements may be made for long distance transfer with 90–125 psig air. The dense phase pulse system of Figure 5.4 may operate at 10–30 psig.

Linear velocities, carrying capacity as cuft of free air per lb of solid and power input as HP/tons per hour (tph) are listed in Table 5.1 as a general guide for a number of substances. These data are for 4-, 5-, and 6-in. lines; for 8-in. lines, both Sat. and HP/tph are reduced by 15%, and for 10-in. by 25%. Roughly, air velocities in low positive pressure systems are 2000 ft/min for light materials,

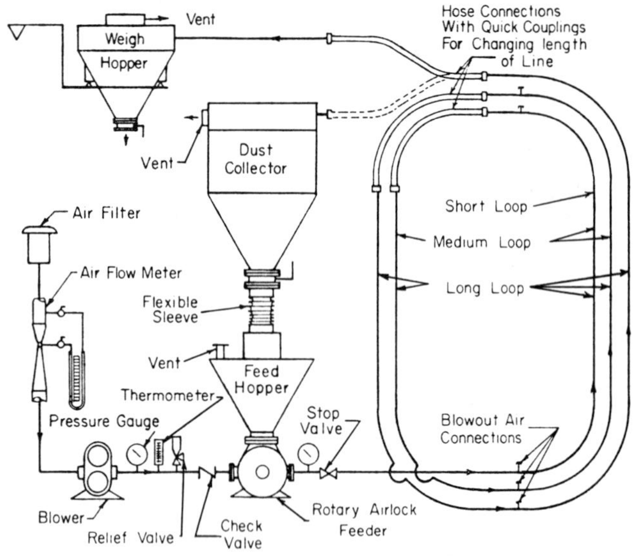

Figure 5.4. Sketch of pilot plant arrangement for testing pneumatic conveying under positive pressure (*Kraus,* Pneumatic Conveying of Bulk Materials, *McGraw-Hill, New York, 1980*).

3000–4000 ft/min for medium densities such as those of grains, and 5000 ft/min and above for dense materials such as fly ash and cement; all of these velocities are of free air, at atmospheric pressure.

Another set of rules for air velocity as a function of line length

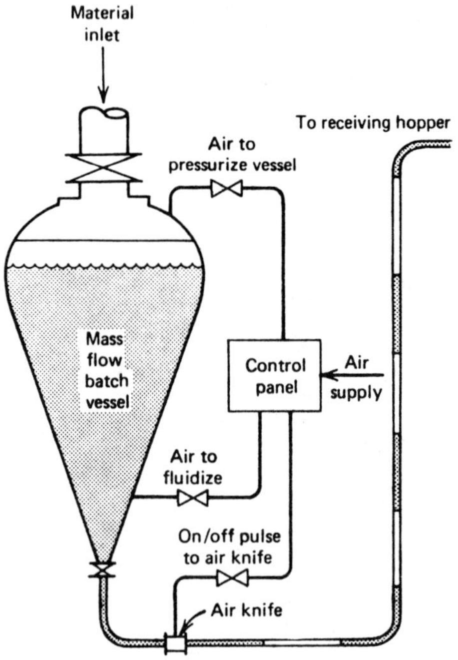

Figure 5.5. Concept of dense phase transfer of friable materials, by intermittent injection of material and air pulses, air pressures normally 10–30 psig and up to 90 psig (*Sturtevant Engineering Co., Boston, MA*).

and bulk density is due to Gerchow (1980) and is

Line length (ft)	ft/min 55 lb/cuft	55–85	85–115
200	4000	5000	6000
500	5000	6000	7000
1000	6000	7000	8000

Conveying capacity expressed as vol % of solids in the stream usually is well under 5 vol %. From Table 5.1, for example, it is about 1.5% for alumina and 6.0% for polystyrene pellets, figured at atmospheric pressure; at 12 psig these percentages will be roughly doubled, and at subatmospheric pressures they will be lower.

POWER CONSUMPTION AND PRESSURE DROP

The power consumption is made up of the work of compression of the air and the frictional losses due to the flows of air and solid through the line. The work of compression of air at a flow rate m'_a and $C_P/C_v = 1.4$ is given by

$$w_c = 3.5(53.3)(T+460)m'_a[(P_2/P_1)^{0.2857} - 1] \quad \text{(ft lbf/sec)} \tag{5.15}$$

with the flow rate in lb/sec.

Frictional losses are evaluated separately for the air and the solid. To each of these, contributions are made by the line itself, the elbows and other fittings, and the receiving equipment. It is conservative to assume that the linear velocities of the air and solid are the same. Since the air flow normally is at a high Reynolds number, the friction factor may be taken constant at $f_a = 0.015$. Accordingly the frictional power loss of the air is given by

$$w_1 = \Delta P_1 m'_a/\rho_a = (u^2/2g)\left[1 + 2n_c + 4n_f + (0.015/D)\left(L + \sum L_i\right)\right]m'_a \quad \text{(ft lbf/sec)}. \tag{5.16}$$

The unity in the bracket accounts for the entrance loss, n_c is the number of cyclones, n_f is the number of filters, L is the line length, and L_i is the equivalent length of an elbow or fitting. For long radius bends one rule is that the equivalent length is 1.6 times the actual length of the bend. Another rule is that the long bend radius is 12 times the nominal size of the pipe. Accordingly,

$$L_i = 1.6(\pi R_i/2) = 2.5R_i = 2.5D''_i \text{ ft}, \quad \text{with } D''_i \text{ in inches.} \tag{5.17}$$

The value of g is 32.2 ft lb m/(lbf sec^2).

The work being done on the solid at the rate of m'_s lb/sec is made up of the kinetic gain at the entrance (w_2), the lift (w_3) through an elevation Δz, friction in the line (w_4), and friction in the elbow (w_5). Accordingly,

$$w_2 = \frac{u^2}{2g}m'_s \quad \text{(ft lbf/sec)}. \tag{5.18}$$

The lift work is

$$w_3 = \Delta z \frac{g}{g_c} m'_s = \Delta z m'_s \quad \text{(ft lbf/sec)}. \tag{5.19}$$

The coefficient of sliding friction f_s of the solid equals the tangent of the angle of repose. For most substances this angle is 30–45° and

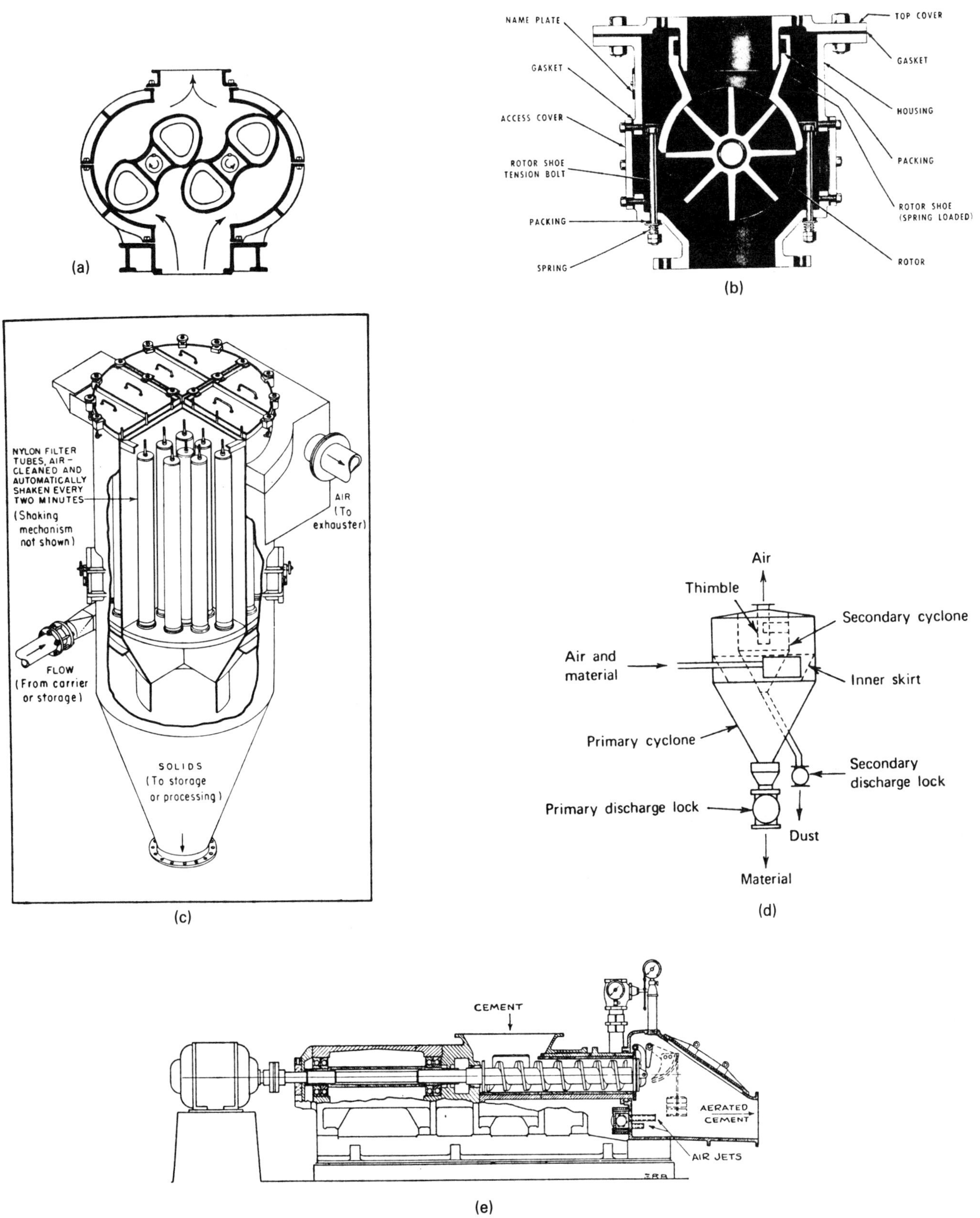

Figure 5.6. Components of pneumatic conveying systems. (a) Rotary positive displacement blower for pressure or vacuum. (b) A rotary airlock feeder for fine materials (*Detroit Stoker Co.*). (c) A four-compartment receiver-filter (*Fuller Co., Bethlehem, PA*). (d) A two-stage cyclone receiver. (e) The Fuller–Kinyon pump for cement and other fine powders. Powder is fed into the aeration chamber with a screw and is fluidized with compressed air (*Fuller Co., Bethlehem, PA*).

the value of f_s is 0.58–1.00. The sliding friction in the line is

$$w_4 = f_s L m_s' \quad \text{(ft lbf/sec)}, \tag{5.20}$$

where L is the line length.

Friction in the curved elbows is enhanced because of centrifugal force so that

$$w_5 = f_s \frac{u^2}{gR}\left(\frac{2\pi R}{4}\right) m_s' = 0.0488 f_s u^2 m_s' \quad \text{(ft lbf/sec)}. \tag{5.21}$$

The total frictional power is

$$w_f = w_1 + w_2 + w_3 + w_4 + w_5, \tag{5.22}$$

and the total power consumption is

$$w = \frac{(w_c + w_f)}{550\eta(1.8 m_s')} \quad \text{[HP/(ton/hr)]}, \tag{5.23}$$

where η is the blower efficiency. Pressure drop in the line is obtained from the frictional power, the total flow rate, and the density of the mixture:

$$\Delta P = \frac{w_f}{144(m_a' + m_s')} \rho_m \quad \text{(psi)}. \tag{5.24}$$

The specific air rate, or saturation, is

$$\text{saturation} = 0.7854(60) D^2 \quad \text{(cuft/min of air)/(lb/min of solid)]}, \tag{5.25}$$

where the velocity of the air is evaluated at atmospheric pressure.

Example 5.2 makes the calculations described here for power and pressure drop, and compares the result with the guidelines of Table 5.1.

5.3. MECHANICAL CONVEYORS AND ELEVATORS

Granular solids are transported mechanically by being pushed along or dragged along or carried. Movement may be horizontal or vertical or both. In the process plant distances may be under a hundred feet or several hundred feet. Distances of several miles may be covered by belts servicing construction sites or mines or power plants. Capacities range up to several hundred tons/hr. The principal kinds of mechanical conveyors are illustrated in Figures 5.7–5.13 and will be described. Many construction features of these machines are arbitrary. Thus manufacturers' catalogs are the ultimate source of information about suitability for particular services, sizes, capacities, power requirements and auxiliaries. Much of the equipment has been made in essentially the present form for about 100 years by a number of manufacturers so that a body of standard practice has developed.

PROPERTIES OF MATERIALS HANDLED

The physical properties of granular materials that bear particularly on their conveying characteristics include size distribution, true and bulk densities, and angle of repose or coefficient of sliding friction, but other less precisely measured or described properties are also of concern. A list of pertinent properties appears in Table 5.2. The elaborate classification given there is applied to about 500 materials in the FMC Corporation Catalog 100 (1983, pp. B.27–B.35) but is too extensive for reproduction here. For each material the table also identifies the most suitable design of screw conveyor of this company's manufacture and a factor for determining the power requirement. An abbreviated table of about 150 substances appears in the *Chemical Engineers Handbook* (1984, p. 7.5). Hudson (1954, pp. 6–9), describes the characteristics of about 100 substances in relation to their behavior in conveyors. Table 5.3 lists bulk densities, angles of respose at rest, and allowable angles of inclination which are angles of repose when a conveyor is in motion; references to more extensive listings of such data are given in this table.

The angle of repose is a measure of the incline at which conveyors such as screws or belts can carry the material. The tangent of the angle of repose is the coefficient of sliding friction. This property is a factor in the power needed to transfer the material by pushing or dragging as in pneumatic, screw, flight, and Redler equipment.

Special provisions need to be made for materials that tend to form bridges; Figure 5.13(a) is an example of a method of breaking up bridges in a storage bin so as to ensure smooth flow out. Materials that tend to pack need to be fluffed up as they are pushed along by a screw; adjustable paddles as in Figure 5.7(d) may be sufficient.

SCREW CONVEYORS

These were invented by Archimedes and assumed essentially their present commercial form a hundred years or so ago. Although the equipment is simple in concept and relatively inexpensive, a body of experience has accumulated whereby the loading, speed, diameter, and length can be tailored to the characteristics of the materials to be handled. Table 5.4, for example, recognizes four classes of materials, ranging from light, freeflowing, and nonabrasive materials such as grains, to those that are abrasive and have poor flowability such as bauxite, cinders, and sand. Only a portion of the available data are reproduced in this table.

Lengths of screw conveyors usually are limited to less than about 150 ft; when the conveying distance is greater than this, a belt or some other kind of machine should be chosen. The limitation of length is due to structural strength of the shaft and coupling. It is expressed in terms of the maximum torque that is allowable. Formulas for torque and power of screw conveyors are given in Table 5.4 and are applied to selection of a conveyor in Example 5.3.

Several designs of screws are shown in Figure 5.7. The basic design is one in which the pitch equals the diameter. Closer spacing is needed for carrying up steep inclines, and in fact very fine pitch screws operating at the relatively high speeds of 350 rpm are used to convey vertically. The capacity of a standard pitch screws drops off sharply with the inclination, for example:

Angle (degrees)	<8	20	30	45
Percent of capacity	100	55	30	0

Allowable loadings as a percentage of the vertical cross section depend on the kind of material being processed; examples are shown in Table 5.4.

BELT CONVEYORS

These are high capacity, relatively low power units for primarily horizontal travel and small inclines. The maximum allowable inclination usually is 5–15° less than the angle of repose; it is shown as "recommended maximum inclination" in Table 5.3 for some substances, and is the effective angle of repose under moving conditions.

The majority of conveyor belts are constructed of fabric, rubber, and wire beads similarly to automobile tires, but they are made also of wire screen or even sheet metal for high temperature

Example 5.2

Size and Power Requirement of a Pneumatic Transfer Line

A pneumatic transfer line has 300 ft of straight pipe, two long radius elbows, and a lift of 50 ft. A two-stage cyclone is at the receiving end. Solid with a density of 125 lb/cuft is at the rate of 10 tons/hr and the free air is at 5000 ft/min. Inlet condition is 27 psia and 100°F. Investigate the relation btween line diameter and power requirement.

On a first pass, the effect of pressure loss on the density of the air will be neglected.

Mass flow rate of solid:

$$m_s' = 20{,}000/3600 = 5.56\ \text{lb/sec}.$$

Mass flow rate of air:

$$m_a' = \frac{5000}{60}\frac{\pi}{4}(0.075)D^2 = 4.91D^2\ \text{lb/sec}.$$

Density of air:

$$\rho_a = 0.075\left(\frac{27}{14.7}\right) = 0.138\ \text{lb/cuft}.$$

Density of mixture:

$$\rho_m = \frac{(m_a' + m_s')}{m_a'/\rho_a + m_s'/\rho_s} = \frac{(m_a' + 5.56)}{m_a'/0.138 + 5.56/125}$$

Linear velocity of air at inlet:

$$u = \frac{5000}{60}\left(\frac{14.7}{27}\right) = 45.37\ \text{fps}.$$

Assume air and solid velocities equal. Elbow radius = $12D$. Elbow equivalent length,

$$L_e = 1.6(\pi/2)(12D) = 30.2D$$

Power for compression from 14.7 psia and 560 R to 27 psia,

$$k/(k-1) = 3.5,$$

$$w_c = 3.5RT_1[(P_2/P_1)^{0.2857} - 1]m_a'$$
$$= 3.5(53.3)(560)[(27/14.7)^{0.2857} - 1]4.91D^2$$
$$= 97305D^2\ \text{ft lbf/sec}.$$

Frictional contribution of air

$$w_1 = \frac{u^2}{2g}[5 + (0.015/D)(300 + 2(30.2)D]m_a'$$
$$= [(45.4)^2/64.4][5.9 + (4.5/D)](4.91D^2)$$
$$= 157.1D^2(5.9 + 4.5/D)$$

For the solid, take the coefficient of sliding friction to be $f_s = 1$. Power loss is made up of four contributions. Assume no slip velocity;

$$w_s = w_2 + w_3 + w_4 + w_5$$
$$= [u^2/2g + \Delta Z + f_sL + 2(0.0488)f_su^2]m_s'$$
$$= 5.56[45.4^2/64.4 + 50 + 300 + 2(0.0488)45.4^2]$$
$$= 3242.5\text{ft lbf/sec}.$$

Total friction power:

$$w_f = 3242.5 + 157.1D^2(5.9 + 4.5/D).$$

Pressure drop:

$$\Delta P = \frac{w_f}{144(m_a' + m_s')}\rho_m\ \text{psi}.$$

Fan power at $\eta = 0.5$:

$$\dot{P} = \frac{w_c + w_f}{550(0.5)(10)} = \frac{w_c + w_f}{2750}\text{HP/tph},$$

$$\text{saturation} = \frac{5000(\pi/4)D^2}{20{,}000/60} = 11.78D^2\ \text{SCFM/(lb/min)}.$$

1 PS	D (ft)	m_a'	ρ_m	w_c	w_f
3	0.2557	0.3210	2.4808	6362	3484
4	0.3356	0.5530	1.5087	10,959	3584
5	0.4206	0.8686	1.0142	17,214	3704
6	0.5054	1.2542	0.7461	24,855	3837

1 PS	ΔP (psi)	HP/TPH	SCFM/ lb/min
3	10.2	3.58	0.77
4	6.1	5.29	1.33
5	4.1	7.60	2.08
6	2.9	10.44	3.00

From Table 5.1, data for pebble lime are

sat = 1.7 SCFM/(lb/min)
power = 3.0 HP/TPH

and for soda ash:

sat = 1.9 SCFM/(lb/min)
power = 3.4 HP/TPH.

The calculated values for a 4 in. line are closest to the recommendations of the table.

services. A related design is the apron conveyor with overlapping pans of various shapes and sizes (Fig. 5.8), used primarily for short travel at elevated temperatures. With pivoted deep pans they are also effective elevators.

Flat belts are used chiefly for moving large objects and cartons. For bulk materials, belts are troughed at angles of 20–45°. Loading of a belt may be accomplished by shovelling or directly from overhead storage or by one of the methods shown on Figure 5.9. Discharge is by throwing over the end of the run or at intermediate points with plows.

TABLE 5.2. Codes for Characteristics of Granular Materials[a]

Major Class	Material Characteristics Included	Code Designation
Density	Bulk Density, Loose	Actual lbs/ft³
Size	Very Fine No. 200 Sieve (.0029") And Under	A_{200}
	No. 100 Sieve (.0059") And Under	A_{100}
	No. 40 Sieve (.016") And Under	A_{40}
	Fine No. 6 Sieve (.132") And Under	B_0
	Granular ½" And Under	$C_{½}$
	Granular 3" And Under	D_3
	(¹)Lumpy Over 3" To Be Special X=Actual Maximum Size	D_x
	Irregular Stringy, Fibrous, Cylindrical, Slabs, etc.	F
Flowability	Very Free Flowing—Flow Function > 10	1
	Free Flowing—Flow Function > 4 But < 10	2
	Average Flowability—Flow Function > 2 But < 4	3
	Sluggish—Flow Function < 2	4
Abrasiveness	Mildly Abrasive —Index 1-17	5
	Moderately Abrasive—Index 18-67	6
	Extremely Abrasive—Index 68-416	7
Miscellaneous Properties Or Hazards	Builds Up and Hardens	F
	Generates Static Electricity	G
	Decomposes—Deteriorates in Storage	H
	Flammability	J
	Becomes Plastic or Tends to Soften	K
	Very Dusty	L
	Aerates and Becomes Fluid	M
	Explosiveness	N
	Stickiness-Adhesion	O
	Contaminable, Affecting Use	P
	Degradable, Affecting Use	Q
	Gives Off Harmful or Toxic Gas or Fumes	R
	Highly Corrosive	S
	Mildly Corrosive	T
	Hygroscopic	U
	Interlocks, Mats or Agglomerates	V
	Oils Present	W
	Packs Under Pressure	X
	Very Light and Fluffy—May Be Windswept	Y
	Elevated Temperature	Z

[a] Example: A fine 100 mesh material with an average density of 50 lb/cuft that has average flowability and is moderately abrasive would have a code designation $50A_{100}36$; if it were dusty and mildly corrosive, it would be $50A_{100}36LT$.

(FMC Corp., Materials Handling Division, Homer City, PA, 1983).

Power is required to run the empty conveyor and to carry the load horizontally and vertically. Table 5.5 gives the equations, and they are applied in Example 5.4. Squirrel-cage ac induction motors are commonly used as drives. Two- and four-speed motors are available. Mechanical efficiencies of speed reducing couplings between motor and conveyor range from 95 to 50%. Details of idlers, belt trippers, cleaners, tension maintaining devices, structures, etc. must be consulted in manufacturers' catalogs. The selection of belt for strength and resistance to abrasion, temperature, and the weather also is a topic for specialists.

BUCKET ELEVATORS AND CARRIERS

Bucket elevators and carriers are endless chains to which are attached buckets for transporting granular materials along vertical, inclined or horizontal paths. Figure 5.10 shows two basic types: spaced buckets that are far apart and continuous which overlap. Spaced buckets self-load by digging the material out of the boot and are operated at speeds of 200–300 fpm; they are discharged centrifugally. Continuous buckets operate at lower speeds, and are used for friable materials and those that would be difficult to pick up in the boot; they are fed directly from a loading chute and are discharged by gravity. Bucket carriers are essentially forms of pan conveyors; they may be used instead of belt conveyors for shorter distances and when they can be made of materials that are

TABLE 5.3. Bulk Densities, Angles of Repose, and Allowable Angles of Inclination

Material	Average Weight (lb/cuft)	Angle of Repose (degrees)	Recommended Maximum Inclination
Alum, fine	45–50	30–45	
Alumina	50–65	22	10–12
Aluminum sulfate	54	32	17
Ammonium chloride	45–52		
Ammonium nitrate	45		
Ammonium sulfate	45–58		
Asbestos shred	20–25		
Ashes, coal, dry, $\frac{1}{2}$ in. max	35–40	40	20–25
Ashes, coal, wet, $\frac{1}{2}$ in. max	45–50	50	23–27
Ashes, fly	40–45	42	20–25
Asphalt, $\frac{1}{2}$ in. max	45		
Baking powder	40–55		18
Barium carbonate	72		
Bauxite, ground	68	35	20
Bentonite, 100 mesh max	50–60		
Bicarbonate of soda	40–50		
Borax, $\frac{1}{2}$ in.	55–60		
Borax, fine	45–55		20–22
Boric acid, fine	55		
Calcium acetate	125		
Carbon, activated, dry, fine	8–20		
Carbon black, pelleted	20–25		
Casein	36		
Cement, Portland	94	39	20–23
Cement, Portland, aerated	60–75		
Cement clinker	75–95	30–40	18–20
Charcoal	18–25	35	20–25
Chips, paper mill	20–25		
Clay, calcined	80–100		
Clay, dry, fine	100–120	35	20–22
Clay, dry, lumpy	60–75	35	18–20
Coal, anthracite, $\frac{1}{2}$ in. max	60	35	18
Coal, bituminous, 50 mesh max	50–54	45	24
Coal, bituminous, $\frac{1}{2}$ in. max	43–50	40	22
Coal, lignite	40–45	38	22
Coke breeze, $\frac{1}{4}$ in. max	25–35	30–45	20–22
Copper sulfate	75–85	31	17
Cottonseed, dry, delinted	35	29	16
Cottonseed, dry, not delinted	18–25	35	19
Cottonseed meal	35–40	35	22
Cryolite dust	75–90		
Diatomaceous earth	11–14		
Dicalcium phosphate	40–50		
Disodium phosphate	25–31		
Earth, as excavated, dry	70–80	35	20
Earth, wet, containing clay	100–110	45	23
Epsom salts	40–50		
Feldspar, $\frac{1}{2}$ in. screenings	70–85	38	18
Ferrous sulfate	60–75		
Flour, wheat	35–40		
Fullers earth, dry	30–35	23	
Fullers earth, oily	60–65		
Grain, distillery, spen, dry	30		
Graphite, flake	40		
Grass seed	10–12		
Gravel, bank run	90–100	38	20
Gravel, dry, sharp	90–100		15–17
Gravel, pebbles	90–100	30	12
Gypsum dust, aerated	60–70	42	23
Gypsum, $\frac{1}{2}$ in. screenings	70–80	40	21
Iron oxide pigment	25	40	25
Kaolin talc, 100 mesh	42–56	45	23
Lactose	32		
Lead arsenate	72		

TABLE 5.3—(*continued*)

Lead oxides	60–150		
Lime, $\frac{1}{4}$ in. max	60–65	43	23
Lime, hydrated, $\frac{1}{4}$ in. max	40	40	21
Lime, hydrated, pulverized	32–40	42	22
Limestone, crushed	85–90	38	18
Limestone dust	80–85		20
Lithopone	45–50		
Magnesium chloride	33		
Magnesium sulfate	70		
Milk, dry powder	36		
Phosphate, triple super, fertilizer	50–55	45	30
Phosphate rock, pulverized	60	40	25
Polystyrene beads	40		
Potassium nitrate	76		
Rubber, pelletized	50–55	35	22
Salt, common, coarse	40–55		
Salt, dry, fine	70–80	25	11
Salt cake, dry, coarse	85	36	21
Salt cake, dry, pulverized	60–85		
Saltpeter	80		
Sand, bank, damp	100–130	45	20–22
Sand, bank, dry	90–110	35	16–18
Sawdust	10–13	36	22
Shale, crushed	85–90	39	22
Soap chips	15–25	30	18
Soap powder	20–25		
Soda ash briquetts	50	22	7
Soda ash, heavy	55–65	32	19
Soda ash, light	20–35	37	22
Sodium bicarbonate	41	42	23
Sodium nitrate	70–80	24	11
Starch	25–50	24	12
Sugar, granulated	50–55		
Sugar, powdered	50–60		
Trisodium phosphate, pulverized	50	40	25
Wood chips	10–30		27
Zinc oxide, heavy	30–35		
Zinc oxide, light	10–15		

Other tables of these properties appear in these publications:
1. Conveyor Equipment Manufacturers Association, *Belt Conveyors for Bulk Materials,* 1966, pp. 25–33.
2. Stephens–Adamson Mfg. Co. Catalog 66, 1954, pp. 634–636.
3. FMC Corporation Material Handling Equipment Division Catalog 100, 1983, pp. B.27–B.35.
4. *Perry's Chemical Engineers Handbook,* 1984, p. 7.5.

TABLE 5.4. Sizing Data for Screw Conveyors[a]

(a) Diameter (rpm and cuft/hr)

Diam. of Conveyor, Inches	△ Max. Lump Size, Inches	Maximum Recommended Speed R.P.M.	Capacities, Cubic Feet Per Hour: At Maximum Recommended Speed	Capacities, Cubic Feet Per Hour: At One R.P.M.	Maximum Recommended Speed R.P.M.	Capacities, Cubic Feet Per Hour: At Maximum Recommended Speed	Capacities, Cubic Feet Per Hour: At One R.P.M.
		Loading of Materials in Trough Class I—45% Full			Loading of Materials in Trough Class II—30% Full		
6	¾	165	875	2.27	120	180	1.5
9	1½	150	1200	8.0	100	560	5.6
12	2	140	2700	19.3	90	1200	13.3
14	2½	130	4000	30.8	85	1790	21.1
16	3	120	5600	46.6	80	2510	31.4
18	3	115	7600	66.1	75	3400	45.4
20	3½	105	9975	95.0	70	4340	62.1

[a] Example 5.3 utilizes these data.
(Stephens–Adamson Co. Catalog, 1954, p. 66).

TABLE 5.4(a)—(*continued*)

Diam. of Conveyor, Inches	△ Max. Lump Size, Inches	Maximum Recommended Speed R.P.M.	Capacities, Cubic Feet Per Hour: At Maximum Recommended Speed	Capacities, Cubic Feet Per Hour: At One R.P.M.	Maximum Recommended Speed R.P.M.	Capacities, Cubic Feet Per Hour: At Maximum Recommended Speed	Capacities, Cubic Feet Per Hour: At One R.P.M.
		Loading of Materials in Trough Class II X—30% Full			Loading of Materials in Trough Class III—15% Full		
6	¾	60	90	1.5	60	45	.75
9	1½	50	280	5.6	50	140	2.8
12	2	50	665	13.3	50	335	6.7
14	2½	45	950	21.1	45	470	10.5
16	3	45	1410	31.4	45	705	15.7
18	3	40	1850	45.4	40	910	22.7
20	3½	40	2485	62.1	40	1240	31.1

(b) Characteristics of Some Materials (A Selection From the Original Table)

Materials	Approx. Weight per Cubic Foot	Capacity Classification	‡Type of Conveyor to Use	Horsepower Factor "F"
Alfalfa meal	17	II	A,B,C	.4
Alum, lumpy	50–60	II	G,H,J	1.5
Alum, pulverized	45–50	II	A,B,C	.8
*Alumina	60	III	K	2.0
Aluminum, hydrate	15–20	II	A,B,C	.8
Ammonium sulphate	52	...	.....	....
Asbestos, shredded	20–25	II	G,H,J	1.0
*Ashes, dry	35–40	IIX	D	2.0
Asphalt, crushed	45	II	A,B,C	.5
Bakelite, powdered	30–40	II	A,B,C	1.4
Baking powder	41	II	A,B,C	.6
Barley	38	I	A,B,C	.4
†Bauxite, crushed	75–85	III	E	1.8
Beans, castor	36	I	A,B,C	.5
Beans, navy, dry	48	I	A,B,C	.4
Bentonite	51	IIX	D	1.0
*Bones, crushed	35–40	IIX	D	2.0
*Bones, granulated or ground	50	IIX	D	1.7
*Bone black	20–25	IIX	D	1.7
Bonechar	40	IIX	D	1.7
*Bone meal	55–60	IIX	D	1.7
Borax, powdered	53	II	A,B,C	.7
Boric acid powder	30–40	II	A,B,C	.8
Bran	16	II	A,B,C	.4

(c) Factor *S* in the Formula for Power *P*

Diameter of Conveyor, Inches	Type of Hanger Bearing: SEALMASTER Ball Bearing	Babbitt, Bronze or Oil-Impregnated Wood	Self-Lubricating Bronze	Hard Iron
4	12	21	33	50
6	18	33	54	80
9	32	54	96	130
10	38	66	114	160
12	55	96	171	250
14	78	135	255	350
16	106	186	336	480
18	140	240	414	600
20	165	285	510	700
24	230	390	690	950

(d) Limits of Horsepower and Torque

Diameter of Conveyor, Inches	Diameter of Coupling, Inches	Maximum Horsepower at 100 R.P.M.	Maximum Torque Capacity in Inch Pounds
4	1	1.5	950
6,9,10	1½	5.0	3200
9,10,12	2	10.0	6300
12,14	2 7/16	15.0	9500
12,14,16,18	3	25.0	16000
20	3 7/16	40.0	25000

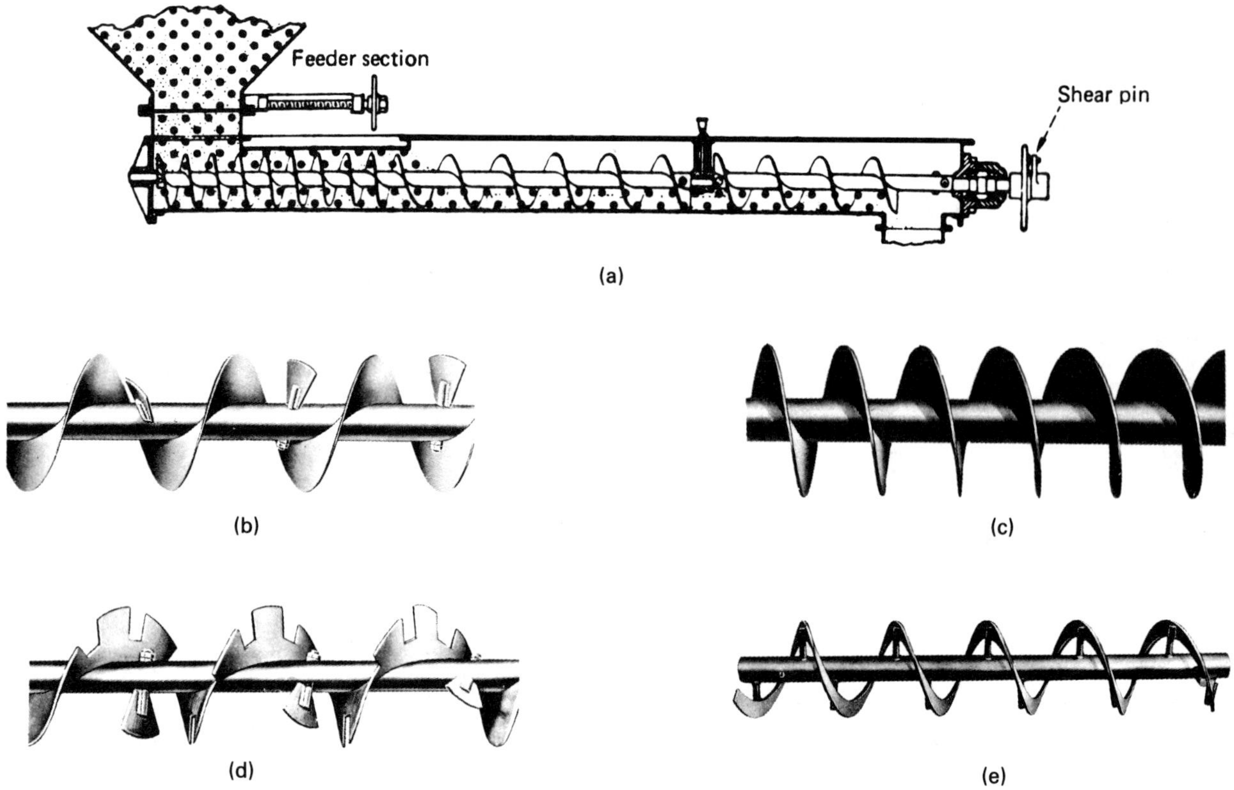

Figure 5.7. A screw conveyor assembly and some of the many kinds of screws in use. (a) Screw conveyor assembly with feed hopper and discharge chute. (b) Standard shape with pitch equal to the diameter, the paddles retard the forward movement and promote mixing. (c) Short pitch suited to transfer of material up inclines of as much as 20°. (d) Cut flight screws combine a moderate mixing action with forward movement, used for light, fine, granular or flaky materials. (e) Ribbon flights are suited to sticky, gummy or viscous substances.

EXAMPLE 5.3
Sizing a Screw Conveyor

Dense soda ash with bulk density 60 lb/cuft is to be conveyed a distance of 100 ft and elevated 12 ft. The material is class II-X with a factor $F = 0.7$. The bearings are self-lubricated bronze and the drive is V-belt with $\eta = 0.93$. The size, speed, and power will be selected for a rate of 15 tons/hr.

$$Q = 15(2000)/60 = 500 \text{ cuft/hr}.$$

According to Table 5.4(a) this capacity can be accommodated by a 12 in. conveyor operating at

$$\omega = (500/665)(50) = 37.6 \text{ rpm}, \quad \text{say 40 rpm}$$

From Table 5.4(c) the bearing factor is

$$s = 171.$$

Accordingly,

$$\dot{P} = [171(40) + 0.7(500)(60)]100 + 0.51(12)(30{,}000)/10^6$$
$$= 2.97 \text{ HP}$$
$$\text{motor HP} = G\dot{P}/\eta = 1.25(2.97)/0.93 = 3.99,$$
$$\text{torque} = 63{,}000(2.97)/40 = 4678 \text{ in. lb}.$$

From Table 5.4(d) the limits for a 12 in. conveyor are 10.0 HP and 6300 in. lb so that the selection is adequate for the required service.

A conveyor 137 ft long would have a shaft power of 4.00 HP and a torque of 6300 in. lbs, which is the limit with a 2 in. coupling; a sturdier construction would be needed at greater lengths.

For comparison, data of Table 5.5 show that a 14 in. troughed belt has an allowable speed of 267 fpm at allowable inclination of 19° (from Table 5.3), and the capacity is

$$2.67(0.6)(38.4) = 61.5 \text{ tons/hr},$$

far more than that of the screw conveyor.

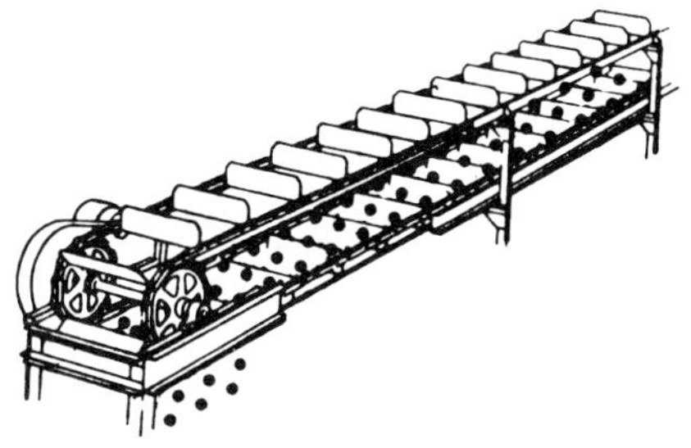

(a)

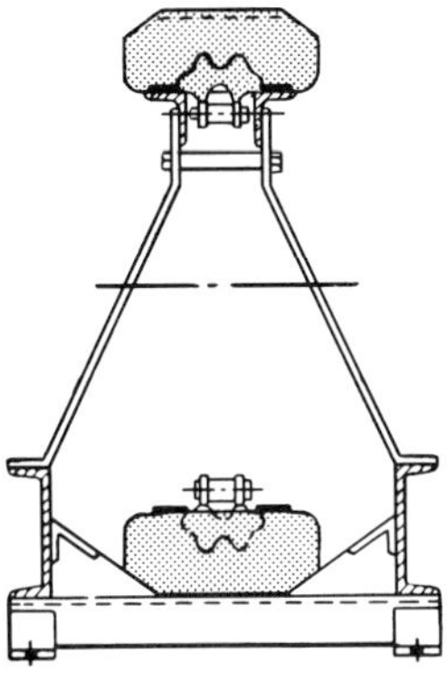

(b)

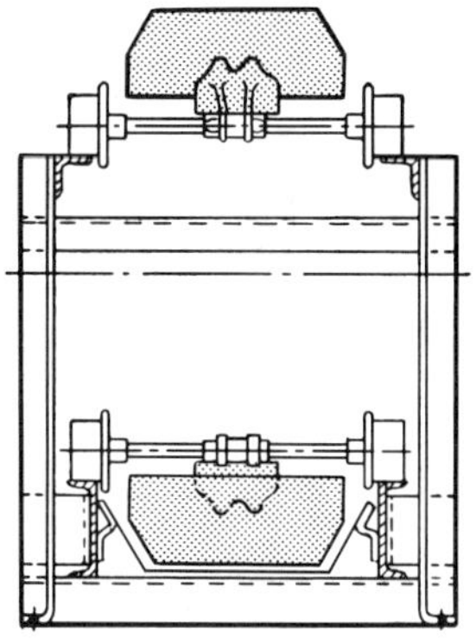

(c)

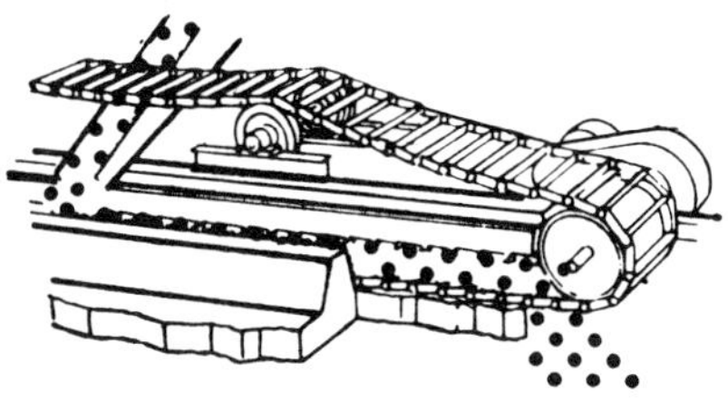

(d)

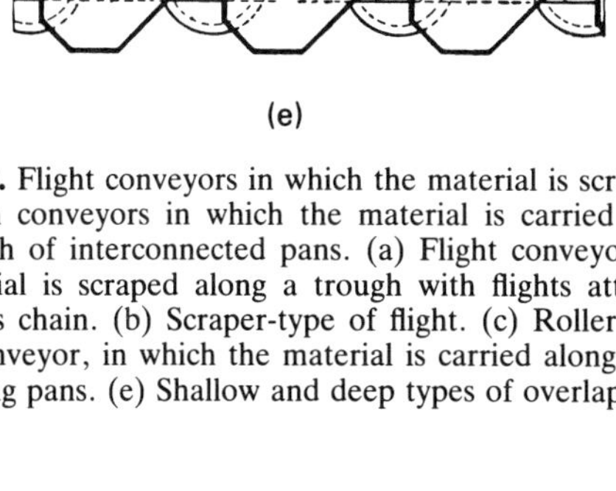

(e)

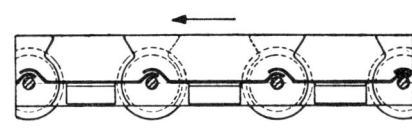

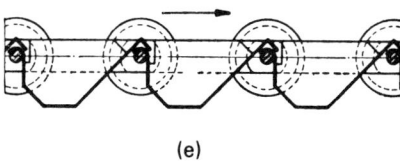

Figure 5.8. Flight conveyors in which the material is scraped along, and apron conveyors in which the material is carried along in a closed path of interconnected pans. (a) Flight conveyor, in which the material is scraped along a trough with flights attached to a continuous chain. (b) Scraper-type of flight. (c) Roller flights. (d) Apron conveyor, in which the material is carried along in moving, overlapping pans. (e) Shallow and deep types of overlapping pans.

TABLE 5.5. Belt Conveyor Data[a]

(a) Capacity (tons/hr) at 100 ft/min, 100 lb/cuft, and Indicated Slope Angle

45° Troughed Belt

Belt Width (inch)	0°	10°	20°	30°
14	27.99	33.00	38.40	43.80
16	38.70	45.60	52.50	60.00
18	51.00	60.00	69.00	78.60
20	65.10	76.20	87.90	100.0
24	98.10	114.9	132.0	149.7
30	160.8	187.5	214.8	243.6
36	238.5	277.8	318.0	360.3
42	331.8	385.8	441.3	499.5
48	440.1	511.5	584.4	661.2
54	563.1	654.6	747.9	845.7
60	702.0	815.4	931.2	1053.0
66	856.5	994.2	1134.9	1282.8
72	1026.0	1190.1	1358.4	1535.4

Flat Belt

Belt width (inches)	5°	10°	20°	30°
14	2.85	6.69	14.01	21.42
16	3.87	9.18	19.05	29.16
18	5.07	11.88	24.90	38.10
20	6.39	15.06	31.50	48.24
24	9.57	22.47	47.10	72.06
30	13.51	36.45	76.32	116.8
36	22.86	53.73	112.6	172.3
42	31.65	74.37	155.9	238.5
48	39.84	90.15	196.2	300.0
54	53.49	125.7	263.4	403.2
60	66.60	156.5	327.9	501.9
66	81.12	190.5	399.6	611.1
72	96.99	238.0	477.9	731.1

[a]Example 5.4 utilizes these data. Power $= P_{horizontal} + P_{vertical} + P_{empty}$ (HP), where $P_{horizontal} = (0.4 + L/300)(W/100)$, $P_{vertical} = 0.001HW$, and P_{empty} obtained from part (c), with $H =$ lift (ft), $L =$ horizontal travel (ft), and $W =$ tons/hr.

(a) From Conveyor Equipment Manufacturers Association, 1979; (b) from Stephens–Adamson Catalog 66, 1954; (c) from Hudson, 1954].

(b) Maximum Recommended Belt Speeds for Nondusting Service

Belt Width, Inches	Belt Speed in Feet per Minute						
	Lump Stone or Ore	Gravel	Lump Coal	Crushed Ore and Stone †	Slack Coal †	Sand	Wood Chips Grain
12	250	300	250	350	350	350	400
18	300	350	300	400	400	400	500
24	350	400	350	450	450	450	600
30	400	450	400	500	500	500	700
36	450	500	450	550	550	550	800
42	500	550	500	550	600	600	800
48	550	600	550	550	650	600	800
54	550	600	600	550	700	600	800
60	550	600	650	550	700	600	800
66	550	600	700	550	700	600	800
72	550	600	700	550	700	600	800

TABLE 5.5—(*continued*)

(c) Power to Drive Empty Conveyor

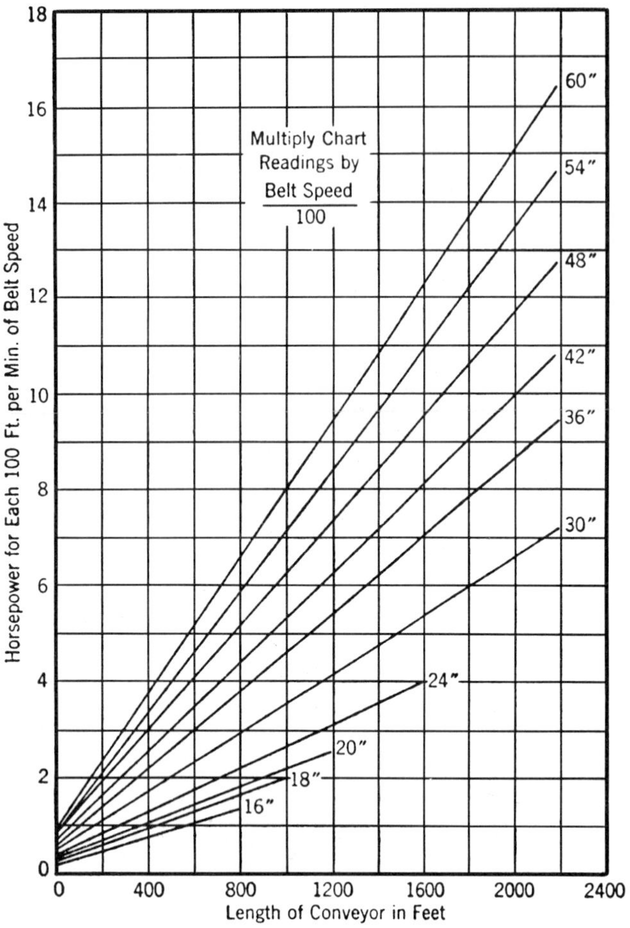

particularly suited to a process. Capacity and power data for bucket machines are given in Table 5.6. Flight and apron conveyors are illustrated in Figure 5.11.

CONTINUOUS FLOW CONVEYOR ELEVATORS

One design of a drag-type of machine is the Redler shown on Figure 5.12. They function because the friction against the flight is greater than that against the wall. Clearly they are versatile in being able to transfer material in any direction and have the often important merit of being entirely covered. Circular cross sections are available but usually they are square, from 3 to 30 in. on a side, and operate at speeds of 30–250 ft/min, depending on the material handled and the construction. Some data are shown in Table 5.7. Most dry granular materials such as wood chips, sugar, salt, and soda ash are handled very well in this kind of conveyor. More difficult to handle are very fine materials such as cement or those that tend to pack such as hot grains or abrasive materials such as sand or crushed stone. Power requirement is dependent on the coefficient of sliding friction. Factors for power calculations of a few substances are shown in Table 5.7.

The closed-belt (zipper) conveyor of Figure 5.13 is a carrier that is not limited by fineness or packing properties or abrasiveness. Of course, it goes in any direction. It is made in a nominal 4-in. size, with a capacity rating by the manufacturer of 0.07 cuft/ft of travel. The power requirement compares favorably with that of open belt conveyors, so that it is appreciably less than that of other types. The formula is

$$\mathrm{HP} = 0.001[(L_1/30 + 5)u + (L_2/16 + 2L_3)T], \tag{5.26}$$

where

u = ft/min,
T = tons/hr,
L_1 = total belt length (ft),
L_2 = length of loaded horizontal section (ft),
L_3 = length of loaded vertical section (ft).

Speeds of 200 ft/min or more are attainable. Example 5.5 shows that the power requirement is much less than that of the Redler conveyor.

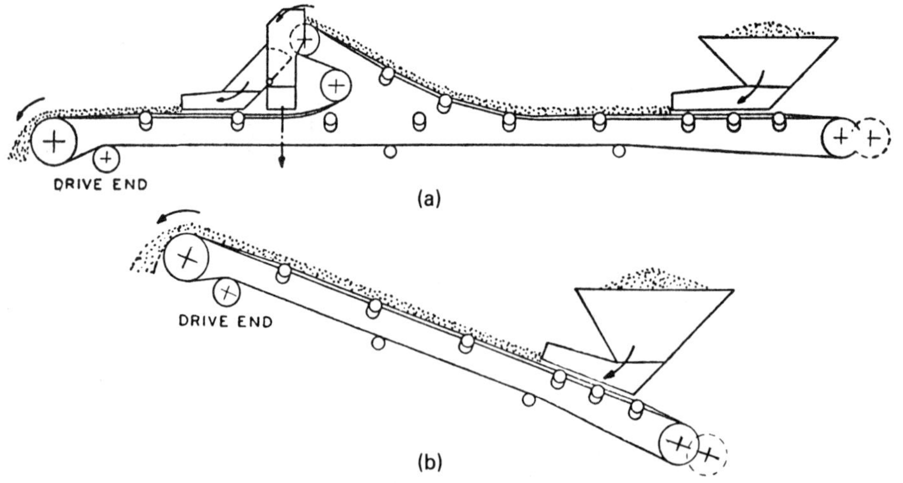

Figure 5.9. Some arrangements of belt conveyors (*Stephens-Adamson Co.*) and types of idlers (*FMC Corp.*). (a) Horizontal conveyor with discharge at an intermediate point as well as at the end. (b) Inclined conveyor, satisfactory up to 20° with some materials. (c) Inclined or retarding conveyor for lowering materials gently down slopes. (d) A flat belt idler, rubber cushion type. (e) Troughed belt idler for high loadings; usually available in 20°, 35°, and 45° side inclinations.

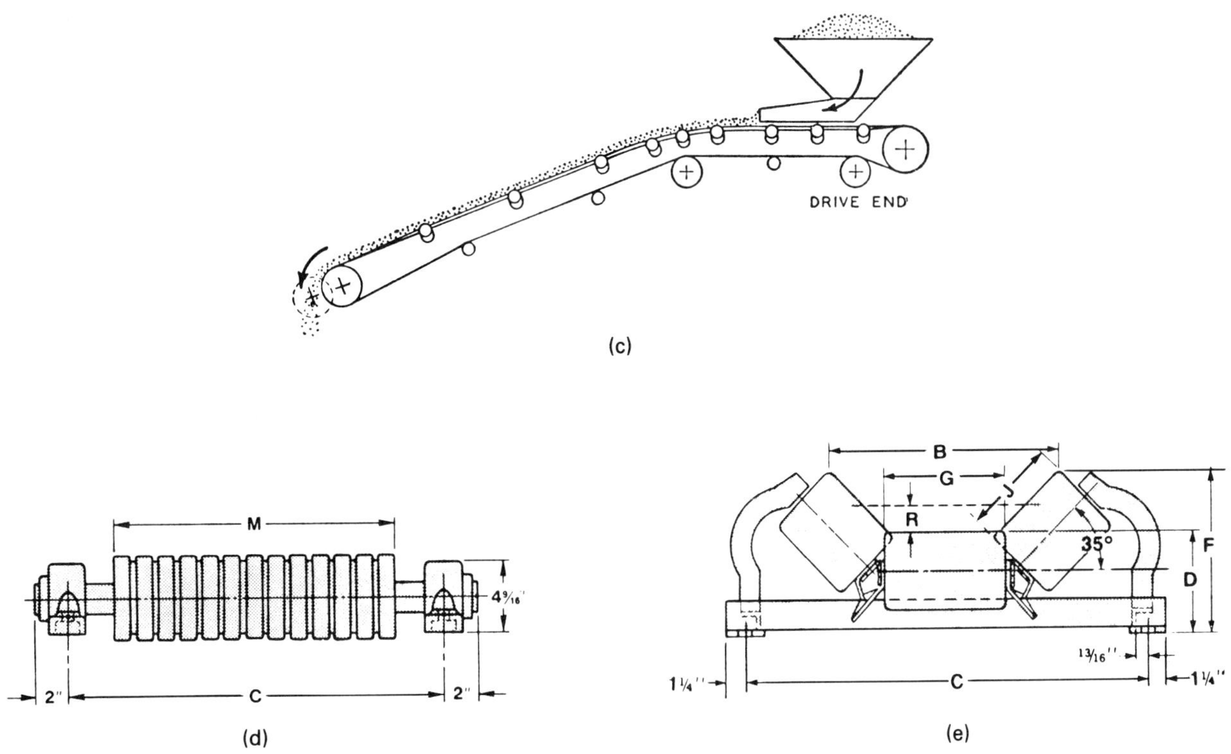

Figure 5.9—*(continued)*

Closing Comments. Most kinds of conveyors and elevators are obtainable from several manufacturers, each of whom builds equipment to individual standards of sturdiness, materials of construction, mechanical details, performance, and price. These differences may be decisive in individual cases. Accordingly, a selection usually must be made from a manufacturer's catalog, and ultimately with the advice of the manufacturer.

5.4. SOLIDS FEEDERS

Several types are illustrated in Figures 5.9 and 3.7. Rates are controlled by adjusting gates or rotation speeds or translation speeds. All of these methods require free flow from a storage bin which may be inhibited by bridging or arching. The device of Figure 5.9(a) provides motion to break up such tendencies.

For the most part the devices shown provide only rough feed rate control. More precise control is achieved by continuous weighing. The equipment of Figure 3.16(l) employs measurements of belt speed and the weight impressed on one or several of the belt idlers to compute and control the weight rate of feed; precision better than 0.5% is achievable. For some batch processes, the feeder discharges into an overhead weighing hopper for accurate measurement of the charge. Similar systems are used to batch feed liquids when integrating flow meters are not sufficiently accurate.

Example 5.4
Sizing a Belt Conveyor

Soda ash of bulk density 60 lb/cuft is to be transported at 400 tons/hr a horizontal distance of 1200 ft up an incline of 5°. The running angle of repose of this material is 19°. The conveyor will be sized with the data of Table 5.5.

Consider a 24 in. belt. From Table 5.5(a) the required speed is

$$u = (400/132)100 = 303 \text{ ft/min}.$$

Since the recommended maximum speed in Table 5.5(b) is 350 fpm, this size is acceptable:

$$\text{conveyor length} = 1200/\cos 5° = 1205 \text{ ft},$$

$$\text{rise} = 1200 \tan 5° = 105 \text{ ft}.$$

With the formulas and graph (c) of Table 5.5, the power requirement becomes

$$\begin{aligned}\text{Power} &= P_{\text{horizontal}} + P_{\text{vertical}} + P_{\text{empty}} \\ &= (0.4 + 1200/300)(400/100) \\ &\quad + 0.001(105)(400) + 303(3.1)/100 \\ &= 69.0 \text{ HP}.\end{aligned}$$

Perhaps 10 to 20% more should be added to compensate for losses in the drive gear and motor.

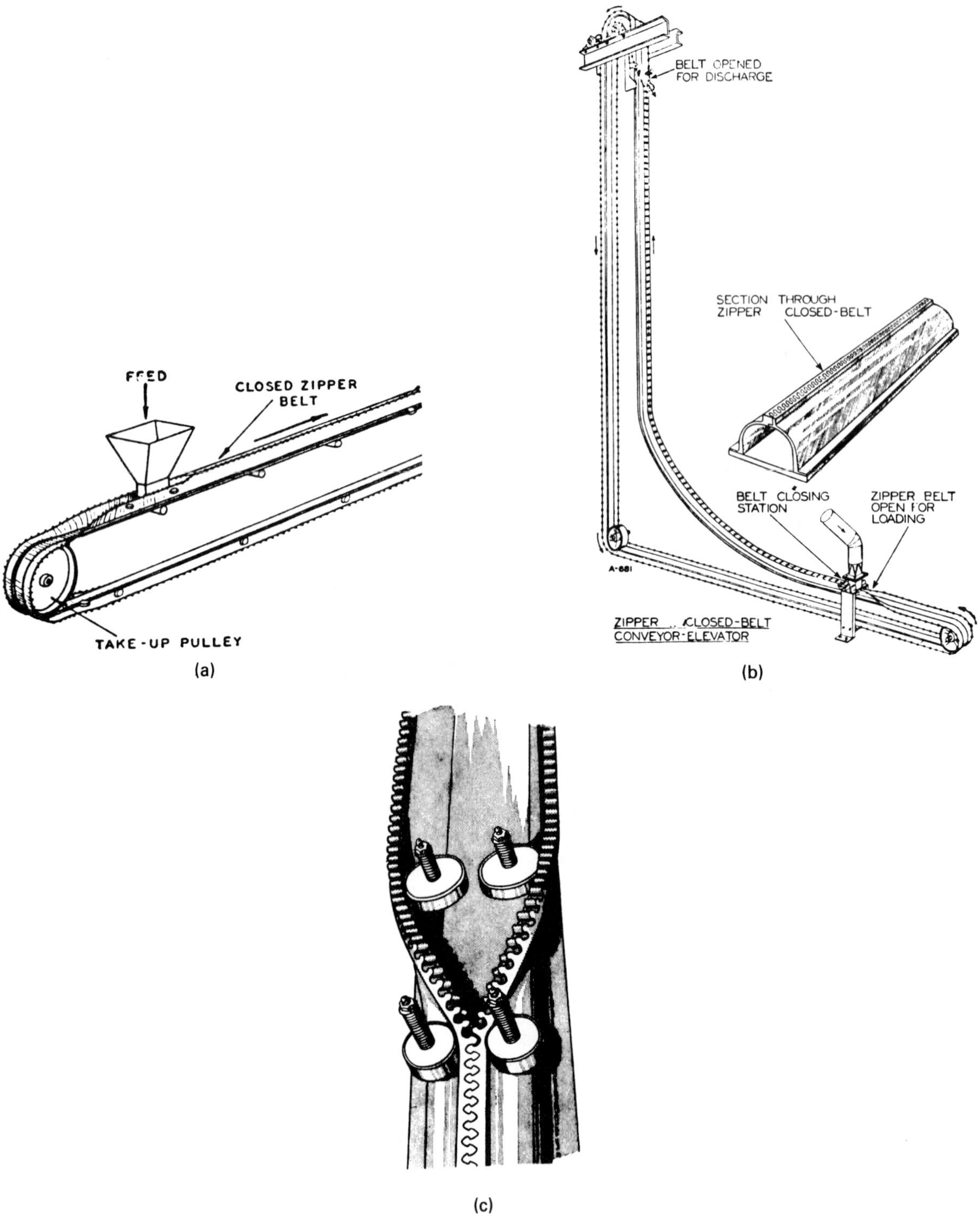

Figure 5.10. Closed belt (zipper) for conveying in any direction (*Stephens-Adamson Co.*). (a) Arrangement of pulley, feed hopper and open and closed belt regions. (b) The tubular belt conveyor for horizontal and vertical transport; a section of the zippered closed belt is shown. (c) Showing how the zipper closes (on downward movement of the belt in this sketch) or opens (on upward movement of the belt).

TABLE 5.6. Capacities and Power Requirements of Bucket Elevator Conveyors

(a) Gravity Discharge Elevators Used Primarily For Coal[a,c]

Size of bucket, in. $L \times W$	Capacity, tons/hr. at 100 ft./min.			Hp.† with material at 50 lb./cu. ft.					
	Bucket spacing, in.			Per 10-ft. vertical lift			Per 100-ft. horizontal run		
				Spacing of buckets, in.					
	18	24	36	18	24	36	18	24	36
16 × 15	46	35	23	0.59	0.44	0.30	5.32	4.24	3.04
20 × 15	58	44	29	.74	.56	.37	6.32	4.97	3.54
24 × 15	70	52	35	.90	.67	.45	7.34	5.74	4.04
20 × 20	104	..	52	1.30	....	.66	9.20	....	4.85
24 × 20	125	..	63	1.60	....	.80	10.92	....	5.74
30 × 20	159	..	79	2.00	....	1.00	13.70	....	7.08
36 × 20	191	..	95	2.42	....	1.21	16.30	....	8.40

(b) Capacities and Maximum Size of Lumps of Centrifugal Discharge Elevators[b,c]

Bucket spacing, in.	Size, length by width, in.	Speed, ft./min.	Max. lumps		Capacity, tons/hr.		
			All lumps	10% lumps	35-lb. material	50-lb. material	100-lb. material
13	6 × 4	225	½	2½	5	7	14
16	8 × 5	230	¾	3	9	13	27
16	10 × 6	230	1	3½	16	23	47
18	12 × 7	268	1¼	4	27	38	77
18	14 × 7	268	1¼	4	32	46	92
19	16 × 8	262	1½	4½	44	63	127

(c) Centrifugal Discharge of Continuous Belt and Bucket Elevators[c]

Bucket spacing, in.	Size, length by width	Speed, ft./min.	Max. lumps		Capacity, tons/hr.		
			All lumps	10% lumps	35 lb. material	50 lb. material	100 lb. material
13	6 × 4	225	½	2½	5	7	14
16	8 × 5	258	3/3	3	11	15	30
16	10 × 6	258	1	3½	18	26	52
18	12 × 7	298	1¼	4	30	42	85
18	14 × 7	298	1¼	4	36	52	103
18	16 × 8	298	1½	4½	53	114	152

[a] Buckets 80% full.
[b] Buckets 75% full.
[c] Horsepower = 0.002 (tons/hr)(lift in feet).
(Link Belt Co.)

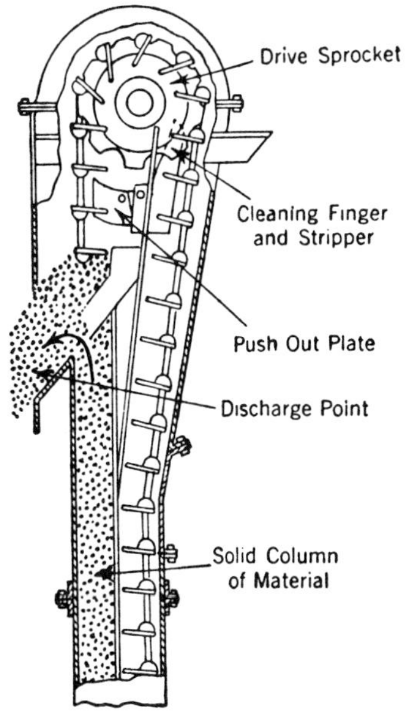

(a)

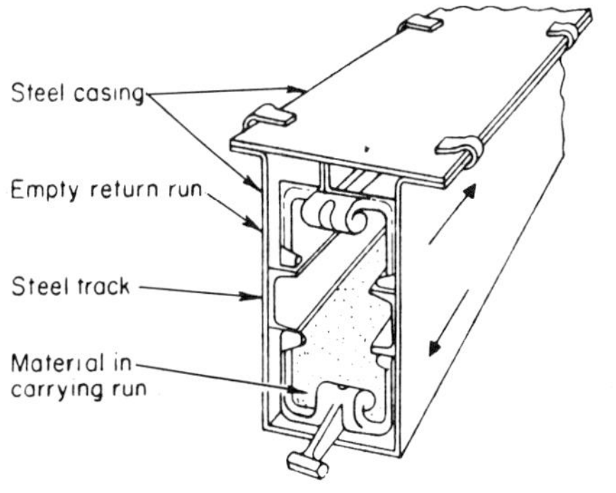

(b)

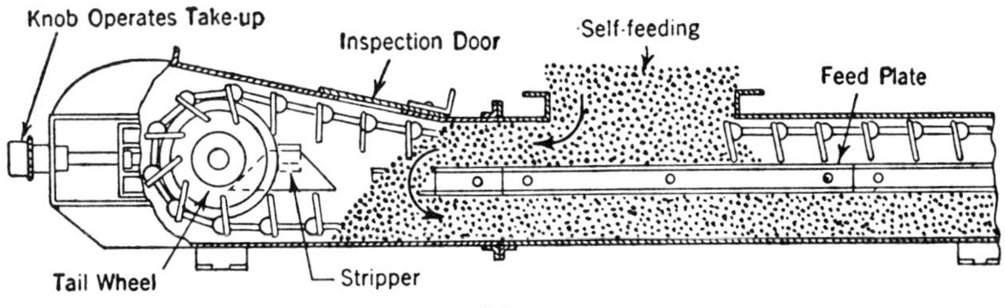

(c)

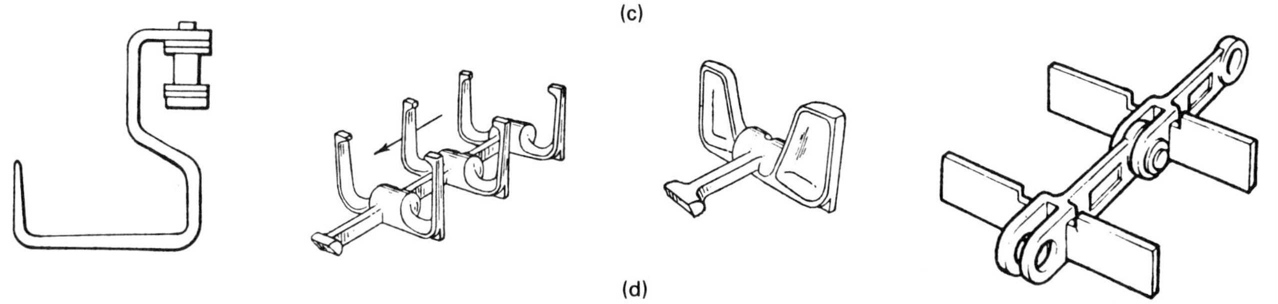

(d)

Figure 5.11. Drag-type enclosed conveyor-elevator (Redler Design) for transfer in any direction (*Stephens-Adamson Mfg. Co.*). (a) Head and discharge end of elevator. (b) Carrying and return runs. (c) Loading end. (d) Some shapes of flights; some are made close-fitting and edged with rubber or plastics to serve as cleanouts.

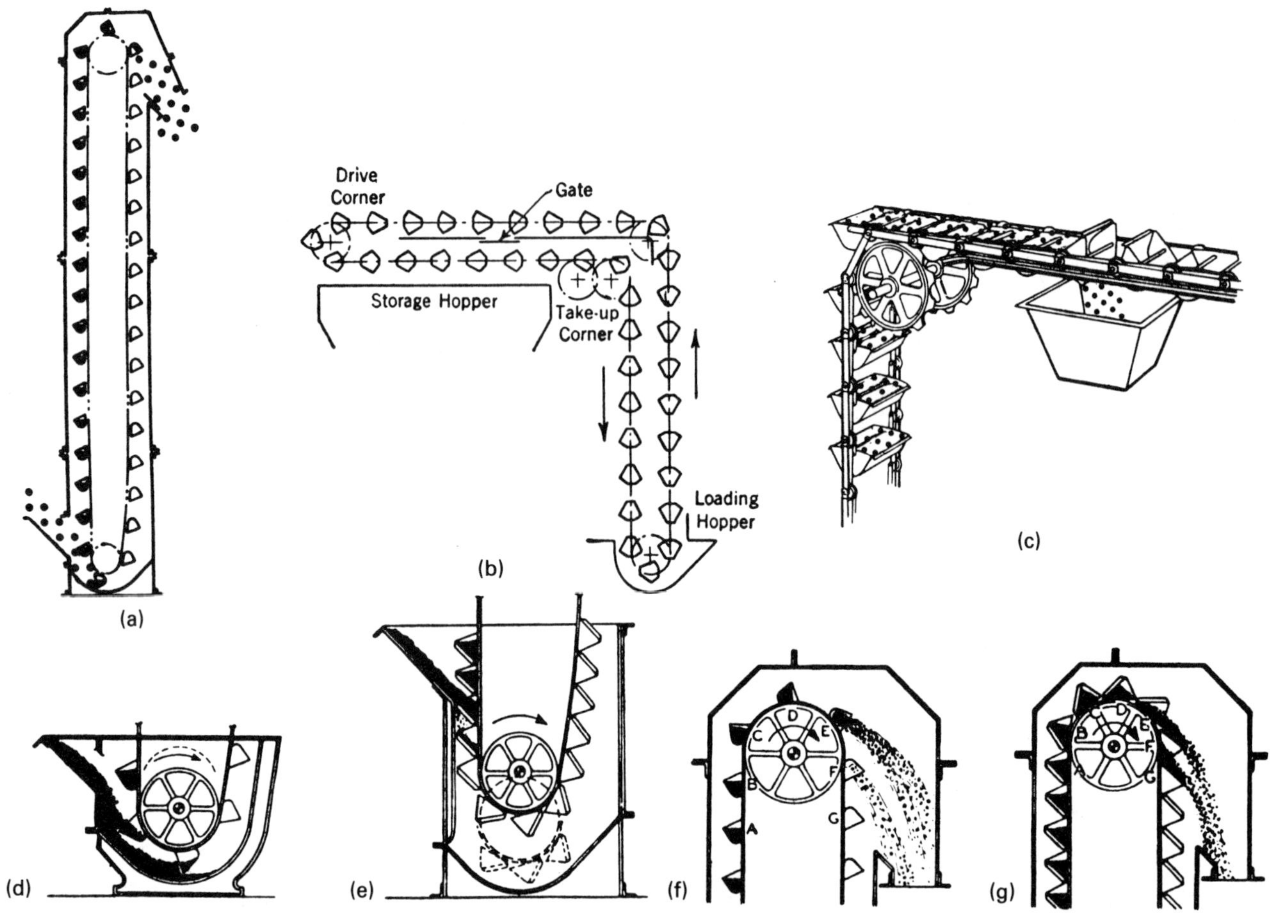

Figure 5.12. Bucket elevators and conveyors. (a) Spaced bucket elevator. (b) Bucket conveyor for vertical and horizontal travel. (c) Discharge of pivoted buckets on horizontal path. (d) Spaced buckets receive part of their load directly and part by scooping the bottom. (e) Continuous buckets are filled as they pass through the loading leg with a feed spout above the tail wheel. (f) Centrifugal discharge of spaced buckets. (g) Discharge mode of continuous buckets.

TABLE 5.7. Speed and Horsepower of Drag-Type Conveyors of Redler Design[a]

(a) Typical Speeds (ft/min)[b]

MATERIAL HANDLED	1000 Conv.	1000 Elev.	2000 Conv.	3000 Conv.
Coal	125	125	80	150
Coke	40	40	40	40
Flyash	30	30	30	30
Grain (Whole)	125	125	80	250
(Processed)	125	100	80	150
Salt	125	100	80	150
Wood (Chips)	100	80	80	150
(Sawdust)	100	100	80	150

[a] HP = 0.001 ($FL + GH + K$) (tons/hr), where H = rise (ft), L = horizontal run (ft), F, G, and K are factors from Table (b); factor E is not used in this formula.

[b] Series 1000, 2000, and 3000 differ in the shapes and sturdiness of the flights.

(Stephens–Adamson Mfg. Co.).

(b) Factors *F, G,* and *K* for Use in the Power Equation for Three Sizes of Units

Material	Weight per Cubic Foot, pounds	K	3″ Units E	3″ Units F	3″ Units G	11″ Units E	11″ Units F	11″ Units G	19″ Units E	19″ Units F	19″ Units G
Beans, dry navy	54	100	1.5	2.9	4.4	1.2	2.3	3.1	1.1	2.0	2.6
Bicarbonate of soda, dry, pulverized	55	0	3.0	6.9	8.1	2.4	5.2	5.4	2.2	4.6	4.3
Bran	26	0	4.1	8.3	3.8	3.0	5.9	2.8	2.6	5.0	2.4
Cellulose acetate dry, coarse granular	10	80	8.0	15.9	4.4	5.3	10.0	3.1	4.4	8.5	2.6
Cement, dry Portland	60–90	0	2.9	7.4	6.0	2.3	5.4	4.1	2.1	4.8	3.4
Clay, dry lumpy	40–100	80	3.1	5.9	4.6	2.4	4.5	3.3	2.1	4.0	2.8
Clay, pulverized	25–80	0	6.0	17.7	6.8	4.3	11.9	4.6	3.8	9.9	3.7
Coal, minus ¼″ slack dry with l'ge proportion fines	40–50	40	2.4	4.6	4.4	1.9	3.5	3.1	1.7	3.1	2.6
Coal, minus ¼″ slack moderately wet	45–55	40	3.3	6.1	5.4	2.6	4.7	3.8	2.3	4.2	3.1
Coal, minus ¼″ slack very wet	50–60	20	2.5	5.4	5.7	2.0	4.1	4.0	1.8	3.6	3.2
Coal, minus 1½″ slack dry or damp	40–50	40				2.0	3.7	3.1	1.8	3.3	2.6
Coal, sized wet or dry	40–50	80	2.2	4.1	3.8	1.7	3.1	2.8	1.5	2.8	2.4
Coconut, shredded	25	20	3.0	6.1	3.2	2.2	4.3	2.4	1.9	3.7	2.1
Coffee, ground	28	20	2.4	4.8	3.2	1.8	3.5	2.4	1.5	3.0	2.1
Corn flakes	12	0	3.8	7.9	2.3	2.6	5.2	1.9	2.1	4.3	1.7
Flour, wheat	30–40	0	3.2	7.1	3.5	2.4	5.0	2.6	2.1	4.3	2.3
Fuller's earth, dry granular	42	80	3.1	6.9	7.3	2.3	5.0	4.9	2.1	4.4	4.0
Lime, burned or "quick" lump or "pebble"	50	200	2.7	5.0	7.0	2.2	3.9	4.8	2.0	3.5	3.8
Lime, dry burned small lumps and dust	50	120	3.5	6.9	6.5	2.8	5.5	4.4	2.5	4.9	3.6
Lime, fine with tendency to pack	40–60	300	4.4	8.8	7.5	3.4	6.6	5.1	3.0	5.8	4.1
Lime, hydrated	10–25	0	11.1	35.5	7.0	7.3	22.5	4.8	6.0	18.1	3.8
Salt, dry granulated	80	80	1.9	3.8	5.7	1.6	3.1	4.0	1.5	2.8	3.2
Salt rock	75	100	1.9	3.5	6.5	1.6	2.9	4.4	1.5	2.6	3.6
Sand, silica coarse dry	90–100	160	2.1	4.2	7.0	1.8	3.4	4.8	1.7	3.2	3.8
Sand very fine, dry	90–100	120	2.4	5.2	7.3	2.0	4.1	4.9	1.9	3.8	4.0
Sawdust, dry	10–30	0	6.2	14.5	4.1	4.1	9.4	3.0	3.4	7.6	2.5
Soda ash, light	25–35	20	4.5	11.4	6.5	3.3	7.8	4.4	2.8	6.6	3.6
Soybean meal	40	20	2.2	4.8	4.6	1.7	3.5	3.3	1.5	3.1	2.8
Starch, lump	30	80	2.1	3.9	3.8	1.6	2.9	2.8	1.4	2.5	2.4
Starch, pulverized	25–45	0	5.4	15.9	6.0	3.9	10.7	4.1	3.4	8.9	3.4
Sugar, dry granulated	50	160	2.7	5.8	9.1	2.2	4.4	6.1	2.0	3.9	4.8
Sugar, brown	40–50	40	4.4	8.5	7.5	3.4	6.4	5.1	3.0	5.6	4.1
Wheat, dry fairly clean	48	40	1.7	3.3	5.2	1.3	2.5	3.6	1.2	2.2	3.0
Wood chips, dry	15–30	40	3.5	6.6	2.8	2.4	4.5	2.2	2.0	3.7	1.9

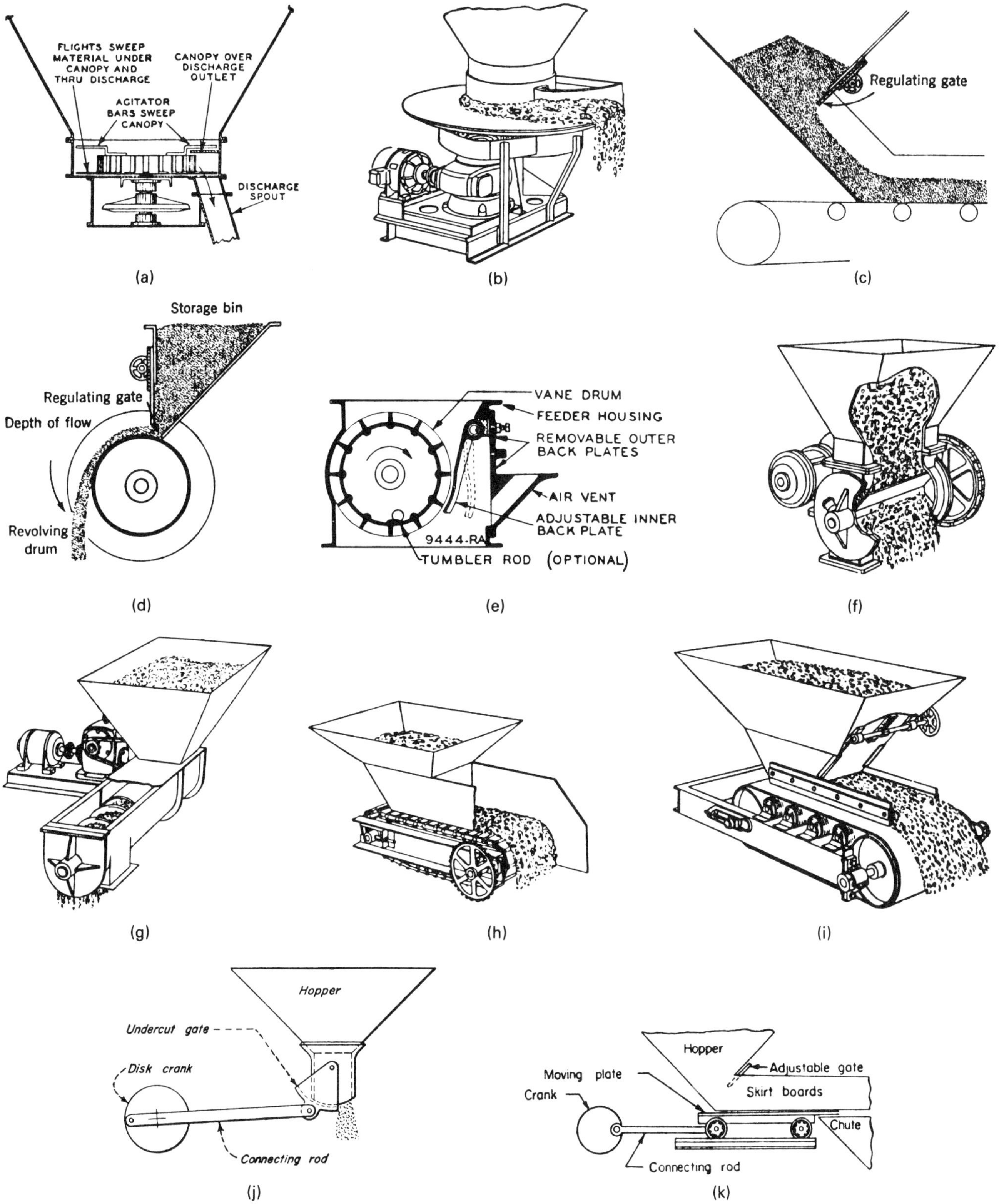

Figure 5.13. Types of feeders for granular solids; also suitable are conveyors such as closed belt, Redler, and bucket types. (a) Bin discharge feeder. (b) Rotary plate feeder with adjustable collar and speed. (c) Flow controlled by an adjustable gate. (d) Rotary drum feeder, regulated by gate and speed. (e) Rotary vane feeder, can be equipped with air lock for fine powders. (f) Vane or pocket feeder. (g) Screw feeder. (h) Apron conveyor feeder. (i) Belt conveyor feeder. (j) Undercut gate feeder. (k) Reciprocating plate feeder. (l) Vibrating feeder, can transfer uphill, downhill, or on the level. (m) "Air-slide" feeder for powders that can be aerated. (n) Weighing belt feeder; unbalance of the weigh beam causes the material flow rate onto the belt to change in the direction of restoring balance.

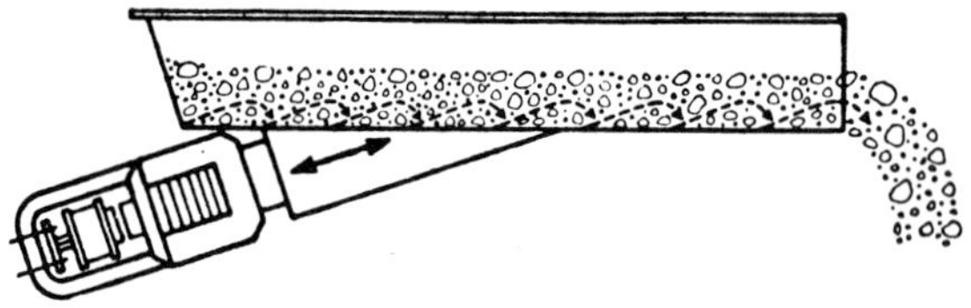

(l)

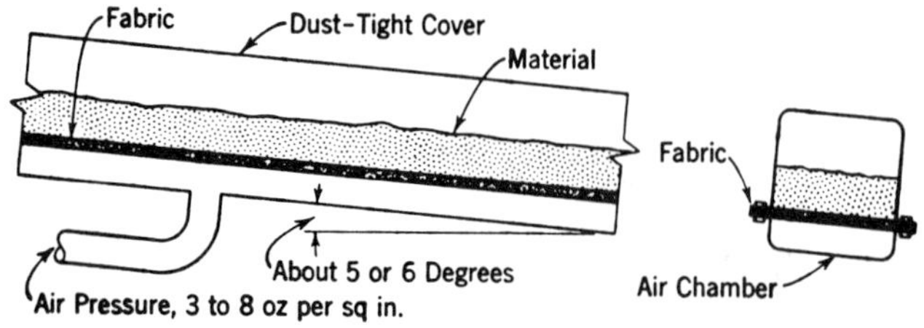

(m)

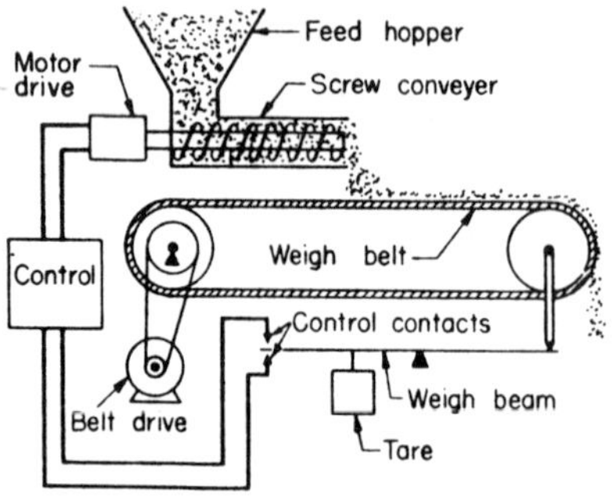

(n)

Figure 5.13—*(continued)*

EXAMPLE 5.5
Comparison of Redler and Zippered Belt Conveyors

Soda ash of bulk density 30 lb/cuft is to be moved 120 ft horizontally and 30 ft vertically at the rate of 350 cuft/hr. Compare power requirements of Redler and zippered belt conveyors for this service.

A 3-in Redler is adequate:

$$u = \frac{350}{60(\pi/4)(3/12)^2} = 118.8 \text{ fpm},$$

which is within the range of Table 5.7(a),

$$\text{tons/hr} = 350(30)/2000 = 5.25$$

Take constants from Table 5.7(b) for a Redler.

$$\text{HP} = \frac{5.25}{1000}[11.4(120) + 6.5(30) + 20] = 8.31.$$

For a closed belt,

$$u = \frac{350}{0.07(60)} = 83.3 \text{ fpm},$$

which is well under the 200 fpm that could be used,

$$L_1 = 300, \quad L_2 = 120, \quad L_3 = 30.$$

Use Eq. (5.26):

$$\begin{aligned} \text{HP} &= 0.001\{(300/30 + 5)83.3 + [120/16 + 2(30)]5.25\} \\ &= 1.60. \end{aligned}$$

REFERENCES

1. T.H. Allegri, *Materials Handling Principles and Practice,* Van Nostrand Reinhold, New York, 1984.
2. A.G. Bain and S.T. Bonnington, *The Hydraulic Transport of Solids by Pipeline,* Pergamon, New York, 1970.
3. M.V. Bhatic and P.N. Cheremisinoff, *Solid and Liquid Conveying Systems,* Technomic, Lancaster, PA, 1982.
4. A.J. Bobkowicz and W.G. Gauvin, The effects of turbulence in the flow characteristics of model fibre suspensions, *Chem. Eng. Sci.* **22,** 229–247 (1967).
5. R. Clift, Conveyors, hydraulic, *Encycl. Chem. Process. Des.* **11,** 262–278 (1980).
6. H. Colijn, *Mechanical Conveyors for Bulk Solids,* Elsevier, New York, 1985.
7. Conveyor Equipment Manufacturers Association, *Belt Conveyors for Bulk Materials,* Van Nostrand Reinhold, New York, 1979.
8. D.W. Dodge and A.B. Metzner, Turbulent flow of non-newtonian systems, *AIChE J.* **5,** 189 (1959).
9. G.H. Ewing, Pipeline transmission, in *Marks' Mechanical Engineers Handbook,* McGraw-Hill, New York, 1978, pp. 11.134–11.135.
10. FMC Corp. Material Handling Equipment Division, Catalog 100, Homer City, PA, 1983.
11. F.J. Gerchow, Conveyors, pneumatic, in *Encycl. Chem. Process. Des.* **11,** 278–319 (1980); *Chem. Eng.,* (17 Feb. 1975, 31 Mar. 1975).
12. H.V. Hawkins, Pneumatic conveyors, in *Marks' Mechanical Engineers Handbook,* McGraw-Hill, New York, 1978, pp. 10.50–10.63.
13. J.W. Hayden and T.E. Stelson, Hydraulic conveyance of solids in pipes, in Zandi, Ref. 27, 1971, pp. 149–163.
14. W.G. Hudson, *Conveyors and Related Equipment,* Wiley, New York, 1954.
15. E. Jacques and J.G. Montfort, Coal transportation by slurry pipeline, in Considine (Ed.), *Energy Technology Handbook,* McGraw-Hill, New York, 1977, pp. 1.178–1.187.

16. M. Kraus, *Pneumatic Conveying of Bulk Materials,* McGraw-Hill, New York, 1980.

17. R.A. Kulwiec (Ed.), *Material Handling Handbook,* Wiley, New York, 1985.

18. D.E. Perkins, and J.E. Wood, Design and Select Pneumatic Conveying Systems, *Hydrocarbon Processing* 75–78 (March 1974).

19. G.J. Raymus, Pneumatic conveyors, in *Perry's Chemical Engineers Handbook,* McGraw-Hill, New York, 1984, pp. 7.17–7.25.

20. P.E. Solt, Conveying, pneumatic troubleshooting, *Encycl. Chem. Process. Des.* **11,** 214–226 (1980).

21. Stephens-Adamson Mfg. Co., General Catalog 66, Aurora, IL, 1954, and updated sections.

22. H.A. Stoess, *Pneumatic Conveying,* Wiley, New York, 1983.

23. E.J. Wasp, T.C. Aude, R.H. Seiter, and T.L. Thompson, in Zandi, Ref. 27, 1971, pp. 199–210.

24. E.J. Wasp, J.P. Kenny, and R.L. Gandhi, *Solid-Liquid Flow in Slurry Pipeline Transportation,* Trans. Tech. Publ., 1977, Gulf, Houston, 1979.

25. E.J. Wasp, T.L. Thompson, and P.E. Snoek, The era of slurry pipelines, *Chem. Technol.,* 552–562 (Sep. 1971).

26. O.A. Williams, *Pneumatic and Hydraulic Conveying of Solids,* Dekker, New York, 1983.

27. I. Zandi (Ed.), *Advances in Solid-Liquid Flow in Pipes and Its Applications,* Pergamon, New York, 1971.

6
FLOW OF FLUIDS

The transfer of fluids through piping and equipment is accompanied by friction and may result in changes in pressure, velocity, and elevation. These effects require input of energy to maintain flow at desired rates. In this chapter, the concepts and theory of fluid mechanics bearing on these topics will be reviewed briefly and practical and empirical methods of sizing lines and auxiliary equipment will be emphasized.

6.1. PROPERTIES AND UNITS

The basis of flow relations is Newton's relation between force, mass, and acceleration, which is

$$F = (m/g_c)a. \tag{6.1}$$

When F and m are in lb units, the numerical value of the coefficient is $g_c = 32.174$ lb ft/lbf sec^2. In some other units,

$$g_c = 1\frac{\text{kg m/sec}^2}{\text{N}} = 1\frac{\text{g cm/sec}^2}{\text{dyn}} = 9.806\frac{\text{kg m/sec}^2}{\text{kg}_\text{f}}.$$

Since the common engineering units for both mass and force are 1 lb, it is essential to retain g_c in all force–mass relations. The interconversions may be illustrated with the example of viscosity whose basic definition is force/(velocity)(distance). Accordingly the viscosity in various units relative to that in SI units is

$$\begin{aligned}1\ \text{Ns/m}^2 &= \frac{1}{9.806}\text{kg}_\text{f}\ \ \text{s/m}^2 = 10\ \text{g/(cm)(s)}\\ &= 10\ \text{P} = 0.0672\ \text{lb/(ft)(sec)}\\ &= \frac{0.0672}{32.174}\text{lbf sec/ft}^2 = 0.002089\ \text{lbf sec/ft}^2.\end{aligned}$$

In data books, viscosity may be recorded either in force or mass units. The particular merit of SI units (kg, m, s, N) is that $g_c = 1$ and much confusion can be avoided by consistent use of that system. Some numbers of frequent use in fluid flow problems are

Viscosity: 1 cPoise = 0.001 N s/m^2 = 0.4134 lb/(ft)(hr).
Density: 1 g m/cm^3 = 1000 kg/m^3 = 62.43 lb/ft^3.
Specific weight: 62.43 lbf/cuft = 1000 kg$_\text{f}$/m^3.
Pressure: 1 atm = 0.10125 MPa = 0.10125(10^6) N/m^2 = 1.0125 bar.

Data of densities of liquids are empirical in nature, but the effects of temperature, pressure, and composition can be estimated; suitable methods are described by Reid et al. (*Properties of Gases and Liquids,* McGraw Hill, New York, 1977), the *API Refining Data Book* (American Petroleum Institute, Washington, DC, 1983), and the *AIChE Data Prediction Manual* (1984–date). The densities of gases are represented by equations of state of which the simplest is that of ideal gases; from this the density is given by:

$$\rho = 1/V = MP/RT, \quad \text{mass/volume} \tag{6.2}$$

where M is the molecular weight. For air, for example, with P in atm and T in °R,

$$\rho = \frac{29P}{0.73T}, \quad \text{lb/cuft.} \tag{6.3}$$

For nonideal gases a general relation is

$$\rho = MP/zRT, \tag{6.4}$$

where the compressibility factor z is correlated empirically in terms of reduced properties T/T_c and P/P_c and the acentric factor. This subject is treated for example by Reid et al. (1977, p. 26) and Walas (1985, pp. 17, 70). Many PVT equations of state are available. That of Redlich and Kwong may be written in the form

$$V = b + RT/(P + a/\sqrt{T}V^2), \tag{6.5}$$

which is suitable for solution by direct iteration as used in Example 6.1.

Flow rates are expressible as linear velocities or in volumetric, mass, or weight units. Symbols for and relations between the several modes are summarized in Table 6.1.

The several variables on which fluid flow depends may be gathered into a smaller number of dimensionless groups, of which the Reynolds number and friction factor are of particular importance. They are defined and written in the common kinds of units also in Table 6.1. Other dimensionless groups occur less frequently and will be mentioned as they occur in this chapter; a long list is given in *Perry's Chemical Engineers Handbook* (McGraw-Hill, New York, 1984, p. 5.62).

EXAMPLE 6.1
Density of a Nonideal Gas from Its Equation of State
The Redlich–Kwong equation of carbon dioxide is

$$(P + 63.72(10^6)/\sqrt{T}V^2)(V - 29.664) = 82.05T$$

with P in atm, V in mL/g mol and T in K. The density will be found at $P = 20$ and $T = 400$. Rearrange the equation to

$$V = 29.664 + (82.05)(400)/(20 + 63.72(10^6)/\sqrt{400}V^2).$$

Substitute the ideal gas volume on the right, $V = 1641$; then find V on the left; substitute that value on the right, and continue. The successive values of V are

$$V = 1641, 1579, 1572.1, 1571.3, 1571.2, \cdots \quad \text{mL/g mol}$$

and converge at 1571.2. Therefore, the density is

$$\rho = 1/V = 1/1571.2, \quad \text{or} \quad 0.6365\ \text{g mol/L} \quad \text{or} \quad 28.00\ \text{g/L}.$$

TABLE 6.1. Flow Quantities, Reynolds Number, and Friction Factor

Flow Quantity	Symbol and Equivalent	Typical Units: Common	Typical Units: SI
Linear	u	ft/sec	m/sec
Volumetric	$Q = uA = \pi D^2 u/4$	cuft/sec	m^3/sec
Mass	$\dot{m} = \rho Q = \rho A u$	lb/sec	kg/sec
Weight	$\dot{w} = \gamma Q = \gamma A u$	lbf/sec	N/sec
Mass/area	$G = \rho u$	lb/(sqft)(sec)	kg/m^2 sec
Weight/area	$G_\gamma = \gamma u$	lbf/(sqft)(sec)	N/m^2 sec

Reynolds Number (with $A = \pi D^2/4$)

$$\mathrm{Re} = \frac{Du\rho}{\mu} = \frac{Du}{\nu} = \frac{DG}{\mu} = \frac{4Q\rho}{\pi D\mu} = \frac{4\dot{m}}{\pi D\mu} \quad (1)$$

Friction Factor

$$f = \frac{\Delta P}{\rho} \Big/ \left(\frac{L}{D}\frac{u^2}{2g_c}\right) = 2g_c D\Delta P/L\rho u^2 = 1.6364 \Big/ \left[\ln\left(\frac{0.135\varepsilon}{D} + \frac{6.5}{\mathrm{Re}}\right)\right]^2 \quad (2)$$

(Round's equation)

$$\frac{\Delta P}{\rho} = \frac{L}{D}\frac{u^2}{2g_c} f = \frac{8LQ^2}{g_c \pi^2 D^5} f = \frac{8L\dot{m}^2}{g_c \pi^2 \rho^2 D^5} f = \frac{LG^2}{2g_c D\rho^2} f \quad (3)$$

In the units

D = in., $\dot{m}$ = lb/hr

Q = cuft/sec, μ = cP

ρ = specific gravity

$$\mathrm{Re} = \frac{6.314\dot{m}}{D\mu} = \frac{1.418(10^6)\rho Q}{D\mu} \quad (4)$$

$$\frac{\Delta P}{L} = \frac{3.663(10^{-9})\dot{m}^2}{\rho D^5} f, \text{ atm/ft} \quad (5)$$

$$= \frac{5.385(10^{-8})\dot{m}^2}{\rho D^5} f, \text{ psi/ft} \quad (6)$$

$$= \frac{0.6979\rho Q^2}{D^5} f, \text{ psi/ft} \quad (7)$$

Laminar Flow

$\mathrm{Re} < 2300$

$$f = 64/\mathrm{Re} \quad (2a)$$

$\Delta P/L = 32\mu u/D^2$

$$= \frac{1.841(10^{-7})\mu\dot{m}}{\rho D^4}, \text{ atm/ft} \quad (5a)$$

$$= \frac{2.707(10^{-6})\mu\dot{m}}{\rho D^4}, \text{ psi/ft} \quad (6a)$$

$$= \frac{35.083\mu Q}{D^4}, \text{ psi/ft} \quad (7a)$$

Gravitation Constant

$g_c = 1$ kg m/N sec^2
$= 1$ g cm/dyn sec^2
$= 9.806$ kg m/kgf sec^2
$= 32.174$ lbm ft/lbf sec^2
$= 1$ slug ft/lbf sec^2
$= 1$ lbm ft/poundal sec^2

6.2. ENERGY BALANCE OF A FLOWING FLUID

The energy terms associated with the flow of a fluid are

1. Elevation potential $(g/g_c)z$,
2. Kinetic energy, $u^2/2g_c$,
3. Internal energy, U,
4. Work done in crossing the boundary, PV,
5. Work transfer across the boundary, W_s,
6. Heat transfer across the boundary, Q.

Figure 6.1 represents the two limiting kinds of regions over which energy balances are of interest: one with uniform conditions throughout (completely mixed), or one in plug flow in which gradients are present. With single inlet and outlet streams of a uniform region, the change in internal energy within the boundary is

$$\begin{aligned} d(mU) &= m\,dU + U\,dm = m\,dU + U(dm_1 - dm_2) \\ &= dQ - dW_s + [H_1 + u_1^2/2g_c + (g/g_c)z_1]\,dm_1 \\ &\quad - [H_2 + u_2^2/2g_c + (g/g_c)z_2]\,dm_2. \end{aligned} \quad (6.6)$$

One kind of application of this equation is to the filling and emptying of vessels, of which Example 6.2 is an instance.

Under steady state conditions, $d(mU) = 0$ and $dm_1 = dm_2 = dm$, so that Eq. (6.6) becomes

$$\Delta H + \Delta u^2/2g_c + (g/g_c)\Delta z = (Q - W_s)/m, \quad (6.7)$$

or

$$\Delta U + \Delta(PV) + \Delta u^2/2g_c + (g/g_c)\Delta z = (Q - W_s)/m, \quad (6.8)$$

or

$$\Delta U + \Delta(P/\rho) + \Delta u^2/2g_c + (g/g_c)\Delta z = (Q - W_s)/m. \quad (6.9)$$

For the plug flow condition of Figure 6.1(b), the balance is made in terms of the differential changes across a differential length dL of the vessel, which is

$$dH + (1/g_c)u\,du + (g/g_c)\,dz = dQ - dW_s, \quad (6.10)$$

where all terms are per unit mass.

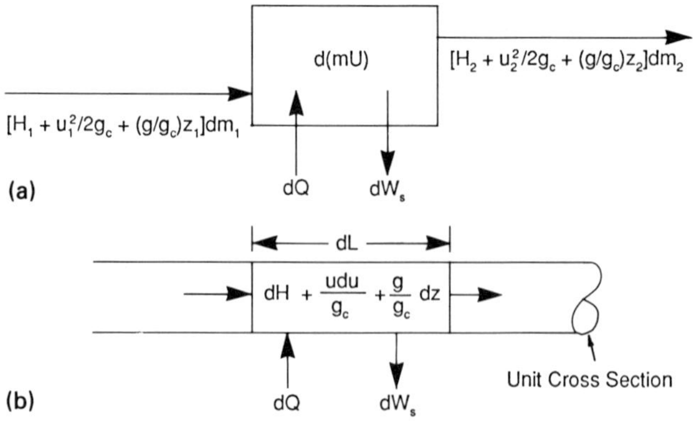

Figure 6.1. Energy balances on fluids in completely mixed and plug flow vessels. (a) Energy balance on a bounded space with uniform conditions throughout, with differential flow quantities dm_1 and dm_2. (b) Differential energy balance on a fluid in plug flow in a tube of unit cross section.

EXAMPLE 6.2
Unsteady Flow of an Ideal Gas through a Vessel

An ideal gas at 350 K is pumped into a 1000 L vessel at the rate of 6 g mol/min and leaves it at the rate of 4 g mol/min. Initially the vessel is at 310 K and 1 atm. Changes in velocity and elevation are negligible. The contents of the vessel are uniform. There is no work transfer.

Thermodynamic data:

$$U = C_v T = 5T,$$
$$H = C_p T = 7T.$$

Heat transfer:

$$dQ = h(300 - T)\,d\theta$$
$$= 15(300 - T)\,d\theta.$$

The temperature will be found as a function of time θ with both $h = 15$ and $h = 0$.

$$dn_1 = 6\,d\theta,$$
$$dn_2 = 4\,d\theta,$$
$$dn = dn_1 - dn_2 = 2\,d\theta,$$
$$n_0 = P_0 V/RT_0 = 1000/(0.08205)(310) = 39.32 \text{ gmol},$$
$$n = n_0 + 2\theta,$$

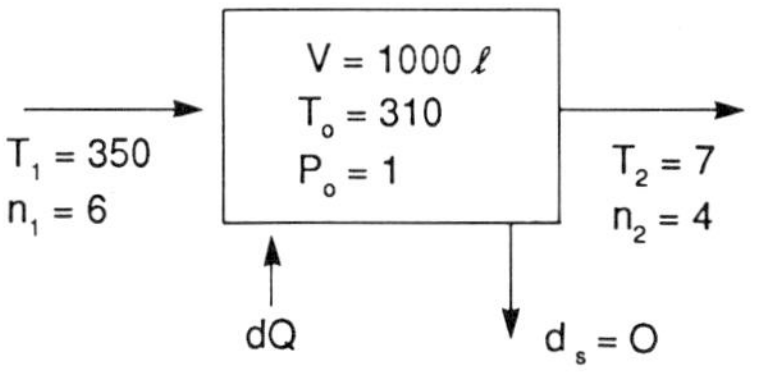

Energy balance

$$d(nU) = n\,dU + U\,dn = nC_v\,dT + C_v T(2\,d\theta)$$
$$= H_1\,dn_1 - H_2\,dn_2 + dQ - dw_s$$
$$= C_p(6T_1 - 4T)\,d\theta + h(300 - T)\,d\theta.$$

This rearranges into

$$\int_0^\theta \frac{d\theta}{n_0 + 2\theta} = \int_{310}^{T_2} \frac{dT}{(1/C_v)[6C_pT_1 + 300h - (4C_p + 2C_v + h)T}$$
$$= \begin{cases} \displaystyle\int_{310}^{T_2} \frac{dT}{3840 - 10.6T}, & h = 15, \\ \displaystyle\int_{310}^{T_2} \frac{dT}{2940 - 7.6T}, & h = 0. \end{cases}$$

The integrals are rearranged to find T,

$$T_2 = \begin{cases} 362.26 - 52.26\left(\dfrac{1}{1 + 0.05090\theta}\right)^{5.3}, & h = 15, \\ 386.84 - 76.84\left(\dfrac{1}{1 + 0.05090\theta}\right)^{3.8}, & h - 0. \end{cases}$$

Some numerical values are

	T_2		P	
θ	$h=15$	$h=0$	$h=15$	$h=0$
0	310	310	1	1
0.2	312.7	312.9	1.02	1.02
0.5	316.5	317.0		
1	322.1	323.2		
5	346.5	354.4		
10	356.4	370.8	1.73	1.80
∞	362.26	386.84	∞	∞

The pressures are calculated from

$$P = \frac{nRT}{V} = \frac{(39.32 + 2\theta)(0.08205)T}{1000}.$$

Friction is introduced into the energy balance by noting that it is a mechanical process, dW_f, whose effect is the same as that of an equivalent amount of heat transfer dQ_f. Moreover, the total effective heat transfer results in a change in entropy of the flowing liquid given by

$$T\,dS = dQ + dW_f. \tag{6.11}$$

When the thermodynamic equivalent

$$dH = V\,dP + T\,dS \tag{6.12}$$

and Eq. (6.11) are substituted into Eq. (6.10), the net result is

$$V\,dP + (1/g_c)u\,du + (g/g_c)\,dz = -(dW_s + dW_f), \tag{6.13}$$

which is known as the mechanical energy balance. With the expression for friction of Eq. (6.18) cited in the next section, the mechanical energy balance becomes

$$V\,dP + (1/g_c)u\,du + (g/g_c)\,dz + \frac{fu^2}{2g_cD}\,dL = -dW_s. \tag{6.13'}$$

For an incompressible fluid, integration may be performed term by term with the result

$$\Delta P/\rho + \Delta u^2/2g_c + (g/g_c)\Delta z = -(W_s + W_f). \tag{6.14}$$

The apparent number of variables in Eq. (6.13) is reduced by the substitution $u = V/A$ for unit flow rate of mass, where A is the cross-sectional area, so that

$$V\,dP + (1/g_cA^2)V\,dV + (g/g_c)\,dz = -(dW_s + dW_f). \tag{6.15}$$

Integration of these energy balances for compressible fluids under several conditions is covered in Section 6.7.

The frictional work loss W_f depends on the geometry of the system and the flow conditions and is an empirical function that will be explained later. When it is known, Eq. (6.13) may be used to find a net work effect W_s for otherwise specified conditions.

The first three terms on the left of Eq. (6.14) may be grouped into a single stored energy terms as

$$\Delta E = \Delta P/\rho + \Delta u^2/2g_c + (g/g_c)\Delta z, \tag{6.16}$$

EXAMPLE 6.3
Units of the Energy Balance

In a certain process the changes in stored energy and the friction are

$$\Delta E = -135 \text{ ft lbf/lb}$$
$$w_f = 13 \text{ ft lbf/lb}.$$

The net work will be found in several kinds of units:

$$w_s = -(\Delta E + w_f) = 122 \text{ ft lbf/lb},$$
$$w_s = 122 \frac{\text{ft lbf}}{\text{lb}} \frac{4.448N}{\text{lbf}} \frac{2.204 \text{ lb}}{\text{kg}} \frac{m}{3.28 \text{ ft}}$$
$$= 364.6 \frac{\text{N m}}{\text{kg}}, \quad 364.6 \frac{\text{J}}{\text{kg}},$$
$$w_s = 364.6 \frac{\text{N m}}{\text{kg}} \frac{\text{kgf}}{9.806 \text{ N}} = 37.19 \frac{\text{m kgf}}{\text{kg}}.$$

At sea level, numerically lbf = lb and kgf = kg. Accordingly,

$$w_s = 122 \frac{\text{ft lbf}}{\text{lb}} \frac{\text{lb}}{\text{lbf}} \frac{\text{kgf}}{\text{kg}} \frac{m}{3.28 \text{ ft}} = 37.19 \frac{\text{kgf m}}{\text{kg}},$$

as before.

---

and the simpler form of the energy balance becomes

$$\Delta E + W_f = -W_s. \tag{6.17}$$

The units of every term in these energy balances are alternately:

ft lb_f/lb with $g_c = 32.174$ and g in ft/sec^2 (32.174 at sea level).
N m/kg = J/kg with $g_c = 1$ and g in m/sec^2 (1.000 at sea level).
kg_f m/kg with $g_c = 9.806$ and g in m/sec^2 (9.806 at sea level).

Example 6.3 is an exercise in conversion of units of the energy balances.

The sign convention is that *work input is a negative quantity* and consequently results in an increase of the terms on the left of Eq. (6.17). Similarly, work is produced by the flowing fluid only if the stored energy ΔE is reduced.

6.3. LIQUIDS

Velocities in pipe lines are limited in practice because of

1. the occurrence of erosion.
2. economic balance between cost of piping and equipment and the cost of power loss because of friction which increases sharply with velocity.

Although erosion is not serious in some cases at velocities as high as 10–15 ft/sec, conservative practice in the absence of specific knowledge limits velocities to 5–6 ft/sec.

Economic optimum design of piping will be touched on later, but the rules of Table 6.2 of typical linear velocities and pressure drops provide a rough guide for many situations.

The correlations of friction in lines that will be presented are for new and clean pipes. Usually a factor of safety of 20–40% is advisable because pitting or deposits may develop over the years. There are no recommended fouling factors for friction as there are for heat transfer, but instances are known of pressure drops to double in water lines over a period of 10 years or so.

In lines of circular cross section, the pressure drop is represented by

$$\Delta P = f\rho \frac{L}{D} \frac{u^2}{2g_c}. \tag{6.18}$$

For other shapes and annular spaces, D is replaced by the hydraulic diameter

$$D_h = 4(\text{cross section})/\text{wetted perimeter}.$$

For an annular space, $D_h = D_2 - D_1$.

In laminar flow the friction is given by the theoretical Poiseuille equation

$$f = 64/N_{re}, \quad N_{Re} < 2100, \quad \text{approximately}. \tag{6.19}$$

At higher Reynolds numbers, the friction factor is affected by the roughness of the surface, measured as the ratio ε/D of projections on the surface to the diameter of the pipe. Values of ε are as follows; glass and plastic pipe essentially have $\varepsilon = 0$.

	ε (ft)	ε (mm)
Riveted steel	0.003–0.03	0.9–9.0
Concrete	0.001–0.01	0.3–3.0
Wood stave	0.0006–0.003	0.18–0.9
Cast iron	0.00085	0.25
Galvanized iron	0.0005	0.15
Asphalted cast iron	0.0004	0.12
Commercial steel or wrought iron	0.00015	0.046
Drawn tubing	0.000005	0.0015

The equation of Colebrook [*J. Inst. Civ. Eng.* London, **11,** pp. 133–156 (1938–1939)] is based on experimental data of Nikuradze [*Ver. Dtsch. Ing. Forschungsh. 356* (1932)].

$$\frac{1}{\sqrt{f}} = 1.14 - 0.869 \ln\left(\frac{\varepsilon}{D} + \frac{9.38}{N_{Re}\sqrt{f}}\right), \quad N_{Re} > 2100. \tag{6.20}$$

Other equations equivalent to this one but explicit in f have been devised. A literature review and comparison with more recent experimental data are made by Olujic [*Chem. Eng.*, 91–94, (14 Dec. 1981)]. Two of the simpler but adequate equations are

$$f = 1.6364\left[\ln\left(\frac{0.135\varepsilon}{D} + \frac{6.5}{N_{Re}}\right)\right]^{-2} \tag{6.21}$$

[Round, *Can. J. Chem. Eng.* **58,** 122 (1980)],

$$f = \left\{-0.8686 \ln\left[\frac{\varepsilon}{3.7D} - 2.1802 \ln\left(\frac{\varepsilon}{3.7D} + \frac{14.5}{N_{Re}}\right)\right]\right\}^{-2} \tag{6.22}$$

[Schacham, *Ind. Eng. Chem. Fundam.* **19**(5), 228 (1980)]. These

TABLE 6.2. Typical Velocities and Pressure Drops in Pipelines

	Liquids (psi/100 ft)		
	Liquids within 50°F of Bubble Point	Light Oils and Water	Viscous Oils
Pump suction	0.15	0.25	0.25
Pump discharge	2.0 (or 5–7 fps)	2.0 (or 5–7 fps)	2.0 (or 3–4 fps)
Gravity flow to or from tankage, maximum	0.05	0.05	0.05
Thermosyphon reboiler inlet and outlet	0.2		

Gases (psi/100 ft)		
Pressure (psig)	0–300 ft Equivalent Length	300–600 ft Equivalent Length
−13.7 (28 in. Vac)	0.06	0.03
−12.2 (25 in. Vac)	0.10	0.05
−7.5 (15 in. Vac)	0.15	0.08
0	0.25	0.13
50	0.35	0.18
100	0.50	0.25
150	0.60	0.30
200	0.70	0.35
500	2.00	1.00

Steam	psi/100 ft	Maximum ft/min
Under 50 psig	0.4	10,000
Over 50 psig	1.0	7000

Steam Condensate

To traps, 0.2 psi/100 ft. *From bucket traps,* size on the basis of 2–3 times normal flow, according to pressure drop available. *From continuous drainers,* size on basis of design flow for 2.0 psi/100 ft

Control Valves

Require a pressure drop of at least 10 psi for good control, but values as low as 5 psi may be used with some loss in control quality

Particular Equipment Lines (ft/sec)	
Reboiler, downcomer (liquid)	3–7
Reboiler, riser (liquid and vapor)	35–45
Overhead condenser	25–100
Two-phase flow	35–75
Compressor, suction	75–200
Compressor, discharge	100–250
Inlet, steam turbine	120–320
Inlet, gas turbine	150–350
Relief valve, discharge	$0.5v_c$ [a]
Relief valve, entry point at silencer	v_c [a]

[a] v_c is sonic velocity.

three equations agree with each other within 1% or so. The Colebrook equation predicts values 1–3% higher than some more recent measurements of Murin (1948), cited by Olujic (*Chemical Engineering*, 91–93, Dec. 14, 1981).

For orientation purposes, the pressure drop in steel pipes may be found by the rapid method of Table 6.3, which is applicable to highly turbulent flow for which the friction factor is given by von Karman's equation

$$f = 1.3251[\ln(D/\varepsilon) + 1.3123)]^{-2}. \tag{6.23}$$

Under some conditions it is necessary to employ Eq. (6.18) in differential form. In terms of mass flow rate,

$$dP = \frac{8\dot{m}^2 f}{g_c \pi^2 \rho D^5} dL. \tag{6.24}$$

Example 6.4 is of a case in which the density and viscosity vary along the length of the line, and consequently the Reynolds number and the friction factor also vary.

FITTINGS AND VALVES

Friction due to fittings, valves and other disturbances of flow in pipe lines is accounted for by the concepts of either their equivalent lengths of pipe or multiples of the velocity head. Accordingly, the pressure drop equation assumes either of the forms

$$\Delta P = f(L + \sum L_i)\rho u^2/2g_c D, \tag{6.25}$$

$$\Delta P = [f(L/D) + \sum K_i]\rho u^2/2g_c. \tag{6.26}$$

Values of equivalent lengths L_i and coefficients K_i are given in Tables 6.4 and 6.5. Another well-documented table of K_i is in the *Chemical Engineering Handbook* (McGraw-Hill, New York, 1984 p. 5.38).

Comparing the two kinds of parameters,

$$K_i = fL_i/D \tag{6.27}$$

so that one or the other or both of these factors depend on the friction factor and consequently on the Reynolds number and possibly ε. Such a dependence was developed by Hooper [*Chem. Eng.*, 96–100, (24 Aug. 1981)] in the equation

$$K = K_1/N_{\text{Re}} + K_2(1 + 1/D), \tag{6.28}$$

where D is in inches and values of K_1 and K_2 are in Table 6.6. Hooper states that the results are applicable to both laminar and turbulent regions and for a wide range of pipe diameters. Example 6.5 compares the several systems of pipe fittings resistances. The K_i method usually is regarded as more accurate.

ORIFICES

In pipe lines, orifices are used primarily for measuring flow rates but sometimes as mixing devices. The volumetric flow rate through a thin plate orifice is

$$Q = C_d A_0 \left(\frac{2\Delta P/\rho}{1 - \beta^4}\right)^{1/2}, \tag{6.29}$$

A_0 = cross sectional area of the orifice,

$\beta = d/D$, ratio of the diameters of orifice and pipe.

For corner taps the coefficient is given by

$$C_d \approx 0.5959 + 0.0312\beta^{2.1} - 0.184\beta^8 + (0.0029\beta^{2.5})(10^6/\text{Re}_D)^{0.75} \tag{6.30}$$

(International Organization for Standards Report DIS 5167, Geneva, 1976). Similar equations are given for other kinds of orifice taps and for nozzles and Venturi meters.

TABLE 6.3. Approximate Computation of Pressure Drop of Liquids and Gases in Highly Turbulent Flow in Steel Pipes[a]

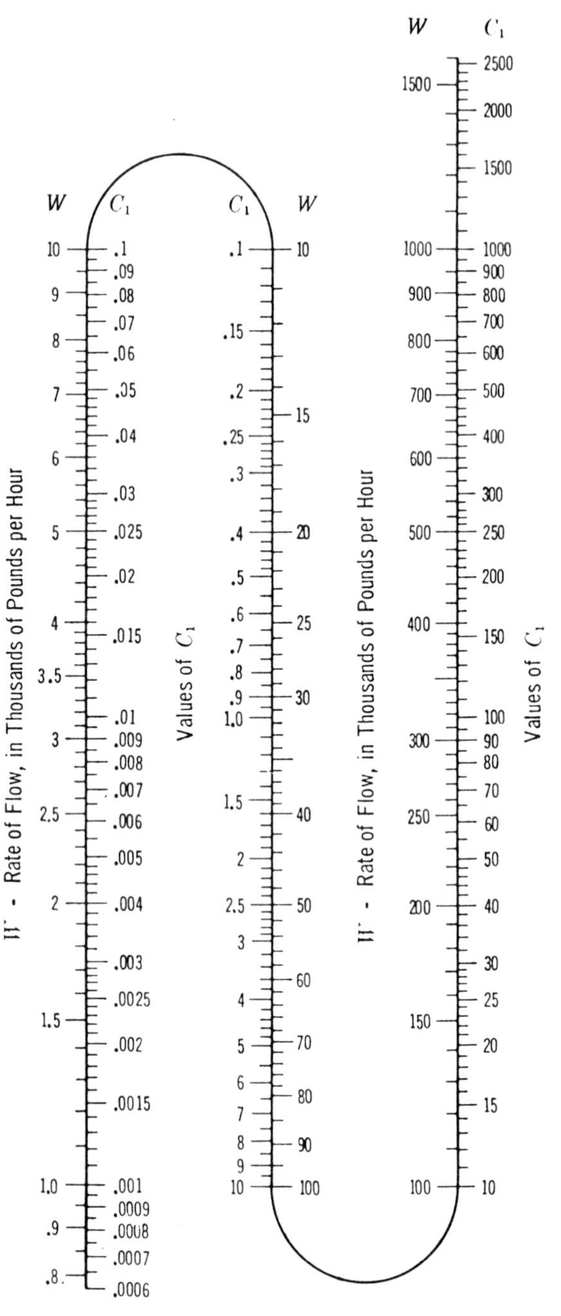

Nominal Pipe Size In.	Schedule Number	Value of C_2
⅛	40 s	7,920,000
	80 x	26,200,000
½	40 s	93,500
	80 x	186,100
	160	4,300,000
	... xx	11,180,000
¾	40 s	21,200
	80 x	36,900
	160	100,100
	... xx	627,000
1	40 s	5,950
	80 x	9,640
	160	22,500
	... xx	114,100
1½	40 s	627
	80 x	904
	160	1,656
	... xx	4,630
2	40 s	169
	80 x	236
	160	488
	... xx	899
2½	40 s	66.7
	80 x	91.8
	160	146.3
	... xx	380.0
3	40 s	21.4
	80 x	28.7
	160	48.3
	... xx	96.6
3½	40 s	10.0
	80 x	13.2
4	40 s	5.17
	80 x	6.75
	120	8.94
	160	11.80
	... xx	18.59
5	40 s	1.59
	80 x	2.04
	120	2.69
	160	3.59
	... xx	4.93
6	40 s	0.610
	80 x	0.798
	120	1.015
	160	1.376
	... xx	1.861
8	20	0.133
	30	0.135
	40 s	0.146
	60	0.163
	80 x	0.185
	100	0.211
	120	0.252
	140	0.289
	... xx	0.317
	160	0.333
10	20	0.0397
	30	0.0421
	40 s	0.0447
	60 x	0.0514
	80	0.0569
	100	0.0661
	120	0.0753
	140	0.0905
	160	0.1052

Nominal Pipe Size In.	Schedule Number	Value of C_2
12	20	0.0157
	30	0.0168
	... s	0.0175
	40	0.0180
	... x	0.0195
	60	0.0206
	80	0.0231
	100	0.0267
	120	0.0310
	140	0.0350
	160	0.0423
14	10	0.00949
	20	0.00996
	30 s	0.01046
	40	0.01099
	... x	0.01155
	60	0.01244
	80	0.01416
	100	0.01657
	120	0.01898
	140	0.0218
	160	0.0252
16	10	0.00463
	20	0.00421
	30 s	0.00504
	40 x	0.00549
	60	0.00612
	80	0.00700
	100	0.00804
	120	0.00926
	140	0.01099
	160	0.01244
18	10	0.00247
	20	0.00256
	.. s	0.00266
	30	0.00276
	.. x	0.00287
	40	0.00298
	60	0.00335
	80	0.00376
	100	0.00435
	120	0.00504
	140	0.00573
	160	0.00669
20	10	0.00141
	20 s	0.00150
	30 x	0.00161
	40	0.00169
	60	0.00191
	80	0.00217
	100	0.00251
	120	0.00287
	140	0.00335
	160	0.00385
24	10	0.000534
	20 s	0.000565
	.. x	0.000597
	30	0.000614
	40	0.000651
	60	0.000741
	80	0.000835
	100	0.000972
	120	0.001119
	140	0.001274
	160	0.001478

Note: The letters s, x, and xx in the columns of Schedule Numbers indicate Standard, Extra Strong, and Double Extra Strong pipe, respectively.

[a] $\Delta P_{100} = C_1 C_2/\rho$ psi/100 ft, with ρ in lb/cuft.
(Crane Co. *Flow of Fluids through Fittings, Valves and Pipes*, Crane Co., New York, 1982).

EXAMPLE 6.4
Pressure Drop in Nonisothermal Liquid Flow

Oil is pumped at the rate of 6000 lb/hr through a reactor made of commercial steel pipe 1.278 in. ID and 2000 ft long. The inlet condition is 400°F and 750 psia. The temperature of the outlet is 930°F and the pressure is to be found. The temperature varies with the distance, L ft, along the reactor according to the equation

$$T = 1500 - 1100\exp(-0.0003287L) \quad (°\text{F})$$

The viscosity and density vary with temperature according to the equations

$$\mu = \exp\left(\frac{7445.3}{T + 459.6} - 6.1076\right), \quad \text{cP},$$
$$\rho = 0.936 - 0.00036T, \quad \text{g/mL}.$$

Round's equation applies for the friction factor:

$$N_{Re} = \frac{4\dot{m}}{\pi D\mu} = \frac{4(6000)}{\pi(1.278/12)2.42\mu} = \frac{29{,}641}{\mu},$$
$$\varepsilon/D = \frac{0.00015(12)}{1.278} = 0.00141,$$

$$f = \frac{1.6364}{[\ln[0.135(0.00141) + 6.5/N_{Re}]^2}.$$

The differential pressure is given by

$$-dP = \frac{8\dot{m}^2}{g_c\pi^2\rho D^5} f\,dL = \frac{8(6000/3600)^2}{32.2\pi^2 62.4\rho(1.278/12)^5(144)} f\,dL$$
$$= \frac{0.568f}{\rho}dL, \quad \text{psi},$$
$$P = 750 - \int_0^L \frac{0.586f}{L}dL = 750 - \int_0^L I\,dL.$$

The pressure profile is found by integration with the trapezoidal rule over 200 ft increments. The computer program and the printout are shown. The outlet pressure is 7000/psi.

For comparison, taking an average temperature of 665°F,

$$\mu = 1.670, \quad \rho = 0.697$$
$$N_{Re} = 17{,}700, \quad f = 0.00291,$$
$$P_{out} = 702.5.$$

```
10 ! Example 6.4:pressure drop
   in nonisothermal flow
20 READ L,P,D ! (D = length inc
   rement
30 DATA 0,750,200
40 GOSUB 180
50 I1=1
60 GOSUB 150
70 I1=1
80 L=L+D
90 GOSUB 180
100 P=P-.5*D*(I1+I)
110 GOSUB 150
120 IF L>1800 THEN 140
130 GOTO 70
140 END
150 DISP USING 160 ; L,T,R1/1000
    ,100*F,P
160 IMAGE DDDD,2X,DDD.D,2X,DDD.D
    ,2X,D.DD,2X,DDD.D
170 RETURN
180 T=1500-1100*EXP(-(.0003287*L
    ))
190 M=EXP(7445.3/(T+459.6)-6.107
    6)
200 R=.936-.00036*T
210 R1=29641/M
220 F=1.6364/LOG(.135*.00141+6.5
    /R1)^2
230 I=.568*F/R
240 RETURN
```

L	T	$N_{Re}/1000$	100f	P
0	400.0	2.3	4.85	750.0
200	470.0	4.4	3.99	743.6
400	535.5	7.5	3.49	737.9
600	596.9	11.6	3.16	732.8
800	654.4	16.7	2.95	727.9
1000	708.2	22.7	2.80	723.2
1200	758.5	29.5	2.69	718.5
1400	805.7	37.1	2.61	713.9
1600	849.9	45.2	2.55	709.3
1800	891.3	53.8	2.51	704.7
2000	930.0	62.7	2.47	700.1

TABLE 6.4. Equivalent Lengths of Pipe Fittings[a]

Pipe size, in.	Standard ell	Medium radius ell	Long-radius ell	45-deg ell	Tee	Gate valve, open	Globe valve, open	Swing check, open
1	2.7	2.3	1.7	1.3	5.8	0.6	27	6.7
2	5.5	4.6	3.5	2.5	11.0	1.2	57	13
3	8.1	6.8	5.1	3.8	17.0	1.7	85	20
4	11.0	9.1	7.0	5.0	22	2.3	110	27
5	14.0	12.0	8.9	6.1	27	2.9	140	33
6	16.0	14.0	11.0	7.7	33	3.5	160	40
8	21	18.0	14.0	10.0	43	4.5	220	53
10	26	22	17.0	13.0	56	5.7	290	67
12	32	26	20.0	15.0	66	6.7	340	80
14	36	31	23	17.0	76	8.0	390	93
16	42	35	27	19.0	87	9.0	430	107
18	46	40	30	21	100	10.2	500	120
20	52	43	34	23	110	12.0	560	134
24	63	53	40	28	140	14.0	680	160
36	94	79	60	43	200	20.0	1,000	240

[a] Length of straight pipe (ft) giving equivalent resistance.
(Hicks and Edwards, *Pump Application Engineering*, McGraw-Hill, New York, 1971).

POWER REQUIREMENTS

A convenient formula in common engineering units for power consumption in the transfer of liquids is

$$\dot{P} = \frac{\text{(volumetric flow rate)(pressure difference)}}{\text{(equipment efficiency)}}$$
$$= \frac{\text{(gals/min)(lb/sq in.)}}{1714\text{(fractional pump eff)(fractional driver eff)}} \text{ horsepower.} \tag{6.30a}$$

Efficiency data of drivers are in Chapter 4 and of pumps in Chapter 7. For example, with 500 gpm, a pressure difference of 75 psi, pump efficiency of 0.7, and driver efficiency of 0.9, the power requirement is 32.9 HP or 24.5 kw.

6.4. PIPELINE NETWORKS

A system for distribution of fluids such as cooling water in a process plant consists of many interconnecting pipes in series, parallel, or branches. For purposes of analysis, a point at which several lines meet is called a node and each is assigned a number as on the figure of Example 6.6. A flow rate from node i to node j is designated as Q_{ij}; the same subscript notation is used for other characteristics of the line such as f, L, D, and N_{Re}.

Three principles are applicable to establishing flow rates, pressures, and dimensions throughout the network:

1. Each node i is characterized by a unique pressure P_i.
2. A material balance is preserved at each node: total flow in equals total flow out, or net flow equals zero.
3. The friction equation $P_i - P_j = (8\rho/g_c\pi^2)f_{ij}L_{ij}Q_{ij}^2/D_{ij}^5$ applies to the line connecting node i with j.

In the usual network problem, the terminal pressures, line lengths, and line diameters are specified and the flow rates throughout are required to be found. The solution can be generalized, however, to determine other unknown quantities equal in number to the number of independent friction equations that describe the network. The procedure is illustrated with the network of Example 6.6.

The three lines in parallel between nodes 2 and 5 have the same pressure drop $P_2 - P_5$. In series lines such as 37 and 76 the flow rate is the same and a single equation represents friction in the series:

$$P_3 - P_6 = kQ_{36}^2(f_{37}L_{37}/D_{37}^5 + f_{76}L_{76}/D_{76}^5).$$

The number of flow rates involved is the same as the number of lines in the network, which is 9, plus the number of supply and destination lines, which is 5, for a total of 14. The number of material balances equals the number of nodes plus one for the overall balance, making a total of 7.

The solution of the problem requires $14 - 7 = 7$ more relations to be established. These are any set of 7 friction equations that involve the pressures at all the nodes. The material balances and pressure drop equations for this example are tabulated.

From Eqs. (4)–(10) of Example 6.6, any combination of seven quantities Q_{ij} and/or L_{ij} and/or D_{ij} can be found. Assuming that the Q_{ij} are to be found, estimates of all seven are made to start, and the corresponding Reynolds numbers and friction factors are found from Eqs. (2) and (3). Improved values of the Q_{ij} then are found

TABLE 6.5. Velocity Head Factors of Pipe Fittings[a]

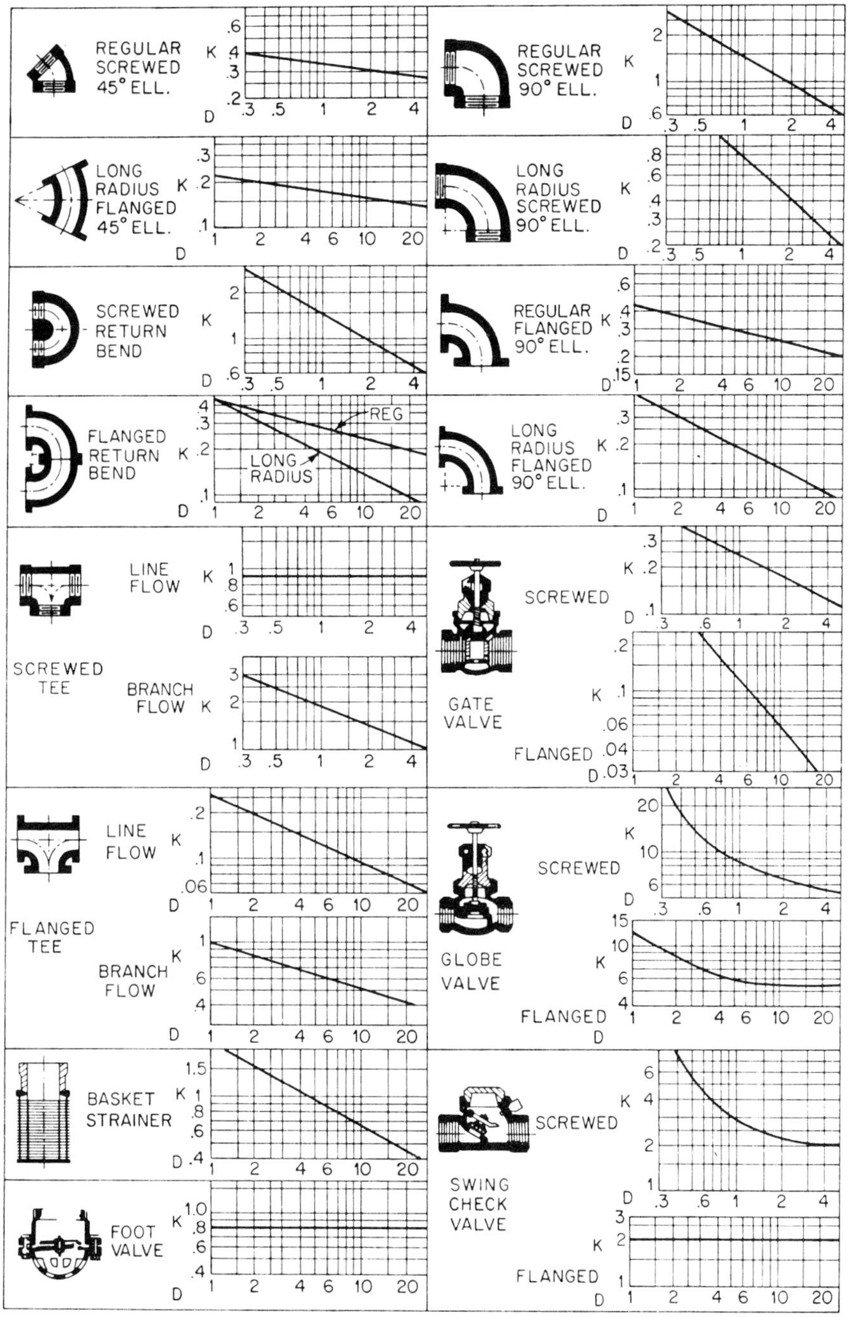

[a] $h = Ku^2/2g_c$, ft of fluid.
(Hydraulic Institute, Cleveland, OH, 1957).

TABLE 6.6. Velocity Head Factors of Pipe Fittings[a]

			Fitting type	K_1	K_∞
Elbows	90°		Standard (R/D = 1), screwed	800	0.40
			Standard (R/D = 1), flanged/welded	800	0.25
			Long-radius (R/D = 1.5), all types	800	0.20
		Mitered elbows (R/D=1.5)	1 Weld (90° angle)	1,000	1.15
			2 Weld (45° angles)	800	0.35
			3 Weld (30° angles)	800	0.30
			4 Weld (22½° angles)	800	0.27
			5 Weld (18° angles)	800	0.25
	45°		Standard (R/D = 1), all types	500	0.20
			Long-radius (R/D = 1.5), all types	500	0.15
			Mitered, 1 weld, 45° angle	500	0.25
			Mitered, 2 weld, 22½° angles	500	0.15
	180°		Standard (R/D = 1), screwed	1,000	0.60
			Standard (R/D = 1), flanged/welded	1,000	0.35
			Long radius (R/D = 1.5), all types	1,000	0.30
Tees	Used as elbow		Standard, screwed	500	0.70
			Long-radius, screwed	800	0.40
			Standard, flanged or welded	800	0.80
			Stub-in-type branch	1,000	1.00
	Run-through tee		Screwed	200	0.10
			Flanged or welded	150	0.50
			Stub-in-type branch	100	0.00
Valves	Gate, ball, plug		Full line size, β = 1.0	300	0.10
			Reduced trim, β = 0.9	500	0.15
			Reduced trim, β = 0.8	1,000	0.25
	Globe, standard			1,500	4.00
	Globe, angle or Y-type			1,000	2.00
	Diaphragm, dam type			1,000	2.00
	Butterfly			800	0.25
	Check		Lift	2,000	10.00
			Swing	1,500	1.50
			Tilting-disk	1,000	0.50

Note: Use R/D = 1.5 values for R/D = 5 pipe bends, 45° to 180°. Use appropriate tee values for flow through crosses.

[a] Inlet, flush, $K = 160/N_{Re} + 0.5$. Inlet, intruding, $K = 160/N_{Re} = 1.0$. Exit, $K = 1.0$. $K = K_1/N_{Re} + K_2(1 + 1/D)$, with D in inches. [Hooper, *Chem. Eng.* 96–100 (24 Aug. 1981)].

from Eqs. (4)–(10) with the aid of the Newton–Raphson method for simultaneous nonlinear equations.

Some simplification is permissible for water distribution systems in metallic pipes. Then the Hazen–Williams formula is adequate, namely

$$\Delta h = \Delta P/\rho = 4.727L(Q/130)^{1.852}/D^{4.8704} \quad (6.31)$$

with linear dimensions in ft and Q in cuft/sec. The iterative solution method for flowrate distribution of Hardy Cross is popular. Examples of that procedure are presented in many books on fluid mechanics, for example, those of Bober and Kenyon (*Fluid Mechanics,* Wiley, New York, 1980) and Streeter and Wylie (*Fluid Mechanics,* McGraw-Hill, New York, 1979).

With particularly simple networks, some rearrangement of equations sometimes can be made to simplify the solution. Example 6.7 is of such a case.

6.5. OPTIMUM PIPE DIAMETER

In a chemical plant the capital investment in process piping is in the range of 25–40% of the total plant investment, and the power consumption for pumping, which depends on the line size, is a substantial fraction of the total cost of utilities. Accordingly, economic optimization of pipe size is a necessary aspect of plant design. As the diameter of a line increases, its cost goes up but is accompanied by decreases in consumption of utilities and costs of pumps and drivers because of reduced friction. Somewhere there is an optimum balance between operating cost and annual capital cost.

For small capacities and short lines, near optimum line sizes may be obtained on the basis of typical velocities or pressure drops such as those of Table 6.2. When large capacities are involved and lines are long and expensive materials of construction are needed, the selection of line diameters may need to be subjected to complete economic analysis. Still another kind of factor may need to be taken into account with highly viscous materials: the possibility that heating the fluid may pay off by reducing the viscosity and consequently the power requirement.

Adequate information must be available for installed costs of piping and pumping equipment. Although suppliers quotations are desirable, published correlations may be adequate. Some data and references to other published sources are given in Chapter 20. A simplification in locating the optimum usually is permissible by ignoring the costs of pumps and drivers since they are essentially insensitive to pipe diameter near the optimum value. This fact is clear in Example 6.8 for instance and in the examples worked out by Happel and Jordan (*Chemical Process Economics,* Dekker, New York, 1975).

Two shortcut rules have been derived by Peters and Timmerhaus (1980; listed in Chapter 1 References) for optimum diameters of steel pipes of 1-in. size or greater, for turbulent and laminar flow:

$$D = 3.9Q^{0.45}\rho^{0.13}, \quad \text{turbulent flow}, \quad (6.32)$$

$$D = 3.0Q^{0.36}\mu^{0.18}, \quad \text{laminar flow}. \quad (6.33)$$

D is in inches, Q in cuft/sec, ρ in lb/cuft, and μ in cP. The factors involved in the derivation are: power cost = 0.055/kWh, friction loss due to fittings is 35% that of the straight length, annual fixed charges are 20% of installation cost, pump efficiency is 50%, and cost of 1-in. IPS schedule 40 pipe is \$0.45/ft. Formulas that take additional factors into account also are developed in that book.

Other detailed studies of line optimization are made by Happel and Jordan (*Chemical Process Economics,* Dekker, New York, 1975) and by Skelland (1967). The latter works out a problem in simultaneous optimization of pipe diameter and pumping temperature in laminar flow.

Example 6.8 takes into account pump costs, alternate kinds of drivers, and alloy construction.

6.6. NON-NEWTONIAN LIQUIDS

Not all classes of fluids conform to the frictional behavior described in Section 6.3. This section will describe the commonly recognized types of liquids, from the point of view of flow behavior, and will summarize the data and techniques that are used for analyzing friction in such lines.

VISCOSITY BEHAVIOR

The distinction in question between different fluids is in their viscosity behavior, or relation between shear stress τ (force per unit area) and the rate of deformation expressed as a lateral velocity

EXAMPLE 6.5
Comparison of Pressure Drops in a Line with Several Sets of Fittings Resistances

The flow considered is in a 12-inch steel line at a Reynolds number of 6000. With $\varepsilon = 0.00015$, Round's equation gives $f = 0.0353$. The line composition and values of fittings resistances are:

	Table 6.4	Table 6.5	Table 6.6		
	L	K	K_1	K_2	K
Line	1000	—	—	—	—
6 LR ells	20	0.25	500	0.15	0.246
4 tees, branched	66	0.5	150	0.15	0.567
2 gate valves, open	7	0.05	300	0.10	0.158
1 globe valve	340	5.4	1500	4.00	4.58
	1738	9.00			8.64

Table 6.4, $\dfrac{\Delta P}{(\rho u^2/2g_c)} = \dfrac{f}{D}(1738) = 61.3,$

Table 6.5, $\dfrac{\Delta P}{(\rho u^2/2g_c)} = f\dfrac{L}{D} + \sum K_i$

$$= \frac{0.0353(1000)}{1} + 9.00 = 44.3,$$

Table 6.6, $\dfrac{\Delta P}{(\rho u^2/2g_c)} = 35.3 + 8.64 = 43.9.$

The value $K = 0.05$ for gate valve from Table 6.5 appears to be low: *Chemical Engineering Handbook*, for example, gives 0.17, more nearly in line with that from Table 6.6. The equivalent length method of Table 6.4 gives high pressure drops; although convenient, it is not widely used.

EXAMPLE 6.6
A Network of Pipelines in Series, Parallel, and Branches: the Sketch, Material Balances, and Pressure Drop Equations

Pressure drop:

$$\Delta P_{ij} = (8\rho/g_c\pi^2) f_{ij} L_{ij} Q_{ij}^2 / D_{ij}^5 = k f_{ij} L_{ij} Q_{ij}^2 / D_{ij}^5. \quad (1)$$

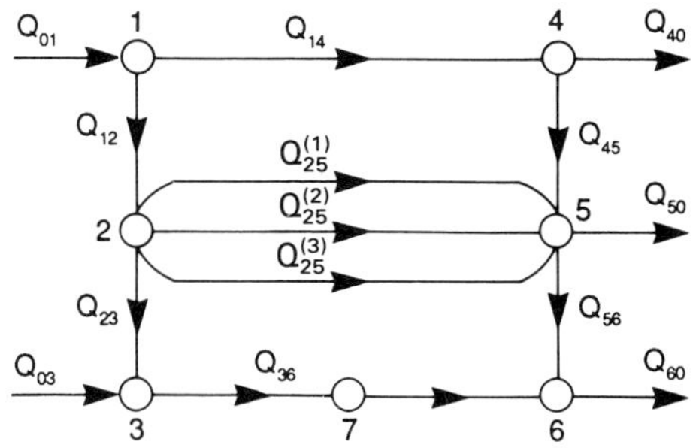

Reynolds number:

$$(N_{Re})_{ij} = 4Q_{ij}\rho/\pi D_{ij}\mu. \quad (2)$$

Friction factor:

$$f_{ij} = 1.6364/[\ln(\varepsilon/D_{ij} + 6.5/(N_{Re})_{ij})]^2. \quad (3)$$

Pressure drops in key lines:

$$\Delta p_{12} = P_1 - P_2 - kf_{12}L_{12}Q_{12}^2/D_{12}^5 = 0, \quad (4)$$
$$\Delta p_{23} = P_2 - P_3 - kf_{23}L_{23}Q_{23}^2/D_{23}^5 = 0, \quad (5)$$
$$\Delta p_{25} = P_2 - P_5 = kf_{25}^{(1)}L_{25}^{(1)}(Q_{25}^{(1)})^2/(D_{25}^{(1)})^5 \quad (6)$$
$$= kf_{25}^{(2)}L_{25}^{(2)}(Q_{25}^{(2)})^2/(D_{25}^{(2)})^5 \quad (7)$$
$$= kf_{25}^{(3)}L_{25}^{(3)}(Q_{25}^{(3)})^2/(D_{25}^{(3)})^5, \quad (8)$$
$$\Delta p_{45} = P_4 - P_5 - kf_{45}L_{45}Q_{45}^2/D_{45}^5 = 0, \quad (9)$$
$$\Delta p_{56} = P_5 - P_6 - kf_{56}L_{56}Q_{56}^2/D_{56}^5 = 0 \quad (10)$$

Node	Material Balance at Node:	
1	$Q_{01} - Q_{12} - Q_{14} = 0$	(11)
2	$Q_{12} - Q_{23} - Q_{25}^{(1)} - Q_{25}^{(2)} - Q_{25}^{(3)} = 0$	(12)
3	$Q_{03} + Q_{23} - Q_{36} = 0$	(13)
4	$Q_{14} - Q_{40} - Q_{45} = 0$	(14)
5	$Q_{45} + Q_{25}^{(1)} + Q_{25}^{(2)} + Q_{25}^{(3)} - Q_{50} - Q_{56} = 0$	(15)
6	$Q_{36} + Q_{56} - Q_{60} = 0$	(16)
Overall	$Q_{01} + Q_{03} - Q_{40} - Q_{50} - Q_{60} = 0$	(17)

EXAMPLE 6.7
Flow of Oil in a Branched Pipeline

The pipeline handles an oil with sp gr = 0.92 and kinematic viscosity of 5 centistokes(cS) at a total rate of 12,000 cuft/hr. All three pumps have the same output pressure. At point 5 the elevation is 100 ft and the pressure is 2 atm gage. Elevations at the other points are zero. Line dimensions are tabulated following. The flow rates in each of the lines and the total power requirement will be found.

Line	L (ft)	D (ft)
14	1000	0.4
24	2000	0.5
34	1500	0.3
45	4000	0.75

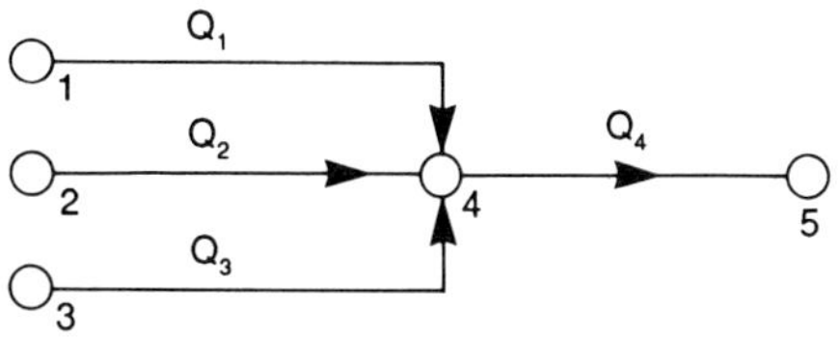

$$Q_1 + Q_2 + Q_3 = Q_4 = 12{,}000/3600 = 3.333 \text{ cfs} \quad (1)$$

$$N_{Re} = \frac{4Q}{\pi DV} = \frac{4Q}{\pi D(5/92{,}900)} = \frac{23{,}657Q}{D}$$
$$= \begin{cases} 59{,}142Q_1 \\ 47{,}313Q_2, \\ 78{,}556Q_3, \\ 31{,}542Q_4, \end{cases} \quad (2)$$

Example 6.7—(*continued*)

$\varepsilon = 0.00015$ ft,

$$h_f = \frac{8fLQ^2}{g_c\pi^2D^5} = 0.0251fLQ^2/D^5 \text{ ft}, \quad (3)$$

$$h_{f1} = h_{f2} = h_{f3}, \quad (4)$$

$$\frac{f_1L_1Q_1^2}{D_1^5} = \frac{f_2L_2Q_2^2}{D_2^5} = \frac{f_3L_3Q_3^2}{D_3^5} \quad (5)$$

$$Q_2 = Q_1\left[\frac{f_1L_1}{f_2L_2}\left(\frac{D_2}{D_1}\right)^5\right]^{1/2} = 1.2352\left(\frac{f_1}{f_2}\right)^{1/2}Q_1, \quad (6)$$

$$Q_3 = Q_1\left[\frac{f_1L_1}{f_3L_3}\left(\frac{D_3}{D_1}\right)^5\right]^{1/2} = 0.3977\left(\frac{f_1}{f_3}\right)^{1/2}Q_1, \quad (7)$$

$$Q_1\left[1 + 1.2352\left(\frac{f_1}{f_2}\right)^{1/2} + 0.3977\left(\frac{f_1}{f_3}\right)^{1/2}\right] = Q_4 = 3.333, \quad (8)$$

$$f = \frac{1.6364}{[\ln(2.03(10^{-5})/D + 6.5/N_{Re}]^2}.$$

For line 45,

$$(N_{Re})_4 = 31542\,(3.333) = 105{,}140,$$

$$f_4 = 0.01881,$$

$$(h_f)_{45} = \frac{0.02517(0.01881)(4000)(3.333)^2}{(0.75)^5} = 88.65 \text{ ft}.$$

Procedure:

1. As a first trial assume $f_1 = f_2 = f_3$, and find $Q_1 = 1.266$ from Eq. (8).
2. Find Q_2 and Q_3 from Eqs. (6) and (7).
3. With these values of the Q_i, find improved values of the f_i and hence improved values of Q_2 and Q_3 from Eqs. (6) and (7).
4. Check how closely $Q_1 + Q_2 + Q_3 - 3.333 \overset{?}{=} 0$.
5. If check is not close enough, adjust the value of Q_1 and repeat the calculations.

The two trials shown following prove to be adequate.

Q_1	Q_2	Q_3	Q_4	$10/3 - Q_4$	f_1
1.2660	1.5757	0.4739	3.3156	0.0023	0.02069
1.2707	1.5554	0.5073	3.3334	0.0001	0.02068

Summary:

Line	N_{Re}	f	Q	h_f
14	75,152	0.02068	1.2707	82.08
24	60,121	0.02106	1.5554	82.08
34	99,821	0.02053	0.5073	82.08
45	105,140	0.01881	3.3333	88.65

$$h_{f14} = h_{f24} = h_{f34} = \frac{0.02517(0.02068)(1000)(1.2707)^2}{(0.4)^5} = 82.08 \text{ ft}.$$

Velocity head at discharge:

$$\frac{u_5^2}{2g_c} = \frac{1}{2g_c}\left(\frac{Q_4}{(\pi/4)D^2}\right)^2 = 0.88 \text{ ft}.$$

Total head at pumps:

$$h_p = \frac{2(2117)}{0.92(62.4)} + 100$$
$$+ 0.88 + 82.08 + 88.65$$
$$= 345.36 \text{ ft}.$$

Power

$$= \gamma Q_4 h_p$$
$$0.92(62.4)(10/3)345.36$$
$$= 66{,}088 \text{ ft lb/sec}$$
120.2 HP, 89.6 kW.

```
10 ! Example 6.7;flow in a bran
   ched pipeline
20 READ D1,D2,D3,L1,L2,L3
30 DATA .4,.5,.3,1000,2000,1500
40 INPUT Q1
50 Q2=1.2352*Q1
60 Q3=.3977*Q1
70 R1=59142*Q1
80 R2=47313*Q1
90 R3=78556*Q1
100 F1=1.6364/LOG(.135*.00015/D1
    +6.5/R1)^2
110 F2=1.6364/LOG(.135*.00015/D2
    +6.5/R2)^2
120 F3=1.6364/LOG(.135*.00015/D3
    +6.5/R3)^2
130 Q2=1.2352*Q1*(F1/F2)^.5 ! im
    proved value
140 Q3=.3977*Q1*(F1/F3)^.5 ! imp
    roved value
150 X=10/3-Q1-Q2-Q3 ! should be
    less than 0.0001
160 DISP X,Q1,Q2,Q3
170 GOTO 40 ! choose another val
    ue of Q1 if condition of ste
    p 150 is not satisfied
180 END
```

Example 6.8
Economic Optimum Pipe Size for Pumping Hot Oil with a Motor or Turbine Drive

A centrifugal pump and its spare handle 1000 gpm of an oil at 500°F. Its specific gravity is 0.81 and its viscosity is 3.0 cP. The length of the line is 600 ft and its equivalent length with valves and other fittings is 900 ft. There are 12 gate valves, two check valves, and one control valve.

Suction pressure at the pumps is atmospheric; the pump head exclusive of line friction is 120 psi. Pump efficiency is 71%. Material of construction of line and pumps is 316 SS. Operation is 8000 hr/yr.

Characteristics of the alternate pump drives are:

a. Turbines are 3600 rpm, exhaust pressure is 0.75 bar, inlet pressure is 20 bar, turbine efficiency is 45%. Value of the high pressure steam is \$5.25/1000 lbs; that of the exhaust is \$0.75/1000 lbs.
b. Motors have efficiency of 90%, cost of electricity is \$0.065/kWh.

Cost data are:

1. Installed cost of pipe is $7.5D$ \$/ft and that of valves is $600D^{0.7}$ \$ each, where D is the nominal pipe size in inches.

EXAMPLE 6.8—*(continued)*

2. Purchase costs of pumps, motors and drives are taken from *Manual of Economic Analysis of Chemical Processes, Institut Francais du Petrole* (McGraw-Hill, New York, 1976).
3. All prices are as of mid-1975. Escalation to the end of 1984 requires a factor of 1.8. However, the location of the optimum will be approximately independent of the escalation if it is assumed that equipment and utility prices escalate approximately uniformly; so the analysis is made in terms of the 1975 prices. Annual capital cost is 50% of the installed price/year.

The summary shows that a 6-in. line is optimum with motor drive, and an 8-in. line with turbine drive. Both optima are insensitive to line sizes in the range of 6–10 in.

$$Q = 1000/(7.48)(60) = 2.2282 \text{ cfs}, \quad 227.2 \text{ m}^3/\text{hr},$$

$$N_{\text{Re}} = \frac{4Q\rho}{\pi D\mu} = \frac{4(2.2282)(0.81)(62.4)}{\pi(0.000672)(3)D} = \frac{71{,}128}{D},$$

$$f = 1.6364\left[\ln\frac{0.135(0.00015)}{D} + \frac{6.5D}{71{,}128}\right]^2.$$

Pump head:

$$h_p = \frac{120(144)}{0.81(62.4)} + \frac{8fLQ^2}{g\pi^2 D^5}$$
$$= 341.88 + 124.98f/D^5 \text{ ft}.$$

Motor power:

$$P_m = \frac{Q\rho}{\eta_p\eta_m}h_p = \frac{2.2282(50.54)}{550(0.71(0.90))}h_p$$
$$= 0.3204h_p, \quad \text{HP}$$

Turbine power:

$$P_t = \frac{2.2282(50.54)}{550(0.71)}h_p = 0.2883h_p, \quad \text{HP}.$$

Steam

$$= 10.14 \text{ kg/HP (from the "manual")}$$
$$= 10.14(0.2883)(2.204)h_p/1000 = 0.006443h_p, \quad 1000 \text{ lb/hr}.$$

Power cost:

$$0.065(8000)(\text{kw}), \$/\text{yr},$$

Steam cost:

$$4.5(8000)(1000 \text{ lb/hr}), \$/\text{yr}.$$

Installed pump cost factors for alloy, temperature, etc (data in the "manual")

$$= 2[2.5(1.8)(1.3)(0.71)] = 8.2.$$

Summary:

IPS	**4**	**6**	**8**	**10**
D (ft)	0.3355	0.5054	0.6651	0.8350
100f	1.89	1.87	1.89	1.93
h_p (ft)	898	413	360	348
Pump efficiency	0.71	0.71	0.71	0.71
motor (kW)	214.6	98.7	86.0	83.2
Steam, 1000 lb/hr	5.97	2.66	2.32	2.25
Pump cost, 2 installed	50,000	28,000	28,000	28,000
Motor cost, 2 installed	36,000	16,000	14,000	14,000
Turbine cost, 2 installed	56,000	32,000	28,000	28,000
Pipe cost	18,000	27,000	36,000	45,000
Valve cost	23,750	31,546	38,584	45,107
Equip cost, motor drive	127,750	93,546	107,584	123,107
Equip cost, turbine drive	147,750	109,546	121,584	137,107
Power cost ($/yr)	111,592	51,324	44,720	43,264
Steam cost ($/yr)	208,440	95,760	83,520	80,834
Annual cost, motor drive	175,467	98,097	98,512	104,817
Annual cost, turbine drive	282,315	150,533	144,312	149,387

gradient, $\dot{\gamma} = du/dx$. The concept is represented on Figure 6.2(a): one of the planes is subjected to a shear stress and is translated parallel to a fixed plane at a constant velocity but a velocity gradient is developed between the planes. The relation between the variables may be written

$$\tau = F/A = \mu(du/dx) = \mu\dot{\gamma}, \tag{6.34}$$

where, by definition, μ is the viscosity. In the simplest case, the viscosity is constant, and the fluid is called Newtonian. In the other cases, more complex relations between τ and $\dot{\gamma}$ involving more than one constant are needed, and dependence on time also may be present. Classifications of non-Newtonian fluids are made according to the relation between τ and $\dot{\gamma}$ by formula or shape of plot, or according to the mechanism of the resistance of the fluid to deformation.

The concept of an apparent viscosity

$$\mu_a = \tau/\dot{\gamma} \tag{6.35}$$

is useful. In the Newtonian case it is constant, but in general it can be a function of τ, $\dot{\gamma}$, and time θ.

Non-Newtonian behavior occurs in solutions or melts of polymers and in suspensions of solids in liquids. Some τ–$\dot{\gamma}$ plots are shown in Figure 6.2, and the main classes are described following.

1. *Pseudoplastic liquids* have a τ–$\dot{\gamma}$ plot that is concave downward. The simplest mathematical representation of such relations is a power law

$$\tau = K\dot{\gamma}^n, \quad n < 1 \tag{6.36}$$

with $n < 1$. This equation has two constants; others with many more than two constants also have been proposed. The apparent viscosity is

$$\mu_a = \tau/\dot{\gamma} = K/\dot{\gamma}^{1-n}. \tag{6.37}$$

Since n is less than unity, the apparent viscosity decreases with the deformation rate. Examples of such materials are some polymeric solutions or melts such as rubbers, cellulose acetate and napalm; suspensions such as paints, mayonnaise, paper pulp, or detergent slurries; and dilute suspensions of inert solids. Pseudoplastic properties of wallpaper paste account for good spreading and adhesion, and those of printing inks prevent their running at low speeds yet allow them to spread easily in high speed machines.

2. *Dilatant liquids* have rheological behavior essentially

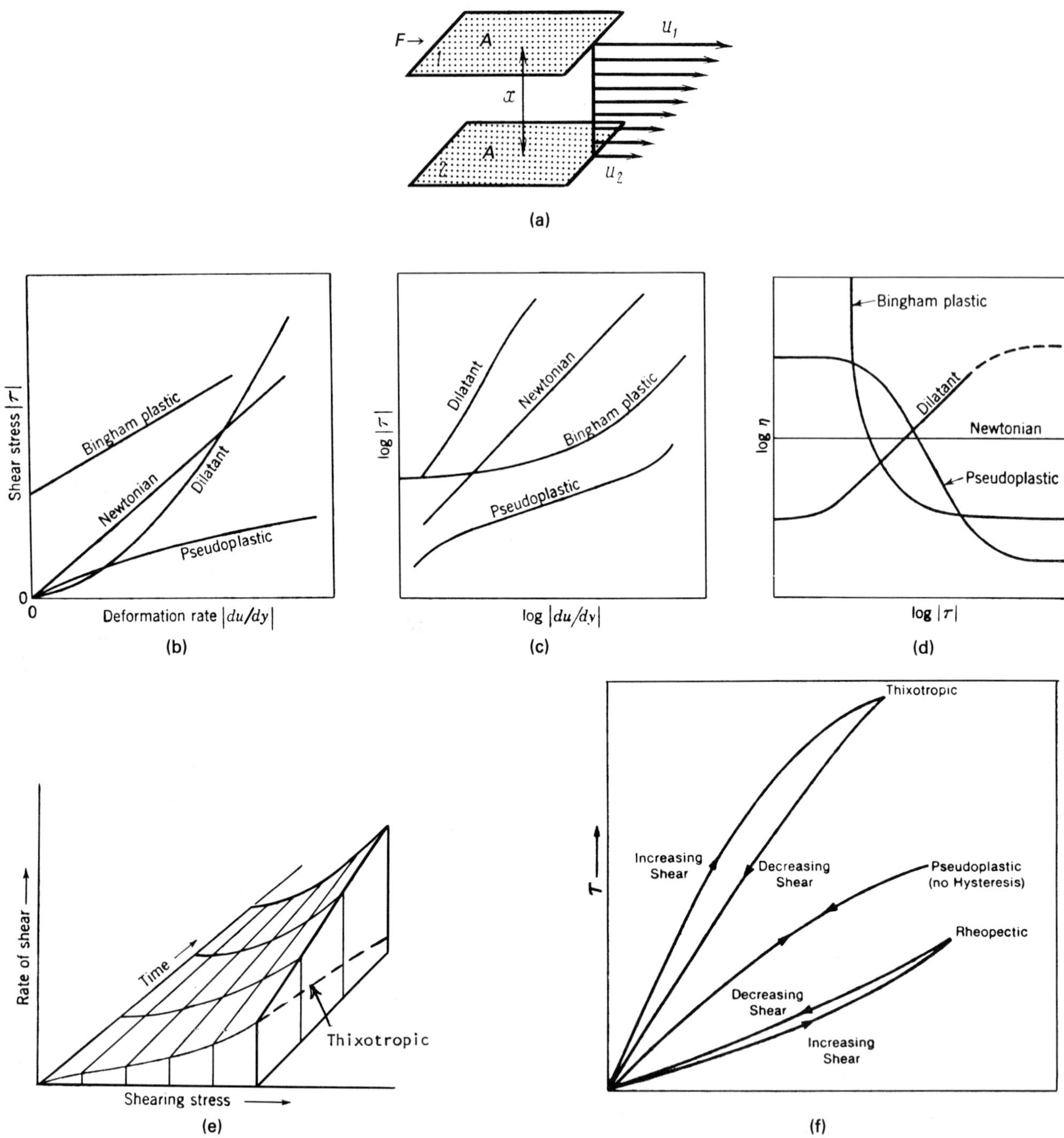

Figure 6.2. Relations between shear stress, deformation rate, and viscosity of several classes of fluids. (a) Distribution of velocities of a fluid between two layers of areas A which are moving relatively to each other at a distance x under influence of a force F. In the simplest case, $F/A = \mu(du/dx)$ with μ constant. (b) Linear plot of shear stress against deformation. (c) Logarithmic plot of shear stress against deformation rate. (d) Viscosity as a function of shear stress. (e) Time-dependent viscosity behavior of a rheopectic fluid (thixotropic behavior is shown by the dashed line). (f) Hysteresis loops of time-dependent fluids (arrows show the chronology of imposed shear stress).

opposite those of pseudoplastics insofar as viscosity behavior is concerned. The τ–$\dot{\gamma}$ plots are concave upward and the power law applies

$$\tau = K\dot{\gamma}^n, \quad n > 1, \tag{6.38}$$

but with n greater than unity; other mathematical relations also have been proposed. The apparent viscosity, $\mu_a = K\dot{\gamma}^{n-1}$, increases with deformation rate. Examples of dilatant materials are pigment-vehicle suspensions such as paints and printing inks of high concentrations; starch, potassium silicate, and gum arabic in water; quicksand or beach sand in water. Dilatant properties of wet cement aggregates permit tamping operations in which small impulses produce more complete settling. Vinyl resin plastisols exhibit pseudoplastic behavior at low deformation rates and dilatant behavior at higher ones.

3. *Bingham plastics* require a finite amount of shear stress before deformation begins, then the deformation rate is linear. Mathematically,

$$\tau = \tau_0 + \mu_B(du/dx) = \tau_0 + \mu_B\dot{\gamma}, \tag{6.39}$$

where μ_B is called the coefficient of plastic viscosity. Examples of materials that approximate Bingham behavior are drilling muds; suspensions of chalk, grains, and thoria; and sewage sludge. Bingham characteristics allow toothpaste to stay on the brush.

4. *Generalized Bingham or yield-power law* fluids are represented by the equation

$$\tau = \tau_0 + K\dot{\gamma}^n. \tag{6.40}$$

Yield-dilatant ($n > 1$) materials are rare but several cases of

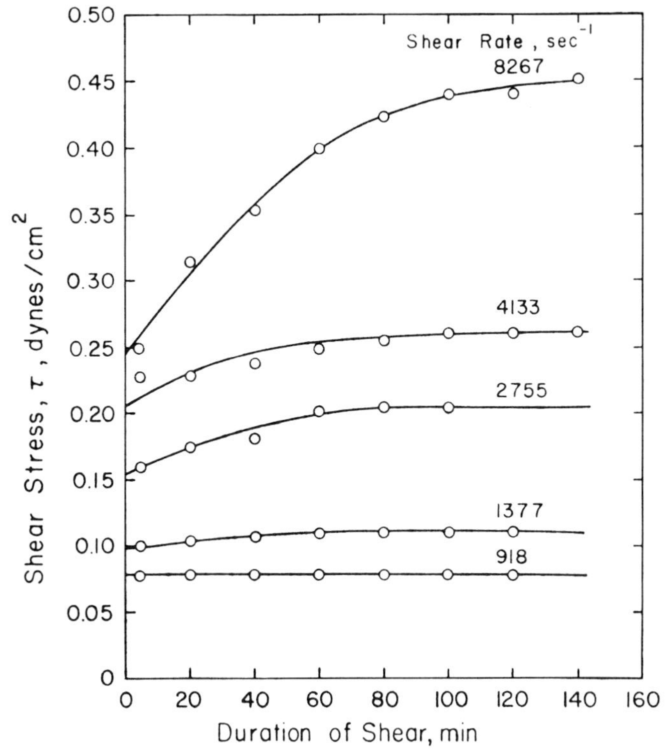

Figure 6.3. Time-dependent rheological behavior of a rheopectic fluid, a 2000 molecular weight polyester [*after Steg and Katz,* J. Appl. Polym. Sci. **9,** *3177 (1965)*].

yield-pseudoplastics exist. For instance, data from the literature of a 20% clay in water suspension are represented by the numbers $\tau_0 = 7.3$ dyn/cm^2, $K = 1.296$ dyn(sec)^n/cm^2 and $n = 0.483$ (Govier and Aziz, 1972, p. 40). Solutions of 0.5–5.0% carboxypolymethylene also exhibit this kind of behavior, but at lower concentrations the yield stress is zero.

5. *Rheopectic fluids* have apparent viscosities that increase with time, particularly at high rates of shear as shown on Figure 6.3. Figure 6.2(f) indicates typical hysteresis effects for such materials. Some examples are suspensions of gypsum in water, bentonite sols, vanadium pentoxide sols, and the polyester of Figure 6.3.

6. *Thixotropic fluids* have a time-dependent rheological behavior in which the shear stress diminishes with time at a constant deformation rate, and exhibits hysteresis [Fig. 6.2(f)]. Among the substances that behave this way are some paints, ketchup, gelatine solutions, mayonnaise, margarine, mustard, honey, and shaving cream. Nondrip paints, for example, are thick in the can but thin on the brush. The time-effect in the case of the thixotropic crude of Figure 6.4(a) diminishes at high rates of deformation. For the same crude, Figure 6.4(b) represents the variation of pressure gradient in a pipe line with time and axial position; the gradient varies fivefold over a distance of about 2 miles after 200 min. A relatively simple relation involving five constants to represent thixotropic behavior is cited by Govier and Aziz (1972, p. 43):

$$\tau = (\mu_0 + c\lambda)\dot{\gamma}, \tag{6.41}$$

$$d\lambda/d\theta = a - (a + b\dot{\gamma})\lambda. \tag{6.42}$$

The constants μ_0, a, b, and c and the structural parameter λ are obtained from rheological measurements in a straightforward manner.

7. *Viscoelastic fluids* have the ability of partially recovering their original states after stress is removed. Essentially all molten polymers are viscoelastic as are solutions of long chain molecules such as polyethylene oxide, polyacrylamides, sodium carboxymethylcellulose, and others. More homely examples are egg whites, dough, jello, and puddings, as well as bitumen and napalm. This property enables eggwhites to entrap air, molten polymers to form threads, and such fluids to climb up rotating shafts whereas purely viscous materials are depressed by the centrifugal force.

Two concepts of deformability that normally are applied only to solids, but appear to have examples of gradation between solids and liquids, are those of shear modulus E, which is

$$E = \text{shear stress/deformation}, \tag{6.43}$$

and relaxation time θ^*, which is defined in the relation between the residual stress and the time after release of an imposed shear stress, namely,

$$\tau = \tau_0 \exp(-\theta/\theta^*). \tag{6.44}$$

A range of values of the shear modulus (in kgf/cm^2) is

Gelatine	
0.5% solution	4×10^{-10}
10% solution (jelly)	5×10^{-2}
Raw rubber	1.7×10^2
Lead	4.8×10^4
Wood (oak)	8×10^4
Steel	8×10^5

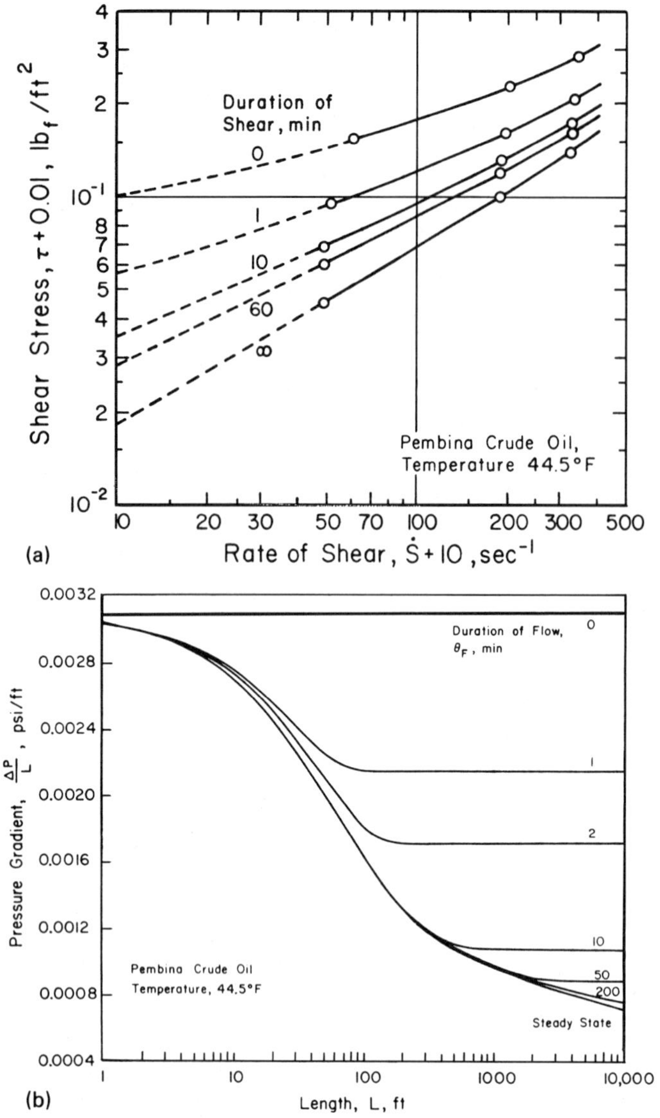

Figure 6.4. Shear and pipeline flow data of a thixotropic Pembina crude oil at 44.5°F. (a) Rheograms relating shear stress and rate of shear at several constant durations of shear (*Ritter and Govier,* Can. J. Chem. Eng. **48,** *505* (*1970*)]. (b) Decay of pressure gradient of the fluid flowing from a condition of rest at 15,000 barrels/day in a 12 in. line [*Ritter and Batycky,* SPE Journal **7,** *369* (*1967*)].

and that of relaxation time (sec) is

Water	3×10^{-6}
Castor oil	2×10^{-3}
Copal varnish	2×10
Colophony (at 55°C)	5×10
Gelatine, 0.5% solution	8×10^{2}
Colophony (at 12°C)	4×10^{6}
Ideal solids	∞

Examples thus appear to exist of gradations between the properties of normally recognized true liquids (water) and true solids.

Elastic properties usually have a negligible effect on resistance to flow in straight pipes, but examples have been noted that the resistances of fittings may be as much as 10 times as great for viscoelastic liquids as for Newtonian ones.

PIPELINE DESIGN

The sizing of pipelines for non-Newtonian liquids may be based on scaleup of tests made under the conditions at which the proposed line is to operate, without prior determination and correlation of rheological properties. A body of theory and some correlations are available for design with four mathematical models:

$$\tau_w = K\dot{\gamma}^n, \quad \text{power law,} \tag{6.45}$$

$$\tau_w = \tau_y + \mu_B\dot{\gamma}, \quad \text{Bingham plastic,} \tag{6.46}$$

$$\tau_w = \tau_y + K\dot{\gamma}^n, \quad \text{Generalized Bingham or yield-power law,} \tag{6.47}$$

$$\tau_w = K'(8\bar{V}/D)^{n'} \quad \text{Generalized power law} \tag{6.48}$$

(Metzner–Reed) (*AIChE J.* **1,** 434, 1955).

In the last model, the parameters may be somewhat dependent on the shear stress and deformation rate, and should be determined at magnitudes of those quantities near those to be applied in the plant.

The shear stress τ_w at the wall is independent of the model and is derived from pressure drop measurements as

$$\tau_w = D\Delta P/4L. \tag{6.49}$$

Friction Factor. In rheological literature the friction factor is defined as

$$f = \frac{D\Delta P}{4L\rho V^2/2g_c} \tag{6.50}$$

$$= \frac{\tau_w}{\rho V^2/2g_c}. \tag{6.51}$$

This value is one-fourth of the friction factor used in Section 6.3. For the sake of consistency with the literature, the definition of Eq. (6.50) will be used with non-Newtonian fluids in the present section.

Table 6.2 lists theoretical equations for friction factors in laminar flows. In terms of the generalized power law, Eq. (6.48),

$$f = \frac{\tau_w}{\rho V^2/2g_c} = \frac{K'(8V/D)^{n'}}{\rho V^2/2g_c}$$

$$= \frac{16}{D^{n'}V^{2-n'}\rho/g_cK'8^{n'-1}}. \tag{6.52}$$

By analogy with the Newtonian relation, $f = 16/\mathrm{Re}$, the denominator of Eq. (6.52) is designated as a modified Reynolds number,

$$\mathrm{Re}_{\mathrm{MR}} = D^{n'}V^{2-n'}\rho/g_cK'8^{n'-1}. \tag{6.53}$$

The subscript MR designates Metzner–Reed, who introduced this form.

Scale Up. The design of pipelines and other equipment for handling non-Newtonian fluids may be based on model equations with parameters obtained on the basis of measurements with viscometers or with pipelines of substantial diameter. The shapes of plots of τ_w against $\dot{\gamma}$ or $8V/D$ may reveal the appropriate model. Examples 6.9 and 6.10 are such analyses.

In critical cases of substantial economic importance, it may be advisable to perform flow tests—Q against ΔP—in lines of moderate size and to scale up the results to plant size, without necessarily trying to fit one of the accepted models. Among the effects that may not be accounted for by such models are time

EXAMPLE 6.9
Analysis of Data Obtained in a Capillary Tube Viscometer

Data were obtained on a paper pulp with specific gravity 1.3, and are given as the first four columns of the table. Shear stress τ_w and deformation rate $\dot{\gamma}$ are derived by the equations applying to this kind of viscometer (Skelland, 1967, p. 31; Van Wazer et al., 1963, p. 197):

$$\tau_w = D\Delta P/4L,$$
$$\dot{\gamma} = \frac{3n'+1}{4n'}\left(\frac{8V}{D}\right)$$
$$n' = \frac{d\ln(\tau_w)}{d\ln(8V/D)}$$

The plot of $\log \tau_w$ against $\log(8V/D)$ shows some scatter but is approximated by a straight line with equation

$$\tau_w = 1.329(8V/D)^{0.51}.$$

Since

$$\dot{\gamma} = (2.53/2.08)(8V/D),$$

the relation between shear stress and deformation is given by the equation

$$\tau_w = 1.203\dot{\gamma}^{0.51}$$

D (cm)	L (cm)	$\dot{m}$ (g/sec)	P (Pa)	8V/D (1/sec)	τ_w (Pa)
0.15	14	0.20	3200	464	8.57
0.15	14	0.02	1200	46.4	3.21
0.30	28	0.46	1950	133.5	5.22
0.30	28	0.10	860	29.0	2.30
0.40	28	1.20	1410	146.9	5.04

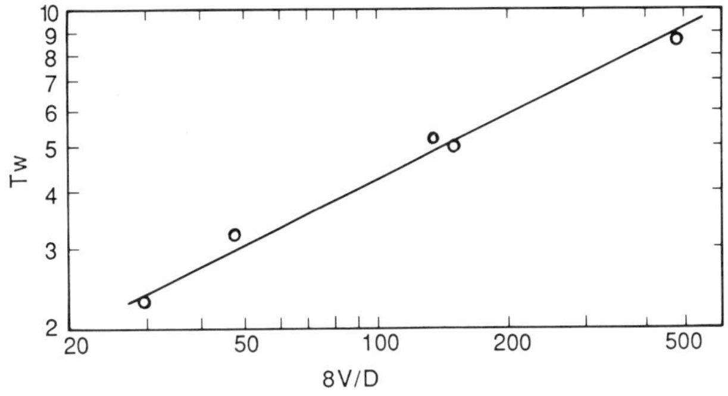

EXAMPLE 6.10
Parameters of the Bingham Model from Measurements of Pressure Drops in a Line

Data of pressure drop in the flow of a 60% limestone slurry of density 1.607 g/ml were taken by Thomas [*Ind. Eng. Chem.* **55,** 18–29 (1963)]. They were converted into data of wall shear stress $\tau_w = D\Delta P/4L$ against the shear rate $8V/D$ and are plotted on the figure for three line sizes.

The Buckingham equation for Bingham flow in the laminar region is

$$\frac{8V}{D} = \frac{\tau_w}{\mu_B}\left[1 - \frac{4}{3}\left(\frac{\tau_0}{\tau_w}\right) + \frac{1}{3}\left(\frac{\tau_0}{\tau_w}\right)^4\right]$$
$$\simeq \frac{1}{\mu_B}\left(\tau_w - \frac{4}{3}\tau_0\right)$$

The second expression is obtained by neglecting the fourth-power term. The Bingham viscosity μ_B is the slope of the plot in the laminar region and is found from the terminal points as

$$\mu_B = (73\text{–}50)/(347\text{–}0) = 0.067\ \text{dyn sec/cm}^2.$$

From the reduced Buckingham equation,

$$\tau_0 = 0.75\tau_w \quad (\text{at } 8V/D = 0)$$
$$= 37.5.$$

Accordingly, the Bingham model is represented by

$$\tau_w = 37.5 + 0.067(8V/D), \quad \text{dyn/cm}^2$$

with time in seconds.

Transitions from laminar to turbulent flow may be identified off the plots:

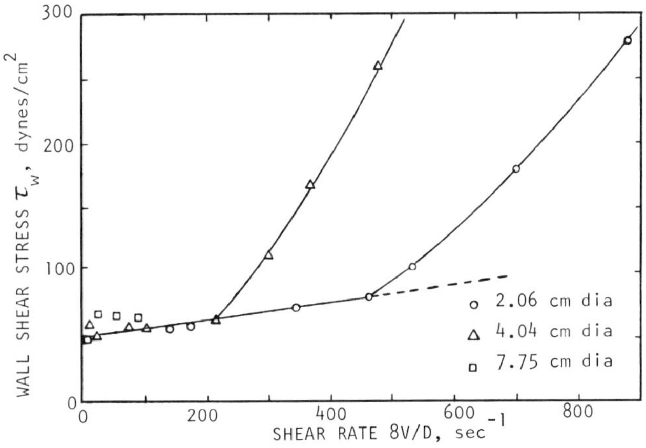

$D = 2.06$ cm, $8V/D = 465$, $V = 120$ cm/sec
4.04 215, 109
7.75 (critical not reached).

The transition points also can be estimated from Hanks' correlation [*AIChE J.* **9**, 45, 306 (1963)] which involves these expressions:

$$x_c = (\tau_0/\tau_w)_c,$$
$$\text{He} = D^2\tau_0\rho/\mu_B^2,$$
$$x_c/(1-x_c)^3 = \text{He}/16{,}800,$$
$$\text{Re}_{Bc} = (1 - \tfrac{4}{3}x_c + \tfrac{1}{3}x_c^4)\text{He}/8x_c.$$

The critical linear velocity finally is evaluated from the critical Reynolds number of the last equation with the following results;

Examples 6.10—*(continued)*

D (cm)	10^{-4} He	x_c	Re_{Bc}	V_c
2.06	5.7	0.479	5635	114 (120)
4.04	22.0	0.635	8945	93 (109)
7.75	81.0	0.750	14,272	77

The numbers in parentheses correspond to the break points on the figure and agree roughly with the calculated values.

The solution of this problem is based on that of Wasp et al. (1977).

---

dependence, pipe roughness, pipe fitting resistance, wall slippage, and viscoelastic behavior. Although some effort has been devoted to them, none of these particular effects has been well correlated. Viscoelasticity has been found to have little effect on friction in straight lines but does have a substantial effect on the resistance of pipe fittings. Pipe roughness often is accounted for by assuming that the relative effects of different roughness ratios ε/D are represented by the Colebrook equation (Eq. 6.20) for Newtonian fluids. Wall slippage due to trace amounts of some polymers in solution is an active field of research (Hoyt, 1972) and is not well predictable.

The scant literature on pipeline scaleup is reviewed by Heywood (1980). Some investigators have assumed a relation of the form

$$\tau_w = D\Delta P/4L = kV^a/D^b$$

and determined the three constants K, a, and b from measurements on several diameters of pipe. The exponent a on the velocity appears to be independent of the diameter if the roughness ratio ε/D is held constant. The exponent b on the diameter has been found to range from 0.2 to 0.25. How much better this kind of analysis is than assuming that $a = b$, as in Eq. (6.48), has not been established. If it can be assumed that the effect of differences in ε/D is small for the data of Examples 6.9 and 6.10, the measurements should plot as separate lines for each diameter, but such a distinction is not obvious on those plots in the laminar region, although it definitely is in the turbulent region of the limestone slurry data.

Observations of the performance of existing large lines, as in the case of Figure 6.4, clearly yields information of value in analyzing the effects of some changes in operating conditions or for the design of new lines for the same system.

Laminar Flow. Theoretically derived equations for volumetric flow rate and friction factor are included for several models in Table 6.7. Each model employs a specially defined Reynolds number, and the Bingham models also involve the Hedstrom number,

$$\mathrm{He} = \tau_0 \rho D^2/\mu_B^2. \tag{6.54}$$

These dimensionless groups also appear in empirical correlations of the turbulent flow region. Although even in the approximate Eq. (9) of Table 6.7, group He appears to affect the friction factor, empirical correlations such as Figure 6.5(b) and the data analysis of Example 6.10 indicate that the friction factor is determined by the Reynolds number alone, in every case by an equation of the form, $f = 16/\mathrm{Re}$, but with Re defined differently for each model. Table 6.7 collects several relations for laminar flows of fluids.

Transitional Flow. Reynolds numbers and friction factors at which the flow changes from laminar to turbulent are indicated by the breaks in the plots of Figures 6.4(a) and (b). For Bingham models, data are shown directly on Figure 6.6. For power-law liquids an equation for the critical Reynolds number is due to Mishra and Triparthi [*Trans. IChE* **51,** T141 (1973)],

$$\mathrm{Re}'_c = \frac{1400(2n+1)(5n+3)}{(3n+1)^2}. \tag{6.55}$$

The Bingham data of Figure 6.6 are represented by the equations of Hanks [*AIChE J.* **9,** 306 (1963)],

$$(\mathrm{Re}_B)_c = \frac{\mathrm{HE}}{8x_c}\left(1 - \frac{4}{3}x_c + \frac{1}{3}x_c^4\right), \tag{6.56}$$

$$\frac{x_c}{(1-x_c)^3} = \frac{\mathrm{He}}{16{,}800}. \tag{6.57}$$

They are employed in Example 6.10.

Turbulent Flow. Correlations have been achieved for all four models, Eqs. (6.45)–(6.48). For power-law flow the correlation of Dodge and Metzner (1959) is shown in Figure 6.5(a) and is represented by the equation

$$\frac{1}{\sqrt{f}} = \frac{4.0}{(n')^{0.75}}\log_{10}[\mathrm{Re}_{n'} f^{(1-n'/2)}] - \frac{0.40}{(n')^{1.2}}. \tag{6.58}$$

These authors and others have demonstrated that these results can represent liquids with a variety of behavior over limited ranges by

TABLE 6.7. Laminar Flow: Volumetric Flow Rate, Friction Factor, Reynolds Number, and Hedstrom Number

Newtonian

$$f = 16/\mathrm{Re}, \quad \mathrm{Re} = DV\rho/\mu \tag{1}$$

Power Law [Eq. (6.45)]

$$Q = \frac{\pi D^3}{32}\left(\frac{4n}{3n+1}\right)\left(\frac{\tau_w}{K}\right)^{1/n} \tag{2}$$

$$f = 16/\mathrm{Re}' \tag{3}$$

$$\mathrm{Re}' = \frac{\rho VD}{K}\left(\frac{4n}{1+3n}\right)^n \cdot \left(\frac{D}{8V}\right)^{n-1} \tag{4}$$

Bingham Plastic [Eq. (6.46)]

$$Q = \frac{\pi D^3 \tau_w}{32\mu_B}\left[1 - \frac{4}{3}\frac{\tau_o}{\tau_w} + \frac{1}{3}\left(\frac{\tau_o}{\tau_w}\right)^4\right] \tag{5}$$

$$\mathrm{Re}_B = DV\rho/\mu_B \tag{6}$$

$$\mathrm{He} = \tau_0 D^2 \rho/\mu_B^2 \tag{7}$$

$$\frac{1}{\mathrm{Re}_B} = \frac{f}{16} - \frac{\mathrm{He}}{6\mathrm{Re}_B^2} + \frac{\mathrm{He}^4}{3f^3\mathrm{Re}_B^8} \quad \text{(solve for } f\text{)} \tag{8}$$

$$f \simeq \frac{96 Re_B^2}{6\mathrm{Re}_B + \mathrm{He}} \quad [\text{neglecting } (\tau_0/\tau_w)^4 \text{ in Eq. (5)}] \tag{9}$$

Generalized Bingham (Yield-Power Law) [Eq. (6.47)]

$$Q = \frac{\pi D^3}{32}\frac{4n}{3n+1}\left(\frac{\tau_w}{K}\right)^{1/n}\left(1 - \frac{\tau_y}{\tau_w}\right) \times \left\{1 - \frac{\tau_y/\tau_w}{2n+1}\left[1 + \frac{2n}{n+1}\left(\frac{\tau_y}{\tau_w}\right)\left(1 + n\frac{\tau_y}{\tau_w}\right)\right]\right\} \tag{10}$$

$$f = \frac{16}{\mathrm{Re}'}\left(1 - \frac{2\mathrm{He}}{f\mathrm{Re}'^2}\right) \times \left\{1 - \frac{1}{(2n-1)}\frac{2\mathrm{He}}{f\mathrm{Re}'^2}\left[1 + \frac{2n}{(n+1)}\frac{2\mathrm{He}}{f\mathrm{Re}'^2}\left(1 + n\cdot\frac{2\mathrm{He}}{f\mathrm{Re}'^2}\right)\right]\right\} \tag{11}$$

[Re′ by Eq. (4) and He by Eq. (7)]

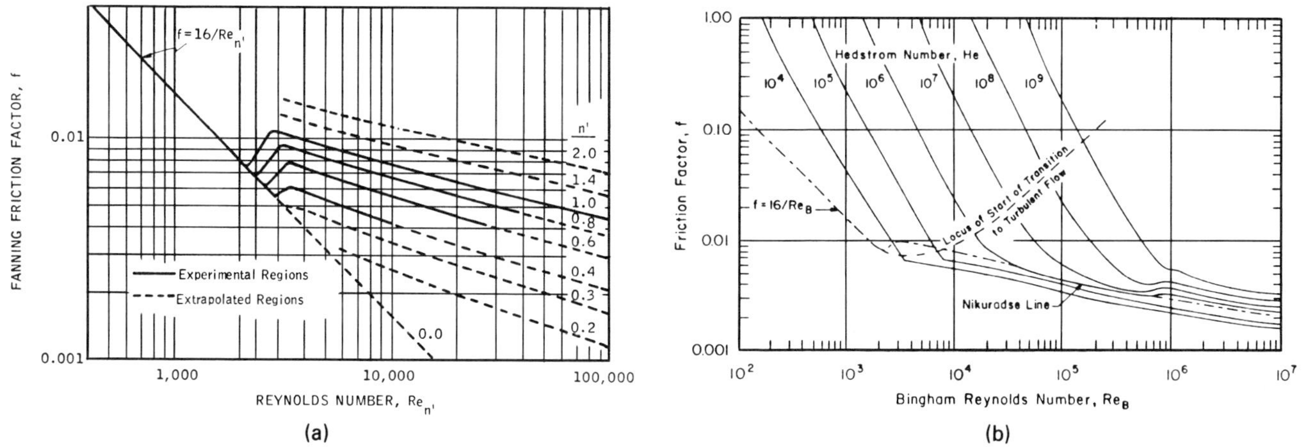

Figure 6.5. Friction factors in laminar and turbulent flows of power-law and Bingham liquids. (a) For pseudoplastic liquids represented by $\tau_w = K'(8V/D)^{n'}$, with K' and n' constant or dependent on τ_w: $1/\sqrt{f} = [4.0/(n')^{0.75}]\log_{10}[\mathrm{Re}_{n'} f^{(1-n'/2)}] - 0.40/(n')^{1.2}$, [*Dodge and Metzner,* AIChE J. **5,** *189 (1959)*]. (b) For Bingham plastics, $\mathrm{Re}_B = DV\rho/\mu_B$, $\mathrm{He} = \tau_0 D^2\rho/\mu_B^2$ [*Hanks and Dadia,* AIChE J. **17,** *554 (1971)*].

evaluating K' and n' in the range of shear stress $\tau_w = D\Delta P/4L$ that will prevail in the required situation.

Bingham flow is represented by Figure 6.5(b) in terms of Reynolds and Hedstrom numbers.

Theoretical relations for generalized Bingham flow [Eq. (6.47)] have been devised by Torrance [*S. Afr. Mech. Eng.* **13,** 89 (1963)]. They are

$$\frac{1}{\sqrt{f}} = \left(\frac{2.69}{n} - 2.95\right) + \frac{1.97}{n}\ln(1-x) + \frac{1.97}{n}\ln(\mathrm{Re}'_T f^{1-n/2}) + \frac{0.68}{n}(5n-8) \quad (6.59)$$

with the Reynolds number

$$\mathrm{Re}_T = D^n V^{2-n}\rho/8^{n-1}K \quad (6.60)$$

and where

$$x = \tau_0/\tau_w. \quad (6.61)$$

In some ranges of operation, materials may be represented approximately equally well by several models, as in Example 6.11 where the power-law and Bingham models are applied.

6.7. GASES

The differential energy balances of Eqs. (6.10) and (6.15) with the friction term of Eq. (6.18) can be integrated for compressible fluid flow under certain restrictions. Three cases of particular importance are of isentropic or isothermal or adiabatic flows. Equations will be developed for them for ideal gases, and the procedure for nonideal gases also will be indicated.

ISENTROPIC FLOW

In short lines, nozzles, and orifices, friction and heat transfer may be neglected, which makes the flow essentially isentropic. Work transfer also is negligible in such equipment. The resulting theory is a basis of design of nozzles that will generate high velocity gases for power production with turbines. With the assumptions indicated, Eq. (6.10) becomes simply

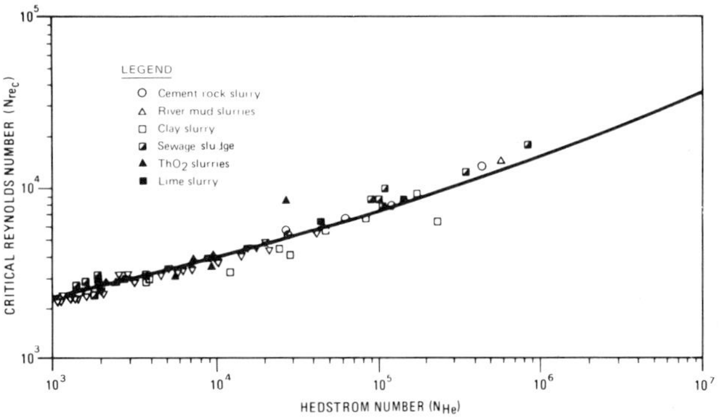

Figure 6.6. Critical Reynolds number for transition from laminar to turbulent flow of Bingham fluids. The data also are represented by Eqs. (6.56) and (6.57): (○) cement rock slurry; (△) river mud slurries; (□) clay slurry; (◪) sewage sludge; (▲) ThO_2 slurries; (■) lime slurry. [*Hanks and Pratt,* SPE Journal, *342–346 (Dec. 1967)*].

$$dH + (1/g_c)u\,du = 0, \quad (6.62)$$

which integrates into

$$H_2 - H_1 + \frac{1}{2g_c}(u_2^2 - u_1^2) \approx 0. \quad (6.63)$$

One of these velocities may be eliminated with the mass balance,

$$\dot{m} = u_2A_2/V_2 = u_1A_1/V_1 \quad (6.64)$$

so that

$$u_2^2 - u_1^2 = (\dot{m}V_2/A_2)^2[1 - (A_2V_1/A_1V_2)^2]. \quad (6.65)$$

For ideal gases substitutions may be made from

$$H_2 - H_1 = C_p(T_2 - T_1) \quad (6.66)$$

and

$$T_2/T_1 = (P_2/P_1)^{(k-1)/k} = (V_1/V_2)^k. \tag{6.67}$$

After these substitutions are made into Eq. (6.63), the results may be solved for the mass rate of flow as

$$\dot{m}/A_2 = \frac{\left(\frac{2g_cP_1}{V_1}\right)^{1/2}\left\{\frac{k}{k-1}\left[\left(\frac{P_2}{P_1}\right)^{2/k} - \left(\frac{P_2}{P_1}\right)^{(k+1)/k}\right]\right\}^{1/2}}{\left[1-\left(\frac{A_2}{A_1}\right)^2\left(\frac{P_2}{P_1}\right)^{2/k}\right]^{1/2}}. \tag{6.68}$$

At specified mass flow rate and inlet conditions P_1 and V_1, Eq. (6.68) predicts a relation between the area ratio A_2/A_1 and the pressure ratio P_2/P_1 when isentropic flow prevails. It turns out that, as the pressure falls, the cross section at first narrows, reaches a minimum at which the velocity becomes sonic; then the cross section increases and the velocity becomes supersonic. In a duct of constant cross section, the velocity remains sonic at and below a critical pressure ratio given by

$$\frac{P_s}{P_1} = \left(\frac{2}{k+1}\right)^{k/(k+1)}. \tag{6.69}$$

The sonic velocity is given by

$$u_s = \sqrt{g_c(\partial P/\partial \rho)_S} \rightarrow \sqrt{g_ckRT/M_w}, \tag{6.70}$$

where the last result applies to ideal gases and M_w is the molecular weight.

ISOTHERMAL FLOW IN UNIFORM DUCTS

When elevation head and work transfer are neglected, the mechanical energy balance equation (6.13) with the friction term of Eq. (6.18) become

$$V\,dP + (1/g_c)u\,du + \frac{fu^2}{2g_cD}dL = 0. \tag{6.71}$$

Make the substitutions

$$u = G/\rho = GV \tag{6.72}$$

and the ideal gas relation

$$V = P_1V_1/P \quad \text{and} \quad dV/V = -dP/P \tag{6.73}$$

so that Eq. (6.71) becomes

$$\frac{P\,dP}{P_1V_1} - \frac{G^2}{g_c}\ln\left(\frac{P_1}{P_2}\right) + \frac{fG^2}{2g_cD}dL = 0. \tag{6.74}$$

This is integrated term-by-term between the inlet and outlet conditions,

$$\frac{P_2^2 - P_1^2}{2P_1V_1} + \frac{G^2}{g_c}\ln\left(\frac{P_1}{P_2}\right) + \frac{fG^2L}{2g_cD} = 0 \tag{6.75}$$

and may be rearranged into

$$P_2^2 = P_1^2 - \frac{2P_1V_1G^2}{g_c}\left[\frac{fL}{2D} + \ln\left(\frac{P_1}{P_2}\right)\right]. \tag{6.76}$$

In terms of a density, ρ_m, at the average pressure in the line,

$$P_2 = P_1 - \frac{fG^2L}{2g_cD\rho_m}. \tag{6.77}$$

The average density may be found with the aid of an approximate evaluation of P_2 based on the inlet density; a second trial is never justified. Eqs. (6.76) and (6.77) and the approximation of Eq. (6.76) obtained by neglecting the logarithmic term are compared in Example 6.12. The restriction to ideal gases is removed in Section 6.7.4.

ADIABATIC FLOW

The starting point for development of the integrated adiabatic flow energy balance is Eq. (6.71), and again ideal gas behavior will be assumed. The equation of condition of a static adiabatic process, $PV^k = \text{const}$, is not applicable to the flow process; the appropriate

---

EXAMPLE 6.11
Pressure Drop in Power-Law and Bingham Flow

A limestone slurry of density 1.693 g/mL is pumped through a 4-in. (152 mm) line at the rate of 4 ft/sec (1.22 m/sec). The pressure drop (psi/mile) will be calculated. The slurry behavior is represented by

a. The power-law with $n = 0.165$ and $K = 34.3\ \text{dyn sec}^{0.165}/\text{cm}^2$ ($3.43\ \text{Pa sec}^{0.165}$).
b. Bingham model with $\tau_0 = 53\ \text{dyn/cm}^2$ (5.3 Pa) and $\mu_B = 22$ cP (0.022 Pa sec).

Power law:

$$\begin{aligned} \text{Re}' &= D^nV^{2-n}\rho/8^{n-1}K \\ &= (0.152)^{0.165}(1.22)^{1.835}(1693)(8)^{0.835}/3.43 \\ &= 2957, \\ f &= 0.0058 \quad [\text{Fig. 6.6(a)}] \end{aligned}$$

$$\begin{aligned} \frac{\Delta P}{L} &= \frac{4f\rho V^2}{2g_cD} = \frac{4(0.0058)(1693)(1.22)^2}{2(0.152)} \\ &= 192.3\ \text{N/(m}^2\text{)(m)} \quad [g_c = \text{kgm/sec}^2/\text{N}], \\ &\rightarrow 192.3(14.7/101{,}250)1610 = 45.0\ \text{psi/mile}. \end{aligned}$$

Bingham:

$$\begin{aligned} \text{Re}_B &= \frac{DV\rho}{\mu_B} = \frac{0.152(1.22)(1693)}{0.022} = 14{,}270, \\ \text{He} &= \tau_0D^2\rho/\mu_B^2 = 5.3(0.152)^2(1693)/(0.022)^2 \\ &= 428{,}000, \\ \text{critical Re}_B &= 12{,}000 \quad (\text{Fig. 6.5}), \\ f &= 0.007 \quad [\text{Fig. 6.6(b)}], \\ \frac{\Delta P}{L} &= \frac{0.007}{0.0058}45.0 = 54.3\ \text{psi/mile}. \end{aligned}$$

one is obtained as follows. Begin with

$$dH = -d\left(\frac{u^2}{2g_c}\right) = \frac{G^2V\,dV}{g_c} \tag{6.78}$$

$$= C_p\,dT = \frac{Rk}{k-1}dT = \frac{k}{k-1}d(PV), \tag{6.79}$$

from which

$$d(PV) = \left(\frac{k-1}{k}\right)\frac{G^2}{g_c}V\,dV, \tag{6.80}$$

and the integral is

$$PV = P_1V_1 - \left(\frac{k-1}{k}\right)\frac{G^2}{2g_c}(V^2 - V_1^2). \tag{6.81}$$

Also

$$V\,dP = d(PV) - (PV)\frac{dV}{V} \tag{6.82}$$

Substitutions into Eq. (6.71) result in

$$d(PV) - PV\frac{dV}{V} + \frac{G^2}{g_c}V\,dV + \frac{fG^2}{2g_cD}dL = 0. \tag{6.83}$$

Further substitutions from Eqs. (6.80) and (6.81) and multiplying through by $2kg_c/G^2V^2$ result in

$$2\frac{dV}{V} - \left[\frac{2kg_cP_1V_1}{G^2} + (k-1)V_1^2\right]\frac{dV}{V^3} + (k-1)\frac{dV}{V} + \frac{kf}{D}dL = 0. \tag{6.84}$$

Integrating from V_1 to V_2 and $L = 0$ to L gives

$$(k+1)\ln\frac{V_2}{V_1} + \frac{1}{2}\left[\frac{2kg_cP_1V_1}{G^2} + (k-1)V_1^2\right]\left(\frac{1}{V_2^2} - \frac{1}{V_1^2}\right) + \frac{kfL}{D} = 0 \tag{6.85}$$

or

$$\frac{fL}{D} = \frac{1}{2k}\left[\frac{2kg_cP_1V_1}{G^2V_1^2} + (k-1)\right]\left[1 - \left(\frac{V_1}{V_2}\right)^2\right] + \frac{k+1}{2k}\ln\left(\frac{V_1}{V_2}\right)^2. \tag{6.86}$$

In terms of the inlet Mach number,

$$M_1 = u_1/\sqrt{g_ckRT/M_w} = GV_1/\sqrt{g_ckRT/M_w}, \tag{6.87}$$

the result becomes

$$\frac{fL}{D} = \frac{1}{2k}\left(k - 1 + \frac{2}{M_1^2}\right)\left[1 - \left(\frac{V_1}{V_2}\right)^2\right] + \frac{k+1}{2k}\ln\left(\frac{V_1}{V_2}\right)^2. \tag{6.88}$$

When everything else is specified, Eqs. (6.86) or (6.88) may be solved for the exit specific volume V_2. Then P_2 may be found from Eq. (6.81) or in the rearrangement

$$\frac{P_2V_2}{P_1V_1} = \frac{T_2}{T_1} = 1 + \left(\frac{k-1}{2k}M_1^2\right)\left[1 - \left(\frac{V_2}{V_1}\right)^2\right], \tag{6.89}$$

from which the outlet temperature likewise may be found.

Although the key equations are transcendental, they are readily solvable with hand calculators, particularly those with root-solving provisions. Several charts to ease the solutions before the age of calculators have been devised: M.B. Powley, *Can. J. Chem. Eng.*, 241–245 (Dec. 1958); C.E. Lapple, reproduced in *Perry's Chemical Engineers' Handbook*, McGraw-Hill, New York, 1973, p. 5.27; O. Levenspiel, reproduced in *Perry's Chemical Engineers' Handbook*, McGraw-Hill, New York, 1984, p. 5.31; Hougen, Watson, and Ragatz, *Thermodynamics*, Wiley, New York, 1959, pp. 710–711.

In all compressible fluid pressure drop calculations it is usually justifiable to evaluate the friction factor at the inlet conditions and to assume it constant. The variation because of the effect of temperature change on the viscosity and hence on the Reynolds number, at the usual high Reynolds numbers, is rarely appreciable.

NONIDEAL GASES

Without the assumption of gas ideality, Eq. (6.71) is

$$\frac{dP}{V} + \frac{G^2}{g_c}\frac{dV}{V} + \frac{fG^2}{2g_cD}dL = 0. \tag{6.90}$$

In the isothermal case, any appropriate PVT equation of state may be used to eliminate either P or V from this equation and thus permit integration. Since most of the useful equations of state are pressure-explicit, it is simpler to eliminate P. Take the example of one of the simplest of the non-ideal equations, that of van der Waals

$$P = \frac{RT}{V-b} - \frac{a}{V^2}, \tag{6.91}$$

of which the differential is

$$dP = \left(-\frac{RT}{(V-b)^2} + \frac{2a}{V^3}\right)dV. \tag{6.92}$$

Substituting into Eq. (6.90),

$$\left(-\frac{RT}{(V-b)^2} + \frac{2a}{V^3} + \frac{G^2}{g_c}\right)\frac{dV}{V} + \frac{fG^2}{2g_cD}dL = 0. \tag{6.93}$$

Although integration is possible in closed form, it may be more convenient to perform the integration numerically. With more accurate and necessarily more complicated equations of state, numerical integration will be mandatory. Example 6.13 employs the van der Waals equation of steam, although this is not a particularly suitable one; the results show a substantial difference between the ideal and the nonideal pressure drops. At the inlet condition, the compressibility factor of steam is $z = PV/RT = 0.88$, a substantial deviation from ideality.

6.8. LIQUID–GAS FLOW IN PIPELINES

In flow of mixtures of the two phases in pipelines, the liquid tends to wet the wall and the gas to concentrate in the center of the channel, but various degrees of dispersion of each phase in the other may exist, depending on operating conditions, particularly the individual flow rates. The main patterns of flow that have been recognized are indicated on Figures 6.7(a) and (b). The ranges of conditions over which individual patterns exist are represented on maps like those of Figures 6.7(c) and (d). Since the concept of a

EXAMPLE 6.12
Adiabatic and Isothermal Flow of a Gas in a Pipeline
Steam at the rate of 7000 kg/hr with an inlet pressure of 23.2 barabs and temperature of 220°C flows in a line that is 77.7 mm dia and 305 m long. Viscosity is $28.5(10^{-6})$N sec/m^2 and specific heat ratio is $k = 1.31$. For the pipe, $\varepsilon/D = 0.0006$. The pressure drop will be found in (a) isothermal flow; (b) adiabatic flow. Also, (c) the line diameter for sonic flow will be found.

$$V_1 = 0.0862\ \mathrm{m^3/kg},$$
$$G = 7000/(3600)(\pi/4)(0.0777)^2 = 410.07\ \mathrm{kg/m^2\,sec},$$
$$\mathrm{Re}_1 = \frac{DG}{\mu} = \frac{0.0777(410.07)}{28.5(10^{-6})} = 1.12(10^6),$$
$$f = 1.6364/[\ln(0.135)(0.0006) + 6.5/1.2(10^6)]^2 = 0.0187.$$

Inlet sonic velocity:

$$u_{s1} = \sqrt{g_c kRT_1/M_w} = \sqrt{1(1.31)(8314)493.2/18.02} = 546\ \mathrm{m/sec}$$
$$M_1 = u_1/u_{s1} = GV_1/u_{s1} = 410.07(0.0862)/546 = 0.0647.$$

As a preliminary calculation, the pressure drop will be found by neglecting any changes in density:

$$\Delta P = \frac{fG^2L}{2g_c D\rho} = \frac{0.0187(410.07)^2(305)}{2(1)(0.0777)(1/0.0862)} = 5.32(10^5)\ \mathrm{N/m^2},$$
$$\therefore P_2 = 23.2 - 5.32 = 17.88\ \mathrm{bar}.$$

(a) Isothermal flow. Use Eq. (6.76):

$$\frac{2P_1V_1G^2}{g_c} = 2(23.2)(10^5)(0.0862)(410.07)^2 = 6.726(10^{10}),$$
$$P_2 = \left[P_1^2 - \frac{2P_1V_{1G}^2}{g_c}\left(\frac{fL}{2D} + \ln\frac{P_1}{P_2}\right)\right]^{1/2}$$
$$= 10^5\sqrt{538.24 - 6.726\left(0.0187(305)/2(0.0777) + \ln\frac{23.2(10^5)}{P_2}\right)}$$
$$= 17.13(10^5)\ \mathrm{N/m^2},$$

and

$$\Delta P = 23.2 - 17.13 = 5.07\ \mathrm{bar}.$$

When the logarithmic term is neglected,

$$P_2 = 17.07(10)^5\ \mathrm{N/m^2}.$$

(b) Adiabatic flow. Use Eq. (6.88):

$$\frac{fL}{D} = \frac{1}{2k}\left(k - 1 + \frac{2}{M_1^2}\right)\left[1 - \left(\frac{V_1}{V_2}\right)^2\right] + \frac{k+1}{2k}\ln\left(\frac{V_1}{V_2}\right)^2, \qquad (1)$$
$$\frac{0.0187(305)}{0.0777} = \frac{1}{2.62}\left(0.31 + \frac{2}{0.0647^2}\right)$$
$$\times\left[1 - \left(\frac{V_1}{V_2}\right)^2\right] + \frac{2.31}{2.62}\ln\left(\frac{V_1}{V_2}\right)^2,$$
$$73.4 = 182.47\left[1 - \left(\frac{V_1}{V_2}\right)^2\right] + 0.8817\ln\left(\frac{V_1}{V_2}\right)^2,$$
$$\therefore \frac{V_1}{V_2} = 0.7715.$$

Equation (6.89) for the pressure:

$$\frac{P_2V_2}{P_1V_1} = \frac{T_2}{T_1} = \left[1 + \frac{(k-1)}{2k}M_1^2\right]\left[1 - \left(\frac{V_2}{V_1}\right)^2\right]$$
$$= 1 + \frac{0.31(0.0647)^2}{2.62}[1 - (1.2962)^2]$$
$$= 0.9997,$$
$$P_2 = 0.9997P_1\left(\frac{V_1}{V_2}\right) = 0.9997(23.2)(10^5)(0.7715)$$
$$= 17.89(10^5)\ \mathrm{N/m^2},$$
$$\Delta P = 23.2 - 17.89 = 5.31\ \mathrm{bar}.$$

(c) Line diameter for sonic flow. The critical pressure ratio is

$$\frac{P_2}{P_1} = \left(\frac{2}{k+1}\right)^{k/(k-1)} = 0.5439, \quad \text{with } k = 1.31,$$
$$G = \frac{7000/3600}{(\pi/4)D^2} = \frac{2.4757}{D^2},$$
$$M_1 = \frac{GV_1}{U_{s_1}} = \frac{2.4757(0.0862)}{546D^2} = \frac{3.909(10^{-4})}{D^2}. \qquad (2)$$

Equation (6.89) becomes

$$0.5439(V_2/V_1) = 1 + 0.1183M_1^2[1 - (V_2/V_1)^2], \qquad (3)$$
$$fL/D = 0.0187(305)/D = 5.7035/D \qquad (4)$$
$$= \text{rhs of Eq. (6.88)}.$$

Procedure

1. Assume D.
2. Find M_1 [Eq. (2)].
3. Find V_2/V_1 from Eq. (6.89) [Eq. (3)].
4. Find rhs of Eq. (6.88) [Eq. (1)].
5. Find $D = 5.7035/$[rhs of Eq. (6.88)] [Eq. (4)].
6. Continue until steps 1 and 5 agree.

Some trials are:

D	M_1	Eq. (6.89) V_1/V_2	Eq. (6.88) rhs	D
0.06	0.1086	0.5457	44.482	0.1282
0.07	0.0798	0.5449	83.344	0.06843
0.0697	0.08046	0.5449	81.908	0.06963

$\therefore D = 0.0697$ m.

```
10  ! Example 6.12. Line dia for
      sonic flow
20  K=1.31
30  INPUT D ! (Trial value)
40  M=.0003909/D^2 ! (Eq 2)
50  INPUT V ! (=V1/V2)
60  GOSUB 130
70  IF ABS(X1)>=.0001 THEN 50
80  F=1/2/K*(K-1+2/M^2)*(1-V^2)+
      (K+1)/2/K*LOG(V^2) ! (Eq 1)
90  D1=5.7035/F
100 DISP D,D1
110 GOTO 30 ! (For another trial
      value of D if it is not clo
      se enough to calculated D1)
120 END
130 X1=-(.5439/V)+1+(K-1)/2/K*M^
      2*(1-1/V^2)
140 DISP X1
150 RETURN
```

particular flow pattern is subjective and all the pertinent variables apparently have not yet been correlated, boundaries between regions are fuzzy, as in (d).

It is to be expected that the kind of phase distribution will affect such phenomena as heat transfer and friction in pipelines. For the most part, however, these operations have not been correlated yet with flow patterns, and the majority of calculations of two-phase flow are made without reference to them. A partial exception is annular flow which tends to exist at high gas flow rates and has been studied in some detail from the point of view of friction and heat transfer.

The usual procedure for evaluating two-phase pressure drop is to combine pressure drops of individual phases in some way. To this end, multipliers ϕ_i are defined by

$$(\Delta P/L)_{\text{two-phase}} = \phi_i^2 (\Delta P/L)_i. \tag{6.94}$$

In the following table, subscript L refers to the liquid phase, G to the gas phase, and $L0$ to the total flow but with properties of the liquid phase; x is the weight fraction of the vapor phase.

Subscript	Re	$\Delta P/L$	ϕ^2
G	DGx/μ_G	$f_G G^2 x^2/2g_c D\rho_G$	$(\Delta P/L)/(\Delta P/L)_G$
L	$DG(1-x)/\mu_L$	$f_L G^2(1-x)^2/2g_c D\rho_L$	$(\Delta P/L)/(\Delta P/L)_L$
LO	DG/μ_L	$f_{LO} G^2/2g_c D\rho_L$	$(\Delta P/L)/(\Delta P/L)_{LO}$

In view of the many other uncertainties of two phase flow correlations, the friction factors are adequately represented by

$$f = \begin{cases} 64/\text{Re}, & \text{Re} < 2000, \text{ Poiseuille equation}, & (6.95) \\ 0.32/\text{Re}^{0.25}, & \text{Re} > 2000, \text{ Blasius equation}. & (6.96) \end{cases}$$

HOMOGENEOUS MODEL

The simplest way to compute line friction in two-phase flow is to adopt some kinds of mean properties of the mixtures and to employ the single phase friction equation. The main problem is the assignment of a two-phase viscosity. Of the number of definitions that have been proposed, that of McAdams et al. [*Trans.* ASME **64,** 193–200 (1942)] is popular:

$$1/\mu_{\text{two-phase}} = x/\mu_G + (1-x)/\mu_L. \tag{6.97}$$

The specific volumes are weight fraction additive,

$$V_{\text{two-phase}} = xV_G + (1-x)V_L \tag{6.98}$$

so that

$$1/\rho_{\text{two-phase}} = x/\rho_G + (1-x)/\rho_L, \tag{6.99}$$

where x is the weight fraction of the gas. Pressure drops by this method tend to be underestimated, but are more nearly accurate at higher pressures and higher flow rates.

With the Blasius equation (6.96), the friction factor and the pressure gradient become, with this model,

$$f = \frac{0.32}{(DG)^{0.25}}\left(\frac{x}{\mu_g} + \frac{1-x}{\mu_L}\right)^{0.25}, \tag{6.100}$$

$$\frac{\Delta P}{L} = \frac{fG^2}{2g_c D[x/\rho_G + (1-x)/\rho_L]}. \tag{6.101}$$

A particularly simple expression is obtained for the multiplier in terms of the Blasius equation:

$$\phi_{L0}^2 = \frac{\Delta P/L}{(\Delta P/L)_{L0}} = \frac{1 - x + x\rho_L/\rho_G}{(1 - x + x\mu_L/\mu_G)^{0.25}}. \tag{6.102}$$

Some values of ϕ_{L0}^2 from this equation for steam are

x	P = 0.689 bar	P = 10.3 bar
0.01	3.40	1.10
0.10	12.18	1.95
0.50	80.2	4.36

High values of multipliers are not uncommon.

EXAMPLE 6.13
Isothermal Flow of a Nonideal Gas

The case of Example 6.12 will be solved with a van der Waals equation of steam. From the *CRC Handbook of Chemistry and Physics* (CRC Press, Boca Raton, FL, 1979),

$$a = 5.464\ \text{atm}(\text{m}^3/\text{kg mol})^2 = 1703.7\ \text{Pa}(\text{m}^3/\text{kg})^2,$$
$$b = 0.03049\ \text{m}^3/\text{kg mol} = 0.001692\ \text{m}^3/\text{kg},$$
$$RT = 8314(493.2)/18.02 = 2.276(10^5)\ \text{N m/kg}.$$

Equation (6.93) becomes

$$\int_{0.0862}^{V_2}\left[\frac{-2.276(10^5)}{(V-0.00169)^2} + \frac{3407.4}{V^3} + (410.07)^2\right]\frac{dV}{V} + \frac{0.0187(410.07)^2(305)}{2(0.0777)} = 0,$$

$$\phi = \int_{0.0862}^{V_2}\left[\frac{-0.0369}{(V-0.00169)^2} + \frac{5.52(10^{-4})}{V^3} + 0.0272\right]\frac{dV}{V} + 1 = 0$$

The integration is performed with Simpson's rule with 20 intervals. Values of V_2 are assumed until one is found that makes $\phi = 0$. Then the pressure is found from the v dW equation:

$$P_2 = \frac{2.276(10^5)}{(V_2 - 0.00169)} - \frac{1703.7}{V_2^2}$$

Two trials and, a linear interpolation are shown. The value $P_2 = 18.44$ bar compares with the ideal gas 17.13.

V_2	ϕ	P_2
0.120	−0.0540	
0.117	+0.0054	
0.1173	0	18.44 bar

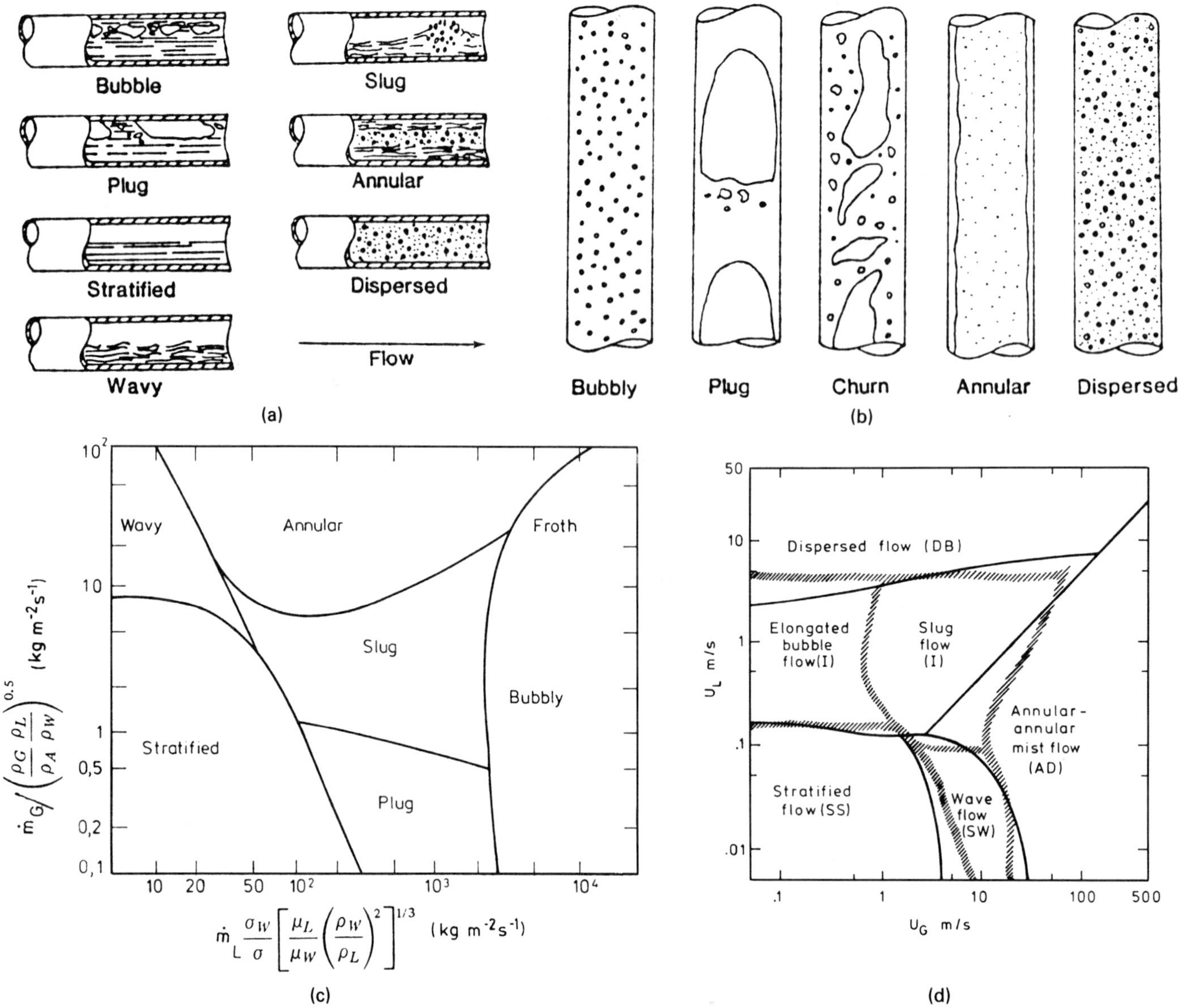

Figure 6.7. Flow patterns and correlations of flow regimes of liquid–gas mixtures in pipelines. (a) Patterns in horizontal liquid–gas flow. (b) Patterns in vertical liquid–gas flow. (c) Correlations of ranges of flow patterns according to Baker [*Oil Gas J.* **53**(*12*), *185* (*1954*)], as replotted by Bell et al. [Chem. Eng. Prog. Symp. Ser. **66,** *159* (*1969*)]; σ is surface tension of the liquid, and σ_w that of water. (d) Flow regimes of water/air at 25°C and 1 atm [*Taitel and Dukler,* AIChE J. **22,** *47* (*1976*)]; the fuzzy boundaries are due to Mandhane et al. [Int. J. Two-Phase Flow **1,** *537* (*1974*)].

SEPARATED FLOW MODELS

Pressure drop in two-phase flow is found in terms of pressure drops of the individual phases with empirical multipliers. The basic relation is

$$(\Delta P/L)_{\text{two-phase}} = \phi_G^2(\Delta P/L)_G = \phi_L^2(\Delta P/L)_L = \phi_{L0}^2(\Delta P/L)_{L0}. \tag{6.103}$$

The last term is the pressure drop calculated on the assumption that the total mass flow has the properties of the liquid phase.

Some correlations of multipliers are listed in Table 6.8. Lockhart and Martinelli distinguish between the various combinations of turbulent and laminar (viscous) flows of the individual phases; in this work the transition Reynolds number is taken as 1000 instead of the usual 2000 or so because the phases are recognized to disturb each other. Item 1 of Table 6.8 is a guide to the applicability of the Lockhart–Martinelli method, which is the oldest, and two more recent methods. An indication of the attention that has been devoted to experimentation with two phase flow is the fact that Friedel (1979) based his correlation on some 25,000 data points.

Example 6.14 compares the homogeneous and Lockhart–Martinelli models for the flow of a mixture of oil and hydrogen.

OTHER ASPECTS

The pattern of annular flow tends to form at higher gas velocities; the substantial amount of work done on this topic is reviewed by

TABLE 6.8. Two-Phase Flow Correlations of Pressure Drop

1. Recommendations

μ_L/μ_G	G (kg/m^2 sec)	Correlation
<1000	all	Friedel
>1000	>100	Chisholm–Baroczy
>1000	<100	Lockhart–Martinelli

2. Lockhart-Martinelli Correlation

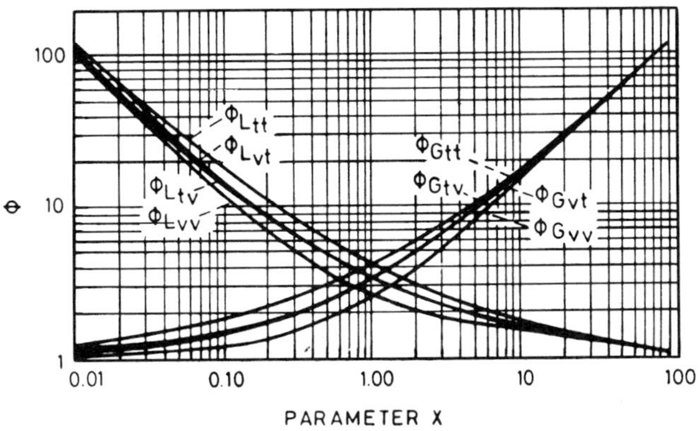

$\phi_L^2 = 1 + C/X + 1/X^2$
$\phi_G^2 = 1 + CX + X^2$
$X^2 = (\Delta P/L)_L/(\Delta P/L)_G$

Liquid	Gas	Subscript	C
Turbulent	Turbulent	*tt*	20
Viscous	Turbulent	*vt*	12
Turbulent	Viscous	*tv*	10
Viscous	Viscous	*vv*	5

3. Chisholm-Baroczy Correlation

$\phi_{LO}^2 = 1 + (Y^2 - 1)[Bx^{(2-n)/2}(1-x)^{(2-n)/2} + x^{2-n}] = (\Delta P/L)/(\Delta P/L)_{LO}$
$n = 0.25$
$Y^2 = (\Delta P/L)_{GO}/(\Delta P/L)_{LO}$
$B = 55/G^{0.5}, \quad 0 < Y < 9.5$
$= 520/YG^{0.5}, \quad 9.5 < Y < 28$
$= 15{,}000/Y^2G^{0.5}, \quad Y > 28$
x = weight fraction gas

4. Friedel Correlation

$$\phi_{LO}^2 = E + \frac{3.24FH}{\mathrm{Fr}^{0.045}\mathrm{We}^{0.035}}, \quad \mathrm{Fr} = G^2/g_c D\rho_{TP}^2$$

$$E = (1-x)^2 + x^2\frac{\rho_L f_{GO}}{\rho_G f_{LO}}, \quad \mathrm{We} = G^2D/\rho_{TP}\sigma$$

$$F = x^{0.78}(1-x)^{0.24}, \quad \rho_{TP} = \left(\frac{x}{\rho_G} + \frac{1-x}{\rho_L}\right)^{-1}$$

$$H = \left(\frac{\rho_L}{\rho_G}\right)^{0.91}\left(\frac{\mu_G}{\mu_L}\right)^{0.19}\left(1 - \frac{\mu_G}{\mu_L}\right)^{0.7}, \quad x = \text{weight fraction gas}$$

1. (P.B. Whalley, cited by G.F. Hewitt, 1982). 2. [Lockhart and Martinelli, *Chem. Eng. Prog.* **45,** 39–48 (1949); Chisholm, *Int. J. Heat Mass Transfer* **10,** 1767–1778 (1967)]. 3. [Chisholm, *Int. J. Heat Mass Transfer* **16,** 347–348 (1973); Baroczy, *Chem. Eng. Prog. Symp. Ser.* **62,** 217–225 (1965)]. 4. (Friedel, European Two Phase Flow Group Meeting, Ispra, Italy, Paper E2, 1979, cited by G.F. Hewitt, 1982).

EXAMPLE 6.14
Pressure Drop and Void Fraction in Liquid–Gas Flow
A mixture of an oil and hydrogen at 500 psia and 200°F enters a 3 in. Schedule 40 steel line. Data are:

Oil: 140,000 lb/hr, 51.85 lb/cuft, 2700 cfh, viscosity 15 cP.

Hydrogen: 800 lb/hr, 0.142 lb/cuft, 5619 cfh, viscosity $2.5(10^{-7})$ lbf sec/sqft.

The pressure drop in 100 ft of line will be found, and also the voidage at the inlet condition.

$$\mathrm{Re}_L = \frac{4\dot{m}}{\pi D g_c \mu} = \frac{4(140{,}000/3600)}{\pi(0.2557)(32.2)0.15},$$

$$\mathrm{Re}_G = \frac{4(800/3600)}{\pi(0.2557)(32.2)(2.5)(10^{-7})} = 137{,}500,$$

$$\frac{\varepsilon}{D} = 0.00059.$$

Round equations:

$$f = \frac{1.6434}{[\ln(0.135\varepsilon/D + 6.5/\mathrm{Re}]^2} = \begin{cases} 0.0272, & \text{liquid,} \\ 0.0204, & \text{gas,} \end{cases}$$

$$(\Delta P/L)_L = \frac{8f\dot{m}^2}{\pi^2 g_c \rho D^5} = \frac{8(0.0272)(38.89)^2}{\pi^2(32.2)(51.85)(0.2557)^5}$$
$$= 18.27 \text{ psf/ft},$$

$$(\Delta P/L)_G = \frac{8(0.0204)(0.222)^2}{\pi^2(32.2)(0.142)(0.2557)^5} = 0.1663 \text{ psf/ft},$$

$$X^2 = 18.27/0.1633 = 111.8.$$

Lockhart–Martinelli–Chisholm:

$$c = 20 \text{ for TT regime (Table 6.8)},$$

$$\phi_L^2 = 1 + \frac{C}{X} + \frac{1}{X^2} = 2.90,$$

$$\therefore (\Delta P/L) \text{ two phase} = \phi_L^2(\Delta P/L)_L = 2.90(18.27)$$
$$= 53.0 \text{ psf/ft}, \quad 36.8 \text{ psi/100 ft}.$$

Check with the *homogeneous model:*

$$x = \frac{800}{140{,}000 + 800} = 0.0057 \text{ wt fraction gas},$$

$$\mu = \left[\frac{0.0057}{2.5(10^{-7})} + \frac{0.9943}{3.13(10^{-4})}\right]^{-1} = 3.85(10^{-5}) \frac{\text{lbf sec}}{\text{sqft}},$$

$$\rho = \left[\frac{0.0057}{0.142} + \frac{0.9943}{51.85}\right]^{-1} = 16.86 \text{ lb/cuft},$$

$$\mathrm{Re} = \frac{4(39.11)}{\pi(32.2)(0.2557)3.85(10^{-5})} = 157{,}100$$

$$f = 0.0202,$$

$$\frac{\Delta P}{L} = \frac{8(0.0202)(39.11)^2}{\pi^2(32.2)(16.86)(0.2557)^5} = 42.2 \text{ psf/ft},$$

compared with 53.0 by the LMC method.
Void fraction by Eq. (6.104):

$$\varepsilon_G = 1 - 1/\phi_L = 1 - 1/\sqrt{2.90} = 0.413,$$

compared with input flow condition of

$$\varepsilon = \frac{Q_G}{Q_G + Q_L} = \frac{5619}{5619 + 2700} = 0.675.$$

Method of Premoli [Eqs. (6.105) and (6.106)]:
Surface tension $\sigma = 20$ dyn/cm, 0.00137 lbf/ft,

$$\mathrm{We} = \frac{DG^2}{g_c \rho_L \sigma} = \frac{16\dot{m}^2}{\pi^2 g_c D^3 \rho_L \sigma}$$
$$= \frac{16(38.89)^2}{\pi^2(32.2)(0.2557)^3(51.85)(0.00137)} = 64{,}118,$$

$$\mathrm{Re} = 19{,}196,$$

$$E_1 = 1.578(19196)^{-0.19}(51.85/0.142)^{0.22} = 0.8872,$$

$$E_2 = 0.0273(6411.8)(19196)^{-0.51}(51.85/0.142)^{-0.08} = 7.140,$$

$$y = 5619/2700 = 2.081,$$

$$yE_2 = 2.081(7.140) = 14.86.$$

Clearly, this term must be less than unity if Eq. (6.105a) for S is to be valid, so that equation is not applicable to this problem as it stands. If yE_2 is replaced by $y/E_2 = 0.2914$, then

$$S = 1 + 0.8872\left(\frac{2.081}{1.2914} - 0.2914\right)^{0.5} = 2.02,$$

and the voidage is

$$\varepsilon = \frac{5619}{5619 + 2.02(2700)} = 0.51,$$

which is a plausible result. However, Eqs. (6.105) and (6.105a) are quoted correctly from the original paper; no numerical examples are given there.

Hewitt (1982). A procedure for stratified flow is given by Cheremisinoff and Davis [*AIChE J.* **25**, 1 (1979)].

Voidage of the holdup in the line is different from that given by the proportions of the incoming volumetric flows of the two phases, but is of course related to it. Lockhart and Martinelli's work indicates that the fractional gas volume is

$$\varepsilon = 1 - 1/\phi_L, \tag{6.104}$$

where ϕ_L is defined in Table 6.8. This relation has been found to give high values. A correlation of Premoli et al. [*Termotecnica* **25,** 17–26 (1971); cited by Hewitt, 1982] gives the void fraction in terms of the incoming volumetric flow rates by the equation

$$\varepsilon_G = Q_G/(Q_G + SQ_L), \tag{6.105}$$

where S is given by the series of equations

$$\begin{aligned} &S = 1 + E_1[y/(1 + yE_2) - yE_2]^{1/2}, \\ &E_1 = 1.578\,\mathrm{Re}^{-0.19}(\rho_L/\rho_G)^{0.22}, \\ &E_2 = 0.0273\,\mathrm{We}\,\mathrm{Re}^{-0.51}(\rho_L/\rho_G)^{-0.08}, \\ &y = Q_G/Q_L, \quad \mathrm{Re} = DG/\mu_L, \quad \mathrm{We} = DG^2/\sigma\rho_L. \end{aligned} \tag{6.105'}$$

Direct application of these equations in Example 6.14 is not successful, but if E_2 is taken as the reciprocal of the given expression, a plausible result is obtained.

6.9. GRANULAR AND PACKED BEDS

Flow through granular and packed beds occurs in reactors with solid catalysts, adsorbers, ion exchangers, filters, and mass transfer equipment. The particles may be more or less rounded or may be shaped into rings, saddles, or other structures that provide a desirable ratio of surface and void volume.

Natural porous media may be consolidated (solids with holes in them), or they may consist of unconsolidated, discrete particles. Passages through the beds may be characterized by the properties of porosity, permeability, tortuosity, and connectivity. The flow of underground water and the production of natural gas and crude oil, for example, are affected by these characteristics. The theory and properties of such structures is described, for instance, in the book of Dullien (*Porous Media, Fluid Transport and Pore Structure,* Academic, New York, 1979). A few examples of porosity and permeability are in Table 6.9. Permeability is the proportionality constant k in the flow equation $u = (k/\mu)\, dP/dL$.

Although consolidated porous media are of importance in chemical engineering, only unconsolidated porous media are incorporated in process equipment, so that further attention will be restricted to them.

Granular beds may consist of mixtures of particles of several sizes. In flow problems, the mean surface diameter is the appropriate mean, given in terms of the weight fraction distribution, x_i, by

$$D_p = 1/(\sum x_i/D_i). \qquad (6.106)$$

When a particle is not spherical, its characteristic diameter is taken as that of a sphere with the same volume, so that

$$D_p = (6V_p/\pi)^{1/3}. \qquad (6.107)$$

SINGLE PHASE FLUIDS

Extensive measurements of flow in and other properties of beds of particles of various shapes, sizes and compositions are reported by Leva et al. (1951). Differences in voidage are pronounced as Figure 6.8(c) shows.

A long-established correlation of the friction factor is that of Ergun (*Chem. Eng. Prog.* **48,** 89–94, 1952). The average deviation from his line is said to be ±20%. The friction factor is

$$f_p = \frac{g_c D_p \varepsilon^3}{u^2(1-\varepsilon)}\left(\frac{\Delta P}{L}\right) \qquad (6.108)$$

$$= 150/\mathrm{Re}_p + 1.75 \qquad (6.109)$$

with

$$\mathrm{Re}_p = D_p G/\mu(1-\varepsilon). \qquad (6.110)$$

TABLE 6.9. Porosity and Permeability of Several Unconsolidated and Consolidated Porous Media

Media	Porosity (%)	Permeability (cm^2)
Berl saddles	68–83	1.3×10^{-3}–3.9×10^{-3}
Wire crimps	68–76	3.8×10^{-5}–1.0×10^{-4}
Black slate powder	57–66	4.9×10^{-10}–1.2×10^{-9}
Silica powder	37–49	1.3×10^{-10}–5.1×10^{-10}
Sand (loose beds)	37–50	2.0×10^{-7}–1.8×10^{-6}
Soil	43–54	2.9×10^{-9}–1.4×10^{-7}
Sandstone (oil sand)	8–38	5.0×10^{-12}–3.0×10^{-8}
Limestone, dolomite	4–10	2.0×10^{-11}–4.5×10^{-10}
Brick	12–34	4.8×10^{-11}–2.2×10^{-9}
Concrete	2–7	1.0×10^{-9}–2.3×10^{-7}
Leather	56–59	9.5×10^{-10}–1.2×10^{-9}
Cork board	—	3.3×10^{-6}–1.5×10^{-5}
Hair felt	—	8.3×10^{-6}–1.2×10^{-5}
Fiberglass	88–93	2.4×10^{-7}–5.1×10^{-7}
Cigarette filters	17–49	1.1×10^{-5}
Agar-agar	—	2.0×10^{-10}–4.4×10^{-9}

(A.E. Scheidegger, *Physics of Flow through Porous Media,* University of Toronto Press, Toronto, Canada, 1974).

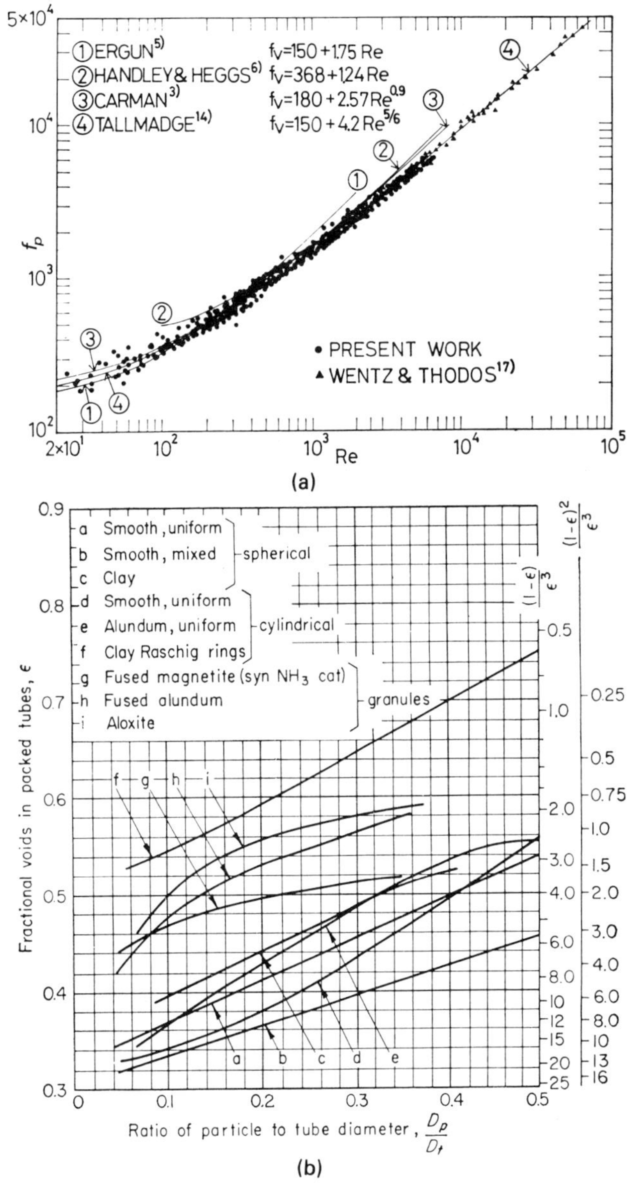

Figure 6.8. Friction factors and void fractions in flow of single phase fluids in granular beds. (a) Correlation of the two-phase friction factor, $\mathrm{Re} = D_p G/(1-\varepsilon)\mu$ and $f_p = [g_c D_p \varepsilon^3/\rho u^2(1-\varepsilon)](\Delta P/L) = 150/\mathrm{Re} + 4.2/(\mathrm{Re})^{1/6}$ [*Sato et al.,* J. Chem. Eng. Jpn. **6,** *147–152 (1973)*]. (b) Void fraction in granular beds as a function of the ratio of particle and tube diameters [*Leva, Weintraub, Grummer, Pollchik, and Storch,* U.S. Bur. Mines Bull. *504 (1951)*].

The pressure gradient accordingly is given by

$$\frac{\Delta P}{L} = \frac{G^2(1-\varepsilon)}{\rho g_c D_p \varepsilon^3}\left[\frac{150(1-\varepsilon)\mu}{D_P G} + 1.75\right]. \tag{6.111}$$

For example, when $D_p = 0.005$ m, $G = 50$ kg/m^2 sec, $g_c = 1$ kgm/N sec^2, $\rho = 800$ kg/m^3, $\mu = 0.010$ N sec/m^2, and $\varepsilon = 0.4$, the gradient is $\Delta P/L = 0.31(10^5)$ Pa/m.

An improved correlation is that of Sato (1973) and Tallmadge (*AIChE J.* **16,** 1092 (1970)] shown on Figure 6.8(a). The friction factor is

$$f_p = 150/\mathrm{Re}_p + 4.2/\mathrm{Re}_p^{1/6} \tag{6.112}$$

with the definitions of Eqs. (6.108) and (6.110). A comparison of Eqs. (6.109) and (6.112) is

Re_p	5	50	500	5000
f_p (Ergun)	31.8	4.80	2.05	1.78
f_p (Sato)	33.2	5.19	1.79	1.05

In the highly turbulent range the disagreement is substantial.

TWO-PHASE FLOW

Operation of packed trickle-bed catalytic reactors is with liquid and gas flow downward together, and of packed mass transfer equipment with gas flow upward and liquid flow down.

Concurrent flow of liquid and gas can be simulated by the homogeneous model of Section 6.8.1 and Eqs. 6.109 or 6.112, but several adequate correlations of separated flows in terms of Lockhart–Martinelli parameters of pipeline flow type are available. A number of them is cited by Shah (*Gas-Liquid-Solid Reactor Design,* McGraw-Hill, New York, 1979, p. 184). The correlation of Sato (1973) is shown on Figure 6.9 and is represented by either

$$\phi = (\Delta P_{LG}/\Delta P_L)^{0.5} = 1.30 + 1.85(X)^{-0.85}, \quad 0.1 < X < 20, \tag{6.113}$$

or

$$\log_{10}\left(\frac{\Delta P_{LG}}{\Delta P_L + \Delta P_G}\right) = \frac{0.70}{[\log_{10}(X/1.2)]^2 + 1.00}, \tag{6.114}$$

where

$$X = \sqrt{(\Delta P/L)_L/(\Delta P/L)_G}. \tag{6.115}$$

The pressure gradients for the liquid and vapor phases are calculated on the assumption of their individual flows through the bed, with the correlations of Eqs. (6.108)–(6.112).

The fraction h_L of the void space occupied by liquid also is of interest. In Sato's work this is given by

$$h_L = 0.40(a_s)^{1/3} X^{0.22}, \tag{6.116}$$

where the specific surface is

$$a_s = 6(1-\varepsilon)/D_p. \tag{6.117}$$

Additional data are included in the friction correlation of Specchia and Baldi [*Chem. Eng. Sci.* **32,** 515–523 (1977)], which is represented by

$$f_{LG} = \frac{g_c D_p \varepsilon}{3\rho_G u_G^2(1-\varepsilon)}\left(\frac{\Delta P}{L}\right), \tag{6.118}$$

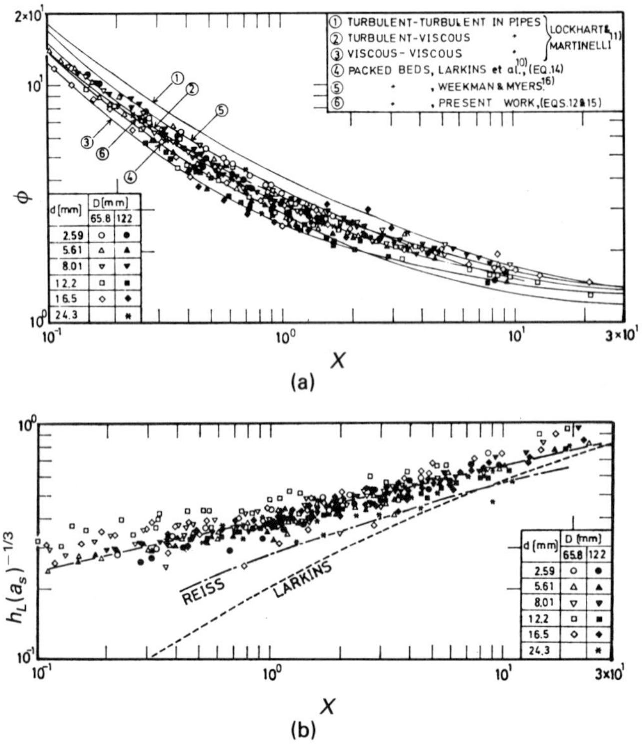

Figure 6.9. Pressure drop gradient and liquid holdup in liquid–gas concurrent flow in granular beds. [*Sato, Hirose, Takahashi, and Toda,* J. Chem. Eng. Jpn. **6,** *147–152 (1973)*]. (a) Correlation of the two phase pressure drop gradient $\Delta P/L$, $\phi = 1.30 + 1.85X^{-0.85}$. (b) Correlation of frictional holdup h_L of liquid in the bed; a_s is the specific surface, 1/mm, d is particle diameter, and D is tube diameter. $h_L = 0.4a_s^{1/3}X^{0.22}$.

$$\ln f_{LG} = 7.82 - 1.30\ln(Z/\psi^{1.1}) - 0.0573[\ln(Z/\psi^{1.1})]^2. \tag{6.119}$$

The parameters in Eq. (6.119) are

$$Z = (\mathrm{Re}_G)^{1.167}/(\mathrm{Re}_L)^{0.767}, \tag{6.120}$$

$$\psi = \frac{\sigma_w}{\sigma_L}\left[\frac{\mu_L}{\mu_w}\left(\frac{\rho_w}{\rho_L}\right)^2\right]^{1/3}. \tag{6.121}$$

Liquid holdup was correlated in this work for both nonfoaming and foaming liquids.

$$\text{Nonfoaming,} \quad h_L = 0.125(Z/\psi^{1.1})^{-0.312}(a_s D_p/\varepsilon)^{0.65}, \tag{6.122}$$

$$\text{Foaming,} \quad h_L = 0.06(Z/\psi^{1.1})^{-0.172}(a_s D_p/\varepsilon)^{0.65}. \tag{6.123}$$

The subscript w in Eq. (6.121) refers to water.

Countercurrent flow data in towers with shaped packings are represented by Figure 13.37. The pressure drop depends on the viscosity of the liquid and on the flow rates and densities of the liquid and gas, as well as on characteristics of the packing which are represented here by the packing factor F. Nominally, the packing factor is a function of the specific surface a_s and the voidage ε, as

$$F = a_s/\varepsilon^3, \tag{6.124}$$

but calculated values are lower than the experimental values shown in the table by factors of 2–5 or so. Clearly the liquid holdup reduces the effective voidage to different extents with different packings. The voidages of the packings in the table range from 70 to

95%, whereas voidages obtained with small spherical or cylindrical packings normally used as catalysts are less than 40% or so, which makes them impractical for countercurrent operation. However, catalysts are made in the forms of rings or saddles when very low pressure drop or countercurrent operation is desirable.

Even when they are nominally the same type and size, packings made by different manufacturers may differ substantially in their pressure drop and mass transfer behavior, so that manufacturers data should be obtained for final design.

Many data on individual packings are given by Billet (*Distillation Engineering*, Chemical Pub. Co., New York), in *Chemical Engineers Handbook* (McGraw-Hill, New York, 1984, p. 18.23) and with Figure 13.37.

The uppermost line of Figure 13.37(a) marks the onset of flooding which is the point at which sharp increase of pressure drop obtains on a plot against liquid rate. Flooding limits also are represented on Figure 13.36; in practice, it is customary to operate at a gas rate that is 70% of that given by the line, although there are many data points below this limit in this correlation.

Mesh or other open structures as vessel packing have attractive pressure drop and other characteristics, but each type has quite individual behavior so that it is best to consult their manufacturer's data.

6.10. GAS–SOLID TRANSFER

Equipment for pneumatic conveying is described in Section 5.2 along with some rules for calculating power requirements. Here the latter topic will be supplemented from a more fundamental point of view.

CHOKING VELOCITY

Although the phenomena are not clearcut, partial settling out of solids from the gas stream and other instabilities may develop below certain linear velocities of the gas called choking velocities. Normal pneumatic transport of solids accordingly is conducted above such a calculated rate by a factor of 2 or more because the best correlations are not more accurate. Above choking velocities the process is called dilute phase transport and, below, dense phase transport.

What appears to be the best correlation of choking velocities is due to Yang [*AIChE J.* **21**, 1013–1015 (1975)], supplemented by Punwani et al. and Yang (cited by Teo and Leung, 1984, pp. 520–521). The choking velocity U_{gc} and voidage ε_c are found by simultaneous solution of the equations

$$G_s/\rho_s = (U_{gc} - U_t)(1 - \varepsilon_c) \tag{6.125}$$

or

$$\varepsilon_c = 1 - G_s/\rho_s(U_{gc} - U_t) \tag{6.126}$$

and

$$gD(\varepsilon_c^{-4.7} - 1) = 3.41(10^5)(\rho_g/\rho_s)^{2.2}(U_{gc} - U_t)^2, \tag{6.127}$$

where G_s is the mass rate of flow of solid per unit cross section and the other terms are defined in Table 6.10. When ε_c from Eq. (6.126) is substituted into Eq. (6.127), the single unknown in that equation is readily found with a root solving routine. For the case of Example 6.15, $G_s = 29.6\ \text{kg/m}^2\ \text{sec}$, $U_t = 0.45\ \text{m/sec}$, $\rho_s = 1282\ \text{kg/m}^3$, and $\rho_g = 1.14\ \text{kg/m}^3$. Accordingly, $U_{gc} = 1.215\ \text{m/sec}$ and $\varepsilon_c = 0.9698$.

TABLE 6.10. Equations for the Calculation of Pressure Drop in Gas–Solid Transport

Solid Friction Factor f_s According to Various Investigators

Investigator	f_s	
Stemerding (1962)	0.003	(1)
Reddy and Pei (1969)	$0.046U_p^{-1}$	(2)
Van Swaaij, Buurman, and van Breugel (1970)	$0.080U_p^{-1}$	(3)
Capes and Nakamura (1973)	$0.048U_p^{-1.22}$	(4)
Konno and Saito (1969)	$0.0285\sqrt{gD}\,U_p^{-1}$	(5)
Yang (1978), vertical	$0.00315\dfrac{1-\varepsilon}{\varepsilon^3}\left[\dfrac{(1-\varepsilon)U_t}{U_f - U_p}\right]^{-0.979}$	(6)
Yang (1976), horizontal	$0.0293\dfrac{1-\varepsilon}{\varepsilon^3}\left[\dfrac{(1-\varepsilon)U_f}{\sqrt{gD}}\right]^{-1.15}$	(7)

Free Setting Velocity

$K = D_p\left[\dfrac{g\rho_f(\rho_p - \rho_f)}{\mu_f^2}\right]^{1/3}$		(8)
$U_{t(\text{Stokes})} = \dfrac{gD_p^2(\rho_p - \rho_f)}{18\mu_f}, \quad K < 3.3$	$K < 3.3$	(9)
$U_{t(\text{intermediate})} = \dfrac{0.153g^{0.71}D_p^{1.14}(\rho_p - \rho_f)^{0.71}}{\rho_f^{0.29}\mu_f^{0.43}},$	$3.3 < K < 43.6$	(10)
$U_{t(\text{Newton})} = 1.75\left(\dfrac{gD_p(\rho_p - \rho_f)}{\rho_f}\right)^{1/2},$	$43.6 < K < 2360$	(11)

Particle Velocity

Investigator	U_p	
Hinkle (1953)	$U_g - U_t$	(12)
IGT (1978)	$U_g(1 - 0.68D_p^{0.92}\rho_p^{0.5}\rho_f^{-0.2}D^{-0.54})$	(13)
Yang (1976)	$U_g - U_t\left[\left(1 + \dfrac{f_sU_p^2}{2gD}\right)^{\varepsilon^{4.7}}\right]^{1/2}$	(14)

Voidage

$$\varepsilon = 1 - 4\dot{m}_p/\pi D^2(\rho_p - \rho_f)U_p \tag{15}$$

Notation: U_f is a fluid velocity, U_p is particle velocity, U_t is particle free settling velocity, $\dot{m}_s$ is mass rate of flow of solid, D = pipe diameter, D_p is particle diameter, $g = 9.806\ \text{m/sec}^2$ at sea level.
(Klinzing, *Gas–Solid Transport*, McGraw-Hill, New York, 1981).

PRESSURE DROP

The relatively sparse data on dense phase transport is described by Klinzing (1981) and Teo and Leung (1984). Here only the more important category of dilute phase transport will be treated.

The pressure drop in simultaneous flow of gas and solid particles is made up of contributions from each of the phases. When the particles do not interact significantly, as in dilute transport, the overall pressure drop is represented by

$$\Delta P = \rho_p(1 - \varepsilon)Lg + \rho_f\varepsilon Lg + \frac{2f_g\rho_fU_f^2L}{D} + \frac{2f_s\rho_p(1 - \varepsilon)U_p^2L}{D} \tag{6.128}$$

for vertical transport; in horizontal transport only the two frictional terms will be present. The friction factor f_g for gas flow is the normal one for pipe flow; except for a factor of 4, it is given by Eq. (6.19) for laminar flow and by the Round equation (6.21) for turbulent flow. For the solid friction factor f_s, many equations of

EXAMPLE 6.15
Pressure Drop in Flow of Nitrogen and Powdered Coal

Powdered coal of 100 μm dia and 1.28 specific gravity is transported vertically through a 1-in. smooth line at the rate of 15 g/sec. The carrying gas is nitrogen at 1 atm and 25°C at a linear velocity of 6.1 m/sec. The density of the gas is 1.14 kg/m^3 and its viscosity is 1.7(10^{-5}) N sec/m^2. The equations of Table 6.10 will be used for the various parameters and ultimately the pressure gradient $\Delta P/L$ will be found:

$$\text{Eq. (8)},\quad K = 10^{-4}\left\{\frac{9.806(1.14)(1282-1.14)}{[1.7(10^{-5})]^2}\right\}^{1/3} = 3.67,$$

$$\text{Eq. (10)},\quad U_t = \frac{0.153(9.806)^{0.71}(0.0001)^{1.14}(1282-1.14)^{0.71}}{1.14^{0.29}[1.7(10^{-5})]^{0.43}}$$

$$= 0.37\text{ m/sec (0.41 m/sec by Stokes' law)},$$

$$\text{Eq. (15)},\quad \varepsilon = 1 - \frac{0.015}{(\pi/4)(0.0254)^2(1282-1.14)U_p} = 1 - \frac{0.0231}{U_p}, \quad \text{(I)}$$

$$\text{Eq. (14)},\quad U_p = 6.1 - 0.45\sqrt{1 + f_s U_p^2/2(9.806)(0.0254)}$$

$$= 6.1 - 0.45\sqrt{1 + 2.007 f_s U_p^2} \quad \text{(II)}$$

$$\text{Eq. (7)},\quad f_s = \frac{0.00315(1-\varepsilon)}{\varepsilon^3}\left[\frac{(1-E)0.45}{6.1-U_p}\right]^{-0.979} \quad \text{(III)}$$

Equations (I), (II), and (III) are solved simultaneously with the results:

$$\varepsilon = 0.9959 \quad \text{and} \quad U_p = 5.608,$$

For the calculation of the pressure drop,

$$f_s = 0.0031 \quad \text{(Yang equation)},$$

$$\mathrm{Re}_f = \frac{DU_f\rho_f}{\mu_f} = \frac{0.0254(6.1)(1.14)}{1.7(10^{-5})} = 10{,}390.$$

Therefore, Round's Eq. (6.21) applies:

$$f_f = \tfrac{1}{4} f_{\text{Round}} = 0.0076,$$

Eq. (6.128),

$$\Delta P/L = 9.806[1282(1-0.9959) + 1.14(0.9959)]$$
$$+ (2/0.0254)[0.0076(1.14)(6.1)^2$$
$$+ 0.0031(1282)(0.0041)(5.608)^2]$$
$$= 51.54 + 11.13 + 25.38 + 40.35 = 128.4\text{ Pa/m}.$$

With Eqs. (5) and (13), no trial calculations are needed.

$$\text{Eq. (13)},\quad U_p = 6.1[1 - 0.68(0.0001)^{0.92}(1282)^{0.5}$$
$$\times (1.14)^{-0.2}(0.0254)^{-0.54}]$$
$$= 5.88\text{ m/sec},$$

$$\text{Eq. (15)},\quad \varepsilon = 1 - 0.0231/5.78 = 0.9960,$$

$$\text{Eq. (5)},\quad f_s = 0.0285\sqrt{9.806(0.0254)}/5.88 = 0.00242.$$

Therefore, the solid frictional gradient is obtained from the calculated value 40.35 in the ratio of the friction factors.

$$(\Delta P/L)_{\text{solid friction}} = 40.35(0.00242/0.0031) = 31.5\text{ Pa/m}.$$

```
10 ! Example 6.15. Pressure dro
   p in flow of nitrogen and po
   wdered coal
20 INPUT U
30 E=1-.0231/U ! (Eq I)
40 F=.003151*(1-E)/E^3*(.45*(1-
   E)/(6.1-U))^-.979 ! (Eq III)
50 G=-U+6.1-.45*(1+2.007*F*U^2)
   ^.5 ! (should = 0)
60 PRINT "U=";U
70 PRINT "G=";G
80 GOTO 20 ! (if G is not suffi
   ciently close to zero)
90 END

U= 5.608
G=-.000089348061
```

varying complexity have been proposed, of which some important ones are listed in Table 6.10.

These equations involve the free settling velocity U_t, for which separate equations also are shown in the table. At lower velocities Stokes' law applies, but corrections must be made at higher ones. The particle velocity U_p is related to other quantities by Eqs. (12)–(14) of the table, and the voidage in turn is represented by Eq. (15). In a review of about 20 correlations, Modi et al. (*Proceedings, Powder and Bulk Solids Handling and Processing Conference,* Powder Advisory Center, Chicago, 1978, cited by Klinzing, 1981) concluded that the correlations of Konno and Saito (1969) and of Yang (1976, 1978) gave adequate representation of pneumatic conveying of coal. They are applied in Example 6.15 and give similar results there.

6.11. FLUIDIZATION OF BEDS OF PARTICLES WITH GASES

As the flow of fluid through a bed of solid particles increases, it eventually reaches a condition at which the particles are lifted out of permanent contact with each other. The onset of that condition is called minimum fluidization. Beyond this point the solid–fluid mass exhibits flow characteristics of ordinary fluids such as definite viscosity and flow through lines under the influence of hydrostatic head difference. The rapid movement of particles at immersed surfaces results in improved rates of heat transfer. Moreover, although heat transfer rate between particles and fluid is only moderate, 1–4 Btu/(hr)(sqft)(°F), the amount of surface is so great, 10,000–150,000 sqft/cuft, that temperature equilibration between phases is attained within a distance of a few particle diameters. Uniformity of temperature, rapid mass transfer, and rapid mixing of solids account for the great utility of fluidized beds in process applications.

As the gas flow rate increases beyond that at minimum fluidization, the bed may continue to expand and remain homogeneous for a time. At a fairly definite velocity, however, bubbles begin to form. Further increases in flow rate distribute themselves between the dense and bubble phases in some ways that are not well correlated. Extensive bubbling is undesirable when intimate contacting between phases is desired, as in drying processes or solid catalytic reactions. In order to permit bubble formation, the

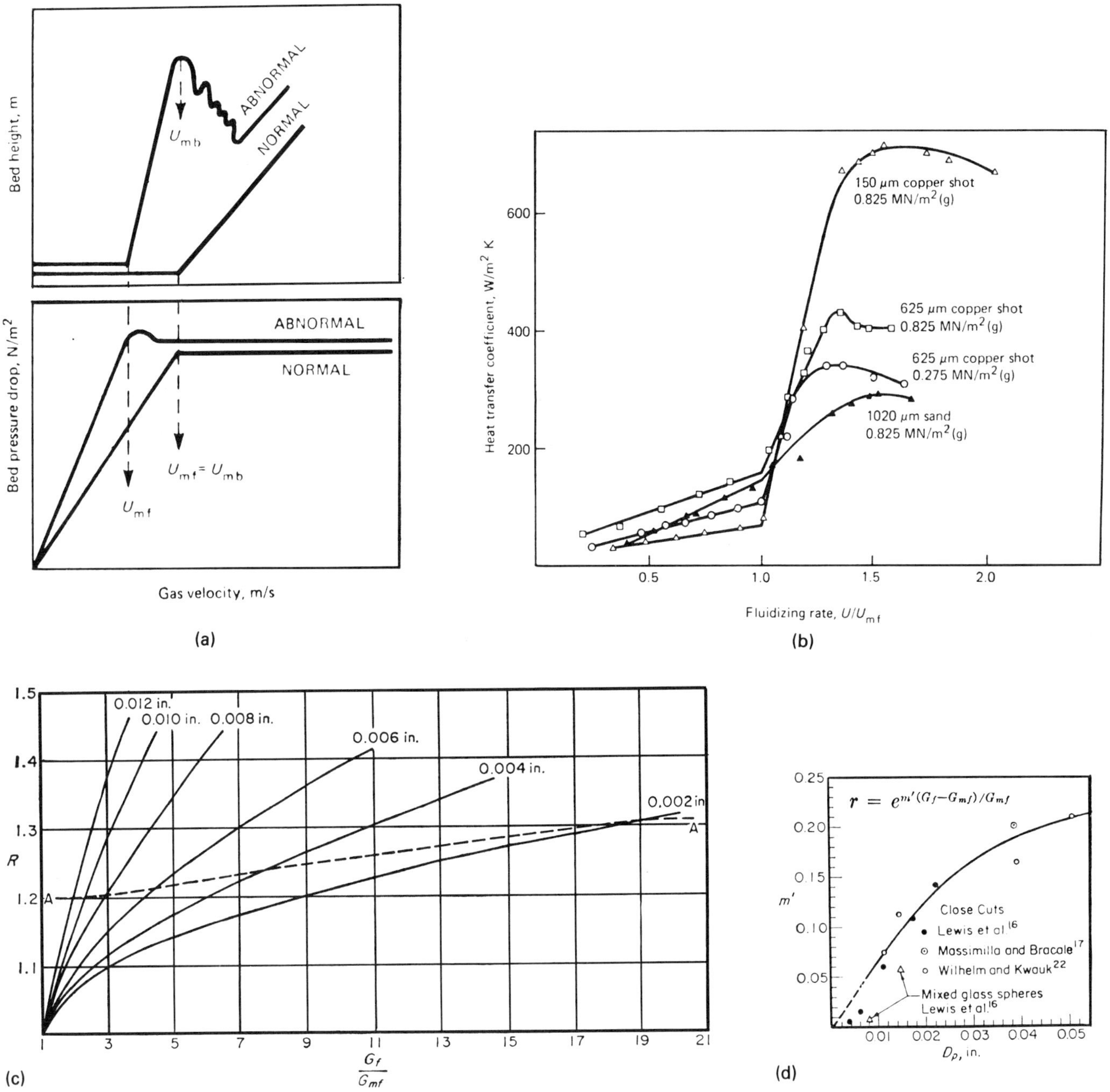

Figure 6.10. Characteristics of gas–solid fluidization. (a) Schematic of the progress of pressure drop and bed height with increasing velocity, for "normal" and "abnormal" behavior. For normal systems, the rates at minimum fluidization and minimum bubbling are the same. (b) Behavior of heat transfer coefficient with gas flow rate analogous to part (a). The peak depends on the density and diameter of the particles (*Botteril,* Fluid Bed Heat Transfer, *Academic, New York, 1975*). (c) Bed expansion ratio as a function of reduced flow rate and particle size. The dashed line is recommended for narrow size range mixtures (*Leva, 1959, p. 102*). (d) Correlation of fluctuations in level, the ratio of the maximum level of disturbed surface to average level (*Leva, 1959, p. 105*). (e) Bed voidage at minimum fluidization (*Leva, 1959*). Agarwal and Storrow: (*a*) soft brick; (*b*) absorption carbon; (*c*) broken Raschig rings; (*d*) coal and glass powder; (*e*) carborundum; (*f*) sand. U.S. Bureau of Mines: (*g*) round sand, $\phi_S = 0.86$; (*h*) sharp sand, $\phi_S = 0.67$; (*i*) Fischer–Tropsch catalyst, $\phi_S = 0.58$; (*j*) anthracite coal, $\phi_S = 0.63$; (*k*) mixed round sand, $\phi_S = 0.86$. Van Heerden et al.: (*l*) coke; (*m*) carborundum. (*f*) Coefficient *C* in the equation for mass flow rate at minimum fluidization (*Leva, 1959*): $G_{mf} = CD_p^2 g_c \rho_F(\rho_S - \rho_F)/\mu$ and $C = 0.0007\,\mathrm{Re}^{-0.063}$. (g) Minimum bubbling and fluidization velocities of cracking catalysts (*Harriott and Simone, in* Cheremisinoff and Gupta, Eds., Handbook of Fluids in Motion, *Ann Arbor Science, Ann Arbor, MI, 1983,* p. *656*). (h) Minimum fluidization and bubbling velocities with air as functions of particle diameter and density [*Geldart,* Powder Technol. **7,** *285* (*1973*)]. (i) Transport disengagement height, TDH, as a function of vessel diameter and superficial linear velocity [*Zenz and Weil,* AIChE J. **4,** *472* (*1958*)]. (j) Good fluidization conditions (*W.V. Battcock and K.K. Pillai, "Particle size in Pressurised Combustors,"* Proc. Fifth International Conference on Fluidised Bed Combustion, Mitre Corp., Washington D.C., *1977*).

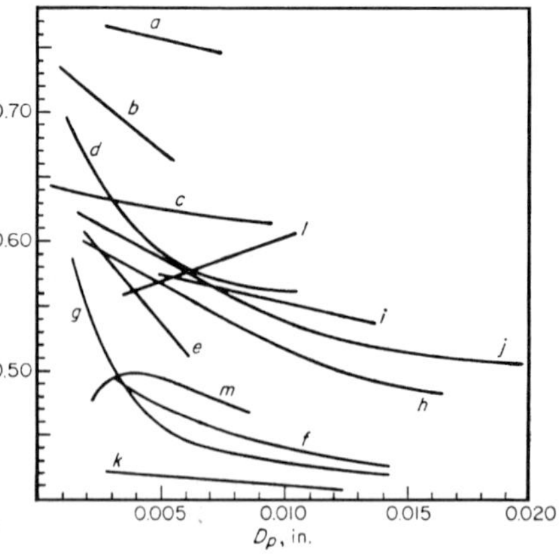

(e)

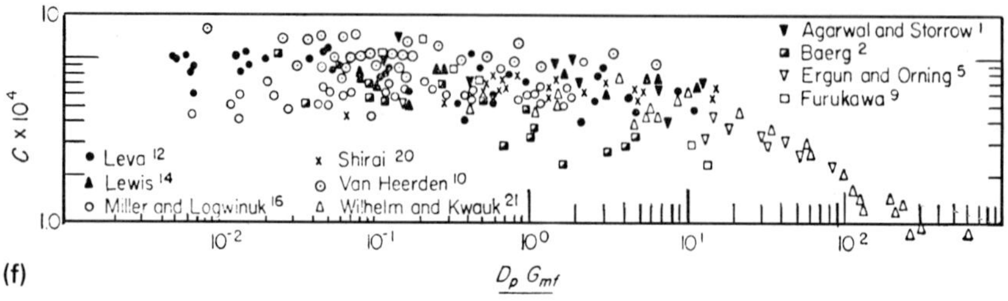

(f)

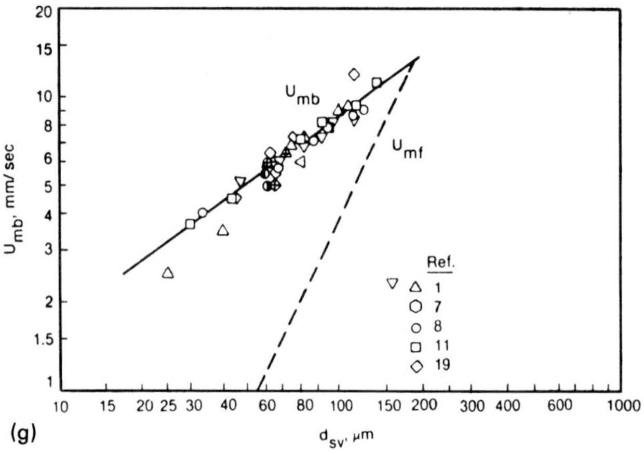

(g)

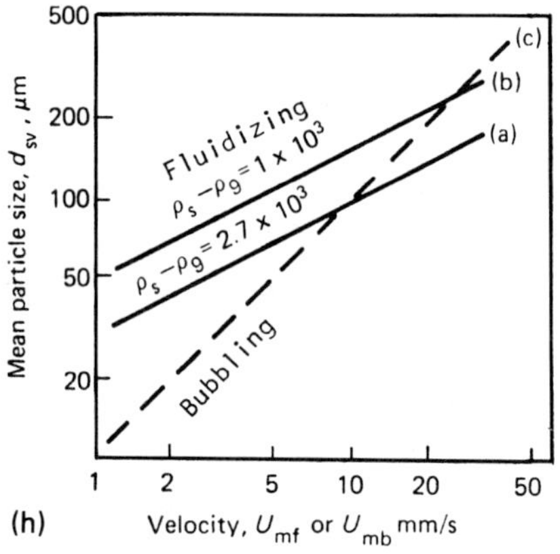

(h)

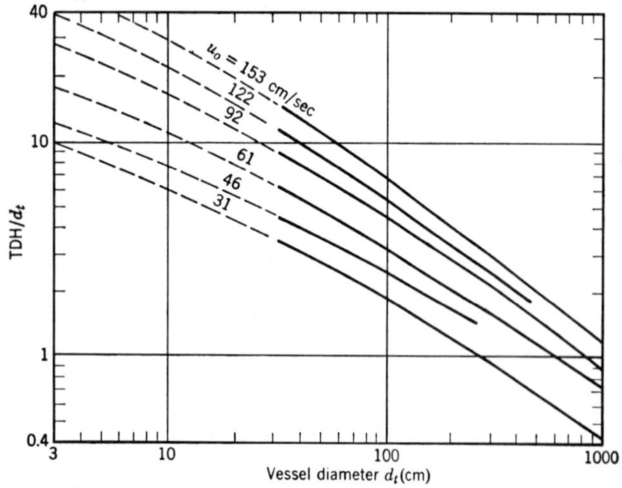

(i)

Figure 6.10—*(continued)*

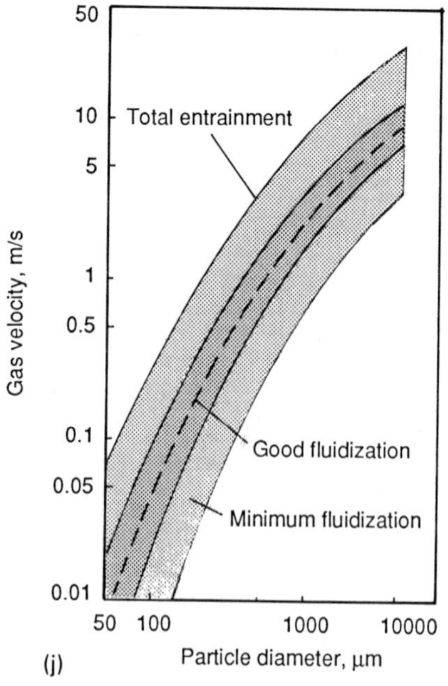

(j)

particles appear to interlock to form a skin around the bubble and thus prevent free particles from raining through those spaces. Bubble sizes become large at high rates of flow and may eventually reach the diameter of the vessel, at which time slugging and severe entrainment will occur.

Onset of fluidization commonly is detected by noting a break in the plot of flow against pressure drop. For a range beyond the minimum fluidizing velocity, the pressure drop remains constant and equal to the weight of the bed but the bed level rises gradually and bubbles are generated at an increasing rate. Not in all cases, however, is the fluidization behavior entirely smooth. Figure 6.10(a) compares "normal" with a case of "abnormal" behavior. Among the reasons for abnormality are aggregation of particles because of stickiness or attractive forces between small particles and interlocking of rough surfaces. It is even possible for bubbling to occur before the onset of fluidization by formation of channels in the bed.

CHARACTERISTICS OF FLUIDIZATION

Six different regimes of fluidization are identified in Figure 6.11 and its legend. Particulate fluidization, class (b) of the figure, is desirable for most processing since it affords intimate contacting of phases. Fluidization depends primarily on the sizes and densities of the particles, but also on their roughness and the temperature, pressure, and humidity of the gas. Especially small particles are subject to electrostatic and interparticle forces.

Four main classes characterized by diameters and differences in densities of the phases are identified in Figure 6.12 and its legend. Groups A and B are most frequently encountered; the boundary between them is defined by the equation given in the legend. Group A particles are relatively small, 30–150 μm dia, with densities below 1.5 g/cc. Their bed behavior is "abnormal" in that the bed expands appreciably before bubbling sets in, and the minimum bubbling velocity always is greater than the minimum fluidization velocity. The bubbles disengage quickly. Cracking catalysts that have been studied extensively for their fluidization behavior are in this class. Group B materials have $d_p = 150$–$500\,\mu$m and are 1.5–4.0 g/mL. The bed expansion is small, and minimum bubbling and fluidization velocities are nearly the same. The bubbles also disengage rapidly. Coarse sand and glass beads that have been favorite study materials fall in this group. Group C comprises small cohesive particles whose behavior is influenced by electrostatic and van der Waals forces. Their beds are difficult to fluidize and subject to channelling. Group D particles are large, 1 mm or more, such as lead shot and grains. They do not fluidize well and are usually handled in spouted beds, such as Figure 9.13(f).

Among the properties of particles most conducive to smooth fluidization are the following:

1. rounded and smooth shape,
2. in the range of 50–500 μm diameter,
3. a broad spectrum of particle sizes, with ratios of largest to smallest sizes in the range of 10 to 25,
4. enough toughness to resist attrition.

Such tailoring of properties is feasible for many catalyst-carrier formulations, but drying processes, for instance, may be restricted by other considerations. Fluidization of difficult materials can be maintained by mechanical or ultrasonic vibration of the vessel, or pulsation of the supply of the fluid, or mechanical agitation of the contents of the vessel, or by addition of fluidization aids such as fine foreign solids.

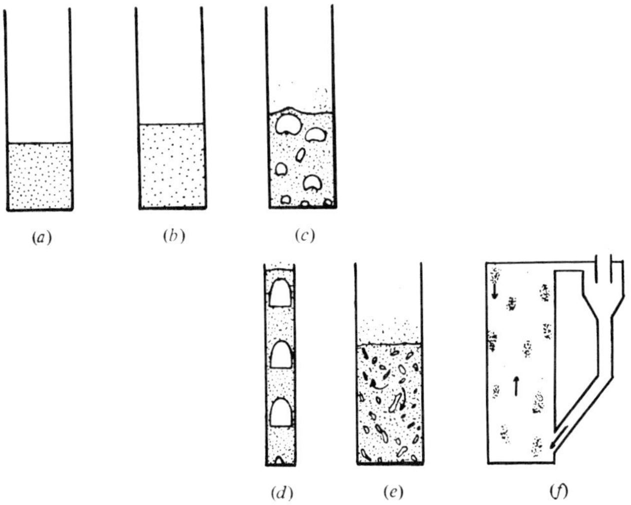

Figure 6.11. Six regimes of fluidization identified with increasing gas superficial velocity (*Grace, 1982*).

	Velocity Range	Regime	Appearance and Principal Features
(a)	$0 \le u < u_{mf}$	fixed bed	particles are quiescent; gas flows through interstices
(b)	$u_{mf} \le u < u_{mb}$	particulate fluidization	bed expands smoothly in a homogeneous manner; top surface is well defined; some small-scale particle motion; little tendency for particles to aggregate; very little fluctuation
(c)	$u_{mb} \le u < u_{ms}$	bubbling fluidization	void regions form near the distributor, grow mostly by coalescence, and rise to the surface; top surface is well defined with bubbles breaking through periodically; irregular pressure fluctuations of appreciable amplitude
(d)	$u_{ms} \le u < u_k$	slugging regime	voids fill most of the column cross section; top surface rises and collapses with reasonably regular frequency; large and regular pressure fluctuations
(e)	$u_k \le u < u_{tr}$	turbulent regime	small voids and particle clusters dart to and fro; top surface difficult to distinguish; small-amplitude pressure fluctuations only
(f)	$u_{tr} \le u$	fast fluidization	no upper surface to bed; particles are transported out the top and must be replaced by adding solids at or near the bottom; clusters or strands of particles move downward, mostly near the wall, while gas, containing widely dispersed particles, moves upward; at fixed solid feed rate, increasingly dilute as u is increased

SIZING EQUIPMENT

Various aspects of the hydrodynamics of gas–solid fluidization have been studied extensively with conclusions that afford guidance to the interpretation and extension of pilot plant data. Some of the leading results bearing on the sizing of vessels will be discussed here. Heat transfer performance is covered in Chapter 17. Example 6.16 applies to some of the cited data.

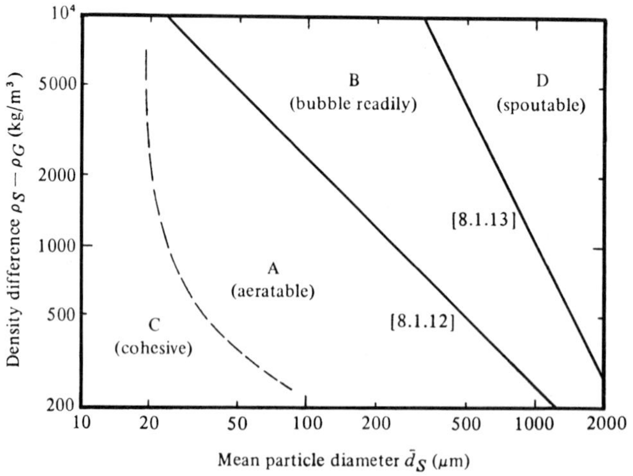

Figure 6.12. Characteristics of four kinds of groups of particles classified by Geldart [Powder Technol. **6,** *201–205* (*1972*); **7,** *285–292* (*1973*)]. The boundary between A and B is represented by the equation $\bar{d}_S = 44{,}000\rho_F^{0.1}\mu_F^{0.9}/g(\rho_S - \rho_F)$ and that between B and D by $(\rho_S - \rho_F)\,\bar{d}_S^2 = 10^{-3}$ kg/m.

Feature	Group C	Group A	Group B	Group D
Distinguishing word or phrase	Cohesive	aeratable	bubble readily	spoutable
Example	Flour	fluid cracking catalyst	sand	wheat
Particle size for $\rho_S = 2.5\ \text{g/cm}^3$	$\leq 20\ \mu m$	$20 < \bar{d}_S \leq 90\ \mu m$	$90 < \bar{d}_S \leq 650\ \mu m$	$>650\ \mu m$
Channeling	Severe	little	negligible	negligible
Spouting	None	none	shallow beds only	readily
Collapse rate	—	slow	rapid	rapid
Expansion	Low because of channeling	high; initially bubble-free	medium	medium
Bubble shape	channels, no bubbles	flat base, spherical cap	rounded with small indentation	rounded
Rheological character of dense phase	high yield stress	apparent viscosity of order 1 poise	apparent viscosity of order 5 poise	apparent viscosity of order 10 poise
Solids mixing	very low	high	medium	low
Gas back mixing	very low	high	medium	low
Slugging mode	flat raining plugs	axisymmetric	mostly axi-symmetric	mostly wall slugs
Effect of $\bar{d}_S$ (within group) on hydrodynamics	unknown	appreciable	minor	unknown
Effect of particle size distribution	unknown	appreciable	negligible	can cause segregation

Solids of practical interest often are mixtures of a range of particle diameters, but, for convenience, correlations are expressed in terms of a single size which is almost invariably taken as the surface average diameter given by

$$d_p = 1/\sum x_i d_i, \tag{6.129}$$

where x_i is the weight fraction of the material having a diameter d_i measured by screen analysis. Particles that deviate substantially from a spherical shape are characterized as having the diameter of a sphere with the same volume as the particle. The sphericity is defined as the ratio

$$\phi = \text{(surface of a sphere)/(surface of the particle with the same volume)} \tag{6.130}$$

and is always less than unity. Accordingly, the relation between the effective particle size d_p and that found by screen analysis is

$$d_p = \phi d_{\text{screen}}. \tag{6.131}$$

EXAMPLE 6.16
Dimensions of a Fluidized Bed Vessel
A fluidized bed is to hold 10,000 kg of a mixture of particles whose true density is 1700 kg/m^3. The fluidizing gas is at 0.3 m^3/sec, has a viscosity of 0.017 cP or $1.7(E-5)$ N sec/m^2 and a density of 1.2 kg/m^3. The size distribution of the particles is

d (μm)	252	178	126	89	70	50	30	10
x (wt fraction)	0.088	0.178	0.293	0.194	0.113	0.078	0.042	0.014
u_t (m/sec)	3.45	1.72	0.86	0.43	0.27	0.14	0.049	0.0054

The terminal velocities are found with Stokes' equation

$$u_t = \frac{g(\rho_p - \rho)}{18\mu} d_p^2 = \frac{9.81(1700-1.2)(E-12)}{18[1.7(E-5)]}[d_p\,(\mu\text{m})]^2.$$

(a) The average particle size is

$$d_p = 1\Big/\sum (x_i/d_i) = 84.5\,\mu\text{m}.$$

(b) With $d_p = 84.5$ and density difference of 1699 kg/m^3, the material appears to be in Group A of Figure 6.12.

(c) Minimum fluidization velocity with Eq. (6.133)

$$u_{mf} = \frac{0.0093[84.5(E-6)]^{1.82}(1700-1.2)^{0.94}}{[1.7(E-5)]^{0.88}(1.2)^{0.06}} = 0.0061\text{ m/sec},$$

and with Eqs. (6.134) and (6.135),

$$\text{Ar} = \frac{1.2(1700-1.2)(9.81)[84.5(E-6)]^3}{[1.7(E-5)]^2} = 41.75,$$

$$\text{Re}_{mf} = \sqrt{(27.2)^2 + 0.0408(41.75)} - 27.2 = 0.0313,$$

$$u_{mb} = \frac{\mu\,\text{Re}_{mf}}{d_p\rho} = \frac{1.7(E-5)(0.0313)}{84.5(E-6)(1.2)} = 0.0052\text{ m/sec}.$$

Use the larger value, $u_{mf} = 0.0061$, as the conservative one.

(d) Minimum bubbling velocity, with Eq. (6.136),

$$u_{mb} = 33(84.5)(E-6)[1.2/1.7(E-5)]^{0.1} = 0.0085\text{ m/sec},$$

$$\therefore u_{mb}/u_{mf} = 0.0085/0.0061 = 1.39.$$

From Eq. (6.139),

$$\frac{u_{mb}}{u_{mf}} = \frac{82[1.7(E-5)]^{0.6}(1.2)^{0.06}}{9.81[84.5(E-6)]^{1.3}(1700-1.2)} = 1.35,$$

which is in rough agreement.

(e) Voidage at minimum bubbling from Eq. (6.138):

$$\frac{\varepsilon_{mb}^3}{1-\varepsilon_{mb}} = 47.4\left\{\frac{[1.7(E-5)]^2}{9.81[84.5(E-6)]^3(1700)^2}\right\}^{0.5} = 0.1948,$$

$$\therefore \varepsilon_{mb} = 0.469.$$

It is not certain how nearly consistent this value is with those at minimum fluidization read off Figure 6.10(e). Only a limited number of characteristics of the solids are accounted for in Eq. (6.138).

(f) Operating gas velocity. The ratios of entraining and minimum fluidizing velocities for the two smallest particle sizes present are

$$0.049/0.0061 = 8.03,\quad \text{for } 30\,\mu\text{m},$$

$$0.0054/0.0061 = 0.89,\quad \text{for } 10\,\mu\text{m}.$$

Entrainment of the smallest particles cannot be avoided, but an appreciable multiple of the minimum fluidizing velocity can be used for operation; say the ratio is 5, so that

$$u_f = 5u_{mf} = 5(0.0061) = 0.0305\text{ m/sec}.$$

(g) Bed expansion ratio. From Figure 6.10(c) with $d_p =$ 84.5 μm or 0.0033 in. and $G_f/G_{mf} = 5$,

$$R = \begin{cases} 1.16, & \text{by interpolation between the full lines,} \\ 1.22, & \text{off the dashed line.} \end{cases}$$

Take $R = 1.22$ as more conservative. From Eq. (6.140) the ratio of voidages is

$$\varepsilon_{mb}/\varepsilon_{mf} = 5^{0.22} = 1.42.$$

From part (e), $\varepsilon_{mb} = 0.469$ so that $\varepsilon_{mf} = 0.469/1.42 = 0.330$. Accordingly, the ratio of bed levels is

$$L_{mb}/L_{mf} = (1-\varepsilon_{mf})/(1-\varepsilon_{mb}) = 0.67/0.531 = 1.262.$$

Although the value of ε_{mf} appears somewhat low, the value of R checks roughly the one from Figure 6.10(c).

(h) Fluctuations in level. From Figure 6.10(d), with $d_p =$ 0.0033 in., the value of $m' = 0.02$, so that

$$r = \exp[0.02(5-1)] = 1.083.$$

(i) TDH from Figure 6.10(i). At $u_f = u_{mf} - 4(0.0061) =$ 0.0244 m/sec, the abscissa is off the plot, but a rough extrapolation and interpolation indicates about 1.5 m for TDH.

(j) Dimensions of the bed and vessel. With a volumetric flow rate of 0.3 m^3/sec, the required diameter is

$$D = \sqrt{0.3/(0.305)(\pi/4)} = 3.54\text{ m}.$$

With a charge of 10,000 kg of solids and a voidage at minimum bubbling of 0.469, the height of the minimum bubbling bed is

$$L = \frac{10000}{1700(1-0.469)(\pi/4)D^2} = 1.13\text{ m}.$$

This value includes the expansion factor which was calculated separately in item (g) but not the fluctuation parameter; with this correction the bed height is

$$L_b = 1.13(1.083) = 1.22\text{ m}.$$

The vessel height is made up of this number plus the TDH of 1.5 m or

$$\text{vessel height} = 1.22 + 1.5 = 2.72\text{ m}.$$

Minimum Fluidization. The fundamental nature of this phenomenon has led to many correlations for its prediction. That of Leva (1959) applies to Reynolds numbers $\mathrm{Re}_{mf} = d_p G_{mf}/\mu < 5$, and is

$$G_{mf} = 688 D_p^{1.82} \frac{[\rho_F(\rho_S - \rho_F)]^{0.94}}{\mu^{0.88}} \tag{6.132}$$

in the common units G_{mf} in lb/(hr)(sqft), D_p in inches, densities in lb/cuft, and viscosity in cP. In SI units it is

$$U_{mf} = \frac{0.0093 d_p^{1.82}(\rho_p - \rho_f)^{0.94}}{\mu^{0.88}\rho_f^{0.06}}. \tag{6.133}$$

The degree of confidence that can be placed in the correlation is indicated by the plot of data on which it is based in Figure 6.10(f). An equation more recently recommended by Grace (1982) covers Reynolds numbers up to 1000:

$$\mathrm{Re}_{mf} = d_p u_{mf}\rho/\mu = \sqrt{(27.2)^2 + 0.0408(\mathrm{Ar})} = 27.2, \tag{6.134}$$

where

$$\mathrm{Ar} = \rho(\rho_p - \rho) g d_p^3/\mu^2. \tag{6.135}$$

Here also the data show much scatter, so that pilot plant determinations of minimum fluidization rates usually are advisable.

Minimum Bubbling Conditions. Minimum bubbling velocities for Group B substances are about the same as the minimum fluidization velocities, but those of Group A substances are substantially greater. For Group A materials the correlation of Geldart and Abrahamsen [*Powder Technol* **19,** 133 (1978)] for minimum bubbling velocity is

$$u_{mb} = 33 d_p(\mu/\rho)^{-0.1}. \tag{6.136}$$

For air at STP this reduces to

$$u_{mb} = 100 d_p. \tag{6.137}$$

For cracking catalysts represented on Figure 6.10(g), Harriott and Simone (1983) present an equation for the ratio of the two kinds of velocities as

$$\frac{u_{mb}}{u_{mf}} = \frac{82\mu^{0.6}\rho^{0.06}}{g d_p^{1.3}(\rho_p - \rho)}. \tag{6.138}$$

The units of this equation are SI; the coefficient given by Cheremisinoff and Cheremisinoff (1984, p. 161) is incorrect. Figures 6.10(g) and (h) compare the two kinds of velocities over a range of particle diameters. Voidage at minimum bubbling is correlated by an equation of Cheremisinoff and Cheremisinoff (1984, p. 163):

$$\varepsilon_{mb}^3/(1 - \varepsilon_{mb}) = 47.4(g d_p^3 \rho_p^2/\mu^2)^{-0.5}. \tag{6.139}$$

Bed Expansion and Fluctuation. The change of bed level with increasing gas rate is represented schematically in Figure 6.10(a). The height remains constant until the condition of minimum fluidization is reached, and the pressure drop tends to level off. Then the bed continues to expand smoothly until some of the gas begins to disengage from the homogeneous dense phase and forms bubbles. The point of onset of bubbling corresponds to a local maximum in level which then collapses and attains a minimum. With increasing gas rate, the bed again continues to expand until entrainment develops and no distinct bed level exists. Beyond the minimum bubbling point, some fraction of the excess gas continues through the dense phase but that behavior cannot be predicted with any accuracy.

Some smoothed data of expansion ratio appear in Figure 6.10(c) as a function of particle size and ratio of flow rates at minimum bubbling and fluidization. The rather arbitrarily drawn dashed line appears to be a conservative estimate for particles in the range of 100 μm.

Ordinarily under practical conditions the flow rate is at most a few multiples of the minimum fluidizing velocity so the local maximum bed level at the minimum bubbling velocity is the one that determines the required vessel size. The simplest adequate equation that has been proposed for the ratio of voidages at minimum bubbling and fluidization is

$$\varepsilon_{mb}/\varepsilon_{mf} = (G_{mb}/G_{mf})^{0.22} \tag{6.140}$$

$$= 2.64\mu^{0.89}\rho^{0.54}/g^{0.22} d_p^{1.06}(\rho_p - \rho)^{0.22} \tag{6.141}$$

The last equation results from substitution of Eq. (6.138) into (6.140). Then the relative bed level is found from

$$L_{mb}/L_{mf} = (1 - \varepsilon_{mf})/(1 - \varepsilon_{mb}). \tag{6.142}$$

Either ε_{mb} or ε_{mf} must be known independently before Eq. (6.141) can be applied, either by application of Eq. (6.139) for ε_{mb} or by reading off a value of ε_{mf} from Figure 6.8(c) or Figure 6.10(e). These values are not necessarily consistent.

At high gas velocities the bed level fluctuates. The ratio of maximum disturbed level to the average level is correlated in terms of G_f/G_{mf} and the particle diameter by the equation

$$r = \exp[m'(G_f - G_{mf})/G_{mf}], \tag{6.143}$$

where the coefficient m' is given in Figure 6.10(d) as a function of particle diameter.

Freeboard. Under normal operating conditions gas rates somewhat in excess of those for minimum fluidization are employed. As a result particles are thrown into the space above the bed. Many of them fall back, but beyond a certain height called the transport disengaging height (TDH), the entrainment remains essentially constant. Recovery of that entrainment must be accomplished in auxiliary equipment. The TDH is shown as a function of excess velocity and the diameter of the vessel in Figure 6.10(i). This correlation was developed for cracking catalyst particles up to 400 μm dia but tends to be somewhat conservative at the larger sizes and for other materials.

Viscosity. Dense phase solid–gas mixtures may be required to flow in transfer line catalytic crackers, between reactors and regenerators and to circulate in dryers such as Figures 9.13(e), (f). In dilute phase pneumatic transport the effective viscosity is nearly that of the fluid, but that of dense phase mixtures is very much greater. Some data are given by Schügerl (in Davidson and Harrison, 1971, p. 261) and by Yates (1983). Apparent viscosities with particles of 50–550 μm range from 700 to 1300 cP, compared with air viscosity of 0.017 cP at room temperature. Such high values of the viscosity place the flow definitely in the laminar flow range. However, information about friction in flow of fluidized mixtures through pipelines is not easy to find in the open literature. Someone must know since many successful transfer lines are in operation.

REFERENCES

General

1. M.M. Denn, *Process Fluid Mechanics,* Prentice-Hall, Englewood Cliffs, NJ, 1980.
2. O. Levenspiel, *Engineering Flow and Heat Exchange,* Plenum, New York, 1984.
3. M. Modell and R.C. Reid, *Thermodynamics and Its Applications,* Prentice-Hall, Englewood Cliffs, NJ, 1983.
4. V.L. Streeter and E.B. Wylie, *Fluid Mechanics,* McGraw-Hill, New York, 1979.

Non-Newtonian Fluids

5. G.W. Govier and K. Aziz, *Flow of Complex Mixtures in Pipes,* Van Nostrand Reinhold, New York, 1972.
6. N.I. Heywood, Pipeline design for non-Newtonian fluids, *Inst. Chem. Eng. Symp. Ser. No.* 60, 33–52 (1980).
7. J.W. Hoyt, The effect of additives on fluid friction, *Trans. ASME J. Basic Eng.,* 258 (June 1972).
8. P.A. Longwell, *Mechanics of Fluid Flow,* McGraw-Hill, New York, 1966.
9. R.D. Patel, Non-Newtonian flow, in *Handbook of Fluids in Motion,* (Cheremisinoff and Gupta, Eds.), Ann Arbor Science, Ann Arbor, MI, 1983, pp. 135–177.
10. A.H.P. Skelland, *Non-Newtonian Flow and Heat Transfer,* Wiley, New York, 1967.
11. J.R. Van Wazer, J.W. Lyons, K.Y. Kim, and R.E. Colwell, *Viscosity and Flow Measurement,* Wiley-Interscience, New York, 1963.
12. E.J. Wasp, J.P. Kenny, and R.L. Gandhi, *Solid Liquid Flow Slurry Pipeline Transportation,* Trans. Tech. Publications, Clausthal, Germany, 1977.

Two-phase Flow

13. D. Chisholm, Gas–liquid flow in pipeline systems, in *Handbook of Fluids in Motion,* (Cheremisinoff and Gupta, Eds.) Ann Arbor Science, Ann Arbor, MI, 1983, pp. 483–513.
14. D. Chisholm, *Two-Phase Flow in Pipelines and Heat Exchangers,* George Godwin, London, 1983.
15. G.W. Govier and K. Aziz, *The Flow of Complex Mixtures in Pipes,* Van Nostrand Reinhold, New York, 1972.
16. G.F. Hewitt, Liquid–gas systems, in *Handbook of Multiphase Systems,* (G. Hetsroni, Ed.), Hemisphere, New York, 1982, pp. 2.1–2.94.

Gas–Solid (Pneumatic) Transport

17. G. Klinzing, *Gas–Solid Transport,* McGraw-Hill, New York, 1981.
18. N.P. Cheremisinoff, and R. Gupta (Eds.), Gas–solid flows, in *Handbook of Fluids in Motion,* Ann Arbor Science, Ann Arbor, MI, 1983, pp. 623–860.
19. C.S. Teo and L.S. Leung, Vertical flow of particulate solids in standpipes and risers, in *Hydrodynamics of Gas–Solids Fluidization,* (N.P. Cheremisinoff and P.N. Cheremisinoff, Eds.), Gulf, Houston, 1984, pp. 471–542.

Fluidization

20. J.S.M. Botteril, *Fluid-Bed Heat Transfer,* Academic, New York, 1975.
21. N.P. Cheremisinoff and P.N. Cheremisinoff, *Hydrodynamics of Gas–Solid Fluidization,* Gulf, Houston, 1984.
22. J.F. Davidson and D. Harrison, Eds., *Fluidization,* Academic, New York, 1971.
23. J.R. Grace, Fluidization, Section 8 of G. Hetsroni, 1982.
24. G. Hetsroni (Ed.), *Handbook of Multiphase Systems,* McGraw-Hill, New York, 1982.
25. M. Leva, *Fluidization,* McGraw-Hill, New York, 1959.
26. J.C. Yates, *Fundamentals of Fluidized-Bed Chemical Processes,* Butterworths, London, 1983.

7

FLUID TRANSPORT EQUIPMENT

Although liquids particularly can be transported by operators carrying buckets, the usual mode of transport of fluids is through pipelines with pumps, blowers, compressors, or ejectors. Those categories of equipment will be considered in this chapter. A few statements will be made at the start about piping, fittings, and valves, although for the most part this is information best gleaned from manufacturers' catalogs. Special problems such as mechanical flexibility of piping at elevated temperatures are beyond the scope here, and special problems associated with sizing of piping for thermosyphon reboilers and the suction side of pumps for handling volatile liquids are deferred to elsewhere in this book.

7.1. PIPING

Standard pipe is made in a discrete number of sizes that are designated by nominal diameters in inches, as "inches IPS (iron pipe size)." Table A5 lists some of these sizes with dimensions in inches. Depending on the size, up to 14 different wall thicknesses are made with the same outside diameter. They are identified by schedule numbers, of which the most common is Schedule 40. Approximately,

$$\text{Schedule number} = 1000\,P/S,$$

where

P = internal pressure, psig
S = allowable working stress in psi.

Tubing for heat exchangers, refrigeration, and general service is made with outside diameters measured in increments of 1/16 or 1/8 in. Standard size pipe is made of various metals, ceramics, glass, and plastics.

Dimensional standards, materials of construction, and pressure ratings of piping for chemical plants and petroleum refineries are covered by ANSI Piping Code B31.3 which is published by the ASME, latest issue 1980. Many details also are given in such sources as Crocker and King, *Piping Handbook* (McGraw-Hill, New York, 1967), *Perry's Chemical Engineers Handbook* (1984), and *Marks Standard Handbook for Mechanical Engineers* (1987).

In sizes 2 in. and less screwed fittings may be used. Larger joints commonly are welded. Connections to equipment and in lines whenever need for disassembly is anticipated utilize flanges. Steel flanges, flanged fittings, and valves are made in pressure ratings of 150, 300, 600, 900, 1500, and 2500 psig. Valves also are made in 125 and 250 psig cast iron. Pressure and temperature ratings of this equipment in various materials of construction are specified in the piping code, and are shown in *Chem. Eng. Handbook* 1984, pp. 6.75–6.78.

VALVES

Control of flow in lines and provision for isolation of equipment when needed are accomplished with valves. The basic types are relatively few, some of which are illustrated in Figure 7.1. In gate valves the flow is straight through and is regulated by raising or lowering the gate. The majority of valves in the plant are of this type. In the wide open position they cause little pressure drop. In globe valves the flow changes direction and results in appreciable friction even in the wide open position. This kind of valve is essential when tight shutoff is needed, particularly of gas flow. Multipass plug cocks, butterfly valves, slide valves, check valves, various quick-opening arrangements, etc. have limited and often indispensable applications, but will not be described here.

The spring in the relief valve of Figure 7.1(c) is adjusted to open when the pressure in the line exceeds a certain value, at which time the plug is raised and overpressure is relieved; the design shown is suitable for pressures of several hundred psig.

More than 100 manufacturers in the United States make valves that may differ substantially from each other even for the same line size and pressure rating. There are, however, independent publications that list essentially equivalent valves of the several manufacturers, for example the books of Zappe (1981) and Lyons (1975).

CONTROL VALVES

Control valves have orifices that can be adjusted to regulate the flow of fluids through them. Four features important to their use are capacity, characteristic, rangeability and recovery.

Capacity is represented by a coefficient

$$C_d = C_v/d^2,$$

where d is the diameter of the valve and C_v is the orifice coefficient in equations such as the following

$$Q = C_v\sqrt{(P_1 - P_2)/\rho_w}, \quad \text{gal/min of liquid},$$
$$Q = 22.7C_v\sqrt{(P_1 - P_2)P_2/\rho_a T}, \quad \text{SCFM of gas when } P_2/P_1 > 0.5,$$
$$Q = 11.3C_v P_1/\sqrt{\rho_a T}, \quad \text{SCFM of gas when } P_2/P_1 < 0.5,$$

where P_i is pressure in psi, ρ_w is specific gravity relative to water, ρ_a is specific gravity relative to air, and T is temperature °R. Values of C_d of commercial valves range from 12 for double-seated globe valves to 32 for open butterflies, and vary somewhat from manufacturer to manufacturer. Chalfin (1980) has a list.

Characteristic is the relation between the valve opening and the flow rate. Figure 7.1(h) represents the three most common forms. The shapes of plugs and ports can be designed to obtain any desired mathematical relation between the pressure on the diaphragm, the travel of the valve stem, and the rate of flow through the port. *Linear* behavior is represented mathematically by $Q = kx$ and *equal percentage* by $Q = k_1 \exp(k_2 x)$, where x is the valve opening. *Quick-opening* is a characteristic of a bevel-seated or flat disk type of plug; over a limited range of 10–25% of the maximum stem travel is approximately linear.

Over a threefold load change, the performances of linear and equal percentage valves are almost identical. When the pressure drop across the valve is less than 25% of the system drop, the equal percentage type is preferred. In fact, a majority of characterized valves currently are equal percentage.

Rangeability is the ratio of maximum to minimum flows over which the valve can give good control. This concept is difficult to quantify and is not used much for valve selection. A valve generally can be designed properly for a suitably wide flow range.

Recovery is a measure of the degree of pressure recovery at the valve outlet from the low pressure at the vena contracta. When flashing occurs at the vena contracta and the pressure recovery is high, the bubbles collapse with resulting cavitation and noise. The more streamlined the valve, the more complete the pressure recovery; thus, from this point of view streamlining seems to be an undesirable quality. A table of recovery factors of a number of valve types is given by Chalfin (1980); such data usually are provided by manufacturers.

These characteristics and other properties of 15 kinds of valves are described by Chalfin (1980).

Pressure drop. Good control requires a substantial pressure drop through the valve. For pumped systems, the drop through the valve should be at least 1/3 of the pressure drop in the system, with a minimum of 15 psi. When the expected variation in flow is small, this rule can be relaxed. In long liquid transportation lines, for instance, a fully open control valve may absorb less than 1% of the system pressure drop. In systems with centrifugal pumps, the variation of head with capacity must be taken into account when sizing the valve. Example 7.2, for instance, illustrates how the valve drop may vary with flow in such a system.

Types of valves. Most flow control valves are operated with adjustable air pressure on a diaphragm, as in Figure 7.1(d), since this arrangement is more rapid, more sensitive and cheaper than

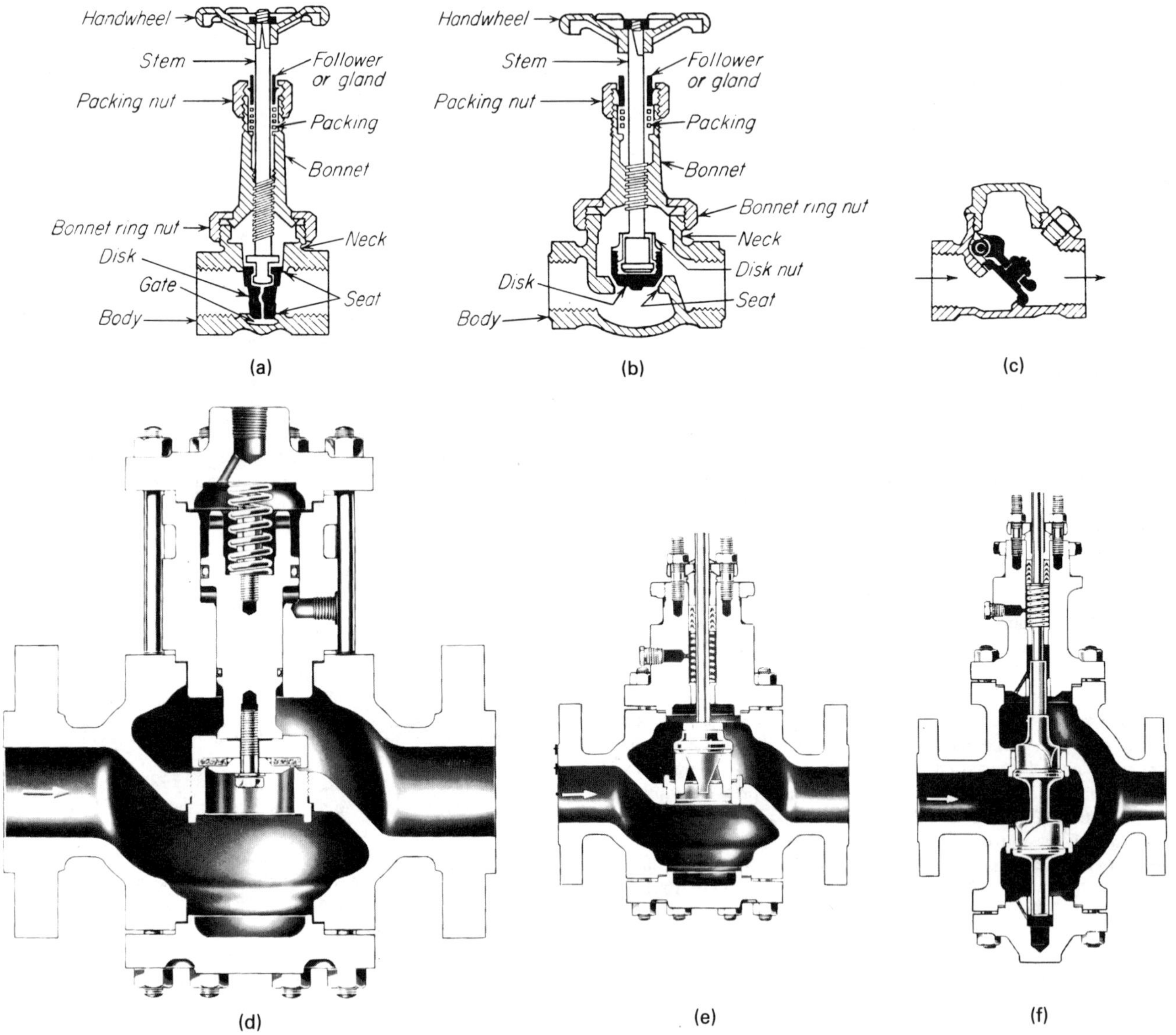

Figure 7.1. Some kinds of manual and automatically controlled valves. (a) Gate valve, for the majority of applications. (b) Globe valve, when tight shutoff is needed. (c) Swing check valve to ensure flow in one direction only. (d) A pressure relief valve, in which the plug is raised on overpressure. (e) A control valve with a single port. (f) A double-port, reverse-acting control valve. (g) A control valve with a double port, in which the correct opening is maintained by air pressure above the diaphragm. (*A*) valve body; (*B*) removable seat; (*C*) discs; (*D*) valve-stem guide; (*E*) guide bushing; (*F*) valve bonnet; (*G*) supporting ring; (*H*) supporting arms; (*J*) diaphragm; (*K*) coupling between diaphragm and valve stem; (*L*) spring-retaining rod; (*M*) spring; (*N*) spring seat; (*O*) pressure connection. (*Fischer.*) (h) Relation between fractional opening and fractional flow of three modes of valve openings.

electrical motor control. Double-ported valve (d) gives better control at large flow rates; the pressures on the upper and lower plugs are balanced so that less force is needed to move the stem. The single port (e) is less expensive but gives a tighter shutoff and is generally satisfactory for noncritical service. The reverse acting valve (f) closes on air failure and is desirable for reasons of safety in some circumstances.

7.2. PUMP THEORY

Pumps are of two main classes: centrifugal and the others. These others mostly have positive displacement action in which the discharge rate is largely independent of the pressure against which they work. Centrifugal pumps have rotating elements that impart high velocity initially and high pressure head ultimately to the liquid. Elements of their theory will be discussed here. A glossary of pump terms and terms relating primarily to centrifugal pumps are defined in the Glossary at the end of this chapter. The chief variables involved in pump theory are listed here with typical units:

D, diameter of impeller (ft or m),
H, output head (ft or m),
n, rotational speed (1/sec),
$\dot{P}$, output power (HP or kW),
Q, volumetric discharge rate (cfs or m^3/sec),
μ, viscosity (lb/ft sec or N sec/m^2),
ρ, density (lb/cuft or kg/m^3),
ε, surface roughness (ft or m).

BASIC RELATIONS

A dimensional analysis with these variables reveals that the functional relations of Eqs. (7.1) and (7.2) must exist:

$$gH/n^2D^2 = \phi_1(Q/nD^3, D^2n\rho/\mu, \varepsilon/D), \quad (7.1)$$
$$\dot{P}/\rho n^3D^5 = \phi_2(Q/nD^3, D^2n\rho/\mu, \varepsilon/D). \quad (7.2)$$

The group $D^2n\rho/\mu$ is the Reynolds number and ε/D is the roughness ratio. Three new groups also have arisen which are named

capacity coefficient, $C_Q = Q/nD^3$, (7.3)
head coefficient, $C_H = gH/n^2D^2$, (7.4)
power coefficient, $C\dot{P} = \dot{P}/\rho n^3D^5$. (7.5)

The hydraulic efficiency is expressed by these coefficients as

$$\eta = gH\rho Q/\dot{P} = C_HC_Q/C_P. \quad (7.6)$$

Although this equation states that the efficiency is independent of the diameter, in practice this is not quite true. An empirical relation is due to Moody [*ASCE Trans.* **89,** 628 (1926)]:

$$\eta_2 = 1 - (1 - \eta_1)(D_1/D_2)^{0.25}. \quad (7.7)$$

Geometrically similar pumps are those that have all the dimensionless groups numerically the same. In such cases, two different sets of operations are related as follows:

$$Q_2/Q_1 = (n_2/n_1)(D_2/D_1)^3, \quad (7.8)$$
$$H_2/H_1 = (n_2D_2/n_1D_1)^2, \quad (7.9)$$
$$\dot{P}_2/\dot{P}_1 = (\rho_2/\rho_1)(n_2/n_1)^3(D_2/D_1)^5. \quad (7.10)$$

The performances of geometrically similar pumps also can be represented in terms of the coefficients C_Q, C_H, C_P, and η. For instance, the data of the pump of Figure 7.2(a) are transformed into the plots of Figure 7.2(b). An application of such generalized curves is made in Example 7.1.

Another dimensionless parameter that is independent of diameter is obtained by eliminating D between C_Q and C_H with the result,

$$N_s = nQ^{0.5}/(gH)^{0.75}. \quad (7.11)$$

This concept is called the specific speed. It is commonly used in the

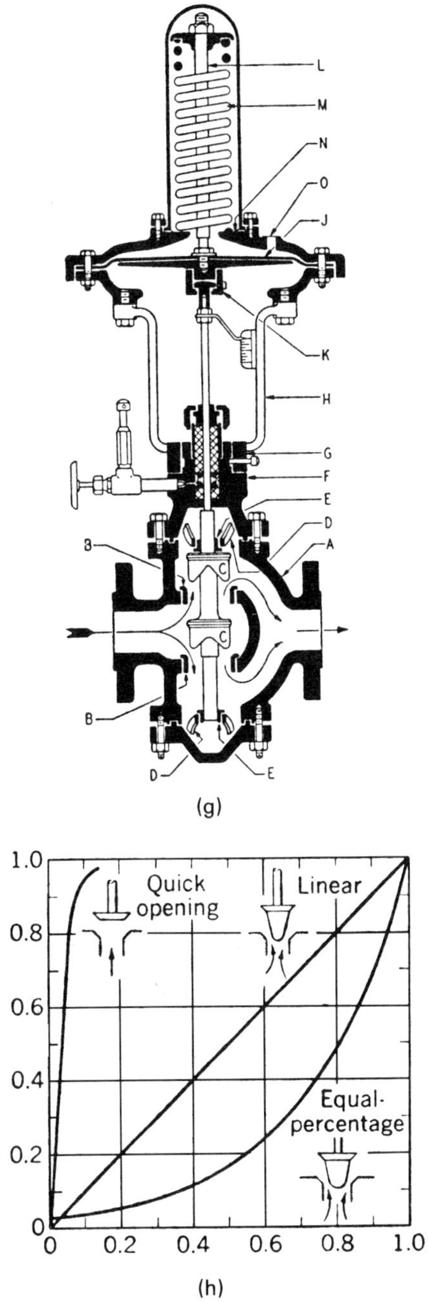

Figure 7.1—(*continued*)

EXAMPLE 7.1
Application of Dimensionless Performance Curves

Model and prototypes are represented by the performance curves of Figure 7.2. Comparisons are to be made at the peak efficiency, assumed to be the same for each. Data off Figure 7.2(b) are:

$$\eta = 0.93,$$
$$C_H = gH/n^2D^2 = 5.2,$$
$$C_P = \dot{P}/\rho n^3D^5 = 0.69,$$
$$C_Q = Q/nD^3 = 0.12.$$

(**a**) The prototype is to develop a head of 76 m:

$$n = \left(\frac{gH}{C_HD^2}\right)^{0.5} = \left(\frac{9.81(76)}{5.2(0.371)^2}\right)^{0.5} = 32.27 \text{ rps},$$
$$Q = nD^3C_Q = 32.27(0.371)^3(0.12) = 0.198 \text{ m}^3/\text{sec},$$
$$\dot{P} = \rho n^3D^5C_p = 1000(32.27)^3(0.371)^5(0.69)$$
$$= 0.163(10^6) \text{ W}, \ 163 \text{ kW}.$$

(**b**) The prototype is to have a diameter of 2 m and to rotate at 400 rpm:

$$Q = nD^3C_Q = (400/60)(2)^3(0.12) = 6.4 \text{ m}^3/\text{sec},$$
$$H = n^2D^2C_H/g = (400/60)^2(2)^2(5.2)/9.81 = 94.2 \text{ m},$$
$$\dot{P} = \rho n^3D^5C_p = 1000(400/60)^3(2)^5(0.69)$$
$$= 6.54(10^6) \text{ kgm}^2/\text{sec}^3,$$
$$6.54(10^6) \text{ N m/sec}, 6540 \text{ kW}.$$

(**c**) Moody's formula for the effect of diameter on efficiency gives

$$\eta_2 = 1 - (1-\eta_1)(D_1/D_2)^{0.25} = 1 - 0.07(0.371/2)^{0.25}$$
$$= 0.954 \quad \text{at } 2 \text{ m},$$

compared with 0.93 at 0.371 m.

(**d**) The results of (**a**) and (**b**) also are obtainable directly from Figure 7.2(a) with the aid of Eqs. (7.7), (7.8), and (7.9). Off the figure at maximum efficiency,

$$\eta = 0.93, \quad Q = 0.22, \quad H = 97, \quad \text{and} \quad P = 218.$$

When the new value of H is to be 76 m and the diameter is to remain the same,

$$n_2 = 35.6(H_2/H_1)^{0.5} = 35.6(76/97)^{0.5} = 31.5 \text{ rps},$$
$$Q_2 = Q_1(n_2/n_1) = 0.22(H_2/H_1)^{0.5} = 0.195 \text{ m}^3/\text{sec},$$
$$\dot{P}_2 = \dot{P}_1(\rho_2/\rho_1)(n_2/n_1)^3(D_2/D_1)^5 = 218(H_2/H_1)^{1.5} = 151.2 \text{ kW}.$$

These values agree with the results of (**a**) within the accuracy of reading the graphs.

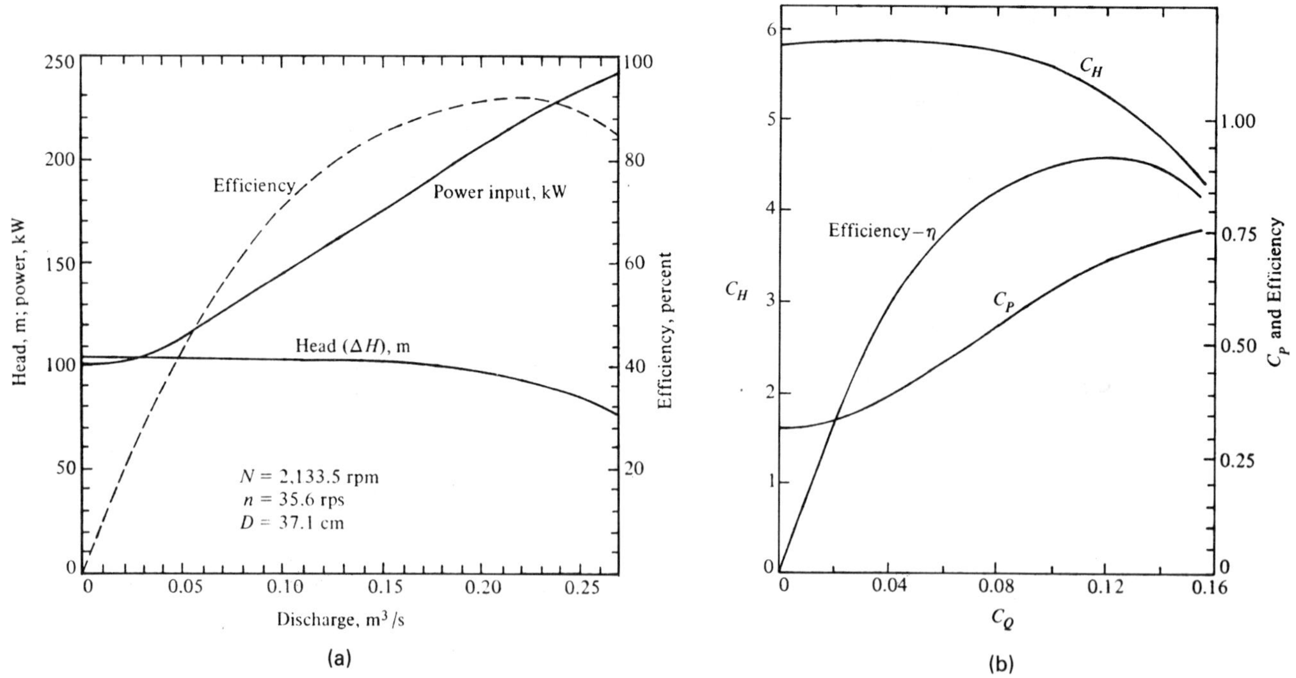

Figure 7.2. Performance curves in dimensional and dimensionless forms: (a) Data of a pump with a specific diameter and rotation speed. (b) Dimensionless performance curves of all pumps geometrically similar to (a). The dashed lines identify the condition of peak efficiency. (*After Daugherty and Franzini,* Fluid Mechanics with Engineering Applications, *McGraw-Hill, New York, 1957*).

EXAMPLE 7.2
Operating Points of Single and Double Pumps in Parallel and Series

The head loss in a piping system is represented by the equation

$$H_s = 50 + 6.0(Q/100)^2 + H_v,$$

where H_v is the head loss in the control valve. The pump to be used has the characteristic curve of the pump of Figure 7.7(b) with an 8 in. impeller; that curve is represented closely by the equation

$$H_p = 68 - 0.5(Q/100) - 4.5(Q/100)^2.$$

The following will be found (see Figure 7.17):

(a) The values of H_v corresponding to various flow rates Q gpm.
(b) The flow rate and head on the pumps when two pumps are connected in parallel and the valve is wide open ($H_v = 0$).
(c) The same as (b) but with the pumps in series.
(d) The required speed of the pump at 80 gpm when no control valve is used in the line.

(a) The operating point is found by equating H_s and H_p from which

$$H_v = 68 - 0.5(Q/100) - 4.5(Q/100)^2 - [50 + 6.0(Q/100)^2].$$

Some values are

$Q/100$	0.8	1.0	1.2	1.286
H_v	10.88	7.00	2.28	0
H_s				59.92

(b) In parallel each pump has half the total flow and the same head H_s:

$$50 + 6.0(Q/100)^2 = 68 - (0.5/2)(Q/100) - (4.5/4)(Q/100)^2,$$
$$\therefore Q = 157.2 \text{ gpm}, \quad H_s = 64.83 \text{ ft}.$$

(c) In series each pump has the same flow and one-half the total head loss:

$$\tfrac{1}{2}(50 + 6.0(Q/100)^2] = 68 - 0.5(Q/100) - 4.5(Q/100)^2,$$
$$\therefore Q = 236.1 \text{ gpm}, \quad H_s = 83.44 \text{ ft}.$$

Series flow allows 50% greater gpm than parallel.

(d) $H_s = 50 + 4.8 = 54.8,$
$$H_p = (68 - 0.4 - 2.88)(n/1750)^2,$$
$$\therefore n = 1750\sqrt{54.8/64.72} = 1610 \text{ rpm}.$$

mixed units

$$N_s = (\text{rpm})(\text{gpm})^{0.5}/(\text{ft})^{0.75}. \tag{7.12}$$

For double suction pumps, Q is one half the pump output.

The net head at the suction of the pump impeller must exceed a certain value in order to prevent formation of vapor and resulting cavitation of the metal. This minimum head is called the net positive suction head and is evaluated as

NPSH = (pressure head at the source)
+ (static suction head)
− (friction head in the suction line)
− (vapor pressure of the liquid). (7.13)

Usually each manufacturer supplies this value for his equipment. (Some data are in Figure 7.7.) A suction specific speed is defined as

$$S = (\text{rpm})(\text{gpm})^{0.5}/(\text{NPSH})^{0.75}. \tag{7.14}$$

Standards for upper limits of specific speeds have been established, like those shown in Figure 7.6 for four kinds of pumps. When these values are exceeded, cavitation and resultant damage to the pump may occur. Characteristic curves corresponding to widely different values of N_s are shown in Figure 7.3 for several kinds of pumps handling clear water. The concept of specific speed is utilized in Example 7.3. Further data are in Figure 7.6.

Recommendations also are made by the Hydraulic Institute of suction specific speeds for multistage boiler feed pumps, with $S = 7900$ for single suction and $S = 6660$ for double suction. Thus the required NPSH can be found by rearrangement of Eq. (7.14) as

$$\text{NPSH} = [(\text{rpm})(\text{gpm})^{0.5}/S]^{4/3}. \tag{7.15}$$

For example, at 3500 rpm, 1000 gpm, and $S = 7900$, the required NPSH is 34 ft.

For common fluids other than water, the required NPSH usually is lower than for cold water; some data are shown in Figure 7.16.

PUMPING SYSTEMS

The relation between the flow rate and the head developed by a centrifugal pump is a result of its mechanical design. Typical curves are shown in Figure 7.7. When a pump is connected to a piping system, its head must match the head loss in the piping system at the prevailing flow rate. The plot of the flow rate against the head loss in a line is called the system curve. The head loss is given by the mechanical energy balance,

$$H_s = \frac{\Delta P}{\rho} + \frac{\Delta u^2}{2g_c} + \Delta z + \frac{fLu^2}{2gD} + H_v, \tag{7.16}$$

where H_v is the head loss of a control valve in the line.

The operating point may be found as the intersection of plots of the pump and system heads as functions of the flow rate. Or an equation may be fitted to the pump characteristic and then solved simultaneously with Eq. (7.16). Figure 7.17 has such plots, and Example 7.2 employs the algebraic method.

In the normal situation, the flow rate is the specified quantity. With a particular pump curve, the head loss of the system may need to be adjusted with a control valve in the line to make the system and pump heads the same. Alternately, the speed of the pump can be adjusted to make the pump head equal to that of the system. From Eq. (7.9) the relation between speeds and pump heads at two

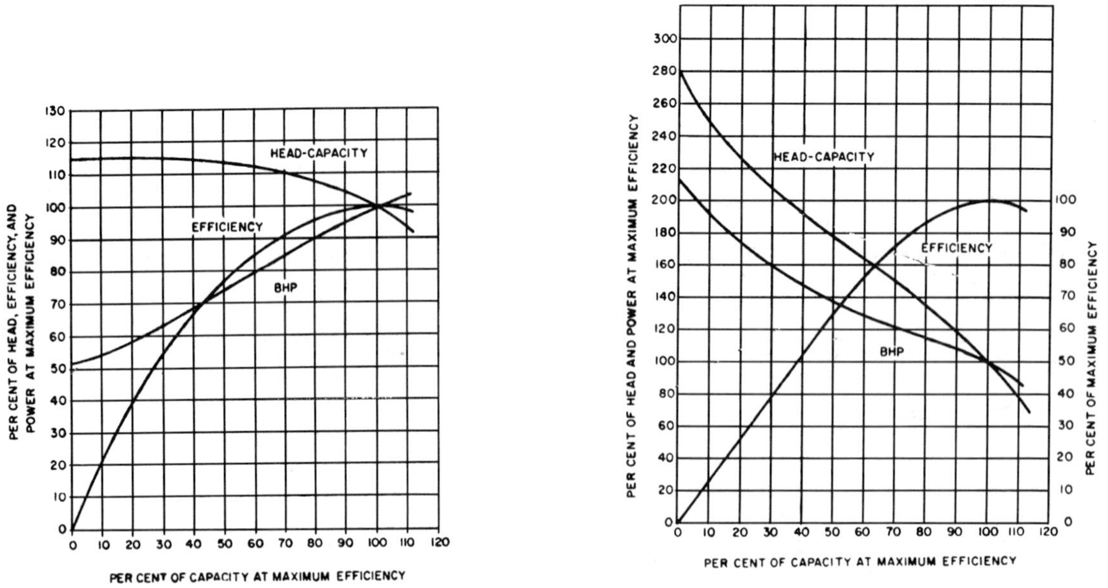

(a) (b)

Figure 7.3. Performance curves of single-suction impellers corresponding to two values of the specific speed. (a) $N_s = 1550$, centrifugal pump. (b) $N_s = 10{,}000$, mixed and axial flow pumps.

conditions is

$$n_2 = n_1(H_2/H_1)^{0.5}. \tag{7.17}$$

Example 7.2 is of cases with control valve throttling and pump speed control. In large systems, the value of power saved can easily overbalance the extra cost of variable speed drives, either motor or steam turbine.

When needed, greater head or greater capacity may be obtained by operating several pumps in series or parallel. In parallel operation, each pump develops the same head (equal to the system head), and the flow is the sum of the flows that each pump delivers at the common head. In series operation, each pump has the same flow rate and the total head is the sum of the heads developed by the individual pumps at the prevailing flow rate, and equal to the system head. Example 7.1 deals with a pair of identical pumps, and corresponding system and head curves are shown in Figure 7.17.

7.3. PUMP CHARACTERISTICS

A centrifugal pump is defined in the glossary at the end of this chapter as a machine in which a rotor in a casing acts on a liquid to give it a high velocity head that is in turn converted to pressure head by the time the liquid leaves the pump. Other common nomenclature relating to the construction and performance of centrifugal and related kinds of pumps also is in that table.

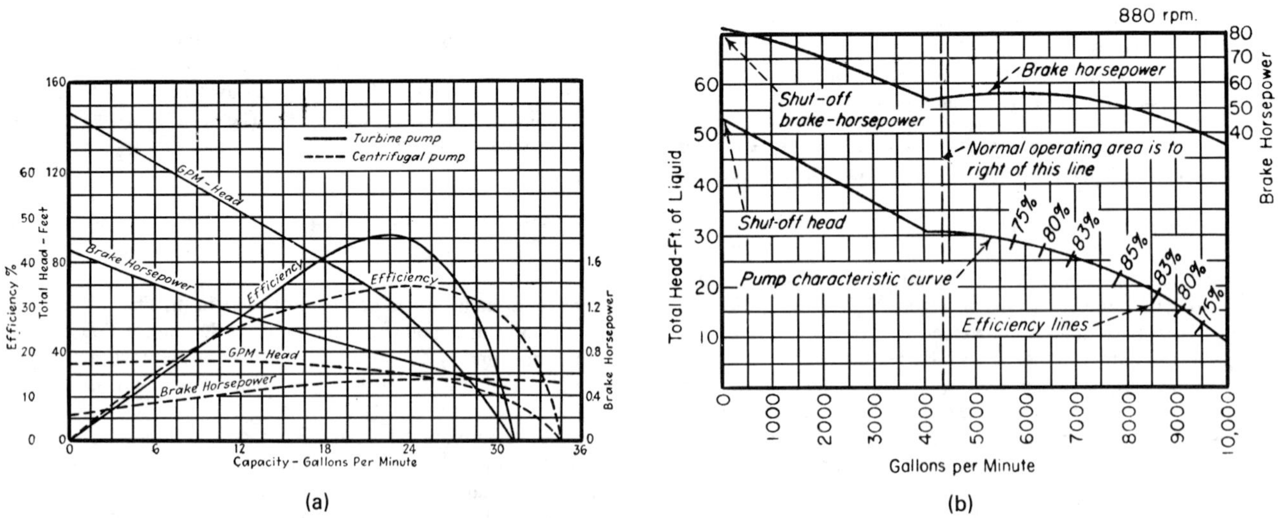

(a) (b)

Figure 7.4. Performance of several kinds of pumps. (a) Comparison of small centrifugal and turbine pumps (*Kristal and Annett, 1940*). (b) An axial flow pump operating at 880 rpm (Chem. Eng. Handbook, *1973*). (c) An external gear pump like that of Figure 7.12(e) (*Viking Pump Co.*). (d) A screw-type positive displacement pump. (e) NPSH of reciprocating positive displacement pumps.

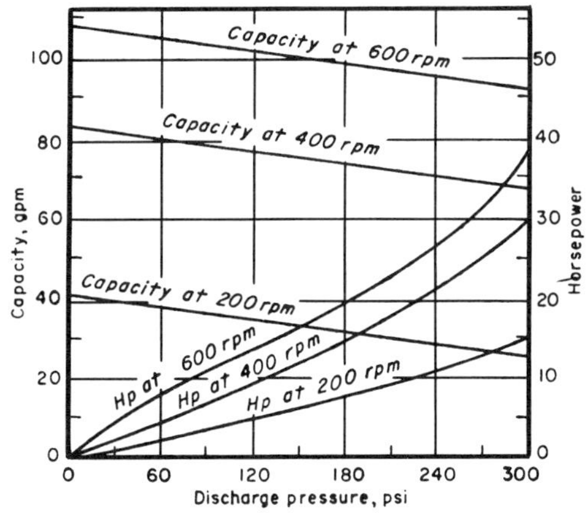

(c)

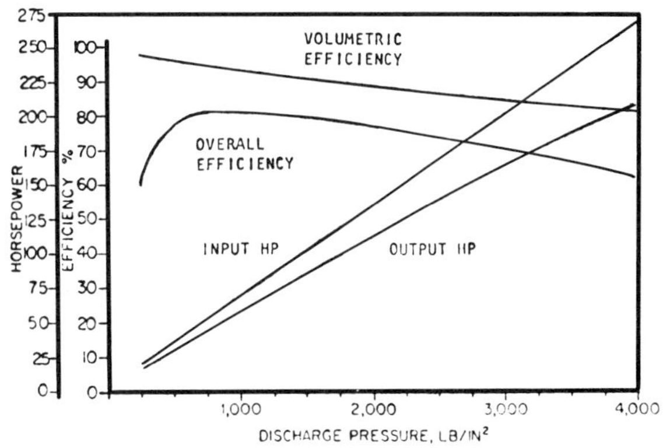

(d)

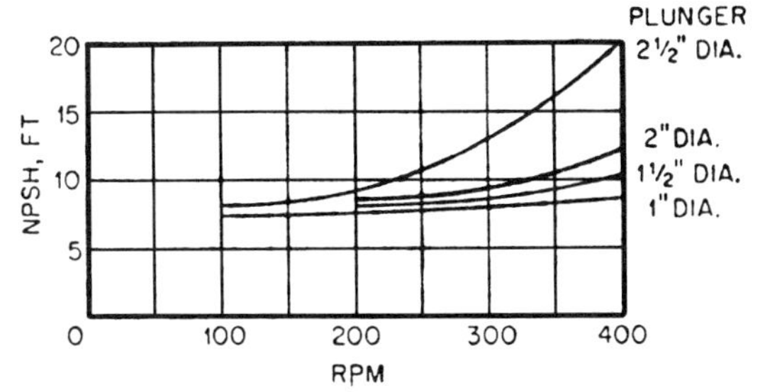

(e)

Figure 7.4—(*continued*)

(a) Efficiencies as % of those with direct piston drive:

Stroke, in	5	8	10	20	30	40	50
Crank-and-flywheel pump				87	88	90	92
Piston pump	60	70	74	84	86	88	90
High-pressure pump	55	64	67	76	78	80	81

(b) Efficiencies of crankshaft-driven pumps of various sizes:

Water HP	3	5	10	20	30	50	75	100	200
Efficiency (%)	55	65	72	77	80	83	85	86	88

(c) % of flow above and below the mean; curve is shown for triplex double-acting:

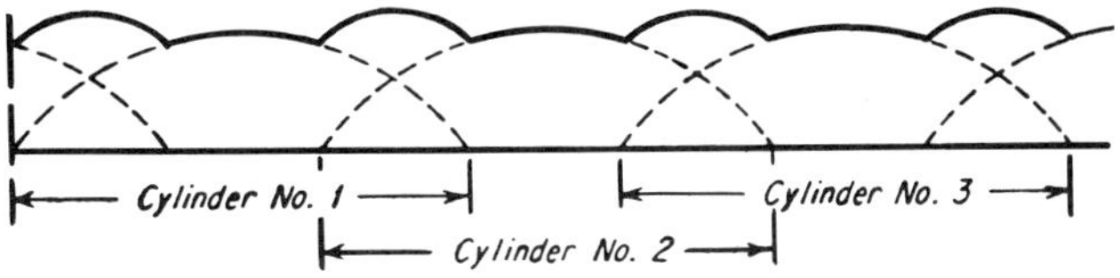

Type	Number of Plungers	% above Mean	% below Mean	Plunger Phase
Duplex (double)	2	24	22	180°
Triplex	3	6	17	120°
Quaduplex	4	11	22	90°
Quintaplex	5	2	5	72°
Sextuplex	6	5	9	60°

(d) Efficiency as a function of % reduced pressure or % reduced speed:

% Full-Load Developed Pressure	Mechanical Efficiency	% Speed	Mechanical Efficiency
20	82	44	93.3
40	88	50	92.5
60	90.5	73	92.5
80	92	100	92.5
100	92.5		

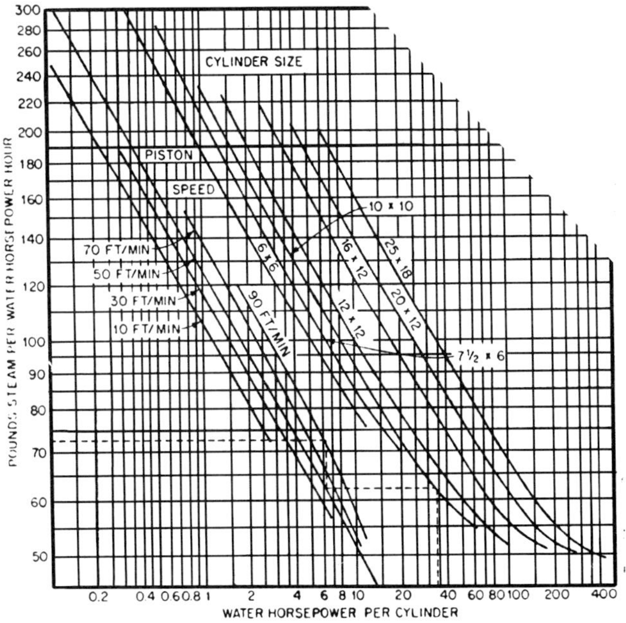

(e) Approximate steam consumption of reciprocating pumps (*Hydraulic Institute, Cleveland, OH, 1957*). *Example:* A pump with a 10 × 10 cylinder and developing 33 HP at 90 ft/min needs 73 lb steam/water HP. The 50 fpm line is a reference line.

Figure 7.5. Data relating to the performance of piston and plunger pumps.

Example 7.3
Check of Some Performance Curves with the Concept of Specific Speed

(**a**) The performance of the pump of Figure 7.7(b) with an 8 in. impeller will be checked by finding its specific speed and comparing with the recommended upper limit from Figure 7.6(b). Use Eq. (7.12) for N_s

Q (gpm)	100	200	300
H (ft)	268	255	225
N_s (calcd)	528	776	1044
N_s [Fig. 7.10(a)]	2050	2150	2500
NPSH	5	7	13

Clearly the performance curves are well within the recommended upper limits of specific speed.

(**b**) The manufacturer's recommended NPSH of the pump of Figure 7.7(c) with an 8 in. impeller will be checked against values from Eq. (7.15) with $S = 7900$:

Q (gpm)	100	150	200
H (ft)	490	440	300
NPSH (mfgr)	10	18	35
NPSH [Eq. (7.15)]	7.4	9.7	11.8

The manufacturer's recommended NPSHs are conservative.

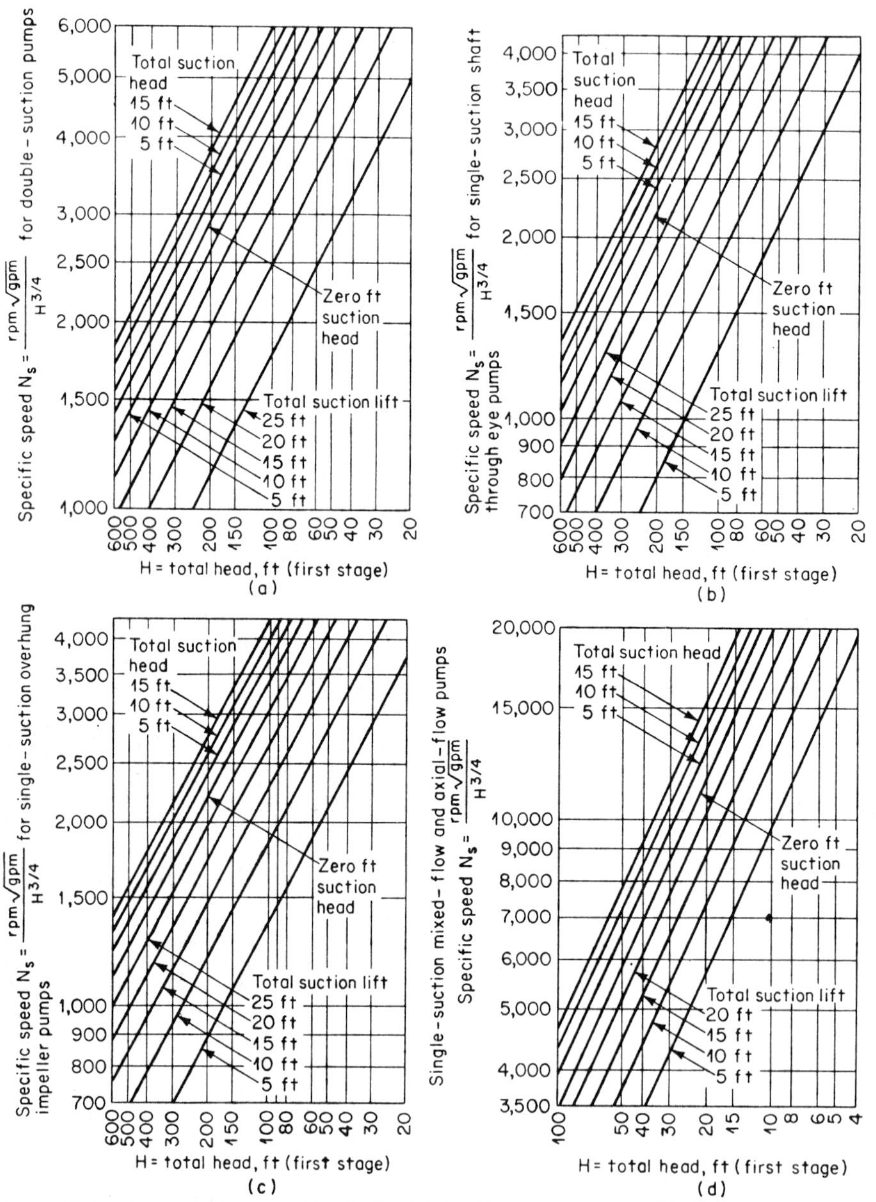

Figure 7.6. Upper specific-speed limits for (a) double-suction pumps (shaft through impeller eye) handling clear water at 85°F at sea level, (b) single-suction pumps (shaft through impeller eye) handling clear water at 85°F at sea level, (c) single-suction pumps (overhung-impeller type) handling clear water at 85°F at sea level, (d) single-suction mixed- and axial-flow pumps (overhung-impeller type) handling clear water at 85°F at sea level. (*Hydraulic Institute, Cleveland, OH, 1957*).

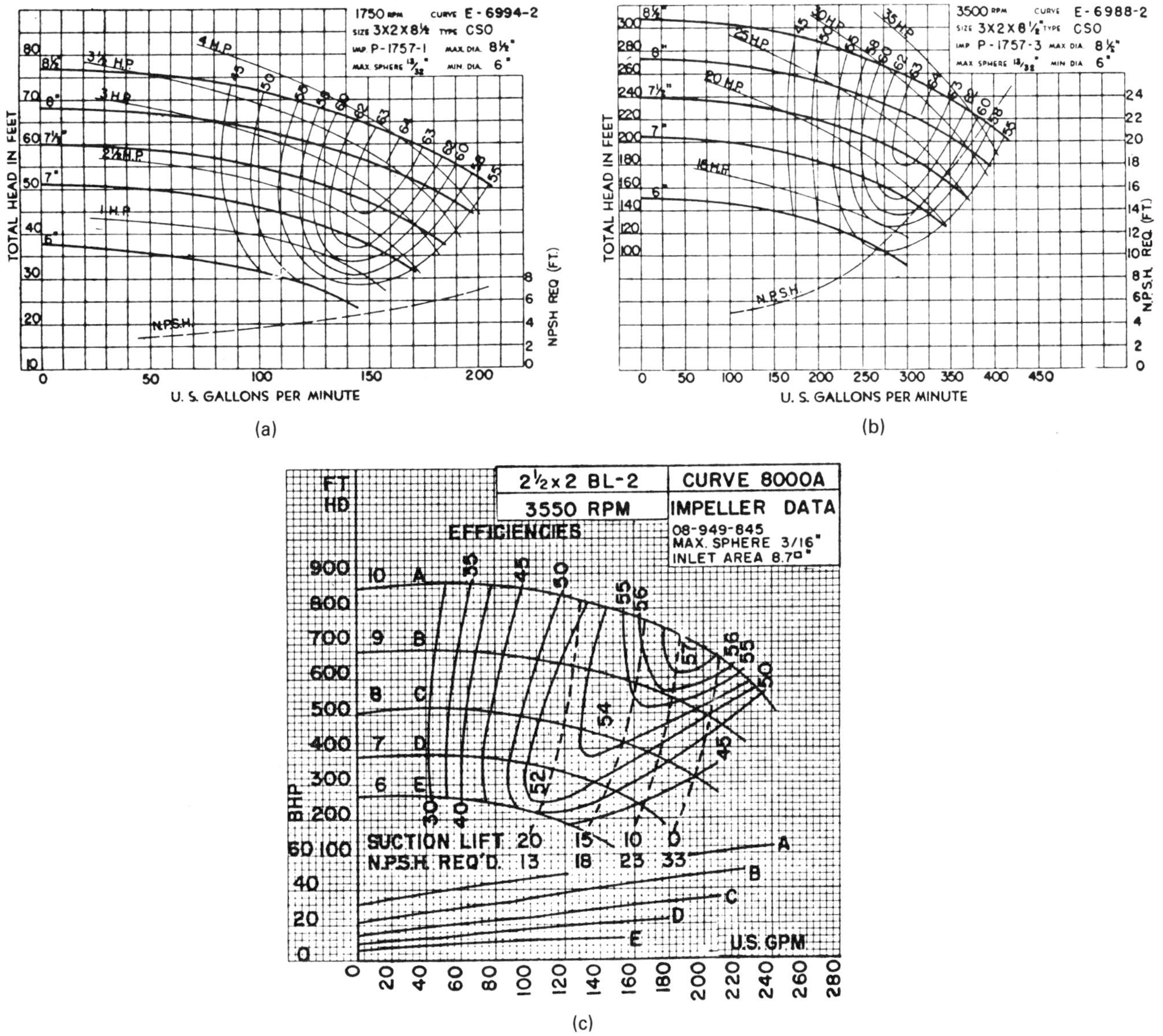

Figure 7.7. Characteristic curves of centrifugal pumps when operating on water at 85°F (*Allis Chalmers Co.*). (a) Single suction, 1750 rpm. (b) The pump of (a) operated at 3500 rpm. (c) Multistage, single suction, 3550 rpm.

The basic types of centrifugals are illustrated in Figure 7.9. A volute is a gradually expanding passage in which velocity is partially converted to pressure head at the outlet. The diffuser vanes of Figures 7.9(b) and 7.10(d) direct the flow smoothly to the periphery. The volute design is less expensive, more amenable to use with impellers of different sizes in the same case, and, as a consequence, by far the most popular construction. Diffuser construction is used to a limited extent in some high pressure, multistage machines. The double suction arrangement of Figure 7.9(d) has balanced axial thrust and is favored particularly for severe duty and where the lowered NPSH is an advantage. Multistage pumps, however, are exclusively single suction.

Some of the many kinds of impellers are shown in Figure 7.10. For clear liquids, some form of closed impeller [Figure 7.10(c)] is favored. They may differ in width and number and curvature of the vanes, and of course in the primary dimension, the diameter. Various extents of openness of impellers, [Figs. 7.10(a) and (b)] are desirable when there is a possibility of clogging as with slurries or pulps. The impeller of Figure 7.10(e) has both axial propeller and centrifugal vane action; the propeller confers high rates of flow but the developed pressure is low. Figure 7.3(b) represents a typical axial pump performance.

The turbine impeller of Figure 7.10(h) rotates in a case of uniform diameter, as in Figure 7.12(j). As Figure 7.4(a) demonstrates, turbine pump performance resembles that of positive displacement types. Like them, turbines are essentially self-priming, that is, they will not vapor bind.

All rotating devices handling fluids require seals to prevent leakage. Figure 7.13 shows the two common methods that are used: stuffing boxes or mechanical seals. Stuffing boxes employ a soft packing that is compressed and may be lubricated with the pump liquid or with an independent source. In mechanical seals, smooth metal surfaces slide on each other, and are lubricated with a very small leakage rate of the pump liquid or with an independent liquid.

Performance capability of a pump is represented on diagrams like those of Figure 7.7. A single point characterization often is made by stating the performance at the peak efficiency. For example, the pump of Figure 7.7(c) with a 9 in. impeller is called a

(a) Single-suction, 1800 rpm standard pumps:

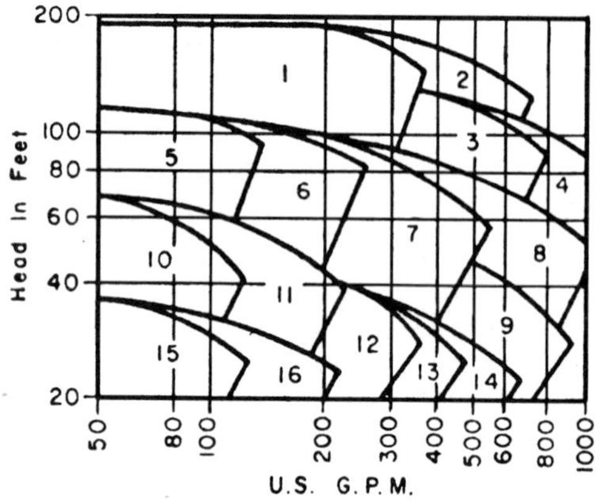

Key No.	Suction and Discharge	Approximate Cost	Horsepower Range at 1.0 Sp Gr
1	4 × 3	$1200	$7\frac{1}{2}$–25
2	6 × 4	1350	20–30
3	4 × 3	1200	15–25
4	5 × 4	1500	15–30
5	2 × $1\frac{1}{2}$	750	2–$7\frac{1}{2}$
6	$2\frac{1}{2}$ × 2	1050	3–10
7	4 × 3	1200	5–15
8	5 × 4	1350	$7\frac{1}{2}$–20
9	5 × 4 (1200 rpm)	1500	3–$7\frac{1}{2}$
10	$1\frac{1}{2}$ × $1\frac{1}{2}$	700	1–2
11	2 × 2	750	$1\frac{1}{2}$–3
12	3 × $2\frac{1}{2}$ (1200 rpm)	1050	$1\frac{1}{2}$–3
13	4 × 3 (1200 rpm)	1200	2–5
14	5 × 4 (1200 rpm)	1350	2–5
15	$2\frac{1}{2}$ × $2\frac{1}{2}$	500	$\frac{3}{4}$–$1\frac{1}{2}$
16	3 × 3	600	1–2

(b) Single-suction, 3600 rpm standard pumps:

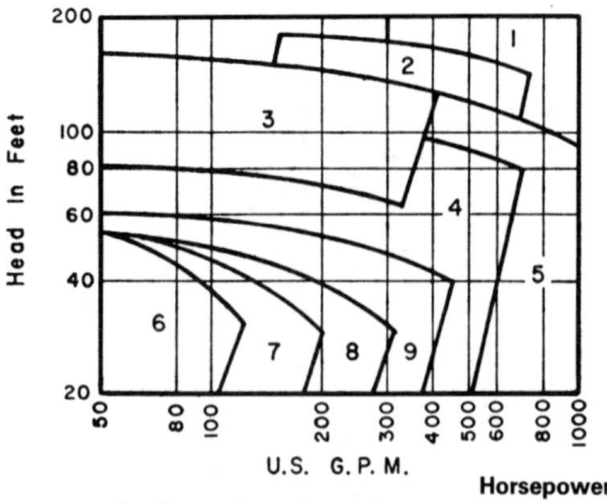

Key No.	Suction and Discharge	Approximate Cost	Horsepower Range at 1.0 Sp. Gr.
1	5 × 4	$2250	40–60
2	4 × 3	1950	20–40
3	3 × $2\frac{1}{2}$	1650	$7\frac{1}{2}$–20
4	4 × 4	1800	3–20
5	5 × 5	1950	5–30
6	2 × $1\frac{1}{2}$	1200	$\frac{3}{4}$–2
7	$2\frac{1}{2}$ × 2	1350	1–3
8	3 × $2\frac{1}{2}$	1500	2–$7\frac{1}{2}$
9	4 × 3	1650	2–$7\frac{1}{2}$

(c) Single-suction 1800 and 3600 rpm refinery pumps for elevated temperatures and pressures:

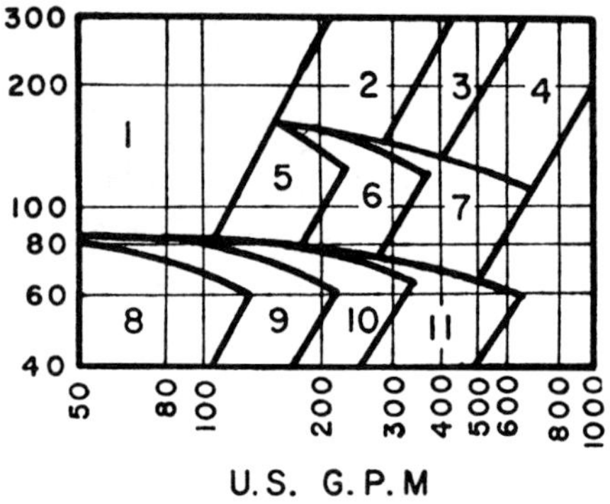

Key No.	Suction and Discharge	Approximate Cost	Horsepower Range at 1.0 Sp Gr
1	2 × $1\frac{1}{2}$	$3400	$7\frac{1}{2}$–30
2	3 × 2 (3600 rpm)	3700	15–50
3	4 × 3	4300	20–75
4	6 × 4	4800	40–125
5	3 × 2	4200	5–15
6	4 × 3 (1800 rpm)	4500	$7\frac{1}{2}$–20
7	6 × 4	5400	15–40
8	2 × $1\frac{1}{2}$	3400	1–5
9	3 × 2 (1800 rpm)	3700	2–$7\frac{1}{2}$
10	4 × 3	4300	3–10
11	6 × 4	4800	5–15

Figure 7.8. Typical capacity-head ranges of some centrifugal pumps, their 1978 costs and power requirements. Suction and discharge are in inches (*Evans, 1979, Vol. 1*).

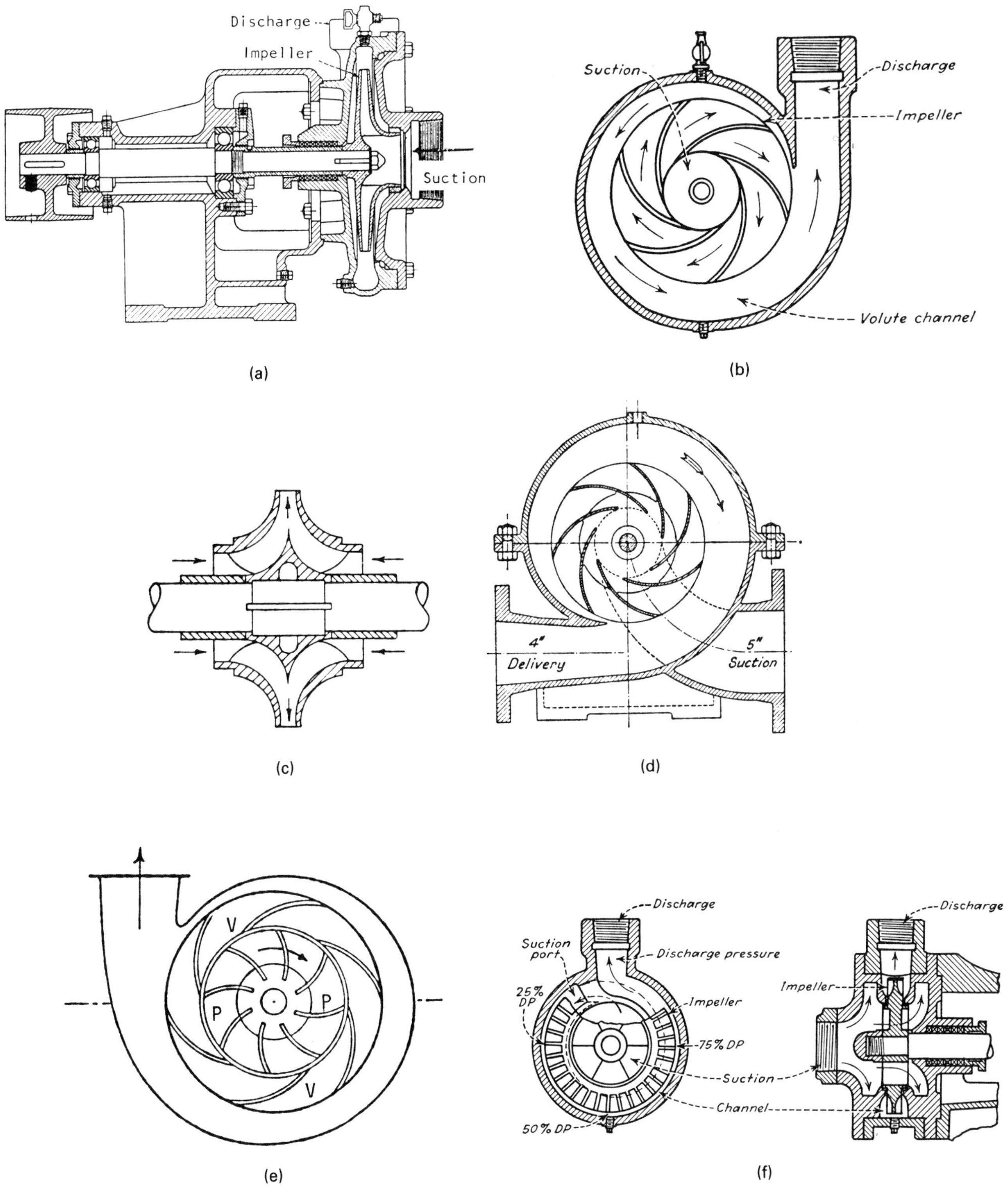

Figure 7.9. Some types of centrifugal pumps. (a) Single-stage, single suction volute pump. (b) Flow path in a volute pump. (c) Double suction for minimizing axial thrust. (d) Horizontally split casing for ease of maintenance. (e) Diffuser pump: vanes V are fixed, impellers P rotate. (f) A related type, the turbine pump.

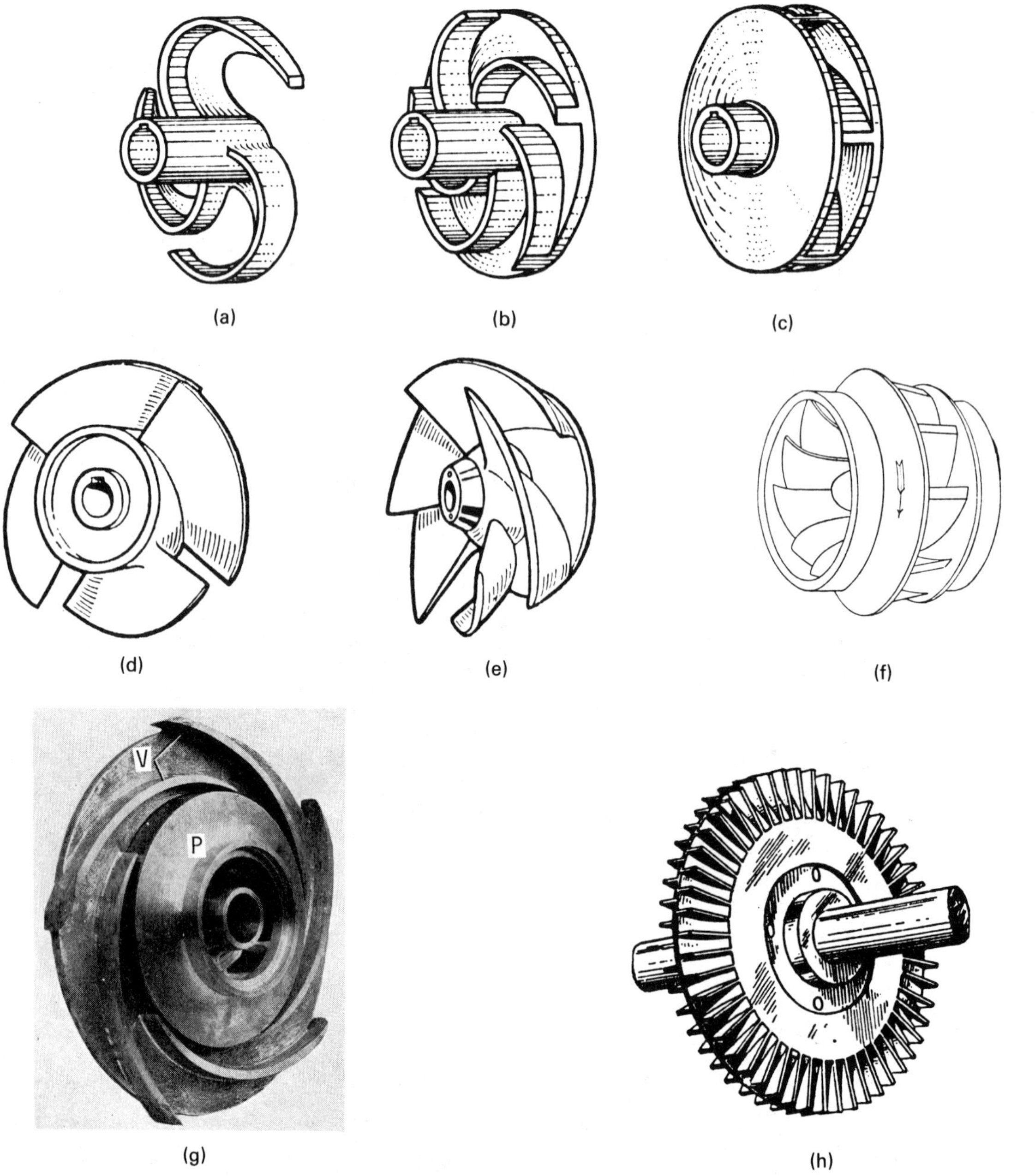

Figure 7.10. Some types of impellers for centrifugal pumps. (a) Open impeller. (b) Semiopen impeller. (c) Shrouded impeller. (d) Axial flow (propeller) type. (e) Combined axial and radial flow, open type. (f) Shrouded mixed-flow impeller. (g) Shrouded impeller (P) in a case with diffuser vanes (V). (h) Turbine impeller.

175 gpm and 560 ft head pump at a peak efficiency of 57%; it requires a 15 ft suction lift, an 18 ft NPSH and 43 BHP. Operating ranges and costs of commercial pumps are given in Figure 7.8. General operating data are in Figure 7.4.

Although centrifugal pumps are the major kinds in use, a great variety of other kinds exist and have limited and sometimes unique applications. Several kinds of positive displacement types are sketched in Figure 7.12. They are essentially self-priming and have a high tolerance for entrained gases but not usually for solids unless they may be crushed. Their characteristics and applications are discussed in the next section.

7.4. CRITERIA FOR SELECTION OF PUMPS

The kind of information needed for the specification of centrifugal, reciprocating and rotary pumps is shown on forms in Appendix B. General characteristics of classes of pumps are listed in Table 7.1 and their ranges of performance in Table 7.2. Figure 7.14 shows recommended kinds of pumps in various ranges of pressure and flow rate. Suitable sizes of particular styles of a manufacturer's pumps are commonly represented on diagrams like those of Figure 7.8. Here pumps are identified partly by the sizes of suction and discharge nozzles in inches and the rpm; the key number also

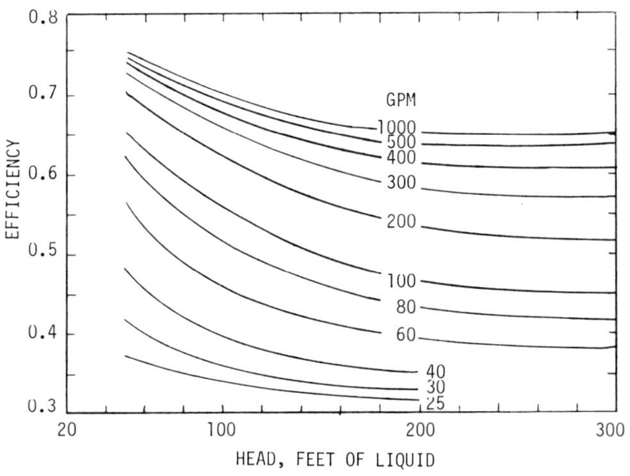

Figure 7.11. Approximate efficiencies of centrifugal pumps in terms of GPM and head in feet of liquid.

identifies impeller and case size and other details which are stated in a catalog. Each combination of head and capacity will have an efficiency near the maximum of that style. Although centrifugal pumps function over a wide range of pressure and flow rates, as represented by characteristic curves like those of Figures 7.2 and 7.7, they are often characterized by their performance at the peak efficiency, as stated in the previous section. Approximate efficiencies of centrifugal pumps as functions of head and capacity are on Figure 7.11 and elsewhere here.

Centrifugal pumps have a number of good qualities:

1. They are simple in construction, are inexpensive, are available in a large variety of materials, and have low maintenance cost.
2. They operate at high speed so that they can be driven directly by electrical motors.
3. They give steady delivery, can handle slurries and take up little floor space.

Some of their drawbacks are

4. Single stage pumps cannot develop high pressures except at very high speeds (10,000 rpm for instance). Multistage pumps for high

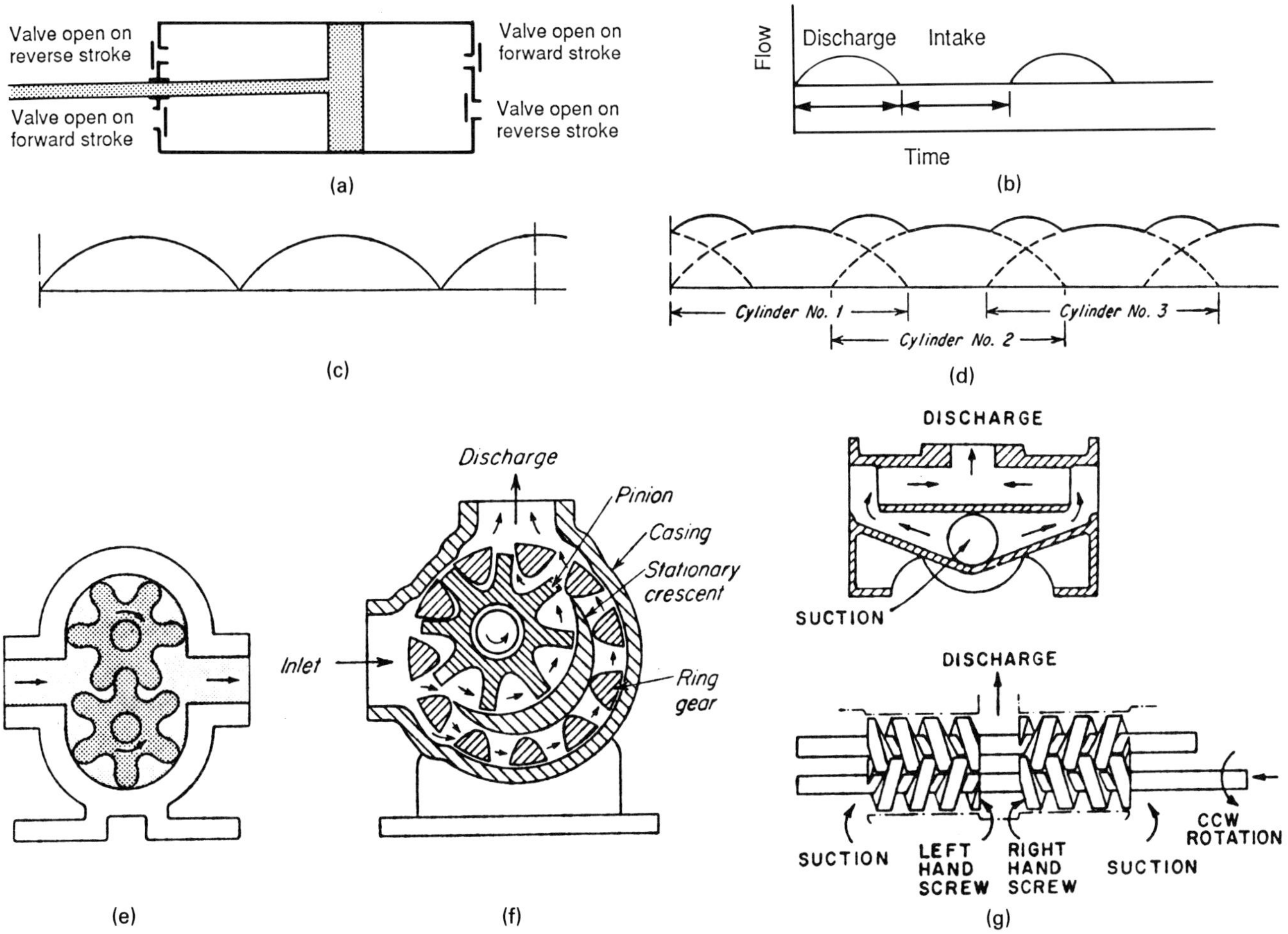

Figure 7.12. Some types of positive displacement pumps. (a) Valve action of a double acting reciprocating piston pump. (b) Discharge curve of a single acting piston pump operated by a crank; half-sine wave. (c) Discharge curve of a simplex double acting pump as in (a). (d) Discharge curve of a duplex, double acting pump. (e) An external gear pump; characteristics are in Figure 7.8(c). (f) Internal gear pump; the outer gear is driven, the inner one follows. (g) A double screw pump. (h) Peristaltic pump in which fluid is squeezed through a flexible tube by the follower. (i) Double diaphragm pump shown in discharge position (BIF unit of General Signal). (j) A turbine pump with essentially positive displacement characteristics (*data on Fig. 7.4(a)*).

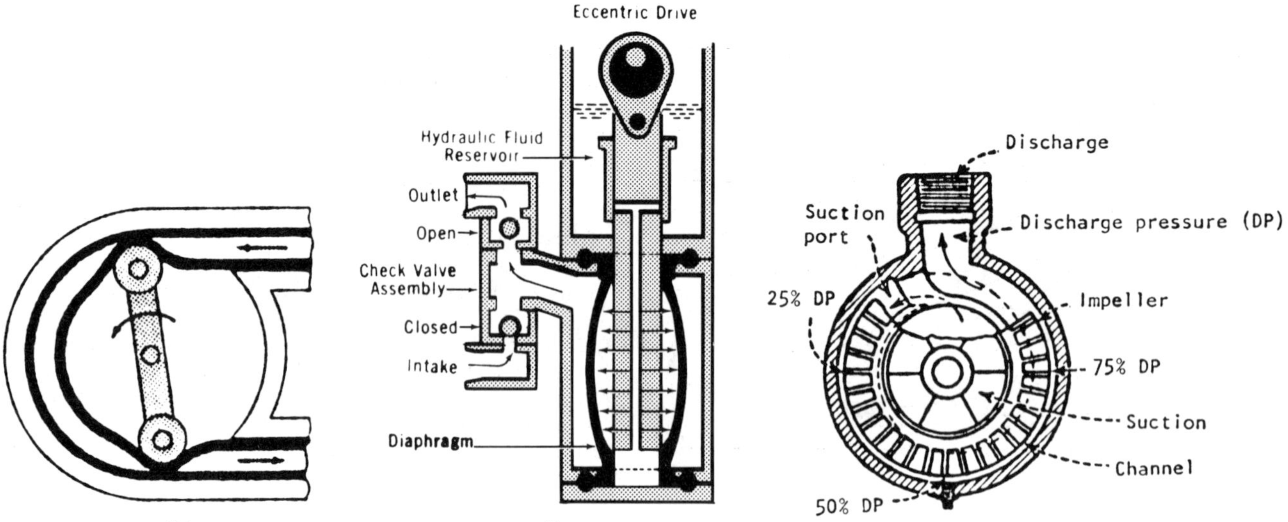

(h) (i) (j)

Figure 7.12—*(continued)*

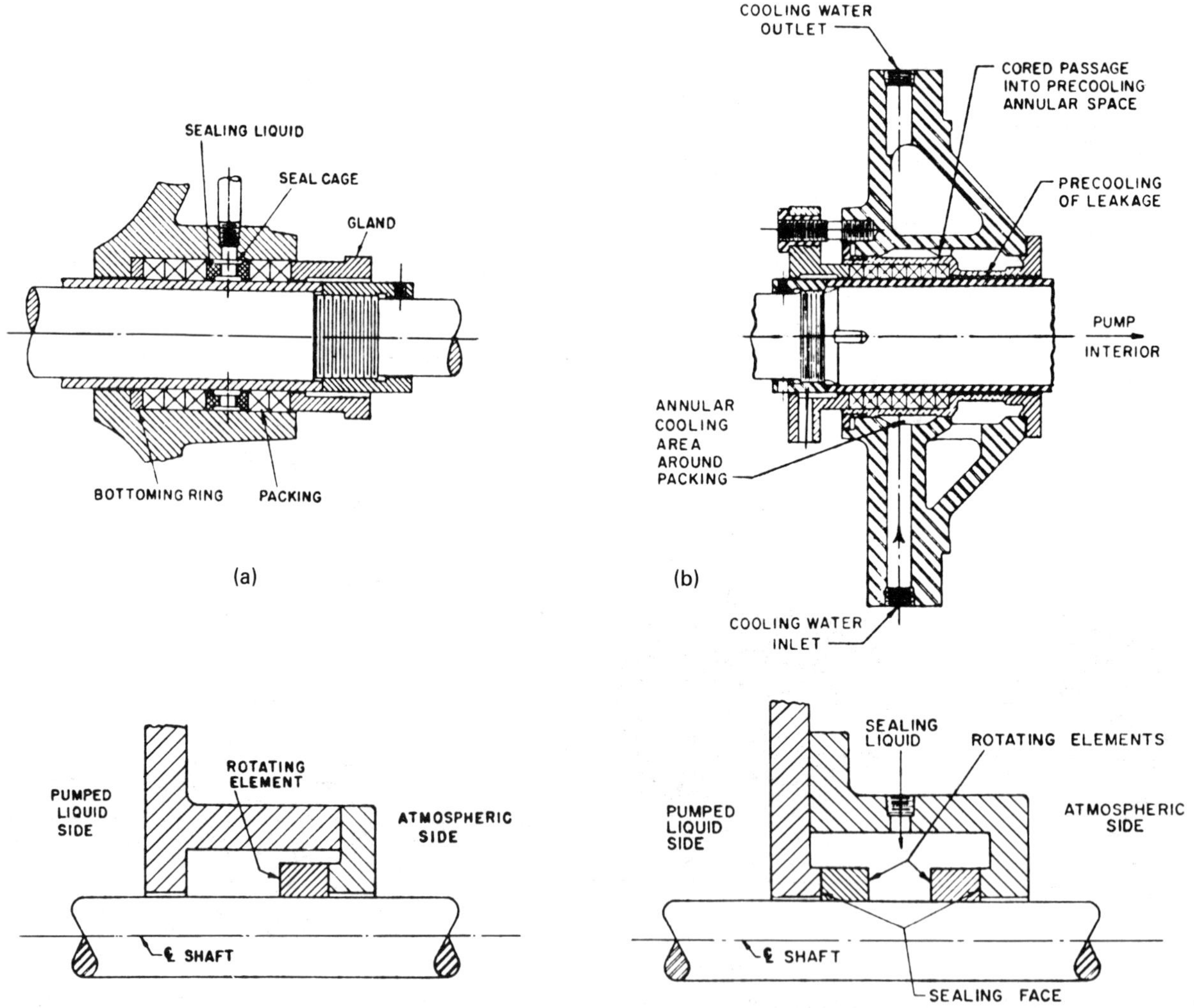

(c) (d)

Figure 7.13. Types of seals for pump shafts. (a) Packed stuffing box; the sealing liquid may be from the pump discharge or from an independent source. (b) Water cooled stuffing box. (c) Internal assembly mechanical seal; the rotating and fixed surfaces are held together by the pressure of the pump liquid which also serves as lubricant; a slight leakage occurs. (d) Double mechanical seal with independent sealing liquid for handling toxic or inflammable liquids.

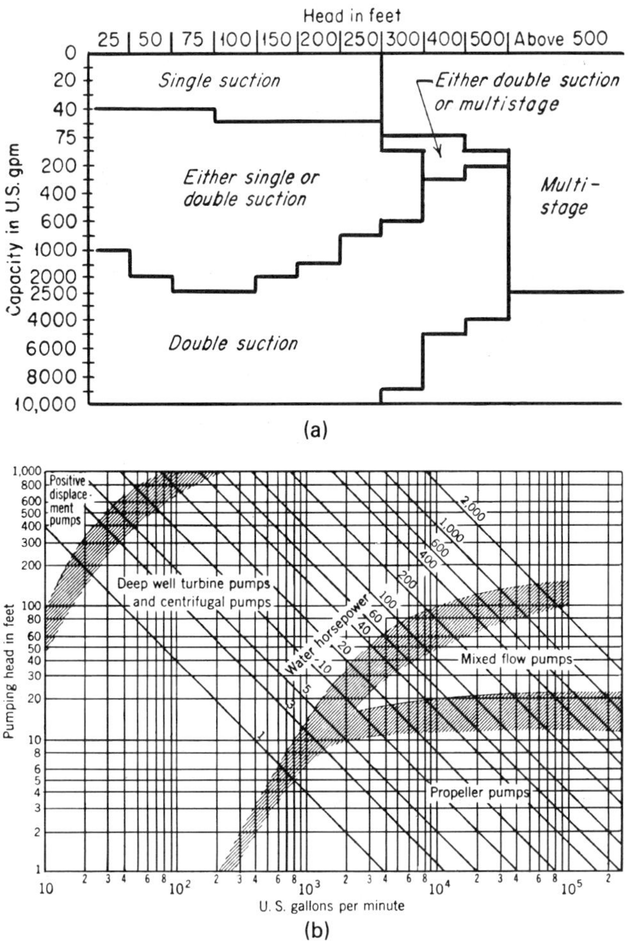

Figure 7.14. Range of applications of various kinds of pumps. (a) Range of applications of single and double suction pumps (*Allis-Chalmers Co.*). (b) Recommended kinds of pumps for various kinds of head and flow rate (*Fairbanks, Morse, and Co.*).

pressures are expensive, particularly in corrosion-resistant materials.

5. Efficiencies drop off rapidly at flow rates much different from those at peak efficiency.
6. They are not self-priming and their performance drops off rapidly with increasing viscosity. Figure 7.15 illustrates this effect.

On balance, centrifugal pumps always should be considered first in comparison with reciprocating or rotary positive displacement types, but those do have their places. Range of applications of various kinds of pumps are identified by Figure 7.14.

Pumps with reciprocating pistons or plungers are operated with steam, motor or gas engine drives, directly or through gears or belts. Their mode of action is indicated on Figure 7.12(a). They are always used with several cylinders in parallel with staggered action to smooth out fluctuations in flow and pressure. Figure 7.5(c) shows that with five cylinders in parallel the fluctuation is reduced to a maximum of 7%. External fluctuation dampers also are used. Although they are self-priming, they do deteriorate as a result of cavitation caused by release of vapors in the cylinders. Figure 7.4(e) shows the NPSH needed to repress cavitation. Application of reciprocating pumps usually is to low capacities and high pressures of 50–1000 atm or more. Some performance data are shown in Figure 7.5.

Diaphragm pumps [Fig. 7.12(i)] also produce pulsating flow. They are applied for small flow rates, less than 100 gpm or so, often for metering service. Their utility in such applications overbalances the drawback of their intrinsic low efficiencies, of the order of 20%.

Screw pumps [Fig. 7.12(g)] are suited for example to high viscosity polymers and dirty liquids at capacities up to 2000 gpm and pressures of 200 atm at speeds up to 3000 rpm. They are compact, quiet, and efficient. Figure 7.4(d) shows typical performance data.

Gear pumps [Figs. 7.12(e) and (f)] are best suited to handling clear liquids at a maximum of about 1000 gpm at 150 atm. Typical performance curves are shown in Figure 7.4(c).

Peristaltic pumps [Fig. 7.12(h)] move the liquid by squeezing a tube behind it with a rotor. Primarily they are used as metering pumps at low capacities and pressures in corrosive and sanitary services when resistant flexible tubes such as those of teflon can be used, and in laboratories.

Turbine pumps [Figs. 7.9(f), 7.12(i), and 7.4(a)] also are called regenerative or peripheral. They are primarily for small capacity and high pressure service. In some ranges they are more efficient than centrifugals. Because of their high suction lifts they are suited to handling volatile liquids. They are not suited to viscous liquids or abrasive slurries.

7.5. EQUIPMENT FOR GAS TRANSPORT

Gas handling equipment is used to transfer materials through pipe lines, during which just enough pressure or head is generated to overcome line friction, or to raise or lower the pressure to some required operating level in connected process equipment. The main classes of this kind of equipment are illustrated in Figures 7.18 and 7.19 and are described as follows.

1. *Fans* accept gases at near atmospheric pressure and raise the pressure by approximately 3% (12 in. of water), usually on air for ventilating or circulating purposes.
2. *Blowers* is a term applied to machines that raise the pressure to an intermediate level, usually to less than 40 psig, but more than accomplished by fans.
3. *Compressors* are any machines that raise the pressure above the levels for which fans are used. Thus, in modern terminology they include blowers.
4. *Jet compressors* utilize a high pressure gas to raise other gases at low pressure to some intermediate value by mixing with them.
5. *Vacuum pumps* produce subatmospheric pressures in process equipment. Often they are compressors operating in reverse but other devices also are employed. Operating ranges of some commercial equipment are stated in Table 7.3.
6. *Steam jet ejectors* are used primarily to evacuate equipment but also as pumps or compressors. They are discussed in Section 7.7.

Application ranges of fans and compressors are indicated on Figures 7.20 and 7.21. Some of these categories of equipment now will be discussed in some detail.

FANS

Fans are made either with axial propellers or with a variety of radial vanes. The merits of different directions of curvature of the vanes are stated in Figure 7.24 where the effect of flow rate of pressure, power, and efficiency also are illustrated. Backward curved vanes are preferable in most respects. The kinds of controls used have a marked effect on fan performance as Figure 7.23 shows. Table 7.4 shows capacity ranges and other characteristics of various kinds of

TABLE 7.1. Characteristics of Various Kinds of Pumps

Pump Type	Construction Style	Construction Characteristics	Notes
Centrifugal (horizontal)	single-stage overhung, process type	impeller cantilevered beyond bearings	capacity varies with head
	two-stage overhung	two impellers cantilevered beyond bearings	used for heads above single-stage capability
	single-stage impeller between bearings	impeller between bearings; casing radially or axially split	used for high flows to 1083 ft (330 m) head
	chemical	casting patterns designed with thin sections for high-cost alloys	have low pressure and temperature ratings
	slurry	designed with large flow passages	low speed and adjustable axial clearance; has erosion control features
	canned	no stuffing box; pump and motor enclosed in a pressure shell	low head capacity limits when used in chemical services
	multistage, horizontally split casing	nozzles located in bottom half of casing	have moderate temperature-pressure ranges
	multistage, barrel type	outer casing contains inner stack of diaphragms	used for high temperature-pressure ratings
Centrifugal (vertical)	single-stage, process type	vertical orientation	used to exploit low net positive section head (NPSH) requirements
	multistage	many stages with low head per stage	low-cost installation
	inline	inline installation, similar to a valve	low-cost installation
	high speed	speeds to 380 rps, heads to 5800 ft (1770 m)	high head/low flow; moderate costs
	slump	casing immersed in sump for easy priming and installation	low cost
	multistage, deep well	long shafts	used for water well service
Axial	propeller	propeller-shaped impeller	vertical orientation
Turbine	regenerative	fluted impeller. Flow path resembles screw around periphery	capacity independent of head; low flow/high head performance
Reciprocating	piston, plunger	slow speeds	driven by steam engine cylinders or motors through crankcases
	metering	consists of small units with precision flow control system	diaphragm and packed plunger types
	diaphragm	no stuffing box	used for chemical slurries; can be pneumatically or hydraulically actuated
Rotary	screw	1, 2, or 3 screw rotors	for high-viscosity, high-flow high-pressure services
	gear	intermeshing gear wheels	for high-viscosity, moderate-pressure/moderate-flow services

(Cheremisinoff, 1981).

TABLE 7.2. Typical Performances of Various Kinds of Pumps[a]

Type	Style	Capacity (gpm)	Max Head (ft)	Max *P* (psi)	NPSH (ft)	Max *T* (°F)	Efficiency (%)
Centrifugal (horizontal)	single-stage overhung	15–5,000	492	600	6.56–19.7	851	20–80
	two-stage overhung	15–1,200	1394	600	6.56–22.0	851	20–75
	single-stage impeller between bearings	15–40,000	1099	980	6.56–24.9	401–851	30–90
	chemical	1000	239	200	3.94–19.7	401	20–75
	slurry	1000	394	600	4.92–24.9	851	20–80
	canned	1–20,000	4921	10,000	6.56–19.7	1004	20–70
	multistage horizontal split	20–11,000	5495	3000	6.56–19.7	401–500	65–90
	multistage, barrel type	20–9,000	5495	6000	6.56–19.7	851	40–75
Centrifugal (vertical)	single stage	20–10,000	804	600	0.98–19.7	653	20–85
	multistage	20–80,000	6004	700	0.98–19.7	500	25–90
	inline	20–12,000	705	500	6.56–19.7	500	20–80
	high speed	5–400	5807	2000	7.87–39.4	500	10–50
	sump	10–700	197	200	0.98–22.0		45–75
	multistage deep well	5–400	6004	2000	0.98–19.7	401	30–75
Axial	propeller	20–100,000	39	150	6.56	149	65–85
Turbine	regenerative	1–2000	2493	1500	6.56–8.20	248	55–85
Reciprocating	piston, plunger	10–10,000	1.13×10^6	>50,000	12.1	554	65–85
	metering	0–10	1.70×10^5	50,000	15.1	572	20
	diaphragm	4–100	1.13×10^5	3500	12.1	500	20
Rotary	screw	1–2000	6.79×10^4	3000	9.84	500	50–80
	gear	1–5000	11,155	500	9.84	653	50–80

[a] 1 m^3/min = 264 gpm, 1 m = 3.28 ft, 1 bar = 14.5 psi, °C = (°F − 32)/1.8.

TABLE 7.3. Operating Ranges of Some Commercial Vacuum Producing Equipment

Type of Pump	Operating Range (mm Hg)
Reciprocating piston	
1-stage	760–10
2-stage	760–1
Rotary piston oil-sealed	
1-stage	$760–10^{-2}$
2-stage	$760–10^{-3}$
Centrifugal multistage (dry) liquid jet	760–200
Mercury Sprengel	$760–10^{-3}$
Water aspirator (18°C)	760–15
Two-lobe rotary blower (Roots type)	$20–10^{-4}$
Turbomolecular	$10^{-1}–10^{-10}$
Zeolite sorption (liquid nitrogen cooled)	$760–10^{-3}$
Vapor jet pumps	
Steam ejector	
1-stage	760–100
2-stage	760–10
3-stage	760–1
4-stage	$760–3 \times 10^{-1}$
5-stage	$760–5 \times 10^{-2}$
Oil ejector (1-stage)	$2–10^{-2}$
Diffusion-ejector	$2–10^{-4}$
Mercury diffusion with trap	
1-stage	$10^{-1}–<10^{-6}$
2-stage	$1–<10^{-6}$
3-stage	$10–<10^{-6}$
Oil diffusion	
1-stage	$10^{-1}–5 \times 10^{-6}$
4-stage fractionating (untrapped)	$5 \times 10^{-1}–10^{-9}$
4-stage fractionating (trapped)	$5 \times 10^{-1}–10^{-12}$
Getter-ion (sputter-ion)	$10^{-3}–10^{-11}$
Sublimation (titanium)	$10^{-4}–10^{-11}$
Cryopumps (20 K)	$10^{-2}–10^{-10}$
Cryosorption (15 K)	$10^{-2}–10^{-12}$

(*Encyclopedia of Chemical Technology,* Wiley-Interscience, New York, 1978–1984).

fans. Figure 7.24 allows exploration of the effects of changes in specific speed or diameter on the efficiencies and other characteristics of fans. The mutual effects of changes in flow rate, pressure, speed, impeller diameter, and density are related by the "fan laws" of Table 7.5, which apply to all rotating propelling equipment.

COMPRESSORS

The several kinds of commercial compressors are identified in this classification:

1. Rotodynamic
   a. Centrifugal (radial flow)
   b. Axial flow
2. Positive displacement
   a. Reciprocating piston
   b. Rotary (screws, blades, lobes, etc.).

Sketches of these several types are shown in Figures 7.19 and 7.20 and their application ranges in Figures 7.20 and 7.21.

CENTRIFUGALS

The head-flow rate curve of a centrifugal compressor often has a maximum as shown on Figure 3.21, similar to the pump curve of Figure 7.7(c). To the left the developed head increases with flow, but to the right the head decreases with increasing flow rate. At the peak the flow pulsates and the machine vibrates. This operating point is called the *surge limit* and is always identified by the manufacturer of the equipment, as shown on Figure 7.25 for those centrifugal and axial machines. Stable operation exists anywhere right of the surge limit. Another kind of flow limitation occurs when the velocity of the gas somewhere in the compressor approaches sonic velocity. The resulting shock waves restrict the flow; a slight increase in flow then causes a sharp decline in the developed pressure.

Table 7.6 shows as many as 12 stages in a single case. These machines are rated at either 10K or 12K ft/stage. The higher value corresponds to about 850 ft/sec impeller tip speed which is near the limit for structural reasons. The limitation of head/stage depends on

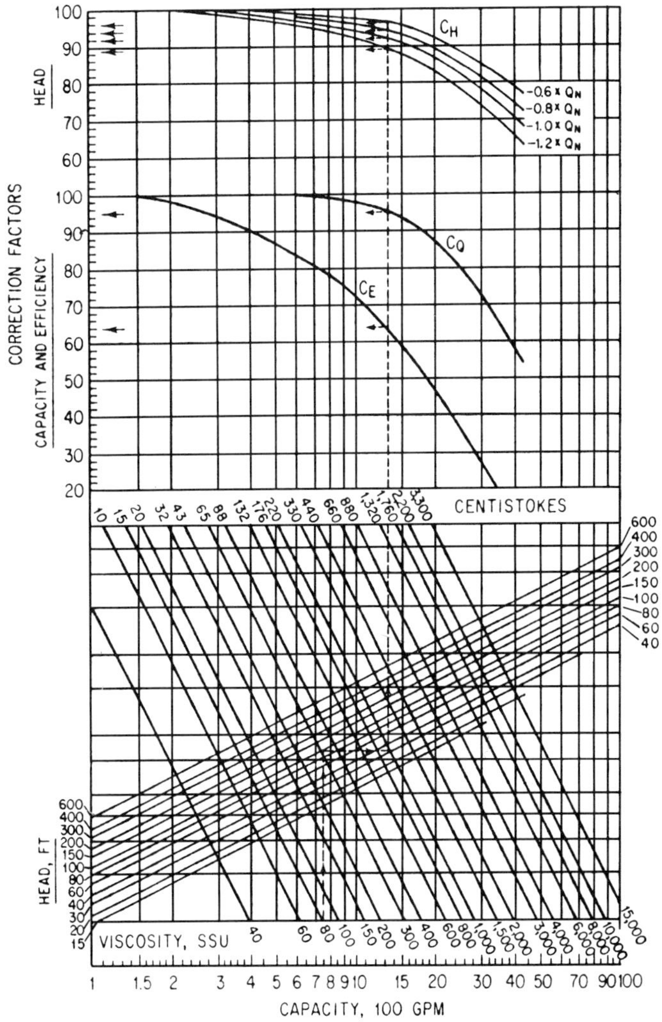

(a)

Figure 7.15. Effects of viscosity on performance of centrifugal pumps: (a) Hydraulic Institute correction chart for pumping liquids. (b) Typical performances of pumps when handling viscous liquids. The dashed lines on the chart on the left refer to a water pump that has a peak efficiency at 750 gpm and 100 ft head; on a liquid with viscosity 1000 SSU (220 CS) the factors relative to water are efficiency 64%, capacity 95% and head 89% that of water at 120% normal capacity ($1.2Q_H$).

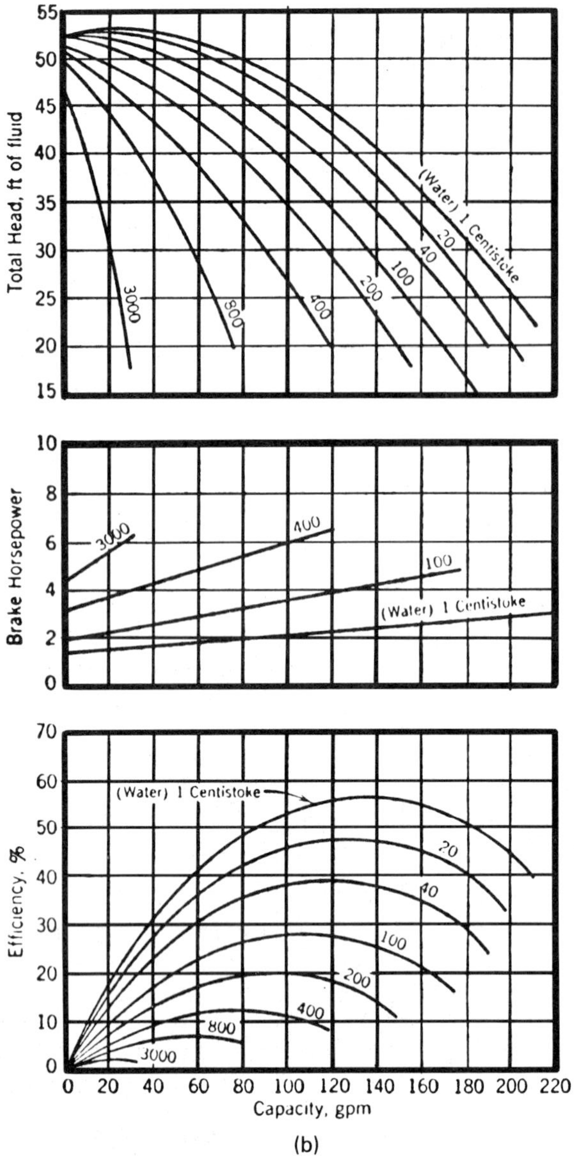

Figure 7.15—(*continued*)

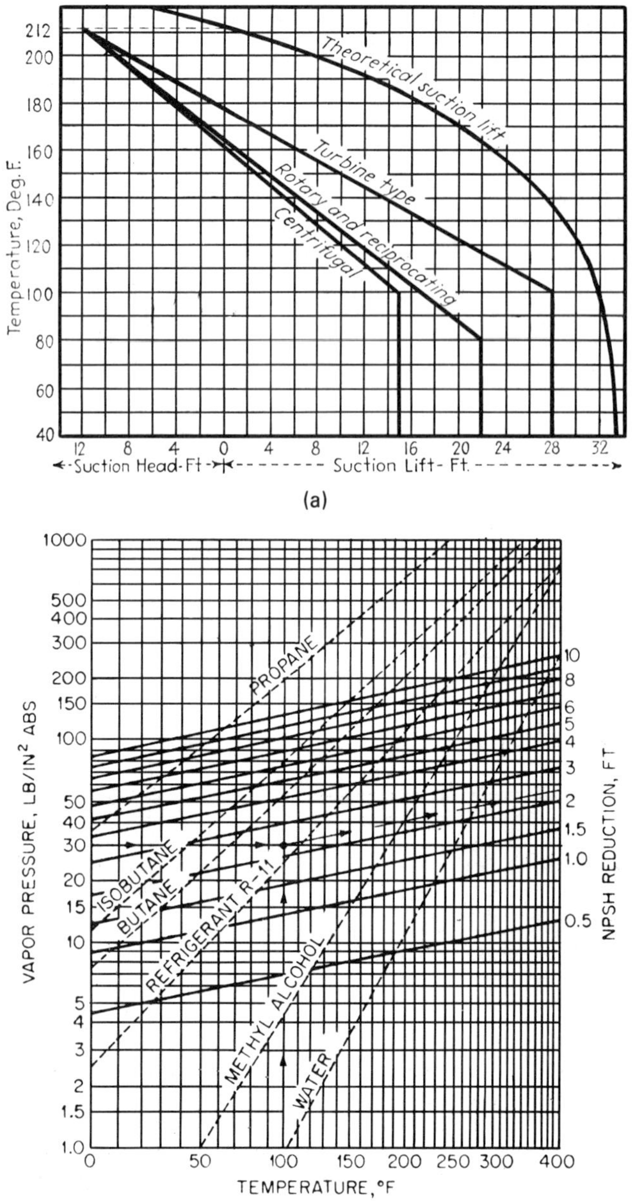

Figure 7.16. Recommended values of net positive suction head (NPSH) at various temperatures or vapor pressures: (a) NPSH of several types of pumps for handling water at various temperatures. (b) Correction of the cold water NPSH for vapor pressure. The maximum recommended correction is one-half of the cold water value. The line with arrows shows that for a liquid with 30 psia vapor pressure at 100°F, the reduction in NPSH is 2.3 ft (*data of Worthington International Inc.*).

the nature of the gas and the temperature, as indicated on Figure 7.26. Maximum compression ratios of 3–4.5 per stage with a maximum of 8–12 per machine are commonly used. Discharge pressures as high as 3000–5000 psia can be developed by centrifugal compressors.

A specification form is included in Appendix B and as Table 4.4. Efficiency data are discussed in Section 7.6, Theory and Calculations of Gas Compression: Efficiency.

AXIAL FLOW COMPRESSORS

Figure 7.18(b) shows the axial flow compressor to possess a large number of blades attached to a rotating drum with stationary but adjustable blades mounted on the case. Typical operating characteristics are shown on Figure 7.22(a). These machines are suited particularly to large gas flow rates at maximum discharge pressures of 80–130 psia. Compression ratios commonly are 1.2–1.5 per stage and 5–6.5 per machine. Other details of range of applications are stated on Figure 7.20. According to Figure 7.21, specific speeds of axial compressors are in the range of 1000–3000 or so.

Efficiencies are 8–10% higher than those of comparable centrifugal compressors.

RECIPROCATING COMPRESSORS

Reciprocating compressors are relatively low flow rate, high pressure machines. Pressures as high as 35,000–50,000 psi are

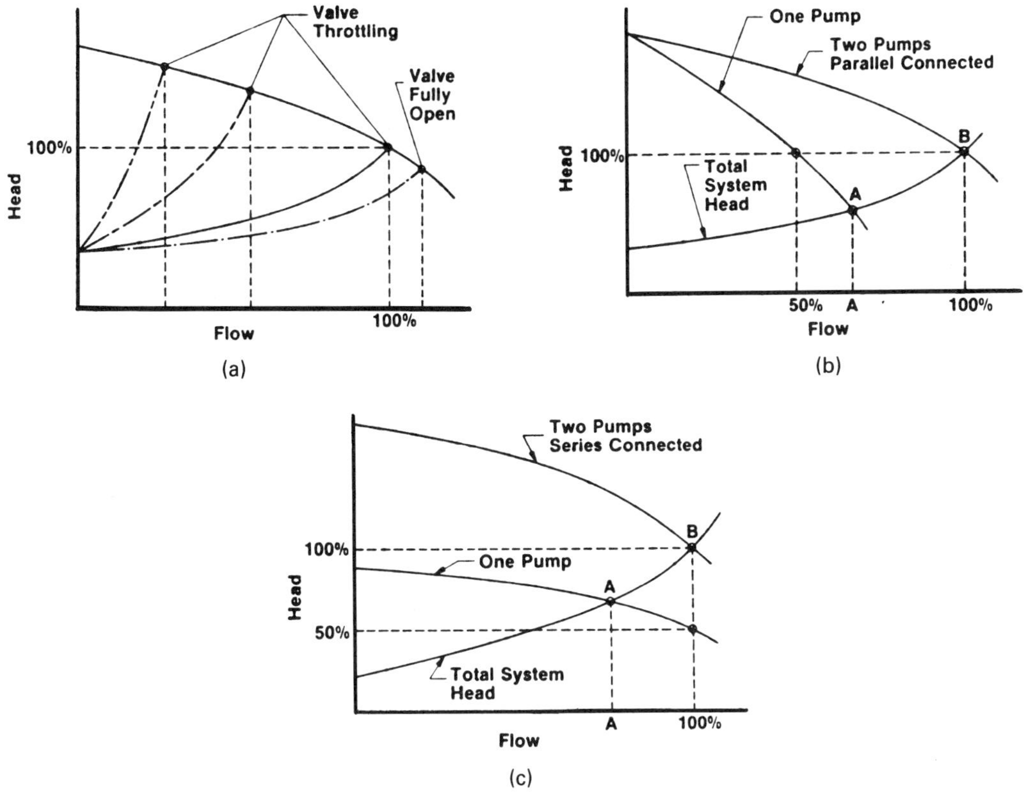

Figure 7.17. Operating points of centrifugal pumps under a variety of conditions. (a) Operating points with a particular pump characteristic and system curves corresponding to various amounts of flow throttling with a control valve. (b) Operating point with two identical pumps in parallel; each pump delivers one-half the flow and each has the same head. (c) Operating point with two identical pumps in series; each pump delivers one-half the head and each has the same flow.

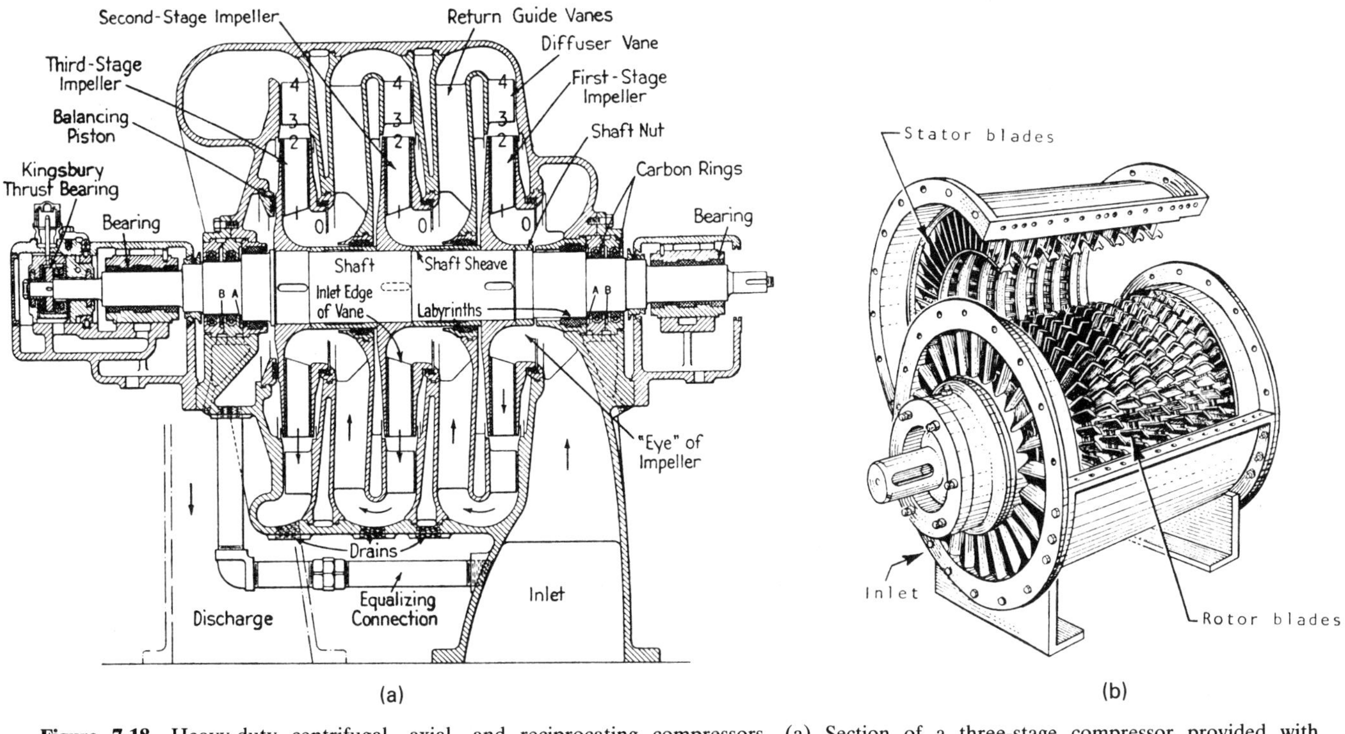

Figure 7.18. Heavy-duty centrifugal, axial, and reciprocating compressors. (a) Section of a three-stage compressor provided with steam-sealed packing boxes (*DeLaval Steam Turbine Co.*). (b) An axial compressor (*Clark Brothers Co.*). (c) Double-acting, two-stage reciprocating compressor with water-cooled jacket and intercooler (*Ingersoll-Rand Co.*).

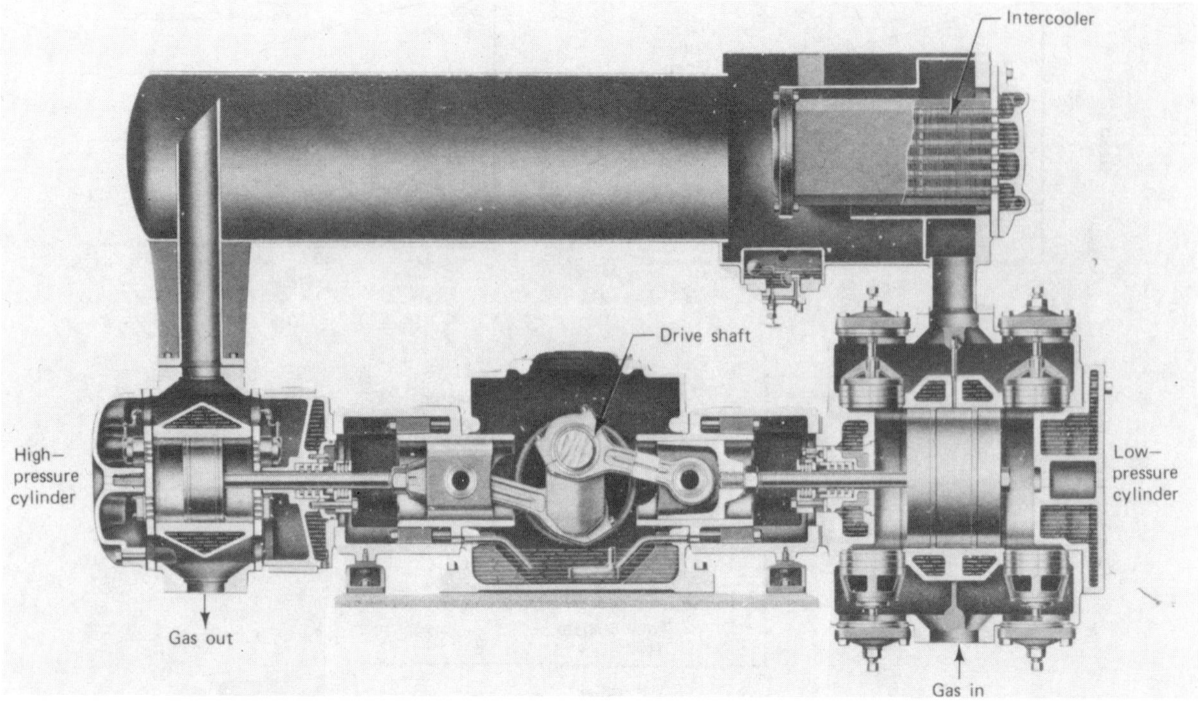

Figure 7.18—(*continued*) (c)

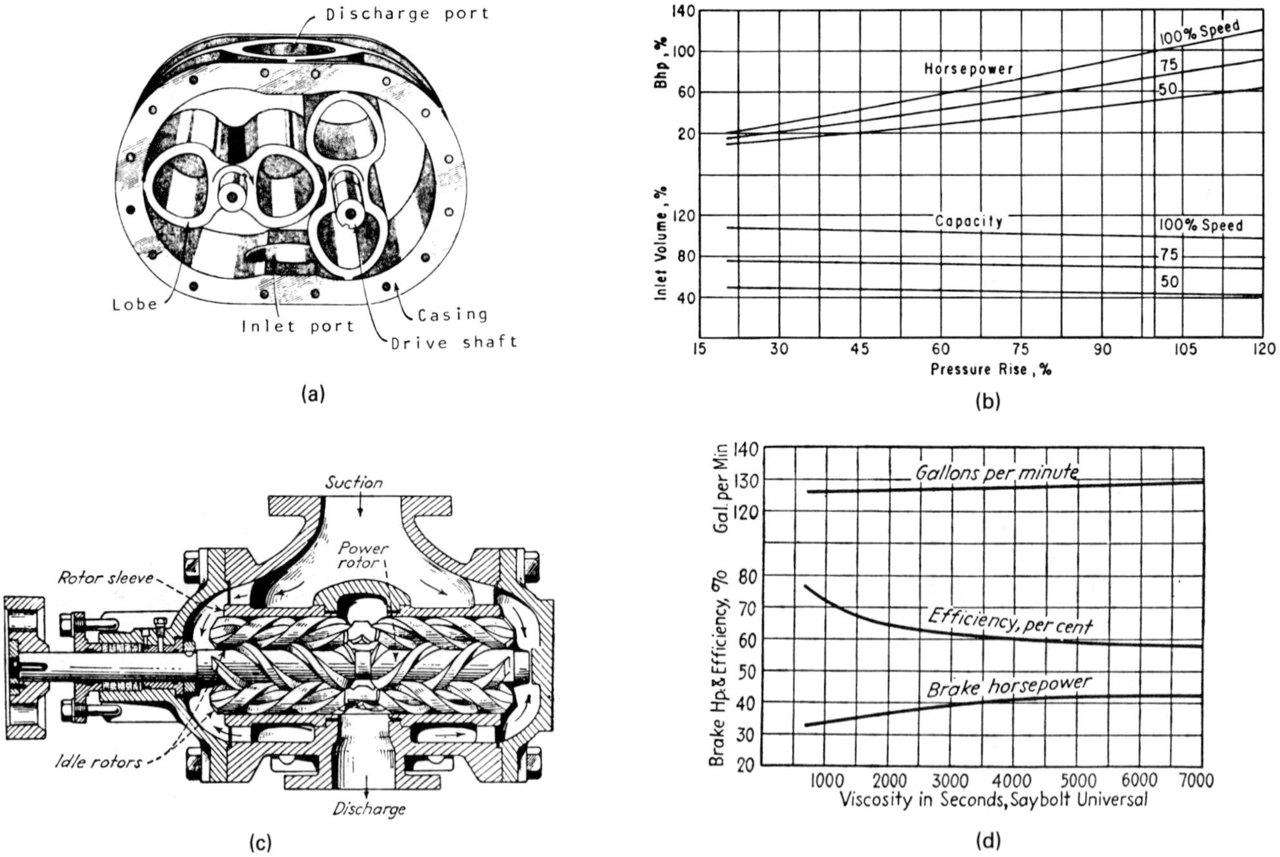

Figure 7.19. Some rotary positive displacement compressors. (a) A two-lobe blower. (b) Performance of a two-lobe blower (*Roots-Connersville Co.*). (c) A screw pump with one power and two idle rotors (*Kristal and Annett, 1940*). (d) Performance of 3.5″ screw pump handling oils at 1150 rpm against 325 psig (*Kristal and Annett, 1940*). (e) Principle of the liquid ring seal compressor (*Nash Engineering Co.*). (f) A sliding vane blower (*Beach-Russ Co.*).

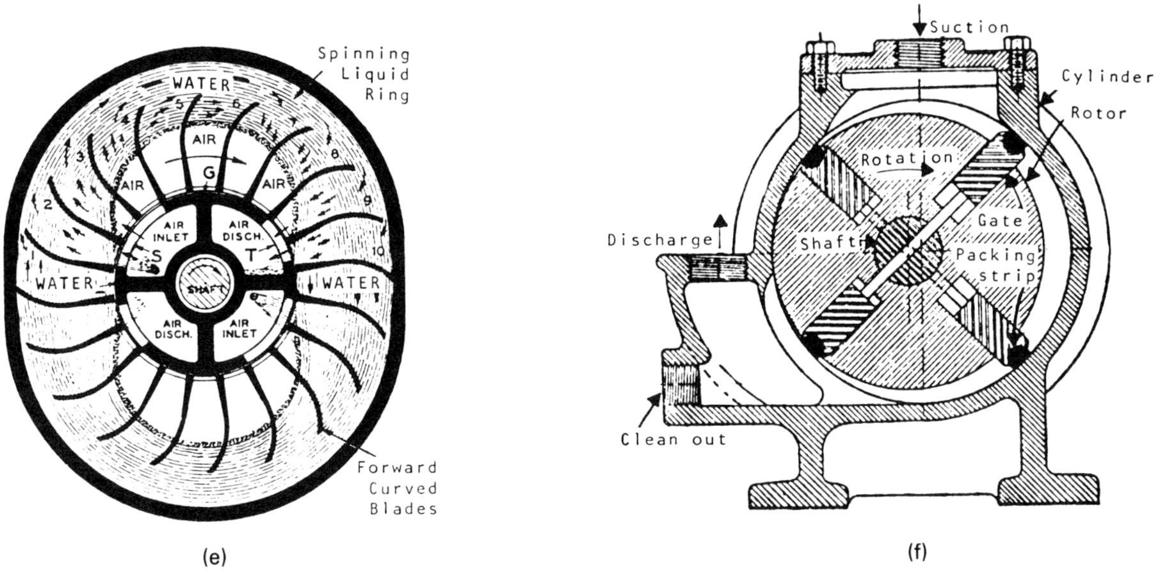

Figure 7.19.—*(continued)*

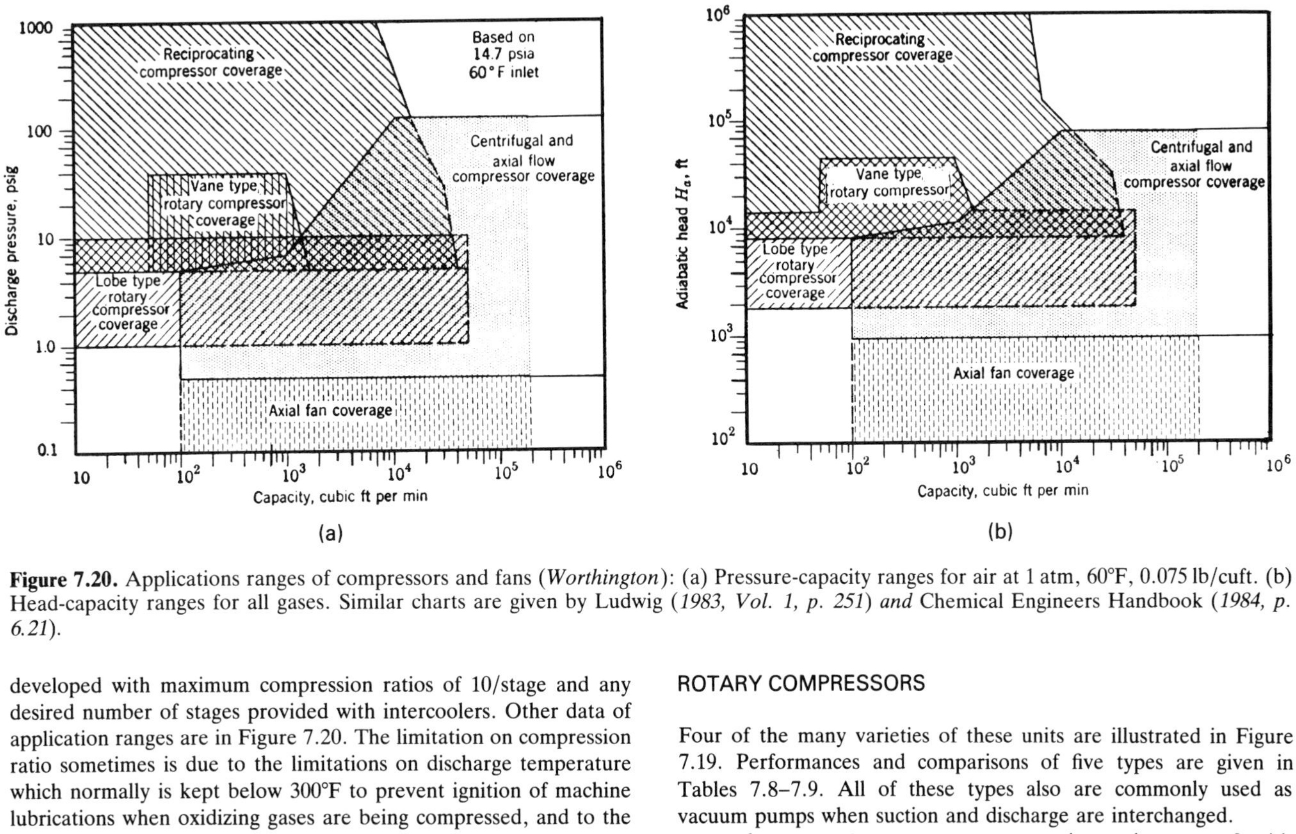

Figure 7.20. Applications ranges of compressors and fans (*Worthington*): (a) Pressure-capacity ranges for air at 1 atm, 60°F, 0.075 lb/cuft. (b) Head-capacity ranges for all gases. Similar charts are given by Ludwig (*1983, Vol. 1, p. 251) and* Chemical Engineers Handbook (*1984, p. 6.21*).

developed with maximum compression ratios of 10/stage and any desired number of stages provided with intercoolers. Other data of application ranges are in Figure 7.20. The limitation on compression ratio sometimes is due to the limitations on discharge temperature which normally is kept below 300°F to prevent ignition of machine lubrications when oxidizing gases are being compressed, and to the fact that power requirements are proportional to the absolute temperature of the suction gas.

A two-stage double-acting compressor with water cooled cylinder jackets and intercooler is shown in Figure 7.18(c). Selected dimensional and performance data are in Table 7.7. Drives may be with steam cylinders, turbines, gas engines or electrical motors. A specification form is included in Appendix B. Efficiency data are discussed in Section 7.6, Theory and Calculations of Gas Compression: Temperature Rise, Compression Ratio, Volumetric Efficiency.

ROTARY COMPRESSORS

Four of the many varieties of these units are illustrated in Figure 7.19. Performances and comparisons of five types are given in Tables 7.8–7.9. All of these types also are commonly used as vacuum pumps when suction and discharge are interchanged.

Lobe type units operate at compression ratios up to 2 with efficiencies in the range of 80–95%. Typical relations between volumetric rate, power, speed, and pressure boost are shown in Figure 7.19(b).

Spiral screws usually run at 1800–3600 rpm. Their capacity ranges up to 12,000 CFM or more. Normal pressure boost is 3–20 psi, but special units can boost pressures by 60–100 psi. In vacuum service they can produce pressures as low as 2 psia. Some other performance data are shown with Figure 7.19(d).

The sliding vane compressor can deliver pressures of 50 psig or

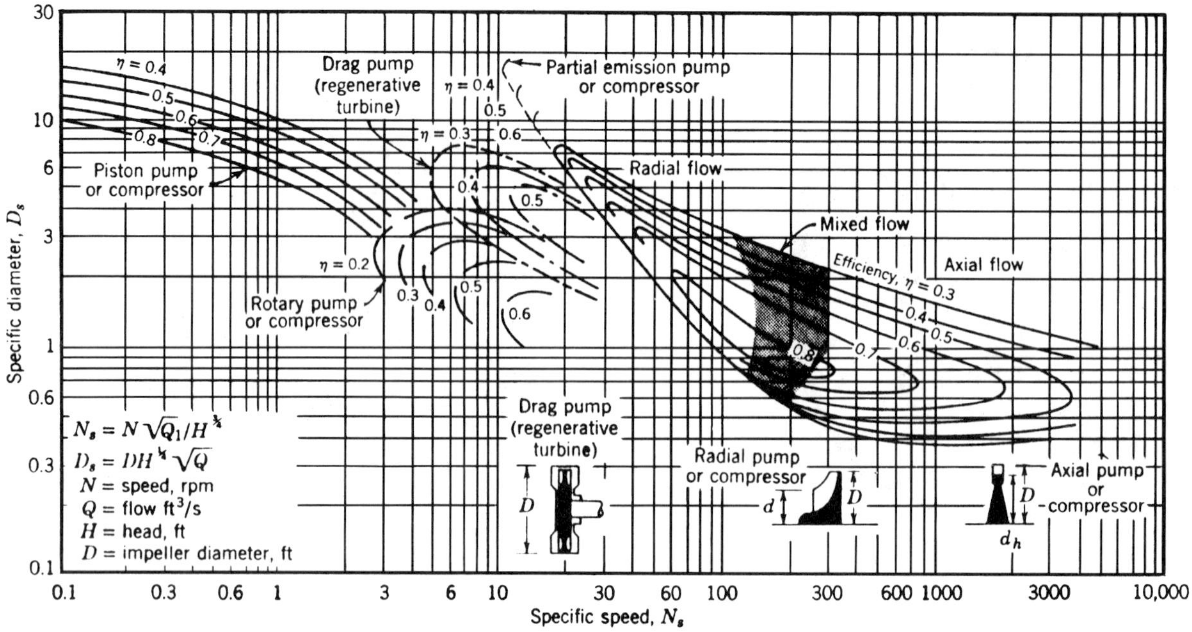

Figure 7.21. Operating ranges of single-stage pumps and compressors [*Balje,* Trans. ASME, J. Eng. Power. **84,** *103* (*1962*)]. *Example:* atmospheric air at the rate of 100,000 SCFM is compressed to 80,000 ft lbf/ft (41.7 psig) at 12,000 rpm; calculated $N_s = 103$; in the radial flow region with about 80% efficiency, $D_s = 1.2–1.6$, so that $D = 2.9–3.9$ ft.

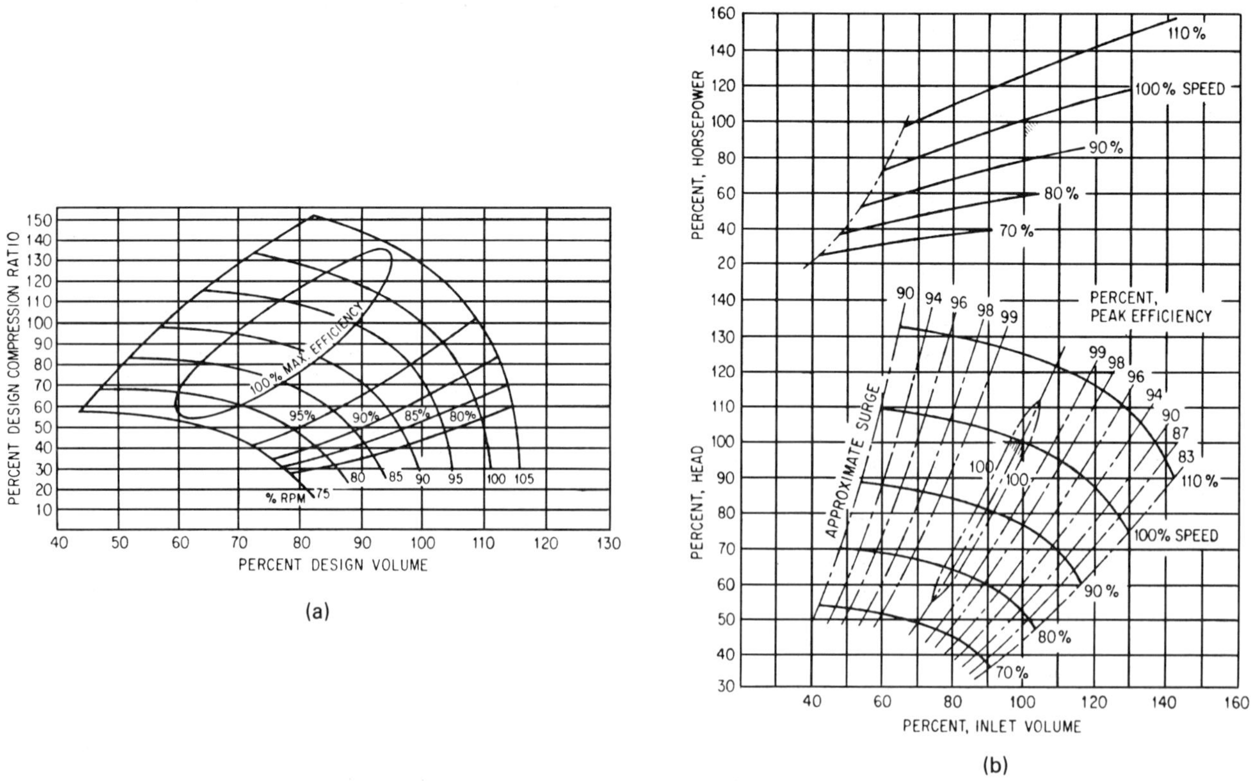

Figure 7.22. Performances of dynamic compressors: (a) Axial compressor. (b) Centrifugal compressor. All quantities are expressed as percentages of those at the design condition which also is the condition of maximum efficiency (De Laval Engineering Handbook, *McGraw-Hill, New York, 1970*).

TABLE 7.4. Performance Characteristics of Fans[a]

Description	Quantity (1000 acfm)		Head Inches Water	Opt. V (fps)	Max q_{ad}	Diameter (in.)		N_s	D_s	Peak Eff.
	Min	Max				Min	Max			
Axial propeller	8	20	10	410	0.13	23	27	470	0.63	77
Axial propeller	20	90	8	360	0.12	27	72	500	0.60	80
Axial propeller	6	120	2.5	315	0.10	27	84	560	0.50	84
Radial air foil	6	100	22	250	0.45	18	90	190	0.85	88
Radial BC	3	35	18	260	0.63	18	90	100	1.35	78
Radial open MH	2	27	18	275	0.55	18	66	97	1.45	56
Radial MH	2	27	18	250	0.55	18	66	86	1.53	71
Radial IS	2	27	18	250	0.55	18	66	86	1.53	66
Vane BI flat	1	10	12	250	0.43	10	30	210	0.81	70
Vane FC	1	10	2	65	1.15	10	30	166	0.65	66

[a] $q_{ad} = 32.2H/V^2$, $N_s = NQ^{0.5}/V^{0.75}$ (specific speed), $D_s = DV^{0.25}/Q^{0.5}$ (specific diameter), where D = diameter (ft), H = head (ft), Q = suction flow rate (cfs), V = impeller tip speed (fps), and N = rotation speed (rpm).
(Evans, 1979).

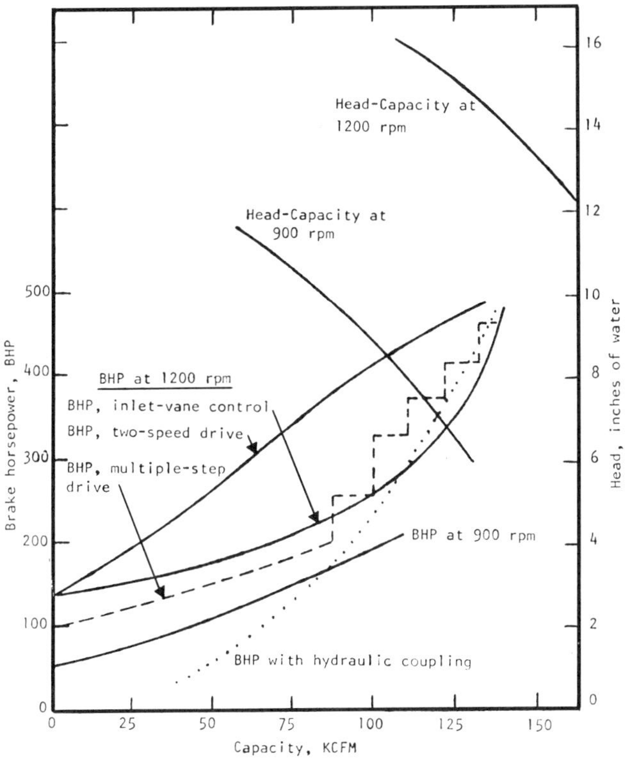

Control Type	Control Cost	Required Power Input	Advantages (A), and Disadvantages (D)
a	low	high	(A) simplicity; (D) high power input
b	moderate	moderate	(A) lower input power; (D) higher cost
c	low	moderate	(A) simplicity; (D) fan erosion
d	moderate	moderate	(D) complex; also needs dampers
e	high	low	(A) simple; no dampers needed

Figure 7.23. Performances of fans with several kinds of controls (*American Standard Co. Inc.*). (a) A damper in the duct with constant-speed fan drive, (b) two-speed fan driver, (c) inlet vanes or inlet louvers with a constant-speed fan drive, (d) multiple-step variable-speed fan drive, and (e) hydraulic or electric coupling with constant-speed driver giving wide control over fan speed.

TABLE 7.5. Fan Laws[a]

Fan Law Number		Ratio of	–	Ratio	×	Ratio		Ratio
1	a	cfm		$size^3$	×	rpm		1
	b	press	–	$size^2$	×	rpm^2	×	δ
	c	HP		$size^5$	×	rpm^3		δ
2	a	cfm		$size^2$	×	$press^{1/2}$		$1/\delta^{1/2}$
	b	rpm	–	$1/size$	×	$press^{1/2}$	×	$1/\delta^{1/2}$
	c	HP		$size^2$	×	$press^{3/2}$		$1/\delta^{1/2}$
3	a	rpm		$1/size^3$	×	cfm		1
	b	press	–	$1/size^4$	×	cfm^2	×	δ
	c	HP		$1/size^4$	×	cfm^3		δ
4	a	cfm		$size^{4/3}$	×	$HP^{1/3}$		$1/\delta^{1/3}$
	b	press	–	$1/size^{4/3}$	×	$HP^{2/3}$	×	$\delta^{1/3}$
	c	rpm		$1/size^{5/3}$	×	$HP^{1/3}$		$1/\delta^{1/3}$
5	a	size		$cfm^{1/2}$	×	$1/press^{1/4}$		$\delta^{1/4}$
	b	rpm	–	$1/cfm^{1/2}$	×	$press^{3/4}$	×	$1/\delta^{3/4}$
	c	HP		cfm	×	press		1
6	a	size		$cfm^{1/3}$	×	$1/rpm^{1/3}$		1
	b	press	–	$cfm^{2/3}$	×	$rpm^{4/3}$	×	δ
	c	HP		$cfm^{5/3}$	×	$rpm^{4/3}$		δ
7	a	size		$press^{1/2}$	×	$1/rpm$		$1/\delta^{1/2}$
	b	cfm	–	$press^{3/2}$	×	$1/rpm^2$	×	$1/\delta^{3/2}$
	c	HP		$press^{5/2}$	×	$1/rpm^2$		$1/\delta^{3/2}$
8	a	size		$1/HP^{1/4}$	×	$cfm^{3/4}$		$\delta^{1/4}$
	b	rpm	–	$HP^{3/4}$	×	$1/cfm^{5/4}$	×	$1/\delta^{3/4}$
	c	press		HP	×	$1/cfm$		1
9	a	size		$HP^{1/2}$	×	$1/press^{3/4}$		$\delta^{1/4}$
	b	rpm	–	$1/HP^{1/2}$	×	$press^{5/4}$	×	$1/\delta^{3/4}$
	c	cfm		HP	×	$1/press$		1
10	a	size		$HP^{1/5}$	×	$1/rpm^{3/5}$		$1/\delta^{1/5}$
	b	cfm	–	$HP^{3/5}$	×	$1/rpm^{4/3}$	×	$1/\delta^{3/5}$
	c	press		$HP^{2/5}$	×	$rpm^{4/5}$		$\delta^{3/5}$

[a] $\delta = \rho/g_c$.
For example, the pressure P varies as $D^2N^2\rho/g_c$ line 1(b), $Q^2(\rho/g_c)/D^4$ line 3(b), $\dot{P}^{2/3}(\rho/g_c)^{1/3}/D^{4/3}$ line 4(b), $Q^{2/3}N^{4/3}\rho/g_c$ line 6(b), $\dot{P}/Q$ line 8(c), and $\dot{P}^{2/5}N^{4/5}(\rho/g_c)^{3/5}$ line 10(c).
(R.D. Madison, 1949).

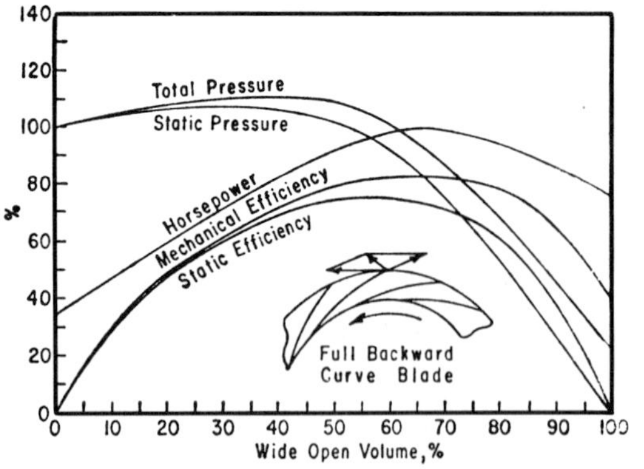

(a)

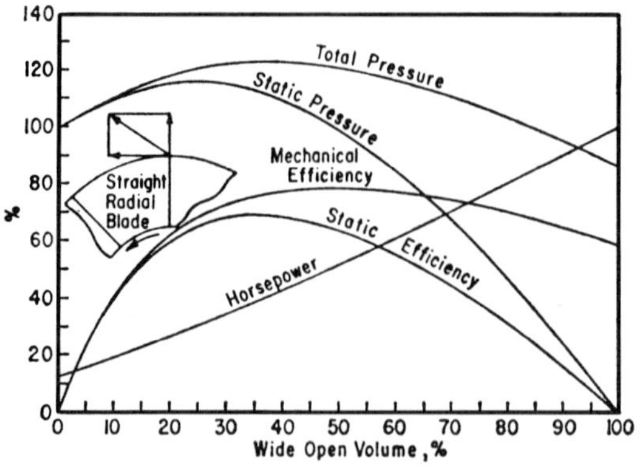

(b)

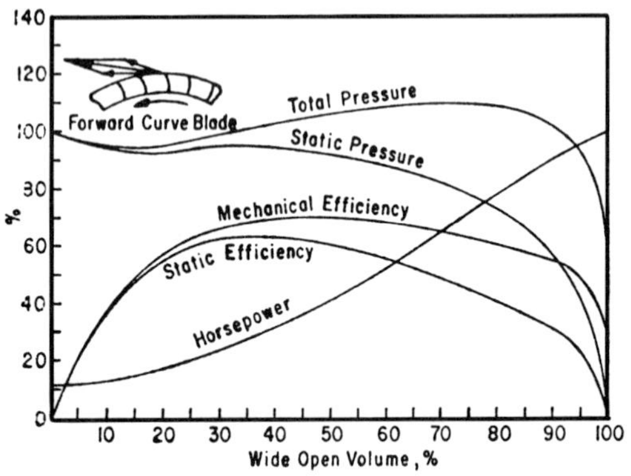

(c)

	Backwardly Curved	Radial Blade	Forwardly Curved
First Cost*	High	Medium	Low
Efficiency	High	Medium	Low
Stability of Operation	Good	Good	Poor
Space Required	Medium	Medium	Small
Tip Speed	High	Medium	Low
Resistance to Abrasion	Medium	Good	Poor
Ability to Handle Sticky Materials	Medium	Good	Poor

(d)

Figure 7.24. Performances of fans with various-shaped blades (*Green Fuel Economizer Co.*): (a) Backward curved blades. (b) Straight radial blades. (c) Forward curved blades. (d) Comparison of characteristics of the several blade types (*Sturtevant*).

TABLE 7.6. Specifications of Centrifugal Compressors

Frame	Normal Inlet Flow Range[a] (ft^3/min)	Nominal Polytropic Head per Stage[b] (H_p)	Nominal Polytropic Efficiency (η_p)	Nominal Maximum No. of Stages[c]	Speed at Nominal Polytropic Head/Stage
29M	500–8000	10,000	0.76	10	11500
38M	6000–23,000	10,000/12,000	0.77	9	8100
46M	20,000–35,000	10,000/12,000	0.77	9	6400
60M	30,000–58,000	10,000/12,000	0.77	8	5000
70M	50,000–85,000	10,000/12,000	0.78	8	4100
88M	75,000–130,000	10,000/12,000	0.78	8	3300
103M	110,000–160,000	10,000	0.78	7	2800
110M	140,000–190,000	10,000	0.78	7	2600
25MB (H) (HH)	500–5000	12,000	0.76	12	11500
32MB (H) (HH)	5000–10,000	12,000	0.78	10	10200
38MB (H)	8000–23,000	10,000/12,000	0.78	9	8100
46MB	20,000–35,000	10,000/12,000	0.78	9	6400
60MB	30,000–58,000	10,000/12,000	0.78	8	5000
70MB	50,000–85,000	10,000/12,000	0.78	8	4100
88MB	75,000–130,000	10,000/12,000	0.78	8	3300

[a] Maximum flow capacity is reduced in direct proportion to speed reduction.
[b] Use either 10,000 or 12,000 ft for each impeller where this option is mentioned.
[c] At reduced speed, impellers can be added.
(Elliott Co.).

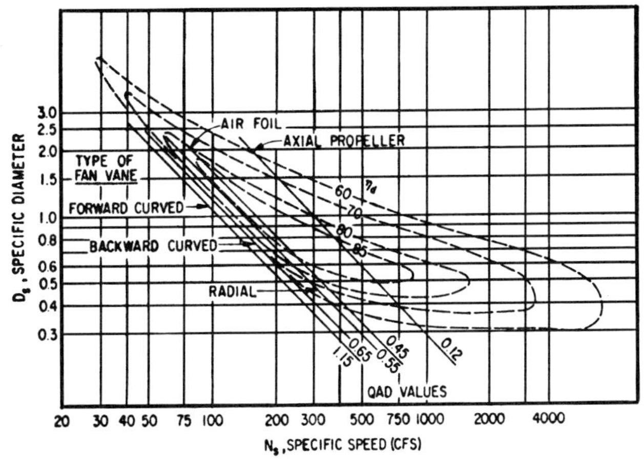

Figure 7.25. Efficiency and head coefficient q_{ad} as functions of specific speeds and specific diameters of various kinds of impellers (*Evans, 1979*). *Example:* An axial propeller has an efficiency of 70% at $N_s = 200$ and $D_s = 1.5$; and 85% at $N_s = 400$ and $D_s = 0.8$. See Table 7.4 for definitions of q_{ad}, N_s, and D_s.

pull a vacuum of 28 in. of mercury. A two-stage unit can deliver 250 psig. A generous supply of lubricant is needed for the sliding vanes. Table 7.9 shows that power requirements are favorable in comparison with other rotaries.

Liquid-liner compressors produce an oil-free discharge of up to 125 psig. The efficiency is relatively low, 50% or so, but high enough to make them superior to steam jet ejectors for vacuum service. The liquid absorbs the considerable heat of compression and must be circulated and cooled; a 200 HP compressor requires 100 gpm of cooling water with a 10°F rise. When water vapor is objectionable in the compressed gas, other sealing liquids are used; for example, sulfuric acid for the compression of chlorine. Figure 7.19(e) shows the principle and Table 7.10 gives specifications of some commercial units.

7.6. THEORY AND CALCULATIONS OF GAS COMPRESSION

The main concern of this section is how to determine the work requirement and the effluent conditions of a compressor for which the inlet conditions and the outlet pressure are specified. Theoretical methods allow making such calculations for ideal and real gases and gas mixtures under isothermal and frictionless adiabatic (isentropic) conditions. In order that results for actual operation can be found it is neecessary to know the efficiency of the equipment. That depends on the construction of the machine, the mode of operation, and the nature of the gas being processed. In the last analysis such information comes from test work and its correlation by manufacturers and other authorities. Some data are cited in this section.

DIMENSIONLESS GROUPS

The theory of dimensionless groups of Section 7.2, Basic Relations, also applies to fans and compressors with rotating elements, for example, Eqs. (7.8)–(7.10) which relate flow rate, head, power, speed, density, and diameter. Equivalent information is embodied in Table 7.5. The concept of specific speed, Eqs. (7.11) and (7.12), also is pertinent. In Figures 7.21 and 7.25 it is the basis for identifying suitable operating ranges of various types of compressors.

IDEAL GASES

The ideal gas or a gas with an equation of state

$$PV = zRT \tag{7.18}$$

is a convenient basis of comparison of work requirements for real gases and sometimes yields an adequate approximation of these work requirements. Two limiting processes are isothermal and isentropic (frictionless adiabatic) flows. Changes in elevation and velocity heads are considered negligible here. With constant compressibility z the isothermal work is

$$W = \int_{P_1}^{P_2} V\,dP = zRT\ln(P_2/P_1). \tag{7.19}$$

Under isentropic conditions and with constant heat capacities, the pressure–volume relation is

$$PV^k = P_1V_1^k = \text{const}, \tag{7.20}$$

where

$$k = C_p/C_v \tag{7.21}$$

TABLE 7.7. Some Sizes of One- and Two-Stage Reciprocating Compressors

(a) Horizontal, One-Stage, Belt-Driven

Diameter Cylinder (in.)	Stroke (in.)	Displacement (cuft/min.)	rpm	Air Pressure (lb/sq in.)	Brake HP at Rated Pressure	Openings (in.)	
						Inlet	Outlet
$7\frac{1}{2}$	6	106	310	80–100–125	15.9–17–18	$2\frac{1}{2}$	$2\frac{1}{2}$
$8\frac{1}{2}$	9	170	300	80–100–125	25–27–29	3	3
10	10	250	285	80–100–125	36–38.5–41	$3\frac{1}{2}$	$3\frac{1}{2}$
11	12	350	270	80–100–125	51–57–60	—	4
$8\frac{1}{2}$	6	136	350	40–60	15–18.5	—	3
10	9	245	300	40–75	27–34	$3\frac{1}{2}$	$3\frac{1}{2}$
11	10	312	285	40–75	34–43	4	4
13	12	495	270	40–75	54–70	5	5
12	9	350	300	20–45	30–42	4	4
13	10	435	285	30–45	42–52	6	6
15	12	660	270	30–50	59–74	7	7

(Worthington Corp.).

TABLE 7.7—*(continued)*

(b) Horizontal, One-Stage, Steam-Driven[a]

Diameter, Steam Cylinder (in.)	Diameter, Air Cylinder, (in.)	Stroke (in.)	Displacement, (cuft/min)	rpm	Air Pressure, (lb/sq in.)
7	$7\frac{1}{2}$	6	106	350	80–100–125
8	$8\frac{1}{2}$	9	170	300	80–100–125
9	10	10	250	285	80–100–125
10	11	12	350	270	80–100–125
7	$8\frac{1}{2}$	6	136	350	40–60
8	10	9	245	300	40–75[b]
9	11	10	312	285	40–75[b]
10	13	12	495	270	40–75[b]
8	12	9	350	300	20–45[c]
9	13	10	435	285	20–45[c]
10	15	12	660	270	20–50[c]

[a] All machines have piston-type steam valves.
[b] 110-lb steam necessary for maximum air pressure.
[c] 125-lb steam necessary for maximum air pressure.
(Worthington Corp.).

(c) Horizontal, Two-Stage, Belt-Driven

Diameter Cylinder (in.)				
Low Pressure	High Pressure	Stroke (in.)	rpm	Piston Displacement (cuft free air/min)
4	$2\frac{1}{8}$	4	500	28
6	$2\frac{7}{8}$	6	350	65
8	$3\frac{8}{4}$	8	300	133
10	$4\frac{7}{8}$	10	275	241

(Ingersoll–Rand Co.).

TABLE 7.8. Summary of Rotary Compressor Performance Data

	Type				
	Helical Screw	Spiral Axial	Straight Lobes	Sliding Vanes	Liquid Liner
Configuration, features (male × female)	4 × 6	2 × 4	2 × 2	8 Blades	16 Sprockets
Max displacement (cfm)	20,000	13,000	30,000	6,000	13,000
Max diameter (in.)	25	16	18	33	48
Min diameter (in.)	4	6	10	5	12
Limiting tip speed (Mach)	0.30	0.12	0.05	0.05	0.06
Normal tip speed (Mach)	0.24	0.09	0.04	0.04	0.05
Max L/d, low pressure	1.62	2.50	2.50	3.00	1.1
Normal L/d, high pressure	1.00	1.50	1.50	2.00	1.00
V factor for volumetric efficiency	7	3	5	3	3
X factor for displacement	0.0612	0.133	0.27	0.046	0.071
Normal overall efficiency	75	70	68	72	50
Normal mech. eff. at ±100 HP (%)	90	93	95	94	90
Normal compression ratio R_c	2/3/4	3	1.7	2/3/4	5
Normal blank-off R_c	6	5	5	7	9
Displacement form factor A_e	0.462	1.00	2.00	0.345	0.535

(Evans, 1979).

TABLE 7.9. Five Rotary Compressors for a Common Service

	Type				
	Helical Screw	Spiral Axial	Straight Lobes	Sliding Vanes	Liquid Liner
Suction loss θ_i	9.35	1.32	0.89	0.90	1.40
Discharge loss θ_e	7.35	1.04	0.70	0.70	1.10
Intrinsic corr. B	1.185	1.023	1.016	1.016	1.025
Adiabatic eff. η_{ad}	85.6	97.7	98.5	98.5	97.9
Slippage W_s (%)	28.5	16.6	11.8	11.8	3.0
Slip eff. η_s (%)	71.5	83.4	88.2	88.2	97.0
Thermal eff. η_t (%)	89.2	93.7	95.8	95.5	42.5
Volumetric eff. E_{vr}	68.0	85.7	89.1	89.9	96.6
Displacement (cfm)	14,700	11,650	11,220	11,120	10,370
Rotor dia. (in.)	26.6	26.2	27.0	65.0	45.5
Commercial size, $d \times L$	25×25	22×33	22×33	46×92[a]	43×48[b]
Speed (rpm)	3,500	1,250	593	284	378
Motor (HP)	1,100	800	750	750	1,400
Service factor	1.09	1.11	1.10	1.12	1.10
Discharge temp °F	309	270	262	263	120

[a] Twin 32.5×65 or triplet 26.5×33 (667 rpm) are more realistic.
[b] Twin 32×32 (613 rpm) alternate where $L = d$.
(Evans, 1979).

is the ratio of heat capacities at constant pressure and constant volume and

$$C_v = R - C_p. \tag{7.22}$$

A related expression of some utility is

$$T_2/T_1 = (P_2/P_1)^{(k-1)/k}. \tag{7.23}$$

Since k ordinarily is a fairly strong function of the temperature, a suitable average value must be used in Eq. (7.20) and related ones.

Under adiabatic conditions the flow work may be written as

$$W = H_2 - H_1 = \int_{P_1}^{P_2} V\, dP. \tag{7.24}$$

Upon substitution of Eq. (7.20) into Eq. (7.24) and integration, the isentropic work becomes

$$W_s = H_2 - H_1 = P_1^{1/k} V_1 \int_{P_1}^{P_2} dP/P^{1/k}$$
$$= \left(\frac{k}{k-1}\right) z_1 R T_1 \left[\left(\frac{P_2}{P_1}\right)^{(k-1)/k} - 1\right]. \tag{7.25}$$

TABLE 7.10. Specifications of Liquid Liner Compressors

Compressor (size)	Pressure (psi)	Capacity (cuft/min)	Motor (HP)	Speed (rpm)
	5	1020	40	
K-6	10	990	60	570
	15	870	75	
	20	650	100	
621		26	$7\frac{1}{2}$	3500
1251	35	120	40	1750
1256		440	100	1750
621		23	10	3500
1251	80	110	50	1750
1256		410	150	1750

(Nash Engineering Co.).

In multistage centrifugal compression it is justifiable to take the average of the inlet and outlet compressibilities so that the work becomes

$$W_s = H_2 - H_1 = \left(\frac{k}{k-1}\right)\left(\frac{z_1 + z_2}{2}\right) R T_1 \left[\left(\frac{P_2}{P_1}\right)^{(k-1)/k} - 1\right]. \tag{7.26}$$

When friction is present, the problem is handled with *empirical* efficiency factors. The isentropic compression efficiency is defined as

$$\eta_s = \frac{\text{isentropic work or enthalpy change}}{\text{actually required work or enthalpy change}}. \tag{7.27}$$

Accordingly,

$$W = \Delta H = W_s/\eta_s = (\Delta H)_s/\eta_s. \tag{7.28}$$

When no other information is available about the process gas, it is justifiable to find the temperature rise from

$$\Delta T = (\Delta T)_s/\eta_s \tag{7.29}$$

so that

$$T_2 = T_1(1 + (1/\eta_s)[(P_2/P_1)^{(k-1)/k} - 1]. \tag{7.30}$$

A case with variable heat capacity is worked out in Example 7.5.

For mixtures, the heat capacity to use is the sum of the mol

EXAMPLE 7.4
Gas Compression, Isentropic and True Final Temperatures
With $k = 1.4$, $P_2/P_1 = 3$ and $\eta_s = 0.71$; the final temperatures are $(T_2)_s = 1.369T_1$ and $T_2 = 1.519T_1$ with Eqs. (7.24) and (7.31).

---

fraction weighted heat capacities of the pure components,

$$C_p = \sum x_i C_{pi}. \tag{7.31}$$

REAL PROCESSES AND GASES

Compression in reciprocating and centrifugal compressors is essentially adiabatic but it is not frictionless. The pressure–volume behavior in such equipment often conforms closely to the equation

$$PV^n = P_1 V_1^n = \text{const.} \tag{7.32}$$

Such a process is called polytropic. The equation is analogous to the isentropic equation (7.20) but the polytropic exponent n is different from the heat capacity ratio k.

Polytropic exponents are deduced from PV measurements on the machine in question. With reciprocating machines, the PV data are recorded directly with engine indicators. With rotary machines other kinds of instruments are used. Such test measurements usually are made with air.

Work in polytropic compression of a gas with equation of state $PV = zRT$ is entirely analogous to Eq. (7.26). The hydrodynamic work or the work absorbed by the gas during the compression is

$$W_{hd} = \int_{P_1}^{P_2} V\,dP = \left(\frac{n}{n-1}\right) z_1 R T_1 \left[\left(\frac{P_2}{P_1}\right)^{(n-1)/n} - 1\right]. \tag{7.33}$$

Manufacturers usually characterize their compressors by their polytropic efficiencies which are defined by

$$\eta_p = \left(\frac{n}{n-1}\right) \Big/ \left(\frac{k}{k-1}\right) = \frac{n(k-1)}{k(n-1)}. \tag{7.34}$$

The polytropic work done on the gas is the ratio of Eqs. (7.33) and (7.34) and comprises the actual mechanical work done on the gas:

$$W_p = W_{hd}/\eta_p = \left(\frac{k}{k-1}\right) z_1 R T_1 \left[\left(\frac{P_2}{P_1}\right)^{(n-1)/n} - 1\right] \tag{7.35}$$

Losses in seals and bearings of the compressor are in addition to W_p; they may amount to 1–3% of the polytropic work, depending on the machine.

The value of the polytropic exponent is deduced from Eq. (7.34) as

$$n = \frac{k\eta_p}{1 - k(1 - \eta_p)}. \tag{7.36}$$

The isentropic efficiency is

$$\eta_s = \frac{\text{isentropic work [Eq. (7.25)]}}{\text{actual work [Eq. (7.35)]}} \tag{7.37}$$

$$= \frac{(P_2/P_1)^{(k-1)/k} - 1}{(P_2/P_1)^{(n-1)/n} - 1} \tag{7.38}$$

$$= \frac{(P_2/P_1)^{(k-1)/k} - 1}{(P_2/P_1)^{(k-1)/k\eta_p} - 1}. \tag{7.39}$$

The last version is obtained with the aid of Eq. (7.34) and relates the isentropic and polytropic efficiencies directly. Figure 7.27(b) is a plot of Eq. (7.39). Example 7.6 is an exercise in the relations between the two kinds of efficiencies.

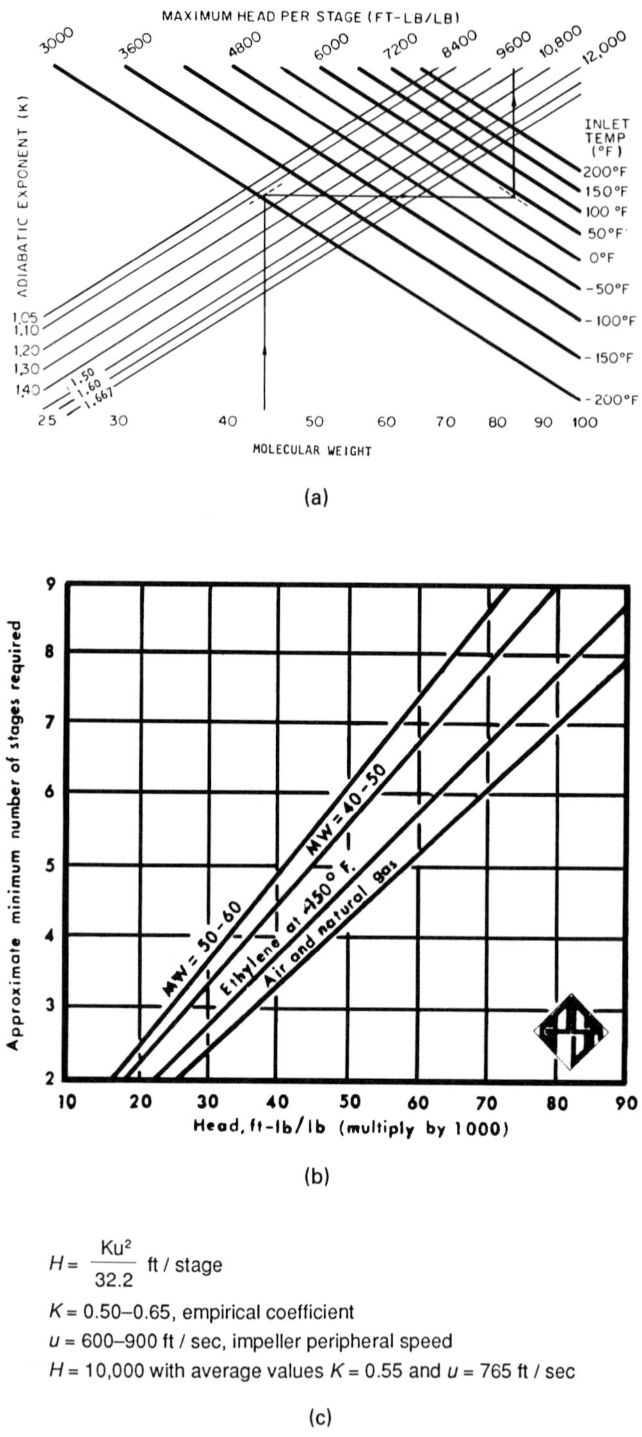

$H = \dfrac{Ku^2}{32.2}$ ft / stage

K = 0.50–0.65, empirical coefficient

u = 600–900 ft / sec, impeller peripheral speed

H = 10,000 with average values K = 0.55 and u = 765 ft / sec

(c)

Figure 7.26. Several ways of estimating allowable polytropic head per stage of a multistage centrifugal compressor. (a) Single-stage head as a function of k, molecular weight, and temperature (*Elliott Co.*). (b) Single-stage head as a function of the nature of the gas (NGPSA Handbook, *Gas Processors Assn, Tulsa, OK, 1972*), obtained by dividing the total head of the compressor by number of stages. $H = Ku^2/32.2$ ft/stage, $K = 0.50$–0.65, empirical coefficient, $u = 600$–900 ft/sec, impeller peripheral speed, and $H = 10{,}000$ with average values $K = 0.55$ and $u = 765$ ft/sec. (c) An equation and parameters for estimation of head.

EXAMPLE 7.5
Compression Work with Variable Heat Capacity
Hydrogen sulfide heat capacity is given by

$$C_p = 7.629 + 3.431(E-4)T + 5.809(E-6)T^2 - 2.81(E-9)T^3, \quad \text{cal/g mol},$$

with T in K. The gas is to be compressed from 100°F (310.9 K) and 14.7 psia to 64.7 psia.

Assuming the heat capacity to be independent of pressure in this low range, the isentropic condition is

$$\Delta S = \int_{T_1}^{T_2} (C_p/T)\,dT - R\ln(P_2/P_1) = \int_{310.9}^{T_2} (C_p/T)\,dT - 1.987\ln(64.7/14.7) = 0.$$

By trial, with a root-solving program,

$T_2 = 441.1$ K, 334.4°F (compared with 345°F from Example 7.7).

The isentropic enthalpy change becomes

$$\Delta H_s = \int_{310.1}^{441.1} C_p\,dT = 1098.1 \text{ cal/g mol} \rightarrow 1098.1(1.8)/34.08 = 58.0 \text{ Btu/lb},$$

compared with 59.0 from Example 7.7. The integration is performed with Simpson's rule on a calculator.

The actual final temperature will vary with the isentropic efficiency. It is found by trial from the equation

$$1098.1/\eta_s = \int_{1098.1}^{T_2} C_p\,dT.$$

Some values are

η_s	1.0	0.75	0.50	0.25
T_2	441.1	482.93	564.29	791.72

WORK ON NONIDEAL GASES

The methods discussed thus far neglect the effect of pressure on enthalpy, entropy, and heat capacity. Although efficiencies often are not known well enough to justify highly refined calculations, they may be worth doing in order to isolate the uncertainties of a design. Compressibility factors are given for example by Figure 7.29. Efficiencies must be known or estimated.

Thermodynamic Diagram Method. When a thermodynamic diagram is available for the substance or mixture in question, the flow work can be found from the enthalpy change,

$$W = \Delta H. \tag{7.40}$$

The procedure is illustrated in Example 7.7 and consists of these steps:

1. Proceed along the line of constant entropy from the initial condition to the final pressure P_2 and enthalpy $(H_2)_s$.
2. Evaluate the isentropic enthalpy change $(\Delta H)_s = (H_2)_s - H_1$.
3. Find the actual enthalpy change as

$$\Delta H = (\Delta H)_s/\eta_s \tag{7.41}$$

and the final enthalpy as

$$H_2 = H_1 + (\Delta H)_s/\eta_s. \tag{7.42}$$

4. At the final condition (P_2, H_2) read off any other desired properties such as temperature, entropy or specific volume.

Thermodynamic diagrams are known for light hydrocarbons, refrigerants, natural gas mixtures, air, and a few other common substances. Unless a substance or mixture has very many applications, it is not worthwhile to construct a thermodynamic diagram for compression calculations but to use other equivalent methods.

General Method. The effects of composition of mixtures and of pressure on key properties such as enthalpy and entropy are deduced from *PVT* equations of state. This process is described in books on thermodynamics, for example, Reid, Prausnitz, and Sherwood (*Properties of Liquids and Gases,* McGraw-Hill, New York, 1977) and Walas (*Phase Equilibria in Chemical Engineering,* Butterworths, Stoneham, MA, 1985). Only the simplest correlations of these effects will be utilized here for illustration.

For ideal gases with heat capacities dependent on temperature, the procedure requires the isentropic final temperature to be found by trial from

$$\Delta S = \int_{T_1}^{T_{2s}} (C_p/T)\,dT - R\ln(P_2/P_1) \rightarrow 0, \tag{7.43}$$

and then the isentropic enthalpy change from

$$\Delta H = \int_{T_1}^{T_{2s}} C_p\,dT. \tag{7.44}$$

The final temperature T_2 is found by trial after applying a known isentropic efficiency,

$$(\Delta H)_s/\eta_s = \int_{T_1}^{T_2} C_p\,dT. \tag{7.45}$$

The fact that heat capacities usually are represented by empirical polynomials of the third or fourth degree in temperature accounts for the necessity of solutions of equations by trial.

Example 7.5 applies this method and checks roughly the calculations of Example 7.7 with the thermodynamic diagram of this substance. The pressures are relatively low and are not expected to generate any appreciable nonideality.

This method of calculation is applied to mixtures by taking a mol fraction weighted heat capacity of the mixture,

$$C_p = \sum x_i C_{pi}. \tag{7.46}$$

When the pressure range is high or the behavior of the gas is

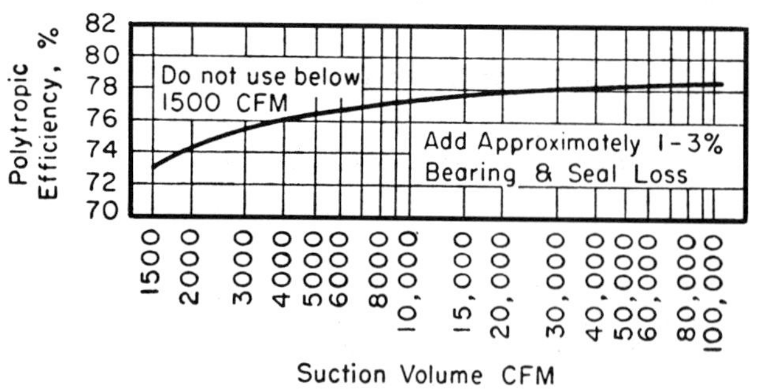

TABLE OF CORRECTION FACTORS DUE TO COMPRESSION RATIO

Compression Ratio	Inlet Volume in CFM 1500	2000	3000	4000	5000	7500	15,000	30,000	60,000 & over
1.35	1.0	1.0	1.0	1.0	1.0	1.0	1.0	1.0	1.0
1.75	.983	.990	.990	.994	.996	.996	.996	.996	1.0
2.25	.976	.977	.986	.985	.991	.995	.995	.995	1.0
2.80	—	.969	.980	.980	.989	.992	.993	.994	1.0
5.00	—	—	.965	.970	.972	.985	.991	.996	.997
10.00	—	—	—	.944*	.955*	.960*	.985*	.989*	.995
15.00	—	—	—	—	.935*	.957*	.971*	.986*	.993

Notes: Asterisk indicates figures applying only to high molecular weight hydrocarbons.
Factors apply on one compressor body with six or less impellers.

(a)

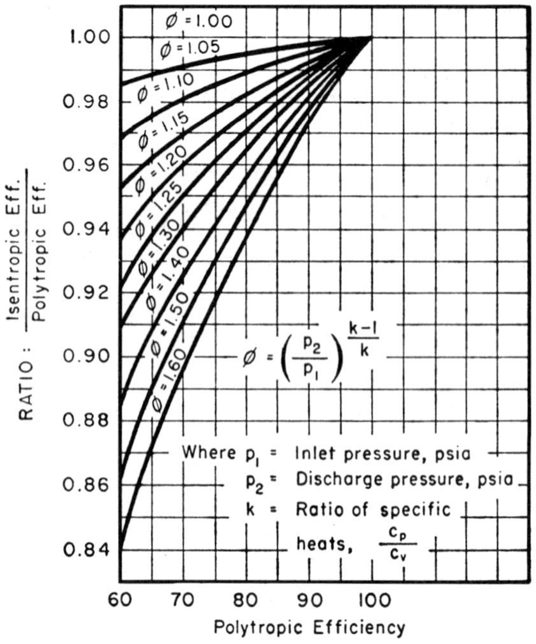

(b)

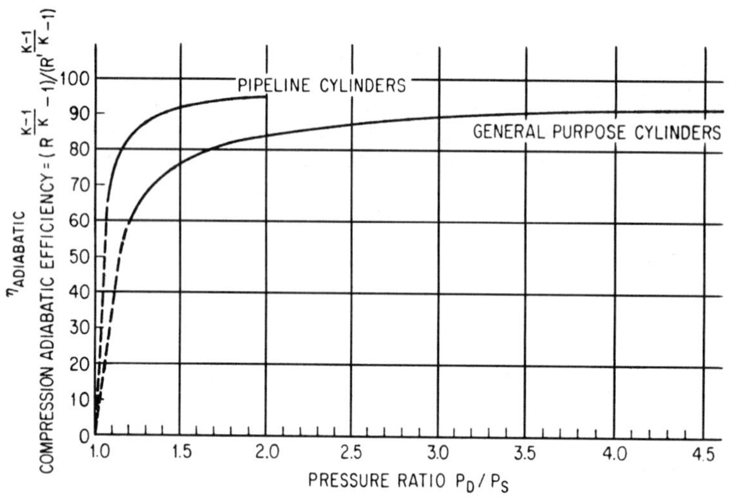

Compression Ratio	Efficiencies (Engine-driven*)
1.1	50–60
1.2	60–70
1.3	65–80
1.5	70–85
2.0	75–88
2.5	80–89
3.0	82–90
4.0	85–90

*Multiply by 0.95 for motor-driven compressors.

(c)

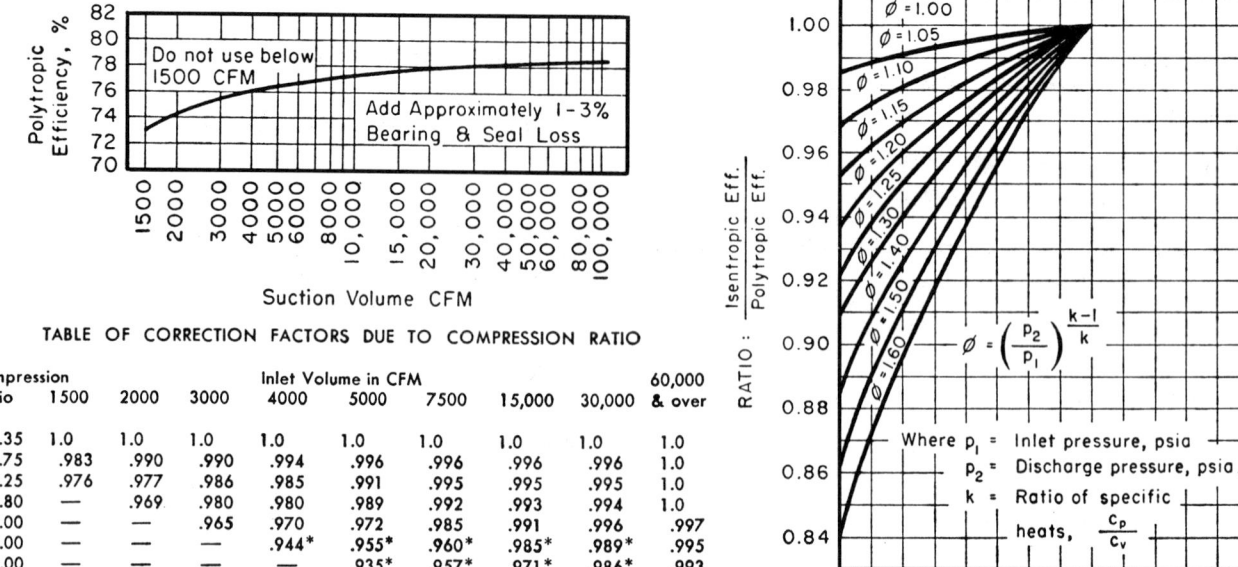

Figure 7.27. Efficiencies of centrifugal and reciprocating compressors. (a) Polytropic efficiencies of centrifugal compressors as a function of suction volume and compression ratio (*Clark Brothers Co.*). (b) Relation between isentropic and polytropic efficiencies, Eqs. (7.22) (7.23). (c) Isentropic efficiencies of reciprocating compressors (De Laval Handbook, *McGraw-Hill, New York, 1970*). Multiply by 0.95 for motor drive. Gas engines require 7000–8000 Btu/HP.

Example 7.6
Polytropic and Isentropic Efficiencies

Take $\eta_p = 0.75$, $k = 1.4$, and $P_2/P_1 = 3$. From Eq. (7.39), $n = 1.6154$ and $\eta_s = 0.7095$. With Figure 7.27(b), $\phi = 3^{0.2857} = 1.3687$, $\eta_s = 0.945\eta_p = 0.709$. The agreement is close.

---

nonideal for any other reason, the isentropic condition becomes

$$\Delta S = \int_{T_1}^{T_{2s}} (C_p'/T)\,dT - R\ln(P_2/P_1) + \Delta S_1' - \Delta S_2' \to 0. \quad (7.47)$$

After the final isentropic temperature T_{2s} has been found by trial, the isentropic enthalpy change is obtained from

$$(\Delta H)_s = \int_{T_1}^{T_{2s}} C_p'\,dT + \Delta H_1' - \Delta H_{2s}'. \quad (7.48)$$

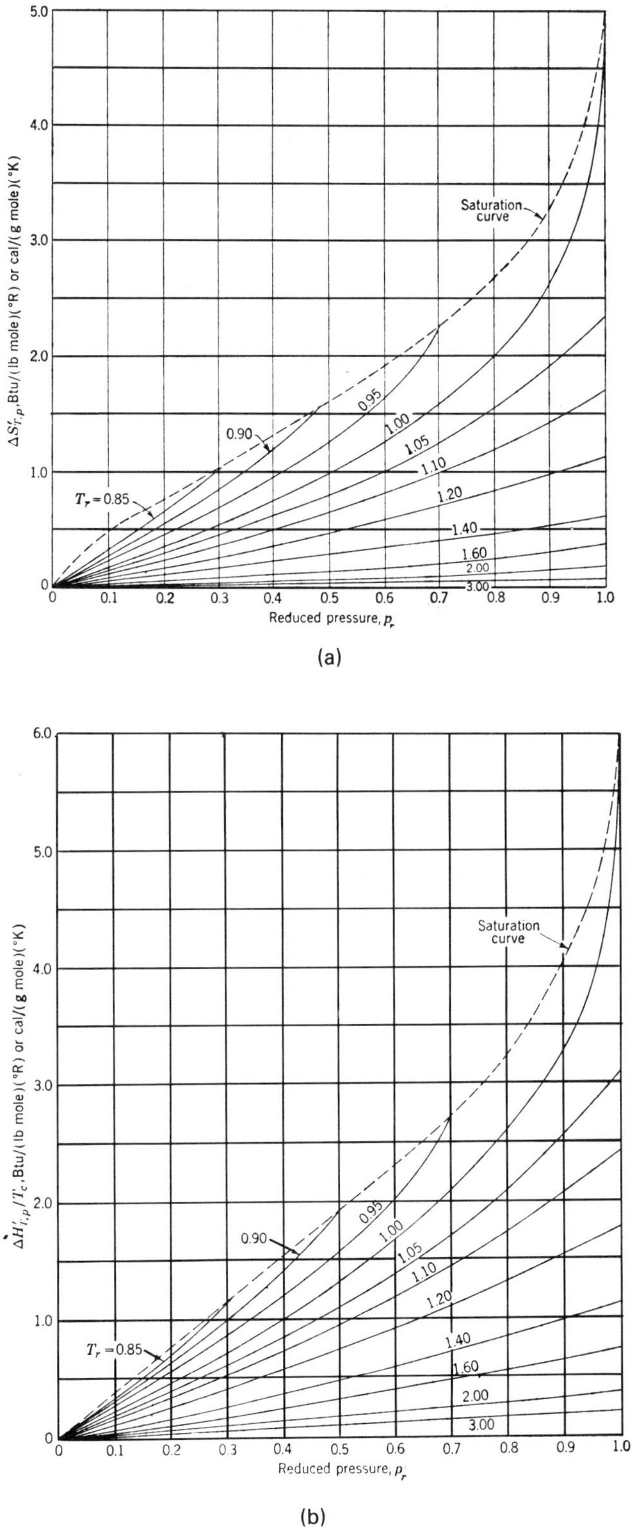

Figure 7.28. Residual entropy and enthalpy as functions of reduced properties. (a) Residual entropy. (b) Residual enthalpy. *Drawn by Smith and Van Ness* (Introduction to Chemical Engineering Thermodynamics, *McGraw-Hill, New York, 1959) from data of Lydersen et al. For illustrative purposes primarily*; *see text for other sources.*]

In terms of a known isentropic efficiency the final temperature T_2 then is found by trial from

$$(\Delta H)_s/\eta_s = \int_{T_1}^{T_2} C'_p \, dT + \Delta H'_1 - \Delta H'_2. \tag{7.49}$$

In these equations the heat capacity C'_p is that of the ideal gas state or that of the real gas near zero or atmospheric pressure. The residual properties $\Delta S'_1$ and $\Delta H'_1$ are evaluated at (P_1, T_1) and $\Delta S'_2$ and $\Delta H'_2$ at (P_2, T_2). Figure 7.28 gives them as functions of reduced temperature T/T_c and reduced pressure P/P_c. More accurate methods and charts for finding residual properties from appropriate equations of state are presented in the cited books of Reid et al. (1977) and Walas (1985).

For mixtures, pseudocritical properties are used for the evaluation of the reduced properties. For use with Figure 7.28, Kay's rules are applicable, namely,

$$(P_c)_{\text{mix}} = \sum x_i P_{ci}, \tag{7.50}$$

$$(T_c)_{\text{mix}} = \sum x_i T_{ci}, \tag{7.51}$$

but many equations of state employ particular combining rules.

Example 7.8 compares a solution by this method with the assumption of ideal behavior.

EFFICIENCY

The efficiencies of fluid handling equipment such as fans and compressors are empirically derived quantities. Each manufacturer will supply either an efficiency or a statement of power requirement for a specified performance. Some general rules have been devised for ranges in which efficiencies of some classes equipment usually fall. Figure 7.27 gives such estimates for reciprocating compressors. Fan efficiencies can be deduced from the power-head curves of Figure 7.24. Power consumption or efficiencies of rotary and reciprocating machines are shown in Tables 7.7, 7.8, and 7.9.

Polytropic efficiencies are obtained from measurements of power consumption of test equipment. They are essentially independent of the nature of the gas. As the data of Figure 7.27 indicate, however, they are somewhat dependent on the suction volumetric rate, particularly at low values, and on the compression ratio. Polytropic efficiencies of some large centrifugal compressors are listed in Table 7.6. These data are used in Example 7.9 in the selection of a machine for a specified duty.

The most nearly correct methods of Section 7.6.4 require knowledge of isentropic efficiencies which are obtainable from the polytropic values. For a given polytropic efficiency, which is independent of the nature of the gas, the isentropic value is obtained with Eq. (7.39) or Figure 7.27(b). Since the heat capacity is involved in this transformation, the isentropic efficiency depends on the nature of the substance and to some extent on the temperature also.

TEMPERATURE RISE, COMPRESSION RATIO, VOLUMETRIC EFFICIENCY

The isentropic temperature in terms of compression ratio is given for ideal gases by

$$(T_2)_s = T_1(P_2/P_1)^{(k-1)/k}. \tag{7.52}$$

For polytropic compression the final temperature is given directly by

$$T_2 = T_1(P_2/P_1)^{(n-1)/n} \tag{7.53}$$

EXAMPLE 7.7
Finding Work of Compression with a Thermodynamic Chart

Hydrogen sulfide is to be compressed from 100°F and atmospheric pressure to 50 psig. The isentropic efficiency is 0.70. A pressure-enthalpy chart is taken from Starling (*Fluid Thermodynamic Properties for Light Petroleum Systems,* Gulf, Houston, TX, 1973). The work and the complete thermodynamic conditions for the process will be found.

The path followed by the calculation is 1–2–3 on the sketch. The initial enthalpy is −86 Btu/lb. Proceed along the isentrop $S = 1.453$ to the final pressure, 64.7 psia, and enthalpy $H_2 = -27$. The isentropic enthalpy change is

$$\Delta H_s = -27 - (-86) = 59 \text{ Btu/lb}.$$

The true enthalpy change is

$$\Delta H = 59/0.70 = 84.3.$$

The final enthalpy is

$$H_3 = -86 + 84.3 = -1.7.$$

Other conditions at points 2 and 3 are shown on the sketch. The work is

$$\dot{W} = \Delta H = 84.3 \text{ Btu/lb}$$
$$\rightarrow 84.3/2.545 = 33.1 \text{ HP hr/(1000 lb)}.$$

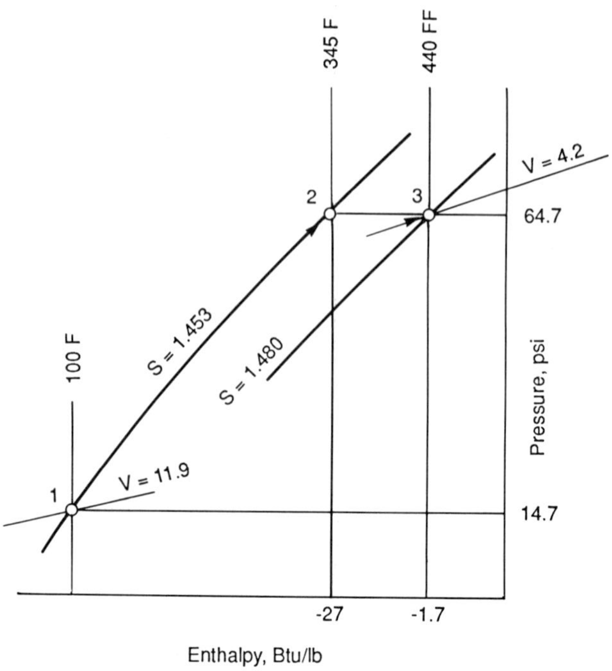

EXAMPLE 7.8
Compression Work on a Nonideal Gas

Hydrogen sulfide at 450 K and 15 atm is to be compressed to 66 atm. The isentropic final temperature and the isentropic enthalpy change will be found with the aid of Figure 7.28 for the residual properties.

The critical properties are $T_c = 373.2$ K and $P_c = 88.2$ atm. The heat capacity is stated in Example 7.5:

$$T_{r1} = 450/373.2 = 1.21,$$
$$P_{r1} = 15/88.2 = 0.17,$$
$$P_{r2} = 66/88.2 = 0.75,$$
$$\therefore \Delta S_1' = 0.15,$$
$$\Delta H_1' = 0.2(373.2) = 75.0,$$
$$\Delta S = \int_{450}^{T_2} \frac{Cp}{T} dT - 1.987 \ln \frac{66}{15} + 0.15 - \Delta S_2' \stackrel{?}{=} 0, \quad (1)$$
$$\Delta H_s = \int_{450}^{T_2} Cp\, dT + 75.0 - \Delta H_2'. \quad (2)$$

1. Assume a value of T_2.
2. Evaluate T_{r2} and $\Delta S_2'$.
3. Integrate Eq. (1) numerically and note the righthand side.
4. Continue with trial values of T_2 until $\Delta S = 0$.
5. Find $\Delta H_2'$ and finally evaluate ΔH_s.

Two trials are shown.

T_2	Tr_2	$\Delta S_2'$	ΔS	$\Delta H_2'$	ΔH_s
600	1.61	0.2	−0.047		
626.6	1.68	0.2	+0.00009	187	1487.7

When the residual properties are neglected,

$$T_2 = 623.33 \text{ K} \quad \text{(compared with 626.6 real)},$$
$$\Delta H_s = 1569.5 \quad \text{(compared with 1487.7 real)}.$$

Real temperature rise:

With $\eta_s = 0.75$, the enthalpy change is 1487.7/0.75 and the enthalpy balance is rearranged to

$$\Delta = -\frac{1487.7}{0.75} + \int_{450}^{T_2} Cp^*\, dT + 75 - \Delta H_2' \stackrel{?}{=} 0$$

Trial T_2	T_r	$\Delta H_2'$	rhs
680	1.82	109	+91.7
670.79	1.80	112	−0.021
670.80	1.80		+0.075

$$\therefore T_2 = 670.79 \text{ K}.$$

For ideal gas

$$\Delta = -\frac{1569.5}{0.75} + \int_{450}^{T_2} Cp^*\, dT \xrightarrow{?} 0$$

By trial:

$$T_2 = 670.49 \text{ K}$$

Nonideality is slight in this example.

EXAMPLE 7.9
Selection of a Centrifugal Compressor

A hydrocarbon mixture with molecular weight 44.23 is raised from 41°F and 20.1 psia to 100.5 psia at the rate of 2400 lb mol/hr. Its specific heat ratio is $k = 1.135$ and its inlet and outlet compressibilities are estimated as $z_1 = 0.97$ and $z_2 = 0.93$. A size of compressor will be selected from Table 7.6 and its expected performance will be calculated:

$$2400 \text{ lb mol/hr} = 1769 \text{ lb/min},$$
$$10{,}260 \text{ cfm}$$

From Table 7.6, the smallest compressor for this gas rate is #38M. Its characteristics are

$$\dot{N} = 8100 \text{ rpm at } 10\text{–}12 \text{ K ft/stage}$$
$$\eta_p = 0.77$$

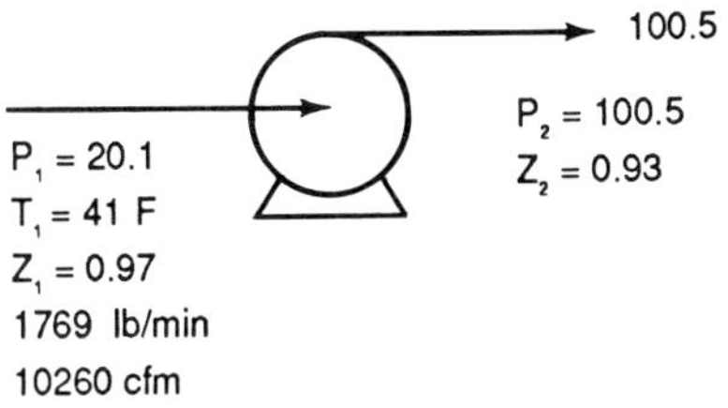

Accordingly,

$$\frac{n-1}{n} = \frac{k-1}{k\eta_p} = \frac{0.135}{1.135(0.77)} = 0.1545.$$

Using Eq. (7.35) for the polytropic head,

$$H_p = \left(\frac{Z_1 + Z_2}{2}\right)\left(\frac{k}{k-1}\right)RT_1\left[\left(\frac{P_2}{P_1}\right)^{(n-1)/n} - 1\right]$$
$$= 0.95\left(\frac{1.135}{0.135}\right)\left(\frac{1544}{44.23}\right)(501)[5^{0.1545} - 1]$$
$$= 39430 \text{ ft}.$$

From Figure 7.26(a), the max head per stage is 9700, and from Figure 7.26(b) the min number of stages is about 4.5. Accordingly, use five stages with standard 10,000 ft/stage impellers. The required speed with the data of Table 7.6 is

$$\text{speed} = 8100\sqrt{39430/10{,}000(5)} = 7190 \text{ rpm}.$$

Power absorbed by the gas is

$$\dot{P}_{\text{gas}} = \frac{\dot{m}H_p}{33{,}000\eta_p} = \frac{1769(39{,}430)}{33{,}000(0.77)} = 2745 \text{ HP}.$$

Friction losses $\cong 3\%$ max;

$$\therefore \text{ total power input} = 2745/0.97 = 2830 \text{ HP max}.$$

---

or alternately in terms of the isentropic efficiency by

$$(\Delta T)_{\text{actual}} = T_2 - T_1 = (\Delta T)_{\text{isentropic}}/\eta_s \quad (7.54)$$

so that

$$T_2 = T_1 + (\Delta T)_s/\eta_s = T_1\{1 + (1/\eta_s)[(P_2/P_1)^{(k-1)/k} - 1]\}. \quad (7.55)$$

The final temperature is read off directly from a thermodynamic diagram when that method is used for the compression calculation, as in Example 7.7. A temperature calculation is made in Example 7.10. Such determinations also are made by the general method for nonideal gases and mixtures as in Example 7.8 and for ideal gases in Example 7.4.

Compression Ratio. In order to save on equipment cost, it is desirable to use as few stages of compression as possible. As a rule, the compression ratio is limited by a practical desirability to keep outlet temperatures below 300°F or so to minimize the possibility of ignition of machine lubricants, as well as the effect that power requirement goes up as outlet temperature goes up. Typical compression ratios of reciprocating equipment are:

Large pipeline compressors	1.2–2.0
Process compressors	1.5–4.0
Small units	up to 6.0

For minimum equipment cost, the work requirement should be the same for each stage. For ideal gases with no friction losses between stages, this implies equal compression ratios. With n stages, accordingly, the compression ratio of each stage is

$$P_{j+1}/P_j = (P_n/P_1)^{1/n}. \quad (7.56)$$

Example 7.11 works out a case involving a nonideal gas and interstage pressure losses.

In centrifugal compressors with all stages in the same shell, the allowable head rise per stage is stated in Table 7.6 or correlated in Figure 7.26. Example 7.9 utilizes these data.

Volumetric Efficiency. For practical reasons, the gas is not completely discharged from a cylinder at each stroke of a reciprocating machine. The clearance of a cylinder is filled with compressed gas which reexpands isentropically on the return stroke. Accordingly, the gas handling capacity of the cylinder is less than the product of the cross section by the length of the stroke. The volumetric efficiency is

$$n_v = \frac{\text{suction gas volume}}{\text{cylinder displacement}}$$
$$= 1 - f_c[(P_2/P_1)^{1/k} - 1], \quad (7.57)$$

where

$$f_c = \frac{\text{clearance volume}}{\text{cylinder displacement volume}}.$$

For a required volumetric suction rate Q (cfm), the required product of cross section A_s (sqft), stroke length L_s (ft), and speed N (rpm) is given by

$$A_sL_sN = Q/\eta_v. \quad (7.58)$$

Example 7.10
Polytropic and Isentropic Temperatures

Take $k = 1.4$, $(P_2/P_1) = 3$, and $\eta_p = 0.75$. From Eq. (7.34),

$$(n-1)/n = (k-1)/k\eta_p = 0.3810$$

and from Eq. (7.39)

$$n_s = \frac{3^{0.2857} - 1}{3^{0.3810} - 1} = 0.7094$$

so that from Eq. (7.53),

$$T_2/T_1 = 3^{0.3810} = 1.5198, \text{ isentropic,}$$

and from Eq. (7.54),

$$T_2/T_1 = 1 + (1/0.7094)(3^{0.2857} - 1) = 1.5197, \text{ polytropic.}$$

7.7. EJECTOR AND VACUUM SYSTEMS

Application ranges of the various kinds of devices for maintenance of subatmospheric pressures in process equipment are shown in Table 7.3. The use of mechanical pumps—compressors in reverse—for such purposes is mentioned earlier in this chapter. Pressures also can be reduced by the action of flowing fluids. For instance, water jets at 40 psig will sustain pressures of 0.5–2.0 psia. For intermediate pressure ranges, down to 0.1 Torr or so, steam jet ejectors are widely favored. They have no moving parts, are quiet, easily installed, simple, and moderately economical to operate, and readily adaptable to handling corrosive vapor mixtures. A specification form is in Appendix B.

EJECTOR ARRANGEMENTS

Several ejectors are used in parallel when the load is variable or because the process system gradually loses tightness between maintenance shutdowns—then some of the units in parallel are cut in or out as needed.

Multistage units in series are needed for low pressures. Sketches are shown in Figure 7.30 of several series arrangements. In Figure 7.30(a), the first stage drives the process vapors, and the second stage drives the mixture of those vapors with the motive steam of the first stage. The other two arrangements employ interstage condensers for the sake of steam economy in subsequent stages. In contact (barometric) condensers the steam and other

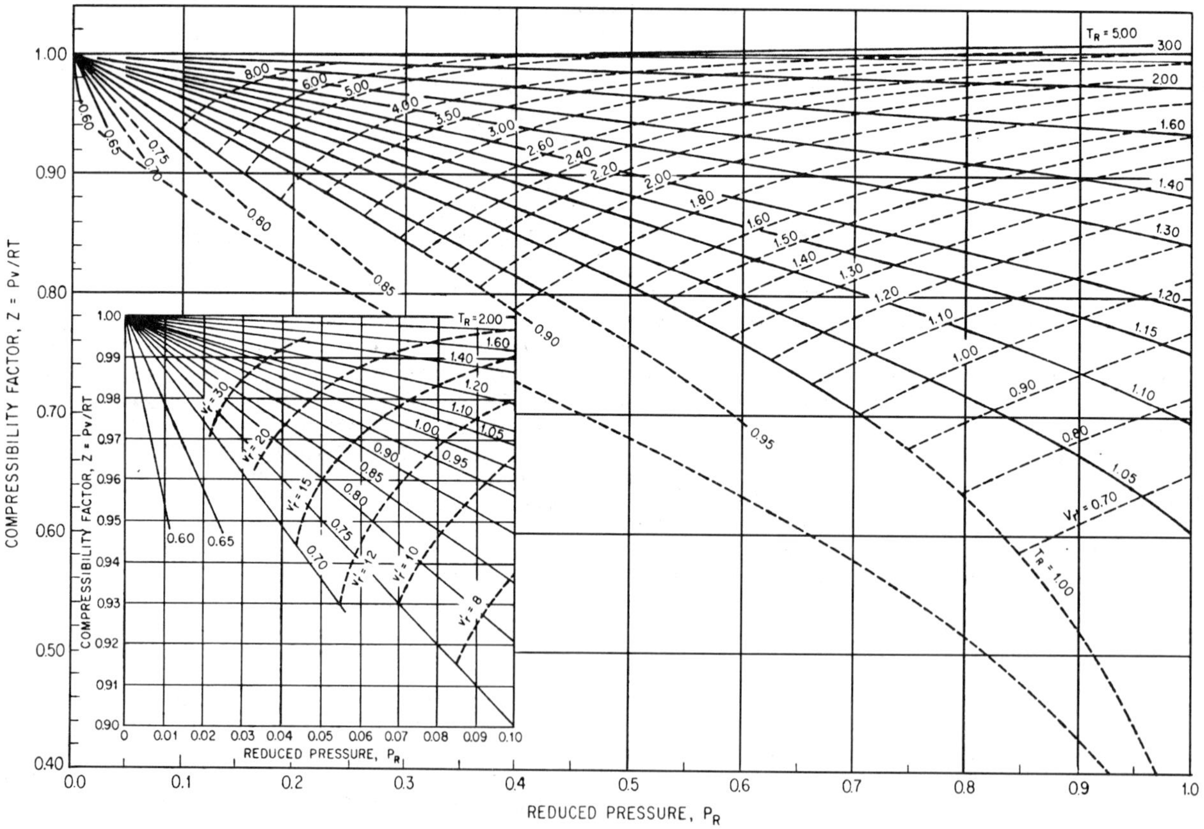

Figure 7.29. Compressibility factors, $z = PV/RT$, of gases. Used for the solution of Example 7.11. $P_R = P/P_c$, $T_R = T/T_c$, and $V_{r'} = P_cV/RT_c$.

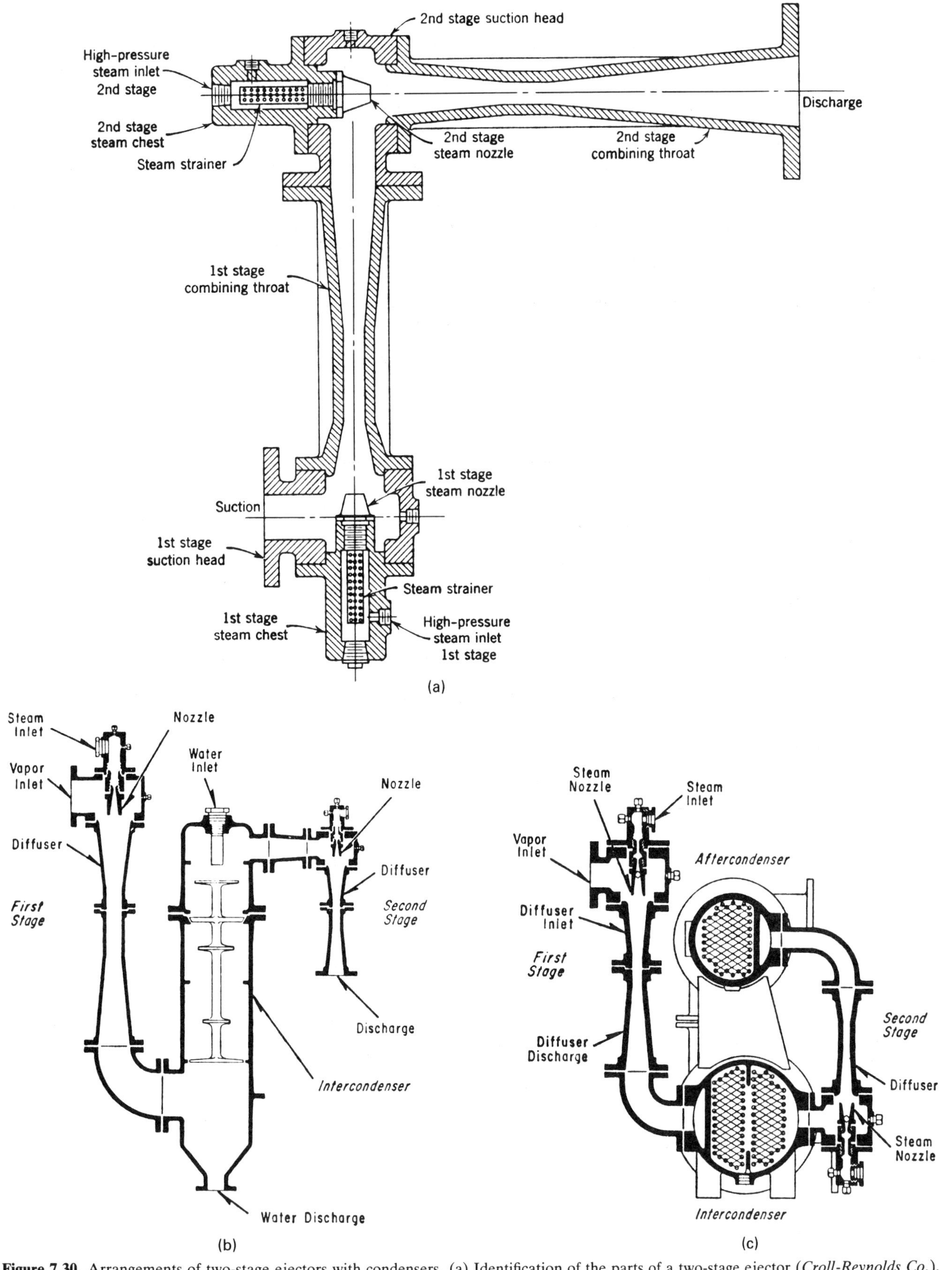

Figure 7.30. Arrangements of two-stage ejectors with condensers. (a) Identification of the parts of a two-stage ejector (*Croll-Reynolds Co.*). (b) A two-stage ejector with interstage barometric condenser (*Elliot Co.*). (c) A two-stage ejector with surface condensers interstage and terminal (*Elliot Co.*).

condensables are removed with a cold water spray. The tail pipes of the condensers are sealed with a 34 ft leg into a sump, or with a condensate pump operating under vacuum. Surface condensers permit recovery of valuable or contaminating condensates or steam condensate for return as boiler feed. They are more expensive than barometrics, and their design is more complex than that of other kinds of condensers because of the large amounts of noncondensables that are present.

As many as six stages are represented on Figure 7.30, combined with interstage condensers in several ways. Barometric condensers are feasible only if the temperature of the water is below its bubblepoint at the prevailing pressure in a particular stage. Common practice requires the temperature to be about 5°F below the bubblepoint. Example 7.13 examines the feasibility of installing intercondensers in that process.

AIR LEAKAGE

The size of ejector and its steam consumption depend on the rate at which gases must be removed from the process. A basic portion of such gases is the air leakage from the atmosphere into the system.

Theoretically, the leakage rate of air through small openings, if they can be regarded as orifices or short nozzles, is constant at vessel pressures below about 53% of atmospheric pressure. However, the openings appear to behave more nearly as conduits with relatively large ratios of lengths to diameters. Accordingly sonic flow is approached only at the low pressure end, and the air mass inleakage rate is determined by that linear velocity and the low density prevailing at the vessel pressure. The content of other gases in the evacuated vessel is determined by each individual process. The content of condensables can be reduced by interposing a refrigerated condenser between process and vacuum pump.

Standards have been developed by the Heat Exchange Institute for rates of air leakage into commercially tight systems. Their chart is represented by the equation

$$m = kV^{2/3}, \tag{7.59}$$

where m is in lb/hr, V is the volume of the system in cuft, and the

---

EXAMPLE 7.11
Three-Stage Compression with Intercooling and Pressure Loss between Stages

Ethylene is to be compressed from 5 to 75 atm in three stages. Temperature to the first stage is 60°F, those to the other stages are 100°F. Pressure loss between stages is 0.34 atm (5 psi). Isentropic efficiency of each stage is 0.87. Compressibilities at the inlets to the stages are estimated from Figure 7.29 under the assumption of equal compression ratios as $z_0 = 0.98$, $z_1 = 0.93$, and $z_2 = 0.83$. The interstage pressures will be determined on the basis of equal power load in each stage. The estimated compressibilities can be corrected after the pressures have been found, but usually this is not found necessary. $k = C_p/C_v = 1.228$ and $(k-1)/k = 0.1857$.

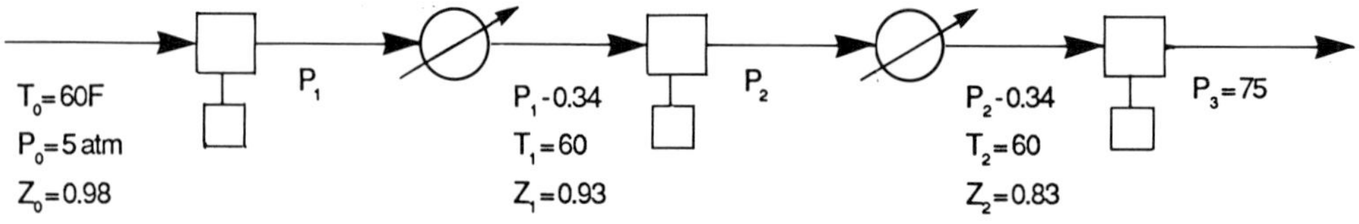

With equal power in each stage

$$\dot{P}_i = \frac{z_i R T_i k}{(k-1)\eta_s}\left[\left(\frac{P}{P_i}\right)^{0.1857} - 1\right]$$
$$= 0.98(520)[(P_1/5)^{0.1857} - 1]$$
$$= 0.93(560)\left\{\left(\frac{P_2}{(P_1 - 0.34)}\right)^{0.1857} - 1\right\}$$
$$= 0.83(560)\{(75/(P_2 - 0.34)]^{0.1857} - 1\}$$

Values of P_1 will be assumed until the value of P_2 calculated by equating the first two terms equals that calculated from the last two terms. The last entries in the table are the interpolated values.

	P_2	
P_1	1+2	2+3
12	27.50	28.31
12.5	29.85	28.94
13.0	32.29	29.56
12.25	—28.60—	

$$\text{Total power} = \frac{3(0.98)(1.987)(520)}{0.1857(2545)0.87}\left[\left(\frac{12.25}{5}\right)^{0.1857} - 1\right]$$
$$= 1.34\ \text{HP/(lb mol/hr)}.$$

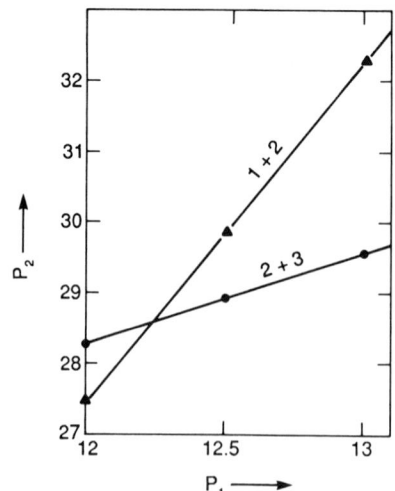

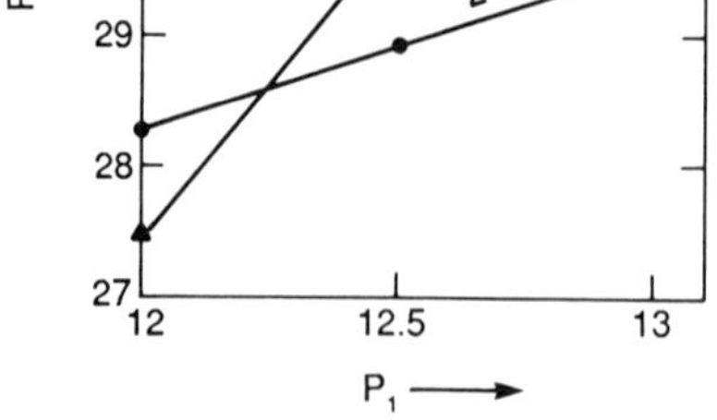

TABLE 7.11. Estimated Air Leakages Through Connections, Valves, Stuffing Boxes Etc. of Process Equipment[a]

Type Fitting	Estimated Average Air Leakage (lb/hr)
Screwed connections in sizes up to 2 in.	0.1
Screwed connections in sizes above 2 in.	0.2
Flanged connections in sizes up to 6 in.	0.5
Flanged connections in sizes 6 in. to 24 in. including manholes	0.8
Flanged connections in sizes 24 in. to 6 ft	1.1
Flanged connections in sizes above 6 ft	2.0
Packed valves up to $\frac{1}{2}$ in. stem diameter	0.5
Packed valves above $\frac{1}{2}$ in. stem diameter	1.0
Lubricated plug valves	0.1
Petcocks	0.2
Sight glasses	1.0
Gage glasses including gage cocks	2.0
Liquid sealed stuffing box for shaft of agitators, pumps, etc. (per in. shaft diameter)	0.3
Ordinary stuffing box (per in. of diameter)	1.5
Safety valves and vacuum breakers (per in. of nominal size	1.0

[a] For conservative practice, these leakages may be taken as supplementary to those from Eq. (7.59). Other practices allow 5 lb/hr for each agitator stuffing box of standard design; special high vacuum mechanical seals with good maintenance can reduce this rate to 1–2 lb/hr.

[From C.D. Jackson, *Chem. Eng. Prog.* **44,** 347 (1948)].

coefficient is a function of the process pressure as follows:

Pressure (Torr)	>90	20–90	3–20	1–3	<1
k	0.194	0.146	0.0825	0.0508	0.0254

For each agitator with a standard stuffing box, 5 lb/hr of air leakage is added. Use of special vacuum mechanical seals can reduce this allowance to 1–2 lb/hr.

For a conservative design, the rate from Eq. (7.59) may be supplemented with values based on Table 7.11. Common practice is to provide oversize ejectors, capable of handling perhaps twice the standard rates of the Heat Exchange Institute.

Other Gases. The gas leakage rate correlations cited are based on air at 70°F. For other conditions, corrections are applied to evaluate an effective air rate. The factor for molecular weight M is

$$f_M = 0.375 \ln(M/2) \quad (7.60)$$

and those for temperature T in °F of predominantly air or predominantly steam are

$$f_A = 1 - 0.00024(T - 70), \quad \text{for air}, \quad (7.61)$$

$$f_S = 1 - 0.00033(T - 70), \quad \text{for steam}. \quad (7.62)$$

An effective or equivalent air rate is found in Example 7.12.

EXAMPLE 7.12
Equivalent Air Rate

Suction gases are at the rate of 120 lb/hr at 300°F and have a molecular weight of 90. The temperature factor is not known as a function of molecular weight so the value for air will be used. Using Eqs. (7.61) and (7.62),

$$m = 120(0.375)\ln(90/2)[1 - 0.00024(300-70)]$$
$$= 161.8\ \text{lb/hr equivalent air.}$$

STEAM CONSUMPTION

The most commonly used steam is 100 psig with 10–15° superheat, the latter characteristic in order to avoid the erosive effect of liquids on the throats of the ejectors. In Figure 7.31 the steam consumptions are given as lb of motive steam per lb of equivalent air to the first stage. Corrections are shown for steam pressures other than 100 psig. When some portion of the initial suction gas is condensable, downward corrections to these rates are to be made for those ejector assemblies that have intercondensers. Such corrections and also the distribution of motive steam to the individual stages are problems best passed on to ejector manufacturers who have experience and a body of test data.

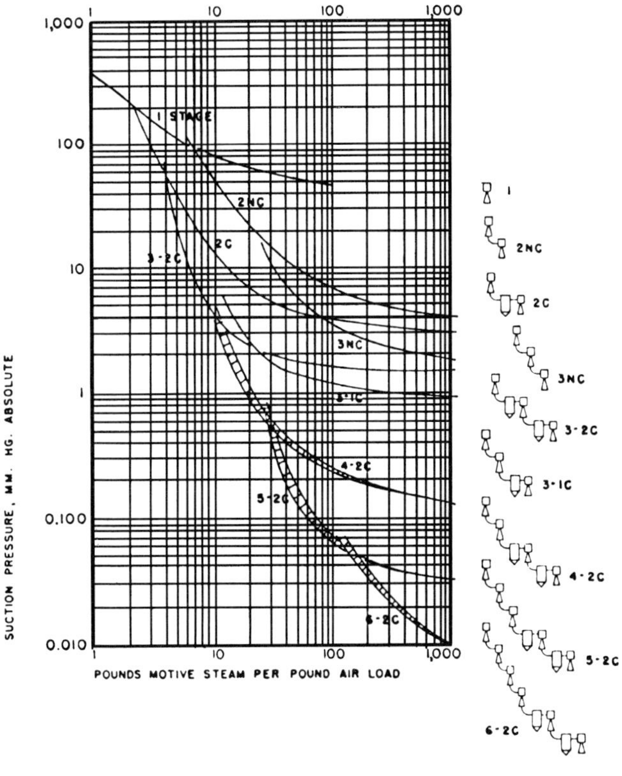

Figure 7.31. Steam requirements of ejectors at various pressure levels with appropriate numbers of stages and contact intercondensers. Steam pressure 100 psig, water temperature 85°F. Factor for 65 psig steam is 1.2 and for 200 psig steam it is 0.80 (*Worthington Corp*).

EXAMPLE 7.13
Interstage Condensers

A four-stage ejector is to evacuate a system to 0.3 Torr. The compression ratio in each stage will be

$$(P_4/P_0)^{1/4} = (760/0.3)^{1/4} = 7.09.$$

The individual stage pressures and corresponding water bubblepoint temperatures from the steam tables are

Discharge of stage	0	1	2	3	4
Torr	0.3	2.1	15.1	107	760
°F		14	63.7	127.4	

The bubblepoint temperature in the second stage is marginal with normal cooling tower water, particularly with the practical restriction to 5°F below the bubblepoint. At the discharge of the third stage, however, either a surface or barometric condenser is quite feasible. At somewhat higher process pressure, two interstage condensers may be practical with a four-stage ejector, as indicated on Figure 7.31.

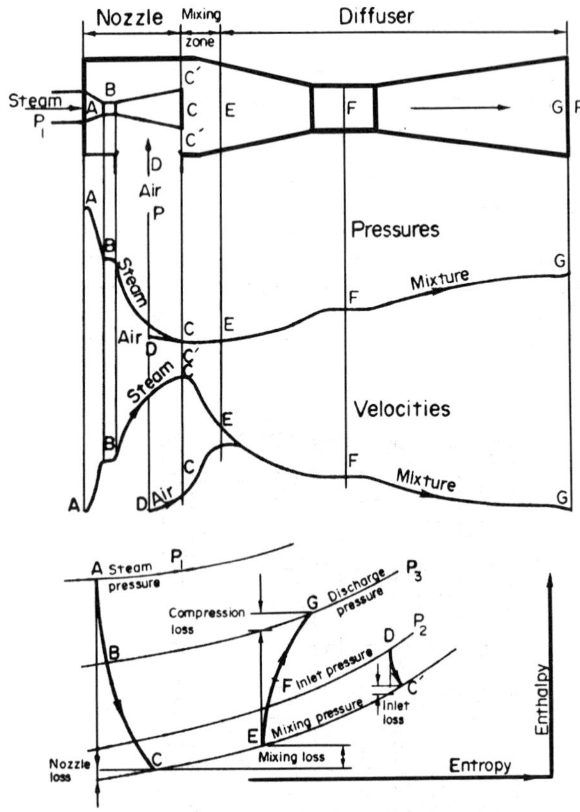

Figure 7.32. Progress of pressures, velocities, enthalpies and entropies in an ejector (*Coulson and Richardson,* Chemical Engineering, *Pergamon, 1977, New York, Vol. 1*).

When barometric condensers are used, the effluent water temperature should be at least 5°F below the bubblepoint at the prevailing pressure. A few bubblepoint temperatures at low pressures are:

Absolute (in. Hg)	0.2	0.5	1.0	2.0
Bubblepoint °F	34.6	58.8	79.0	101.1

Interstage pressures can be estimated on the assumption that compression ratios will be the same in each stage, with the suction to the first stage at the system pressure and the discharge of the last stage at atmospheric pressure. Example 7.13 examines at what stages it is feasible to employ condensers so as to minimize steam usage in subsequent stages.

EJECTOR THEORY

The progress of pressure, velocity, and energy along an ejector is illustrated in Figure 7.32. The initial expansion of the steam to point C and recompression of the mixture beyond point E proceed adiabatically with isentropic efficiencies of the order of 0.8. Mixing in the region from C to E proceeds with approximate conservation of momenta of the two streams, with an efficiency of the order of 0.65. In an example worked out by Dodge (1944, pp. 289–293), the compounding of these three efficiencies leads to a steam rate five times theoretical. Other studies of single-stage ejectors have been made by Work and Haedrich (1939) and DeFrate and Hoerl (1959), where other references to theory and data are made.

The theory is in principle amenable to the prediction of steam distribution to individual stages of a series, but no detailed procedures are readily available. Manufacturers charts such as Figure 7.31 state only the consumption of all the stages together.

GLOSSARY FOR CHAPTER 7

PUMP TERMS

Head has the dimensions $[F][L]/[M]$; for example, ft lbf/lb or ft; or N m/kg or m:

a. pressure head $= \Delta P/\rho$;
b. velocity head $= \Delta u^2/2g_c$;
c. elevation head $= \Delta z(g/g_c)$, or commonly Δz;
d. friction head in line, $H_f = f(L/D)u^2/2g_c$;
e. system head H_s is made up of the preceding four items;
f. pump head equals system head, $H_p = H_s$, under operating conditions;
g. static suction head equals the difference in levels of suction liquid and the centerline of the pump;
h. static suction lift is the static suction head when the suction level is below the centerline of the pump; numerically a negative number.

NPSH (net positive suction head) = (pressure head of source) + (static suction head) − (friction head of the suction line) − (vapor pressure of the flowing liquid).

Hydraulic horsepower is obtained by multiplying the weight rate of flow by the head difference across the pump and converting to horsepower. For example, HHP = (gpm)(psi)/1714 = (gpm)(sp gr)(ft)/3960.

Brake horsepower is the driver power output needed to operate the pump. BHP = HHP/(pump efficiency).

Driver horsepower, HP = BHP/(driver efficiency) = HHP/(pump efficiency)(driver efficiency).

TERMS CONCERNING CENTRIFUGAL AND RELATED PUMPS

Axial flow is flow developed by axial thrust of a propeller blade, practically limited to heads under 50 ft or so.

Centrifugal pump consists of a rotor (impeller) in a casing in which a liquid is given a high velocity head that is largely converted to pressure head by the time the liquid reaches the outlet.

Characteristic curves are plots or equations relating the volumetric flow rate through a pump to the developed head or efficiency or power or NPSH.

Diffuser type: the impeller is surrounded by gradually expanding passages formed by stationary guide vanes [Figs. 7.2(b) and 7.3(d)].

Double suction: two incoming streams enter at the eye of the impeller on opposite sides, minimizing axial thrust and worthwhile for large, high head pumps [Fig. 7.2(b)].

Double volute: the liquid leaving the impeller is collected in two similar volutes displaced 180° with a common outlet; radial thrust is counterbalanced and shaft deflection is minimized, resulting in lower maintenance and repair, used in high speed pumps producing above 500 ft per stage.

Impeller: the rotor that accelerates the liquid.

a. Open impellers consist of vanes attached to a shaft without any form of supporting sidewall and are suited to handling slurries without clogging [Fig. 7.2(a)].

b. Semienclosed impellers have a complete shroud on one side [Fig. 7.3(c)]; they are essentially nonclogging, used primarily in small size pumps; clearance of the open face to the wall is typically 0.02 in. for 10 in. diameters.

c. Closed impellers have shrouds on both sides of the vanes from the eye to the periphery, used for clear liquids [Fig. 7.3(b)].

Mechanical seals prevent leakage at the rotating shaft by sliding metal on metal lubricated by a slight flow of pump liquid or an independent liquid [Figs. 7.4(c) and (d)].

Mixed flow: develops head by combined centrifugal action and propeller action in the axial direction, suited to high flow rates at moderate heads [Fig. 7.3(e)].

Multistage: several pumps in series in a single casing with the objective of developing high heads. Figure 7.6(c) is of characteristic curves.

Performance curves (see characteristic curves).

Single suction: the liquid enters on one side at the eye of the impeller; most pumps are of this lower cost style [Fig. 7.2(c)].

Split case: constructed so that the internals can be accessed without disconnecting the piping [Fig. 7.2(a)].

Stuffing box: prevent leakage at the rotating shaft with compressed soft packing that may be wetted with the pump liquid or from an independent source [Figs. 7.4(a) and (b)].

Volute type: the impeller discharges the liquid into a progressively expanding spiral [Fig. 7.2(a)].

REFERENCES

Compressors

1. Compressors in *Encyclopedia of Chemical Processing and Design,* Dekker, New York, 1979, Vol. 10, pp. 157–409.
2. F.L. Evans, Compressors and fans, in *Equipment Design Handbook for Refineries and Chemical Plants,* Gulf, Houston, 1979, Vol. 1, pp. 54–104.
3. H. Gartmann, *DeLaval Engineering Handbook,* McGraw-Hill, New York, 1970, pp. 6.61–6.93.
4. R. James, Compressor calculation procedures, in *Encyclopedia of Chemical Processing and Design,* Dekker, New York, Vol. 10, pp. 264–313.
5. E.E. Ludwig, Compressors, in *Applied Process Design for Chemical and Petrochemical Plants,* Gulf, Houston, 1983, Vol. 3, pp. 251–396.
6. R.D. Madison, *Fan Engineering,* Buffalo Forge Co., Buffalo, NY, 1949.
7. H.F. Rase and M.H. Barrow, *Project Engineering of Process Plants,* Wiley, New York, pp. 297–347.

Ejectors

1. L.A. DeFrate and V.W. Haedrich, *Chem. Eng. Prog. Symp. Ser.* **21,** 43–51 (1959).
2. B.F. Dodge, *Chemical Engineering Thermodynamics,* McGraw-Hill, New York, 1944, pp. 289–293.
3. F.I. Evans, *Equipment Design Handbook for Refineries and Chemical Plants,* Gulf, Houston, 1979, Vol. 1, pp. 105–117.
4. E.E. Ludwig, *loc. cit.,* Vol. 1, pp. 206–239.
5. R.E. Richenberg and J.J. Bawden, Ejectors, steam jet, in *Encyclopedia of Chemical Processing and Design,* Dekker, New York, Vol. 17, pp. 167–194.
6. L.T. Work and V.W. Haedrich, *Ind. Eng. Chem.* **31,** 464–477 (1939).

Piping

1. ANSI Piping Code, ASME, New York, 1980.
2. S. Chalfin, Control valves, *Encyclopedia of Chemical Processing and Design,* Dekker, New York, 1980, Vol. 11, pp. 187–213.
3. F.L. Evans, *Equipment Design Handbook for Refineries and Chemical Plants,* Gulf, Houston, 1979, Vol. 2; piping, pp. 188–304; valves, pp. 315–332.
4. J.W. Hutchinson, *ISA Handbook of Control Valves,* Inst. Soc. America, Research Triangle Park, NC, 1976.
5. R.C. King, *Piping Handbook,* McGraw-Hill, New York, 1967.
6. J.L. Lyons, *Encyclopedia of Valves,* Van Nostrand Reinhold, New York, 1975.
7. *Marks' Standard Handbook for Mechanical Engineers,* McGraw-Hill, New York, 1987.
8. *Perry's Chemical Engineers' Handbook,* McGraw-Hill, New York, 1984.
9. R. Weaver, *Process Piping Design,* Gulf, Houston, 1973, 2 Vols.
10. P. Wing, Control valves, in *Process Instruments and Controls Handbook,* (D.M. Considine, Ed.), McGraw-Hill, New York, 1974.
11. R.W. Zappe, *Valve Selection Handbook,* Gulf, Houston, pp. 19.1–19.60, 1981.

Pumps

1. D. Azbel and N.P. Cheremisinoff, *Fluid Mechanics and Fluid Operations,* Ann Arbor Science, Ann Arbor, MI, 1983.
2. N.P. Cheremisinoff, *Fluid Flow: Pumps, Pipes and Channels,* Ann Arbor Science, Ann Arbor, MI, 1981.
3. F.L. Evans, *loc. cit.,* Vol. 1, pp. 118–171.
4. H. Gartmann, *DeLaval Engineering Handbook,* McGraw-Hill, New York, 1970, pp. 6.1–6.60.
5. I.J. Karassik and R. Carter, *Centrifugal Pump Selection Operation and Maintenance,* F.W. Dodge Corp., New York, 1960.
6. I.J. Karassik, W.C. Krutsch, W.H. Fraser, and Y.J.P. Messina, *Pump Handbook,* McGraw-Hill, New York, 1976.
7. F.A. Kristal and F.A. Annett, *Pumps,* McGraw-Hill, New York, 1940.
8. E.E. Ludwig, *loc. cit.,* Vol. 1, pp. 104–143.
9. S. Yedidiah, *Centrifugal Pump Problems,* Petroleum Publishing, Tulsa, OK, 1980.

8

HEAT TRANSFER AND HEAT EXCHANGERS

Basic concepts of heat transfer are reviewed in this chapter and applied primarily to heat exchangers, which are equipment for the transfer of heat between two fluids through a separating wall. Heat transfer also is a key process in other specialized equipment, some of which are treated in the next and other chapters. The three recognized modes of heat transfer are by conduction, convection, and radiation, and may occur simultaneously in some equipment.

8.1. CONDUCTION OF HEAT

In a solid wall such as Figure 8.1(a), the variation of temperature with time and position is represented by the one-dimensional Fourier equation

$$\frac{\partial T}{\partial \theta} = kA \frac{\partial^2 T}{\partial x^2} \tag{8.1}$$

For the most part, only the steady state condition will be of concern here, in which the case the partial integral of Eq. (8.1) becomes

$$Q = -kA \frac{dT}{dx}, \tag{8.2}$$

assuming the thermal conductivity k to be independent of temperature. Furthermore, when both k and A are independent of position,

$$Q = -kA \frac{\Delta T}{\Delta x} = \frac{kA}{L}(T_0 - T_L), \tag{8.3}$$

in the notation of Figure 8.1(a).

Equation (8.3) is the basic form into which more complex situations often are cast. For example,

$$Q = kA_{\text{mean}} \frac{\Delta T}{L} \tag{8.4}$$

when the area is variable and

$$Q = UA(\Delta T)_{\text{mean}} \tag{8.5}$$

in certain kinds of heat exchangers with variable temperature difference.

THERMAL CONDUCTIVITY

Thermal conductivity is a fundamental property of substances that basically is obtained experimentally although some estimation methods also are available. It varies somewhat with temperature. In many heat transfer situations an average value over the prevailing temperature range often is adequate. When the variation is linear with

$$k = k_0(1 + \alpha T), \tag{8.6}$$

the integral of Eq. (8.2) becomes

$$\begin{aligned} Q(L/A) &= k_0[T_1 - T_2 + 0.5\alpha(T_1^2 - T_2^2)] \\ &= k_0(T_1 - T_2)[1 + 0.5\alpha(T_1 + T_2)], \end{aligned} \tag{8.7}$$

which demonstrates that use of a value at the average temperature gives an exact result. Thermal conductivity data at several temperatures of some metals used in heat exchangers are in Table 8.1. The order of magnitude of the temperature effect on k is illustrated in Example 8.1.

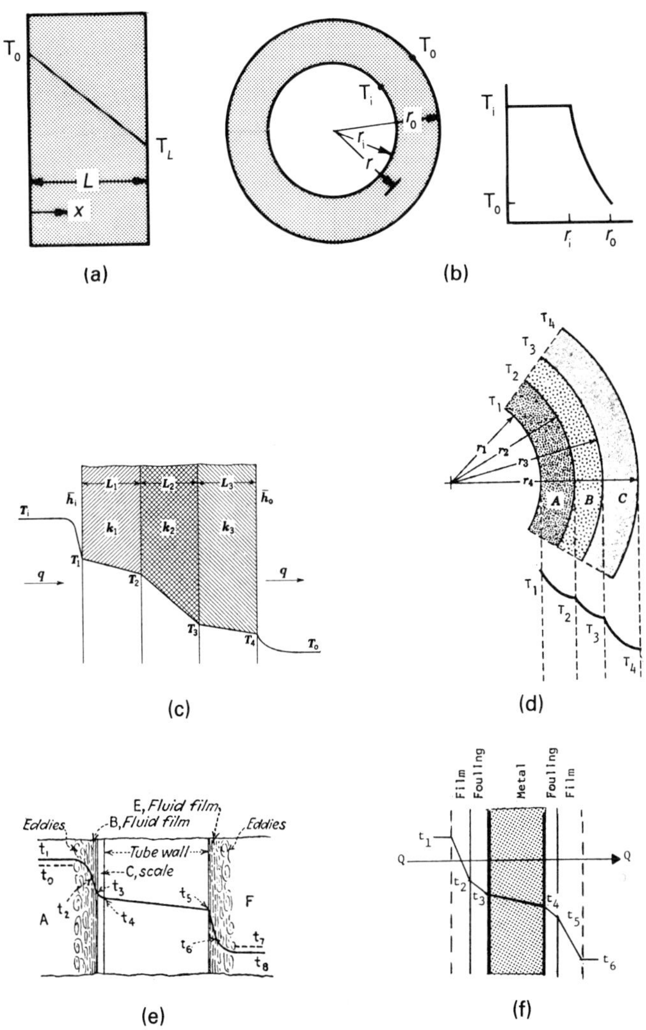

Figure 8.1. Temperature profiles in one-dimensional conduction of heat. (a) Constant cross section. (b) Hollow cylinder. (c) Composite flat wall. (d) Composite hollow cylindrical wall. (e) From fluid A to fluid F through a wall and fouling resistance in the presence of eddies. (f) Through equivalent fluid films, fouling resistances, and metal wall.

TABLE 8.1. Thermal Conductivities of Some Metals Commonly Used in Heat Exchangers [kBtu/(hr)(sqft)(°F/ft)]

	Temperature (°F)			
Metal or Alloy	**−100**	**70**	**200**	**1000**
Steels				
Carbon	—	30.0	27.6	22.2
1 Cr $\frac{1}{4}$ Mo	—	19.2	19.1	18.0
410	—	13.0	14.4	—
304	—	9.4	10.0	13.7
316	8.1	9.4	—	13.0
Monel 400	11.6	12.6	13.8	22.0
Nickel 200	—	32.5	31.9	30.6
Inconel 600	—	8.6	9.1	14.3
Hastelloy C	—	7.3	5.6	10.2
Aluminum	—	131	133	—
Titanium	11.8	11.5	10.9	12.1
Tantalum	—	31.8	—	—
Copper	225	225	222	209
Yellow brass	56	69	—	—
Admiralty	55	64	—	—

HOLLOW CYLINDER

As it appears on Figure 8.1(b), as the heat flows from the inside to the outside the area changes constantly. Accordingly the equivalent of Eq. (8.2) becomes, for a cylinder of length N,

$$Q = -kN(2\pi r)\frac{dT}{dr}, \tag{8.8}$$

of which the integral is

$$Q = \frac{2\pi kN(T_1 - T_2)}{\ln(r_2/r_1)}. \tag{8.9}$$

This may be written in the standard form of Eq. (8.4) by taking

$$A_m = 2\pi LNr_{\rm lm} \tag{8.10}$$

and

$$L = r_2 - r_1, \tag{8.11}$$

where

$$r_{\rm lm} = (r_2 - r_1)/\ln(r_2/r_1) \tag{8.12}$$

is the logarithmic mean radius of the hollow cylinder. This concept is not particularly useful here, but logarithmic means also occur in other more important heat transfer situations.

COMPOSITE WALLS

The flow rate of heat is the same through each wall of Figure 8.1(c). In terms of the overall temperature difference,

$$Q = UA(T_1 - T_4), \tag{8.13}$$

where U is the overall heat transfer coefficient and is given by

$$\frac{1}{U} = \frac{1}{k_a/L_a} + \frac{1}{k_b/L_b} + \frac{1}{k_c/L_c}. \tag{8.14}$$

The reciprocals in Eq. (8.14) may be interpreted as resistances to heat transfer, and so it appears that thermal resistances in series are additive.

For the composite hollow cylinder of Figure 8.1(d), with length N,

$$Q = \frac{2\pi N(T_1 - T_4)}{\ln(r_2/r_1)/k_a + \ln(r_3/r_2)/k_b + \ln(r_4/r_3)/k_c}. \tag{8.15}$$

With an overall coefficient U_i based on the inside area, for example,

$$Q = 2\pi Nr_iU_i(T_1 - T_4) = \frac{2\pi N(T_1 - T_4)}{1/U_ir_i}. \tag{8.16}$$

On comparison of Eqs. (8.15) and (8.16), an expression for the inside overall coefficient appears to be

$$\frac{1}{U_i} = r_i\left[\frac{\ln(r_2/r_1)}{k_a} + \frac{\ln(r_3/r_2)}{k_b} + \frac{\ln(r_4/r_3)}{k_c}\right]. \tag{8.17}$$

In terms of the logarithmic mean radii of the individual cylinders,

$$\frac{1}{U_i} = r_i\left[\frac{1}{k_ar_{ma}/(r_2 - r_1)} + \frac{1}{k_br_{mb}/(r_3 - r_2)} + \frac{1}{k_cr_{mc}/(r_4 - r_3)}\right],$$

which is similar to Eq. (8.14) for flat walls, but includes a ratio of radii as a correction for each cylinder.

FLUID FILMS

Heat transfer between a fluid and a solid wall can be represented by conduction equations. It is assumed that the difference in temperature between fluid and wall is due entirely to a stagnant film of liquid adhering to the wall and in which the temperature profile is linear. Figure 8.1(e) is a somewhat realistic representation of a temperature profile in the transfer of heat from one fluid to another through a wall and fouling scale, whereas the more nearly ideal Figure 8.1(f) concentrates the temperature drops in stagnant fluid and fouling films.

Since the film thicknesses are not definite quantities, they are best combined with the conductivities into single coefficients

$$h = k/L \tag{8.18}$$

so that the rate of heat transfer through the film becomes

$$Q = hA\Delta T. \tag{8.19}$$

Through the five resistances of Figure 8.1(f), the overall heat

---

EXAMPLE 8.1
Conduction through a Furnace Wall

A furnace wall made of fire clay has an inside temperature of 1500°F and an outside one of 300°F. The equation of the thermal conductivity is $k = 0.48[1 + 5.15(E - 4)T]$ Btu/(hr)(sqft)(°F/ft). Accordingly,

$$Q(L/A) = 0.48(1500 - 300)[1 + 5.15(E - 4)(900)] = 0.703.$$

If the conductivity at 300°F had been used, $Q(L/A) = 0.554$.

EXAMPLE 8.2
Effect of Ignoring the Radius Correction of the Overall Heat Transfer Coefficient

The two film coefficients are 100 each, the two fouling coefficients are 2000 each, the tube outside diameter is 0.1 ft, wall thickness is 0.01 ft, and thermal conductivity of the metal is 30:

$$r_i/r_o = 0.04/0.05 = 0.8,$$
$$r_m = (0.05 - 0.04)/\ln 1.25 = 0.0448,$$
$$r_m/r_o = 0.8963,$$
$$U_o = [1/100(0.8) + 1/2000(0.8) + 1/(30/0.01)(0.8963) + 1/100 + 1/2000]^{-1} = 41.6721.$$

Basing on the inside area,

$$U_i = [1/100 + 1/2000 + [(30/0.01)(0.0448/0.04)]^{-1} + 0.8/100 + 0.8/2000]^{-1} = 52.0898.$$

Ignoring the corrections,

$$U = (2/100 + 2/2000 + 1/30/0.01)^{-1} = 46.8750.$$

The last value is very nearly the average of the other two.

transfer coefficient is given by

$$\frac{1}{U} = \frac{1}{h_1} + \frac{1}{h_2} + \frac{1}{k_3/L_3} + \frac{1}{h_4} + \frac{1}{h_5}, \tag{8.20}$$

where L_3 is the thickness of the metal.

If the wall is that of hollow cylinder with radii r_i and r_o, the overall heat transfer coefficient based on the outside surface is

$$\frac{1}{U_o} = \frac{1}{h_1(r_i/r_o)} + \frac{1}{h_2(r_i/r_0)} + \frac{1}{(k_3/L_3)(r_m/r_0)} + \frac{1}{h_4} + \frac{1}{h_5}, \tag{8.21}$$

where r_m is the mean radius of the cylinder, given by Eq. (8.12).

Since wall thicknesses of heat exchangers are relatively small and the accuracy of heat transfer coefficients may not be great, the ratio of radii in Eq. (8.21) often is ignored, so that the equation for the overall coefficient becomes simply

$$\frac{1}{U} = \frac{1}{h_1} + \frac{1}{h_2} + \frac{1}{k_3/L_3} + \frac{1}{h_4} + \frac{1}{h_5}. \tag{8.22}$$

The results of the typical case of Example 8.2, however, indicate that the correction may be significant. A case with two films and two solid cylindrical walls is examined in Example 8.3.

EXAMPLE 8.3
A Case of a Composite Wall: Optimum Insulation Thickness for a Steam Line

A 3 in. IPS Sched 40 steel line carries steam at 500°F. Ambient air is at 70°F. Steam side coefficient is 1000 and air side is 3 Btu/(hr)(sqft)(°F). Conductivity of the metal is 30 and that of insulation is 0.05 Btu/(hr)(sqft)(°F/ft). Value of the steam is $5.00/MBtu. cost of the insulation is $1.5/(yr)(cuft). Operation is 8760 hr/yr. The optimum diameter d of insulation thickness will be found.

Pipe:

$d_o = 0.2917$ ft,
$d_i = 0.2557$ ft,
$\ln(d_o/d_i) = 0.1317.$

Insulation:

$$\ln(d_o/d_i) = \ln(d/0.2917). \tag{1}$$

Heat transfer coefficient based on inside area:

$$U_i = d_i\left[\frac{1}{1000d_i} + \frac{0.1317}{30} + \frac{\ln(d/0.2917)}{0.05} + \frac{1}{3d}\right]^{-1}, \tag{2}$$
$$Q/A_i = U_i\Delta T = 430U_i.$$

Steam cost:

$$C_1 = 5(10^{-6})(8760)Q/A_i = 0.0438Q/A_i, \quad \$\,(\text{yr})(\text{sqft inside}). \tag{3}$$

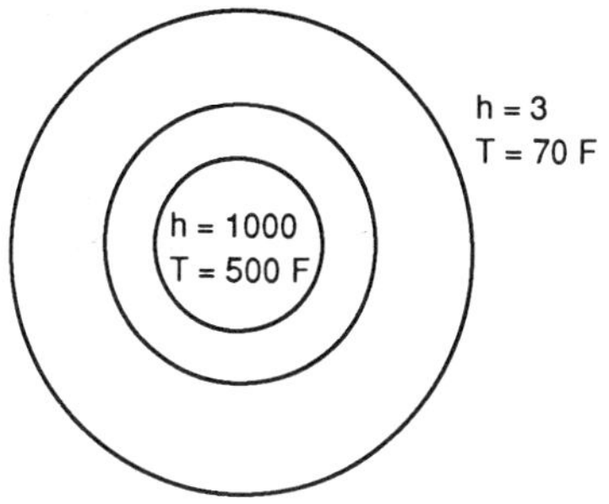


```
 10 ! Example 8.3. Optimum insul
    ation thickness
 20 READ D1,D2
 30 DATA .2917,.2557
 40 INPUT D
 50 U1=.001/D2+.1307/30+LOG(D/D1
    )/.05+1/3/D
 60 U=1/U1/D2
 70 Q=430*U
 80 C1=.0438*Q
 90 C2=1.5*(D^2-D1^2)/D2^2
100 C=C1+C2 ! (req'd to be minim
    um)
110 PRINT USING 120 ; D,U,C1,C2,
    C
120 IMAGE .DDD,X,.DDD,X,DD.DD,X,
    DD.DD,X,DD.DDDD
130 GOTO 40
140 END
```

Example 8.3—*(continued)*

D	U	C_1	C_2	$C_1 + C_2$
.490	.354	6.66	3.56	10.2147
.494	.349	6.57	3.65	10.2118
.495	.347	6.54	3.67	10.2117 *
.496	.346	6.52	3.69	10.2118
.500	.341	6.43	3.78	10.2148

Insulation cost:

$$C_2 = 1.5V_{ins}/A_i = \frac{1.5(d^2 - 0.2917^2)}{(0.2557)^2}, \quad \$/(\text{yr})(\text{sqft inside}). \quad (4)$$

Total cost:

$$C = C_1 + C_2 \rightarrow \text{minimum}. \quad (5)$$

Substitute Eqs. (2)–(4) into Eq. (5). The outside diameter is the key unknown.

The cost curve is fairly flat, with a minimum at $d = 0.50$ ft, corresponding to 1.25 in. thickness of insulation. Some trials are shown with the computer program. A more detailed analysis of insulation optima is made by Happel and Jordan [*Chem. Process Econ.*, 380 (1975)], although their prices are dated. Section 8.12 also discusses insulation.

Heat transfer coefficients are empirical data and derived correlations. They are in the form of overall coefficients U for frequently occurring operations, or as individual film coefficients and fouling factors.

8.2. MEAN TEMPERATURE DIFFERENCE

In a heat exchanger, heat is transferred between hot and cold fluids through a solid wall. The fluids may be process streams or independent sources of heat such as the fluids of Table 8.2 or sources of refrigeration. Figure 8.2 shows such a process with inlet and outlet streams, but with the internal flow pattern unidentified because it varies from case to case. At any cross section, the differential rate of heat transfer is

$$dQ = U(T - T')\,dA = -mc\,dT = m'c'\,dT'. \quad (8.23)$$

The overall heat transfer rate is represented formally by

$$Q = UA(\Delta T)_m. \quad (8.24)$$

The mean temperature difference $(\Delta T)_m$ depends on the terminal temperatures, the thermal properties of the two fluids and on the flow pattern through the exchanger.

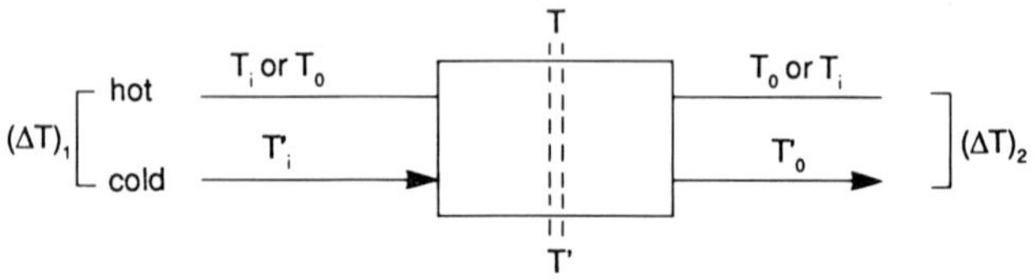

Figure 8.2. Terminal temperatures and temperature differences of a heat exchanger, with unidentified internal flow pattern.

SINGLE PASS EXCHANGER

The simplest flow patterns are single pass of each fluid, in either the same or opposite directions. Temperature profiles of the main kinds of thermal behavior are indicated on Figure 8.3(a). When the unbroken lines [cases (a)–(e)] are substantially straight, the mean temperature is expressed in terms of the terminal differences by

$$(\Delta T)_m = (\Delta T)_{\text{log mean}} = \frac{(\Delta T)_2 - (\Delta T)_1}{\ln[(\Delta T)_2/(\Delta T)_1]}. \quad (8.25)$$

This is called the logarithmic mean temperature difference. The temperature profiles are straight when the heat capacities are

TABLE 8.2. Properties of Heat Transfer Media

Medium	Trade Name	Phase	°F	atm, gage	Remarks
Electricity	—		100–4500	—	—
Water	—	vapor	200–1100	0–300	—
Water	—	liquid	300–400	6–15	—
Flue gas	—	gas	100–2000	0–7	—
Diphenyl–diphenyl oxide eutectic	Dowtherm A	liquid or vapor	450–750	0–9	nontoxic, carbonizes at high temp
Di + triaryl cpds	Dowtherm G	liquid	20–700	0–3	sensitive to oxygen
Ethylene glycol, inhibited	Dow SR-1	liquid	−40–250	0	acceptable in food industry
Dimethyl silicones	Dow Syltherm 800	liquid	−40–750	0	low toxicity
Mixed silanes	Hydrotherm	liquid	−50–675	0	react with oxygen and moisture
Aromatic mineral oil	Mobiltherm, Mobil	liquid	100–600	0	not used with copper based materials
Chlorinated biphenyls	Therminol, Monsanto	liquid	50–600	0	toxic decomposition products
Molten nitrites and nitrates of K and Na	Hi-Tec, DuPont	liquid	300–1100	0	resistant alloys needed above 850°F
Sodium–potassium eutectic		liquid	100–1400	0	stainless steel needed above 1000°F
Mercury		vapor	600–1000	0–12	low pressure vapor, toxic, and expensive

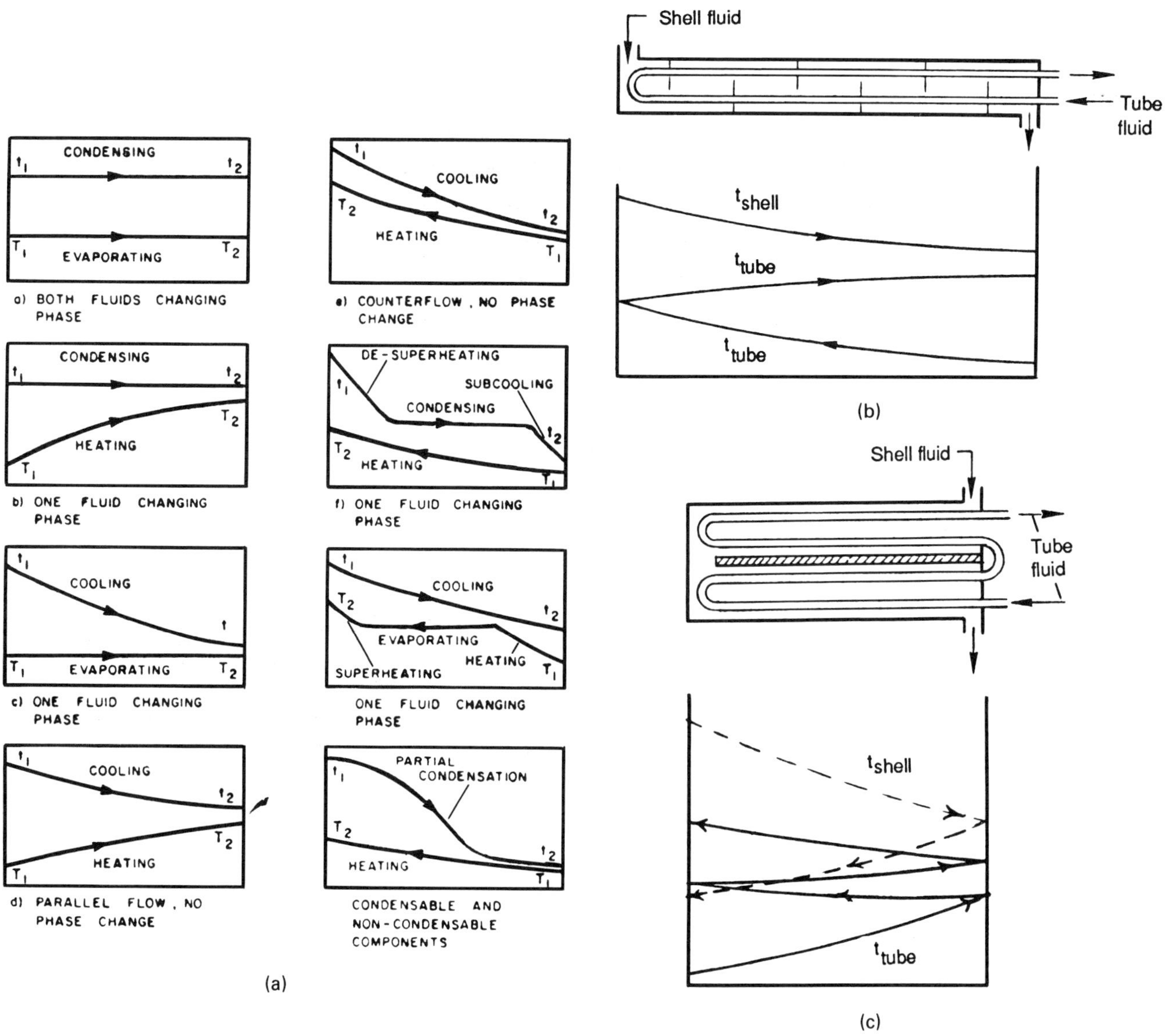

Figure 8.3. Temperature profiles in heat exchangers. (a) In parallel or countercurrent flow, with one or two phases. (b) One shell pass, two tube passes. (c) Two shell passes, four tube passes.

substantially independent of temperature over the range of the process, or when a phase change occurs at constant temperature.

When the profiles consist of linear sections, as in cases (f) and (g), the exchanger can be treated as a three-section assembly, each characterized by its own log mean temperature difference, for which intermediate temperatures may be found by direct calculation or by trial. Heat transfer for a case such as (h) with continuously curved profile must be evaluated by integration of Eq. (8.23).

MULTIPASS EXCHANGERS

For reasons of compactness of equipment, the paths of both fluids may require several reversals of direction. Two of the simpler cases of Figure 8.3 are (b) one pass on the shell side and two passes on the tube side and (c) two passes on the shell side and four on the tube side. On a baffled shell side, as on Figure 8.4(c), the dominant flow is in the axial direction, so this pattern still is regarded as single pass on the shell side. In the cross flow pattern of Figure 8.5(c), each stream flows without lateral mixing, for instance in equipment like Figure 8.6(h). In Figure 8.6(i) considerable lateral mixing would occur on the gas side. Lateral mixing could occur on both sides of the plate exchanger of Figure 8.6(h) if the fins were absent.

Mean temperature differences in such flow patterns are obtained by solving the differential equation. Analytical solutions have been found for the simpler cases, and numerical ones for many important complex patterns, whose results sometimes are available in generalized graphical form.

F-METHOD

When all of the terminal temperatures are known, the mean temperature difference is found directly from

$$(\Delta T)_m = F(\Delta T)_{\text{log mean}}, \qquad (8.26)$$

where the correction factor F depends on the flow pattern and is

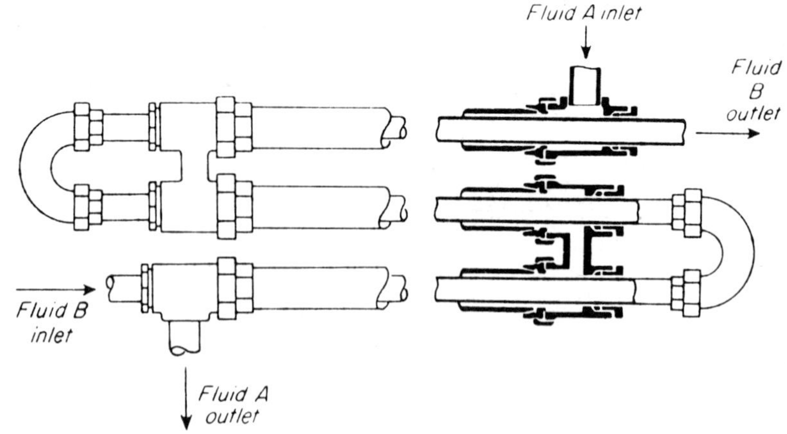

(a)

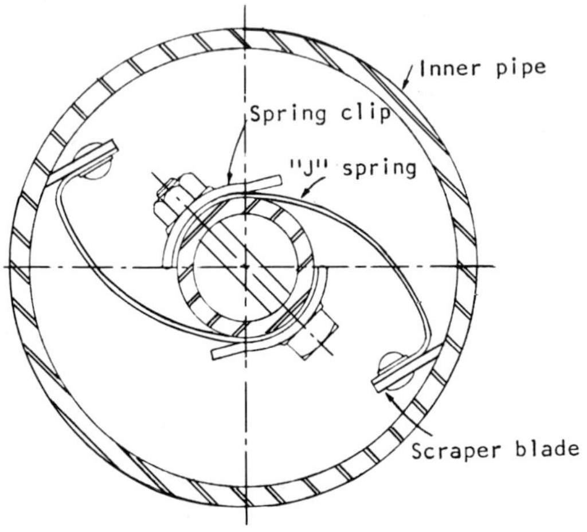

(b)

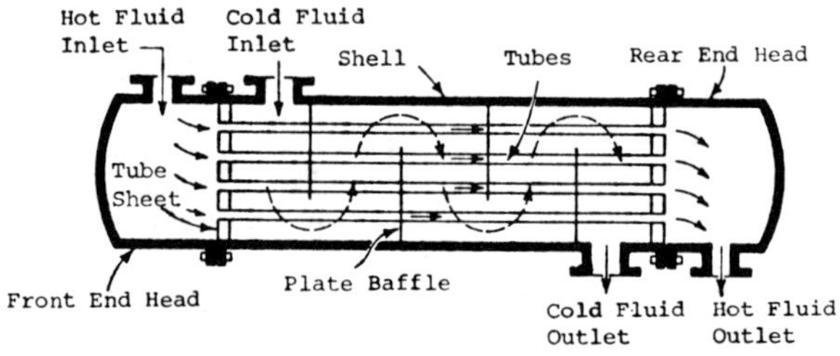

(c)

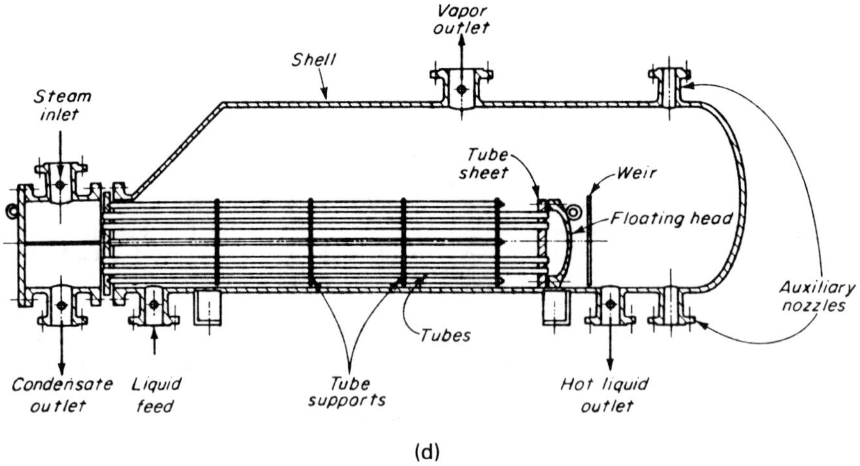

(d)

Figure 8.4. Example of tubular heat exchangers (see also Fig. 8.14). (a) Double-pipe exchanger. (b) Scraped inner surface of a double-pipe exchanger. (c) Shell-and-tube exchanger with fixed tube sheets. (d) Kettle-type reboiler. (e) Horizontal shell side thermosiphon reboiler. (f) Vertical tube side thermosiphon reboiler. (g) Internal reboiler in a tower. (h) Air cooler with induced draft fan above the tube bank. (i) Air cooler with forced draft fan below the tube bank.

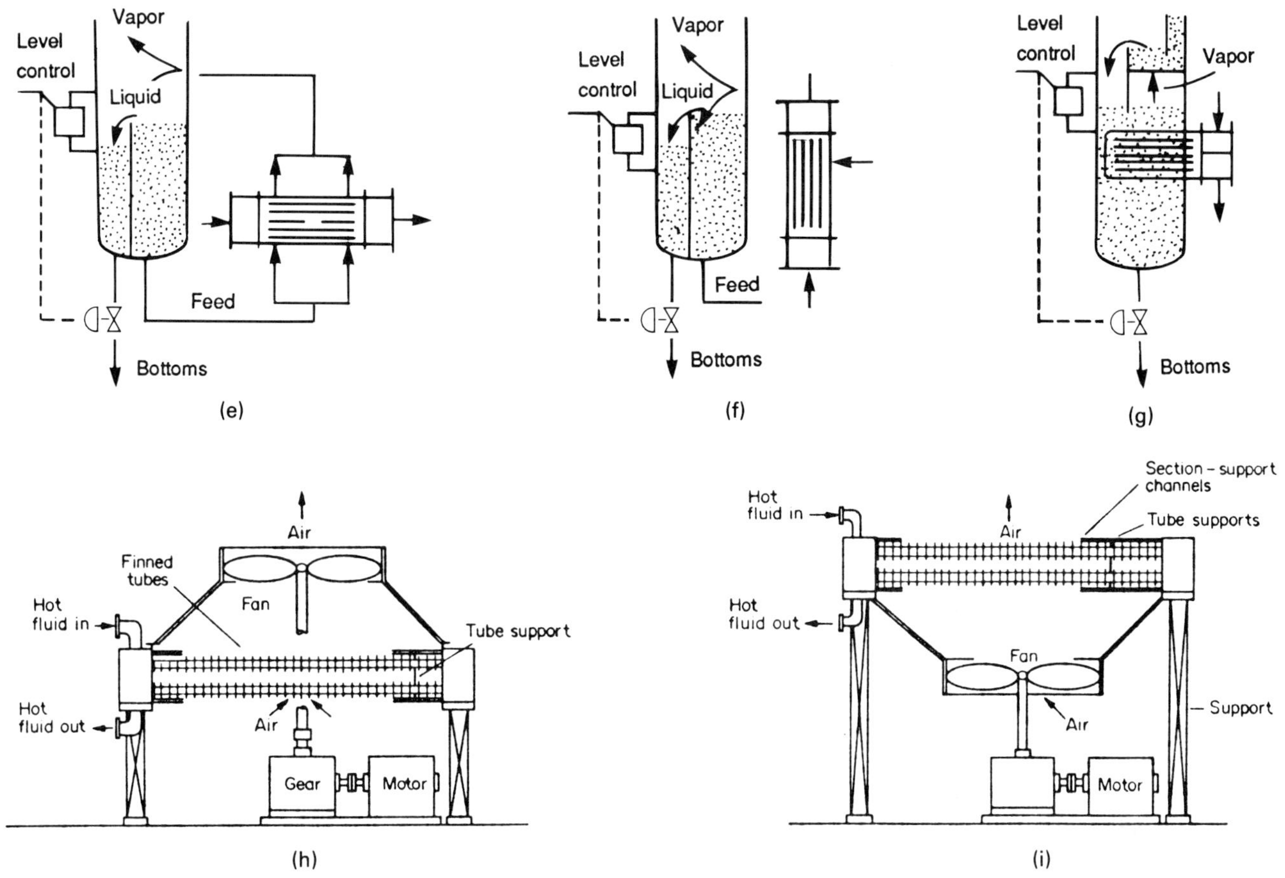

Figure 8.4.—(*continued*)

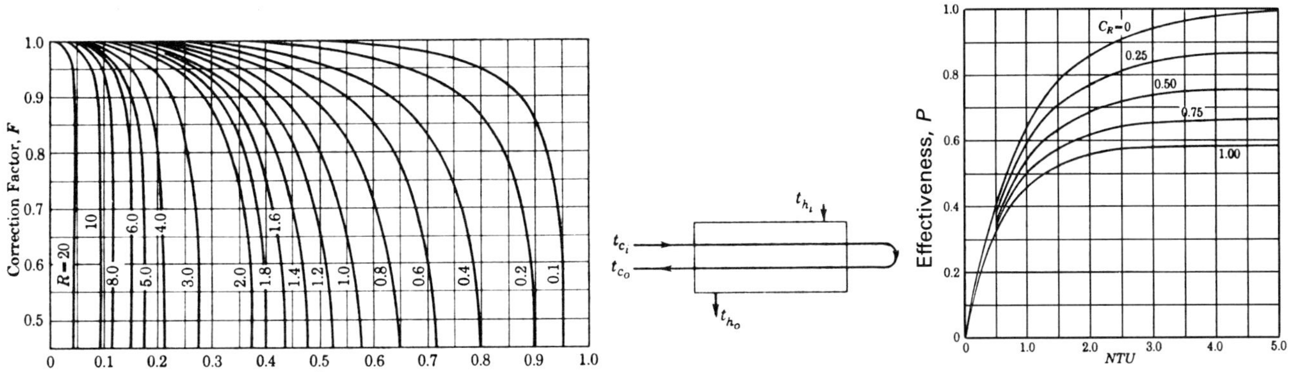

(a)

Figure 8.5. Correction factor *F*, effectiveness and number of transfer units in multipass and cross flow heat exchangers (*Bowman et al.*, Trans ASME **283,** *1940; Kays and London, 1984*):

$$P = \frac{T_i - T_o}{T_i - T_i'}, \quad R = \frac{T_i' - T_o'}{T_i - T_o},$$

T on the tubeside, *T'* on the shellside. *i* = input, *o* = output. (a) One pass on shellside, any multiple of two passes on tubeside. (b) Two passes on shell side, any multiple of four on tubeside. (c) Cross flow, both streams unmixed laterally. (d) Cross flow, one stream mixed laterally. (e) Cross flow, both streams mixed laterally. (f) Effectiveness and number of transfer units in parallel and countercurrent flows. (g) Three shell passes, multiples of six on tubeside. (h) Four shell passes, multiples of eight on tubeside. (i) Five shell passes, multiples of ten on tubeside. (j) Six shell passes, multiples of 12 on tubeside.

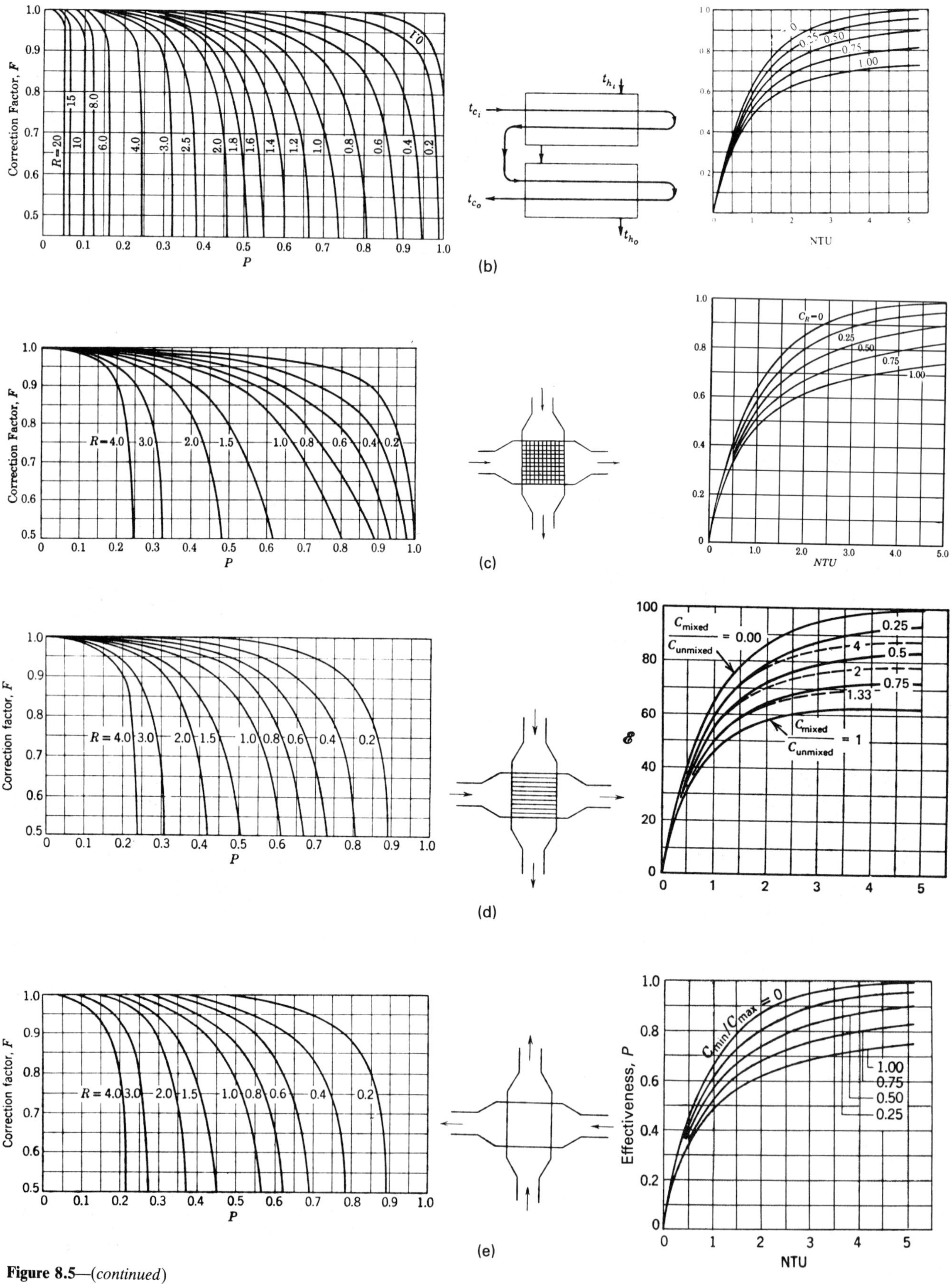

Figure 8.5—(*continued*)

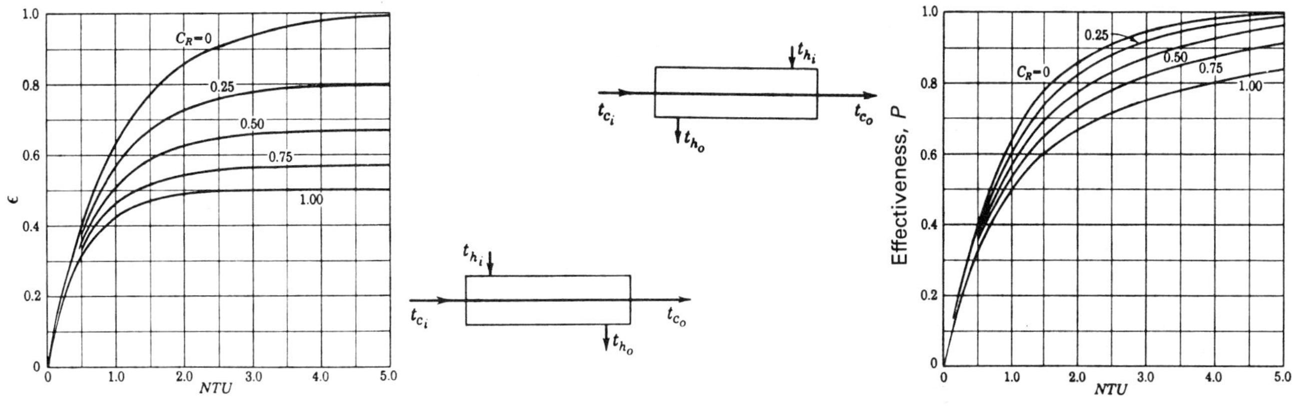

(f)

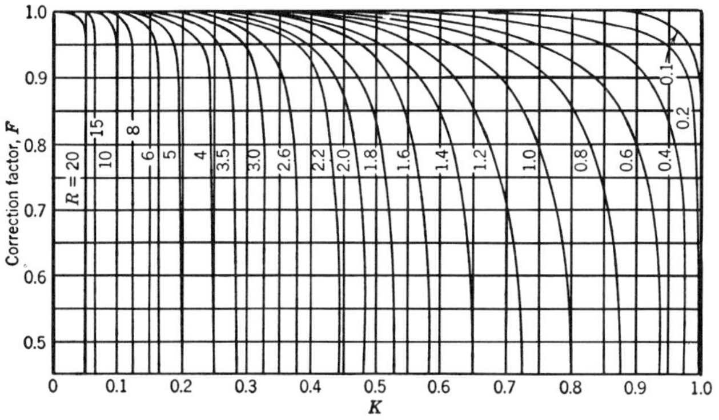

(g)

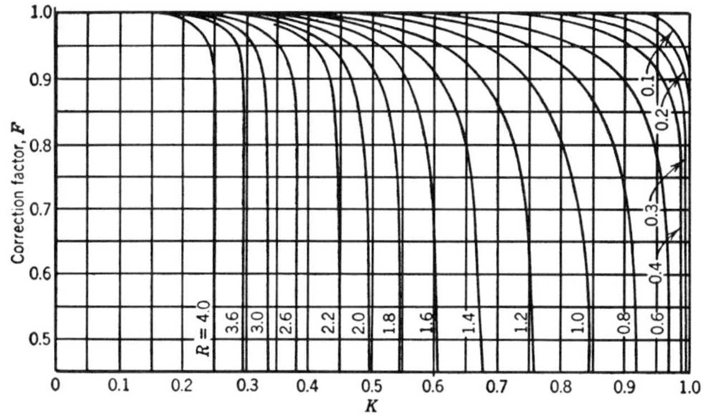

(h)

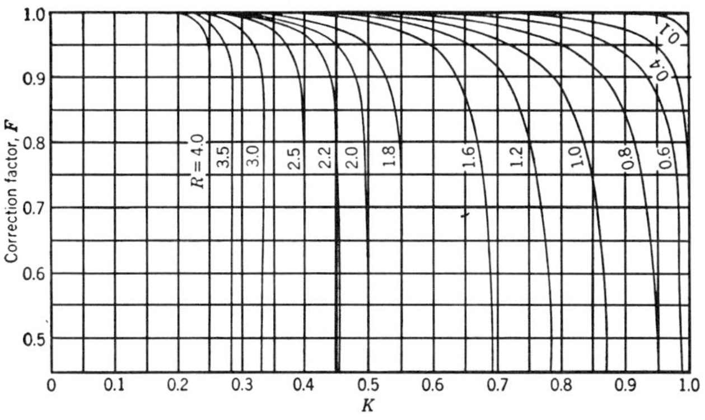

(i)

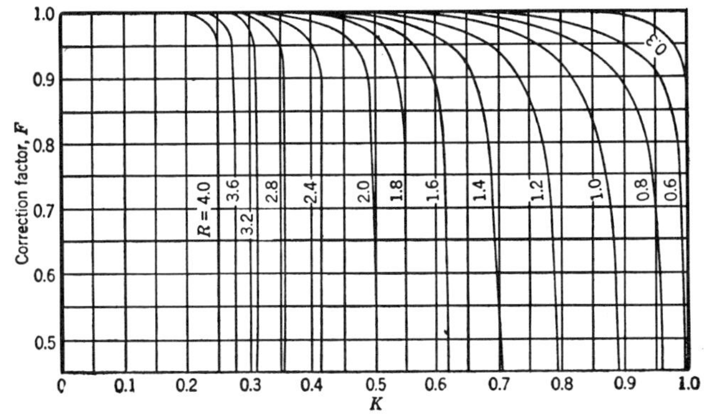

(j)

Figure 8.5—*(continued)*

expressed in terms of these functions of the terminal temperatures:

$$P = \frac{T_o - T_i}{T_i' - T_i} = \frac{\text{actual heat transfer}}{\text{maximum possible heat transfer}}, \tag{8.27}$$

$$R = \frac{T_i - T_o}{T_o' - T_i'} = \frac{mc}{m'c'}. \tag{8.28}$$

Some analytical expressions for F are shown in Table 8.3, and more graphical solutions in Figure 8.5.

This method is especially easy to apply when the terminal temperatures are all known, because then F and $(\Delta T)_{\text{log mean}}$ are immediately determinable for a particular flow pattern. Then in the heat transfer equation

$$Q = UAF(\Delta T)_{\text{lm}} \tag{8.29}$$

any one of the quantities Q, U, or A may be found in terms of the others. A solution by trial is needed when one of the terminal temperatures is unknown, as shown in Example 8.4. The next

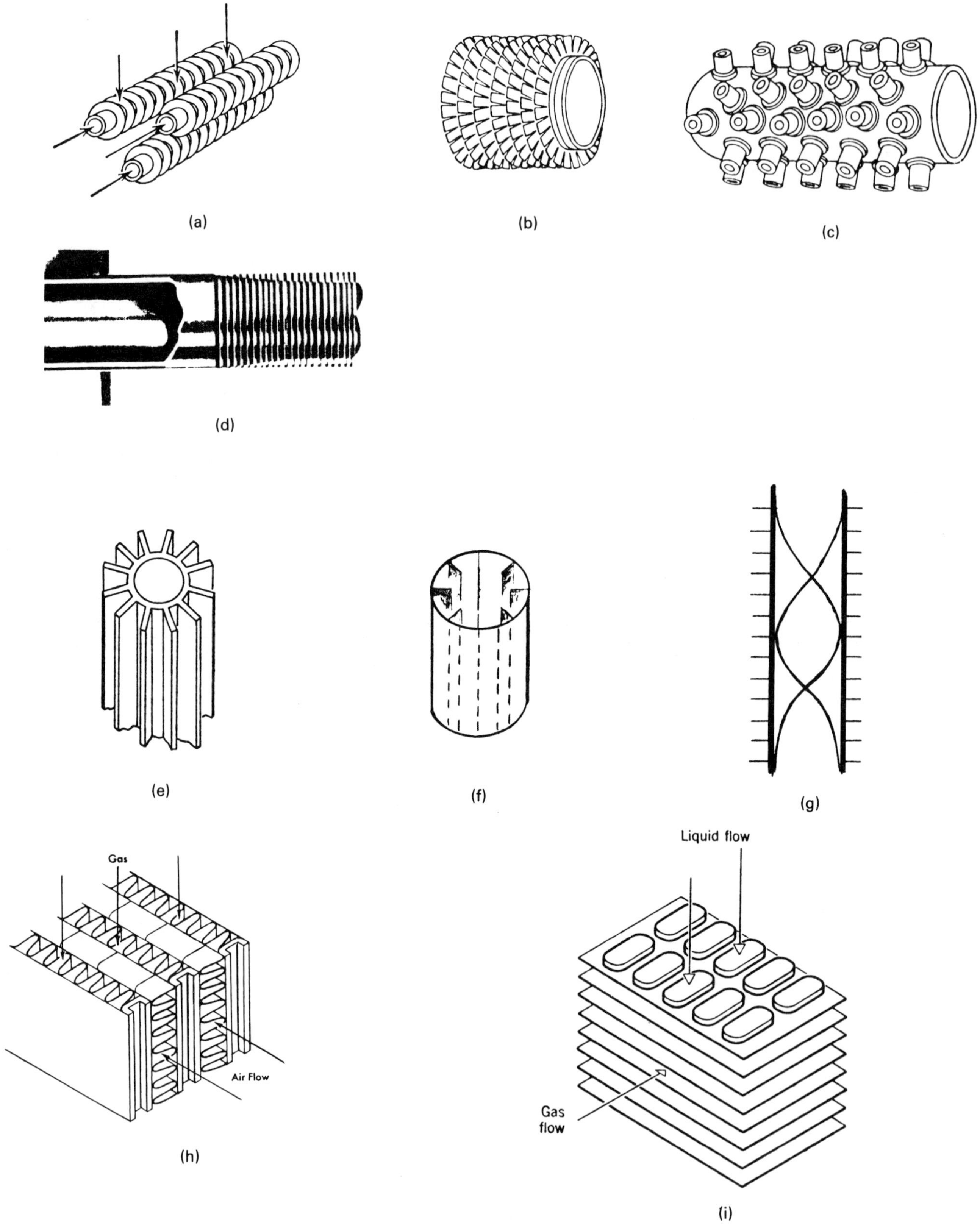

Figure 8.6. Examples of extended surfaces on one or both sides. (a) Radial fins. (b) Serrated radial fins. (c) Studded surface. (d) Joint between tubesheet and low fin tube with three times bare surface. (e) External axial fins. (f) Internal axial fins. (g) Finned surface with internal spiral to promote turbulence. (h) Plate fins on both sides. (i) Tubes and plate fins.

TABLE 8.3. Formulas for Mean Temperature Difference and Effectiveness in Heat Exchangers

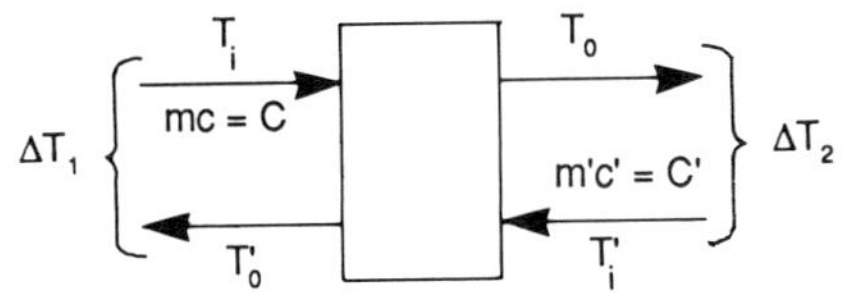

1. Parallel or countercurrent flow,

$$(\Delta T)_m = (\Delta T)_{\text{logmean}} = (\Delta T_1 - \Delta T_2)/\ln(\Delta T_1/\Delta T_2).$$

2. In general,

$$(\Delta T)_m = F(\Delta T)_{\text{logmean}}$$

or

$$(\Delta T)_m = \theta(T_i - T_i'),$$

where F and θ depend on the actual flow paths on the shell and tube sides and are expressed in terms of these quantities:

$C = C_{\min}/C_{\max}$,
$P = (T_o - T_i)/(T_i' - T_i)$ = actual heat transfer/(maximum possible heat transfer),
$R = (T_i - T_o)/(T_o' - T_i') = m'c'/mc$.

3. Number of transfer units, N or NTU, is

$$N = UA/C_{\min} = P/\theta,$$

where $C_{\min}$ is the smaller of the two values mc or $m'c'$ of the products of mass rate of flow times the heat capacity.
4. In parallel flow, $P = N\theta = \{1 - \exp[-N(1+C)]\}/(1+C)$.
5. In countercurrent flow, $P = N\theta = \{1 - \exp[-N(1-C)]\}/\{1 - C\exp[-N(1-C)]\}$.
6. One shell pass and any multiple of two tube passes,

$$F = \frac{\sqrt{R^2+1}}{R-1}\cdot\ln\left(\frac{1-P}{1-PR}\right)\Bigg/\ln\left[\frac{2-P(R+1-\sqrt{R^2+1})}{2-P(R+1+\sqrt{R^2+1})}\right],\quad R\neq 1,$$

$$F = \frac{P}{1-P}\cdot\sqrt{2}\Bigg/\ln\left[\frac{2-P(2-\sqrt{2})}{2-P(2+\sqrt{2})}\right],\quad R=1,$$

$$P = 2\left\{1 + C + (1+C^2)^{1/2}\frac{1+\exp[-N(1+C^2)^{1/2}]}{1-\exp[-N(1+C^2)^{1/2}]}\right\}^{-1}.$$

7. Two shell passes and any multiple of four tube passes,

$$F = \left[\frac{\sqrt{R^2+1}}{2(R-1)}\ln\frac{1-P}{1-PR}\right]\Bigg/\ln\left[\frac{2/P - 1 - R + (2/P)\sqrt{(1-P)(1-PR)} + \sqrt{R^2+1}}{2/P - 1 - R + (2/P)\sqrt{(1-P)(1-PR)} - \sqrt{R^2+1}}\right].$$

8. Cross flow,
   (a) Both streams laterally unmixed, $P = 1 - \exp\{[\exp(-NCn) - 1]/Cn\}$, where $n = N^{-0.22}$.
   (b) Both streams mixed, $P = \{1/[1 - \exp(-N)] + C/[1 - \exp(-NC)] - 1/N\}^{-1}$.
   (c) $C_{\max}$ mixed, $C_{\min}$ unmixed, $P = (1/C)\{1 - \exp[-C(1 - e^{-N})]\}$.
   (d) $C_{\min}$ mixed, $C_{\max}$ unmixed, $P = 1 - \exp\{-(1/C)[1 - \exp(-NC)]\}$.
9. For more complicated patterns only numerical solutions have been made. Graphs of these appear in sources such as Heat Exchanger Design Handbook (HEDH, 1983) and Kays and London (1984).

method to be described, however, may be more convenient in such a case.

θ-METHOD

One measure of the size of heat transfer equipment is the number of transfer units N defined by

$$N = UA/C_{\min}, \tag{8.30}$$

where $C_{\min}$ is the smaller of the two products of mass flow rate and heat capacity, mc or $m'c'$. N is so named because of a loose analogy with the corresponding measure of the size of mass transfer equipment.

A useful combination of P and N is their ratio

$$\theta = \frac{P}{N} = \frac{C_{\min}(T_o - T_i)}{UA(T_i' - T_i)} = \frac{Q}{UA(T_i' - T_i)} = \frac{(\Delta T)_m}{(T_i' - T_i)}, \tag{8.31}$$

where $(T_o - T_i)$ is the temperature change of the stream with the smaller value of mc. Thus θ is a factor for obtaining the mean temperature difference by the formula:

$$(\Delta T)_m = \theta(T_i' - T_i) \tag{8.32}$$

when the two inlet temperatures are known.

The term P often is called the exchanger effectiveness. Equations and graphs are in Table 8.3 and Figure 8.4. Many graphs for θ, like those of Figure 8.7, may be found in the *Heat Exchanger Design Handbook* (HEDH, 1983). When sufficient other data are known about a heat exchange process, an unknown outlet temperature can be found by this method directly without requiring trial calculations as with the F-method. Example 8.5 solves such a problem.

SELECTION OF SHELL-AND-TUBE NUMBERS OF PASSES

A low value of F means, of course, a large surface requirement for a given heat load. Performance is improved in such cases by using several shells in series, or by increasing the numbers of passes in the same shell. Thus, two 1–2 exchangers in series are equivalent to one large 2–4 exchanger, with two passes on the shell side and four passes on the tube side. Usually the single shell arrangement is more economical, even with the more complex internals. For economy, F usually should be greater than 0.7.

EXAMPLE

A shell side fluid is required to go from 200 to 140°F and the tube side from 80 to 158°F. The charts of Figure 8.5 will be used:

$$P = (200 - 140)/(200 - 80) = 0.5,$$
$$R = (158 - 80)/(200 - 140) = 1.30.$$

For a 1–2 exchanger,	$F = 0.485$:
2–4	0.92
4–8	0.98

The 1–2 exchanger is not acceptable, but the 2–4 is acceptable. If the tube side outlet were at 160 instead of 158, F would be zero for the 1–2 exchanger but substantially unchanged for the others.

8.3. HEAT TRANSFER COEFFICIENTS

Data are available as overall coefficients, individual film coefficients, fouling factors, and correlations of film coefficients in terms of

EXAMPLE 8.4
Performance of a Heat Exchanger with the F-Method
Operation of an exchanger is represented by the sketch and the equation

$$Q/UA = 50 = F(\Delta T)_{lm}$$

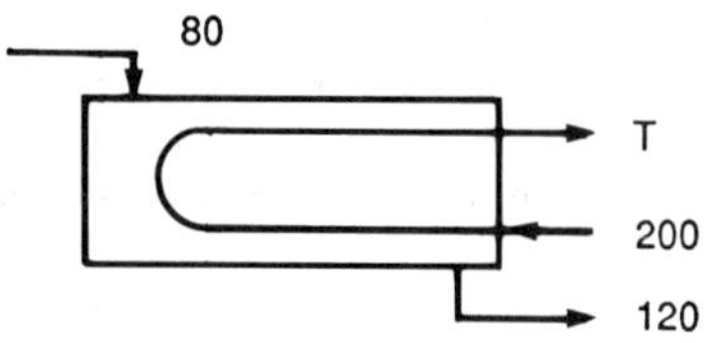

The outlet temperature of the hot fluid is unknown and designated by T. These quantities are formulated as follows:

$$P = \frac{200 - T}{200 - 80},$$

$$R = \frac{200 - T}{120 - 80},$$

$$(\Delta T)_{lm} = \frac{T - 80 - (200 - 120)}{\ln[(T - 80)/(200 - 120)]}$$

F is represented by the equation of Item 6 of Table 8.3, or by Figure 8.4(a). Values of T are tried until one is found that satisfies $G \equiv 50 - F(\Delta T)_{lm} \cong 0$. The printout shows that

$$T = 145.197.$$

The sensitivity of the calculation is shown in the following tabulation:

T	P	R	$(\Delta T)_{lm}$	F	G
145.0	0.458	1.375	72.24	0.679	0.94
145.197	0.457	1.370	72.35	0.691	0.00061
145.5	0.454	1.363	72.51	0.708	−1.34

```
10 ! Example 8.4. The F-method
20 SHORT P,R,F,T1
30 INPUT T
40 P=(200-T)/120
50 R=(200-T)/40
60 T1=(T-160)/LOG((T-80)/80)
70 E=(R^2+1)^.5
80 F=E/(R-1)*LOG((1-P)/(1-P*R))
90 F=F/LOG((2-P*(R+1-E))/(2-P*(
   R+1+E)))
95 G=50-F*T1
100 PRINT "T=";T
110 PRINT "G=";G
120 PRINT "P=";P
130 PRINT "R=";R
140 PRINT "F=";F
150 PRINT "T1=";T1
160 GOTO 30
170 END

T= 145.197
G= .00240286
P= .45669
R= 1.3701
F= .69109
T1= 72.346
```

physical properties and operating conditions. The reliabilities of these classes of data increase in the order of this listing, but also the ease of use of the data diminishes in the same sequence.

OVERALL COEFFICIENTS

The range of overall heat transfer coefficients is approximately 10–200 Btu/(hr)(sqft)(°F). Several compilations of data are available, notably in the *Chemical Engineers Handbook* (McGraw-Hill, New York, 1984, pp. 10.41–10.46) and in Ludwig (1983, pp. 70–73). Table 8.4 qualifies each listing to some extent, with respect to the kind of heat transfer, the kind of equipment, kind of process stream, and temperature range. Even so, the range of values of U usually is two- to three-fold, and consequently only a rough measure of equipment size can be obtained in many cases with such data. Ranges of the coefficients in various kinds of equipment are compared in Table 8.5.

FOULING FACTORS

Heat transfer may be degraded in time by corrosion, deposits of reaction products, organic growths, etc. These effects are accounted for quantitatively by fouling resistances, $1/h_f$. They are listed separately in Tables 8.4 and 8.6, but the listed values of coefficients include these resistances. For instance, with a clean surface the first listed value of U in Table 8.4 would correspond to a clean value of $U = 1/(1/12 - 0.04) = 23.1$. How long a clean value could be maintained in a particular plant is not certain. Sometimes fouling develops slowly; in other cases it develops quickly as a result of process upset and may level off. A high coefficient often is desirable, but sometimes is harmful in that excessive subcooling may occur or film boiling may develop. The most complete list of fouling factors with some degree of general acceptance is in the TEMA (1978) standards. The applicability of these data to any particular situation, however, is questionable and the values probably not better than ±50%. Moreover, the magnitudes and uncertainties of arbitrary fouling factors may take the edge off the importance of precise calculations of heat transfer coefficients. A brief discussion of fouling is by Walker (1982). A symposium on this important topic is edited by Somerscales and Knudsen (1981).

INDIVIDUAL FILM COEFFICIENTS

Combining individual film coefficients into an overall coefficient of heat transfer allows taking into account a greater variety and range of conditions, and should provide a better estimate. Such individual coefficients are listed in Tables 8.6 and 8.7. The first of these is a very cautious compilation with a value range of 1.5- to 2-fold. Values of the fouling factors are included in the coefficient listings of both tables but are not identified in Table 8.7. For clean service, for example, involving sensible heat transfer from a medium organic to heating a heavy organic,

$$U = 10{,}000/(57 - 16 + 50 - 34) = 175$$

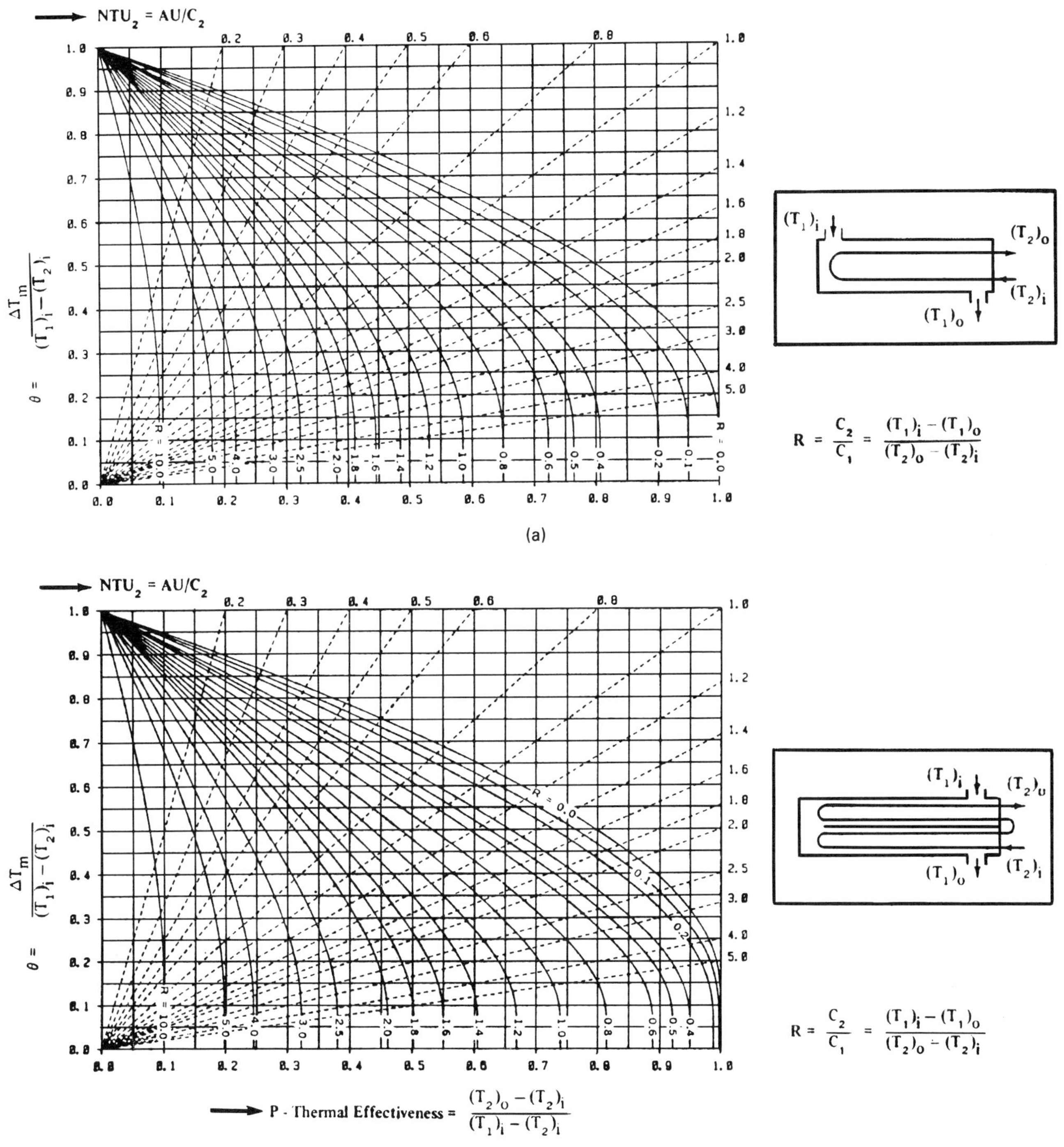

Figure 8.7. θ correction charts for mean temperature difference: (a) One shell pass and any multiple of two tube passes. (b) Two shell passes and any multiple of four tube passes. [(*HEDH, 1983*); *after Mueller in Rohsenow and Hartnett,* Handbook of Heat Transfer, *Section 18, McGraw-Hill, New York, 1973. Other cases also are covered in these references.*]

EXAMPLE 8.5
Application of the Effectiveness and the θ Method
Operating data of an exchanger are shown on the sketch. These data include

$$UA = 2000,$$
$$m'c' = 1000, \quad mc = 800,$$
$$C = C_{min}/C_{max} = 0.8.$$

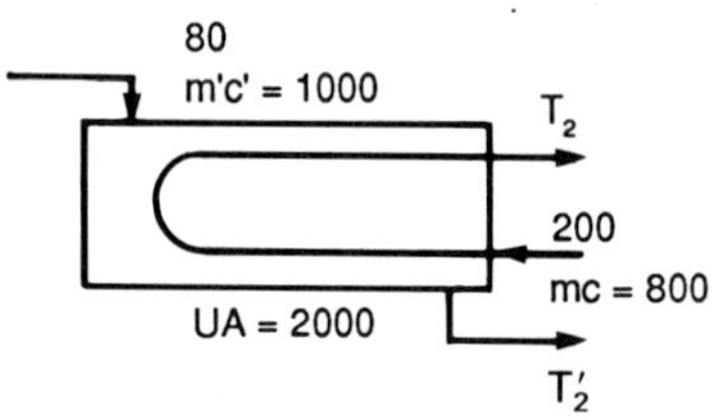

The equation for effectiveness P is given by item 6 of Table 8.3 or it can be read off Figure 8.4(a). Both P and θ also can be read off Figure 8.4(a) at known N and $R = C_2/C_1 = 0.8$. The number of transfer units is

$$N = UA/C_{min} = 2000/800 = 2.5,$$
$$C = C_{min}/C_{max} = 0.8,$$
$$D = \sqrt{1 + C^2} = 1.2806,$$
$$P = \frac{2}{1 + C + D[1 + \exp(-ND)]/1 - \exp(-ND)} = 0.6271,$$
$$\theta = P/N = 0.2508,$$
$$\Delta T_m = \theta(200 - 80) = 30.1,$$
$$Q = UA(\Delta T)_m = 2000(30.1) = 60{,}200,$$
$$= 800(200 - T_2) = 1000(T_2' - 80),$$
$$\therefore T_2 = 124.75,$$
$$T_2' = 140.2.$$

T_2 also may be found from the definition of P:

$$P = \frac{\text{actual } \Delta T}{\text{max possible } \Delta T} = \frac{200 - T_2}{200 - 80} = 0.6271,$$
$$\therefore T_2 = 124.78.$$

With this method, unknown terminal temperatures are found without trial calculations.

compared with a normal value of

$$U = 10{,}000/(57 + 50) = 93,$$

where the averages of the listed numbers in Table 8.6 are taken in each case.

METAL WALL RESISTANCE

With the usual materials of construction of heat transfer surfaces, the magnitudes of their thermal resistances may be comparable with the other prevailing resistances. For example, heat exchanger tubing of 1/16 in. wall thickness has these values of $1/h_w = L/k$ for several common materials:

Carbon steel	$1/h_w = 1.76 \times 10^{-4}$
Stainless steel	5.54×10^{-4}
Aluminum	0.40×10^{-4}
Glass	79.0×10^{-4}

which are in the range of the given film and fouling resistances, and should not be neglected in evaluating the overall coefficient. For example, with the data of this list a coefficient of 93 with carbon steel tubing is reduced to 88.9 when stainless steel tubing is substituted.

DIMENSIONLESS GROUPS

The effects of the many variables that bear on the magnitudes of individual heat transfer coefficients are represented most logically and compactly in terms of dimensionless groups. The ones most pertinent to heat transfer are listed in Table 8.8. Some groups have ready physical interpretations that may assist in selecting the ones appropriate to particular heat transfer processes. Such interpretations are discussed for example by Gröber et al. (1961, pp. 193–198). A few are given here.

The Reynolds number, $Du\rho/\mu = \rho u^2/(\mu u/D)$, is a measure of the ratio of inertial to viscous forces.

The Nusselt number, $hL/k = h/(k/L)$, is the ratio of effective heat transfer to that which would take place by conduction through a film of thickness L.

The Peclet number, $DGC/k = GC/(k/D)$ and its modification, the Graetz number wC/kL, are ratios of sensible heat change of the flowing fluid to the rate of heat conduction through a film of thickness D or L.

The Prandtl number, $C\mu/k = (\mu/\rho)/(k/\rho C)$, compares the rate of momentum transfer through friction to the thermal diffusivity or the transport of heat by conduction.

The Grashof number is interpreted as the ratio of the product of the buoyancy and inertial forces to the square of the viscous forces.

The Stanton number is a ratio of the temperature change of a fluid to the temperature drop between fluid and wall. Also, St = (Nu)/(Re)(Pr).

An analogy exists between the transfers of heat and mass in moving fluids, such that correlations of heat transfer involving the Prandtl number are valid for mass transfer when the Prandtl number $C\mu/k$ is replaced by the Schmidt number $\mu/\rho k_d$. This is of particular value in correlating heat transfer from small particles to fluids where particle temperatures are hard to measure but measurement of mass transfer may be feasible, for example, in vaporization of naphthalene.

8.4. DATA OF HEAT TRANSFER COEFFICIENTS

Specific correlations of individual film coefficients necessarily are restricted in scope. Among the distinctions that are made are those of geometry, whether inside or outside of tubes for instance, or the shapes of the heat transfer surfaces; free or forced convection; laminar or turbulent flow; liquids, gases, liquid metals, non-Newtonian fluids; pure substances or mixtures; completely or partially condensable; air, water, refrigerants, or other specific substances; fluidized or fixed particles; combined convection and radiation; and others. In spite of such qualifications, it should be

TABLE 8.4. Overall Heat Transfer Coefficients in Some Petrochemical Applications, U Btu/(hr)(sqft)(°F)[a]

In Tubes	Outside Tubes	Type Equipment	Velocities (ft/sec) Tube	Velocities (ft/sec) Shell	Overall Coefficient	Temp. Range (°F)	Estimated Fouling Tube	Estimated Fouling Shell	Estimated Fouling Overall
A. Heating-cooling									
Butadiene mix. (Super-heating)	steam	H	25–35	—	12	400–100	—	—	0.04
Solvent	solvent	H	—	1.0–1.8	35–40	110–30	—	—	0.0065
Solvent	propylene (vaporization)	K	1–2	—	30–40	40–0	—	—	0.006
C_4 unsaturates	propylene (vaporization)	K	20–40	—	13–18	100–35	—	—	0.005
Solvent	chilled water	H	—	—	35–75	115–40	0.003	0.001	—
Oil	oil	H	—	—	60–85	150–100	0.0015	0.0015	—
Ethylene-vapor	condensate and vapor	K	—	—	90–125	600–200	0.002	0.001	—
Ethylene vapor	chilled water	H	—	—	50–80	270–100	0.001	0.001	—
Condensate	propylene (refrigerant)	K-U	—	—	60–135	60–30	0.001	0.001	—
Chilled water	transformer oil	H	—	—	40–75	75–50	0.001	0.001	—
Calcium brine-25%	chlorinated C_1	H	1–2	0.5–1.0	40–60	−20–+10	0.002	0.005	—
Ethylene liquid	ethylene vapor	K-U	—	—	10–20	−170–(−100)	—	—	0.002
Propane vapor	propane liquid	H	—	—	6–15	−25–100	—	—	0.002
Lights and chlor. HC	steam	U	—	—	12–30	−30–260	0.001	0.001	—
Unsat. light HC, CO, CO_2, H_2	steam	H	—	—	10–2	400–100	—	—	0.3
Ethonolamine	steam	H	—	—	15–25	400–40	0.001	0.001	—
Steam	air mixture	U	—	—	10–20	−30–220	0.0005	0.0015	—
Steam	styrene and tars	U (in tank)	—	—	50–60	190–230	0.001	0.002	—
Chilled water	freon-12	H	4–7	—	100–130	90–25	0.001	0.001	—
Water[b]	lean copper solvent	H	4–5	—	100–120	180–90	—	—	0.004
Water	treated water	H	3–5	1–2	100–125	90–110	—	—	0.005
Water	C_2-chlor. HC, lights	H	2–3	—	6–10	360–100	0.002	0.001	—
Water	hydrogen chloride	H	—	—	7–15	230–90	0.002	0.001	—
Water	heavy C_2-chlor.	H	—	—	45–30	300–90	0.001	0.001	—
Water	perchlorethylene	H	—	—	55–35	150–90	0.001	0.001	—
Water	air and water vapor	H	—	—	20–35	370–90	0.0015	0.0015	—
Water	engine jacket water	H	—	—	230–160	175–90	0.0015	0.001	—
Water	absorption oil	H	—	—	80–115	130–90	0.0015	0.001	—
Water	air-chlorine	U	4–7	—	8–18	250–90	—	—	0.005
Water	treated water	H	5–7	—	170–225	200–90	0.001	0.001	—
B. Condensing									
C_4 unsat.	propylene refrig.	K	v	—	58–68	60–35	—	—	0.005
HC unsat. lights	propylene refrig.	K	v	—	50–60	45–3	—	—	0.0055
Butadiene	propylene refrig.	K	v	—	65–80	20–35	—	—	0.004
Hydrogen chloride	propylene refrig.	H	—	—	110–60	0–15	0.012	0.001	—

(continued)

TABLE 8.4—*(continued)*

In Tubes	Outside Tubes	Type Equipment	Velocities (ft/sec) Tube	Velocities (ft/sec) Shell	Overall Coefficient	Temp. Range (°F)	Estimated Fouling Tube	Estimated Fouling Shell	Estimated Fouling Overall
Lights and chloro-ethanes	propylene refrig.	KU	—	—	15–25	130–(−20)	0.002	0.001	—
Ethylene	propylene refrig.	KU	—	—	60–90	120–(−10)	0.001	0.001	—
Unsat. chloro HC	water	H	7–8	—	90–120	145–90	0.002	0.001	—
Unsat. chloro HC	water	H	3–8	—	180–140	110–90	0.001	0.001	—
Unsat. chloro HC	water	H	6	—	15–25	130–(−20)	0.002	0.001	—
Chloro-HC	water	KU	—	—	20–30	110–(−10)	0.001	0.001	—
Solvent and non cond.	water	H	—	—	25–15	260–90	0.0015	0.004	—
Water	propylene vapor	H	2–3	—	130–150	200–90	—	—	0.003
Water	propylene	H	—	—	60–100	130–90	0.0015	0.001	—
Water	steam	H	—	—	225–110	300–90	0.002	0.0001	—
Water	steam	H	—	—	190–235	230–130	0.0015	0.001	—
Treated water	steam (exhaust)	H	—	—	20–30	220–130	0.0001	0.0001	—
Oil	steam	H	—	—	70–110	375–130	0.003	0.001	—
Water	propylene cooling and cond.	H	—	—	{25–50 110–150}	{30–45 (C) 15–20 (Co)}	0.0015	0.001	—
Chilled water	air-chlorine (part and cond.)	U	—	—	{8–15 20–30}	{8–15 (C) 10–15 (Co)}	0.0015	0.005	—
Water	light HC, cool and cond.	H	—	—	35–90	270–90	0.0015	0.003	—
Water	ammonia	H	—	—	140–165	120–90	0.001	0.001	—
Water	ammonia	U	—	—	280–300	110–90	0.001	0.001	—
Air-water vapor	freon	KU	—	—	{10–50 10–20}	60–10	—	—	0.01
C. Reboiling									
Solvent, Copper-NH_3	steam	H	7–8	—	130–150	180–160	—	—	0.005
C_4 unsat.	steam	H	—	—	95–115	95–150	—	—	0.0065
Chloro. HC	steam	VT	—	—	35–25	300–350	0.001	0.001	—
Chloro. unsat. HC	steam	VT	—	—	100–140	230–130	0.001	0.001	—
Chloro. ethane	steam	VT	—	—	90–135	300–350	0.001	0.001	—
Chloro. ethane	steam	U	—	—	50–70	30–190	0.002	0.001	—
Solvent (heavy)	steam	H	—	—	70–115	375–300	0.004	0.0005	—
Mono-di-ethanolamines	steam	VT	—	—	210–155	450–350	0.002	0.001	—
Organics, acid, water	steam	VT	—	—	60–100	450–300	0.003	0.0005	—
Amines and water	steam	VT	—	—	120–140	360–250	0.002	0.0015	—
Steam	naphtha frac.	Annulus Long. F.N.	—	—	15–20	270–220	0.0035	0.0005	—
Propylene	C_2, C_2^-	KU	—	—	120–140	150–40	0.001	0.001	—
Propylene-butadiene	butadiene, unsat.	H	—	25–35	15–18	400–100	—	—	0.02

[a] Fouling resistances are included in the listed values of *U*.
[b] Unless specified, all water is untreated, brackish, bay or sea. Notes: H = horizontal, fixed or floating tube sheet, U = U—tube horizontal bundle, K = kettle type, V = vertical, R = reboiler, T = thermosiphon, v = variable, HC = hydrocarbon, (C) = cooling range Δt, (Co) = condensing range Δt.
(Ludwig, 1983).

TABLE 8.5. Ranges of Overall Heat Transfer Coefficients in Various Types of Exchangers [U Btu/(hr)(sqft)(°F)][a]

Equipment	Process	U
Shell-and-tube exchanger [Fig. 8.4(c)]	gas (1 atm)–gas (1 atm)	1–6
	gas (250 atm)–gas (250 atm)	25–50
	liquid–gas (1 atm)	2–12
	liquid–gas (250 atm)	35–70
	liquid–liquid	25–200
	liquid–condensing vapor	50–200
Double-pipe exchanger [Fig. 8.4(a)]	gas (1 atm)–gas (1 atm)	2–6
	gas (250 atm)–gas (250 atm)	25–90
	liquid–gas (250 atm)	35–100
	liquid–liquid	50–250
Irrigated tube bank	water–gas (1 atm)	3–10
	water–gas (250 atm)	25–60
	water–liquid	50–160
	water–condensing vapor	50–200
Plate exchanger [Fig. 8.8(a)]	water–gas (1 atm)	3–10
	water–liquid	60–200
Spiral exchanger [Fig. 8.8(c)]	liquid–liquid	120–440
	liquid–condensing steam	160–600
Compact [Fig. 8.6(h)]	gas (1 atm)–gas (1 atm)	2–6
	gas (1 atm)–liquid	3–10
Stirred tank, jacketed	liquid–condensing steam	90–260
	boiling liquid–condensing steam	120–300
	water–liquid	25–60
Stirred tank, coil inside	liquid–condensing steam	120–440
	water–liquid	90–210

[a] 1 Btu/(hr)(sqft)(°F) = 5.6745 W/m^2 K.
Data from (HEDH, 1983).

borne in mind that very few proposed correlations are more accurate than ±20% or so.

Along with rate of heat transfer, the economics of practical exchanger design requires that pumping costs for overcoming friction be taken into account.

DIRECT CONTACT OF HOT AND COLD STREAMS

Transfer of heat by direct contact is accomplished in spray towers, in towers with a multiplicity of segmented baffles or plates (called shower decks), and in a variety of packed towers. In some processes heat and mass transfer occur simultaneously between phases; for example, in water cooling towers, in gas quenching with water, and in spray or rotary dryers. Quenching of pyrolysis gases in transfer lines or towers and contacting on some trays in fractionators may involve primarily heat transfer. One or the other, heat or mass transfer, may be the dominant process in particular cases.

Data of direct contact heat transfer are not abundant. The literature has been reviewed by Fair (1972) from whom specific data will be cited.

One rational measure of a heat exchange process is the number of transfer units. In terms of gas temperatures this is defined by

$$N_g = \frac{T_{g,\text{in}} - T_{g,\text{out}}}{(T_g - T_L)_{\text{mean}}}. \tag{8.33}$$

The logarithmic mean temperature difference usually is applicable. For example, if the gas goes from 1200 to 150°F and the liquid countercurrently from 120 to 400°F, the mean temperature difference is 234.5 and $N_g = 4.48$. The height of a contact zone then is obtained as the product of the number of transfer units and the height H_g of a transfer unit. Several correlations have been made of the latter quantity, for example, by Cornell, Knapp, and Fair (1960) and modified in the *Chemical Engineers Handbook* (1973, pp. 18.33, 18.37). A table by McAdams (1954, p. 361) shows that in spray towers the range of H_g may be 2.5–10 ft and in various kinds of packed towers, 0.4–4 ft or so.

Heat transfer coefficients also have been measured on a volumetric or cross section basis. In heavy hydrocarbon fractionators, Neeld and O'Bara (1970) found overall coefficients of 1360–3480 Btu/(hr)(°F)(sqft of tower cross section). Much higher values have been found in less viscous systems.

Data on small packed columns were correlated by Fair (1972) in the form

$$Ua = CG^mL^n, \quad \text{Btu/(hr)(cuft)(°F)}, \tag{8.34}$$

where the constants depend on the kind of packing and the natures of the fluids. For example, with air–oil, 1 in. Raschig rings, in an 8 in. column

$$Ua = 0.083G^{0.94}L^{0.25}. \tag{8.35}$$

When G and L are both 5000 lb/(hr)(sqft), for instance, this formula gives $Ua = 2093$ Btu/(hr)(cuft)(°F).

In spray towers, one correlation by Fair (1972) is

$$h_ga = 0.043G^{0.8}L^{0.4}/Z^{0.5} \quad \text{Btu/(hr)(cuft)(°F)}. \tag{8.36}$$

TABLE 8.6. Typical Ranges of Individual Film and Fouling Coefficients [h Btu/(hr)(sqft)(°F)]

Fluid and Process	Conditions	P (atm)	$(\Delta T)_{max}$ (°F)	$10^4 h$	$10^4 h_f$
Sensible					
Water	liquid			7.6–11.4	6–14
Ammonia	liquid			7.1–9.5	0–6
Light organics	liquid			28–38	6–11
Medium organics	liquid			38–76	9–23
Heavy organics	liquid heating			23–76	11–57
Heavy organics	liquid cooling			142–378	11–57
Very heavy organics	liquid heating			189–568	23–170
Very heavy organics	liquid cooling			378–946	23–170
Gas		1–2		450–700	0–6
Gas		10		140–230	0–6
Gas		100		57–113	0–6
Condensing transfer					
Steam ammonia	all condensable	0.1		4.7–7.1	0–6
Steam ammonia	1% noncondensable	0.1		9.5–14.2	0–6
Steam ammonia	4% noncondensable	0.1		19–28	0–6
Steam ammonia	all condensable	1		3.8–5.7	0–6
Steam ammonia	all condensable	10		2.3–3.8	0–6
Light organics	pure	0.1		28–38	0–6
Light organics	4% noncondensable	0.1		57–76	0–6
Light organics	pure	10		8–19	0–6
Medium organics	narrow range	1		14–38	6–30
Heavy organics	narrow range	1		28–95	11–28
Light condensable mixes	narrow range	1		23–57	0–11
Medium condensable mixes	narrow range	1		38–95	6–23
Heavy condensable mixes	medium range	1		95–190	11–45
Vaporizing transfer					
Water		<5	45	5.7–19	6–12
Water		<100	36	3.8–14	6–12
Ammonia		<30	36	11–19	6–12
Light organics	pure	20	36	14–57	6–12
Light organics	narrow range	20	27	19–76	6–17
Medium organics	pure	20	36	16–57	6–17
Medium organics	narrow range	20	27	23–95	6–17
Heavy organics	pure	20	36	23–95	11–28
Heavy organics	narrow range	20	27	38–142	11–45
Very heavy organics	narrow range	20	27	57–189	11–57

Light organics have viscosity <1 cP, typically similar to octane and lighter hydrocarbons.
Medium organics have viscosities in the range 1–5 cP, like kerosene, hot gas oil, light crudes, etc.
Heavy organics have viscosities in the range 5–100 cP, cold gas oil, lube oils, heavy and reduced crudes, etc.
Very heavy organics have viscosities above 100 cP, asphalts, molten polymers, greases, etc.
Gases are all noncondensables except hydrogen and helium which have higher coefficients.
Conversion factor: 1 Btu/(hr)(sqft)(°F) = 5.6745 W/m^2 K.
(After HEDH, 1983, 3.1.4-4).

In a tower with height $Z = 30$ ft and with both G and L at 5000 lb/(hr)(cuft), for example, this formula gives $h_g a = 215$.

In liquid–liquid contacting towers, data cited by Fair (1972) range from 100–12,000 Btu/(hr)(cuft)(°F) and heights of transfer units in the range of 5 ft or so. In pipeline contactors, transfer rates of 6000–60,000 Btu/(hr)(cuft)(°F) have been found, in some cases as high as 200,000.

In some kinds of equipment, data only on mass transfer rates may be known. From these, on the basis of the Chilton–Colburn analogy, corresponding values of heat transfer rates can be estimated.

NATURAL CONVECTION

Coefficients of heat transfer by natural convection from bodies of various shapes, chiefly plates and cylinders, are correlated in terms of Grashof, Prandtl, and Nusselt numbers. Table 8.9 covers the most usual situations, of which heat losses to ambient air are the most common process. Simplified equations are shown for air. Transfer of heat by radiation is appreciable even at modest temperatures; such data are presented in combination with convective coefficients in item 16 of this table.

FORCED CONVECTION

Since the rate of heat transfer is enhanced by rapid movement of fluid past the surface, heat transfer processes are conducted under such conditions whenever possible. A selection from the many available correlations of forced convective heat transfer involving single phase fluids, including flow inside and outside bare and extended surfaces, is presented in Table 8.10. Heat transfer resulting in phase change, as in condensation and vaporization, also is covered in this table. Some special problems that arise in interpreting phase change behavior will be mentioned following.

TABLE 8.7. Individual Film Resistances (1/*h*) Including Fouling Effects, with *h* in Btu/(hr)(sqft)(°F)

Fluid	Kind of Heat Transfer		
	Sensible	Boiling	Condensing
Aromatic liquids			
Benzene, toluene, ethylbenzene, styrene	0.007	0.011	0.007
Dowtherm	0.007	—	—
Inorganic solutions			
$CaCl_2$ Brine (25%)	0.004	—	—
Heavy acids	0.013	—	—
NaCl Brine (20%)	0.0035	—	—
Misc. dilute solutions	0.005	—	—
Light hydrocarbon liquids			
C_3, C_4, C_5	0.004	0.007	0.004
Chlorinated hydrocarbons	0.004	0.009	0.007
Miscellaneous organic liquids			
Acetone	0.007	—	—
Amine solutions			
Saturated diethanolamine and mono-ethanolamine (CO_2 and H_2S)	0.007	—	—
Lean amine solutions	0.005	—	—
Oils			
Crude oil	0.015	—	—
Diesel oil	0.011	—	—
Fuel oil (bunker C)	0.018	—	—
Gas oil			
Light	0.0125	—	0.015
Heavy (typical of cat. cracker feed)	0.014	—	0.018
Gasoline (400° EP)	0.008	0.010	0.008
Heating oil (domestic 30° API)	0.010	—	—
Hydroformate	0.006	—	—
Kerosine	0.009	—	0.013
Lube oil stock	0.018	—	—
Naphthas			
Absorption	0.008	0.010	0.006
Light virgin	0.007	0.010	0.007
Light catalytic	0.006	0.010	0.007
Heavy	0.008	0.011	0.0085
Polymer (C_8's)	0.008	0.010	0.008
Reduced crude	0.018	—	—
Slurry oil (fluid cat. cracker)	0.015	—	—
Steam (no noncondensables)			0.001
Water			
Boiler water	0.003	—	—
Cooling tower (untreated)	0.007	—	—
Condensate (flashed)	0.002	—	—
River and well	0.007	—	—
Sea water (clean and below 125°F)	0.004	—	—
Gases in turbulent flow			
Air, CO, CO_2, and N_2	0.045		
Hydrocarbons (light through naphthas)	0.035		

(Fair and Rase, Pet Refiner *33* (7), 121, 1854; Rase and Barrow, Project Engineering of Process Plants, 224, Wiley, 1957.)

CONDENSATION

Depending largely on the nature of the surface, condensate may form either a continuous film or droplets. Since a fluid film is a partial insulator, dropwise condensation results in higher rates of condensation. Promoters are substances that make surfaces nonwetting, and may be effective as additives in trace amounts to the vapor. Special shapes of condensing surfaces also are effective in developing dropwise condensation. None of these effects has been generally correlated, but many examples are cited in HEDH and elsewhere. Condensation rates of mixtures are influenced by both heat and mass transfer rates; techniques for making such calculations have been developed and are a favorite problem for implementation on computers. Condensation rates of mixtures that form immiscible liquids also are reported on in HEDH. Generally, mixtures have lower heat transfer coefficients in condensation than do pure substances.

BOILING

This process can be nuclear or film type. In nuclear boiling, bubbles detach themselves quickly from the heat transfer surface. In film boiling the rate of heat transfer is retarded by an adherent vapor film through which heat supply must be by conduction. Either mode

TABLE 8.8. Dimensionless Groups and Units of Quantities Pertaining to Heat Transfer

Symbol	Number	Group
Bi	Biot	hL/k
Fo	Fourier	$k\theta/\rho CL^2$
Gz	Graetz	wC/kL
Gr	Grashof	$D^3\rho^2 g\beta\Delta T/\mu^2$
Nu	Nusselt	hD/k
Pe	Peclet	$DGC/k = (\text{Re})(\text{Pr})$
Pr	Prandtl	$C\mu/k$
Re	Reynolds	DG/μ, $Du\rho/\mu$
Sc	Schmidt	$\mu/\rho k_d$
St	Stanton	$hC/G = (\text{Nu})/(\text{Re})(\text{Pr})$

Notation	Name and Typical Units
C	heat capacity [Btu/(lb)(°F), cal/(g)(°C)]
D	diameter (ft, m)
g	acceleration of gravity [ft/(hr)^2, m/sec^2]
G	mass velocity [lb/(hr)(ft)^2, kg/sec)(m)^2]
h	heat transfer coefficient [Btu/(hr)(sqft)(°F), W/(m)^2(sec)]
k	thermal conductivity [Btu/(hr)(sqft)(°F/ft), cal/(sec)(cm^2)(C/cm)]
k_d	diffusivity (volumetric) [ft^2/hr, cm^2/sec]
L	length (ft, cm)
T, ΔT	temperature, temperature difference (°F or °R, °C or K)
u	linear velocity (ft/hr, cm/sec)
U	overall heat coefficient (same as units of h)
w	mass rate of flow (lb/hr, g/sec)
β	Thermal expansion coefficient (1/°F, 1/°C)
θ	time (hr, sec)
μ	viscosity [lb/(ft)(hr), g/(cm)(sec)]
ρ	density [lb/(ft)^3, g/(cm)^3]

can exist in any particular case. Transition between modes corresponds to a maximum heat flux and the associated critical temperature difference. A table of such data by McAdams (*Heat Transmission*, McGraw-Hill, New York, 1954, p. 386) shows the critical temperature differences to range from 42–90°F and the maximum fluxes from 42–126 KBtu/(hr)(sqft) for organic substances and up to 410 KBtu/(hr)(sqft) for water; the nature of the surface and any promoters are identified. Equations (40) and (41) of Table 8.10 are for critical heat fluxes in kettle and thermosyphon reboilers. Beyond the maximum rate, film boiling develops and the rate of heat transfer drops off very sharply.

Evaluation of the boiling heat transfer coefficient in vertical tubes, as in thermosyphon reboilers, is based on a group of equations, (42)–(48), of Table 8.10. A suitable procedure is listed following these equations in that table.

EXTENDED SURFACES

When a film coefficient is low as in the cases of low pressure gases and viscous liquids, heat transfer can be improved economically by employing extended surfaces. Figure 8.6 illustrates a variety of extended surfaces. Since the temperature of a fin necessarily averages less than that of the bare surface, the effectiveness likewise is less than that of bare surface. For many designs, the extended surface may be taken to be 60% as effective as bare surface, but this factor depends on the heat transfer coefficient and thermal conductivity of the fin as well as its geometry. Equations and corresponding charts have been developed for the common geometries and are shown, for example, in HEDH (1983, Sec. 2.5.3) and elsewhere. One chart is given with Example 8.6. The efficiency η of the extended surface is defined as the ratio of a realized heat transfer to the heat transfer that would be obtained if the fin were at the bare tube temperature throughout. The total heat transfer is the sum of the heat transfers through the bare and the extended surfaces:

$$Q = Q_b + Q_e = U_b A_b (1 + \eta A_e/A_b)(T_b - T_{\text{fluid}}).$$

A_b is the tube surface that is not occupied by fins. Example 8.6 performs an analysis of this kind of problem.

8.5. PRESSURE DROP IN HEAT EXCHANGERS

Although the rate of heat transfer to or from fluids is improved by increase of linear velocity, such improvements are limited by the economic balance between value of equipment saving and cost of pumping. A practical rule is that pressure drop in vacuum condensers be limited to 0.5–1.0 psi (25–50 Torr) or less, depending on the required upstream process pressure. In liquid service, pressure drops of 5–10 psi are employed as a minimum, and up to 15% or so of the upstream pressure.

Calculation of tube-side pressure drop is straightforward, even of vapor–liquid mixtures when their proportions can be estimated. Example 8.7 employs the methods of Chapter 6 for pressure drop in a thermosiphon reboiler.

The shell side with a number of segmental baffles presents more of a problem. It may be treated as a series of ideal tube banks connected by window zones, but also accompanied by some bypassing of the tube bundles and leakage through the baffles. A hand calculation based on this mechanism (ascribed to K.J. Bell) is illustrated by Ganapathy (1982, pp. 292–302), but the calculation usually is made with proprietary computer programs, that of HTRI for instance.

A simpler method due to Kern (1950, pp. 147–152) nominally considers only the drop across the tube banks, but actually takes account of the added pressure drop through baffle windows by employing a higher than normal friction factor to evaluate pressure drop across the tube banks. Example 8.8 employs this procedure. According to Taborek (HEDH, 1983, 3.3.2), the Kern predictions usually are high, and therefore considered safe, by a factor as high as 2, except in laminar flow where the results are uncertain. In the case worked out by Ganapathy (1982, pp. 292–302), however, the Bell and Kern results are essentially the same.

8.6. TYPES OF HEAT EXCHANGERS

Heat exchangers are equipment primarily for transferring heat between hot and cold streams. They have separate passages for the two streams and operate continuously. They also are called recuperators to distinguish them from regenerators, in which hot and cold streams pass alternately through the same passages and exchange heat with the mass of the equipment, which is intentionally made with large heat capacity. Recuperators are used mostly in cryogenic services, and at the other extreme of temperature, as high temperature air preheaters. They will not be discussed here; a detailed treatment of their theory is by Hausen (1983).

Being the most widely used kind of process equipment is a claim that is made easily for heat exchangers. A classified directory of manufacturers of heat exchangers by Walker (1982) has several hundred items, including about 200 manufacturers of shell-and-tube equipment. The most versatile and widely used exchangers are the shell-and-tube types, but various plate and other types are valuable and economically competitive or superior in some applications. These other types will be discussed briefly, but most of the space following will be devoted to the shell-and-tube types, primarily

TABLE 8.9. Equations for Heat Transfer Coefficients of Natural Convection

Vertical plates and cylinders, length L

$$X_L = (\mathrm{Gr})(\mathrm{Pr}) = \left(\frac{L^3\rho_f^2 g\beta_f\Delta t}{\mu_f^2}\right)\left(\frac{c_p\mu}{k}\right)_f \quad (1)$$

$$hL/k = 0.13X_L^{1/3}, \quad \text{turbulent}, \quad 10^9 < X_L < 10^{12} \quad (2)$$

$$h = 0.19(\Delta t)^{1/3}, \quad \text{for air}, \quad \Delta t \text{ in °F}, h \text{ in Btu/(hr)(sqft)(°F)} \quad (3)$$

$$hL/k = 0.59X_L^{1/4}, \quad \text{laminar}, \quad 10^4 < X_L < 10^9 \quad (4)$$

$$h = 0.29(\Delta t/L)^{1/4}, \quad \text{for air}, \quad L \text{ in ft} \quad (5)$$

Single horizontal cylinder, diameter D_0

$$X_D = \frac{D_0^3\rho_s^2 g\beta_s\Delta t}{\mu_s^2}\left(\frac{c_p\mu}{k}\right) \quad (6)$$

$$hD_0/k = 0.53X_D^{1/4}, \quad 10^3 < X_D < 10^9 \quad (7)$$

$$h = 0.18(\Delta t)^{1/3}, \quad \text{for air}, \quad 10^9 < X_D < 10^{12} \quad (8)$$

$$h = 0.27(\Delta t/D_0)^{1/4}, \quad 10^4 < X_D < 10^9 \quad (9)$$

Horizontal plates, rectangular, L the smaller dimension

$$X_L = \frac{L^3\rho_f^2 g\beta_f\Delta t}{\mu_f^2}\left(\frac{c_p\mu}{k}\right)_f \quad (10)$$

Heated plates facing up or cooled facing down

$$hL/k = 0.14X_L^{1/3}, \quad 2(10^7) < X_L < 3(10^{10}), \quad \text{turbulent}$$

$$h = 0.22(\Delta t)^{1/3}, \quad \text{for air} \quad (11)$$

$$hL/k = 0.54X_L^{1/4}, \quad 10^5 < X_L < 2(10^7), \quad \text{laminar} \quad (12)$$

$$h = 0.27(\Delta t/L)^{1/4} \quad (13)$$

Heated plates facing down, or cooled facing up

$$hL/k = 0.27X_L^{1/4}, \quad 3(10^5) < X_L < 3(10^{10}), \quad \text{laminar} \quad (14)$$

$$h = 0.12(\Delta t/L)^{1/4}, \quad \text{for air} \quad (15)$$

Combined convection and radiation coefficients, $h_c + h_r$, for horizontal steel or insulated pipes in a room at 80°F (16)

Nominal Pipe Dia (in.)	$(\Delta t)_s$, Temperature Difference (°F) from Surface to Room														
	50	100	150	200	250	300	400	500	600	700	800	900	1000	1100	1200
$\frac{1}{2}$	2.12	2.48	2.76	3.10	3.41	3.75	4.47	5.30	6.21	7.25	8.40	9.73	11.20	12.81	14.65
1	2.03	2.38	2.65	2.98	3.29	3.62	4.33	5.16	6.07	7.11	8.25	9.57	11.04	12.65	14.48
2	1.93	2.27	2.52	2.85	3.14	3.47	4.18	4.99	5.89	6.92	8.07	9.38	10.85	12.46	14.28
4	1.84	2.16	2.41	2.72	3.01	3.33	4.02	4.83	5.72	6.75	7.89	9.21	10.66	12.27	14.09
8	1.76	2.06	2.29	2.60	2.89	3.20	3.88	4.68	5.57	6.60	7.73	9.05	10.50	12.10	13.93
12	1.71	2.01	2.24	2.54	2.82	3.13	3.83	4.61	5.50	6.52	7.65	8.96	10.42	12.03	13.84
24	1.64	1.93	2.15	2.45	2.72	3.03	3.70	4.48	5.37	6.39	7.52	8.83	10.28	11.90	13.70

(McAdams, *Heat Transmission,* McGraw-Hill, New York, 1954).

because of their importance, but also because they are most completely documented in the literature. Thus they can be designed with a degree of confidence to fit into a process. The other types are largely proprietary and for the most part must be process designed by their manufacturers.

PLATE-AND-FRAME EXCHANGERS

Plate-and-frame exchangers are assemblies of pressed corrugated plates on a frame, as shown on Figure 8.8(a). Gaskets in grooves around the periphery contain the fluids and direct the flows into and out of the spaces between the plates. Hot and cold flows are on opposite sides of the plates. Figure 8.8(b) shows a few of the many combinations of parallel and countercurrent flows that can be maintained. Close spacing and the presence of the corrugations result in high coefficients on both sides—several times those of shell-and-tube equipment—and fouling factors are low, of the order of $1\text{–}5 \times 10^{-5}$ Btu/(hr)(sqft)(°F). The accessibility of the heat exchange surface for cleaning makes them particularly suitable for fouling services and where a high degree of sanitation is required, as in food and pharmaceutical processing. Operating pressures and temperatures are limited by the natures of the available gasketing materials, with usual maxima of 300 psig and 400°F.

Since plate-and-frame exchangers are made by comparatively few concerns, most process design information about them is proprietary but may be made available to serious enquirers. Friction factors and heat transfer coefficients vary with the plate spacing and the kinds of corrugations; a few data are cited in HEDH (1983, 3.7.4–3.7.5). Pumping costs per unit of heat transfer are said to be lower than for shell-and-tube equipment. In stainless steel

TABLE 8.10. Recommended Individual Heat Transfer Coefficient Correlations[a]

A. Single Phase Streams

a. Laminar Flow, Re < 2300

Inside tubes

$$Nu_T = \sqrt[3]{3.66^3 + 1.61^3\, Pe(d/L)}, \quad 0.1 < Pe(d/L) < 10^4 \tag{1}$$

Between parallel plates of length L and separation distance s

$$Nu_T = 3.78 + \frac{0.0156[Pe(s/L)]^{1.14}}{1 + 0.058[Pe(s/L)]^{0.64}\, Pr^{0.17}}, \quad 0.1 < Pe(s/L) < 10^3 \tag{2}$$

In concentric annuli with d_i inside, d_o outside, and hydraulic diameter $d_h = d_o - d_i$. I, heat transfer at inside wall; II, at outside wall; III, at both walls at equal temperatures

$$Nu_T = Nu_\infty + f\left(\frac{d_i}{d_o}\right)\frac{0.19[Pe(d_h/L)]^{0.8}}{1 + 0.117[Pe(d_h/L)]^{0.467}} \tag{3}$$

$$\text{Case I:} \quad Nu_{i\infty} = 3.66 + 1.2\left(\frac{d_i}{d_o}\right)^{-0.8} \tag{4}$$

$$\text{Case II:} \quad Nu_{o\infty} = 3.66 + 1.2\left(\frac{d_i}{d_o}\right)^{0.5} \tag{5}$$

$$\text{Case III:} \quad Nu_b = 3.66 + \left[4 - \frac{0.102}{(d_i/d_o) + 0.2}\right]\left(\frac{d_i}{d_o}\right)^{0.04} \tag{6}$$

$$\text{Case I:} \quad f\left(\frac{d_i}{d_o}\right) = 1 + 0.14\left(\frac{d_i}{d_o}\right)^{-0.5} \tag{7}$$

$$\text{Case II:} \quad f\left(\frac{d_i}{d_o}\right) = 1 + 0.14\left(\frac{d_i}{d_o}\right)^{1/3} \tag{8}$$

$$\text{Case III:} \quad f\left(\frac{d_i}{d_o}\right) = 1 + 0.14\left(\frac{d_i}{d_o}\right)^{0.1} \tag{9}$$

b. Turbulent Flow, Re > 2300

Inside tubes

$$Nu = 0.0214(Re^{0.8} - 100)\, Pr^{0.4}\left[1 + \left(\frac{d}{L}\right)^{2/3}\right], \quad 0.5 < Pr < 1.5 \tag{10}$$

$$Nu = 0.012(Re^{0.87} - 280)\, Pr^{0.4}\left[1 + \left(\frac{d}{L}\right)^{2/3}\right], \quad 1.5 < Pr < 500 \tag{11}$$

Concentric annuli: Use d_h for both Re and Nu. Nu_{tube} from Eqs. (10) or (11)

$$\text{Case I:} \quad \frac{Nu_i}{Nu_{tube}} = 0.86\left(\frac{d_i}{d_o}\right)^{-0.16}$$

$$\text{Case II:} \quad \frac{Nu_o}{Nu_{tube}} = 1 - 0.14\left(\frac{d_i}{d_o}\right)^{0.6} \tag{13}$$

$$\text{Case III:} \quad \frac{Nu_b}{Nu_{tube}} = \frac{0.86(d_i/d_o)^{0.84} + [1 - 0.14(d_i/d_o)^{0.6}]}{1 + d_i/d_o} \tag{14}$$

Across one row of long tubes: d = diameter, s = center-to-center distance, $a = s/d$, $\psi = 1 - \pi/4a$, $L = \pi d/2$

$$Re_{\psi,L} = wL/\psi\nu \tag{15}$$

$$Nu_{o,row} = 0.3 + \sqrt{Nu_{L,lam}^2 + Nu_{L,turb}^2} \tag{16}$$

$$Nu_{L,lam} = 0.664\sqrt{Re_{\psi,L}}\, Pr^{1/3} \tag{17}$$

$$Nu_{L,turb} = 0.037\, Re_{\psi,L}^{0.8}\, Pr/[1 + 2.443\, Re_{\psi,L}^{-0.1}(Pr^{2/3} - 1)] \tag{18}$$

$$Nu_{L,row} = \alpha L/\lambda \tag{19}$$

[a] Special notation used in this table: α = heat transfer coefficient (W/m^2 K) (instead of h), η = viscosity (instead of μ), and α = thermal conductivity (instead of k).
(Based on HEDH, 1983).

TABLE 8.10—*(continued)*

Across a bank of n tubes deep:

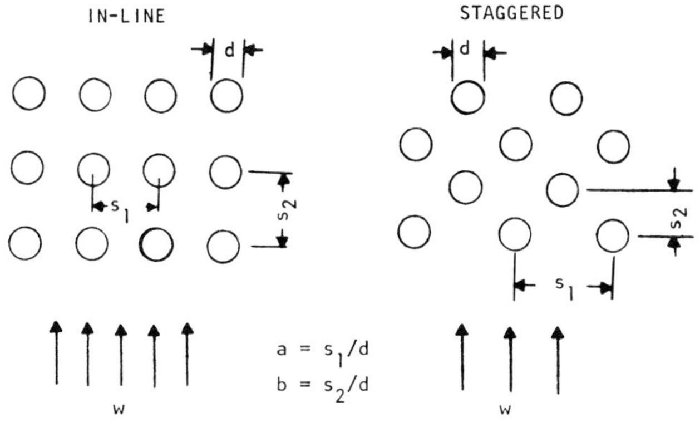

$$\psi = 1 - \pi/4a \quad \text{if } b \geq 1 \tag{20}$$
$$\psi = 1 - \pi/4ab \quad \text{if } b < 1 \tag{21}$$

$$\mathrm{Nu}_{o,\mathrm{bank}} = \alpha L/\lambda = f_A\,\mathrm{Nu}_{o,\mathrm{row}}/K, \quad n \geq 10 \tag{22}$$
$$\mathrm{Nu}_{o,\mathrm{bank}} = [1 + (n-1)f_A]\,\mathrm{Nu}_{o,\mathrm{row}}/Kn, \quad n < 10 \tag{23}$$

[$\mathrm{Nu}_{o,\mathrm{row}}$ from Eq. (16)]

$$f_{A,\text{in-line}} = 1 + (0.7/\psi^{1.5})[(b/a - 0.3)/(b/a + 0.7)^2] \tag{24}$$
$$f_{A,\text{stag}} = 1 + 2/3b \tag{25}$$
$$K = (\mathrm{Pr}/\mathrm{Pr}_w)^{0.25}, \quad \text{for liquid heating} \tag{26}$$
$$K = (\mathrm{Pr}/\mathrm{Pr}_w)^{0.11}, \quad \text{for liquid cooling} \tag{27}$$
$$K = (T/T_w)^{0.12}, \quad \text{for gases} \tag{28}$$

Subscript w designates wall condition

Banks of radial high-fin tubes: ε = (bare tube surface)/(total surface of finned tube)

In line:

$$\mathrm{Nu} = 0.30\,\mathrm{Re}^{0.625}\varepsilon^{-0.375}\,\mathrm{Pr}^{0.333}, \quad 5 < \varepsilon < 12, \quad 5000 < \mathrm{Re} < 10^5 \tag{29}$$

Staggered: $a = s_1/d$, $b = s_2/d$, s = spacing of fins

$$\mathrm{Nu} = 0.19(a/b)^{0.2}(s/d)^{0.18}(h/d)^{-0.14}\,\mathrm{Re}^{0.65}\mathrm{Pr}^{0.33}, \quad 100 < \mathrm{Re} < 20{,}000 \tag{30}$$

Banks of radial low-fin tubes: D = diameter of finned tube, s = distance between fins, h = height of fin; following correlation for D = 22.2 mm, s = 1.25 mm, and h = 1.4 mm

$$\mathrm{Nu} = 0.0729\,\mathrm{Re}^{0.74}\,\mathrm{Pr}^{0.36}, \quad 5000 < \mathrm{Re} < 35{,}000 \tag{31}$$
$$\mathrm{Nu} = 0.137\,\mathrm{Re}^{0.68}\,\mathrm{Pr}^{0.36}, \quad 35{,}000 < \mathrm{Re} < 235{,}000 \tag{32}$$
$$\mathrm{Nu} = 0.0511\,\mathrm{Re}^{0.76}\,\mathrm{Pr}^{0.36}, \quad 235{,}000 < \mathrm{Re} < 10^6 \tag{33}$$

B. Condensation of Pure Vapors

On vertical tubes and other surfaces; $\dot{\Gamma}$ = condensation rate per unit of periphery

$$\frac{\bar{\alpha}}{\lambda_l}\left[\frac{\eta_l^2}{\rho_l(\rho_l - \rho_g)g_n}\right]^{1/3} = 1.47\left(\frac{4\dot{\Gamma}}{\eta_l}\right)^{-1/3} \tag{34}$$

On a single horizontal tube: $\dot{\Gamma}$ = condensation rate per unit length of tube

$$\frac{\alpha}{\lambda_l}\left[\frac{\eta_l^2}{\rho_l(\rho_l - \rho_g)g_n}\right]^{1/3} = 1.51\left(\frac{4\dot{\Gamma}}{\eta_l}\right)^{-1/3} \tag{35}$$

(continued)

TABLE 8.10—*(continued)*

On a bank of N horizontal tubes: Γ = condensation rate per unit length from the bottom tube

$$\frac{\alpha}{\lambda_l}\left[\frac{\eta_l^2}{\rho_l(\rho_l-\rho_g)g_n}\right]^{1/3} = 1.51\left(\frac{4\dot{\Gamma}}{\eta_l}\right)^{-1/3} N^{-1/6} \quad (36)$$

C. Boiling

Single immersed tube: $\dot{q}$ heat flux (W/m^2), p_c = critical pressure, bars, $p_r = p/p_c$

$$\alpha = 0.1000\dot{q}^{0.7}p^{0.69}[1.8p_r^{0.17} + 4p_r^{1.2} + 10p_r^{10}], \quad \text{W/m}^2\,\text{K} \quad (37)$$

Kettle and horizontal thermosiphon reboilers

$$\alpha = 0.27\exp(-0.027BR)\dot{q}^{0.7}p_c^{0.69}p_r^{0.17} + \alpha_{nc} \quad (38)$$

BR = difference between dew and bubblepoints (°K); if more than 85, use 85

$$\alpha_{nc} = \begin{cases} 250\ \text{W/m}^2\,\text{K}, & \text{for hydrocarbons} \\ 1000\ \text{W/m K}, & \text{for water} \end{cases} \quad (39)$$

Critical heat flux in kettle and horizontal thermosiphon reboilers

$$\dot{q}_{\max} = 80{,}700p_cp_r^{0.35}(1-p_r)^{0.9}\psi_b, \quad \text{W/m}^2 \quad (40)$$

ψ_b = (external peripheral surface of tube bundle)/
(total tube area); if >0.45, use 0.45

Boiling in vertical tubes: thermosiphon reboilers

Critical heat flux: p_c critical pressure, bars; D_i tube ID, m; L tube length, m

$$\dot{q} = 393{,}000(D_i^2/L)^{0.35}p_c^{0.61}p_r^{0.25}(1-p_r), \quad \text{W/m}^2 \quad (41)$$

Heat transfer coefficient with Eqs. (42)–(48) and following procedure

$$\alpha_{tp} = \alpha_{nb} + \alpha_c \quad (42)$$

$$\alpha_c = 0.023\left(\frac{\dot{m}(1-x)D}{\eta_l}\right)^{0.8}\left(\frac{\eta c_p}{\lambda}\right)_l^{0.4}\frac{\lambda_l}{D}F \quad (43)$$

$$\alpha_{nb} = 0.00122\left(\frac{\lambda_l^{0.79}C_{pl}^{0.45}\rho_l^{0.49}}{\sigma^{0.5}\eta_l^{0.29}\Delta h_v^{0.24}\rho_g^{0.24}}\right)\Delta T_{\text{sat}}^{0.24}\Delta p_{\text{sat}}^{0.75}S \quad (44)$$

$$F = 1 \quad \text{for } 1/X_{tt} \le 0.1 \quad (45)$$

$$F = 2.35(1/X_{tt} + 0.213)^{0.736} \quad \text{for } 1/X_{tt} > 0.1 \quad (46)$$

$$S = 1/(1 + 2.53\times10^{-6}\,\text{Re}_{tp}^{1.17}) \quad (47)$$

$$X_{tt} \cong [(1-x)/x]^{0.9}(\rho_g/\rho_l)^{0.5}(\eta_l/\eta_g)^{0.1} \quad (48)$$

Procedure for finding the heat transfer coefficient and required temperature difference when the heat flux $\dot{q}$, mass rate of flow $\dot{m}$ and fraction vapor x are specified

1. Find X_{tt}, Eq. (48)
2. Evaluate F from Eqs. (45), (46)
3. Calculate α_c, Eq. (43)
4. Calculate $\text{Re}_{tp} = \dot{m}F^{1.25}(1-x)D/\eta_l$
5. Evaluate S from Eq. (47)
6. Calculate α_{nb} for a range of values of ΔT_{sat}
7. Calculate α_{tp} from Eq. (42) for this range of ΔT_{sat} values
8. On a plot of calculated $\dot{q} = \alpha_{tp}\Delta T_{\text{sat}}$ against α_{tp}, find the values of α_{tp} and ΔT_{sat} corresponding to the specified $\dot{q}$

Example 8.6
Sizing an Exchanger with Radial Finned Tubes

A liquid is heated from 150 to 190°F with a gas that goes from 250 to 200°F. The duty is 1.25 MBtu/hr. The inside film coefficient is 200, the bare tube outside coefficient is $h_b = 20$ Btu/(hr)(sqft)(°F). The tubes are 1 in. OD, the fins are $\frac{5}{8}$ in. high, 0.038 in. thick, and number 72/ft. The total tube length will be found with fins of steel, brass, or aluminum:

$$\text{LMTD} = (60 - 50)/\ln(60/50) = 54.8,$$
$$U_b = (1/20 + 1/200)^{-1} = 18.18.$$

Fin surface:

$$A_e = 72(2)(\pi/4)[(2.25^2 - 1)/144] = 3.191 \text{ sqft/ft}.$$

Uncovered tube surface:

$$A_b = (\pi/12)[1 - 72(0.038/12)] = 0.2021 \text{ sqft/ft},$$
$$A_e/A_b = 3.191/0.2021 = 15.79,$$
$$y_b = \text{half-fin thickness} = 0.038/2(12) = 0.00158 \text{ ft}.$$

Abscissa of the chart:

$$x = (r_e - r_b)\sqrt{h_b/y_b k} = [(2.25 - 1)/24]\sqrt{20/0.00158k}$$
$$= 5.86/\sqrt{k},$$
$$r_e/r_b = 2.25,$$
$$A_b = Q/U_b \Delta T(1 + \eta A_e/A_b)$$
$$= 1.25(10^6)/18.18(54.8)(1 + 15.79\eta) \text{ sq ft}.$$

Find η from the chart. Tube length, $L = A_b/0.2021$ ft.

	k	x	η	A_b	L
Steel	26	1.149	0.59	121.6	602
Brass	60	0.756	0.76	96.5	477
Al	120	0.535	0.86	86.1	426

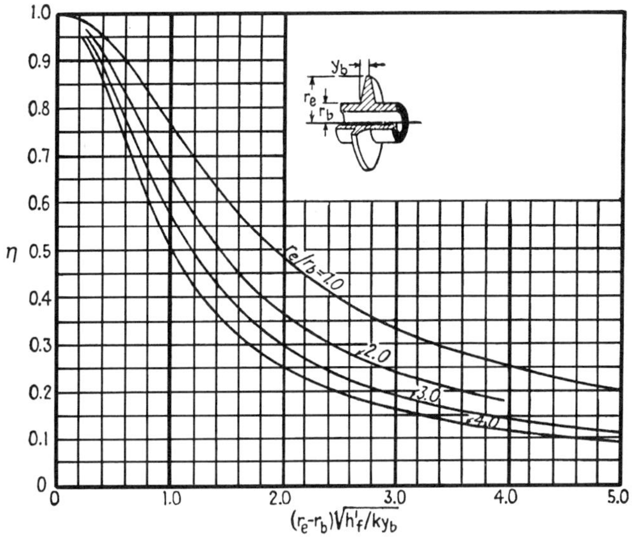

Example 8.7
Pressure Drop on the Tube Side of a Vertical Thermosiphon Reboiler

Liquid with the properties of water at 5 atm and 307°F is reboiled at a feed rate of 2800 lb/(hr)(tube) with 30 wt % vaporization. The tubes are 0.1 ft ID and 12 ft long. The pressure drop will be figured at an average vaporization of 15%. The Lockhart–Martinelli, method will be used, following Example 6.14, and the formulas of Tables 6.1 and 6.8:

	Liquid	Vapor
$\dot{m}$ (lb/hr)	2380	420
μ (lb/ft hr)	0.45	0.036
ρ (lb/cuft)	57.0	0.172
Re	67340	148544
f	0.0220	0.0203
$\Delta P/L$ (psi/ft)	0.00295	0.0281

$$X^2 = 0.00295/0.0281 = 0.1051,$$
$$C = 20,$$
$$\phi_L^2 = 1 + 20/X + 1/X^2 = 72.21,$$
$$(\Delta P/L) \text{ two phase} = 72.21(0.00295) = 0.2130,$$
$$\Delta P = 0.2130(12) = 2.56 \text{ psi}, \quad 5.90 \text{ ft water}.$$

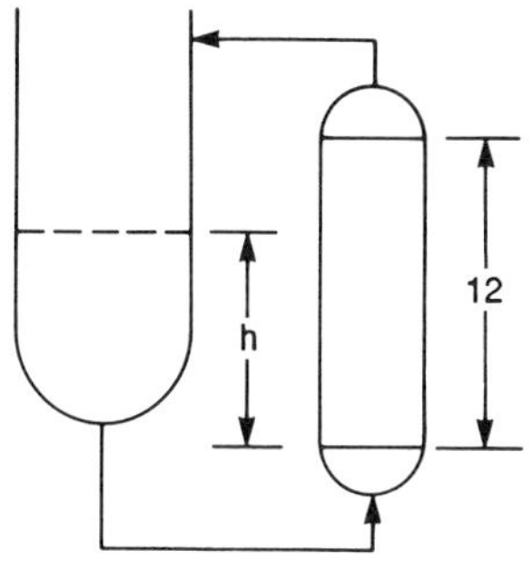

Average density in reboiler tubes is

$$\rho_m = \frac{2800}{2380/57 + 420/0.172} = 1.13 \text{ lb/cuft}.$$

Required height of liquid in tower above bottom of tube sheet

$$\rho_L h = 2.56(144) + 1.13(12),$$
$$h = 382.2/57 = 6.7 \text{ ft}.$$

EXAMPLE 8.8
Pressure Drop on the Shell Side with 25% Open Segmental Baffles, by Kern's Method (1950, p. 147)

Nomenclature and formulas:

$$\text{hydraulic diameter } D_h = \begin{cases} 1.1028P_t^2/D_t - D_t, & \text{triangular pitch,} \\ 1.2732P_t^2/D_t - D_t, & \text{square pitch,} \end{cases}$$

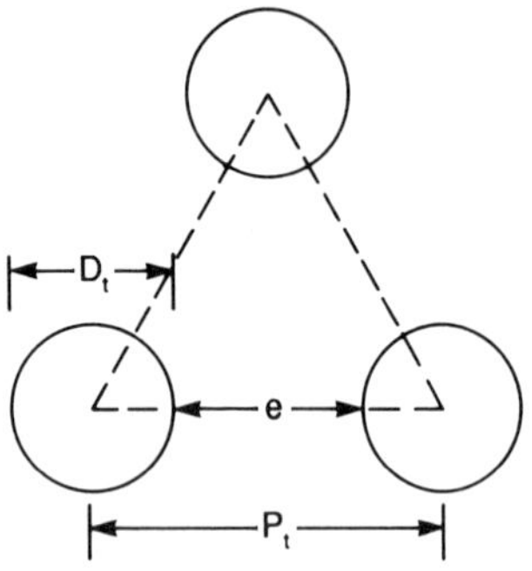

D_s = shell diameter,
B = distance between baffles,
N = number of baffles,
A_s = flow area = D_sBC/P_t,
$G_s = \dot{m}/A_s$, lb/(hr)(sqft),

$$\text{Re} = D_hG_s/\mu,$$
$$f = 0.0121\text{Re}^{-0.19}, \quad 300 < \text{Re} < 10^6, \quad 25\% \text{ segmental baffles,}$$
$$\Delta P = \frac{fG_s^2D_s(N+1)}{2g\rho D_h} = \frac{fG_s^2D_s(N+1)}{5.22(10^{10})sD_h}, \quad \text{psi,}$$
s = specific gravity.

Numerical example:

$\dot{m}$ = 43,800 lb/hr,
s = 0.73 sp gr,
μ = 0.097 lb/ft hr,
D_t = 1 in.,
P_t = 1.25 in., triangular pitch,
$C = 1.25 - 1.00 = 0.25$ in.,
D_s = 21.25 in., 1.77 ft.,
D_h = 0.723 in., 0.0603 ft.,
B = 5 in.,
N = 38 baffles,
$A_s = 21.25(0.25)(5)/1.25(144) = 0.1476$ sqft,
$G_s = 43{,}800/0.1476 = 296{,}810$ lb/(hr)(sqft),
$\text{Re} = 0.0603(296{,}810)/0.97 = 18{,}450$,
$f = 0.0121(18{,}450)^{-0.19} = 0.00187$,

$$\Delta P = \frac{0.00187(296{,}810)^2(1.77)(39)}{5.22(10^{10})(0.73)(0.0603)} = 4.95 \text{ psi.}$$

construction, the plate-and-frame construction cost is 50–70% that of shell-and-tube, according to Marriott (*Chem. Eng.*, April 5, 1971).

A process design of a plate-and-frame exchanger is worked out by Ganapathy (1982, p. 368).

SPIRAL HEAT EXCHANGERS

As appears on Figure 8.8(c), the hot fluid enters at the center of the spiral element and flows to the periphery; flow of the cold fluid is countercurrent, entering at the periphery and leaving at the center. Heat transfer coefficients are high on both sides, and there is no correction to the log mean temperature difference because of the true countercurrent action. These factors may lead to surface requirements 20% or so less than those of shell-and-tube exchangers. Spiral types generally may be superior with highly viscous fluids at moderate pressures. Design procedures for spiral plate and the related spiral tube exchangers are presented by Minton (1970). Walker (1982) lists 24 manufacturers of this kind of equipment.

COMPACT (PLATE-FIN) EXCHANGERS

Units like Figure 8.6(h), with similar kinds of passages for the hot and cold fluids, are used primarily for gas service. Typically they have surfaces of the order of 1200 m^2/m^3 (353 sqft/cuft), corrugation height 3.8–11.8 mm, corrugation thickness 0.2–0.6 mm, and fin density 230–700 fins/m. The large extended surface permits about four times the heat transfer rate per unit volume that can be achieved with shell-and-tube construction. Units have been designed for pressures up to 80 atm or so. The close spacings militate against fouling service. Commercially, compact exchangers are used in cryogenic services, and also for heat recovery at high temperatures in connection with gas turbines. For mobile units, as in motor vehicles, the designs of Figures 8.6(h) and (i) have the great merits of compactness and light weight. Any kind of arrangement of cross and countercurrent flows is feasible, and three or more different streams can be accommodated in the same equipment. Pressure drop, heat transfer relations, and other aspects of design are well documented, particularly by Kays and London (1984) and in HEDH (1983, Sec. 3.9).

AIR COOLERS

In such equipment the process fluid flows through finned tubes and cooling air is blown across them with fans. Figures 8.4(g) and (h) show the two possible arrangements. The economics of application of air coolers favors services that allow 25–40°F temperature difference between ambient air and process outlet. In the range above 10 MBtu/(hr), air coolers can be economically competitive with water coolers when water of adequate quality is available in sufficient amount.

Tubes are 0.75–1.00 in. OD, with 7–11 fins/in. and 0.5–0.625 in. high, with a total surface 15–20 times bare surface of the tube. Fans are 4–12 ft/dia, develop pressures of 0.5–1.5 in. water, and require power inputs of 2–5 HP/MBtu/hr or about 7.5 HP/100 sqft of exchanger cross section. Spacings of fans along the length of the equipment do not exceed 1.8 times the width of the cooler. Face velocities are about 10 ft/sec at a depth of three rows and 8 ft/sec at a depth of six rows.

Standard air coolers come in widths of 8, 10, 12, 16, or 20 ft, lengths of 4–40 ft, and stacks of 3–6 rows of tubes. Example 8.8 employs typical spacings.

Three modes of control of air flow are shown in Figure 3.3(e). Precautions may need to be taken against subcooling to the freezing point in winter.

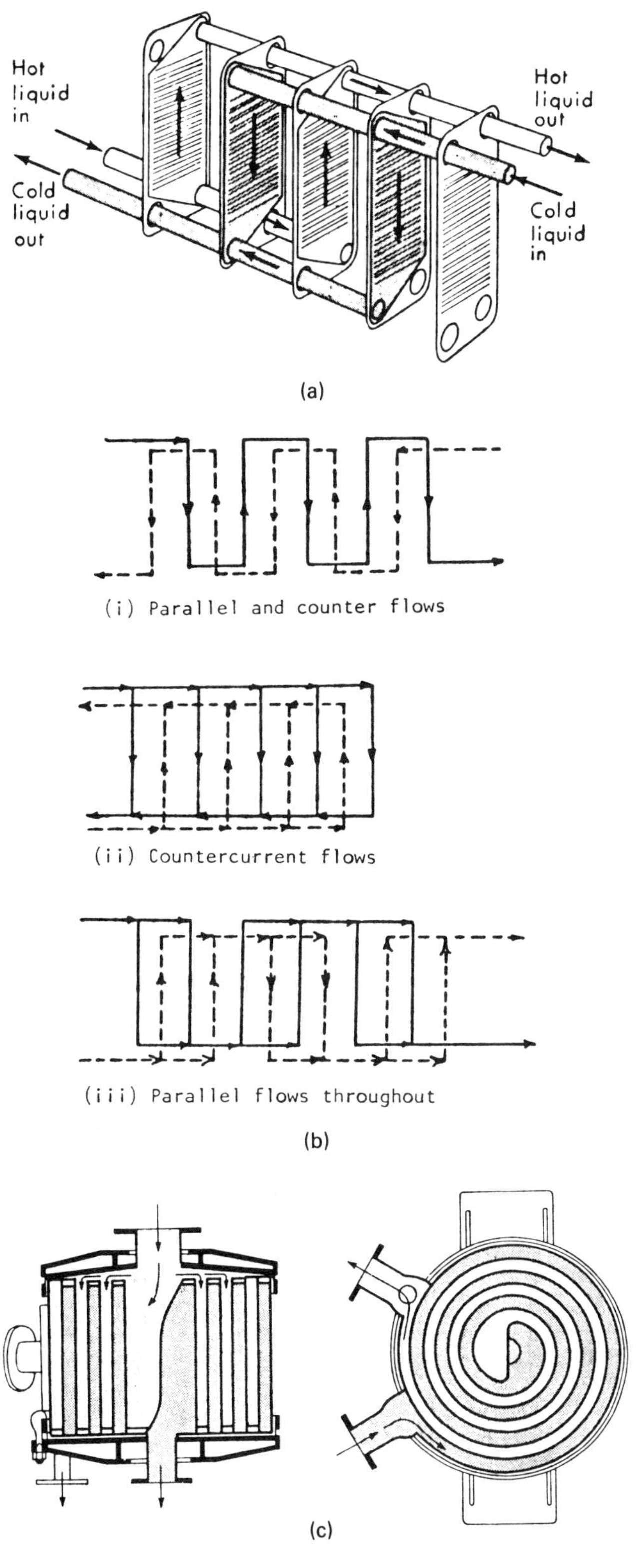

Figure 8.8. Plate and spiral compact exchangers. (a) Plate heat exchanger with corrugated plates, gaskets, frame, and corner portals to control flow paths. (b) Flow patterns in plate exchangers, (i) parallel-counter flows; (ii) countercurrent flows; (iii) parallel flows throughout. (c) Spiral exchanger, vertical, and horizontal cross sections.

Forced draft arrangement, from below the tubes, Figure 8.4(h), develops high turbulence and consequently high heat transfer coefficients. Escape velocities, however, are low, 3 m/sec or so, and as a result poor distribution, backmixing and sensitivity to cross currents can occur. With induced draft from above the tubes, Figure 8.4(g), escape velocities may be of the order of 10 m/sec and better flow distribution results. This kind of installation is more expensive, the pressure drops are higher, and the equipment is bathed in hot air which can be deteriorating. The less solid mounting also can result in noisier operation.

Correlations for friction factors and heat transfer coefficients are cited in HEDH. Some overall coefficients based on external bare tube surfaces are in Tables 8.11 and 8.12. For single passes in cross flow, temperature correction factors are represented by Figure 8.5(c) for example; charts for multipass flow on the tube side are given in HEDH and by Kays and London (1984), for example. Preliminary estimates of air cooler surface requirements can be made with the aid of Figures 8.9 and 8.10, which are applied in Example 8.9.

DOUBLE-PIPES

This kind of exchanger consists of a central pipe supported within a larger one by packing glands [Fig. 8.4(a)]. The straight length is limited to a maximum of about 20 ft; otherwise the center pipe will sag and cause poor distribution in the annulus. It is customary to operate with the high pressure, high temperature, high density, and corrosive fluid in the inner pipe and the less demanding one in the annulus. The inner surface can be provided with scrapers [Fig. 8.4(b)] as in dewaxing of oils or crystallization from solutions. External longitudinal fins in the annular space can be used to improve heat transfer with gases or viscous fluids. When greater heat transfer surfaces are needed, several double-pipes can be stacked in any combination of series or parallel.

Double-pipe exchangers have largely lost out to shell-and-tube units in recent years, although Walker (1982) lists 70 manufacturers of them. They may be worth considering in these situations:

1. When the shell-side coefficient is less than half that of the tube side; the annular side coefficient can be made comparable to the tube side.
2. Temperature crosses that require multishell shell-and-tube units can be avoided by the inherent true countercurrent flow in double pipes.
3. High pressures can be accommodated more economically in the annulus than they can in a larger diameter shell.
4. At duties requiring only 100–200 sqft of surface the double-pipe may be more economical, even in comparison with off-the-shelf units.

The process design of double-pipe exchangers is practically the simplest heat exchanger problem. Pressure drop calculation is straightforward. Heat transfer coefficients in annular spaces have been investigated and equations are cited in Table 8.10. A chapter is devoted to this equipment by Kern (1950).

8.7. SHELL-AND-TUBE HEAT EXCHANGERS

Such exchangers are made up of a number of tubes in parallel and series through which one fluid travels and enclosed in a shell through which the other fluid is conducted.

CONSTRUCTION

The shell side is provided with a number of baffles to promote high velocities and largely more efficient cross flow on the outsides of the

TABLE 8.11. Overall Heat Transfer Coefficients in Air Coolers [U Btu/(hr)(°F)(sqft of outside bare tube surface)]

Liquid Coolers				Condensers	
Material	Heat-Transfer Coefficient, [Btu/(hr) (ft²)(°F)]	Material	Heat-Transfer Coefficient, [Btu/(hr)(ft²)(°F)]	Material	Heat-Transfer Coefficient, [Btu/(hr)(ft²)(°F)]
Oils, 20° API	10–16	Heavy oils, 8–14° API		Steam	140–150
200°F avg. temp	10–16	300°F avg. temp	6–10	Steam	
300°F avg. temp	13–22	400°F avg. temp	10–16	10% noncondensibles	100–110
400°F avg. temp	30–40	Diesel oil	45–55	20% noncondensibles	95–100
		Kerosene	55–60	40% noncondensibles	70–75
Oils, 30° API		Heavy naphtha	60–65	Pure light hydrocarbons	80–85
150°F avg. temp	12–23	Light naphtha	65–70	Mixed light hydrocarbons	65–75
200°F avg. temp	25–35	Gasoline	70–75	Gasoline	60–75
300°F avg. temp	45–55	Light hydrocarbons	75–80	Gasoline-steam mixtures	70–75
400°F avg. temp	50–60	Alcohols and most organic solvents	70–75	Medium hydrocarbons	45–50
Oils, 40° API				Medium hydrocarbons w/steam	55–60
150°F avg. temp	25–35	Ammonia	100–120	Pure organic solvents	75–80
200°F avg. temp	50–60	Brine, 75% water	90–110	Ammonia	100–110
300°F avg. temp	55–65	Water	120–140		
400°F avg. temp	60–70	50% ethylene glycol and water	100–120		

Vapor Coolers

Material	Heat-Transfer Coefficient [Btu/(hr)(ft²)(°F)]				
	10 psig	50 psig	100 psig	300 psig	500 psig
Light hydrocarbons	15–20	30–35	45–50	65–70	70–75
Medium hydrocarbons and organic solvents	15–20	35–40	45–50	65–70	70–75
Light inorganic vapors	10–15	15–20	30–35	45–50	50–55
Air	8–10	15–20	25–30	40–45	45–50
Ammonia	10–15	15–20	30–35	45–50	50–55
Steam	10–15	15–20	25–30	45–50	55–60
Hydrogen					
100%	20–30	45–50	65–70	85–95	95–100
75% vol	17–28	40–45	60–65	80–85	85–90
50% vol	15–25	35–40	55–60	75–80	85–90
25% vol	12–23	30–35	45–50	65–70	80–85

[Brown, *Chem. Eng.* (27 Mar. 1978)].

TABLE 8.12. Overall Heat Transfer Coefficients in Condensers, Btu/(hr)(sqft)(°F)[a]

Liquid Coolants

Vapor	Coolant	Btu/(hr)(sqft)(°F)
Alcohol	water	100–200
Dowtherm	tall oil	60–80
Dowtherm	Dowtherm	80–120
Hydrocarbons		
high boiling under vacuum	water	18–50
low boiling	water	80–200
intermediate	oil	25–40
kerosene	water	30–65
kerosene	oil	20–30
naphtha	water	50–75
naphtha	oil	20–40
Organic solvents	water	100–200
Steam	water	400–1000
Steam-organic azeotrope	water	40–80
Vegetable oils	water	20–50

Air Coolers

Vapor	Btu/(hr)(bare sqft)(°F)
Ammonia	100–120
Freons	60–80
Hydrocarbons, light	80–100
Naphtha, heavy	60–70
Naphtha, light	70–80
Steam	130–140

[a] Air cooler data are based on 50 mm tubes with aluminum fins 16–18 mm high spaced 2.5–3 mm apart; coefficients based on bare tube surface. Excerpted from HEDH, 1983.

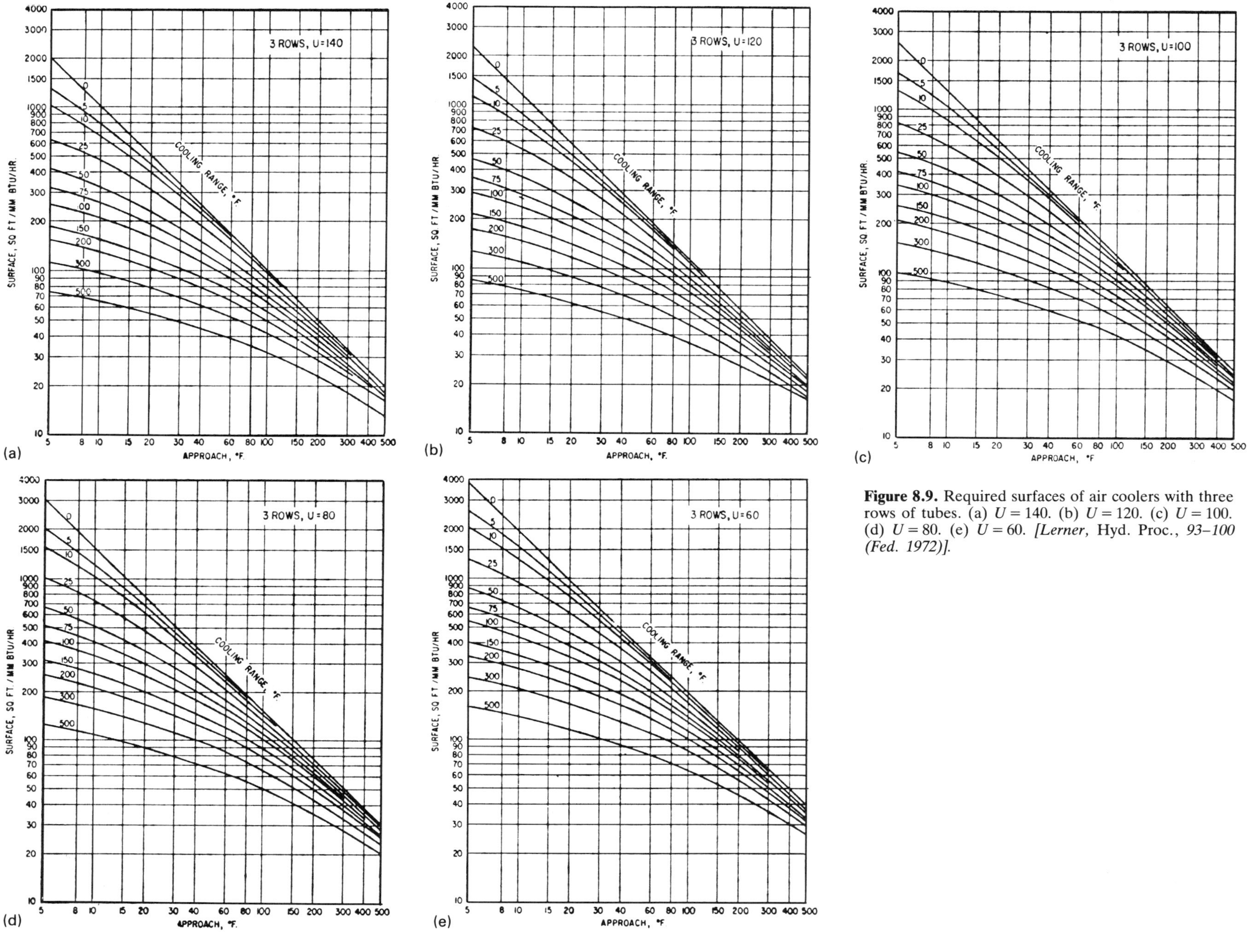

Figure 8.9. Required surfaces of air coolers with three rows of tubes. (a) $U = 140$. (b) $U = 120$. (c) $U = 100$. (d) $U = 80$. (e) $U = 60$. *[Lerner,* Hyd. Proc., *93–100 (Fed. 1972)].*

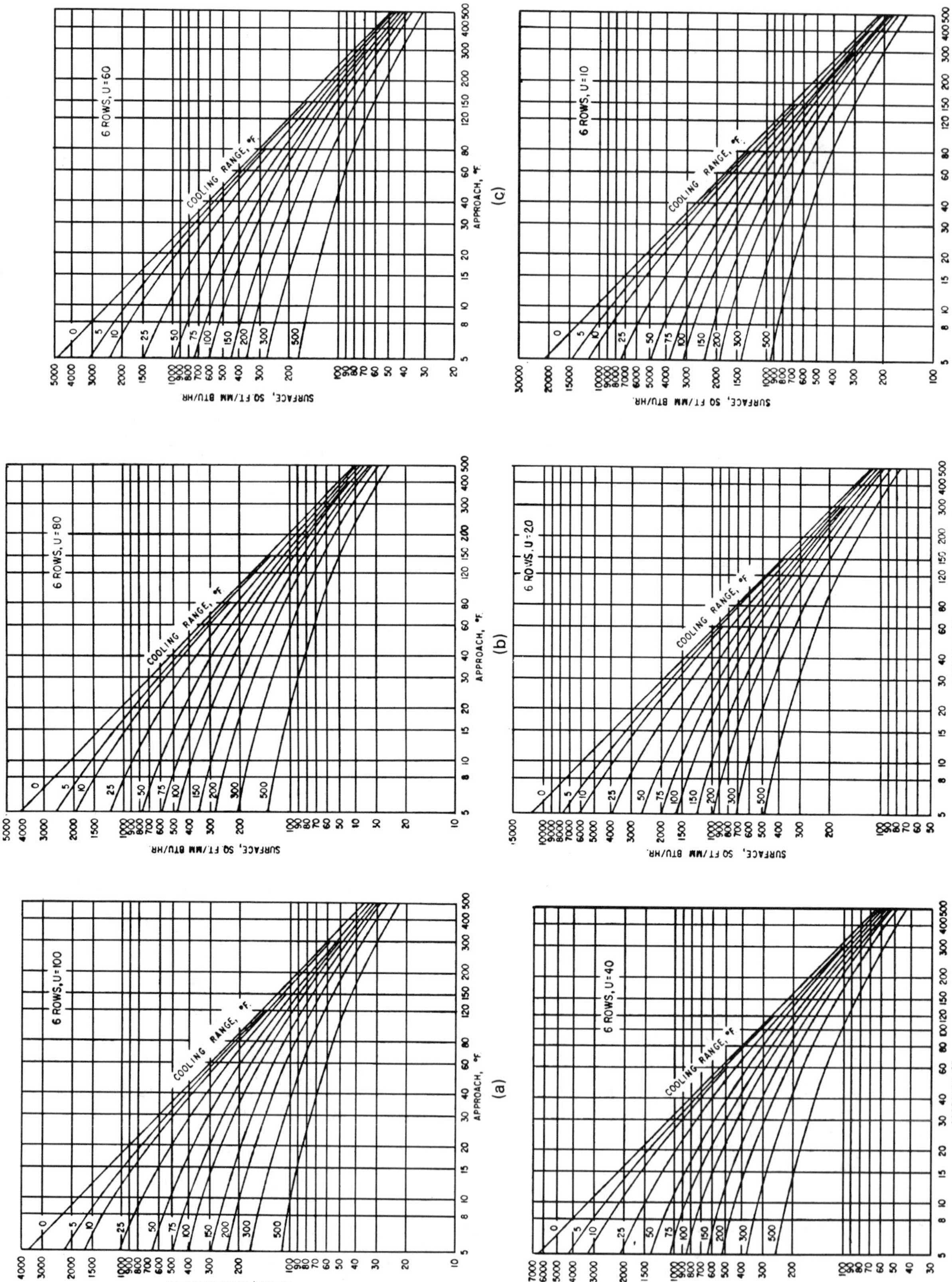

Figure 8.10. Required surfaces of air coolers with six rows of tubes. (a) $U = 100$ Btu/(hr)(sqft)(°F). (b) $U = 80$. (c) $U = 60$. (d) $U = 40$. (e) $U = 20$. (f) $U = 10$. [*Lerner*, Hyd. Proc., *93–100 (Feb. 1972)*].

Example 8.9
Estimation of the Surface Requirements of an Air Cooler
An oil is to be cooled from 300 to 150°F with ambient air at 90°F, with a total duty of 20 MBtu/hr. The tubes have 5/8 in. fins on 1 in. OD and 2–5/16 in. triangular spacing. The tube surface is given by

$A = 1.33NWL$, sqft of bare tube surface,
N = number of rows of tubes, from 3 to 6,
W = width of tube bank, ft,
L = length of tubes, ft.

According to the data of Table 8.12, the overall coefficient may be taken as $U = 60$ Btu/(hr)(°F)(sqft of bare tube surface). Exchangers with 3 rows and with 6 rows will be examined.

Approach = 150 − 90 = 60°F,
Cooling range = 300 − 150 = 150°F,

From Figure 8.9(f), 3 rows,

$A = 160$ sqft/MBtu/hr)
→ 160(20) = 3200 sqft
= 1.33(3)WL.

When $W = 16$ ft, $L = 50$ ft.

Two fans will make the ratio of section length to width, 25/16 = 1.56 which is less than the max allowable of 1.8. At 7.5 HP/100 sqft,

$$\text{Power} = \frac{16(50)}{100}7.5 = 60\ \text{HP}.$$

From Figure 8.10(c), 6 rows,

$A = 185$ sqft/(MBtu/hr)
→ 185(20) = 3700 sqft.
= 1.33(6)WL.

When $W = 16$ ft, $L = 29$ ft.

Since $L/W = 1.81$, one fan is marginal and two should be used:

$$\text{Power} = [16(29)/100]7.5 = 34.8\ \text{HP}.$$

The 6-row construction has more tube surface but takes less power and less space.

tubes. Figure 8.4(c) shows a typical construction and flow paths. The versatility and widespread use of this equipment has given rise to the development of industrywide standards of which the most widely observed are the TEMA standards. Classifications of equipment and terminology of these standards are summarized on Figure 8.11.

Baffle pitch, or distance between baffles, normally is 0.2–1.0 times the inside diameter of the shell. Both the heat transfer coefficient and the pressure drop depend on the baffle pitch, so that its selection is part of the optimization of the heat exchanger. The window of segmental baffles commonly is about 25%, but it also is a parameter in the thermal-hydraulic design of the equipment.

In order to simplify external piping, exchangers mostly are built with even numbers of tube passes. Figure 8.12(c) shows some possible arrangements, where the full lines represent partitions in one head of the exchanger and the dashed lines partitions in the opposite head. Partitioning reduces the number of tubes that can be accommodated in a shell of a given size. Table 8.12 is of such data. Square tube pitch in comparison with triangular pitch accommodates fewer tubes but is preferable when the shell side must be cleaned by brushing.

Two shell passes are obtained with a longitudinal baffle, type *F* in Figures 8.11(a) or 8.3(c). More than two shell passes normally are not provided in a single shell, but a 4–8 arrangement is thermally equivalent to two 2–4 shells in series, and higher combinations are obtained with more shells in series.

ADVANTAGES

A wide range of design alternates and operating conditions is obtainable with shell-and-tube exchangers, in particular:

- Single phases, condensation or boiling can be accommodated in either the tubes or the shell, in vertical or horizontal positions.
- Pressure range and pressure drop are virtually unlimited, and can be adjusted independently for the two fluids.
- Thermal stresses can be accommodated inexpensively.
- A great variety of materials of construction can be used and may be different for the shell and tubes.
- Extended surfaces for improved heat transfer can be used on either side.
- A great range of thermal capacities is obtainable.
- The equipment is readily dismantled for cleaning or repair.

TUBE SIDE OR SHELL SIDE

Several considerations may influence which fluid goes on the tube side or the shell side.

The tube side is preferable for the fluid that has the higher pressure, or the higher temperature or is more corrosive. The tube side is less likely to leak expensive or hazardous fluids and is more easily cleaned. Both pressure drop and laminar heat transfer can be predicted more accurately for the tube side. Accordingly, when these factors are critical, the tube side should be selected for that fluid.

Turbulent flow is obtained at lower Reynolds numbers on the shell sidc, so that the fluid with the lower mass flow preferably goes on that side. High Reynolds numbers are obtained by multipassing the tube side, but at a price.

DESIGN OF A HEAT EXCHANGER

A substantial number of parameters is involved in the design of a shell-and-tube heat exchanger for specified thermal and hydraulic conditions and desired economics, including: tube diameter, thickness, length, number of passes, pitch, square or triangular; size of shell, number of shell baffles, baffle type, baffle windows, baffle spacing, and so on. For even a modest sized design program, Bell (in HEDH, 1983, 3.1.3) estimates that 40 separate logical designs may need to be made which lead to $2^{40} = 1.10 \times 10^{12}$ different paths through the logic. Since such a number is entirely too large for normal computer processing, the problem must be simplified with

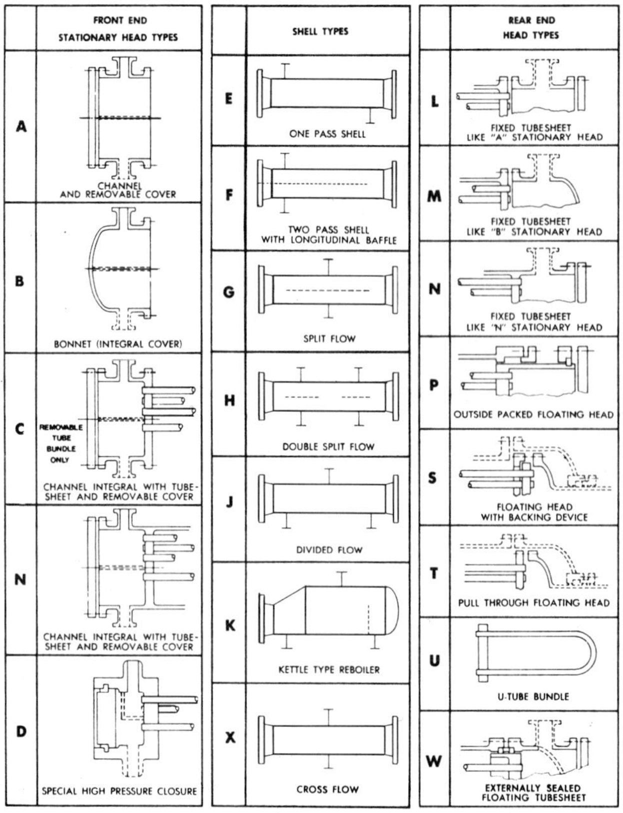

(a)

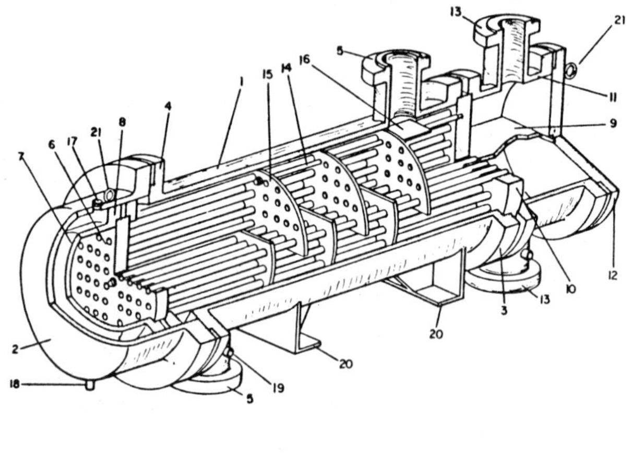

1. SHELL
2. SHELL COVER
3. SHELL CHANNEL
4. SHELL COVER END FLANGE
5. SHELL NOZZLE
6. FLOATING TUBESHEET
7. FLOATING HEAD
8. FLOATING HEAD FLANGE
9. CHANNEL PARTITION
10. STATIONARY TUBESHEET
11. CHANNEL
12. CHANNEL COVER
13. CHANNEL NOZZLE
14. TIE RODS AND SPACERS
15. TRANSVERSE BAFFLES OR SUPPORT PLATES
16. IMPINGEMENT BAFFLE
17. VENT CONNECTION
18. DRAIN CONNECTION
19. TEST CONNECTION
20. SUPPORT SADDLES
21. LIFTING RING

(b)

Figure 8.11. Tubular Exchanger Manufacturers Association classification and terminology for heat exchangers. (a) TEMA terminology for shells and heads of heat exchangers. (b) Terminology for parts of a TEMA type AES heat exchanger. The three letters A, E, and S come from part (a).

some arbitrary decisions based on as much current practice as possible.

A logic diagram of a heat exchanger design procedure appears in Figure 8.13. The key elements are:

1. Selection of a tentative set of design parameters, Box 3 of Figure 8.13(a).
2. Rating of the tentative design, Figure 8.13(b), which means evaluating the performance with the best correlations and calculation methods that are feasible.
3. Modification of some design parameters, Figure 8.13(c), then rerating the design to meet thermal and hydraulic specifications and economic requirements.

A procedure for a tentative selection of exchanger will be described following. With the exercise of some judgement, it is feasible to perform simpler exchanger ratings by hand, but the present state of the art utilizes computer rating, with in-house programs, or those of HTRI or HTFS, or those of commercial services. More than 50 detailed numerical by hand rating examples are in the book of Kern (1950) and several comprehensive ones in the book of Ganapathy (1982).

TENTATIVE DESIGN

The stepwise procedure includes statements of some rules based on common practice.

1. Specify the flow rates, terminal temperatures and physical properties.
2. Calculate the LMTD and the temperature correction factor F from Table 8.3 or Figure 8.5.
3. Choose the simplest combination of shell and tube passes or number of shells in series that will have a value of F above 0.8 or so. The basic shell is 1–2, one shell pass and two tube passes.
4. Make an estimate of the overall heat transfer coefficient from Tables 8.4–8.7.
5. Choose a tube length, normally 8, 12, 16, or 20 ft. The 8 ft long exchanger costs about 1.4 times as much as the 20 ft one per unit of surface.
6. Standard exchanger tube diameters are 0.75 or 1 in. OD, with pitches shown in Table 8.13.
7. Find a shell diameter from Table 8.13 corresponding to the selections of tube diameter, length, pitch, and number of passes made thus far for the required surface. As a guide, many heat exchangers have length to shell diameter ratios between 6 and 8.
8. Select the kinds and number of baffles on the shell side.

The tentative exchanger design now is ready for detailed evaluation with the best feasible heat transfer and pressure drop data. The results of such a rating will suggest what changes may be needed to satisfy the thermal, hydraulic, and economic requirements for the equipment. Example 8.10 goes through the main part of such a design.

8.8. CONDENSERS

Condensation may be performed inside or outside tubes, in horizontal or vertical positions. In addition to the statements made in the previous section about the merits of tube side or shell side: When freezing can occur, shell side is preferable because it is less likely to clog. When condensing mixtures whose lighter components are soluble in the condensate, tube side should be adopted since drainage is less complete and allows condensation (and dissolution) to occur at higher temperatures. Venting of noncondensables is more positive from tube side.

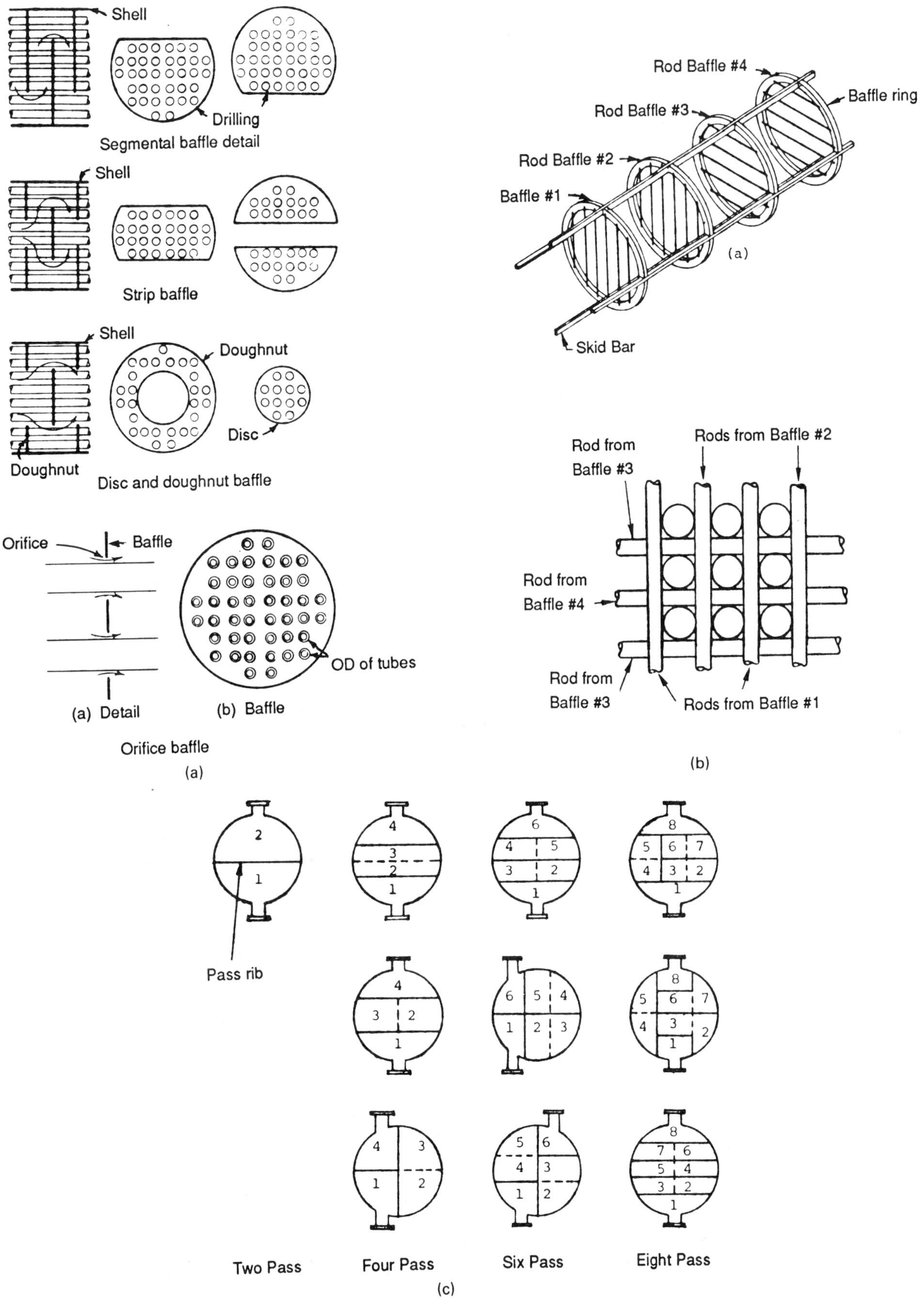

Figure 8.12. Arrangements of cross baffles and tube-side passes. (a) Types of cross baffles. (b) Rod baffles for minimizing tube vibrations; each tube is supported by four rods. (c) Tube-side multipass arrangements.

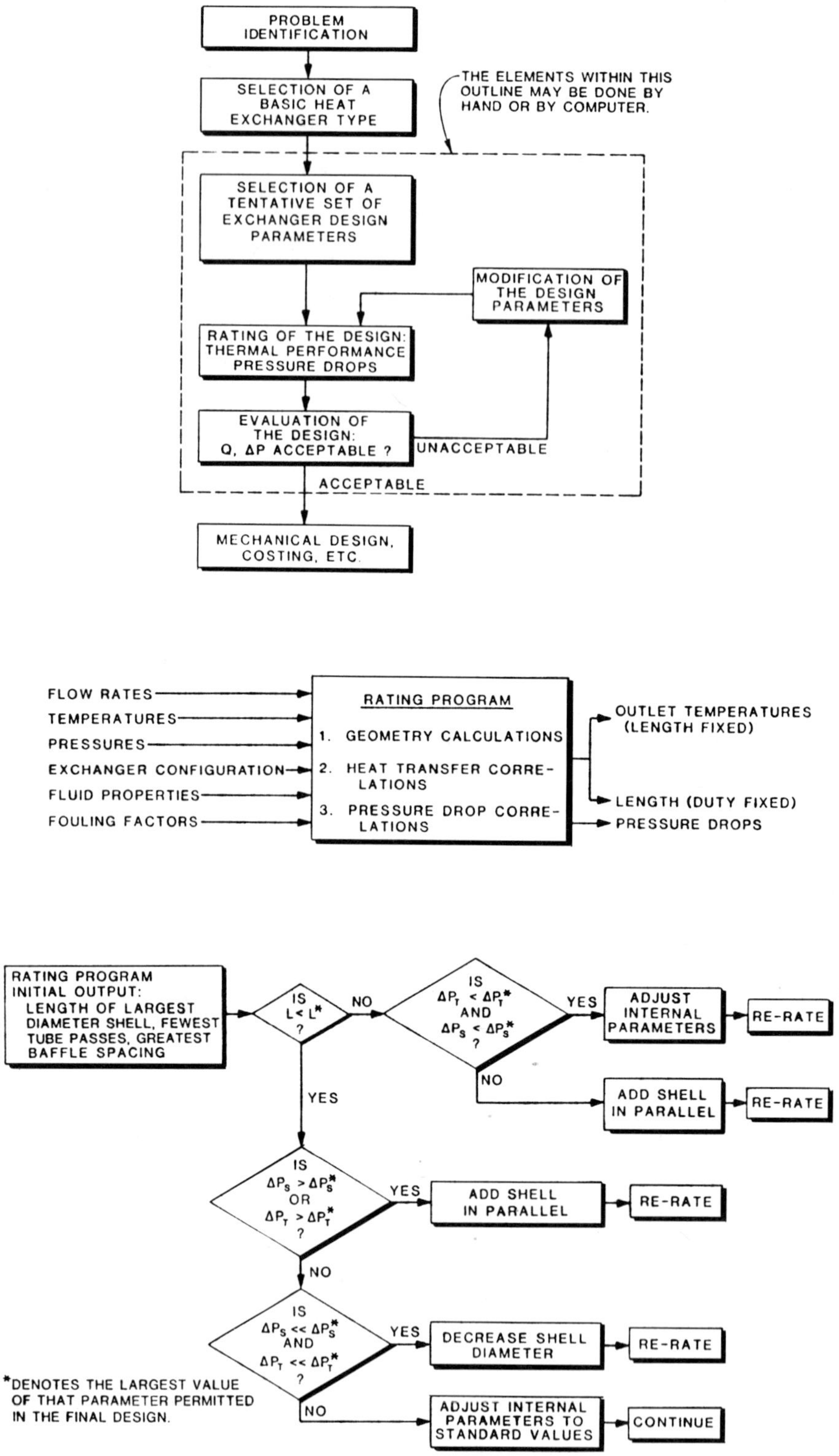

Figure 8.13. A procedure for the design of a heat exchanger, comprising a tentative selection of design parameters, rating of the performance, modification of this design if necessary, and re-rating to meet specifications (*see also Bell, in* Heat Exchanger Design Handbook, *Section 3.1.3, Hemisphere Publishing Company, 1983*).

TABLE 8.13. Tube Counts of Shell-and-Tube Heat Exchangers[a]

Heat Exchanger Tube Sheet Layout Count Table

37	35	33	31	29	27	25	23¼	21¼	19¼	17¼	15¼	13¼	12	10	8	I.D. of Shell (In.)		
1269	1143	1019	881	763	663	553	481	391	307	247	193	135	105	69	33	¾" on 1⁵⁄₁₆" Δ	Fixed Tubes	One-Pass
1127	1007	889	765	667	577	493	423	343	277	217	157	117	91	57	33	¾" on 1" Δ		
965	865	765	665	587	495	419	355	287	235	183	139	101	85	53	33	¾" on 1" □		
699	633	551	481	427	361	307	247	205	163	133	103	73	57	33	15	1" on 1¼" Δ		
595	545	477	413	359	303	255	215	179	139	111	83	65	45	33	17	1" on 1¼" □		
1242	1088	964	846	734	626	528	452	370	300	228	166	124	94	58	32	¾" on 1⁵⁄₁₆" Δ	Fixed Tubes	Two-Pass
1088	972	858	746	646	556	468	398	326	264	208	154	110	90	56	28	¾" on 1" Δ		
946	840	746	644	560	486	408	346	280	222	172	126	94	78	48	26	¾" on 1" □		
688	608	530	462	410	346	292	244	204	162	126	92	62	52	32	16	1" on 1¼" Δ		
584	522	460	402	348	298	248	218	172	136	106	76	56	40	26	12	1" on 1¼" □		
1126	1008	882	768	648	558	460	398	304	234	180	134	94	64	34	8	¾" on 1⁵⁄₁₆" Δ	U Tubes[2]	
1000	882	772	674	566	484	406	336	270	212	158	108	72	60	26	8	¾" on 1" Δ		
884	778	688	586	506	436	362	304	242	188	142	100	72	52	30	12	¾" on 1" □		
610	532	466	396	340	284	234	192	154	120	84	58	42	26	8	XX	1" on 1¼" Δ		
526	464	406	356	304	256	214	180	134	100	76	58	38	22	12	XX	1" on 1¼" □		
1172	1024	904	788	680	576	484	412	332	266	196	154	108	84	48	XX	¾" on 1⁵⁄₁₆" Δ	Fixed Tubes	Four-Pass
1024	912	802	692	596	508	424	360	292	232	180	134	96	72	44	XX	¾" on 1" Δ		
880	778	688	590	510	440	366	308	242	192	142	126	88	72	48	XX	¾" on 1" □		
638	560	486	422	368	308	258	212	176	138	104	78	60	44	24	XX	1" on 1¼" Δ		
534	476	414	360	310	260	214	188	142	110	84	74	48	40	24	XX	1" on 1¼" □		
1092	976	852	740	622	534	438	378	286	218	166	122	84	56	28	XX	¾" on 1⁵⁄₁₆" Δ	U Tubes[2]	
968	852	744	648	542	462	386	318	254	198	146	98	64	52	20	XX	¾" on 1" Δ		
852	748	660	560	482	414	342	286	226	174	130	90	64	44	24	XX	¾" on 1" □		
584	508	444	376	322	266	218	178	142	110	74	50	36	20	XX	XX	1" on 1¼" Δ		
500	440	384	336	286	238	198	166	122	90	66	50	32	16	XX	XX	1" on 1¼" □		
1106	964	844	732	632	532	440	372	294	230	174	116	80	XX	XX	XX	¾" on 1⁵⁄₁₆" Δ	Fixed Tubes	Six-Pass
964	852	744	640	548	464	388	322	258	202	156	104	66	XX	XX	XX	¾" on 1" Δ		
818	224	634	536	460	394	324	266	212	158	116	78	54	XX	XX	XX	¾" on 1" □		
586	514	442	382	338	274	226	182	150	112	82	56	34	XX	XX	XX	1" on 1¼" Δ		
484	430	368	318	268	226	184	154	116	88	66	44	XX	XX	XX	XX	1" on 1¼" □		
1058	944	826	716	596	510	416	358	272	206	156	110	74	XX	XX	XX	¾" on 1⁵⁄₁₆" Δ	U Tubes[2]	
940	826	720	626	518	440	366	300	238	184	134	88	56	XX	XX	XX	¾" on 1" Δ		
820	718	632	534	458	392	322	268	210	160	118	80	56	XX	XX	XX	¾" on 1" □		
562	488	426	356	304	252	206	168	130	100	68	42	30	XX	XX	XX	1" on 1¼" Δ		
478	420	362	316	268	224	182	152	110	80	60	42	XX	XX	XX	XX	1" on 1¼" □		
1040	902	790	682	576	484	398	332	258	198	140	94	XX	XX	XX	XX	¾" on 1⁵⁄₁₆" Δ	Fixed Tubes	Eight-Pass
902	798	694	588	496	422	344	286	224	170	124	82	XX	XX	XX	XX	¾" on 1" Δ		
760	662	576	490	414	352	286	228	174	132	94	XX	XX	XX	XX	XX	¾" on 1" □		
542	466	400	342	298	240	190	154	120	90	66	XX	XX	XX	XX	XX	1" on 1¼" Δ		
438	388	334	280	230	192	150	128	94	74	XX	XX	XX	XX	XX	XX	1" on 1¼" □		
1032	916	796	688	578	490	398	342	254	190	142	102	68	XX	XX	XX	¾" on 1⁵⁄₁₆" Δ	U Tubes[2]	
908	796	692	600	498	422	350	286	226	170	122	82	52	XX	XX	XX	¾" on 1" Δ		
792	692	608	512	438	374	306	254	194	146	106	70	48	XX	XX	XX	¾" on 1" □		
540	464	404	340	290	238	190	154	118	90	58	38	24	XX	XX	XX	1" on 1¼" Δ		
456	396	344	300	254	206	170	142	98	70	50	34	XX	XX	XX	XX	1" on 1¼" □		
37	**35**	**33**	**31**	**29**	**27**	**25**	**23¼**	**21¼**	**19¼**	**17¼**	**15¼**	**13¼**	**12**	**10**	**8**	**I.D. of Shell (in.)**		

[1] Allowance made for Tie Rods.
[2] R.O.B. = $2\frac{1}{2} \times$ Tube Dia. Actual Number of "U" Tubes is one-half the above figures.
[a] A 3/4 in. tube has 0.1963 sqft/ft, a 1 in. OD has 0.2618 sqft/ft. Allowance made for tie rods.
[b] R.O.B. = $2\frac{1}{2} \times$ tube dia. Actual number of "U" tubes is one-half the above figures.

EXAMPLE 8.10
Process Design of a Shell-and-Tube Heat Exchanger

An oil at the rate of 490,000 lb/hr is to be heated from 100 to 170°F with 145,000 lb/hr of kerosene initially at 390°F. Physical properties are

Oil	0.85 sp gr, 3.5 cP at 135°F, 0.49 sp ht
Kerosene	0.82 sp gr, 0.4 cP at 200°F, 0.61 sp ht

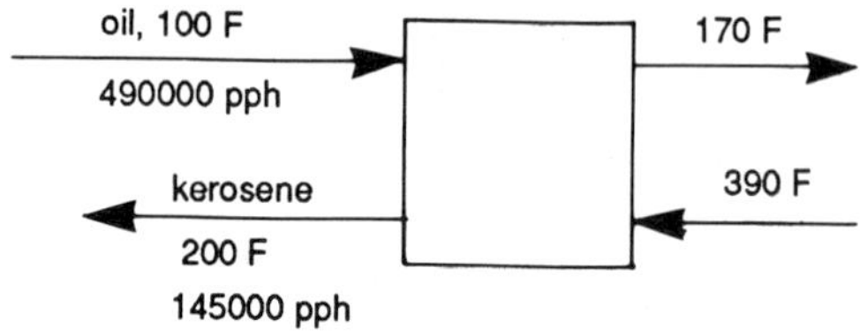

Kerosene outlet:

$$T = 390 - (490{,}000/145{,}000)(0.49/0.61)(170 - 100)$$
$$= 200°F,$$
$$\text{LMTD} = (220 - 100)/\ln 2.2 = 152.2,$$
$$P = (170 - 100)/(390 - 100) = 0.241,$$
$$R = (390 - 200)/(170 - 100) = 2.71.$$

From Figure 8.5(a), $F = 0.88$, so a 1–2 exchanger is satisfactory:

$$\Delta T = 152.2(0.88) = 133.9.$$

From Table 8.6, with average values for medium and heavy organics,

$$U = 10^4/(57 + 16 + 50 + 34) = 63.7,$$
$$Q = 490{,}000(0.49)(170 - 100) = 1.681(10^7)\ \text{Btu 1 hr},$$
$$A = Q/U\Delta T = 1.681(10^7)/63.7(133.9) = 1970\ \text{sqft},$$
$$1970/0.2618 = 7524.8\ \text{ft of 1 in. OD tubing}.$$

Use $1\frac{1}{4}$ in. pitch, two tube pass. From Table 8.13,

L (ft)	Required No. Tubes	D_{shell} (number of tubes) Triangular	Square
8	940	—	—
12	627	35 (608)	37 (584)
16	470	31 (462)	33 (460)
20	376	29 (410)	31 (402)

Use 16 ft tubes on $1\frac{1}{4}$ in. square pitch, two pass, 33 in. shell

$$L/D = 16/(33/12) = 5.82,$$

which is near standard practice. The 20 ft length also is acceptable but will not be taken.

The pressure drops on the tube and shell sides are to be calculated.

Tube side: 0.875 in. ID, 230 tubes, 32 ft long: Take one velocity head per inlet or outlet, for a total of 4, in addition to friction in the tubes. The oil is the larger flow so it will be placed in the tubes.

$$\dot{m} = 490{,}000/230 = 2130.4\ \text{lb/(hr)(tube)}.$$

Use formulas from Table 6.1

$$\text{Re} = 6.314(2130.4)/0.875(3.5) = 4392,$$
$$f = 1.6364/[\ln(5(10^{-7})/0.875 + 6.5/4392)]^2 = 0.0385,$$
$$\Delta P_f = 5.385(10^{-8})(2130)^2(32)(0.0385)/0.85(0.875)^5$$
$$= 0.691\ \text{psi}.$$

Expansion and contraction:

$$\Delta P_e = 4\rho(u^2/2g_e) = 4(53.04)(3.26)^2/(64.4)(144) = 0.243\ \text{psi},$$
$$\therefore \Delta P_{\text{tube}} = 0.691 + 0.243 = 0.934\ \text{psi}.$$

Shellside. Follow Example 8.8:

$$D_h = 1.2732(1.25/12)^2/(1/12) - 1/12 = 0.0824\ \text{ft},$$
$$B = 1.25\ \text{ft between baffles},$$
$$E = 0.25/12\ \text{ft between tubes},$$
$$D_s = 33/12 = 2.75\ \text{ft shell diameter},$$
$$A_s = 2.75(1.25)(0.25/12)/(1.25/12) = 0.6875\ \text{sqft},$$
$$G_s = 145{,}000/0.6875 = 210{,}909\ \text{lb/(hr)(sqft)},$$
$$\text{Re} = 0.0824(210{,}909)/0.4(2.42) = 17{,}952,$$
$$f = 0.0121(17{,}952)^{-0.19} = 0.00188,$$
$$\Delta P_{\text{shell}} = 0.00188(210{,}909)^2(2.75)(13)/5.22(10^{10})(0.82)(0.0824)$$
$$= 0.85\ \text{psi}.$$

The pressure drops on each side are acceptable. Now it remains to check the heat transfer with the equations of Table 8.10 and the fouling factors of Table 8.6.

CONDENSER CONFIGURATIONS

The several possible condenser configurations will be described. They are shown on Figure 8.14.

Condensation Inside Tubes: Vertical Downflow. Tube diameters normally are 19–25 mm, and up to 50 mm to minimize critical pressure drops. The tubes remain wetted with condensate which assists in retaining light soluble components of the vapor. Venting of noncondensables is positive. At low operating pressures, larger tubes may be required to minimize pressure drop; this may have the effect of substantially increasing the required heat transfer surface. A disadvantage exists with this configuration when the coolant is fouling since the shell side is more difficult to clean.

Condensation Inside Tubes: Vertical Upflow. This mode is used primarily for refluxing purposes when return of a hot condensate is required. Such units usually function as partial condensers, with the lighter components passing on through. Reflux condensers usually are no more than 6–10 ft long with tube diameters of 25 mm or more. A possible disadvantage is the likelihood of flooding with condensate at the lower ends of the tubes.

Condensation Outside Vertical Tubes. This arrangement requires careful distribution of coolant to each tube, and requires a sump and a pump for return to a cooling tower or other source of coolant. Advantages are the high coolant side heat transfer

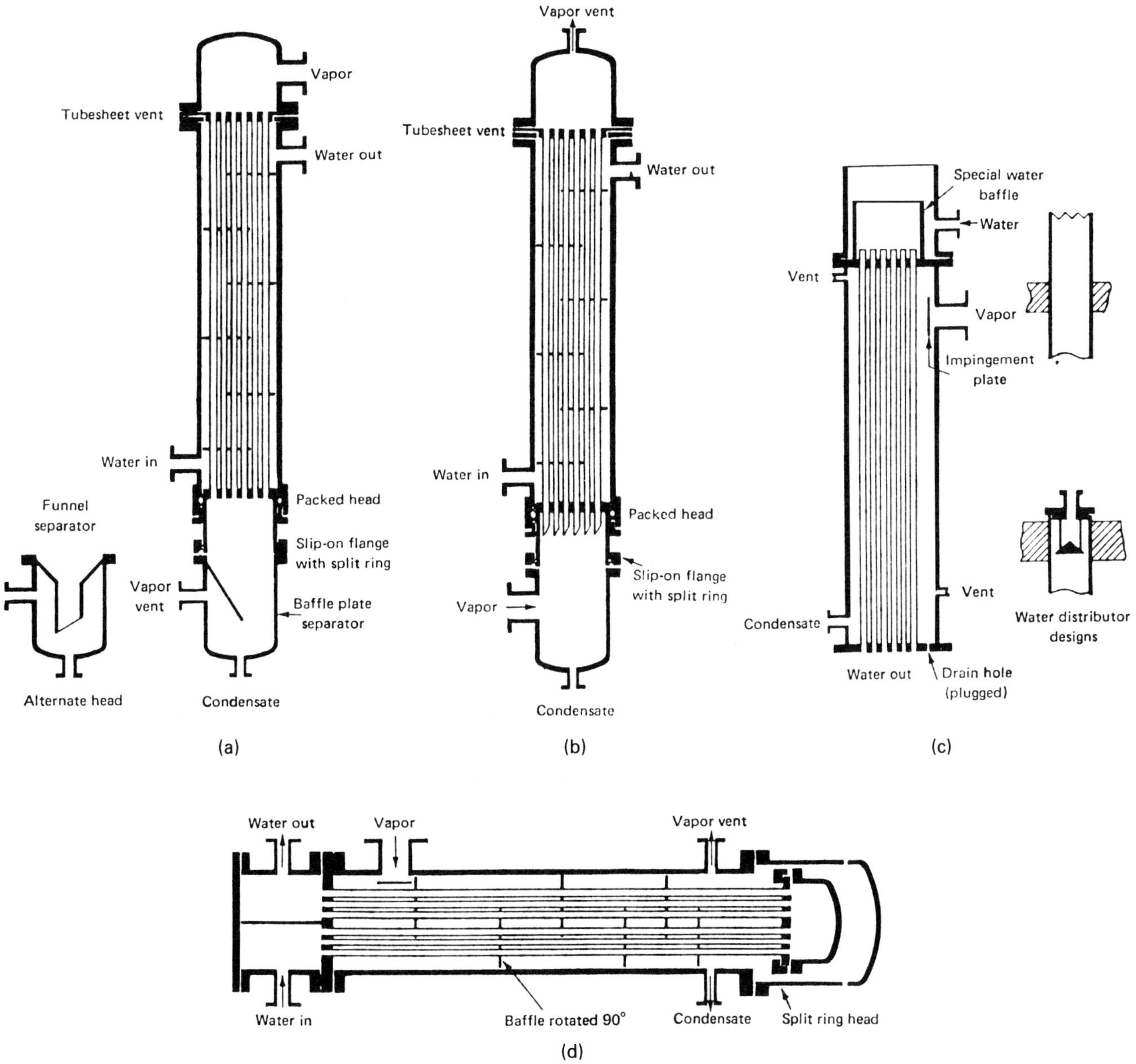

Figure 8.14. Some arrangements of shell-and-tube condensers. (a) Condensate inside tubes, vertical upflow. (b) Inside tubes, vertical downflow. (c) Outside tubes, vertical downflow. (d) Condensate outside horizontal tubes. (HEDH, *1983*, *3.4.3*).

coefficient and the ease of cleaning. The free draining of condensate is a disadvantage with wide range mixtures.

Condensation Inside Horizontal Tubes. This mode is employed chiefly in air coolers where it is the only feasible mode. As condensation proceeds, liquid tends to build up in the tubes, then slugging and oscillating flow can occur.

Condensation Outside Horizontal Tubes. Figure 8.14(d) shows a condenser with two tube passes and a shell side provided with vertically cut baffles that promote side to side flow of vapor. The tubes may be controlled partially flooded to ensure desired subcooling of the condensate or for control of upstream pressure by regulating the rate of condensation. Low-fin tubes often are advantageous, except when the surface tension of the condensates exceeds about 40 dyn/cm in which event the fins fill up with stagnant liquid. The free draining characteristic of the outsides of the tubes is a disadvantage with wide condensing range mixtures, as mentioned. Other disadvantages are those generally associated with shell side fluids, namely at high pressures or high temperatures or corrosiveness. To counteract such factors, there is ease of cleaning if the coolant is corrosive or fouling. Many cooling waters are scale forming; thus they are preferably placed on the tube side. On balance, the advantages often outweigh the disadvantages and this type of condenser is the most widely used.

DESIGN CALCULATION METHOD

Data for condensation are described in Section 8.4 and given in Tables 8.4–8.7, and a few additional overall coefficients are in Table

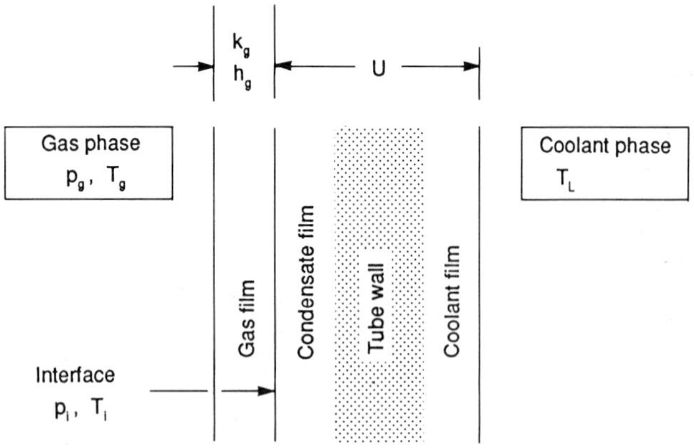

Figure 8.15. Model for partial condensation in the presence of uncondensed material: $U(T_i - T_L) = h_g(T_g - T_i) + \lambda k_g(p_g - p_i)$. [*A.P. Colburn and O.A. Hougen,* Ind. Eng. Chem. **26,** *1178–1182 (1934)*].

8.12. The calculation of condensation of pure vapors is straightforward. That of mixtures occurs over a range of temperatures and involves mass transfer resistance through a gas film as well as heat transfer resistance by liquid and fouling films. A model due to Colburn and Hougen (1934) is represented by Figure 8.15. The overall rate of heat transfer is regarded as the sum of the sensible heat transfer through a gas film and the heat of condensation of the material transferred by diffusion from the gas phase to the interface. The equation of this heat balance is, in terms of the notation of Figure 8.15,

$$U(T_i - T_L) = h_g(T_g - T_i) + \lambda k_g(p_g - p_i). \tag{8.37}$$

The temperature T_L of the coolant is related to the heat transfer Q by

$$dQ = \dot{m}_L C_L \, dT_L$$

or the integrated form

$$T_L = T_{L0} + \Delta Q/\dot{m}_L C_L. \tag{8.38}$$

A procedure will be described for taking the vapor from its initial dewpoint T_{g0} to its final dewpoint corresponding to the required amount of condensation. Gas temperatures are specified at intermediate points and the heat balance is applied over one interval at a time.

1. Prepare the condensing curve, a plot of the vapor temperature T_g against the amount of heat removed Q, by a series of isothermal flashes and enthalpy balances.
2. Starting at the inlet temperature T_{g0}, specify a temperature T_g a few degrees less, and note the heat transfer ΔQ corresponding to this temperature difference from the condensing curve.
3. Find the temperature T_L of the coolant with Eq. (8.38).
4. Assume an interfacial temperature T_i, then find the corresponding vapor pressure p_i and latent heat λ.
5. From available correlations, find values of the coefficients h_g, k_g, and U which are temperature- and composition-dependent, although they sometimes may be taken as constant over some ranges.
6. Check if these values satisfy the heat balance of Eq. (8.37). If not, repeat the process with other estimates of T_i until one is found that does satisfy the heat balance.
7. Continue with other specifications of the vapor temperature T_g, one interval at a time, until the required outlet temperature is reached.
8. The heat transfer area will be found by numerical integration of

$$A = \int_0^Q \frac{dQ}{U(T_i - T_L)}. \tag{8.39}$$

Examples of numerical applications of this method are in the original paper of Colburn and Hougen (1934), in the book of Kern (1950, p. 346) and in the book of Ludwig (1983, Vol. 3, p. 116).

The Silver–Bell–Ghaly Method

This method takes advantage of the rough proportionality between heat and mass transfer coefficients according to the Chilton–Colburn analogy, and employs only heat transfer coefficients for the process of condensation from a mixture. The sensible heat Q_{sv} of the vapor is transferred through the gas film

$$dQ_{sv} = h_g(T_g - T_i)\, dA. \tag{8.40}$$

In terms of an overall heat transfer coefficient U that does not include the gas film, the total heat transfer Q_T that is made up of the latent heat and the sensible heats of both vapor and liquid is represented by

$$dQ_T = U(T_i - T_L)\, dA. \tag{8.41}$$

When the unknown interfacial temperature T_i is eliminated and the ratio Z of sensible and total heat transfers

$$Z = dQ_{sv}/dQ_T \tag{8.42}$$

is introduced, the result is

$$dQ_T = \frac{U(T_g - T_L)}{1 + ZU/h_g}\, dA, \tag{8.43}$$

which is solved for the heat transfer area as

$$A = \int_0^{Q_T} \frac{1 + ZU/h_g}{U(T_g - T_L)}\, dQ_T. \tag{8.44}$$

Since the heat ratio Z, the temperatures and the heat transfer coefficients vary with the amount of heat transfer Q_T up to a position in the condenser, integration must be done numerically. The coolant temperature is evaluated from Eq. (8.38). Bell and Ghaly (1973) examine cases with multiple tube passes.

The basis of the method was stated by Silver (1947). A numerical solution of a condenser for mixed hydrocarbons was carried out by Webb and McNaught (in Chisholm, 1980, p. 98); comparison of the Silver–Bell–Ghaly result with a Colburn–Hougen calculation showed close agreement in this case. Bell and Ghaly (1973) claim only that their method predicts values from 0 to 100% over the correct values, always conservative. A solution with constant heat transfer coefficients is made in Example 8.11: A recent review of the subject has been presented by McNaught (in Taborek et al., 1983, p. 35).

8.9. REBOILERS

Reboilers are heat exchangers that are used primarily to provide boilup for distillation and similar towers. All types perform partial vaporization of a stream flowing under natural or forced circulation

EXAMPLE 8.11
Sizing a Condenser for a Mixture by the Silver–Bell–Ghatly Method

A mixture with initial dewpoint 139.9°C and final bubblepoint 48.4°C is to be condensed with coolant at a constant temperature of 27°C. The gas film heat transfer coefficient is 40 W/m^2 K and the overall coefficient is 450. Results of the calculation of the condensing curve are

T (°C)	139.9	121.6	103.3	85.0	66.7	48.4
Q (W)	0	2154	3403	4325	5153	5995

In the following tabulation, over each temperature interval are shown the average gas temperature, the value of Z, and the value of the integrand of Eq. (8.44). The integrand is plotted following.

Interval	1	2	3	4	5
$(T_g)_m$	130.75	112.45	94.15	75.85	57.4
Z	0.1708	0.1613	0.1303	0.0814	0.0261
Integrand × (10^5)	6.26	7.32	8.31	8.71	9.41

The heat transfer surface is the area under the stepped curve, which is $a = 0.454\ \text{m}^2$. A solution that takes into account the substantial variation of the heat transfer coefficients along the condenser gives the result $A = 0.385\ \text{m}^2$ (Webb and McNaught, in Chisholm, 1980, p. 98).

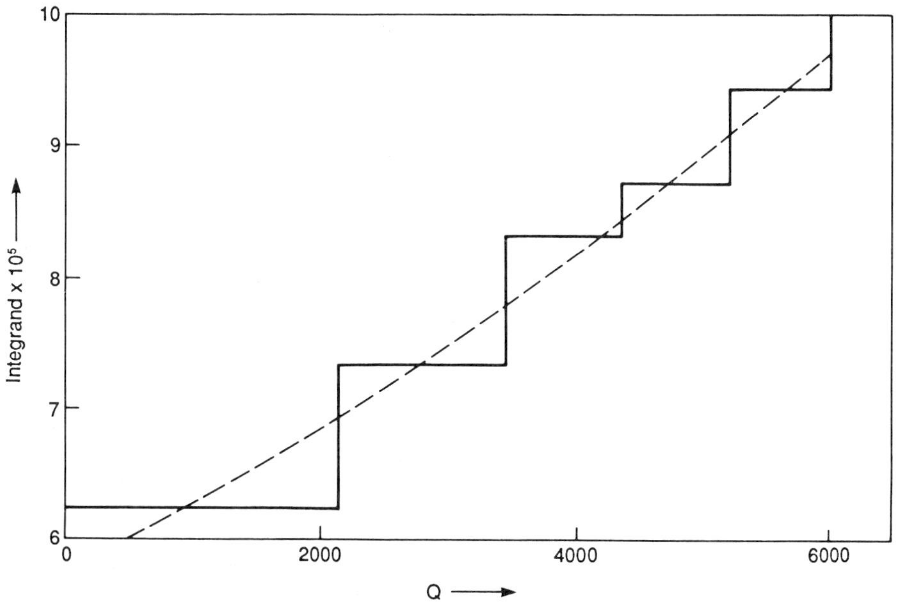

conditions. Sketches of a kettle and two types of thermosiphon reboilers are in Figure 8.4. Internal reboilers, with a tube bundle built into the tower bottom, also have some application. Flow through a vertical unit like that of Figure 8.4(f) may be forced with a pump in order to improve heat transfer of viscous or fouling materials, or when the vaporization is too low to provide enough static head difference, or when the tower skirt height is too low. A summary guide to the several types of reboilers is in Table 8.14.

KETTLE REBOILERS

Kettle reboilers consist of a bundle of tubes in an oversize shell. Submergence of the tubes is assured by an overflow weir, typically 5–15 cm higher than the topmost tubes. An open tube bundle is preferred, with pitch to diameter ratios in the range of 1.5–2. Temperature in the kettle is substantially uniform. Residence time is high so that kettles are not favored for thermally sensitive materials. The large shell diameters make kettles uneconomic for high pressure operation. Deentraining mesh pads often are incorporated. Tube bundles installed directly in the tower bottom are inexpensive but the amount of surface that can be installed is limited.

HORIZONTAL SHELL SIDE THERMOSIPHONS

The fraction vaporized in thermosiphon reboilers usually can be made less than in kettles, and the holdup is much less. Less static head difference is needed as driving force for recirculation in comparison with vertical units. Circulation rate can be controlled by throttling the inlet line. Because of the forced flow, there is a temperature gradient, from the inlet bubblepoint to the exit bubblepoint, whereas in a kettle the boiling temperature is more nearly uniform, at the exit bubblepoint. Consequently, for the same percentage vaporization, the mean temperature difference between shell and tube sides will be greater for thermosiphons than for kettles. Or for the same mean temperature difference, the percentage vaporization can be made less. Large surface requirements favor horizontal over vertical thermosiphons. Horizontal tube bundles are easier to maintain. The usual arguments for tube side versus shell side also are applicable.

VERTICAL THERMOSIPHONS

Circulation is promoted by the difference in static heads of supply liquid and the column of partially vaporized material. The exit

TABLE 8.14. A Guide to the Selection of Reboilers

	Reboiler Type			
Process Conditions	Kettle or Internal	Horizontal Shell-Side Thermosiphon	Vertical Tube-Side Thermosiphon	Forced Flow
Operating pressure				
Moderate	E	G	B	E
Near critical	B-E	R	Rd	E
Deep vacuum	B	R	Rd	E
Design ΔT				
Moderate	E	G	B	E
Large	B	R	G-Rd	E
Small (mixture)	F	F	Rd	P
Very small (pure component)	B	F	P	P
Fouling				
Clean	G	G	G	E
Moderate	Rd	G	B	E
Heavy	P	Rd	B	G
Very heavy	P	P	Rd	B
Mixture boiling range				
Pure component	G	G	G	E
Narrow	G	G	B	E
Wide	F	G	B	E
Very wide, with viscous liquid	F-P	G-Rd	P	B

[a] Category abbreviations: B, best; G, good operation; F, fair operation, but better choice is possible; Rd, risky unless carefully designed, but could be best choice in some cases; R, risky because of insufficient data; P, poor operation; E, operable but unnecessarily expensive.
(HEDH, 1983, 3.6.1).

weight fraction vaporized should be in the range of 0.1–0.35 for hydrocarbons and 0.02–0.10 for aqueous solutions. Circulation may be controlled with a valve in the supply line. The top tube sheet often is placed at the level of the liquid in the tower. The flow area of the outlet piping commonly is made the same as that of all the tubes. Tube diameters of 19–25 mm diameter are used, lengths up to 12 ft or so, but some 20 ft tubes are used. Greater tube lengths make for less ground space but necessitate taller tower skirts.

Maximum heat fluxes are lower than in kettle reboilers. Because of boiling point elevations imposed by static head, vertical thermosiphons are not suitable for low temperature difference services.

Shell side vertical thermosiphons sometimes are applied when the heating medium cannot be placed on the shell side.

FORCED CIRCULATION REBOILERS

Forced circulation reboilers may be either horizontal or vertical. Since the feed liquid is at its bubblepoint, adequate NPSH must be assured for the pump if it is a centrifugal type. Linear velocities in the tubes of 15–20 ft/sec usually are adequate. The main disadvantages are the costs of pump and power, and possibly severe maintenance. This mode of operation is a last resort with viscous or fouling materials, or when the fraction vaporized must be kept low.

CALCULATION PROCEDURES

Equations for boiling heat transfer coefficients and maximum heat fluxes are Eqs. (37) through (48) of Table 8.10. Estimating values are in Tables 8.4–8.7. Roughly, boiling coefficients for organics are 300 Btu/(hr)(sqft)(°F), or 1700 W/m^2 K; and for aqueous solutions, 1000 Btu/(hr)(sqft)(°F), or 5700 W/m^2 K. Similarly, maximum fluxes are of the order of 20,000 Btu/(hr)(sqft), or 63,000 W/m^2, for organics; and 35,000 Btu/(hr)(sqft) or 110,000 W/m^2, for aqueous systems.

The design procedure must start with a specific geometry and heat transfer surface and a specific percentage vaporization. Then the heat transfer coefficient is found, and finally the required area is calculated. When the agreement between the assumed and calculated surfaces is not close enough, the procedure is repeated with another assumed design. The calculations are long and tedious and nowadays are done by computer.

Example 8.12 summarizes the results of such calculations made on the basis of data in *Heat Exchanger Design Handbook* (1983). Procedures for the design of kettle, thermosiphon and forced circulation reboilers also are outlined by Polley (in Chisholm, 1980, Chap. 3).

8.10. EVAPORATORS

Evaporators employ heat to concentrate solutions or to recover dissolved solids by precipitating them from saturated solutions. They are reboilers with special provisions for separating liquid and vapor phases and for removal of solids when they are precipitated or crystallized out. Simple kettle-type reboilers [Fig. 8.4(d)] may be adequate in some applications, especially if enough freeboard is provided. Some of the many specialized types of evaporators that are in use are represented on Figure 8.16. The tubes may be horizontal or vertical, long or short; the liquid may be outside or inside the tubes, circulation may be natural or forced with pumps or propellers.

Natural circulation evaporators [Figs. 18.16(a)–(e)] are the most popular. The forced circulation type of Figure 18.16(f) is most versatile, for viscous and fouling services especially, but also the most expensive to buy and maintain. In the long tube vertical design, Figure 8.16(d), because of vaporization the liquid is in annular or film flow for a substantial portion of the tube length, and accordingly is called a rising film evaporator. In falling film

EXAMPLE 8.12
Comparison of Three Kinds of Reboilers for the Same Service

The service is reboiling a medium boiling range hydrocarbon mixture at 10 atm with a duty of 14,600 kW. The designs are calculated in HEDH (1983, 3.6.5) and are summarized here.

In each case a specific geometry and surface are assumed; then the heat transfer coefficients are evaluated, and the area is checked. When agreement between assumed and calculated areas is not close, another design is assumed and checked.

Of the three sets of calculations summarized here, only that for the kettle need not be repeated. Both the others should be repeated since the assumed designs are too conservative to be economical.

Quantity	Kettle	Horizontal TS	Vertical TS
Rated area (m^2)	930	930	480
Tube length (m)	6.1	6.1	4.9
Tube OD (mm)	19	19	—
Tube ID (mm)	—	—	21.2
Vaporization (%)	30	25	25
U (W/m^2 K)	674	674	928
$(\Delta T)_m$	25	44.8	44.8
Calculated area (m^2)	866	483	350
Calculated $\dot{q}$ (W/m^2)	16,859	30,227	41,174
$\dot{q}_{max}$ (W/m^2)	—	—	67,760

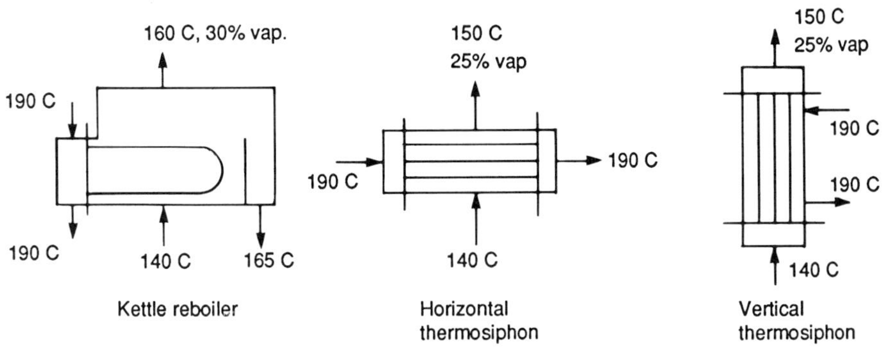

evaporators, liquid is distributed to the tops of the individual tubes and flows down as a film. The hydrostatic head is eliminated, the pressure drop is little more than the friction of the vapor flow, and heat transfer is excellent. Since the contact time is short and separation of liquid and vapor is virtually complete, falling film evaporation is suitable for thermally sensitive materials.

Long tube vertical evaporators, with either natural or forced circulation are the most widely used. Tubes range from 19 to 63 mm diameter, and 12–30 ft in length. The calandria of Figure 8.16(b) has tubes 3–5 ft long, and the central downtake has an area about equal to the cross section of the tubes. Sometimes circulation in calandrias is forced with built in propellors. In some types of evaporators, the solids are recirculated until they reach a desired size. In Figure 8.16(f), fresh feed is mixed with the circulating slurry. In Figure 8.16(g) only the clear liquid is recirculated, and small more nearly uniform crystals are formed.

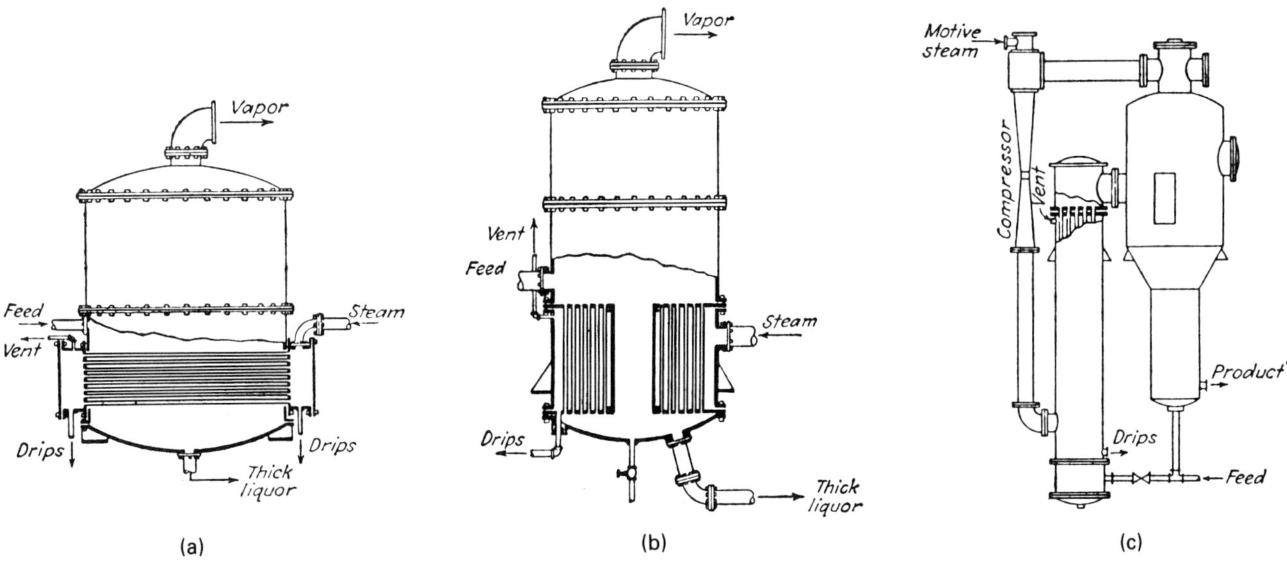

Figure 8.16. Some types of evaporators. (a) Horizontal tube. (b) Calandria type. (c) Thermocompressor evaporator. (d) Long tube vertical. (e) Falling film. (f) Forced circulation evaporator-crystallizer. (g) Three types of "Oslo/Krystal" circulating liquid evaporator-crystallizers.

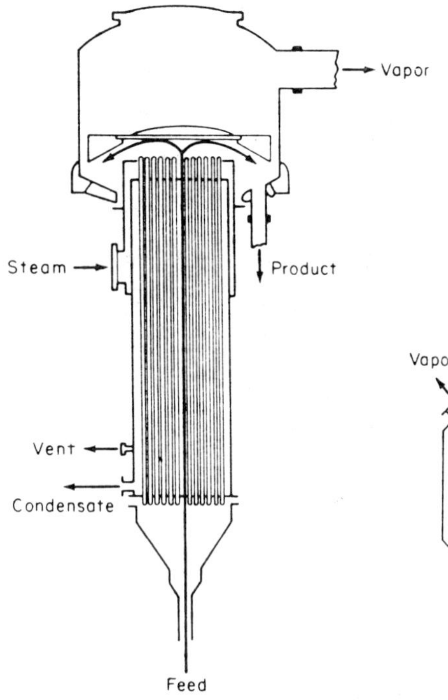

(d)

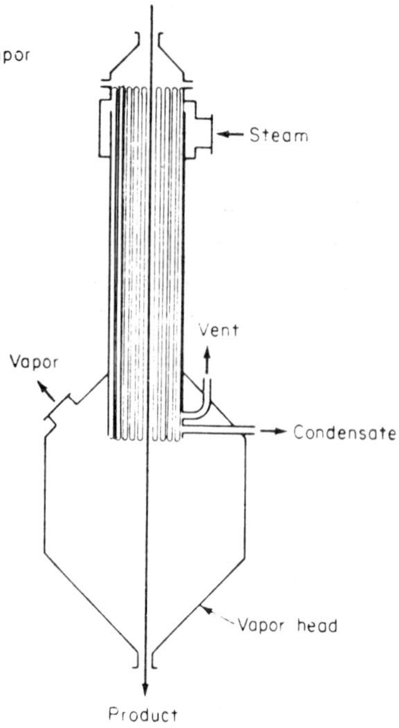

(e)

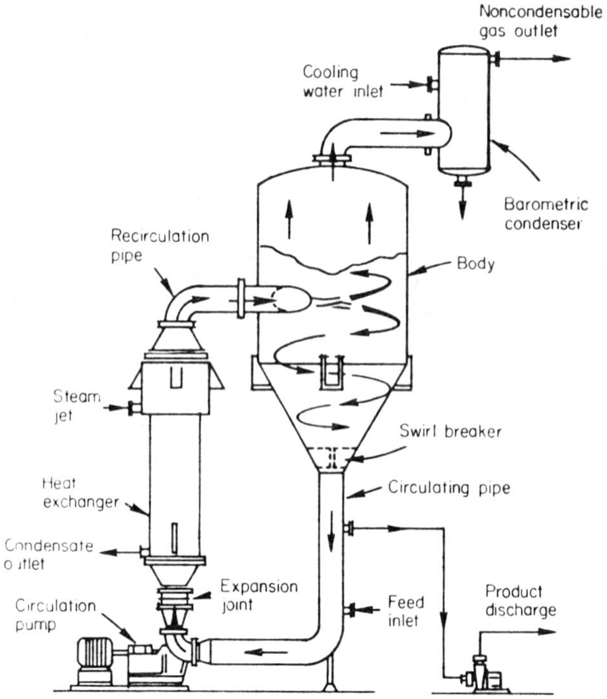

(f)

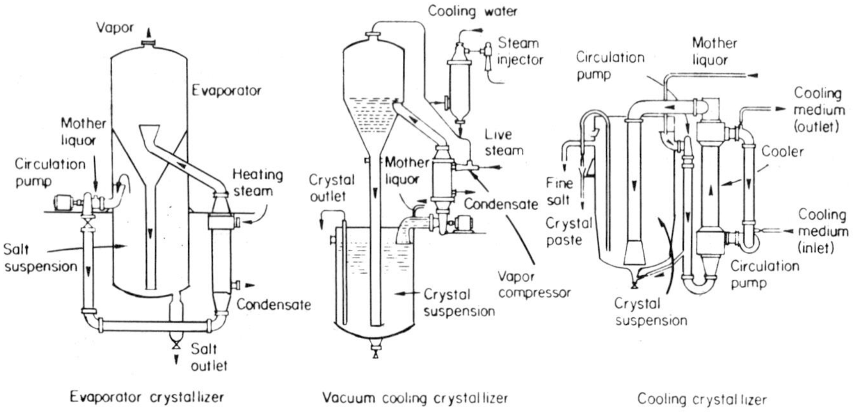

(g)

Figure 8.16—(*continued*)

THERMAL ECONOMY

Thermal economy is a major consideration in the design and operation of evaporators. This is improved by operating several vessels in series at successively lower pressures and utilizing vapors from upstream units to reboil the contents of downstream units. Figure 8.17 shows such arrangements. Thermal economy is expressed as a ratio of the amount of water evaporated in the complete unit to the amount of external steam that is supplied. For a single effect, the thermal economy is about 0.8, for two effects it is 1.6, for three effects it is 2.4, and so on. Minimum cost usually is obtained with eight or more effects. When high pressure steam is available, the pressure of the vapor can be boosted with a steam jet compressor [Fig. 8.16(c)] to a usable value; in this way savings of one-half to two-thirds in the amount of external steam can be achieved. Jet compressor thermal efficiencies are 20–30%. A possible drawback is the contamination of condensate with entrainment from the evaporator. When electricity is affordable, the pressure of the vapor can be boosted mechanically, in compressors with efficiencies of 70–75%.

Because of the elevation of boiling point by dissolved solids, the difference in temperatures of saturated vapor and boiling solution may be 3–10°F which reduces the driving force available for heat transfer. In backward feed [Fig. 8.17(b)] the more concentrated solution is heated with steam at higher pressure which makes for lesser heating surface requirements. Forward feed under the influence of pressure differences in the several vessels requires more surface but avoids the complications of operating pumps under severe conditions.

Several comprehensive examples of heat balances and surface

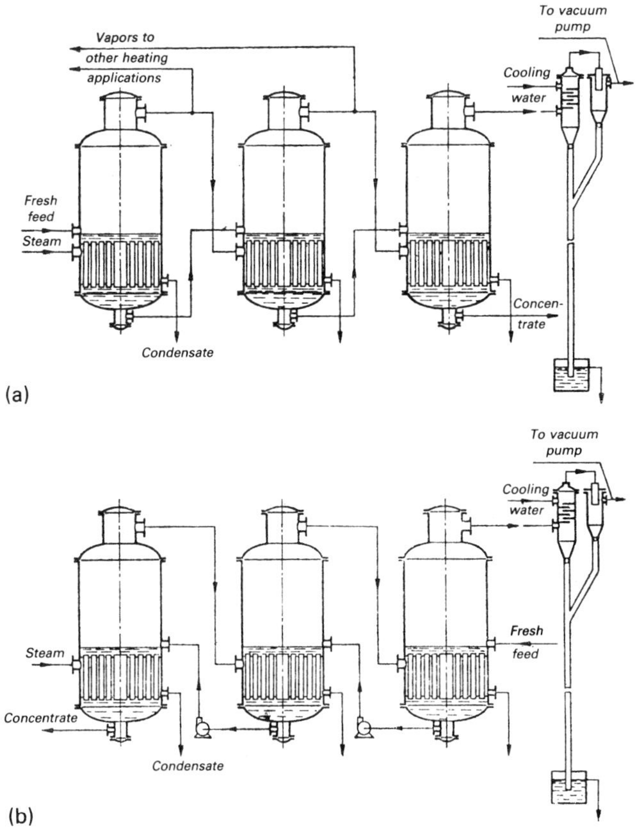

Figure 8.17. Forward and backward of liquid flow with respect to steam flow in triple-effect evaporators. (a) Forward flow of liquid by action of pressure differences in the vessels. (b) Backward-pumped flow of liquid through the vessels.

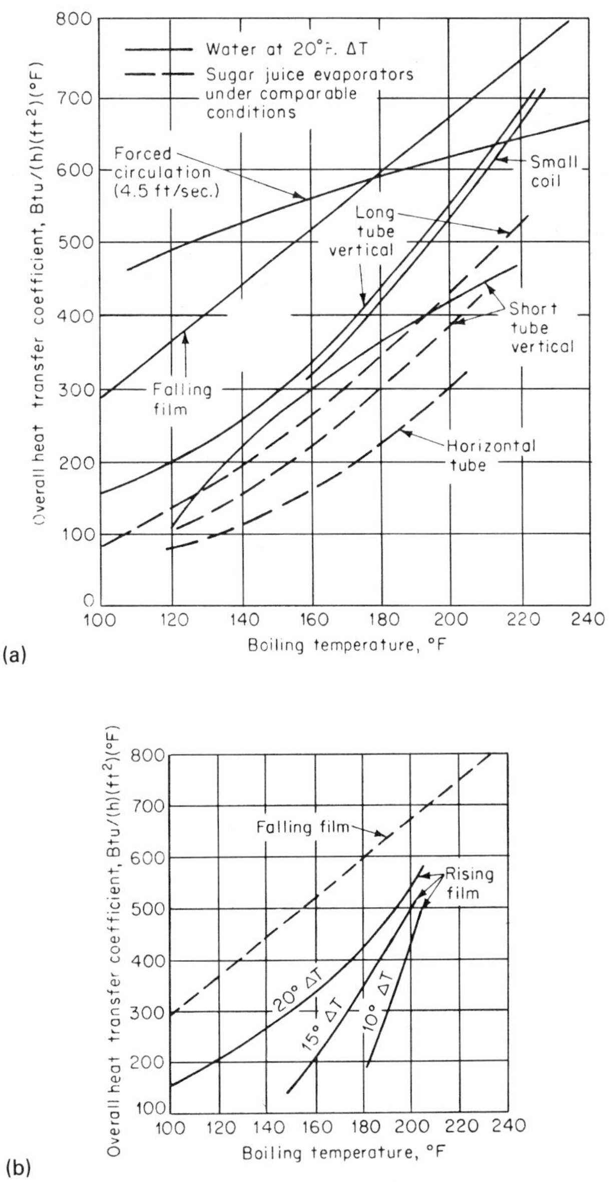

Figure 8.18. Overall heat transfer coefficients in some types of evaporations. (a) Water and sugar juice evaporators; (b) Sea water evaporators. *[F.C. Standiford,* Chem. Eng., *157–176 (9 Dec. 1963)].*

requirements of multiple effect evaporation are worked out by Kern (1950).

SURFACE REQUIREMENTS

The data of Tables 8.4–8.7 and particularly 8.10 for boiling liquids are applicable to evaporators when due regard is given the more severe fouling that can occur. For example, cases have been cited in which fouling presents fully half the resistance to heat transfer in evaporators. Some heat transfer data specifically for evaporators are in Figure 8.18. Forced circulation and falling film evaporators have the higher coefficients, and the popular long tube vertical, somewhat poorer performance.

With such data, an estimate can be made of a possible evaporator configuration for a required duty, that is, the diameter, length, and number of tubes can be specified. Then heat transfer correlations can be applied for this geometry and the surface recalculated. Comparison of the estimated and calculated surfaces will establish if another geometry must be estimated and checked. This procedure is described in Example 8.12.

8.11. FIRED HEATERS

High process temperatures are obtained by direct transfer of heat from the products of combustion of fuels. Maximum flame temperatures of hydrocarbons burned with stoichiometric air are about 3500°F. Specific data are cited by Hougen, Watson, and Ragatz (*Chemical Process Principles,* Vol. I, Wiley, New York, 1954, p. 409) and in *Marks Mechanical Engineers Handbook,* (1978, p. 4.57). With excess air to ensure complete combustion the temperatures are lower, but still adequate for the attainment of process temperatures above 2000°F when necessary. Lower temperatures are obtained with heat transfer media such as those of Table 8.2 which are in turn serviced in direct-fired heaters.

DESCRIPTION OF EQUIPMENT

In fired heaters and furnaces, heat is released by combustion of fuels into an open space and transferred to fluids inside tubes which are ranged along the walls and roof of the combustion chamber.

The heat is transferred by direct radiation and convection and also by reflection from refractory walls lining the chamber.

Three zones are identified in a typical heater such as that of Figure 8.19(a). In the *radiant zone,* heat transfer is predominantly (about 90%) by radiation. The *convection zone* is "out of sight" of the burners; although some transfer occurs by radiation because the temperature still is high enough, most of the transfer here is by convection. The application of extended surfaces permits attainment of heat fluxes per unit of bare surface comparable to those in the radiant zone. *Shield section* is the name given to the first two rows or so leading into the convection section. On balance these tubes receive approximately the same heat flux as the radiant tubes because the higher convection transfer counteracts the lesser radiation due to lack of refractory wall backing. Accordingly, shield tubes are never finned.

The usual temperature of flue gas entering the shield section is 1300–1650°F and should be 200–300°F above the process temperature at this point. The proportions of heat transferred in the radiant and convection zones can be regulated by recirculation of hot flue gases into the radiant zone, as sketched on Figure 8.19(b). Such an operation is desirable in the thermal cracking of hydrocarbons, for instance, to maintain a proper temperature profile; a negative gradient may cause condensation of polymeric products that make coke on the tubes. Multiple chambers as in

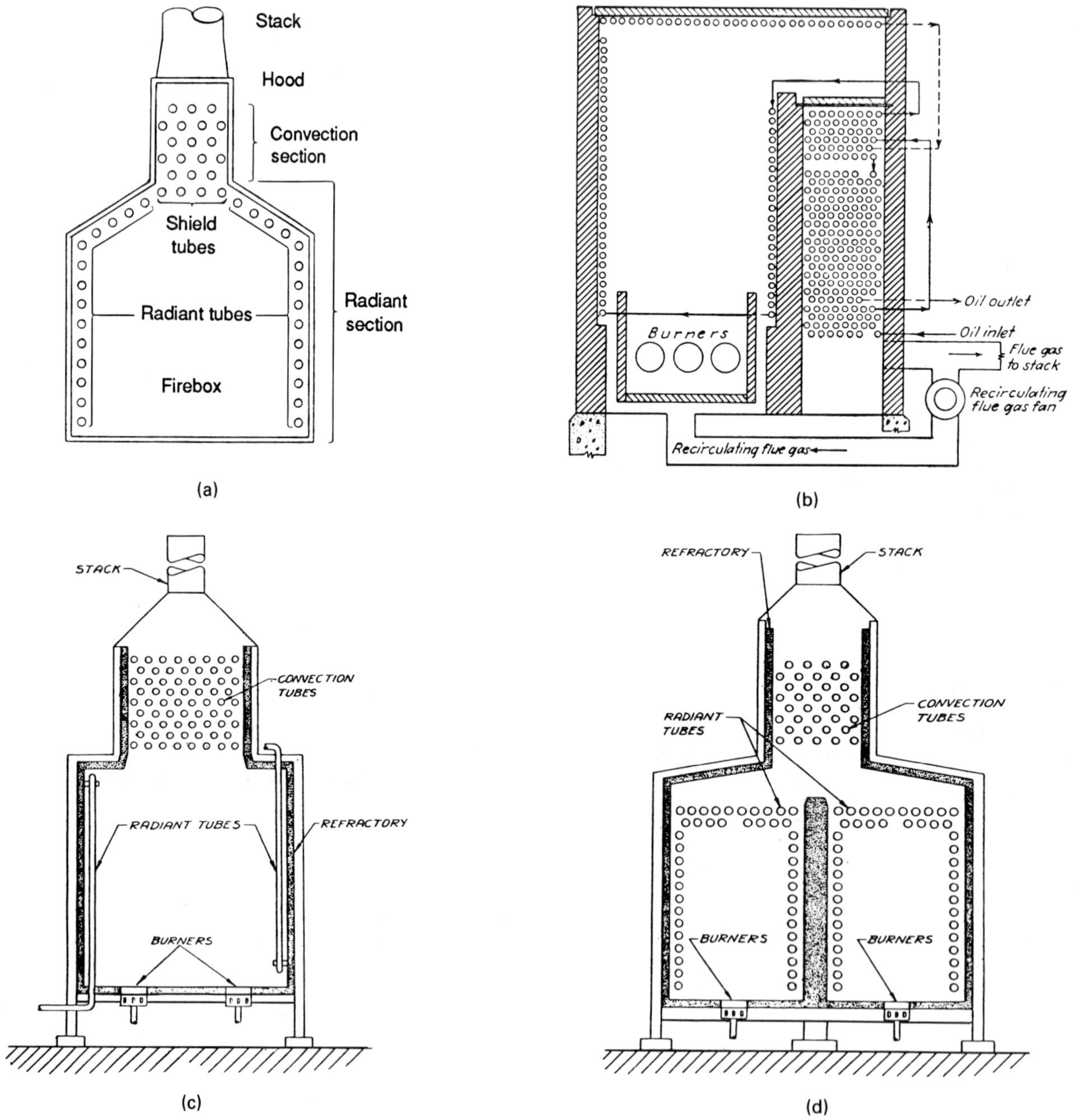

Figure 8.19. Some types of process fired heaters (See also Fig. 17.16 for a radiation panel heater). (a) Radiant, shield, and convection sections of a box-type heater. (b) Heater with a split convection section for preheating before and soaking after the radiant section (*Lobo and Evans, 1939*). (c) Vertical radiant tubes in a cylindrical shell. (d) Two radiant chambers with a common convection section.

Figure 8.19(d) also provide some flexibility. In many operations, about 75% of the heat is absorbed in the radiant zone of a fired heater.

Horizontal tube supports are made of refractory steel to withstand the high temperatures. Hangers for vertical tubes make for a less expensive construction per unit of tube surface. Furnaces are lined with shaped light weight refractory brick 5–8 in. thick. A 1 in. layer of insulating brick is placed between the lining and the metal shell.

Differences of opinion exist among designers with respect to housing shapes and tube arrangements. Nelson (*Petroleum Refinery Engineering,* McGraw-Hill, New York, 1958, p. 587), for example, describes a dozen types. The most common are cylindrical shells with vertical tubes and cabin or box types with horizontal tubes. Figures 8.19 and 17.16 are of typical constructions. Convection zones are most commonly at the top. Process fluid goes first through the convection section and usually leaves the radiant tubes at the top, particularly when vaporization occurs in them. In the more complex flow pattern of Figure 8.19(b), some of the convection tubes are used for preheat and the remainder to maintain the process fluid at a suitable reaction temperature that was attained in the radiant tubes. Some of the convection zone also may be used for steam generation or superheating or for other heat recovery services in the plant.

Capacities of 10–200 MBtu/hr can be accommodated in heaters with single radiant chambers, and three to four chambers with a common convection section are feasible. Stoichiometric combustion air requirements of typical fuels are tabulated:

Fuel	LHV (Btu/lb)	Combustion Air lb/lb	Combustion Air lb/1000 Btu
Methane	21,500	17.2	0.800
Propane	19,920	15.2	0.763
Light fuel oil	17,680	14.0	0.792
Heavy fuel oil	17,420	13.8	0.792
Anthracite	12,500	4.5	0.360

Burners may be located in the floor or on the ends of the heaters. Liquid fuels are atomized with steam or air or mechanically. A particularly effective heater design is equipped with radiant panel (surface combustion) burners, illustrated in Figure 17.16(a), (b). The incandescent walls are located 2–3 ft from the tubes. The furnace side of the panel may reach 2200°F whereas the outer side remains at 120°F because of continual cooling by the air–gas mixture. Radiant panel burners require only 2–5% excess air compared with 10–20% for conventional burners. Heaters equipped with radiant panels cost more but provide better control of temperatures of reactions such as pyrolysis of hydrocarbons to ethylene for instance.

Distances between tube banks are of the order of 20 ft or so. A rough guide to box size is about 4 cuft/sqft of radiant transfer surface, but the ultimate criterion is sufficient space to avoid impingement of flames on the tubes. Some additional notes on dimensions are stated with the design procedure of Table 8.18.

Tubes are mounted approximately one tube diameter from the refractory walls. Usual center-to-center spacing is twice the outside tube diameter. Wider spacings may be employed to lower the ratio of peak flux at the front of the tube to the average flux. For single rows of tubes, some values of these ratios are

Center-to-center/diameter	1	1.5	2	2.5	3
Max flux/avg flux	3.1	2.2	1.8	1.5	1.2

Less is gained by extending the ratio beyond 2.0. Excessive fluxes may damage the metal or result in skin temperatures that are harmful to the process fluid.

A second row of tubes on triangular spacing contributes only about 25% of the heat transfer of the front row. Accordingly, new furnaces employ only the more economical one-row construction. Second rows sometimes are justifiable on revamp of existing equipment to marginally greater duty.

HEAT TRANSFER

Performance of a heater is characterized by the average heat flux in the radiant zone and the overall thermal efficiency. Heat fluxes of representative processes are listed in Table 8.15. Higher fluxes make for a less expensive heater but can generate high skin temperatures inside and out. Thermal sensitivity of the process fluid, the strength of the metal and its resistance to corrosion at elevated temperatures are factors to be taken into account in limiting the peak flux. Because of the refractory nature of water, however, allowable fluxes in steam boilers may reach 130,000 Btu/(hr)(sqft), in comparison with a maximum of about 20,000 in hydrocarbon service. Example 8.13 is a study of the effect of tube spacing on inside film peak temperatures.

A certain amount of excess air is needed to ensure complete combustion. Typical minimum excess requirements are 10% for gaseous fuels and 15–20% for liquids. Radiant panel burners may get by with 2–5% excess air.

Efficiency is the ratio of total heat absorbed in radiant,

TABLE 8.15. Typical Radiant Fluxes and Process Temperatures

Service	Average Radiant Rate ($Btu/hr/ft^2$) (Based on OD)	Temperature (°F)
Atmospheric crude heaters	10,000–14,000	400–700
Reboilers	10,000–12,000	400–550
Circulating oil heaters	8000–11,000	600
Catalytic reformer change and reheat	7500–12,000	800–1000
Delayed coking heater	10,000–11,000	925
Visbreaker heaters—heating section	9000–10,000	700–950
Soaking section	6000–7000	950
Lube vacuum heaters	7500–8500	850
Hydrotreater and hydrocracker charge heaters	10,000	700–850
Catalytic-cracker feed heaters	10,000–11,000	900–1050
Steam superheaters	9000–13,000	700–1500
Natural gasoline plant heaters	10,000–12,000	—
Ethylene and propylene synthesis	10,000–15,000	1300–1650

EXAMPLE 8.13
Peak Temperatures

An average flux rate is 12,000 Btu/(hr)(sqft) and the inside film coefficient is 200 Btu/(hr)(sqft)(°F). At the position where the average process temperature is 850°F, the peak inside film temperature is given by $T = 850 + 12{,}000°R/200$. At the several tube spacings the peak temperatures are:

Center-to-center/diameter	1	1.5	2	2.5	3
Peak (°F)	1036	982	958	948	9.22

For heavy liquid hydrocarbons the upper limit of 950°F often is adopted.

convection, and heat recovery sections of the heater to the heat released by combustion. The released heat is based on the lower heating value of the fuel and ambient temperature. With standard burners, efficiencies may be in the range 60–80%; with radiant panels, 80–82%. Within broad limits, any specified efficiency can be attained by controlling excess air and the extent of recovery of waste heat.

An economical apportionment of heat absorption between the radiant and convection zones is about 75% in the radiant zone. This can be controlled in part by recirculation of flue gases into the radiant chamber, as shown in Figure 8.19(b).

Because of practical limitations on numbers and possible locations of burners and because of variations in process temperatures, the distribution of radiant flux in a combustion chamber is not uniform. In many cases, the effect of such nonuniformity is not important, but for sensitive and chemically reacting systems it may need to be taken into account. A method of estimating quickly a flux distribution in a heater of known configuration is illustrated by Nelson (1958, p. 610). A desired pattern can be achieved best in a long narrow heater with a multiplicity of burners, as on Figure 17.16 for instance, or with a multiplicity of chambers. A procedure for design of a plug flow heater is outlined in the *Heat Exchanger Design Handbook* (1983, 3.11.5). For most practical purposes, however, it is adequate to assume that the gas temperature and the heat flux are constant throughout the radiant chamber. Since the heat transfer is predominantly radiative and varies with the fourth power of the absolute temperature, the effect of even substantial variation in stock temperature on flux distribution is not significant. Example 8.14 studies this problem.

DESIGN OF FIRED HEATERS

The design and rating of a fired heater is a moderately complex operation. Here only the completely mixed model will be treated. For this reason and because of other generalizations, the method to be described affords only an approximation of equipment size and performance. Just what the accuracy is, it is hard to say. Even the relatively elaborate method of Lobo and Evans (1939) is able to predict actual performance only within a maximum deviation of 16%.

EXAMPLE 8.14
Effect of Stock Temperature Variation

A combustion chamber is at 2260°R, a stock enters at 1060°R and leaves at 1360°R. Accordingly, the heat fluxes at the inlet and outlet are approximately in the ratio $(2.26^4 - 1.06^4)/2.26^4 - 1.36^4) = 1.095$. The small effect of even greater variation in flux on a mild cracking operation is illustrated in Figure 8.22.

Pertinent equations and other relations are summarized in Table 8.16, and a detailed stepwise procedure is listed in Table 8.17. A specific case is worked out in detail in Example 8.15. Basically, a heater configuration and size and some aspects of the performance are assumed in advance. Then calculations are made of the heat transfer that can be realized in such equipment. Adjustments to the design are made as needed and the process calculations repeated. Details are given in the introduction to Example 8.16. Figures 8.20, 8.21, and 8.22 pertain to this example. Some of the approximations used here were developed by Wimpress (1963); his graphs were converted to equation form for convenience. Background and more accurate methods are treated notably by Lobo and Evans (1939) and more briefly by Kern (1950) and Ganapathy (1982). Charts of gas emissivity more elaborate than Figure 8.23 appear in these references.

An early relation between the heat absorption Q in a radiant zone of a heater, the heat release Q_f, the effective surface A_{cp} and the air/fuel ratio R lb/lb is due to Wilson, Lobo, and Hottel [*Ind. Eng. Chem.* **24,** 486, (1932)]:

$$Q_f/Q = 1 + (R/4200)\sqrt{Q_f/A_{cp}}. \tag{8.45}$$

Although it is a great simplification, this equation has some utility in appraising directional effects of changes in the variables. Example 8.16 considers changes in performance with changes in excess air.

Heat transfer in the radiant zone of a fired heater occurs largely by radiation from the flue gas (90% or so) but also significantly by convection. The combined effect is represented by

$$Q/A = h_r(T_g^4 - T_s^4) + h_c(T_g - T_s), \tag{8.46}$$

where T_g and T_s are absolute temperatures of the gas and the receiving surface. The radiative properties of a gas depend on its chemical nature, its concentration, and the temperature. In the thermal range, radiation of flue gas is significant only from the triatomic molecules H_2O, CO_2, and SO_2, although the amount of the last is small and usually neglected. With fuels having the composition C_xH_{2x}, the ratio of partial pressures is $p_{H_2O}/p_{CO_2} = 1$. In Figure 8.23, the emissivity of such a gas is represented as a function of temperature and the product PL of the partial pressures of water and carbon dioxide and the path of travel defined by the mean beam length. Item 8 of Table 8.16 is a curve fit of such data.

When other pertinent factors are included and an approximation is introduced for the relatively minor convection term, the heat transfer equation may be written

$$Q/\alpha A_{cp}F = 1730[(T_g/1000)^4 - (T_s/1000)^4] + 7(T_g - T_s). \tag{8.47}$$

Here the absorptivity depends on the spacing of the tubes and is given by item 5 of Table 8.16. The cold plane area A_{cp} is the product of the number of tubes by their lengths and by the center-to-center spacing. The combination αA_{cp} is equal to the area of an ideal black plane that has the same absorptivity as the tube

TABLE 8.16. Equations and Other Relations for Fired Heater Design

1. Radiant zone heat transfer

$$\frac{Q_R}{\alpha A_R F} = 1730\left[\left(\frac{T_g + 460}{1000}\right)^4 - \left(\frac{T_t + 460}{100}\right)^4\right] + 7(T_g - T_t)$$

2. Radiant zone heat balance

$$\frac{Q_R}{\alpha A_R F} = \frac{Q_n}{\alpha A_R F}\left(1 + \frac{Q_a}{Q_n} + \frac{Q_f}{Q_n} - \frac{Q_L}{Q_n} - \frac{Q_g}{Q_n}\right)$$

Q_R is the enthalpy absorbed in the radiant zone, Q_a is the enthalpy of the entering air, Q_f that of the entering fuel, Q_L is the enthalpy loss to the surroundings, Q_g is the enthalpy of the gas leaving the radiant zone; Q_a and Q_f are neglected if there is no preheat, and Q_L/Q_n is about 0.02–0.03; Q_n is the total enthalpy released in the furnace

3. Enthalpy Q_s, of the stack gas, given by the overall heat balance

$$Q_s/Q_n = 1 + (1/Q_n)(Q_a + Q_f - Q_L - Q_R - Q_{convection})$$

4. Enthalpy Q_g, of the flue gas as a function of temperature, °F

$$Q_g/Q_n = [a + b(T/1000 - 0.1)](T/1000 - 0.1)$$

z = fraction excess air

$a = 0.22048 - 0.35027z + 0.92344z^2$

$b = 0.016086 + 0.29393z - 0.48139z^2$

5. Absorptivity, α, of the tube surface with a single row of tubes

$\alpha = 1 - [0.0277 + 0.0927(x - 1)](x - 1)$

x = (center-to-center spacing)/(outside tube diameter)

6. Partial pressure of $CO_2 + H_2O$

$P = 0.288 - 0.229x + 0.090x^2$

x = fraction excess air

7. Mean beam lengths L of radiant chambers

Dimensional Ratio[a] Rectangular Furnaces	Mean Length L (ft)
1. 1-1-1 to 1-1-3 1-2-1 to 1-2-4	2/3 $\sqrt[3]{\text{furnace volume}}$, (ft^3)
2. 1-1-4 to 1-1-∞	1.0 × smallest dimension
3. 1-2-5 to 1-2-8	1.3 × smallest dimension
4. 1-3-3 to 1-∞-∞	1.8 × smallest dimension
Cylindrical Furnaces	
5. $d \times d$	2/3 diameter
6. $d \times 2d$ to $d \times \infty d$	1 × diameter

[a] Length, width, height in any order.

8. Emissivity ϕ of the gas (see also Fig. 8.20).

$\phi = a + b(PL) + c(PL)^2$

PL = product of the partial pressure (6) and the mean beam length (7)

$z = (T_g + 460)/1000$

$a = 0.47916 - 0.19847z + 0.022569z^2$

$b = 0.047029 + 0.0699z - 0.01528z^2$

$c = 0.000803 - 0.00726z + 0.001597z^2$

9. Exchange factor F

$F = a + b\phi + c\phi^2$

ϕ = gas emissivity, (8)

$z = A_w/\alpha A_R$

$a = 0.00064 + 0.0591z + 0.00101z^2$

$b = 1.0256 + 0.4908z - 0.058z^2$

$c = -0.144 - 0.552z + 0.040z^2$

TABLE 8.16—(*continued*)

10. Overall heat transfer coefficient U_c in the convection zone

$$U_c = (a + bG + cG^2)(4.5/d)^{0.25}$$

G = flue gas flow rate, lb/(sec)(sqft open cross section)
d = tube outside diameter, (in.)
$z = T_f/1000$, average outside film temperature
$a = 2.461 - 0.759z + 1.625z^2$
$b = 0.7655 + 21.373z - 9.6625z^2$
$c = 9.7938 - 30.809z + 14.333z^2$

11. Flue gas mass rate G_f

$$\frac{10^6 G_f}{Q_n} = \begin{bmatrix} 840 + 8.0x, & \text{with fuel oil} \\ 822 + 7.78x, & \text{with fuel gas} \end{bmatrix} \quad \text{lb/MBtu heat release}$$

x = fraction excess air

TABLE 8.17. Procedure for the Rating of a Fired Heater, Utilizing the Equations of Table 8.16

1. Choose a tube diameter corresponding to a cold oil velocity of 5–6 ft/sec
2. Find the ratio of center-to-center spacing to the outside tube diameter. Usually this is determined by the dimensions of available return bends, either short or long radius
3. Specify the desired thermal efficiency. This number may need modification after the corresponding numbers of tubes have been found
4. Specify the excess combustion air
5. Calculate the total heat absorbed, given the enthalpies of the inlet and outlet process streams and the heat of reaction
6. Calculate the corresponding heat release, (heat absorbed)/efficiency
7. Assume that 75% of the heat absorption occurs in the radiant zone. This may need to be modified later if the design is not entirely satisfactory
8. Specify the average radiant heat flux, which may be in the range of 8000–20,000 Btu/(hr)(sqft). This value may need modification after the calculation of Step 28 has been made
9. Find the needed tube surface area from the heat absorbed and the radiant flux. When a process-side calculation has been made, the required number of tubes will be known and will not be recalculated as stated here
10. Take a distance of about 20 ft between tube banks. A rough guide to furnace dimensions is a requirement of about 4 cuft/sqft of radiant transfer surface, but the ultimate criterion is sufficient space to avoid flame impingement
11. Choose a tube length between 30 and 60 ft or so, so as to make the box dimensions roughly comparable. The exposed length of the tube, and the inside length of the furnace shell, is 1.5 ft shorter than the actual length
12. Select the number of shield tubes between the radiant and convection zones so that the mass velocity of the flue gas will be about 0.3–0.4 lb/(sec)(sqft free cross section). Usually this will be also the number of convection tubes per row
13. The convection tubes usually are finned
14. The cold plane area is

    $$A_{cp} = \text{(exposed tube length)(center-to-center spacing)} \\ \text{(number of tubes exclusive of the shield tubes)}$$

15. The refractory area A_w is the inside surface of the shell minus the cold plane area A_{cp} of Step 14

    $$A_w = 2[W(H + L) + H \times L)] - A_{cp}$$

    where W, H, and L are the inside dimensions of the shell
16. The absorptivity α is obtained from Eq. (5) when only single rows of tubes are used. For the shield tubes, $\alpha = 1$
17. The sum of the products of the areas and the absorptivities in the radiant zone is

    $$\alpha A_R = A_{\text{shield}} + \alpha A_{cp}$$

18. For the box-shaped shell, the mean beam length L is approximated by

    $$L = \tfrac{2}{3}(\text{furnace volume})^{1/3}$$

TABLE 8.17—*(continued)*

19. The partial pressure P of $CO_2 + H_2O$ is given in terms of the excess air by Eq. (6)
20. The product PL is found with the results of Steps 18 and 19
21. The mean tube wall temperature T_t in the radiant zone is given in terms of the inlet and outlet process stream temperatures by

$$T_t = 100 + 0.5(T_1 + T_2)$$

22. The temperature T_g of the gas leaving the radiant zone is found by combining the equations of the radiant zone heat transfer [Eq. (1)] and the radiant zone heat balance [Eq. (2)]. With the approximation usually satisfactory, the equality is

$$\frac{Q_n}{\alpha A_R F}\left(1 - 0.02 - \frac{Q_g}{Q_n}\right) = 1730\left[\left(\frac{T_g + 460}{1000}\right)^4 - \left(\frac{T_t + 460}{1000}\right)^4\right] + 7(T_g - T_t)$$

The solution of this equation involves other functions of T_g, namely, the emissivity ϕ by Eq. (8), the exchange factor F by Eq. (9) and the exit enthalpy ratio Q_g/Q_n by Eq. (4)
23. The four relations cited in Step 22 are solved simultaneously by trial to find the temperature of the gas. Usually it is in the range 1500–1800°F. The Newton–Raphson method is used in the program of Table 8.18. Alternately, the result can be obtained by interpolation of a series of hand calculations
24. After T_g has been found, calculate the heat absorbed Q_R by Eq. (1)
25. Find the heat flux

$$Q/A = Q_R/A_{radiant}$$

and compare with value specified in Step 8. If there is too much disagreement, repeat the calculations with an adjusted radiant surface area
26. By heat balance over the convection zone, find the inlet and outlet temperatures of the process stream
27. The enthalpy of the flue gas is given as a function of temperature by Eq. (4). The temperature of the inlet to the convection zone was found in Step 23. The enthalpy of the stack gas is given by the heat balance [Eq. (3)], where all the terms on the right-hand side are known. Q_s/Q_n is given as a function of the stack temperature T_s by Eq. (4). That temperature is found from this equation by trial
28. The average temperature of the gas film in the convection zone is given in terms of the inlet and outlet temperatures of the process stream and the flue gas approximately by

$$T_f = 0.5\left[T_{L1} + T_{LO} + \frac{(T_{g1} - T_{L1}) - (T_s - T_{LO})}{\ln[(T_{g1} - T_{L1})/(T_s - T_{LO})]}\right]$$

The flow is countercurrent
29. Choose the spacing of the convection tubes so that the mass velocity is $G = 0.3$–0.4 lb/(sec)(sqft free cross section). Usually this spacing is the same as that of the shield tubes, but the value of G will not be the same if the tubes are finned
30. The overall heat transfer coefficient is found with Eq. (10)
31. The convection tube surface area is found by

$$A_c = Q_c/U_c \text{ (LMTD)}$$

and the total length of bare of finned tubes, as desired, by dividing A_c by the effective area per foot
32. Procedures for finding the pressure drop on the flue gas side, the draft requirements and other aspects of stack design are presented briefly by Wimpress.

[Based partly on the graphs of Wimpress, *Hydrocarbon Process.* **42**(10), 115–126 (1963)].

EXAMPLE 8.15
Design of a Fired Heater

The fuel side of a heater used for mild pyrolysis of a fuel oil will be analyzed. The flowsketch of the process is shown in Figure 8.20, and the tube arrangement finally decided upon is in Figure 8.21. Only the temperatures and enthalpies of the process fluid are pertinent to this aspect of the design, but the effect of variation of heat flux along the length of the tubes on the process temperature and conversion is shown in Figure 8.22. In this case, the substantial differences in heat flux have only a minor effect on the process performance.

Basic specifications on the process are the total heat release (102.86 MBtu/hr), overall thermal efficiency (75%), excess air (25%), the fraction of the heat release that is absorbed in the radiant section (75%), and the heat flux (10,000 Btu/(hr)(sqft).

In the present example, the estimated split of 75% and a

Example 8.15—(*continued*)

radiant rate of 10,000 lead to an initial specification of 87 tubes, but 90 were taken. The final results are quite close to the estimates, being 77.1% to the radiant zone and 9900 Btu/(hr)(sqft) with 90 tubes. If the radiant rate comes out much different from the desired value, the number of tubes is changed accordingly.

Because of the changing temperature of the process stream, the heat flux also deviates from the average value. This variation is estimated roughly from the variation of the quantity

$$\beta = 1730(T_g^4 - T_L^4) + 7.0(T_g - T_L),$$

where the gas temperature T_g, in the radiant zone is constant and T_L is the temperature of the process stream, both in °R. In comparison with the average flux, the effect is a slightly increased preheat rate and a reduced flux in the reaction zone. The inside skin temperature also can be estimated on the reasonable assumptions of heat transfer film coefficients of more than 100 before cracking starts and more than 200 at the outlet. For the conditions of this example, with $Q/A = 9900$ and $T_g = 2011$°R, these results are obtained:

T_L(°F)	β/β_{724}	h	T_{skin}(°F)
547	1.093	>100	<655
724	1	>100	<823
900	1.878	>200	<943

The equation numbers cited following are from Table 8.16. The step numbers used following are the same as those in Table 8.17:

1. Flow rate = 195,394/3600(0.9455)(62.4) = 0.9200 cfs, velocity = 5.08 fps in 6–5/8 in. OD Schedule 80 pipe.
2. Short radius return bends have 12 in. center-to-center.
3. $\eta = 0.75$.
4. Fraction excess air = 0.25.
5. From the API data book and a heat of cracking of 332 Btu/(lb gas + gasoline):

$$H_{900} = 0.9(590) + 0.08(770) + 0.02(855) = 609.6 \text{ Btu/lb},$$
$$Q_{total} = 195{,}394(609.6 - 248) + 19{,}539(332) = 77.14(\text{E}6).$$

6. Heat released:

$$Q_n = 77.14/0.75 = 102.86(\text{E}6) \text{ Btu/lb}.$$

7. Radiant heat absorption:

$$Q_R = 0.75(77.14)(\text{E}6) = 57.86(\text{E}6).$$

8. (Q/A) rad = 10,000 Btu/(hr)(sqft), average.
9. Radiant surface:

$$A = 57.86(\text{E}6)/10{,}000 = 5786 \text{ sqft}.$$

11. Tube length = 5786/1.7344 = 3336 ft; 40 foot tubes have an exposed length of 38.5 ft; $N = 3336/38.5 = 86.6$, say 92 radiant tubes.
12. From Eq. (11) the flue gas rate is

$$G_f = 102.85(1020) = 104{,}907 \text{ lb/hr}.$$

With four shield tubes, equilateral spacing and 3 in. distance to walls,

$$G = \frac{104{,}907(12)}{3600(38.5)(27.98)} = 0.325 \text{ lb/sec sqft}.$$

13. The 90 radiant tubes are arranged as shown on Figure 8.22: 4 shields, 14 at the ceiling, and 36 on each wall. Dimensions of the shell are shown.
14. $A_{cp} = (38.5)(1)(90 - 4) = 3311$ sqft.
15. Inside surface of the shell is

$$A_s = 2[20(37 + 38.5) + 37(38.5)] = 5869 \text{ sqft}.$$

Refractory surface,

$$A_w = 5869 - 3311 = 2558 \text{ sqft}.$$

16. (Center-to-center)/OD = 12/6.625 = 1.81,

$$\alpha = 0.917, \text{ single rows of tubes} \quad [\text{Eq. (5)}].$$

17. Effective absorptivity:

$$\alpha A_R = 4(38.5)(1) + 0.917(3311) = 3190 \text{ sqft},$$
$$A_w/\alpha A_r = 2558/3190 = 0.8018.$$

18. Mean beam length:

$$L = (2/3)(20 \times 37 \times 38.5)^{1/3} = 20.36.$$

19. From Eq. (6), with 25% excess air,

$$P = 0.23.$$

20. $PL = 0.23(20.36) = 4.68$ atm ft.
21. Mean tube wall temp: The stream entering the radiant section has absorbed 25% of the total heat.

$$H_1 = 248 + 0.25(77.14)(\text{E}6)/195{,}394 = 346.7,$$
$$T_1 = 565\text{°F},$$
$$T_t = 100 + (565 + 900)/2 = 832.5.$$

22–24. Input data are summarized as:

$$PL = 4.68,$$
$$D_1 = 0.8018,$$
$$D_2 = 0.25,$$
$$T_1 = 832.5,$$
$$Q_1 = Q_n/\alpha A_R = 102.86(\text{E}6)/3190 = 32{,}245.$$

From program "FRN-1",

$$T_g = 1553.7,$$
$$F = 0.6496 \quad [\text{Eq. (9)}],$$
$$Q_R = \alpha A_R F\left\{1730\left[\left(\frac{T_g + 460}{1000}\right)^4 - \left(\frac{T_t + 460}{1000}\right)^4\right] + 7(T_g - T_t)\right\}$$
$$= 3190(0.6496)(28{,}679) = 59.43(\text{E}6).$$

Compared with estimated 57.86(E6) at 75% heat absorption

EXAMPLE 8.15—*(continued)*

in the radiant section. Repeat the calculation with an estimate of 60(E6)

$$H_1 = 248 + (77.14 - 60)(E6)/195{,}394 = 335.7,$$
$$T_1 = 542,$$
$$T_t = 100 + 0.5(542 + 900) = 821,$$
$$T_g = 1550.5,$$
$$F = 0.6498,$$
$$Q_R = 3190(0.6498)(28{,}727) = 59.55(E6).$$

Interpolating,

Q_{assumed}	T_1	T_t	T_g	Q_{calcd}	Q/A
57.86	565	832.5	1553.7	59.43	
60.00	542	821	1550.5	59.55	
Interpolation	[547		1551.2	59.50	9900]

26–27.

$$Q_{\text{conv}} = (77.14 - 59.50)(E6) = 17.64(E6).$$

Fraction lost in stack gas

$$Q_s/Q_n = 1 - 0.02 - 0.75 = 0.23.$$

From (Eq. (4),

$$T_s = 920°\text{F}.$$

28–31.

$$\text{LMTD} = 735.6$$

mean gas film temp is

$$T_f = 0.5(400 + 547 + 735.6) = 841.3.$$

Since $G = 0.325$ lb/(sec)(sqft),

$$V_c = 5.6\ \text{Btu/(hr)(sqft)(°F)} \quad [(\text{Eq. (10)}],$$
$$A_{\text{conv}} = \frac{17.64(E6)}{735.6(5.6)} = 4282\ \text{sqft},$$
$$\frac{4282}{1.7344(38.5)} = 64.1\ \text{bare tubes}$$

or 16 rows of 4 tubes each. Spacing the same as of the shield tubes.

Beyond the first two rows, extended surfaces can be installed.

$$\text{Total rows} = 2 + 14/2 = 9.$$

bank, and is called the equivalent cold plane area. Evaluation of the exchange factor F is explained in item 9 of Table 8.16. It depends on the emissivity of the gas and the ratio of refractory area A_w to the equivalent cold plane area αA_{cp}. In turn, $A_w = A - A_{cp}$, where A is the area of the inside walls, roof, and floor that are covered by refractory.

In the convection zone of the heater, some heat also is transferred by direct radiation and reflection. The several contributions to overall heat transfer specifically in the convection zone of fired heaters were correlated by Monrad [*Ind. Eng. Chem.* **24,** 505 (1932)]. The combined effects are approximated by item 10 of Table 8.16, which is adequate for estimating purposes. The relation depends on the temperature of the gas film which is taken to be the sum of the average process temperature and one-half of the log mean temperature difference between process and flue gas over the entire tube bank. The temperature of the gas entering the convection zone is found with the trial calculation described in Steps 22–23 of Table 8.17 and may utilize the computer program of Table 8.18.

8.12. INSULATION OF EQUIPMENT

Equipment at high or low temperatures is insulated to conserve energy, to keep process conditions from fluctuating with ambient conditions, and to protect personnel who have occasion to approach the equipment. A measure of protection of the equipment metal against atmospheric corrosion also may be a benefit. Application of insulation is a skilled trade. Its cost runs to 8–9% of purchased equipment cost.

In figuring heat transfer between equipment and surroundings, it is adequate to take account of the resistances of only the insulation and the outside film. Coefficients of natural convection are in Table 8.9 and properties of insulating materials at several

EXAMPLE 8.16
Application of the Wilson–Lobo–Hottel Equation

In the case of Example 8.15, 25% excess air was employed, corresponding to 19.0 lb/air/lb fuel, the heat release was $Q_f = 102.86(10^6)$ Btu/hr, and $\alpha A_{cp} = 3036$. The effect will be found of changing the excess air to 10% (16.72 lb air/lb fuel) on the amount of fuel to be fired while maintaining the same heat absorption.

Ratioing Eq. (8.45) to yield the ratio of the releases at the two conditions,

$$\frac{Q_{f2}}{102.86(106)} = \frac{1 + (16.72/4200)\sqrt{Q_{f2}/3036}}{1 + (19.0/4200)\sqrt{102.86(10^6)/3036}}$$

$$= \frac{1 + 0.0722\sqrt{Q_{f2}}(10^{-6})}{1.8327}$$

$$\therefore Q_{f2} = 95.82(10^6)\ \text{Btu/hr},$$

which is the heat release with 10% excess air.

$$\text{With 25\% excess air,} \quad Q/Q_f = 1/1.8327 = 0.5456,$$
$$\text{With 10\% excess air,} \quad Q/Q_f = 0.5456(102.86/95.82) = 0.5857,$$

which shows that approximately 7% more of the released heat is absorbed when the excess air is cut from 25% down to 10%.

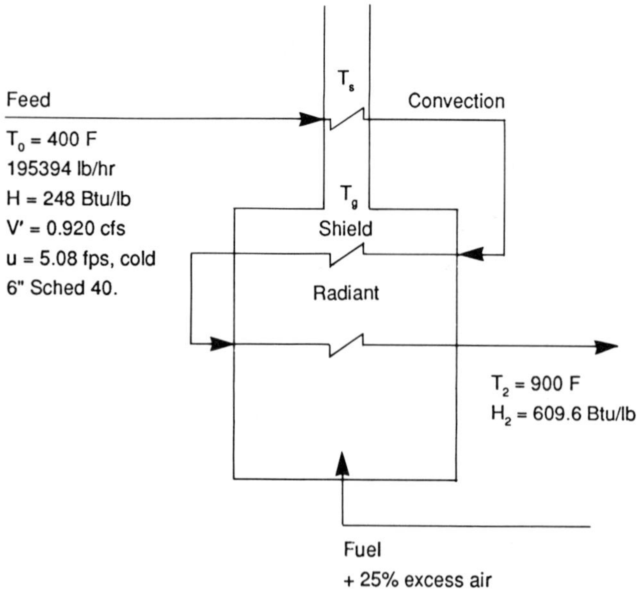

Figure 8.20. Flowsketch of process of Example 8.16.

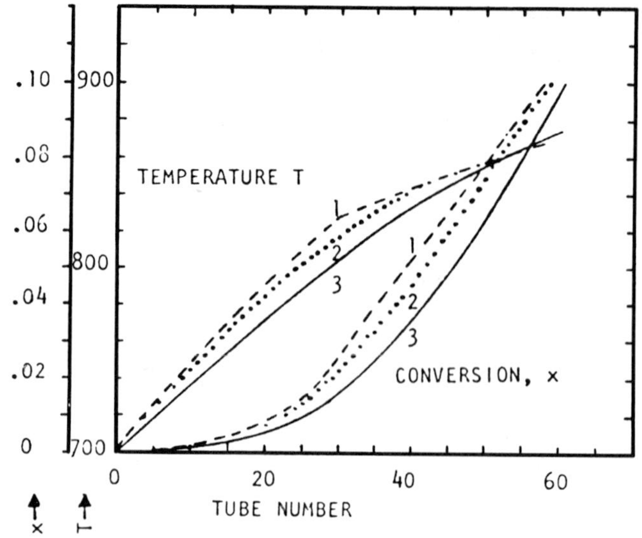

Figure 8.22. Effects of three modes of heat flux distribution on temperature and conversion in pyrolysis of a fuel oil: (1) two levels, 12,500 and 7500; (2) linear variation between the same limits; (3) constant at 10,000 Btu/(hr)(sqft). Obtained by method of Example 8.16.

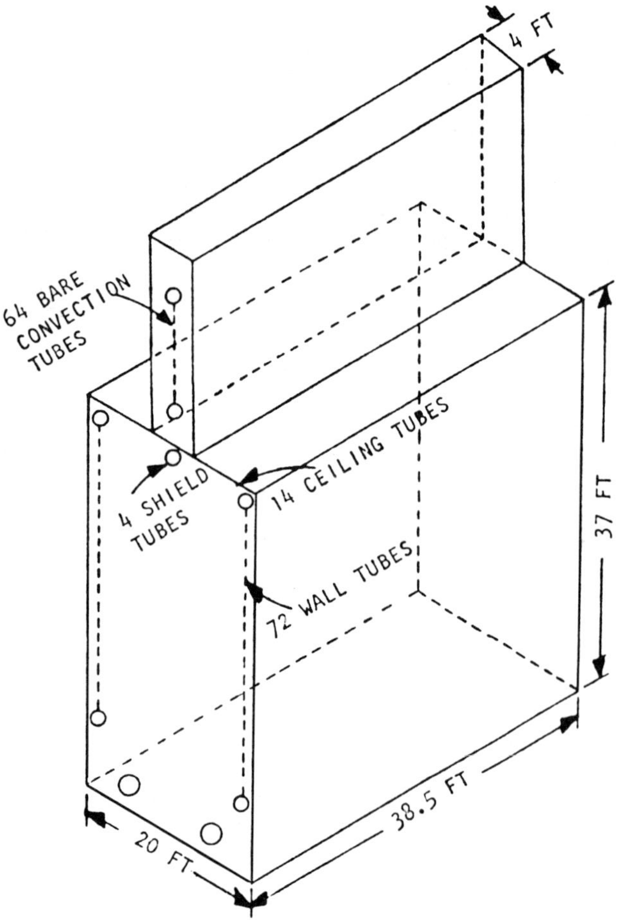

Figure 8.21. Tube and box configuration of the fired heater of Example 8.16.

temperature levels are in Tables 8.19–8.21. Outdoors under windy conditions, heat losses are somewhat greater than indoors at natural convections. Tabulations of economic thicknesses in *Chemical Engineers Handbook* (McGraw-Hill, New York, 1984, 11.55–11.58) suggest that 10–20% greater thickness of insulation is justified at wind velocity of 7.5 miles/hr.

The optimum thickness of insulation can be established by economic analysis when all of the cost data are available, but in practice a rather limited range of thicknesses is employed. Table 8.22 of piping insulation practice in one instance is an example.

The procedure for optimum selection of insulation thicknesses is exemplified by Happel and Jordan [*Chem. Process Economics,* 380 (1975)]. They take into account the costs of insulation and fuel, payout time, and some minor factors. Although their costs of fuel are off by a factor of 10 or more, their conclusions have some validity if it is recognized that material costs likewise have gone up by roughly the same factor. They conclude that with energy cost of \$2.5/million Btu (adjusted by a factor of 10), a payout time of 2 years, for pipe sizes of 2–8 in., the optimum thicknesses in

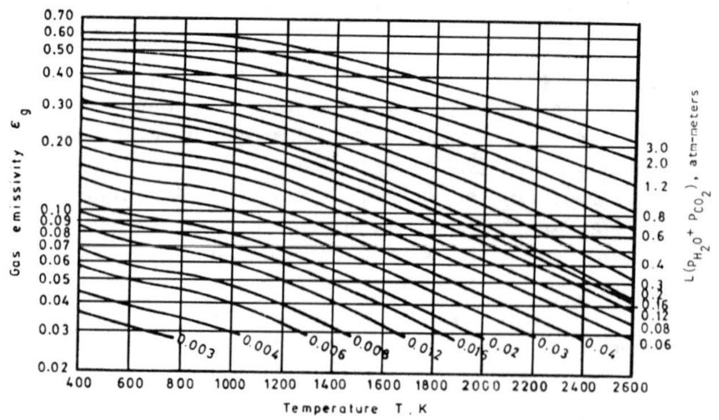

Figure 8.23. Total emissivity of carbon dioxide and water with $P_{H_2O}/P_{CO_2} = 1$ and a total pressure of 1 atm [*Hadvig,* J. Inst. Fuel **43,** *129 (1970)].*

TABLE 8.18. Program for Finding the Radiant Gas Temperature by Steps 22 and 23 of Table 8.17

```
 10 ! Example 8.16. Design of a
    fired heater.Radiant gas tem
    p by step 22. Program "FRN-1
    ",tape 2
 20 ! P=PL, product of partial p
    ressures of CO2+H20 and mean
     beam length
 30 ! D1=Aw/αAr
 40 ! D2=fraction excess air
 50 ! T1=tube surface temperatur
    e
 60 ! Q1=Qn/αAr
 70 ! Q2=Qg/Qn, Eq.8
 80 ! F1=emissivity, Eq.6
 90 ! F=exchange factor, Eq. 7
100 ! J = RHS-LHS of step 22
110 SHORT T
120 READ P,D1,D2,T1,Q1
130 DATA 4.1,.9605,.25,672,42828
140 INPUT T
150 GOSUB 270
160 J1=J
170 T=1.0001*T
180 GOSUB 270
190 J2=J
200 DISP T
210 H=.0001*T*J1/(J2-J1)
220 T=T/1.0001-H
230 IF ABS(H/T)<=.0001 THEN 250
240 GOTO 150
250 PRINT "RADIANT GAS TEMP=";T
260 END
270 Z1=(T+460)/1000
280 A1=.47916-.19847*Z1+.022569*
    Z1^2
290 B1=.047029+.0699*Z1-.01528*Z
    1^2
300 C1=-.000803-.00726*Z1+.00159
    7*Z1^2
310 F1=A1+B1*P+C1*P^2
320 Z2=D1
330 A2=.00064+.0591*Z2+.00101*Z2
    ^2
340 B2=1.0256+.4908*Z2-.058*Z2^2
350 C2=-.144-.552*Z2+.04*Z2^2
360 F=A2+B2*F1+C2*F1^2
370 Z3=D2
380 A3=.22048-.35057*Z3+.92344*Z
    3^2
390 B3=.016086+.29393*Z3-.48139*
    Z3^2
400 Q2=(A3+B3*(T/1000-.1))*(T/10
    00-.1)
410 J=-(Q1/F*(.98-Q2))+1730*(((T
    +460)/1000)^4-((T1+460)/1000
    )^4)+7*(T-T1)
420 RETURN
```

insulation depend on the process temperature according to:

T (°F)	200	400	600
Thickness (in.)	0.5	1.0	1.25

The data of Table 8.22 are roughly in agreement with these calculations. Optimum thicknesses of pipe insulation also are tabulated in *Chemical Engineers Handbook* (1984, 11.56); they cover both indoor and outdoor conditions, temperature ranges of 150–1200°F and energy costs of 1–8 dollars/million Btu.

For very large tanks storing volatile liquids and subject to pressure buildup and breathing losses, it is advisable to find economic thickness of insulation by economic analysis. The influence of solar radiation should be taken into account; a brief treatment of this topic is in the book of Threlkeld (*Thermal Environmental Engineering,* Prentice-Hall, Englewood Cliffs, NJ, 1970). In at least one application, rigid urethane foam sprayed onto storage tanks in 2 in. thickness and covered with a 4 mil thickness of neoprene rubber for weather proofing was economically attractive.

Although resistance to heat transfer goes up as the thickness of pipe insulation is increased, the external surface also increases; a thickness may be reached at which the heat transfer becomes a minimum and then becomes larger. In accordance with this kind of behavior, heat pickup by insulated refrigerated lines of small diameters can be greater than that of bare lines. In another instance, electrical transmission lines often are lagged to increase the rate of heat loss. An example worked out by Kreith (*Principles of Heat Transfer,* Intext, New York, 1973, p. 44) reveals that an insulated 0.5 in. OD cable has a 45% greater heat loss than a bare one.

LOW TEMPERATURES

Insulants suited to cryogenic equipment are characterized by multiple small spaces or pores that occlude more or less stagnant air of comparatively low thermal conductivity. Table 8.19 lists the most common of these materials. In application, vapor barriers are provided in the insulating structure to prevent inward diffusion of atmospheric moisture and freezing on the cold surface with resulting increase in thermal conductivity and deterioration of the insulation. Sealing compounds of an asphalt base are applied to the surface of the insulation which then is covered with a weatherproof jacket or cement coating. For truly cryogenic operations such as air liquefaction and rectification in which temperatures as low as −300°F are encountered, all of the equipment is enclosed in a box, and then the interstices are filled with ground cork.

MEDIUM TEMPERATURES

Up to about 600°F, 85% magnesia has been the most popular material. It is a mixture of magnesia and asbestos fibers so constructed that about 90% of the total volume is dead air space. Equivalents are available for situations where asbestos is undesirable. Such insulants are applied to the equipment in the form of slabs or blankets which are held in place with supports and clips spotwelded to the equipment. They are covered with cement to seal gaps and finished off with a canvas cover that is treated for resistance to the weather. A galvanized metal outer cover may be preferred because of its resistance to mechanical damage of the insulation.

A mixture of diatomaceous earth and an asbestos binder is suitable for temperatures up to the range of 1600–1900°F. Johns-Manville "Superex" is one brand. Since this material is more expensive than 85% magnesia, a composite may be used to save money: sufficient thickness of the high temperature resistant material to bring its external surface to below 600°F, finished off with 85% magnesia in appropriate thickness. Table 8.22(c) is one standard specification of this type.

REFRACTORIES

Equipment made of metal and subject to high temperatures or abrasive or corrosive conditions often is lined with ceramic material.

TABLE 8.19. Thermal Conductivities of Insulating Materials for Low Temperatures [*k* Btu/(hr)(sqft)(°F/ft)]

Material	Bulk/ Density, (lb/cuft)	Temp (°F)	*h*	Material	Bulk Density, (lb/cuft)	Temp (°F)	*h*
Corkboard	6.9	100	0.022	Rubber board,			
		−100	0.018	expanded,			
		−300	0.010	"Rubatex"	4.9	100	0.018
Fibreglas with						−100	0.015
asphalt coating						−300	0.004
(board)	11.0	100	0.023	Silica aerogel,			
		−100	0.014	powder	5.3	100	0.013
		−300	0.007	"Santocel"		0	0.012
Glass blocks,						−100	0.010
expanded,							
"Foamglas"	10.6	100	0.036	Vegetable fiber-			
		−100	0.033	board, asphalt	14.4	100	0.028
		−300	0.018	coating		−100	0.021
Mineral wool						−300	0.013
board,							
"Rockcork"	14.3	100	0.024	Foams:	2.9	−100	0.015
		−100	0.017	Polystyrene[a]	5.0	−100	0.019
		−300	0.008	Polyurethane[b]			

[a]Test space pressure, 1.0 atm; $k = 0.0047$ at 10^{-3} mm Hg.
[b]Test space pressure, 1.0 atm; $k = 0.007$ at 10^{-3} mm Hg.
(*Marks Mechanical Engineers Handbook,* 1978, p. 4.64).

When the pressure is moderate and no condensation is likely, brick construction is satisfactory. Some of the materials suited to this purpose are listed in Table 8.21. Bricks are available to withstand 3000°F. Composites of insulating brick next to the wall and stronger brick inside are practical. Continuous coats of insulants are formed by plastering the walls with a several inch thickness of concretes of various compositions. "Gunite" for instance is a mixture of 1 part cement and 3 parts sand that is sprayed onto walls and even irregular surfaces. Castable refractories of lower density and greater insulating powers also are common. With both brickwork and castables, an inner shell of thin metal may be provided to guard against leakage through cracks that can develop in the refractory lining. For instance, a catalytic reformer 4 ft OD designed for 650 psig and 1100°F has a shell 1.5 in. thick, a light weight castable lining 4-5/8 in. thick and an inner shell of metal 1/8 in. thick. A catalytic cracker 10 ft dia designed for 75 psig and 1100°F has a 3 in. monolithic concrete liner and 3 in. of blanket insulation on the outside. Ammonia synthesis reactors that operate at 250 atm and 1000°F are insulated on the inside to keep the wall below about 700°F, the temperature at which steels begin to decline in strength, and also to prevent access of hydrogen to the shell since that causes embrittlement. An air gap of about 0.75 in. between the outer shell and the insulating liner contributes significantly to the overall insulating quality.

TABLE 8.20. Thermal Conductivities of Insulating Materials for High Temperatures [*k* Btu/(hr)(sqft)°F/ft)]

Material	Bulk Density, lb/cuft	Max Temp (°F)	100°F	300°F	500°F	1000°F	1500°F	2000°F
Asbestos paper, laminated	22	400	0.038	0.042				
Asbestos paper, corrugated	16	300	0.031	0.042				
Diatomaceous earth, silica, powder	18.7	1500	0.037	0.045	0.053	0.074		
Diatomaceous earth, asbestos and bonding material	18	1600	0.045	0.049	0.053	0.065		
Fiberglas block, PF612	2.5	500	0.023	0.039				
Fiberglas block, PF614	4.25	500	0.021	0.033				
Fiberglas block, PF617	9	500	0.020	0.033				
Fiberglas, metal mesh blanket, #900	—	1000	0.020	0.030	0.040			
Glass blocks, average values	14–24	1600	—	0.046	0.053	0.074		
Hydrous calcium silicate, "Kaylo"	11	1200	0.032	0.038	0.045			
85% magnesia	12	600	0.029	0.035				
Micro-quartz fiber, blanket	3	3000	0.021	0.028	0.042	0.075	0.108	0.142
Potassium titanate, fibers	71.5	—	—	0.022	0.024	0.030		
Rock wool, loose	8–12	—	0.027	0.038	0.049	0.078		
Zirconia grain	113	3000	—	—	0.108	0.129	0.163	0.217

(Marks, *Mechanical Engineers Handbook,* 1978, p. 4.65).

TABLE 8.21. Properties of Refractories and Insulating Ceramics[a]

(a) Chemical Composition of Typical Refractories

No.	Refractory Type	SiO_2	Al_2O_3	Fe_2O_3	TiO_2	CaO	MgO	Cr_2O_3	SiC	Alkalies	Resistance to: Siliceous Steel-Slag	High-lime Steel-Slag	Fused Mill-Scale	Coal-Ash Slag
1	Alumina (fused)	8–10	85–90	1–1.5	1.5–2.2	—	—	—	—	0.8–1.3[a]	E	G	F	G
2	Chrome	6	23	15[b]	—	—	17	38	—	—	G	E	E	G
3	Chrome (unburned)	5	18	12[b]	—	—	32	30	—	—	G	E	E	G
4	Fire clay (high-heat duty)	50–57	36–42	1.5–2.5	1.5–2.5	—	—	—	—	1–3.5[c]	F	P	P	F
5	Fire clay (super-duty)	52	43	1	2	—	—	—	—	2[c]	F	P	F	F
6	Forsterite	34.6	0.9	7.0	—	1.3	55.4							
7	High-alumina	22–26	68–72	1–1.5	3.5	—	—	—	—	1–1.5[c]	G	F	F	F
8	Kaolin	52	45.4	0.6	1.7	0.1	0.2	—	—	—	F	P	G[d]	F
9	Magnesite	3	2	6	—	3	86	—	—	—	P	E	E	E
10	Magnesite (unburned)	5	7.5	8.5	—	2	64	10	—	—	P	E	E	E
11	Magnesite (fused)	—	—	—	—	—	—	—	—	—	F	E	E	E
12	Refractory porcelain	25–70	25–60	—	—	—	—	—	—	1–5	G	F	F	F
13	Silica	96	1	1	—	2	—	—	—	—	E	P	F	P
14	Silicon carbide (clay bonded)	7–9	2–4	0.3–1	1	—	—	—	85–90	—	E	G	F	E
15	Sillimanite (mullite)	35	62	0.5	1.5	—	—	—	—	0.5[c]	G	F	F	F
16	Insulating fire-brick (2600°F)	57.7	36.8	2.4	1.5	0.6	0.5	—	—	—	P	P	G[e]	P

(b) Physical Properties of Typical Refractories[g]

Refractory No.	Fusion Point °F	Fusion Point Pyrometric Cone	Deformation under Load (% at °F and lb/in.)	Spalling Resistance[f]	Repeat Shrinkage after 5 hr (% °F)	Wt. of Straight 9 in. Brick (lb)
1	3390+	39+	1 at 2730 and 50	G	+0.5 (2910)	9–10.6
2	3580+	41+	shears 2740 and 28	P	−0.5–1.0 (3000)	11.0
3	3580+	41+	shears 2955 and 28	F	−0.5–1.0 (3000)	11.3
4	3060–3170	31–33	2.5–10 at 2460 and 25	G	±0–1.5 (2550)	7.5
5	3170–3200	33–34	2–4 at 2640 and 25	E	±0–1.5 (2910)	8.5
6	3430	40	10 at 2950	F	—	9.0
7	3290	36	1–4 at 2640 and 25	E	−2–4 (2910)	7.5
8	3200	34	0.5 at 2640 and 25	E	−0.7–1.0 (2910)	7.7
9	3580+	41+	shears 2765 and 28	P	−1–2 (3000)	10.0
10	3580+	41+	shear 2940 and 28	F	−0.5–1.5 (3000)	10.7
11	3580+	41+		F	—	10.5
12	2640–3000	16 + 30		G		
13	3060–3090	31–32	shears 2900 and 25	P	+0.5–0.8 (2640)	6.5
14	3390	39	0–1 at 2730 and 50	E	+2 (2910)	8–9.3
15	3310–3340	37–38	0–0.5 at 2640 and 25	E	−0–0.8 (2910)	8.5
16	2980–3000	29–30	0.3 at 2200 and 10	G	−0.2 (2600)	2.25

[a] Divide by 12 to obtain the units *k* Btu/(hr)(sqft)(°F/ft).
[b] As FeO.
[c] Includes lime and magnesia.
[d] Excellent if left above 1200°F.
[e] Oxidizing atmosphere.
[f] E = Excellent. G = Good. F = Fair. P = Poor.
[g] [Some data from Trostel, *Chem. Met. Eng.* (Nov. 1938)].
Marks, *Mechanical Engineers Handbook,* McGraw-Hill, New York, 1978, pp. 6.172–6.173.

TABLE 8.22. Specifications of Thicknesses of Pipe Insulation for Moderate and High Temperatures, in Single or Double Strength as May Be Needed

(a) Insulation of 85% Magnesia or Equivalent up to 600°F

Pipe Size (in.)	Standard Thick (in.)	Double Standard Thick (in.)
1–1/2 or less	7/8	1–15/16
2	1–1/32	2–5/32
2–1/2	1–1/32	2–5/32
3	1–1/32	2–5/32
4	1–1/8	2–1/4
5	1–1/8	2–5/16
6	1–1/8	2–5/16
8	1–1/4	2–1/2
10	1–1/4	2–1/2
12–33	1–1/2	3

(b) Molded Diatomaceous Earth Base Insulation, to 1900°F, Single or Double Thickness as Needed

Pipe Size (in.)	Thickness (in.)	
1–1/2	2	2
2	1–1/4	2–1/8
2–1/2	1–5/16	1–13/16
3	1–9/16	2–1/16
4	1–9/16	2–1/16
5	1–1/2	2
6	1–1/2	2–1/16
8	1–1/2	2
10	1–9/16	2–1/8
12	1–9/16	2–1/8
14–33	1–1/2	2

(c) Combination Insulation, Inner Layer of Diatomaceous Earth Base, and Outer of 85% Magnesia or Equivalent, for High Level Insulation to 1900°F

Inner Layer		Outer Layer	
Pipe Size (in.)	Thickness (in.)	Nominal Pipe Size (in.)	Thickness (in.)
1–1/2 or less	2	no outer layer	–
2	1–1/4	4–1/2	1–1/2
2–1/2	1–5/16	5	1–1/2
3	1–9/16	6	1–1/2
4	1–9/16	7	1–1/2
5	1–1/2	8	2
6	1–1/2	9	2
8	1–1/2	11	2
10	1–9/16	14	2
12	1–9/16	16	2
14–33	1–1/2	17–36	2

Data of an engineering contractor.

8.13. REFRIGERATION

Process temperatures below those attainable with cooling water or air are attained through refrigerants whose low temperatures are obtained by several means:

1. Vapor compression refrigeration in which a vapor is compressed, then condensed with water or air, and expanded to a low pressure and correspondingly low temperature through a valve or an engine with power takeoff.
2. Absorption refrigeration in which condensation is effected by absorption of vapor in a liquid at high pressure, then cooling and expanding to a low pressure at which the solution becomes cold and flashed.
3. Steam jet action in which water is chilled by evaporation in a chamber maintained at low pressure by means of a steam jet ejector. A temperature is 55°F or so is commonly attained, but down to 40°F may be feasible. Brines also can be chilled by evaporation to below 32°F.

The unit of refrigeration is the ton which is approximately the removal of the heat of fusion of a ton of ice in one day, or 288,000 Btu/day, 12,000 Btu/hr, 200 Btu/min. The reciprocal of the efficiency, called the coefficient of performance (COP) is the term employed to characterize the performances of refrigerating processes:

$$\text{COP} = \frac{\text{energy absorbed by the refrigerant at the low temperature}}{\text{energy input to the refrigerant}}.$$

A commonly used unit of COP is (tons of refrigeration)/(horsepower input). Some of the refrigerants suited to particular temperature ranges are listed in Tables 1.10, 8.23, and 8.24.

COMPRESSION REFRIGERATION

A basic circuit of vapor compression refrigeration is in Figure 8.24(a). After compression, vapor is condensed with water cooling and then expanded to a low temperature through a valve in which the process is essentially at constant enthalpy. In large scale installations or when the objective is liquefaction of the "permanent" gases, expansion to lower temperatures is achieved in turboexpanders from which power is recovered; such expansions are approximately isentropic. The process with expansion through a valve is represented on a pressure–enthalpy diagram in Figure 8.24(b).

A process employing a circulating brine is illustrated in Figure 8.24(c); it is employed when cooling is required at several points distant from the refrigeration unit because of the lower cost of circulation of the brine, and when leakage between refrigerant and process fluids is harmful.

For an overall compression ratio much in excess of four or so, multistage compression is more economic. Figure 8.24(d) shows two stages with intercooling to improve the capacity and efficiency of the process.

Many variations of the simple circuits are employed in the interest of better performance. The case of Example 8.17 has two stages of compression but also two stages of expansion, a scheme due originally to Windhausen (in 1901). The flashed vapor of the intermediate stage is recycled to the high pressure compressor. The numerical example shows that an improved COP is attained with the modified circuit. In the circuit with a centrifugal compressor of Figure 8.25, the functions of several intermediate expansion valves and flash drums are combined in a single vessel with appropriate internals called an economizer. This refrigeration unit is used with a fractionating unit for recovering ethane and ethylene from a mixture with lighter substances.

Low temperatures with the possibility of still using water for final condensation are attained with cascade systems employing coupled circuits with different refrigerants. Refrigerants with higher vapor pressures effect condensation of those with lower vapor pressures. Figure 8.26 employs ethylene and propylene in a cascade for servicing the condenser of a demethanizer which must be cooled to −145°F. A similar process is represented on a flowsketch in the book of Ludwig (1983, Vol. 1, p. 249). A three element cascade with methane, ethylene and propylene refrigerants is calculated by

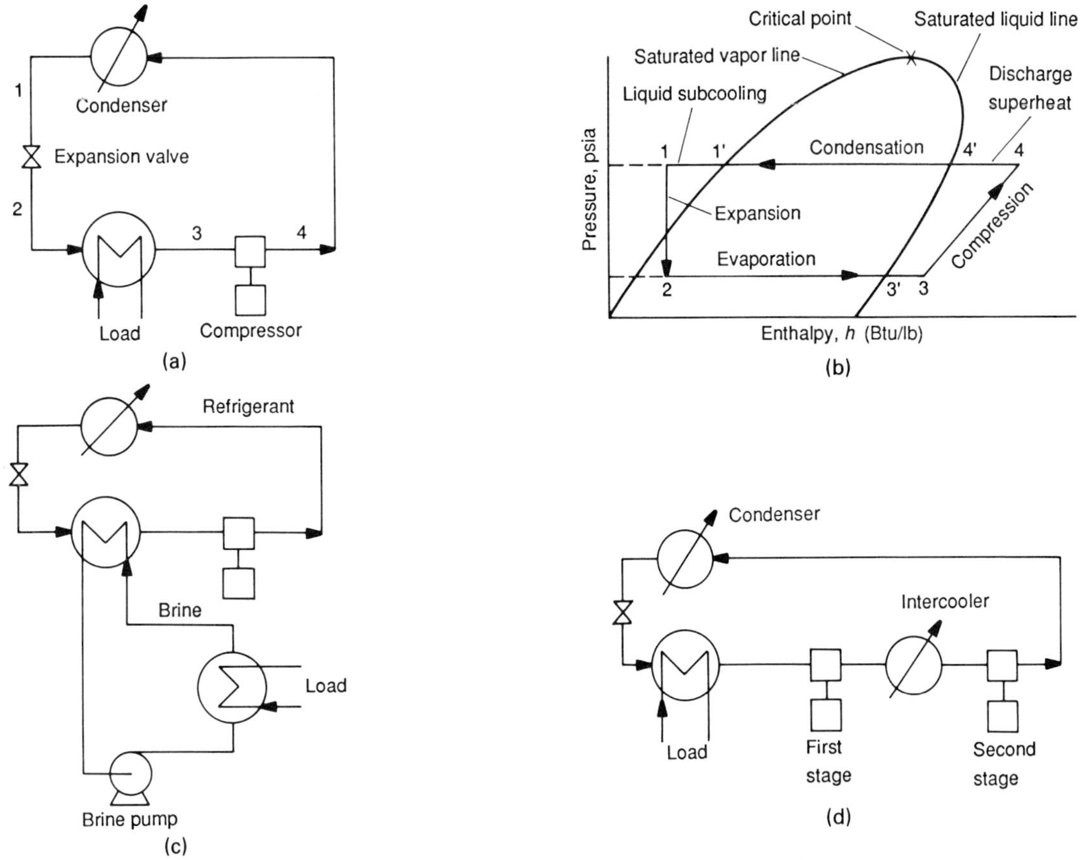

Figure 8.24. Simpler circuits of compression refrigeration (see also Example 8.17). (a) Basic circuit consisting of a compressor, condenser, expansion valve and evaporator (load). (b) Conditions of the basic circuit as they appear on a pressure–enthalpy diagram; the primed points are on the vapor–liquid boundary curve. (c) Circuit with circulation of refrigerated brine to process loads. (d) Circuit with two-stage compression and intercooling.

EXAMPLE 8.17
Two-Stage Propylene Compression Refrigeration with Interstage Recycle

A propylene refrigeration cycle operates with pressures of 256, 64, and 16 psia. Upon expansion to 64 psia, the flashed vapor is recycled to the suction of the high pressure stage while the liquid is expanded to 16 psia to provide the needed refrigeration at −9°F. The ratios of refrigeration to power input will be compared without and with interstage recycle.

Basis: 1 lb of propylene to the high pressure stage. Conditions

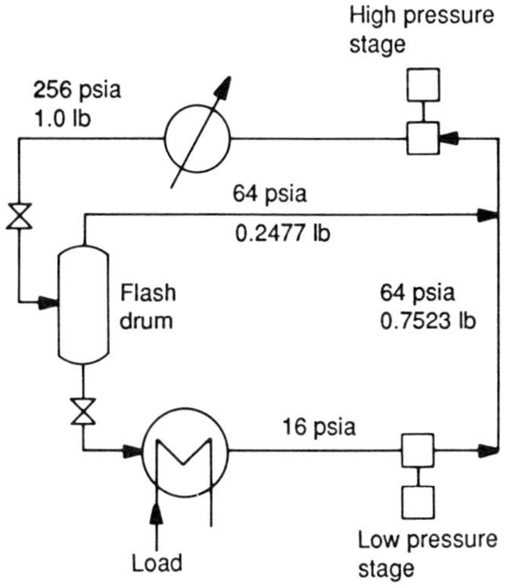

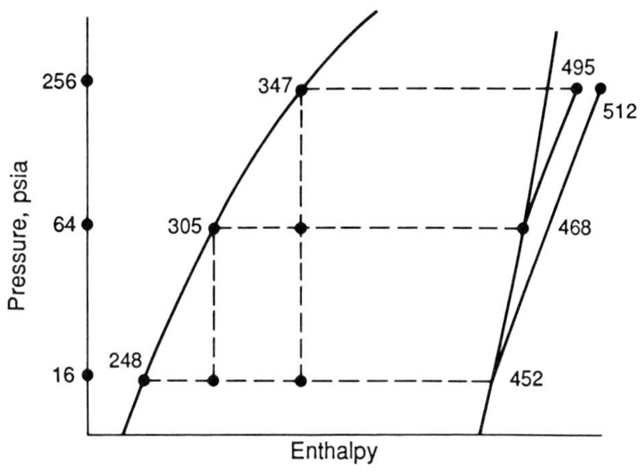

Example 8.17—*(continued)*

are shown on the pressure–enthalpy and flow diagrams. Isentropic compression and isenthalpic expansion are taken. Without recycle,

refrigeration = 452–347 = 105 Btu/lb,
work = 512–452 = 60 Btu/lb,
COP = 105/60 = 1.75.

With recycle,

interstage vapor = (347–305)/(468–305) = 0.2577 lb/lb,
refrigeration = (452–305)0.7423 = 109.1 Btu/lb,
work = (495–468)0.2577 + (512–452)0.7423 = 51.5 Btu/lb,
COP = 109.1/51.5 = 2.12,

which points out the improvement in coefficient of performance by the interstage recycle.

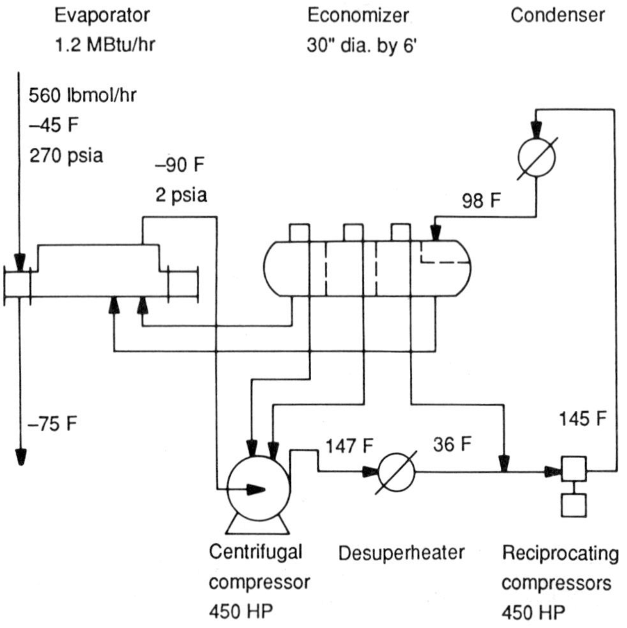

Figure 8.25. A refrigeration system for the overhead condenser of a fractionator for recovering ethane and ethylene. Freon-12 is the refrigerant. The economizer combines the functions of several expansion valves and flash drums for intermediate recycle of flashed vapors.

Bogart (1981, pp. 44–47); it attains −240°F with a maximum pressure of 527 psia.

REFRIGERANTS

Several refrigerants commonly used above −80°F or so are compared in Table 8.23. Ethylene and butane also are in use, particularly in refineries where they are recoverable from the process streams. Properties of the freons (also known by the trade name genetrons) are listed in Table 8.24. Freon 12 is listed in both tables so some comparisons of all of these refrigerants is possible. The refrigerants of Table 8.23 have similar performance. When ammonia or some hydrocarbons are made in the plant, their election as refrigerants is logical. Usually it is preferred to operate at suction pressures above atmospheric to avoid inleakage of air. The nonflammability and nontoxicity of the freons is an attractive quality. Relatively dense vapors such as Ref-12, -22, and -500 are preferred with reciprocating compressors which then may have smaller cylinders. For most equipment sizes, Ref-12 or -114 can be adopted for greater capacity with the same equipment. Ref-22 and -500 are used with specially built centrifugals to obtain highest capacities.

Ammonia absorption refrigeration is particularly applicable when low level heat is available for operation of the stripper reboiler and power costs are high. Steam jet refrigeration is the large scale system of choice when chilled water is cold enough, that is above 40°F or so.

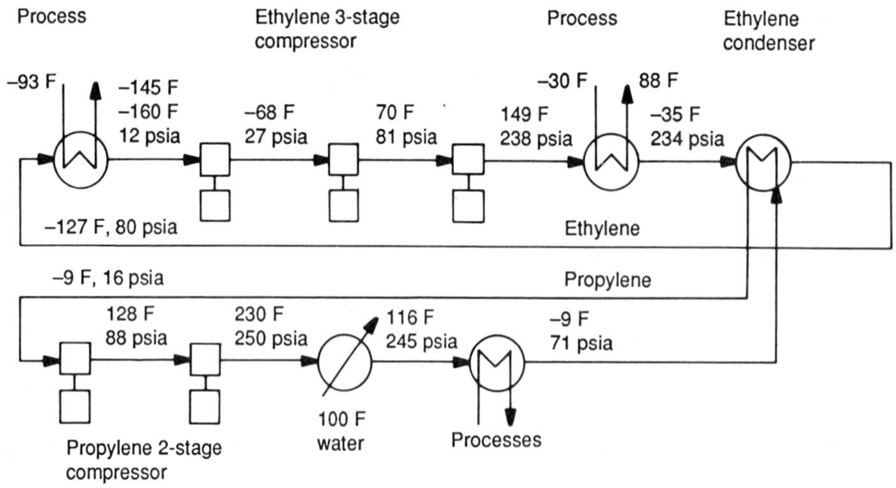

Figure 8.26. A cascade refrigeration system employing ethylene and propylene for condensing the overhead of a demethanizer at −145°F. The diagram is somewhat simplified.

TABLE 8.23. Comparative Data of Refrigerants[a]

Evaporator Temp (°F)		−80	−60	−40	−20	0	20	40	60
Evaporator pressure (psia)	ammonia		5.55	10.4	18.3	30.4	48.0	73.0	107.5
	propylene	7.20	12.5	20.7	32.1	48.0	70.0	96.0	131
	propane	5.55	9.78	16.2	25.5	38.1	56.0	80.0	110
	freon 12	2.88	5.36	9.3	15.3	23.8	35.7	51.7	72.4
Condensed Liquid Temperature 95°F; Condenser Pressure in psia: Ammonia 197; Propylene 212; Propane 177; 12 123									
lb refrigerant/ min/ton refrigeration	ammonia		0.454	0.446	0.438	0.432	0.426	0.422	0.418
	propylene	2.07	1.96	1.87	1.79	1.72	1.66	1.60	1.54
	propane	2.18	2.04	1.93	1.83	1.74	1.67	1.59	1.53
	freon 12	5.18	4.89	4.65	4.42	4.22	4.05	3.88	3.74
CMF of refrigerant/ min/ton refrigeration	ammonia		20.4	11.1	6.45	3.96	2.52	1.69	1.14
	propylene	27.1	15.7	9.18	5.85	3.84	2.53	1.80	1.28
	propane	37.4	20.0	12.0	7.29	4.77	3.12	2.13	1.50
	freon 12	59.9	31.7	18.0	10.8	6.79	4.44	3.00	2.09
Brake horsepower/ ton refrigeration	ammonia		4.31	3.23	2.41	1.78	1.26	0.835	0.483
	propylene	5.00	3.96	3.10	2.35	1.74	1.20	0.830	0.485
	propane	4.98	3.87	3.03	2.32	1.75	1.24	0.800	0.458
	freon 12	5.70	4.33	3.31	2.47	1.83	1.30	0.848	0.490
Condensed Liquid Temperature 125°F; Condenser Pressure in psia: Ammonia 303; Propylene 314; Propane 260; Freon 12 184									
lb refrigerant min/ton refrigeration	ammonia		0.492	0.483	0.474	0.466	0.460	0.454	0.450
	propylene	2.67	2.50	2.35	2.22	2.11	2.01	1.93	1.86
	propane	2.86	2.63	2.44	2.29	2.16	2.04	1.94	1.84
	freon 12	6.42	5.98	5.61	5.28	5.00	4.75	4.53	4.33
CFM of refrigerant/ ton/ refrigeration	ammonia		22.0	12.0	6.97	4.26	2.72	1.82	1.23
	propylene	35.2	20.0	11.5	7.32	4.72	3.08	2.18	1.56
	propane	50.0	25.8	15.4	9.16	5.94	3.79	2.63	1.80
	freon 12	74.0	38.8	21.7	12.9	8.05	5.21	3.50	2.42
Brake horsepower/ ton refrigeration	ammonia		5.68	4.38	3.33	2.54	1.90	1.38	0.952
	propylene	7.49	5.96	4.71	3.66	2.79	2.03	1.55	1.10
	propane	7.47	5.85	4.60	3.59	2.81	2.07	1.50	1.03
	freon 12	8.09	6.25	4.78	3.67	2.78	2.07	1.49	1.02

[a] The horsepowers are based on centrifugal compressor efficiencies without economizers.

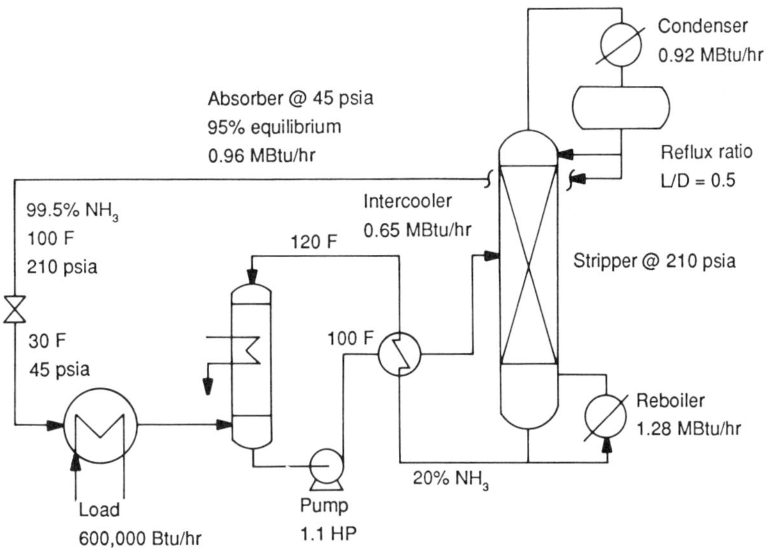

Figure 8.27. An ammonia absorption refrigeration process for a load of 50 tons at 30°F. The conditions were established by Hougen, Watson, and Ragatz (Thermodynamics, *Wiley, New York, 1959, pp. 835–842).*

TABLE 8.24. Comparative Data of Freon Refrigerants

Refrigerant Number (ARI Designation)	11	12	22	113	114	500	502
Chemical name	trichloromono-fluoromethane	dichlorodi-fluoromethane	monochlorodi-fluoromethane	trichlorotri-fluoroethane	dichlorotetra-fluoroethane	azeotrope of dichlorodi-fluoromethane and difluoroethane	azeotrope of monochlorodi-fluoromethane and monochloropenta-fluoroethane
Chemical formula	CCl_3F	CCl_2F_2	$CHClF_2$	$CCl_2F{-}CClF_2$	$C_2Cl_2F_4$	73.8% CCl_2F_2 26.2% CH_3CHF_2	48.8% $CHCLF_2$ 51.2% $CClF_2{-}CF_3$
Molecular w1	137.38	120.93	86.48	187.39	170.93	99.29	111.64
Gas constant R [(ft lb/lb R)]	11.25	12.78	17.87	8.25	9.04	15.57	13.87
Boiling point at 1 atm (°F)	74.7	−21.62	−41.4	117.6	38.4	−28.0	−50.1
Freezing point at 1 atm (°F)	−168	−252	−256	−31	−137	−254	*a*
Critical temperature (°F)	388.0	233.6	204.8	417.4	294.3	221.1	194.1
Critical pressure (psia)	635.0	597.0	716.0	495.0	474.0	631.0	618.7
Specific heat of liquid, 86°F	0.220	0.235	0.335	0.218	0.238	0.300	0.305
Specific heat of vapor, C_p 60°F at 1 atm	*a*	0.146	0.149	*a*	0.156	0.171	0.164
Specific heat at vapor, C_v 60°F at 1 atm	*a*	0.130	0.127	*a*	0.145	0.151	0.161
Ratio C_p/C_v = K (86°F at 1 atm)	1.11	1.14	1.18	1.12	1.09	1.13	1.023
Saturation pressure (psia) at							
−50°F	0.52	7.12	11.74	*a*	1.35	8.395	14.74
0°F	2.55	23.85	38.79	0.84	5.96	27.96	45.94
40°F	7.03	51.67	83.72	2.66	15.22	60.94	94.90
105°F	25.7	141.25	227.65	11.58	50.29	167.85	244.40
Net refrigerating effect (Btu/lb) 40–105°F (no subcooling)	67.56	49.13	66.44	54.54	43.46	59.82	43.72
Cycle efficiency (% Carnot cycle) 40–105°F	90.5	83.2	81.8	87.5	84.9	82.0	76.1
Liquid circulated 40–105°F [(lb/min/ton)]	2.96	4.07	3.02	3.66	4.62	3.35	4.58
Theoretical displacement 40–105°F (cuft/min/ton)]	16.1	3.14	1.98	39.5	9.16	2.69	2.04
Theoretical horsepower per ton 40–105°F	0.676	0.736	0.75	0.70	0.722	0.747	0.806
Coefficient at performance 40–105°F (4.71/HP per ton)	6.95	6.39	6.29	6.74	6.52	6.31	5.86
Cost compared with R 11	1.00	1.57	2.77	2.15	2.97	2.00	5.54

[a] Data not available or not applicable.
(Carrier Air Conditioning Co.).

ABSORPTION REFRIGERATION

The most widely used is ammonia absorption in water. A flowsketch of the process is in Figure 8.27. Liquid ammonia at a high pressure is obtained overhead in a stripper, and then is expanded through a valve and becomes the low temperature vapor–liquid mixture that functions as the refrigerant. The low pressure vapor is absorbed in weak liquor from the bottom of the stripper. Energy input to the refrigeration system is primarily that of the steam to the stripper reboiler and a minor amount of power to the pump and the cooling water circulation.

This kind of system has a useful range down to the atmospheric boiling point of ammonia, −28°F or −33°C, or even lower. Two or three stage units are proposed for down to −94°F. Sizing of equipment is treated by Bogart (1981).

Another kind of absorption refrigerant system employs aqueous lithium bromide as absorbent and circulating water as the refrigerant. It is used widely for air conditioning systems, in units of 600–700 tons producing water at 45°F.

CRYOGENICS

This term is applied to the production and utilization of temperatures in the range of liquid air, −200°F and lower. A great deal of information is available on this subject of special interest, for instance in *Chemical Engineers Handbook* (1984, 12.47–12.58) and in the book of Arkhanov et al. (1981).

REFERENCES

1. K.J. Bell and M.A. Ghaly, An approximate generalized design method for multicomponent partial condensers, *Chem. Eng. Prog. Symp. Ser.* **131,** 72–79 (1973).
2. V. Cavaseno et al. (Eds.), *Process Heat Exchange,* McGraw-Hill, New York, 1979.
3. D. Chisholm (Ed.), *Developments in Heat Exchange Technology I,* Applied Science, London, 1980.
4. J.R. Fair, Process heat transfer by direct fluid-phase contact, *Chem. Eng. Prog. Symp. Ser.* **118,** 1–11 (1972); *Chem. Eng.,* (12 June 1972).
5. V. Ganapathy, *Applied Heat Transfer,* PennWell Books, Tulsa, OK, 1982.
6. H. Gröber, S. Erk, and U. Grigull, *Fundamentals of Heat Transfer,* McGraw-Hill, New York, 1961.
7. H. Hausen, *Heat Transfer in Counterflow, Parallel Flow and Cross Flow,* McGraw-Hill, New York, 1983.
8. HEDH, *Heat Exchanger Design Handbook* (E.U. Schlünder et al., Eds.), Hemisphere, New York, 1983–date, 5 vols.
9. M. Jakob, *Heat Transfer,* Wiley, New York, 1957, Vol. 2.
10. S. Kakac, A.E. Bergles, and F. Mayinger (Eds.), *Heat Exchangers: Thermal-Hydraulic Fundamentals and Design,* Hemisphere, New York, 1981.
11. W.M. Kays and A.L. London, *Compact Heat Exchangers,* McGraw-Hill, New York, 1984.
12. D.Q. Kern, *Process Heat Transfer,* McGraw-Hill, New York, 1950.
13. S.K. Kutateladze and V.M. Borishanskii, *Concise Encyclopedia of Heat Transfer,* Pergamon, New York, 1966.
14. E.E. Ludwig, *Applied Process Design for Chemical and Petrochemical Plants,* Gulf, Houston, 1983, Vol. 3, pp. 1–200.
15. P.E. Minton, Designing spiral plate and spiral tube exchangers, *Chem. Eng.,* (4 May 1970); (18 May 1970).
16. R.K. Neeld and J.T. O'Bara, Jet trays in heat transfer service, *Chem. Eng. Prog.* **66**(7), 53 1970.
17. P.A. Schweitzer (Ed.), *Handbook of Separation Techniques for Chemical Engineers,* McGraw-Hill, New York, 1979, Sec. 2.3, Evaporators, Sec. 2.4, Crystallizers.
18. L. Silver, Gas cooling with aqueous condensation, *Trans. Inst. Chem. Eng.* **25,** 30–42 (1947).
19. E.F.C. Somerscales and J.G. Knudsen (Eds.), *Fouling of Heat Transfer Equipment,* Hemisphere, New York, 1981.
20. J. Taborek, G.F. Hewitt, and N. Afgan (Eds.), *Heat Exchangers Theory and Practice,* Hemisphere, New York, 1983.
21. TEMA Standards, Tubular Exchanger Manufacturers Association, Tarrytown, NY, 1978.
22. G. Walker, *Industrial Heat Exchangers,* Hemisphere, New York, 1982.*

* The book by Walker (Appendix D, 1982) has a guide to the literature of heat transfer in book form and describes the proprietary services HTFS (Heat Transfer and Fluid Services) and HTRI (Heat Transfer Research Inc.).

Fired Heaters (see also Ganapathy, HEDH, and Kern above)

23. F.A. Holland, R.M. Moores, F.A. Watson, and J.K. Wilkinson, *Heat Transfer,* Heinemann, London, 1970.
24. H.C. Hottel, in *McAdams Heat Transmission,* McGraw-Hill, New York, 1954.
25. W.E. Lobo and J.E. Evans, Heat transfer in the radiant section of petroleum heaters, *Trans. AIChE* **35,** 743 (1939).
26. C.C. Monrad, Heat transmission in the convection section of pipe stills, *Ind. Eng. Chem.* **24,** 505 (1932).
27. D.W. Wilson, W.E. Lobo, and H.C. Hottel, Heat transmission in the radiant section of tube stills, *Ind. Eng. Chem.* **24,** 486 (1932).
28. R.N. Wimpress, Rating fired heaters, *Hydrocarbon Process.* **42**(10), 115–126 (1963); Generalized method predicts fired-heater performance, *Chem. Eng.,* 95–102 (22 May 1978).

Selected American Petroleum Institute Standards (API, Washington, D.C.)

29. Std. 660, Shell-and-Tube Heat Exchangers for General Refinery Services, 1982.
30. Std. 661, Air-Cooled Heat Exchangers for General Refinery Services, 1978.
31. Std. 665, API Fired Heater Data Sheet, 1966, 1973.

Insulation

32. *Marks Mechanical Engineers Handbook,* McGraw-Hill, New York, 1978, pp. 6.169–6.177.
33. H.F. Rase and M.H. Barrow, *Project Engineering of Process Plants,* Wiley, New York, 1957, Chap. 19.
34. G.B. Wilkes, *Heat Insulation,* Wiley, New York, 1950.

Refrigeration

35. A. Arkhanov, I. Marfenina, Ye. Mikulin, *Theory and Design of Cryogenic Systems,* Mir Publishers, Moscow, 1981.
36. ASHRE, *Thermophysical Properties of Refrigerants,* American Society of Heating, Refrigeration and Air-Conditioning Engineers, Atlanta, GA, 1976.
37. M. Bogart, *Ammonia Absorption Refrigeration in Industrial Processes,* Gulf, Houston, 1981.
38. *Carrier System Design Manual,* Carrier Air Conditioning Co., Syracuse, NY, 1964, Part 4, Refrigerants, brines and oils.
39. F.L. Evans, *Equipment Design Handbook for Refineries and Chemical Plants,* Gulf, Houston, 1979, Vol. 1, pp. 172–196.
40. T.M. Flynn and K.D. Timmerhaus, Cryogenic processes, in *Chemical Engineers Handbook,* 1984, pp. 12.46–12.58.
41. W.B. Gosney, *Principles of Refrigeration,* Cambridge University Press, Cambridge, 1982.
42. E.E. Ludwig, *Applied Process Design for Chemical and Petroleum Plants,* Gulf, Houston, 1983, Vol. 1, pp. 201–250.
43. Y.R. Mehra, Refrigerating properties of ethylene, ethane, propylene and propane, *Chem. Eng.,* 97 (18 Dec. 1978); 131 (15 Jan. 1979); 95 (12 Feb. 1979); 165 (26 Mar. 1979).

9

DRYERS AND COOLING TOWERS

The processes of the drying of solids and the evaporative cooling of process water with air have a common foundation in that both deal with interaction of water and air and involve simultaneous heat and mass transfer. Water cooling is accomplished primarily in packed towers and also in spray ponds or in vacuum spray chambers, the latter for exceptionally low temperatures. Although such equipment is comparatively simple in concept, it is usually large and expensive, so that efficiencies and other aspects are considered proprietary by the small number of manufacturers in this field.

In contrast, a great variety of equipment is used for the drying of solids. Thomas Register lists about 35 pages of U.S. manufacturers of drying equipment, classified with respect to type or the nature of the material being dried. In a major respect, dryers are solids handling and transporting equipment, notable examples being perforated belt conveyors and pneumatic conveyors through which hot air is blown. Solids being dried cover a range of sizes from micron-sized particles to large slabs and may have varied and distinctive drying behaviors. As in some other long-established industries, drying practices of necessity have outpaced drying theory. In the present state of the art, it is not possible to design a dryer by theory without experience, but a reasonably satisfactory design is possible from experience plus a little theory.

Performances of dryers with simple flow patterns can be described with the aid of laboratory drying rate data. In other cases, theoretical principles and correlations of rate data are of value largely for appraisal of the effects of changes in some operating conditions when a basic operation is known. The essential required information is the residence time in the particular kind of dryer under consideration. Along with application of possible available rules for vessel proportions and internals to assure adequate contacting of solids and air, heat and material balances then complete a process design of a dryer.

In order to aid in the design of dryers by analogy, examples of dimensions and performances of the most common types of dryers are cited in this chapter. Theory and correlation of heat and mass transfer are treated in detail elsewhere in this book, but their use in the description of drying behavior will be indicated here.

9.1. INTERACTION OF AIR AND WATER

Besides the obvious processes of humidification and dehumidification of air for control of environment, interaction of air and water is a major aspect of the drying of wet solids and the cooling of water for process needs. Heat and mass transfer then occur simultaneously. For equilibrium under adiabatic conditions, the energy balance is

$$k_g\lambda(p_s - p) = h(T - T_w), \tag{9.1}$$

where p_s is the vapor pressure at the wet bulb temperature T_w. The moisture ratio, H lb water/lb dry air, is related to the partial pressure of the water in the air by

$$H = \frac{18}{29}\frac{p}{P - p} \simeq \frac{18}{29}\frac{p}{P}, \tag{9.2}$$

the approximation being valid for relatively small partial pressures. Accordingly, the equation of the adiabatic saturation line may be written

$$H_s - H = (h/\lambda k)(T - T_w) \tag{9.3}$$

$$= (C/\lambda)(T - T_s). \tag{9.4}$$

For water, numerically $C \simeq h/k$, so that the wet bulb and adiabatic saturation temperatures are identical. For other vapors this conclusion is not correct.

For practical purposes, the properties of humid air are recorded on psychrometric (or humidity) charts such as those of Figures 9.1 and 9.2, but tabulated data and equations also are available for greater accuracy. A computer version is available (Wiley Professional Software, Wiley, New York). The terminal properties of a particular adiabatic humification of air are located on the same saturation line, one of those sloping upwards to the left on the charts. For example, all of these points are on the same saturation line: $(T, H) = (250, 0.008)$, $(170, 0.026)$ and $(100, 0.043)$; the saturation enthalpy is 72 Btu/lb dry, but the individual enthalpies are less by the amounts 2.5, 1.2, and 0, respectively.

Properties such as moisture content, specific volume, and enthalpy are referred to unit mass of dry air. The units employed on Figure 9.1 are lb, cuft, °F, and Btu; those on Figure 9.2 are SI. The data are for standard atmospheric pressure. How to correct them for minor deviations from standard pressure is explained for example in *Chemical Engineers' Handbook* (McGraw-Hill, New York, 1984, 12.10). An example of reading the charts is with the legend of Figure 9.1. Definitions of common humidity terms and their units are given following.

1. Humidity is the ratio of mass of water to the mass of dry air,

$$H = W_w/W_a. \tag{9.5}$$

2. Relative humidity or relative saturation is the ratio of the prevailing humidity to the saturation humidity at the same temperature, or the ratio of the partial pressure to the vapor pressure expressed as a percentage,

$$\%RH = 100H/H_s = 100p/p_s. \tag{9.6}$$

3. The relative absolute humidity is

$$(H/H_s)_{\text{absolute}} = \left(\frac{p}{P - p}\right) \Big/ \left(\frac{p_s}{P - p_s}\right) \tag{9.7}$$

4. Vapor pressure of water is given as a function of temperature by

$$p_s = \exp(11.9176 - 7173.9/(T + 389.5)), \quad \text{atm}, \quad °\text{F}. \tag{9.8}$$

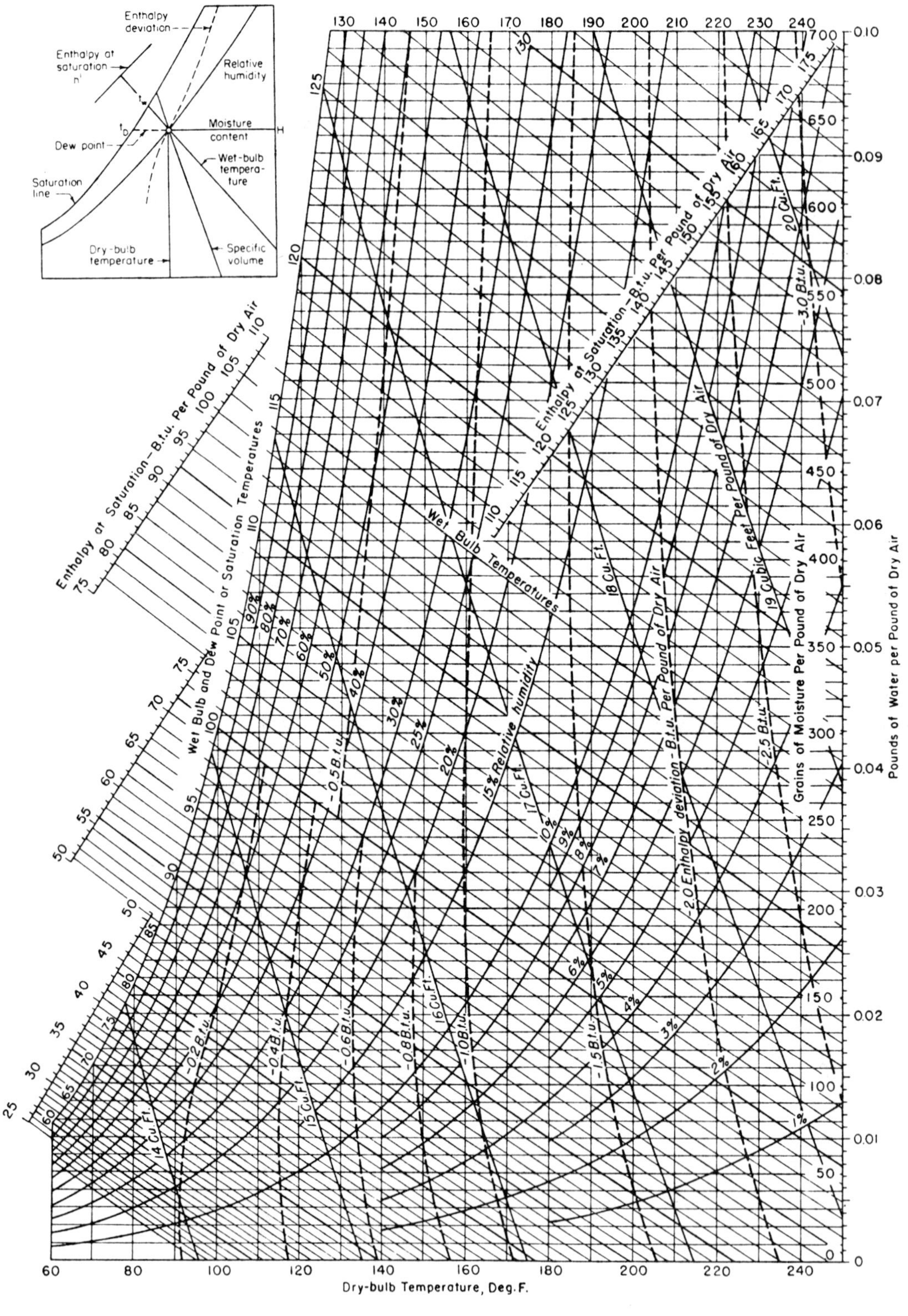

Figure 9.1. Psychrometric chart in English units (*Carrier Corp. Syracuse, NY*). Example: For air at 200°F with $H = 0.03$ lb/lb: $T_s = 106.5$°F, $V_h = 17.4$ cuft/lb dry, $100H/H_s = 5.9\%$, $h = h_s + D = 84 - 1.7 = 82.3$ Btu/lb dry.

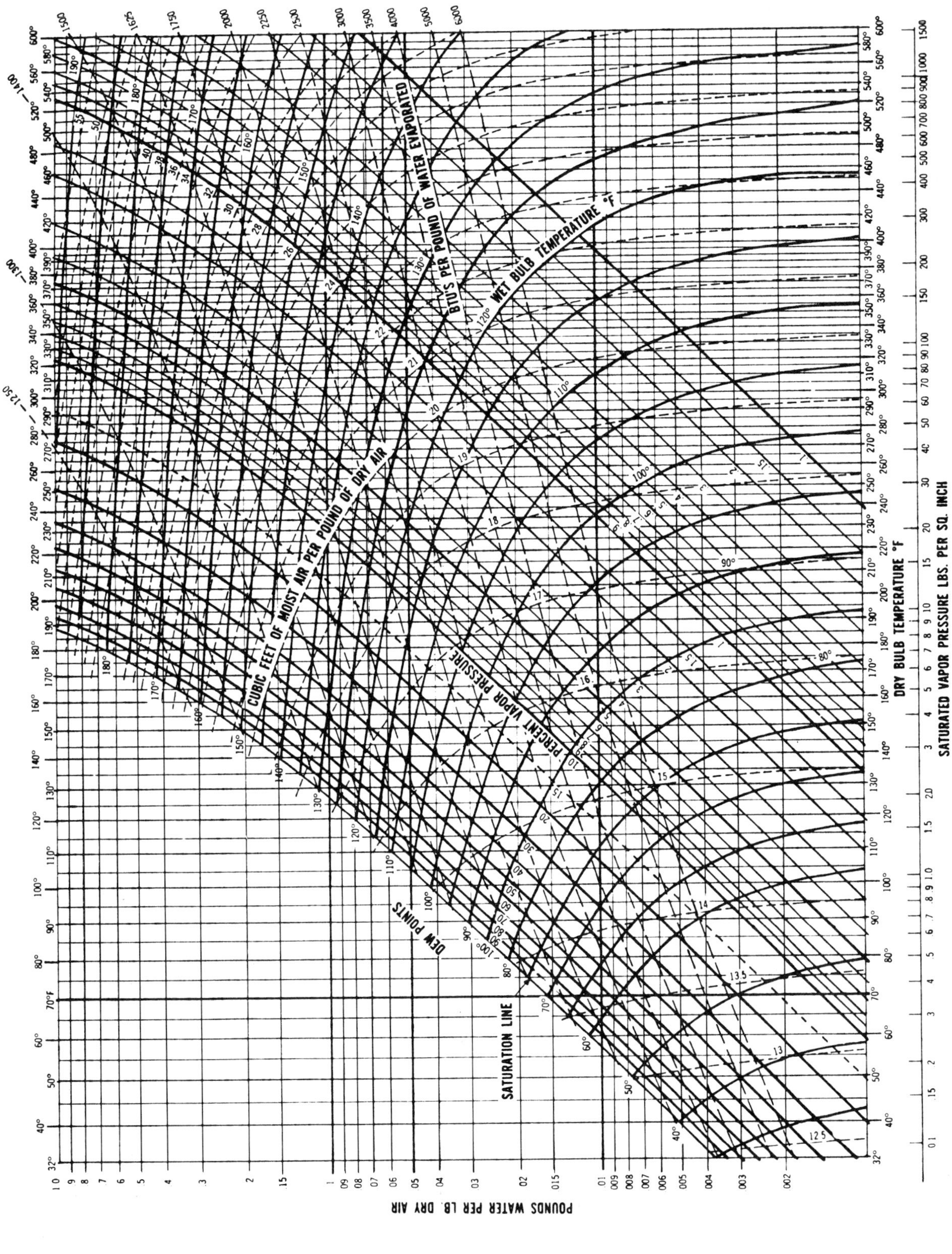

Figure 9.2. Psychrometric chart for a wide temperature range, 32–600°F *(Proctor and Schwartz, Inc., Horsham, PA).*

EXAMPLE 9.1
Conditions in an Adiabatic Dryer

The air to a dryer has a temperature of 250°F and a wet bulb temperature of 101.5°F and leaves the process at 110°F. Water is evaporated off the surface of the solid at the rate of 1500 lb/hr. Linear velocity of the gas is limited to a maximum of 15 ft/sec. The diameter of the vessel will be found.

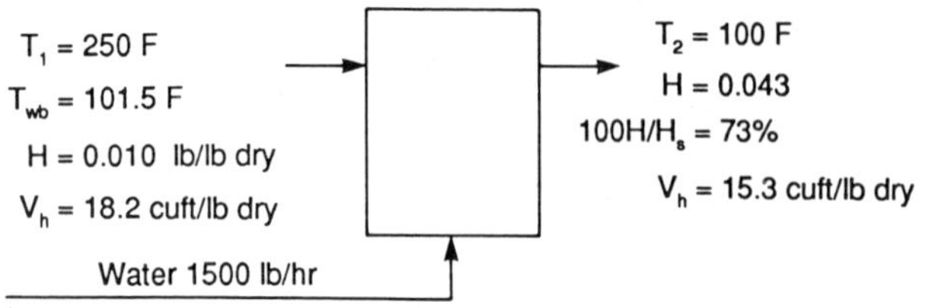

Terminal conditions of the air are read off the adiabatic saturation line and appear on the sketch:

$$\text{Dry air} = \frac{1500}{0.043 - 0.010} = 45{,}455 \text{ lb/hr}$$

$$\rightarrow \frac{45{,}455(18.2)}{3600} = 229.8 \text{ cfs},$$

$$D = \sqrt{229.8/15(\pi/4)} = 4.4 \text{ ft}.$$

5. The humid volume is the volume of 1 lb of dry air plus the volume of its associated water vapor,

$$V_h = 0.73(1/29 + h/18)(T + 459.6)/P, \quad \text{cuft/(lb dry air)}. \tag{9.9}$$

6. Humid specific heat is

$$C_h = C_a + C_w H = 0.24 + 0.45H, \quad \text{Btu/(F)(lb dry air)}. \tag{9.10}$$

7. The wet bulb temperature T_w is attained by measurement under standardized conditions. For water, T_w is numerically nearly the same as the adiabatic saturation temperature T_s.
8. The adiabatic saturation temperature T_s is the temperature attained if the gas were saturated by an adiabatic process.
9. With heat capacity given by item 6, the enthalpy of humid air is

$$h = 0.24T + (0.45T + 1100)H. \tag{9.11}$$

On the psychrometric chart of Figure 9.1, values of the saturation enthalpy h_s and a correction factor D are plotted. In these terms the enthalpy is

$$h = h_s + D. \tag{9.12}$$

In Figure 9.2, the enthalpy may be found by interpolation between the lines for saturated and dry air.

In some periods of drying certain kinds of solids, water is brought to the surface quickly so that the drying process is essentially evaporation of water from the free surface. In the absence of intentional heat exchange with the surrounding or substantial heat losses, the condition of the air will vary along the adiabatic saturation line. Such a process is analyzed in Example 9.1.

For economic reasons, equilibrium conditions cannot be approached closely. In a cooling tower, for instance, the effluent air is not quite saturated, and the water temperature is not quite at the wet bulb temperature. Percent saturation in the vicinity of 90% often is feasible. Approach is the difference between the temperatures of the water and the wet bulb. It is a significant determinant of cooling tower size as these selected data indicate:

Approach (°F)	5	10	15	20	25
Relative tower volume	2.4	1.6	1.0	0.7	0.55

Other criteria for dryers and cooling towers will be cited later.

9.2. RATE OF DRYING

In a typical drying experiment, the moisture content and possibly the temperature of the material are measured as functions of the time. The inlet and outlet rates and compositions of the gas also are noted. From such data, the variation of the rate of drying with either the moisture content or the time is obtained by mathematical differentiation. Figure 9.3(d) is an example. The advantage of expressing drying data in the form of rates is that their dependence on thermal and mass transfer driving forces is more simply correlated. Thus, the general drying equation may be written

$$-\frac{1}{A}\frac{dW}{d\theta} = h(T_g - T) = k_p(P - P_g) = k_H(H - H_g), \tag{9.13}$$

where subscript g refers to the gas phase and H is the moisture content, (kg/kg dry material), corresponding to a partial or vapor pressure P. Since many correlations of heat and mass transfer coefficients are known, the effects of many changes in operating conditions on drying rates may be ascertainable. Figures 9.3(g) and (h) are experimental evidence of the effect of humidity of the air and (i) of the effect of air velocity on drying rates.

Other factors, however, often complicate drying behavior. Although in some ranges of moisture contents the drying process may be simply evaporation off a surface, the surface may not dry uniformly and consequently the effective amount of surface may change as time goes on. Also, resistance to diffusion and capillary flow of moisture may develop for which phenomena no adequate correlations are known. Furthermore, shrinkage may occur on drying, particularly near the surface, which hinders further movement of moisture outwards. In other instances, agglomerates of particles may disintegrate on partial drying.

Some examples of drying data appear in Figure 9.3. Commonly recognized zones of drying behavior are represented in Figure 9.3(a). Equilibrium moisture contents assumed by various materials in contact with air of particular humidities is represented by (b). The shapes of drying rate curves vary widely with operating conditions and the physical state of the solid; (b) and others are some examples. No correlations have been developed or appear possible whereby such data can be predicted. In higher ranges of moisture content of some materials, the process of drying is essentially evaporation of moisture off the surface, and its rate remains constant until the surface moisture is depleted as long as the condition of the air remains the same. During this period, the rate is independent of the nature of the solid. The temperature of

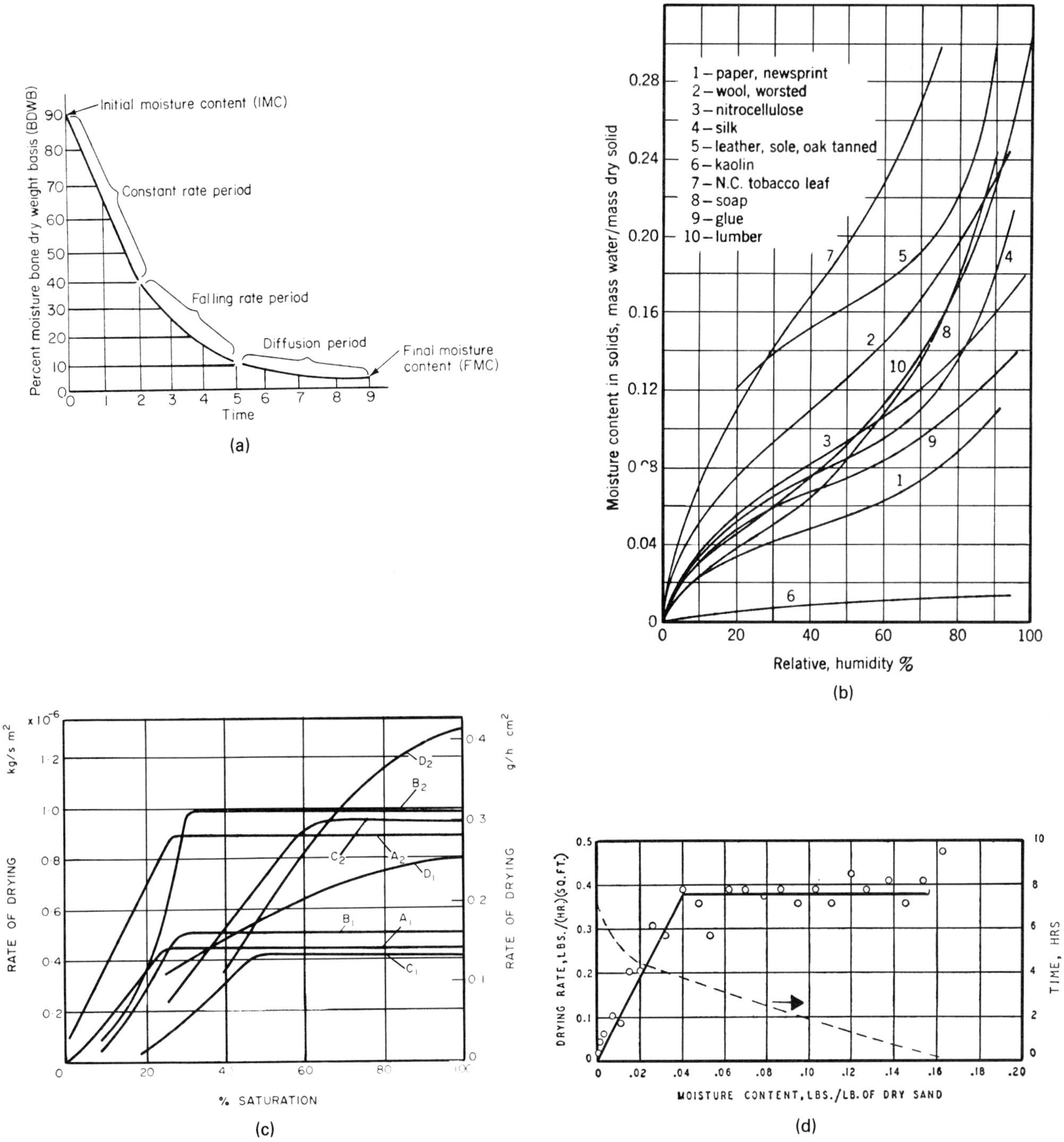

Figure 9.3. (a) Classic drying curve of moisture content against time; a heat-up period in which no drying occurs also is usually present *(Proctor and Schwartz, Inc.; Schweitzer, p. 4.144)*. (b) Equilibrium moisture content as a function of relative humidity; many other data are tabulated in Chemical Engineers Handbook *(McGraw-Hill, New York, 1984, 20.12). (These data are from National Academy of Science, copyright 1926.)* (c) Rate of drying as a function of % saturation at low (subscript 1) and high (subscript 2) drying rates: (A) glass spheres, 60 μm, bed 51 mm deep; (B) silica flour, 23.5 μm, 51 mm deep; (C) silica flour, 7.5 μm, 51 mm bed; (D) silica flour, 2.5 μ, 65 mm deep *(data of Newitt et al.,* Trans. Inst. Chem. Eng. **27,** *1 (1949)*. (d) Moisture content, time and drying rates in the drying of a tray of sand with superheated steam; surface 2.35 sqft, weight 27.125 lb. The scatter in the rate data is due to the rough numerical differentiation *(Wenzel, Ph.D. thesis, University of Michigan, 1949)*. (e) Temperature and drying rate in the drying of sand in a tray by blowing air across it. Dry bulb 76.1°C, wet bulb 36.0°C *(Ceaglske and Hougen,* Trans. AIChE **33,** *283 (1937)*. (f) Drying rates of slabs of paper pulp of several thicknesses *[after McCready and McCabe,* Trans. AIChE **29,** *131 (1933)]*. (g) Drying of asbestos pulp with air of various humidities *[McCready and McCabe,* Trans. AIChE **29,** *131 (1933)]*. (h) Effect of temperature difference on the coefficient K of the falling rate equation $-dW/d\theta = KW$ *[Sherwood and Comings,* Trans. AIChE **27,** *118 (1932)]*. (i) Effect of air velocity on drying of clay slabs. The data are represented by $R = 2.0u^{0.74}(H_w - H)$. The dashed line is for evaporation in a wetted wall tower *(Walker, Lewis, McAdams, and Gilliland,* Principles of Chemical Engineering, *McGraw-Hill, New York, 1937)*.

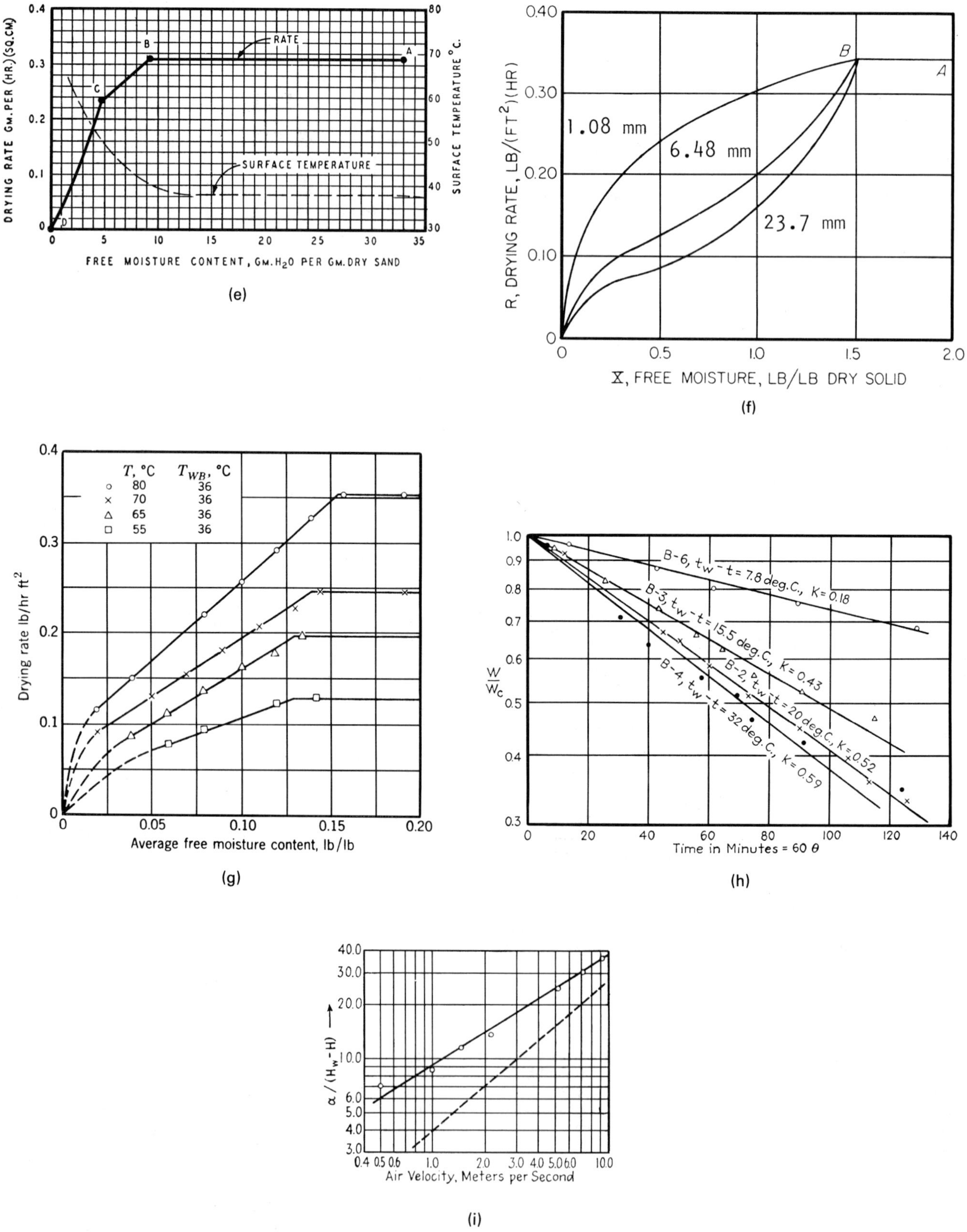

Figure 9.3—(*continued*)

EXAMPLE 9.2
Drying Time over Constant and Falling Rate Periods with Constant Gas Conditions

The data of Figure 9.3(d) were obtained on a sample that contained 27.125 lb dry sand and had an exposed drying surface of 2.35 sqft. Take the case of a sample that initially contained 0.168 lb moisture/lb dry material and is to be dried to $W = 0.005$ lb/lb. In these units, the constant rate shown on the graph is transformed to

$$-\frac{1}{2.35}\frac{dW}{d\theta} = \frac{0.38}{27.125} \quad \text{(lb/lb)/(hr)(sqft)},$$

which applies down to the critical moisture content $W_c = 0.04$ lb/lb. The rate behavior over the whole moisture range is

$$-\frac{dW}{d\theta} = \begin{cases} 0.03292, & 0.04 < W < 0.168, \\ 0.823W, & W < 0.04. \end{cases}$$

Accordingly, the drying time is

$$\begin{aligned}\theta &= \frac{W - W_c}{0.03292} + \frac{1}{0.823}\ln\left(\frac{W_c}{W}\right) \\ &= \frac{0.168 - 0.04}{0.03292} + \frac{1}{0.823}\ln\left(\frac{0.04}{0.005}\right) \\ &= 6.42 \text{ hr}.\end{aligned}$$

This checks the reading off the plot of the original data on Figure 9.3(d).

---

the evaporate assumes the wet bulb temperature of the air. Constant rate zones are shown in (d) and (e), and (e) reports that temperatures are truly constant in such a zone.

The moisture content at which the drying rate begins to decline is called critical. Some of the variables on which the transition point depends are indicated in Figures 9.3(c) and (g). The shape of the falling rate curve sometimes may be approximated by a straight line, with equation

$$-\frac{dW}{d\theta} = k(W - W_e), \tag{9.14}$$

where W_e is the equilibrium moisture content. When W_e is zero as it often is of nonporous granular materials, the straight line goes through the origin. (d) and (h) illustrate this kind of behavior. The drying time is found by integration of the rate plots or equations. The process is illustrated in Example 9.2 for straight line behavior. Other cases require numerical integration. Each of the examples of Figure 9.3 corresponds to a particular substantially constant gas condition. This is true of shallow bed drying without recirculation of humid gas, but in other kinds of drying equipment the variation of the rate with time and position in the equipment, as well as with the moisture content, must be taken into account.

An approximation that may be justifiable is that the critical moisture content is roughly independent of the drying conditions and that the falling rate curve is linear. Then the rate equations may be written

$$-\frac{1}{A}\frac{dW}{d\theta} = \begin{cases} k(H_s - H_g), & W_c < W < W_0, \\ \dfrac{k(H_s - H_g)(W - W_e)}{W_c - W_e}, & W_e < W < W_c. \end{cases} \tag{9.15}$$

Examples 9.3 and 9.4 apply these relations to a countercurrent dryer in which the humidity driving force and the equilibrium moisture content vary throughout the equipment.

LABORATORY AND PILOT PLANT TESTING

The techniques of measuring drying of stationary products, as on trays, are relatively straightforward. Details may be found in the references made with the data of Figure 9.3. Mass transfer resistances were eliminated by Wenzel through use of superheated steam as the drying medium.

In some practical kinds of dryers, the flow patterns of gas and solid are so complex that the kind of rate equation discussed in this section cannot be applied readily. The sizing of such equipment is essentially a scale-up of pilot plant tests in similar equipment. Some manufacturers make such test equipment available. The tests may establish the residence time and the terminal conditions of the gas and solid. Dusting behavior and possible need for recycling of gas or of dried material are among the other factors that may be noted.

Such pilot plant data are cited for the rotary dryer of Example 9.6. For the pneumatic conveying dryer of Example 9.8, the tests establish heat and mass transfer coefficients which can be used to calculate residence time under full scale operation.

Scale-up factors as small as 2 may be required in critical cases, but factors of 5 or more often are practicable, particularly when the tests are analyzed by experienced persons. The minimum dimensions of a test rotary dryer are 1 ft dia by 6 ft long. A common criterion is that the product of diameter and rpm be in the range 25–35. A laboratory pneumatic conveying dryer is described by Nonhebel and Moss (1971). The veseel is 8 cm dia by about 1.5 m long. Feed rate suggested is 100 g/min and the air velocity about 1 m/sec. They suggest that 6–12 passes of the solid through this equipment may be needed to obtain the requisite dryness because of limitations in its length.

The smallest pilot spray dryer supplied by Bowen Engineering Co. is 30 in. dia by 2.5–6.0 ft high. Atomization is with 15 SCFM of air at 100 psig. Air rate is 250 actual cfm at 150–1000°F. Evaporation rates of 15–80 lb/hr are attained, and particles of product range from 5 to 40 μm.

A pilot continuous multitray dryer is available from the Wyssmont Co. It is 4 ft dia by 5 ft high with 9 trays and can handle 25–200 lb/hr of feed.

Batch fluidized bed dryers are made in quite small sizes, of the order of 100 lb/hr of feed as the data of Table 9.14(a) show, and are suitable for pilot plant work.

9.3. CLASSIFICATION AND GENERAL CHARACTERISTICS OF DRYERS

Removal of water from solids is most often accomplished by contacting them with air of low humidity and elevated temperature. Less common, although locally important, drying processes apply heat radiatively or dielectrically; in these operations as in freeze drying, the role of any gas supply is that of entrainer of the humidity.

The nature, size, and shape of the solids, the scale of the operation, the method of transporting the stock and contacting it with gas, the heating mode, etc. are some of the many factors that

EXAMPLE 9.3
Drying with Changing Humidity of Air in a Tunnel Dryer
A granular material deposited on trays or a belt is moved through a tunnel dryer countercurrently to air that is maintained at 170°F with steam-heated tubes. The stock enters at 1400 lb dry/hr with $W = 1.16$ lb/lb and leaves with 0.1 lb/lb. The air enters at 5% relative humidity ($H_g = 0.0125$ lb/lb) and leaves at 60% relative humidity at 170°F ($H_g = 0.203$ lb/lb). The air rate found by moisture balance is 7790 lb dry/hr:

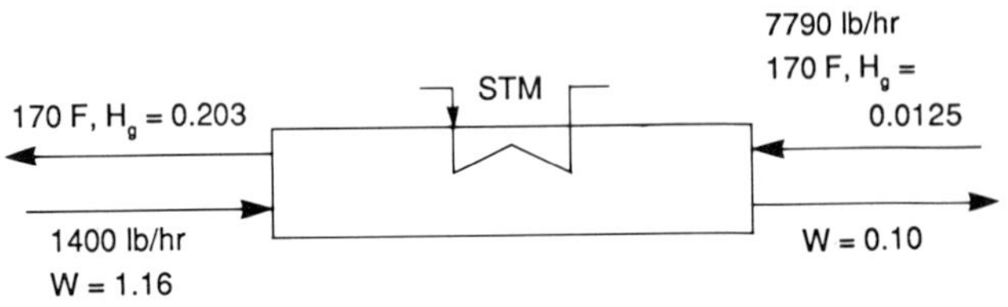

Drying tests reported by Walker, Lewis, McAdams, and Gilliland, *Principles of Chemical Engineering*, McGraw-Hill, New York, (1937, p. 671) may be represented by the rate equation

$$-100\frac{dW}{d\theta} = \begin{cases} 0.28\ (\text{lb/lb})/\text{hr}, & 0.58 < W < 1.16, \\ 0.28(W - W_e)/(0.58 - W_e), & W_e < W < 0.58. \end{cases} \quad (1)$$

The air was at 95°F and 7% relative humidity, corresponding to a humidity driving force of $H_s - H_g = 0.0082$. Equilibrium moisture content as a function of the fraction relative humidity (RH), and assumed independent of temperature, is represented by

$$W_e = 0.0036 + 0.1539(\text{RH}) - 0.097(\text{RH})^2. \quad (2)$$

The critical moisture content is assumed indpendent of the drying rate. Accordingly, under the proposed operating conditions, the rate of drying will be

$$-100\frac{dW}{d\theta} = \begin{cases} \dfrac{0.28(H_s - H_g)}{0.0082}, & 0.58 < W < 1.16, \\ \dfrac{0.28(H_s - H_g)(W - W_e)}{0.0082(0.58 - 0.014)}, & W_e < W < 0.58. \end{cases} \quad (3)$$

With moisture content of the stock as a parameter, the humidity of the air is calculated by moisture balance from

$$H_g = 0.0125 + (1400/7790)(W - 0.1). \quad (4)$$

The corresponding relative humidities and wet bulb temperatures and corresponding humidities H_s are read off a psychrometric chart. The equilibrium moisture is found from the relative humidity by Eq. (2). The various corrections to the rate are applied in Eq. (3). The results are tabulated, and the time is found by integration of the rate data over the range $0.1 < W < 1.16$.

W	H_g	H_s	RH	W_e	Rate	1/Rate
1.16	0.203	0.210			0.239	4.184
1.00	0.174	0.182			0.273	3.663
0.9	0.156	0.165			0.303	3.257
0.8	0.138	0.148			0.341	2.933
0.7	0.120	0.130			0.341	2.933
0.58	0.099	0.110	0.335	0.044	0.356	2.809
0.50	0.084	0.096	0.29	0.040	0.333	3.003
0.4	0.066	0.080	0.24	0.035	0.308	3.247
0.3	0.048	0.061	0.18	0.028	0.213	4.695
0.2	0.030	0.045	0.119	0.021	0.162	6.173
0.1	0.0125	0.0315	0.050	0.011	0.102	9.804

The drying time is

$$\theta = \int_{1.16}^{0.10} \frac{dw}{\text{rate}} = 4.21\ \text{hr}, \quad \text{by trapezoidal rule.}$$

The length of tunnel needed depends on the space needed to ensure proper circulation of air through the granular bed. If the bed moves through the dryer at 10 ft/hr, the length of the dryer must be at least 42 ft.

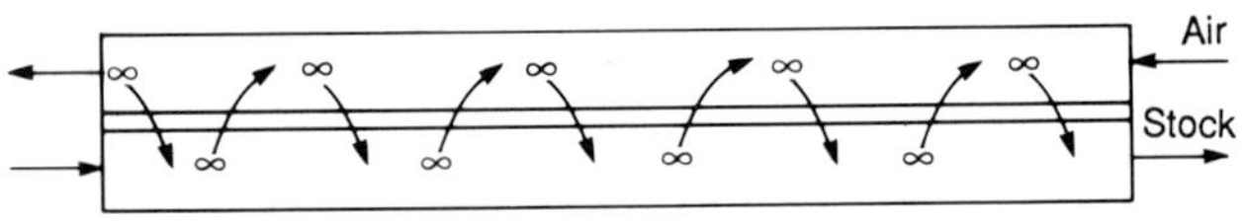

have led to the development of a considerable variety of equipment. The most elaborate classification of dryers is that of Kröll (1978) which assigns one of 10 letters for the kind of solid and one of seven numbers for the kind of operation. As modified by Keey (1972), it comprises 39 main classes and a total of 70 with subclasses. Less comprehensive but perhaps more practical classifications are shown in Table 9.1. They take into account the method of operation, the physical form of the stock, special features, scale of production, and drying time.

In a later section, the characteristics and performances of the most widely used equipment will be described in some detail. Many types are shown in Figure 9.4. Here some comparisons are made. Evaporation rates and thermal efficiencies are compared in Table 9.2, while similar and other data appear in Table 9.3. The wide spreads of these numbers reflect the diversity of individual designs of the same general kind of equipment, differences in moisture contents, and differences in drying properties of various materials. Fluidized bed dryers, for example, are operated as batch or continuous, for pharmaceuticals or asphalt, at rates of hundreds or many thousands of pounds per hour.

An important characteristic of a dryer is the residence time distribution of solids in it. Dryers in which the particles do not move relatively to each other provide uniform time distribution. In spray, pneumatic conveying, fluidized bed, and other equipment in which the particles tumble about, a substantial variation in residence time develops. Accordingly, some particles may overdry and some remain wet. Figure 9.5 shows some data. Spray and pneumatic conveyors have wide time distributions; rotary and fluidized bed units have narrower but far from uniform ones. Differences in particle size also lead to nonuniform drying. In pneumatic conveying dryers particularly, it is common practice to recycle a portion of the product continuously to ensure adequate overall drying. In other cases recycling may be performed to improve the handling characteristics when the feed material is very wet.

EXAMPLE 9.4
Effects of Moist Air Recycle and Increase of Fresh Air Rate in Belt Conveyor Drying

The conditions of Example 9.3 are taken except that recycle of moist air is employed and the equilibrium moisture content is assumed constant at $W_e = 0.014$. The material balance in terms of the recycle ratio R appears on the sketch:

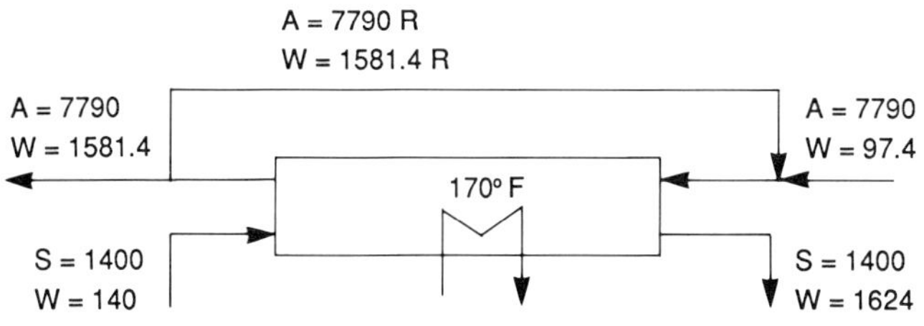

A = air, W = water, S = dry solid

Humidity of the air at any point is obtained from the water balance

$$H_g = \frac{1581.4R + 97.4 + 1400(W - 0.1)}{7790(R+1)}. \qquad (1)$$

The vapor pressure is

$$p_s = \exp[11.9176 - 7173.9/(T_s + 389.5)] \text{ atm}. \qquad (2)$$

The saturation humidity is

$$H_s = (18/29)p_s/(1 - p_s). \qquad (3)$$

The heat capacity is

$$C = 0.24 + 0.45H_g. \qquad (4)$$

With constant air temperature of 170°F, the equation of the adiabatic saturation line is

$$170 - T_s = \frac{\lambda}{C}(H_s - H_g) \simeq \frac{900}{C}(H_s - H_g). \qquad (5)$$

The drying rate equations above and below the critical moisture content of 0.58 are

$$-100\frac{dW}{d\theta} = \begin{cases} 34.15(R+1)^{0.8}(H_s - H_g), & 0.58 < W < 1.16, \quad (6) \\ 60.33(R+1)^{0.8}(H_s - H_g)(W - 0.014), & W < 0.58. \quad (7) \end{cases}$$

When fresh air supply is simply increased by a factor $R + 1$ and no recycle is employed, Eq. (1) is replaced by

$$H_g = \frac{97.4(R+1) + 1400(W - 0.1)}{7790(R+1)}. \qquad (8)$$

The solution procedure is:

1. Specify the recycle ratio R (lbs recycle/lb fresh air, dry air basis).
2. Take a number of discrete values of W between 1.16 and 0.1. For each of these find the saturation temperature T_s and the drying rates by the following steps.
3. Assume a value of T_s.
4. Find H_g, P_s, H_s, and C from Eqs. (1)–(4).
5. Find the value of T_s from Eq. (5) and compare with the assumed value. Apply the Newton–Raphson method with numerical derivatives to ultimately find the correct value of T_s and the corresponding value of H_s.
6. Find the rate of drying from Eqs. (6), (7).
7. Find the drying time by integration of the reciprocal rate as in Example 9.3, with the trapezoidal rule.

The printout shows saturation temperatures and reciprocal rates for $R = 0$, 1, and 5 with recycle; and for $R = 1$ with only the fresh air rate increased, using Eq. (8). The residence times for the four cases are

$R = 0$,	moist air,	$\theta = 3.667$ hrs
$= 1$,	moist air,	$= 2.841$
$= 5$,	moist air,	$= 1.442$
$= 1$,	fresh air,	$= 1.699$.

Although recycling of moist air does reduce the drying time because of the increased linear velocity, an equivalent amount of fresh air is much more effective because of its lower humidity. The points in favor of moist air recycle, however, are saving in fuel when the fresh air is much colder than 170°F and possible avoidance of case hardening or other undesirable phenomena resulting from contact with very dry air.

R = 0

W	T_s	1 / Rate
1.16	150.21	3.9627
1.00	145.92	3.4018
.90	142.86	3.1043
.80	139.45	2.8365
.70	135.62	2.5918
.60	131.24	2.3680
.50	126.19	2.5187
.40	120.25	2.8795
.30	113.08	3.5079
.20	104.15	4.8223
.10	92.45	9.2092

R = 1, fresh air

W	T_s	1 / Rate
1.16	132.62	1.3978
1.00	128.81	1.2985
.90	126.19	1.2395
.80	123.35	1.1815
.70	120.25	1.1248
.60	116.85	1.0693
.58	116.12	1.0582
.50	113.08	1.1839
.40	108.89	1.4112
.30	104.15	1.7979
.20	98.74	2.6014
.10	92.45	5.2893

Example 9.4—(*continued*)

R = 1, moist air

W	T_s	1 / Rate
1.16	150.21	2.2760
1.00	148.15	2.1043
.90	146.77	2.0088
.80	145.33	1.9181
.70	143.81	1.8323
.60	142.21	1.7509
.58	141.88	1.7351
.50	140.52	1.9534
.40	138.72	2.3526
.30	136.82	3.0385
.20	134.79	4.4741
.10	132.62	9.3083

R = 5, moist air

W	T_s	1 / Rate
1.16	150.21	.9451
1.00	149.54	.9208
.90	149.11	.9060
.80	148.68	.8916
.70	148.24	.8776
.60	147.79	.8630
.58	147.70	.8604
.50	147.33	.9918
.40	146.87	1.2302
.30	146.40	1.6364
.20	145.92	2.4824
.10	145.43	5.3205

```
10 ! Example 9.4. Belt conveyor
    drying
20 R=1 ! change for other cases
30 INPUT W
40 H1=(1581.4*(R+1)+97.4+1400*(
    W-.1))/7790/(R+1)
45 ! H1=(97.4*(R+1)+1400*(W-.1)
    )/7790/(R+1) ! Replace line
    40 with this when no recycle
    is used.
50 C=.24+.45*H1
60 T=120 ! Trial sat temp
70 GOSUB 200
80 Y1=Y
90 T=1.0001*T
100 GOSUB 200
110 Y2=Y
120 K=.0001*T*Y1/(Y2-Y1)
130 T=T/1.0001-K ! Newton-Raphso
    n
140 IF ABS(K/T)<=.00001 THEN 160
150 GOTO 70
160 PRINT USING 170 ; W,T,1/R1 !
     Subst R2 for R1 when W<.58
170 IMAGE D.DD,2X,DDD.DD,2X,D.DD
    DD
175 GOTO 30
180 END
200 ! SR for sat temp
210 P=EXP(11.9176-7173.9/(T+389.
    5))
220 H=.621*P/(1-P) ! sat humidit
    y
230 Y=170-T-900*(H-H1)/C
240 R1=34.15*(R+1)^.8*(H-H1)
250 R2=60.33*(R+1)^.8*(H-H1)*(W-
    .014)
260 RETURN
270 END
```

PRODUCTS

More than one kind of dryer may be applicable to a particular product, or the shape and size may be altered to facilitate handling in a preferred kind of machine. Thus, application of through-circulation drying on tray or belt conveyors may require prior extrusion, pelleting, or briquetting. Equipment manufacturers know the capabilities of their equipment, but they are not always reliable guides to comparison with competitive kinds since they tend to favor what they know best. Industry practices occasionally change over a period of time. For example, at one time rotary kilns were used to dry and prepare fertilizer granules of a desired size range by accretion from concentrated solutions onto the mass of drying particles. Now this operation is performed almost exclusively in fluidized bed units because of economy and controlability of dust problems.

Typical examples of products that have been handled successfully in particular kinds of dryers are listed in Table 9.4. The performance data of later tables list other examples.

COSTS

Differences in thermal economies are stated in the comparisons of Table 9.2 and other tables. Some equipment cost data are in Chapter 20. When the capacity is large enough, continuous dryers are less expensive than batch units. Those operating at atmospheric pressure cost about 1/3 as much as those at vacuum. Once-through air dryers are one-half as expensive as recirculating gas equipment. Dielectric and freeze driers are the most expensive and are justifiable only for sensitive and specialty products. In the range of 1–50 Mtons/yr, rotary, fluidized bed and pneumatic conveying dryers cost about the same, although there are few instances where they are equally applicable.

SPECIFICATION FORMS

A listing of key information relating to dryer selection and design is in Table 9.5. Questionnaires of manufacturers of several kinds of dryers are in Appendix C.

9.4. BATCH DRYERS

Materials that require more than a few minutes drying time or are in small quantity are treated on a batch basis. If it is granular, the material is loaded on trays to a depth of 1–2 in. with spaces of approximately 3 in. between them. Perforated metal bottoms allow drying from both sides with improved heat transfer. Hot air is blown across or through the trays. Cross velocities of 1000 ft/min are feasible if dusting is not a problem. Since the rate of evaporation increases roughly with the 0.8 power of the linear velocity, high velocities are desirable and are usually achieved by internal recirculation with fans. In order to maintain humidity at operable levels, venting and fresh air makeup are provided at rates of 5–50% of the internal circulation rate. Rates of evaporation of 0.05–0.4 lb/(hr)(sqft tray area) and steam requirements of 1.5–2.3 lbs/lb evaporation are realized.

Drying under vacuum is commonly practiced for sensitive materials. Figure 9.6 shows cross and through circulation tray

TABLE 9.1. Classification of Dryers by Several Criteria[a]

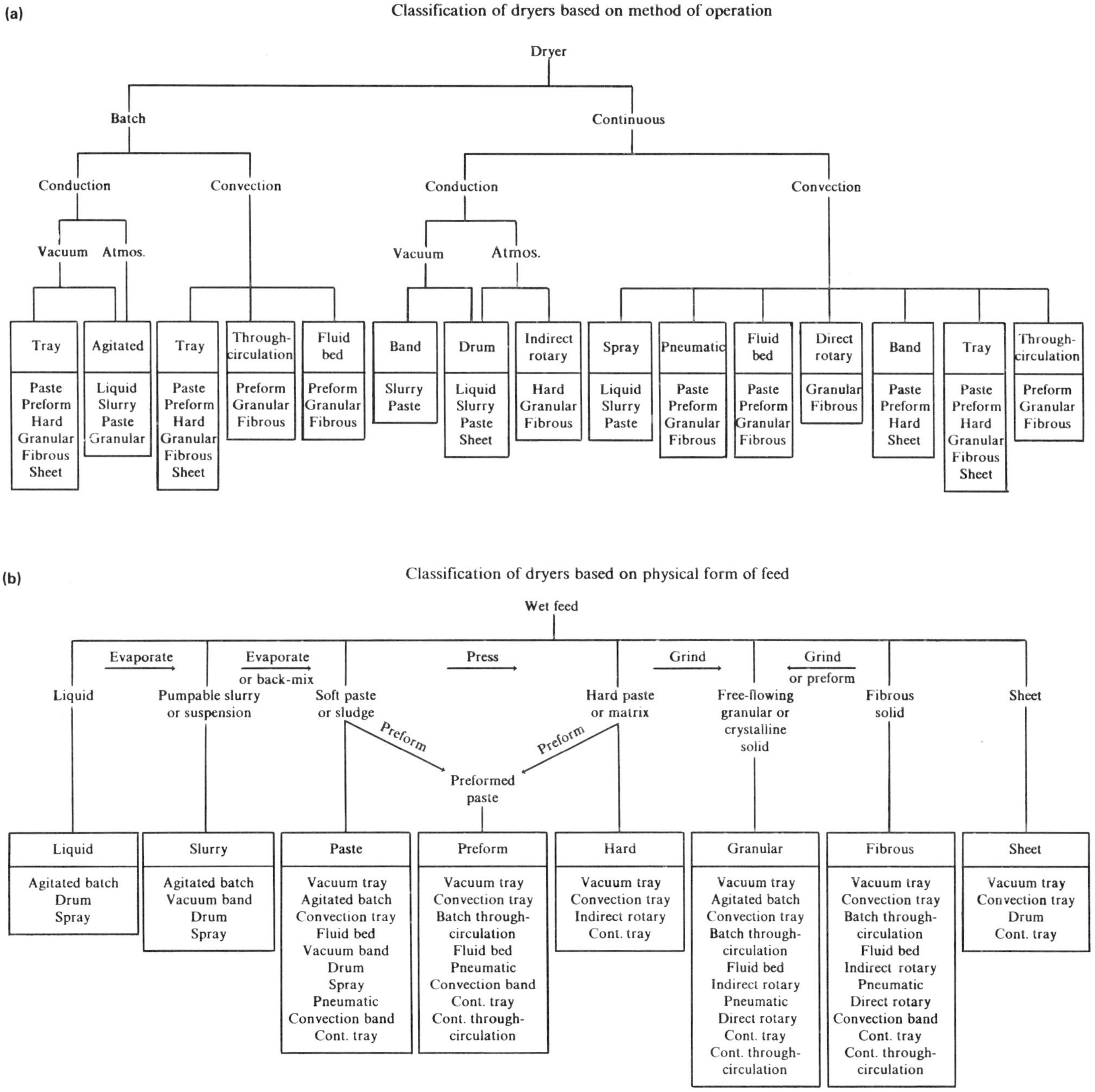

[a] See Figure 9.4 for sketches of dryer types.
[Items (a)–(d) by Nonhebel and Moss, 1971, pp. 45, 48–50].

TABLE 9.1—(*continued*)

(c)

Classification of dryers by scale of production

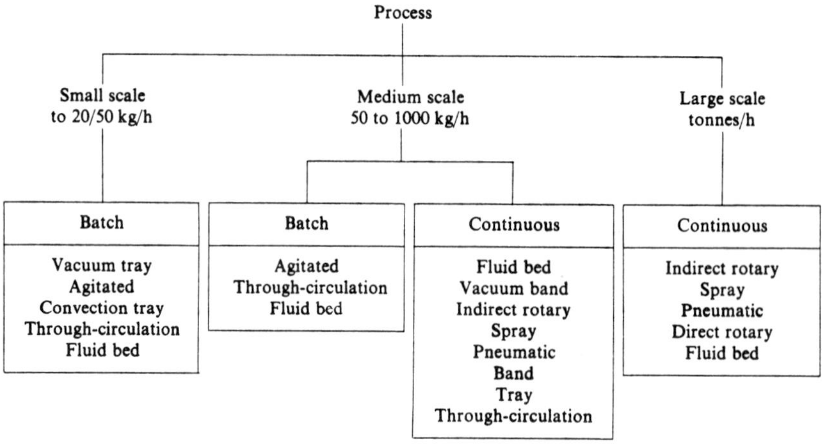

(d)

Classification of dryers by suitability for special features

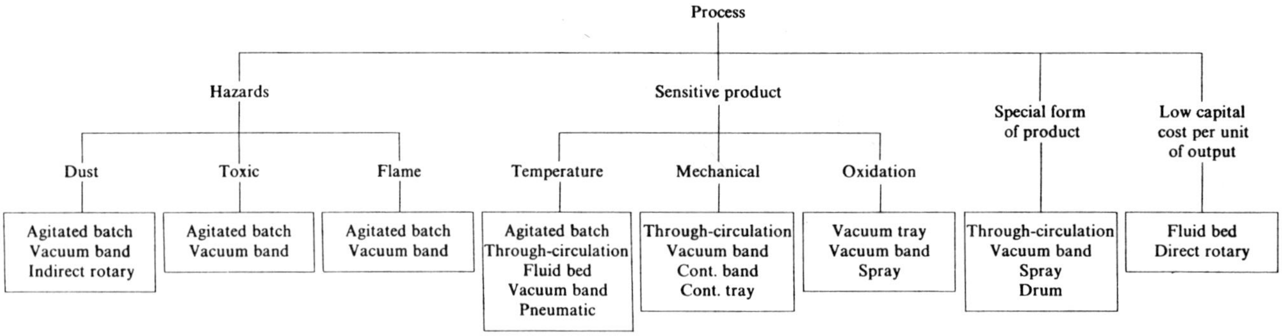

arrangements. The typical operating data of Table 9.6 cover a wide range of drying times, from a fraction of an hour to many hours. Charging, unloading, and cleaning are labor-intensive and time-consuming, as much as 5–6 hr for a 200-tray dryer, with trays about 5 sqft and 1–1.5 in. deep, a size that is readily handled manually. They are used primarily for small productions of valuable and thermally sensitive materials. Performance data are in Tables 9.6(b) and (c).

Through circulation dryers employ perforated or open screen bottom tray construction and have baffles that force the air through the bed. Superficial velocities of 150 ft/min are usual, with pressure drops of 1 in. or so of water. If it is not naturally granular, the material may be preformed by extrusion, pelleting, or briquetting so that it can be dried in this way. Drying rates are greater than in cross flow. Rates of 0.2–2 lb/(hr)(sqft tray area) and thermal efficiencies of 50% are realized. Table 9.7(d) has performance data.

Several types of devices that are used primarily for mixing of granular materials have been adapted to batch drying. Examples appear in Figure 9.8. They are suited to materials that do not stick to the walls and do not agglomerate during drying. They may be jacketed or provided with heating surfaces in the form of tubes or platecoils, and are readily arranged for operation under vacuum when handling sensitive materials. The double-cone tumbler has been long established. Some operating data are shown in Table 9.7. It and V-shaped dryers have a gentle action that is kind to fragile materials, and are discharged more easily than stationary cylinders or agitated pans. The fill proportion is 50–70%. When heated with 2 atm steam and operating at 10 Torr or so, the evaporation rate is 0.8–1.0 lb/(hr)(sqft of heating surface).

Fixed cylinders with rotating ribbons or paddles for agitation and pans with vertical agitators are used to a limited extent in batch operation. Pans are used primarily for materials that become sticky during drying. Table 9.7 and Figure 9.7 are concerned with this kind of equipment.

A detailed example of capital and operating costs of a jacketed vacuum dryer for a paste on which they have laboratory drying data is worked out by Nonhebel and Moss (1971, p. 110).

Fluidized bed dryers are used in the batch mode on a small scale. Table 9.14(a) has some such performance data.

9.5. CONTINUOUS TRAY AND CONVEYOR BELT DRYERS

Trays of wet material loaded on trucks may be moved slowly through a drying tunnel: When a truck is dry, it is removed at one end of the tunnel, and a fresh one is introduced at the other end. Figure 9.8(c) represents such equipment. Fresh air inlets and humid

TABLE 9.2. Evaporation Rates and Thermal Efficiencies of Dryers

Equipment	Figure 9.4	(lb/hr)/sqft	(lb/hr)/cuft	Efficiency[a] (%)
Belt conveyor	e			46–58
Shelf				
Flow through	a	0.02–2.5		18–41
Flow past	a	0.02–3.1		18–41
Rotary				
Roto-louvre		7.2–15.4		23–66
Parallel current direct fired		6.1–16.4		65
Parallel current warm air	f	6.1–16.4		50
Countercurrent direct fired		6.1–16.4		60
Countercurrent warm air	f	6.1–16.4		45
Steam tube	h	6.1–16.4		85
Indirect fired	g	6.1–16.4		25
Tunnel				36–42
Pneumatic				
0.5 mm dia granules	o		6.2	26–63
1.0 mm			1.2	26–63
5 mm			0.25	26–63
Spray	m		0.1–3	21–50
Fluidized bed	n		50–160	20–55
Drum	l	1.4–5.1		36–73
Spiral agitated				
High moisture	i	1–3.1		36–63
Low moisture	i	0.1–0.5		36–63
Splash paddle	k		5.6	65–70
Scraped multitray	d	0.8–1.6		

[a] Efficiency is the ratio of the heat of evaporation to the heat input to the dryer.

TABLE 9.3. Comparative Performances of Basic Dryer Types

	Basic Dryer Type					
	Tray	Conveyor	Rotary	Spray	Flash	Fluid Bed
Product	filter cake	clay	sand	TiO_2	spent grain	coal
Drying time (min)	1320	9.5	12	<1.0	<1.0	2.0
Inlet gas temperature (°F)	300	420	1650	490	1200	1000
Initial moisture (% dry basis)	233	25	6	100	150	16
Final moisture (% dry basis)	1	5.3	0.045	0	14	7.5
Product loading (lb dry/ft^2)	3.25	16.60	N.A.	N.A.	N.A.	21 in. deep
Gas velocity (ft/min)	500	295	700	50	2000	1000
Product dispersion in gas	slab	packed bed	gravity flow	spray	dispersed	fluid bed
Characteristic product shape	thin slab	extrusion	granules	spherical drops	grains	$\frac{1}{2}$-in. particles
Capacity [lb evap./(h)(dryer area)]	0.34	20.63	1.35[a]	0.27[a]	10[a]	285
Energy consumed (Btu/lb evap.)	3000	1700	2500	1300	1900	2000
Fan [hp/(lb evap./h)]	0.042	0.0049	0.0071	0.019	0.017	0.105

[a] lb evap./(h)(dryer, volume).
(Wentz and Thygeson, 1979: tray column from Perry, *Chemical Engineers' Handbook,* 4th ed., p. 20–7; conveyor and spray columns from Proctor and Schwartz, Division of SCM; rotary, flash, and fluid bed columns from Williams–Gardner, 1971, pp. 75, 149, 168, 193).

air outlets are spaced along the length of the tunnel to suit the rate of evaporation over the drying curve. This mode of operation is suited particularly to long drying times, from 20 to 96 hr for the materials of Table 9.6(e).

In the rotating tray assembly of Figure 9.8(a), material enters at the top and is scraped onto successive lower trays after complete revolutions. A leveler on each tray, shown in Figure 9.8(b), ensures uniform drying. Although the air flow is largely across the surface of the bed, the turnover of the material as it progresses downward makes the operation more nearly through-circulation. A cooling zone is readily incorporated in the equipment. The contacting process is complex enough that laboratory tray drying tests are of little value. A pilot plant size unit is cited in Section 9.2. Some industrial data on rotary tray drying are in Table 9.9, and some other substances that have been handled successfully in this equipment are listed in Table 9.4.

An alternate design has fixed jacketed trays for indirect heating. Scrapers attached to the central shaft drop the material from tray to tray. Like the rotating tray equipment, this equipment is limited to free flowing materials, but has the advantage of being essentially dust free.

Equipment developed essentially for movement of granular solids has been adapted to drying. Screw conveyors, for instance, have been used but are rarely competitive with belt conveyors,

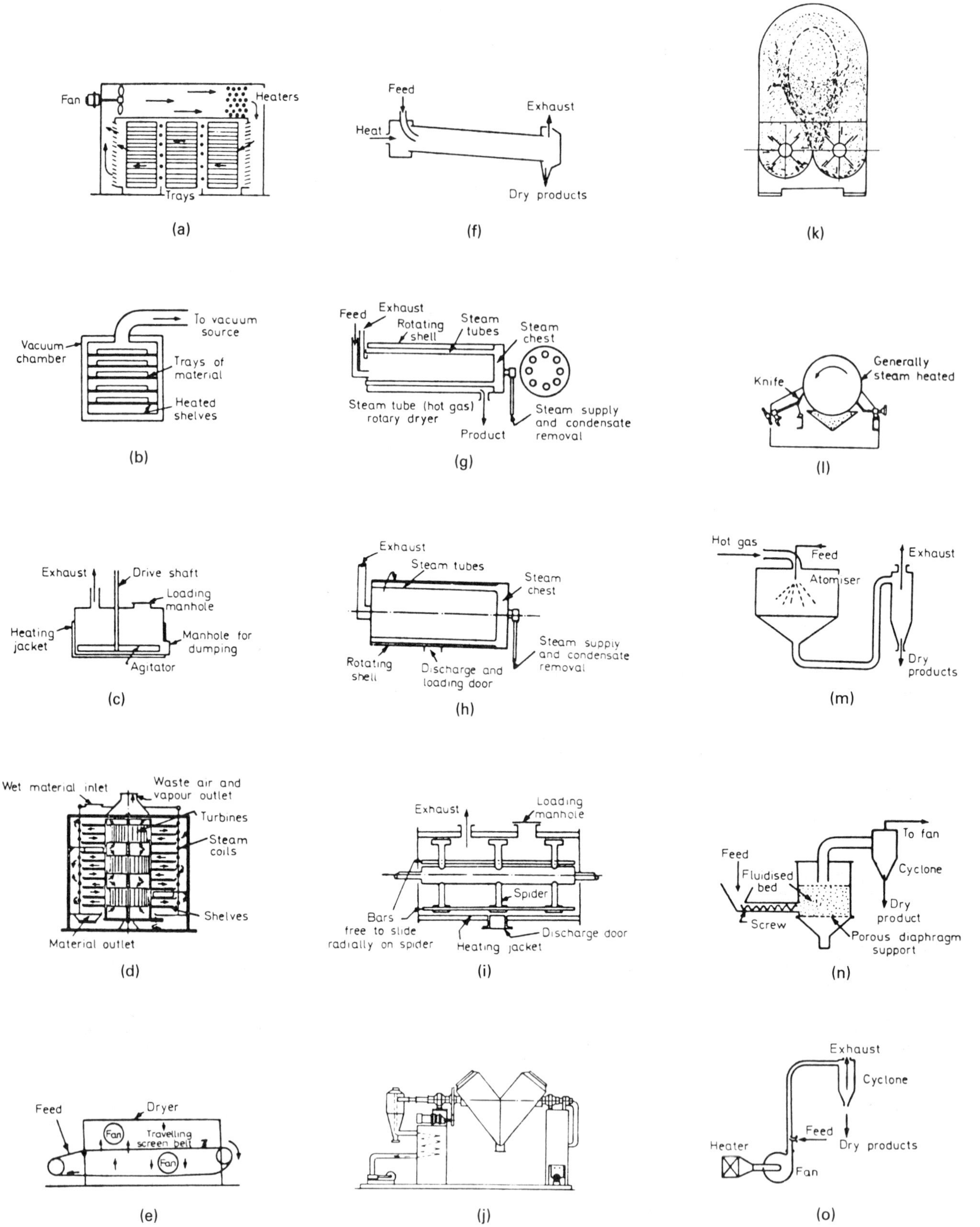

Figure 9.4. Types of dryers cited in Tables 9.1 and 9.2. (a) Tray or compartment. (b) Vacuum tray. (c) Vertical agitated batch vacuum drier. (d) Continuous agitated tray vertical turbo. (e) Continuous through circulation. (f) Direct rotary. (g) Indirect rotary. (h) Agitated batch rotary (atmos or vacuum). (i) Horizontal agitated batch vacuum drier. (j) Tumble batch dryer. (k) Splash dryer. (l) Single drum. (m) Spray. (n) Fluidized bed dryer. (o) Pneumatic conveying *(mostly after Nonhebel and Moss, 1971).*

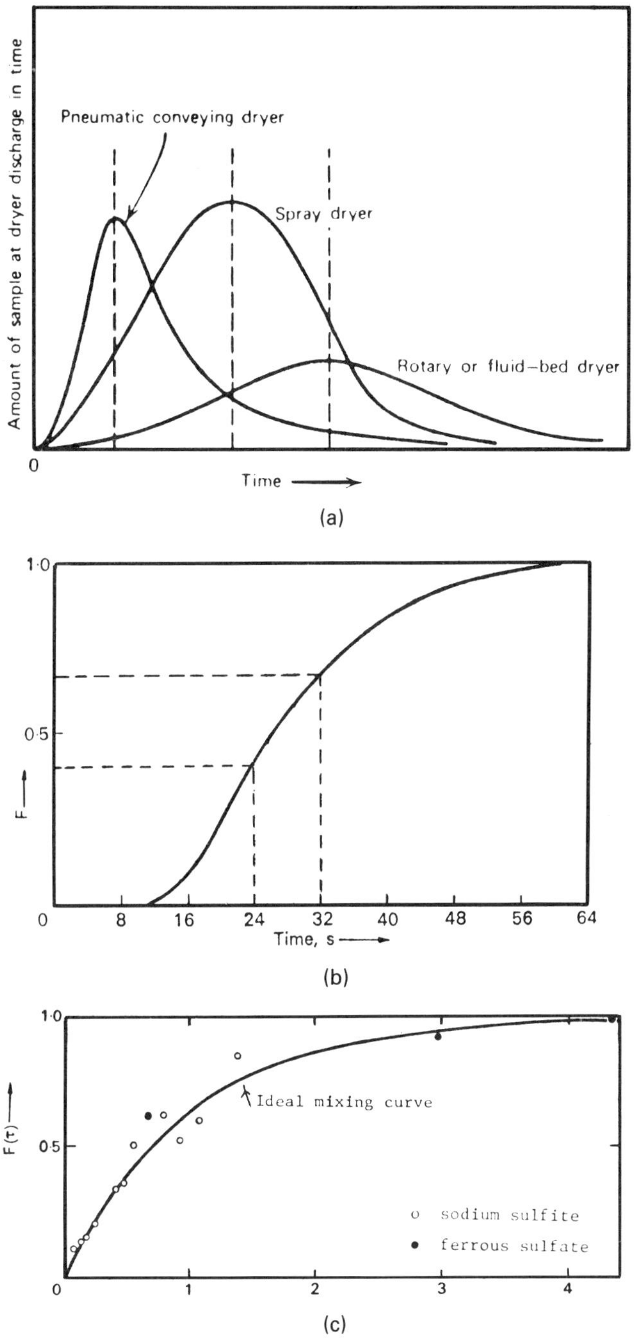

Figure 9.5. Residence time distribution in particle dryers. (a) Four types of dryers (*McCormick, 1979*). (b) Residence time distribution of air in a detergent spray tower; example shows that 27% (difference between the ordinates) has a residence time between 24 and 32 sec [*Place et al.*, Trans. Inst. Chem. Eng. **37,** *268* (*1959*)]. (c) Fluidized bed drying of two materials (*Vanacek et al.*, Fluidized Bed Drying, *1966*).

particularly for materials that tend to degrade when they are moved. From the point of view of drying, belt conveyors are of two types: with solid belts and air flow across the top of the bed, called convection drying, or with perforated belts and through circulation of the air. The screw conveyor of Figure 9.8(f) has indirect heating.

Solid belts are used for pastes and fine powders. Through

TABLE 9.4. Examples of Products Dried in Specific Kinds of Equipment

1. *Spray dryers*: rubber chemicals, sulfonates, inorganic phosphates, ceramics, kaolin, coffee, detergents, pharmaceuticals, pigments, inks, lignosulfonate wood waste, melamine and urea formaldehyde resins, polyvinyl chloride, microspheres, skim milk, eggs, starch, yeast, silica gel, urea, salts
2. *Drum dryers*: potatoes, cereals, buttermilk, skim milk, dextrins, yeasts, instant oat meal, polyacylamides, sodium benzoate, propionates, acetates, phosphates, chelates, aluminum oxide, *m*-disulfuric acid, barium sulfate, calcium acetate–arsenate-carbonate–hydrate-phosphate, caustic, ferrous sulfate, glue, lead arsenate, sodium benzene sulfonate, and sodium chloride
3. *Vacuum drum dryers*: syrups, malted milk, skim milk, coffee, malt extract, and glue
4. *Vacuum rotary dryers*: plastics, organic polymers, nylon chips, chemicals of all kinds, plastic fillers, plasticizers, organic thickeners, cellulose acetate, starch, and sulfur flakes
4. *Belt conveyor dryers*: yeast, charcoal briquettes, synthetic rubber, catalysts, soap, glue, silica gel, titanium dioxide, urea formaldehyde, clays, white lead, chrome yellow, and metallic stearates
6. *Pneumatic conveyor dryers*: yeast filter cake, starch, whey, sewage sludge, gypsum, fruit pulp, copper sulfate, clay, chrome green, synthetic casein, and potassium sulfate
7. *Rotary multitray dryer*: pulverized coal, pectin, penicillin, zinc sulfide, waste slude, pyrophoric zinc powder, zinc oxide pellets, calcium carbonate, boric acid, fragile cereal products, calcium chloride flakes, caffein, inorganic fluorides, crystals melting near 100°F, prilled pitch, electronic grade phosphors, and solvent-wet organic solids
8. *Fluidized bed dryer*: lactose base granules, pharmaceutical crystals, weed killer, coal, sand, limestone, iron ore, polyvinyl chloride, asphalt, clay granules, granular desiccant, abrasive grit, and salt
9. *Freeze dryers*: meat, seafood, vegetables, fruits, coffee, concentrated beverages, pharmaceuticals, veterinary medicines, and blood plasma
10. *Dielectric drying*: baked goods, breakfast cereals, furniture timber blanks, veneers, plyboard, plasterboard, water-based foam plastic slabs, and some textile products
11. *Infrared drying*: sheets of textiles, paper and films, surface finishes of paints and enamels, and surface drying of bulky nonporous articles.

circulation belts are applied to granules more than about 3 mm in narrowest dimension. When the feed is not in suitable granular form, it is converted in a preformer to a size range usually of 3–15 mm. Belts are made of chain mail mesh or metal with 2 mm perforations or slots of this width.

Several arrangements of belt dryers are shown in Figures 9.8(c)–(e). In the wet zone, air flow usually is upward, whereas in the drier and cooling zones it is downward in order to minimize dusting. The depth of material on the belt is 1–8 in. Superficial air velocities of 5 ft/sec usually are allowable. The multizone arrangement of Figure 9.8(e) takes advantage of the fact that the material becomes lighter and stronger and hence can be loaded more deeply as it dries. Each zone also can be controlled separately for air flow and temperature. The performance data of Table 9.9 cover a range of drying times from 11 to 200 min, and thermal efficiencies are about 50%.

Laboratory drying rate data of materials on trays are best obtained with constant air conditions. Along a belt conveyor or in a tray-truck tunnel, the moisture contents of air and stock change with position. Example 9.3 shows how constant condition drying tests can be adapted to belt conveyor operation. The effects of recycling moist air and of increasing the air velocity beyond that studied in the laboratory tests are studied in Example 9.4. Recycling does reduce drying time because of the increased air velocity, but it

TABLE 9.5. Specification Form for a Dryer[a]

1. Operation	mode	batch/continuous
	operating cycle	—— h
2. Feed	(a) material to be dried	——
	(b) feed rate	—— kg/h
	(c) nature of feed	solution/slurry/sludge/granular/ fibrous/sheet/bulky
	(d) physical properties of solids:	
	initial moisture content	—— kg/kg
	hygroscopic-moisture content	—— kg/kg
	heat capacity	—— kJ/kg°C
	bulk density, wet	—— kg/m^3
	particle size	—— mm
	(e) moisture to be removed:	
	chemical composition	——
	boiling point at 1 bar	—— °C
	heat of vaporization	—— MJ/kg
	heat capacity	—— kJ/kg°C
	(f) feed material is	scaling/corrosive/toxic/abrasive/ explosive
	(g) source of feed	——
3. Product	(a) final moisture content	—— kg/kg
	(b) equilibrium-moisture content at 60% r.h.	—— kg/kg
	(c) bulky density	—— kg/m^3
	(d) physical characteristics	granular/flaky/fibrous/powdery/ sheet/bulky
4. Design restraints	(a) maximum temperature when wet	—— °C
	when dry	—— °C
	(b) manner of degradation	——
	(c) material-handling problems,	
	when wet	——
	when dry	——
	(d) will flue-gases contaminate product?	——
	(e) space limitations	——
5. Utilities	(a) steam available at	—— bar pressure (10^6 N/m^2)
	maximum quantity	—— kg/h
	costing	—— $/kg
	(b) other fuel	——
	at	—— kg/h
	with heating value	—— MJ/kg
	costing	—— $/kg
	(c) electric power	—— V
	frequency	—— hz
	phases	——
	costing	—— $/kWh
6. Present method of drying		——
7. Rate-of-drying data under constant external conditions:		——
		——
or data from existing plant		——
		——
8. Recommended materials of construction		
	(a) parts in contact with wet material	——
	(b) parts in contact with vapors	——

[a] Questionnaires of several manufacturers are in Appendix C.
(Keey, 1972, p. 325).

is not as effective in this regard as the same increase in the amount of fresh air. Recycling is practiced, however, to reduce heat consumption when the fresh air is cold and to minimize possible undesirable effects from over-rapid drying with low humidity air. Parallel current operation also avoids overrapid drying near the end. For parallel flow, the moisture balance of Example 9.4 becomes

$$H_g = \frac{97.4(R+1) + 1400(1.16 - W)}{7790(R+1)} \tag{9.16}$$

and replaces line 30 of the computer program.

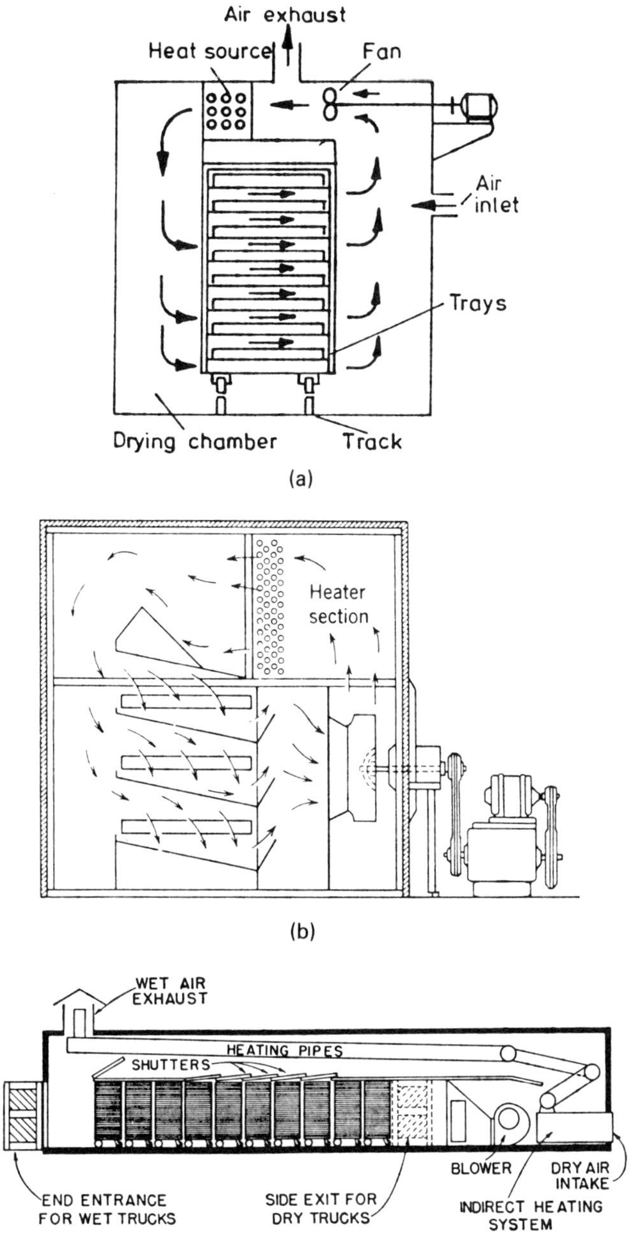

Figure 9.6. Tray dryer arrangements, batch and continuous. Performance data are in Table 9.5. (a) Air flow across the surfaces of the trays. (b) Air circulation forced through the beds on the trays (*Proctor and Schwartz Inc.*). (c) Continuous drying of trays mounted on trucks that move through the tunnel; air flow may be in parallel or countercurrent (*P.W. Kilpatrick, E. Lowe, and W.B. Van Arsdel,* Advances in Food Research, *Academic, New York, 1955, Vol. VI, p. 342*).

The kind of data desirable in the design of through-circulation drying are presented for a particular case by Nonhebel and Moss (1971, p. 147). They report on effects of extrusion diameters of the original paste, the bed depth, air linear velocity, and air inlet humidity, and apply these data to a design problem.

9.6. ROTARY CYLINDRICAL DRYERS

Rotating cylindrical dryers are suited for free-flowing granular materials that require drying times of the order of 1 hr or less. Materials that tend to agglomerate because of wetness may be preconditioned by mixing with recycled dry product.

Such equipment consists of a cylindrical shell into which the wet material is charged at one end and dry material leaves at the other end. Figure 9.9 shows some examples. Drying is accomplished by contact with hot gases in parallel or countercurrent flow or with heat transfer through heated tubes or double shells. Designs are available in which the tubes rotate with the shell or are fixed in space.

Diameters typically are 4–10 ft and lengths are 4–15 diameters. The product of rpm and diameter is typically between 25 and 35. Superficial gas velocities are 5–10 ft/sec; but lower values may be needed for fine products, and rates up to 35 ft/sec may be allowable for coarse materials. To promote longitudinal travel of the solid, the shell is mounted on a slope of 1 in 40 or 20.

In a countercurrent dryer the exit temperature of the solid approaches that of the inlet gas. In a parallel current dryer, the exit gas is 10–20°C above that of the solid. For design purposes the temperature of the exit solid in parallel flow may be taken as 100°C.

Flights attached to the shell lift up the material and shower it as a curtain through which the gas flows. Cross sections of some dryers are shown in Figure 9.10. The shape of flights is a compromise between effectiveness and ease of cleaning. The number is between 2 and 4 times the diameter of the shell in feet, and their depth is between $\frac{1}{12}$ and $\frac{1}{8}$ of the diameter. Holdup in the dryer depends on details of design and operation, but 7–8% is a usual figure. Cross-sectional holdup is larger at the wet end than at the dry end. An 85% free cross section commonly is adopted for design purposes; the rest is taken up by flights and settled and cascading solids.

Residence time depends on the nature of the material and mechanical features of the dryer. The performance data of Table 9.10 show a range of 7–90 min. A formula cited by Williams-Gardner (1971, p. 133) for the geometrical residence time is

$$\theta = kL/nDS, \tag{9.17}$$

where L is the length, D is the diameter, n is rpm, and S is the slope (in./ft). The coefficient k varies from 3 to 12 for various countercurrent single shell dryers. The formula may be of some value in predicting roughly the effects of changes in the quantities included in it.

The only safe way of designing a rotary dryer is based on pilot plant tests or by comparison with known performance of similar operations. Example 9.5 utilizes pilot plant data for upscaling a dryer. The design of Example 9.6 also is based on residence time and terminal conditions of solid and air established in a pilot plant.

When heating by direct contact with hot gases is not feasible because of contamination or excessive dusting, dryers with jacketed shells or other kinds of heat transfer surfaces are employed. Only enough air to entrain away the moisture is employed. The temperature of the solid approaches the boiling temperature of the water in the constant rate period. Figure 9.10 shows designs in which the heating tubes are fixed in space or are attached to the rotating shell. Table 9.10 gives some performance data.

Combined indirect and indirect dryers pass the hot gases first through a jacket or tubes, and then wholly or in part through the open dryer. Efficiencies of such units are higher than of direct units, being in the range 60–80%. Table 9.10(d) shows performance data. Since the surfaces are hot, this equipment is not suitable for

TABLE 9.6. Performance Data of Batch Tray and Tray–Truck Dryers

(a) Cross-Flow Operation

	Coated Tablets	PTFE	Aspirin Base Granules	Stearates	Chalk	Filter Cake	Filter Cake	Filter Cake
Capacity, wet charge (lb)	120	80	56	20,000	1800	3000	2800	4300
Number of trays	40	20	20	320	72	80	80	80
Tray area (ft^2)	140	70	70	4800	1130	280	280	280
Depth of loading (in.)	0.5	1.0	0.5	2.0	2.0	1.0	1.0	1.5
Initial moisture (% w/w basis)	25	25–30	15	71	46	70	70	80
Final moisture (% w/w basis)	nil	0.4	0.5	0.5	2.0	1.0	1.0	0.25
Maximum air temperature (°F)	113	284	122	200	180	300	200	200
Loading (lb/ft^2)	0.9	1.2	0.4	0.9	0.91	3.25	3.04	11.7
Drying time (hr)	12	5.5	14	24	4.5	22	45	12
Overall drying rate (lb/hr)	2.6	5.3	0.84	62.5	185	96.6	43.2	90
Evaporative rate ($lb/hr/ft^2$)	0.0186	0.05	0.008	0.013	0.327	0.341	0.184	0.317
Total installed HP	1	1	1	45	6	4	2	2

(Williams–Gardner, 1971, p. 75, Table 12: first three columns courtesy Calmic Engineering Co.; last five columns courtesy A.P.V.—Mitchell (Dryers) Ltd.)

(b) Vacuum Dryers with Steam Heated Shelves

	Soluble Aspirin	Paint Pigment	Ferrous Glutinate	Ferrous Succinate	Lithium Hydroxide	Tungsten Alloy	Stabilized Diazamin
Capacity, wet product (lb/h)	44	30.5	41.6	52.5	36.8	12.8	4.6
Tray area (ft^2)	108	108	108	108	54	215	172
Depth of loading (in.)	1	2	0.5	1	1	0.5	0.75
Initial moisture (% w/w basis)	72.4	49.3	25	37.4	59	1.6	22.2
Final moisture (% w/w basis)	1.25	0.75	0.5	18.8	0.9	nil	0.5
Max temp (°F)	104	158	203	203	122	239	95
Loading [lb charge (wet) ft^2]	6.1	102	2.3	1.94	3.08	7.16	1.22
Drying time (hr)	15	36	6	4	4.5	12	48
Overall drying rate (lb moisture evaporated/ft^2/hr)	0.293	0.14	0.11	0.11	0.034	0.013	0.0058
Total installed HP	6	6	6	6	3	2	5
Vacuum (in. Hg)	29.5	28	27	27	27	29	22–23

(Williams–Gardner, 1971, p. 88, Table 15: courtesy Calmic Engineering Co.).

(c) Vacuum Dryers with Steam-Heated Shelves

Material	Sulfur Black	Calcium Carbonate	Calcium Phosphate
Loading (kg dry material/m^2)	25	17	33
Steam pressure (kPa gauge)	410	410	205
Vacuum (mm Hg)	685–710	685–710	685–710
Initial moisture content (%, wet basis)	50	50.3	30.6
Final moisture content (%, wet basis)	1	1.15	4.3
Drying time (hr)	8	7	6
Evaporation rates (kg/sec m^2)	8.9×10^{-4}	7.9×10^{-4}	6.6×10^{-4}

(*Chemical Engineers' Handbook*, McGraw-Hill, New York, 1984, p. 20.23, Table 20.8).

(d) Through Circulation Dryers

Kind of Material	Granular Polymer	Vegetable	Vegetable Seeds
Capacity (kg product/hr)	122	42.5	27.7
Number of trays	16	24	24
Tray spacing (cm)	43	43	43
Tray size (cm)	91.4 × 104	91.4 × 104	85 × 98
Depth of loading (cm)	7.0	6	4
Physical form of product	crumbs	0.6-cm diced cubes	washed seeds
Initial moisture content (%, dry basis)	11.1	669.0	100.0
Final moisture content (%, dry basis)	0.1	5.0	9.9
Air temperature (°C)	88	77 dry-bulb	36
Air velocity, superficial (m/sec)	1.0	0.6–1.0	1.0
Tray loading (kg product/m^2)	16.1	5.2	6.7
Drying time (hr)	2.0	8.5	5.5
Overall drying rate (kg water evaporated/hr m^2)	0.89	11.86	1.14
Steam consumption (kg/kg water evaporated)	4.0	2.42	6.8
Installed power (kW)	7.5	19	19

(Proctor and Schwartz Co.).

TABLE 9.6—(*continued*)

(e) Tray and Tray–Truck Dryers

Material	Color	Chrome Yellow	Toluidine Red	Half-Finished Titone	Color
Type of dryer	2-truck	16-tray dryer	16-tray	3-truck	2-truck
Capacity (kg product/hr)	11.2	16.1	1.9	56.7	4.8
Number of trays	80	16	16	180	120
Tray spacing (cm)	10	10	10	7.5	9
Tray size (cm)	60 × 75 × 4	65 × 100 × 2.2	65 × 100 × 2	60 × 70 × 3.8	60 × 70 × 2.5
Depth of loading (cm)	2.5–5	3	3.5	3	
Initial moisture (%, bone-dry basis)	207	46	220	223	116
Final moisture (%, bone-dry basis)	4.5	0.25	0.1	25	0.5
Air temperature (°C)	85–74	100	50	95	99
Loading (kg product/m^2)	10.0	33.7	7.8	14.9	9.28
Drying time (hr)	33	21	41	20	96
Air velocity (m/sec)	1.0	2.3	2.3	3.0	2.5
Drying (kg water evaporated/hr m^2)	0.59	65	0.41	1.17	0.11
Steam consumption (kg/kg water evaporated)	2.5	3.0	—	2.75	
Total installed power (kW)	1.5	0.75	0.75	2.25	1.5

(Proctor and Schwartz Co.).

TABLE 9.7. Performance of Agitated Batch Dryers (See Fig. 9.7)

(a) Double-Cone Tumbler

	Tungsten Carbide	Polyester Resin	Penicillin	Hydroquinone	Prussian Blue Pigment
Volatile ingredient	naphtha	water	acetone	water	water
Physical nature of charge	heavy slurry	pellets	powder	powder	filtercake
Dryer dia (ft)	2	2	2	2	2
Dryer capacity (ft^3)	2.5	2.5	2.5	2.5	2.5
Method of heating	hot water	steam	hot water	hot water	steam
Heating medium temperature (°F)	180	240	140	150	225
Vacuum (mm Hg abs)	40–84	12–18	40	50–100	40–110
Initial volatile content (% w/w basis)	18.0	0.34	27.9	5.0	83
Final volatile content (% w/w basis)	nil	0.01	nil	0.25	4.8
Weight of charge (lb)	640	130	55	61	142.5
Bulk density of charge (lb/ft^3)	256	51.5	21.5	26.5	58.5
Drying time (min)	155	215	90	50	480

(Courtesy Patterson Division, Banner Industries Inc.; Williams–Gardner, 1971).

(b) Paddle, Ribbon, and Pan[a]

Material	Type of Dryer	Size of Dryer (mm) Length	Size of Dryer (mm) Dia	Driving Motor (HP)	Wet Charge (kg)	Filling Ratio	Initial Moisture Content (%, Wet Basis)	Absolute Pressure in Dryer (mb)	Jacket Temp (°C)	Drying Time (hr)	Mean Overall Coeff. U_c (W/m °C)
Organic paste	HCRP	5500	1200		4000	0.36	30	200	80	15	35
Different fine aromatic organic compound crystals	HCRP	3800	1350	15	2260	0.2	68	265	125	6	45
	HCRP	3800	1350	15	4660	0.4	75	265	125	8	60
	HCRP	5500	1200		2100	0.2	6	200	125	4	25
Anthracene (water and pyridine)	HCRP	8900	1800	35	37000	0.72	76	665–1000	170	16	75
Dyestuff paste	HCSB	2750	1200	10	2000	0.3	70	265	105	14	30
Different organic pastes	PVP		1800	15	1080	0.4	41	1000	125	32	35
	PVP		2450	25	800	0.4	35	665	125	$7\frac{1}{2}$	25
Different dyestuff pastes	PVP		1800	15	1035	0.4	61	1000	125	11	135
	PVP		2450	20–30	2400	0.7	64	470	125	12	115

[a] HCRP = paddle agitator; HCSP = ribbon agitator; PVP = pan with vertical paddles.
(Nonhebel and Moss, 1971).

TABLE 9.7—(*continued*)

(c) Pan Dryer

	Sodium Thiosulphate	Potassium Zeolite	Arsenic Pentoxide
Dryer diameter	6 ft 0 in.	2 ft 3 in.	8 ft 0 in.
Dryer depth	2 ft 0 in.	1 ft 0 in.	2 ft 0 in.
Capacity (lb product)	12 cwt	14 lb	$2\frac{1}{2}$ ton/day
Initial moisture (% w/w basis)	37	40	35
Final moisture (% w/w basis)	0	1	2–3
Method of heating	steam	steam	steam
Atmospheric (a) or vacuum (b)	(b) 26 Hg	(a) 60 lb/in.^{2}/gauge	(b)
Drying temperature: material (°F)			
Drying temperature: shelf (°F)		153 C	
Bulk density product (lb/ft^3)			
Drying time (hr/batch)	5	3	8
Material of construction	SS	MS	SS

[Courtesy A.P.V.—Mitchell (Dryers) Ltd., Williams–Gardner, 1971].

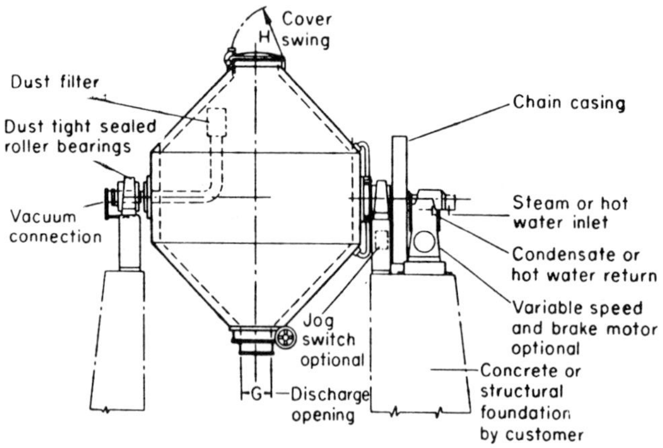

(a)

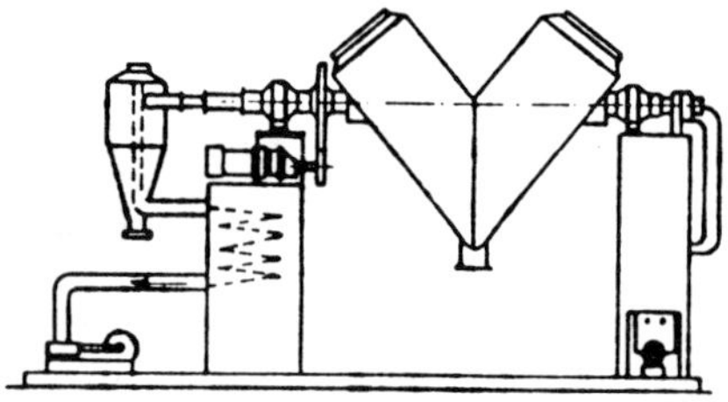

(b)

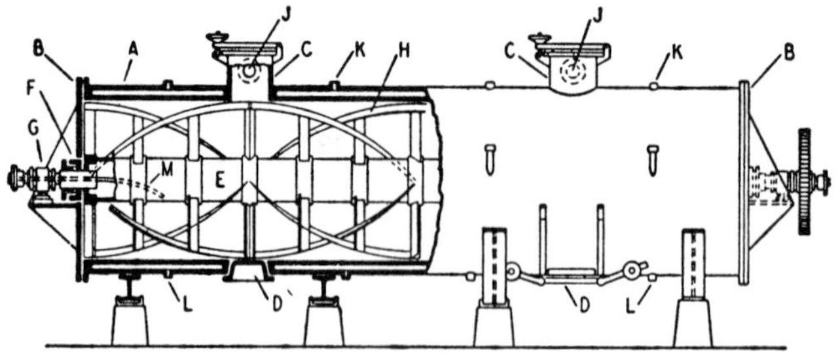

(c)

Figure 9.7. Tumbling and agitated heated dryers for atmospheric and vacuum batch operation. (a) Double cone tumbler; performance data in Table 9.6(a) (*Pennsalt Chem. Co.*). (b) V-shaped tumbler. (c) Ribbon agitated cylinder; performance data in Table 9.6(b). (*A*) jacketed shell; (*B*) heads; (*C*) charging connections; (*D*) discharge doors; (*E*) agitator shaft; (*F*) stuffing box; (*G*) shaft bearings; (*H*) agitator blades; (*J*) vapor outlets; (*K*) steam inlets; (*L*) condensate outlets; (*M*) discharge siphon for shaft condensate (*Buflovak Equip. Div., Blaw Knox Co.*) (d) Paddle agitated cylinder. Performance data in Table 9.6(b). (e) Horizontal pan with agitator blades. Data are Table 9.6(b).

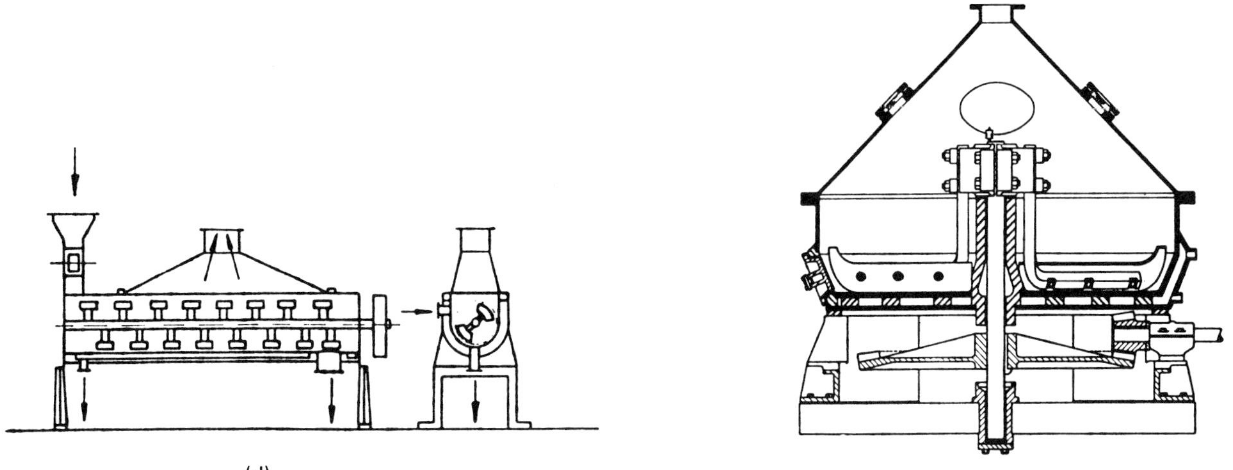

(d) (e)

Figure 9.7—*(continued)*

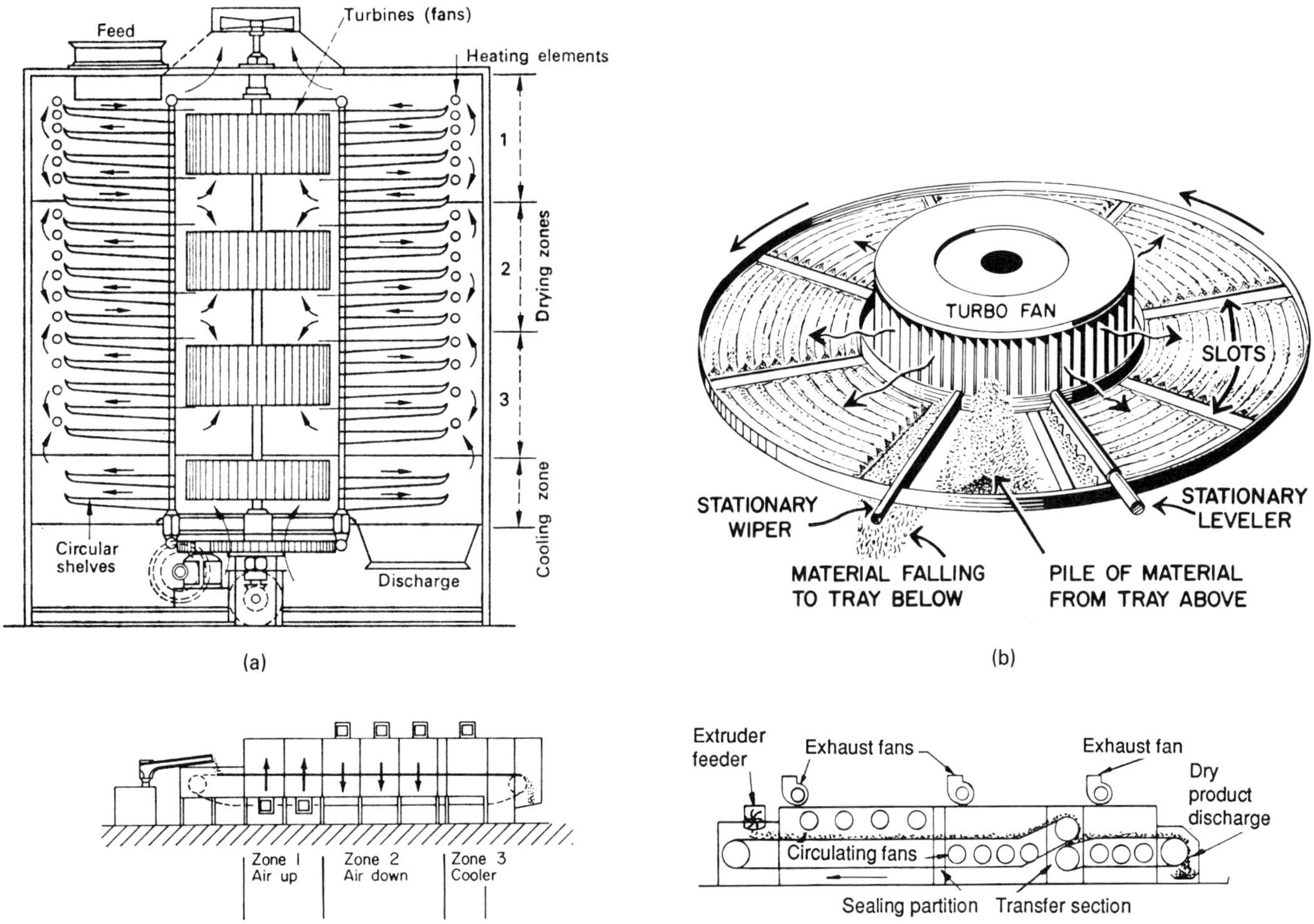

(a) (b) (c) (d)

Figure 9.8. Rotary tray, through-circulation belt conveyor, and heated screw conveyor dryers. (a) Rotary tray dryer (*Wyssmont Co.*). (b) Action of a rotating tray and wiper assembly (*Wyssmont Co.*). (c) A single conveyor belt with air upflow in wet zone and downflow in dry (*Proctor and Schwartz Inc.*). (d) A two-stage straight-through belt conveyor dryer. (e) A three-belt conveyor dryer; as the material becomes dryer, the loading becomes deeper and the belt longer (*Proctor and Schwartz Inc.*). (f) Screw conveyor dryer with heated hollow screw (*Bepex Corp.*).

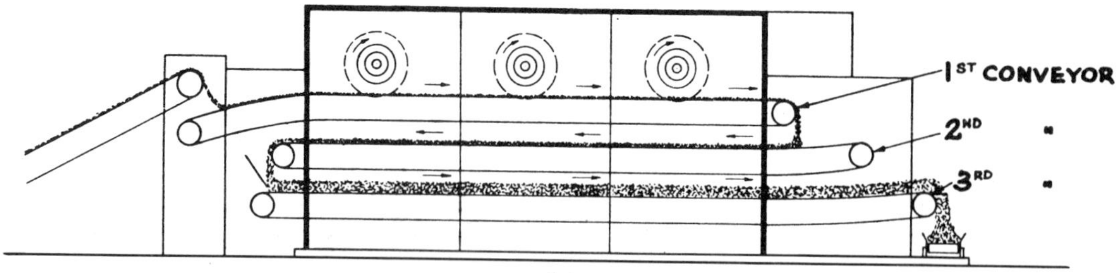

(e)

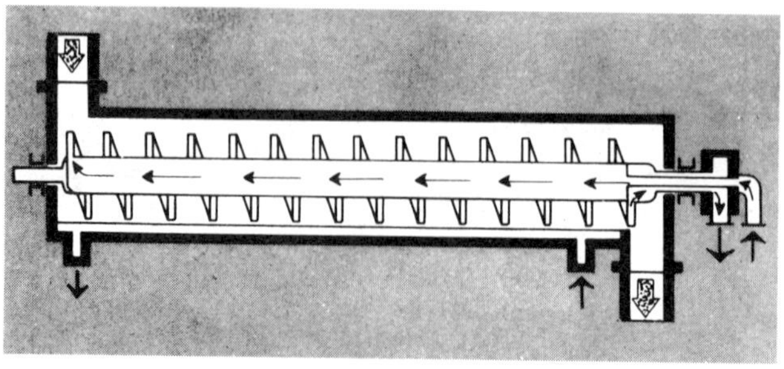

(f)

Figure 9.8—*(continued)*

TABLE 9.8. Performance of Rotary Tray and Pan Dryers

(a) Multitray Dryers at Atmospheric Pressure

	China Clay	Bread Crumbs	Cu–Ni Concentrate	Catalyst Pellets	Kaolin	Calcium Chloride	Urea	Vitamin Powder
Dryer height	—	—	—	23 ft	23 ft	47 ft	47 ft	12 ft
Dryer diameter	—	—	—	19 ft	19 ft	31 ft	15 ft	9 ft
Tray area (ft^2)	7000	2000 (drying) 1000 (cooling)	2900	—	—	—	—	—
Capacity (lb/product/hr)	31,000	1680	19,000	4200	10,000	24,000	5000	200
Initial moisture (% w/w basis)	30	36	22	45	35	25	20	20
Final moisture (% w/w basis)	10	5	5	18	5	1	0.2	5
Product temperature (°F)	160	100	200	—	—	—	—	—
Residence time (min)	40	40 (drying) 20 (cooling)	25	—	—	—	—	—
Evaporation rate (lb/ft^2/hr)	9.100	804	4060	2050	4600	11,000	100	37
Method of heating	external oil	steam	external oil	external gas	external oil	internal gas	external steam	external steam
Heat consumption (Btu/lb moisture evaporated)	1750	—	2200	1750	1850	1800	3500	2700
Installed HP	80	25	60	23	47	65	75	$2\frac{1}{2}$

Williams–Gardner, 1971).
(First three columns courtesy Buell Ltd.; last five columns courtesy The Wyssmont Co., Inc.).

(b) Multiple Vacuum Pan Dryer

	Sodium Hydrosulphite	Maneb	Melamine	Activated Carbon
Dryer diameter (pans) (m)	2	2	2	2
Number of pans	5	17	11	17
Area (approx)(m^2)	12.4	42.8	27.6	42.8
Dry product (lb/hr)	1100	660	1870	440
Initial moisture (% w/w)	4	23	11	62
Final moisture (% w/w)	0.1	0.5	0.03	3
Heating	hot water	steam 1.3 atm	steam 2.5 atm	steam 2.5 atm
Pan temperature (°C)	98	105	125	125
Evaporation rate (lb/ft^2/hr)	0.325	0.325	0.79	0.78
Drying time (min)	15	170	12	30

(Data of Krauss–Maffei-Imperial GmbH).

TABLE 9.9. Performance of Through-Circulation Belt Conveyor Dryers [See Figs. 9.8(c)–(e)]

(a) Data of A.P.V.—Mitchell (Dryers) Ltd.

	Fertilizers	Bentonite	Pigment	Nickel Hydroxide	Metallic Stearate
Effective dryer length	42 ft 6 in.	60 ft 0 in.	24 ft 0 in.	24 ft 0 in.	41 ft 3 in.
Effective band width	8 ft 6 in.	8 ft 6 in.	4 ft 0 in.	4 ft 0 in.	6 ft 0 in.
Capacity (lb product/hr)	2290	8512	100	125	125
Method of feeding / Feedstock preforming	oscillator	oscillator	extruder	extruder	extruder
Initial moisture (% w/w basis)	45.0	30	58.9	75	75
Final moisture (% w/w basis)	2.0	10.0	0.2	0.5	0.2
Drying time (min)	16	14	60	70	60
Drying rate (lb evaporated/ft^2/hr)	7.0	6.5	2.0	7.5	1.5
Air temperature range (°F)	—	—	—	—	—
Superficial air velocity (ft/min)	200	200	180	180	125
Heat consumption (Btu/lb evaporated)	—	—	—	—	—
Method of heating	direct oil	direct oil	steam	steam	steam
Fan installed HP	35	50	14	14	28

(Williams–Gardner, 1971).

(b) Data of Krauss–Maffei–Imperial GmbH

	Aluminium Hydrate	Polyacrylic Nitrile	Sulfur	Calcium Carbonate	Titanium Dioxide
Effective dryer length	32 ft 9 in.	43 ft 0 in.	28 ft 0 in.	50 ft 0 in.	108 ft 0 in.
Effective band width	6 ft 6 in.	6 ft 6 in.	6 ft 6 in.	6 ft 3 in.	9 ft 6 in.
Capacity (lb product/hr)	615	2070	660	1800	6000
Method of feeding / Feedstock preforming	grooved drum	extruder	extruder	extruder	extruder
Initial moisture (% w/w basis)	38.0	55.0	45.0	60.0	50.0
Final moisture (% w/w basis)	0.2	1.0	1.0	0.5	0.5
Drying time (min)	26	52	110	40	45
Drying rate (lb evaporated/hr/ft^2)	2.88	3.37	3.57	5.73	6.0
Air temperature range (°F)	233	186/130	194/230	320	314/392
Superficial air velocity (ft/min)	140	100/216	140	160	150
Heat consumption (lb steam/lb evaporated)	1.7–1.8	1.8–1.9	1.8–1.9	1.7–1.8	1.8–1.9
Method of heating	50 lb/$in.^2$ steam	25 lb/$in.^2$ steam	90 lb/$in.^2$ steam	160 lb/$in.^2$ steam	260 lb/$in.^2$ steam
Fan installed hp (approx.)	25	65	20	35	80

(Williams–Gardner, 1971).

(c) Data of Proctor and Schwartz Inc.

Kind of Material	Inorganic Pigment	Cornstarch	Fiber Staple		Charcoal Briquettes	Gelatin	Inorganic Chemical
Capacity (kg dry product/hr)	712	4536	1724		5443	295	862
			Stage A	Stage B			
Approximate dryer area (m^2)	22.11	66.42	57.04	35.12	52.02	104.05	30.19
Depth of loading (cm)	3	4	—	—	16	5	4
Air temperature (°C)	120	115–140	130–100	100	135–120	32–52	121–82
Loading (kg product/m^2)	18.8	27.3	3.5	3.3	182.0	9.1	33
Type of conveyor (mm)	1.59 by 6.35 slots	1.19 by 4.76 slots	2.57-diameter holes, perforated plate		8.5 × 8.5 mesh screen	4.23 × 4.23 mesh screen	1.59 × 6.35 slot
Preforming method or feed	rolling extruder	filtered and scored	fiber feed		pressed	extrusion	rolling extruder
Type and size of preformed particle (mm)	6.35-diameter extrusions	scored filter cake	cut fiber		64 × 51 × 25	2-diameter extrusions	6.35-diameter extrusions
Initial moisture content (% bone-dry basis)	120	85.2	110		37.3	300	111.2
Final moisture content (% bone-dry basis)	0.5	13.6	9		5.3	11.1	1.0
Drying time (min)	35	24	11		105	192	70
Drying rate [kg water evaporated/(hr m^2)]	38.39	42.97	17.09		22.95	9.91	31.25
Air velocity (superficial)(m/sec)	1.27	1.12	0.66		1.12	1.27	1.27
Heat source per kg water evaporated [steam kg/kg gas (m^3/kg)]	gas 0.11	steam 2.0	steam 1.73		waste heat	steam 2.83	gas 0.13
Installed power (kW)	29.8	119.3	194.0		82.06	179.0	41.03

(*Perrys Chemical Engineers Handbook*, McGraw-Hill, New York, 1984).

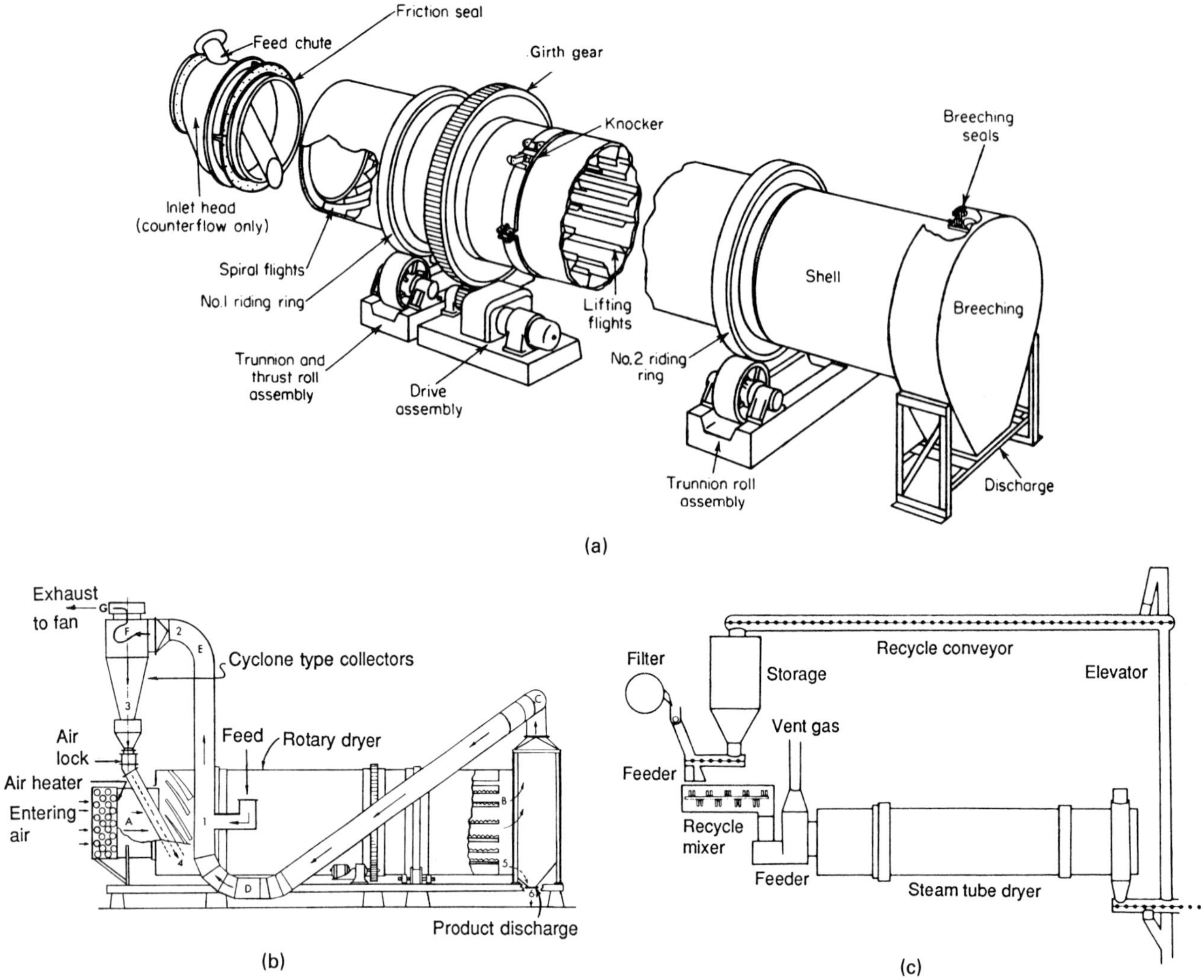

Figure 9.9. Rotary dryer assemblies. (a) Parts of the shell of a direct fired rotary dryer (*C.E. Raymond Bartlett Snow Co.*). (b) Assembly of a rotary dryer with pneumatic recycle of fines (*Standard Steel Corp.*). (c) Steam tube dryer with mechanical conveyor for partial recycle of product for conditioning of the feed.

thermally sensitive materials and, of course, may generate dust if the gas rate through the open dryer is high.

In the Roto-Louvre design of Figure 9.10(b) the gas enters at the wall, flows first through the bed of particles and subsequently through the shower of particles. Performance data are in Tables 9.10(b) and (c).

A formula for the power required to rotate the shell is given by Wentz and Thygeson (1979):

$$P = 0.45 W_t v_r + 0.12 BDNf, \qquad (9.18)$$

where P is in watts, W_t is the weight (kg) of the rotating parts, v_r is the peripheral speed of the carrying rollers (m/sec), B is the holdup of solids (kg), D is diameter of the shell (m), N is rpm, and f is the number of flights along the periphery of the shell. Information about weights may be obtained from manufacturers catalogs or may be estimated by the usual methods for sizing vessels. Fan and driver horsepower are stated for the examples of Tables 9.10(a)–(c). The data of Table 9.10(a) are represented roughly by

$$P = 5 + 0.11 DL, \qquad (9.19)$$

where P is in HP and the diameter D and length L are in feet.

9.7. DRUM DRYERS FOR SOLUTIONS AND SLURRIES

Solutions, slurries and pastes may be spread as thin films and dried on steam heated rotating drums. Some of the usual arrangements are shown on Figure 9.11. Twin drums commonly rotate in opposite directions inward to nip the feed, but when lumps are present that could damage the drums, rotations are in the same direction. Top feed with an axial travelling distributor is most common. Dip feed is shown in Figure 9.11(d) where an agitator also is provided to keep solids in suspension. When undesirable boiling of the slurry in the

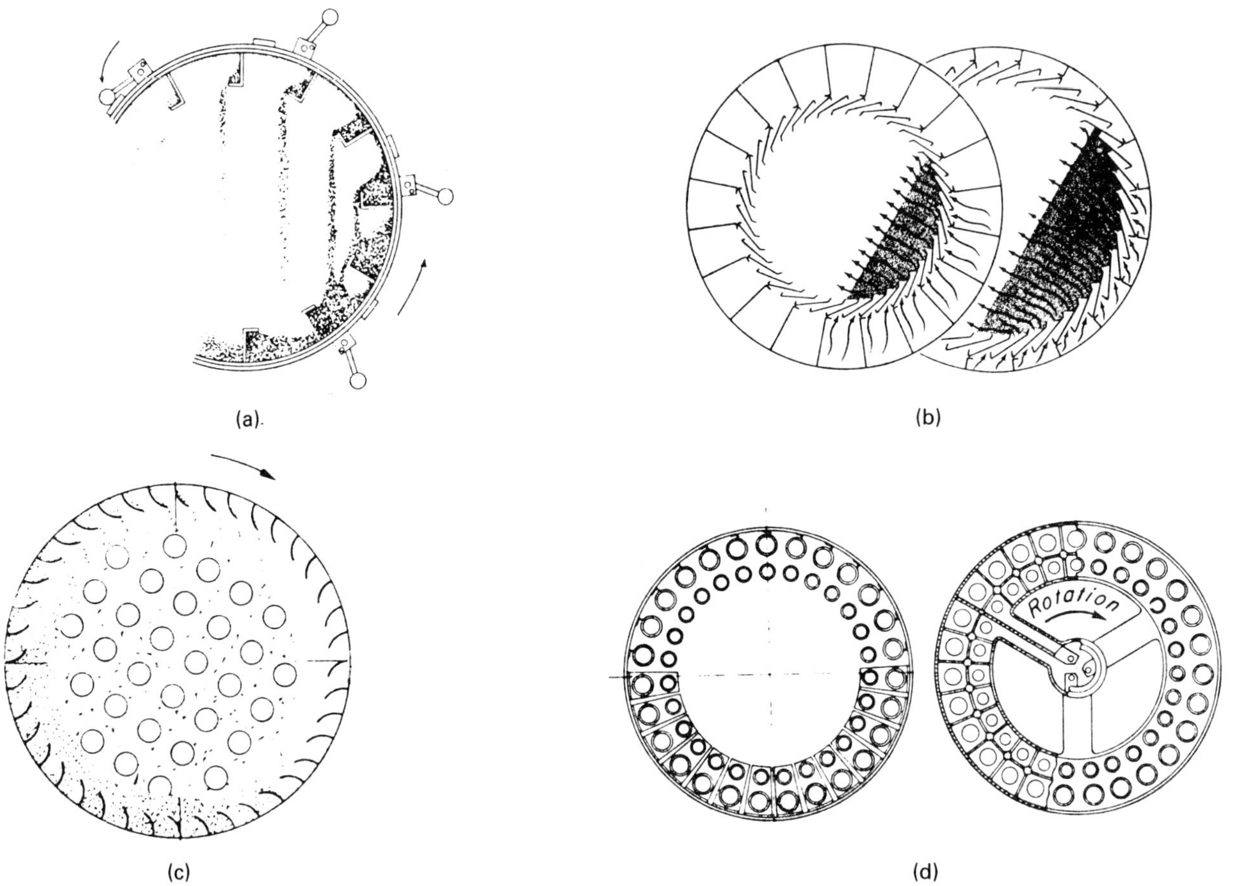

Figure 9.10. Cross sections of rotary dryers. (a) Action of the flights in cascading the drying material. The knockers are for dislodging material that tends to cling to the walls. (b) Cross section of chamber of rotolouvre dryer showing product depths and air flows at feed and discharge ends. The air enters at the wall and flows through the bed as well as through the cloud of showered particles (*Link-Belt Co.*). (c) Showering action in a dryer with fixed steam tubes and rotating shell. (d) Section and steam manifold at the end of a dryer in which the steam tubes rotate with the dryer.

pan could occur, splash feed as in Figure 9.11(c) is employed. Example 9.7 describes some aspects of an actual installation.

For mechanical reasons the largest drum made is 5 ft dia by 12 ft with 188 sqft of curved surface. A 2 × 2 ft drum also is listed in manufacturers' catalogs. Performance data are in Tables 9.11 and 9.12.

The material comes off as flakes 1–3 mm or less thick. They are broken up to standard size of about $\frac{1}{4}$ in. square. That process makes fines that are recycled to the dryer feed. Drying times fall in the range of 3–12 sec. Many laboratory investigations have been made of drying rates and heat transfer coefficients, but it appears that the only satisfactory basis for sizing plant equipment is pilot plant data obtained with a drum of a foot or more in diameter. Usually plant performance is superior to that of pilot plant units because of steadier long time operation.

Rotation speeds of the examples in Table 9.12 show a range of 1–24 rpm. Thin liquids allow a high speed, thick pastes a low one. In Table 9.13(c) the evaporation rates group in the range 15–30 kg/m^2 hr, but a few of the data are far out of this range. The few data in Table 9.13(a) show that efficiencies are comparatively high, 1.3 lb steam/lb water evaporated.

A safe estimate of power requirement for double drum dryers is approx 0.67 HP/(rpm)(100 sqft of surface). Maintenance can be as high as 10%/yr of the installed cost. Knives last from 1 to 6 months depending on abrasiveness of the slurry. Competitors for drum dryers are solid belt conveyors that can can handle greater thicknesses of pasty materials, and primarily spray dryers that have largely taken over the field.

9.8. PNEUMATIC CONVEYING DRYERS

Free-flowing powders and granules may be dried while being conveyed in a high velocity air stream. The necessary equipment is variously called pneumatic conveying dryer, pneumatic dryer, air lift dryer, or flash dryer. The basic system consists of an air heater, solids feeding device, vertical or inclined drying leg, cyclone or other collector and an exhaust fan. Figure 9.12 shows some of the many commercial equipment. Provision for recycling some of the product generally is included. Some of the materials being handled successfully in pneumatic dryers are listed in Table 9.5.

Readily handled particles are in the size range 1–3 mm. When the moisture is mostly on the surface, particles up to 10 mm have been processed. Large particles are brought down to size in dispersion devices such as knife, hammer or roller mills.

Typical performance data are summarized in Table 9.13. In practice air velocities are 10–30 m/sec. The minimum upward velocity should be 2.5–3 m/sec greater than the free fall velocity of the largest particles. Particles in the range of 1–2 mm correspond to an air velocity of 25 m/sec. Since agglomerates may exist under drying conditions, the safest design is that based on pilot plant tests or prior experience.

EXAMPLE 9.5
Scale-Up of a Rotary Dryer

Tests on a laboratory unit come up with the stated conditions for drying a pelleted material at the rate of 1000 lb dry/hr:

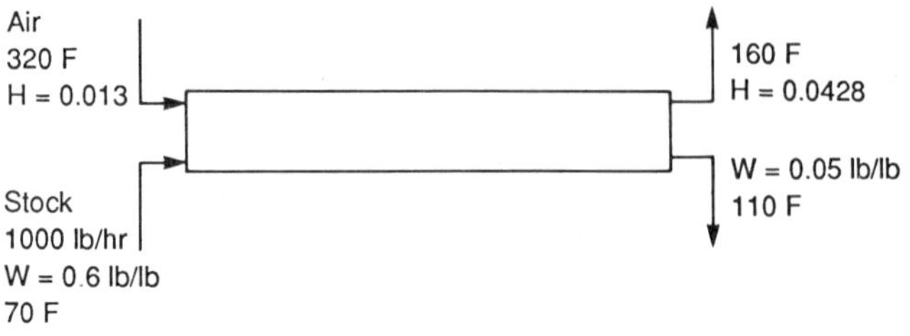

The residence time is 20 min. The speed is 3–4 rpm. On the average, 7.5% of the cross section is occupied by solid. Because of dusting problems, the linear velocity of the air is limited to 12 ft/sec. The diameter and length will be found. Since the inlet and outlet conditions are specified and the moisture transfer is known, the heat balance can be made. The heat capacity of the solid is 0.24:

$$\text{moisture evap} = 1000(0.6\text{–}0.05) = 550\ \text{lb/hr}$$
$$\text{air rate} = 550/(0.0428 - 0.013) = 18{,}456\ \text{lb/hr}$$

Off a psychrometric chart, the sp vol of the air is 15.9 cuft/(lb dry). The diameter is

$$D = \left(\frac{18{,}456(15.9)}{3600(12)(1-0.075)\pi/4}\right)^{1/2} = 3.06\ \text{ft}, \quad \text{say } 3.0\ \text{ft}.$$

The length is

$$L = \frac{30(20/60)}{0.075\pi D^2/4} = 18.9\ \text{ft}.$$

EXAMPLE 9.6
Design Details of a Countercurrent Rotary Dryer

Pilot plants indicate that a residence time of 3 hr is needed to accomplish a drying with the conditions indicated on the sketch. For reasons of entrainment, the air rate is limited to 750 lbs dry/(hr)(sqft cross section). Properties of the solid are 50 lb/cuft and 0.22 Btu/(lb)(°F). Symbols on the sketch are A = dry air, S = dry solid, W = water:

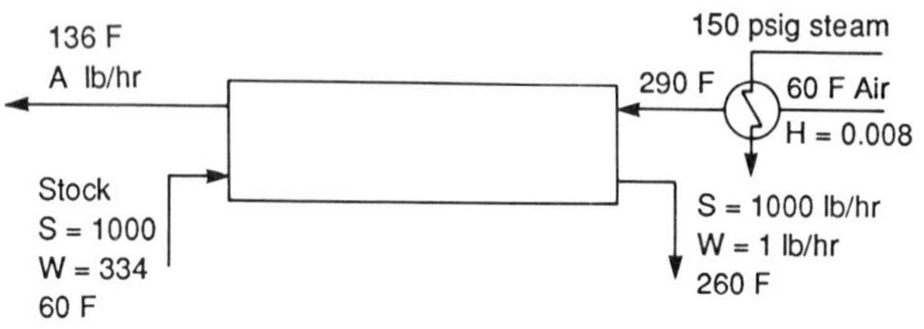

In terms of the dry air rate, A lb/hr, the average moist heat capacity is

$$C = 0.24 + 0.45[0.008 + \tfrac{1}{2}(333/A)] = 0.2436 + 74.93/A.$$

In the dryer, the enthalpy change of the moist air equals the sum of the enthalpy changes of the moisture and of the solid. Add 7% for heat losses. With steam table data,

$$\begin{aligned}(0.2436 + 74.93/A)A(290 - 136) &= 1.07[333(1120.3) + 1(228)\\ &\quad + 1000(0.22)(260 - 60) - 334(28)]\\ &= 1.07(407{,}936) = 43{,}649],\\ \therefore A &= 11{,}633 \quad \text{lb/hr}.\end{aligned}$$

The exit humidity is

$$H = 0.008 + 333/11{,}633 = 0.0366\ \text{lb/lb},$$

which corresponds to an exit dewpoint of 96°F, an acceptable value.

With the allowable air rate of 750 lb/hr sqft, the diameter of the dryer is

$$D = \sqrt{11{,}633/750\pi/4} = 4.44\ \text{ft}, \quad \text{say } 4.5\ \text{ft}.$$

Say the solid occupies 8% of the cross section. With a solids density of 50 lb/cuft, the dryer volume,

$$V = 3(1000/50)/0.08 = 750\ \text{cuft},$$

and the length is

$$L = 750/(4.5)^2\pi/4 = 47.2\ \text{ft}.$$

The standard number of flights is 2–4 times the diameter, or

$$\text{number} = (2\text{–}4)4.5 = 9\text{–}18, \quad \text{say } 12.$$

The product of rpm and diameter is 25–35

$$\therefore \text{rpm} = (25\text{–}35)/4.5 = 5.5\text{–}7.8, \quad \text{say } 6.7.$$

The stm heater duty is

$$Q_s = 11{,}633(0.2436)(290 - 60) = 651{,}733\ \text{Btu/hr},$$

150 psig stm,

$$\text{stm} = 651{,}733/857 = 760.5\ \text{lb/hr}.$$

Evaporation efficiency is

$$\eta = 333/760.5 = 0.438\ \text{lb water/lb stm}.$$

The efficiency of the dryer itself is

$$\eta_d = 407{,}936/651{,}733 = 0.626\ \text{Btu/Btu}.$$

TABLE 9.10. Performance Data of Rotary Dryers

(a) Direct Heated Dryers

	Sugar Beet Pulp[a]	Calcium Carbonate[a]	Blast Furnace Slag[a]	Lead Concentrate[b]	Sand[b]	Zinc Concentrate[b]	Ammonium Sulphate[c]	Fine Salt[c]	Crystals[d]	Chemicals[d]
Air flow	parallel	parallel	parallel	parallel	parallel	parallel	counter	counter	counter	indirect counter
Dryer length	9 ft 2 in.	6 ft 3 in.	7 ft 2 in.	4 ft 6 in.	4 ft 6 in.	7 ft 6 in.	9 ft 0 in.	5 ft 0 in.	10 ft 0 in.	4 ft 6 in.
Dryer length	46 ft 0 in.	34 ft 0 in.	40 ft 0 in.	35 ft 0 in.	32 ft 6 in.	60 ft 0 in.	40 ft 0 in.	40 ft 0 in.	60 ft 0 in.	27 ft 0 in.
Method of heating	oil	oil	oil	oil	gas	oil	gas	steam	steam	Louisville steam tube
Method of feed	screw	belt	belt	screw	chute	screw	conveyor	feeder	screw	screw
Initial moisture (% w/w)	82	13.5	33	14	5.65	18	2.5	5.0	7.0	1.5
Final moisture (% w/w)	10	0.5	nil	8	0.043	8	0.2	0.1	8.99	0.1
Evaporation (lb/hr)	34,000	6000	11,600	1393	701	8060	1120	400	1150	63
Capacity (lb evaporated/ft^3 dryer volume)	11	6	7	2.5	1.35	2.3	0.5	0.52	0.245	—
Efficiency (Btu supplied/water evaporated)	1420	1940	1710	2100	2550	1850	1920	2100	1650	—
Inlet air temperature (°F)	1560	1560	1560	1300	1650	1500	400	280	302	—
Outlet air temperature (°F)	230	220	248	200	222	200	180	170	144	—
Residence time (av. min)	20	25	30	20	12	20	15	40	70	30/40
Fan HP	70	40	50	20	5	75	25	8	—	10
Motive HP	15	20	25	10	10	55	60	15	60	—
Fan capacity (std. air ft^3/min)	45000	8500	18,000	2750	2100	12,000	18,500	6500	—	—

[a] Courtesy Buell Ltd.
[b] Courtesy Head Wrightson (Stockton) Ltd.
[c] Courtesy Edgar Allen Aerex Ltd.
[d] Courtesy Constantin Engineers Ltd.—Louisville Dryers; Williams–Gardner, 1971.

(b) Roto–Louvre Dryers

	Bone Meal	Sugar[a]	Sulfate of Ammonia[a]	Bread Crumbs	Bentonite
Dryer diameter	7 ft 6 in.	7 ft 6 in.	7 ft 6 in.	4 ft 6 in.	8 ft 10 in.
Dryer length	12 ft 0 in.	25 ft 0 in.	25 ft 0 in.	20 ft 0 in.	30 ft 0 in.
Initial moisture (% w/w basis)	17.0	1.5	1.0	37	45
Final moisture (% w/w basis)	7.0	0.03	0.2	2.5	11
Method of feed	screw	screw	chute	chute	chute
Evaporation rate (lb/hr)	1660	500	400	920	7100
Efficiency (Btu supplied/lb evaporation)	74.3	40	—	55	62.5
Method of heating	steam	steam	steam	gas	oil
Inlet air temperature (°F)	203	194	248	572	842
Outlet air temperature (°F)	122	104	149	158	176
Residence time, min	9.3	12.5	9.0	25.7	37.3
Fan HP (absorbed)	49.3	52.2	55	13.7	54.3
Motive HP (absorbed)	8	12.5	15	2.3	20.0
Fan capacity (ft^3/min)					
Inlet	9560	18,000	16,000	5380	20,000
Outlet	14,000	22,300	21,000	5100	25,000

[a] Combined two-stage dryer-cooler.
(Courtesy Dunford and Elliott Process Engineering Ltd.; Williams–Gardner, 1971).

(continued)

TABLE 9.10—(*continued*)

(c) Roto–Louvre Dryers

Material Dried	Ammonium Sulfate	Foundry Sand	Metallurgical Coke
Dryer diameter	2 ft 7 in.	6 ft 4 in.	10 ft 3 in.
Dryer length	10 ft	24 ft	30 ft
Moisture in feed (% wet basis)	2.0	6.0	18.0
Moisture in product (% wet basis)	0.1	0.5	0.5
Production rate (lb/hr)	2500	32,000	38,000
Evaporation rate (lb/hr)	50	2130	8110
Type of fuel	steam	gas	oil
Fuel consumption	255 lb/hr	4630 ft^3/hr	115 gal/hr
Calorific value of fuel	837 Btu/lb	1000 Btu/ft^3	150,000 Btu/gal
Efficiency (Btu supplied per lb evaporation)	4370	2170	2135
Total power required (HP)	4	41	78

(FMC Corp.; *Chemical Engineers' Handbook,* 1984, p. 20.20).

(d) Indirect–Direct Double Shell Dryers

	Indirect–Direct Double Shell		
	Coal	Anhydrite	Coke
Dryer diameter	7 ft 6 in.	5 ft 10 in.	5 ft 10 in.
Dryer length	46 ft 0 in.	35 ft 0 in.	35 ft 0 in.
Initial moisture content (% w/w basis)	22	6.0	15
Final moisture (% w/w basis)	6	1.0	1.0
Evaporation rate (lb/hr)	5800	2300	1600
Evaporation—volume ratio (lb/ft^3/hr)	3.5	3.15	2.2
Heat source	coal	oil	oil
Efficiency (Btu supplied/lb water evaporated)	1250	1250	1340
Inlet air temperature (°F)	1200	1350	1350
Outlet air temperature (°F)	160	160	200

(Courtesy Edgar Allen Aerex Ltd.; Williams–Gardner, 1971).

(e) Steam Tube Dryers

	Class 1	Class 2	Class 3
Class of materials	high moisture organic, distillers' grains, brewers' grains, citrus pulp	pigment filter cakes, blanc fixe, barium carbonate, precipitated chalk	finely divided inorganic solids, water-ground mica, water-ground silica, flotation concentrates
Description of class	wet feed is granular and damp but not sticky or muddy and dries to granular meal	wet feed is pasty, muddy, or sloppy, product is mostly hard pellets	wet feed is crumbly and friable, product is powder with very few lumps
Normal moisture content of wet feed (% dry basis)	233	100	54
Normal moisture content of product (% dry basis)	11	0.15	0.5
Normal temperature of wet feed (K)	310–320	280–290	280–290
Normal temperature of product (K)	350–355	380–410	365–375
Evaporation per product (kg)	2	1	0.53
Heat load per lb product (kJ)	2250	1190	625
Steam pressure normally used (kPa gauge)	860	860	860
Heating surface required per kg product (m^2)	0.34	0.4	0.072
Steam consumption per kg product (kg)	3.33	1.72	0.85

(*Chemical Engineers' Handbook,* 1984).

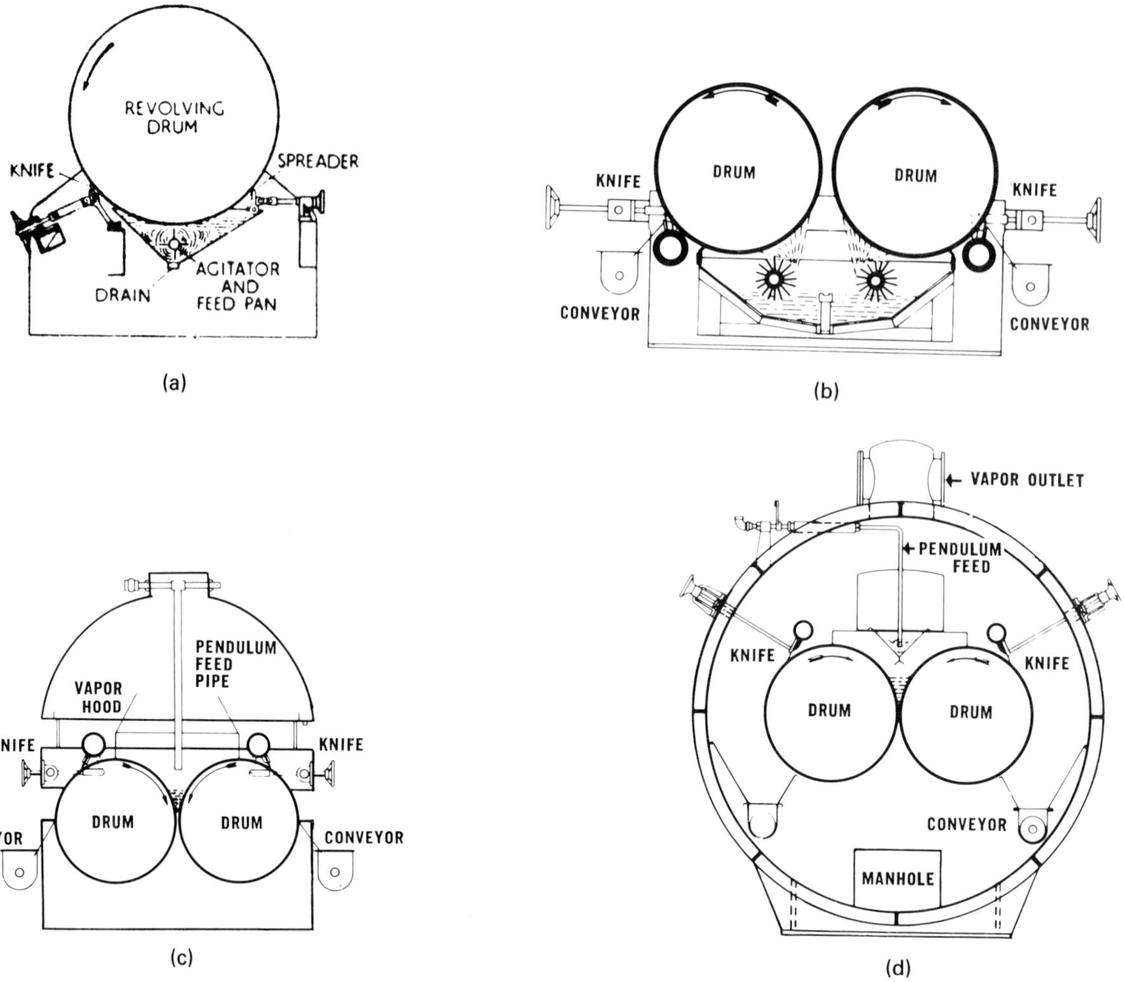

Figure 9.11. Drum dryers for solutions and thin slurries (*Buflovak Equip. Div., Blaw Knox Co.*). (a) Single drum dryer with dip feed and spreader. (b) Double drum dryer with splash feed. (c) Double drum dryer with top feed, vapor hood, knives and conveyor. (d) Double drum dryer with pendulum feed, enclosed for vacuum operation.

Single pass residence times are 0.5–3 sec, but most commercial operations employ some recycling of the product so that average residence times are brought up to 60 sec. Recycling also serves to condition the feed if it is very wet. The spread of residence times in pneumatic dryers, as indicated by Figure 9.5(a), is broad, so feed that has a particularly wide size distribution may not dry uniformly. Recycling, however, assists uniformity, or several dryers in series or preclassification of particle sizes may be employed.

Since the contact time is short, heat-sensitive materials with good drying characteristics are particularly suited to this kind of dryer, but sticky materials obviously are not. Moreover, since attrition may be severe, fragile granules cannot be handled safely. Other kinds of dryers should be considered for materials that have substantial falling rate drying periods.

Pilot plant work is essential as a basis for full scale design. It may be directed to finding suitable velocities, temperatures and drying times, or it may employ more basic approaches. The data provided for Example 9.8, for instance, are of particle size distribution, partial pressure of water in the solution, and heat and mass transfer coefficients. These data are sufficient for the

---

Example 9.7
Description of a Drum Drying System

A detergent drying plant handles 86,722 lb/day of a slurry containing 52% solids and makes 45,923 lb/day of product containing 2% water. The dryers are two sets of steam-heated double drums, each 3.5 ft dia by 10 ft, with a total surface of 440 sqft. Each drum is driven with a 10 HP motor with a variable speed transmission. Each trolley top spreader has a 0.5 HP motor. Each side conveyor has a 1 HP motor and discharges to a common belt conveyor that in turn discharges to a bucket elevator that feeds a flaker where the product is reduced to flakes less than 0.25 in. square. Fines are removed in an air grader and recycled to the dryer feed tank.

TABLE 9.11. Performance Data of Drum Dryers

(a) Drum Dryers

	Yeast Cream	Stone Slop	Starch Solutions	Glaze	Zirconium Silicate	Brewers Yeast	Clay Slip
Feed solids (% by weight)	16	40	36	64	70	25	75
Product moisture (% w/w basis)	5.7	0.2	5	0.2	0.2	5	9
Capacity (lb prod./hr)	168	420	300–400	225	1120	146	4000
Dryer type (a) single, (b) twin, (c) double	(a)	(a)	(a)	(a)	(a)	(a)	(a)
Drum							
diameter	4 ft 0 in.	2 ft 6 in.	48 in.	18 in.	36 in.	28 in.	48 in.
length	10 ft 0 in.	5 ft 0 in.	120 in.	36 in.	72 in.	60 in.	120 in.
Type of feed method	top roller	dip	top roller	side	dip	center nip	side
Steam pressure (lb/in^2 gauge)	80	60	80	—	80	40	40
Atmospheric or vacuum	atmos.	atmos.	atmos.	atmos.	atmos.	atmos.	atmos.
Steam consumption (lb/lb evaporated)	—	—	1.3	1.3	—	—	1.35
Average effective area (%)	—	—	86	—	—	—	65
Evaporation/ft^2/hr	6.5	4	5	9	8.4	6	8.4

(Courtesy A.P.V. Mitchell Dryers, Ltd.; Williams–Gardner, 1971).

(b) Drum Dryers in the Size Range 0.4 × 0.4–0.8 × 2.25 m[a]

Type of Dryer and Feed Size by Letter, A, B, or C	Drum Speed (rev/min)	Steam Press (bar, g)	Type of Material	Physical Form of Feed	Solids in Feed (%)	H_2O in Product (%)	Output of Dried Product ($g/sec\,m^2$)	Evaporation Rate of Water ($g/sec\,m^2$)
			inorganic salts					
Single (dip)	4.4	3.5[b]	alk. carbs	—	50	8–12	5.5	4.9
Single (splash)	1	3.0	$Mg(OH)_2$	thick slurry	35	0.5	1.9	1.5
Twin (splash) A	3	3.0	$Fe(OH)_3$	thin slurry	22	3.0	4.3	1.3
Double	3–8	5.0	Na Acetate	solution	20	0.4–10	2.0–7.0	8–24
Double and twin	7–9	2–3	Na_2SO_4	solution	24	0.15–5.5	4.7–6.1	11–12
Double and twin	5–9	4–6	Na_2HPO_4	solution	44	0.8–0.9	8.2–11.1	9–14
Twin (dip) A	5	5.5	organic salts	solution	27	2.8	1.9	5.2
Twin A	3	5.5	organic salts	solution	33	13.0	1.4	2.6
Twin B	2[c]	3.5	organic salts	solution	20	1.0	1.0	3.8
Twin C	5	5.5	organic salts	solution	39	0.4	3.9	6.1
Twin C	$5\frac{1}{2}$	5.5	organic salts	solution	42	1.0	2.1	4.6
Twin C	6	5.5	organic salts	solution	35	5.0	4.1	7.2
Twin (splash) A	3–5	5.0	organic salts	thin slurry	20	1.7–3.1	1.0–1.9	3.7–7.3
Double A	$5\frac{1}{2}$	6.0	organic salts	solution	11	—	1.1	9
Double B	$6\frac{1}{2}$	5–6	organic salts	solution	40	3	3.4	4.9
Twin (dip) A	5	5.5		thin slurry	30	1.2	2.4	5.5
Twin A	5	5.0	organic	viscous soln.	28	10.5	1.9	4.2
Double	2	3.0	compounds	viscous soln.	—	6.0	0.7	—
Double	$4\frac{1}{2}$	3.5		thin slurry	25	1.0	0.4–1.9	3.5–5.0
Twin (dip)	5	5.0	organic	(a) solution	25	0.5	0.3	0.8
Twin	10	5.5	compounds of	(b) thick slurry	30	2.5	2.0	4.6
Twin	10	5.5	low surface tension	(c) thick slurry	35	—	3.1	—
Double	11	5.5	similar letters	(b) thin paste	46	—	6.4	7.3
Double	12	5.5	for same	(c) thick paste	58	—	6.0	4.3
Double	11	5.5	compound	(a) solution	20	0.5	0.24	1.0

[a] Dryer dia and width (m): (A) 0.457 × 0.457; (B) 0.71 × 1.52; (C) 0.91 × 2.54.
[b] Plus external hot air flow.
[c] Stainless steel drum.
(Nonhebel and Moss, 1971).

TABLE 9.12. Performance of Drum Dryers

(a) Single, Double drum and Vacuum Drums

Material	Method of Feed	Moisture Content, (% Wet Basis) Feed	Product	Steam Pressure, (lb/sq in.)	Drum Speed (rpm)	Feed Temp. (°F)	Capacity [lb product/(hr)(sq ft)]	Vacuum (in. Hg)
Double-drum dryer								
Sodium sulfonate	trough	53.6	6.4	63	$8\frac{1}{2}$	164	7.75	
Sodium sulfate	trough	76.0	0.06	56	7	150	3.08	
Sodium phosphate	trough	57.0	0.9	90	9	180	8.23	
Sodium acetate	trough	39.5	0.44	70	3	205	1.51	
Sodium acetate	trough	40.5	10.03	67	8	200	5.16	
Sodium acetate	trough	63.5	9.53	67	8	170	3.26	
Single-drum dryer								
Chromium sulfate	spray film	48.5	5.47	50	5	—	3.69	
Chromium sulfate	dip	48.0	8.06	50	4	—	1.30	
Chromium sulfate	pan	59.5	5.26	24	$2\frac{1}{2}$	158	1.53	
Chromium sulfate	splash	59.5	4.93	55	$1\frac{3}{4}$	150	2.31	
Chromium sulfate	splash	59.5	5.35	53	$4\frac{3}{4}$	154	3.76	
Chromium sulfate	dip	59.5	4.57	53	$5\frac{3}{4}$	153	3.36	
Vegetable glue	pan	60–70	10–12	20–30	6–7	—	1–1.6	
Calcium arsenate	slurry	75–77	0.5–1.0	45–50	3–4	—	2–3	
Calcium carbonate	slurry	70	0.5	45	2–3	—	1.5–3	
Twin-drum dryer								
Sodium sulfate	dip	76	0.85	55	7	110	3.54	
Sodium sulfate	top	69	0.14	60	$9\frac{1}{2}$	162	4.27	
Sodium sulfate	top	69	5.47	32	$9\frac{1}{2}$	116	3.56	
Sodium sulfate	splash	71	0.10	60	6	130	4.30	
Sodium sulfate	splash	71.5	0.17	60	12	140	5.35	
Sodium sulfate	splash	71.5	0.09	60	10	145	5.33	
Sodium phosphate	splash	52.5	0.59	58	$5\frac{1}{2}$	208	8.69	
Sodium phosphate	dip	55	0.77	60	$5\frac{1}{2}$	200	6.05	
Sodium sulfonate	top	53.5	8–10	63	$8\frac{1}{2}$	172	10.43	
Vacuum single-drum dryer								
Extract	pan	59	7.75	35	8	—	4.76	27.9
Extract	pan	59	2.76	35	6	—	1.92	27.9
Extract	pan	59	2.09	36	4	—	1.01	atmos.
Extract	pan	56.5	1.95	35	$7\frac{1}{2}$	—	3.19	22.7
Extract	pan	56.5	1.16	50	$2\frac{1}{2}$	—	0.75	atmos.
Skim milk	pan	65	2–3	10–12	4–5	—	2.5–3.2	
Malted milk	pan	60	2	30–35	4–5	—	2.6	
Coffee	pan	65	2–3	5–10	$1–1\frac{1}{2}$	—	1.6–2.1	
Malt extract	spray film	65	3–4	3–5	0.5–1.0	—	1.3–1.6	
Tanning extract	pan	50–55	8–10	30–35	8–10	—	5.3–6.4	
Vegetable glue	pan	60–70	10–12	15–30	5–7	—	2–4	

(*Perrys Chemical Engineers Handbook*, McGraw-Hill, 1950 edition).

calculation of residence time when assumptions are made about terminal temperatures.

9.9. FLUIDIZED BED DRYERS

Free flowing granular materials that require relatively short drying times are particularly suited to fluidized bed drying. When longer drying times are necessary, multistaging, recirculation or batch operation of fluidized beds still may have advantages over other modes.

A fluidized bed is made up of a mass of particles buoyed up out of permanent contact with each other by a flowing fluid. Turbulent activity in such a bed promotes high rates of heat and mass transfer and uniformity of temperature and composition throughout. The basic system includes a solids feeding device, the fluidizing chamber with a perforated distributing plate for the gas, an overflow duct for removal of the dry product, a cyclone and other equipment for collecting fines, and a heater and blower for the gaseous drying medium.

Much ingenuity has been applied to the design of fluidized bed drying. Many different arrangements of equipment are illustrated and described in the comprehensive book of Kröll (1978) for instance. Figure 9.13(a) depicts the basic kind of unit and the other items are a few of the many variants. Tables 9.14 and 9.15 are selected performance data.

Shallow beds are easier to maintain in stable fluidization and of course exert a smaller load on the air blower. Pressure drop in the air distributor is approximately 1 psi and that through the bed equals the weight of the bed per unit cross section. Some pressure drop data are shown in Table 9.14. The cross section is determined by the gas velocity needed for fluidization as will be described. It is usual to allow 3–6 ft of clear height between the top of the bed and the air exhaust duct. Fines that are entrained are collected in a cyclone and blended with the main stream since they are very dry

TABLE 9.12—(*continued*)

(b) Single and Double Drum with Various Feed Arrangements

Kind of Dryer, Kind of Stock	Moisture Content In (%)	Moisture Content Out (%)	Vapor Pressure Absolute (bar)	Rotation Speed (1/min)	Unit Product Capacity (kg/m² hr)	Drying Rate (kg/m² hr)
Single drum, dip feed						
Alkali carbonate	50	8 bis 12	3.5	4.4	20	17.8
Double drum, dip feed						
Organic salt solution	73	2.8	5.5	5	6.8	18.6
Organic compound, dilute slurry	70	1.2	5.5	5	8.6	19.6
Organic compound, solution	75	0.5	5.0	5	1.1	1.9
Single drum with spreading rolls						
Skim milk concentrate	50	4	3.8	24	15.8	14.2
Whey concentrate	45	4.3	5.0	16	10 bis 11.8	7.4 bis 8.8
Cuprous oxide	58	0.5	5.2	10	11.0	14.3
Single drum, splash feed						
Magnesium hydroxide, dense slurry	65	0.5	3.0	1	6.8	5.4
Double drum, splash feed						
Iron hydroxide, dilute slurry	78	3.0	3.0	3	15.4	4.7
Organic salt, dilute slurry	80	1.7 bis 3.1	5.0	3 bis 5	3.6 bis 6.8	13.3 bis 26.2
Sodium acetate	50	4.0	6.0	5	10.0	9.3
Sodium sulfate	70	2.3	7.8	5	18.0	40.4
Double drum, top feed						
Beer yeast	80	8.0	6.0	5	10.0	36.2
Skim milk, fresh	91.2	4.0	6.4	12	6.2	61.5
Organic salt solution	89	—	6.0	5.5	4	32.3
Organic salt solution	60	3	5 bis 6	6.5	12.2	17.7
Organic compound, dilute slurry	75	1	3.5	4.5	1.4 bis 6.8	12.6 bis 18
Double drum with spreading rolls						
Potato pulp	76.2	11.4	8	5	22.5	61.1

(Kröll, 1978, p. 348).

due to their small size. Normally entrainment is 5–10% but can be higher if the size distribution is very wide. It is not regarded as feasible to permit high entrainment and recycle back to the drying chamber, although this is common practice in the operation of catalytic cracking equipment.

Mixing in shallow beds is essentially complete; Figure 9.5(c) shows some test data in confirmation. The corresponding wide distribution of residence times can result in nonuniform drying, an effect that is accentuated by the presence of a wide distribution of particle sizes. Multiple beds in series assure more nearly constant residence time for all particles and consequently more nearly uniform drying. The data of Table 9.14(b) are for multiple zone dryers. Figures 9.13(c) and (d) have additional zones for cooling the product before it leaves the equipment. Another way of assuring

TABLE 9.13. Performance Data of Pneumatic Conveying Dryers (Sketches in Fig. 9.12)

(a) Raymond Flash Dryer

	Fine Mineral	Spent Grain	Organic Chemical	Chicken Droppings	Fine Coal Filter Cake
Method of feed	pump	belt	screw	pump	screw
Material size, mesh	−100	—	−30	—	−30
Product rate (lb/hr)	27,000	9000	900	2300	2000
Initial moisture content (% w/w basis)	25	60	37	70	30
Final moisture content (% w/w basis)	nil	12	3	12	8.5
Air inlet temperature (°F)	1200	1200	450	1300	1200
Air outlet temperature (°F)	200/300	200/300	200/300	200/300	200/300
Method of heating	direct oil	direct oil	direct oil	direct oil	direct oil
Heat consumption (Btu/lb water evaporated)	1.6×10^3	1.9×10^3	3.1×10^3	1.9×10^3	1.4×10^3
Air recirculation	no	no	no	no	no
Material recirculation	yes	yes	no	yes	no
Material of construction	MS	MS/SS	MS	MS	MS
Fan capacity (std. ft³/min)	18,000	22,000	4300	8500	1500
Installed fan HP	110	180	30	50	10
Product exit temperature (°F)	200	—	200	—	135

(Courtesy International Combustion Products Ltd.; Williams–Gardner, 1971).

TABLE 9.13—*(continued)*

(b) Buttner–Rosin Pneumatic Dryer

	Metallic Stearate	Starch	Adipic Acid	Fiber	Coal Filter Cake
Method of feed	sling	sling	screw	distributor	distributor
Material size	fine	fine	−30 mesh	$-\frac{1}{4}$ in.	−30 mesh
Product rate (lb/hr)	280	13,236	10,000	2610	67,200
Initial moisture (% w/w basis)	40	34	10	62.4	32
Final moisture (% w/w basis)	0.5	13	0.2	10	6
Air inlet temperature (°F)	284	302	320	752	1292
Air outlet temperature (°F)	130	122	149	230	212
Method of heating	steam	steam	steam	oil	PF
Heat consumption (Btu/lb/water evaporated)	2170	1825	2400	1720	1590
Air recirculation	no	no	no	no	yes
Material recirculation	yes	no	yes	yes	yes
Fan capacity (std. ft^3/min)	1440	26,500	9500	12,500	27,000
Installed fan HP	15	220	65	60	250
Product exit temperature (°F)	104	95	120	140	158

(Courtesy Rosin Engineering Ltd.; Williams–Gardner, 1971).

(c) Pennsalt–Berks Ring Dryer

	Metal[a] Stearates	Spent[a] Grains	Sewage[b] Sludge	Starches	Polystyrene Beads
Method of Feed	**belt feeder rotary valve**	**back mixer rotary valve**	**vibratory feeder rotary valve**	**cascading rotary valve screen**	**vibratory feeder rotary valve**
Product rate (lb/hr)	240	1120	4300	5000	1000
Initial moisture (% w/w basis)	55	80	45	35	2.0
Final moisture (% w/w basis)	1	5	12	10	0.2
Air inlet temperature (°F)	250	500	600	300	175
Air outlet temperature (°F)	150	170	170	130	115
Method of heating	steam	gas	oil	steam	steam
Heat consumption (Btu/lb water evaporated)	2900	1800	1750	2000	5000
Air recirculation	no	no	no	no	no
Material recirculation	yes	yes	yes	no	no
Material of construction	SS	MSG	MS	MSG	SS
Fan capacity (std ft^3/min)	3750	16,500	8250	15,000	900
Installed fan HP	20	75	60	60	7.5

[a] Ring dryer application.
(Courtesy Pennsalt Ltd.; Williams–Gardner, 1971).

(d) Various Pneumatic Dryers

Material	Location	Tube Dia (cm)	Tube Height (m)	Gas Rate (m^3/hr) (NTP)	Gas Temp (°C) In	Gas Temp (°C) Out	Solid Rate (kg/hr)	Solid Temp (°C) In	Solid Temp (°C) Out	Moisture (%) In	Moisture (%) Out	Air/Solid Ratio (m^3/kg) (NTP)	Air/Solid Ratio (kg/kg)	Water Evaporated (kg/hr)
Ammonium sulphate	Japan	18	1	1100	215	76	950	38.5	63	2.75	0.28	1.2	1.5	23.5
Sewage sludge filter cake	U.S.A.	—	—	1200	700	121	2270	15	71	80	10	5.3	7.2	1590
Coal 6 mm	U.S.A.	—	—	50,000	371	80	51,000	15	57	9	3	1.0	1.3	4350
Hexamethylene tetramine	Germany	30	38[a]	3600	93	50	2500	—	48	6–10	0.08–0.15	1.4	1.9	18.1

[a] 23 m vertical, 15 m horizontal.
(Nonhebel and Moss, 1971).

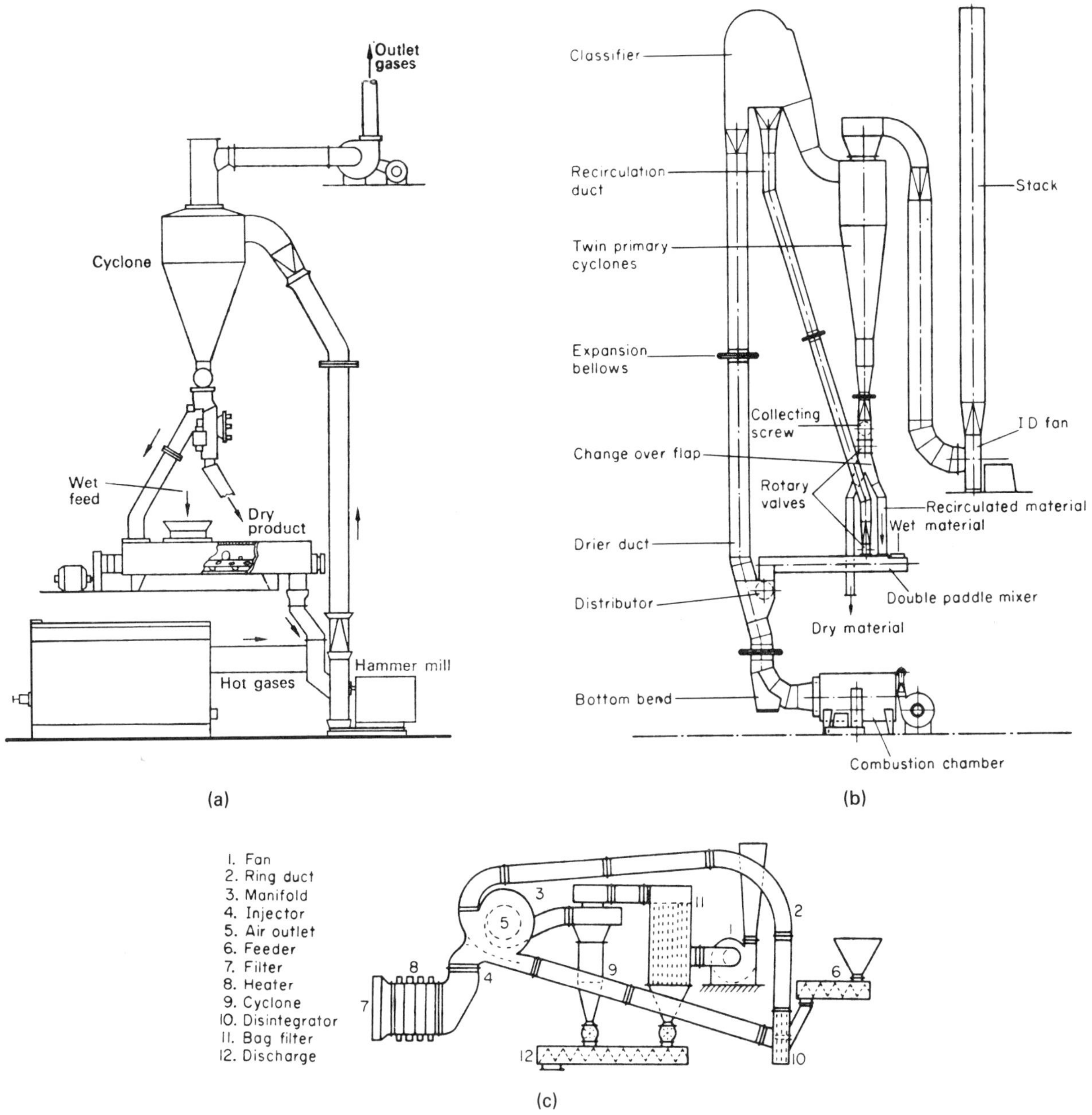

Figure 9.12. Examples of pneumatic conveying dryers; corresponding performance data are in Table 9.13. (a) Raymond flash dryer, with a hammer mill for disintegrating the feed and with partial recycle of product (*Raymond Division, Combustion Engineering*). (b) Buttner–Rosin pneumatic dryer with separate recycle and disintegration of large particles (*Rosin Engineering Ltd.*). (c) Berks ring dryer; the material circulates through the ring-shaped path, product is withdrawn through the cyclone and bag filter (*Pennsalt Chemical Co.*).

complete drying is a recirculation scheme like that of Figure 9.13(e). In batch operation the time can be made as long as necessary.

Stable fluidization requires a distribution of particle sizes, preferably in the range of a few hundred microns. Normally a size of 4 mm or so is considered an upper limit, but the coal dryers of Tables 9.15(a) and (b) accommodate sizes up to 0.5 in. Large and uniformly sized particles, such as grains, are dried successfully in spouted beds [Fig. 9.13(f)]: Here a high velocity gas stream entrains the solid upward at the axis and releases it at the top for flow back through the annulus. Some operations do without the mechanical draft tube shown but employ a naturally formed central channel.

One way of drying solutions or pastes under fluidizing conditions is that of Figure 9.13(g). Here the fluidized mass is of auxiliary spheres, commonly of plastic such as polypropylene, into which the solution is sprayed. The feed material deposits uniformly on the spheres, dries there, and then is knocked off automatically as it leaves the drier and leaves the auxiliary spheres behind. When a mass of dry particles can be provided to start a fluidized bed drying process, solutions or pastes can be dried after deposition on the seed material as on the auxiliary spheres. Such a process is employed, for instance, for growing fertilizer granules of desired larger sizes, and has largely replaced rotary dryers for this purpose.

A few performance data of batch fluid dryers are in Table

EXAMPLE 9.8
Sizing a Pneumatic Conveying Dryer

A granular solid has a moisture content of 0.035 kg/kg dry material which is to be reduced to 0.001 kg/kg. The charge is at the rate of 9.72 kg/sec, is at 60°C and may not be heated above 90°C. Inlet air is at 450°C and has a moisture content of 0.013 kg/kg dry air.

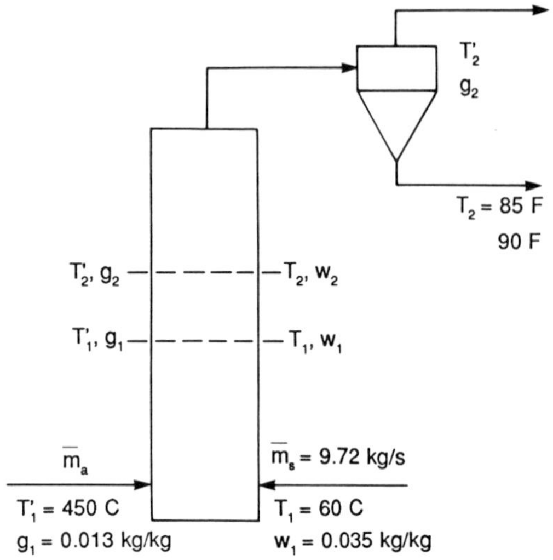

Specific gravity of the solid is 1.77 and its heat capacity is 0.39 cal/g °C. The settling velocity of the largest particle present, 2.5 mm dia, is 10 m/sec. Heat capacity of the air is taken as 0.25 cal/g °C and the latent heat at 60°C as 563 cal/g. Experimental data for this system are reported by Nonhebel and Moss (1971, pp. 240ff) and are represented by the expressions:

Heat transfer coefficient:

$$ha = 0.47 \text{ cal/(kg solid)(°C)}.$$

Vapor pressure:

$$P = \exp(13.7419 - 5237.0/T), \quad \text{atm, K}.$$

Mass transfer coefficient:

$$k_g a = \exp(-3.1811 - 1.7388 \ln w - 0.2553(\ln w)^2),$$

where w is the moisture content of the solid (kg/kg) in the units kg water/(kg solid)(atm)(sec).

In view of the strong dependence of the mass transfer coefficient on moisture content and the 35-fold range of that property, the required residence time and other conditions will be found by analyzing the performance over small decrements of the moisture content.

An air rate is selected on the assumption that the exit of the solid is at 85°C and that of the air is 120°C. These temperatures need not be realized exactly, as long as the moisture content of the exit air is below saturation and corresponds to a partial pressure less than the vapor pressure of the liquid on the solid. The amount of heat transferred equals the sum of the sensible heat of the wet solid and the latent heat of the lost moisture. The enthalpy balance is based on water evaporating at 60°C:

$$\bar{m}_s[(0.39 + 0.001)(85 - 60) + (0.035 - 0.001)(85 - 60 + 563)]$$
$$= \bar{m}_a[(0.25 + 0.480(0.001))(450 - 120) + 0.48(0.034)(120 - 60)],$$

$$\bar{m}_a = \frac{29.77\bar{m}_s}{83.64} = \frac{29.77(9.72)}{83.64} = \begin{cases} 3.46 \text{ kg/sec}, \\ 7.08 \text{ m}^3/\text{sec} \quad \text{at } 450°\text{C}, \\ 3.85 \text{ m}^3/\text{sec} \quad \text{at } 120°\text{C}. \end{cases}$$

At a tower diameter of 0.6 m,

$$U = \frac{Q}{0.36\pi/4} = \begin{cases} 25.0 \text{ m/sec} \quad \text{at } 450°\text{C}, \\ 13.6 \text{ m/sec} \quad \text{at } 120°\text{C}. \end{cases}$$

These velocities are great enough to carry the largest particles with settling velocity of 10 m/sec.

Equations are developed over intervals in which $W_1 \to W_2$, $T_1 \to T_2$, and $T'_1 \to T'_2$.

The procedure will be:

1. Start with known W_1, T_1, and T'_1.
2. Specify a moisture content W_2.
3. Assume a value T_2 of the solid temperature.
4. Calculate T'_2 from the heat balance.
5. Check the correctness of T_2 by noting if the times for heat and mass transfers in the interval are equal.

$$\theta_h = \frac{Q}{ha(\Delta T)_{\text{lm}}} = \frac{Q}{0.47(\Delta T)_{\text{lm}}}$$

$$\theta_m = \frac{w_1 - w_2}{k_g a(\Delta P)_{\text{lm}}}$$

Heat balance:

$$\bar{m}_s[0.391(T_2 - T_1) + (W_1 - W_2)(T_2 - T_1 + 563)]$$
$$= \bar{m}_a\{[0.25 + 0.48(0.001)](T'_1 - T'_2)$$
$$+ 0.48(W_1 - W_2)(T'_2 - 60)\}.$$

Substitute $\bar{m}_s/\bar{m}_a = 9.72/3.46 = 2.81$ and solve for T'_2.

$$T'_2 = \frac{-0.25048T'_1 + 28.8(W_1 - W_2) + 2.81 \times [0.39(T_2 - T_1) + (W_1 - W_2)(T_2 - T_1 + 563)]}{0.48(W_1 - W_2) - 0.25048}. \quad (1)$$

$$g_1 = 0.013 + \frac{\bar{m}_s}{\bar{m}_a}(W_1 - 0.013) = 0.013 + 2.81(W_1 - 0.013). \quad (2)$$

$$P_1 = \frac{g_1}{18/29 + g_1} = \frac{g_1}{0.6207 + g_1} \quad \text{(partial pressure in air)}. \quad (3)$$

$$g_2 = 0.013 + 2.81(W_2 - 0.013). \quad (4)$$

$$P_2 = \frac{g_2}{0.6207 + g_2}. \quad (5)$$

$$Pa_1 = \exp[13.7419 - 5237.9/(T_1 + 273.2)], \quad \text{vapor pressure}. \quad (6)$$

$$Pa_2 = \exp[13.7419 - 5237.9/(T_2 + 273.2)]. \quad (7)$$

$$(\Delta P)_{\text{lm}} = \frac{(Pa_1 - P_1) - (Pa_2 - P_2)}{\ln[(Pa_1 - P_1)/(Pa_2 - P_2)]}. \quad (8)$$

$$(\Delta T)_{\text{lm}} = \frac{T'_1 - T_1 - (T'_2 - T_2)}{\ln[(T'_1 - T_1)/(T'_2 - T_2)]}. \quad (9)$$

$$\Delta Q = 0.391(T_2 - T_1) + (W_1 - W_2)(T_2 - T_1 + 563),$$
per kg of solid. (10)

$$\bar{W} = 0.5(W_1 + W_2). \quad (11)$$

$$k_g a = \exp[-3.1811 - 1.7388 \ln \bar{W} - 0.2553(\ln \bar{W})^2]. \quad (12)$$

$$\theta_h = \Delta Q/ha(\Delta T)_{\text{lm}} = \Delta Q/0.43(\Delta T)_{\text{lm}}, \quad \text{heating time}. \quad (13)$$

$$\theta_m = (W_1 - W_2)/k_g a(\Delta P)_{\text{lm}}, \quad \text{mass transfer time}. \quad (14)$$

$Z = \theta_h - \theta_m \to 0$ when the correct value of T_2 has been selected. (15)

After the correct value of T_2 has been found for a particular interval, make $W_2 \to W_1$, $T_2 \to T_1$, and $T'_2 \to T'_1$. Specify a

Example 9.8—(*continued*)

decremented value of W_2, assume a value of T_2, and proceed. The solution is tabulated.

W	T	T′	θ(sec)
0.035	60	450	0
0.0325	73.04	378.2	0.0402
0.03	75.66	352.2	0.0581
0.025	77.41	315.3	0.0872
0.02	77.23	286.7	0.1133
0.015	76.28	261.3	0.1396
0.01	75.15	236.4	0.1687
0.005	74.67	208.4	0.2067
0.003	75.55	192.4	0.2317
0.001	79.00	165.0	0.2841

When going directly from 0.035 to 0.001,

$$T_2 = 80.28,$$
$$T_2' = 144.04,$$
$$\theta = 0.3279 \text{ sec}.$$

The calculation could be repeated with a smaller air rate in order to reduce its exit temperature to nearer 120°C, thus improving thermal efficiency.

In the vessel with diameter = 0.6 m, the air velocities are

$$u_a = \begin{cases} 25.0 \text{ m/sec} & \text{at } 450°\text{C inlet} \\ 5.15 \text{ m/sec} & \text{at } 165°\text{C outlet} \end{cases}$$

$$20.1 \text{ m/sec} \quad \text{average.}$$

The vessel height that will provide the needed residence time is

$$H = \bar{u}_a\theta = 20.1(0.2841) = 5.70 \text{ m}.$$

Very fine particles with zero slip velocity will have the same holdup time as the air. The coarsest with settling velocity of 10 m/sec will have a net forward velocity of

$$\bar{u}_s = 20.1 - 10 = 10.1 \text{ m/sec},$$

which corresponds to a holdup time of

$$\theta = 5.7/10.1 = 0.56 \text{ sec},$$

which is desirable since they dry more slowly.

After the assumption of T_2, other quantities are evaluated in the order shown in this program.

```
10 ! Example 9.8. Pneumatic con
   veying dryer
20 ! Finding the exit solids te
   mp T2 by trial, then all dep
   endent quantities
30 INPUT W1,W2,T1,A1 ! A1 is th
   e inlet air temp T1'
40 INPUT T2 ! Trial value
50 A2=(2.81*(.391*(T2-T1)+(W1-W
   2)*(T2-T1+563))-.25048*A1+28
   .8*(W1-W2))/(.48*(W1-W2)-.25
   048)
60 G1=.013+2.81*(W1-.013)
70 P1=G1/(.6207+G1)
80 G2=.013+2.81*(W2-.013)
90 P2=G2/(.6207+G2)
100 Q1=EXP(13.7419-5237.9/(T1+27
   3.2)) ! vapor pressure
110 Q2=EXP(13.7419-5237.9/(T2+27
   3.2))
120 P3=(Q1-P1-Q2+P2)/LOG((Q1-P1)
   /(Q2-P2)) ! (ΔP)lm
130 T3=(A1-T1-A2+T2)/LOG((A1-T1)
   /(A2-T2)) ! (ΔT)lm
140 Q=.391*(T2-T1)+(W1-W2)*(T2-T
   1+563)
150 H1=Q/.47/T3 ! heating time
160 W=.5*(W1+W2)
170 K=EXP(-3.1811-1.7388*LOG(W)-
   .2533*LOG(W)^2)
180 H2=(W1-W2)/K/P3 ! vaporizati
   on time
190 Z=H1-H2 ! time difference sh
   ould be zero
200 DISP Z
210 DISP A2,H1
220 GOTO 40 ! if Z is not near e
   nough to zero; otherwise the
    correct value of T2 has bee
   n found
230 END
```

Data for the first interval

```
W1= .035
W2= .0325
T1= 73.04
T1'= 450

T2= 73.04
T2'= 378.16969111
Time= 4.02283660795E-2
```

9.14(a). This process is faster and much less labor-intensive than tray drying and has largely replaced tray drying in the pharmaceutical industry which deals with small production rates. Drying rates of 2–10 lb/(hr)(cuft) are reported in this table, with drying times of a fraction of an hour to several hours. In the continuous operations of Table 9.15, the residence times are at most a few minutes.

Thermal efficiency of fluidized bed dryers is superior to that of many other types, generally less than twice the latent heat of the water evaporated being required as heat input. Power requirements are a major cost factor. The easily dried materials of Table 9.15(a) show evaporation rates of 58–103 lb/(hr)(HP installed) but the more difficult materials of Table 9.15(d) show only 5–18 lb/(hr) (HP installed). The relatively large power requirements of fluidized bed dryers are counterbalanced by their greater mechanical simplicity and lower floor space requirements.

Air rates in Table 9.15 range from 13 to 793 SCFM/sqft, which is hardly a guide to the selection of an air rate for a particular case. A gas velocity twice the minimum fluidization velocity may be taken as a safe prescription. None of the published correlations of minimum fluidizing velocity is of high accuracy. The equation of Leva (*Fluidization*, McGraw-Hill, New York, 1959) appears to be as good as any of the later ones. It is

$$G_{mf} = 688D_p^{1.83}[\rho_g(\rho_s - \rho_g)]^{0.94}/\mu^{0.88}, \tag{9.20}$$

where G_{mf} is in lb/(hr)(sqft), ρ_g and ρ_s are densities of the gas and solid (lb/cuft), D_p is the particle diameter (in.), and μ is the gas viscosity (cP). In view of the wide scatter of the data on which this correlation is based, shown on Figure 6.14(f), it appears advisable to find the fluidization velocity experimentally for the case in hand.

Although it is embarrassing again to admit the fact, unfortunately all aspects of fluidized bed drying must be established with pilot plant tests. The wide ranges of performance parameters in Tables 9.14 and 9.15 certainly emphasize this conclusion. A limited exploration of air rates and equipment size can be made on the basis of a drying rate equation and fluidization correlations from the literature. This is done in Example 9.9. A rough approximation of a drying rate equation can be based on through circulation drying of the granular material on a tray, with gas flow downward.

9.10. SPRAY DRYERS

Suitable feeds to a spray dryer are solutions or pumpable pastes and slurries. Such a material is atomized in a nozzle or spray wheel, contacted with heated air or flue gas and conveyed out of the equipment with a pneumatic or mechanical type of conveyor. Collection of fines with a cyclone separator or filter is a major aspect of spray dryer operation. Typical equipment arrangements and flow patterns are shown in Figure 9.14.

The action of a high speed spray wheel is represented by Figure 9.14(e); the throw is lateral so that a large diameter vessel is required with this form of atomization, as shown in Figure 9.14(a). The flow from nozzles is largely downward so that the dryer is slimmer and taller. Parallel flow of air and spray downward is the most common

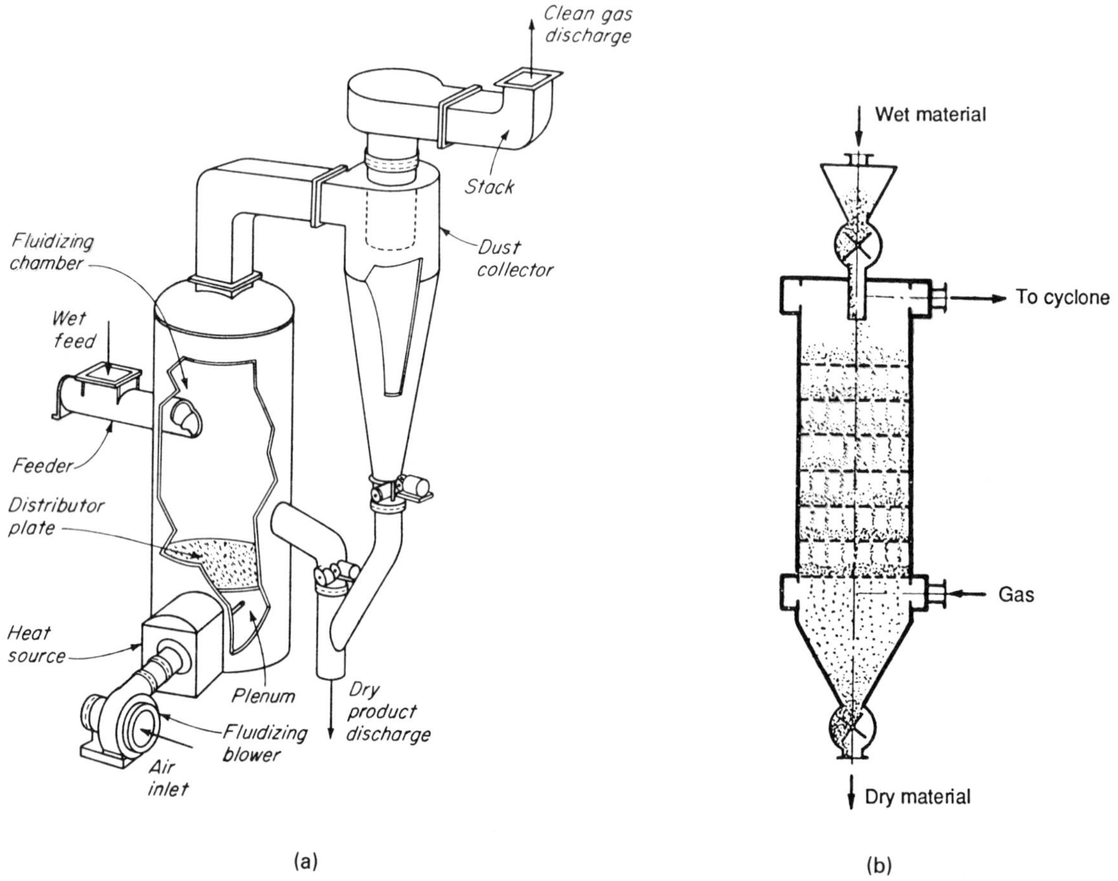

Figure 9.13. Fluidized bed dryers. (a) Basic equipment arrangement (*McCabe and Smith,* Unit Operations in Chemical Engineering, *McGraw-Hill, New York, 1984*). (b) Multiple bed dryer with dualflow distributors; performance data are in Table 9.14(b) (*Romankov, in Davidson and Harrison,* Fluidisation, *Academic, New York, 1971*). (c) A two-bed dryer with the lower one used as cooler: (a, b, c) rotary valves; (d) drying bed; (e) cooling bed; (f, g) air distributors; (h, i) air blowers; (k) air filter; (l) air heater; (m) overflow pipe; (n) product collector (*Kroll, 1978*). (d) Horizontal multizone dryer: (a) feeder; (b) air distributor; (c) fluidized bed; (d) partitions; (e) dust guard; (f) solids exit; (g) drying zone; (h) cooling zone; (i, k) blowers; (l, m) air plenums; (n) air duct; (o) dust collector; (p) exhaust fan (*Kroll, 1978*). (e) Circulating fluidized bed used for removal of combined water from aluminum hydroxide: (a) feed; (b) fluidized bed; (c) solids exit; (d) fuel oil inlet; (e) primary air inlet; (f) secondary air inlet; (g) gas exit (*Kroll, 1978*). (f) Spouted bed with draft tube for drying coarse, uniform-sized granular materials such as grains [*Yang and Keairns,* AIChE Symp. Ser. **176,** *218* (*1978*), *Fig. 1*]. (g) Fluidized bed dryer for sludges and pastes. The fluidized solids are fine spheres of materials such as polypropylene. The wet material is sprayed in, deposits on the spheres and dries there. At the outlet the spheres strike a plate where the dried material is knocked off and leaves the dryer as flakes. The auxiliary spheres remain in the equipment: (a) feed; (b) distributor; (c) spheres loaded with wet material; (d) returning spheres; (e) striking plate; (f) hot air inlet; (g) air and solids exit (*Kroll, 1978*).

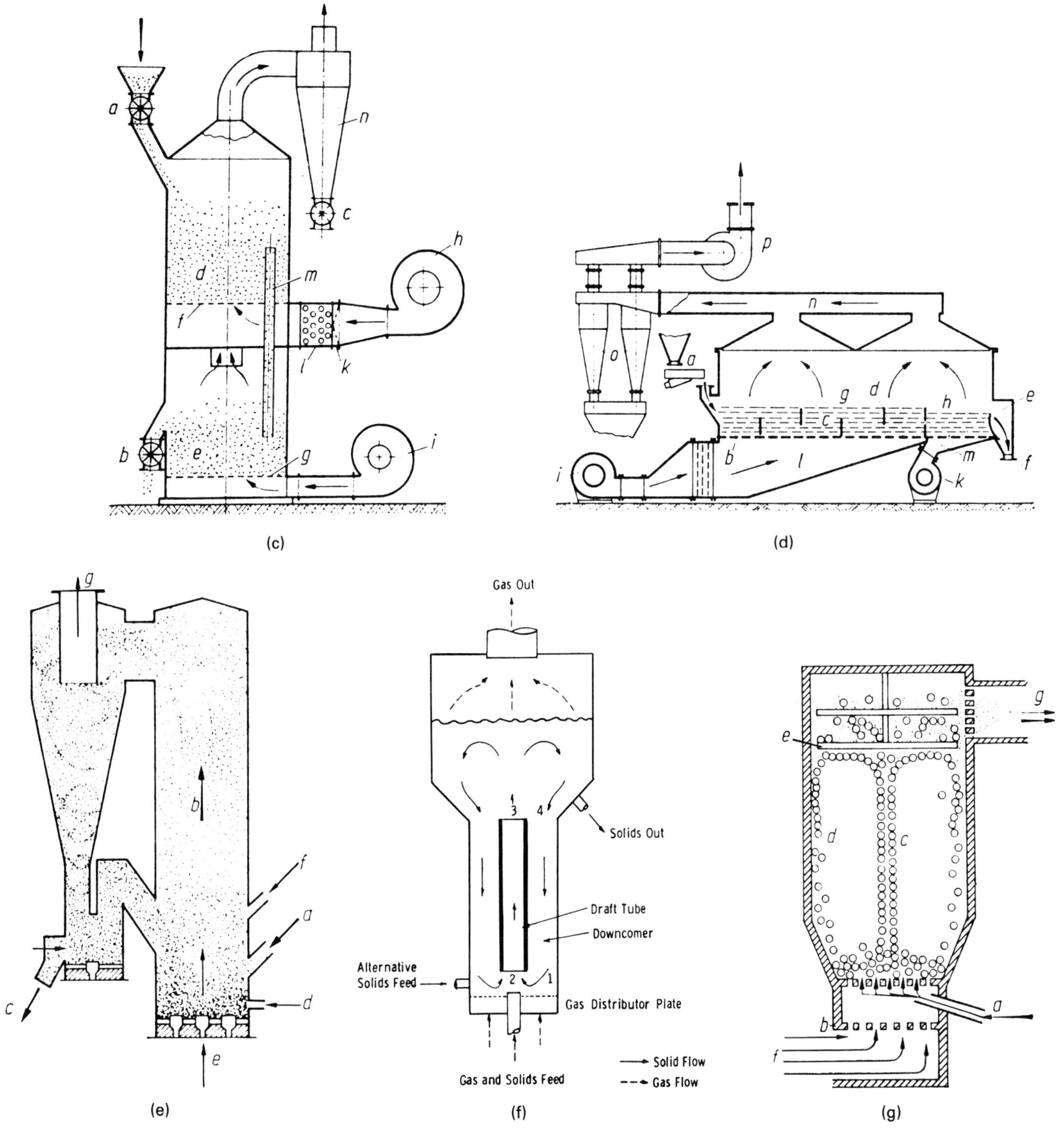

Figure 9.13—*(continued)*

arrangement, but the left-hand figure of Figure 9.14(d) is in counterflow. Figure 9.14(c) has tangential input of cooling air. In some operations, the heated air is introduced tangentially; then the process is called mixed flow. Most of the entries in Table 9.16(a) are parallel flow; but the heavy duty detergent is in counterflow, and titanium dioxide is either parallel or mixed flow. Counterflow is thermally more efficient, results in less expansion of the product particles, but may be harmful to thermally sensitive products because they are exposed to high air temperatures as they leave the dryer. The flat bottomed dryer of Figure 9.14(c) contacts the exiting solids with cooling air and is thus adapted to thermally sensitive materials.

Two main characteristics of spray drying are the short drying time and the porosity and small, rounded particles of product. Short

TABLE 9.14. Performance Data of Fluidized Bed Dryers: Batch and Multistage Equipment

(a) Batch Dryers

	Ammonium Bromide	Lactose Base Granules	Pharmaceutical Crystals	Liver Residue	Weed Killer
Holding capacity (lb wet product)	100	104	160	280	250
Bulk density, dry (lb/ft^3)	75	30	20	30	35
Initial moisture (% w/w basis)	6	10	65	50	20–25
Final moisture (% w/w basis)	1	2	0.4	5.0	1.0
Final drying temperature (°F)	212	158	248	140	140
Drying time (min)	20	90	120	75	210
Fan capacity (ft^3/min at 11 in. w.g.)	750	1500	3000	4000	3000
Fan HP	5	10	20	25	20
Evaporation rate (lb H_2O/hr)	15	5.7	52	100	17

(Courtesy Calmic Engineering Co. Ltd.; Williams–Gardner, 1971).

(b) Multistage Dryers with Dual-flow Distributors [Equipment Sketch in Fig. 9.13(b)]

Function	Heater	Cooler	Drier	Cooler
Material	Wheat Grains	Wheat Grains	Slag	Quartz Sand
Particle size (diameter)(mm)	5 × 3	5 × 3	0.95	1.4
Material feed rate (metric tons/hr)	1.5	1.5	7.0	4.0
Column diameter (m)	0.90	0.83	1.60	1.70
Perforated trays (shelves):				
Hole diameter (mm)	20	20	20; 10	20
Proportion of active section	0.4	0.4	0.4; 0.4	0.4
Number of trays	10	6	1; 2	20
Distance between trays (mm)	20	20	25; 40	15
Total pressure drop on fluidized bed (kgf/m^2)	113	64	70[a]	40
Hydraulic resistance of material on one tray (kgf/m^2)	7.8	9.2	20; 10	1.8
Inlet gas temperature (°C)	265	38	300	20
Gas inlet velocity (m/sec)	8.02	3.22	4.60	0.74
Material inlet temperature (°C)	68	175	20	350
Material discharge temperature (°C)	175	54	170	22
Initial humidity (% on wet material)	25	—	8	—
Final humidity (% on wet material)	2.8	—	0.5	—
Blower conditions				
Pressure (kgf/m^2)	450	250	420	250
Throughput (m^3/min)	180 (80°C)	130 (50°C)	360 (70°C)	100 (35°C)
Power consumption (HP)	50	20	75	7.5

[a] With grids and two distributor plates.
(Romankov, in Davidson and Harrison, *Fluidisation,* Academic, New York, 1971).

drying time is a particular advantage with heat sensitive materials. Porosity and small size are desirable when the material subsequently is to be dissolved (as foods or detergents) or dispersed (as pigments, inks, etc.). Table 9.17 has some data on size distributions, bulk density, and power requirements of the several types of atomizers.

The mean residence time of the gas in a spray dryer is the ratio of vessel volume to the volumetric flow rate. These statements are made in the literature regarding residence times for spray drying:

Source	Time (sec)
Heat Exchanger Design Handbook (1983)	5–60
McCormick (1979)	20
Masters (1976)	20–40 (parallel flow)
Nonhebel and Moss (1971)	<60
Peck (1983)	5–30
Wentz and Thygeson (1979)	<60
Williams–Gardner (1971)	4–10 (<15 ft dia); 10–20 (>15 ft dia)

Residence times of air and particles are far from uniform; Figure 9.5(a) and (b) is a sample of such data.

Because of slip and turbulence, the average residence times of particles are substantially greater than the mean time of the air, definitely so in the case of countercurrent or mixed flow. Surface moisture is removed rapidly, in less than 5 sec as a rule, but falling rate drying takes much longer. Nevertheless, the usual drying operation is completed in 5–30 sec. The residence time distribution of particles is dependent on the mixing behavior and on the size distribution. The coarsest particles fall most rapidly and take longest for complete drying. If the material is heat-sensitive, very tall towers in parallel flow must be employed; otherwise, countercurrent or mixed flows with high air temperatures may suffice. In some cases it may be feasible to follow up incomplete spray drying with a pneumatic dryer.

Drying must be essentially completed in the straight sided zones of Figures 9.14(a) and (b). The conical section is for gathering and efficient discharge of the dried product. The lateral throw of spray wheels requires a vessel of large diameter to avoid

TABLE 9.15. Performance Data of Continuous Fluidized Bed Dryers

(a) Data of Fluosatatic Ltd.

	Coal	Sand	Silica Sand	Limestone	Iron Ore
Material size, mesh	$\frac{1}{2}$–0	−25–0	−18–0	$\frac{3}{16}$–0	−$\frac{3}{8}$–0
Method of feed	twin screw	bucket elev.	conv.	conv.	conv.
Product rate (lb product/hr)	448,000	22,400	112,000	67,000	896,000
Initial moisture (% w/w basis)	11	6	6	15	3
Final moisture (% w/w basis)	5.5	0.1	0.1	0.1	0.75
Residence time (min)	1	1.25	1.5	1.25	0.5
Dryer diameter (ft)	10	3.0	7.25	5.5	8.5
Fluid bed height (in.)	18	12	12	12	18
Air inlet temperature (°F)	1000	1200	1200	1200	1200
Air outlet temperature (°F)	170	212	212	212	212
Air quantity (ft^3/min std.)	40,000	2000	9000	13,000	45,000
Material exit temperature (°F)	140	220	220	220	220
Evaporation (lb/hr)	24,640	1430	6720	11,880	20,400
Method of heating	coal	gas	oil	oil	oil
Heat consumption (Btu/lb water evaporated)	1830	1620	1730	1220	2300
Fan installed HP	240	20	80	115	350

(Williams–Gardner, 1971).

(b) Data of Head Wrightston Stockton Ltd.

	Coal	Silicious Grit	Glass Sand	Sand	Asphalt
Method of feed	screw feeder	chute	chute	chute	chute
Material size	−$\frac{1}{2}$ in.	−$\frac{1}{16}$ in.	−36 mesh	−$\frac{1}{16}$ in.	−$\frac{3}{16}$ in.
Product rate (lb product/hr)	190,000	17,920	15,680	33,600	22,400
Initial moisture (% w/w basis)	14	5	7	5	5
Final moisture (% w/w basis)	7	0	0	0	0.5
Residence time (min)	2	$1\frac{1}{2}$	3	3	10
Dryer diameter	7 ft 3 in.	3 ft 0 in.	4 ft 6 in.	6 ft 6 in.	8 ft 0 in.
Fluid bed height (in.)	21	12	12	12	24
Air inlet temperature (°F)	1000	1400	1400	1400	470
Air outlet temperature (°F)	135	230	230	230	220
Air quantity (ft^3/min std)	20,000	2000	2000	3500	7000
Material exit temp (°F)	140	230	230	230	220
Evaporated rate (lb/hr)	11,200	896	1097	1680	1120
Method of heating	coke-oven gas	gas oil	town gas	gas oil	gas oil
Heat consumption (Btu/lb water evaporated)	2000	2250	2000	2200	1800
Fan installed HP	210	$32\frac{1}{2}$	18	30	90

(Williams–Gardner, 1971).

(c) Data of Pennsalt Ltd.

	Abrasive Grit	Clay Granules	Sand	Granular Desiccant	Household Salt
Product rate (lb/hr)	2200	1000	14,000	150	13,500
Initial moisture (% w/w basis)	9	22	6	25	4
Final moisture (% w/w basis)	dry	3	dry	7	0.03
Air inlet temperature (°F)	580	160	325	300	390
Air outlet temperature (°F)	210	120	140	205	230
Method of heating	gas	steam	gas	gas	steam
Heat consumption (Btu/lb water evaporated)	2700	3800	2700	3600	5100
Bulk density (lb/ft^3)	120	60	90	30	60
Average drying time (min)	2.5	30	3	24	4
Fan capacity (ft^3/min std.)	2.5	1.35	1.05	0.84	1.05
Installed fan HP	10	45	25	5	50

(Williams–Gardner, 1971).

TABLE 9.15—*(continued)*

(d) Data of Rosin Engineering Ltd.

	Sodium Perborate	Weed Killer	PVC	Coal	Sand
Method of feed	screw	vibrator	screw	vibrator	vibrator
Material size	30–200 mesh	5–1 mm flake	60–120 mesh	3 mesh–zero	30–120 mesh
Product rate (lb product/hr)	11,400	5100	10,075	440,000	112,000
Initial moisture (% w/w basis)	3.5	14	2.0	8	8
Final moisture (% w/w basis)	0.0	0.2	0.2	1	0.2
Residence time (min)	1.5	11	30	0.3	0.45
Drier bed size (ft × ft)	22.5 × 5.5	18 × 4.5	23 × 6	16 × 6.6	12.5 × 3.2
Fluid bed height (in.)	4	3	18	5	6
Air inlet temperature (°F)	176	212	167	932	1202
Air outlet temperature (°F)	104	150	122	180	221
Air quantity (ft^3/min std)	6600	14,200	5400	67,330	8000
Material exit temperature (°F)	104	205	122	180	212
Evaporation (lb/hr)	400	720	183	33,440	9750
Method of heating	steam	steam	steam	coke-oven gas	oil
Heat consumption (Btu/lb water evaporated)	2100	3060	4640	1970	2200
Fan installed HP	33	40	34	600	70

(Williams–Gardner, 1971).

EXAMPLE 9.9
Sizing a Fluidized Bed Dryer

A wet solid at 100°F contains $W = 0.3$ lb water/lb dry and is to be dried to $W = 0.01$. Its feed rate is 100 lb/hr dry. The air is at 350°F and has $H_{g0} = 0.015$ lb water/lb dry. The rate of drying is represented by the equation

$$-\frac{dW}{d\theta} = 60(H_s - H_g), \quad \text{(lb/lb)/min.}$$

The solid has a heat capacity 0.35 Btu/(lb)(°F), density 150 lb/cuft, and average particle size 0.2 μm (0.00787 in.). The air has a viscosity of 0.023 cP and a density of 0.048 lb/cuft. The fluidized bed may be taken as a uniform mixture. A suitable air rate and dimensions of the bed will be found:

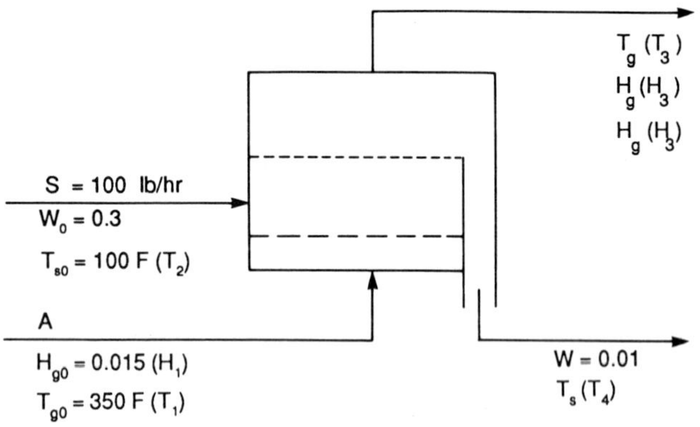

Symbols used in the computer program are in parentheses.

Minimum fluidizing rate by Leva's formula:

$$G_{mf} = \frac{688 D_p^{1.83}[0.048(150 - 0.048)]^{0.94}}{\mu^{0.88}}$$

$$= \frac{688(0.00787)^{1.83}[0.048(150 - 0.048)]^{0.94}}{(0.023)^{0.88}}$$

$$= 17.17 \text{ lb/(hr)(sqft)}.$$

Let $G_f = 2G_{mf} = 34.34$ lb/(hr)(sqft).
Expanded bed ratio

$$(L/L_0) = (G_f/G_{mf})^{0.22} = 2^{0.22} = 1.16.$$

Take voidage at minimum fluidization as

$$\varepsilon_{mf} = 0.40,$$
$$\therefore \varepsilon_f = 0.464.$$

Drying time:

$$\theta = \frac{W_0 - W}{60(H_s - H_g)} = \frac{0.3 - 0.01}{60(H_s - H_g)}. \quad (1)$$

Since complete mixing is assumed, H_s and H_g are exit conditions of the fluidized bed.

Humidity balance:

$$\dot{A}(H_g - H_{g0}) = \dot{S}(W_0 - W),$$
$$H_g = 0.015 + 0.29\dot{S}/\dot{A}. \quad (2)$$

Average heat capacity:

$$C_g = \tfrac{1}{2}(C_{g0} + C_g) = 0.24 + 0.45[(0.015 + H_g)/2]$$
$$= 0.2434 + 0.225H_g. \quad (3)$$

Heat balance:

$$\dot{A}C_g(T_{g0} - T_g) = \dot{S}[(C_s + W)(T_s - T_{s0}) + \lambda(W_0 - W)],$$
$$(\dot{A}/\dot{S})C_g(350 - T_g) = 0.36(T_s - 100) + 900(0.29). \quad (4)$$

EXAMPLE 9.9—*(continued)*

Adiabatic saturation line:

$$T_g - T_s = \frac{\lambda}{C_g}(H_s - H_g) = \frac{900}{C_g}(H_s - H_g). \qquad (5)$$

Vapor pressure:

$$P_s = \exp[11.9176 - 7173.9/(T_s + 389.5)]. \qquad (6)$$

Saturation humidity:

$$H_s = \frac{18}{29}\frac{P_s}{1 - P_s}. \qquad (7)$$

Eliminate T_3 between Eqs. (4) and (5):

$$T_s = 350 - \frac{0.36(T_4 - 100) + 261}{RC_g}$$
$$= T_4 + \frac{900(H_4 - H_3)}{C_g}, \quad [T_3 \equiv T_g,\ T_4 \equiv T_s]. \qquad (8)$$

Procedure: For a specified value of $R = \dot{A}/\dot{S}$, solve Eqs. (6), (7), and (8) simultaneously.

R	T_g	T_s	H_g	H_s	θ (min)
5	145.14	119.84	0.0730	0.0803	0.662
6	178.11	119.74	0.0633	0.0800	0.289
8	220.09	119.60	0.0513	0.0797	0.170
10	245.72	119.52	0.0440	0.0795	0.136
12	262.98	119.47	0.0392	0.0794	0.120

Take

$R = 10$ lb air/lb solid,
$\dot{A} = 10(100) = 1000$ lb/hr,
$\theta = 0.136$ min.

Cross section:

$\dot{A}/G_f = 1000/34.34 = 29.12$ sqft, 6.09 ft dia.

Avg density:

$\frac{1}{2}(1/20.96 + 1/19.03) = 0.0501$ lb/cuft.

Linear velocity:

$$u = \frac{G_f}{\rho\varepsilon(60)} = \frac{34.34}{0.0501(0.464)(60)} = 24.62 \text{ fpm}.$$

Bed depth:

$$L = u\theta = 24.62(0.136) = 3.35 \text{ ft}.$$

Note: In a completely mixed fluidized bed, the drying time is determined by the final moisture contents of the air and solid. When drying is entirely in the falling rate period with rate equation

$$-\frac{dW}{d\theta} = \frac{k(H_s - H_g)}{W_c}W, \quad W \le W_c,$$

the drying time will be

$$\theta = \frac{W_c}{k(H_s - H_g)W}$$

where H_s, H_g, and W are final conditions. When the final W is small, 0.01 in the present numerical example, the single stage drying time will be prohibitive. In such cases, multistaging, batch drying, or some other kind of drying equipment must be resorted to.

```
10 ! Example 9.9. Fluidized bed
   dryer
20 INPUT R ! =A/S, ratio of rat
   es of flow of air and solid
30 H3=.015+.29/R ! =Hg
40 C1=.2434+.225*H3
50 INPUT T4 ! Trial value of Ts
60 GOSUB 200
70 Y1=Y
80 T4=1.0001*T4
90 GOSUB 200
100 Y2=Y
110 K=.0001*Y1/(Y2-Y1)
120 T4=T4/1.0001-K
130 DISP T4
140 IF ABS(K/T4)<=.00001 THEN 16
    0
150 GOTO 60
160 DISP USING 170 ; R,T3,T4,H3,
    H4,T5
170 IMAGE DD,X,DDD.D,X,DDD.D,X,.
    DDDD ,X,.DDDD,X, .DDD
180 END
200 ! SR for T4
210 P=EXP(11.9176-7173.9/(T4+389
    .5))
220 H4=18*P/29/(1-P) ! = Hs
230 T3=T4+900*(H4-H3)/C1 ! = Tg
240 Y=-T3+350-(.36*(T4-100)+261)
    /R/C1
250 T5=.29/(H4-H3)/60 ! = time
260 RETURN
270 END
```

R	T_g	T_s	H_g	H_s	Time
5	145.1	119.84	.0730	.0803	.662
6	178.1	119.74	.0633	.0800	.289
8	220.1	119.61	.0513	.0797	.170
10	245.7	119.53	.0440	.0795	.136
12	263.0	119.47	.0392	.0794	.120
15	280.4	119.42	.0343	.0792	.108

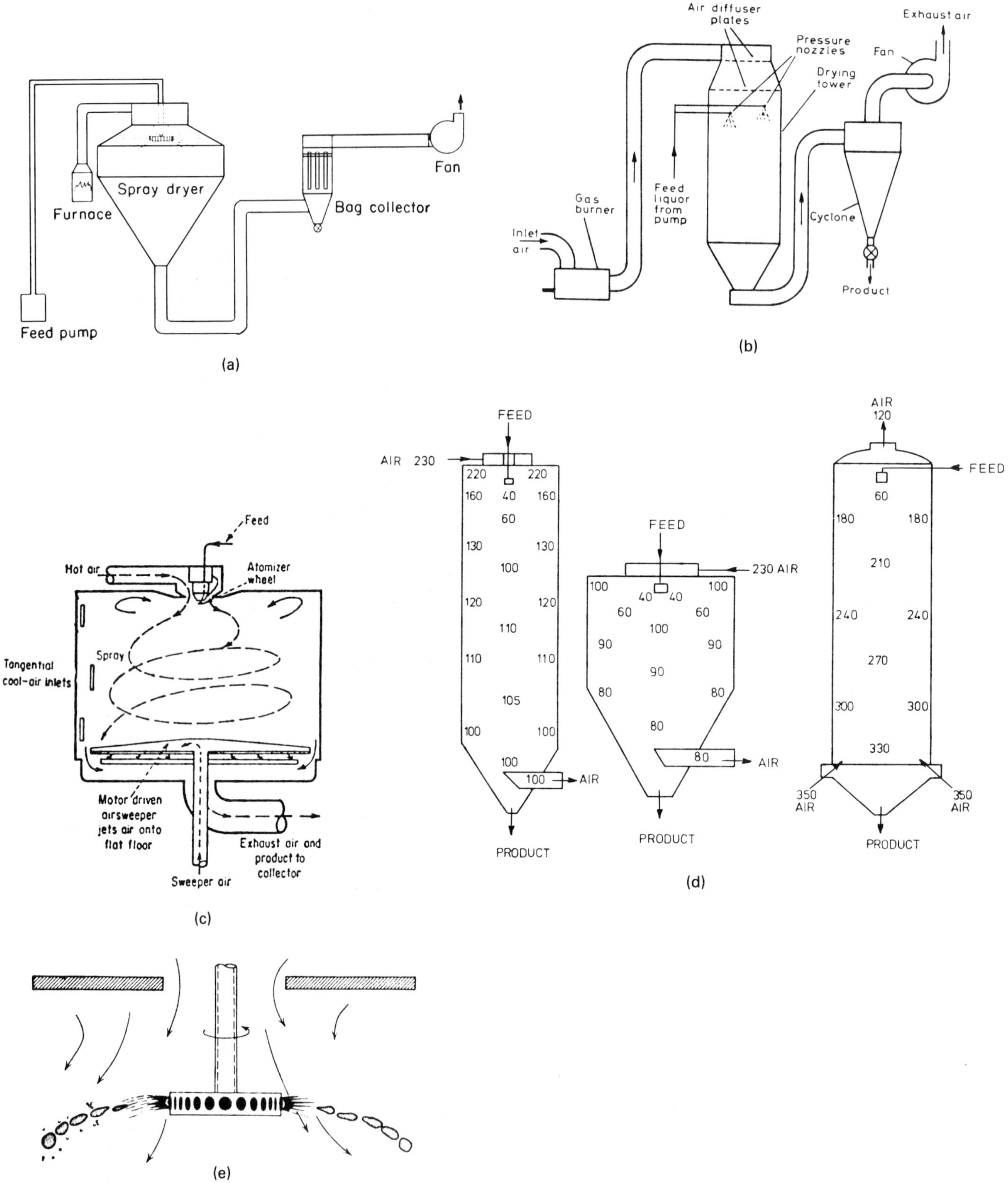

Figure 9.14. Spray dryer arrangements and behavior. (a) Spray dryer equipped with spray wheel; straight section $L/D = 0.5$–1.0 (*Proctor and Schwartz Inc.*). (b) Spray dryer equipped with spray nozzle; straight section $L/D = 4$–5 (*Nonhebel and Moss, 1971*). (c) Spray dryer for very heat sensitive products; flat bottom, side air ports and air sweeper to cool leaving particles. (d) Distribution of air temperatures in parallel and countercurrent flows (*Masters, 1976, p. 18, Fig. 1.5*). (e) Droplet-forming action of a spray wheel (*Stork–Bowen Engineering Co.*).

TABLE 9.16. Performance Data of Spray Dryers

(a) Data of Kröll (1978)

Kind of Stock	Moisture Content In (%)	Moisture Content Out (%)	Spray Device	Flow Pattern	Air Temperature In (°C)	Air Temperature Out (°C)
Skim milk, $d = 60\ \mu m$	48–55	4	wheel or	parallel	250	95–100
	50–60	4	nozzle			
			170–200 bar	parallel	250	95–100
Whole milk	50–60	2.5	wheel or nozzle			
			100–140 bar	parallel	170–200	
Eggs, whole	74–76	2–4	wheel or nozzle	parallel	140–200	50–80
Eggs, yolks	50–55	2–4	wheel or nozzle	parallel	140–200	50–80
Eggs, whites	87–90	7–9	wheel or nozzle	parallel	140–200	50–80
Coffee, instant, $300\ \mu m$	75–85	3–3.5	nozzle	parallel	270	110
Tea, instant	60	2	nozzle, 27 bar	parallel	190–250	
Tomatoes	65–75	3–3.5	wheel	parallel	140–150	
Food yeast	76–78	8	wheel	parallel	300–350	100
Tannin	50–55	4	wheel	parallel	250	90
PVC emulsion, 90% > 80 μm < 60 μm	40–70	0.01–0.1	wheel or nozzle or pneumatic	parallel	165–300	
Melamine–urethane–formaldehyde resins	30–50	0	wheel 140–160 m/sec	parallel	200–275	65–75
Heavy duty detergents	35–50	8–13	nozzle, 30–60 bar	counter	350–400	90–110
Kaolin	35–40	1	wheel	parallel	600	120

(b) Performance of a Dryer 18 ft Dia by 18 ft High with a Spray Wheel and a Fan Capacity of 11,000 cfm at the Outlet[a]

Material	Air Temp (°F) In	Air Temp (°F) Out	% Water in Feed	Evaporation Rate (lb/hr)
Blood, animal	330	160	65	780
Yeast	440	140	86	1080
Zinc sulfate	620	230	55	1320
Lignin	400	195	63	910
Aluminum hydroxide	600	130	93	2560
Silica gel	600	170	95	2225
Magnesium carbonate	600	120	92	2400
Tanning extract	330	150	46	680
Coffee extract *A*	300	180	70	500
Coffee extract *B*	500	240	47	735
Magnesium chloride	810	305	53	1140 (to dihydrate)
Detergent *A*	450	250	50	660
Detergent *B*	460	240	63	820
Detergent *C*	450	250	40	340
Manganese sulfate	600	290	50	720
Aluminum sulfate	290	170	70	230
Urea resin *A*	500	180	60	505
Urea resin *B*	450	190	70	250
Sodium sulfide	440	150	50	270
Pigment	470	140	73	1750

[a] The fan on this dryer handles about 11,000 cuft/min at outlet conditions. The outlet-air temperature includes cold air in-leakage, and the true temperature drop caused by evaporation must therefore be estimated from a heat balance.
(Bowen Engineering Inc.).

TABLE 9.17. Particle Diameters, Densities, and Energy Requirements

(a) Atomizer Performance

Type	Size Range (μm)	Power Input (kWh/1000 L)
Single fluid nozzle	8–800	0.3–0.5
Pneumatic nozzle	3–250	
Spray wheel	2–550	0.8–1.0
Rotating cup	25–950	

(b) Dry Product Size Range

Product	μm
Skim milk	20–250
Coffee	50–600
Eggs	5–500
Egg white	1–40
Color pigments	1–50
Detergents	20–2000
Ceramics	15–500

(c) Bulk Density of Sprayed Product as Affected by Air Inlet Temperature and Solids Content of Feed[a]

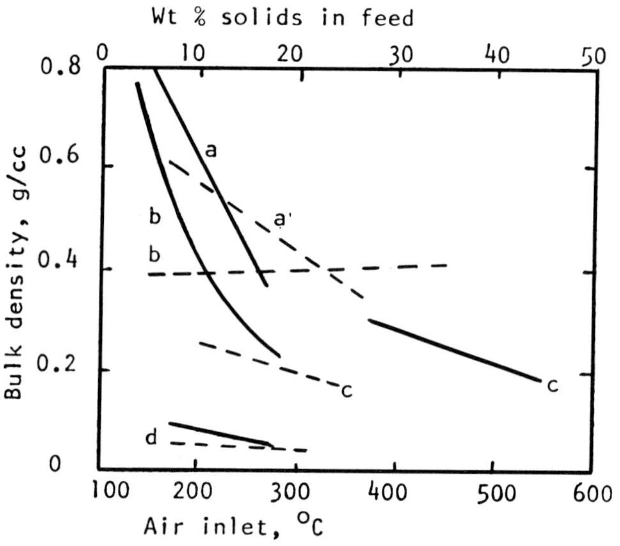

[a] The full lines are against temperature, the dashed ones against concentration: (a) sodium silicate; (b) coffee extract, 22%; (c) water dispersible dye, 19.5%; (d) gelatin.

[Data of Duffie and Marshall, *Chem. Eng. Prog.* **49,** 417 480 (1953)].

accumulation of wet material on the walls; length to diameter ratios of 0.5–1.0 are used in such cases. The downward throw of nozzles permits small diameters but greater depths for a given residence time; L/D ratios of 4–5 or more are used.

ATOMIZATION

Proper atomization of feed is the key to successful spray drying. The three devices of commercial value are pressure nozzles, pneumatic nozzles, and rotating wheels of various designs. Usual pressures employed in nozzles range from 300 to 4000 psi, and orifice diameters are 0.012–0.15 in. An acceptably narrow range of droplet sizes can be made for a feed of particular physical properties by adjustment of pressure and diameter. Multiple nozzles are used for atomization in large diameter towers. Because of the expense of motive air or steam, pneumatic nozzles are used mostly in small installations such as pilot plants, but they are most suitable for dispersion of stringy materials such as polymers and fibers. The droplet size increases as the motive pressure is lessened, the range of 60–100 psi being usual. The action of a rotating wheel is indicated in Figure 9.14(e). Many different shapes of orifices and vanes are used for feeds of various viscosities, erosiveness, and clogging tendencies. Operating conditions are up to 60,000 lb/hr per atomizer, speeds up to 20,000 rpm, and peripheral speeds of 250–600 ft/sec.

The main variables in the operation of atomizers are feed pressure, orifice diameter, flow rate and motive pressure for nozzles and geometry and rotation speed of wheels. Enough is known about these factors to enable prediction of size distribution and throw of droplets in specific equipment. Effects of some atomizer characteristics and other operating variables on spray dryer performance are summarized in Table 9.18. A detailed survey of theory, design and performance of atomizers is made by Masters (1976), but the conclusion is that experience and pilot plant work still are essential guides to selection of atomizers. A clear choice between nozzles and spray wheels is rarely possible and may be arbitrary. Milk dryers in the United States, for example, are equipped with nozzles, but those in Europe usually with spray wheels. Pneumatic nozzles may be favored for polymeric solutions, although data for PVC emulsions in Table 9.16(a) show that spray wheels and pressure nozzles also are used. Both pressure nozzles and spray wheels are shown to be in use for several of the applications of Table 9.16(a).

APPLICATIONS

For direct drying of liquids, slurries, and pastes, drum dryers are the only competition for spray dryers, although fluidized bed dryers sometimes can be adapted to the purpose. Spray dryers are capable of large evaporation rates, 12,000–15,000 lb/hr or so, whereas a 300 sqft drum dryer for instance may have a capacity of only 3000 lb/hr. The spherelike sprayed particles often are preferable to drum dryer flakes. Dust control is intrinsic to spray dryer construction but will be an extra for drum dryers. The completely enclosed operation of spray dryers also is an advantage when toxic or noxious materials are handled.

THERMAL EFFICIENCY

Exit air usually is maintained far from saturated with moisture and at a high temperature in order to prevent recondensation of moisture in parallel current operation, with a consequent lowering of thermal efficiency. With steam heating of air the overall efficiency is about 40%. Direct fired dryers may have efficiencies of 80–85% with inlet temperatures of 500–550°C and outlet of 65–70°C. Steam consumption of spray dryers may be 1.2–1.8 lb steam/lb evaporated, but the small unit of Table 9.19(b) is naturally less efficient. A 10% heat loss through the walls of the dryer often is taken for design purposes. Pressure drop in a dryer is 15–50 in. of water, depending on duct sizes and the kind of separation equipment used.

DESIGN

The design of spray dryers is based on experience and pilot plant determinations of residence time, air conditions, and air flow rate. Example 9.10 utilizes such data for the sizing of a commercial scale spray dryer.

TABLE 9.18. Effects of Variables on Operation of Spray Dryers

Variable Increased	Factors Increased	Factors Decreased
Chamber inlet temperature	*Feed rate* and thus: product rate, particle size (*b*), product moisture content, chamber wall build-up (*a*)	bulk density (*b*)
Chamber outlet temperature	product thermal degradation (*a*)	*feed rate* and thus: product rate particle size (*b*) product moisture content chamber wall build-up
Gas volume rate	*feed rate* and thus: product rate, particle size (*b*), product moisture content, chamber wall build-up (*a*)	residence time
Feed concentration	product rate, bulk density (*b*), particle size (*b*)	
Atomizer speed Atomizer disc diameter		
For stable lattices	bulk density	*particle size* and thus: product moisture content chamber wall build-up
For unstable lattices	coagulation (*a*) and thus: particle size, product moisture content, chamber wall build-up	
Atomizer vane depth Atomizer vane number	bulk density (*b*)	particle size (*b*) and thus: product moisture content, chamber wall build-up
Atomizer vane radial length		*For unstable lattices* particle size chamber wall build-up
Feed surface tension	bulk density (*b*)	particle size (*b*)
Chamber inlet gas humidity	product moisture content, chamber wall build-up (*a*)	

[a] This factor will only occur if a critical value of the variable is exceeded.
[b] Not for suspensions.
(Nonhebel and Moss, 1971).

The smallest pilot unit supplied by Bowen Engineering has a diameter of 30 in. and straight side of 29 in., employs parallel flow, up to 25 ACFM, 150–1000°F, particle sizes 30–40 μm average, either pneumatic nozzle or spray wheel. The performance of this unit is given in Table 9.19. The magnitude of the "product number" is arrived at by pilot plant work and experience; it increases with increased difficulty of drying or thermal sensitivity or both. Although much useful information can be obtained on this small scale, Williams-Gardner (1971) states that data on at least a 7 ft dia dryer be obtained for final design of large capacity units.

9.11. THEORY OF AIR–WATER INTERACTION IN PACKED TOWERS

The key properties of mixtures of air and water vapor are described in Section 9.1. Here the interactions of air and water in packed towers under steady flow conditions will be analyzed. The primary objectives of such operations may be to humidify or dehumidify the air as needed for particular drying processes or other processes, or to cool process water used for heat transfer elsewhere in the plant. Humidification–dehumidification usually is accomplished in spray towers, whereas cooling towers almost invariably are filled with some type of packing of open structure to improve contacting but with minimum pressure drop of air.

Analysis of the interaction of air and water involves the making of material and enthalpy balances. These are made over a differential section of the tower shown on Figure 9.15(a) and are subsequently integrated to establish the size of equipment for a given performance. In terms of empirical heat, k_h, and mass, k_m, transfer coefficients, these balances are

$$G\,dh = LC_L\,dT = L\,dT \quad (9.21)$$

$$= k_m(h_s - h)\,dz \quad (9.22)$$

$$= k_h(T - T_s)\,dz. \quad (9.23)$$

In Eq. (9.21) the heat capacity of water has been taken as unity. The approximations that are involved in making an enthalpy difference a driving force are discussed for example by Foust et al. (1980). Rearrangement and integration leads to the results

$$\text{NTU} = \frac{k_m Z}{L} \quad (9.24)$$

$$= \int_{T_1}^{T_2} \frac{dT}{h_s - h} \quad (9.25)$$

$$= \frac{G}{L}\int_{h_1}^{h_2} \frac{dh}{h_s - h}. \quad (9.26)$$

TABLE 9.19. Product Numbers and Performance of a 30 × 29 in. Pilot Plant Spray Dryer

(a) Product Numbers of Selected Materials

Material	*Product number*
1. COLOURS	
Reactive dyes	5–6
Pigments	5–11
Dispersed dyes	16–26
2. FOODSTUFFS	
Carbohydrates	14–20
Milk	17
Proteins	16–28
3. PHARMACEUTICALS	
Blood insoluble/soluble	11–22
Hydroxide gels	6–10
Riboflavin	15
Tannin	16–20
4. RESINS	
Acrylics	10–11
Formaldehyde resin	18–28
Polystyrene	12–15
5. CERAMICS	
Alumina	11–15
Ceramic colours	10

(Bowen Engineering Inc.).

(b) Performance of the Pilot Unit as a Function of Product Number[a]

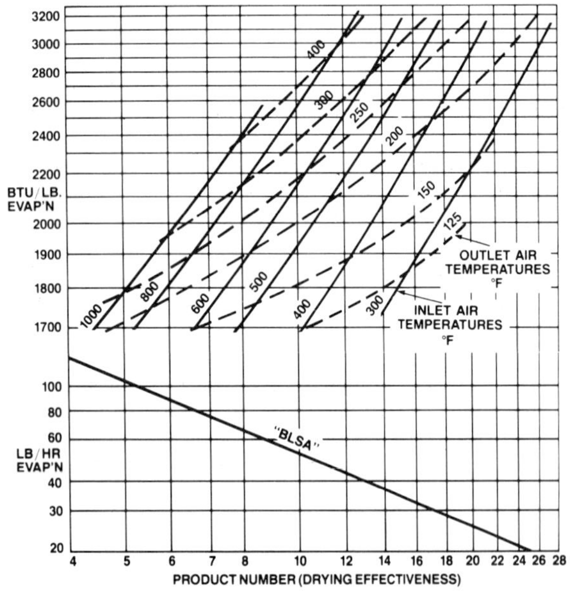

[a] Example: For a material with product number = 10 and air inlet temperature of 500°F, the evaporation rate is 53 lb/hr, input Btu/lb evaporated = 1930, and the air outlet temperature is 180°F.
(Bowen Engineering).

Both forms of the integral are employed in the literature to define the number of transfer units. The relation between them is

$$k_m Z/G = (L/G)(\mathrm{NTU}). \tag{9.27}$$

The height of a transfer unit is

$$\mathrm{HTU} = Z/(\mathrm{NTU}) = L/k_m = (L/G)(G/k_m). \tag{9.28}$$

The quantity G/k_m sometimes is called the height of a transfer unit expressed in terms of enthalpy driving force, as in Figure 9.16, for example:

$$G/k_m = (G/L)(\mathrm{HTU}). \tag{9.29}$$

Integration of Eq. (9.21) provides the enthalpy balance around one end of the tower,

$$L(T - T_1) + G(h - h_1). \tag{9.30}$$

Combining Eqs. (9.22) and (9.23) relates the saturation enthalpy and temperature,

$$h_s = h + (k_m/k_h)(T - T_s). \tag{9.31}$$

In Figure 9.15(c), Eq. (9.31) is represented by the line sloping upwards to the left. The few data that apparently exist suggest that the coefficient ratio is a comparatively large number. In the absence of information to the contrary, the ratio commonly is taken infinite, which leads to the conclusion that the liquid film resistance is negligible and that the interface is at the bulk temperature of the water. For a given value of T, therefore, the value of h_s in Eq. (9.25) is found from the equilibrium relation (h_s, T_s) of water and the corresponding value of h from the balance Eq. (9.30). When the coefficient ratio is finite, a more involved approach is needed to find the integrand which will be described.

The equilibrium relation between T_s and h_s is represented on the psychrometric charts Figures 9.1 and 9.2, but an analytical representation also is convenient. From Section 9.1,

$$h_s = 0.24T_s + (18/29)(0.45T_s + 1100)[p_s/(1 - p_s)], \tag{9.32}$$

where the vapor pressure is represented by

$$p_s = \exp[11.9176 - 7173.9/(T_s + 389.5)]. \tag{9.33}$$

Over the limited ranges of temperature that normally prevail in cooling towers a quadratic fit to the data,

$$h_s = a + bT_s + cT_s^2$$

may be adequate. Then an analytical integration becomes possible for the case of infinite k_m/k_h. This is done by Foust et al. (1980) for example.

The Cooling Tower Institute (1967) standardized their work in terms of a Chebyshev numerical integration of Eq. (9.25). In this method, integrands are evaluated at four temperatures in the interval, namely,

$$\begin{aligned} &T_2 + 0.1(T_2 - T_1), && \text{corresponding integrand } I_1,\\ &T_2 + 0.4(T_2 - T_1), && \text{corresponding integrand } I_2,\\ &T_1 - 0.4(T_2 - T_1), && \text{corresponding integrand } I_3,\\ &T_1 - 0.1(T_2 - T_1), && \text{corresponding integrand } I_4. \end{aligned} \tag{9.34}$$

Then the integral is

$$\int_{T_1}^{T_2} \frac{dT}{h_s - h} = 0.25(T_2 - T_1)(I_1 + I_2 + I_3 + I_4). \tag{9.35}$$

EXAMPLE 9.10
Sizing a Spray Dryer on the Basis of Pilot Plant Data

Feed to a spray dryer contains 20% solids and is to be dried to 5% moisture at the rate of 500 lb/hr of product. Pilot plant data show that a residence time of 6 sec is needed with inlet air of 230°F, $H = 0.008$ lb/lb, and exit at 100°F. Ambient air is at 70°F and is heated with steam. Enthalpy loss to the surroundings is 10% of the heat load on the steam heater. The vessel is to have a 60° cone. Air rate and vessel dimensions will be found.

Enthalpy, humidity, and temperatures of the air are read off the psychrometric chart and recorded on the sketch.

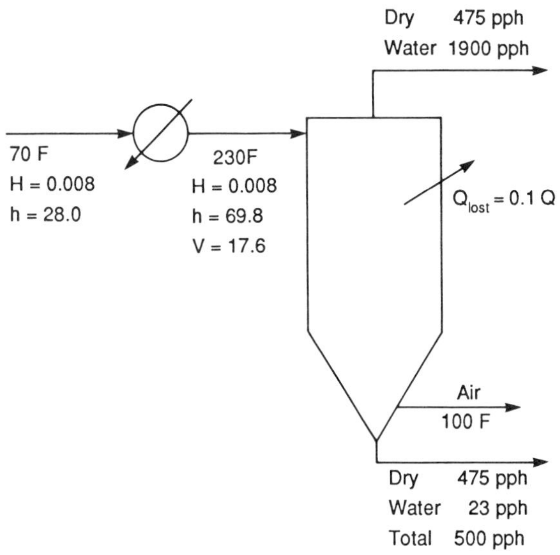

Enthalpy loss of air is

$$0.1(69.8 - 28.0) = 4.2 \text{ Btu/lb}.$$

Exit enthalpy of air is

$$h = 69.8 - 4.2 = 65.6.$$

At 100°F and this enthalpy, other properties are read off the psychrometric chart as

$$H = 0.0375 \text{ lb/lb},$$
$$V = 14.9 \text{ cuft/lb}.$$

Air rate is

$$A = \frac{1900 - 25}{0.0375 - 0.008} = 63{,}559 \text{ lb/hr}$$
$$\rightarrow \frac{63{,}559}{3600}\left(\frac{17.6 + 14.9}{2}\right) = 287 \text{ cfs}.$$

With a residence time of 6 sec, the dryer volume is

$$V_d = 287(6) = 1721.4 \text{ cuft}.$$

Make the straight side four times the diameter and the cone 60°:

$$1721.4 = 4D(\pi D^2/4) + \frac{0.866\pi D^3}{12} = 3.3683D^3,$$
$$\therefore D = 8.0 \text{ ft}.$$

When $k_m/k_h \rightarrow \infty$, evaluation of the integrands is straightforward. When the coefficient ratio is finite and known, this procedure may be followed:

1. For each of the four values of T, find h from Eq. (9.30).
2. Eliminate h_s between Eqs. (9.31) and (9.32) with the result

$$h + (k_m/k_h)(T - T_s)$$
$$= 0.24T_s + (18/29)(0.45T_s + 1100)[p_s/(1 - p_s)]. \quad (9.36)$$

3. Substitution of Eq. (9.33) into (9.36) will result in an equation that has T_s as the only unknown. This is solved for with the Newton–Raphson method.
4. Substitution of this value of T_s back into Eq. (9.31) will evaluate h_s.
5. The integrand $1/(h_s - h)$ now may be evaluated at each temperature and the integration performed with Eq. (9.35).

Example 9.11 employs this method for finding the number of transfer units as a function of liquid to gas ratio, both with finite and infinite values of k_m/k_h. The computer programs for the solution of this example are short but highly desirable. Graphical methods have been widely used and are described for example by Foust et al. (1980).

TOWER HEIGHT

The information that is ultimately needed about a cooling tower design is the height of packing for a prescribed performance. This equals the product of the number of transfer units by the height of each one,

$$Z = (\text{NTU})(\text{HTU}). \quad (9.37)$$

Some HTU data for cooling tower packing have been published, for example, those summarized on Figure 9.16. Other data appear in the additional literature cited for this chapter. Several kinds of tower fill made of redwood slats are illustrated in Figure 9.17. The numbers N of such decks corresponding to particular NTUs and (L/G)s are given by the equation

$$N = \frac{[(\text{NTU}) - 0.07](L/G)^b}{a}. \quad (9.38)$$

Values of a and b are given for each type of fill with Figure 9.17. These data are stated to be for 120°F inlet water. Although the authors state that corrections should be estimated for other temperatures, they do not indicate how this is to be done. For example, with deck type C, NTU = 2 and $L/G = 1.2$: $N = (2 - 0.07)(1.2)^{0.60}/0.092 = 23.4$ decks, or a total of 31.2 ft since the deck spacing is 16 in. The data of Figure 9.16 are used in Example 9.11.

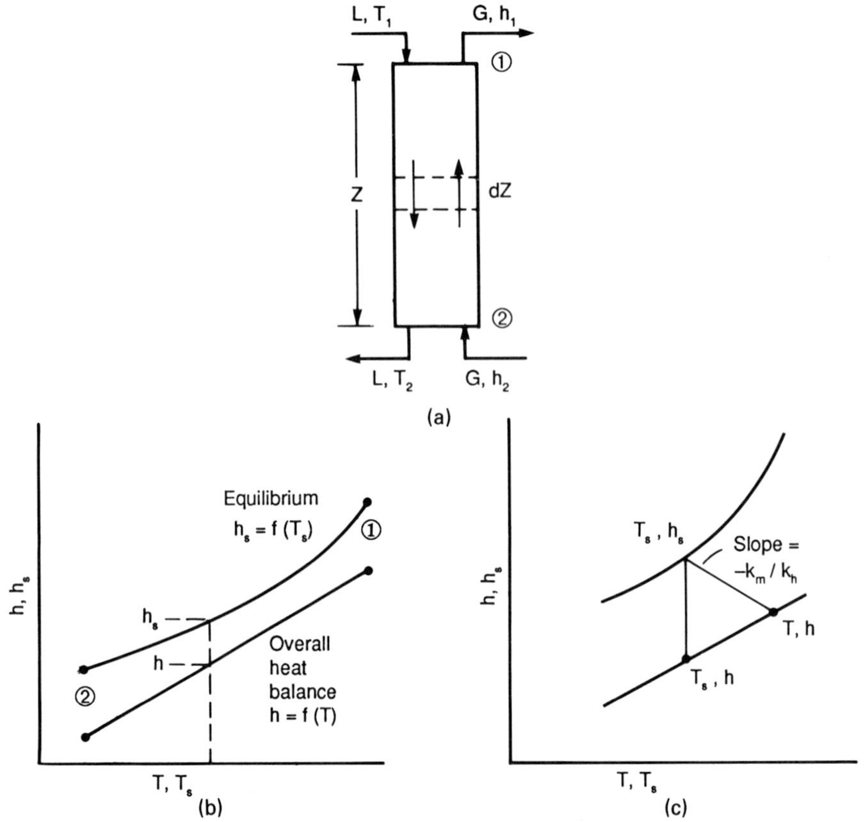

Figure 9.15. Relations in a packed continuous flow air–water contactor. (a) Sketch of the tower with differential zone over which the enthalpy and material balances are made. (b) Showing equilibrium and operating lines from which the integrand $1/(h_s - h)$ can be found as a function of liquid temperature T. (c) Showing interfacial conditions as determined by the coefficient ratio k_m/k_h; when this value is large, interfacial and saturation temperatures are identical.

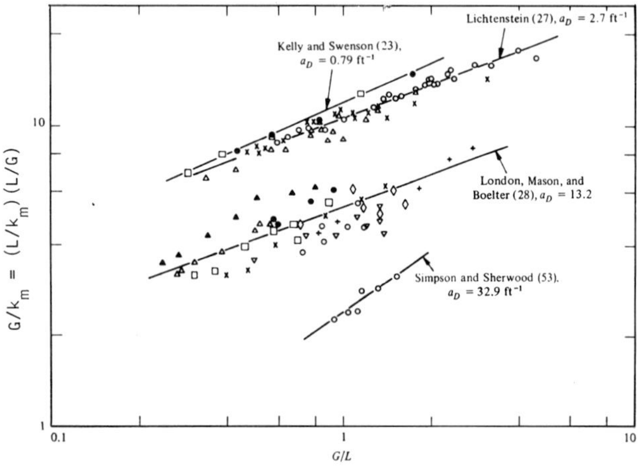

Figure 9.16. Data of heights of transfer units of packings characterized by the specific surface a_d (sqft/cuft). The ordinate is $G/k_m = Z/\int dh/(h_s - h)$, which is related to the form of NTU used in this chapter by

$$\mathrm{HTU} = Z/\mathrm{NTU} = Z\bigg/\int \frac{dT}{h_s - h} = L/k_m = (G/k_m)(L/G).$$

The equation of the London line is equivalent to

$$\mathrm{HTU} = 5.51(L/G)^{0.59}.$$

(*Sherwood et al., 1975*).

9.12. COOLING TOWERS

Cooling of water in process plants is accomplished most economically on a large scale by contacting it with air in packed towers. For reasons of economy, the tower fill is of a highly open structure. Efficient ring and structured packings of the sort used for distillation and other mass transfer processes are too expensive and exert too high a power load on the fans. Standard cooling tower practice allows a maximum of 2 in. of water pressure drop of the air. Water loadings range 500–2000 lb/(hr)(sqft) or 1–4 gpm/sqft. Gas loadings range 1300–1800 lb/(hr)(sqft) or between 300 and 400 ft/min. The liquid to gas ratio L/G normally is in the range 0.75–1.50 and the number of transfer units or the tower characteristic, $\mathrm{HTU} = k_m Z/L$, vary from 0.5 to 2.50.

The most common fill is of wooden slats of rectangular or triangular cross section arranged as in Figure 9.17. Corrugated sheets of asbestos–concrete have some application and also PVC construction unless the temperatures are above 160°F.

Fan power consumption is the major operating cost and can be counterbalanced in part by greater investment in natural draft construction. In the majority of process applications, fan-operated towers are preferred. Very large installations such as those in power plants employ chimney assisted natural draft installations. A limited use of atmospheric towers is made in areas where power costs are especially high.

The main types of cooling towers are represented on Figure 9.18. Their chief characteristics and some pros and cons will be discussed in order.

EXAMPLE 9.11
Sizing of a Cooling Tower: Number of Transfer Units and Height of Packing

Water is to be cooled from 110 to 75°F by contact with air that enters countercurrently at 90°F with a dewpoint of 60°F. The data of London et al. (1940) of Figure 9.16 for height of transfer unit are applicable. Calculations will be made for two values of the coefficient ratio k_m/k_h, namely, 25 and ∞ Btu/(°F) (lb dry air), of Eq. (9.31). The effect of the ratio of liquid to gas rates, L/G, will be explored.

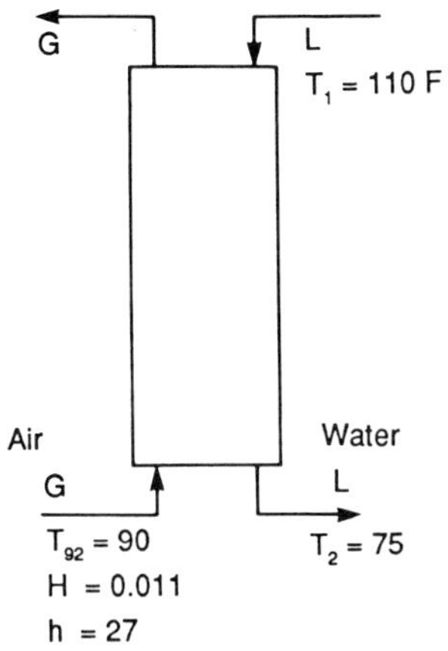

The maximum allowable L/G corresponds to equilibrium between exit air and entering water at 110. The saturation enthalpy at 110°F is 92, so that Eq. (9.30) becomes

$$\left(\frac{L}{G}\right)_{max} = \frac{92-27}{110-75} = 1.857.$$

The several trials will be made at $L/G = (0.6, 1.0, 1.4, 1.7)$.

The applicable equations with numerical substitutions are listed here and incorporated in the computer program for solution of this problem [Eqs. (9.30)–(9.33)]:

$$h = 27 + (L/G)(T - 75),$$
$$h_s = h + 25(T - 75),$$
$$h_s = 0.24T + (18/29)(0.45T + 1100)P_s/(1 - P_s),$$
$$P_s = \exp[11.9176 - 7173.9/(T_s + 389.5)].$$

When $k_m/k_h \to \infty$, T_s in Eq. (9.33) is replaced by T.

The four temperatures at which the integrands are evaluated for the Chebyshev integration are found with Eq. (9.34) and tabulated in the calculation summary following.

Equations (9.30) and (9.31) are solved simultaneously for h and h_s with the aid of the Newton–Raphson method as used in the computer program; the integrands are evaluated and the integration are completed with Eq. (9.35).

The number of transfer units is sensitive to the value of L/G, but the effect of k_m/k_h is more modest, at least over the high range used; data for this ratio do not appear to be prominently recorded. Figure 9.16 shows a wide range of heights of transfer units for the different kinds of packings, here characterized by the surface a_d (sqft/cuft) and substantial variation with L/G. The last line of the calculation summary shows variation of the tower height with L/G.

Data of London et al. (1940) of Figure 9.16:

$$(G/L)(\text{HTU}) = 5.51(G/L)^{0.41}$$

or

$$\text{HTU} = 5.51(L/G)^{0.59}.$$

Tower height:

$$Z = (\text{HTU})(\text{NTU}).$$

For several values of L/G:

L/G	0.6	1	1.4	1.7
HTU (ft)	4.08	5.51	6.72	7.54

Evaluation of interfacial temp and the NTU for $L/G = 1$ with $k_m/k_h = 25$:

T	h	T_s	$1/(h_s - h)$
78.5	30.5	78.099	0.0864
89	41	88.517	0.0709
96	48	95.400	0.0575
106.5	58.5	105.581	0.0385
			0.2533

$\therefore$ NTU = (110 − 75(0.2533)/4 = 2.217.

For other values of L/G:

T	h	$1/(h_s - h)$ $L/G = 0.6$	1	1.4	1.7
78.5	30.5	0.0751	0.0864	0.0943	0.1043
89	41	0.0518	0.0709	0.1167	0.2200
96	48	0.0398	0.0575	0.1089	0.3120
106.5	58.5	0.0265	0.0385	0.0724	0.1987
		0.1933	0.2533	0.3923	0.8350
NTU →		1.691	2.217	3.433	7.306

With $k_m/k_h \to \infty$:

T	h	$1/(h_s - h)$ $L/G = 0.6$	1	1.4	1.7
78.5	30.5	0.0725	0.0807	0.90	0.1006
89	41	0.0494	0.0683	0.1107	0.2070
96	48	0.0376	0.0549	0.1020	0.2854
106.5	106.5	0.0248	0.0361	0.0663	0.1778
		0.1844	0.2400	0.3700	0.7708
NTU →		1.613	2.100	3.238	6.745
Z →		6.58	11.57	21.76	50.86

```
10 ! Example 9.11 with Km/Kh =
   25
20 READ T1,L
30 DATA 106.5,1.7
40 H2=27+L*(T1-75)
50 INPUT T ! = Ts
60 GOSUB 180
70 Y1=Y
80 T=1.0001*T
90 GOSUB 180
100 Y2=Y
```

Example 9.11—*(continued)*

```
110 K=.0001*Y1/(Y2-Y1)
120 T=T/1.0001-K
130 DISP T
140 IF ABS(K/T)<=.00001 THEN 160
150 GOTO 60
160 DISP T,H,1/(H-H2)
170 END
180 ! SR FOR T, H
190 P=EXP(11.9176-7173.9/(T+389
    5))
200 M=18*P/29/(1-P)
210 H=.24*T1+(.45*T1+1100)*M
220 H1=H2-25*(T-T1)
230 Y=H1-H ! = 0 ?
240 RETURN
250 END
```

```
 10 ! Example 9.11 with infinite
     Km/Kh
 20 L=1.4
 30 INPUT T1
 40 H2=27+L*(T1-75)
 50 P=EXP(11.9176-7173.9/(T1+389
     .5))
 60 M=18*P/29/(1-P) ! = Hs
 70 H=.24*T1+(.45*T1+1100)*M
 80 DISP 1/(H-H2)
 90 GOTO 30
100 END
```

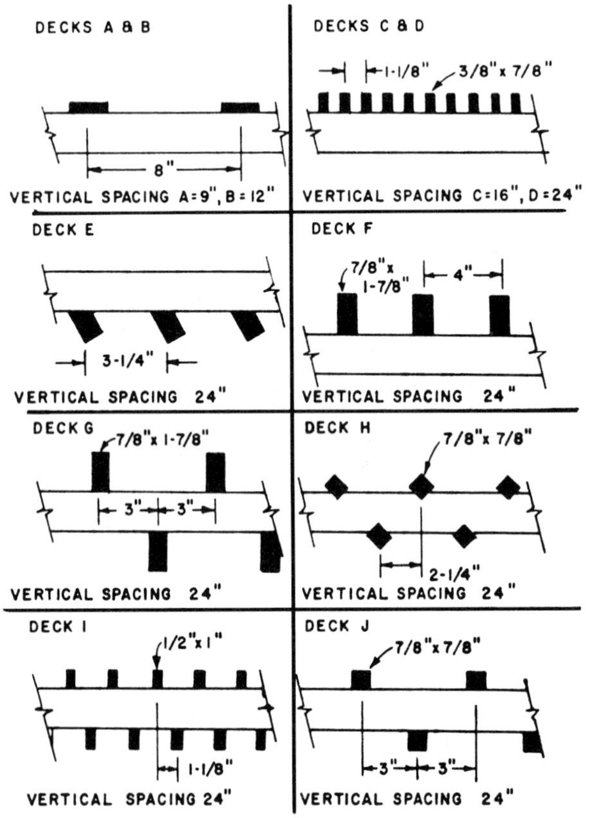

Factors in Eq. 9.38 for the Number of Decks

Deck Type	*a*	*b*
A	0.060	0.62
B	0.070	0.62
C	0.092	0.60
D	0.119	0.58
E	0.110	0.46
F	0.100	0.51
G	0.104	0.57
H	0.127	0.47
I	0.135	0.57
J	0.103	0.54

Figure 9.17. Kinds of fill made of redwood slats for cooling towers, and factors for determining the required number of decks with inlet water at 120°F (*Cheremisinoff and Cheremisinoff, 1981*).

a. Atmospheric towers are effective when prevailing wind velocities are 5 miles/hr or more. For access to the wind they are narrow but long, lengths of 2000 ft having been constructed. Water drift losses are relatively large. The savings because of elimination of tall chimney or fan power is counterbalanced by increased size because of less efficient cross flow and variations in wind velocity.

b. Chimney assisted natural draft towers also eliminate fans. Most of the structure is the chimney, the fill occupying only 10–12% of the tower height at the bottom. The temperature and humidity of the air increase as the air flows upward so that its buoyancy increases and results in rapid movement through the chimney. Smaller units are made as circular cylinders since these can be built rapidly. The hyperboloidal shape has greater strength for a given wall thickness. In towers as large as 250 ft dia and 450 ft high, wall thicknesses of 5–8 in. of reinforced concrete are adequate. The enlarged cross section at the top converts some kinetic energy into pressure energy which assists in dispelling the exit humid air into the atmosphere.

The ratio of base diameter to height is 0.75–0.85, the ratio of throat and base diameters is 0.55–0.65, and the ratio of vertical depth of air opening to base diameter is 0.1–0.12. Air velocity through the tower is 3–6 ft/sec, water flow rates range from 600 to 1800 lb/(hr)(sqft). Two towers each 375 ft high are able to service a 500 MW power plant. Natural draft towers are uneconomical below heights of 70 ft. The upper limit is imposed principally by environmental visual considerations; towers 500 ft high are in existence. A cost comparison is made with item d.

c. Hyperbolic fan assisted towers can have as much as three times the capacity of the same size natural draft towers. The fans provide greater control than the natural draft systems; for example, they may be turned on only at peak loads. Rules of thumb cited by Cheremisinoff and Cheremisinoff (1981) for relative sizing is that fan assisted hyperbolic towers may have diameters 2/3 and heights 1/2 those of purely natural draft designs.

d. Countercurrent-induced draft construction is the most widely used type in process industries. Mechanical draft is capable of a greater degree of control than natural draft and such towers are able in some cases to cool the water within 2°F of the wet bulb temperature of the air. The elevated fan location introduces some structural and noise problems. The flow of air is quite uniform across the cross section and its discharge is positive and at high velocity so that there is little backflow of humid air into the tower. A cost comparison (dated 1978) with hyperbolic towers is made by Singham (1983, Sec. 3.12.4.1). The case is for a water rate of 6.1 m^3/sec, cooling range of 8.5°C, approach of 10°C, and wet bulb of 17°C. The cost of the natural draft tower

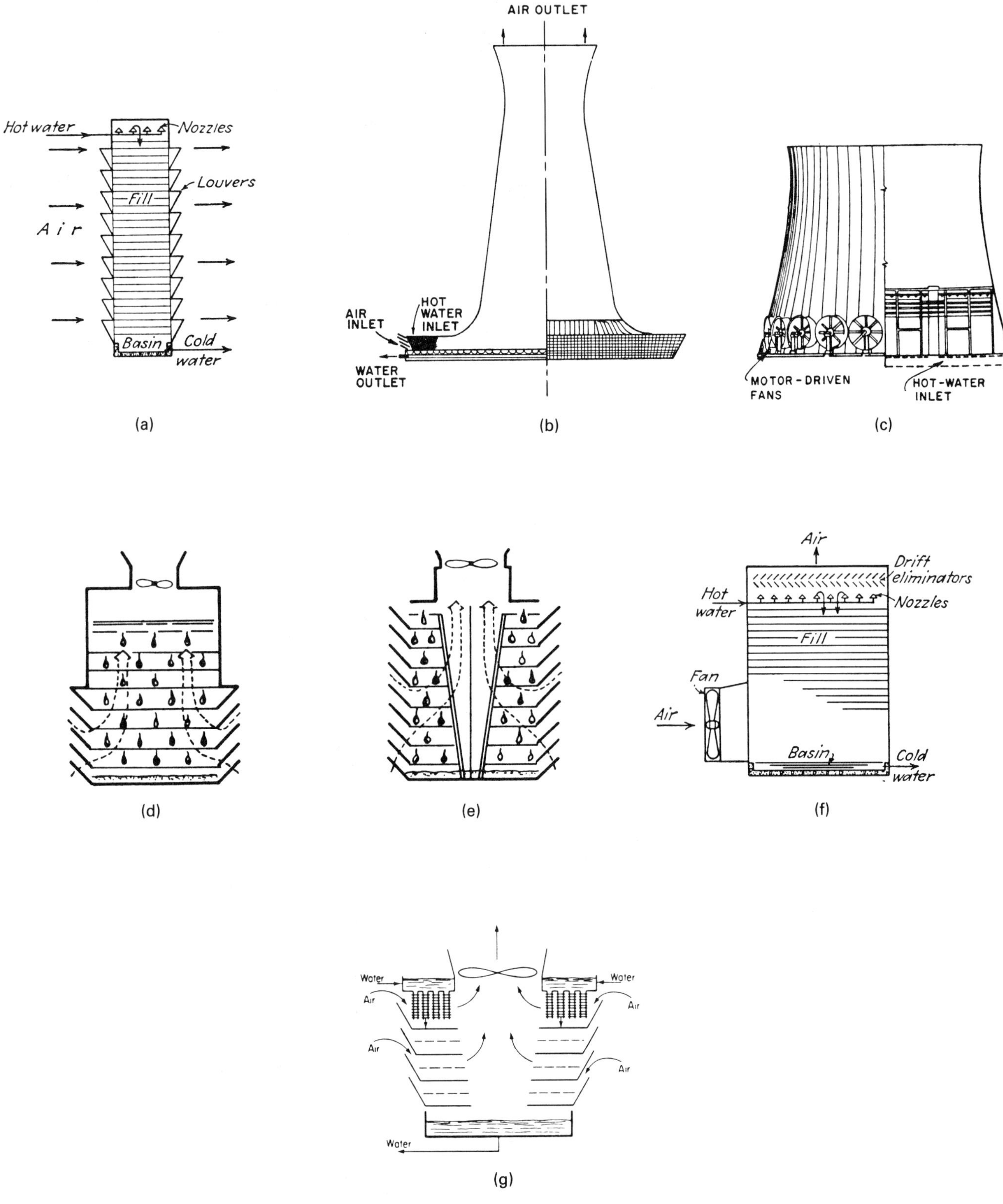

Figure 9.18. Main types of cooling towers. (a) Atmospheric, dependent on wind velocity. (b) Hyperbolic stack natural draft. (c) Hyperbolic assisted with forced draft fans. (d) Counterflow-induced draft. (e) Crossflow-induced draft. (f) Forced draft. (g) Induced draft with surface precooler for very hot water; also called wet/dry tower. [(*b*)–(*e*) *from Cheremisinoff and Cheremisinoff, 1981*).

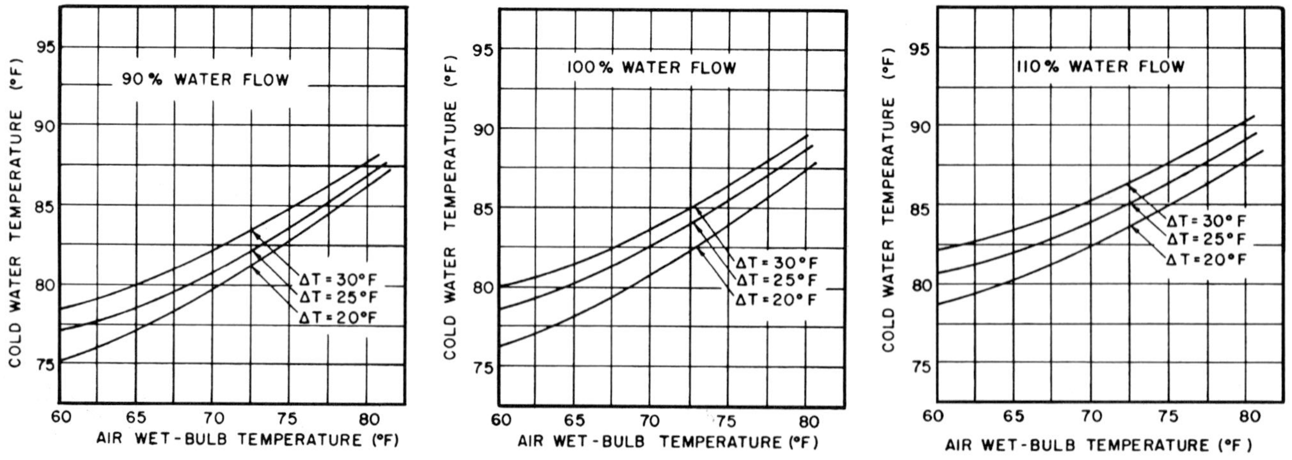

Figure 9.19. Typical cooling tower performance curves (*Cheremisinoff and Cheremisinoff, 1981*).

TABLE 9.20. Selected Data Required with Bids of Cooling Towers

A. *Cooling Tower*
1. Number of cells
2. Cell dimensions, ft, in.
3. Tower length, ft, in.
4. Tower width, ft, in.
5. Tower height, ft, in.
6. Casing, material and dimensions
7. Structure, material and dimensions
8. Fill decks, material and dimensions
9. Partitions and baffles, materials and dimensions
10. Drift eliminators, material and dimensions
11. Fan stacks, material and dimensions
12. Fan deck, material and dimensions
13. Louvers, material and dimensions
14. Board feet of fill
15. Board feet total tower
16. Height of fan stacks, ft, in.
17. Post extension below curb, ft, in.
18. Total shipping weight, lb
19. Total operating weight, lb

B. *Fans*
1. Number of units
2. Type and manufacturer
3. Diameter, ft, in.
4. Number of blades per fan
5. Blade material
6. Hub material
7. rpm
8. Tip speed, fpm
9. Mechanical efficiency, %
10. Static efficiency, %
11. Weight, lb

C. *Motors*
1. Number of units
2. Size, HP
3. Type and manufacturer
4. Full load speed, rpm
5. Frame size
6. Full load current, amps
7. Locked rotor current amps
8. Weight, lb

H. *Distribution System*
1. Number and size of inlet flanges
2. Height of water inlet above curb, ft, in.
3. Header material
4. Lateral material
5. Nozzle, or downspout material

J. *Design Performance*
1. Pumping head from top of basin curb, ft
2. Spray loss, max %
3. Evaporation loss, max %
4. Fill wetted surface, ft^2
5. Total wetted surface, ft^2
6. Effective splash surface, ft^2
7. Effective cooling volume, ft^3 (from eliminators to water level)
8. Air volume per fan, cfm
9. Static pressure, inches of water
10. Output horsepower/motor/(turbine)
11. Tower loading, gpm/ft^2

K. *Drawings and Performance Curves*
1. Tower outline elevation
2. Foundation outline
3. Fill rack details
4. Drift eliminator details
5. Tower sheeting arrangement
6. A series of guaranteed performance curves within limits of CTI Test Procedure ATP-105, latest revision

(Excerpted from Cheremisinoff and Cheremisinoff, 1981).

was 1.2 M pounds and that of the mechanical draft was 0.75 M pounds, but the fan power was 775 kW. The opinion was expressed that mechanical draft towers are more economical at water rates below 1.25 m^3/sec (19,800 gpm).

e. Crossflow induced draft offer less resistance to air flow and can operate at higher velocities, which means that less power and smaller cell sizes are needed than for counterflows. The shorter travel path of the air makes them less efficient thermally. The cross flow towers are made wider and less high, consequently with some saving in water pumping cost.

f. Forced draft towers locate the fans near ground level which requires simpler support structures and possibly lower noise levels. A large space must be provided at the bottom as air inlet. Air distribution is poor because it must make a 90° turn. The humid air is discharged at low velocity from the top of the tower and tends to return to the tower, but at the same time the drift loss of water is less. The pressure drop is on the discharge side of the fan which is less power-demanding than that on the intake side of induced draft towers.

g. Wet–dry towers employ heat transfer surface as well as direct contact between water and air. Air coolers by themselves are used widely for removal of sensible heat from cooling water on a comparatively small scale when cooling tower capacity is limited. Since dry towers cost about twice as much as wet ones, combinations of wet and dry sometimes are applied, particularly when the water temperatures are high, of the order of 160°F, so that evaporation losses are prohibitive and the plumes are environmentally undesirable. The warm water flows first through tubes across which air is passed and then enters a conventional packed section where it is cooled further by direct contact with air. Separate dampers for air to the dry and wet sections can throw greater load on the wet section in summer months.

WATER FACTORS

Evaporation losses are about 1% of the circulation for every 10°F of cooling range. Windage or drift losses are 0.3–1.0% for natural draft towers and 0.1–0.3% for mechanical draft. Usually the salt content of the circulating water is limited to 3–7 times that of the makeup. Blowdown of 2.5–3% of the circulation accordingly is needed to maintain the limiting salt concentration.

TESTING AND ACCEPTANCE

At the time of completion of an installation, the water and air conditions and the loads may not be exactly the same as those of the design specification. Acceptance tests performed then must be analyzed to determine if the performance is equivalent to that under the design specifications. Such tests usually are performed in accordance with recommendations of the Cooling Tower Institute.

The supplier generally provides a set of performance curves covering a modest range of variation from the design condition, of which Figure 9.19 is a sample. Some of the data commonly required with bids of cooling tower equipment are listed in Table 9.20, which is excerpted from a 10-page example of a cooling tower requisition by Cheremisinoff and Cheremisinoff (1981).

REFERENCES

Drying

1. W.L. Badger and J.T. Banchero, *Introduction to Chemical Engineering,* McGraw-Hill, New York, 1955.
2. C.W. Hall, *Dictionary of Drying,* Dekker, New York, 1979.
3. R.B. Keey, *Drying Principles and Practice,* Pergamon, New York, 1972.
4. R.B. Keey, *Introduction to Industrial Drying Operations,* Pergamon, New York, 1978.
5. K. Kröll, *Trockner und Trocknungsverfahren,* Springer-Verlag, Berlin, 1978.
6. P.Y. McCormick, Drying, in *Encyclopedia of Chemical Technology,* Wiley, New York, 1979, Vol. 8, pp. 75–113.
7. K. Masters, *Spray Drying,* George Godwin, London, 1976.
8. A.S. Mujumdar (Ed.), *Advances in Drying,* Hemisphere, New York, 1980–1984, 3 vols.
9. G. Nonhebel and A.A.H. Moss, *Drying of Solids in the Chemical Industry,* Butterworths, London, 1971.
10. R.E. Peck, Drying solids, in *Encyclopedia of Chemical Processing and Design,* Dekker, New York, 1983, Vol. 17, pp. 1–29.
11. E.U. Schlünder, Dryers, in *Heat Exchanger Design Handbook,* Hemisphere, New York, 1983, Sec. 3.13.
12. G.A. Schurr, Solids drying, in *Chemical Engineers Handbook,* McGraw-Hill, New York, 1984, pp. 20.4–20.8.
13. T.H. Wentz and J.R. Thygeson, Drying of wet solids, in *Handbook of Separation Techniques for Chemical Engineers,* (Schweitzer, Ed.), McGraw-Hill, New York, 1979.
14. A. Williams-Gardner, *Industrial Drying,* Leonard Hill, Glasgow, 1971.

Cooling Towers

1. N.P. Cheremisinoff and P.N. Cheremisinoff, *Cooling Towers: Selection, Design and Practice,* Ann Arbor Science, Ann Arbor, MI, 1981.
2. Cooling Tower Institute, *Performance Curves,* CTI, Spring, TX, 1967.
3. A.S. Foust et al., *Principles of Unit Operations,* Wiley, New York, 1980.
4. D.Q. Kern, *Process Heat Transfer,* McGraw-Hill, New York, 1950.
5. T.K. Sherwood, R.L. Pigford, and C.R. Wilke, *Mass Transfer,* McGraw-Hill, New York, 1975.
6. J.R. Singham, Cooling towers, in *Heat Exchanger Design Handbook,* Hemisphere, New York, 1983, Sec. 3.12.

Data on Performance of Cooling Tower Packing

1. Hayashi, Hirai, and Okubo, *Heat Transfer Jpn. Res.* **2**(2) 1–6 (1973).
2. Kelly and Swenson, *Chem. Eng. Prog.* **52,** 263 (1956), cited in Figure 9.16.
3. Lichtenstein, *Trans. ASME* **66,** 779 (1943), cited in Figure 9.16.
4. London, Mason, and Boelter, *Trans. ASME* **62,** 41 (1940), cited in Figure 9.16.
5. Lowe and Christie, *Proceedings, International Heat Transfer Conference,* Boulder, CO, 1961, Part V, pp. 933–950.
6. Simpson and Sherwood, *Refrig. Eng.* **52,** 535 (1946), cited in Figure 9.16.
7. Tezuka, *Heat Transfer Jpn. Res.* **2**(3), 40–52 (1973).

10

MIXING AND AGITATION

Agitation is a means whereby mixing of phases can be accomplished and by which mass and heat transfer can be enhanced between phases or with external surfaces. In its most general sense, the process of mixing is concerned with all combinations of phases of which the most frequently occurring ones are

1. *gases with gases.*
2. *gases into liquids: dispersion.*
3. *gases with granular solids: fluidization, pneumatic conveying, drying.*
4. *liquids into gases: spraying and atomization.*
5. *liquids with liquids: dissolution, emulsification, dispersion.*
6. *liquids with granular solids: suspension.*
7. *pastes with each other and with solids.*
8. *solids with solids: mixing of powders.*

Interaction of gases, liquids, and solids also may take place, as in hydrogenation of liquids in the presence of a slurried solid catalyst where the gas must be dispersed as bubbles and the solid particles must be kept in suspension.

Three of the processes involving liquids, numbers 2, 5, and 6, employ the same kind of equipment; namely, tanks in which the liquid is circulated and subjected to a certain amount of shear. This kind of equipment has been studied most extensively. Although some unusual cases of liquid mixing may require pilot plant testing, general rules have been developed with which mixing equipment can be designed somewhat satisfactorily. This topic will be emphasized in this chapter.

The other mixing operations of the list require individual kinds of equipment whose design in some cases is less quantified and is based largely on experience and pilot plant work. Typical equipment for such purposes will be illustrated later in this chapter. Phase mixing equipment which accomplishes primarily mass transfer between phases, such as distillation and extraction towers, also are covered elsewhere. Stirred reactors are discussed in Chapter 17.

Circulation and shear of the liquid in a vessel can be accomplished with external pumps and appropriate location of suction and discharge nozzles, but a satisfactory combination of vertical and lateral flows is obtained more economically by internal impellers, baffles, and draft tubes. Some general statements about dimensions, proportions, and internals of a liquid mixing vessel can be made.

10.1. A BASIC STIRRED TANK DESIGN

The dimensions of the liquid content of a vessel and the dimensions and arrangement of impellers, baffles and other internals are factors that influence the amount of energy required for achieving a needed amount of agitation or quality of mixing. The internal arrangements depend on the objectives of the operation: whether it is to maintain homogeneity of a reacting mixture or to keep a solid suspended or a gas dispersed or to enhance heat or mass transfer. A basic range of design factors, however, can be defined to cover the majority of cases, for example as in Figure 10.1.

THE VESSEL

A dished bottom requires less power than a flat one. When a single impeller is to be used, a liquid level equal to the diameter is optimum, with the impeller located at the center for an all-liquid system. Economic and manufacturing considerations, however, often dictate higher ratios of depth to diameter.

BAFFLES

Except at very high Reynolds numbers, baffles are needed to prevent vortexing and rotation of the liquid mass as a whole. A baffle width one-twelfth the tank diameter, $w = D_t/12$; a length extending from one half the impeller diameter, $d/2$, from the tangent line at the bottom to the liquid level, but sometimes terminated just above the level of the eye of the uppermost impeller. When solids are present or when a heat transfer jacket is used, the baffles are offset from the wall a distance equal to one-sixth the baffle width. Four radial baffles at equal spacing are standard; six are only slightly more effective, and three appreciably less so. When the mixer shaft is located off center (one-fourth to one-half the tank radius), the resulting flow pattern has less swirl, and baffles may not be needed, particularly at low viscosities.

DRAFT TUBES

A draft tube is a cylindrical housing around and slightly larger in diameter than the impeller. Its height may be little more than the diameter of the impeller or it may extend the full depth of the liquid, depending on the flow pattern that is required. Usually draft tubes are used with axial impellers to direct suction and discharge streams. An impeller-draft tube system behaves as an axial flow pump of somewhat low efficiency. Its top to bottom circulation behavior is of particular value in deep tanks for suspension of solids and for dispersion of gases. About a dozen applications are illustrated by Sterbacek and Tausk (1965, pp. 283ff) and a chapter is devoted to their use by Oldshue (1983, 469ff).

IMPELLER TYPES

A basic classification is into those that circulate the liquid axially and those that achieve primarily radial circulation. Some of the many shapes that are being used will be described shortly.

IMPELLER SIZE

This depends on the kind of impeller and operating conditions described by the Reynolds, Froude, and Power numbers as well as individual characteristics whose effects have been correlated. For the popular turbine impeller, the ratio of diameters of impeller and vessel falls in the range, $d/D_t = 0.3\text{–}0.6$, the lower values at high rpm, in gas dispersion, for example.

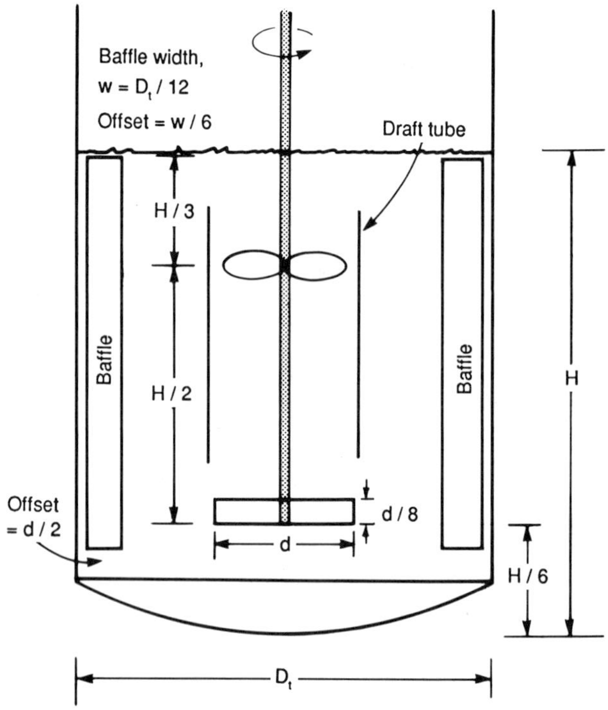

Figure 10.1. A basic stirred tank design, not to scale, showing a lower radial impeller and an upper axial impeller housed in a draft tube. Four equally spaced baffles are standard. H = height of liquid level, D_t = tank diameter, d = impeller diameter. For radial impellers, $0.3 \le d/D_t \le 0.6$.

IMPELLER SPEED

With commercially available motors and speed reducers, standard speeds are 37, 45, 56, 68, 84, 100, 125, 155, 190, and 320 rpm. Power requirements usually are not great enough to justify the use of continously adjustable steam turbine drives. Two-speed drives may be required when starting torques are high, as with a settled slurry.

IMPELLER LOCATION

Expert opinions differ somewhat on this factor. As a first approximation, the impeller can be placed at 1/6 the liquid level off the bottom. In some cases there is provision for changing the position of the impeller on the shaft. For off-bottom suspension of solids, an impeller location of 1/3 the impeller diameter off the bottom may be satisfactory. Criteria developed by Dickey (1984) are based on the viscosity of the liquid and the ratio of the liquid depth to the tank diameter, h/D_t. Whether one or two impellers are needed and their distances above the bottom of the tank are identified in this table:

Viscosity [cP (Pa sec)]	Maximum level h/D_t	Number of Impellers	Impeller Clearance	
			Lower	Upper
<25,000 (<25)	1.4	1	$h/3$	—
<25,000 (<25)	2.1	2	$D_t/3$	$(2/3)h$
>25,000 (>25)	0.8	1	$h/3$	—
>25,000 (>25)	1.6	2	$D_t/3$	$(2/3)h$

Another rule is that a second impeller is needed when the liquid must travel more than 4 ft before deflection.

Side entering propellors are placed 18–24 in. above a flat tank floor with the shaft horizontal and at a 10° horizontal angle with the centerline of the tank; such mixers are used only for viscosities below 500 cP or so.

In dispersing gases, the gas should be fed directly below the impeller or at the periphery of the impeller. Such arrangements also are desirable for mixing liquids.

10.2. KINDS OF IMPELLERS

A rotating impeller in a fluid imparts flow and shear to it, the shear resulting from the flow of one portion of the fluid past another. Limiting cases of flow are in the axial or radial directions so that impellers are classified conveniently according to which of these flows is dominant. By reason of reflections from vessel surfaces and obstruction by baffles and other internals, however, flow patterns in most cases are mixed. When a close approach to axial flow is particularly desirable, as for suspension of the solids of a slurry, the impeller may be housed in a draft tube; and when radial flow is needed, a shrouded turbine consisting of a rotor and a stator may be employed.

Because the performance of a particular shape of impeller usually cannot be predicted quantitatively, impeller design is largely an exercise of judgment so a considerable variety has been put forth by various manufacturers. A few common types are illustrated on Figure 10.2 and are described as follows:

a. The three-bladed mixing propeller is modelled on the marine propeller but has a pitch selected for maximum turbulence. They are used at relatively high speeds (up to 1800 rpm) with low viscosity fluids, up to about 4000 cP. Many versions are available: with cutout or perforated blades for shredding and breaking up lumps, with sawtooth edges as on Figure 10.2(g) for cutting and tearing action, and with other than three blades. The stabilizing ring shown in the illustration sometimes is included to minimize shaft flutter and vibration particularly at low liquid levels.

b. The turbine with flat vertical blades extending to the shaft is suited to the vast majority of mixing duties up to 100,000 cP or so at high pumping capacity. The simple geometry of this design and of the turbines of Figures 10.2(c) and (d) has inspired extensive testing so that prediction of their performance is on a more rational basis than that of any other kind of impeller.

c. The horizontal plate to which the impeller blades of this turbine are attached has a stabilizing effect. Backward curved blades may be used for the same reason as for type e.

d. Turbine with blades are inclined 45° (usually). Constructions with two to eight blades are used, six being most common. Combined axial and radial flow are achieved. Especially effective for heat exchange with vessel walls or internal coils.

e. Curved blade turbines effectively disperse fibrous materials without fouling. The swept back blades have a lower starting torque than straight ones, which is important when starting up settled slurries.

f. Shrouded turbines consisting of a rotor and a stator ensure a high degree of radial flow and shearing action, and are well adapted to emulsification and dispersion.

g. Flat plate impellers with sawtooth edges are suited to emulsification and dispersion. Since the shearing action is localized, baffles are not required. Propellers and turbines also are sometimes provided with sawtooth edges to improve shear.

h. Cage beaters impart a cutting and beating action. Usually they are mounted on the same shaft with a standard propeller. More violent action may be obtained with spined blades.

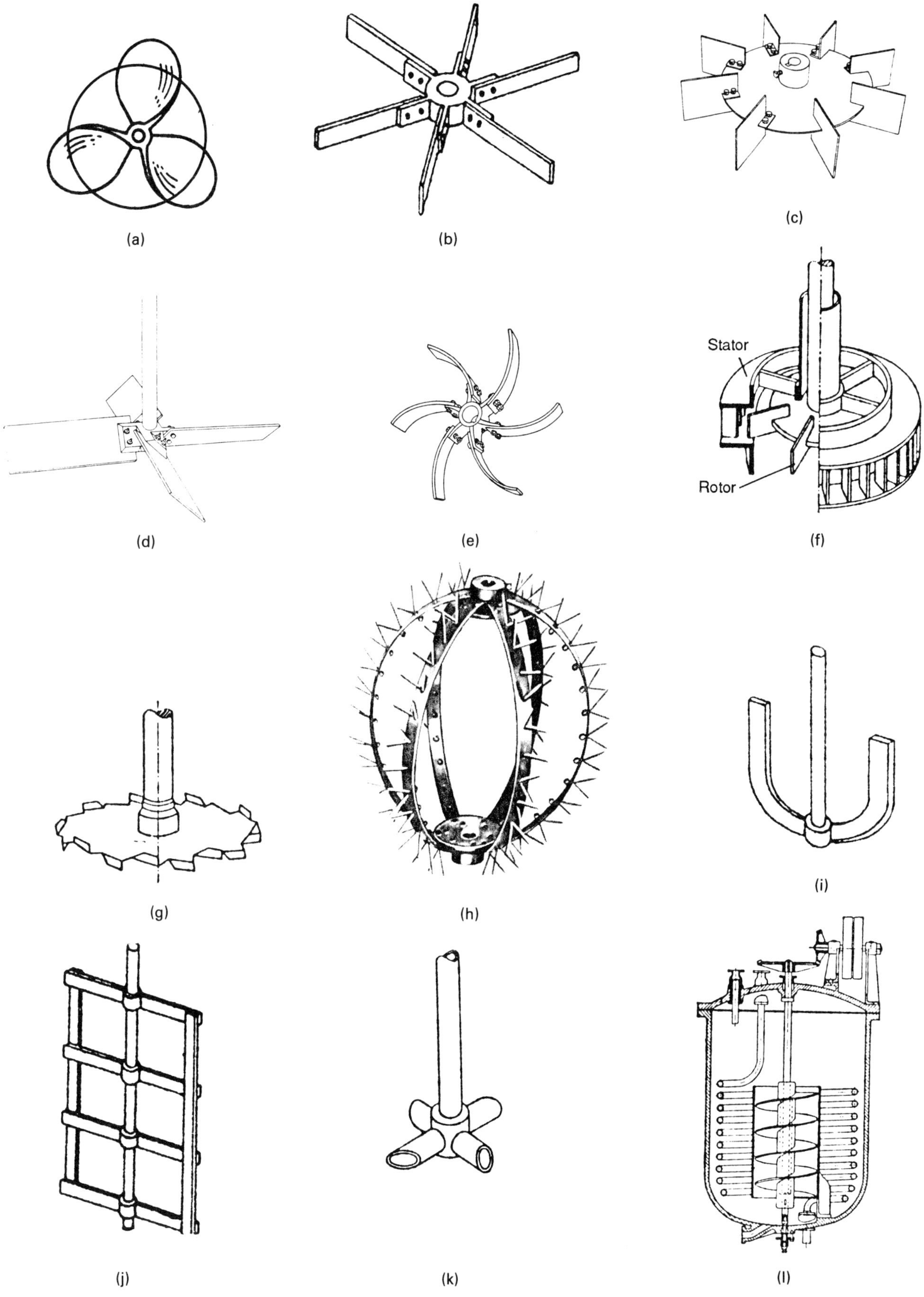

Figure 10.2. Representative kinds of impellers (descriptions in the text).

i. Anchor paddles fit the contour of the container, prevent sticking of pasty materials, and promote good heat transfer with the wall.
j. Gate paddles are used in wide, shallow tanks and for materials of high viscosity when low shear is adequate. Shaft speeds are low. Some designs include hinged scrapers to clean the sides and bottom of the tank.
k. Hollow shaft and hollow impeller assemblies are operated at high tip speeds for recirculating gases. The gas enters the shaft above the liquid level and is expelled centrifugally at the impeller. Circulation rates are relatively low, but satisfactory for some hydrogenations for instance.
l. This arrangement of a shrouded screw impeller and heat exchange coil for viscous liquids is perhaps representative of the many designs that serve special applications in chemical processing.

10.3. CHARACTERIZATION OF MIXING QUALITY

Agitation and mixing may be performed with several objectives:

1. Blending of miscible liquids.
2. Dispersion of immiscible liquids.
3. Dispersion of gases in liquids.
4. Suspension of solid particles in a slurry.
5. Enhancement of heat exchange between the fluid and the boundary of a container.
6. Enhancement of mass transfer between dispersed phases.

When the ultimate objective of these operations is the carrying out of a chemical reaction, the achieved specific rate is a suitable measure of the quality of the mixing. Similarly the achieved heat transfer or mass transfer coefficients are measures of their respective operations. These aspects of the subject are covered in other appropriate sections of this book. Here other criteria will be considered.

The uniformity of a multiphase mixture can be measured by sampling of several regions in the agitated mixture. The time to bring composition or some property within a specified range (say within 95 or 99% of uniformity) or spread in values—which is the blend time—may be taken as a measure of mixing performance. Various kinds of tracer techniques may be employed, for example:

1. A dye is introduced and the time for attainment of uniform color is noted.
2. A concentrated salt solution is added as tracer and the measured electrical conductivity tells when the composition is uniform.
3. The color change of an indicator when neutralization is complete when injection of an acid or base tracer is employed.
4. The residence time distribution is measured by monitoring the outlet concentration of an inert tracer that can be analyzed for accuracy. The shape of response curve is compared with that of a thoroughly (ideally) mixed tank.

The last of these methods has been applied particularly to chemical reaction vessels. It is covered in detail in Chapter 17. In most cases, however, the RTDs have not been correlated with impeller characteristics or other mixing parameters. Largely this also is true of most mixing investigations, but Figure 10.3 is an uncommon example of correlation of blend time in terms of Reynolds number for the popular pitched blade turbine impeller. As expected, the blend time levels off beyond a certain mixing intensity, in this case beyond Reynolds numbers of 30,000 or so. The acid–base indicator technique was used. Other details of the test work and the scatter of the data are not revealed in the published information. Another practical solution of the problem is typified by Table 10.1 which relates blend time to power input to

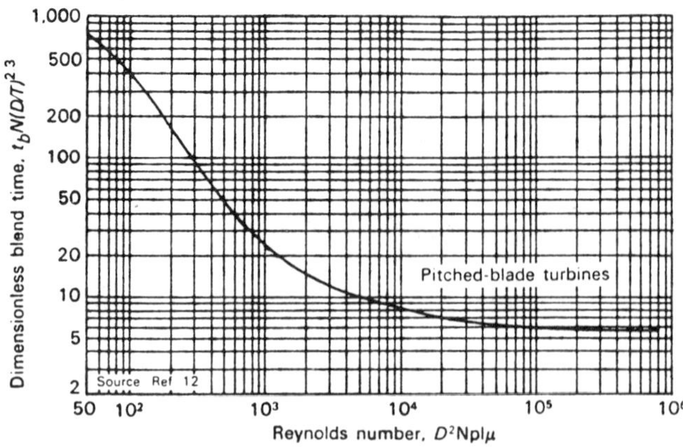

Figure 10.3. Dimensionless blend time as a function of Reynolds number for pitched turbine impellers with six blades whose $W/D = 1/5.66$ [*Dickey and Fenic,* Chem. Eng. **145,** (*5 Jan. 1976*)].

vessels of different sizes and liquids of various viscosities. A review of the literature on blend times with turbine impellers has been made by Brennan and Lehrer [*Trans. Inst. Chem. Eng.* **54,** 139–152 (1975)], who also did some work in the range $10^4 < N_{Re} < 10^5$ but did not achieve a particularly useable correlation.

An impeller in a tank functions as a pump that delivers a certain volumetric rate at each rotational speed and corresponding power input. The power input is influenced also by the geometry of the equipment and the properties of the fluid. The flow pattern and the degree of turbulence are key aspects of the quality of mixing. Basic impeller actions are either axial or radial, but, as Figure 10.4 shows, radial action results in some axial movement by reason of deflection from the vessel walls and baffles. Baffles contribute to turbulence by preventing swirl of the contents as a whole and elimination of vortexes; offset location of the impeller has similar effects but on a reduced scale.

Power input and other factors are interrelated in terms of certain dimensionless groups. The most pertinent ones are, in common units:

$N_{Re} = 10.75Nd^2S/\mu,$	Reynolds number,	(10.1)
$N_P = 1.523(10^{13})P/N^3d^5S,$	Power number,	(10.2)
$N_Q = 1.037(10^5)Q/Nd^3,$	Flow number,	(10.3)
$t_bN,$	Dimensionless blend time,	(10.4)

TABLE 10.1. Blending Data for Four-Bladed 45° Turbines[a]

V→	1000 gal			5000 gal			10,000 gal			20,000 gal		
μ↓ θ→	6	12	30	6	12	30	6	12	30	6	12	30
100	1†	½†	¼†	3	2	1	3	2	1	7½	5	2
250	1½†	¾†	⅓†	3	2	1	5	3	1½	10	7½	3
500	3†	1½†	¾†	3	2	1	7½	5	2	15	10	5
1000	1½*	2†	1†	5	3	1½	15	7½	3	20	15	7½

[a] Motor horsepowers for various batch volumes, viscosities in cP, blend times in minutes.
* Denotes single four-bladed, 45° axial-flow impeller (unshaded selections).
† Denotes portable geardrive mixer with single 1.5-pitch propeller ("shaded" selections).
(Oldshue, 1983, p. 91).

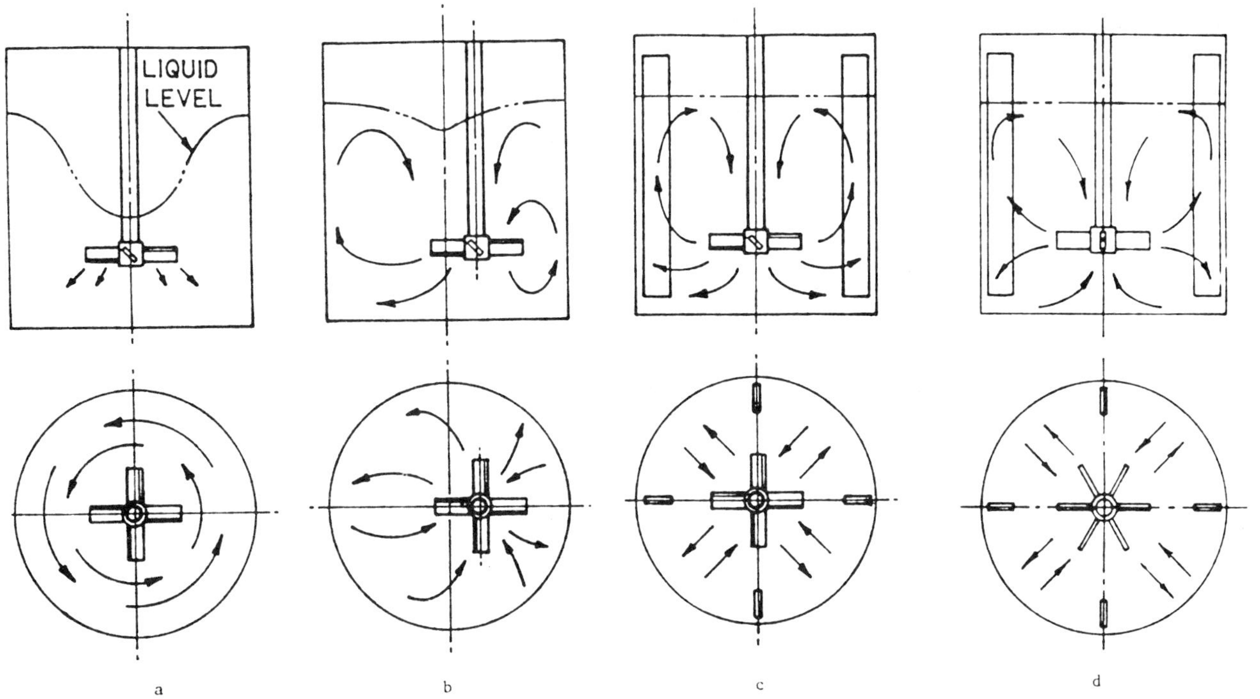

Figure 10.4. Agitator flow patterns. (a) Axial or radial impellers without baffles produce vortexes. (b) Offcenter location reduces the vortex. (c) Axial impeller with baffles. (d) Radial impeller with baffles.

$N_{Fr} = 7.454(10^{-4})N^2d,$ Froude number, (10.5)

d = impeller diameter (in.),
D = vessel diameter (in.),
N = rpm of impeller shaft,
P = horsepower input,
Q = volumetric pumping rate (cuft/sec),
S = specific gravity,
t_b = blend time (min),
μ = viscosity (cP).

The Froude number is pertinent when gravitational effects are significant, as in vortex formation; in baffled tanks its influence is hardly detectable. The power, flow, and blend time numbers change with Reynolds numbers in the low range, but tend to level off above $N_{Re} = 10{,}000$ or so at values characteristic of the kind of impeller. Sometimes impellers are characterized by their limiting N_p, as an $N_p = 1.37$ of a turbine, for instance. The dependencies on Reynolds number are shown on Figures 10.5 and 10.6 for power, in Figure 10.3 for flow and in Figure 10.7 for blend time.

Rough rules for mixing quality can be based on correlations of power input and pumping rate when the agitation system is otherwise properly designed with a suitable impeller (predominantly either axial or radial depending on the process) in a correct location, with appropriate baffling and the correct shape of vessel. The power input per unit volume or the superficial linear velocity can be used as measures of mixing intensity. For continuous flow reactors, for instance, a rule of thumb is that the contents of the vessel should be turned over in 5–10% of the residence time. Specifications of superficial linear velocities for different kinds of operations are stated later in this chapter. For baffled turbine agitation of reactors, power inputs and impeller tip speeds such as

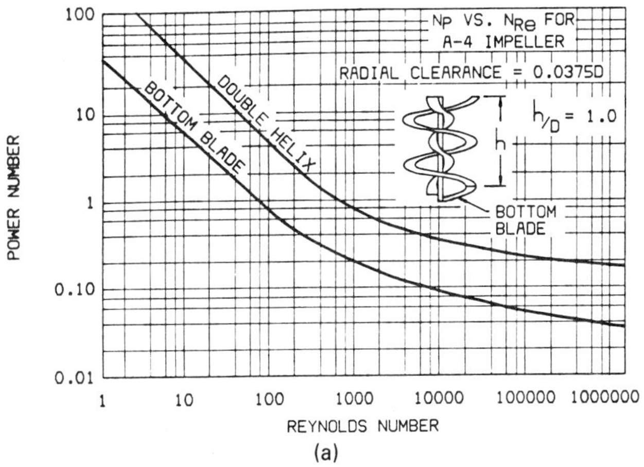

Figure 10.5. Power number, $N_p = Pg_c/N^3D^5\rho$, against Reynolds number, $N_{Re} = ND^2\rho/\mu$, for several kinds of impellers: (a) helical shape (*Oldshue, 1983*); (b) anchor shape (*Oldshue, 1983*); (c) several shapes: (1) propeller, pitch equalling diameter, without baffles; (2) propeller, $s = d$, four baffles; (3) propeller, $s = 2d$, without baffles; (4) propeller, $s = 2d$, four baffles; (5) turbine impeller, six straight blades, without baffles; (6) turbine impeller, six blades, four baffles; (7) turbine impeller, six curved blades, four baffles; (8) arrowhead turbine, four baffles; (9) turbine impeller, inclined curved blades, four baffles; (10) two-blade paddle, four baffles; (11) turbine impeller, six blades, four baffles; (12) turbine impeller with stator ring; (13) paddle without baffles (data of Miller and Mann); (14) paddle without baffles (data of White and Summerford). All baffles are of width $0.1D$ [*after Rushton, Costich, and Everett,* Chem. Eng. Prog. **46**(*9*), *467* (*1950*)].

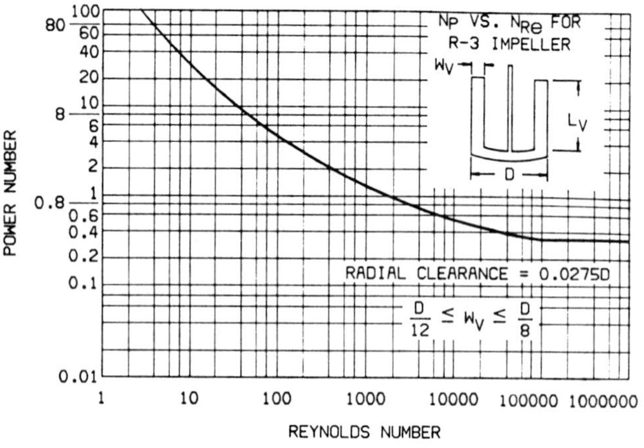

(b)

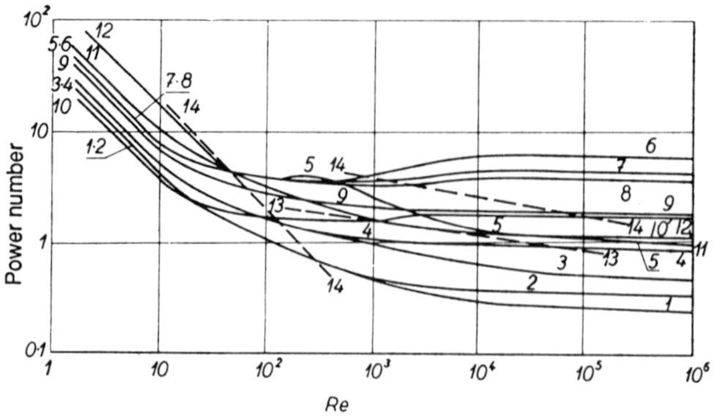

(c)

Figure 10.5—(*continued*)

the following may serve as rough guides:

Operation	HP/1000 gal	Tip Speed (ft/sec)
Blending	0.2–0.5	
Homogeneous reaction	0.5–1.5	7.5–10
Reaction with heat transfer	1.5–5.0	10–15
Liquid-liquid mixtures	5	15–20
Liquid-gas mixtures	5–10	15–20
Slurries	10	

The low figure shown for blending is for operations such as incorporation of TEL into gasoline where several hours may be allowed for the operation.

Example 10.1 deals with the design and performance of an agitation system to which the power input is specified. Some degree of consistency is found between the several rules that have been cited.

10.4. POWER CONSUMPTION AND PUMPING RATE

These basic characteristics of agitation systems are of paramount importance and have been investigated extensively. The literature is

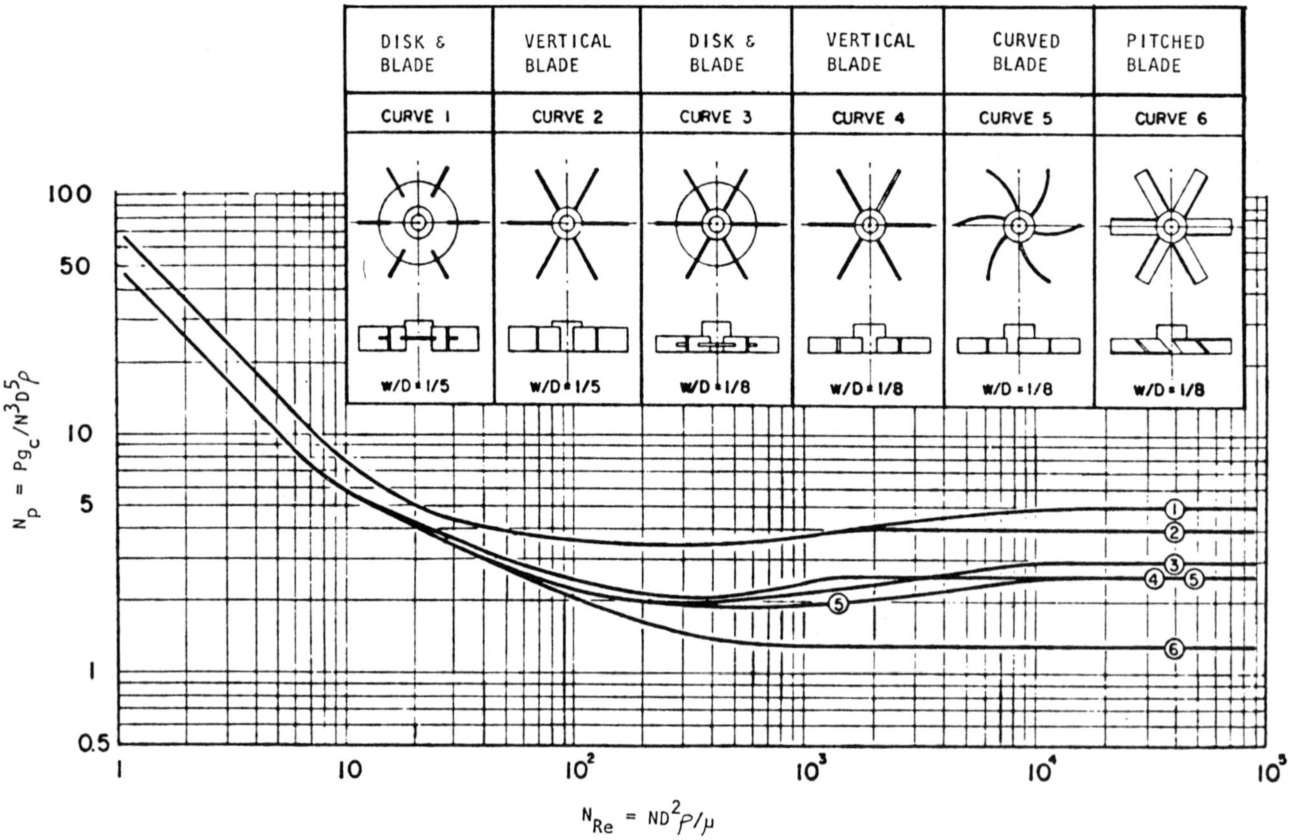

Figure 10.6. Power number against Reynolds number of some turbine impellers [*Bates, Fondy, and Corpstein,* Ind. Eng. Chem. Process. Des. Dev. **2**(*4*) *311* (*1963*)].

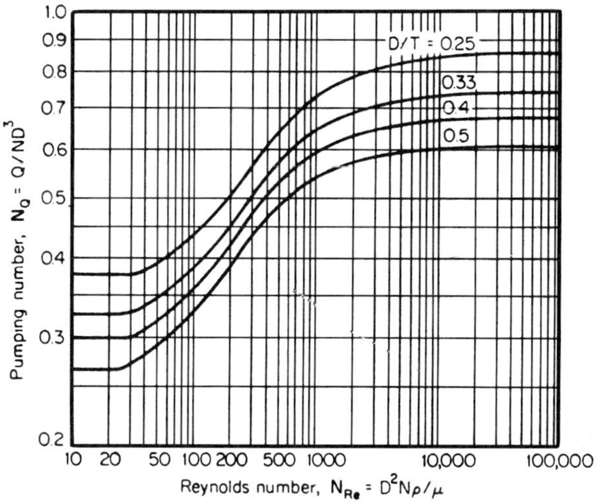

Figure 10.7. Flow number as a function of impeller Reynolds number for a pitched blade turbine with $N_p = 1.37$. D/T is the ratio of impeller and tank diameters. [*Dickey, 1984,* **12,** *7;* Chem. Eng., *102–110* (*26 Apr. 1976*)].

reviewed, for example, by Oldshue (1983, pp. 155–191), Uhl and Gray (1966, Vol. 1), and Nagata (1975). Among the effects studied are those of type and dimensions and locations of impellers, numbers and sizes of baffles, and dimensions of the vessel. A few of the data are summarized on Figures 10.5–10.7. Often it is convenient to characterize impeller performance by single numbers; suitable ones are the limiting values of the power and flow numbers at high Reynolds numbers, above 10,000–30,000 or so, for example:

Type	No. baffles	N_p	N_Q
Propeller	0	0.3	
Propeller	3–8	0.33–0.37	0.40–0.55
Turbine, vertical blade	0	0.93–1.08	0.33–0.34
Turbine, vertical blade	4	3–5	0.70–0.85
Pitched turbine, 45°	0	0.7	0.3
Pitched turbine, 45°	4	1.30–1.40	0.60–0.87
Anchor	0	0.28	

A correlation of pumping rate of pitched turbines is shown as Figure 10.7.

Power input per unit volume as a measure of mixing intensity or quality was cited in Section 10.3 and in Chapter 17. From the correlations cited in this section, it is clear that power input and Reynolds number together determine also the pumping rate of a given design of impeller. This fact has been made the basis of a method of agitator system design by the staff of Chemineer. The superficial linear velocity—the volumetric pumping rate per unit cross section of the tank—is adopted as a measure of quality of mixing. Table 10.2 relates the velocity to performance of three main categories of mixing: mixing of liquids, suspension of solids in slurries, and dispersion of gases. A specification of a superficial velocity will enable selection of appropriate impeller size, rotation speed, and power input with the aid of charts such as Figures 10.6 and 10.7. Examples 10.1 and 10.2 are along these lines.

The combination of HP and rpm that corresponds to a particular superficial velocity depends on the size of the tank, the size of the impeller, and certain characteristics of the system. Tables 10.3, 10.4, and 10.5 are abbreviated combinations of horsepower and rpm that are suitable at particular pumping rates for the three main categories of mixing. More complete data may be found in the literature cited with the tables.

1. For mixing of liquids, data are shown for a viscosity of 5000 cP, but data also have been developed for 25,000 cP, which allow for

---

EXAMPLE 10.1
Impeller Size and Speed at a Specified Power Input

For a vessel containing 5000 gal of liquid with specific gravity = 0.9 and viscosity of 100 cP, find size and speed of a pitched turbine impeller to deliver 2 HP/1000 gal. Check also the superficial linear velocity and the blend time.

The dimensions of the liquid content are 9.5 ft high by 9.5 ft dia. Take

$$d = 0.4D = 0.4(9.5)(12) = 45.6\text{ in., say 46 in., impeller,}$$
$$P = 2V = 2(5) = 10\text{ HP},$$
$$N_{Re} = \frac{10.75SNd^2}{\mu} = \frac{10.75(0.9)(46)^2N}{1000} = 20.47N,$$
$$N_P = \frac{1.523(10^{13})P}{N^3D^5S} = \frac{1523(10^{13})(10)}{0.9(46)^5N^3} = \frac{821,600}{N^3}.$$

Solve for N by trial with the aid of curve 6 of Figure 10.6.

Trial N	N_{Re}	N_p	N (Eq. (2))
56	1146	1.3	85.8
84	1720	1.3	85.8

Take $N = 84$ rpm.
According to Figure 10.7 at $d/D = 0.4$,

$$N_Q = 0.61,$$
$$Q = N_QNd^3 = 0.61(84/60)(46/12)^3 = 48.1\text{ cfs},$$
$$u_s = 48.1/[(\pi/4)(9.5)^2] = 0.68\text{ fps}.$$

This value corresponds to moderate to high mixing intensity according to Table 10.2.

From Figure 10.3, at $N_{Re} = 1720$, blend time is given by

$$t_bN(d/D)^{2.3} = 17.0$$

or

$$t_b = \frac{17}{84(0.4)^{2.3}} = 1.67\text{ min}.$$

According to Table 10.1, the blend time is less than 6 min, which agrees qualitatively.

TABLE 10.2. Agitation Results Corresponding to Specific Superficial Velocities

ft/sec	Description
Liquid Systems	
0.1–0.2	low degree of agitation; a velocity of 0.2 ft/sec will a. blend miscible liquids to uniformity when specific gravity differences are less than 0.1 b. blend miscible liquids to uniformity if the ratio of viscosities is less than 100 c. establish liquid movement throughout the vessel d. produce a flat but moving surface
0.3–0.6	characteristic of most agitation used in chemical processing; a velocity of 0.6 ft/sec will e. blend miscible liquids to uniformity if the specific gravity differences are less than 0.6 f. blend miscible liquids to uniformity if the ratio of viscosities is less than 10,000 g. suspend trace solids (less than 2%) with settling rates of 2–4 ft/min h. produce surface rippling at low viscosities
0.7–1.0	high degree of agitation; a velocity of 1.0 ft/sec will i. blend miscible liquids to uniformity if the specific gravity differences are less than 1.0 j. blend miscible liquids to uniformity if the ratio of viscosities is less than 100,000 k. suspend trace solids (less than 2%) with settling rates of 4–6 ft/min l. produce surging surface at low viscosities
Solids Suspension	
0.1–0.2	minimal solids suspension; a velocity of 0.1 ft/sec will a. produce motion of all solids with the design settling velocity b. move fillets of solids on the tank bottom and suspend them intermittently
0.3–0.5	characteristic of most applications of solids suspension and dissolution; a velocity of 0.3 ft/sec will c. suspend all solids with the design settling velocity completely off the bottom of the vessel d. provide slurry uniformity to at least one-third of the liquid level e. be suitable for slurry drawoff at low exit nozzle locations
0.6–0.8	when uniform solids distribution must be approached; a velocity of 0.6 ft/sec will f. provide uniform distribution to within 95% of liquid level g. be suitable for slurry drawoff up to 80% of liquid level
0.9–1.0	when the maximum feasible uniformity is needed. A velocity of 0.9 ft/sec will h. provide slurry uniformity to 98% of the liquid level i. be suitable for slurry drawoff by means of overflow
Gas Dispersion	
0.1–0.2	used when degree of dispersion is not critical to the process; a velocity of 0.2 ft/sec will a. provide nonflooded impeller conditions for coarse dispersion b. be typical of situations that are not mass transfer limited
0.3–0.5	used where moderate degree of dispersion is needed; a velocity of 0.5 ft/sec will c. drive fine bubbles completely to the wall of the vessel d. provide recirculation of dispersed bubbles back into the impeller
0.6–1.0	used where rapid mass transfer is needed; a velocity of 1.0 ft/sec will e. maximize interfacial area and recirculation of dispersed bubbles through the impeller

[Chemineer, Co. Staff, *Chem. Eng.*, 102–110 (26 April 1976); 144–150 (24 May 1976); 141–148 (19 July 1976)].

EXAMPLE 10.2
Effects of the Ratios of Impeller and Tank Diameters

Power and rpm requirements will be investigated and compared with the data of Table 10.3. The superficial velocity is 0.6 ft/sec, $V = 5000$ gals, Sp Gr = 1.0. Viscosities of 100 cP and 5000 cP will be considered.

With $h/D = 1$, $D = h = 9.47$ ft,

pumping rate $Q = 0.6(\pi/4)(9.47)^2 = 42.23$ cfs,

$$N_Q = 1.037(10^5)Q/Nd^3 = 4.3793/Nd^3 \quad (1)$$

$$N_{Re} = 10.7Nd^2S/\mu = 0.00214Nd^2, \quad \mu = 5000, \quad (2)$$

$$P = N_pN^3d^5S/1.523(10^{13}), \quad (3)$$

N_p from Figure 10.6.

For several choices of d/D, solve Eqs. (1) and (2) simultaneously with Figure 10.7. With $\mu = 5000$ cP;

d/D	d	N	N_Q [Eq. (1)]	N_{Re} [Eq. (2)]	N_Q (Fig. 10.7)	N_p	P (HP)
0.25	28.4	300	0.637	518	0.64	1.4	45.9
0.33	37.5	145	0.573	436	0.57	1.45	21.5
0.50	56.8	52	0.460	359	0.45	1.5	8.2

With $\mu = 100$ cP, turbulence is fully developed.

d/D	d	N	N_Q	N_{Re}	N_Q (Fig. 10.7)	N_p	P
0.25	28.4	228	0.839	18,990	0.84	1.3	18.7
0.33	37.5	112	0.742	16,850	0.74	1.3	8.9
0.50	56.8	40	0.597	13,800	0.60	1.3	3.2

Table 10.3 gives these combinations of HP/rpm as suitable: 25/125, 20/100, 10/56, 7.5/37. The combination 10/56 checks roughly the last entry at 5000 cP. Table 10.3 also has data for viscosities of 25,000 cP, thus allowing for interpolation and possibly extrapolation.

TABLE 10.3. Mixing of Liquids; Power and Impeller Speed (hp/rpm) for Two Viscosities, as a Function of the Liquid Superficial Velocity; Pitched Blade Turbine Impeller

	Volume (gal)					
	5000 cP			25,000 cP		
ft/sec	1000	2000	5000	1000	2000	5000
0.1	2/280	2/190	2/100	2/125	2/84	7.5/125
	1/190	1/100		1.5/84	1.5/56	5/100
						5/84
						3/56
0.2	2/190	2/125	5/125	3/84	5/125	10/84
	1/100	2/84	3/84	2/84	3/84	7.5/68
		1.5/84	3/68	1.5/56	3/68	5/45
			2/45		2/45	3/37
0.3	2/125	3/84	7.5/125	5/125	15/155	20/100
	1.5/84	1.5/56	5/100	5/84	7.5/68	15/68
			5/84	3/68	5/45	10/45
			3/56	2/45	3/37	7.5/37
0.4	2/84	5/125	10/84	7.5/84	10/84	30/100
	1.5/56	3/68	7.5/68	5/56	7.5/45	25/84
		3/56	5/45			20/68
		2/45	3/37			10/37
0.5	5/125	7.5/125	15/100	15/155	25/125	75/190
	3/84	5/84	10/68	10/100	20/100	60/155
			7.5/45	10/84	15/84	40/100
				7.5/68	10/56	15/45
0.6	5/100	15/155	25/125	20/155	25/100	40/84
	3/68	10/100	20/100	15/125	15/68	30/68
	3/56	7.5/84	10/56		15/56	25/56
	2/45	3/37	7.5/37		10/45	20/37
0.7	7.5/125	10/84	15/68	25/155	40/155	75/125
	5/84	7.5/68	15/56	15/84	30/100	50/84
		5/45	10/45		25/84	30/45
			10/37		20/68	
0.8	10/125	10/68	30/100	30/155	50/155	75/100
	7.5/100	7.5/56	25/84	25/125	40/125	60/84
			20/68	20/100		50/68
			15/45			40/56
0.9	15/155	15/84	60/155	40/155	75/190	75/84
	10/100	10/56	40/100	30/125	60/155	60/68
	7.5/84	7.5/45		25/100	40/100	50/56
					30/68	
1.0	10/84	30/155	50/100	40/125	60/125	125/125
	7.5/68	25/125	40/84	30/100	50/100	100/100
		20/100	30/68		50/84	75/68
		15/68	25/56		40/84	60/56

[Hicks, Morton, and Fenic, *Chem. Eng.*, 102–110 (26 April 1976)].

interpolation and possibly extrapolation. The impeller is a pitched-blade turbine.

2. For suspension of solids, the tables pertain to particles with settling velocities of 10 ft/min, but data are available for 25 ft/min. The impeller is a pitched-blade turbine.
3. For gas dispersion the performance depends on the gas rate. Data are shown for a superficial inlet gas rate of 0.07 ft/sec, but data are available up to 0.2 ft/sec. Four baffles are specified and the impeller is a vertical blade turbine.

Example 10.2 compares data of Table 10.4 with calculations based on Figures 10.6 and 10.7 for all-liquid mixing. Power and rpm requirements at a given superficial liquid velocity are seen to be very sensitive to impeller diameter. When alternate combinations of HP/rpm are shown in the table for a particular performance, the design of the agitator shaft may be a discriminant between them. The shaft must allow for the torque and bending moment caused by the hydraulic forces acting on the impeller and shaft. Also, the impeller and shaft must not rotate near their resonant frequency. Such mechanical details are analyzed by Ramsey and Zoller [*Chem. Eng.*, 101–108 (30 Aug. 1976)].

10.5. SUSPENSION OF SOLIDS

Besides the dimensions of the vessel, the impeller, and baffles, certain physical data are needed for complete description of a slurry mixing problem, primarily:

1. Specific gravities of the solid and liquid.
2. Solids content of the slurry (wt %).
3. Settling velocity of the particles (ft/min).

The last of these may be obtained from correlations when the mesh size or particle size distribution is known, or preferably experimentally. Taking into account these factors in their effect on suspension quality is at present a highly empirical process. Tables

TABLE 10.4. Suspension of Solids; Power and Impeller Speed (hp/rpm) for Two Settling Velocities, as a Function of the Superficial Velocity of the Liquid; Pitched Blade Turbine Impeller

	Volume (gal)					
	10 ft/min			25 ft/min		
ft/sec	1000	2000	5000	1000	2000	5000
0.1	1/190	2/190	5/125	2/190	2/125	5/125
		1/100	3/84	1/190	2/84	3/84
			3/68	1/100	1.5/84	3/68
			2/45		1.5/56	2/45
0.2	1/100	2/125	7.5/125	2/125	3/84	15/155
		1.5/84	5/100			10/100
			5/84			7.5/68
			3/56			5/45
0.3	2/190	2/84	3/37	1.5/84	5/125	10/84
		1.5/56			3/68	
					2/45	
0.4	2/155	5/155	7.5/84	2/84	7.5/155	7.5/45
	1.5/100		5/56	1.5/56	5/100	
					3/56	
0.5	1.5/84	3/84	15/155	2/68	7.5/125	15/84
	2/125		10/100	2/56	5/84	10/56
			7.5/68			7.5/37
			5/45			
0.6	2/100	5/125	10/84	3/84	5/56	25/125
	1.5/68	3/68				20/100
		3/56				15/68
		2/45				10/45
0.7	2/84	7.5/155	15/84	7.5/155	15/155	30/100
	1.5/56	7.5/125	10/56	5/125	10/100	25/84
		5/84	7.5/45	5/100	7.5/84	20/68
			7.5/37	3/68	7.5/68	15/56
0.8	3/84	7.5/84	25/125	7.5/125	10/84	60/155
		5/56	20/100	5/84		40/100
			15/68			30/68
			10/45			25/56
0.9	7.5/155	15/155	40/155	10/125	15/84	75/190
	5/125	10/100	30/100	7.5/100		60/125
	5/100	7.5/68	25/84			50/100
	3/68		20/68			40/84
1.0	7.5/125	20/100	50/100	15/155	30/155	75/125
	5/84	15/84	40/84	10/100	25/125	75/100
		10/84	30/68		20/100	60/84
			25/56			50/84

[Gates, Morton, and Fondy, *Chem. Eng.*, 144–150 (24 May 1976)].

10.2–10.5 are one such process; the one developed by Oldshue (1983) will be examined shortly.

Suspension of solids is maintained by upward movement of the liquid. In principle, use of a draft tube and an axial flow impeller will accomplish this flow pattern most readily. It turns out, however, that such arrangements are suitable only for low solids contents and moderate power levels. In order to be effective, the cross section of the draft tube must be appreciably smaller than that of the vessel, so that the solids concentration in the draft tube may become impractically high. The usually practical arrangement for solids suspension employs a pitched blade turbine which gives both axial and radial flow.

For a given tank size, the ultimate design objective is the relation between power input and impeller size at a specified uniformity. The factors governing such information are the slurry volume, the slurry level, and the required uniformity. The method of Oldshue has corrections for these factors, as F_1, F_2, and F_3. When multiplied together, they make up the factor F_4 which is the ordinate of Figure 10.8(d) and which determines what combinations of horsepower and ratio of impeller and vessel diameters will do the required task. Example 10.3 employs this method, and makes a comparison with the Chemineer method of Tables 10.2 and 10.3.

10.6. GAS DISPERSION

Gases are dispersed in liquids usually to facilitate mass transfer between the phases or mass transfer to be followed by chemical reaction. In some situations gases are dispersed adequately with spargers or porous distributors, but the main concern here is with the more intense effects achievable with impeller driven agitators.

SPARGERS

Mixing of liquids and suspension of solids may be accomplished by bubbling with an inert gas introduced uniformly at the bottom of the tank. For mild agitation a superficial gas velocity of 1 ft/min is used, and for severe, one of about 4 ft/min.

TABLE 10.5. Dispersion of Gases; Power and Impeller Speed (hr/rpm) for Two Gas Inlet Superficial Velocities, as a Function of the Liquid Superficial Velocity; Vertical Blade Turbine Impeller

	Volume (gal)					
	0.07 ft/sec			0.20 ft/sec		
ft/sec	1500	3000	5000	1500	3000	5000
0.1	2/56	5/84	7.5/68	3/56	7.5/68	10/45
			15/155		10/100	
0.2	2/45	7.5/125	10/84	3/45	15/155	15/68
			7.5/45			20/100
				5/100	10/84	25/125
0.3	3/84	7.5/68	10/45		7.5/45	
	3/68	5/45	10/56			
	3/56	7.5/84				
		5/56				
0.4	5/125	10/84	15/68	5/84	10/45	30/155
	5/84	10/100	20/100	7.5/155	10/56	20/68
	5/100	10/45	15/84	5/56		15/45
	5/45	10/56	20/68			15/56
0.5	7.5/125	15/155	25/125	7.5/125	15/68	25/84
	7.5/155	15/68	25/84	7.5/68	15/84	25/100
	7.5/68	15/84	25/100	7.5/84	15/45	25/56
	7.5/84	15/45	25/56		15/56	
0.6	10/84	20/100	30/155	10/84	20/100	30/100
	10/100	20/68	30/100	10/100	20/68	30/125
		20/45	30/125			30/68
			30/68			30/45
0.7	10/56	25/125	40/155	10/56	25/125	40/155
		25/84	40/84		25/84	40/84
		25/100	40/100		25/100	40/100
		25/56	40/56		25/56	40/56
0.8	15/155	30/155	50/100	15/155	30/155	50/100
	15/84	30/100	50/68	15/84	30/100	50/68
		30/125	50/84		30/125	50/84
			50/45			50/56
0.9	15/68	30/68	60/125	15/68	30/68	60/125
			60/155			60/155
			60/84			60/84
			60/56			60/56
1.0	20/100	40/155	75/190	25/125	40/155	75/190
	20/68	40/84	75/100	25/84	40/84	75/100
			75/125			75/125

[Hicks and Gates, *Chem. Eng.*, 141–148 (19 July 1976)].

MASS TRANSFER

The starting point of agitator design is properly a mass transfer coefficient known empirically or from some correlation in terms of parameters such as impeller size and rotation, power input, and gas flow rate. Few such correlations are in the open literature, but some have come from two of the industries that employ aerated stirred tanks on a large scale, namely liquid waste treating and fermentation processes. A favored method of studying the absorption of oxygen is to measure the rate of oxidation of aqueous sodium sulfite solutions. Figure 10.9 summarizes one such investigation of the effects of power input and gas rate on the mass transfer coefficients. A correlation for fermentation air is given by Dickey (1984, **12**-17):

$$k_L a = \text{rate}/(\text{concentration driving force}) = 0.064(P_g/V)^{0.7}u_g^{0.2}, \quad 1/\text{sec}, \tag{10.6}$$

with P_g/V in HP/1000 gal and superficial gas velocity u_g in ft/sec. A general correlation of mass transfer coefficient that does not have power input as a factor is given by Treybal (*Mass Transfer Operations*, McGraw-Hill, New York, 1980, 156); presumably this is applicable only below the minimum power input here represented by Figure 10.11.

When mass transfer coefficients are not determinable, agitator design may be based on superficial liquid velocities with the criteria of Table 10.2.

SYSTEM DESIGN

The impeller commonly used for gas dispersion is a radial turbine with six vertical blades. For a liquid height to diameter ratio $h/D \leq 1$, a single impeller is adequate; in the range $1 \leq h/D \leq 1.8$ two are needed, and more than two are rarely used. The lower and upper impellers are located at distances of 1/6 and 2/3 of the liquid level above the bottom. Baffling is essential, commonly with four baffles of width 1/12 that of the tank diameter, offset from the wall at 1/6 the width of the baffle and extending from the tangent line of the wall to the liquid level. The best position for inlet of the gas is below and at the center of the lower impeller; an open pipe is commonly used, but a sparger often helps. Since ungassed power is significantly larger than gassed, a two-speed motor is desirable to prevent overloading, the lower speed to cut in automatically when the gas supply is interrupted and rotation still is needed.

MINIMUM POWER

Below a critical power input the gas bubbles are not affected laterally but move upward with their natural buoyancy. This condition is called gas flooding of the impeller. At higher power inputs the gas is dispersed radially, bubbles impinge on the walls and are broken up, consequently with improvement of mass transfer. A correlation of the critical power input is shown as Figure 10.10.

POWER CONSUMPTION OF GASSED LIQUIDS

At least partly because of its lower density and viscosity, the power to drive a mixture of gas and liquid is less than that to drive a liquid. Figure 10.11(a) is a correlation of this effect, and other data at low values of the flow number Q/Nd^3 are on Figure 10.11(b). The latter data for Newtonian fluids are correlated by the equation

$$P_g/P = 0.497(Q/Nd^3)^{-0.38}(N^2d^3\rho_L/\sigma)^{-0.18}, \tag{10.7}$$

where the last group of terms is the Weber number, ρ_L is the density of the liquid, and σ is its surface tension.

SUPERFICIAL LIQUID VELOCITY

When mass transfer data are not known or are not strictly pertinent, a quality of mixing may be selected by an exercise of judgment in terms of the superficial liquid velocity on the basis of the rules of Table 10.2. For gas dispersion, this quantity is related to the power input, HP/1000 gal, the superficial gas velocity and the ratio d/D in Figure 10.12.

DESIGN PROCEDURES

On the basis of the information gathered here, three methods are possible for the design of agitated gas dispersion. In all cases the size of the tank, the ratio of impeller and tank diameters and the gas feed rate are specified. The data are for radial turbine impellers with six vertical blades.

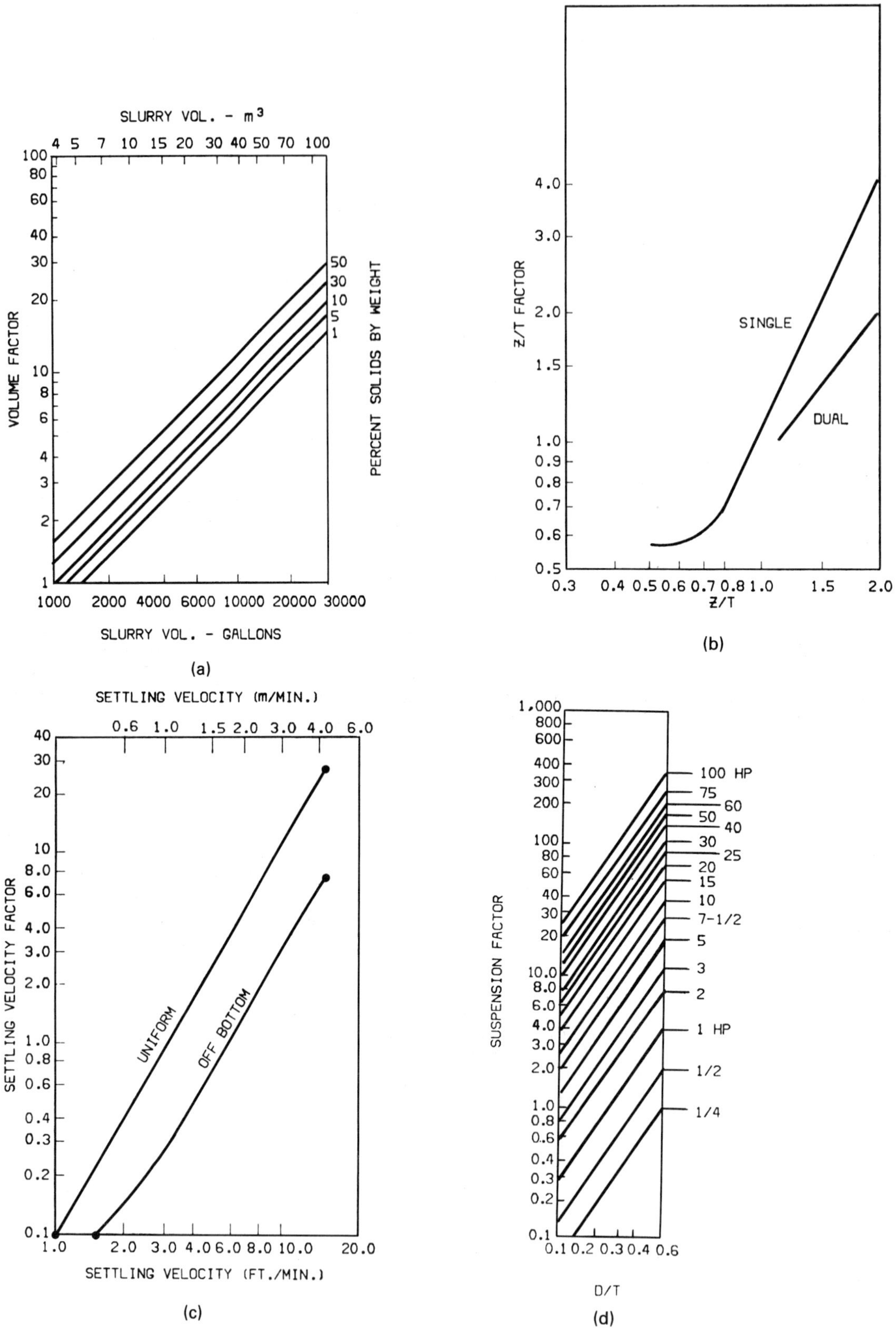

Figure 10.8. Suspension of solids. Power and ratio of diameters of impeller and tank, with four-bladed 45° impeller, width/diameter = 0.2. [*method of Oldshue (1983)*]. (a) The factor on power consumption for slurry volume, F_1. (b) The factor on power requirement for single and dual impellers at various h/D ratios, F_2. (c) The effect of settling velocity on power consumption, F_3. (d) Suspension factor for various horsepowers: $F_4 = F_1F_2F_3$.

EXAMPLE 10.3
Design of the Agitation System for Maintenance of a Slurry
These conditions are taken:

$V = 5000$ gal,
$h/D = 1$,
settling velocity = 10 ft/min,
solids content = 10 wt %

Reading from Figure 10.8,

$$F_1 = 4,$$
$$F_2 = 1.1,$$
$$F_3 = \begin{cases} 3.0, & \text{off bottom,} \\ 10.0, & \text{uniform,} \end{cases}$$
$$F_4 = F_1 F_2 F_3 = \begin{cases} 13.2, & \text{off bottom,} \\ 44, & \text{uniform.} \end{cases}$$

The relation between the ratio of impeller and vessel diameters, d/D and HP is read off Figure 10.8(d).

	HP	
d/D	Off btm	Uniform
0.2	20	65
0.4	7.5	25
0.6	4	12

Comparing with readings from Tables 10.2 and 10.3,

Superficial liq. velocity	HP/rpm
0.3 (off btm)	10/45, 10/56
0.6 (uniform)	30/155, 30/125, 30/100, 30/68

These results correspond roughly to those of the Oldshue method at $d/D = 0.4$. The impeller sizes can be determined with Figures 10.6 and 10.7.

1. Start with a known required mass transfer coefficient. From a correlation such as Figure 10.9 or Eq. (10.6) the gassed power per unit volume will become known, and the total gassed power to the tank will be P_g. The ratio of gassed power to ungassed power is represented by Figure 10.11(a) and the equations given there; at this stage the rotation speed N is not yet known. This value is found by trial by simultaneous solution with Figure 10.6 which relates the Reynolds and power numbers; the power here is the ungassed power. The value of N that results in the precalculated P_g will be the correct one. Curve 2 of Figure 10.6 is the one applicable to gas dispersion with the data of this section.
2. Start with a choice of superficial liquid velocity u_L made in accordance with the criteria of Table 10.2. With the aid of the known gas velocity u_s and d/D, find P_g/V from Figure 10.12. Then proceed to find N by trial with Figures 10.11(a) and 10.6 as in method 1.

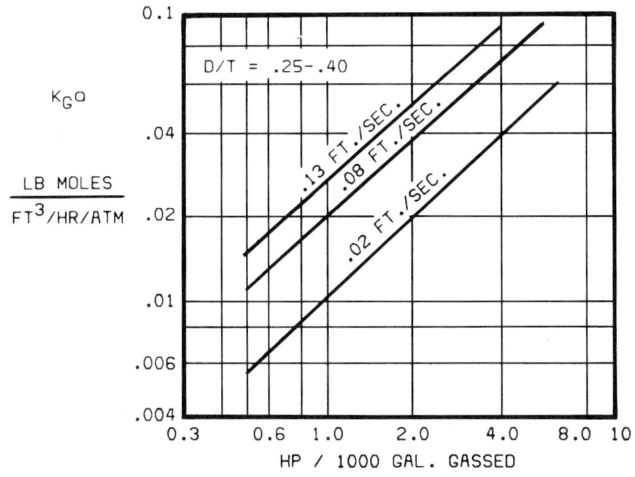

Figure 10.9. Typical data of mass transfer coefficients at various power levels and superficial gas rates for oxidation of sodium sulfite in aqueous solution. $d/D = 0.25–0.40$ (*Oldshue, 1983*).

3. As soon as a superficial liquid velocity has been selected, a suitable combination of HP/rpm can be taken from Table 10.5.

These procedures are applied in Example 10.4.

As general rules, levels of 5–12 HP/1000 gal are typical of aerobic fermentation vessels, and 1–3 HP/1000 gal of aerobic waste treatment; concentrations and oxygen requirements of the microorganisms are different in the two kinds of processes.

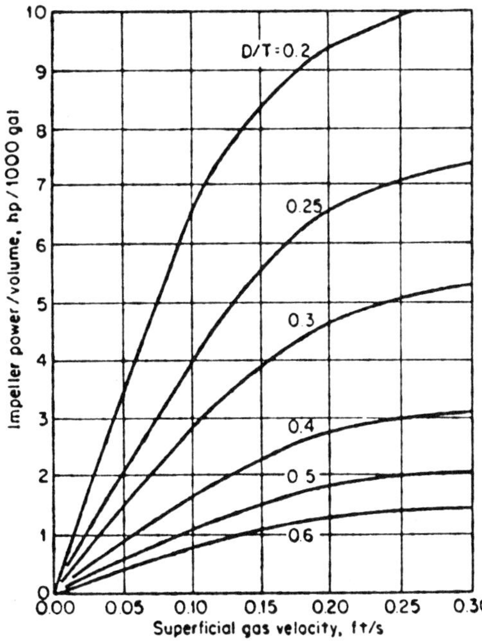

Figure 10.10. Minimum power requirement to overcome flooding as a function of superficial gas velocity and ratio of impeller and tank diameters, d/D. [*Hicks and Gates,* Chem. Eng., *141–148* (*19 July 1976*)].

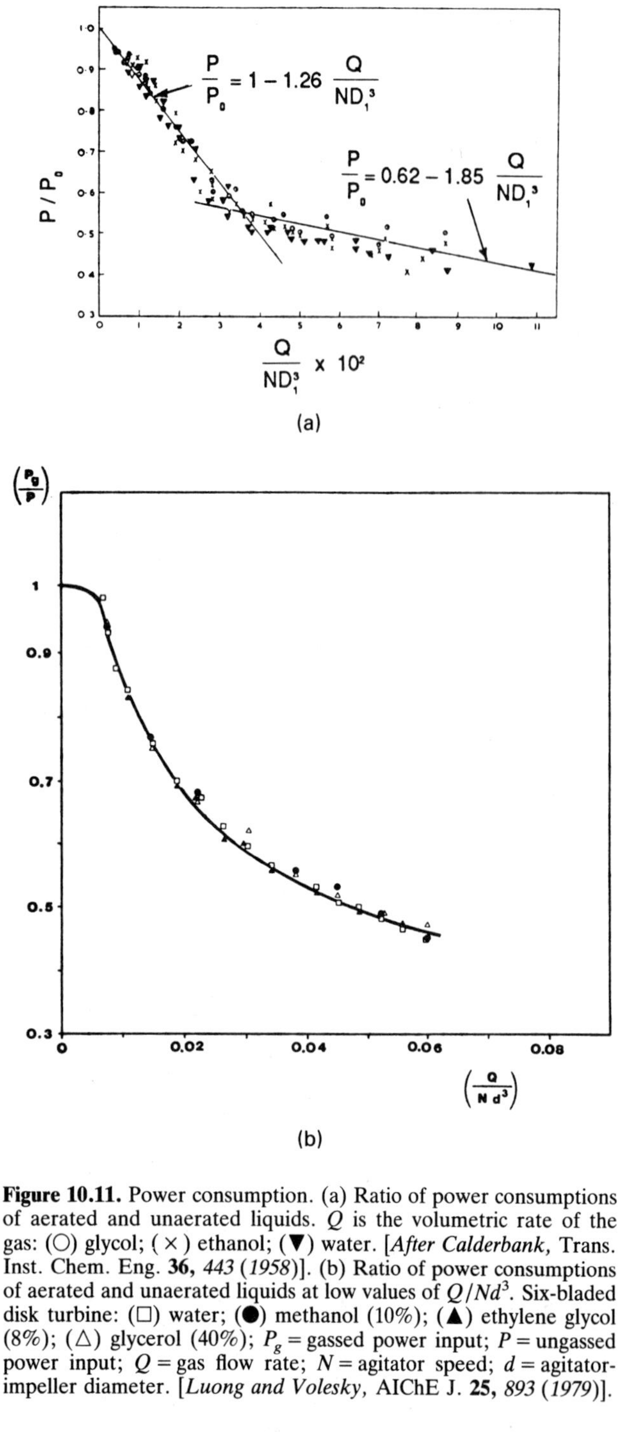

Figure 10.11. Power consumption. (a) Ratio of power consumptions of aerated and unaerated liquids. Q is the volumetric rate of the gas: (○) glycol; (×) ethanol; (▼) water. [*After Calderbank,* Trans. Inst. Chem. Eng. **36,** *443* (*1958*)]. (b) Ratio of power consumptions of aerated and unaerated liquids at low values of Q/Nd^3. Six-bladed disk turbine: (□) water; (●) methanol (10%); (▲) ethylene glycol (8%); (△) glycerol (40%); P_g = gassed power input; P = ungassed power input; Q = gas flow rate; N = agitator speed; d = agitator-impeller diameter. [*Luong and Volesky,* AIChE J. **25,** *893* (*1979*)].

10.7. IN-LINE BLENDERS AND MIXERS

When long residence time is not needed for chemical reaction or other purposes, small highly powered tank mixers may be suitable, with energy inputs measured in HP/gal rather than HP/1000 gal. They bring together several streams continuously for a short contact time (at most a second or two) and may be used whenever the effluent remains naturally blended for a sufficiently long time, that is, when a true solution is formed or a stable emulsion-like mixture. When it is essential that the mixing be immediate each stream will

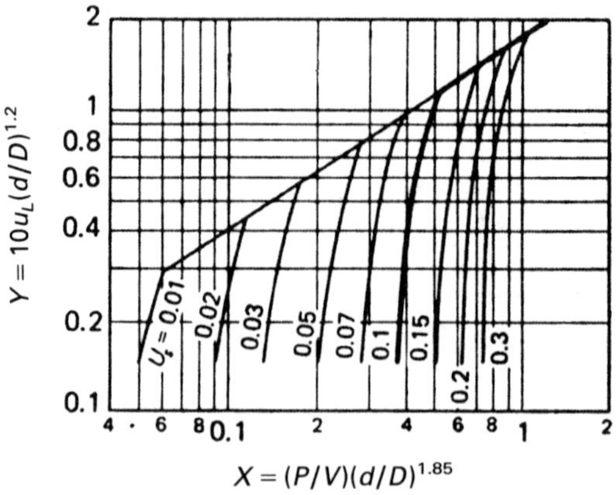

Figure 10.12. Relation between power input, P/V HP/1000 gal, superficial liquid velocity u_L ft/sec, ratio of impeller and tank diameters, d/D, and superficial gas velocity u_s ft/sec. [*Hicks and Gates,* Chem. Eng., *141–148* (*19 July 1976*)].

have its own feed nozzle, as in Figure 10.13(b), but usually the streams may be combined externally near the blender and then given the works, as in Figure 10.13(a).

One manufacturer gives these power ratings:

Tank size (gal)	1	5	10	30
Motor HP	0.5	1	2	3

Another ties in the line and motor sizes:

Line size, (in.)	1–4	6–8	10–12
Motor HP	0.5	1	2

But above viscosities of 10 cP a body one size larger than the line size is recommended.

Other devices utilize the energy of the flowing fluid to do the mixing. They are inserts to the pipeline that force continual changes of direction and mixing. Loading a section of piping with tower packing is an example but special assemblies of greater convenience have been developed, some of which are shown in Figure 10.14. In each case manufacturer's literature recommends the sizes and pressure drops needed for particular services.

The Kenics mixer, Figure 10.14(a), for example, consists of a succession of helical elements twisted alternately in opposite directions. In laminar flow for instance, the flow is split in two at each element so that after n elements the number of striations becomes 2^n. The effect of this geometrical progression is illustrated in Figure 10.14(b) and points out how effective the mixing becomes after only a few elements. The Reynolds number in a corresponding empty pipe is the major discriminant for the size of mixer, one manufacturer's recommendations being

N_{Re}	Number of Elements
Less than 10	24
10–2000	12–18
More than 2000	6

Besides liquid blending applications, static mixers have been used for mixing gases, pH control, dispersion of gases into liquids, and dispersion of dyes and solids in viscous liquids. They have the advantages of small size, ease of operation, and relatively low cost. The strong mixing effect enhances the rate of heat transfer from viscous streams. Complete heat exchangers are built with such

Example 10.4
HP and rpm Requirements of an Aerated Agitated Tank

A tank contains 5000 gal of liquid with sp gr = 1.0 and viscosity 100 cP that is aerated and agitated. The ratio of impeller to tank diameters is $d/D = 0.4$. Two sets of conditions are to be examined.

a. The air rate is 972 SCFM or 872 ACFM at an average submergence of 4 ft. The corresponding superficial gas velocity is 0.206 ft/sec or 0.063 m/sec. A mass transfer coefficient $k_La = 0.2$/sec is required; Dickey's equation (10.6) applies. Find the power and rpm needed.

b. The air rate is 296 ACFM, 0.07 ft/sec, 0.0213 m/sec. The required intensity of mixing corresponds to a liquid superficial velocity of 0.5 ft/sec. Find the power, rotation speed, and mass transfer coefficients for sulfite oxidation and for fermentation.

a. $d = 0.4(9.47) = 3.79\ \text{ft},\ 45.46\ \text{in.},$

$$k_La = 0.064(P_g/V)^{0.7}u_g^{0.2} = 0.2,$$
$$P_g/V = [0.2/0.064(0.206)^{0.2}]^{1/0.7} = 8.00\ \text{HP},$$
$$P_g = 5(8.0) = 40.0\ \text{HP/5000 gal},$$
$$Q/Nd^3 = 872/(379)^3N = 16.02/N,$$
$$N_{\text{Re}} = 10.75Nd^2S/\mu = 10.75(45.46)^2N/100 = 222N.$$

Equation (10.2),

$$N_p = 1.523(10^{13})P/N^3d^5S = 78{,}442P/N^3.$$

Curve 2 of Figure (10.6) applies. P_g/P from Figure 10.10(a). Solve by trial.

N	Q/Nd^3	P_g/P	N_{Re}	N_p	P	P_g
100	0.160	0.324	22,200	4	51	16.5
150	0.107	0.422	33,300	4	172	72.6
127	0.1261	0.3866	28,194	4	104.5	40.4 ≅ 40.0

The last entry of P_g checks the required value 40.0. Find the corresponding superficial liquid velocity with Figure 10.12:

$$X = (P/V)(d/D)^{1.85} = 8.04(0.4)^{1.85} = 1.48,$$

at $u_G = 0.206$ ft/sec, $Y = 2.0$,

$$\therefore u_L = 2/10(0.4)^{1.2} = 0.60\ \text{ft/sec}.$$

From Table 10.2, a liquid velocity of 0.6–0.7 ft/sec will give moderate to high dispersion. Table 10.5 gives possible HP/rpm combination of 30/125, somewhat less than the value found here.

b. With liquid circulation velocity specified,

$$u_L = 0.5\ \text{ft/sec}.$$

Use Figure 10.12:

$$Y = iou_L(d/D)^{1.2} = 10(0.5)(0.4)^{1.2} = 1.67,$$
$$X = 0.8,$$
$$P_g/V = 0.8/(0.4)^{1.85} = 4.36\ \text{HP/1000 gal}$$

(this does exceed the minimum of 1.6 from Figure 10.11),

$$P_g = 5(4.36) = 21.8,$$
$$\frac{Q}{Nd^3} = 296/(3.79)^3N = 5.437/N,$$
$$N_{\text{Re}} = 222N \quad \text{(part a)},$$
$$N_{=}\frac{78{,}442P}{N} \quad \text{(part a)}.$$

Solve by trial, using Figure 10.10(a) and curve 2 of Figure 10.6.

N	Q/Nd^3	P_g/P	N_{Re}	N_p	P	P_g
100	0.0544	0.5194	22,200	4	51	26.5
94	0.0576	0.5130		4	42.35	21.7 ≅ 2.8

The closest reading from Table 10.5 is HP/rpm = 25/100 which is a good check.

For sulfite oxidation, at $u_g = 0.07$ ft/sec,

$$P_g/V = 4.36\ \text{HP/1000 gal}, \quad \text{from Figure 10.9},$$
$$k_ga = 0.07\ \text{lb mol/(cuft)/(hr)(atm)}.$$

For fermentation, Eq. 10.6 gives

$$k_La = 0.064(4.36)^{0.7}(0.07)^{0.2}$$
$$= 0.105\frac{\text{lb mol/(cuft)(sec)}}{\text{lb mol/cuft}}.$$

mixing inserts in the tubes and are then claimed to have 3–5 times normal capability in some cases.

10.8. MIXING OF POWDERS AND PASTES

Industries such as foods, cosmetics, pharmaceuticals, plastics, rubbers, and also some others have to do with mixing of high viscosity liquids or pastes, of powders together and of powders with pastes. Much of this kind of work is in batch mode. The processes are so diverse and the criteria for uniformity of the final product are so imprecise that the nonspecialist can do little in the way of equipment design, or in checking on the recommendations of equipment manufacturers. Direct experience is the main guide to selection of the best kind of equipment, predicting how well and quickly it will perform, and what power consumption will be. For projects somewhat out of direct experience and where design by analogy may not suffice, testing in pilot plant equipment is a service provided by many equipment suppliers.

A few examples of mixers and blenders for powders and pastes are illustrated in Figure 10.15. For descriptions of available equipment—their construction, capacity, performance, power consumption, etc.—the primary sources are catalogs of manufacturers and contact with their offices. Classified lists of manufacturers, and some of their catalog information, appear in the *Chemical Engineering Catalog* (Reinhold, New York, annually) and in the *Chemical Engineering Equipment Buyers Guide* (McGraw-Hill, New York, annually). Brief descriptions of some types of equipment are in *Perry's Chemical Engineers Handbook* (McGraw-Hill, New York, 1984 and earlier editions). Well-classified descriptions, with figures, of paste mixers are in Ullmann (1972,

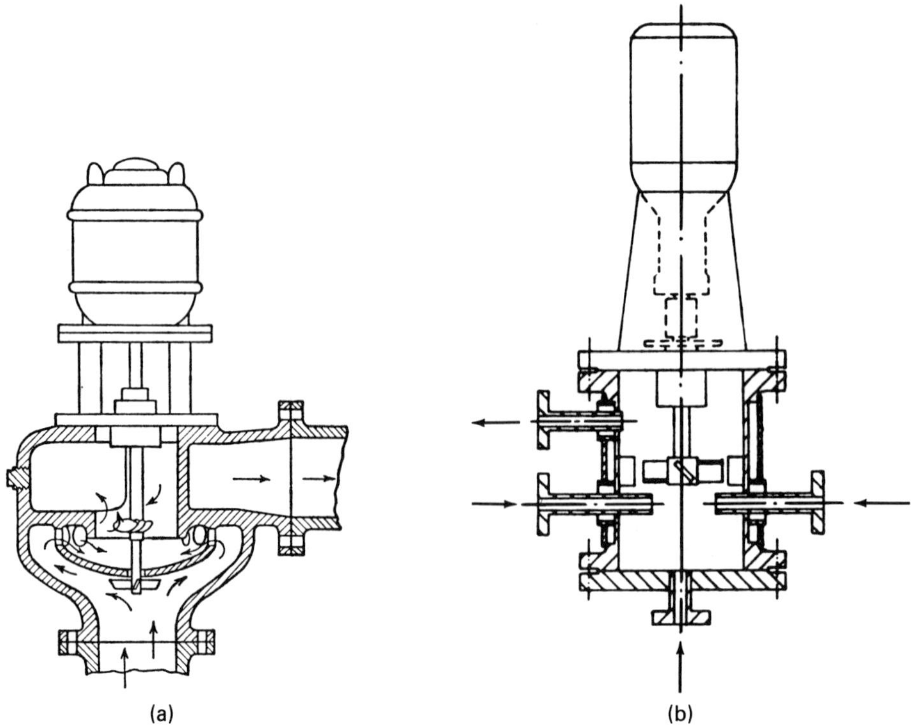

Figure 10.13. Motor-driven in-line blenders: (a) Double impeller made by Nettco Corp.; (b) three-inlet model made by Cleveland Mixer Co.

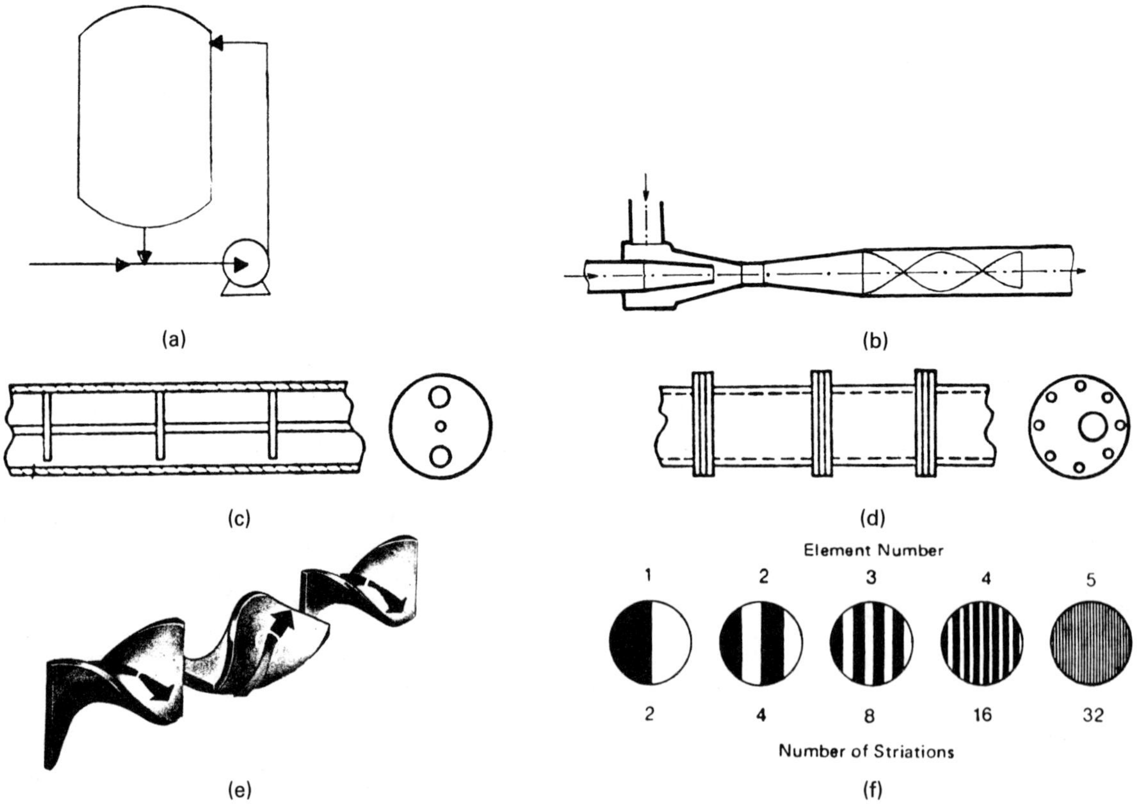

Figure 10.14. Some kinds of in-line mixers and blenders. (a) Mixing and blending with a recirculating pump. (b) Injector mixer with a helical baffle. (c) Several perforated plates (orifices) supported on a rod. (d) Several perforated plates flanged in. (e) Hellical mixing elements with alternating directions (Kenics Corp.). (f) Showing progressive striations of the flow channels with Kenics mixing elements.

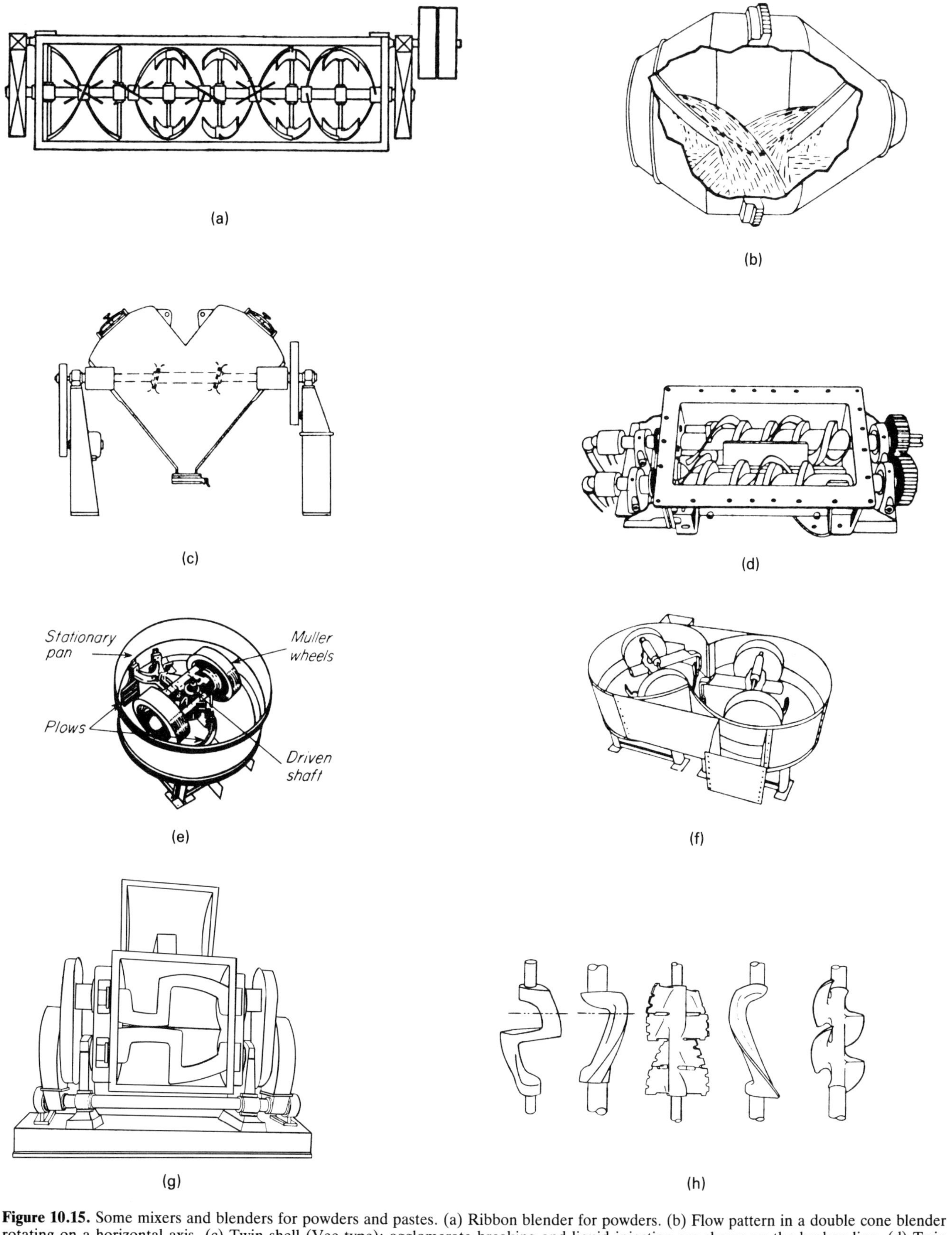

Figure 10.15. Some mixers and blenders for powders and pastes. (a) Ribbon blender for powders. (b) Flow pattern in a double cone blender rotating on a horizontal axis. (c) Twin shell (Vee-type); agglomerate breaking and liquid injection are shown on the broken line. (d) Twin rotor; available with jacket and hollow screws for heat transfer. (e) Batch muller. (f) Twin mullers operated continuously. (g) Double-arm mixer and kneader (*Baker–Perkins Inc.*). (h) Some types of blades for the double-arm kneader (*Baker–Perkins Inc.*).

Vol. 2, pp. 282–300) and a similar one for powder mixers (*loc. cit.*, pp. 301–311). Since this equipment industry has been quite stable, older books are still useful, notably those of Riegel (1953), Mead (1964), and particularly Kieser (1934–1939).

REFERENCES

1. R.S. Brodkey (Ed.), *Turbulence in Mixing Operations,* Academic, New York, 1975.
2. Chemineer Co. Staff, *Liquid Agitation,* Reprint of 12 articles from *Chemical Engineering,* 8 Dec. 1975–6 Dec. 1976.
3. D.S. Dickey, In *Handbook of Chemical Engineering Calculations,* (N.P. Chopey and T.G. Hicks Eds.), McGraw-Hill, New York, 1984.
4. S. Harnby, M.F. Edwards, and A.W. Nienow, *Mixing in the Process Industries,* Butterworths, Stoneham, MA, 1985.
5. A.J. Kieser, *Handbuch der chemisch-technischen Apparate,* Springer-Verlag, Berlin, 1934–1939.
6. W.J. Mead, *Encyclopedia of Chemical Process Equipment,* Reinhold, New York, 1964.
7. S. Nagata, *Mixing Principles and Applications,* Wiley, New York, 1975.
8. J.Y. Oldshue, *Fluid Mixing Technology,* McGraw-Hill, New York, 1983.
9. E.R. Riegel, *Chemical Process Machinery,* Reinhold, New York, 1953.
10. Z. Sterbacek and P. Tausk, *Mixing in the Chemical Industry,* Pergamon, New York, 1965.
11. J.J. Ulbrecht and G.K. Patterson, *Mixing of Liquids by Mechanical Agitation,* Gordon & Breach, New York, 1985.
12. V. Uhl and J.B. Gray (Eds.), *Mixing Theory and Practice,* Academic, New York, 1966, 1967, 2 vols.
13. *Ullmann's Encyclopedia of Chemical Technology,* Verlag Chemie, Weinheim, Germany, 1972, Vol. 2, pp. 249–311.

11

SOLID–LIQUID SEPARATION

Solid–liquid separation is concerned with mechanical processes for the separation of liquids and finely divided insoluble solids.

11.1. PROCESSES AND EQUIPMENT

Much equipment for the separation of liquids and finely divided solids was invented independently in a number of industries and is of diverse character. These developments have occurred without benefit of any but the most general theoretical considerations. Even at present, the selection of equipment for specific solid–liquid separation applications is largely a process of scale-up based on direct experimentation with the process material.

The nature and sizing of equipment depends on the economic values and proportions of the phases as well as certain physical properties that influence relative movements of liquids and particles. Pressure often is the main operating variable so its effect on physical properties should be known. Table 11.1 is a broad classification of mechanical processes of solid–liquid separation. Clarification is the removal of small contents of worthless solids from a valuable liquid. Filtration is applied to the recovery of valuable solids from slurries. Expression is the removal of relatively small contents of liquids from compressible sludges by mechanical means.

Whenever feasible, solids are settled out by gravity or with the aid of centrifugation. In dense media separation, an essentially homogeneous liquid phase is made by mixing in finely divided solids (less than 100 mesh) of high density; specific gravity of 2.5 can be attained with magnetite and 3.3 with ferrosilicon. Valuable ores and coal are floated away from gangue by such means. In flotation, surface active agents induce valuable solids to adhere to gas bubbles which are skimmed off. Magnetic separation also is practiced when feasible. Thickeners are vessels that provide sufficient residence time for settling to take place. Classifiers incorporate a mild raking action to prevent the entrapment of fine particles by the coarser ones that are to be settled out. Classification also is accomplished in hydrocyclones with moderate centrifugal action.

TABLE 11.1. Chief Mechanical Means of Solid–Liquid Separation

1. Settling
   - a. by gravity
     - i. in thickeners
     - ii. in classifiers
   - b. by centrifugal force
   - c. by air flotation
   - d. by dense media flotation
   - e. by magnetic properties
2. Filtration
   - a. on screens, by gravity
   - b. on filters
     - i. by vacuum
     - ii. by pressure
     - iii. by centrifugation
3. Expression
   - a. with batch presses
   - b. with continuous presses
     - i. screw presses
     - ii. rolls
     - iii. discs

Freely draining solids may be filtered by gravity with horizontal screens, but often filtration requires a substantial pressure difference across a filtering surface. An indication of the kind of equipment that may be suitable can be obtained by observations of sedimentation behavior or of rates of filtration in laboratory vacuum equipment. Figure 11.1 illustrates typical progress of sedimentation. Such tests are particularly used to evaluate possible flocculating processes or agents. Table 11.2 is a classification of equipment based on laboratory tests; test rates of cake formation range from several cm/sec to fractions of a cm/hr.

Characteristics of the performance of the main types of commercial SLS equipment are summarized in Table 11.3. The completeness of the removal of liquid from the solid and of solid from the liquid may be important factors. In some kinds of equipment residual liquid can be removed by blowing air or other gas through the cake. When the liquid contains dissolved substances that are undesirable in the filter cake, the slurry may be followed by

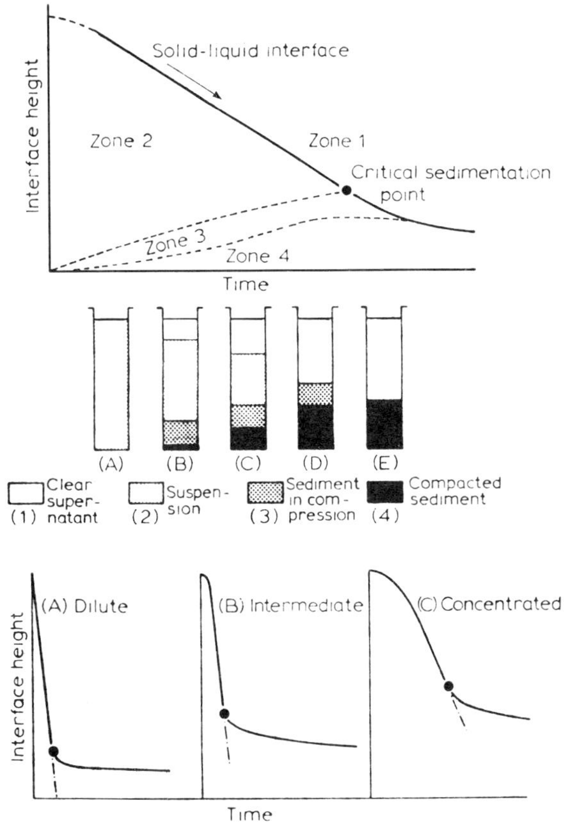

Figure 11.1. Sedimentation behavior of a slurry, showing loose and compacted zones (*Osborne, 1981*).

TABLE 11.2. Equipment Selection on the Basis of Rate of Cake Buildup

Process Type	Rate of Cake Buildup	Suitable Equipment
Rapid filtering	0.1–10 cm/sec	gravity pans; horizontal belt or top feed drum; continuous pusher type centrifuge
Medium filtering	0.1–10 cm/min	vacuum drum or disk or pan or belt; peeler type centrifuge
Slow filtering	0.1–10 cm/hr	pressure filters; disc and tubular centrifuges; sedimenting centrifuges
Clarification	negligible cake	cartridges; precoat drums; filter aid systems; sand deep bed filters

(Tiller and Crump, 1977; Flood, Parker, and Rennie, 1966).

pure water to displace the residual filtrate. Qualitative cost comparisons also are shown in this table. Similar comparisons of filtering and sedimentation types of centrifuges are in Table 11.19.

Final selection of filtering equipment is inadvisable without some testing in the laboratory and pilot plant. A few details of such work are mentioned later in this chapter. Figure 11.2 is an outline of a procedure for the selection of filter types on the basis of appropriate test work. Vendors need a certain amount of information before they can specify and price equipment; typical inquiry forms are in Appendix C. Briefly, the desirable information includes the following.

1. Flowsketch of the process of which the filtration is a part, with the expected qualities and quantities of the filtrate and cake.
2. Properties of the feed: amounts, size distribution, densities and chemical analyses.
3. Laboratory observations of sedimentation and leaf filtering rates.
4. Pretreatment options that may be used.
5. Washing and blowing requirements.
6. Materials of construction.

A major aspect of an SLS process may be conditioning of the slurry to improve its filterability. Table 11.4 summarizes common pretreatment techniques, and Table 11.5 lists a number of flocculants and their applications. Some discussion of pretreatment is in Section 11.3.

11.2. THEORY OF FILTRATION

Filterability of slurries depends so markedly on small and unidentified differences in conditions of formation and aging that no correlations of this behavior have been made. In fact, the situation is so discouraging that some practitioners have dismissed existing filtration theory as virtually worthless for representing filtration behavior. Qualitatively, however, simple filtration theory is directionally valid for modest scale-up and it may provide a structure on which more complete theory and data can be assembled in the future.

As filtration proceeds, a porous cake of solid particles is built up on a porous medium, usually a supported cloth. Because of the fineness of the pores the flow of liquid is laminar so it is represented by the equation

$$Q = \frac{dV}{dt} = \frac{A\Delta P}{\mu R}. \tag{11.1}$$

The resistance R is made up of those of the filter cloth R_f and that of the cake R_c which may be assumed proportional to the weight of the cake. Accordingly,

$$Q = \frac{dV}{dt} = \frac{A\Delta P}{\mu(R_f + R_c)} = \frac{A\Delta P}{\mu(R_f + \alpha cV/A)}, \tag{11.2}$$

α = specific resistance of the cake (m/kg),
c = wt of solids/volume of liquid (kg/m^3),
μ = viscosity (N sec/m^2)
P = pressure difference (N/m^2)
A = filtering surface (m^2)
V = volume of filtrate (m^3)
Q = rate of filtrate accumulation (m^3/sec).

R_f and α are constants of the equipment and slurry and must be evaluated from experimental data. The simplest data to analyze are those obtained from constant pressure or constant rate tests for which the equations will be developed. At constant pressure Eq. (11.2) is integrated as

$$\frac{A\Delta P}{\mu} t = R_f V + \frac{\alpha c}{2A} V^2 \tag{11.3}$$

and is recast into linear form as

$$\frac{t}{V/A} = \frac{\mu}{\Delta P} R_f + \frac{\mu \alpha c}{2\Delta P} \frac{V}{A}. \tag{11.4}$$

The constants R_f and α are derivable from the intercept and slope of the plot of t/V against V. Example 11.1 does this. If the constant pressure period sets in when $t = t_0$ and $V = V_0$, Eq. (11.4) becomes

$$\frac{t - t_0}{V - V_0} = \frac{\mu}{A\Delta P} R_f + \frac{\mu \alpha c}{2A^2 \Delta P}(V + V_0). \tag{11.5}$$

A plot of the left hand side against $V + V_0$ should be linear.

At constant rate of filtration, Eq. (11.2) can be written

$$Q = \frac{V}{t} = \frac{A\Delta P}{\mu(R_f + \alpha cV/A)} \tag{11.6}$$

and rearranged into the linear form

$$\frac{\Delta P}{Q} = \frac{\Delta P}{V/t} = \frac{\mu}{A} R_f + \frac{\mu \alpha c}{A^2} V. \tag{11.7}$$

The constants again are found from the intercept and slope of the linear plot of $\Delta P/Q$ against V.

After the constants have been determined, Eq. (11.7) can be employed to predict filtration performance under a variety of constant rate conditions. For instance, the slurry may be charged with a centrifugal pump with a known characteristic curve of output pressure against flow rate. Such curves often may be represented by parabolic relations, as in Example 11.2, where the data are fitted by an equation of the form

$$P = a - Q(b + cQ). \tag{11.8}$$

The time required for a specified amount of filtrate is found by integration of

$$t = \int_0^V dV/Q. \tag{11.9}$$

TABLE 11.3. Comparative Performance of SLS Equipment[a]

	Product Parameters			Feed Conditions Favoring Use			Equipment Characteristics			Direct Costs		
	Solids in Liquid Product	Liquid in Solid Product	Wash* Possibilities	Solids Concentration	Solids Density	Particle Size	Power	Space	Holdup	Initial	Operating	Maintenance
Filtration												
Vacuum drum filter	F	G	E[d]	high to med.	—	medium	high	medium	medium	high	high	medium
Disc filters	F	G	P to F	medium	—	fine	high	medium	medium	med. to high	high	medium
Horizontal filter	F	G	G to E[d]	high to med.	—	coarse	high	medium	medium	medium	high	medium
Precoat filter	E	P**	P to F**	very low	—	slimy	high to med.	medium	medium	high	very high	medium
Leaf (Kelly) filter	G to E[d]	F	F to G	low	—	fine, slimy	med. to low	medium	medium	medium	very high	medium
Sedimentation												
Thickener	G to E	P	P	medium	dense	medium	low	very high	very high	med. to low	low	very low
Clarifier	G	P	very P	low	med. dense	fine	very low	very high	very high	med. to low	low	very low
Classifier	P	P	P to F	medium	dense	coarse	low	high	high	med. to low	low	low
Centrifugation												
Disc	F to G	P	P	low to med.	medium	fine	high	low	low	high	high	high
Solid bowl	P	F	P to F	med. to high	medium	med. to fine	high	low	low	med. to high	high	high
Basket	P to F	E	E[d]	med. to high	—	coarse	high	low	low	medium	high	high
Liquid cyclones												
Large	P	P to F	P	low to med.	high	medium	med. to low	low	low	very low	medium	high
Small multiple	P to F	P	very P	low	med. to high	fine	med. to low	low	low	low	medium	medium
Screens	P	P to F	P	med. to high	—	coarse to med.	low	very low	very low	very low	medium	med. to high
Ultrafiltration	E	P to F	P	low	—	very fine	med. to high	high	high	high	high	very high

[a] P = Poor. F = Fair. G = Good. E = Excellent. * Decantation wash always possible. [d] Displacement wash feasible. ** Solids product contaminated by precoat material. (Purchas, 1981).

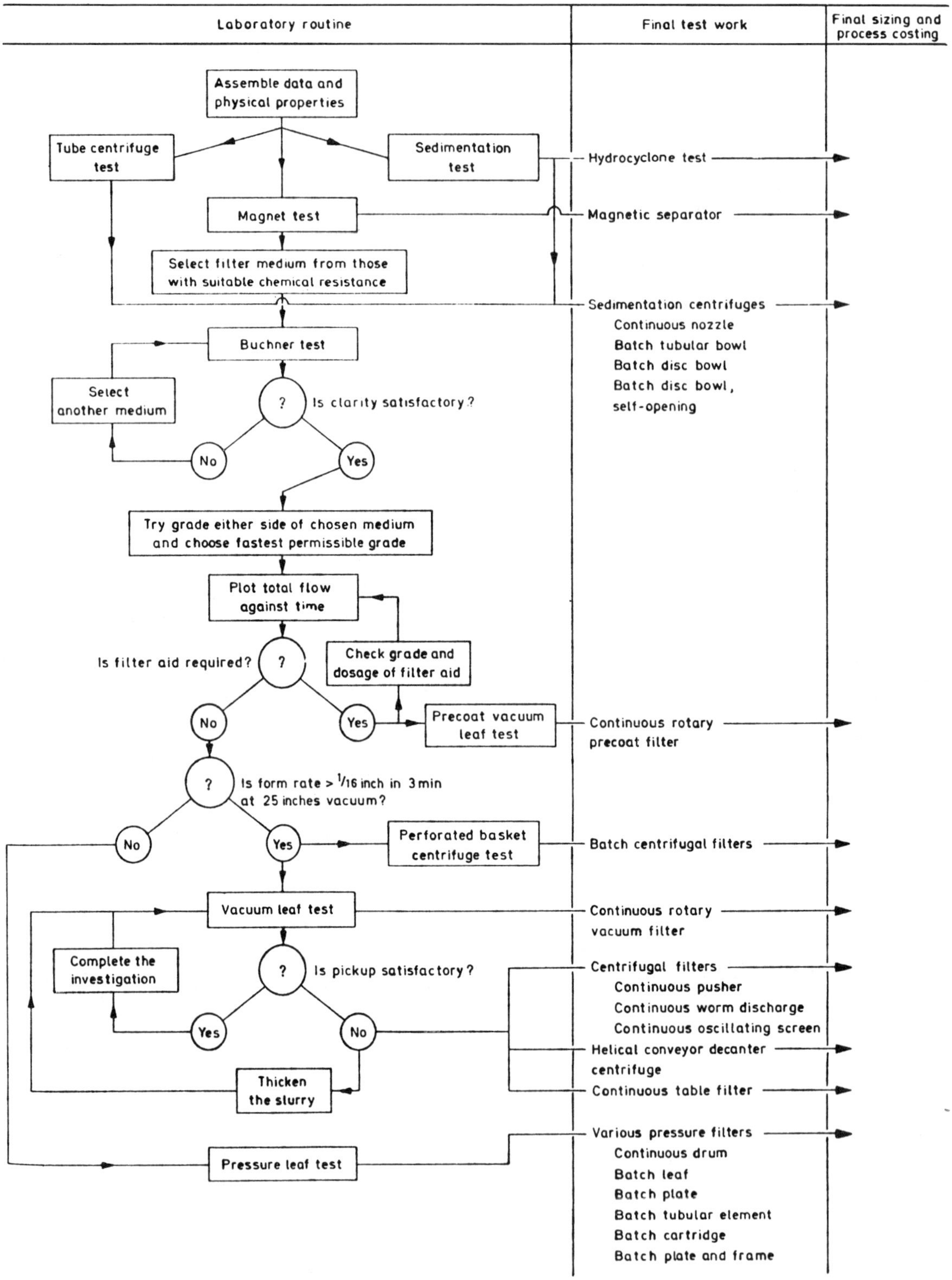

Figure 11.2. Experimental routine for aiding the selection of solid-liquid separation equipment (*Davies, 1965*).

TABLE 11.4. Action and Effects of Slurry Pretreatments

Action On	Technique	Effects
1. Liquid	1. heating 2. dilution with solvent	reduction of viscosity, thereby speeding filtration and settling rates and reducing cake moisture content
	3. degassing and stripping	prevents gas bubbles forming within the medium or cake and impeding filtration
2. Solid particles	1. coagulation by chemical additives	destabilizes colloidal suspensions, allowing particles to agglomerate into microflocs
	2. flocculation by natural or forced convection	microflocs are brought into contact with each other to permit further agglomeration into large flocs
	3. aging	size of individual particles increases, e.g., by crystal growth
3. Concentration of solids	1. increase by appropriate first-stage device such as settling tank, cyclone flotation cell or filter/thickener	rate of filtration increased, especially if initial concentration <2%
	2. classify to eliminate fines, using sedimentation or cyclone	rate of filtration increased and cake moisture content reduced
	3. add filter powder (e.g., diatomite) or other solids to act as 'body aid'	rate of filtration increased by more porous cake and possibly by high total solid concentration
4. Solid/liquid interaction	1. heat treatment, e.g., Porteus process involving pressure cooking 2. freeze/thaw 3. ultrasonics 4. ionized radiation	physical methods which condition sludge and induce coagulation and/or flocculation
	5. addition of wetting agents	reduces the interfacial surface tension, improves the draining characteristics of the cake, and decreases the residual moisture content

(Purchas, 1981).

TABLE 11.5. Natures and Applications of Typical Flocculants

Trade Name	Composition	Type or Mechanism	Typical Application	Normal Range of pH Effectiveness	Normal Effective Concentration	Approx. Price per lb[a]	Manufacturer
Alum	$Al_2(SO_4)_3 \cdot XH_2O$	electrolytic and coagulation	water treatment	5–10	15 ppm	2¢	inorganic chemical manufacturers
Ferric sulfate	$Fe_3(SO_4)XH_2O$	electrolytic coagulation	water treatment and chemical processing	any	5–100 ppm	2¢	inorganic chemical manufacturers
Sodium CMC	sodium carboxymethylcellulose	coagulation and bridging	mineral processing	3–9	0.03–0.5 lb/ton	50¢	Hercules, DuPont
Kelgin W	algins	coagulation and bridging	water treatment	4–11	up to 5 ppm	\$1.50	Kelco Co.
Separan	acrylamide polymer	bridging	chemical processing	2–10	0.2–10 ppm	\$1.00–\$2.00	Dow Chemical Co.
Fibrefloc	animal glue	electrolytic	waste treatment	1–9	5–30 ppm	18¢	Armour and Co.
Corn starch	corn starch	bridging	mineral processing	2–10	10 lb/ton	7¢	—
Polynox	polyethylene oxide	bridging	chemical processing	2–10	1–50 ppm	\$2.00	Union Carbide
Silica sol	activated silica sol	electrolytic coagulation	waste treatment	4–6	1–20 ppm	1.5¢ as sodium silicate	inorganic chemical manufacturers
Sodium aluminate	sodium aluminate	coagulation	water treatment	3–12	2–10 ppm	10¢	National Aluminate
Guar gum	guar gum	bridging	mineral processing	2–12	0.02–0.3 lb/ton	35¢	General Mills
Sulfuric acid	H_2SO_4	electrolytic	waste treatment	1–5	highly variable	1¢	inorganic chemical manufacturers

[a] 1966 prices, for comparison only.
(Purchas, 1981).

EXAMPLE 11.1
Constants of the Filtration Equation from Test Data

Filtration tests were performed on a $CaCO_3$ slurry with these properties:

$$C = 135 \text{ kg solid/m}^3 \text{ liquid},$$
$$\mu = 0.001 \text{ N sec/m}^2.$$

The area of the filter leaf was 500 cm^2. Data were taken of the volume of the filtrate (L) against time (sec) at pressures of 0.5 and 0.8 bar. The results will be analyzed for the filtration parameters:

		0.5 bar		0.8 bar	
(L)	V/A	t	t/(V/A)	t	t/(V/A)
0.5	0.01	6.8	680	4.8	480
1	0.02	19.0	950	12.6	630
1.5	0.03	36.4	1213	22.8	760
2	0.04	53.4	1335	35.6	890
2.5	0.05	76.0	1520	50.5	1010
3	0.06	102.0	1700	69.0	1150
3.5	0.07	131.2	1874	88.2	1260
4	0.08	163.0	2038	112.0	1400
4.5	0.09			—	—
5	0.10			165.0	1650

The units of V/A are m^3/m^2. Equation (11.2) is

$$\frac{d(V/A)}{dt} = \frac{\Delta P}{\mu(R_f + \alpha CV/A)},$$

whose integral may be written

$$\frac{R_f}{\Delta P/\mu} + \frac{\alpha C}{2(\Delta P/\mu)} \frac{V}{A} = \frac{t}{V/A}.$$

Intercepts and slopes are read off the linear plots. At 0.5 bar,

$$\Delta P/\mu = 0.5(10^5)/0.001 = 0.5(10^8),$$
$$R_f = 600\Delta P/\mu = 3.0(10^{10}) \quad \text{m}^{-1},$$
$$\alpha = [18{,}000(2)/C]\Delta P/\mu = 36{,}000(0.5)(10^8)/135$$
$$= 1.333(10^{10}) \quad \text{m/kg}.$$

At 0.8 bar,

$$\Delta P/\mu = 0.8(10^8),$$
$$R_f = 375(0.8)(10^8) = 3(10^{10}) \quad \text{m}^{-1},$$
$$\alpha = 12{,}750(2)(0.8)(10^8)/135 = 1.511(10^{10}) \quad \text{m/kg}.$$

Fit the data with Almy–Lewis equation, Eq. (11.24),

$$\alpha = kp^n,$$
$$n = \frac{\ln(\alpha_1/\alpha_2)}{\ln(P_1/P_2)} = \frac{\ln(1.511/1.333)}{\ln(0.8/0.5)} = 0.2664,$$
$$k = 1.511(10^{10})/0.8^{0.2664} = 1.604(10^{10}),$$
$$\therefore \alpha = 1.604(10^{10})P^{0.2664}, \quad \text{m/kg, } P \text{ in bar}.$$

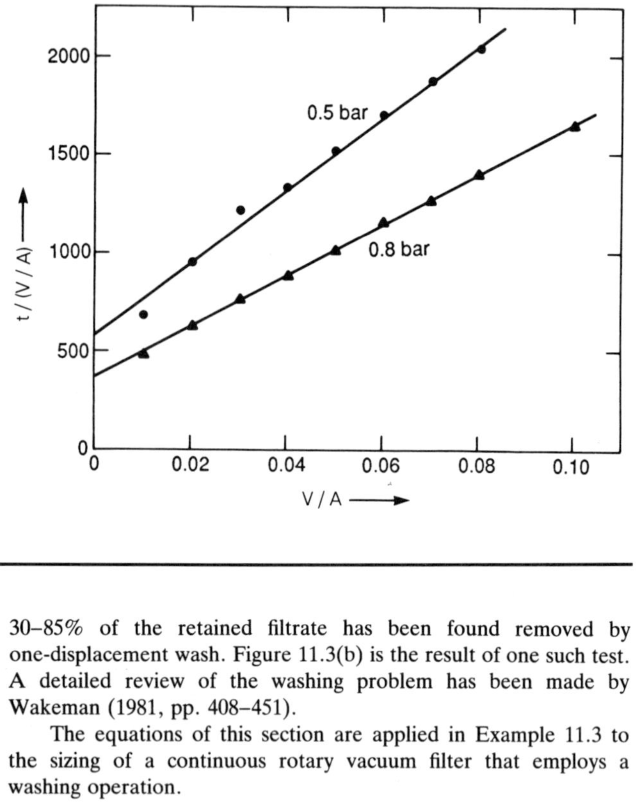

Basic filtration Eq. (11.2) is solved for the amount of filtrate,

$$V = \frac{A}{\mu c\alpha}\left(\frac{A\Delta P}{Q} - \mu R_f\right). \tag{11.10}$$

Equations (11.8) and (11.10) are solved simultaneously for ΔP and Q at specified values of V and the results tabulated so:

V	ΔP	Q	1/Q	t
0	—	—	—	0
—	—	—	—	—
V_{final}	—	—	—	t_{final}

Integration is accomplished numerically with the Simpson or trapezoidal rules. This method is applied in Example 11.2.

When the filtrate contains dissolved substances that should not remain in the filter cake, the occluded filtrate is blown out; then the cake is washed by pumping water through it. Theoretically, an amount of wash equal to the volume of the pores should be sufficient, even without blowing with air. In practice, however, only 30–85% of the retained filtrate has been found removed by one-displacement wash. Figure 11.3(b) is the result of one such test. A detailed review of the washing problem has been made by Wakeman (1981, pp. 408–451).

The equations of this section are applied in Example 11.3 to the sizing of a continuous rotary vacuum filter that employs a washing operation.

COMPRESSIBLE CAKES

Resistivity of filter cakes depends on the conditions of formation of which the pressure is the major one that has been investigated at length. The background of this topic is discussed in Section 11.3, but here the pressure dependence will be incorporated in the filtration equations. Either of two forms of pressure usually is taken,

$$\alpha = \alpha_0 P^n \tag{11.11}$$

or

$$\alpha = \alpha_0(1 + kP)^n. \tag{11.12}$$

EXAMPLE 11.2
Filtration Process with a Centrifugal Charge Pump
A filter press with a surface of $50\,\text{m}^2$ handles a slurry with these properties:

$$\mu = 0.001\,\text{N sec/m}^2,$$
$$C = 10\,\text{kg/m}^3,$$
$$\alpha = 1.1(10^{11})\,\text{m/kg},$$
$$R_f = 6.5(10^{10})\,\text{m}^{-1}.$$

The feed pump is a centrifugal with a characteristic curve represented by the equation

$$\Delta P = 2 - Q(0.00163Q - 0.02889), \quad \text{bar} \tag{1}$$

with Q in m^3hr. Find (a) the time required to obtain $50\,\text{m}^3$ of filtrate; (b) the volume, flow rate, and pressure profiles. Equation (11.2) of the text solved for V becomes

$$V = \frac{A}{\alpha\mu C}\left(\frac{A\Delta P}{Q} - \mu R_f\right) = \frac{50}{1.1(10^9)}\left[\frac{50(10^5)\Delta P}{Q/3600} - 6.5(10^7)\right]$$
$$= 818.1\left(\frac{\Delta P}{Q} - 0.0036\right). \tag{2}$$

Equations (1) and (2) are solved simultaneously to obtain the tabulated data. The time is found by integration with the trapezoidal rule:

$$t = \int_0^{50} \frac{dV}{Q}$$

V	ΔP	Q	t (hr)
0	0.1576	43.64	0
10	0.6208	39.27	0.24
20	0.9896	35.29	0.51
30	1.2771	31.71	0.81
40	1.4975	28.53	1.14
50	1.6648	25.72	1.51

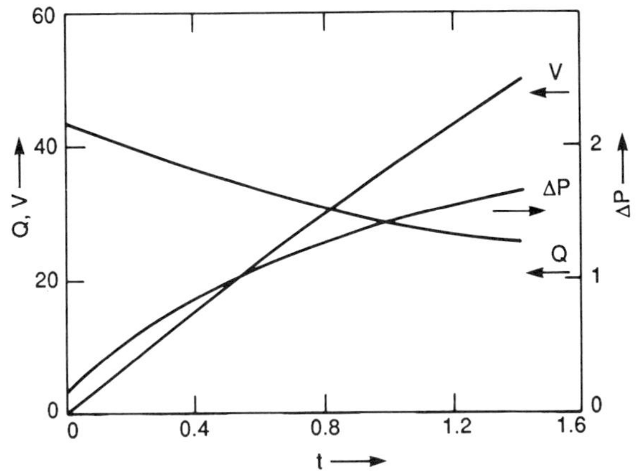

The first of these does not extrapolate properly to resistivity at low pressures, but often it is as adequate as the more complex one over practical ranges of pressure.

Since the drag pressure acting on the particles of the cake varies from zero at the face to the full hydraulic pressure at the filter cloth, the resistivity as a function of pressure likewise varies along the cake. A mean value is defined by

$$\frac{1}{\bar{\alpha}} \simeq \left(\frac{1}{\alpha}\right)_{\text{mean}} = \frac{1}{\Delta P_c}\int_0^{\Delta P_c} \frac{dP}{\alpha} \simeq \frac{1}{\Delta P}\int_0^{\Delta P} \frac{dP}{\alpha}, \tag{11.13}$$

where ΔP_c is the pressure drop through the cake alone. In view of the roughness of the usual correlations, it is adequate to use the overall pressure drop as the upper limit instead of the drop through the cake alone.

With Eq. (11.12) the mean value becomes

$$\bar{\alpha} = \frac{\alpha_0 k(1-n)\Delta P}{(1+k\Delta P)^{1-n} - 1}. \tag{11.14}$$

The constants α_0, k, and n are determined most simply in compression-permeability cells as explained in Section 11.4, but those found from filtration data may be more appropriate because the mode of formation of a cake also affects its resistivity. Equations (11.14) and (11.2) together become

$$\frac{d(V/A)}{dt} = \frac{\Delta P}{\mu}\left[R_f + \frac{\alpha_0 ck(1-n)\Delta P}{(1+k\Delta P)^{n-1} - 1}\frac{V}{A}\right]^{-1}, \tag{11.15}$$

which integrates at constant pressure into

$$\frac{2t}{V/A} = \frac{2\mu}{\Delta P}R_f + \frac{\alpha_0 ck\mu(1-n)}{(1+k\Delta P)^{1-n} - 1}(V/A) \tag{11.16}$$

The four unknown parameters are α_0, k, n, and R_f. The left-hand side should vary linearly with V/A. Data obtained with at least three different pressures are needed for evaluation of the parameters, but the solution is not direct because the first three parameters are involved nonlinearly in the coefficient of V/A. The analysis of constant rate data likewise is not simple.

The mean resistivity at a particular pressure difference can be evaluated from a constant pressure run. From three such runs—ΔP_1, ΔP_2, and ΔP_3—three values of the mean resistivity—$\bar{\alpha}_1$, $\bar{\alpha}_2$, and $\bar{\alpha}_3$—can be determined with Eq. (11.2) and used to find the three constants of the expression for an overall mean value,

$$\bar{\alpha} = \alpha_0(1+k\Delta P)^n, \tag{11.17}$$

which is not the same as Eq. (11.12) but often is as satisfactory a representation of resistivity under practical filtration conditions. Substituting Eq. (11.17) into Eq. (11.2), the result is

$$\frac{d(V/A)}{dt} = \frac{\Delta P}{\mu[R_f + \alpha_0 c(1+k\Delta P)^n(V/A)]}. \tag{11.18}$$

Integration at constant pressure gives the result

$$\frac{\alpha_0 c\mu(1+k\Delta P)^n}{2\Delta P}\frac{V}{A} + \mu R_f/\Delta P = \frac{t}{V/A}. \tag{11.19}$$

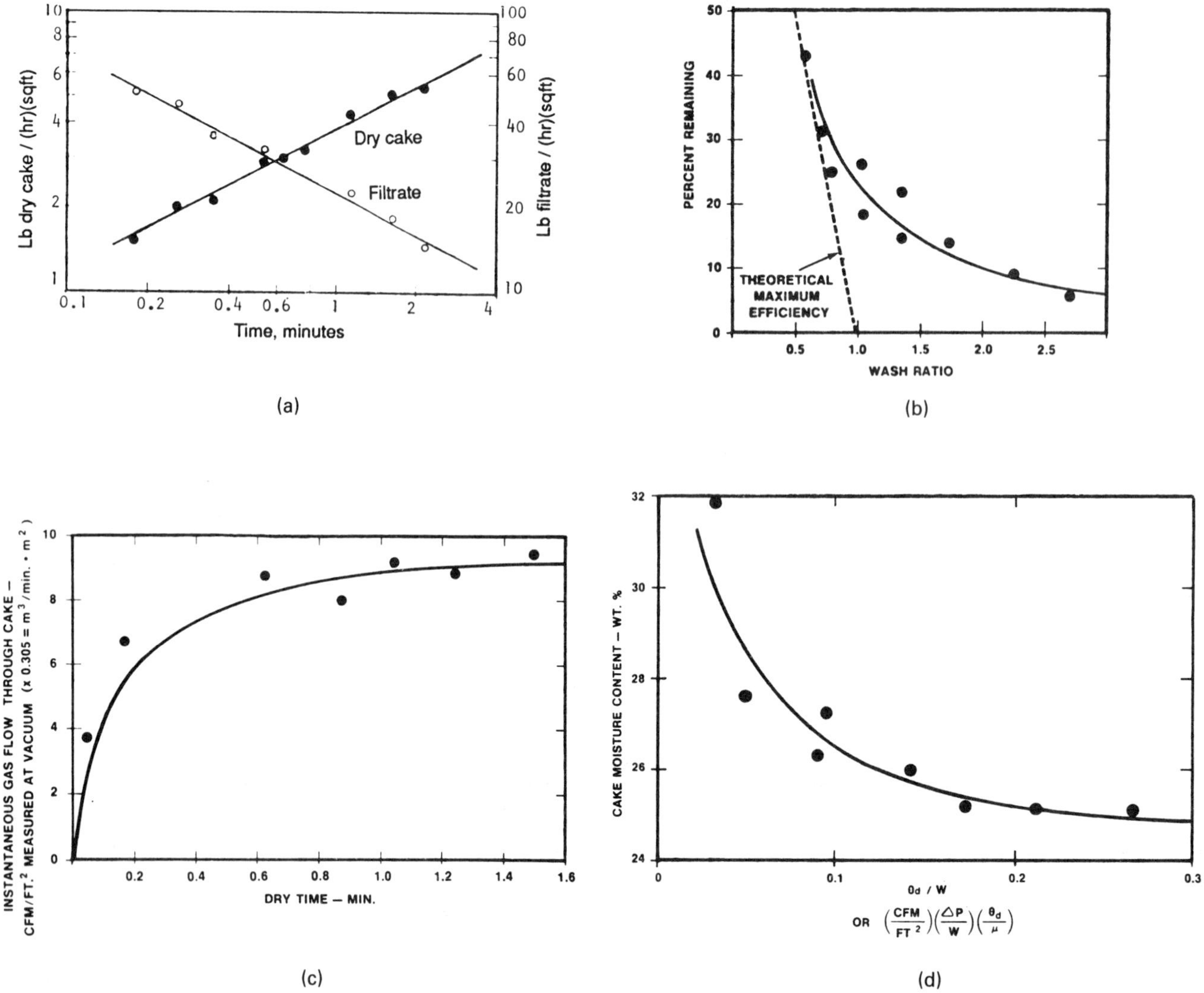

Figure 11.3. Laboratory test data with a vacuum leaf filter. (a) Rates of formation of dry cake and filtrate. (b) Washing efficiency. (c) Air flow rate vs. drying time. (d) Correlation of moisture content with the air rate, pressure difference ΔP, cake amount W lb/sqft, drying time θ_d min and viscosity of liquid (*Dahlstrom and Silverblatt, 1977*).

Example 11.3
Rotary Vacuum Filter Operation

A TiO_2 slurry has the properties

$$c = 200\ \text{kg solid/m}^2\ \text{liquid},$$
$$\rho_s = 4270\ \text{kg/m}^3,$$
$$\mu = 0.001/3600\ \text{N hr/m}^2,$$
$$\alpha = 1.6(E12)\ \text{m/kg (item 4 of Fig. 11.2)},$$
$$\varepsilon = 0.6.$$

Cloth resistance is $R_f = 1(E10)\ \text{m}^{-1}$. Normal peripheral speed is about 1 m/min. Filtering surface is 1/3 of the drum surface and washing surface is 1/6 of the drum surface. The amount of wash equals the pore space of the cake. The cake thickness is to be limited to 1 cm. At suitable operating pressures, find the drum speed in rph and the drum diameter:

$$\text{cake thickness} = 0.01\ \text{m} = \frac{C}{\rho_s(1-\varepsilon)}\frac{V_f}{A}$$
$$= \frac{200}{4270(0.4)}\frac{V_f}{A},$$
$$\frac{V_f}{A} = \frac{0.01(4270)(0.4)}{200} = 0.0854\ \text{m}^3/\text{m}^2,$$
$$\text{wash liquid} = \text{pore volume}$$
$$= 0.01(0.6) = 0.006\ \text{m}^2\ \text{m}^2. \quad (1)$$

With the pressure difference in bar,

$$\frac{d(V/A)}{dt} = \frac{10^5 \Delta P_b}{(0.001/3600)[10^{10} + 160(10^{10})V/A]}$$
$$= \frac{36\Delta P_b}{1 + 160V/A}. \quad (2)$$

Example 11.3—*(continued)*

The integral at constant pressure is

$$80(V_f/A)^2 + V_f/A = 36\Delta P_b t_f. \tag{3}$$

With $V_f/A = 0.0854$,

$$\Delta P_b t_f = 0.01858,$$
$$t_f = 0.01858/\Delta P_b = 1/3\dot{n}_f \tag{4}$$
$$\dot{n}_f = 17.94\Delta P_b, \tag{5}$$

where $\dot{n}_f$ is the rph speed needed to make the 1 cm thick cake.

From Eq. (2) the washing rate is

$$r_w = \frac{36\Delta P_b}{1 + 160(0.0854)} = 2.455\Delta P_b. \tag{6}$$

Washing time:

$$t_w = \frac{0.006}{2.455\Delta P_b} = \frac{0.00244}{\Delta P_b} \geq \frac{1}{\dot{n}_w}, \tag{7}$$

$$\dot{n}_w \leq 68.3\Delta P_b \tag{8}$$

Comparing (5) and (8), it appears that an rph to meet the filtering requirements is 68.3/17.94 = 3.8 times that for washing and is the controlling speed.

With a peripheral speed of 60 m/hr

$$60 = \pi D n,$$
$$D = 60/\pi n = 19.1/\dot{n}. \tag{9}$$

The parameters at several pressures are

ΔP_b (bar)	0.2	0.4	0.6	0.8
$\dot{n}$ (rph)	3.59	7.18	10.76	14.35
D (m)	5.3	2.66	1.78	1.33

If the peripheral speed were made 1.22 m/min, a drum 1.0 m dia would meet the requirements with $\Delta P = 0.8$ bar. Another controllable feature is the extent of immersion which can be made greater or less than 1/3. Sketches of a rotary vacuum filter are in Figure 11.12.

Eq. (11.19) could be written in terms of $\bar{\alpha}$ from Eq. (11.17) and would then have the same form as Eq. (11.2), but with only R_f as a parameter to be found from a single run at constant pressure. In Example 11.1, the mean resistivity is found from the simpler equation

$$\bar{\alpha} = \alpha_0(\Delta P)^n. \tag{11.20}$$

Analysis of the filtration of a compressible material is treated in Example 11.4.

11.3. RESISTANCE TO FILTRATION

The filtration equation

$$\frac{Q}{A} = \frac{\Delta P}{\mu(R_f + \alpha c V/A)} \tag{11.2}$$

considers the overall resistance to flow of filtrate to be made up of contributions from the filter medium R_f, and from the cake with specific resistance α.

FILTER MEDIUM

In practice, a measured R_f includes the effects of all factors that are independent of the amount of the cake; in a plate-and-frame press, for instance, piping and entrance and exit losses will be included, although most of the resistance usually is due to the medium itself. Aging and the resulting increase in resistance is a recognized behavior, particularly of media made of fibers. Particles are gradually occluded in the media so thoroughly that periodic cleaning cannot restore the original condition. The degree of penetration of the medium depends on the porosity, the pore sizes, particles sizes, and velocity. Normally R_f is found to depend on the operating pressure; on plots like those of Example 11.1, the two intercepts may correspond to different values of R_f at the two pressures.

Data for some filter media are shown in Table 11.6. Although these porosities and permeabilities are of unused materials, the relative values may be useful for comparing behaviors under filtration conditions. Permeability K_p normally is the property reported rather than the resistivity that has been discussed here. It is defined by the equation

$$Q/A = K_p \Delta P/\mu L, \tag{11.21}$$

where L is the thickness. The relation to the resistivity is

$$R_f = L/K_p. \tag{11.22}$$

Thus the filtration resistivity of the medium includes its thickness. Typical measured values of R_f are of the order of $10^{10}\,\text{m}^{-1}$; for comparison, the fine filter sheet of Table 1.6, assuming it to be 1 mm thick, has $L/K_p = 0.001/0.15(10^{-12}) = 0.7(10^{10})\,\text{m}^{-1}$.

CAKE RESISTIVITY

A fundamental relation for the flow resistance of a bed of particles is due to Kozeny (*Ber. Wien. Akad.* **135a,** 1927, 271–278):

$$\alpha = K s_0^2(1-\varepsilon)/\varepsilon^3, \tag{11.23}$$

K = approximately 5 at low porosities,
s_0 = specific surface of the particles,
ρ_s = density of the particles,
ε = porosity, volume voids/volume of cake.

Because the structure of a cake is highly dependent on operating conditions and its history, the Kozeny equation is only of qualitative value to filtration theory by giving directional effects.

At increasing pressures, the particles or aggregates may be distorted and brought closer together. The rate of flow also may affect the structure of a cake: at low rates a loose structure is formed, at higher ones fine particles are dragged into the previously formed bed. The drag pressure at a point in a cake is the difference between the pressure at the filter medium and the pressure loss due to friction up to that point. As the drag pressure at a distance from the filter cloth increases, even at constant filtering pressure, the porosity and resistance adjust themselves continuously. Figure 11.4(a) shows such effects of slurry concentration and filtering rates

EXAMPLE 11.4
Filtration and Washing of a Compressible Material
A kaolin slurry has the properties

$$c = 200 \text{ kg solid/m}^3 \text{ filtrate},$$
$$\mu = 0.001 \text{ N sec/m}^2,\ 2.78(E-7) \text{ N hr/m}^2,$$
$$\rho_s = 200 \text{ kg/m}^3,$$
$$\alpha = 87(E10)(1 + P/3.45)^{0.7} \text{ m/kg with } P \text{ in bar},$$
$$\varepsilon = 1 - 0.460(1 + P/3.45)^{0.12}.$$

The equations for α and ε are taken from Table 11.8.

Filtration will proceed at a constant rate for 15 min, the pressure will rise to 8 bar and filtration will continue at this pressure until the end of the operation. Filter cloth resistance is $R_f = 1(10^{10})\ \text{m}^{-1}$. The down time per batch is 1 hr.

a. Find the maximum daily production of filtrate.
b. The filtrate will be blown and then washed with a volume of water equal to the pore space of the cake. Find the maximum daily production of filtrate under these conditions.

Part (a)
Basis 1 m^2 of filtering surface. At $P = 8$ bar, or $8(10^5)$ Pa

$$\alpha = 87(10^{10})(1 + 8/3.45)^{0.7} = 2.015(10^{12}) \text{ m/kg},$$
$$\varepsilon = 1 - 0.46(1 + 8/3.45)^{0.12} = 0.47,$$
$$\mu c\alpha = (0.001/3600)(200)(2.015)(10^{12}) = 1.12(10^8) \text{ N hr/m}^4.$$

The filtration equation (11.2) is

$$\frac{dV}{dt} = \frac{A\Delta P}{\mu(R_f + \alpha CV/A)} = \frac{\Delta P}{(0.001/3600)[10^{10} + 2.015(10^{12})(200)V]}$$
$$= \frac{\Delta P}{2780 + 1.12(10^8)V}.$$

The rate when $t = 0.25h$ and $\Delta P = 8(10^5)$ Pa,

$$Q = \frac{8(10^5)}{2780 + 1.12(10^8)Qt} = \frac{8(10^5)}{2780 + 0.28(10^8)Q}$$
$$= 0.1691 \text{ m}^3/\text{m}^2 \text{ hr}.$$

The amount of filtrate at this time is

$$V_0 = Qt = 0.1691(0.25) = 0.0423 \text{ m}^3.$$

The integral of the rate equation at constant P is

$$2780(V_f - 0.0423) + 0.56(10^8)(V_f^2 - (0.00423)^2]$$
$$= 8(10^5)(t_f - 0.25).$$

Filtering period is

$$t_f = 0.25 + 0.0035(V_f - 0.0423) + 70.0(V_f^2 - 0.0018).$$

Daily production rate,

$$R_d = (\text{no of batches/day})(\text{filtrate/batch})$$
$$= \frac{24V_f}{t_d + t_f} = \frac{24V_f}{1 + t_f},\ \text{m}^3/(\text{m}^2)(\text{day})$$
$$= \frac{24V_f}{1.25 + 0.0035(V_f - 0.0423) + 70(V_f^2 - 0.0018)}.$$

The tabulation shows that R_d is a max when $V_f = 0.127$.

V_f	t_f	R_d
0.12		1.3507
0.126		1.3526
0.127	1.2533	1.3527 (max)
0.128		1.3526
0.129		1.3525
0.130		1.3522

Part (b)

$$\text{Amount of wash liquid} = \frac{cV_f\varepsilon}{\rho_s(1-\varepsilon)} = \frac{200(0.47)}{2500(0.53)} = 0.0709V_f,$$

wash rate = filtering rate at the conclusion of the filtration

$$= \frac{\Delta p}{\mu(R_f + \alpha cV_f)} = \frac{8(10^5)}{2780 + 1.12(10^8)V_f},\quad \text{m}^3/\text{hr},$$
$$t_w = \text{wash time} = \frac{0.709V_f[2780 + 1.12(10^8)V_f]}{8(10^5)}$$
$$= V_f(0.000246 + 9.926V_f),$$
$$R_d = \frac{24V_f}{1 + t_f + t_w}$$
$$= \frac{24V_f}{[1 + 0.0035(V - 0.0423) + 7010(V_f^2 - 0.0018) + V_f(0.000246 + 9.926V_f)]}.$$

The optimum operation is found by trial:

$$V_f = 0.105,$$
$$t_f = 1.0805,$$
$$t_w = 0.1095,$$
$$R_d = 1.1507 \text{ (max), daily production rate.}$$

on the parameters of the correlating equation

$$\alpha = \alpha_0(\Delta P)^n. \tag{11.24}$$

The measurements were obtained with a small filter press. Clearly, the resistivity measured at a particular rate is hardly applicable to predicting performance at another rate or at constant pressure.

COMPRESSIBILITY–PERMEABILITY (CP) CELL MEASUREMENTS

The probable success of correlation of cake resistivity in terms of all the factors that have been mentioned has not been great enough to have induced any serious attempts of this nature, but the effect of pressure has been explored. Although the α's can be deduced from

TABLE 11.6. Porosities and Permeabilities of Some Filter Media

Porosity (%)	
Wedge wire screen	5–10
Perforated sheet	20
Wire mesh:	
Twill weave	15–25
Square	30–35
Porous plastics, metals, ceramics	30–50
Crude kieselguhr	50–60
Porous ceramic, special	70
Membranes, plastic foam	80
Asbestos/cellulose sheets	80
Refined filter aids (diatomaceous earth expanded perlite)	80–90
Paper	60–95
Scott plastic foam	97
Permeability, $10^{12}K_p$ (m^2) (compare Eq. (11.22))	
Filter aids	
Fine	0.05–0.5
Medium	1–2
Coarse	4–5
Cellulose fibre pulp	1.86
Cellulose fibre + 5% asbestos	0.34
Filter sheets	
Polishing	0.017
Fine	0.15
Clarifying	1.13
Sintered metal	
3 μm pore size	0.20
8 μm pore size	1.0
28 μm pore size	7.5
75 μm pore size	70

(Purchas, 1981).

filtration experiments, as done in Example 11.1, a simpler method is to measure them in a CP cell as described briefly later in this chapter. Equation (11.24) for the effect of pressure was proposed by Almy and Lewis (1912). For the materials of Figure 1.2(b), for instance, it seems to be applicable over at least moderate stretches of pressure. Incidentally, these resistances are not represented well by the Kozeny porosity function $(1-\varepsilon)/\varepsilon^3$; for substance 6, the ratio of resistivities at 100 and 1 psia is 22 and the ratio of the porosity functions is 2.6. The data of Table 11.7 also show a substantial effect of pressure on resistivity.

Since the drag pressure varies along the cake as a result of friction, porosity and resistivity also will vary with position. Figure 11.5 shows such data at three different overall pressures. The axial profile of the normalized pressure, P_{local}/P_{face}, appears to be a unique function of fractional distance along the cake, independent of the filtering pressure. The resistivity will vary along the cake just as the porosity does. As the cake builds up, moreover, the drag pressure, porosity, and resistivity at a particular distance from the filter medium also will vary. Consequently, since the resistivity does not necessarily change linearly with position, any mean value also is likely to vary as the cake builds up. Thus, in the filtration equation even a mean value of α has to be expressed as a function of P and V. The proper mathematical representation of a filtration process is by means of an integro-differential equation with a moving boundary (the face of the cake). Such an analysis was made by Wakeman (1978) and a similar one by Tiller, Crump, and Ville (1979). At present, unfortunately, such a mathematical approach to filtration problems is more of academic than practical value. One of the factors that is not taken into account is the effect of flow rate on the formation and stability of loose cake structures; such behavior normally is not reproducible.

ANOTHER FORM OF PRESSURE DEPENDENCE

Equation (11.24) cannot be entirely valid because it predicts zero resistivity at zero pressure, whereas cakes do have structures and significant resistivities even at minimal operating pressures. Modified Eq. (11.12) is extrapolatable, and is rewritten here as

$$\alpha = \alpha_0(1+kP)^n \tag{11.25}$$

with a similar one for porosity

$$\varepsilon = 1-(1-\varepsilon_0)(1+kP)^n. \tag{11.26}$$

Some data fitted to these equations by Tiller et al. (1979) are in Table 11.10; here the constant k is the same for both α and ε, although this is not necessarily generally the case. Unfortunately, these data show that the parameters are not independent of the pressure range. Apparently the correlation problem has not been solved. Perhaps it can be concluded that insofar as the existing filtration theory is applicable to real filtering behavior, the approximation of Almy and Lewis may be adequate over the moderate ranges or pressures that are used commonly, somewhere between 0.5 and 5 atm.

PRETREATMENT OF SLURRIES

Since the sizes of particles and agglomerates of the slurry are a main determinant of a rate of filtration, any methods of influencing these sizes are of great practical value. For example, Figures 1.2(b) and (c) show $CaCO_3$ and TiO_2 each to be precipitated at two different values of pH with resultant great differences in resistivity and porosity. At 10 psia, for instance, the resistivities of the two $CaCO_3$'s are in the ratio of 5, with corresponding differences in rate of filtration. Pretreatment of a slurry to enhance coagulation and particle growth is an important aspect of filter process design. Another method of long standing for improving filtration behavior is the formation of an open cake structure by addition of relatively large and rigid particles of a filter aid. The common methods of pretreatment are listed in Table 11.4, and some chemical flocculants that are of practical value are described in Table 11.5. These effects cannot be predicted safely and must be measured.

11.4. THICKENING AND CLARIFYING

When dilute slurries are encountered on a large scale, it is more economical to concentrate them before filtering. This is accomplished by sedimentation or thickening in tanks for an appropriate period. Typical designs of thickeners are sketched in Figure 11.6. The slurry is introduced at the top center, clear liquid overflows the top edge, whereas the solids settle out and are worked gradually towards the center with slowly rotating rakes towards the discharge port at the bottom center. The concentrated slurry then is suitable for filtration or other further processing. Clarifiers are similar devices, primarily for recovering clear liquids from dilute suspensions. Some characteristics of sedimentation equipment are given in Table 11.3 and typical applications are listed in Table 11.9 and 14.7. Sedimentation rates often are assisted by addition of flocculating agents, some of which are listed in Table 11.5. Specifically, pilot plant testing is advisable when

1. The expecting filtering area is expected to be substantial, measured in tens of m^2.
2. Cake washing is critical.

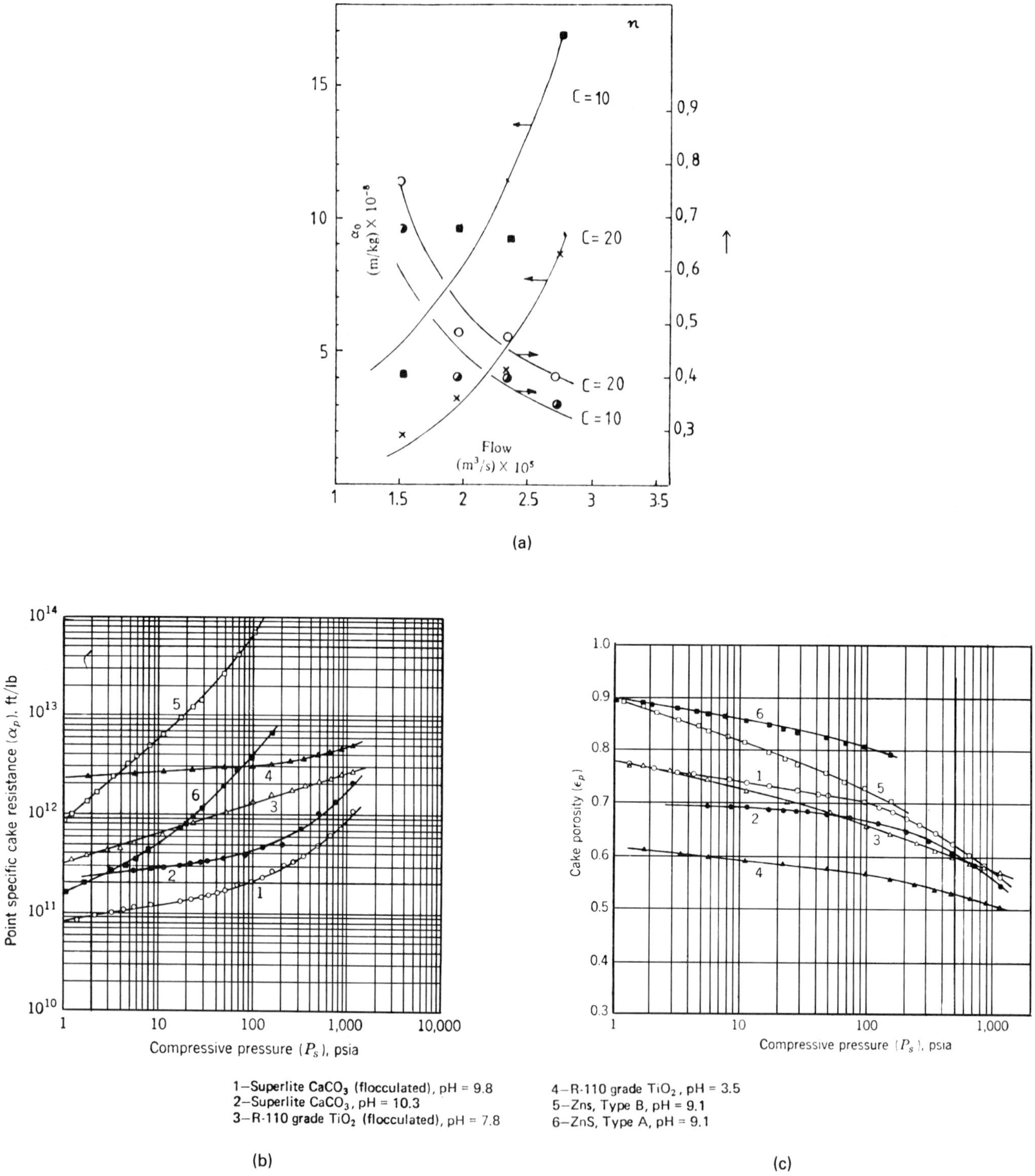

Figure 11.4. Data of compressibilities and porosities of filter cakes. (a) Parameters of the correlation $\alpha = \alpha_0(\Delta P)^n$ for resistivity of $CaSiO_3$ filter cakes at two rates and two concentrations (*Rushton and Katsoulas, 1984*). (b) Resistivity as a function of pressure measured in a compressibility-permeability (CP) cell [*Grace,* Chem. Eng. Prog. **49,** *303, 367, 427 (1953)*]. (c) Porosity as a function of pressure for the same six materials (*Grace, loc. cit.*).

3. Cake drying is critical.
4. Cake removal may be a problem.
5. Precoating may be needed.

11.5. LABORATORY TESTING AND SCALE-UP

Laboratory filtration investigations are of three main kinds:

1. observation of sedimentation rates;
2. with small vacuum or pressure leaf filters;
3. with pilot plant equipment of the types expected to be suitable for the plant.

Sedimentation tests are of value particularly for rapid evaluation of the effects of aging, flocculants, vibration, and any other variables that conceivably could affect a rate of filtration. The results may suggest what kinds of equipment to exclude from further consideration and what kind is likely to be worth investigating. For instance, if sedimentation is very rapid, vertical leaves are excluded, and top feed drums or horizontal belts are indicated; or it may be indicated that the slurry should be preconcentrated in a thickener before going to filtration. If the settling is very slow, the use of filter aids may be required, etc. Figure 11.1 illustrates typical sedimentation behavior. Figure 11.2 summarizes an experimental routine.

Vacuum and pressure laboratory filtration assemblies are shown in Figure 11.7. Mild agitation with air sometimes may be preferable to the mechanical stirrer shown, but it is important that any agglomerates of particles be kept merely in suspension and not broken up. The test record sheet of Figure 11.8 shows the kind of data that normally are of interest. Besides measurements of filtrate and cake amounts as functions of time and pressure, it is desirable to test washing rates and efficiencies and rates of moisture removal with air blowing. Typical data of these kinds are shown in Figure 11.3. Detailed laboratory procedures are explained by Bosley (1977) and Dahlstrom and Silverblatt (1977). Test and scale-up procedures for all kinds of SLS equipment are treated in the book edited by Purchas (1977).

Before any SLS equipment of substantial size is finally selected, it is essential to use the results of pilot plant tests for guidance. Although many vendors are in a position to do such work, pilot equipment should be used at the plant site where the slurry is made. Because slurries often are unstable, tests on shipments of slurry to the vendors pilot plant may give misleading results. It may be possible to condition a test slurry to have a maximum possible resistivity, but a plant design based on such data will have an unknown safety factor and may prove uneconomical.

TABLE 11.7. Specific Resistances of Some Filter Cakes

Material	Filtration Pressure psi	Resistance SI Units, m/kg
High grade kieselguhr	—	1.64×10^{9}
Ordinary kieselguhr	25	1.15×10^{11}
	100	1.31×10^{11}
Carboraffin charcoal	1.4	3.14×10^{10}
	10	5.84×10^{10}
Calcium carbonate	25	2.21×10^{11}
(precipitated)	100	2.68×10^{11}
Ferric oxide (pigment)	25	8.04×10^{11}
	100	14.12×10^{11}
Mica clay	25	4.81×10^{11}
	100	8.63×10^{11}
Colloidal clay	25	5.10×10^{12}
	100	6.47×10^{12}
Magnesium hydroxide	25	3.24×10^{12}
(gelatinous)	100	6.97×10^{12}
Aluminium hydroxide	25	2.16×10^{13}
(gelatinous)	100	4.02×10^{13}
Ferric hydroxide	25	1.47×10^{13}
(gelatinous)	100	4.51×10^{13}
Thixotropic mud	80	6.77×10^{14}
Theoretical figures for rigid spheres:		
$d = 10\ \mu m$	—	6.37×10^{9}
$d = 1\ \mu m$	—	6.37×10^{11}
$d = 0.1\ \mu m$	—	6.37×10^{13}

(Carman, 1938).

COMPRESSION–PERMEABILITY CELL

Such equipment consists of a hollow cylinder fitted with a permeable bottom and a permeable piston under controlled pressure. Slurry is charged to the slurry, cake is formed with gentle suction, and the piston is lowered to the cake level. The rate of flow of filtrate at low head through the compressed cake is measured at a series of pressures on the piston. From the results the resistivity of the cake becomes known as a function of pressure. The data of Figures 11.4(b) and (c) were obtained this way; those of Figure 11.4(a) by filtration tests.

There is much evidence, however, that the resistivity behavior of a cake under filtration conditions may be different from that measured in a CP cell. The literature is reviewed by Wakeman (1978). CP cell data are easily obtained and may be of value in a qualitative sense as an indication of the sensitivity of resistivity to pressure, but apparently are not of acceptable engineering accuracy for the design of filtration equipment. The deduction of resistivities from filtration tests is illustrated in Example 11.1.

THE SCFT CONCEPT

No serious attempt has yet been made to standardize filtration tests and to categorize filtration behavior in generally accepted terms. A possibly useful measure of filterability, however, has been proposed by Purchas (1977; 1981). The time in minutes required to form a cake 1 cm thick when the cell is operated with a differential of 500 Torr (0.67 bar) is called the Standard Cake Formation Time (SCFT), t_F. The pressure of 500 Torr is selected because it is obtained easily with common laboratory equipment. The procedure suggested is to make a series of tests at several cake thicknesses and to obtain the SCFT by interpolation, rather than to interrupt a single test to make observations of cake thickness. A direct relation exists, of course, between the SCFT and resistivity α; some examples are

Material	α (m/kg)	SCFT t_F (min)
Filter aid	1.64(E9)	0.26
$CaCO_3$	2.21(E11)	34.6
Colloidal clay	5.10(E12)	798

Full scale filtration equipment requirements can be estimated quickly in terms of t_F. For instance, when the resistance of the filter medium is neglected, the constant pressure Eq. (11.3) may be written as

$$\Delta Pt = \frac{\alpha c}{2}\left(\frac{V}{A}\right)^2 = \frac{\alpha c}{2}\left[\frac{(1-\varepsilon)L}{c}\right]^2, \tag{11.27}$$

where L is the thickness of the cake in meters. Upon rationing in

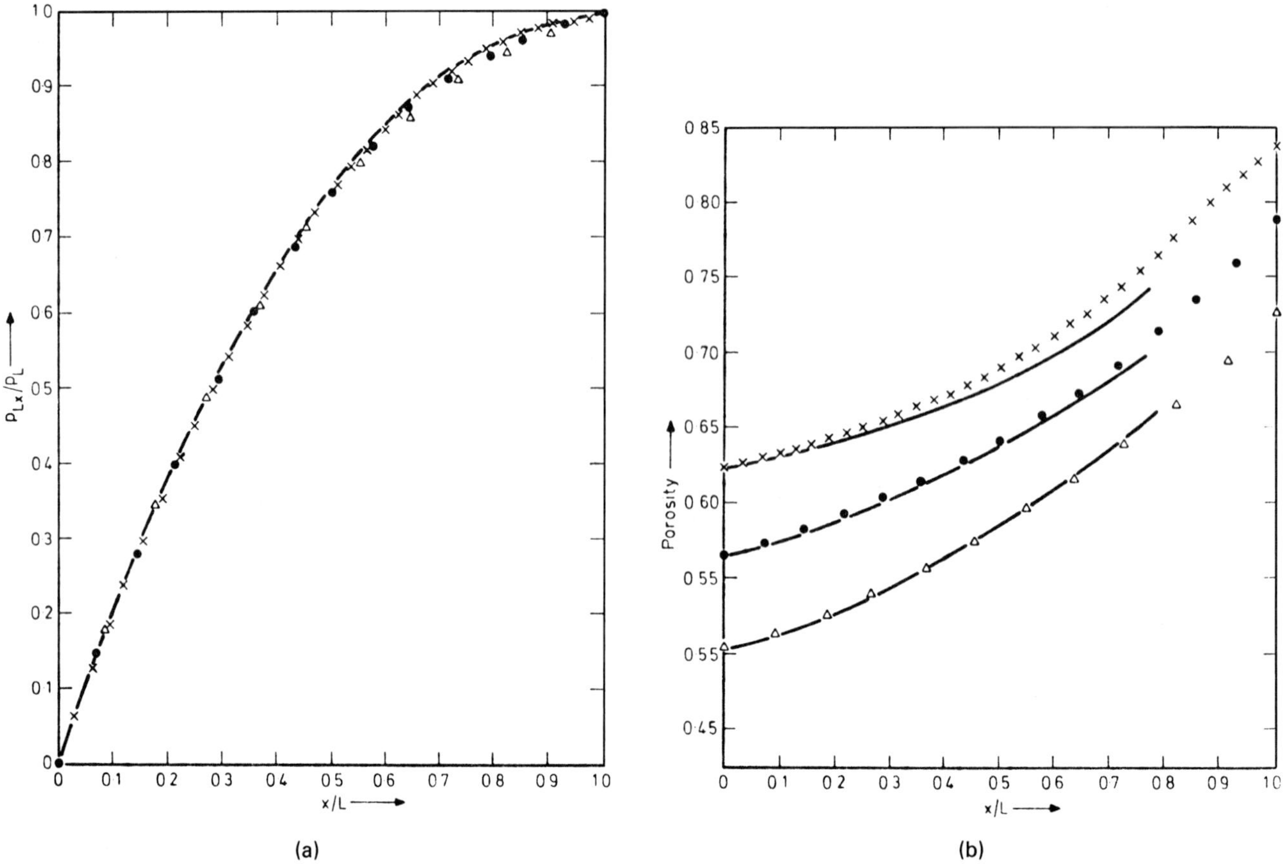

Figure 11.5. Axial distribution of pressure and porosity of an ignition-plug clay measured in a CP cell. (a) Normalized pressure distribution as a function of normalized distance [(– – –) experimental filtration data; theoretical curves: (×) $\Delta P = 98\ \mathrm{kN\ m^{-2}}$; (●) $\Delta P = 294\ \mathrm{kN\ m^{-2}}$; (△) $\Delta P = 883\ \mathrm{kN\ m^{-2}}$]. (b) Porosity distributions at three pressures. The curves are by Wakeman (1978).

the SCFT data for 0.01 m,

$$\frac{\Delta Pt}{0.67t_F} = (100L)^2, \tag{11.28}$$

with ΔP in bar. From this relation the filtering time can be found at a specified pressure and cake thickness and when t_F is known.

SCALE-UP

Sizing of full scale equipment on the basis of small scale tests requires a consideration of possible ranges of at least the following variables:

1. filterability as measured by cake and medium resistivity;
2. feed rate and concentration;
3. operating conditions, particularly pressure and high initial rates;
4. behavior of the filter cloth with time.

Safety factors for scale up from laboratory leaf tests are difficult to generalize. On the basis of pilot plant work, adjustments of 11–21% are made to plate-and-frame filter areas or rates, and 14–20% to continuous rotary filters, according to Table 1.4.

The performance of solid–liquid separation equipment is difficult to predict by the engineer without some specific experience in this area. Unfortunately, it must be again recommended that the advice of experienced vendors should be sought, as well as that of expert consultants.

11.6. ILLUSTRATIONS OF EQUIPMENT

Equipment for solid–liquid separation is available commercially from many sources. About 150 names and addresses of suppliers in the United States and abroad are listed by Purchas (1981). Classifications of vendors with respect to the kind of equipment are given, for instance, in *Chemical Engineering Catalog* (Reinhold, New York, annual) and in *Chemical Engineering Equipment Buyers Guide* (McGraw-Hill, New York, annual).

The variety of solid–liquid separation equipment is so great that only a brief selection can be presented here. The most extensive modern picture gallery is in the book of Purchas (1981). The older encyclopedia of Kieser (Spamer-Springer, Berlin, 1937, Vol. 2) has 250 illustrations in 130 pages of descriptions; the pictures do not appear to have aged particularly. Illustrations in manufacturers catalogs are definitive and often reveal the functioning as well as aspect of the equipment. The selected figures of this chapter are primarily line drawings that best reveal the functioning modes of the equipment.

Figure 11.9 shows two models of sand filters whose purpose is to remove small contents of solids from large quantities of liquids. The solids deposit both on the surface of the bed and throughout the bed. They are removed intermittently by shutting off the main

TABLE 11.8. Parameters of Equations for Resistivity α and Porosity ε of Some Filter Cakes

$$\alpha = \alpha_0\left(1 + \frac{p_s}{p_a}\right)^n$$

$$(1-\varepsilon) = (1-\varepsilon_0)\left(1 + \frac{p_s}{p_a}\right)^{\beta^*}$$

Material	Pressure range, kPa	p_a, kPa	α_0, m kg^{-1} × 10^{-10}	n	$(1-\varepsilon_0)$	β^*
$CaCO_3$ (ref. 7)	3–480	1	11	0.15	0.209	0.06
$CaCO_3$ (ref. 8)	7–550	7	5.1	0.2	0.225	0.06
	550–7000	790	8.1	0.9	0.263	0.22
Darco-B (ref. 8)	7–275	1.7	1.1	0.4	0.129	0.08
	275–7000	520	4.7	1.8	0.180	0.18
Kaolin-Al_2SO_4 (ref. 8)	7–415	7	43	0.3	0.417	0.04
	415–7000	345	87	0.7	0.460	0.12
Solka-Floc (ref. 8)	7–275	2.75	0.00058	1.0	0.132	0.16
	275–7000	260	0.13	2.0	0.237	0.26
Talc-C (ref. 8)	7–1400	5.5	4.7	0.55	0.155	0.16
	1400–7000	1400	35	1.8	0.339	0.25
TiO_2 (ref. 8)	7–7000	7	18	0.35	0.214	0.1
Tungsten (ref. 8)	7–480	7	0.39	0.15	0.182	0.05
	480–7000	520	0.38	0.9	0.207	0.22
Hong Kong	1–15	1	42	0.35	0.275	0.09
pink kaolin (ref. 9)	15–1000	12	70	0.55	0.335	0.1
Gairome clay (ref. 10)	4–1000	3.4	370	0.55	0.309	0.09

(Tiller et al, 1979)

flow and backwashing with liquid. The concentrated sludge then must be disposed of in some way. Beds of charcoal are employed similarly for clarification of some organic liquids; they combine adsorption and mechanical separation.

Clarification of a large variety of liquids is accomplished with cartridge filters which come in a large variety of designs. Usually the cartridges are small, but liquid rates in excess of 5000 gpm have been designed for. The filtering surface may be a fine metal screen or an assembly of closely spaced disks whose edge face functions as the filtering surface, or woven or matted fibers. The operation is intermittent, with either flushing back of the accumulated solids or replacement of the filtering elements in the body of the cartridge, or in some instances the solids are scraped off the filtering surface with a built-in mechanism and then flushed out in concentrated form. The variety of cartridge filters are described in detail in books by Warring (1981), Purchas (1981), and Cheremisinoff and Azbel (1983). Table 11.10 is a selected list of some of their applications and the minimum sizes of particles that are removed.

Figure 11.6 is of two types of sedimentation equipment, and Figure 12.2(e) of another. They are used for clarifying a valuable liquid or for preparing a concentrated slurry for subsequent filtration. They depend on gravitational sedimentation. Removal is assisted by rake action, or by the conical sides of the vessel of Figure 11.6(b).

Figure 11.10 is of the main kinds of filters that can be operated at superatmospheric pressures which may be necessary with otherwise slow filtering slurries. Commercial sizes are listed in Table 11.11. They all operate on intermittent cycles of cake formation, washing, dewatering with air blowing and cake removal. The plate-and-frame design of Figure 11.10(a) is the most widely recognized type. In it, cake removal is effected after separating the plates. The horizontal plate design of Figure 11.10(b) is popular in smaller sizes under, 2 ft dia or so; the plates are lifted out of the casing for cake removal. The other units all have fixed spacings between the leaves. From them the cakes may be blown back with air or flushed back or scraped off manually. The Vallez unit of Figure 11.10(f) ordinarily does not require the case to be opened for cleaning.

Figure 11.11 is of continuous horizontal filtering equipment that operate primarily with vacuum, although they could be housed in pressure-tight casings for operation at superatmospheric pressure or with volatile liquids. Both the belt and the rotary units are well suited to rapidly settling and free draining slurries. In comparison with rotary drum vacuum filters, the horizontal equipment of Figure 11.11(c) has the merit of more readily accessible piping, a real advantage from a servicing point of view.

Figure 11.12 represents the main kinds of rotary drum filters. Commercial sizes are listed in Table 11.14. The flowsketch of Figure 11.12(a) identifies the main auxiliaries required for this kind of filtration process. Feed to the drum may be dip-type as in Figure 11.12(b), but top feed designs also are widely used. The unit with internal filtering surface of Figure 11.12(c) is suited particularly to rapidly settling solids and has been adapted to pressure operation.

Cake removal usually is with a scraper into a screw or belt conveyor, but Figure 11.12(d) depicts the use of a drum with a filtering belt that is subject to a continual cleaning process. Some filters have a multi parallel string discharge assembly whose path follows that of the belt shown.

The double drum filter of Figure 11.12(e) has obvious merit particularly when top feeding is desirable but it is not used widely nowadays. Disk filters of the type of Figure 11.12(f) are the most widely used rotary type when washing of the cake is not necessary.

Figure 11.13 is of a variety of devices that utilize centrifugal force to aid in the separation of solid and liquid mixtures. Figure

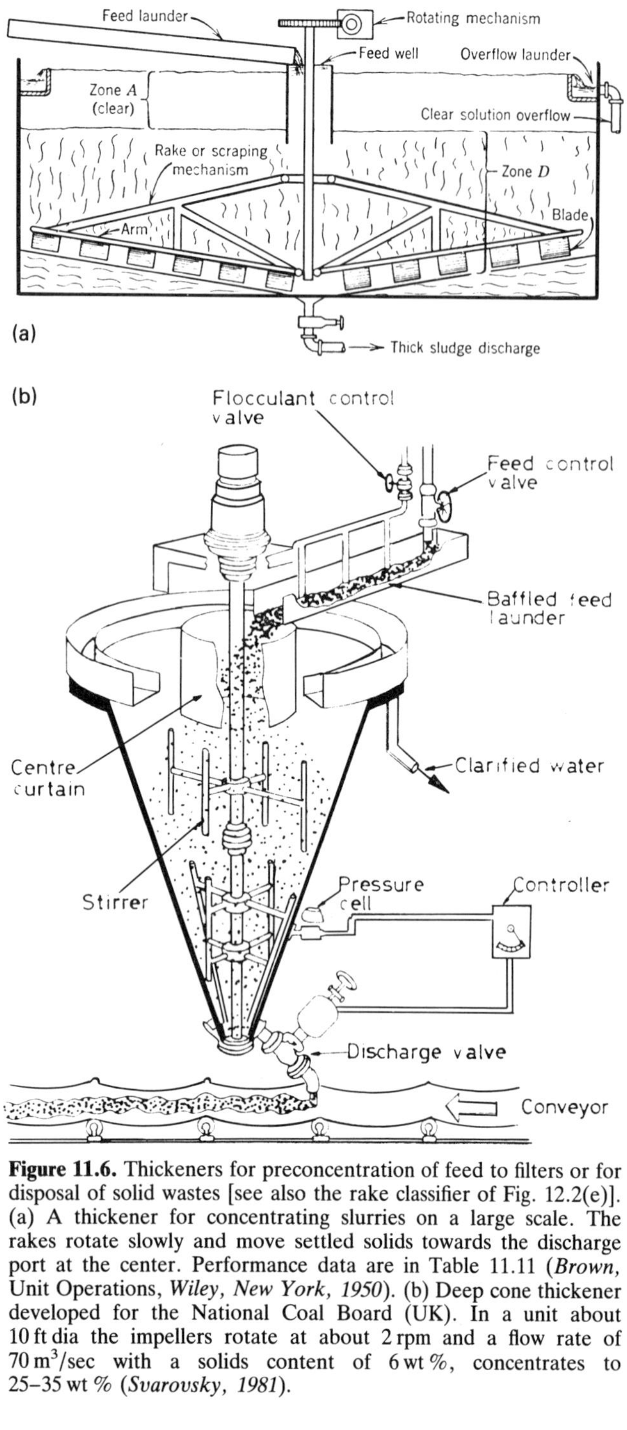

Figure 11.6. Thickeners for preconcentration of feed to filters or for disposal of solid wastes [see also the rake classifier of Fig. 12.2(e)]. (a) A thickener for concentrating slurries on a large scale. The rakes rotate slowly and move settled solids towards the discharge port at the center. Performance data are in Table 11.11 (*Brown, Unit Operations, Wiley, New York, 1950*). (b) Deep cone thickener developed for the National Coal Board (UK). In a unit about 10 ft dia the impellers rotate at about 2 rpm and a flow rate of 70 m^3/sec with a solids content of 6 wt %, concentrates to 25–35 wt % (*Svarovsky, 1981*).

11.13(a) performs cake removal at reduced rotating speed, whereas the design of Figure 11.13(d) accomplishes this operation without slowing down. The clarifying centrifuge of Figure 11.13(e) is employed for small contents of solids and is cleaned after shutdown. The units of Figures 11.13(b) and (c) operate continuously, the former with discharge of cake by a continuous helical screw, the latter by a reciprocating pusher mechanism that operates at 30–70 strokes/min and is thus substantially continuous.

Hydrocyclones generate their own, mild centrifugal forces. Since the acceleration drops off rapidly with diameter, hydrocyclones are made only a few inches in diameter. For larger capacities, many units are used in parallel. The flow pattern is shown schematically in Figure 11.13(f). The shapes suited to different applications are indicated in Figure 11.13(g). In Figure 11.13(h), the centrifugal action in a hydrocyclone is assisted by a high speed impeller. This assistance, for example, allows handling of 6% paper pulp slurries in comparison with only 1% in unassisted units. Hydrocyclones are perhaps used much more widely for dust separation than for slurries.

TABLE 11.9. Performances of Sedimentation Equipment

(a) Thickeners[a]

	% solids		Unit area, sq. ft./ton, day
	Feed	Underflow	
Alumina, Bayer process:			
Red-mud primary settlers	3–4	10–25	20–30
Red-mud washers	6–8	15–20	10–15
Red-mud final thickener	6–8	20–35	10–15
Trihydrate seed thickener	2–8	30–50	12–30
Cement, West process	16–20	60–70	15–25
Cement kiln dust	9–10	45–55	3–18
Coral	12–18	45–55	15–25
Cyanide slimes	16–33	40–55	5–13
Lime mud:			
Acetylene generator	12–15	30–40	15–33
Lime-soda process	9–11	35–45	15–25
Paper industry	8–10	32–45	14–18
Magnesium hydroxide from brine	8–10	25–50	60–100
Metallurgical (flotation or gravity concentration):			
Copper concentrates	14–50	40–75	2–20
Copper tailings	10–30	45–65	4–10
Lead concentrates	20–25	60–80	7–18
Zinc concentrates	10–20	50–60	3–7
Nickel:			
Leached residue	20	60	8
Sulfide concentrate	3–5	65	25
Potash slimes	1–5	6–25	40–125
Uranium:			
Acid leached ore	10–30	25–65	2–10
Alkaline leached ore	20	60	10
Uranium precipitate	1–2	10–25	50–125

(b) Clarifiers

Application	Overflow rate, gal./min., sq. ft.	Detention time, hr.
Primary sewage treatment (settleable-solids removal)	0.4	2
Secondary sewage treatment (final clarifiers—activated sludge and trickling filters)	0.55–0.7	1.5–2
Water clarification (following 30-min. flocculation)	0.4–0.55	3
Lime and lime-soda softening (high rate—upflow units)	1.5	2
Industrial wastes	Must be tested for each application	

[a] See also Table 14.7.

(*Perry's Chemical Engineers Handbook, McGraw-Hill, New York,* 1963, *pp.* 19.49, 19.52).

11.7. APPLICATIONS AND PERFORMANCE OF EQUIPMENT

Data of commercially available sizes of filtration equipment, their typical applications, and specific performances are available only to a limited extent in the general literature, but more completely in

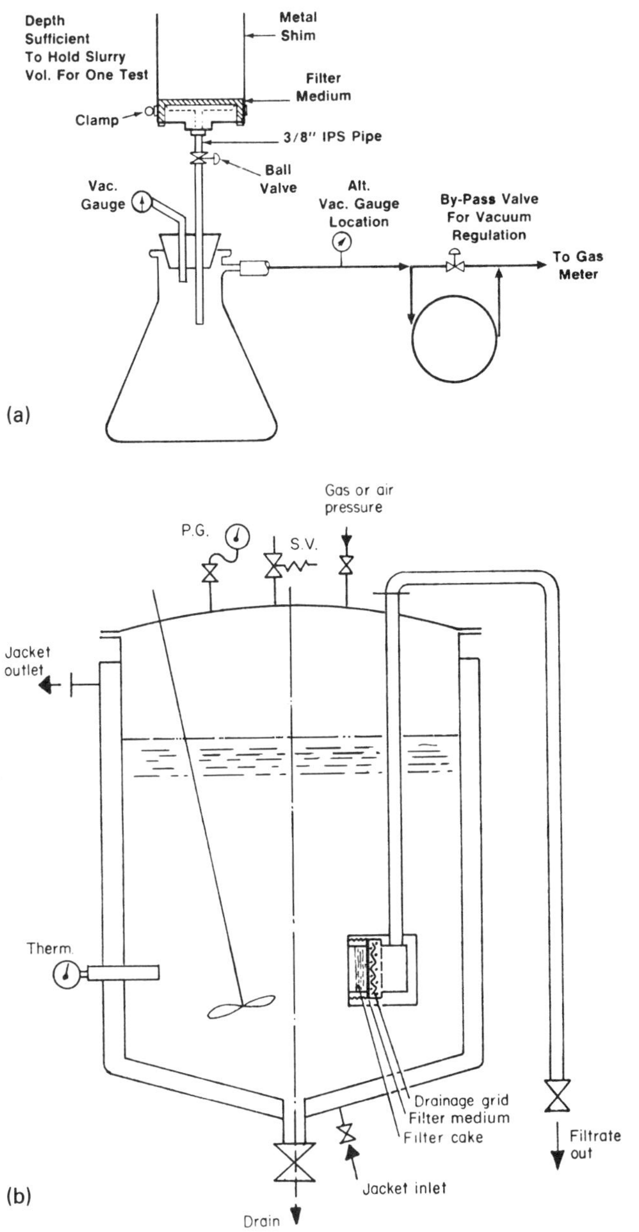

Figure 11.7. Two types of laboratory filter arrangements. (a) Vacuum test filter arrangement; standard sizes are 0.1, 0.05, or 0.025 sqft (*Dahlstrom and Silverblatt, 1977*). (b) Laboratory pressure filter with a vertical filtering surface and a mechanical agitator; mild air agitation may be preferred (*Bosley, 1977*).

manufacturers' literature. Representative data are collected in this section and summarized in tabular form. One of the reasons why more performance data have not been published is the difficulty of describing each system concisely in adequate detail. Nevertheless, the limited listings here should afford some perspective of the nature and magnitude of some actual and possibly potential applications.

Performance often is improved by appropriate pretreatment of the slurry with flocculants or other means. An operating practice that is finding increasing acceptance is the delaying of cake deposition by some mechanical means such as scraping, brushing, severe agitation, or vibration. In these ways most of the filtrate is expelled before the bulk of the cake is deposited. Moreover, when the cake is finally deposited from a thickened slurry, it does so with an open structure that allows rapid filtration. A similar factor is operative in belt or top feed drum filters in which the coarse particles drop out first and thus form the desirable open structure. A review of such methods of enhancement of filtration rates is by Svarovsky (1981).

The relative suitability of the common kinds of solid–liquid separation equipment is summarized in Table 11.3. Filtration is the most frequently used operation, but sedimentation as a method of pretreatment and centrifugation for difficulty filterable materials has many applications. Table 11.15 gives more detail about the kinds of filters appropriate to particular services.

Representative commercial sizes of some types of pressure filters for operation in batch modes are reported in Table 11.11. Some of these data are quite old, and not all of the equipment is currently popular; thus manufacturers should be consulted for the latest information. Commercially available size ranges of continuous belt, rotary drum, rotary disk, and horizontal rotary filters are listed in Table 11.12. For the most part these devices operate with vacua of 500 Torr or less.

Sedimentation equipment is employed on a large scale for mineral and ore processing. These and other applications are listed in Table 11.9(a). The clarification operations of Table 11.9(b) are of water cleaning and sewage treatment. The sludges that are formed often are concentrated further by filtration. Such applications are listed in Table 11.16 along with other common applications of plate-and-frame filter presses. Sludge filter cakes are compressible and have high resistivity so that the elevated pressures at which presses can be operated are necessary for them. Among the kinds of data given here are modes of conditioning the slurries, slurry concentrations, cake characteristics, and cycle times.

Clarification of a great variety of industrial liquids is accomplished on smaller scales than in tank clarifiers by application of cartridge filters; some of these applications are listed in Table 11.10.

Cycle times, air rates, and minimum cake thicknesses in operation of rotary drum filters are stated in Table 11.13. A few special applications of horizontal belt filters are given in Table 11.14, but in recent times this kind of equipment is taking over many of the traditional functions of rotary drum filters. Belt filters are favored particularly for freely filtering slurries with wide range of particle sizes.

The applications listed in Table 11.17 and 11.18 are a few of those of rotary drum, rotary disk, and tipping or tilting pan filters. The last type employs a number of vacuum pans on a rotating circular track; after the cake is formed, the pans are blown back with air and then tipped to discharge the cake. The data of these tables include particle size range, moisture content of the cake, filtering rate, solids handling rate, vacuum pump load and degree of vacuum. Clearly a wide range of some of these variables occurs in practice.

Characteristics of centrifugal filters and sedimentation centrifuges are in Table 11.19. The filtering types are made to handle from less than 5 tons/hr to more than 100 tons/hr of solids, with g-levels ranging from 30 to 3000. For sedimentation types, the g-levels listed range up to 18,000, but high values can be used only with small diameter equipment because of metal strength limitations. Capacity of sedimentation types is measured in terms of liquid rates, the maximum listed here being 100,000 L/hr. An outstanding feature of centrifugal separators is the small sizes of particles that can be handled satisfactorily; the values in the table cover the range 1–400 μm. Short retention time is a feature of centrifuge operation that may be of interest when unstable materials need to be processed.

FILTRATION LEAF TEST DATA SHEET – VACUUM AND PRESSURE

Company ________ Mat'l as Received: Date ________ Test No. ________
Address ________ Solids: ________ % Date Tested ________
________ Analysis ________ By ________
________ Liquid: ________ % Location ________
Filter Type ________ Leaf Size ________ $Ft.^2$ Analysis ________
Used Shim: No ________ Yes ________ Precoat Forming Liquid ________ Temp. ________ °F/°C

Run No.	Filter Media and/or Precoat Type	Feed Temp., °F/°C	% Solids in Feed		Vacuum - in. Hg. Pressure - PSI.				TIME MIN.					Air Flow		Filtrate		Precoat Penetration	Wash		Cake/Precoat Thickness, In.	Dia. of Shaved Area, in.	Cake Weights			
			As Prepared	Back Calculated	Form	Wash	Dry	After Cake Cracks	Form	Dewater	Wash	Dry	To Crack or Gas Breakthrough After Form/Wash	(1)	Temp., °F °C	ML.	Clarity		ML.	Temp., °F/°C			Tare GMS.	Wet & Tare GMS.	Dry & Tare GMS.	Dish No.

CAKE DISCHARGE		REAGENT TREATMENT	
RUNS	COMMENT	RUNS	COMMENT

REMARKS: (1) Record Basis of Observation in Space Provided.

Figure 11.8. A filtration leaf test data sheet (*Dahlstrom and Silverblatt, 1977*).

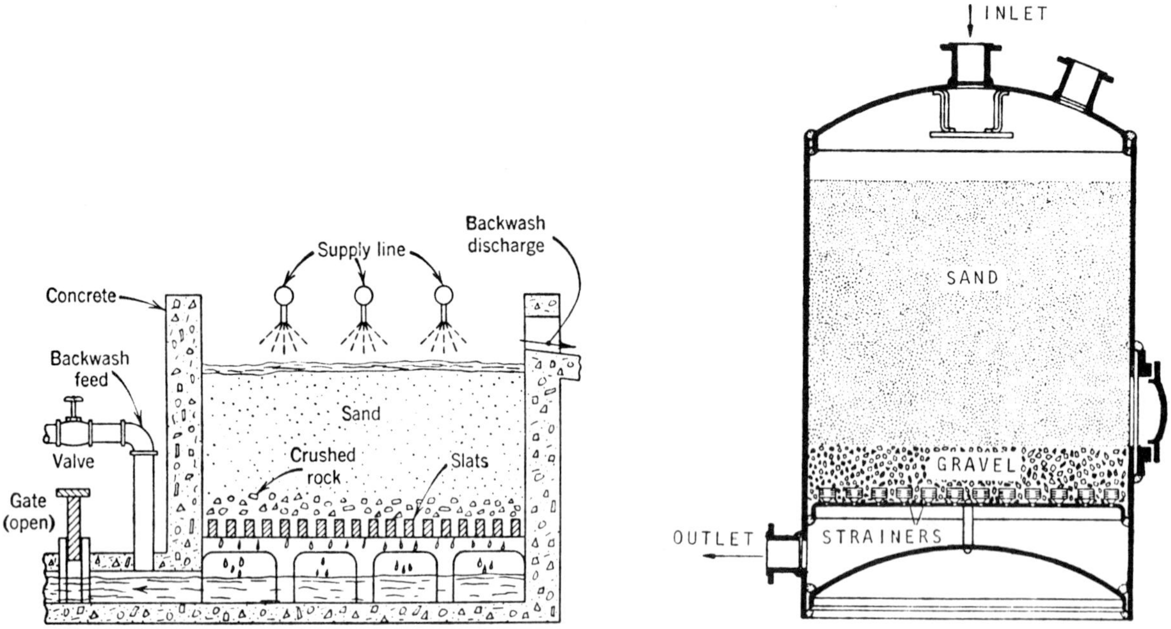

(a) (b)

Figure 11.9. Deep bed sand filters for removal of small contents of solids from large quantities of liquids. Accumulations from the top and within the bed are removed by intermittent backwashing. Charcoal may be used instead of sand for clarifying organic liquids. (a) Gravity operation. (b) Pressure operation.

TABLE 11.10. Application of Cartridge Filters in Industry and Typical Particle Size Ranges Removed

Industry and Liquid	Typical Filtration Range
Chemical Industry	
Alum	60 mesh–60 μm
Brine	100–400 mesh
Ethyl Alcohol	5–10 μm
Ferric Chloride	30–250 mesh
Herbicides/Pesticides	100–700 mesh
Hydrochloric Acid	100 mesh to 5–10 μm
Mineral Oil	400 mesh
Nitric Acid	40 mesh to 5–10 μm
Phosphoric Acid	100 mesh to 5–10 μm
Sodium Hydroxide	1–3 to 5–10 μm
Sodium Hypochlorite	1–3 to 5–10 μm
Sodium Sulfate	5–10 μm
Sulfuric Acid	250 mesh to 1–3 μm
Synthetic Oils	25–30 μm
Petroleum Industry	
Atmospheric Reduced Crude	25–75 μm
Completion Fluids	200 mesh to 1–3 μm
DEA	250 mesh to 5–10 μm
Deasphalted Oil	200 mesh
Decant Oil	60 mesh
Diesel Fuel	100 mesh
Gas Oil	25–75 μm
Gasoline	1–3 μm
Hydrocarbon Wax	25–30 μm
Isobutane	250 mesh
MEA	200 mesh to 5–10 μm
Naphtha	25–30 μm
Produced Water for Injection	1–3 to 15–20 μm
Residual Oil	25–50 μm
Seawater	5–10 μm
Steam Injection	5–10 μm
Vacuum Gas Oil	25–75 μm
All Industries	
Adhesives	30–150 mesh
Boiler Feed Water	5–10 μm
Caustic Soda	250 mesh
Chiller Water	200 mesh
City Water	500 mesh to 1–3 μm
Clay Slip (ceramic and china)	20–700 mesh
Coal-Based Synfuel	60 mesh
Condensate	200 mesh to 5–10 μm
Coolant Water	500 mesh
Cooling Tower Water	150–250 mesh
Deionized Water	100–250 mesh
Ethylene Glycol	100 mesh to 1–3 μm
Floor Polish	250 mesh
Glycerine	5–10 μm
Inks	40–150 mesh
Liquid Detergent	40 mesh
Machine Oil	150 mesh
Pelletizer Water	250 mesh
Phenolic Resin Binder	60 mesh
Photographic Chemicals	25–30 μm
Pump Seal Water	200 mesh to 5–10 μm
Quench Water	250 mesh
Resins	30–150 mesh
Scrubber Water	40–100 mesh
Wax	20–200 mesh
Wellwater	60 mesh to 1–3 μm

(Courtesy of Ronningen-Petter Division, Dover Corporation, Portage, MI; Cheremisinoff and Azbel, 1983).

TABLE 11.11. Sizes of Commercial Discontinuous Pressure Filters

(a) Approximate Area and Cake Capacity for Various Sizes of Plate and Frame Filters[a]

Size of filter plate (mm)	Effective Filtration area per Chamber (m^2)		Cake-Holding Capacity per Chamber per 25 mm of Chamber Thickness l	
	Cast Iron	Wood	Cast Iron	Wood
250	0.096	0.054	1.2	0.6
360	0.2	0.123	2.5	1.43
470	0.35	0.21	4.4	2.5
630	0.66	0.45	8.3	5.4
800	1.1	0.765	13.7	9.3
1000	1.74	1.2	21.62	14.6
1200	2.5	1.76	31.4	21.36
1450	3.7	2.46	46.24	30.2

(b) Sizes of Kelly Filters (in.)

	30 × 49	40 × 108	48 × 120	60 × 108
Number of frames	6	8	10	12
Spacing between frames (in.)	4	4	4	4
Filter area (sqft)	50	250	450	650

(c) Standard Sweetland Filter

No.	ID[b] (in.)	Length of Shell (in.)	No. Leaves 2 in. Space	No. Leaves 4 in. Space	Filter Area 2 in. Spacing (sqft)	Filter Area 4 in. Spacing (sqft)	Total Weight[c] of Filter (lbs)
1	10	$20\frac{1}{2}$	9	5	8	$4\frac{1}{2}$	550
2	16	$36\frac{1}{2}$	18	9	46	23	2150
5	25	61	30	15	185	92	7300
7	25	82	41	20	252	123	9350
10	31	109	54	27	523	262	16500
12	37	145	72	36	1004	502	29600

(d) Vallez Filter (Largest Size Only, 20 ft Long, 7 ft high, 7 ft wide)[d]

Spacing of Leaves (in.)	No. of Leaves	O.D. of Leaf (in.)	Filter Area (sqft)	Cake Capacity (cuft)
3	52	52	1232	65
4	39	52	924	72
5	31	52	734	79
6	23	52	646	92

(e) Characteristics of Typical Vertical-Tank Pressure Leaf Filters[e]

Tank Diam (in.)	Filter Area (sqft)	No. of Leaves	Leaf Spacing (in.)	Max. Cake Capacity (cuft)	Tank Volume (gal)	Approx. Overall Height (ft)	Approx. Shipping Weight (lb)
18	19	5	3	1.8	38	5.5	625
18	24	5	3	2.3	45	6.0	650
18	27	7	2	1.7	38	5.5	650
18	35	7	2	2.2	45	6.0	675
30	80	9	3	7.2	128	6.5	1125
30	95	9	3	8.7	132	7.0	1200
30	110	12	2	6.6	128	6.5	1180
30	125	12	2	8.0	132	7.0	1275
48	320	16	3	30.0	435	8.8	2900
48	370	16	3	35.0	500	9.3	3050
48	440	21	2	28.0	435	8.8	3125
48	510	21	2	32.0	500	9.3	3325

[a] F. H. Schule, Ltd.
[b] Diameter of leaf 1 in. less.
[c] Filled with water.
[d] There are smaller sizes with leaves the outside diameters of which are $44\frac{1}{2}$, 36, 30, and 22 in.; for the 30 in. leaves, four lengths of shell are available.
[e] T. Shriver & Co., Inc.

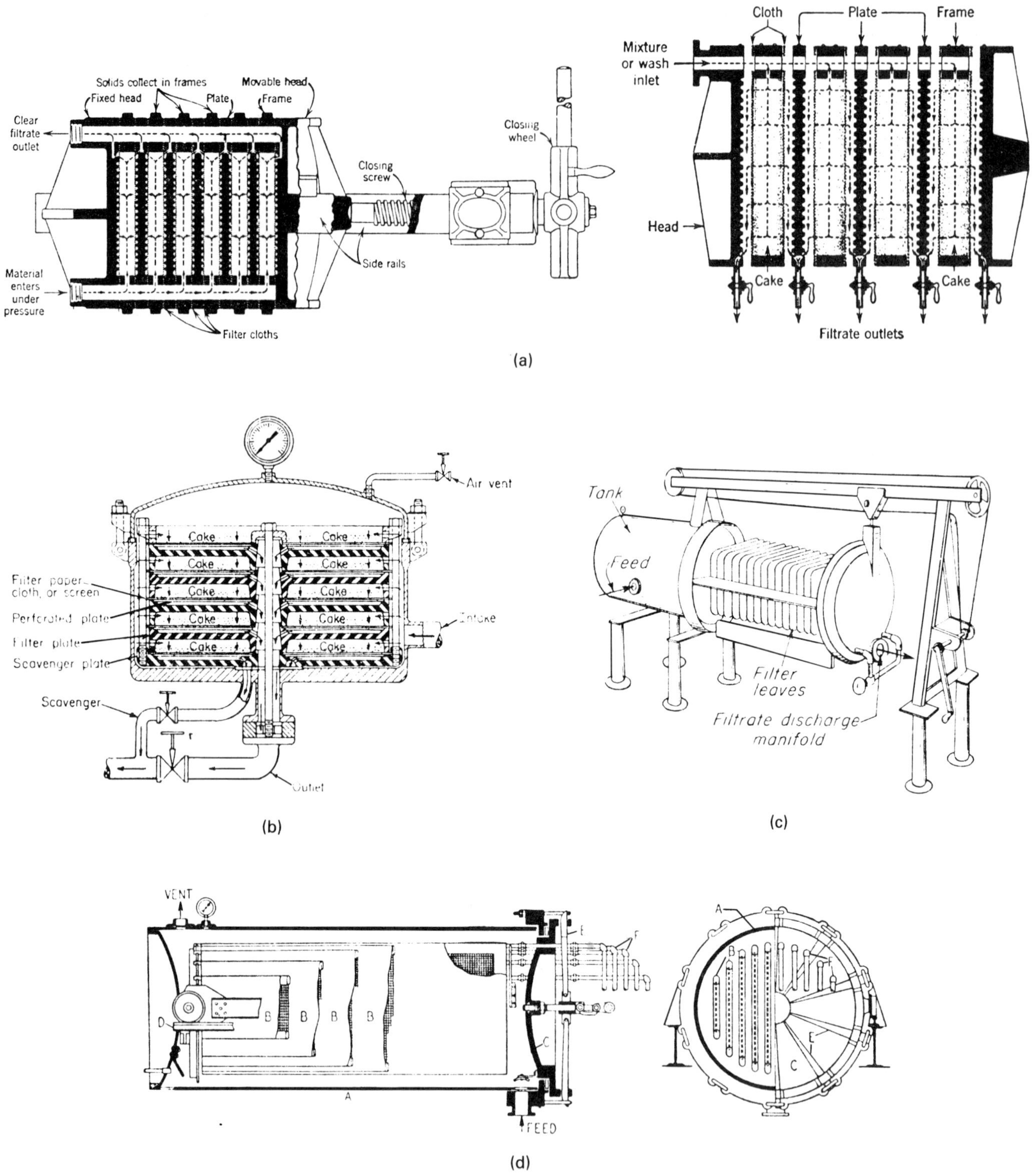

Figure 11.10. Pressure filters for primarily discontinuous operation. (a) Classic plate-and-frame filter press and details; the plates are separated for manual removal of the cake (*T. Shriver Co.*). (b) Horizontal plate filter; for cleaning, the head is removed and the plates are lifted out of the vessel (*Sparkler Mfg. Co.*). (c) Pressure leaf filter; the leaf assembly is removed from the shell and the cake is scraped off without separating the leaves (*Ametek Inc.*). (d) The Kelly filter has longitudinal leaves mounted on a carriage; for cleaning, the assembly is slid out of the shell (*Oliver United Filters*). (e) The Sweetland filter has circular leaves and a split casing; the lower half of the casing is dropped to allow access for removal of the cake (*Oliver United Filters*). (f) The Vallez filter has circular leaves rotating at about 1 rpm to promote cake uniformity when the solids have a wide size range; removal of blown-back or washed back cake is accomplished with a built-in screw conveyor without requiring the shell to be opened (*Goslin–Birmingham Co.*).

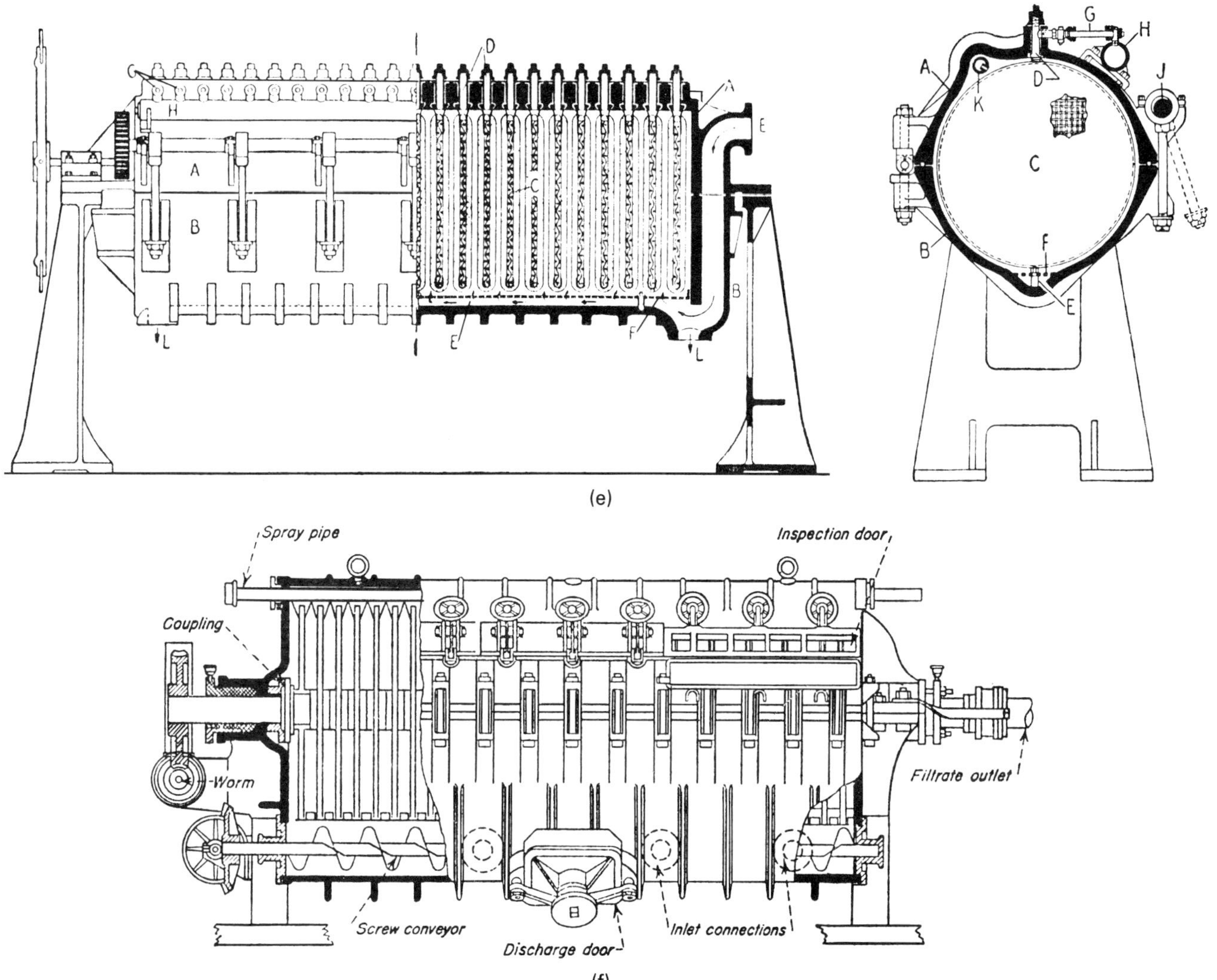

Figure 11.10.—(*continued*)

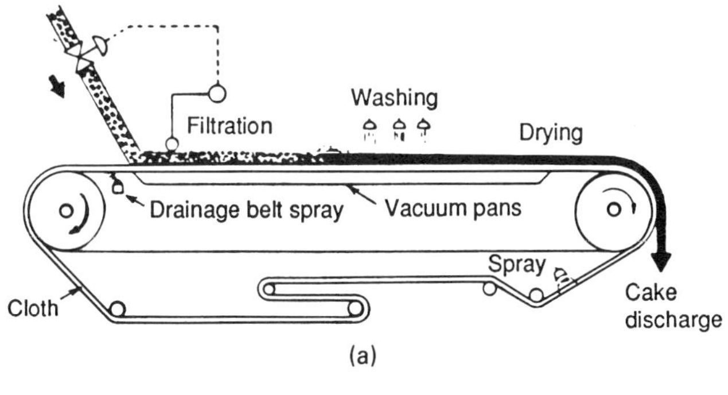

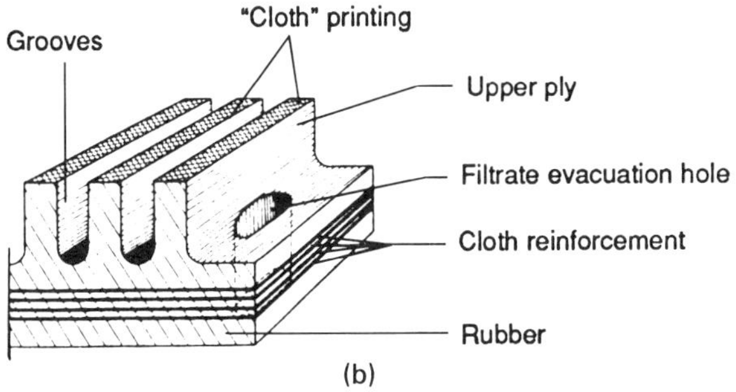

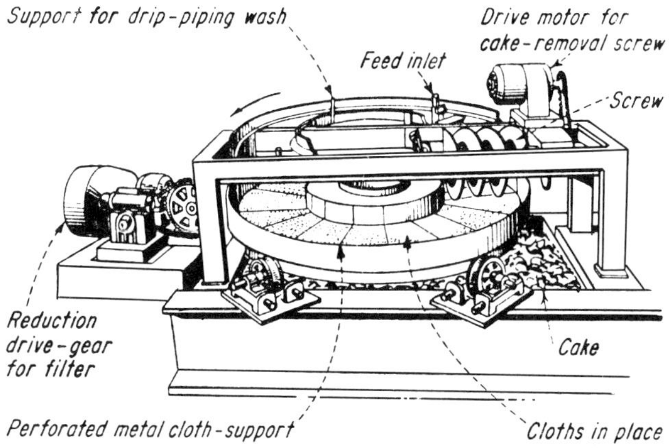

Figure 11.11. Continuous horizontal vacuum filters especially suited to free settling and draining solids. (a) Principle of the conveyor belt filter; units may operate up to 0.5 m/sec with a cycle time up to 10 min and produce cake thicknesses up to 15 cm. (b) Showing the construction of a grooved rubber belt support for the filter cloth of the belt filter (*Purchas, 1981*). (c) Rotating horizontal vacuum filter; the unit has readily accessible piping and is amenable to thorough washing of free draining solids (*Dorr–Oliver Inc.*).

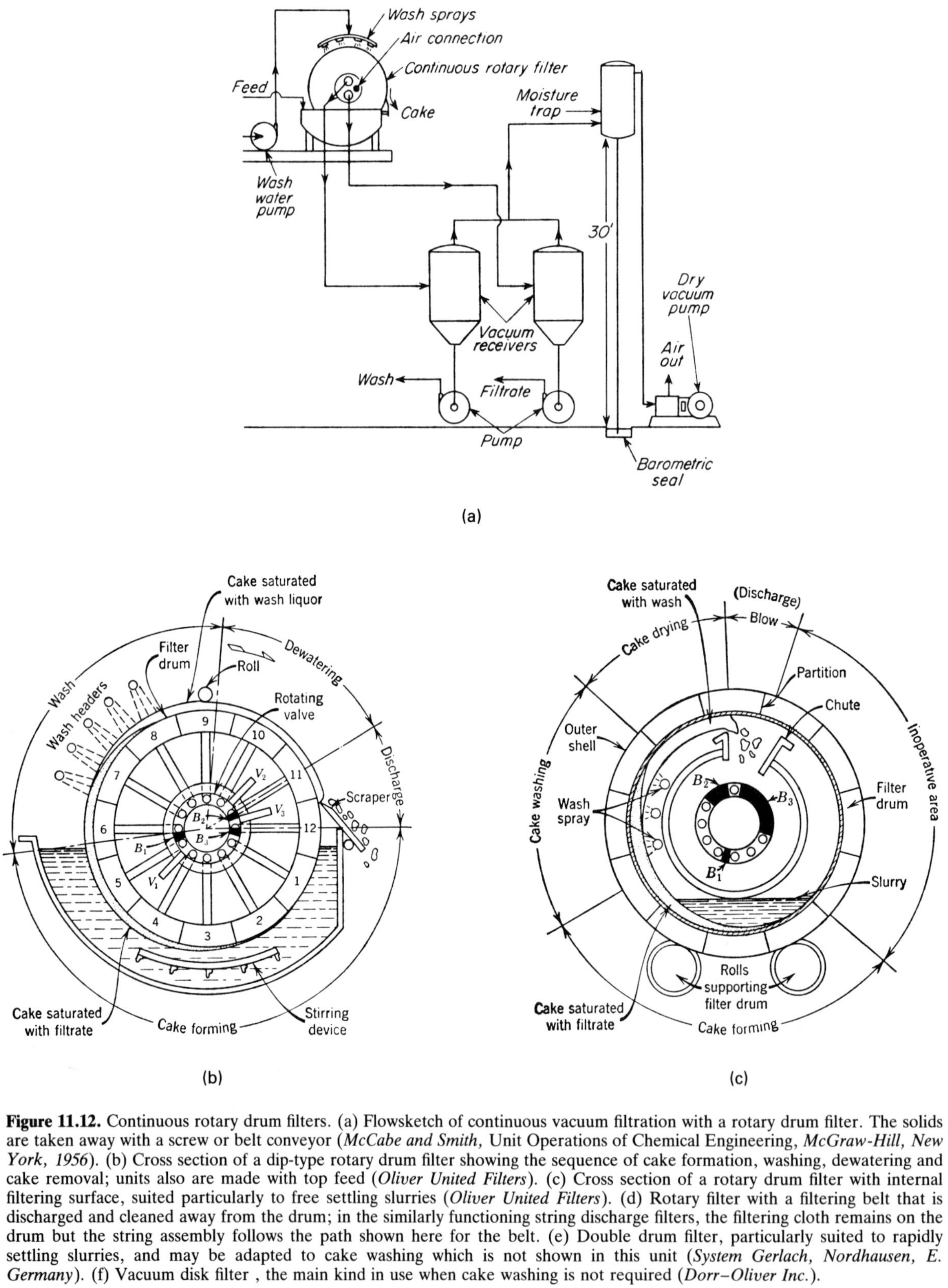

Figure 11.12. Continuous rotary drum filters. (a) Flowsketch of continuous vacuum filtration with a rotary drum filter. The solids are taken away with a screw or belt conveyor (*McCabe and Smith,* Unit Operations of Chemical Engineering, *McGraw-Hill, New York, 1956*). (b) Cross section of a dip-type rotary drum filter showing the sequence of cake formation, washing, dewatering and cake removal; units also are made with top feed (*Oliver United Filters*). (c) Cross section of a rotary drum filter with internal filtering surface, suited particularly to free settling slurries (*Oliver United Filters*). (d) Rotary filter with a filtering belt that is discharged and cleaned away from the drum; in the similarly functioning string discharge filters, the filtering cloth remains on the drum but the string assembly follows the path shown here for the belt. (e) Double drum filter, particularly suited to rapidly settling slurries, and may be adapted to cake washing which is not shown in this unit (*System Gerlach, Nordhausen, E. Germany*). (f) Vacuum disk filter , the main kind in use when cake washing is not required (*Dorr–Oliver Inc.*).

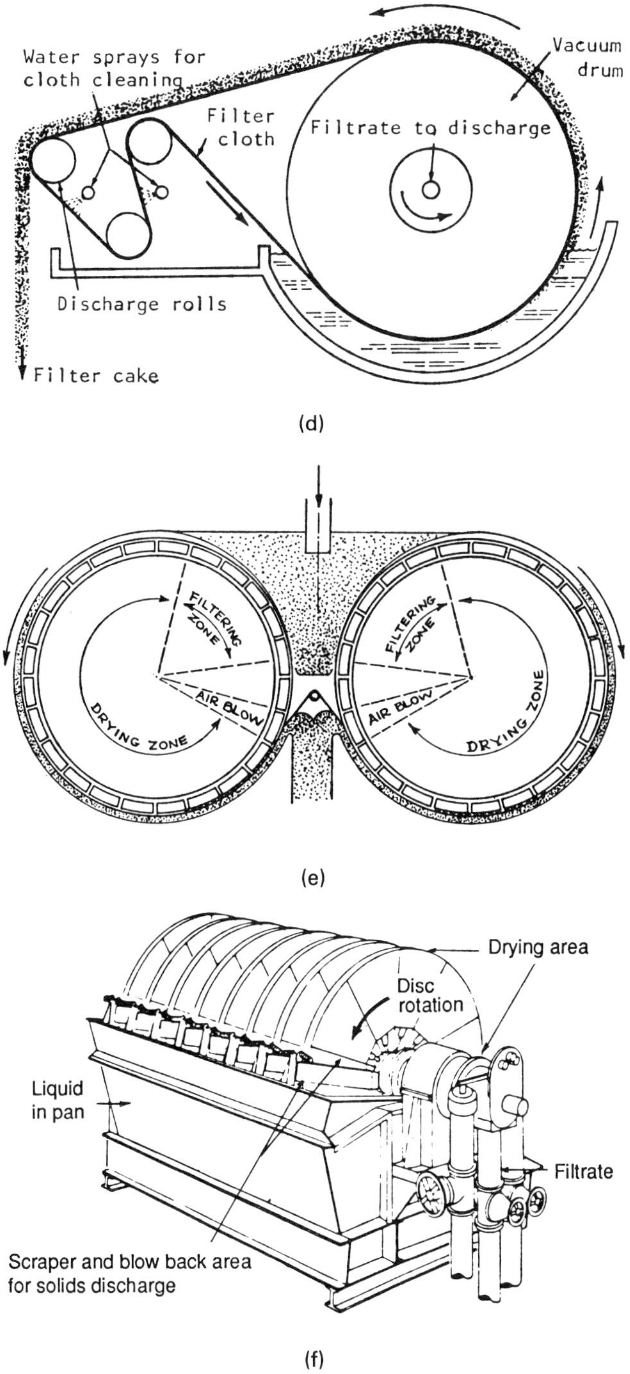

Figure 11.12—(*continued*)

TABLE 11.12. Sizes of Commercial Continuous Vacuum Filters

(a) Horizontal Belt Filters[a]

Series	Ft^2 Range	No. Vac. Pans
2600	10–45	1
4600	45–200	1
6900	150–700	1
9600	130–500	2
13,600	600–1200	2

(Eimco).

(b) Rotary Drum, Disk, and Horizontal Filters

Rotary Drum Component Filters[b]

Drum[c] Diam (ft)	Filter Surface Area (sqft), Length (ft): 4	6	8	10	12	14	16	18	20	22	24
6	76	113	151	189	226						
8			200	250	300	350	400				
10				310	372	434	496	558	620		
12					456	532	608	684	760	836	912

Disk Component Filters[d]

Disk diam (ft)[e]	6	7	8	9	10	11
Number of disks						
Min.	2	3	4	5	6	7
Max.	8	9	10	11	12	13
Filtering area per disk (sqft)	47	67	90	117	147	180

Horizontal Filters

Dia (ft)[f]	6	8	10	13	15	16	17	18	19	20	22	24
Area (sqft)												
Nom	28	50	78	133	177	201	227	254	283	314	380	452
Eff	25	45	65	120	165	191	217	244	273	304	372	444

[a] Filtrate 10–1600 lb/(hr)(sqft).
[b] Adaptable to knife, wire, string, belt, or roll discharge.
[c] All-plastic construction filters also available in 3 and 4 ft drum dia, providing filter areas of 9 to 100 sqft.
[d] All disks are composed of 10 sectors. Disk spacing is 16 in.
[e] The American filter, a similar disk filter, also available in 4 ft diameter, with 20 sqft disk.
[f] Also available in 3, 4, and 11.5 ft diameter.
(Dorr–Oliver Inc.).

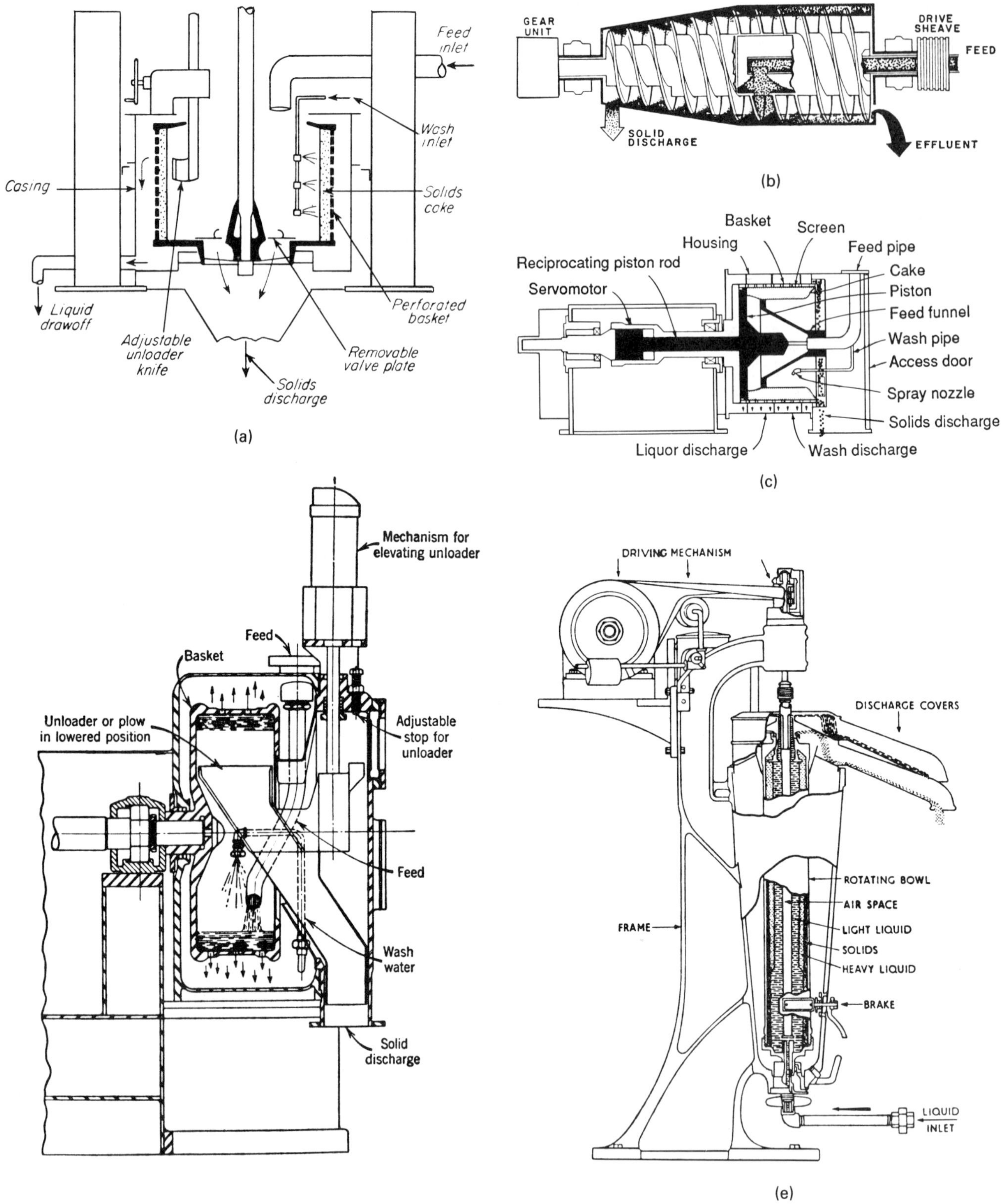

Figure 11.13. Filtering centrifuges. (a) Top suspended batch centrifugal filter; the cake is scraped off the screen intermittently at lowered rotation speeds of 50 rpm or so, cake thicknesses of 2–6 in., cycle time per load 2–3 min (*McCabe and Smith,* Unit Operations of Chemical Engineering, *McGraw-Hill, New York, 1956*). (b) A solid bowl centrifugal filter with continuous helical screw discharge of the cake (*Bird Machine Co.*). (c) Pusher type of centrifuge in which the cake is discharged with a reciprocating pusher mechanism that operates while the machine is at full speed (*Baker–Perkins Co.*). (d) Horizontal centrifugal with automatic controls for shutting off the feed, washing the cake and scraping it off, all without slowing down the rotation (*Baker–Perkin Co.*). (e) Supercentrifuge for removing small contents of solids from liquids; dimensions 3–6 in. by 5 ft, speed 1000 rps, acceleration 50,000 *g*, 50–500 gal/hr, cleaned after shutdown. (f) Pattern of flow in a hydrocyclone. (g) The shape of hydrocyclone adapted to the kind of service. (h) Centrifugal action of a cyclone assisted by a high speed impeller (*Voight Gmbh*).

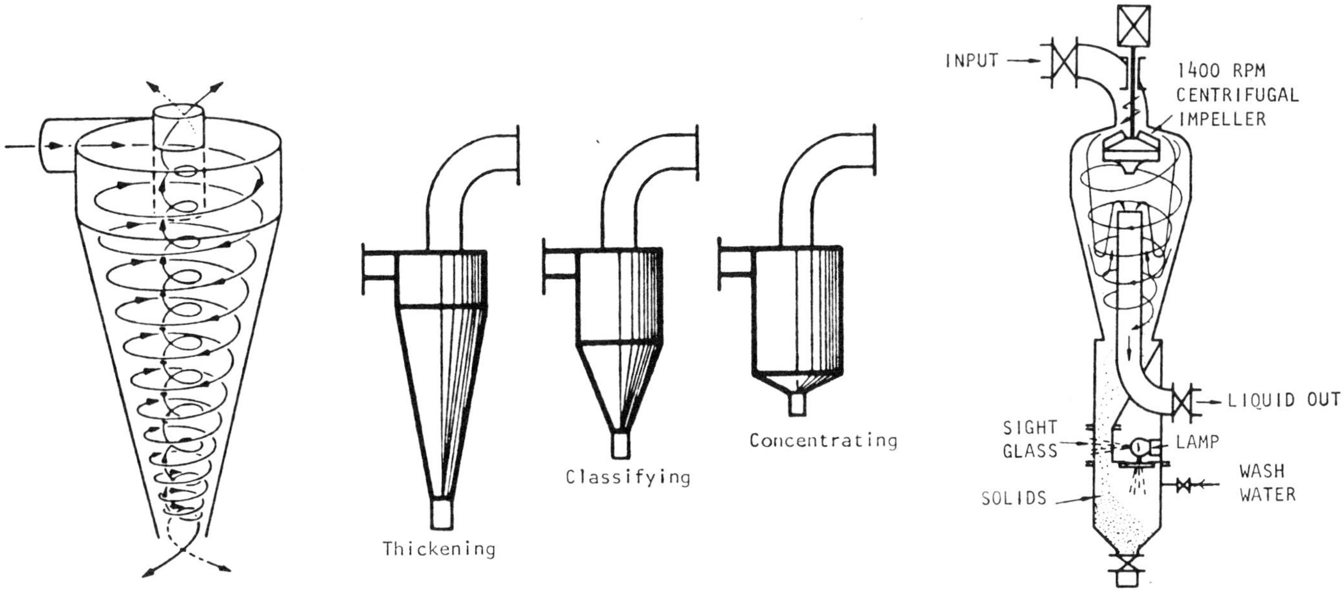

(f) (g) (h)

Figure 11.13—(*continued*)

TABLE 11.13. Typical Applications of Industrial Filters

Material	Characteristics	Filtrate Rate kg/(m²)(hr)	Equipment Type[a] A	B	C	D	E	Vacuum (Torr)	Pressure (atm)
Flotation concentrates	minerals, <0.3 m	300–1000	—	—	x	—	x	450–600	—
Sedimentation concentrates	>0.3 mm	6000–42,000	—	—	x	—	x	50–150	—
Crystals and granules	0.05–0.3 mm	600–2000	—	—	x	—	x	100–300	—
Beverages, juices	worthless solids, use filter aids	150–5000	x	x	—	—	—	—	2.5–3.5
Pigments	smeary, sticky, 0.06 mm	120–300	—	—	x	x	—	500–680	—
		batch mode	x	x	—	—	—		2.5–4
Limestone, oxide minerals	fine, high density	200–1000	—	—	x	—	—	450–600	—
		batch mode	x	x	—	x	—	—	2.5–4
Cane sugar mud	fibrous, viscous		x	x	x	—	—	—	—
Mineral oils	high viscosity, 1–20% bleaching clays	100–1000	—	x	—	—	—	—	4
Liquid fuels	low viscosity, bleaching clays	800–2500	—	x	—	—	—	—	<4
Varnishes, lacquers	cloudy, viscous, solid adsorbents	15–18	x	—	—	—	—	—	1
Fats, oils, waxes	worthless solids, 50–70°C	500–800	x	x	—	—	—	—	—
Sewage sludge	colloidal, slimy	15–150	—	—	x	—	—	550–600	—
Pulp and paper	fibrous, free filtering	150–500	—	—	x	—	—	150–500	—
Cement	fine limestone, shale, clay, etc	300–1000	—	—	x	—	—	450–630	—

[a] Equipment type: (A) filter press; (B) leaf pressure filters, such as Kelly, Sweetland, etc.; (C) continuous vacuum filter; (D) batch rotary filter; (E) continuous rotary filter.

TABLE 11.14. Design and Operating Factors for Continuous Vacuum Filters

(a) Typical Factors for Cycle Design

	Submergence[a]			% of Cycle		
Filter type	Apparent	Effective Maximum	Total under[b] Active Vac or Pressure	Max[c] for Washing	Max for[d] Dewatering Only	Required for Cake Discharge
Drum						
Standard scraper	35	30	80	29	50–60	20
Roll discharge	35	30	80	29	50–60	20
Belt	35	30	75	29	45–50	25
Coil or string	35	30	75	29	45–50	25
Precoat	35, 55, 85	35, 55, 85	93	30	10	5
Horizontal belt	as req'd	as req'd	lengthen as req'd	as req'd	as reqd	0
Horizontal table	as req'd	as req'd	80	as req'd	as req'd	20
Tilting pan	as req'd	as req'd	75	as req'd	as req'd	25
Disc	35	28	75	none	45–50	25

[a] Total available for effective subm., cake washing, drying, etc.

[b] Value for bottom feed filters assume no trunnion stuffing boxes, except for precoat. Consult manufacturers for availability of higher submergences.

[c] Maximum washing on a drum filter starts at horizontal centerline on rising side and extends to 15 past top dead center.

[d] Dewatering means drainage of liquor from cake formed during submergence.

(b) Typical Air Flow Rates

Type of Filter	Air Flow at 500 Torr Vacuum $[m^3/(h)(m^2)]$
Rotary drum	50–80
Precoat drum	100–150
Nutsche	30–60
Horizontal belt or pan	100–150

(c) Minimum Cake Thickness for Effective Discharge

	Minimum Design Thickness	
Filter Type	(in.)	(mm)
Drum		
Belt	1/8–3/16	3–5
Roll discharge	1/32	1
Std scrapter	1/4	6
Coil	1/8–3/16	3–5
String discharge	1/4	6
Precoat	0–1/8 max	0–3 max
Horizontal belt	1/8–3/16	3–5
Horizontal table	3/4	20
Tilting pan	3/4–1	20–25
Disc	3/8–1/2	10–13

[(a, b) Purchas, 1981; (c) Purchas, 1977].

TABLE 11.15. Typical Performance Data for Horizontal Belt Filters

Application	Filter area, m^2	Slurry feed characteristics % solids	pH	t/hr	Wash ratio (wt/wt based on dry solids)	Solubles recovery %	Final cake moisture %
Dewatering metallic concentrates	8	40	—	20	—	—	7
Brine precipitate sludge	25	12	—	1	8	90	50
Calcine leach	60	45	10	78	1	99.7	14
Uranium leach pulp	120	50	1–2	300	0.4	99.3	18
Cyanide leach gold pulp	120	50	10–11	80	0.6	99.6	20

(Delfilt Ltd.; Purchas, 1981).

TABLE 11.16. Examples of Filter Press Performance for Dewatering of Wastes in Municipal, Potable Water and Industrial Effluents

Type of Material	Nature and level of conditioning	Filtration cycle time (Hr)	Solids feed Wt/Wt (%)	Cake Wt/Wt (%)	Cake thickness (mm)	Remarks
Fine waste slurry	Polyelectrolytes 0.05–0.3 lb/ton	0.5–2	15–35	75–82	25–40	More than 80% below
Frothed tailings	Polyelectrolytes 0.05–0.3 lb/ton	1–2.5	15–35	73–80	25–40	240BS mesh
Primary sewage sludge	5–25% lime with 5–15% copperas, 5–25% lime and 3–6% ferric chloride	3–7 1.5–2	4–7 3–6	40–55 35–50	25–32 25–32	
Digested sewage sludge	or 1-2% ACH(Al_2O_3)	2–3				
Heat treated sludge		1–2	12–15	50–70	32	
Mixed sewage sludge including surplus activated	Up to 3% aluminium chlorohydrate (Al_2O_3 basic) or 30% lime with 30% copperas or 3–8% $FECl_3$	3–6 2–4	up to 4 up to 4	30–45 30–40	32 25	Proportion of surplus activ. sludge is 40% by weight
Paper Mill Humus sludge	1% ACH	8	0.5–1.5	30–45	25	
Paper Mill pool effluent sludge	10% lime, 10% copperas of 1% $FECl_3$	1–3	1–1.5	40–55	25	
Pickling and plating sludge	Up to 10% lime if required	1.5–3	2–3	30–45	25–32	
Potable water treatment sludge	In some instances no conditioning is required 0.2–1.5% polyelectrolyte (Frequently it is possible to decant large quantities of clarified water after conditioning and before filtration).	3–8	0.5–3	25–35	19–25	
Brine sludge		1.5–3	10–25	60–70	20–25	
Hydroxide sludge	1 mg/l polyelectrolyte or 10% lime	1.5–3	0.5–1.5	35–45	25–32	
Lead hydroxide sludge		0.5	45	80	32	

(Edwards and Jones Ltd.).

TABLE 11.17. Operating Data of Some Vacuum Filter Applications

Application	*Type of vacuum filter frequency used*[b]	*Solids content of feed,* wt/wt	*Solids handling rate, kg dry solids* h^{-1} m^{-2} *filter surface*[c]	*Moisure content of cake,* wt/wt	*Air flow* m^3 h^{-1} m^{-2} *filter surface*[d]	*Vacuum,* mmHg
Chemicals						
Alumina hydrate	Top feed drum	40	450–750	15	90	125
Barium nitrate	Top feed drum	80	1250	5	450	250
Barium sulphate	Drum	40	50	30	18	500
Bicarbonate of soda	Drum	50	1750	12	540	300
Calcium carbonate	Drum	50	125	22	36	500
Calcium carbonate (precipitated)	Drum	30	150	40	36	550
Calcium sulphate	Tipping pan	35	600	30	90	450
Caustic lime mud	Drum	30	750	50	108	375
Sodium hypochlorite	Belt discharge drum	12	150	30	54	500
Titanium dioxide	Drum	30	125	40	36	500
Zinc stearate	Drum	5	25	65	54	500
Minerals						
Frothed coal (coarse)	Top feed drum	30	750	18	72	300
Frothed coal (fine)	Drum or disc	35	400	22	54	375
Frothed coal tailings	Drum	40	200	30	36	550
Copper concentrates	Drum	50	300	10	36	525
Lead concentrates	Drum	70	1000	12	54	550
Zinc concentrates	Drum	70	750	10	36	500
Flue dust (blast furnace)	Drum	40	150	20	54	500
Fluorspar	Drum	50	1000	12	90	375

Notes:

[a] The information given should only be used as a general guide, for slight differences in the nature, size range and concentration of solids, and in the nature and temperature of liquor in which they are suspended, can significantly affect the performance of any filter.

[b] It should not be assumed that the type of filter stated is the only suitable unit for each application. Other types may be suitable, and the ultimate selection will normally be a compromise based on consideration of many factors regarding the process and the design features of the filter.

[c] The handling rate (in kg h^{-1} m^{-2}) generally refers to dry solids except where specifically referred to as filtrate.

[d] The air volumes stated are measured at the operating vacuum (i.e. they refer to attenuated air).

(Osborne, 1981).

TABLE 11.18. Typical Performance Data of Rotary Vacuum Filters

Material	Approximate particle size	Feed solids conc. wt %	Filtration rate (9) kg/(m²)(hr)	Vacuum Pump (9) $m^3/(m^2)(min)$	mm Hg
Disc filter					
Flotation coal	33–43%–200 mesh	22–26	300–600	1.5	500
Copper concentrates	90%–200 mesh	60–70	250–450	0.5	500
Magnetic concentrates	80–95%–325 mesh	55–65	1000–2000	2.5–3.0	600–650
Coal refuse	35–50%–250 mesh	35–40	100–125	0.6	500
Magnesium hydroxide	15 microns av. size	10–15	40–60	0.6	500
Drum filter					
(1) Sugar cane mud	Limed for flocculation	7–18 by vol.	25–75	0.2	500
$CaCO_3$ mud recausticising	—	35–40	500–600	1.8–2	250–380
(2) Corn starch	15–18 microns, av. size	32–42	110–150	0.9–1	560
Sewage sludge					
Primary	Flocculated	5–8	15–30	0.5	500
Primary digested	Flocculated	4–7	10–20	0.5	509
(3) Leached uranium ore	50–60%–200 mesh Flocculated	50–60	150–220	0.5–	500
Kraft pulp	Long fibre	1–1½	220–300	Barometric leg	
(4) Kaolin clay	98–75%–2 micron	25–35	30–75	0.5	600
Belt drum filter					
(5) Sugar cane mud	Seperan flocculated	7–18 by vol.	90–250	0.2	500
Sewage sludge					
Primary	Flocculated	5–8	30–50	0.5	500
Primary digested	Flocculated	4–7	15–35	0.5	500
Corn gluten	Self flocculating	16–20 oz/U.S. gal	15–30	0.6	500
Corn starch	15–18 microns, av. size	32–42	180–250	0.9–1	500
(3) Gold cyanide leached off	65%–200 mesh	50–60	300–600	0.5	500
(3) Spent vegetable carbon	98%–325 mesh	100–130 gm/litres	30–50	1.5	500
Dextrose processing					
Steel mill dust	20–40%–2 microns	40–50	170–300	0.6–1.2	500
(3) Sodium hypochlorite	Fine	12	150	0.9	500
Top feed drum					
Iron ore concentrates	2–4%–200 mesh 8 mesh top size	35	6300–7300	15	150
(6) Sodium Chloride	5–10%–100 mesh	25–35	1000–1500	30	150
Bone char	1%–70 mesh	8–20	1200–1700	40	90
(6) Ammonium sulphate	5–15%–35 mesh	35–40% by vol.	1000–1700	45–60	75
Tilting pan filter					
(7) Gypsum from digested phosphate rock	40–50 micron av.	35–40	600–900	1.2–1.5	500
(8) Leached cobalt residue	—200 mesh	45–50	250	3	380
(8) Alumina-silica gel catalyst	—	12	270	0.9	500
(7) Pentaerythritol	—	30–40	75–100	3.6	500

Notes: (1) Filtrate very dirty—must be recirculated back to clarifier—cake washed.
(2) String discharge filter.
(3) Cake washed.
(4) Roller discharge drum filter.
(5) Filtrate very clean—goes directly to evaporation—cake washed.
(6) top feed filter drier.
(7) Two or three stages of counter-current washing.
(8) Three stages of counter-current washing.
(9) Based on total filter area.

(Data of Envirotech Corp.).

TABLE 11.19. Data of Centrifugal Filters and Sedimentation Centrifuges (*Purchas, 1977*)

(a) Operating Ranges of Main Types of Centrifugal Filters

Type of Centrifuge	Continuous	Automatically Discharged at Full Speed	Automatically Discharged at Reduced Speed	g-Factor Range (F_c)	Minimum Solid Concentration in Feed [% by Volume (C_v)]	Possibility of Washing	Minimum Particle Size, mm	Minimum Filtrability Coefficient (k)(m/sec)	Maximum Retention Time (Sec)
Oscillating	x			30–120	40	no	0.3	5×10^{-4}	6
Tumbler	x			50–300	40	no	0.2	2×10^{-4}	6
Worm Screen	x			500–3000	20	poor	0.06	1×10^{-5}	15
Pusher	x			300–2000	30	good	0.08	5×10^{-5}	60
Peeler		x	x	300–1600	5	very good	0.01	2×10^{-7}	as wanted
Pendulum			x	200–1200	5	very good	0.005	1×10^{-7}	as wanted

(Hultsch and Wilkesmann; Purchas, 1977).

(continued)

TABLE 11.19—*(continued)*

(b) Criteria for Selection of Sedimentation Centrifuges

Parameter	Tubular Bowl	Skimmer Pipe	Disc	Scroll
Solids concentration. vol./vol.	<1%	up to about 40%	up to about 20%	any as long as it remains pumpable
Particle size range processable for density difference under 1 g/cc and liquor viscosity 1 cP	$\frac{1}{2}$–50 μm	10 μm–6 mm	1–400 μm	5 μm–6 mm
Settling time of 1 litre under 1 g	Few hours to infinity	$\frac{1}{2}$ hr to days	several hours	$\frac{1}{2}$–1 hr
Settling time of 50 cc at 2000 g	5–15 min	1–5 min	5–10 min	1–5 min
Approximation maximum throughput for largest machine	5000 litre/hr	15,000 litre/hr	100,000 litre/hr	70,000 litre/hr
Approximate nominal throughput for largest machine	1250 litre/hr	12,000 litre/hr	40,000 litre/hr	30,000 litre/hr
Nature of bottle spun solids	Can be any consistency	Must be fluid to pasty	Must not be too cohesive	Preferably compact and cohesive
Batch or continuous	Batch	Semi	Semi or continuous	Continuous
Floc applicable	Possibly but not usual	Yes	No	Yes
g levels used	Up to 18,000. 60,000 Laboratory model	Up to 1600	4500–12,000	500–4000
Maximum sigma value × 10^7 cm^2	5	4	10	14

(F.A. Records).

REFERENCES

1. C. Almy and W.K. Lewis, Factors determining the capacity of a filter press, *Ind. Eng. Chem.* **4,** 528 (1912).
2. N.P. Cheremisinoff and D. Azbel, *Liquid Filtration,* Ann Arbor Science, Ann Arbor, MI, 1981.
3. R. Bosley, Pressure vessel filters, in Purchas, Ref. 14, 1977, pp. 367–401.
4. D.A. Dahlstrom and C.E. Silverblatt, Continuous filters, in Purchas, Ref. 14, 1977, pp. 445–492.
5. E. Davies, Filtration equipment for solid–liquid separation, *Trans. Inst. Chem. Eng.* **43**(8), 256–259 (1965).
6. J.E. Flood, H.E. Parker, and F.W. Rennie, Solid-liquid separation, *Chem. Eng.* 163–181 (30 June 1966).
7. M.P. Freeman and J.A. FitzPatrick (Eds.), Theory, practice and process principles for physical separations, Proceedings of the Engineering Foundation Conference, Pacific Grove California, Oct.–Nov. 1977, Engineering Foundation or AIChE, 1981.
8. C. Gelman, H. Green, and T.H. Meltzer, Microporous membrane filtration, in Azbel and Cheremisinoff, Ref. 3, 1981, pp. 343–376.
9. C. Gelman and R.E. Williams, Ultrafiltration, in Cheremisinoff and Azbel, Ref. 3, 1981, pp. 323–342.
10. J. Gregory (Ed.), *Solid–Liquid Separation,* Ellis Horwood, Chichester, England, 1984.
11. K.J. Ives, Deep bed filtration, in Svarovsky, Ref. 17, 1981, pp. 284–301.
12. D.G. Osborne, Gravity thickening, in Svarovsky, Ref. 17, 1981, pp. 120–161.
13. D.G. Osborne, Vacuum filtration, in Svarovsky, Ref. 17, 1981, pp. 321–357.
14. D.B. Purchas, (Ed.), *Solid–Liquid Separation Equipment Scale-Up,* Uplands Press, London, 1977.
15. D.B. Purchas, *Solid–Liquid Separation Technology,* Uplands Press, London, 1981.
16. A. Rushton and C. Katsoulas, Practical and theoretical aspects of constant pressure and constant rate filtration, in Gregory, Ref. 10, 1984, pp. 261–272.
17. L. Svarovsky (Ed.), *Solid–Liquid Separation,* Butterworths, London, 1981.
18. F.M. Tiller (Ed.), *Theory and Practice of Solid–Liquid Separation,* University of Houston, Houston, 1978.
19. F.M. Tiller and J.R. Crump, Solid–liquid separation: an overview, *Chem. Eng. Prog.,* **73**(10), 65–75 (1977).
20. F.M. Tiller, J.R. Crump, and C. Ville, Filtration theory in its historical perspective; a revised approach with surprises, Second World Filtration Congress, The Filtration Society, London, 1979.
21. R.J. Wakeman, A numerical integration of the differential equations describing the formation of and flow in compressible filter cakes. *Trans. Inst. Chem. Eng.* **56,** 258–265 (1978).
22. R.J. Wakeman, Filter cake washing, in Svarovsky, Ref. 17, 1981, pp. 408–451.
23. R.H. Warring, *Filters and Filtration Handbook,* Gulf, Houston, 1981.
24. Solids Separation Processes, International Symposium, Dublin, April 1980, EFCE Publication Series No. 9, Institution of Chemical Engineers, Symposium Series No. 59, Rugby, England, 1980.

12

DISINTEGRATION, AGGLOMERATION, AND SIZE SEPARATION OF PARTICULATE SOLIDS

From the standpoint of chemical processing, size reduction of solids is most often performed to make them more reactive chemically or to permit recovery of valuable constituents. Common examples of comminution are of ores for separation of valuable minerals from gangue, of limestone and shale for the manufacture of cement, of coal for combustion and hydrogenation to liquid fuels, of cane and beets for recovery of sugar, of grains for recovery of oils and flour, of wood for the manufacture of paper, of some flora for recovery of natural drugs, and so on.

Since the process of disintegration ordinarily is not highly selective with respect to size, the product usually requires separation into size ranges that are most suitable to their subsequent processing. Very small sizes are necessary for some applications, but in other cases intermediate sizes are preferred. Thus the byproduct fines from the crushing of coal are briquetted with pitch binder into 3–4–in. cubes when there is a demand for coal in lump form. Agglomeration in general is practiced when larger sizes are required for ease of handling, or to reduce dust nuisances, or to densify the product for convenient storage or shipping, or to prepare products in final form as tablets, granules, or prills.

Comminution and size separation are characterized by the variety of equipment devised for them. Examples of the main types can be described here with a few case studies. For real, it is essential to consult manufacturers' catalogs for details of construction, sizes, capacities, space, and power requirements. They are properly the textbooks for these operations, since there are few generalizations in this area for prediction of characteristics of equipment. A list of about 90 U.S. and Canadian manufacturers of size separation equipment is given in the Encyclopedia of Chemical Technology *[**21,** 137 (1983)], together with identification of nine equipment types. The* Chemical Engineering Equipment Buyers Guide *(McGraw-Hill, New York) and* Chemical Engineering Catalog *(Penton/Reinhold, New York) also provide listings of manufacturers according to kind of equipment.*

12.1. SCREENING

Separation of mixtures of particulate solids according to size may be accomplished with a series of screens with openings of standard sizes. Table 12.1 compares several such sets of standards. Sizes smaller than the 38 μm in these tables are determined by elutriation, microscopic examination, pressure drop measurements, and other indirect means. The distribution of sizes of a given mixture often is of importance. Some ways of recording such data are illustrated in Figure 16.4 and discussed in Section 16.2.

The distribution of sizes of a product varies with the kind of disintegration equipment. Typical distribution curves in normalized form are presented in Figure 12.1, where the size is given as a percentage of the maximum size normally made in that equipment. The more concave the curves, the greater the proportion of fine material. According to these correlations, for example, the percentages of material greater than 50% of the maximum size are 50% from rolls, 15% from tumbling mills, and only 5% from closed circuit conical ball mills. Generalization of these curves may have led to some loss of accuracy since the RRS plots of the data shown in Figure 12.1(c) deviate much more than normally from linearity.

In order to handle large lumps, separators are made of sturdy parallel bars called grizzlies. Punched plates are used for intermediate sizes and woven screens for the smallest sizes. Screening is best performed dry, unless the feed is the product of wet grinding or is overly dusty and an equipment cover is not feasible. Wetting sometimes is used to prevent particles from sticking together. Types of screens and other classifiers to cover a range of sizes are shown in Figure 12.2. Usually some kind of movement of the stock or equipment is employed to facilitate the separations.

REVOLVING SCREENS OR TROMMELS

One type is shown in Figure 12.2(a). They are perforated cylinders rotating at 15–20 rpm, below the critical velocity. The different-sized perforations may be in series as shown or they may be on concentric surfaces. They are suitable for wet or dry separation in the range of 60–10 mm. Vertically mounted centrifugal screens run at 60–80 rpm and are suitable for the range of 12–0.4 mm.

Examples of performance are: (1) a screen 3 ft dia by 8 ft long with 5-mesh screen at 2 rpm and an inclination of 2° has a capacity of 600 cuft/hr of sand; (2) a screen 9 ft dia by 8 ft long at 10 rpm and an inclination of 7° can handle 4000 cuft/hr of coke.

Flat screens are vibrated or shaken to force circulation of the bed of particles and to prevent binding of the openings by oversize particles. Usually several sizes are arranged vertically as in Figures 12.2(b) and (c), but sometimes they are placed in line as in the cylindrical screen of Figure 12.2(a). Inclined screens vibrate at 600–7000 strokes/min. They are applicable down to 38 μm or so, but even down to 200 mesh at greatly reduced capacity. Horizontal screens have a vibration component in the horizintal direction to convey the material along; they operate in the range of 300–3000 strokes/min.

Shaking or reciprocating screens are inclined slightly. Speeds are in the range of 30–1000 strokes/min; the lower speeds are used for coal and nonmetallic minerals down to 12 mm, and higher speeds may size down to 0.25 mm. The bouncing rubber balls of Figure 12.2(c) prevent permanent blinding of the perforations.

Rotary sifters are of either gyratory or reciprocating types. They operate at 500–600 rpm and are used for sizes of 12 mm–50 μm, but have low capacity for fine sizes.

CAPACITY OF SCREENS

For coarse screening, the required area per unit of hourly rate may be taken off Figure 12.3. More elaborate calculation procedures that take into account smaller sizes and design features of the equipment appear in the following references:

Mathews, *Chem. Eng.* 76 (10 July 1972) and presented in *Chemical*

TABLE 12.1. Comparison Table of United States, Tyler, Canadian, British, French, and German Standard Sieve Series

U.S.A. (1)		TYLER (2)	CANADIAN (3)		BRITISH (4)		FRENCH (5)		GERMAN (6)
*Standard	Alternate	Mesh Designation	Standard	Alternate	Nominal Aperture	Nominal Mesh No.	Opg. M.M.	No.	Opg.
125 mm	5"		125 mm	5"					
106 mm	4.24"		106 mm	4.24"					
100 mm	4"		100 mm	4"					
90 mm	3½"		90 mm	3½"					
75 mm	3"		75 mm	3"					
63 mm	2½"		63 mm	2½"					
53 mm	2.12"		53 mm	2.12"					
50 mm	2"		50 mm	2"					
45 mm	1¾"		45 mm	1¾"					
37.5 mm	1½"		37.5 mm	1½"					
31.5 mm	1¼"		31.5 mm	1¼"					
26.5 mm	1.06"	1.05"	26.5 mm	1.06"					
25.0 mm	1"		25.0 mm	1"					25.0 mm
22.4 mm	7/8"	.883"	22.4 mm	7/8"					
19.0 mm	3/4"	.742"	19.0 mm	3/4"					20.0 mm
									18.0 mm
16.0 mm	5/8"	.624"	16.0 mm	5/8"					16.0 mm
13.2 mm	.530"	.525"	13.2 mm	.530"					
12.5 mm	1/2"		12.5 mm	1/2"					12.5 mm
11.2 mm	7/16"	.441"	11.2 mm	7/16"					
									10.0 mm
9.5 mm	3/8"	.371"	9.5 mm	3/8"					
8.0 mm	5/16"	2½	8.0 mm	5/16"					8.0 mm
6.7 mm	.265"	3	6.7 mm	.265"					
6.3 mm	1/4"		6.3 mm	1/4"					6.3 mm
5.6 mm	No. 3½	3½	5.6 mm	No. 3½					
							5.000	38	5.0 mm
4.75 mm	4	4	4.75 mm	4					
4.00 mm	5	5	4.00 mm	5			4.000	37	4.0 mm
3.35 mm	6	6	3.35 mm	6	3.35 mm	5			
							3.150	36	3.15 mm
2.80 mm	7	7	2.80 mm	7	2.80 mm	6			
2.36 mm	8	8	2.36 mm	8	2.40 mm	7	2.500	35	2.5 mm
2.00 mm	10	9	2.00 mm	10	2.00 mm	8	2.000	34	2.0 mm
1.70 mm	12	10	1.70 mm	12	1.68 mm	10	1.600	33	1.6 mm
1.40 mm	14	12	1.40 mm	14	1.40 mm	12			
							1.250	32	1.25 mm
1.18 mm	16	14	1.18 mm	16	1.20 mm	14			
1.00 mm	18	16	1.00 mm	18	1.00 mm	16	1.000	31	1.0 mm
850 µm	20	20	850 µm	20	850 µm	18			
							.800	30	800 µm
710 µm	25	24	710 µm	25	710 µm	22			
							.630	29	630 µm
600 µm	30	28	600 µm	30	600 µm	25			
500 µm	35	32	500 µm	35	500 µm	30	.500	28	500 µm
425 µm	40	35	425 µm	40	420 µm	36			
							.400	27	400 µm
355 µm	45	42	355 µm	45	355 µm	44			
							.315	26	315 µm
300 µm	50	48	300 µm	50	300 µm	52			
250 µm	60	60	250 µm	60	250 µm	60	.250	25	250 µm
212 µm	70	65	212 µm	70	210 µm	72			
							.200	24	200 µm
180 µm	80	80	180 µm	80	180 µm	85			
							.160	23	160 µm
150 µm	100	100	150 µm	100	150 µm	100			
125 µm	120	115	125 µm	120	125 µm	120	.125	22	125 µm
106 µm	140	150	106 µm	140	105 µm	150			
							.100	21	100 µm
90 µm	170	170	90 µm	170	90 µm	170			90 µm
							.080	20	80 µm
75 µm	200	200	75 µm	200	75 µm	200			
									71 µm
63 µm	230	250	63 µm	230	63 µm	240	.063	19	63 µm
									56 µm
53 µm	270	270	53 µm	270	53 µm	300			
							.050	18	50 µm
45 µm	325	325	45 µm	325	45 µm	350			45 µm
							.040	17	40 µm
38 µm	400	400	38 µm	400					

(1) U.S.A. Sieve Series—ASTM Specification E-11-70
(2) Tyler Standard Screen Scale Sieve Series.
(3) Canadian Standard Sieve Series 8-GP-1d.
(4) British Standards Institution, London BS-410-62.
(5) French Standard Specifications, AFNOR X-11-501.
(6) German Standard Specification D1N 4188.

* These sieves correspond to those recommended by ISO (International Standards Organization) as an International Standard and this designation should be used when reporting sieve analysis intended for international publication.

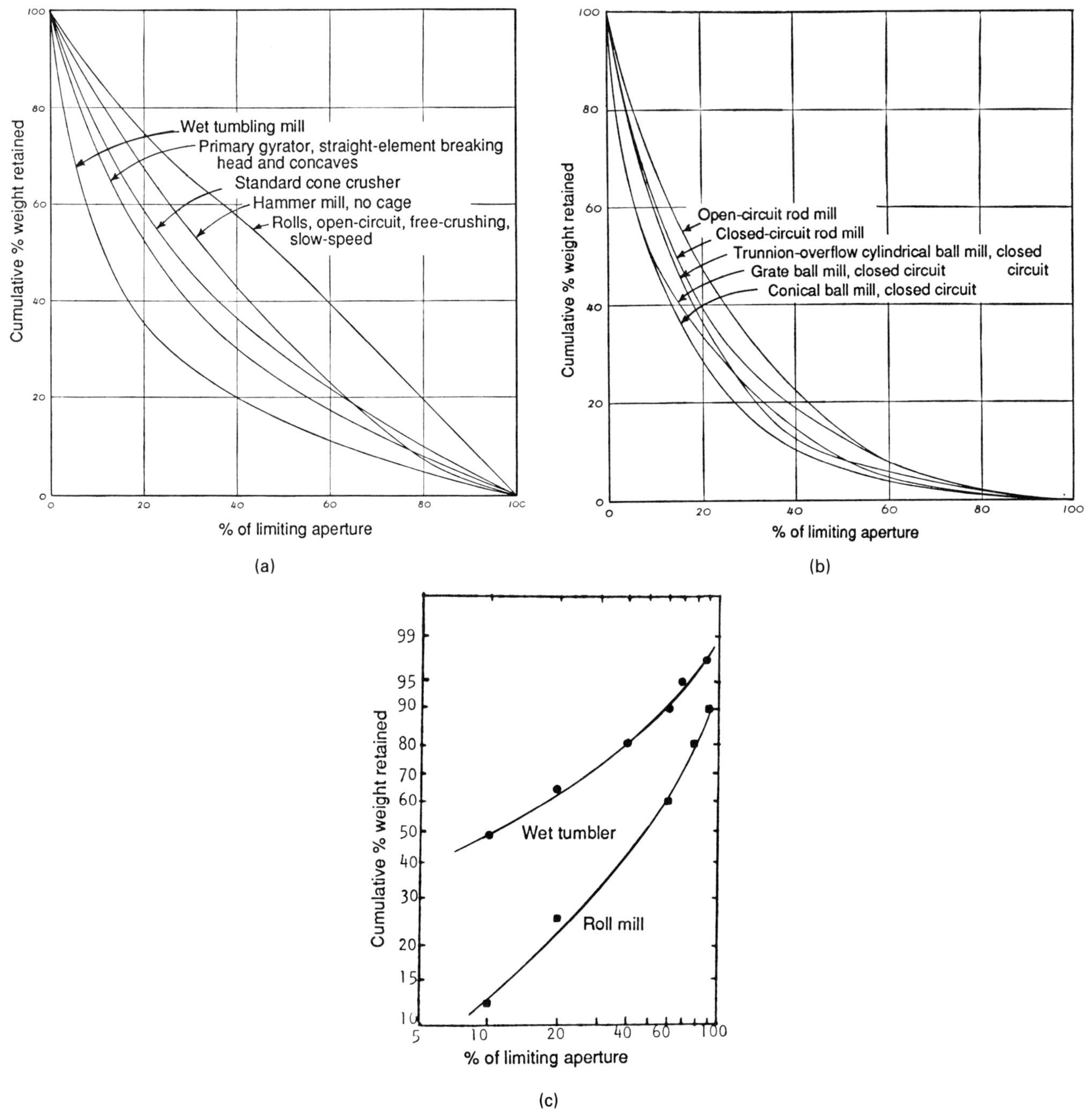

Figure 12.1. Normalized cumulative size distribution curves of comminuted products. (a) From various kinds of crushing equipment. (b) From rod and ball mills. (c) RRS plots of two curves (*Taggart, 1951*).

Engineers' Handbook, McGraw-Hill, New York, 1984, p. 21.17.

Kelly and Spottiswood, *Introduction to Mineral Processing,* 1982, p. 193.

V.K. Karra, Development of a model for predicting the screening performance of a vibrating screen, *CIM Bull.* **72,** 167–171 (Apr. 1979).

The last of these procedures is in the form of equations suitable for use on a computer.

12.2. CLASSIFICATION WITH STREAMS OF AIR OR WATER

Entrainment of particles with streams of air or water is particularly suitable for removal of small particles from mixtures. Complete distribution curves can be development by employing several stages operating at suitable conditions in series.

AIR CLASSIFIERS

Although screens of 150 mesh and finer are made, they are fragile and slow, so that it is often preferable to employ air elutriation to

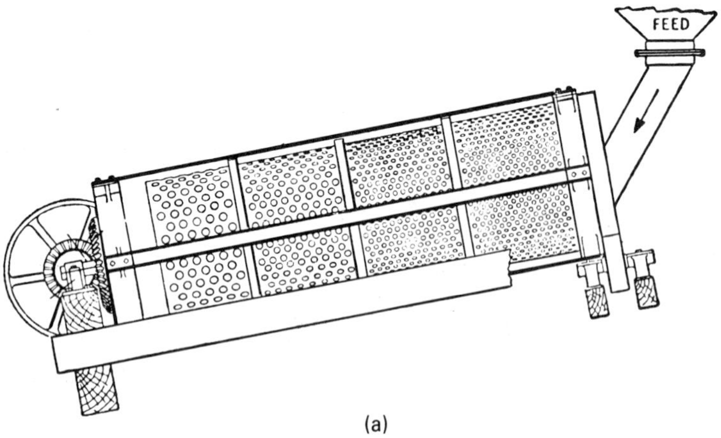

(a)

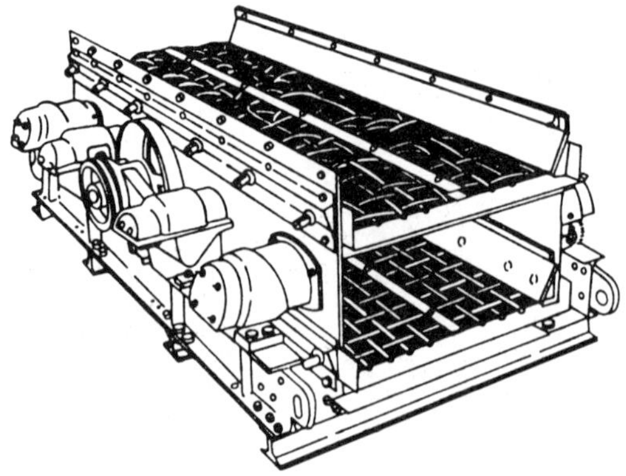

(b)

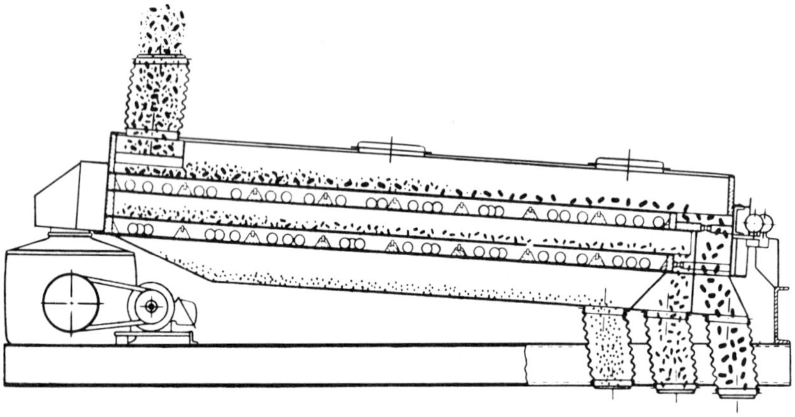

Model Number	Screen Size per Surface	Capacity in Tons per Hour		Motor	
		Pellets	Crumbles	HP	RPM
203	30″ x 60″	13	8	2	1800
43*	40″ x 84″	25	15	2	1800
83*	40″ x 120″	35	23	3	1800
523	60″ x 120″	50	34	7.5	1200
73	80″ x 144″	85	55	10	1200

(c)

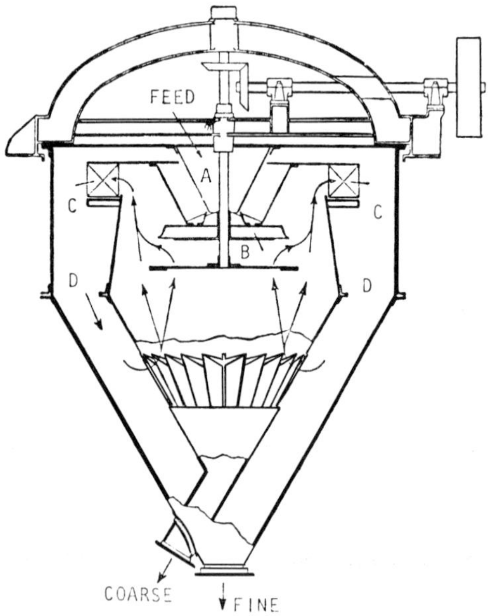

(d)

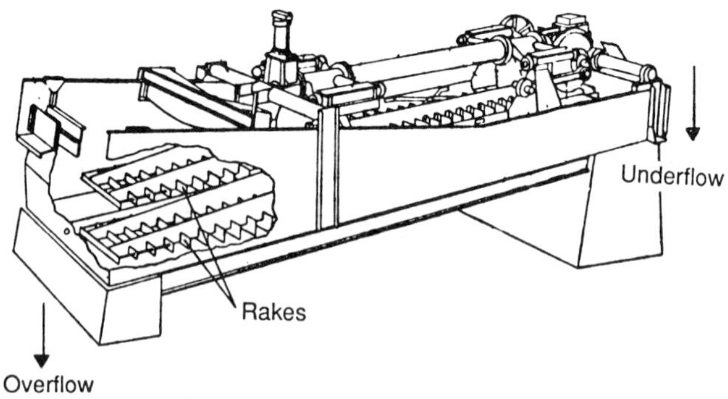

(e)

Figure 12.2. Equipment for classifying particulate solids by size from more than 0.5 in. to less than 150 mesh. (a) Rotating cylinder (trommel) for sizing particles greater than 0.5 in., 2–10 rpm, 10–20° inclination. (b) Heavy duty vibrating screen, 1200–1800 vib/min (*Tyler–Niagara, Combustion Engineering Inc.*). (c) Three-product reciprocating flat screen, 500–600 rpm, with bouncing rubber balls to unbind the openings, dry products to 100 mesh (*Rotex Inc.*). (d) Air classifier for products less than 150 mesh. Feed enters at A, falls on the rotating plate B, fines are picked up by air suction fans C, transferred to zone D where they separate out and fall to the discharge, and air recirculates back to fans C (*Sturtevant Mill Co.*). (e) Dorr drag rake wet classifier. (f) Hydrocyclone.

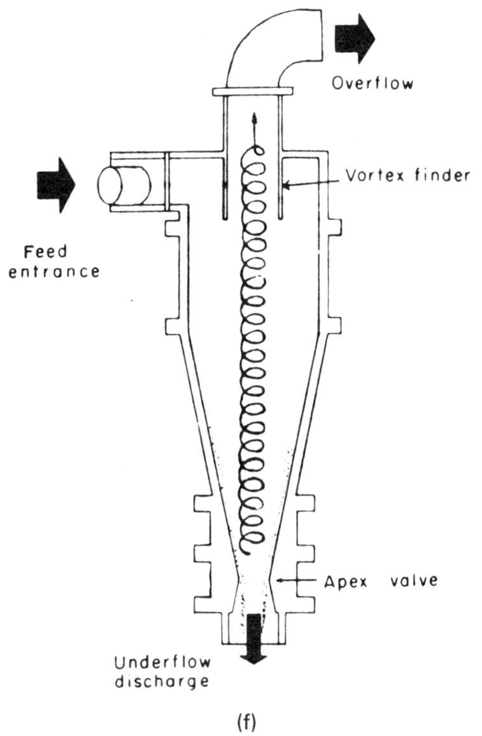

Figure 12.2—*(continued)*

remove fine particles. The equipment of Figure 12.2(d) employs a rotating plate that throws the particles into the air space from which the finer particles are removed and subsequently recovered.

WET CLASSIFIERS

These are used to make two product size ranges, oversize and undersize, with some overlap. The break commonly is between 28 and 200 mesh. A considerable variety of equipment of this nature is available, and some 15 kinds are described by Kelly and Spottiswood (1982, pp. 200–201). Two of the most important kinds, the drag rake classifier and the hydrocyclone, will be described here.

The classifier of Figure 12.2(e) employs two set of rakes that alternately raise, lower, and move the settled solids up the incline to the discharge. Movement of the rakes is sufficient to keep the finer particles in suspension and discharge them at the lower end. More construction detail of the Dorr classifier may be found in older books, for example, the 1950 edition of the *Chemical Engineers Handbook* (McGraw-Hill, New York). The stroke rate may be 9/min when making separation at 200 mesh and up to 32/min for 28 mesh rapid settling sands. Widths range from 1 to 20 ft, lengths to 40 ft, capacity of 5–850 tons slurry/hr, loads from 0.5 to 150 HP. The solids content of the feed is not critical, and that of the overflow may be 2–20% or more.

Hydrocyclones, also called hydroclones, employ self-generated mild centrifugal forces to separate the particles into groups of predominantly small and predominantly large ones. Because of bypassing, the split of sizes is not sharp. The characteristic diameter of the product is taken as d_{50}, the diameter than which 50 wt % of the material is greater or less. The key elements of a hydrocyclone are identified on Figure 12.2(f). A typical commercial unit made by Krebs Engineers has an inlet area about 7% of the cross-sectional area between the vessel wall and the vortex finder, a vortex finder with diameter 35–40% that of the vessel, and an apex diameter not less than 25% that of the vortex finder. For such a unit, the equation for the cut point is

$$d_{50} = \frac{13.2D^{0.675}\exp(-0.301 + 0.0945V - 0.00356V^2 + 0.0000684V^3)}{(\Delta P)^{0.3}(S-1)^{0.5}} \tag{12.1}$$

and the slurry flow rate is

$$Q = 0.7(\Delta P)^{0.5}D^2 \tag{12.2}$$

in the units d_{50} μm, vessel diameter D in inches, V = vol % of solids in the feed, ΔP is the pressure drop in psi, S = specific gravity, and Q is the flow rate in gpm (Mular and Jull, in Mular and Bhappu 1978, p. 397). Performance characteristics of one line of commercial hydrocyclones are shown in Figure 12.3(b). Comparison of the chart and equations is made in Example 12.1.

Hydrocyclones are small and inexpensive separators for handling feeds up to about 600 cuft/min and removing particles in the range of 300–5 μm from dilute suspensions. Large diameters (up to about 24 in.) have greater volumetric capacity but also a greater cutpoint on particle diameter. Series and parallel arrangements may be made for any desired compromise between these quantities. In comparison with drag rake classifiers, hydrocyclones are smaller, cost about the same to operate but have lower costs for capital and installation. They are preferred in closed circuit grinding.

12.3. SIZE REDUCTION

Crushing is applied to large lumps of feed stock and grinding to smaller lumps, often the products of crushing, but the size distinction is not overly sharp. The process of size reduction results in a range of product sizes whose proper description is with the complete cumulative size distribution, but for convenience a characteristic diameter corresponding to 80% pass in the cumulative distribution curve is commonly quoted.

Some devices employ impact (hammers) and others employ crushing by nipping (rolls or jaws). Within limits, kinetic energy and dimensions of crushing elements can be selected to give a desired reduction ratio. Because of the deformability of solid materials, however, a theoretical limit does exist to the size of particles that can be crushed. These limits are 1 μm for quartz and 3–5 μm for limestone. The products of crushing these sizes, of course, can be very much smaller, so that really there is no practical lower limit to grinding.

In practical operations, only about 1% of the input energy to the mill appears as new surface energy of the product. Nevertheless, empirical relations for power consumption based on the extent of size reduction have been developed. One such relation is

$$W = 10W_i(1/\sqrt{d} - 1/\sqrt{d_i}), \quad \text{kWh/ton}, \tag{12.3}$$

where d and d_i are the final and initial diameters (μm) corresponding to 80% cumulative passing. The work index W_i is related to the crushing strength of the material; typical values appear in Table 12.2. Example 12.2 compares a result from this formula with direct data from a manufacturer's catalog.

Characteristics of the main common types of size reduction equipment are listed in Table 12.3, including size of feed, size of product, capacity, power consumption, and average reduction ratio. Coarse comminuters perform with reduction ratios less than 10, fine ones with ratios of 100 or more. From very large to very fine may require several operations in series, as in the flowsketch of Figure 12.4(b), where three stages of crushing and two of classification are shown.

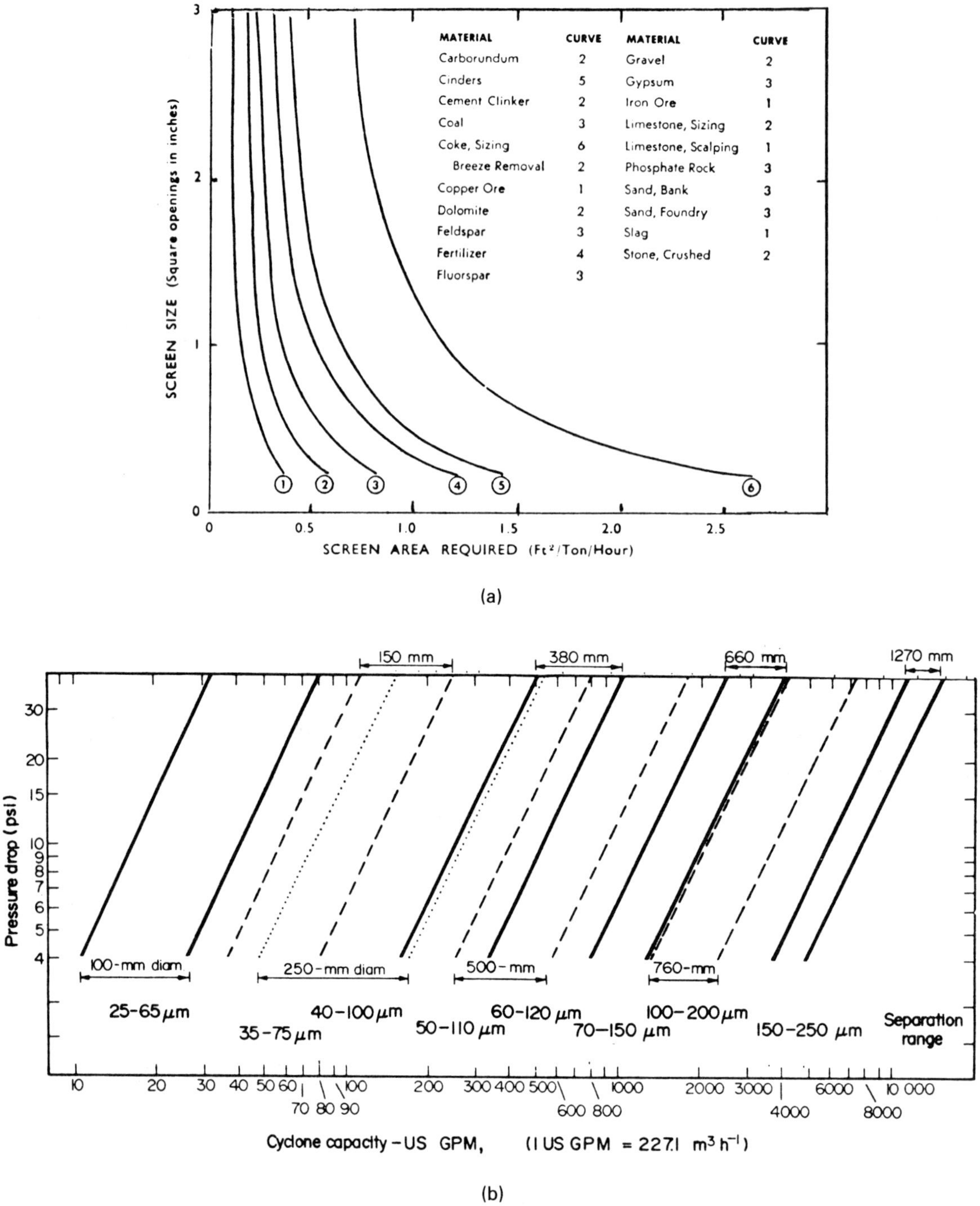

Figure 12.3. Performances of screens and hydrocyclones. (a) Capacities of screens for various products (*Denver Equipment Co.*). (b) Capacity, separation range and pressure drop of hydrocyclones (*Krebs Engineers*). Example: A 380 mm dia vessel has a separation range of 50–110 μm, and can handle between 200 and 450 gpm at a pressure drop of 7.5 psi.

Toughness, hardness, and temperature sensitivity are some of the properties that influence choice of equipment and operating conditions. Fibrous materials require cutting rather than crushing action. Temperature-sensitive materials such as plastics and rubber need to be cooled with ambient or refrigerated air. Cryogenic processing that involves immersion of the material in liquid nitrogen is employed even for such prosaic materials as scrap automobiles and rubber tires; the low temperatures enhance brittleness and result in lowered power consumptions.

The kinds of equipment used for certain materials are identified in Table 12.4. Usually several kinds are more or less equally suited. Then the choice may be arbitrary and based on experience or on marginal considerations. Table 12.5 presents a broader range of materials that are being ground in four of the principal kinds of fine

Example 12.1
Sizing a Hydrocyclone

A hydrocyclone assembly is required to handle 10,000 gpm of slurries of a solid with specific gravity 2.9 with a cutoff point of $d_{50} = 100\ \mu m$. The allowable pressure is $\Delta P = 5$ psi. Several slurry concentrations V will be examined. Substituting into Eq. (12.1), with z the function of V in parentheses,

$$100 = \frac{13.2D^{0.675}e^z}{5^{0.3}1.9^{0.5}},$$

whence

$$D = (16.92/e^z)^{1.4815}.$$

The corresponding capacity of one hydrocyclone is

$$Q = 0.7(5)^{0.5}D^2.$$

The results are tabulated following at several values of V:

V	e^z	D in.	D mm	Q	No. Units in Parallel
5	1.0953	57.7	1466	5214	2
10	1.4282	39.0	989	2375	4
20	2.0385	23.0	584	828	12
30	3.2440	11.6	293	209	48

From Figure 12.3(b), with 5 psi a 660 mm unit will handle 1000 gpm and have a cutoff between 50 and 150 μm. This corresponds to the calculated data with V about 19 vol %. For a more detailed study of hydrocyclone sizing, the article of Mular and Jull (in Mular and Bhappu, 1980, pp. 376–403) may be consulted. The pressure drop can be adjusted to compensate for changes in slurry concentration.

grinders. Performances of attrition, cutter and jet mills with some materials are given with Table 12.7. Additional operating data arranged by material are referred to in Table 12.10.

Closed circuit grinding employs a means whereby only material smaller than a specified size appears in the product. A less precise mode of operation employs an air stream through the equipment at such a rate that only the appropriately fine material is withdrawn and the rest remains until it is crushed to size. Ball mills sometimes are operated in this fashion, and also the ring-roller mill of Figure 12.4(a). For closer size control, all of the crushed material is withdrawn as it is formed and classified externally into product and recycle. The other examples of Figure 12.4 illustrate several such schemes.

Wet grinding with water is practiced when dusting is a problem, or when subsequent processing is to be done wet, as of ores that are later subjected to separation by flotation or sink-float processes. Removal of a slurry from a ball mill is easier than of dry material; there are cases where this advantage is controlling. Because of the lubricating effect of the water, power consumption of wet milling is less per ton, but this advantage may be outweighed by corrosion of the equipment.

12.4. EQUIPMENT FOR SIZE REDUCTION

Some of the many available kinds of size reduction equipment can be described here. Manufacturers' catalogs have the most complete descriptions of the equipment and almost always provide typical or expected performance data. Useful compilations of such information are by Taggart (1945) and the *Chemical Engineers Handbook* (1984, Section 8, as well as older editions).

CRUSHERS

Lumps as large as several feet in diameter are crushed in gyratory or jaw crushers. Figure 12.5(a) shows a type of crusher that is made in widths from about 5 to 70 in. and with gaps from 4 to 60 in. Stroke rates vary from 300 to 100/min. The average feed is subjected to 8–10 strokes before it becomes small enough to escape. The jaw crusher is suited to hard, abrasive, and also sticky feeds; it makes minimum fines but the product may be slabby because of the long, narrow exit. Gyratory crushers are more suited to slabby feeds and make a more rounded product.

ROLL CRUSHERS

Toothed rolls such as those of Table 12.8(b) can handle relatively large lumps, for example, 14 in. maximum with 24 in. rolls according to the table. To smooth rolls, the feed size is limited by the angle of nip which depends on the surface conditions but often is approximately 16° or arccos 0.961. Accordingly the relation between the diameters of the roll d_r and feed d_f and the gap d_0 between the rolls is given by

$$d_r = (0.961d_f - d_0)/0.039. \qquad (12.4)$$

For example, with $d_f = 1$ in. and $d_0 = 0.25$ in., the roll diameter is figured as 18 in. Table 12.8(b) lists 16 in. as the smallest size suitable for this service, which appears to be somewhat marginal in comparison with the calculated result. According to the formula, 1 in. lumps could be nipped by 16 in. rolls with a spacing of 0.34 in. It is not possible to state who is smarter, the formula or the manufacturer.

Figure 12.5(b) shows a smooth roll assembly. Usually only one of the rolls is driven and one is spring mounted to prevent damage by uncrushable material in the feed. Reduction ratios shown in Table 12.8(c) range only between 2:1 and 4:1. The proportion of fines is comparatively small. Sets of rolls in series with decreasing settings are used to achieve overall high reduction ratios. The rolls of a pair can be driven at the same or different speeds, within a range of 50–900 rpm. The capacity generally is about 25% of the maximum corresponding to a continuous ribbon of material passing between the rolls. A sample listing of materials that are ground in roll mills is in Table 12.5(a). In the arrangement of Figure 12.4(c), the upper pair of rolls is the primary crusher whereas the lower pair works on recycle of the oversize.

Hammer mills employ rotating elements that beat the material until it is small enough to fall through the screen at the bottom of the casing. Product size is determined by the speed of the hammers and the size of the screen openings. Table 12.9(a) shows the former effect. The units of this table operate at speeds up to 900 rpm and make size reductions of 40:1 or so. The smaller units of Table 12.9(b) operate at speeds to 16,000 rpm and make very fine powders. Because of the heating effect, they often are operated with a stream of ambient or refrigerated air for cooling. Under these conditions even heat softening materials such as natural resins or chicle can be ground satisfactorily. Hammer mills are the

TABLE 12.2. Typical Values of the Work Index W_i kWh/ton, of Eq. (12.3)

Material	Work Index W_i	Material	Work Index W_i
All materials tested	13.81	Kyanite	18.87
Andesite	22.13	Lead ore	11.40
Barite	6.24	Lead-zinc ore	11.35
Basalt	20.41	Limestone	11.61
Bauxite	9.45	Limestone for cement	10.18
Cement clinker	13.49	Manganese ore	12.46
Cement raw material	10.57	Magnesite, dead burned	16.80
Chrome ore	9.60	Mica	134.50
Clay	7.10	Molybdenum	12.97
Clay, calcined	1.43	Nickel ore	11.88
Coal	11.37	Oil shale	18.10
Coke	20.70	Phosphate fertilizer	13.03
Coke, fluid petroleum	38.60	Phosphate rock	10.13
Coke, petroleum	73.80	Potash ore	8.88
Copper ore	13.13	Potash salt	8.23
Coral	10.16	Pumice	11.93
Diorite	19.40	Pyrite ore	8.90
Dolomite	11.31	Pyrrhotite ore	9.57
Emery	58.18	Quartzite	12.18
Feldspar	11.67	Quartz	12.77
Ferro-chrome	8.87	Rutile ore	12.12
Ferro-manganese	7.77	Sandstone	11.53
Ferro-silicon	12.83	Shale	16.40
Flint	26.16	Silica	13.53
Fluorspar	9.76	Silica sand	16.46
Gabbro	18.45	Silicon carbide	26.17
Galena	10.19	Silver ore	17.30
Garnet	12.37	Sinter	8.77
Glass	3.08	Slag	15.76
Gneiss	20.13	Slag, iron blast furnace	12.16
Gold ore	14.83	Slate	13.83
Granite	14.39	Sodium silicate	13.00
Graphite	45.03	Spodumene ore	13.70
Gravel	25.17	Syenite	14.90
Gypsum rock	8.16	Tile	15.53
Ilmenite	13.11	Tin ore	10.81
Iron ore	15.44	Titanium ore	11.88
Hematite	12.68	Trap rock	21.10
Hematite—Specular	15.40	Uranium ore	17.93
Oolitic	11.33	Zinc ore	12.42
Limanite	8.45		
Magnetite	10.21		
Taconite	14.87		

[F.C. Bond, Bri. Chem. Eng. **6**, 378–385, 543–548 (1961)].

EXAMPLE 12.2
Power Requirement for Grinding

Cement clinker is to be reduced from an initial $d_{80} = 1500\ \mu m$ to a final d_{80} of 75 μm. From Table 12.2 the work index is $W_i = 13.49$. Substituting into Eq. (12.3),

$$W = 10(13.49)(1/\sqrt{75} - 1/\sqrt{1500}) = 12.1\ \text{kW/(ton/hr)}.$$

According to Table 12.7(b), a 3 ft × 24 in. ball mill requires 10 HP for a rate of 0.5 tons/hr, or 14.9 kW/(ton/hr), a rough check of the result from the equation.

principal equipment for cryogenic processing when products of 50–100 mesh are adequate. Scrap automobiles and rubber tires are chilled with liquid nitrogen and are thereby made highly brittle to facilitate grinding. Nitrogen consumption runs about 0.25 kg/kg steel and up to 0.65 kg/kg rubber [Biddulph, *Chem. Eng.*, (11 Feb. 1980)].

This equipment is particularly suited to crushing of soft, friable materials to cube-shaped products with small proportions of fines. For fibrous materials, the screen is provided with cutting edges. Some data are in Table 12.7(c). A list of materials that are handled in hammer mills is in Table 12.5(a), and other products are referred to in Table 12.10.

Tumbling mills consist of vessels rotating about the horizontal and charged with a mass of relatively small elements that tumble and crush the process material as they fall. Their function may be to mix as well as grind, in batch or continuous operation, in open or closed circuit. Figure 12.4(d) shows a closed circuit arrangement with a ball mill. The crushing elements most commonly are steel balls of several sizes, or ceramic pebbles, or rods the length of the shell, or a range of sizes of the process material that is thus made to grind itself. In processing of minerals, tumbling mills often operate wet with slurries of about 80% solids, just thin enough to permit flow in and out of the equipment.

The mode in which the material grinds itself is called *autogenous grinding*. Such operation can achieve size reduction from 25 cm to 0.1 mm in one step. Autogeneous mills operate at 80–85% of the critical speed, which is the speed at which the grinding media are thrown to the wall and cling to it. They are desirable for mineral treatment since they release the mineral content without overgrinding which could complicate a subsequent flotation process, for instance. Materials for which the process is used are friable and grainy, such as silica rock, asbestos, basic slag, bauxite, cement clinker, dolomite, ferrosilicon, limestone, specular hematite, and taconite. In comparison with ball milling, steel consumption is largely eliminated but energy costs are greater by between 25 and 100% because of lower impacting forces with low density materials.

Rod mills [Fig. 12.5(f)] are capable of taking feed as large as 50 mm and reducing it to 300 mesh, but ordinarily the cutoff point is larger. The performance data of Table 12.6(e) shows a product range from 8 to 65 mesh. Rods in use range from 25–150 mm dia; smaller ones tend to bend and break. The ratio of rod length to vessel diameter is kept in the range of 1.4–1.6. Ratios below 1.25 tend to result in tangling. Maximum usable rod length is about 6 m; above this they tend to bend. About 45% of the bulk volume of the mill is occupied by rods. Rotation is at 50–65% of critical speed. Rod consumption normally is in the range of 0.1–1.0 kg steel/ton of ore for wet grinding, and about 10–20% less for dry grinding. Because the coarse feed tends to spread the rods at the feed end, grinding takes place preferentially on the large particles and results in a product of relatively narrow size range. Accordingly, rod mills are nearly always run in open circuit.

Ball mills serve as a final stage of comminution. Balls have a greater ratio of surface area to weight than rods so they are better suited to fine grinding. The length to diameter ratio ranges from less than 1 to about 1.5. Rotation speed is greater than that of rod mills, being 70–80% of critical. Mills that are subjected to vibration can operate above the critical speed. The bulk volume of balls is about 50% of the mill volume.

The Denver ball mills for which operating data are shown in Table 12.6(a) normally are charged with equal weights of 2-, 3-, and 4-in. balls; or for finer grinding, with equal weights of 1.5-, 2-, and 3-in. balls. Figure 12.5(d) is of the widely used conical shape of mill in which a range of sizes of balls group themselves axially during operation. The balls range from 5 in. down, the large ones for crushing the large lumps and the small ones acting on the small

TABLE 12.3. Operating Ranges for Commonly Used Size Reduction Equipment

Equipment	Size of Feed (mm)	Size of Product (mm)	Reduction Ratio	Capacity (tons/hr)	Power Consumption (kW)
Gyratory crushers	200–2000	25–250	8	100–500	100–700
Jaw crushers	100–1000	25–100	8	10–1000	5–200
Cone crushers	50–300	5–50	8	10–1000	20–250
Impact breakers	50–300	1–10	40	10–1000	100–2000
Rod mills	5–20	0.5–2	10	20–500	100–4000
Ball mills	1–10	0.01–0.1	100	10–300	50–5000
Hammer mills	5–30	0.01–0.1	400	0.1–5	1–100
Jet mills	1–10	0.003–0.05	300	0.1–2	2–100

lumps. The performance data of Table 12.6(b) are for wet grinding; dry grinding capacities are 10–20% less. Segregation of balls by size also is achieved in cylindrical shapes with spiral twists in the liner profile (Trelleborg AB).

Tube mill is the term applied to a mill of uniform diameter with greater ratio of length to diameter, in the range of 4–5. Because of the greater length, and correspondingly greater residence time, a finer product can be obtained. Figure 12.5(e) shows a tube mill with three compartments.

Pebble mills are single compartment tube mills with ceramic balls as the grinding medium. They are used when contamination with iron must be avoided, as for grinding and mixing of light colored pigments, food products, and pharmaceuticals. Since the grinding rate is roughly proportional to the weight of the balls, the grinding rate with pebbles is only about 1/3 that with steel balls of the same volume. This is clear from the data of Table 12.6(b). The batch operating data of Table 12.6(c) are for grinding sand of density 100 lb/cuft with a charge of 50 vol % of pebbles or 33 vol % of steel balls. The obvious advantage of batch grinding is that any degree of fineness can be obtained by allowing sufficient time.

TABLE 12.4. Size Reduction Equipment Commonly Used in the Chemical Process Industries

Material	Equipment
Asbestos and mica	roll crushers, hammer, and jet mills
Cement	gyratory, jaw and roll crushers, roller, and ball mills
Clays	pan crushers, ring-rollers, and bead mills
Coal	roll crushers, pulverizers, ball, ring-roller, and bowl mills
Coke	rod, ball, and ring-roller mills
Colors and pigments	hammer, jet, and ring-roller mills
Cosmetics and pharmaceuticals	dispersion and colloid mills
Cotton and leather	rotary cutters
Flour and feed meal	roller, attrition, hammer, and pin mills
Graphite	ball, tub, ring-roller, and jet mills
Hard rubber	roller mills
Lime and Shells	hammer and ring-roller mills
Metallic minerals	gyratory and jaw crushers, tumbling mills
Paper and plastics	cutters and slitters
Phosphates	ball and ring-roller mills
Polymers	pulverizers, attrition mills
Pressed cakes	hammer and attrition mills
Refractories	gyratory and jaw crushers, pan and ball mills
Salts	cage and hammer mills
Soaps	hammer, multicage, and screen mills
Starch	hammer and pin mills
Stone and aggregate	gyratory, jaw, and roll crushers
Sulfur	ring-roller mills
Talc and soapstones	roll crushers, ring-roller, pebble, and jet mills

Roller mills [Figs. 12.4(a) and 12.6(f)]. Such equipment employs cylindrical or tapered surfaces that roll along flatter surfaces and crush nipped particles. In the ring-roller mill of Figure 12.4(a), spring-loaded rolls are forced against a rotating ring and crush the material that is thrown between them with ploughs. In another design the ring is stationary, and the assembly of rolls is rotated and maintained in contact with the ring by centrifugal force. The unit shown is equipped with built-in air classification: as fines are formed, they are removed with an air stream. For closer control of size distribution, ring-roller mills often are operated with external air classification similar to the process of Figure 12.4(d). The performance data of Table 12.6(c) are for products ranging from 20 to 200 mesh, with appropriate control of air rates. Ring roller mills are used widely for grinding of materials from coal to hard rock and cement clinker; some applications are cited in Tables 12.5(b) and 12.10.

Some special equipment is shown in Figure 12.6.

a. *Gyratory crusher* was described earlier. The sketch shows the cone-shaped element rotating in a tapered passage. Gyratories are made to handle even larger lumps than jaw crushers. Very large lumps are broken first by explosives or pneumatic hammers before feeding to crushers.
b. *The squirrel-cage disintegrator* illustrated has four concentric cages with horizontal impact bars. The cages rotate in alternately opposite directions, strike the feed and disintegrate or tear it apart. Coal, other friable materials, and fibrous materials are handled in this equipment.
c. *Disc-type attrition mills* have surfaces that rotate past each other with close clearance and high speed, usually in opposite directions but sometimes in the same direction at different speeds. Clearances are adjustable with spring loading in increments of 0.001 in. Maximum feed sizes are 10–25 mm, diameters 12–48 in., and speeds 1200–7000 rpm. Table 12.5(b) is a list of materials that have been ground in disc mills. Some data are in Table 12.7(b).
d. *Colloid mills* are used to grind and disperse solids in liquids and to prepare emulsions. Adjustable clearances are between 0.001 and 0.050 in., and peripheral speeds to 10,000 ft/min. They are used, for example, to make lubricating greases by dispersion of calcium stearate in hydrocarbon oils.
e. *Buhrstone mills* are an ancient example of an attrition mill. Nowadays they are used mostly for fine grinding of paints, inks, and pharmaceuticals.
f. *Roller mills,* also called spindle mills, act by crushing nipped materials between a rotating cylindrical or tapered surface and a

TABLE 12.5. Materials that Have Been Ground in Particular Kinds of Mills

(a) Crusher Rolls

ammonium nitrate
asbestos
barley malt
bauxite
beet pulp
bone
casein
catalyst beads
cereals
charcoal
cheese
chemicals
coal
cocoa cakes
coconut shells
coffee
cork
corn
corn cobs
corn meal
cottonseed
cracker meal
crackings
crimping grains
dog food cakes
DDT
dolomite lime
dried biscuits
dried apple pulp
feedstuffs
flaxseed
floor tile
flour
fuller's earth
glue
grains
gun powder
insulating materials
iron oxide
lumpy chemicals and flour
magnesium oxide
malt
malted milk
meat scraps
mustard seed
oil bearing seeds
pelletized feeds
pepper
pharamaceuticals
plastics
reclaim rubber
resin
salt
soy beans
spices
sponge iron
starch
uranium concentrates

(b) Disc Attrition Mills

alloy powders
alum
aluminum chips
apples, dried
asbestos
bark
borax
brake lining scrap
brass chips
caustic soda
cereals
chalk
charcoal
chemical salts
chips
cloves
cocoa
coconut shells
copper powders
copra
cork
corn
cottonseed and hulls
drugs
dye stuffs
egg shells
feathers
fertilizers
fish meal
glue
gum arabic
hops
leather
metal powder
mica
nuts and shells
oil cake
paris green
peanuts and hulls
pepper
phosphates
plaster
potash
potatoes
pumice
rice and hulls
roots
rosin
rubber
sawdust
salt
suds
soy beans
spices
starch
shavings
tankage
tobacco stems
wood pulp

(c) Roller Mills

alum
barytes
hematite
insecticide roots
bentonite
clay
cement clinker
chalk
cocoa
DDT
dolomite
feldspar
graphite
gypsum
kaolin
lime
limestone
mica
phosphate rock
resins
soy bean cake
sulfur
talc
titanium dioxide

(d) Hammer Mills

aluminum tristearate
animal glue
antioxidants
asbestos
asphalt
aspirin
bagasse
barley
bentonite, dried
bone char
brewer's yeast
calcium carbonate
calcium phosphate
carbon black
cellulose acetate
cinnamon
clay
coal
cocoa cake
cocoa-sugar mixtures
coconut shells
corn meal
cottonseed cake
diatomaceous earth
dyestuffs
etching powder
ginger
graphite
guar gum
gum acacia
gypsum
irish moss
lactose
lead, red
licorice root
lime, hydrated
mica
milo grain
oats, rolled
oyster shells
pentaerythritol
perlite
pigments
plastic molding compounds
potato flour
pyrethrum
saccharin
sage
soya flour
sugar
talc
tobacco stems
vermiculite

(e) Fluid Jet Mills

aluminum
aluminum oxide
antibiotics
asbestos
barytes
benzene hexachloride
carbon
carborundum
coal
cocoa
cryolite
DDT
dieldrin
fatty acids
feldspar
ferrochrome
frits
fuller's earth
graphite
iron oxide
lead oxide
mica
molybdenum disulfide
nephelene syenite
phenolics
PVC
pyrethrum
resins
rotenone
salts
shellacs
silica gel
silicon
silicon carbide
sugar
sulfa drugs
sulfur
talc
titanium dioxide
toluidine red
vanilla beans
vitamins
waxes
yeast

(After Mead, "Encyclopedia of Chemical Process Equipment," Reinhold, N.Y., 1964.)

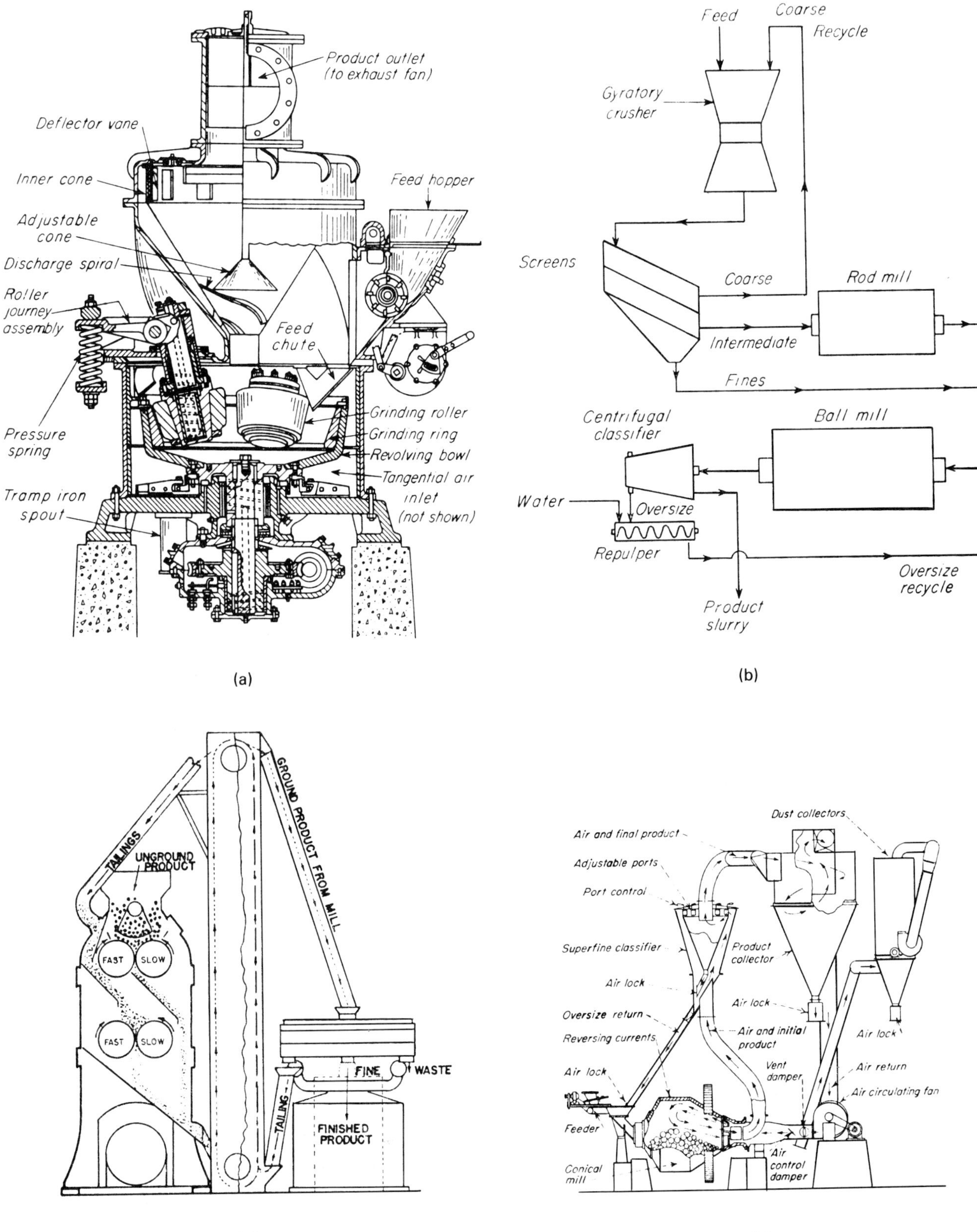

Figure 12.4. Closed-circuit grinding processes, in which coarse products are captured and recirculated until they are brought down to size. (a) Ring-roller mill (Raymond) with built-in air classification; crushing action is by rotating vertical rolls acting on a revolving bowl ring. (b) Flowsketch of closed-circuit grinding with three stages of grinding and two of classifying (*McCabe and Smith,* Unit Operations, *McGraw-Hill, New York, 1976*). (c) A two-pair high roller mill (Schutz–O'Neil Co.); recycle is reground in the lower rolls; Table 12.5(c) lists materials ground by this equipment. (d) A Hardinge conical ball mill in a closed circuit with an air classifier and dust collectors (*Hardinge Co.*).

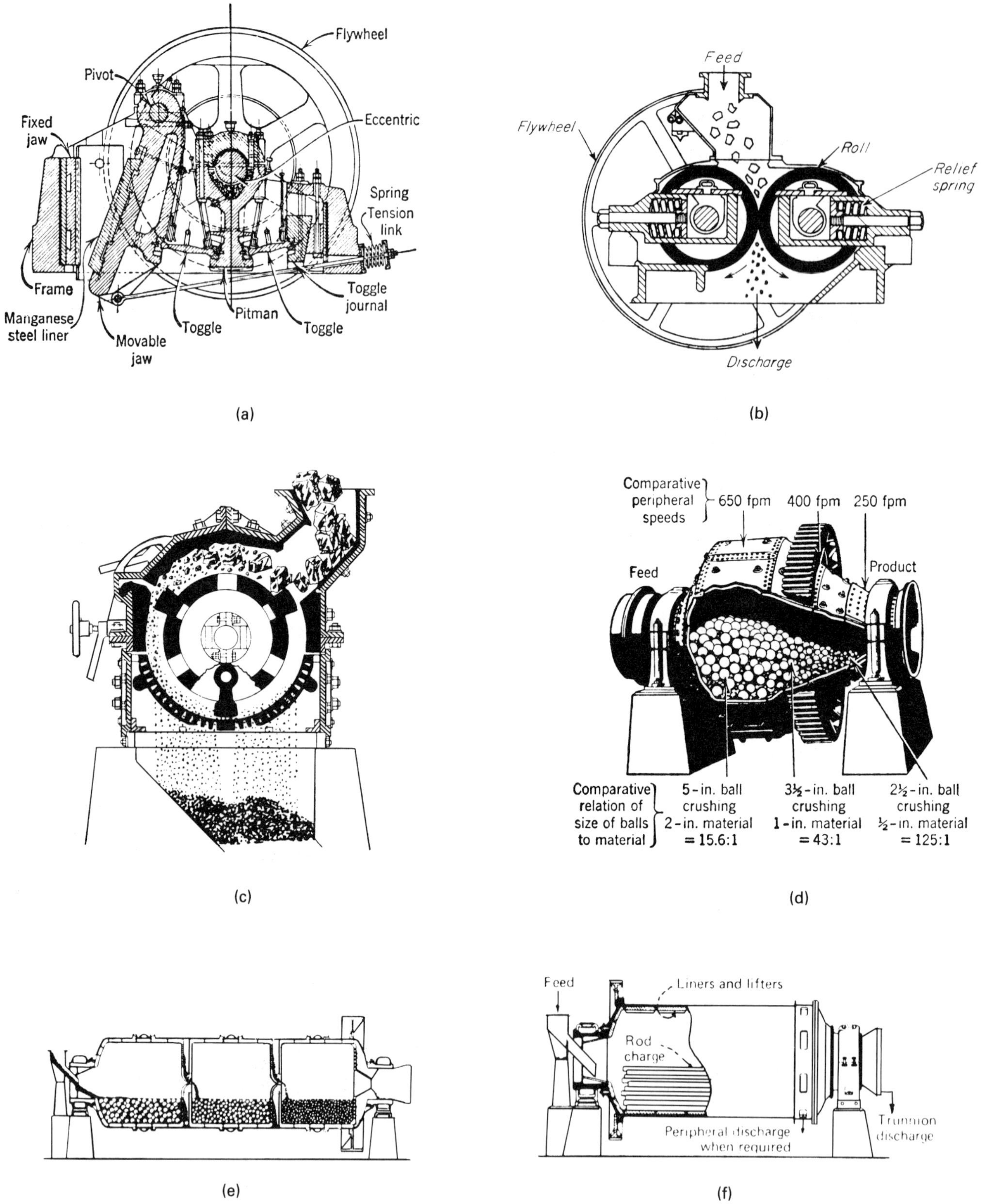

Figure 12.5. Jaw, roll, impact, and tumbling equipment for size reduction. (a) Blake-type jaw crusher operates at 200–300 strokes/min (*Allis–Chalmers Co.*). (b) Smooth roll crusher, for which operating data are in Table 12.8(b). (c) Swing hammer mill; operating data in Table 12.7(a). (d) View of a conical ball mill, showing distributions of balls and material and crushing ranges; data in Table 12.6(b) (*Hardinge Co.*). (e) Tube mill with three compartments, length to diameter ratio 3–5. (f) Rod mill in a cylindrical tumbler, $L/D = 1.2$–1.6; data in Table 12.6(d).

TABLE 12.6. Performance of Ball, Pebble, and Rod Mills in Continuous and Batch Modes

(a) Capacities of Some Straight-Sided Ball Mills on Quartz to Various Meshes

Denver Ball Mill Size Dia. × Lgt. (ft)	Capacity (tons per 24 hrs) Medium-hard Quartz: 2-in to 35 mesh	1-in to 48 mesh	½-in to 65 mesh	½-in to 100 mesh	¼-in to 200 mesh	Rpm Mill	Horsepower: To Run	Of Motor
3 × 2	15	11	9	6		33	7½	10
3 × 3	20	16	14	9		33	10	15
3 × 4	25	21	19	12	7	33	12	15
3 × 6	35	31	29	18	9	33	17½	20
3 × 9	50	46	44	27	13	33	24	25
4 × 3	42	34	30	22	12	28	17	20
4 × 5	63	55	50	31	16	28	28	30
4 × 10	116	108	103	62	26	28	49	50
5 × 3	77	63	55	40	22	26	34	40
5 × 6	130	116	110	67	33	26	57	60
5 × 12	250	236	224	136	54	26	103	125

(Denver Equipment Co.).

(b) Hardinge Conical Ball Mills in Continuous Wet Grinding; Dry Grinding Rates Are 10–20% Less

Size	Weight of Mill	Weight of Lining	Weight of Balls Maximum (lbs)	Rpm*	Motor (max. hp)	Capacity (tons per 24 hrs): 1½-in to 10 mesh	½-in to 100 mesh	¼-in to 98% −325 mesh
3′ × 24″	3,050	2,400	2,400	39.8	10	32	12	4
5′ × 22″	10,200	8,000	8,300	30.4	40	140	49	19
6′ × 36″	17,100	11,700	17,500	27.7	75	282	97	38
8′ × 48″	29,000	23,000	43,500	23.8	200	820	274	108
10′ × 66″	50,600	35,000	83,500	21.2	450	1,900	632	249

(Hardinge Co.).

(c) Hardinge Conical Pebble Mills in Continuous Wet Grinding; Dry Grinding Rates Are 10–20% Less

Size	Weight of Mill	Weight of Lining	Weight of Balls Maximum (lbs)	Rpm*	Motor (max. hp)	Capacity (tons per 24 hrs): 1½-in to 10 mesh	½-in to 100 mesh	¼-in to 98% −325 mesh
3′ × 24″	3,000	1,300	700	40.4	5	15	5.5	2.1
5′ × 22″	9,600	4,000	2,300	31.2	15	54	19	7.5
6′ × 36″	16,500	6,500	4,800	28.2	30	117	42	17
8′ × 48″	19,400	12,300	12,700	24.1	75	326	117	45
10′ × 66″	35,900	16,800	25,500	21.4	150	675	242	95

(Hardinge Co.).

(d) Pebble and Balls Mills for Batch Grinding of Sand of 100 lb/cuft; Pebble Charge 50 vol %, Steel Ball Charge 33 vol %

Mill No.	I.D. of Steel Cylinder	Pebble Mills: Capacity, Dry Grinding (lbs) Porcelain	Buhrstone	Approx. rpm	Ball Mills: Capacity, Dry Grinding (lbs)	Rpm
8½A	24 × 24″	104	9C	40	188	36
8½C	24 × 36″	164	145	40	280	36
6 A	30 × 36″	273	245	36	440	32
5	36 × 42″	440	440	32	740	29
4½	42 × 48″	720	720	29	1152	27
2 A	60 × 60″	1916	1890	19	2936	17
1 A	72 × 72″	3456	3410	16	5076	14
1 C	72 × 120″	5900	5850	16	8460	14

(Paul Abbe Co.).

(e) Performance of Marcy Rod Mills

Size, ft.	Rod charge, tons	Hp. to run	Mill speed, r.p.m.	Capacity, tons/24 hr.: No. 8 sieve	No. 20 sieve	No. 35 sieve	No. 48 sieve	No. 65 sieve
2 × 4	0.9	4– 6	38	20	15	12	10	7
3 × 6	3.6	18– 22	30	105	80	65	50	40
4 × 8	7.6	44– 48	25	240	180	145	120	90
5 × 10	14.5	85– 95	21	525	390	315	260	195
6 × 12	24.1	135–150	17½	855	640	510	425	320
7 × 15	42.1	225–250	15	1600	1200	965	800	600
8 × 12	43.4	230–250	13.2	1675	1250	1000	830	625
9 × 12	54.7	310–340	12.5	2240	1680	1350	1115	835

(Mine and Smelter Division, Kennedy Van Saun Co).

TABLE 12.7. Some Other Kinds of Disintegrators

(a) Ring Roller Mills to Make Down to 100 Mesh

Barytes, 8 to 10 tons/hr to 40 mesh
Coal, 5 to 6 tons/hr to 40 mesh
Coke (96 hour) $3\frac{1}{2}$ to 4 tons/hr to 20 mesh
Fire clay, 8 to 11 tons/hr
Florida pebble, 7 t. to 85%—60 mesh
Florida pebble, 3 t. to 95%—100 mesh
Gannister, 10 to 12 tons/hr to 14 mesh
Iron borings, 8 to 10 tons/hr to 20 mesh
Limestone, 8 to 12 tons/hr to 20 mesh
Limestone, 3 to 4 tons/hr to 85%—200 mesh
Manganese, 2 to 4 tons/hr to 80 mesh
Marble, 3 to 4 tons/hr to 95%—100 mesh
Oyster shells, 4 to 5 tons/hr to 60 mesh

The No. 2 mill has 50 per cent larger capacities
The No. 0 capacity is approximately 35 per cent of the figures for No. $1\frac{1}{2}$ in this table.
Size of feed: 1″ to $1\frac{1}{2}$″.

DIMENSIONS AND SPEEDS FOR STURTEVANT RING ROLL MILLS

Size	Ring Diam. × Face	Rolls Diam. × Face	Ring Speed (rpm)	Horsepower
NO. 0	24″ × 7″	14″ × 7″	125	8 to 15
No. $1\frac{1}{2}$	45″ × 8″	16″ × $10\frac{1}{2}$″	64	45 to 50
No. 2	44″ × 14″	18″ × 14″	70	75

(Sturtevant Mill Co.).

(b) Attrition Mills for Tough Organic Materials

Material	Size-reduction details	Unit*	Capacity lb./hr.	Hp.
Alkali cellulose	Shredding for xanthation	B	4,860	5
Asbestos	Fluffing and shredding	C	1,500	50
Bagasse	Shredding	B	1,826	5
Bronze chips	$\frac{1}{8}$ in. to No. 100 sieve size	A	50	10
Carnauba wax	No. 4 sieve to 65% < No. 60 sieve	D	1,800	20
Cast-iron borings	$\frac{1}{4}$ in. to No. 100 sieve	A	100	10
Cast-iron turnings	$\frac{1}{4}$ in. to No. 100 sieve	E	500	50
Cocoanut shells	2 × 2 × $\frac{1}{4}$ in. to 5/100 sieve	B	1,560	17
	5/100 sieve to 43% < No. 200 sieve	D	337	20
Cork	2/20† sieve to 20/120 < No. 200 sieve	D	145	15
Corn cobs	1 in. to No. 10 sieve	F	1,500	150
Cotton seed oil and solvent	Oil release from 10/200 sieve product	B	2,400	30
Mica	4 × 4 × $\frac{1}{4}$ in. to 3/60 sieve	B	2,800	6
	8/60 to 75% < 60/200 sieve	D	510	7.5
Oil-seed cakes (hydraulic)	1-$\frac{1}{2}$ in. to No. 16 sieve	F	15,000	100
Oil-seed residue (screw press)	1 in. to No. 16 sieve size	F	25,000	100
Oil-seed residue (solvent)	$\frac{1}{4}$ in. to No. 16 sieve	F	35,000	100
Rags	Shredding for paper stock	B	1,440	11
Ramie	Shredding	B	820	10
Sodium sulfate	35/200 sieve to 80/325 sieve	B	11,880	10
Sulfite pulp sheet	Fluffing for acetylation, etc.	C	1,500	50
Wood flour	10/50 sieve to 35% < 100 sieve	D	130	15
Wood rosin	4 in. max. to 45% < 100 sieve	B	7,200	15

* A—8 in. single-runner mill
B—24 in. single-runner mill
C—36 in. single-runner mill
D—20 in. double-runner mill
E—24 in. double-runner mill
F—36 in. double-runner mill
† 2/20, or smaller than No. 2 and larger than No. 20 sieve size.
(Sprout-Waldron Co.).

TABLE 12.7—(*continued*)

(c) Rotary Cutters for Fibrous Materials

Material	Screen opening	Feed rate, lb./hr.	Hp.	Air	Remarks on product
Amosite asbestos pencils	$1\frac{1}{2}''$	1000	11	Yes	Finer fiber bundles average length 2″
Cellophane bags	$\frac{11}{32}''$	200	10	Yes	Finer than $\frac{5}{16}''$
Cork	$\frac{3}{16}''$	525	16	Yes	90% 4/24″ sieve
Chemical cotton	60 mesh	120	15	Yes	Flock; 35% under No. 100 sieve
Leather scrap	$\frac{3}{4}''$	600	20	Yes	Precutting before shredding
Fiberglass	$\frac{3}{16}''$	300	18	Yes	1″ (approx.) lengths
Waste paper	$\frac{5}{16}''$	338	13	Yes	Through No. 4 sieve and finer
Sheet pulp	40 mesh	150	15	Yes	Flock; 85%, 40/100 sieve
Tenite scrap	$\frac{5}{16}''$	340	12	No	Granulated for reuse
Vinylite scrap	$\frac{7}{32}''$	300	15	Yes	35%, 6/10 sieve; granular
$\frac{1}{3}''$ Geon sheet	$\frac{5}{16}''$	540	11	No	99%, 4/20 sieve; for molding granules
Cotton rags	$\frac{3}{4}''$	500	11	Yes	No linting
Buna scrap	10 mesh	264	12	Yes	Granular
Neoprene scrap	30 mesh	90	14	Yes	20°F, temperature rise
Soft-wood chips	$\frac{1}{8}''$	960	12	Yes	90%, 10/50 sieve
Hard-wood chips	$\frac{1}{16}''$	290	11	Yes	83%, 20/100 sieve

* 90 per cent 4/24 sieve, *i.e.*, 90 per cent is through No. 4 and on No. 24 sieve.
(Sprout-Waldron Co.).

TABLE 12.8. Performance of Jaw and Roll Crushers

(a) Capacities and Data on Blake Type Jaw Crushers (Selected Items)[a]

Size of Jaw Opening (in.)	Capacity (tons/hr) (1 ton to 20 cuft Capacity) Open Side Setting (in.)							Jaw Motion (in.)	Horse Power Req.	rpm
	1	$1\frac{1}{2}$	2	$2\frac{1}{2}$	3	4				
10 × 7	7	9	12				A	$\frac{5}{8}$	$7\frac{1}{2}$	300
	8	12	16				B			
20 × 10		15	20	24	31		A	$\frac{5}{8}$	15	275
		24	32	40	49		B			
30 × 18				38	45	61	A	$\frac{11}{16}$	40	250
			48	60	74	102	B			

[a] A-straight jaw plates; B-nonchoking jaw plates.
* (Data supplied by Allis–Chalmers Mfg. Co., Milwaukee, WI).

(b) Double Toothed-Roll Crushers on Coal

Roll Size (in.) Dia	Face	Maximum Size Lump (in.)	Roll (rpm)	Capacity (TPH) Reducing to $1\frac{1}{4}$ to 2	Minimum Motor (HP)
18	18	4	150	39–67	8
18	20	4	150	46–75	8–10
18	24	4	150	52–88	10–12
24	18	14	125	46–74	12–18
24	20	14	125	54–82	15–20
24	24	14	125	62–98	15–20

(Stephens–Adamson Co.)

(c) Relation of Capacity, Size of Feed, Roll Setting, and Speed of Rolls for Sturtevant Balanced Crushing Rolls; Screening in Closed Circuit (Average Rock, Which Can Be Nipped at Speeds Named)

Size of Roll Dia × Face (in.)	Feed Cubes (in.)	Roll Setting (in.)	Speed (rpm)	Capacity (tons/hr)
16 × 10	1.25	0.61	200	26.6
	1	0.25	212	11.6
	0.75	0.2	225	9.8
	0.50	0.125	245	6.67
	0.25	0.065	272	3.86
24 × 15	2	1	115	56.4
	1.5	0.54	130	34.4
	1	0.25	140	17.15
	0.75	0.2	150	14.7
	0.5	0.125	163	10
36 × 20	3	1.5	59	87
	2.5	1	62	61
	2	0.5	70	34.2
	1.5	0.37	78	29.2
	1	0.25	85	20.9

(Sturtevant Mill Co.).

TABLE 12.9. Performance of Impact Disintegrators

(a) Hammer Mills

	tons/hr		
	Limestone, $\frac{1}{8}$ in. Slots	Limestone, $\frac{1}{4}$ in. Slots	Burnt lime, $\frac{1}{4}$ in. Slots
0 Swing-sledge	2–4	4–7	7–9
1 Swing-sledge	6–10	12–15	18–20
2 Swing-sledge	12–15	20–30	60–70
00 Hinged-hammer pulverizer	1–2	2–4	4–6

Approximate Screen Analysis of Product, Reducing 3 in. Limestone

Grate Spacing (in.)	Passing through Mesh Stated			
	$\frac{1}{4}$ in.	10 mesh	50 mesh	100 mesh
$\frac{1}{4}$	99.8%	85%	50%	40%
$\frac{1}{8}$		99	70	60

$\frac{1}{8}$ in. slots means that the grating space was $\frac{1}{8}$ in.

Dimensions and Speeds

	Length	Width	Inside Diameter	Inside Width	Feed Opening	Pulley Speed (rpm)	Approx. HP
0 Swing-sledge	4ft 3in.	4ft 1in.	24 in.	10 in	13 × 11 in.	1200–1500	12
1 Swing-sledge	5ft 1in.	5ft 8in.	30 in.	20 in.	17 × 20 in.	1000–1300	40
2 Swing-sledge	6ft	7ft	36 in.	30 in.	20 × 30 in.	1000–1200	75
00 Hinged-hammer pulverizer	2ft 5in.	3ft	16 in.	11 in.	12 × 12 in.	1200–3600	5–20
0–24 in. Hinged hammer pulverizer	3ft 7in.	5ft 8in.	24 in.	24 in.	$12\frac{1}{2}$ × 24 in.	1000–1200	15–20

(Data supplied by Sturtevant Mill Co., Boston, MA).

(b) High Speed "Mikro-Pulverizer"

Material	Mesh Fineness	No.1 (5 HP)	No. 2 (15 HP)	No. 3 (40 HP)	No. 4 (75 HP)
Aluminum Hydrate	99.8% through 200	600	1,800	4,800	9,000
Ball Clay	98% through 325	600	1,800	4,800	9,000
Calcium Arsenate	99% through 300	1,250	3,750	10,000	18,750
Bituminous Coal	70% through 200	500	1,500	4,000	7,500
Carbon Black	99.99% through 325	450	1,350	3,600	6,750
Cellulose Acetate (Pulp)	94% through 40	200	600	1,600	3,000
Chrome Yellow	99.9% through 200	1,250	3,750	10,000	18,750
Dry Color Slurry	Smooth Slurry	800	2,400	6,400	12,000
Face Powder Mixture	Good Blend	600	1,800	4,800	9,000
Gypsum, Raw	88% through 100	1,650	5,000	13,200	24,750
Iron Blue	95% through 325	750	2,250	6,000	11,250
Kaolin	99.9% through 325	750	2,250	6,000	11,250
Malted Milk	99% through 20	625	1,875	5,000	9,400
Molding Compound	90% through 16	750	2,250	6,000	11,250
Soap Powder	96% through 20	1,500	4,500	12,000	22,500
Soybean Flake	94% through 100	300	900	2,400	4,500
Sugar	99% through 100	600	1,800	4,800	9,000
Tile Clay Body	100% through 16	1,650	5,000	13,200	24,750
Titanium Dioxide	99.8% through 325	600	1,800	4,800	9,000
White Lead	99.99% through 325	1,000	3,000	8,000	15,000
Zinc Oxide	99.9% through 325	600	1,800	4,800	9,000

Top Rotor Speeds—Approximate Idle Loads

Unit	Speed	HP
No. 1	9,600 RPM	1½
No. 2	6,900 RPM	4
No. 3	4,600 RPM	12
No. 4	3,450 RPM	18

(Pulverizing Machinery Co.).

(c) Steam- or Air-Operated Jet Mills

Material	Mill Diameter	Grinding Type	Medium Flow	Solid Feed Rate	Approx. Avg. Particle Size (μ)
Titanium Dioxide	30"	steam	4000 lbs/hr	2250 lbs/hr	less than 1
Sulfur	24"	air	1000 cfm	1300 lbs/hr	3-4
Talc (varies)	30"	steam	4000 lbs/hr	2000 lbs/hr	2
Iron Oxide Pigment	30"	steam	4000 lbs/hr	1000 lbs/hr	2-3
Cryolite	30"	steam	4000 lbs/hr	1000 lbs/hr	3
Barytes	30"	steam	4000 lbs/hr	1800 lbs/hr	3-4
Fuller's Earth	20"	steam	1200 lbs/hr	600 lbs/hr	3-4, 5 top
Anthracite Coal	20"	air	1000 cfm	1000 lbs/hr	5-6
DDT (50%)	24"	air	1000 cfm	1400 lbs/hr	3-4
Procaine-Penicillin	8"	air	100 cfm	25 lbs/hr	5, 20 top

(Sturtevant Mill Co.).

TABLE 12.10. Mill Performance Data for Grinding of Specific Products

Material	Equipment	Handbook Table No.
Anthracite	ball mill CC	46
Barite	wet Hardinge ball mill	35
Cement clinker	three-compartment wet tube mill	42
Fertilizers	hammer mill	41
Fuller's earth	roller	48
Grain	attrition	32
Gypsum rock	ring-roller	45
Iron oxide	ring-roller	47
Limestone	ring-roller	34
Limestone	wet Hardinge ball mill	35
Metal stearates	hammer mill	50
Oyster shells	hammer mill	38
Phosphates	ball mill	39–40
Quicklime	ball mill CC	44
Rubber	roller mill	51
Seed cake	hammer mill	33
Siliceous refractories	pebble mill	36
Slate	three-compartment wet tube mill	43
Sodium carbonate	roller	48
Sulfur	ring-roller	49

Note: CC is closed circuit grinding; the ring-roller mill has built-in air classification.
(From *Chemical Engineers' Handbook*, McGraw-Hill, New York, 1984, pp. 8.48–8.60).

flatter surface. When the rotating surface is cylindrical and the flat surface is horizontal, the equipment is called a dry pan mill. The equipment shown throws the crushed material outwards where it is picked up and removed with an air stream. Table 12.5(c) is a list of materials that are being ground in roller mills. The ring-roller mill of Figure 12.4(a) is in this class.

g. *Fluid jet pulverizers* have opposed high speed gas jets that cause collision and disintegration of the particles. A size classifier and fan return larger sizes to the jet stream. The "Majac" jet mill of Figure 12.6(g) is a related kind of device; it has a horizontal section in which high speed gas jets act on the particles. These mills are used primarily for specialty fine grinding of high-value materials. Performance data of Micronizers are in Table 12.9(c); those of the Majac pulverizer are expected to be similar.

12.5. PARTICLE SIZE ENLARGEMENT

For many purposes, lumps of materials of intermediate sizes are the most desirable forms, neither too small nor too large. For instance, beds of overly small granules of catalysts exhibit too great resistance to flow of reacting fluids, and too small particles in suspensions settle out or filter too slowly. Other situations that benefit from size enlargement of particles are listed in Table 12.11.

Because of adhesive forces, particles tend to stick together, particularly small particles that have a large ratio of surface to mass. If a mass is vibrated or shaken lightly, for instance, smaller particles penetrate the interstices between larger ones with increase of contact area and adhesion of the mass. Substances differ naturally in their tendency to agglutinate; as examples, the following groups of materials are listed in the order of increasing tendency to agglutination:

1. Superphosphate, sulfates, and NPK fertilizers.
2. Carbamide and diammophosphate.
3. Ammonium phosphate, potassium chloride, potassium bicarbonate, and salt.

Adhesion of any mass of particles can be developed by sufficiently high pressure, but lower pressures suffice upon addition of liquid or syrupy binders. Table 12.12 is a list of some commercial agglomerations and the binders that they employ, and Table 12.13 shows how much moisture is needed.

The main types of processes used industrially for particle size enlargement are five in number, defined as follows:

1. *Compaction* is achieved either by compression or extrusion. *Compression* is done either into a mold to give a final shape or into a sheet or block that is later broken up to proper sizes. *Extrudates* are formed under pressure in dies of a variety of cross sections; as they leave the die they are broken up or cut to size.
2. *Agglomeration* is accomplished under tumbling or otherwise agitated conditions, with or without binding agents. Size is controlled by adjusting the residence time and by gradual addition of feed and binder, slurry or solution.
3. *Globulation* is the formation of droplets of solution, slurry, or melt followed by solidification by prilling, spray drying, or fluidized bed operation. Control of particle size is best achieved in fluidized beds.
4. *Heat bonding* is of two types: *nodulization* in which material is tumbled while heated to give hard rounded granules and *sintering* in which the product is an integrated mass that is subsequently broken to size.
5. *Flocculation,* coagulation and growth of particles in dilute slurries, to assist in subsequent sedimentation and filtration.

A particular industry may employ more than one of these techniques, for instance the manufacture of solid catalysts. Spherical catalysts are made in rotating pan granulators (Fig. 12.7). If the rheological properties are suitable, the material can be extruded (Fig. 12.8), then cut into short cylinders, and subsequently tumbled (Fig. 12.9) into rounded shapes. Smaller spherical beads, for instance, of catalysts for moving bed processes, are made by precipitation or coagulation in an immiscible fluid. Pellets or rings are made on tabletting machines (Fig. 12.10). Although the process is more expensive than extrusion, the product is more nearly uniform. Both extrusion and tabletting result in diffusion resistant skins that, however, usually are eliminated on drying or calcination of the catalyst. Ammonia synthesis catalyst is made by sintering (Fig. 12.11) or fusion of the several ingredients, then crushed and used as irregular lumps of size ranges 1.5–3, 6–10, and 12–21 mm.

In the following, the main equipment for particle size enlargement will be illustrated and discussed.

TUMBLERS

The particles of a granular mass will cohere when they are tumbled and sprayed lightly with a liquid binder which often is water or a concentrated solution of the material being agglomerated. The growth may be due to agglomeration of small particles or to layering of material evaporated from the sprayed solution. Rotary kilns of the kind used for drying or chemical reaction (cement or lime burning, for instance) are adapted to size enlarging service. Usually the tumbling action is less intense, only enough to expose the material to sprays. The sprays are fine and are applied to the surface of the bed of particles. The tumbling action then distributes the liquid uniformly through the mass.

The disk granulator of Figure 12.7 is a shallow pan, inclined 45–70° to the horizontal and rotating at speeds of 10–30 rpm. The ratio of pan diameter to collar height is 3–5. The variety of materials to which this equipment is applied is indicated by the listing of part (e) with this figure and in Table 12.14. As the rotation proceeds, fresh solids and spray are charged continually. The finer particles settle to the bottom, the largest remain at the top and then overflow the collar and constitute the product. Because of the size stratification, the product of disk granulation is more uniform in size than of drum granulators which discharge a mixed product. Some performance data in addition to those shown with Figure 12.7 are:

Material	Diameter (mm)	kg/(min)(m^2)
Iron ore	10–25	11.4
Cement flour		18
Fertilizer	1.6–3.3	14.3

The data of Figure 12.7(c) for cement kiln feed are 42–44 kg/(min)(m^2).

Pans also are made with height more nearly equal to the diameter. In one such device the material is continually lifted onto

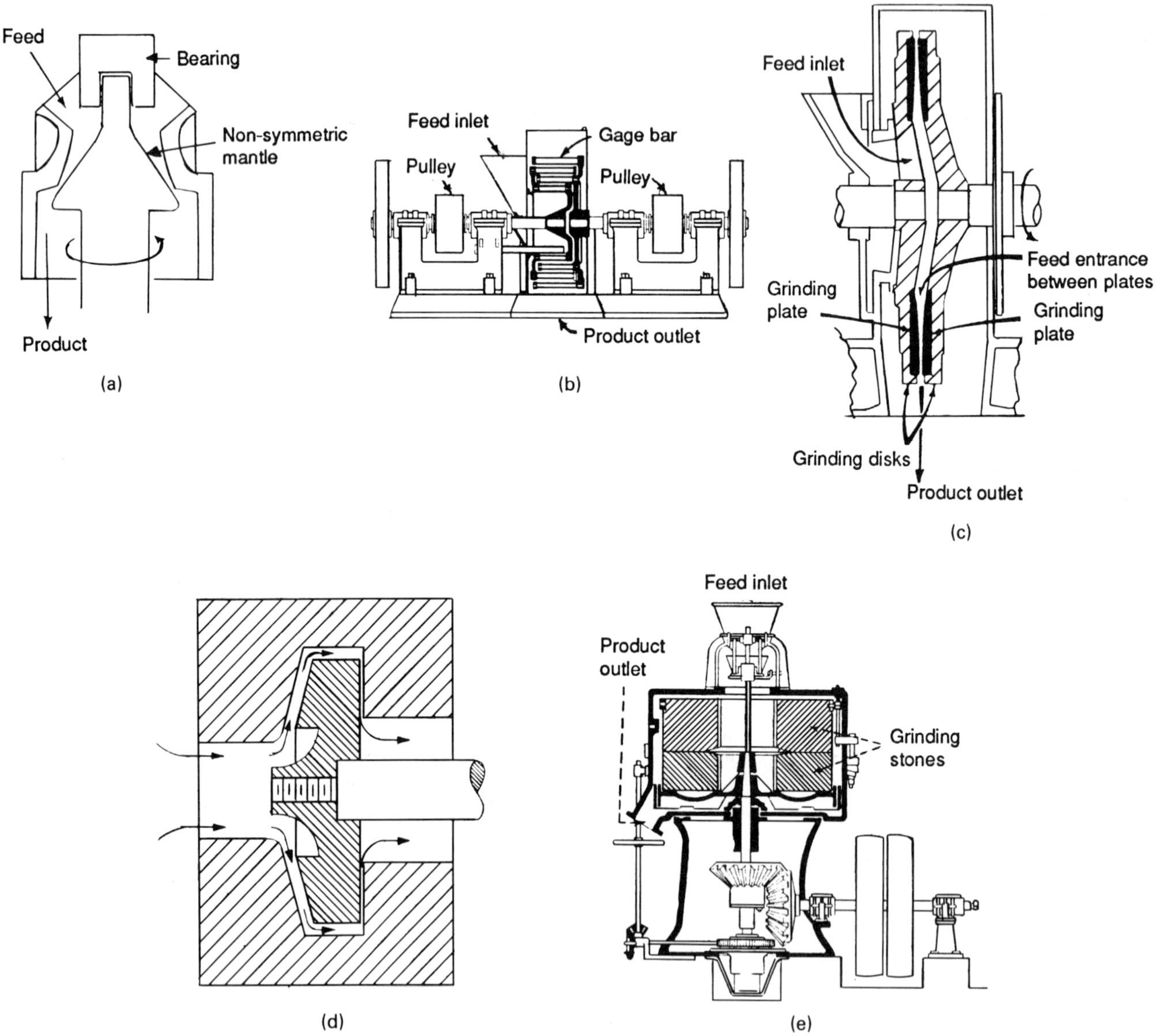

Figure 12.6. Examples of mostly less common devices for size reduction. (a) Schematic of a gyratory crusher for very large lumps. (b) Squirrel-cage disintegrator with four cages. (c) Disc-type attrition mill, rotating at 1200–7000 rpm, clearances adjustable by increments of 0.001 in. (d) Schematic of colloid mill, clearance adjustable between 0.001 and 0.050 in., peripheral speeds to 10,000 ft/min. (e) Buhrstone attrition mill, used for making flour and grinding paints, printers inks and pharmaceuticals. (f) Roller or spindle mill; the crushed material is thrown outwards and removed with an air stream. (g) Majac fluid energy mill making a −200 mesh product; opposed air jets cause high speed collisions and disintegration of the material.

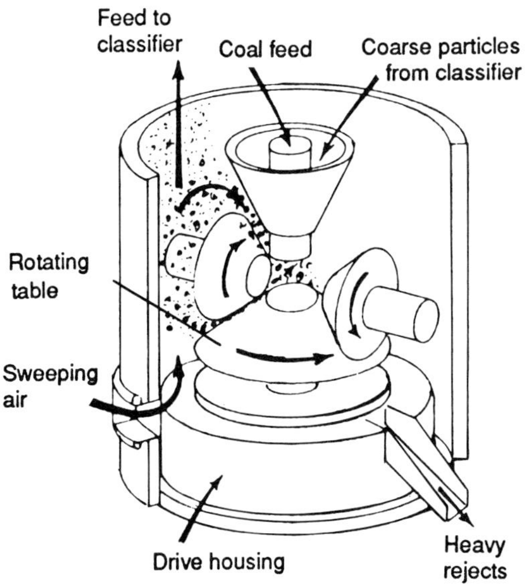

(f)

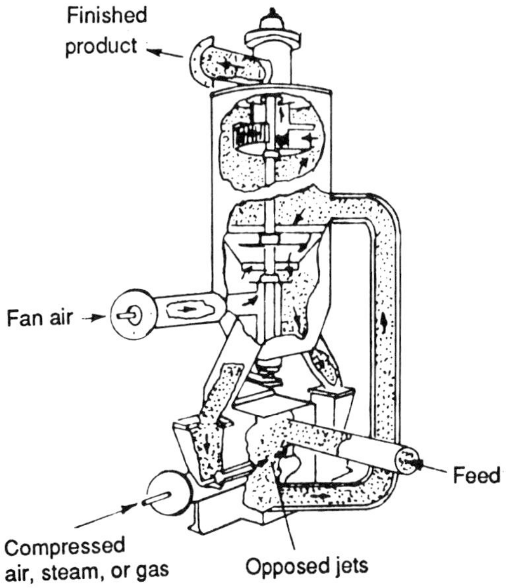

(g)

Figure 12.6—(*continued*)

TABLE 12.11. Benefits of Size Enlargement and Examples of Such Applications

Benefit	Examples of Application
1. Production of useful structural forms and shapes	pressing of intricate shapes in powder metallurgy; manufacture of spheres by planetary rolling
2. Preparation of definite quantity units	metering, dispensing, and administering of drugs in pharmaceutical tablets
3. Reduced dusting losses	briquetting of waste fines
4. Creation of uniform, nonsegregating blends of fine materials	sintering of fines in the steel industry
5. Better product appearance	manufacture of fuel briquets
6. Prevention of caking and lump formation	granulation of fertilizers
7. Improvement of flow properties	granulation of ceramic clay for pressing operations
8. Greater bulk density to improve storage and shipping of particulates	pelleting of carbon black
9. Reduction of handling hazards with irritating and obnoxious materials	flaking of caustic
10. Control of solubility	production of instant food products
11. Control of porosity and surface-to-volume ratio	pelleting of catalyst supports
12. Increased heat transfer rates	agglomeration of ores and glass batch for furnace feed
13. Removal of particles from liquids	pellet flocculation of clays in water using polymeric bridging agents
14. Fractionation of particle mixtures in liquids	selective oil agglomeration of coal particles from dirt in water
15. Lower pressure drop in packed beds	reactors with granular catalysts

an internal sizing screen from which the oversize is taken off as product (Sherrington and Oliver, 1981, p. 69).

Rotating Drum Granulators. The equipment of Figure 12.9 is largely free of internals to promote mixing, but provides just sufficient turnover to effect redistribution of the spray throughout the mass. With heavy sprays and little tumbling action, excess and nonuniform agglomeration will occur. Granules 4–6 mm dia commonly are made by layering from the sprayed solutions. Fertilizer granules made this way are larger, more dense, and harder than those made by prilling. The trend in the industry has been for prilling towers to be replaced by drum granulators and for those in turn to be replaced by fluidized bed granulators in which dusting problems are most controllable.

A pitch of as much as 10° is used to assist material transport over the length of the drum. Length to diameter ratios of 2–3 are used and speeds of 10–20 rpm. Recommended speeds are about 50% of the critical speed for the dry material; then adequate cascading occurs and the range of particle size distribution is narrowed. Figure 12.9(b) shows the results of such tests on small scale granulation of a fertilizer.

Another application of tumblers is to the manufacture of mixed fertilizers, in which solid ammonium nitrate, liquid ammonia, liquid phosphoric acid, and liquid sulfuric acid are charged separately and reacted. The incidental agglomeration is excessive, however, and the process must be followed by appropriate crushing and size classification.

Various designs of powder blenders can be equipped with sprays and used as granulators. Figure 12.12 is of a trough equipped with two sets of paddles that rotate in opposite directions and throw the particles to the center where they are wetted. Since the mixing is predominantly lateral rather than axial, a measure of plug flow exists and results in a narrower distribution of sizes than from an empty drum, and may approach that from inclined disk granulators. The shape is not quite as rounded, but can be improved if desired by tumbling subsequently in a dry drum. Dimensions and performance of commercial units are shown with Figure 12.12.

TABLE 12.12. List of Agglomerated Products and Their Binders

Material	Binder	Agglomeration Equipment
Activated Charcoal	Lignosulfonate	Turbulator*
Alumina	Water	Turbulator*/Disc
Animal Feed	Molasses	Ring Extruder
Boric Acid	Water	Disc Pelletizer
Carbon Black & Iron Powder	Alcohol-Carbowax	Turbulator*
Carbon, Synthetic Graphite	Sodium Silicate	Turbulator*/Disc
Cement, Raw Mix	Water	Disc Pelletizer
Cement Kiln Dust	Water	Turbulator*/Disc
Charcoal	Starch Gel	Briquetter
Chrome Carbide	Alcohol	Disc Pelletizer
Clay, Attapulgite	Water	Turbulator*/Disc
Clay, Bentonite	Water	Turbulator*
Coal, Anthracite	Pitch	Briquetter
Coal, Bituminous	Lignosulfonate	Disc Pelletizer
Coal Dust	Water	Turbulator*
Coke, Petroleum	Pitch	Briquetter
Continuous Casting Flux	Water	Turbulator*/Disc
Copper Smelter Dust	Sodium Silicate	Turbulator*
Copper Sulphite Concentrate	Sodium Silicate	Disc Pelletizer
Detergent Dust	Water	Disc Pelletizer
Dolomite Kiln Dust	Water	Turbulator*/Disc
Dye Pigment	Lignosulfonate	Turbulator*/Disc
Electric Furnace Dust	Water	Turbulator*/Disc
Fertilizer	Ammonia	Drum
Flourspar	Sodium Silicate	Disc Pelletizer
Flourspar	Lime-Molasses	Briquetter
Flyash (boiler)	Water	Turbulator*/Disc
Flyash (high carbon)	Lignosulfonate	Briquetter
Glass Batch	Caustic Soda	Disc Pelletizer
Glass Batch	Water	Briquetter
Herbicide	Lignosulfonate-Water	Turbulator*/Disc
Herbicide	Clay-Carbowax	Briquetter
Iron Ore	Bentonite-water	Drum
Lignite	Gilsonite-Water	Turbulator*/Disc
Limestone	Clay-Water	Turbulator*
Manganese Ore	Lime-Molasses	Briquetter
Manganese Oxide	Sulfuric Acid	Turbulator*/Disc
Phosphate Rock	Phosphoric Acid	Turbulator*/Disc
Plastic Powder	Alcohol	Disc Pelletizer
Potash Fines	Water	Disc Pelletizer
Sodium Borate	Sulfuric Acid	Turbulator*/Disc
Sulfur Powder	Clay	Compactor
Tungsten Carbide	Alcohol	Disc Pelletizer
Zeolite	Clay-Water	Turbulator*/Disc

(Koerner and MacDougal, 1983).

Sticky, very fine, and highly aerated materials can be granulated in drums with high speed impellers with pegs or pins instead of paddles. In Figure 12.13, the material enters at one end, is immediately wetted, and emerges as pellets at the other end. Residence times are under a minute. The data with this figure show that the bulk density of carbon black is increased by a factor of 11, although with about 50% binder in the product.

As mentioned, other powder blending devices can be adapted to granulation, but unless most of the equipment is on hand, it is best to adopt a proven design with which some manufacturer has experience. If the stakes are high enough, the cost of a development program with other equipment may be justifiable.

ROLL COMPACTING AND BRIQUETTING

Agglomeration of finely divided materials is accomplished at high rates and low costs by compression between rolls. The form of product may be continuous sheets that subsequently are broken up to desired size or it may be lumps or briquets of finished form and size. A few shapes are shown in Figure 12.14, but a great variety of simple geometrical shapes can be made with readily available rolls. Briquets are a low cost product, rough in shape, and not of highly uniform weight. When smooth appearance and weight uniformity are demanded, tabletting is the process to be used. The great variety of materials that have been compacted with rolls is indicated

TABLE 12.13. Moisture Requirements for Successful Granulation in Tumbling Machines

Raw material	Approximate size analyses of raw material, less than indicated mesh	Moisture content of balled product (% H_2O)
Precipitated calcium carbonate	200	29.5–32.1
Hydrated lime	325	25.7–26.6
Pulverized coal	48	20.8–22.1
Calcined ammonium metavananiate	200	20.9–21.8
Lead–zinc concentrate	20	6.9–7.2
Iron pyrite calcine	100	12.2–12.8
Specular hematite concentrate	150	9.4–9.9
Taconite concentrate	150	9.2–10.1
Magnetic concentrate	325	9.8–10.2
Direct shipping open pit ores	10	10.3–10.9
Underground iron ore	0.25 in.	10.4–10.7
Basic oxygen converter fume	1 μm	9.2–9.6
Row cement meal	150	13.0–13.9
Utilities–fly ash	150	24.9–25.8
Fly ash–sewage sludge composite	150	25.7–27.1
Fly ash–clay slurry composite	150	22.4–24.9
Coal–limestone composite	100	21.3–22.8
Coal–iron ore composite	48	12.8–13.9
Iron ore–limestone composite	100	9.7–10.9
Coal–iron ore–limestone composite	14	13.3–14.8

Courtesy McDowell Wellman Company.

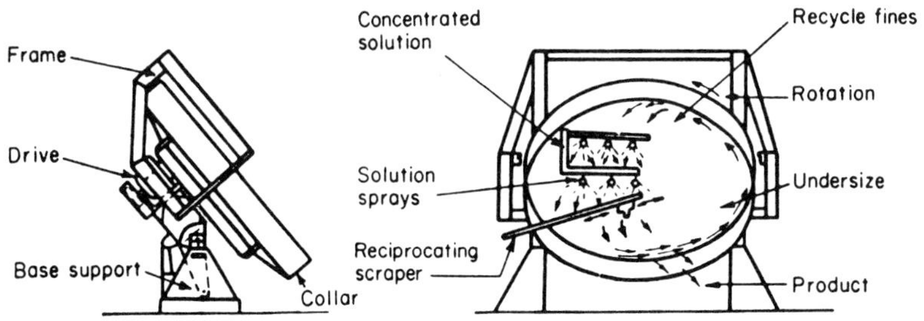

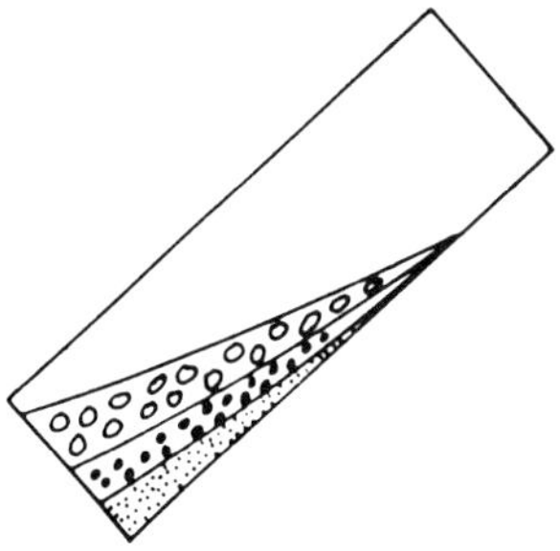

Dish size (m)	Motor (kW)	Capacity (kg s^{-1})	Material	Remarks
0·36	0·18	0·013	Tungsten carbide	16 x 60 mesh micropellets
0·36	0·18	0·0044	Alumina	
0·99	0·55	0·13	Phosphate rock	85% 4 x 30 mesh product
0·99	0·55	0·076	Bituminous coal filter cake	Feed to pan dryer
0·99	0·55	0·076	Beryl ore mix	Feed to sinter belt
0·99	0·55	0·15	Copper precipitate	
1·37	2·2	0·28	Frit enamel mix	Feed to furnace
2·59	11	8·5	Zinc concentrate sinter mix	Micropelletized sinter machine feed
2·59	11	0·85	Chromate	For electric ore furnace
2·59	7·3	0·93	Bituminous coal fines	For coking furnace
3·05	15	1·7–2·3	Raw shale fines	For expanding in rotary kiln
3·05	18	2·8	Bituminous coal filter cake	
3·66	22	3·4	Zinc sulphide ore	For fluid bed roasting of 4 x 30 mesh pellets
4·27	37	11	Nitrogen fertilizer material	Feed: hot melt and recycle
5·49	44	11	Magnetite ore	Feed to travelling grate – indurating section

(c)

Figure 12.7. Rotating dish granulator applications and performance (*Sherrington and Oliver, 1981*). (a) Edge and face view of a dish granulator, diameters to 25 ft, Froude no. $n^2D/g_c = 0.5$–0.8. (b) Stratification of particle sizes during rotation. (c) Typical applications of dish granulation (*Dravo Corp.*). (d) Capacity and power (*Dravo Corp.*). (e) Performance on cement kiln feed.

Disk size, ft.	70 lb./cu. ft. material Pelletizing		125 lb./cu. ft. Pelletizing	
	Approx. capacity, tons/hr.	Horse-power	Approx. capacity, tons/hr.	Horse-power
18	30	40	40	50
15	18	25	25	30
12	10	12	15	16
9	5	6	10	7½
6	3	3	5	5
3¼	½	1	1	1

(d)

Diameter (m)	3·6	4
Depth (cm)	91	91
Speed (rpm)	17·5	14·0
Drive (kW)		
Installed	30	37
Used	26	25–30
Feed rate (kg s^{-1})	7·1	8·5–10·1
Moisture (%)	12·5–13·5	12·5–13·5
Granule porosity (%)	26	26
Granule compressive strength (kg)	2·7–6·7	2·7–6·7
Powder feed position	Bottom centre	
Water feed positions		
Main	Jets above powder feed	
Secondary	Fine sprays in top section of pan	

(e)

Figure 12.7—(*continued*)

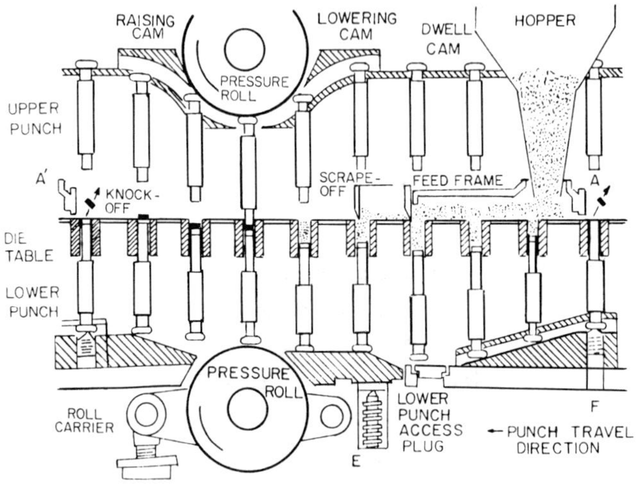

(a)

Number of tooling stations	27	33	45
Output (tablets/min)	1,000–2,700	1,200–3,300	1,600–4,500
Max. tablet diameter (in.)	$1\frac{3}{16}$	$1\frac{1}{16}$	$\frac{3}{4}$
Fill depth (in.):			
Standard	$0-\frac{11}{16}$	$0-\frac{11}{16}$	$0-\frac{11}{16}$
Optional	$\frac{11}{16}-1\frac{3}{8}$	$\frac{11}{16}-1\frac{3}{8}$	$\frac{11}{16}-1\frac{3}{8}$
Max. operating pressure (tons)	10	10	10
Pressure release adjustment (tons)	0–10	0–10	0–10
Upper punch entrance (in.)	$\frac{3}{16}-\frac{7}{16}$	$\frac{3}{16}-\frac{7}{16}$	$\frac{3}{16}-\frac{7}{16}$

(b)

Series	37	45	55	61
Number of stations	37	45	55	61
Max. operating pressure (tons)	10	6.5	6.5	6.5
Max. depth of fill (in.)	$\frac{13}{16}$	$\frac{11}{16}$	$\frac{11}{16}$	$\frac{11}{16}$
Max. tablet diameter (in.)	1	$\frac{5}{8}$	$\frac{7}{16}$	$\frac{7}{16}$
Output (tablets/min)	888–3,552	2,050–8,200	2,500–10,000	2,775–11,100

(c)

Figure 12.8. Operation and specifications of rotary tabletting machines (*Carstensen, 1984*). (a) Action of the punches of a rotary tabletting machine. (b) Specifications of a Sharples Model 328 (*Stokes–Pennwalt Co.*). (c) Specifications of a Manesty Rotapress Mk 11 (*Manesty Machines Ltd. and Thomas Engineering Inc.*).

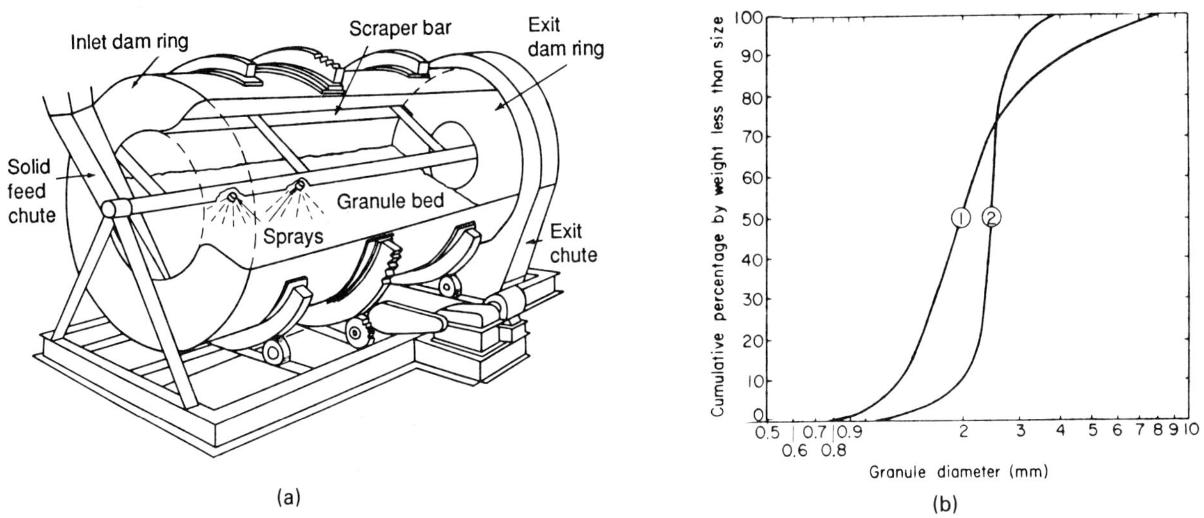

Application	Diameter (ft)	Length (ft)	Installed Power (HP)	rpm	Approximate[a] Capacity (tons/hr)
Fertilizer granulation	5	10	15	10–17	7.5
	8[b]	16	75	8–14	40
Iron ore balling	9	31	60	12–14	54
	12	33	75	10	98

[a] Capacity excludes recycle; actual drum throughput may be much higher.
[b] Inclination 2°.

(c)

Figure 12.9. Rolling drum granulator sketch and performance. (a) Sketch of a rolling drum granulator (*Sherrington and Oliver, 1981*). (b) Effect of rotational speed on size distribution: (1) at 20% of critical speed; (2) at 50%. (c) Performance data on commercial units (*Capes and Fouda, 1984*).

by the listing of Tables 12.15 and 12.16. A survey of equipment currently marketed worldwide is made by Pietsch (1976); an excerpt is in Table 12.16.

Compacting of specific materials can be facilitated with certain kinds of additives. Binders are additives that confer strength to the agglomerates, and lubricants reduce friction during the operation. Some additives may function both ways. A few of the hundreds of binders that have been tried or proposed are listed in Table 12.12. Lubricants include the liquids water, glycerine, and lubricating oils; and typical solids are waxes, stearic acid, metallic stearates, starch, and talc.

Successful compacting has been accomplished at temperatures as high as 1000°C. Extrudates are 1–10 mm thick. The information of commercially available equipment of Table 12.17 is representative. Rolls range in size from 130 mm dia by 50 mm wide to 910 mm dia by 550 mm wide. Capacities are 10–6000 kg/h, and energy requirements, 2–16 kWh/ton. Compacting of mixed fertilizers and similar materials is accomplished by pressures of 30–1200 atm, of plastics and resins by 1200–2500 atm, and of metal powders above 5000 atm. Feed supply may be in the vertical or horizontal direction, by gravity or forced feed. Horizontal feeding is less bothered by entrapped air.

Figures 12.10(a) and (b) show product in sheet form which is subsequently broken down to size. Pellets of large size also may be made for subsequent crushing. For instance, pellets 5 cm dia by 1 cm thick are made for the pharmaceutical trade for breaking up to serve as coarse granular feed to tabletting.

TABLETTING

Rotary compression machines convert powders and granules into hard tablets of quite uniform weight, notably of pharmaceuticals, but also of some solid catalyst formulations. The process is illustrated in Figure 12.8(a). A powder is loaded into a die where it is retained by a lower punch; then it is compressed with an upper punch, and the tablet is ejected by raising both punches.

Most tablets are small, the largest shown in Figure 12.8 is 1-3/16 in. dia and the greatest depth of fill is 1-3/8 in., but other machines make tablets 4 in. dia and exert forces of 100 tons. The degree of weight uniformity normally aimed at is indicated by the specifications of the U.S. Pharmacopeia. This states that, of a sample of 20 tablets, only two may differ from the mean by the percentage stated following and only one may deviate by twice the percentage stated:

Weight of Tablet (mg)	% Deviation
Equal to or less than 13	15
13–130	10
130–324	7.5
More than 324	5

Greater weight uniformity is achieved with coarse powders or granules as feed. Too large a proportion of fines may cause the tablets to come apart upon ejection. Satisfactory feed can be prepared by first making large tablets in another machine and then

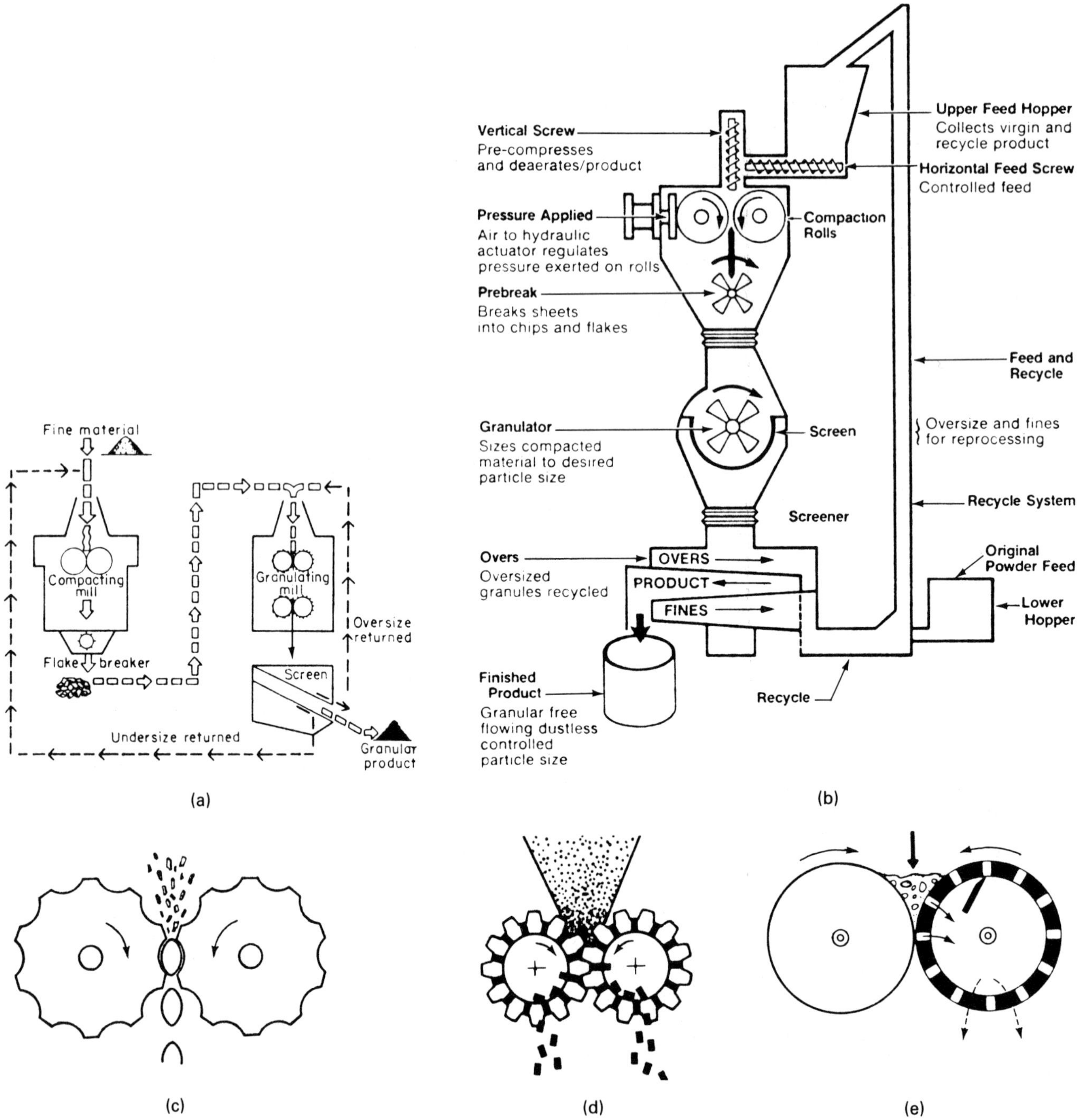

Figure 12.10. Equipment for compacting, briquetting, and pelleting. (a) Flowsketch of a process for compacting fine powders, then granulating the mass (*Allis–Chalmers Co.*). (b) Integrated equipment for roll compacting and granulating (*Fitzpatrick Co.*). (c) A type of briquetting rolls. (d) A gear pelleter. (e) A double roll extruder.

breaking them into coarse granules, or by batch-fluidized bed granulation. For pharmaceuticals the range of allowable additives to facilitate tabletting is limited. Magnesium stearate is a common lubricant to the extent of 0–2%, and corn starch is a common binder in the range of 0–5%. Disintegrants and fillers also are used. Preparation of such mixes is accomplished in powder blenders and fixed by granulation.

The Stokes machines of Figure 12.8 operate at 35–100 rpm and the large Manesty at 45–180 rpm. Maximum forces for these sizes of tablets are 10 tons, but up to 100 tons may be needed for tablets 2.5–4 in. dia. The largest machines shown can be driven with about 50 HP.

EXTRUSION PROCESSES

Powders, pastes and melts are pelleted by extrusion through a die followed by cutting. Binders and lubricants may need to be incorporated in the feed, but the process usually is not feasible for

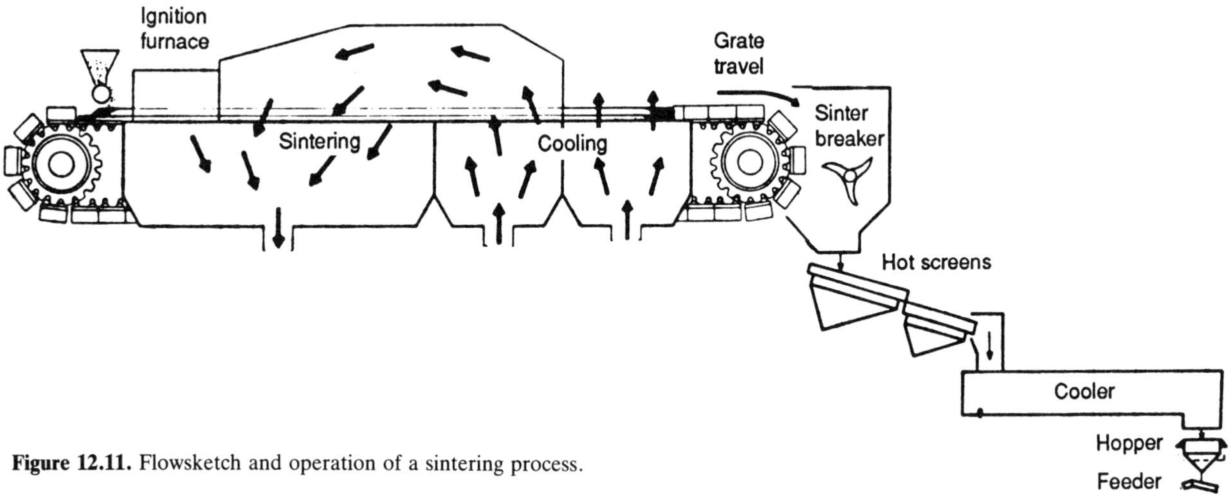

Figure 12.11. Flowsketch and operation of a sintering process.

TABLE 12.14. Industries that Employ Disk Granulators and Some of the Products They Process

Industry	Typical Application
Steel	Electric Furnace Baghouse Dust, BOF Dust, OH Dust, Coke Fines, Raw Materials, Iron Ore Pelletizing
Foundry	Baghouse Dust, Mold Sand Fines
Ferroalloy	Silicon, Ferrosilicon, Ferromanganese, Ferrochrome
Copper	Concentrates, Smelter Dust, Precipitates
Lead/Zinc	Concentrates, Sinter Mix, Flue Dust, Drosses
Other Metals	Tungsten, Molybdenum, Antimony, Brass, Tin, Berrylium, Precious Metals, Aluminum, Silicon, Nickel
Glass	Glass Rawmix, Furnace Dust, Glass Powder
Ceramics	Alumina, Catalyst, Molecular Sieves, Substrates, Insulator Body, Tilemix, Press Feed, Proppants, Frits, Colors
Refractories	Bauxite, Alumina, Kiln Dust, Blends
Cement/Lime	Raw Meal, Kiln Dust
Chemicals	Soda Ash, Sodium Sulfate, Detergents, Cleaners, Zinc Oxide, Pigments, Dyes, Pharmaceutical Compounds, Industrial Carbons, Carbon Black
Ag-Chemicals	Fertilizers, Pesticides, Herbicides, Insecticides, Soil Conditioners, Aglime, Dolomite, Trace Minerals, N-P-K raw Materials
Foods	Instant Drink Mix, Powdered Process Foods, Sugar, Sweetners, Confectionary Mix
Coal	Coal Fines
Power	Coal Fines, Fly Ash, FGD Sludge, Boiler Ash, Wood Ash
Nonmetallic Minerals	Clay, Talc, Kaolin, Fluorspar, Feldspar, Diatomaceous Earth, Fullers Earth, Perlite
Pulp, Paper, Wood	Paper Dust, Wood Fines, Sander Dust, Boiler Ash
Solid Waste	Incinerator Ash, Refuse Fines, Mixed Refuse, Dried Sludge

(Koerner and MacDougal, 1983).

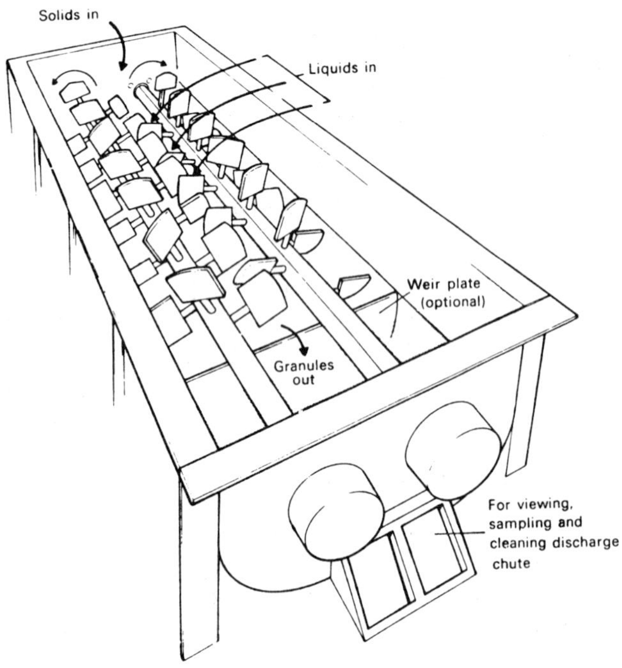

Blunger dimensions

Length (m)	4·5
Width (m)	1·4
Height (m)	1·07
Screw diameter (m)	0·8
Pitch (m)	1·7
Shaft speed (rpm)	55
Capacity ($kg\ s^{-1}$)	23–25
Installed power (kW)	104

(a)

MODEL	MATERIAL BULK DENSITY LB/FT^3	APPROXIMATE CAPACITY (TONS/HR)	SIZE (WIDTH × LENGTH) (FT)	SPEED (RPM)	DRIVE (HP)
A	25	8	2 × 8	56	15
	50	15	2 × 8	56	20
	75	22	2 × 8	56	25
	100	30	2 × 8	56	30
B	25	30	4 × 8	56	30
	50	60	4 × 8	56	50
	75	90	4 × 8	56	75
	100	120	4 × 8	56	100
C	25	30	4 × 12	56	50
	50	60	4 × 12	56	100
	75	90	4 × 12	56	150
	100	120	4 × 12	56	200
	125	180	4 × 12	56	300

(b)

Figure 12.12. Paddle blending granulator and typical performance. (a) Sketch of a double paddle trough granulator (*Sherrington and Oliver, 1984*). (b) Performance in granulation of fertilizers (*Feeco International*).

abrasive materials. Economically feasible power requirements correspond to the range of 100–200 lb/HP hr. The main types of machines are illustrated in Figures 12.10(e) and 12.15.

Screw extruders usually are built with a single screw as shown, but may have as many as four screws and the die may have multiple holes of various cross sections. An 8 in. dia screw can have a capacity of 2000 lb/hr of molten plastics. Tubing can be extruded at 150–300 ft/min. To make pellets, the extrudate goes to cutting machines in which parts as small as washers can be made at rates as high as 8000/min. The extrusion of plastics is described at length by Schwartz and Goodman (*Plastics Materials and Processes,* Van Nostrand Reinhold, New York, 1982); such equipment is applicable to other situations.

Ring pellet mills consist of a power driven rotating ring with radial holes, friction driven rolls to force the material through the holes and knives to cut the extrudate to desired lengths. The feed is charged with screw feeders into the spaces between the rolls and the feed distributor flights. Hole diameters range from 1.6 to 32 mm. The force of compaction is due to flow friction through the die. Differing flow and compression characteristics are accommodated by varying the thickness of the ring. A production rate of 200 lb/HP hr has been quoted for a normal material through 0.25 in.

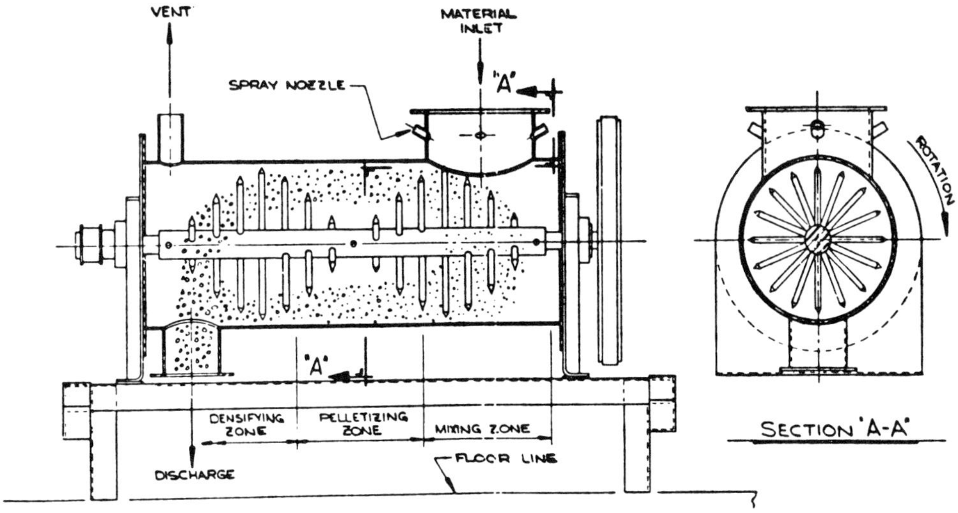

(a)

Carbon black feed	
Rate, Mg/day	26.3[a]
Bulk density, kg/m^3	51.3
Pellets produced	
Wet basis	
Production rate, kg/h	2108.3
Bulk density, kg/m^3	562.3
Densification ratio	11.0
Dry basis	
Production rate, kg/h	1096.3
Bulk density, kg/m^3	394.1
Binder	
Specific gravity	1.05
Injection rate, kg/h	1011.5
Use ratio, weight of binder to weight of wet pellets	0.92
Power Consumption[b]	
Rate, kW	18.5
Per Mg of wet pellets, kWh	15.0[c]
Production quality	
Rotap test (5 min), %	1.4 (avg. of 45 samples)
Crushing strength, g	25 (avg. of 73 samples)

[a]Average from 5-day test, plus subsequent production.
[b]Ammeter readings.
[c]Cold shell.

(b)

Figure 12.13. Pin mixers which operate at high speed for granulation of fine and aerated powders. (a) Pinmixer for the granulation of wetted fine powders. (b) Performance of a pinmixer, dimensions 0.67 m dia by 2.54 m, for pelleting a furnace oil carbon black (*Capes*, 1980).

holes. The life of a die is measured in hours. Units of 300 HP are made. Some applications are cited in Table 12.17.

The large tonnage application is to the preparation of animal feeds but many smaller scale applications also are being made. A survey of this literature is made by Sherrington and Oliver (1981).

PRILLING

In the process a molten material is disintegrated into droplets which are allowed to fall and to solidify in contact with an air stream. The equipment is similar to that for spray drying, but the mechanism is simpler in that no evaporation occurs and as one consequence the product is less porous and stronger. A sketch of a prilling process is in Figure 12.16. At one time the chief application was to fertilizers, but a list of many other prillable materials is given in Table 12.18.

Dimensional and some operating data for prilling of urea and ammonium nitrate also are in Table 12.18. Towers as high as 60 m have been installed. Because of the expense of towers, prilling is not competitive with other granulation processes until capacities of 200–400 tons/day are reached.

(a)

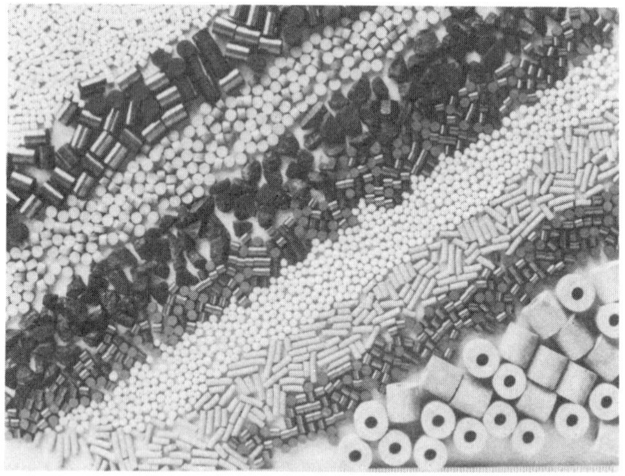

(b)

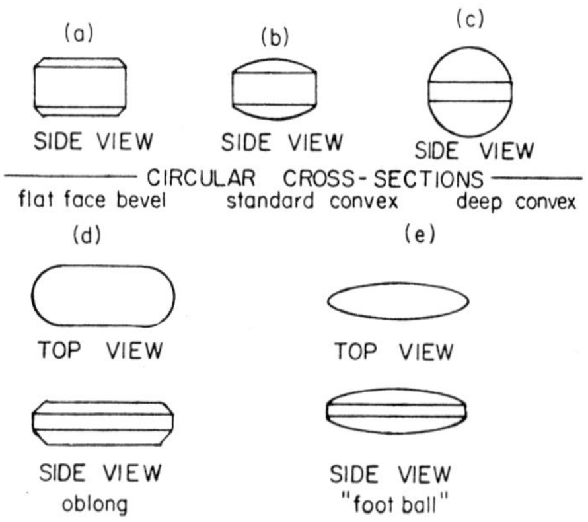

(c)

Figure 12.14. Common shapes and sizes of pellets made by some agglomeration techniques. (a) Sizes and shapes of briquets made on roll-type machines. (b) Catalyst pellets made primarily by extrusion and cutting (*Imperial Chemical Industries*). (c) Some of the shapes made with tabletting machines.

Materials suitable for prilling are those that melt without decomposition, preferably have a low heat of solidification and a high enough melting point to permit the use of ambient air for cooling. Because of high viscosities, spray wheels are preferred to spray nozzles. The wheels often are equipped with scrapers to prevent clogging. Since several wheels are needed for capacity, they are often arranged in line and the cross section of the tower is made rectangular.

The density of the prills is reduced substantially when much evaporation occurs: with 0.2–0.5% water in the feed, ammonium nitrate prills have a specific gravity of 0.95, but with 3–5% water it falls to 0.75. Prilled granules usually are less dense than those made by layering growth in drum or fluidized bed granulators. The latter processes also can make larger prills economically. To make large prills, a tall tower is needed to ensure solidification before the bottom is reached. The size distribution depends very much on the character of the atomization but can be made moderately uniform. Some commercial data of cumulative % less than size are:

% Less than size	0	5	50	95	100
Dia (mm)	1.2	1.6	2.4	3.5	4.8

Cooling of the prills can be accomplished more economically in either rotary drums of fluidized beds than in additional tower height. Fluidized bed coolers are cheaper and better because more easily dust controllable, and also because they can be incorporated in the lower section of the tower. After cooling, the product is screened, and the fines are returned to the melter and recycled.

FLUIDIZED AND SPOUTED BEDS

In fluidized bed granulation, liquid or solution is sprayed onto or into the bed, and growth occurs by agglomeration as a result of binding of small particles by the liquid or by layering as a result of evaporation of solution on the surfaces of the particles. The granules grown by layering are smoother and harder. Some attrition also occurs and tends to widen the size distribution range of the product. Larger agglomerates are obtained when the ratio of droplet/granule diameters decreases. Increase in the rate of the fluidizing gas and in the temperature of the bed decreases penetration and wetting of the bed and hence leads to smaller granule sizes. A narrower and more concentrated spray wets a smaller proportion of the particles and thus leads to larger size product. The bed is often made conical so that the larger particles are lifted off the bottom and recirculated more thoroughly.

Initial particle size distribution often is in the range of 50–250 μm. The product of Table 12.19(a) is 0.7–2.4 mm dia.

A wide range of operating conditions is used commercially. Performance data are in Table 12.19. Gas velocities cover a range of 3–20 times the minimum fluidizing velocity or 0.1–2.5 m/sec. Bed expansion ratios are up to 3 or so. As in fluidized bed drying, bed depths are low, usually between 12 and 24 in. Evaporation rates are in the range 0.005–1.0 kg/(sec)(m^2).

Batch fluidized bed granulation is practiced for small production rates or when the residence time must be long. Figure 12.17(a) and Table 12.19(a) are of an arrangement to make granules as feed to pharmaceutical tabletting. A feature of this equipment is the elaborate filter for preventing escape of fine particles and assuring their eventual growth. A continuous process for recovery of pellets of sodium sulfate from incineration of paper mill wastes is the subject of Figure 12.17(b) and Table 12.19(b).

TABLE 12.15. Alphabetical List of Some of the Materials that Have Been Successfully Compacted by Roll Presses

Acrylic resins, activated carbon, adipic acid, alfalfa, alga powder, alumina, aluminium, ammonium chloride, animal feed, anthracite, asbestos
Barium chloride, barium sulfate, battery masses, bauxite, bentonite, bitumen, bone meal, borax, brass turnings
Cadmium oxide, calcined dolomite, calcium chloride, calcium oxide, carbomethylcellulose (CMC), carbonates, catalysts, cellulose acetate, ceramics, charcoal, clay, coal, cocoa powder, coffee powder, coke, copper, corn starch
Detergents, dextrine, dimethylterephthalate (DMT), dolomite, ductile metals, dusts, dyes
Earthy ores, eggshells, elastomers, emulsifiers, epoxy resins
Feldspar, ferroalloys, ferrosilicium, fertilizers, flue dusts, fluorspar, fly ash, foodstuffs, fruit powders, fruit wastes, fungicides
Gipsum, glass making mixtures, glass powder, grain waste, graphite, gray iron chips and turnings
Herb teas, herbicides, hops, hydrated lime
Ice, inorganic salts, iron oxide, iron powder, insecticides
Kaolin, kieselgur, kieserite
Lead, lead oxide, leather wastes, LD-dust, lignite, lime, limestone, lithium carbonate, lithium fluoride, lithium hydroxide
Magnesia, magnesium carbonates, magnetite, maleic anhydrate, manganese dioxide, metal powders, molding compounds, molybdenum, monocalciumphosphate (MCP)
Naphthalene, nickel powders, nickel ores, niobium oxide
Ores, organic chlorides, organic silicates, oil shale, oyster shells
Pancreas powder, penicillin, pharmaceuticals, phosphate ores, plastics, polyvinylchloride (PVC), potash, potassium compounds, protein pigments, pyrites, pyrocatechol
Raisin seeds, reduced ores, refractory materials, rice starch, rock salt
Salts, sawdust, scrap metals, shales, silicates, soda ash, sodium chloride, sodium compounds, sodium cyanide, sponge iron, steel turnings, stone wool, sugar, sulfur
Teas, tin, titanium sponge, turnings
Urea, urea formaldehyde
Vanadium, vermiculite, vitamins
Waxes, welding powder, wood dust, wood shavings
Yeast (dry)
Zinc oxide, zirconium sand

(Pietsch, 1976).

The multicompartment equipment of Figure 12.17(c) permits improved control of process conditions and may assure a narrower size distribution because of the approach to plug flow.

Some of the fluidized bed dryers of Figure 9.13 could be equipped with sprays and adapted to granulation. The dryer performance data of Tables 9.14 and 9.15 may afford some concept of the sizes and capacities of suitable granulators, particularly when the sprays are somewhat dilute and evaporation is a substantial aspect of the process.

Spouted beds are applicable when granule sizes larger than those that can be fluidized smoothly. Above 1 mm or so, large bubbles begin to form in the bed and contacting deteriorates. Two arrangements of spraying into a spouted bed are shown in Figure 12.17(d). Particles grow primarily by deposition from the evaporated liquid that wets them as they flow up the spout and down the annulus. As the performance data of Table 12.19(c) indicate, particles up to 5 mm dia can be made and even quite dilute solutions can be processed. The diameter of the spout can be deduced from the given gas rates and the entraining velocities of the particles being made. Figure 9.13(f) is a more complete sketch of a spouted bed arrangement. Example 9.9 is devoted to sizing a fluidized bed dryer, but many aspects of that design are applicable to a granulation process.

SINTERING AND CRUSHING

This process originally was developed to salvage iron ore fines that could not be charged directly to the blast furnace. Although other applications appear to be feasible, the original application to iron ore fines seems to be still the only one, possibly because the scale of the operation is so great. Figure 12.11 is a sketch of the process. A mixture of ore fines, some recycle, 14–25% of calcite or dolomite as fluxes, and 2.5–5% solid fuel is placed on a conveyor to a depth of 12 in. or so. It is conveyed into an ignition furnace, burned and fused together, and then cooled and crushed to size. Fines under 6 mm are recycled. Sinter feed to blast furnaces is about 40–50 mm in the major dimension. The equipment is very large. One with a conveyor 5 m wide and 120 m long has a capacity of 27,000 tons/day.

Nodulizing is another process of size enlargement by fusion. This employs a rotary kiln like those used for cement manufacture. The product is uniform, about 0.5 in. dia, and more dense than sinter.

Sintering of powdered metals such as aluminum, beryllium, tungsten, and zinc as well as ceramics under pressure is widely practiced as a shaping process, but that is different from the sintering process described here.

TABLE 12.16. Roll Pressing Equipment Offered by Two Manufacturers

(a)

Model	Roll dia./mm	Max. roll width/mm	Position of rolls/feeder	Max. force/ metric tons	Overload system	Approx. capacity/ kg^{-1}	Feeder type	Press drive/ kW	Feeder drive / kW	Roll shapes	Max. feed temp./ °C
L 200/50	200	50	horiz./vert. or vert./horiz.	~10	None	10–100	screw	3/4	0·5	smooth/corrugated/ pocketed	80
K 26/100	200	100	horizontal/vertical	~20	"	100–200	"	11	3	"	80
K 27/200	300	200	"	~40	"	200–500	"	22	7·5	"	80
K 27/300	300	300	"	~80	hydraulic	500–1 000	"	30	7·5	"	80
CS 25	230	65	"	25	"	100–300	"	7	3	"	120
CS 50	406	119	"	50	"	300–1 000	"	15	5	"	150
MS 75	500	230	"	75	"	1 000–10 000	"	22	7·5	"	120
MS 150	500	280	"	150	"	3 000–15 000	"	75	11	"	1 000
MS 200	710	460	"	200	"	up to 50 000	"	300	15	"	1 000
MS 300	710	550	"	300	"	up to 60 000	screw(s)	400	15	"	1 000
MS 350	910	250	"	350		up to 40 000	screw	250	15	"	1 000

(b)

Model	Roll dia./mm	Max. roll width/mm	Position of rolls/feeder	Max. force/ metric tons	Overload system	Approx. capacity/ kg^{-1}	Feeder type	Press drive/ kW	Feeder drive / kW	Roll shapes	Max. feed temp./ °C
B 100	130	50	vertical/horizontal	10	hydraulic	20	screw	as required	as required	smooth/corrugated/ pocketed	ambient
B 150	200	75	"	20	"	200	"			"	"
B 220	300	75	"	30	"	1 500	"			"	"
B 300	380	100	"	60	"	3 000	"			"	"
B 400	460	150	"	125	"	5 000	"			"	"
B 500	610	200	"	250	"	15 000	"			"	"
D 100	130		"	20	"	50	"			"	"
D 150	200		"	40	"	200	"			"	"
D 300	330		"	70	"	3 000	"			"	"
DH 400	520		horizontal/vertical	140	"	6 000	"			"	800
DH 500	710		"	270	"	20 000	"			"	800
DH 600	920		"	500	"	50 000	"			"	800

[(a) Hutt Gmbh; (b) K.R. Komarek, Inc. From Pietsch, 1976].

TABLE 12.17. Some Applications of Rotating Ring Pelletizers (see Figure 12.15(b))

Material	Reason To Pellet	LB/HP/HR (KG/KW/HR)	Pellet Size (Inches Diameter) (Millimeter Dia.)
Asbestos Shorts	Densify, Reduce Dust In at 20lb/ft3 (320kg/m3) Out at 65lb/ft3 (1041kg/m3)	45 (27)	3/8" (9.5mm)
Acrylamide-Dry Wet	Handling	80 (49) 170 (103)	1/4" (6.4mm)
Bagasse	Densify, Reduce Dust In at 5lb/ft3 (80kg/m3) Out at 30lb/ft3 (480kg/m3)	80 (49)	3/8" (9.5mm)
Bauxite	Handling	300 (182)	1/2" (12.8mm)
Brewers Grain (Spent)	Densify, Handling In at 13lb/ft3 (208kg/m3) Out at 36lb/ft3 (577kg/m3)	150 (91)	1/4" (6.4mm)
Clay Base Material	Handling, Densify, Calcine	100-300 (61-182)	1/8" to 3/4" (3.2mm to 19mm)
Cryolite Filter Cake	Handling	100 (61)	3/8" (9.5mm)
Domolite	Handling	200 (122)	1/4" to 3/16" (6.4mm to 4.8mm)
Herbicide	Handling, Control, Solubility	150 (91)	12/64" (4.8mm)
Insecticide	Defined Form, Reduce Dust	120 (73)	1/8" (3.2mm)
Iron Oxide	Calcining, Reduce Dust	50-100 (30- 61)	1/8" to 1/4" (3.2mm to 6.4mm)
Lignite	Eliminate Fines	100 (61)	1/8" to 1/4" (3.2mm to 6.4mm)
Nylon Film Scrap	Densify	60 (36)	1/8" (3.2mm)
Paper Scrap	Densify	83 (50)	1/2" (12.7mm)
Phenolic Molding Compound	Reduce Dust, Handling	60 (36)	1/8" (3.2mm)
Polyethelyene Film	Densify from 5lb/ft3 (80kg/m3) to 20 lb/ft3 (320kg/m3)	30 (18)	1/8" to 3/16" (3.2mm to 4.8mm)
Polystyrene Foam	Densify from 4lb/ft3 (64kg/m3) to 24lb/ft3 (384kg/m3)	164 (100)	1/8" (3.2mm)
Polypropylene Film	Densify	40 (24)	1/8" (3.2mm)
Rubber Accelerator	Reduce Dust, Handling	192 (117)	12/64" (4.8mm)
Starch	Handling	75 (46)	12/64" (4.8mm)
Sawdust	Burn	60 (36)	1/4" (6.4mm)
Salt	Handling, Reduce Dust	70 (43)	1/8" (3.2mm)

(Sprout Waldron Co.).

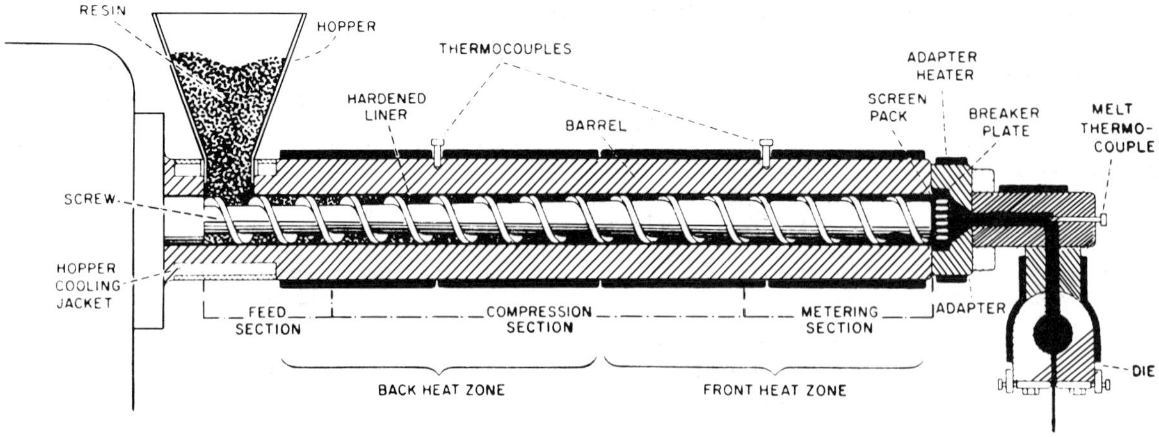

(a)

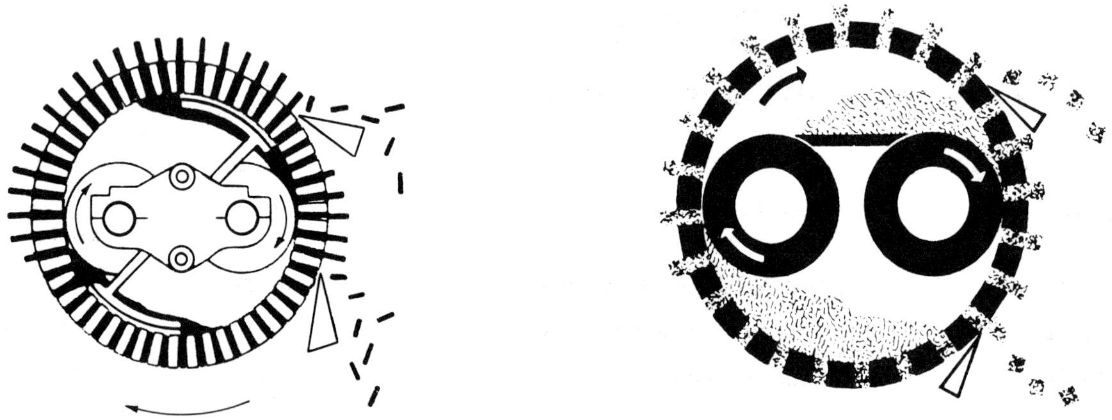

(b)

Figure 12.15. Two types of extrusion pelleting equipment. (a) Screw-type extruder for molten plastics: The die is turned 90° in the illustration from its normal position for viewing purposes. The extruded material is cooled and chopped subsequently as needed (*U.S. Industrial Chemical Co.*). (b) Ring extruders: material is charged with screw conveyors to the spaces between the inner rolls and the outer perforated ring, the ring rotates, material is forced through the dies and cut off with knives.

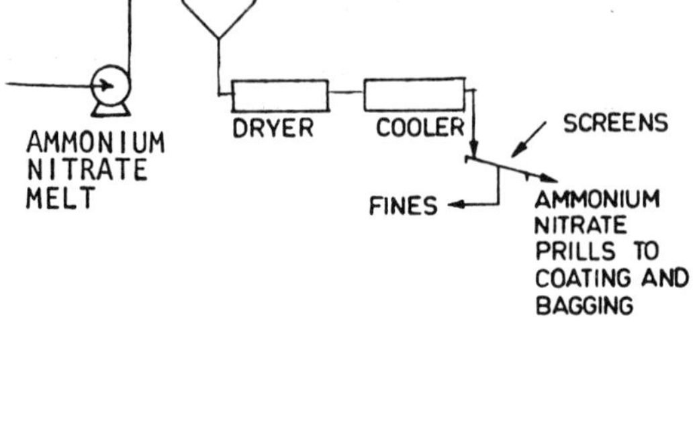

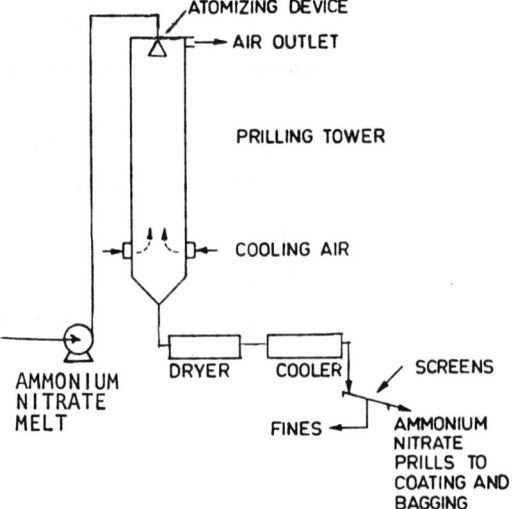

Figure 12.16. A prilling tower for ammonium nitrate, product size range 0.4–2.0 mm. The dryer is not needed if the moisture content of the melt is less than about 0.5%.

TABLE 12.18. List of Typical Prillable Materials and Performances of Some Prilling Operations

(a) List of Typical Prillable Materials

Adhesives	Pentachlorophenol
Adipic Acid	Petroleum wax
Alpha naphthol	Phenolic resins–Novalak resin
Ammonium nitrate and additives	Pine rosin
Asphalt	Polyethylene resins
Bisphenol-A	Polystyrene resins
Bitumen	Polypropylene–maleic anhydride
Carbon pitch	Potassium nitrate
Caustic soda	Resins
Cetyl alcohol	Sodium glycols
Coal-derived waxes	Sodium nitrate
Coal tar pitch	Sodium nitrite
Dichloro-benzidine	Sodium sulphate
Fatty acids	Stearic acid
Fatty alcohols	Stearyl alcohol
Epoxy resins	Substituted aliphatics
Hydrocarbon resins	Substituted amides
High-melting inorganic salts	Sulphur
Ink formulations	Urea and additives
Lauric acid	Urea–sulphur mix
Myristic acid	Wax–resin blends
Myristyl alcohol	
Paraffins	

(b) Data for the Prilling of Urea and Ammonium Nitrate

Tower size			
Prill tube height, ft		130	
Rectangular cross section, ft		11 by 21.4	
Cooling air			
rate, lb/h		360,000	
inlet temperature		ambient	
temperature rise, °F		15	
Melt			
Type	Urea		Ammonium Nitrate
rate, lb/h	35,200 (190 lb H_2O)		43,720 (90 lb H_2O)
inlet temperature, °F	275		365
Prills			
outlet temperature, °F	120		225
size, mm	approximately 1 to 3		

(HPD Inc.).

TABLE 12.19. Performance of Fluidized Bed and Spouted Bed Granulators

(a) Batch Fluidized Bed Granulator to Make Feed to Pharmaceutical Tablets; the Sketch Is in Figure 12.17(a)

	APPROXIMATE RANGE
Batch load, dry basis, lb	20 to 400[a]
Volume of container for static bed, ft^3	2 to 15
Fluidizing air fan, hp	5 to 25
Air (Steam) heating capacity, Btu/h	70,000 to 600,000
Drying air temperature, °C	40 to 80
Granulating liquid spray[b]	Two fluid nozzle
Air volume	½ to 2 SCFM
Liquid volume	500 to 1500 cm^3/min
Batch processing time, min	30 to 50
Average granule size	24 to 8 mesh

[a]Batch capacity exceeds 1500 lb in the largest modern units.
[b]Typical granulating liquids are gelatin or sodium carboxymethyl cellulose solutions.

(Capes and Fouda, 1984).

(b) Performance of Fluidized Bed Granulation of Two Waste Products; Sketch Is in Figure 12.17(b) for Paper Mill Waste

TYPE OF SLUDGE	INCINERATOR SIZE	BED TEMPERATURE	CAPACITY	GRANULAR PRODUCT COMPOSITION
Oil refinery waste sludge (85–95% water)	40 ft high; 20 ft ID at base increasing to 28 ft at top	1330°F	31 × 10^3 lb/hr of sludge	Start-up material was silica sand; replaced by nodules of various ash components such as $CaSO_4$, Na, Ca, Mg silicates, Al_2O_3 after operation of incinerator.
Paper mill waste liquor[a] (40% solids)	20 ft ID at top	1350°F	31 × 10^3 lb/hr	Sulfur added to produce 90–95% Na_2SO_4 and some Na_2CO_3

(Capes and Fouda, 1984).

(c) Applications of Spouted Bed Granulations

	Feed solution		Product		Gas temperature			
Material	*Moisture content (%)*	*Temperature (°C)*	*Size (mm)*	*Moisture (%)*	*Inlet (°C)*	*Outlet (°C)*	*Gas flow rate ($m^3 s^{-1}$)*	*Capacity ($kg h^{-1}$)*
Complex fertilizer	27	15	3–3·5	2·4	170	70	13·9	4000
Potassium chloride	68	15	4–5	–	200	60	13·9	1000
Ammonium nitrate	4	175	2·5–4	0·2	15	55	13·9	9500
Sulphur	–	135	2–5	–	15	–	$1·1 \times 10^{-2}$	40
Inorganic pigments, e.g. natural sienna	45	–	3–5	–	280	100		
Organic dyes, e.g. acid blue black	63	–	1–3	6·5	226	154		
Ammonium sulphate	60	70	~2	–	190	83	$\sim 1·3 \times 10^{-2}$	~2·7
Sodium chloride	77	–	~4·5	–	120	70	$\sim 1·8 \times 10^{-2}$	~1·2

(Sherrington and Oliver, 1981).

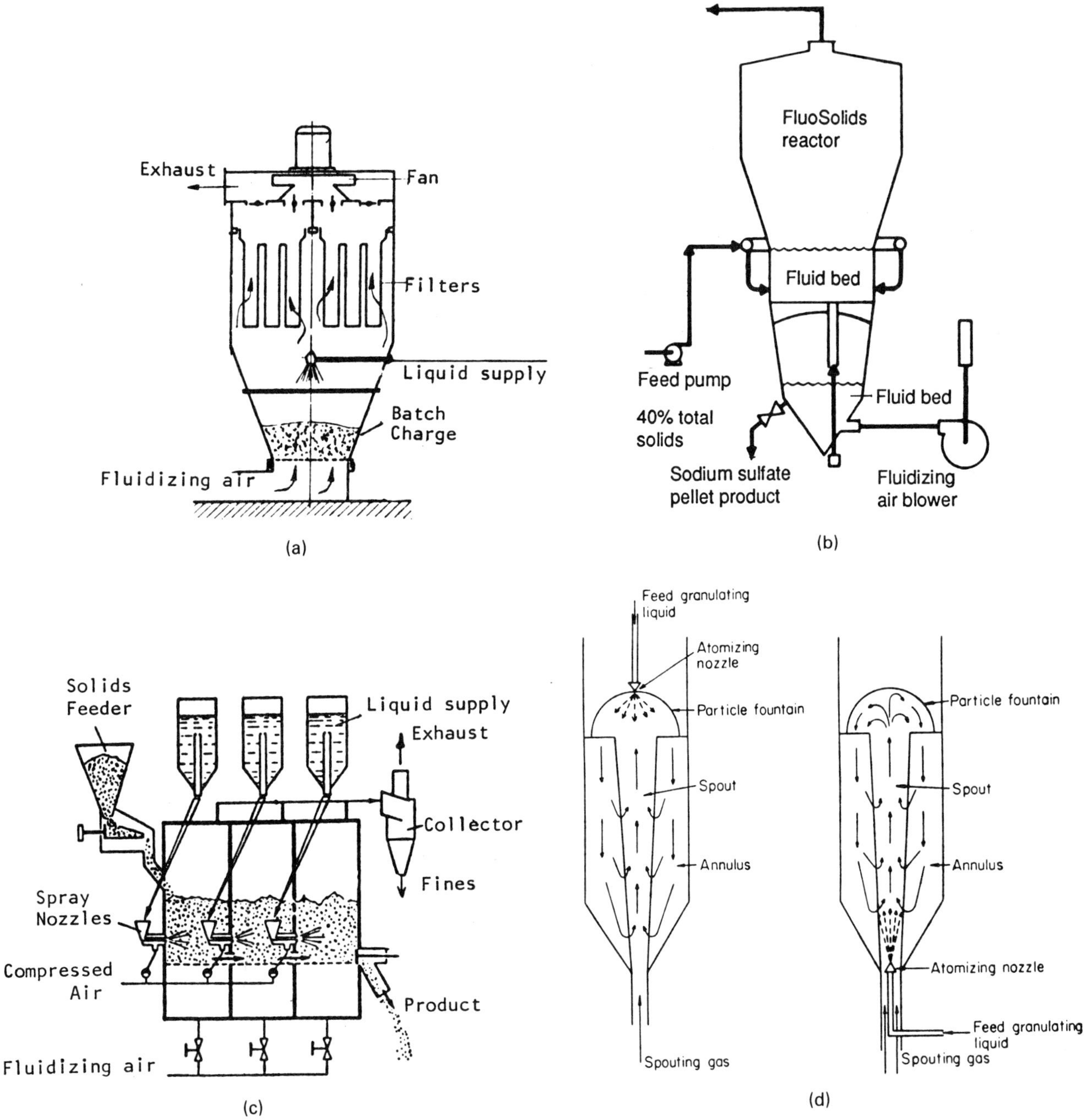

Figure 12.17. Fluidized bed and spouted bed granulators. (a) A batch fluidized bed granulator used in the pharmaceutical industry; performance data in Table 12.19(a). (b) Part of a fluidized bed incineration process for paper mill waste recovering sodium sulfate pellets; performance data in Table 12.19(b). (c) A three-stage fluidized bed granulator for more complete control of process conditions and more nearly uniform size distribution. (d) Two modes of injection of spray to spouted beds, into the body on the left and at the top on the right; performance data in Table 12.19(c).

REFERENCES

Size Reduction and Separation

1. W.L. Badger and J.T. Banchero, *Introduction to Chemical Engineering,* McGraw-Hill, New York, 1955.
2. W.M. Goldberger, Solid–solid systems, in *Chemical Engineers' Handbook,* McGraw-Hill, New York, 1984.
3. E.G. Kelly and D.J. Spottiswood, *Introduction to Mineral Processing,* Wiley, New York, 1982.
4. W.J. Mead (Ed.), Balling devices, Briquet machines, Grinding, Mills colloid, Mills roller, Screening, in *Encyclopedia of Chemical Process Equipment,* Reinhold, New York, 1964.
5. A.L. Mular and R.B. Bhappu (Eds.), *Mineral Processing Plant Design,* AIMME, New York, 1980.
6. E.J. Pryor, *Mineral Processing,* Elsevier, New York, 1965.
7. E.R. Riegel, *Chemical Process Machinery,* Reinhold, New York, 1953.
8. G.C. Sresty, Crushing and grinding equipment, in *Chemical Engineers' Handbook,* McGraw-Hill, New York, 1984, pp. 8.9–8.59.
9. A.F. Taggart (Ed.), *Handbook of Mineral Dressing,* Wiley, New York, 1945.
10. A.F. Taggart, *Elements of Ore Dressing,* Wiley, New York, 1951.
11. B.A. Wills, *Mineral Processing Technology,* Pergamon, New York, 1985.

Size Enlargement

12. C.E. Capes, *Particle Size Enlargement,* Elsevier, New York, 1980.
13. C.E. Capes, Size englargement, in *Chemical Engineers Handbook,* McGraw-Hill, New York, 1984, pp. 8.60–8.72.
14. C.E. Capes and A.E. Fouda, Agitation methods, in Ref. 17, pp. 286–294.
15. A.E. Capes and A.E. Fouda, Prilling and other spray methods, in Ref. 17, pp. 294–307.
16. J.T. Carstensen, Tabletting and pelletization in the pharmaceutical industry, in Ref. 17, pp. 252–268.
17. M.E. Fayed and L. Otten, (Eds.), *Handbook of Powder Science and Technology,* Van Nostrand Reinhold, New York, 1984.
18. R.M. Koerner and J.A. MacDougal (Eds.), *Briquetting and Agglomeration,* Institute for Briquetting and Agglomeration, Erie, PA, 1983.
19. R.A. Limons, Sintering iron ore, in Ref. 17, pp. 307–331.
20. W. Pietsch, *Roll Pressing,* Heyden, London, 1976.
21. P.J. Sherrington and R. Oliver, *Granulation,* Heyden, London, 1981.
22. N.E. Stanley-Wood (Ed.), *Enlargement and Compaction of Particulate Solids,* Butterworths, London, 1983.

13

DISTILLATION AND GAS ABSORPTION

The feasibility of separation of mixtures by distillation, absorption, or stripping depends on the fact that the compositions of vapor and liquid phases are different from each other at equilibrium. The vapor or gas phase is said to be richer in the more volatile or lighter or less soluble components of the mixture. Distillation employs heat to generate vapors and cooling to effect partial or total condensation as needed. Gas absorption employs a liquid of which the major components are essentially nonvolatile and which exerts a differential solvent effect on the components of the gas. In a complete plant, gas absorption is followed by a stripping operation for regeneration and recycle of the absorbent and for recovering the preferentially absorbed substances. In reboiled absorbers, partial stripping of the lighter components is performed in the lower part of the equipment. In distillation, absorption, or rectification and stripping are performed in the same equipment. Figures 13.1 and 13.2 show the basic types of equipment.

These distinctions between the two operations are partly traditional. The equipment is similar, and the mathematical treatment, which consists of material and energy balances and phase equilibrium relations, also is the same for both. The fact, however, that the bulk of the liquid phase in absorption-stripping plants is nonvolatile permits some simplifications in design and operation.

Equipment types are of two kinds, tray-type or packed, stagewise or continuous. The trays function as individual stages and produce stepwise changes in concentration. In packed towers concentration changes occur gradually. Until recently packed towers were used only in small equipment and where their construction was an advantage under corrosive conditions or when low pressure drop was mandatory. The picture now has changed somewhat and both types often are competitive over a wide range of sizes.

13.1. VAPOR–LIQUID EQUILIBRIA

This topic is concerned with the relations between vapor and liquid compositions over a range of temperature and pressure. Functionally, the dependence of the mol fraction y_i of component i in the vapor phase depends on other variables as

$$y_i = f(T, P, x_1, x_2, \ldots, x_n). \tag{13.1}$$

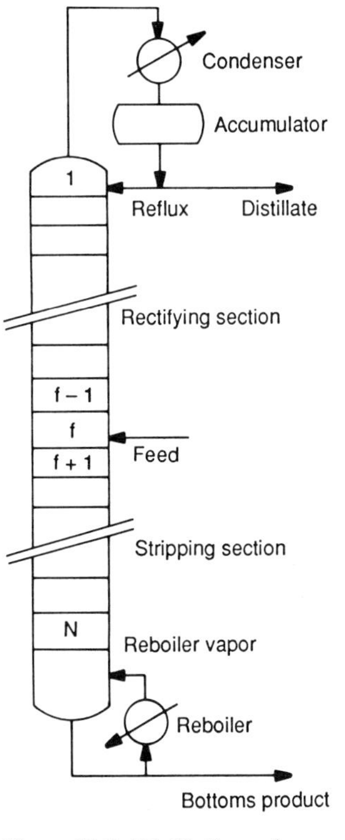

Figure 13.1. Distillation column assembly.

The dependence on composition alone often is approximated by

$$y_i = K_i x_i, \tag{13.2}$$

where K_i, the vaporization equilibrium ratio (VER) is assumed to depend primarily on the temperature and pressure and the nature of the substance, and only secondarily on the composition. Equation (13.2) can be viewed as suggested by Raoult's law,

$$y_i = (P_i^{sat}/P)x_i \tag{13.3}$$

with

$$(K_i)_{ideal} = P_i^{sat}/P, \tag{13.4}$$

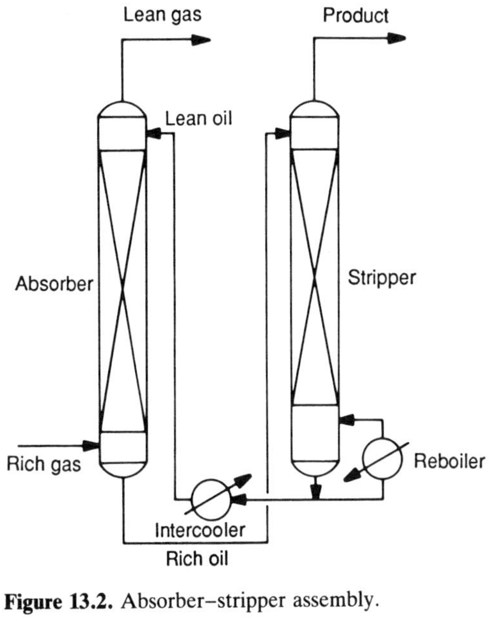

Figure 13.2. Absorber–stripper assembly.

where P_i^{sat} is the vapor pressure of component i and P is the system pressure. Several correlations have been developed for VERs, chiefly for hydrocarbon systems, for example, the one in Figure 13.3. The effect of composition is expressed in terms of a convergence pressure, which is explained for instance in the *API Data Book* (1969–date). The correction is small for system pressures under 10 atm or so and is neglected in this book.

A more nearly complete expression of K_i is derived upon noting that at equilibrium partial fugacities of each component are the same in each phase, that is

$$\hat{f}_i^v = \hat{f}_i^L \tag{13.5}$$

or in terms of fugacity and activity coefficients,

$$y_i \hat{\phi}_i^v P = x_i \gamma_i \phi_i^{sat} P_i^{sat}. \tag{13.6}$$

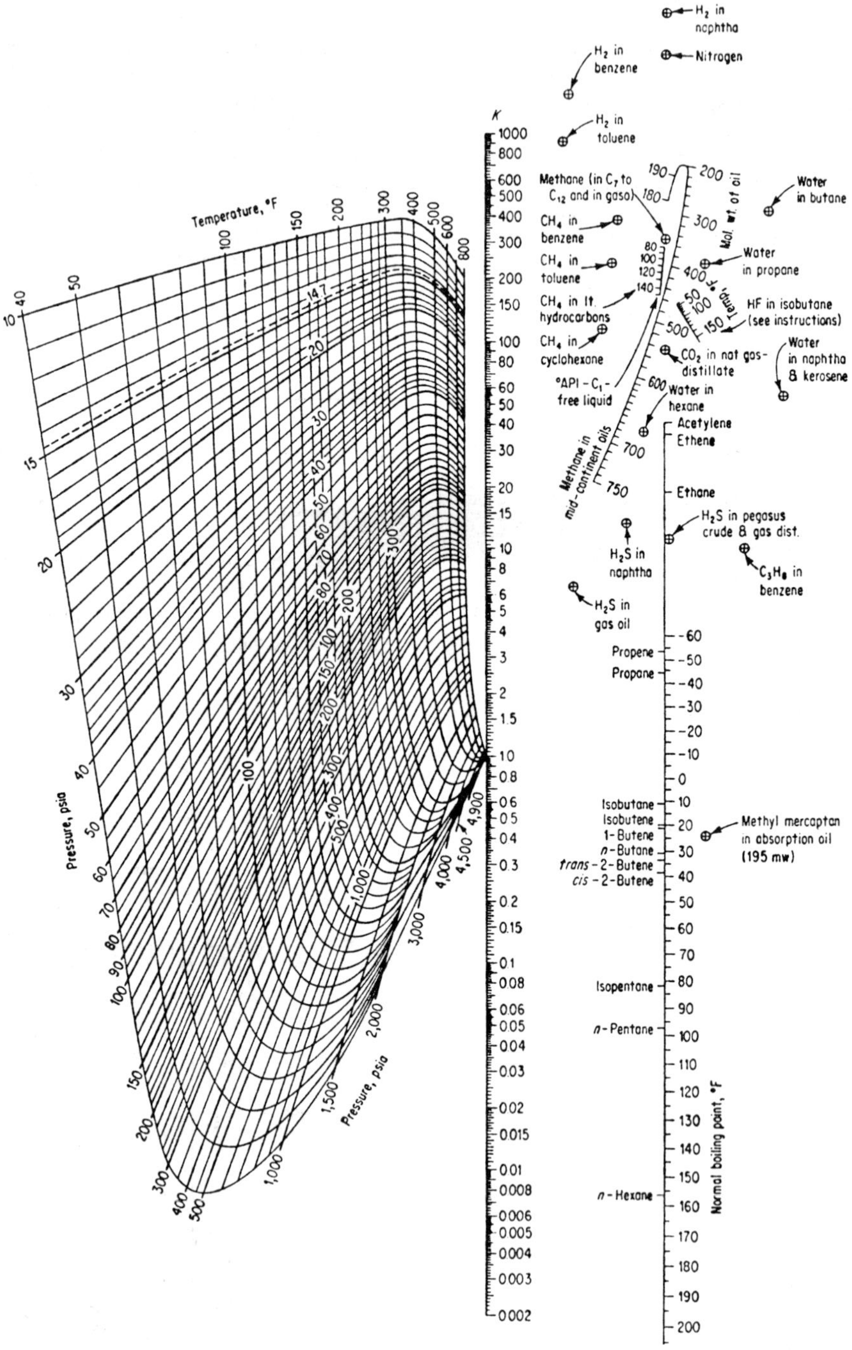

Figure 13.3. Vaporization equilibrium ratios (*Hadden and Grayson, 1961; Courtesy Mobil Oil Corp., New York*).

The relation between mol fractions becomes

$$y_i = \frac{\gamma_i \phi_i^{sat} P_i^{sat}}{\hat{\phi}_i^v P} x_i, \tag{13.7}$$

which makes the vaporization equilibrium ratio

$$K_i = \frac{\gamma_i \phi_i^{sat} P_i^{sat}}{\hat{\phi}_i^v P}. \tag{13.8}$$

Additionally usually small corrections for pressure, called Poynting factors, also belong in Eq. (13.6) and following but are omitted here. The new terms are

γ_i = activity coefficient in the liquid phase,
ϕ_i^{sat} = fugacity coefficient of the pure component at its vapor pressure,
$\hat{\phi}_i^v$ = partial fugacity coefficient in the vapor phase.

Equations for fugacity coefficients are derived from equations of state. Table 13.1 has them for the popular Soave equation of state. At pressures below 5–6 atm, the ratio of fugacity coefficients in Eq. (13.8) often is near unity. Then the VER may be written

$$K_i = \gamma_i P_i^{sat}/P \tag{13.9}$$

and is independent of the nature of the vapor phase.

Values of the activity coefficients are deduced from experimental data of vapor–liquid equilibria and correlated or extended by any one of several available equations. Values also may be calculated approximately from structural group contributions by methods called UNIFAC and ASOG. For more than two components, the correlating equations favored nowadays are the Wilson, the NRTL, and UNIQUAC, and for some applications a solubility parameter method. The first and last of these are given in Table 13.2. Calculations from measured equilibrium compositions are made with the rearranged equation

$$\gamma_i = \frac{\phi_i^v P}{\phi_i^{sat} P_i^{sat}} \frac{y_i}{x_i} \tag{13.10}$$

$$\simeq \frac{P}{P_i^{sat}} \frac{y_i}{x_i}. \tag{13.11}$$

The last approximation usually may be made at pressures below 5–6 atm. Then the activity coefficient is determined by the vapor pressure, the system pressure, and the measured equilibrium compositions.

Since the fugacity and activity coefficients are mathematically complex functions of the compositions, finding corresponding compositions of the two phases at equilibrium when the equations are known requires solutions by trial. Suitable procedures for making flash calculations are presented in the next section, and in greater detail in some books on thermodynamics, for instance, the one by Walas (1985). In making such calculations, it is usual to start by assuming ideal behavior, that is,

$$\hat{\phi}_i^v/\phi_i^{sat} = \gamma_i = 1. \tag{13.12}$$

After the ideal equilibrium compositions have been found, they are used to find improved values of the fugacity and activity coefficients. The process is continued to convergence.

TABLE 13.1. The Soave Equation of State and Fugacity Coefficients

Equation of State

$$P = \frac{RT}{V-b} - \frac{a\alpha}{V(V+b)}$$

$$z^3 - z^2 + (A - B - B^2)z - AB = 0$$

Parameters

$a = 0.42747 R^2 T_c^2/P_c,$
$b = 0.08664 RT_c/P_c$
$\alpha = [1 + (0.48508 + 1.55171\omega - 0.15613\omega^2)(1 - T_r^{0.5})]^2$

$\alpha = 1.202 \exp(-0.30288 T_r)$ and for hydrogen (Graboski and Daubert, 1979)

$A = a\alpha P/R^2T^2 = 0.42747 \alpha P_r/T_r^2$

$B = bP/RT = 0.08664 P_r/T_r$

Mixtures

$$a\alpha = \sum\sum y_i y_j (a\alpha)_{ij}$$

$$b = \sum y_i b_i$$

$$A = \sum\sum y_i y_j A_{ij}$$

$$B = \sum y_i B_i$$

Cross parameters

$(a\alpha)_{ij} = (1 - k_{ij})\sqrt{(a\alpha)_i (a\alpha)_j}$

k_{ij} in table

$k_{ij} = 0$ for hydrocarbon pairs and hydrogen

Correlations in Terms of Absolute Differences between Solubility Parameters of the Hydrocarbon, δ_{HC} and of the Inorganic Gas

Gas	k_{ij}
H_2S	$0.0178 + 0.0244\,\lvert\delta_{HC} - 8.80\rvert$
CO_2	$0.1294 - 0.0292\,\lvert\delta_{HC} - 7.12\rvert - 0.0222\,\lvert\delta_{HC} - 7.12\rvert^2$
N_2	$-0.0836 + 0.1055\,\lvert\delta_{HC} - 4.44\rvert - 0.0100\,\lvert\delta_{HC} - 4.44\rvert^2$

Fugacity Coefficient of a Pure Substance

$$\ln\phi = z - 1 - \ln\left[z\left(1 - \frac{b}{V}\right)\right] - \frac{a\alpha}{bRT}\ln\left(1 + \frac{b}{V}\right)$$

$$= z - 1 - \ln(z - B) - \frac{A}{B}\ln\left(1 + \frac{B}{z}\right)$$

Fugacity Coefficients in Mixtures

$$\ln\phi_i = \frac{b_i}{b}(z-1) - \ln\left[z\left(1 - \frac{b}{V}\right)\right] + \frac{a\alpha}{bRT}\left[\frac{b_i}{b} - \frac{2}{a\alpha}\sum_j y_j (a\alpha)_{ij}\right]\ln\left(1 + \frac{b}{V}\right)$$

$$= \frac{B_i}{B}(z-1) - \ln(z - B) + \frac{A}{B}\left[\frac{B_i}{B} - \frac{2}{a\alpha}\sum_j y_j (a\alpha)_{ij}\right]\ln\left(1 + \frac{B}{z}\right)$$

(*Walas*, 1985).

TABLE 13.2. Activity Coefficients from Solubility Parameters and from the Wilson Equation

Binary Mixtures

Name	Parameters	$\ln\gamma_1$ and $\ln\gamma_2$
Scatchard–Hildebrand	δ_1, δ_2	$\frac{V_1}{RT}(1-\phi_1)^2(\delta_1-\delta_2)^2$
		$\frac{V_2}{RT}\phi_1^2(\delta_1-\delta_2)^2$
	$\phi_1 = V_1x_1/(V_1x_1+V_2x_2)$	
Wilson	$\lambda_{12}, \lambda_{21}$	$-\ln(x_1+\Lambda_{12}x_2)+x_2\left(\frac{\Lambda_{12}}{x_1+\Lambda_{12}x_2}-\frac{\Lambda_{21}}{\Lambda_{21}x_1+x_2}\right)$
		$-\ln(x_2+\Lambda_{21}x_1)-x_1\left(\frac{\Lambda_{12}}{x_1+\Lambda_{12}x_2}-\frac{\Lambda_{21}}{\Lambda_{21}x_1+x_2}\right)$
	$\Lambda_{12}=\frac{V_2^L}{V_1^L}\exp\left(-\frac{\lambda_{12}}{RT}\right)$	$\Lambda_{21}=\frac{V_1^L}{V_2^L}\exp\left(-\frac{\lambda_{21}}{RT}\right)$
	V_i^L molar volume of pure liquid component i.	

Ternary Mixtures

$$\ln\gamma_1 = 1-\ln(x_1\Lambda_{i1}+x_2\Lambda_{i2}+x_3\Lambda_{i3})-\frac{x_1\Lambda_{1i}}{x_1+x_2\Lambda_{12}+x_3\Lambda_{13}}-\frac{x_2\Lambda_{2i}}{x_1\Lambda_{21}+x_2+x_3\Lambda_{23}}-\frac{x_3\Lambda_{3i}}{x_1\Lambda_{31}+x_2\Lambda_{32}+x_3}$$

$$\Lambda_{ii}=1$$

Multicomponent Mixtures

Equation	Parameters	$\ln\gamma_i$
Scatchard–Hildebrand	δ_i	$\frac{V_i}{RT}\left[\delta_i-\sum_j \frac{x_jV_j\delta_j}{\sum_k x_kV_k}\right]^2$
Wilson	$\Lambda_{ij}=\frac{V_j^L}{V_i^L}\exp\left(-\frac{\lambda_{ij}}{RT}\right)$	$-\ln\left(\sum_{j=1}^m x_j\Lambda_{ij}\right)+1-\sum_{k=1}^m \frac{x_k\Lambda_{ki}}{\sum_{j=1}^m x_j\Lambda_{kj}}$
	$\Lambda_{ij}=\Lambda_{jj}=1$	

RELATIVE VOLATILITY

The compositions of vapor and liquid phases of two components at equilibrium sometimes can be related by a constant relative volatility which is defined as

$$\alpha_{12}=\frac{y_1}{x_1}\Big/\frac{y_2}{x_2}=\left(\frac{y_1}{1-y_1}\right)\Big/\left(\frac{x_1}{1-x_1}\right). \tag{13.13}$$

Then

$$\frac{y_1}{1-y_1}=\alpha_{12}\frac{x_1}{1-x_1}. \tag{13.14}$$

In terms of vaporization equilibrium ratios,

$$\alpha_{12}=K_1/K_2=\gamma_1P_1^{\text{sat}}/\gamma_2P_2^{\text{sat}}, \tag{13.15}$$

and when Raoult's law applies the relative volatility is the ideal value,

$$\alpha_{\text{ideal}}=P_1^{\text{sat}}/P_2^{\text{sat}}. \tag{13.16}$$

Usually the relative volatility is not truly constant but is found to depend on the composition, for example,

$$\alpha_{12}=k_1+k_2x_1. \tag{13.17}$$

Other relations that have been proposed are

$$\frac{y_1}{1-y_1}=k_1+k_2\left(\frac{x_1}{1-x_1}\right) \tag{13.18}$$

and

$$\frac{y_1}{1-y_1}=k_1\left(\frac{x_1}{1-x_1}\right)^{k_2}. \tag{13.19}$$

A variety of such relations is discussed by Hala (*Vapor–Liquid*

Example 13.1
Correlation of Relative Volatility

Data for the system ethanol + butanol at 1 atm are taken from the collection of Kogan et al. (1966, #1038). The values of $x/(100-x)$, $y/(100-y)$, and α are calculated and plotted. The plot on linear coordinates shows that relative volatility does not plot linearly with x, but from the linear log–log plot it appears that

$$\frac{y}{100-y} = 4.364\left(\frac{x}{100-x}\right)^{1.045} \quad \text{or} \quad \alpha = 4.364\left(\frac{x}{100-x}\right)^{0.045}.$$

x	y	x	y
0	0	39.9	74.95
3.45	12.5	53.65	84.3
6.85	22.85	61.6	88.3
10.55	32.7	70.3	91.69
14.5	41.6	79.95	95.08
18.3	49.6	90.8	97.98
28.4	63.45	100.0	100.0

x	α	x/100 − x	y/100 − y
3.5	4.00	0.04	0.14
6.9	4.03	0.07	0.30
10.6	4.12	0.12	0.49
14.5	4.20	0.17	0.71
18.8	4.25	0.23	0.98
26.4	4.38	0.40	1.74
39.9	4.51	0.66	2.99
53.7	4.64	1.16	5.37
61.6	4.70	1.60	7.55
70.3	4.66	2.37	11.03
80.0	4.85	3.99	19.33
90.8	4.91	9.87	48.50

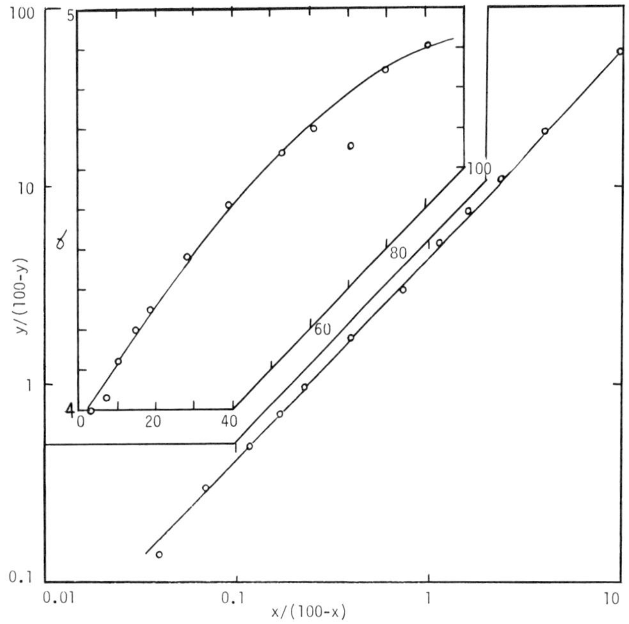

Equilibria, Pergamon, London, 1967). Other expressions can be deduced from Eq. (13.15) and some of the equations for activity coefficients, for instance, the Scatchard–Hildebrand of Table 13.2. Then

$$\alpha_{12} = \frac{y_1}{x_1}\bigg/\frac{y_2}{x_2} = \frac{P_1^{\text{sat}}}{P_2^{\text{sat}}}\exp\left\{\frac{(\delta_1-\delta_2)^2}{RT}[V_1(1-\phi_1)^2 - V_2\phi_1^2]\right\}, \tag{13.20}$$

where

$$\phi_1 = \frac{V_1x_1}{V_1x_1 + V_2x_2} \tag{13.21}$$

is the volume fraction of component 1 in the mixture.

Beyond a certain complexity these analytical relations between vapor and liquid compositions lose their utility. The simplest one, Eq. (13.14), is of value in the analysis of multistage separating equipment. When the relative volatility varies modestly from stage to stage, a geometric mean often is an adequate value to use. Applications are made later. Example 13.1 examines two ways of interpreting dependence of relative volatility on composition.

BINARY x–y DIAGRAMS

Equilibria between the components of a binary mixture are expressed as a functional relation between the mol fractions of the usually more volatile component in the vapor and liquid phases,

$$y = f(x). \tag{13.22}$$

The definition of relative volatility, Eq. (13.14), is rearranged into this form:

$$y = \frac{\alpha x}{1+(\alpha-1)x} \tag{13.23}$$

Representative x–y diagrams appear in Figure 13.4. Generally they are plots of direct experimental data, but they can be calculated from fundamental data of vapor pressure and activity coefficients. The basis is the bubblepoint condition:

$$y_1 + y_2 = \frac{\gamma_1 P_1^{\text{sat}}}{P}x_1 + \frac{\gamma_2 P_2^{\text{sat}}}{P}(1-x_1) = 1. \tag{13.24}$$

In order to relate y_1 and x_1, the bubblepoint temperatures are found over a series of values of x_1. Since the activity coefficients depend on the composition of the liquid and both activity coefficients and vapor pressures depend on the temperature, the calculation requires a respectable effort. Moreover, some vapor–liquid measurements must have been made for evaluation of a correlation of activity coefficients. The method does permit calculation of equilibria at several pressures since activity coefficients are substantially independent of pressure. A useful application is to determine the effect of pressure on azeotropic composition (Walas, 1985, p. 227).

13.2. SINGLE-STAGE FLASH CALCULATIONS

The problems of interest are finding the conditions for onset of vaporization, the bubblepoint; for the onset of condensation, the

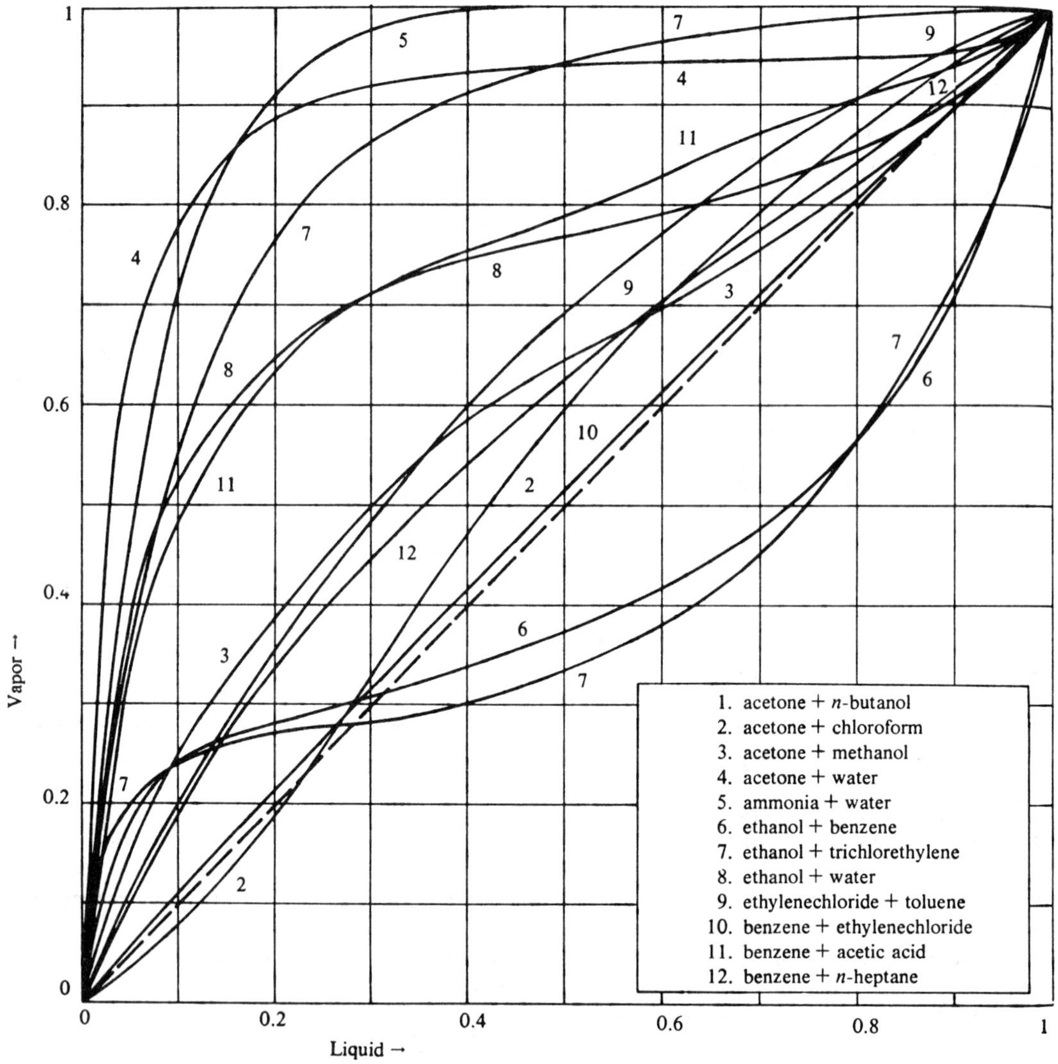

Figure 13.4. Some vapor-liquid composition diagrams at essentially atmospheric pressure. This is one of four such diagrams in the original reference (*Kirschbaum,* Destillier und Rektifiziertechnik, *Springer, Berlin, 1969*). Compositions are in weight fractions of the first-named.

dewpoint; and the compositions and the relative amounts of vapor and liquid phases at equilibrium under specified conditions of temperature and pressure or enthalpy and pressure. The first cases examined will take the K_i to be independent of composition. These problems usually must be solved by iteration, for which the Newton–Raphson method is suitable. The dependence of K on temperature may be represented adequately by

$$K_i = \exp[A_i - B_i/(T + C_i)]. \tag{13.25}$$

An approximate relation for the third constant is

$$C_i = 18 - 0.19T_{b_i}, \tag{13.26}$$

where T_{b_i} is the normal boiling point in °K. The dependence of K on pressure may be written simply as

$$K_i = a_i P^{b_i}. \tag{13.27}$$

Linear expressions for the enthalpies of the two phases are

$$h_i = a_i + b_i T, \tag{13.28}$$

$$H_i = c_i + d_i T, \tag{13.29}$$

assuming negligible heats of mixing. The coefficients are evaluated by readings off Figure 13.3, for example, and tabulations of pure component enthalpies. First derivatives are needed for application of the Newton–Raphson method:

$$\partial K_i/\partial T = B_i K_i/(T + C_i)^2, \tag{13.30}$$

$$\partial K_i/\partial P = b_i K_i/P. \tag{13.31}$$

BUBBLEPOINT TEMPERATURE AND PRESSURE

The temperature at which a liquid of known composition first begins to boil is found from the equation

$$f(T) = \sum K_i x_i - 1 = 0, \tag{13.32}$$

where the K_i are known functions of the temperature. In terms of Eq. (13.25), the Newton–Raphson algorithm is

$$T = T - \frac{-1 + \sum K_i x_i}{\sum [B_i K_i x_i/(T + C_i)^2]}. \tag{13.33}$$

Similarly, when Eq. (13.27) represents the effect of pressure, the bubblepoint pressure is found with the N–R algorithm:

$$f(P) = \sum K_i x_i - 1 = 0, \tag{13.34}$$

$$P = P - \frac{-1 + \sum a_i P^{b_i} x_i}{\sum a_i b_i P_i^{b_i - 1} x_i}. \tag{13.35}$$

DEWPOINT TEMPERATURE AND PRESSURE

The temperature or pressure at which a vapor of known composition first begins to condense is given by solution of the appropriate equation,

$$f(T) = \sum y_i/K_i - 1 = 0, \tag{13.36}$$

$$f(P) = \sum y_i/K_i - 1 = 0. \tag{13.37}$$

In terms of Eqs. (13.25) and (13.27) the N–R algorithms are

$$T = T + \frac{-1 + \sum y_i/K_i}{\sum [(y_i/K_i^2)\partial K_i/\partial T]} = T + \frac{-1 + \sum y_i/K_i}{\sum [B_i y_i/K_i(T + C_i)^2]}, \tag{13.38}$$

$$P = P + \frac{-1 + \sum y_i/K_i}{\sum [(y_i/K_i^2)\partial K_i/\partial P]} = P + \frac{(-1 + \sum y_i/K_i)P}{\sum (b_i y_i/K_i)}. \tag{13.39}$$

FLASH AT FIXED TEMPERATURE AND PRESSURE

At temperatures and pressures between those of the bubblepoint and dewpoint, a mixture of two phases exists whose amounts and compositions depend on the conditions that are imposed on the system. The most common sets of such conditions are fixed T and P, or fixed H and P, or fixed S and P. Fixed T and P will be considered first.

For each component the material balances and equilibria are:

$$Fz_i = Lx_i + Vy_i, \tag{13.40}$$

$$y_i = K_i x_i. \tag{13.41}$$

On combining these equations and introducing $\beta = V/F$, the fraction vaporized, the flash condition becomes

$$f(\beta) = -1 + \sum x_i = -1 + \sum \frac{z_i}{1 + \beta(K_i - 1)} = 0, \tag{13.42}$$

and the corresponding N–R algorithm is

$$\beta = \beta + \frac{-1 + \sum [z_i/(1 + \beta(K_i - 1))]}{\sum \{(K_i - 1)z_i/[1 + \beta(K_i - 1)]^2\}}. \tag{13.43}$$

After β has been found by successive approximation, the phase compositions are obtained with

$$x_i = \frac{z_i}{1 + \beta(K_i - 1)}, \tag{13.44}$$

$$y_i = K_i x_i. \tag{13.45}$$

A starting value of $\beta = 1$ always leads to a converged solution by this method.

FLASH AT FIXED ENTHALPY AND PRESSURE

The problem will be formulated for a specified final pressure and enthalpy, and under the assumption that the enthalpies are additive (that is, with zero enthalpy of mixing) and are known functions of temperature at the given pressure. The enthalpy balance is

$$H_F = (1 - \beta) \sum x_i H_{iL} + \beta \sum y_i H_{iV} \tag{13.46}$$

$$= (1 - \beta) \sum \frac{z_i H_{iL}}{1 + \beta(K_i - 1)} + \beta \sum \frac{K_i z_i H_{iV}}{1 + \beta(K_i - 1)}. \tag{13.47}$$

This equation and the flash Eq. (13.42) constitute a set:

$$f(\beta, T) = -1 + \sum \frac{z_i}{1 + \beta(K_i - 1)} = 0, \tag{13.48}$$

$$g(\beta, T) = H_F - (1 - \beta) \sum \frac{z_i H_{iL}}{1 + \beta(K_i - 1)} - \beta \sum \frac{K_i z_i H_{iV}}{1 + \beta(K_i - 1)} = 0, \tag{13.49}$$

from which the phase split β and temperature can be found when the enthalpies and the vaporization equilibrium ratios are known functions of temperature. The N–R method applied to Eqs. (13.48) and (13.49) finds corrections to initial estimates of β and T by solving the linear equations

$$h\frac{\partial f}{\partial \beta} + k\frac{\partial f}{\partial T} + f = 0, \tag{13.50}$$

$$h\frac{\partial g}{\partial \beta} + k\frac{\partial g}{\partial T} + g = 0, \tag{13.51}$$

where all terms are evaluated at the assumed values (β_0, T_0) of the two unknowns. The corrected values, suitable for the next trial if that is necessary, are

$$\beta = \beta_0 + h, \tag{13.52}$$

$$T = T_0 + K. \tag{13.53}$$

Example 13.2 applies these equations for dewpoint, bubblepoint, and flashes.

EQUILIBRIA WITH Ks DEPENDENT ON COMPOSITION

The procedure will be described only for the case of bubblepoint temperature for which the calculation sequence is represented on Figure 13.5. Equations (13.8) and (13.32) are combined as

$$f(T) = \sum \frac{\gamma_i \phi_i^{\text{sat}} P_i^{\text{sat}}}{\hat{\phi}_i P} x_i - 1 = 0. \tag{13.54}$$

The liquid composition is known for a bubblepoint determination, but the temperature is not at the start, so that starting estimates must be made for both activity and fugacity coefficients. In the flow diagram, the starting values are proposed to be unity for all the variables. After a trial value of the temperature is chosen, subsequent calculations on the diagram can be made directly. The correct value of T has been chosen when $\sum y_i = 1$.

Since the equations for fugacity and activity coefficients are complex, solution of this kind of problem is feasible only by computer. Reference is made in Example 13.3 to such programs. There also are given the results of such a calculation which reveals the magnitude of deviations from ideality of a common organic system at moderate pressure.

EXAMPLE 13.2
Vaporization and Condensation of a Ternary Mixture

For a mixture of ethane, n-butane, and n-pentane, the bubblepoint and dewpoint temperatures at 100 psia, a flash at 100°F and 100 psia, and an adiabatic flash at 100 psia of a mixture initially liquid at 100°F will be determined. The overall composition z_i, the coefficients A, B, and C of Eq. (13.22) and the coefficients a, b, c, and d of Eqs. (13.28) and (13.29) are tabulated:

Coefficients

	z	A	B	C	a	b	c	d
C_2	0.3	5.7799	2167.12	−30.6	122	0.73	290	0.45
nC_4	0.3	6.1418	3382.90	−60.8	96	0.56	267	0.34
nC_5	0.4	6.4610	3978.36	−73.4	90	0.55	260	0.40

The bubblepoint temperature algorithm is

$$T = T - \frac{-1 + \sum K_i x_i}{\sum [B_i K_i x_i/(T + C_i)^2]}, \tag{13.33}$$

and the dewpoint temperature algorithm is

$$T = T + \frac{-1 + \sum y_i/K_i}{\sum [B_i y_i/K_i(T + C_i)^2]}. \tag{13.38}$$

Results of successive iterations are

Bubblepoint	Dewpoint
1000.0000	700.0000
695.1614	597.8363
560.1387	625.9790
506.5023	635.3072
496.1742	636.0697
495.7968	636.0743
495.7963	636.0743

The algorithm for the fraction vapor at specified T and P is

$$\beta = \frac{V}{F} = \beta + \frac{-1 + \sum z_i/(1 + \beta(K_i - 1))}{\sum (K_i - 1)z_i/(1 + \beta(K_i - 1))^2}, \tag{13.43}$$

and the equations for the vapor and liquid compositions are

$$x_i = z_i/(1 + \beta(K_i - 1)), \tag{13.44}$$
$$y_i = K_i x_i. \tag{13.45}$$

Results for successive iterations for β and the final phase compositions are

β		z_i	x_i	y_i
1.0000	C_2	0.3	0.1339	0.7231
0.8257	nC_4	0.3	0.3458	0.1833
0.5964	nC_5	0.4	0.5203	0.0936
0.3986				
0.3038				
0.2830				
0.2819				

Adiabatic flash calculation: Liquid and vapor enthalpies off charts in the API data book are fitted with linear equations

$$h = a + bT \quad (°F), \tag{13.28}$$
$$H = c + dT \ (°F). \tag{13.29}$$

The inlet material to the flash drum is liquid at 100°F, with $H_0 = 8{,}575.8$ Btu/lb mol. The flash Eq. (13.43) applies to this part of the example. The enthalpy balance is

$$H_0 = 8575.8$$
$$= (1 - \beta) \sum M_i x_i h_i + \beta \sum M_i y_i H_i \tag{13.46}$$
$$= (1 - \beta) \sum \frac{M_i z_i h_i}{1 + \beta(K_i - 1)} + \beta \sum \frac{K_i M_i z_i H_i}{1 + \beta(K_i - 1)}. \tag{13.47}$$

The procedure consists of the steps.

1. Assume T.
2. Find the K_i, h_i, and H_i.
3. Find β from the flash equation (13.43).
4. Evaluate the enthalpy of the mixture and compare with H_0, Eq. (13.47).

The results of several trials are shown:

T (°R)	β	H
530.00	0.1601	8475.70
532.00	0.1681	8585.46
531.82	0.1674	8575.58~ 8575.8, check.

The final VERs and the liquid and vapor compositions are:

	K	x	y
C_2	4.2897	0.1935	0.8299
nC_4	0.3534	0.3364	0.1189
nC_5	0.1089	0.4701	0.0512

The numerical results were obtained with short computer programs which are given in Walas (1985, p. 317).

13.3. EVAPORATION OR SIMPLE DISTILLATION

As a mixture of substances is evaporated, the residue becomes relatively depleted in the more volatile constituents. A relation for binary mixtures due to Rayleigh is developed as follows: The differential material balance for a change dL in the amount of liquid remaining is

$$-y\,dL = d(LX) = L\,dx + X\,dL. \tag{13.55}$$

Upon rearrangement and integration, the result is

$$\ln\left(\frac{L}{L_0}\right) = \int_{x_0}^{x} \frac{dx}{x - y}. \tag{13.56}$$

In terms of a constant relative volatility

$$y = \frac{\alpha x}{1 + (\alpha - 1)x}, \tag{13.57}$$

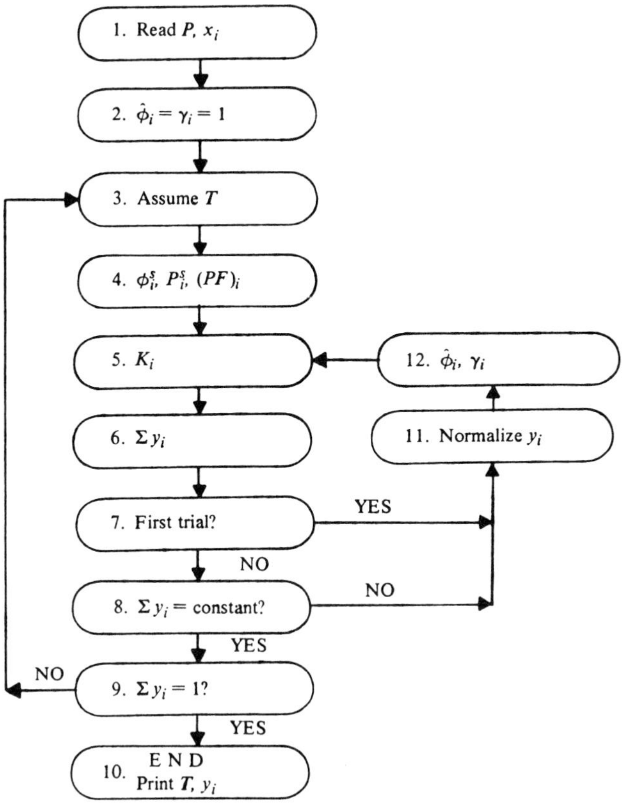

Figure 13.5. Calculation diagram for bubblepoint temperature (*Walas,* Phase Equilibria in Chemical Engineering, *Butterworths, Stoneham, MA, 1985*).

the integral becomes

$$\ln\frac{L}{L_0} = \frac{1}{x-1}\ln\frac{x(1-x_0)}{x_0(1-x)} + \ln\frac{1-x_0}{1-x}. \tag{13.58}$$

MULTICOMPONENT MIXTURES

Simple distillation is not the same as flashing because the vapor is removed out of contact with the liquid as soon as it forms, but the process can be simulated by a succession of small flashes of residual liquid, say 1% of the original amount each time. After n intervals, the amount of residual liquid F is

$$F = L_0(1-0.01n) \tag{13.59}$$

and

$$\beta = \frac{V}{F} = \frac{0.01L_0}{(1-0.01n)L_0} = \frac{0.01}{1-0.01n}. \tag{13.60}$$

Then the flash equation (13.42) becomes a function of temperature,

$$f(T_n) = -1 + \sum \frac{z_i}{1+0.01(K_i-1)/(1-0.01n)} = 0. \tag{13.61}$$

Here z_i is the composition at the end of interval n and K_i also may be taken at the temperature after interval n. The composition is found by material balance as

$$Lz_i = L_0(1-0.01n)z_i = L_0\left[z_{i0} - 0.01\sum_{k=1}^{n} y_{ik}\right], \tag{13.62}$$

where each composition y_{ik} of the flashed vapor is found from Eqs. (13.44) and (13.45)

$$y_i = K_i x_i = \frac{K_i z_i}{1+0.01(K_i-1)/(1-0.01n)} \tag{13.63}$$

and is obtained during the process of evaluating the temperature with Eq. (13.61). The VERs must be known as functions of temperature, say with Eq. (13.25).

13.4. BINARY DISTILLATION

Key concepts of the calculation of distillation are well illustrated by analysis of the distillation of binary mixtures. Moreover, many real systems are essentially binary or can be treated as binaries made up of two pseudo components, for which it is possible to calculate upper and lower limits to the equipment size for a desired separation.

The calculational base consists of equilibrium relations and material and energy balances. Equilibrium data for many binary systems are available as tabulations of x vs. y at constant temperature or pressure or in graphical form as on Figure 13.4. Often they can be extended to other pressures or temperatures or expressed in mathematical form as explained in Section 13.1. Sources of equilibrium data are listed in the references. Graphical calculation of distillation problems often is the most convenient

---

Example 13.3
Bubblepoint Temperature with the Virial and Wilson Equations

A mixture of acetone (1) + butanone (2) + ethylacetate (3) with the composition $x_1 = x_2 = 0.3$ and $x_3 = 0.4$ is at 20 atm. Data for the system such as vapor pressures, critical properties, and Wilson coefficients are given with a computer program in Walas (1985, p. 325). The bubblepoint temperature was found to be 468.7 K. Here only the properties at this temperature will be quoted to show deviations from ideality of a common system. The ideal and real K_i differ substantially.

Component	ϕ^{sat}	$\hat{\phi}^{v}$	$\phi^{sat}/\hat{\phi}^{v}$	γ
1	0.84363	0.84353	1.00111	1.00320
2	0.79219	0.79071	1.00186	1.35567
3	0.79152	0.78356	1.00785	1.04995

Component	K_{ideal}	K_{real}	y
1	1.25576	1.25591	0.3779
2	0.72452	0.98405	0.2951
3	0.77266	0.81762	0.3270

---

method, but numerical procedures may be needed for highest accuracy.

MATERIAL AND ENERGY BALANCES

In terms of the nomenclature of Figure 13.6, the balances between stage n and the top of the column are

$$V_{n+1}y_{n+1} = L_n x_n + Dx_D, \tag{13.64}$$
$$V_{n+1}H_{n+1} = L_n h_n + Dh_D + Q_c \tag{13.65}$$
$$= L_n h_n + DQ', \tag{13.66}$$

where

$$Q' = h_D + Q_c/D \tag{13.67}$$

is the enthalpy removed at the top of the column per unit of overhead product. These balances may be solved for the liquid/vapor ratio as

$$\frac{L_n}{V_{n+1}} = \frac{y_{n+1} - x_D}{x_n - x_D} = \frac{Q' - H_{n+1}}{Q' - h_n} \tag{13.68}$$

and rearranged as a combined material and energy balance as

$$\frac{L_n}{V_{n+1}} = \frac{y_{n+1} - x_D}{x_n - x_D} x_n + \frac{H_{n+1} - h_n}{Q' - h_n} x_D. \tag{13.69}$$

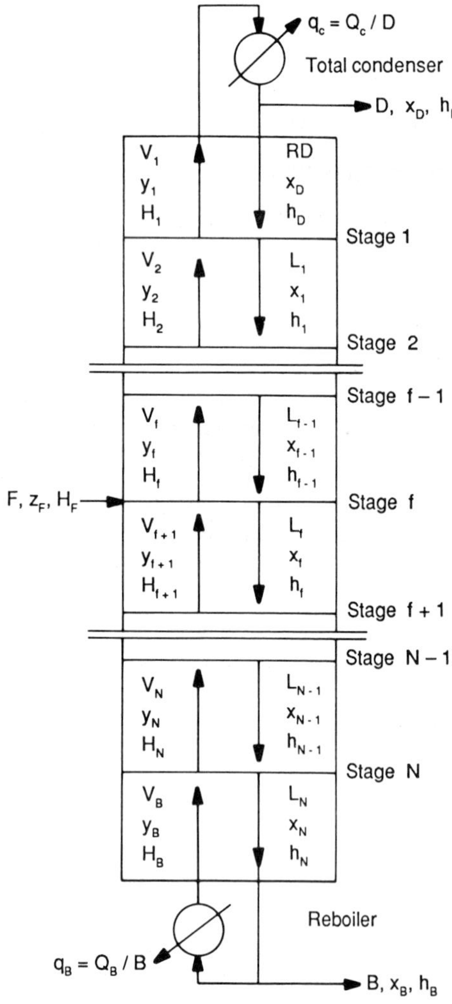

Figure 13.6. Model of a fractionating tower.

Similarly the balance between plate m below the feed and the bottom of the column can be put in the form

$$y_m = \frac{Q'' - H_m}{Q'' - h_{m+1}} x_{m+1} + \frac{h_{m+1} - H_m}{h_{m+1} - Q''} x_B, \tag{13.70}$$

where

$$Q'' = h_B - Q_b/B \tag{13.71}$$

is the enthalpy removed at the bottom of the column per unit of bottoms product.

For the problem to be tractable, the enthalpies of the two phases must be known as functions of the respective phase compositions. When heats of mixing and heat capacity effects are small, the enthalpies of mixtures may be compounded of those of the pure components; thus

$$H = yH_a + (1 - y)H_b, \tag{13.72}$$
$$h = xh_a + (1 - x)h_b, \tag{13.73}$$

where H_a and H_b are vapor enthalpies of the pure components at their dewpoints and h_a and h_b are corresponding liquid enthalpies at their bubblepoints.

Overall balances are

$$F = D + B, \tag{13.74}$$
$$Fz_F = Dx_D + Bx_B, \tag{13.75}$$
$$FH_F = Dh_D + Bh_B. \tag{13.76}$$

In the usual distillation problem, the operating pressure, the feed composition and thermal condition, and the desired product compositions are specified. Then the relations between the reflux rates and the number of trays above and below the feed can be found by solution of the material and energy balance equations together with a vapor–liquid equilibrium relation, which may be written in the general form

$$f(x_n, y_n) = 0. \tag{13.77}$$

The procedure starts with the specified terminal compositions and applies the material and energy balances such as Eqs. (13.64) and (13.65) and equilibrium relations alternately stage by stage. When the compositions from the top and from the bottom agree closely, the correct numbers of stages have been found. Such procedures will be illustrated first with a graphical method based on constant molal overflow.

CONSTANT MOLAL OVERFLOW

When the molal heats of vaporization of the two components are equal and the tower is essentially isothermal throughout, the molal flow rates L_n and V_n remain constant above the feed tray, and L_m and V_m likewise below the feed. The material balances in the two sections are

$$y_{n+1} = \frac{L_n}{V_{n+1}} x_n + \frac{D}{V_{n+1}} x_D, \tag{13.64}$$
$$y_m = \frac{L_{m+1}}{V_m} x_{m+1} - \frac{B}{V_m} x_B. \tag{13.78}$$

The flow rates above and below the feed stage are related by the liquid–vapor proportions of the feed stream, or more generally by the thermal condition of the feed, q, which is the ratio of the heat required to convert the feed to saturated vapor and the heat of vaporization, that is,

$$q = (H_F^{sat} - H_F)/(\Delta H)_{vap}. \tag{13.79}$$

For instance, for subcooled feed $q > 1$, for saturated liquid $q = 1$, and for saturated vapor $q = 0$. Upon introducing also the reflux ratio

$$R = L_n/D, \tag{13.80}$$

the relations between the flow rates become

$$L_m = L_n + qF = RD + qF, \tag{13.81}$$

$$V_m = L_m - B = RD + qF - B. \tag{13.82}$$

Accordingly, the material balances may be written

$$y = \frac{R}{R+1}x_n + \frac{1}{R+1}x_D, \tag{13.83}$$

$$y_m = \frac{RD+qF}{RD+qF-B}x_{m+1} - \frac{B}{RD+qF-B}x_B. \tag{13.84}$$

The coordinates of the point of intersection of the material balance lines, Eqs. (13.83) and (13.84), are located on a "q-line" whose equation is

$$y = \frac{q}{q-1}x + \frac{1}{q-1}x_F. \tag{13.85}$$

Figure 13.7(b) shows these relations.

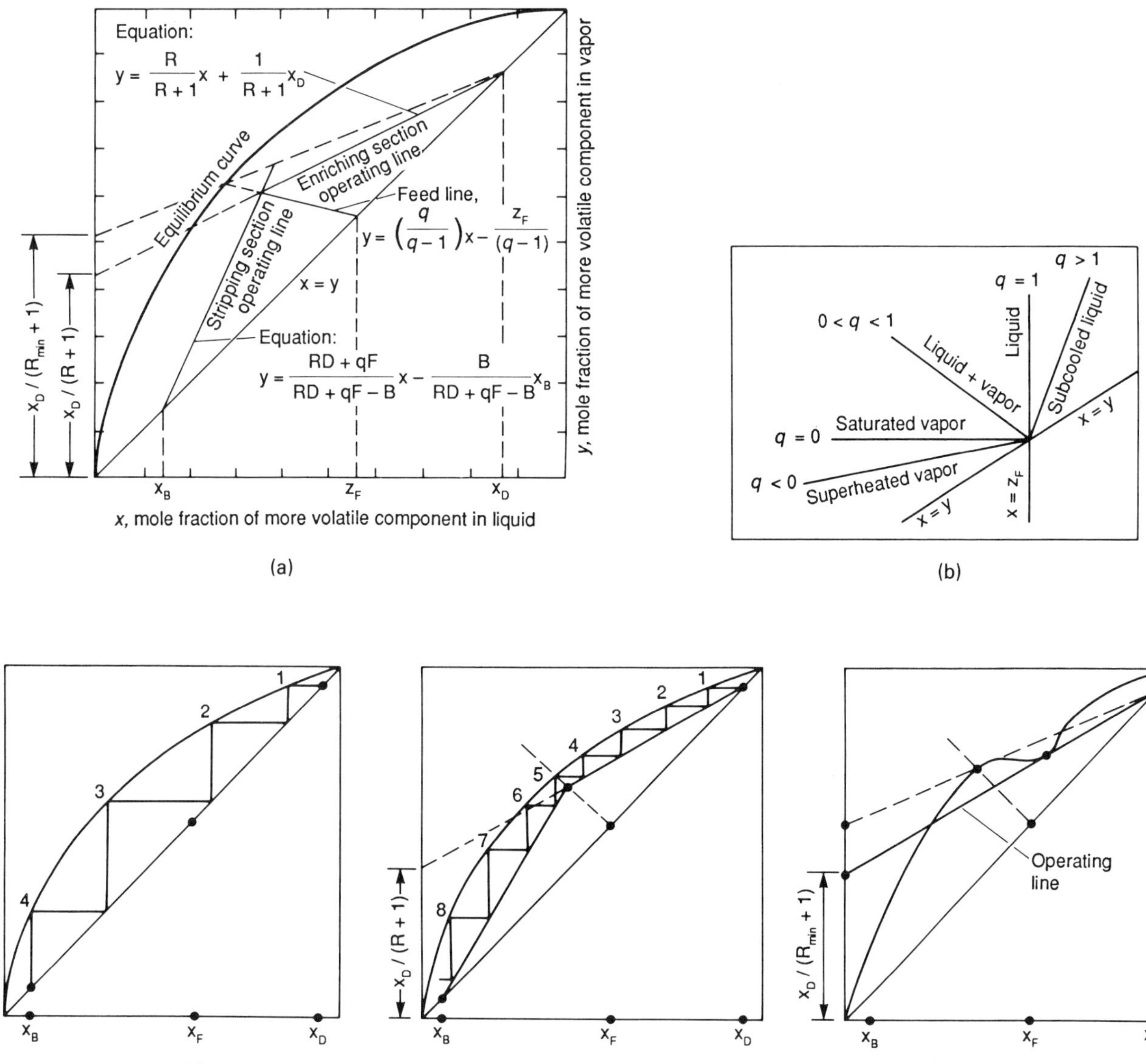

Figure 13.7. Features of McCabe–Thiele diagrams for constant molal overflow. (a) Operating line equations and construction and minimum reflux construction. (b) Orientations of q-lines, with slope $= q/(q-1)$, for various thermal conditions of the feed. (c) Minimum trays, total reflux. (d) Operating trays and reflux. (e) Minimum reflux determined by point of contact nearest x_D.

BASIC DISTILLATION PROBLEM

The basic problem of separation by distillation is to find the numbers of stages below and above the feed stage when the quantities x_F, x_D, x_B, F, D, B, and R are known together with the phase equilibrium relations. This means that all the terms in Eqs. (13.83) and (13.84) are to be known except the running x's and y's. The problem is solved by starting with the known compositions, x_D and x_B, at each end and working one stage at a time towards the feed stage until close agreement is reached between the pairs (x_n, y_n) and (x_m, y_m). The procedure is readily implemented on a programmable calculator; a suitable program for the enriching section is included in the solution of Example 13.4. A graphical solution is convenient and rapid when the number of stages is not excessive, which depends on the scale of the graph attempted.

Figure 13.7 illustrates various aspects of the graphical method. A minimum number of trays is needed at total reflux, that is, with no product takeoff. Minimum reflux corresponds to a separation requiring an infinite number of stages, which is the case when the equilibrium curve and the operating lines touch somewhere. Often this can occur on the q-line, but another possibility is shown on Figure 13.7(e). The upper operating line passes through point (x_D, x_D) and $x_D/(R+1)$ on the left ordinate. The lower operating line passes through the intersection of the upper with the q-line and point (x_B, x_B). The feed tray is the one that crosses the intersection of the operating lines on the q-line. The construction is shown with Example 13.5. Constructions for cases with two feeds and with two products above the feed plate are shown in Figure 13.8.

Optimum Reflux Ratio. The reflux ratio affects the cost of the tower, both in the number of trays and the diameter, as well as the cost of operation which consists of costs of heat and cooling supply and power for the reflux pump. Accordingly, the proper basis for choice of an optimum reflux ratio is an economic balance. The sizing and economic factors are considered in a later section, but reference may be made now to the results of such balances summarized in Table 13.3. The general conclusion may be drawn that the optimum reflux ratio is about 1.2 times the minimum, and also that the number of trays is about 2.0 times the minimum. Although these conclusions are based on studies of systems with nearly ideal vapor–liquid equilibria near atmospheric pressure, they often are applied more generally, sometimes as a starting basis for more detailed analysis of reflux and tray requirements.

Azeotropic and Partially Miscible Systems. Azeotropic mixtures are those whose vapor and liquid equilibrium compositions are identical. Their x–y lines cross or touch the diagonal. Partially miscible substances form a vapor phase of constant composition over the entire range of two-phase liquid compositions; usually the horizontal portion of the x–y plot intersects the diagonal, but those of a few mixtures do not, notably those of mixtures of methylethylketone and phenol with water. Separation of azeotropic mixtures sometimes can be effected in several towers at different pressures, as illustrated by Example 13.6 for ethanol-water mixtures. Partially miscible constant boiling mixtures usually can be separated with two towers and a condensate phase separator, as done in Example 13.7 for n-butanol and water.

UNEQUAL MOLAL HEATS OF VAPORIZATION

Molal heats of vaporization often differ substantially, as the few data of Table 13.4 suggest. When sensible heat effects are small, however, the condition of constant molal overflow still can be preserved by adjusting the molecular weight of one of the components, thus making it a pseudocomponent with the same molal heat of vaporization as the other substance. The x–y diagram and all of the compositions also must be converted to the adjusted molecular weight. Example 13.5 compares tray requirements on the basis of true and adjusted molecular weights for the separation of ethanol and acetic acid whose molal heats of vaporization are in the ratio 1.63. In this case, the assumption of constant molal overflow with the true molecular weight overestimates the tray requirements. A more satisfactory, but also more laborious, solution of the problem takes the enthalpy balance into account, as in the next section.

MATERIAL AND ENERGY BALANCE BASIS

The enthalpies of mixtures depend on their compositions as well as the temperature. Enthalpy–concentration diagrams of binary mixtures, have been prepared in general form for a few important systems. The most comprehensive collection is in Landolt-Börnstein [**IV4b**, 188, (1972)] and a few diagrams are in *Chemical Engineers Handbook* (1984), for instance, of ammonia and water, of ethanol and water, of oxygen and nitrogen, and some others. Such diagrams are named after Merkel.

For purposes of distillation calculations, a rough diagram of saturated vapor and liquid enthalpy concentration lines can be drawn on the basis of pure component enthalpies. Even with such a rough diagram, the accuracy of distillation calculation can be much superior to those neglecting enthalpy balances entirely. Example 13.8 deals with preparing such a Merkel diagram.

A schematic Merkel diagram and its application to distillation calculations is shown in Figure 13.9. Equilibrium compositions of vapor and liquid can be indicated on these diagrams by tielines, but are more conveniently used with associated x–y diagrams as shown with this figure. Lines passing through point P with coordinates (x_D, Q') are represented by Eq. (13.69) and those through point Q with coordinates (x_B, Q'') by Eq. (13.70). Accordingly, any line through P to the right of PQ intersects the vapor and liquid enthalpy lines in corresponding (x_n, y_{n+1}) and similarly the intersections of random lines through Q determine corresponding (x_{m+1}, y_m). When these coordinates are transferred to the x–y diagram, they determine usually curved operating lines. Figure 13.9(b) illustrates the stepping off process for finding the number of stages. Points P, F, and Q are collinear.

The construction for the minimum number of trays is independent of the heat balance. The minimum reflux corresponds to a minimum condenser load Q and hence to a minimum value of $Q' = h_D + Q_c/D$. It can be found by trial location of point P until an operating curve is found that touches the equilibrium curve.

ALGEBRAIC METHOD

Binary systems of course can be handled by the computer programs devised for multicomponent mixtures that are mentioned later. Constant molal overflow cases are handled by binary computer programs such as the one used in Example 13.4 for the enriching section which employ repeated alternate application of material balance and equilibrium stage-by-stage. Methods also are available that employ closed form equations that can give desired results quickly for the special case of constant or suitable average relative volatility.

Minimum Trays. This is found with the Fenske–Underwood equation,

$$N_{\min} = \frac{\ln[x_D(1-x_B)/x_B(1-x_D)]}{\ln \alpha} \tag{13.86}$$

EXAMPLE 13.4
Batch Distillation of Chlorinated Phenols

A mixture of chlorinated phenols can be represented as an equivalent binary with 90% 2,4-dichlorphenol (DCP) and the balance 2,4,6-trichlorphenol with a relative volatility of 3.268. Product purity is required to be 97.5% of the lighter material, and the residue must be below 20% of 2,4-DCP. It is proposed to use a batch distillation with 10 theoretical stages. Vaporization rate will be maintained constant.

a. For operation at constant overhead composition, the variations of reflux ratio and distillate yield with time will be found.
b. The constant reflux ratio will be found to meet the overhead and bottoms specifications.

a. At constant overhead composition, $y_D = 0.975$: The composition of the residue, x_{10}, is found at a series of reflux ratios between the minimum and the value that gives a residue composition of 0.2.

```
10 ! Example 13.9. Distillation
     at constant Yd
20 A=3.268
30 OPTION BASE 1
40 DIM X(10),Y(12)
50 Y(1)=.975
60 INPUT R
70 FOR N=1 TO 10
80 X(N)=1/(A/Y(N)-A+1)
90 Y(N+1)=1/(R+1)*(R*X(N)+Y(1))
100 NEXT N
110 Z=(Y(1)-.9)/(Y(1)-X(10)) ! =
     L/Lo
120 I=(R+1)/(Y(1)-X(10))^2 ! Int
     egrand of Eq 4
130 PRINT USING 140 ; R,X(10),Z,
     I
140 IMAGE D.DDDD,2X,.DDDD,2X,D.D
     DDD,2X,DDD.DDDDD
150 GOTO 60
160 END
```

With $q = 1$ and $x_n = 0.9$,

$$y_n = \frac{\alpha x_n}{1+(\alpha-1)x_n} = \frac{3.268(0.9)}{1+2.268(0.9)} = 0.9671,$$

$$R_m/(R_m+1) = \frac{0.975-0.9671}{0.975-0.9} = 0.1051,$$

$$\therefore R_m = 0.1174.$$

The btms compositions at a particular value of R are found by successive applications of the equations

$$x_n = \frac{y_n}{\alpha-(\alpha-1)y_n}, \tag{1}$$

$$y_{n+1} = \frac{R}{R+1}x_n + \frac{1}{R+1}y_D. \tag{2}$$

Start with $y_1 = y_D = 0.975$. The calculations are performed with the given computer program and the results are tabulated. The values of L/L_0 are found by material balance:

$$L/L_0 = (0.975-0.900)/(0.975-x_L) \tag{3}$$

The values of V/L_0 are found with Eq. (13.111).

$$\frac{V}{L_0} = (y_D - x_{L0})\int_{x_{L_0}}^{x_L} \frac{R+1}{(y_D-x_L)^2}\,dx_L$$
$$= (0.975-0.900)\int_{0.9}^{x_L} \frac{R+1}{(0.975-x_L)^2}\,dx_L. \tag{4}$$

From the tabulation, the cumulative vaporization is

$$V/L_0 = 1.2566.$$

The average reflux ratio is

$$\bar{R} = \frac{V-D}{D} = \frac{V}{D} - 1 = \frac{V}{L_0 - L} - 1 = \frac{V/L_0}{1-L/L_0} - 1$$
$$= \frac{1.2566}{1-0.0968} - 1 = 0.3913.$$

R	x_L	L/L_o	Integrand	V/L_o	$t/\bar{t}$
.1174	.9000	1.0001	198.69073	0.0000	0.000
.1500	.8916	.8989	165.17980	.1146	.091
.2000	.8761	.7585	122.74013	.2820	.224
.2500	.8571	.6362	89.94213	.4335	.345
.3000	.8341	.5321	65.43739	.5675	.452
.3500	.8069	.4461	47.75229	.6830	.544
.4000	.7760	.3768	35.33950	.7793	.620
.4500	.7422	.3222	26.76596	.8580	.683
.5000	.7069	.2797	20.86428	.9210	.733
.6000	.6357	.2210	13.89632	1.0138	.807
.7000	.5694	.1849	10.33322	1.0741	.855
.8000	.5111	.1617	8.36592	1.1150	.887
.9000	.4613	.1460	7.20138	1 1440	.910
1.0000	.4191	.1349	6.47313	1.1657	.928
1.2000	.3529	.1206	5.68386	1.1959	.952
1.4000	.3040	.1118	5.32979	1.2160	.968
1.6000	.2667	.1059	5.18287	1.2308	.979
1.8000	.2375	.1017	5.14847	1.2421	.988
2.0000	.2141	.0986	5.18132	1.2511	.996
2.1400	.2002	.0968	5.23097	1.2566	1.000

Example 13.4—(*continued*)

This is less than the constant reflux, $R = 0.647$, to be found in part b.

At constant vaporization rate, the time is proportional to the cumulative vapor amount:

$$\frac{t}{\bar{t}} = \frac{V}{V_{\text{final}}} = \frac{V/L_0}{1.2566}. \tag{5}$$

Also

$$D/L_0 = 1 - L/L_0. \tag{6}$$

From these relations and the tabulated data, D/L_0 and R are plotted against reduced time $t/\bar{t}$.

b. At constant reflux: A reflux ratio is found by trial to give an average overhead composition $\bar{y}_D = 0.975$ and a residue composition $x_L = 0.2$. The average overhead composition is found with material balance

$$\bar{y}_D = [x_{L0} - (L/L_0)x_L]/(1 - L/L_0). \tag{7}$$

The value of L/L_0 is calculated as a function of y_D from

$$\ln \frac{L}{L_0} = \int_{0.9}^{x_L} \frac{1}{y_D - x_L}\, dx_L. \tag{8}$$

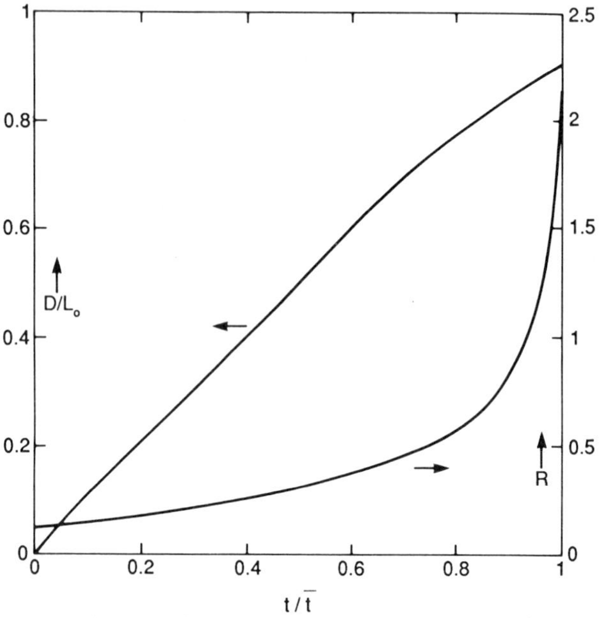


```
10 ! Example 13.9. Distillation
     at constant reflux
20 A=3.268
30 OPTION BASE 1
40 DIM X(10),Y(11)
50 INPUT R ! reflux ratio
60 INPUT Y(1)
70 FOR N=1 TO 10
80 X(N)=1/(A/Y(N)-A+1)
90 Y(N+1)=1/(R+1)*(R*X(N)+Y(1))
100 NEXT N
110 I=1/(Y(1)-X(10))
120 DISP USING 130 ; Y(1),X(10),
    I
130 IMAGE .DDDDD,2X,.DDDD,2X,DD.
    DDDD
140 GOTO 60
150 END
```

At a trial value of R, values of x_{10} are found for a series of assumed y_D's until x_{10} equals or is less than 0.20. The given computer program is based on Eqs. (1) and (2). The results of two trials and interpolation to the desired bottoms composition, $x_L = 0.200$, are

R	0.6	0.7	0.647
x_L	0.2305	0.1662	0.200

y_D	x_L	$1/(y_D - x_L)$	L/L_0	$\bar{y}_D$
Reflux ratio $R = 0.6$				
0.99805	0.9000	10.2035		
0.99800	0.8981	10.0150	0.9810	
0.99750	0.8800	8.5127	0.8295	
0.99700	0.8638	7.5096	0.7286	
0.99650	0.8493	6.7917	0.6568	
0.99600	0.8361	6.2521	0.6026	
0.99550	0.8240	5.8314	0.5602	
0.99500	0.8130	5.4939	0.5263	
0.99400	0.7934	4.9855	0.4750	
0.99300	0.7765	4.6199	0.4379	
0.99200	0.7618	4.3436	0.4100	
0.99100	0.7487	4.1270	0.3879	
0.99000	0.7370	3.9522	0.3700	
0.98500	0.6920	3.4135	0.3135	
0.98000	0.6604	3.1285	0.2827	
0.97500	0.6357	2.9471	0.2623	
0.97000	0.6152	2.8187	0.2472	
0.96500	0.5976	2.7217	0.2354	
0.96000	0.5819	2.6450	0.2257	
0.95500	0.5678	2.5824	0.2176	
0.95000	0.5548	2.5301	0.2104	
0.90000	0.4587	2.2662	0.1671	
0.85000	0.3923	2.1848	0.1441	
0.80000	0.3402	2.1751	0.1286	
0.75000	0.2972	2.2086	0.1171	
0.70000	0.2606	2.2756	0.1079	0.9773
0.65000	0.2286	2.3730	0.1001	0.9746
0.60000	0.2003	2.5019	0.0933	0.9720
Reflux ratio $R = 0.7$				
0.99895	0.9000	10.1061		
0.99890	0.8963	9.7466	0.9639	
0.99885	0.8927	9.4206	0.9312	
0.99880	0.8892	9.1241	0.9015	
0.99870	0.8824	8.5985	0.8488	
0.99860	0.8758	8.1433	0.8032	
0.99840	0.8633	7.4019	0.7288	
0.99820	0.8518	6.8306	0.6716	
0.99800	0.8410	6.3694	0.6254	
0.99700	0.7965	4.9875	0.4857	
0.99600	0.7631	4.2937	0.4160	
0.99500	0.7370	3.8760	0.3739	
0.99400	0.7159	3.5958	0.3456	
0.99300	0.6983	3.3933	0.3249	
0.99200	0.6835	3.2415	0.3094	
0.99100	0.6076	2.6082	0.2969	
0.99000	0.6594	3.0248	0.2869	
0.98000	0.5905	2.5674	0.2366	
0.97000	0.5521	2.3929	0.2151	
0.96000	0.5242	2.2946	0.2015	
0.95000	0.5013	2.2287	0.1913	
0.94000	0.4816	2.1815	0.1832	
0.93000	0.4639	2.1455	0.1763	
0.92000	0.4479	2.1182	0.1704	
0.91000	0.4334	2.0982	0.1652	
0.90000	0.4193	2.0803	0.1605	
0.85000	0.3611	2.0454	0.1423	
0.80000	0.3148	2.0610	0.1294	
0.75000	0.2761	2.1101	0.1194	
0.70000	0.2429	2.1877	0.1112	
0.65000	0.2137	2.2920	0.1041	
0.60000	0.1877	2.4254	0.0979	0.9773
0.55000	0.1643	2.5927	0.0923	0.9748
0.50000	0.1431	2.8019	0.0872	0.9723

Example 13.5.
Distillation of Substances with Widely Different Molal Heats of Vaporization

The molal heats of vaporization of ethanol and acetic acid are 9225 and 5663 cal/g mol. A mixture with ethanol content of $x_F = 0.50$ is to be separated into products with $x_B = 0.05$ and $x_D = 0.95$. Pressure is 1 atm, feed is liquid at the boiling point, and the reflux ratio is to be 1.3 times the minimum. The calculation of tray requirements is to be made with the true molecular weight, 60.05, of acetic acid and with adjustment to make the apparent molal heat of vaporization the same as that of ethanol, which becomes

$$60.05(9225/5663) = 98.14.$$

The adjusted mol fractions, x' and y', are related to the true ones by

$$x' = \frac{x}{x + 0.6119(1-x)}, \quad y' = \frac{y}{y + 0.6119(1-y)}.$$

The experimental and converted data are tabulated following and plotted on McCabe–Thiele diagrams. The corresponding compositions involved in this distillation are:

$$x_B = 0.05, \quad x'_B = 0.0792$$
$$x_F = 0.50, \quad x'_F = 0.6204$$
$$x_D = 0.95, \quad x'_D = 0.9688$$

x	y	x'	y'
0.0550	0.1070	0.0869	0.1638
0.0730	0.1440	0.1140	0.2156
0.1030	0.1970	0.1580	0.2862
0.1330	0.2740	0.2004	0.3815
0.1660	0.3120	0.2454	0.4257
0.2070	0.3930	0.2990	0.5141
0.2330	0.4370	0.3318	0.5592
0.2820	0.5260	0.3909	0.6446
0.3470	0.5970	0.4648	0.7077
0.4600	0.7500	0.5820	0.8306
0.5160	0.7930	0.6353	0.8623
0.5870	0.8540	0.6990	0.9053
0.6590	0.9000	0.7595	0.9363
0.7280	0.9340	0.8139	0.9586
0.6160	0.9660	0.8788	0.9789
0.9240	0.9900	0.9521	0.9939

In terms of the true molecular weight, minimum reflux is given by

$$x_D/(R_{\min} + 1) = 0.58,$$

whence

$$R_m = 0.6379,$$
$$R = 1.3(0.6379) = 0.8293,$$
$$x_D/(R+1) = 0.5193,$$
$$x'_D/(R+1) = 0.5296.$$

Taking straight operating lines in each case, the numbers of trays are

$N = 11.0$ with true molecular weight of acetic acid,
$N' = 9.8$ with adjusted molecular weight.

In this case it appears that assuming straight operating lines, even though the molal heats of vaporization are markedly different, results in overestimation of the number of trays needed for the separation.

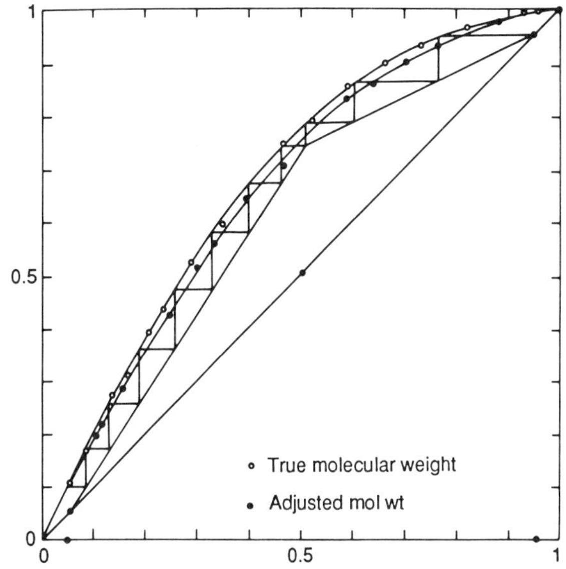

a. Construction with true molecular weight, $N = 11$.

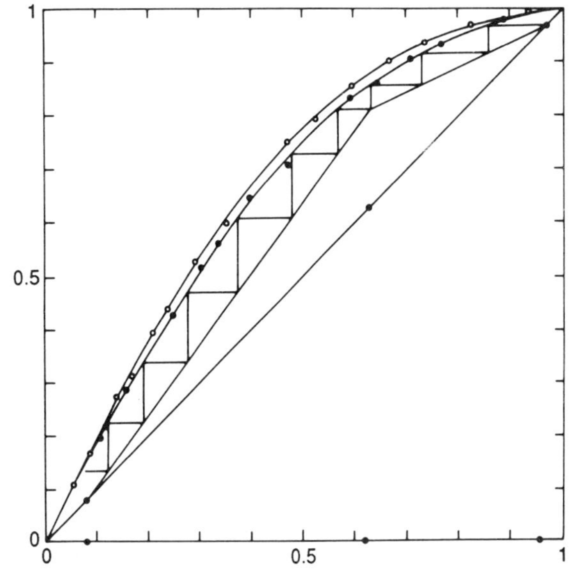

b. Construction with adjusted molecular weight, $N = 9.8$.

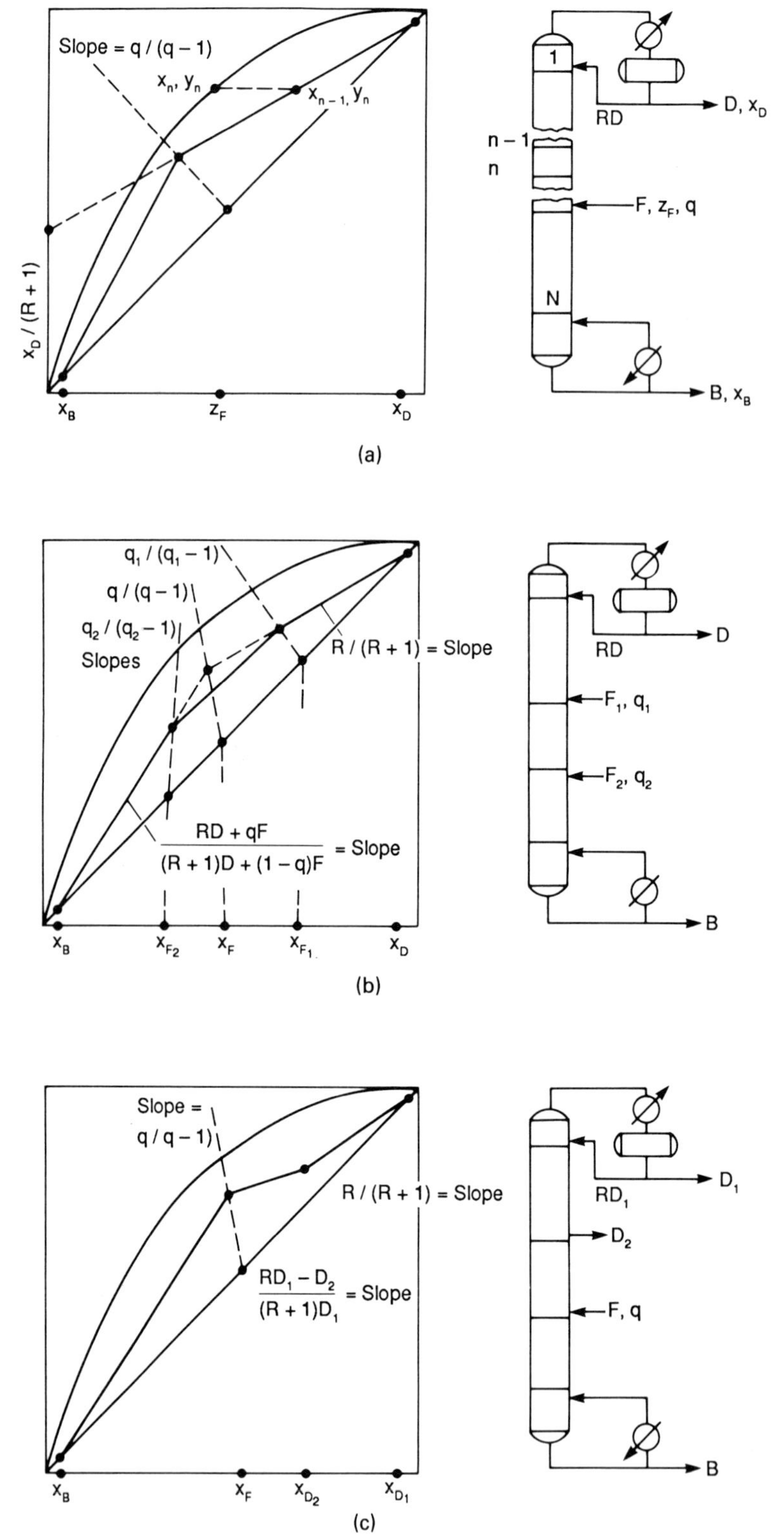

Figure 13.8. Operating and q-line construction with several feeds and top products. (a) One feed and one overhead product. (b) Two feeds and one overhead product. (c) One feed and two products from above the feed point.

TABLE 13.3. Economic Optimum Reflux Ratio for Typical Petroleum Fraction Distillation near 1 atm[a]

	Factor for optimum reflux $f=(R_{opt}/R_m)-1$, $R_{opt}=(1+f)R_m$								Factor for optimum trays N_{opt}/N_m		
	$N_m=10$, R_m			$N_m=20$, R_m			$N_m=50$, R_m		$N_m=10$, R_m	$N_m=20$, R_m	$N_m=50$, R_m
	1	3	10	1	3	10	1	10	1 to 10	1 to 10	1 to 10
Base case	0.20	0.12	0.10	0.24	0.17	0.16	0.31	0.21	2.4	2.3	2.1
Payout time 1 yr	0.24	0.14	0.12	0.28	0.20	0.17	0.37	0.24	2.2	2.1	2.0
Payout time 5 yr	0.13	0.09	0.07	0.17	0.13	0.10	0.22	0.15	2.7	2.5	2.2
Steam cost \$0.30/M lb	0.22	0.13	0.11	0.27	0.16	0.14	0.35	0.22	2.3	2.1	2.0
Steam cost \$0.75/M lb	0.18	0.11	0.09	0.21	0.13	0.11	0.29	0.19	2.5	2.3	2.1
$G_a=50$ lb mole/(hr)(sqft)	0.06	0.04	0.03	0.08	0.06	0.05	0.13	0.08	3.1	2.8	2.4

[a] The "base case" is for payout time of 2 yr, steam cost of \$0.50/1000 lb, vapor flow rate $G_a=15$ lb mol/(hr)(sqft). Although the capital and utility costs are prior to 1975 and are individually far out of date, the relative costs are roughly the same so the conclusions of this analysis are not far out of line. Conclusion: For systems with nearly ideal VLE, R is approx. $1.2R_{min}$ and N is approx. $2.0N_{min}$.
(Happel and Jordan, *Chemical Process Economics*, Dekker, New York, 1975).

Minimum Reflux. Underwood's method employs two relations. First an auxiliary parameter θ is found in the range $1<\theta<\alpha$ by solving

$$\frac{\alpha x_F}{\alpha-\theta}+\frac{1-x_F}{1-\theta}=1-q \tag{13.87}$$

or

$$(1-q)\theta^2+[(\alpha-1)x_F+q(\alpha+1)-\alpha]\theta-\alpha q=0, \tag{13.88}$$

or in two important special cases:

$$\text{when } q=0,\quad \theta=\alpha-(\alpha-1)x_F, \tag{13.89}$$

$$\text{when } q=1,\quad \theta=\frac{\alpha}{(\alpha-1)x_F+1}. \tag{13.90}$$

Then R_m is found by substitution into

$$R_m=-1+\frac{\alpha x_D}{\alpha-\theta}+\frac{1-x_D}{1-\theta}. \tag{13.91}$$

Formulas for the numbers of trays in the enriching and stripping sections at operating reflux also are due to Underwood (*Trans. Inst. Chem. Eng.* **10,** 112–152, 1932). For above the feed, these groups of terms are defined:

$$K_1=L_n/V_n=R/(R+1), \tag{13.92}$$

$$\phi_1=K_1(\alpha-1)/(K_1\alpha-1). \tag{13.93}$$

Then the relation between the compositions of the liquid on tray 1

Example 13.6
Separation of an Azeotropic Mixture by Operation at Two Pressure Levels

At atmospheric pressure, ethanol and water form an azetrope with composition $x=0.846$, whereas at 95 Torr the composition is about $x=0.94$. As the diagram shows, even at the lower pressure the equilibrium curve hugs the $x=y$ line. Accordingly, a possibly feasible separation scheme may require three columns, two operating at 760 Torr and the middle one at 95 Torr, as shown on the sketch. The basis for the material balance used is that 99% of the ethanol fed to any column is recovered, and that the ethanol-rich products from the columns have $x=0.8$, 0.9, and 0.995, resp.

Although these specifications lead to only moderate tray and reflux requirements, in practice distillation with only two towers and the assistance of an azeotropic separating agent such as benzene is found more economical. Calculation of such a process is made by Robinson and Gilliland (1950, p. 313).

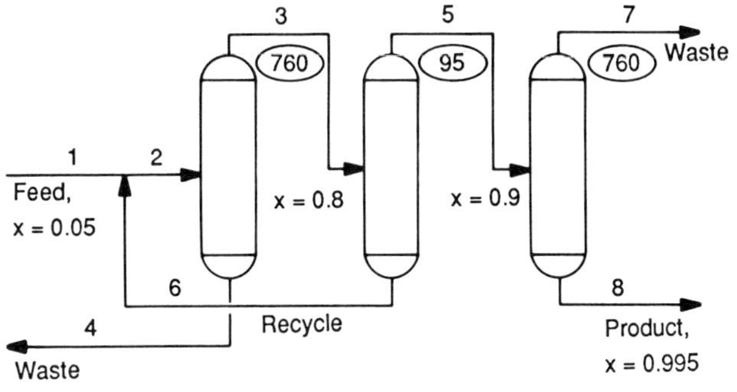

	1	2	3	4	5	6	7	8
Ethanol	5	5.05000	4.9995	0.05050	4.949500	0.049995	0.04950	4.90000
Water	95	95.69992	1.24987	94.45005	0.54994	0.69993	0.52532	0.02462

EXAMPLE 13.6—(*continued*)

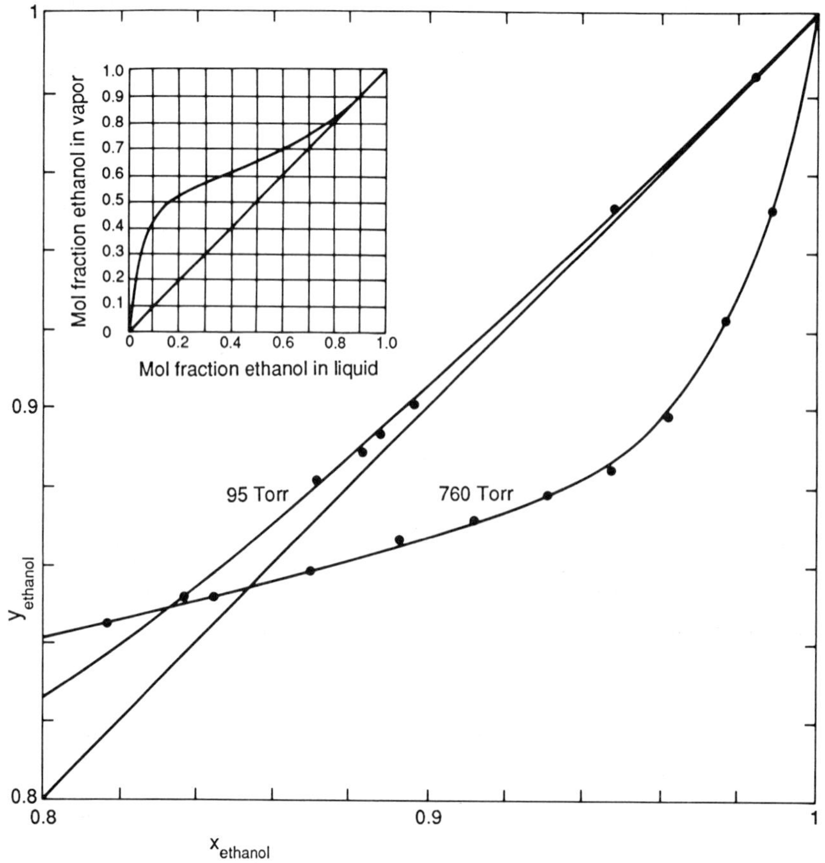

Ethanol-water vapor-liquid equilibria at 95 and 760 Torr.

EXAMPLE 13.7
Separation of a Partially Miscible Mixture

Water and n-butanol in the concentration range of about 50–98.1 mol % water form two liquid phases that boil at 92.7°C at one atm. On cooling to 40°C, the hetero-azeotrope separates into phases containing 53 and 98 mol % water.

A mixture containing 12 mol % water is to be separated by distillation into products with 99.5 and 0.5 mol % butanol. The accompanying flowsketch of a suitable process utilizes two columns with condensing-subcooling to 40°C. The 53% saturated solution is refluxed to the first column, and the 98% is fed to the second column. The overhead of the second column contains a small amount of butanol that is recycled to the condenser for recovery. The recycle material balance is shown with the sketch.

The three sets of vapor–liquid equilibrium data appearing on the x–y diagram show some disagreement, so that great accuracy cannot be expected from determination of tray requirements, particularly at the low water concentrations. The upper operating line in the first column is determined by the overall material balance so it passes through point (0.995, 0.995), but the initial point on the operating line is at $x = 0.53$, which is the composition of the reflux. The construction is shown for 50% vaporized feed. That result and those for other feed conditions are summarized:

q	R_m	$R_m = 1.3R_m$	N
1	2.02	2.62	12
0.5	5.72	7.44	8
0	9.70	12.61	6

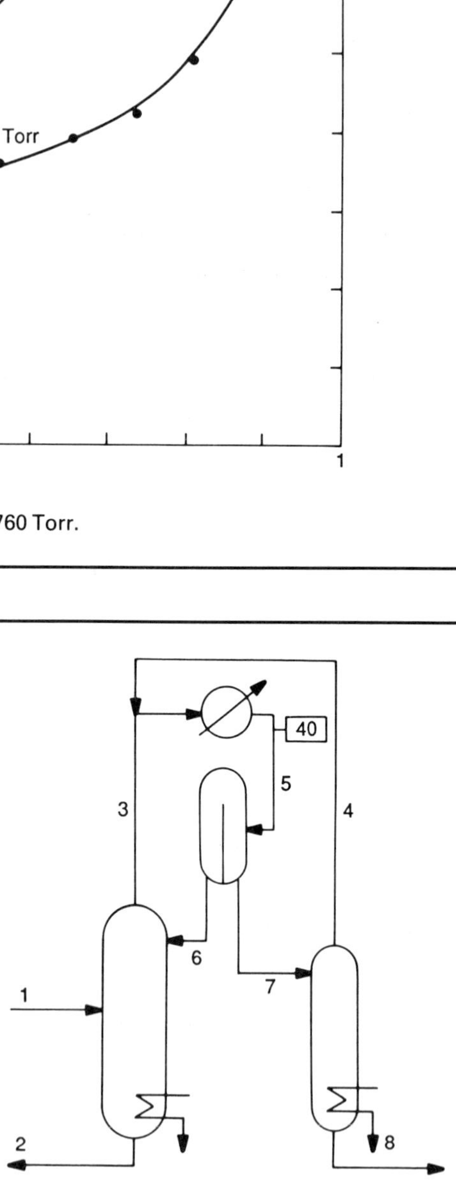

	1	2	3	4	5	6	7	8
Water	12	0.44	18.4139	0.7662	19.1801	6.8539	12.3262	11.56
Butanol	88	87.94	6.1379	0.1916	6.3295	6.0779	0.2516	0.06
	100	88.38	24.5518	0.9578	25.5096	12.9318	12.5778	11.62
% Water	12	0.5	75	80	75.19	53	98	99.5

Example 13.7—*(continued)*

In the second column, two theoretical trays are provided and are able to make a 99.6 mol % water waste, slightly better than the 99.5 specified. The required L/V is calculated from compositions read off the diagram:

$$L/V = (0.966 - 0.790)/(0.996 - 0.981) = 13.67.$$

If live steam were used instead of indirect heat, the bottoms concentration would be higher in water. This distillation is studied by Billet (1979, p. 216). Stream compositions are given below the flowsketch.

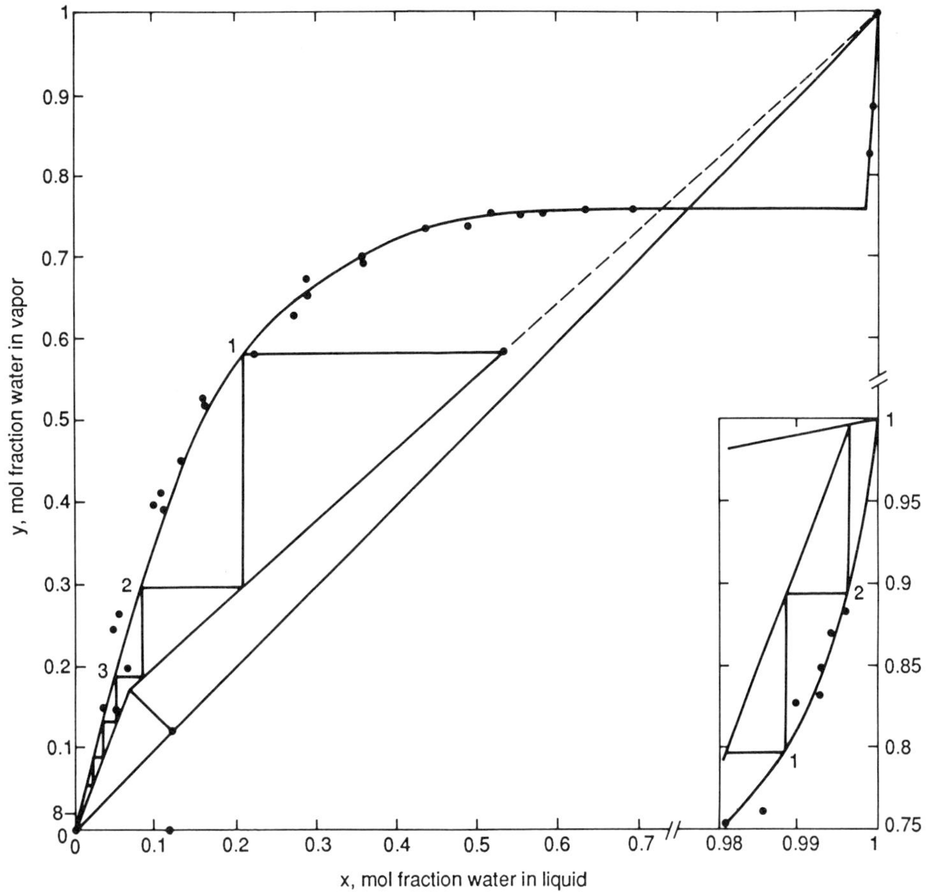

Equilibrium stage requirements for the separation of water and n-butanol.

and that on tray n is

$$(K_1\alpha)^{n-1} = \frac{1/(1-x_1) - \phi_1}{1/(1-x_n) - \phi_1}. \tag{13.94}$$

Since the overhead composition x_D is the one that is specified rather than that of the liquid on the top tray, x_1, the latter is eliminated from Eq. (13.94). The relative volatility definition is applied

$$\frac{\alpha x_1}{1-x_1} = \frac{x_D}{1-x_D}, \tag{13.95}$$

from which

$$\frac{1}{1-x_1} = \frac{x_D + \alpha(1-x_D)}{\alpha(1-x_D)}. \tag{13.96}$$

With this substitution, Eq. (13.94) becomes

$$(K_1\alpha)^{n-1} = \frac{[x_D + \alpha(1-x_D)]/\alpha(1-x_D) - \phi_1}{1/(1-x_n) - \phi_1}. \tag{13.97}$$

The number of trays above the feed plus the feed tray is obtained after substituting the feed composition x_F for x_n.

Below the feed,

$$K_2 = V_m/L_m = (RD + qF - B)/(RD + qF), \tag{13.98}$$

$$\phi_2 = (\alpha - 1)/(K_2\alpha - 1). \tag{13.99}$$

The relation between the compositions at the bottom and at tray m is

$$(K_2\alpha)^m = \frac{1/x_B - \phi_2}{1/x_m - \phi_2}. \tag{13.100}$$

The number of trays below the feed plus the feed tray is found after replacing x_m by x_F. The number of trays in the whole column then is

$$N = m + n - 1. \tag{13.101}$$

Example 13.9 applies these formulas.

TABLE 13.4. Molal Heats of Vaporization at Their Normal Boiling Points of Some Organic Compounds That May Need To Be Separated from Water

			Molecular Weight	
Compound	NBP (°C)	cal/g mol	True	Adjusted[a]
Water	100	9717	18.02	18.02
Acetic acid	118.3	5663	60.05	103.04
Acetone	56.5	6952	58.08	81.18
Ethylene glycol	197	11860	62.07	50.85
Phenol	181.4	9730	94.11	94.0
n-Propanol	97.8	9982	60.09	58.49
Ethanol	78.4	9255	46.07	48.37

[a] The adjustment of molecular weight is to make the molal heat of vaporization the same as that of water.

13.5. BATCH DISTILLATION

A batch distillation plant consists of a still or reboiler, a column with several trays, and provisions for reflux and for product collection. Figure 13.10(c) is a typical equipment arrangement with controls. The process is applied most often to the separation of mixtures of several components at production rates that are too small for a continuous plant of several columns equipped with individual reboilers, condensers, pumps, and control equipment. The number of continuous columns required is one less than the number of components or fractions to be separated. Operating conditions of a typical batch distillation making five cuts on an 8-hr cycle are in Figure 13.11.

Operation of a batch distillation is an unsteady state process whose mathematical formulation is in terms of differential equations since the compositions in the still and of the holdups on individual trays change with time. This problem and methods of solution are treated at length in the literature, for instance, by Holland and Liapis (*Computer Methods for Solving Dynamic Separation Problems,* 1983, pp. 177–213). In the present section, a simplified analysis will be made of batch distillation of binary mixtures in columns with negligible holdup on the trays. Two principal modes of operating batch distillation columns may be employed:

1. With constant overhead composition. The reflux ratio is adjusted continuously and the process is discontinued when the concentration in the still falls to a desired value.
2. With constant reflux. A reflux ratio is chosen that will eventually produce an overhead of desired average composition and a still residue also of desired composition.

Both modes usually are conducted with constant vaporization rate at an optimum value for the particular type of column construction. Figure 13.10 represents these modes on McCabe–Thiele diagrams. Small scale distillations often are controlled

Example 13.8
Enthalpy–Concentration Lines of Saturated Vapor and Liquid of Mixtures of Methanol and Water at a Pressure of 2 atm

A basis of 0°C is taken. Enthalpy data for methanol are in *Chemical Engineers' Handbook* (McGraw-Hill, New York, 1984, p. 3.204) and for water in Keenan et al. (*Steam Tables: SI Units,* Wiley, New York, 1978).

Methanol: $T = 82.8°C$
$H_v = 10{,}010$ cal/g mol,
$h_L = 1882$ cal/g mol,
$\Delta H_v = 8128$ cal/g mol,
$C_p = 22.7$ cal/g mol °C.

Water: $T = 120.6°C$
$H_v = 11{,}652$ cal/g mol,
$h_L = 2180$ cal/g mol,
$\Delta H_v = 9472$ cal/g mol,

Experimental x–y data are available at 1 and 3 atm (Hirata, 1976, #517, #519). Values at 2 atm can be interpolated by eye. The lines show some overlap. Straight lines are drawn connecting enthalpies of pure vapors and enthalpies of pure liquids. Shown is the tie line for $x = 0.5$, $y = 0.77$.

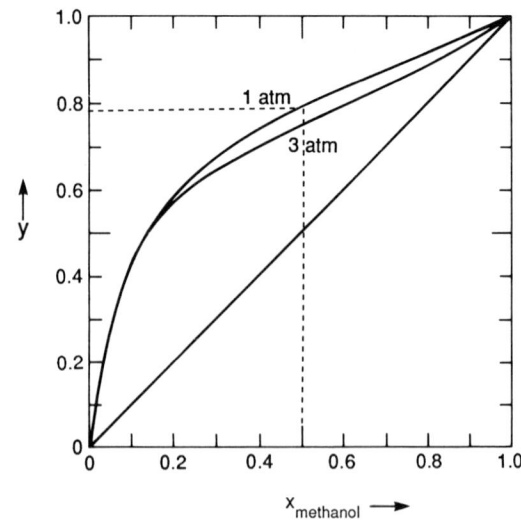

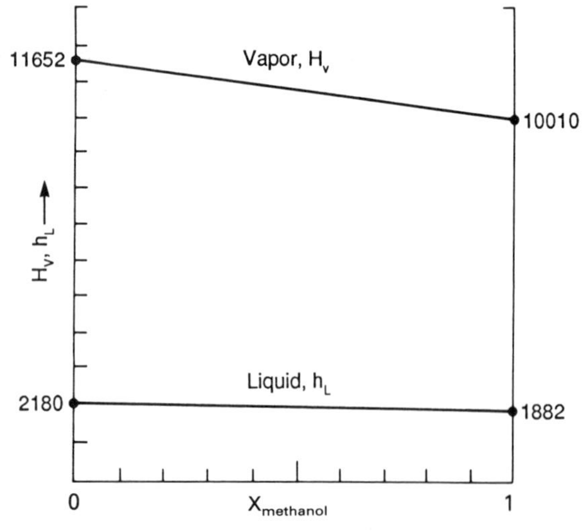

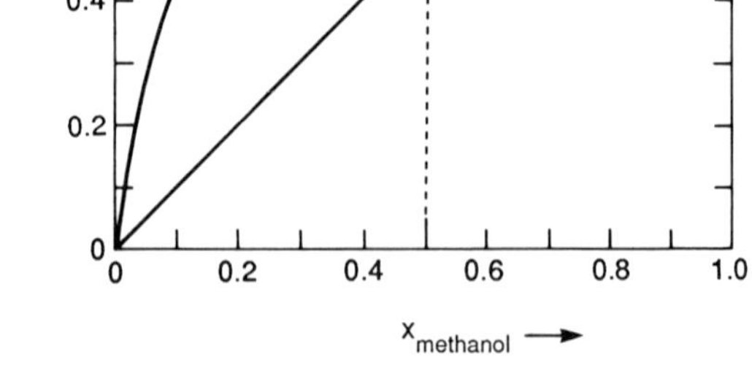

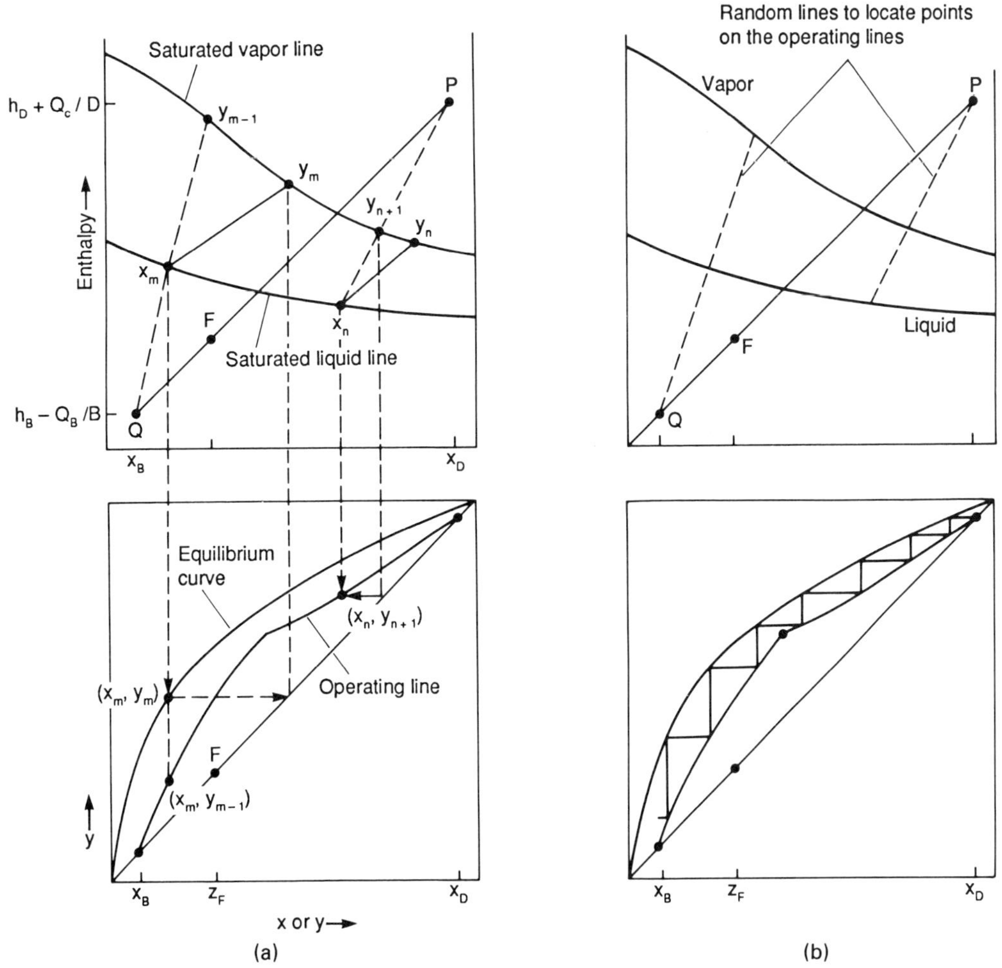

Figure 13.9. Combined McCabe–Thiele and Merkel enthalpy–concentration diagrams for binary distillation with heat balances. (a) Showing key lines and location of representative points on the operating lines. (b) Completed construction showing determination of the number of trays by stepping off between the equilibrium and operating lines.

manually, but an automatic control scheme is shown in Figure 13.10(c). Constant overhead composition can be assured by control of temperature or directly of composition at the top of the column. Constant reflux is assured by flow control on that stream. Sometimes there is an advantage in operating at several different reflux rates at different times during the process, particularly with multicomponent mixtures as on Figure 13.11.

MATERIAL BALANCES

Assuming negligible holdup on the trays, the differential balance between the amount of overhead, dD, and the amount L remaining in the still is

$$y_D\,dD = -y_D\,dL = -d(Lx_L) = -L\,dx_L - x_L\,dL, \tag{13.102}$$

which is integrated as

$$\ln(L/L_0) = \int_{x_{L_0}}^{x_L} \frac{1}{y_D - x_L}\,dx_L. \tag{13.103}$$

The differences $y_D - x_L$ depend on the number of trays in the column, the reflux ratio, and the vapor–liquid equilibrium relationship. For constant molal overflow these relations may be taken as

$$y_{n+1} = \frac{R}{R+1}x_n + \frac{1}{R+1}y_D, \tag{13.104}$$

$$y_n = f(x_n). \tag{13.105}$$

When the overhead composition is constant, Eq. 13.103 is integrable directly, but the same result is obtained by material balance,

$$\frac{L}{L_0} = \frac{y_D - x_{L_0}}{y_D - x_L}. \tag{13.106}$$

With variable overhead composition, the average value is represented by the same overall balance,

$$\bar{y}_D = \frac{x_{L_0} - (L/L_0)x_L}{1 - (L/L_0)}, \tag{13.107}$$

Example 13.9
Algebraic Method for Binary Distillation Calculation

An equimolal binary mixture which is half vaporized is to be separated with an overhead product of 99% purity and 95% recovery. The relative volatility is 1.3. The reflux is to be selected and the number of trays above and below the feed are to be found with the equations of Section 13.4.6.

The material balance is

Component	F	D	x_D	B	X_B
1	50	49.50	0.99	0.50	0.0100
2	50	0.48	0.01	49.52	0.9900
Total	100	49.98		50.02	

Minimum no. of trays,

$$N_m = \frac{\ln(0.99/0.01)(0.99/0.01)}{\ln 1.3} = 35.03.$$

For minimum reflux, by Eqs. (13.88) and (13.91),

$$0.5\theta^2 + [0.3(0.5) + 0.5(2.3) - 1.3]\theta - 1.3(0.5) = 0,$$

$$\theta^2 = 1.3,$$

$$\theta = 1.1402,$$

$$R_m = -1 + \frac{1.3(0.99)}{1.3 - 1.1402} + \frac{0.01}{1 - 0.1402} = 6.9813,$$

$$R = 1.2R_m = 8.3775,$$

$$K_1 = \frac{R}{R+1} = 0.8934,$$

$$\phi_1 = \frac{0.8934(1.3-1)}{0.8934(1.3) - 1} = 1.6608,$$

$$\frac{1}{1-x_1} = \frac{0.99 + 1.3(0.01)}{1.3(0.01)} = 77.1538,$$

$$(K_1\alpha)^{n-1} = (1.1614)^{n-1} = \frac{77.1538 - 1.6608}{1/(1-0.5) - 1.6608} = 222.56,$$

$$\therefore n = 37.12,$$

$$K_2 = \frac{8.3775(49.98) + 0.5(100) - 50.02}{468.708} = 0.8933,$$

$$\phi_2 = \frac{1.3 - 1}{0.8933(1.3) - 1} = 1.8600,$$

$$[0.8933(1.3)]^m = \frac{1/0.01 - 1.8600}{1/0.5 - 1.8600} = 701.00,$$

$$\therefore m = 43.82,$$

$$\therefore N = m + n - 1 = 37.12 + 43.82 - 1 = 79.94 \text{ trays}.$$

but it is also necessary to know what reflux will result in the desired overhead and residue compositions.

For constant overhead composition at continuously varied reflux ratios, the total vaporization is found as follows. The differential balance is

$$dD = dV - dL = (1 - dL/dV)\,dV \tag{13.108}$$

The derivative dL/dV is the slope of the operating line so that

$$1 - \frac{dL}{dV} = 1 - \frac{R}{R+1} = \frac{1}{R+1}. \tag{13.109}$$

Substitution from Eqs. (13.103), (13.106), and (13.109) into Eq. (13.108) converts this into

$$dV = L_0(x_{L_0} - \bar{y}_D)\frac{R+1}{(x_L - \bar{y}_D)^2}\,dx_L, \tag{13.110}$$

from which the total amount of vapor generated up to the time the residue composition becomes x_L is

$$V = L_0(x_{L_0} - \bar{y}_D)\int_{x_{L_0}}^{x_L} \frac{R+1}{(x_L - \bar{y}_D)^2}\,dx_L. \tag{13.111}$$

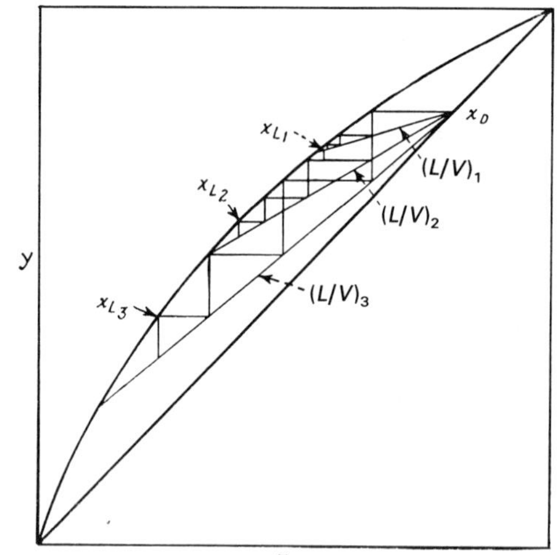

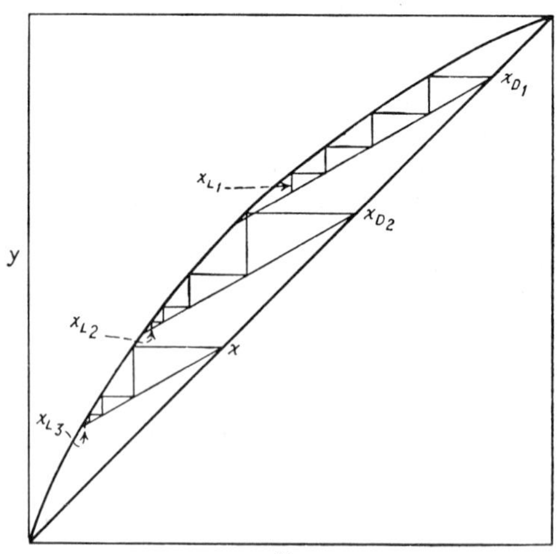

Figure 13.10. Batch distillation: McCabe–Thiele constructions and control modes. (a) Construction for constant overhead composition with continuously adjusted reflux rate. (b) Construction at constant reflux at a series of overhead compositions with an objective of specified average overhead composition. (c) Instrumentation for constant vaporization rate and constant overhead composition. For constant reflux rate, the temperature or composition controller is replaced by a flow controller.

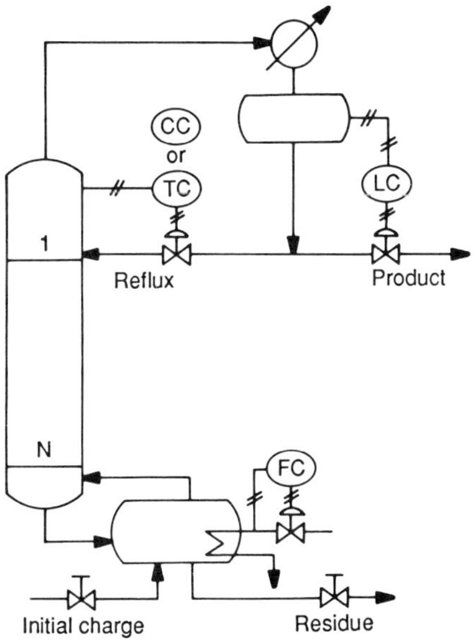

(c)

Figure 13.10—(*continued*)

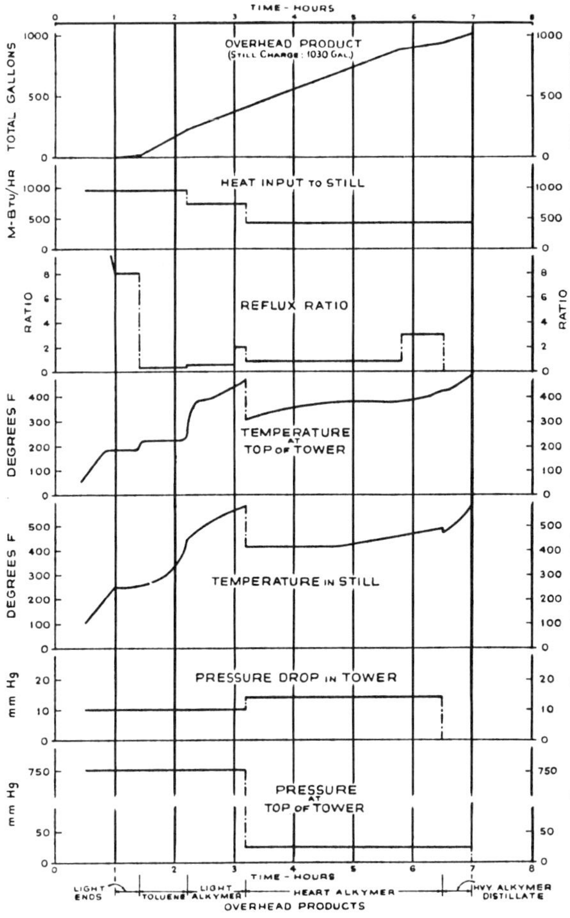

Figure 13.11. Operation of a batch distillation with five cuts.

At constant vaporization rate the time is proportional to the amount of vapor generated, or

$$t/\bar{t} = V/V_{\text{total}}. \tag{13.112}$$

Hence the reflux ratio, the amount of distillate, and the bottoms composition can be related to the fractional distillation time. This is done in Example 13.4, which studies batch distillations at constant overhead composition and also finds the suitable constant reflux ratio that enables meeting required overhead and residue specifications. Although the variable reflux operation is slightly more difficult to control, this example shows that it is substantially more efficient thermally—the average reflux ratio is much lower—than the other type of operation.

Equation (13.97) can be used to find the still composition—x_n in that equation—at a particular reflux ratio in a column-reboiler combination with n stages. Example 13.4 employs instead a computer program with Equations (13.104) and (13.105). That procedure is more general in that a constant relative volatility need not be assumed, although that is done in this particular example.

13.6 MULTICOMPONENT SEPARATION: GENERAL CONSIDERATIONS

A tower comprised of rectifying (above the feed) and stripping (below the feed) sections is capable of making a more or less sharp separation between two products or pure components of the mixture, that is, between the light and heavy key components. The *light key* is the most volatile component whose concentration is to be controlled in the bottom product and the *heavy key* is the least volatile component whose concentration is to be controlled in the overhead product. Components of intermediate volatilities whose distribution between top and bottom products is not critical are called *distributed keys.* When more than two sharply separated products are needed, say n top and bottom products, the number of columns required will be $n - 1$.

In some cases it is desirable to withdraw sidestreams of intermediate compositions from a particular column. For instance, in petroleum fractionation, such streams may be mixtures of suitable boiling ranges or which can be made of suitable boiling range by stripping in small auxiliary columns. Other cases where intermediate streams may be withdrawn are those with minor but critical impurities that develop peak concentrations at these locations in the column because of inversion of volatility as a result of concentration gradient. Thus, pentyne-1 in the presence of n-pentane in an isoprene-rich C_5 cracked mixture exhibits this kind of behavior and can be drawn off as a relative concentrate at an intermediate point. In the rectification of fermentation alcohol, whose column profile is shown in Figure 13.12(a), undesirable esters and higher alcohols concentrate at certain positions because their solubilities are markedly different in high and low concentrations of ethanol in water, and are consequently withdrawn at these points.

Most distillations, however, do not develop substantial concentration peaks at intermediate positions. Figure 13.12(b) is of normal behavior.

SEQUENCING OF COLUMNS

The number n of top and bottom products from a battery of $n - 1$ columns can be made in several different ways. In a direct method, the most volatile components are removed one-by-one as overheads in successive columns with the heaviest product as the bottoms of the last column. The number of possible ways of separating components goes up sharply with the number of products, from two arrangements with three products to more than 100 with seven products. Table 13.5 identifies the five possible arrangements for

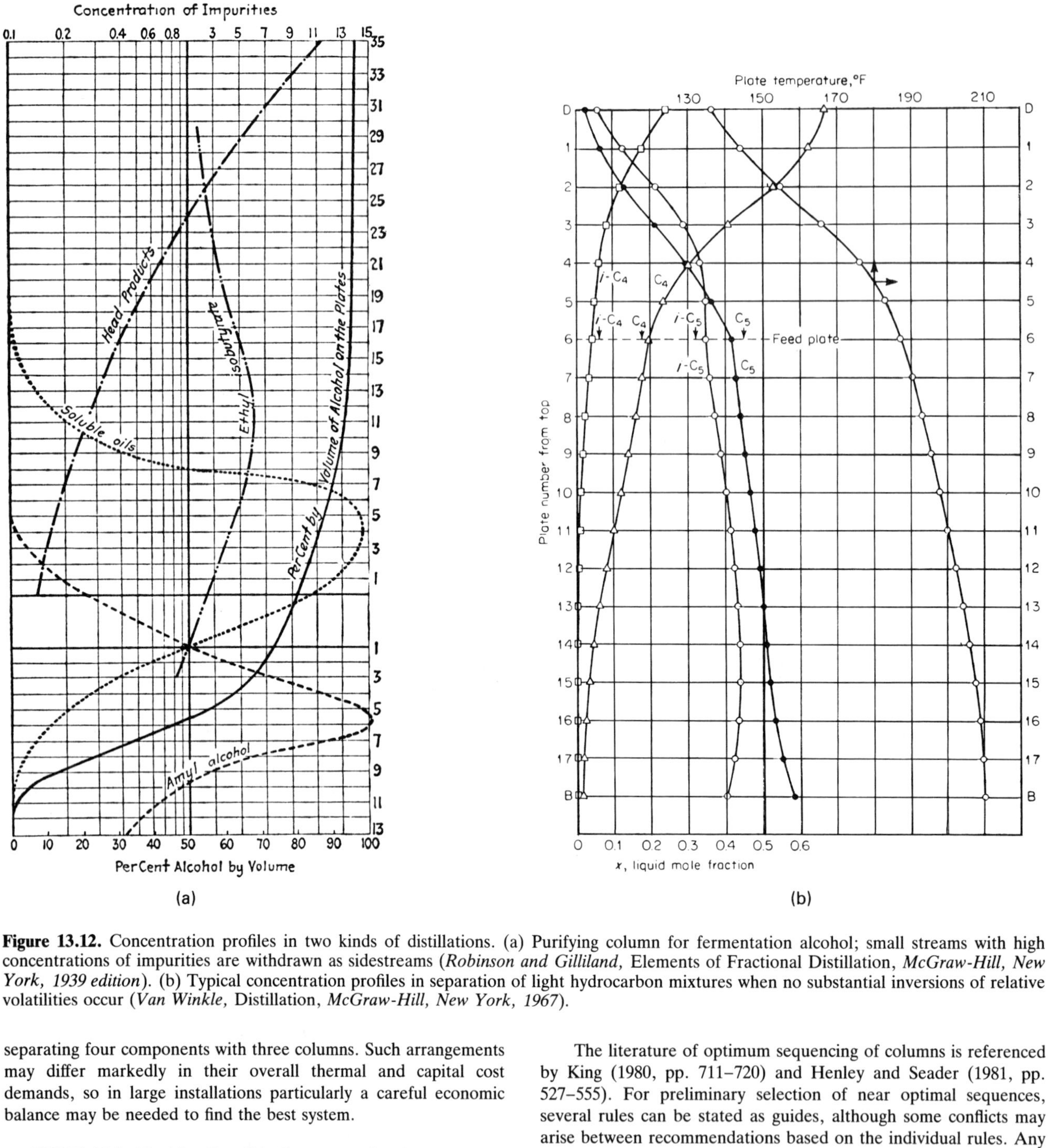

Figure 13.12. Concentration profiles in two kinds of distillations. (a) Purifying column for fermentation alcohol; small streams with high concentrations of impurities are withdrawn as sidestreams (*Robinson and Gilliland,* Elements of Fractional Distillation, *McGraw-Hill, New York, 1939 edition*). (b) Typical concentration profiles in separation of light hydrocarbon mixtures when no substantial inversions of relative volatilities occur (*Van Winkle,* Distillation, *McGraw-Hill, New York, 1967*).

separating four components with three columns. Such arrangements may differ markedly in their overall thermal and capital cost demands, so in large installations particularly a careful economic balance may be needed to find the best system.

TABLE 13.5. The Five Possible Sequences for the Separation of Four Components ABCD by Three Columns

Column 1		Column 2		Column 3	
Ovhd	Btms	Ovhd	Btms	Ovhd	Btms
A	BCD	B	CD	C	D
A	BCD	BC	D	B	C
AB	CD	A	B	C	D
ABC	D	A	BC	B	C
ABC	D	AB	C	A	B

The literature of optimum sequencing of columns is referenced by King (1980, pp. 711–720) and Henley and Seader (1981, pp. 527–555). For preliminary selection of near optimal sequences, several rules can be stated as guides, although some conflicts may arise between recommendations based on the individual rules. Any recommended cases then may need economic evaluations.

1. Perform the easiest separation first, that is, the one least demanding of trays and reflux, and leave the most difficult to the last.
2. When neither relative volatility nor concentration in the feed varies widely, remove the components one-by-one as overhead products.
3. When the adjacent ordered components in the process feed vary widely in relative volatility, sequence the splits in the order of decreasing relative volatility.
4. When the concentrations in the feed vary widely but the relative

volatilities do not, sequence the splits to remove components in the order of decreasing concentration in the feed.

NUMBER OF FREE VARIABLES

The performance of a given column or the equipment requirements for a given separation are established by solution of certain mathematical relations. These relations comprise, at every tray, heat and material balances, vapor–liquid equilibrium relations, and mol fraction constraints. In a later section, these equations will be stated in detail. For now, it can be said that for a separation of C components in a column of n trays, there still remain a number, $C+6$, of variables besides those involved in the cited equations. These must be fixed in order to define the separation problem completely. Several different combinations of these $C+6$ variables may be feasible, but the ones commonly fixed in column operation are the following:

Item	Name	Number of Variables
1	feed rate	1
2	feed composition	$C-1$
3	feed enthalpy	1
4	ratio of overhead and feed rates	1
5	reflux enthalpy	1
6	reflux ratio, L/D or L/V	1
7	number of trays	1
8	column pressure	1
		$C+6$

A common alternate specification is of the overhead and bottoms compositions expressed through distribution of the keys (two variables) as a replacement of items 4 and 7.

13.7. ESTIMATION OF REFLUX AND NUMBER OF TRAYS (FENSKE–UNDERWOOD–GILLILAND METHOD)

The first step in the design of distillation equipment is specification of the required distribution of light and heavy key components. Then the specific operating conditions and equipment size are established, ultimately on the basis of an economic balance or simply by exercise of judgment derived from experience. The design parameters that need to be determined include intermediate ones such as limiting reflux and trays that are needed for establishing a working design. These design parameters are the following:

1. Minimum number of theoretical trays,
2. Distribution of nonkeys between the overhead and bottoms products,
3. Minimum reflux,
4. Operating reflux,
5. Number of theoretical trays,
6. Location of the feed tray,
7. Tray efficiencies.

In packed towers, the variation of conditions from top to bottom is continuous and not interrupted as at trays. Nevertheless, it is convenient to speak of packing heights equivalent to a theoretical tray (HETU), so that tray tower theory can be applied to the design of packed towers.

All of the values of this list can be established at least approximately by rapid shortcut methods. In some instances such values may be useful as final ones, but ordinarily they are for exploratory purposes or as a starting basis for a computer design. Computer design of fractionation is an iterative process which depends for rapid convergence on good starting estimates of the principal quantities.

The background of shortcut methods is well treated in the books of King (1980) and Henley and Seader (1981). Here attention will be directed to application of the techniques. These shortcut methods assume constant molal overflow in the rectifying and stripping zones and constant relative volatilities, which may be taken at the conditions of the feed tray or as a geometric mean of the values at the top and bottom of the column. Since the top conditions are not known completely in advance, evaluation of a mean relative volatility is an iterative process that can be started with the value at the feed tray or at the feed condition. Particular modes of variation of α sometimes are assumed. The method of Winn assumes that the vaporization equilibrium ratios vary as

$$K_{1k} = \beta K_{hk}^{\delta} \tag{13.113}$$

or

$$\alpha = K_{1k}/K_{hk} = \beta K_{hk}^{\delta-1} \tag{13.114}$$

The constants β and δ for the conditions of the tower are deduced from log–log plots of K's, which usually are available for hydrocarbons and natural gas constituents but can be evaluated from

$$K = \gamma P^{\text{sat}}/P, \tag{13.115}$$

with activity coefficient γ of unity if no better information is known.

MINIMUM TRAYS

This is found from the relative volatility and the distribution of the keys between the overhead and bottoms by the Underwood–Fenske equation

$$N_m = \frac{\ln[(x_D/x_B)_{1k}/(x_D/x_B)_{hk}]}{\ln(\alpha_{1k}/\alpha_{hk})} = \frac{\ln[(d/b)_{1k}/(d/b)_{hk}]}{\ln(\alpha_{1k}/\alpha_{hk})}. \tag{13.116}$$

In terms of the variation of VERs according to Eq. (13.113),

$$N_m = \frac{\ln[(d/b)_{1k}/(d/b)_{hk}^{\delta}]}{\ln \beta} \tag{13.117}$$

DISTRIBUTION OF NONKEYS

A convenient approximation is that the distributions of nonkeys require the minimum number of trays as given by Eq. (13.116). Designating the nonkey by subscript nk, that equation becomes

$$\ln(d/b)_{nk} = \ln(d/b)_{1k} + N_m \ln(\alpha_{nk}/\alpha_{1k}) \tag{13.118}$$

or

$$(d/b)_{nk} = (d/b)_{1k}(\alpha_{nk}/\alpha_{1k})^{N_m}. \tag{13.119}$$

The distribution of nonkeys actually depends somewhat on the reflux ratio. For instance, in the case of Example 13.10, the distributions at minimum trays (total reflux) and minimum reflux are substantially different. Often it turns out, however, that the distributions predicted by Eq. (13.119) are close to those at finite reflux whenever R is near $1.2R_m$, which is often near the economic value for the reflux ratio. Further discussion of this topic is by Hengstebeck (Distillation, 1961) and Stupin and Lockhart (1968) whose work is summarized by King (1980, p. 434). Knowledge of the complete distribution is needed for estimation of top and bottom temperatures and for determination of the minimum reflux by the method to be cited.

EXAMPLE 13.10
Shortcut Design of Multicomponent Fractionation

A mixture of the given composition and relative volatilities has a thermal condition $q = 0.8$ and a pressure of 10 atm. It is to be fractionated so that 98% of component C and 1% of component E will appear in the overhead. The tray and reflux requirements are to be found. In the following table, the quantities in brackets are calculated in the course of the solution. f_i, d_i, and b_i are the mols of component i per mol of total feed.

	α	f	d	b
A	3.1	0.03	[0.0300]	[1.5(E − 5)]
B	2.6	0.07	[0.0698]	[0.0002]
C *lk*	2.2	0.15	0.147	0.0030
D	1.3	0.33	[0.0481][a]	[0.2819][a]
D *hk*	1.0	0.30	0.003	0.297
F	0.8	0.12	[0.0000]	[0.1200]

[a] The corrected distribution of component D will be found along with the minimum reflux.

The minimum number of trays is

$$N_m = \frac{\ln\left[\dfrac{0.147}{0.003}\Big/\dfrac{0.003}{0.297}\right]}{\ln 2.2} = 10.76$$

The distribution of component A is found as

$$\left(\frac{d}{b}\right)_i = \left(\frac{f-b}{b}\right)_i = \left(\frac{d}{b}\right)_{lk}\left(\frac{\alpha_i}{\alpha_{lk}}\right)^{N_m}$$
$$= \frac{0.147}{0.003}\left(\frac{3.1}{2.2}\right)^{10.76} = 1962,$$
$$b_i = \frac{f_i}{1+(d/b)_i} = \frac{0.03}{1+1962} = 1.5(E-5),$$
$$d_i = f_2 - b_i = 0.03 - 1.5(E-5) = 0.300.$$

Distributions of the other components are found in the same way.

Since component D is distributed, two values of θ are found from Eq. (13.120):

$$\frac{3.1(0.03)}{3.1-\theta} + \frac{2.6(0.7)}{2.6-\theta} + \frac{2.2(0.15)}{2.2-\theta} + \frac{1.3(0.33)}{1.3-\theta}$$
$$+ \frac{1(0.3)}{1-\theta} + \frac{0.8(0.12)}{0.8-\theta} = 1 - 0.8,$$
$$\therefore\ \theta_1 = 1.8817, \quad \theta_2 = 1.12403.$$

The overhead content d_D of component D and the minimum reflux are found from the two equations

$$(R_m+1)D = (R_m+1)(0.2498 + d_D)$$
$$= \frac{3.1(0.03)}{3.1-\theta_1} + \frac{2.6(0.07)}{2.6-\theta_1} + \frac{2.2(0.147)}{2.2-\theta_1}$$
$$+ \frac{1.3d_D}{1.3-\theta_1} + \frac{0.003}{1-\theta_1}$$
$$= \frac{3.1(0.03)}{3.1-\theta_2} + \frac{2.6(0.007)}{2.6-\theta_2} + \frac{2.2(0.147)}{2.2-\theta_2}$$
$$+ \frac{1.3d_D}{1.3-\theta_2} + \frac{0.003}{1-\theta_2}.$$

Upon substituting $\theta_1 = 1.8817$, $\theta_2 = 1.12403$,

$$d_D = 0.09311,$$
$$D = 0.2498 + 0.09311 = 0.3429,$$
$$(R_m+1)D = 1.1342,$$
$$R_m = 2.3077.$$

Let $R = 1.2\ R_m = 1.2(2.3077) = 2.7692$. Apply Eq. (13.124):

$$X = \frac{R - R_m}{R+1} = \frac{0.2(2.3077)}{3.7692} = 0.1225,$$
$$Y = 0.5313$$
$$N = \frac{N_m + Y}{1-Y} = \frac{10.76 + 0.5313}{1-0.5313} = 24.1.$$

Feed plate location:

$$\frac{N_{\text{above}}}{N_{\text{below}}} = \frac{\ln\left(\dfrac{0.147}{0.15}\Big/\dfrac{0.003}{0.300}\right)}{\ln\left(\dfrac{0.15}{0.003}\Big/\dfrac{0.3}{0.297}\right)} = 1.175.$$

Since $N_{\text{above}} + N_{\text{below}} = 24.1$,

$$\text{feed tray} = \frac{24.1}{1+1/1.175} = 13 \text{ from the top.}$$

For comparison, apply Eqs. (13.129) and (13.130):

$$\frac{N_r^*}{24 - N_r^*} = \left[\frac{0.6572}{0.3428}\left(\frac{0.30}{0.15}\right)\left(\frac{0.003/0.6572}{0.003/0.3428}\right)^2\right]^{0.206}$$
$$= 1.0088,$$
$$N_r^* = 12.05,$$
$$N_r = 12.05 - 0.5 \log 24 = 10.46 \text{ from the top.}$$

Presumably 10.46 from the top is more accurate than 13.0, but it also may be in error because of the approximate fashion in which the distributions of nonkeys were found.

Note that the predicted distributions of component D do not agree closely.

	d	b
From minimum trays	0.0481	0.2819
From minimum reflux	0.09303	0.2370

MINIMUM REFLUX

The method of Underwood employs auxiliary parameters θ derived from the equation

$$\sum_{i=1}^{C} \frac{\alpha_i x_{Fi}}{\alpha_i - \theta} = 1 - q, \qquad (13.120)$$

where q is the thermal condition of the feed and the summation extends over all the components in the feed. The only roots required are those in numerical value between the relative volatilities of the light and heavy keys. For instance, if there is one distributed component, subscript $_{dk}$, the required roots θ_1 and θ_2 are in the ranges

$$\alpha_{1k} > \theta_1 > \alpha_{dk},$$
$$\alpha_{dk} > \theta_2 > \alpha_{hk}.$$

Then the minimum reflux and the distribution of the intermediate component are found from the two equations that result from substitution of the two values of θ into Underwood's second equation

$$R_m + 1 = \frac{1}{D} \sum \frac{\alpha_i d_i}{\alpha_i - \theta}. \qquad (13.121)$$

The number of values of θ and the number of Eqs. (13.121) is equal to 1 plus the number of components with relative volatilities between those of the light and heavy keys. When there is no distributed component, Eq. (13.121) may be used in terms of mol fractions and only a single form is needed for finding the minimum reflux,

$$R_m + 1 = \sum \frac{\alpha_i x_{iD}}{\alpha_i - \theta}. \qquad (13.122)$$

Occasionally the minimum reflux calculated by this method comes out a negative number. That, of course, is a signal that some other method should be tried, or it may mean that the separation between feed and overhead can be accomplished in less than one equilibrium stage.

OPERATING REFLUX

As discussed briefly in Section 13.4, the operating reflux is an amount in excess of the minimum that ultimately should be established by an economic balance between operating and capital costs for the operation. In many cases, however, as stated there the assumptions $R = 1.2R_m$ often is close to the optimum and is used without further study unless the installation is quite a large one.

ACTUAL NUMBER OF THEORETICAL TRAYS

An early observation by Underwood (*Trans. Inst. Chem. Eng.* **10,** pp. 112–152, 1932) of the plate–reflux relation was

$$(R - R_m)(N - N_m) = \text{const}, \qquad (13.123)$$

but no general value for the constant was possible. Several correlations of calculated data between these same variables have since been made. A graphical correlation made by Gilliland (*Ind. Eng. Chem.* **32,** 1101, 1940) has found wide acceptance because of its fair accuracy and simplicity of use. Of the several representations of the plot by equations, that of Molokanov et al. [*Int. Chem. Eng.* **12,** 209–212 (1972)] is accurate and easy to use:

$$Y = \frac{N - N_{\min}}{N + 1} = 1 - \exp\left[\left(\frac{1 + 54.4X}{11 + 117.2X}\right)\left(\frac{X - 1}{X^{0.5}}\right)\right], \qquad (13.124)$$

where

$$X = \frac{R - R_{\min}}{R + 1}, \qquad (13.125)$$

from which the number of theoretical trays is

$$N = \frac{N_m + Y}{1 - Y}. \qquad (13.126)$$

The Gilliland correlation appears to be conservative for feeds with low values of q (the thermal condition of the feed), and can be in error when there is a large difference in tray requirements above and below the feed. The principal value of the correlation appears to be for preliminary exploration of design variables which can be refined by computer calculations. Although it is often used for final design, that should be done with caution. Other possibly superior but more difficult to use correlations have been proposed and are described in standard textbooks; for example, Hines and Maddox (1985).

FEED TRAY LOCATION

Particularly when the number of trays is small, the location of the feed tray has a marked effect on the separation in the column. An estimate of the optimum location can be made with the Underwood–Fenske equation (13.116), by applying it twice, between the overhead and the feed and between the feed and the bottoms. The ratio of the numbers of rectifying N_r and stripping N_s trays is

$$\frac{N_r}{N_s} = \frac{\ln[(d/f)_{1k}/(d/f)_{hk}]}{\ln[(f/b)_{1k}/(f/b)_{hk}]} \qquad (13.127)$$

$$= \frac{\ln[(x_d/x_f)_{1k}/(x_d/x_f)_{hk}]}{\ln[(x_f/x_b)_{1k}/(x_f/x_b)_{hk}]}. \qquad (13.128)$$

An improved relation that, however, requires more information is due to Akashah, Erbar, and Maddox [*Chem. Eng. Commun.* **3,** 461 (1979)]. It is

$$N_r = N_r^* - 0.5\log(N_t), \qquad (13.129)$$

where N_t is the total number of trays in the column and N_r^* is given by the empirical Kirkbride (*Petrol. Refiner* **23** (9), 321, 1944) equation,

$$\frac{N_r^*}{N_t - N_r^*} = \left[\frac{B}{D}\left(\frac{x_{1k}}{x_{hk}}\right)_f \left(\frac{x_{B1k}}{x_{Dhk}}\right)^2\right]^{0.206}. \qquad (13.130)$$

TRAY EFFICIENCIES

The calculations made thus far are of theoretical trays, that is, trays on which vapor–liquid equilibrium is attained for all components. Actual tray efficiencies vary widely with the kind of system, the flow rates, and the tray construction. The range can be from less than 10% to more than 100% and constitutes perhaps the greatest uncertainty in the design of distillation equipment. For hydrocarbon fractionation a commonly used efficiency is about 60%. Section 13.14 discusses this topic more fully.

13.8. ABSORPTION FACTOR SHORTCUT METHOD OF EDMISTER

This method finds the product distribution ratio b/d for each component in a column with known numbers of trays above and below the feed and with a known reflux ratio. The flowsketch and nomenclature appear on Figure 13.13.

An absorption factor for each component i on each tray j is defined as

$$A_{ij} = L_j/V_jK_{ij}, \tag{13.131}$$

but usually it is understood to apply to a specific component so the subscript i is dropped and the absorption factors on tray j become

$$A_j = L_j/V_jK_j. \tag{13.132}$$

Similarly a stripping factor for each component is defined as

$$S_j = K_jV_j/L_j. \tag{13.133}$$

The ratio of bottom and overhead flow rates for each component is

$$\frac{b}{d} = \frac{\phi_1 + (L_d/DK_d)\phi_2 - (1-q)F}{\psi_1 + (V_b/B)\psi_2 - 1}, \tag{13.134}$$

with which the individual flow rates of each component are found

$$b_i = \frac{f_i}{1 + (b/d)_i}, \tag{13.135}$$

$$d_i = f_i - b_i. \tag{13.136}$$

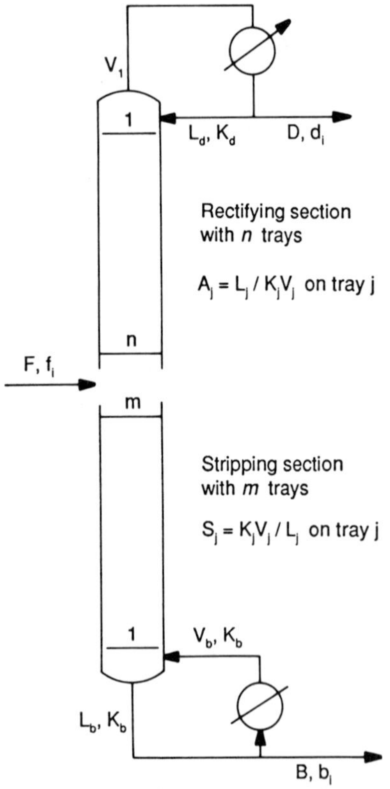

Figure 13.13. Sketch and nomenclature for the absorption factor method.

The function ϕ and ψ are defined as

$$\phi_1 = \frac{A_e^{n+1} - 1}{A_e - 1}, \tag{13.137}$$

$$\phi_2 = (A_1A_n)^{n/2}, \tag{13.138}$$

$$\psi_1 = \frac{S_e^{m+1} - 1}{S_e - 1}, \tag{13.139}$$

$$\psi_2 = (S_1S_m)^{m/2}. \tag{13.140}$$

The effective absorption and stripping factors in each zone are approximately

$$A_e = -0.5 + \sqrt{A_n(A_1 + 1) + 0.25}, \tag{13.141}$$

$$S_e = -0.5 + \sqrt{S_m(S_1 + 1) + 0.25}. \tag{13.142}$$

A certain number of initial estimates must be made when applying Edmister's method which are improved by iteration.

1. Initial estimates must be made of the top and bottom temperatures so that the A_1 and S_1 can be estimated. These estimates will be adjusted by bubblepoint calculations after b and d have been found by the first iteration.
2. The temperature at the feed zone may be found by taking a linear temperature gradient.
3. Estimates must be made of V/L at the top and bottom and the feed zone. In distillation problems, assumption of constant molal overflow in each zone probably is within the accuracy of the method. In stripping or absorption columns, first iteration evaluations of the amounts of stripping or absorption will provide improved estimates of V/L at the key points in the columns.

A distillation problem is worked out by this method by Edmister [*Pet. Eng.*, 128–142 (Sept. 1948)]. The method is developed there.

For independent absorbers and strippers, the Kremser–Brown formulas apply. The fraction absorbed is

$$\phi_a = \frac{A_e^{n+1} - A_e}{A_e^{n+1} - 1}, \tag{13.143}$$

and the fraction stripped is

$$\phi_s = \frac{S_e^{m+1} - S_e}{S_e^{m+1} - 1}. \tag{13.144}$$

An absorber is calculated by this method in Example 13.11.

13.9. SEPARATIONS IN PACKED TOWERS

Continuous changes in compositions of phases flowing in contact with each other are characteristic of packed towers, spray or wetted wall columns, and some novel equipment such as the HIGEE contactor (Fig. 13.14). The theory of mass transfer between phases and separation of mixtures under such conditions is based on a two-film theory. The concept is illustrated in Figure 13.15(a).

In its simplest form, the rate of mass transfer per unit area across these films is

$$N/A = k_G(y - y^*) = k_L(x^* - x). \tag{13.145}$$

Two special cases are commonly recognized.

EXAMPLE 13.11
Calculation of an Absorber by the Absorption Factor Method

A mixture of a given composition is to have 60% of its n-butane removed by scrubbing with an oil in a 4-tray tower operating essentially isothermally at a pressure of 4 atm. The oil feed rate per 100 mol of feed gas will be found. The data are

	z_f	K	ϕ
C_1	0.253	54	
C_2	0.179	14	
C_3	0.222	3.5	
nC_4	0.240	0.5	0.600
nC_5	0.105	0.2	
	1.000		

The Kremser–Brown formula (Eq. (13.143)) for the fraction absorbed is applied to nC_4:

$$\phi = (A_e^5 - A_e)/(A_e^5 - 1) = 0.6,$$
$$\therefore A_e = 0.644, \quad \text{by trial.}$$

Estimate that 27 mol of gas is absorbed. Let L_d represent the lean oil rate: For nC_4

$$A_1 = \frac{L_d}{KV_1} = \frac{L_d}{0.5(73)}, \quad A_n = \frac{L_d + 27}{0.5(100)}.$$

Substitute into Eq. (13.141),

$$A_e = -0.5 + \left[\frac{(L_d + 27)}{50}\left(\frac{L_d}{36.5} + 1\right) + 0.25\right]^{1/2} = 0.644,$$
$$\therefore L_d = 12.46, \quad \text{by trial.}$$

For the other components,

$$A_e = -0.5 + \left[\frac{12.46 + 27}{100K}\left(\frac{12.46}{73K} + 1\right) + 0.25\right]^{1/2},$$
$$\phi = \frac{A_e^5 - A_e}{A_e^5 - 1},$$
$$b = 100 z_f \phi.$$

The results are tabulated and show that the calculated value, 27.12, is close to the assumed, 27.00.

	z_f	K	A_e	ϕ	b
C_1	0.253	54	0.00728	0.00728	0.18
C_2	0.179	14	0.02776	0.02776	0.50
C_3	0.222	3.5	0.1068	0.1068	2.37
nC_4	0.240	0.5	0.644	0.600	14.40
nC_5	0.105	0.2	1.4766	0.9208	9.67
	1.000				27.12

1. Equimolal counterdiffusion between the phases, as in distillation with McCabe–Thiele approximations.
2. Diffusion through a stagnant film, as in absorption or stripping processes involving the transfer of a single component between liquid and vapor phases. Since there is a concentration gradient of the diffusing substance in the films, a correction is applied to the mass transfer coefficient. It is shown in books on mass transfer that the effective coefficient of a stagnant film is

$$(k_G)_{\text{effective}} = k_G/(y - y^*)_{\text{log mean}}, \tag{13.146}$$

where

$$(y - y^*)_{\text{log mean}} = \frac{(1-y) - (1-y^*)}{\ln[(1-y)/(1-y^*)]} = \frac{(y^* - y)}{\ln[(1-y)/(1-y^*)]}. \tag{13.147}$$

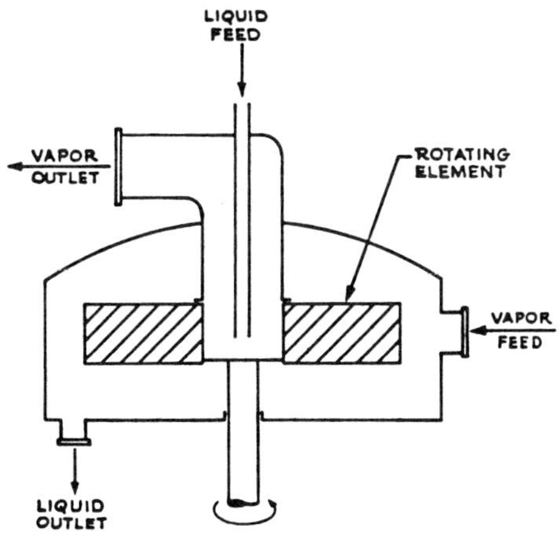

Figure 13.14. A centrifugal packed fractionator, trade name HIGEE, Imperial Chemical Industries. Units have been operated with 500 times gravitational acceleration, with 3–18 theoretical stages, up to 36 in. dia, employing perforated metal packing. For distillation, one unit is needed for rectification and one for stripping. Units have been used primarily for gas stripping and on offshore platforms because of compactness [*Ramshaw,* Chem. Eng., *13–14 (Feb. 1983)].*

MASS TRANSFER COEFFICIENTS

Numerous investigations have been conducted of mass transfer coefficients in vessels with a variety of kinds of packings. Many of the more acceptable results are cited in recent books on mass transfer, for instance, those of Sherwood et al. (*Mass Transfer,* McGraw-Hill, New York, 1975), Cussler (*Diffusion,* Cambridge, 1984), and Hines and Maddox (1985). A convenient correlation of mass transfer coefficients in granular beds covering both liquid and vapor films is that of Dwivedi and Upadhyay [*Ind. Eng. Chem. Process Des. Dev.* **16,** 157 (1977)], namely,

$$\varepsilon j_d = \frac{0.765}{\text{Re}^{0.82}} + \frac{0.365}{\text{Re}^{0.386}} \tag{13.148}$$

$$j_d = (\text{Sh})/(\text{Re})(\text{Sc})^{2/3} \quad \text{(Chilton–Colburn factor)}, \tag{13.149}$$
$$\text{Sh} = kd/\mathcal{D} \quad \text{(Sherwood number)}, \tag{13.150}$$
$$\text{Sc} = \mu/\rho\mathcal{D} \quad \text{(Schmidt number)}, \tag{13.151}$$
$$\text{Re} = du\rho/\mu = 4w/\pi d^2\mu \quad \text{(Reynolds number)},$$

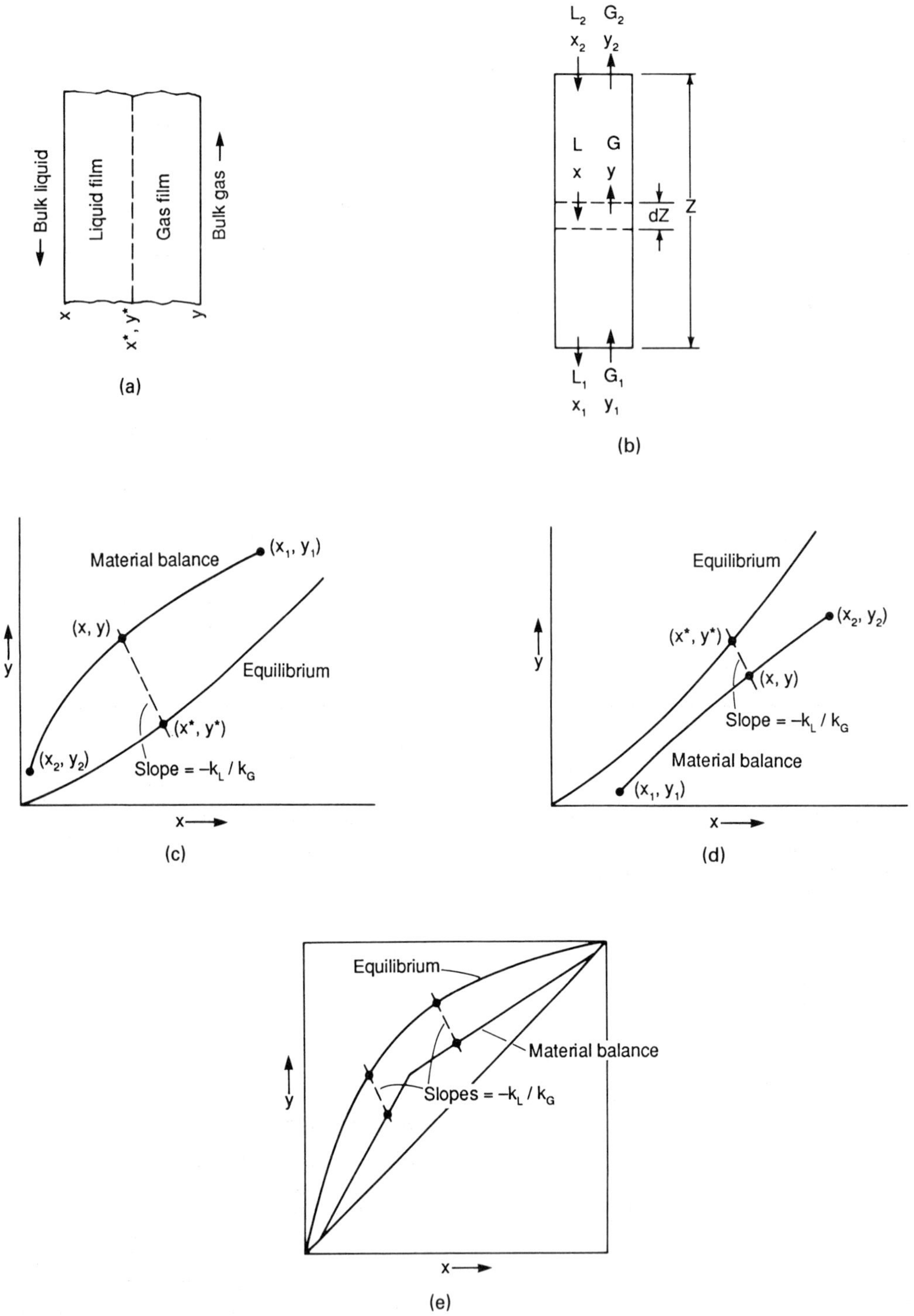

Figure 13.15. Mechanism, nomenclature, and constructions for absorption, stripping and distillation in packed towers. (a) Two-film mechanism with equilibrium at the interface. (b) Sketch and nomenclature for countercurrent absorption or stripping in a packed tower. (c) Equilibrium and material balance lines in absorption, showing how interfacial concentrations are found. (d) Equilibrium and material balance lines in stripping, showing how interfacial concentrations are found. (e) Equilibrium and material balance lines in distillation, showing how interfacial concentrations are found.

where

d = particle diameter,
$\mathscr{D}$ = diffusivity of the substance being transferred,
k = mass transfer coefficient,
u = linear velocity of the fluid,
w = mass rate of flow of the fluid,
ε = fractional voidage between particles,
ρ = density of the fluid,
μ = viscosity of the fluid. (13.152)

Most of the properties change somewhat from one end to the other of industrial columns for effecting separations, so that the mass transfer coefficients likewise vary. Perhaps the property that has the most effect is the mass rate of flow which appears in the Reynolds number. Certainly it changes when there is a substantial transfer of material between the two phases in absorption or stripping; and even under conditions of constant molal overflow in distillation processes, the mass rate of flow changes because of differences of the molecular weights of the substances being separated. As a practical expedient, however, mass transfer coefficients are evaluated at mean conditions in a column.

DISTILLATION

Only the important case of constant molal overflow will be considered. The material balance around the lower end of the column of Figure 13.15(b) is

$$Gy + L_1x_1 = G_1y_1 + Lx, \qquad (13.153)$$

which becomes at constant molal overflow

$$y = \frac{L}{G}x + \left(y_1 - \frac{L}{G}x_1\right). \qquad (13.154)$$

The rate balance on an element of height dz of a column of unit cross section is

$$-dN = d(Gy) = G\,dy = k_Ga(y - y^*)\,dz \qquad (13.155)$$
$$= d(Lx) = L\,dx = k_La(x^* - x)\,dz, \qquad (13.156)$$

where a is the interfacial surface per unit volume of the packed bed.

These equations relate the interfacial concentrations (x^*, y^*) to those in the bulks of the liquid and gas phases (x, y); thus

$$\frac{y^* - y}{x^* - x} = -\frac{k_L}{k_G}. \qquad (13.157)$$

The bulk concentrations (x, y) are related by the material balance Eq. (13.144), and the equilibrium concentrations (x^*, y^*) from experimental data in graphical, tabular, or equation form,

$$y^* = f(x^*) \qquad (13.158)$$

for instance, at constant relative volatility,

$$y^* = \frac{\alpha x^*}{1 + (\alpha - 1)x^*}. \qquad (13.159)$$

Corresponding points (y, y^*) in a column where the ratio k_L/k_G is known are found as follows: At a particular composition x, the value of y is known from Eq. (13.154). Then corresponding values (x^*, y^*) are related linearly by Eq. (13.157). Substitution into Eq. (13.158) then will establish the value of y^* corresponding to the selected y. Figures 13.14(c), (d), (e) display graphical procedures for this operation.

By rearrangements of Eqs. (13.155) and (13.156) the height of the column is given by

$$Z = \frac{G}{k_Ga}\int_{y_1}^{y_2} \frac{dy}{y^* - y} \qquad (13.160)$$
$$= \frac{L}{k_La}\int_{x_1}^{x_2} \frac{dx}{x - x^*}. \qquad (13.161)$$

The integrals in these equations are measures of the difficulty of the separation. Under some conditions they are roughly equal to the number of theoretical trays for the same change in concentration (y_1, y_2) or (x_1, x_2). Accordingly, they are called numbers of transfer units.

$$\mathrm{NTU}_G = \int_{y_1}^{y_2} \frac{dy}{y^* - y}, \qquad (13.162)$$

$$\mathrm{NTU}_L = \int_{x_1}^{x_2} \frac{dx}{x - x^*}. \qquad (13.163)$$

Consequently, it is natural to call the coefficients of the integrals the height of a transfer unit,

$$\mathrm{HTU}_G = G/k_ga, \qquad (13.164)$$
$$\mathrm{HTU}_L = L/k_La. \qquad (13.165)$$

These terms sometimes are used interchangeably with height equivalent to a theoretical stage (HETS), but they are nearly the same only when the ratio k_L/k_G is a large number in the case of HTU_G. Example 13.12 studies this difference.

The concepts NTU and HTU are defined only for binary distillations and the transfer of a single substance in absorption or stripping. Since most processes of industrial interest involve multicomponents, the HETS of packed towers is the more useful concept, and may be evaluated readily from test data and tray calculations.

ABSORPTION OR STRIPPING

Neither mass nor molal flow rates are constant in these operations. In cases where essentially only one component is being transferred between phases, it is sometimes convenient to recognize the flow rates G' and L' of solute-free phases. They are related to the total flow rates by

$$G' = G(1 - y) = G_1(1 - y_1), \qquad (13.166)$$
$$L' = L(1 - x) = L_1(1 - x_1). \qquad (13.167)$$

The material balance around the lower end of the column of Figure 13.15(b),

$$Gy + L_1x_1 = G_1y_1 + Lx \qquad (13.168)$$

can be written

$$\frac{y}{1 - y} = \frac{L'}{G'}\left(\frac{x}{1 - x}\right) + \left(\frac{y_1}{1 - y_1} - \frac{L'}{G'}\frac{x_1}{1 - x_1}\right) \qquad (13.169)$$

or in the linear form

$$Y = \frac{L'}{G'} X + \left(Y_1 - \frac{L'}{G'} X_1 \right) \tag{13.170}$$

with the substitutions

$$X = \frac{x}{1 - x}, \tag{13.171}$$

$$Y = \frac{y}{1 - y}. \tag{13.172}$$

The equilibrium curve also can be transformed into these coordinates. These transformations are useful for graphical determinations of numbers of theoretical trays rather than for determination of numbers of transfer units. Example 13.13 employs both sets of units.

Example 13.12
Numbers of Theoretical Trays and of Transfer Units with Two Values of k_L/k_G for a Distillation Process

An equimolal mixture at its boiling point is to be separated into 95 and 5% contents of the lighter component in the top and bottoms products. The relative volatility is $\alpha = 2$, the minimum reflux is 1.714, and the operating reflux is 50% greater. The two values of k_L/k_G to be examined are -1 and ∞.

The relation between interfacial and bulk concentrations is that of Eq. (13.157), $(y^* - y)/(x^* - x) = -k_L/k_G$. At a series of values of x, corresponding values of y^* and y may be read off with the graphical constructions shown on Figures (b) and (c) of this example. The values for slope $= -1$ are tabulated, but those for slope $= \infty$ are calculated from the equations of the equilibrium and operating lines and are not recorded. The integrands of Eq. (13.160) also are tabulated for both cases, and the numbers of transfer units are obtained by integration with the trapezoidal rule:

$$\text{NTU} = \int_{y_1}^{y_2} \frac{dy}{y^* - y}$$

a. The number of theoretical trays stepped off on the McCabe–Thiele diagram is 16.2.
b. With $k_L/k_G = 1$, the number of transfer units is 30.7.
c. With $k_L/k_G = \infty$, the number of transfer units is 15.4.

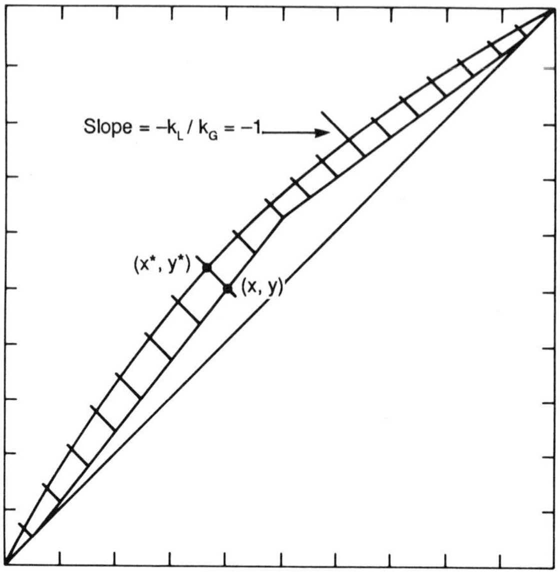

(b) Construction with $k_L/k_G = 1$, showing takeoff of vapor concentrations in the bulk, y, and at the interface, y^*. Number of transfer found by integration = 15.4.

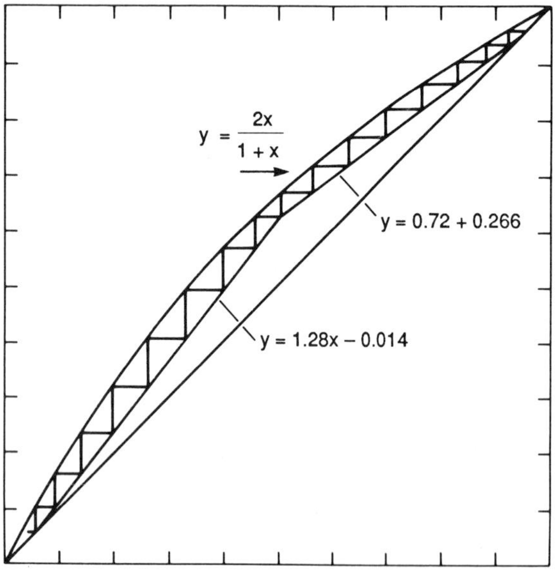

(a) McCabe–Thiele construction showing that 16.2 trays are needed to contain 95 and 5% of the lighter substance in the products from a 50% boiling liquid feed.

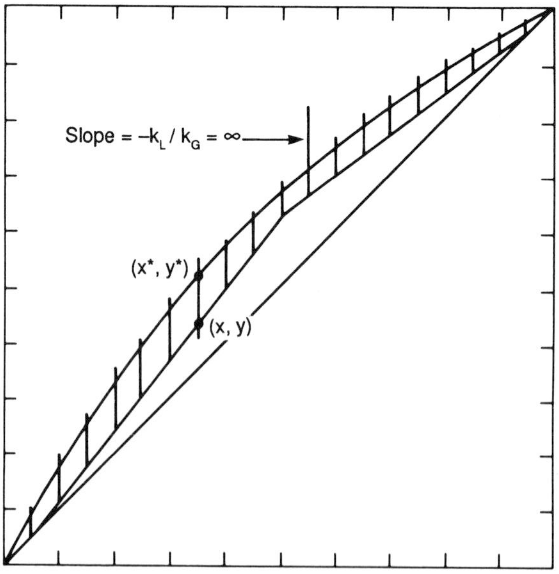

(c) Construction with $k_L/k_G = \infty$. Number of transfer units found by integration = 30.6.

Example 13.12—*(continued)*

Within the accuracy of the trapezoidal rule integration and of the graphical determination of the number of trays, the numbers 16.2 and 15.4 are substantially the same. The infinite value of the ratio of mass transfer coefficients k_L/k_G means that all of the resistance to mass transfer is in the gas film:

x	y	y_1^*	$1/(y_\infty^* - y)$	$1/(y_1^* - y)$
0.05	0.05	0.068	22.105	55.56
0.10	0.114	0.149	14.745	28.57
0.15	0.178	0.209	12.067	32.26
0.2	0.242	0.279	10.949	27.03
0.25	0.306	0.345	10.638	25.64
0.3	0.370	0.411	10.924	24.39
0.35	0.434	0.474	11.832	25.00
0.4	0.498	0.536	13.619	26.31
0.45	0.526	0.593	17.039	32.26
0.5	0.626	0.648	24.590	45.45
0.55	0.662	0.687	20.974	40.00
0.6	0.698	0.728	19.231	33.33
0.65	0.734	0.763	18.560	34.48
0.7	0.770	0.798	18.681	35.71
0.75	0.806	0.832	19.533	38.46
0.8	0.842	0.870	21.327	35.71
0.85	0.878	0.902	24.439	41.67
0.9	0.914	0.933	29.969	52.63
0.95	0.950	0.965	41.053	66.67

Example 13.13
Trays and Transfer Units for an Absorption Process

The solute content of a gas with $y_1 = 0.40$ is to be reduced to $y_2 = 0.05$. The entering solvent is solute-free, $x_1 = 0$, and is to leave with $x_2 = 0.19$. The equilibrium relationship is represented by the equation

$$y^* = x^*(1 + 5x^*),$$

and the ratio of mass transfer coefficients is $k_L/k_G = 1$.

In terms of solute-free coordinates, the equation of the material balance line is

$$Y = 2.6176X + 0.0526,$$

calculated with the given terminal concentrations. In terms of mol fractions the material balance line is curved, with equation

$$y = \frac{2.6176x/(1-x) + 0.0526}{2.6176x/(1-x) + 1.0526}.$$

The equation of the equilibrium curve in solute-free coordinates is

$$\frac{Y}{1+Y} = \frac{X}{1+X}\left(1 + \frac{5X}{1+X}\right).$$

Constructions for the numbers of trays in both sets of coordinates are made. They agree within the accuracy of graphical constructions on this scale, $N = 4.7$ with (x, y) and $N = 4.5$ with (X, Y).

For the transfer unit determination with the given ratio of mass transfer coefficients, corresponding values of (y, y^*) are found by intersections of the material balance and equilibrium lines with lines whose slopes are $-k_L/k_G = -1$ as indicated on Figure (a) and in detail with Example 13.12. These values are tabulated together with the corresponding integrands. The number of transfer units is found by trapezoidal rule integration of

$$\text{NTU} = \int_{0.05}^{0.40} \frac{dy}{(1-y)\ln[(1-y^*)/(1-y)]}$$
$$= 6.52.$$

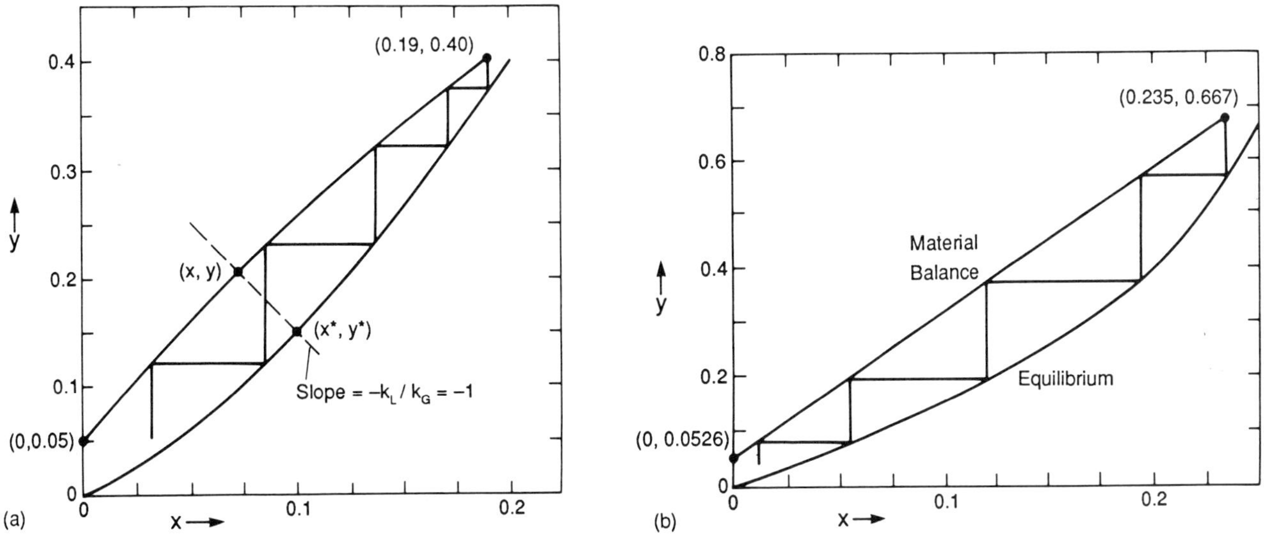

The two values of N should be the same, but there is a small disagreement because of construction inaccuracies on this scale: (a) construction with mol fraction coordinates, $N = 4.7$; (b) construction with solute-free coordinates, $N = 4.5$.

EXAMPLE 13.13—*(continued)*

x	y	y^*	Integrand
0	0.05	0.009	24.913
0.01	0.0733	0.020	19.296
0.02	0.0959	0.036	17.242
0.03	0.1178	0.052	15.757
0.04	0.1392	0.069	14.818
0.05	0.1599	0.086	14.119
0.06	0.1801	0.102	13.405
0.07	0.1998	0.122	13.469
0.08	0.2189	0.141	13.467
0.09	0.2375	0.160	13.548
0.10	0.2556	0.180	13.888
0.11	0.2733	0.202	14.703
0.12	0.2906	0.224	15.709
0.13	0.3074	0.246	16.998
0.14	0.3237	0.268	18.683
0.15	0.3397	0.290	20.869
0.16	0.3553	0.312	23.862
0.17	0.3706	0.335	28.877
0.18	0.3854	0.358	37.304
0.19	0.4000	0.381	53.462

The rate balance on an element of height dz of a column of unit cross section, as in Figure 13.15(b), is

$$-dN = d(Gy) = (k_G)_{\text{eff}}a(y - y^*)\,dz \tag{13.173}$$
$$= d(Lx) = (k_L)_{\text{eff}}a(x^* - x)\,dz. \tag{13.174}$$

Expanding the differential of Eq. (13.163),

$$d(Gy) = d\left(\frac{G'}{1-y}\right) = \frac{G'}{(1-y)^2}\,dy = \frac{G}{1-y}\,dy. \tag{13.175}$$

Introducing Eqs. (13.146) and (13.175) into Eq. (13.173) and integrating, the height becomes

$$Z = \left(\frac{G}{k_G a}\right)_{\text{mean}} \int_{y_1}^{y_2} \frac{(y-y^*)_{\text{lm}}}{(1-y)(y-y^*)}\,dy. \tag{13.176}$$

On replacing the log mean term by Eq. (13.147), the result becomes

$$Z = \left(\frac{G}{k_g a}\right)_m \int_{y_1}^{y_2} \frac{1}{(1-y)\ln[(1-y)/(1-y^*)]}\,dy. \tag{13.177}$$

The variable flow rate G is used here instead of the constant G' because the mass transfer coefficient k_G depends more directly on G. As used in Eqs. (13.176) and (13.177), a mean value of the coefficient is preferred in practice in preference to accounting for its variation within the integral.

The integrals are defined as numbers of transfer units for absorption or stripping,

$$\text{NTU}_G = \int_{y_1}^{y_2} \frac{1}{(1-y)\ln[(1-y)/(1-y^*)]}\,dy, \tag{13.178}$$

$$\text{NTU}_L = \int_{x_1}^{x_2} \frac{1}{(1-x)\ln[(1-x)/(1-x^*)]}\,dx, \tag{13.179}$$

and the heights of transfer units are

$$\text{HTU}_G = (G/k_G a)_{\text{mean}}, \tag{13.180}$$
$$\text{HTU}_L = (L/k_L a)_{\text{mean}}. \tag{13.181}$$

HTUs vary with the kind of packing, the flow rates, the distribution of flow across the cross section, and sometimes with the packing height and column diameter. They are necessarily experimental data. Some of these data are discussed at the end of this chapter.

The way in which interfacial concentrations y^* are related to the bulk concentrations y required for evaluation of the integrand of Eq. (13.176) is explained on Figure 13.14(c), (d), and in Example 13.13, which finds trays and transfer units for an absorption problem.

13.10. BASIS FOR COMPUTER EVALUATION OF MULTICOMPONENT SEPARATIONS

Until the advent of computers, multicomponent distillation problems were solved manually by making tray-by-tray calculations of heat and material balances and vapor–liquid equilibria. Even a partially complete solution of such a problem required a week or more of steady work with a mechanical desk calculator. The alternatives were approximate methods such as those mentioned in Sections 13.7 and 13.8 and pseudobinary analysis. Approximate methods still are used to provide feed data to iterative computer procedures or to provide results for exploratory studies.

The two principal tray-by-tray procedures that were performed manually are the Lewis and Matheson and Thiele and Geddes. The former started with estimates of the terminal compositions and worked plate-by-plate towards the feed tray until a match in compositions was obtained. Invariably adjustments of the amounts of the components that appeared in trace or small amounts in the end compositions had to be made until they appeared in the significant amounts of the feed zone. The method of Thiele and Geddes fixed the number of trays above and below the feed, the reflux ratio, and temperature and liquid flow rates at each tray. If the calculated terminal compositions are not satisfactory, further trials with revised conditions are performed. The twisting of temperature and flow profiles is the feature that requires most judgement. The Thiele–Geddes method in some modification or other is the basis of most current computer methods. These two forerunners of current methods of calculating multicomponent phase separations are discussed briefly with calculation flowsketches by Hines and Maddox (1985).

Computer programs for multistage operations embodying heat and material balances and sophisticated phase equilibrium relations are best left to professionals. Most such work is done by service organizations that specialize in chemical engineering process calculations or by specialists in engineering organizations. A few valuable programs appear in the open literature:

1. A Wang–Henke program appears in J. Christensen (Ed.) (*Stagewise Computations—Computer Programs for Chemical Engineering Education,* Sterling Swift Publishing, Manchaia, TX, 1972).
2. A Naphthali–Sandholm program appears in Fredenslund, Gmehling, and Rasmussen (*Vapor–Liquid Equilibria Using UNIFAC,* Elsevier, New York, 1977).
3. A Newton–Raphson SC (simultaneous correction) program of Newman is reproduced by King (*Separation Processes,* McGraw-Hill, New York, Appendix E).

Abundant descriptions of the theoretical basis and procedures for computer methods appear in recent literature and are summarized in books by Holland (1981), King (1980, Chap. 10), and Henley and Seader (1981, Chap. 15). The present chapter will be devoted to the basic equations, the kinds of process specifications that can be made and met, and convergence criteria applicable to iterative calculations of problems of distillation, absorption, and stripping. To a certain extent, the same methods are applicable to liquid–liquid extraction and other phase separation processes.

SPECIFICATIONS

The variables most commonly fixed in operations of distillation columns are listed in Section 13.6. Detailed calculation processes of column performance may require other intermediate or tentative specifications whose nature depends on the particular computer algorithm used. These specifications are identified with the descriptions of the three chief methods of this section.

THE MESH EQUATIONS

The letters of this acronym refer to *M*aterial balances, *E*quilibria between vapor and liquid, *S*ummations of mol fractions to unity, and *H*eat or enthalpy balances. The quantities and notation pertaining to a single equilibrium stage and to an assembly of them are represented on Figure 13.16. In the simplest case a distillation stage exchanges two inlet and two outlet streams with adjacent stages. In addition, some stages will have in or out material or heat flows. Computer programs can be written in general form to include these factors on each stage to accommodate multiple feeds, side streams, and intermediate condensing or boiling. Enthalpy transfers sometimes are effected with hollow trays through which a heat transfer medium is circulated, or commonly by pumping a sidestream through an external heat exchanger and returning it to the column. The latter practice is particularly common for

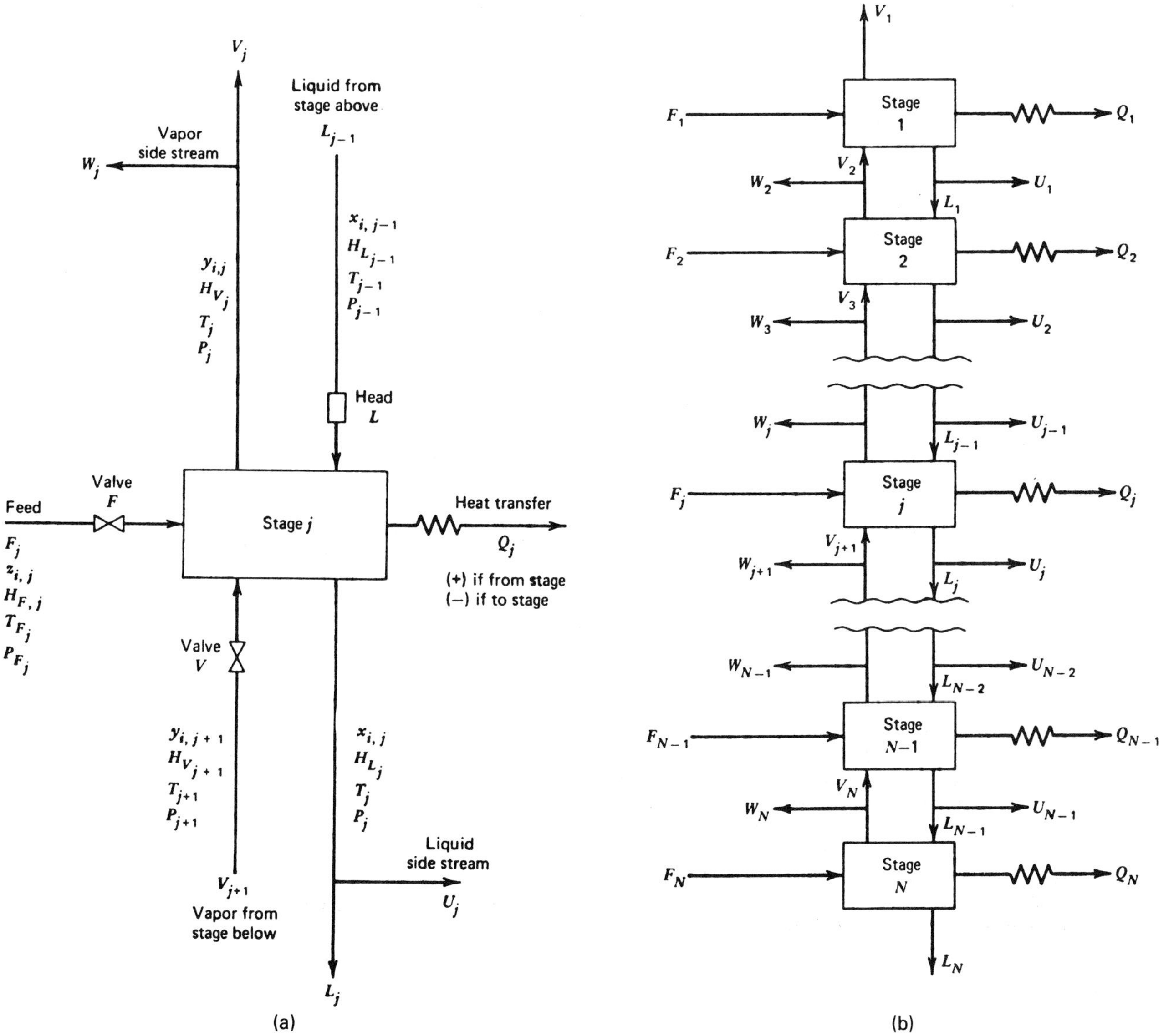

Figure 13.16. Flow patterns and nomenclature of a single equilibrium stage and a cascade of them (*after Henley and Seader, 1981*). (a) A single equilibrium stage. (b) An assembly of N stages.

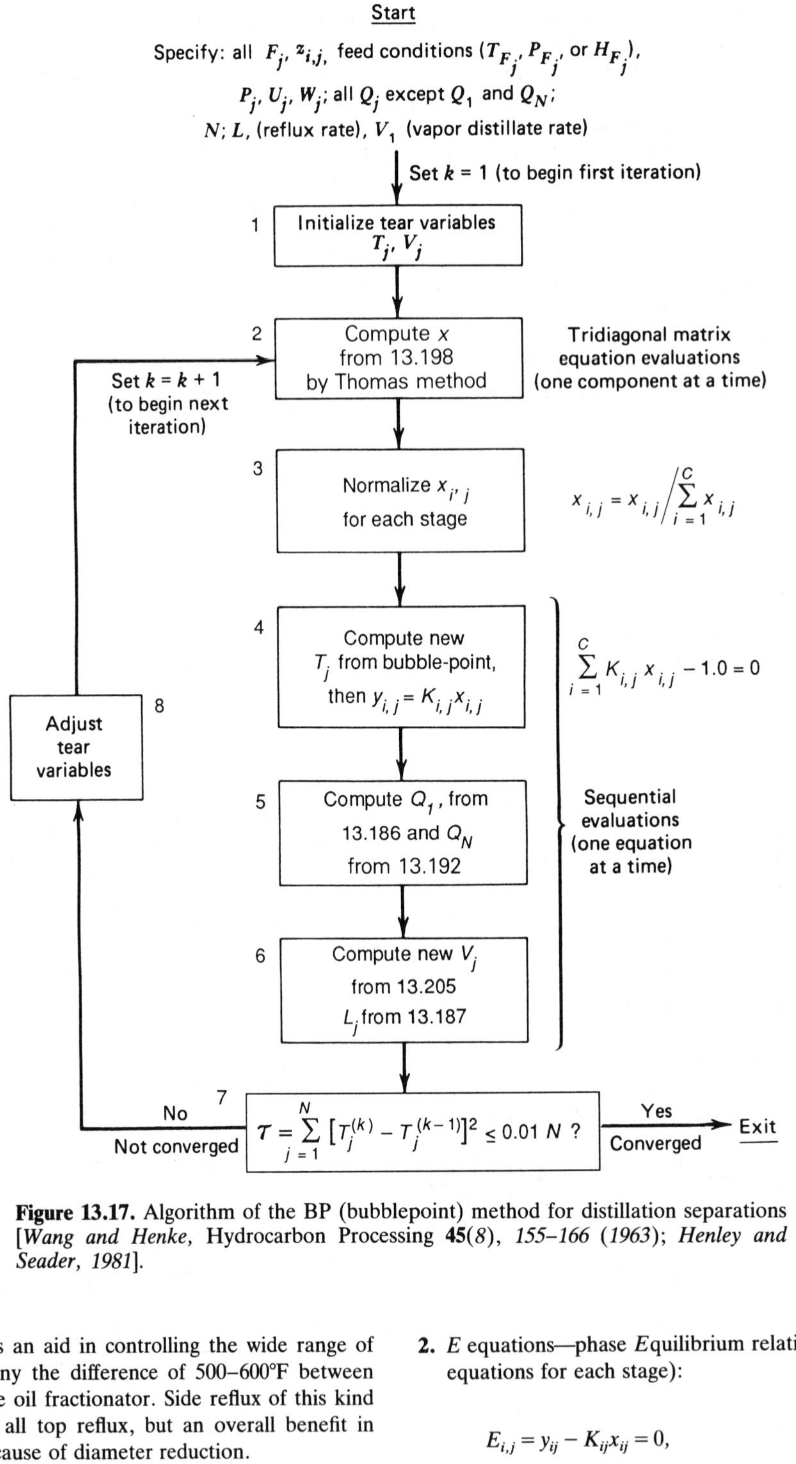

Figure 13.17. Algorithm of the BP (bubblepoint) method for distillation separations [*Wang and Henke,* Hydrocarbon Processing **45**(*8*), *155–166* (*1963*); *Henley and Seader, 1981*].

petroleum fractionation as an aid in controlling the wide range of vapor rates that accompany the difference of 500–600°F between top and bottom of a crude oil fractionator. Side reflux of this kind requires more trays than all top reflux, but an overall benefit in equipment cost results because of diameter reduction.

For every component, C in number, on every stage, N in number, there are material, equilibrium, and energy balances, and the requirement that the mol fractions of liquid and vapor phases on each tray sum to unity. The four sets of these equations are:

1. M equations—*M*aterial balance for each component (C equations for each stage):

$$M_{ij} = L_{j-1}x_{i,j-1} + V_{j+1}y_{i,j+1} + F_j z_{ij} - (L_j + U_j)x_{ij} - (V_j + W_i)y_{ij} = 0. \tag{13.182}$$

2. E equations—phase *E*quilibrium relation for each component (C equations for each stage):

$$E_{i,j} = y_{ij} - K_{ij}x_{ij} = 0, \tag{13.183}$$

where K_{ij} is the phase equilibrium ratio.

3. S equations—mole fraction *S*ummations (one for each stage):

$$(S_y)_j = \sum_{i=1}^{C} y_{ij} - 1.0 = 0, \tag{13.184}$$

$$(S_x)_j = \sum_{i=1}^{C} x_{ij} - 1.0 = 0. \tag{13.185}$$

4. *H* equation—energy balance (one for each stage):

$$H_j = L_{j-1}H_{L_{j-1}} + V_{j+1}H_{V_{j+1}} + F_jH_{F_j} - (L_j + U_j)H_{L_j} - (V_j + W_j)H_{V_j} - Q_j = 0, \quad (13.186)$$

where kinetic and potential energy changes are ignored.

In order to simplify these equations, the liquid rate at each stage is eliminated with the substitutions

$$L_j = V_{j+1} + \sum_{m=1}^{j} (F_m - U_m - W_m) - V_1, \quad (13.187)$$

and the vapor compositions by the equilibrium relations

$$y_{ij} = K_{ij}x_{ij}. \quad (13.188)$$

Three other variables occurring in the MESH equations are functions of more fundamental variables, namely,

$$K_{ij} = K(T_j, P_j, x_{ij}, y_{ij}), \quad (13.189)$$
$$H_{LJ} = H_L(T_j, P_j, x_j), \quad (13.190)$$
$$H_{Vj} = H_V(T_j, P_j, y_j). \quad (13.191)$$

The reboiler load is determined by the overall energy balance,

$$Q_N = \sum_{j=1}^{N} (F_jH_{F_j} - U_jH_{L_j} - W_jH_{V_j}) - \sum_{j=1}^{N-1} Q_j - V_1H_{V_1} - L_NH_{L_N}. \quad (13.192)$$

When all of the following variables are specified,

$N, F_j, z_{ij}, T_j, P_j, U_j, W_j,$ and Q_j (except Q_1 and Q_N), for $i = 1$ to C and $j = 1$ to N,

the MESH equations reduce in number to $N(2C + 3)$ in the same number of variables, and are hence in principle solvable. The equations are nonlinear, however, and require solution by some iterative technique, invariably involving linearization at some stage in the calculation process.

Almost all computer programs employed currently adopt the Thiele–Geddes basis; that is, they evaluate the performance of a column with a specified feed, bottoms/overhead ratio, reflux ratio, and numbers of trays above and below the feed. Specific desired product distributions must be found by interpolation between an appropriate range of exploratory runs. The speed and even the possibility of convergence of an iterative process depends on the values of starting estimates of the variables to be established eventually. Accordingly, the best possible starting estimates should be made by methods such as those of Sections 13.7 and 13.8, or on the basis of experience.

After values of the variables T_j and V_j, called tear variables, are specified, Eqs. (182)ff become a linear set in the x_{ij} variables. Initial estimates of the vapor flows are made by assuming constant molal overflow modified by taking account of external inputs and outputs, and those of the temperatures by assuming a linear gradient between estimated top and bottom temperatures. Initially, also, the K_{ij} are taken as ideal values, independent of composition, and for later iterations the compositions derived from the preceding one may be used to evaluate corrected values of K_{ij}. With appropriate substitutions,

$$A_jx_{i,j-1} + B_jx_{ij} + C_jx_{i,j+1} = D_j, \quad (13.193)$$

where

$$A_j = V_j + \sum_{m=1}^{j-1} (F_m - W_m - U_m) - V_1, \quad 2 \le j \le N, \quad (13.194)$$

$$B_j = -\left[V_{j+1} + \sum_{m=1}^{j} (F_m - W_m - U_m) - V_1 + U_j + (V_j + W_j)K_{i,j}\right], \quad 1 \le j \le N \quad (13.195)$$

$$C_j = V_{j+1}K_{i,j+1}, \quad 1 \le j \le N-1, \quad (13.196)$$

$$D_j = -F_jz_{ij}, \quad 1 \le j \le N, \quad (13.197)$$

the modified MESH equations can be written as a tridiagonal matrix, thus

$$\begin{bmatrix} B_1 & C_1 & 0 & 0 & 0 & \cdots & \cdots & & \cdots & 0 \\ A_2 & B_2 & C_2 & 0 & 0 & \cdots & \cdots & & \cdots & 0 \\ 0 & A_3 & B_3 & C_3 & 0 & \cdots & \cdots & & \cdots & 0 \\ \cdots & \cdots & & & & & & & & \cdots \\ \cdots & & & & & & & & & \cdots \\ \cdots & & & & & & & & & \cdots \\ \cdots & & & & & & & & & \cdots \\ \cdots & & & & & & & & & \cdots \\ \cdots & & & & & & & & \cdots & \cdots \\ 0 & \cdots & & \cdots & \cdots & 0 & A_{N-2} & B_{N-2} & C_{N-2} & 0 \\ 0 & \cdots & & \cdots & \cdots & 0 & 0 & A_{N-1} & B_{N-1} & C_{N-1} \\ 0 & \cdots & & \cdots & \cdots & 0 & 0 & 0 & A_N & B_N \end{bmatrix} \times \begin{bmatrix} x_{i,1} \\ x_{i,2} \\ x_{i,3} \\ \cdots \\ \cdots \\ \cdots \\ \cdots \\ \cdots \\ \cdots \\ x_{i,N-2} \\ x_{i,N-1} \\ x_{i,N} \end{bmatrix} = \begin{bmatrix} D_1 \\ D_2 \\ D_3 \\ \cdots \\ \cdots \\ \cdots \\ \cdots \\ \cdots \\ \cdots \\ D_{N-2} \\ D_{N-1} \\ D_N \end{bmatrix}. \quad (13.198)$$

The tridiagonal matrix is readily solved by computer by a method due to Thomas which is explained by Wang and Henke [*Hydrocarbon Proc.* **45**(8), 155–163 (1966)] and by Henley and Seader (1981). A FORTRAN program is given by Gerald and Wheatley [*Appl. Numer. Anal.*, 146, (1984)] and King (1980, Appendix E). A program in BASIC language is by Pachner (*Handbook of Numerical Analysis Applications,* McGraw-Hill, New York, 1984, Program P103).

After solution of the matrix for the liquid phase mol fractions x_{ij}, the next step is to make improved estimates of T_j and V_j for the next iteration. Three different procedures have been commonly employed for proceeding to succeeding trials, differing in simplicity or particular merit for certain kinds of problems.

1. BP (bubblepoint) methods. Temperatures are corrected iteratively by determinations of bubblepoints. The method is satisfactory for mixtures with relatively narrow ranges of volatilities. The parent program of this type is that of Wang and Henke (1966) which is flowsketched on Figure 13.17 and described in the next section. The availability of a FORTRAN program was cited earlier in this section.
2. SR (sum-rates) method. The new liquid flow rates are taken

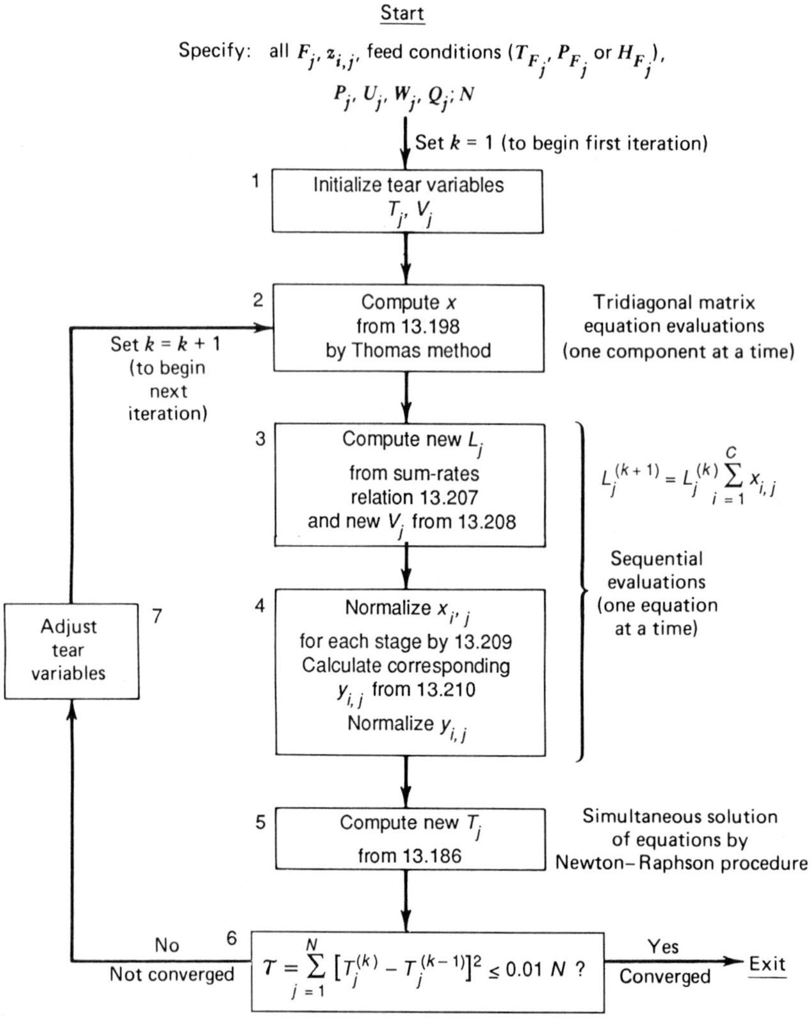

Figure 13.18. Algorithm for the SR (sum rates) method for absorbers and strippers [*Birningham and Otto,* Hydrocarbon Processing **46**(*10*), *163–170* (*1967*); *Henley and Seader, 1981*].

proportional to the nonnormalized sums of mol fractions, the vapor rates by subsequent material balances, and the new temperatures by enthalpy balances. A flowsketch of the calculation process is in Figure 13.18, and a brief description also is given subsequently. This method is particularly suited to separations involving substances with widely differing volatilities, as in absorbers and strippers, where the bubblepoint method breaks down.

3. SC (simultaneous correction) method. The MESH equations are reduced to a set of $N(2C+1)$ nonlinear equations in the mass flow rates of liquid components l_{ij} and vapor components v_{ij} and the temperatures T_j. The enthalpies and equilibrium constants K_{ij} are determined by the primary variables l_{ij}, v_{ij}, and T_j. The nonlinear equations are solved by the Newton–Raphson method. A convergence criterion is made up of deviations from material, equilibrium, and enthalpy balances simultaneously, and corrections for the next iterations are made automatically. The method is applicable to distillation, absorption and stripping in single and multiple columns. The calculation flowsketch is in Figure 13.19. A brief description of the method also will be given. The availability of computer programs in the open literature was cited earlier in this section.

THE WANG–HENKE BUBBLEPOINT METHOD

The procedure is outlined in Figure 13.17. The input data are listed above Box 1 and include all external material and enthalpy flows except condenser and reboiler loads, the number of trays, the reflux rate, and the reboiler load. The process is iterative, starting with estimates of temperature and vapor flow rates on each tray and making successive improvements in these values until a convergence criterion on temperatures is satisfied.

Box 1. Initial estimates of the temperature are made by taking linear variation between estimated overhead dewpoint and bottoms bubblepoint. The vapor rates are estimated on the basis of constant molal overflow with due regard to input or output sidestreams.

Box 2. The system represented by the matrix Eq. (13.198) consists of linear equations that are solved for the liquid mol fractions x_{ij}.

Box 3. In general the mol fractions will not sum to unity, so that they are normalized as

$$(x_{ij})_{\text{normalized}} = x_{ij} \Big/ \sum_{i=1}^{C} x_{ij}. \tag{13.199}$$

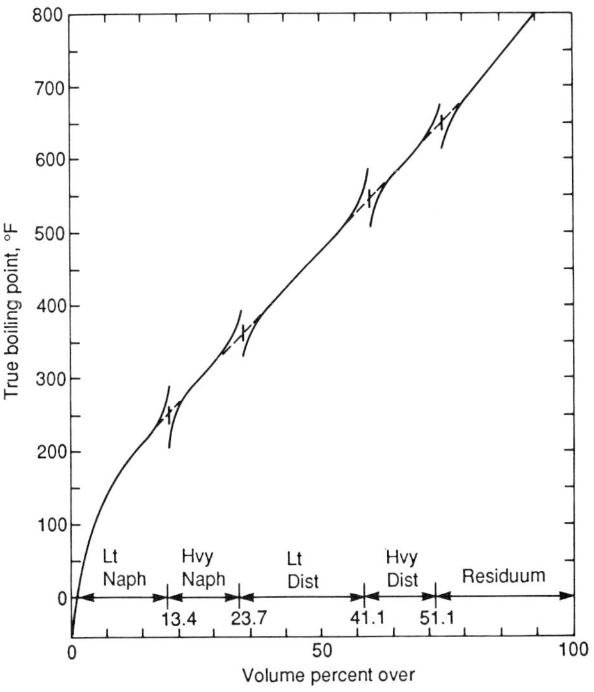

Figure 13.19. True boiling point (TBP) curve of a crude oil, with superimposed TBP curves of five fractions into which it is separated by a typical fractionating system like that of Figure 15.20. The separations are not sharp, with as much as 50°F difference between the end point of a light product and the initial of the next heavier one. It is common to speak of the gap between the 95 and 5% points rather than the end points.

Box 4. New values of the stage temperatures T_j are calculated as bubblepoints with the normalized x_{ij}. Initially the effect of vapor compositions y_{ij} on K_{ij} is ignored and the vapor compositions are found with

$$y_{ij} = K_{ij}x_{ij}. \qquad (13.200)$$

Subsequently, the values of y_{ij} from the previous iteration can be used in the evaluation of K_{ij}.

Box 5. The enthalpies H_{Vj} and H_{Lj} can be evaluated with Eqs. (13.190) and (13.191) since T_j, P_j, x_{ij}, and y_{ij} have been estimated. The condenser load Q_1 is figured with Eq. (13.186) and the reboiler load Q_N with Eq. (13.192).

Box 6. The new vapor rates V_j are found with the heat balances, Eqs. (13.201)–(13.205), and the new liquid rates with Eq. (13.187):

$$\alpha_j V_j + \beta_j V_{j+1} = \gamma_j, \qquad (13.201)$$

where

$$\alpha_j = H_{L_{j-1}} - H_{V_j}, \qquad (13.202)$$

$$\beta_j = H_{V_{j+1}} - H_{L_j}, \qquad (13.203)$$

$$\gamma_j = \left[\sum_{m=1}^{j-1} (F_m - W_m - U_m) - V_1\right](H_{L_j} - H_{L_{j-1}}) + F_j(H_{L_j} - H_{F_j}) + W_j(H_{V_j} - H_{L_j}) + Q_j, \qquad (13.204)$$

$$V_j = \frac{\gamma_{j-1} - \alpha_{j-1}V_{j-1}}{\beta_{j-1}}. \qquad (13.205)$$

Box 7. The convergence criterion imposes a tolerance on the differences between successive iterations of the temperatures

$$\tau = \sum_1^N (T_j^{(k)} - T_j^{(k-1)})^2 \le 0.01N. \qquad (13.206)$$

Box 8. If the criterion is not satisfied, the values of T_j found in Box 4 and the vapor rates V_j of Box 6 are the new starting values to be input to Box 2.

THE SR (SUM-RATES) METHOD

In this method, temperatures for succeeding iterations are found by enthalpy balances rather than by bubblepoint determinations, after new values of the liquid and vapor flow rates have been estimated from solution of the equations for the liquid mol fractions. This procedure is suited to absorption and stripping problems for which the BP method breaks down because of the wide range of relative volatilities involved. The algorithm appears in Figure 13.18. Input data are the same as for the BP method.

Box 1. Initial temperatures and vapor flow rates are estimated in the same way as in the BP method.

Box 2. The mol fractions are found by solution of the tridiagonal matrix as in the BP method.

Box 3. At this point the x_{ij} are not normalized but their sum is applied to estimate new liquid flow rates from the relation

$$L_j^{(k+1)} = L_j^{(k)} \sum_{i=1}^{C} x_{ij}. \qquad (13.207)$$

The corresponding vapor rates are obtained by the material balance, which is a rearrangement of Eq. (13.187),

$$V_j = L_{j-1} - L_N + \sum_{m=j}^{N} (F_m - W_m - U_m). \qquad (13.208)$$

Box 4. Then the x_{ij} are normalized by

$$(x_{ij})_{\text{normalized}} = x_{ij} \Big/ \sum_{i=1}^{C} x_{ij}; \qquad (13.209)$$

the values of y_{ij} are obtained by

$$y_{ij} = K_{ij}x_{ij} \qquad (13.210)$$

and also normalized,

$$y_{ij} = y_{ij} \Big/ \sum_{i=1}^{C} y_{ij}. \qquad (13.211)$$

When the K_{ij} depend on the vapor phase compositions, values of y_{ij} from the previous iteration are used.

Box 5. New temperatures are calculated from the enthalpy balances Eq. (13.186). The temperature is implicit in these equations because of its involvement in the enthalpies and the K_{ij}. Accordingly, the temperature must be found by the Newton–Raphson method for simultaneous nonlinear equations.

Box 6. The convergence criterion is

$$\tau = \sum (T_j^{(k)} - T_j^{(k-1)})^2 \le 0.01N? \qquad (13.212)$$

Box 7. If the convergence criterion is not satisfied, the values of V_j

from Box 3 and the temperatures from Box 5 are input to Box 2.

SC (SIMULTANEOUS CORRECTION) METHOD

A brief description of this procedure is abstracted from the fuller treatment of Henley and Seader (1981). The MESH equations (13.182)–(13.186) in terms of mol fractions are transformed into equations with molal flow rates of individual components in the liquid phase l_{ij} and vapor phase v_{ij} as the primary variables. The relations between the transformed variables are in this list:

$$L_j = \sum_{i=1}^{C} l_{ij}, \quad V_j = \sum_{i=1}^{C} v_{ij}, \quad x_{i,j} = \frac{l_{i,j}}{L_i}, \quad y_{ij} = \frac{v_{ij}}{V_i}, \quad f_{ij} = F_j z_{ij}, \quad s_j = U_j/L_j, \quad S_j = W_j/V_j. \quad (13.213)$$

The balance equations become three groups totalling $N(2C+1)$ in number:

Material balance:

$$M_{i,j} = l_{i,j}(1+s_j) + v_{ij}(1+S_j) - l_{ij-1} - v_{ij+1} - f_{ij} = 0. \quad (13.214)$$

Phase equilibria:

$$E_{i,j} = K_{ij} l_{ij} \frac{\sum_{\kappa=1}^{C} v_{\kappa j}}{\sum_{\kappa=1}^{C} l_{\kappa j}} - v_{ij} = 0. \quad (13.215)$$

Energy balance:

$$H_j = H_{L_j}(1+s_j)\sum_{i=1}^{C} l_{ij} + H_{V_j}(1+S_j)\sum_{i=1}^{C} v_{ij} - H_{L_{j-1}}\sum_{i=1}^{C} l_{ij-1} - H_{V_{j+1}}\sum_{i=1}^{C} v_{ij+1} - H_{F_j}\sum_{i=1}^{C} f_{ij} - Q_j = 0. \quad (13.216)$$

When N and all f_{ij}, P_F, P_j, s_j, S_j, and Q_i are specified, there remain $N(2C+1)$ unknowns, the same as the number of MEH equations (13.214)–(13.216). They are nonlinear equations in the primary variables l_{ij}, v_{ij}, and T_j for $i = 1$ to C and $j = 1$ to N. The T_j are involved implicitly in equations for the enthalpies and equilibrium constants.

The convergence criterion adopted is

$$\tau_3 = \sum_{j=1}^{N}\left\{(H_j)^2 + \sum_{i=1}^{C}[(M_{ij})^2 + (E_{ij})^2]\right\} \le \varepsilon_3 = N(2C+1)\left(\sum_{j=1}^{N} F_j^2\right)10^{-10}. \quad (13.217)$$

It will ensure that the converged variables will be accurate to generally at least four significant figures.

The algorithm of the procedure is in Figure 13.20.

Box 1. Initial estimates of the stage temperatures are taken from linear variations between estimated overhead dewpoint and bottoms bubblepoint temperatures. Those of the vapor rates are based on the assumption of constant molal overflow with due regard to sidestreams, and those of the liquid rates are made consistent with the material flow balances.

Box 2. With the initializations of Box 1, the matrix of the MEH equations is tridiagonal like Eq. (13.198) and may be solved for the l_{ij} and v_{ij} by the Thomas algorithm.

Box 3. Evaluate the discrepancy function made up of deviations from zero of the mass M, equilibrium E, and enthalpy H functions of Eqs. (13.214)–(13.216):

$$\tau_3 = \sum_{j=1}^{N}\left\{(H_j)^2 + \sum_{i=1}^{C}[(M_{ij})^2 + (E_{ij})^2]\right\}. \quad (13.218)$$

Box 4. The discrepancy function τ_3 is compared with the tolerance ε_3

$$\varepsilon_3 = N(2C+1)\left(\sum_{j=1}^{N} F_j^2\right)10^{-10}. \quad (13.219)$$

If $\tau_3 \le \varepsilon_3$, the process has converged and final data are evaluated in Boxes 5 and 6. If $\tau_3 > \varepsilon_3$, proceed to the next iteration by way of Box 7.

Box 5. The total flow rates are found by summing up the component flow rates

$$L_j = \sum_{i=1}^{C} l_{ij} \quad (13.220)$$

and

$$V_j = \sum_{i=1}^{C} v_{ij}. \quad (13.221)$$

Box 6. Evaluate condenser and reboiler loads by heat balances if they have not been specified.

Box 7. When $\tau_3 > \varepsilon_3$, corrections to the l_{ij}, v_{ij}, and T_j are calculated from the nonlinear MEH equations by the Newton–Raphson method. In these equations the enthalpies and equilibrium constants usually are nonlinear functions of the temperatures.

Box 8. Employ a process for evaluating the optimum fraction of a calculated correction of each variable to be applied to the next trial. That is,

$$(\Delta a)_{\text{optimum}} = t(\Delta a)_{\text{calculated}}, \quad 0 < t \le 1. \quad (13.222)$$

The selection process is described by Henley and Seader. The optimally corrected values of l_{ij}, v_{ij}, and T_j are input to Box 4 for the next iteration.

13.11. SPECIAL KINDS OF DISTILLATION PROCESSES

Conditions sometimes exist that may make separations by distillation difficult or impractical or may require special techniques. Natural products such as petroleum or products derived from vegetable or animal matter are mixtures of very many chemically unidentified substances. Thermal instability sometimes is a problem. In other cases, vapor–liquid phase equilibria are unfavorable. It is true that distillations have been practiced successfully in some natural product industries, notably petroleum, long before a scientific basis was established, but the designs based on empirical rules are being improved by modern calculation techniques. Even unfavorable vapor–liquid equilibria sometimes can be ameliorated by changes of operating conditions or by chemical additives. Still, it must be recognized that there may be superior separation techniques in some cases, for instance, crystallization, liquid–liquid extraction, supercritical extraction, foam fractionation, dialysis, reverse osmosis, membrane separation, and others. The special distillations exemplified in this section are petroleum, azeotropic, extractive, and molecular distillations,

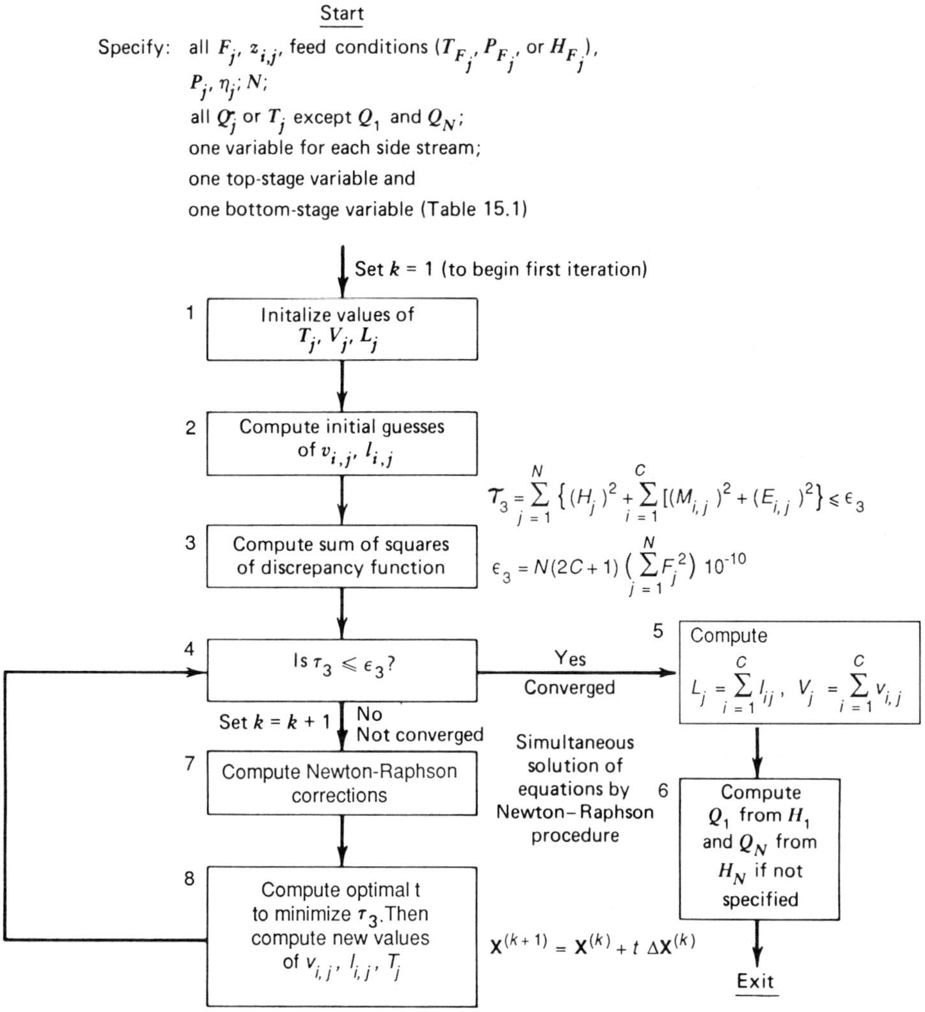

Figure 13.20. Algorithm of the SC (simultaneous correction) method for all multistage separations of fluid mixtures [*Naphthali and Sandholm,* AIChE J. **17,** *148* (*1971*); *Henley and Seader, 1981*].

PETROLEUM FRACTIONATION

Crude oils are mixtures of many substances, mostly unidentified chemically, that cover a boiling range of less than 0°F to more than 1000°F. Lower molecular weight substances are identifiable and may be recovered as pure substances, but the usual products of petroleum fractionation are mixtures with relatively narrow boiling ranges that have found consumer acceptance as final products or are suitable for further processing in the plant. On the typical refinery flow diagram of Figure 13.21, several of the processes represented as blocks either involve or are followed by distillation.

Important properties of petroleum and its fractions are measured by standardized procedures according to the API or ASTM. A particularly distinctive property is the true boiling point (TBP) curve as a function of the volume percent distilled under standardized conditions. Figure 13.19 is the TBP curve of a whole crude on which are superimposed curves of products that can be taken off sidestreams from a main distillation column, as in Figure 19.21. As samples of the distillate are collected, their densities and other properties of interest also are measured. The figure with Example 13.14 is of such measurements.

A representative petroleum fractionation process is summarized on Figure 13.22. Steam stripping of the sidestreams removes light ends and narrows the 95–5% temperature gap discussed in Example 13.14. The only source of heat supply to the column is at the feed point. A sufficient portion of the feed must be vaporized to be equivalent to the sum of all the products removed from the column above the feed point. Usually an additional amount of 2–5%, called overflash, may be needed to cover heat losses and reflux requirements. Because of the large temperature gradient and the high temperatures, the vapor volumes are large and also change greatly as the temperature falls along the column and sidestreams are withdrawn. Optimization of the size and cost of the fractionator usually requires removal of heat and provision of reflux at intermediate points rather than exclusively at the top as in most distillations, despite the need for additional trays to maintain efficient fractionation. The vapor rates at sidestream drawoffs usually are critical ones so they are checked by heat balances. Empirical rules have been developed for reflux ratios at drawoffs that ensure quality of these products.

The older, empirical practices for the design and operation of petroleum fractionators are stated in books such as that of Nelson

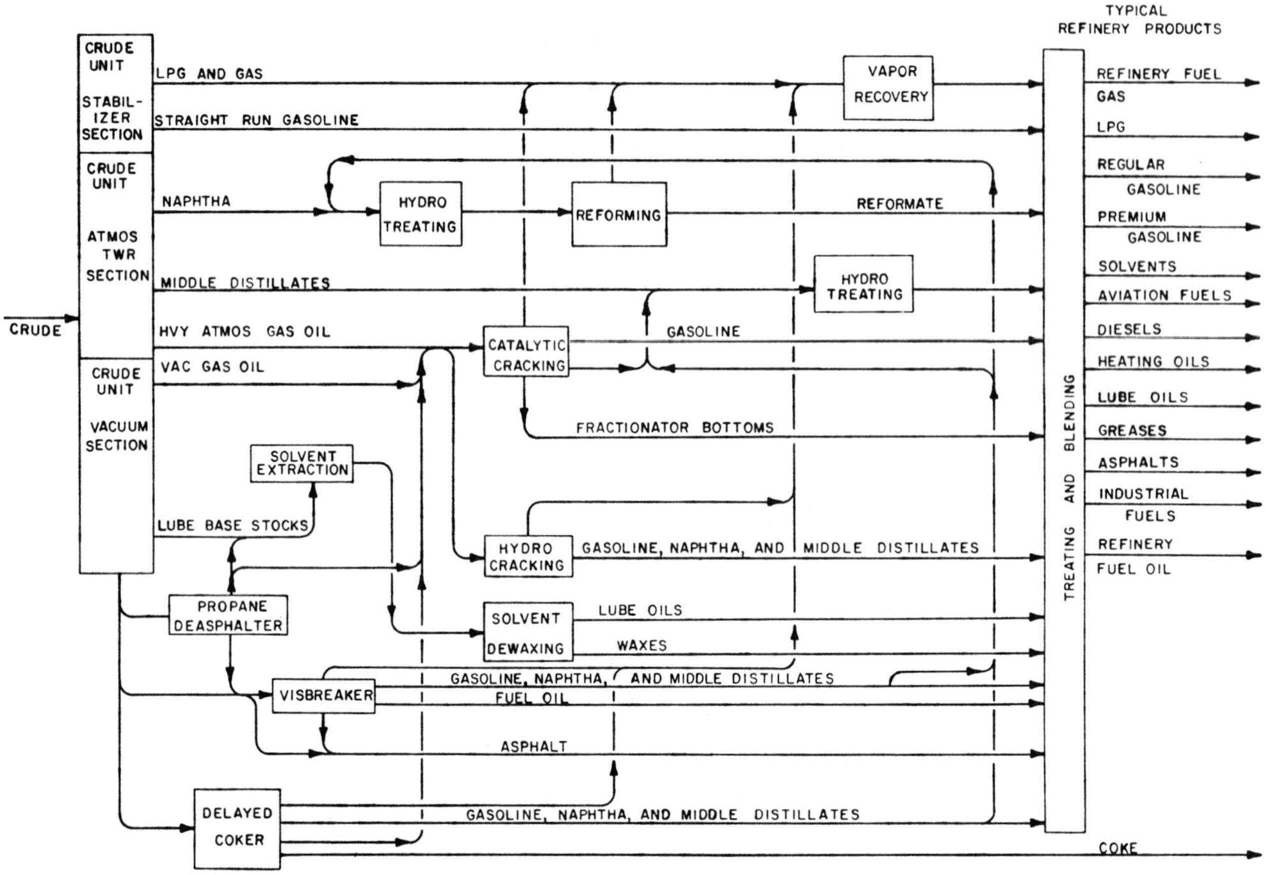

Figure 13.21. Petroleum refinery block diagram. Several of the processes identified by blocks include distillation or are followed by distillation (*Gary and Handwerk,* Petroleum Refining, *Dekker, New York, 1975*).

(*Petroleum Refinery Engineering,* McGraw-Hill, New York, 1958). Some such rules are collected in Table 13.6. A recent coverage of this subject is by Watkins (*Petroleum Refinery Distillation,* 1979), and an estimation procedure for distillations of naphthas without sidestreams is described by Broughton and Uitti [*Encycl. Chem. Process. Des.* **16,** 186–198 (1982)]. An engineer versed in these techniques can prepare a near optimum design in a few days. For the most part, nowadays, only rough estimates of tray numbers and heat balances need be made as starting estimates for eventual computer design of the process.

The basis of the fractionation design is the true boiling point curve. This is replaced by a stepped curve made up of fractions boiling over ranges of 10–25°F. The lighter components up to pentanes or hexanes are treated as such, but the other components are pseudocomponents characterized by their average boiling points, specific gravities, molecular weights, and other properties necessary to calculation of the distillation behavior. For full range crude oil fractionation, as many as 50 pseudocomponents may be required to represent the real TBP curve. In the case of naphtha fractionators without sidestreams, 20 pseudocomponents may be sufficient. Calculated compositions of products in terms of pseudocomponents can be reconstituted into smooth TBP curves to ensure that conventional specifications such as initial and final boiling points are met. The operation of converting a mixture characterized by TBP and specific gravity curves into a mixture of a discrete number of components with compositions expressed in mol fractions is performed in Example 13.14.

EXTRACTIVE DISTILLATION

In such a process an additive or solvent of low volatility is introduced in the separation of mixtures of low relative volatilities or for concentrating a mixture beyond the azeotropic point. From an extractive distillation tower, the overhead is a finished product and the bottoms is an extract which is separated down the line into a product and the additive for recycle. The key property of the additive is that it enhance the relative volatilities of the substances to be separated. From a practical point of view, the additive should be stable, of low cost, require moderate reboiler temperatures particularly for mixtures subject to polymerization or thermal degradation, effective in low to moderate concentrations, and easily recoverable from the extract. Some common additives have boiling points 50–100°C higher than those of the products.

Selection of an Additive. Ultimately the choice of an extractive distillation solvent will require a certain amount of experimental work, but some screening process should be employed to limit its scope. Examination of solvents that are being used or have been studied for successful commercial operations is a starting point. Some rules involving similarities or differences in polarities or hydrogen bonding have been proposed. The less soluble of a pair of substances usually will have the enhanced volatility. Accordingly, a comparison of solubility parameters may be a guide: A good additive should have a solubility parameter appreciably different from one of the components and closer to that of the other. Such an

Example 13.14
Representation of a Petroleum Fraction by an Equivalent Number of Discrete Components

The true boiling point and specific gravity variation with the volume percent distilled are found by standard ASTM procedures. In the present case, the smooth TBP curve is replaced by a stepped curve of eleven pseudo components characterized by their 50% boiling points and specific gravities. Their molecular weights are obtained with the general correlation of Figure (c); then the mol fractions are calculated. Vaporization equilibrium ratios and relative volatilities can be read off charts such as Figure 13.3, which are available for higher boiling ranges than this one. Then any required distillation can be calculated by any suitable standard method.

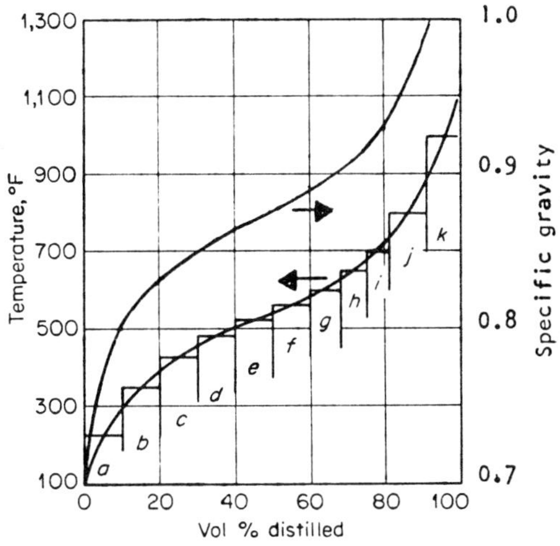

(a)

Component	Vol %	TBP 50% t, F	ρ, g/cu cm	Mol. weight	Mol. frac.
a	10	225	0.745	102	0.1730
b	10	350	0.815	141	0.1370
c	10	430	0.842	165	0.1215
d	10	485	0.860	192	0.0990
e	10	528	0.870	210	0.0987
f	10	565	0.880	227	0.0920
g	8	600	0.896	242	0.0705
h	7	650	0.913	270	0.0562
i	6	700	0.930	300	0.0434
j	10	800	0.955	353	0.0641
k	9	1000	1.030	485	0.0450
					1.0000

(b)

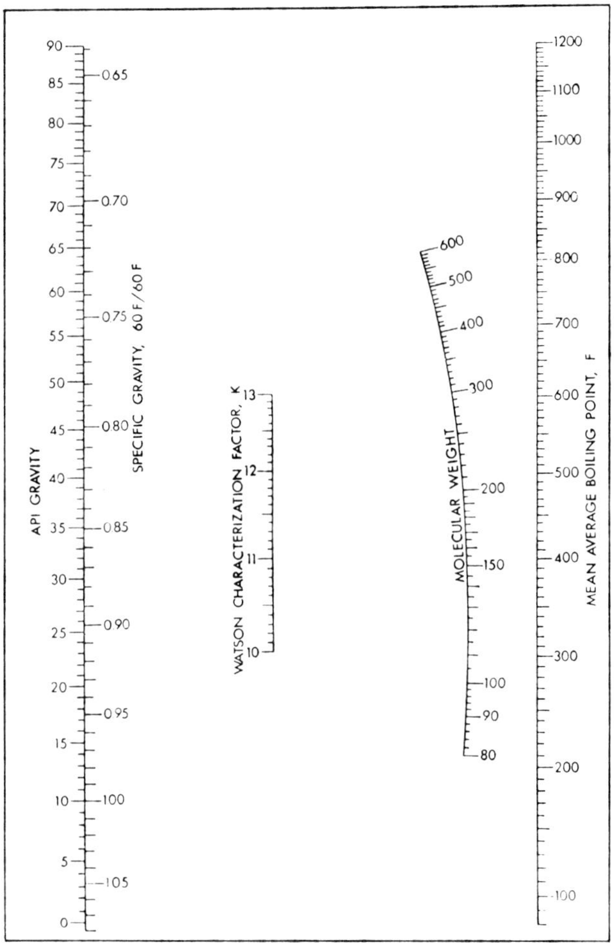

(c)

(a) Experimental true boiling point and specific gravity curves, and the equivalent stepped curve. (b) Mol fraction composition of the 11 pseudo components with equivalent vaporization behavior. (c) Standard correlations of properties of petroleum fractions.

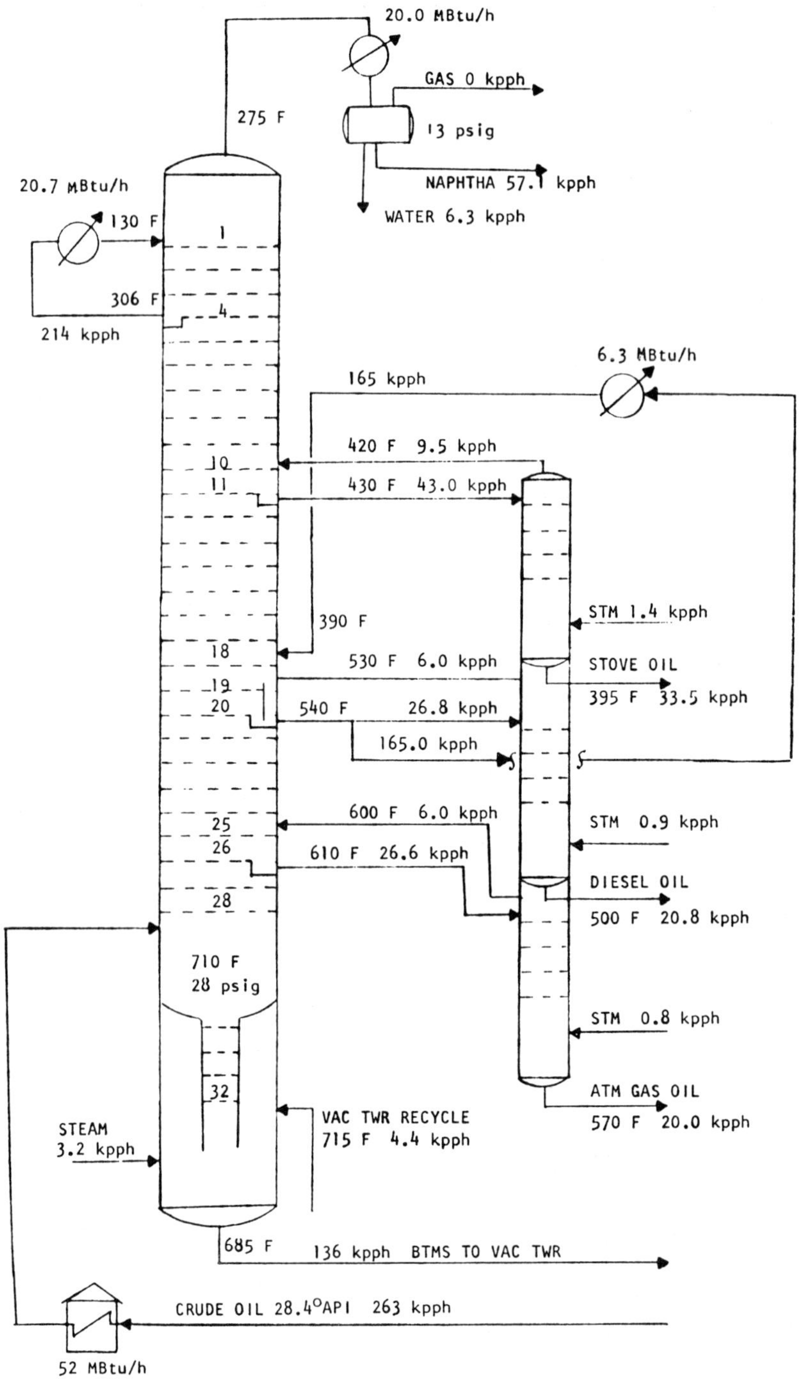

Figure 13.22. Material and energy flows in distillation of 20,000 BPSD (263,000 lb/hr) of 28.4° API crude oil into five products. The main tower is 11 ft dia by 94 ft TT, and the stripper is 3 ft dia by 54 ft TT.

explanation may be correct for the enhancement of the volatility of isooctane (7.55) relative to that of toluene (8.91) in the presence of phenol (12.1) or aniline (11.5), both of which are commercially feasible additives. The data of Figure 13.23(a) do show that the volatility of isooctane is enhanced by the presence of phenol. The numbers in parentheses are the solubility parameters. In the case of acetone (9.8), chloroform (9.3), and methyl-isobutylketone (8.3), the data of Figure 13.23(b) show that chloroform has the enhanced volatility, although its solubility parameter is closer to that of the solvent. A possible interpretation of the data is that association of the ketones as a consequence of their hydrogen bonding capabilities reduces the volatility of the acetone. Explanations of the effects of dissolved solids, as in Figures 13.23(c) and (d), are more obscure, although a substantial number of other cases also is known.

TABLE 13.6. Some Rules for Design and Operation of Petroleum Fractionators

(a) Draw Tray Temperature T_{dt} as a Function of the Bubblepoint T_{bp} of the Stream:

$$T_{dt} = \begin{cases} \exp(0.0040 T_{bp} + 4.404), & 200 \le T_{bp} \le 325°F \\ \exp(0.002952 T_{bp} + 4.744), & 325 \le T_{bp} \le 600°F \end{cases}$$

(b) Gap and Overlap of Top and Sidestream Products in Terms of Reflux and Plates

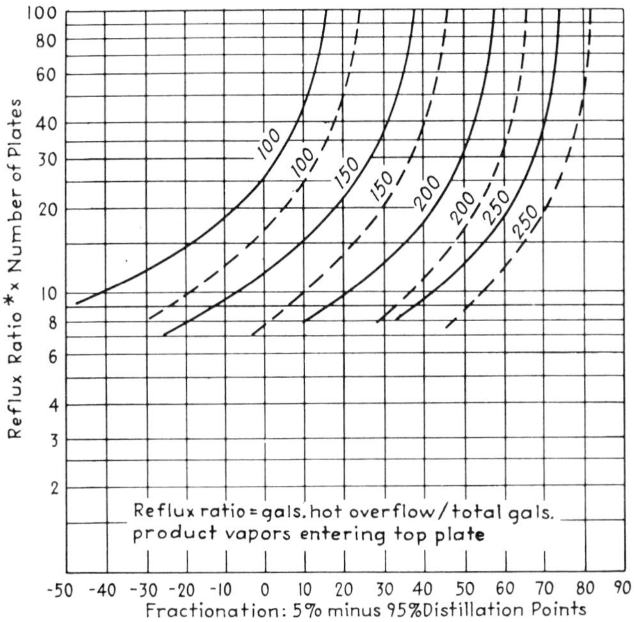

(c) Gap and Overlap between Sidestream Products in Terms of Reflux and Plates

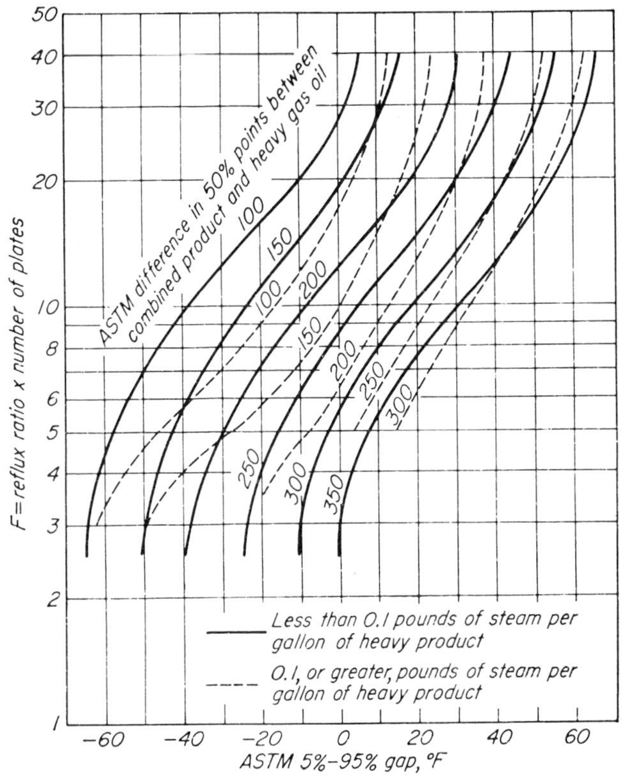

Numbers on the streams are °F differences between the 50% points of the streams. Dashed lines are with stripping steam, full ones without [Packie, *Trans. AIChE* **37,** 51 (1941)].

(d) Number of Trays between Drawoffs

Separation	Number of Trays
Light naphtha to heavy naphtha	6 to 8
Heavy naphtha to light distillate	6 to 8
Light distillate to heavy distillate	4 to 6
Heavy distillate to atmospheric gas oil	4 to 6
Flash zone to first draw tray	3 to 4
Steam and reboiled stripping sections	4

(e) Normal Stripping Steam Usage

Product	lb Steam/gal
Naphtha	0.2–0.5
Kerosene or diesel fuel	0.2–0.6
Gas oil	0.1–0.5
Neutral oils	0.4–0.9
Topped crude oil	0.4–1.2
Residual cylinder stock	1.0 up

(f) Superficial Linear Velocities in Towers

Operation	Pressure (psia or mm)	Tray Spacing (in.)	Superficial Tower Velocity (ft/sec)
Topping	17 lb	22	2.6–3.3
Cracking	40 lb	22	1.5–2.2
Pressure dist. rerun	20 lb	22	2.8–3.7[a]
Solution rerun	25 lb	22	2.8–3.5
Pressed dist. rerun	25 lb	22	2.8–3.9[a]
Pressed dist. rerun	60 mm	24	6.0–9.0
Vacuum	30 mm	30	9.0–12.0
Vaccum	90 mm	24	5.0–8.0
Stabilizer	160 lb	18	2.2–2.8
Nat. gaso. absorber	50 lb	14	1.0–1.3
Nat. gaso. absorber	400 lb	18	0.5–0.8

[a] Greatly dependent on quantity of steam.

(g) Pressure Drop 0.1–0.2 psi/tray

(h) Overflash into Tower Feed Zone is 2–5%

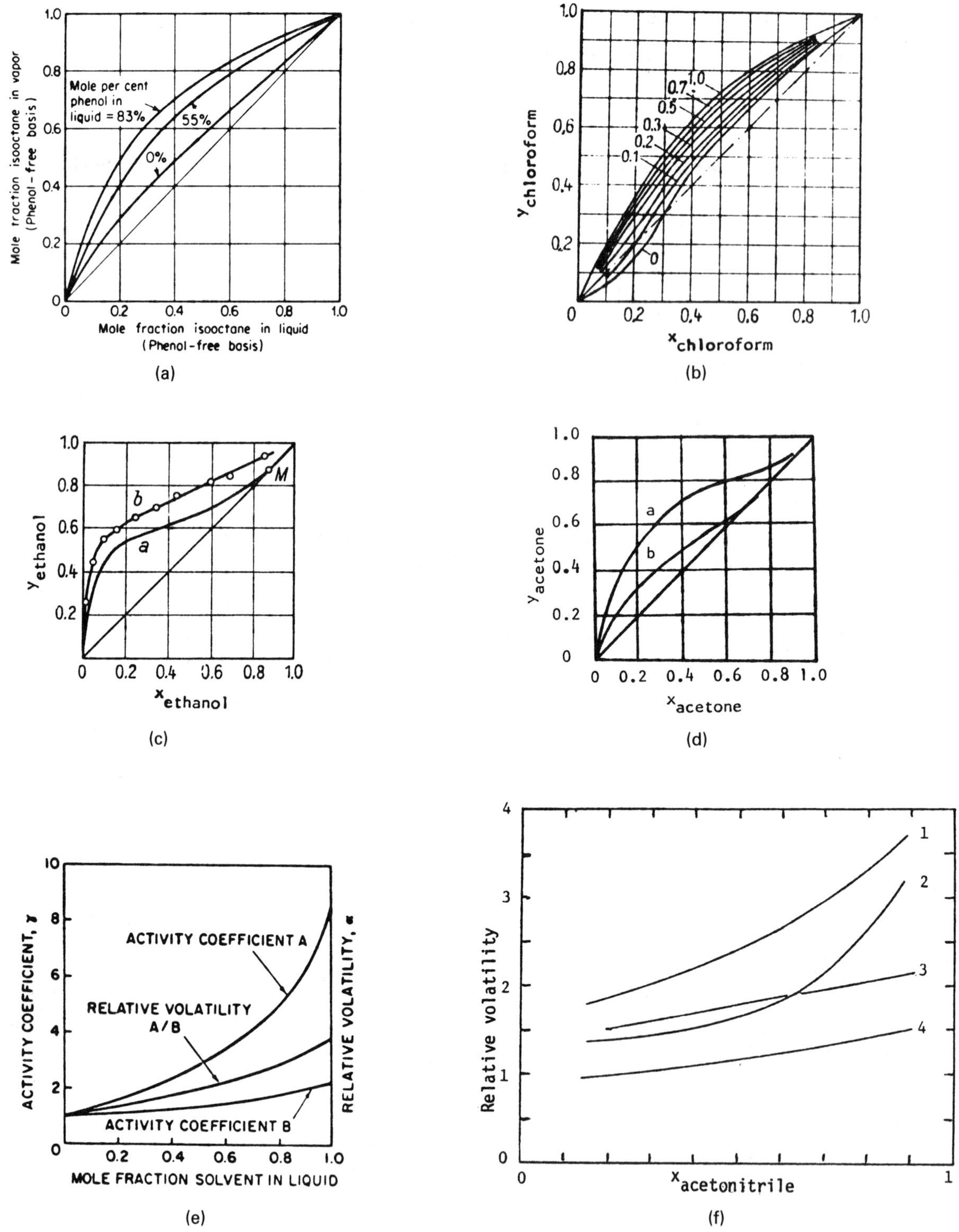

Figure 13.23. Examples of vapor–liquid equilibria in presence of solvents. (a) Mixture of *i*-octane and toluene in the presence of phenol. (b) Mixtures of chloroform and acetone in the presence of methylisobutylketone. The mole fraction of solvent is indicated. (c) Mixture of ethanol and water: (*a*) without additive; (*b*) with 10 g $CaCl_2$ in 100 mL of mix. (d) Mixture of acetone and methanol: (*a*) in 2.3*M* $CaCl_2$; (*b*) salt-free. (e) Effect of solvent concentration on the activity coefficients and relative volatility of an equimolal mixture of acetone and water (*Carlson and Stewart, in Weissbergers Technique of Organic Chemistry IV, Distillation, 1965*). (f) Relative volatilities in the presence of acetonitrile. Compositions of hydrocarbons in liquid phase on solvent-free basis: (1) 0.76 isopentane + 0.24 isoprene; (2) 0.24 iC_5 + 0.76 IP; (3) 0.5 iC_5 + 0.5 2-methylbutene-2; (4) 0.25–0.76 2MB2 + 0.75–0.24 IP [*Ogorodnikov et al.*, Zh. Prikl. Kh. **34,** *1096–1102 (1961)*].

Measurements of binary vapor–liquid equilibria can be expressed in terms of activity coefficients, and then correlated by the Wilson or other suitable equation. Data on all possible pairs of components can be combined to represent the vapor–liquid behavior of the complete mixture. For exploratory purposes, several rapid experimental techniques are applicable. For example, differential ebulliometry can obtain data for several systems in one laboratory day, from which infinite dilution activity coefficients can be calculated and then used to evaluate the parameters of correlating equations. Chromatography also is a well-developed rapid technique for vapor–liquid equilibrium measurement of extractive distillation systems. The low-boiling solvent is deposited on an inert carrier to serve as the adsorbent. The mathematics is known from which the relative volatility of a pair of substances can be calculated from the effluent trace of the elutriated stream. Some of the literature of these two techniques is cited by Walas (1985, pp. 216–217).

Some Available Data. A brief list of extractive distillation processes of actual or potential commercial value is in Table 13.7; the column of remarks explains why this mode of separation is adopted. The leading applications are to the separation of close-boiling aromatic, naphthenic, and aliphatic hydrocarbons and of olefins from diolefins such as butadiene and isoprene. Miscellaneous separations include propane from propylene with acrylonitrile as solvent (DuPont, U.S. Pat. 2,980,727) and ethanol from propanol with water as solvent [Fig. 13.24(b)].

Earlier explorations for appropriate solvents may have been conducted by the Edisonian technique of trying whatever was on the laboratory shelves. An extensive list of mixtures and the extractive distillation solvents that have been studied is in the book of Kogan (*Azeotropic and Extractive Distillation,* Leningrad, 1971, pp. 340–430, in Russian).

Some of the many solvents that have been examined for certain hydrocarbon separations are listed in Table 13.8; part (c) for *n*-butane and butene-2 separations includes data showing that addition of some water to the solvent enhances the selectivity. The diolefins butadiene and isoprene are available commercially as byproducts of cracking operations and are mixed with other close-boiling saturated, olefinic and acetylenic hydrocarbons, often as many as 10–20 different ones. The most widely used extractive solvents are *n*-methylpyrrolidone (NMP), dimethylformamide (DMF), furfural and acetonitrile (ACN), usually with 10–20% water to improve selectivity, although at the expense of reduced solvent power and the consequent need for a greater proportion of solvent. A few of the many available data for these important separations appear in Figure 13.23(f) and Table 13.9. They show the effects of hydrocarbon proportions, the content of solvent, and the concentration of water in the solvent. Sufficient data are available for the major pairs of commercial mixtures to permit evaluation of parameters of the Wilson or other equations for activity coefficients in multicomponent mixtures, and thus to place the design of the equipment on a rational basis. Another distinction between possible additives is their solvent power. Table 13.10, for example, shows that diolefins are much more soluble in DMF than in ACN, and thus DMF circulation need be less.

Calculation Methods. An often satisfactory approximation is to take the mixture in the presence of the solvent to be a pseudobinary of the keys on a solvent-free basis, and to employ the McCabe–Thiele or other binary distillation method to find tray and reflux demands. Since the relative volatility varies with concentration of the solvent, different equilibrium curves are used for above and below the feed based on average loads in those zones. Figure 13.25 is of such a construction.

When data of activity coefficients of all pairs of components are known, including those with the solvent, any of the standard calculation procedures for multicomponent distillation, which include ternaries, may be used. Composition profiles found by tray-by-tray calculations in two cases appear in Figure 13.24.

To the number of trays found by approximate methods, a few trays are added above the solvent feed point in order to wash back any volatilized solvent. Nonvolatility is a desirable property, but most otherwise suitable solvents do have appreciable volatilities.

Extractive Distillation Recovery of Isoprene. A typical flowsketch and material balance of distillation and solvent recovery towers for extracting isoprene from a mixture of cracked products with aqueous acetonitrile appears in Figure 13.26. A description of the flowsheet of a complete plant is given in Example 2.10. In spite of the fact that several trays for washing by reflux are provided, some volatilization of solvent still occurs so that the complete plant

TABLE 13.7. Examples of Extractive Distillation Processes for the Separation of Ideal, Nonideal, and Azeotropic Systems

Additive	Mixture To Be Separated	ΔT (°C)[a]	Remarks
Aniline	*n*-heptane–methylcyclohexane	2.7	ideal mixture ($\alpha = 1.07$)
	benzene–cyclohexane	0.7	azeotrope
	n-heptane–toluene	12.8	non-ideal mixtures; asymptotic approach of equilibrium curve to diagonal
Phenol	*n*-heptane–toluene	12.8	
	iso-octane–toluene	11.4	
	methylcyclohexane–toluene	9.5	
Ethyleneglycol monobutylether	methylethylketone–water	20.4	azetrope
Diethylether	ethanol–water	21.6	azeotrope
Higher ketones and alcohols	acetone–methanol	8.5	azeotrope
Higher esters and alcohols	ethylacetate–ethanol	1.3	azeotrope
Higher ketones and chloro compounds	acetone–chloroform	5.0	azetrope

[a] ΔT is the difference in atmospheric boiling points, °C.

TABLE 13.8. Relative Volatilities of Three Binary Systems and Their Enhancement in the Presence of Several Solvents

(a) *n*-Heptane/Methylcyclohexane with Relative Volatility of 1.07

Solvent	Mole per cent in liquid phase	T, °C. (av.)	Av. rel. volatility, α_S	Improvement factor, α_S/α	Ref. No.
Aniline	92	139	1.52	1.42	1
	78	121	1.40	1.31	1
	70	110	1.27	1.19	2
	58	113	1.26	1.18	1
Furfural	79	—	1.35	1.26	1
Phenol	81	—	1.31	1.24	1
Nitrobenzene	82	—	1.31	1.24	1
Dichlorodiethyl ether	81	—	1.28	1.20	1
Aminocyclohexane	76	—	1.16	1.08	1
Pyridine	70	—	1.4	1.31	2
Ethanol	70	—	1.3	1.21	2
n-Butanol	70	—	1.3	1.21	2
tert-Butanol	70	—	1.25	1.17	2
Acetic acid	70	—	1.27	1.19	2
None	—	—	1.07	1.00	2

[1] Griswold, Andres, Van Berg, and Kasch, *Ind. Eng. Chem.*, **38**, 66 (1946).
[2] Fenske, Carlson, and Quiggle, *Ind. Eng. Chem.*, **39**, 1322 (1947).

(b) Cyclohexane/Benzene with Relative Volatility of 1.02

Solvent	Mole per cent in charge	T, °C.	Relative volatility, α_S	Improvement factor, α_S/α
Acetic acid	69.0	84	1.75	1.78
Methanol	67.3	53	1.58	1.61
Ethanol	67.3	65	1.36	1.38
n-Propanol	70.5	79	1.26	1.28
Isopropanol	67.9	70	1.22	1.24
Dioxane	67.4	86	1.75	1.78
Chlorex (dichlorodiethyl ether)	67.5	105	2.31	2.36
Methyl Cellosolve	66.7	85	1.84	1.88
Cellosolve	67.5	95	1.58	1.61
Carbitol	66.8	87	1.99	2.03
Acetone	66.3	55	2.03	2.07
Methyl ethyl ketone	65.1	72	1.78	1.81
Diacetone	67.3	89	1.82	1.85
Pyridine	66.9	93	1.83	1.86
Aniline	66.8	93	2.11	2.16
Nitromethane	67.8	74	3.00	3.06
Nitrobenzene	68.2	102	2.25	2.30
Acetonitrile	67.3	65	2.85	2.92
Furfural	67.1	79	3.10	3.16
Phenol	66.8	92	2.01	2.05

Updike, Langdon, and Keyes, *Trans. Am. Inst. Chem. Engrs.*, **41**, 717 (1945).

TABLE 13.8—*(continued)*

(c) Butane/2-Butene with Relative Volatility of about 1.08[a] The asterisks denote that data are included for both dry and wet solvents

Solvent	Vol/vol HC	Temperature, °F	$\alpha = \gamma_{C_4}/\gamma_{C_4=}$
Hydroxyethylacetate	4	133	1.54
Methylsalicylate	2	156	1.46
Dimethylphthalate	2	142	1.41
Ethyl oxalate	2	154	1.38
Carbitolacetate	2	160	1.35
Diethyl carbonate	2	172	1.28
Amylacetate	2	180	1.21
* Acetonitrile	2	137	1.49
Butyronitrile	2	161	1.42
Acrylonitrile	2	156	1.23
Acetonyl acetone	2	141	1.43
Cyclohexanone	2	171	1.32
Acetophorone	2	145	1.31
Methylhexyl ketone	2	166	1.27
Methylamyl ketone	2	173	1.23
Methylisobutyl ketone	2	171	1.23
Methyldiisobutyl ketone	2	177	1.18
Nitromethane	1.8	134	1.60
Nitroethane	2	146	1.46
1-Chloro-1-nitropropane	2	155	1.46
Nitrobenzene	2	150	1.41
o-Nitrotoluene	2	155	1.38
o-Nitroanisole	2	130	1.30
n-Formylmorpholine	4.6	133	1.60
Morpholine	2	160	1.41
Pyridine	2	176	1.35
Quinoline	2	148	1.33
Picoline	2	188	1.29
Benzyl alcohol	3	144	1.48
* Phenol	2	138	1.47
Diacetone alcohol	2	146	1.32
Butyl alcohol	2	152	1.21
2-Ethyl butyl alcohol	2	161	1.20
o-Hexanol	2	159	1.18
tert-Butyl alcohol	2	154	1.16
Benzaldehyde	2	145	1.42
* Furfural	3	158	1.40
3,4-Diethoxybenzaldehyde	2	125	1.11
Butyraldehyde	2	165	1.09
* Aniline	3	130	1.65
o-Chloroaniline	2	152	1.44
Methylaniline	2	146	1.42
o-Toluidine	2	148	1.38
Dimethyl aniline	2	169	1.37
n-Tributyl amine	2	176	1.09
Cellosolve	2	152	1.40
Dichloroethylether	2	152	1.39
Anisole	2	175	1.28
Butyl Cellosolve	2	163	1.24
Diethyl Cellosolve	2	179	1.23
Diethyl carbitol	2	173	1.23
n-Butylether	2	197	1.10
Solvents with water			
* Furfural, 96 wt %	3.7	128	1.78
* Aniline, 96.5 wt %	4.4	132	1.77
Methylacetoacetate (90 vol %)	3.6	134	1.67
* Phenol (90 vol %)	2.5	133	1.66
Acetonylacetone (95 vol %)	3	128	1.58
* Acetonitrile (90 vol %)	4	133	1.58
Benzyl alcohol (95 vol %)	2.5	133	1.51
o-Chlorophenol (90 vol %)	5.0	180	1.50

[Data from Hess, Narragon, and Coghlin, *Chem. Eng. Prog. Symp. Ser.* **2**, 72–96 (1952)].

TABLE 13.9. Relative Volatilities of C_4 and C_5 Hydrocarbons in Various Solvents
(a) Volatilities of Butenes Relative to Butadiene at 40°C

Mole Fraction Hydrocarbons in Solution	DMF with 12% H_2O			Acetonitrile					
				Dry			13% H_2O		
	1-Butene	*trans*-2-Butene	*cis*-2-Butene	1-Butene	*trans*-2-Butene	*cis*-2-Butene	1-Butene	*trans*-2-Butene	*cis*-2-Butene
0.05	2.10	1.66	1.6	—	—	—	1.86	1.47	1.41
0.10	1.90	1.62	1.43	1.81	1.44	1.37	1.73	1.38	1.31
0.15	2.15	1.65	1.40	1.83	1.50	1.35	1.73	1.43	1.20

[Galata, Kofman, and Matveeva, *Chem. Chem. Tech.* (in Russian) **2,** 242–255 (1962)].

(b) Volatilities at Infinite Dilution and 20°C Relative to Butadiene

Compound	Solvent				
	None	Furfural	DMF	NMP	ACN
n-Butane	0.88	3.0	3.04	3.84	3.41
Isobutylene	1.03	2.03	2.00	2.45	2.20
Butene-1	1.01	1.97	1.95	2.44	2.16
t-Butene-2	0.86	1.42	1.54	2.02	1.70
c-Butene-2	—	1.29	1.40	1.76	1.56

(Evans and Sarno, Shell Development Co.).

(c) Enhancement of Relative Volatilities of C_4 and C_5 Hydrocarbons

Class and Name of Hydrocarbons	No. Solvent α_0	α/α_0 in Presence of Solvent		
		Furfural	DMF	ACN
Alkanes/alkenes				
n-Butane/1-butene	0.83	1.53	1.83	1.70
i-Butene/1-butene	1.14	1.52	1.83	1.70
i-Pentane/2-methyl-1-butene	1.08	1.66	1.85	—
n-Pentane/2-methyl-1-butene	0.84	1.61	1.80	—
Alkanes/dienes				
1-Butene/butadiene	1.04	1.57	1.91	1.69
Trans-2-butene/butadiene	0.83	1.44	1.85	1.71
cis-2-butene/butadiene	0.76	1.41	1.84	1.71
2-Methyl-2-butene/isoprene	0.87	1.45	1.84	—
Isoprene/3-methyl-1-butene	0.88	—	1.37	—

(Galata et al., loc cit.]

TABLE 13.10. Solubilities (wt %) of Classes of C_4 and C_5 Hydrocarbons in Various Solvents

Compounds	DMF with		ACN with		Furfural with
	8% Water	12% Water	8% Water	12% Water	4% Water
Saturated	7	5	14	9	7
Olefins	18	13	32	22	18
Diolefins	58	41	32	22	24
Acetylenes	69	50	—	—	—

[Galata, Kofman, and Matveeva, *Chem. Chem. Tech.* (in Russian) **2,** 242–255 (1962)].

also has water wash columns on both hydrocarbon product streams. A further complication is that acetonitrile and water form an azeotrope containing about 69 mol % solvent. Excess water enters the process in the form of a solution to control polymerization of the unsaturates in the hotter parts of the towers and reboilers.

Two feasible methods for removal of as much water as desired from the azeotrope are depicted on Figure 13.27. The dual pressure process takes advantage of the fact that the azeotropic composition is shifted by change of pressure; operations at 100 and 760 Torr result in the desired concentration of the mixture. In the other method, trichlorethylene serves as an entrainer for the water. A ternary azeotrope is formed that separates into two phases upon condensation. The aqueous layer is rejected, and the solvent layer is recycled to the tower. For economic reasons, some processing beyond that shown will be necessary since the aqueous layer contains some acetonitrile that is worth recovering or may be regarded as a pollutant.

AZEOTROPIC DISTILLATION

The objective of azeotropic distillation is the separation or concentration beyond the azeotropic point of mixtures with the aid of an entrainer to carry some of the components overhead in a column. An azeotrope is a constant boiling mixture with vapor and liquid phases of the same composition. A related class of systems is that of partially miscible liquids that also boil at constant temperature. The two phases exert their individual vapor pressures so that the boiling temperature and vapor composition remain constant over the full range of immiscibility, but the compositions of vapor and overall liquid phases are only accidentally the same. In most cases of immiscible liquids, the horizontal portion of an x–y diagram crosses the $x = y$ line, for instance, the system of n-butanol and water of Example 13.5. The system of methylethylketone and water is one of the few known exceptions for which the immiscible boiling range does not cross the $x = y$ line of Figure 13.28(b). Artificial systems can be constructed with this behavior. Thus Figures 13.28(c) and (d) are of diagrams synthesized with two different sets of parameters of the Margules equation for the activity coefficients; one of the x–y lines crosses the diagonal and the other does not. Figure 13.28(a) for acetone and water is representative of the most common kind of homogeneous azeotropic behavior.

The overhead stream of the distillation column may be a low-boiling binary azeotrope of one of the keys with the entrainer or more often a ternary azeotrope containing both keys. The latter kind of operation is feasible only if condensation results in two liquid phases, one of which contains the bulk of one of the key components and the other contains virtually all of the entrainer which can be returned to the column. Figure 13.29(a) is of such a flow scheme. When the separation resulting from the phase split is not complete, some further processing may make the operation technically as well as economically feasible.

Data of Azeotropes. The choice of azeotropic entrainer for a desired separation is much more restricted than that of solvents for extractive distillation, although many azeotropic data are known. The most extensive compilation is that of Ogorodnikov, Lesteva, and Kogan (*Handbook of Azeotropic Mixtures* (in Russian), 1971). It contains data of 21,069 systems, of which 1274 are ternary, 60 multicomponent, and the rest binary. Another compilation (*Handbook of Chemistry and Physics,* 60th ed., CRC Press, Boca Raton, FL, 1979) has data of 685 binary and 119 ternary azeotropes. Shorter lists with grouping according to the major substances also are available in *Lange's Handbook of Chemistry* (12th ed., McGraw-Hill, New York, 1979). Data of some ternary systems are in Table 13.11.

Commercial Examples. The small but often undesirable contents of water dissolved in hydrocarbons may be removed by distillation. In drying benzene, for instance, the water is removed overhead in the azeotrope, and the residual benzene becomes dry enough for processing such as chlorination for which the presence of water is harmful. The benzene phase from the condenser is refluxed to the tower. Water can be removed from heavy liquids by addition of some light hydrocarbon which then is cooked out of the liquid as an azeotrope containing the water content of the original heavy liquid. Such a scheme also is applicable to the breaking of aqueous emulsions in crude oils from tar sands. After the water is removed

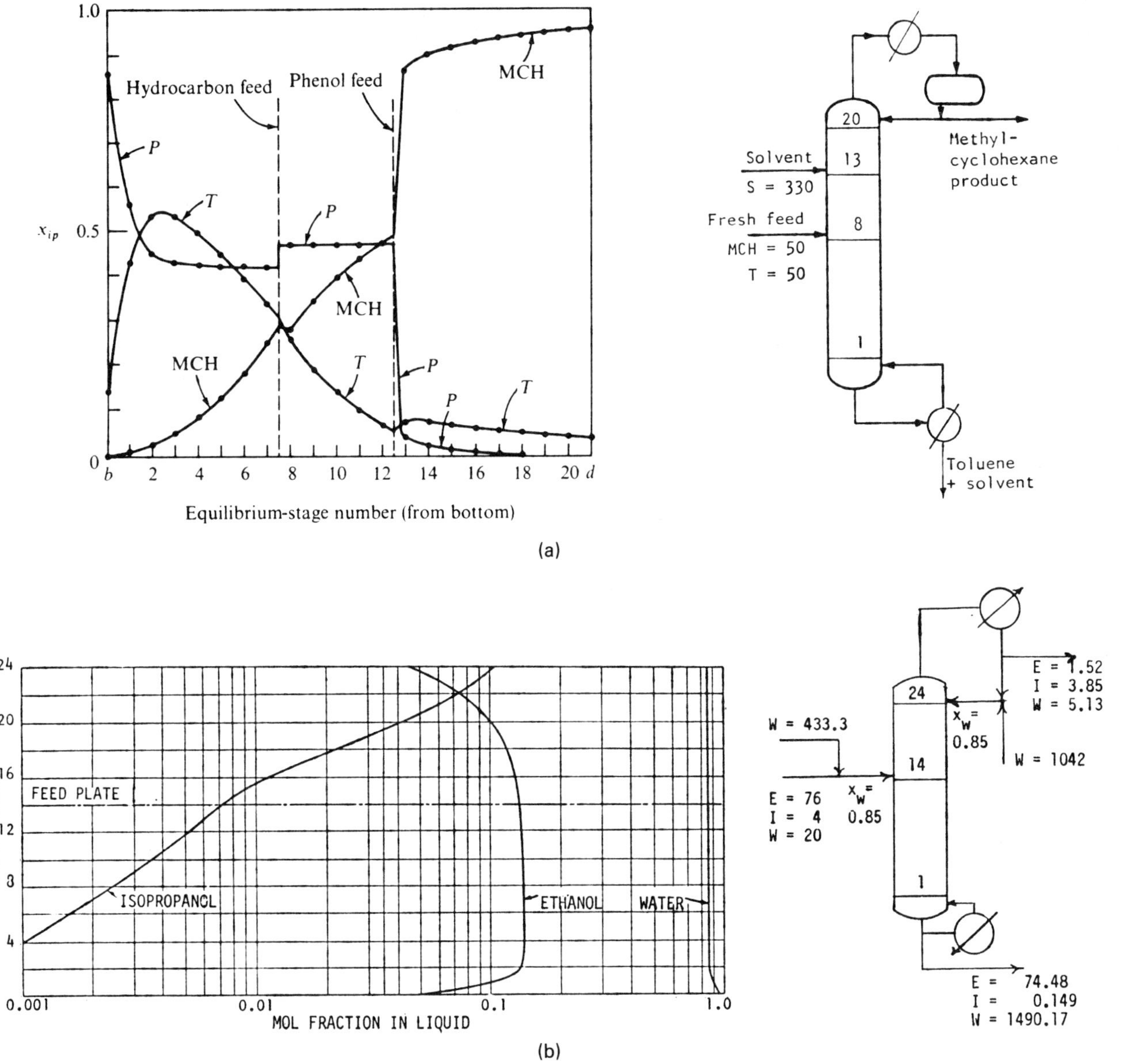

Figure 13.24. Composition profiles and flowsketches of two extractive distillation processes. (a) Separation of methylcyclohexane and toluene with phenol as solvent (*data calculated by Smith, 1963*). (b) Separation of aqueous ethanol and isopropanol, recovering 98% of the ethanol containing 0.2 mol % isopropanol, employing water as the solvent. Flow rates are in mols/hr (*data calculated by Robinson and Gilliland, 1950*).

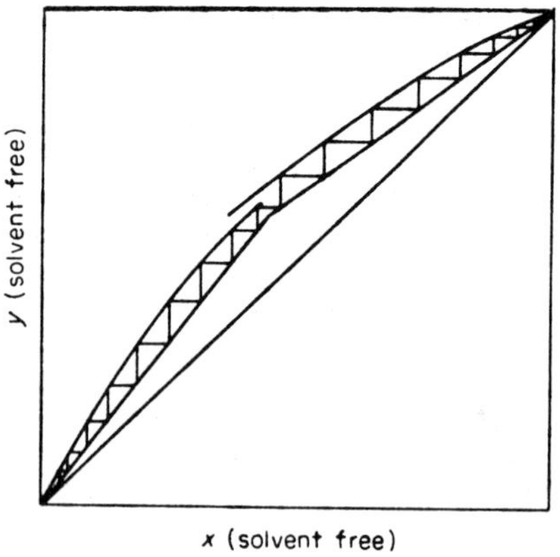

Figure 13.25. Illustrating McCabe–Thiele construction of pseudobinary extractive distillation with smaller relative volatility below the feed plate.

azeotropically, solids originally dissolved or entrained in the aqueous phase settle out readily from the dry hydrocarbon phase. Even in the evaporation of water from caustic, the addition of kerosene facilitates the removal of water by reducing the temperature to which the pot must be heated.

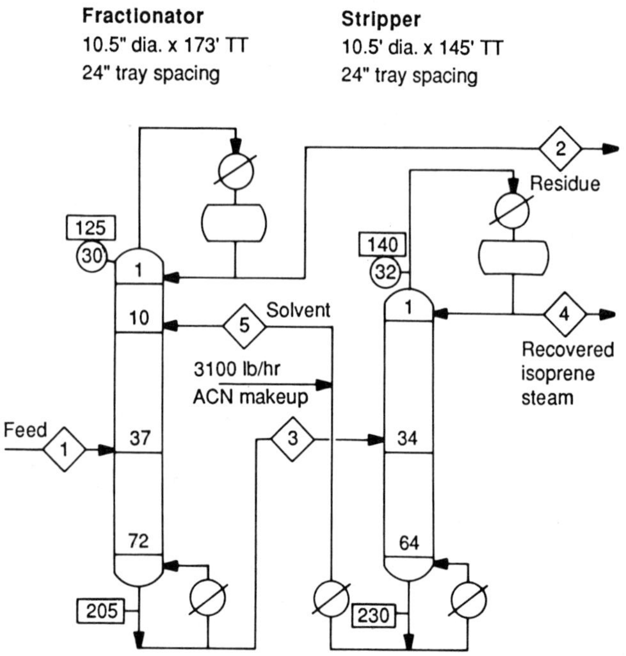

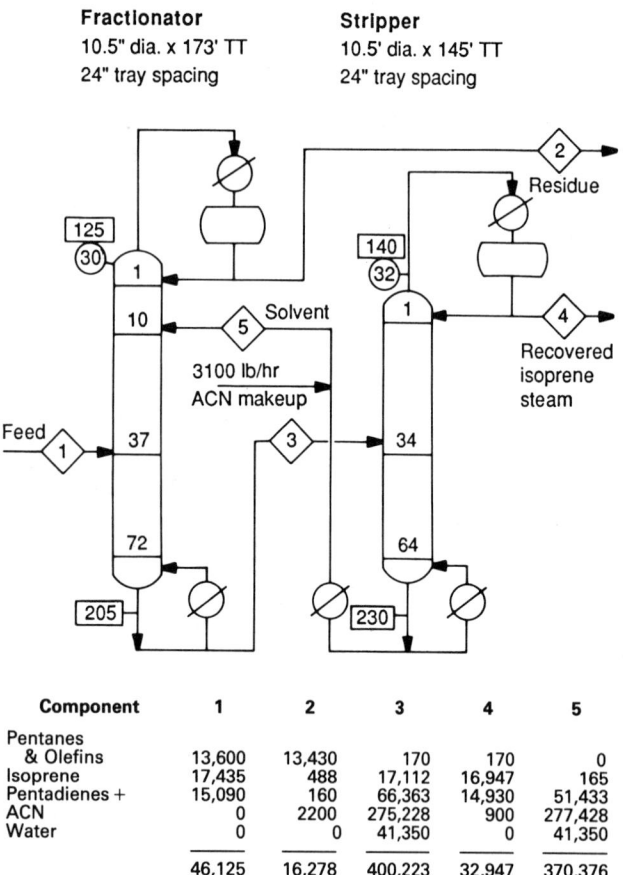

Component	1	2	3	4	5
Pentanes & Olefins	13,600	13,430	170	170	0
Isoprene	17,435	488	17,112	16,947	165
Pentadienes +	15,090	160	66,363	14,930	51,433
ACN	0	2200	275,228	900	277,428
Water	0	0	41,350	0	41,350
	46,125	16,278	400,223	32,947	370,376

Figure 13.26. Flowsketch for the recovery of isoprene from a mixture of C_5s with aqueous acetonitrile. Flow quantities in lb/hr, pressures in psia, and temperatures in °F. Conditions are approximate. (Data of The C. W. Nofsinger Co.)

Ordinary rectification for the dehydration of acetic acid requires many trays if the losses of acid overhead are to be restricted, so that azeotropic processes are used exclusively. Among the entrainers that have been found effective are ethylene dichloride, *n*-propyl acetate, and *n*-butyl acetate. Water contents of these azeotropes are 8, 14, and 28.7 wt %, respectively. Accordingly, the *n*-butyl acetate is the most thermally efficient of these agents. The *n*-propyl acetate has been used in large installations, in the first stage as solvent for extraction of acetic acid and then as azeotropic entrainer to remove the accompanying

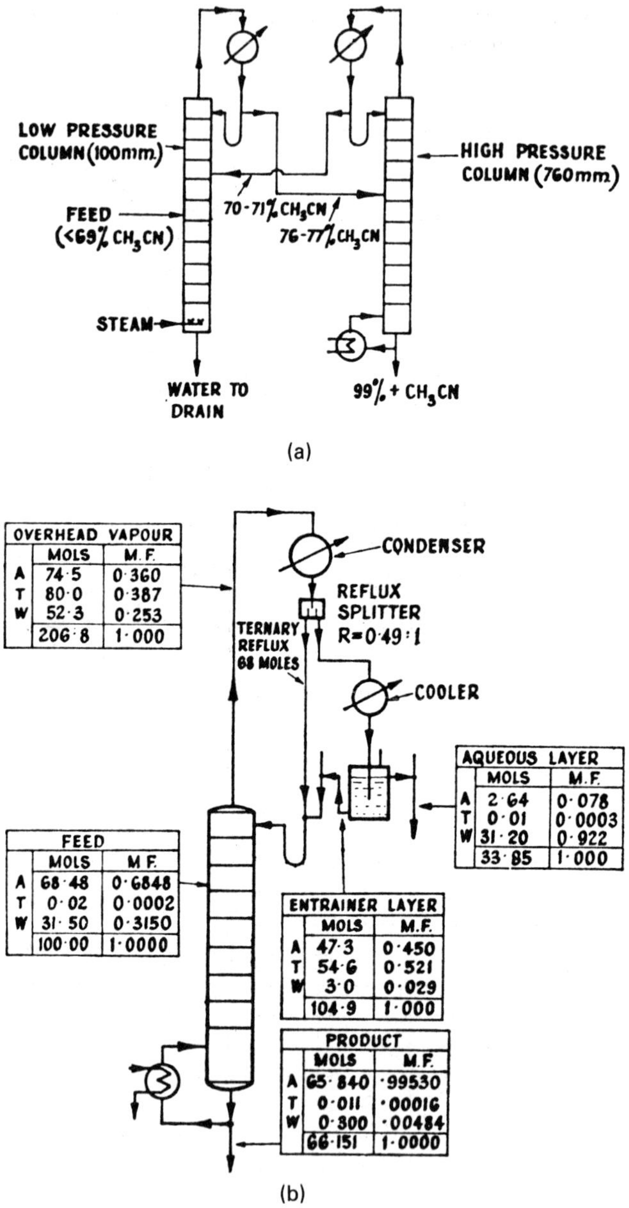

Figure 13.27. Separation of the azeotropic mixture of acetonitrile and water which contains approximately 69 mol % or 79.3 wt % of acetonitrile. (*Pratt,* Countercurrent Separation Processes, *Elsevier, New York, 1967, pp. 194, 497*). (a) A dual pressure process with the first column at 100 Torr and the second at 760 Torr. (b) Process employing trichlorethylene as entrainer which carries over the water in a ternary azeotrope that in turn separates into two phases upon condensation.

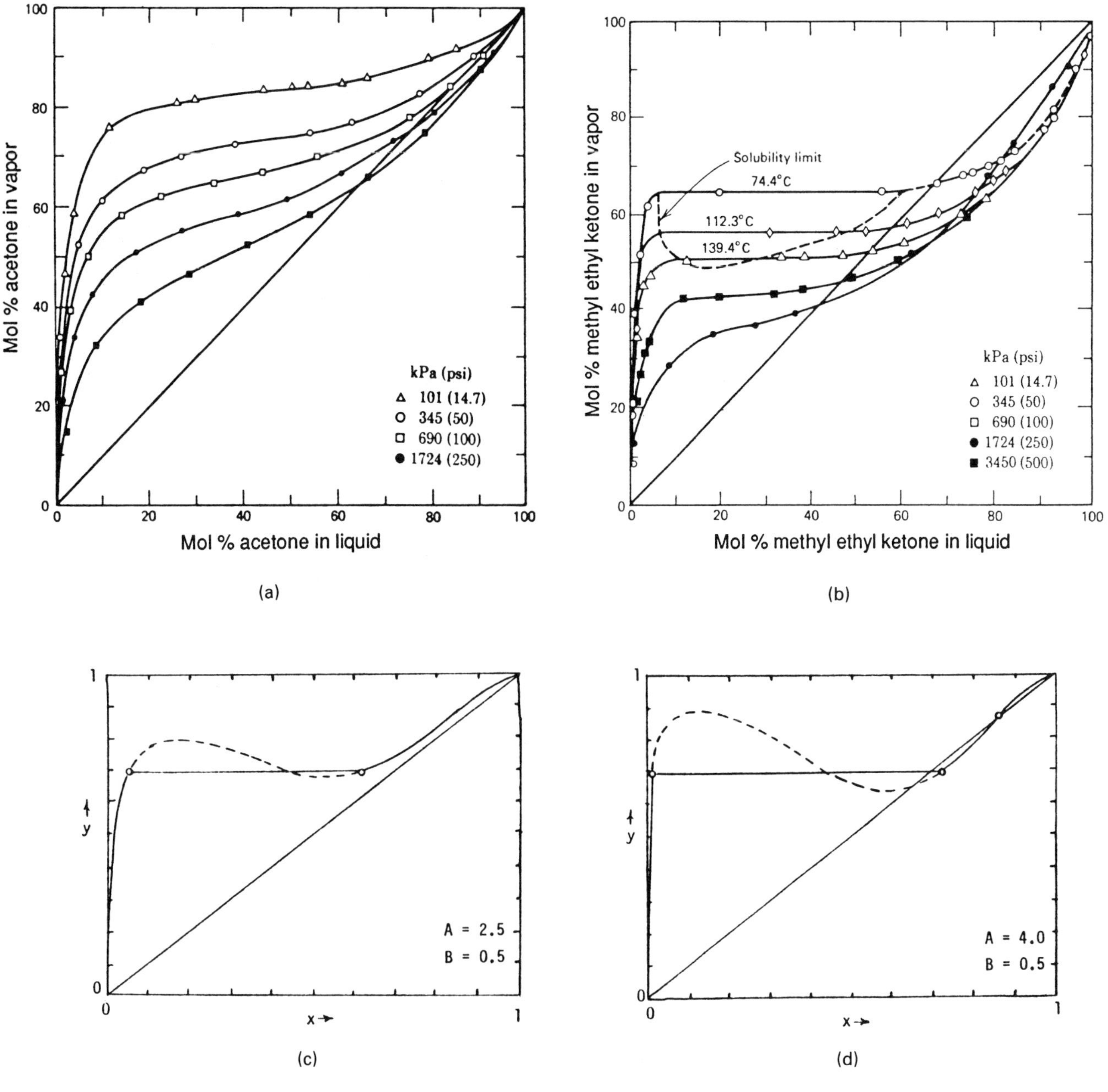

Figure 13.28. Vapor–liquid equilibria of some azeotropic and partially miscible liquids. (a) Effect of pressure on vapor–liquid equilibria of a typical homogeneous azeotropic mixture, acetone and water. (b) Uncommon behavior of the partially miscible system of methylethylketone and water whose two-phase boundary does not extend byond the $y = x$ line. (c) x–y diagram of a partially miscible system represented by the Margules equation with the given parameters and vapor pressures $P_1^0 = 3$, $P_2^0 = 1$ atm; the broken line is not physically significant but is represented by the equation. (d) The same as (c) but with different values of the parameters; here the two-phase boundary extends beyond the $y = x$ line.

water. Extractive distillation with a high boiling solvent that is immiscible with water upon condensation is technically feasible for acetic acid drying but is a more expensive process.

Ethanol forms an azeotrope containing 5 wt % water. In older installations, dissolved salts were employed to break the azeotrope. Typical data are in Figure 13.28(c). Several substances form ternary azeotropes with ethanol and water, including benzene, gasoline, and trichlorethylene. The first is not satisfactory because of slight decomposition under distillation conditions. A flowsketch of a process employing benzene is in Figure 13.29(a). In a modernization of the benzene process (Raphael Katzen Associates, Cincinnati, OH), a high purity ethanol is made by controlling the distillation so that the lower 10 trays or so are free of benzene. Another entrainer, diethyl ether, has the desirable property of forming an azeotrope with water but not with ethanol. The water content of the azeotrope is small so that the operation is conducted at 8 atm to shift the composition to a higher value of 3% water. In small installations, drying with molecular sieves is a competitive

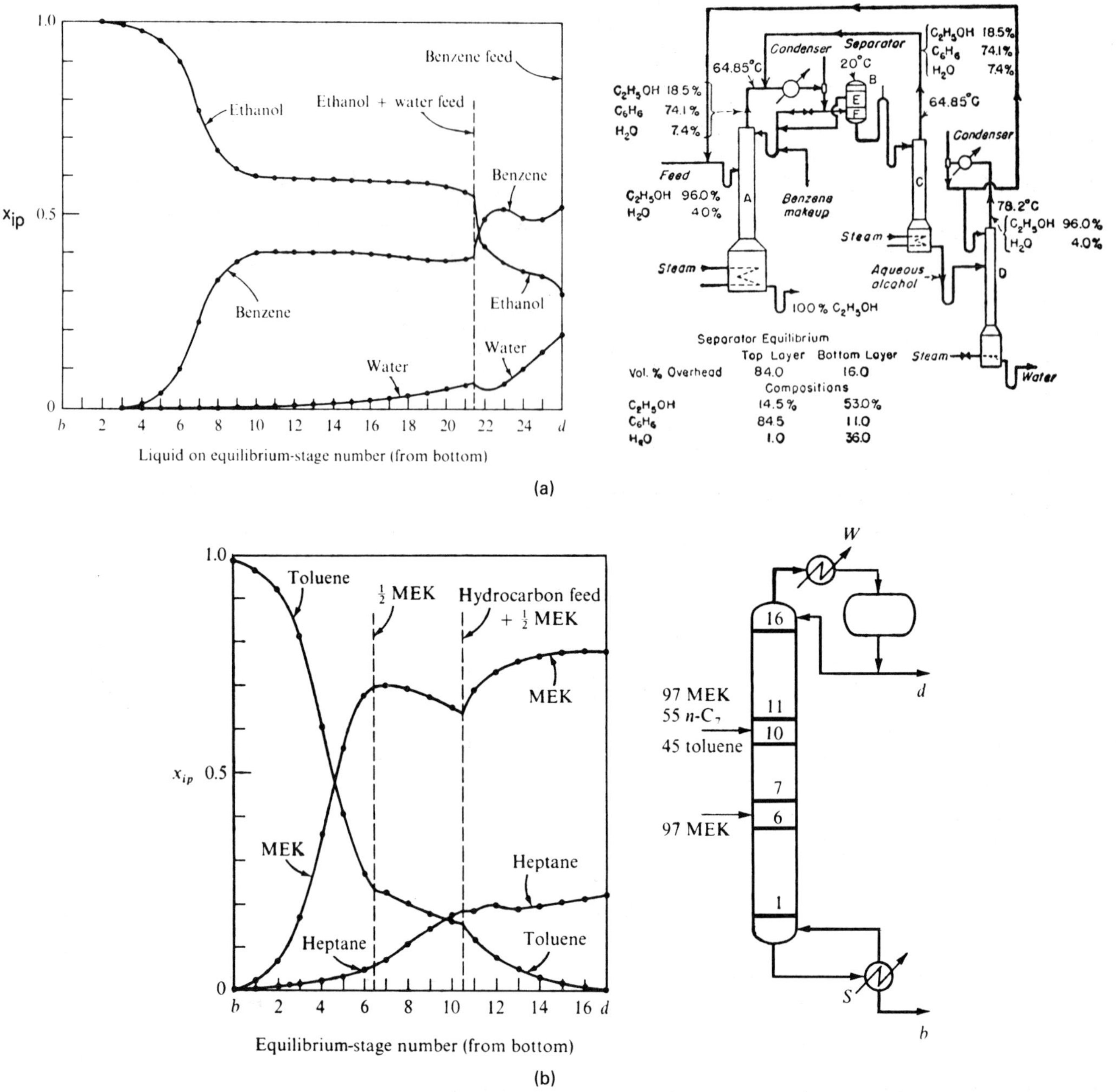

Figure 13.29. Composition profiles and flowsketches of two azeotropic distillation processes (*adapted by King, 1980*). (a) Separation of ethanol and water with benzene as entrainer. Data of the composition profiles in the first column were calculated by Robinson and Gilliland, (1950); the flowsketch is after Zdonik and Woodfield (*in* Chemical Engineers Handbook, *McGraw-Hill, New York, 1950, p. 652*). (b) Separation of n-heptane and toluene with methylethylketone entrainer which is introduced in this case at two points in the column (*data calculated by Smith, 1963*).

process. Separations with membranes of both vapor and liquid phases, supercritical extraction with carbon dioxide and many other techniques have been proposed for removal of water from ethanol.

Formic acid can be dehydrated with propyl formate as entrainer. Small contents of formic acid and water in acetic acid can be entrained away with chloroform which forms binary azeotropes with water and formic acid but no other azeotropes in this system.

Some hydrocarbon separations can be effected azeotropically. Figure 13.29(b) shows an operation with methylethylketone which entrains n-heptane away from toluene. Hexane in turn is an effective entrainer for the purification of methylethylketone by distilling the latter away from certain oxide impurities that arise during the synthesis process.

Design. When the vapor–liquid equilibria are known, in the form of UNIQUAC parameters for instance, the calculation of azeotropic distillation may be accomplished with any of the standard multicomponent distillation procedures. The Naphthali–

Sandholm algorithm (Fig. 13.20) and the θ-method of Holland (1981) are satisfactory. Another tray-by-tray algorithm is illustrated for azeotropic distillation by Black, Golding, and Ditsler [*Adv. Chem. Ser.* **115,** 64 (1972)]. A procedure coupling the tower, decanter, and stripper of Figure 13.29(a) is due to Prokopakis and Seider [*AIChE J.* **29,** 49 (1983)]. Two sets of composition profiles obtained by tray-by-tray calculations appear in Figures 13.29(a) and (b).

MOLECULAR DISTILLATION

This process is an evaporation that is conducted at such low pressures that the distance between the hot and condensing surfaces is less than the mean free path of the molecules. Each unit is a single stage, but several units in series are commonly employed. Molecular distillation is applied to thermally sensitive high molecular weight materials in the range of 250–1200 molecular

TABLE 13.11. Selected Ternary Aezotropic Systems at Atmospheric Pressure

(a) Systems with Water and Alcohols

	B.P. 760 mm.		% By weight		
	Other component	Azeotrope	Water	Alcohol	Other component
A—Ethyl Alcohol **(B.P. 78.3°)**					
Ethyl acetate (6)	77.1	70.3	7.8	9.0	83.2
Diethyl formal (7)	87.5	73.2	12.1	18.4	69.5
Diethyl acetal (8)	103.6	77.8	11.4	27.6	61.0
Cyclohexane	80.8	62.1	7	17	76
Benzene (4)	80.2	64.9	7.4	18.5	74.1
Chloroform	61.2	55.5	3.5	4.0	92.5
Carbon tetrachloride	76.8	61.8	4.3	9.7	86.0
Ethyl iodide	72.3	61	5	9	86
Ethylene chloride	83.7	66.7	5	17	78
B—n-Propyl Alcohol **(B.P. 97.2°)**					
n-Propyl formate (6)	80.9	70.8	13	5	82
n-Propyl acetate (6)	101.6	82.2	21.0	19.5	59.5
Di-*n*-propyl formal (7)	137.4	86.4	8.0	44.8	47.2
Di-*n*-propyl acetal (8)	147.7	87.6	27.4	51.6	21.0
Di-*n*-propyl ether (7)	91.0	74.8	11.7	20.2	68.1
Cyclohexane	80.8	66.6	8.5	10.0	81.5
Benzene (4)	80.2	68.5	8.6	9.0	82.4
Carbon tetrachloride	76.8	65.4	5	11	84
Diethyl ketone	102.2	81.2	20	20	60
C—Isopropyl Alcohol (B.P. 82.5°)					
Cyclohexane	80.8	64.3	7.5	18.5	74.0
Benzene (4)	80.2	66.5	7.5	18.7	73.8
D—n-Butyl Alcohol **(B.P. 117.8°)**					
n-Butyl formate (6)	106.6	83.6	21.3	10.0	68.7
n-Butyl acetate (6)	126.2	89.4	37.3	27.4	35.3
Di-*n*-butyl ether (7)	141.9	91	29.3	42.9	27.7
E—Isobutyl Alcohol (B.P. 108.0°)					
Isobutyl formate (6)	94.4	80.2	17.3	6.7	76.0
Isobutyl acetate	117.2	86.8	30.4	23.1	46.5
F—tert-Butyl Alcohol **(B.P. 82.6°)**					
Benzene (4)	80.2	67.3	8.1	21.4	70.5
Carbon tetrachloride (9)	76.8	64.7	3.1	11.9	85.0
G—n-Amyl Alcohol **(B.P. 137.8°)**					
n-Amyl formate (6)	131.0	91.4	37.6	21.2	41.2
n-Amyl acetate (6)	148.8	94.8	56.2	33.3	10.5
H—Isoamyl Alcohol (B.P. 131.4°)					
Isoamyl formate (6)	124.2	89.8	32.4	19.6	48.0
Isoamyl acetate (6)	142.0	93.6	44.8	31.2	24.0
I—Allyl Alcohol **(B.P. 97.0°)**					
n-Hexane	69.0	59.7	5	5	90
Cyclohexane	80.8	66.2	8	11	81
Benzene	80.2	68.2	8.6	9.2	82.2
Carbon tetrachloride	76.8	65.2	5	11	84

(Lange, *Handbook of Chemistry*, McGraw-Hill, New York, 1979).

TABLE 13.11—*(continued)*

(b) Other Systems

Component *A* Mole % *A* = 100 − (*B* + *C*)	Components *B* and *C*	Mole % *B* and *C*	Temp., °C.
Water	Carbon tetrachloride Ethanol	57.6 23.0	61.8 2 phase
	Trichloroethylene Ethanol	38.4 41.2	67.25 2 phase
	Trichloroethylene Allyl alcohol	49.2 17.3	71.4 2 phase
	Trichloroethylene Propyl alcohol (n)	51.1 16.6	71.55 2 phase
	Ethanol Ethyl acetate	12.4 60.1	70.3
	Ethanol Benzene	22.8 53.9	64.86
	Allyl alcohol Benzene	9.5 62.2	68.3
	Propyl alcohol (n) Benzene	8.9 62.8	68.48
Carbon disulfide	Methanol Ethyl bromide	24.1 35.4	33.92
Methyl formate	Ethyl bromide Isopentane	23.8 31.0	16.95
	Ethyl ether Pentane (n)	7.2 48.2	20.4
Propyl lactate (n)	Phenetol Menthene	35.2 34.1	163.0

weights, such as oils, fats, waxes, essential oils and scents, vitamins and hormone concentrates, and to the deodorization of high molecular weight materials.

Operating pressures are in the range of 1 m Torr. For example, the mean free paths of normal triglycerides of 800 molecular weight are

P (m Torr)	Path (mm)
8	7
3	25
1	50

The theoretical Langmuir equation for the rate of evaporation is

$$w' = 2100P^0\sqrt{M/T} \quad \mathrm{kg/m^2\,hr} \tag{13.223}$$

with the vapor pressure P^0 in Torr, the temperature in K, and with M as the molecular weight. Industrial apparatus may have 80–90% of these rates because of inefficiencies. Some numerical values at 120°C are:

Compound	*M*	P^0 (Torr)	w' (kg/m^2 hr)
Stearic acid	284	35.0	1.87
Cholesterin	387	0.5	2.02
Tristearin	891	0.0001	2.74

From Langmuir's equation it is clear that it is possible to separate substances of the same vapor pressure but different molecular weights by molecular distillation.

Apparatus and Operating Conditions. The main kinds of commercial units are illustrated in Figure 13.30. In the falling film type, the material flows by gravity as a thin film on a vertical heated cylinder, evaporates there, and is condensed on a concentric cooled surface. Diameters range from 2 to 50 cm, heights 2 to 10 m, and feed rates from 1 to 60 L/hr. In order to prevent channelling, the surface of the evaporator is made rough or other means are employed. The cross section of a wiped film commercial still is shown in Figure 13.30(b). Contact times in commercial apparatus may be as low as 0.001 sec.

In the centrifugal still, the material that is charged to the bottom creeps up the heated, rotating conical surface as a thin film, is evaporated, and then condensed and discharged. The film thickness is 0.05–0.1 mm. Rotors are up to about 1.5 m dia and turn at 400–500 rpm. Evaporating areas are up to 4.5 m^2 per unit, feed rates range from 200 to 700 L/hr, and distillates range from 2 to 400 L/hr, depending on the service. From three to seven stills in series are used for multiple redistillation of some products. Two stills 1.5 m dia can process a tank car of oil in 24 hr. A typical pumping train for a large still may comprise a three-stage steam ejector, two oil boosters and a diffusion pump, of capacity 1000–5000L/sec, next to the still. Equivalent mechanical pumps may be employed instead of the ejectors, depending on the economic requirements.

The evaporator of Figure 13.30(d) is for service intermediate between those of ordinary film evaporators and molecular stills, with greater clearances and higher operating pressures than in the latter equipment. The rotating action permits handling much more viscous materials than possible in film evaporators.

13.12. TRAY TOWERS

Contacting of vapor and liquid phases on trays is either in countercurrent flow or with cross flow of liquid against vapor flow upward. The spacing of trays is determined partly by the necessity of limiting carryover of entrainment from one tray to another, and is thus related to the vapor velocity and the diameter of the vessel. For reasons of accessibility of trays to periodic servicing, however, their spacing commonly is 20–24 in. Then workmen can go up or down the tower through removable sections of the trays and have enough room to work in. For the same reason, tray diameters are restricted to a minimum of 30 in. When a smaller size is adequate, cartridge trays that can be lifted out of the vessel as a group, as in Figure 13.31(b), or packed towers are adopted. A data sheet for recording key data of a tray tower is in Table 13.12. The tedious calculations of many mechanical details of tray construction usually are relegated to computers.

COUNTERCURRENT TRAYS

The three main kinds of trays with countercurrent flow of liquid and gas are:

Dualflow, with round holes in the 1/8–1/2 in. range, extensively tested by Fractionation Research Inc. (FRI).

Turbogrid, with slots 1/4–1/2 in. wide, developed by Shell Development Co.

Ripple trays, made of perforated corrugated sheets, with vapor flow predominantly through the peaks and liquid through the valleys, developed by Stone and Webster.

Although some of the vapor and liquid flows through the openings are continuous, the bulk of the flows pass alternately with a surging action. The absence of downcomers means a greater bubbling area and consequently a greater vapor handling capacity, and also allows a close spacing to be used, as little as 9 in. in some applications. The action in such cases approaches those of towers filled with structured packings. Their turndown ratio is low, that is, the liquid drains completely off the tray at lowered vapor rates. Consequently, countercurrent trays have never found widespread use.

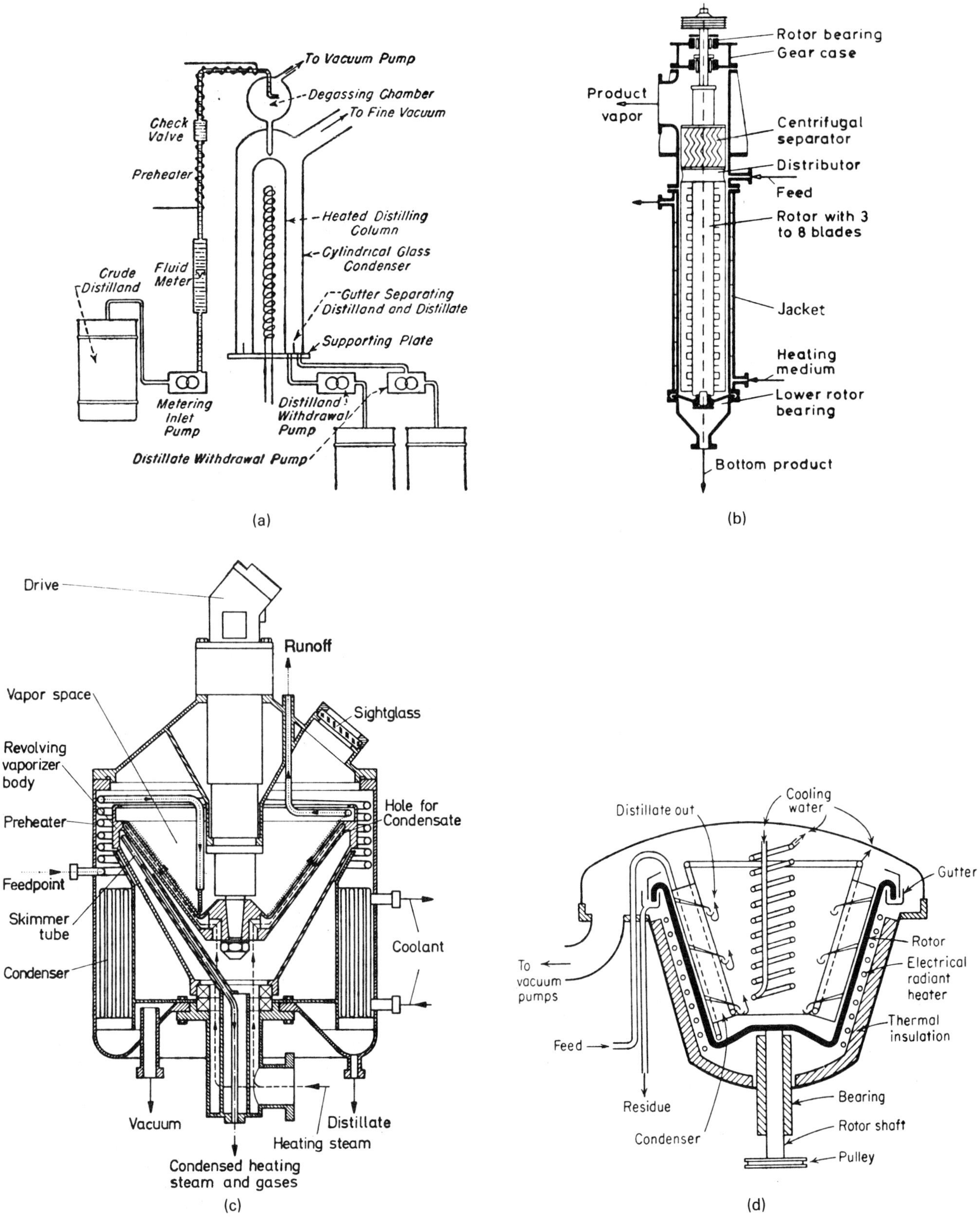

Figure 13.30. Molecular distillation and related kinds of equipment. (a) Principle of the operation of the falling film still (Chemical Engineers Handbook, *McGraw-Hill, New York, 1973*). (b) Thin-layer evaporator with rigid wiper blades (*Luwa Co., Switzerland*). (c) The Liprotherm rotating thin film evaporator, for performance intermediate to those of film evaporators and molecular stills (*Sibtec Co., Stockholm*). (d) Centrifugal molecular still [*Hickman,* Ind. Eng. Chem. **39,** *686* (*1947*)].

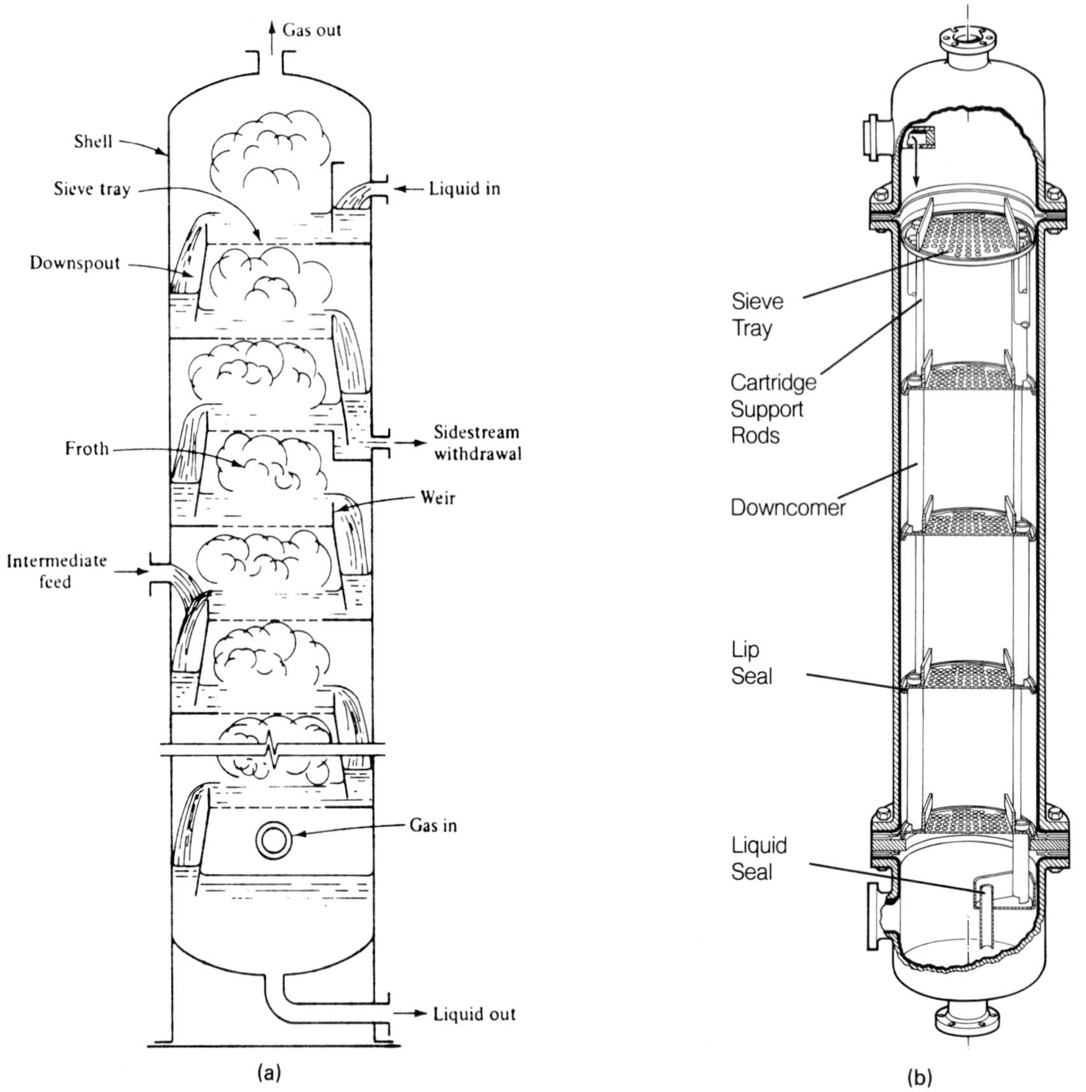

Figure 13.31. Assembled sieve tray towers. (a) Flowsketch of a sieve tray tower (*Treybal, 1980*). (b) Cartridge type sieve tray tower in small diameters (*Pfaudler Co.*).

On crossflow trays, the path of liquid is horizontal from downcomer to weir and in contact with vapor dispersed through openings in the tray floor. Such flows are illustrated in Figures 13.31 and 13.32. Depending on the rate and on the diameter, the liquid flow may be single, double, or four-pass. A common rule for dividing up the flow path is a restriction of the liquid rate to a maximum of about 8 gpm/in. of weir length. Usually towers 5 ft dia and less are made single pass. Since efficiency falls off as the flow path is shortened, a maximum of two passes sometimes is specified, in which cash flow rates may approach 20 gpm/in. of weir.

The main kinds of cross flow trays with downcomers in use are sieve, valve, and bubblecap.

SIEVE TRAYS

A liquid level is maintained with an overflow weir while the vapor comes up through the perforated floor at sufficient velocity to keep most of the liquid from weeping through. Hole sizes may range from 1/8 to 1 in., but are mostly 1/4–1/2 in. Hole area as a percentage of the active cross section is 5–15%, commonly 10%. The precise choice of these measurements is based on considerations of pressure drop, entrainment, weeping, and mass transfer efficiency. The range of conditions over which tray operation is satisfactory and the kinds of malfunctions that can occur are indicated roughly in Figure 13.33(a) and the behavior is shown schematically on Figure 13.32(e).

The required tower diameter depends primarily on the vapor rate and density and the tray spacing, with a possibly overriding restriction of accommodating sufficient weir length to keep the gpm/in. of weir below about 8. Figure 13.33(b) is a correlation for the flooding velocity. Allowable velocity usually is taken as 80% of the flooding value. Corrections are indicated with the figure for the fractional hole area other than 10% and for surface tension other than 20 dyn/cm. Moreover, the correction for the kind of operation given with Figure 13.34 for valve trays is applicable to sieve trays.

Weir heights of 2 in. are fairly standard and weir lengths about 75% of the tray diameter. For normal conditions downcomers are

TABLE 13.12. Tray Design Data Sheet

Item No. or Service				
Tower diameter, I.D.				
Tray spacing, inches				
Total trays in section				
Max. △ P, mm Hg				
Conditions at Tray No.				
Vapor to tray, °F				
Pressure,				
Compressibility				
°Density, lb./cu. ft.				
°Rate, lb./hr.				
cu. ft./sec. (cfs)				
cfs $\sqrt{D_V/(D_L - D_V)}$				
Liquid from tray, °F				
Surface tension				
Viscosity, cp				
°Density, lb./cu. ft.				
°Rate, lb./hr.				
GPM hot liquid				
Foaming tendency	None____	Moderate____	High____	Severe____

°These values are required in this form for direct computer input.

Notes

1. Tray numbering: top to bottom_____; bottom to top_____.
2. Number of flow paths or passes_____.
3. Minimum rate as a % of design rate_____.
4. Allowable downcomer velocity_______ft/sec.
5. Bottom tray downcomer: total drawoff_____other_________.
6. Adjustable weir required: yes_____; no_____.
7. Tray material and thickness_______________.
8. Valve or cap material__________________.

(International Critical Tables, McGraw-Hill, New York, 1929).

sized so that the depth of liquid in them is less than 50% and the residence time more than 3 sec. For foaming and foam-stable systems, the residence time may be two to three times this value. The topic of tray efficiency is covered in detail in Section 13.6, but here it can be stated that they are 80–90% in the vicinity of $F = u_v\sqrt{\rho_v} = 1.0$ (ft/sec)(lb/cuft)$^{1/2}$ for mixtures similar to water with alcohols and to C_6–C_7 hydrocarbons.

A detailed design of a tray includes specification of these items:

1. Hole dia, area, pitch and pattern.
2. Blanking of holes for less than eventual load.
3. Downcomer type, size, clearance, and weir height.
4. Tray thickness and material.
5. Pressure drop.
6. Turndown ratio before weeping begins.
7. Liquid gradient.

Correlations for checking all of these specifications are known. An example is worked out by Fair (in Smith, 1963, Chap. 15). The basis is holes 3/16 in. dia, fractional open area of 0.10, weir height of 2 in. and tray spacing 24 in.

The correlation of Figure 13.33 has no provision for multipass liquid flow. Corrections could be made by analogy with the valve tray correlation, as suggested at the close of Example 13.15.

VALVE TRAYS

The openings in valve trays are covered with liftable caps that adjust themselves to vapor flow. Illustrations of two kinds of valves are in Figure 13.32(b). The caps rest about 0.1 in. above the floor and rise to a maximum clearance of 0.32 in. The commonest hole diameter is 1.5 in. but sizes to 6 in. are available. Spacing of the standard diameter is 3–6 in. With 3 in. spacing, the number of valves is 12–14/sqft of free area. Some of the tray cross section is taken up by the downcomer, by supports, and by some of the central manway structure.

In spite of their apparent complexity of construction in comparison with sieve trays, they usually are less expensive than sieve trays because of their larger holes and thicker plates which need less support. They are more subject to fouling and defer to sieves for such services.

Tray diameters may be approximated with Figure 13.34 which is for "normal" systems, 24 in. tray spacing, and 80% of flooding. For other tray spacings, corrections may be approximated with the

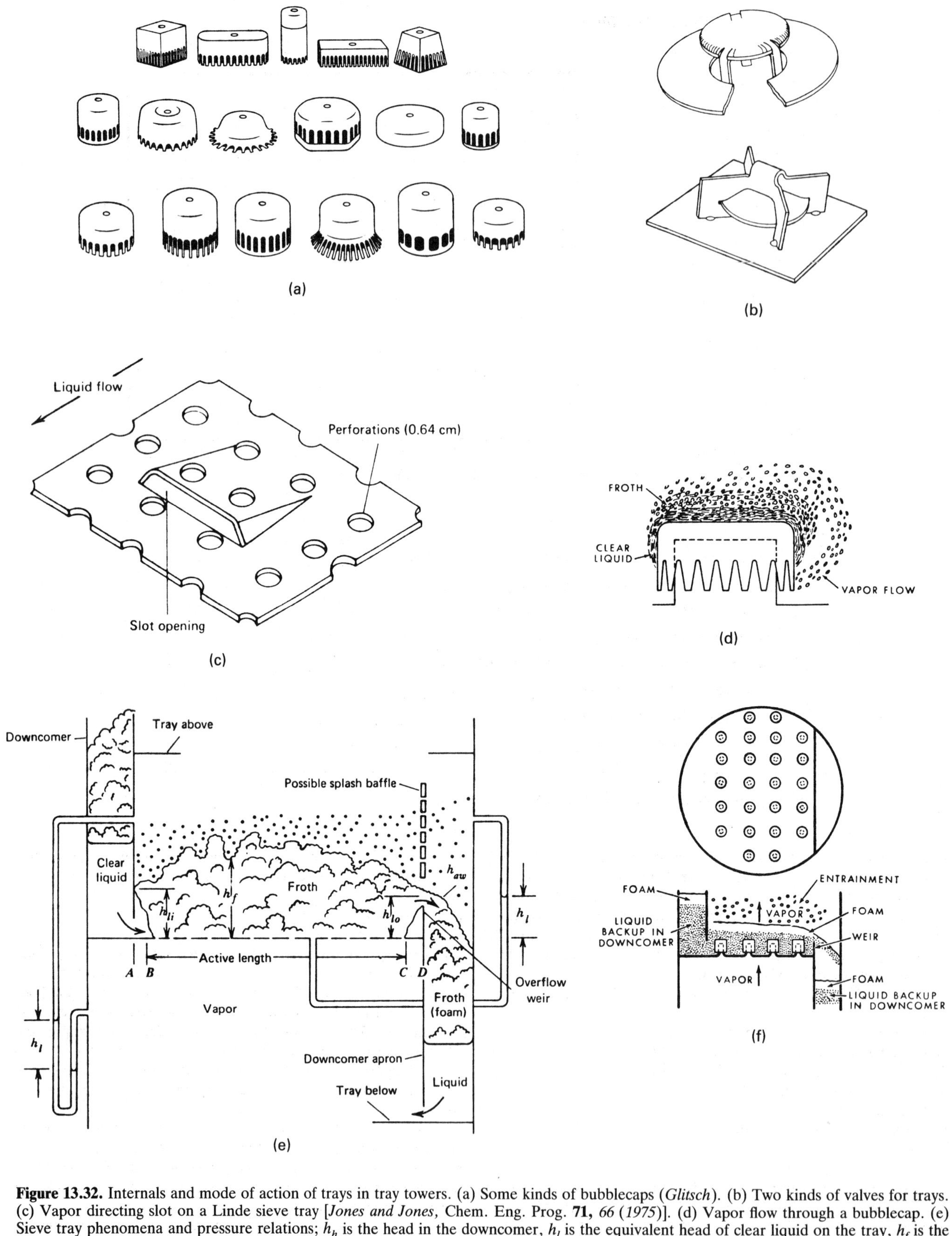

Figure 13.32. Internals and mode of action of trays in tray towers. (a) Some kinds of bubblecaps (*Glitsch*). (b) Two kinds of valves for trays. (c) Vapor directing slot on a Linde sieve tray [*Jones and Jones,* Chem. Eng. Prog. **71,** *66 (1975)*]. (d) Vapor flow through a bubblecap. (e) Sieve tray phenomena and pressure relations; h_h is the head in the downcomer, h_l is the equivalent head of clear liquid on the tray, h_f is the visible height of froth on the tray, and h_t is the pressure drop across the tray (*Bolles, in Smith, 1963*). (f) Assembly of and action of vapor and liquid on a bubblecap tray.

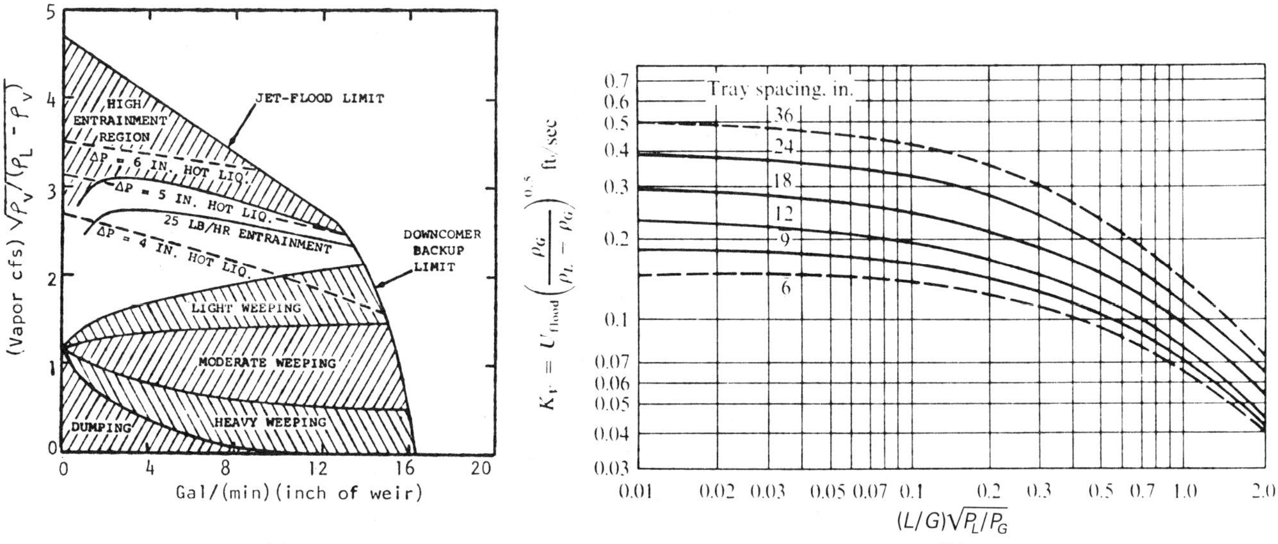

(a) (b)

Figure 13.33. Operating ranges of malfunctions and flooding velocity correlation of sieve trays. (a) Performance of a typical sieve tray, showing ranges of weeping, dumping, entrainment, and flooding. (b) Correlation of flooding at various tray spacings. For normal operation, take 80% of the flooding rate as a design condition. To correct for surface tension (dyn/cm), multiply the ordinate by $(\sigma/20)^{0.2}$. To correct for other than 10% hole area, multiply the ordinate by 0.9 for 8% and by 0.8 for 6% [*after Fair and Matthews,* Pet. Refin. **37**(*4*), *153* (*1958*)].

sieve tray correlation of Figure 13.33(c). Factors for correcting the allowable volumetric rate for various degrees of foaming are given with the figure.

Formulas and procedures for calculation of detailed tray specifications are presented, for example, by Glitsch Inc. (Bulletin 4900, *Ballast Tray Design Manual,* Dallas, TX, 1974), and illustrated with a completely solved numerical problem.

BUBBLECAP TRAYS

Bubblecap assemblies serve to disperse the vapor on the tray and to maintain a minimum level of liquid. A few of the many kinds that have been used are in Figure 13.32, together with illustrations of their mode of action and assembly on a tray. The most used kinds are 4 or 6 in. dia round caps. Because of their greater cost and

---

Example 13.15
Comparison of Diameters of Sieve, Valve, and Bubblecap Trays for the Same Service

A C_3 splitter has 24 in. tray spacing and will operate at 80% of flooding. These data are applicable:

$W_v = 271{,}500$ lb/hr of vapor,
$Q_v = 27.52$ cfs of vapor,
$W_L = 259{,}100$ lb/hr of liquid,
$Q_L = 1100$ gpm,
$\rho_V = 2.75$ lb/cuft,
$\rho_L = 29.3$ lb/cuft.

Sieve tray: Use Figure 13.33(b):

$$\text{abscissa} = (259{,}100/271{,}500)\sqrt{2.75/29.3} = 0.2924,$$
$$\text{ordinate, } C = 0.24,$$
$$u_G = C\sqrt{(\rho_L - \rho_V)/\rho_V} = 0.24\sqrt{29.3/2.75 - 1} = 0.746 \text{ fps}.$$

Allowable velocity at 80% of flooding,

$$u_G = 0.8(0.746) = 0.597 \text{ fps},$$
$$\therefore D = \sqrt{Q_v/(\pi/4)u_G} = \sqrt{27.52/(\pi/4)(0.597)} = 7.67 \text{ ft}.$$

Valve tray: Use Figure 13.34:

$$(\text{cfs})\sqrt{\rho_V/(\rho_L - \rho_V)} = 27.52\sqrt{2.75/(29.3 - 2.75)} = 8.86,$$
$$\therefore D = \begin{cases} 9.4 \text{ ft}, & \text{one pass}, \\ 7.6 \text{ ft}, & \text{two passes}. \end{cases}$$

Bubblecap tray: Use Eq. (13.224):

$$K = 4.2 \text{ for 24 in. tray spacing},$$
$$\therefore D = 0.0956[271{,}500/4.2\sqrt{29.3(2.75)}]^{1/2} = 8.11 \text{ ft}.$$

The correlations for sieve and bubblecap trays have no provision for multipass flow of liquid. Their basic data may have been obtained on smaller towers with liquid flow equivalent to two-pass arrangement in towers 8 ft dia. The sieve tray correlation should be adapted to multipass flow by comparison with results obtained by the valve tray correlation in specific cases.

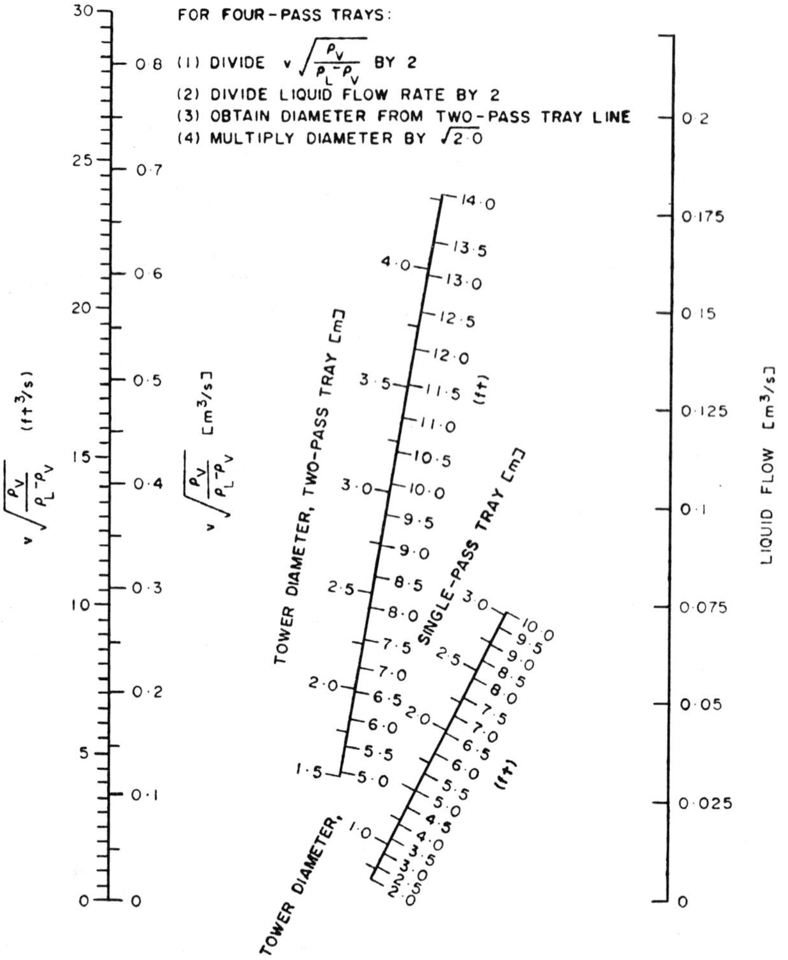

Service	System Factor
Nonfoaming, regular systems	1.00
Fluorine systems, e.g., BF_3, Freon	0.90
Moderate foaming, e.g., oil absorbers, amine and glycol regenerators	0.85
Heavy foaming, e.g., amine and glycol absorbers	0.73
Severe foaming, e.g., MEK units	0.60
Foam-stable systems, e.g., caustic regenerators	0.30–0.60

Figure 13.34. Chart for finding the diameters of valve trays. Basis of 24 in. tray spacing and 80% of flood for nonfoaming services. Use Figure 13.32(b) for approximate adjustment to other tray spacings, and divide the $V_{\text{load}} = V\sqrt{\frac{\rho_V}{\rho_L - \rho_V}}$ by the given "system factor" for other services (*Glitsch Inc., Bulletin 4900, Dallas, TX, 1974*).

problems with hydraulic gradient, bubblecap trays are rarely installed nowadays, having lost out since about 1950 to the other two kinds. Since they do have a positive liquid seal and will not run dry, they are used sometimes in low liquid flow rate situations such as crude vacuum towers, but even there they have lost out largely to structured tower packings which have much lower pressure drop.

The allowable vapor velocity and the corresponding tray diameter are represented by the work of Souders and Brown, which is cited in standard textbooks, for example Treybal (1980). Its equivalent is the "Jersey Critical" formula,

$$D = 0.0956(W_v/K\sqrt{\rho_L\rho_v})^{1/2}, \quad \text{ft} \qquad (13.224)$$

with the factor K dependent on the tray spacing as follows:

Tray spacing (in.)	18	24	30	30+
K	3.4	4.2	4.7	5.0

Here

W_v = vapor flow rate (lb/hr),
ρ_v = vapor density (lb/cuft),
ρ_L = liquid density (lb/cuft).

Example 13.15 compares diameters of sieve, valve, and

bubblecap trays calculated with the relations cited in this section. All of these relations presumably are based on limiting the amount of entrainment to a level that does not affect efficiency appreciably. Accordingly, the differences in diameters found in that example are due less, perhaps, to differences in performances of the different kinds of trays than to the particular data on which the correlations are based.

A factor that is of concern with bubblecap trays is the development of a liquid gradient from inlet to outlet which results in corresponding variation in vapor flow across the cross section and usually to degradation of the efficiency. With other kinds of trays this effect rarely is serious. Data and procedures for analysis of this behavior are summarized by Bolles (in Smith, 1963, Chap. 14). There also are formulas and a numerical example of the design of all features of bubblecap trays. Although, as mentioned, new installations of such trays are infrequent, many older ones still are in operation and may need to be studied for changed conditions.

13.13. PACKED TOWERS

In comparison with tray towers, packed towers are suited to small diameters (24 in. or less), whenever low pressure is desirable, whenever low holdup is necessary, and whenever plastic or ceramic construction is required. Applications unfavorable to packings are large diameter towers, especially those with low liquid and high vapor rates, because of problems with liquid distribution, and whenever high turndown is required. In large towers, random packing may cost more than twice as much as sieve or valve trays.

Depth of packing without intermediate supports is limited by its deformability; metal construction is limited to depths of 20–25 ft, and plastic to 10–15 ft. Intermediate supports and liquid redistributors are supplied for deeper beds and at sidestream withdrawal or feed points. Liquid redistributors usually are needed every $2\frac{1}{2}$–3 tower diameters for Raschig rings and every 5–10 diameters for pall rings, but at least every 20 ft.

The various kinds of internals of packed towers are represented in Figure 13.35 whose individual parts may be described one-by-one:

(a) is an example column showing the inlet and outlet connections and some of the kinds of internals in place.
(b) is a combination packing support and redistributor that can also serve as a sump for withdrawal of liquid from the tower.
(c) is a trough-type distributor that is suitable for liquid rates in excess of 2 gpm/sqft in towers two feet and more in diameter. They can be made in ceramics or plastics.
(d) is an example of a perforated pipe distributor which is available in a variety of shapes, and is the most efficient type over a wide range of liquid rates; in large towers and where distribution is especially critical, they are fitted with nozzles instead of perforations.
(e) is a redistribution device, the rosette, that provides adequate redistribution in small diameter towers; it diverts the liquid away from the wall towards which it tends to go.
(f) is a holddown plate to keep low density packings in place and to prevent fragile packings such as those made of carbon, for instance, from disintegrating because of mechanical disturbances at the top of the bed.

KINDS OF PACKINGS

The broad classes of packings for vapor–liquid contacting are either random or structured. The former are small, hollow structures with large surface per unit volume that are loaded at random into the vessel. Structured packings may be layers of large rings or grids, but are most commonly made of expanded metal or woven wire screen that are stacked in layers or as spiral windings.

The first of the widely used random packings were Raschig rings which are hollow cylinders of ceramics, plastics, or metal. They were an economical replacement for the crushed rock often used then. Because of their simplicity and their early introduction, Raschig rings have been investigated thoroughly and many data of their performance have been obtained which are still useful, for example, in defining the lower limits of mass transfer efficiency that can be realized with improved packings.

Several kinds of rings are shown in Figure 13.36. They are being made in a variety of internal structure and surface modifications. Pall rings in metal and plastics are perhaps the most widely used packings. One brand, "Hy-Pak," has corrugated walls and more intrusions than the standard designs shown in the figure. Cascade minirings, with height less than the diameter, appear to have improved efficiency in comparison with some other pall rings. Saddles are more efficient because of greater surface and improved hydrodynamics. In plastic construction, Figure 13.36(h), they are made with a variety of holes and protrusions to enlarge the specific surface. When ceramic construction is necessary, saddles are the preferred packings. A survey of efficiencies of packed beds is in Table 13.13.

Whenever possible, the ratio of tower and packing diameters should exceed 15. As a rough guide, 1 in. packing is used for gas rates of about 500 cfm and 2 in. for gas rates of 2000 cfm or more.

Structured packings are employed particularly in vacuum service where pressure drops must be kept low. Because of their open structure and large specific surfaces, their mass transfer efficiency is high when proper distribution of liquid over the cross section can be maintained. Table 13.14 is a comparison of various features of five commercial makes of structured packings. The HIGEE centrifugal fractionator of Figure 13.14 employs structured packing in the form of perforated metal.

Ultimately, the choice of packing is based on pressure drop and mass transfer efficiency. Since packings of individual manufacturers differ in detail, the manufacturers pressure drop data should be used. A few such data are in Figures 13.37 and 13.38. Mass transfer efficiency is discussed in the next section.

FLOODING AND ALLOWABLE LOADS

The main operating limitation of the operation of a packed bed is the onset of flooding. Then the interstices tend to fill with liquid, the gas becomes unable to flow smoothly, and the pressure drop begins to rise sharply. The classic correlation of the flood point is due to Sherwood and Lobo et al. It is shown in Figure 13.39. Clearly, there is much scatter and many more recent kinds of packings are not covered. Nevertheless, it is fairly standard practice to design for a flow rate of 70–80% of that given by the correlation. In case the liquid is a foaming type, the factor is 40% of the flooding rate, or some means of eliminating the foam is found.

The correlation of Eckert (Fig. 13.37) combines a pressure drop relation and safe flow rates insofar as staying away from the flooding point is concerned. A flooding line corresponds to pressure drops in excess of 2 in. water/ft. In use, a pressure drop is selected, and the correlation is applied to find the corresponding mass velocity G from which the tower diameter then is calculated. Another correlation recommended by a manufacturer of packings appears in Figure 13.40. Example 13.16 compares these correlations for a specific case; they do not compare any more closely than could be expected from the scatter of flooding data.

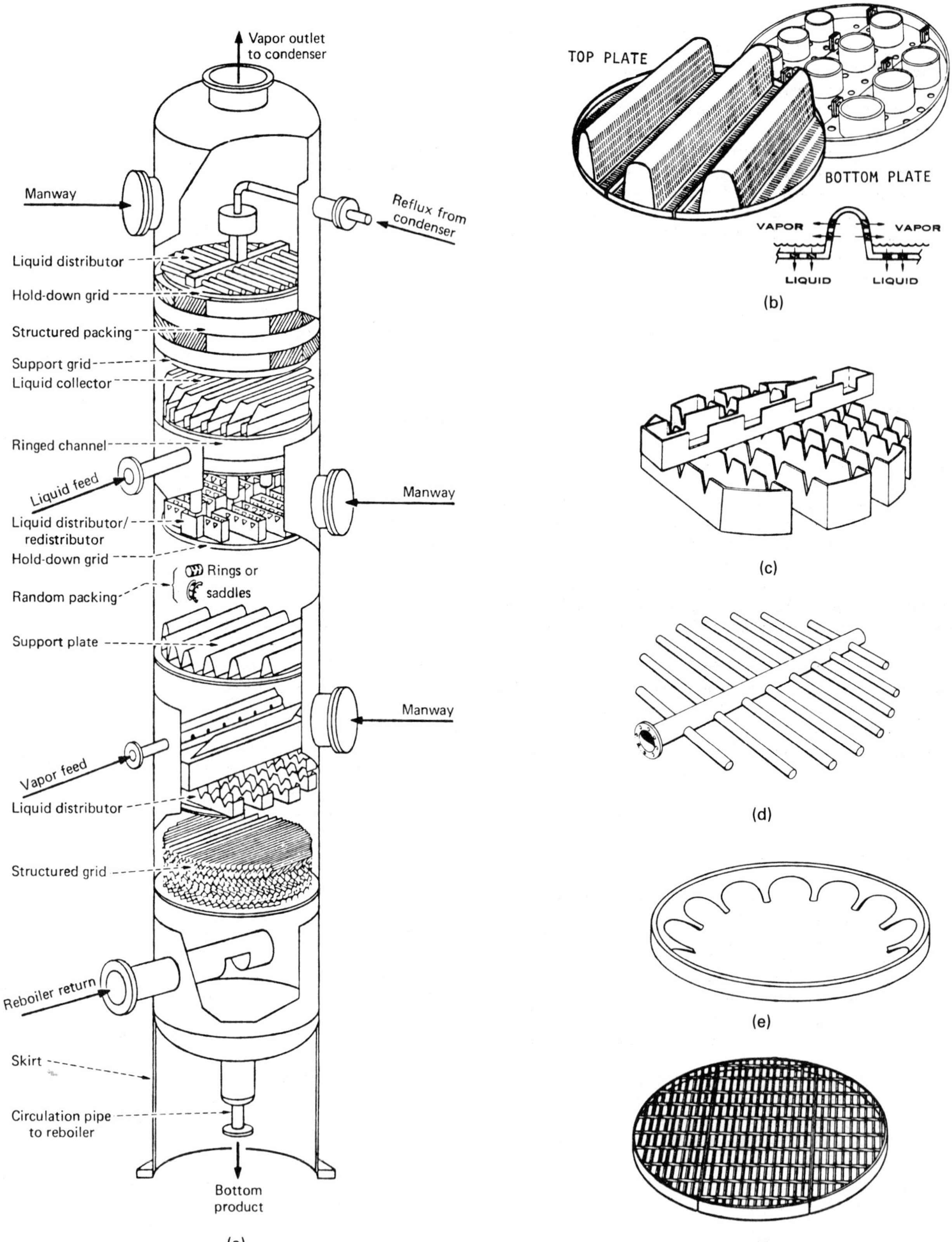

Figure 13.35. Packed column and internals. (a) Example packed column with a variety of internals [*Chen,* Chem. Eng. **40,** (*5 Mar. 1984*)]. (b) Packing support and redistributor assembly. (c) Trough-type liquid distributor. (d) Perforated pipe distributor. (e) Rosette redistributor for small towers. (f) Hold-down plate, particularly for low density packing.

Figure 13.36. Some kinds of tower packings: (a) Raschig ring; (b) partition or Lessing ring; (c) double spiral ring; (d) metal pall ring; (e) plastic pall ring; (f) ceramic Berl saddle (*Maurice A. Knight Co.*); (g) ceramic intalox saddle (*Norton Co.*); (h) plastic intalox saddle (*Norton Co.*); (i) metal intalox saddle (*Norton Co.*); (j) Tellerette (*Chem-Pro Co.*); (k) plastic tripak (*Polymer Piping and Metals Co.*); (l) metal tripak (*Polymer Piping and Metals Co.*); (m) wood grid; (n) section through expanded metal packing; (o) sections of expanded metal packings placed alternatively at right angles (*Denholme Co.*); (p) GEM structured packing (*Glitsch Inc.*).

TABLE 13.13. Survey of Efficiencies of Packed Beds

System	Dia. m	Packing Type	Size, m	Bed Depth m	HETP, m	HTU, m	System press., kPa
Hydrocarbons							
Absorber	0.91	Pall rings	0.05	7.0	0.85	—	5,964
L. O. top fractionator	0.91	Pall rings	0.05	5.2	0.76	—	1,083
L. O. bottom fractionator	1.22	Pall rings	0.05	5.2	0.85	—	1,083
Deethanizer top	0.46	Pall rings	0.038	6.1	0.88	—	2,069
Deethanizer bottom	0.76	Pall rings	0.05	5.5	1.01	—	2,069
Depropanizer top	0.59	Pall rings	0.038	4.88	0.98	—	1,862
Depropanizer bottom	0.59	Pall rings	0.038	7.32	0.73	—	1,862
Debutanizer top	0.50	Pall rings	0.038	3.66	0.73	—	621
Debutanizer bottom	0.50	Pall rings	0.038	5.49	0.61	—	621
Pentane-isopentane	0.46	Pall rings	0.025	2.74	0.46	—	101
Light/heavy naphtha	0.38	Pall rings	0.025	3.05	0.62	0.54	13
	0.38	Intalox	0.025	3.05	0.76	0.61	13
	0.38	Raschig rings	0.025	3.05	0.71	0.52	13
Iso-octane/toluene	0.38	Pall rings	0.025	3.05	0.43	0.45	13
	0.38	Pall rings	0.025	3.05	0.53	0.51	13
Gas plant absorber	1.22	Pall rings	0.025	7.0	0.88	—	6,206
2,2,4-trimethyl-pentane/ methylcyclo-hexane	0.91	Stedman		7.6	0.88	—	101
	3.35	Stedman		2.1	0.13	—	101
Hydrocarbons/water							
Acetone	0.36	Intalox	0.025	3.96	0.46	—	101
	0.46	Pall rings	0.025	8.38	0.37	—	101
Methanol (batch)	0.61	Intalox	0.038	4.27	—	0.76	101
Methanol	0.41	Pall rings	0.025	4.27	—	0.52	101
	0.30	Intalox	0.025	8.23	0.46	—	101
Isopropanol	0.53	Plastic pall	0.038	4.88	—	0.84	101
	0.33	Intalox	0.025	6.40	—	0.76	101
	0.46	Intalox	0.025	3.35	0.48	—	101
Ethylene glycol	1.07	Pall rings	0.038	4.88	0.91	—	31
Propylene glycol	0.25	Intalox	0.013	1.83	—	0.86	101
Furfural	0.51	Intalox	0.038	5.49	0.61	—	101
Formic acid	0.91	Pall rings	0.05	10.67	0.76	—	101
Acetone (absorption)	0.61	Intalox	0.038	5.49	—	0.46	101
Benzolchloride/ benzene/steam	0.61	Pall rings	0.025	5.18	—	1.07	101
Tall oil/steam	3.66	Intalox	0.05	10.46	0.76	—	101
Methylisobutyl ketone/steam	1.07	Intalox	0.038	8.53	1.22	—	101
Acetone	0.38	Intalox	0.05	2.90	0.53	0.47	101
	0.38	Pall rings	0.038	2.90	0.46	0.55	101
	0.38	Pall rings	0.025	2.90	0.44	0.34	101
	0.38	Intalox	0.025	2.90	0.52	0.32	101
	0.38	Berl	0.025	2.90	0.52	0.34	101
	0.38	Cer. Raschig	0.025	2.90	1.05	0.36	101
	0.38	Raschig rings	0.025	2.90	0.52	0.36	101
	0.38	Pall rings	0.016	2.90	0.40	0.32	101
Methanol	0.38	Pall rings	0.025	2.90	0.66	0.67	101
Polar hydrocarbons							
Methyl furan/ methyl tetra-hydrofuran	0.61	Intalox	0.038	14.63	0.53	—	101
Benzoic acid/ toluene	0.61	Intalox	0.038	6.40	0.46	—	101
Methone (5, 5 dimethyl 1, 3 cyclohexanedione), batch	0.56	Pall rings	0.038	9.75	0.49	—	101
Monochloro acetic acid/ acetic anhydride	0.25	Intalox	0.025	2.74	—	0.86	8.0
Tar acid distillation (batch)	0.46	Pall rings	0.038	9.14	0.49	—	13
Cresols (batch)	0.46	Pall rings	0.038	9.14	0.85	—	13
Benzene/ monochloro-benzene	0.38	Intalox	0.038	2.90	1.80	0.52	101
	0.38	Intalox	0.025	2.90	1.13	0.76	101
Methylethylketone/ toluene	0.38	Pall rings	0.025	2.90	0.35	0.29	101
	0.38	Raschig rings	0.025	2.90	0.30	0.31	101
	0.38	Intalox	0.025	2.90	0.23	0.27	101
	0.38	Berl	0.025	2.90	0.31	0.31	101
	0.38	Cer. Raschig	0.025	2.90	0.46	0.30	101
	0.38	Pall rings	0.025	2.90	0.40	0.28	101
	0.38	Raschig rings	0.025	2.90	0.35	0.30	101
	0.38	Intalox	0.025	2.90	0.29	0.26	101
	0.38	Berl	0.025	2.90	0.34	0.29	101
	0.38	Cer. Raschig	0.025	2.90	0.31	0.27	101
Phenol/ortho-creosol	0.46	Pall rings	0.038	9.14	0.49	—	13
Fatty acid	0.76	Pall rings	0.038	12.19	0.85	—	4.94
Benzene/mono-chloro-benzene	1.83	Intalox	0.038	9.75	1.07	—	13
DMPC/ DMPC cresols/ DBOC batch	0.46	Pall rings	0.038	9.14	0.49	—	2.67
CH_3Cl/ CH_2Cl_2/ $CHCl_3$/ CCl_4	0.48	Intalox	0.025	20.73	0.46	—	101
Methylene/ light ends	0.46	Intalox	0.025	13.41	0.46	—	101
Methylene/ product	0.64	Intalox	0.038	13.72	0.46	—	101
Chloroform/ product	0.56	Intalox	0.038	27.43	0.46	—	101

[Eckert, *Chem. Eng. Prog.* **59**(5) 76 (1963)].

TABLE 13.14. Comparison of Structured Tower Packings

	Goodloe packing	Hyperfil packing	Koch-Sulzer packing	Neo-Kloss packing	Leva film trays
General information:					
Type	Knitted multifilament	Knitted multifilament	Corrugated woven-wire fabric	Rolled screen with spacers	Multiple unsealed downcomer trays on close tray spacing
Approximate number of units 12 in and larger sold through 1975	610	90	500	43	120
Largest diameter sold to date	5 ft, 8 in	5 ft	11 ft	6 ft	14 ft, 6 in
Materials in which available	[a]	[a]	[a]	[a]	[b]
Process and system considerations:					
Minimum head pressure, torr	1	1	0.5	0.5[i]	5
Liquid considerations:					
Minimum rates, gal/(min)(ft²)[c]	0.016	0.1[i]	0.08	0.1[i]	0.05
Maximum rate, gal/(min)(ft²)[c]	>4.9	4.7[i]	>8	3.8[i]	5
Maximum viscosity, cP	200				>100
Holdup, fraction of total volume, typical	0.07–0.12	0.1	0.04	0.03	
Sensitivity to uneven initial liquid distribution	Moderate	Moderate	Fairly low	High[d]	Moderate
Vapor F factor, based on internal cross-sectional area of shell, typical	0.5–1.5	1.0	1.8	2.5–3.0	
Maximum	1.7	1.4	3.3	4.0	2.0
Minimum	low	0.14		0.16	0.25
HETS, in:					
Range	3½–8½	3½–9	4–10	4–18	12–24
Typical	5	5	7	8	18
Pressure drop, mmHg per foot of packed height	[e]	[e]	See Fig. 3	See Fig. 2	See Fig. 4
Pressure drop per theoretical stage	[f]	[f]	[g]	[g]	[g]
Fouling considerations:					
Sensitive to particulate solids?	Yes	Yes	Moderately[h]	No	No
Sensitive to fouling by tarry substances?	Yes	Yes	Moderately[h]	Yes	No
Sensitive to fouling from polymer formation?	Yes	Yes	Moderately[h]	Yes	No
Mechanical considerations:					
Is the device furnished as a package including shell and internals?	Optional	Optional	Optional	Yes	Optional
Can it be installed through shell manholes?	Yes[j]	Yes	Yes	No	No
Can it be installed in an existing shell with only minor modifications?	Yes[j]	Yes	Yes		Yes[k]
Test facilities:					
Are pilot test facilities available?	Yes	Yes	Yes	Yes	Yes

[a]Any metal capable of being drawn into wire.
[b]Any metal which can be fabricated into the required shapes.
[c]These liquid rates are generally those claimed by the manufacturers. Very low liquid loadings [below 0.2 gal/(min)(ft²)] always require special attention to the design of the liquid-distribution system.
[d]Neo-Kloss packing requires highly precise initial liquid distribution because liquid cannot spread from one layer of screen to the next. Care is needed in the design of the distribution system (provided by vendor), in its installation, and in prevention of fouling.
[e]No general curves are available for estimating from F factor. See vendor bulletins for calculational methods.
[f]Vendor bulletins indicate that pressure drop per theoretical stage will be about 0.5 mmHg or less.
[g]For Koch-Sulzer packing, Neo-Kloss packing, and Leva film trays, a preliminary estimate can be made by dividing the pressure drop per foot of packing at a typical or expected F factor loading from Fig. 2, 3, or 4 by an assumed HETS (in feet).
[h]Relatively good irrigation properties minimize the potential for dry spots which promote fouling.
[i]Vendor considers those values to be extremes normally used but not absolute limits.
[j]Techniques have been developed to permit installation through a manhole. However, it is preferred and usually less costly to provide full shell flanges on either new or existing columns.
[k]Full shell opening required.

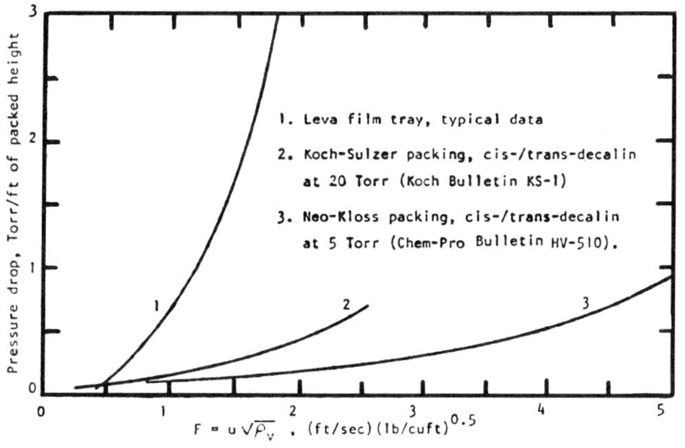

(P.G. Nygren, in Schweitzer, 1979).

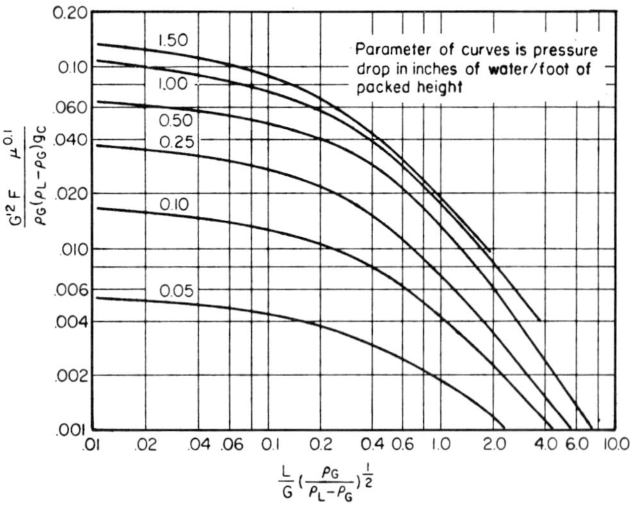

L = Liquid rate, lb/(s)(ft^2)
G' = Gas rate, lb/(s)(ft^2)
ρ_L = Liquid density, lb/ft^3
ρ_G = Gas density, lb/ft^3
F = Packing factor
μ = Viscosity of liquid, centipoise
g_c = Gravitational constant = 32.2

(a)

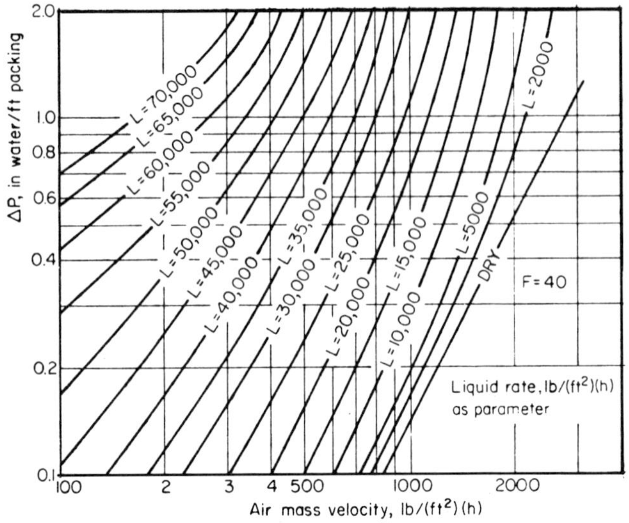

(b)

Type of Packing	Mat'l.	Nominal Packing Size, In										
		¼	⅜	½	⅝	¾	1	1¼	1½	2	3	3½
Super Intalox	Ceramic	—	—	—	—	—	60	—	—	30	—	—
Super Intalox	Plastic	—	—	—	—	—	33	—	—	21	16	—
Intalox saddles	Ceramic	725	330	200	—	145	98	—	52	40	22	—
Hy-Pak rings	Metal	—	—	—	—	—	42	—	—	18	15	—
Pall rings	Plastic	—	—	—	97	—	52	—	40	25	—	16
Pall rings	Metal	—	—	—	70	—	48	—	28	20	—	16
Berl saddles	Ceramic	900^{x}	—	240^{x}	—	170^{h}	110^{h}	—	65^{h}	45^{x}	—	—
Raschig rings	Ceramic	1,600^{b,x}	1,000^{b,x}	580^{c}	380^{c}	255^{c}	155^{d}	125^{e,x}	95^{e}	65^{f}	37^{g,x}	—
Raschig rings $\frac{1}{32}$-in wall	Metal	700^{x}	390^{x}	300^{x}	170	155	115^{x}	—	—	—	—	—
Raschig rings $\frac{1}{16}$-in wall	Metal	—	—	410	290	220	137	110^{x}	83	57	32^{x}	—
Tellerettes	Plastic	—	—	—	—	—	40	—	—	20	—	—
Maspak	Plastic	—	—	—	—	—	—	—	—	32	20	—
Lessing exp.	Metal	—	—	—	—	—	—	—	30	—	—	—
Cross partition	Ceramic	—	—	—	—	—	—	—	—	—	70	—

[a] $\frac{1}{32}$ Wall [b] $\frac{1}{16}$ Wall [c] $\frac{3}{32}$ Wall [d] ⅛ Wall [e] $\frac{3}{16}$ Wall [f] ¼ Wall [g] ⅜ Wall [h] Packing factors obtained in 16- and 30-in. I.D. towers. [x] Extrapolated

(c)

Figure 13.37. Corrrelation of flow rates, typical pressure drop behavior, and packing factors of random packed beds. [*Eckert, Foote, and Walter,* Chem. Eng. Prog. **62**(*1*), *59* (*1966*); *Eckert,* Chem. Eng. (*14 Apr. 1975*)]. (a) Correlation of flow rate and pressure drop in packed towers. (b) Typical pressure drop data: 2 in. porcelain intalox saddles, with $F = 40$, in a bed 30 in. dia by 10 ft high. (c) Packing factors, F, of wet random packings.

LIQUID DISTRIBUTION

Liquid introduced at a single point at the top of a packed bed migrates towards the walls. Relatively high liquid rates, as in distillation operations where the molal flow rates of both phases are roughly comparable, tend to retard this migration. When liquid rates are low, the maldistribution is more serious. In any event good distribution must be provided initially. A common rule is that the number of liquid streams should be 3–5/sqft in towers larger than 3 ft dia, and several times this number in smaller towers. Some statements about redistribution are made at the beginning of this section.

LIQUID HOLDUP

The amount of liquid holdup in the packing is of interest when the liquid is unstable or when a desirable reaction is to be carried out in the vessel. A correlation for Raschig rings, Berl saddles, and intalox saddles is due to Leva (*Tower Packings and Packed Tower Design*, U.S. Stoneware Co., Akron, OH, (1953):

$$L_w = 0.004(L/D_p)^{0.6}, \quad \text{cuft liquid/cuft bed} \tag{13.225}$$

with L in lb liquid/(hr)(sqft) and D_p is packing size (in.). For instance, when $L = 10{,}000$ and $D_p = 2$, then $L_w = 0.066$ cuft/cuft.

PRESSURE DROP

Although several attempts have been made to correlate data of pressure drop in packed beds in accordance with the general theory of granular beds, no useful generalization has been achieved. In any event, all manufacturers make available such data measured for their packings, usually only for the air-water system. Samples of such data are in Figures 13.37, 13.38, and 13.40.

13.14. EFFICIENCIES OF TRAYS AND PACKINGS

The numbers of theoretical or equilibrium stages needed for a given vapor–liquid separation process can be evaluated quite precisely when the equilibrium data are known, but in practice equilibrium is not attained completely on trays, and the height of packing equivalent to a theoretical stage is a highly variable quantity. In a few instances, such as in large diameter towers (10 ft or so), a significant concentration gradient exists along the path of liquid flow, so that the amount of mass transfer may correspond to more than that calculated from the average terminal compositions. Mass transfer performance of packed beds is most conveniently expressed in terms of HETS (height equivalent to a theoretical stage), particularly when dealing with multicomponent mixtures to which the concept of HTU (height of a transfer unit) is difficult to apply. In addition to the geometrical configuration of the tray or packing, the main factors that affect their efficiencies are flow rates, viscosities, relative volatilities, surface tension, dispersion, submergence, and others that are combined in dimensionless groups such as Reynolds and Schmidt.

TRAYS

In spite of all the effort that has been expended on this topic, the prediction of mass transfer efficiency still is not on a satisfactory basis. The relatively elaborate method of the *AIChE Bubble-Tray Manual* (AIChE, New York, 1958) is based on the two-film theory but has not had a distinguished career. A number of simpler correlations have been proposed and have some value as general guidance. That literature has been surveyed recently by Vital, Grossel, and Olsen [*Hyd. Proc.*, 55–56 (Oct. 1984); 147–153 (Nov. 1984); 75–78 (Dec. 1984)].

Efficiency of mass transfer is expressed as the ratio of the actual change in mol fraction to the change that could occur if equilibrium were attained

$$E = \Delta y/(\Delta y)_{\text{equilibrium}}. \tag{13.226}$$

Because of concentration gradients along the tray, primarily in the liquid phase, the overall efficiency is different from a point efficiency. Since the hydraulics of the tray usually cannot be known accurately, point and overall efficiencies are difficult to relate. In Table 13.15, for instance, three kinds of efficiencies are shown:

E_{0G} is an overall efficiency based on average changes in the vapor phase mol fraction.

E_{mv} is the Murphree efficiency, in which $(\Delta y)_{\text{equilib}}$ is used as in equilibrium with the liquid leaving the tray.

E_0 is the ratio of theoretical trays needed for a given separation to the actual number required, and is called the overall efficiency.

Since the efficiency may vary with the position on an individual tray and on the position of the tray in the tower, the three kinds are not the same. When more than one value is shown in Table 13.15 or other literature, the smallest value should be taken as the overall efficiency when that number is needed.

The values of Tables 13.15 and 13.16 probably are not the optima in all cases. The graphs of Figure 13.41 indicate that efficiencies depend markedly on the vapor flow factor, $F = u\sqrt{\rho}$, and there often is a peak in the efficiency curve. Figure 13.42 shows the effect of liquid flow rate across the tray and through the downcomer, measured as a percentage of the flow required to fill the downcomer of this particular tray.

Some of the available methods for estimating tray efficiencies will be described. A useful summary of the AIChE bubble-cap tray method is in the book of King (1980, pp. 621–626). Some of the literature that has found fault with this method is cited by Vital et al. (1984).

The method of O'Connell is popular because of its simplicity and the fact that predicted values are conservative (low). It expresses the efficiency in terms of the product of viscosity and relative volatility, $\mu\alpha$, for fractionators and the equivalent term HP/μ for absorbers and strippers. The data on which it is based are shown in Figure 13.43. For convenience of use with computer programs, for instance, for the Underwood–Fenske–Gilliland method which is all in terms in equations not graphs, the data have been replotted and fitted with equations by Negahban (University of Kansas, 1985). For fractionators,

$$E = 53.977 - 22.527(\log x) + 3.0700(\log x)^2 - 11.000(\log x)^3, \tag{13.227}$$

where $x = \mu\alpha$, μ is viscosity (cP) and α is relative volatility. For absorbers and strippers,

$$E = 39.425 + 20.034(\log x) + 1.3480(\log x)^2 - 0.3528(\log x)^3, \tag{13.228}$$

where $x = \mathrm{HP}/\mu$, H is Henry's law constant [lbmol/(cuft)(atm)], P is in atm, and μ is in cP.

The equation of McFarland, Sigmund, and Van Winkle [*Hyd. Proc.*, 111–114 (Jul. 1972)] is based primarily on data obtained in pilot plant and laboratory units. It shows a weak dependence on several dimensionless groups. About 800 data points were correlated. The absolute average deviation was 10.6%, and 90% of the calculated values were within 24% of the experimental ones.

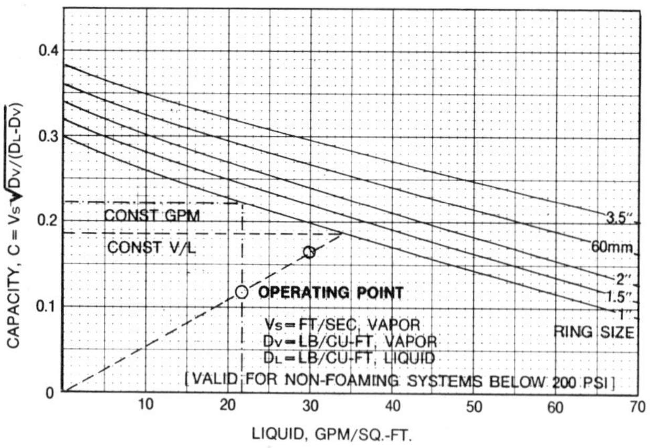

(a)

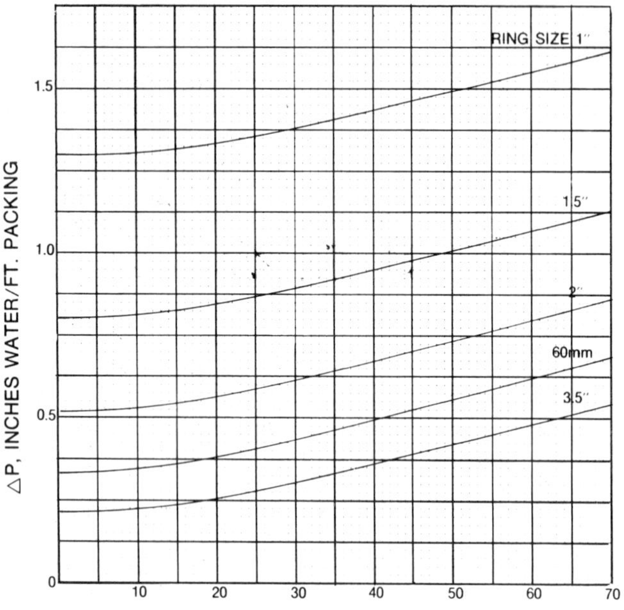

(b)

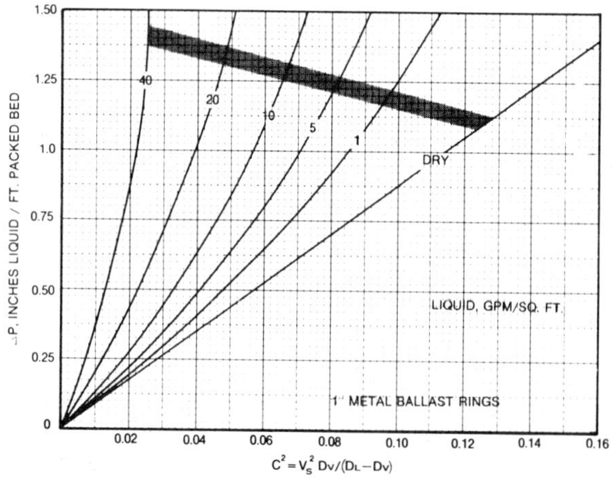

(c)

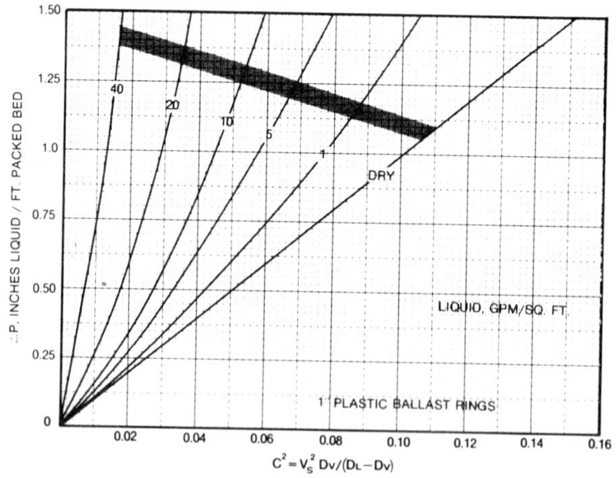

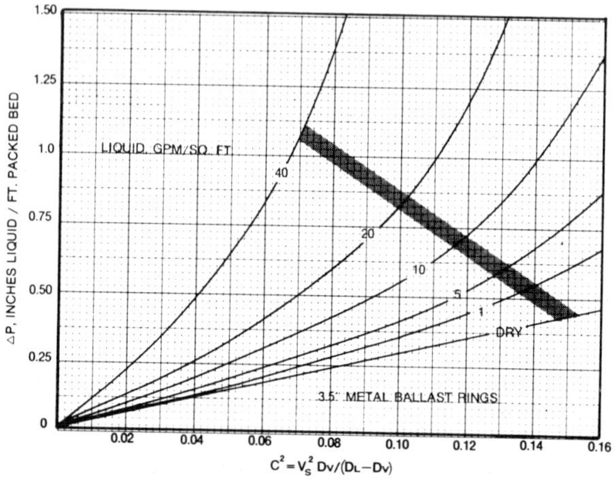

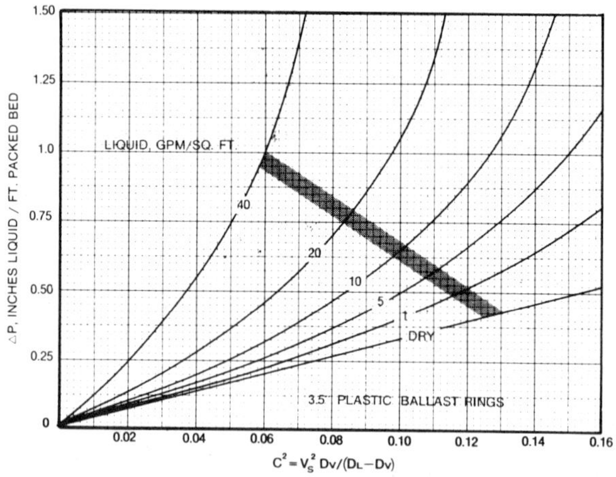

(c)

Figure 13.38. Capacity and pressure drop in beds of pall ("ballast") rings (*Glitsch, Inc.*). (a) Capacity chart for pall rings. V_s = vapor velocity (ft/sec). Example with 1 in. rings,

$$\text{fractional loading} = C/C_F = \begin{cases} 0.161/0.188 = 0.856 \text{ at constant } V/L \\ 0.161/0.200 = 0.805 \text{ at constant gpm.} \end{cases}$$

(b) Pressure drop at 85% of flooding. (c) Pressure drops with 1 and 3.5 in. metal and plastic rings at a range of flow rates.

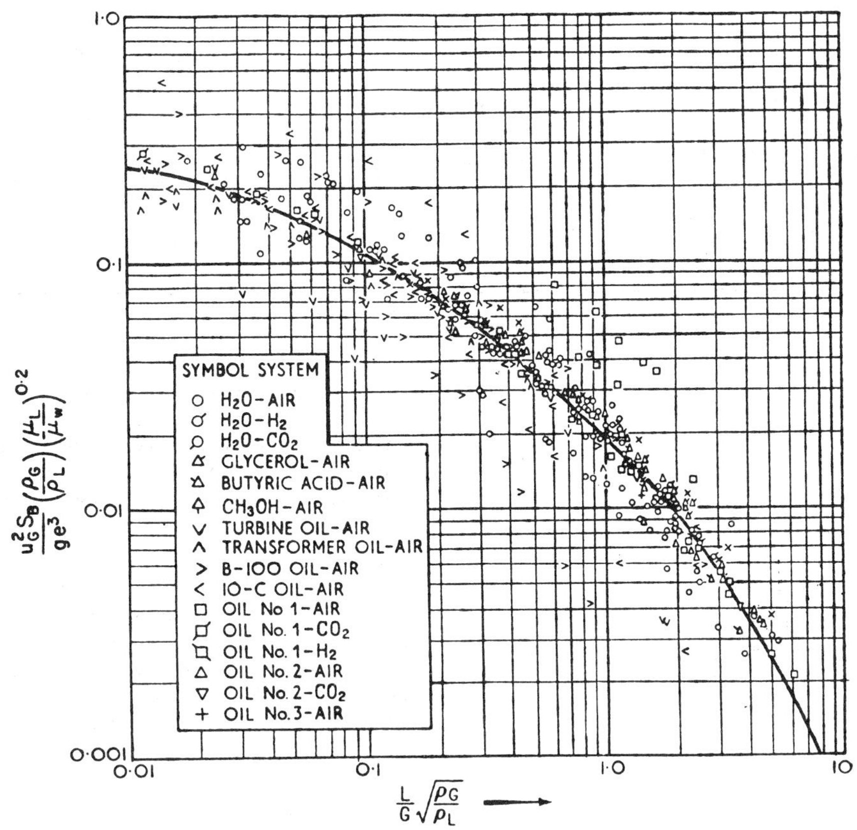

Figure 13.39. The Sherwood–Lobo correlation of flooding limit in random packed beds. μ_G is the superficial linear velocity of the gas, μ_L/μ_w is the ratio of viscosities of the liquid and water, S_B is the specific surface of the packing (sqft/cuft), ρ_G and ρ_L are densities of gas and liquid, and ε is the fraction voids; the ratio S_B/ε^3 is the factor F of the table with Figure 13.37 [*Sherwood, Shipley and Holloway,* Ind. Eng. Chem. **30,** *765* (*1938*); *Lobo, Friend, and Hashmall,* Trans. AIChE **41,** *693* (*1945*)].

Example 13.16
Performance of a Packed Tower by Three Methods

A packed tower with 3 in. metal pall rings will be analyzed for the system of Example 13.15. The packing factor is $F = 15$ sqft/cuft.

(a) Use the correlation of Figure 13.37:

$$\text{abscissa} = (L/G)\sqrt{\rho_V/(\rho_L-\rho_V)}$$
$$= (259{,}100/271{,}500)\sqrt{2.75/(29.3-2.75)} = 0.3071.$$

The ordinate y is read off the figure for several values of $\Delta P/L$; then the flow rate G' and the cross sectional areas are calculated from

$$G' = \sqrt{g_c\rho_V(\rho_L-\rho_V)/F} = 12.52\sqrt{y},\quad \text{lb/(sec)(sqft)},$$
$$A = 271{,}500/3600G' = 75.417/G',\quad \text{sqft},$$

$\Delta P/L$	y	G'	A	D
0.25	0.019	1.726	43.69	7.45
0.50	0.035	2.34	32.23	6.41
1.00	0.048	2.74	27.52	5.92

(b) Apply the method of Figure 13.38 for a tower 7.5 ft dia, 44.18 sqft.

$$\text{abscissa} = 1100/A = 24.9\ \text{gpm/sqft},$$
$$\text{ordinate}\quad C = (27.52/A)\sqrt{2.75/(29.3-2.75)}$$
$$= 8.857/44.18 = 0.2004.$$

The intersection of the line through the origin and the operating point (24.9, 0.2004) with the 3 in. ring line (interpolated) is at

$$C_F = 0.27.$$

Therefore,

$$\%\ \text{flooding} = 100(0.2004/0.27) = 74\%.$$

Check the pressure drop by this method.

$$C^2 = (0.2004)^2 = 0.040.$$

Interpolating on Figure 13.38(c) to 3 in. pall rings, at 1100/44.18 = 24.9 gpm/sqft,

$$\Delta P/L = 0.35\ \text{in. water/ft},$$

Example 13.16—*(continued)*

which is a rough check of the value $\Delta P/L = 0.25$ by method (a). Method (b), however, predicts that flooding would occur when $A = 32.23$, whereas method (a) says this size is acceptable if the pressure drop of 0.50 can be tolerated.

(c) Check the flooding by the Sherwood–Lobo correlation (Figure 13.39):

$$(L/G)\sqrt{\rho_G/\rho_L} = (259{,}100/271{,}500)\sqrt{2.75/29.3} = 0.2924,$$

$$y = 0.53 = [(Q/\pi D^2)^2/16g_c](\rho_G/\rho_L)(u_L/u_w)^{0.1}$$
$$= (27.52/\pi D^2)^2 2.75/16(32.2)(29.3),$$
$$D = \begin{cases} 5.64\text{ ft}, & \text{at flooding,} \\ 6.31\text{ ft}, & \text{at 80\% of flooding.} \end{cases}$$

These values are more nearly consistent with the data of Figure 13.37.

The equation is

$$E_{MV} = 7.0(D_g)^{0.14}(\text{Sc})^{0.25}(\text{Re})^{0.08}, \tag{13.229}$$

where

E_{MV} = percent efficiency,
$D_g = \sigma_L/\mu_L U_V$,
$\text{Sc} = \mu_L/\rho_L D_{LK}$,
$\text{Re} = h_W U_V \rho_V/[(\mu_L)(\text{FA})]$,
σ_L = surface tension (lb/hr²),
μ_L = liquid viscosity (lb/ft hr),
U_V = superficial vapor velocity (ft/hr),
D_{LK} = diffusivity of light key component (ft²/hr),
h_W = height of weir (ft),
ρ_L = liquid density (lb/ft³),
ρ_V = vapor density (lb/ft³),
FA = fractional free area available for vapor flow.

The equation of Bakowski [*Br. Chem. Eng.* **14**, 945 (1969); **8**, 384, 472 (1963)] is

$$E_{oc} = \frac{1}{1 + 3.7(10^4)KM/h'\rho_l T} \tag{13.230}$$

where

E_{oc} = overall column efficiency (fractional),
K = vapor–liquid equilibrium ratio, y^*/x,
y^* = gas-phase concentration at equilibrium (mole fraction),
x = liquid-phase concentration (mole fraction),
M = molecular weight,
h' = effective liquid depth (mm),
ρ_l = liquid density (kg/m³),
T = temperature (K).

The equation of Chu, Donovan, Bosewell, and Furmeister [*J. Appl. Chem.* **1**, 529 (1951)] is

$$\log_{10} E = 1.67 + 0.30\log_{10}(L/V) - 0.25\log_{10}(\mu_L \alpha) + 0.30h_L, \tag{13.231}$$

where

L, V = the liquid and vapor flow rates (kmol/sec),
μ_L = the viscosity of the liquid feed (mN sec/m²),
α = the relative volatility of the key components,
h_L = the effective submergence (m), taken as the distance from the top of the slot to the weir lip plus half the slot height.

Four of these relations are applied in Example 13.17. The McFarland and Bakowski methods bracket experimental values, whereas the other methods give low values. This comparison, of course, probably is not generally valid. Even experimental values are not exact, unless they are found for exactly the desired operating conditions, the same tray design, and for the same key components. Nevertheless, the collected experimental data and the several correlations that have been cited supply a background on which judgement can be applied to specific problems.

PACKED TOWERS

The most useful measure of the separating power of packed towers is the HETP, the height equivalent to a theoretical plate or stage. It is evaluated simply as the ratio of packed height used for a certain degree of separation to the theoretical number of stages. Its relation to the fundamental quantity, HTU, or the height of a transfer unit, is

$$\text{HETP} = \text{HTU}\frac{\ln(mV/L)}{mV/L - 1}, \tag{13.232}$$

where m is the slope of the equilibrium curve. In distillation, the equilibrium and operating lines diverge below the feed point and converge above it. As a result the value of mV/L averages approximately unity for distillation so that HETP and HTU become essentially equal. Usually this is not true in absorption–stripping processes.

Data also are reported as mass transfer coefficients. For the gas phase, the relation to the HTU is

$$(\text{HTU})_G = G/k_G aP, \tag{13.233}$$

where G is the molal flow rate of the gas, say in the units lbmol/(hr)(sqft), P is the total pressure and $k_G a$ has the units lbmol/(hr)(cuft)(unit of pressure). The liquid phase relation is

$$(\text{HTU})_L = Lk_L a\rho_L, \tag{13.234}$$

where L is the molal flow rate of the liquid [lbmol/(hr)(sqft)], $k_L a$ has the units lbmol/(hr)(cuft)(unit of concentration difference), and ρ_L is the liquid density. The individual HTUs are combined into overall expressions by

$$(\text{HTU})_{OG} = (\text{HTU})_G + (m'V/L)H_L, \tag{13.235}$$
$$(\text{HTU})_{OL} = (\text{HTU})_L + (L/m''V)H_G. \tag{13.236}$$

The positions of the slopes m' and m'' of the equilibrium curve are identified in Figure 13.44(a).

Selected data of HETP, $k_G a$, HTU and pressure drop are in Figures 13.45 and 13.46.

Mass Transfer Coefficients. A relation covering liquid and vapor phase mass transfer coefficients is cited in Section 13.9.

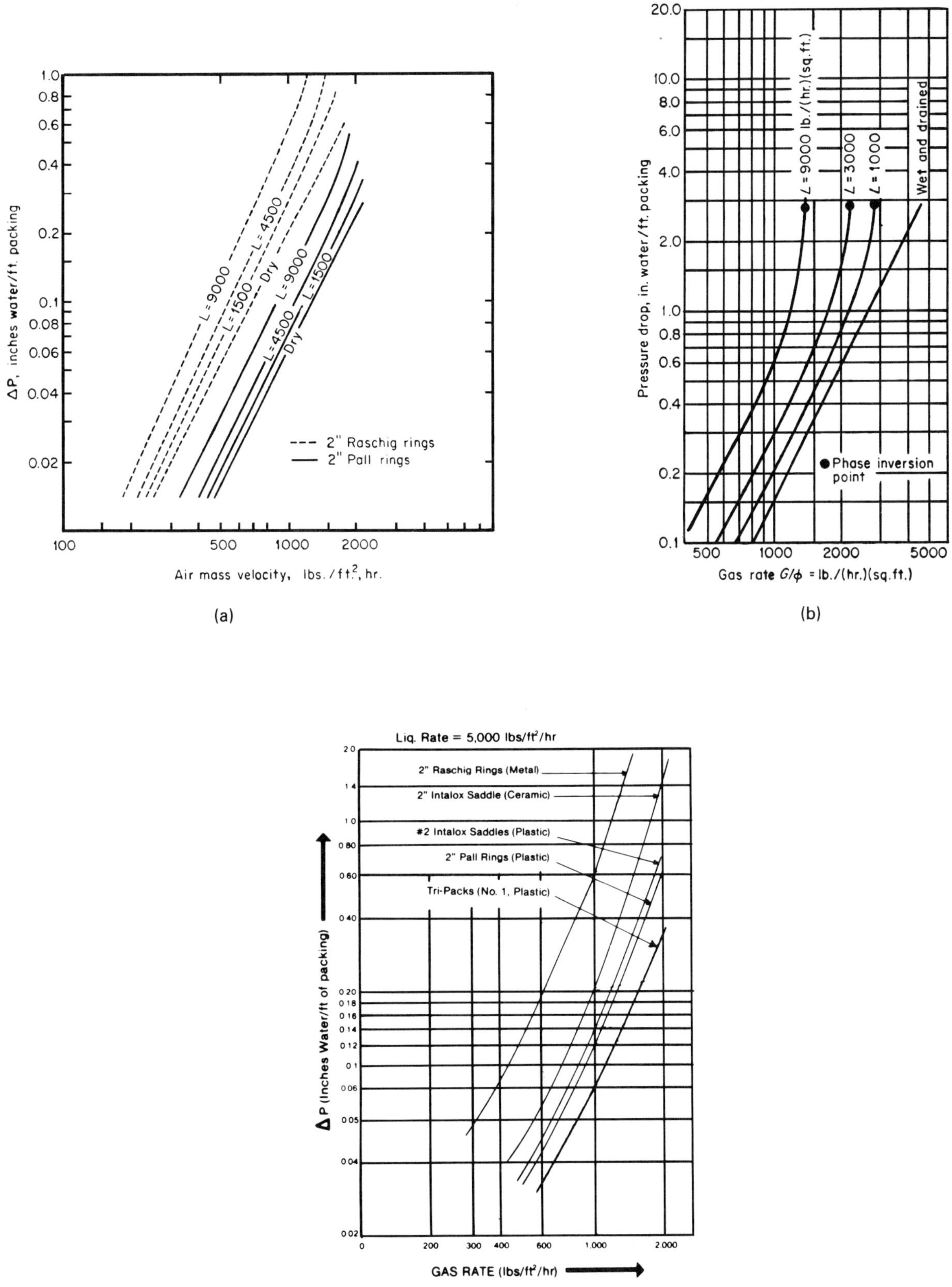

Figure 13.40. Comparison of pressure drops through several kinds of packed beds. (a) 2 in. Raschig and pall rings [*Eckert et al.,* Chem. Eng. Prog. **54**(*1*), *70* (*1958*)]. (b) 1 in. Tellerettes [*Teller and Ford,* Ind. Eng. Chem. **50,** *1201* (*1958*)]. (c) 2 in. plastic Tripack and other 2 in. packings (*Polymer Piping and Metals Co.*).

TABLE 13.15. Survey of Tray Efficiencies

System	Col. dia., m	Press., kPa	Temp., °K	Weir height, m	Tray efficiency, % E_{oG}	E_{mv}	E_o	Remarks
Bubble-cap								
Ethanol/Water	0.46					98		
	—					70	95	
	0.11					64.6		
	0.46	101			80–90	90–130		with splash baffle
	—					95		
	—				81			±11%
	0.48			0.81		85		±5%
	0.15					99.8		Carey (1934)
	0.196	101				80		
Methylcyclohexane/ toluene	0.11					64.6		
Air/water	1.52		290				83	
Carbon dioxide/water	0.076					80	125	no mixing
	—						100	50% mixing
Acetic acid/water	0.46	101				65		
	0.59	58				54		±5%
Deuterium/hydrogen	0.027		20			50	44	±3%
Oxygen/nitrogen	—						76	±9%
Acetone/water	5.49				91			
	—				83			±2%
Ethylene dichloride/ toluene	—			0.032		95		
Sugar/water	1.52					80		
$CHCl_3/CCl_4$	—				90			±5%
Ammonia/water	0.305	101	283			77		
Methanol/water	1.0					90		
	—					70		
Acetone/benzene	—				79			±1%
Methanol/Isopropanol/ water	0.45				68	68		
	—				70			
	—				68			
Acetone/methanol/water	0.45				60			
	—				60			
	—				80			
Gasoline stabilizers	1.44	1,820			75	100		Brown (1936)
Benzene/toluene	0.2					60		Lewis (1930)
	0.2					70		Carey (1934)
	0.15					58		Carey (1934)
Aniline/water	0.2					58		Carey (1934)
Naphtha/water	2.74					65		Lewis (1928)
Isopropanol/water	0.45					78		
Methanol/isopropanol	0.45					64		
Acetone/methanol	0.45					61		
Benzene/toluene/xylene	0.2					75		Lewis (1930)
Naphtha/pinene/aniline	0.2					90		Lewis (1930)
Sieve								
Ethanol/Water	0.076	101			45.5			
	0.127					85		
	0.196	101				90		
	—					71.4		CA-100455C (1970)
	Lab						120	Brown (1936)
Methylethylketone/Water	0.08				41			
Acetone/water	0.05	101			25.5			
	0.11				43.5			±11.5%
	0.15		373			80		
Benzene/water	0.05	101			9.6			
Toluene/water	0.05	101			7.1			
n-Heptane/ methylcyclohexane	0.04					77.6		±2.07%
n-Heptane/cyclohexane	1.2	165		0.05			85	
	2.44	165		0.05			75	
	1.2	164		0.05		90		60% flood
Toluene/ methylcyclohexane	0.15	101					54.6	±15%
	—	27					55.5	±5%

TABLE 13.15—(*continued*)

System	Col. dia., m	Press., kPa	Temp., °K	Weir height, m	Tray efficiency, % E_{oG}	E_{mv}	E_o	Remarks
Sieve, cont.								
Methylcyclohexane/	0.05	101					91	
toluene	—					88		
Propane/butane	—						100	±5%
Carbon dioxide/water	0.08					80	125	no mixing
	—						100	50% mixing
	0.15	111	298			89.7		
n-Octane/toluene	0.15	101		0.025			38	
Air/water/ammonia	0.08		298	0.08	85.7	96.4		±0.1
	—			0.03	70			
	—					65	50	
Oxygen/water/ammonia	0.15					75		±7%
Ammonia/water	0.3	101	283			89		
Methylisobutylketone/ water	0.08		298	0.08	41.5	64		±1.7
Ethylene dichloride/ toluene	0.05	101					75	
Methylethylketone/toluene	0.15			0.05			88	
Air/ethanol	—			0.1	80			±20%
Air/propanol	—			0.03	77			±5%
Methanol/CCl_4	0.11				25.7			±0.8
Methanol/water	0.11				56			±13%
	—				79.2			
	1.0					93		
	—					90		CA-100455C (1970)
Acetone/CCl_4	0.11				50			±9%
Isopropanol/water	—						72.9	CA-100455C (1970)
Benzene/toluene	0.127					75		
	—				76.5			±7.1%
	10.7	101	353	0.08		80.5		1 pass
	—					80.5		2 pass
	—			0.14		85.2		4 pass
Benzene/methanol	0.18	690		0.05	86.4	94.2	85	
Cyclohexane/toluene	2.4	103					70	
Ethylbenzene/styrene	—	13						
	0.5	13		0.038		75		
Helium/	—				90			±5%
methylisobutylketone				0.038				
Nitrogen/isobutanol	—				80			
Nitrogen/cyclohexanol	—			0.051	70			±20%
Acetic acid/water	0.46	101				75		
Benzene/propanol	0.46					58.6		96.6% flood
Cyclohexane/n-heptane/	0.038				78			
toluene	—				110			
	—				93			
n-Heptane/toluene	—			0.02		45		
	—					62		
n-Heptane/benzene	—			0.02		55		
	—					68		
CCl_4/benzene	0.032						71	no vapor pulsing
	—						73	vapor pulsing
Isobutane/n-butane	—	2,068				110		
Ethanol/water/furfural	—					80		CA-165714Y (1980)
n-Hexane/ethanol/	0.1	101	333		70			+0.9
methylcyclopentane	—				70.3			−2.5
	—				71			−4.4
n-Hexane/ethanol/	0.1	101	333		55			+14.3
methylcyclopentane/	—				60			−1.5
benzene	—				6			−1.9
Benzene/n-propanol	0.46	101	366	0.08		54		±5%, 60% flood
	0.03	101	366			57		±5%, 60% flood
Toluene/n-propanol	0.46	101	366	0.08		61	65	±5%, 60% flood
	0.03	101	366			57		±8%, 60% flood
Beer/water	Comm.						120	Brown (1936)

(*continued*)

TABLE 13.15—(*continued*)

System	Col. dia., m	Press., kPa	Temp., °K	Weir height, m	Tray efficiency, % E_{oG}	E_{mv}	E_o	Remarks
Sieve, cont.								
Air/water triethyleneglycol	—	101	297	0.05		62		
APV-West								
Methanol/water	1.0					81		
Kascade								
Ethanol/water	0.2					54.1		
	0.2						70	
Methylcyclohexane/toluene	0.2					44.6		
							72	
Oxygen/water	0.2					84		
Tunnel								
Furfural/isobutane & butylene	4.0	593	318			25		±4%, ΔP = 12.5
Furfural/n-butane & butylene	4.0	593	318			25		±7%, ΔP = 12.5
Turbogrid								
Ammonia/water	0.3	101	283			75		
Ethanol/water	—	0.24				85		
	—	101					85	±11%
Methanol/water	—					87		±2%
	0.1	101	298			95		±3.9%
	Small					86		
Methanol/isopropanol/ water	0.15					66.4		
Methanol/isopropanol	Small					65		
V-Grid								
Air/water/ammonia	—					70	60	
Combination valve-sieve								
Benzene/propanol	0.46				32			
Ethylenebenzene/styrene	—	13					80	
Wyatt Perfavalve								
Propanol/toluene	0.46					76.7		74.7% flood
Propanol/benzene	0.46					55.5		44.4% flood
Valve								
Benzene/toluene/xylene	2.43					69		±3%
Ethanol/water	0.032						70	
	0.06					56		CA-107562W (1975)
n-Propanol/benzene	0.46					73		±5%, 60% flood
n-Propanol/toluene	0.46					51		±5%, 60% flood
Ethylbenzene/styrene	0.5	13				85		
Benzene/C_8 aromatics	2.43				44		88	top
					66			bottom
Round valve								
Ethanol/water	0.196	101				92		
L-Type valve								
Ethanol/water	0.196	101				68		
Nutter valve								
C_6/C_7	—	165					96	float valve
iC_4/nC_4		1,131					121	float valve
Propanol/benzene	0.5					63.8		29.6% flood
Propanol/toluene	0.5		303			75.3		
Cyclohexane/heptane	1.2						96	20% flood
Koch valve								
Cyclohexane/heptane	1.2						93.8	50% flood
Benzene/toluene/xylene	2.43	92	346	0.08		48	88	
Glitsch								
n-Butane/isobutane	1.3	1,138		0.05			122	V-1 Ballast
Cyclohexane/n-heptane	1.3	165		0.05		99.5		V-1 Ballast
	1.2						97	21% flood, Valve
Methanol/water	0.1	101					High	A-1 Ballast CA-31366g (1967)
	—			0.05		88		±2%, Valve (downcomers)

TABLE 13.15—*(continued)*

System	Col. dia., m	Press., kPa	Temp., °K	Weir height, m	Tray efficiency, % E_{oG}	E_{mv}	E_o	Remarks
Ripple								
Methanol/water	1.0					73		
Ammonia/water	0.3	101				82		
Light gasoline	2.0						48	CA-68739C (1968)
Uniflex								
Methanol/water	1.0					87		
Baffle								
Toluene/methylcyclohexane	—	101				87		
Angle								
Methanol/water	—	101					70	CA-18097g (1972)
Crossflow plate								
Benzene/toluene	—					60		
Ethanol/water	—					70		
Jet								
Air/water/ triethyleneglycol	—	101	297	0.05		65		
Methylethylketone/toluene	0.15			0.03			93	

[References given in the original: Vital, Grossel, and Olsen, *Hyd. Proc.*, 147–153 (Nov. 1984)].

TABLE 13.16. Efficiency Data of Some Operations with Bubblecap, Ripple, and Turbogrid Trays

Disperser	System	Column diameter, ft	Tray spacing, in	Pressure, psia	Static submergence, in	Efficiency, % E_{mv}*	E_{oc}†	Remarks	Ref.
Bubble cap	Ethanol-water	1.31	10.6	14.7	1.18	83–87	......	......	1
		1.31	16.3	14.7	1.18	84–97			
		2.5	14	14.7	1.2	80–85	......	......	2
	Methanol-water	3.2	15.7	14.7	1.0	90–95	......	......	3
	Ethyl benzene–styrene	2.6	19.7	1.9	0.2	55–68	......	......	4
	Cyclohexane–*n*-heptane	4.0	24	14.7	0.25	65–90	......	......	5
				24	4.25	65–90	......		
				50		65–90			
	Cyclohexane–*n*-heptane	4.0	24	5	0.6	65–85	......	Tunnel caps	6
				24		75–100	......		
	Benzene-toluene	1.5	15.7	14.7	1.5	70–80	......	......	7
	Toluene-isooctane	5.0	24	14.7	0.4	......	60–80	......	8
Ripple sieve	Methanol-water	3.2	15.7	14.7	...	70–90	......	10.8% open	3
	Ethanol-water	2.5	14	14.7	1.0	75–85		10.4%	2
	Methanol-water	3.2	15.7	14.7	1.57	90–100		4.8% open	3
	Ethyl benzene–styrene	2.6	19.7	1.9	0.75	70		12.3% open	9
	Benzene-toluene	1.5	15.7	14.7	3.0	60–80		18% open	7
	Methyl alcohol–*n*-propyl alcohol–*sec*-butyl alcohol	6.0	18	18	1.38	......	64	......	10
	Mixed xylenes + C_8-C_{10} paraffins and naphthenes	13.0	21	25	1.25	......	86	......	5
	Cyclohexane–*n*-heptane	4.0	24	5	2.0	60–70	......	14% open	13
				24		80	......	14% open	13
		4.0	24	5	2.0	70–80	......	8% open	12
	Isobutane-*n*-butane	4.0	24	165	2.0	110	......	14% open	13
		4.0	24	165	2.0	120	......	8% open	12
		4.0	24	300	2.0	110	......	8% open	12
		4.0	24	400	2.0	100	......	8% open	12
Turbogrid valve	Methanol-water	3.2	15.7	14.7	...	70–80	......	14.7% open	3
	Ethanol-water	2.5	14	14.7	1.0	75–85	......	......	2
	Ethyl benzene–styrene	2.6	19.7	1.9	0.75	75–85	......	......	4
	Cyclohexane–*n*-heptane	4.0	24	20	3.0	......	50–96	Rect. valves	11
	n-Butane-isobutene	4.0	24	165	3.0	......	104–121	Rect. valves	11
	Benzene-toluene	1.5	15.7	14.7	3.0	75–80	......	......	7

References
1. Kirschbaum, *Z. Ver. Dtsch. Ing. Beth. Verfahrenstech.*, (5), 131 (1938); (3), 69 (1940).
2. Kirschbaum, *Distillier-Rektifiziertechnik*, 4th ed., Springer-Verlag, Berlin and Heidelberg, 1969.
3. Kastanek and Standart, *Sep. Sci*, **2**, 439 (1967).
4. Billet and Raichle, *Chem. Ing. Tech.*, **38**, 825 (1966); **40**, 377 (1968).
5. AIChE Research Committee, *Tray Efficiency in Distillation Columns*, final report, University of Delaware, Newark, 1958.
6. Raichle and Billet, *Chem. Ing. Tech.*, **35**, 831 (1963).
7. Zuiderweg, Verburg, and Gilissen, *Proc. Intn. Symp.*, Brighton, England, 1960.
8. Manning, Marple, and Hinds, *Ind. Eng. Chem.*, **49**, 2051 (1957).
9. Billet, *Proc. Intn. Symp.*, Brighton, England, 1970.
10. Mayfield, Church, Green, Lee, and Rasmussen, *Ind. Eng. Chem.*, **44**, 2238 (1952).
11. Fractionation Research, Inc., "Report of Tests of Nutter Type B Float Valve Tray," July 2, 1964, from Nutter Engineering Co., Tulsa.
12. Sakata and Yanagi, *Inst. Chem. Eng. Symp. Ser.*, no. 56, 3.2/21 (1979).
13. Yanagi and Sakata, *Ind. Eng. Chem. Process Des. Dev.* **21**, 712 (1982).

*See Eq. (18-32).
†See Eq. (18-28).
NOTE: To convert feet to meters, multiply by 0.3048; to convert inches to centimeters, multiply by 2.54; and to convert pounds-force per square inch to kilopascals, multiply by 6.895.

(*Chemical Engineers' Handbook*, McGraw-Hill, New York, 1984).

HETP Correlations. Most of the data available for correlation are laboratory data and not indicative of large scale behavior except perhaps on a comparative basis. Some guidelines for full scale tray behavior are stated by Frank [*Chem. Eng.* **111,** (14 Mar. 1977)] in this table:

Type of Packing/Application	HETP (m)
25 mm dia packing	0.46
38 mm dia packing	0.66
50 mm dia packing	0.9
Absorption duty	1.5–1.8
Small diameter columns (<0.6 m dia.)	column diameter
Vacuum columns	values as above +0.1 m

A correlation for Raschig rings and Berl saddles by Murch [*Ind. Eng. Chem.* **45,** 2616 (1953)] covers columns up to 30 in. dia and 10 ft high. His relations are

$$\text{HETP} = C_1 G'^{C_2} d_c^{C_3} Z^{1/3} \alpha \mu_L / \rho_L, \tag{13.237}$$

where

G' = mass velocity of vapor (kg/m^2 sec) of tower area,
d_c = column diameter (m),
Z = packed height (m),
α = relative volatility,
μ_L = liquid viscosity (N sec/m^2),
ρ_L = liquid density (kg/m^3).

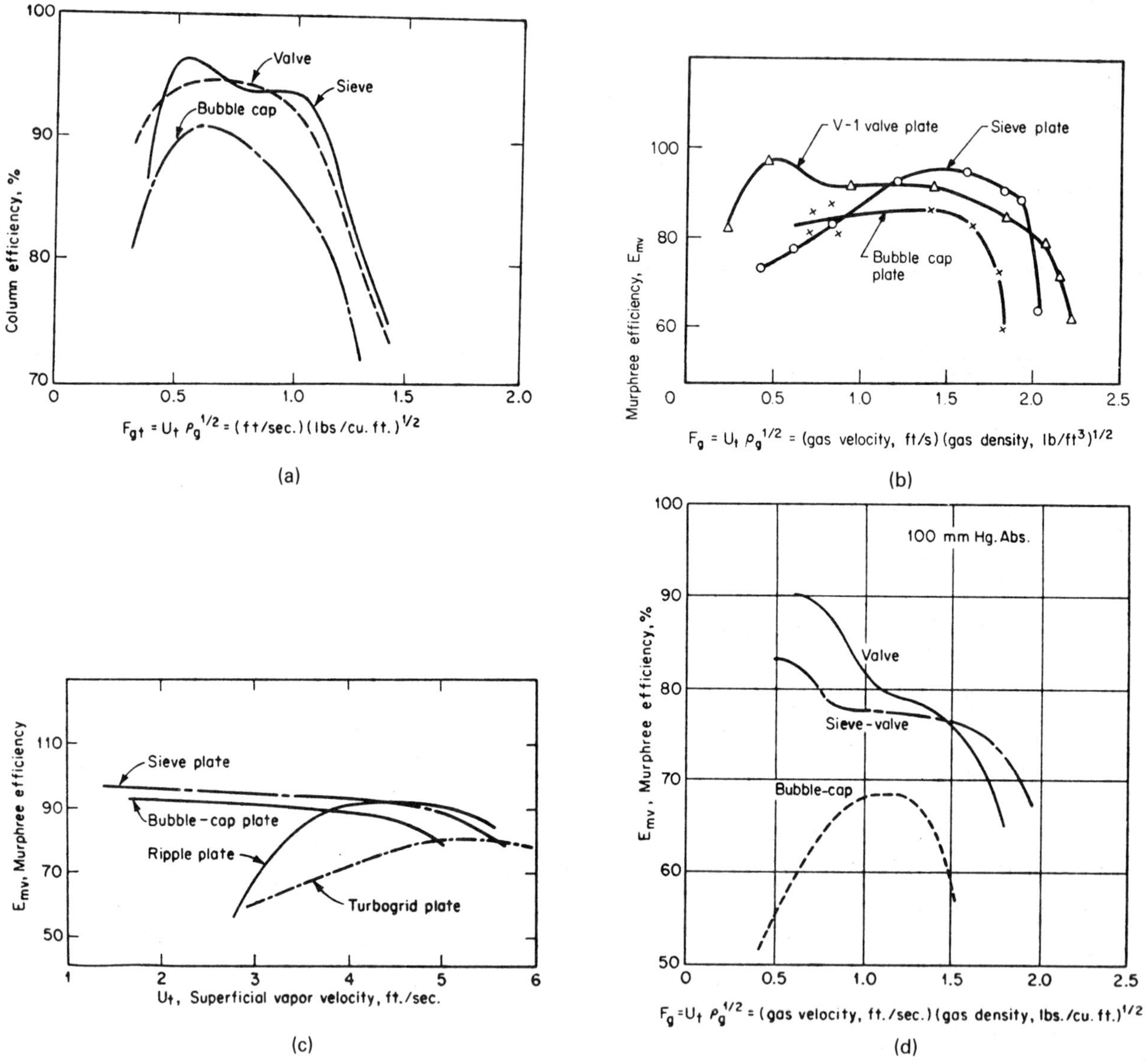

Figure 13.41. Efficiencies of some fractionations with several types of trays as a function of vapor factor $F = u\sqrt{\rho}$ or linear velocity. (a) Data of methanol/water in a column 3.2 ft dia [*data of Kastanek, Huml, and Braun,* Inst. Chem. Eng. Symp. Ser. **32**(*5*), *100* (*1969*)]. (b) System cyclohexane/n-heptane in a 4 ft dia sieve column [*Sakata and Yanagi,* Inst. Chem. Eng. Symp. Ser. **56,** *3.2/21* (*1979*)]; valve tray data (*Bulletin 160, Glitsch Inc., 1967*). (c) Methanol/water [*Standart et al.,* Br. Chem. Eng. **11,** *1370* (*1966*); Sep. Sci. **2,** *439* (*1967*). (d) Styrene/ethylbenzene at 100 Torr [*Billet and Raichle,* Chem. Ing. Tech. **38,** *825* (*1966*); **40,** *377* (*1968*)]. (e) Ethanol/water (*Kirschbaum,* Destillier und Rektifiziertechnik, *Springer, Berlin, 1969*). (f) Methanol/water [*Kastanek, Huml, and Braun,* Inst. Chem. Eng. Symp. Ser. **32,** *5.100*, (*1969*)].

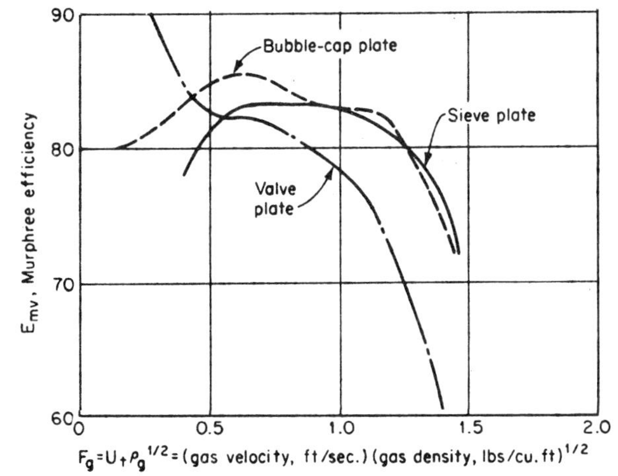

(e)

Figure 13.41—(*continued*)

Values of the constants C_i appear in this tabulation:

Type of Packing	Size (mm)	C_1 ($\times 10^{-5}$)	C_2	C_3
Rings	6			1.24
	9	0.77	−0.37	1.24
	12.5	7.43	−0.24	1.24
	25	1.26	−0.10	1.24
	50	1.80	0	1.24
Saddles	12.5	0.75	−0.45	1.11
	25	0.80	−0.14	1.11

A correlation for 25 and 70 mm Raschig rings by Ellis [*Birmingham Univ. Chem. Eng.* **5**(1), 21 (1953)] with HETP (m) is

$$\text{HETP} = 18d_r + 12m[(G'/L' - 1)], \tag{13.238}$$

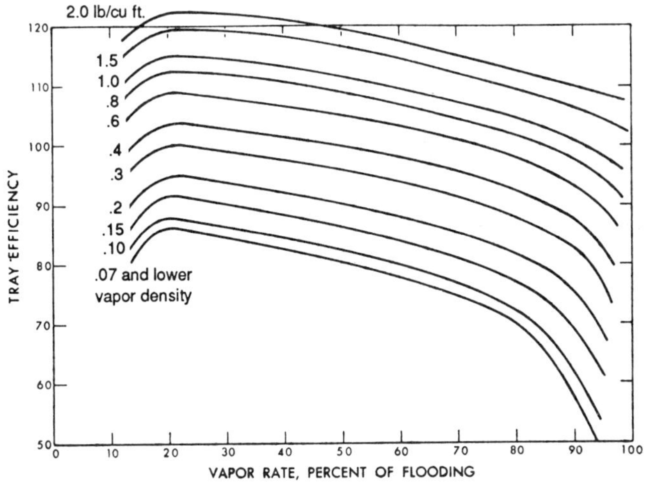

Figure 13.42. Efficiency of Glitsch V-1 valve trays on isobutane/butane and cyclohexane/n-heptane as a function of vapor density and percent of flood, measured by Fractionation Research Inc. (*Glitsch Inc., Bulletin 160, Dallas, TX, 1958*).

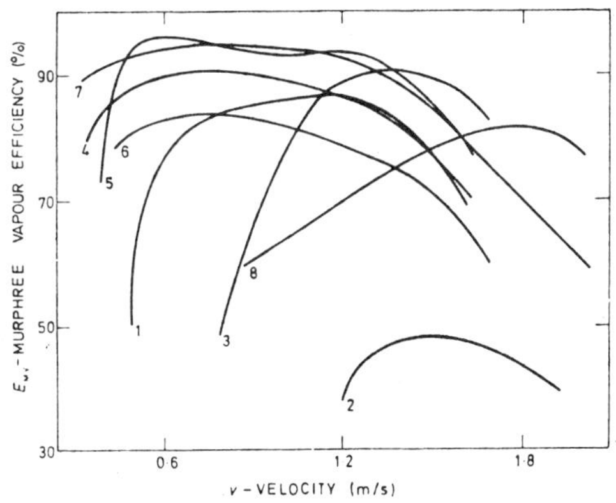

1: Uniflux plate, $F_c = 0{\cdot}1455$, $h_w = 70$ mm
2: Kittel plate, $F_c = 0{\cdot}213$, elliptic slot $14{\cdot}5 \times 8$ mm
3: Ripple plate, $F_c = 0{\cdot}1082$, $d = 3{\cdot}0$ mm
4: Bubble-cap plate, $F_c = 0{\cdot}099$, $h_w = 60$ mm
5: Sieve plate, $F_c = 0{\cdot}042$, $h_w = 40$ mm, $d = 4{\cdot}0$ mm
6: APV–West plate, $F_c = 0{\cdot}1072$, $h_w = 45$ mm
7: Glitsch valve plate, A1, $F_c = 0{\cdot}14$, $h_w = 50$ mm
8: Turbogrid, set IV, $F_c = 0{\cdot}147$, $d = 4{\cdot}5$ mm

(f)

where

d_r is the diameter of the rings (m),
m is the average slope of equilibrium curve,
G' is the vapor mass flow rate,
L' is the liquor mass flow rate.

HTU data have been correlated by Cornell et al. (1960) and updated by Bolles and Fair [*Inst. Chem. Eng. Symp. Ser.* **56**(2), 3.3/35 (1979)]. Pall rings, Raschig rings, and saddles are covered in the original article, but only the pall ring results are quoted here. Separate relations for the liquid and vapor phases are represented by Eqs. (13.239) and (13.240) and Figure 13.44.

$$H_L = \phi(\text{Sc}_L)^{0.5}(C)(Z/10)^{0.15} \tag{13.239}$$

where

H_L = height of liquid phase transfer unit (ft),
ϕ = parameter from figure,
Sc_L = liquid phase Schmidt number, $\mu_L/\rho_L D_L$,
C = flooding correction factor from figure,
Z = height of column packing (ft);

$$H_V = \frac{\psi(\text{Sc}_V)^{0.5}}{(L_m f_1 f_2 f_3)^{0.6}}\left(\frac{d_c}{12}\right)^{1.24}\left(\frac{Z}{10}\right)^{1/3}, \tag{13.240}$$

where

H_V = height of gas film transfer unit (ft),
Sc_V = vapor-phase Schmidt number,
d_c = column diameter (in.),
Z = packing height (ft),
ψ = parameter from figure,
L_m = liquid flow rate (lb/hr) ft^2),
$f_1 = (\mu_L/1.005)^{0.16}$, μ_L (cP),
$f_2 = (1/\rho_L)^{1.25}$, ρ_L (g/cm^3),
$f_3 = (72.8/\sigma)^{0.8}$, σ (dyn/cm).

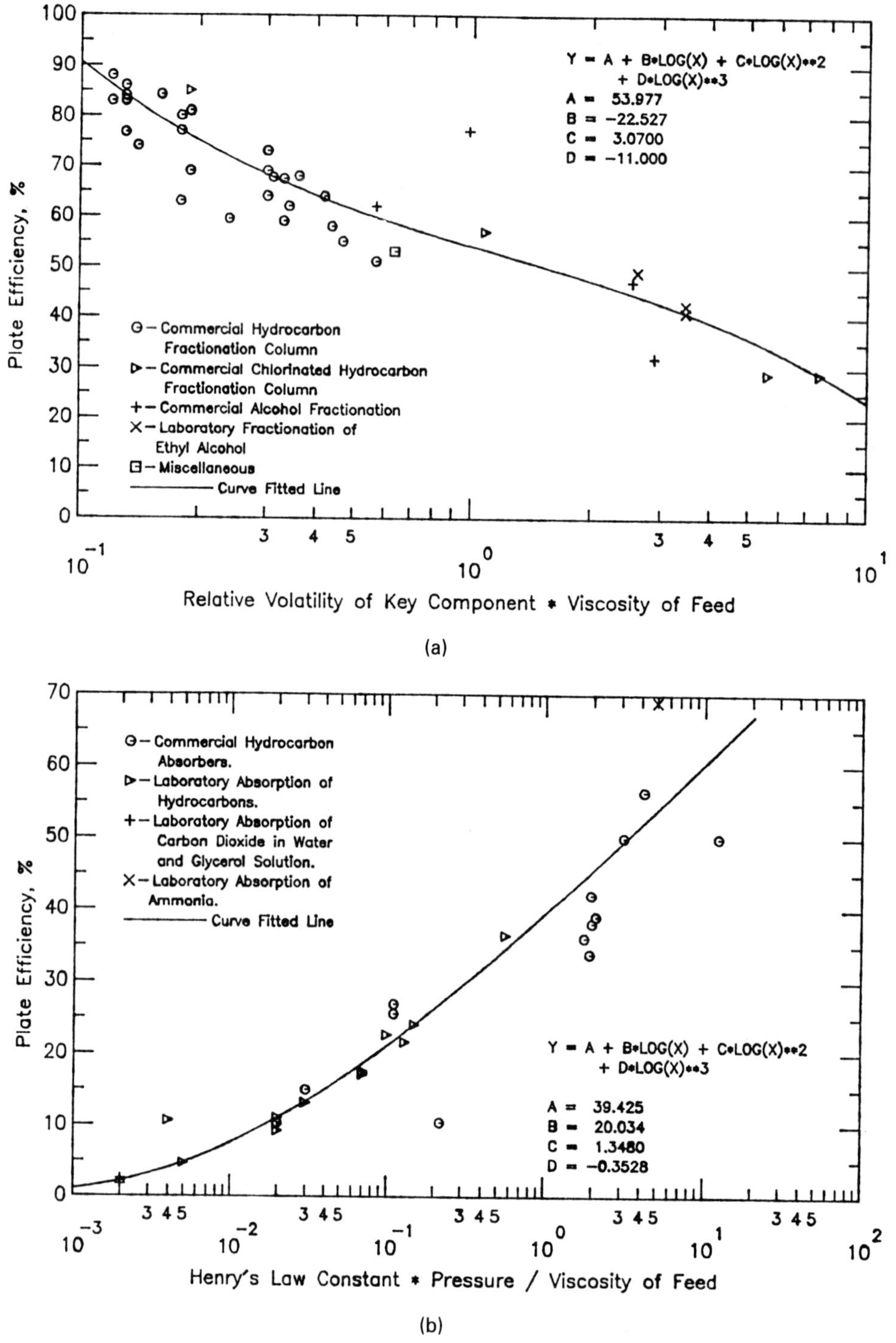

Figure 13.43. Efficiencies of fractionators and absorber-strippers. The original curves of O'Connell [Trans. AIChE **42,** *741 1946*)] have been replotted and fitted with equations, as shown on the figures, by S. Negahban (*University of Kansas, 1985*). (a) Fractionators (the viscosity μ is in cP). (b) Absorbers and strippers; H = Henry's law constant in lb mol/(cuft)(atm), P is in atm, and μ is in cP.

Example 13.17
Tray Efficiency for the Separation of Acetone and Benzene

(a) Method of McFarland et al.: The operating data are taken from their article, as follows:

Acetone mole fraction, $x_1 = 0.637$,
Benzene mole fraction, $x_2 = 0.363$,
Temperature T (°F) = 166,
Superficial vapor mass velocity G (lb/hr sqft) = 3820,
Vapor velocity u_v (ft/hr) = 24,096,
Weir height, h_w (ft) = 0.2082,
Fraction free area FA = 0.063.

The pertinent physical properties of the mixture are

$\mu_L = 0.609$ lb/ft hr, 0.252 cP,
$\sigma_L = 5.417(10^5)$ lb/sqft hr, 18.96 dyn/cm,
$D_{\text{light key}} = 2.32(10^{-4})$ sqft/hr.

The dimensionless groups appearing in the correlation are

$$N_{Dg} = \sigma_L/\mu_L u_V = 37,$$
$$N_{Sc} = \mu_L/\rho_L D_{LK} = 55,$$
$$N_{Re} = h_w G/\mu_L(\text{FA}) = 2.07(10^4).$$

The tray efficiency is found with Eq. (13.229):

$$E = 7.0(N_{Dg})^{0.14}(N_{Sc})^{0.25}(N_{Re})^{0.08} = 71\%.$$

(b) Method of O'Connell: The relative volatility is 3.24 at $x = 0.05$ and 1.63 at $x = 0.95$, or a geometrical mean value of $\alpha = 2.30$. Accordingly,

$$\mu\alpha = 0.252(2.3) = 0.58,$$

and, from Figure 13.43 or Eq. (13.227),

$$E = 56\%.$$

(c) Method of Bakowski, Eq. (13.230):

$K = y/x = 1.20$ at $x = 0.637$,
$M = 58$,
$h' = 50$ mm,
$\rho_t = 820\ \text{kg/m}^3$,
$T = 348$ K,

$$E = \frac{100}{1 + 37{,}000(1.20)(58)/50(820)(348)} = 84.7\%.$$

(d) Method of Chu et al., Eq. (13.231):

$$\text{L/V} = \begin{cases} 0.8, & \text{above the feed,} \\ 1.2, & \text{below the feed,} \end{cases}$$

$$\log E = 1.67 - 0.25\log(0.58) + 0.3(0.05) + 0.3\log(L/V)$$
$$= 1.744 + 0.3\log(L/V),$$

$$E = \begin{cases} 51.9\%, & \text{above the feed,} \\ 58.6\%, & \text{below the feed.} \end{cases}$$

(e) Experimental data: Table 13.15 shows $E = 79\%$ for acetone/benzene in bubblecap tower and $E = 85\%$ for methanol/benzene with sieve trays. Figure 13.41 shows that efficiencies above 80% are readily attainable near $u\sqrt{\rho} = 1.0$.

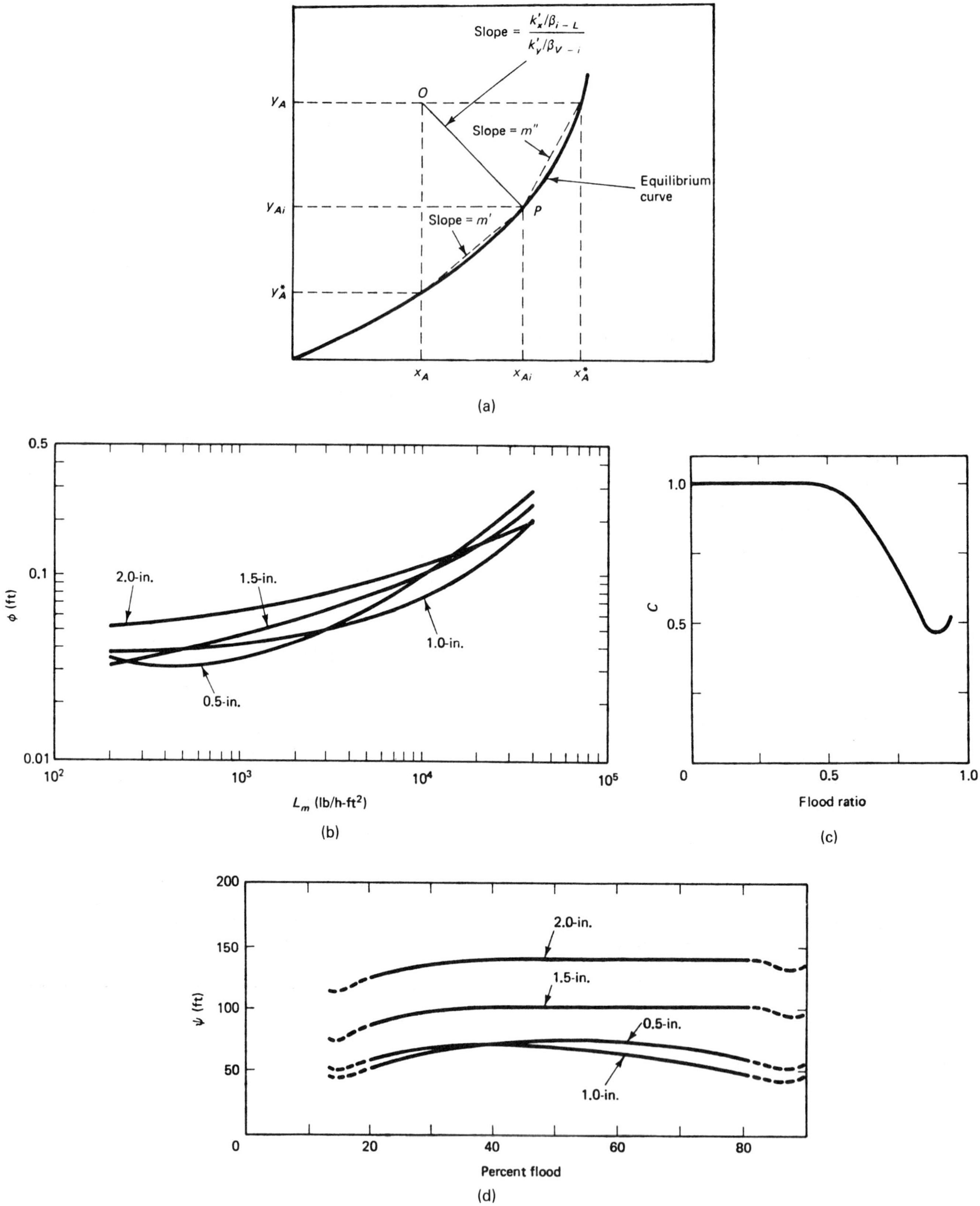

Figure 13.44. Factors in Eqs. (13.239) and (13.240) for HTUs of liquid and vapor films; and slopes m' and m'' of the combining Eqs. (13.235) and (13.236): [*Bolles and Fair,* Inst. Chem. Eng. Symp. Ser. **56**(*2*), *3.3/3.5*, (*1979*)]. (a) Definitions of slopes m' and m'' in Eqs. (13.235) and (13.236) for combining liquid and gas film HTUs; $\beta = 1$ for equimolal counter diffusion; $\beta = (x_B)_{\text{mean}}$ for diffusion through a stagnant film. (b) Factor ϕ of the liquid phase Eq. (13.239). (c) Factor C of the liquid phase, Eq. (13.239). (d) Factor ψ of the gas phase, Eq. (13.240), for metal pall rings.

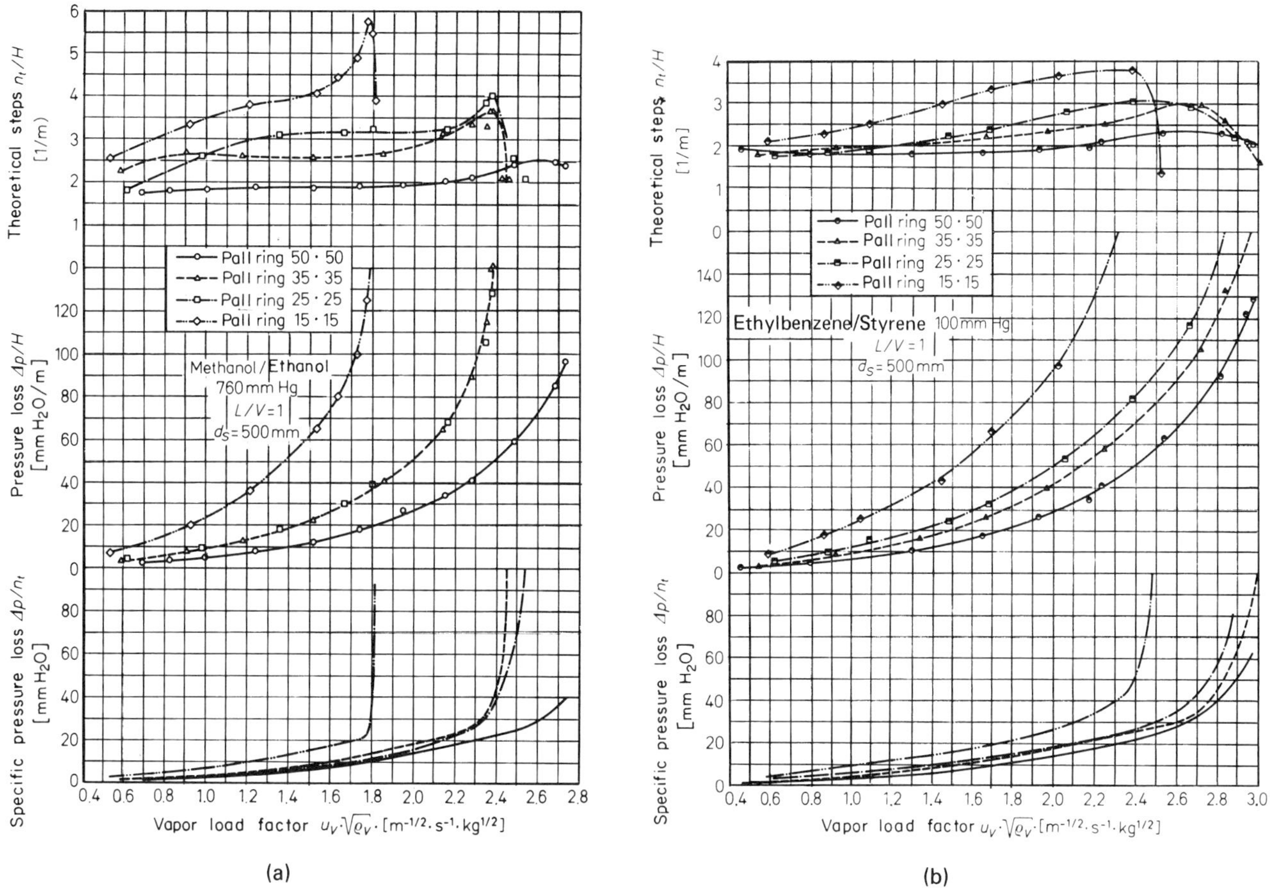

Figure 13.45. Number of stages per meter (reciprocal of HETP), pressure loss per meter and pressure loss per theoretical stage in a 500 mm dia column filled with metal pall rings. Other charts in the original show the effects of packing height and column diameter, as well as similar data for Raschig rings (*Billet, 1979*). (a) Methanol/ethanol at 760 Torr and total reflux in a column 500 mm dia. (b) Ethylbenzene/styrene at 100 Torr and total reflux in a column 500 mm dia.

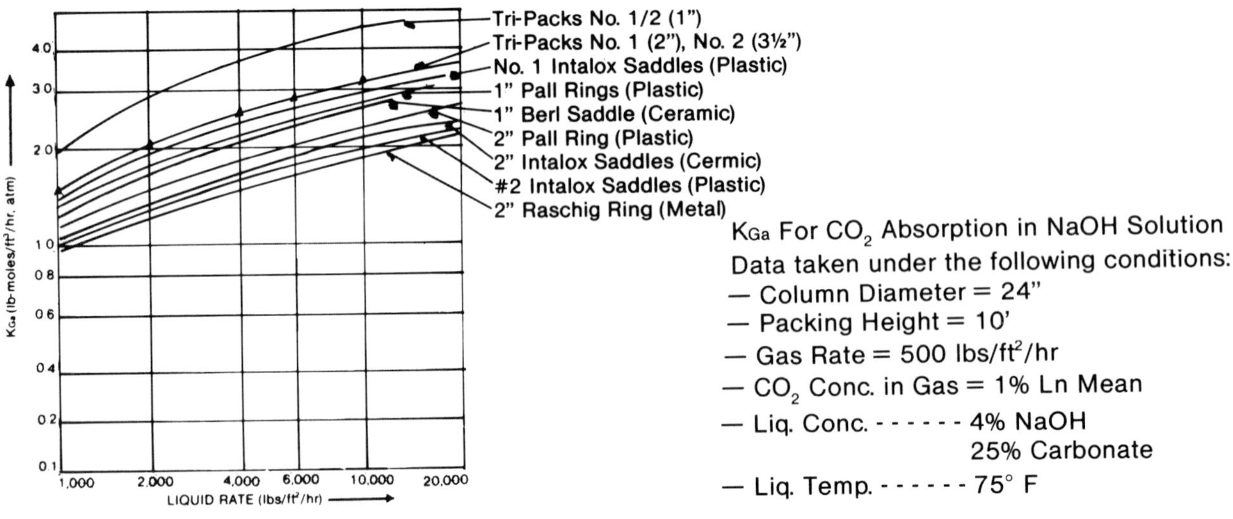

(a)

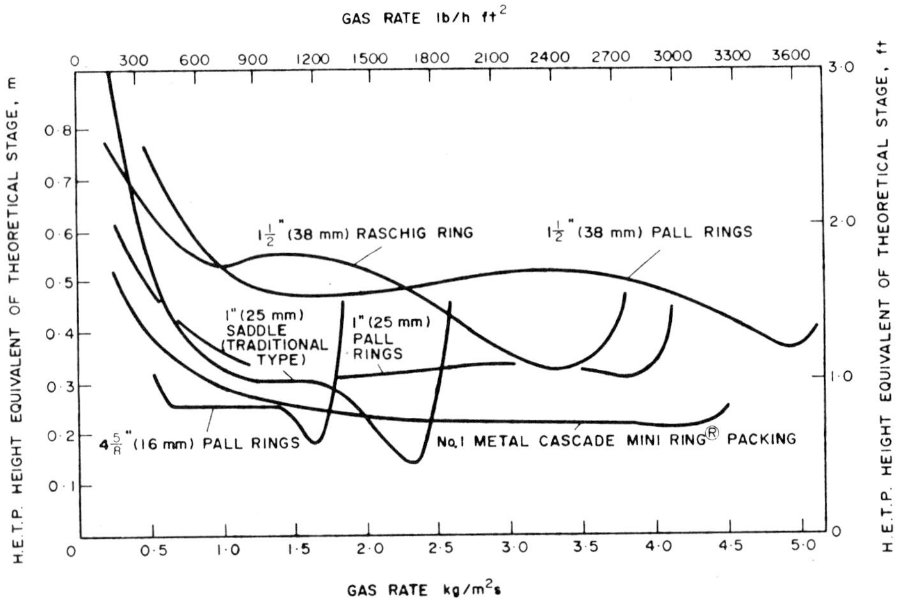

(b)

Figure 13.46. Data of HETP, HTU, and $K_g a$ for several systems and kinds of tower packings. (a) $K_g a$ for absorption of ammonia in NaOH with various packings (*Polymer Piping and Materials Co.*). (b) HETP of several packings as functions of the gas rate (*I. Eastham, cited by Coulson and Richardson, 1978,* Vol. *2*, p. *515*). (c) HETP for ethylbenzene/styrene at 100 Torr; curve 1 for 2 in. metal pall rings [*Billet,* Chem. Eng. Prog. **63**(*9*), *55* (*1967*)]; curve 2 for 1 in. metal pall rings (*Billet,* loc. cit.); curve 3 for Sulzer packing (*Koch Engineering Co.*). (d) $K_g a$ for absorption of CO_2 in NaOH with various packings; CMR are cascade mini rings which are pall rings with heights about one-half the diameters (*Mass Transfer International*). (e) HTU for absorption of ammonia from air with water [*Teller,* Chem. Eng. Prog. **50**, *70*, (*1954*). (f) HTU for absorption of ammonia from air with water (*Wen, S.M. Thesis, University of West Virginia, 1953*). (g) Typical HETP for distillation with 2 in. metal pall ("ballast") rings (*redrawn from Bulletin 217, Glitsch Inc.*).

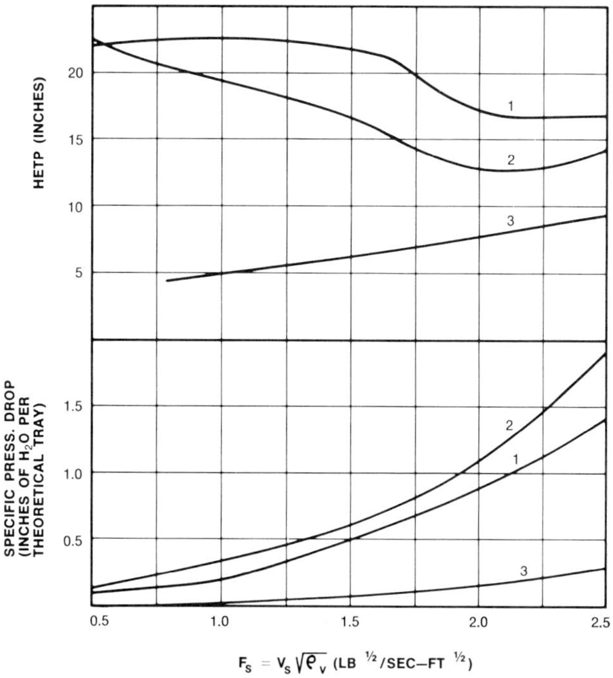
HETP (INCHES)
20
15
10
5
1
2
3
SPECIFIC PRESS. DROP (INCHES OF H_2O PER THEORETICAL TRAY)
1.5
1.0
0.5
2
1
3
0.5
1.0
1.5
2.0
2.5
$F_S = V_S\sqrt{\rho_V}$ (LB $^{1/2}$/SEC—FT $^{1/2}$)

(c)

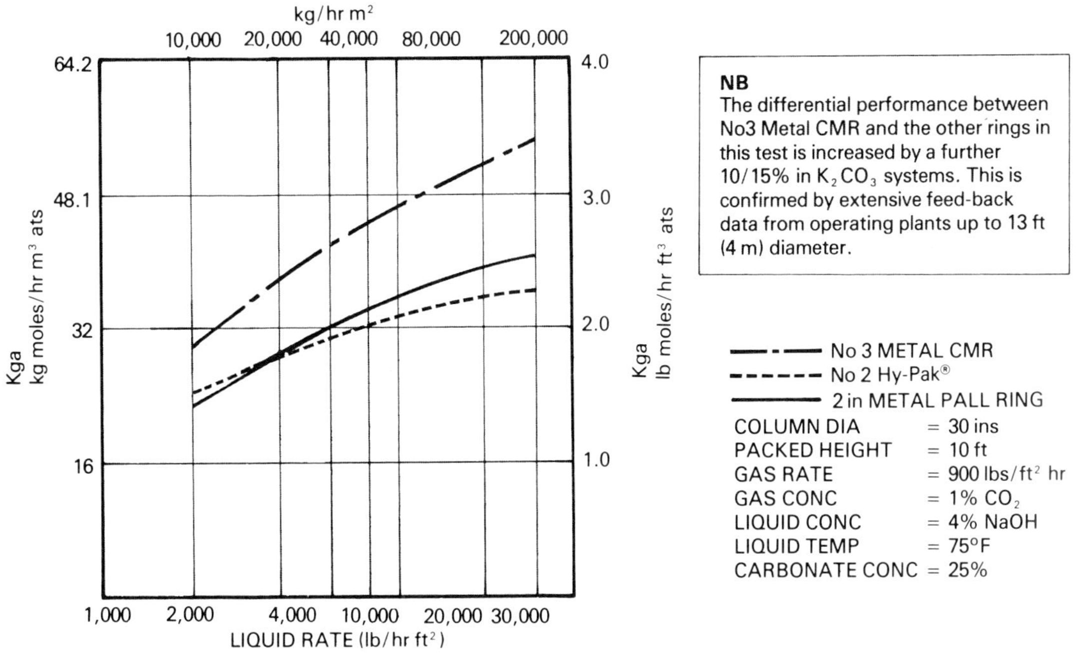
kg/hr m^2
10,000 20,000 40,000 80,000 200,000
64.2
48.1
32
16
4.0
3.0
2.0
1.0
Kga kg moles/hr m^3 ats
Kga lb moles/hr ft^3 ats
1,000 2,000 4,000 10,000 20,000 30,000
LIQUID RATE (lb/hr ft^2)
NB
The differential performance between No3 Metal CMR and the other rings in this test is increased by a further 10/15% in K_2CO_3 systems. This is confirmed by extensive feed-back data from operating plants up to 13 ft (4 m) diameter.
No 3 METAL CMR
No 2 Hy-Pak®
2 in METAL PALL RING
COLUMN DIA = 30 ins
PACKED HEIGHT = 10 ft
GAS RATE = 900 lbs/ft^2 hr
GAS CONC = 1% CO_2
LIQUID CONC = 4% NaOH
LIQUID TEMP = 75°F
CARBONATE CONC = 25%

(d)

Figure 13.46—(*continued*)

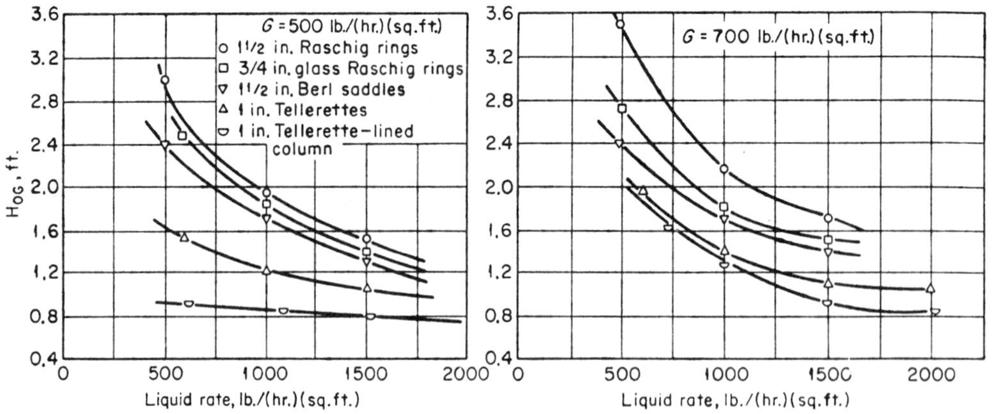

(e)

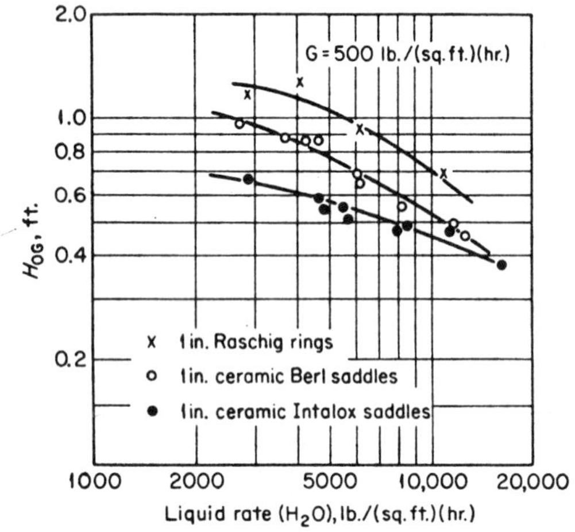

(f)

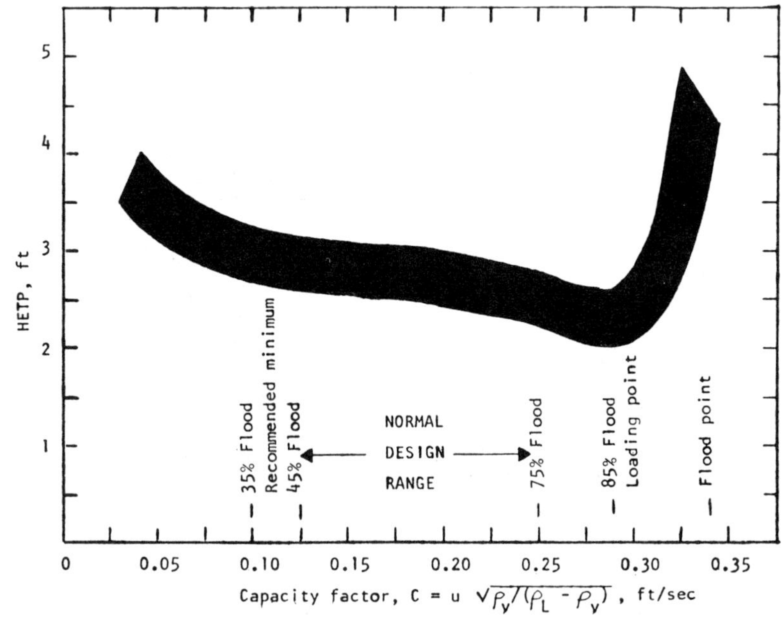

(g)

Figure 13.46—*(continued)*

REFERENCES

General

1. R. Billet, *Distillation Engineering,* Chemical Publishing Co., New York, 1979.
2. J.M. Coulson and J.F. Richardson, *Chemical Engineering,* Pergamon, New York, 1978, Vol. 2.
3. J.R. Fair, Liquid–gas systems, in *Chemical Engineers Handbook,* McGraw-Hill, New York, 1984, Section 18.
4. R.J. Hengstebeck, *Distillation,* Reinhold, New York, 1961.
5. E.J. Henley and J.D. Seader, *Equilibrium-Stage Processes in Chemical Engineering,* Wiley, New York, 1981.
6. A.L. Hines and R.N. Maddox, *Mass Transfer Fundamentals and Applications,* Prentice-Hall, Englewood Cliffs, NJ, 1985.
7. C.D. Holland, *Fundamentals of Multicomponent Distillation,* McGraw-Hill, New York, 1981.
8. V. Kafarov, *Mass Transfer,* Mir Publishers, Moscow, 1985.
9. C.J. King, *Separation Processes,* McGraw-Hill, New York, 1980.
10. C.S. Robinson and E.R. Gilliland, *Elements of Fractional Distillation,* McGraw-Hill, New York, 1950.
11. J.D. Seader, Distillation, in *Chemical Engineers Handbook,* McGraw-Hill, New York, 1984, Section 13.
12. B.D. Smith, *Design of Equilibrium Stage Processes,* McGraw-Hill, New York, 1963.
13. R.E. Treybal, *Mass Transfer Operations,* McGraw-Hill, New York, 1980.
14. S.M. Walas, *Phase Equilibria in Chemical Engineering,* Butterworths, Stoneham, MA, 1985.

Special Topics

1. D.B. Broughton and K.D. Uitti, Distillation estimates for naphtha cuts, in *Encyclopedia of Chemical Processing and Design,* Dekker, New York, 1982, Vol. 16, pp. 186–198.
2. J.S. Eckert, Design of packed columns, in Reference 8, 1979.
3. R.W. Ellerbe, Batch distillation, in Reference 8, 1979.
4. J. Hollo, et al., *Applications of Molecular Distillation,* Budapest, 1971.

5. P.G. Nygren, High efficiency low pressure drop packings, in Ref. 8, 1979, pp. 1.241–1.253.
6. D.F. Othmer, Azeotropic and extractive distillation, *Encycl. Chem. Tech.* **3,** 352–377 (1978).
7. G. Prokopakis, Azeotropic and extractive distillation, *Encycl. Chem. Technol. Suppl.,* 145–158 (1984).
8. P.A. Schweitzer, Editor, *Handbook of Separation Techniques for Chemical Engineers,* McGraw-Hill, New York, 1979.
9. W.J. Stupin and F.J. Lockhart, Distillation, thermally coupled, in *Encyclopedia of Chemical Processing and Design,* Dekker, New York, 1982, Vol. 16, pp. 279–299.
10. T.J. Vital, S.S. Grossel, and P.I. Olsen, Estimating tray efficiency, *Hyd. Proc.,* 55–56 (Oct. 1984); 147–153 (Nov. 1984); 75–78 (Dec. 1984).

Vapor–Liquid Equilibrium Data Collections

1. *API Technical Data Book-Petroleum Refining,* American Petroleum Institute, Washington, D.C., 1983–date, Chaps. 8 and 9.
2. J. Gmehling, U. Onken et al., *Vapor–Liquid Equilibrium Data Collection,* DECHEMA, Frankfurt/Main, Germany, 1979–date.
3. M. Hirata, et al., *Computer Aided Data Book of Vapor Liquid Equilibria,* Elsevier, New York, 1976.
4. V.B. Kogan, et al., *Equilibria between Vapor and Liquid* (in Russian), Izdatelstvo Nauka, Moscow, 1966.
5. Landolt-Boernstein Zahlenwerte und Funktionen, II2a, 1960; IV4b, 1972. New Series Group IV, Vol. 3, 1975, Springer, Berlin.
6. *NGPSA Engineering Data Book,* Natural Gas Processors Suppliers Associations, Tulsa, OK, 1972, Chap. 18, and later editions.

14

EXTRACTION AND LEACHING

Extraction is a process whereby a mixture of several substances in the liquid phase is at least partially separated upon addition of a liquid solvent in which the original substances have different solubilities. When some of the original substances are solids, the process is called leaching. In a sense, the role of solvent in extraction is analogous to the role of enthalpy in distillation. The solvent-rich phase is called the extract, and the solvent-poor phase is called the raffinate. A high degree of separation may be achieved with several extraction stages in series, particularly in countercurrent flow.

Processes of separation by extraction, distillation, crystallization, or adsorption sometimes are equally possible. Differences in solubility, and hence of separability by extraction, are associated with differences in chemical structure, whereas differences in vapor pressure are the basis of separation by distillation. Extraction often is effective at near-ambient temperatures, a valuable feature in the separation of thermally unstable natural mixtures or pharmaceutical substances such as penicillin.

The simplest separation by extraction involves two substances and a solvent. Equilibria in such cases are represented conveniently on triangular diagrams, either equilateral or right-angled, as for example on Figures 14.1 and 14.2. Equivalent representations on rectangular coordinates also are shown. Equilibria between any number of substances are representable in terms of activity coefficient correlations such as the UNIQUAC or NRTL. In theory, these correlations involve only parameters that are derivable from measurements on binary mixtures, but in practice the resulting accuracy may be poor and some multicomponent equilibrium measurements also should be used to find the parameters. Finding the parameters of these equations is a complex enough operation to require the use of a computer. An extensive compilation of equilibrium diagrams and UNIQUAC and NRTL parameters is that of Sorensen and Arlt (1979–1980). Extensive bibliographies have been compiled by Wisniak and Tamir (1980–1981).

The highest degree of separation with a minimum of solvent is attained with a series of countercurrent stages. Such an assembly of mixing and separating equipment is represented in Figure 14.3(a), and more schematically in Figure 14.3(b). In the laboratory, the performance of a continuous countercurrent extractor can be simulated with a series of batch operations in separatory funnels, as in Figure 14.3(c). As the number of operations increases horizontally, the terminal concentrations E_1 and R_3 approach asymptotically those obtained in continuous equipment. Various kinds of more sophisticated continuous equipment also are widely used in laboratories; some are described by Lo et al. (1983, pp. 497–506). Laboratory work is of particular importance for complex mixtures whose equilibrium relations are not known and for which stage requirements cannot be calculated.

In mixer-separators the contact times can be made long enough for any desired approach to equilibrium, but 80–90% efficiencies are economically justifiable. If five stages are required to duplicate the performance of four equilibrium stages, the stage efficiency is 80%. Since mixer-separator assemblies take much floor space, they usually are employed in batteries of at most four or five units. A large variety of more compact equipment is being used. The simplest in concept are various kinds of tower arrangements. The relations between their dimensions, the operating conditions, and the equivalent number of stages are the key information.

Calculations of the relations between the input and output amounts and compositions and the number of extraction stages are based on material balances and equilibrium relations. Knowledge of efficiencies and capacities of the equipment then is applied to find its actual size and configuration. Since extraction processes usually are performed under adiabatic and isothermal conditions, in this respect the design problem is simpler than for thermal separations where enthalpy balances also are involved. On the other hand, the design is complicated by the fact that extraction is feasible only of nonideal liquid mixtures. Consequently, the activity coefficient behaviors of two liquid phases must be taken into account or direct equilibrium data must be available.

14.1. EQUILIBRIUM RELATIONS

On a ternary equilibrium diagram like that of Figure 14.1, the limits of mutual solubilities are marked by the binodal curve and the compositions of phases in equilibrium by tielines. The region within the dome is two-phase and that outside is one-phase. The most common systems are those with one pair (Type I, Fig. 14.1) and two pairs (Type II, Fig. 14.4) of partially miscible substances. For instance, of the approximately 1000 sets of data collected and analyzed by Sorensen and Arlt (1979), 75% are Type I and 20% are Type II. The remaining small percentage of systems exhibit a considerable variety of behaviors, a few of which appear in Figure 14.4. As some of these examples show, the effect of temperature on phase behavior of liquids often is very pronounced.

Both equilateral and right triangular diagrams have the property that the compositions of mixtures of all proportions of two mixtures appear on the straight line connecting the original mixtures. Moreover, the relative amounts of the original mixtures corresponding to an overall composition may be found from ratios of line segments. Thus, on the figure of Example 14.2, the amounts of extract and raffinate corresponding to an overall composition M are in the ratio $E_1/R_N = MR_N/E_1M$.

Experimental data on only 26 quaternary systems were found by Sorensen and Arlt (1979), and none of more complex systems, although a few scattered measurements do appear in the literature. Graphical representation of quaternary systems is possible but awkward, so that their behavior usually is analyzed with equations. To a limited degree of accuracy, the phase behavior of complex mixtures can be predicted from measurements on binary mixtures, and considerably better when some ternary measurements also are available. The data are correlated as activity coefficients by means of the UNIQUAC or NRTL equations. The basic principle of application is that at equilibrium the activity of each component is the same in both phases. In terms of activity coefficients this

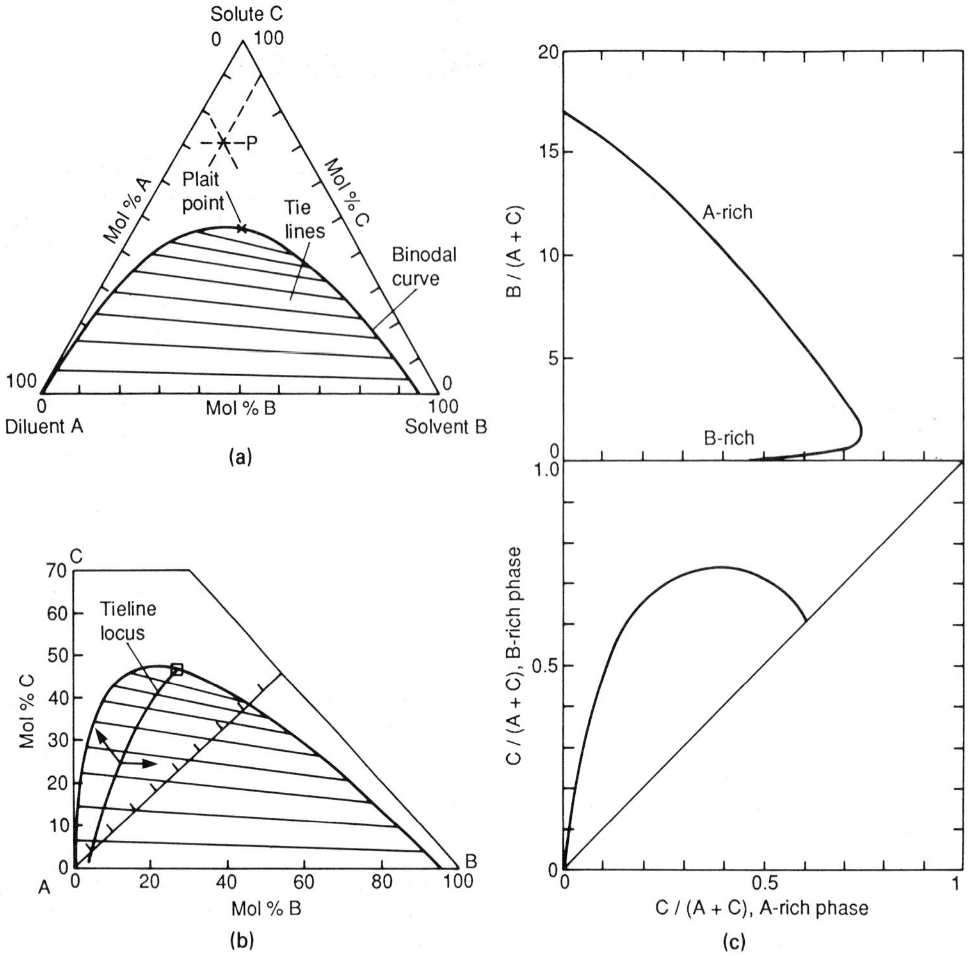

Figure 14.1. Equilibria in a ternary system, type 1, with one pair of partially miscible liquids; A = 1-hexene, B = tetramethylene sulfone, C = benzene, at 50°C (*R.M. De Fre, thesis, Gent, 1976*). (a) Equilateral triangular plot; point *P* is at 20% A, 10% B, and 70% C. (b) Right triangular plot with tielines and tieline locus, the amount of A can be read off along the perpendicular to the hypotenuse or by difference. (c) Rectangular coordinate plot with tieline correlation below, also called Janecke and solvent-free coordinates.

condition is for component *i*,

$$\gamma_i x_i = \gamma_i^* x_i^*, \tag{14.1}$$

where * designates the second phase. This may be rearranged into a relation of distributions of compositions between the phases,

$$x_i^* = (\gamma_i / \gamma_i^*) x_i = K_i x_i, \tag{14.2}$$

where K_i is the distribution coefficient. The activity coefficients are functions of the composition of the mixture and the temperature. Applications to the calculation of stage requirements for extraction are described later.

Extraction behavior of highly complex mixtures usually can be known only from experiment. The simplest equipment for that purpose is the separatory funnel, but complex operations can be simulated with proper procedures, for instance, as in Figure 14.3(c). Elaborate automatic laboratory equipment is in use. One of them employs a 10,000–25,000 rpm mixer with a residence time of 0.3–5.0 sec, followed by a highly efficient centrifuge and two chromatographs for analysis of the two phases (Lo et al., 1983, pp. 507).

Compositions of petroleum mixtures sometimes are represented adequately in terms of some physical property. Three examples appear in Figure 14.5. Straight line combining of mixtures still is valid on such diagrams.

Basically, compositions of phases in equilibrium are indicated with tielines. For convenience of interpolation and to reduce the clutter, however, various kinds of tieline loci may be constructed, usually as loci of intersections of projections from the two ends of the tielines. In Figure 14.1 the projections are parallel to the base and to the hypotenuse, whereas in Figures 14.2 and 14.6 they are horizontal and vertical.

Several tieline correlations in equation form have been proposed, of which three may be presented. They are expressed in weight fractions identified with these subscripts:

CA solute C in diluent phase A
CS solute C in solvent phase S
SS solvent S in solvent phase S
AA diluent A in diluent phase A
AS diluent A in solvent phase S
SA solvent S in diluent phase A.

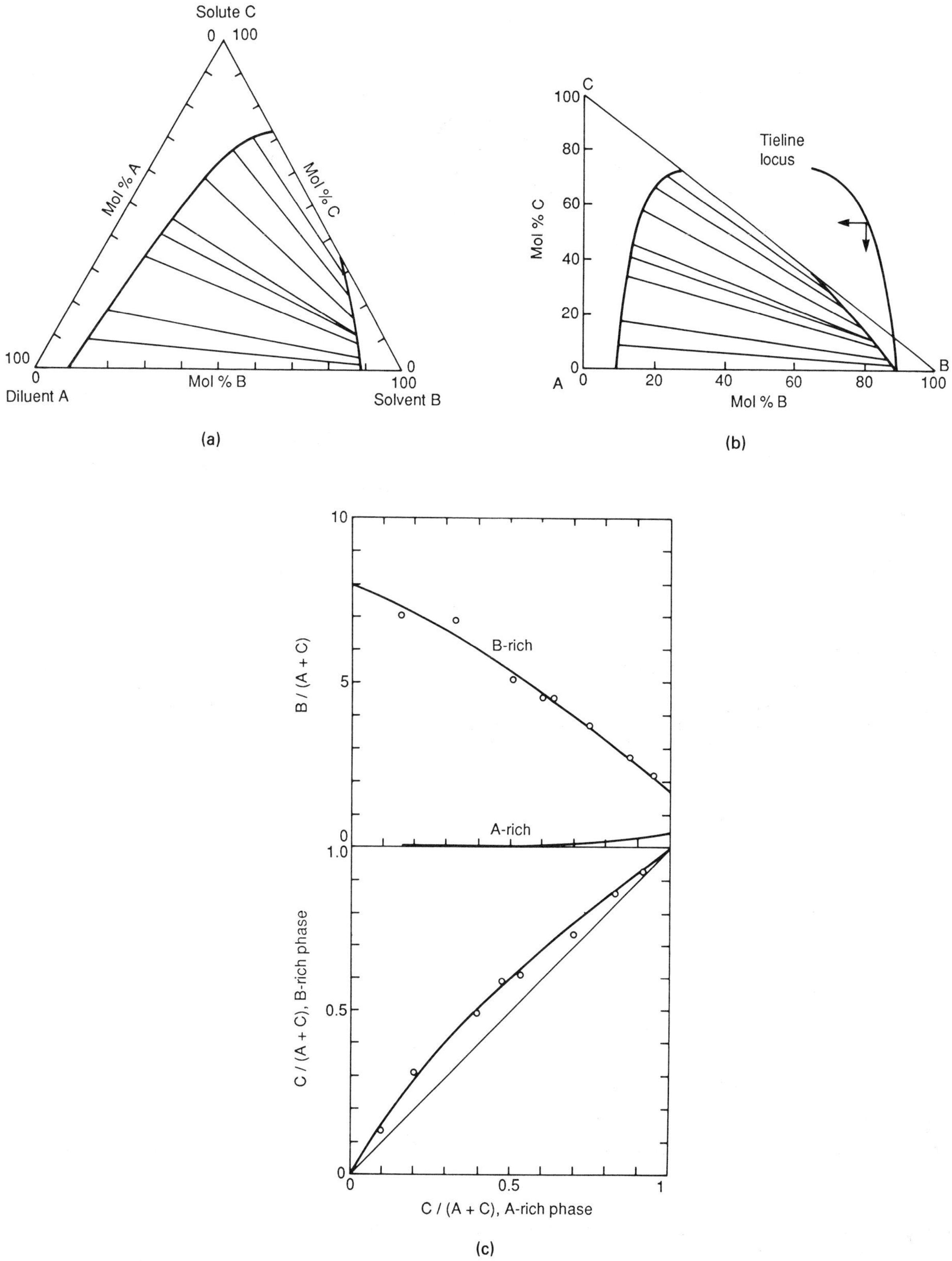

Figure 14.2. Equilibria in a ternary system, type II, with two pairs of partially miscible liquids; A = hexane, B = aniline, C = methylcyclopentane, at 34.5°C [*Darwent and Winkler,* J. Phys. Chem. **47,** *442* (*1943*)]. (a) Equilateral triangular plot. (b) Right triangular plot with tielines and tieline locus. (c) Rectangular coordinate plot with tieline correlation below, also called Janecke and solvent-free coordinates.

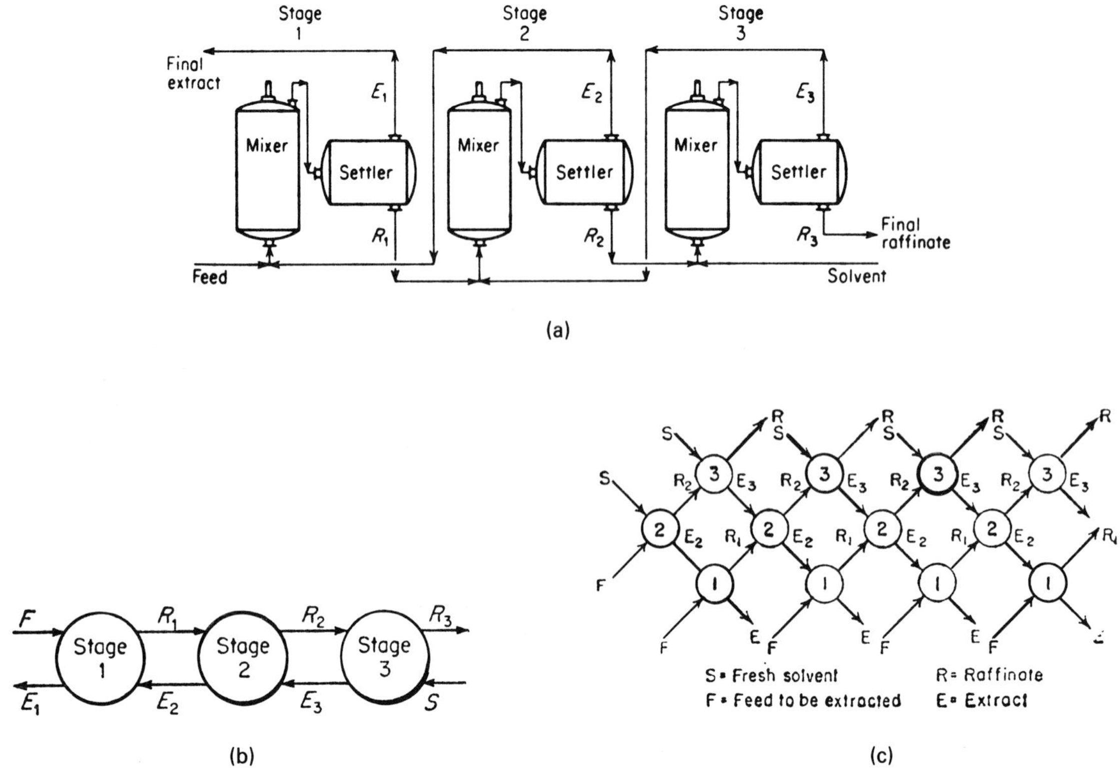

Figure 14.3. Representation of countercurrent extraction batteries. (a) A battery of mixers and settlers (or separators). (b) Schematic of a three-stage countercurrent battery. (c) Simulation of the performance of a three-stage continuous countercurrent extraction battery with a series of batch extractions in separatory funnels which are designated by circles on the sketch. The numbers in the circles are those of the stages. Constant amounts of feed F and solvent S are mixed at the indicated points. As the number of operations is increased horizontally, the terminal compositions E_1 and R_3 approach asymptotically the values obtained in continuous countercurrent extraction (*Treybal, 1963, p. 360*).

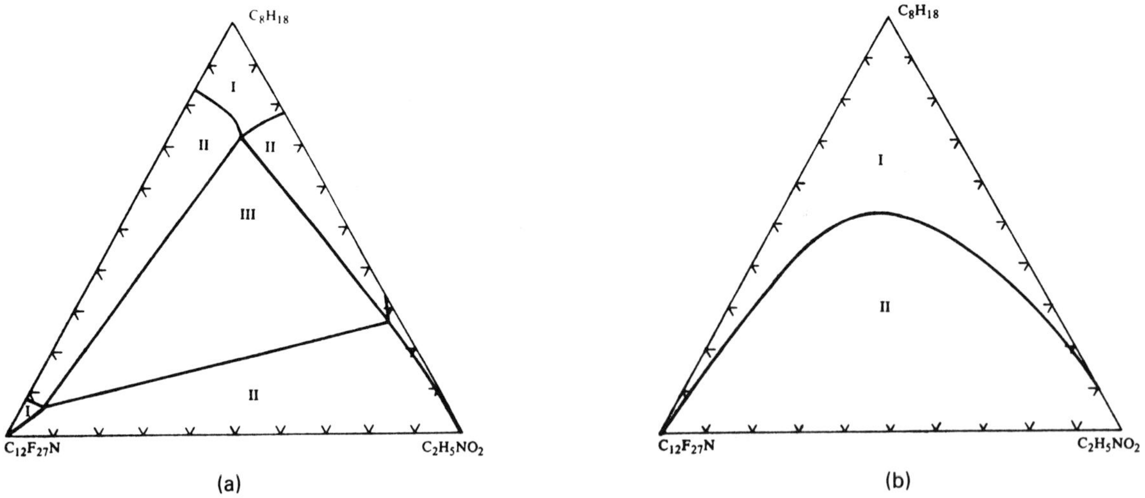

Figure 14.4. Less common examples of ternary equilibria and some temperature effects. (a) The system 2,2,4-trimethylpentane + nitroethane + perfluorobutylamine at 25°C; the Roman numerals designate the number of phases in that region [*Vreeland and Dunlap,* J. Phys. Chem. **61,** *329* (*1957*)]. (b) Same as (a) but at 51.3°C. (c) Glycol + dodecanol + nitroethane at 24°C; 12 different regions exist at 14°C [*Francis,* J. Phys. Chem. **60,** *20* (*1956*)]. (d) Docosane + furfural + diphenylhexane at several temperatures [*Varteressian and Fenske,* Ind. Eng. Chem. **29,** *270* (*1937*)]. (e) Formic acid + benzene + tribromomethane at 70°C; the pair formic acid/benzene is partially miscible with 15 and 90% of the former at equilibrium at 25°C, 43 and 80% at 70°C, but completely miscible at some higher temperature. (f) Methylcyclohexane + water + -picoline at 20°C, exhibiting positive and negative tieline slopes; the horizontal tieline is called solutropic (*Landolt-Börnstein II2b*).

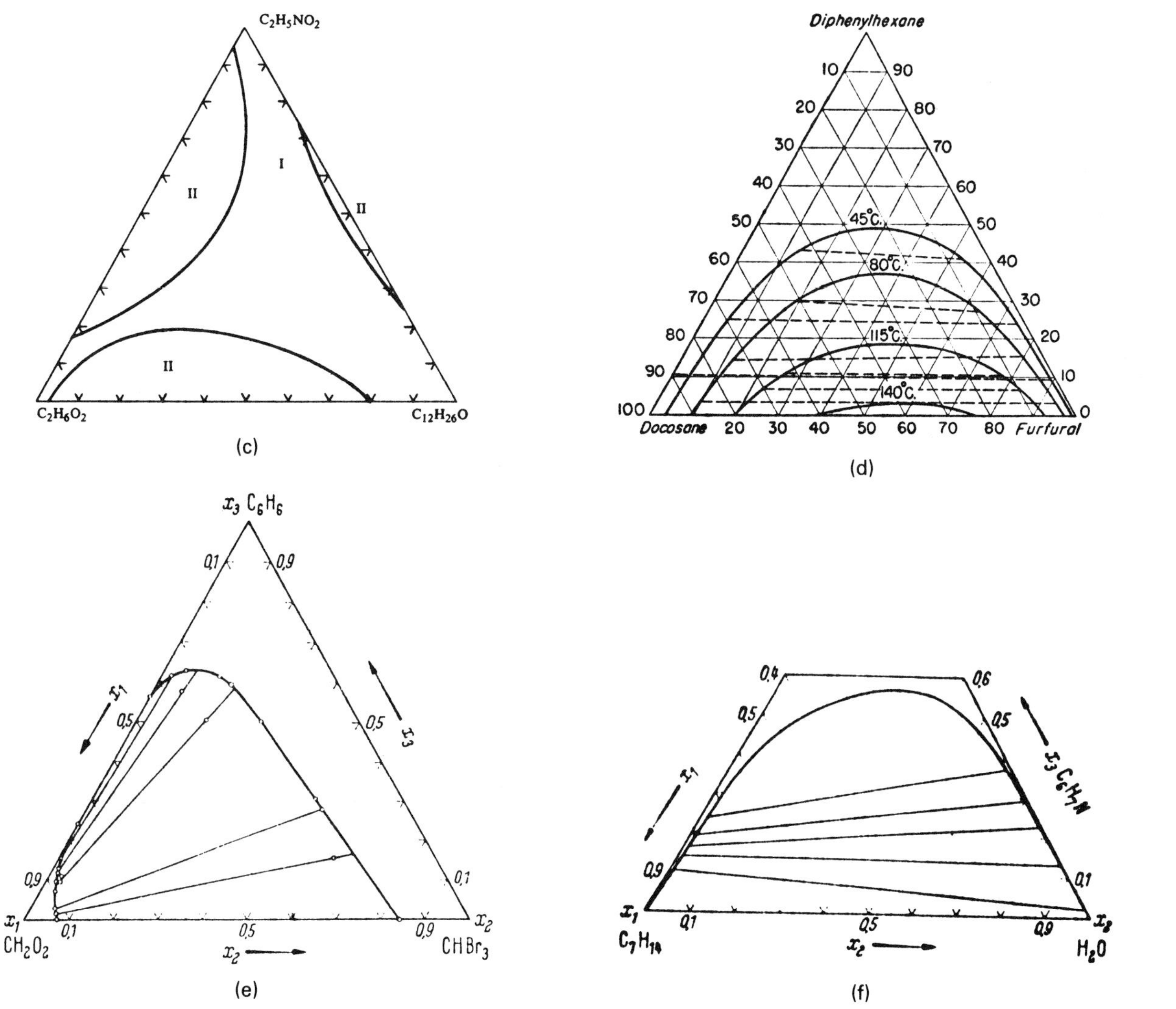

Figure 14.4—(*continued*)

Ishida, *Bull. Chem. Soc. Jpn.* **33,** 693 (1960):

$$X_{CS}X_{SA}/X_{CA}X_{SS} = K(X_{AS}X_{SA}/X_{AA}X_{SS})^n. \tag{14.3}$$

Othmer and Tobias, *Ind. Eng. Chem.* **34,** 693 (1942):

$$(1 - X_{SS})/X_{SS} = K[(1 - X_{AA})/X_{AA}]^n. \tag{14.4}$$

Hand, *J. Phys. Chem.* **34,** 1961 (1930):

$$X_{CS}/X_{SS} = K(X_{CA}/X_{AA})^n. \tag{14.5}$$

These equations should plot linearly on log–log coordinates; they are tested in Example 14.1.

A system of plotting both binodal and tieline data in terms of certain ratios of concentrations was devised by Janecke and is illustrated in Figure 14.1(c). It is analogous to the enthalpy–concentration or Merkel diagram that is useful in solving distillation problems. Straight line combining of mixture compositions is valid in this mode. Calculations for the transformation of data are made most conveniently from tabulated tieline data. Those for Figure 14.1 are made in Example 14.2. The x–y construction shown in Figure 14.2 is the basis for a McCabe–Thiele construction for finding the number of extraction stages, as applied in Figure 14.7.

14.2. CALCULATION OF STAGE REQUIREMENTS

Although the most useful extraction process is with countercurrent flow in a multistage battery, other modes have some application. Calculations may be performed analytically or graphically. On flowsketches like those of Example 14.1 and elsewhere, a single box represents an extraction stage that may be made up of an individual mixer and separator. The performance of differential contactors such as packed or spray towers is commonly described as the height equivalent to a theoretical stage (HETS) in ft or m.

SINGLE STAGE EXTRACTION

The material balance is

$$\begin{aligned}&\text{feed} + \text{solvent} = \text{extract} + \text{raffinate},\\ &F + S = E + R.\end{aligned} \tag{14.6}$$

This nomenclature is shown with Example 14.3. On the triangular diagram, the proportions of feed and solvent locate the mix point

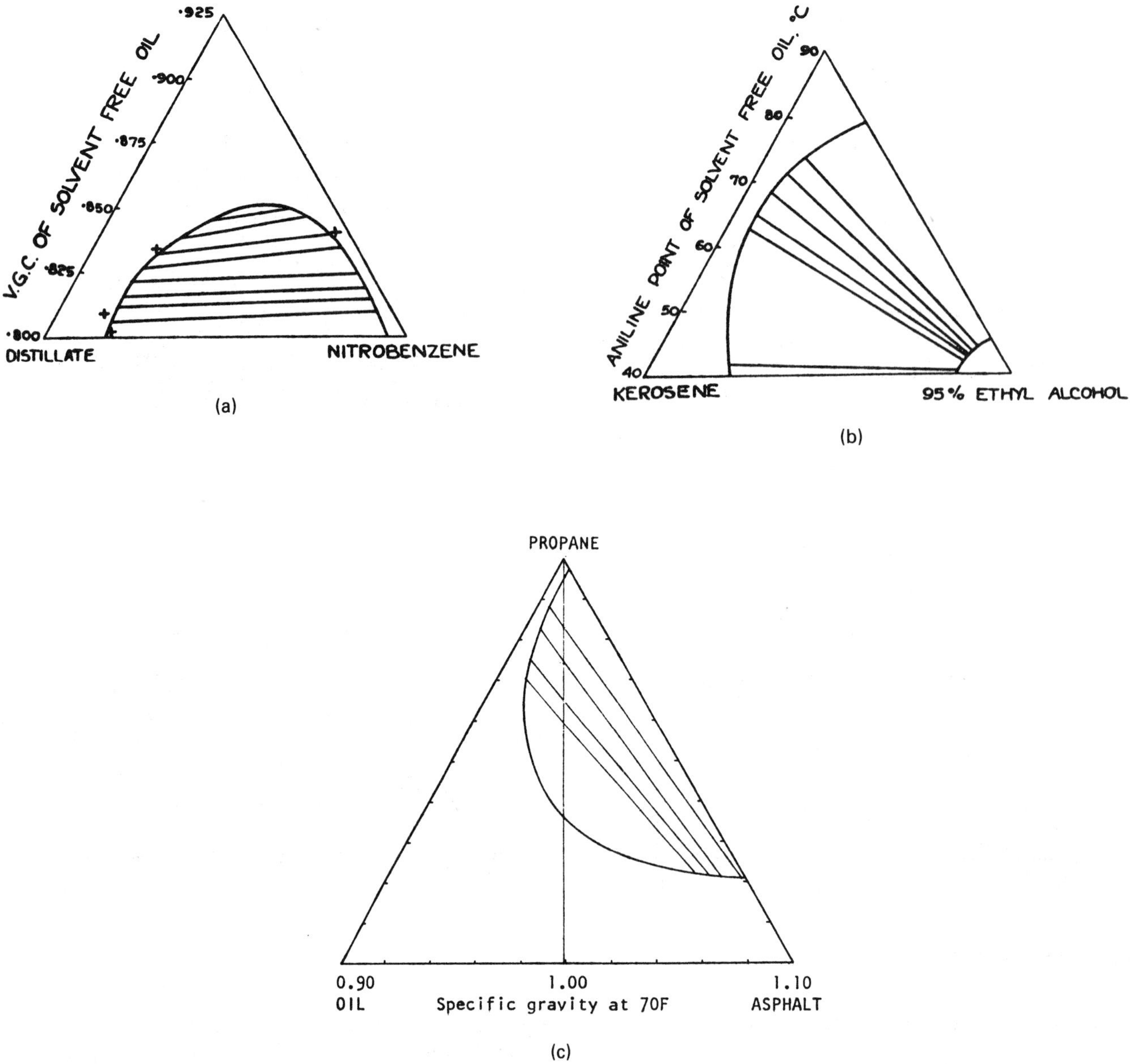

Figure 14.5. Representation of solvent extraction behavior in terms of certain properties rather than direct compositions [*Dunstan et al.,* Sci. Pet., *1825–1855* (*1938*)]. (a) Behavior of a naphthenic distillate of VGC = 0.874 with nitrobenzene at 10°C. The viscosity-gravity constant is low for paraffins and high for naphthenes. (b) Behavior of a kerosene with 95% ethanol at 17°C. The aniline point is low for aromatics and naphthenes and high for paraffins. (c) Behavior of a dewaxed crude oil with liquid propane at 70°F, with composition expressed in terms of specific gravity.

M. The extract *E* and raffinate *R* are located on opposite ends of the tieline that goes through *M*.

CROSSCURRENT EXTRACTION

In this process the feed and subsequently the raffinate are treated in successive stages with fresh solvent. The sketch is with Example 14.3. With a fixed overall amount of solvent the most efficient process is with equal solvent flow to each stage. The solution of Example 14.3 shows that crosscurrent two stage operation is superior to one stage with the same total amount of solvent.

IMMISCIBLE SOLVENTS

The distribution of a solute between two mutually immiscible solvents can be represented by the simple equation,

$$Y = K'X, \tag{14.7}$$

where

X = mass of solute/mass of diluent,
Y = mass of solute/mass of solvent.

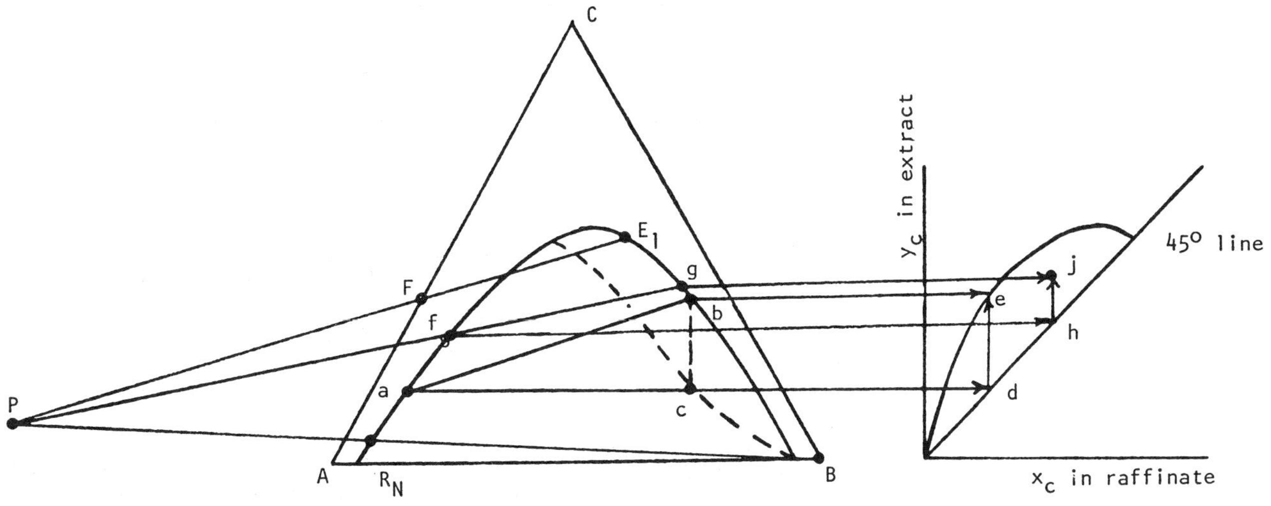

Figure 14.6. Construction of points on the distribution and operating curves: Line *ab* is a tieline. The dashed line is the tieline locus. Point *e* is on the equilibrium distribution curve, obtained as the intersection of paths *be* and *ade*. Line *Pfg* is a random line from the difference point *P* and intersecting the binodal curve in *f* and *g*. Point *j* is on the operating curve, obtained as the intersection of paths *gj* and *fhj*.

When K' is not truly constant, some kind of mean value may be applicable, for instance, a geometric mean, or the performance of the extraction battery may be calculated stage by stage with a different value of K' for each. The material balance around the first stage where the raffinate leaves and the feed enters and an intermediate stage k (as in Fig. 14.8, for instance) is

$$EY_F + RX_{k-1} = EY_k + RX_n. \tag{14.8}$$

In terms of the extraction ratio,

$$A = K(E/R), \tag{14.9}$$

Example 14.1
The Equations for Tieline Data

The tieline data of the system of Example 14.1 are plotted according to the groups of variables in the equations of Ishida, Hand, and Othmer and Tobias with these results:

Ishida: $y = 1.00x^{0.67}$ [Eq. (14.3)],
Hand: $y = 0.078x^{1.11}$ [Eq. (14.5)],
Othmer and Tobias: $y = 0.88x^{0.90}$ [Eq. (14.4)].

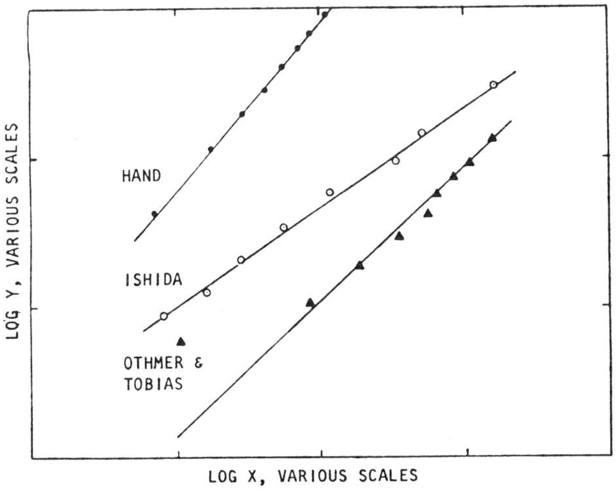

The last correlation is inferior for this particular example as the plots show.

x_{AA}	x_{CA}	x_{SA}	x_{AS}	x_{CS}	x_{SS}
98.945	0.0	1.055	5.615	0.0	94.385
92.197	6.471	1.332	5.811	3.875	90.313
83.572	14.612	1.816	6.354	9.758	83.889
75.356	22.277	2.367	7.131	15.365	77.504
68.283	28.376	3.341	8.376	20.686	70.939
60.771	34.345	4.884	9.545	26.248	64.207
54.034	39.239	6.727	11.375	31.230	57.394
47.748	42.849	9.403	13.505	35.020	51.475
39.225	45.594	15.181	18.134	39.073	42.793

$\frac{10^4 x_{AS}x_{SA}}{x_{AA}x_{SS}}$	$\frac{x_{CS}x_{SA}}{x_{CA}x_{SS}}$	$\frac{1}{x_{AA}}-1$	$\frac{1}{x_{SS}}-1$	x_{CA}/x_{AA}	x_{CS}/x_{SS}
6.34	0	0.0107	0.0595	0	0
9.30	0.0088	0.0846	0.1073	0.070	0.043
16.46	0.0129	0.1966	0.1928	0.178	0.116
28.90	0.0211	0.3270	0.2903	0.296	0.198
58.22	0.0343	0.4645	0.4097	0.416	0.292
119.47	0.0581	0.6455	0.5575	0.565	0.409
339.77	0.0933	0.8507	0.7423	0.726	0.544
516.67	0.1493	1.0943	0.9427	0.897	0.680
1640	0.3040	1.5494	1.3368	1.162	0.913

Example 14.2
Tabulated Tieline and Distribution Data for the System A = 1-Hexene, B = Tetramethylene Sulfone, C = Benzene, Represented in Figure 14.1

Experimental tieline data in mol %:

Left Phase			Right Phase		
A	C	B	A	C	B
98.945	0.0	1.055	5.615	0.0	94.385
92.197	6.471	1.332	5.811	3.875	90.313
83.572	14.612	1.816	6.354	9.758	83.888
75.356	22.277	2.367	7.131	15.365	77.504
68.283	28.376	3.341	8.376	20.686	70.938
60.771	34.345	4.884	9.545	26.248	64.207
54.034	39.239	6.727	11.375	31.230	57.394
47.748	42.849	9.403	13.505	35.020	51.475
39.225	45.594	15.181	18.134	39.073	42.793

Calculated ratios for the Jänecke coordinate plot of Figure 14.1:

Left Phase		Right Phase	
$\frac{B}{A+C}$	$\frac{C}{A+C}$	$\frac{B}{A+C}$	$\frac{C}{A+C}$
0.0108	0	16.809	0
0.0135	0.0656	9.932	0.4000
0.0185	0.1488	5.190	0.6041
0.0248	0.2329	3.445	0.6830
0.0346	0.2936	2.441	0.7118
0.0513	0.3625	1.794	0.7333
0.0721	0.4207	1.347	0.7330
0.1038	0.4730	1.061	0.7217
0.1790	0.5375	0.748	0.6830

The x–y plot like that of Figure 14.6 may be made with the tieline data of columns 5 and 2 expressed as fractions or by projection from the triangular diagram as shown.

the material balance becomes

$$(A/K)Y_F + X_{k-1} = AX_k + X_n. \tag{14.10}$$

When these balances are made stage-by-stage and intermediate compositions are eliminated, assuming constant A throughout, the result relates the terminal compositions and the number of stages. The expression for the fraction extracted is

$$\phi = \frac{X_F - X_n}{X_F - Y_S/K} = \frac{A^{n+1} - A}{A^{n+1} - 1}. \tag{14.11}$$

This is of the same form as the Kremser–Brown equation for gas absorption and stripping and the Turner equation for leaching. The solution for the number of stages is

$$n = -1 + \frac{\ln[(A - \phi)/(1 - \phi)]}{\ln A}. \tag{14.12}$$

When A is the only unknown, it may be found by trial solution of these equations, or the Kremser–Brown stripping chart may be used. Example 14.4 applies these results.

14.3. COUNTERCURRENT OPERATION

In countercurrent operation of several stages in series, feed enters the first stage and final extract leaves it, and fresh solvent enters the last stage and final raffinate leaves it. Several representations of

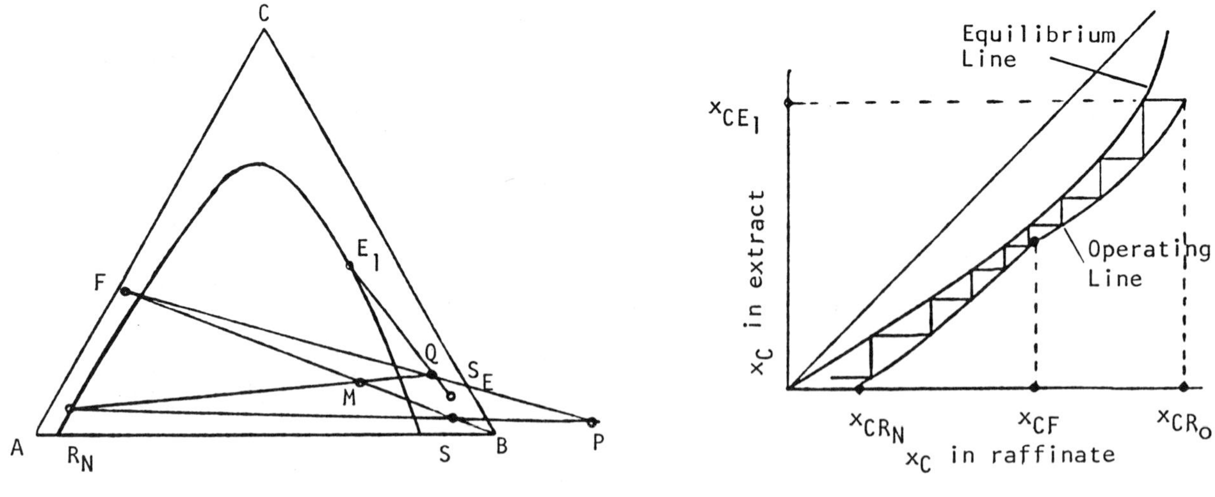

(a)

Figure 14.7. Locations of operating points P and Q for feasible, total, and minimum extract reflux on triangular diagrams, and stage requirements determined on rectangular distribution diagrams. (a) Stages required with feasible extract reflux. (b) Operation at total reflux and minimum number of stages. (c) Operation at minimum reflux and infinite stages.

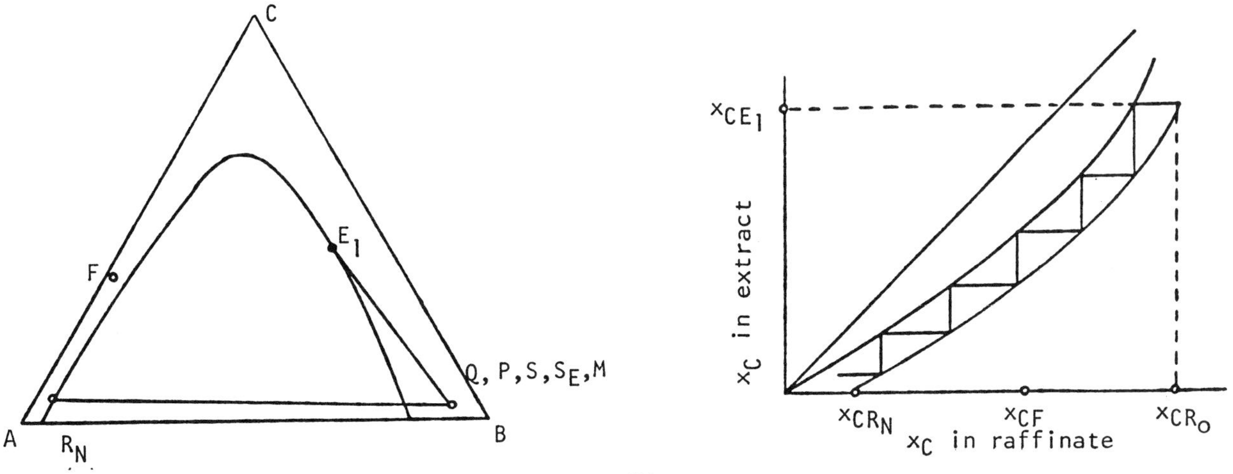

(b)

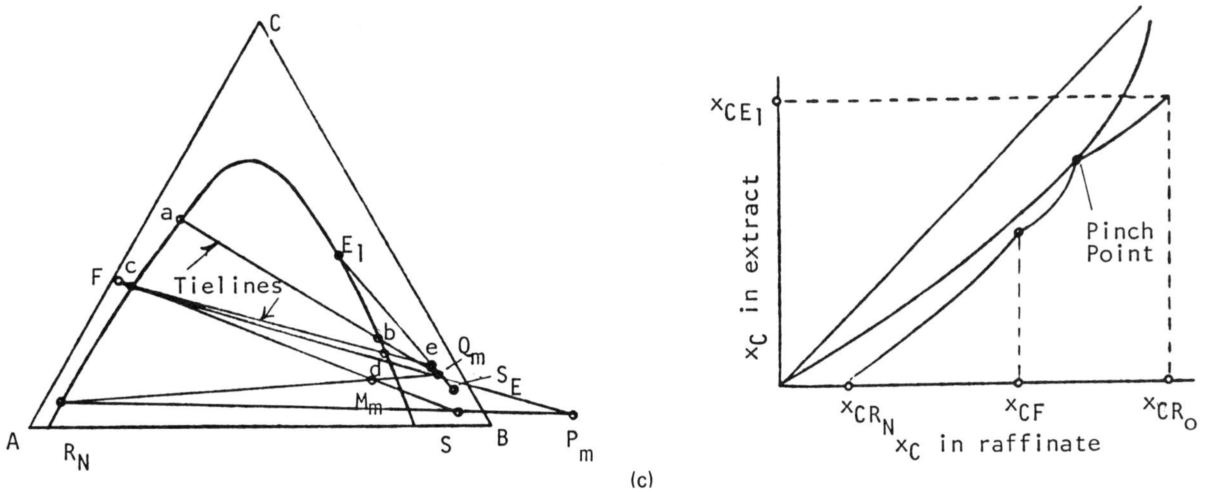

(c)

Figure 14.7—(*continued*)

such processes are in Figure 14.3. A flowsketch of the process together with nomenclature is shown with Example 14.5. The overall material balance is

$$F + S = E_1 + R_N = M \tag{14.13}$$

or

$$F - E_1 = R_N - S = P. \tag{14.14}$$

The intersection of extended lines FE_1 and R_NS locates the operating point P. The material balance from stage 1 through k is

$$F + E_{k+1} = E_1 + R_k \tag{14.15}$$

or

$$F - E_1 = R_k - E_{k+1} = P. \tag{14.16}$$

Accordingly, the raffinate from a particular stage and the extract from a succeeding one are on a line through the operating point P. Raffinate R_k and extract E_k streams from the same stage are located at opposite ends of the same tieline.

The operation of finding the number of stages consists of a number of steps:

1. Either the solvent feed ratio or the compositions E_1 and R_N serve to locate the mix point M.
2. The operating point P is located as the intersection of lines FE_1 and R_NS.
3. When starting with E_1, the raffinate R_1 is located at the other end of the tieline.
4. The line PR_1 is drawn to intersect the binodal curve in E_2.

The process is continued with the succeeding values R_2, E_3, R_3, E_4, . . . until the final raffinate composition is reached.

When number of stages and only one of the terminal compositions are fixed, the other terminal composition is selected by trial until the stepwise calculation finds the prescribed number of stages. Example 14.6 applies this kind of calculation to find the stage requirements for systems with Types I and II equilibria.

Evaluation of the numbers of stages also can be made on rectangular distribution diagrams, with a McCabe–Thiele kind of construction. Example 14.5 does this. The Janecke coordinate plots like those of Figures 14.1 and 14.2 also are convenient when many stages are needed, since then the triangular construction may

EXAMPLE 14.3
Single Stage and Cross Current Extraction of Acetic Acid from Methylisobutyl Ketone with Water
The original mixture contains 35% acetic acid and 65% MIBK. It is charged at 100 kg/hr and extracted with water.

a. In a single stage extractor water is mixed in at 100 kg/hr. On the triangular diagram, mix point M is midway between F and S. Extract and raffinate compositions are on the tieline through M. Results read off the diagram and calculated with material balance are

	E	*R*
Acetic acid	0.185	0.16
MIBK	0.035	0.751
Water	0.78	0.089
kg/hr	120	80

b. The flowsketch of the crosscurrent process is shown. Feed to the first stage and water to both stages are at 100 kg/hr. The extract and raffinate compositions are on the tielines passing through mix points M_1 and M_2. Point M is for one stage with the same total amount of solvent. Two stage results are:

	E_1	R_1	E_2	R_2
Acetic acid	0.185	0.160	0.077	0.058
MIBK	0.035	0.751	0.018	0.902
Water	0.780	0.089	0.905	0.040
kg/hr	120	80	113.4	66.6

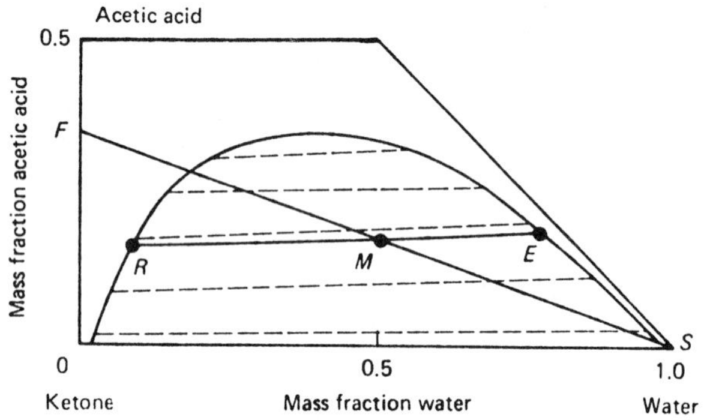

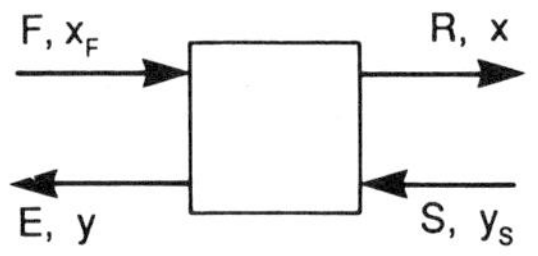

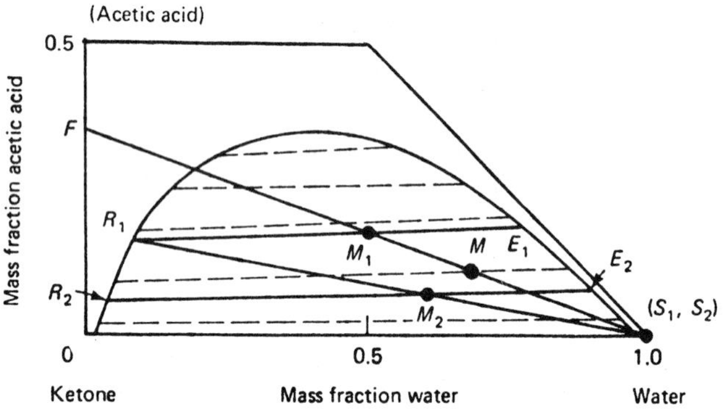

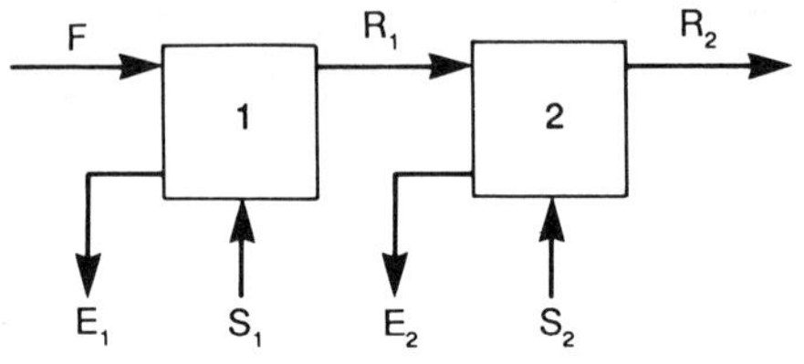

become crowded and difficult to execute accurately unless a very large scale is adopted. The Janecke method was developed by Maloney and Schubert [*Trans. AIChE* **36,** 741 (1940)]. Several detailed examples of this kind of calculation are worked by Treybal (1963), Oliver (*Diffusional Separation Processes,* Wiley, New York, 1966), and Laddha and Degaleesan (1978).

MINIMUM SOLVENT/FEED RATIO

Both maximum and minimum limits exist of the solvent/feed ratio. The maximum is the value that locates the mix point M on the binodal curve near the solvent vertex, such as point M_{max} on Figure 14.7(b). When an operating line coincides with a tieline, the number of stages will be infinite and will correspond to the minimum solvent/feed ratio. The pinch point is determined by the intersection of some tieline with line R_NS. Depending on whether the slopes of the tielines are negative or positive, the intersection that is closest or farthest from the solvent vertex locates the operating point for minimum solvent. Figure 14.9 shows the two cases. Frequently, the tieline through the feed point determines the minimum solvent quantity, but not for the two cases shown.

EXTRACT REFLUX

Normally, the concentration of solute in the final extract is limited to the value in equilibrium with the feed, but a countercurrent stream that is richer than the feed is available for enrichment of the extract. This is essentially solvent-free extract as reflux. A flowsketch and nomenclature of such a process are given with Example 14.7. Now there are two operating points, one for above the feed and one for below. These points are located by the following procedure:

1. The mix point is located by establishing the solvent/feed ratio.
2. Point Q is at the intersection of lines R_NM and E_1S_E, where S_E refers to the solvent that is removed from the final extract, and may or may not be of the same composition as the fresh solvent S. Depending on the shape of the curve, point Q may be inside

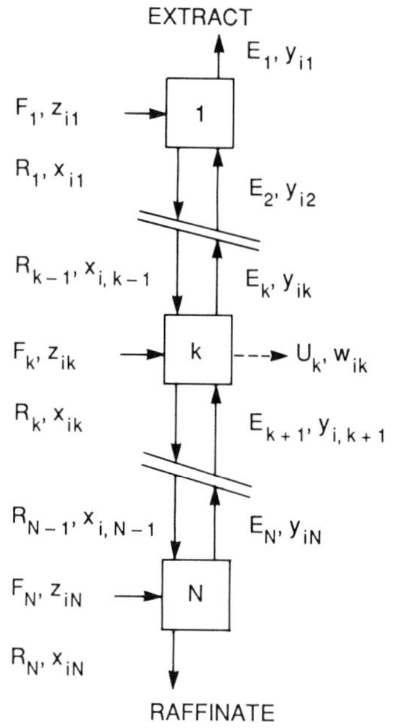

Figure 14.8. Model for liquid–liquid extraction. Subscript i refers to a component: $i = 1, 2, \ldots, c$. In the commonest case, F_1 is the only feed stream and F_N is the solvent, or F_k may be a reflux stream. Withdrawal streams U_k can be provided at any stage; they are not incorporated in the material balances written here.

the binodal curve as in Example 14.7, or outside as in Figure 14.7.

3. Point P is at the intersection of lines $R_N M$ and $E_1 S_E$, where S_E refers to the solvent removed from the extract and may or may not be the same composition as the fresh solvent S.

Determination of the stages uses Q as the operating point until the raffinate composition R_k falls below line FQ. Then the operation is continued with operating point P until R_N is reached.

MINIMUM REFLUX

For a given extract composition E_1, a pinch point develops when an operating line through either P or Q coincides with a tieline. Frequently, the tieline that passes through the feed point F determines the reflux ratio, but not on Figure 14.7(c). The tieline that intersects line FS_E nearest point S_e locates the operating point Q_m for minimum reflux. In Figure 14.7(c), intersection with tieline $Fcde$ is further away from point S_E than that with tieline abQ_m, which is the one that locates the operating point for minimum reflux in this case.

MINIMUM STAGES

As the solvent/feed ratio is increased, the mix point M approaches the solvent point S, and poles P and Q likewise do so. At total reflux all of the points P, Q, S, S_E, and M coincide; this is shown in Figure 14.7(b).

Examples of triangular and McCabe–Thiele constructions for feasible, total, and minimum reflux are shown in Figure 14.7.

Example 14.4
Extraction with an Immiscible Solvent

A feed containing 30 wt % of propionic acid and 70 wt % trichlorethylene is to be extracted with water. Equilibrium distribution of the acid between water (Y) and TCE (X) is represented by $Y = K'X$, with $K' = 0.38$. Section 14.3 is used.

a. The ratio E/R of water to TCE needed to recover 95% of the acid in four countercurrent stages will be found:

$$X_F = 30/70,$$
$$X_n = 1.5/70,$$
$$Y_S = 0,$$
$$\phi = (30 - 1.5)/(30 - 0) = 0.95 = (A^5 - A/(A^5 - 1).$$

By trial,

$$A = 1.734,$$
$$E/R = A/K' = 1.734/0.38 = 4.563.$$

b. The number of stages needed to recover 95% of the acid with $E/R = 3.5$ is found with Eq. 14.12.

$$A = K'E/R = 0.38(3.5) = 1.330, \quad \phi = 0.95,$$
$$n = -1 + \frac{\ln[(A-\phi)/(1-\phi)]}{\ln A} = -1$$
$$+ \ln[(1.330 - 0.95)/(1 - 0.95)]/\ln(1.330) = 6.11$$

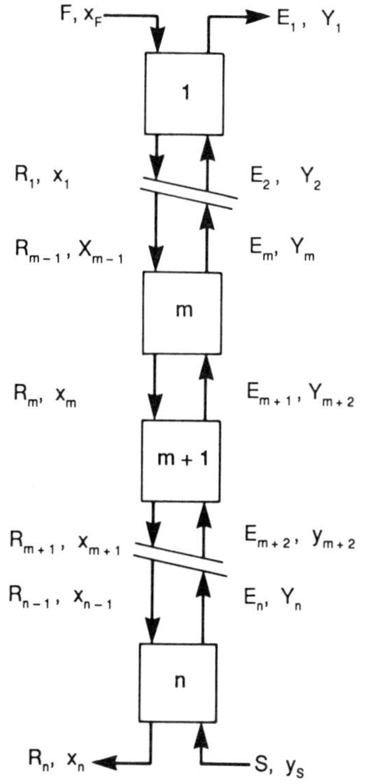

EXAMPLE 14.5
Countercurrent Extraction Represented on Triangular and Rectangular Distribution Diagrams
The specified feed F and the desired extract E_1 and raffinate R_N compositions are shown. The solvent/feed ratio is in the ratio of the line segments MS/MF, where the location of point M is shown as the intersection of lines E_1R_N and FS.

Phase equilibrium is represented by the tieline locus. The equilibrium distribution curve is constructed as the locus of intersections of horizontal lines drawn from the right-hand end of a tieline with horizontals from the left-hand end of the tielines and reflected from the 45° line.

The operating curve is drawn similarly with horizontal projections from pairs of random points of intersection of the binodal curve by lines drawn through the difference point P. Construction of these curves also is explained with Figure 14.6.

The rectangular construction shows that slightly less than eight stages are needed and the triangular that slightly more than eight are needed. A larger scale and greater care in construction could bring these results closer together.

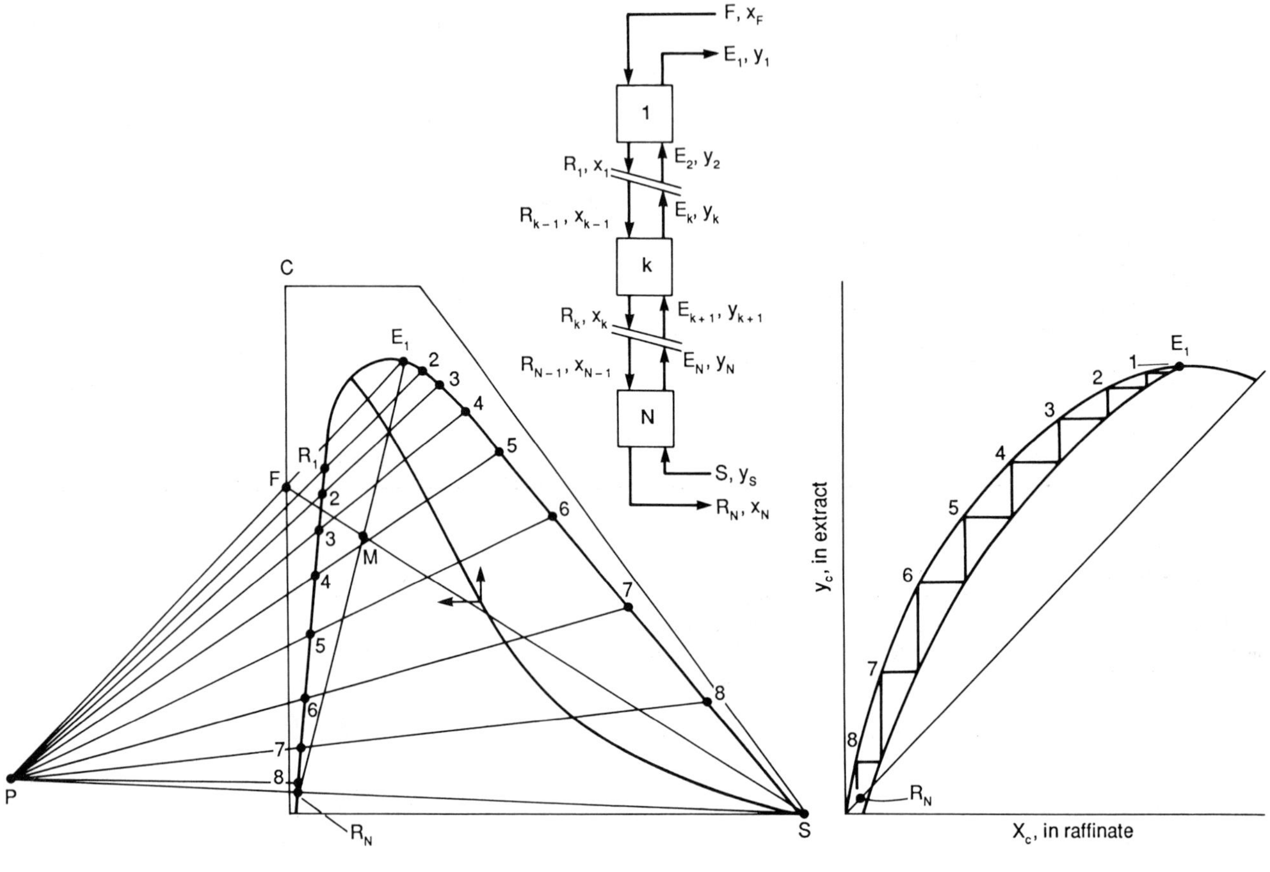

Naturally, the latter constructions are analogous to those for distillation since their forms of equilibrium and material balances are the same. References to the literature where similar calculations are performed with Janecke coordinates were given earlier in this section.

Use of reflux is most effective with Type II systems since then essentially pure products on a solvent-free basis can be made. In contrast to distillation, however, extraction with reflux rarely is beneficial, and few if any practical examples are known. A related kind of process employs a second solvent to wash the extract countercurrently. The requirements for this solvent are that it be only slighly soluble in the extract and easily removable from the extract and raffinate. The sulfolane process is of this type; it is described, for example, by Treybal (1980) and in more detail by Lo et al. (1983, pp. 541–545).

14.4. LEACHING OF SOLIDS

Leaching is the removal of solutes from admixture with a solid by contracting it with a solvent. The solution phase sometimes is called the overflow, but here it will be called extract. The term underflow or raffinate is applied to the solid phase plus its entrained or occluded solution.

Equilibrium relations in leaching usually are simpler than in liquid–liquid equilibria, or perhaps only appear so because few measurements have been published. The solution phase normally contains no entrained solids so its composition appears on the hypotenuse of a triangular diagram like that of Example 14.8. Data for the raffinate phase may be measured as the holdup of solution by the solid, K lb solution/lb dry (oil-free) solid, as a function of the concentration of the solution, y lb oil/lb solution. The correspond-

EXAMPLE 14.6
Stage Requirements for the Separation of a Type I and a Type II System

a. The system with A = heptane, B = tetramethylene sulfone, and C = toluene at 50°C [Triparthi, Ram, and Bhimeshwara, *J. Chem. Eng. Data* **20,** 261 (1975)]: The feed contains 40% C, the extract 70% C on a TMS-free basis or 60% overall, and raffinate 5% C. The construction shows that slightly more than two equilibrium stages are needed for this separation. The compositions of the streams are read off the diagram:

	Feed	Extract	Raffinate
Heptane	60	27	2
TMS	0	13	93
Toluene	40	60	5

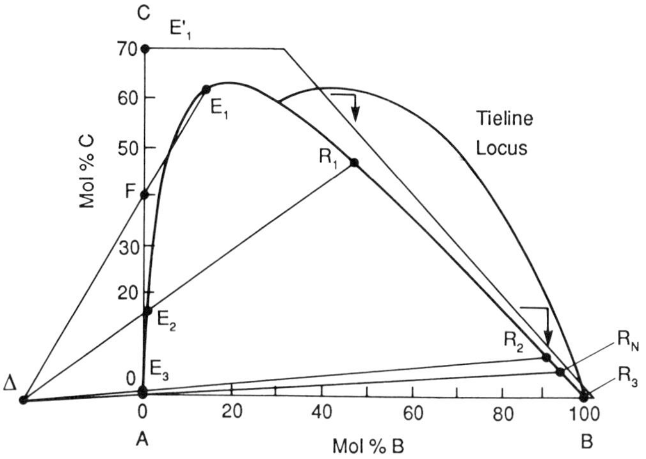

The material balance on heptane is

$$40 = 0.6E + 0.05(100 - E),$$

whence $E = 63.6$ lb/100 lb feed, and the TMS/feed ratio is

$$0.13(63.6) + 0.93(36.4) = 42 \text{ lb}/100 \text{ lb feed}.$$

b. The type II system with A = octane, B = nitroethane, and C = 2,2,4-trimethylpentane at 25°C [Hwa, Techo, and Ziegler, *J. Chem. Eng. Data* **8,** 409 (1963)]: The feed contains 40% TMP, the extract 60% TMP, and the raffinate 5% TMP. Again, slightly more than two stages are adequate.

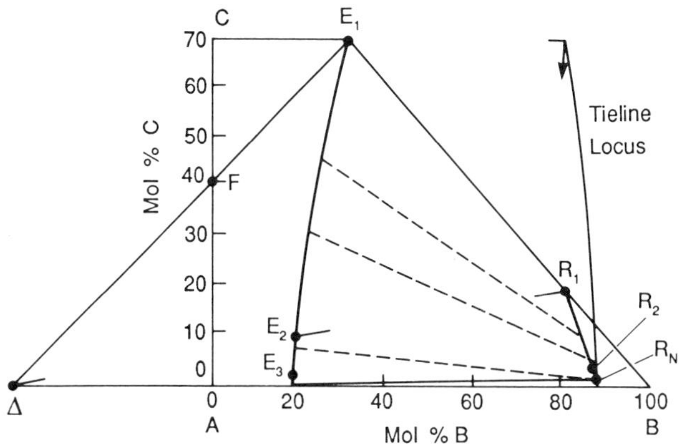

ing weight fraction of oil in the raffinate or underflow is

$$x = Ky/(K + 1). \tag{14.17}$$

Since the raffinate is a mixture of the solution and dry solid, the equilibrium value in the raffinate is on the line connecting the origin with the corresponding solution composition y, at the value of x given by Eq. (14.17). Such a raffinate line is constructed in Example 14.8.

Material balance in countercurrent leaching still is represented by Eqs. (14.14) and (14.16). Compositions R_k and E_{k+1} are on a line through the operating point P, which is at the intersection of

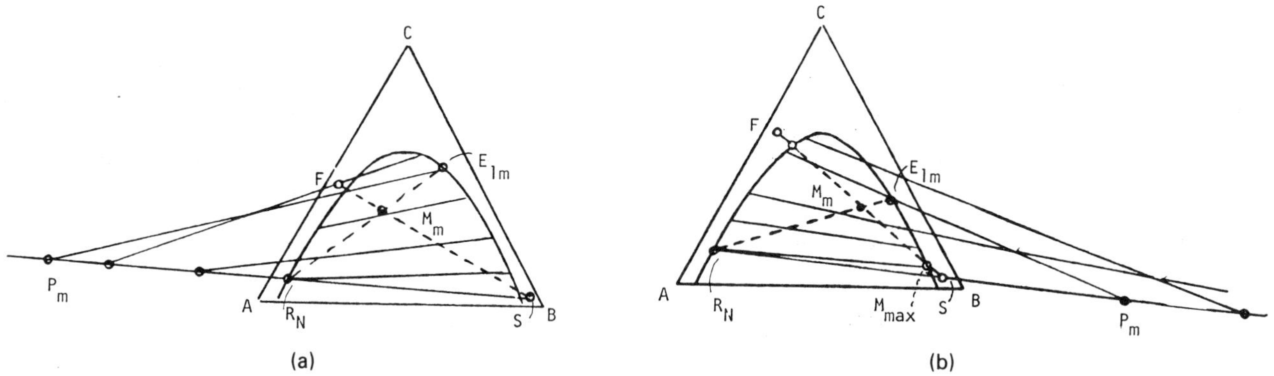

Figure 14.9. Minimum solvent amount and maximum extract concentration. Determined by location of the intersection of extended tielines with extended line R_NS. (a) When the tielines slope down to the left, the furthest intersection is the correct one. (b) When the tielines slope down to the right, the nearest intersection is the correct one. At *maximum* solvent amount, the mix point M_m is on the binodal curve.

Example 14.7

Countercurrent Extraction Employing Extract Reflux

The feed F, extract E_1, and raffinate R_N are located on the triangular diagram. The ratio of solvent/feed is specified by the location of the point M on line SF.

Other nomenclature is identified on the flowsketch. The solvent-free reflux point R_0 is located on the extension of line SE_1. Operating point Q is located at the intersection of lines SR_0 and R_NM. Lines through Q intersect the binodal curve in compositions of raffinate and reflux related by material balance: for instance, R_n and E_{n+1}. When the line QF is crossed, further constructions are made with operating point P, which is the intersection of lines FQ and SR_N.

In this example, only one stage is needed above the feed F and five to six stages below the feed. The ratio of solvent to feed is

$$S/F = FM/MS = 0.196,$$

and the external reflux ratio is

$$r = E_{1R}/E_{1P} = (R_0S/R_0E_1)(QE_1/SQ) = 1.32.$$

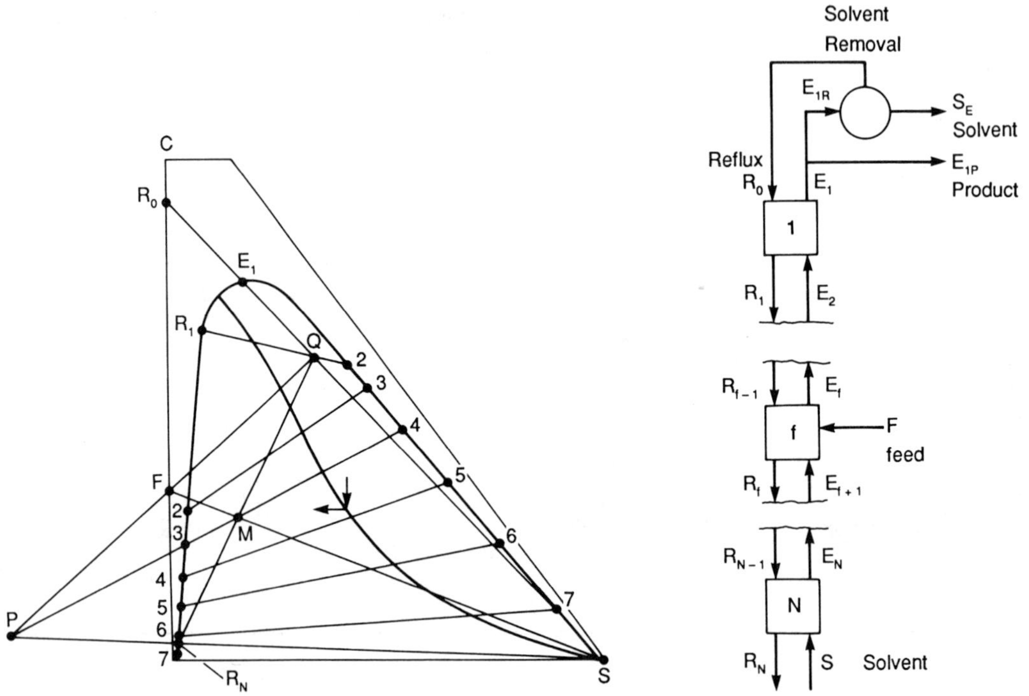

Flowsketch and triangular diagram construction with extract reflux.

Example 14.8

Leaching of an Oil-Bearing Solid in a Countercurrent Battery

Oil is to be leached from granulated halibut livers with pure ether as solvent. Content of oil in the feed is 0.32 lb/lb dry (oil-free) solids and 95% is to be recovered. The economic upper limit to extract concentration is 70% oil. Ravenscroft [*Ind. Eng. Chem.* **28,** 851 (1934)] measured the relation between the concentration of oil in the solution, y, and the entrainment or occlusion of solution by the solid phase, K lb solution/lb dry solid, which is represented by the equation

$$K = 0.19 + 0.126y + 0.810y^2.$$

The oil content in the entrained solution then is given by

$$x = K/(K+1)y, \quad \text{wt fraction},$$

and some calculated values are

y	0	0.1	0.2	0.3	0.4	0.5	0.6	0.7	0.8
x	0	0.0174	0.0397	0.0694	0.1080	0.1565	0.2147	0.2821	0.3578

Points on the raffinate line of the triangular diagram are located on lines connecting values of y on the hypotenuse (solids-free) with the origin, at the values of x and corresponding y from the preceding tabulation.

Feed composition is $x_F = 0.32/1.32 = 0.2424$.

Oil content of extract is $y_1 = 0.7$.

Oil content of solvent is $y_S = 0$.

Amount of oil in the raffinate is $0.32(0.05) = 0.016$ lb/lb dry, and the corresponding entrainment ratio is

$$K_N = 0.016/y_N = 0.19 + 0.126y_N + 0.81y_N^2.$$

EXAMPLE 14.8—(*continued*)

Solving by trial,

$y_N = 0.0781,$

$K_N = 0.2049,$

$x_N = 0.0133$ (final raffinate composition).

The operating point P is at the intersection of lines FE_1 and SR_N. The triangular diagram construction shows that six stages are needed.

The equilibrium line of the rectangular diagram is constructed with the preceding tabulation. Points on the material balance line are located as intersections of random lines through P with these results:

y	0	0.1	0.2	0.3	0.4	0.5	0.6	0.7
x	0.013	0.043	0.079	0.120	0.171	0.229	0.295	0.368

The McCabe–Thiele construction also shows that six stages are needed.

Point P is at the intersection of lines E_1F and SR_N. Equilibrium compositions are related on lines through the origin, point A. Material balance compositions are related on lines through the operating point P.

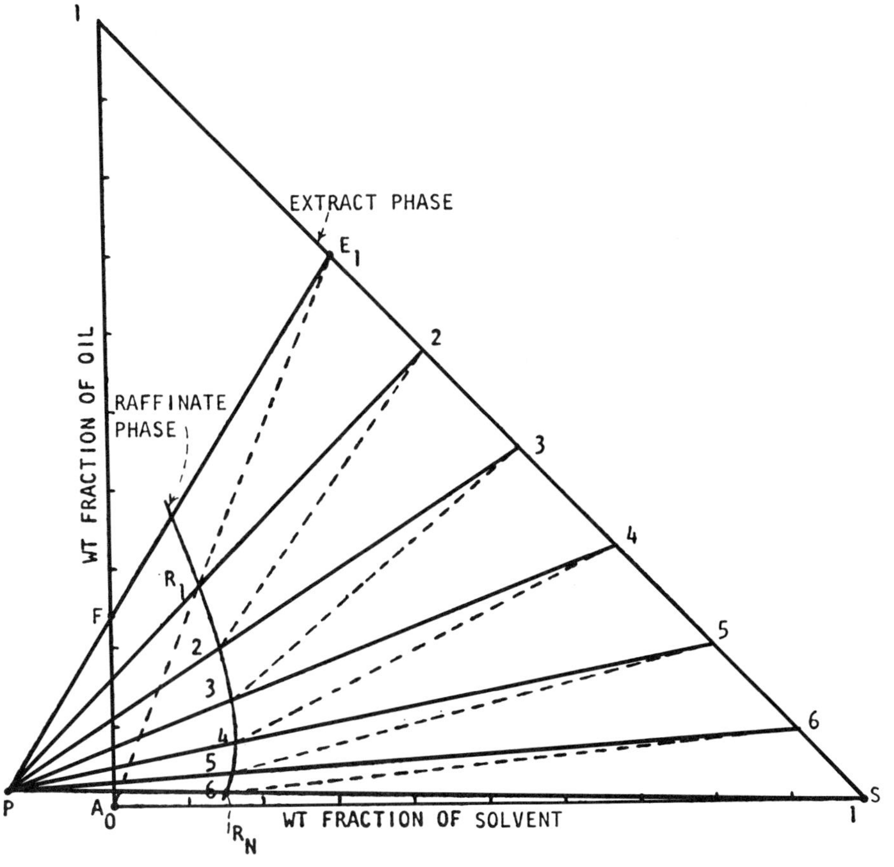

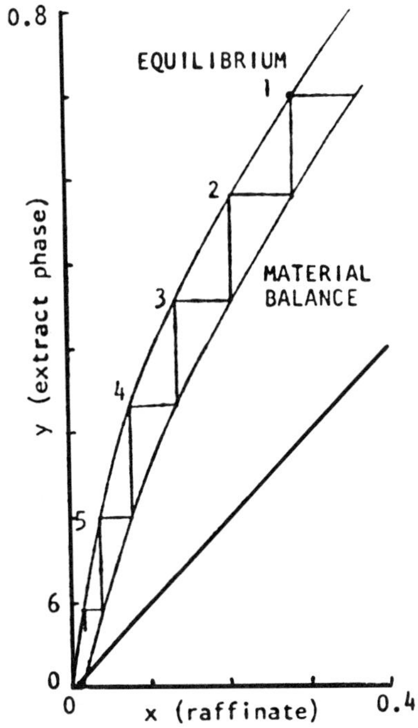

lines FE_1 and SR_N. Similarly, equilibrium compositions R_k and E_k are on a line through the origin. Example 14.8 evaluates stage requirements with both triangular diagram and McCabe–Thiele constructions. The mode of construction of the McCabe–Thiele diagram is described there.

These calculations are of equilibrium stages. The assumption is made that the oil retained by the solids appears only as entrained solution of the same composition as the bulk of the liquid phase. In some cases the solute may be adsorbed or retained within the interstices of the solid as solution of different concentrations. Such deviations from the kind of equilibrium assumed will result in stage efficiencies less than 100% and must be found experimentally.

14.5. NUMERICAL CALCULATION OF MULTICOMPONENT EXTRACTION

Extraction calculations involving more than three components cannot be done graphically but must be done by numerical solution of equations representing the phase equilibria and material balances over all the stages. Since extraction processes usually are adiabatic and nearly isothermal, enthalpy balances need not be made. The solution of the resulting set of equations and of the prior determination of the parameters of activity coefficient correlations requires computer implementation. Once such programs have been developed, they also may be advantageous for ternary extractions, particularly when the number of stages is large or several cases must be worked out. Ternary graphical calculations also could be done on a computer screen with a little effort and some available software.

The notation to be used in making material balances is shown on Figure 14.8. For generality, a feed stream F_k is shown at every stage, and a withdrawal stream U_k also could be shown but is not incorporated in the balances written here. The first of the double subscripts identifies the component i and the second the stage number k; a single subscript refers to a stage.

For each component, the condition of equilibrium is that its activity is the same in every phase in contact. In terms of activity coefficients and concentrations, this condition on stage k is written:

$$\gamma_{ik}^E y_{ik} = \gamma_{ik}^R x_{ik} \tag{14.18}$$

or

$$y_{ik} = K_{ik} x_{ik}, \tag{14.19}$$

where

$$K_{ik} = \gamma_{ik}^R / \gamma_{ik}^E \tag{14.20}$$

is the distribution ratio. The activity coefficients are functions of the temperature and the composition of their respective phases:

$$\gamma_{ik}^E = f(T_k, y_{1k}, y_{2k}, \dots, y_{ck}), \tag{14.21}$$

$$\gamma_{ik}^R = f(T_k, x_{1k}, x_{2k}, \dots, x_{ck}). \tag{14.22}$$

The most useful relations of this type are the NRTL and UNIQUAC which are shown in Table 14.1.

Around the kth stage, the material balance is

$$R_{k-1}x_{i,k-1} + E_{k+1}y_{i,k+1} + F_k z_{ik} - R_k x_{ik} - E_k y_{ik} = 0. \tag{14.23}$$

When combined with Eq. (14.19), the material balance becomes

$$R_{k-1}x_{i,k-1} - (R_k + E_k K_{ik})x_{ik} + E_{k+1}K_{i,k+1}x_{i,k+1} = -F_k z_{ik}. \tag{14.24}$$

In the top stage, $k = 1$ and $R_0 = 0$ so that

$$-(R_1 + V_1 K_{i1})x_{i1} + E_2 K_{i2} x_{i2} = -F_1 z_{i1}. \tag{14.25}$$

In the bottom stage, $k = N$ and $E_{N+1} = 0$ so that

$$R_{N-1}x_{i,N-1} - (R_N + E_N K_{iN})x_{iN} = -F_N z_{iN}. \tag{14.26}$$

The overall balance from stage 1 through stage k is

$$R_k = E_{k+1} - E_1 + \sum_1^k F_k, \tag{14.27}$$

which is used to find raffinate flows when values of the extract flows have been estimated.

For all stages for a component i, Eqs. (14.24)–(14.26) constitute a tridiagonal matrix which is written

$$\begin{bmatrix} B_1 & C_1 & & & & \\ A_2 & B_2 & C_2 & & & \\ & A_j & B_j & C_j & & \\ & & A_{N-1} & B_{N-1} & C_{N-1} & \\ & & & A_N & B_N \end{bmatrix} \begin{bmatrix} x_{i1} \\ x_{i2} \\ x_{ij} \\ x_{iN-1} \\ x_{iN} \end{bmatrix} = \begin{bmatrix} D_1 \\ D_2 \\ D_j \\ D_{N-1} \\ D_N \end{bmatrix} \tag{14.28}$$

When all of the coefficients are known, this can be solved for the concentrations of component i in every stage. A straightforward method for solving a tridiagonal matrix is known as the Thomas algorithm to which references are made in Sec. 13.10, "Basis for Computer Evaluation of Multicomponent Separations: Specifications."

INITIAL ESTIMATES

Solution of the equations is a process in which the coefficients of Eq. (14.28) are iteratively improved. To start, estimates must be made of the flow rates of all components in every stage. One procedure is to assume complete removal of a "light" key into the extract and of the "heavy" key into the raffinate, and to keep the solvent in the extract phase throughout the system. The distribution of the keys in the intermediate stages is assumed to vary linearly, and they must be made consistent with the overall balance, Eq. (14.27), for each component. With these estimated flowrates, the values of x_{ik} and y_{ik} are evaluated and may be used to find the activity coefficients and distribution ratios, K_{ik}. This procedure is used in Example 14.9.

PROCEDURE

The iterative calculation procedure is outlined in Figure 14.10. The method is an adaptation to extraction by Tsuboka and Katayama (1976) of the distillation calculation procedure of Wang and Henke [*Hydrocarb. Proc.* **45**(8), 155–163 (1967)]. It is also presented by Henley and Seader (1981, pp. 586–594).

1. The initial values of the flowrates and compositions x_{ik} and y_{ik} are estimated as explained earlier.
2. The values of activity coefficients and distribution ratios are evaluated.
3. The coefficients in the tridiagonal matrix are evaluated from Eqs. (14.24)–(14.26). The matrix is solved once for each component.
4. The computed values of iteration $(r+1)$ are compared with those of the preceding iteration as

$$\tau_1 = \sum_{i=1}^{C} \sum_{k=1}^{N} |x_{ik}^{(r+1)} - x_{ik}^{(r)}| \le \varepsilon_1 = 0.01NC. \tag{14.29}$$

   The magnitude, $0.01NC$, of the convergence criterion is arbitrary.
5. For succeeding evaluations of activity coefficients, the values of the mol fractions are normalized as

$$(x_{ik})_{\text{normalized}} = x_{ik} \Big/ \sum_{i=1}^{C} x_{ik}, \qquad (y_{ik})_{\text{normalized}} = y_{ik} \Big/ \sum_{i=1}^{C} y_{ik}. \tag{14.30}$$

6. When the values of x_{ik} have converged, a new set of y_{ik} is calculated with

$$y_{ik} = K_{ik} x_{ik}. \tag{14.19}$$

7. A new set of extract flow rates is calculated from

$$E_k^{(s+1)} = E_k^{(s)} \sum_{i=1}^{C} y_{ik}, \tag{14.31}$$

   where s is the outer loop index number.

TABLE 14.1. NRTL and UNIQUAC Correlations for Activity Coefficients of Three-Component Mixtures[a]

NRTL

$$\ln \gamma_i = \frac{\tau_{1i}G_{1i}x_1 + \tau_{2i}G_{2i}x_2 + \tau_{3i}G_{3i}x_3}{G_{1i}x_1 + G_{2i}x_2 + G_{3i}x_3}$$

$$+ \frac{x_1 G_{i1}}{x_1 + G_{12}x_2 + G_{13}x_3}\left[\tau_{i1} - \frac{x_2\tau_{21}G_{21} + x_3\tau_{31}G_{31}}{x_1 + x_2 G_{21} + x_3 G_{31}}\right]$$

$$+ \frac{x_2 G_{i2}}{G_{12}x_1 + x_2 + G_{32}x_3}\left[\tau_{i2} - \frac{x_1\tau_{12}G_{12} + x_3\tau_{32}G_{32}}{x_1 G_{12} + x_2 + x_3 G_{32}}\right]$$

$$+ \frac{x_3 G_{i3}}{G_{13}x_1 + G_{23}x_2 + x_3}\left[\tau_{i3} - \frac{x_1\tau_{13}G_{13} + x_2\tau_{23}G_{23}}{G_{13}x_1 + G_{23}x_2 + x_3}\right]$$

$\tau_{ii} = 0$

$G_{ii} = 1$

UNIQUAC

$$\ln \gamma_i = \ln\frac{\phi_i}{x_i} + 5q_i \ln\frac{\theta_i}{\phi_i} + l_i - \frac{\phi_i}{x_i}(x_1 l_1 + x_2 l_2 + x_3 l_3) + q_i[1 - \ln(\theta_1\tau_{1i} + \theta_2\tau_{2i} + \theta_3\tau_{3i})]$$

$$- \frac{\theta_1\tau_{i1}}{\theta_1 + \theta_2\tau_{21} + \theta_3\tau_{31}} - \frac{\theta_2\tau_{i2}}{\theta_1\tau_{12} + \theta_2 + \theta_3\tau_{32}} - \frac{\theta_3\tau_{i3}}{\theta_1\tau_{13} + \theta_2\tau_{23} + \theta_3}$$

$\tau_{ii} = 1$

$$\phi_i = \frac{r_i x_i}{r_1 x_1 + r_2 x_2 + r_3 x_3}$$

$$\theta_i = \frac{q_i x_i}{q_1 x_1 + q_2 x_2 + q_3 x_3}$$

$l_i = 5(r_i - q_i) - r_i + 1$

[a] NRTL equation: There is a pair of parameters g_{jk} and g_{kj} for each pair of substances in the mixture; for three substances, there are three pairs. The other terms of the equations are related to the basic ones by

$\tau_{jk} = g_{jk}/RT,$

$G_{jk} = \exp(-\alpha_{jk}\tau_{jk}).$

For liquid–liquid systems usually, $\alpha_{jk} = 0.4$.

UNIQUAC equation: There is a pair of parameters u_{jk} and u_{kj} for each pair of substances in the mixture:

$\tau_{jk} = \exp(-u_{jk}/RT).$

The terms with single subscripts are properties of the pure materials which are usually known or can be estimated.

The equations are extended readily to more components.

(See, for example, Walas, *Phase Equilibria in Chemical Engineering,* Butterworths, 1985).

8. The criterion for convergence is

$$\tau_2 = \sum_{k=1}^{N} (1 - E_k^{(s)}/E_k^{(s+1)})^2 \le \varepsilon_2 = 0.01N. \quad (14.32)$$

The magnitude, $0.01N$, of the convergence criterion is arbitrary.

9. If convergence has not been attained, new values of R_k are calculated from Eq. (14.27).

10. Distribution ratios K_{ik} are based on normalized values of x_{ik} and y_{ik}.

11. The iteration process continues through the inner and outer loops.

EXAMPLE 14.9
Trial Estimates and Converged Flow Rates and Compositions in All Stages of an Extraction Battery for a Four-Component Mixture

Benzene is to be recovered from a mixture with hexane using aqueous dimethylformamide as solvent in a five-stage extraction battery. Trial estimates of flow rates for starting a numerical solution are made by first assuming that all of the benzene and all of the solvent ultimately appear in the extract and all of the hexane appears in the raffinate. Then flow rates throughout the battery are assumed to vary linearly with stage number. Table 1 shows these estimated flowrates and Table 2 shows the corresponding mol fractions. Tables 3 and 4 shows the converged solution made by Henley and Seader (1981, p. 592); they do not give any details of the solution but the algorithm of Figure 14.10 was followed.

TABLE 1. Estimated mol/hr

	Extract					Raffinate				
Stage	Total	*H*	*B*	*D*	*W*	Total	*H*	*B*	*D*	*W*
0	—									
1	1100	0	100	750	250	400	300	100	0	0
2	1080	0	80	750	250	380	300	80	0	0
3	1060	0	60	750	250	360	300	60	0	0
4	1040	0	40	750	250	340	300	40	0	0
5	1020	0	20	750	250	320	300	20	0	0
$N+1$	1000	0	0	750	250	300	300	0	0	0

TABLE 2. Estimated Mol Fractions

Stage	y_{ij}				x_{ij}			
j	*H*	*B*	*D*	*W*	*H*	*B*	*D*	*W*
1	0.0	0.0909	0.6818	0.2273	0.7895	0.2105	0.0	0.0
2	0.0	0.0741	0.6944	0.2315	0.8333	0.1667	0.0	0.0
3	0.0	0.0566	0.7076	0.2359	0.8824	0.1176	0.0	0.0
4	0.0	0.0385	0.7211	0.2404	0.9375	0.0625	0.0	0.0
5	0.0	0.0196	0.7353	0.2451	1.0000	0.0	0.0	0.0

TABLE 3. Converged Mol Fractions

Stage	y_{ij}				x_{ij}			
j	*H*	*B*	*D*	*W*	*H*	*B*	*D*	*W*
1	0.0263	0.0866	0.6626	0.2245	0.7586	0.1628	0.0777	0.0009
2	0.0238	0.0545	0.6952	0.2265	0.8326	0.1035	0.0633	0.0006
3	0.0213	0.0309	0.7131	0.2347	0.8858	0.0606	0.0532	0.0004
4	0.0198	0.0157	0.7246	0.2399	0.9211	0.0315	0.0471	0.0003
5	0.0190	0.0062	0.7316	0.2432	0.9438	0.0125	0.0434	0.0003

TABLE 4. Converged mol/hr

	Extract	Raffinate
Hexane	29.3	270.7
Benzene	96.4	3.6
DMF	737.5	12.5
Water	249.0	0.1
Total	1113.1	286.9

Solutions of four cases of three- and four-component systems are presented by Tsuboka and Katayama (1976); the number of outer loop iterations ranged from 7 to 41. The four component case worked out by Henley and Seader (1981) is summarized in Example 14.9; they solved two cases with different water contents of the solvent, dimethylformamide.

14.6. EQUIPMENT FOR EXTRACTION

Equipment for extraction and leaching must be capable of providing intimate contact between two phases so as to effect transfer of solute between them and also of ultimately effecting a complete separation of the phases. For so general an operation, naturally a substantial variety of equipment has been devised. A very general classification of equipment, their main characteristics and industrial applications is in Table 14.2. A detailed table of comparisons and ratings of 20 kinds of equipment on 14 characteristics has been prepared by Pratt and Hanson (in Lo et al., 1983, p. 476). Some comparisons of required sizes and costs are in Table 14.3.

Selected examples of the main categories of extractors are represented in Figures 14.11–14.15. Their capacities and performance will be described in general terms insofar as possible, but sizing of liquid–liquid extraction equipment always requires some pilot plant data or acquaintance with analogous cases. Little detailed information about such analogous situations appears in the open literature. Engineers familiar with particular kinds of equipment, such as their manufacturers, usually can predict performance with a minimum amount of pilot plant data.

Literature data is almost entirely for small equipment whose capacity and efficiency cannot be scaled up to commercial sizes, although it is of qualitative value. Extraction processes are sensitive because they operate with small density differences that are sensitive to temperature and the amount of solute transfer. They also are affected by interfacial tensions, the large changes in phase flow rates that commonly occur, and even by the direction of mass transfer. For comparison, none of these factors is of major significance in vapor-liquid contacting.

CHOICE OF DISPERSE PHASE

Customarily the phase with the highest volumetric rate is dispersed since a larger interfacial area results in this way with a given droplet size. In equipment that is subject to backmixing, such as spray and packed towers but not sieve tray towers, the disperse phase is made the one with the smaller volumetric rate. When a substantial difference in resistances of extract and raffinate films to mass transfer exists, the high phase resistance should be compensated for with increased surface by dispersion. From this point of view, Laddha and Degaleesan (1978, pp. 194) point out that water should be the dispersed phase in the system water + diethylamine + toluene. The dispersed phase should be the one that wets the material of construction less well. Since the holdup of continuous phase usually is greater, the phase that is less hazardous or less expensive should be continuous. It is best usually to disperse a highly viscous phase.

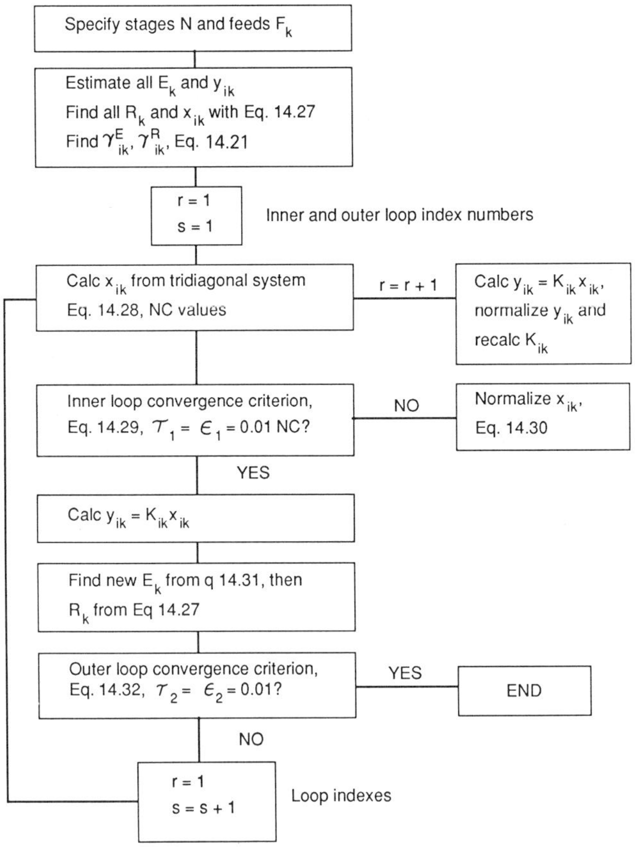

Figure 14.10. Algorithm for computing flows and compositions in an extraction battery of a specified number of stages (*after Henley and Seader, 1981*).

MIXER-SETTLERS

The original and in concept the simplest way of accomplishing extractions is to mix the two phases thoroughly in one vessel and then to allow the phases to separate in another vessel. A series of such operations performed with series or countercurrent flows of the phases can accomplish any desired degree of separation. Mixer-settlers have several advantages and disadvantages, for instance:

Pros. The stages are independent, can be added to or removed as needed, are easy to start up and shut down, are not bothered by suspended solids, and can be sized for high (normally 80%) efficiencies.

Cons. Emulsions can be formed by severe mixing which are hard to break up, pumping of one or both phases between tanks may be required, independent agitation equipment and large floor space needs are expensive, and high holdup of valuable or hazardous solvents exists particularly in the settlers.

Some examples of more or less compact arrangements of mixers and settlers are in Figures 14.11 and 14.14(c). Mixing equipment is described in Chapter 10 where rules for sizing, blending, mixing intensity, and power requirements are covered, for instance Figure 10.3 for blend times in stirred tanks. Mixing with impellers in tanks is most common, but also is accomplished with pumps, jet mixers [Fig. 14.11(b)], line mixers and static mixers.

TABLE 14.2. Features and Industrial Applications of Liquid–Liquid Extractors

Types of extractor	General features	Fields of industrial application
Unagitated columns	Low capital cost Low operating and maintenance cost Simplicity in construction Handles corrosive material	Petrochemical Chemical
Mixer-settlers	High-stage efficiency Handles wide solvent ratios High capacity Good flexibility Reliable scale-up Handles liquids with high viscosity	Petrochemical Nuclear Fertilizer Metallurgical
Pulsed columns	Low HETS No internal moving parts Many stages possible	Nuclear Petrochemical Metallurgical
Rotary-agitation columns	Reasonable capacity Reasonable HETS Many stages possible Reasonable construction cost Low operating and maintenance cost	Petrochemical Metallurgical Pharmaceutical Fertilizer
Reciprocating-plate columns	High throughput Low HETS Great versatility and flexibility Simplicity in construction Handles liquids containing suspended solids Handles mixtures with emulsifying tendencies	Pharmaceutical Petrochemical Metallurgical Chemical
Centrifugal extractors	Short contacting time for unstable material Limited space required Handles easily emulsified material Handles systems with little liquid density difference	Pharmaceutical Nuclear Petrochemical

(Reprinted by permission from T. C. Lo, Recent Developments in Commercial Extractors, Engineering Foundation Conference on Mixing Research, Rindge, N.H., 1975).

Capacities of line mixers such as those of Figure 10.13 and of static mixers such as those of Figure 10.14 are stated in manufacturers catalogs. A procedure for estimating mixing efficiencies from basic correlations is illustrated by Laddha and Degaleesan (1978, p. 424).

Separation of the mixed phases is accomplished by gravity settling or less commonly by centrifugation. It can be enhanced by inducing coalescence with packing or electrically, or by shortening the distance of fall to a coalesced phase. Figures 14.11(d), 18.2, and 18.3 are some examples. Chapter 18 deals with some aspects of the separation of liquid phases.

A common basis for the design of settlers is an assumed droplet size of 150 μm, which is the basis of the standard API design method for oil–water separators. Stokes law is applied to find the settling time. In open vessels, residence times of 30–60 min or superficial velocities of 0.5–1.5 ft/min commonly are provided. Longitudinal baffles can cut the residence time to 5–10 min. Coalescence with packing or wire mesh or electrically cut these

TABLE 14.3. Comparisons of Performance and Costs of Extraction Equipment

(a) Some Comparisons and Other Performance Data

System	Equipment	Total Flow Capacity (Imp. gal/hrft^2) Pilot Plant	Plant
Co–Ni–D2EHPA H_2SO_4	Mixco agitated	(4 in.) 300	(60 in.) 170
	Karr reciprocating	(3 in.) 900	
	sieve plate pulse	(2 in.) 900	
Zr–Hf–TBP HNO_3	Mixco agitated		(30 in.) 184
	sieve plate pulse (steel)	(2 in.) 500	
	sieve plate pulse (Teflon)	(2 in.) 1345	(10 in.) 1345
	RDC		(30 in.) 135
Hf–Zr–MIBK SCN^-	spray column		(4 in.) 2450
Rare earths–D2EHPA H_2SO_4	Podbielniak centrifuge		(4 feed dia) 30,000 gal/hr
U–amine-solvent-in-pulp H_2SO_4	sieve plate pulse	(2 in.) 600	(10 in.) 900
Cu–Lix 64N H_2SO_4	mixer settlers		60–120
Cu–Ni–amine HCl	mixer settlers		60–120

(b) Cost Comparison, 1970 Prices, for Extraction of 150 gpm of Aqueous Feed Containing 5 g/L of Cu with 100 gpm Solvent, Recovering 99% of the Copper

Contactor	Equipment Required No.	Dia. (ft)	Length (ft)	Equip. Cost $ × 1000	Total cost $ × 1000
Mixer settler	2	—	—	60	151.2
Mixco	3	5	16	100	246.7
Pulse	1	5	60	160	261.5
Kenics	3	2	28	230	336.1
Podbielniak		3–D36	—	300	378.0
Graesser	15	5	3.0	88	308.0

[a] Mixers have 150 gal capacity, settlers are 150 sqft by 4 ft deep with 9 in. solvent layer.

(G.M. Ritcey and A.W. Ashbrook, *Solvent Extraction*, 1979, Vol. II).

times substantially. A chart for determining separation of droplets of water with a plate pack of 3/4 in. spacing is reproduced by Hooper and Jacobs (in Schweitzer, 1979, 1.343–1.358). Numerical examples of settler design also are given in that work. For especially difficult separations or for space saving, centrifuges are applied. Liquid hydrocyclones individually have low efficiencies, but a number in series can attain 80–85% efficiency overall. Electrical coalescence is used commonly for separation of brine from crude oil; the subject is treated by Waterman (*Chem. Eng. Prog.* **61**(10), 51 1965).

A control system for a mixer-settler is represented by Figure 3.19.

SPRAY TOWERS

These are empty vessels with provisions for introducing the liquids as dispersed or continuous phases and for removing them. Figure 14.12(a) shows both phases dispersed, which may be demanded when substantial changes in volumetric or physical properties result from solute transfer. Capacities of spray towers are high because of their openness, and they are not bothered by suspended solids. Backmixing is severe in towers of more than a few inches in diameter. Without operating experience to the contrary, even towers 20–40 ft high cannot be depended upon to function as more than single stages. The cross section is determined by the flooding velocity; that of the continuous phase is correlated by the equation

$$V_{CF} = \frac{4000\Delta\rho^{0.28}}{[0.483\mu_C^{0.075}\rho_C^{0.5} + d_P^{0.056}\rho_D^{0.5}(V_D/V_C)^{0.5}]^2}\ \text{ft/hr}, \quad (14.33)$$

where a factor of 0.4 suggested by Treybal has been incorporated for safe design. The units are ft lb hr; the viscosity μ_C lb/ft hr = 2.42 cP. For large capacities, several parallel towers of at most 2 ft dia should be used. Commercially, spray towers are suitable for liquid–liquid processes in which rapid, irreversible chemical reactions occur, as in neutralization of waste acids. The substantial literature of flooding, holdup, mass transfer and axial mixing in small spray towers is reviewed by Laddha and Degaleesan (1978, pp. 221–255) and more briefly by Cavers (in Lo et al., 1983, pp. 320–328).

PACKED TOWERS

Since mass transfer in packed or spray towers occurs differentially rather than stagewise, their performance should be expressed in terms of the number of transfer units (NTU) rather than the number of theoretical stages (NTS). For dilute systems, the number of transfer units is given in terms of the terminal concentrations and the equilibrium relation by

$$\text{NTU} = \int_{x_1}^{x_2} \frac{dx}{x - x_{\text{equilib}}}. \quad (14.34)$$

In order to permit sizing a tower, data must be available of the height of a transfer unit (HTU). This term often is used interchangeably with the height equivalent to a theoretical stage (HETS), but strictly they are equal only for dilute solutions when the ratio of the extract and raffinate flow rates, E/R, equals the distribution coefficient, $K = x_E/x_R$ (Treybal, 1963, p. 350). Extractor performance also is expressible in terms of mass transfer coefficients, for instance, K_Ea, which is related to the number and height of transfer units by

$$\frac{K_Ea\Delta C}{E/S} = \frac{\text{NTU}}{Z} = \frac{1}{\text{HTU}}, \quad (14.35)$$

where E/S is the extract flow rate per unit cross section and ΔC is mean concentration difference of the solute. Correlations of this quantity based on data from towers of 1–2 in. dia have been made, for example, by Laddha and Degaleesan (1978). They may be of qualitative value in predicting performance of commercial equipment when combined with some direct pilot plant information. In commercial size towers, HETS of 2–5 ft may be realized. Mass transfer drops off sharply with axial distance, so that the dispersed phase is redistributed every 5–7 ft. A sketch of a redistributor is in Figure 14.12(e). Extractors with three or more beds are not uncommon.

Packed towers may be employed when 5–10 stages suffice. They are not satisfactory at interfacial tensions above 10 dyn/cm. Even at this condition, sieve trays have greater efficiency, and at much higher interfacial tensions some form of agitated tower is required.

Metal and ceramic packings tend to remain wetted with the

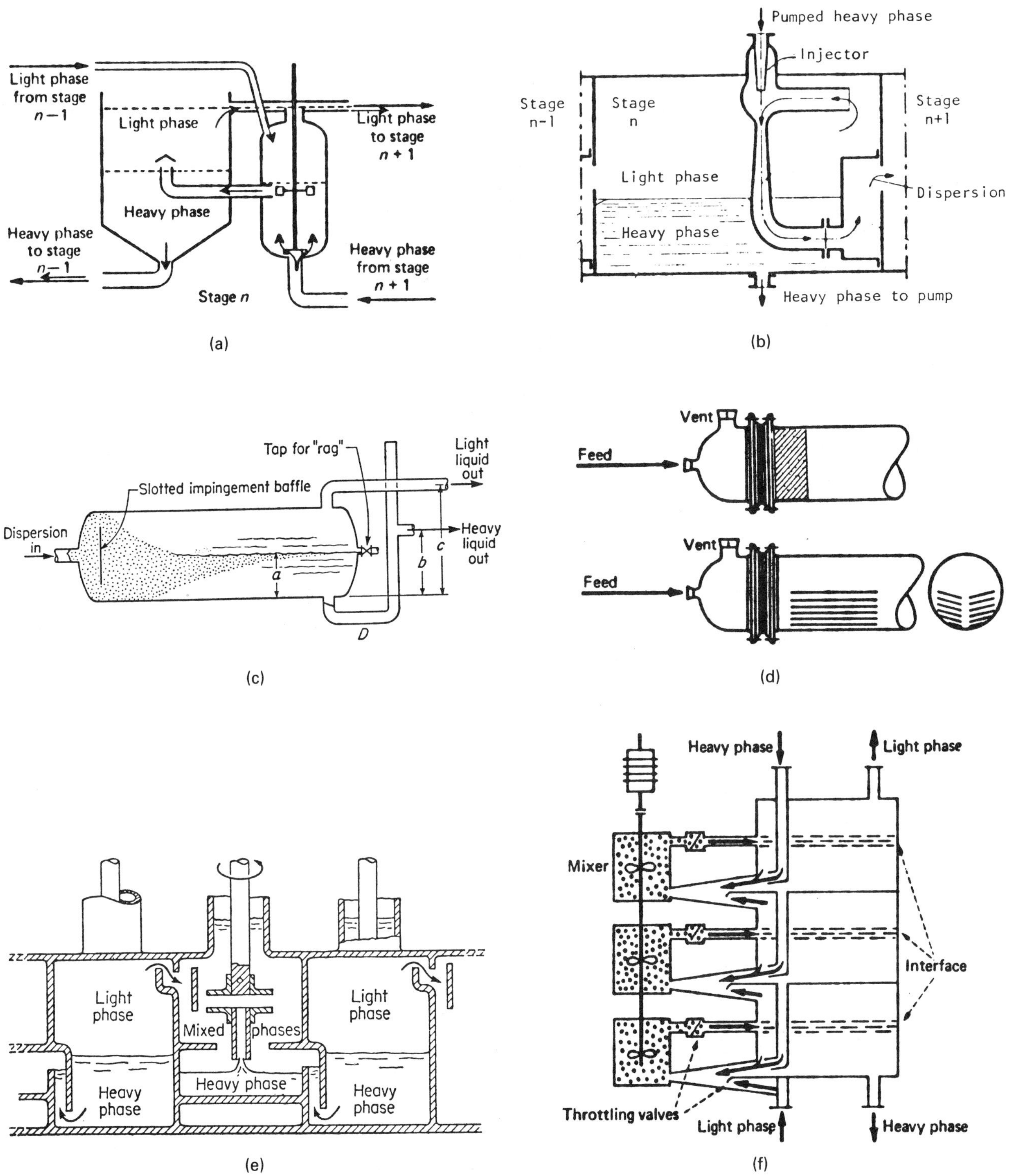

Figure 14.11. Some types and arrangements of mixers and settlers. (a) Kemira mixer-settler (*Mattila,* Proc. Solvent Extraction Conference, *ISEC 74, Inst. Chem. Eng., London, 1974*); (b) Injection mixer and settler (*Ziolkowski, 1961*). (c) Gravity settler; "rag" is foreign material that collects at the interface. (d) Provisions for improving rate of settling: (top) with packing or wire mesh; (bottom) with a nest of plates. (e) Compact arrangement of pump mixers and settlers [*Coplan et al.,* Chem. Eng. Prog. **50,** *403* (*1954*)]. (f) Vertical arrangement of a battery of settlers and external mixers (*Lurgi Gesellschaften*).

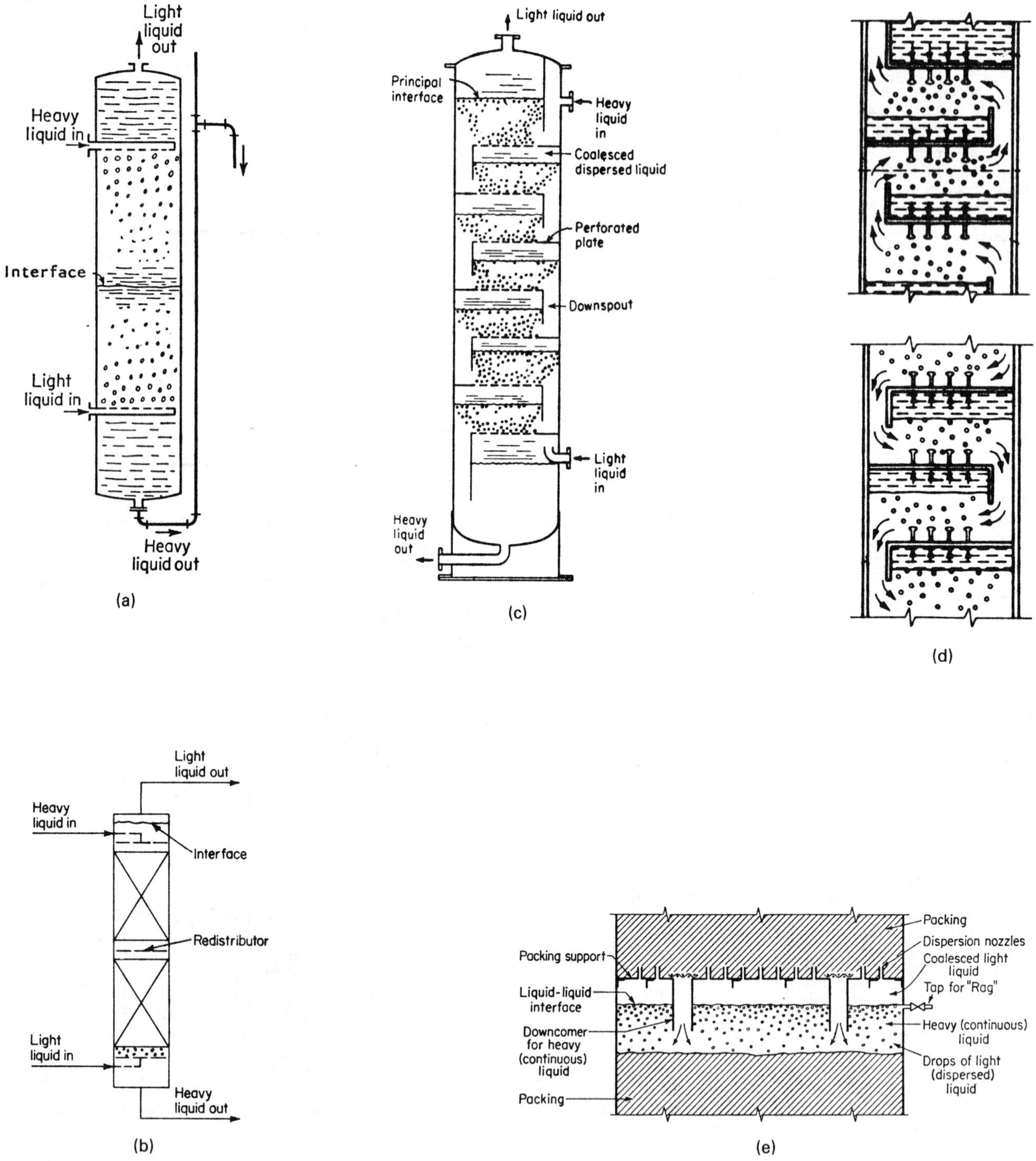

Figure 14.12. Tower extractors without agitation. (a) Spray tower with both phases dispersed. (b) Two-section packed tower with light phase dispersed. (c) Sieve tray tower with light phase dispersed. (d) Sieve tray construction for light phase dispersed (left) and heavy phase dispersed (right). (e) Redistributor for packed tower with light phase dispersed (*Treybal, 1963*).

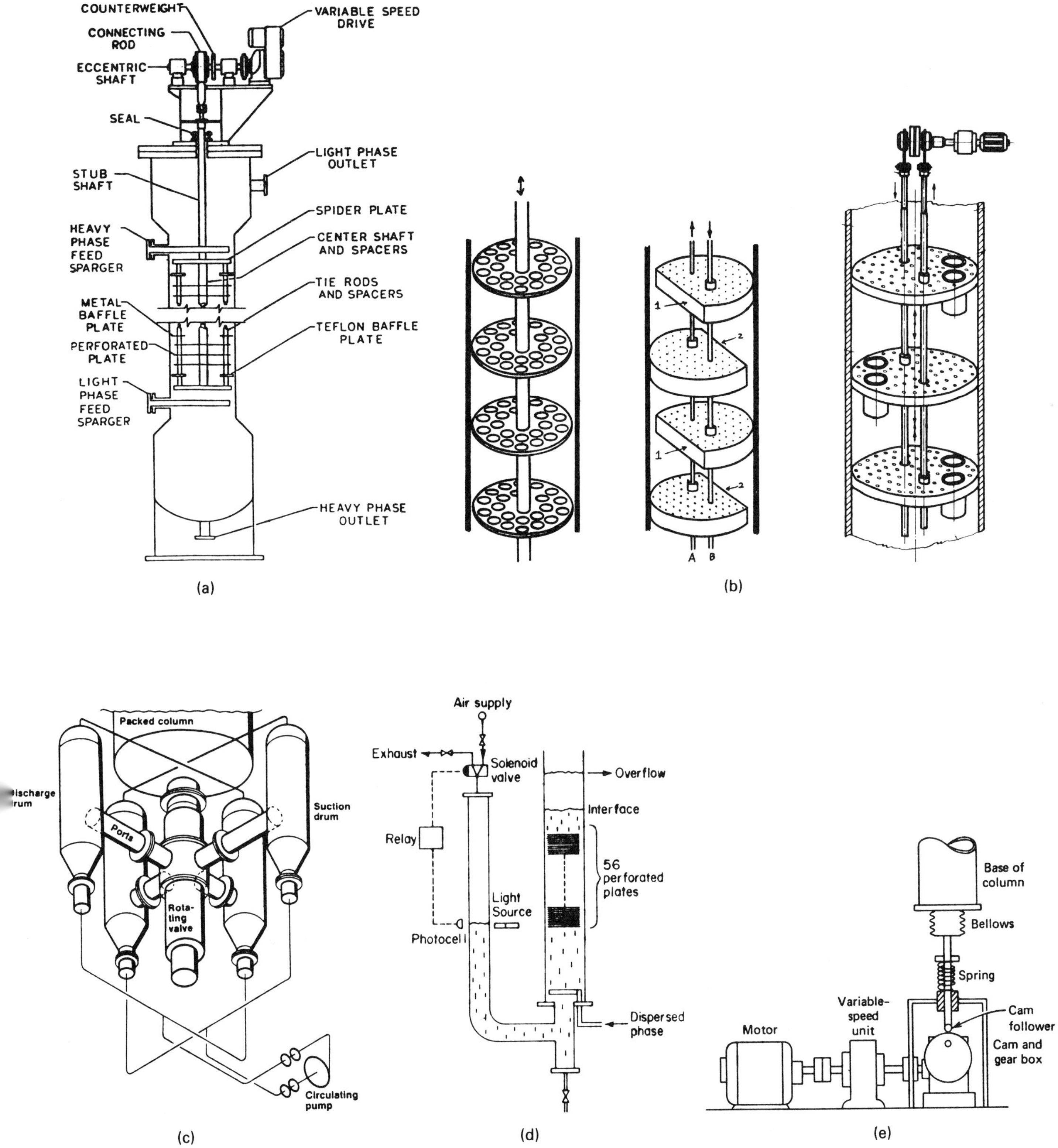

Figure 14.13. Towers with reciprocating trays or with pulsing action. (a) Assembly of a 36 in. Karr reciprocating tray column (*Chem. Pro. Co.*). (b) Sieve trays used in reciprocating trays columns; (left) large opening trays for the Karr column; (middle) countermotion trays with cutouts; (right) countermotion trays with downpipes for heavy phase. (c) Rotary valve pulsator, consisting of a variable speed pump and a rotary valve that alternately links the column with pairs of suction and discharge vessels. (d) Sieve tray tower with a pneumatic pulser [Proc. Int. Solv. Extr. Conf. **2,** *1571* (*1974*)]. (e) A pulser with a cam-operated bellows.

liquid that first wets them, so that the tower should be charged first with the continuous phase. Thermoplastics tend to be preferentially oil wetted, but they can be wetted by aqueous phase if immersed in it for several days.

Intalox saddles and pall rings of 1–1.5-in. size are the most commonly used packings. Smaller sizes tend to be less effective since their voids are of the same order of magnitude as drop sizes. The flooding correlation of Figure 14.16 is recommended by Eckert (1984); a safe design is about 70% of the value obtained with this correlation. Dispersed phase loadings should not exceed 25 gal/(min)(sqft). Dispersion is best accomplished with perforated plates in which hole sizes are 3/16–1/4 in. Velocities through the holes should not exceed 0.8 ft/sec, but if short riser tubes are employed the velocities can be as high as 1.5 ft/sec.

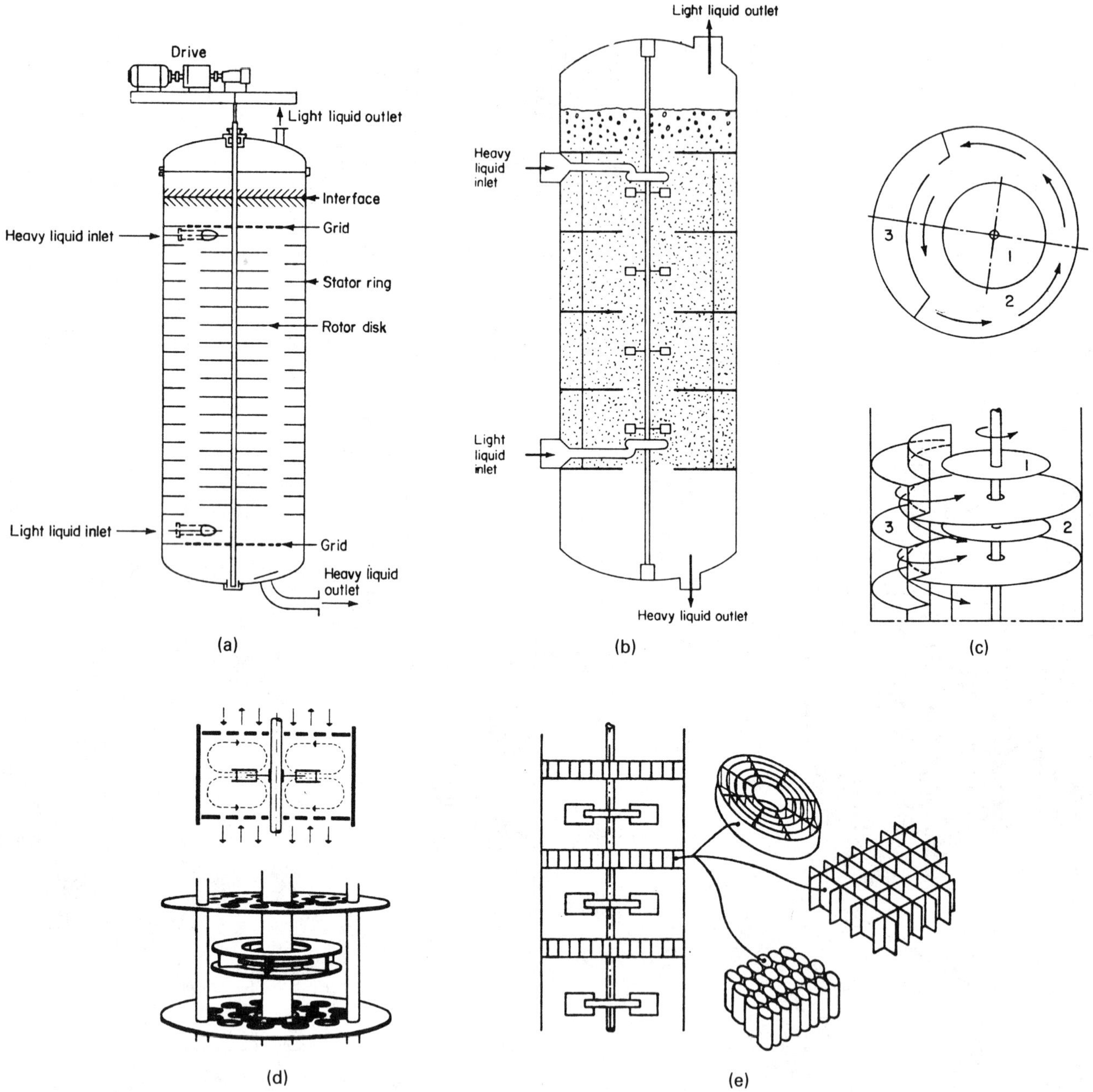

Figure 14.14. Tower extractors with rotary agitators. (a) RDC (rotating disk contactor) extraction tower (*Escher B.V., Holland*). (b) Oldshue–Rushton extractor with turbine impellers and stator rings (*Mixing Equip. Co.*). (c) ARD (asymmetric rotating disk) extractor: (1) rotating disk rotor; (2) mixing zone; (3) settling zone (*Luwa A.G.*). (d) Kuhni extractor, employing turbine impellers and perforated partitions (*Kühni Ltd.*). (e) EC (enhanced coalescing) extractor [*Fischer et al.*, Chem. Ing. Tech., *228* (*Mar. 1983*)]. (f) Model of Scheibel extractor employing baffled mixing stages and wire mesh separating zones (*E.G. Scheibel Inc.*). (g) Model of Scheibel extractor employing shrouded turbine impellers and flat stators, suited for larger diameter columns (*E.G. Scheibel Inc.*).

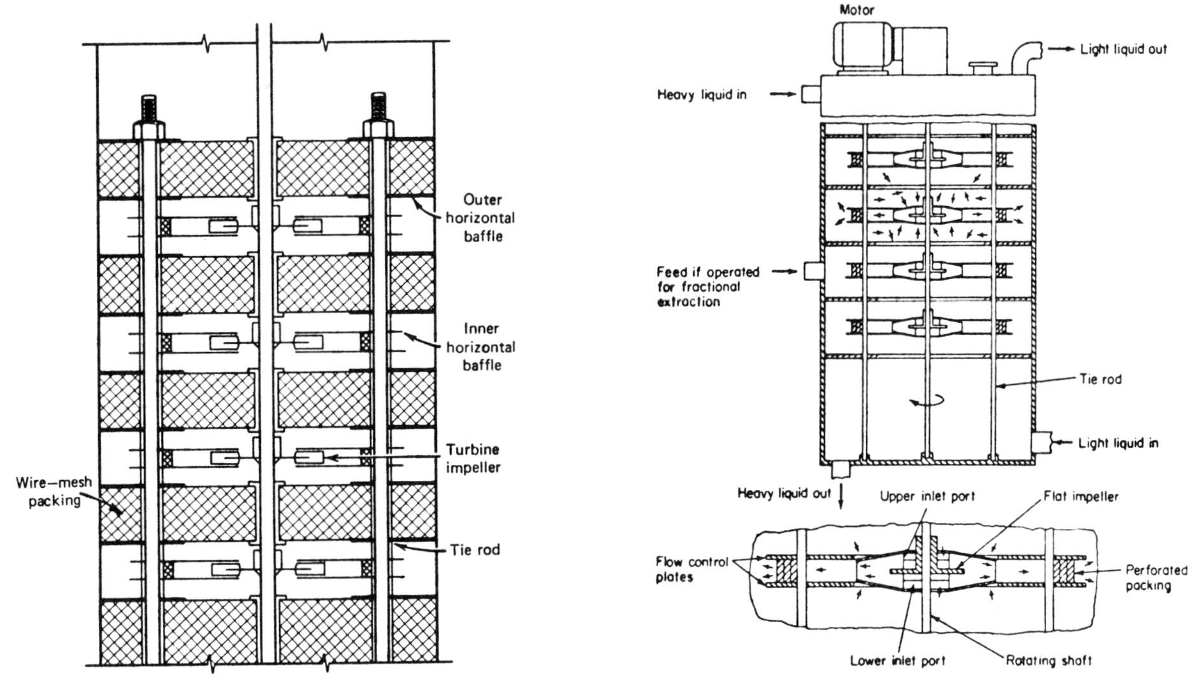

(f) (g)

Figure 14.14—(*continued*)

SIEVE TRAY TOWERS

In sieve or perforated tray towers, the continuous phase runs across each tray and proceeds to the next one through a downcomer or riser. The dispersed phase is trapped as a coalesced layer at each tray and redispersed. The designs for light phase or heavy phase dispersion are shown in Figure 14.12(d). Either phase may be the dispersed one, but usually it is the raffinate. Both the reduced axial mixing because of the presence of the trays and the repeated dispersion tend to improve the efficiency over the other kinds of unagitated towers.

Hole diameters are much smaller than for vapor–liquid contacting, being 3–8 mm, usually on triangular spacing of 2–3 dia, and occupy from 15 to 25% of the available tray area. The area at a downcomer or riser is not perforated, nor is the area at the support ring which may be an inch or two wide. Velocities through the holes are kept below about 0.8 ft/sec to avoid formation of very small droplets. The head available for flow of the continuous pulse is the tray spacing. It is estimated as 4.5 velocity heads and thus is given by the equation

$$h = 4.5V_d^2\rho_C/2g_c \tag{14.36}$$

where V_d is the linear velocity in the downcomer. It is usual to fill the downcomer with packing to coalesce entrainment; then the downcomer cross section must be made correspondingly larger.

Diameter of the Tower. The cross section of the tower must be made large enough to accommodate the downcomer and the perforated zone. Diameters of 12 ft or more are common.

Tray spacing is from 6–24 in., the larger dimension to facilitate servicing the trays in place when necessary. Both the downcomer cross section and the depth of coalesced layer are factors related to the spacing, and so is the efficiency. The depth of coalesced layer at each tray must be sufficient to force the liquid through the holes. In the range of 1 ft/sec through the holes, surface tension does not affect the flow significantly, so that the head–velocity relationship is the common one through orifices, namely,

$$V = 0.67\sqrt{2g_c h\Delta\rho/\rho_D}. \tag{14.37}$$

A correction also can be applied for the ratio of perforated and total tray areas. For the case of Example 14.10, the depth of coalesced layer is 1.6 in. according to this equation.

Tray Efficiency. A rough correlation for tray efficiency is due to Treybal (1963); as modified by Krishnamurty and Rao [*Ind. Eng. Chem. Process. Des. Dev.* **7,** 166 (1968)] it has the form

$$E = (0.35Z_T^{0.5}/\sigma d_0^{0.35})(V_D/V_C)^{0.42}, \tag{14.38}$$

where the interfacial tension σ is in dyn/cm and the tray spacing Z_T and hole diameter d_0 are in ft. Efficiencies and capacities of several kinds of extractors are summarized in Figure 14.17.

Application of the rules given here for sizing extraction towers without mechanical agitation is made in Example 14.10. The results probably are valid within only about 25%. The need for some pilot plant information of the particular system is essential.

PULSED PACKED AND SIEVE TRAY TOWERS

A rapid reciprocating motion imparted to the liquid in a tower results in improved mass transfer. This action can be accomplished without parts and bearings in contact with the process liquids and consequently has found favor for handling hazardous and corrosive liquids as in nuclear energy applications. Most of the applications still are in that industry, but several other installations are listed by Lo et al. (1983, pp. 345, 366). Packed columns up to 3 m dia and 10 m high with throughputs in excess of 200 m^3/hr are in use.

Both packed and perforated plate towers are in use. The most

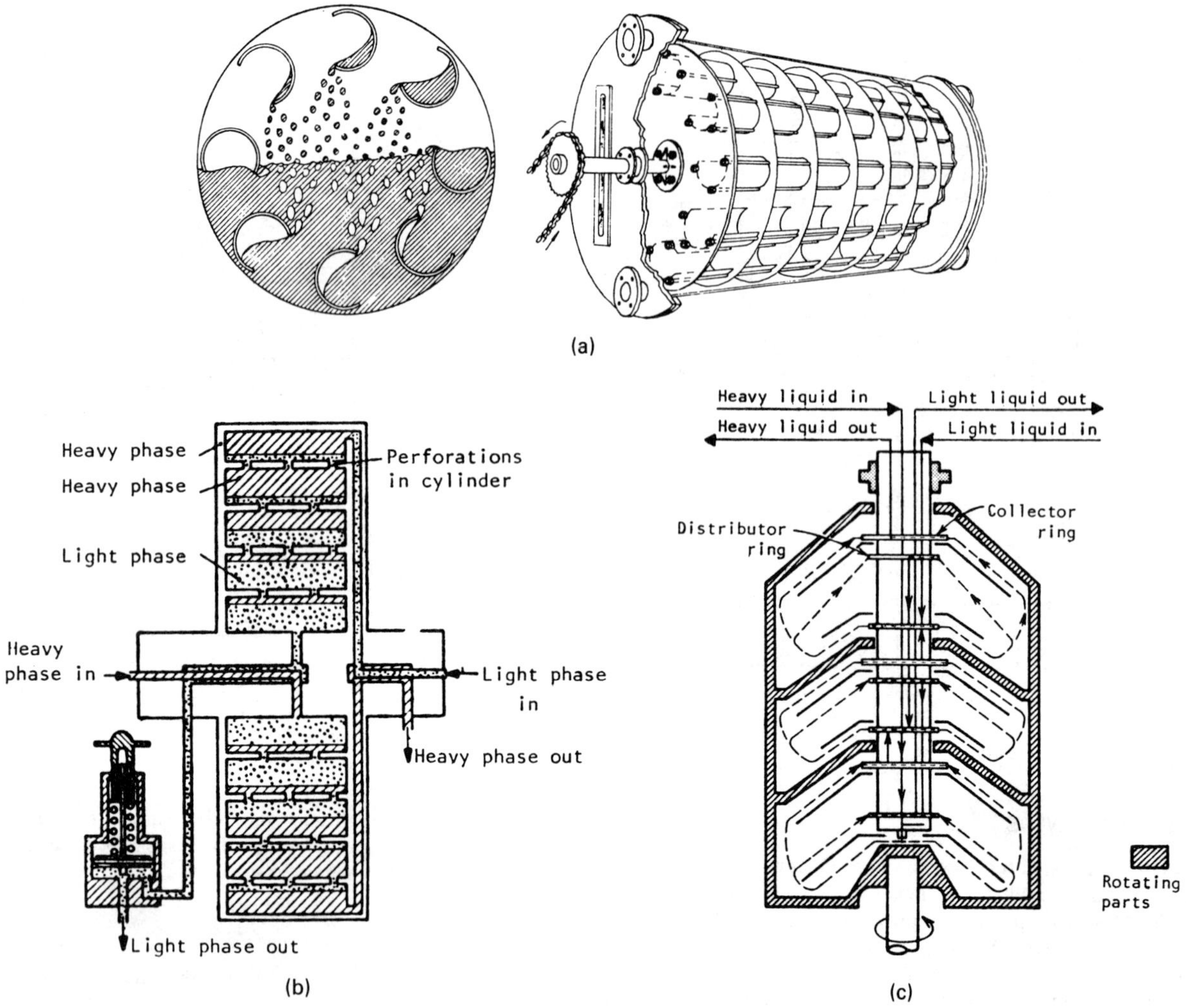

Figure 14.15. A horizontal rotating extractor and two kinds of centrifugal extractors. (a) The RTL (formerly Graesser raining bucket) horizontal rotating extractor; both phases are dispersed at some portion of the rotation (*RTL S. A., London*). (b) Operating principle of the Podbielniak centrifugal extractor; it is made up of several concentric perforated cylinders (*Baker–Perkins Co.*). (c) The Luwesta centrifugal extractor (schematic diagram) (*Luwa Corp.*).

commonly used packing is 1 in. Raschig rings. A "standard" geometry for the plates is 3 mm dia holes on triangular spacing to give 23% open area, plate thickness of 2 mm, and plate spacing of 50 mm. Reissinger and Schröter (1978) favor 2 mm holes and 100 mm plate spacing. The action of the plates is to disperse the heavy phase on the upstroke and the light phase on the down stroke.

Pulsing is uniform across the cross section, and accordingly the height needed to achieve a required extraction is substantially independent of the diameter as long as hydrodynamic similarity is preserved. Although correlations for flooding, holdup, and HTU are not well generalized, a major correlating factor is the product of frequency f and amplitude A_p; in practical applications fA_p is in the range of 20–60 mm/sec.

One large user has standardized on a frequency of 90 cycles/min and amplitudes of vibration of 6–25 mm. Three kinds of pulsing modes are shown in Figures 14.13(c)–(e). The rotary valve pulsator consists of two reservoirs each on the suction and discharge of a variable speed centifugal pump and hooked to a rotating valve. Pneumatic and reciprocating pump pulsers also are popular.

Extraction efficiency can be preserved over a wide range of throughputs by adjusting the product fA_p. A comparison of several correlations of HTU made by Logsdail and Slater (in Lo et al., 1983, pp. 364) shows a four- to five-fold range, but a rough conservative rule can be deduced from these data, namely

$$\text{HTU} = 3.7/(fA_p)^{1/3}, \quad 20 \le fA_0 \le 60\ \text{mm/sec}, \tag{14.39}$$

which gives an HTU of 1 m at $fP_p = 50$ mm/sec. In small diameter extractors, data for HETS of 0.2–0.5 m or less have been found, as appear in Figure 14.17.

Flooding, holdup, and mass transfer rates are highly interdependent and are not simply related. Reissinger and Schröter (1978) state that tray towers in comparison with other types have good efficiencies at 60 m^3/m^2 hr at frequencies of 60–90/min and amplitudes of 10 mm. Packed towers have about 2/3 the capacities of tray towers. Also in comparison with unagitated towers, which are limited to interfacial tensions below 10 dyn/cm, pulsed towers are not limited by interfacial tension up to 30–40 dyn/cm. Some

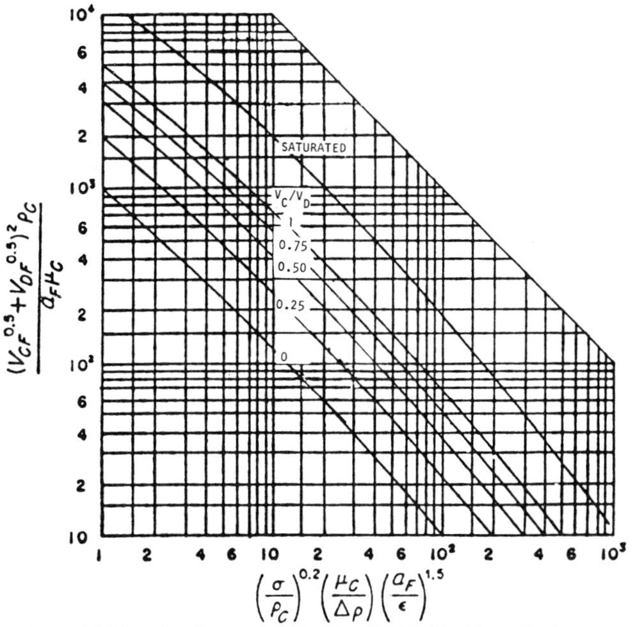

Figure 14.16. Flooding velocities in liquid–liquid packed towers [*J.S. Eckert*, Encycl. Chem. Process. Des. **21,** *149–165 (1984)*]. V = ft/hr (superficial velocity); C = continuous phase; D = disperse phase; a = sqft area of packing/cuft; Δ = difference; ε = void fraction in packing; μ = viscosity centipoise continuous phase; ρ = lb/cuft; σ = (dynes/cm) interfacial surface tension; F = packing factor.

further comparisons are made in Tables 14.3 and 14.4 and Figure 14.17.

RECIPROCATING TRAY TOWERS

Desirable motion can be imparted to the liquids by reciprocating motion of the plates rather than by pulsing the entire liquid mass. This mode employs much less power and provides equally good extraction efficiency. A 30 in. dia tower 20 ft high is sufficiently agitated with a 1.5 HP motor. Some arrangements of such extractors are shown in Figure 14.13.

The holes of reciprocating plates are much larger than those of pulsed ones. Typical specifications of such extractors are: Holes are 9/16 in. dia, open area is 50–60%, stroke length 0.5–1.0 in., 100-150 strokes/min at 0.75 in. stroke length, plate spacing normally 2 in. but may vary from 1–6 in. when the physical properties vary significantly in different parts of the tower. In towers about 30 in. dia, HETS is 20–25 in. and throughputs are up to 40 m^3/m^2 hr (2000 gal/hr sqft). Scaleup formulas for HETS and reciprocating speed, fA_p, are stated by the manufacturer, Chem Pro Corp.:

$$(\mathrm{HETS})_2/(\mathrm{HETS})_1 = (D_2/D_1)^{0.36}, \qquad (14.40)$$

$$(fA_p)_2/(fA_p)_1 = (D_1/D_2)^{0.14}. \qquad (14.41)$$

The performance of a reciprocating tower is compared with several other small extractors in Figure 14.17.

An extractor with countermotion of alternate plates is known as the VPE (vibrating plate extractor). Figure 14.13(b) shows the arrangement. This model also is constructed with segmented plates or with downcomers for passage of the continuous phase. At least during some portion of the cycle, the light phase coalesces and is trapped below the tray, just as in static tray extractors. The capacity of these units is greater than of those with full trays and the efficiency remains high. Some data (Lo et al., 1984, p. 386) indicate that some commercial extractions are completed satisfactorily in towers 4–8 m high at rates of 35–100 m^3/m^2 hr.

ROTATING DISK CONTACTOR (RDC)

The concept of arranging a battery of mixer-settlers in a vertical line in a single shell has been implemented in a variety of ways. In the RDC (Rotary Disk Contactor) extractor, the impellers are flat disks, the mixing zones are separated by partial diametral baffles called stators, but distinct settling zones are not provided. Figure 14.14(a) is a sketch. Because of its geometrical simplicity and its effectiveness, the RDC is one of the most widely employed of agitated extractors. The situations in which it may not be suitable are when only a few stages are needed, in which case mixer-settlers will be satisfactory and cheaper; or when their large holdup and long residence times may be harmful to unstable substances; or for systems with low interfacial tensions and low density differences because then stable emulsions may be formed by the intense agitation.

According to the comparisons of small units in Figure 14.17, the RDC is intermediate in stage efficiency and throughput. The value of HETS = 0.3 m from this figure compares roughly with the HTU = 0.4 or 0.75, depending on which phase is dispersed, of the pilot plant data of Example 14.11.

The design procedure used by Kosters, of Shell Oil Co., who developed this equipment, requires pilot plant measurements on the particular system of HTU and slip velocity as functions of power input. The procedure for scaleup is summarized in Table 14.5, and results of a typical design worked out by Kosters (in Lo et al., 1983, pp. 391–405) are summarized in Example 14.11. Scaleup by this method is said to be reliable in going from 64 mm dia to 4–4.5 m dia. The data of Figure 14.18 are used in this study.

OTHER ROTARY AGITATED TOWERS

One of the first agitated tower extractors was developed by Scheibel (*AIChE. J.* **44,** 681, 1948). The original design, like Figure 14.14(f), employed settling zones packed with wire mesh, but these were found unnecessary in most cases and now flat partitions between mixing zones are used. The Mixco [Fig. 14.14(b)] and Scheibel–York [Fig. 14.14(g)] units differ primarily in the turbine impellers, the Mixco being open and the other shrouded. In spite of the similarity of their equipment, the manufacturers have possibly different ranges of experience. Since extractor selection is not on an entirely rational basis, a particular body of experience may be critical for fine tuning.

Enhanced coalescing between stages is provided in the designs of Figure 14.14(e). The Kühni extractor of Figure 14.14(d) employs shrouded turbine impellers and perforated plate partitions between compartments and extending over the entire cross section. The ARD (asymmetric rotating disk) extractor has lateral spaces for settling between agitation zones.

Some performance data are cited for the Kühni by Ritcey and Ashbrook (1979, p. 102):

% Free Cross Section	m^3/m^2 hr	HETS (m)
10	10	0.08
40	50	0.20

Although not all equipment is compared, Figure 14.17 shows the Kühni to have a high efficiency but somewhat lower capacity than the RDC and other units.

Most of these types of equipment have at least several hundred installations. The sizing of full scale equipment still requires pilot planting of particular systems. The scaleup procedures require geometrical and hydrodynamic similarities between the pilot and full scale plants. Hydrodynamic similarity implies equalities of

Example 14.10
Sizing of Spray, Packed, or Sieve Tray Towers
Five theoretical stages are needed for liquid–liquid extraction of a system with these properties:

$$Q_D = 600\ \text{cuft/hr},$$
$$Q_c = 500\ \text{cuft/hr},$$
$$\rho_D = 50\ \text{lb/cuft},$$
$$\rho_c = 60\ \text{lb/cuft},$$
$$\mu_D = 0.5\ \text{cP},$$
$$\mu_c = 1.0\ \text{cP, } 2.42\ \text{lb/ft hr},$$
$$\sigma = 10\ \text{dyn/cm, interfacial tension},$$
$$d_0 = 0.0208\ \text{ft } (0.25''),\ \text{hole size},$$
$$d_p = 0.02\ \text{ft (droplet diameter)}.$$

Spray tower: The flooding velocity is found with Eq. (14.33):

$$V_D = \frac{4000(10)^{0.28}}{[0.483(2.42)^{0.075}(60)^{0.5} + (0.02)^{0.056}(50)^{0.5}(1.2)^{0.5}]^2}$$
$$= 73.0\ \text{ft/hr},$$
$$A_c = 600/73.0 = 8.22\ \text{sqft},$$
$$D = 3.24\ \text{ft}.$$

To accommodate five stages, a total height of 100 ft or so would be needed. Two towers each 3.5 ft dia by 50 ft high would be suitable.

Packed tower: Flooding velocity is obtained with Figure 14.17. For 1 in. metal pall rings,

$$F\varepsilon^2 = a_p/\varepsilon = 63/0.94 = 67.02,$$
$$(\mu_c/\Delta\rho)(\sigma/\rho_c)^{0.2}(a_p/\varepsilon)^{1.5} = (1/10)(10/60)^{0.2}(67.02)^{1.5} = 38.34,$$
$$\therefore 200 = V_c[1 + (V_D/V_c)^{0.5}]^2\rho_c/a_p\mu_c$$
$$= V_c[1 + 1.2^{0.5}]^2 60/63(1),$$
$$V_c = 47.83\ \text{ft/hr, at the flooding point.}$$

Take 70% of flooding:

$$A_c = 500/0.7(47.83) = 14.93\ \text{sqft}$$
$$D = 4.36\ \text{ft}.$$

Take a conservative HETS = 5 ft. Then the tower will be 4.5 ft dia with 25 ft of packing and two redistributors, a total of about 35 ft.

Sieve tray tower: Take 1.5 ft tray spacing, 0.25 in. holes on 0.75 in. triangular spacing. The downcomer area is found with Eq. (14.36):

$$\Delta h = 1.5 = \frac{4.5V_D^2\rho_c}{2g_c\Delta\rho} = \frac{4.5(60)}{2(4.18)(10^8)(10)}V_D^2,$$

$$V_D = 6815\ \text{ft/hr},$$
$$A_d = 600/6815 = 0.088\ \text{sqft},$$
$$D_d = 4.02\ \text{in., downcomer diameter.}$$

Take hole velocity = 0.8 ft/sec, 2880 ft/hr:

$$\text{total hole area} = \frac{600}{2880} = 0.2083\ \text{sqft}.$$

$$\frac{\text{tray area}}{\text{hole area}} = \frac{0.866d_s^2}{\frac{1}{2}(\pi/4)d_0^2} = 2.21\left(\frac{ds}{d}\right)^2$$
$$= 2.21(3)^2 = 19.89,$$
$$\text{tray area} = 19.89(0.2083) = 4.14\ \text{sqft}.$$

Add area of two 4 in. pipes, 4.5 in. OD = 0.11 sqft.

$$\text{area} = 4.14 + 0.11 = 4.25\ \text{sqft},$$
$$\text{dia} = 27.9\ \text{in.}$$

Add 2 in. for support rings, making the diameter 30 in.

Tray efficiency from Eq. (14.38):

$$E = \frac{0.35(Z_t)^{0.5}}{d_0^{0.35}}\left(\frac{V_D}{V_C}\right)^{0.42} = \frac{0.35(1.5)^{0.5}}{10.(0.0208)^{0.35}}(1.2)^{0.42} = 0.18,$$
$$\text{number of trays} = 5/0.18 = 27.8,$$
$$\text{tower height} = 1.5(28) + 6 = 48\ \text{ft, including 3 ft at each end.}$$

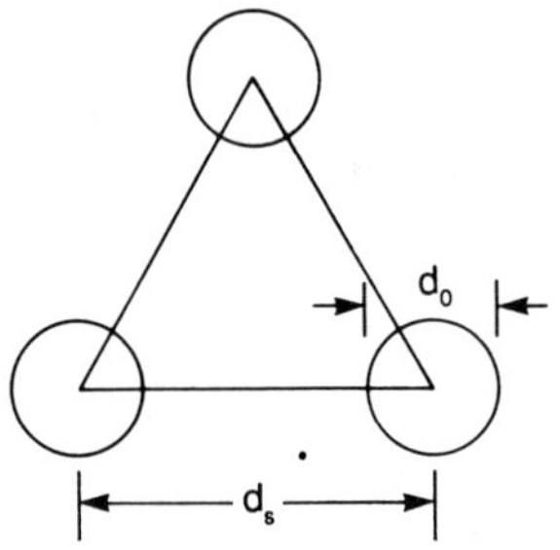

Summary:

	Height	Diameter
Spray	100	3.2
Packed	35	4.5
Sieve tray	48	2.5

droplet diameters, fractional holdups, and linear superficial velocities. Also preserved are the specific radial discharge rates, defined by Q/DH = (volumetric flow rate)/(vessel dia) (compartment height).

A detailed design of an ARD extractor based on pilot plant work is presented by Misek and Marek (in Lo et al., 1983, pp. 407–417). The design and operating parameters of the ARD extractor are related to the vessel diameter D (mm); thus:

Free cross section = 25%.
Disk diameter = $0.49D$.
Chamber height = $1.3D^{0.67}$.
Agitator rpm = $15{,}000/D^{0.78}$.

A manufacturer's bulletin on a 150 mm dia ARD extractor gives HETS = 0.4 m and capacity 15 m^3/m^2 hr.

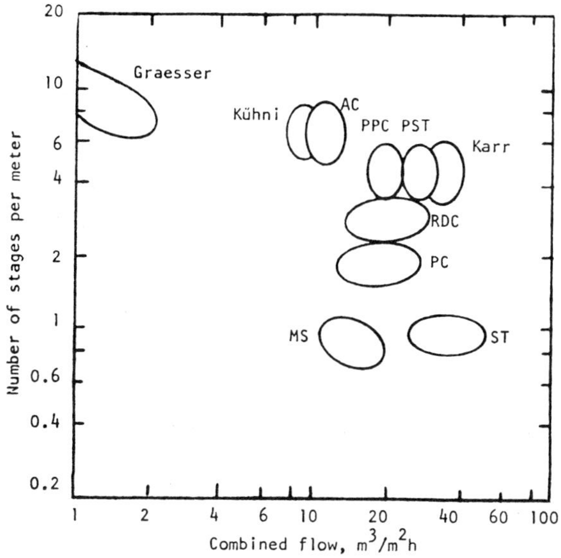

Figure 14.17. Efficiency and capacity range of small diameter extractors, 50–150 mm dia. Acetone extracted from water with toluene as the disperse phase, $V_d/V_c = 1.5$. Code: AC = agitated cell; PPC = pulsed packed column; PST = pulsed sieve tray; RDC = rotating disk contactor; PC = packed column; MS = mixer-settler; ST = sieve tray [*Stichlmair,* Chem. Ing. Tech. **52**(*3*), *253–255* (*1980*)].

Less specific information about the other kinds of extractors mentioned here is presented by Lo et al. (1983, pp. 419–448) but no integrated examples. The information perhaps could be run down in the abundant literature cited there, or best from the manufacturers.

OTHER KINDS OF EXTRACTORS

Some novel types and variations of basic types of extractors have been developed, most of which have not found wide acceptance, for instance pulsed rotary towers. The literature of a few of them is listed by Baird (in Lo et al., 1983, pp. 453–457). Here the extractors illustrated in Figure 14.15 will be described.

Graesser Raining Bucket Contactor. The Graesser "raining bucket" contactor consists of a horizontal rotating shell with a shaft that carries a number of diametral partitions extending to the wall. Between the partitions are buckets that carry the liquid and cascade it through each phase. No attempt is made to effect dispersion beyond simply emptying the buckets. The light and heavy phases are alternately both dispersed. They are introduced and withdrawn at opposite ends. The speed of rotation is 1–40 rpm, depending on the diameter, and is gauged to effect proper mass transfer and yet avoid emulsification. An approximately linear relation exists between mass transfer rate and diameter up to about 2 m. Production contactors usually are designed to provide about 0.3 theoretical stages per compartment. Phase flow ratios of 6:1 are accommodated readily and ratios of up to 20:1 also have been designed for. Solids can be leached in this equipment; for instance, bitumen is dissolved away from tar sands with kerosene.

A commercial unit 5 ft dia by 18 ft long has 26 × 7-in. wide compartments each with 16 × 8-in. buckets and provides six theoretical stages. A unit 12 in. dia by 3 ft long has a capacity of 30 gal/hr at 8 rpm. A unit 6 ft dia has a capacity of 6000 gal/hr at 1.4 rpm.

Centrifugal Contactors. These devices have large capacities per unit, short residence times, and small holdup. They can handle systems that emulsify easily or have small density differences or large interfacial tensions or need large ratios of solvent to feed. Some types are employed as separators of mixtures made in other equipment, others as both mixers and settlers, and some as differential contactors.

The Podbielniak contactor is a differential type. It is constructed of several perforated concentric cylinders and is shown schematically in Figure 14.15(b). Input and removal of the phases at each section are accomplished through radial tubes. The flow is countercurrent with alternate mixing and separating occurring respectively at the perforations and between the bands. The position of the interface is controlled by the back pressure applied on the light phase outlet.

Residence time can be as short as 10 sec. One 750 gpm unit is said to have a total liquid holdup of 200 gal. From 3–10 stages per unit have been reported, although Table 14.6 shows a range of 1.8–7.7. A 65 in. dia casing can accommodate throughputs up to 25,000 gal/hr. An economic comparison of a Podbielniak with other

TABLE 14.4. Maximum Loads and Diameters of Extractors

Column Type	Maximum Load ($m^3/(m^2)(h)$)	Maximum Column Diameter (m)	Maximum Throughput (m^3/h)
Graesser contactor	<10	7.0	380
Scheibel	<20	1.0	16
Asymmetric rotating-disk	≈25	3.2–5.0	200
Lurgi tower	≈30	8.0	1500
Pulsed packed	≈40	2.8	250
Rotating-disk contactor	≈40	4.0	500
Kühni	≈50	3.0	350
Pulsed sieve-tray extractor	≈60	3.0	420
Karr	80–100	1.0	<80

These data apply at a high interfacial tension (30–40 dyn/cm), a viscosity similar to water, an inlet ratio of the phases of 1:1 parts by volume, and a density difference of approximately 0.6 g/cm^3.

(Reissinger and Schröter, 1978).

EXAMPLE 14.11
Design of a Rotating Disk Contactor

A hydrocarbon mixture containing 10% aromatics and at the rate of 55.5 m^3/hr is to be treated with a solvent at the rate of 173.6 m^3/hr. Ten stages are needed for the extraction. Pilot plant data are available for the HTU and the slip velocity; they are shown on the graphs for solvent either continuous or dispersed. The procedure of Table 14.5 was applied by Kosters (in Lo et al., 1983, pp. 391–405) with the following results:

	Solvent Continuous	Solvent Dispersed
Vessel dia (m)	2.1	1.7
Stator dia (m)	1.47	1.19
Rotor dia (m)	1.26	1.02
HTU (m)	0.41	0.75
$(HTU)_{eff}$ (m)	0.663	1.107
Number of compartments	40	81
Compartment height (m)	0.20	0.17
Total height (m)	10.4	15.7
Rotor speed (rpm)	15–60	15–70
Power (theoretical kW)	4.6	2.8

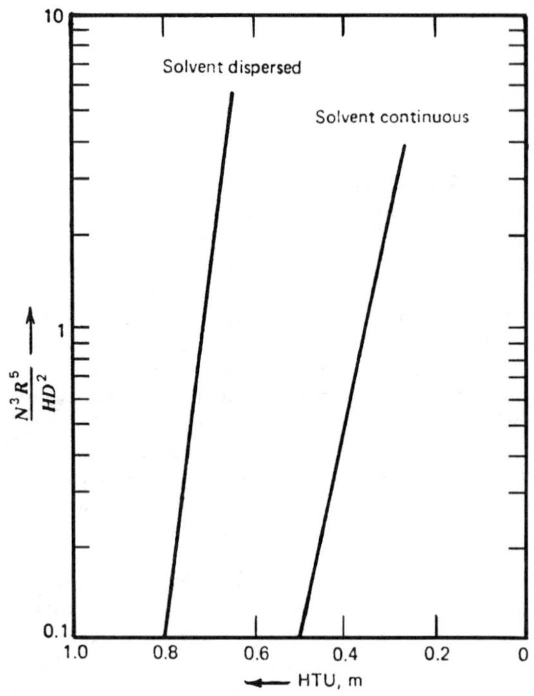

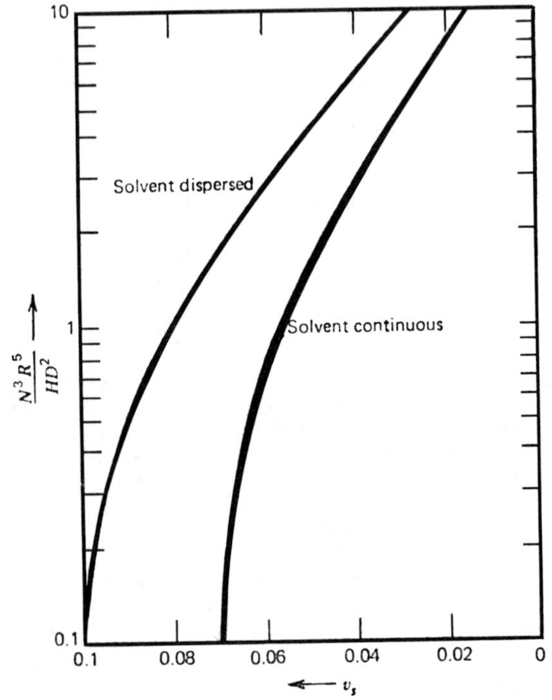

extractors is made in Table 14.3(b). Although its basic cost is high, it requires few auxiliaries so that the overall cost of an extraction plant is not drastically out of line in every instance. Nevertheless, this equipment is used primarily when short residence time and other characteristic features are indispensable.

Other kinds of centrifugals also are used widely. Some are described by Hafez (in Lo et al., 1983, pp. 459–474) and performance data are presented in Table 14.6. Characteristics of centrifugals that are used primarily for removal of solids from slurries are summarized in Table 11.18.

LEACHING EQUIPMENT

In leaching processes, finely divided solids are contacted with solvents to remove soluble constituents. Usually some kind of multistage and countercurrent operation is desirable. The most bothersome aspect is handling of the wet solids.

In the leaching battery of Figure 14.19(a), the solids are transported between vessels with slurry pumps and are mixed in line with countercurrent solution from the next stage. For the process to be effective, the solids must settle freely. The tanks have sloped bottoms and slowly moving rakes that scrape the solids towards the center discharge. Units employed for treating ores, for example, are very large, 100–200 ft dia. A few performance data of settlers are in Table 14.7.

Solids being extracted remain fixed in the cells of the battery of Figure 14.20(b). Fresh solvent is charged to the cell that is most nearly exhausted and next to be taken off stream, then solution proceeds through the other cells in series and leaves as finished extract from the cell that has been charged most recently. For sugar beet extraction a battery normally consists of 10–14 cells. Cells have volumes ranging from 4–12 m^3 and height to diameter ratios as high as 1.5. Since leaching is faster at elevated temperatures, the solutions are heated between cells. Leaching time is 60–100 min. The amount of solution made is 110 kg/100 kg beets and contains 13–16% sugar. Various kinds of barks and seeds also are extracted in this kind of equipment. Further details of the equipment

TABLE 14.5. Formulas for Sizing an RDC

1. Stator opening diameter, $S = 0.7D$, where D is vessel diameter
2. Rotor diameter, $R = 0.6D$
3. Height to diameter ratio of a compartment:

D (m)	0.5–1.0	1.0–1.5	1.5–2.5	2.5
H/D	0.15	0.12	0.1	0.08–0.1

4. Power input, Figure 14.20(a)
5. Fractional holdup at flooding, h_f, from Figure 14.20(b)
6. Slip velocity V_s preferably is obtained experimentally, but is given approximately by Figure 14.20(c)
7. Superficial velocity of the continuous phase at flooding,

$$V_{Cf} = \frac{V_s \exp(-h_f)}{V_D/V_C h_f + 1/(1-h_f)},$$

where V_C and V_D are the superficial velocities of the continuous and dispersed phases

8. Holdup h at an operating velocity V_C, say 70–80% of flooding,

$$V_C = \frac{V_s \exp(-h)}{V_D/V_C h + 1/(1-h)};$$

solve by trial for h when other quantities are specified

9. Effective height of a transfer unit,

$$(\text{HTU})_{\text{eff}} = (\text{HTU})_{\text{pilot plant}} + \text{HDU}$$

in terms of a value obtained in a pilot plant and a calculated height of a diffusion unit (HDU)

10. Height of a diffusion unit, $\text{HDU} = H(1/\text{Pe}_C + 1/\text{Pe}_D)$
11. Factors E_C and E_D for evaluating the Peclet numbers,

$$E_C = 0.5V_cH + 0.012RNH(S/D)^2$$
$$E_D = E_C[4.2(10^5)(V_D/h)^{3.3}/D^2];$$

when the correction in brackets is less than unity, make $E_{D_E} = E_C$

12. The Peclet numbers are

$$1/\text{Pe}_C = E_C(1-h)/HV_C,$$
$$1/\text{Pe}_D = E_Dh/HV_D$$

13. Final expression for height of a diffusion unit is

$$\text{HDU} = E_c(1-h)/V_C + E_dh/V_D$$

(W.C.G. Kosters, Shell Oil Co.).

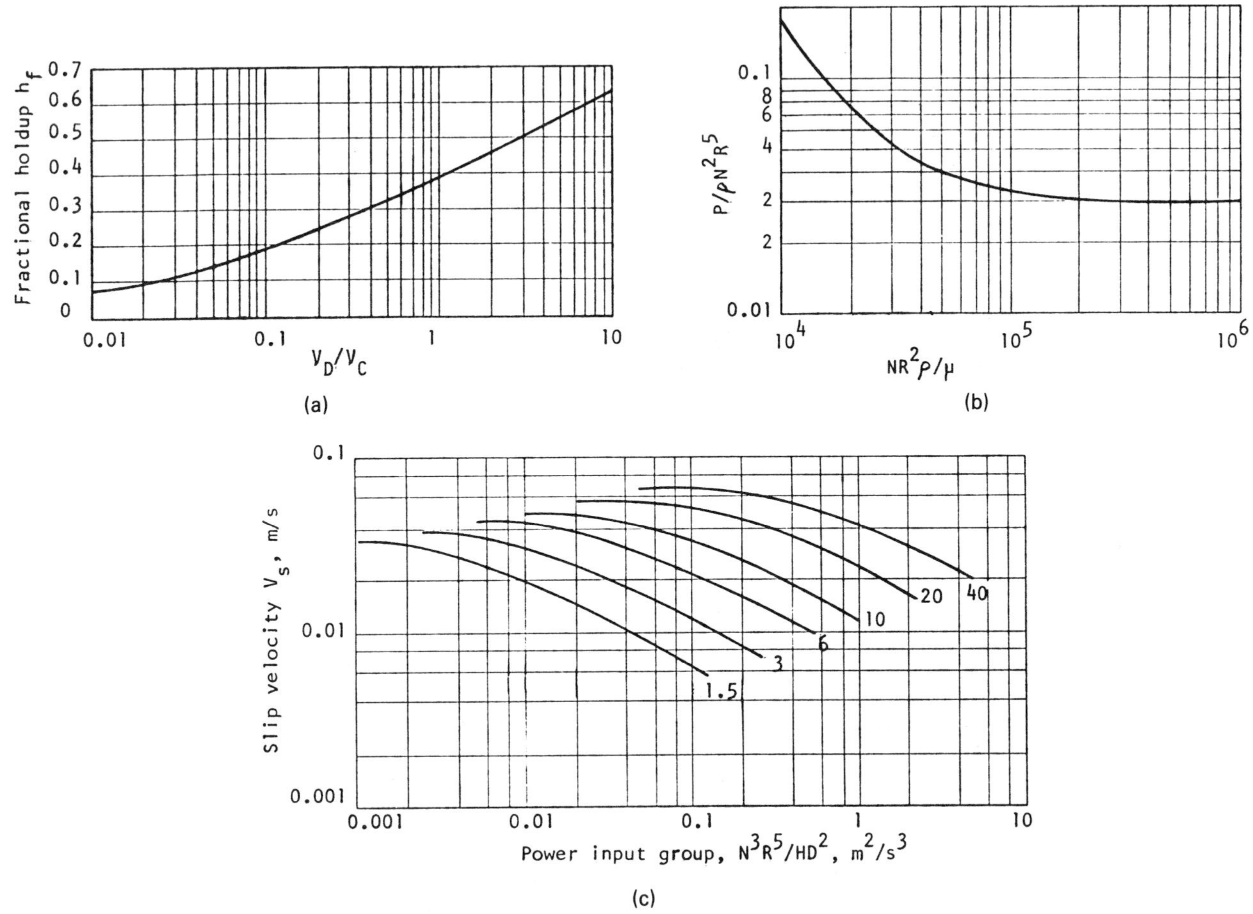

Figure 14.18. Holdup at flooding, power input, and slip velocity in an RDC (*Kosters, in Lo, Baird, and Hanson, 1983*). (a) Fractional holdup at flooding, h_f, as a function of flow ratio of the phases. (b) Power input to one rotor as a function of rotation speed N and radius R. (c) Slip velocity versus power input group for density difference of 0.15 g/mL, at the indicated surface tensions (dyn/cm).

TABLE 14.6. Performance of Centrifugal Extractors

SPECIFICATIONS[a]

Extractor	Model	Volume, m^3	Capacity, m^3/hr	rpm	Motor Mounting	Motor Power, kW	Diameter, m
Podbielniak	E 48	0.925	113.5	1,600	Side	24	1.2
Quadronic	Hiatchi 4848	0.9	72	1,500	Side	55	1.2
α-Laval	ABE 216	0.07	21	6,000	Top	30	
UPV			6	1,400	Bottom	14	
Luwesta	EG 10006		5	4,500	Bottom		
Robatel SGN	LX6 70NL	0.072	3.5	1,600	Top, side		1.3
Robatel BXP	BXP 800	0.220	50	1,000	Top	15	0.8
Westfalia	TA 15007	0.028	30	3,500	Top	63	0.7
SRL/ANL		0.003	0.05	3,500	Top		0.1
MEAB	SMCS-10	0.00012	0.3	22,000	Bottom		

[a]Operating pressures are in the range 300–1750 kPa; operating temperatures cover a very wide range; operating flow ratios cover the range $\frac{10}{1}$ - $\frac{1}{10}$ easily.

PERFORMANCE

Extractor	System	Operating Variables: rpm	$R = Q_h/Q_l$	Q_t, m^3/hr	Flooding, %	Number of Theoretical Stages
Podbielniak						
B-10	Kerosene-NBA[a]-water	3000	0.5	5.1	73	6–6.5
D-18	Kerosene-NBA-water	2000	0.5	11.1	58	5–5.5
A-1	Oil-aromatics-phenol[b]	5000	3.5	0.01–0.02	33–66	5–7.7
9000	Broth-penicillin B-pentacetate	2900	4.4	7.5		1.8
		2900	3.4	7.5		2.04
		2900	2.4	7.5		2.21
9500	Some system	2900	3.5	7.5		2.04
		2700	3.5	7.5		2.19
		2500	3.5	7.5		2.30
		2300	3.5	7.5		2.36
	Oil-aromatics-furfural	2000	4.0	12.0	90	3–6
A-1	IAA[c]-boric acid-water	5000	1–0.3	0.01–0.03	44–95	3.5–7.7
		3000	1.0	0.01	44	2.3
		4075	1.0	0.01	44	2.8
		4600	1.0	0.01	44	2.96
UPV	Oil-aromatics-phenol[b]	1400	0.8–1.2	6	75	2–5.8
Robatel SGN						
LX-168N	Uranyl nitrate-30% TBP	1500	1–0.2	2.1–4.5		7
LX-324	Some system	3100	1.6	24–63		3.4–3.9
SRL single stage	Uranyl nitrate-Ultrasene	1790	0.5–1.5	6.4–12	33–96	0.92–0.99
ANL single stage	Uranyl nitrate-TBP/dodacane	3500	0.3–4	0.8–1.6	50	0.97–1

[a]Normal butyl amine.
[b]Containing 1.7–5% water.
[c]Isoamyl alcohol.
[d]Number of theoretical and actual stages.

(M. Hafez, in Lo et al., 1983, pp. 459–474).

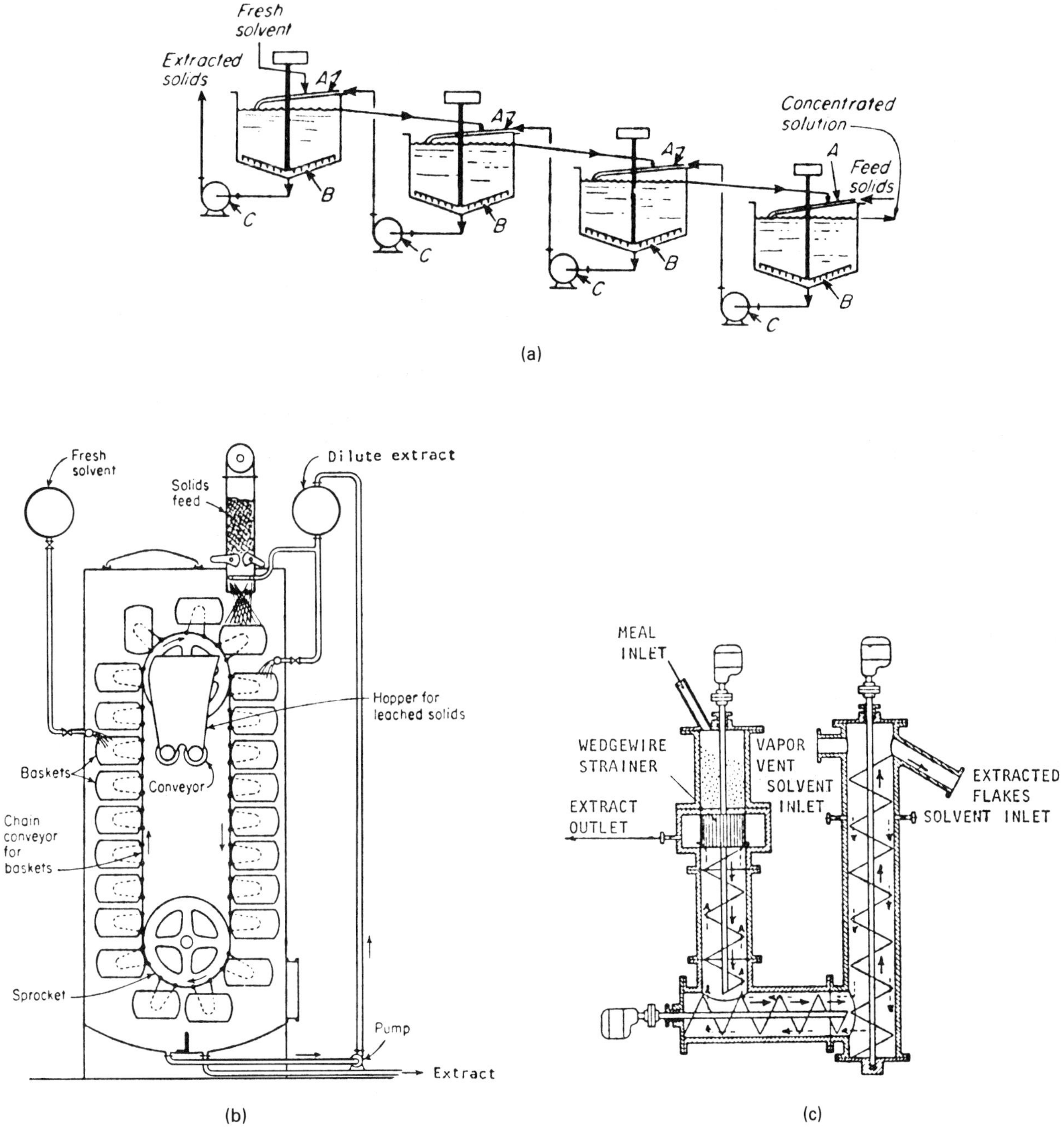

Figure 14.19. Continuous leaching equipment. (a) A battery of thickeners of the type shown, for example, in Figure 13.9(a), used in countercurrent leaching. The slurry is pumped between stages counter to the liquid flow: (*A*) mixing line for slurry and solution; (*B*) scraper arms; (C) = slurry pumps. (b) A bucket elevator with perforated buckets used for continuous extraction, named the Bollmann or Hansa–Muehle system [*Goss,* J. Am. Oil Chem. Soc. **23,** *348* (*1946*)]. (c) A countercurrent leaching system in which the solid transport is with screw conveyors; a similar system is named Hildebrandt. (d) The Bonotto multi-tray tower extractor. The trays rotate while the solid is scraped and discharged from tray to tray. The solid transport action is similar to that of the rotary tray dryer of Figure 9.8(a) [*Goss,* J. Am. Oil Chem. Soc. **23,** *348* (*1946*)]. (e) Rotocel extractor, which consists of about 18 wedge-shaped cells in a rotating shell. Fresh solvent is charged to the last cell and the drained solutions are pumped countercurrently to each cell in series (*Blaw-Knox Co.*).

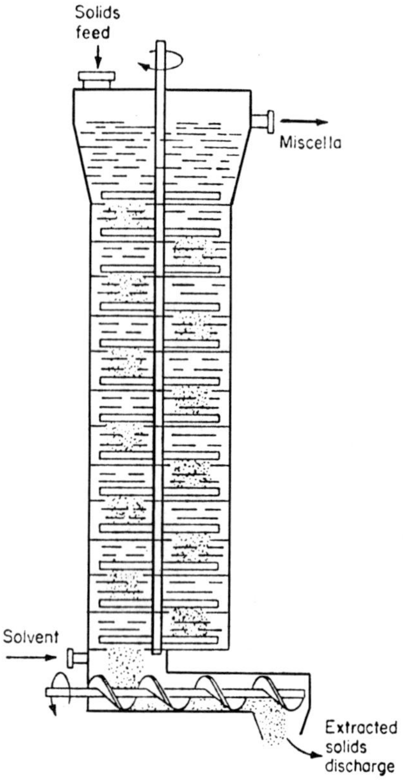

(d)

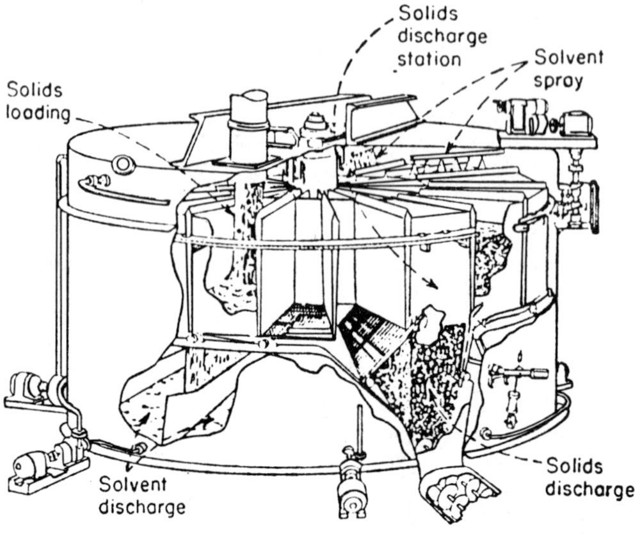

(e)

Figure 14.19—*(continued)*

arrangement are given by Badger and McCabe (*Elements of Chemical Engineering,* McGraw-Hill, New York, 1936).

Continuous transport of the solids against the solution is employed in several kinds of equipment, including screw, perforated belt, and bucket conveyors. One operation carries a bed of seeds 3–4 ft thick on a perforated belt that moves only a few feet per minute. Fresh solvent is applied 1/5 to 1/3 of the distance from the discharge, percolates downward, is collected in pans, and is redistributed by pumps countercurrently to the travel of the material.

The vertical bucket elevator extractor of Figure 14.19(b) stands 40–60 ft high and can handle as much as 50 tons/hr with 1–2 HP. The buckets have perforated bottoms. As they start to descend, they are filled with fresh flaked material and sprayed with dilute intermediate extract. The solution percolates downward from bucket to bucket. As the travel turns upward, the buckets are subjected to countercurrent extraction with solution from fresh solvent that is charged about 1/3 the distance from the top. There is sufficient travel time for drainage before discharge of the spent flakes.

Countercurrent action is obtained in the Bonotto extractor of Figure 14.19(d). It has a number of trays arranged in vertical line and provided with scrapers to discharge solids through staggered openings in the trays. The principle of this mode of solid transport

TABLE 14.7. Performance of Settling Tanks

No.	Size (ft)	Slurry	Mesh	Rate of Feed (tons/day)	Solids in Feed	Solids in Underflow (%)	Remarks
4	6 × 5	Paint pigment	300	39	5.7%	33	Solubles washed out
3	16 × 8	Iron oxide	300	162	10	33	C.C.D. washing
1	20 × 8	Zinc, copper, lead ore	99.5% −200	400	20	40	
2	25 × 10	Calcium carbonate	200	450 each	10	38	Feed is 14° Bé caustic liquor
1	40 × 10	Flotation tailings	65% −200	800	20	55	To recover the water
1	40 × 12	Flotation mill concentrates		1050	25	56	

(Hardinge Co.).

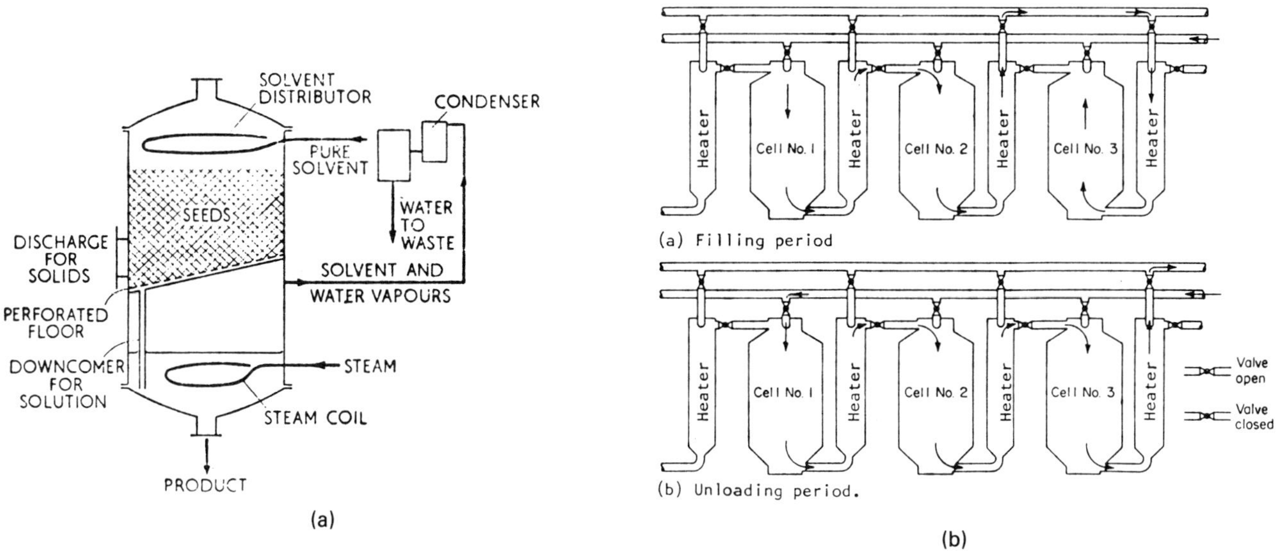

Figure 14.20. Single tank and battery of tanks as equipment for batch leaching. (a) A single tank extractor of the type used for recovering the oil from seeds. (b) Principle of the leaching battery. Cells are charged with solid and solvent is pumped through heaters and cells in series. In the figure, cell 1 has been exhausted and is being taken off stream and cell 3 has just been charged. (*Badger and McCabe,* Elements of Chemical Engineering, *McGraw-Hill, New York, 1936*).

is similar to that of Figure 9.8(b). Solvent is charged at the bottom of the tower and leaves at the top, and the spent solid is removed with a screw conveyor.

Few performance data of leaching equipment have found their way into the open literature, but since these processes have long been exploited, a large body of information must be in the files of manufacturers and users of such equipment.

REFERENCES

1. P.J. Bailes, C. Hanson, M.A. Hughes, and M.W.T. Pratt, Extraction, liquid–liquid, *Encycl. Chem. Process. Des.* **21,** 19–125 (1984).
2. A.E. Dunstan et al., Eds., in *Science of Petroleum,* Solvent extraction methods of refining, 1817–1929, Oxford University Press, Oxford, 1938, Section 28.
3. J.S. Eckert, Extraction, liquid–liquid, packed tower design, *Encycl. Chem. Process. Des.* **21,** 149–166 (1984).
4. C. Hanson Ed., *Recent Advances in Liquid–Liquid Extraction,* Pergamon, New York, 1971.
5. E.J. Henley and J.D. Seader, *Equilibrium-Stage Separation Operations in Chemical Engineering,* Wiley, New York, 1981.
6. A.E. Karr, Design scale up and application of the reciprocating plate extraction column, *Sep. Sci. Technol.* **15,** 877–905 (1980).
7. G.S. Laddha and T.E. Degaleesan, *Transport Phenomena in Liquid Extraction,* Tata McGraw-Hill, New York, 1978.
8. T.C. Lo, M.H.I. Baird, and C. Hanson, Eds., *Handbook of Solvent Extraction,* Wiley, New York, 1983.
9. K.H. Reissinger and J. Schröter, Selection criteria for liquid–liquid extractors, *Chem. Eng.,* 109–118 (6 Nov. 1978); also *Encycl. Chem. Process. Des.* **21,** 125–149 (1984).
10. G.M. Ritcey and A.W. Ashbrook, *Solvent Extraction with Applications to Process Metallurgy,* Elsevier, New York, 1979, Parts I, II.
11. L.A. Robbins, Liquid–liquid extraction, in Ref. 13, pp. 1.256–1.282; in *Chemical Engineers' Handbook,* McGraw-Hill, New York, 1984, pp. 15.1–15.20, 21.56–21.83.
12. H. Sawistowski and W. Smith, *Mass Transfer Process Calculations,* Wiley, New York, 1963.
13. P.A. Schweitzer, Ed., *Handbook of Separation Techniques for Chemical Engineers,* McGraw-Hill, New York, 1979.
14. J.M. Sorensen and W. Arlt, *Liquid–Liquid Equilibrium Data Collection,* DECHEMA, Frankfurt/Main, Germany, 1979–1980.
15. R.E. Treybal, *Liquid Extraction,* McGraw-Hill, New York, 1951, 1963.
16. R.E. Treybal, *Mass Transfer Operations,* McGraw-Hill, New York, 1980.
17. T. Tsuboka and T. Katayama, Design algorithm for liquid-liquid separation processes, *J. Chem. Eng. Jpn.* **9,** 40–45 (1976).
18. S.M. Walas, *Phase Equilibria in Chemical Engineering,* Butterworths, Stoneham, Mass., 1985.
19. J. Wisniak and A. Tamir, *Liquid–Liquid Equilibrium and Extraction Bibliography,* Elsevier, New York, 1980.
20. J. Wisniak and A. Tamir, *Phase Diagrams: A Literature Source Book,* Elsevier, New York, 1981.
21. Z. Ziolkowski, *Liquid Extraction in the Chemical Industry* (in Polish), PWT, Warsaw, 1961.

15

ADSORPTION AND ION EXCHANGE

Separation of the components of a fluid can be effected by contacting them with a solid that has a preferential attraction for some of them. Such processes are quantitatively significant when the specific surfaces of the solids are measured in hundreds of m^2/g. Suitable materials are masses of numerous fine pores that were generated by expulsion of volatile substances. The most important adsorbents are activated carbon, prepared by partial volatilization or combustion of a carbonaceous body, and activated alumina, silica gel, and molecular sieves which are all formed by expulsion of water vapor from a solid. The starting material for silica gel is a coagulated silicic acid and that for molecular sieves is hydrated aluminum silicate crystals that end up as porous crystal structures. Porous glasses made by leaching with alkai have some application in chromatography. Physical properties of common adsorbents are listed in Tables 15.1 and 15.2. Representative manufacturing processes are represented on Figure 15.1.

The amount of adsorption is limited by the available surface and pore volume, and depends also on the chemical natures of the fluid and solid. The rate of adsorption also depends on the amount of exposed surface but, in addition, on the rate of diffusion to the external surface and through the pores of the solid for accessing the internal surface which comprises the bulk of the surface. Diffusion rates depend on temperature and differences in concentration or partial pressures. The smaller the particle size, the greater is the utilization of the internal surface, but also the greater the pressure drop for flow of bulk fluid through a mass of the particles.

In ion exchange equipment, cations or anions from the fluid deposit in the solid and displace equivalent amounts of other ions from the solid. Suitable solids are not necessarily porous; the ions are able to diffuse through the solid material. A typical exchange is that of H^+ or OH^- ions from the solid for some undesirable ions in the solution, such as Ca^{++} or SO_4^{--}. Eventually all of the ions in the solid are replaced, but the activity is restored by contacting the exhausted solid with a high concentration of the desired ion, for example, a strong acid to replace lost hydrogen ions.

For economic reasons, saturated adsorbents and exhausted ion exchangers must be regenerated. Most commonly, saturation and regeneration are performed alternately and intermittently, but equipment can be devised in which these processes are accomplished continuously by countercurrent movement of the solid and fluid streams. Only a few such operations have proved economically feasible. The UOP and Toray processes for liquid adsorption are not true continuous processes but are effectively such.

Desorption is accomplished by elevating the temperature, or reducing the pressure, or by washing with a suitable reagent. The desorbed material may be recovered as valuable product in concentrated form or as a waste in easily disposable form. Adsorbent carbons used for water treating often must be regenerated by ignition in a furnace. Relatively small amounts of adsorbents that are difficult to regenerate are simply discarded.

15.1. ADSORPTION EQUILIBRIA

The amount of adsorbate that can be held depends on the concentration or partial pressure and temperature, on the chemical nature of the fluid, and on the nature, specific surface, method of preparation, and regeneration history of the solid. For single adsorbable components of gases, the relations between amount adsorbed and the partial pressure have been classified into the six types shown in Figure 15.2. Many common systems conform to Type I, for example, some of the curves of Figure 15.3. Adsorption data are not highly reproducible because small contents of impurities and the history of the adsorbent have strong influences on their behavior.

One of the simplest equations relating amount of adsorption and pressure with some range of applicability is that of Freundlich,

$$w = aP^n \tag{15.1}$$

and its generalization for the effect of temperature

$$\omega = aP^n \exp(-b/T). \tag{15.2}$$

The exponent n usually is less than unity. Both gas and liquid adsorption data are fitted by the Freundlich isotherm. Many liquid data are fitted thus in a compilation of Landolt-Börnstein (II/3, *Numerical Data and Functional Relationships in Science and Technology,* Springer, New York, 1956, pp. 525–528), but their gas data are presented in graphical form only (LB IV 4/b, 1972, pp. 121–187). The effect of temperature also is correlated by a theory of Polanyi, whereby all data of a particular system fall on the same curve; Figure 15.4 is an example. For isothermal data, a combination of the Freundlich and Langmuir equations was developed by Yon and Turnock (*Chem. Eng. Prog. Symposium Series* **117,** 67, 1971):

$$w = kP^n/(1 + kP^n). \tag{15.3}$$

Individuals of multicomponent mixtures compete for the limited space on the adsorbent. Equilibrium curves of binary mixtures, when plotted as x vs. y diagrams, resemble those of vapor–liquid mixtures, either for gases (Fig. 15.5) or liquids (Fig. 15.6). The shapes of adsorption curves of binary mixtures, Figure 15.7, are varied; the total adsorptions of the components of the pairs of Figure 15.7 would be more nearly constant over the whole range of compositions in terms of liquid volume fractions rather than the mol fractions shown.

Higher molecular weight members of homologous series adsorb preferentially on some adsorbents. The desorption data of Figure 15.8 attest to this, the hydrogen coming off first and the pentane last. In practical cases it is not always feasible to allow sufficient time for complete removal of heavy constituents so that the capacity of regenerated adsorbent becomes less than that of fresh, as Figure 15.9 indicates. Repeated regeneration causes gradual deterioration

TABLE 15.1. Physical Properties of Adsorbents

	Particle Form*	Mesh Size	Effective Diameter D_p, ft.	Bulk Density ρ_b, Lb/cu.ft.	External Void Fraction F_a	External Surface a_v, sq.ft.	Specific Heat C_x, Btu/lb °F	Reactivation Temperature °F	Examples
Activated Carbon....	P	4 × 6	0.0128	30	0.34	310	0.25	200–1000	Columbia L
	P	6 × 8	0.0092	30	0.34	446	"	" "	" "
	P	8 × 10	0.0064	30	0.34	645	"	" "	" "
	G	4 × 10	0.0110	30	0.40	460	0.25	" "	Pittsburgh BPL
	G	6 × 16	0.0062	30	0.40	720	"	" "	"
	G	4 × 10	0.0105	28	0.44	450	"	" "	Witco 256
Silica Gel..........	G	3 × 8	0.0127	45	0.35	230	0.22	250–450	Davison 03
	G	6 × 16	0.0062	45	0.35	720	"	" "	" "
	S	4 × 8	0.0130	50	0.36	300	0.25	300–450	Mobil Sorbead R
Activated Alumina..	G	4 × 8	0.0130	52	0.25	380	0.22	350–600	Alcoa Type F
	G	8 × 14	0.0058	52	0.25	480	"	" "	" " "
	G	14 × 28	0.0027	54	0.25	970	"	" "	" " "
	S	(1/4")	0.0208	52	0.30	200	0.22	350–1000	Alcoa Type H
	S	(1/8")	0.0104	54	0.30	400	"	" "	" " "
Molecular Sieves...	G	14 × 28	0.0027	30	0.25	970	0.23	300–600	Davison, Linde
	P	(1/16")	0.0060	45	0.34	650	"	" "	"
	P	(1/8")	0.0104	45	0.34	400	"	" "	"
	S	4 × 8	0.0109	45	0.37	347	"	" "	"
	S	8 × 12	0.0067	45	0.37	565	"	" "	"

* P = pellets; G = granules; S = spheroids

(Fair, 1969).

TABLE 15.2. Data of Molecular Sieves
(a) Structures and Applications

Framework	Cationic Form	Formula of Typical Unit Cell	Window	Effective Channel Diameter (Å)	Application
A	Na	$Na_{12}[(AlO_2)_{12}(SiO_2)_{12}]$	8-ring (obstructed)	3.8	Desiccant. CO_2 removal from natural gas
	Ca	$Ca_5Na_2[(AlO_2)_{12}(SiO_2)_{12}]$	8-ring (free)	4.4	Linear paraffin separation. Air separation
	K	$K_{12}[(AlO_2)_{12}(SiO_2)_{12}]$	8-ring (obstructed)	2.9	Drying of cracked gas containing C_2H_4, etc.
X	Na	$Na_{86}[(AlO_2)_{86}(SiO_2)_{106}]$	12-ring	8.4	Pressure swing H_2 purification
	Ca	$Ca_{40}Na_6[(AlO_2)_{86}(SiO_2)_{106}]$	12-ring	8.0	Removal of mercaptans from natural gas
	Sr, Ba[a]	$Sr_{21}Ba_{22}[(AlO_2)_{86}(SiO_2)_{106}]$	12-ring	8.0	Xylene separation
Y	Na	$Na_{56}[(AlO_2)_{56}(SiO_2)_{136}]$	12-ring	8.0	Xylene separation
	K	$K_{56}[(AlO_2)_{56}(SiO_2)_{136}]$	12-ring	8.0	Xylene separation
Mordenite	Ag	$Ag_8[(AlO_2)_8(SiO_2)_{40}]$	12-ring	7.0	I and Kr removal from nuclear off-gases[32–34]
	H	$H_8[(AlO_2)_8(SiO_2)_{40}]$			
Silicalite	—	$(SiO_2)_{96}$	10-ring	6.0	Removal of organics from water
ZSM-5	Na	$Na_3[(AlO_2)_3(SiO_2)_{93}]$	10-ring	6.0	Xylene separation[31]

[a] Also K-BaX.

(Ruthven, 1984).

TABLE 15.2—(continued)

(b) Typical Properties of Union Carbide Type X Molecular Sieves

Basic type	Nominal pore diameter, angstroms	Available form	Bulk density, lb/ft³	Heat of adsorption (max), Btu/lb H_2O	Equilibrium H_2O capacity, % wt	Molecules adsorbed	Molecules excluded	Applications
3A	3	Powder 1/16-in pellets 1/8-in pellets	30 44 44	1800	23 20 20	Molecules with an effective diameter < 3 Å, including H_2O and NH_3	Molecules with an effective diameter > 3 Å, e.g., ethane	The preferred molecular sieve adsorbent for the commerical dehydration of unsaturated hydrocarbon streams such as cracked gas, propylene, butadiene, and acetylene. It is also used for drying polar liquids such as methanol and ethanol.
4A	4	Powder 1/16-in pellets 1/8-in pellets 8 × 12 beads 4 × 8 beads 14 × 30 mesh	30 45 45 45 45 44	1800	28.5 22 22 22 22 22	Molecules with an effective diameter < 4 Å, including ethanol, H_2S, CO_2, SO_2, C_2H_4, C_2H_6, and C_3H_6	Molecules with an effective diameter > 4 Å, e.g., propane	The preferred molecular sieve adsorbent for static dehydration in a closed gas or liquid system. It is used as a static desiccant in household refrigeration systems; in packaging of drugs, electronic components and perishable chemicals; and as a water scavenger in paint and plastic systems. Also used commercially in drying saturated hydrocarbon streams.
5A	5	Powder 1/16-in pellets 1/8-in pellets	30 43 43	1800	28 21.5 21.5	Molecules with an effective diameter < 5 Å, including n-C_4H_9OH,† n-C_4H_{10},† C_3H_8 to $C_{22}H_{46}$, R-12	Molecules with an effective diameter > 5 Å, e.g., iso compounds and all 4-carbon rings	Separates normal paraffins from branched-chain and cyclic hydrocarbons through a selective adsorption process.
10X	8	Powder 1/16-in pellets 1/8-in pellets	30 36 36	1800	36 28 28	Iso paraffins and olefins, C_6H_6, molecules with an effective diameter < 8 Å	Di-n-butylamine and larger	Aromatic hydrocarbon separation.
13X	10	Powder 1/16-in pellets 1/8-in pellets 8 × 12 beads 4 × 8 beads 14 × 30 mesh	30 38 38 42 42 38	1800	36 28.5 28.5 28.5 28.5 28.5	Molecules with an effective diameter < 10 Å	Molecules with an effective diameter > 10 Å, e.g., $(C_4F_9)_3N$	Used commercially for general gas drying, air plant feed purification (simultaneous removal of H_2O and CO_2), and liquid hydrocarbon and natural gas sweetening (H_2S and mercaptan removal).

(Kovach, 1978).

of adsorbent; Figure 15.10 reports this for a molecular sieve operation.

Representation and generalization of adsorption equilibria of binary and higher mixtures by equation is desirable, but less progress has been made for such systems than for vapor–liquid or liquid–liquid equilibria. The Yon and Turnock equations (1971) applied to components 1 and 2 of binary mixtures are

$$w_1/w_{1,\text{sat}} = k_1 P_1^{n_1} \theta, \tag{15.4}$$

$$w_2/w_{2,\text{sat}} = k_2 P_2^{n_2} \theta, \tag{15.5}$$

$$\theta = 1/(1 + k_1 P_1^{n_1} + k_2 P_2^{n_2}). \tag{15.6}$$

They have been found useful as an empirical correlation method for adsorption on molecular sieves [Maurer, *Am. Chem. Soc. Symp. Ser.* **135,** 73 (1980)]. Other attempts at prediction or correlation of multicomponent adsorption data are reviewed by Ruthven (1984). In general, however, multicomponent equilibria are not well correlatable in general form so that design of equipment is best based on direct laboratory data with the exact mixture and the exact adsorbent at anticipated pressure and temperature.

Adsorption processes are sensitive to temperature, as the data of Figures 15.3, 15.5, and 15.11 show. Thus practical adsorption processes are complicated by the substantial heats of adsorption that necessarily develop. These are of the same order of magnitude as heats of condensation. Some data are in Figure 15.4

15.2. ION EXCHANGE EQUILIBRIA

Ion exchange is a chemical process that can be represented by a stoichiometric equation, for example, when ion A in solution replaces ion B in the solid phase,

$$\text{A (solution)} + \text{B (solid)} \rightleftharpoons \text{A (solid)} + \text{B (solution)} \tag{15.7}$$

or

$$\text{A} + \bar{\text{B}} / \bar{\text{A}} + \text{B}, \tag{15.8}$$

where the overstrike designates a component in the solid phase. The equilibrium constant is called the selectivity, designated by K_{AB},

$$K_{\text{AB}} = C_{\bar{\text{A}}} C_{\text{B}} / C_{\text{A}} C_{\bar{\text{B}}} \tag{15.9}$$

$$= x_{\bar{\text{A}}} x_{\text{B}} / x_{\text{A}} x_{\bar{\text{B}}} \tag{15.10}$$

$$= \left[\frac{x_{\bar{\text{A}}}}{1 - x_{\bar{\text{A}}}}\right] \Big/ \left[\frac{x_{\text{A}}}{1 - x_{\text{A}}}\right]. \tag{15.11}$$

The last equation relates the mol fractions of the ion originally in the solution at equilibrium in the liquid (x_{A}) and solid ($x_{\bar{\text{A}}}$) phases.

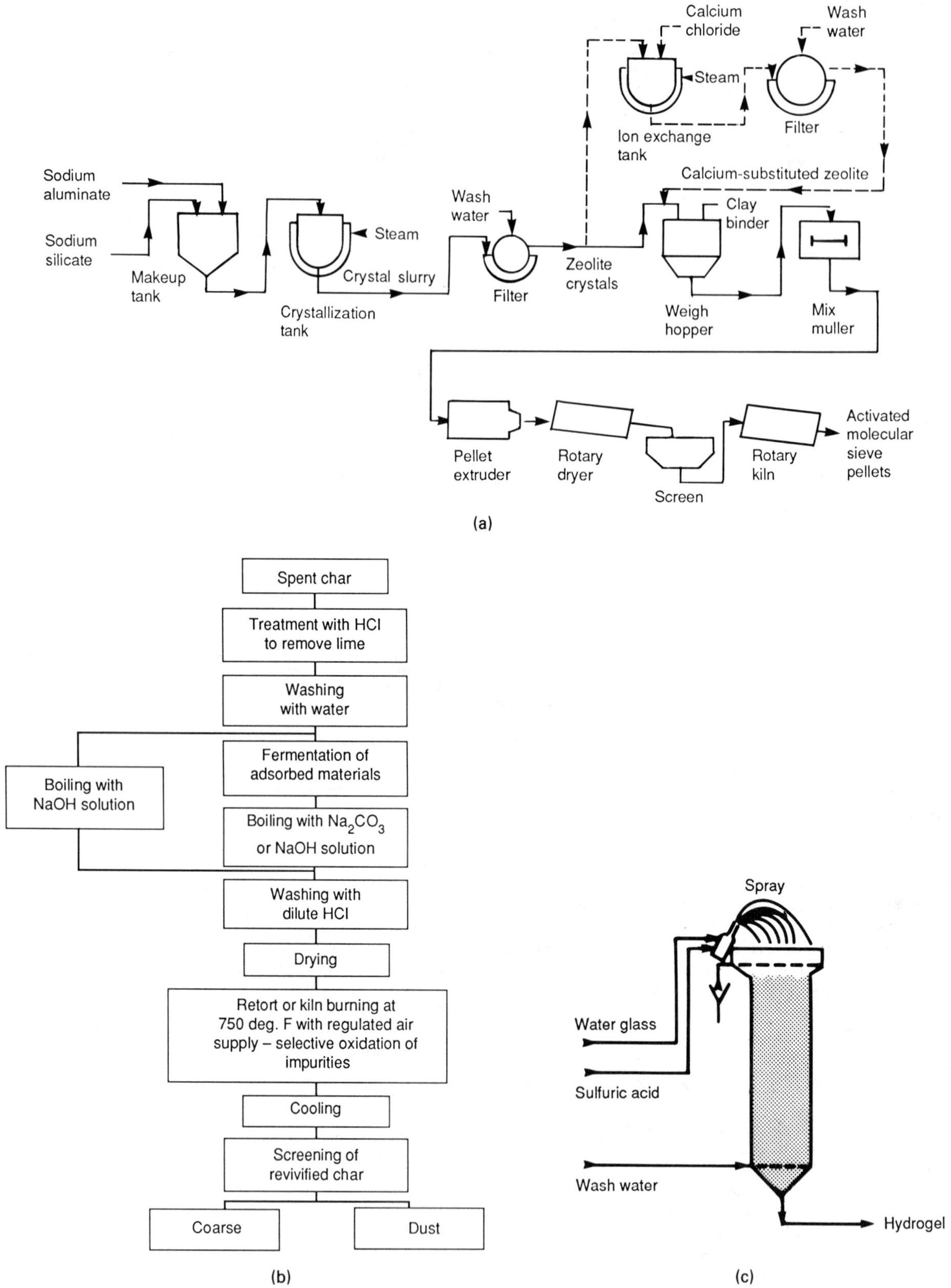

Figure 15.1. Processes for making adsorbents. (a) Flowsketch of a process for making molecular sieve adsorbents. (b) Process for reactivation of bone char. (c) Silica gel by the BASF process. The gel is formed and solidifies in air from sodium silicate and sulfuric acid, then is washed free of sodium sulfate with water (*Ullmann,* Encyclopedia of Chemical Technology, *Verlag Chemie, Weinheim, Germany*).

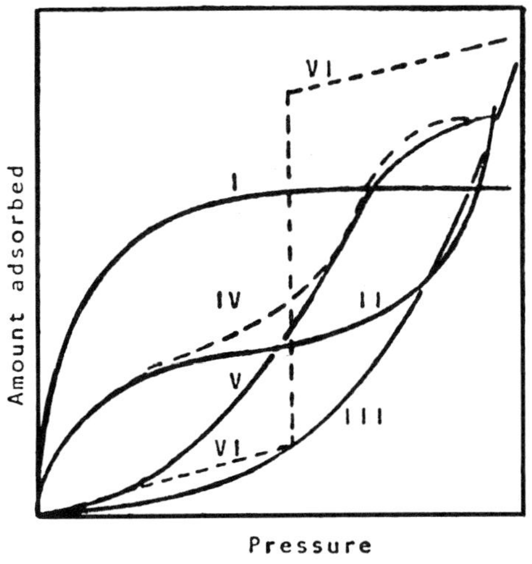

Figure 15.2. Types of adsorption isotherms: (I) monomolecular layer; (II and III) multimolecular layers; (IV and V) multimolecular layers and condensation in pores; (VI) phase transition of a monomolecular layer on the surface (*after Brunauer,* Physical Adsorption, *Princeton Univ. Press, 1945*).

The residual mol fraction in the liquid phase corresponding to a given mol fraction or degree of saturation in the solid phase is

$$x_A = \frac{1}{1 + K_{AB}(1 - x_{\bar{A}})/x_{\bar{A}}}. \quad (15.12)$$

Approximate values of the selectivity of various ions are shown in Table 15.3; for a particular pair, K_{AB} is the ratio of tabulated values for each.

When the exchanged ion D is divalent, the reaction is

$$D + 2\bar{B} \rightleftharpoons \bar{D} + 2B, \quad (15.13)$$

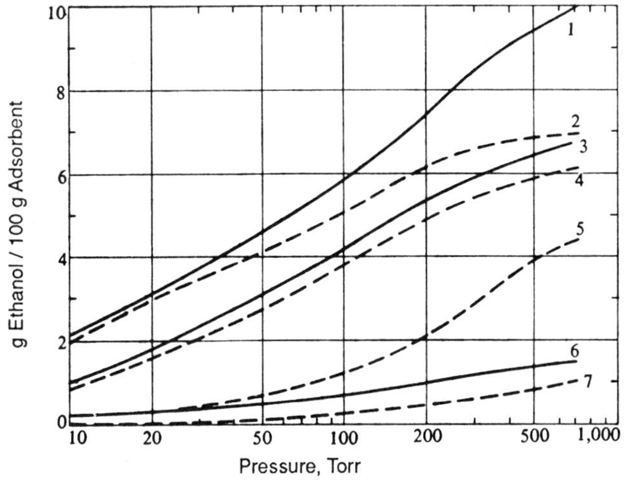

Figure 15.3. Effects of temperature, pressure, and kind of adsorbent on the amount of ethane adsorbed: (1) activated carbon at 25°C; (2) type 4A molecular sieve (MS) at 0°C; (3) type 5A MS at 25°C; (4) type 4A MS at 25°C; (5) type 4A MS at 75°C; (6) silica gel at 25°C; (7) type 4A MS at 150°C. (*Data from Union Carbide Corp.*)

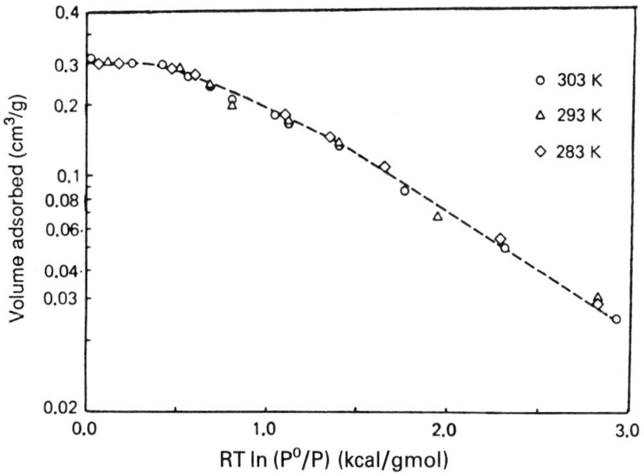

Figure 15.4. Polanyi characteristic curve for effect of temperature on adsorption of n-butane on silica gel [*Al-Sahhat et al.,* Ind. Eng. Chem. Process. Des. Dev. **20,** *658* (*1981*)].

and the equilibrium constant or selectivity is given by

$$(\bar{C}/C)K_{DB} = x_B^2 x_{\bar{D}} / x_{\bar{B}}^2 x_D \quad (15.14)$$

$$= (x_{\bar{D}}/x_D)[(1 - x_D)/(1 - x_{\bar{D}})]^2, \quad (15.15)$$

where C and $\bar{C}$ are the total concentrations of the two kinds of ions in the solution and in the solid, respectively.

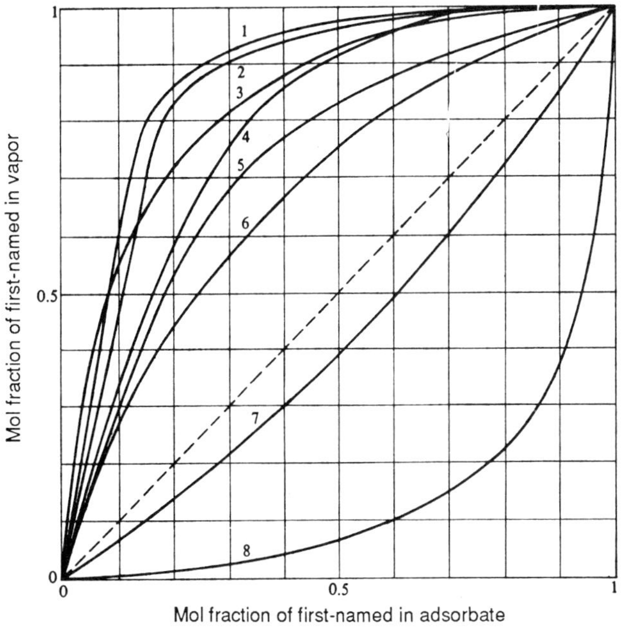

Figure 15.5. Adsorption of binary mixtures: (1) ethane + ethylene. Type 4A MS 25°C, 250 Torr; (2) ethane + ethylene. Type 4A MS, 25°C, 730 Torr; (3) ethane + ethylene. Type 4A MS, 75°C, 730 Torr; (4) carbon dioxide + hydrogen sulfide. Type 5A MS, 27°C, 760 Torr; (5) n-pentane + n-hexane, type 5A MS, 100°C, 760 Torr; (6) ethane + ethylene, silica gel, 25°C, 760 Torr; (7) ethane + ethylene, Columbia G carbon, 25°C, 760 Torr; (8) acetylene + ethylene. Type 4A MS, 31°C, 740 Torr. (*Data from Union Carbide Corp.*)

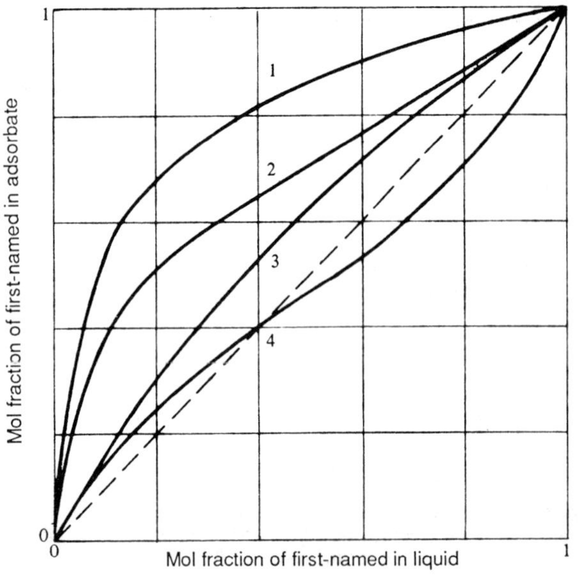

Figure 15.6. Binary liquid adsorption equilibria on X–Y diagrams: (1) toluene + iso-octane on silica gel (Eagle and Scott, 1950); (2) toluene + iso-octane on charcoal (Eagle and Scott, 1950); (3) ethylene dichloride + benzene on boehmite (Kipling); (4) ethylene dichloride + benzene on charcoal (Kipling). (*Kipling in* Proceedings of the Second International Congress of Surface Activity. (*1957*), *Vol. III, p. 462.*)

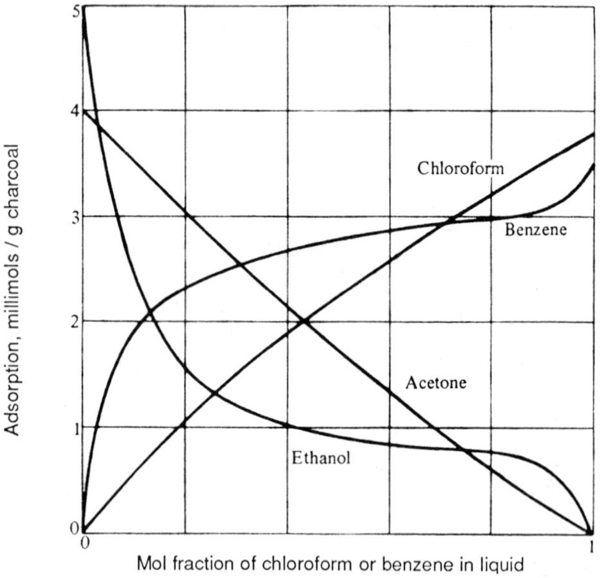

Figure 15.7. Adsorption of liquid mixtures on charcoal. Chloroform + acetone and benzene + ethanol. The ordinate gives the amount of each individual substance that is adsorbed, the abscissa the mol fraction of chloroform (mixed with acetone) or the mol fraction of benzene (mixed with ethanol). (*Data gathered by Kipling.* Adsorption from Solutions of Non-Electrolytes, *1965*).

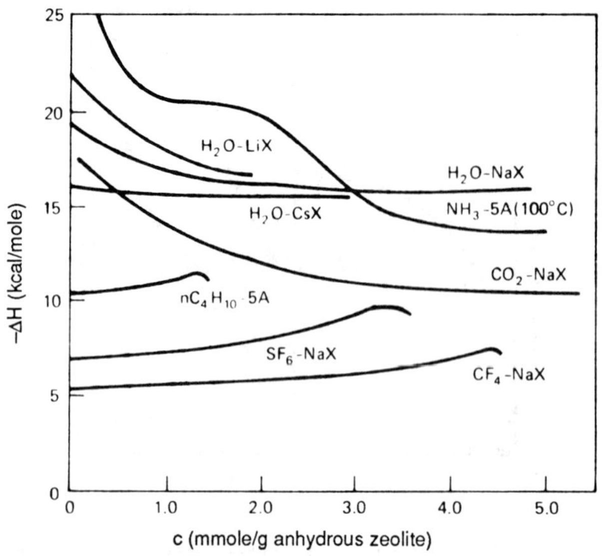

Figure 15.8. Variation of isosteric heat of adsorption with coverage showing the difference in trends between polar and nonpolar sorbates. $nC_4H_{10}-5A$ (data of Schirmer et al.); CF_4–NaX, SF_6–NaX (data of Barrer and Reucroft); CO_2–NaX, (data of Huang and Zwiebel), NH_3–5A (data of Schirmer et al.); H_2O–LiX, NaX, and CsX, (data of Avgul et al.). [*Ruthven,* Sep Purification Methods **5**(*2*), *189* (*1976*)].

Example 15.1 is concerned with such an exchange and regeneration process.

15.3. ADSORPTION BEHAVIOR IN PACKED BEDS

Adsorption is performed most commonly in fixed vertical beds of porous granular adsorbents. Flow of adsorbing fluid usually is down through the bed, that of regenerant usually is upward. Moving and fluidized beds have only a limited application in the field.

If the time is sufficient, the adsorbent nearer the inlet of the fluid becomes saturated at the prevailing inlet fluid concentration but a concentration gradient develops beyond the saturation zone. Figure 5.12 depicts this behavior. The region of falling concentration is called the mass transfer zone (MTZ). The gradient is called the adsorption wave front and is usually S-shaped. When its leading edge reaches the exit, breakthrough is said to have been attained. Practically, the breakthrough is not regarded as necessarily at zero concentration but at some low value such as 1% or 5% of the inlet that is acceptable in the effluent. A hypothetical position, to the left of which in Figure 15.12(b) the average adsorbate content equals the saturation value, is called the stoichiometric front. The distance between this position and the exit of the bed is called the length of unused bed (LUB). The exhaustion time is attained when the effluent concentration becomes the same as that of the inlet, or some practical high percentage of it, such as 95 or 99%.

The shape of the adsorption front, the width of the MTZ, and the profile of the effluent concentration depend on the nature of the adsorption isotherm and the rate of mass transfer. Practical bed depths may be expressed as multiples of MTZ, values of 5–10 multiples being economically feasible. Systems that have linear adsorption isotherms develop constant MTZs whereas MTZs of convex ones (such as Type I of Figure 15.1) become narrower, and those of concave systems become wider as they progress through

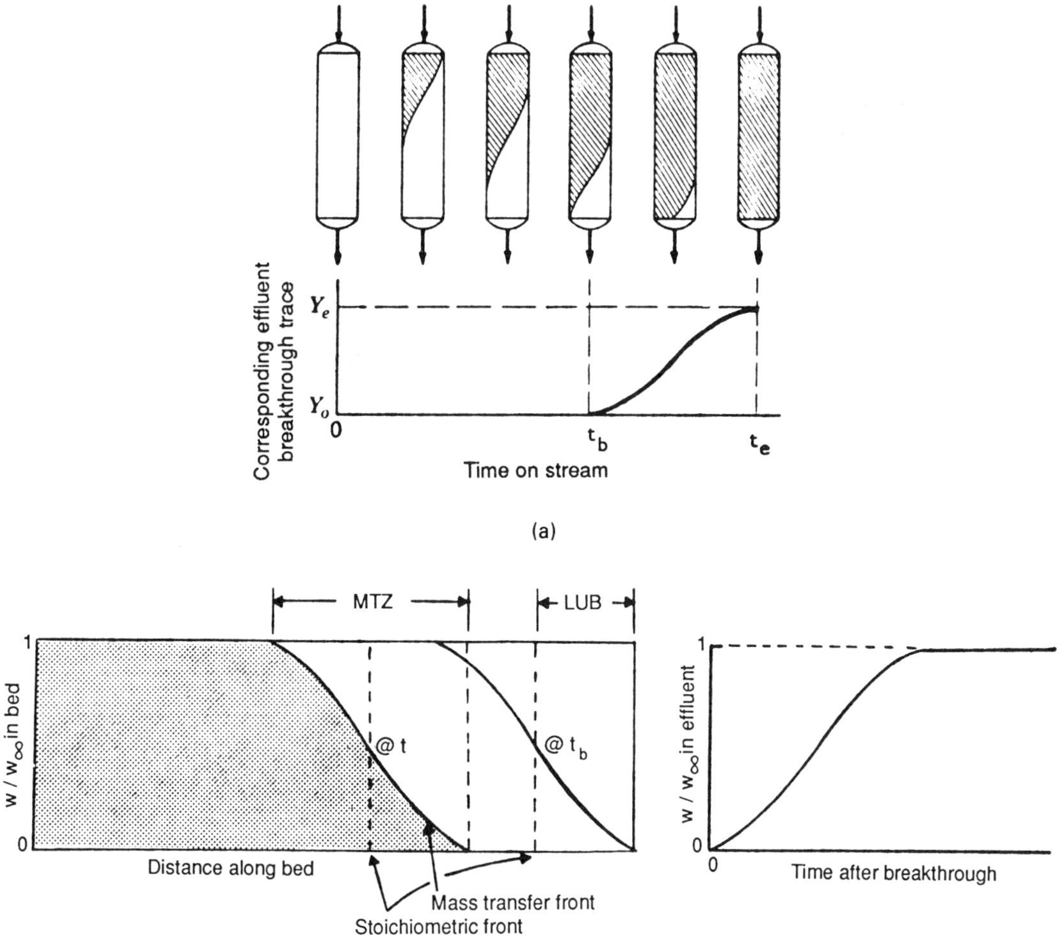

Figure 15.9. Concentrations in adsorption beds as a function of position and of effluent as a function of time. (a) Progress of a stable mass transfer front through an adsorption bed and of the effluent concentration (*Lukchis, 1973*). (b) The mass transfer zone (MTZ), the length of unused bed (LUB), stoichiometric front, and profile of effluent concentration after breakthrough.

the bed. The last types are called unfavorable isotherms; separations in such cases usually are accomplished more economically by some other kind of process. The narrower the MTZ, the greater the degree of utilization of the bed.

The rate of mass transfer from fluid to solid in a bed of porous granular adsorbent is made up of several factors in series:

1. Diffusion to the external surface.
2. Deposition on the surface.
3. Diffusion in the pores.
4. Diffusion along the surface.

Various combination of shapes of isotherms and mass transfer factors have been taken into account by solutions of the problem in the literature. One of the simpler cases was adopted by Hougen and Marshall (1947, see Figure 15.13), who took a linear isotherm and diffusion to the external surface as controlling the rate. They developed the solution in analytical form, of which several approximations that are easier to use are mentioned for instance by Vermeulen et al. (1984, p. 16.28). A graphical form of the solution appears in Figure 15.13. This shows the effluent concentration ratio, C/C_0, in terms of a time parameter τ, at a number of values of a parameter Z', which involves the bed length Z. In Example 15.2, this chart is used to find the concentration profile of the effluent, the break and exhaustion times, and the % utilization of the adsorbent bed. In this case, the model affords a fair comparison with experimental data.

Many investigations have been conducted of the mass transfer coefficient at the external surfaces of particles and of other diffusional mechanisms. Some of the correlations are discussed in Chapters 13 and 17. A model developed by Rosen [*Ind. Eng. Chem.* **46,** 1590 (1954)] takes into account both external film and pore diffusional resistance to mass transfer together with a linear isotherm. A numerical example is worked out by Hines and Maddox (1985, p. 485).

In the model developed by Thomas [*J. Am. Chem. Soc.* **66,** 1664 (1944)], the controlling mechanism is the surface kinetics represented by the Langmuir isotherm. Extensions of this work by Vermeulen et al. (1984) incorporate external surface and pore diffusional resistances.

No comprehensive comparisons of the several models with each other and with experimental data appear to have been published.

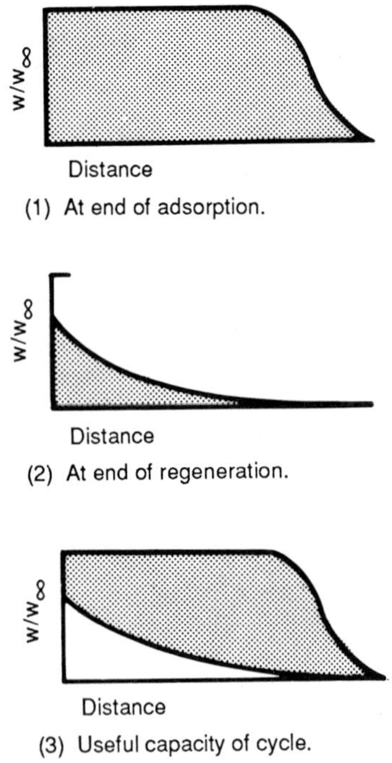

Figure 15.10. Incomplete regeneration of adsorbent bed by a thermal-swing cycle.

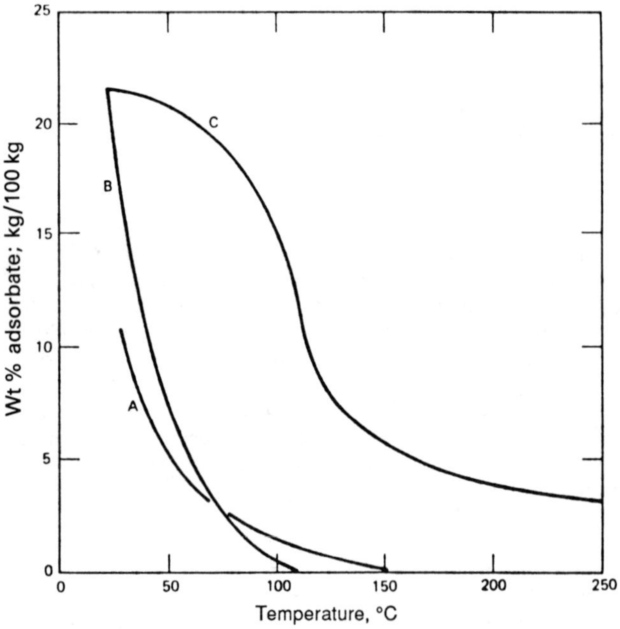

Figure 15.11. Effect of temperature on molecular sieve type 5A, silica gel and activated alumina at water vapor pressure, 13.3 kPa (100 mm Hg). A, molecular sieve type 5A; B, silica-type adsorbent; C, alumina-type adsorbent. (*Chi and Cummings, 1978*).

TABLE 15.3. Gas Phase Adsorption Cycles, Steam Requirements, and Operation

(a) Typical Cycle Times for Adsorber Operation

	High Pressure Gas Dryer		Organic Solvent Recovery Unit	
	A	B	A	B
Onstream.................	24	24	2.00	1.00
Depressure/purge.........	2	1	.....	.....
Hot gas....................	10	13	.....	.....
Steam.....................	...	...	0.75	0.67
Hot gas....................	...	...	0.33	.....
Cold gas...................	5	8	0.42	0.33
Pressure/standby.........	7	2	0.50	.....
	24	24	2.00	1.00

(Fair, 1969).

(b) Steam/Solvent Ratios and Amount of Adsorbate for a Coconut-Shell Carbon 6–12 Mesh, 1200 m²/g

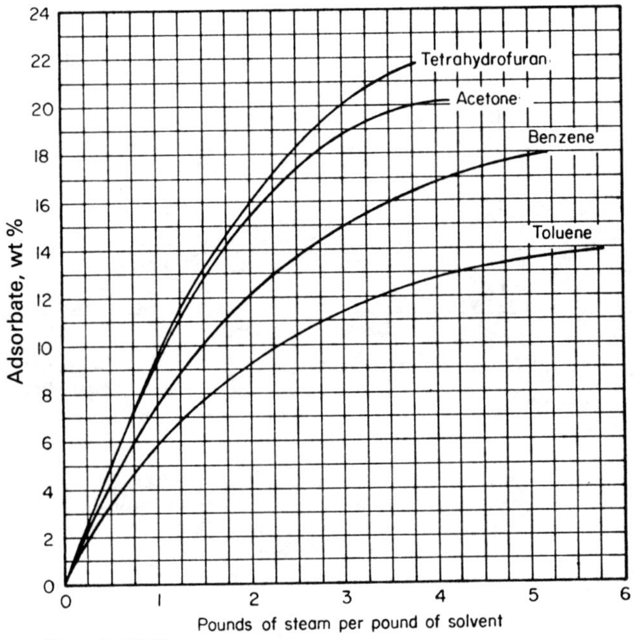

(Kovach, 1979).

(c) Typical Operating Parameters for Gas Phase Adsorption

	Range	*Design*
Superficial gas velocity	20 to 50 cm/s (40 to 100 ft/min)	40 cm/s (80 ft/min)
Adsorbent bed depth	3 to 10 MTZ	5 MTZ
Adsorption time	0.5 to 8 h	4 h
Temperature	−200 to 50°C	
Inlet concentration		
Adsorption base	100 to 5000 vppm	
LEL base	40%	
Adsorbent particle size	0.5 to 10 mm	4 to 8 mm
Working charge	5 to 20% wt	10%
Steam solvent ratio	2:1 to 8:1	4:1
Adsorbent void volume	38 to 50%	45%
Steam regeneration temperature	105 to 110°C	
Inert gas regenerant temperature	100 to 300°C	
Regeneration time	½ adsorption time	
Number of adsorbers	1 to 6	2 to 3

(Kovach, 1979).

EXAMPLE 15.1
Application of Ion Exchange Selectivity Data

The $SO_4^=$ ion of an aqueous solution containing $C = 0.018$ eq/L is to be replaced with Cl^- ion from a resin with $\bar{C} = 1.2$ eq/L. The reaction is

$$SO_4^=(\text{solution}) + 2Cl^-(\text{resin}) \rightleftarrows SO_4^=(\text{resin}) + 2Cl^-(\text{solution}),$$
$$D + 2\bar{B} \rightleftarrows \bar{D} + 2B.$$

From Table 15.3, the selectivity ratio $K_{DB} = 0.15/1.0 = 0.15$, and

$$K_{DB}\bar{C}/C = 0.15(1.2)/0.018 = 10.$$

Then Eq. (15.15) becomes

$$x_{\bar{D}}/(1 - x_{\bar{D}})^2 = 10x_D/(1 - x_D)^2.$$

For several values of mol fraction x_D of $SO_4^=$ in solution, the corresponding mol fractions $x_{\bar{D}}$ in the resin are calculated and tabulated:

$x_{SO_4^=}$	
In Solution	In Resin
1	1
0.1	0.418
0.05	0.284
0.01	0.0853

For regeneration of the resin, a 12% solution of NaCl will be used; its ion concentration is 2.23 eq/L. Other values for the system remain at $\bar{C} = 1.2$ eq/L and $K_{DB} = 0.15$. Accordingly,

$$K_{DB}\bar{C}/C = 0.15(1.2)/2.23 = 0.0807$$

and Eq. (15.15) becomes

$$x_{\bar{D}}/(1 - x_{\bar{D}})^2 = 0.0807x_D/(1 - x_D)^2.$$

The values of $x_{SO_4^=}$ in the liquid phase will be calculated for several values in the resin. Those results will be used to find the minimum amount of regenerant solution needed for each degree of regeneration

$x_{SO_4^=}$		
In Resin	In Solution	L regenerant/ L resin
0.1	0.455	1.06
0.05	0.319	1.60
0.01	0.102	5.22

Sample calculation for the last entry of the table: The equivalents of $SO_4^=$ transferred from the resin to the solution are

$$0.99(1.2) = 1.188 \text{ eq/L}.$$

The minimum amount of solution needed for this regeneration is

$$\frac{1.188}{0.102(2.23)} = 5.22 \text{ L solution/Liter}.$$

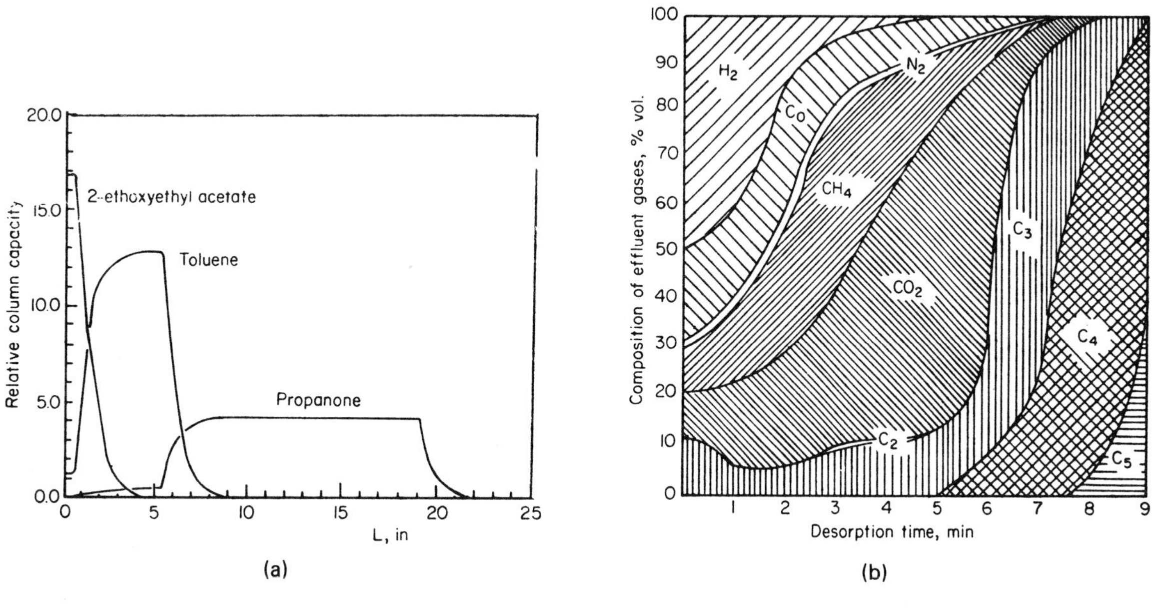

Figure 15.12. Multicomponent mixtures, adsorption, and desorption. (a) Concentrations of the components of a ternary mixture in continuous adsorption, as in a moving bed unit (*Kovach, 1979*). (b) Composition of a desorbed stream consisting of several components as a function of time.

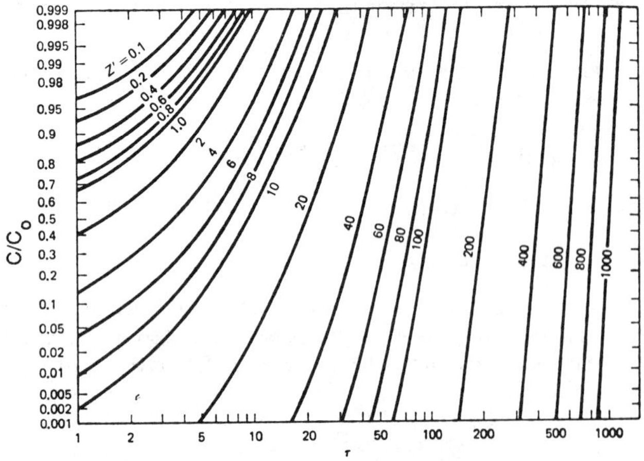

Figure 15.13. Dependence of the concentration ratio, C/C_0 of the effluent from an adsorber on parameters of bed length and time; for the case of a linear isotherm, zero initial adsorbate content and constant inlet composition C_0

$Z' = (K_f a/\varepsilon u_i)Z$, bed length parameter,

$\tau = (K_f a/K_D \rho_b)(t - Z/u_i)$, time parameter,

$K_D = q/C$, coefficient of linear adsorption isotherm,

u_i = interstitial velocity in the bed,

ε = voidage of the bed,

Z = length of the bed,

$K_f a$ = mass transfer coefficient, $(L^3 \text{ fluid})/(L^3 \text{ bed})(\text{time})$.

(*Hougen and Watson,* Chemical Process Principles, *Wiley, New York, 1947, p. 1086*; *Hougen and Marshall,* Chem. Eng. Prog. **43,** *197 (1947)*; *Vermeulen et al.,* Chemical Engineers' Handbook, *McGraw-Hill, New York, 1984, p. 16.29.*)

Moreover, they are all based on isothermal behavior and approximations of adsorption isotherms and have not been applied to multicomponent mixtures. The greatest value of these calculation methods may lie in the prediction of effects of changes in basic data such as flow rates and slopes of adsorption isotherms after experimental data have been measured of breakthroughs and effluent concentration profiles. In a multicomponent system, each substance has a different breakthrough which is affected by the presence of the other substances. Experimental curves such as those of Figure 15.14 must be the basis for sizing an adsorber.

Since taking samples of adsorbent from various positions in the bed for analysis is difficult, it is usual to deduce the shape of the adsorption front and the width of the MTZ from the effluent concentration profile which may be monitored with a continuous analyzer-recorder or by sampling. The overall width of the MTZ, for instance, is given in terms of the exhaustion and breakthrough times and the superficial velocity as

$$\text{width} = u_s(t_e - t_b)/\varepsilon.$$

REGENERATION

Adsorbents are restored to essentially their original condition for reuse by desorption. Many hundreds of cycles usually are feasible, but eventually some degradation occurs, as in Figure 15.15 for instance, and the adsorbent must be discarded.

The most common method of regeneration is by purging the bed with a hot gas. Operating temperatures are characteristic of the adsorbent; suitable values at atmospheric pressure are shown in Table 15.1. The exit temperature of the gas usually is about 50°F higher than that of the end of the bed. Typical cycle times for adsorption and regeneration and steam/adsorbent ratios are given in Table 15.4. Effluent composition traces of a multicomponent system are in Figure 15.9. Complete removal of adsorbate is not always economically feasible, as suggested by Table 15.4(b). The effect of incomplete removal on capacity is shown schematically by Figure 15.10. Sufficient heat must be supplied to warm up the adsorbent and the vessel, to provide heat of desorption and enthalpy absorption of the adsorbate, and to provide for heat losses to the surroundings. Table 15.4(c) suggests that regeneration times be about one-half the adsorption times. For large vessels, it may be worthwhile to make the unsteady heating calculation by the general methods applicable to regenerators, as presented, for instance, by Hausen (*Heat Transfer in Counterflow, Parallel Flow and Cross-flow,* McGraw-Hill, New York, 1983).

Purging of the adsorbate with an inert gas at much reduced pressure is feasible in high pressure adsorption plants. The adsorption of Example 15.2, for instance, is conducted at 55 atm, so that regeneration could be accomplished at a pressure of only a few atmospheres without heating. If the adsorbate is valuable, some provision must be made for recovering it from the desorbing gas.

Ignition of adsorbents in external furnaces is practiced to remove some high molecular weight materials that are difficult to volatilize. This is done, for example, for reactivation of carbon from water treating for trace removal of impurities such as phenol. Caustic solution can convert the phenol into soluble sodium phenate in readily disposable concentrated form as an alternate process for regeneration.

Displacement of the adsorbate with another substance that is in turn displaced in process is practiced, for instance, in liquid phase recovery of paraxylene from other C_8 aromatics. In the Sorbex process, suitable desorbents are toluene and paradiethylbenzene. This process is described later.

15.4. ADSORPTION DESIGN AND OPERATING PRACTICES

When continuous operation is necessary, at least two adsorbers are employed, one on adsorption and the other alternately on regeneration and cooling. In cases where breakthrough is especially harmful, three vessels are used, one being regenerated, the other two onstream with the more recently regenerated vessel downstream, as in Figure 15.16.

Beds usually are vertical; adsorbers 45 ft high and 8–10 ft dia are in use. When pressure drop must be minimized, as in the recovery of solvents from atmospheric air, horizontal vessels with shallow beds are in common use. Process gas flow most often is downward and regenerant gas flow is upward to take advantage of counterflow effects. Upflow rates are at most about one-half the fluidizing velocity of the particles. Vertical and horizontal types are represented on Figure 15.17.

A major feature of adsorber design is the support for the granular adsorbent, preferably one with a low pressure drop. The combination of Figure 15.18(a) of grid, screens, and support beams is inexpensive to fabricate and maintain, has a low heat capacity and a low pressure drop. The construction of Figure 15.18(b) is suited to adsorbers that must be dumped frequently. Supports of layers of ceramic balls or gravel or anthracite, resting on the bottom of the vessel, are suited to large vessels and when corrosion-resistant construction is required. Typical arrangements are shown in Figures 17.26 and 17.27. The successive layers increase in diameter by factors of 2–4 up to 1 in. or so. Holddown balls also may be

EXAMPLE 15.2
Adsorption of *n*-Hexane from a Natural Gas with Silica Gel
Hexane is to be recovered from a natural gas with silica gel. Molecular weight of the gas is 17.85, the pressure is 55.4 atm, temperature is 94°F, and the content of *n*-hexane is 0.853 mol % or 0.0982 lb/cuft. The bed is 43 in. deep and the superficial velocity is 11.4 ft/min. Other data are shown with the sketch:

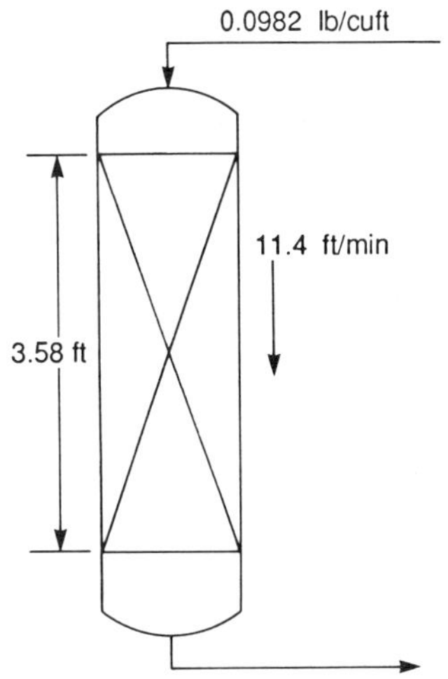

$Z = 3.58$ ft, bed depth,
$u_s = 11.4$ ft/min, superficial velocity,
$D_p = 0.01$ ft, particle diameter,
$a = 284$ sqft/cuft, packing external surface,
$\rho_b = 52$ lb/cuft, bed density,
$\varepsilon = 0.35$ bed voidage.

From these and physical property data, the Schmidt and Reynolds numbers are calculated as

$$\mathrm{Sc} = 1.87, \quad \mathrm{Re} = 644.$$

The equation of Dwivedi and Upadhyay, Eq. (13.148), is applicable:

$$J_d = \frac{k_g}{u_s}\mathrm{Sc}^{2/3} = \frac{1}{\varepsilon}\left(\frac{0.765}{\mathrm{Re}^{0.82}} + \frac{0.365}{\mathrm{Re}^{0.386}}\right),$$

$$\therefore k_g = \frac{11.4}{0.35(1.87)^{2/3}}(0.0038 + 0.0301) = 0.7268 \text{ ft/min},$$

$$k_g a = 0.7268(284) = 206.4 \text{ cuft gas/(cuft solid)(min)}.$$

Saturation content of adsorbate is 0.17 lb/lb solid. Accordingly, the coefficient of the linear adsorption isotherm is

$$k_d = \frac{0.17}{0.0982} = 1.731 \frac{\text{lb hexane/lb solid}}{\text{lb hexane/cuft gas}}.$$

Use the Hougen–Marshall chart (Fig. 15.13):

$$Z' = \frac{k_g a Z}{u_s} = \frac{206.4(3.58)}{11.4} = 64.82,$$

$$t = \frac{k_d \rho_b}{k_g a}\tau + \frac{z}{u_s/\varepsilon} = \frac{1.731(52)}{206.4}\tau + \frac{3.58}{11.4/0.35}$$
$$= 0.436\tau + 0.11 \quad \text{min.}$$

Values of τ are read off Figure 15.13 and converted into values of t:

C/C_0	τ	t (min)
0.01	40	17.56
0.05	45	19.74
0.1	50	21.92
0.2	53	23.23
0.4	60	26.28
0.6	65	28.46
0.8	73	31.95
0.9	79	34.57
0.95	82	35.87
0.99	92	40.24

The total amount adsorbed to the breakpoint, at $C/C_0 = 0.01$, per sqft of bed cross section is

$$0.0982(11.4)(17.56) = 19.66 \text{ lb/sqft cross section.}$$

The saturation amount for the whole bed is

$$3.58(0.17)(52) = 31.65 \text{ lb/sqft cross section.}$$

Accordingly,

$$\text{utilization of bed} = (19.66/31.65)(100\%) = 62.1\%.$$

The calculated concentration profile is compared in the figure with experimental data, Run 117, of McLeod and Campbell, *Soc. Pet. Eng. J.*, 166 (June 1966):

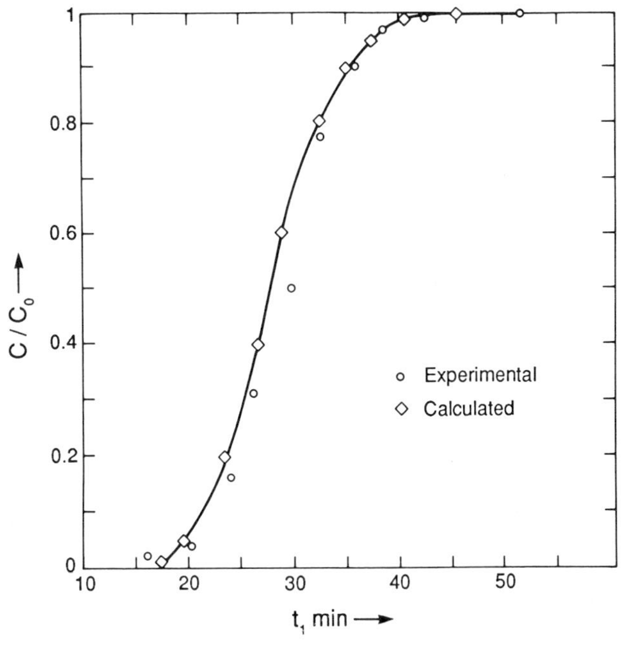

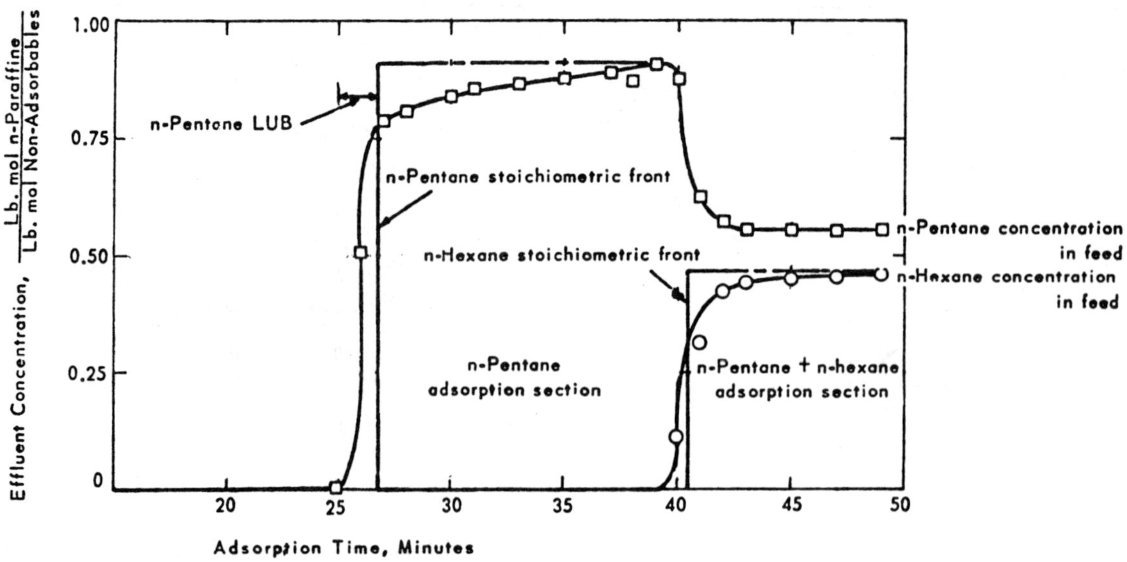

Figure 15.14. Breakthrough curves in the adsorption of a mixture of hydrocarbons with composition n-butane 0.4 mol %, n-pentane 25.9, n-hexane 23.9 iso and cyclic hydrocarbons 49.8 mol % (*Lee, in* Recent Advances in Separation Science, *CRC Press, Boca Raton, FL, 1972, Vol. II, pp. 75–110*).

provided at the top to prevent disturbance of the top layer of adsorbent by incoming high velocity gas or entrainment by upflowing gases. When regeneration is by heating, a drawback of the ball support arrangement is their substantial heat capacity, which slows up the heating rate and subsequent cooling to process temperature.

Representative values and ranges of operating parameters are summarized in Table 15.4. Cycle times for some adsorptions are adjusted to work shift length, usually multiples of 8 hr, with valve adjustments made by hand. When cycle times are short, as for solvent recovery, automatic opening and closing of valves is necessary.

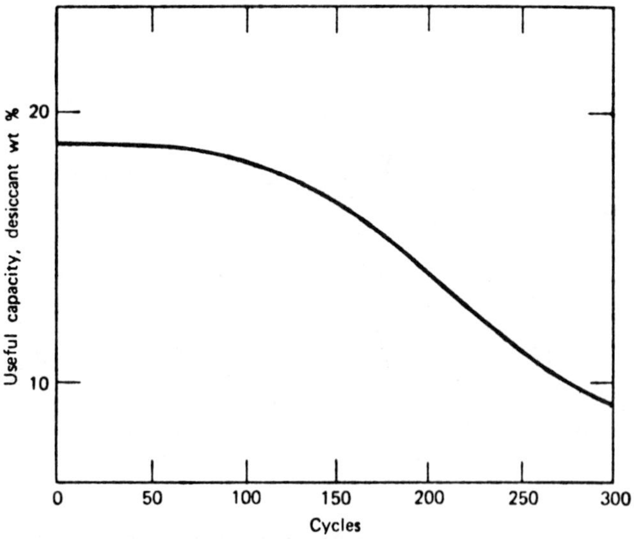

Figure 15.15. Capacity decline with service of a molecular sieve (*plant data, Davison Sieve 562*). Flow, 8150 kg mol; pressure, 3600 kPa (36 atm); temperature, 15°C; water content, 96 kg/hr; minimum cycle time, 24 hr. (*Chi and Cummings, 1978*).

Steam rates for regeneration of a particular adsorbent carbon are shown in Table 15.4(b). Steam/solvent ratios as high as 8 sometimes are necessary.

Data for liquid phase adsorption are typified by water treating for removal of small but harmful amounts of impurities. Some conditions are stated by Bernardin [*Chem. Eng.*, (18 Oct. 1976)]. Water flow rates are 5–10 gpm/sqft. When suspended solids are present, the accumulation on the top of the bed is backwashed at 15–20 gpm/sqft for 10–20 min/day. The adsorbent usually is not regenerated in place but is removed and treated in a furnace. Accordingly, a continuous operation is desirable, and one is simulated by periodic removal of spent adsorbent from the bottom of the vessel with a design like that of Figure 15.18(b) and replenishing of fresh adsorbent at the top. The pulses of spent and fresh carbon are 2–10% of the total bed. Height to diameter ratio in such units is about 3.

A carbon adsorber for handling 100,000 gal/day of water consists of two vessels in series, each 10 ft dia by 11 ft sidewall and containing 20,000 lb of activated carbon. Total organic carbon is reduced from 650 mg/L to 25 mg/L, and phenol from 130 mg/L to less than 0.1 mg/L.

The capacity of regeneration furnaces is selected so that they operate 80–90% of the time. In multiple-hearth furnaces the loading is 70–80 lb/(sqft)(day). In countercurrent direct fired rotary kilns, a 6% volumetric loading is used with 45 min at activation temperature.

Details of the design and performance of other liquid phase adsorptions such as the Sorbex processes are proprietary.

15.5. ION EXCHANGE DESIGN AND OPERATING PRACTICES

Ion exchange processes function by replacing undesirable ions of a liquid with ions such as H^+ or OH^- from a solid material in which the ions are sufficiently mobile, usually some synthetic resin. Eventually the resin becomes exhausted and may be regenerated by contact with a small amount of solution with a high content of the desired ion. Resins can be tailored to have selective affinities for

TABLE 15.4. Properties of Ion-Exchange Materials

(a) Physical Properties

Material	Shape* of particles	Bulk wet density (drained), kg/L	Moisture content (drained), % by weight	Swelling due to exchange, %	Maximum operating temperature,† °C	Operating pH range	Exchange capacity: Dry, equivalent/kg	Exchange capacity: Wet, equivalent/L
Cation exchangers: strongly acidic								
Polystyrene sulfonate								
Homogeneous (gel) resin	S				120–150	0–14		
4% cross-linked		0.75–0.85	64–70	10–12			5.0–5.5	1.2–1.6
6% cross-linked		0.76–0.86	58–65	8–10			4.8–5.4	1.3–1.8
8–10% cross-linked		0.77–0.87	48–60	6–8			4.6–5.2	1.4–1.9
12% cross-linked		0.78–0.88	44–48	5			4.4–4.9	1.5–2.0
16% cross-linked		0.79–0.89	42–46	4			4.2–4.6	1.7–2.1
20% cross-linked		0.80–0.90	40–45	3			3.9–4.2	1.8–2.0
Macroporous structure								
10–12% cross-linked	S	0.81	50–55	4–6	120–150	0–14	4.5–5.0	1.5–1.9
Sulfonated phenolic resin	G	0.74–0.85	50–60	7	50–90	0–14	2.0–2.5	0.7–0.9
Sulfonated coal	G							
Cation exchangers: weakly acidic								
Acrylic (pK 5) or methacrylic (pK 6)								
Homogeneous (gel) resin	S	0.70–0.75	45–50	20–80	120	4–14	8.3–10	3.3–4.0
Macroporous	S	0.67–0.74	50–55	10–100	120	...	~8.0	2.5–3.5
Phenolic resin	G	0.70–0.80	~50	10–25	45–65	0–14	2.5	1.0–1.4
Polystyrene phosphonate	G, S	0.74	50–70	<40	120	3–14	6.6	3.0
Polystyrene aminodiacetate	S	0.75	68–75	<100	75	3–14	2.9	0.7
Polystyrene amidoxime	S	~0.75	58	10	50	1–11	2.8	0.8–0.9
Polystyrene thiol	S	~0.75	45–50	...	60	1–13	~5	2.0
Cellulose								
Phosphonate	F	...	...	...	...	...	~7.0	
Methylene carboxylate	F, P, G	...	...	...	...	...	~0.7	
Greensand (Fe silicate)	G	1.3	1–5	0	60	6–8	0.14	0.18
Zeolite (Al silicate)	G	0.85–0.95	40–45	0	60	6–8	1.4	0.75
Zirconium tungstate	G	1.15–1.25	~5	0	>150	2–10	1.2	1.0
Anion exchangers: strongly basic								
Polystyrene-based								
Trimethyl benzyl ammonium (type I)								
Homogeneous, 8% CL	S	0.70	46–50	~20	60–80	0–14	3.4–3.8	1.3–1.5
Macroporous, 11% CL	S	0.67	57–60	15–20	60–80	0–14	3.4	1.0
Dimethyl hydroxyethyl ammonium (type II)								
Homogeneous, 8% CL	S	0.71	~42	15–20	40–80	0–14	3.8–4.0	1.2
Macroporous, 10% CL	S	0.67	~55	12–15	40–80	0–14	3.8	1.1
Acrylic-based								
Homogeneous (gel)	S	0.72	~70	~15	40–80	0–14	~5.0	1.0–1.2
Macroporous	S	0.67	~60	~12	40–80	0–14	3.0–3.3	0.8–0.9
Cellulose-based								
Ethyl trimethyl ammonium	F	...	...	...	100	4–10	0.62	
Triethyl hydroxypropyl ammonium		...	...	...	100	4–10	0.57	
Anion exchangers: intermediately basic (pK 11)								
Polystyrene-based	S	0.75	~50	15–25	65	0–10	4.8	1.8
Epoxy-polyamine	S	0.72	~64	8–10	75	0–7	6.5	1.7
Anion exchangers: weakly basic (pK 9)								
Aminopolystyrene								
Homogeneous (gel)	S	0.67	~45	8–12	100	0–7	5.5	1.8
Macroporous	S	0.61	55–60	~25	100	0–9	4.9	1.2
Acrylic-based amine								
Homogeneous (gel)	S	0.72	~63	8–10	80	0–7	6.5	1.7
Macroporous	S	0.72	~68	12–15	60	0–9	5.0	1.1
Cellulose-based								
Aminoethyl	P	...	...	...	...		1.0	
Diethyl aminoethyl	P	...	...	...	...		~0.9	

*Shapes: C, cylindrical pellets; G, granules; P, powder; S, spheres.
†When two temperatures are shown, the first applies to H form for cation, or OH form for anion, exchanger; the second, to salt ion.
NOTE: To convert kilograms per liter to pounds per cubic foot, multiply by 6.238×10^1; $°F = 9/5\ °C + 32$.

(*Chemical Engineers' Handbook*, McGraw-Hill, New York, 1984; a larger table complete with trade names is in the 5th edition, 1973).

(b) Selectivity Scale for Cations on 8% Crosslinked Resin

Li^+	1.0	Zn^{2+}	3.5
H^+	1.3	Co^{2+}	3.7
Na^+	2.0	Cu^{2+}	3.8
NH_4^+	2.6	Cd^{2+}	3.9
K^+	2.9	Be^{2+}	4.0
Rb^+	3.2	Mn^{2+}	4.1
Cs^+	3.3	Ni^{2+}	3.9
Ag^+	8.5	Ca^{2+}	5.2
UO_2^{2+}	2.5	Sr^{2+}	6.5
Mg^{2+}	3.3	Pb^{2+}	9.9
		Ba^{2+}	11.5

(Bonner and Smith, *J. Phys. Chem.* **61**, 1957, p. 326).

(c) Approximate Selectivity Scale for Anions on Strong-Base Resins

I^-	8	HCO_3^-	0.4
NO_3^-	4	CH_3COO^-	0.2
Br^-	3	F^-	0.1
HSO_4^-	1.6	OH^-(Type I)	0.05–0.07
NO_2^-	1.3		
CN^-	1.3	SO_4^{2-}	0.15
Cl^-	1.0	CO_3^{2-}	0.03
BrO_3^-	1.0	HPO_4^{2-}	0.01
OH^-(Type II)	0.65		

(Bonner and Smith, *J. Phys. Chem.* **61**, 1957, p. 326).

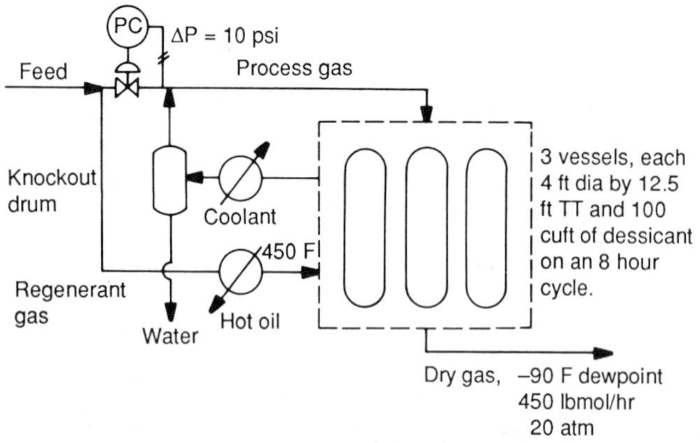

Figure 15.16. A three-vessel drying system for a cracked light hydrocarbon stream. Valve operation usually is on automatic timer control. Recycled process gas serves as regenerant.

particular kinds of ions, for instance, mercury, boron, ferrous iron, or copper in the presence of iron. Physical properties of some commercial ion exchange resins are listed in Table 15.3 together with their ion exchange capacities. The most commonly used sizes are $-20+50$ mesh (0.8–0.3 mm) and $-40+80$ mesh (0.4–0.18 mm).

Rates of ion exchange processes are affected by diffusional resistances of ions into and out of the solid particles as well as resistance to external surface diffusion. The particles are not really solid since their volume expands by 50% or more by imbibition of water. For monovalent exchanges in strongly ionized resins, half times with intraparticle diffusion controlling are measured in seconds or minutes. For film diffusion, half times range from a few minutes with $0.1N$ solutions up to several hours with $0.001N$ solutions. Film diffusion rates also vary inversely with particle diameter. A rough rule is that film diffusion is the controlling mechanism when concentrations are below 0.1–$1.0N$, which is the situation in many commercial instances. Then the design methods can be same as for conventional adsorbers.

Ion exchange materials have equilibrium exchange capacities of about 5 meq/g or 2.27 g eq/lb. The percentage of equilibrium exchange that can be achieved practically depends on contact time, the concentration of the solution, and the selectivity or equilibrium constant of the particular system. The latter factor is discussed in Section 15.2 with a numerical example.

Commercial columns range up to 6 m dia and bed heights from 1 to 6 m, most commonly 1–3 m. Freeboard of 50–100% is provided to accommodate bed expansion when regenerant flow is upward. The liquid must be distributed and withdrawn uniformly over the cross section. Perforated spiders like those of Figure 15.19 are suitable. The usual support for the bed of resin is a bed of gravel or layers of ceramic balls of graded sizes as in Figure 17.27. Balls sometimes are placed on top of the bed to aid in distribution or to prevent disturbance of the top level. Since the specific volume of the material can change 50% or more as a result of water absorption and ion–ion exchange, the distributor must be located well above the initial charge level of fresh resin.

Liquid flow rates may range from 1 to 12 gpm/sqft, commonly 6–8 gpm/sqft. When the concentration of the exchange ion is less than 50 meq/L, flow rates are in the range of 15–80 bed volumes (BV)/hr. For demineralizing water with low mineral content, rates as high as 400 BV/hr are used. Regenerant flow rates are kept low, in the range of 0.5–5.0 BV/hr, in order to allow attainment of equilibrium with minimum amounts of solution.

The ranges of possible operating conditions that have been stated are very broad, and averages cannot be depended upon. If the proposed process is similar to known commercial technology, a new design can be made with confidence. Otherwise laboratory work must be performed. Experts claim that tests on columns 2.5 cm dia and 1 m bed depth can be scaled up safely to commercial diameters. The laboratory work preferably is done with the same bed depth as in the commercial unit, but since the active exchange zone occupies only a small part of a normal column height, the exchange capacity will be roughly proportional to the bed height, and tests with columns 1 m high can be dependably scaled up. The laboratory work will establish process flow rates, regenerant quantities and flow rates, rinsing operations, and even deterioration of performance with repeated cycles.

Operating cycles for liquid contacting processes such as ion exchange are somewhat more complex than those for gas adsorption. They consist of these steps:

1. Process stream flow for a proper period.
2. A rinse for recovering possibly valuable occluded process solution.
3. A backwash to remove accumulated foreign solids from the top of the bed and possibly to reclassify the particle size distribution.
4. The flow of regenerant for a proper period.
5. Rinse to remove occluded regenerant.

As complex a cyclic process as this may demand cycle times of more than a few hours. Very high ion concentrations or high volumetric rates may require batteries of vessels and automatic switching of the several streams, or continuously operating equipment. Several continuous ion exchange plants are being operated successfully. The equipment of Figure 15.20 employs pulsed transfer of solid between exchange and regenerant zones as often as every 4 min to every 20 or 30 min. Attrition of the resin may require replacement of as much as 30% of the resin each year in water conditioning applications.

Fluidized bed units such as the multistage unit of Figure 15.20 suffer from some loss of efficiency because the intense mixing eliminates axial concentration gradients. They do have the merit, however, of not being bothered by the presence of foreign solid particles.

The economic break between fixed bed and continuous operation has been estimated as ion concentrations of $0.5N$, or flow rates above 300 gpm, or when three or more parallel beds are required to maintain continuous operation. The original application of continuous ion exchange was to treatment of radioactive wastes, but some installations of ordinary water treating have been made.

Resin requirements for two extremes of ion concentration are analyzed in Example 15.3. The high concentration stream clearly is a candidate for continuous ion exchange.

ELECTRODIALYSIS

In this process, dissolved electrolytes are removed by application of electromotive force across a battery of semipermeable membranes constructed from cation and anion exchange resins. The cation membrane passes only cations and the anion membrane only anions. The two kinds of membranes are stacked alternately and separated about 1 mm by sheets of plastic mesh that are still provided with flow passages. When the membranes and spacers are compressed together, holes in the corners form appropriate conduits for inflow and outflow. Membranes are 0.15–0.6 mm thick. A commercial stack may contain several hundred compartments or pairs of membranes in parallel. A schematic of a stack assembly is

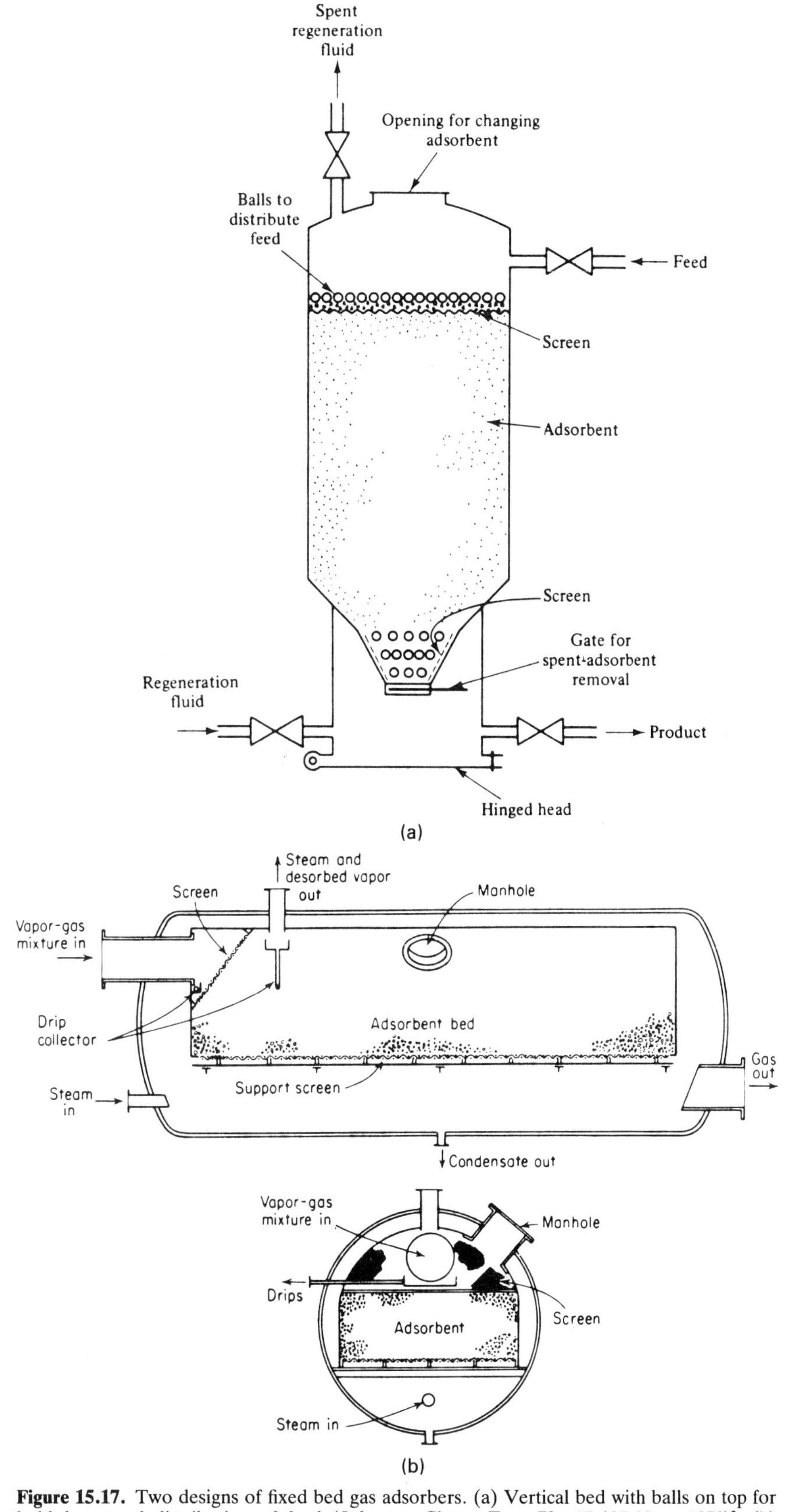

Figure 15.17. Two designs of fixed bed gas adsorbers. (a) Vertical bed with balls on top for hold-down and distribution of feed (*Johnson,* Chem. Eng. **79,** *87* (*27 Nov. 1972*)]. (b) Horizontal fixed bed for low pressure drop operation [*Treybal,* Mass Transfer Operations, *McGraw-Hill, New York, 1980*; *Logan, U.S. Pat. 2,180,712* (*1939*)].

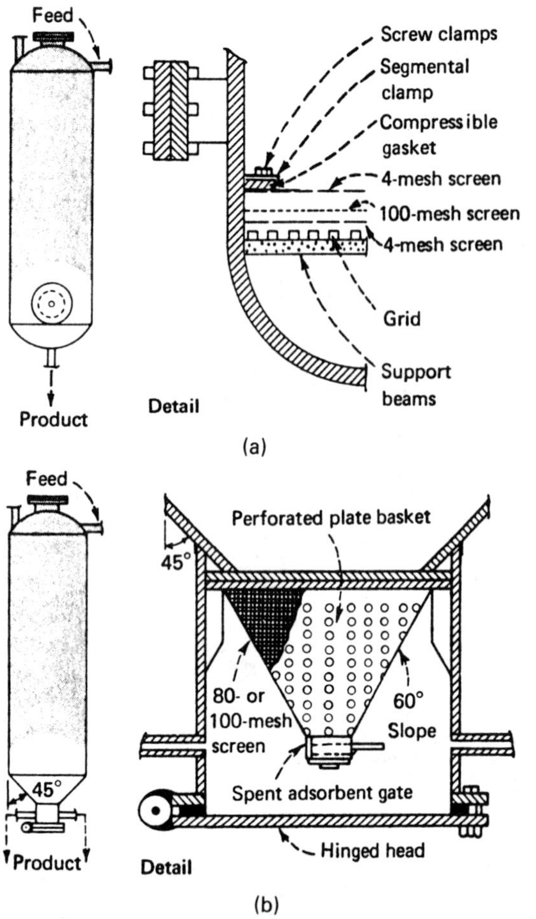

Figure 15.18. Two types of supports for adsorbent beds [*Johnston,* Chem. Eng., (*27 Nov. 1972*)]. See also Figures 17.23 and 17.24. (a) Common type of flat screen support. (b) Conical-type of support suited to frequent removal of adsorbent.

in Figure 15.21. Properties of commercially produced membranes are in Table 15.5 and performance data are in Table 15.6.

Membranes may be manufactured by mixing powdered ion exchange resin with a solution of binder polymer and pouring the heated mixture under pressure onto a plastic mesh or cloth. The concentration of the ion exchanger is normally 50–70%. They are chiefly copolymers of styrene and divinylbenzene, sulfonated with sulfuric acid for introduction of the cation exchange group.

Standard cell sizes are up to 30 by 45 in. In an individual stack the compartments are in parallel, but several stacks in series are employed to achieve a high degree of ion exchange. The ion exchange membrane is not depleted and does not need regeneration. The mechanism is that an entering cation under the influence of an emf replaces an H^+ ion from the resin and H^+ from solution on the opposite face of membrane replaces the migrating cation.

Table 15.6 shows that pressures drops may be as high as 900 psi. Flow rates in a single stage are about 1 gal/(hr)(sqft of available membrane surface). The process is distinguished by very low power requirements: the desalination of sea water, for instance, consumes 11–12 kWh/1000 gal. One stage effects a reduction of about 50% in salt content, so several stages in series are used for high performance. A flow sketch of a three-stage electrodialysis plant is in Figure 15.21(c).

Like many other specialities, electrodialysis plants are purchased as complete packages from a few available suppliers. Membrane replacement is about 10% per year. Even with prefiltering the feed, cleaning of membranes may be required at intervals of a few months. The comparative economics of electrodialysis for desalting brackish waters is discussed by Belfort (1984): for lower salinities, electrodialysis and reverse osmosis are competitive, but for higher ones electrodialysis is inferior. Electrodialysis has a number of important unique applications, for removal of high contents of minerals from foods and pharmaceuticals, for recovery of radioactive and other substances from dilute solutions, in electro-oxidation reduction processes and others.

15.6. PRODUCTION SCALE CHROMATOGRAPHY

When a mixture of two substances is charged to a chromatographic column, one of them may be held more strongly than the other. Elution with an inert fluid will remove the more lightly held substance first, then the other. Separations even between very similar substances can be very sharp. Figure 15.22(a) is an example of a chromatogram. Only fluid–solid chromatography is an adsorptive process, but gas–liquid and liquid–liquid are used more frequently since liquids with suitable absorption properties are easier to find than solid adsorbents. The active sorbent is a high-boiling solvent deposited on a finely divided inert solid carrier. The process is one of absorption, but the behavior is much like that of adsorption. The principal application is to chemical analysis. Relative retention times on various sorbents are key data which are extensively tabulated, for instance in Meites (*Handbook of Analytical Chemistry,* McGraw-Hill, New York, 1963).

Chromatographic separations are necessarily intermittent with alternate injections and elutions, although a measure of continuity can be achieved with an assembly of several units, or with suitably sized surge tanks. A process flowsketch appears in Figure 15.22(b). Information on production scale chromatography is provided by Conder (1973). Only separations difficult to achieve by other means are economical with chromatography.

Individual drums are provided for each product fraction. A detector monitors the separation and provides signals for controlling the injection and collection sequence. The operation of partial condensers for the dilute eluted streams presents challenges because of aerosol formations. When a valuable carrier such as nitrogen is used, it must be cleaned up and recycled.

A 1968 estimate of the cost breakdown for a plant with a column 4 ft dia by 15 ft high and a throughput of 200–920 tons/yr has been converted to a percentage basis in Table 15.7 because of its age. The costs are said to not vary greatly with throughput or the nature of the separation, although this analysis has been made specifically for the separation of α- and β-pinenes. The temperature was 165°C and the solvent was Carbowax 20M. The design was based on data in a 4 in. dia column which had a capacity of 200–1500 mL/hr.

Some of the materials for which chromatographic separation should be considered are essential oils, terpenoids, steroids, alkaloids, pharmaceuticals, metal chelates, isotopes, and close-boiling isomers. For easy separations, vacuum distillation, liquid–liquid extraction, and fractional crystallization are less expensive.

15.7. EQUIPMENT AND PROCESSES

Adsorbents are made in pellet form by extrusion or pressing or in granular form by crushing and classification of larger masses or in spherical or globular form by precipitation in an inert gaseous or liquid medium. Typical processes for some adsorbent preparations are represented in Figure 15.1. The BASF process of Figure 15.1(c)

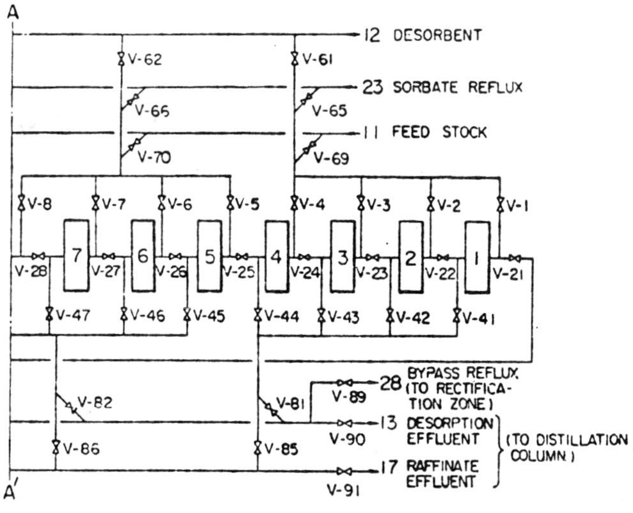

(a)

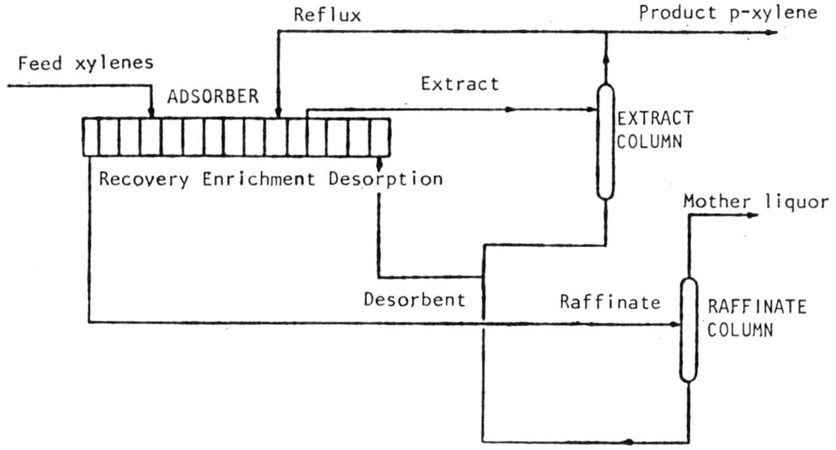

(b)

Figure 15.19. The Toray simulated continuous adsorption process. (a) Showing the main valving for a seven-chamber adsorption system [*Otani et al., U.S. Pat. 3,761,533, (25 Sep. 1973)*]. (b) Flowsketch for recovery of paraxylene by continuous adsorption [*Otani et al.,* Chem. Economy Eng. Rev. **3**(*6*), *56–59* (*1971*)].

for silica gel employs a special mixing nozzle in which water glass is mixed intermittently with a stream of sulfuric acid to form an unstable sol that is sprayed directly into air where the globules solidify immediately. Mixing time in the nozzle is 0.1–1.0 sec. The sodium sulfate is washed out in a tower to which the particles fall. The process of initial activation or reactivation is shown for bone char. In general a heating process, sometimes combined with oxidation, is required to drive gases out of the solid mass and thus to make it porous.

GAS ADSORPTION

The usual equipment for gas adsorption is a number of vessels containing fixed beds of the adsorbent, at least two vessels for achieving overall continuous operation. Figure 15.17 shows suitable vertical and horizontal vessels. The vertical ones are less likely to form channels and usually are favored. Bed depths as high as 45 ft are in use. Horizontal vessels are preferred when pressure drops must be kept low, as in recovery of solvents from air in printing or paint establishments. Modes of support of granular beds are shown in Figures 15.18, 15.24, and 17.27.

A three-bed adsorption unit is illustrated in Figure 15.16. It is used to dry the feed to a distillation column with a top temperature of −70°F; thus a water dewpoint of −90°F is required. One of the vessels always is on regeneration and cooling down, and the other two in series on adsorption, with the more recently reactivated one downstream. A bleed off the process stream is diverted to use as regenerant. After the gas leaves the vessel being regenerated, the water is condensed out by cooling and the gas returns to the process downstream of a control valve that maintains a 10 psi differential.

Normally adsorption is conducted at or as near ambient temperature as possible and regeneration is at 350–500°F. A new process developed by Union Carbide Corp, the AHR process (adsorption heat recovery), allows the temperature to rise to 200–400°F as a result of the heat of adsorption and effects the desorption at comparable temperatures with a stream of inert gas

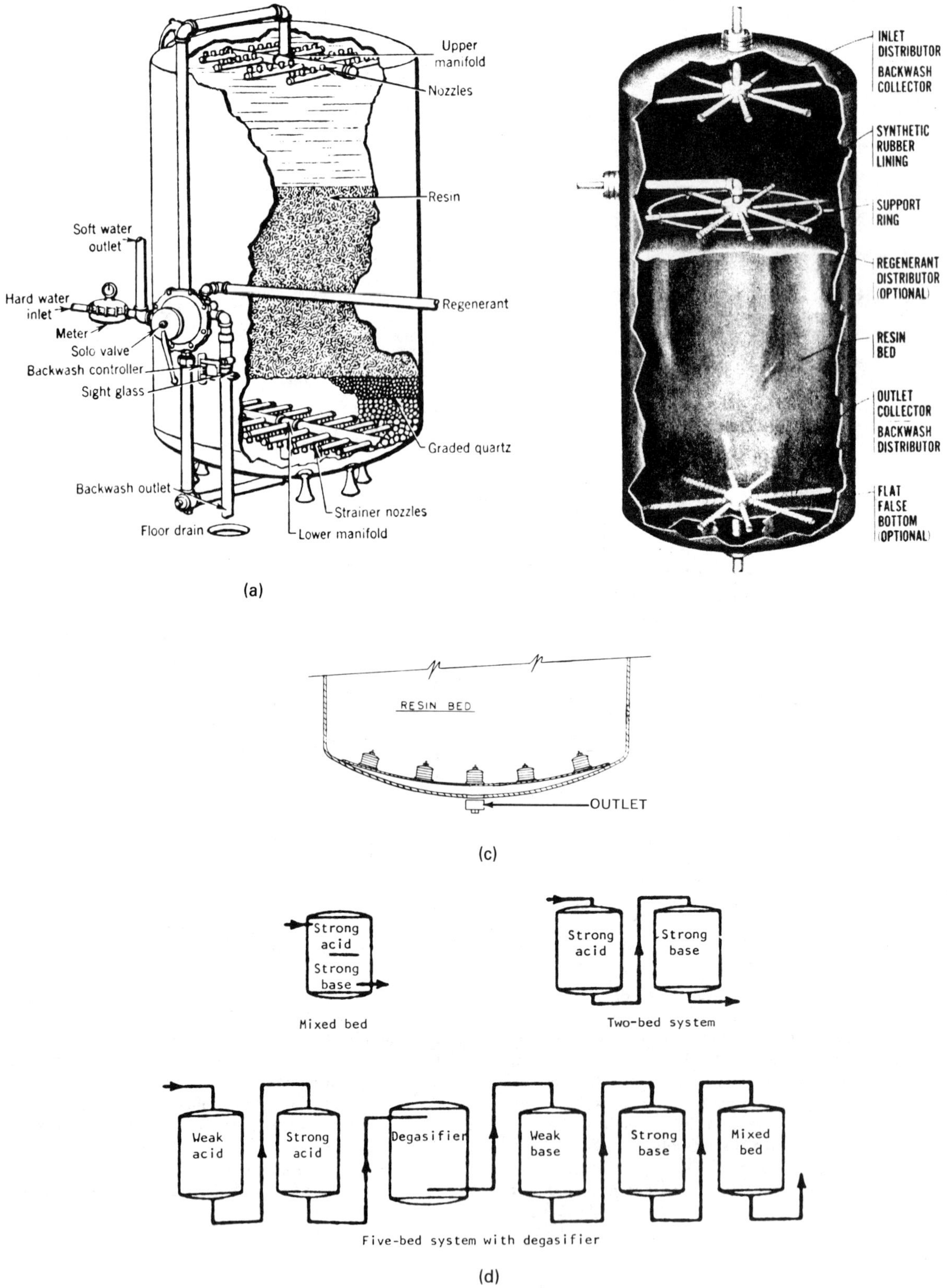

Figure 15.20. Fixed bed ion exchange vessels and arrangements. (a) Typical design of a water softener, showing bed support, distributor, and effluent collector. (b) Vessel with radial-type distributors and collectors (*Illinois Water Treatment Co.*). (c) A double-dish underdrain system (*Permutit Co.*). (d) Some arrangements of vessels for cation and anion exchange.

Example 15.3
Size of an Ion Exchanger for Hard Water
A hard water contains 120 ppm of $CaCO_3$, 90% of which is to be removed with a hydrogen exchange resin of capacity 5 meq/g. By the method of Example 15.1 it is ascertained that under these conditions 98% of H^+ ion of the resin will be replaced by the Ca^{++} at equilibrium. The minimum amount of resin will correspond to the equilibrium value. That amount will be calculated for treating 100 gpm of water on a 24 hr cycle. The mol wt of $CaCO_3 = 100.06$.

$$\text{resin capacity} = 0.98(0.005)(100.06) = 0.490 \text{ lb } CaCO_3/\text{lb resin},$$
$$CaCO_3 \text{ removed} = 0.9(8.34)(100)(1440)(120)(10^{-6}) = 129.7 \text{ lb}/24 \text{ hr},$$
$$\text{resin needed} = 129.7/0.49 = 264.7 \text{ lb, or } 4.71 \text{ cuft of resin with sp gr} = 0.9.$$

For comparison, the amount of resin needed to remove the Na^+ from a 3.5% solution of NaCl at the rate of 100 gpm in 24 hr will be found:

$$\text{resin capacity} = 5 \text{ meq/g} = 0.005 \text{ lb mol/lb},$$
$$Na^+ \text{ removed} = 0.035(8.34)(100)(1440)/58.5 = 718.5 \text{ lb mol/day}.$$

Accordingly,

$$\text{resin} = 718.5/0.005 = 142{,}700 \text{ lb},$$

pointing out that a fixed bed unit on such a long cycle may not be practical for such a high concentration of ion to be exchanged.

such as nitrogen or carbon dioxide which is recycled after the water is condensed out. The process is applied to removal of as much as 20% of water from ethanol with cycle times of the order of 1 hr, instead of the common 24 hr, even for much smaller contents of water (Anon, *Chem. Eng.*, April 15, 1985, p. 17).

Continuous fluidized bed equipment has been utilized for gas adsorption, but usually attrition losses of comparatively expensive adsorbents have been prohibitive and the loss of efficiency because of axial mixing has been a serious handicap. Drying equipment such as those of Figure 9.13 presumably can be operated in reverse to recover valuable substances from a vapor phase, and the forward mode applied for regeneration in associated equipment. Other possibly suitable fluidized bed configurations are those of the reactors of Figures 17.32(a), (c), and (d).

Moving bed gas adsorbers also have been proposed and used, patterned after moving bed gas oil crackers. In the Hypersorber of Figure 15.23, flows of gas and solids are countercurrent in a single vessel. After saturation, the solid is stripped with steam and removed at the bottom of the tower, and gas is lifted to cooling and adsorption zones. The control mechanism for solids flow and typical performance for ethylene recovery from cracked gases also are shown with the figure. Partly because of attrition losses and the advent of competitive processes for ethylene recovery, the Hypersorber was abandoned after a few years. The simpler Nofsinger moving bed adsorber of Figure 15.24 also has not proved commercially attractive.

LIQUID PHASE ADSORPTION

A major application of liquid phase adsorption is to the removal of relatively small amounts of impurities or color bodies in water treating, sugar refining, and other processes. Both batch and

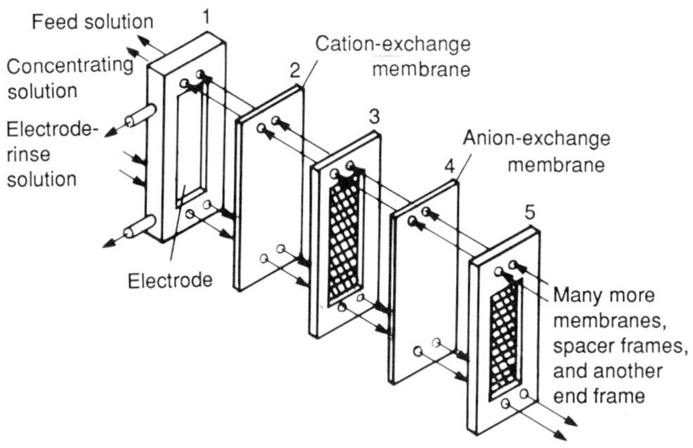

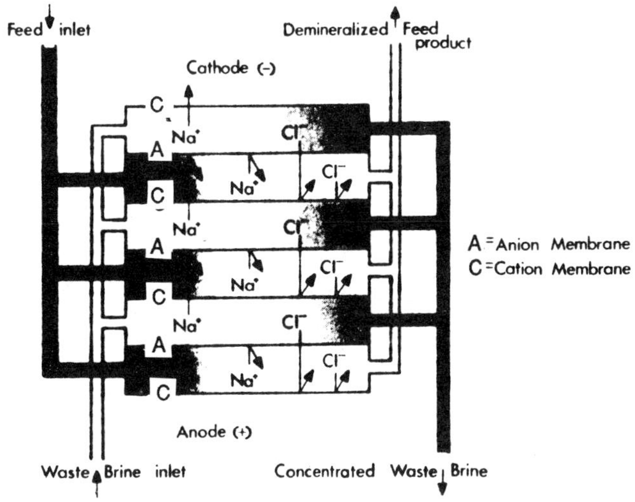

Figure 15.21. Electrodialysis equipment and processes. (a) View of the components of an electrodialysis stack (*Lacey, 1978*). (b) Flow pattern through an electrodialyzer for removal of NaCl from water (*Ionics Inc.*). (c) Electroreduction with the use of an ion exchange diaphragm. (d) Flowsketch of a three-stage electrodialysis for treatment of brackish water (*Rogers, in Belfort, 1984*).

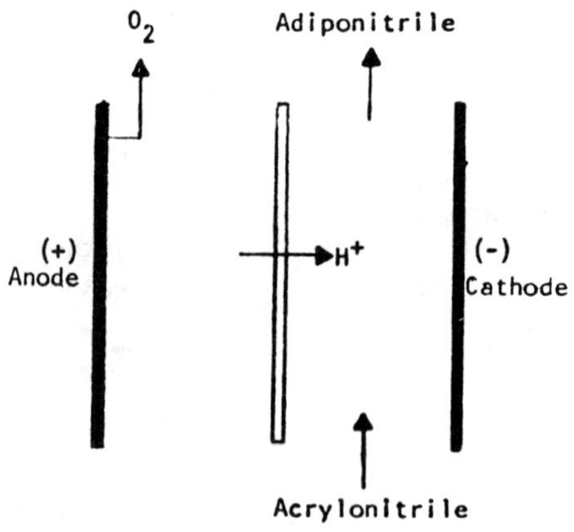

(c)

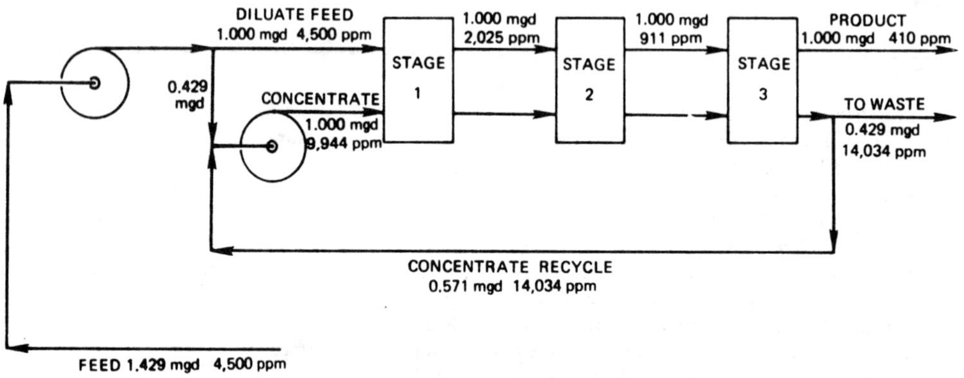

(d)

Figure 15.21—*(continued)*

TABLE 15.5. Properties of Membranes for Electrodialysis

Manufacturer	Name of membranes	Membrane	Thickness (mm)	Capacity (meq/gm)	Electrical resistance (Ω cm² in 0.1 N NaCl)	Reinforcement
Ionac Chemical Co. New Jersey	Ionac	MC-3142	0.15	1.06	9.1	Yes
		MC-3470	0.35	1.05	10.5	Yes
		MA-3148	0.17	0.93	10.1	Yes
		MA-3475	0.40	1.13	23	Yes
		IM-12	0.13	—	4	Yes
American Machine and Foundry Connecticut	A.M.F.	C-60	0.30	1.5	6	No
		A-60	0.30	1.6	5	No
Ionics Inc. Massachusetts	Nepton	CR61 AZL 183	0.60	2.7	9	Yes
		AR 111 BZL 183	0.60	1.8	14	Yes
Asahi Glass Co. Ltd. Tokyo, Japan	Selemion	CMV	0.15	1.4	6.1	Yes
		AMV	0.14		4.0	Yes
Tokuyama Soda Ltd. Tokyo, Japan	Neosepta	CL 25 T	0.16	1.8–2.0	3.5	Yes
		AV 4 T	0.15	1.5–2.0	4.0	Yes
Asahi Chemical Industry Co. Ltd. Tokyo, Japan	A.C.I. or Acipex	DK 1	0.23	2.6	6.5	Yes
		DA 1	0.21	1.5	4.5	Yes
Ben-Gurion University of the Negev, Research & Development Authority Beersheva, Israel	Neginst	NEGINST-HD	0.35	0.8	12	Yes
		NEGINST-HD	0.35	0.8	10	Yes
		NEGINST-HC	0.2	1.6	6	No
		NEGINST-HC	0.2	1.7	8	No

(Belfort, 1984).

TABLE 15.6. Performance of Electrodialysis Equipment on Treatment of 3000 ppm Brackish Water

	Single stack, MK II, four stages	Single stack, MK III, three stages	Single stack, MK III, one stage	Three stacks in series, MK III
Typical hydraulic flow rate				
U.S. gal/24-h day	16,700	55,600	166,700	166,700
U.S. gal/min	11.6	38.6	116	116
Pressure drop at typical flow, lb/in^2	47	44	14	42
Number of membranes	540	900	900	2,700
Size of membranes, in × in	18 × 20	18 × 40	18 × 40	18 × 40
Total area of membranes, ft^2	1,350	4,500	4,500	13,500
% total area available for transfer	62	64	64	64
Approximate weight, lb	1,300	2,800	2,800	8,400
Approximate overall height, including legs	4'6"	6'10"	6'8"	6'8"
Demineralization per pass (25°C, high-Cl water, typical flow), %	88.5	88.3	52	90.0
Current required for 3000-ppm feed, A	Stages 1 and 2: 19 Stages 3 and 4: 8	Stages 1 and 2: 36 Stage 3: 12	46	Stage 1: 46 Stage 2: 24 Stage 3: 12
Voltage required for 3000-ppm feed†	Stages 1 and 2: 180 Stages 3 and 4: 150	Stages 1 and 2: 350 Stage 3: 150	640	Stage 1: 640 Stage 2: 500 Stage 3: 420
Direct-current kW/stack for 3000-ppm feed†	4.6	14.1	29	Stage 1: 29 Stage 2: 12 Stage 3: 5
Direct-current kWh/1000 gal product‡ for 3000-ppm feed†	7.4	6.8	4.7	7.4

*Ionics, Incorporated, Watertown, Mass, 1979. These units use the EDR process, in which polarity and fluid flow are periodically reversed. In general, addition of acid and antiprecipitant to the feed is not necessary in this process.

†For typical brackish water containing a high proportion of sodium chloride.

‡Approximately 10% of flow wasted during reversal.

(Spiegler, 1984).

continuous equipment are illustrated in Figure 15.25. The batch process consists of slurrying the liquid with powdered adsorbent and then separating the two phases by filtration. The saturated adsorbent—carbon from water treating or fullers earth from oil treating—is regenerated by ignition as in the block diagram of Figure 15.1(b), or sometimes by treatment with suitable reactive solvents such as sodium hydroxide for adsorbed phenol from water. In the semicontinuous process of Figure 15.25(b), pulses of adsorbent are withdrawn periodically from the bottom and fresh material is charged in at the top. The pulses are 2–10% of the volume of the bed. Some data of adsorbent treating of water were given in Section 15.4. Attrition losses in moving beds for liquid treating are less than for gas treating. In the similar process of ion exchange of Figure 15.25(a), ion exchange losses of 30% per year are mentioned.

The successful simulation of continuous moving bed adsorption process developed by UOP (Universal Oil Products) is illustrated in Figure 15.26. For the process being simulated, part (a) of the figure

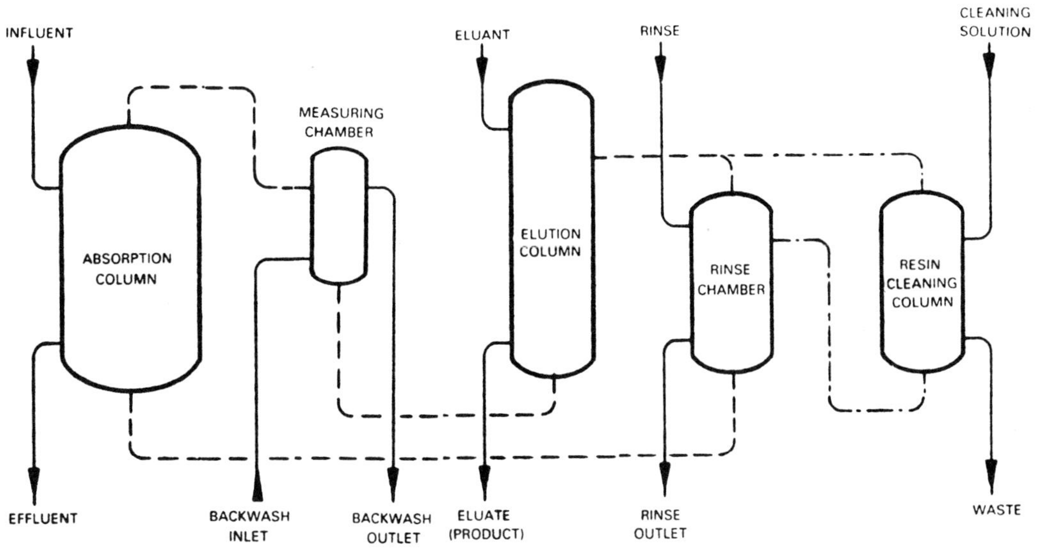

Figure 15.22. A process for recovering uranium from mine waters. The absorption column is 2.16 m dia, water flow rate is 28.5 m/hr, resin transfer off the top of the absorption column is 87 L every 3 hr, inlet concentration 3–6 mg U/L, outlet 0.002–0.009 mg U/L (*Himsley and Bennett, in Naden and Streat, 1984, pp. 144–168; U.S. Pat. 4,018,677*).

TABLE 15.7. Economic Data for a Chromatographic Process with Throughput of 400–920 tons/yr, with Column 4 ft dia by 15 ft high[a]

Equipment Cost	Percent	Annual Operating Cost	Percent
Feed preparation and injection	9.4	Maintenance and taxes	19.9
Column	13.1	Operating labor	13.7
Detection and control	2.6	Utilities and supplies	5.7
Fraction collection and heat exchange	18.7	Packing replacement	40.8
Carrier recycle	11.2	Depreciation (10 year)	19.9
Process piping and building	16.5		100
Engineering and construction	28.5		
	100		

[a] Data are given only on percentage bases because of their age.
(Abcor Inc., 1968).

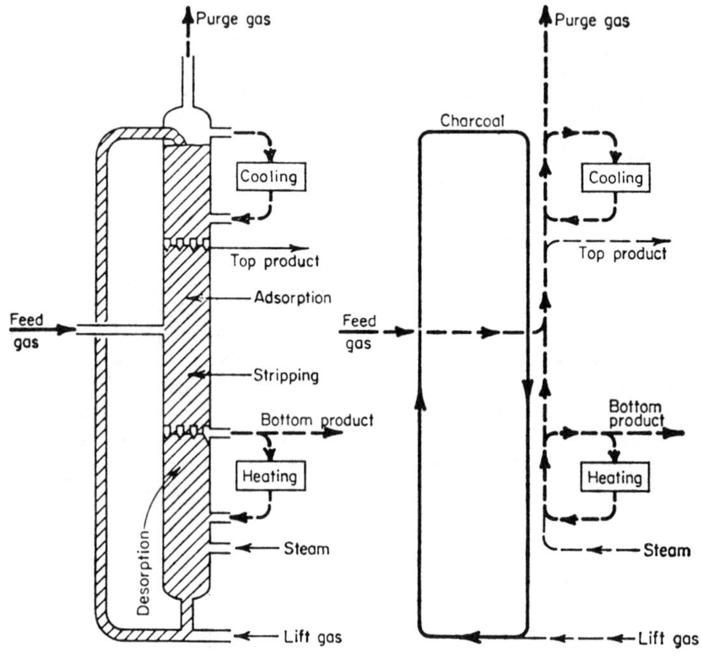

(a)

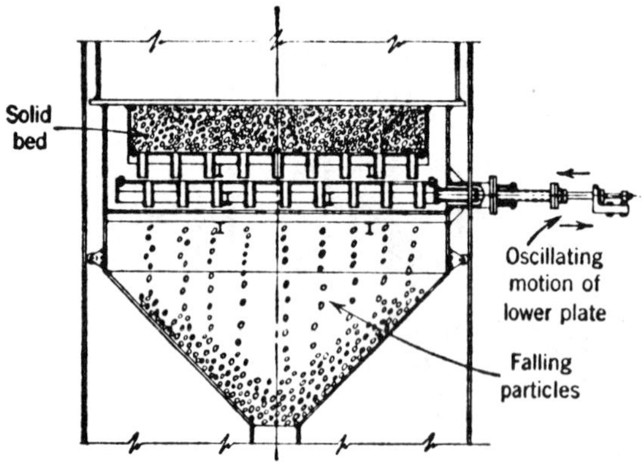

(b)

Component	Compositions, vol %			
	Feed gas	Bottoms	Overhead	Purge gas
Hydrogen	39.8	.....	31.6	61.8
Nitrogen	1.7	.....	1.4	2.5
Carbon monoxide	0.9	.....	0.8	1.3
Oxygen	0.1	0.1	0.1	0.2
Methane	51.3	.....	66.1	33.7
Carbon dioxide	0.2	2.9	.....	0.1
Acetylene	0.2	3.6		
Ethylene	5.8	92.7	.....	0.4
Ethane	Tr.	0.7		
Total	100.0	100.0	100.0	100.0

(c)

Figure 15.23. Hypersorber continuous moving bed gas phase adsorption system (*See Mantell,* Adsorption, *McGraw-Hill, New York, 1951*). (a) Schematic pattern of flows of gas and solid adsorbent (*Hengstebeck,* Petroleum Processing, *McGraw-Hill, New York, 1959*). (b) Solids flow rate control mechanism. (c) Typical separation performance.

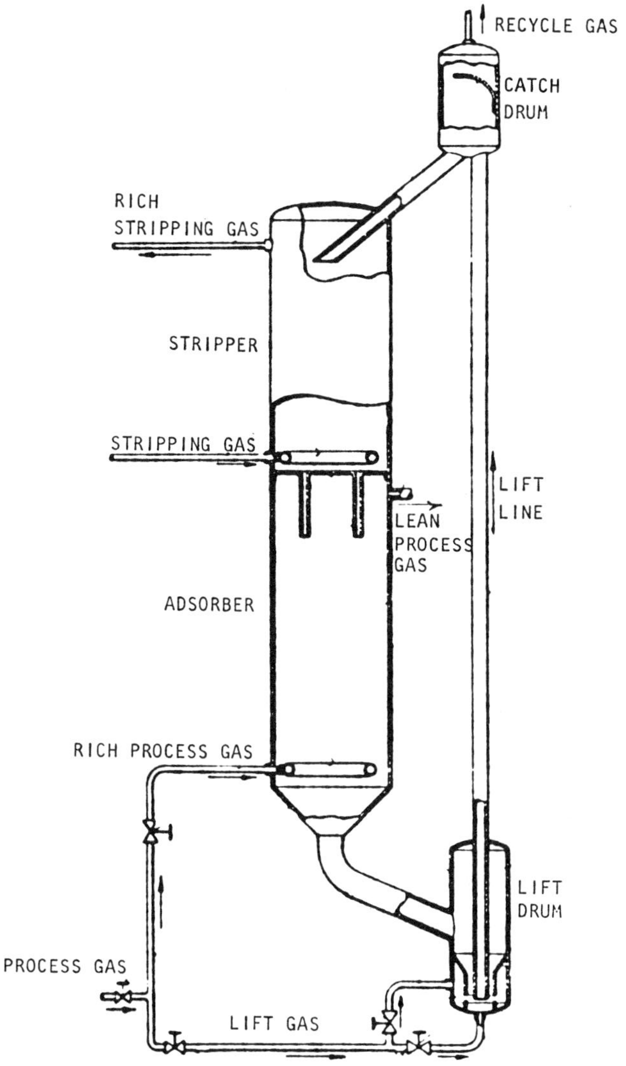

Figure 15.24. Nofsinger continuous moving bed adsorber [*Spangler and Price, U.S. Pat. 3,442,066 (6 May 1969)*].

shows flows of adsorbent and fluids and the composition profiles along the tower. The simulated process employs 12 fixed beds in a single vessel, in which input and output streams are individually controlled. The points of entry and withdrawal of the four external streams—feed, extract, raffinate, adsorbent—are controlled with a single special rotary valve. Periodically each stream is switched to the adjacent bed so that the four liquid access positions are always maintained the same distance apart. Satisfactory operation is assured by uniform feeds and withdrawals and flushing of lines between their uses for regeneration and other purposes. The internal constructions of the tower, such as the mechanism of feed and withdrawal at individual beds, are not revealed in the literature. As of 1984, some 60 large capacity installations for various hydrocarbon isomer separations with molecular sieves have been made. The largest column mentioned is 22 ft dia. The distribution across the cross section has been worked out so that scale-up from 3 in. to commercial size is reliable. The process is described briefly in articles by Broughton (1978; 1984) and in several patents listed in the first of these articles.

A variant of this process developed by Toray industries is illustrated in Figure 15.19. The beds are individual rather than in a single shell and utilize individual on-off valves. One application, to the separation of aromatics, is mentioned in the literature [Otani, *Chem. Eng.* **80**(9), 106 (17 Sep. 1973)].

ION EXCHANGE

Because of the large volumes of dilute electrolytes that sometimes need to be treated, continuous processing with ion exchange materials is more common than liquid phase adsorption, although fixed bed processes still are predominant. Typical arrangements of fixed beds appear in Figure 15.20. Any particular ion exchange resin is capable of exchanging only cations or anions. The two kinds of resins may be mixed and incorporated in the same vessel or they may be used separately in their own vessels. Cation exchange resins may be strongly or weakly acid, and anion exchange resins, strongly or weakly basic. The choice of an ion exchange system depends on the composition of the feed, the product quality required, the scale of the operation, and the economics of the process. Three of the many possible arrangements of vessels are sketched in Figure 15.20(d). Series combinations of vessels are employed when leakage is highly undesirable. The inlet to the last stage is monitored and the information is taken as a guide to transfer of the first vessel in line to regeneration.

All of the continuous processes of Figures 15.27 and 15.22 employ intermittent transfer of spent resin out of the primary vessel to regeneration facilities. Although all of the operations of exchange, rinsing, and regeneration can be performed in elegantly designed equipment such as Figure 15.27(a), greater flexibility is inherent in a multivessel plant such as Figure 15.22. This is of an operating plant for which some size and operating data are given here, and more appear in the reference article.

Performances of four fluidized bed ion exchange plants are described by Cloete (in Naden and Streat, 1984, pp. 661–667). One of the exchange columns is 4.85 m dia, has 12 stages each 1 m high, with perforated trays having holes 12 mm dia with a capacity of 640 m^3/hr of uranium mine waters.

Ion exchange resins are used widely as heterogeneous catalysts of processes that require acid or base catalysis, for example, hydration of propylene to isopropanol, reaction of isobutylene with acetonitrile, and many others. The same kind of equipment is suitable as for ion exchange, but usually regeneration is not necessary, although some degradation of the resin naturally occurs over a period of time.

ION EXCHANGE MEMBRANES AND ELECTRODIALYSIS

Only one basic design of electrodialysis equipment for demineralization appears to be in use. This is an assembly of alternate cation and anion ion exchange sheets separated by spacers in groups of several hundred clamped together between electrodes. The assembly physically resembles a plate-and-frame filter press. Figures 15.21(a) and (b) show such assemblies, and some dimensional data were stated in Section 15.5, Electrodialysis.

The concentrates of salt solutions made by electrodialysis of seawater are suited as feed to the evaporators of salt manufacturing plants with considerable savings in overall energy requirements. Other applications also are based on the concentrating effects of electrodialysis, for instance, tenfold increases of concentrations of depleted streams from nickel and copper plating plants are made routinely.

In applications such as the electroreduction of Figure 15.21(c) and electrolysis of brine to caustic and chlorine, single membranes serve as diaphragms between electrodes, permitting passage of ions but retarding diffusion and mixing of feed and product solutions.

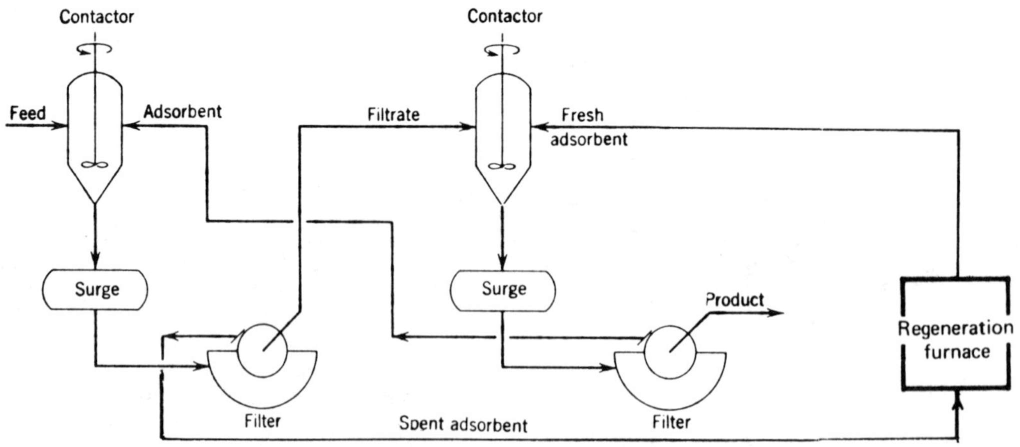

(a)

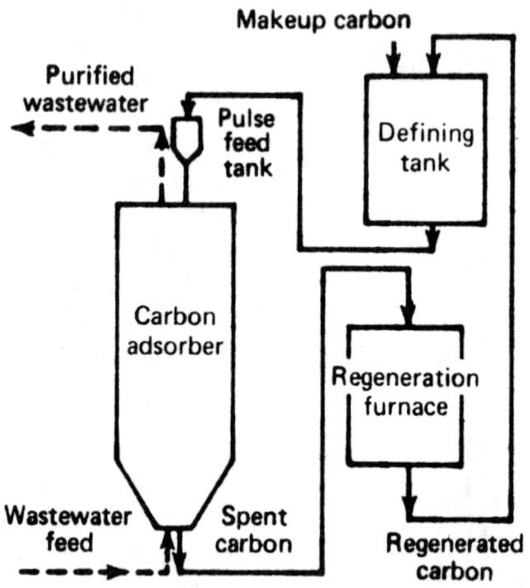

(b)

Figure 15.25. Liquid phase adsorption processes for water treated with activated carbon and petroleum treated with clay adsorbents. (a) A two-stage slurry tank and filter process. (b) Continuous pulsed bed operation, individual pulses 2–10% of bed volume as needed.

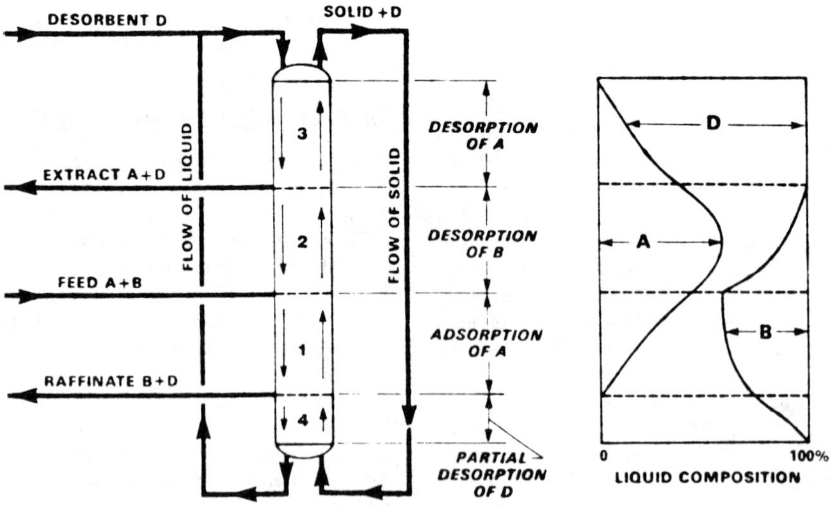

(a)

Figure 15.26. Continuous and UOP simulated continuous moving bed liquid adsorption processes [*Broughton,* Sep. Sci. Technol. **19,** *723–736 1984–1985*)]. (a) Continuous moving bed liquid adsorption process flows and composition profiles. (b) UOP Sorbex simulated moving bed adsorption process.

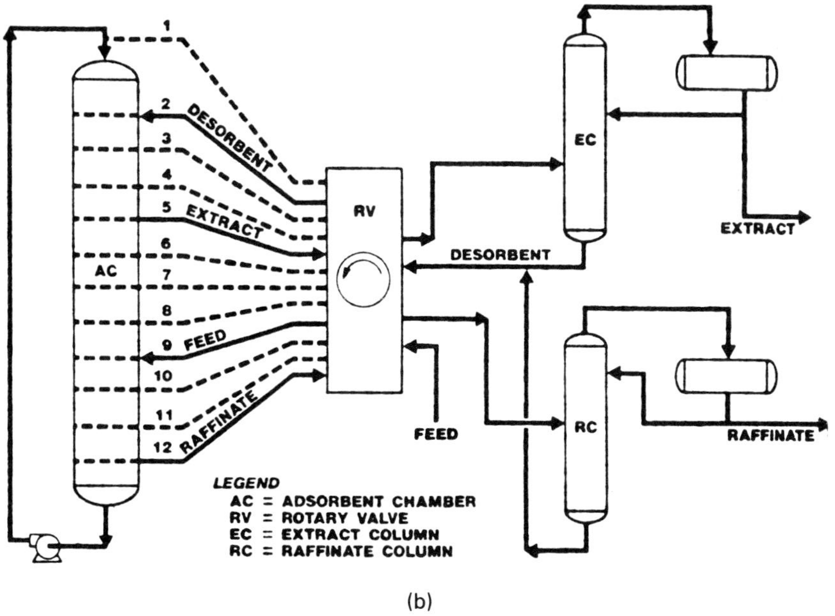

Figure 15.26—(*continued*)

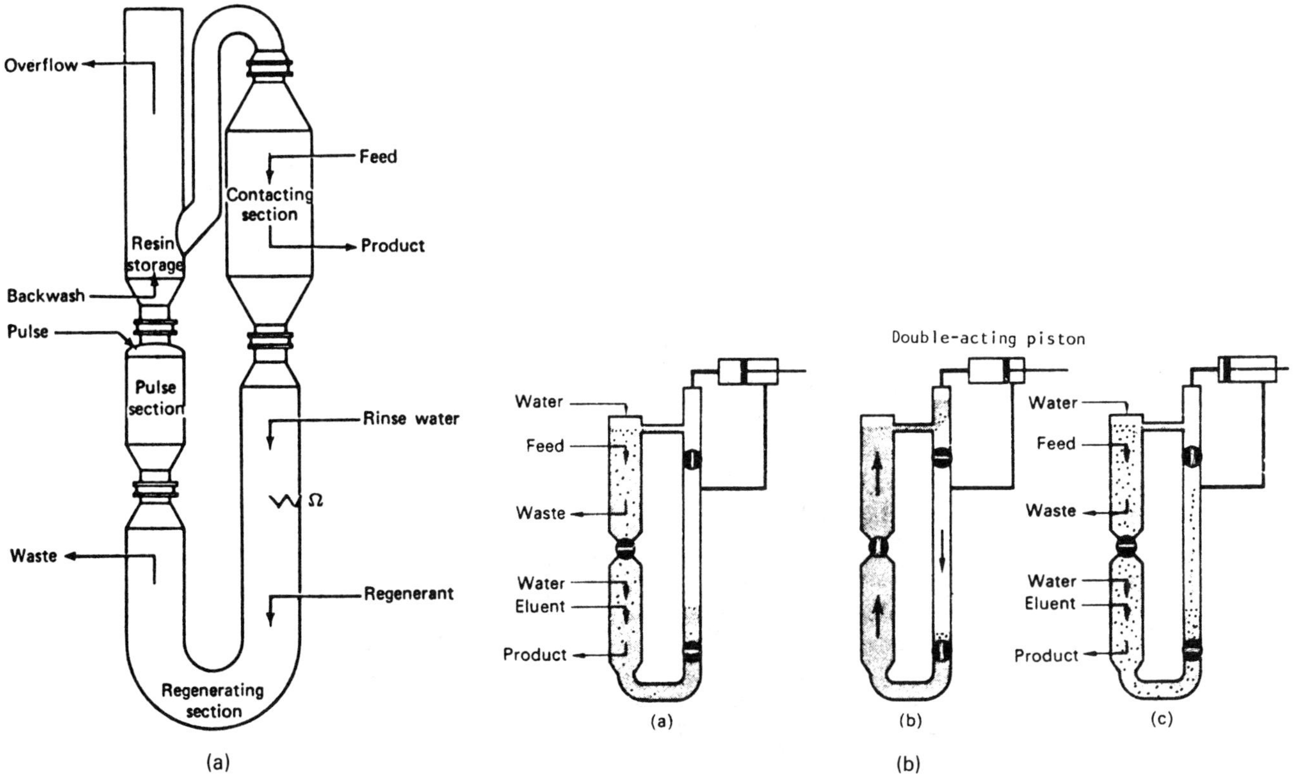

Figure 15.27. Continuous ion exchange equipment. (a) The Higgins moving bed unit; the consolidated resin bed is recirculated upwards with a hydraulic pulse. [*U.S. Pat. 3,580,842, (25 May 1942)*]. (b) Operating sequence in the Higgins contactor. The double-acting piston simultaneously sucks liquid from the top of the column and delivers it to the bottom: (a) solution pumping for several minutes; (b) resin movement for 3–5 sec; (c) solution pumping for several minutes (*Coulson and Richardson,* Chemical Engineering, *Pergamon, New York, 1979, Vol. 3, p. 520*). (c) In the Asahi system, resin is transferred between the adsorber and the regenerator at 10–60 min intervals (*Asahi Chem. Ind., U.S. Pat. 3,152,072*). (d) The Himsley multistage slurry adsorber; the resin is pumped as a slurry, one stage at a time into and from the regenerator [*Himsley, Can. Pat. 980,467, (23 Dec. 1975)*].

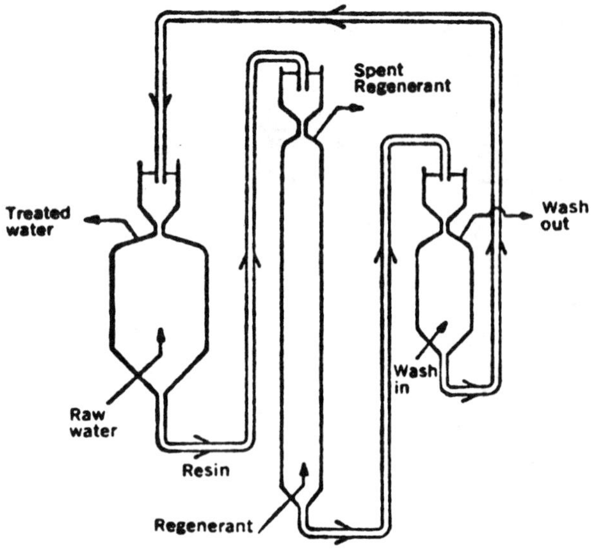

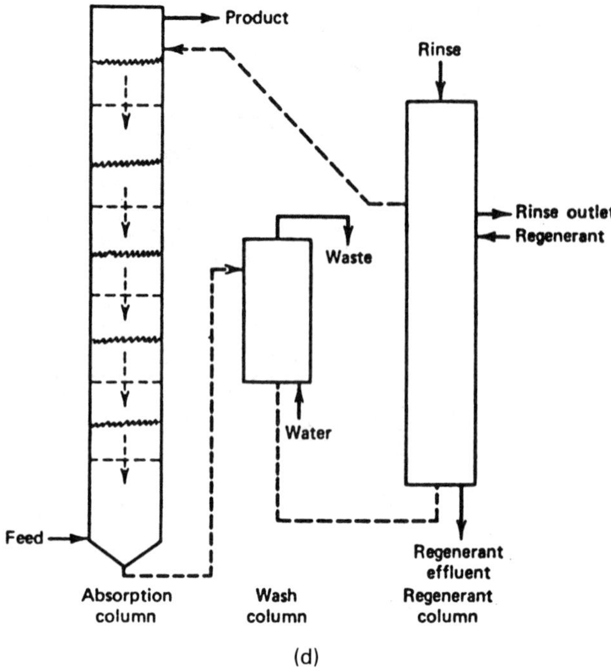

Figure 15.27—(*continued*)

CHROMATOGRAPHIC EQUIPMENT

The components of a commercial process employing chromatographic separation are represented on the flowsketch of Figure 15.28. The process is intermittent with very short cycles. The cost breakdown of a plant for the separation of α- and β-pinenes is given in Table 15.7, which is based on pilot plant work in a 4 in. dia column. That company is no longer in that business; thus the test data are not available, and the operating conditions are not known. Other data for the same separation, however, are presented in Table 15.8. The largest column considered there, 0.4 m dia, has a production rate of 4300 kg/(m^2)(day), with cycle times of about 1 min. Valentin (1981) refers to literature where columns of 1 and 2 m dia are described and mentions that diameters up to 5 m are feasible and may have throughputs of 10,000 metric tons/yr.

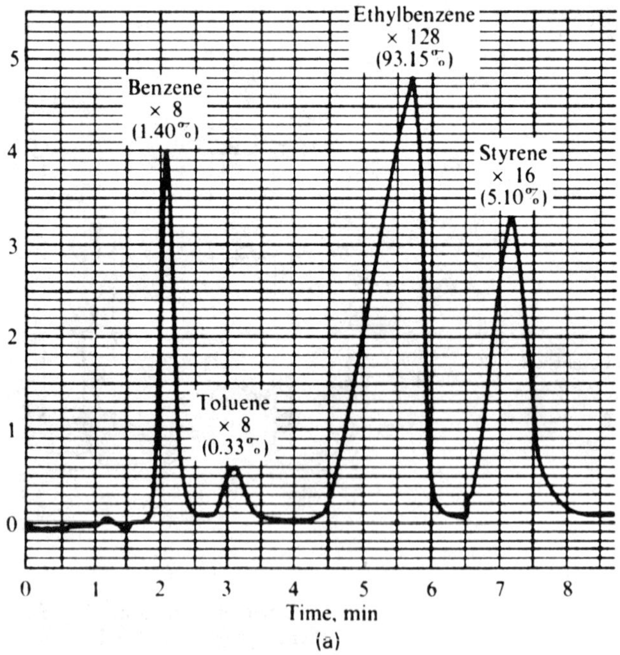

Figure 15.28. Chromatographic separations. (a) Typical chromatogram produced by gas–liquid chromatography. (b) Flowsketch of a production scale chromatographic unit [*Ryan, Timmins, and O'Donnell,* Chem. Eng. Prog. **64,** *53* (*Aug. 1968*)].

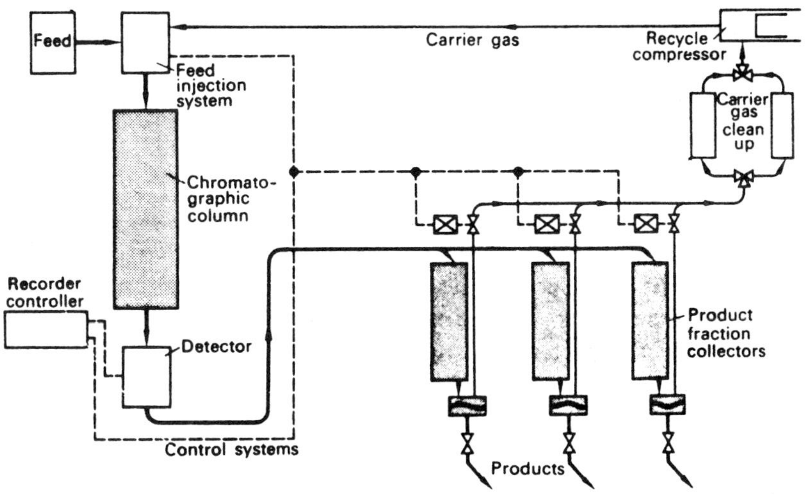

Figure 15.28—*(continued)*

TABLE 15.8. Examples of Chromatographic Separations in a Column 125 mm dia
(a) Five Separations

Case	Mixture	Charge (mol %)	Product (mol %)	Temperature (°C)/ Pressure (Torr)	Remarks
1	Pentane	99.3	99.995	225/760	extreme purification
	Isopentane	0.7	50 ppm		
2	Benzyl alcohol	99.5	99.6	150/50	selective removal of small impurities
	Benzyl aldehyde	0.1			
	Others	0.4			
3	α-Pinene	70.0	99/2	160/760	simultaneous purification of two materials
	β-Pinene	30.0	1/98		
4	Essence of cloves			220/760	recovery of a thermally sensitive natural product
	Eugenol	75	99.8		
5	Bromo-3-thiophene	91.6	99.0	150/760	a difficult separation of position isomers
	Bromo-2-thiophene	4.6	1.0		
	Others	3.8			

(Bonmati and Guiochon, Parfums, Cosmetiques, Aromes, 37–59, Sept/Oct 1976).

(b) Details of the Separation of α-Pinene and β-Pinene

Operating Conditions	Other Data		ELF–SRTI Process	
Column				
diameter (mm)	100	305	125	400
Length (m)	2.7	unknown	1.5	1.5
Baffles	yes	yes	no	no
Programmer	peak deflection		time-based	
Carrier gas	He	He	He	H_2
Velocity (cm/sec)	9.2	—	9.5	9
Temperature (°C)	160	160	160	160
Cycle (sec)	80	—	80	65
Purity	98.5–98.6		99.1–97.8	
Productivity (kg/day)	19	160	40	540

REFERENCES

Adsorption

1. R.A. Anderson, Adsorption (general), *Encycl. Chem. Process. Des.* **2,** 174–213 (1977).
2. G.B. Broughton, Adsorptive separation (liquids), *Encycl. Chem. Technol.* **1,** 563–581 (1978).
3. D.B. Broughton, Production scale adsorption of liquid mixtures by simulated moving bed technology, *Sep. Sci. Technol.* **19,** 723–736 (1984–1985).
4. C.H. Chi and W.P. Cummings, Adsorptive separation (gases), *Encycl. Chem. Technol.* **1,** 544–563 (1978).
5. J.R. Fair, Sorption processes, *Chem. Eng.,* 90–110 (14 Jul. 1969).
6. A.L. Hines and R.N. Maddox, *Mass Transfer Fundamentals and Applications,* Prentice-Hall, Englewood Cliffs, NJ, 1985.
7. J.L. Kovach, Gas adsorption, in *Handbook of Separation Processes for Chemical Engineers,* Schweitzer, (Ed.), McGraw-Hill, New York, 1979, pp. 3.3–3.47.
8. M.N.Y. Lee, Novel separations with molecular sieve adsorption, in *Recent Developments in Separation Science II,* CRC Press, Boca Raton, FL, 1972, pp. 75–110.
9. C.M. Lukchis, Adsorption systems, *Chem. Eng.,* (11 June 1973); (9 July 1973); (6 Aug. 1973).
10. D.M. Ruthven, *Principles of Adsorption and Adsorption Processes,* Wiley, New York, 1984.
11. T. Vermeulen et al., Adsorption design, *Encycl. Chem. Process. Des.* **2,** 162–174 (1977); Adsorption separation, *Encycl. Chem. Technol.* **1,** 531–544 (1978); Adsorption and ion exchange, in *Chemical Engineer's Handbook,* McGraw-Hill, New York, 1984, Sec. 16.

Ion Exchange

1. R.E. Anderson, Ion exchange separation, in *Handbook of Separation Techniques for Chemical Engineers,* (Schweitzer, Ed.), pp. 1.359–1.414.
2. D. Naden and M. Streat, Eds., *Ion Exchange Technology,* Horwood, Chichester, England, 1984.
3. Vermeulen et al., Ref. A.11.
4. R.M. Wheaton and E.J. Lefevre, Ion exchange, *Encycl. Chem. Technol.* **13,** 678–705 (1981).

Electrodialysis and Ion Exchange Membranes

1. G. Belfort, Ed., *Synthetic Membrane Processes,* Academic, New York, 1984.
2. D.S. Flett, Ed., *Ion Exchange Membranes,* Horwood, Chichester, England, 1983.
3. R.E. Lacey, Ion exchange separations, in Schweitzer, Ed., *Handbook of Separation Techniques for Chemical Engineers,* McGraw-Hill, New York, 1979, pp. 1.449–1.465.
4. K.S. Spiegler, Electrodialysis, in *Chemical Engineers' Handbook,* McGraw-Hill, New York, 1984, pp. 17.37–17.45.
5. H. Strathmann, Electrodialysis and its application in the chemical process industry, *Sep. Purification Methods* **14**(1), 41–66 (1985).

Chromatography

1. R. Bonmati and G. Guiochon, *Parfums Cosmetiques Aromes,* 37–59 (Sep./Oct. 1976).
2. J.R. Conder, Production Scale Gas Chromatography, in *New Developments in Gas Chromatography,* H. Purnell, (Ed.), Wiley, New York, 1973.
3. A.E. Rodrigues and D. Tondeur, Eds., *Percolation Processes: Theory and Applications,* Sijthoff and Noordhoff, Alphen aan den Rijn, Netherlands, 1981.
4. P. Valentin, in Rodrigues and Tondeur, loc cit., pp. 141–196.

16

CRYSTALLIZATION FROM SOLUTIONS AND MELTS

Dissolved or molten substances are recoverable in solid form by precipitation upon cooling or upon removal of the solvent or by addition of precipitating agents. For convenience a distinction is made between two kinds of processes:

1. *In solution crystallization, the crystals are separated away from a solvent, often water. In the case of inorganic solids particularly, the operating temperature is far below their melting points.*
2. *In melt crystallization, two or more substances of comparable melting points are separated by some degree of cooling. The degree of completeness of such separations depends on the phase equilibrium relations. When the crystals must be refined to remove occluded substances, the recovered material may leave the process in molten form. Subsequently, it may be solidified as flakes or sprayed granules.*

The design of crystallizers is based on knowledge of phase equilibria, solubilities, rates and amounts of nuclei generation, and rates of crystal growth. Each system is unique in most of these respects and not often predictable. The kind of information needed for design of a continuous crystallizer is indicated by the data supplied for Example 16.1.

Although theoretical advances are being made, the current state of the art of crystallization requires pilot plant evaluation of parameters of equations and of such operating variables as

1. *crystal size distribution,*
2. *effects of impurities and additives,*
3. *residence time,*
4. *circulation rate,*
5. *mixing efficiency,*
6. *allowable degrees of supersaturation or subcooling,*
7. *heat transfer characteristics,*

and others peculiar to the particular kind of equipment under consideration for the full scale plant.

This chapter will discuss the main concepts associated with crystallization practice, and will describe the main types of equipment used nowadays, together with some indications of their performance and applicability.

16.1. SOLUBILITIES AND EQUILIBRIA

The variation of the solubilities of most substances with temperature is fairly regular, and usually increases with temperature. When water is the solvent, breaks may occur in solubility curves because of formation of hydrates. Figure 16.1(a) shows such breaks, and they can be also discerned in Figures 16.2(b) and (c). Unbroken lines usually are well enough represented by second degree polynomials in temperature, but the Clapeyron-type equation with only two constants, $\ln x = A + B/T$, is of good accuracy, as appears for some cases on Figure 16.1(b).

A convenient unit of solubility is the mass of solute per unit mass of solvent, or commonly g solute/100 g solvent. Interconversions with molal units and mol fractions are made readily when densities of the solutions are known.

Under quiescent conditions a concentration substantially in excess of normal solubility or a temperature lower than the normal saturation temperature can be maintained. The maximum supersaturation appears to be a fairly reproducible quantity, but is reduced or even eliminated by stirring or by the introduction of dust or seed crystals. Some data are shown in Figure 16.1(c) and in Table 16.1. They are expressed as $\Delta C = C - C_{sat}$ or as $\Delta C/C_{sat}$ or as $\Delta T = T - T_{sat}$. According to the data of Table 16.1(d), subcooling correlates roughly with the heat of solution. The increments ΔC and ΔT can be quite substantial quantities.

The several regions of varying stability are represented by Figure 16.1(d). At concentrations above or temperatures below those represented by the supersaturation line, nuclei form and crystals grow spontaneously, although the rates of these processes do depend on the depth of penetration of the unstable region. Little control can be exercised on behavior in this region. In the metastable region, growth of crystals will occur even under quiescent conditions when dust or seeds are introduced and nuclei can be generated by agitation. Behavior in the metastable region is largely controllable so that it is the practical operating region for production of crystals of significant sizes.

Practically feasible extents of supersaturation or subcooling are fairly small and depend on the substance and the temperature. Some data appear in Table 16.2. Since the recommended values are one-half the maxima listed, they rarely are more than 2°C or so. This means that very high circulation rates through heat exchangers are needed. Thus, in the urea process of Example 16.1, the temperature rise is 2°F, and the volumetric circulation rate is about 150 times the fresh feed rate.

PHASE DIAGRAMS

Equilibria between liquid and solid phases over wide ranges of temperature are represented compactly on phase diagrams. The effect of moderate pressure on condensed phases is negligible. Aqueous systems often are complicated by the formation of hydrates, and other substances also may form intermolecular compounds. Of the substances of Figure 16.2, KCl does not form a hydrate, but NaCl and $MgSO_4$ do. Mixtures always have lower melting points than those of the pure components. The lowest temperature and the corresponding composition at which a liquid phase can be present identify the eutectic ("easy melting"), for example, point *C* on Figure 16.2(a) and point *B* on Figure 16.2(b). Binary and ternary eutectics also are identified on the ternary diagram [Fig. 16.2(f)].

The effects of evaporation or chilling on the amounts and compositions of the liquid and solid phases can be followed on the diagrams. Example 16.2 does this. Mixtures that form eutectics cannot be separated completely by chilling. The amount and nature of a separated solid phase depends on the temperature and the overall composition. Examples 16.2(c) and (d) make such calculations. Mixtures that are completely miscible in both liquid and solid phases, such as Figure 16.2(d), can be separated

EXAMPLE 16.1
Design of a Crystallizing Plant

A plant is to make 10,000 lb/hr of urea crystals from a solution that contains 75% dissolved salt. The material balance and operating conditions are shown on the sketch. Key crystallization data are given by Bennett (1981, p. 452) as

1. The residence time is 3.4 hr.
2. The temperature rise across the heater is 2°F.

Other information deduced from pilot plant work is:

3. The feed contains 75% solids, but 1200 lb/hr of wash water from the centrifuge also is returned to the crystallizer.
4. The liquor contains 66.8% dissolved urea and has a specific gravity of 1.17 at the operating temperature of 130°F.
5. The slurry contains 28 lb crystals/cuft and has a specific gravity of 1.354.
6. A purge stream amounting to 7% of the feed liquor is withdrawn as shown on the sketch.
7. The pressure is 60 Torr, at which the saturation temperature of steam is 106°F. The superheat of 24°F is neglected in figuring the vapor density and velocity.
8. Depth of liquid in the vessel should not exceed 10 ft and the vapor velocity should not exceed that given by the equation, $u = 0.06\sqrt{\rho_L/\rho_g - 1}$. A wire mesh deentrainer is not feasible because of encrustation.
9. Heat capacity of the solid is 0.62 Btu/(lb)(°F) and the heat of crystallization is 104 Btu/lb.
10. For sizing the vacuum ejector, air leakage is estimated at 25 lb/hr and carbon dioxide is 34 lb/hr.
11. The coefficient of heat transfer in the exchanger is 200 Btu/(hr)(sqft)(°F).

Calculations:

$$\text{vapor rate} = 4471(296)/3600 = 367.6 \text{ cfs},$$

$$\text{slurry holdup} = 10{,}000(3.4)/28 = 1214.3 \text{ cuft},$$

$$u_{max} = 0.06\sqrt{84.5/0.0034 - 1} = 9.46 \text{ fps},$$

$$D_{min} = \sqrt{367.6/9.46(\pi/4)} = 7.03 \text{ ft}.$$

The corresponding liquid depth is

$$h = 1214.3/(\pi/4)D^2 = 31.0 \text{ ft},$$

which is too great a value.

Try $D = 12.5$ ft:

$$\text{dished head capacity} = 152 \text{ cuft} \quad \text{(Figure 18.5)},$$

$$\text{straight side} = (1214.3 - 152)/(12.5)^2(\pi/4) = 8.66 \text{ ft}, \quad \text{say } 9.0 \text{ ft}.$$

Together with the depth of liquid in the dished head, the total depth will be close to the 10 ft specified as the maximum. From Figure 18.5, a free board of 5.5 ft is adequate in the absence of a deentraining pad. Accordingly, the vessel will have a diameter of 12.5 ft, a straight side of 14.5 ft, and dished heads designed for full vacuum. The sketch is to scale.

Sufficient data are given for finding the heat balance and the liquor circulation rate, and for sizing the auxiliaries such as lines, pump, heat exchanger and vacuum system, but those calculations will not be made.

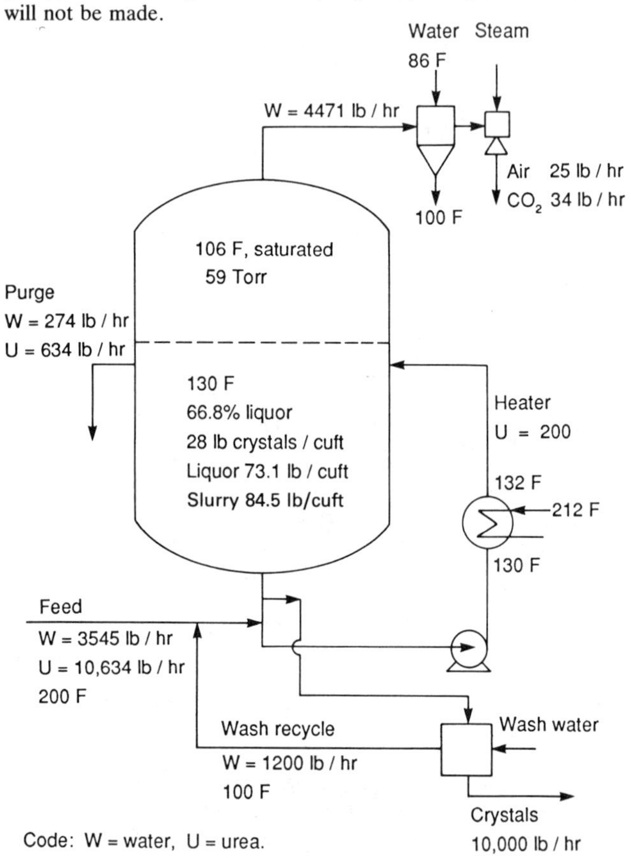

essentially completely in multistage equipment, although such processes are not often feasible. The possible extent of separation of multicomponent mixtures can be interpreted with a phase diagram like those of Figure 16.2(f) and Example 16.3. Phase diagrams are fairly plentiful, but published ones usually seem to be of the system they were interested in and not of the one you are interested in. Fortunately, nowadays phase diagrams can be developed at moderate cost and expenditure of time with differential scanning calorimeters.

Estimates of phase diagrams can be made on the assumption of ideal behavior or with activity coefficient data based on binary measurements that are more easily obtained. In such cases, clearly, it should be known that intermolecular compounds do not form. The freezing behaviors of ideal mixtures over the entire range of temperatures can be calculated readily. The method is explained for example by Walas (*Phase Equilibria in Chemical Engineering*, Butterworths, Stoneham, MA, 1985, Example 8.9).

ENTHALPY BALANCES

Although the thermal demands of crystallization processes are small compared with those of possibly competitive separation processes such as distillation or adsorption; nevertheless, they must be known. For some important systems, enthalpy–composition diagrams have been prepared, like those of Figure 16.3, for instance. Calculations also may be performed with the more widely

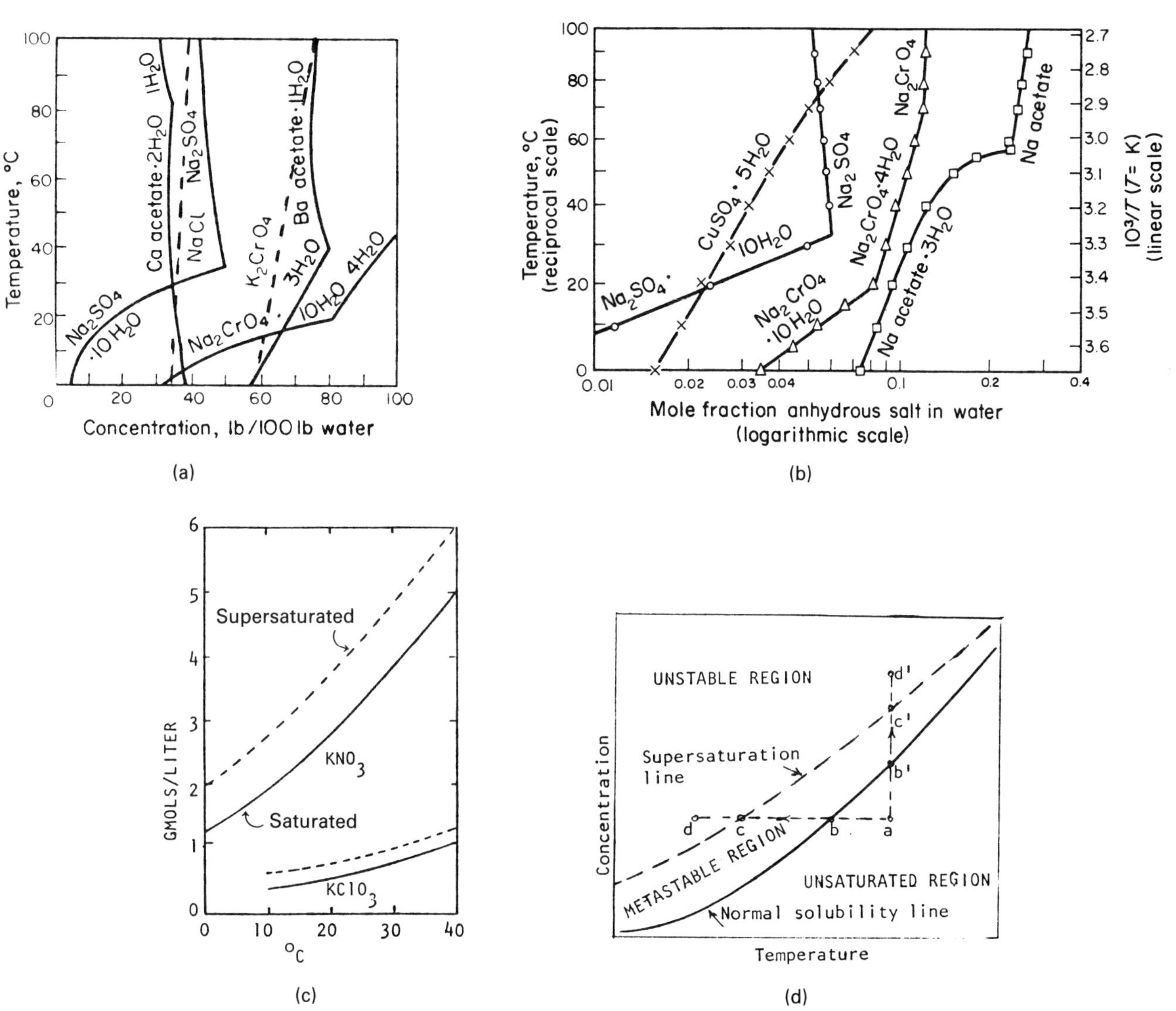

Figure 16.1. Solubility relations. (a) Linear plot of solubilities against temperature (*Mullin, 1972*). (b) Solubility against temperature plotted according to the equation $z = \exp(A + B/T)$ (*Mullin, 1972*). (c) Normal and supersolubilities of two salts (*data collected by Khamskii, 1969*). (d) Identification of regions on solubility plots. In the unstable region, nucleation and growth are spontaneous. In the metastable region growth can occur on externally introduced particles. Along a–d to the left or along a–d' upwards, nucleation and growth can start at c or c', but a substantial nuclei growth rate will not be achieved until d or d' are reached.

available data of heat capacities and heats of solution. The latter are most often recorded for infinite dilution, so that their utilization will result in a conservative heat balance. For the case of Example 16.3, calculations with the enthalpy–concentration diagram and with heat of solution and heat capacity data are not far apart.

16.2. CRYSTAL SIZE DISTRIBUTION

Crystal size distribution (CSD) is measured with a series of standard screens. The openings of the various mesh sizes according to the U.S. Standard are listed in Example 6.6, and according to the British Standard in Figure 16.4. Table 12.1 is a complete listing. The size of a crystal is taken to be the average of the screen openings of successive sizes that just pass and just retain the crystal.

The cumulative wt % either greater or less than a specified screen opening is recorded. The amount of a size less than a particular screen opening and greater than the next smaller size is called the differential amount. Typical size distribution data on Figure 16.4 are plotted in two cumulative modes, greater than or less than, and as differential polygons or histograms. For some purposes the polygon may be smoothed and often is shown that way. Some theoretical cumulative and differential distribution curves of similar nature are shown in Figure 16.5; the abscissas are proportional to the crystal length.

Cumulative data often are represented closely by the Rosin–Rammler–Sperling (RRS) equation

$$y = 100 \exp[-(d/d_m)^n], \qquad (16.1)$$

where d is the diameter, d_m is a mean diameter corresponding to $y = 100/e = 36.8\%$ and n is called the uniformity factor. The greater n, the more nearly uniform the distribution. The log–log plot of this

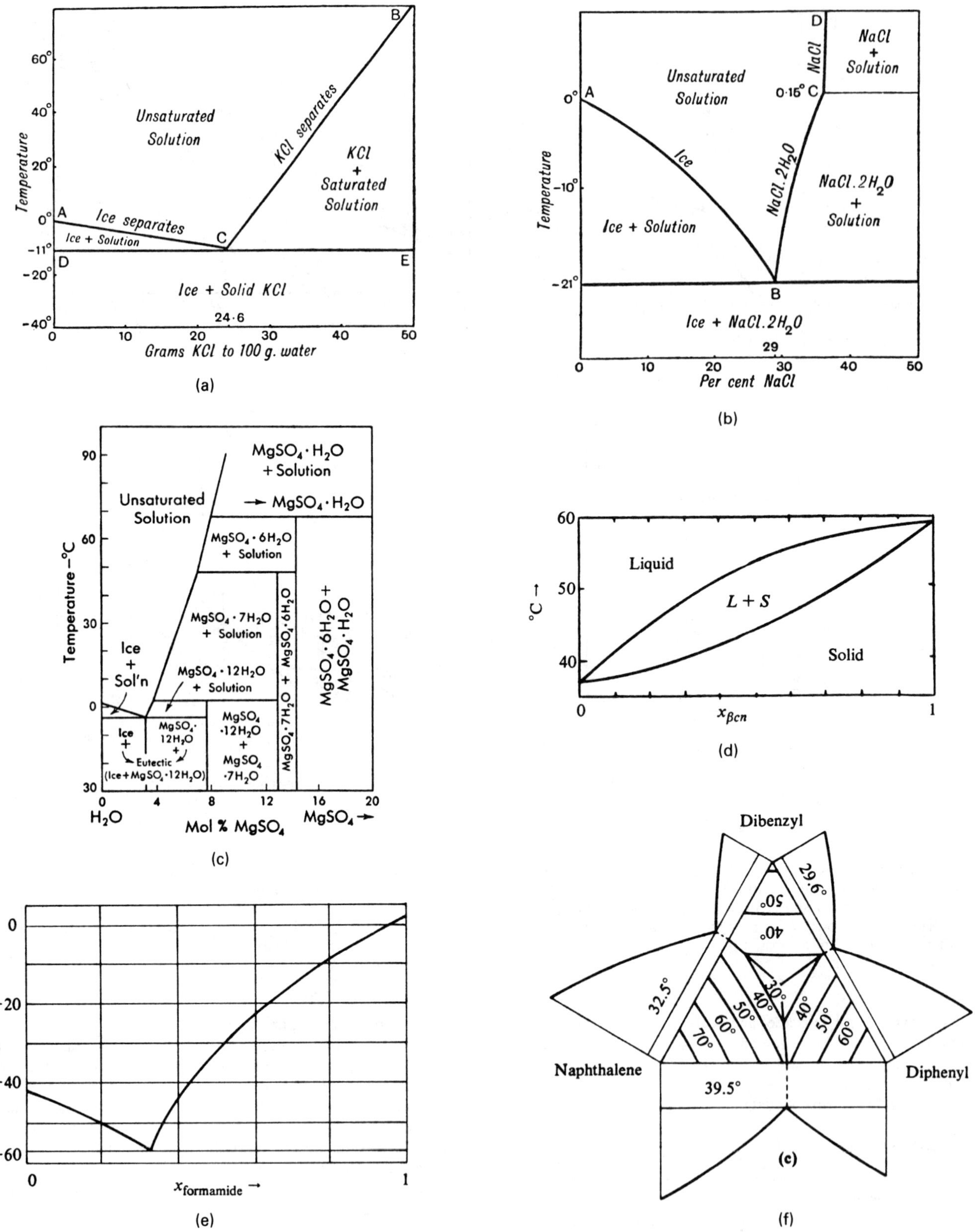

Figure 16.2. Some phase diagrams. (a) The water end of the system potassium chloride and water. (b) The water end of the system sodium chloride and water. (c) The water end of the system magnesium sulfate and water; the heptahydrate goes to the mono at 150°C, and to anhydrous at 200°C. (d) β-methylnaphthalene and β-chloronaphthalene form solid solutions. (e) Mixtures of formamide and pyridine form a simple eutectic. (f) These mixtures form binary eutectics at the indicated temperatures and a ternary eutectic at mol fractions 0.392 dibenzyl, 0.338 diphenyl, and 0.27 naphthalene.

TABLE 16.1. Data of Supersaturation and Subcooling of Solutions

(a) Maximum Supersaturation of Solutions at 20°C, $\beta = \Delta C / C_0$

Solute	Tovbin and Krasnova's data [144]	Gorbachev and Shlykov's data [34]	Fisher's data [152]*
KCl	0.095	0.39	—
KBr	0.056	0.102	—
KI	0.029	—	—
$KClO_3$	0.41	—	—
KNO_3	0.36	1.08	—
NH_4NO_3	0.10	—	—
$NaNO_3$	0.064	—	—
$Mg(NO_3)_2$	0.93	—	—
K_2SO_4	0.37	—	0.34
$K_2C_2O_4$	0.41	—	—
K_2CrO_4	0.093	—	—
$K_2Cr_2O_7$	0.62	—	0.32
$Ba(NO_3)_2$	0.40	—	—
$CuSO_4$	1.50	—	—
$HgCl_2$	0.43	—	—
$K_3Fe(CN)_6$	0.13	—	—
$K_4Fe(CN)_6$	0.54	—	—
$KBrO_3$	—	2.71	—
KIO_3	—	1.60	—
$Na_2C_2O_4$	—	—	0.86
$(NH_4)_2C_2O_4 \cdot H_2O$	—	—	0.36

*Fisher's results were obtained at 25°C.

(b) Temperature Dependence of the Maximum Supersaturation of Salt Solutions

Solute	t, °C	C_0, moles/liter	C, moles/liter	$\alpha = C - C_0$	β
KNO_3	0	1.25	2.03	0.78	0.62
	10	1.96	2.78	0.81	0.41
	20	2.76	3.75	0.99	0.36
	30	3.83	4.84	1.01	0.26
	40	4.97	6.00	1.03	0.20
KCl	0	3.33	3.88	0.55	0.16
	10	3.72	4.12	0.40	0.11
	20	4·03	4.42	0.39	0.095
	30	4.29	4.45	0.16	0.037
	40	4.45	4.58	0.13	0.029
$KClO_3$	10	0.40	0.65	0.25	0.62
	20	0.58	0.82	0.24	0.41
	30	0.80	1.05	0.25	0.32
	40	1.11	1.32	0.21	0.19
K_2CrO_4	10	2.68	2.96	0.28	0.11
	20	2.74	3.00	0.26	0.093
	30	2.82	3.03	0.21	0.073
	40	2.98	3.07	0.19	0.065

(c) Maximum Supercooling of Salt Solutions at Various Temperatures

Solute	Heat of solution λ, cal/mole	$t_0 - t$, °C	$\lambda(t_0 - t)$, cal·mole⁻¹·deg⁻¹
KCl	4046	19.6	78897
KBr	5080	16.3	80804
KI	5110	15.5	79205
$KBrO_3$	9760	8.8	84788
KIO_3	6780	13.5	91490
$KClO_3$	9950	6.6	65670
KNO_3	8800	13.0	114400
$KClO_4$	12100	6.3	76230
KCNS	6100	13.0	79300
$NaNO_3$	5030	13.0	65399
$NaClO_3$	5600	12.0	67200
NaCl	12200	51.0	62220
NH_4Cl	3880	20.0	77600
$(NH_4)_2SO_4$	2370	24.0	56880
NH_4NO_3	6320	10.3	65016
$HgCl_2$	3300	25.0	82500
$CuSO_4$	2750	36.7	80925
$NaClO_4$	3600	20.0	72000
NH_4ClO_4	6360	12.0	76320
$Ba(ClO_4)_2$	9400	9.0	84600

(d) Dependence of the Maximum Supercooling of Solutions on Heat of Solution

Solute	t_0, °C	t, °C	$\theta = t_0 - t$, °C	Solute	t_0, °C	t, °C	$\theta = t_0 - t$, °C
KNO_3	20	−1.0	21.0	KBr	10.5	−1.8	12.3
	30	8.9	21.1		20	8.0	12.0
	40	18.9	21.1		30	17.8	12.2
	50	28.8	21.2		40	28.0	12.0
	60	38.8	21.2		50	37.9	12.1
	70	48.9	21.1		60	47.7	12.3
KCl	50	6.7	43.3		80	67.7	12.3
	60	16.6	43.4	K_2SO_4	90	3.0	87.0
	70	26.7	43.3		100	13.0	87.0
	80	36.7	43.3				
	90	46.6	43.4				

(Khamskii, 1969).

equation should be linear. On Figure 16.4(c) the scatter about the straight line is small, but several of the plots of commercial data of Figure 16.6 deviate somewhat from linearity at the larger diameters.

Two other single numbers are used to characterize size distributions. The median aperture, MA or d_{50}, is the screen opening through which 50% of the material passes. The coefficient of variation is defined by the equation

$$CV = 100(d_{16} - d_{84})/2d_{50}. \tag{16.2}$$

The origin of this concept is that the fraction of the total area under a normal distribution curve between the 16 and 84% points is twice the standard deviation. Thc smaller CV, the more nearly uniform the crystal sizes. Products of DTB crystallizers, for instance, often have CVs of 30–50%. The number is useful as a measure of consistency of operation of a crystallizer. Some details are given by Mullin (1972, pp. 349, 389).

TABLE 16.2. Maximum Allowable Supercooling ΔT (°C) and Corresponding Supersaturation ΔC (g/100 g water) at 25°C[a]

Substance	ΔT	ΔC
NH_4 alum	3.0	1.0
NH_4Cl	0.7	0.3
NH_4NO_3	0.6	3.0
$(NH_4)_2SO_4$	1.8	0.5
$NH_4H_2PO_4$	2.5	2.3
$CuSO_4.5H_2O$	1.4	1.0
$FeSO_4.7H_2O$	0.5	0.6
K alum	4.0	1.0
KBr	1.1	0.6
KCl	1.1	0.3
KI	0.6	0.4
KH_2PO_4	9.0	4.6
KNO_3	0.4	0.6
KNO_2	0.8	0.8
K_2SO_4	6.0	1.3
$MgSO_4.7H_2O$	1.0	1.3
$NiSO_4.7H_2O$	4.0	4.4
$NaBr.2H_2O$	0.9	0.9
$Na_2CO_3.10H_2O$	0.6	2.8
$Na_2CrO_4.10H_2O$	1.6	0
NaCl	4.0	0.2
$Na_2B_4O_7.10H_2O$	4.0	0.9
NaI	1.0	1.7
$NaHPO_4.12H_2O$	0.4	1.5
$NaNO_3$	0.9	0.7
$NaNO_2$	0.9	0.6
$Na_2SO_4.10H_2O$	0.3	0.7
$Na_2S_2O_3.5H_2O$	1.0	2.2
Urea	2.0	

[a] Working values usually are not more than one-half the maxima.
(After Mullin, 1972).

16.3. THE PROCESS OF CRYSTALLIZATION

The questions of interest are how to precipitate the crystals and how to make them grow to suitable sizes and size distributions. Required sizes and size distributions are established by the need for subsequent recovery in pure form and ease of handling, and by traditional commercial practices or consumer preferences.

CONDITIONS OF PRECIPITATION

The most common methods of precipitating a solid from a solution are by evaporation of the solvent or by changing to a temperature at which the solubility is lower. Usually solubility is decreased by lowering the temperature. Some examples are in Figure 16.1. The limit of removal is determined by the eutectic composition. According to the data of Figure 16.2, for instance, a 24.6% solution of KCl will solidify completely at −11°C and a 3.5% solution of $MgSO_4$ will do so at 4°C; these values represent the limits to which salt is recoverable by chilling. Complete recovery, however, is accomplished by evaporation.

A precipitate may be formed as a result of chemical reaction between separately soluble gases or liquids. Commercial examples are productions of sodium sulfate, ammonium sulfate, and ammonium phosphate.

Precipitation also can be induced by additives, a process generally called salting out because salts with ions common to those whose precipitation is desired are often used for this purpose. For instance, ammonium chloride is recovered from spent Solvay liquors by addition of sodium chloride and the solubility of $BaCl_2$ can be reduced from 32% to 0.1% by addition of 32% of $CaCl_2$. Other kinds of precipitants also are used, for instance, alcohol to precipitate aluminum sulfate from aqueous solutions.

Foreign substances even in minute amounts may have other kinds of effects on crystallization: They may inhibit or accelerate growth rate or change the shape of crystals, say from rounded to needlelike, or otherwise. One of the problems sometimes encountered with translating laboratory experience to full scale operation is that the synthetic liquors used in the laboratory may not contain the actually occurring impurities, and thus give quite different performance. Substances that modify crystal formation are very important industrially and many such materials have been the subject of patents.

SUPERSATURATION

A saturated solution is one that is in equilibrium with the solid phase and will remain unchanged indefinitely at a particular

---

EXAMPLE 16.2
Using the Phase Diagrams of Figure 16.2

a. Evaporation of a solution of $MgSO_4$ at 30°C: As water is removed, the composition moves along the horizontal. When the salt concentration reaches about 6%, precipitation of heptahydrate begins and is completed at about 13%. Between 13 and 14% salt, the precipitate is a mixture of solid hepta and solid hexa hydrates. Beyond 14%, the mixture consists of mixtures of solid hepta and mono hydrates in proportions determined by the amount of water present overall.

b. Chilling of a 6% solution of $MgSO_4$: precipitation of heptahydrate begins at about 35°C. At about 2°C, the mixture consists of solid dodecahydrate and unsaturated solution. Below −4°C complete solidification exists; the product is a mixture of pure dodecahydrate and an intimate eutectic mixture of ice and dodeca crystals.

c. Recovery of pyridine: As appears on the diagram of Figure 16.2(e), the eutectic contains 33% formamide and 67% pyridine. When the mixture contains 80% pyridine, the maximum possible recovery of pure pyridine is

$$P = (0.8 - 0.67)/(1 - 0.67) = 0.39, \text{ or } 39\%.$$

d. Recovery of formamide: When the mixture of Figure 16.2(e) contains 80% formamide, the maximum recovery of the pure material is

$$F = (0.8 - 0.33)/(1 - 0.33) = 0.70, \text{ or } 70\%.$$

e. At 50°C, the liquid phase of Figure 16.2(d) contains 35% and the solid phase 74% of β-chloronaphthalene.

f. The progress of crystallization of a ternary mixture such as that of Figure 16.2(f) is described in Example 16.8.

EXAMPLE 16.3

Heat Effect Accompanying the Cooling of a Solution of $MgSO_4$

A 30% solution of $MgSO_4$ is cooled from 150F to 50F. Data of the initial and final conditions are taken off the equilibrium diagram, Fig. 16.3(b). At the lower temperature, 27% of the mixture crystallizes out as the heptahydrate.

	Original at 150°F	Final (at 50°F) Total	Liquid	Solid
Water (lb)	38.6	38.6	38.6	—
$MgSO_4{\cdot}7H_2O$ (lb)	61.4	61.4	34.4	27.0
Total (lb)	100	100	73.0	27.0
H (Btu/lb)	−3	−82	−53	−161

Accordingly the change in enthalpy is

$$\Delta H = -82 - (-3) = -79 \text{ Btu/lb}.$$

This value will be compared with a calculation using data of heat capacities and heat of solution. From *Perry's Chemical Engineers' Handbook* (1984), the heat solution of the heptahydrate is −39.2 Btu/lb and its heat capacity is 0.36 Btu/(lb)(°F). The enthalpy change of the cooling and crystallization process is

$$\Delta H = [0.386 + 0.614(0.36)](50 - 150) + 0.27(-39.2) = -71.3 \text{ Btu/lb},$$

which is a poor check of the value found with the aid of the equilibrium diagram. Possible sources of error of the second method include the use of heat of solution at infinite dilution instead of the prevailing concentration and the assumption that the heat capacities are additive.

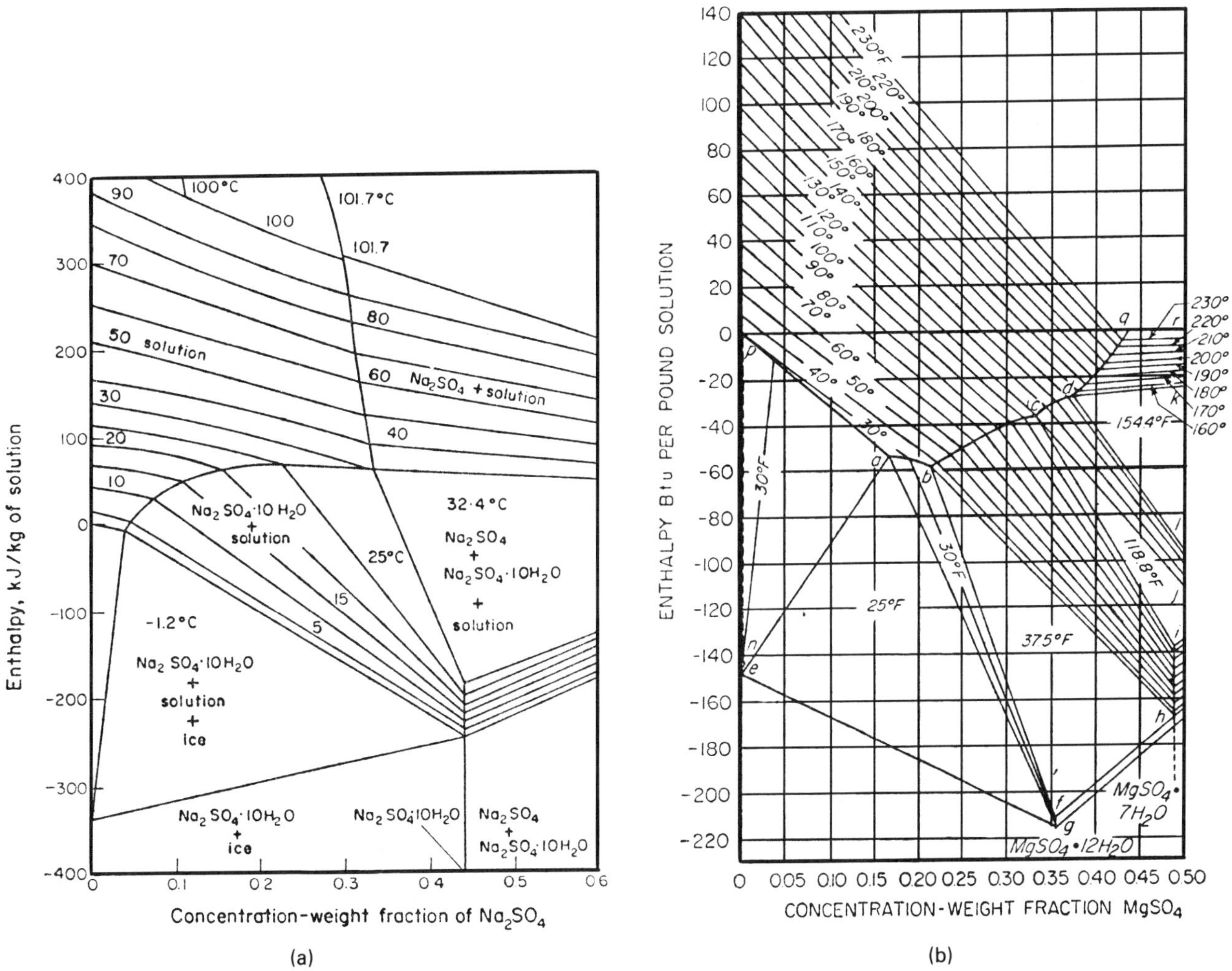

Figure 16.3. Enthalpy–composition diagrams of some salt solutions. Several other diagrams are in the compilation of Landolt-Börnstein, IV 4b, 1972, pp. 188–224. (a) sodium sulfate/water; (b) magnesium sulfate/water (*after* Chemical Engineers' Handbook, *1963 edition, McGraw-Hill, New York*); (c) sodium carbonate/water.

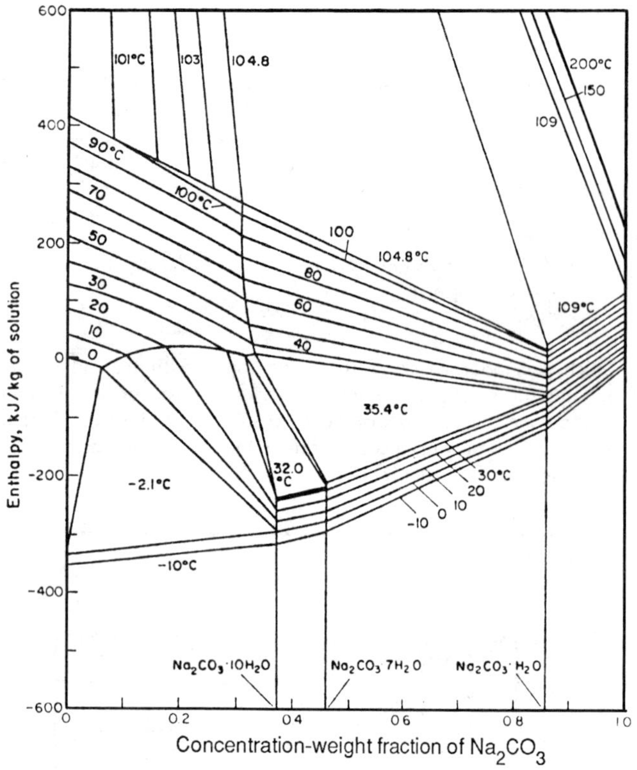

(c)

Figure 16.3—(*continued*)

temperature and composition of other constituents. Greater than normal concentrations also can be maintained in what is called a supersaturated condition which is metastable. Metastability is sensitive to mechanical disturbances such as agitation, ultrasonics, and friction and the introduction of solid particles. Under those conditions, solids will separate out until normal saturation is obtained. When great care is taken, the metastable state is reproducible. A thermodynamic interpretation of metastability can be made in terms of the Gibbs energy of mixtures. In Figure 16.5(a), the solid line a–b is of unsaturated solution and the straight line b–e is of mixtures of all proportions of pure solid and saturated solution represented by point b. Points c and d are at the points of inflection of the plot and represent the limits of metastability. Thus line b–c represents the range of concentrations between the saturated and supersaturated values.

Several measures of supersaturation are being used in terms of the saturation concentration C_0; thus

$\alpha = \Delta C_s = C - C_0$, the difference in concentrations,

$\beta = \Delta C_s / C$, the relative difference,

$\gamma = C/C_0 = \beta + 1$, the concentration ratio,

with similar definitions for subcooling or superheating. The data of Figure 16.1(c) and Table 16.1 show that excess concentration and metastable cooling can be quite substantial amounts.

GROWTH RATES

Crystallization can occur only from supersaturated solutions. Growth occurs first by formation of nuclei and then by their gradual growth. At concentrations above supersaturation, as at point d' on Figure 16.1(d), nucleation is conceived to be spontaneous and rapid. In the metastable region, nucleation is caused by mechanical shock or friction and secondary nucleation can result from the breakup of already formed crystals. It has been observed that the rate depends on the extent of supersaturation; thus

$$\frac{dC_N}{d\theta} = k(C - C_0)^m. \tag{16.3}$$

Values of the exponent m have been found to range from 2 to 9, but have not been correlated to be of quantitative value for prediction.

B.S. mesh number	*Sieve aperture, μm*	*Fractional weight per cent retained*	*Cumulative weight per cent oversize*	*Cumulative weight per cent undersize*
7	2360	1·2	1·2	98·8
10	1700	2·9	4·1	95·9
14	1180	18·8	22·9	77·1
18	850	28·8	51·7	48·3
25	600	22·0	73·7	26·3
36	425	11·1	84·8	15·2
52	300	6·0	90·8	9·2
72	212	3·9	94·7	5·3
100	150	1·8	96·5	3·5
150	106	1·3	97·8	2·2
>150	—	2·2	—	—

(a)

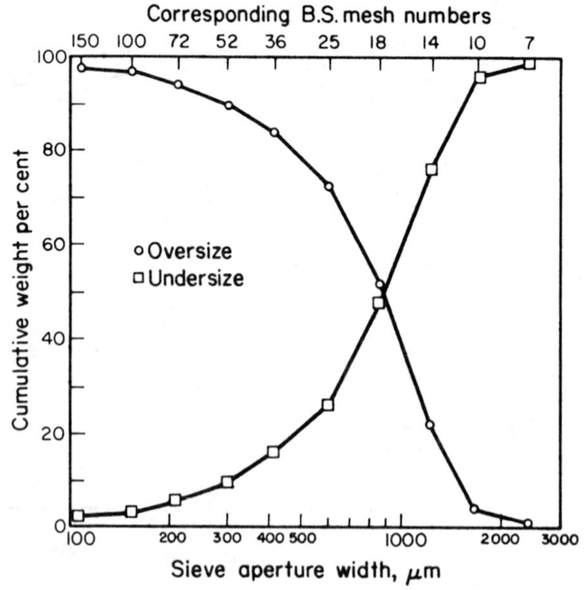

(b)

Figure 16.4. Several ways of recording the same data of crystal size distribution (CSD) (*Mullin, 1972*). (a) The data. (b) Cumulative wt % retained or passed, against sieve aperture. (c) Log–log plot according to the RRS equation $P = \exp[(-d/d_m)^n]$; off this plot, $d_{50\%} = 850$, $d_m = 1000$, $n = 1.8$. (d) Differential polygon. (e) Differential histogram.

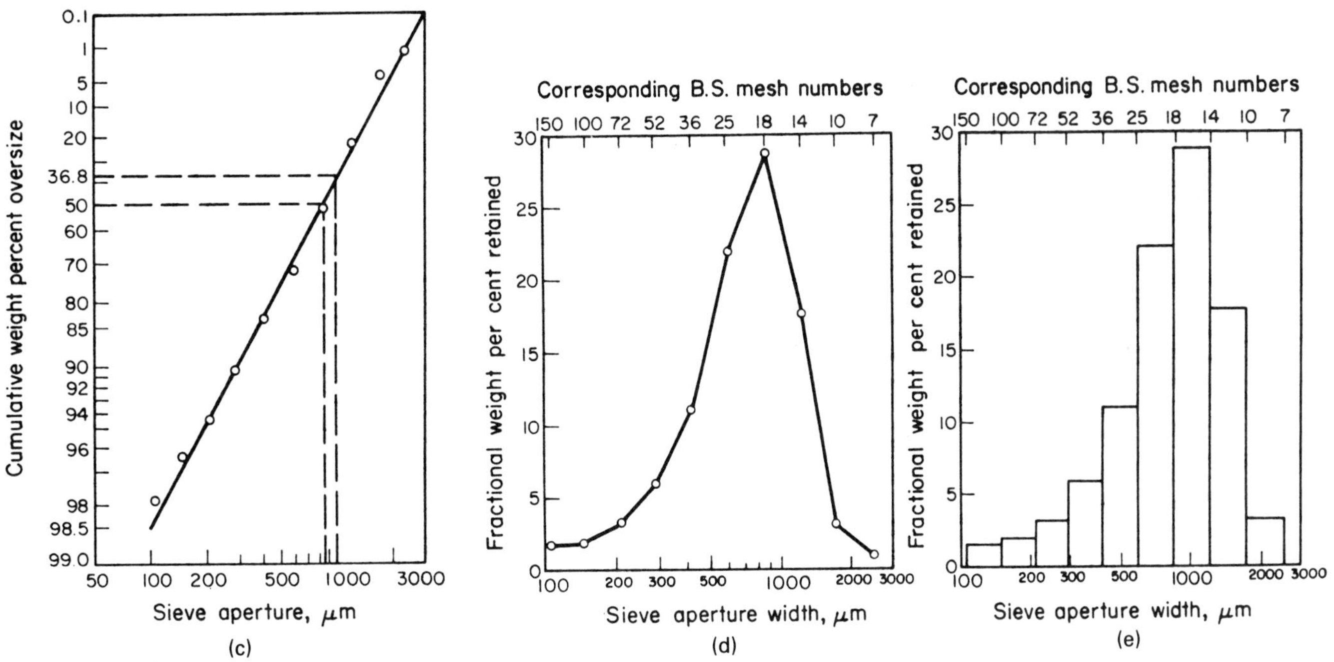

Figure 16.4—(*continued*)

Nucleation rates are measured by counting the numbers of crystals formed over periods of time.

The growth rates of crystals depend on their instantaneous surface and the linear velocity of solution past the surface as well as the extent of supersaturation, and are thus represented by the equation

$$\frac{dW_c}{d\theta} = kuA(C - C_0)^n. \tag{16.3'}$$

Values of the exponent have been found of the order of 1.5, but

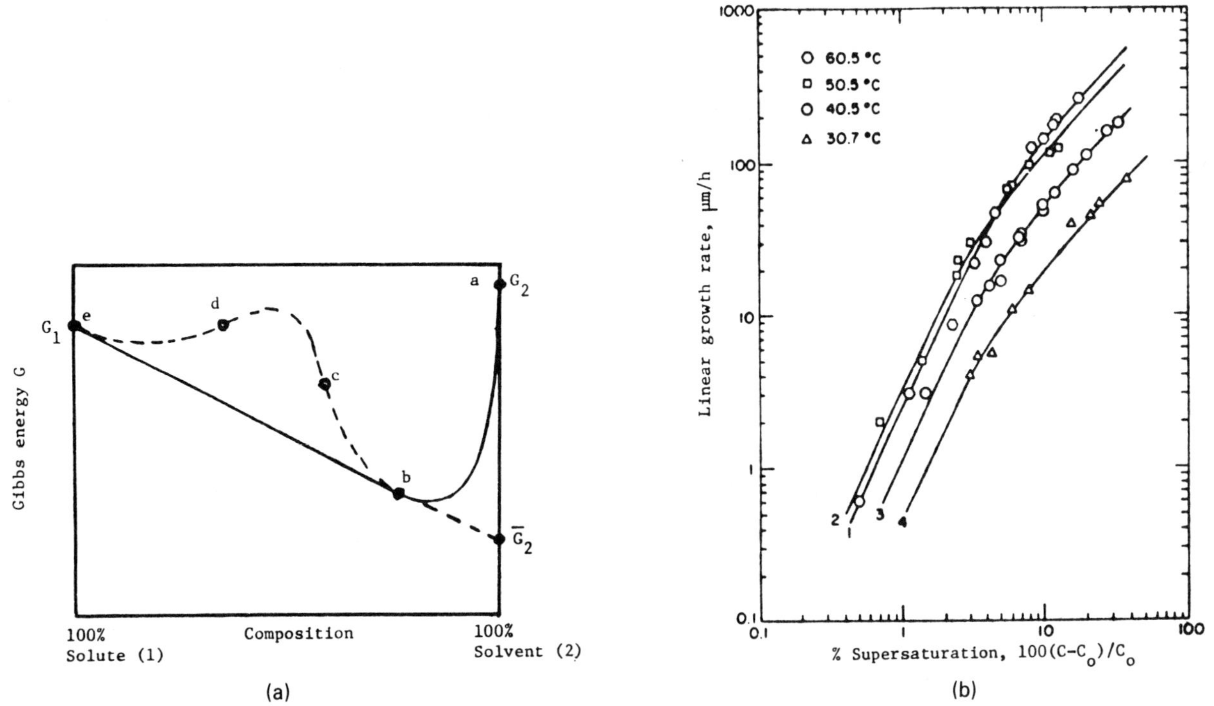

Figure 16.5. Supersaturation behavior. (a) Schematic plot of the Gibbs energy of a solid solute and solvent mixture at a fixed temperature. The true equilibrium compositions are given by points *b* and *e*, the limits of metastability by the inflection points *c* and *d*. For a salt-water system, point *d* virtually coincides with the 100% salt point *e*, with water contents of the order of 10^{-6} mol fraction with common salts. (b) Effects of supersaturation and temperature on the linear growth rate of sucrose crystals [*data of Smythe (1967) analyzed by Ohara and Reid, 1973*].

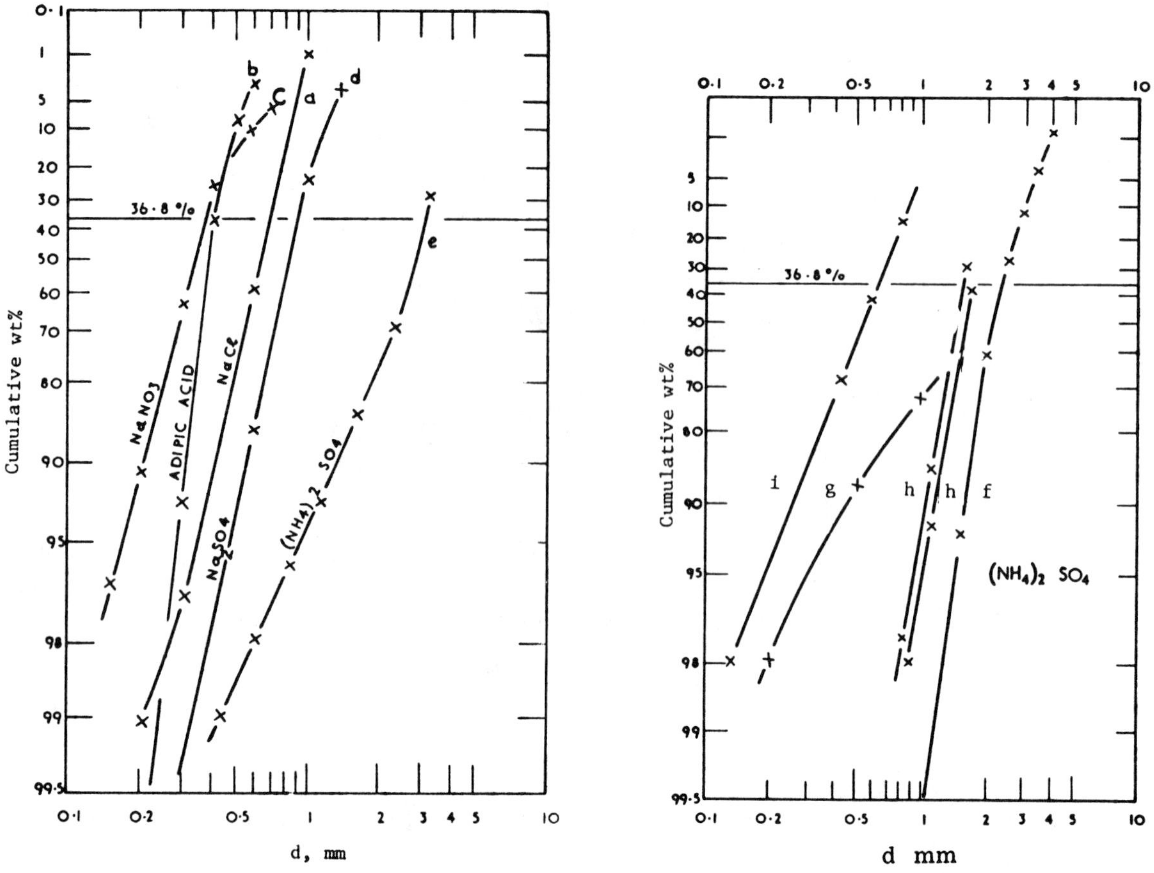

Figure 16.6. Crystal size distributions of several materials in several kinds of crystallizers (*Bamforth, 1965*).

Code	Crystallizer	Substance	d_m	n
a	Escher-Wyss	NaCl	0.7	4.7
b	Giavanola	adipic acid	0.4	8.1
c	Matusevich	$NaNO_3$	0.37	4.0
d	Kestner	Na_2SO_4	0.92	4.7
e	Oslo-Krystal	$(NH_4)_2SO_4$	3.2	2.1
f	Oslo-Krystal	$(NH_4)_2SO_4$	2.35	6.0
g	Sergeev	$(NH_4)_2SO_4$	—	1.5(?)
h	DTB	$(NH_4)_2SO_4$	1.6	5.7
i	Standard saturator	$(NH_4)_2SO_4$	0.62	2.6

The parameters are those of the RRS equation, Eq. 16.1.

again no correlation of direct use to the design of crystallizers has been achieved. The sucrose growth data of Figure 16.5(b) are not quite log–log linear as predicted by this equation.

In laboratory and commercial crystallizations, wide size distributions usually are the rule, because nuclei continue to form throughout the process, either spontaneously or by breakage of already formed crystals. Large crystals of more or less uniform size are desirable. This condition is favored by operating at relatively low extents of supersaturation at which the nucleation rate is low but the crystals already started can continue to grow. The optimum extent of supersaturation is strictly a matter for direct experimentation in each case. As a rough guide, the data for allowable subcooling and corresponding supersaturation of Table 16.2 may serve. Since the recommended values are one-half the maxima shown, it appears that most crystallizations under commercial conditions should operate with less than about 2°C subcooling or the corresponding supersaturation. The urea plant design of Example 16.4 is based on 2°F heating.

Growth rates of crystals also must be measured in the laboratory or pilot plant, although the suitable condition may be expressed simply as a residence time. Table 16.3 gives a few growth rate data at several temperatures and several extents of supersaturation for each substance. In most instances the recommended supersaturation measured as the ratio of operating to saturation concentrations is less than 1.1. It may be noted that at a typical rate of increase of diameter of 10^{-7} m/sec, the units used in this table, the time required for an increase of 1 mm is 2.8 hr.

Batch crystallizers often are seeded with small crystals of a known range of sizes. The resulting crystal size distribution for a given overall weight gain can be estimated by an approximate

EXAMPLE 16.4
Deductions from a Differential Distribution Obtained at a Known Residence Time
The peak of the differential distribution obtained with a residence time of $\bar{t} = 2$ hr corresponds to $L_{pr} = 1.2$ mm. Assuming ideal mixing, $L_{pr}/G\bar{t} = 1.2/2G = 3$, and $G = 0.2$ mm/h. With this knowledge of G, crystal size distributions could be found at other residence times.

relation known as the McCabe Delta-L Law, which states that each original crystal grows by the same amount ΔL. The relation between the relative masses of the original and final size distributions is given in terms of the incremental ΔL by

$$R = \frac{\sum w_i(L_{0i} + \Delta L)^3}{\sum w_i L_{0i}^3}. \qquad (16.4)$$

When R is specified, ΔL is found by trial, and then the size distribution is evaluated. Example 16.5 does this.

Some common substances for which crystallization data are reported in the literature and in patents are listed in Table 16.4.

16.4. THE IDEAL STIRRED TANK

All continuous crystallizers are operated with some degree of mixing, supplied by internal agitators or by pumparound. The important limiting case is that of ideal mixing in which conditions are uniform throughout the vessel and the composition of the effluent is the same as that of the vessel content. In crystallization literature, this model carries the awkward name MSMPR (mixed suspension mixed product removal). By analogy with the terminology of chemical reactors it could be called CSTC (continuous stirred tank crystallizer). Several such tanks in series would be called a CSTC battery. A large number of tanks in series would approach plug flow, but the crystal size distribution still would not be uniform if nucleation continued along the length of the crystallizer.

The process to be analyzed is represented by Figure 16.4. What will be found are equations for the cumulative and differential size distributions in terms of residence time and growth rate. The principal notation is summarized here.

Q = volumetric feed rate,
V_c = volume of holdup in the tank
n = number of crystals per unit volume
L = length of the crystal
G = linear growth rate of the crystal
t = time
$\bar{t} = V_c/Q$, mean residence time
$x = L/G\bar{t}$, reduced time
ϕ_m = cumulative mass distribution
n^0 = zero side nuclei concentration, also called zero size population density
B^0 = nucleation rate
a_v = volume shape factor = volume of crystal/(length)$^3 = \pi/6$ for spheres, = 1 for cubes.

The case being considered is that in which the feed contains no

TABLE 16.3. Mean Overall Growth Rates of Crystals (m/sec) at Each Face[a]

Crystallising substance	°C	*S*	$\bar{v}$ (m/s)
$(NH_4)_2SO_4 \cdot Al_2(SO_4)_3 \cdot 24H_2O$	15	1·03	$1{\cdot}1 \times 10^{-8}$*
	30	1·03	$1{\cdot}3 \times 10^{-8}$*
	30	1·09	$1{\cdot}0 \times 10^{-7}$*
	40	1·08	$1{\cdot}2 \times 10^{-7}$*
NH_4NO_3	40	1·05	$8{\cdot}5 \times 10^{-7}$
$(NH_4)_2SO_4$	30	1·05	$2{\cdot}5 \times 10^{-7}$*
	60	1·05	$4{\cdot}0 \times 10^{-7}$
	90	1·01	$3{\cdot}0 \times 10^{-8}$
$NH_4H_2PO_4$	20	1·06	$6{\cdot}5 \times 10^{-8}$
	30	1·02	$3{\cdot}0 \times 10^{-8}$
	30	1·05	$1{\cdot}1 \times 10^{-7}$
	40	1·02	$7{\cdot}0 \times 10^{-8}$
$MgSO_4 \cdot 7H_2O$	20	1·02	$4{\cdot}5 \times 10^{-8}$*
	30	1·01	$8{\cdot}0 \times 10^{-8}$*
	30	1·02	$1{\cdot}5 \times 10^{-7}$*
$NiSO_4 \cdot (NH_4)_2SO_4 \cdot 6H_2O$	25	1·03	$5{\cdot}2 \times 10^{-9}$
	25	1·09	$2{\cdot}6 \times 10^{-8}$
	25	1·20	$4{\cdot}0 \times 10^{-8}$
$K_2SO_4 \cdot Al_2(SO_4)_3 \cdot 24H_2O$	15	1·04	$1{\cdot}4 \times 10^{-8}$*
	30	1·04	$2{\cdot}8 \times 10^{-8}$*
	30	1·09	$1{\cdot}4 \times 10^{-7}$*
	40	1·03	$5{\cdot}6 \times 10^{-8}$*
KCl	20	1·02	$2{\cdot}0 \times 10^{-7}$
	40	1·01	$6{\cdot}0 \times 10^{-7}$
KNO_3	20	1·05	$4{\cdot}5 \times 10^{-8}$
	40	1·05	$1{\cdot}5 \times 10^{-7}$
K_2SO_4	20	1·09	$2{\cdot}8 \times 10^{-8}$*
	20	1·18	$1{\cdot}4 \times 10^{-7}$*
	30	1·07	$4{\cdot}2 \times 10^{-8}$*
	50	1·06	$7{\cdot}0 \times 10^{-8}$*
	50	1·12	$3{\cdot}2 \times 10^{-7}$*
KH_2PO_4	30	1·07	$3{\cdot}0 \times 10^{-8}$
	30	1·21	$2{\cdot}9 \times 10^{-7}$
	40	1·06	$5{\cdot}0 \times 10^{-8}$
	40	1·18	$4{\cdot}8 \times 10^{-7}$
NaCl	50	1·002	$2{\cdot}5 \times 10^{-8}$
	50	1·003	$6{\cdot}5 \times 10^{-8}$
	70	1·002	$9{\cdot}0 \times 10^{-8}$
	70	1·003	$1{\cdot}5 \times 10^{-7}$
$Na_2S_2O_3 \cdot 5H_2O$	30	1·02	$1{\cdot}1 \times 10^{-7}$
	30	1·08	$5{\cdot}0 \times 10^{-7}$
Citric acid monohydrate	25	1·05	$3{\cdot}0 \times 10^{-8}$
	30	1·01	$1{\cdot}0 \times 10^{-8}$
	30	1·05	$4{\cdot}0 \times 10^{-8}$
Sucrose	30	1·13	$1{\cdot}1 \times 10^{-8}$*
	30	1·27	$2{\cdot}1 \times 10^{-8}$*
	70	1·09	$9{\cdot}5 \times 10^{-8}$
	70	1·15	$1{\cdot}5 \times 10^{-7}$

[a] The supersaturation is expressed by $S = C/C_0$, with C the amount dissolved and C_0 the normal solubility (kg crystals/kg water). The mean growth velocity is that at one face of the crystal; the length increase is $G = 2\bar{v}$ (m/sec). Data are for crystals in the size range 0.5–1.0 mm in the presence of other crystals. The asterisk denotes that the growth rate probably is size-dependent.
(Mullin, 1972).

EXAMPLE 16.5
Batch Crystallization with Seeded Liquor

Seed crystals with this size distribution are charged to a batch crystallizer:

L_o, length (mm)	0.251	0.178	0.127	0.089	0.064
w (wt fraction)	0.09	0.26	0.45	0.16	0.04

On the basis of the McCabe ΔL law, these results will be found:

a. The length increment that will result in a 20-fold increase in mass of the crystals.
b. The mass growth corresponding to the maximum crystal length of 1.0 mm.

When L is the increment in crystal length, the mass ratio is

$$R = \frac{\sum w_i(L_{0i}+L)^3}{\sum w_i L_{0i}^3} = \frac{\sum w_i(L_{0i}+L)^3}{0.09346} = 20$$

a. By trial, the value of $L = 0.2804$ mm.
b. When $L = 1 - 0.251 = 0.749$, $R = 181.79$.

The size distributions and the computer program are tabulated.

```
10 ! Example 16.4. Batch crysta
     llization with seeded liquor
20 OPTION BASE 1
25 DIM L0(5),W(5),S(5)
30 MAT READ L0,W
40 DATA .251,.178,.127,.089,.06
     4,.09,.26,.45,.16,.04
50 INPUT L
60 S=0
70 FOR I=1 TO 5
80 S(I)=(L0(I)+L)^3*W(I)
100 S=S(I)+S
110 PRINT USING 120 ; W(I),L0(I)
     ,L0(I)+L
120 IMAGE .DDD,2X,.DDD,2X,D.DDDD
130 NEXT I
135 PRINT
140 PRINT "INCREMENT L=";L
150 PRINT "SUMMATION=";S
160 PRINT "WEIGHT RATIO=";S/.003
     93458
170 END
```

```
 W        Lo       Lo+L
.090     .251     .2510
.260     .178     .1780
.450     .127     .1270
.160     .089     .0890
.040     .064     .0640

INCREMENT L= 0
SUMMATION= .00393458126
WEIGHT RATIO= 1.00000032024


.090     .251     .5314
.260     .178     .4584
.450     .127     .4074
.160     .089     .3694
.040     .064     .3444

INCREMENT L= .2804
SUMMATION= 7.86768511336E-2
WEIGHT RATIO= 19.9962514763


.090     .251    1.0000
.260     .178     .9270
.450     .127     .8760
.160     .089     .8380
.040     .064     .8130

INCREMENT L= .749
SUMMATION= .71526668218
WEIGHT RATIO= 181.789843434
```

nuclei but they are generated in the tank. The balance on the number of crystals is

rate of generation = rate of efflux

or

$$V_c \frac{dn}{dt} = Qn. \tag{16.5}$$

Upon substituting for the linear growth rate

$$G = dL/dt \tag{16.6}$$

and rearranging,

$$\frac{dn}{n} = \frac{Q}{V_c}dt = \frac{Q}{V_c G}dL = \frac{dL}{\bar{t}G} = dx \tag{16.7}$$

where

$$\bar{t} = V_c/Q \tag{16.8}$$

is the mean residence time and

$$x = L/G\bar{t} = t/\bar{t} \tag{16.9}$$

is the dimensionless time. Integration of the equation

$$\int_0^n \frac{dn}{n} = \frac{1}{G\bar{t}}\int_0^L dL \tag{16.10}$$

is

$$n = n^0 \exp(-L/G\bar{t}) = n^0 \exp(-x), \tag{16.11}$$

TABLE 16.4. Some Common Substances for which Crystallization Data Are Reported in the Literature and in Patents[a]

Compound	*Remark or aspect referred to*
Ag–halides	
Ag_2CrO_4	growth kinetics
AlF_3	
Al_2O_3–corundum	
$Al\,NH_4(SO_4)_2$	
$Al\,K(SO_4)_2$	influence of supersaturation
$Al(OH)_3$	
H_3BO_3	
$Na_2B_4O_7$	oleic acid conducive
$BaSO_4$	nucleation
	growth
	habit
$BaCO_3$	
$BaTiO_4$	
$CaSO_4$	citrates, SO_4'', elevated temp.
$CaCO_3$	metaphosphate conducive
$CaCl_2$	
$Ca(NO_3)_2$	
$K_2Cr_2O_7$	rhythmic crystallisation
$CuSO_4$	excess H_2SO_4 detrimental
$CuCl_2$	
$FeSO_4$	
H_2O	nucleation
	growth
NH_4J	nucleation
K-halides	Pb^{2+}, Zn^{2+} conducive
KH_2PO_4	
KNO_3	
K_2SO_4	
K_2CrO_4	
$MgSO_4$	t = 45°C, borax conducive
$MgCl_2$	
$MnCl_2$	
LiF	
LiCl	
Li_2SO_4	
NaCl	Pb, Fe, Al, Zn conducive; caking inhibited by ferrocyanides; urea leads to octahedral prisms
Na_2CO_3	Na_2SO_4 conducive
$NaHCO_3$	
Na_2SO_4	wetting agents conducive
$Na_2S_2O_3$	
$NaClO_3$	
NaCN	
NH_4NO_3	paraffin, urea, dyes
	methods of crystallising
	effect of additives: conducive
$(NH_4)_2SO_4$	urea, Fe^{2+}, Mg^{2+}, tannin, pH5; Al^{3+} and Fe^{3+} lead to needle formation
	removal of admixtures
	crystal growth
	methods of crystallising
$(NH_4)_2S_2O_5$	
NH_4HCO_3	coarse grained, stabilisation
NH_4Cl	Zn^{++}, Pb^{++}, NH_4^+, wood extract
H_3PO_4	
$NH_4H_2PO_4$	Fe^{3+} and NH_4^+ conducive
$(NH_4)_2HPO_4$	
$NiSO_4$	
$Pb(NO_3)_2$	
$PbCO_3$	
$SrSO_4$	
$ZnSO_4$	
anthracene	
adipic acid	
sugars	
citric acid	
phenols	
xylenes	
naphthalene	
paraffin	
urea	methods and parameters of crystallisation
	NH_4Cl, $MgCO_3$
	glyoxal, cyanuric acid
	surface-active agents
Na-acetate	
NaK-tartarate	
pentaerythrite	
pepsine	
terephthalic acid	

[a] (The references, some 400 in number, are given by Nyvlt, 1971 Appendix A).

where

$$n^0 = \lim_{L \to 0} \frac{dn}{dL} \tag{16.12}$$

is the concentration of crystals of zero length which are the nuclei; it also is called the zero size population density.

The nucleation rate is

$$B^0 = \lim_{L \to 0} \frac{dn}{dt} = \lim_{L \to 0} \left(\frac{dL}{dt} \frac{dn}{dL} \right) \tag{16.13}$$

$$= Gn^0. \tag{16.14}$$

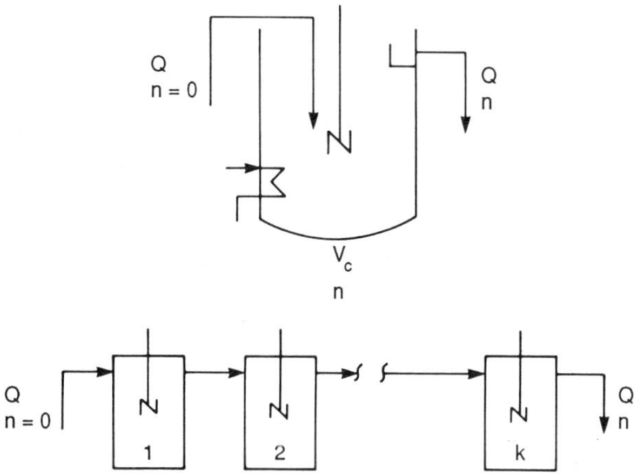

Figure 16.7. Material balancing of continuous stirred tank crystallizers (CSTC). (a) The single stage CSTC. (b) Multistage battery with overall residence time $\bar{t} = (1/Q) \sum_1^k V_{ci}$.

The number of crystals per unit volume is

$$n_c = \int_0^\infty n\,dL = \int_0^\infty n^0 \exp(-L/G\bar{t})\,dL = n^0 G\bar{t}. \tag{16.15}$$

The total mass of crystals per unit volume is

$$m_c = \int_0^\infty mn\,dL = \int_0^\infty a_v \rho_c L^3 n^0 \exp(-L/G\bar{t})\,dL$$
$$= 6a_v \rho_c n^0 (G\bar{t})^4, \tag{16.16}$$

where a_v is the volumetric shape factor and ρ_c the crystal density. Accordingly, the number of crystals per unit mass is

$$n_c/m_c = 1/6a_v \rho_c (G\bar{t})^3. \tag{16.17}$$

The mass of crystals per unit volume with length less than L or with dimensionless residence time less than x is

$$m_L = \int_0^L mn\,dL = a_v \rho_c (G\bar{t})^4 n^0 \int_0^x x^3 e^{-x}\,dx. \tag{16.18}$$

The value of the integral is

$$\int_0^x x^3 e^{-x}\,dx = 6[1 - e^{-x}(1 + x + x^2/2 + x^3/6)]. \tag{16.19}$$

This expression has a maximum value at $x = 3$ and the corresponding length L_{pr} is called the predominant length

$$L_{pr} = 3G\bar{t}. \tag{16.20}$$

The cumulative mass distribution is

$$\phi_m = m_L/m_c = 1 - e^{-x}(1 + x + x^2/2 + x^3/6), \tag{16.21}$$

and the differential mass distribution is

$$d\phi_m/dx = x^3 e^{-x}/6, \tag{16.22}$$

which has a maximum value of 0.224 at $x = 3$.

The nucleation rate must generate one nucleus for every crystal present in the product. In terms of M', the total mass rate of production of crystals,

$$B^0 = \frac{M'}{n_c/m_c} = \frac{M'}{6a_v \rho_c (G\bar{t})^3} = \frac{1.5M'}{a_v \rho_c L_{pr}^3}. \tag{16.23}$$

The principal quantities related by these equations are ϕ_m, $d\phi_m/dx$, L, L_{pr}, $\bar{t}$, n^0, and B^0. Fixing a certain number of these will fix the remaining one. Size distribution data from a CSTC are analyzed in Example 16.6. In Example 16.7, the values of the predominant length L_{pr} and the linear growth rate G are fixed. From these values, the residence time and the cumulative and differential mass distributions are found. The effect of some variation in residence time also is found. The values of n^0 and B^0 were found, but they are ends in themselves. Another kind of condition is analyzed in Example 16.4.

MULTIPLE STIRRED TANKS IN SERIES

Operation in several tanks in series will provide narrower size distributions. Equations were developed by Nyvlt (1971) for two main cases. With generation of nuclei in the first stage only, the cumulative and differential distributions for k stages are

$$\phi_m = 1 - e^{-kx} \sum_{j=0}^{k+2} \frac{n^j x^j}{j!}, \tag{16.24}$$

$$\frac{d\phi_m}{dx} = \frac{k(kx)^{k+2}}{(k+2)!} e^{-kx}. \tag{16.25}$$

The multistage distributions are plotted in Figure 16.8 for several values of the number of stages. Maxima of the differential distributions occur at

$$x_{max} = 1 + 2/k, \tag{16.26}$$

and the values of those maxima are represented by

$$\left(\frac{d\phi_m}{dx}\right)_{max} = \frac{k^{k+3}(1 + 2/k)^{k+2}}{(k+2)!} \exp[-(k+2)]. \tag{16.27}$$

Some numerical values are:

k	1	2	3	4	5	10
x_{max}	3	2	1.67	1.5	1.4	1.2
$(d\phi_m/dx)_{max}$	0.224	0.391	0.526	0.643	0.745	1.144

Nyvlt (1971) also develops equations for multistage crystallizers in which nuclei form at the same rate in all stages. For two such stages, the cumulative distribution is represented by

$$\phi_m = 1 - 0.5e^{-x}[1 + x + x^2/2 + x^3/6]$$
$$- 0.5e^{-2x}[1 + 2x + 2x^2 + (4/3)x^3 + (2/3)x^4]. \tag{16.28}$$

A comparison of two-stage crystallizers with nucleation in the first stage only and with nucleation in both stages appears in Figure 16.9. The uniformity of crystal size is not as good with nucleation proceeding in every stage; the difference is especially pronounced at larger numbers of stages, which are not shown here but are by Nyvlt (1971).

As in the operation of chemical reactors, multistaging requires shorter residence time for the same performance. For the same L/G ratio, the relative crystallization times of k stages and one stage to reach the peaks are given by Eq. (16.26) as

$$\bar{t}_k/\bar{t}_1 = (1 + k/2)/3, \tag{16.29}$$

which is numerically 0.4 for five stages. Not only is the time shortened, but the size distribution is narrowed. What remains is how to maintain substantial nucleation in only the first stage. This could be done by seeding the first stage and then operating at such low supersaturation that spontaneous nucleation is effectively retarded throughout the battery. Temperature control also may be feasible.

APPLICABILITY OF THE CSTC MODEL

Complete mixing, of course, is not practically realizable and in any event may have a drawback in that intense agitation will cause much secondary nucleation. Some rules for design of agitation of solid suspensions are discussed in Chapter 10, notably in Table 10.2; internal velocities as high as 1.0 ft/sec may be desirable.

Equations can be formulated for many complex patterns, combinations of mixed and plug flow, with decanting of supernatant liquor that contains the smaller crystals and so on. A modification to the CSTC model by Jancic and Garside (1976) recognizes that linear crystal growth rate may be size-dependent; in one instance

Example 16.6
Analysis of Size Distribution Data Obtained in a CSTC
Differential distribution data obtained from a continuous stirred tank crystallizer are tabulated.

w	L	$\sum w/L^3$
0.02	0.340	0.5089
0.05	0.430	1.1377
0.06	0.490	1.6477
0.08	0.580	2.0577
0.10	0.700	2.3493
0.13	0.820	2.5851
0.13	1.010	2.7112
0.13	1.160	2.7945
0.10	1.400	2.8310
0.09	1.650	2.8510
0.04	1.980	2.8562
0.03	2.370	2.8584

The last column is of the summation $\sum_0^L w_i/L_i^3$ at corresponding values of crystal length L. The volumetric shape factor is $a_v = 0.866$, the density is 1.5 g/mL, and the mean residence time was 2.0 hr. The linear growth rate G and the nucleation rate B^0 will be found.

The number of crystals per unit mass smaller than size L is

$$N = \frac{1}{a_v \rho} \sum_0^L \frac{w_i}{L_i^3}. \tag{1}$$

It is also related to the CSTC material balance by

$$dN/dL = n = n^0 \exp(-L/G\bar{t}). \tag{2}$$

Integration of Eq. (2) is

$$N = \int_0^L n^0 \exp(-L/G\bar{t})\, dL = G\bar{t}n^0[1 - \exp(-L/G\bar{t})]. \tag{3}$$

Combining Eqs. (1) and (3),

$$\sum w_i/L_i^3 = a_v \rho G\bar{t}n^0[1 - \exp(-L/G\bar{t})]. \tag{4}$$

The two unknowns G and n^0 may be found by nonlinear regression with the 12 available data for L_i. However, two representative values of L_i are taken here, and the unknowns are solved for by simultaneous solution of two equations. When

$$L = 0.58, \quad \sum = 2.0577,$$
$$L = 1.40, \quad \sum = 2.8310.$$

Substituting into Eq. (4) and ratioing,

$$\frac{2.8310}{2.0577} = \frac{1 - \exp(-1.4/G\bar{t})}{1 - \exp(-0.58/G\bar{t})};$$

by trial,

$$G\bar{t} = 0.5082$$
$$G = 0.5082/2 = 0.2541.$$

With $L = 1.4$ in Eq. (4),

$$2.8310 = 0.866(1.5)(0.5082)n^0[1 - \exp(-1.4/0.5082)],$$

from which

$$n^0 = 4.58 \text{ nuclei/mm}^4$$
$$= 4.58(10)^{12} \text{ nuclei/m}^4.$$

Accordingly,

$$B^0 = Gn^0 = 0.2541(10)^{-3}(4.58)(10)^{12} = 1.16(10)^9 \text{ nuclei/m}^3\text{ hr}.$$

The cumulative mass size distribution is represented by

$$\phi_m = 1 - e^{-x}(1 + x + x^2/2 + x^3/6)$$

with

$$x = L/G\bar{t} = L/0.5082.$$

This distribution should be equivalent to the original one, but may not check closely because the two points selected may not have been entirely representative. Moreover, although the data were purportedly obtained in a CSTC, the mixing may not have been close to ideal.

they find that

$$G = G^0(1 + L/G^0\bar{t})^{0.65}.$$

Other studies have tried to relate sizes of draft tubes, locations and sizes of baffles, circulation rate, and so on to crystallization behavior. So far the conclusions are not general enough to do a designer much good. A possibly useful concept, the separation index (SI), is mentioned by Mullin (1976, p. 293):

$$\text{SI} = (\text{kg of 1 mm equivalent crystals})/\text{m}^3\text{ hr}.$$

For inorganic salts in water at near ambient temperature, a value of SI in the range of 100–150 kg/m^3/hr may be expected. An illustration of the utilization of pilot plant data and plant experience in the design of a urea crystallizer is in Example 16.1.

In general, the design policy to be followed is to utilize as much laboratory and pilot plant information as possible, to work it into whatever theoretical pattern is applicable, and to finish off with a comfortable safety factor. There may be people who know how; they should be consulted.

16.5. KINDS OF CRYSTALLIZERS

The main kinds of crystallizers are represented in Figure 16.10. They will be commented on in order. Purification of products of melt crystallization is treated separately.

Batch crystallizers are used primarily for production of fine chemicals and pharmaceuticals at the rate of 1–100 tons/week. The one exception is the sugar industry that still employs batch vacuum crystallization on a very large scale. In that industry, the syrup is concentrated in triple- or quadruple-effect evaporators, and crystallization is completed in batch vacuum pans that may or may not be equipped with stirrers [Fig. 16.11(g)].

EXAMPLE 16.7
Crystallization in a Continuous Stirred Tank with Specified Predominant Crystal Size

Crystals of citric acid monohydrate are to made in a CSTC at 30°C with predominant size $L_{pr} = 0.833$ mm (20 mesh). The density is 1.54 g/mL, the shape factor $a_v = 1$ and the solubility is 39.0 wt %. A supersaturation ratio $C/C_0 = 1.05$ is to be used.

Take the growth rate, $G = 2\bar{v}$, to be one-half of the value given in Table 16.3:

$$G = dL/d\theta = 4(10^{-8})\ \text{m/sec},\ 0.144\ \text{mm/hr}.$$

The predominant size is related to other quantities by

$$L_{pr} = 0.833 = 3G\bar{t},$$

from which

$$\bar{t} = 0.833/(3)(0.144) = 1.93\ \text{hr}.$$

For a mass production rate of 15 kg/hr of crystals, $C = 15$, the nucleation rate is

$$B^0 = \frac{1.5C}{a_v \rho_c L_{pr}^3} = \frac{1.5(15)}{1(1.5)[0.833(E-3)]^3}$$
$$= 2.595(10)^{10}\ \text{nuclei/m}^3\ \text{hr}.$$

The zero size concentration of nuclei is

$$n^0 = B^0/G = 2.595(10)^{10}/4(10)^{-8} = 6.49(10)^{17}\ \text{nuclei/m}^4.$$

Accordingly, the equation of the population density is

$$n = n^0 \exp(-L/G\bar{t}) = \exp(41.01 - 360L).$$

The cumulative mass distribution is

$$\phi_m = 1 - e^{-x}(1 + x + x^2/2 + x^3/6),$$

where

$$x = L/G\bar{t} = 3.60L,\quad \text{with } L \text{ in mm}.$$

The differential distributions are differences between values of ϕ_m at successive values of crystal length L. The tabulation shows cumulative and differential distributions at the key $\bar{t} = 1.93$ hr, and also at 1.5 and 3.0 hr. The differential distributions are plotted and show the shift to larger sizes as residence time is increased, but the heights of the peaks are little affected.

		$\bar{t}$ = 1.5 h		$\bar{t}$ = 1.93 h		$\bar{t}$ = 3.0 h	
Mesh	mm	Cum	Diff	Cum	Diff	Cum	Diff
0	0	1.0000	.0002	1.0000	.0020	1.0000	.0517
6	3.327	.9998	.0010	.9980	.0068	.9483	.0623
7	2.794	.9989	.0041	.9912	.0185	.8859	.0913
8	2.362	.9948	.0136	.9728	.0424	.7947	.1226
9	1.981	.9812	.0350	.9304	.0778	.6720	.1410
10	1.651	.9462	.0603	.8526	.1021	.5310	.1260
12	1.397	.8859	.0983	.7505	.1322	.4051	.1183
14	1.168	.7876	.1152	.6183	.1278	.2868	.0873
16	.991	.6724	.1343	.4905	.1268	.1995	.0693
20	.833	.5381	.1304	.3637	.1071	.1302	.0482
24	.701	.4076	.1157	.2565	.0845	.0820	.0323
28	.589	.2919	.0929	.1720	.0614	.0497	.0204
32	.495	.1990	.0684	.1106	.0416	.0293	.0123
35	.417	.1306	.0483	.0690	.0274	.0169	.0074
42	.351	.0823	.0324	.0416	.0173	.0096	.0043
48	.295	.0499	.0312	.0243	.0108	.0053	.0025
60	.246	.0287	.0119	.0135	.0058	.0028	.0013
65	.208	.0168	.0073	.0077	.0034	.0015	.0007
80	.175	.0095	.0043	.0042	.0019	.0008	.0004
100	.147	.0052	.0037	.0023	.0016	.0004	.0003
150	.104	.0015	.0011	.0006	.0005	.0001	.0001
200	.074	.0004	.0004	.0002	.0002	.0000	.0000

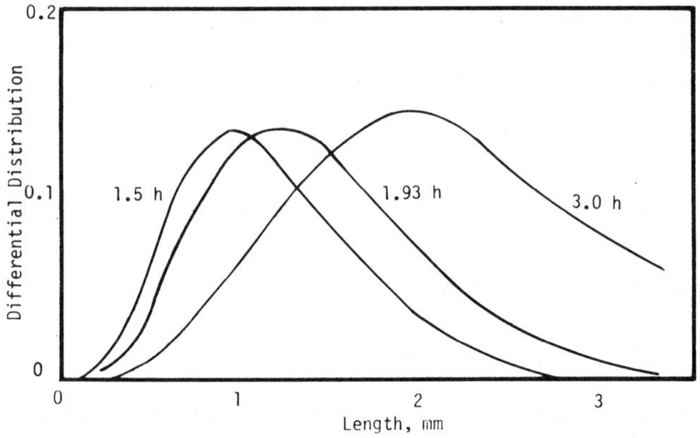

Natural circulation evaporators like those shown on Figure 8.16 may be equipped for continuous salt removal and thus adapted to crystallization service. For large production rates, however, forced circulation types such as the DTB crystallizer of Figure 16.10(g), with some control of crystal size, are the most often used. The lower limit for economic continuous operation is 1–4 tons/day of crystals, and the upper limit in a single vessel is 100–300 tons/day, but units in parallel can be used for unlimited capacity.

Many special types of equipment have been developed for particular industries, possibly extreme examples being the simple open ponds for solar evaporation of brines and recovery of salt, and the specialized vacuum pans of the sugar industry that operate with syrup on the tubeside of calandrias and elaborate internals to eliminate entrainment. Some modifications of basic types of crystallizers often carry the inventor's or manufacturer's name. For their identification, the book of Bamforth (1965) may be consulted.

The basic equipment descriptions following carry the letter designations of Figure 16.10.

(a) Jacketed pipe scraped crystallizers. These are made with inner pipe 6–12 in. dia and 20–40 ft long, often arranged in tiers of three or more connected in series. Scraper blades rotate at 15–30 rpm. Temperatures of −75 to +100°F have been used and viscosities in excess of 10,000 cP present no problems. Although the action is plug flow with tendency to uniform crystal size, the larger particles settle to the bottom and grow at the expense of the smaller ones that remain suspended, with the result that a wide range of sizes is made. Capacity is limited by rates of heat transfer; coefficients of 10–25 Btu/(hr)(sqft)(°F) usually are attainable. Higher coefficients are obtainable in Votators (Cherry Burrell Co.) that have more intense scraping action. Pilot units of 4 in. by 4 ft and larger are made.

(b) Swenson–Walker type. In comparison with jacketed pipes, they have the advantage of being more accessible for cleaning. The standard unit is 24 in. wide, 26 in. high, and 10 ft long. Four units in line may be driven off one shaft. Capacity is limited by heat transfer rates which may be in the range of 10–25 Btu/(hr)(sqft)(°F), with an

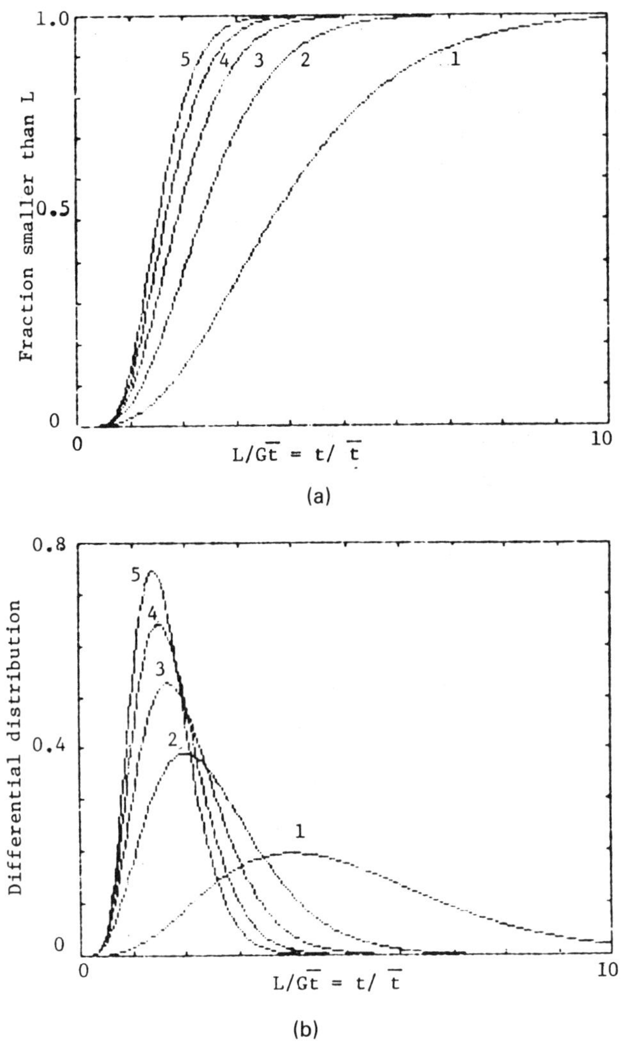

Figure 16.8. Theoretical crystal size distributions from an ideal stirred tank and from a series of tanks with generation of nuclei only in the first tank. Equations of the curves and for the peak values are in the text. (a) Cumulative distributions. (b) Differential distributions.

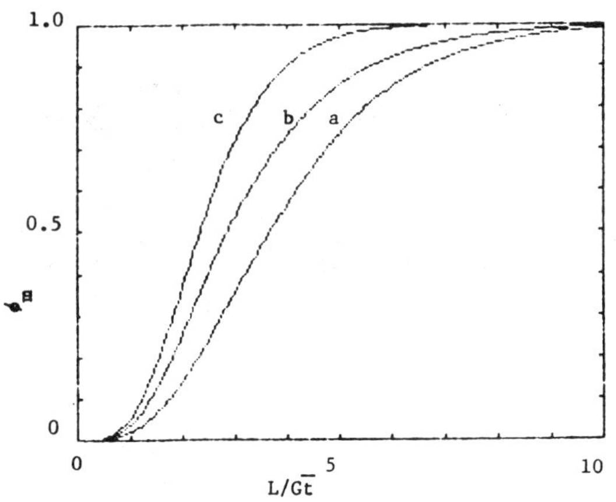

Figure 16.9. Cumulative size distribution in continuous stirred tanks. (a) one tank; (b) two tanks in series, nucleation in both; (c) two tanks in series, nucleation in only the first.

effective area of 3 sqft/ft of length. According to data in *Chemical Engineers' Handbook* (3rd ed., McGraw-Hill, New York, 1950, p. 1071), a 40 ft unit is able to produce 15 tons/day of trisodium phosphate, and a 50 ft unit can make 8 tons/day of Glaubers salt. The remarks about crystal size distribution made under item (a) apply here also.

(c) Batch stirred and cooled types. Without agitation, crystallization time can be 2–4 days; an example is given in *Chemical Engineer's Handbook* (1950, p. 1062). With agitation, times of 2–8 hr are sometimes cited. The limitation is due to attainable rates of heat transfer. Without encrustation of surfaces by crystals, coefficients of 50–200 Btu/(hr)(sqft)(°F) are realizable, but temperature differences are maintained as low as 5–10°F in order to keep supersaturation at a level that prevents overnucleation. Stirring breaks corners off crystals and results in secondary nucleation so that crystal size is smaller than in unagitated tanks. Larger crystal sizes are obtained by the standard practice of seeding with an appropriate range of fine crystals. Calculation of the performance of such an operation is made in Example 16.5. Teflon heat transfer tubes that are thin enough to flex under the influence of circulating liquid cause a continual descaling that maintains good heat transfer consistently, 20–65 Btu/(hr)(sqft)(°F). Circulating types such as Figures (d) and (e) often are operated in batch mode, the former under vacuum if needed. High labor costs keep application of batch crystallizers to small or specialty production.

(d) Circulating evaporators. Some units are built with internal coils or calandrias and are simply conventional evaporators with provisions for continual removal of crystals. Forced circulation and external heat exchangers provide better temperature control. High velocities in the tubes keep the surfaces scoured. Temperature rise is limited to 3–10°F per pass in order to control supersaturation and nucleation. Operation under vacuum often is practiced. When the boiling point elevation is not excessive, the off vapors may be recompressed and used again for heating purposes. Multiple effect units in series for thermal economy may be used for crystallizing evaporators as they are for conventional evaporation. Pilot units of 2 ft dia are made, and commercial units up to 40 ft dia or so.

(e) Circulating cooling crystallizers. Such operations are feasible when the solubility falls sharply with decreasing temperature. Coolers usually are applied to smaller production rates than the evaporative types. Cooling is 1–2°F per pass and temperature differences across the tubes are 5–15°F.

The special designs of Figure 16.11 mostly feature some control of crystal size. They are discussed in order.

(a) Draft tube baffle (DTB) crystallizer. The growing crystals are circulated from the bottom to the boiling surface with a slow moving propeller. Fine crystals are withdrawn from an annular space, redissolved by heating to destroy unwanted nuclei and returned with the feed liquor. The temperature rise caused by mixing of heated feed and circulating slurry is 1–2°F. The fluidized bed of large crystals at the bottom occupies 25–50% of the vessel volume. Holdup time is kept sufficient for crystal growth to the desired size. Products such as KCl, $(NH_4)_2SO_4$, and $(NH_4)H_2PO_4$ can be made in this equipment in the range of 6–20 mesh. Reaction and crystallization can be accomplished simultaneously in DTB units. The reactants can be charged into the recirculation line or into the draft tube. Examples are the production of ammonium sulfate from ammonia and sulfuric acid and the neutralization of waste acids with lime. The heat of reaction is removed by evaporation of water.

(b) Direct contact refrigeration. Such equipment is operated as

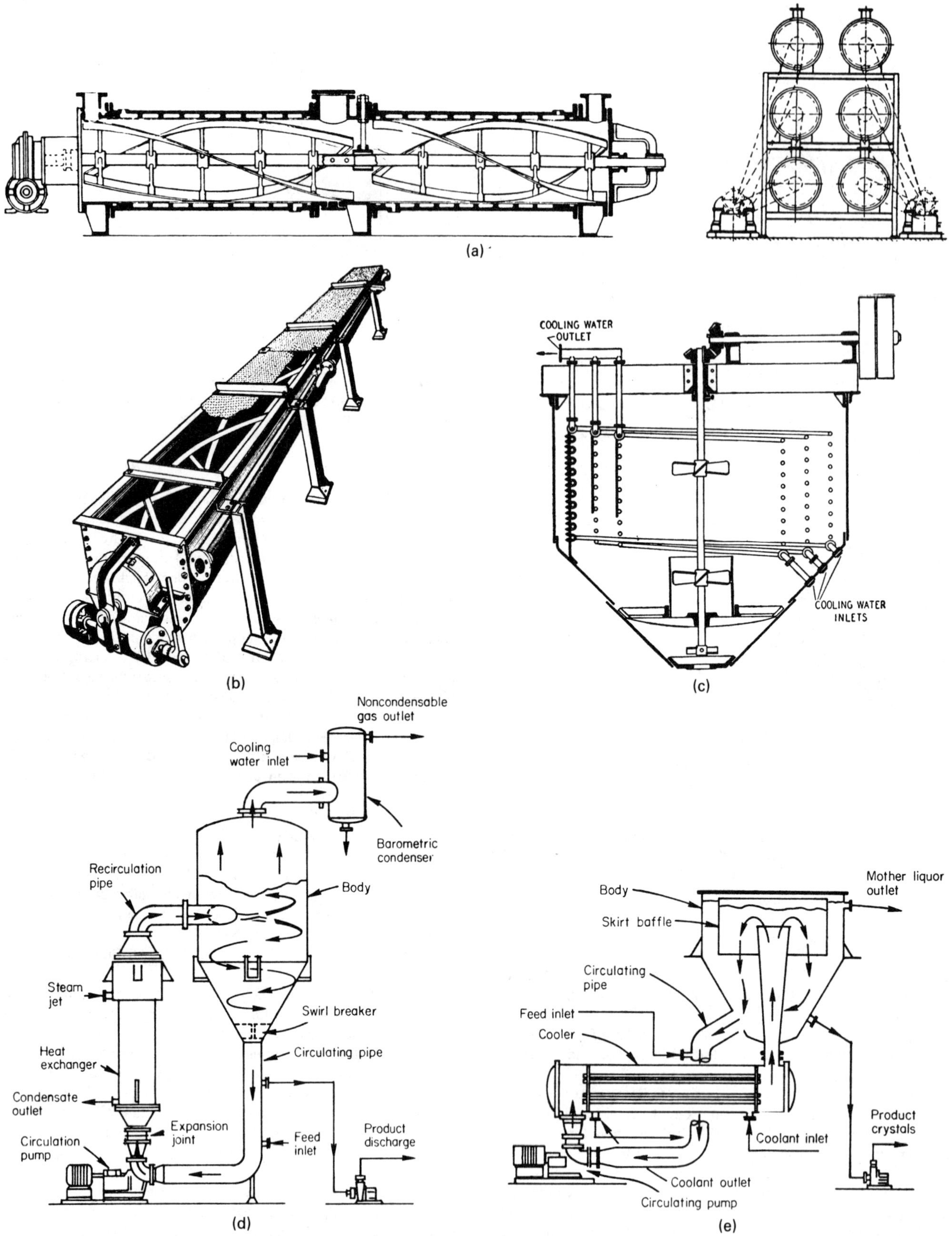

Figure 16.10. Basic types of batch and continuous crystallizers. (a) Jacketed scraped pipe and assembly of six units (*Riegel,* Chemical Process Machinery, *Reinhold, N.Y., 1953*). (b) Swenson–Walker jacketed scraped trough (*Swenson Evaporator Co., Riegel, 1953*). (c) Batch stirred tank with internal cooling coil (*Badger, and McCabe,* Elements of Chemical Engineering, *McGraw-Hill, New York, 1936*). (d) Crystallization by evaporation, with circulation through an external heater (*Schweitzer,* loc. cit., *p. 2.170*). (e) Crystallization by chilling, with circulation through an external cooler. (*P.A. Schweitzer, Ed.,* Handbook of Separation Techniques for Chemical Engineers, *McGraw-Hill, New York, 1979, p. 2.166*).

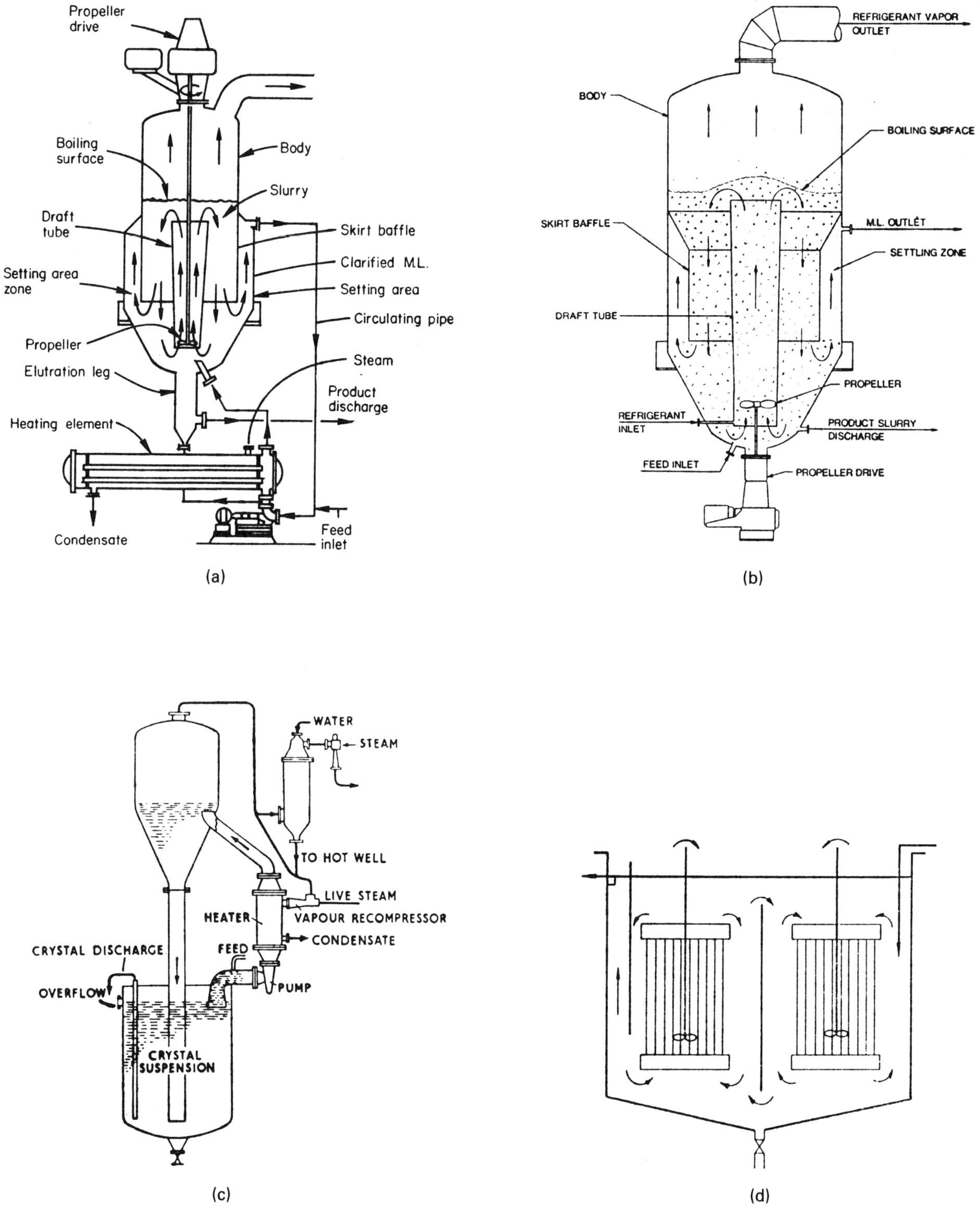

Figure 16.11. Examples of special kinds of crystallizers. (a) Swenson draft tube baffle (DTB) crystallizer; crystals are brought to the surface where growth is most rapid, the baffle permits separation of unwanted fine crystals, resulting in control of size. (b) Direct chilling by contact with immiscible refrigerant, attains very low temperatures and avoids encrustation of heat transfer surfaces. Freons and propane are in common use. (c) Oslo "Krystal" evaporative classifying crystallizer. Circulation is off the top, the fine crystals are destroyed by heating, large crystals grow in the body of the vessel. (d) Twinned crystallizer. When one chamber is maintained slightly supersaturated and the other slightly subsaturated, coarse crystals can be made. (*Nyvlt, 1971*). (e) APV–Kestner long tube salting evaporator; large crystals (0.5 mm or so) settle out. (f) Escher-Wyss or Tsukushima DP (double propeller) crystallizer. The double propeller maintains upward flow in the draft tube and downward flow in the annulus, resulting in highly stable suspensions. (g) A vacuum pan for crystallization of sugar (*Honolulu Iron Works*).

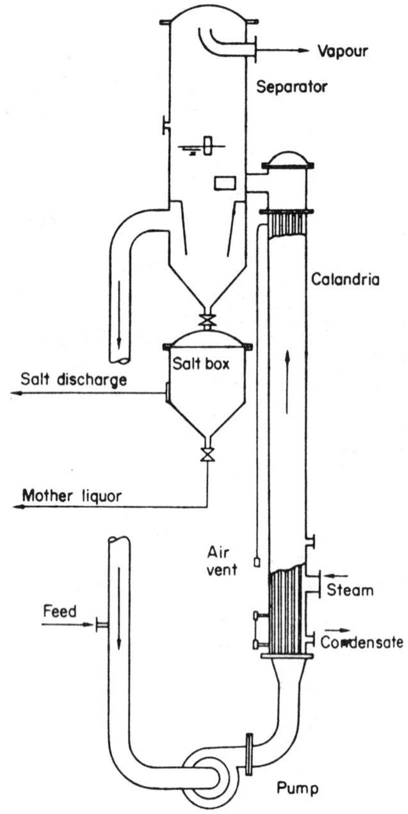
Vapour
Separator
Calandria
Salt box
Salt discharge
Mother liquor
Air vent
Steam
Feed
Condensate
Pump

(e)

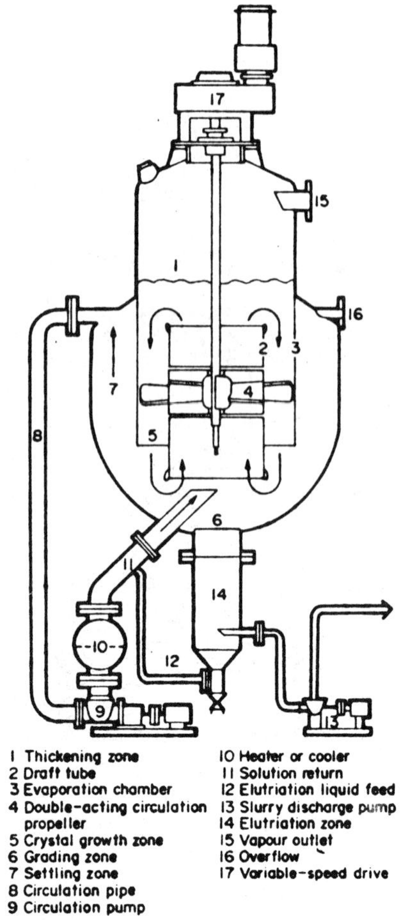
17
15
16
1
2
3
4
5
6
7
8
9
10
11
12
13
14
1 Thickening zone
2 Draft tube
3 Evaporation chamber
4 Double-acting circulation propeller
5 Crystal growth zone
6 Grading zone
7 Settling zone
8 Circulation pipe
9 Circulation pump
10 Heater or cooler
11 Solution return
12 Elutriation liquid feed
13 Slurry discharge pump
14 Elutriation zone
15 Vapour outlet
16 Overflow
17 Variable-speed drive

(f)

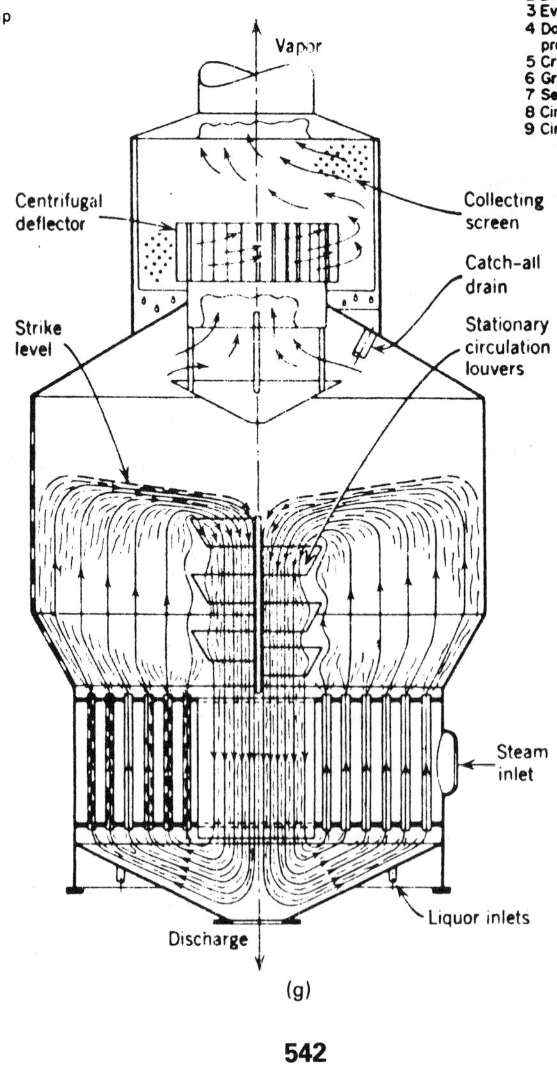
Vapor
Centrifugal deflector
Collecting screen
Catch-all drain
Strike level
Stationary circulation louvers
Steam inlet
Liquor inlets
Discharge

(g)

Figure 16.11—(*continued*)

low as −75°F. Essentially immiscible refrigerant is mixed with the liquor and cools it by evaporation. The effluent refrigerant is recovered, recompressed, and recycled. Direct contacting eliminates the need for temperature difference across a heat transfer tube which can be economically more than 5–15°F, and also avoids scaling problems since the liquor must be on the outside of the tubes when refrigerant is used. Examples are crystallization of caustic with freon or propane and of *p*-xylene with propane refrigerant.

(c) Oslo "Krystal" evaporative classifying crystallizer. The supernatant liquid containing the fines is circulated through the external heater where some of the fines are redissolved because of the temperature rise. The settled large crystals are withdrawn at the bottom. The recirculation rate is much greater than the fresh feed rate. In one operation of $MgSO_4.7H_2O$ crystallization, fresh feed saturated at 120°C is charged at 2000 kg/hr to the vessel maintained at 40°C and is mixed with a recirculated rate of 50,000 kg/hr to produce a mixture that is temporarily at 43°C, which then evaporates and cools. Vessel sizes as large as 15 ft dia and 20 ft high are mentioned in the literature. The same principle is employed with cooling type crystallization operations.

(d) Twinned crystallizer. Feed is to the right chamber. The rates of recirculation and forward feed are regulated by the position of the center baffle. Improved degree of uniformity of crystal size is achieved by operating one zone above saturation temperature and the other below. Fine particles are dissolved and the larger ones grow at their expense. Even with both zones at the same temperature, the series operation of two units in series gives more nearly uniform crystal size distribution than can be made in a single stirred tank. It is not stated if any such crystallizers are operated outside Nyvlt's native land, Czechoslovakia, that also produces very fine tennis players (Lendl, Mandlikova, Navratilova, Smid, and Sukova).

(e) APV-Kestner long tube vertical evaporative crystallizers are used to make small crystals, generally less than 0.5 mm, of a variety of substances such as NaCl, Na_2SO_4, citric acid, and others; fine crystals recirculate through the pump and heater.

(f) Escher-Wyss (Tsukushima) double propeller maintains flow through the draft tube and then annulua and maintains highly stable suspension characteristics.

(g) Sugar vacuum pan. This is an example of the highly specialized designs developed in some long-established industries. Preconcentration is effected in multiple effect evaporators; then crystallization is accomplished in the pans.

16.6. MELT CRYSTALLIZATION AND PURIFICATION

Some mixtures of organic substances may be separated advantageously by cooling and partial crystallization. The extent of such recovery is limited by the occurrence of eutectic behavior. Examples 16.2 and 16.8 consider such limitations. Sometimes these limitations can be circumvented by additions of other substances that change the phase equilibria or may form easily separated compounds with one of the constituents that are subsequently decomposed for recovery of its constituents.

Thus the addition of *n*-pentane to mixtures of *p*-xylene and *m*-xylene permits complete separation of the xylenes which form a binary eutectic with 11.8% para. Without the *n*-pentane, much para is lost in the eutectic, and none of the meta is recoverable in pure form. A detailed description of this process is given by Dale (1981), who calls it extractive crystallization. Other separation processes depend on the formation of high melting molecular compounds or clathrates with one of the constituents of the mixture. One example is carbon tetrachloride that forms a compound with *p*-xylene and alters the equilibrium so that its separation from *m*-xylene is facilitated. Hydrocarbons form high-melting hydrates with water; application of propane hydrate formation for the desalination of water has been considered. Urea forms crystalline complexes with straight chain paraffins such as the waxy ingredients of lubricating oils. After separation, the complex may be decomposed at 75–80°C for recovery of its constituents. This process also is described by Dale (1981). Similarly thiourea forms crystalline complexes with isoparaffins and some cyclic compounds.

Production rates of melt crystallization of organic materials usually are low enough to warrant the use of scraped surface crystallizers like that of Figure 16.10(a). A major difficulty in the production of crystals is the occlusion of residual liquor on them which cuts the overall purity of the product, especially so because of low temperatures near the eutectic and the consequent high viscosities. Completeness of removal of occluded liquor by centrifugation or filtration often is limited because of the fragility and fineness of the organic crystals.

MULTISTAGE PROCESSING

In order to obtain higher purity, the first product can be remelted and recrystallized, usually at much higher temperatures than the eutectic so that occlusion will be less, and of course at higher concentration. In the plant of Figure 16.12, for instance, occlusion from the first stage is 22% with a content of 8% *p*-xylene and an overall purity of 80%; from the second stage, occlusion is 9% with a PX content of 42% but the overall purity is 95% PX; one more crystallization could bring the overall purity above 98% or so.

Because the handling of solids is difficult, particularly that of soft organic crystals, several crystallization processes have been developed in which solids do not appear outside the crystallizing equipment, and the product leaves the equipment in molten form. For organic substances, crystalline form and size usually are not of great importance as for products of crystallization from aqueous solutions. If needed, the molten products can be converted into flakes or sprayed powder, or in extreme cases they can be recrystallized out of a solvent.

THE METALLWERK BUCHS PROCESS

The Metallwerk Buchs (MWB) process is an example of a batch crystallization that makes a molten product and can be adapted to multistaging when high purities are needed. Only liquids are transferred between stages; no filters or centrifuges are needed. As appears on Figure 16.13, the basic equipment is a vertical thin film shell-and-tube heat exchanger. In the first phase, liquor is recirculated through the tubes as a film and crystals gradually freeze out on the cooled surface. After an appropriate thickness of solid has accumulated, the recirculation is stopped. Then the solid is melted and taken off as product or transferred to a second stage for recrystallization to higher purity.

PURIFICATION PROCESSES

As an alternative to multistage batch crystallization processes with their attendant problems of material handling and losses, several types of continuous column crystallizers have been developed, in which the product crystals are washed with their own melts in countercurrent flow. Those illustrated in Figures 16.14–16.17 will be described. Capacities of column purifiers as high as 500 gal/(hr)(sqft) have been reported but they can be less than one-tenth as much. Lengths of laboratory size purifiers usually are less than three feet.

Schildknecht Column [Fig. 16.14(a)]. This employs a rotating spiral or screw to move the solids in the direction against the flow of

Example 16.8
Crystallization from a Ternary Mixture

The case is that of mixtures of the three isomeric nitrotoluenes for which the equilibrium diagram is shown. Point P on the diagram has the composition 0.885 para, 0.085 meta, and 0.030 ortho. The temperature at which crystals begin to form must be found experimentally or it may be calculated quite closely from the heats and temperatures of fusion by a method described for instance by Walas (Example 8.9, *Phase Equilibria in Chemical Engineering,* Butterworths, Stoneham, MA, 1985). It cannot be found with the data shown on the diagram. In the present case, incipient freezing is at 46°C, with para coming out at point P on the diagram. As cooling continues, more and more pure para crystals form. The path is along straight line PS which corresponds to constant proportions of the other two isomers since they remain in the liquid phase. At point S, −13°C, which is on the eutectic trough of meta and ortho, the meta also begins to precipitate. Para and meta continue to precipitate along the trough until the ternary eutectic E is reached at −40°C when complete solidification occurs. The cooling path is shown on the phase diagram. The recovery of pure para at equilibrium at various temperatures and the composition of the liquid phase are tabulated. (Coulson and Warner, *A Problem in Chemical Engineering Design: The Manufacture of Mononitrotoluene,* Inst. of Chem. Eng., Rugby, England, 1949).

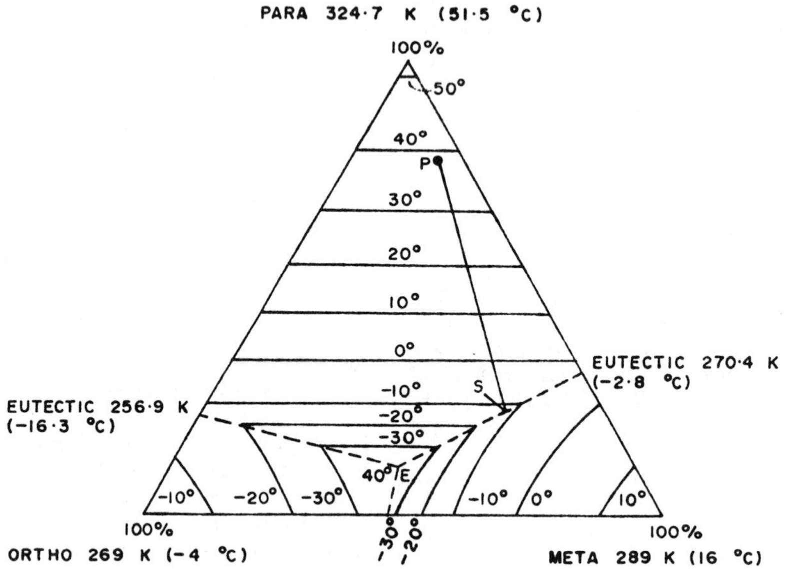

Temperature, °C	Para-deposited, kg	Mother liquor, kg	Composition of mother liquor (per cent by weight) O	M	P
46	0	100	3·0	8·5	88·5
40	39·6	60·4	5·0	14·0	81·0
30	66·7	33·3	8·7	24·8	66·5
20	75·0	25·0	12·0	34·0	54·0
10	79·6	20·4	14·7	41·6	43·7
0	82·3	17·7	16·7	48·0	35·3
−10	84·8	15·2	17·0	53·9	27·1

the fluid. The conveyor is of open construction so that the liquid can flow through it but the openings are small enough to carry the solids. Throughputs of 50 L/hr have been obtained in a 50 mm dia column. Because of the close dimensional tolerances that are needed, however, columns larger than 200 mm dia have not been successful. Figure 16.14(a) shows a section for the formation of the crystals, but columns often are used only as purifiers with feed of crystals from some external source.

***Philips Crystallization Process* [*Fig. 16.14(b)*].** The purifying equipment consists of a vessel with a wall filter and a heater at the bottom. Crystals are charged from an external crystallizer and forced downwards with a reciprocating piston or with pulses from a pump. The washing liquid reflux flows from the melting zone where it is formed upward through the crystal bed and out through the wall filter. Pulse displacement is 0.3–0.6 cm/sqcm of column cross section, with a frequency of 200–250/min. For many applications reflux ratios of 0.05–0.60 are suitable. Evaluation of the proper combination of reflux and length of purifier must be made empirically.

From a feed containing 65% p-xylene, a column 1000 sqcm in cross section can make 99% PX at the rate of 550 kg/hr, and 99.8%

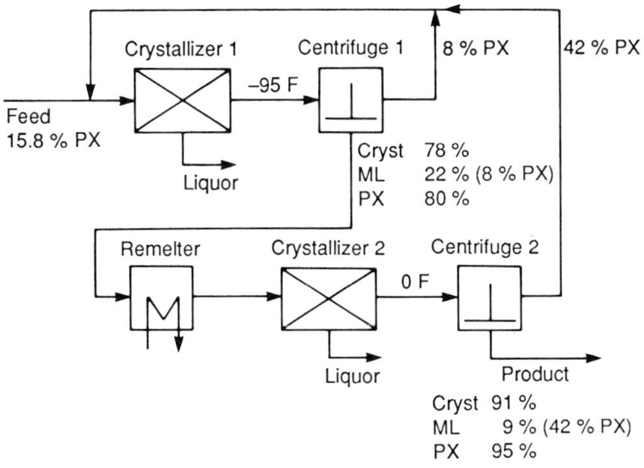

Figure 16.12. Humble two-stage process for recovery of p-xylene by crystallization. Yield is 82.5% of theoretical. ML = mother liquor, PX = p-xylene (*Haines, Powers and Bennett, 1955*).

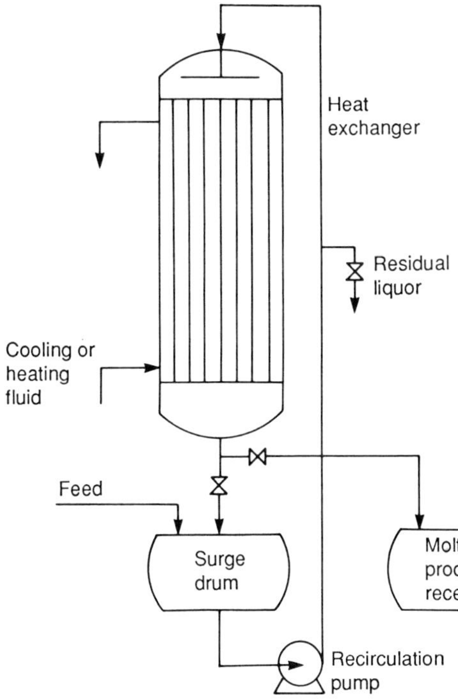

Figure 16.13. MWB (Metallwerk Buchs) batch recirculating crystallizer, with freezing on and melting off insides of thin film heat exchanger tubes; adaptable to multistage processing without external solids handling (*Mützenberg and Saxer, 1971*).

PX at 100 kg/hr; this process has been made obsolete, however, by continuous adsorption with molecular sieves. Similarly, a feed of 83 mol % of 2-methyl-5-vinyl pyridine has been purified to 95% at the rate of 550 g/hr cm^2 and 99.7% at 155 g/hr cm^2. At one time, columns of more than 60 cm dia were in operation.

***Brodie Crystallizer-Purifier* [*Fig. 16.14(c)*].** This equipment combines a horizontal scraped surface crystallizer with a vertical purifying section. The capacity and performance of the purifier depends strongly on the sizes of the crystals that enter that zone. In order to ensure adequate crystal size, residence times in the crystallizing zone as long as 24 hr may be needed. No data of residence times are stated in the original article. Some operating data on the recovery of para-dichlorbenzene from a mixture containing 75% of this material are reported for a purifier that is 1.14 sqft cross section as follows, as well as data for some other materials.

Reflux ratio	2	0.5	0.25
Feed rate (gal/hr)	29	60	90
Residue rate (gal/hr)		20	30
Product rate (gal/hr)	20	40	60
PDCB in residue (%)		25	25
Product purity (%)	99.997	99.99	99.5

***TNO Bouncing Ball Purifier* (*Fig. 16.15*).** The basis for this design is the observation that small crystals melt more readily and have a greater solubility than large ones. The purifier is a column with a number of sieve trays attached to a central shaft that oscillates up and down. As the slurry flows through the tower, bouncing balls on each tray impact the crystals and break up some of them. The resulting small crystals melt and enrich the liquid phase, thus providing an upward refluxing action on the large crystals that continue downward to the melting zone at the bottom. Reflux is returned from the melting zone and product is taken off.

Specifications of a pilot plant column are:

diameter, 80 mm,
hole size, 0.6 × 0.6 mm,
number of balls/tray, 30,
diameter of balls, 12 mm,
amplitude of vibration, 0.3 mm,
frequency, 50/sec,
number of trays, 13,
tray spacing, 100 mm.

For the separation of benzene and thiophene that form a solid solution, a tray efficiency of more than 40% could be realized. Flow rates of 100–1000 kg/m^2 hr have been tested. The residence time of crystals was about 30 min per stage. Eutectic systems also have been handled satisfactorily. A column 500 mm dia and 3 m long with 19 trays has been built; it is expected to have a capacity of 300 tons/yr.

***Kureha Double-Screw Purifier* (*Fig. 16.16*).** This unit employs a double screw with intermeshing blades that express the liquid from the crystal mass as it is conveyed upward. The melt is formed at the top, washes the rising crystals countercurrently, and leaves as residue at the bottom. A commercial unit has an effective height of 2.6 m and a cross section of 0.31 m^2. When recovering 99.97% p-dichlorbenzene from an 87% feed, the capacity is 7000 metric tons/yr. The feed stock comes from a tank crystallizer and filter. Data on other eutectic systems are shown, and also on separation of naphthalene and thiophene that form a solid solution; a purity of 99.87% naphthalene is obtained in this equipment.

***Brennan–Koppers Purifier* (*Fig. 16.17*).** This equipment employs top melting like the Kureha and wall filters like the Philips. Upward movement of the crystals is caused by drag of the flowing fluid. The crystal bed is held compact with a rotating top plate or piston that is called a harvester. It has a corrugated surface that scrapes off the top of the top of the bed and openings that permit the crystals to enter the melting zone at any desired rate. The melt flows downward through the openings in the harvester, washes the

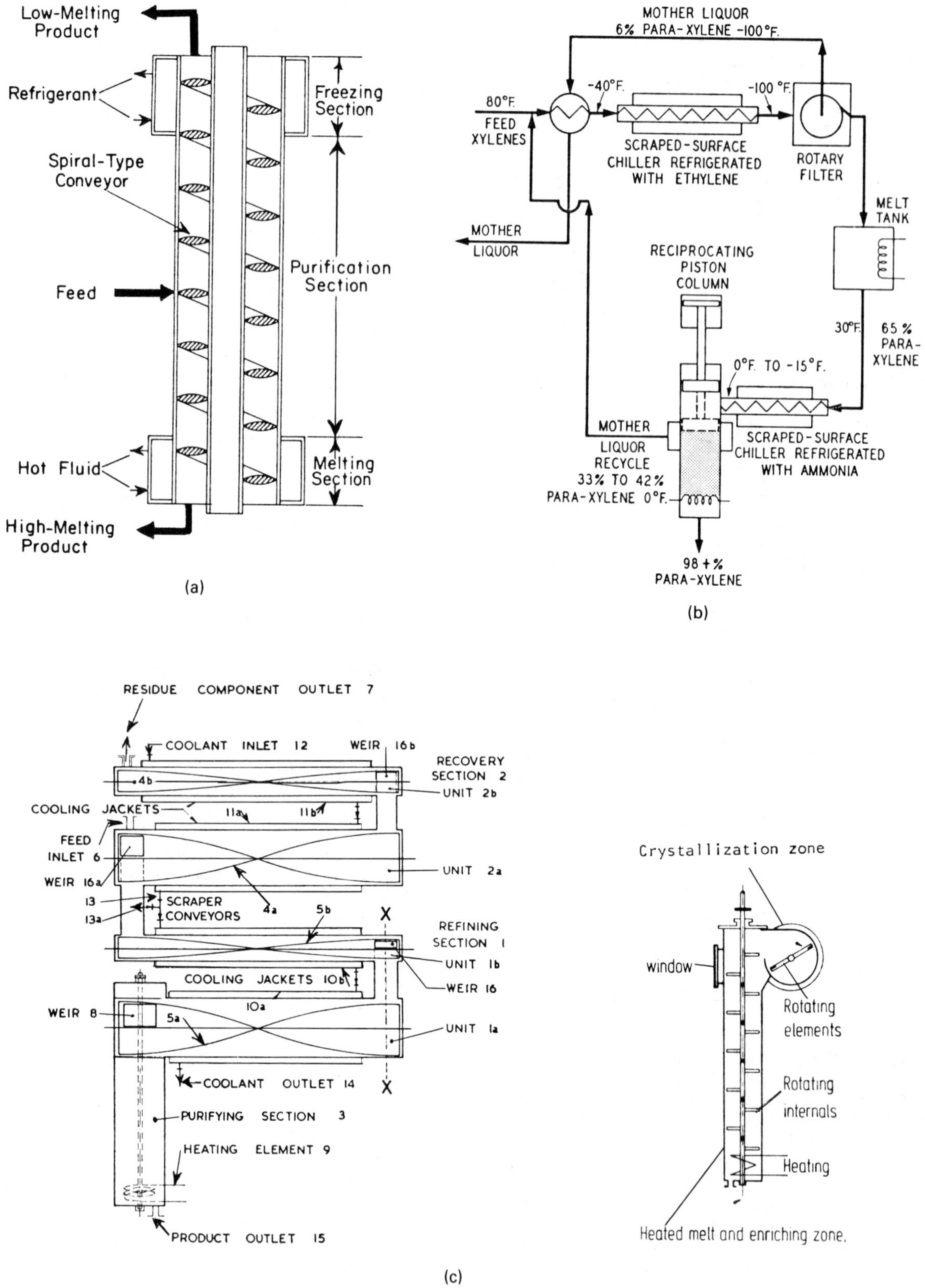

Figure 16.14. Three types of crystal purifiers with different ways of transporting the crystals. (a) Spiral or screw conveyor type, laboratory scale, but successful up to 200 mm dia [*Schildknecht,* Z. Anal. Chem. **181,** *254 (1961)*]. (b) Philips purifier with reciprocating piston or pulse pump drive [*McKay, Dale, and Weedman,* Ind. Eng. Chem. **52,** *197 (1960)*]. (c) Combined crystallizer and purifier, gravity flow of the crystals; purifier details on the right (*Brodie, 1971*).

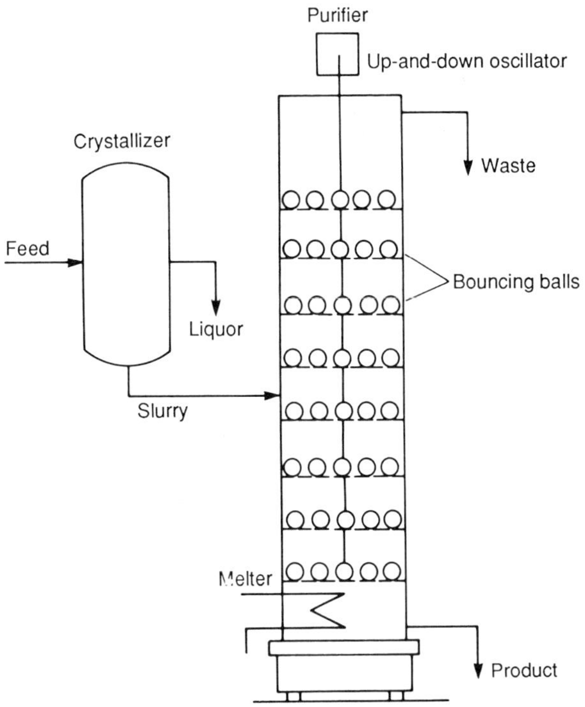

Figure 16.15. TNO Crystal Purifier (*Arkenbout et al., 1976, 1978*).

upwardly moving crystals, and leaves through the sidewall filter as residue. The movement of crystals is quite positive and not as dependent on particle size as in some other kinds of purifiers. Data are given in the patent (U.S. Pat. 4,309,878) about purification of 2,6-ditertiary butyl para cresol; the harvester was operated at 40–60 rpm and filtration rates of 100 lb/(hr)(sqft) were obtained. Other information supplied directly by E.D. Brennan are that a 24 in. dia unit stands 9 ft high without the mixer and that the following performances have been achieved:

	Diameter (in.)	Purity (wt %) Feed	Purity (wt %) Product	Prod. Rate (lb/hr/ft^2)
A. Pilot plant tests				
Acetic acid	3	83	99.85	100
p-Dichlorbenzene	6	70	99.6	380
Naphthalene (high sulfur)	6	68	98	220
Di-*t*-butyl-*p*-cresol	3	85	99.1	210
	6	85	99.1	230
B. Commercial operation				
Di-*t*-butyl-*p*-cresol	24	90	99.5	340

All feeds were prepared in Armstrong scraped surface crystallizers.

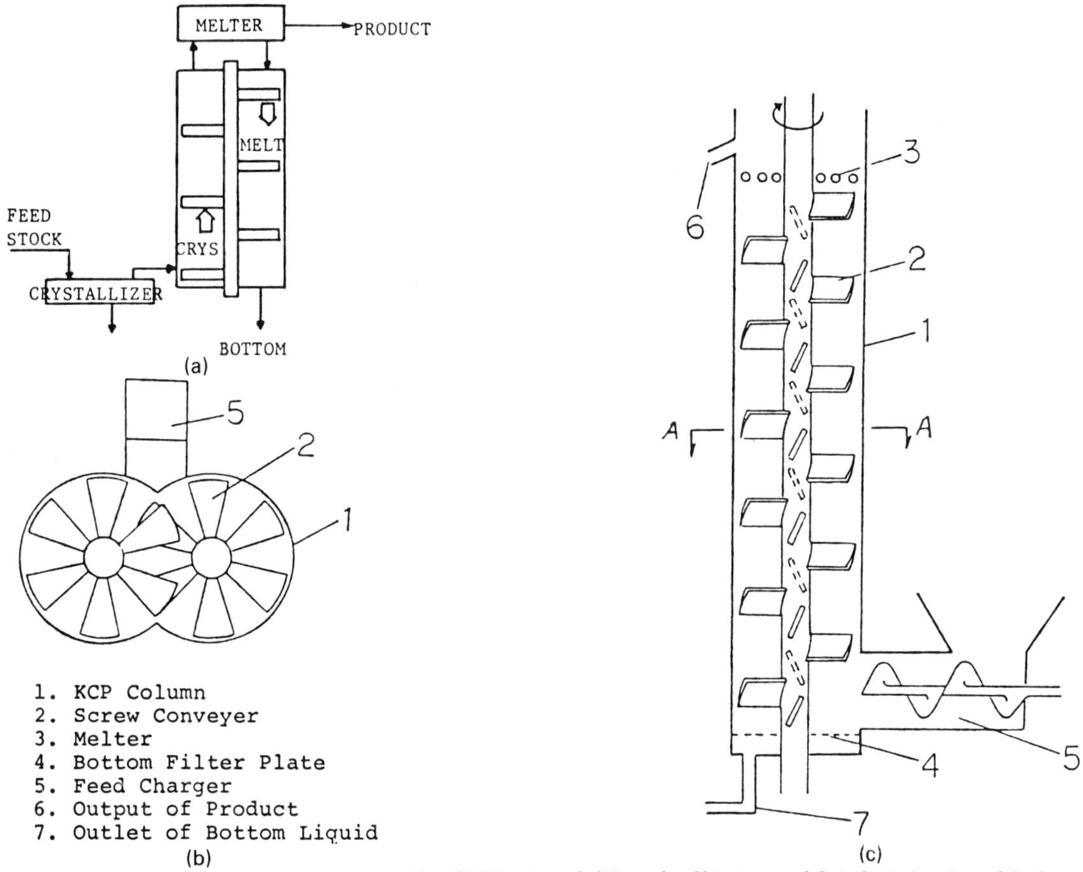

Figure 16.16. Kureha continuous crystal purifier (KCP column) (*Yamada, Shimizu, and Saitoh, in Jancic and DeJong, 1982, pp. 265–270*). (a) Flowsketch. (b) Dumbbell-shaped cross section at AA. (c) Details of column and screw conveyor.

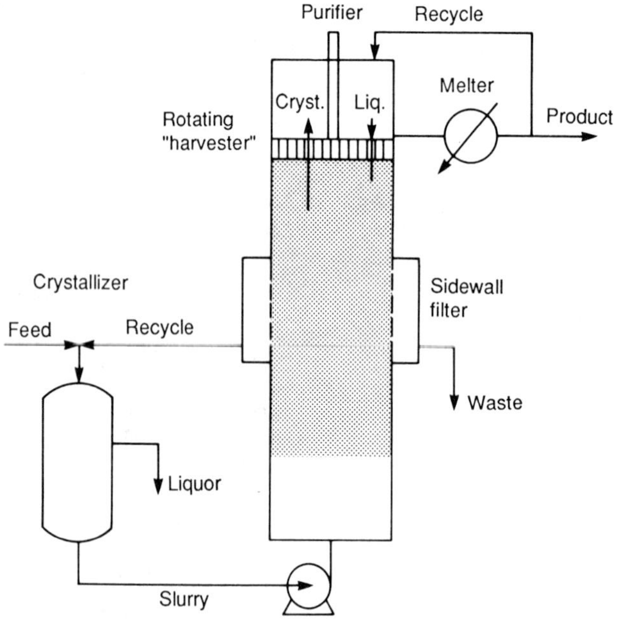

Figure 16.17. Brennan–Koppers crystal purifier (*Brennan, 1982*).

REFERENCES

Crystallization from Solutions

1. A.W. Bamforth, *Industrial Crystallization,* Leonard Hill, London, 1965.
2. R.C. Bennett, Crystallization design, *Encycl. Chem. Process. Des.* **13,** 421–455 (1981).
3. R.C. Bennett, Crystallization from solution, in *Chemical Engineers' Handbook,* McGraw-Hill, New York, 1984, pp. 19.24–19.40.
4. E.D. DeJong and S.I. Jancic, *Industrial Crystallization 1978,* North-Holland, Amsterdam, 1979.
5. Industrial Crystallization, Symposium of Inst. Chem. Eng., Inst. Chem. Eng, London, 1969.
6. S.I. Jancic and E.J. DeJong, Eds., *Industrial Crystallization 1981,* North-Holland, Amsterdam, 1982.
7. S.I. Jancic and J. Garside, in Ref. 10, 1976, p. 363.
8. E.V. Khamskii, *Crystallization from Solution,* Consultants Bureau, New York, 1969.
9. J.W. Mullin, *Crystallization,* Butterworths, London, 1972.
10. J.W. Mullin, Ed., *Symposium on Industrial Crystallization,* Plenum, New York, 1976.
11. J.W. Mullin, Crystallization, *Encycl. Chem. Technol.* **7,** 1978, pp. 243–285.
12. J.W. Mullin, Bulk crystallization, in *Crystal Growth* (Pamplin, Ed.), Pergamon, New York, 1980, pp. 521–565.
13. J. Nyvlt, Crystallization as a unit operation in chemical engineering, in Ref. 5, 1969, pp. 1–23.
14. J. Nyvlt, *Industrial Crystallization from Solutions,* Butterworths, London, 1971.
15. J. Nyvlt, *Industrial Crystallization: The Present State of the Art,* Verlag Chemie, Weinheim, 1978.
16. M. Ohara and R.C. Reid, *Modelling Crystal Growth Rates from Solution,* Prentice-Hall, Englewood Cliffs, NJ, 1973.
17. A.D. Randolph and M.A. Larson, *Theory of Particulate Processes,* Academic, New York, 1971.
18. G. Singh, Crystallization from Solution, in *Handbook of Separation Techniques for Chemical Engineers* (Schweitzer, Ed.), McGraw-Hill, New York, 1979, pp. 2.151–2.182.

Melt Crystallization

19. R. Albertins, W.C. Gates, and J.E. Powers, Column crystallization, in Ref. 32, 1967, pp. 343–367.
20. G.J. Arkenbout, Progress in continuous fractional crystallization, *Sep. Purification Methods* **7**(1), 99–134 (1978).
21. G.J. Arkenbout, A. vanKujik, and W.M. Smit, Progress in continuous fractional crystallization, in Ref. 10, 1976, pp. 431–435.
22. E.D. Brennan (Koppers Co.), Process and Apparatus for Separating and Purifying a Crystalline Material, U.S. Pat. 4,309,878 (12 Jan. 1982).
23. J.A. Brodie, A continuous multistage melt purification process, *Mech. Chem. Eng. Trans., Inst. Eng. Australia,* 37–44 (May 1971).
24. G.H. Dale, Crystallization: extractive and adductive, *Encycl. Chem. Process. Des.* **13,** 456–506 (1981).
25. R.A. Findlay, Adductive crystallization, in *New Chemical Engineering Separation Techniques* (Schoen, Ed.), Wiley-Interscience, New York, 1958.
26. R.A. Findlay and J.A. Weedman, Separation and purification by crystallization, in *Advances in Petroleum Chemistry and Refining,* Wiley-Interscience, New York, 1958, Vol. 1, pp. 118–209.
27. H.W. Haines, J.M. Powers, and R.B. Bennett, Separation of xylenes, *Ind. Eng. Chem.* **47,** 1096 (1955).
28. J.D. Henry and C.C. Moyers, Crystallization from the melt, in *Chemical Engineers' Handbook,* McGraw-Hill, New York, 1984, pp. 17.2–17.12.
29. D.L. McKay, Phillips fractional solidification process, in Ref. 32, 1967, pp. 427–439.
30. A.B. Mützenberg and K. Saxer, The MWB crystallizer, *Dechema Monographien* **66,** 313–320 (1971).
31. J. Yamada, C. Shimizu, and S. Saitoh, Purification of organic chemicals by the Kureha Continuous Crystal Purifier, in Ref. 6, 1982, pp. 265–270.
32. M. Zief and W.R. Wilcox, *Fractional Soldification,* Dekker, New York, 1967, Vol. 1.

17

CHEMICAL REACTORS

17.1. DESIGN BASIS AND SPACE VELOCITY

This chapter summarizes the main principles of chemical kinetics and catalysis; also it classifies and describes some of the variety of equipment that is suitable as chemical reactors. Because of the diversity of the behavior of chemical reactions, few rules are generally applicable to the design of equipment for such purposes. Reactors may be stirred tanks, empty or packed tubes or vessels, shell-and-tube devices or highly specialized configurations, in any of which heat transfer may be provided. These factors are balanced in individual cases to achieve economic optima. The general rules of other chapters for design of pressure vessels, heat exchangers, agitators, and so on naturally apply to reactors.

DESIGN BASIS

Although the intent of this chapter is not detailed design, it is in order to state what is included in a proper design basis, for example at least these items:

1. Stoichiometry of the participating reactions.
2. Thermal and other physical properties.
3. Heats of reaction and equilibrium data.
4. Rate of reaction, preferably in equation form, relating it to composition, temperature, pressure, impurities, catalysts and so on. Alternately tabular or graphical data relating compositions to time and the other variables listed.
5. Activity of the catalyst as a function of onstream time.
6. Mode of catalyst reactivation or replacement.
7. Stability and controllability of the process.
8. Special considerations of heat and mass transfer.
9. Corrosion and safety hazards.

REACTION TIMES

In practical cases reaction times vary from fractions of a second to many hours. The compilation of Table 17.1 of some commercial practices may be a basis for choosing by analogy an order of magnitude of reactor sizes for other processes.

For ease of evaluation and comparison, an apparent residence time often is used instead of the true one; it is defined as the ratio of the reactor volume to the inlet volumetric flow rate,

$$\bar{t}_{\text{app}} = V_r/V'_0.$$

On the other hand, the true residence time must be found by integration,

$$\bar{t} = \int dV_r/V' = \int dn'/rV'.$$

Since the rate of reaction r and the volumetric flow rate V' at each position depend on T, P, and local molal flow rate n' of the key component of the reacting mixture, finding the true residence time is an involved process requiring many data. The easily evaluated apparent residence time usually is taken as adequate for rating sizes of reactors and for making comparisons.

A related concept is that of space velocity which is the ratio of a flow rate at STP (60°F, 1 atm usually) to the size of the reactor. The most common versions of space velocities in typical units are:

GHSV (gas hourly space velocity) = (volumes of feed as gas at STP/hr)/(volume of the reactor or its content of catalyst) = (SCFH gas feed)/cuft.

LHSV (liquid hourly space velocity) = (Volume of liquid feed at 60°F/hr)/volume of reactor or catalyst) = (SCFH liquid feed)/cuft.

WHSV (weight hourly space velocity) = (lb of feed/hr)/(lb of catalyst). Other combinations of units of the flow rate and reactor size often are used in practice, for instance.

BPSD/lb = (barrels of liquid feed at 60°F per stream day)/(lb catalyst), but it is advisable to write out such units in each case to avoid confusion with the standard meanings of the given acronyms. Since the apparent residence time is defined in terms of the actual inlet conditions rather than at standard T and P, it is not the reciprocal of GHSV or LHSV, although the units are the same.

17.2. RATE EQUATIONS AND OPERATING MODES

The equations of this section are summarized and extended in Table 17.2. The term "rate of reaction" used here is the rate of decomposition per unit volume,

$$r_a = -\frac{1}{V}\frac{dn_a}{dt}, \quad \text{mol/(unit time)(unit volume)}. \tag{17.1}$$

A rate of formation will have the opposite sign. When the volume is constant, the rate is the derivative of the concentration

$$r_a = -\frac{dC_a}{dt}, \quad \text{at constant volume}. \tag{17.2}$$

In homogeneous environments the rate is expressed by the law of mass action in terms of powers of the concentrations of the reacting substances

$$r_a = -\frac{1}{V}\frac{dn_a}{dt} = kC_a^{\alpha}C_b^{\beta}\cdots. \tag{17.3}$$

When the reaction mechanism truly follows the stoichiometric equation

$$\nu_a A + \nu_b B + \cdots \rightarrow \text{products}, \tag{17.41}$$

the exponents are the stoichiometric coefficients; thus,

$$r_a = k(C_a)^{\nu_a}(C_b)^{\nu_b}\cdots, \tag{17.5}$$

but α, β, $\cdots$ often are purely empirical values—integral or nonintegral, sometimes even negative.

The coefficient k is called the specific rate. It is taken to be independent of the concentrations of the reactants but does depend primarily on temperature and the nature and concentration of

TABLE 17.1. Residence Times and/or Space Velocities in Industrial Chemical Reactors

Product (raw materials)	Type	Reactor phase	Catalyst	Conditions T,°C	Conditions P, atm	Residence time or space velocity	Source and page
1. Acetaldehyde (ethylene, air)	FB	L	Cu and Pd chlorides	50–100	8	6–40 min	[2] 1, [7] 3
2. Acetic anhydride (acetic acid)	TO	L	Triethyl phosphate	700–800	0.3	0.25–5 s	[2]
3. Acetone (*i*-propanol)	MT	LG	Ni	300	1	2.5 h	[1] *1* 314
4. Acrolein (formaldehyde, acetaldehyde)	FL	G	MnO, silica gel	280–320	1	0.6 s	[1] *1* 384, [7] 33
5. Acrylonitrile (air, propylene, ammonia)	FL	G	Bi phosphomolybdate	400	1	4.3 s	[3] 684, [2] 47
6. Adipic acid (nitration of cyclohexanol)	TO	L	Co naphthenate	125–160	4–20	2 h	[2] 51, [7] 49
7. Adiponitrile (adipic acid)	FB	G	H_3BO_3 H_3PO_4	370–410	1	3.5–5 s 350–500 GHSV	[1] *2* 152, [7] 52
8. Alkylate (*i*-C_4, butenes)	CST	L	H_2SO_4	5–10	2–3	5–40 min	[4] 223
9. Alkylate (*i*-C_4, butenes)	CST	L	HF	25–38	8–11	5–25 min	[4] 223
10. Allyl chloride (propylene, Cl_2)	TO	G	N.A.	500	3	0.3–1.5 s	[1] *2* 416, [7] 67
11. Ammonia (H_2, N_2)	FB	G	Fe	450	150	28 s 7,800 GHSV	[6] 61
12. Ammonia (H_2, N_2)	FB	G	Fe	450	225	33 s 10,000 GHSV	[6] 61
13. Ammonia oxidation	Flame	G	Pt gauze	900	8	0.0026 s	[6] 115
14. Aniline (nitrobenzene, H_2)	B	L	$FeCl_2$ in H_2O	95–100	1	8 h	[1] *3* 289
15. Aniline (nitrobenzene, H_2)	FB	G	Cu on silica	250–300	1	0.5–100 s	[7] 82
16. Aspirin (salicylic acid, acetic anhydride)	B	L	None	90	1	>1 h	[7] 89
17. Benzene (toluene)	TU	G	None	740	38	48 s 815 GHSV	[6] 36, [9] 109
18. Benzene (toluene)	TU	G	None	650	35	128 s	[1] *4* 183, [7] 98
19. Benzoic acid (toluene, air)	SCST	LG	None	125–175	9–13	0.2–2 h	[7] 101
20. Butadiene (butane)	FB	G	Cr_2O_3, Al_2O_3	750	1	0.1–1 s	[7] 118
21. Butadiene (1-butene)	FB	G	None	600	0.25	0.001 s 34,000 GHSV	[3] 572
22. Butadiene sulfone (butadiene, SO_2)	CST	L	*t*-butyl catechol	34	12	0.2 LHSV	[1] *5* 192
23. *i*-Butane (*n*-butane)	FB	L	$AlCl_3$ on bauxite	40–120	18–36	0.5–1 LHSV	[4] 239, [7] 683
24. *i*-Butane (*n*-butane)	FB	L	Ni	370–500	20–50	1–6 WHSV	[4] 239
25. Butanols (propylene hydroformylation)	FB	L	PH_3-modified Co carbonyls	150–200	1,000	100 g/L-h	[1] *5* 373
26. Butanols (propylene hydroformylation)	FB	L	Fe penta-carbonyl	110	10	1 h	[7] 125
27. Calcium stearate	B	L	None	180	5	1–2 h	[7] 135
28. Caprolactam (cyclohexane oxime)	CST	L	Polyphos-phoric acid	80–110	1	0.25–2 h	[1] *6* 73, [7] 139
29. Carbon disulfide (methane, sulfur)	Furn.	G	None	500–700	1	1.0 s	[1] *6* 322, [7] 144
30. Carbon monoxide oxidation (shift)	TU	G	Cu-Zn or Fe_2O_3	390–220	26	4.5 s 7,000 GHSV	[6] 44
30′. Port. cement	Kiln	S		1400–1700	1	10 h	[11]

TABLE 17.1—(*continued*)

Product (raw materials)	Type	Reactor phase	Catalyst	Conditions T,°C	P, atm	Residence time or space velocity	Source and page
31. Chloral (Cl_2, acetaldehyde)	CST	LG	None	20–90	1	140 h	[7] 158
32. Chlorobenzenes (benzene, Cl_2)	SCST	LG	Fe	40	1	24 h	[1] *8* 122
33. Coking, delayed (heater)	TU	LG	None	490–500	15–4	250 s	[1] *10* 8
34. Coking, delayed (drum, 100 ft max.)	B	LG	None	500–440	4	0.3–0.5 ft/s vapor	[1] *10* 8
35. Cracking, fluid-catalytic	FL	G	SiO_2, Al_2O_3	470–540	2–3	0.5–3 WHSV	[4] 162
36. Cracking, hydro- (gas oils)	FB	LG	Ni, SiO_2, Al_2O_3	350–420	100–150	1–2 LHSV	[11]
37. Cracking (visbreaking residual oils)	TU	LG	None	470–495	10–30	450 s 8 LHSV	[11]
38. Cumene (benzene, propylene)	FB	G	H_3PO_4	260	35	23 LHSV	[11]
39. Cumene hydroperoxide (cumene, air)	CST	L	Metal porphyrins	95–120	2–15	1–3 h	[7] 191
40. Cyclohexane (benzene, H_2)	FB	G	Ni on Al_2O_3	150–250	25–55	0.75–2 LHSV	[7] 201
41. Cyclohexanol (cyclohexane, air)	SCST	LG	None	185–200	48	2–10 min	[7] 203
42. Cyclohexanone (cyclohexanol)	CST	L	N.A.	107	1	0.75 h	[8] (1963)
43. Cyclohexanone (cyclohexanol)	MT	G	Cu on pumice	250–350	1	4–12 s	[8] (1963)
44. Cyclopentadiene (dicyclopentadiene)	TU	G	None	220–300	1–2	0.1–0.5 LHSV	[7] 212
45. DDT (chloral, chlorobenzene)	B	L	Oleum	0–15	1	8 h	[7] 233
46. Dextrose (starch)	CST	L	H_2SO_4	165	1	20 min	[8] (1951)
47. Dextrose (starch)	CST	L	Enzyme	60	1	100 min	[7] 217
48. Dibutylphthalate (phthalic anhydride, butanol)	B	L	H_2SO_4	150–200	1	1–3 h	[7] 227
49. Diethylketone (ethylene, CO)	TO	L	Co oleate	150–300	200–500	0.1–10 h	[7] 243
50. Dimethylsulfide (methanol, CS_2)	FB	G	Al_2O_3	375–535	5	150 GHSV	[7] 266
51. Diphenyl (benzene)	MT	G	None	730	2	0.6 s 3.3 LHSV	[7] 275, [8] (1938)
52. Dodecylbenzene (benzene, propylene tetramer)	CST	L	$AlCl_3$	15–20	1	1–30 min	[7] 283
53. Ethanol (ethylene, H_2O)	FB	G	H_3PO_4	300	82	1,800 GHSV	[2] 356, [7] 297
54. Ethyl acetate (ethanol, acetic acid)	TU, CST	L	H_2SO_4	100	1	0.5–0.8 LHSV	[10] 45, 52, 58
55. Ethyl chloride (ethylene, HCl)	TO	G	$ZnCl_2$	150–250	6–20	2 s	[7] 305
56. Ethylene (ethane)	TU	G	None	860	2	1.03 s 1,880 GHSV	[3] 411, [6] 13
57. Ethylene (naphtha)	TU	G	None	550–750	2–7	0.5–3 s	[7] 254
58. Ethylene, propylene chlorohydrins (Cl_2, H_2O)	CST	LG	None	30–40	3–10	0.5–5 min	[7] 310, 580

(continued)

TABLE 17.1—(*continued*)

Product (raw materials)	Type	Reactor phase	Catalyst	Conditions T,°C	P, atm	Residence time or space velocity	Source and page
59. Ethylene glycol (ethylene oxide, H_2O)	TO	LG	1% H_2SO_4	50–70	1	30 min	[2] 398
60. Ethylene glycol (ethylene oxide, H_2O)	TO	LG	None	195	13	1 h	[2] 398
61. Ethylene oxide (ethylene, air)	FL	G	Ag	270–290	1	1 s	[2] 409, [7] 322
62. Ethyl ether (ethanol)	FB	G	WO_3	120–375	2–100	30 min	[7] 326
63. Fatty alcohols (coconut oil)	B	L	Na, solvent	142	1	2 h	[8] (1953)
64. Formaldehyde (methanol, air)	FB	G	Ag gauze	450–600	1	0.01 s	[2] 423
65. Glycerol (allyl alcohol, H_2O_2)	CST	L	H_2WO_4	40–60	1	3 h	[7] 347
66. Hydrogen (methane, steam)	MT	G	Ni	790	13	5.4 s 3,000 GHSV	[6] 133
67. Hydrodesulfurization of naphtha	TO	LG	Co-Mo	315–500	20–70	1.5–8 LHSV 125 WHSV	[4] 285, [6] 179, [9] 201
68. Hydrogenation of cottonseed oil	SCST	LG	Ni	130	5	6 h	[6] 161
69. Isoprene (*i*-butene, formaldehyde)	FB	G	HCl, silica gel	250–350	1	1 h	[7] 389
70. Maleic anhydride (butenes, air)	FL	G	V_2O_5	300–450	2–10	0.1–5 s	[7] 406
71. Melamine (urea)	B	L	None	340–400	40–150	5–60 min	[7] 410
72. Methanol (CO, H_2)	FB	G	ZnO, Cr_2O_3	350–400	340	5,000 GHSV	[7] 421
73. Methanol (CO, H_2)	FB	G	ZnO,Cr_2O_3	350–400	254	28,000 GHSV	[3] 562
74. *o*-Methyl benzoic acid (xylene, air)	CST	L	None	160	14	0.32 h 3.1 LHSV	[3] 732
75. Methyl chloride (methanol, Cl_2)	FB	G	Al_2O_3 gel	340–350	1	275 GHSV	[2] 533
76. Methyl ethyl ketone (2-butanol)	FB	G	ZnO	425–475	2–4	0.5–10 min	[7] 437
77. Methyl ethyl ketone (2-butanol)	FB	G	Brass spheres	450	5	2.1 s 13 LHSV	[10] 284
78. Nitrobenzene (benzene, HNO_3)	CST	L	H_2SO_4	45–95	1	3–40 min	[7] 468
79. Nitromethane (methane, HNO_3)	TO	G	None	450–700	5–40	0.07–0.35 s	[7] 474
80. Nylon-6 (caprolactam)	TU	L	Na	260	1	12 h	[7] 480
81. Phenol (cumene hydroperoxide)	CST	L	SO_2	45–65	2–3	15 min	[7] 520
82. Phenol (chloro-benzene, steam)	FB	G	Cu, Ca phosphate	430–450	1–2	2 WHSV	[7] 522
83. Phosgene (CO, Cl_2)	MT	G	Activated carbon	50	5–10	16 s 900 GHSV	[11]
84. Phthalic anhydride (*o*-xylene, air)	MT	G	V_2O_5	350	1	1.5 s	[3] 482, 539, [7] 529
85. Phthalic anhydride (naphthalene, air)	FL	G	V_2O_5	350	1	5 s	[9] 136, [10] 335
86. Polycarbonate resin (bisphenol-A, phosgene)	B	L	Benzyltri-ethylammonium chloride	30–40	1	0.25–4 h	[7] 452
87. Polyethylene	TU	L	Organic peroxides	180–200	1,000–1,700	0.5–50 min	[7] 547
88. Polyethylene	TU	L	Cr_2O_3, Al_2O_3, SiO_2	70–200	20–50	0.1–1,000 s	[7] 549

TABLE 17.1—(*continued*)

Product (raw materials)	Type	Reactor phase	Catalyst	Conditions T,°C	P, atm	Residence time or space velocity	Source and page
89. Polypropylene	TO	L	R_2AlCl, $TiCl_4$	15–65	10–20	15–100 min	[7] 559
90. Polyvinyl chloride	B	L	Organic peroxides	60	10	5.3–10 h	[6] 139
91. *i*-Propanol (propylene, H_2O)	TO	L	H_2SO_4	70–110	2–14	0.5–4 h	[7] 393
92. Propionitrile (propylene, NH_3)	TU	G	CoO	350–425	70–200	0.3–2 LHSV	[7] 578
93. Reforming of naphtha (H_2/hydrocarbon = 6)	FB	G	Pt	490	30–35	3 LHSV 8,000 GHSV	[6] 99
94. Starch (corn, H_2O)	B	L	SO_2	25–60	1	18–72 h	[7] 607
95. Styrene (ethylbenzene)	MT	G	Metal oxides	600–650	1	0.2 s 7,500 GHSV	[5] 424
96. Sulfur dioxide oxidation	FB	G	V_2O_5	475	1	2.4 s 700 GHSV	[6] 86
97. *t*-Butyl methacrylate (methacrylic acid, *i*-butene)	CST	L	H_2SO_4	25	3	0.3 LHSV	[1] 5 328
98. Thiophene (butane, S)	TU	G	None	600–700	1	0.01–1 s	[7] 652
99. Toluene diisocyanate (toluene diamine, phosgene)	B	LG	None	200–210	1	7 h	[7] 657
100. Toluene diamine (dinitrotoluene, H_2)	B	LG	Pd	80	6	10 h	[7] 656
101. Tricresyl phosphate (cresyl, $POCl_3$)	TO	L	$MgCl_2$	150–300	1	0.5–2.5 h	[2] 850, [7] 673
102. Vinyl chloride (ethylene, Cl_2)	FL	G	None	450–550	2–10	0.5–5 s	[7] 699

Abbreviations
Reactors: batch (B), continuous stirred tank (CST), fixed bed of catalyst (FB), fluidized bed of catalyst (FL), furnace (Furn.), multitubular (MT), semicontinuous stirred tank (SCST), tower (TO), tubular (TU).
Phases: liquid (L), gas (G), both (LG).
Space velocities (hourly): gas (GHSV), liquid (LHSV), weight (WHSV).
Not available (N.A.)

REFERENCES

1. J.J. McKetta, ed., "Encyclopedia of Chemical Processing and Design," Marcel Dekker, New York, 1976 to date (referenced by volume).
2. W.L. Faith, D.B. Keyes, and R.L. Clark, "Industrial Chemicals," revised by F.A. Lowenstein and M.K. Moran, John Wiley & Sons, New York, 1975.
3. G.F. Froment and K.B. Bischoff, "Chemical Reactor Analysis and Design," John Wiley & Sons, New York, 1979.
4. R.J. Hengstebeck, "Petroleum Processing," McGraw-Hill, New York, 1959.
5. V.G. Jenson and G.V. Jeffreys, "Mathematical Methods in Chemical Engineering," 2nd ed., Academic Press, New York, 1977.
6. H.F. Rase, "Chemical Reactor Design for Process Plants: Vol. 2, Case Studies," John Wiley & Sons, New York, 1977.
7. M. Sittig, "Organic Chemical Process Encyclopedia," Noyes, Park Ridge, N.J., 1969 (patent literature exclusively).
8. Student Contest Problems, published annually by AIChE, New York (referenced by year).
9. M.O. Tarhan, "Catalytic Reactor Design," McGraw-Hill, New York, 1983.
10. K.R. Westerterp, W.P.M. van Swaaij, and A.A.C.M. Beenackers, "Chemical Reactor Design and Operation," John Wiley & Sons, New York, 1984.
11. Personal communication (Walas, 1985).

catalysts. Temperature dependence usually is represented by

$$k = k_\infty \exp(-E/RT) = \exp(a' - b'/T), \tag{17.6}$$

where E is the energy of activation.

Specific rates of reactions of practical interest cannot be found by theoretical methods of calculation nor from correlations in terms of the properties of the reactants. They must be found empirically in every case together with the complete dependence of the rate of reaction on concentrations, temperature, and other pertinent factors. The analysis of experimental data will be ignored here since the emphasis is placed on the use of known rate equations.

Integration of the rate equation is performed to relate the composition to the reaction time and the size of the equipment. From a rate equation such as

$$-\frac{dC_a}{dt} = kC_a^\alpha C_b^\beta C_c^\gamma, \tag{17.7}$$

TABLE 17.2. Basic Rate Equations

1. *The reference reaction* is

$$v_aA + v_bB + \cdots \rightarrow v_rR + v_sS + \cdots$$
$$\Delta v = v_r + v_s + \cdots - (v_a + v_b + \cdots)$$

2. *Stoichiometric balance* for any component *i*,

$$n_i = n_{i0} \pm (v_i/v_a)(n_{a0} - n_a)$$
$$\begin{cases} + \text{ for product (right-hand side, RHS)} \\ - \text{ for reactant (left-hand side, LHS)} \end{cases}$$
$$C_i = C_{i0} \pm (v_i/v_a)(C_{a0} - C_a), \quad \text{at constant } T \text{ and } V \text{ only}$$
$$n_t = n_{t0} + (\Delta v/v_a)(n_{a0} - n_a)$$

3. *Law of mass action*

$$r_a = -\frac{1}{V_r}\frac{dn_a}{dt} = kC_a^{v_a}C_b^{v_b}\cdots$$
$$= kC_a^{v_a}[C_{b0} - (v_b/v_a)(C_{a0} - C_a)]^{v_b}\cdots$$
$$r_a = kC_a^{\alpha}[C_{b0} - (v_b/v_a)(C_{a0} - C_a)]^{\beta}\cdots$$

where it is not necessarily true that $\alpha = v_a$, $\beta = v_b, \cdots$

4. *At constant volume,* $C_a = n_a/V_r$

$$kt = \int_{C_a}^{C_{a0}} \frac{1}{C_a^{\alpha}[C_{b0} - (v_b/v_a)(C_{a0} - C_a)]^{\beta}\cdots}\,dC_a$$
$$kt = \int_{n_a}^{n_{a0}} \frac{V_{\cdots}^{-1+\alpha+\beta}}{n_a^{\alpha}[n_{b0} + (v_b/v_a)(n_{a0} - n_a)]^{\beta}\cdots}\,dn_a$$

Completed integrals for some values of α and β are in Table 17.3

5. *Ideal gases at constant pressure:*

$$V_r = \frac{n_tRT}{P} = \frac{RT}{P}\left[n_{t0} + \frac{\Delta v}{v_a}(n_{a0} - n_a)\right]$$
$$r_a = kC_a^{\alpha}$$
$$kt = \left(\frac{RT}{P}\right)^{\alpha-1}\int_{n_a}^{n_{a0}} \frac{[n_{t0} + (\Delta v/v_a)(n_{a0} - n_a)]^{\alpha-1}}{n_a^{\alpha}}\,dn_a$$

6. *Temperature effect* on the specific rate:

$$k = k_{\infty}\exp(-E/RT) = \exp(a' - b'/T)$$
$$E = \text{energy of activation}$$

7. *Simultaneous reactions:* The overall rate is the algebraic sum of the rates of the individual reactions. For example, take the three reactions:

1. $A + B \xrightarrow{k_1} C + D.$
2. $C + D \xrightarrow{k_2} A + B.$
3. $A + C \xrightarrow{k_3} E.$

The rates are related by:

$$r_a = r_{a1} + r_{a2} + r_{a3} = k_1C_aC_b - k_2C_cC_d + k_3C_aC_c$$
$$r_b = -r_d = k_1C_aC_b - k_2C_cC_d$$
$$r_c = -k_1C_aC_b + k_2C_cC_d + k_3C_aC_c$$
$$r_e = -k_3C_aC_c$$

The number of independent rate equations is the same as the number of independent stoichiometric relations. In the present example, reactions 1 and 2 are a reversible reaction and are not independent. Accordingly, C_c and C_d, for example, can be eliminated from the equations for r_a and r_b which then become an integrable system. Usually only systems of linear differential equations with constant coefficients are solvable analytically. Many such cases are treated by Rodiguin and Rodiguina (1964) *Consecutive Chemical Reactions,* Van Nostrand, N.Y.

8. *Mass transfer resistance:*

$$C_{ai} = \text{interfacial concentration of reactant A}$$
$$r_a = -\frac{dC_a}{dt} = k_d(C_a - C_{ai}) = kC_{ai}^{\alpha} = k\left(C_a - \frac{r_a}{k_d}\right)^{\alpha}$$
$$kt = \int_{C_a}^{C_{a0}} \frac{1}{(C_a - r_a/k_d)^{\alpha}}\,dC_a$$

The relation between r_a and C_a must be established (numerically if need be) from the second line before the integration can be completed

9. *Solid-catalyzed reactions,* some Langmuir-Hinshelwood mechanisms for the reference reaction $A + B \rightarrow R + S$.

1. Adsorption rate of A controlling

$$r_a = -\frac{1}{V}\frac{dn_a}{dt} = kP_a\theta_v$$
$$\theta_v = 1\Big/\left[1 + \frac{K_a}{K_e}\frac{P_rP_s}{P_b} + K_bP_b + K_rP_r + K_sP_s + K_IP_I\right]$$
$$K_e = P_rP_s/P_aP_b \quad \text{(equilibrium constant)}$$

I is an adsorbed substance that is chemically inert

2. Surface reaction rate controlling:

$$r = kP_aP_b\theta_v^2$$
$$\theta_v = 1\Big/\left(1 + \sum K_jP_j\right),$$

summation over all substances absorbed

3. Reaction $A_2 + B \rightarrow R + S$, with A_2 dissociated upon adsorption and with surface reaction rate controlling:

$$r_a = kP_aP_b\theta_v^3$$
$$\theta_v = 1/(1 + \sqrt{K_aP_a} + K_bP_b + \cdots)$$

4. At constant *P* and *T* the P_i are eliminated in favor of n_i and the total pressure by

$$P_a = \frac{n_a}{n_t}P$$
$$P_i = \frac{n_i}{n_t}P = \frac{n_{i0} \pm (v_i/v_a)(n_{a0} - n_a)}{n_{t0} + (\Delta v/v_a)(n_{a0} - n_a)}P$$
$$\begin{cases} + \text{ for products, RHS} \\ - \text{for reactants, LHS} \end{cases}$$
$$V = \frac{n_tRT}{P}$$
$$kt = \int_{n_a}^{n_{a0}} \frac{1}{VP_aP_b\theta_v^2}, \quad \text{for a case (2) batch reaction}$$

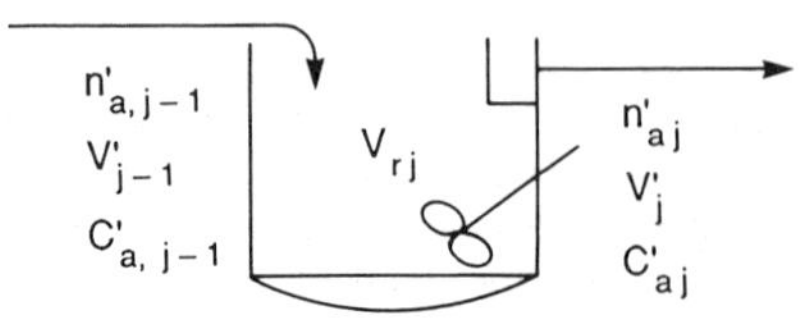

TABLE 17.2—(*continued*)

10. *A continuous stirred tank reactor battery* (*CSTR*)
Material balances:

$$n'_{a0} = n'_a + r_{a1}V_{r1}$$
$$\vdots$$
$$n'_{a,j-1} = n'_{aj} + r_{aj}V_{rj}, \quad \text{for the } j\text{th stage}$$

For a first order reaction, with $r_a = kC_a$,

$$\frac{C_{aj}}{C_{a0}} = \frac{1}{(1+k_1\bar{t}_1)(1+k_2\bar{t}_2)\cdots(1+k_j\bar{t}_j)}$$
$$= \frac{1}{(1+k\bar{t}_i)^j},$$

for j tanks in series with the same temperatures and residence times $\bar{t}_i = V_{ri}/V'_i$, where V' is the volumetric flow rate

11. *Plug flow reactor* (PFR):

$$r_a = -\frac{dn'_a}{dV_r} = kC_a^{\alpha}C_b^{\beta}\cdots$$
$$= k(n'_a/V')^{\alpha}(n'_b/V')^{\beta}\cdots$$

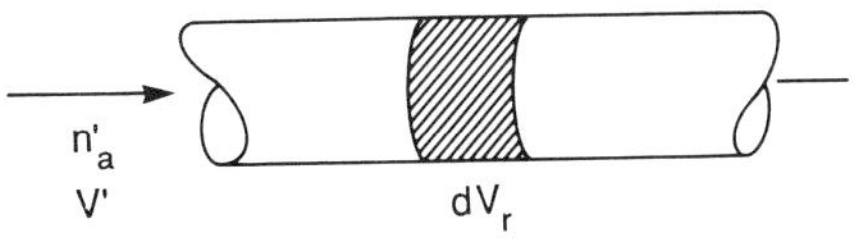

12. *Material and energy balances* for batch, CSTR and PFR are in Tables 17.4, 17.5, and 17.6
13. *Notation*
A, B, R, S are participants in the reaction; the letters also are used to represent concentrations
$C_i = n_i/V_r$ or n'_i/V', concentration
n_i = mols of component i in the reactor
n'_i = molal flow rate of component i
V_r = volume of reactor
V' = volumetric flow rate
v_i = stoichiometric coefficient
r_i = rate of reaction of substance i [mol/(unit time)(unit volume)]
α, β = empirical exponents in a rate equation

the concentrations C_b and C_c first must be eliminated with the aid of the stoichiometric equation of the process. Item 4 of Table 17.2 is an example. When several reactions occur simultaneously, the overall rate of a particular participant is the algebraic sum of its rates in individual reactions. Item 7 of Table 17.2 is an example. The number of differential equations representing the reacting system is the same as the number of independent stoichiometric equations; appropriate concentrations are eliminated with stoichiometry to develop an integrable set of equations. Integrals of common isothermal, constant volume rate equations are summarized in Table 17.3, and a simple case of a process at constant pressure is item 5 of Table 17.3.

An overall conversion rate may depend on rates of mass transfer between phases as well as chemical rates. In the simplest case, mass transfer and chemical transformation occur in series; advantage is taken of the equality of these two rates at steady state conditions to eliminate interfacial concentrations from the rate equations and thus to permit integration. Item 8 of Table 17.2 is an example.

Rates of fluid phase reactions catalyzed by solids also can be represented at least approximately by powers of the concentrations. A more fundamental approach, however, takes into account mechanisms of adsorption and of reaction on the catalyst surface. A few examples of resulting equations are in item 9 of Table 17.2.

Practical solid-catalyzed rate processes also may be influenced by rates of diffusion to the external and internal surfaces. In the latter case the rate equation is modified by inclusion of a catalyst effectiveness to become

$$r_a = k\eta f(C_a). \tag{17.8}$$

The effectiveness is a measure of the utilization of the internal surface of the catalyst. It depends on the dimensions of the catalyst particle and its pores, on the diffusivity, specific rate, and heat of reaction. With a given kind of catalyst, the only control is particle size to which the effectiveness is proportional; a compromise must be made between effectiveness and pressure drop. In simple cases η can be related mathematically to its parameters, but in such important practical cases as ammonia synthesis its dependence on parameters is complex and strictly empirical. Section 17.5 deals with this topic.

Reaction processes may be conducted under nonflow or steady flow conditions. One mode of the latter is tubular flow or, in the limiting case, plug flow, in which all molecules have substantially the same residence time. The rate equation for a plug flow reactor (PFR) is

$$r_a = -\frac{dn'_a}{dV_r} = kC_a^{\alpha}C_b^{\beta}\cdots = k\left(\frac{n'_a}{V'}\right)^{\alpha}\left(\frac{n'_b}{V'}\right)^{\beta}\cdots, \tag{17.9}$$

where V_r is the reactor volume and the primes (′) designate flow rates. Flow reactions of gases take place at substantially constant pressure so that V' will depend on the extent of conversion if there is a change in the number of mols. Item 11 of Table 17.2 is an example of the rate equation for such conditions.

The other mode of flow reaction employs one or more stirred tanks in series, which is called a continuous stirred tank (CSTR) battery. The rate of reaction in a single tank is

$$r_a = \frac{n'_{a0} - n'_a}{V_r} \simeq \frac{C_{a0} - C_a}{V_r/V'} = \frac{C_{a0} - C_a}{\bar{t}} = kC_a^{\alpha}C_b^{\beta}\cdots. \tag{17.10}$$

The relation in terms of concentrations is valid if the volumetric rates into and out of the tank are substantially the same. Stirring is assumed sufficient to maintain uniform composition and temperature in the tank; then the effluent conditions are the same as those of the tank. Relations for several tanks in series are in item 10 of Table 17.2.

17.3. MATERIAL AND ENERGY BALANCES OF REACTORS

All chemical reactions are accompanied by some heat effects so that the temperature will tend to change, a serious result in view of the sensitivity of most reaction rates to temperature. Factors of equipment size, controllability, and possibly unfavorable product distribution of complex reactions often necessitate provision of means of heat transfer to keep the temperature within bounds. In practical operation of nonflow or tubular flow reactors, truly isothermal conditions are not feasible even if they were desirable. Individual continuous stirred tanks, however, do maintain substantially uniform temperatures at steady state when the mixing is intense enough; the level is determined by the heat of reaction as well as the rate of heat transfer provided.

In many instances the heat transfer aspect of a reactor is

TABLE 17.3. Some Isothermal Rate Equations and Their Integrals

1. A → products:

$$-\frac{dA}{dt} = kA$$

$$\frac{A}{A_0} = \begin{cases} \exp[-k(t-t_0)], & \alpha = 1 \\ \left[\dfrac{1}{1+kA_0^{\alpha-1}(t-t_0)}\right]^{1/(\alpha-1)}, & \alpha \neq 1 \end{cases}$$

2. A + B → products:

$$-\frac{dA}{dt} = kAB = kA(A+B_0-A_0)$$

$$k(t-t_0) = \frac{1}{B_0-A_0}\ln\frac{A_0(A+B_0-A_0)}{AB_0}$$

3. Reversible reaction $A \underset{k_2}{\overset{k_1}{\rightleftharpoons}} B$:

$$-\frac{dA}{dt} = k_1A - k_2(A_0+B_0-A) = (k_1+k_2)A - k_2(A_0+B_0)$$

$$(k_1+k_2)(t-t_0) = \ln\frac{k_1A_0-k_2B_0}{(k_1+k_2)A-k_2(A_0+B_0)}$$

4. Reversible reaction, second order, $A + B \underset{k_2}{\overset{k_1}{\rightleftharpoons}} R + S$

$$-\frac{dA}{dt} = k_1AB - k_2RS = k_1A(A+B_0-A_0)$$

$$- k_2(A_0+R_0-A)(A_0+S_0-A)$$

$$= \alpha A^2 + \beta A - \gamma$$

$$\alpha = k_1 - k_2$$

$$\beta = k_1(B_0-A_0) + k_2(2A_0+R_0+S_0)$$

$$\gamma = k_2(A_0+R_0)(A_0+S_0)$$

$$q = \sqrt{\beta^2+4\alpha\gamma}$$

$$k(t-t_0) = \begin{cases} \dfrac{2\alpha A_0+\beta}{2\alpha A+\beta}, & q = 0 \\ \dfrac{1}{q}\ln\left[\left(\dfrac{2\alpha A_0+\beta-q}{2\alpha A_0+\beta+q}\right)\left(\dfrac{2\alpha A+\beta+q}{2\alpha A+\beta-q}\right)\right], & q \neq 0 \end{cases}$$

5. The reaction $\nu_aA + \nu_bB \rightarrow \nu_rR + \nu_sS$ *between ideal gases at constant T and P*

$$-\frac{dn_a}{dt} = \frac{kn_a^\alpha}{V^{\alpha-1}}$$

$$V = n_t\frac{RT}{P} = \left[n_{t0} + \frac{\Delta\nu}{\nu_a}(n_{a0}-n_a)\right]\frac{RT}{P}$$

$$k(t-t_0) = \begin{cases} \displaystyle\int_{n_a}^{n_{a0}} \frac{V^{\alpha-1}}{n_a^\alpha}\,dn_a, & \text{in general} \\ \dfrac{RT}{P}\left[n_{b0} + \dfrac{\Delta\nu}{\nu_a}\left(\dfrac{1}{n_a}-\dfrac{1}{n_{a0}}\right) - \dfrac{\Delta\nu}{\nu_a}\ln\left(\dfrac{n_{a0}}{n_a}\right)\right], & \text{when } \alpha = 2 \end{cases}$$

6. Equations readily solvable by Laplace transforms. For example:

$$A \underset{k_3}{\overset{k_1}{\rightleftharpoons}} B \overset{k_2}{\longrightarrow} C$$

Rate equations are

$$-\frac{dA}{dt} = k_1A - k_2B$$

$$-\frac{dB}{dt} = -k_1A + (k_2+k_3)B$$

$$-\frac{dC}{dt} = -k_2B$$

Laplace transformations are made and rearranged to

$$(s+k_1)\bar{A} + k_3\bar{B} = A_0$$

$$-k_1\bar{A} + (s+k_2+k_3)\bar{B} = B_0$$

$$-k_2\bar{B} + s\bar{C} = C_0$$

These linear equations are solved for the transforms as

$$D = s^2 + (k_1+k_2+k_3)s + k_1k_2$$

$$\bar{A} = [A_0s + (k_2+k_3)A_0 + K_3B_0]/D$$

$$\bar{B} = [B_0s + k_1(A_0+B_0)]/D$$

$$\bar{C} = (k_2\bar{B} + C_0)/s$$

Inversion of the transforms can be made to find the concentrations *A*, *B*, and *C* as functions of the time *t*. Many such examples are solved by Rodiguin and Rodiguina (*Consecutive Chemical Reactions*, Van Nostrand, New York, 1964).

---

paramount. Many different modes have been and are being employed, a few of which are illustrated in Section 6. The design of such equipment is based on material and energy balances that incorporate rates and heats of reaction together with heat transfer coefficients. Solution of these balances relates the time, composition, temperature, and rate of heat transfer. Such balances are presented in Tables 17.4–17.7 for four processes:

1. Nonflow reactors.
2. Plug flow reactors.
3. Continuous stirred tanks.
4. Flow reactor packed with solid catalyst.

The data needed are the rate equation, energy of activation, heat of reaction, densities, heat capacities, thermal conductivity, diffusivity, heat transfer coefficients, and usually the stoichiometry of the process. Simplified numerical examples are given for some of these cases. Item 4 requires the solution of a system of partial differential equations that cannot be made understandable in concise form, but some suggestions as to the procedure are made.

17.4. NONIDEAL FLOW PATTERNS

The CSTR with complete mixing and the PFR with no axial mixing are limiting behaviors that can be only approached in practice. Residence time distributions in real reactors can be found with tracer tests.

RESIDENCE TIME DISTRIBUTION

In the most useful form the test consists of a momentary injection of a known amount of inert tracer at the inlet of the operating vessel and monitoring of its concentration at the outlet. The data are used most conveniently in reduced form, as $E = C/\bar{C}_0$ in terms of $t_r = t/\bar{t}$, where

C = concentration of tracer at the outlet,
$\bar{C}_0$ = initial average concentration of tracer in the vessel,
$\bar{t} = V_r/V'$ = average residence time.

The plotted data usually are somewhat skewed bell-shapes. Some

TABLE 17.4. Material and Energy Balances of a Nonflow Reaction

Rate equations:

$$r_a = -\frac{1}{V_r}\frac{dn_a}{d\theta} = kC_a^{\alpha} = k\left(\frac{n_a}{V_r}\right)^{\alpha} \quad (1)$$

$$k = \exp(a' - b'/T) \quad (2)$$

Heat of reaction:

$$\Delta H_r = \Delta H_{r298} + \int_{298}^{T} \Delta C_p \, dT \quad (3)$$

Rate of heat transfer:

$$Q' = UA(T_s - T) \quad (4)$$

(the simplest case is when UA and T_s are constant)
Enthalpy balance:

$$\frac{dT}{dn_a} = \frac{1}{\rho V_r \bar{C}_p}\left[\Delta H_r + \frac{UA(T_s - T)}{V_r k(n_a/V_r)}\right] \quad (5)$$

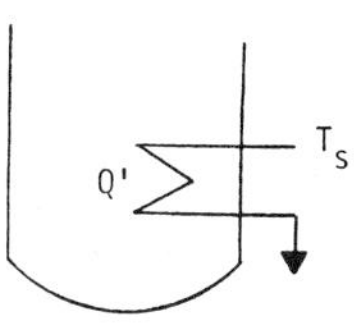

$$\frac{dT}{dC_a} = \frac{1}{\rho \bar{C}_p}\left[\Delta H_r + \frac{UA(T_s - T)}{V_r k C_a}\right] \quad (6)$$

$$T = T_0 \quad \text{when} \quad C_a = C_{a0} \quad (7)$$

$$\bar{C}_p = \frac{1}{\rho V_r}\sum n_i C_{pi} \quad (8)$$

Solve Eq. (6) to find $T = f(C_a)$; combine Eqs. (1) and (2) and integrate as

$$\theta = \int_{C_a}^{C_{a0}} \frac{1}{C_a^{\alpha}\exp[a' - b'/f(C_a)]}\, dC_a \quad (9)$$

Temperature and time as a function of composition are shown for two values of UA/V_r for a particular case represented by

$$\frac{dT}{dC_a} = \frac{1}{50}\left[5000 + 5T + \frac{UA(300 - T)}{V_r k C_a^2}\right]$$

$$k = \exp(16\text{–}5000/T)$$

$$T_0 = 350$$

$$C_{a0} = 1$$

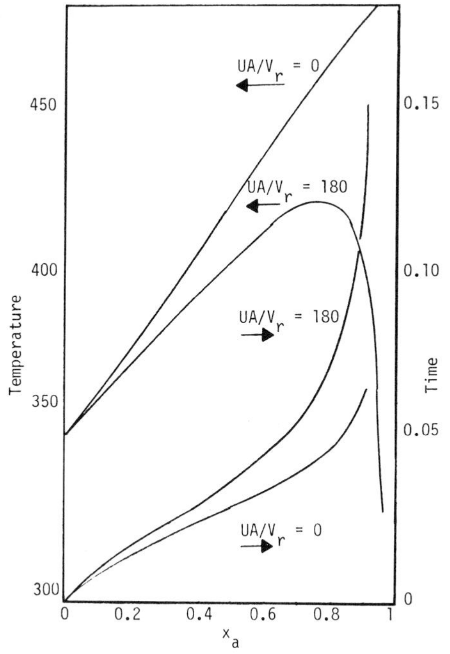

TABLE 17.5. Material and Energy Balance of a CSTR

The sketch identifies the nomenclature
Mean residence time:

$$\bar{t} = V_r/V' \quad (1)$$

Temperature dependence:

$$k = \exp(a' - b'/T) \quad (2)$$

Rate equation:

$$r_a = kC_a^{\alpha} = kC_{a0}^{\alpha}(1 - x)^{\alpha}, \; x = (C_{a0} - C_a)/C_{a0} \quad (3)$$

Material balance:

$$C_{a0} = C_a + k\bar{t}C_a \quad (4)$$

$$x = k\bar{t}C_{a0}^{\alpha-1}(1 - x)^{\alpha} \quad (5)$$

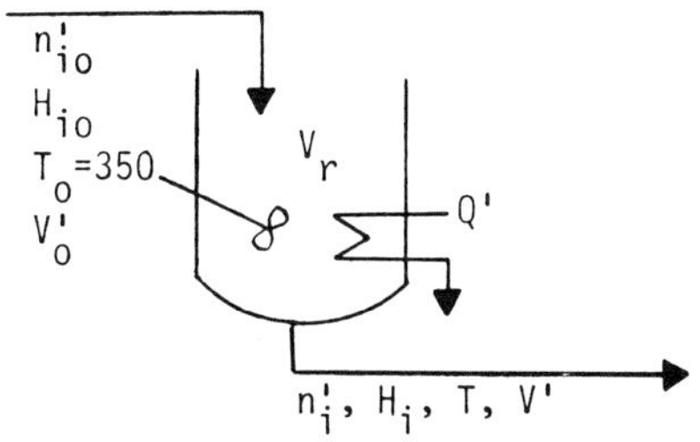

Enthalpy balance:

$$\sum n_i' H_i - \sum n_{i0}' H_{i0} = Q' - \Delta H_r(n_{a0}' - n_a') \quad (6)$$

$$H_i = \int_{298}^{T} C_{pi}\, dT \quad (7)$$

$$\Delta H_r = \Delta H_{r298} + \int_{298}^{T} \Delta C_p \, dT \quad (8)$$

For the reaction $aA + bB \rightarrow rR + sS$,

$$\Delta C_p = rC_{pr} + sC_{ps} - aC_{pa} - bC_{pb} \quad (9)$$

When the heat capacities are equal and constant, the heat balance is

$$\bar{C}_p \rho V'(T - T_0) = Q' - \Delta H_{r298} V'(C_{a0} - C_a) \quad (10)$$

Example:

$k = \exp(16\text{–}5500/T)$
$C_{a0} = 5$ g mol/L
$V' = 2000$ L/hr
$\Delta H_r = -5$ kcal/g mol
$\rho C_p = 0.9$ kcal/(L)(K)
$\alpha = 2$
$T_0 = 350$

	x = 0.90		x = 0.95	
$\bar{t}$	T	Q'	T	Q'
1	419.5	80	471.3	171
2	398.5	42	444.9	123
3	387.1	22	430.8	98
4	379.4	8	421.3	81
5	373.7	- 2	414.2	68
6	369.1	-11	408.6	58
7	365.3	-17	404.0	50
8	362.1	-23	400.0	43
9	359.3	-28	396.6	36
10	356.9	-33	393.6	31

(continued)

TABLE 17.5—(continued)

Eqs. (2) and (5) combine to

$$T = \frac{5500}{16 - \ln[x/5\bar{t}(1-x)^2]}$$

and Eq. (10) becomes

$$Q' = 2[0.9(T - 350) - 25x], \quad \text{Mcal/hr}$$

The temperature and the rate of heat input Q' are tabulated as functions of the residence time for conversions of 90 and 95%

---

actual data are shown in Figure 17.1 together with lines for ideal CSTR and PFR. Such shapes often are represented approximately by the Erlang statistical distribution which also is the result for an n-stage stirred tank battery,

$$E(t_r) = \frac{C}{\bar{C}_0} = \frac{n^n t_r^{n-1}}{(n-1)!}\exp(-nt_r), \tag{17.11}$$

where n is the characterizing parameter; when n is not integral, $(n-1)!$ is replaced by the gamma function $\Gamma(n)$. C_0 is the initial average concentration. The variance,

$$\sigma^2 = \int_0^\infty E(t_r - 1)^2\, dt_r = 1/n \tag{17.12}$$

of this distribution is a convenient single parameter characterization of the spread of residence times. This quantity also is related to the Peclet number, $\mathrm{Pe} = uL/D_e$, by

$$\sigma^2 = 2/\mathrm{Pe} - [1 - \exp(-\mathrm{Pe})]/\mathrm{Pe}^2, \tag{17.13}$$

where

u = linear velocity in the axial direction,
L = distance in the axial direction,
D_e = axial eddy diffusivity or dispersion coefficient.

TABLE 17.6. Material and Energy Balances of a Plug Flow Reactor (PFR)

The balances are made over a differential volume dV_r of the reactor
Rate equation:

$$dV_r = \frac{-dn'_a}{r_a} \tag{1}$$

$$= -\frac{1}{k}\left(\frac{V'}{n'_a}\right)^\alpha dn'_a \tag{2}$$

$$= -\exp\left(\frac{-a' + b'}{T}\right)\left(\frac{n'_t RT}{Pn'_a}\right)^\alpha dn'_a \tag{3}$$

Enthalpy balance:

$$\Delta H_r = \Delta H_{r298} + \int_{298}^{T} \Delta C_p\, dT \tag{4}$$

$$dQ = U(T_s - T)\, dA_p = \frac{4U}{D}(T_s - T)\, dV_r$$

$$= -\frac{4U(T_s - T)}{Dr_a} dn'_a \tag{5}$$

$$dQ + \Delta H_r\, dn'_a = \sum n_i\, dH_i = \sum n_i C_{pi}\, dT \tag{6}$$

$$\frac{dT}{dn'_a} = \frac{\Delta H_r - 4U(T_s - T)/Dr_a}{\sum n_i C_{pi}} = f(T, T_s, n'_a) \tag{7}$$

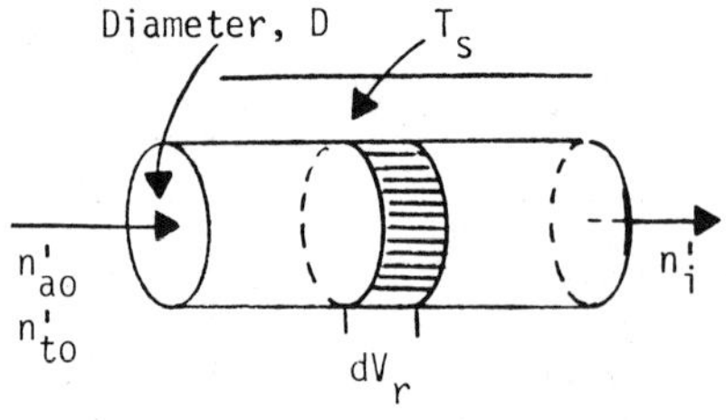

At constant T_s, Eq. (7) may be integrated numerically to yield the temperature as a function of the number of mols

$$T = \phi(n'_a) \tag{8}$$

Then the reactor volume is found by integration

$$V_r = \int_{n'_a}^{n'_{a0}} \frac{1}{\exp[a' - b'/\phi(n'_a)][Pn'_a/n'_t R\phi(n'_a)]^\alpha} dn'_a \tag{9}$$

Adiabatic process:

$$dQ = 0 \tag{10}$$

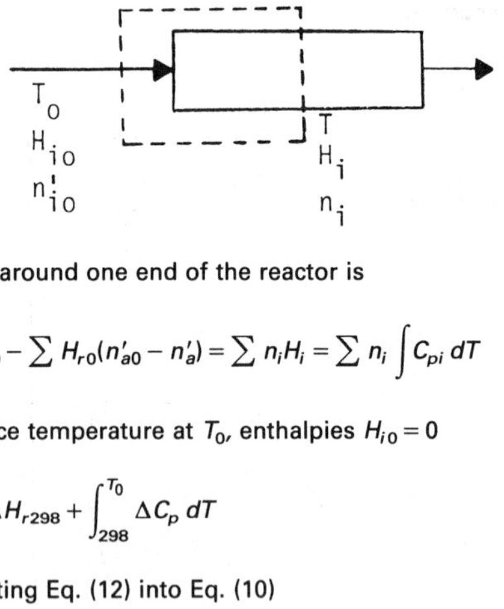

The balance around one end of the reactor is

$$\sum n_{i0}H_{i0} - \sum H_{r0}(n'_{a0} - n'_a) = \sum n_i H_i = \sum n_i \int C_{pi}\, dT \tag{11}$$

With reference temperature at T_0, enthalpies $H_{i0} = 0$

$$\Delta H_{r0} = \Delta H_{r298} + \int_{298}^{T_0} \Delta C_p\, dT \tag{12}$$

Substituting Eq. (12) into Eq. (10)

$$\left[-\Delta H_{r298} + \int_{298}^{T_0} \Delta C_p\, dT\right](n'_{a0} - n'_a) = \sum n_i \int_{T_0}^{T} C_{pi}\, dT \tag{13}$$

Adiabatic process with $\Delta C_p = 0$ and with constant heat capacities

$$T = T_0 - \frac{\Delta H_{r298}(n'_{a0} - n'_a)}{\sum n_i C_{pi}} \tag{14}$$

This expression is substituted instead of Eq. (8) to find the volume with Eq. (9)

TABLE 17.7. Material and Energy Balances of a Packed Bed Reactor

Diffusivity and thermal conductivity are taken appreciable only in the radial direction

Material balance equation:

$$\frac{\partial x}{\partial z}-\frac{D}{u}\left(\frac{\partial^2 x}{\partial r^2}+\frac{1}{r}\frac{\partial x}{\partial r}\right)-\frac{\rho}{u_0C_0}r_c=0 \qquad (1)$$

Energy balance equation:

$$\frac{\partial T}{\partial z}-\frac{k}{GC_p}\left(\frac{\partial^2 T}{\partial r^2}+\frac{1}{r}\frac{\partial T}{\partial r}\right)+\frac{\Delta H_r\rho}{GC_p}r_c=0 \qquad (2)$$

At the inlet:

$$x(0,r)=x_0 \qquad (3)$$

$$T(0,r)=T_0 \qquad (4)$$

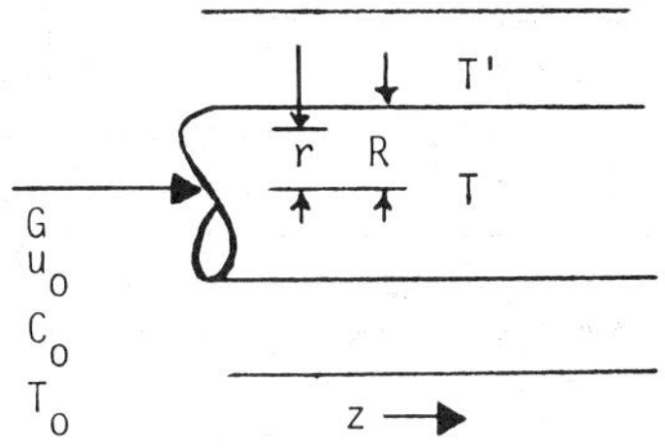

At the center:

$$r=0,\quad \frac{\partial x}{\partial r}=\frac{\partial T}{\partial r}=0 \qquad (5)$$

At the wall:

$$r=R,\quad \frac{\partial x}{\partial r}=0 \qquad (6)$$

$$\frac{\partial T}{\partial r}=\frac{U}{k}(T'-T) \qquad (7)$$

When the temperature T' of the heat transfer medium is not constant, another enthalpy balance must be formulated to relate T' with the process temperature T.

A numerical solution of these equations may be obtained in terms of finite difference equivalents, taking m radial increments and n axial ones. With the following equivalents for the derivatives, the solution may be carried out by direct iteration:

$$r=m(\Delta r)$$
$$z=n(\Delta z) \qquad (8)$$

$$\frac{\partial T}{\partial z}=\frac{T_{m,n+1}-T_{m,n}}{\Delta z} \qquad (9)$$

$$\frac{\partial T}{\partial r}=\frac{T_{m+1,n}-T_{m,n}}{\Delta r} \qquad (10)$$

$$\frac{\partial^2 T}{\partial r^2}=\frac{T_{m+1,n}-2T_{m,n}+T_{m-1,n}}{(\Delta r)^2} \qquad (11)$$

Expressions for the x-derivatives are of the same form:

r_c = rate of reaction, a function of s and T
G = mass flow rate, mass/(time)(superficial cross section)
u = linear velocity
D = diffusivity
k = thermal conductivity

At large values of Pe, the ratio Pe/n approaches 2.

The superficial Peclet number in packed beds,

$$\mathrm{Pe}=u_0d_p/D_e$$

is very roughly correlated (Wen and Fan, *Models for Flow Systems and Chemical Reactors*, Dekker, NY, 1975) in terms of the dimensionless groups $\mathrm{Re}=u_0d_p\rho/\mu$ and $\mathrm{Sc}=\mu/\rho D_m$, where

d_p = particle diameter,
D_m = molecular diffusivity,
ε = fraction voids in the bed.

The correlations are

$$\varepsilon\mathrm{Pe}=0.20+0.011\mathrm{Re}^{0.48},\quad \text{for liquids, standard deviation 46\%,} \qquad (17.14)$$

$$\frac{1}{\mathrm{Pe}}=\frac{0.3}{\mathrm{Re\,Sc}}+\frac{0.5}{1+3.8/\mathrm{Re\,Sc}},\quad \text{for gases.} \qquad (17.15)$$

There are no direct correlations of the variance (or the corresponding parameter n) in terms of the geometry and operating conditions of a vessel. For this reason the RTD is not yet a design tool, but it does have value as a diagnostic tool for the performance of existing equipment on which tracer tests can be made. RTDs obtained from tracer tests or perhaps estimated from dispersion coefficient data or correlations sometimes are applicable to the prediction of the limits between which a chemical conversion can take place in the vessel.

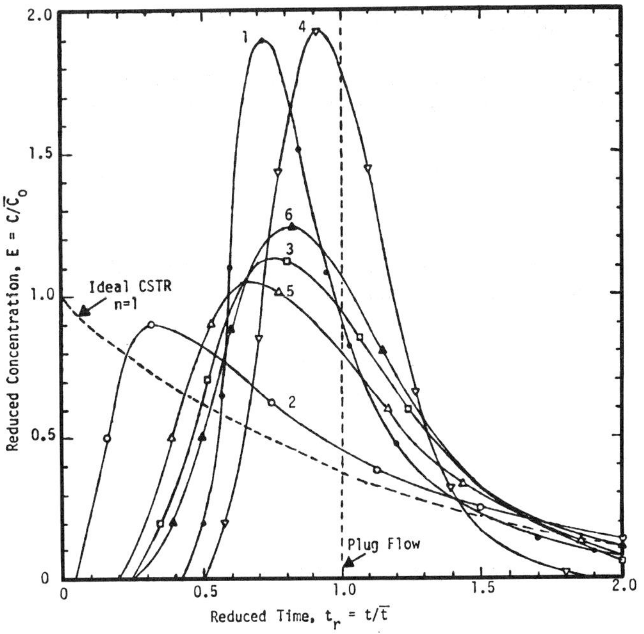

No.	Code	Process	σ^2	n	Pe
1	○	aldolization of butyraldehyde	0.050	20.0	39.0
2	●	olefin oxonation pilot plant	0.663	1.5	1.4
3	□	hydrodesulfurization pilot plant	0.181	5.5	9.9
4	▽	low temp hydroisomerization pilot	0.046	21.6	42.2
5	△	commercial hydrofiner	0.251	4.0	6.8
6	▲	pilot plant hydrofiner	0.140	7.2	13.2

Figure 17.1. Residence time distributions of some commercial and pilot fixed bed reactors. The variance, the equivalent number of CSTR stages, and the Peclet number are given for each.

CONVERSION IN SEGREGATED AND MAXIMUM MIXED FLOWS

In some important cases, limiting models for chemical conversion are the segregated flow model represented by the equation

$$\bar{C}/C_0 = 1 - x = \int_0^\infty (C/C_0)_{\text{batch}} E(t_r)\, dt_r = \int_0^\infty (C/C_0)_{\text{batch}} E(t)\, dt \tag{17.16}$$

and the maximum-mixedness model represented by Zwietering's equation. For a rate equation $r_c = kC^\alpha$ this equation is

$$\frac{dx}{dt_r} - kC_0^{\alpha-1}x^\alpha + \frac{E(t_r)}{1 - \int_0^{t_r} E(t_r)\, dt_r} x = 0, \tag{17.17}$$

with the boundary condition

$$dx/dt_r = 0 \quad \text{when } t_r \to \infty, \tag{17.18}$$

which is used to find the starting value x_∞ from

$$kC_0^{\alpha-1}x_\infty - \frac{E(t_r)}{1 - \int_0^\infty E(t_r)\, dt_r} x_\infty = 0. \tag{17.19}$$

Numerical integration of the equation is sufficiently accurate by starting at $(x_\infty, t_r \simeq 4)$ and proceeding to $t_r = 0$ at which time the value of x is the conversion in the reactor with residence time distribution $E(t_r)$.

With a given RTD the two models may correspond to upper and lower limits of conversion or reactor sizes for simple rate equations; thus

	Conversion Limit	
Reaction Order	**Segrated**	**Max-Mix**
More than 1	upper	lower
Less than 1	lower	upper
Complex	?	?

Relative sizes of reactors based on the two models are given in Figure 17.2 for second- and half-order reactions at several conversions. For first order reactions the ratio is unity. At small values of the parameter n and high conversions, the spread in reactor sizes is very large. In many packed bed operations, however, with proper initial distribution and redistribution the value of the parameter n is of the order of 20 or so, and the corresponding spread in reactor sizes is modest near conversions of about 90%. In such cases the larger predicted vessel size can be selected without undue economic hardship.

The data also can be rearranged to show the conversion limits for a reactor of a given size.

When the rate equation is complex, the values predicted by the two models are not necessarily limiting. Complexities can arise from multiple reactions, variation of density or pressure or temperature, incomplete mixing of feed streams, minimax rate behavior as in autocatalytic processes, and possibly other behaviors. Sensitivity of the reaction to the mixing pattern can be established in such cases, but the nature of the conversion limits will not be ascertained. Some other, possibly more realistic models will have to be devised to represent the reaction behavior. The literature has many examples of models but not really any correlations (Naumann and Buffham, 1983; Wen and Fan; Westerterp et al., 1984).

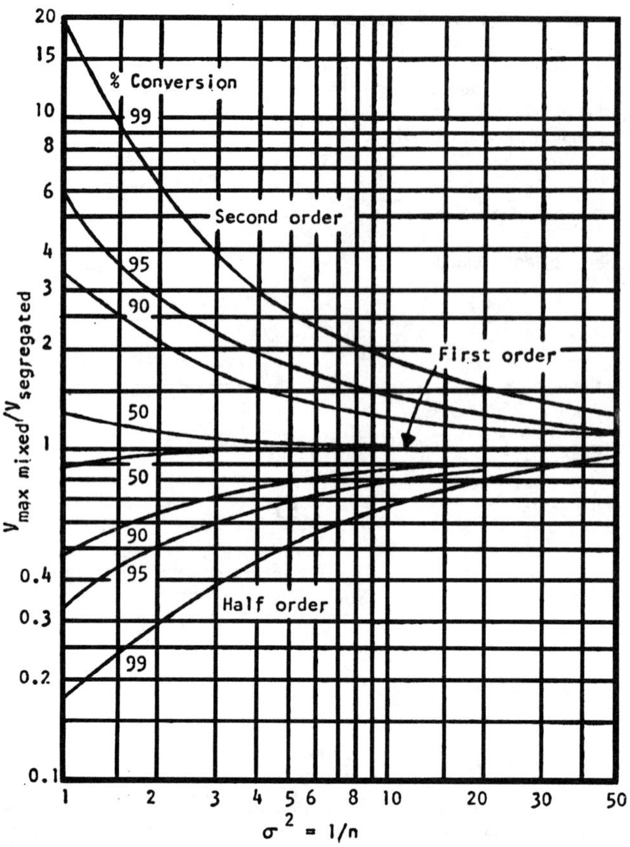

Figure 17.2. Relative volumes of maximum-mixed and segregated flow reactors with the same RTDs identified by $n = 1/\sigma^2$, as a function of conversion for second- and half-order reactions. For first-order reactions the ratio is unity throughout.

CONVERSION IN SEGREGATED FLOW AND CSTR BATTERIES

The mixing pattern in an n-stage CSTR battery is intermediate between segregated and maximum mixed flow and is characterized by residence time distribution with variance $\sigma^2 = 1/n$. Conversion in the CSTR battery is found by solving n successive equations

$$\frac{C_{j-1}}{C_0} = \frac{C_j}{C_0} + \frac{k\bar{t}}{n} C_0^{\alpha-1}\left(\frac{C_j}{C_0}\right)^\alpha \quad \text{for } j = 1\text{–}n \tag{17.20}$$

for $C_n/C_0 = 1 - x$. The ratio of required volumes of CSTR batteries and segregated flow reactors is represented by Figure 17.3 for several values of n over a range of conversions for a second order reaction. Comparison with the maximum mixed/segregated flow relation of Figure 17.2 shows a distinct difference between the two sets of ratios.

DISPERSION MODEL

Although it also is subject to the limitations of a single characterizing parameter which is not well correlated, the Peclet number, the dispersion model predicts conversions or residence times unambiguously. For a reaction with rate equation $r_c = kC^\alpha$, this model is represented by the differential equation

$$\frac{1}{\text{Pe}}\frac{d^2x}{dz^2} - \frac{dx}{dz} + k\bar{t}C_0^{\alpha-1}(1-x)^\alpha = 0 \tag{17.21}$$

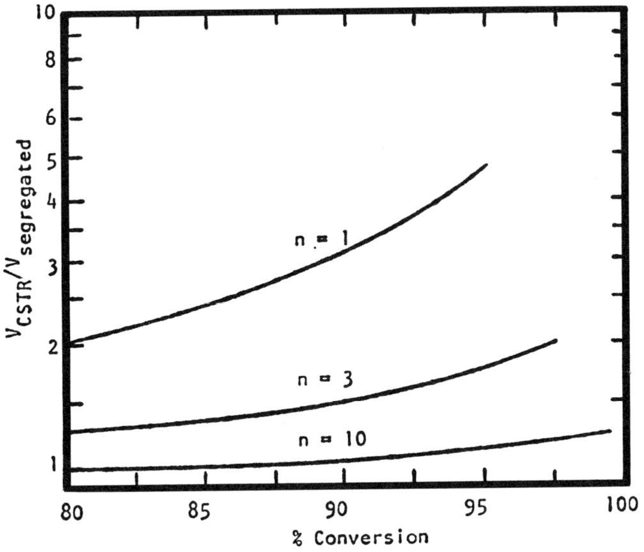

Figure 17.3. Ratio of volumes of an n-stage CSTR battery and a segregated flow reactor characterized by a residence time distribution with variance $\sigma^2 = 1/n$. Second-order reaction.

with the boundary conditions

$$\text{at } z = 0, \quad \left(1 - x + \frac{1}{\text{Pe}}\frac{dx}{dz}\right)_0 = 1, \tag{17.22}$$

$$\text{at } z = 1, \quad \frac{dx}{dz} = 0, \tag{17.23}$$

where

$x = 1 - C/C_0$, fractional conversion,

z = axial distance/length of reactor.

An analytical solution can be found only for a first-order reaction. The two-point boundary condition requires a special numerical procedure. Plots of solutions for first and second order reactions are shown in Figures 17.4 and 17.5.

LAMINAR AND RELATED FLOW PATTERNS

A tubular reactor model that may apply to viscous fluids such as polymers has a radial distribution of linear velocities represented by

$$u = (1 + 2/m)\bar{u}(1 - \beta^m), \tag{17.24}$$

where $\beta = r/R$. When $m = 2$, the pattern is Poiseuille or laminar flow, and, when m is infinite, it is plug flow. The residence time along a streamline is

$$t = \bar{t}/(1 + 2/m)(1 - \beta^m). \tag{17.25}$$

The average conversion over all the stream lines is

$$\frac{\bar{C}}{C_0} = \frac{1}{\pi R^2}\int \left(\frac{C}{C_0}\right)_{\text{streamline}} d(\pi r^2) = 2\int_0^1 \left(\frac{C}{C_0}\right)_{\text{streamline}} \beta\, d\beta. \tag{17.26}$$

For first-order reaction, for example

$$\frac{\bar{C}}{C_0} = 2\int_0^1 \exp\left[\frac{-k\bar{t}}{(1 + 2/m)(1 - \beta^m)}\right]\beta\, d\beta \tag{17.27}$$

and for second-order

$$\frac{\bar{C}}{C_0} = 2\int_0^1 \frac{1}{1 + kC_0\bar{t}/(1 + 2/m)(1 - \beta^m)}\beta\, d\beta. \tag{17.28}$$

These integrals must be evaluated numerically. Variation in residence time will contribute, for example, to the spread in molecular weight distribution of polymerizations.

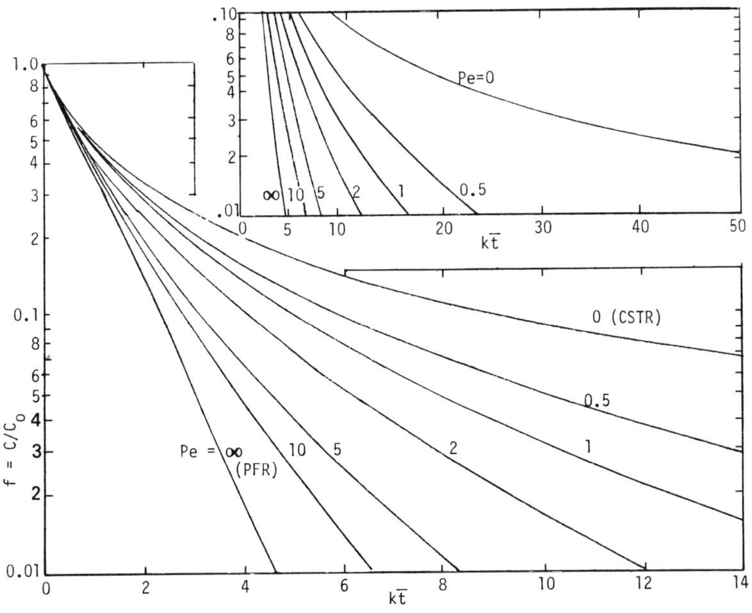

Figure 17.4. Dispersion model. Conversion of first-order reaction as function of the Peclet number.

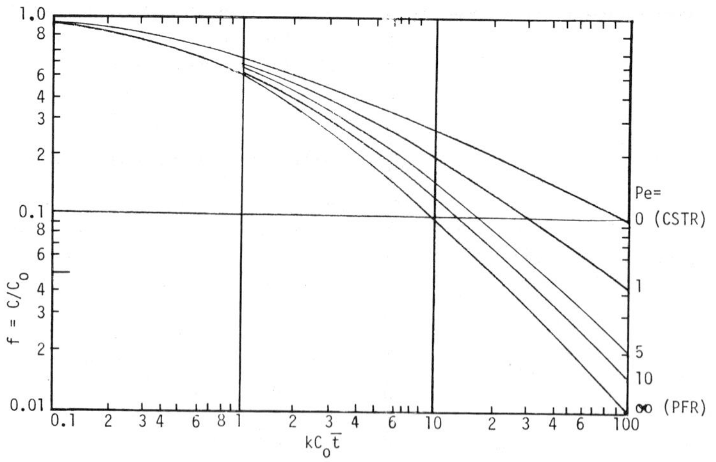

Figure 17.5. Second-order reaction with dispersion identified by the Peclet number, $Pe = uL/D_L$.

17.5. SELECTION OF CATALYSTS

A catalyst is a substance that increases a rate of reaction by participating chemically in intermediate stages of reaction and is liberated near the end in a chemically unchanged form. Over a period of time, however, permanent changes in the catalyst—deactivation—may occur. Inhibitors are substances that retard rates of reaction. Many catalysts have specific actions in that they influence only one reaction or group of definite reactions. An outstanding example is the living cell in which there are several hundred different catalysts, called enzymes, each one favoring a specific chemical process.

The mechanism of a catalyzed reaction—the sequence of reactions leading from the initial reactants to the final products—is changed from that of the uncatalyzed process and results in a lower overall energy of activation, thus permitting a reduction in the temperature at which the process can proceed favorably. The equilibrium condition is not changed since both forward and reverse rates are accelerated equally. For example, a good hydrogenation catalyst also is a suitable dehydrogenation accelerator; the most favorable temperature will be different for each process, of course.

A convenient classification is into homogeneous and heterogeneous catalysts. The former types often are metal complexes that are soluble in the reaction medium, but acids and bases likewise have a long known history of catalytic action. The specific action of a particular metal complex can be altered by varying the ligands or coordination number of the complex or the oxidation state of the central metal atom. Advantages of homogeneous catalysts are their specificity and low temperature and pressure requirements. Their main drawbacks are difficulty of recovery from the process fluid, often rapid degradation, and relatively high cost. Classic examples of homogeneous catalysis are the inversion of sugar with mineral acids, olefin alkylation with hydrofluoric acid, and the use of ammonia in the Solvay process and of nitrogen oxides in the Chamber process. A modern development is the synthesis of acetic acid from methanol and CO in the presence of homogeneous rhodium complexes.

The problem of separating the catalyst at the end of the operation can be eased in some cases by attaching the catalyst to a solid support, for instance, liquid phosphoric acid in the pores of a solid carrier for the vapor phase synthesis of cumene and the fairly wide application of enzymes that are attached (immobilized) by various means to solid polymers. Some metal ligands also are being combined with solid polymers.

HETEROGENEOUS CATALYSTS

By far the greatest tonnages of synthetic chemicals are manufactured in fluid phases with solid catalysts. Such materials are cheap, are easily separated from the reaction medium, and are adaptable to either flow or nonflow reactors. Their drawbacks are a lack of specificity and often high temperature and pressure requirements. The principal components of most heterogeneous catalysts are three in number:

1. A catalytically active substance or mixture.
2. A carrier of more or less large specific surface on which the catalyst proper is deposited as a thin film, either for economy when the catalyst is expensive or when the catalyst itself cannot be prepared with a suitable specific surface.
3. Promoters, usually present in relatively small amount, which enhance the activity or retard degradation.

Some composite catalysts are designed to promote several reactions of a sequence leading to the final products. A basic catalyst often can be selected with general principles, but subsequent fine tuning of a commercially attractive design must be done in a pilot plant or sometimes on a plant scale.

Analogy to what is known to be effective in chemically similar problems usually provides a start for catalyst design, although a scientific basis for selection is being developed. This involves a study in detail of the main possible intermediate reactions that could occur and of the proton and electron receptivity of the catalyst and possible promoters, as well as reactant bond lengths and crystals lattice dimensions. Several designs are made from this fundamental point of view in the book of Trimm (1980). A thorough coverage of catalytic reactions and catalysts arranged according to the periodic table is underway in a series edited by Roiter (1968–date). Industrial catalyst practice is summarized by Thomas (1970) who names manufacturers of specific catalysts. Specific processes and general aspects of catalysis are covered in three books edited by Leach (1983–1985): In a chapter by Wagner, there are lists of 40 catalysts with the kinds of processes for which they are effective and of 49 catalytic processes with the homogeneous or heterogeneous catalysts that have been used with them. Many industrial processes are described by Satterfield (1980). Cracking, reforming, partial oxidation, hydrodesulfurization, and catalysis by transition-metal complexes are treated in detail by Gates et al. (1979) and the catalytic conversion of hydrocarbons by Pines (1981). The mechanisms and other aspects of organic catalysis are described in one of the volumes of the series edited by Bamford and Tipper (1978). A vast literature exists for enzyme processes; that technology is well reviewed in two articles in *Ullmann's Encyclopedia* (Biotechnologie, Enzyme) as well as by Bailey and Ollis (1986). In the present text, Table 17.1 identifies the catalyst used in most of the 100 processes listed.

Intermediate processes of catalyzed organic reactions may involve neutral free radicals $R^{\cdot}$, positive ions R^+, or negative ions R^- as short-lived reactants. A classification of catalysts and processes from the point of view of elementary reactions between reagents and catalysts is logically desirable but has not yet been worked out. However, there is a wealth of practice more or less completely documented, some proprietary but available at a price. The ensuing discussions are classified into kinds of catalysts and into kinds of processes.

KINDS OF CATALYSTS

To a certain extent, it is known what kinds of reactions are speeded up by certain classes of catalysts, but individual members of the same class may differ greatly in activity, selectivity, resistance to degradation, and cost. Even small differences in these properties can mean large sums of money on the commercial scale. Solid catalysts, the most usual kind, are not particularly specific or selective, so that there is a considerable crossing of lines in classifications between kinds of catalysts and kinds of reactions they favor. Nevertheless, leading relations can be brought out.

Strong acids are able to donate protons to a reactant and to take them back. Into this class fall the common acids, aluminum halides, and boron trifluoride. Also acid in nature are silica, alumina, aluminosilicates, metal sulfates and phosphates, and sulfonated ion exchange resins. The nature of the active sites on these kinds of solids still is not completely understood. The majority of reactions listed subsequently are catalytically influenced to some extent by acidic substances. Zeolites are dehydrated aluminosilicates with small pores of narrow size distribution, to which is due their highly selective catalytic action since only molecules small enough to enter the pores can react. In cracking operations they are diluted to 10–15% in silica–alumina to restrain their great activity; the composite catalyst still is very active but makes less carbon, makes lower amounts of C_3–C_4 products, and has a longer life. Their greater activity has led to the supplanting of fluidized bed crackers by riser-tube reactors. When zeolites are incorporated in reforming catalysts, they crack isoparaffins into straight chains that enter the pores and convert into higher octane substances.

Base catalysis is most effective with alkali metals dispersed on solid supports or in the homogeneous form as aldoxides, amides, and so on. Small amounts of promoters may be added to form organoalkali compounds that really have the catalytic power. Basic ion exchange resins also are useful. Some base-catalyzed processes are isomerization and oligomerization of olefins, reaction of olefins with aromatics, and hydrogenation of polynuclear aromatics.

Metal oxides, sulfides, and hydrides form a transition between acid–base and metal catalysts. They catalyze hydrogenation–dehydrogenation as well as many of the reactions catalyzed by acids such as cracking and isomerization. Their oxidation activity is related to the possibility of two valence states which allow oxygen to be released and reabsorbed alternately. Common examples are oxides of cobalt, iron, zinc, and chromium; and hydrides of precious metals which can release hydrogen readily. Sulfide catalysts are more resistant than metallic catalysts to formation of coke deposits and to poisoning by sulfur compounds; their main application is to hydrodesulfurization.

Metals and alloys. The principal industrial metallic catalysts are found in periodic group VIII which are transition elements with almost completed 3d, 4d, and 5d electron orbits. According to one theory, electrons from adsorbed molecules can fill the vacancies in the incomplete shells and thus make a chemical bond. What happens subsequently will depend on the operating conditions. Platinum, palladium, and nickel, for example, form both hydrides and oxides; they are effective in hydrogenation (vegetable oils, for instance) and oxidation (ammonia or sulfur dioxide, for instance). Alloys do not always have catalytic properties intermediate between those of the pure metals since the surface condition may be different from the bulk and the activity is a property of the surface. Addition of small amounts of rhenium to Pt/Al_2O_3 results in a smaller decline of activity with higher temperature and slower deactivation rate. The mechanism of catalysis by alloys is in many instances still controversial.

Transition-metal organometallic catalysts in solution are effective for hydrogenation at much lower temperatures than metals such as platinum. They are used for the reactions of carbon monoxide with olefins (hydroformylation) and for some oligomerizations. The problem of separating the catalyst from solution sometimes is avoided by anchoring or immobilizing the catalyst on a polymer support containing pendant phosphine groups and in other ways.

KINDS OF CATALYZED ORGANIC REACTIONS

A fundamental classification of organic reactions is possible on the basis of the kinds of bonds that are formed and destroyed and the natures of eliminations, substitutions, and additions of groups. Here a more pragmatic list of 20 commercially important individual kinds or classes of reactions will be discussed.

1. Alkylations, for example, of olefins with aromatics or isoparaffins, are catalyzed by sulfuric acid, hydrofluoric acid, BF_3, and $AlCl_3$.
2. Condensations of aldehydes and ketones are catalyzed homogeneously by acids and bases, but solid bases are preferred, such as anion exchange resins and alkali or alkaline earth hydroxides or phosphates.
3. Cracking, a rupturing of carbon—carbon bonds, for example, of gas oils to gasoline, is favored by silica–alumina, zeolites, and acid types generally.
4. Dehydration and dehydrogenation combined utilizes dehydration agents combined with mild dehydrogenation agents. Included in this class of catalysts are phosphoric acid, silica–magnesia, silica–alumina, alumina derived from aluminum chloride, and various metal oxides.
5. Esterification and etherification may be accomplished by catalysis with mineral acids of BF_3; the reaction of isobutylene with methanol to make MTBE is catalyzed by a sulfonated ion exchange resin.
6. Fischer–Tropsch oligomerization of CO + hydrogen to make hydrocarbons and oxygenated compounds. Iron promoted by potassium is favored, but the original catalyst was cobalt which formed a carbonyl in process.
7. Halogenation and dehalogenation are catalyzed by substances that exist in more than one valence state and are able to accept and donate halogens freely. Silver and copper halides are used for gas-phase reactions, and ferric chloride commonly for liquid phase. Hydrochlorination (the absorption of HCl) is promoted

by $BiCl_3$ or $SbCl_3$ and hydrofluorination by sodium fluoride or chromia catalysts that fluoride under reaction conditions. Mercuric chloride promotes addition of HCl to acetylene to make vinyl chloride.

8. Hydration and dehydration employ catalysts that have a strong affinity for water. Alumina is the principal catalyst, but also used are aluminosilicates, metal salts, and phosphoric acid or its metal salts on carriers and cation exchange resins.
9. Hydrocracking is catalyzed by substances that promote cracking and hydrogenation together. Nickel and tungsten sulfides on acid supports and zeolites loaded with palladium are used commercially.
10. Hydrodealkylation, for example, of toluene to benzene, is promoted by chromia–alumina with a low sodium content.
11. Hydrodesulfurization uses sulfided cobalt/molybdena/alumina, or alternately with nickel and tungsten substituted for Co and Mo.
12. Hydroformylation, or the oxo process, is the reaction of olefins with CO and hydrogen to make aldehydes. The catalyst base is cobalt naphthenate which transforms to cobalt hydrocarbonyl in place. A rhodium complex that is more stable and functions at a lower temperature also is used.
13. Hydrogenation and dehydrogenation employ catalysts that form unstable surface hydrides. Transition-group and bordering metals such as Ni, Fe, Co, and Pt are suitable, as well as transition group oxides or sulfides. This class of reactions includes the important examples of ammonia and methanol syntheses, the Fischer–Tropsch and oxo and synthol processes and the production of alcohols, aldehydes, ketones, amines, and edible oils.
14. Hydrolysis of esters is speeded up by both acids and bases. Soluble alkylaryl sulfonic acids or sulfonated ion exchange resins are satisfactory.
15. Isomerization is promoted by either acids or bases. Higher alkylbenzenes are isomerized in the presence of $AlCl_3/HCl$ or BF_3/HF; olefins with most mineral acids, acid salts, and silica alumina; saturated hydrocarbons with $AlCl_3$ or $AlBr_3$ promoted by 0.1% of olefins.
16. Metathesis is the rupture and reformation of carbon—carbon bonds, for example of propylene into ethylene plus butene. Catalysts are oxides, carbonyls or sulfides of Mo, W, or rhenium.
17. Oxidation catalysts are either metals that chemisorb oxygen readily such as platinum or silver, or transition metal oxides that are able to give and take oxygen by reason of their having several possible oxidation states. Ethylene oxide is formed with silver, ammonia is oxidized with platinum, and silver or copper in the form of metal screens catalyze the oxidation of methanol to formaldehyde.
18. Polymerization of olefins such as styrene is promoted by acid or base catalysts or sodium; polyethylene is made with homogeneous peroxides.
19. Reforming is the conversion primarily of naphthenes and alkanes to aromatics, but other chemical reactions also occur under commercial conditions. Platinum or platinum/rhenium are the hydrogenation–dehydrogenation component of the catalyst and alumina is the acid component responsible for skeletal rearrangements.
20. Steam reforming is the reaction of steam with hydrocarbons to make town gas or hydrogen. For town gas a representative catalyst composition contains 13 wt % Ni, 12.1% U, and 0.3 wt % K; it is particularly resistant to poisoning by sulfur. To make hydrogen, the catalyst contains oxides of Ni, Ca, Si, Al, Mg, and K. Specific formulations are given by Satterfield (1980).

PHYSICAL CHARACTERISTICS OF SOLID CATALYSTS

Although a few very active solid catalysts are used as fine wire mesh or other finely divided form, catalysts are mostly porous bodies whose total surface is measured in m^2/g. These and other data of some commercial catalysts are shown in Table 17.8. The physical characteristics of major importance are as follows.

1. *Particle size.* In gas fluidized beds the particle diameters average less than 0.1 mm; smaller sizes impose too severe loading on entrainment recovery equipment. In slurry beds the particles can be about 1 mm dia. In fixed beds the range is 2–5 mm dia. The competing factors are that the pressure drop increases with diminishing diameter and the accessibility of the internal surface decreases with increasing diameter. With poorly thermally conducting materials, severe temperature gradients or peaks arise with large particles that may lead to poor control of the reaction and the development of undesirable side reactions like carbonization.
2. *Specific surface.* Solid spheres of 0.1 mm dia have a specific surface of $0.06\,m^2/mL$ and an activated alumina one of about $600\,m^2/mL$. Other considerations aside, a large surface is desirable because the rate of reaction is proportional to the amount of accessible surface. Large specific surfaces are associated with pores of small diameters and are substantially all internal surface.
3. *Pore diameters and their distribution.* Small pores limit accessibility of internal surface because of increased resistance to diffusion of reactants inwards. Diffusion of products outward also is slowed down and may result in degradation of those products. When the catalyst is expensive, the inaccessible internal surface is a liability. A more or less uniform pore diameter is desirable, but the distribution usually is statistical and only molecular sieves have nearly uniform pores. Those catalyst granules that are extrudates of compacted masses of smaller particles have bimodal pore size distribution, between the particles and within them. Clearly a compromise between large specific surface and its accessibility as measured by pore diameter is required in some situations.
4. *Effective diffusivity.* Resistance to diffusion in a catalyst pore is due to collisions with other molecules and with the walls of the pore. The corresponding diffusivities are called bulk diffusivity and Knudsen diffusivity D_K. Many data and correlations of the former type exist; the latter is calculable from the following formula (Satterfield, 1970, p. 42):

$$D_K = \frac{19{,}400\theta^2}{S_g\rho_p}\left(\frac{T}{M}\right)^{1/2},$$

where

θ = fraction porosity,
S_g = specific surface per unit mass,
ρ_p = density,
T = temperature (K),
M = molecular weight.

This equation applies to uniform cylindrical pores whose length equals the thickness of the catalyst through which the diffusion takes place. The actual diffusivity in common porous catalysts usually is intermediate between bulk and Knudsen. Moreover, it depends on the pore size distribution and on the true length of

TABLE 17.8. Physical Properties of Some Commercial Catalysts and Carriers[a]

Designation	Nominal Size	Surface Area (m^2/g)	Total Void Fraction	$D^{*}_{eff} \times 10^3$ (cm^2/sec)	Average Tortuosity Factor τ_p Parallel-Path Pore Model	$r_e = 2V_g/S_g$ (Å)	τ_m Based on Average Pore Radius
T-126	3/16 × 1/8 in.	197	0.384	29.3	3.7 ± 0.2	29	0.45
T-1258		302	0.478	33.1	3.8 ± 0.2	23.6	0.41
T-826		232	0.389	37.7	3.9 ± 0.1	21.4	0.26
T-314		142	0.488	20.0	7.1 ± 0.9	41.5	1.2
T-310		154	0.410	16.6	3.8 + 0.1	34.3	0.67
G-39	3/16 × 3/16 in.	190	0.354	17.5	4.8 ± 0.3	22.4	0.53
G-35		—	0.354	18.2	4.9 ± 0.1		—
T-606		—	0.115	27.7	2.9 ± 0.2		—
G-58		6.4	0.389	87.0	2.8 ± 0.3	543.	2.87
T-126	1/4 × 1/4 in.	165	0.527	38.8	3.6 ± 0.3	49.0	0.79
T-606		—	0.092	0.71	79 ± 28		—
G-41		—	0.447	21.9	4.4 ± 0.1		—
G-52		—	0.436	27.4	3.9 + 0.2		—
G-56	1/2 × 1/2 in.	42	0.304	8.1	11.1 ± 1.1	84.	3.74
BASF	5 × 5 mm	87.3	0.500	11.8	7.3 + 0.7	41.	2.05
Harshaw	1/4 × 1/4 in.	44	0.489	13.3	7.2 ± 0.1	91.	3.95
Haldor Topsøe	1/4 × 1/4 in.	143	0.433	15.8[e]	2.8	25.8	0.83

Catalyst	Description
T-126	Activated γ-alumina
T-1258	Activated γ-alumina
T-826	3% CoO, 10% MoO_3, and 3% NiO on alumina
T-314	About 8–10% Ni and Cr in the form of oxides on an activated alumina
T-310	About 10–12% nickel as the oxide on an activated alumina
T-606	Specially compounded refractory oxide support
G-39	A cobalt-molybdenum catalyst, used for simultaneous hydrodesulfurization of sulfur compounds and hydrogenation of olefins
G-35	A cobalt-molybdenum catalyst supported on high-purity alumina, used for hydrodesulfurization of organic sulfur compounds
G-41	A chromia-alumina catalyst, used for hydrodealkylation and dehydrogenation reactions
G-58	Palladium-on-alumina catalyst, for selective hydrogenation of acetylene in ethylene
G-52	Approximately 33 wt % nickel on a refractory oxide support, prereduced. Used for oxygen removal from hydrogen and inert gas streams
G-56	A nickel-base catalyst used for steam reforming of hydrocarbons
BASF	A methanol synthesis catalyst, prereduced
Harshaw	A methanol synthesis catalyst, prereduced
Haldor Topsøe	A methanol synthesis catalyst, prereduced

[a] The measured effective diffusivities are those of hydrogen in nitrogen at room temperature and pressure except that of Haldor Topsoe which is of helium in nitrogen.

[Satterfield and Cadle, *Ind. Eng. Chem. Process Design and Development*, **7**, 256 (1968)].

path. Two tortuosity factors are defined:

τ_p = ratio of measured diffusivity to that calculated with the known pore size distribution and bulk diffusivity and the thickness of the catalyst mass.

τ_m = ratio of measured diffusivity to that calculated from the Knudsen formula with a mean pore diameter.

The data of Table 17.8 exhibit a fairly narrow range of τ_p, an average of about 4, but there seems to be no pattern to τ_m, which is not surprising since the diffusions actually are intermediate between bulk and Knudsen in these cases. In order to be able to calculate the effective diffusivity, it is necessary to know the pore size distribution, the specific surface, the porosity, and bulk diffusivity in the reaction mixture under reaction conditions. Such a calculation is primarily of theoretical interest. Practically it is more useful to simply measure the diffusivity directly, or even better to measure the really pertinent property of catalyst effectiveness as defined next.

CATALYST EFFECTIVENESS

Catalyst effectiveness is a measure of the extent of utilization of internal surface; it is the ratio of a rate of reaction actually achieved with the catalyst particle to the rate that would prevail if all of the internal surface were exposed to the reactant concentration at the external surface of the particle. The rate equation accordingly is modified to

$$r = k\eta f(C_s), \tag{17.30}$$

where η is the catalyst effectiveness and C_s is the concentration of the reactant at the external surface. For isothermal reactions, η

always is less than unity, but very large values can develop for exothermic reactions in poorly conducting catalysts.

A great deal of attention has been devoted to this topic because of the interesting and often solvable mathematical problems that it presents. Results of such calculations for isothermal zero-, first-, and second-order reactions in uniform cylindrical pores are summarized in Figure 17.6. The abscissa is a modified Thiele modulus whose basic definition is

$$\phi = R/k_v C_s^{n-1}/D_{\text{eff}}, \tag{17.31}$$

where R is a linear dimension (the radius of a sphere, for example), k_v the specific rate on a volumetric basis, C_s the surface concentration, n the order of the reaction, and D_{eff} the effective diffusivity. For nonisothermal reactions, those with variable volume and with rate equations of the Langmuir–Hinshelwood or other complex types, additional parameters are involved. Although such calculations can be made, they still require measurements of effective diffusivity as well as a number of unverifiable assumptions. Accordingly in practical cases it is preferable to make direct measurements of catalyst effectiveness and to correlate them with operating parameters. The effectiveness is deduced by comparing conversion with the reference particle size with those with successively small particle sizes until the effect disappears. Two examples are presented to illustrate the variables that are taken into account and the magnitudes of the effects.

For synthesis of ammonia the effectiveness has been measured by Dyson and Simon [*Ind. Eng. Chem. Fundam.* **7,** 605–610 (1968)] and correlated by the equation

$$\eta = b_0 + b_1 T + b_2 x + b_3 T^2 + b_4 x^2 + b_5 T^3 + b_6 x^3, \tag{17.32}$$

where T is in K, x is fractional conversion of nitrogen, and the b_i depend on pressure as given in this table:

Pressure (atm)	b_0	b_1	b_2	b_3	b_4	b_5	b_6
150	−17.539096	0.07697849	6.900548	-1.082790×10^{-4}	−26.42469	4.927648×10^{-3}	38.93727
225	−8.2125534	0.03774149	6.190112	-5.354571×10^{-5}	−20.86963	2.379142×10^{-3}	27.88403
300	−4.6757259	0.02354872	4.687353	-3.463308×10^{-5}	−11.28031	1.540881×10^{-3}	10.46627

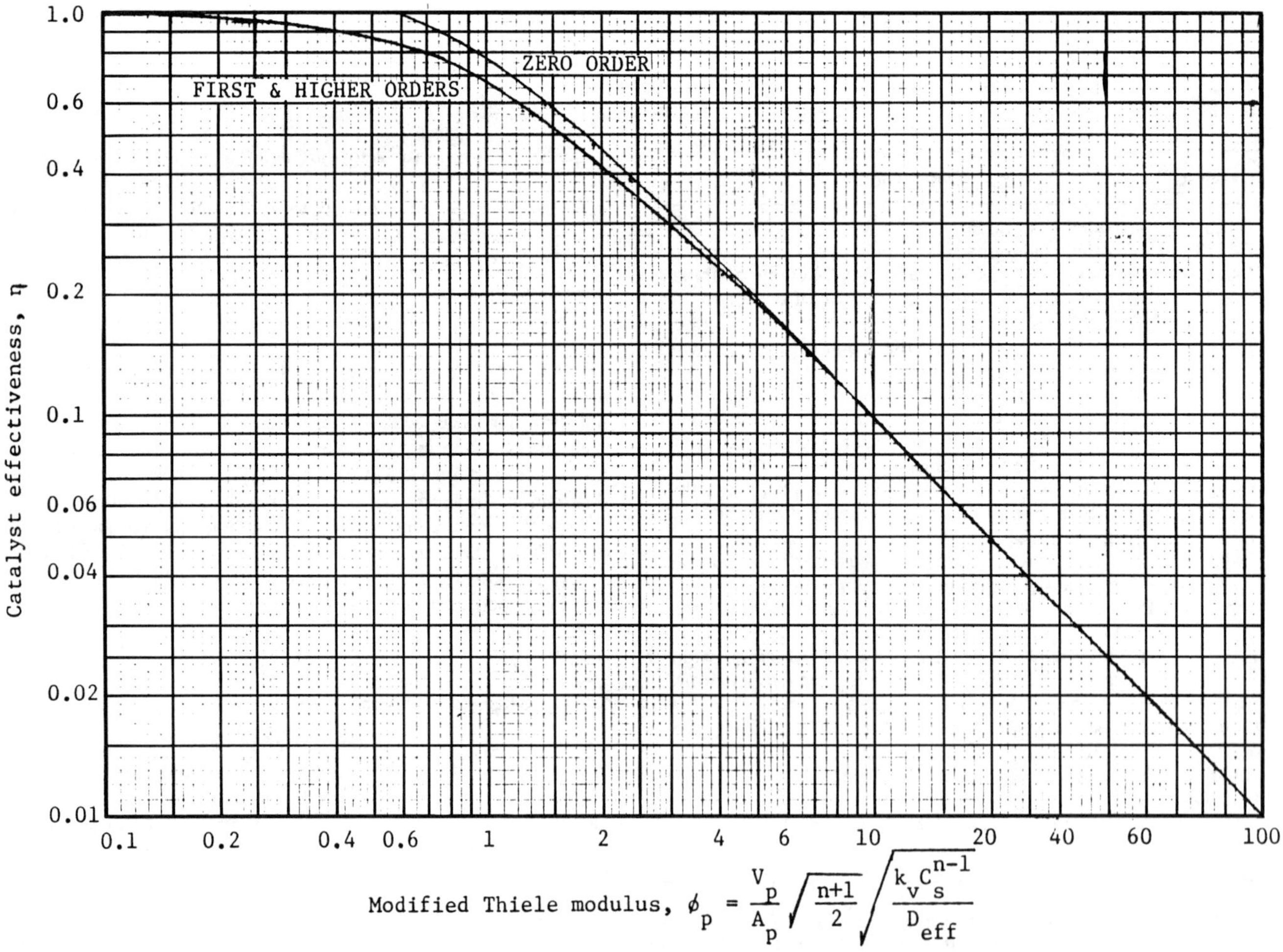

Figure 17.6. Generalized chart of catalyst effectiveness for reactions of order n in particles with external surface A_p and volume V_p. The upper curve applies exactly to zero-order reaction in spheres, and the lower one closely for first- and second-order reactions in spheres.

The reference mixture has $H_2/N_2 = 3$ and contains 12.7% inert; other ratios had slightly different effectiveness. The particle diameters are 6–10 mm. Some calculations from this equation at 225 atm are:

T	x	η
700	0.25	0.81
700	0.10	0.57
650	0.25	0.91

For oxidation of sulfur dioxide, measurements of effectiveness were made by Kadlec, Pour, and Regner [*Coll. Czech. Chem. Commun.* **33,** 2388, 2526 (1968)] whose data are shown following. They are at atmospheric pressure. The initial content of SO_2 and the conversion have little effect on the result. Both increase in size of granule and temperature lower the effectiveness, although the effect of temperature is somewhat erratic.

Experimentally Determined Effectiveness Factors

		Conversion					
°C	% SO_2	0.4	0.5	0.6	0.7	0.8	0.9
Irregular grain shape, fraction 5–6 mm							
460	7	0.84	0.84	0.82	0.83	0.82	0.81
480	7	0.60	0.62	0.62	0.62	0.60	0.60
500	7	—	0.54	0.51	0.50	0.50	0.52
520	7	—	0.35	0.35	0.35	0.38	0.38
Cylindrical granules of 6 mm diameter and 12 mm length							
460	7	0.57	0.57	0.59	0.60	0.60	0.60
	10	0.58	0.62	0.63	0.63	0.62	0.62
480	7	0.53	0.54	0.56	0.57	0.56	0.57
	10	0.44	0.45	0.45	0.46	0.45	0.47
500	7	0.25	0.25	0.27	0.28	0.27	0.31
	10	0.26	0.27	0.30	0.30	0.31	0.30
520	7	—	0.21	0.21	0.22	0.22	0.23
	10	—	0.20	0.21	0.21	0.22	0.24

The rate equations of both these processes are quite complex, and there is little likelihood that the effectiveness could be deduced mathematically from fundamental data as functions of temperature, pressure, conversion, and composition, which is the kind of information needed for practical purposes. Perhaps the only estimate that can be made safely is that, in the particle size range below 1 mm or so, the effectiveness probably is unity. The penetration of small pores by liquids is slight so that the catalysts used in liquid slurry systems are of the low specific surface type or even nonporous.

17.6. TYPES AND EXAMPLES OF REACTORS

Almost every kind of holding or contacting equipment has been used as a chemical reactor at some time, from mixing nozzles and centrifugal pumps to the most elaborate towers and tube assemblies. This section is devoted to the general characteristics of the main kinds of reactors, and also provides a gallery of selected examples of working reactors.

The most obvious distinctions are between nonflow (batch) and continuous operating modes and between the kinds of phases that are being contacted. A classification of appropriate kinds of reactors on the basis of these two sets of distinctions is in Figure 17.7.

When heterogeneous mixtures are involved, the conversion rate often is limited by the rate of interphase mass transfer, so that a large interfacial surface is desirable. Thus, solid reactants or catalysts are finely divided, and fluid contacting is forced with mechanical agitation or in packed or tray towers or in centrifugal pumps. The rapid transfer of reactants past heat transfer surfaces by agitation or pumping enhances also heat transfer and reduces harmful temperature gradients.

Batch processing is used primarily when the reaction time is long or the required daily production is small. The same batch equipment often is used to make a variety of products at different times. Otherwise, it is not possible to generalize as to the economical transition point from batch to continuous operation. One or more batch reactors together with appropriate surge tanks may be used to simulate continuous operation on a daily or longer basis.

STIRRED TANKS

Stirred tanks are the most common type of batch reactor. Typical proportions are shown on Figures 17.8 and 10.1, and modes of level control on Figure 3.6. Stirring is used to mix the ingredients initially, to maintain homogeneity during reaction, and to enhance heat transfer at a jacket wall or internal surfaces. The reactor of Figure 17.9(b) employs a pumparound for mixing of the tank contents and for heat transfer in an outside exchanger. Pumparound or recycle in general may be used to adapt other kinds of vessels to service as batch mode reactors; for example, any of the packed vessels of Figure 17.10(a)–(e). A pumparound tubular flow reactor is employed for the polymerization of ethylene on Figure 17.11(c); as the polymer is formed, it is bled off at a much lower rate than that of the recirculation, so that in a sense the action of this equipment approaches batch operation.

Some special industrial stirred reactors are illustrated in Figure 17.10: (b) is suitable for pasty materials, (c) for viscous materials, and the high recirculation rate of (d) is suited to intimate contacting of immiscible liquids such as hydrocarbons with aqueous solutions.

Many applications of stirred tank reactors are to continuous processing, either with single tanks or multiple arrangements as in Figures 17.9(c)–(d). Knowledge of the extent to which a stirred tank does approach complete mixing is essential to being able to predict its performance as a reactor. The other limiting case is that of plug flow, in which all nonreacting molecules have the same residence time. Deviations from the limiting cases of complete mixing (in a CISTR) and no axial mixing (in a PFR) are evaluated with residence time distributions (RTDs) based on analyses of tracer tests.

At present, RTD behavior has not been well correlated with operating or design factors, but the technique is of diagnostic value with existing equipment. CISTR (continuous ideal stirred tank reactor) behavior is approached when the mean residence time is 5–10 times the mixing time, which is in turn the length of time needed to achieve homogeneity of a mixture of several inputs. Often this is achieved by 50–200 revolutions of a properly designed stirrer. Although mixing times have been the subject of many studies in the literature (Westerterp et al., 1984, p. 254), no useful generalizations have been deduced. The mixing time depends on the geometry and the speed and power of the agitator. A propeller above and a turbine below on the same shaft, baffles attached to the wall of the tank, and possibly a draft tube around the shaft for effective recirculation of the contents constitute a basic design. However, no completely rational design of mixing equipment is possible at this time, so that in critical cases experts should be consulted. Chapter 10 also deals with this topic.

Power input per unit volume and impeller tip speeds are often used measures of the intensity of stirring, assuming correct proportions of the vessel and proper baffling. Appropriate ranges

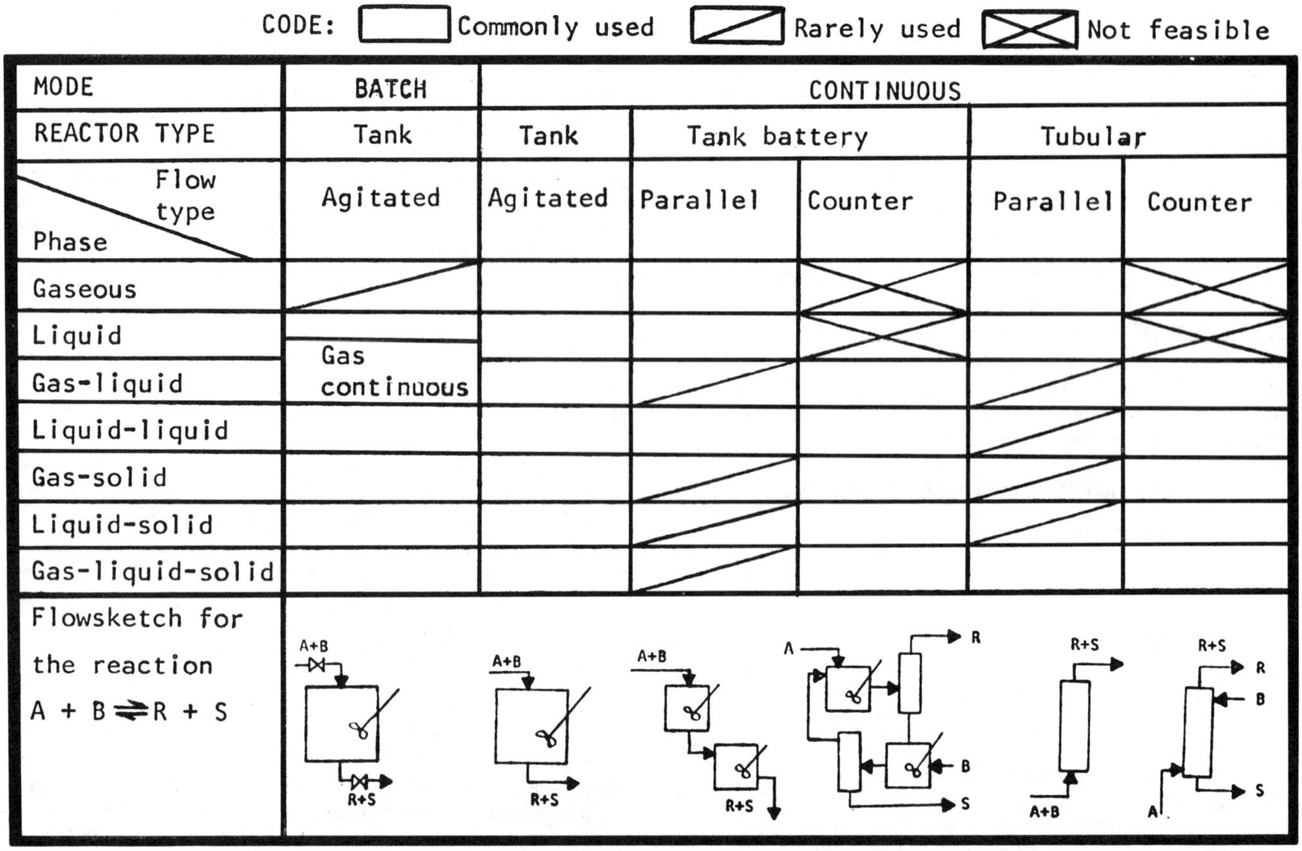

CODE:
Commonly used
Rarely used
Not feasible
MODE
BATCH
CONTINUOUS
REACTOR TYPE
Tank
Tank
Tank battery
Tubular
Flow type
Phase
Agitated
Agitated
Parallel
Counter
Parallel
Counter
Gaseous
Liquid
Gas-liquid
Gas continuous
Liquid-liquid
Gas-solid
Liquid-solid
Gas-liquid-solid
Flowsketch for the reaction A + B ⇌ R + S
A+B
R+S
A+B
R+S
A+B
R+S
A
R
B
S
R+S
A+B
R+S
R
B
S
A

Figure 17.7. Classification of reactors according to the mode of operation and the kinds of phases involved.

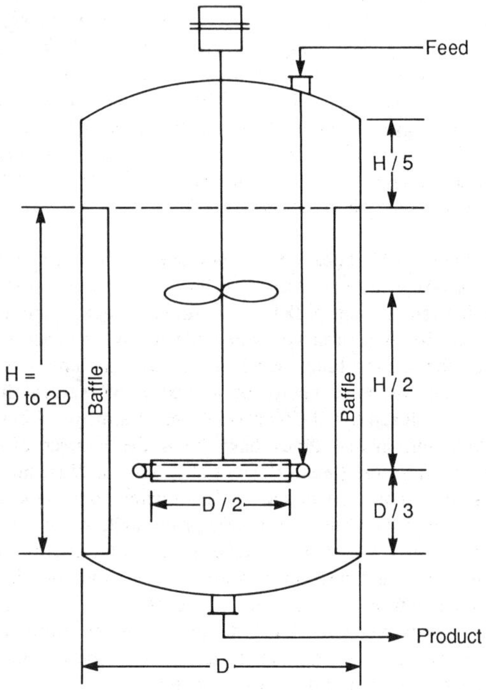

Figure 17.8. Typical proportions of a stirred tank reactor with radial and axial impellers, four baffles, and a sparger feed inlet.

for some reaction conditions are as tabulated:

Operation	kW/m^3 [a]	Tip speed (m/sec)
Blending	0.05–0.1	
Homogeneous reaction	0.1–0.3	2.5–3.3
Reaction with heat transfer	0.3–1.0	3.5–5.0
Gas–liquid, liquid–liquid	1–2	5–6
Slurries	2–5	

[a] $1\ kW/m^3 = 5.08$ HP/1000 gal

Heat transfer coefficients in stirred tank operations are discussed in Section 17.7.

For a given load and conversion, the total volume of a CSTR (continuous stirred tank reactor) battery decreases with the number of stages, sharply at first and then more slowly. When the reaction is first order, for example, $r = kC$, the ratio of total reactor volume V_r of n stages to the volumetric feed rate V'_0 is represented by

$$kV_r/V'_0 = n[(C_0/C)^{1/n} - 1]. \tag{17.33}$$

At conversions of 95 and 99%, some values from this equation are

n	1	2	3	4	5	10
kV_r/V'_0 at 95%	19	6.9	5.1	4.5	5.1	3.5
kV_r/V'_0 at 99%	99	18.0	10.9	9.7	7.6	5.9

Since the cost of additional controls, agitators, and pumps can counterbalance the savings in volume, four or five tanks in a battery normally prove to be an optimum number, but a larger number of stages may be economical with a single shell design like Figure

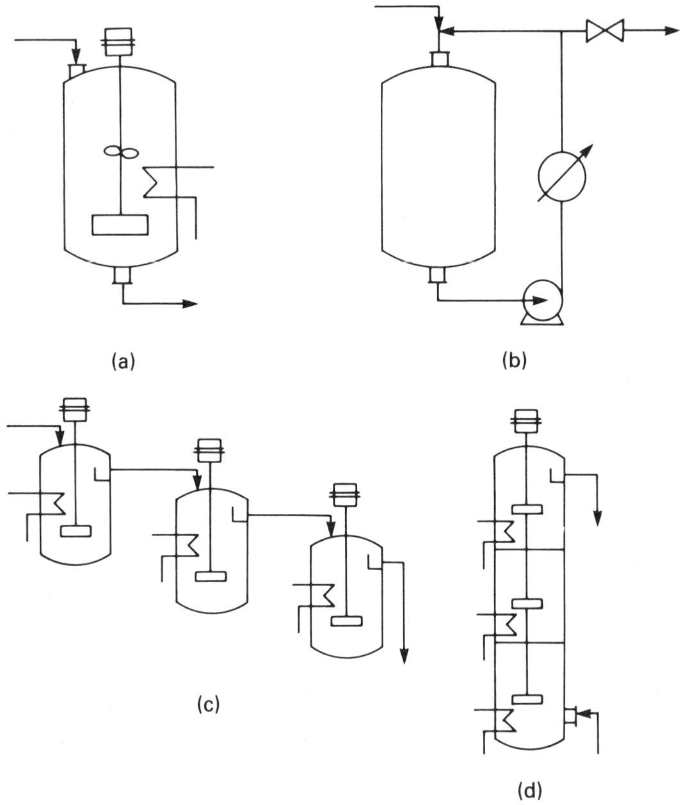

Figure 17.9. Stirred tank reactors, batch and continuous. (a) With agitator and internal heat transfer surface, batch or continuous. (b) With pumparound mixing and external heat transfer surface, batch or continuous. (c) Three-stage continuous stirred tank reactor battery. (d) Three-stage continuous stirred tank battery in a single shell.

17.9(d), particularly when the stages are much less efficient than ideal ones.

For some purposes it is adequate to assume that a battery of five or so CSTRs is a close enough approximation to a plug flow reactor. The tubular flow reactor is smaller and cheaper than any comparable tank battery, even a single shell arrangement. For a first order reaction the ratio of volumes of an n-stage CSTR and a PFR is represented by

$$(V_r)_{\mathrm{CSTR}}/(V_r)_{\mathrm{PFR}} = n[(C_0/C)^{1/n} - 1]/\ln(C_0/C). \qquad (17.34)$$

For example, when $n = 5$ and conversion is 99%, the ratio is 1.64. For second-order and other-order reactions a numerical solution for the ratio is needed, one of which is represented by Figure 17.12. For a second order reaction the ratio is 1.51 at 99% conversion with five stages.

A further difference between CSTR batteries and PFRs is that of product distributions with complex reactions. In the simple case, $A \rightarrow B \rightarrow C$ for example, a higher yield of intermediate product B is obtained in a PFR than in a single CSTR. It is not possible to generalize the results completely, so that the algebra of each individual reacting system must be worked out to find the best mode.

TUBULAR FLOW REACTORS

The ideal behavior of tubular flow reactors (TFR) is plug flow, in which all nonreacting molecules have equal residence times. Any backmixing that occurs is incidental, the result of natural turbulence or that induced by obstructions to flow by catalyst granules or tower packing or necessary internals of the vessels. The action of such obstructions can be two-edged, however, in that some local backmixing may occur, but on the whole a good approach to plug flow is developed because large scale turbulence is inhibited. Any required initial blending of reactants is accomplished in mixing nozzles or by in-line mixers such as those of Figures 10.13–10.14. As a result of chemical reaction, gradients of concentration and temperature are developed in the axial direction of TFRs.

TFRs may be of pipe diameters ranging from 1 to 15 cm or so, or they may be vessels of diameters measured in meters. Figure 17.13 is of a variety of vessel configurations. Single tube reactors more than 1000 m long are used, in which case they are trombone-shaped as on Figures 17.14(f) and 17.15(c). The selection of diameter is a result of compromise between construction cost, pumping cost, and required heat transfer. In some cases it may be necessary to avoid the laminar flow region, which is below Reynolds numbers of 2300–4000 or so, if the reaction is complex and a spread of residence times is harmful.

When many tubes in parallel are needed, a shell-and-tube construction like that of heat exchangers is employed; the vessel then may be regarded as a heat exchanger in which a reaction occurs incidentally. Heat transfer to single tubes is accomplished with jackets in Figure 17.14(f) and in a fired heater in Figure 17.15(c). Some of the many designs of fired heaters that are suitable for pyrolysis and other high temperature reactions are illustrated on

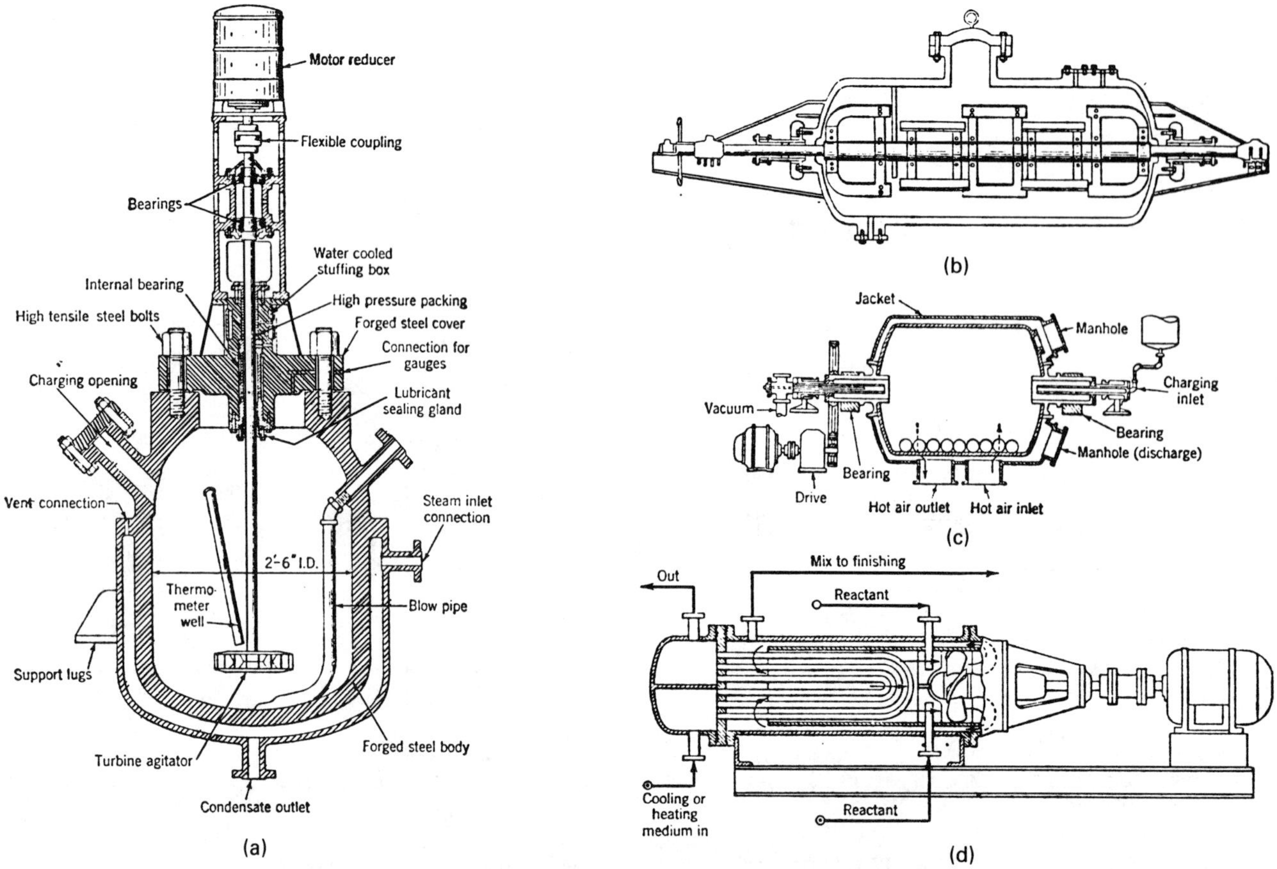

Figure 17.10. Several modes of mixing in commercial tank reactors. (a) Steam-jacketed autoclave, 120 gal, 200 psig, 300°F (*courtesy Blaw-Knox Co.*). (b) Horizontal autoclave, 650 gal, 100 psig (*courtesy Blaw-Knox Co.*). (c) Ball-mill sulfonator [*Groggins. Courtesy McGraw-Hill, New York*]. (d) Horizontal heat-exchange reactor (*courtesy Stratford Engineering Corp. patents issued and pending*).

Figure 17.16. In the process for making phenol, monochlorbenzene, and aqueous caustic are reacted at 320°C and 200 atm in multipass tubes of 10 cm dia or so in a fired heater.

In general, the construction of TFRs is dictated by the need for accommodation of granular catalysts as well as for heat transfer. Some of the many possible arrangements are illustrated on Figure 17.13 and elsewhere in this section.

Some unusual flow reactors are shown in Figure 17.14. The residence times in the units for high temperature pyrolysis to make acetylene and ethylene and for the oxidation of ammonia are measured in fractions of a second; acetic anhydride is made by mixing reactants quickly in a centrifugal pump; NO is formed at very high temperature in an electric furnace; and ethylene is polymerized at high or low pressures in the two units shown.

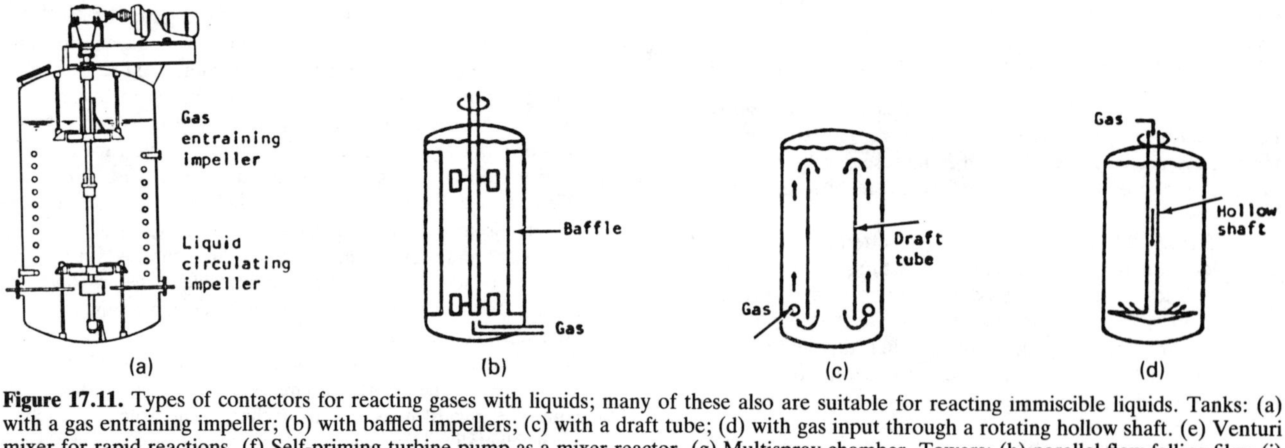

Figure 17.11. Types of contactors for reacting gases with liquids; many of these also are suitable for reacting immiscible liquids. Tanks: (a) with a gas entraining impeller; (b) with baffled impellers; (c) with a draft tube; (d) with gas input through a rotating hollow shaft. (e) Venturi mixer for rapid reactions. (f) Self-priming turbine pump as a mixer-reactor. (g) Multispray chamber. Towers: (h) parallel flow falling film; (i) spray tower with gas as continuous phase; (j) parallel flow packed tower; (k) counter flow tray tower. (l) A doublepipe heat exchanger used as a tubular reactor.

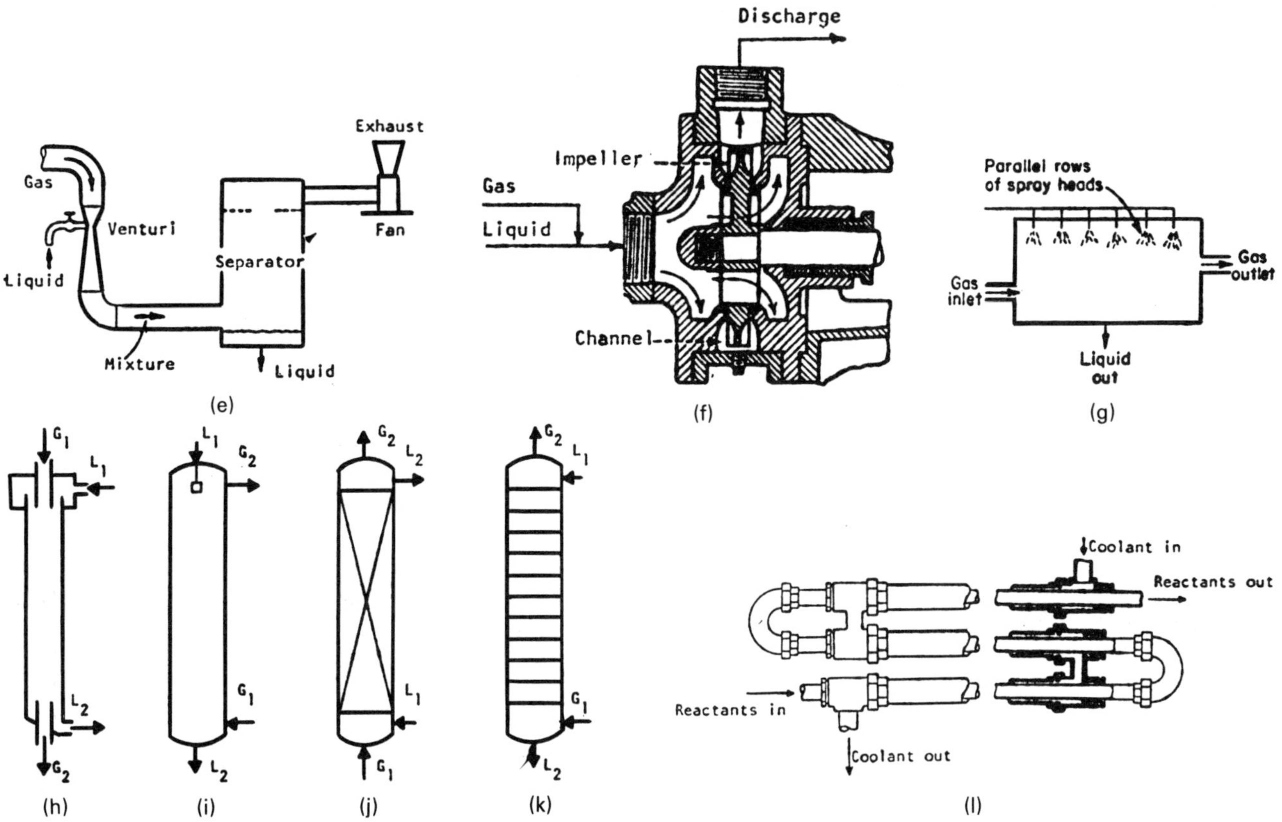

Figure 17.11—(*continued*)

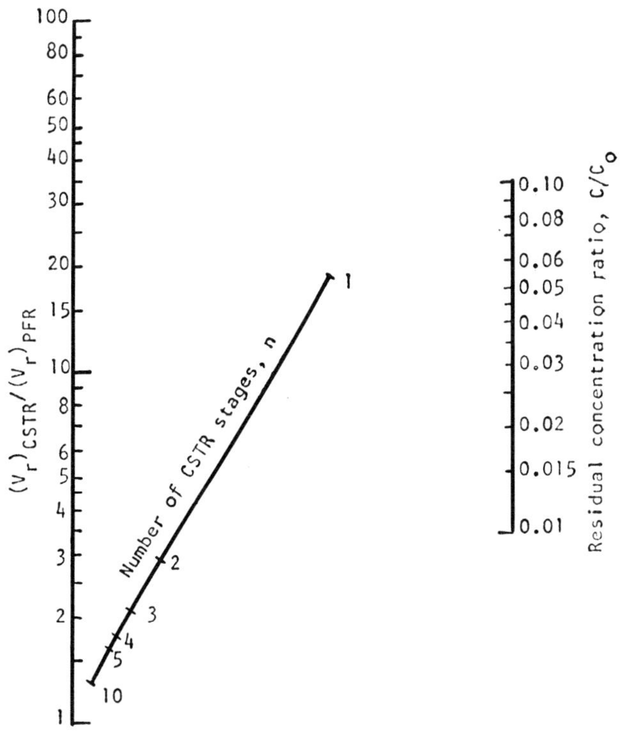

Figure 17.12. Ratio of volumes of an n-stage CSTR battery and a plug flow reactor as a function of residual concentration ratio C/C_0 with a rate equation $r = kC^2$.

GAS–LIQUID REACTIONS

Except with highly volatile liquids, reactions between gases and liquids occur in the liquid phase, following a transfer of gaseous participants through gas and liquid films. The rate of mass transfer always is a major or limiting factor in the overall transformation process. Naturally the equipment for such reactions is similar to that for the absorption of chemically inert gases, namely towers and stirred tanks. Figure 17.11 illustrates schematically types of gas–liquid reactors. Figure 17.17 shows specific examples of such reactors: In the synthesis of butynediol, acetylene at high pressure is bubbled into aqueous formaldehyde at several positions along a tower in (a). The heat of absorption of nitrogen oxides in water to make nitric acid is removed in two ways in the equipment of (b) and (e). Fats are hydrogenated in a continuous multistage stirred reactor in (c) and under batch conditions in a coil-cooled stirred tank in (d). A thin film reactor is used for the sulfonation of dodecylbenzene with SO_3 in (f). Hydrogen is recirculated with a hollow-shaft agitator to convert nitrocaprolactam in (g). A shell-and-tube design is used for the reaction of ammonia and adipic acid in (h).

Reactions between gases and liquids may involve solids also, either as reactants or as catalysts. Table 17.9 lists a number of examples. The lime/limestone slurry process is the predominant one for removal of SO_2 from power plant flue gases. In this case it is known that the rate of the reaction is controlled by the rate of mass transfer through the gas film.

Some gases present in waste gases are recovered by scrubbing with absorbent chemicals that form loose compounds; the absorbent then may be recovered for reuse by elevating the temperature or lowering the pressure in a regenerator. Such loose compounds may exert appreciable back pressure in the absorber, which must be taken into account when that equipment is to be sized.

In all cases, a limiting reactor size may be found on the basis of mass transfer coefficients and zero back pressure, but a size determined this way may be too large in some cases to be economically acceptable. Design procedures for mass transfer equipment are in other chapters of this book. Data for the design of gas–liquid reactors or chemical absorbers may be found in books such as those by Astarita, Savage, and Bisio (*Gas Treating with Chemical Solvents,* Wiley, New York, 1983) and Kohl and Riesenfeld (*Gas Purification,* Gulf, Houston, TX, 1979).

FIXED BED REACTORS

The fixed beds of concern here are made up of catalyst particles in the range of 2–5 mm dia. Vessels that contain inert solids with the sole purpose of improving mass transfer between phases and developing plug flow behavior are not in this category. Other uses of inert packings are for purposes of heat transfer, as in pebble heaters and induction heated granular beds—these also are covered elsewhere.

The catalyst in a reactor may be loaded in several ways, as:

1. a single large bed,
2. several horizontal beds,
3. several packed tubes in a single shell,
4. a single bed with imbedded tubes,
5. beds in separate shells.

Some of the possibilities are illustrated in Figures 17.13 and 17.18. Variations from a single large bed are primarily because of a need for control of temperature by appropriate heat transfer, but also for redistribution of the flow or for control of pressure drop. There are few fixed bed units that do not have some provision for heat transfer. Only when the heat of reaction is small is it possible to regulate the inlet temperature so as to make adiabatic operation feasible; butane dehydrogenation, for example, is done this way.

Because of their long industrial histories and worldwide practice, the sulfuric acid and ammonia industries have been particularly inventive with regard to reactors. A few designs for SO_2 oxidation are illustrated in Figure 17.19. Their dominant differences are in modes of temperature control to take advantage of high rates of reaction at high temperature and favorable equilibrium conversion at lower temperatures. Figure 17.19(g) shows the temperature profile achieved in that equipment, and Figure 17.20 presents patterns of temperature control in the production of SO_3, ammonia, and methanol.

A selection of ammonia reactors is illustrated in Figures 17.21 and 17.22. These vessels incorporate particularly elaborate means for temperature regulation. The basic flow pattern is indicated in

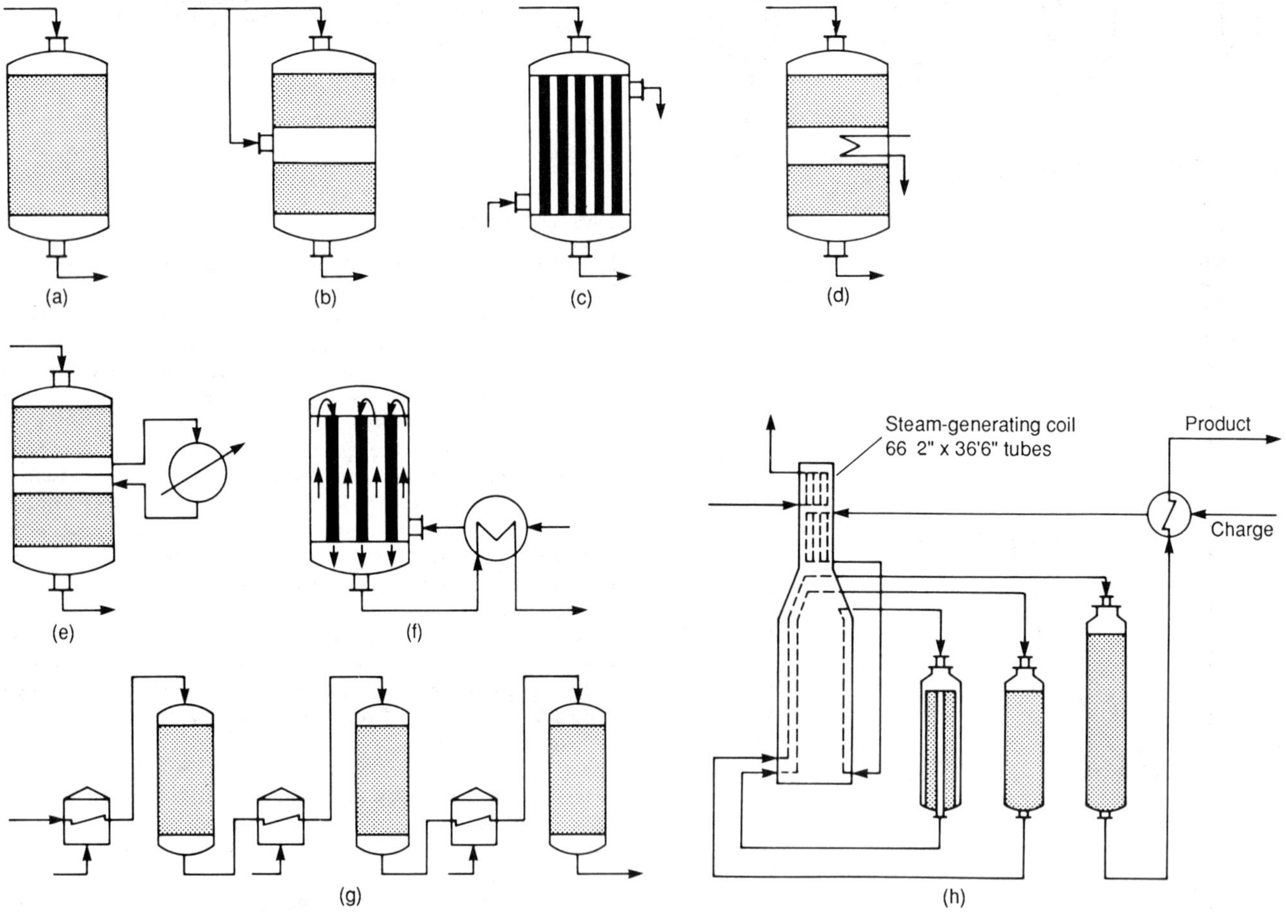

Figure 17.13. Multibed catalytic reactors: (a) adiabatic; (b) interbed coldshot injection; (c) shell and tube; (d) built-in interbed heat exchanger; (e) external interbed exchanger; (f) autothermal shell, outside influent-effluent heat exchanger; (g) multishell adiabatic reactor with interstage fired heaters; (h) platinum-catalyst, fixed bed reformer for 5000 bpsd charge rate; reactors 1 and 2 are 5.5 ft dia by 9.5 ft high and reactor 3 is 6.5 × 12.0 ft.

Figure 17.21(a), and some temperature profiles in Figures 17.22(d) and 17.23(e). For modern high capacity performance in single units, reactors with short travel paths through the catalyst and pressures below 200 atm are favored. Comparative performance data over a range of conditions appear in Figure 17.22.

Thermal effects also are major factors in the design of reactors for making synthetic fuels. The units of Figure 17.24 for synthesis of methanol and gasoline are typical fixed bed types.

Catalytic reformers upgrade low octane naphthas into gasoline in the presence of hydrogen to retard deposition of carbon on the catalyst. Temperatures to 500°C and pressures to 35 atm are necessary. Representative reactors are shown in Figure 17.25.

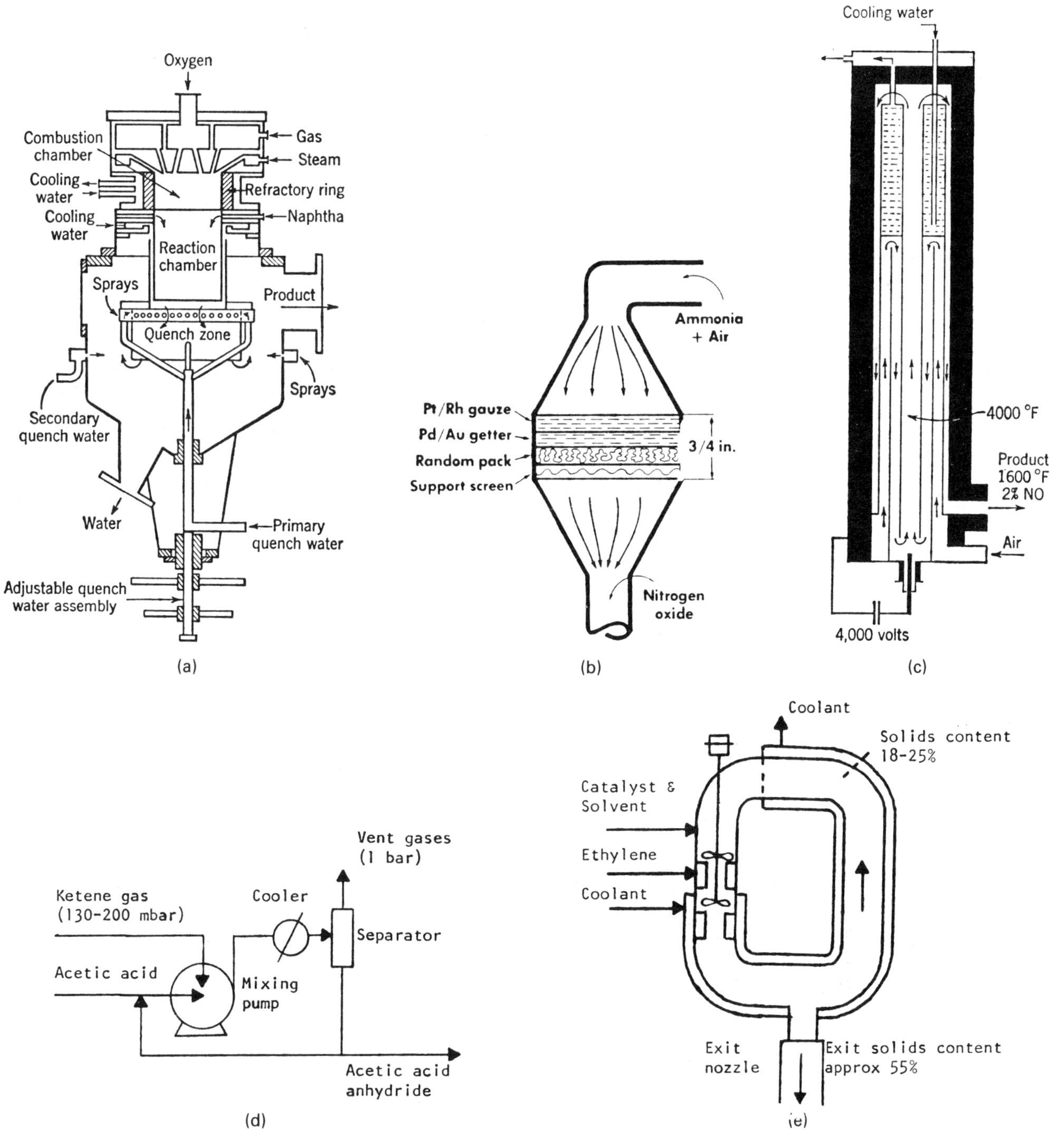

Figure 17.14. Some unusual reactor configurations. (a) Flame reactor for making ethylene and acetylene from liquid hydrocarbons [*Patton et al.,* Pet Refin **37**(*11*) *180*, (*1958*)]. (b) Shallow bed reactor for oxidation of ammonia, using Pt-Rh gauze [*Gillespie and Kenson,* Chemtech, *625* (*Oct. 1971*)]. (c) Schoenherr furnace for fixation of atmospheric nitrogen. (d) Production of acetic acid anhydride from acetic acid and gaseous ketene in a mixing pump. (e) Phillips reactor for low pressure polymerization of ethylene (closed loop tubular reactor). (f) Polymerization of ethylene at high pressure.

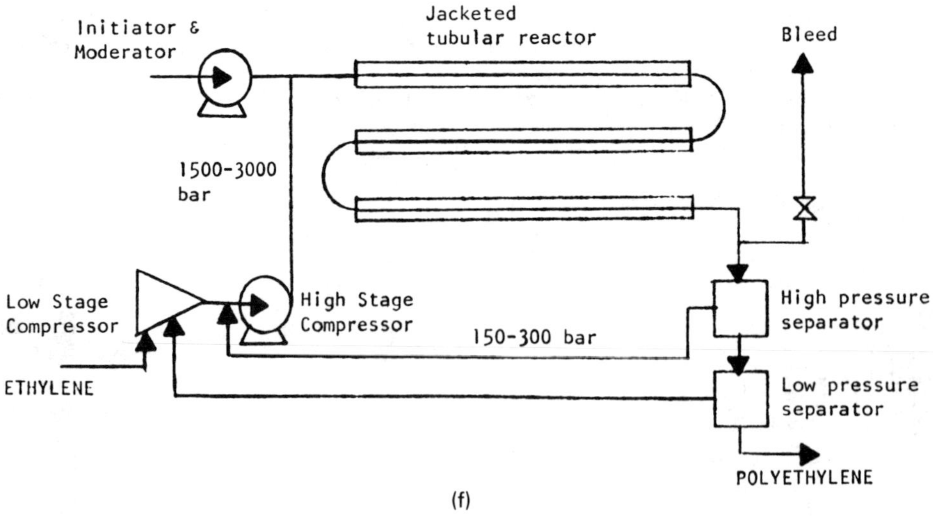

Figure 17.14—(*continued*)

Feedstocks to such units usually must be desulfurized; a reactor like that of Figure 17.26 hydrogenates sulfur compounds to hydrogen sulfide, which is readily removed.

Fluid flow through fixed bed reactors usually is downward. Instead of screens for supporting catalyst in the vessel, a support of graduated sizes of inert material is used, as illustrated in Figure 17.27. Screens become blinded by the small particles of catalyst. A similar arrangement is used at the top to prevent disturbance of the catalyst level by the high velocity fluids.

MOVING BEDS

In such vessels granular or lumpy material moves vertically downward as a mass. The solid may be a reactant or a catalyst or a heat carrier. The reactor of Figure 17.28(a) was used for the fixation of nitrogen in air at about 4000°F. The heat-carrying pebbles are heated by direct contact with combustion gases, dropped into a reaction zone supplied with reacting air, and then recycled with elevators to the reheating zone. The treated air must be quenched

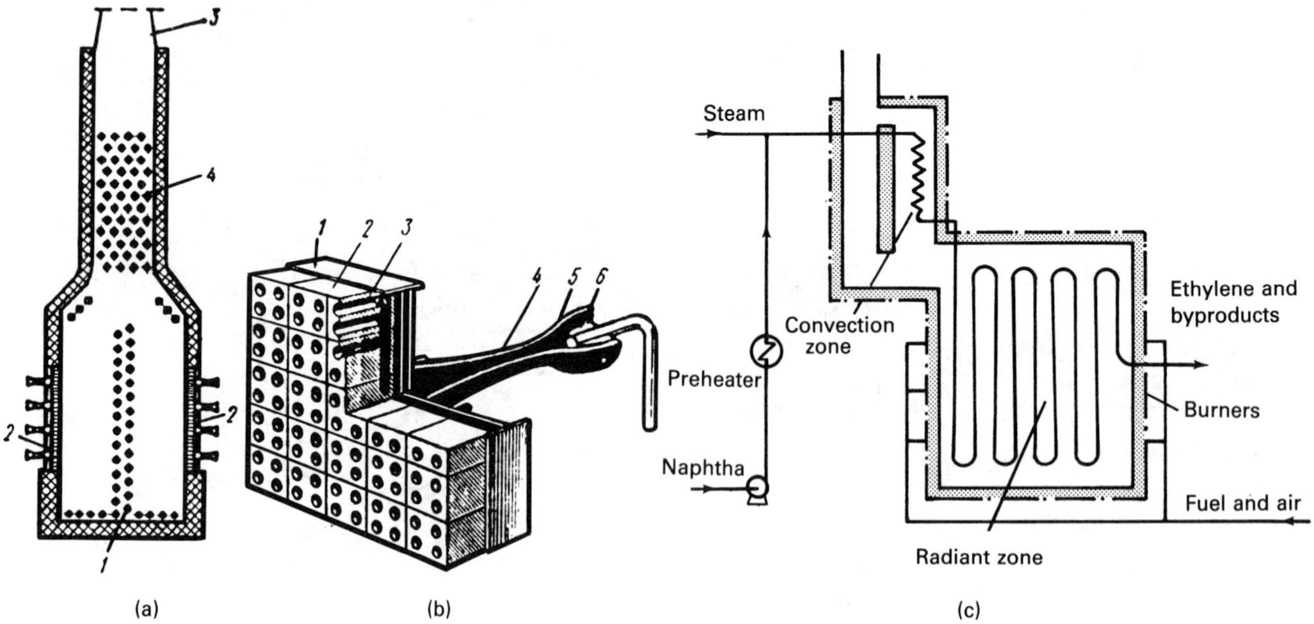

Figure 17.15. A fired heater as a high temperature reactor. (a) Arrangement of tubes and burners: (1) radiant tubes; (2) radiant panel burners; (3) stack; (4) convection chamber tubes (*Sukhanov,* Petroleum Processing, *Mir, Moscow, 1982*). (b) Radiant (surface-combustion) panel burner: (1) housing; (2) ceramic perforated prism; (3) tube; (4) injector; (5) fuel gas nozzle; (6) air throttle (*Sukhanov,* Petroleum Processing, *Mir, Moscow, 1982*). (c) Fired tubular cracking furnace for the preparation of ethylene from naphtha.

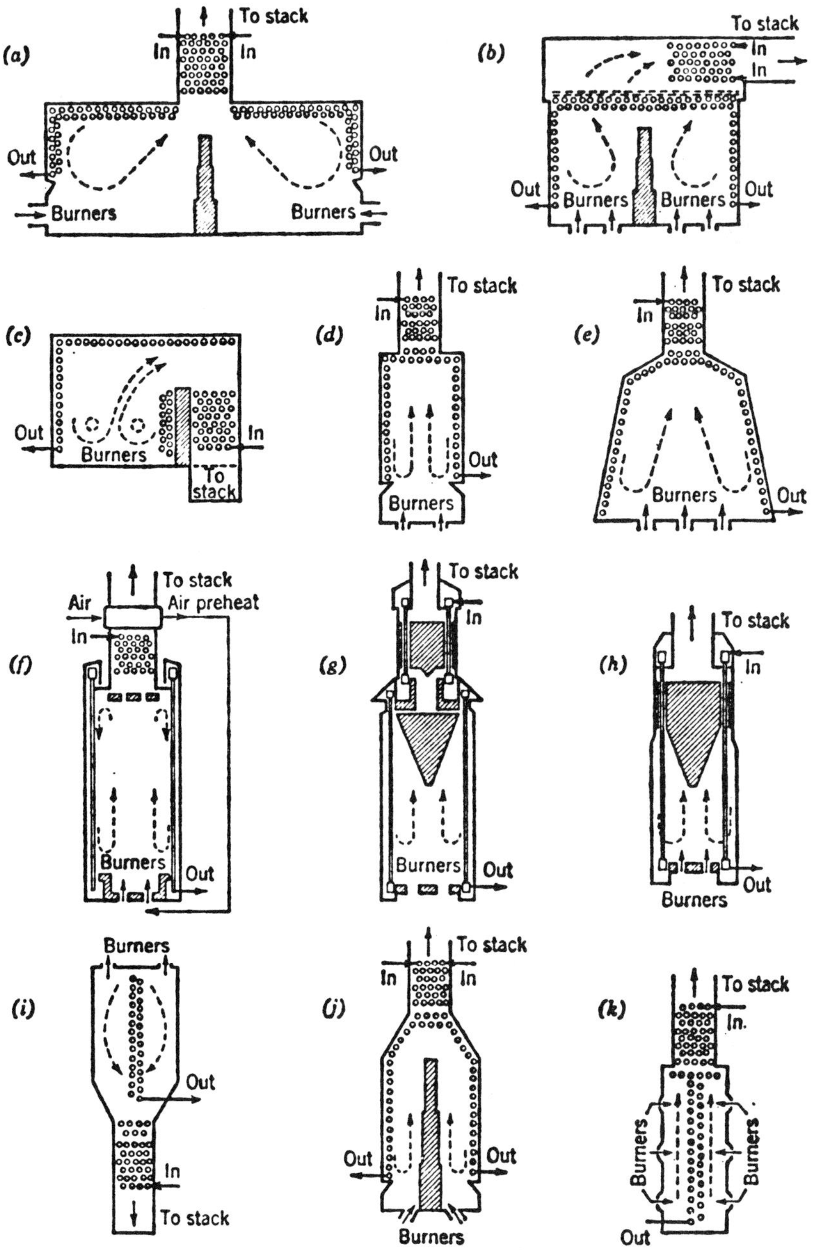

Figure 17.16. Basic types of tubular furnaces [*Nelson,* Petroleum Refinery Engineering, *McGraw-Hill, 1958. Courtesy McGraw-Hill, New York*].

at the rate of about 25,000°F/sec to retain a concentration of about 1% nitrogen oxides. In another such operation, two units are used in parallel, one being heated while the other is reacting.

The pebble heater, Figure 17.28(b), is used in the same manner; its application to the pyrolysis of oils to make ethylene also did not prove competitive and has been abandoned.

Units like that of Figure 17.28(c) were employed at one time in the catalytic cracking of gas oils. The catalyst is transferred between regenerating and reacting zones with bucket elevators or air lifts. Some data for this equipment are given with the figure.

Two examples in which the solid itself is reactive are the shale oil retorts of Figure 17.29. Crushed oil shale is charged at the top, air and gaseous fuel at the bottom. When the shale moving downward reaches a temperature of 900°F, the kerogen decomposes into oil vapor, gas, and carbonaceous residue. There are many designs of pilot plant retorts, but the only commercial units at present are in the USSR and China.

KILNS AND HEARTH FURNACES

These units are primarily for high temperature services, the kilns up to 2500°F and the furnaces up to 4000°F. Usual construction is steel-lined with ceramics, sometimes up to several feet in thickness.

Vertical kilns are used for materials that do not fuse or soften, as for the burning of limestone or dolomite. Many such operations are batch: the fresh solid is loaded into the kiln, heated with combustion products until reaction is complete, and then dumped. The lime kiln of Figure 17.30(c), however, operates continuously as a moving bed reactor. These vessels range in size from 8 to 15 ft dia and are 50–80 ft high. For calcination of lime the peak temperatures are about 2200°F, although decomposition proceeds freely at 1850°F. Fuel supply may be coke mixed with the limestone if the finished lime can tolerate the additional ash, or gaseous or liquid fuels. Space velocity is 0.8–1.5 lb CaO/(hr)(cuft of kiln), or 45–100 lb CaO/(hr)(sqft of kiln cross section), depending on the size and modernity of the kiln, the method of firing, and the lump size which is in the range of 4–10 in.

Rotary kilns have many applications as reactors: between finely divided solids (cement), between liquids and solids (salt cake from salt and sulfuric acid), between gases and solids, and for the decomposition of solids (SO_3 and lime from $CaSO_4$). The kiln is a long narrow cylinder with a length-to-diameter ratio of 10–20. General purpose kilns are 100–125 ft long, but cement kilns as large as 12 ft dia by 425 ft long are operated. An inclination to the horizontal of 2–5 deg is sufficient to move the solid along. Speed of rotation is 0.25–2 rpm. Lumps up to 1 in. dia or fine powders are

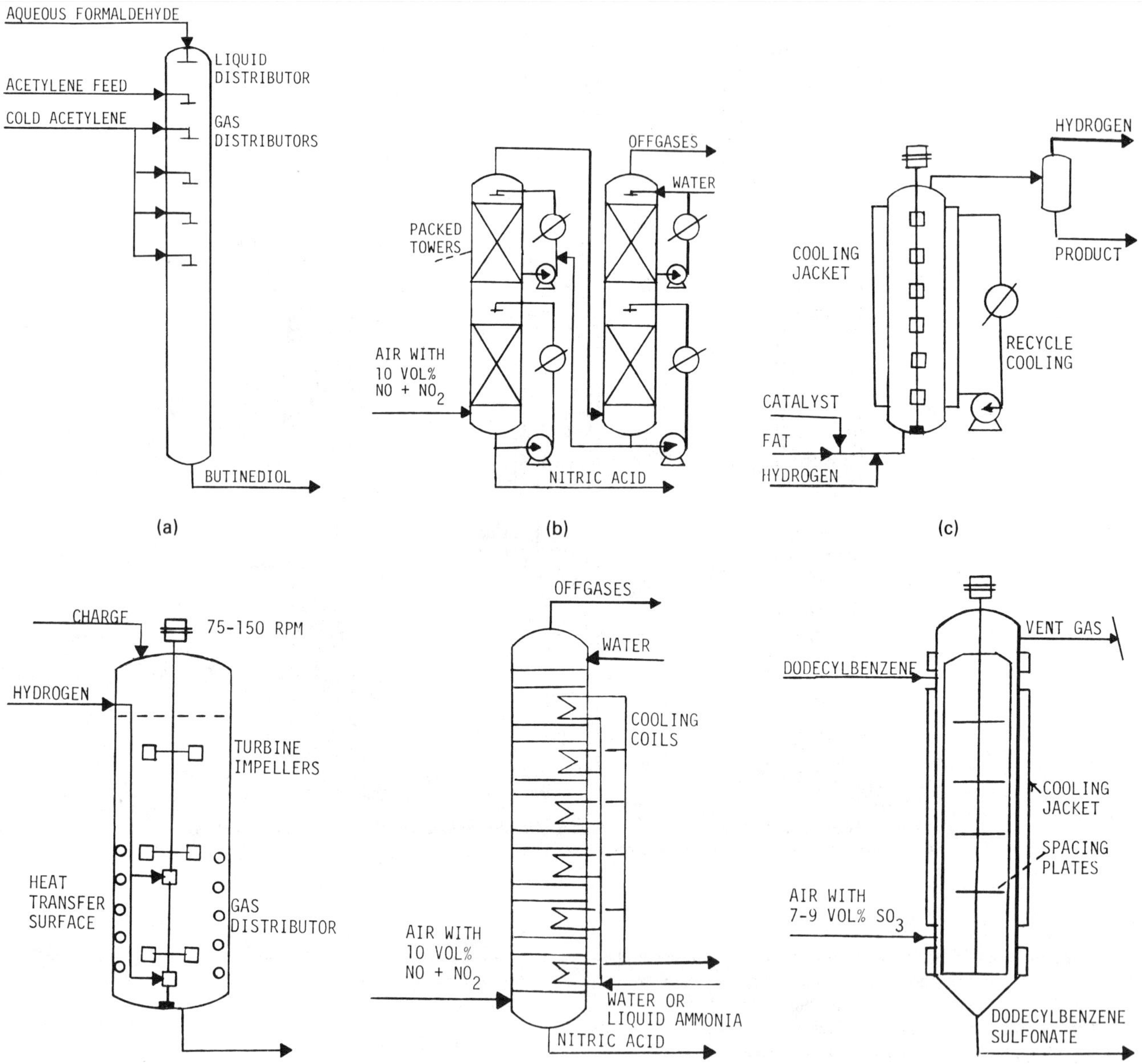

Figure 17.17. Examples of reactors for specific liquid–gas processes. (a) Trickle reactor for synthesis of butinediol 1.5 m dia by 18 m high. (b) Nitrogen oxide absorption in packed columns. (c) Continuous hydrogenation of fats. (d) Stirred tank reactor for batch hydrogenation of fats. (e) Nitrogen oxide absorption in a plate column. (f) A thin film reactor for making dodecylbenzene sulfonate with SO_3. (g) Stirred tank reactor for the hydrogenation of caprolactam. (h) Tubular reactor for making adiponitrile from adipic acid in the presence of phosphoric acid.

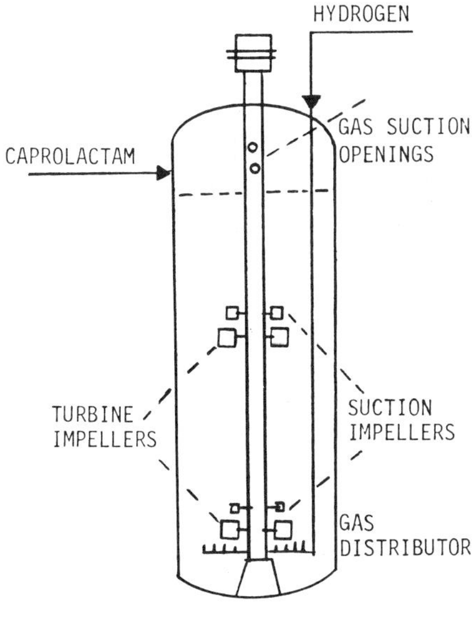

(g)

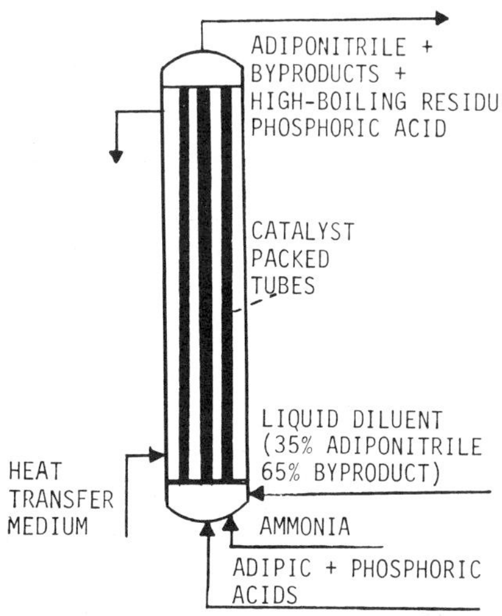

(h)

Figure 17.17—(*continued*)

TABLE 17.9. Examples of Fluidized Bed Processes

A. Catalytic Processes

1. Oil cracking and reforming
2. Recovery of high concentrations of benzene from gas oils
3. Olefin production from crude oil
4. Chlorine by oxidation of HCl
5. Acetylene from methane
6. Preparation of unsaturated aldehydes
7. Reduction of nitro compounds to amines
8. Oxidation of SO_2 to SO_3
9. Phthalic anhydride from naphthalene or *o*-xylene
10. Maleic acid anhydride from benzene
11. Formaldehyde from methanol
12. Chlorination of methane and ethylene
13. Fischer–Tropsch synthesis of gasoline
14. Hydrogenation of ethylene
15. Oxidation of ammonia
16. Ethylene oxide from ethylene
17. Butadiene from ethanol
18. Dehydrogenation of isopropanol
19. Isomerization of *n*-butane
20. Post-chlorination of PVC
21. Decomposition of ozone
22. Preparation of chlorinated hydrocarbons
23. Preparation of melamine resins
24. Isoprene synthesis
25. Reduction of vinyl acetate
26. Preparation of acrylonitrile

B. Noncatalytic Processes

1. Gasification of coal
2. Fluid bed coking
3. Pyrolytic cracking of methane
4. Preparation of activated carbon
5. Ethylene by cracking of petroleum fractions
6. Combustion of coal
7. Burning of oil shale
8. Combustion of municipal and industrial wastes
9. Burning of black liquor (paper industry)
10. Roasting of sulfides of iron, copper, and zinc
11. Combustion of sulfur in a sand bed
12. Decomposition of waste sulfuric acid and sulfates
13. Cracking of chlorides such as $FeCl_2$, $NiCl_3$, and $AlCl_3$
14. Volatilization of rhenium
15. Burning of limestone and dolomite
16. Cement burning
17. Reduction of iron ores and metallic oxides
18. Chlorination of ores of aluminum, titanium, nickel, cobalt, and tin
19. Chlorination of roasted pyrites and iron ores
20. Chlorination of lime
21. Calcination of aluminum hydroxide to alumina
22. Preparation of aluminum sulfate from bauxite
23. Preparation of fluorides aluminum trifluoride, uranium tetra- and hexafluorides
24. Preparation of pure tungsten from the fluoride
25. Calcination of phosphates
26. Preparation of phosphorus oxychloride
27. Preparation of carbon disulfide
28. Preparation of hydrazine
29. Preparation of nitric acid
30. Preparation of nitrates of ammonia and sodium
31. Preparation of sodium carbonate
32. Preparation of hydrogen cyanide
33. Hydrochlorination of uranium fuel elements
34. Preparation of uranium trioxide from the nitrate
35. Recovery of uranium from nuclear fuels
36. Removal of fluorine from offgases of aluminum electrolysis
37. Heating of heat transfer media such as sand
38. Cooling of granular masses such as fertilizers
39. Drying of finely divided materials such as flotation ores and raw phosphates
40. Coating of fuel elements by pyrolytic cracking of chlormethylsilanes

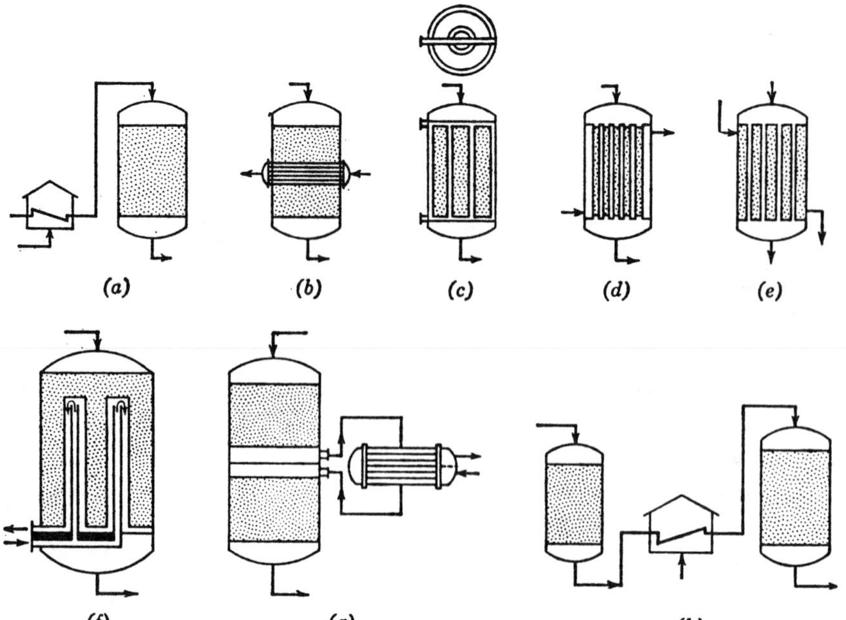

Figure 17.18. Heat transfer in fixed-bed reactors: (a) adequate preheat; (b) internal heat exchanger; (c) annular cooling spaces; (d) packed tubes; (e) packed shell; (f) tube and thimble; (g) external heat exchanger; (h) multiple shell, with external heat transfer (*Walas, 1959*).

usual. Heating mostly is with combustion gases, but some low temperature heating may be accomplished through heated jackets. Figures 17.30(a) and (b) show the temperature profiles of gas and stock in a cement kiln and space velocities of a number of kiln processes.

Multiple-hearth furnaces are suited to continuous handling of solids that exhibit a limited amount of fusion or sintering. In the kind shown on Figure 17.30(e), the scrapers rotate, in other kinds the plates rotate, and in still others the scrapers oscillate and discharge the plates at each stroke. Material is charged at the top, moves along as rotation proceeds, and drops onto successively lower plates while combustion gases or gaseous reactants flow upward. This equipment is used to roast ores, burn calcium sulfate or bauxite, and reactivate the absorbent clays of the petroleum industry. A reactor with nine trays, 16 ft dia and 35 ft high can roast about 1,250 lb/hr of iron pyrite, at a residence time of about 4–5 hr.

Hearth furnaces consist of one or more flat or concave pans, either moving or stationary, usually equipped with scraper-stirrers. Although such equipment is used mostly for ore treating and metallurgical purposes, a few inorganic chemical processes utilize them, for example, Leblanc soda ash, sodium sulfide from salt cake and coal, sodium sulfate and hydrogen chloride from salt and sulfuric acid, and sodium silicate from sand and soda ash. A kind of salt-cake furnace is shown in Figure 17.30(d). Salt and sulfuric acid are charged continuously to the center of the pan, and the rotating scrapers gradually work the reacting mass towards the periphery where the sodium sulfate is discharged over the edge. The products leave at about 1000°F. Pans of 11–18 ft dia can handle 6–10 tons/24 hr of salt. For comparison, a Laury horizontal rotating cylindrical kiln, 5 × 22 ft, has a capacity of 1 ton/hr of salt cake.

Very high temperature operations such as those of production of glass or metals utilize single-hearth furnaces, often with heat

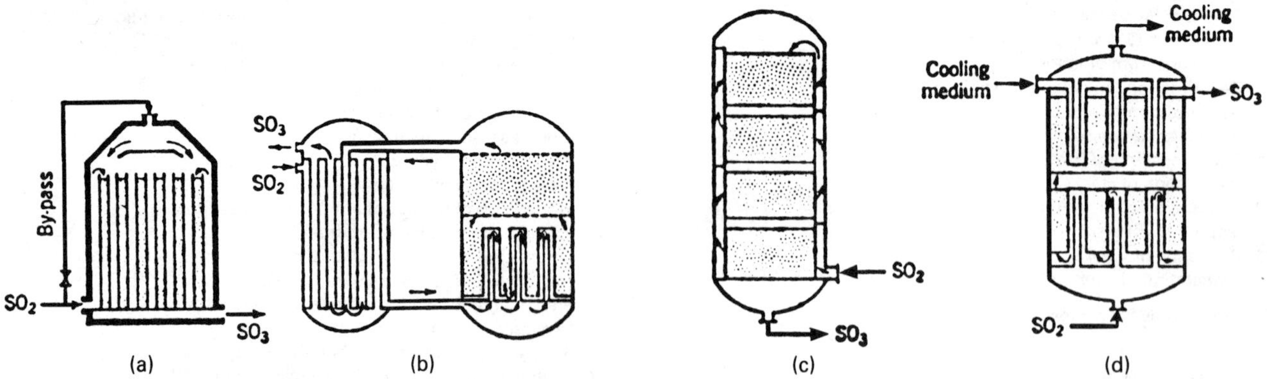

Figure 17.19. Reactors for the oxidation of sulfur dioxide: (a) Feed–product heat exchange. (b) External heat exchanger and internal tube and thimble. (c) Multibed reactor, cooling with charge gas in a spiral jacket. (d) Tube and thimble for feed against product and for heat transfer medium. (e) BASF-Knietsch, with autothermal packed tubes and external exchanger. (f) Sper reactor with internal heat transfer surface. (g) Zieren-Chemiebau reactor assembly and the temperature profile (*Winnacker-Weingartner,* Chemische Technologie, *Carl Hanser Verlag, Munich, 1950–1954*).

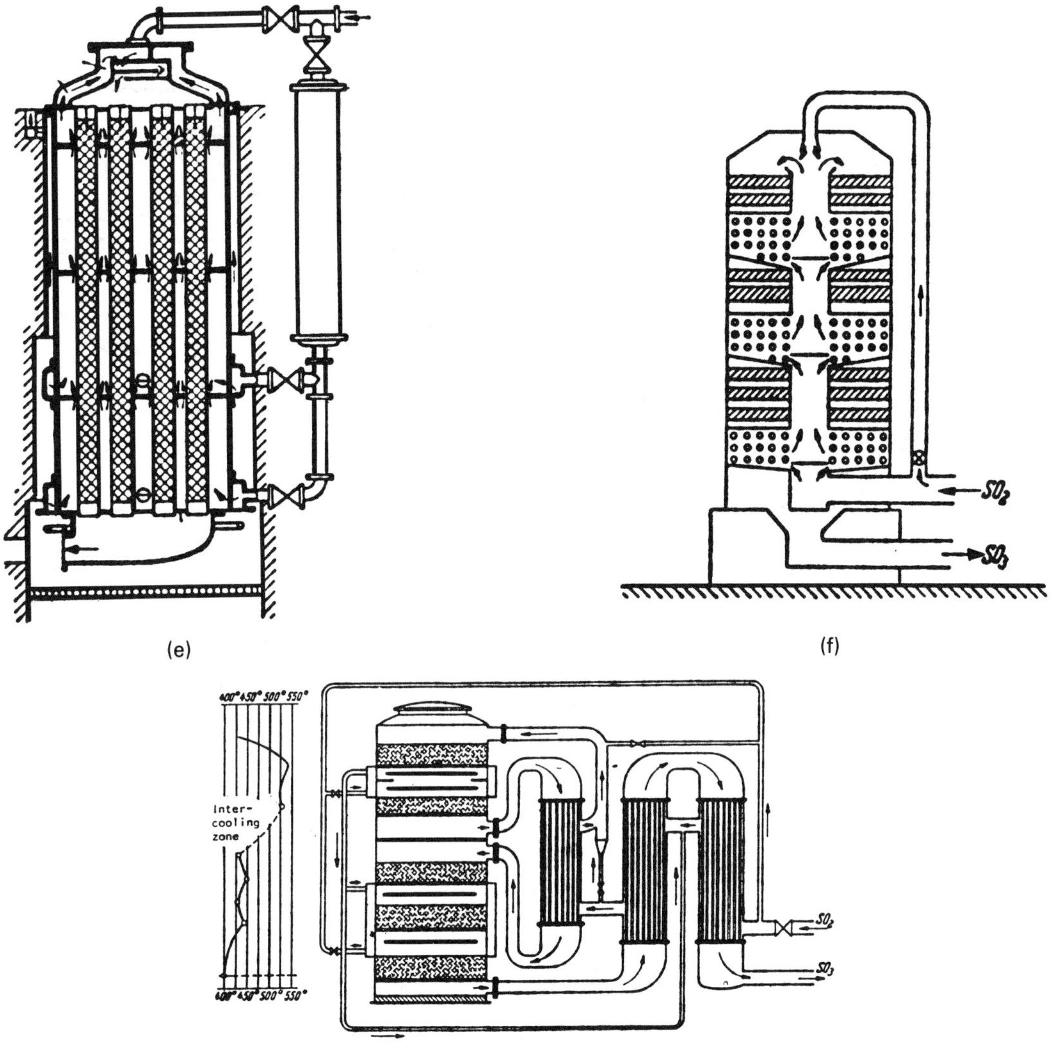

(e) (f) (g)

Figure 17.19—(*continued*)

regenerators for fuel economy. The Siemens-Martin furnace of Figure 17.30(d) with a hearth 13 ft wide and 40 ft long has a production rate of 10 tons/hr of steel with a residence time of 10 hr. The hearth volume is about 5000 cuft and the total regenerator volume is about 25,000 cuft.

FLUIDIZED BED REACTORS

This term is restricted here to equipment in which finely divided solids in suspension interact with gases. Solids fluidized by liquids are called slurries. Three phase fluidized mixtures occur in some coal liquefaction and petroleum treating processes. In dense phase gas–solid fluidization, a fairly definite bed level is maintained; in dilute phase systems the solid is entrained continuously through the reaction zone and is separated out in a subsequent zone.

The most extensive application of fluidization has been to catalytic cracking of petroleum fractions. Because the catalyst degrades in a few minutes, it is circulated continuously between reaction and regeneration zones. Figure 17.31(a) is a version of such equipment. The steam stripper is for the removal of occluded oil before the catalyst is to be burned. The main control instrumentation of a side-by-side system is shown in Figure 3.6(h).

Fluid catalytic vessels are very large. Dimensions and performance of a medium capacity unit (about 50,000 BPSD, 60 kg/sec) are shown with the figure. Other data for a reactor to handle 15,000 BPSD are a diameter of 25 ft and a height of 50 ft. Catalyst holdup and other data of such a reactor are given by Kraft, Ulrich, and O'Connor (in Othmer (Ed.), *Fluidization*, Reinhold, New York, 1956) as follows:

Item	Quantity
Unit charge, nominal	15,000 BPSD
Catalyst inventory, total	250 tons
Catalyst inventory, regenerator bed	100 tons
Superficial velocity, regenerator	2.5 fps
Bed density, regenerator	28.0 lb/cuft
Flue gas plus solids density, cyclone inlet	0.5 lb/cuft
Catalyst circulation rate, unit	24.0 tons/min
Catalyst circulation rate, to cyclones	7.0 tons/min
Catalyst loss rate, design expectation	2.0 tons/day

Figure 17.31(b) is of a unit in which most of the cracking occurs

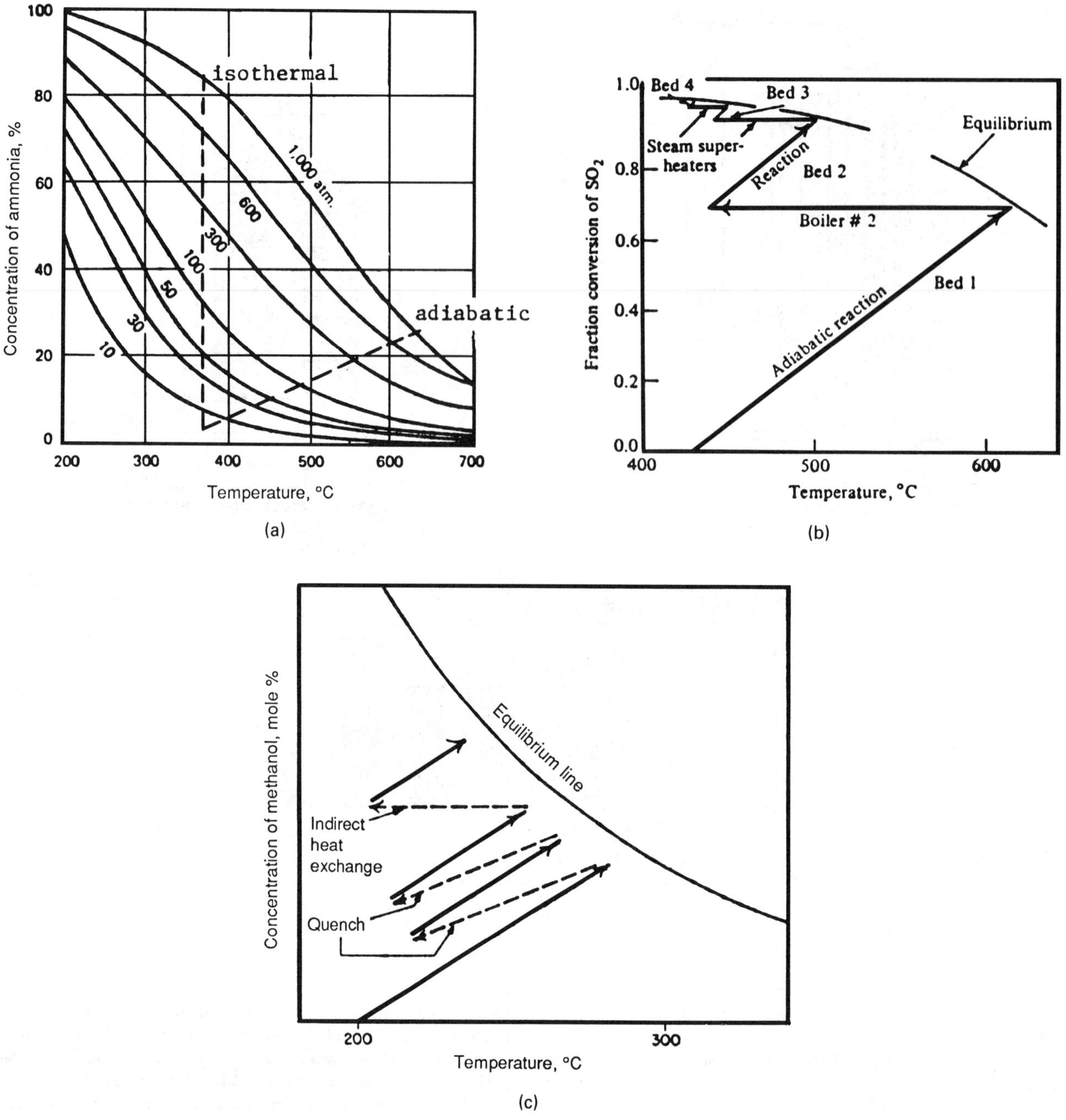

Figure 17.20. Control of temperature in multibed reactors so as to utilize the high rates of reaction at high temperatures and the more favorable equilibrium conversion at lower temperatures. (a) Adiabatic and isothermal reaction lines on the equilibrium diagram for ammonia synthesis. (b) Oxidation of SO_2 in a four-bed reactor at essentially atmospheric pressure. (c) Methanol synthesis in a four bed reactor by the ICI process at 50 atm; not to scale; 35% methanol at 250°C, 8.2% at 300°C, equilibrium concentrations.

in a transfer line, an operation that became feasible with the development of highly active zeolite catalysts. The reaction is completed in the upper zone, but the main function of that zone is to separate product and spent catalyst. In contrast to the dense-phase bed of a large reactor, in which mixing can approach ideality, the dilute phase transfer line is more nearly in plug flow. Accordingly, a much smaller reaction zone suffices; moreover, superior product distribution and greater gasoline yield result. Similar reactor configurations are shown in Figures 17.31(c) and (d) of other petroleum processes.

The mechanism of interaction between catalyst and gas in a large fluidized bed is complex and is not well correlated with design factors. In the bed itself, large bubbles of a foot or more in diameter form and are irrigated with a rain of catalyst particles. This process occurs in parallel with a well-mixed fluidized bed. Above the bed level and before the entrained catalyst is recovered in cyclones, the reaction continues in dilute phase plug flow. Since even the physical behavior of fluidized beds is not well understood, the design of such reactors is done largely on the basis of fairly large pilot plants and by analogy with earlier experience in this area.

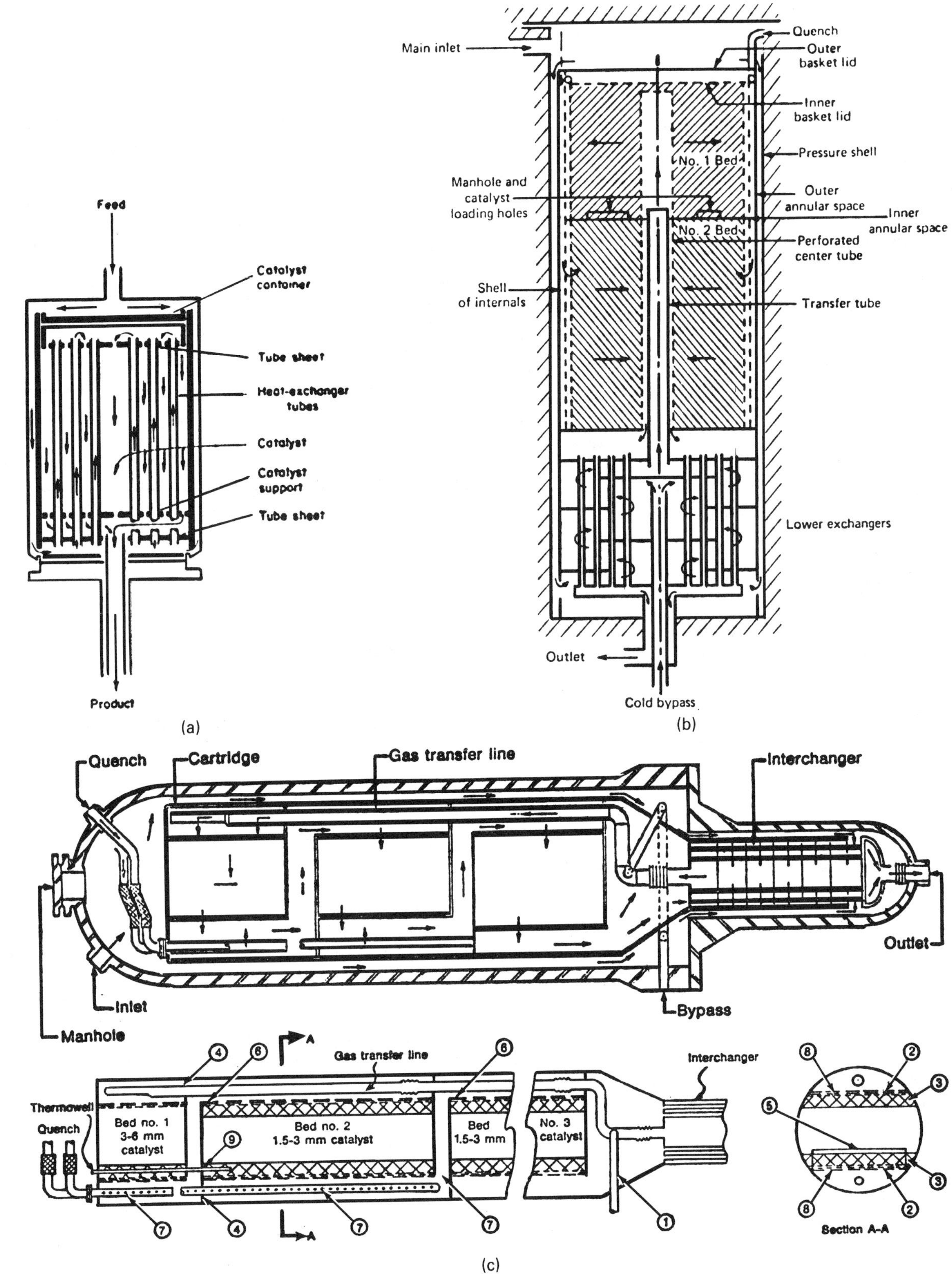

Figure 17.21. Some recent designs of ammonia synthesis converters. (a) Principle of the autothermal ammonia synthesis reactor. Flow is downwards along the wall to keep it cool, up through tubes imbedded in the catalyst, down through the catalyst, through the effluent-influent exchanger and out. (b) Radial flow converter with capacities to 1800 tons/day (*Haldor Topsoe Co., Hellerup, Denmark*). (c) Horizontal three-bed converter and detail of the catalyst cartridge. Without the exchanger the dimensions are 8 × 85 ft, pressure 170 atm, capacity to 2000 tons/day (*Pullman Kellogg*). (d) Vessel sketch, typical temperature profile and typical data of the ICI quench-type converter. The process gas follows a path like that of part (a) of this figure. Quench is supplied at two points (*Imperial Chemical Industries*).

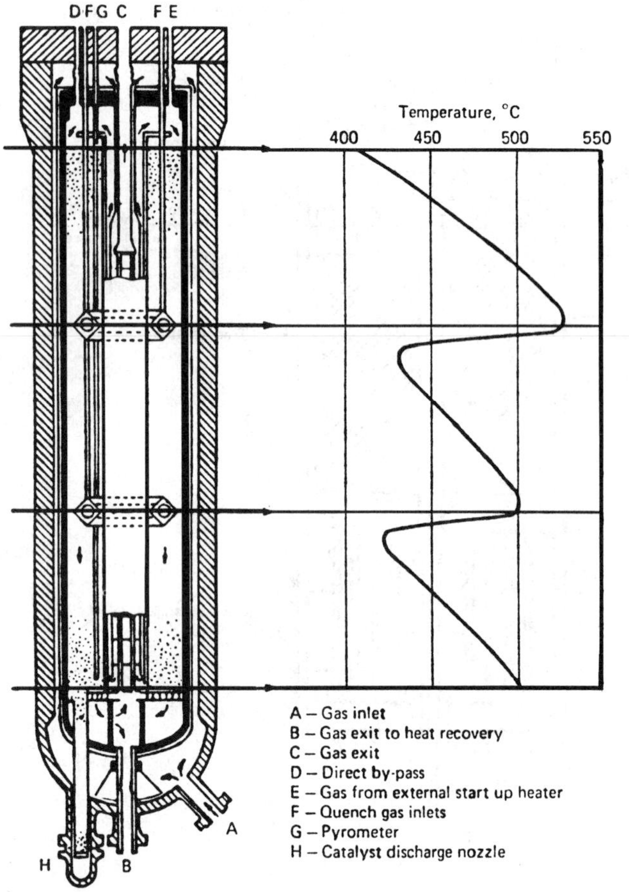

Typical Data for ICI Quench Converters of Various Sizes

Capacity (short tpd)	660	990	1100	1650
Pressure (psig)	4700	3200	4250	3220
Inlet gas composition (%)				
Ammonia	4.0	3.0	3.2	1.4
Inerts	15.0	12.0	15.0	12.0
Inlet gas flow (MM scfh)	10.6	18.0	18.5	24.5
Catalyst volume (ft^3)	740	1170	1100	2400
Pressure vessel				
Internal diameter (in.)	80	96	95	109
Length (in.)	437	493	472	700
Weight (short ton)				
Cartridge shell	14.2	34.2	22.8	56.4
Heat exchanger	15.5	30.0	25.4	23.8
Pressure vessel (less cover)	130	128	182	240

Typical Data for an ICI Quench Converter of 1300 Short Tons/Day Capacity

Pressure (psig)	2200	3200	4000	4700
Inlet gas flow (MM scfh)[a]	25.8	21.2	19.8	19.0
Catalyst volume (ft^3)	2600	1730	1320	1030
Pressure vessel				
Internal diameter (in.)	120	102	96	89
Length (in.)	663	606	528	488
Weight (short ton)				
Cartridge shell	68.5	40.8	29.2	23.6
Heat exchanger	37.1	25.4	20.7	17.9
Pressure vessel (less cover)	186	184	187	189
Converter pressure drop (psi)	140	104	87	91

[a]Composition: 2% NH_3, 12% inerts (CH_4 + A), 21.5% N_2, 64.5% H_2 by vol.

(d)

Figure 17.21—(*continued*)

The earliest fluidized process was the noncatalytic Winkler process for gasification of coal in 1921. Other noncatalytic processes, and some catalytic ones, are listed in Table 17.9. A few noncatalytic reactors are shown in Figure 17.32. Cracking of naphthas to ethylene with circulating hot sand as the heat carrier is shown in part (a); at the operating temperature of 720–850°C, much carbon deposits on the sand but is not at all harmful as it would be on the surfaces of tubular cracking units. In the dilute phase process of calcination of alumina, part (b), the circulating solid is the product itself; combustion products from sprays of oil and auxiliary air furnish the motive power. The calcining unit for lime of part (c) is an example of a successful multistage reactor; residence time in the calcining zone is 2 hr, in the cooling zone 0.5 hr, and in each of the preheating zones 1 hr. Multibed units for petroleum operations have not been feasible, but some units have been built with a degree of baffling that simulates staging in a rough fashion. The catalyst of the phthalic anhydride reactor of part (d) does not need to be regenerated so the fluidized bed remains in place; since the reaction is highly sensitive to temperature, the oxidation is kept under control with much imbedded heat transfer surface and by cold injections. A modern coal gasifier appears in part (e); a thirtyfold circulation of spent char is employed along with the fresh feed to counteract the agglomeration tendency of many coals. The H-Coal reactor of part (f) operates with a three-phase mixture. The catalyst does not circulate but bubbles in place. Activity is maintained by bleeding off and replenishing 1–2% of the catalyst holdup per day. Operating conditions are 450°C and 3000 psig. Both coal and heavy petroleum residua are handled successfully. The unit is known as an "ebullating bed."

The literature of fluidization phenomena and technology is extensive. A good although dated bibliography is in *Ullmann's Encyclopedia* (1973, Vol. 3, pp. 458–460). The book by Cheremisinoff and Cheremisinoff (1984) has more than 500 abstracts of articles on fluidization hydrodynamics, mixing and heat transfer, but little on reactor technology. Other literature on fluidization is cited in the References of Chapter 6.

17.7. HEAT TRANSFER IN REACTORS

Maintenance of proper temperature is a major aspect of reactor operation. The illustrations of several reactors in this chapter depict a number of provisions for heat transfer. The magnitude of required heat transfer is determined by heat and material balances as described in Section 17.3. The data needed are thermal conductivities and coefficients of heat transfer. Some of the factors influencing these quantities are associated in the usual groups for heat transfer; namely, the Nusselt, Stanton, Prandtl, and Reynolds dimensionless groups. Other characteristics of particular kinds of reactors also are brought into correlations. A selection of practical results from the abundant literature will be assembled here. Some modes of heat transfer to stirred and fixed bed reactors are represented in Figures 17.33 and 17.18, and temperature profiles in

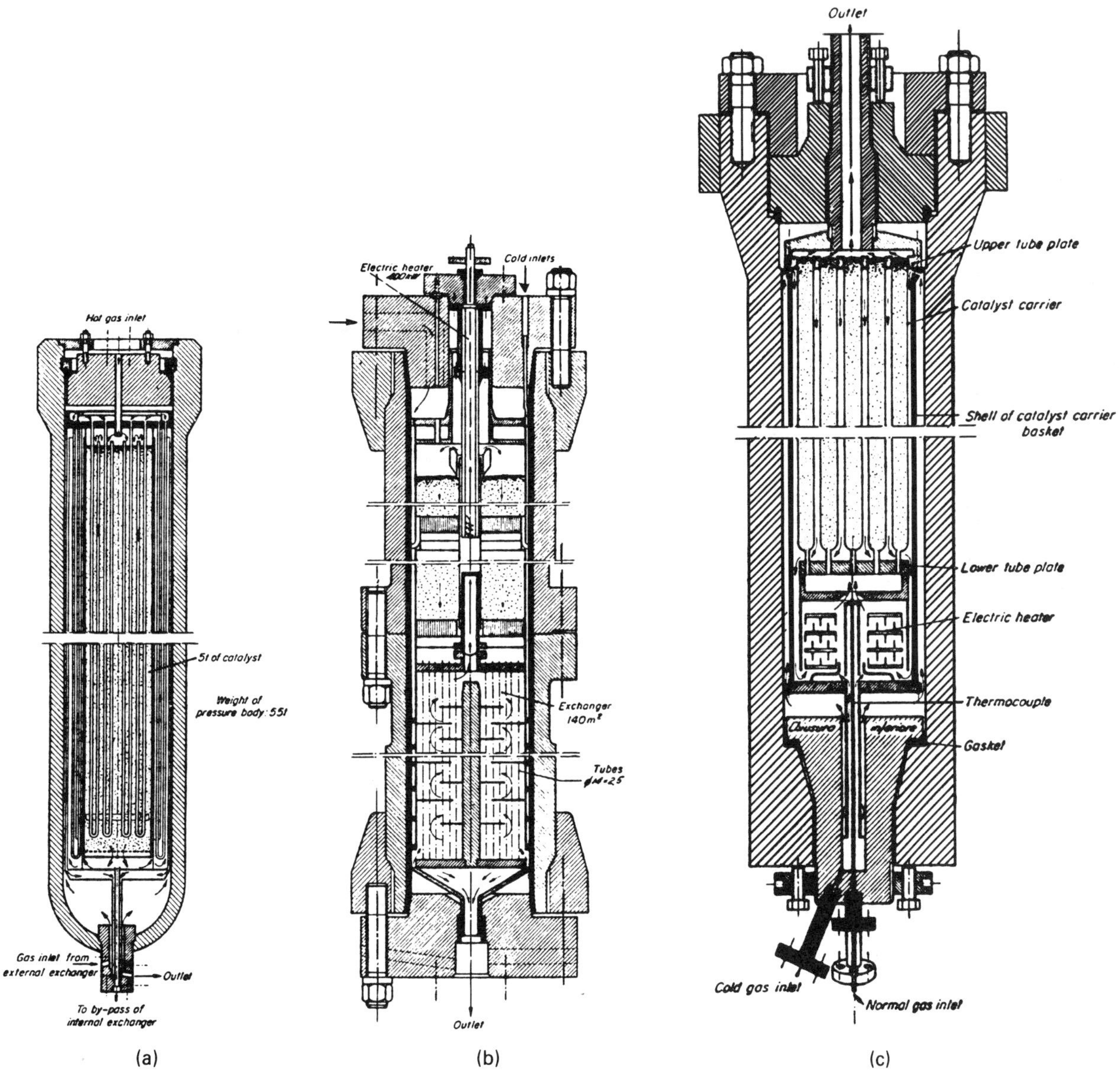

Figure 17.22. Representative ammonia converters operating at various pressures and effluent concentrations (*Vancini, 1971*). (a) Original Uhde design operating at 125 atm; typical dimensions, 1.4 × 7 m. (b) Haber–Bosch–Mittasch converter operating at 300 atm; typical dimensions, 1.1 × 12.8 m. (c) Claude converter operating at 1000 atm; typical dimensions 1.2 × 7 m. (d) Fauser–Montecatini (old style) converter operating at 300 atm with external heat exchange, showing axial profiles of temperature and ammonia concentration.

Comparison of Performance

Process	Pressure (bar)	Effluent ammonia (%)	TPD/m^3	Catalyst life (yr)
Uhde	125	7–8	10	>2
Haber-Bosch	300	13–15	25	2
Claude	1000	22–24	120	0.25
Fauser	300	12–17	25	2

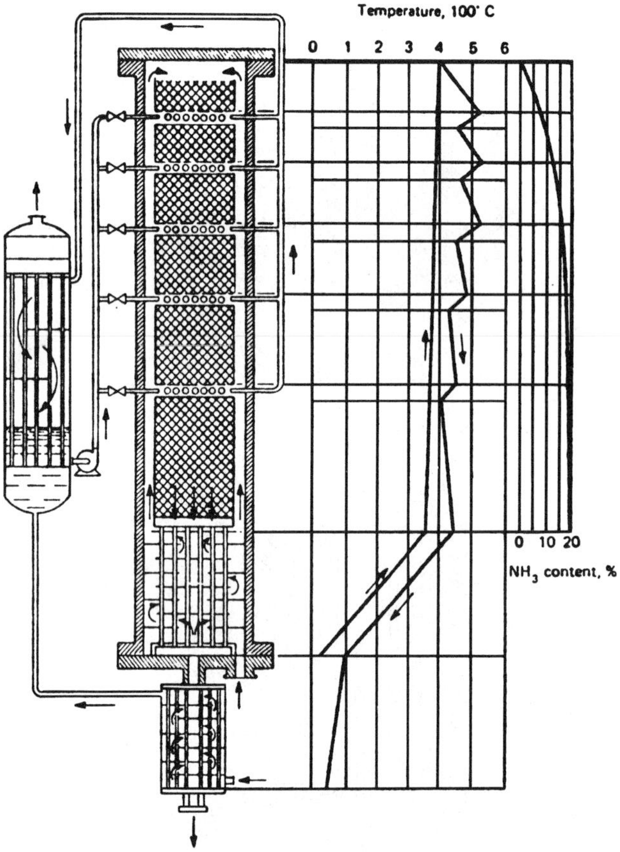

(d)

Figure 17.22—(*continued*)

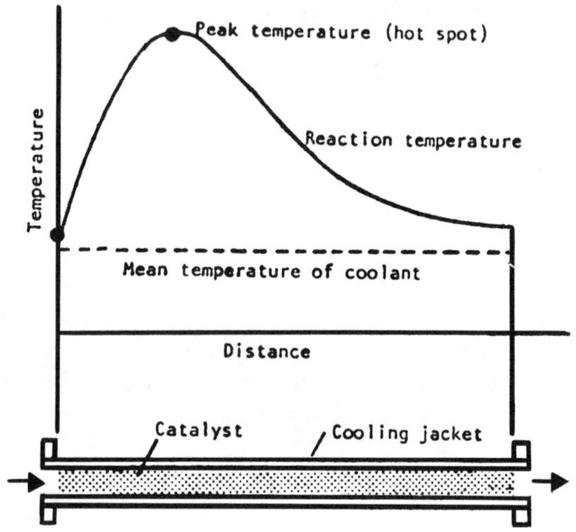

(a)

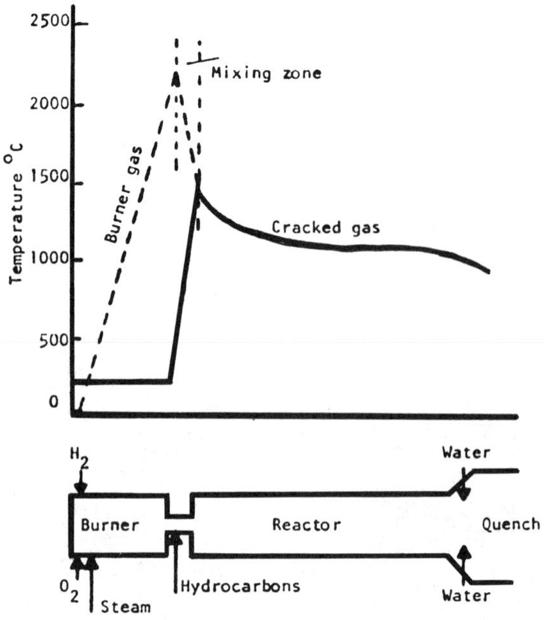

(b)

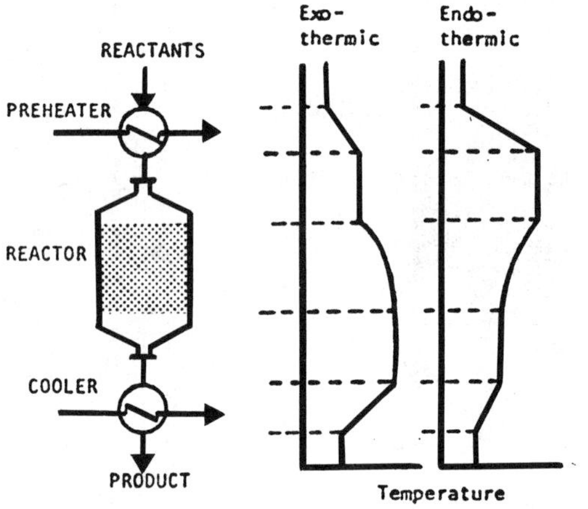

(c)

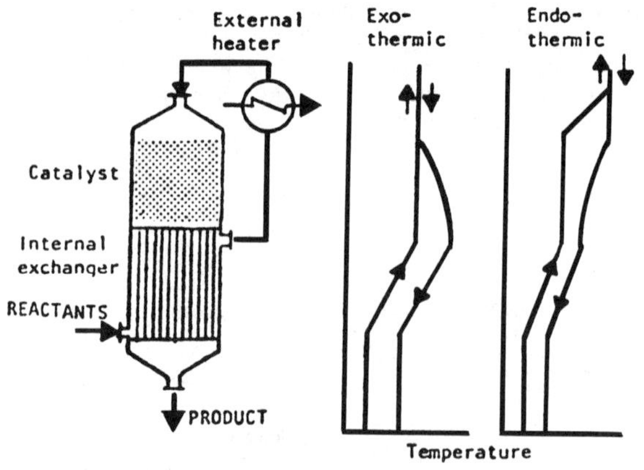

(d)

Figure 17.23. Representative temperature profiles in reaction systems (see also Figs. 17.20, 17.21(d), 17.22(d), 17.30(c), 17.34, and 17.35). (a) A jacketed tubular reactor. (b) Burner and reactor for high temperature pyrolysis of hydrocarbons (*Ullmann, 1973, Vol. 3, p. 355*); (c) A catalytic reactor system in which the feed is preheated to starting temperature and product is properly adjusted; exo- and endothermic profiles. (d) Reactor with built-in heat exchange between feed and product and with external temperature adjustment; exo- and endothermic profiles.

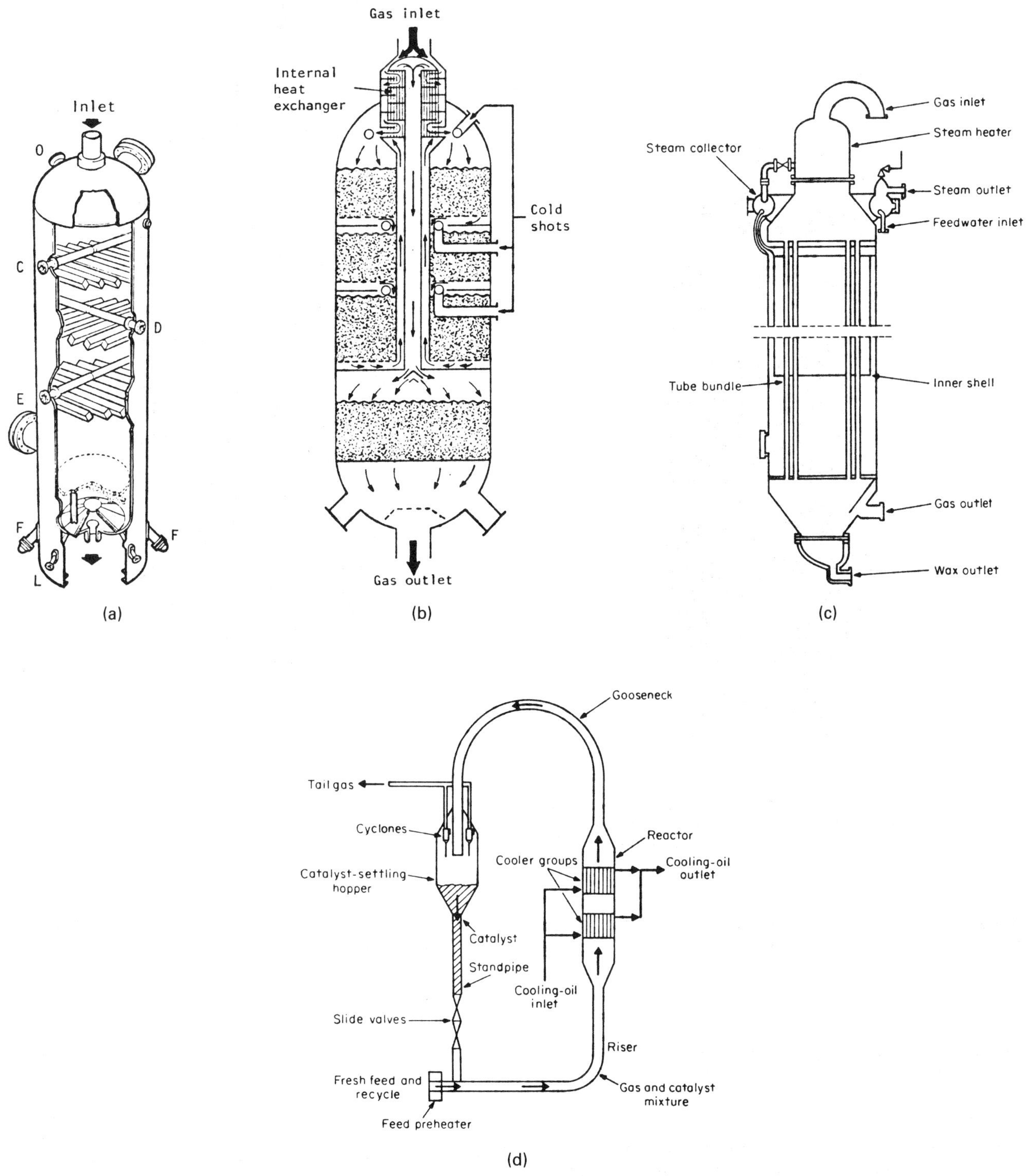

Figure 17.24. Types of reactors for synthetic fuels [*Meyers (Ed.),* Handbook of Synfuels Technology, *McGraw-Hill, New York, 1984*]. (a) ICI methanol reactor, showing internal distributors. C, D and E are cold shot nozzles, F = catalyst dropout, L = thermocouple, and O = catalyst input. (b) ICI methanol reactor with internal heat exchange and cold shots. (c) Fixed bed reactor for gasoline from coal synthesis gas; dimensions 10 × 42 ft, 2000 2-in. dia tubes packed with promoted iron catalyst, production rate 5 tons/day per reactor. (d) Synthol fluidized bed continuous reactor system for gasoline from coal synthesis gas.

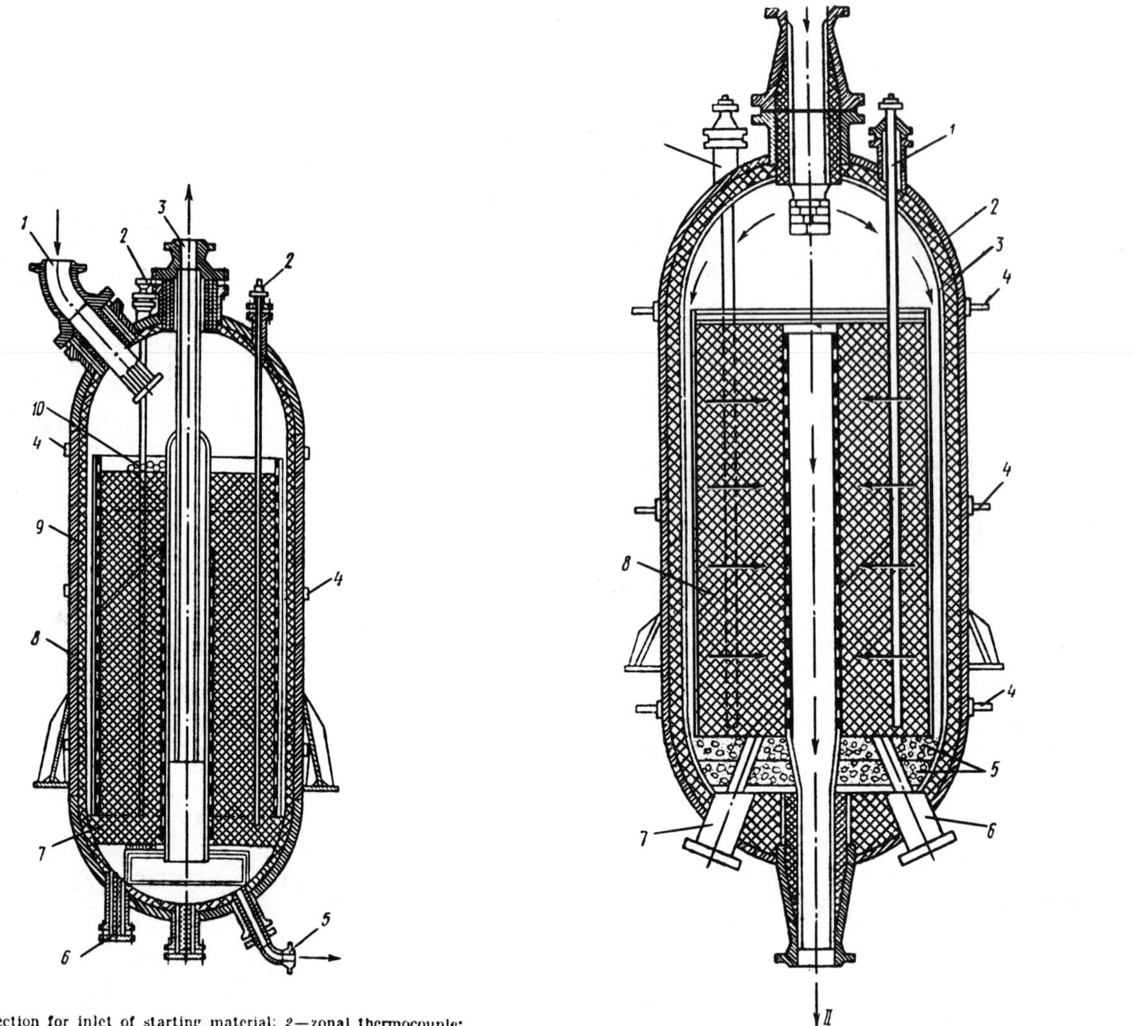

1—pipe connection for inlet of starting material; 2—zonal thermocouple; 3—pipe connection for product discharge; 4—external thermocouple; 5—pipe connection for discharge of product during ejecting of system (in catalyst regeneration); 6—pipe connection for catalyst discharge; 7—light-weight fireclay; 8—reactor housing; 9—gunite lining; 10—porcelain balls

(a)

1—multizonal thermocouple; 2—reactor housing; 3—lining; 4—surface thermocouple; 5—porcelain balls; 6—pipe connection for discharge of product during ejecting of system (in catalyst regeneration); 7—pipe connection for catalyst discharge; 8—catalyst. Lines: I—gas-product mixture; II—reaction products

(b)

Figure 17.25. Catalytic reforming reactors of axial and radial flow types. The latter is favored because of lower pressure drop (*Sukhanov,* Petroleum Processing, *Mir, Moscow, 1982*). (a) Axial flow pattern. (b) Radial flow pattern.

several industrial reactors appear in Figures 17.20–17.23, 17.30, 17.34, and 17.35.

STIRRED TANKS

Values of overall coefficients of heat transfer are collected in Tables 17.10–17.12. Two sets of formulas for tank-side film coefficients are in Tables 17.13 and 17.14. They relate the Nusselt number to the Reynolds and Prandtl numbers and several other factors. In the equation for jacketed tanks, for example,

$$h_0(\text{jacket})\frac{T}{k} = 0.85\left(\frac{D^2N\rho}{\mu}\right)^{0.66}\left(\frac{C_p\mu}{k}\right)^{0.33}\left(\frac{\mu}{\mu_s}\right)^{0.14} \times \left(\frac{Z}{T}\right)^{-0.56}\left(\frac{D}{T}\right)^{0.13} \tag{17.35}$$

the rpm, the tank and impeller diameters, and the liquid depth as well as a viscosity ratio are involved. Table 17.14 identifies the kind of impeller that was used in the investigation, but in general test results have shown that approximately the same heat transfer coefficient is obtained with flat-blade turbines, pitched-blade turbines, or propellers. Axial flow turbines produce the most circulation for a given power input and heat transfer is related directly to the flow, so that this kind of impeller usually is favored. From Eq. (17.35), the coefficient is proportional to the 0.66 power of the rpm, $N^{0.66}$, and from Chapter 10, the power input at high Reynolds numbers varies as the cube root of N. Accordingly it appears that the coefficient is proportional to the 0.22 exponent of the power input to the stirred tank,

$$h \propto P^{0.22}$$

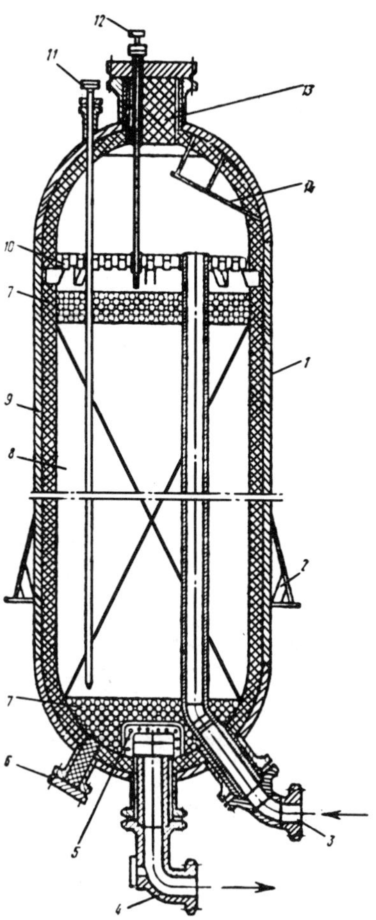

1—housing; 2—support ring; 3—inlet pipe connection; 4—discharge pipe connection; 5—sieve; 6—catalyst discharge pipe connection; 7—porcelain balls; 8—catalyst space; 9—reactor lining (mesh-reinforced gunite); 10—distribution tray (for uniform distribution of catalyst during charging); 11—union for zonal thermocouple; 12—union for thermocouple; 13—lightweight fireclay brick; 14—baffle

Figure 17.26. Reactor for hydrofining diesel oils, with ceramic lining (*Sukhanov*, Petroleum Processing, *Mir, Moscow, 1982*).

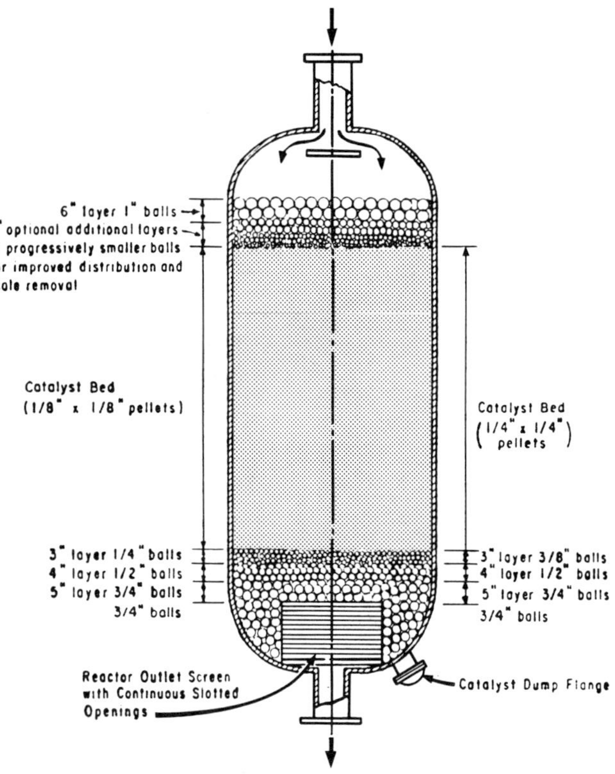

Figure 17.27. Catalyst packed adiabatic reactor, showing application of ceramic balls of graduated sizes for support at the bottom and hold-down at the top (*Rase,* Chemical Reactor Design for Process Plants, *Wiley, New York, 1977).*

and consequently that the coefficient of heat transfer is little affected by large increases of power input.

Since most of the literature in this area is relatively old, practitioners apparently believe that what has been found out is adequate or is kept confidential. Table 17.14 has the recommended formulas.

PACKED BED THERMAL CONDUCTIVITY

The presence of particles makes the effective conductivity of a gas greater than the molecular conductivity by a factor of 10 or more. The nature of the solid has little effect at Reynolds numbers above 100 or so; although the effect is noticeable at the lower values of Re, it has not been completely studied. Besides the Reynolds, Prandtl, and Peclet numbers, the effective diffusivity depends on the molecular conductivity, porosity, particle size, and flow conditions. Plots in terms of Re, Pr, and Pe (without showing actual data points) are made by Beek (1962, Fig. 3), but the simpler plots obtained by a number of investigators in terms of the Reynolds number alone appear on Figure 17.36(a). As Table 17.15 shows, most of the data were obtained with air whose $\mathrm{Pr} = 0.72$ and $k_f = 0.026\,\mathrm{kcal/(m)(hr)(°C)}$ at about 100°C. Accordingly, the data could be generalized to present the ratio of effective and molecular conductivities as

$$k_e/k_f = 38.5k_e. \tag{17.36}$$

Equations of the highest and lowest lines on this figure then may be written

$$k_e/k_f = 8.08 + 0.1027\mathrm{Re} \quad \text{(Kwong and Smith)}, \tag{17.37}$$

$$k_e/k_f = 13.85 + 0.0623\mathrm{Re} \quad \text{(Quinton and Storrow)}. \tag{17.38}$$

At higher temperatures, above 300°C or so, radiation must contribute to the effective conductivity, but there are so many other uncertainties that the radiation effect has not been studied at length.

HEAT TRANSFER AT WALLS, TO PARTICLES, AND OVERALL

The correlations cited in Tables 17.16 and 17.18 are of the Nusselt number in terms of the Reynolds and Prandtl numbers, or of the Reynolds alone. They are applicable only above specified Reynolds numbers, about 40 in most cases; clearly they do not predict correctly the coefficient of natural convection, at $\mathrm{Re} = 0$.

Wall coefficients are obtainable from particle–fluid data by a rule of Beek (1962),

$$h_w = 0.8h_p. \tag{17.39}$$

This is how Eq. (8) of Table 17.18 is deduced; Eq. (9) represents

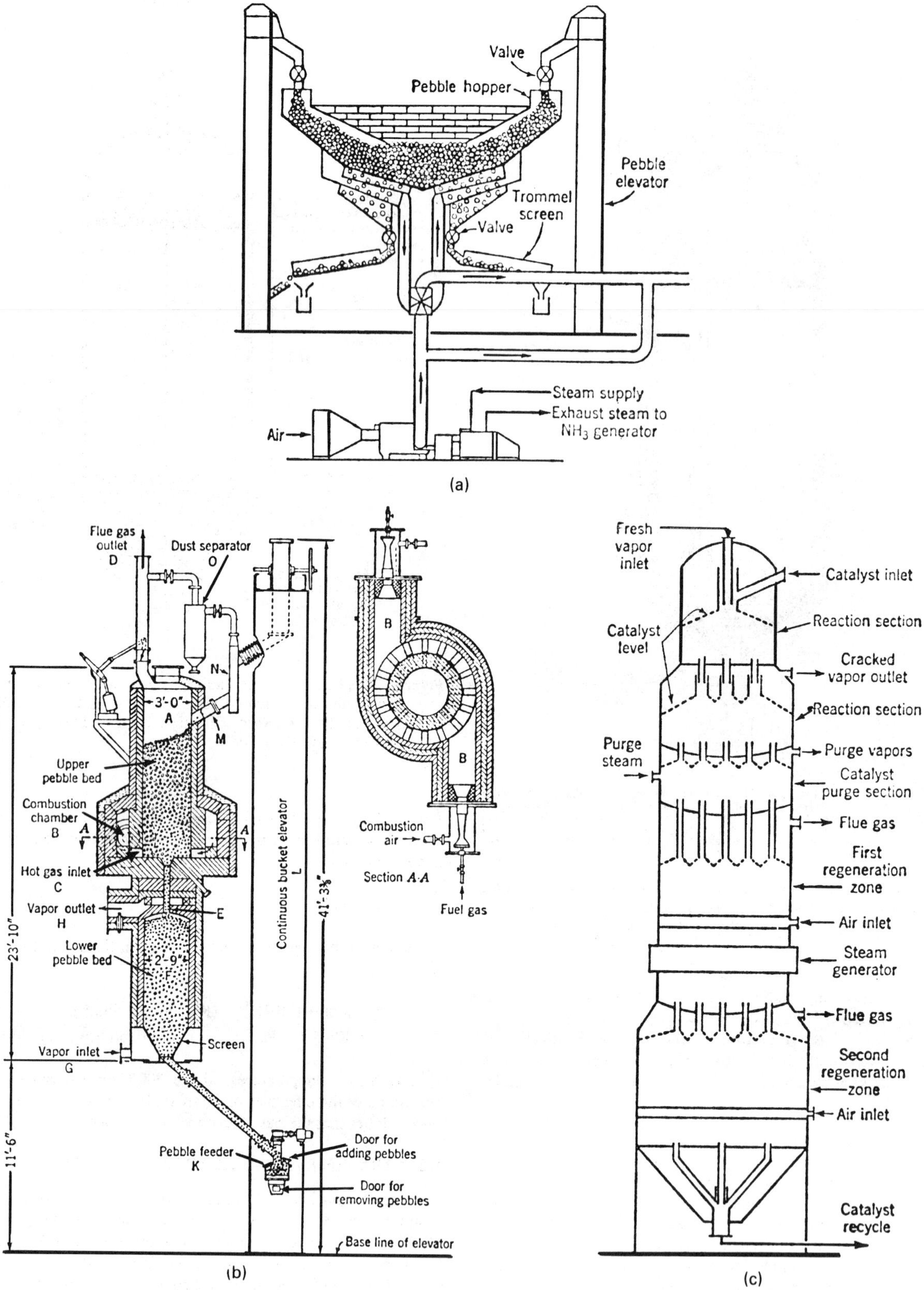

Figure 17.28. Reactors with moving beds of catalyst or solids for heat supply. (a) Pebble reactor for direct oxidation of atmospheric nitrogen; two units in parallel, one being heated with combustion gases and the other used as the reactor [*Ermenc,* Chem. Eng. Prog. **52,** *149* (*1956*)]. (b) Pebble heater which has been used for making ethylene from heavier hydrocarbons (*Batchelder and Ingols, U.S. Bureau of Mines Report Invest. No. 4781, 1951*). (c) Moving bed catalytic cracker and regenerator; for 20,000 bpsd the reactor is 16 ft dia, catalyst circulation rate 2–7 lbs/lb oil, attrition rate of catalyst 0.1–0.5 lb/ton circulated, pressure drop across air lift line is about 2 psi (*L. Berg in Othmer* (*Ed.*), Fluidization, *Reinhold, New York, 1956*).

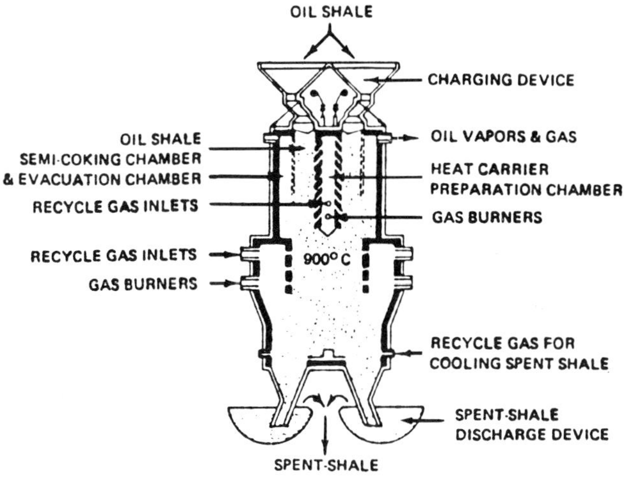

(a)

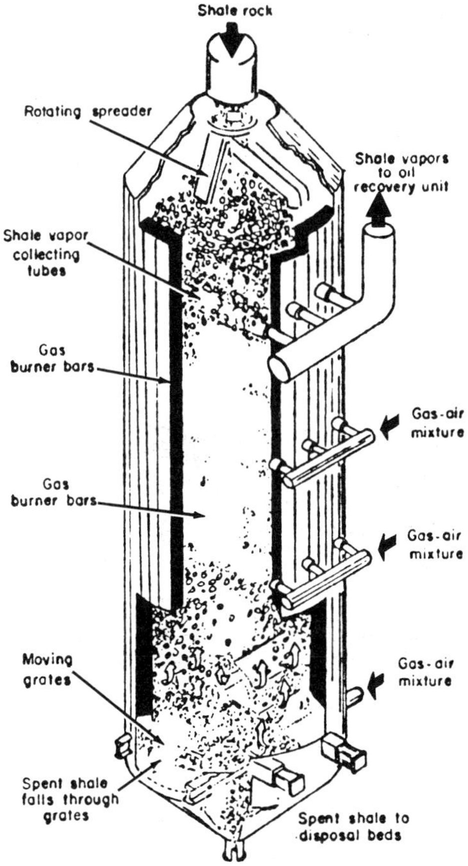

(b)

Figure 17.29. Moving bed reactors for cracking and recovery of shale oil. (a) Kiviter retort, USSR 200–300 tons/day [*J.W. Smith, in Meyers (Ed.),* Handbook of Synfuels Technology, *McGraw-Hill, New York, 1984*]. (b) Paraho retort for shale oil recovery (*Paraho Oil Shale Demonstration, Grand Junction, CO*).

the same data but is simply a curve fit of Figure 17.36(c) at an average value Pr = 0.65.

Data of heat transfer between particle and fluid usually are not measured directly because of the experimental difficulties, but are deduced from measurements of mass transfer coefficients assuming the Colburn analogy to apply,

$$\begin{aligned}(\text{Sherwood})(\text{Schmidt})^{2/3} &= (\text{Nusselt})(\text{Prandtl})^{2/3}\\ &= \text{function of Reynolds.} \qquad (17.40)\end{aligned}$$

Thus, in Figure 17.36(c), if the Nusselt number is replaced by the Sherwood and the Prandtl by the Schmidt, the relation will be equally valid for mass transfer.

The ratio, L/D, of length to diameter of a packed tube or vessel has been found to affect the coefficient of heat transfer. This is a dispersion phenomenon in which the Peclet number, uL/D_{disp}, is involved, where D_{disp} is the dispersion coefficient. Some 5000 data points were examined by Schlünder (1978) from this point of view; although the effect of L/D is quite pronounced, no clear pattern was deduced. Industrial reactors have L/D above 50 or so; Eqs. (6) and (7) of Table 17.18 are asymptotic values of the heat transfer coefficient for such situations. They are plotted in Figure 17.36(b).

Most investigators have been content with correlations of the Nusselt with the Reynolds and Prandtl numbers, or with the Reynolds number alone. The range of numerical values of the Prandtl number of gases is small, and most of the investigations have been conducted with air whose Pr = 0.72 at 100°C. The effect of Pr is small on Figure 17.36(c), and is ignored on Figure 17.36(b) and in some of the equations of Tables 17.17 and 17.18.

The equations of Table 17.18 are the ones recommended for coefficients of heat transfer between wall and fluid in packed vessels.

For design of equipment like those of Figure 17.28, coefficients of heat transfer between particle and fluid should be known. Direct measurements with this objective have been made with metallic packings heated by electrical induction or current. Some correlations are given in Table 17.17. Glaser and Thodos [*AIChE J.* **4,** 63 (1958)] correlated such data with the equation

$$(h_p/C_pG)(C_p\mu/k)^{2/3} = \frac{0.535}{(\text{Re}')^{0.3} - 1.6}, \quad 100 < \text{Re}' < 9200, \qquad (17.41)$$

where

$$\text{Re}' = \phi\sqrt{a_p}\, G/\mu(1-\varepsilon),$$
$$a_p = \text{surface of a single particle},$$
$$\phi = \text{sphericity of the particle}.$$

The formulas of Table 17.18 also could be used by taking $h_p = 1.25h_w$. In moving bed catalyst regenerators, heat fluxes of the order of 25,000 Btu/(hr)(cuft) have been estimated to occur between fluid and particle. Fluid and particle temperatures consequently differ very little.

FLUIDIZED BEDS

A distinctive feature of fluidized beds is a high rate of heat transfer between the fluid and immersed surfaces. Some numerical values are shown on Figure 17.37. For comparison, air in turbulent flow in pipelines has a coefficient of about 25 Btu/(hr)(sqft)(°F). (a) is of calculations from several correlations of data for the conditions identified in Table 17.19; (b) shows the effect of diameters of quartz particles; and (c) pertains to 0.38 mm particles of several substances.

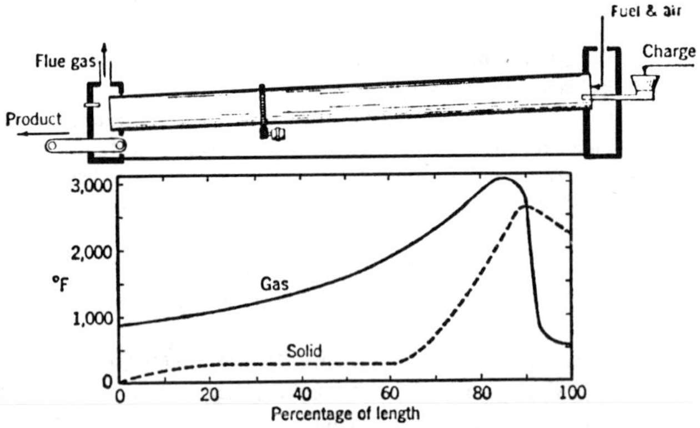

(a)

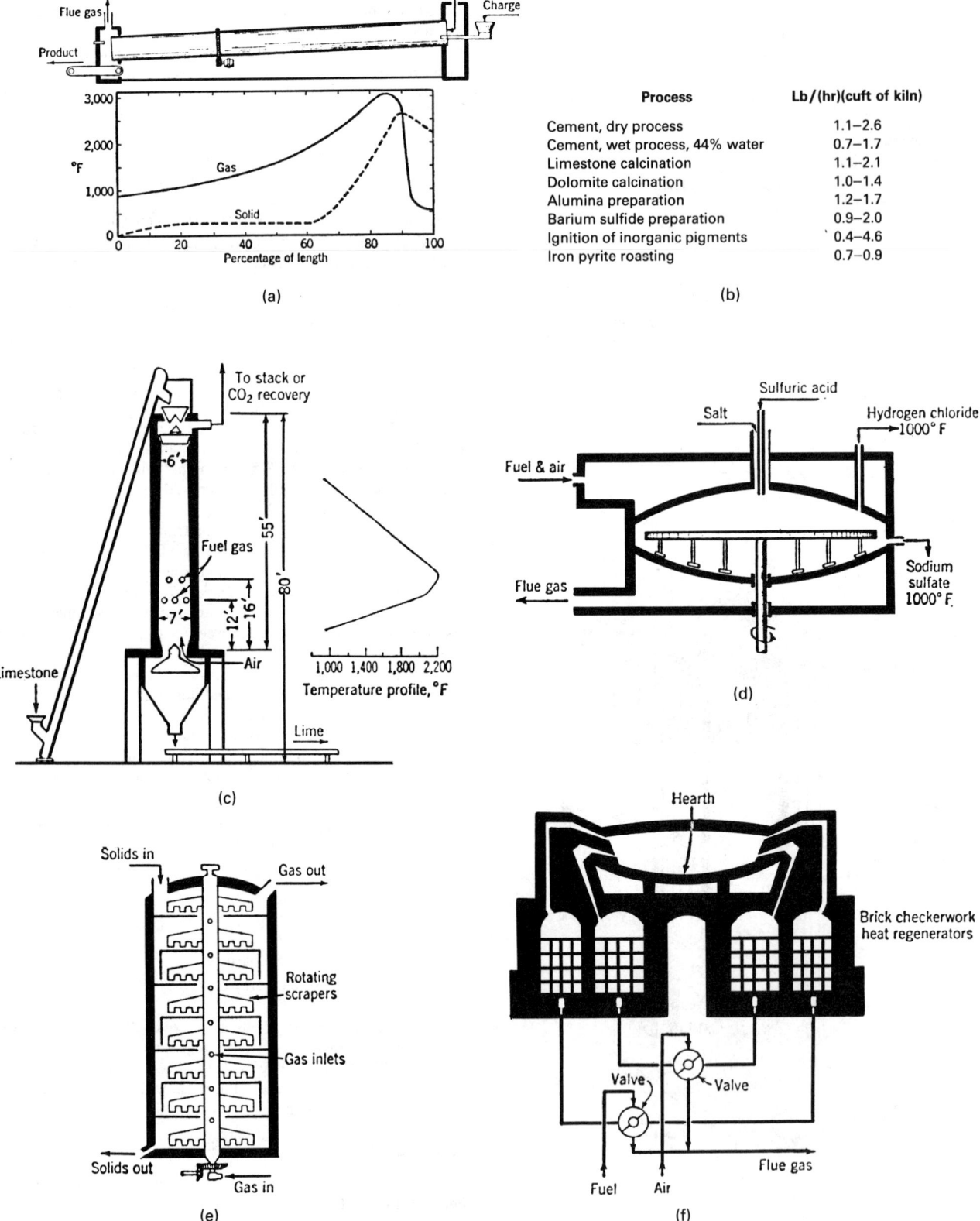

Process	Lb/(hr)(cuft of kiln)
Cement, dry process	1.1–2.6
Cement, wet process, 44% water	0.7–1.7
Limestone calcination	1.1–2.1
Dolomite calcination	1.0–1.4
Alumina preparation	1.2–1.7
Barium sulfide preparation	0.9–2.0
Ignition of inorganic pigments	0.4–4.6
Iron pyrite roasting	0.7–0.9

(b)

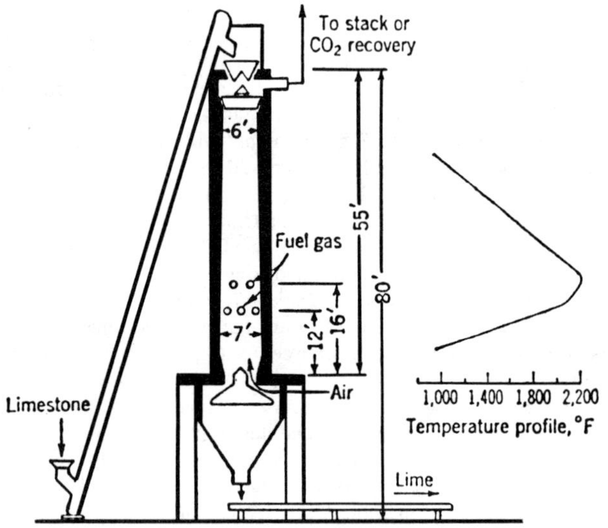

(c)

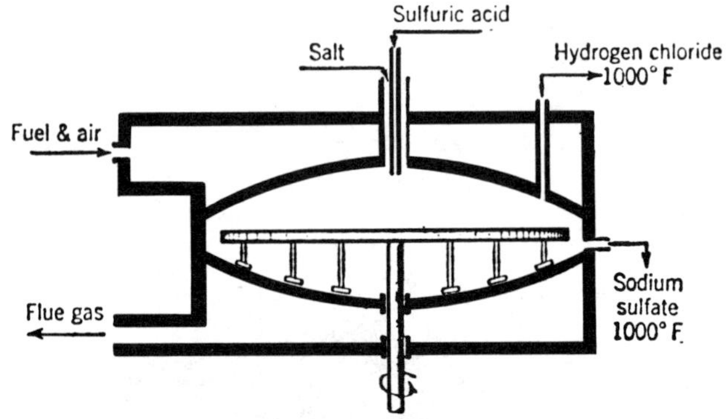

(d)

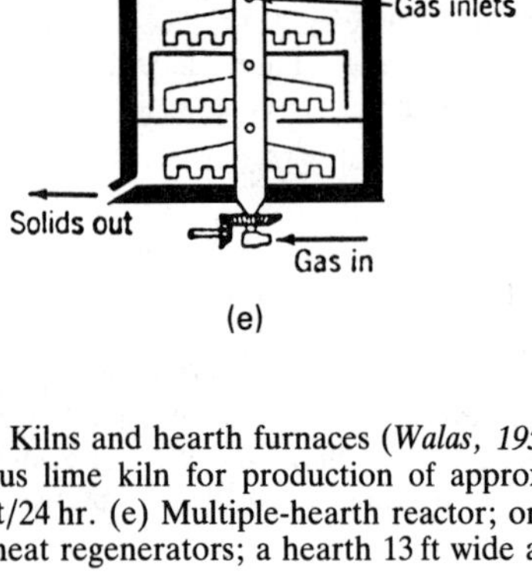

(e)

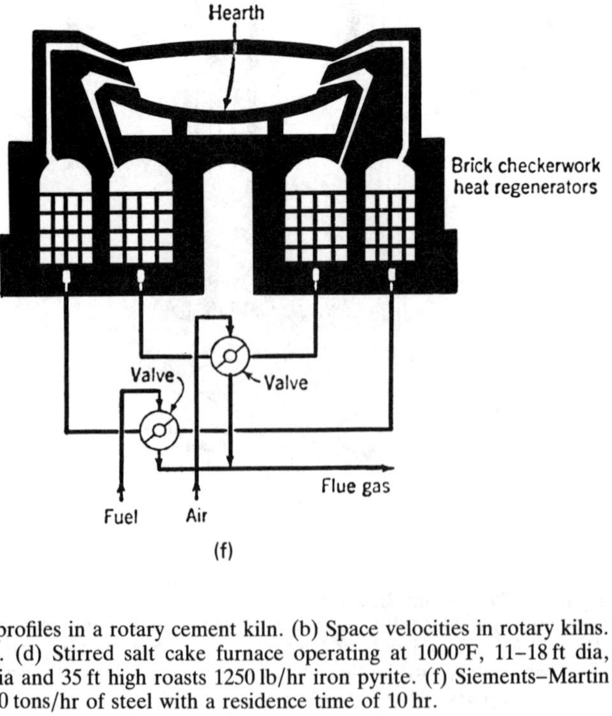

(f)

Figure 17.30. Kilns and hearth furnaces (*Walas, 1959*). (a) Temperature profiles in a rotary cement kiln. (b) Space velocities in rotary kilns. (c) Continuous lime kiln for production of approximately 55 tons/24 hr. (d) Stirred salt cake furnace operating at 1000°F, 11–18 ft dia, 6–10 tons salt/24 hr. (e) Multiple-hearth reactor; one with 9 trays, 16 ft dia and 35 ft high roasts 1250 lb/hr iron pyrite. (f) Siements–Martin furnace and heat regenerators; a hearth 13 ft wide and 40 ft long makes 10 tons/hr of steel with a residence time of 10 hr.

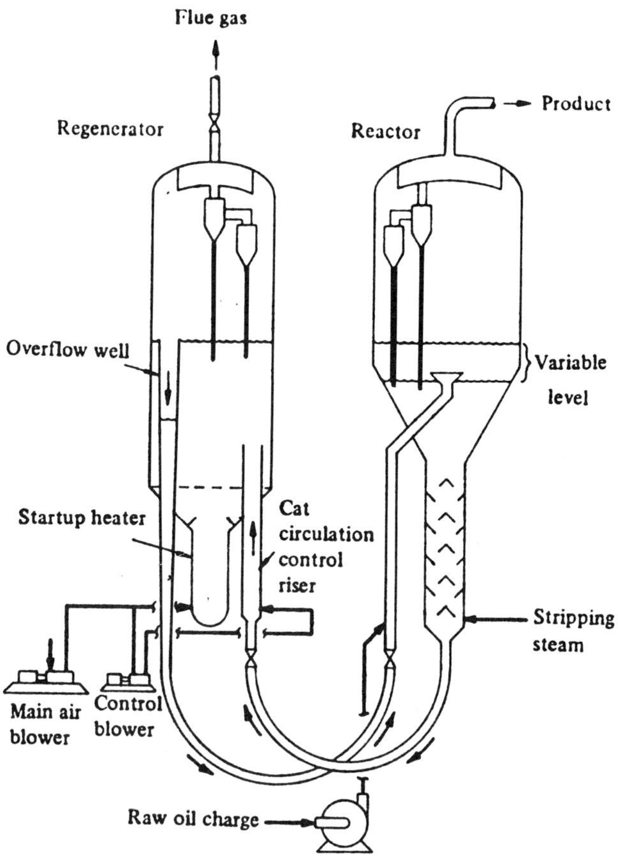

Typical FCC Operating Parameters

Parameter	Value
Feed capacity, kg/s	60 fresh; 95 total
Reactor diameter, m	6.7
Bed depth of reactor, m	Variable 0–1.6
Reactor dilute-phase velocity, m/s	0.75
Reactor temperature, °C	520
Stripper diameter, m	3.3
Stripper height, m	11
U-bend diameter, m	0.86
Catalyst circulation rate, kg/s	450
Regenerator diameter, m	10.7
Regenerator bed depth, m	6.64
Regenerator gas velocity, m/s	0.64
Height of regenerator cyclones inlet above bed, m	10.3
Regenerator temperature, °C	670
Regenerator pressure, kN/m^3 gauge	170
Catalyst inventory in regenerator, Mg	190
Entrainment to regenerator cyclones (estimated), kg/s	260
Catalyst losses in regenerator flue gas, g/s	15

(a)

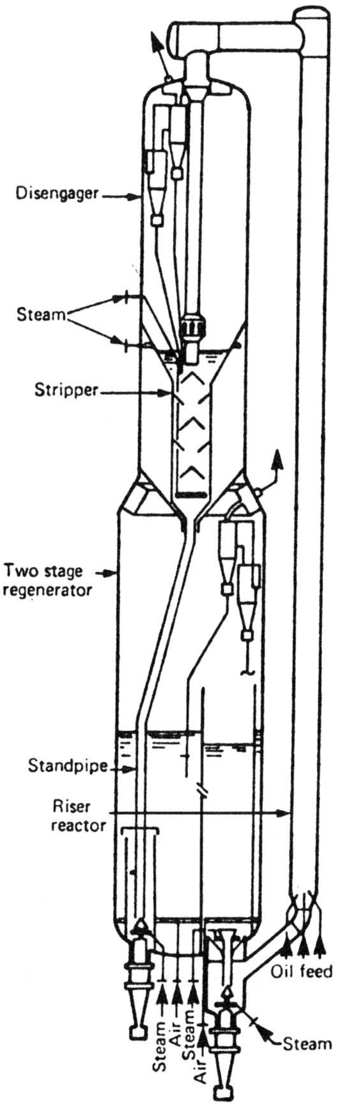

(b)

Figure 17.31. Fluidized bed reactor processes for the conversion of petroleum fractions. (a) Exxon Model IV fluid catalytic cracking (FCC) unit sketch and operating parameters. (*Hetsroni,* Handbook of Multiphase Systems, *McGraw-Hill, New York, 1982*). (b) A modern FCC unit utilizing active zeolite catalysts; the reaction occurs primarily in the riser which can be as high as 45 m. (c) Fluidized bed hydroformer in which straight chain molecules are converted into branched ones in the presence of hydrogen at a pressure of 1500 atm. The process has been largely superseded by fixed bed units employing precious metal catalysts (*Hetsroni,* loc. cit.). (d) A fluidized bed coking process; units have been built with capacities of 400–12,000 tons/day.

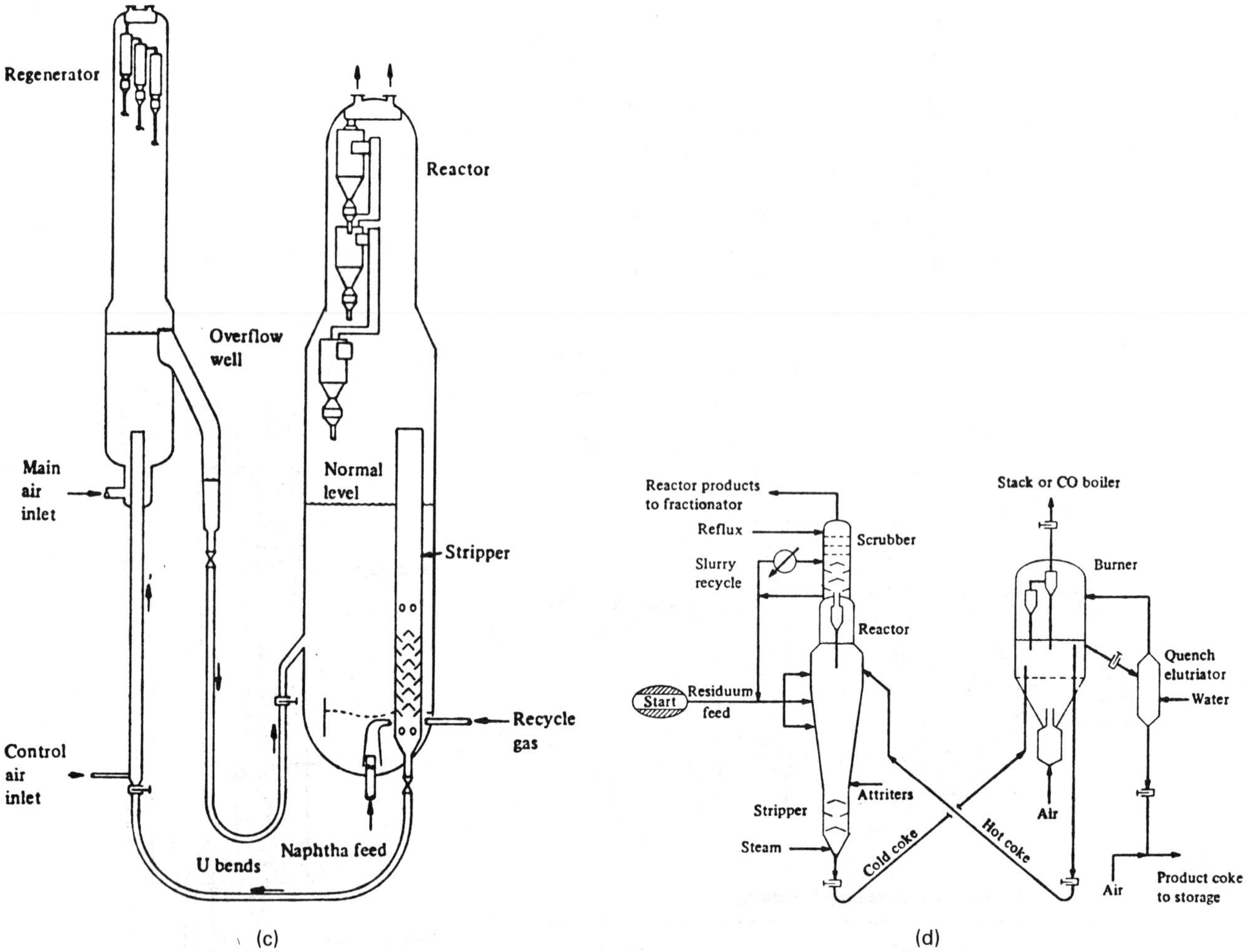

Figure 17.31—(*continued*)

Temperature in a fluidized bed is uniform unless particle circulation is impeded. Gas to particle heat flow is so rapid that it is a minor consideration. Heat transfer at points of contact of particles is negligible and radiative transfer also is small below 600°C. The mechanisms of heat transfer and thermal conductivity have been widely studied; the results and literature are reviewed, for example, by Zabrodsky (1966) and by Grace (1982, pp. 8.65–8.83).

Heat transfer behavior is of importance at the walls of the vessel where it determines magnitudes of heat losses to the surroundings and at internal surfaces used for regulation of the operating temperature. The old correlations for heat transfer coefficients of Wender and Cooper (1958) (shown on Fig. 17.38) and those of Vreedenburg (1960) (shown in Table 17.17) still are regarded as perhaps the best. (See Table 17.20.) A fair amount of scatter of the data obtained by various investigators is evident in Figure 17.38. Vreedenburg utilized additional data in his correlating, and, consequently, his figures show even more scatter. On Figure 17.37(a) also there is much disagreement; but if lines 8 and 9 by the same investigators and part of line 3 are ignored, the agreement becomes fair.

Some opinion is that the correlations for vertical tubes should be taken as standard. Coefficients at the wall appear to be about 10% less than at vertical tubes on the axis of the vessel and those for horizontal tubes perhaps 5–6% less (Korotjanskaja et al., 1984, p. 315).

As appears on Figure 17.37, a peak rate of heat transfer is attained. It has been correlated by Zabrodsky et al. (1976) for particles smaller than 1 mm by the equation

$$h_{\max} d_s / k_g = 0.88 \mathrm{Ar}^{0.213}, \quad 100 < \mathrm{Ar} < 1.4 \times 10^5, \tag{17.42}$$

$$\mathrm{Ar} = g \rho_g (\rho_s - \rho_g) d_s^3 / \mu_g^2. \tag{17.43}$$

17.8. CLASSES OF REACTION PROCESSES AND THEIR EQUIPMENT

In this section, industrial reaction processes are classified primarily with respect to the kinds of phases participating, and instances are given of the kind of equipment that has been found suitable. As always, there is much variation in practice because of local or historical or personal circumstances which suggests that a certain latitude in new plant design is possible.

HOMOGENEOUS GAS REACTIONS

Ethylene is made by pyrolysis of hydrocarbon vapors in tubes of 50–100 mm dia and several hundred meters long with a reaction time of several seconds; heat is supplied by mixing with superheated steam and by direct contact of the tube with combustion gases.

Reactors to make polyethylene are 34–50 mm dia with 10–20 m long turns totalling 400–900 m in length. The tube is jacketed and heated or cooled at different positions with pressurized water.

A flow reactor is used for the production of synthesis gas, $CO + H_2$, by direct oxidation of methane and other hydrocarbons in the presence of steam. Preheated streams are mixed and react in a

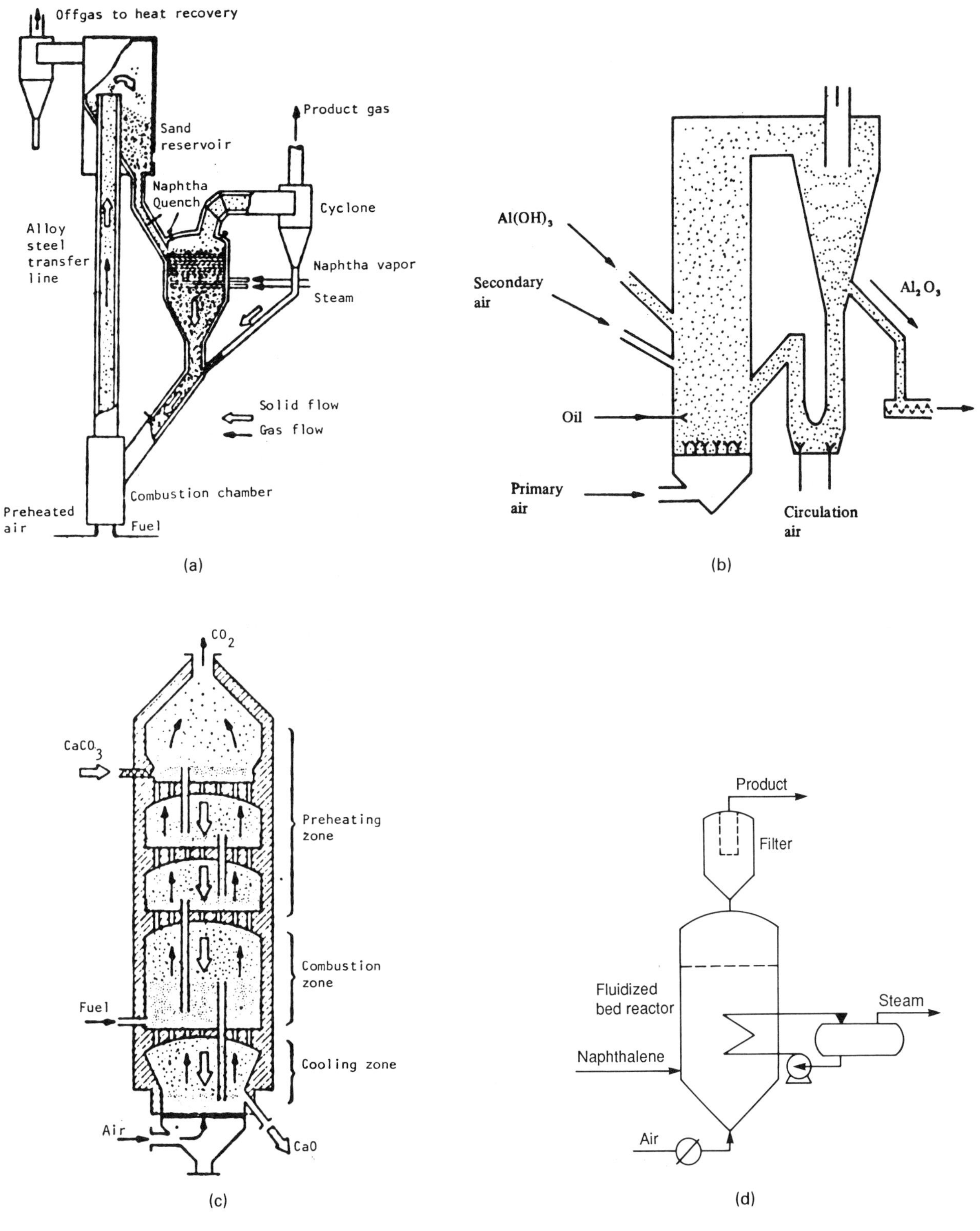

Figure 17.32. Other fluidized bed reaction systems. (a) Cracking of naphtha to ethylene with circulating hot sand at 720–850°C (*Lurgi*). (b) Circulating fluidized bed process for production of alumina by calcination (*Lurgi*). (c) Multibed reactor for calcination of limestone (*Dorr-Oliver*). (d) Synthesis of phthalic anhydride; cooling surface is in the bed (*Badger-Sherwin-Williams*). (e) Coal gasifier with two beds to counteract agglomeration, with spent char recirculating at 20–30 times the fresh feed rate (*Westinghouse*). (f) Ebbulating bed reactor of the H-Coal and H-Oil process for converting these materials at high temperature and pressure into gas and lighter oils (*Meyer,* Handbook of Synfuels Technology, *McGraw-Hill, 1984*).

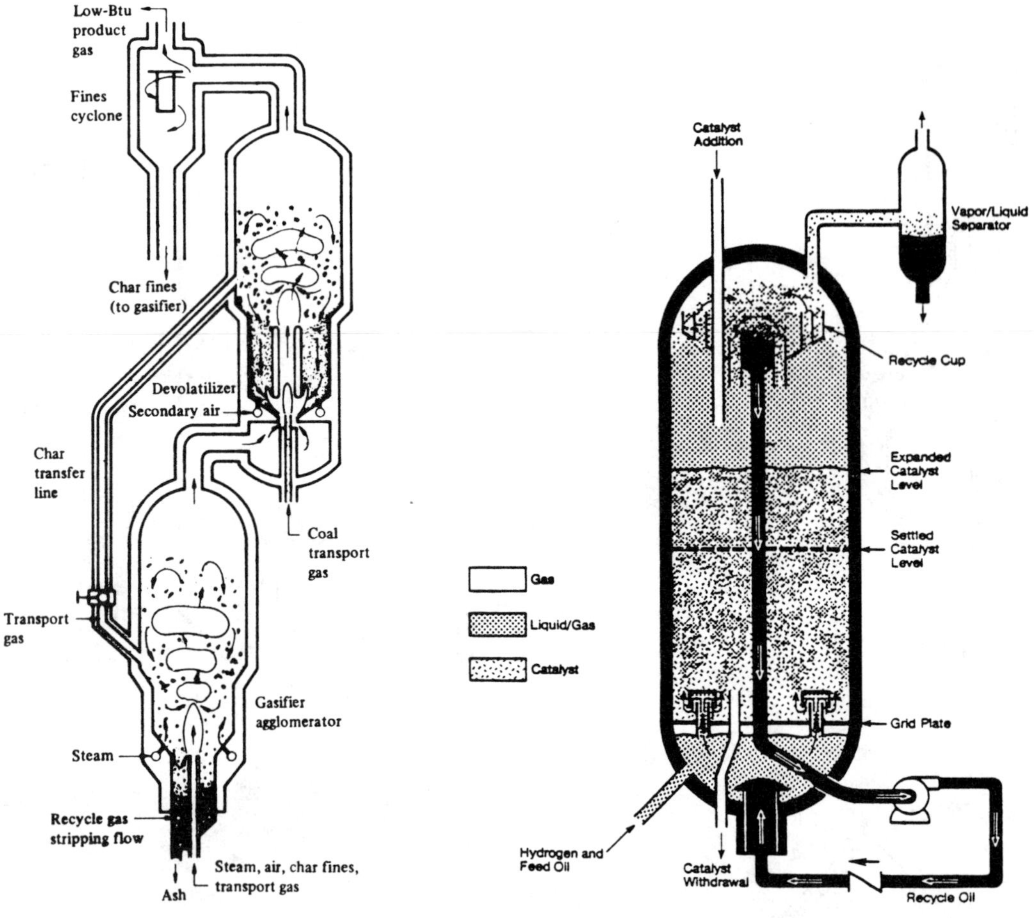

Figure 17.32—(*continued*)

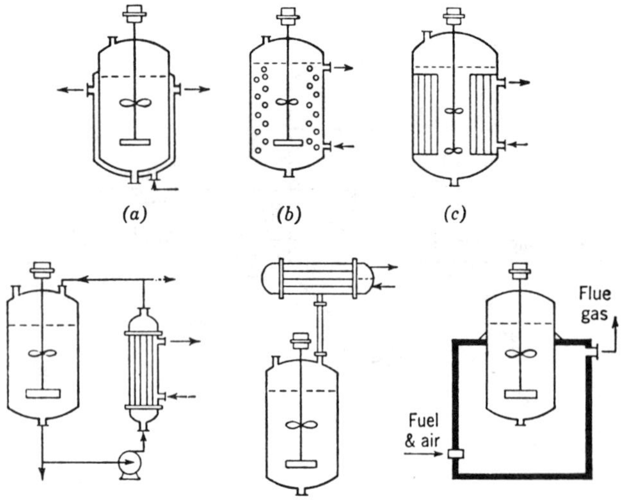

Figure 17.33. Heat transfer to stirred-tank reactors: (a) jacket; (b) internal coils; (c) internal tubes; (d) external heat exchanger; (e) external reflux condenser; (f) fired heater (*Walas, 1959*).

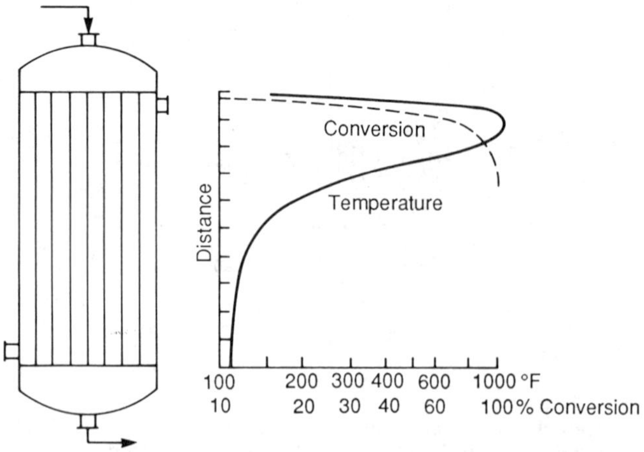

Figure 17.34. Temperature and conversion profiles in a water-cooled shell-and-tube phosgene reactor, 2-in. tubes loaded with carbon catalyst, equimolal CO and Cl_2.

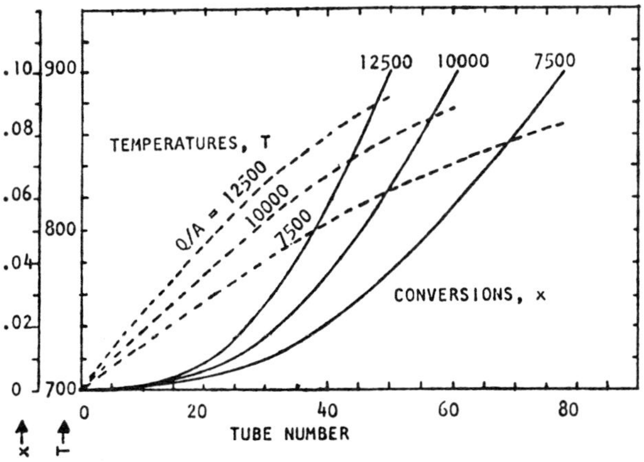

Figure 17.35. Temperature and conversion profiles of mild thermal cracking of a heavy oil in a tubular furnace with a back pressure of 250 psig and at several heat fluxes [Btu/hr(sqft)].

flow nozzle. Burning and quenching are performed in different zones of a ceramic-lined tower.

HOMOGENEOUS LIQUID REACTIONS

Almost innumerable instances of such reactions are practiced. Single-batch stirred tanks, CSTR batteries, and tubular flow reactors are all used. Many examples are given in Table 17.1. As already pointed out, the size of equipment for a given purpose depends on its type. A comparison has been made of the production of ethyl acetate from a mixture initially with 23% acid and 46% ethanol; these sizes were found for 35% conversion of the acid (Westerterp, 1984, pp. 41–58):

Reactor	V_r/V'_0 [m^3/(kg/day)]
Batch (1/3 downtime)	1.04
PFR	0.70
CSTR	1.22
3-stage CSTR	0.85

Some of the homogeneous liquid systems of Table 17.1 are numbers 2, 16, 22, 28, 42, 53, 54, and 96, some in batch, mostly continuous.

LIQUID-LIQUID REACTIONS

Such reactions can take place predominantly in either the continuous or disperse phase or in both phases or mainly at the interface. Mutual solubilities, distribution coefficients, and the amount of interfacial surface are factors that determine the overall rate of conversion. Stirred tanks with power inputs of 5–10 HP/1000 gal or extraction-type equipment of various kinds are used to enhance mass transfer. Horizontal TFRs usually are impractical unless sufficiently stable emulsions can be formed, but mixing baffles at intervals are helpful if there are strong reasons for using such equipment. Multistage stirred chambers in a single shell are used for example in butene–isobutane alkylation with sulfuric acid catalyst. Other liquid–liquid processes listed in Table 17.1 are numbers 8, 27, 45, 78, and 90.

GAS-LIQUID REACTIONS

Intimate contacting between chemically reacting gas and liquid phases is achieved in a variety of equipment, some examples of which follow

a. Tanks equipped with turbine agitators with or without internal gas recirculation. An example is air oxidation of cyclohexane to cyclohexanol and cyclohexanone.
b. Bubble towers with parallel flow of the phases, gas dispersed, with or without trays or packing. In such equipment isobutene from a mixture of C_4 hydrocarbons forms tertiary butanol in contact with aqueous sulfuric acid.
c. Countercurrent flow of the two phases in tray or packed towers, as in ordinary absorption processes. Absorption of nitrogen oxides in water to make nitric acid is a prime example.
d. Tubular or multitubular reactors are usable when the volumetric rate of the gas is so much greater than that of the liquid that substantial mixing of phases exists. Adipic acid nitrile is made from gaseous ammonia and liquid adipic acid with a volumetric ratio of 1500; the residence time of the gas phase is about 1 sec, and that of the liquid 180–300 sec.
e. Liquid ejector for entraining the gas. This is used to remove dilute acid or other impurities from waste air by scrubbing with aqueous solutions; the liquid is recirculated so that a gas/liquid volumetric ratio of 100–200 is maintained.
f. Pumps of centrifugal or turbine types are effective mixing devices and can constitute a reactor when the needed residence time is short. Such a device is used, for instance, to make acetic anhydride from acetic acid and ketene [Spes, *Chem. Ing. Tech.* **38,** 963 (1966)].
g. Thin film reactors are desirable when the liquid viscosity is high, the reaction is highly exothermic and short reaction times are adequate. Such a process is the sulfonation of dodecylbenzene with dilute SO_3 [Ujhidy et al., *Chem. Tech.* **18,** 625 (1966)].
h. Packed tower reactors in parallel flow are operated either top-to-bottom or bottom-to-top. Distribution, holdup, and pressure drop behavior can be predicted from mass transfer correlations. Downflow towers have lower pressure drop, but upflow of liquid assures greater liquid holdup and longer contact time which often are advantages.

NONCATALYTIC REACTIONS WITH SOLIDS

The chief examples are smelting for the recovery of metals from ores, cement manufacture, and lime burning. The converters, roasters, and kilns for these purposes are huge special devices, not usually adaptable to other chemical applications. Shale oil is recovered from crushed rock in a vertical kiln on a batch or continuous basis—moving bed in the latter case—sometimes in a hydrogen-rich atmosphere for simultaneous denitrification and desulfurization. The capacity of ore roasters is of the order of 300–700 tons/(day)(m^3 of reactor volume). Rotary kilns for cement have capacities of 0.4–1.1 tons/(day)(m^3); for other purposes the range is 0.1–2.

FLUIDIZED BEDS OF NONCATALYTIC SOLIDS

Fluidized bed operations sometimes are alternates to those with fixed beds. Some of the successful processes are fluid bed combustion of coal, cracking of petroleum oils, ethylene production from gas oils in the presence of fluidized sand as a heat carrier, fluidized bed coking, water–gas production from coal (the original fluidized bed operation), recovery of shale oil from rock, reduction of iron ore with hydrogen at 30 atm pressure, lime burning, HCN from coke + ammonia + propane in a fluidized electric furnace, and many others. Many of these processes have distinct equipment configurations and space velocities that cannot be generalized, except insofar as general relations apply to fluidized bed stability,

TABLE 17.10. Overall Heat Transfer Coefficients in Agitated Tanks [U Btu/(hr)(sqft)(°F)]

Fluid Inside Jacket	Fluid In Vessel	Wall Material	Agitation	U
Steam	water	enameled cast iron	0–400 rpm	96–120
Steam	milk	enameled C.I.	none	200
Steam	milk	enameled C.I.	stirring	300
Steam	milk boiling	enameled C.I.	none	500
Steam	milk	enameled C.I.	200 rpm	86
Steam	fruit slurry	enameled C.I.	none	33–90
Steam	fruit slurry	enameled C.I.	stirring	154
Steam	water	C.I. and loose lead lining	agitated	4–9
Steam	water	C.I. and loose lead lining	none	3
Steam	boiling SO_2	steel	none	60
Steam	boiling water	steel	none	187
Hot water	warm water	enameled C.I.	none	70
Cold water	cold water	enameled C.I.	none	43
Ice water	cold water	stoneware	agitated	7
Ice water	cold water	stoneware	none	5
Brine, low velocity	nitration slurry	—	35–58 rpm	32–60
Water	sodium alcoholate solution	"Frederking" (cast-incoil)	agitated, baffled	80
Steam	evaporating water	copper	—	381
Steam	evaporating water	enamelware	—	36.7
Steam	water	copper	none	148
Steam	water	copper	simple stirring	244
Steam	boiling water	copper	none	250
Steam	paraffin wax	copper	none	27.4
Steam	paraffin wax	cast iron	scraper	107
Water	paraffin wax	copper	none	24.4
Water	paraffin wax	cast iron	scraper	72.3
Steam	solution	cast iron	double scrapers	175–210
Steam	slurry	cast iron	double scrapers	160–175
Steam	paste	cast iron	double scrapers	125–150
Steam	lumpy mass	cast iron	double scrapers	75–96
Steam	powder (5% moisture)	cast iron	double scrapers	41–51

(LIGHTNIN Technology Seminar, Mixing Equipment Co., 1982).

particle size distribution, heat transfer, multistaging, and possibly other factors.

CIRCULATING GAS OR SOLIDS

High temperatures are generated by direct or indirect contact with combustion gases. A circulating bed of granular solids heated in this way has been used for the fixation of nitrogen from air in the range of 2300°C. Pebble heaters originally were developed as pyrolysis reactors to make ethylene, but are no longer used for this purpose. Pebbles are 5–10 mm dia, temperatures of 1700°C are readily attained, heat fluxes are in the vicinity of 15,000 Btu/(hr)(°F)(cuft of pebbles) and contact times are fractions of a second. These characteristics should be borne in mind for new processes, although there are no current examples. Induction heating of fluidized particles has been used to transfer heat to a reacting fluid; in this process the solid remains in the reactor and need not circulate through a heating zone.

FIXED BED SOLID CATALYSIS

This kind of process is used when the catalyst maintains its activity sufficiently long, for several months or a year or two as in the cases of some catalytic reforming or ammonia synthesis processes. A few processes have operated on cycles of reaction and regeneration of less than an hour or a few hours. Cycle timers on automatic valves make such operations completely automatic. A minimum of three vessels usually is needed: One on-stream, one being regenerated, and the last being purged and prepared for the next cycle. Adsorption processes are conducted this way. The original Houdry cracking process employed 10 min on-stream. One catalytic reforming process employs seven or so reactors with one of them

TABLE 17.11. Jacketed Vessels Overall Heat Transfer Coefficients

Jacket fluid	Fluid in vessel	Wall material	Overall U* Btu/(h·ft²·°F)	Overall U* J/(m²·s·K)
Steam	Water	Stainless steel	150–300	850–1700
Steam	Aqueous solution	Stainless steel	80–200	450–1140
Steam	Organics	Stainless steel	50–150	285– 850
Steam	Light oil	Stainless steel	60–160	340– 910
Steam	Heavy oil	Stainless steel	10– 50	57– 285
Brine	Water	Stainless steel	40–180	230–1625
Brine	Aqueous solution	Stainless steel	35–150	200– 850
Brine	Organics	Stainless steel	30–120	170– 680
Brine	Light oil	Stainless steel	35–130	200– 740
Brine	Heavy oil	Stainless steel	10– 30	57– 170
Heat-transfer oil	Water	Stainless steel	50–200	285–1140
Heat-transfer oil	Aqueous solution	Stainless steel	40–170	230– 965
Heat-transfer oil	Organics	Stainless steel	30–120	170– 680
Heat-transfer oil	Light oil	Stainless steel	35–130	200– 740
Heat-transfer oil	Heavy oil	Stainless steel	10– 40	57– 230
Steam	Water	Glass-lined CS	70–100	400– 570
Steam	Aqueous solution	Glass-lined CS	50– 85	285– 480
Steam	Organics	Glass-lined CS	30– 70	170– 400
Steam	Light oil	Glass-lined CS	40– 75	230– 425
Steam	Heavy oil	Glass-lined CS	10– 40	57– 230
Brine	Water	Glass-lined CS	30– 80	170– 450
Brine	Aqueous solution	Glass-lined CS	25– 70	140– 400
Brine	Organics	Glass-lined CS	20– 60	115– 340
Brine	Light oil	Glass-lined CS	25– 65	140– 370
Brine	Heavy oil	Glass-lined CS	10– 30	57– 170
Heat-transfer oil	Water	Glass-lined CS	30– 80	170– 450
Heat-transfer oil	Aqueous solution	Glass-lined CS	25– 70	140– 400
Heat-transfer oil	Organics	Glass-lined CS	25– 65	140– 370
Heat-transfer oil	Light oil	Glass-lined CS	20– 70	115– 400
Heat-transfer oil	Heavy oil	Glass-lined CS	10– 35	57– 200

*Values listed are for moderate nonproximity agitation. CS = carbon steel.

(*Perry's Chemical Engineers Handbook,* McGraw-Hill, New York, 1984).

down every week. Regeneration usually is done in place, but eventually the catalyst must be removed and replaced. Platinum and other precious metals are recovered from the catalyst carriers in the factory.

A granular catalyst sometimes serves simultaneously as tower packing for reaction and separation of the participants by distillation, particularly when the process is reversible and removal of the product is necessary for complete conversion to take place. This is the case of the reaction of methanol and isobutene to make methyl tertiary-butyl ether (MTBE) in the presence of granular acid ion exchange resin catalyst. MTBE is drawn off the bottom of the tower and excess methanol off the top. Such a process is applicable

TABLE 17.12. Overall Heat Transfer Coefficients with Immersed Coils [U expressed in Btu/(h · ft² · °F)]

Type of coil	Coil spacing, in.†	Fluid in coil	Fluid in vessel	Temp. range, °F.	U‡ without cement	U with heat-transfer cement
⅜ in. o.d. copper tubing attached with bands at 24-in. spacing	2	5 to 50 lb./sq. in. gage steam	Water under light agitation	158–210	1-5	42–46
	3⅛			158–210	1-5	50-53
	6¼			158–210	1-5	60–64
	12½ or greater			158–210	1-5	69-72
⅜ in. o.d. copper tubing attached with bands at 24-in. spacing	2	50 lb./sq. in. gage steam	No. 6 fuel oil under light agitation	158–258	1-5	20–30
	3⅛			158–258	1-5	25–38
	6¼			158–240	1-5	30–40
	12½ or greater			158–238	1-5	35–46
Panel coils		50 lb./sq. in. gage steam	Boiling water	212	29	48–54
		Water	Water	158–212	8–30	19–48
		Water	No. 6 fuel oil	228–278	6-15	24–56
			Water	130–150	7	15
			No. 6 fuel oil	130–150	4	9–19

*Data courtesy of Thermon Manufacturing Co.
†External surface of tubing or side of panel coil facing tank.

TABLE 17.13. Summary of Heat-Transfer Coefficients on the Agitated Side

General Equation: $$\frac{h(L)}{\lambda_f} = \alpha\left(\frac{\rho N D_I^2}{\mu}\right)^m\left(\frac{c_p\mu}{\lambda_f}\right)^b\left(\frac{\mu_b}{\mu_w}\right)^c \text{(other terms)}$$

Agitator Type	Transfer Surface	Approx. Reynolds Number Range	L	α	m	b	c	Other Terms	Additional Comments	Ref.
Turbine										
6-blade, flat (baffled)	jacket	10–10^5	D	0.73	0.65	0.33	0.24	—	Use for standard configuration. See p. 357 for details	1,25
	coil	400–1.5×10^6	d_{ct}	0.17	0.67	0.37	See Note 1	$\left(\frac{D_I}{D}\right)^{0.1}\left(\frac{d_{ct}}{D}\right)^{0.5}$	See Note 2. Applies for standard configuration with $D_c/D = 0.7$ and $S_c/d_{ct} = 2$–4; $Z_c/D = 0.15$	29
	vertical baffle-type	10^3–2×10^6	d_{ct}	0.09	0.65	0.33	0.4	$\left(\frac{D_I}{D}\right)^{0.33}\left(\frac{2}{n_{bv}}\right)^{0.2}$		32
6-blade, retreating blade (curved blade)	jacket	10^3–10^6	D	0.68	0.67	0.33	0.14		Revised. See Note 3	27,30
no baffles	coil	10^3–10^6	d_{ct}	1.40	0.62	0.33	0.14	—	Revised	27,30
6-blade, 45° pitched	jacket	20–200	D	0.44[a]	0.67	0.33	0.24	—	Baffles have no effect in Reynolds number range studied in 12-in diameter vessel	33
3-blade retreating	jacket	2×10^4–2×10^6	D	0.37[b]	0.67	0.33	0.14	—	For glass-lined vessels with finger-type baffle	27
Propeller	jacket	2×10^3	D	0.54	0.67	0.25	0.14	—	Limited data, but a large 5 ft diameter tank used, marine-type impeller used at 45° pitch and located at the midpoint of tank. No baffles used	34
Paddle	jacket	600–5×10^5	D	0.112	0.75	0.44	0.25	$\left(\frac{D}{D_I}\right)^{0.40}\left(\frac{w_I}{D_I}\right)^{0.13}$		26,35
	coil	3×10^2–2.6×10^5	d_{ct}	0.87	0.62	0.33	0.14	—		37
Anchor	jacket	10–300	D	1.0	0.5	0.33	0.18	—		34,36
		300–40,000	D	0.36	0.67	0.33	0.18	—		34,36

Notes.

1. $\ln c = -0.202 \ln \mu - 0.357$, with μ in cp.
2. For unbaffled case with coils use 0.65 of h calculated for baffled case (29).
3. With baffles and $N_{Re} < 400$ use value calculated. In fully developed turbulent region baffles increase calculated h by approximately 37% (1)

New nomenclature: d_{ct} is outside tube diameter of coil, D_c is coil diameter, n_{bv} is number of vertical baffle-type coils, S_c is coil spacing, w_I is impeller blade width, and Z_c is height of coil from tank bottom.

[a] For impeller $4\frac{1}{2}$-in. from bottom, 0.535 for impeller 11-in from bottom

[b] For steel impeller, 0.33 for glassed-steel impeller

(Rase, 1977, Vol. 1).

TABLE 17.14. Equations for Heat Transfer Coefficients inside Stirred Tanks[a]

1. To jackets, with paddles, axial flow, and flat blade turbines[1,6,7]

$$h_0(\text{jacket})\frac{T}{k} = 0.85\left(\frac{D^2N\rho}{\mu}\right)^{0.66}\left(\frac{C_p\mu}{k}\right)^{0.33} \times \left(\frac{\mu}{\mu_s}\right)^{0.14}\left(\frac{Z}{T}\right)^{-0.56}\left(\frac{D}{T}\right)^{0.13}$$

2. To helical coils[3,5]

$$h_0(\text{coil})\frac{D}{k} = 0.17\left(\frac{D^2N\rho}{\mu}\right)^{0.67}\left(\frac{C_p\mu}{k}\right)^{0.37} \times \left(\frac{D}{T}\right)^{0.1}\left(\frac{d}{T}\right)^{0.5}\left(\frac{\mu}{\mu_s}\right)^{m}$$

$m = 0.714/\mu^{0.21}$, μ in cP

3. To vertical tubes[2]

$$h_0(\text{tubes})\frac{D}{K} = 0.09\left(\frac{D^2N\rho}{\mu}\right)^{0.65}\left(\frac{C_p\mu}{k}\right)^{0.3} \times \left(\frac{D}{T}\right)^{0.33}\left(\frac{2}{n_b}\right)^{0.2}\left(\frac{\mu}{\mu_s}\right)^{0.14}$$

4. To plate coils[4]

$$h_0(\text{plate coil})\frac{L}{K} = 0.1788\left(\frac{ND^2\rho}{\mu}\right)^{0.448}\left(\frac{C_p\mu}{k}\right)^{0.33}\left(\frac{\mu}{\mu_f}\right)^{0.50}$$

for $N_{Re} < 1.4 \times 10^3$

$$h_0(\text{plate coil})\frac{L}{K} = 0.0317\left(\frac{ND^2\rho}{\mu}\right)^{0.658}\left(\frac{C_p\mu}{k}\right)^{0.33}\left(\frac{\mu}{\mu_f}\right)^{0.50}$$

for $N_{Re} > 4 \times 10^3$

[a] Nomenclature: d = tube diameter, D = impeller diameter, L = plate coil height, N = impeller rotational speed, n_b = number of baffles or of vertical tubes acting as baffles, T = tank diameter, Z = liquid height.

REFERENCES

1. G. Brooks and G.-J. Su, *Chem. Eng. Prog.* **54,** (Oct. 1959).
2. I.R. Dunlap and J.H. Rushton, *Chem. Eng. Prog. Symp. Ser.* **49**(5), 137 (1953).
3. J.Y. Oldshue and A.T. Gretton, *Chem. Eng Prog.* **50**(12) 615 (1954).
4. D.K. Petree and W.M. Small, *AIChE Symp. Series.* **74**(174), (1978).
5. A.H.P. Skelland, W.K. Blake, J.W. Dabrowski, J.A. Ulrich, and T.F. Mach, *AIChE. J.* **11**(9), (1965).
6. F. Strek, *Int. Chem. Eng.* **5,** 533 (1963).
7. V.W. Uhl and J.B. Gray, *Mixing Theory and Practice,* Academic, New York, 1966, Vol. 1.

(Recommended by Oldshue, 1983).

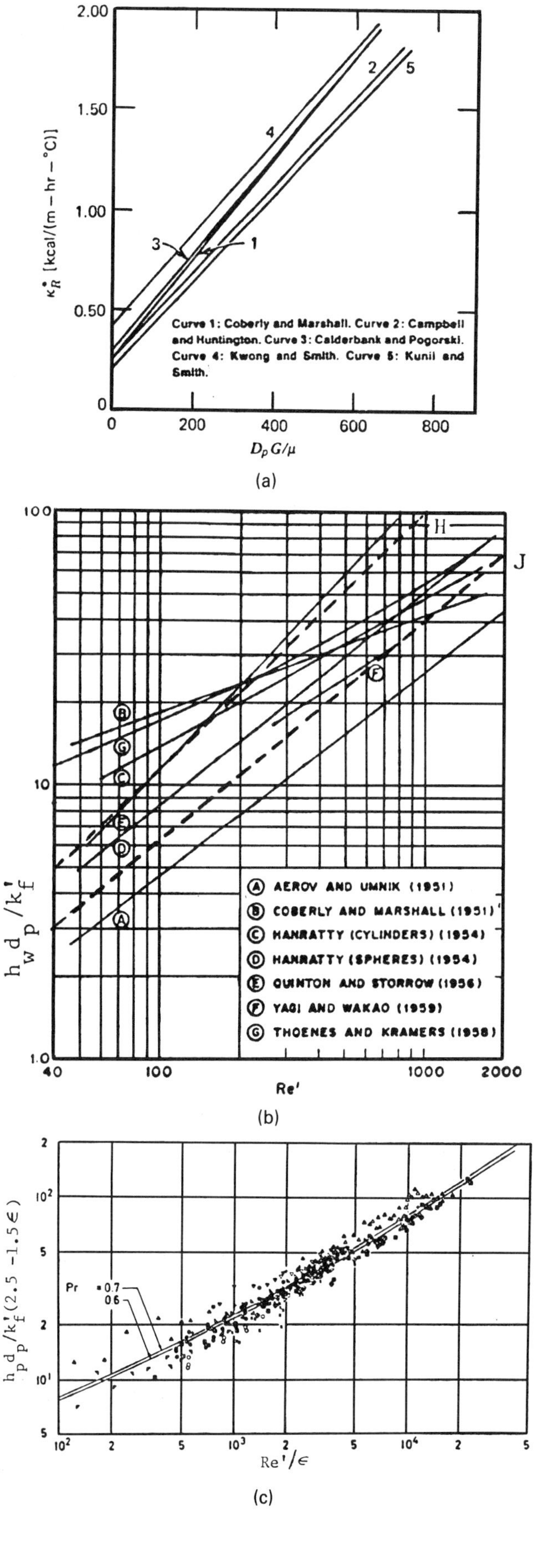

Figure 17.36. Effective thermal conductivity and wall heat transfer coefficient of packed beds. $Re' = d_pG/\mu$, $d_p = 6V_p/A_p$, ε = porosity. (a) Effective thermal conductivity in terms of particle Reynolds number. Most of the investigations were with air of approx. $k'_f = 0.026$, so that in general $k'_e/k'_f = 38.5k'_e$ [*Froment,* Adv. Chem. Ser. **109,** (*1970*)]. (b) Heat transfer coefficient at the wall. Recommendations for L/d_p above 50 by Doraiswamy and Sharma are line H for cylinders, line J for spheres. (c) Correlation of Gnielinski (*cited by Schlünder, 1978*) of coefficient of heat transfer between particle and fluid. The wall coefficient may be taken as $h_w = 0.8h_p$.

TABLE 17.15. Data for the Effective Thermal Conductivity, K_r (kcal/mh°C), and the Tube Wall Film Coefficient, h_w (kcal/m²h°C), in Packed Beds[a]

Authors	Method of Measurement	Heating or Cooling of Gas	Gas	Particles		
				Material	Shape	Diameter D_p, mm
Bakhurov and Boreskov (1947)	Radial temperature and concentration profiles	C	Air	Glass, porcelain, metals, etc.	Spheres, rings, cylinders, granules	3–19
Brötz (1951)	”	H	N_2, CO_2, H_2	Glass, catalyst	Spheres, granules	2–10
Bunnell, Irvin, Olson, and Smith (1949)	Radial temperature profiles	C	Air	Alumina	Cylinders	3
Campbell and Huntington (1952)	”	H, C	Air, natural gas (82% CH_4)	Glass, alumina, aluminum	Spheres, cylinders	5–25
Coberly and Marshall (1951)	”	H	Air	Celite	Cylinders	3–12
Maeda (1952)	”	C	Air	Catalyst	Cylinders	3–10
Quinton and Storrow (1956)	”	H	Air	Glass	Spheres	4.4
Aerov and Umnik (1951b, c)	Packed bed heat exchanger. Single radial temperature	C	Air, CO_2, H_2	Glass, catalyst, porcelain, sand	Spheres, tablets, rings	0.4–10
Hougen and Piret (1951)	Packed bed heat exchanger	C	Air	Celite	Spheres, cylinders	2–12
Molino and Hougen (1952)		H				
Kling (1938b)	”	H	Air	Steel, glass	Spheres	3–4
Verschoor and Schuit (1950)	”	H	Air, H_2	Lead, glass, etc.	Spheres, cylinders, granules	3–10
Bernard and Wilhelm (1950)	Mass diffusion		(Water)	Glass, lead, alumina, etc.	Spheres, cylinders, granules	1–8

when the reaction can be conducted satisfactorily at boiling temperatures; these can be adjusted by pressure.

A variety of provisions for temperature control of fixed beds is described in Section 17.6 and following.

a. Single beds are used when the thermal effects are small. Jacketed walls usually are inadequate for heat transfer to beds, but embedded heat transfer tubes sometimes are used.

b. Multitubular units with catalyst in tubes and heat transfer medium on the shell side are popular. A reactor for making phosgene from carbon monoxide and chloride has 2-in. dia tubes 8 ft long filled with activated carbon catalyst and cooling water on the shell side.

c. Multibed units with built-in interstage heat transfer surface. Economical when the amount of surface is not large. In comparison with type (d), this design may have more difficult maintenance, less flexibility and higher cost because of the shortness of the tubes that may have to be used. The Sper-Rashka converter for SO_2 oxidation has three beds and three large internal exchangers in a single shell (Ullmann, 3rd ed., 1964, Vol. 15, p. 456).

d. Multibed units with external heat exchangers. Several variations

TABLE 17.15—(*continued*)

Tube		Formula (D_p and D_t in m)	Range of N_{Re}' or other variable	Remarks
Diameter, D_t, mm	Length L, m			
100 square	1	$N_{Pe}' = 2.4$ to 6.3	$G \simeq 1000$ kg/m²h	
100	0.2–0.4	Graphs in paper	Gas velocity: 2500–7200 m/h	Special apparatus at low velocities
50	0.05–0.2	$\frac{K_r}{k_g} = 5.0+0.061\left(\frac{D_pG}{\mu}\right)$	30–100	
50–100		$\frac{K_r}{k_g} = 10.0+0.267\left(\frac{G}{\mu a_v}\right)$; $h_w = 2.07\left(\frac{G}{\mu a_v}\right)^{0.47}$	20–500	
127	up to 1	$K_r = 0.27+0.00146\frac{\sqrt{a_p}G}{\mu}$; $h_w = 8.51\ G^{0.33}$	G (kg/m²h): 850–6000	
25–100	0.1–0.4	$\frac{K_r}{k_g} = \begin{cases} 5.5+0.05\left(\frac{D_pG}{\mu}\right) \\ 1.72\left(\frac{D_pG}{\mu}\right)^{0.41} \\ 0.209\left(\frac{D_pG}{\mu}\right)^{0.87} \end{cases}$	0–30 30–100 100–1000	See also Hatta and Maeda (1948 a, b, 1949), Maeda (1950), and Maeda and Kawazoe (1953)
41	0.75	$K_r = 0.36+0.00162\left(\frac{D_pG}{\mu}\right)$; $h_w = 0.04\ G$	30–1100 $500<G<17000$	
65	0.3	$\frac{K_r}{k_g} = 10.5+0.076\ N_{Pr}\left(\frac{4G}{\mu a_v}\right)$ $\frac{h_wD_p}{k_g} = 0.155\ N_{Pr}^{\frac{1}{3}}\left(\frac{4G}{\mu a_v}\right)^{0.75}$	10–3500 150–4000	For rings see paper
35–95	0.16–0.32	$\frac{K_r}{k_g} = 1.23\left(\frac{\sqrt{a_p}G}{\mu}\right)^{0.43}$ No correction for h_w	100–3000	The first paper gives a slightly different formula for cooling
50	0.3	See graph in paper. No correction for h_w	400–3500	
30–50	0.2–0.3	$\frac{K_r}{k_g} = \frac{K_{r,0}}{k_g}+0.10(a_v\ D_t)^{0.50}\left(\frac{G}{\mu a_v}\right)^{0.69}$	10–1000	No correction for h_w
50		Graph in paper of N_{Pe}' vs. N_{Re}'	5–2400	Diffusion of methylene blue in water

[a] a_p = surface of a particle, a_v = surface/volume in the bed. (Kjaer, *Measurement and Calculation of Temperature and Conversion in Fixed-Bed Catalytic Converters,* Haldor Topsoe, Copenhagen, 1958).

of this design with steam generation or feed gas as means of cooling are used for the catalytic oxidation of SO_2.

e. Multibed unit with interstage injection of temperature controlled process fluid or inert fluid for temperature control of the process. In the synthesis of cumene from propylene and benzene in the presence of supported phosphoric acid catalyst, interstage injection of cold process gas and water is used for temperature control and maintenance of catalyst activity (Figure 13.18(g)).

f. Autothermal multitubular unit with heat interchange between feed on the shell side and reacting gas in the packed tubes and between feed and reacted gas in an external or built-in exchanger. Many complex variations of this design have been or are being used for ammonia synthesis.

g. Multibed units in individual shells with interstage heat transfer. From three to seven stages are adopted by different processes for the catalytic reforming of naphthas to gasoline.

FLUIDIZED BED CATALYSIS

Such processes may be conducted to take advantage of the substantial degree of uniformity of temperature and composition and high rates of heat transfer to embedded surfaces. Orthophthalic

TABLE 17.16. Data for the Overall Heat Transfer Coefficient, *u* (kcal/m^{2}h°C), in Packed Beds

Authors	Method of Measurement	Heating or Cooling of Gas	Gas	Particles		
				Material	Shape	Diameter D_p, mm
Campbell and Huntington (1952)	Packed bed heat exchanger	H, C	Air, natural gas (82% CH_4)	Glass, alumina, aluminum	Spheres, cylinders	5–25
Chu and Storrow (1952)	,,	H	Air	Glass, steel, lead, Socony-Vacuum catalyst beads	Spheres	1–6
Colburn (1931)	,,	H	Air	Porcelain, zinc, etc.	Spheres, granules	5–25
Kling (1938b)	,,	H	Air	Steel, glass	Spheres	3–4
Leva (1947)	,,	H	Air, CO_2	Glass, clay, porcelain	Spheres	3–13
Leva and Grummer (1948)	,,	H	Air	Glass, clay, metals, etc.	Spheres, cylinders, granules, etc.	2–25
Leva, Weintraub, Grummer, and Clark (1948)	,,	C	Air, CO_2	Glass, porcelain	Spheres	3–13
Leva (1950)	,,	H	Air	Glass, clay, porcelain, metal	Spheres, rings, cylinders	4–18
Maeda (1952)	,,	C	Air	Catalyst	Cylinders	3–10
Maeda and Kawazoe (1953)	,,	C	Air		Granules, rings, saddles	3–25
Verschoor and Schuit (1950)	,,	H	Air, H_2	Lead, glass, etc.	Spheres, cylinders, granules	3–10
Tasker (1946)	Phthalic anhydride synthesis	C	Air	Catalyst on quartz (?)	Granules	1.7–2.0

anhydride is made by oxidation of naphthalene in a fluidized bed of V_2O_5 deposited on silica gel with a size range of 0.1–0.3 mm with a contact time of 10–20 sec at 350–380°C. Heat of reaction is removed by generation of steam in embedded coils. No continuous regeneration of catalyst is needed. Acrylonitrile and ethylene dichloride also are made under conditions without the need for catalyst regeneration.

From the standpoint of daily capacity, the greatest application of fluidized bed catalysis is to the cracking of petroleum fractions into the gasoline range. In this process the catalyst deactivates in a few minutes, so that advantage is taken of the mobility of fluidized catalyst to transport it continuously between reaction and regeneration zones in order to maintain its activity; some catalyst also must be bled off continuously to maintain permanent poisons such as heavy metal deposits at an acceptable level.

Several configurations of reactor and regenerator have been in

TABLE 17.16—*(continued)*

Tube Diameter D_t, mm	Tube Length L, m	Formula (D_p and D_t in m)	Range of N_{Re}' or other variable	Remarks
50–150		$\frac{U}{GC_p} = 0.76\, e^{-0.0225\, a_v D_t} \left(\frac{G}{\mu a_v}\right)^{-0.42}$	30–1000	U refers to tube axis temperature
25	0.3–1.2	$\frac{UD_t}{k_g} = 0.134 \left(\frac{D_p}{D_t}\right)^{-1.13} \left(\frac{L}{D_t}\right)^{-0.90} \left(\frac{D_pG}{\mu}\right)^{1.17}$ $\frac{UD_t}{k_g} = 15 \left(\frac{D_p}{D_t}\right)^{-0.90} \left(\frac{L}{D_t}\right)^{-1.82} \left(\frac{D_pG}{\mu}\right)^{n}$ $n = 0.55 \left(\frac{L}{D_t}\right)^{0.165}$	$\frac{D_tG}{\mu} < 1600$ $1600 < \frac{D_tG}{\mu} < 3500$	
35–80	0.5–1.2	$U = f\left(\frac{D_p}{D_t}\right) G^{0.83}$	Range of G (kg/m²h): 4500–45000	Function f given in paper. Maximum 0.045 for $\frac{D_p}{D_t} = 0.15$
50	0.3	Graph given in paper	400–3500	
15–52	0.3–0.9	$\frac{UD_t}{k_g} = 0.813\, e^{-6\frac{D_p}{D_t}} \left(\frac{D_pG}{\mu}\right)^{0.90}$	50–3500	
21–52	0.3–0.9		100–4500	Correction factor used for metallic packings
21–52	0.3–0.9	$\frac{UD_t}{k_g} = 3.50\, e^{-4.6\frac{D_p}{D_t}} \left(\frac{D_pG}{\mu}\right)^{0.7}$	150–3000	
15–52	0.3–0.9	$\frac{UD_p}{k_g} = 0.125 \left(\frac{D_pG}{\mu}\right)^{0.75}$	500–12000	Correlation valid for high values of $\frac{D_p}{D_t}$
25–100	0.1–0.4	$\frac{UD_t}{k_g} = 4.9\, e^{-2.2\frac{D_p}{D_t}} \left(\frac{D_pG}{\mu}\right)^{0.60}$	100–600	
52–154		See original paper	30–900	Formula varies with shape of material
30–50	0.2–0.3	$\frac{UD_t}{k_g} = 5.783 \frac{K_{r,0}}{k_g} + 0.085 \left(\frac{D_p}{D_t}\right)^{-0.50} \left(\frac{D_pG}{\mu}\right)^{0.69}$ $+ 0.066 \left(\frac{D_t}{L}\right)\left(\frac{D_p}{D_t}\right)^{-1}\left(\frac{D_pG}{\mu}\right)$	40–4000	$K_{r,0}$ is thermal conductivity of bed with stagnant gas
38	0.4–0.7	$U^{-1} = 0.00123 + 0.54\, G^{-0.83}$	Range of G (kg/m²h): 3000–12000	U refers to tube axis temperature and is corrected for radiation

use, two of which are illustrated in Figures 17.31(a) and (b). Part (a) shows the original arrangement with separate vessels side by side for the two operations. The steam stripper is for removal of occluded oil from the catalyst before it is burned. In other designs the two vessels are in vertical line, often in a single shell with a partition. Part (b) is the most recent design of transfer line cracking which employs highly active zeolite catalysts that are effective at short contact times. The upper vessel is primarily a catalyst disengaging zone. A substantial gradient develops in the transfer line and results in an improvement in product distribution compared with that from mixed reactors such as part (a).

Hundreds of fluidized bed crackers are in operation. The

TABLE 17.17. Heat Transfer Coefficient between Particle and Gas

Authors	Method of Measurement	Heating or Cooling of Gas	Gas	Particles		
				Material	Shape	Diameter, D_p, mm
Furnas (1930a, b, c, 1932)	Unsteady heat transfer	H, C	Air, flue gas	Iron ore, limestone, coke, etc.	Granules	4–70
Löf and Hawley (1948)	"	C	Air	Granitic gravel	Granules	8–34
Saunders and Ford (1940)	"	C	Air	Steel, lead, glass	Spheres	1.6–6.4
Tsukhanova and Shapatina (1943) Chukhanov and Shapatina (1946)	"	C	Air	Steel, chamotte, copper	Spheres, cylinders, granules	2–7
Dayton *et al.* (1952)	Cyclic variations		Air	Glass	Spheres	3–6
Glaser (1955)	"		Air	Stoneware	Raschig rings	5–17
Gamson, Thodos, and Hougen (1943)	Drying		Air	Porous celite	Spheres, cylinders	2–19
Wilke and Hougen (1945)	"		Air	Porous celite	Cylinders	2–19
Taecker and Hougen (1949)	"		Air	Porous clay-kieselguhr	Raschig rings, Berl saddles	6–50
Eichhorn and White (1952)	Dielectrical heating		Air	Plastic	Spheres	0.1–0.7
Satterfield and Resnick (1954)	Decomposition of H_2O_2		Vapors of H_2O and H_2O_2	Catalyst	Spheres	5

vessels are large, as much as 10 m or so in diameter and perhaps twice as high. Such high linear velocities of vapors are maintained that the entire catalyst content of the vessels circulates through the cyclone collectors in an hour or so. Electrical precipitators after the cyclone collectors have been found unnecessary.

Two other fluidized bed petroleum reactors are illustrated as Figures 17.31(c) and (d) and several nonpetroleum applications in Figure 17.32.

GAS–LIQUID REACTIONS WITH SOLID CATALYSTS

The number of commercial processes of this type is substantial. A list of 74 is given by Shah (1979). A briefer list arranged according to the kind of reactor is in Table 17.21. Depending on the circumstances, however, it should be noted that some reactions are conducted industrially in more than one kind of reactor.

Leading characteristics of five main kinds of reactors are described following. Stirred tanks, fixed beds, slurries, and three-phase fluidized beds are used. Catalyst particle sizes are a compromise between pressure drop, ease of separation from the fluids, and ease of fluidization. For particles above about 0.04 mm dia, diffusion of liquid into the pores and, consequently, accessibility of the internal surface of the catalyst have a minor effect on the overall conversion rate, so that catalysts with small specific surfaces, of the order of 1 m^2/g, are adequate with liquid systems. Except in trickle beds the gas phase is the discontinuous one. Except in some operations of bubble towers, the catalyst remains in the vessel, although minor amounts of catalyst entrainment may occur.

TABLE 17.17—(*continued*)

Tube		Formula	Range of	Remarks
Diameter, D_t, mm	Length L, m	(D_p in m)	N_{Re}' or other variable	
150–230	0.5–1	$h_v = A\dfrac{G^{0.7}}{D_p^{\,0.9}}T^{0.3}\,10^{1.68\varepsilon-3.56\varepsilon^2}$	Range of G (kg/m²h): 2300–9200	A given in paper. Corrections for temperature and voids not very reliable
ca. 300 square	0.9	$h_v = 1.82\left(\dfrac{G}{D_p}\right)^{0.7}$	G: 300–1600	
50–200	0.09–0.34	Graphs in paper	G: 2670–5340	Correlated by Löf and Hawley (1948) as: $h_v = 0.152\dfrac{G}{D_p}$
	0.01–0.25	Graphs and $\dfrac{hD_p}{k_g} = 0.24\left(\dfrac{D_pG}{\mu}\right)^{0.83}$	Gas velocity: 0.7–2 m/sec.	
	0.05	Graphs in paper	100–1000	
350		$\dfrac{hD_p}{k_g} = A\left(\dfrac{D_pG}{\mu}\right)^{0.61}$	130–2000	A varies from 0.590 to 0.713. See also Glaser (1938)
		$j_h = \dfrac{h}{GC_p}N_{Pr}^{\,2/3} = 1.064\left(\dfrac{D_pG}{\mu}\right)^{-0.41}$	350–4000	
		$j_h = \dfrac{h}{GC_p}N_{Pr}^{\,2/3} = 1.96\left(\dfrac{D_pG}{\mu}\right)^{-0.51}$	50–350	
		Rings: $j_h = 1.148\left(\dfrac{\sqrt{a_p}\,G}{\mu}\right)^{-0.41}$ Saddles: $j_h = 0.920\left(\dfrac{\sqrt{a_p}\,G}{\mu}\right)^{-0.34}$	100–20000 70–3000	
38 square		Graphs in paper	1–18	
47–75	0.024	$j_h = 0.992\left(\dfrac{D_pG}{\mu}\right)^{-0.34}$	15–160	

[a] a_p = surface of a particle, a_v = surface/volume in the bed.
(Kjaer, *Measurement and Calculation of Temperature and Conversion in Fixed-Bed Catalytic Converters,* Haldor Topsoe, Copenhagen, 1958).

1. Stirred tanks with suspended catalyst are used both in batch and continuously. Hydrogenation of fats or oils with Raney nickel or of caprolactam usually are in batch. Continuous processes include some hydrogenations of fats, some fermentation processes with cellular enzymes and air and the hydrogenation of nitrogen monoxide to hydroxylamine. The gas is distributed with spargers or introduced at the eye of a high-speed impeller in a draft tube. Internal recirculation of the gas also is practiced. The power input depends on the settling tendency of the particles and the required intimacy of gas–liquid mixing. It is greater than in the absence of solids; for example, the solid catalyzed hydrogenation of nitrogen monoxide employs a power input of about $10\,kW/m^3$ (51 HP/1000 gal) compared with 5–10 HP/1000 gal for ordinary liquid–liquid mixing.

2. In ebullated (liquid fluidized) beds the particles are much larger (0.2–1 mm) than in gas fluidization (0–0.1 mm). Little expansion of the bed occurs beyond that at minimum fluidization, so that the bed density is essentially the same as that of the fixed bed. Because substantial internal circulation of the liquid is needed to maintain fluidization, the fluids throughout the reactor are substantially uniform. In the hydrodesulfurization and hydrocracking of petroleum fractions and residua at 100 atm and 400°C, a temperature variation of only 2°C or so obtains in the reactor.

3. Slurry reactors (bubble towers) are fluidized with continuous flow of gas. The particles are smaller (less than 0.1 mm) than in the liquid fluidized systems (0.2–1 mm). In some operations the liquid and solid phases are stationary, but in others they circulate through the vessel. Such equipment has been used in Fischer–Tropsch plants and for hydrogenation of fatty esters to alcohols, furfural to furfuryl alcohol, and of glucose to sorbitol. Hydrogenation of benzene to cyclohexane is done at 50 bar and 220–225°C with Raney nickel of 0.01–0.1 mm dia. The relations between gas velocities, solids

TABLE 17.18. Formulas for the Heat Transfer Coefficient at the Walls of Packed Vessels[a]

Name	Geometry	Formula
1. Beek (1962)	spheres	$Nu = 0.203\,Re^{1/3}\,Pr^{1/3} + 0.220\,Re^{0.8}\,Pr^{0.4}$, $Re < 40$
2. Beek (1962)	cylinders	$Nu = 2.58\,Re^{1/3}\,Pr^{1/3} + 0.094\,Re^{0.8}\,Pr^{0.4}$, $Re < 40$
3. Yagi-Wakao (*Chem. Eng. Eng. Sci.* **5**, 79, 1959)	spheres	$Nu = 0.186\,Re^{0.8}$
4. Hanratty (*Chem. Eng. Sci.* **3**, 209, 1954)	cylinders	$Nu = 0.95\,Re^{0.5}$
5. Hawthorn (*AIChE J* **14**, 69 1968)		$Nu = 0.28\,Re^{0.77}\,Pr^{0.4}$
6. Doraiswamy and Sharma (1984)	spheres	$Nu = 0.17\,Re^{0.79}$, $L/d_t > 50$, $20 < Re < 7600$, $0.05 < d_p/d_t < 0.30$
7. Doraiswamy and Sharma (1984)	cylinders	$Nu = 0.16\,Re^{0.93}$, $L/d_t > 50$, $20 < Re < 800$, $0.03 < d_p/d_t < 0.2$
8. Gnielinski–Martin, Schlünder (1978)		$Nu/(2.5-1.5\varepsilon) = 0.8[2 + F(Re/\varepsilon)^{1/2}(Pr)^{1/3}]$ $F = 0.664\left\{1 + \left[\dfrac{0.0577(Re/\varepsilon)^{0.3}\,Pr^{0.67}}{1 + 2.44(Pr^{2/3} - 1)(Re/\varepsilon)^{-0.1}}\right]^2\right\}^{1/2}$
9. Gnielinski–Martin–Schlünder (1978)		$\ln\dfrac{Nu}{2.5-1.5\varepsilon} \simeq 0.750 + 0.1061\ln(Re/\varepsilon) + 0.0281[\ln(Re/\varepsilon)]^2$

[a] Definitions: $Nu = h_w d_p/k_f$, $Pr = (C_p\mu/k)_f$, h_w = wall coefficient, d_p = particle diameter $= 6V_p/A_p$, k_f = fluid molecular conductivity, ε = porosity, $Re = d_pG/\mu$, G = superficial mass velocity per unit cross section.

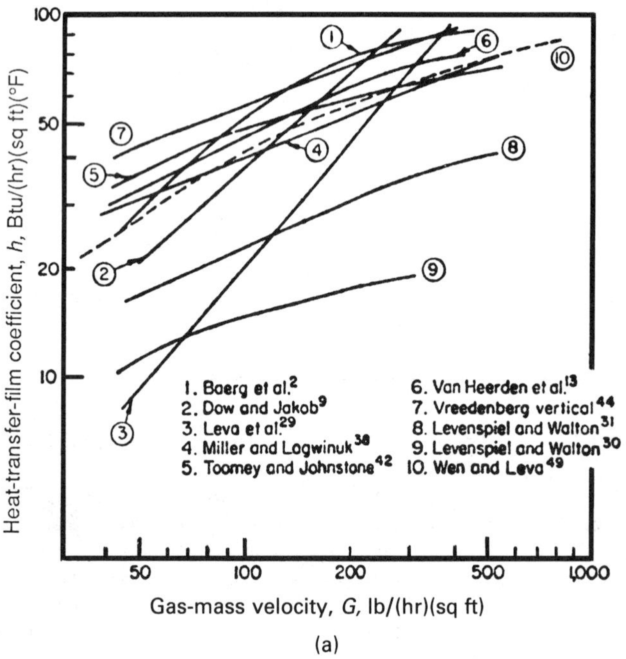

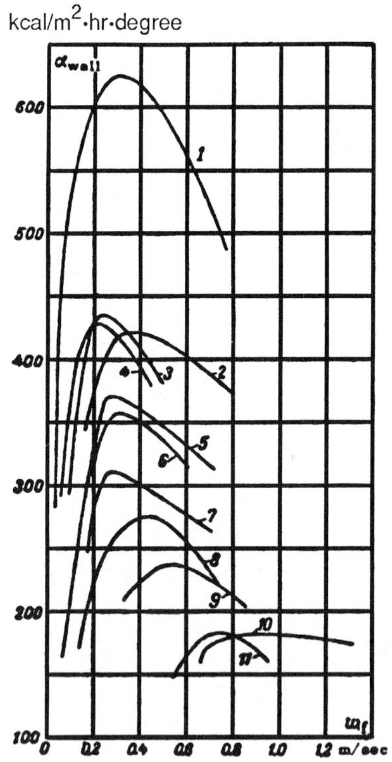

Figure 17.37. Some measured and predicted values of heat transfer coefficients in fluidized beds. 1 Btu/hr(sqft)(°F) = 4.88 kcal/(hr)(m^2)(°C) = 5.678 W/(m^2)(°C). (a) Comparison of correlations for heat transfer from silica sand with particle size 0.15 mm dia fluidized in air. Conditions are identified in Table 17.19 (*Leva, 1959*). (b) Wall heat transfer coefficients as function of the superficial fluid velocity, data of Varygin and Martyushin. Particle sizes in microns: (1) ferrosilicon, $d = 82.5$; (2) hematite, $d = 173$; (3) carborundum, $d = 137$; (4) quartz sand, $d = 140$; (5) quartz sand, $d = 198$; (6) quartz sand, $d = 216$; (7) quartz sand, $d = 428$; (8) quartz sand, $d = 515$; (9) quartz sand, $d = 650$; (10) quartz sand, $d = 1110$; (11) glass spheres, $d = 1160$. (*Zabrodsky et al., 1976, Fig. 10.17*). (c) Effect of air velocity and particle physical properties on heat transfer between a fluidized bed and a submerged coil. Mean particle diameter 0.38 mm: (I) BAV catalyst; (II) iron-chromium catalyst; (III) silica gel; (IV) quartz; (V) marble (*Zabrodsky et al., 1976, Fig. 10.20*).

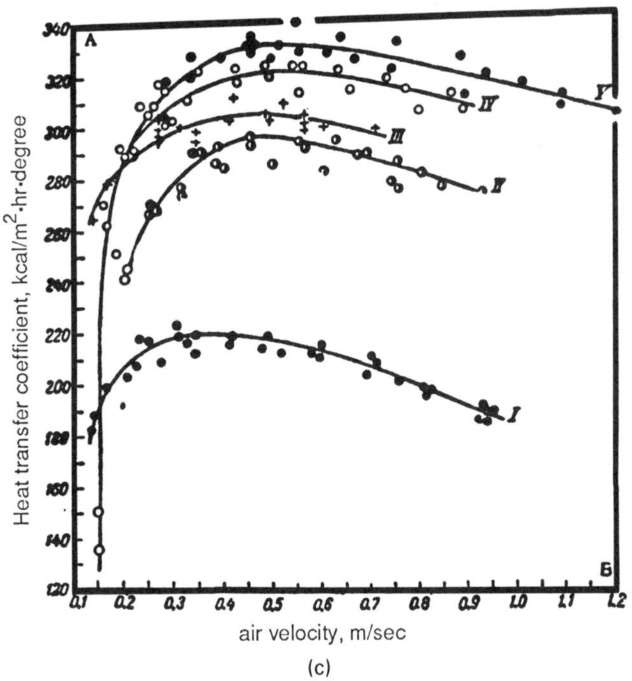

Figure 17.37—(*continued*)

concentrations, bubble sizes, and rates of heat transfer are extensively documented in the literature.

4. In trickle bed reactors the gas and liquid both flow downward through a fixed bed of catalyst. The gas phase is continuous, and the liquid also is continuous as a film on the particles. Provided that the initial distribution is good, liquid distribution remains substantially uniform at rates of 10–30 m^3/m^2 hr superficially, but channelling and hot spots may develop at lower rates. Redistributors sometimes are used. The many correlations that have been developed for packed bed mass transfer are applicable to trickle bed operation. Commercial reactors are 1–4 m dia and 10–30 m long. Hydrocracking and hydrodesulfurization of petroleum and hydration of olefins are commonly practiced in trickle beds at superficial liquid velocities of 3–90 m/hr.

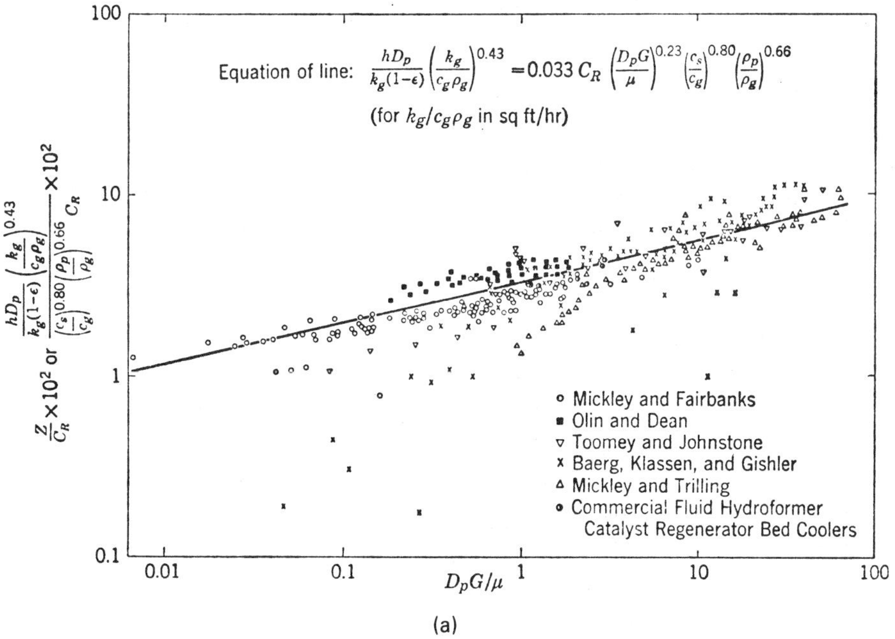

Figure 17.38. Heat transfer coefficient in fluidized beds [*Wender and Cooper,* AIChE J. **4,** *15* (*1958*)]. (a) Heat transfer at immersed vertical tubes. All groups are dimensionless except $k_g/C_g\rho_g$, which is sqft/hr. The constant C_R is given in terms of the fractional distance from the center of the vessel by $C_R = 1 + 3.175(r/R) - 3.188(r/R)^2$. (b) Heat transfer at the wall of a vessel. L_H is bed depth, D_T is vessel diameter.

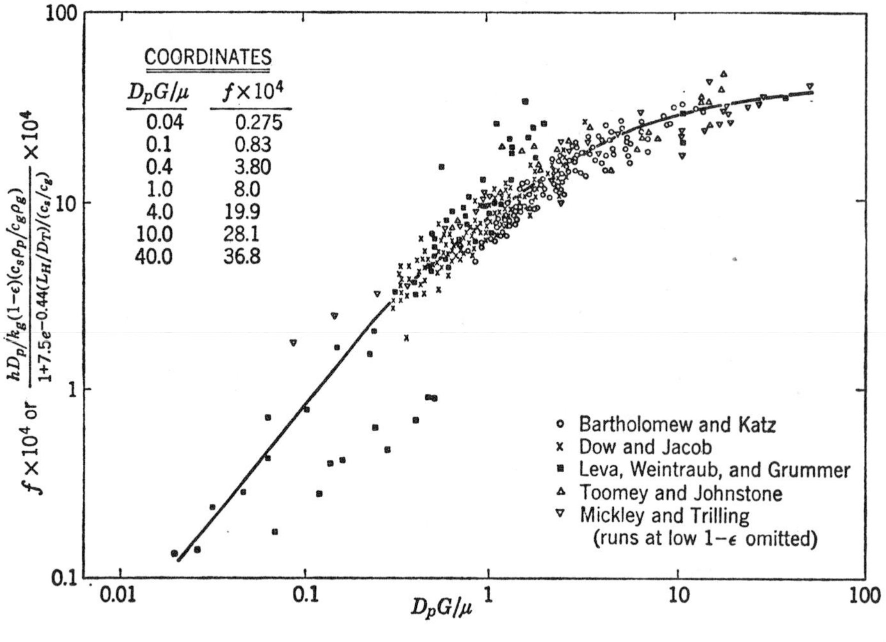

(b)

Figure 17.38—*(continued)*

TABLE 17.19. Experimental Investigations of Heat Transfer in Fluidized Beds[a]

Reference	Solids	Voidage range	Absolute density, lb per cu ft	Particle-size range, ft	Type of apparatus and operation	Vessel diam., in.	Height of heat-transfer area, in.	Bed height, in.	Fluids	Flow range, lb/(hr) (sq ft)	Temp, °F
1	Sands, graphite, soft brick	Dense phase	83–166	8–14 mesh to 36–72 mesh	Steam-jacketed column	1.5	14.5	....	Air	150–1,200	
2	Iron powder, sands, glass beads catalyst	38.8–75	119–434	0.000198–0.00288	Central electric heat	1.25 in. 5.5	4	10	Air	1.85–605	23.5–65.0
3	Sand, aluminum, calcium carbonate	54–95	160–167	0.000277–0.000822	Wall electric heat	4.0	30	30	Air	96–935	300–450
4	Glass beads	Dilute phase	....	0.00023–0.0036	Electric heat from outside	1.959	12	....	Air	95–3,780	
6	Sand, aluminum, graphite, copper catalyst	Dense phase	24.6–27.2	0.00079–0.0126	Central cooling	2.31	Immersed cooling coil	....	Air	40–100	87–145
9	Aerocat, coke, iron powder	52–69	121–466	0.000363–0.000560	Wall steam heating	2.06 and 3.07	23 and 26.5	2–13	Air	50–300	200–220
13	Carborundum, iron oxide, coke, lead fly ash, alloy	Dense phase	37.5–694	0.000262–0.00213	Wall water cooling	3.4	4	16	Air, CH_4, CO_2, town gas, H_2 and N_2 mixtures	44–779	Approx 10–30°C
17	Glass beads	Dense phase	154	0.00010–0.0011	Internal heating by electric wire	3.0	....	....	Air, CO_2, Freon-12, He, H_2, H_2 and N_2 mixtures		
26, 29	Sand, iron catalyst, silica gel	35–75	80–500	0.000129–0.00149	Wall steam heat	2.0 and 4.0	25 and 26	12–25	Air, CO_2, He, N_2	1.47–1,095	258–413
30	Coal	Dense phase	....	0.000432–0.00386	Wall cooling (air)	4.0	24	....	Air	50–1,100	
31	Glass beads catalyst, coal	41.7–86.2	63.6–180	0.000250–0.0142	Wall electric heating	4.0	3 sections, 2, 5, and 2 in.	10–30	Air	79–4,350	
36	Glass beads, microspheres	Dense phase	138–153	0.00022–0.00027	Small electric heater probe	....	....	Approx 18–20	He, air, CH_4, argon	10–150	
37	Glass beads	Dilute phase	151–177	0.000133–0.00149	Internal and external heating	2.875 and 1.00	....	....	Air	2,700	500
38	Silicon carbide, Al_2O_3, silica gel	Dense phase	70–243	0.000287–0.000817	Center wall cooling	2.0	22	....	Air, He, CO_2	6.4–200	120–414
42	Glass beads	Dense phase	167–179	0.000179–0.00278	Wall water cooling	4.73	7 sections, each 5 in. high	13.2–24.6	Air	23.7–1,542	
43, 44	Sand, iron ore	Dense phase	165–330	0.000766–0.00197	Internal cooling	1.35 in. 22.2	....	47, 68	Air	65–300	Approx 100–400
50	Carborundum sand, aluminum powder, lead powder, glass beads	Dense phase	160–700	0.00020–0.010	Internal heating by small cylindrical element	3.94	....	....	Air, CO_2, H_2		

[a] Another list of 29 sources is given by Zabrodsky (1966). (Leva, 1959).

TABLE 17.20. Heat Transfer Coefficients in Fluidized Beds[a]

1. *At vertical tubes* [Vreedenburg, *Chem. Eng. Sci.* **11**, 274 (1960)]:

$$[h(D-d_t)/k_g](d_t/D)^{1/3}(k_g/C_s\mu_g)^{1/2} = C[u(D-d_t)\rho_s/\mu_g]^n$$

Conditions:

$$\rho_s d_s u/\mu_g < 2050,$$
$$\begin{cases} \rho_s u(D-d_t)/\mu_g < 2.4\times 10^5, & C = 2.7\times 10^{-16}, \quad n = 3.4 \\ \rho_s u(D-d_t)/\mu_g > 2.4\times 10^5, & C = 2.2, \quad n = 0.44 \end{cases}$$
$$[h(D-d_t)/k_g](d_t d_s k_g/[(D-d_t)C_s\mu_g]^{1/3} = C[u(D-d_t)g^{0.5}d_s^{1.5}]^n$$

Conditions:

$$\rho_s d_s u/\mu_g > 2550,$$
$$\begin{cases} u(D-d_t)g^{0.5}d_s^{1.5} < 1070, & C = 1.05\times 10^{-4}, \quad n = 2.0 \\ u(D-d_t)g^{0.5}d_s^{1.5} > 1070, & C = 240, \quad n = 0.8 \end{cases}$$

For off-center locations, the factor C is multiplied by C_R which is given in terms of the fractional distance from the center by

$$C_R = 1 + 3.175(r/R) - 3.188(r/R)^2$$

2. *At vertical tubes,* see the correlation of Wender and Cooper on Figure 17.17(a)

3. *At horizontal tubes* (Vreedenburg, *loc. cit.*; Andeen and Glicksman, ASME Paper 76-HT-67, 1976):

$$(hd_t/k_g)(k_g/C_s\mu_g)^{0.3} = 0.66[\rho_s d_t u(1-\varepsilon)/\mu_g\varepsilon]^{0.44}, \quad \rho_s d_s u/\mu_g < 2500$$
$$(hd_t/k_g)(k_g/C_s\mu_g)^{0.3} = 900(1-\varepsilon)(d_t u\mu_g/d_s^3\rho_s g)^{0.326}, \quad \rho_s d_s u/\mu_g > 2550$$

4. *At vessel walls,* see Figure 17.17(b) for the correlation of Wender and Cooper.

[a] Notation: Subscript s for solid, subscript g for gas, d_t = tube diameter, D = vessel diameter, g = acceleration of gravity.

5. Upflow fixed beds. The liquid phase is continuous and the gas phase dispersed. This mode of operation has the advantages of better mixing, higher rates of mass and heat transfer, better distribution of liquid flow across the cross section, and better scouring of deactivating deposits from the surface of the catalyst. The disadvantages relative to trickle beds are higher pressure drop, the possibility of occurrence of flooding, and the need for mechanical restraint to prevent fluidization and entrainment of the catalyst. The most prominent example of upflow operation is the SYNTHOIL coal liquefaction process, but this mode of operation is competitive in other cases with the trickle bed, depending on the balance of advantages and disadvantages in particular situations.

TABLE 17.21. Examples of Industrial Gas–Liquid–Solid Reaction Processes

A. Fixed-bed reactors
1. Trickle beds (downflow)
   a. Catalytic hydrodesulfurization, hydrocracking and hydrogenation
   b. Butynediol from acetylene and aqueous formaldehyde
   c. Sorbitol from glycerol
   d. Oxidation of SO_2 in the presence of activated carbon
   e. Hydrogenation of aniline to cyclohexylaniline
2. Upflow (bubble) reactors
   a. Coal liquefaction by SYNTHOIL process
   b. Fischer–Tropsch process
   c. Selective hydrogenation of phenylacetylene and styrene

B. Suspended solid reactors
1. Stirred tanks
   a. Catalytic hydrogenation of fats and oils
   b. Hydrogenation of acetone and nitrocaprolactam
   c. Aerated fermentation with cellular enzymes
   d. Reaction between methanol and hydrogen chloride with $ZnCl_2$ catalyst
2. Slurry towers
   a. Fischer–Tropsch process
   b. Hydrogenation of methyl styrene and carboxy acids
   c. Oxidation and hydration of olefins
   d. Polymerization of ethylene
   e. Calcium hydrophosphite from white phosphorous and lime slurry
   f. Lime/limestone process for removal of SO_2 from flue gases
3. Fluidized bed of catalyst
   a. Calcium acid sulfite from $CaCO_3 + SO_2 + H_2O$
   b. Coal liquefaction
   c. Hydrocracking and hydrodesulfurization

REFERENCES

General

1. J. Beek, Design of packed catalytic reactors, *Adv. Chem. Eng.* **3**, 203–271 (1962).
2. L.K. Doraiswamy and M.M. Sharma, *Heterogeneous Reactions: Analysis, Examples and Reactor Design,* Wiley, New York, 1984, 2 vols.
3. E.B. Nauman and B.A. Buffham, *Mixing in Continuous Flow Systems,* Wiley, New York, 1983.
4. H.F. Rase, *Chemical Reactor Design of Process Plants,* Wiley, New York, 1977, 2 vols.
5. Y.T. Shah, *Gas–Liquid–Solid Reactor Design,* McGraw-Hill, New York, 1979.
6. M.O. Tarhan, *Catalytic Reactor Design,* McGraw-Hill, New York, 1983.
7. S.M. Walas, *Reaction Kinetics for Chemical Engineers,* McGraw-Hill, New York, 1959.
8. S.M. Walas, Chemical reactor data, *Chem. Eng.,* 79–83 (14 Oct. 1985).
9. K.R. Westerterp, W.P.M. Van Swaaij, and A.A.C.M. Beenackers, *Chemical Reactor Design and Operation,* Wiley, New York, 1984.

Types of Reactors

1. Ullmann, Reaktionsapparate, in *Encyclopedia of Chemical Technology,* Verlag Chemie, Weinheim, Germany, 1973, Vol. 3, pp. 320–518.

Catalysts and Chemical Processes

1. J.E. Bailey and D.F. Ollis, *Biochemical Engineering Fundamentals,* McGraw-Hill, New York, 1986.
2. C.H. Bamford and C.F.H. Tipper (Eds.), *Complex Catalytic Processes,* Comprehensive Chemical Kinetics Vol. 20, Elsevier, New York, 1978.
3. B.C. Gates, J.R. Katzer, and G.C.A. Schuit, *Chemistry of Catalytic Processes,* McGraw-Hill, New York, 1979.
4. B.E. Leach (Ed.), *Applied Industrial Catalysis,* Academic, New York, 1983–1985, 3 vols.
5. H. Pines, *Chemistry of Catalytic Conversions of Hydrocarbons,* Academic, New York, 1981.
6. V.A. Roiter (Ed.), *Handbook of Catalytic Properties of Substances* (in Russian), Academy of Sciences, Ukrainian SSR, Kiev, USSR, 1968–date, 4 vols. to date.
7. C.N. Satterfield, *Mass Transfer in Heterogeneous Catalysis,* MIT Press, Cambridge, MA, 1970.
8. C.N. Satterfield, *Heterogeneous Catalysis in Practice,* McGraw-Hill, New York, 1980.
9. S. Strelzoff, *Technology and Manufacture of Ammonia,* Wiley, New York, 1981.
10. C.L. Thomas, *Catalytic Processes and Proven Catalysts,* Academic, New York, 1970.
11. D.L. Trimm, *Design of Industrial Catalysts,* Elsevier, New York, 1980.
12. *Ullmann's Encyclopedia of Chemical Technology:* Biotechnologie, Vol. 8, pp. 497–526, 1972; Enzyme, Vol. 10, pp. 471–561, 1975, Verlag Chemie Weinheim, Germany,
13. C.A. Vancini, *Synthesis of Ammonia,* Macmillan, New York, 1971.

Heat Transfer in Reactors

1. N.P. Cheremisinoff and P.N. Cheremisinoff, *Hydrodynamics of Gas–Solids Fluidization,* Gulf, Houston, 1984, abstract section.
2. M.B. Glaser and G. Thodos, Heat and momentum transfer in flow of gases through packed beds, *AIChE J.* **4,** 63–74 (1958).
3. J.R. Grace, Fluidized bed heat transfer, in *Handbook of Multiphase Systems* (Hetsroni, Ed.), Hemisphere, New York, 1982.
4. L.A. Korotjanskaja et al. cited by Doraiswamy and Sharma, *loc. cit.,* 1984, p. 323.
5. M. Leva, *Fluidization,* McGraw-Hill, New York, 1959.
6. J.Y. Oldshue, *Fluid Mixing Technology,* McGraw-Hill, New York, 1983.
7. H.F. Rase, *Chemical Reactor Design for Process Plants,* Wiley, New York, 1977, 2 vols.
8. E.U. Schlünder, Transport phenomena in packed bed reactors, in *Chemical Reactor Engineering Reviews—Houston,* ACS Symposium 72, American Chemical Society, Washington, DC, 1978.
9. H.A. Vreedenberg, Heat transfer between a fluidized bed and a horizontal tube, *Chem. Eng. Sci.* **9,** 52–60 (1958); Vertical tubes, *Chem. Eng. Sci.* **11,** 274–285 (1960).
10. L. Wender and G.T. Cooper, Heat transfer between fluidized beds and bounding surfaces—correlation of data, *AIChE J.* **4,** 15–23 (1958).
11. S.S. Zabrodsky, *Hydrodynamics and Heat Transfer in Fluidized Beds,* MIT Press, Cambridge, MA, 1966.
12. S.S. Zabrodsky, N.V. Antonishin, and A.L. Parnas, On fluidized bed to surface heat transfer, *Can. J. Chem. Eng.* **54,** 52–58 (1976).

18

PROCESS VESSELS

Vessels in chemical processing service are of two kinds: those substantially without internals and those with internals. The main functions of the first kinds, called drums or tanks, are intermediate storage or surge of a process stream for a limited or extended period or to provide a phase separation by settling. Their sizes may be established by definite process calculations or by general rules based on experience. The second category comprises the shells of equipment such as heat exchangers, reactors, mixers, fractionators, and other equipment whose housing can be designed and constructed largely independently of whatever internals are necessary. Their major dimensions are established by process requirements described in other chapters, but considerations of adequate strength of vessels at operating pressures and temperatures will be treated in this chapter.

The distinction between drums and tanks is that of size and is not sharp. Usually they are cylindrical vessels with flat or curved ends, depending on the pressure, and either horizontal or vertical. In a continuous plant, drums have a holdup of a few minutes. They are located between major equipmenr or supply feed or accumulate product. Surge drums between equipment provide a measure of stability in that fluctuations are not transmitted freely along a chain, including those fluctuations that are characteristic of control instruments of normal sensitivity. For example, reflux drums provide surge between a condenser and its tower and downstream equipment; a drum ahead of a compressor will ensure freedom from liquid entrainment and one ahead of a fired heater will protect the tubes from running dry; a drum following a reciprocating compressor will smooth out pressure surges, etc. Tanks are larger vessels, of several hours holdup usually. For instance, the feed tank to a batch distillation may hold a day's supply, and rundown tanks between equipment may provide several hours holdup as protection of the main storage from possible off-specification product and opportunity for local repair and servicing without disrupting the entire process.

Storage tanks are regarded as outside the process battery limits, on tank farms. Their sizes are measured in units of the capacities of connecting transportation equipment: 34,500 gal tank cars, 8000 gal tank trucks, etc., usually at least 1.5 times these sizes. Time variations in the supply of raw materials and the demand for the products influence the sizes and numbers of storage tanks.

Liquid storage tanks are provided with a certain amount of vapor space or freeboard, commonly 15% below 500 gal and 10% above 500 gal. Common erection practices for liquid storage tanks are:

a. *For less than 1000 gal, use vertical tanks mounted on legs.*
b. *Between 1000 and 10,000 gal, use horizontal tanks mounted on concrete foundation.*
c. *Beyond 10,000 gal, use vertical tanks mounted on concrete foundations.*

Liquids with high vapor pressures, liquefied gases, and gases at high pressure are stored in elongated horizontal vessels, less often in spherical ones. Gases are stored at substantially atmospheric pressure in gas holders with floating roofs that are sealed with liquid in a double wall. Liquefied gases are maintained at subatmospheric temperatures with external refrigeration or autorefrigeration whereby evolved vapors are compressed, condensed, cooled, and returned to storage.

Liquids stored at near atmospheric pressure are subject to breathing losses: As the tank cools during the night air is drawn in, then vaporization occurs to saturation, and the vapor mixture is expelled as the tank warms up during the day. Volatile liquids such as gasoline consequently suffer a material loss and also a change in composition because of the selective loss of lighter constituents.

In order to minimize such effects, several provisions are made, for example:

1. *A floating roof is a pad which floats on the surface of the stored liquid with a diameter of about a foot less than that of the tank. The annular space between the float and the shell may be sealed by one of several available methods.*
2. *An expansion roof allows thermal expansion of the vapor space. It rides with the changing vapor and is sealed with liquid in a double wall.*
3. *A bag of vapor resistant fabric is allowed to expand into a housing of much smaller diameter than that of the storage tank. This is a lower cost construction than either of the other two.*

Weather resistant solids such as coal or sulfur or ores are stored in uncovered piles from which they are retrieved with power shovels and conveyors. Other solids are stored in silos. For short-time storage for process use, solids are stored in bins that are usually of rectangular cross section with cone bottoms and hooked up to process with conveyors. All aspects of the design of such equipment are covered in books by Reisner and Rothe (1971) and Stepanoff (1969).

18.1. DRUMS

Liquid drums usually are placed horizontal and gas–liquid separators vertical, although reflux drums with gas as an overhead product commonly are horizontal. The length to diameter ratio is in the range 2.5–5.0, the smaller diameters at higher pressures and for liquid–liquid settling. A rough dependence on pressure is

P (psig)	0–250	251–500	501+
L/D	3	4	5

The volume of a drum is related to the flow rate through it, but it depends also on the kinds of controls and on how harmful would be the consequences of downstream equipment running dry. Conventionally, the volume often is expressed in terms of the number of minutes of flow on a half-full basis. For many services, 5–10 min half-full is adequate but two notable exceptions are:

1. Fired heater feed surge drum for which the size is 10–30 min half-full.
2. Compressor feed liquid knockout drum which is made large enough to hold 10–20 min of liquid flow, with a minimum volume of 10 min worth of gas flow rate.

Other major services require more detailed consideration, as follows.

18.2. FRACTIONATOR REFLUX DRUMS

Commonly their orientation is horizontal. When a small amount of a second liquid phase (for example, water in an immiscible organic) is present, it is collected in and drawn off a pot at the bottom of the drum. The diameter of the pot is sized on a linear velocity of 0.5 ft/sec, is a minimum of 16 in dia in drums of 4–8 ft dia, and 24 in. in larger sizes. The minimum vapor space above the high level is 20% of the drum diameter or 10 in (Sigales, 1975).

A method of sizing reflux drums proposed by Watkins (1967) is based on several factors itemized in Table 18.1. A factor F_3 is applied to the net overhead product going downstream, then instrument factors F_1 and labor factors F_2 which are added together and applied to the weighted overhead stream, and finally a factor F_4 is applied, which depends on the kind and location of level indicators. When L is the reflux flow rate and D the overhead net product rate, both in gpm, the volume of the drum (gal) is given by

$$V_d = 2F_4(F_1 + F_2)(L + F_3 D) \text{ gal, full.} \quad (18.1)$$

For example, with $L = 400$ gpm and $D = 200$ gpm, at average conditions $F_1 = 1$, $F_2 = 1.5$, $F_3 = 3$, $F_4 = 1.5$, and

$$V_d = 2(1.5)(1 + 1.5)(400 + 3(200)) = 7500 \text{ gal, full}$$

TABLE 18.1 Factors for Sizing Reflux Accumulators

a. Factors F_1 and F_2 on the Reflux Flow Rate

	Instrument Factor F_1		Labor Factor F_2		
Operation	w/ Alarm	w/o Alarm	Good	Fair	Poor
FRC	$\frac{1}{2}$	1	1	1.5	2
LRC	1	$1\frac{1}{2}$	1	1.5	2
TRC	$1\frac{1}{2}$	2	1	1.5	2

b. Factor F_3 on the Net Overhead Product Flow to External Equipment

Operating Characteristics	F_3
Under good control	2.0
Under fair control	3.0
Under poor control	4.0
Feed to or from storage	1.25

c. Factor F_4 for Level Control

	F_4
Board-mounted level recorder	1.0
Level indicator on board	1.5
Gage glass at equipment only	2.0

(Watkins, 1967).

or, 6.25 min half-full. With the best of everything, $F_1 = 0.5$, $F_2 = 1$, $F_3 = 2$, $F_4 = 1$, and

$$V_d = 2(0.5 + 1)(400 + 2(200)) = 2400 \text{ gal, full}$$

or 2.0 min half-full. The sizes figured this way are overruled when the destination of the net product is to a fired heater or a compressor; then the numbers cited in Section 18.1 are applicable.

Although this method seems to take in a number of pertinent factors, it is not rigorous and may not dissuade some practitioners from continuing to size these drums on the basis of 5 min half-full.

18.3. LIQUID–LIQUID SEPARATORS

Vessels for the separation of two immiscible liquids usually are made horizontal and operate full, although some low rate operations are handled conveniently in vertical vessels with an overflow weir for the lighter phase. The latter mode also is used for particularly large flows at near atmospheric pressures, as in the mixer–settler equipment of Figure 3.19. With the usual L/D ratio of three or more, the travel distance of droplets to the separated phase is appreciably shorter in horizontal vessels.

Since the rise or fall of liquid droplets is interfered with by lateral flow of the liquid, the diameter of the drum should be made large enough to minimize this adverse effect. A rule based on the Reynolds number of the phase through which the movement of the liquid drops occurs is proposed by Hooper and Jacobs (1979). The Reynolds number is $D_h u\rho/\mu$, where D_h is the hydraulic diameter and u is the linear velocity of the continuous phase. The rules are:

N_{Re}	Effect
Less than 5000	little problem
5000–20,000	some hindrance
20,000–50,000	major problem may exist
Above 50,000	expect poor separation

The jet effect of an inlet nozzle also may interfere with the phase separation. Ideally the liquid should be introduced uniformly over the cross section, but a baffle at the inlet nozzle may reduce such a disturbance adequately. More elaborate feed diffusers sometimes may be worthwhile. Figure 18.1 shows a perforated baffle.

Fall or rise of droplets of one liquid in another is represented closely by Stokes law,

$$u = g_c(\rho_2 - \rho_1)d^2/18\mu. \quad (18.2)$$

In common units,

$$u = 9.97(10^6)(\rho_2 - \rho_1)d^2/\mu^2, \quad \text{ft/min}, \quad (18.3)$$

where the ρ_i are specific gravities, d is the droplet diameter (ft), and μ is the viscosity of the continuous phase (cP).

The key property is the droplet diameter, of which many studies have been made under a variety of conditions. In agitated vessels, experience shows that the minimum droplet diameters are in the range of 500–5000 μm. In turbulent pipeline flow, Middleman (1974) found that very few droplets were smaller than 500 μm. Accordingly, for separator design a conservative value is 150 μm, which also has been taken as a standard in the *API Manual on Disposal of Refinery Wastes* (1969). With this diameter,

$$u = 2.415(\rho_2 - \rho_1)/\mu, \quad \text{ft/min}. \quad (18.4)$$

Which phase is the dispersed one can be identified with the

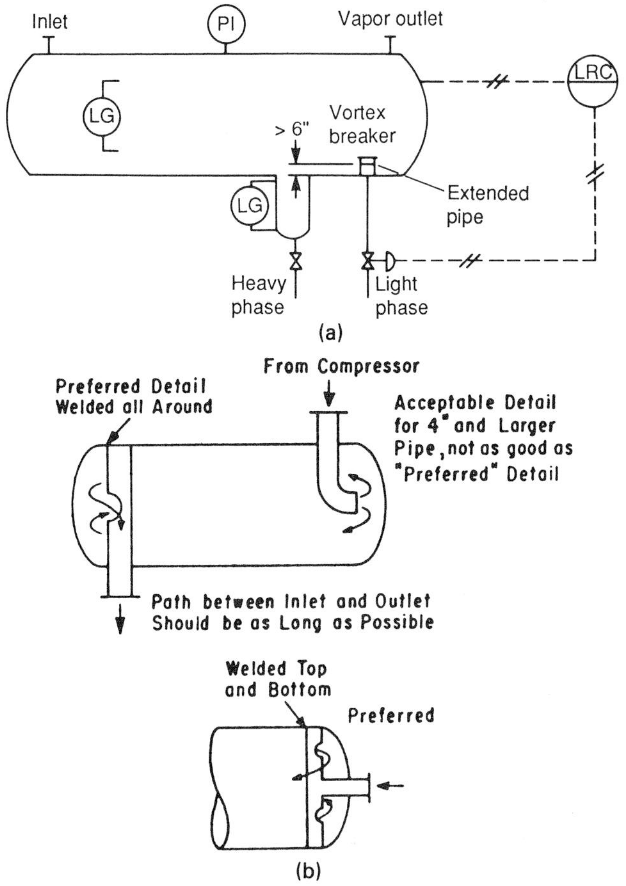

Figure 18.1. Drums for distillation tower reflux and for reciprocating compressor surge. (a) A reflux drum with a pot for accumulation and removal of a heavy phase. The main liquid is removed on level control through a vortex breaker. When the pot is large enough, it can accommodate an interface control for automatic drainage; otherwise the drain valve is hand set and monitored by an operator. (b) Arrangement of a surge drum for eliminating the high frequency response of a reciprocating compressor. Details are given by Ludwig (Applied Process Design for Chemical and Petrochemical Plants, *Gulf, Houston, 1983, Vol. 3*).

factor

$$\psi = \frac{Q_L}{Q_H}\left(\frac{\rho_L u_H}{\rho_H u_L}\right)r^{0.3} \quad (18.5)$$

with the statements of this table (Selker and Schleicher, 1965):

ψ	Result
<0.3	light phase always dispersed
0.3–0.5	light phase probably dispersed
0.5–2.0	phase inversion probable, design for worst case
2.0–3.3	heavy phase probably dispersed
>3.3	heavy phase always dispersed

These relations are utilized in Example 18.1 and the resulting design is represented on Figure 18.2.

COALESCENCE

The rate of separation of liquid phases can be enhanced by shortening the path through which the droplets need rise or fall or by increasing their diameters. Both effects are achieved by forcing the flow between parallel flat or crimped plates or through tower packing or through a mass of packed fibers. The materials should be wetted by the disperse phase and preferably rough. Fine droplets will impinge on the surfaces and will grow by accretion of other droplets. The separator in such cases will consist of a coalescing section and an open section where the now enlarged droplets can separate freely. Figure 18.3 is of a separator equipped with a coalescer that is especially suited to the removal of relatively small quantities of dispersed liquid. Cartridge-type coalescers are described by Redmon (1963). Packed separators have been studied by Davies, Jeffrys, and Azfal (1972) and the subject is reviewed by Laddha and Degaleesan (1983). Coalescence also can be induced electrically, a process that is used widely for the precipitation of brine from crude oils. Proprietary equipment is available for this purpose. The subject is discussed by Waterman (1965) and in detail by Fronczak (1983).

OTHER METHODS

Very fine dispersions can be separated effectively with disk-type centrifuges. Commercial units have capacities of 5–500 gpm and are capable of removing water from hydrocarbons down to the ppm range. A mild centrifugal action is achieved in hydrocyclones. They have been studied for liquid–liquid separation by Sheng, Welker, and Sliepcevich (1974), but their effectiveness was found only modest. The use of hydrocyclones primarily for the recovery of solid particles from liquids is described in the book of Bradley (1965). A symposium on coalescence has papers by Belk (1965), Jordan (1965), Landis (1965), and Waterman (1965).

18.4. GAS–LIQUID SEPARATORS

Droplets of liquid are removed from a gas phase by three chief methods:

1. Settling out under the influence of gravity.
2. Settling out under centrifugal action.
3. Impingement and coalescence on solid surfaces followed by settling.

Available methods for the design of liquid separators are arbitrary in some respects but can be made safe economically. Figure 18.4 illustrates some of these methods.

DROPLET SIZES

The period of time needed for settling out depends on the size distribution of droplets and the required completeness of removal. Under most conditions the droplet diameter is an elusive quantity. A few observations are mentioned by York (1983). Garner et al. (1954) found 95% of evaporator entrainment to be smaller than 18–25 μm. From spray nozzles the droplets are 90 wt % greater than 20 μm. Spray disks made droplet diameters in the range 100–1000 μm. Sprays resulting from splashing and pickup by vapors off condensed liquid films are as large as 5000 μm. Some mists are very fine, however; those in sulfuric acid plants are mostly less than 10 μm, and in some equipment 50 wt % are less than 1 μm (Duros and Kennedy, 1978). On the whole, sprays in process equipment usually are greater than 20 μm, mostly greater than 10 μm.

The amount of entrainment has been studied mostly in distillation equipment. Figure 18.5 summarizes some of these data, and they are applied in Example 18.2. Equation 18.11 incorporates entrainment data indirectly.

A common belief is that 95% of entrainment can be removed in economically sized gravity separators, in excess of 99% with wire

EXAMPLE 18.1
Separation of Oil and Water
Find the dimensions of a drum for the separation of oil and water at these conditions:

Oil at 180 cfh, sp gr = 0.90, viscosity = 10 cP.
water at 640 cfh, sp gr = 1.00, viscosity = 0.7 cP.

Take a droplet size to be 150 μm (0.0005 ft) and that the holdup in the tank is in the same proportions as in the feed. The geometry of the cross section:

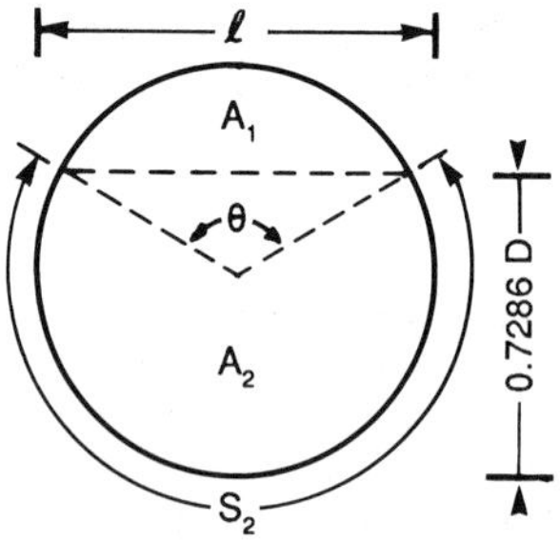

$$A_1 = \frac{180}{120}\frac{\pi}{4}D^2 = \frac{D_2}{8}(\theta - \sin\theta),$$

$$\therefore\ \theta = 2.192\ \text{rad},$$

$$A_2 = 0.7805\frac{\pi}{4}D^2 = 0.6130D^2,$$

$$L = D\sin(\theta/2) = 0.8894D,$$

$$S_2 = D\left(\pi - \frac{\theta}{2}\right) = 2.0456D.$$

Hydraulic diameter of heavy liquid

$$D_h = \frac{4A_2}{L + S_2} = \frac{4(0.6130)D}{0.8894 + 2.0456} = 0.8354D.$$

The dispersion discriminant is

$$\psi = \frac{Q_L}{Q_H}\left(\frac{\rho_L \mu_H}{\rho_H \mu_L}\right)^{0.3} = \frac{180}{640}\left(\frac{0.9(0.7)}{10}\right)^{0.3} = 0.123$$
$$< 0.30.$$

Therefore, oil is the dispersed phase:

$$N_{Re} = \frac{D_h u \rho}{\mu} = \frac{D_h \rho}{\mu}\frac{Q}{\frac{1}{4}\pi D^2} = \frac{(62.4)640\,(0.8354)}{42(0.7)\pi\quad D}$$
$$= \frac{25{,}076}{D}.$$

Velocity of rise:

$$u_r = \frac{2.492(1.00 - 0.90)}{0.7} = 0.356\ \text{ft/min}.$$

Time of rise:

$$t = \frac{0.7286D}{0.356} = 2.0466D\ \text{min}.$$

Forward velocity:

$$u_H = \frac{Q_H}{A_2} = \frac{640}{60(0.6130D^2)} = \frac{17.40}{D^2}\ \text{ft/min}.$$

Flow distance:

$$L_f = tu_H = 2.0466D\left(\frac{17.40}{D^2}\right) = \frac{35.60}{D}\ \text{ft}.$$

The tangent to tangent length of the drum will be approximately 24 in. greater than L_f to accommodate inlet and outlet nozzles and baffles.

The Reynolds number identifies the quality of the separation, $N_{Re} < 5000$ being good.

Some trials are

D (ft)	N_{Re}	*t*	u_H	L_f (ft)
5	5015	10.23	0.696	7.12
3.5	7165	7.16	1.420	10.17
3	8358	6.14	1.933	11.87

A vessel 5×9 ft would give excellent separation; 3×14 ft might be acceptable. A sketch of the proposed drum is in Figure 18.1.

mesh pads and other solid surfaces on which impingement and coalescence are forced, and approaching 100% in scrubbers and high speed centrifuges.

RATE OF SETTLING

The terminal or maximum settling velocity of a small droplet or particle in a gas is governed by one of Newton's equations.

$$u = f\sqrt{g_c D(\rho/\rho_g - 1)}. \qquad (18.6)$$

In laminar flow the friction factor becomes a simple function of the Reynolds number,

$$f = 18/(Du\rho_g/\mu_g). \qquad (18.7)$$

When this substitution is made, the falling velocity becomes

$$u = g_c(\rho - \rho_g)D^2/18\mu, \qquad (18.2')$$

which is Stokes' equation. In view of the uncertainties with which droplet sizes are known in practical situations, Stokes equation usually is regarded as sufficiently descriptive of settling behavior. For example, it predicts that 100 μm droplets of water fall at the rate of 1.0 ft/sec in atmospheric air.

Another approximation of Newton's equation is written

$$u = K\sqrt{\rho/\rho_g - 1}, \qquad (18.8)$$

where the coefficient K depends on the system. For the 100 μm droplets of water in air just cited, the coefficient becomes

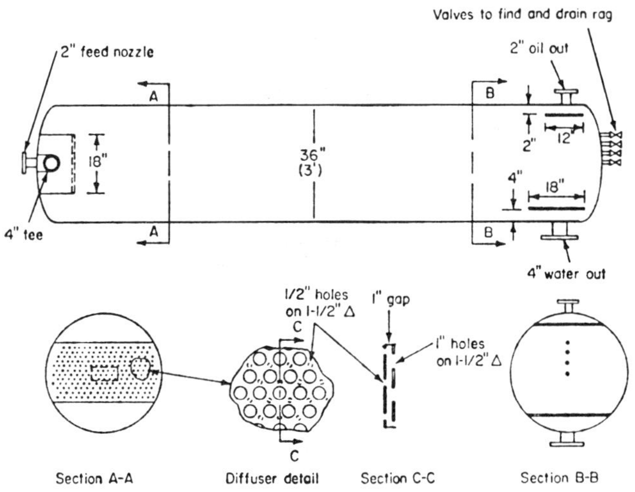

Figure 18.2. A design of an oil-water separator for the conditions of Example 18.1, showing particularly the diffuser at the inlet nozzle and baffles at the outlets. (*Hooper and Jacobs, 1979*).

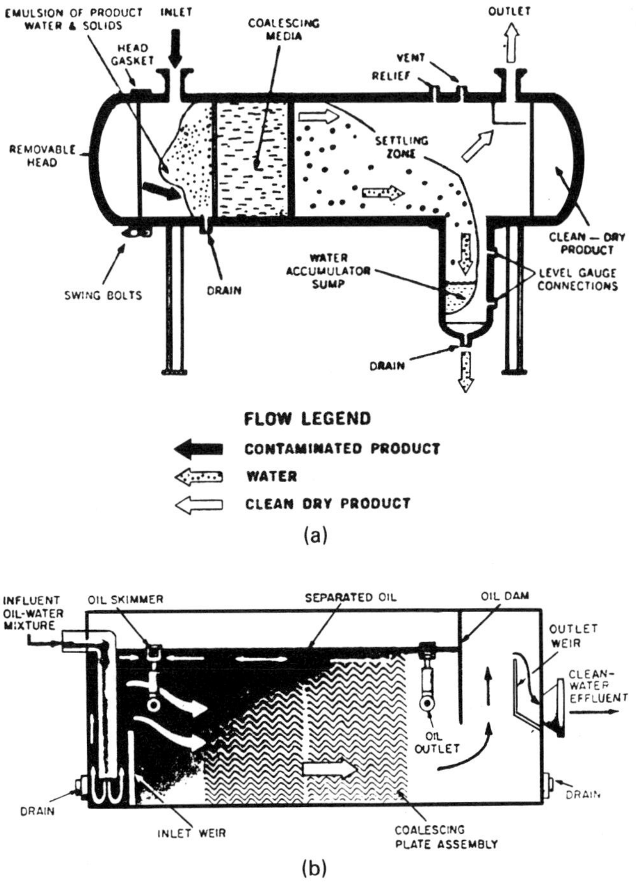

Figure 18.3. Drums with coalescers for assisting in the separation of small amounts of entrained liquid. (a) A liquid–liquid separating drum equipped with a coalescer for the removal of small amounts of dispersed phase. In water–hydrocarbon systems, the pot may be designed for 0.5 ft/sec (*Facet Enterprises, Industrial Division*). (b) An oil–water separator with corrugated plate coalescers (*General Electric Co.*).

$K = 0.035$, and for other sizes it varies as the square of the diameter.

EMPTY DRUMS

The cross section of a vertical settling drum is found from the vapor rate and the allowable linear velocity with the equation

$$u = 0.14\sqrt{\rho/\rho_g - 1}, \quad \text{ft/sec}, \tag{18.9}$$

in which the coefficient of Eq. (18.8) has been evaluated for 200 μm. The vertical dimension is more arbitrarily established. The liquid holdup is determined as in Section 18.2 and Table 18.1. For the vapor space, Watkins (1967) proposes the rules illustrated in Figure 18.6. When the calculated length to diameter ratio comes out less than 3, the length is increased arbitrarily to make the ratio 3; when the ratio comes out more than 5, a horizontal drum is preferably employed. Rules for horizontal drums also are shown on Figure 18.6. The vapor space is made a minimum of 20% of the drum volume which corresponds to a minimum height of the vapor space of 25% of the diameter, but with the further restriction that this never is made less than 12 in. When a relatively large amount of liquid must be held up in the drum, it may be advisable to increase the fraction of the cross section open to the vapor.

The diameter again is figured from the volumetric rate of the vapor and the linear velocity from Eq. (18.9). Since the upward drag of the vapor is largely absent in a horizontal drum, however, the coefficient K often is raised by a factor of 1.25. Example 18.3 deals with the design of both kinds of drums.

WIRE MESH PAD DEENTRAINERS

Pads of fine wire mesh induce coalescence of impinging droplets into larger ones, which then separate freely from the gas phase. Tower packings function similarly but are less effective and more difficult to install. The pads are made of metal wires or plastic strands or fiber glass. These data apply to stainless steel construction:

Efficiency Type	Efficiency (%)	lbs/cuft	sqft/cuft	K Pressure	K Vacuum
Low	99.0	5–7	65	0.40	0.20–0.27
Standard	99.5	9	85	0.35	
High	99.9	12	115	0.35	
Very high	99.9	13–14	120	0.25	

A pad thickness of 4 in. is minimum, 6 in. is popular, and up to 12 in. may be required for fine mists.

The values of K in the preceding table are with a standard disengaging height of 10 in. The effect of other heights h is given by the equation

$$K = 0.021 + 0.0325h, \quad 3 \le h \le 12, \tag{18.10}$$

with a maximum value of 0.40. This relation is for standard efficiency pads. Lower values can be expected in aqueous systems where the surface tension has been reduced by surfactants.

When the pad is installed in a vertical or inclined position, the values of K should be taken 2/3 of the horizontal ones.

At high liquid rates droplets tend to be reentrained and the pad may become flooded. Some data obtained by Poppele (1958) are cited by York (1983, p. 194). A graphical correlation credited to the Fluor Co. is represented by Branan (1983, p. 67) by the equation

$$K = -0.0073 + \frac{0.263}{x^{1.294} + 0.573}, \quad 0.04 \le x \le 6.0, \tag{18.11}$$

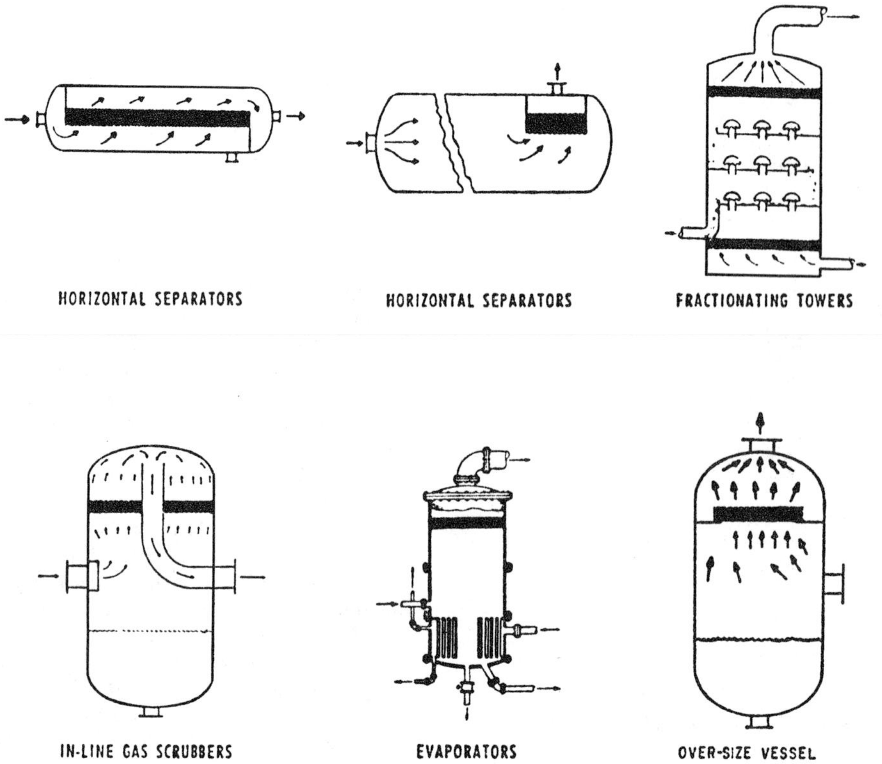

Figure 18.4. Typical installations of mesh pads in equipment (*Metal Textile Corp, Bulletin ME-7; from Ludwig, 1977, Vol. I, p. 159*).

where x is a function of the weight flow rates and densities of the phases

$$x = (W_L/W_V)\sqrt{\rho_V/\rho_L}\,. \tag{18.12}$$

Good performance can be expected at velocities of 30–100% of those calculated with the given Ks. Flooding velocities are at 120–140% of the design rates. At low velocities the droplets drift through the mesh without coalescing. A popular design velocity is about 75% of the allowable. Some actual data of the harmful effect of low velocities were obtained by Carpenter and Othmer (1955); they found, for example, that 99% of 6 μm droplets were removed at 6.8 ft/sec, but 99% of 8 μm at the lower velocity of 3.5 ft/sec.

Pressure drop in pads usually is small and negligible except at flooding; the topic is discussed by York (1983).

In existing drums or when the drum size is determined primarily by the required amount of liquid holdup, the pad dimensions must conform to the superficial velocities given by the design equation. This may necessitate making the pad smaller than the available cross section of the drum. Figure 18.7 shows typical installations. On the other hand, when the pad size is calculated to be greater than the available cross section and there develops a possibility of reentrainment of large droplets from the exit surface of the pad, a downstream settling drum or a high space above the mesh can be provided.

Good design practice is a disengaging space of 6–18 in., the more the better, ahead of the pad and 12 in. above the pad. Other details are shown on Figure 18.8. A design is provided in Example 18.4.

18.5. CYCLONE SEPARATORS

In addition to those already discussed, a variety of proprietary and home-made devices can remove entrainment more or less effectively. Some of them are represented on Figure 18.8.

a. A simple change of direction and impingement on the walls of the drum.
b. Impingement on a baffle.
c. Tangential inlet at high velocity and change of direction.
d. Multiple baffles, without or with coarse spray irrigation.
e. A pipeline deentrainer.

The capacity and effectiveness of proprietary devices such as items c to e cannot be estimated from general knowledge, but manufacturers usually claim that they can be sized to remove 99% of 8 μm droplets or particles. "Separators, entrainment" is an entry in the index of the *Chemical Engineering Catalog* which is a guide to manufacturers who may be consulted about the performance of their equipment.

In cyclone separators the gas enters tangentially at a high velocity, rotates several times, and leaves through a central pipe. Such equipment has been studied widely, particularly for the removal of dusts and catalyst fines in fluidized bed systems. The literature is reviewed by Rietema and Verver (1961), Maas (1979), Zenz (1982), and Semrau (1984).

Typical cyclone dimension ratios are indicated on Figure 18.9. For liquid knockout the bottom head often is made dished as on Figure 18.10 which also shows standard dimensions. Inlet velocities

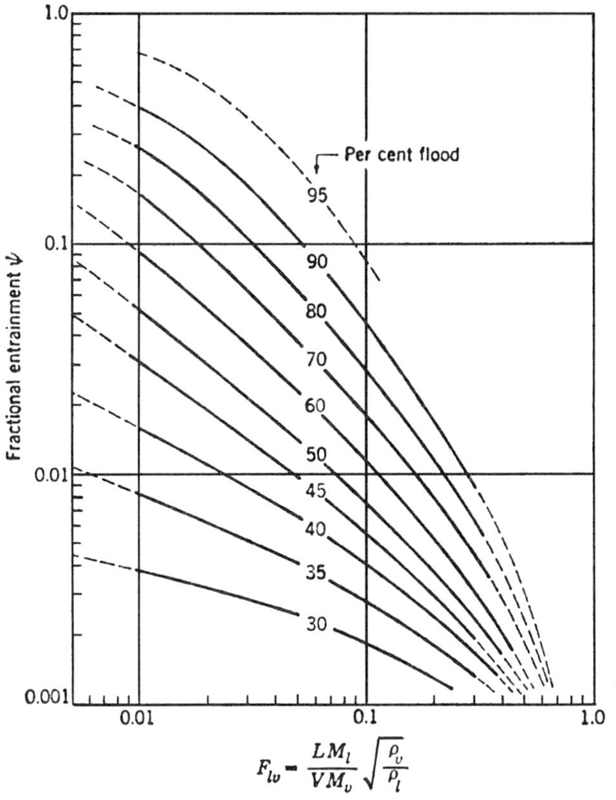

Figure 18.5. Entrainment from sieve trays in the units mols liquid entrained/mol of liquid downflow; LM_L is the weight rate of flow of liquid and VM_V is the weight rate of flow of vapor. The flooding correlation is Figure 13.32(b). [*Fair and Matthews,* Pet. Refiner **37**(*4*), *153* (*1958*)].

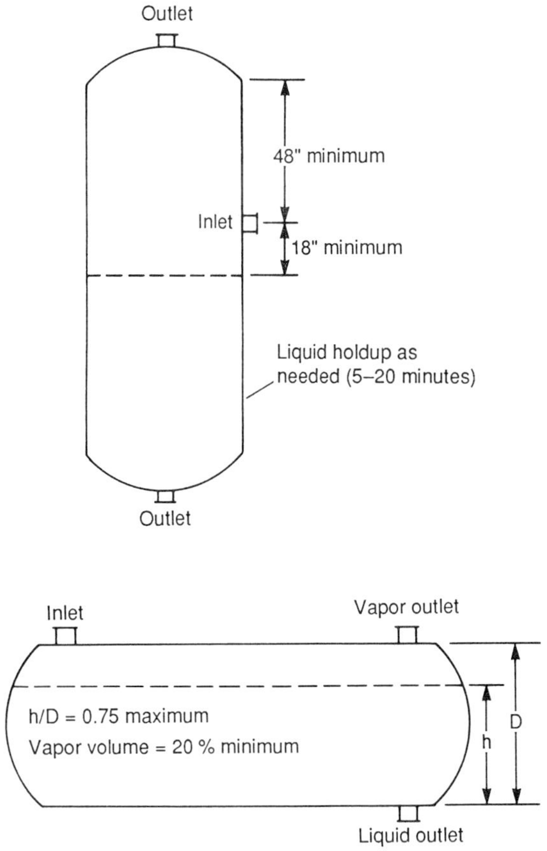

Figure 18.6. Knockout drums. Key dimensions of vertical and horizontal types.

should be in the range 100–150 ft/sec, the higher the better, but may be limited by the occurrence of reentrainment and unacceptable pressure drop. The pressure drop is estimated in terms of velocity heads, a value of four being commonly taken. Accordingly,

$$\Delta P = 4\rho V^2/2g = 4.313\rho(\text{ft/sec}/100)^2, \quad \text{psi.} \qquad (18.13)$$

For atmospheric air, for instance, this becomes

$$\Delta P = 0.323(\text{ft/sec}/100)^2, \quad \text{psi.} \qquad (18.14)$$

For the design of Figure 18.10, the size of the inlet is selected at a specified inlet velocity and required volumetric rate; the other dimensions then are fixed as given for this standard.

Very high velocities tend to skim the liquid film off the vessel wall and off the liquid at the bottom. The liquid also tends to creep up the wall and down the exit pipe where it is picked up by the exit gas. The skirt shown on Figure 18.9 is designed to prevent reentrainment of the creeping liquid, and the horizontal plate in Figure 18.10 prevents vortexing of the accumulated liquid and pickup off its surface.

Efficiencies of 95% for collection of 5 μm droplets can be

EXAMPLE 18.2
Quantity of Entrainment on the Basis of Sieve Tray Correlations

The conditions of Example 13.15 will be used. This is the case of a standard sieve tray with 24 in. spacing and to operate at 80% of flooding. The entrainment correlation is Figure 18.4 for which the value of the abscissa was found to be

$$w_l/w_G\sqrt{\rho_G/\rho_L} = 0.2924.$$

At 80% of flooding the ordinate of Figure 18.4 is

$$\psi = 0.008 \text{ mol entrained liquid/mol liquid downflow.}$$

Since $w_L/w_G = 259{,}100/271{,}500 = 0.954$ mol liquid/mol vapor (assuming the same molecular weights), the entrainment expressed with reference to the vapor flow is

$$\psi = 0.008(0.954) = 0.0076 \text{ mol liquid/mol vapor flow.}$$

The linear velocity of the vapor was found in Example 13.15 to be 0.597 ft/sec for this condition.

Example 18.3
Liquid Knockout Drum (Empty)

Gas at the rate of 3000 cfm and liquid at 25 cfm enter a drum in which entrainment is to be removed. Holdup of liquid in the drum is 10 min. The properties are those of air and water at atmospheric conditions. Find the size of the drum needed to remove droplets greater than 200 μm dia.

Vertical drum, with Eq. (18.9):

$$u = 0.14\sqrt{62.4/0.075 - 1} = 4.04 \text{ ft/sec},$$
$$D = \sqrt{3000/60(\pi/4)(4.04)} = 3.97 \text{ ft}, \quad \text{say } 4.0 \text{ ft}.$$

From Figure 18.5, the vapor space is a minimum of 5.5 ft. The liquid depth is

$$L_{\text{liq}} = \frac{250}{(\pi/4)D^2} = 19.9 \text{ ft} \quad 10 \text{ min holdup},$$
$$L = 19.9 + 5.5 = 25.4 \text{ ft},$$
$$L/D = 25.4/4 = 6.35.$$

With $D = 4.5$ ft, $L = 15.7 + 5.5 = 21.2$, and $L/D = 4.71$.

Horizontal drum:

The allowable velocity is 25% greater:

$$u = 1.25(4.04) = 5.05 \text{ ft/sec}.$$

Try several fractional vapor cross sections ϕ:

$$D = \sqrt{50/5.05(\pi/4)\phi} = \sqrt{12.61/\phi},$$
$$L = 250/(1-\phi)(\pi/4)D^2 = 25.24\phi/(1-\phi),$$
$$h = \text{depth of liquid}.$$

ϕ	h/D	D	L	L/D
0.2	0.75	7.94 (8.0)	6.31 (6.2)	0.78
0.3	0.66	6.48 (6.5)	10.82 (10.8)	1.66
0.4	0.58	5.61 (5.5)	16.83 (17.5)	3.18
0.5	0.50	5.02 (5.0)	25.24 (25.5)	5.10

Accordingly, a horizontal vessel between 5.0 and 5.5 ft dia with a liquid depth between 58 and 50% of the diameter falls in the usual economic range.

achieved by proper design of cyclone separators. For applications such as knockout drums on the suction of compressors, however, it is sufficient to remove only droplets greater than 40–50 μm.

Capacity and efficiency depend on the inlet velocity and the dimensions of the vessel. Correlated studies have been made chiefly for the design of Figure 18.9 with a rectangular inlet whose width is $D/4$ (one-fourth of the vessel diameter) and whose height is 2–3 times the width. A key concept is a critical particle diameter which is the one that is removed to the extent of 50%. The corresponding % removal of other droplet sizes is correlated by Figure 18.11. The equation for the critical particle diameter is

$$(D_p)_{\text{crit}} = \left[\frac{9\mu D}{4\pi N_t V(\rho - \rho_g)}\right]^{0.5}, \tag{18.15}$$

where D is the diameter of the vessel and V is the inlet linear velocity. The quantity N_t is the number of turns made by the gas in the vessel. A graphical correlation given by Zenz (1982) can be

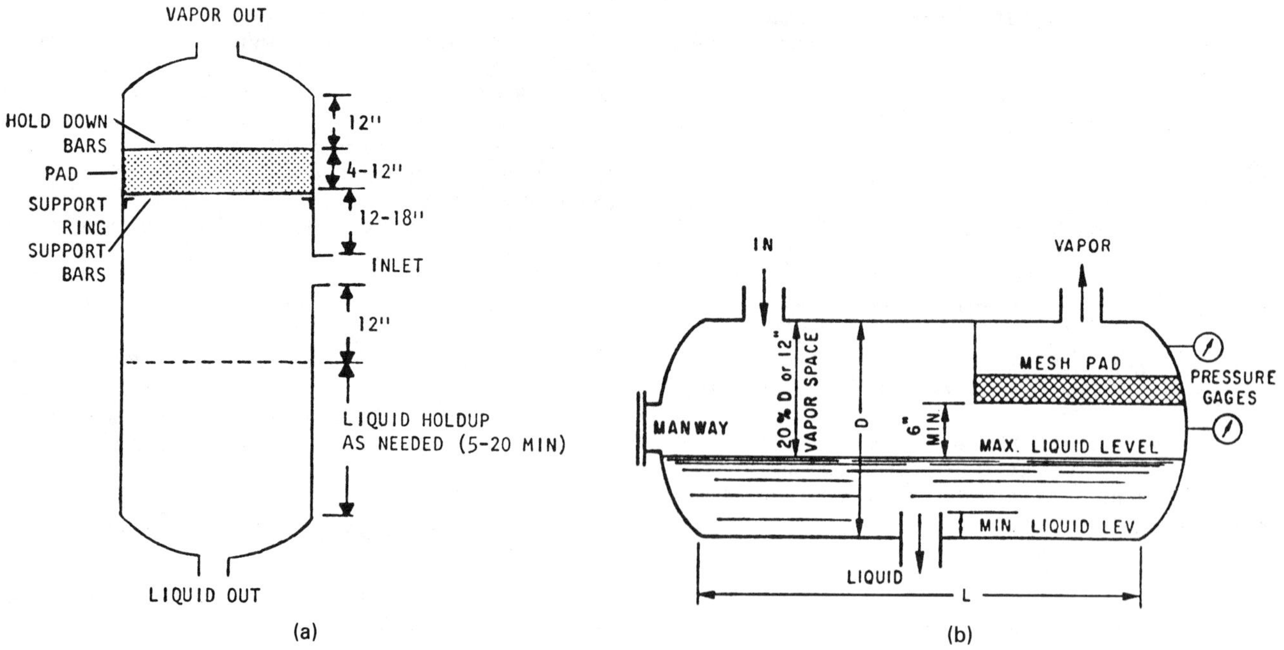

Figure 18.7. Key dimensions of knockout drums equipped with mesh pads. (a) Vertical knockout drum. (b) Horizontal knockout drum.

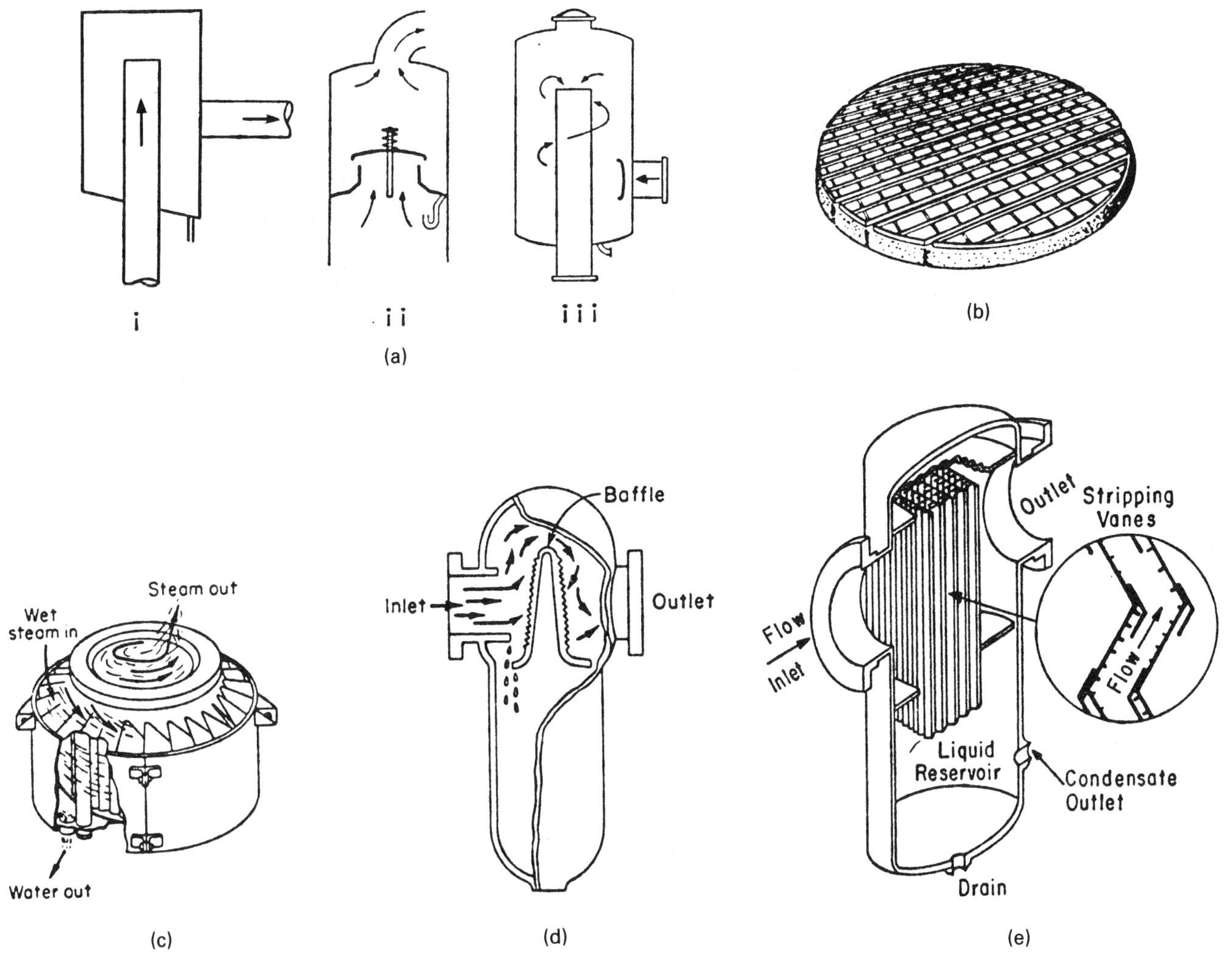

Figure 18.8. Principles of entrainment separation and some commercial types of equipment. (a) Basic principles of entrainment separating equipment: (i) change of direction; (ii) impingement on a baffle; (iii) tangential inlet resulting in centrifugal force. (b) Wire or fiber mesh pad, typical installations as in Figure 18.7. (c) A separator combining impingement and centrifugal force (*V.D. Anderson Co.*). (d) Equipment with impingement and change of direction (*Wright-Austin Co.*). (e) Multiple zig-zag baffle arrangement (*Peerless Mfg. Co.*).

represented by the equation

$$N_t = [0.1079 - 0.00077V + 1.924(10^{-6})V^2]V \qquad (18.16)$$

with V in ft/sec. With a height of opening equal to 2.5 times the width, the volumetric rate is

$$Q = AV = 2.5D^2V/16. \qquad (18.17)$$

These relations are used in Example 18.5 to find the size of a separator corresponding to a specified critical particle diameter, and to the reverse problem of finding the extent of removal of particles when the diameter of the vessel and the velocity are specified.

To obtain a high efficiency, the vessel diameter must be small, but in order to accommodate a required volumetric rate, many units in parallel may be needed. These units, called multicyclones, may be incorporated in a single shell at a cost that may be justifiable in view of greater efficiency and lower pressure drop.

18.6. STORAGE TANKS

Cylindrical tanks for the storage of inflammable liquids above or under ground at near atmospheric pressure are subject to standards of Underwriter Laboratories or of the API. Underwriters covers some smaller sizes. Both sets of standards are restricted to steel construction for essentially noncorrosive service. Various manufacturers supply Underwriter or API tanks as a matter of course.

Standard tanks are made in discrete sizes with some latitude in combinations of diameter and length. For example, Table 18.2 shows the several heights of 30 ft diameter tanks among API standard sizes. The major specification is that of metal wall thickness. In smaller sizes the thickness is determined by requirements of rigidity rather than strength. Some general statements about metal thickness of tanks may be given.

Horizontal tanks. Above ground they are limited to 35,000 gal. Normally they are supported on steel structures or concrete saddles at elevations of 6 to 10 ft. The minimum thickness of shell and heads is 3/16 in. in diameters of 48–72 in. and 1/4 in. in diameters of 73–132 in.

EXAMPLE 18.4
Knockout Drum with Wire Mesh Deentrainer

For the flow conditions of Example 18.2, design a drum with a standard efficiency stainless steel wire mesh pad. For this condition, $k = 0.35$, so that

$$u = 0.35\sqrt{62.4/0.075 - 1} = 10.09 \text{ ft/sec},$$
$$D = \sqrt{50/(\pi/4)u} = 2.51 \text{ ft}.$$

With 2 in. support rings the pad will have a diameter of 34 in.

The size of the drum is set largely by the required liquid holdup of 250 cuft. On the basis of Figure 18.7, the height of vessel above the liquid level is 4 ft. As in Example 18.2, take the diameter to be 4.5 ft. Then

$$L_{\text{liq}} = 25[10/(\pi/4)(4.5)]^2 = 15.7 \text{ ft},$$
$$L = 15.7 + 4.0 = 19.7 \text{ ft},$$
$$L/D = 19.7/4.5 = 4.38.$$

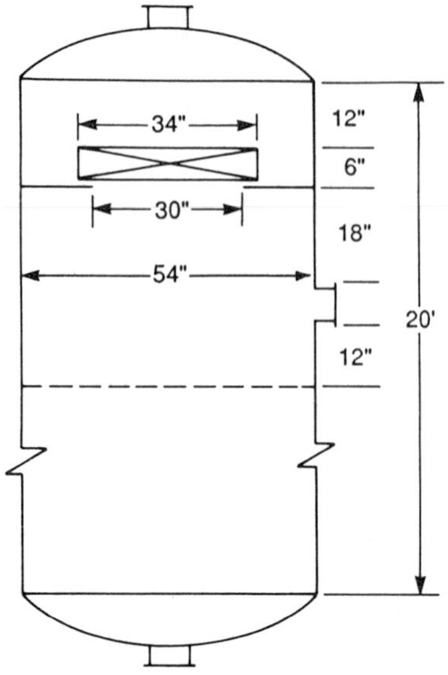

This ratio is acceptable. As a check, use Eqs. (18.11) and (18.12):

$$x \equiv (W_L/W_v)\sqrt{P_v/P_L} = V_L/V_v\sqrt{P_L/P_v}$$
$$= 25/3000\sqrt{62.4/0.075} = 0.24,$$
$$k = -0.0073 + 0.263/[(0.24)^{1.294} + 0.573]$$
$$= 0.353,$$

which is close to the assumed value, $k = 0.35$.

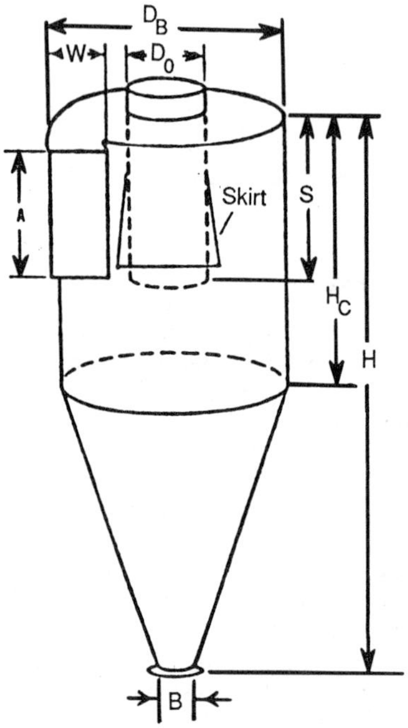

Figure 18.9. Typical dimension ratios of a cyclone separator.

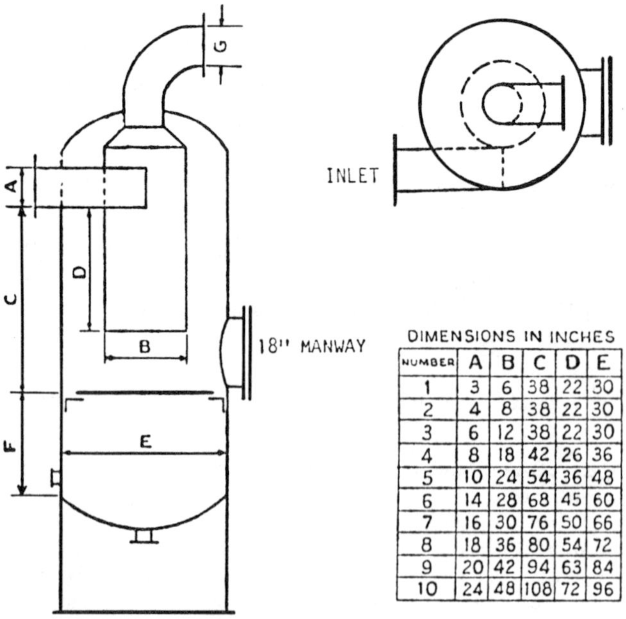

NUMBER	A	B	C	D	E
1	3	6	38	22	30
2	4	8	38	22	30
3	6	12	38	22	30
4	8	18	42	26	36
5	10	24	54	36	48
6	14	28	68	45	60
7	16	30	76	50	66
8	18	36	80	54	72
9	20	42	94	63	84
10	24	48	108	72	96

Figure 18.10. Dimensions of standard liquid knockout drums with tangential inlets.

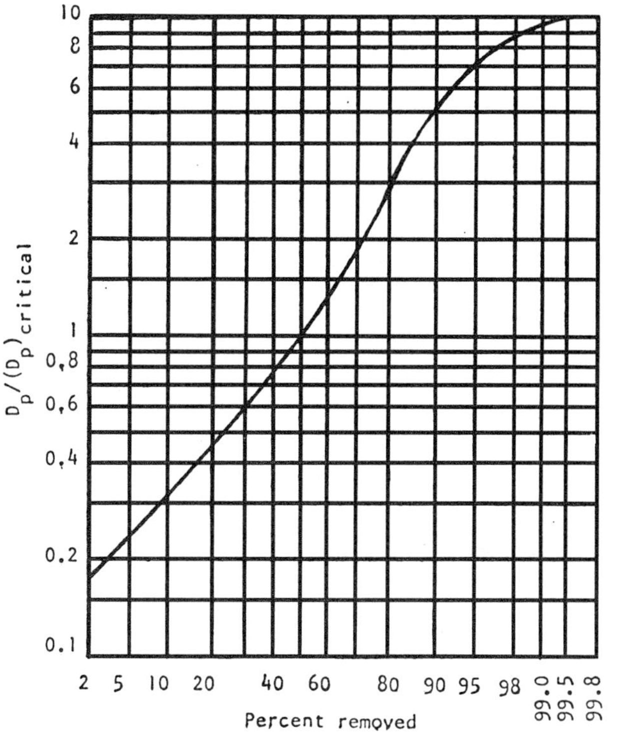

Figure 18.11. Percent removal of particles in a cyclone as a function of their diameters relative to the critical diameter given by Eqs. (18.15) and (18.16) (*Zenz, 1982*).

Vertical tanks. Those supported above ground are made with dished or conical bottoms. Flat bottomed tanks rest on firm foundations of oiled sand or concrete. Supported flat bottoms usually are 1/4 in. thick. Roof plates are 3/16 in. thick. Special roof constructions that minimize vaporization losses were mentioned earlier in this chapter; they are illustrated by Mead (1964), by Riegel (1953), and in manufacturers catalogs. The curved sides are made of several courses of plate with thicknesses graduated to meet requirements of strength. The data of the selected API tanks of Table 18.2 include this information. Figure 18.12 illustrates the facilities that normally are provided for a large storage tank.

In order to minimize hazards, storage tanks for inflammable or toxic materials may be buried. Then they are provided with an overburden of 1.3 times the weight of water that the tank could hold in order to prevent floating after heavy rainfalls.

Cylinders with curved heads are used for pressure storage at 5–230 psig. In the range of 5–10 psig, spheroids and other constructions made up with curved surfaces, as in Figure 18.12(c) are being used in quite large sizes, often with refrigeration to maintain sufficiently low pressures. More illustrations of such equipment appear in manufacturers' catalogs and in the books of Mead (1964) and Riegel (1953).

Mention of vessels for the storage of gases was made at the beginning of this chapter, and Figure 18.12(d) shows the principles of some suitable designs. Design for storage of granular solids includes provisions for handling and withdrawal, as in the case of Figure 18.13.

18.7. MECHANICAL DESIGN OF PROCESS VESSELS

Process design of vessels establishes the pressure and temperature ratings, the length and diameter of the shell, the sizes and locations of nozzles and other openings, all internals, and possibly the material of construction and corrosion allowances. This information must be supplemented with many mechanical details before fabrication can proceed, notably wall thicknesses.

Large storage tanks are supported on a concrete pad on the ground. Other vessels are supported off the ground by various means, as in Figure 18.14.

For safety reasons, the design and construction of pressure vessels are subject to legal and insurance standards. The ASME

Example 18.5
Size and Capacity of Cyclone Separators

Air at 1000 cuft/sec and density of 0.075 lb/cuft contains particles with density 75 lb/cuft. 50% of the 10 μm diameter particles are to be recovered. Find the sizes and numbers of cyclones needed with inlet velocities in the range of 50–150 ft/sec. The inlet is rectangular with width $D/4$ and height $2.5D/4$, where D is the diameter of the vessel.

Equation (18.15) becomes

$$\frac{D}{N_t V} = \frac{4\pi(\rho - \rho_g)D_p^2}{9\mu} = \frac{4\pi(75 - 0.075)}{9(1.285)(10^{-5})}\left(\frac{10}{304,800}\right)^2 = 0.00876,$$

where N_t is given by Eq. (18.16). The number of vessels in parallel is

$$n = \frac{Q'}{AV} = \frac{100}{(2.5/16)D^2V} = \frac{6400}{D^2V}.$$

The results at several velocities are summarized.

V (cfs)	N_t	D (ft)	n
50	3.71	1.62	48.8
100	5.01	4.39	3.32
144	5.32	6.71	1.0

From Figure 18.11, the percentage recoveries of other-sized particles are:

$D_p/(D_p)_{crit}$	% Recovered
0.3	9
0.5	22
0.6	30
1	50
2	70
6	90
9	98.5

When the smallest of these cyclones, 1.62 ft dia, is operated at 150 cuft/sec,

$$N_t = 5.35,$$

$$(D_p)_{crit} = \left[\frac{9(1.285)(10^{-5})(1.62)}{4\pi(5.35)(150)(75 - 0.075)}\right]^{0.5} = 1.574(10^{-5})\ \text{ft},\quad 4.80\ \mu\text{m}.$$

TABLE 18.2. Storage Tanks, Underwriter or API Standard, Selected Sizes

a. Small Horizontal Underwriter Label

Capacity Gallons	Dimensions: Diameter	Length	Thickness	Weight in pounds
280	42″	4′–0″	$\frac{3}{16}$″	540
550	48″	6′–0″	$\frac{3}{16}$″	800
1000	48″	10′–8″	$\frac{3}{16}$″	1260
1000	64″	6′–0″	$\frac{3}{16}$″	1160
1500	64″	9′–0″	$\frac{3}{16}$″	1550
2000	64″	12′–0″	$\frac{3}{16}$″	1950
3000	64″	18′–0″	$\frac{3}{16}$″	2730
4000	64″	24′–0″	$\frac{3}{16}$″	3510

b. Horizontal or Vertical with Underwriter Label

Nominal Capacity Gallons	Dimensions: Diameter	Approx. Length	Thickness	Weight	No. of Supports
5,000	6′–0″	23′–9″	$\frac{1}{4}$″	5,440	3
5,000	7′–0″	17′–6″	$\frac{1}{4}$″	5,130	2
6,000	8′–0″	16′–1″	$\frac{1}{4}$″	5,920	2
6,000	8′–0″	16′–1″	$\frac{5}{16}$″	6,720	2
8,000	8′–0″	21′–4″	$\frac{1}{4}$″	7,280	2
8,000	8′–0″	21′–4″	$\frac{5}{16}$″	8,330	2
10,000	8′–0″	26′–7″	$\frac{1}{4}$″	8,860	3
10,000	8′–0″	26′–7″	$\frac{5}{16}$″	10,510	3
10,000	10′–0″	17′–2″	$\frac{1}{4}$″	8,030	2
10,000	10′–0″	17′–2″	$\frac{5}{16}$″	9,130	2
10,000	10′–6″	15′–8″	$\frac{1}{4}$″	8,160	2
10,000	10′–6″	15′–8″	$\frac{5}{16}$″	9,020	2
15,000	8′–0″	39′–11″	$\frac{1}{4}$″	13,210	4
15,000	8′–0″	39′–11″	$\frac{5}{16}$″	14,620	4
20,000	10′–0″	34′–1″	$\frac{1}{4}$″	14,130	3
20,000	10′–0″	34′–1″	$\frac{5}{16}$″	16,330	3
25,000	10′–6″	38′–9″	$\frac{1}{4}$″	17,040	4
25,000	10′–6″	38′–9″	$\frac{5}{16}$″	19,010	4

c. Large Vertical, API Standard

Dimensions: Diameter	Height	Capacity: 42 gal per bbl	U.S. Gal	Bottom Plates	Shell Plates (Butt Welded): Ring 1	Ring 2	Ring 3	Ring 4	Ring 5	Ring 6	Ring 7	Top Angle	Roof Plates
21′0″	18′0$\frac{3}{4}$″	1,114	46,788	$\frac{1}{4}$″	$\frac{3}{16}$″	$\frac{3}{16}$″	$\frac{3}{16}$″					3″ × 3″ × $\frac{1}{4}$″	$\frac{3}{16}$″
24′0″	24′0″	1,933	81,186	$\frac{1}{4}$″	$\frac{3}{16}$″	$\frac{3}{16}$″	$\frac{3}{16}$″	$\frac{3}{16}$″				3″ × 3″ × $\frac{1}{4}$″	$\frac{3}{16}$″
30′0″	24′0″	3,024	127,008	$\frac{1}{4}$″	$\frac{3}{16}$″	$\frac{3}{16}$″	$\frac{3}{16}$″	$\frac{3}{16}$″				3″ × 3″ × $\frac{1}{4}$″	$\frac{3}{16}$″
30′0″	29′11$\frac{1}{4}$″	3,769	158,300	$\frac{1}{4}$″	$\frac{3}{16}$″	$\frac{3}{16}$″	$\frac{3}{16}$″	$\frac{3}{16}$″	$\frac{3}{16}$″			3″ × 3″ × $\frac{1}{4}$″	$\frac{3}{16}$″
30′0″	35′10$\frac{1}{2}$″	4,510	189,420	$\frac{1}{4}$″	$\frac{3}{16}$″	$\frac{3}{16}$″	$\frac{3}{16}$″	$\frac{3}{16}$″	$\frac{3}{16}$″	$\frac{3}{16}$″		3″ × 3″ × $\frac{1}{4}$″	$\frac{3}{16}$″
30′0′	37′10$\frac{1}{4}$″	4,766	200,161	$\frac{1}{4}$″	$\frac{1}{4}$″	$\frac{3}{16}$″	$\frac{3}{16}$″	$\frac{3}{16}$″	$\frac{3}{16}$″	$\frac{3}{16}$″		3″ × 3″ × $\frac{1}{4}$″	$\frac{3}{16}$″
30′0″	41′9$\frac{3}{4}$″	5,264	221,088	$\frac{1}{4}$″	$\frac{3}{16}$″	$\frac{3}{16}$″	$\frac{3}{16}$″	$\frac{3}{16}$″	$\frac{3}{16}$″	$\frac{3}{16}$″	$\frac{3}{16}$″	3″ × 3″ × $\frac{1}{4}$″	$\frac{3}{16}$″
40′0″	33′10$\frac{3}{4}$″	7,586	318,612	$\frac{1}{4}$″	$\frac{1}{4}$″	$\frac{1}{4}$″	$\frac{3}{16}$″	$\frac{3}{16}$″	$\frac{3}{16}$″			3″ × 3″ × $\frac{1}{4}$″	$\frac{3}{16}$″
50′0″	47′9″	16,700	701,400	$\frac{1}{4}$″	0.35″	0.29″	0.25″	$\frac{1}{4}$″	$\frac{1}{4}$″	$\frac{1}{4}$″		3″ × 3″ × $\frac{1}{4}$″	$\frac{3}{16}$″
60′0″	39′10″	20,054	842,268	$\frac{1}{4}$″	0.34″	0.27″	$\frac{1}{4}$″	$\frac{1}{4}$″	$\frac{1}{4}$″			3″ × 3″ × $\frac{1}{4}$″	$\frac{3}{16}$″
70′0″	40′1″	27,472	1,153,824	$\frac{1}{4}$″	0.40″	0.32″	0.25″	$\frac{1}{4}$″	$\frac{1}{4}$″			3″ × 3″ × $\frac{3}{8}$″	$\frac{3}{16}$″
100′0″	40′0″	55,960	2,350,320	$\frac{1}{4}$″	0.57″	0.45″	0.33″	$\frac{1}{4}$″	$\frac{1}{4}$″			3″ × 3″ × $\frac{3}{8}$″	$\frac{3}{16}$″
150′0″	48′0″	151,076	6,345,192	$\frac{1}{4}$″	1.03″	0.85″	0.68″	0.50″	0.33″	$\frac{1}{4}$″		3″ × 3″ × $\frac{3}{8}$″	$\frac{3}{16}$″

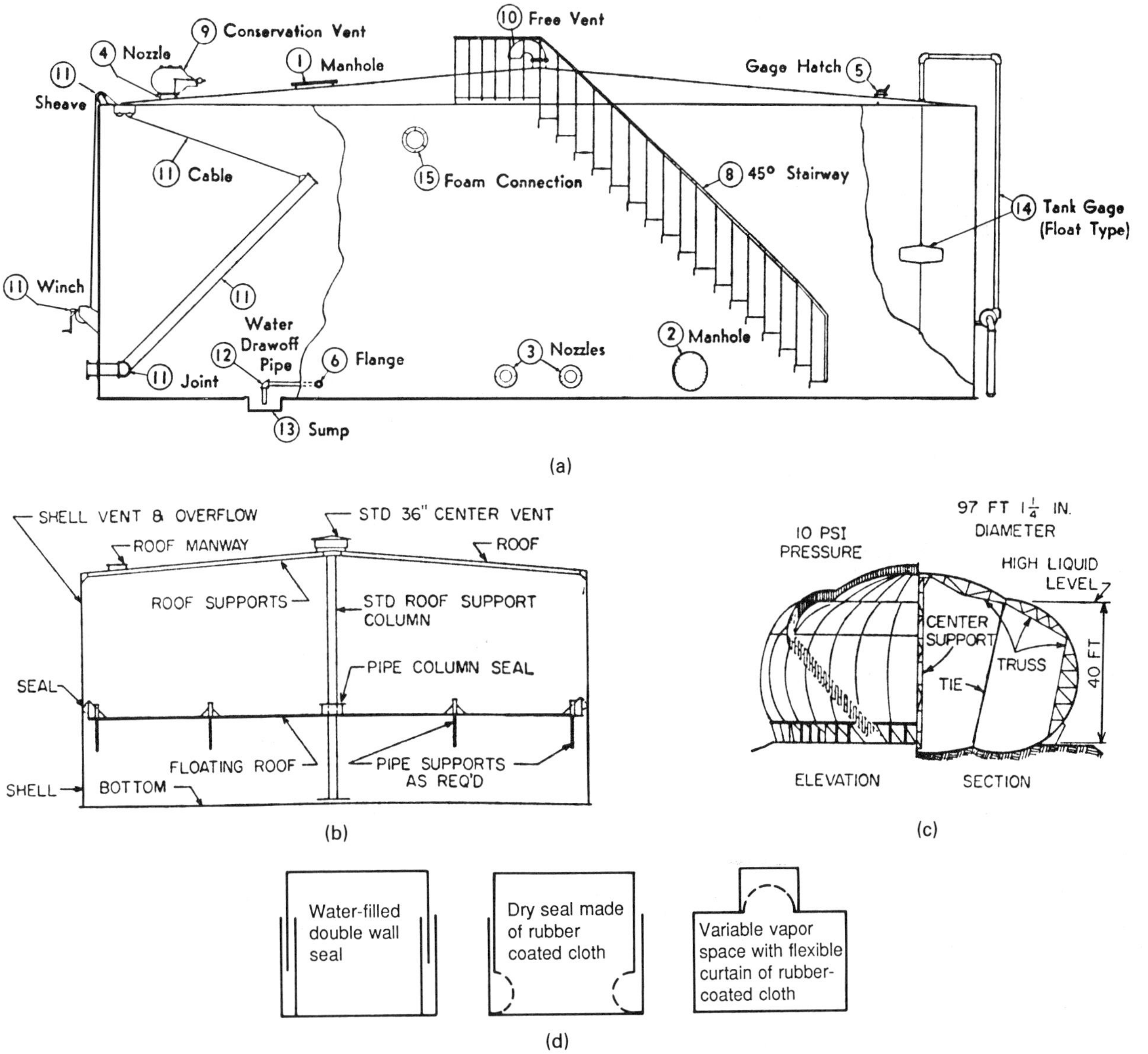

Figure 18.12. Examples of equipment for storage of liquids and gases in large quantities. (a) A large tank and its appurtenances, but with no provision for conservation of breathing losses (*Graver Tank and Mfg. Co.*). (b) Schematic of a covered floating roof tank in which the floating roof rides on the surface of the liquid. They also are made without the fixed roof [*R. Martin,* Petro/Chem. Eng., **23,** (*Aug. 1965*)]. (c) Cutaway of a 40,000 Bbl spheroid for operation at 10 psig (*Chicago Bridge and Iron Co.*). (d) Design principles of tanks for storage of gases or liquids subject to breathing losses at atmospheric pressure: water seal, dry seal with flexible curtain, and variable vapor space controlled by a flexible curtain.

Codes apply to vessels greater than 6 in. dia operating above 15 psig. Section VIII Division 1 applies to pressures below 3000 psig and is the one most often applicable to process work. Above 3000 psig some further restrictions are imposed. Division 2 is not pressure limited but has other severe restrictions. Some of the many details covered by Division 1 are indicated by the references to parts of the code on Figure 18.15.

DESIGN PRESSURE AND TEMPERATURE

In order to allow for possible surges in operation, it is customary to raise the maximum operating pressure by 10% or 10–25 psi, whichever is greater. The maximum operating pressure in turn may be taken as 25 psi greater than the normal. The design pressure of vessels operating at 0–10 psig and 600–1000°F is 40 psig. Vacuum systems are designed for 15 psig and full vacuum. Between −20 and 650°F, 50°F is added to the operating temperature, but higher margins of safety may be advisable in critical situations. When subzero temperatures have an adverse effect on the materials of construction, the working temperature is reduced appropriately for safety.

Allowable tensile stresses are one-fourth the ultimate tensile strength of the material of construction. Values at different temperatures are given in Table 18.4 for some steels of which shells and heads are made. Welded joint efficiencies vary from 100% for double-welded butt joints that are fully radiographed to 60% for

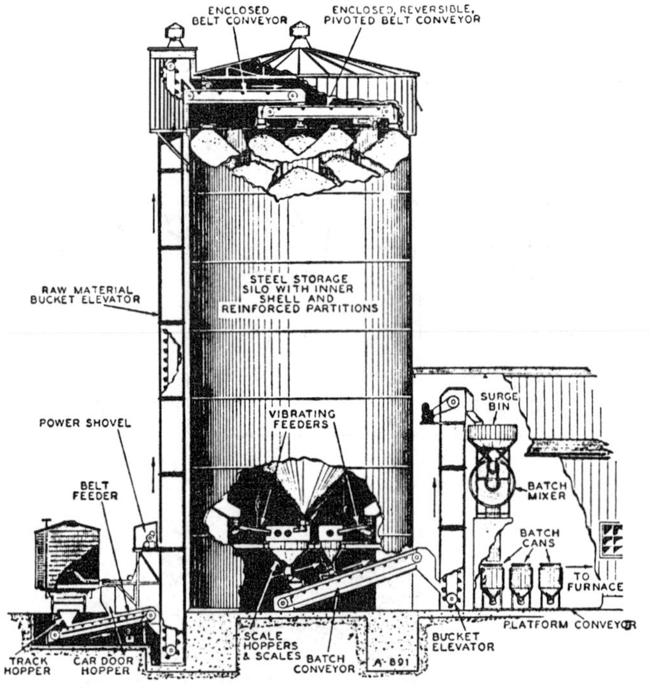

Figure 18.13. Equipment for handling, storing and withdrawing of granular solids in a glass manufacturing plant (*Stephens-Adamson Mfg. Co.*).

single-welded butt joints without backing strips and without radiographing. The Code has details.

SHELLS AND HEADS

Although spherical vessels have a limited process application, the majority of pressure vessels are made with cylindrical shells. The heads may be flat if they are suitably buttressed, but preferably they are some curved shape. The more common types of heads are illustrated on Figure 18.16. Formulas for wall thicknesses are in Table 18.3. Other data relating to heads and shells are collected in Table 18.5. Included are the full volume V_0 and surface S as well as the volume fraction V/V_0 corresponding to a fractional depth H/D in a horizontal vessel. Figure 18.17 graphs this last relationship. For ellipsoidal and dished heads the formulas for V/V_0 are not exact but are within 2% over the whole range.

FORMULAS FOR STRENGTH CALCULATIONS

The ASME Code provides formulas that relate the wall thickness to the diameter, pressure, allowable stress, and weld efficiency. Since they are theoretically sound only for relatively thin shells, some restrictions are placed on their application. Table 18.3 lists these

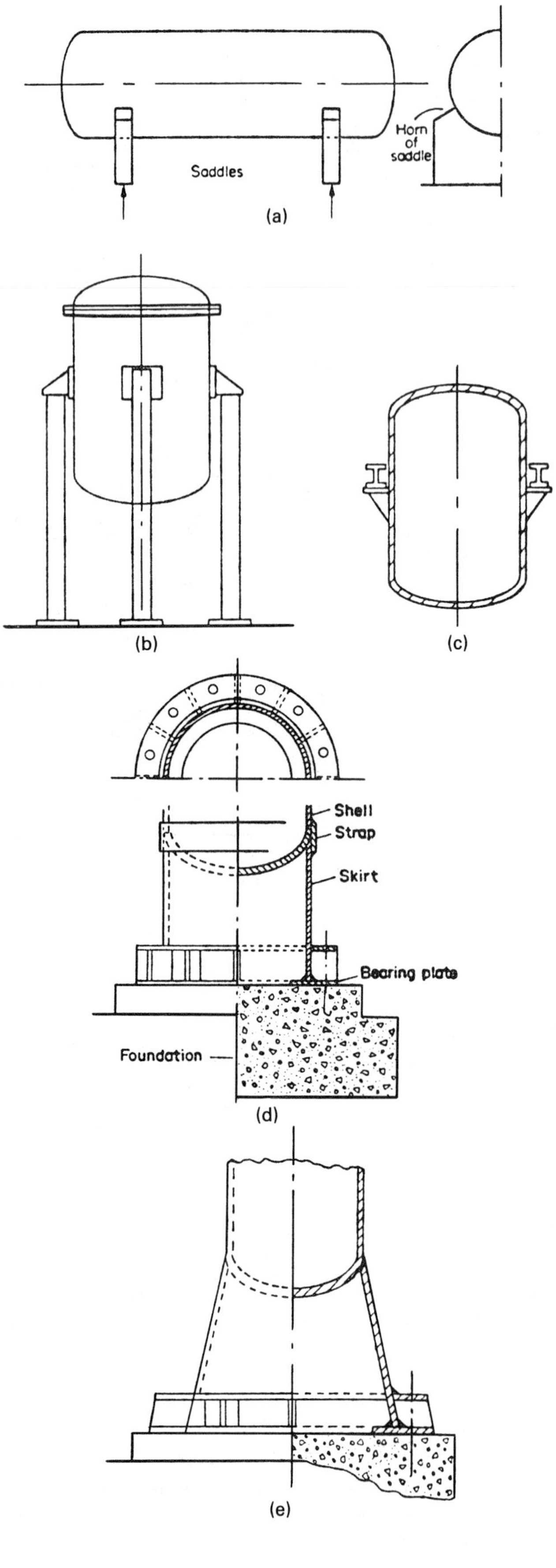

Figure 18.14. Methods of supporting vessels. (a) Saddle supports for horizontal vessels, usually of concrete. (b) Bracket or lug supports resting on legs, for either vertical or horizontal vessels. (c) Bracket or lug supports resting on steel structures, for either vertical or horizontal vessels. (d) Straight skirt support for towers and other tall vessels; the bearing plate is bolted to the foundation. (e) Flared skirt for towers and other tall vessels, used when the required number of bolts is such that the bolt spacing becomes less than the desirable 2 ft.

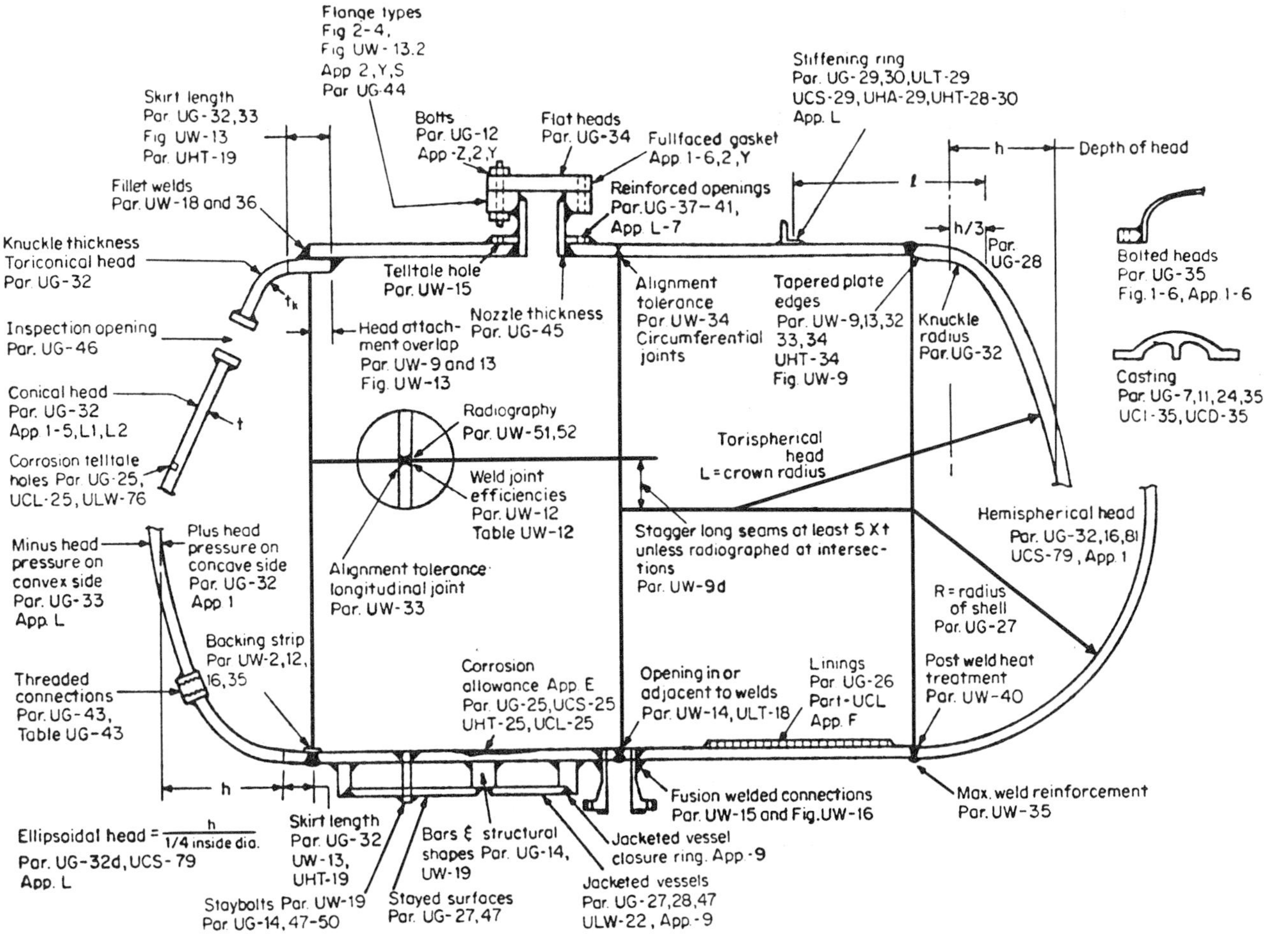

Figure 18.15. References to items covered in the ASME Code for Unfired Pressure Vessels, Section VIII Division 1 (*Chuse and Eber, 1984*).

TABLE 18.3. Formulas for Design of Vessels under Internal Pressure[a]

Item	Thickness t(in.)	Pressure P(psi)	Stress S(psi)	Notes
Cylindrical shell	$\frac{PR}{SE-0.6P}$	$\frac{SEt}{R+0.6t}$	$\frac{P(R+0.6t)}{t}$	$t \le 0.25D,\quad P \le 0.385SE$
Flat flanged head (a)	$D\sqrt{0.3P/S}$	$t^2S/0.3D^2$	$0.3D^2P/t^2$	
Torispherical head (b)	$\frac{0.885PL}{SE-0.1P}$	$\frac{SEt}{0.885L+0.1t}$	$\frac{P(0.885L+0.1t)}{t}$	$r/L=0.06,\quad L \le D+2t$
Torispherical head (b)	$\frac{PLM}{2SE-0.2P}$	$\frac{2SEt}{LM+0.2t}$	$\frac{P(LM+0.2t)}{2t}$	$M=\frac{3+(L/r)^{1/2}}{4}$
Ellipsoidal head (c)	$\frac{PD}{2SE-0.2P}$	$\frac{2SEt}{D+0.2t}$	$\frac{P(D+0.2t)}{2t}$	$h/D=4$
Ellipsoidal head (c)	$\frac{PDK}{2SE-0.2P}$	$\frac{2SEt}{DK+0.2t}$	$\frac{P(DK+0.2t)}{2Et}$	$K=[2+(D/2h)^2]/6,\quad 2 \le D/h \le 6$
Hemispherical head (d) or shell	$\frac{PR}{2SE-0.2P}$	$\frac{2SEt}{R+0.2t}$	$\frac{P(R+0.2t)}{2t}$	$t \le 0.178D,\quad P \le 0.685SE$
Toriconical head (e)	$\frac{PD}{2(SE-0.6P)\cos\alpha}$	$\frac{2SEt\cos\alpha}{D+1.2t\cos\alpha}$	$\frac{P(D+1.2t\cos\alpha)}{2t\cos\alpha}$	$\alpha \le 30^\circ$

[a] Nomenclature: D = diameter (in.), E = joint efficiency (0.6–1.0), L = crown radius (in.), P = pressure (psig), h = inside depth of ellipsoidal head (in.), r = knuckle radius (in.), R = radius (in.), S = allowable stress (psi), t = shell or head thickness (in.).
Note: Letters in parentheses in the first column refer to Figure 18.16.

TABLE 18.4. Maximum Allowable Tensile Stresses (psi) of Plate Steels

(a) Carbon and Low Alloy Steels

A.S.M.E. Specification No.	Grade	Nominal composition	Spec. min. tensile strength	For temperatures not exceeding °F.						
				−20 to 650	700	800	900	1000	1100	1200
Carbon Steel										
SA515	55	C-Si	55,000	13,700	13,200	10,200	6,500	2,500		
SA515	70	C-Si	70,000	17,500	16,600	12,000	6,500	2,500		
SA516	55	C-Si	55,000	13,700	13,200	10,200	6,500	2,500		
SA516	70	C-Si	70,000	17,500	16,600	12,000	6,500	2,500		
SA285	A	...—...	45,000	11,200	11,000	9,000	6,500			
SA285	B	...—...	50,000	12,500	12,100	9,600	6,500			
SA285	C	...—...	55,000	13,700	13,200	10,200	6,500			
Low-Alloy Steel										
SA202	A	Cr-Mn-Si	75,000	18,700	17,700	12,600	6,500	2,500		
SA202	B	Cr-Mn-Si	85,000	21,200	19,800	12,800	6,500	2,500		
SA387	D*	$2\frac{1}{4}$ Cr-I Mo	60,000	15,000	15,000	15,000	13,100	2,800	4,200	1,600

(b) High Alloy Steels

A.S.M.E. Specification No.	Grade	Nominal composition	Specified minimum tensile strength	For temperatures not exceeding °F.										
				−20 to 100	200	400	700	900	1000	1100	1200	1300	1400	1500
SA-240	304	18 Cr-8 Ni	75,000	18,700	15,600	12,900	11,000	10,100	9,700	8,800	6,000	3,700	2,300	1,400
SA-240	304L†	18 Cr-8 Ni	70,000	15,600	13,300	10,000	9,300							
SA-240	310S	25 Cr-20 Ni	75,000	18,700	16,900	14,900	12,700	11,600	9,800	5,000	2,500	700	300	200
SA-240	316	16 Cr-12 Ni-2 Mo	75,000	18,700	16,100	13,300	11,300	10,800	10,600	10,300	7,400	4,100	2,200	1,700
SA-240	410	13 Cr	65,000	16,200	15,400	14,400	13,100	10,400	6,400	2,900	1,000			

(ASME Publications).

relations for cylindrical and spherical shells and for all but the last of the heads of Figure 18.16. For unusual shapes there are no simple methods of design; experience and testing may need to be resorted to if such shapes are required.

The formulas are expressed in terms of inside dimensions.

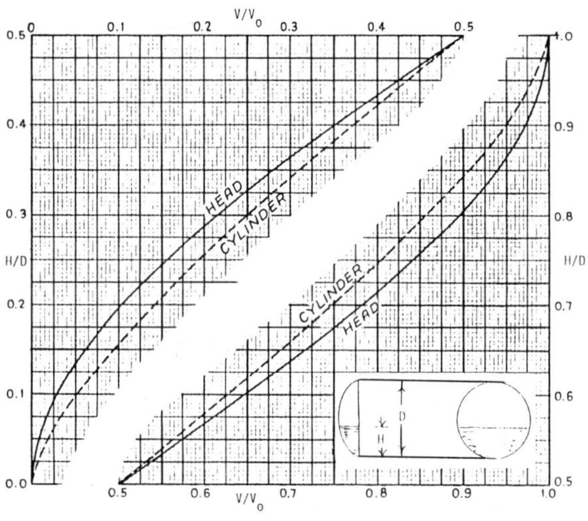

Figure 18.16. Fractional volumes of horizontal cylinders and curved heads at corresponding fractional depths, H/D.

Although they are rarely needed, formulas in terms of outside dimensions, say D_o, may be derived from the given ones by substitution of $D_o - 2t$ for D. For the 2:1 ellipsoidal head, for instance,

$$t = \frac{PD}{2SE - 0.2P} = \frac{P(D_o - 2t)}{2SE - 0.2P} = \frac{PD_o}{2SE + 1.8P}. \quad (18.18)$$

Example 18.6 investigates the dimensions and weight of a vessel to meet specifications. It is brought out that pressure vessels with large L/D ratios are lighter and presumably cheaper. A drawback may be the greater ground space needed by the slimmer and longer construction.

In addition to the shell and heads, contributions to the weight of a vessel include nozzles, manways, any needed internals, and supporting structures such as lugs for horizontal vessels and skirts for vertical ones. Nozzles and manways are standardized for discrete pressure ratings; their dimensions and weights are listed in manufacturers' catalogs. Accounting for these items may contribute 10–20% to the calculated weight of the vessel.

Mechanical design specification sheets (Appendix B) summarize the information that a fabricator needs in addition to the general specifications of the vessel codes. Not all of the data on the specification summary are necessarily in the province of the process engineer; it may depend on the stage of the design and on who else in the organization is available to do the work.

TABLE 18.5. Heads and Horizontal Cylinders: Formulas for Partially Filled Volumes and Other Data

---

Nomenclature

D = diameter of cylinder
H = depth of liquid
S = surface of head
V_0 = volume of full head
θ = angle subtended by liquid level or angle of cone

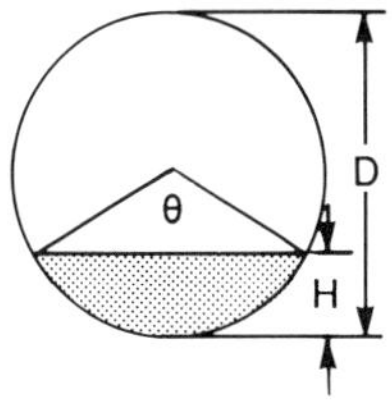

Cylinder

$\theta = 2\arccos(1 - 2H/D)$
$\theta(\text{rad}) = \theta°/57.3$
$V/V_0 = (1/2\pi)(\theta - \sin\theta)$

Hemispherical head

$S = 1.571D^2$
$V = (\pi/3)H^2(1.5D - H)$
$V_0 = (\pi/12)D^3$
$V/V_0 = 2(H/D)^2(1.5 - H/D)$

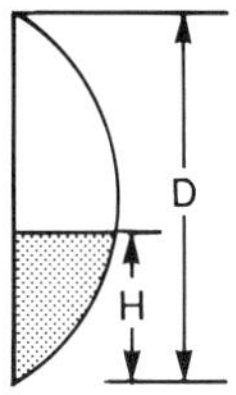

Ellipsoidal head ($h = D/4$)

$S = 1.09D^2$
$V_0 = 0.1309D^3$
$V/V_0 \simeq 2(H/D)^2(1.5 - H/D)$

Torispherical ($L = D$)

$S = 0.842D^2$
$V_0 = 0.0778D^3$
$V/V_0 \simeq 2(H/D)^2(1.5 - H/D)$

Conical

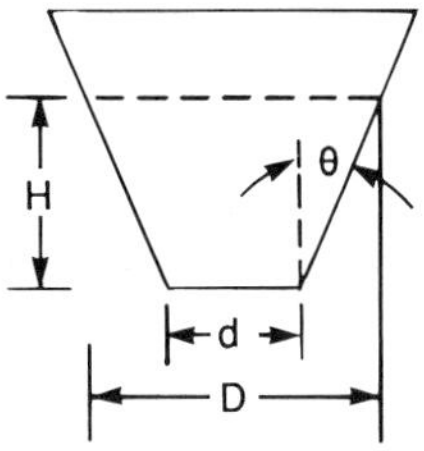

$H = [(D - d)/2]\tan\theta$

$= \begin{cases} 0.5(D - d), & \theta = 45° \\ 0.2887(D - d), & \theta = 30° \end{cases}$

$S = 0.785(D + d)\sqrt{4H^2 + (D - d)^2}$, curved surface
$V = 0.262H(D^2 + Dd + d^2)$

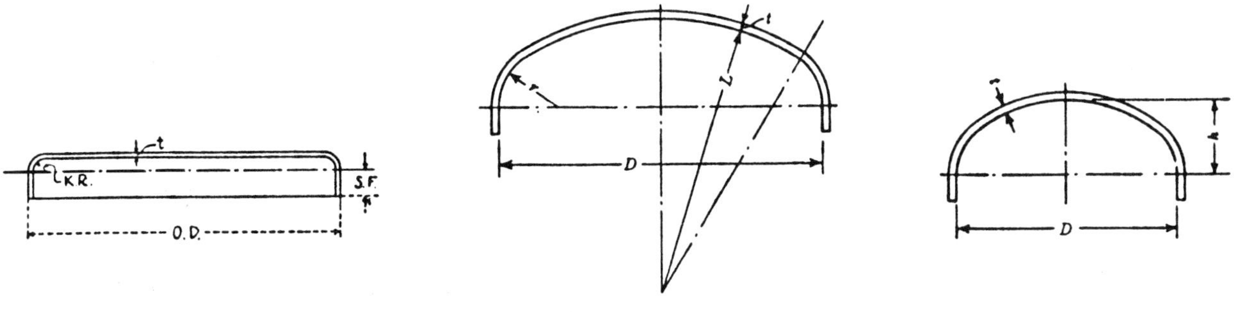

(a) (b) (c)

Figure 18.17. Types of heads for cylindrical pressure vessels. (a) Flat flanged: KR = knuckle radius, SF = straight flange. (b) Torispherical (dished). (c) Ellipsoidal. (d) Spherical. (e) Conical, without knuckle. (f) Conical, with knuckle. (g) Nonstandard, one of many possible types in use.

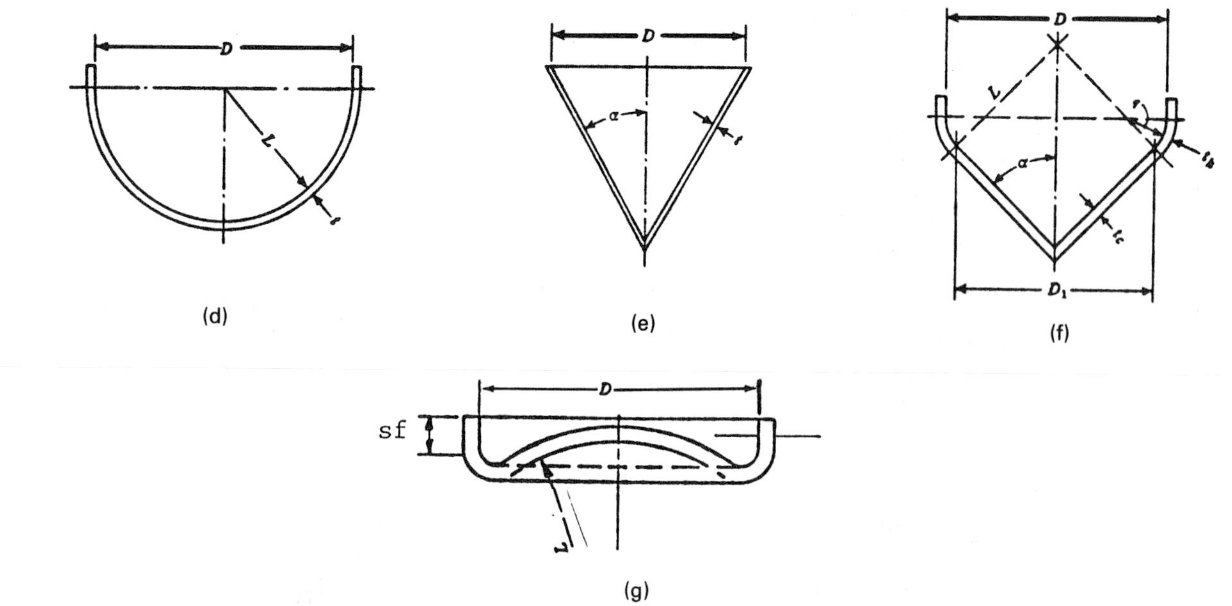

Figure 18.17—(*continued*)

Example 18.6
Dimensions and Weight of a Horizontal Pressure Drum

A drum is to operate at 500°F and 350 psig and to hold 5000 gal at a depth $H/D = 0.8$. Dished heads are to be used. The material is SA285A. Examine the proportions $L/D = 3$ and 5. Formulas are in Table 18.5:

$$V_{\text{tank}} = 5000/7.48 = 668.4 \text{ cuft.}$$

Two heads, capacity with $H/D = 0.8$,

$$V_h = V_0(V/V_0) = 2[0.0778D^3(2)(H/D)^2(1.5 - H/D)]$$
$$= 0.1394D^3.$$

Shell capacity with $H/D = 0.8$,

$$\theta = 2\arccos(1 - 1.6) = 4.4286 \text{ rad},$$
$$V_s = V_0(V/V_0) = (\pi/4)D^2L(1/2\pi)(\theta - \sin\theta)$$
$$= 0.6736D^2L$$
$$V_{\text{liquid}} = 668.4 = 0.1394D^3 + 0.6736D^2L$$

with $L/D = 3$,

$$D = \left(\frac{668.4}{2.1601}\right)^{1/3} = 6.76 \text{ ft}, \quad \text{say } 6.5 \text{ ft},$$

$$L = \frac{668.4 - 0.1394D^3}{0.6736D^2} = 22.1 \text{ ft}, \quad \text{say } 22.0.$$

Allowable stress $S = 11{,}200$ psi.

Say joint efficiency is $E = 0.9$:

$$t_{\text{shell}} = \frac{PR}{SE - 0.6P} = \frac{350(39)}{0.9(11{,}200) - 0.6(350)} = 1.38 \text{ in.}$$

Dished head with $L = D$ and $r/L = 0.06$:

$$t_h = \frac{0.885(350)(78)}{0.9(11{,}200) - 0.1(350)} = 2.41 \text{ in.}$$

Surfaces:

$$\text{shell}, \quad S = \pi DL = 449.3 \text{ sqft},$$
$$\text{heads}, \quad S = 2(0.842)D^2 = 71.2 \text{ sqft},$$
$$\text{Weight} = [449.3(1.4) + 71.2(2.4)]491/12$$
$$= 32{,}730 \text{ lbs.}$$

The results for $L/D = 3$ and 5 are summarized.

Item	$L/D = 3$	$L/D = 5$
D (ft)	6.5	5.5
L (ft)	22.0	32.0
t_{shell} (in.)	1.38 (1.4)	0.957 (1.0)
t_{head} (in.)	2.41 (2.4)	1.67 (1.7)
Weight (lb)	32,730	26,170

The completed vessel will include the weights of nozzles, a manway and reinforcing around the openings, which may total another 10–20%. The weights of these auxiliaries are stated in manufacturers' catalogs.

REFERENCES

1. *API Manual on Refinery Wastes,* Am. Pet. Inst., Washington, D.C., 1969.
2. D. Azbel and N.P. Cheremisinoff, *Chemical and Process Equipment Design: Vessel Design and Selection,* Butterworths, London, 1982.
3. T.E. Belk, Effect of physical chemical parameters on coalescence, *Chem. Eng. Prog.* **61**(10), 72–76 (1965).
4. D. Bradley, *The Hydrocyclone,* Pergamon, New York, 1965.
5. C. Branan, *The Process Engineers Pocket Handbook,* Gulf, Houston, 1976, 1983, Vol. 1, pp. 101–110, Vol. 2, p. 67.
6. C.L. Carpenter and D.F. Othmer, Entrainment removal by a wire mesh separator, *AIChE J.* **1,** 549–557 (1955).
7. R. Chuse, and S.M. Eber, *Pressure Vessels: The ASME Code Simplified,* McGraw-Hill, New York, 1984.
8. G.A. Davies, G.V. Jeffrys, and M. Azfal, A new packing for coalescence and separation of dispersions, *Br. Chem. Eng.* **17,** 709–714 (1972).
9. D.R. Duros and E.D. Kennedy, Acid mist removal, *Chem. Eng. Prog.* **74**(9), 70–77 (1978).
10. F.L. Evans, *Equipment Design Handbook for Refineries and Chemical Plants,* Gulf, Houston, 1980, Vol. 2, pp. 153–165.
11. R.V. Fronczak, Electrical desalting, in *Encyclopedia of Chemical Processing and Design,* Dekker, New York, 1983, Vol. 17, pp. 223–251.
12. F.H. Garner, S.R.M. Ellis, and J.A. Lacey, Size distribution and entrainment of droplets, *Trans. Inst. Chem. Eng.* **32,** 222–235 (1954).
13. W.B. Hooper and L.J. Jacobs, Decantation, in *Handbook of Separation Methods for Chemical Engineers* (P. A. Schweitzer, Ed.), McGraw-Hill, New York, 1979, pp. 1.343–1.358.
14. M.H. Jawad and J.R. Farr, *Structural Analysis and Design of Process Equipment,* Wiley, New York, 1984.
15. G.V. Jordan, Coalescence with porous materials, *Chem. Eng. Prog.* **61**(10), 64–71 (1965).
16. G.S. Laddha and T.E. Degaleesan, in *Handbook of Solvent Extraction* (Lo, Baird, and Hanson, Eds.) Wiley, New York, 1983, p. 125.
17. D.M. Landis, Centrifugal coalescers, *Chem. Eng. Prog.* **61**(10), 58–63 (1965).
18. E.E. Ludwig, *Applied Process Design for Chemical and Petrochemical Plants,* Gulf, Houston, 1977, Vol. 1, pp. 144–180.
19. J.H. Maas, Cyclone separators, in *Handbook of Separation Methods for Chemical Engineers* (P. A. Schweitzer, Ed.), McGraw-Hill, New York, 1979, pp. 6.10–6.17.
20. W.J. Mead, Hoppers and bins and Tanks, in *Encyclopedia of Chemical Process Equipment,* Reinhold, New York, 1964, pp. 546–559 and 941–957.
21. S. Middleman, Drop size distribution produced by turbulent pipe flow of immiscible liquids through a static mixer, *Ind. Eng. Chem., Process. Des. Dev.* **13**(1), 78–83 (1974).
22. H.L. O'Brien, *Petroleum Tankage and Transmission,* Graver Tank and Mfg. Co., East Chicago, IN, 1951.
23. O.C. Redmon, Cartridge type coalescers, *Chem. Eng. Prog.* **59**(9), 87–89 (1963).
24. W. Reisner and M.E. Rothe, *Bins and Bunkers for Handling Bulk Materials,* Trans Tech Publication, Clausthal, Germany, 1971.
25. E.R. Riegel, Tanks, in *Chemical Process Machinery,* Reinhold, New York, 1953, pp. 112–131.
26. K. Rietema and C.G. Verver, *Cyclones in Industry,* Elsevier, New York, 1961.
27. S.S. Safarian and E.C. Harris, *Design and Construction of Silos and Bunkers,* Van Nostrand Reinhold, New York, 1985.
28. A.H. Selker and C.A. Schleicher, Factors affecting which phase will disperse when immiscible liquids are stirred together, *Can. J. Chem. Eng.* **43,** 298–301 (1965).
29. K.T. Semrau, Gas solid separation—cyclones, in *Chemical Engineers Handbook,* McGraw-Hill, New York, 1984, pp. 20.77–20.121.
30. H.P. Sheng, J.R. Welker, and C.M. Sliepcevich, Liquid–liquid separations in a conventional hydroclone, *Can. J. Chem. Eng.* **52,** 487–491 (1974).
31. B. Sigales, How to design reflux drums, *Chem. Eng.* 157–160 (3 Mar. 1975); How to design settling drums, *Chem. Eng.,* 141–143 (23 June 1975); More on how to design reflux drums, *Chem. Eng.,* 87–90 (29 Sep. 1975).
32. A.J. Stepanoff, *Gravity Flow of Bulk Solids and Transport of Solids in Suspension,* Wiley, New York, 1969.
33. L.C. Waterman, Electrical coalescers, *Chem. Eng. Prog.* **61**(10), 51–57 (1965).
34. R.N. Watkins, Sizing separators and accumulators, *Hydrocarbon Proc.* **46**(11), 253–256 (1967).
35. O.H. York, Entrainment separation, in *Encyclopedia of Chemical Processing and Design,* Dekker, New York, 1983, Vol. 19, pp. 168–206.
36. F.A. Zenz, Cyclones, in *Encyclopedia of Chemical Processing and Design,* Dekker, New York, 1982, Vol. 14, pp. 82–97.

19

OTHER TOPICS

Several kinds of processes and equipment that do not fit readily into the categories of earlier chapters have achieved limited but nevertheless valuable applications in chemical processing. Many of them are offered by only a few manufacturers and the performance and economic data are highly proprietary. Here the objective is primarily to describe the principles involved, to point out the main applications, and to refer to sources of more information. Specific manufacturers can be identified by reference to the usual directories of chemical process equipment (References, Chapter 1).

In addition, a number of other topics that also have been ignored in earlier chapters but are of interest to chemical processing will be covered in similarly brief fashion.

19.1. MEMBRANE PROCESSES

Membranes, usually of organic polymers, can be constructed for separation of liquids from dissolved or suspended substances of a wide range of sizes smaller than those normally processed by the kind of filtration equipment described in Chapter 11. The full range of sizes of molecules and particles is illustrated in Figure 19.1. For small dissolved molecules, a phenomenon known as osmosis is the basis for a means of separation. Osmosis becomes manifest when two solutions at the same temperature and pressure but of different concentrations are separated by a semipermeable membrane, namely one that allows passage of the solvent but not the solute. Figure 19.2 illustrates this process. One theory of the action of a semipermeable membrane is that the solvent dissolves in the membrane at the face of higher concentration or higher partial pressure and is released at the other face where the concentration is lower.

The natural tendency is for the solvent to flow in the direction that will equalize the concentrations. It turns out, however, that if a certain pressure, called the osmotic pressure, is imposed on the more concentrated solution, flow the solvent can be forced in the direction from the more concentrated to the more dilute solution. For the case of pure solvent on the low pressure side of the membrane, the osmotic relation is

$$\ln \gamma_w x_w = -\frac{1}{RT}\int_0^{P_{\text{osm}}} \bar{V}_w \, dP = -\frac{\bar{V}_w}{RT} P_{\text{osm}}, \tag{19.1}$$

where γ is the activity coefficient, x is the mol fraction, and $\bar{V}$ is the partial molal volume; subscript w identifies the solvent. For ideal solutions, the activity coefficient is unity. Since nonideality is of common occurrence, this equation may be used to find activity coefficients from measurements of osmotic pressure. Example 19.1 illustrates this process, and points out incidentally how rapidly the osmotic pressure falls off with increasing molecular weight of the solute.

Size ranges for membrane processing by reverse osmosis, ultrafiltration and microfiltration are shown in Figure 19.1. Reverse osmosis is effective in removing solvents away from dissolved molecules. Because of limitations in crushing strengths of membranes, pressures are limited to maxima of about 1000 psi

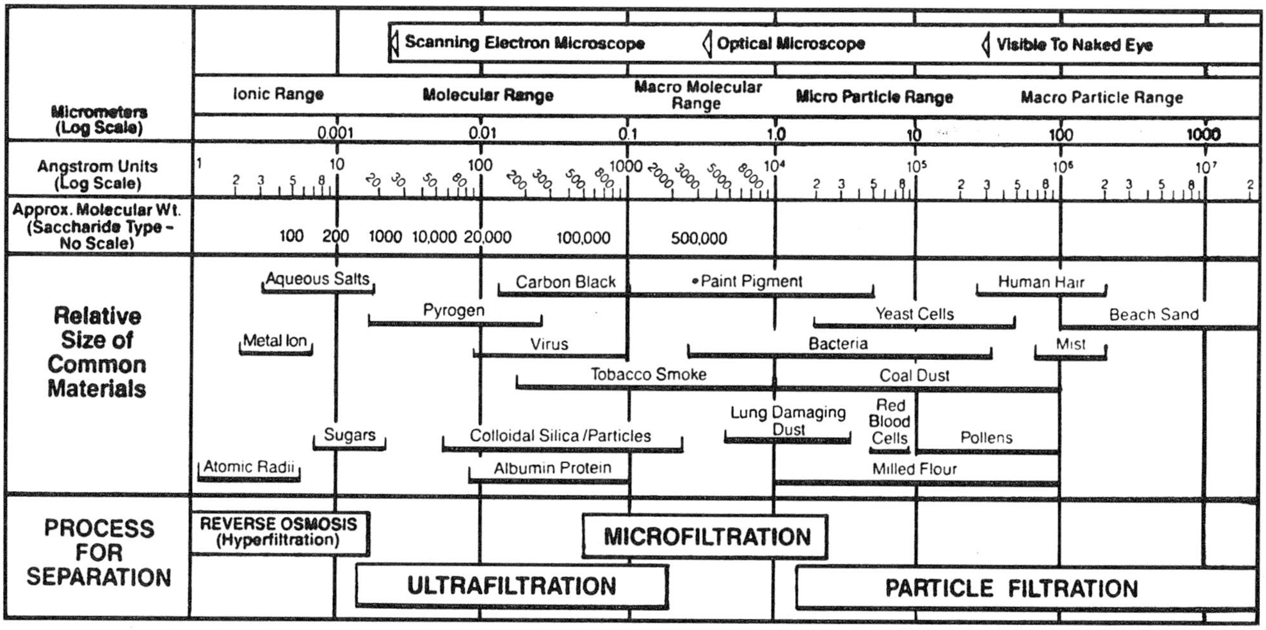

Figure 19.1. Range of molecular weights and particle or droplet sizes of common materials, how they are measured, and the methods employed for their removal from fluids (*Osmonics Inc.*).

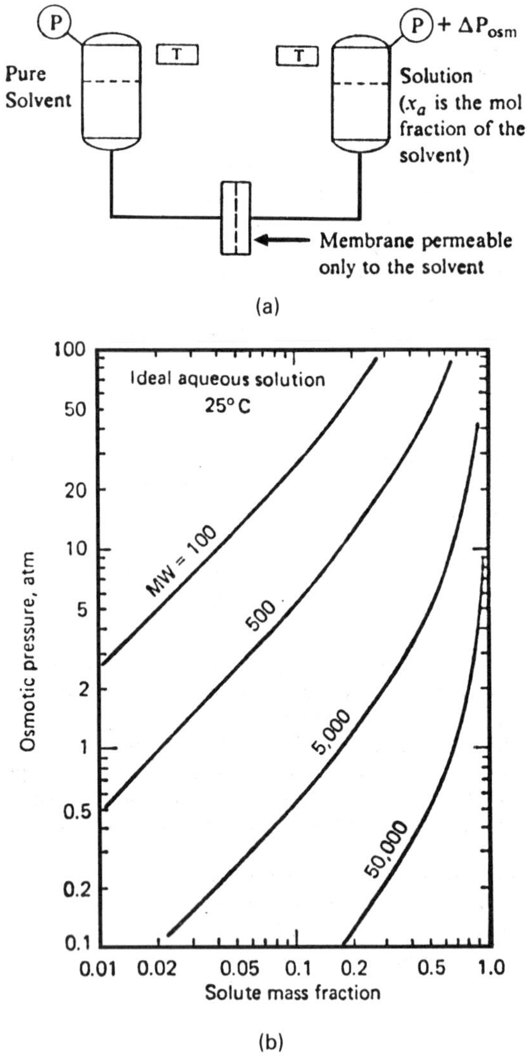

Figure 19.2. Diagram of osmotic behavior and the effect of solute concentration and molecular weight on osmotic pressure. (a) Osmotic-pressure behavior of solutions; ΔP_{osm} is the excess pressure on the solution required to stop flow of solvent through the semipermeable membrane. (b) Effects of solute concentration and molecular weight on osmotic pressure.

(68 atm). Flow rates of 2–200 gal/(sqft)(day) or 0.001–0.1 kg water/m^2 sec are attained in various units. Ultrafiltration operates at 1–10 atm differential and is effective for the molecular weight range of 1000–200,000 which includes many proteins, viruses, and bacteria. Ultra and micro filtrations somewhat overlap. Pressures for microfiltration are about 1 atm differential. Since these processes are relatively expensive, their applications are limited largely to analytical purposes and in water treatment for pharmaceutical manufacturing. Some specific applications are listed in Table 19.1.

MEMBRANES

The first commercially successful membrane was the anisotropic or asymmetric structure invented by Loeb and Sourirajan (1960; cited by Sourirajan, *Reverse Osmosis,* Academic, N.Y., 1970). It is made of cellulose acetate and consists of a dense layer 0.2–0.5 μm thick deposited on a porous structure 50–100 μm thick with pores 0.1–1.0 μm dia. The thin film has the desired solute retention property while offering little resistance to flow and the porous substructure offers little resistance to flow but provides support for the skin. The characteristics of available membranes for reverse osmosis and ultrafiltration are listed in Tables 19.2–19.4.

Hollow fiber membranes are primarily homogeneous. In use, their lower permeability is compensated for by large surface per unit volume of vessel. Fibers are 25–250 μm outside dia, wall thickness 5–50 μm. The cross section of a vessel for reverse osmosis may have 20–35 million fibers/sqft and a surface of 5500-9000 sqft/cuft of vessel. Recently developed hollow fibers for gas permeation processes have anisotropic structures.

EQUIPMENT CONFIGURATIONS

Four principal kinds of membrane assemblies are in use:

a. Tubular, in which the membrane is deposited either on the inside or the outside of porous tubes, most commonly inside for reverse osmosis and outside for ultrafiltration. Figure 19.3(a) shows a single-tube construction but units with 7 or 19 tubes in a single shell are made as standard items. Table 19.5 lists some available sizes. "Dynamic membranes" may be deposited on porous stainless steel tubes from a feed solution containing 50–100 mg/L of membrane-forming solution which consists of polyacrylic acid and hydrous zirconium oxide. Such a membrane can be deposited in 1 hr and replaced as quickly. Fluxes are very high; 100 gal/(sqft)(day) is shown in Table 19.6(a). Some applications are described by Turbak (1981, Vol. II, pp. 434–453).

b. Plate-and-frame construction is shown in Figures 19.3(b) and (c). It is used more commonly for ultrafiltration. A related kind of equipment is the electrodialysis plate-and-frame equipment of Figure 15.21.

c. Spiral wound assemblies are illustrated in Figure 19.4. They consist of a long envelope of membrane sealed on the edges and enclosing a porous material which serves as a channel for the flow of the permeate. The spacer for the feed solution flow channel is a meshlike material through which the solution is forced under pressure. The modules listed in Table 19.5 are 2–8 in. dia, up to 3 ft long, and provide about 250 sqft of membrane surface/cuft of vessel. Dimensions are shown on Figure 19.4(c). According to Table 19.6, reverse osmosis rates of 2500 gal/(sqft)(day) are attained.

d. Hollow fiber assemblies function as one-ended shell-and-tube devices. At one end the fibers are embedded in an epoxy tubesheet and at the other end they are sealed. Overall flows of feed solution and permeate thus are in counterflow. Flow of permeate is into the tubes which takes advantage of the great crushing strengths of the small diameter fibers, and constitutes a "fail-safe" operation since collapse of fibers results in closure whereas bursting would result in leakage. The most serious drawback is some difficulty of cleaning. A widely used equipment of this type is illustrated in Figure 19.5.

APPLICATIONS

The greatest use of membranes is for reverse osmosis desalination of seawater and purification of brackish waters. Spiral wound and hollow fiber equipment primarily are applied to this service. Table 19.6 has some operating data, but the literature is very extensive and reference should be made there for details of performance and economics.

Example 19.1
Applications of the Equation for Osmotic Pressure

a. The osmotic pressure of a sucrose solution is 148.5 atm at 20°C. The concentration is 1.43 kg sucrose/kg water, corresponding to a mol fraction 0.0700 of sucrose. The partial molal volume of water is approx 0.018 L/g mol. Accordingly, the activity coefficient of the water is

$$\gamma = \frac{1}{1-0.07}\exp\left[-\frac{0.018(148.5)}{0.082(293.2)}\right] = 0.9622.$$

The difference from unity appears to be small but is nevertheless significant. At this concentration, if the activity coefficient were unity, the osmotic pressure would be

$$P = -\frac{0.082(293.2)}{0.018}\ln(1-0.07) = 97\text{ atm},$$

which is considerably in error.

b. The effect of molecular weight on ideal osmotic pressures of a variety of solutions containing 0.1 kg solute/kg water is demonstrated in this tabulation:

Mol Weight	Mol fraction x	Ideal Osmotic P (atm)
58.5 (NaCl)	0.0374	50.9
100	0.0177	23.9
342 (sucrose)	0.00524	7.0
1000	0.0018	2.4
10,000	1.8($E-4$)	0.24
100,000 (virus and protein)	1.8($E-5$)	0.024

Figure 19.1 identifies sizes of common molecules and particles. Clearly, osmotic pressures are essentially negligible for molecular weights above 10,000 or so.

Because of the low energy requirements of separations by reverse osmosis, much attention has been devoted to other separations of aqueous solutions, at least on a laboratory scale, for instance, of ethanol/water. Membranes have been found that are moderately effective, but the main obstacle to the process is the very high pressures needed to remove water from high concentrations of ethanol against pure water on the low pressure side. A practical method of circumventing this problem is to replace the water on the low pressure side by a solution of sufficiently high concentration to allow the application of only moderate pressure. The case examined in Example 19.2 utilizes a solution of ethylene glycol on the low pressure side of such concentration that concentration of ethanol above the azeotropic composition can be achieved with a pressure of only 1000 psig. The glycol is easily separated from water by distillation.

GAS PERMEATION

Differences in rates of permeation of membranes by various gases are utilized for the separation of mixtures, for instance, of hydrogen from ammonia plant gas, of carbon dioxide from natural gas, and of helium from natural gas. The successful "Prism" process of Monsanto Co. employs hollow fibers of a porous polysulfone base coated with possibly a thin film of silicone rubber. The fibers are about 800 μm outside and 400 μm inside dia. They are housed in vessels 4–8 in. dia and 10–20 ft long and may contain 10,000–100,000 fibers per bundle. A schematic of such a unit is in Figure 19.6(a). Pressures up to 150 atm are allowable. A unit 4 in. dia by 10 ft long was able to upgrade 290,000 SCFD of ammonia plant purge gas, making a product with 90% hydrogen and a waste of 20% hydrogen from a feed with 37% hydrogen.

TABLE 19.1. Examples of Applications of Ultrafiltration

(a) Applications Involving Retained Colloidal Particles

Material	Application
Pigments and dispersed dyes	Concn/purification of organic pigment slurries; separation of solvents, etc. from pigment/resin in electropaints; concn of pigments in printing effluents
Oil-in-water emulsion globules	Concn of waste oils from metal working/textile scouring; concn of lanolin/dirt from wool scouring
Polymer lattices and dispersions	Concn of emulsion polymers from reactors and washings
Metals/nonmetals/oxides/salts	Concn of silver from photographic wastes; concn of activated carbon slurries; concn of inorganic sludges
Dirt, soils, and clays	Retention of particulates and colloids in turbid water supplies; concn of fines in kaolin processing
Microorganisms	Retention of microbiological solids in activated sludge processing; concn
Plant/animal cellular materials	of viral/bacterial cell cultures; separation of fermentation products from broth; retention of cell debris in fruit juices, etc.; retention of cellular matter in brewery/distillery wastes

(b) Applications Involving Soluble Macromolecules

Material	Application
Proteins and polypeptides	Concn/purification of enzymes; concn/purification of casein and whey proteins; concn/purification of gluten/zein; concn/purification of gelatin; concn/purification of animal blood; retention of haze precursors in clear beverages; retention of antigens in antibiotics solutions; concn/purification of vegetable protein extracts; concn/purification of egg albumen; concn/purification of fish protein extracts; retention of proteins in sugar diffusion juice
Polysaccharides and oligosaccharides	Concn of starch effluents; concn of pectin extracts
Polyphenolics	Concn/purification of lignosulphonates
Synthetic water-soluble polymers	Concn of PVA/CMC desize wastes

[From N. C. Beaton and H. Steadly, in *Recent Developments in Separation Science* (N. N. Li, Ed.), 1981, Vol. VII, pp. 2–29].

TABLE 19.2. Data of Commercial Equipment for Reverse Osmosis and Ultrafiltration

(a) Equipment of Amicon Corp.

Diaflo®	Nominal mol wt cutoff	Apparent pore diam, Å	Water flux, gal/ft²/day at 55 lb/in²
UM 05	500	21	10
UM 2	1,000	24	20
UM 10	10,000	30	60
PM 10	10,000	38	550
PM 30	30,000	47	500
XM 50	50,000	66	250
XM 100A	100,000	110	650
XM 300	300,000	480	1300

(b) Equipment of Nucleopore Corp.

Specified pore size, μm	Pore-size range, μm	Nominal pore density, pores/cm²	Nominal thickness, μm	Typical flow rates at 10 lb/in² (gage), ΔP, 70°F: Water, gal/(min)(ft²)	N_2, ft³/(min)(ft²)
8.0	6.9–8.0	1×10^5	8.0	144.0	138.0
5.0	4.3–5.0	4×10^5	8.6	148.0	148.0
3.0	2.5–3.0	2×10^6	11.0	121.0	128.0
1.0	0.8–1.0	2×10^7	11.5	67.5	95.0
0.8	0.64–0.80	3×10^7	11.6	48.3	76.0
0.6	0.48–0.60	3×10^7	11.6	16.3	33.0
0.4	0.32–0.40	1×10^8	11.6	17.0	33.0
0.2	0.16–0.20	3×10^8	12.0	3.1	8.9
0.1	0.08–0.10	3×10^8	5.3	1.9	5.3
0.08	0.064–0.080	3×10^8	5.4	0.37	2.6
0.05	0.040–0.050	6×10^8	5.4	1.12	1.3
0.03	0.024–0.030	6×10^8	5.4	0.006	0.19

(c) Equipment of Koch Membrane Systems (Formerly Abcor)

Membrane Type (1)	*Nominal M.W. Cutoff (2)*	*Max. Temp (°C) (3)*	*Range pH (4)*	*Maximum Pressure PSI (Kg/cm²)*	*Configuration (5)*
MSD-324*	1,500	90	1-13	150 (10.5)	S
HFK-132	3,500	90	1-13	150 (10.5)	S
HFK-131	5,000	90	1-13	150 (10.5)	S
HFD-300	8,000	80	2-12	150 (10.5)	T
HFM-100	10,000	90	1-13	150 (10.5)	S,T
HFA-251*	15,000	50	2-8	150 (10.5)	T
HFM-180	18,000	90	1-13	150 (10.5)	S,T
HFM-163	18,000	60	2-12	150 (10.5)	S,T
HFP-276	35,000	90	1-13	150 (10.5)	S,T
MSD-181	200,000	90	1-13	150 (10.5)	S,T
MSD-400*	100,000	90	1-13	150 (10.5)	S,T
MSD-405*	250,000	90	1-13	150 (10.5)	S,T
MMP-406*	0.2 Microns	90	1-13	150 (10.5)	S,T
MMP-404*	0.4 Microns	90	1-13	150 (10.5)	S,T
MMP-516*	2 Microns	90	1-13	150 (10.5)	S,T
MMP-407*	2-3 Microns	90	1-13	50 (3.5)	S,T
MMP-600	1-2 Microns	90	1-13	50 (3.5)	S,T
MMP-602	2-3 Microns	90	1-13	50 (3.5)	S

(1) Membranes beginning with "H" designation are stock items.
(= hydrophyllic)*
(2) The nominal molecular weight (M.W.) cutoff is provided as a guide to the relative pore size for these membranes. Since many factors influence the actual MW cutoff, tests must be run to confirm retention for any specific application.
(3) At pH – 6.
(4) At 25°C.
(5) F = Flat Sheet S = Spiral T = Tubular.

Because of the long, narrow configuration, the equipment appears to function in countercurrent mode. Other data of experiments with gas permeators as continuous columns appear in Figures 19.6(b) and (c); the original paper has data on other binary and some complex mixtures.

Permeability of a membrane is determined partly by gas diffusivity, but adsorption phenomena can exist at higher pressures. Separation factors of two substances are approximately in the ratios of their permeabilities, $\alpha_{AB} = P_{oA}/P_{oB}$. Some data of permeabilities and separation factors are in Table 19.7, together with a list of membranes that have been used commercially for particular separations. Similar but not entirely consistent data are tabulated in the *Chemical Engineers Handbook* (McGraw-Hill, New York, 1984, pp. 17.16, 17.18).

19.2. FOAM SEPARATION AND FROTH FLOTATION

Foams are dispersions of gas in a relatively small amount of liquid. When they are still on the surface of the liquid from which they were formed, they also are called froths. Bubbles range in size from about 50 μm to several mm. The data of Table 19.8 show densities of water/air foams to range from 0.8 to 24 g/L. Some dissolved or finely divided substances may concentrate on the bubble surfaces. Beer froth, for instance, has been found to contain 73% protein and 10% water. Surface active substances attach themselves to dissolved materials and accumulate in the bubbles whose formation they facilitate and stabilize. Foam separation is most effective for removal of small contents of dissolved impurities. In the treatment of waste waters for instance, impurities may be reduced from a content measured in parts per million to one measured in parts per billion. High contents of suspended solids or liquids are removed selectively from suspension by a process of froth flotation.

FOAM FRACTIONATION

Some dissolved substances are attracted to surfactants and thus are concentrated and removed with a foam. Such operations are performed in batch or continuous stirred tanks, or in continuous towers as in the flowsketch of Figure 19.7. Compressed air may be supplied through a sparger or ambient air may be drawn into a high speed rotating gas disperser. Improved separation is achieved by staged operation, so that a packed tower is desirable. Moreover, packing assists in the formation of a stable foam since that is difficult to do in an empty tower of several feet in diameter. Larger contents of surfactant usually are needed in large towers than in laboratory units. In pilot plant work associated with the laboratory data of Table 19.8, a tower 2 ft square by 8 ft high was able to treat 120 gal/hr of feed. The laboratory unit was 1 in. dia, so that the gas rate of 154 cm^3/min of Table 19.8 corresponds to a superficial gas velocity of 1.1 ft/min.

Most of the work on foam fractionation reported in the literature is exploratory and on a laboratory scale. A selected list of about 150 topics has been prepared with literature references by Okamoto and Chou (1979). They are grouped into separation of metallic ions, anions, colloids, dyes and organic acids, proteins, and others.

Stable foams that leave the fractionator are condensed for further processing or for refluxing. Condensation may be effected

TABLE 19.3. Properties of Membranes for Reverse Osmosis

(a) Cellulose Acetate Membranes

Membrane	Manufacturer	Volumetric flow rate (l/m²d)	NaCl retention (%)	Pressure (bar)	NaCl concentration (%)
CA*					
(Tubing & Flat)	KALLE	500–2500	98,6–60	40	0,5
CA* (Flat)	DDS***	350–220	99–78	42	0,05
CA* (Tubing)	PATERSON CANDY INT.		97–50	40	0,5
CA* („In-situ cast"; pipe)	ABCOR	450–900	98–90	42	0,5
CA* (Wound module)	UNIV. OIL PROD. (GULF)	650	98	70	3,5
CA* (Hollow fiber)	MONSANTO	130	94	18	3,5
CTA** (Hollow fiber)	DOW CHEM. COMP.	50	99.5	70	3,5
CTA** Ultra-thin wound module	UNIV. OIL PROD. (GULF)	550	99,3	70	3,5

(A. Walch, *Proceedings, Membrane Conference, Lund, Sweden,* 1976; *Ullmann's Encyclopedia of Chemical Technology,* Verlag Chemie, Weinheim, 1978, Vol. 16, pp. 515–535).

* CA: Cellulose-2,5-acetate, ** CTA: Cellulosetriacetate, *** De Danske Sukkerfabriken

TABLE 19.3—*(continued)*

(b) Other Kinds of Membranes

Membrane	Type	Mfgr	Volumetric flow rate/pressure (ml / min • cm² • bar)	Molecular weight range	Conditions
Polyelectrolyte (composite)	UM 05 – UM 10	AMICON	0,005 – 0,08	500 – 10 000	max. 4 bar; 50°C; pH 4 – 10
Polysulfone (asymmetric hollow fibers)	PM 10 – PM 30	AMICON	0,4 – 13	10 000 – 30 000	max. 4 bar; 115°C; pH 0 – 12
Mod acrylics (asymmetric hollow fibers)	XM 50 – XM 300	AMICON	0,6 – 2,0	50 000 – 300 000	max. 2 bar; 70°C; pH 0 – 12
Polyelectrolyte (composite)	PSAC, PSDM	MILLIPORE	0,18 – 0,15	1000 – 40 000	max. 7 bar; 50°C; pH 4 – 10
Cellulose triacetate (60μm)	PEM	GELMAN	0,004	50 000	max. 80°C; pH 1 – 10
Regenerated cellulose (symmetric hollow fibers)	BF 50	DOW CHEM. COMP.	–	5000	max. 70°C; pH 1 – 12
Cellulose acetate (symmetric hollow fibers)	BF 80	DOW CHEM. COMP.	–	30 000	max. 50°C; pH 2 – 8
Regenerated cellulose (100 um)	115	SARTORIUS	0,001 – 0,1	20 000 – 160 000	max. 80°C; pH 1 – 12
Cellulose acetate (100 μm)	117	SARTORIUS	0,001 – 0,1	20 000 – 160 000	max. 80°C; pH 2 – 8
Cellulose nitrate (100 μm)	121	SARTORIUS	0,005 – 0,02	10 000 – 50 000	max. 80°C; pH 1 – 10
ZrO_2/Carbon (asymmetric tube bundle)	Ucarsep	UNION CARBIDE	0,02	30 000 etc.	max. 10 bar; 100°C; pH 1 – 14
Polyamide, polyimide (asym, composite, hollow)	BM 10 – BM 500	BERGHOF	0,004 – 0,7	1000 – 50 000	max. 100°C; pH 2 – 10
Co-polyacrylonitrile (sym)		RHÔNE-POULENC	0,017	70 000	max. 2 bar; pH 2 – 12
Cellulose acetate (asym)	800 – 500	DDS***	0,01 – 0,04	6000 – 60 000	max. 20 – 100 bar; 30 – 50°C; pH 3 – 7
Polysulfone (asym)	GR 5, 6, 8	DDS***		10 000 – 20 000	max. 80°C; pH 0 – 14
Polyacrylonitrile (asym)	FPB – GPA	DORR-OLIVER	0,1 – 0,35	1200 – 100 000	max. 2 – 4 bar; 70 – 80°C; pH 1 – 13
Cellulose acetate (asym)	T2/A – T5/A	PCI⁽⁰⁾	–	1000 – 20 000	max. 10 – 25 bar; 30 – 50°C; pH 3 – 7
Cellulose derivative	T6/B	PCI⁽⁰⁾	–	120 000	max. 10 bar; 60°C; pH 2 – 11
Cellulose acetate (asym)	HFA/100 – HFA/300	ABCOR	0,005 – 0,3	15 000 – 50 000	max. 14 – 100 bar; 30 – 50°C; pH 3 – 7
Polyamide (asym)		ABCOR		50 000	
Cellulose acetate (asym)	UF 6 – UF 100	KALLE	0,01 – 0,1	6000 – 100 000	3 – 10 bar; 30 – 50°C; pH 3 – 7
Polyamide (asym)	PA 40 – PA 100	KALLE	0,06 – 0,1	40 000 – 100 000	3 – 6 bar; max. 70°C; pH 1 – 11

** BM: Berghof-Membran *** DDS: DE DANSKE SUKKERFABRIKEN
(0) PCI: PATERSON CANDY INTERNATIONAL

by a blast of steam, by contact with a hot surface, by chemical antifoaming agents, sonically or ultrasonically, or by contact with a high speed rotating disk as appears in the flowsketch, Figure 19.7.

FROTH FLOTATION

Finely divided solids or immiscible liquids can be made to adhere to gas bubbles and then can be removed from the main liquid. Affinity of a solid for an air bubble can be enhanced with surfactants which adhere to the surface of the solid and make it nonwetting. The main application of froth flotation is to the separation of valuable minerals from gangue. Ores of Cu, Zn, Mo, Pb, and Ni are among those commercially preconcentrated in this way. Reagent requirements of each ore are unique and are established by test. A large amount of experience exists, however, and information is supplied freely by reagent manufacturers. Some recipes are given with descriptions of flotation processes in books on mineral dressing, for example, that of Wills (1985).

Promoters or collectors give the mineral the water-repellent coating that will adhere to an air bubble. Frothers enhance the formation and stability of the air bubbles. Other additives are used to control the pH, to prevent unwanted substances from floating, or to control formation of slimes that may interfere with selectivity.

Air is most commonly dispersed with mechanical agitation. Figure 19.8 illustrates a popular kind of flotation cell in which the gas is dispersed and the pulp is circulated with impellers. Such vessels have capacities of 300–400 cuft. Usually several are connected in series as in Figure 19.8(b). The froth is removed from

TABLE 19.4. Properties of Membranes for Ultrafiltration

	Membrane	Mfgr	Volumetric flow rate (l / m²d)	NaCl retention (%)	Press (bar)	NaCl concen (%)	Stability
Polyamide	Aromatic polyamide (asym hollow fiber B9)	DuPont	50	95	28	0.15	max. 35°C pH 4–11
	Copolyamide asym hollow fiber B10	DuPont	30	98.5	56	3.0	max. 30°C pH 5–9
	Polyamide hydrazide asym wound module	DuPont	500	99	70	3.5	–
	Polypiperazinamide asym fibers	Montecatini	600	98	80	1	chlor-resistent
	Aromatic polyamide ultrathin wound modules	Univ. Oil Prod.	1700	98.9	70	3.5	–
Polyurea	Ethyleneimine, toluylene-diisocyanate NS 100 ultrathin wound module	North Star Res. Inst. (Univ. Oil Prod.)	500	99.6	70	3.5	pH 2–12
Polyfuran	Furfuryl alcohol. H_2SO_4 ; NS 100 ultra thin	North Star Res. Inst.	1000	99.6	70	3.5	pH 2–12
Polyether	Polyphenylene oxide sulfone (5 μm on polypropylene tube	General Electric	1500	84	77	0.1	max. 54°C
	Polysulfone	Rhône-Poulenc	90	99	60	3.5	max. 60°C
Polyhetero-aromatics	Polybenzimidazole asym tube	Celanese	800	95	41	0.5	
	Polybenzimidazole asym hollow fiber	Celanese	50	99.4	70	3.5	
	Polyimide, methoxyl 10 μm	Battelle (BRD)	20	99.7	100	3.5	
Fluoropolymer	Naflon (sulf, 250μm)	DuPont	3	85	100	3.5	
	Permion (pyridine, 25 μm)	RAI-Res. Corp.	1	98.8	100	3.5	
Inorganic membranes	Glass hollow fiber	Stanford Res. Inst.	15	83	102	1	
	Glass hollow fiber	Schott & Gen.		98	120	0.5	
	Graphite, oxidized	Westinghouse (Union Carbide Comp.)	50	80	41	0.5	
Dynamically formed membranes	ZrO_2/polyacrylic acid	Oak Ridge Nat. Lab.	5000	90	70	0.3	
	Polyacrylic acid	Univ. Oil Prod. (Gulf)	2000	80	102	0.3	
Vinyl polymers	Polyvinylpyrrolidone cross linked 75 μm	Univ. Oil Prod. (Gulf)	0.5	99	34	0.8	
	Polyvinyl alcohol cross lined 29 μm	Princeton University	3	93	42	0.6	
	Polyvinyl carbonate 94 μm	Aerojet-Gen. Corp.	3	94	102	3.5	
	Vinyl copolymer 10 μm	Battelle (BRD)	70	96	100	0.5	

(A. Walch, *Proceedings, Membrane Conference, Lund, Sweden,* 1976; *Ullmann's Encyclopedia of Chemical Technology,* Verlag Chemie, Weinheim, 1978, Vol. 16, pp. 515–535).

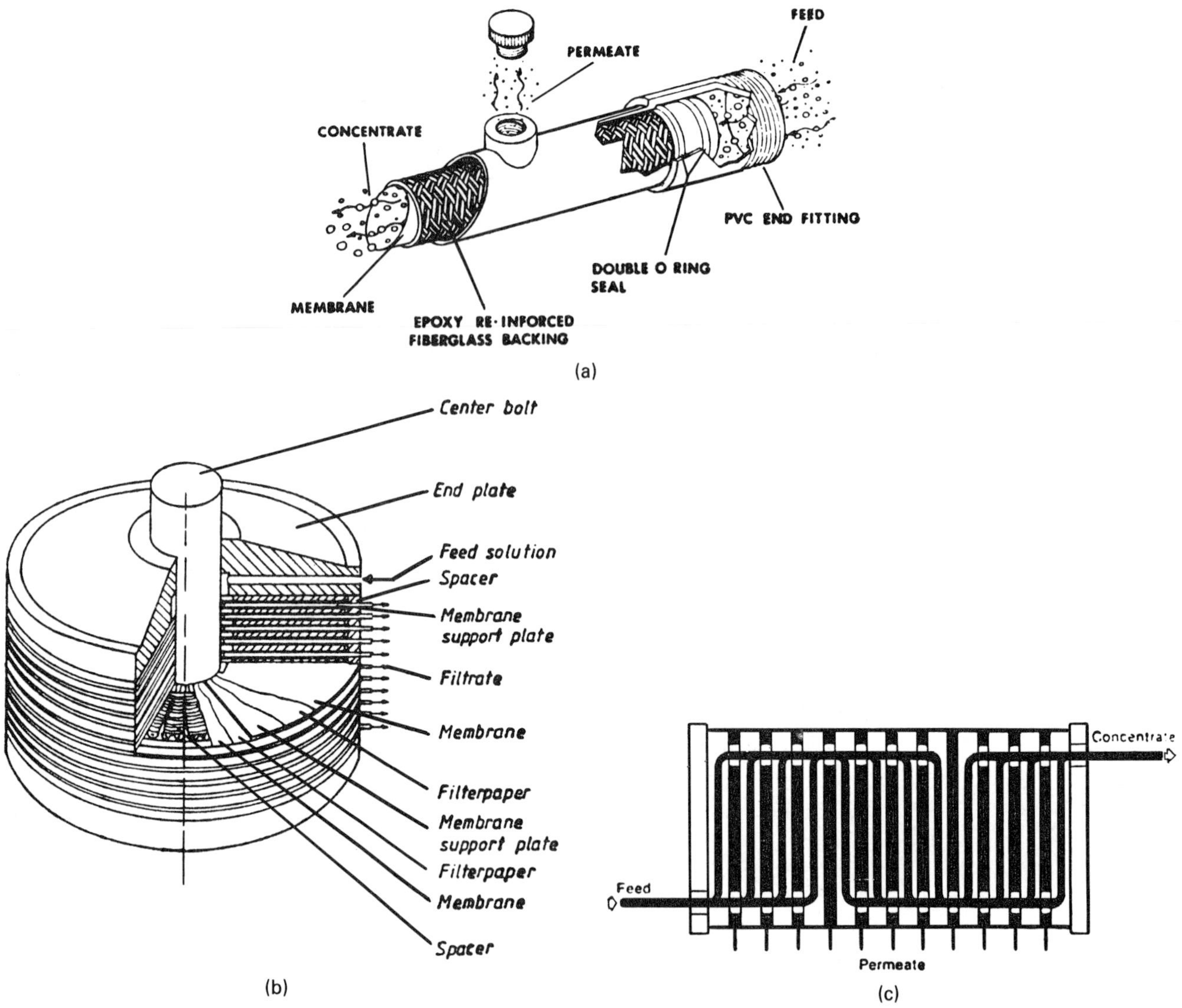

Figure 19.3. Tubular and plate-and-frame membrane modules for reverse osmosis and ultrafiltration. (a) Construction and flow pattern of a single 1 in. dia tube with membrane coating on the inside; in Table 19.4, the "Ultracor" model has seven tubes in a shell and the "Supercor" has 19 [*Koch Membrane Systems* (*Abcor*)]. (b) Assembly of a plate-and-frame ultrafiltration module (*Danish Sugar Co.*). (c) Flow in a plate-and-frame ultrafiltration module.

each cell as it is formed, but the pulp goes through the battery in series. The froth is not highly stable and condenses readily without special provisions as it overflows. Since some entrainment of gangue occurs, usually it is desirable to reprocess the first froth. The flowsketch of Figure 19.9 illustrates such reprocessing. The solids to the first stage are ground here to −65 mesh, which normally is fine enough to release the mineral, and to −200 mesh in the final stage.

Total residence time in a bank of cells may range from 4 to 14 min. A table of approximate capacities of several makes of flotation cells for a pulp with 33% solids of specific gravity = 3 is given in the *Chemical Engineers' Handbook* (McGraw-Hill, New York, 1984, p. 21.49); on an average, an 8-cell bank with 4-min holdup has a capacity of about 1.5 tons solid/(hr) (cuft of cell) and a power requirement of about 0.6 HP/(cuft of cell).

The chief nonmineral application of froth flotation is to the removal or oil or grease or fibrous materials from waste waters of refineries or food processing plants. Oil droplets, for instance, attach themselves to air bubbles which rise to the surface and are skimmed off. Coagulant aids and frothers often are desirable. In one kind of system, the water is saturated with air under pressure and then is pumped into a chamber maintained under a partial vacuum. Bubbles form uniformly throughout the mass and carry out the impurities. The unit illustrated in Figure 19.10 operates at 9 in. mercury vacuum and removes both skimmed and settled sludges. Because of the flocculation effect it is able to process waste water at an enhanced rate of about 5000 gal/(sqft)(day) instead of the usual rate of 800–1000.

In another application, particles of plastics in waste stream are chopped to diameters of 5 mm or less, passed through flotation cells containing proprietary surfactants, and removed as an air froth.

19.3. SUBLIMATION AND FREEZE DRYING

Sublimation is the transformation of a solid directly into vapor and desublimation is the reverse process of condensing the vapor as a

TABLE 19.5. Specifications of Spiral and Tubular Equipment for Reverse Osmosis and Ultrafiltration[a]

Module	Length	Membrane Area/Module
Tubular UF 1″ (2.5 cm) dia.	5 ft. (1.5 m)	1.1 sq. ft. (.10 m²)
Tubular UF 1″ (2.5 cm) dia.	10 ft. (3 m)	2.2 sq. ft. (.20 m²)
ULTRA-COR™ UF Tubes .5″ (1.27 cm) dia.	10 ft. (3 m)	7.4 sq. ft. (.68 m²)
SUPER-COR™ UF Tubes	10 ft. (3 m)	24 ft.² (2.2 m²)
Tubular RO ½″ (1.27 cm) dia.	12 ft. (3.6 m)	48 sq. ft. (4.4 m²)
Spiral UF 2″ (5 cm) dia.	1.2 ft. (.36 m)	2.5 sq. ft. (.23 m²)
Spiral UF 4″ (10 cm) dia.	3 ft. (.9 m)	35-60 sq. ft. (3.2-5.5 m²)
Spiral UF 8″ (20 cm) dia.	3 ft. (.9 m)	150-250 sq. ft. (13.9-23 m²)
Spiral RO 4″ (10 cm) dia.	3 ft. (.9 m)	60 sq. ft. (5.5 m²)

[a] The "Ultracor" model has 7 and the "Supercor" has 19 tubes/shell.
(Koch Membrane Systems, formerly Abcor).

solid. The term pseudosublimation is applied to the recovery of solid condensate from the vaporization of a liquid.

The goal of a commercial sublimation is the separation of a valuable material from nonvolatile ones at temperatures low enough to avoid thermal degradation. The preservation of cell structure (and taste) is a deciding factor in the choice of freeze drying, a special instance of sublimation, foods, pharmaceuticals, and medical products.

Only a few solids have vapor pressures near atmospheric at safe temperatures, among them CO_2, UF_6, $ZrCl_4$, and about 30 organics. Ammonium chloride sublimes at 1 atm and 350°C with decomposition into NH_3 and HCl, but these recombine into pure NH_4Cl upon cooling. Iodine has a triple point 113.5°C and 90.5 Torr; it can be sublimed out of aqueous salt solutions at atmospheric pressure because of the entraining effect of vaporized water.

Sublimation pressures down to 0.001 bar are considered feasible. At lower pressures and in some instances at higher ones, entrainer gas is used, usually air or nitrogen or steam. By such means, for instance, salicyclic acid is purified by sublimation at 150°C with an entrainer of air with sufficient CO_2 to prevent decarboxylation of the acid. At the operating temperature, the vapor pressure is only 0.0144 bar. Operating conditions corresponding to equilibrium in the sublimer appear in Figure 19.11. Equilibrium may be approached in equipment where contact between phases is intimate, as in fluidized beds, but in tray types percent saturation may be as low as 10%.

Among substances that are sublimed under vacuum are anthranilic acid, hydroxyanthraquinone, naphthalene, and β-naphthol. Pyrogallol and *d*-camphor distill from the liquid state but condense as solids. Several metals are purified by sublimation, for instance, magnesium at 600°C and 0.01–0.15 Torr.

The common carrier gases are air or nitrogen or steam. Condensate from a carrier usually is finely divided, snowlike in character, which is sometimes undesirable. Substances which are sublimed in the presence of a carrier gas include anthracene, anthraquinone, benzoic acid, phthalic anhydride, and the formerly mentioned salicylic acid.

A partial list of substances amenable to sublimation is in Table 19.9.

EQUIPMENT

The process of sublimation is analogous to the drying of solids so much the same kind of equipment is usable, including tray dryers (Fig. 9.6), rotary tray dryers (Fig. 9.8), drum dryers [Fig. 9.11(b)], pneumatic conveying dryers (Fig. 9.12), and fluidized beds (Fig. 9.13). The last of these requires the subliming material to be deposited on an inert carrier which is the fluidized material proper.

Condensers usually are large air-cooled chambers whose walls are kept clear with brushes or scrapers or even swinging weights. Scraped or brushed surface crystallizers such as Figure 16.10(a) should have some application as condensers. When a large rate of entrainer gas is employed, a subsequent collecting chamber will be needed. One of the hazards of entrainer sublimation with air is the possibility of explosions even of substances that are considered safe in their normal states.

FREEZE DRYING

Preservation of cell structure, food taste, and avoidance of thermal degradation are reasons for the removal of moisture from such materials by sublimation. The process is preceded by quick freezing which forms small crystals and thus minimum damage to cell walls, and is likely to destroy bacteria. Some of the materials that are being freeze dried commercially are listed in Table 19.9(b).

The most advanced technique of quick freezing is by pouring the material onto a freezing belt. Before drying, the material is granulated or sliced to improve heat and diffusional mass transfer. These operations are conducted in cold rooms at about −46°C.

Sublimation temperatures are in the range of −10 to −40°C and corresponding vapor pressures of water are 2.6–0.13 mbar. Cabinet tray dryers are the most commonly used type. The trays are lifted out of contact with hot surfaces so the heat transfer is entirely by radiation. Loading of 2.5 lb/sqft is usual for foodstuffs. Drying capacity of shelf-type freeze dryers is 0.1–1.0 kg/(hr)(m^2 exposed surface). Another estimate is 0.5–1.6 lb/(hr)(sqft). The ice surface has been found to recede at the rate of 1 mm/hr. Freeze drying also is carried out to a limited extent in vacuum pans, vibrating conveyors, and fluidized beds. Condensers operate as low as −70°C.

Typical lengths of cycles for food stuffs are 5–10 hr, for bacterial pellets 2–20 hr, and for biological fluids 20–50 hr. A production unit with capacity of 500 L may have 75 kW for refrigeration and 50 kW for heating. Conditions for the preparation of freeze dried coffee are preparation of an extract with 20–25% solids, freezing at −25–−43°C, sublimation at approx. 200 Torr to a final final moisture content of 1–3%, total batch processing time of 6–8 hrs.

19.4. PARAMETRIC PUMPING

A class of operations has been devised in which the process fluid is pumped through a particular kind of packed bed in one direction for a while, then in the reverse direction. Each flow direction is at a different level of an operating condition such as temperature, pressure, or pH to which the transfer process is sensitive. Such a periodic and synchronized variation of the flow direction and some operating parameter was given the name of parametric pumping by Wilhelm (1966). A difference in concentrations of an adsorbable–desorbable component, for instance, may develop at the two ends of the equipment as the number of cycles progresses.

TABLE 19.6. Performance Data of Reverse Osmosis Membrane Modules
(a) Data of Belfort (1984).

Module design	Packing density (ft²/ft³)	Water flux at 600 psi (gal/ft² day)	Salt rejection	Water output per unit volume (gal/ft³ day)	Flow channel size (in.)	Ease of cleaning
Tubular						
Brine flow inside tube	30–50	10	Good	300–500	0.5–1.0	Very good
Brine flow outside tube[a]	140	10	Good	1400	0.0–0.125[a]	Good
Spiral wrap[b]	250	10	Good	2500	0.1	Fair
Fiber						
Brine flow inside fiber[c]	1000	5	Fair	5000	0.254	Fair
Brine flow outside fiber	5000–2500	1–3	Fair	5000–7500	0.002	Poor
Flat plate[d]	35	10	Good	350	0.01–0.02	Good
Dynamic membrane[e]	50	100	Poor	5000	~0.25	Good

(b) Data of Crits [*Ind. Water Eng.*, 20–23 (Dec. 1976–Jan. 1977)]

	Tri-acetate hollow fibers	Polyamide hollow fibers	Cellulose acetate spiral-wound
Module sizes and flow, gal/day at 400 lb/in²	5 × 48 in, 4000 gal/day; 10 × 48 in, 20,000 gal/day	4 × 48 in, 4200 gal/day (1); 8 × 48 in, 14,000 gal/day (1)	4 in × 21 ft (6), 4200 gal/day; 8 in × 21 ft (6) 24,000 gal/day
Recommended operating pressure, lb/in²	400	400	400
Flux, permeate rate, gal/day/ft²	1.5	2	15–18
Seals, pressure	2	2	12
Recommended max operating temp, °F	86	95	85
Effluent quality (guaranteed % rejection)	90	90	90
pH range	4–7.5	4–11	4–6.5
Chlorine tolerance	0.5–1.0	0.1 > pH 8.0 0.25 > pH 8.0	0.5–1.0
Influent quality (relative—FI*)	FI–<4	FI<3	FI<15
Recommended influent quality	FI–<3	FI<3	FI<3
Permeate back pressure (static), lb/in²	75	75	0
Biological attack resistance	Resistant	Most resistant	Least resistant
Flushing cleaning	Not effective	Not effective	Effective
Module casing	Epoxy-coated steel	FRP	Epoxy-coated steel and FRP
Field membrane replacement	Yes	No (future yes)	Yes

* FI = fouling index.[2] (1) Initial flow. (6) Six modules per 4200 gal/day.

A schematic of a batch parametric pumped adsorption process is sketched in Figure 19.12(a), whereas Figure 19.12(b) shows the synchronized temperature levels and flow directions. At the start, the interstices of the bed and the lower reservoir are filled with liquid of the initial composition and with the same amount in both. The upper reservoir is empty. The bed is kept cold while the liquid is displaced from the interstices into the upper reservoir by liquid pumped from the lower reservoir. Then the temperature of the bed is raised and liquid is pumped down through the bed. Adsorption occurs from cold liquid and desorption to the hot liquid. For the system of Figure 19.12(c), the separation factor is defined as the ratio of concentrations of the aromatic component in the upper and lower reservoirs; very substantial values were obtained in this case. Data of partial desalination of a solution with an ion exchange resin are in Figure 19.12(d), but here the maximum separation ratio is only about 10.

An intermittent—simulated continuous—operation is described with Figure 19.13. Feed input and withdrawals of products are

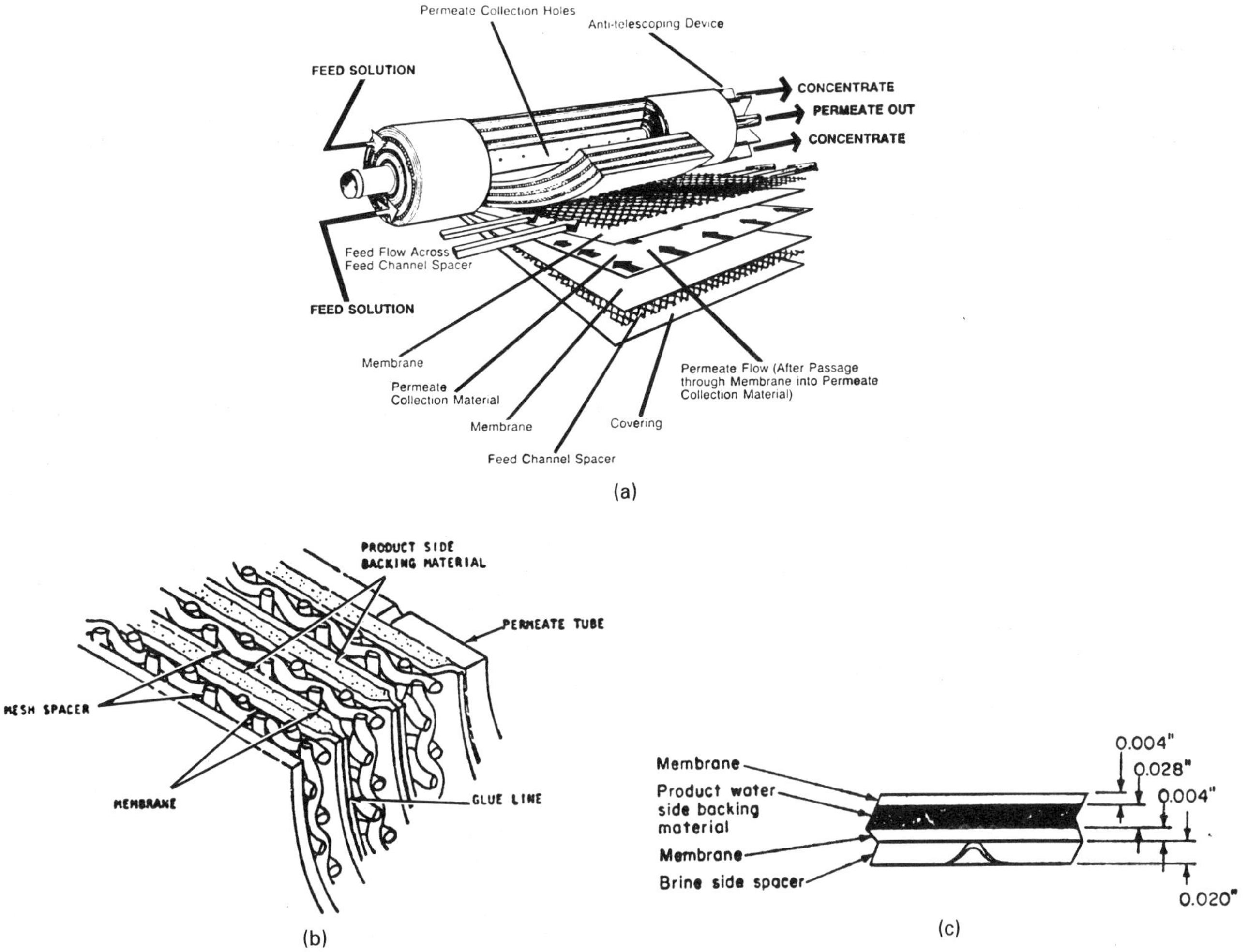

Figure 19.4. The spiral wound membrane module for reverse osmosis. (a) Cutaway view of a spiral wound membrane permeator, consisting of two membranes sealed at the edges and enclosing a porous structure that serves as a passage for the permeate flow, and with mesh spacers outside each membrane for passage of feed solution, then wound into a spiral. A spiral 4 in. dia by 3 ft long has about 60 sqft of membrane surface. (b) Detail, showing particularly the sealing of the permeate flow channel. (c) Thickness of membranes and depths of channels for flows of permeate and feed solutions.

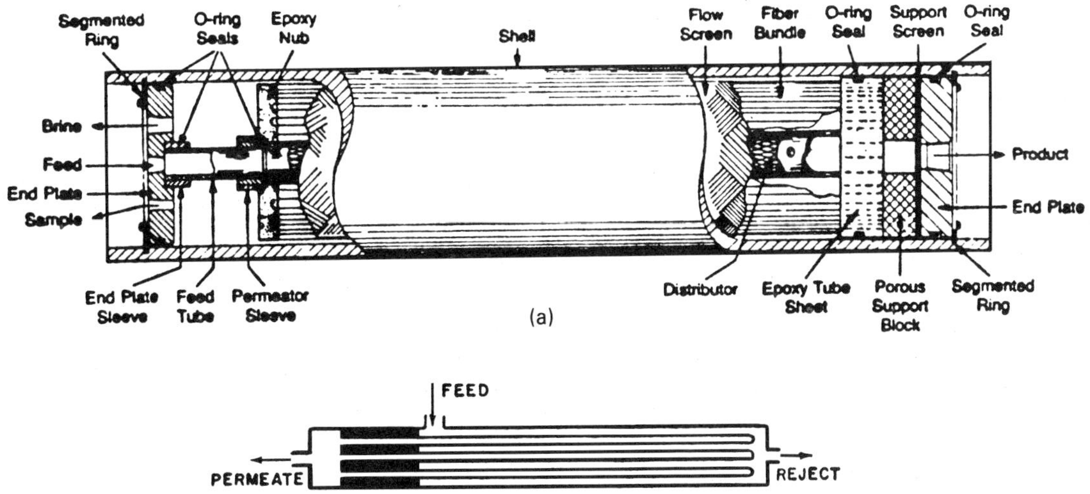

Figure 19.5. The "Permasep" hollow fiber module for reverse osmosis. (a) Cutaway of a DuPont "Permasep" hollow fiber membrane module for reverse osmosis; a unit 1 ft dia and 7 ft active length contains 15–30 million fibers with a surface area of 50,000–80,000 sqft; fibers are 25–250 μm outside dia with wall thickness of 5–50 μm (*DuPont Co.*). (b) The countercurrent flow pattern of a "Permasep" module.

EXAMPLE 19.2
Concentration of a Water/Ethanol Mixture by Reverse Osmosis

The pressure required to drive water out of mixtures of various concentrations of alcohol against pure water at 30°C is calculated from the osmotic equation

$$1 - x_{alc} = \exp\left[-\frac{0.018P}{0.082(303.2)}\right]$$

with the results:

Wt % Alcohol	Mol Fraction Alcohol	P (atm)
10	0.0417	59
50	0.281	456
90	0.779	2085
95.5 (azeotrope)	0.8925	3081
96	0.9038	3234

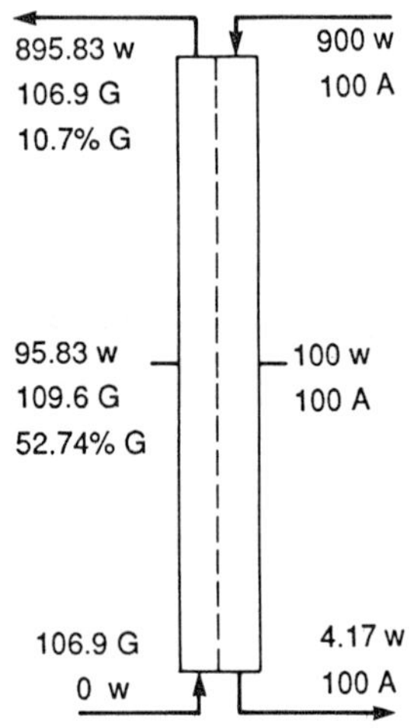

It appears that the pressures needed to make higher than azeotropic composition are beyond the strength of available membranes. A pressure of 1000 psi (68 atm) is feasible. With this pressure the concentrations of solute on the two sides of the membrane are related by

$$\frac{1-x_1}{1-x_2} = \exp\left[-\frac{0.018(68)}{0.082(303.2)}\right] = 0.9520$$

whence

$$x_2 = 1 - (1 - x_1)/0.9520$$

As long as the mol fraction of solute on the low pressure side is kept above the value given by this equation, water can be driven from the side with mol fraction x_1 across the membrane. The solute on the low pressure side should be one that is easily separated from water and any alcohol that may bleed through. Ethylene glycol is such a material which also has the advantage of a relatively low molecular weight, 62. The required minimum concentrations of glycol corresponding to various alcohol concentrations on the high pressure side (68 atm) are tabulated.

Wt % Alcohol	Mol Fraction Alcohol	Mol Fraction Glycol	Wt % Glycol
10	0.0417	≥0	≥0
50	0.281	≥0.2447	≥52.74
90	0.779	≥0.7679	≥91.93
96	0.9038	≥0.8989	≥96.84

The flowsketch shows a feed stream consisting of 100 kg/hr alcohol and 900 kg/hr of water, and making a stream with 96% alcohol. If pure glycol is charged countercurrently at the rate of 106.9 kg/hr, the % glycol at the point in the column where the alcohol content is 50% will be 52.74%, which is high enough to ensure that water can be driven out by a pressure of 68 atm. Beyond this point also, the content of glycol will be high enough to ensure transfer of water out of the alcohol solution. The aqueous glycol will be distilled and recycled. A small increase in its amount will permit some water to be present in the recycle stream.

accomplished with periodic openings and closings of valves without shutting down the equipment at any time. Other modes of operation also can be devised.

Theoretical studies also have applied this cycling principle to liquid–liquid extraction processes with immobilized solvents, and to reversible chemical reactions. Quite comprehensive reviews of the literature of cycling zone separations have been made by Sweed (1972), Wankat (1974), Wankat et al., and (1976).

Although parametric pumping appeared on the academic scene in 1966, no commercial installations appear to have been made, at least no widely publicized ones. Periodic heating and cooling of beds of granules or even of periodic compression and decompression of the process gas appear to be serious economic obstacles.

19.5. SEPARATIONS BY THERMAL DIFFUSION

Separation of mixtures based on differences in thermal diffusivity at present are feasible only for analytical purposes or for production on a very small scale of substances not otherwise recovered easily. Nevertheless, the topic is of some interest to the process engineer as a technique of last resort.

In a vessel with a temperature gradient between a hot and cold surface, a corresponding concentration gradient of a fluid likewise can develop. The substance with the smaller molecular volume usually concentrates in the high temperature region, but other factors including that of molecular shape also affect the relative migrations of components of mixtures. Thus, the sequence of separation of hydrocarbons from hot to cold regions generally is: light normal paraffins, heavy normal paraffins, naphthenes and monocyclic aromatics, and bicyclic aromatics. Isotopes with small differences in molecular weights were the first substances separated by thermal diffusion, but isomers which have identical molecular weights also are being separated.

The basic construction of a horizontal thermal diffusion cell is sketched in Figure 19.14(a). When gases are to be separated, the distance between the plates can be several mm; for liquids it is a fraction of a mm. The separation effects of thermal diffusion and convection currents are superimposed in the equipment of Figure 19.14(b), which is called a thermogravitational or Clusius–Dickel column after the inventors in 1938. A commercially available column used for analytical purposes is in Figure 19.14(c). Several such columns in series are needed for a high degree of separation.

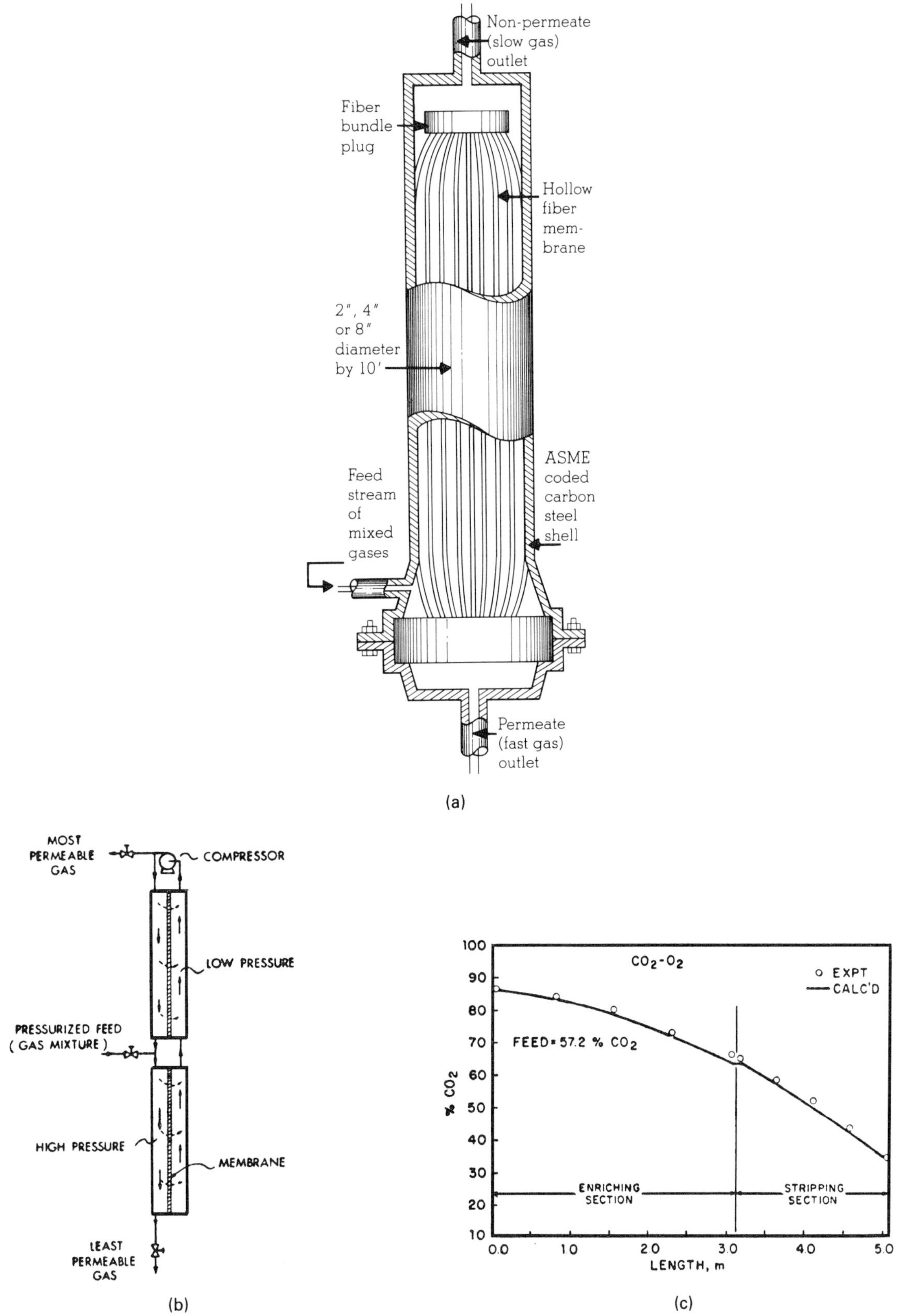

Figure 19.6. Gas permeation equipment and performance. (a) Cutaway of a Monsanto "Prism" hollow fiber module for gas separation by permeation. (b) Flowsketch of a continuous column membrane gas separator. (c) Composition profiles of a mixture of CO_2 and O_2 in a column 5 m long operated at total reflux [Thorman and Hwang in (*Turbak, Ed.*), Synthetic Membranes II, *American Chemical Society, Washington DC, 1981, pp. 259–279*].

TABLE 19.7. Data of Membranes for Gas Permeation Separation

(a) Permeabilities of Helium, Nitrogen, and Methane in Several Membranes at 20°C

Membrane	$P_0 \cdot 10^7$[cm²/s bar] He	N_2	CH_4
Silicon rubber	17.25	11.25	44.25
Polycarbonate	5.03	0.35	0.27
Teflon FEP	4.65	0.19	0.11
Natural rubber	2.70	0.79	—
Polystyrene	2.63	0.17	0.17
Ethyl cellulose	2.33	0.21	0.83
Polyvinyl chloride (plasticized)	1.05	—	0.15
Polyethylene	0.75	0.14	—
Polyvinylfluoride	0.14	0.0014	0.00048

(b) Separation Factors $\alpha_{AB} = P_{0A}/P_{0B}$ for Three Mixtures

Membrane	He/CH_4	He/O_2	H_2/CH_4
Polyacrylonitrile	60 000	—	10 000
Polyethylene	264	35.5	162
Polytetrafluoroethylene	166	45	68.5
Regenerated cellulose	400	48	—
Polyamide 66	214	39	—
Polystyrene	14.6	5.5	21.2
Ethylcellulose	48	3.2	6.6

(c) Examples of Commercial Separations and the Kinds of Membranes Used

Separation process	Membrane
O_2 from air	Ethylcellulose, silicon rubber
He from natural gas	Teflon FEP, asymmetric cellulose acetate
H_2 from refinery gas	Polyimide, polyethylene-terephthalate, polyamide 6
CO_2 from air	Silicon rubber
NH_3 from synthesis gas	Polyethylene terephthalate
H_2S from natural and refinery gas	Silicon rubber, polyvinylidene fluoride
H_2 purification	Pd/Ag alloys

[Membranen, in *Ullmann's Encyclopedia of Chemical Technology*, Verlag Chemie, Weinheim, 1978, Vol. 16, p. 515. Many more data are collected by Hwang, Choi, and Kammermeyer, *Separation Sci.* **9**(6), 461–478 (1974)].

Clusius and Dickel used a column 36 m long to make 99+% pure isotopes of chlorine in HCl. The cascade of Figure 19.15 has a total length of 14 m; most of the annular diameter is 25.4 mm, and the annular widths range from 0.18 to 0.3 mm. The cascade is used to recover the heavy isotope of sulfur in carbon disulfide; a production rate of a 90% concentrate of the heavy isotope of 0.3 g/day was achieved.

Separation of the hydrocarbon isomers of Table 19.10(a) was accomplished in 48 hr in the column of Figure 19.14(c) with 50°C hot wall and 20°C cold wall. The concentration gradient that develops in such a column is shown in Figure 19.14(d). The equilibrium terminal compositions depend on the overall composition, as indicated in Figure 19.14(e). Other kinds of behaviors also occur. Thus mixtures of benzene and cyclohexane are not separated, nor can mixtures of benzene and octadecane when the latter is in excess.

TABLE 19.8. Data of Foam Separation Experiments Made in a 1 in. Dia Column on a Waste Water Containing Radioactive Components and Utilizing Several Different Surfactants

Surfactant	Surf. conc. (gm/liter)	Flow rates (cm³/min) Gas, V	Foam, Q	Foam cond., F	Foam density, ρ_f (gm/liter)	Average bubble diameter, $\bar{D}$ (cm)
Aerosol AY	6.5	154	176	0.197	1.12	0.06
Alipal CO-436	0.375	154	186	0.950	5.10	0.05
Alipal LO-529	0.4	154	174	0.415	2.40	0.06
Deriphat 170C	0.5	154	60	4.92	74	0.025
Igepon CN-42	0.12	154	72	1.6	24	0.038
Tergitol 7	2.0	154	202	0.763	3.77	0.05
Ultrawet SK	0.08	154	173	0.137	0.79	0.10

(Davis and Haas, in *Adsorptive Bubble Separation Techniques* (Lemlich, Ed.), Academic, New York, 1972, pp. 279–297).

Examples of separations of isotopes are in Table 19.10(b). The concentration of U-235 listed there was accomplished in a cascade of 2100 columns, each with an effective height of 14.6 m, inner tube 5 cm dia, gap 0.25 mm, hot surface 87–143°C, and cold surface 63°C, just above the condensation temperature at the operating pressure of 6.7 MPa. Although the process was a technical success,

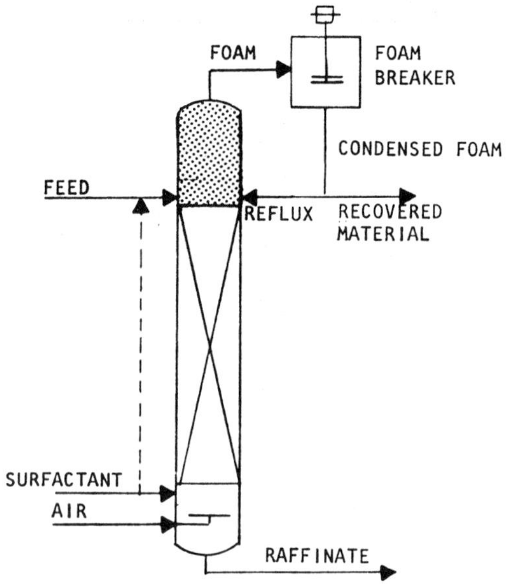

Figure 19.7. Sketch of a foam fractionating column. Surfactants or other foaming agents may be introduced with the feed or separately at a lower feed point. Packing may be employed to minimize axial mixing.

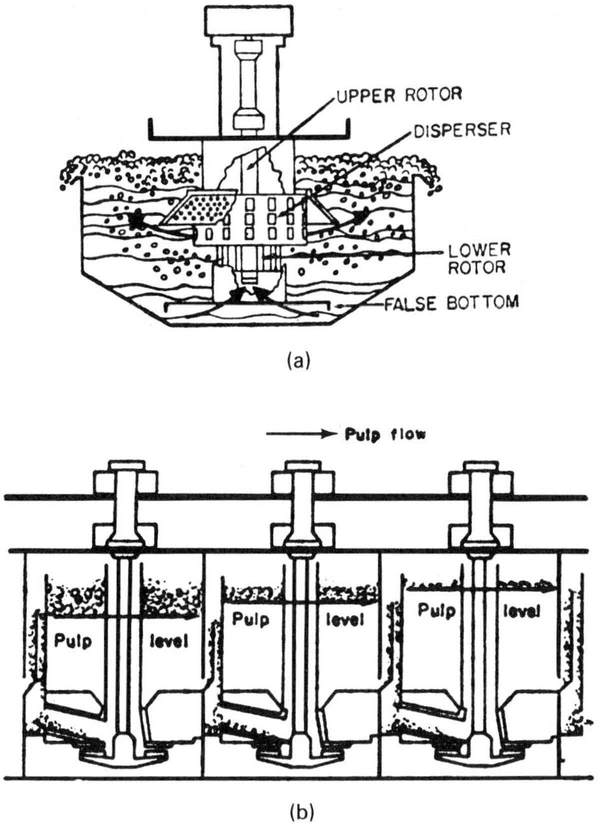

Figure 19.8. The interaction of air and pulp in a froth flotation cell and a series arrangement of such cells: (a) Sectional schematic of flotation cell. Upper portion of rotor draws air down the standpipe for thorough mixing with pulp. Lower portion of rotor draws pulp upward through rotor. Disperser breaks air into minute bubbles. Larger flotation units include false bottom to aid pulp flow. (*WEMCO Division, Envirotech Corp.*). (b) A bank of three flotation cells. The floating concentrate is withdrawn continuously from each stage but the remaining pulp flows in series through the cells.

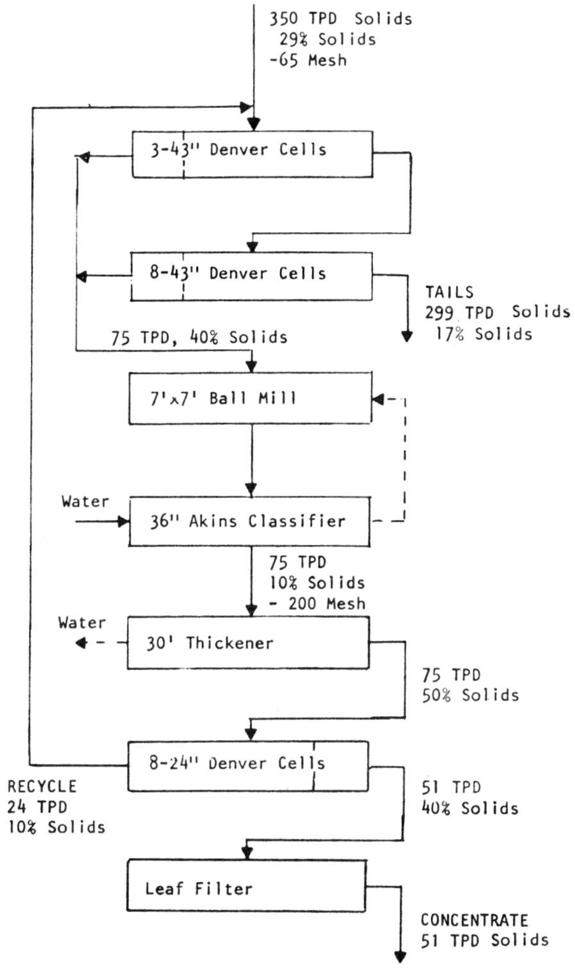

Figure 19.9. Flotation section of a flowsheet for concentration of 350 tons/day of a copper ore (*data of Pima Mining Co., Tucson, AZ*).

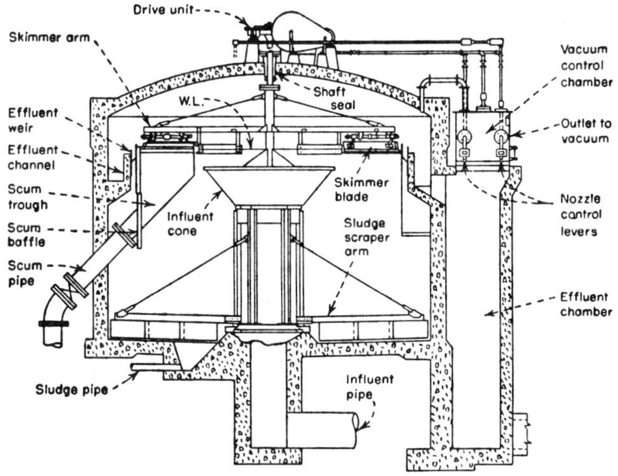

Figure 19.10. Vacuator of the "constant-level" type. The cylindrical tank with a dome-shaped cover is under a constant vacuum of about 9 in. of mercury. Sewage enters a central draft tube from which it is distributed by means of a flared-top section. Floating solids, buoyed up by fine air bubbles, are skimmed from the liquid surface and carried to a trough. Settled solids are removed from the bottom with a scraper mechanism. (*Courtesy of* Engineering News-Record).

it was abandoned in favor of separation by gaseous diffusion which had only 0.7% of the energy consumption.

For separation of hydrocarbons, thermal requirements are estimated to range from 70,000 to 350,000 Btu/lb, compared with heats of vaporization of 150 Btu/lb.

Although thermal diffusion equipment is simple in construction and operation, the thermal requirements are so high that this method of separation is useful only for laboratory investigations or for recovery of isotopes on a small scale, which is being done currently.

19.6. ELECTROCHEMICAL SYNTHESES

Electrolysis plays a role in the manufacture of some key inorganic chemicals on an industrial scale, but rather a minor one in the manufacture of organic chemicals. Chlorine, alkalis, metals, hydrogen, oxygen, and strong oxidizing agents such as $KMnO_4$, F_2, and Cu_2O are made this way. Electroorganic processes of commercial or potentially commercial scale are listed in Table 19.11, which implies that much research is being done in pilot plants and may pay off in the near future. In the United States, the four large tonnage applications are to the manufacture of adiponitrile,

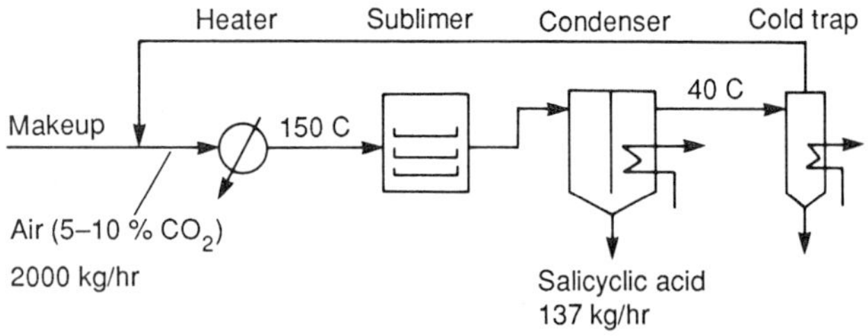

Figure 19.11. Sublimation of salicyclic acid at 1 bar. Vapor pressures are 14.4 mbar at 150°C and 0.023 mbar at 40°C. The air rate shown corresponds to equilibrium in the sublimer, but in some kinds of vessels percent saturation may be as low as 10%. The conditions are those of Mullin [Crystallisation, *Butterworths, London,* **288** (*1972*)].

the Nalco process for lead tetraethyl, which is being phased out from the gasoline industry, the Miles process for dialdehyde starch, which is on standby until the demand picks up, and the 3M electrofluorination process for a variety of products.

Pros and cons of electrochemical processes are not always clear cut. In a few cases, they have lower energy requirements than conventional chemical methods but not usually according to the survey of Table 19.12. The Monsanto process for adiponitrile by electrochemical reduction of acetonitrile is an outstanding example; moreover, comparison of the performances of the original and improved cells [sketched on Figs. 19.16(e) and (f)] suggests the often great leeway in cell design. Small scale electrode processes frequently are handicapped because of the expense of developing

TABLE 19.9. Materials That May Be Purified by Sublimation or Are Being Freeze-Dried

(a) Substances Amenable to Purification by Sublimation[a]

Aluminium chloride	Naphthalene
Anthracene	β-Naphthol
Anthranilic acid	Phthalic anhydride
Anthraquinone	α-Phthalimide
Benzanthrone	Pyrogallol
Benzoic acid	Salicylic acid
Calcium	Sulphur
Camphor	Terephthalic acid
Chromium chloride	Titanium tetrachloride
Ferric chloride	Thymol
Iodine	Uranium hexafluoride
Magnesium	Zirconium tetrachloride

[a] Some others are mentioned in the text.

(b) Products which Are Being Freeze-Dried Commercially.

Foodstuffs	Pharmaceuticals	Animal Tissues and Extracts
Coffee extract	Antibiotics	Arteries
Fish and seafood	Bacterial cultures	Blood
Fruits	Serums	Bones
Fruit juices	Virus solutions	Hormones
Meat		Skin
Milk		Tumors
Tea extract		
Vegetables		

efficient components of cells such as electrodes, diaphragms, membranes, and electrolytes which usually can be justified only for large scale operation.

In comparison with chemical oxidations and reductions, however, electrode reactions are nonpolluting and nonhazardous because of low pressure and usually low temperature. Although electricity usually is more expensive than thermal energy, it is clean and easy to use. Electrolytic processes will become more attractive when inexpensive sources of electricity become developed.

ELECTROCHEMICAL REACTIONS

An equilibrium electrical potential is associated with a Gibbs energy of formation by the equation

$$E^0 = -\Delta G^0/23.06n,$$

where n is the number of gram equivalents involved in the stoichiometric equation of the reaction, ΔG^0 is in kcal/g mol, and E^0 is the potential developed by the reaction in volts. Thus, for the reaction $H_2O \rightleftarrows H_2 + \frac{1}{2}O_2$ at 25°C,

$$E^0 = 54.63/(2)(23.06) = 1.18\ \text{V}$$

and for $HCl \rightleftarrows \frac{1}{2}H_2 + \frac{1}{2}Cl_2$ at 25°C,

$$E^0 = 22.78/23.06 = 0.99\ \text{V}.$$

Practically, reactions are not conducted at equilibrium so that amounts greater than equilibrium potentials are needed to drive a reaction. Major contributions to inefficiency are friction in the electrolyte and other elements of a cell and particularly the overvoltages at the electrodes. The latter are due to adsorption or buildup of electrolysis products such as hydrogen at the electrode surfaces. Figure 19.17(a) shows magnitudes of hydrogen overvoltages at several metals and several currents. The several contributions to voltage drops in a cell are identified in Figures 19.17(b) and (c), whereas Figure 19.17(d) indicates schematically the potential gradient in a cell comprised of five pairs of electrodes in series.

Electrochemical cells are used to supply electrical energy to chemical reactions, or for the reverse process of generating electrical energy from chemical reactions. The first of these applications is of current economic importance, and the other has significant promise for the near future.

FUEL CELLS

A few chemical reactions can be conducted and controlled readily in cells for the production of significant amounts of electrical energy at high efficiency, notably the oxidations of hydrogen or carbon monoxide. Some data of such processes are in Figure 19.18. The basic processes that occur in hydrogen/air cells are in Figure 19.18(a). Equilibrium voltage of such a cell is in excess of 1.0 V at moderate temperatures, but under practical conditions this drops off rapidly and efficiency may become less than 40%, as Figure 19.18(b) shows. Theoretical cell potentials for several reactions of fuel cell interest are in Figure 19.18(c); in theory at least, the oxidations of hydrogen and carbon monoxide are competitive. High temperatures may be adopted to speed up the electrode processes, but they have adverse effects on the equilibria of these particular reactions. Figure 19.18(d) shows the characteristics of major electrochemical fuel systems that have been emphasized thus far. Most of the development effort has been for use in artificial satellites where cost has not been a primary consideration, but

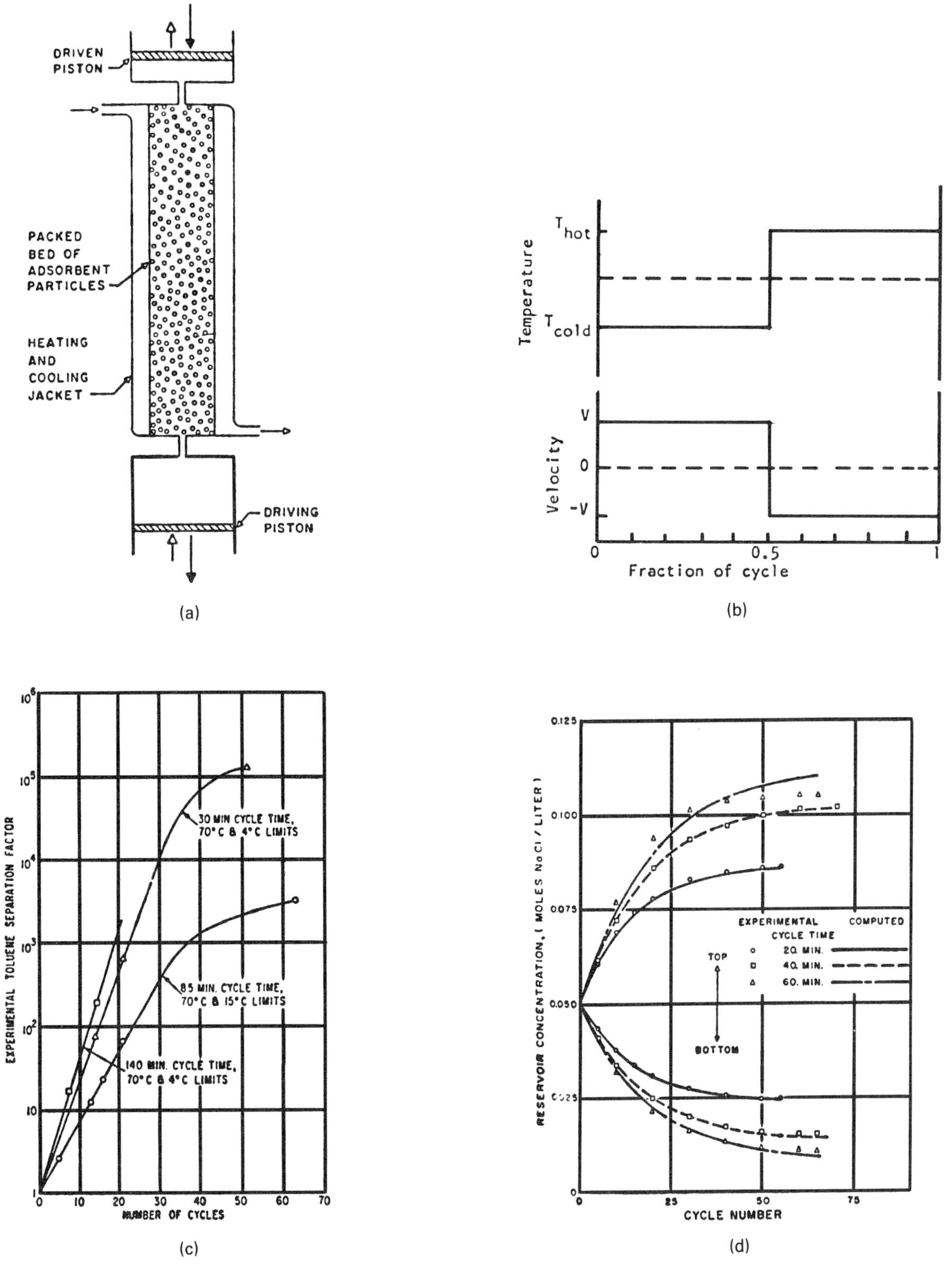

Figure 19.12. Batch parametric processing of solid–liquid interactions such as adsorption or ion exchange. The bottom reservoir and the bed interstices are filled with the initial concentration before pumping is started. (a) Arrangement of adsorbent bed and upper and lower reservoirs for batch separation. (b) Synchronization of temperature levels and directions of flow (positive upward). (c) Experimental separation of a toluene and n-heptane liquid mixture with silica gel adsorbent using a batch parametric pump. (*Reprinted from Wilhelm, 1968, with permission of the American Chemical Society*). (d) Effect of cycle time τ on reservoir concentrations of a closed system for an $NaCl–H_2O$ solution with an ion retardation resin adsorbent. The column is initially at equilibrium with $0.05M$ NaCl at 25°C and $\alpha = 0.8$. The system operates at 5° and 55°C. [*Sweed and Gregory,* AIChE J. **17,** *171* (*1971*)].

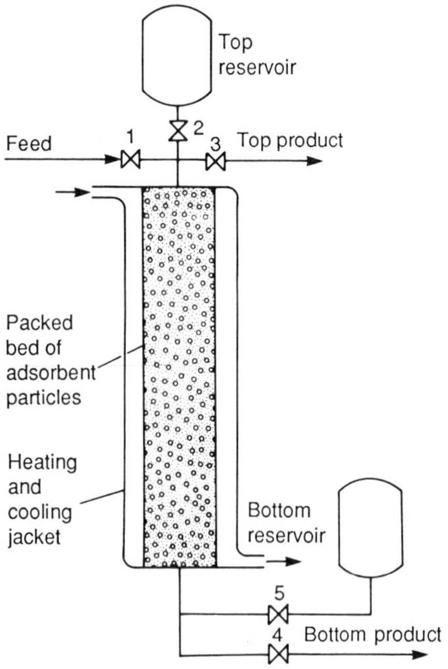

Valve 1	2	3	4	5	T_h	T_c	Process
V_1	—	—	V_1	—		*	feed supplied, btm product removed cold
V_2	—	—	—	V_2		*	feed supplied, btm reservoir chgd. cold
—	V_3	—	—	V_3		*	top reservoir dschgd., btm reservoir chgd. cold
—	—	V_4	—	V_4	*		top product dschgd. hot, btm reservoir dschgd. hot
—	V_5	—	—	V_5	*		top reservoir chgd., btm reservoir dschgd. hot

Figure 19.13. Parametric cycle operating intermittently in five periods; valves that are open each time are identified with their flow rates V_i, and the low and high temperature levels are identified with an asterisk in the proper column.

spinoff to industrial applications has some potential for the near future.

CELLS FOR SYNTHESIS OF CHEMICALS

Cells in which desired chemical reactions can be conducted and controlled are assemblages of pairs of anodes and cathodes between which the necessary potential difference is impressed. The regions near the electrodes may be separated by porous diaphragms to minimize convective mixing of the products formed at the individual electrodes. In recent years, semipermeable or ion-exchange membranes have been employed as diaphragms. In Figure 19.16(a), the membrane allows only Na^+ ions to pass so that the caustic that is made in the cell is essentially free of NaCl. In the mercury cell of Figure 19.16(b), no partition is necessary because the released Na dissolves in the mercury; the amalgam is reacted with water in an electrically neutral zone of the cell to make salt-free caustic. Because of pollution by escaped mercury, such cells have been largely phased out for production of salt-free caustic.

The same process sometimes can be performed efficiently in cells either with or without diaphragms. Figures 19.16(e) and (f) are for making adiponitrile by reduction of acetonitrile. In the newer design, Figure 19.16(f), the flow rate of the electrolyte is high enough to sweep out the generated oxygen quickly enough to prevent reverse oxidation of the product.

Either parallel, called monopolar, or series, called bipolar, electrical connections can be made to the pairs of electrodes in a complete cell. The monopolar types have individual connections to each electrode and thus require only individual pair potential to be applied to the cell assembly. The bipolar mode has electrical connections only to the terminal electrodes. One design such as Figure 19.16(f) has 48 pairs of electrodes in series and requires 600 V. The equipment of Figure 19.19(a) also has bipolar connections. The voltage profile in such equipment is indicated schematically in Figure 19.17(d). Bipolar equipment is favored because of its compactness and, of course, the simplicity of the electrical connections. No adverse comments appear to be made about the high voltages needed.

Although the basic cell design shown schematically in Figures 19.16(a) and 19.19(d) is effective for many applications when dimensions and materials of construction are properly chosen, many special designs have been developed and used, of which only a few can be described here. For the cracking of heavy hydrocarbons to olefins and acetylenes, for instance, the main electrodes may be immersed in a slurry of finely divided coke; the current discharges from particle to particle generate the unsaturates. Only 100–200 V appears to be sufficient.

The most widely used brine electrolytic cells are the Hooker and Diamond Shamrock which are both monopolar, but bipolar designs like that of Figure 19.19(a) also are popular. That figure does not indicate the presence of a diaphragm but one must be used.

Rotating electrodes characterize the BASF cell of Figure 19.19(b), which is used for making adiponitrile. The cell described in the literature has 100 pairs of electrodes 40 cm dia spaced 0.2 mm apart. The rapid flow rate eliminates the need for diaphragms by sweeping out the oxygen as it is formed.

Lead alkyls are made by the action of Grignard reagents on lead anodes in the equipment of Figure 19.19(c). Lead pellets serve as the anode and are replenished as they are consumed. Several tubes 5 cm dia are housed in a single shell for temperature control and as required for capacity.

The simplest kind of cell construction, shown in Figure 19.19(d), suffices for the production of hydrogen by electrolysis of water and for the recovery of chlorine from waste HCl. The term filter-press cell is applied to this kind of equipment because of the layered construction. These two electrolyses are economically feasible under some conditions. Some details are given by Hine (1985).

It has been mentioned already that only a few inorganic and organic electrochemical processes have made it to commercial scale, but the potential may be there and should not be ignored. Recent surveys of the field and of the literature have been made by Hine (1985), Pletcher (1982), and Roberts et al. (1982).

19.7. FERMENTATION PROCESSING

Industrial fermentation is any process involving microorganisms that results in useful products. Among the useful microorganisms are molds, yeasts, algae, and bacteria. They are distinguished from plants and animals by being made of cells of only one kind. Although some kinds are grown as food, yeast or algae, for instance, the main interest here is in chemical manufacture with their assistance. This they accomplish by creating enzymes which catalyze specific reactions. In many respects biochemical processing is like ordinary chemical processing. The recovery and purification of biochemical products, however, often is a more demanding task and offers opportunities for the exercise of ingenuity and the

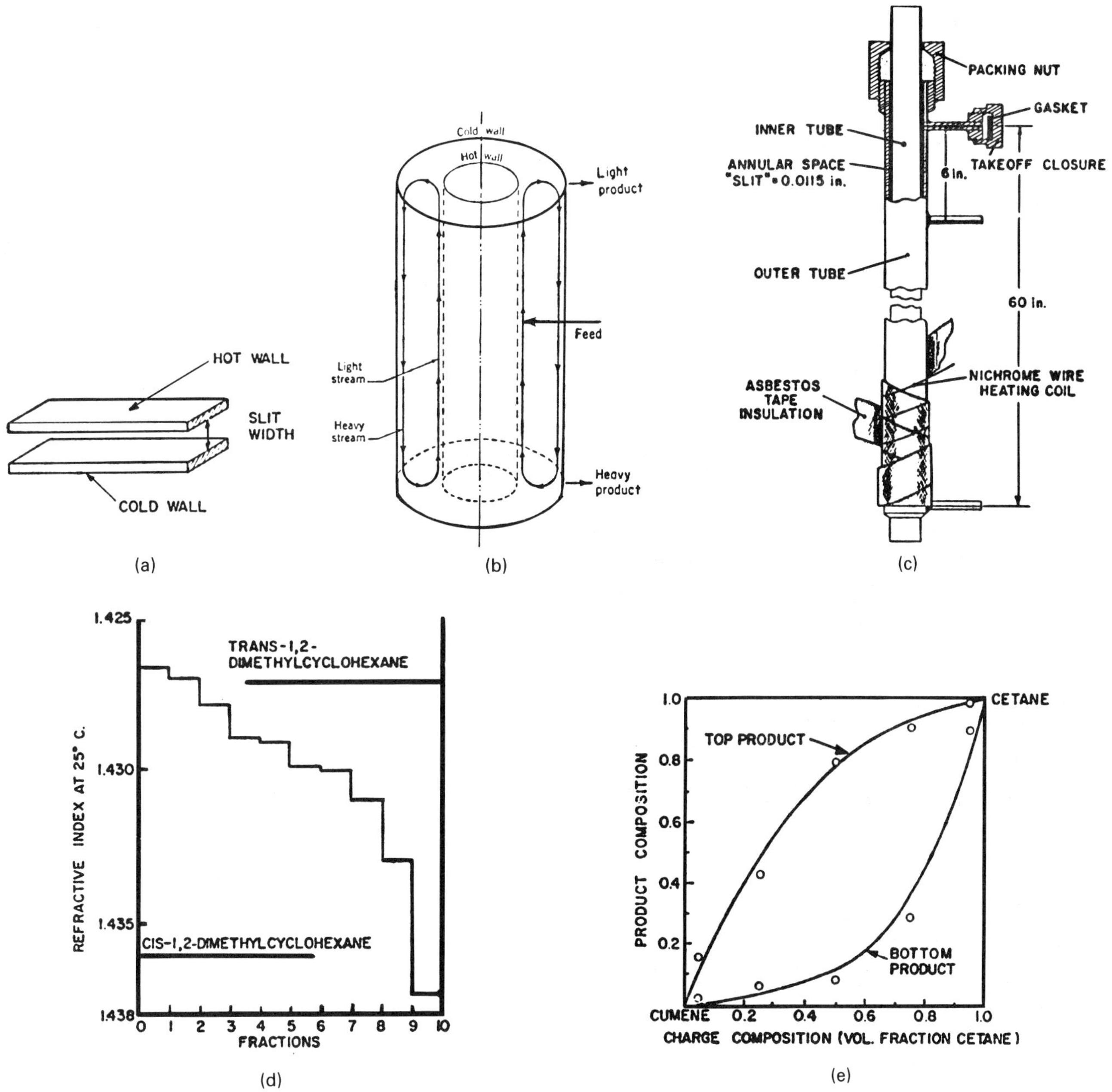

Figure 19.14. Construction and performance of thermal diffusion columns. (a) Basic construction of a thermal diffusion cell. (b) Action in a thermogravitational column. (c) A commercial column with 10 takeoff points at 6 in. intervals; the mean dia of the annulus is 16 mm, width 0.3 mm, volume 22.5 mL (*Jones and Brown, 1960*). (d) Concentration gradients in the separation of cis and trans isomers of 1,2-dimethylcyclohexane (*Jones and Brown, 1960*). (e) Terminal compositions as a function of charge composition of mixtures of cetane and cumene; time 48 hr, 50°C hot wall, 29°C cold wall (*Jones and Brown, 1960*).

application of techniques that are exotic from the point of view of conventional processing. A distinction also is drawn between processes that involve whole cells and those that utilize their metabolic products, enzymes, as catalysts for further processing. A brief glossary of biochemical terms is in Table 19.13.

Major characteristics of microbial processes are:

a. The reaction medium is aqueous.
b. The products are made in low concentration, rarely more than 5–10% for chemicals and much less for enzyme recovery.
c. Reaction temperatures with microorganisms or isolated enzymes are low, usually in the range of 10–60°C, but the optimum spread in individual cases may be 5°C or less.
d. With only a few exceptions, such as potable ethanol or glucose isomerate, the scale of commercial processes is modest, and for enzymes it is measured only in kilograms per day.
e. Batch processing is used preponderantly, but so many conditions must be regulated carefully that computer control is common.

Because of the small scale of enzyme production, laboratory

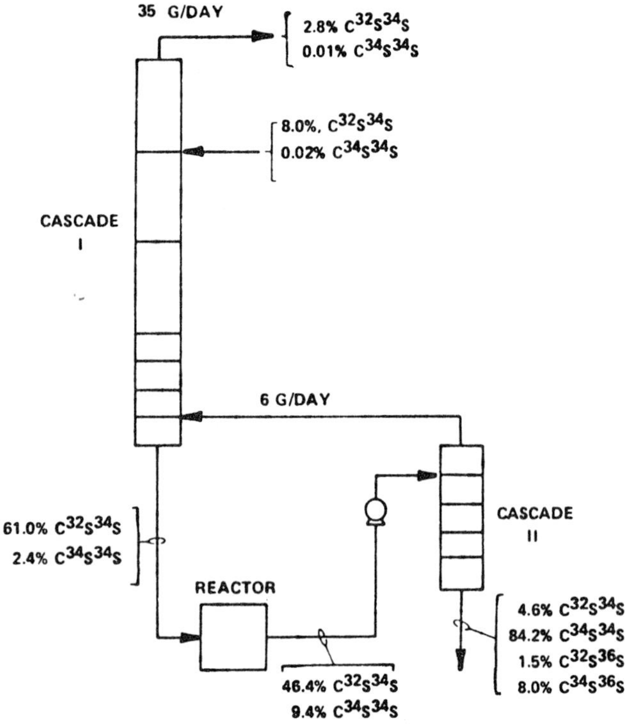

Figure 19.15. The liquid thermal diffusion system for the recovery of heavy sulfur isotope in carbon disulfide. The conditions prevailing at the time after 90% ^{34}S is reached. Each rectangle in the cascades represents a column, each height being proportional to the length of the column. The two cascades have a combined height of 14 m, annular dia 25.4 mm, and annular width 0.18–0.3 mm. Production rate of 90% concentrate of ^{34}S was 0.3 g/day [*W.M. Rutherford,* Ind. Eng. Chem. Proc. Des. Dev. **17,** *17–81 (1978)*].

types of separation and purification operations are often feasible, including: dialysis to remove salts and some low molecular weight substances, ion exchange to remove heavy metals, ultrafiltration with pore sizes under 0.5 μm and pressures of 1–10 atm to remove substances with molecular weights in the range of 15,000–1 million, reverse osmosis to remove water and to concentrate low molecular weight products, and gel permeation chromatography to fractionate a range of high molecular weight substances. Conventional processes of filtration and centrifugation, of drying by freezing or vacuum or spraying, and colloid milling also are used for processing enzymes.

PROCESSING

The three main kinds of fermentation processes are:

a. Growth of microorganisms such as bacteria, fungi, yeasts and others as end products.
b. Recovery of enzymes from cell metabolism, either intracellularly or as secretions, mostly the latter.
c. Production of relatively low molecular weight substances by enzyme catalysis, either with isolated enzymes or with the whole cell.

Some industrial products are listed in Table 19.14. Chemical and fermentation syntheses sometimes are competitive, for instance, of ethanol, acetone, and butanol.

Enzymes are proteins with molecular weights in the range of 15,000–1,000,000 or so. In 1968, for instance, about 1300 were known, but only a few are of industrial significance. They are named after the kinds of reactions that they promote rather than to identify the structure which often is still unknown. Some kinds of enzymes are:

Amylase, which converts polysaccharides (starch or cellulose) to sugars.
Cellulase, which digests cellulose.
Glucose oxidase, which converts glucose to dextrose and levulose.
Isomerase, which converts glucose to fructose.
Lipase, which splits fats to glycerine and fatty acids.
Protease, which breaks down proteins into simpler structures.

Biochemical manufacturing processes consist of the familiar steps of feed preparation, reaction, separation, and purification. The classic mode handles the microorganisms in slurry form in a stirred reactor. Enzyme-catalyzed processes also are performed primarily in stirred tanks, but when the enzymes can be suitably immobilized, that is, attached to solid structures, other kinds of reactor configurations may be preferred. Microbes also are grown in pans or rotating drums under moistened conditions, processes known as solid culture processing. Figure 19.20(a) shows the three modes of microbe culture. Processes that demand extensive handling of moist solids are practiced only on a small scale or when stirred tank action is harmful to cell structures. The process of Figure 19.20(b) consists largely of feed preparation steps.

OPERATING CONDITIONS

The optimum ranges of conditions for microbe growth or enzyme activity are quite narrow and must be controlled closely.

Concentration. A major characteristic of microbial growth and enzymatic conversion processes is low concentrations. The rates of these processes are inhibited by even moderate concentrations of most low molecular weight organic substances, even 1 g/L often being harmful. Nutrients also must be limited, for instance, the following in g/L:

Ammonia	5
Phosphates	10
Nitrates	5
Ethanol	100
Glucose	100

In the fermentation for ethanol, the concentration limit normally is about 8 wt % ethanol, but newer processes have been claimed to function at 10% or so. The search is on for microorganisms, or for creating them, that tolerate high concentrations of reaction products and higher temperatures.

Temperature. Most microbe metabolisms and enzymatic processes function well only in the range of 10–60°C, but in particular cases the active spread of temperatures is only 5–10°C. A classification of microorganisms that is sometimes made is with respect to peak activities near 15°C or near 35°C or near 55°C. The maximum heat effects of metabolic processes can be estimated from heats of formation when the principal chemical participants are known, for instance:

glucose → ethanol, heat of reaction 0.10 kcal/g glucose,

glucose → $CO_2 + H_2O$, heat of reaction 3.74 kcal/g glucose.

TABLE 19.10. Examples of Separations by Thermal Diffusion
(a) Hydrocarbon Isomers

Components	Vol. %	Mol. wt.	Density	Final composition, vol. % Top	Bottom	Separation, %
n-Heptane	50	100	0.6837	95	10	75.4
Triptane	50	100	0.6900	5	90	
Isoöctane	50	114	0.6919	58	40	11.4
n-Octane	50	114	0.7029	42	60	
2-Methylnaphthalene	50	142	0.9905	55.5	42.5	13.1
1-Methylnaphthalene	50	142	1.0163	44.5	57.5	
trans-1,2-Dimethylcyclohexane	40	112	0.7756	100	0	100
cis-1,2-Dimethylcyclohexane	60	112	0.7963	0	100	
p-Xylene	50	106	0.8609	92	0	92
o-Xylene[a]	50	106	0.8799	8	100	
m-Xylene	50	106	0.8639	100	19	80
o-Xylene[a]	50	106	0.8799	0	81	
p-Xylene	50	106	0.8609	50	50	0
m-Xylene	50	106	0.8639	50	50	

(A. L. Jones and G. R. Brown, in *Advances in Petroleum Chemistry and Refining* (McKetta and Kobe, Eds.), Wiley, New York, 1960, Vol. III, pp. 43–76).

[a] *o*-Xylene contains paraffinic impurity (see Fig. 9 in McKetta and Kobe).

(b) Isotopes

Working fluid	Isotope separated	mol % product	Phase	Single column (S) or cascade (C)	Investigator	Year
HCl	^{35}Cl	99.6	Gas	S	Clusius and Dickel	1939
	^{37}Cl	99.4				
Kr	^{84}Kr	98.2	Gas	S	Clusius and Dickel	1941
	^{86}Kr	99.5				
O_2	^{17}O	0.5	Gas	C	Clusius and Dickel	1944
	^{18}O	99.5				
UF_6	^{235}U	0.86	Liquid	C	Manhattan Dist.	1945
N_2	^{15}N	99.8	Gas	S	Clusius & Dickel	1950
Xe	^{134}Xe	1	Gas	C	Clusius et al.	1956
	^{136}Xe	99				
He	^{3}He	10	Gas	C	Bowring and Davies	1958
A	^{36}A	99.8	Gas	C	ORNL†	1961
	^{38}A	23.2				•
Ne	^{20}Ne	99.99	Gas	C	ORNL	1961
	^{22}Ne	99.99				
Kr	^{78}Kr	10	Gas	C	ORNL	1961
	^{86}Kr	96.1				
He§	^{3}He	99	Gas	C	Mound Lab.‡	1962
Ne	^{21}Ne	33.9	Gas	C	ORNL	1963
CH_4	^{13}C	90	Gas	C	Mound Lab.	1963
Xe	^{124}Xe	4.4	Gas	C	ORNL	1964

† Oak Ridge National Laboratory, U.S. AEC, Oak Ridge, Tennessee.
‡ Mound Laboratory, U.S. AEC, Miamisburg, Ohio.
§ Feed not of normal abundance, contained 1 percent ^{3}He from nuclear reaction.
(Benedict et al., 1981).

TABLE 19.11. Electroorganic Synthesis Processes Now Applied Commercially or Past the Pilot Plant Stage

Product[a]	Raw Material[a]	Company (country)	Scale	Type of Process
Commercialized				
Adiponitrile	Acrylonitrile	Monsanto (US) Monsanto (UK) Asahi (Japan)	10^8 kg/yr 10^8 kg/yr 2×10^7 kg/yr	Reductive coupling
p-Aminophenol	Nitrobenzene	(Japan) Holliday (UK)	Not available Not available	Reductive rearrangement
Anthraquinone	Anthracene	Holliday (UK)	Not available	Indirect oxidation
2,5-Dimethoxydihydrofuran	Furan	(Japan) BASF (West Germany)	Not available Not available	Oxidative addition
Fluorinated Organics	Hydrocarbons, aliphatic carboxylic acids, sulfonic acids, amines, etc.	Dia Nippon (Japan) 3M (US)[b]	Not available Not available	Anodic substitution
Gluconic Acid	Glucose	(India)	3×10^5 kg/yr	Oxidation of functional group
Glyoxylic Acid	Oxalic acid	(Japan)	Not available	Reduction of functional group
Hexahydrocarbazole	Tetrahydrocarbazole	BASF (West Germany)	Not available	Reduction
Piperidine	Pyridine	Robinson Bros. (UK)	1.2×10^5 kg/yr	Reduction
Succinic Acid	Maleic acid	(India)	6×10^4 kg/yr	Reduction
Hexadecanedioic Acid	Monomethylazelate	Soda Aromatic Co. (Japan)	Not available	Crum Brown-Walker[c]
Tetraethyl Lead	Ethylmagnesium halide	Nalco (US)	Not available	Anodic
Propylene Oxide	Propylene	BASF (West Germany) others in UK and West Germany	Past pilot-plant Past pilot-plant	Paired synthesis
4,4′-*bis*-Pyridinium Salts	Pyridinium salts	(Japan)	Past pilot-plant	Paired synthesis
Salicylaldehyde	Salicylic acid	(India)	Past pilot-plant	Reduction of functional group
Sebacid Acid Diesters	Adipic acid half esters	BASF (West Germany) (Japan) (USSR)	Past pilot-plant Past pilot-plant Commercial?	Crum Brown-Walker[c]
Benzaldehyde	Toluene	(India)	Past pilot-plant	Indirect oxidation [Mn(III)]
Dihydrophthalic Acid	Phthalic acid	BASF (West Germany)	Commercial?	Reduction
Hydroquinone or Quinone	Benzene	Several	Past pilot-plant	Paired synthesis or anodic oxidation + chemical reduction
Maltol	Furfuryl alcohol	Otsuka (Japan)	Past pilot-plant	Oxidation
Pinacol	Acetone	(Japan) BASF (West Germany)	Past pilot-plant	Reductive coupling

[a] *Formulas are given in Appendix A.*
[b] *Added by author.*
[c] *Oxidative coupling.*
[M. M. Baizer, *J. Appl. Electrochem.* **10,** 285 (1980)].

Some of the energy is used to form the cell structure. Reactions catalyzed by enzymes may be either endo- or exothermic depending on the particular stoichiometry. Because of the diluteness of the solutions normally handled, temperature control is achieved readily. Stirred fermenters are provided with cooling jackets. Internal cooling oils are undesirable because of the difficulty of cleaning them. Fixed beds of immobilized enzymes do not lend themselves readily to jacket cooling, but in many instances the heat effect is so low that the temperature travel can be maintained within the required limits by adjustment of the feed temperature. Multitubular reactors with cooling medium on the shell side may be practical with enzymes immobilized on granules.

Sterilization. This is necessary to prevent the growth of foreign microorganisms. Air is sterilized adequately by the heat of compression. Filters at the inlet remove oil and any microbes that may be present, and filters at the air outlet prevent backflow of foreign microorganisms. The inoculum is prepared under sterile conditions in the laboratory. The substrate is sterilized in an external vessel by holding it at 120°C or so for 1 hr or so.

Aeration. Since metabolism of microorganisms is an oxidative process, the substrate should be kept as nearly saturated as possible. At usual fermenter operating conditions the solubility of oxygen is about 0.03 mmol/L. When the content falls to 0.01 mmol/L, the growth rate falls to about one-half the maximum. Compressed air is introduced through spargers. Dispersion with high-speed agitators rarely is feasible because of possible mechanical destruction of cells. In some sensitive systems, all of the necessary agitation may be provided with an adequate air flow.

TABLE 19.12. Comparative Energy Requirements of Electrochemical and Chemical Processes

Chemical		kcal/kg Electrochemical[a]	kcal/kg Chemical	
Adiponitrile	b	43,177 (10,520)*	65,808	c
Aniline				
Nitrobenzene route	b	36,172	13,919	
Phenol route	b	–	16,736	c
Sorbitol		9,649	958	
Terephthalic Acid		17,382	700	
Phenol	b	35,592	12,251	c
Methyl Ethyl Ketone		6,187	6,690	
			3,233	c
Melamine	b	30,159	15,472	
Hydroquinone	b	52,739	30,814	
Dichloroethane				
HCl route	b	17,773	6,131	
Cl_2 route	b	–	14,819	c

[a] Electrochemical energy adjusted for generating plant efficiency.
[b] Improved Monsanto process.
[c] Energy charged is for hydrocarbon raw materials (different compounds); other compounds begin with the same raw materials.
[d] Chemical route energy given by Rudd et al.; others estimated by Beck et al.
T. Beck et al., *A Survey of Organic Electrolytic Processes,* ANL/OEPM 79–5, Electrochemical Technology Corp., 1979.

Agitation. The purpose of agitation is to keep the microorganisms in suspension, to maintain uniformity to eliminate concentration gradients and hot spots, and to improve heat transfer to the cooling jacket. Rules for the design of agitation systems are covered in Chapter 10. In vessels of a 1000 gal or more, a power input of about 10 HP/1000 gal and impeller tip speeds of 15–20 ft/sec are adequate, but the standard fermenter described in Table 19.15 is supplied with about four times this power.

pH. Biochemical processes are highly sensitive to hydrogen-ion concentration. Most enzymes function best in the range of pH from 5 to 7, but some extremes are pepsin at pH of 1.5 and araginase at pH of 10. For classes of microorganisms, these ranges are common:

Complex cells	6.5–7.5
Bacteria	4–8
Molds	3–7
Yeasts	3–6

Control of pH is accomplished by additions of dilute acid or alkali.

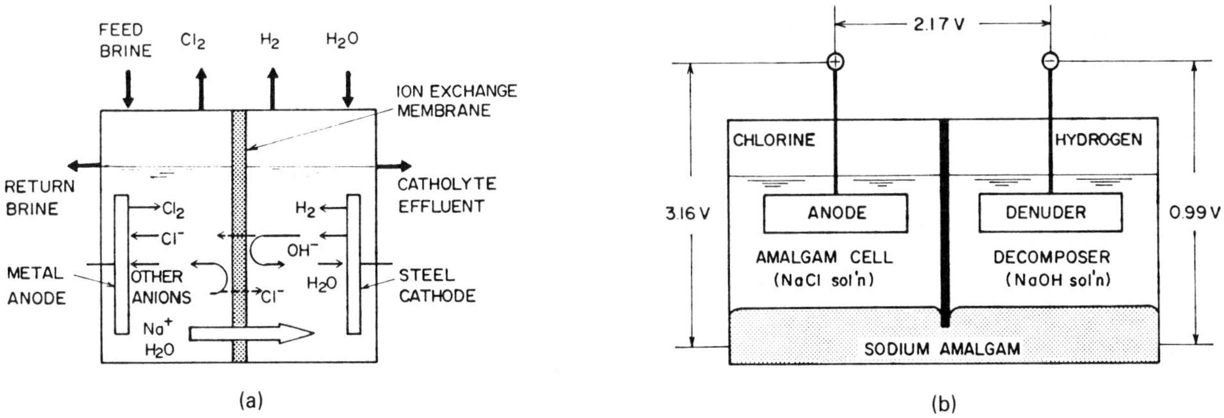

Figure 19.16. Basic designs of electrolytic cells. (a) Basic type of two-compartment cell used when mixing of anolyte and catholyte is to be minimized; the partition may be a porous diaphragm or an ion exchange membrane that allows only selected ions to pass. (b) Mercury cell for brine electrolysis. The released Na dissolves in the Hg and is withdrawn to another zone where it forms salt-free NaOH with water. (c) Monopolar electrical connections; each cell is connected separately to the power supply so they are in parallel at low voltage. (d) Bipolar electrical connections; 50 or more cells may be series and may require supply at several hundred volts. (e) Bipolar-connected cells for the Monsanto adiponitrile process. Spacings between electrodes and membrane are 0.8–3.2 mm. (f) New type of cell for the Monsanto adiponitrile process, without partitions; the stack consists of 50–200 steel plates with 0.0–0.2 mm coating of Cd. Electrolyte velocity of 1–2 m/sec sweeps out generated O_2.

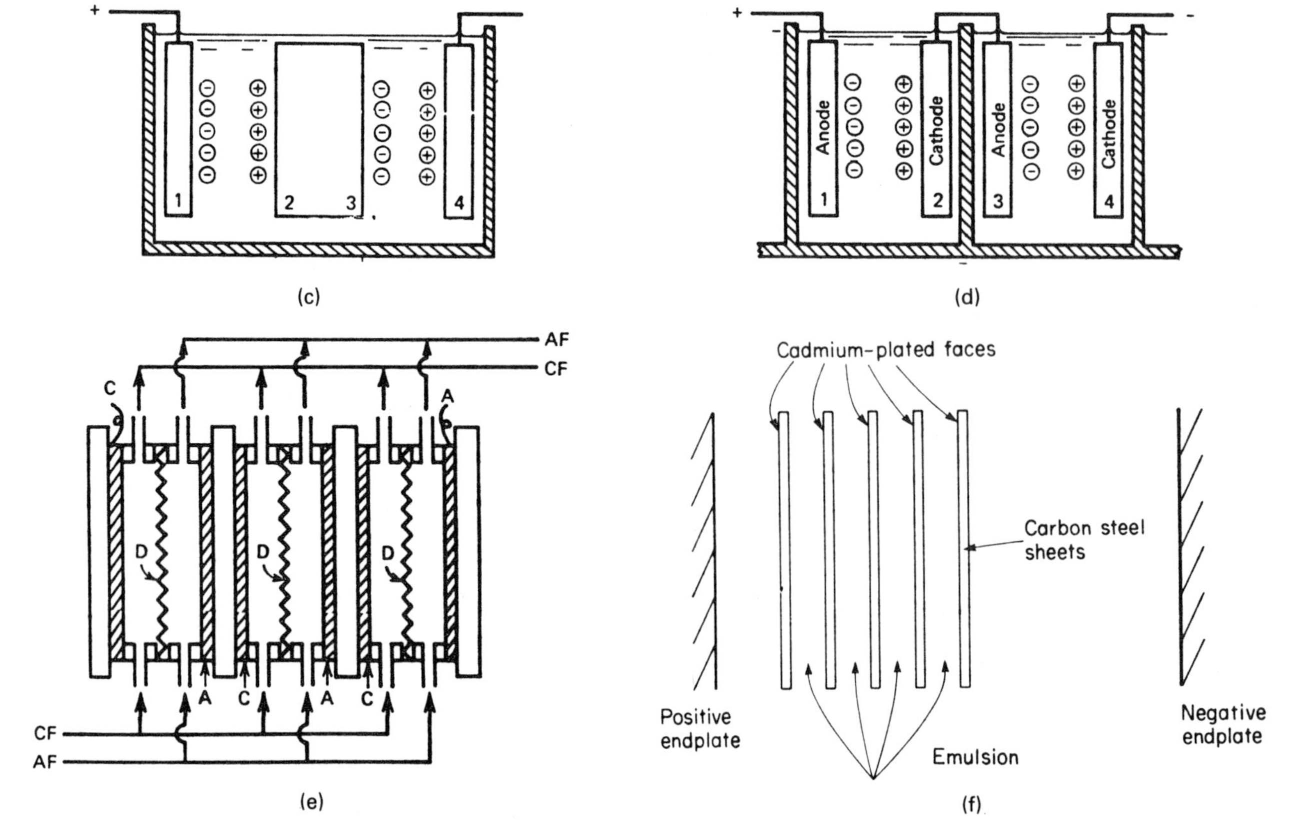

Figure 19.16—(*continued*)

Ion Concentration. Heavy metals, particularly calcium, inhibit enzyme activity. The only feasible method of removing them is with ion exchange resins.

Foam Control. Fermentations tend to froth because metabolites have surfactant properties. Prevention commonly is by addition of antifoam agents such as oils, heavy alcohols, fatty acids, or silicones. High-speed rotating impellers destroy bubbles by direct impact and by throwing them against the wall of the vessel.

REACTORS

Stirred tanks are the chief kind of reactors for handling microorganisms or dissolved isolated enzymes, either as batch units or as continuous stirred tank batteries. When the enzymes are immobilized, a variety of reactor configurations is possible and continuous operation is easily implemented. The immobilization may be on granules or on sheets, and has the further advantage of making the enzymes reusable since recovery of dissolved enzymes rarely is feasible.

	Current Density ($amp\ cm^{-2}$)		
Metal	0	0.01	0.10
Platinized Platinum	0.005	0.035	0.055
Gold	0.02	0.56	0.77
Iron	0.08	0.56	0.82
Smooth Platinum	0.09	---	0.39
Silver	0.15	0.76	0.90
Nickel	0.21	0.65	0.89
Copper	0.23	0.58	0.82
Lead	0.64	1.09	1.20
Zinc	0.70	0.75	1.06
Mercury	0.78	1.10	1.18

(a)

	Amalgam cell at 100 A/dm²	Diaphragm cell at 25 A/dm²
Decomposition voltage	3.16 V	2.17 V
Anode overvoltage, DSA	0.20 V	0.03 V
Cathode overvoltage	0.05 V	0.30 V
Solution *IR*, including bubble effects	0.44 V	0.35 V
Diaphragm		0.60 V
Metal hardware	0.05 V	0.20 V
(Sum) Terminal voltage	3.91 V	3.65 V

(b) (continued)

Figure 19.17. Overvoltage and distribution of voltage drops in cells (*Hine, 1985*). (a) Overvoltage of hydrogen on some metals. (b) Voltage distribution in two kinds of cells for electrolysis of brine. (c) Variation of voltage distribution with current density in the electrolysis of HCl. (d) Schematic of voltage profile in a bipolar cell with five pairs of electrodes.

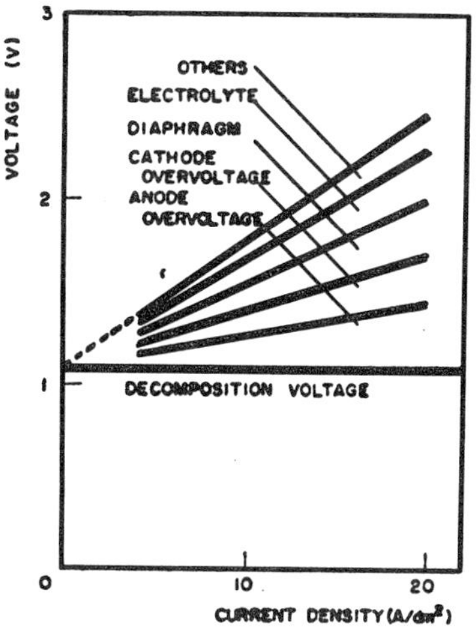

(c)

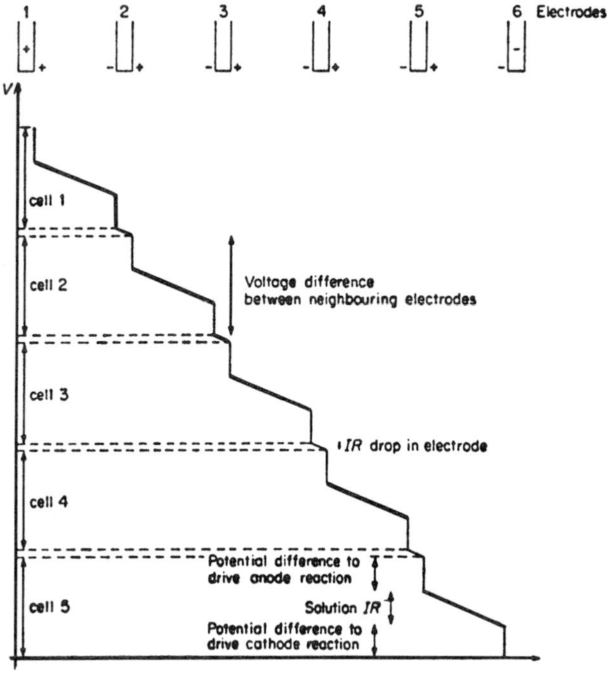

(d)

Figure 19.17—(*continued*)

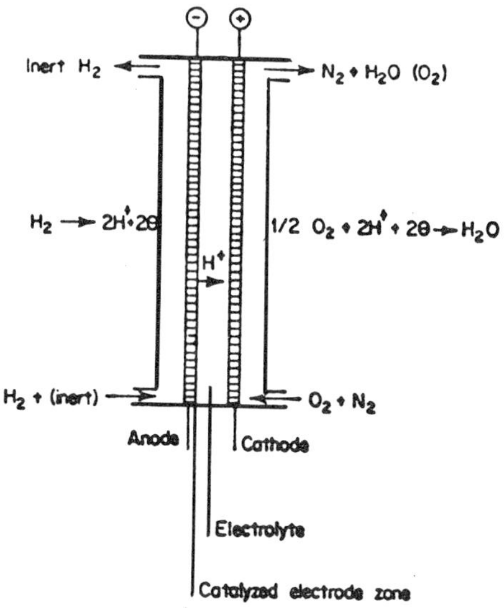

(a)

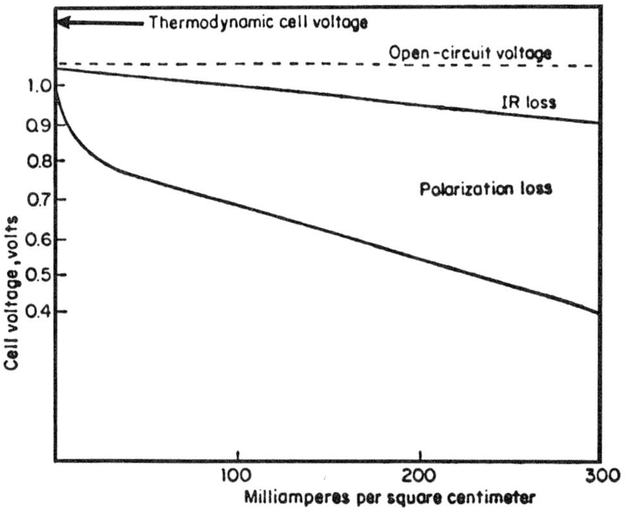

(b)

Reaction	Cell potential, volts					
	25°C	100°C	250°C	500°C	750°C	1000°C
$C + O_2 \rightarrow CO_2$	1.02	1.02	1.02	1.02	1.02	1.01
$2C + O_2 \rightarrow 2CO$	0.71	0.75	0.82	0.93	1.04	1.15
$2CO + O_2 \rightarrow 2CO_2$	1.33	1.30	1.23	1.11	1.00	0.88
$2H_2 + O_2 \rightarrow 2H_2O$	1.23	1.18	1.12	1.05	0.97	0.90

(c)

Figure 19.18. Data of electrochemical fuel cells. (a) Processes in a fuel cell based on the reaction between hydrogen and oxygen. (b) Voltage–current characteristic of a hydrogen–air fuel cell operating at 125°C with phosphoric acid electrolyte [*Adlhart, in* Energy Technology Handbook (*Considine, Ed.*), *1977, p. 4.61*). (c) Theoretical voltages of fuel cell reactions over a range of temperatures. (d) Major electrochemical systems for fuel cells (*Adlhart, in Considine,* loc. cit., *1977, p. 4.62*).

Electrolyte	Current transport	Operating temperature, °C	Electrode catalyst	Reactants: Fuel	Reactants: Oxidant	State of development
Aqueous potassium hydroxide (KOH)	OH^-	20–90	Nickel, silver, platinum metals	Hydrogen, hydrazine	Oxygen, scrubbed air, H_2O_2	Multikilowatt systems developed by several manufacturers.
Aqueous sulfuric acid (H_2SO_4)	H^+	20–80	Tungsten carbide, platinum metals, carbon	Impure hydrogen*	Air	Long life demonstrated in laboratory cells.
Concentrated phosphoric acid (H_3PO_4)	H^+	70–175	Platinum metals	Impure hydrogen*	Air	Multikilowatt system developed and larger systems in development. Long life has been demonstrated.
Fused alkali carbonate	CO_3^{--}	600–800	Nickel, silver	Impure hydrogen*	Air	Several months' life for small cells demonstrated.
Stabilized zirconium oxide	O^{--}	700–1000	Base metal oxides	Impure hydrogen*	Air	10,000 hours' life demonstrated in single cells. Multikilowatt systems in design.

(d)

Figure 19.18—(*continued*)

Many aspects of the design of biochemical reactors are like those of ordinary chemical reactors. The information needed for design are the kinetic data and the dependence of enzyme activity on time and temperature. Many such data are available in the literature, but usually a plant design is based on laboratory data obtained with small fermenters. Standard sizes of such units range from 50 to 1000 L capacity.

A sketch of a plant size fermenter and some of its auxiliaries is in Figure 19.21. Although not shown here, a bottom drive mechanical agitator usually is provided. The standard specification, Table 19.15, of one make of commercial fermenter includes a listing of the many openings that are required, as well as other general information.

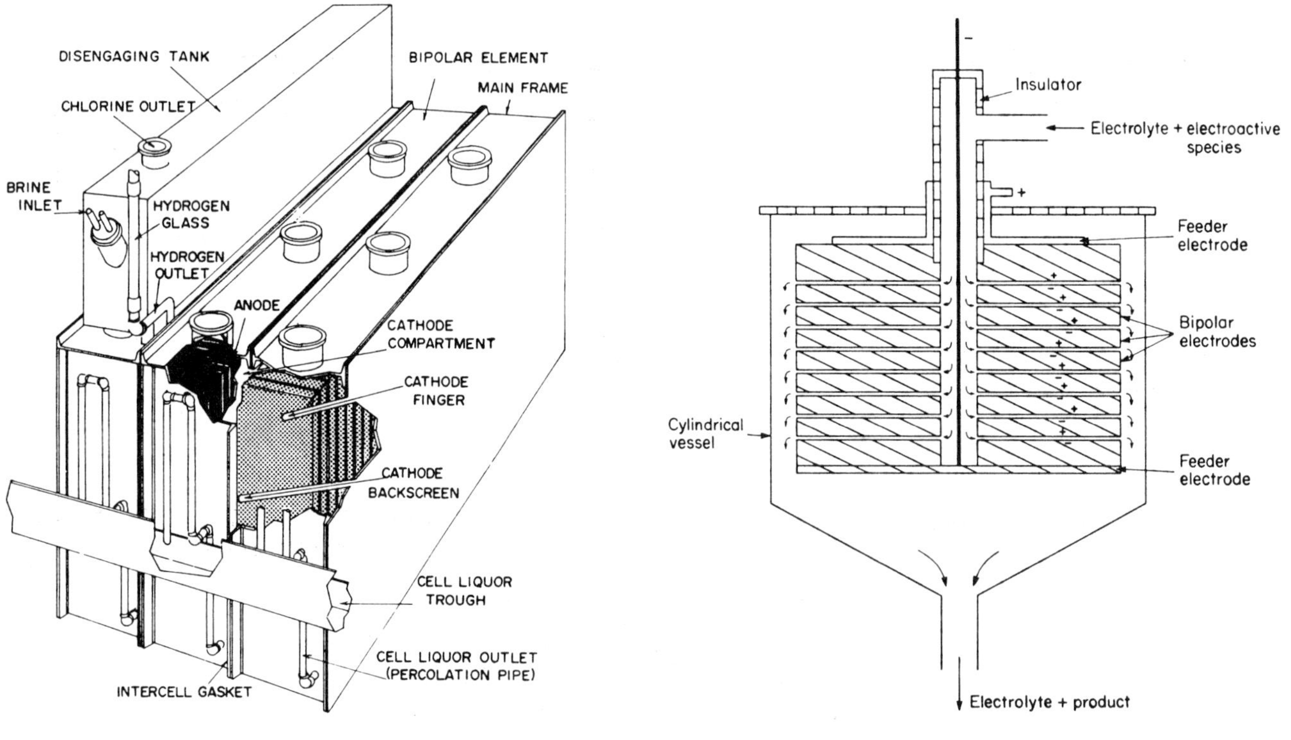

(a) (b)

Figure 19.19. Some special designs of electrolytic cells. (a) Glanor bipolar diaphragm-type cell assembly for chlor-alkali production (*PPG Industries*). (b) BASF capillary gap cell has 100 pairs of graphite plates with gaps of 0.2 mm used for adiponitrile synthesis; anodes are electroplated with lead dioxide [*Beck and Guthke,* Chem. Ing. Tech. **41,** *943* (*1969*)]. (c) Principle of the shell-and-tube reactor for electrolytic oxidation of Grignard reagents to lead alkyls. Lead shot serves as consumable anode which is replenished continuously. Individual tubes are 5 cm dia by 75 cm long [*Danly,* Encycl. Chem. Technol. **8,** *702* (*1979*)]. (d) Simple cells of the type used for electrolysis of HCl and water; voltage breakdown is shown in Figure 19.16(c).

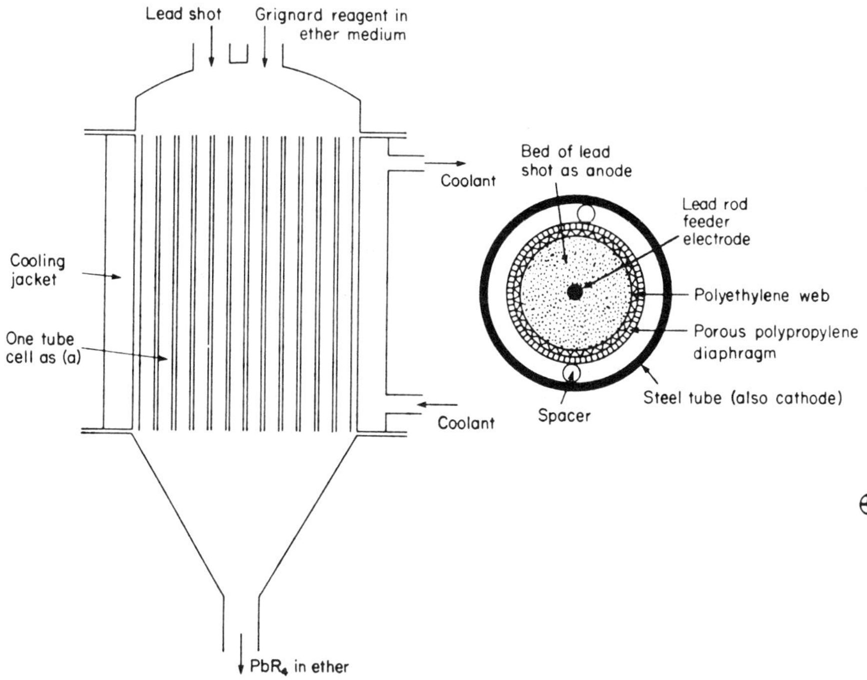

(c)

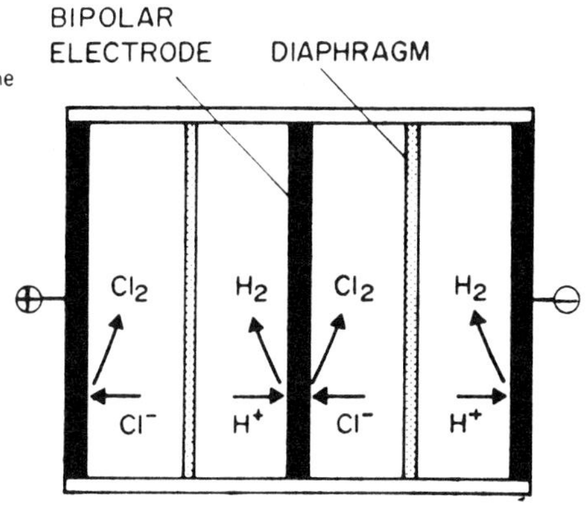

(d)

Figure 19.19—(*continued*)

TABLE 19.13. A Biochemical Glossary

Microorganisms (microbes) are living cells, single or in multiples of the same kind, including bacteria, yeasts, fungi, molds, algae and protozoa. Their metabolic products may be of simple or complex structure

Fermentation is a metabolic process whereby microorganisms grow in the presence of nutrients and oxygen, sometimes in the absence of oxygen. The terms used are aerobic (in the presence of oxygen) and anaerobic (in the absence of oxygen)

Substrate consists of the nutrients on which a microorganism subsists or the chemicals upon which an enzyme acts

Enzymes are made by living cells, and are proteins with molecular weights ranging from about 15,000 to 1,000,000. They are able to catalyze specific reactions

Enzymes, immobilized, are attached to a solid support by adsorption or chemical binding or mechanical entrapment in the pores of a gel structure, yet retain most of their catalytic powers

-ase is a suffix identifying that the substance is an enzyme. The main part of the name describes the nature of the chemical reaction that can be catalyzed, as in cellulase, an enzyme that catalyzes the decomposition of cellulose

TABLE 19.14. Industrial Products of Microbial and Related Processes

A. Significant or marginal products
    - Acetic acid
    - Amino acids
    - Butyric acid
    - Citric acid
    - Ethanol
    - Fructose from glucose
    - Glucose from starch
    - Gluconic acid
    - Methane
    - Nucleotides (glutamic acid, guanyl acid, xanthyllic acid)

B. Products under development or absolesced from microbial synthesis
    - Acids: fumaric, lactic, malic, oxalic and some others
    - Acetone
    - Butanol
    - Butanediol
    - Glycerine
    - Lipids
    - Polyalcohols and other substances

C. Enzymes (extensive lists with properties and industrial suppliers are in the book by Godfrey and Reichelt, 1983).

D. Antibiotics (Lists with major characteristics, sources and manufacturing methods are in, for example, the book of Bailey and Ollis, *Biochemical Engineering Fundamentals,* McGraw-Hill, New York, 1985).

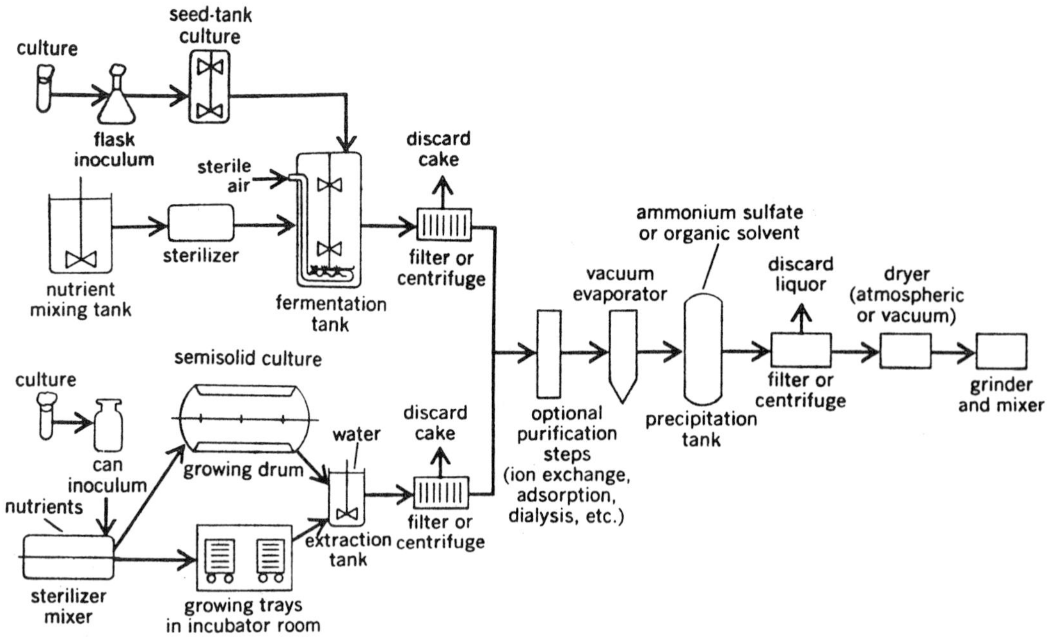

(a)

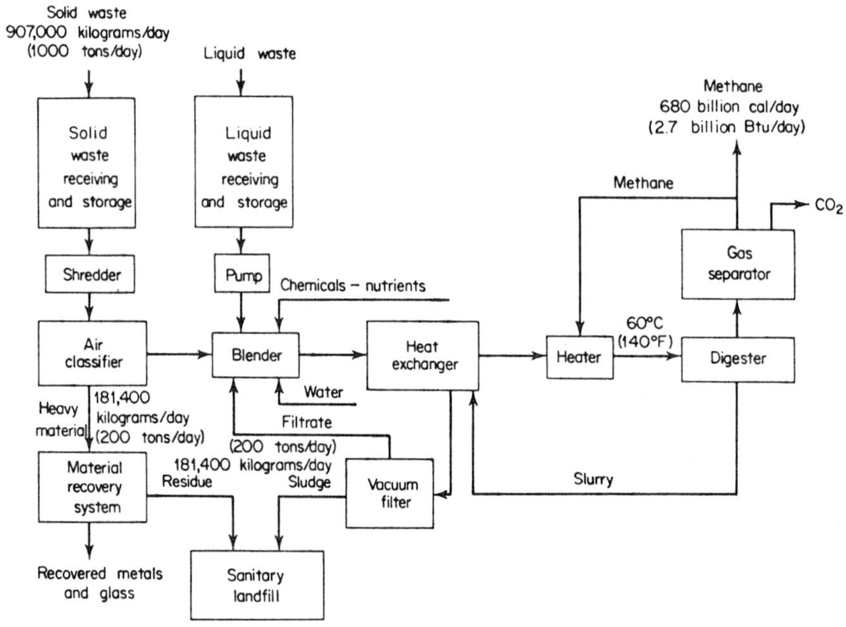

(b)

Figure 19.20. Flowsketches of two processes employing fermentation. (a) Process for enzyme production, showing the use of growing trays, growing drums and stirred tank. Purification steps are the same for all three modes of culture growth. (b) Production of methane-rich gas by anaerobic digestion of finely divided waste solids in a 10–20% slurry. Residence time in the digester is five days [*D.M. Considine (Ed.),* Energy Technology Handbook, *McGraw-Hill, New York, 1977*].

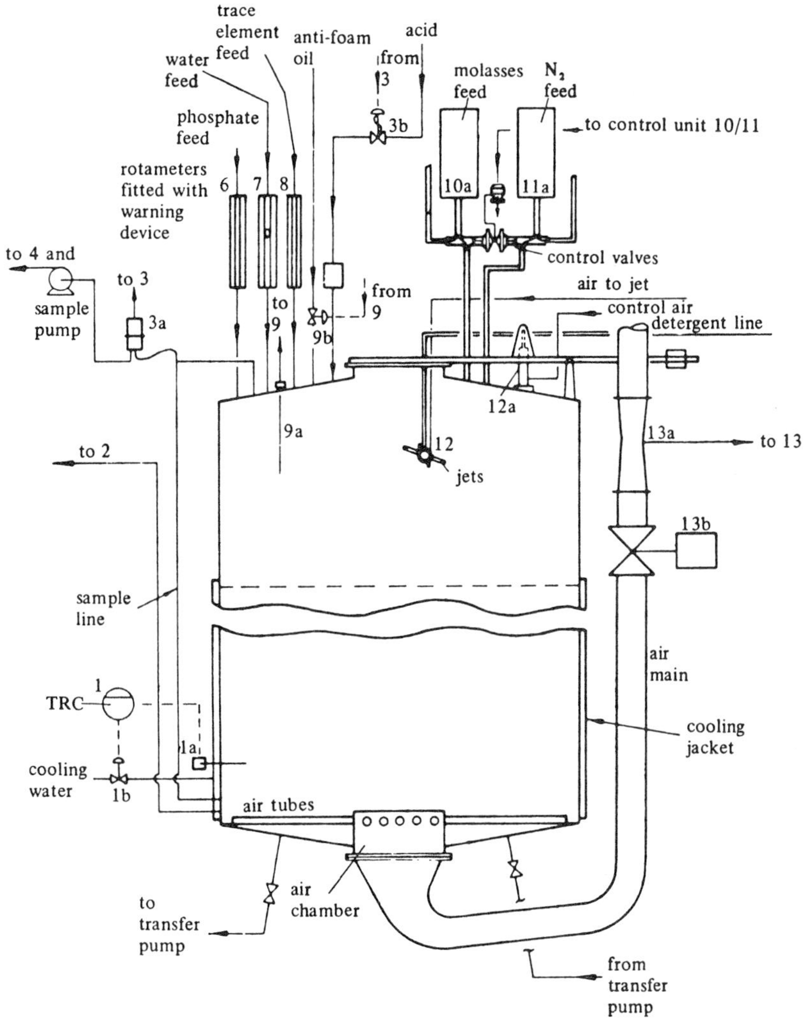

*1 Temperature controller and recorder
1a Resistance thermometer
1b Control valve
*2 Fermenter level
*3 pH recorder and controller
3a pH electrode system
3b Control valve
*4 Yeast concentration recorder
*5 Recorder controller
6 Phosphate feed rotameter
7 Water feed rotameter
8 Trace element feed rotameter

* 9 Foam controller
9a Foam detector
9b Control valve
10/11 Dosage control unit
10a Molasses feed
11a Nitrogen feed
12 Rotor jet
12a Power unit
*13 Air controller recorder
13a Venturi
13b Power operated air control valve
*Indicates panel-mounted instruments.

Figure 19.21. Sketch of a fermenter with its auxiliary equipment. In most cases supplemental agitation by mechanical stirrers is common (*A.J.C. Olsen,* Chem. Ind., *416* (*1960*)].

TABLE 19.15. Standard Specifications of a Fermenter

1. Surfaces in contact with culture are 316 SS, all others 304 SS; free of crevices, mechanically ground and polished to approx 220 grit
2. Approx proportions: height/diameter = 2, impeller/vessel diameter = 0.35, baffle width/vessel diameter = 0.1
3. Maximum working volume = 75–80%, minimum = 25%
4. Ports and penetrations are 20 in number, namely
   A. Steam-sterilizable inoculation/addition port
   B. Combination viewing window/filling port on headplate
   C. Light entrance window and lamp on headplate
   D. Air inlet line
   E. Air exhaust line
   F. Well for temperature control sensor and temperature recorder sensor
   G. Well for thermometer
   H. Water inlet line to jacket of vessel
   I. Water outlet line from jacket of vessel
   J. Rupture disc on headplate and pressure relief valve on jacket
   K. Diaphragm-type pressure gauge
   L. Steam-sterilizable sample port
   M. Steam-sterilizable bottom drain port, discharge valve is flush-bottom
   N. Side-entry port for pH electrode
   O. Top-entering or side-entering (size-dependent) port for installation of the dissolved oxygen electrode
   P. Top-entering port for foam sensor
   Q. Side-entering ports for acid, base, and antifoam addition (valved and piped as required)
   R. Spare penetrations on headplate for insertion of additional sensors i.e., 1 1/8 in. NPT, 1 3/8 in. NPT, 1 3/4 in. NPT
5. Foam breaking: Injection port provided for chemical breaking; mechanical breaker optional, consists of a double disk rotated at high speed with its own drive
6. Agitation system has three six-bladed turbine impellers adjustable along the shaft, maximum tip speed of 1200 ft/min, standard drive of 40 HP for a 5000-L vessel, bottom drive standard, top drive optional
7. Controls and monitors: liquid level, pH, dissolved oxygen, reduction–oxidation (Redox) potential, air rate, temperature, optional automatic sterilization cycle control, rupture disk on vessel, relief valve on jacket

(New Brunswick Scientific Co.).

REFERENCES

Membrane Processes

1. G. Belfort, *Materials Science of Synthetic Membranes: Fundamentals and Water Applications,* Academic, New York, 1984.
2. T.D. Brock, *Membrane Filtration: A User's Guide and Reference Manual,* Science Tech, 1983.
3. S.T. Hwang, C.K. Choi, and K. Kammermeyer, Gaseous transfer coefficients in membranes, *Separation Sci.* **3,** 461–478 (1974).
4. N.N. Li and W.S.W. Ho, Membrane processes, in *Chemical Engineers' Handbook,* McGraw-Hill, New York, 1984, pp. 17.14–17.34.
5. D.R. Lloyd (Ed.), *Materials Science of Synthetic Membranes,* ACS Symposium Series **269,** American Chemical Society, Washington, DC, 1985.
6. M.C. Porter, Membrane filtration, in *Handbook of Separation Techniques for Chemical Engineers* (P. A. Schweitzer, Ed.), McGraw-Hill, New York, 1979, pp. 2.3–2.103.
7. S. Sourirajan, *Reverse Osmosis,* Academic, New York, 1970.
8. A.F. Turbak (Ed.), *Synthetic Membranes: Vol. I, Desalination; Vol. 2, Hyper and Ultrafiltration Uses,* ACS Symposium Series **153** and **154**, American Chemical Society, Washington, DC, 1981.

Foam Separation and Flotation

1. R. Lemlich, *Adsorptive Bubble Separation Techniques,* Academic, New York, 1972.
2. Y. Okamoto and E.J. Chou, Foam separation processes, in *Handbook of Separation Techniques for Chemical Engineers* (P. A. Schweitzer, Ed.), McGraw-Hill, New York, 1979, pp. 2.183–2.197.
3. P. Somasundaran, Foam separation methods, a review, In *Separation and Purification Methods* (Perry and van Oss, Eds.), 1972, Vol. 1, pp. 117–199.
4. T.C. Sorensen, Flotation, in *Chemical Engineers' Handbook,* McGraw-Hill, New York, 1984, pp. 21.46–21.52.
5. B.A. Wills, *Mineral Processing Technology,* Pergamon, New York, 1985.

Sublimation and Freeze Drying

1. W. Corder, Sublimation, in *Chemical Engineers' Handbook,* McGraw-Hill, New York, 1984, pp. 17.12–17.14.
2. N. Ganiaris, Freeze drying, in *Chemical and Process Technology Encyclopedia* (Considine, Ed.), McGraw-Hill, New York, 1974, pp. 523–527.
3. C.A. Holden and H.S. Bryant, Sublimation, *Separation Sci.* **4**(1), 1 (1969).
4. C.J. Major, Freeze Drying, *Chemical Engineers' Handbook,* McGraw-Hill, New York, 1973 edition, pp. 17.26–17.28.
5. G. Matz, Sublimation, in *Ullmann's Encyclopedia of Chemical Technology,* Verlag Chemie, Weinheim, 1972, Vol. 2, pp. 664–671.
6. J.W. Mullin, Sublimation, in *Crystallization,* Butterworths, London, 1972, pp. 284–290.

Parametric Pumping and Cycling Zone Separation

1. N.H. Sweed, in *Recent Developments in Separation Science* (N. N. Li, Ed.), Vol. I, pp. 59–74.
2. P.C. Wankat, Cycling separation processes, *Separation Sci.* **9**(2), 85–116 (1974).
3. P.C. Wankat, J.C.D. Ore, and W.C. Nelson, Cycling zone separations, in *Separation and Purification Methods,* CRC Press, Boca Raton, FL, 1976, Vol. 4, pp. 215–266.
4. R.H. Wilhelm et al., *Ind. Eng. Chem. Fundam.* **5,** 141–144 (1966); **7,** 337–349 (1968).

Thermal Diffusion

1. M. Benedict, T.H. Pigford, and H.W. Levi, *Nuclear Chemical Engineering,* McGraw-Hill, New York, 1981.
2. A.L. Jones and G.B. Brown, Liquid thermal diffusion, in *Advances in Petroleum Chemistry and Refining* (McKetta and Kobe, Eds.), Wiley, New York, 1960, Vol. III, pp. 43–76.
3. G. Vasaru et al., *The Thermal Diffusion Column,* VEB Deutscher Verlag der Wissenschaften, Berlin, 1969.

Electrochemical Syntheses

1. F. Hine, *Electrode Processes and Electrochemical Engineering,* Plenum, New York, 1985.

2. D. Pletcher, *Industrial Electrochemistry,* Chapman and Hall, London, 1982.
3. R. Roberts, R.P. Ouellete, and P.N. Cheremisinoff, *Industrial Applications of Electroorganic Synthesis,* Ann Arbor Science, Ann Arbor, MI, 1982.

Fermentation Processing

1. B. Atkinson, *Biochemical Reactors,* Pion Ltd., London, 1974.
2. B. Atkinson and F. Mativuna, *Biochemical Engineering and Biotechnology Handbook,* Macmillan, Surrey, England, 1983.
3. J.E. Bailey and D.F. Ollis, *Biochemical Engineering Fundamentals,* McGraw-Hill, New York, 1986.
4. T. Godfrey and J. Reichelt, *Industrial Enzymology,* Macmillan, Surrey, England, 1983.
5. P.F. Stanbury and A. Whitaker, *Principles of Fermentation Technology,* Pergamon, New York, 1984.

20
COSTS OF INDIVIDUAL EQUIPMENT

The choice of appropriate equipment often is influenced by considerations of price. A lower efficiency or a shorter life may be compensated for by a lower price. Funds may be low at the time of purchase and expected to be more abundant later, or the economic life of the process is expected to be limited. Alternate kinds of equipment for the same service may need to be considered: water-cooled exchangers vs. air coolers, concrete cooling towers vs. redwood, filters vs. centrifuges, pneumatic conveyors vs. screw or bucket elevators, and so on.

In this chapter, the prices of classes of the most frequently used equipment are collected in the form of correlating equations. The prices are given in terms of appropriate key characteristics of the equipment, such as sqft, gpm, lb/hr, etc. Factors for materials of construction and performance characteristics other than the basic ones also are provided. Although graphs are easily read and can bring out clearly desirable comparisons between related types of equipment, algebraic representation has been adopted here. Equations are capable of consistent reading, particularly in comparison with interpolation on logarithmic scales, and are amenable to incorporation in computer programs.

Unless otherwise indicated, the unit price is \$1000, \$K. Except where indicated, notably for fired heaters, refrigeration systems, and cooling towers (which are installed prices), the prices are purchase prices, FOB, with delivery charges extra. In the United States delivery charges are of the order of 5% of the purchase price, but, of course, dependent on the unit value, as cost per lb or per

Example 20.1
Installed Cost of a Distillation Tower

Shell and trays are made of AISI 304 stainless steel. Dimensional data are:

$D = 4$ ft,
$L = 120$ ft,
$N = 58$ sieve trays,
wall thickness $t_p = 0.50$ in. for pressure,
$t_b = 0.75$ in. at the bottom,
flanged and dished heads weigh 325 lb each,

$$\text{weight } W = (\pi/4)(16)(120(0.5/12)(501) + 2(325) = 32{,}129 \text{ lb}$$

$$C_b = \exp[7.123 + 0.1478(10.38) + 0.02488(10.38)^2 + 0.158(120/4)\ln(0.75/0.50)] = 572{,}686,$$

$f_1 = 1.7,$
$f_2 = 1.189 + 0.0577(4) = 1.420,$
$f_3 = 0.85,$
$f_4 = 1,$

$$C_t = 375.8\exp[0.1739(4)] = 753.4,$$

$$C_{p1} = 204.9(4)^{0.6332}(120)^{0.8016} = 22{,}879,$$

$$\text{purchase price } C = 1.7(572{,}686) + 58(1.42)(0.85)(753.4) + 22{,}879 = \$1{,}049{,}188$$

From Table 20.1, the installation factor is 2.1 so that the installed price is

$$C_{\text{installed}} = 2.1(1{,}049{,}188) = \$2{,}203{,}294.$$

A tower packed with 2 in. pall rings instead of trays:

$$\text{packing volume } V_p = (\pi/4)(4)^2(120) = 1508 \text{ cuft},$$

$$C_{\text{installed}} = 2.1[1.7(572{,}686) + 1508(23.0) + 22{,}879)] = \$2{,}165{,}372.$$

Example 20.2
Purchased and Installed Prices of Some Equipment

a. A box type fired heater with CrMo tubes for pyrolysis at 1500 psig with a duty of 40 million Btu/hr. From Item No. 10 (Table 20.1), the installed price is

$$C_{\text{installed}} = 33.8(1.0 + 0.10 + 0.15)(40)^{0.86} = 1008.32 \text{ K\$, \$1,008,320}.$$

b. A 225 HP-reciprocating compressor with motor drive and belt drive coupling. Items Nos. 2 and 13 (Table 20.1). The installation factor is 1.3.

$$\text{compressor } C = 5960(225)^{0.61} = 162{,}210,$$

$$\text{motor, 1800 rpm, TEFC, } C = 1.2 \times \exp[4.5347 + 0.57065(5.42) + 0.04069(5.42)^2] = \$8113,$$

$$\text{belt drive coupling, } C = 1.2\exp[3.689 + 0.8917(5.42)] = \$6008,$$

$$\text{total installed cost, } C_{\text{total}} = 1.3(162{,}210 + 8113 + 6008) = \$229{,}230.$$

c. A two-stage steam ejector with one surface condenser to handle 200 lb/hr of air at 25 Torr, in carbon steel construction. From Table 20.3 the installation factor is 1.7.

$X = 200/25 = 8,$
$f_1 = 1.6, \quad f_2 = 1.8, \quad f_3 = 1.0$

$$\text{purchase } C = 11(1.6)(1.8(1.0)(8))^{0.41} = 74.31 \text{ K\$, \$74,310},$$

$$\text{installed } C = 1.7C_p = \$126{,}330.$$

cuft. Multipliers have been developed whereby the installed cost of various kinds of equipment may be found. Such multipliers range from 1.2 to 3.0, but details are shown in Table 20.1.

Data are taken from a number of published sources and are updated to the beginning of 1985 with the cost indexes of *Chemical Engineering Magazine,* a selection of which is in Table 20.2. The main sources and the dates of their prices are Hall et al. (1981), Institut Francais du Petrole (1975), and Evans et al. (1979). References also are made to price data of some equipment not covered here. Many data as of mid-1982 have been collected by Ulrich (1984). Perry's *Chemical Engineers Handbook* (1984) has many data scattered throughout; the page numbers having such data are listed in the reference (Green, 1984).

Material of construction is a major factor in the price of equipment so that multipliers for prices relative to carbon steel or other standard materials are given for many of the items covered here. Usually only the parts in contact with process substances need be of special construction, so that, in general, the multipliers are not always as great as they are for vessels that are made entirely of special materials. Thus, when the tube side of an exchanger is special and the shell is carbon steel, the multiplier will vary with the amount of tube surface, as shown in that section.

As with most collections of data, the price data correlated here exhibit a certain amount of scatter. This is due in part to the incomplete characterizations in terms of which the correlations are made, but also to variations among manufacturers, qualities of construction, design differences, market situations, and other factors. Accordingly, the accuracy of the correlations cannot be claimed to be better than ±25% or so.

TABLE 20.1. Index of Equipment

1. Agitators
2. Compressors, turbines, fans
    - Centrifugal compressors
    - Reciprocating compressors
    - Screw compressors
    - Turbines
        - Pressure discharge
        - Vacuum discharge
    - Fans
3. Conveyors
    - Troughed belt
    - Flat belt
    - Screw, steel
    - Screw, stainless
    - Bucket elevator
    - Pneumatic
4. Cooling towers
    - Concrete
    - Wooden
5. Crushers and grinders
    - Cone crusher
    - Gyratory crusher
    - Jaw crusher
    - Hammer mill
    - Ball mill
    - Pulverizer
6. Crystallizers
    - External forced circulation
    - Internal draft tube
    - Batch vacuum
7. Distillation and absorption towers
    - Distillation tray towers
    - Absorption tray towers
    - Packed towers
8. Dryers
    - Rotary, combustion gas heated
    - Rotary, hot air heated
    - Rotary, steam tube heated
    - Cabinet dryers
    - Spray dryers
    - Multiple hearth furnace
9. Evaporators
    - Forced circulation
    - Long tube
    - Falling film
10. Fired heaters
    - Box types
    - Cylindrical types
11. Heat exchangers
    - Shell-and-tube
    - Double pipe
    - Air coolers
12. Mechanical separators
    - Centrifuges
    - Cyclone separators
        - Heavy duty
        - Standard duty
        - Multiclone
    - Disk separators
    - Filters
        - Rotary vacuum belt discharge
        - Rotary vacuum scraper discharge
        - Rotary vacuum disk
        - Horizontal vacuum belt
        - Pressure leaf
        - Plate-and-frame
    - Vibrating screens
13. Motors and couplings
    - Motors
    - Belt drive coupling
    - Chain drive coupling
    - Variable speed drive coupling
14. Pumps
    - Centrifugal
    - Vertical mixed flow
    - Vertical axial flow
    - Gear pumps
    - Reciprocating pumps
15. Refrigeration
16. Steam ejectors and vacuum pumps
    - Ejectors
    - Vacuum pumps
17. Vessels
    - Horizontal pressure vessels
    - Vertical pressure vessels
    - Storage tanks, shop fabricated
    - Storage tanks, field erected

TABLE 20.2. Purchase Prices of Process Equipment (Basic: CE Plant Cost Index = 325, Middle 1985)

1. Agitators
[Meyers and Kime, *Chem. Eng.*, 109–112 (27 Sep. 1976)]

$$C = \exp[a + b \ln HP + c(\ln HP)^2] \quad K\$, \quad 1 < HP < 400$$

		Single Impeller			Dual Impeller		
		Speed 1	2	3	1	2	3
Carbon steel	a	8.57	8.43	8.31	8.80	8.50	8.43
	b	0.1195	−0.0880	−0.1368	0.1603	0.0257	−0.1981
	c	0.0819	0.1123	0.1015	0.0659	0.0878	0.1239
Type 316	a	8.82	8.55	8.52	9.25	8.82	8.72
	b	0.2474	0.0308	−0.1802	0.2801	0.1235	−0.1225
	c	0.0654	0.0943	0.1158	0.0542	0.0818	0.1075

Speeds 1: 30, 37, and 45 rpm
2: 56, 68, 84, and 100 rpm
3: 125, 155, 190, and 230 rpm

2. Compressors, turbines, and fans (K$)

Centrifugal compressors, without drivers (IFP, 1981):

$$C = 6.49(HP)^{0.62} \quad K\$, \quad 200 < HP < 30{,}000$$

Reciprocating compressors without drivers (IFP):

$$C = 5.96(HP)^{0.61} \quad K\$, \quad 100 < HP < 20{,}000$$

Screw compressors with drivers (IFP):

$$C = 1.49(HP)^{0.71} \quad K\$, \quad 10 < HP < 800$$

Turbines (IFP):

Pressure discharge, $C = 0.31(HP)^{0.81}$ K$, $20 < HP < 5000$
vacuum discharge, $C = 0.69(HP)^{0.81}$ K$, $200 < HP < 8000$

Fans with motors (Ulrich)

$$C = f_m f_p \exp[a + b \ln Q + c(\ln Q)^2] \text{ installed cost, K\$}, \quad Q \text{ in KSCFM}$$

	a	*b*	*c*	*Q*
Radial blades	0.4692	0.1203	0.0931	2–500
Backward curved	0.0400	0.1821	0.0786	2–900
Propeller	−0.4456	0.2211	0.0820	2–300
Propeller, with guide vanes	−1.0181	0.3332	0.0647	2–500

Installation factor, f_m

Carbon steel	2.2
Fiberglass	4.0
Stainless steel	5.5
Nickel alloy	11.0

Pressure Factors, F_p

	Centrifugal		Axial	
Pressure (kPa[gage])	Radial	Backward Curved	Prop.	Vane
1	1.0	1.0	1.0	1.00
2	1.15	1.15	—	1.15
4	1.30	1.30	—	1.30
8	1.45	1.45	—	—
16	1.60	—	—	—

3. Conveyors (IFP) K$

Troughed belt: $C = 1.40L^{0.66}$, $10 < L < 1300$ ft
Flat belt: $C = 0.90L^{0.66}$, $10 < L < 1300$ ft
Screw (steel): $C = 0.40L^{0.78}$, $7 < L < 100$ ft
Screw (stainless steel): $C = 0.70L^{0.78}$, $7 < L < 100$ ft
Bucket elevator: $C = 4.22L^{0.63}$, $10 < L < 100$ ft
Pneumatic conveyor (*Chemical Engineers' Handbook*, McGraw-Hill, New York, 1984), 600 ft length

$$C = \exp[3.5612 - 0.0048 \ln W + 0.0913(\ln W)^2], \quad 10 < W < 100 \text{ klb/hr}$$

4. Cooling towers, installed K$

Concrete (IFP) $C = 135fQ^{0.61}$, $1 < Q < 60$ K gal/min:

Δt (°C)	10	12	15
f	1.0	1.5	2.0

Redwood, without basin (Hall): $C = 33.9Q^{0.85}$, $1.5 < Q < 20$ K gal/min

5. Crushers and grinders (IFP) K$

Cone crusher: $C = 1.55W^{1.05}$, $20 < W < 300$ tons/hr
Gyratory crusher: $C = 8.0W^{0.60}$, $25 < W < 200$ tons/hr
Jaw crusher: $C = 6.3W^{0.57}$, $10 < W < 200$ tons/hr
Hammer mill: $C = 2.44W^{0.78}$, $2 < W < 200$ tons/hr
Ball mill: $C = 50.0W^{0.69}$, $1 < W < 30$ tons/hr
Pulverizer: $C = 22.6W^{0.39}$, $1 < W < 5$ tons/hr

6. Crystallizers (IFP, *Chemical Engineers' Handbook*, p. 19.40)

External forced circulation:

$$C = f \exp[4.868 + 0.3092 \ln W + 0.0548(\ln W)^2],$$
$$10 < W < 100 \text{ klb/hr of crystals}$$

Internal draft tube: $C = 178fW^{0.58}$, $15 < W < 100$ klb/hr of crystals
Batch vacuum: $C = 8.16fV^{0.47}$, $50 < V < 1000$ cuft of vessel

Type	Material	*f*
Forced circulation	Mild steel	1.0
	Stainless type 304	2.5
Vacuum batch	Mild steel	1.0
	Rubber-lined	1.3
	Stainless type 304	2.0

7. Distillation and absorption towers, tray and packed (Evans et al., 1984) prices in $

Tray towers:

$$C_t = f_1 C_b + N f_2 f_3 f_4 C_t + C_{p1}$$

Distillation:

$$C_b = \exp[7.123 + 0.1478(\ln W) + 0.02488(\ln W)^2 + 0.01580(L/D)\ln(T_b/T_p)],$$

$9020 < W < 2{,}470{,}000$ lbs of shell exclusive of nozzles and skirt

$C_t = 375.8 \exp(0.1739D)$, $2 < D < 16$ ft tray diameter

N = number of trays

$C_{p1} = 204.9D^{0.6332}L^{0.8016}$, $2 < D < 24$, $57 < L < 170$ ft (platforms and ladders)

Material	f_1	f_2
Stainless steel, 304	1.7	$1.189 + 0.0577D$
Stainless steel, 316	2.1	$1.401 + 0.0724D$
Carpenter 20CB-3	3.2	$1.525 + 0.0788D$
Nickel-200	5.4	
Monel-400	3.6	$2.306 + 0.1120D$
Inconel-600	3.9	
Incoloy-825	3.7	
Titanium	7.7	

(continued)

TABLE 20.2—(*continued*)

Tray Types	f_3
Valve	1.00
Grid	0.80
Bubble cap	1.59
Sieve (with downcomer)	0.95

$f_4 = 2.25/(1.0414)^N$, when the number of trays N is less than 20

T_b is the thickness of the shell at the bottom, T_p is thickness required for the operating pressure, D is the diameter of the shell and tray, L is tangent-to-tangent length of the shell

Absorption:

$C_b = \exp[6.629 + 0.1826(\ln W) + 0.02297(\ln W)^2]$, $4250 < W < 980{,}000$ lb shell

$C_{p1} = 246.4D^{0.7396}L^{0.7068}$, $3 < D < 21$, $27 < L < 40$ ft (platforms and ladders),

f_1, f_2, f_3, and f_4 as for distillation

Packed towers:

$C = f_1C_b + V_pC_p + C_{p1}$

V_p is volume of packing, C_p is cost of packing \$/cuft

Packing Type	C_p (\$/cuft)
Ceramic Raschig rings, 1 in.	19.6
Metal Raschig rings, 1 in.	32.3
Intalox saddles, 1 in.	19.6
Ceramic Raschig rings, 2 in.	13.6
Metal Raschig rings, 2 in.	23.0
Metal Pall rings, 1 in.	32.3
Intalox saddles, 2 in.	13.6
Metal Pall rings, 2 in.	23.0

8. Dryers (IFP)

Rotary combustion gas heated: $C = (1 + f_g + f_m)\exp[4.9504 - 0.5827(\ln A) + 0.0925(\ln A)^2]$, $200 < A < 30{,}000$ sqft lateral surface

Rotary hot air heated: $C = 2.38(1 + f_g + f_m)A^{0.63}$, $200 < A < 4000$ sqft lateral surface

Rotary steam tube: $C = 1.83FA_t^{0.60}$, $500 < A_t < 18{,}000$ sqft tube surface, $F = 1$ for carbon steel, $F = 1.75$ for 304 stainless

Cabinet dryer: $C = 1.15f_pA^{0.77}$, $10 < A < 50$ sqft tray surface

Pressure	f_p
Atmospheric pressure	1.0
Vacuum	2.0

Material	f_m
Mild steel	1.0
Stainless type 304	1.4

Drying Gas	f_g
Hot air	0.00
Combustion gas (direct contact)	0.12
Combustion gas (indirect contact)	0.35

Materials	f_m
Mild steel	0.00
Lined with stainless 304–20%	0.25
Lined with stainless 316–20%	0.50

Spray dryers:

$C = F\exp(0.8403 + 0.8526(\ln x) - 0.0229(\ln x)^2$, $30 < x < 3000$ lb/hr evaporation

Material	F
Carbon steel	0.33
304, 321	1.00
316	1.13
Monel	3.0
Inconel	3.67

Multiple hearth furnaces (Hall et al., 1984)

$C = \exp(a + 0.88N)$, $4 < N < 14$ number of hearths

Diameter (ft)	6.0	10.0	14.25	16.75	18.75	22.25	26.75
Sqft/hearth, approx	12	36	89	119	172	244	342
a	5.071	5.295	5.521	5.719	5.853	6.014	6.094

9. Evaporators (IFP; also Chemical Engineers Handbook, p. 11.42)

Forced circulation: $C = f_m\exp[5.9785 - 0.6056(\ln A) + 0.08514(\ln A)^2]$, $150 < A < 8000$ sqft heat transfer surface

Long tube: $C = 0.36f_mA^{0.85}$, $300 < A < 20{,}000$ sqft

Falling film (316 internals, carbon steel shell)

$C = \exp[3.2362 - 0.0126(\ln A) + 0.0244(\ln A)^2]$, $150 < A < 4000$ sqft

Forced-Circulation Evaporators

Construction Material: Shell/Tube	f_m
Steel/copper	1.00
Monel/cupronickel	1.35
Nickel/nickel	1.80

Long-Tube Evaporators

Construction Material: Shell/Tube	f_m
Steel/copper	1.0
Steel/steel	0.6
Steel/aluminum	0.7
Nickel/nickel	3.3

10. Fired heaters, installed (Hall) K\$

Box type: $C = k(1 + f_d + f_p)Q^{0.86}$, $20 < Q < 200$ M Btu/hr

Tube Material	k
Carbon steel	25.5
CrMo steel	33.8
Stainless	45.0

Design Type	f_d
Process heater	0
Pyrolysis	0.10
Reformer (without catalyst)	0.35

Design Pressure, (psi)	f_p
Up to 500	0
1,000	0.10
1,500	0.15
2,000	0.25
2,500	0.40
3,000	0.60

Cylindrical type: $C = k(1 + f_d + f_p)Q^{0.82}$, $2 < Q < 30$ M Btu/hr

Tube Material	k
Carbon steel	27.3
CrMo steel	40.2
Stainless	42.0

(*continued*)

Design Type	f_d
Cylindrical	0
Dowtherm	0.33

Design Pressure (psi)	f_p
Up to 500	0
1,000	0.15
1,500	0.20

11. Heat exchangers

Shell-and-tube (Evans): $C = f_d f_m f_p C_b$, price in \$

$$C_b = \exp[8.821 - 0.30863(\ln A) + 0.0681(\ln A)^2], \quad 150 < A < 12{,}000 \text{ sqft}$$

Type	f_d
Fixed-head	$\exp[-1.1156 + 0.0906(\ln A)]$
Kettle reboiler	1.35
U-tube	$\exp[-0.9816 + 0.0830(\ln A)]$

Pressure Range (psig)	f_p
100–300	$0.7771 + 0.04981(\ln A)$
300–600	$1.0305 + 0.07140(\ln A)$
600–900	$1.1400 + 0.12088(\ln A)$

$$f_m = g_1 + g_2(\ln A)$$

Material	g_1	g_2
Stainless steel 316	0.8603	0.23296
Stainless steel 304	0.8193	0.15984
Stainless steel 347	0.6116	0.22186
Nickel 200	1.5092	0.60859
Monel 400	1.2989	0.43377
Inconel 600	1.2040	0.50764
Incoloy 825	1.1854	0.49706
Titanium	1.5420	0.42913
Hastelloy	0.1549	0.51774

Double pipe (IFP): $C = 900 f_m f_p A^{0.18}$, $2 < A < 60$ sqft, price in \$

Material: Shell/Tube	f_m
cs/cs	1.0
cs/304L stainless	1.9
cs/316 stainless	2.2

Pressure (bar)	f_p
≤4	1.00
4–6	1.10
6–7	1.25

Air coolers (Hall): $C = 24.6A^{0.40}$, $0.05 < A < 200$ K sqft, price in K\$

12. Mechanical separators

Centrifuges: solid bowl, screen bowl or pusher types

$C = a + bW$, K\$

	Inorganic Process		Organic Process	
Material	a	b	a	b
Carbon steel	42	1.63	—	—
316	65	3.50	98	5.06
Monel	70	5.50	114	7.14
Nickel	84.4	6.56	143	9.43
Hastelloy	—	—	300	10.0
	$10 < W < 90$		$5 < W < 40$ tons/hr	

Disk separators, 316 stainless (IFP):

$C = 8.0Q^{0.52}$, $15 < Q < 150$ gpm, K\$

Cyclone separators (IFP): K\$

heavy duty: $C = 1.39Q^{0.98}$, $2 < Q < 40$ K SCFM

standard duty: $C = 0.65Q^{0.91}$, $2 < Q < 40$ K SCFM

multiclone: $C = 1.56Q^{0.68}$, $9 < Q < 180$ K SCFM

Filters (Hall), prices in \$/sqft:

rotary vacuum belt discharge: $C = \exp[11.20 - 1.2252(\ln A) + 0.0587(\ln A)^2]$, $10 < A < 800$ sqft

rotary vacuum drum scraper discharge: $C = \exp[11.27 - 1.3408(\ln A) + 0.0709(\ln A)^2]$ \$/sqft, $10 < A < 1500$ sqft

rotary vacuum disk: $C = \exp[10.50 - 1.008(\ln A) + 0.0344(\ln A)^2]$ \$/sqft, $100 < A < 4000$ sqft

horizontal vacuum belt: $C = 28300/A^{0.5}$ \$/sqft, $10 < A < 1200$ sqft

pressure leaf: $C = 695/A^{0.29}$ \$/sqft, $30 < A < 2500$ sqft

plate-and-frame: (*Chemical Engineers' Handbook*): $C = 460/A^{0.45}$ \$/sqft, $10 < A < 1000$ sqft

vibrating screen (IFP): $C = 3.1A^{0.59}$ K\$, $0.5 < A < 35$ sqft

13. Motors and couplings, prices in \$

Motors: $C = 1.2\exp[a_1 + a_2(\ln \text{HP}) + a_3(\ln \text{HP})^2]$

Belt drive coupling: $C = 1.2\exp[3.689 + 0.8917(\ln \text{HP})]$

Chain drive coupling: $C = 1.2\exp[5.329 + 0.5048(\ln \text{HP})]$

Variable speed drive coupling:

$C = 12{,}000/(1.562 + 7.877/\text{HP})$, HP < 75

	Coefficients			
Type	a_1	a_2	a_3	HP limit
Open, drip-proof				
3600 rpm	4.8314	0.09666	0.10960	1–7.5
	4.1514	0.53470	0.05252	7.5–250
	4.2432	1.03251	−0.03595	250–700
1800 rpm	4.7075	−0.01511	0.22888	1–7.5
	4.5212	0.47242	0.04820	7.5–250
	7.4044	−0.06464	0.05448	250–600
1200 rpm	4.9298	0.30118	0.12630	1–7.5
	5.0999	0.35861	0.06052	7.5–250
	4.6163	0.88531	−0.02188	250–500
Totally enclosed, fan-cooled				
3600 rpm	5.1058	0.03316	0.15374	1–7.5
	3.8544	0.83311	0.02399	7.5–250
	5.3182	1.08470	−0.05695	250–400
1800 rpm	4.9687	−0.00930	0.22616	7.5–250
	4.5347	0.57065	0.04609	250–400
1200 rpm	5.1532	0.28931	0.14357	1–7.5
	5.3858	0.31004	0.07406	7.5–350
Explosion-proof				
3600 rpm	5.3934	−0.00333	0.15475	1–7.5
	4.4442	0.60820	0.05202	7.5–200
1800 rpm	5.2851	0.00048	0.19949	1–7.5
	4.8178	0.51086	0.05293	7.5–250
1200 rpm	5.4166	0.31216	0.10573	1–7.5
	5.5655	0.31284	0.07212	7.5–200

14. Pumps

Centrifugal (Evans) prices in \$: $C = F_M F_T C_b$, base cast-iron, 3550 rpm VSC

$$C_b = 1.55\exp[8.833 - 0.6019(\ln Q\sqrt{H}) + 0.0519(\ln Q\sqrt{H})^2], \quad Q \text{ in gpm}, \quad H \text{ in ft head}$$

Material	Cost Factor F_M
Cast steel	1.35
304 or 316 fittings	1.15
Stainless steel, 304 or 316	2.00
Cast Gould's alloy no. 20	2.00
Nickel	3.50
Monel	3.30
ISO B	4.95
ISO C	4.60
Titanium	9.70
Hastelloy C	2.95
Ductile iron	1.15
Bronze	1.90

$$F_T = \exp[b_1 + b_2(\ln Q\sqrt{H}) + b_3(\ln Q\sqrt{H})^2]$$

(continued)

TABLE 20.2—*(continued)*

Type	b_1	b_2	b_3
One-stage, 1750 rpm, VSC	5.1029	−1.2217	0.0771
One-stage, 3550 rpm, HSC	0.0632	0.2744	−0.0253
One-stage, 1750 rpm, HSC	2.0290	−0.2371	0.0102
Two-stage, 3550 rpm, HSC	13.7321	−2.8304	0.1542
Multistage, 3550 rpm, HSC	9.8849	−1.6164	0.0834

Type	Flow Range (gpm)	Head Range (ft)	HP (max)
One-stage, 3550 rpm, VSC	50–900	50–400	75
One-stage, 1750 rpm, VSC	50–3500	50–200	200
One-stage. 3550 rpm, HSC	100–1500	100–450	150
One-stage, 1750 rpm, HSC	250–5000	50–500	250
Two-stage, 3550 rpm, HSC	50–1100	300–1100	250
Two-stage, 3550 rpm, HSC	100–1500	650–3200	1450

Vertical mixed flow (IFP): $C = 0.036(\text{gpm})^{0.82}$ K\$, $500 < \text{gpm} < 130{,}000$

Vertical axial flow (IFP): $C = 0.020(\text{gpm})^{0.78}$ K\$, $1000 < \text{gpm} < 130{,}000$

Gear pumps (IFP): $C = \exp[-0.0881 + 0.1986(\ln Q) + 0.0291(\ln Q)^2]$ K\$, $10 < Q < 900$ gpm

Reciprocating (Pikulik and Diaz, 1979):

Cast iron: $C = 63.1Q^{0.81}$ K\$, $15 < Q < 400$ gpm

Others: $C = 653FQ^{0.52}$ K\$, $1 < Q < 400$ gpm

316 stainless	$F = 1.00$
Al bronze	1.40
Nickel	1.86
Monel	2.20

15. **Refrigeration (IFP):** $C = 146FQ^{0.65}$ K\$, $0.5 < Q < 400$ M Btu/hr, installed prices

Temperature Level (°C)	F
0	1.00
−10	1.55
−20	2.10
−30	2.65
−40	3.20
−50	4.00

16. **Steam ejectors and vacuum pumps (Pikulik and Diaz, 1979):**

Ejectors: $C = 11.0f_1f_2f_3X^{0.41}$ K\$, $0.1 < X < 100$

X = (lb air/hr)/(suction pressure in Torr)

Type	f_1	No. Stages	f_2	Material	f_3
No condenser	1.0	1	1.0	carbon steel	1.0
1 surface condenser	1.6	2	1.8	stainless steel	2.0
1 barometric condenser	1.7	3	2.1	hastelloy	3.0
2 surface condensers	2.3	4	2.6		
2 barometric condensers	1.9	5	4.0		

Vacuum pumps: $C = 8.15X^{1.03}$ K\$, $0.3 < X < 15$ (lbs air/hr)/(suction Torr).

17. **Vessels (Evans) prices in \$**

Horizontal pressure vessels: $C = F_M C_b + C_a$

$C_b = \exp[8.571 - 0.2330(\ln W) + 0.04333(\ln W)^2]$, $800 < W < 914{,}000$ lb shell weight

$C_a = 1370D^{0.2029}$, $3 < D < 12$ ft diameter (platforms and ladders)

Vertical vessels: $C = F_M C_b + C_a$

$C_b = \exp[9.100 - 0.2889(\ln W) + 0.04576(\ln W)^2]$, $5000 < W < 226{,}000$ lb

$C_a = 246D^{0.7396}L^{0.7068}$, $6 < D < 10$, $12 < L < 20$ ft tangent-to-tangent

Material	Cost Factor F_M
Stainless steel, 304	1.7
Stainless steel, 316	2.1
Carpenter 20CB-3	3.2
Nickel-200	5.4
Monel-400	3.6
Inconel-600	3.9
Incoloy-825	3.7
Titanium	7.7

Storage tanks, shop fabricated: $C = F_M \exp[2.631 + 1.3673(\ln V) - 0.06309(\ln V)^2]$, $1300 < V < 21{,}000$ gal

Storage tanks, field erected: $C = F_M \exp[11.662 - 0.6104(\ln V) + 0.04536(\ln V)^2]$, $21{,}000 < V < 11{,}000{,}000$ gal

Material of Construction	Cost Factor F_M
Stainless steel 316	2.7
Stainless steel 304	2.4
Stainless steel 347	3.0
Nickel	3.5
Monel	3.3
Inconel	3.8
Zirconium	11.0
Titanium	11.0
Brick-and-rubber-or brick-and-polyester-lined steel	2.75
Rubber- or lead-lined steel	1.9
Polyester, fiberglass-reinforced	0.32
Aluminum	2.7
Copper	2.3
Concrete	0.55

TABLE 20.3. Multipliers for Installed Costs of Process Equipment[a]

Equipment	Multiplier	Equipment	Multiplier
Agitators, carbon steel	1.3	Chimneys and stacks	1.2
stainless steel	1.2	Columns, distillation, carbon steel	3.0
Air heaters, all types	1.5	distillation, stainless steel	2.1
Beaters	1.4	Compressors, motor driven	1.3
Blenders	1.3	steam on gas driven	1.5
Blowers	1.4	Conveyors and elevators	1.4
Boilers	1.5	Cooling tower, concrete	1.2
Centrifuges, carbon steel	1.3	Crushers, classifiers and mills	1.3
stainless steel	1.2	Crystallizers	1.9

(continued)

TABLE 20.3—*(continued)*

Equipment	Multiplier	Equipment	Multiplier
Cyclones	1.4	Pumps, centrifugal, carbon steel	2.8
Dryers, spray and air	1.6	centrifugal, stainless steel	2.0
other	1.4	centrifugal, Hastelloy trim	1.4
Ejectors	1.7	centrifugal, nickel trim	1.7
Evaporators, calandria	1.5	centrifugal, Monel trim	1.7
thin film, carbon steel	2.5	centrifugal, titanium trim	1.4
thin film, stainless steel	1.9	all others, stainless steel	1.4
Extruders, compounding	1.5	all others, carbon steel	1.6
Fans	1.4	Reactor kettles, carbon steel	1.9
Filters, all types	1.4	kettles, glass lined	2.1
Furnaces, direct fired	1.3	kettles, carbon steel	1.9
Gas holders	1.3	Reactors, multitubular, stainless steel	1.6
Granulators for plastic	1.5	multitubular, copper	1.8
Heat exchangers, air cooled, carbon steel	2.5	multitubular, carbon steel	2.2
coil in shell, stainless steel	1.7	Refrigeration plant	1.5
glass	2.2	Steam drums	2.0
graphite	2.0	Sum of equipment costs, stainless steel	1.8
plate, stainless steel	1.5	Sum of equipment costs, carbon steel	2.0
plate, carbon steel	1.7	Tanks, process, stainless steel	1.8
shell and tube, stainless/stainless steel	1.9	Tanks, process, copper	1.9
shell and tube, carbon/stainless steel	2.1	process, aluminum	2.0
Heat exchangers, shell and tube, carbon steel/aluminum	2.2	storage, stainless steel	1.5
shell and tube, carbon steel/copper	2.0	storage, aluminum	1.7
shell and tube, carbon steel /Monel	1.8	storage, carbon steel	2.3
shell and tube, Monel/Monel	1.6	field erected, stainless steel	1.2
shell and tube, carbon steel/Hastelloy	1.4	field erected, carbon steel	1.4
Instruments, all types	2.5	Turbines	1.5
Miscellaneous, carbon steel	2.0	Vessels, pressure, stainless steel	1.7
stainless steel	1.5	pressure, carbon steel	2.8

[a] [J. Gran, *Chem. Eng.*, (6 Apr. 1981)].
Installed Cost = (purchase price)(multiplier).

TABLE 20.4. Chemical Engineering Magazine Cost Indexes

Year	1970	1975	1980	Oct. 1985
CE Plant Cost Index	125.7	182.4	261.2	325.8
Equipment costs	123.8	194.7		347.5
Fabricated equipment	122.7	192.2		335.5
Process machinery	122.9	184.7		333.3
Piping, valves, and fittings	132.0	217.0		385.3
Process instruments and controls	132.1	181.4		323.9
Pumps and compressors	125.6	208.3		421.1
Electrical equipment	99.8	142.1		251.9

REFERENCES

1. *Chemical Engineering Magazine, Modern Cost Engineering,* McGraw-Hill, New York, 1979.
2. *Chemical Engineering Magazine, Modern Cost Engineering II,* McGraw-Hill, New York, 1984.
3. L.B. Evans, A. Mulet, A.B. Corripio, and K.S. Chretien, Costs of pressure vessels, storage tanks, centrifugal pumps, motors, distillation and absorption towers, in Ref. 2, pp. 140–146, 177–183.
4. J. Gran, Improved factor method gives better preliminary cost estimates, in Ref. 2, pp. 76–90.
5. D.W. Green and J.O. Maloney (Eds.), *Perry's Chemical Engineers' Handbook,* McGraw-Hill, New York, 1984, cost data on pp. 6.7, 6.22, 6.112, 6.113, 6.121, 7.19, 11.19, 11.20, 11.21, 11.29, 11.42, 17.27, 17.33, 18.45, 18.46, 18.47, 19.13, 19.40, 19.45, 19.65, 19.89, 19.101, 19.102, 20.37, 20.38, 21.22, 21.45, 22.134, 22.135, 25.69, 25.73–25.75.
6. R.S. Hall, J. Matley, and K.J. McNaughton, Current costs of process equipment, in Ref. 2, pp. 102–137.
7. Institut Francaise du Petrole (IFP), *Manual of Economic Analysis of Chemical Processes,* Technip 1976, McGraw-Hill, New York, 1981.
8. B.G. Liptak, Costs of process instruments, in Ref. 1, pp. 343–375.
9. A. Pikulik and H.E. Diaz, Costs of process equipment and other items, in Ref. 1, pp. 302–317.
10. G.P. Purohit, costs of shell-and-tube heat exchangers, *Chem. Eng.,* (22 Aug. 1983, 4 Mar. 1985, 18 Mar. 1985).
11. G.D. Ulrich, *A Guide to Chemical Engineering Process Design and Economics,* Wiley, New York, 1984.
12. W.M. Vatavuk and R.B. Neveril, Costs of baghouses, electrostatic precipitators, venturi scrubbers, fanc carbon adsorbers, flares and incinerators, in Ref. 2, pp. 184–207.

Appendix A

UNITS, NOTATION, AND GENERAL DATA

TABLE A1. Units and Conversions

Prefixes for Unit Multiples and Submultiples:

10^{-18}	atto	a	10^{1}	deca	da
10^{-15}	femto	f	10^{2}	hecto	h
10^{-12}	pico	p	10^{3}	kilo	k
10^{-9}	nano	n	10^{6}	mega	M
10^{-6}	micro	μ	10^{9}	giga	G
10^{-3}	milli	m	10^{12}	tera	T
10^{-2}	centi	c			
10^{-1}	deci	d			

Length:

1 ft = 0.3048 m = 30.48 cm = 304.8 mm

Volume:

1 cuft = 0.0283 cum = 7.481 U.S. gal
1 cum = 35.34 cuft = 1000 L

Standard gas volume:

22.414 L/g mol at 0°C and 1 atm
359.05 cuft/lb mol at 32°F and 1 atm

Gas constant R:

Energy	Temperature	Mole	R
lb ft^2/sec^2	°Rankine	lb	4.969×10^4
ft lbf	°Rankine	lb	1544
cuft atm	°Rankine	lb	0.7302
cuft (lbf/sqin.)	°Rankine	lb	10.73
Btu	°Rankine	lb	1.987
hP hr	°Rankine	lb	7.805×10^{-4}
kW hr	°Rankine	lb	5.819×10^{-4}
J (abs)	Kelvin	g	8.314
kg m^2/sec^2	Kelvin	kg	8.314×10^3
kgf m	Kelvin	kg	8.478×10^2
cucm atm	Kelvin	g	82.0562
calorie	Kelvin	g	1.987

Gravitational constant:

$$g_c = 1 \text{ kg mass/N sec}^2 = 1 \text{ g cm/dyn sec}^2 = 9.806 \text{ kg mass/kg force sec}^2 = 32.174 \text{ lb mass/lb force sec}^2$$

Mass:

1 lb = 0.4536 kg
1 kg = 2.2046 lb

Density:

1 lb/cuft = 16.018 kg/cum
1 gm/cucm = 62.43 lb/cuft
°API = 141.5/(specific gravity) − 131.5
specific gravity = 141.5/(°API + 131.5)

Force:

1 lb force = 0.4536 kg force = 4.448 Newtons

Pressure:

1 atm = 760 Torr = 760 mm Hg = 101,325 N/sqm
= 1.01325 bar = 10,330 kg/sqm = 14.696 lbf/sq in
= 2,116.2 lbf/sqft
1 bar = 100,000 N/sqm
1 Pa = 1 N/sqm

Energy, work, and heat:

1 Btu = 252.16 cal = 1055.06 J = 0.2930 W hrs
= 10.41 L atm
1 HP hr = 0.7457 kWh = 778 ft lbf = 2545 cal
1 cal = 4.1868 J
1 J = 1 N m = 1 W sec = 0.2388 cal = 0.000948 Btu

Power:

1 ft lbf/sec = 0.0018182 HP = 1.356 W = 0.0012856
Btu/sec = 0.3238 cal/sec
1 W = 1 J/sec = 1 N m/sec

Temperature:

K (Kelvin) = °C (centigrade) + 273.16 = [°F (Fahrenheit) + 459.6]/1.8 = °R (Rankine)/1.8
°R = 1.8 K = °F + 459.6
°C = (°F − 32)/1.8

Temperature difference:

1°C = 1°K = 1.8°R = 1.8°F

Heat capacity and entropy:

1 cal/(g)(°K) = 4.1868 J/(g)(°K) = Btu/(lb)(°R)

Specific energy:

1 cal/g = 4.1868 J/g = 1.8 Btu/lb

TABLE A1—(*continued*)

Volumetric flow:

1 cuft/sec = 0.028316 cum/sec = 28.316 L/sec

Heat flux:

$1\ \text{Btu/(hr)(ft}^2) = 3.1546\ \text{W/m}^2$
$= 2.172\ \text{kcal/(hr)(m}^2)$

Heat transfer coefficient:

$1\ \text{Btu/(hr)(ft}^2\text{)(F)} = 5.6783\ \text{W/m}^2\text{K}$

Surface tension:

$1\ \text{dyn/cm} = 1\ \text{erg/cm}^2$
$= 0.001\ \text{N sec/m}^2$

Viscosity, dynamic:

$1\ \text{cP} = 0.001\ \text{N sec/m}^2$
$= 0.001\ \text{Pa sec}$
$= 0.000672\ \text{lb}_m\text{/ft sec}$
$= 2.42\ \text{lb}_m\text{/ft hr}$
$= 0.0752\ \text{lb}_f\ \text{hr/ft}^2$

Viscosity, kinematic:

$1\ \text{centistoke} = 0.00360\ \text{m}^2\text{/hr}$
$= 0.0388\ \text{ft}^2\text{/hr}$

TABLE A2. Notation[a]

C_p = heat capacity at constant pressure
C_v = heat capacity at constant volume
g_c = gravitational constant (numerical values in Table A1)
h = individual heat transfer coefficient
H = enthalpy
k = thermal conductivity
$k = C_p/C_v$
$K = y/x$ = vaporization equilibrium ratio, VER
m_i = mass fraction of component i of a mixture
M = molecular weight
P = pressure
Q = volumetric flow rate
Q = heat transfer rate
R = gas constant (numerical values in Table A1)
S = entropy
T = temperature, usually °R or °K
u = linear velocity
U = overall heat transfer coefficient
V = volume
x_i = mol fraction of component i in the liquid phase
y_i = mol fraction of component i in the vapor phase
z_i = mol fraction of component i in a mixture
z = PV/RT, compressibility
μ = viscosity
ρ = density
σ = surface tension

[a] Most symbols are defined near where they are used in equations. Unless defined otherwise locally, certain notations have the meanings in this list.

TABLE A3. Properties of Steam and Water

Temp., °F	Absolute pressure, lb/sq in.	Latent heat of evaporation, Btu/lb	Specific volume of steam, cu ft/lb	Density of liquid water, lb/cu ft	Viscosity of liquid water, centipoises	Thermal conductivity of liquid water, (Btu)(ft)/(°F)(ft²)(hr)
32	0.0885	1075.8	3306	62.42	1.786	0.320
35	0.1000	1074.1	2947	62.42	1.689	0.322
40	0.1217	1071.3	2444	62.42	1.543	0.326
45	0.1475	1068.4	2036.4	62.42	1.417	0.329
50	0.1781	1065.6	1703.2	62.39	1.306	0.333
55	0.2141	1062.7	1430.7	62.39	1.208	0.336
60	0.2563	1059.9	1206.7	62.35	1.121	0.340
65	0.3056	1057.1	1021.4	62.30	1.044	0.343
70	0.3631	1054.3	867.9	62.28	0.975	0.346
75	0.4298	1051.5	740.0	62.23	0.913	0.349
80	0.5069	1048.6	633.1	62.19	0.857	0.352
85	0.5959	1045.8	543.5	62.14	0.807	0.355
90	0.6982	1042.9	468.0	62.12	0.761	0.358
95	0.8153	1040.1	404.3	62.03	0.719	0.360
100	0.9492	1037.2	350.4	62.00	0.681	0.362
105	1.1016	1034.3	304.5	61.92	0.646	0.364
110	1.275	1031.6	265.4	61.85	0.614	0.367
115	1.471	1028.7	231.9	61.80	0.585	0.369
120	1.692	1025.8	203.27	61.73	0.557	0.371
125	1.942	1022.9	178.61	61.66	0.532	0.373
130	2.222	1020.0	157.34	61.55	0.509	0.375
135	2.537	1017.0	138.95	61.46	0.487	0.376
140	2.889	1014.1	123.01	61.39	0.467	0.378
145	3.281	1011.2	109.15	61.28	0.448	0.379
150	3.718	1008.2	97.07	61.21	0.430	0.381
155	4.203	1005.2	86.52	61.10	0.414	0.382
160	4.741	1002.3	77.29	61.01	0.398	0.384
165	5.335	999.3	69.19	60.90	0.384	0.385
170	5.992	996.3	62.06	60.79	0.370	0.386
175	6.715	993.3	55.78	60.68	0.357	0.387
180	7.510	990.2	50.23	60.58	0.345	0.388
185	8.383	987.2	45.31	60.47	0.334	0.389
190	9.339	984.1	40.96	60.36	0.333	0.390
195	10.385	981.0	37.09	60.25	0.312	0.391
200	11.526	977.9	33.64	60.13	0.303	0.392
205	12.777	974.8	30.57	60.02	0.293	0.392
210	14.123	971.6	27.82	59.88	0.284	0.393
212	14.696	970.3	26.80	59.75	0.281	0.393
215	15.595	968.4	25.37	59.70	0.277	0.393
220	17.186	965.2	23.15	59.64	0.270	0.394
225	18.93	962.0	21.17	59.48	0.262	0.394
230	20.78	958.8	19.382	59.39	0.255	0.395
235	22.80	955.5	17.779	59.24	0.248	0.395
240	24.97	952.2	16.323	59.10	0.242	0.396
245	27.31	948.9	15.012	58.93	0.236	0.396
250	29.82	945.5	13.821	58.83	0.229	0.396
260	35.43	938.7	11.763	58.52	0.218	0.396
270	41.86	931.8	10.061	58.24	0.208	0.396
280	49.20	924.7	8.645	57.94	0.199	0.396
290	57.55	917.5	7.461	57.64	0.191	0.396
300	67.01	910.1	6.466	57.31	0.185	0.396
310	77.68	902.6	5.626	56.98		0.396
320	89.66	894.9	4.914	56.55		0.395
330	103.06	887.0	4.307	56.31		0.393
340	118.01	879.0	3.788	55.96		0.392
350	134.62	870.7	3.342	55.59		0.390
360	153.04	862.2	2.957	55.22		0.388
370	173.37	853.5	2.625	54.85		0.387
380	195.77	844.6	2.335	54.46		0.385
390	220.37	835.4	2.0836	54.05		0.383
400	247.31	826.0	1.8633	53.65		0.382

Source: Condensed from Keenan and Keyes, *Thermodynamic Properties of Steam,* Wiley, New York, 1936).

TABLE A4. Properties of Air and Steam at Atmospheric Pressure

T (F)	ρ (lbm/cu ft)	c_p (Btu/ lbm F)	$\mu \times 10^5$ (lbm/ ft sec)	$\nu \times 10^3$ (sq ft/ sec)	k (Btu/ hr ft F)
Air					
0	0.086	0.239	1.110	0.130	0.0133
32	0.081	0.240	1.165	0.145	0.0140
100	0.071	0.240	1.285	0.180	0.0154
200	0.060	0.241	1.440	0.239	0.0174
300	0.052	0.243	1.610	0.306	0.0193
400	0.046	0.245	1.750	0.378	0.0212
500	0.0412	0.247	1.890	0.455	0.0231
600	0.0373	0.250	2.000	0.540	0.0250
700	0.0341	0.253	2.14	0.625	0.0268
800	0.0314	0.256	2.25	0.717	0.0286
900	0.0291	0.259	2.36	0.815	0.0303
1000	0.0271	0.262	2.47	0.917	0.0319
1500	0.0202	0.276	3.00	1.47	0.0400
2000	0.0161	0.286	3.45	2.14	0.0471
2500	0.0133	0.292	3.69	2.80	0.051
3000	0.0114	0.297	3.86	3.39	0.054
Steam					
212	0.0372	0.451	0.870	0.234	0.0145
300	0.0328	0.456	1.000	0.303	0.0171
400	0.0288	0.462	1.130	0.395	0.0200
500	0.0258	0.470	1.265	0.490	0.0228
600	0.0233	0.477	1.420	0.610	0.0257
700	0.0213	0.485	1.555	0.725	0.0288
800	0.0196	0.494	1.700	0.855	0.0321
900	0.0181	0.50	1.810	0.987	0.0355
1000	0.0169	0.51	1.920	1.13	0.0388
1200	0.0149	0.53	2.14	1.44	0.0457
1400	0.0133	0.55	2.36	1.78	0.053
1600	0.0120	0.56	2.58	2.14	0.061
1800	0.0109	0.58	2.81	2.58	0.068
2000	0.0100	0.60	3.03	3.03	0.076
2500	0.0083	0.64	3.58	4.30	0.096
3000	0.0071	0.67	4.00	5.75	0.114

TABLE A5. Properties of Steel Pipe

Nominal pipe size, in.	OD, in.	Schedule No.	ID, in.	Flow area per pipe, in.^{2}	Surface per lin ft, ft^2 Outside	Surface per lin ft, ft^2 Inside	Weight per lin ft, lb steel
1/8	0.405	40†	0.269	0.058	0.106	0.070	0.25
		80‡	0.215	0.036	0.106	0.056	0.32
1/4	0.540	40	0.364	0.104	0.141	0.095	0.43
		80	0.302	0.072	0.141	0.079	0.54
3/8	0.675	40	0.493	0.192	0.177	0.129	0.57
		80	0.423	0.141	0.177	0.111	0.74
1/2	0.840	40	0.622	0.304	0.220	0.163	0.85
		80	0.546	0.235	0.220	0.143	1.09
3/4	1.05	40	0.824	0.534	0.275	0.216	1.13
		80	0.742	0.432	0.275	0.194	1.48
1	1.32	40	1.049	0.864	0.344	0.274	1.68
		80	0.957	0.718	0.344	0.250	2.17
1 1/4	1.66	40	1.380	1.50	0.435	0.362	2.28
		80	1.278	1.28	0.435	0.335	3.00
1 1/2	1.90	40	1.610	2.04	0.498	0.422	2.72
		80	1.500	1.76	0.498	0.393	3.64
2	2.38	40	2.067	3.35	0.622	0.542	3.66
		80	1.939	2.95	0.622	0.508	5.03
2 1/2	2.88	40	2.469	4.79	0.753	0.647	5.80
		80	2.323	4.23	0.753	0.609	7.67
3	3.50	40	3.068	7.38	0.917	0.804	7.58
		80	2.900	6.61	0.917	0.760	10.3
4	4.50	40	4.026	12.7	1.178	1.055	10.8
		80	3.826	11.5	1.178	1.002	15.0
6	6.625	40	6.065	28.9	1.734	1.590	19.0
		80	5.761	26.1	1.734	1.510	28.6
8	8.625	40	7.981	50.0	2.258	2.090	28.6
		80	7.625	45.7	2.258	2.000	43.4
10	10.75	40	10.02	78.8	2.814	2.62	40.5
		60	9.75	74.6	2.814	2.55	54.8
12	12.75	30	12.09	115	3.338	3.17	43.8
16	16.0	30	15.25	183	4.189	4.00	62.6
20	20.0	20	19.25	291	5.236	5.05	78.6
24	24.0	20	23.25	425	6.283	6.09	94.7

† Schedule 40 designates former "standard" pipe.
‡ Schedule 80 designates former "extra-strong" pipe.

TABLE A6. Standard Gauges of Sheets, Plates, and Wires

Sheet mills roll steel sheets to U. S. gauge unless otherwise ordered. Plate mills usually roll heavy plates, 3/16 and heavier, and light plate No. 8 to No. 12, to Birmingham gauge. In figuring weights of steel plates add to above the allowance for overweight, adopted by Association American Steel Manufacturers. All steel sheets in our stock are rolled to the U. S. Standard Gauge. Brass is rolled to thickness by Brown & Sharpe's American Gauge. Copper is rolled to thickness by Stubs' or Birmingham Gauge.

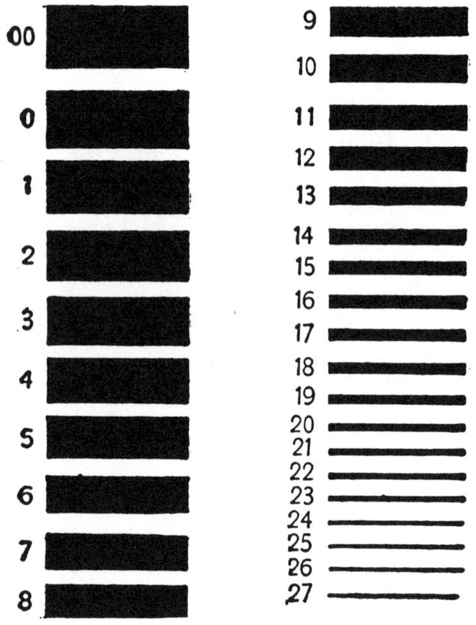

No. of Gauge	THICKNESS AND WEIGHT OF SHEETS AND PLATES							WIRE
	U. S. Standard Gauge Adopted by U. S. Government March 1, 1937			Birmingham or Stubs' Gauge		American or Brown & Sharpe's Gauge		Washburn & Moen Gauge
	Thickness Inches		Weight Lbs. per Sq. Ft.	Thickness, Inches	Weight Lbs. per Sq. Ft.	Thickness, Inches	Weight Lbs. per Sq. Ft.	Thickness, Inches
000	3/8	......	15.00	.425	17.28	.410	16.71	.363
00	11/32	......	13.75	.380	15.45	.365	14.88	.331
0	21/64	......	12.50	.340	13.82	.325	13.26	.307
1	19/64	......	11.25	.300	12.20	.289	11.80	.283
2	9/32	......	10.625	.284	11.55	.258	10.51	.263
3	1/4	.239	10.00	.259	10.53	.229	9.36	.244
4	15/64	.224	9.375	.238	9.68	.204	8.34	.225
5	7/32	.209	8.75	.220	8.95	.182	7.42	.207
6	3/16	.194	8.125	.203	8.25	.162	6.61	.192
7	11/64	.179	7.50	.180	7.32	.144	5.89	.177
8	5/32	.164	6.875	.165	6.71	.128	5.24	.162
9	9/64	.149	6.25	.148	6.02	.114	4.67	.148
10	1/8	.134	5.625	.134	5.45	.102	4.16	.135
11	7/64	.120	5.00	.120	4.88	.091	3.70	.120
12	1/10	.105	4.375	.109	4.43	.08	3.30	.105
13	3/32	.09	3.75	.095	3.86	.072	2.94	.092
14	5/64	.075	3.125	.083	3.37	.064	2.62	.080
15	......	.067	2.813	.072	2.93	.057	2.33	.072
16	1/16	.060	2.50	.065	2.64	.05	2.07	.063
17	......	.054	2.25	.058	2.36	.045	1.85	.054
18	3/64	.048	2.00	.049	1.99	.04	1.64	.047
19	......	.042	1.75	.042	1.71	.036	1.46	.041
20	......	.036	1.50	.035	1.42	.032	1.31	.035
21	1/32	.033	1.375	.032	1.30	.028	1.16	.032
22	......	.030	1.25	.028	1.14	.025	1.03	.027
23	......	.027	1.125	.025	1.02	.023	.922	.026
24	......	.024	1.00	.022	.895	.020	.82	.023
25	......	.021	.875	.020	.813	.018	.73	.020
26	......	.018	.750	.018	.732	.016	.649	.018
27	......	.016	.687	.016	.651	.014	.579	.017
28	......	.015	.625	.014	.569	.012	.514	.016
29	......	.014	.563	.013	....	.011	.461	.015
30	......	.012	.500	.012	....	.01	.408	.014

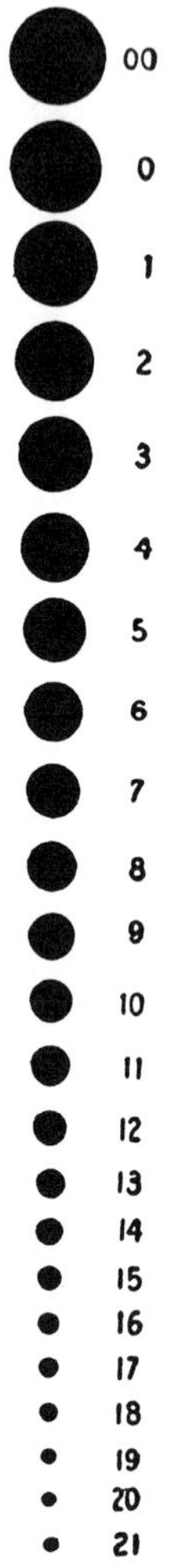

TABLE A7. Weights and Angles of Slide of Various Materials

***Weights of Materials**—The following list gives weights in pounds per cubic foot. Unless otherwise noted, weights are for material in loose, least compacted form. **In figuring Horse Powers,** weights should be increased in proportion to their compressibility.

†Angles of Slide—The angles given are the **minimum** at which the various materials will slide on a steel plate, under **best** condition, for determination of friction. The minimum angle will **increase** as size of particles decrease and with higher moisture content. For definite recommendations refer to S-A Engineers. The inclination of **chutes must be steeper** than minimum angle of slide and S-A Engineers should be consulted for minimum chute slopes.

Friction Factors—The moving-friction factor for any material listed, sliding on steel plate, equals the natural tangent of the "angle of slide" given for that material. See table of natural functions of angles—listed in data section of book. **For example,** the friction factor of cement equals .809 (the natural tangent of 39°, which is the angle of slide for cement).

Specific Gravity—The specific gravity of a material is its weight (in a solid block) compared with that of water at 62° F. Example: As water weighs 62.4 pounds per cubic foot and sulphur weighs 125 pounds, the specific gravity of sulphur is twice that of water or 2.0.

Green Timber—Usually weighs from one-fifth to nearly one-half more than dry. Ordinary building timbers, tolerably seasoned, weigh about one-sixth more.

▲ Solid Cube of material—weights of broken or crushed material decrease, for example, see figures given for coal and for limestone.
** Figures listed are for best conditions (dry, sized and without dust)—The minimum angle will **increase** as size of particles decrease and with higher moisture content. For other conditions refer to S-A Engineers for definite recommendations.

DESCRIPTION	Average Wt. per cu. ft. pounds*	Minimum Angle of Slide†
Air (Atmospheric at 60°F., Under pressure of one atmosphere, 14.7 lbs. per square in., weighs 1/815 as much as water)	.0765	....
Alabaster (marble) ▲	168	....
Alabaster (Real, a compact white plaster of Paris) ▲	144	....
Alfalfa, Ground	15	31°**
" Coarse	9	31°
Alumina, sized or briquette	65	22°
Alumina, Fine, Granulated	55	35°**
Alum, ground	62	35°**
Aluminum Hydrate, Ground	13½	34°*
Aluminum, Sulphate, Granular	54	32°
Ammonium, Sulphate, Damp Granulated	40	45°
Argols, Roasted, Pulverized	63	32°
Ash, Black, Ground	105	27°**
" Fly, Powdered	45	40-45°**
" Volcanic, Powdered	45	45°**
Ashes of Bituminous Coal	35-45	40°
" Damp	40-50	50°
Asphaltum	81	....
Babassu Nuts, ground	45	40°**
Bagasse, Wet Sawdust	4	45°**
Baking Powder	47	38°
Barytes, Granular	144	30°**
Basalt (Similar to Marble) ▲	184	....
Bathstone, Oolite ▲	131	....
Bauxite, Calcined (Granular without dust)	70	26°
" Ground dried	68	35°
" Mine run	85	31°**
Bentonite, pulverized	50	42°
Bismuth ▲	612.4	....
Bones, Animal, pulverized	50	50°
Bone Char, ground	40	35°
Borax, Dehydrated, powdered	75	40°
Bran	16-26	36°
Brick, Best Pressed ▲	140	....
" Common ▲	120	....
" Fire ▲	137 to 144	....
" Soft inferior ▲	100	....
Brickwork, Press Brick, fine joints ▲	128	....
" Medium Quality	112	....
" Coarse, Inferior or Soft ▲	103	....
Calcium Bichromate, Granular	62	25°
" Carbide ▲	139	....
" Carbide, Spent, Powdered	40	45°
" Carbonate (See Limestone)	....	....
" Oxide, Powdered (See Lime)	27	43°
" Phosphate, See Phosphates	....	....
Carbon, ground	50	21°
" Coke, crushed, sized	30	28°
Casein, granular	38-43	30°

DESCRIPTION	Average Wt. per cu. ft. pounds*	Minimum Angle of Slide†
Cellulose Acetate, granular	10	33°
Cement, Portland (per Bbl. net 376 lbs.) (per bag 94 lbs.)	90–100	39°
" Mortar, Portland, 1:2½	135	....
Chalk, ▲	137	....
" Precipitated, Powdered	18	45°
Charcoal, Bone, (Carbonated), granular	60	27°
" Wood Pulp, granular	26.4	35°
Chips, Wood	15–30	22°
Chocolate, Powder	40	45°
Chromic Acid, Flake	75	25°
Cinders, (Coal, Ashes and Clinkers)	40–45	35°**
" Blast Furnace	57	35°**
Clay, Dry in Lump Loose	40–100	35°
" Blended for Tile 11% Moist, Powdered	45	45°
" Ground	60–100	35°
" Fire, Powdered	25–80	45°
" Gray, Granular	50–95	35°
" Pulverized Fire Brick	100	35°
Coal, Anthracite, (Solid) ▲	97	....
" " broken of any Size, Loose	52 to 57	22°
" " " Moderately Shaken	56 to 60	....
" " " , one ton, loose, occupies 40 to 43 cu. ft.	....	....
" " Chestnut	46	22°
Coal, Bituminous, (Solid) ▲	84	....
" " minus ¼", slack, dry	40–50	29°
" " " " " moderately wet	45–55	40°
" " " " " very wet	50–60	33°
" " broken, 1 ton, loose, occupies 43 to 48 cu. ft.	....	....
" " sized, wet or dry	40–50	27°
" " pulverized	30	40°
Cocoanut (see Grains, Seeds and Cereals)	....	....
Coke, pulverized	25	34°
" Loose, 1 ton occupies 80 to 97 cu. ft.	23 to 32	....
" Petroleum, crushed	33-36	27°
Concrete, Cinder with Portland Cement ▲	112	....
" Gravel and Sand with Portland Cement ▲	150	....
" Trap with Portland Cement ▲	155	....
Copper, Oxide, Powdered	190	40°
" Sulphate, Ground	75	31°**
Cork, ¾" to 0"	6	30°
Cryolite ½" to 200 mesh, crushed	52	32°
Cullet (Scrap Glass Sized)	80–120	25°
Dolomite, pulverized	46	41°
" Solid ▲	181	....
Dust, flue, Blast Furnace	40	45°
" Foundry	55–65	38°
" Limestone	55–65	38°
Earth, Common Loam, Perfectly Dry	72 to 80	....
" " " Moist	75–85	....
" Fullers, Raw	42	35°

▲ * † ** See notes preceding table.

(continued on following page)

TABLE A7—*(continued)*

DESCRIPTION	Average Wt. per cu. ft. pounds*	Minimum Angle of Slide†
Eggs	48	...
Feldspar, Pulverized	50–60	40°
" (Crushed)	100	32°
Ferric Sulphate ▲	194	...
Ferrous Sulphate (Copperas) ▲	119	...
Flint ▲	162	...
Fuller's Earth (See Earth)	...	...
Gelatin, Granulated	38	38°
Glass, Common and Plate ▲	161	...
" Batch, Average Mix	100	45°
Glue (Pellet)	45	25°
Gneiss, Solid ▲	168	...
" in Loose Piles	90–100	...
Grains, Seeds and Cereals		
Barley (48 lb. per bushel)	38	...
Beans, Cocoa	37	25°
" Navy	54	22°
Beans, Soy—Cake	45	32°
" " Meal	40	27°
" " —Flour	27	40°
Beans, Soy—Crushed	34	35°
" " Whole	45–50	22°
" " Split	44	25°
Bran	16–26	36°
Brewers Grits	33	24°
Buckwheat (46 lbs., per bu.)	34.5	25°
Clover Seed (60 lbs., per bu.)	48	28°
Cocoa Nibs	32	26°
Cocoanut Meal	32	38°
" Shredded	25	27°
Coffee Beans, Green	42	25°
" Steel Cut	28	23°
Corn, Field (on cob)	45	20°
" Shelled (56 lbs. per bu.)	45	21°
Corn Flakes	12	22°
" Germ	25	25°
" " Flakes	25	36°
" Grits	40	24°
Cornmeal (50 lbs. per bu.)	40	35°
" Muffin Mixture	28	45°
Cotton Seed	25	29°
" " Meal	33	35°
Farina	44	29°
Feed Gluten	31	34°
" Molasses	25	40°
Flax Seed Ground	28	35°
" " (56 lbs. per bu.)	45	21°
Flour (196 lbs. per Barrel)	35 to 40	31°
" Prepared Biscuit	26	40°
Grain, Brewers Spent (Sloppy Wet)	84	...
Grass, Blue, Seed (14 lbs. per bu.)	11.5	30°
Hay in Bales	20	...
Hemp Seed (48 lbs. per bu.)	36	...
Hominy	45	21°
Linseed Meal	27	34°
Linseed Rolled	25	34°
Malt, Dry	32	21°
" Spent Dry	10	28°
Malt Sugar (Ground) Hygroscopic	35	31°
" " (Unground) "	30	31°
Oats, 32 Lbs. Per Bushel	26	21°
" Rolled	18	28°
Pablum	9	32°
" Waste	14	38°
Rice	50	20°
Rye (56 Lbs. Per Bu.)	45	23°
Timothy Seed, (45 Lbs. Per Bu.)	36	28°
Wheat (60 Lbs. Per Bu.)	48	23°
" Germ	32	30°
" " Ground	32	36°
Granite Solid ▲	159	...
Granite, Gneiss ▲	175	...
Gravel	120	30°
Green Stone, Trap, Solid ▲	187	...
" " " Quarried in Loose Piles	107	35°

DESCRIPTION	Average Wt. per cu. ft. pounds*	Minimum Angle of Slide†
Gypsum	142	45°
" In Irregular Lumps	82	30°
" Ground, see Plaster of Paris	56	40°
Hornblende, Solid ▲	187	...
Humus	30–40	...
Ice ▲	57.4	...
Ilmenite, granulated	148	31°
" Fine Ground	120	40°
Iron Oxide Pigment	25	40°
Ivory ▲	114	...
Kalsomine (powder)	32	42°
Kaolin, Green Crushed	64	35°
" Pulverized	22	45°
Kieselguhr, crushed	15	40°
Lead, #70 Red	230	40°
" Silicate, Granulated	230	30°
" Sulphate, Basic Pulverized	184	45°**
Leather ▲	59	...
Lime, Briquette	60	26°
" Burned, Pebble (sized)	53	30°
Lime, Burned, Pulverized	27	43°
" Fine (Spent Dry Carbide)	45	40°
" Mason	17	40°
" Burned or Quick, crushed	50	40°
" Hydrated	10–25	42°
Limestone, solid ▲	166	...
" Pulverized	85	42°
" Mixed Sizes	105	35°
" Coarse, Sized	98	25°
Liquids		
Alcohol	49.3	...
Benzine	53.1	...
Milk	64.3	...
Naptha	53.1	...
Oils, Vegetable	58.7	...
Oils, Mineral	57.1	...
Petroleum	54.8	...
Tallow	58.6	...
Turpentine	54	...
Water Pure Rain Distilled @ 32° F. Bar. 30 in.	62.417	...
" " " " @ 62° F. " "	62.355	...
" " " " @ 212° F. " "	59.7	...
Water, Sea	64.08	...
Wood, Spirit	49.9	...
Lithapone, Granulated	70	40°
" Pulverized	73	40°
Magnesium Carbonate Powdered	9	36°
Masonry of Granite or Limestone, Well Dressed	165	...
" " Brickwork (See Brickwork)	...	...
Metals		
Aluminum ▲	165	...
Babbitt ▲	500–650	...
Brass (7.8 Copper to 8.4 Zinc) ▲	504	...
" Rolled ▲	534	...
Bronze (8 Copper to 1 Tin) (Gunmetal) ▲	552	...
Cadmium ▲	539	...
Chromium ▲	432	...
Copper Cast ▲	542	...
Copper Rolled ▲	555	...
Gold, Cast Pure or 24 Karat ▲	1204	...
Iron, Cast ▲	446	...
Iron Ore, Hematite, Magnetite	135	...
" " , Taconite	115	...
Lead, Commercial ▲	710	...
Magnesium ▲	109	...
Magnesium Sulfate (Epsom Salts) Crystals	55	25°
Manganese ▲	475	...
Mercury at 32° F. ▲	849	...
Nickel ▲	537	...
Silver ▲	655	...
Steel ▲	489.6	...
Tin, Cast ▲	459	...
Zinc or Spelter ▲	437.5	...
Mica, Solid ▲	183	...
" Ground	13.5	36°
Milk, Powdered	40	45°

▲ * † ** See notes preceding table

(Continued on following page)

TABLE A7—(*continued*)

DESCRIPTION	Average Wt. per cu. ft. pounds*	Minimum Angle of Slide†
Molasses, Powdered	21	45°
Molybdenumite Ore, Powdered	107	40°
Mortar, Hardened	103	....
Mud Wet Fluid	108	....
Nails and Spikes, 106 Lbs. Per Keg	....	....
Nitrate of Soda, Pellet Type Granular	68	24°
Nuchar, Granular	22	30°
Paste, Dried, Flaky	10	36°
Phosphate, Powdered	60	40°
" Dicalcium, Granular	60	30°
" Super, Ground	51	45°
" Tri-Sodium, Granulated	60	26°
" Tri-Sodium Pulverized	50	40°
" Rock ▲	200	....
" Florida #20 Mesh, Air Cleaned	93	27°
" Mono Calcium, Powdered	61	40°
Phthalic Anhydride, Flaky	42	24°
Pigment (For Rubber Tires) Powdered	52	45°
Pitch ▲	70	....
" Flake	42	27°
Plaster of Paris, Powdered	50	40°
Potassium, Chloride ▲	124	....
" Sulphate ▲	167	....
Powder, Face	36	45°
Powder, Pudding	40–45	36–40°
Pumice, Pulverized	40	45°
Pyrethrum flowers, coarse ground	20	30°
" spent flowers	32	40°
Pyro, Powdered	56	33°
Quartz ▲	165	....
Resin, Synthetic (from plant), Crushed	40	30°
Resin and Wood Flour, Powdered	19	40°
Rock, Phosphate Pulv.	60	40°
" " Solid ▲	200	....
" Florida Phosphate #20 Mesh Air Cleaned	93	27°
Rosin (From Crude Turpentine) ▲	67	....
Rubber, Scrap (Ground)	23	35°
Rutile, Red Oxide of Lead, Fine Ground	132	32°
" Powdered	107	40°
Salt, Granulated	81	31°
" Rock Crushed, sized	75	25°
Sand, Mine run	90–120	35°**
Sand, Coarse sized	90–100	30°
" Very fine	90–110	32°
" Core	65	39°
" Voids Full of Water	110 to 130	45°**
Sandstone, Solid ▲	147	....
" Quarried & Piled	82	....
Serpentine (Talc) ▲	169	....
Shales, Slate ▲	172	....
Silex	70	44°
Silica, Flour	80	45°
Slag ▲	160 to 180	....
" Furnace Granulated	122	25°
" Birmingham	82	25°
Slate, Solid ▲	175	....
" Fine Ground	82	35°
" Granules, Flaky	87	28°
" Flour	45	45°
Snow (Fresh Fallen)	5 to 12	....
" (Moistened) Compacted By Rain	15 to 50	....
Soap Chips	5–15	28–32°
Soapstone, Talc ▲	169	....
" Fine Ground	60	40°
Soda Ash, Light	25–35	37°
" " Dense	66	32°
" " Briquette	50	22°
" " and Silica Sand (1 to 1.85)	80	29°
" Bicarbonate	30–55	42°
Sodium, Aluminate Ground	72	25–50°
" Antimonate Crushed	49	31°**
" Carbonate, Powdered (Soda Ash)	....	....
" Nitrate, Granular	68	24°
" Phosphates, See Phosphates		
" Sulfite, Powdered	96	40°
" Sulfate (fine and lumps)	88	31°

DESCRIPTION	Average Wt. per cu. ft. pounds*	Minimum Angle of Slide†
Starch, (Powdered)	25–45	45°
" (Lump & Pelleted)	30	28°
" Tablet, Granular Crystals	40	24°
Straw in Bales	25	....
Stucco (Tubed and Untubed), Powdered	50	36° to 38°
Sulphur, Pulverized	50	45°
" Coarse	76	32°
Sugar, Brown	40–50	45°
" Powdered	45	45°
" Granulated	50	35°
" Tailings	57	38°
Talc, Solid (Soapstone) ▲	169	....
" Micaceous, Granulated	62	36°
Tankage, Ground	49	32°
Tar ▲	75	....
Tartaric Acid (Cream of Tartar), Granular	60	35°
Tin, Oxide, Ground	100	35°
Titanium Dioxide (paint pigment), powdered	25	45°
Tobacco Stems, Chopped, Coarse	16	23°
Trap, Rock ▲	187	....
" Quarried, in Piles	107	....
Tripoli (Powdered)	80	40°
Vegetables and Fruits		
Apples (56# per Bushel)	45	....
" Dried (22# per Bushel)	17	....
Beets, Shredded Sugar	7.6	31°
Onions (60# per Bushel)	48	....
Peas (64 Lbs. Per Bushel)	51	....
Potatoes (60 Lbs. Per Bushel)	48	....
Copra, Medium Sized Pieces	33	20°
" Meal, Ground	40	39°
" Expeller Cake Ground	32	30°
" Expeller Cake Chopped	29	20°
Walnut Shells, Ground (320 Mesh)	21	55°
Wax, Bees ▲	60	....
Woods		
Cedar	22	....
Cherry (Perfectly Dry)	44	....
Chestnut (Dry)	30	....
Chips (Dry)	15–32	22°
Cypress	32	....
Elm (Perfectly Dry)	35	....
Fir, Eastern	25	....
Hemlock (Perfectly Dry)	29	....
Hickory (Perfectly Dry)	48	....
Lignumvitae, Dry	41 to 83	....
Locust, Dry	46	....
Mahogany	35–53	....
Maple, Dry	33–40	....
Oak, Live, Perfectly Dry	59	....
" Red, Perfectly Dry	41 to 45	....
Pine, White Perfectly Dry	27	....
" Yellow Perfectly Dry Short leaf	38	....
" " " " Long leaf	44	....
Poplar, Dry	29	....
Red Wood, California Dry	26 to 30	....
Sawdust, Dry	10–30	36°
" Ground	20	45°
Shingles Per 1000, Short 900 Lbs. Long 1400 Lbs.	....	....
Spruce, Dry, California	25	....
Sycamore (Perfectly Dry)	37	....
Walnut, Black (Perfectly Dry)	42	....
Wool Mineral	10	30°
Zinc calcines, Powdered	85	54°
" Ore, roasted, Granular	110	38°
" Oxide	20	45°
" " Leaded, Ground	25–40	50°
" Sulphate, Powdered	72	44°
Zonolite, Fine, Granular	7	30°

▲ * † ** See notes preceding table.

Data of Stephens-Adamson Co., Catalog 66, Aurora, IL, 1954. See also Table 5.3.

TABLE A8. Petroleum Products, Typical Compositions

Summary of Product Types Produced From Petroleum																						
Number of Carbon Atoms		C_1	C_2	C_3	C_4	C_5	C_6	C_7	C_8	C_9	C_{10}	C_{11}	C_{12}	C_{13}	C_{14}	C_{15}	C_{16}	C_{17}	C_{18}	C_{19}	C_{20}	$> C_{20}$
Boiling Point of Normal Paraffin at 760 mm	°C	−161	−89	−42	−0.5	+36	69	98	126	151	174	196	216	235	253	270	287	302	316	329	343	
	°F	−259	−127	−44	+31	97	156	209	258	303	345	384	421	456	488	519	548	575	601	625	649	
Liquefied Petroleum Gas				←	→																	
Precipitation Naphtha							←	—	→													
VM&P Naphtha									←	→												
Mineral Spirits										←	—	—	→									
Reformate						←	—	—	—	—	→											
Gasoline					←	—	—	—	—	—	—	—	→									
Kerosene, Diesel Fuel										←	—	—	—	—	—	—	—	→				
Aviation Turbine Fuel										←	—	—	—	—	—	—	→					
Gas Oil, Fuel Oil													←	—	—	—	—	—	—	—	—	→
Transformer Oil																←	—	—	—	—	—	→
Lubricating Oil																						→
Asphalt, Pitch																						→
Wax																			←	—	—	→

Source: Humble Oil Co.

Appendix B

EQUIPMENT SPECIFICATION FORMS

Specification forms make provision for key data of the equipment and its position in the process, including design characteristics and required performance. A major purpose is a basis for price quotations by suppliers of the equipment.

Although a few forms have been prepared by industry-wide organizations such as TEMA, Hydraulic Institute, and API, most have been prepared for their own use by engineering contractors and design divisions of chemical process organizations. Most private firms have their own libraries, but the merits of some forms have been recognized and copied widely so that much uniformity exists.

The selection of forms in this Appendix is made available through the courtesy of individual companies identified on each form, although one major contributor did not wish to be identified. They have been collected over a period of time and consequently may not be the latest forms in use. For the most part, only the most often used kind of equipment is represented, except for the few items in category K.

A related kind of information is the suppliers inquiry forms of Appendix C.

Following is a listing of the equipment specification forms, included in this Appendix. The number appearing in italic is the page number on which the form appears.

CC-532, 2-56

CATALYTIC CONSTRUCTION CO.
ENGINEERING DIVISION

AGITATOR DATA SHEET

JOB NO. ______ SPEC NO. ______ PAGE ____ OF ____
CLIENT ______ ITEM NO. ______ REQ. NO. ______
LOCATION ______ BY ______ DATE ______
NUMBER REQD. ______ REVISION ____ DATE ____ BY ____

PERFORMANCE DATA

AGITATOR APPLICATION ______
TYPE OF AGITATION ______
FOAMING TENDENCY ______
WORKING CAPACITY-(BATCH OR CONTINUOUS, MIN. & MAX.) ______
MIXING CYCLE ______

MATERIALS TO BE MIXED

COMPONENT	VOLUME %	WEIGHT %	SPECIFIC GRAVITY	VISCOSITY @ CENTIPOISE	TEMP. °F
FINISHED MIX					
SOLIDS IN FINISHED MIX					

MIXING TEMP. ______ °F
ALTITUDE ______
DUTY ______ HRS/DAY
INSTALLATION
INDOOR ______
OUTDOOR ______
CORROSION OR EROSION
DUE TO ______

SETTLING VELOCITY ______ FT/MIN
PARTICLE SIZE ______
AGITATOR IN OPERATION DURING FILLING ______, DURING DRAW-OFF ______
VERTICAL DISTANCE: VESSEL BOTTOM TO DRAW-OFF ______
ADDITIONAL NOTES & DATA ______

VESSEL DATA

VESSEL ______ ITEM NO. ______ SKETCH OR DWG. NO. ______
CAPACITY ______ GAL., DESIGN PRESSURE ______ PSIG @ TEMP. ______ °F, MATERIAL ______
CLEARANCE AVAILABLE FROM MOUNTING FLANGE FOR INSTALLING OR REMOVING AGITATOR ______
DIMENSIONS OF LARGEST OPENING TO PASS IMPELLER ______
NO. & DIMENSIONS OF BAFFLES REQ'D ______

VENDOR'S UNIT CHARACTERISTICS

MANUFACTURER ______
TYPE OR MODEL ______ B.H.P. ______
DRIVE TYPE (DIRECT OR GEARED) ______
______ RATIO ______
IMPELLER: TYPE ______ QUANTITY ______
R.P.M. ______ DIA. ______ SPACING ______
SHAFT LENGTH BELOW MOUNTING FLANGE ______
IMPELLER SHAFT SEAL TYPE ______
BEARINGS TYPE ______
COUPLING TYPE ______
LUBRICATION TYPE ______
MOUNTING TYPE & OPER. ANGLE ______
NO. & TYPE STEADY BEARINGS ______
ELEC. MOTOR: MANUFACTURER ______
TYPE ______ FRAME NO. ______ B.H.P. ______
______ VOLTS, ______ PHASE, ______ CYCLE, R.P.M. ______
STARTING CURRENT ______ FULL LOAD CURRENT ______
POWER FACTOR ______

OUTLINE DWG. NO. ______
SECTIONAL DWG. NO. ______
MATERIALS
IMPELLER ______
SCRAPER BLADE ______
STABILIZER RING ______
IMPELLER SHAFT ______
MIXER DRIVE HOUSING ______
MOUNTING FLANGE ______
PACKING ______
GLAND ______
LANTERN RING ______
COUPLING ______
GASKET ______
MECH SEAL ______
STEADY BEARING ______
WEIGHT: AGITATOR ______ LBS.
WEIGHT: MOTOR ______ LBS.

REMARKS: ______

AGITATOR SPECIFICATION SHEET

Page 1 of 2

REV	BY	CHKD	SUPRV.	DATE	APP'R'V	REMARKS
◇						

PROJ. ENG. / DES SUPRV. / CHECKED BY / DRAWN BY / INITIATOR / DATE

Project ______
1 Service of Unit ______ Pos. No. ______
2 Number Required ______ Location ______
3 Type ______ Mounting ______ Manufacturer ______
4 Model No. ______
5 PROCESS CONDITIONS
6 Class of Operation ______
7 Type of Agitation ______ Period of Agitation ______
C Cycle: Batch: Smallest ______ gal. Normal ______ gal.
Max. ______ gal.
9 Continuous: Rate of flow ______ GPM

10	Liquid	Liquid	Solids or Gas
11 Material			
12 Quantity			
13 Viscosity	____ cp @ ____ °F	____ cp @ ____ °F	____ Mcp @ ____ °F
14 Specific Gravity	____ @ ____ °F	____ @ ____ °F	____ @ ____ °F
15 Components			

16 Solids Characteristics: Size ______ Description ______
17 Mixture Characteristics: Sp.G. ______ @ ______ °F Viscosity ______ cp @ ______ °F
18 Foaming Tendency ______ Separation Time ______ hr.
19 Operating Conditions: Normal Temp. ______ °F Press. ______ psig
Max. Temp. ______ °F Press. ______ psig
20 Vessel Volume: ______ Gals; Diam: ______; Str. Shell: ______
21 MECHANICAL DESIGN DATA
22 Vessel Ref. Drawing: ______ (Attached)
23 Impeller Type: ______; No. ______; Size ______ in; Speed ______ rpm; Spacing ______ in.
24 Shaft Length ______ in. Diameter ______ in.
25 Bearing Type: Shaft ______ Foot ______
26 Seal Type ______ Seal Flow ______
27 Seal Mfr. ______
28 Lubricator ______ Type Lubricant ______
29 Coupling Type: Motor ______ Reducer ______ Shaft ______
30 Stuffing Box Type ______
31 Stabilizer ______ Baffles: No.: ______ Width: ______
32 Mounting Nozzle; ______ Rating: ______ Facing: ______
33 Size of Nozzle for Impeller Installation: ______

TITLE ______ ISSUE DATE SCALE BLDG. NO.
DRAWING NO. ______ ◇

E100G1 (4-71)

AGITATOR SPECIFICATION SHEET

Page 2 of 2

34 DRIVE DATA

35 Driver Type: ______ rpm: ______ hp: ______ Manufacturer: ______
36 Current Characteristics: ______ v ______ ph ______ cycles; Type Encl. ______
37 Starting Torque: ______ Full Load Amps: ______ NEMA Code Letter: ______ Frame: ______
38 Speed Reducer Type: ______
39 Expected Noise Level (Measured 3 ft. from Source) ______ dB
40 ______

41 MATERIALS OF CONSTRUCTION

42 Impeller ______ Lantern Ring ______
43 Impeller Shaft ______ Packing Gland ______
44 Impeller Shaft Coupling(s) ______ Packing ______
45 Mounting Flange ______ Seal ______
46 Mounting Flange Facing ______
47 Seal Faces: Rotating ______ Stationary ______
48 Stuffing Box ______

49 REMARKS

50 ______
51 ______
52 ______
53 ______
54 ______
55 ______

REV	BY	CHKD	SUPRV.	DATE	APP'R'V	REMARKS

INITIATOR	DRAWN BY	CHECKED BY	DES SUPRV.

TITLE	ISSUE DATE	SCALE
		BLDG. NO.
	DRAWING NO.	

E100G1 (4-71)

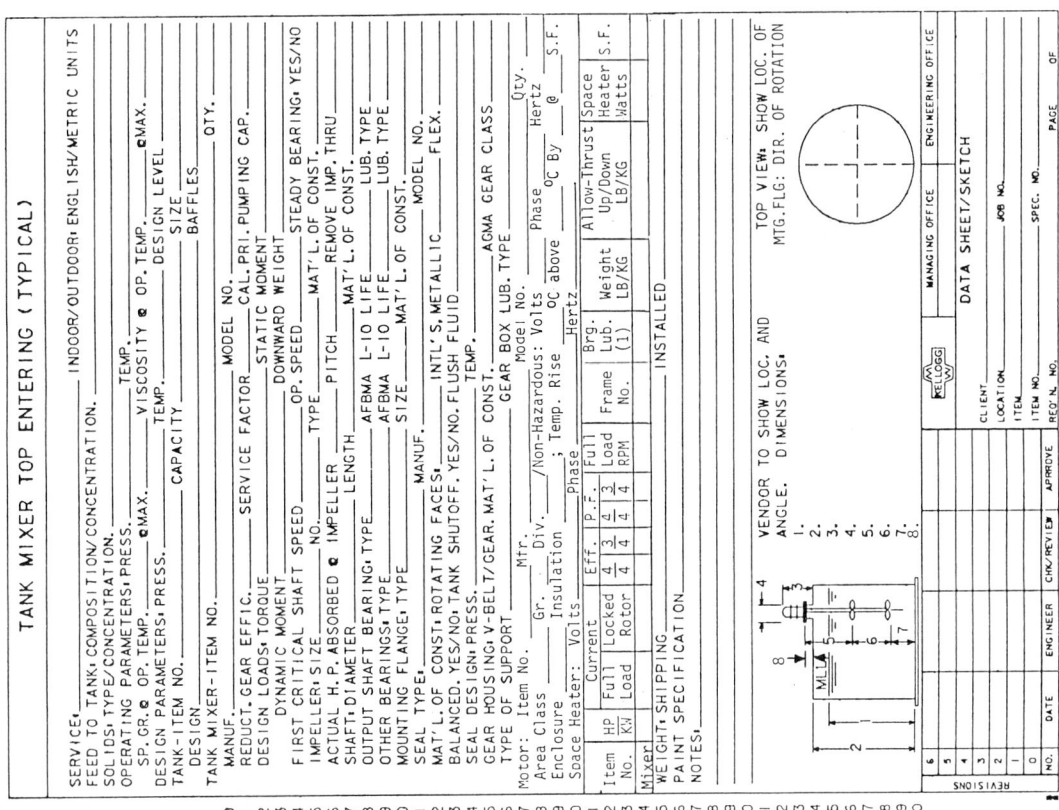

TANK MIXER TOP ENTERING (TYPICAL)

1 SERVICE: ______ INDOOR/OUTDOOR: ENGLISH/METRIC UNITS
2 FEED TO TANK: COMPOSITION/CONCENTRATION. ______
3 SOLIDS: TYPE/CONCENTRATION. ______
4 OPERATING PARAMETERS: PRESS. ______ TEMP. ______
5 SP. GR. @ OP. TEMP. ______ @MAX. ______ VISCOSITY @ OP. TEMP. ______ @MAX. ______
6 DESIGN PARAMETERS: PRESS. ______ TEMP. ______ DESIGN LEVEL ______
7 TANK-ITEM NO. ______ CAPACITY ______ SIZE ______
8 DESIGN ______ BAFFLES ______
9 TANK MIXER-ITEM NO. ______ QTY. ______
10 MANUF. ______ MODEL NO. ______
11 REDUCT. GEAR EFFIC. ______ SERVICE FACTOR ______ CAL. PRI. PUMPING CAP. ______
12 DESIGN LOADS: TORQUE ______ STATIC MOMENT ______
13 DYNAMIC MOMENT ______ DOWNWARD WEIGHT ______
14 FIRST CRITICAL SHAFT SPEED ______ OP. SPEED ______ STEADY BEARING: YES/NO
15 IMPELLER: SIZE ______ NO. ______ TYPE ______ MAT'L. OF CONST. ______
16 ACTUAL H.P. ABSORBED @ IMPELLER ______ PITCH ______ REMOVE IMP. THRU ______
17 SHAFT: DIAMETER ______ LENGTH ______ MAT'L. OF CONST. ______
18 OUTPUT SHAFT BEARING: TYPE ______ AFBMA L-10 LIFE ______ LUB. TYPE ______
19 OTHER BEARINGS: TYPE ______ AFBMA L-10 LIFE ______ LUB. TYPE ______
20 MOUNTING FLANGE: TYPE ______ SIZE ______ MAT'L. OF CONST. ______
21 SEAL TYPE: ______ MANUF. ______ MODEL NO. ______
22 MAT'L. OF CONST: ROTATING FACES: ______ INTL'S, METALLIC ______ FLEX. ______
23 BALANCED. YES/NO: TANK SHUTOFF. YES/NO. FLUSH FLUID ______
24 SEAL DESIGN: PRESS. ______ TEMP. ______
25 GEAR HOUSING: V-BELT/GEAR. MAT'L. OF CONST. ______ AGMA GEAR CLASS ______
26 TYPE OF SUPPORT ______ GEAR BOX LUB. TYPE ______
27 Motor: Item No. ______ Mfr. ______ Model No. ______ Qty. ______
28 Area Class ______ Gr. ______ Div. ______ /Non-Hazardous: Volts ______ Phase ______ Hertz ______
29 Enclosure ______ Insulation ______ ; Temp. Rise ______ °C above ______ °C By ______ @ ______ S.F.
30 Space Heater: Volts ______ Phase ______ Hertz ______

31–33 Item No.	HP / KW	Current Full Load	Current Locked Rotor	Eff. 4/4	Eff. 3/4	P.F. 4/4	P.F. 3/4	Full Load RPM	Frame No.	Brg. Lub. (1)	Weight LB/KG	Allow-Thrust Up/Down LB/KG	Space Heater Watts	S.F.
34 Mixer														

35 WEIGHT: SHIPPING ______ INSTALLED ______
36 PAINT SPECIFICATION ______
37 NOTES: ______
38 ______
39 ______
40 ______

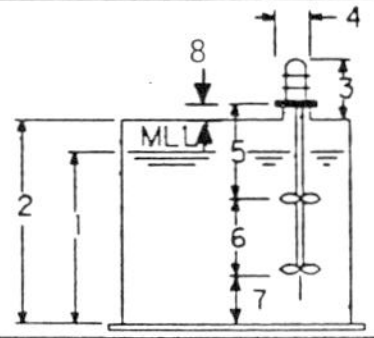

41 VENDOR TO SHOW LOC. AND
42 ANGLE. DIMENSIONS:
43 1.
44 2.
45 3.
46 4.
47 5.
48 6.
49 7.
50 8.

TOP VIEW: SHOW LOC. OF MTG.FLG: DIR. OF ROTATION

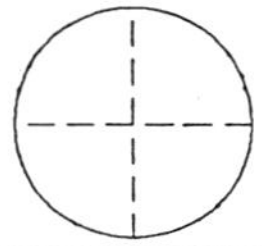

REVISIONS NO.	DATE	ENGINEER	CHK/REVIEW	APPROVE
6				
5				
4				
3				
2				
1				
0				

KELLOGG | MANAGING OFFICE | ENGINEERING OFFICE

DATA SHEET/SKETCH

CLIENT ______
LOCATION ______ JOB NO. ______
ITEM ______
ITEM NO. ______ SPEC. NO. ______
REQ'N. NO. ______ PAGE ______ OF ______

Agitator Specification Sheet—*(continued)*

TANK MIXER SIDE ENTERING (TYPICAL)

SERVICE: ______ INDOOR/OUTDOOR: ENGLISH/METRIC UNITS
FEED TO TANK: COMPOSITION/CONCENTRATION. ______
SOLIDS: TYPE/CONCENTRATION. ______
OPERATING PARAMETERS: PRESS. ______ TEMP. ______
SP. GR. @ OP. TEMP. ______ @MAX. ______ VISCOSITY @ OP. TEMP. ______ @MAX. ______
DESIGN PARAMETERS: PRESS. ______ TEMP. ______ DESIGN LEVEL ______
TANK-ITEM NO. ______ CAPACITY ______ SIZE ______
DESIGN ______ BAFFLES ______
TANK MIXER-ITEM NO. ______ QTY. ______
MANUF. ______ MODEL NO. ______
REDUCT. GEAR EFFIC. ______ SERVICE FACTOR ______ CAL. PRI. PUMPING CAP. ______
DESIGN LOADS: TORQUE ______ STATIC MOMENT ______
DYNAMIC MOMENT ______ DOWNWARD WEIGHT ______
FIRST CRITICAL SHAFT SPEED ______ OPER. SPEED ______
PROPELLER: SIZE ______ TYPE ______ MAT'L. OF CONST. ______
ACTUAL H.P. ABSORBED @ PROPELLER ______ PITCH ______ REMOVE PROP. THRU ______
SHAFT: DIAMETER ______ LENGTH ______ MAT'L. OF CONST. ______
OUTPUT SHAFT BEARING: TYPE ______ AFBMA L-10 LIFE ______ LUB. TYPE ______
OTHER BEARINGS: TYPE ______ AFBMA L-10 LIFE ______ LUB. TYPE ______
MOUNTING FLANGE: TYPE ______ SIZE ______ MAT'L. OF CONST. ______
ANGLE θ ______ °: RIGHT/LEFT HAND MOUNT: DIR. OF ROTAT. ______ FROM MOTOR END
SEAL TYPE: ______ MANUF. ______ MODEL NO. ______
MAT'L. OF CONST: ROTATING FACES: ______ INTL'S. METALIC ______ FLEX. ______
BALANCED. YES/NO: TANK SHUTOFF. YES/NO. FLUSH FLUID ______
SEAL DESIGN: PRESS. ______ TEMP. ______
GEAR HOUSING: V-BELT/GEAR. MAT'L. OF CONST. ______ AGMA GEAR CLASS ______
TYPE OF SUPPORT ______ GEAR BOX LUB. TYPE ______
Motor: Item No. ______ Mfr. ______ Model No. ______ Qty. ______
Area Class ______ Gr. ___ Div. ___ /Non-Hazardous: Volts ______ Phase ______ Hertz ______
Enclosure ______ Insulation ______ ; Temp. Rise ___ °C above ___ °C By ___ @ ___ S.F.
Space Heater: Volts ______ Phase ______ Hertz ______

Item No.	HP KW	Current Full Load	Current Locked Rotor	Eff. 4/4	Eff. 3/4	P.F. 4/4	P.F. 3/4	Full Load RPM	Frame No.	Brg. Lub. (1)	Weight LB/KG	Allow-Thrust Up/Down LB/KG	Space Heater Watts	S.F.
Mixer														

WEIGHT: SHIPPING ______ INSTALLED ______
PAINT SPECIFICATION ______
NOTES: ______

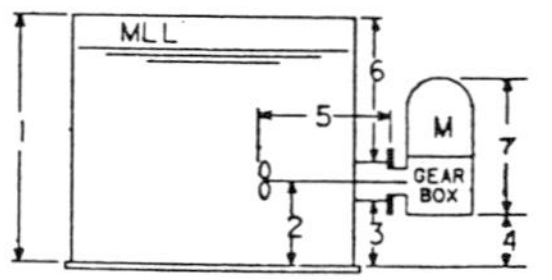

VENDOR TO SHOW LOCATIONS AND ANGLE.
DIMENSION:
1.
2.
3.
4.
5.
6.
7.

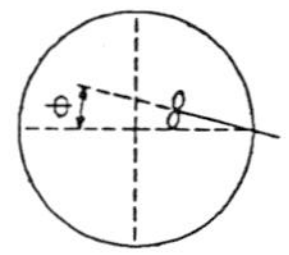

REVISIONS NO.	DATE	ENGINEER	CHK/REVIEW	APPROVE
6				
5				
4				
3				
2				
1				
0				

KELLOGG | MANAGING OFFICE | ENGINEERING OFFICE

DATA SHEET/SKETCH

CLIENT ______
LOCATION ______ JOB NO. ______
ITEM ______
ITEM NO. ______ SPEC. NO. ______
REQ'N. NO. ______ PAGE ___ OF ___

PROCESS DESIGN SECTION
CORPORATE ENGINEERING

SYPHON PUMP (MIXING TEE)
DATA SHEET

EQUIP. POSITION NO.				
SYPHON PUMP NO.				
FLOWSHEET NO.				
PROCESS DATA				
INLET FLUID #1				
DESCRIPTION				
SPECIFIC GRAVITY				
VISCOSITY; CP				
PRESSURE; PSIG				
TEMPERATURE; °F				
FLOW				
INLET FLUID #2				
DESCRIPTION				
SPECIFIC GRAVITY				
VISCOSITY; CP				
PRESSURE; PSIG				
TEMPERATURE; °F				
FLOW				
OUTLET				
TEMPERATURE; °F				
MATERIALS of CONST'N.				
MANUFACTURER				
SIZE NO.				
PIPE CONN'S. - INCHES				
PRESS. INLET				
SUCT. & DELIVERY				

NOTES:

REV	BY	CHKD	SUPRV.	DATE	APP'R'V	REMARKS
◇						

PROJ. ENG. | DES SUPRV. | CHECKED BY | DRAWN BY | INITIATOR | DATE

JOB NO. ______ PLANT LOCATION ______
PROJECT NO. ______ REF. DWG. NO. ______ SEE ABOVE LIST ______
TITLE ______
POSITION NO. - SEE ABOVE LIST ______ SERIES ______ STEP NO. ______
ISSUE DATE | SCALE | BLDG. NO. | DRAWING NO.

E 14089

GENERAL ENGINEERING DIVISION
FORM 66-55 SM 7-54 G.S.

CENTRIFUGAL COMPRESSOR SPECIFICATION

THE FLUOR CORPORATION LTD.

VENDOR MUST FURNISH ALL PERTINENT DATA FOR THIS SPECIFICATION SHEET BEFORE RETURNING

SHEET NO. ______ REV. ______
DATE ______
BY ______ CHK'D. ______
JOB NO. ______

ITEM NO. ______ , SERVICE ______

NO. REQ'D. ______ , DRIVE ______ MANUFACTURER ______

GAS HANDLED				SIZE & TYPE ______ , STAGES
CORROSION FACTORS				DESCRIPTION
OPERATING CONDITION	1:	2:	3:	COMPRESSOR MATERIALS
CAPACITY				CASE
CFM @ INLET CONDS				DIAPHRAGMS
WT. FLOW, LBS/HR.				GUIDE VANES
(M^3SCFD) (SCFM)*				INTER STAGE LABYRINTHS
TEMP., INLET, °F				IMPELLER HUBS & COVERS
DISCH, °F				IMPELLER VANES
PRESS., INLET, PSIA				SHAFT
DISCH, PSIA				SHAFT SLEEVES
DIFFER, PSI				BALANCE DRUM OR DISC
COMP. RATIO				CONSTRUCTION DETAILS
MOLECULAR WT.				CASE: SWP ______ PSIG, MAX. HYDTEST ______ PSIG
SP. GR. (AIR=1)				SPLIT: (HORIZONTAL: VERTICAL: BARREL)
REL. HUMIDITY, %				SUPPORT: (FOOT: PEDESTAL: BRACKET)
"K" VALUE, CP/CV				IMPELLER: TYPE: (OPEN: SEMI-ENCLOSED: ENCLOSED)
COMPRESSIBILITY, "Z"				CONSTR: (CAST: FORGED: RIVETED: WELDED)
ELEV. ABOVE SEA LEVEL				DIAMETER ______ ", VANE THICKNESS ______ "
BHP (INCL. GEAR LOSS)				SHAFT: DIAM. AT IMP ______ ", AT BRGS ______ "
RPM				SPAN: C-C BRGS ______ ", IMP OVERHANG ______ "
IMP. TIP VEL., FPS				CRITICAL SPEED ______ RPM
WATER RATE, #/HP/HR				BRGS: LOCATION (INTERNAL: EXTERNAL)
				RADIAL: TYPE ______ ; PROJ. AREA ______ SQ "
MAX. CAP., CFM @ INL. ______ ; MAX. BHP				THRUST: TYPE ______ ; EFF. AREA ______ SQ "
MIN. CFM, (SURGE POINT)				SEALS:
CONTROL: (SPEED: SUCTION VALVE: INLET VANES)				
SOURCE: (MANUAL: FC: PC: TC)				COUPLING ______ CPLG GRD: YES, NO
STEAM: PRESS. ______ PSIG: TEMP. ______ °F: EXH.				BASE PLATE:
POWER: ______ VOLTS: ______ PH. ______ CYC., COOLING WATER ______ °F				

MFGR SHALL SUPPLY THE FOLLOWING ______ DRAWING NO.	NOZZLES	SIZE	RATING	FACING	LOCATION
PERFORMANCE CURVE	INLET				
SECTIONAL DRAWING	DISCHARGE				
OUTLINE DRAWING	DRAINS				
ROTATION FACING CPLG: ______ CW: CCW	TURB. INLET				
	TURB. EXHAUST				

LUBRICATION SYSTEM	COMPRESSOR WEIGHTS		
COMBINED WITH DRIVER: YES, NO	WT., TOP HALF ______ LBS		
TWIN OIL COOLERS	WT., ROTOR ______ LBS	TESTS: MECHANICAL RUN IN: YES, NO	
TWIN OIL FILTERS	WT., COMPR. BARE ______ LBS	WITNESSED PERFORMANCE: YES, NO	
MAIN OIL PUMP	WT., BASEPLATE ______ LBS	OVERSPEED IMPELLERS TO ______ RPM	
AUX. OIL PUMP	WT., ACCESSORIES ______ LBS	OVERSPEED ROTOR TO ______ RPM	
OIL PRESS. SHUT DOWN	SHIPG WT. LESS DRIVE ______ LBS	HYDROSTATIC TEST PRESS. ______ PSIG	
OIL TEMP. ALARM	FLOOR SPACE	SHOP INSPECTION: YES, NO	

				REDUCTION GEAR
DRIVER: MAKE				MFGR:
TYPE				TYPE:
RATED HP				ARRANGEMENT: (COUPLED) (INTEGRAL)
RPM				LOW SPEED SHAFT, ______ RPM
SHIPPING WEIGHT				RATING ______ HP
SHIPPING POINT				SERVICE FACTOR
SHIPMENT, MONTHS				LUBRICATION:

REMARKS:

*"STANDARD" CONDITIONS ARE 60°F & 14.7 PSIA

THE C.W. NOFSINGER COMPANY
KANSAS CITY, MISSOURI

RECIPROCATING COMPRESSOR SPECIFICATION SHEET

SHEET NO. ______
ITEM NO. ______
CWN JOB NO. ______
CUST. JOB NO. ______
DATE ______
BY ______
REVISED ______

CUSTOMER ______
LOCATION ______
SERVICE ______ NO. REQUIRED ______
MFR. ______ MODEL NO. ______ SERIAL NO. ______

	OPERATING CONDITIONS				CONSTRUCTION DETAILS	
1	VAPOR - GAS	MOL. WT. (INCL H_2O)			FRAME RATING: HP	RPM
2	ELEVATION, FT.	BAROMETER, PSIA			LUBE: TYPE	RESERVOIR CAP. MIN
3	INDOOR - OUTDOOR	HEATED: YES - NO			FLOW, GPM	TEMP. IN/OUT, °F
4	CORR/EROS PROPERTIES				WATER: GPM	TEMP. IN/OUT, °F
5			NORM	RATED	WEIGHTS: COMPR	DRIVER
6	MMSCFD/SCFM (14.7 PSIA-60 °F)				FLYWHEEL	SHIPPING
7	WT. FLOW, LB/HR				TESTS: SHOP INSPECT	PERFORMANCE
8	SUCTION, PSIA AT °F				HYDROSTATIC, PSIG	WITNESS
9	DISCHARGE PRESS, PSIA				ACCESS: SUCT. VA. UNLOADING, MANUAL-AUTO	
10	CP/CV, K (SUCT)				CLEARANCE POCKETS, FIXED-VARIABLE	
11	COMPRESSIBILITY, Z (SUCT)				OUTLINE DWG. NO.	API 618: YES-NO

	PERFORMANCE (EACH STAGE)					
12	STAGE - CYLINDERS/STAGE	1ST	2ND	3RD	4TH	5TH
13	DISCHARGE PSIA/°F					
14	RPM					
15	BORE & STROKE					
16	CYL. DISPLACEMENT, ACFM					
17	CLEARANCE, %					
18	COMPRESSION RATIO					
19	VOLUMETRIC EFF.					
20	CAPACITY, MMCFD					
21	BHP/STAGE					
22	VALVE: TYPE					
23	LIFT (IN/OUT), MILS					
24	GAS VELOCITY, FPM					
25	TEST PRESS (WORKING P x 1.5), PSIG					
26	ROD LOAD: TENSION, PSI					
27	COMPRESSION, PSI					
28	INTAKE NOZZLE: SIZE-RATING-FACING					
29	DISCH. NOZZLE: SIZE-RATING-FACING					
	MATERIALS					
30	CYLINDER/LINER					
31	PISTON/RINGS					
32	RODS					
33	VALVES/SPRINGS					
34	ROD PACKING					

	MOTOR DRIVER		TURBINE DRIVER					
35	SUPPLIED BY	MOUNTED BY	SUPPLIED BY		MOUNTED BY			
36	MFR.	TYPE	MFR.		TYPE			
37	ENCLOSURE	RPM	HP	RPM	WATER RATE, LB/HP-HR			
38	HP	SF	INLET STM. PRESS, PSIG:	NORM	MAX			
39	FRAME	INSULATION	INLET STM. TEMP, °F:	NORM	MAX			
40	VOLTS/PH/HZ	TEMP. RISE, °F	EXHAUST STM. PRESS, PSIG:	NORM	MAX			
41	BEARINGS	LUBE	BEARINGS		LUBE			
42	DRIVE TRAIN: BY	MFR.	NOZZLES	SIZE	RATING	FACING	LOCATION	
43	MODEL	RATIO	INLET					
44	TORSIONAL STUDY: YES-NO	BY	EXHAUST					
45	SEPARATE SPEC IF: DIESEL, GAS		API-611: YES - NO					

FORM 21-A6, 4-76

VENDOR TO SUPPLY INFORMATION MARKED ______

PROCESS DESIGN SECTION — CORPORATE ENGINEERING
FAN OR BLOWER DATA SHEET

REMARKS | APP'R'V | DATE | CHKD | SUPRV | BY | REV

OPERATING CONDITIONS

GAS: ______ CFM AT FLOW TEMP: ______
SP. GR. (AIR = 1.0) ______ REL. HUMIDITY AT FLOW TEMP: ______ %
POUNDS/HR: ______ SUCTION PRESS: ______ PSIA
STD. CFM: ______ DISCH. PRESS: ______ PSIA
FLOW TEMP: ______ °F DIFFERENTIAL PRESS: ______ PSIA

MECHANICAL DATA

MFG: ______ BEARINGS – TYPE: ______
SIZE & TYPE: ______ TYPE OF CLOSURE: ______

NOZZLES	SIZE	ASA RATING	FACING	POSITION
SUCTION				
DISCH.				
DRAIN				

METHOD OF SEALING: ______
LUBRICATION: ______
COOLING: ______
COUPLING MFG: ______
TYPE ______
IMPELLER DIAM: RATED: ______ MAX: ______
TYPE: ______ SHAFT DIAM: ______

MATERIALS OF CONSTRUCTION

CASE: ______ WEAR RINGS: ______
IMPELLER: ______ COUPLING: ______
SHAFT: ______ GASKETS: ______
SHAFT SLEEVE: ______

PERFORMANCE

PROPOSAL CURVE NO: ______ TIP SPEED: ______
NO. OF STAGES: ______ RPM ______ OUTLET VELOCITY: ______
RATED EFF: ______ RATED BHP: ______ ROTATION FACING COUPLING END:
MAX. BHP-RATED IMP. ______ X CW ☐ CCW ☐

MOTOR DRIVE DATA

ITEM NO. ______ MTD. BY. ______ VOLTS/PHASE/CYCLE ______ / ______ / ______
HP ______ RPM ______ FRAME ______ BEARINGS: ______ LUBE: ______
MFG. ______ FULL LOAD AMPS: ______ S.F. ______
TYPE: ______ INSUL. ______ L.R. AMPS: ______
ENCLOSURE: CLASS I GROUP D

MISCELLANEOUS

OUTLINE DWG. NO. ______ ACTUAL IMPELLER DIA. ______
SECTION DWG. NO. ______ TEST CURVE NO. ______
FAN OR BLOWER SERIAL NO. ______ WEIGHT POUNDS NET. ______

REMARKS

INITIATOR | DRAWN BY | CHECKED BY | DES SUPRV.

JOB NO. PLANT LOCATION
PROJECT NO. REF. DWG. NO.
TITLE
POSITION NO. SERIES STEP NO.
ISSUE DATE
SCALE
BLDG. NO.
DRAWING NO.

E223B39

PROCESS DESIGN SECTION — CORPORATE ENGINEERING
VACUUM PUMP DATA SHEET

REMARKS | APP'R'V | DATE | CHKD | SUPRV | BY | REV

OPERATING CONDITIONS

GAS HANDLED: ______
MOLECULAR WGT: ______
SUCTION PRESS: ______ "Hg VACUUM
DISCHARGE PRESS: ______ PSIA
SUCTION TEMP: ______ °F
DISCHARGE TEMP: ______ °F
CAPACITY: ______ LB./HR.
CAPACITY: ______ SCFM
CAPACITY INTAKE: ______ ACFM
"K" VALUE: ______
RELATIVE HUMIDITY: ______ %
ALTITUDE: ______ FT.

SEAL FLUID:
COMPOSITION: ______
TEMPERATURE ______ °F
DENSITY: ______ LB./CU. FT.
VISCOSITY: ______ CPS
VAPOR PRESS: ______ PSIA
HEAT EXCHANGER COOLING WATER:
INLET TEMP: ______ °F
OUTLET TEMP: ______ °F
INLET PRESS: ______ PSIG
ALLOWABLE PRESS. DROP: ______ PSI

MECHANICAL DATA

MFG. ______
SIZE & TYPE ______

NOZZLES	SIZE	ASA RATING	FACING	POSITION
SUCTION				
DISCH.				
HEAT EXCH. COOLING				
DRAIN				
SEAL LIQUID				

CPLG. MFG. ______
TYPE: ______

RECEIVER
CAPACITY: ______ GALS.
SIZE: ______
DESIGN PRESS: ______ PSIG-TEMP. ______ °F
CODED: ☐ NON CODED: ☐
HEAT EXCHANGER:
SURFACE AREA ______ SQ. FT.
DUTY: ______ BTU./HR.
SHELL DESIGN: ______ PSIG-TEMP. ______ °F
TUBE DESIGN: ______ PSIG-TEMP. ______ °F
MAX. CASING PRESS: ______ PSIG.

MATERIALS OF CONSTRUCTION

PUMP:
CASING: ______ IMPELLER: ______
SHAFT: ______
SHAFT SEAL: ______
RECEIVER: ______

HEAT EXCHANGER:
SHELL: ______
TUBES: ______
CHANNEL: ______
BAFFLES: ______

PERFORMANCE

RATED CAPACITY: ______ CFM @ ______ "Hg. VAC.
PUMP SPEED: ______ RPM
NO. OF STAGES: ______
RATED BHP ______

SEAL FLUID FLOW RATE: ______ GPM
COOLING WATER FLOW RATE: ______ GPM
ROTATION FACING PUMP END:
CLOCKWISE ☐ COUNTERCLOCKWISE ☐

MOTOR DRIVE DATA

ITEM NO. ______ MTD. BY ______
VOLTS/PHASE/CYCLE/ ______ / ______ / ______
HP ______ RPM ______ FRAME ______ BEARINGS: ______ LUBE ______
MFG. ______ FULL LOAD AMPS: ______ SF ______
TYPE: ______ INSUL: ______ LR AMPS: ______
ENCLOSURE: CLASS I GROUP D

ACCESSORIES

SILENCER: ______
CONTROL VALVE: ______
SOLENOID VALVE: ______
SEPARATOR: ______

INITIATOR | DRAWN BY | CHECKED BY | DES SUPRV.

JOB NO. PLANT LOCATION
PROJECT NO. REF. DWG. NO.
TITLE
POSITION NO. SERIES STEP NO.
ISSUE DATE
SCALE
BLDG. NO.
DRAWING NO.

E217B33

PROCESS DESIGN SECTION
CORPORATE ENGINEERING

STEAM EJECTOR
DATA SHEET SHEET 1 of 2

Service ______ Manufacturer ______
No.of Units ______ Model No. ______
Type ______

PROCESS CONDITIONS

Non-Condensables ______ lb./hr. M.W. ______
Condensables Vapors (Air etc.) ______ lb./hr. M.W. ______
Steam ______ lb./hr.
Suction Conditions at Ejector: Abs.Press. ______ mmHg. Temp. ______ °F.
Discharge Conditions at Ejector: ______ psig Temp. ______ °F.
Volume of Evacuated System ______ cu.ft. Max. Evacuating Time: ______ Minutes
Expected Air Leakage ______ lb/hr.
Duty ______ hrs./day ______ days/week
Steam Conditions Normal: Press ______ psig Temp. ______ °F. Min: Press. ______ psig Temp. ______ °F.
Consumption at Rated Load ______ lb./hr. at Press. ______ psig Temp. ______ °F.
Water Pressure Normal: ______ psig Min: ______ psig Temp. ______ °F
Consumption at Rated Load: ______ gpm

MECHANICAL DESIGN DATA

EJECTOR	1st Stage	2nd Stage	3rd Stage	4th Stage	5th Stage	6th Stage
Size						
Suction Conn. Size						
Discharge Conn. Size						
Steam Conn. Size						
Design Pressure, psig						
Test Pressure, psig						
MATERIALS:						
Suction Chamber						
Steam Chest						
Steam Nozzle						
Diffuser						

BAROMETRIC CONDENSER

Location: (Precondenser, Intercondenser etc.)			
Size			
Condenser Water Rate gpm			
Outlet Water Temp. °F.			
Inlet Water Pressure Req'd; psig			
Vapor Inlet Nozzle Size			
Vapor Outlet Nozzle Size			
Water Inlet Nozzle Size			
Water Outlet Nozzle Size			
Design Pressure, psig			
Test Pressure, psig			
MATERIALS:			
Shell			
Baffles			
Spray Nozzle			

REMARKS | APP'R'V | DATE | SUPRV. | CHKD | BY | REV

PROJ. ENG. | DES SUPRV. | CHECKED BY | DRAWN BY | INITIATOR | DATE

JOB NO. ______ PLANT LOCATION ______
PROJECT NO. ______ REF. DWG. NO. ______
TITLE ______
POSITION NO. ______ SERIES ______ STEP NO. ______
ISSUE DATE
SCALE
BLDG. NO.
DRAWING NO.

E 162 B 20

PROCESS DESIGN SECTION
CORPORATE ENGINEERING

STEAM EJECTOR
DATA SHEET SHEET 2 of 2

MECHANICAL DESIGN CONT'D.

SURFACE CONDENSER			
Location Precondenser, Intercondenser, etc.			
Size			
Surface Area; sq.ft.			
Tubes; No. x O.D. x BWG x Length	x x x	x x x	x x x
Water Flow Rate; gpm			
Water Temp; In/Out °F	/	/	/
Water Side ΔP; psig			
Tubeside Nozzle Sizes; In/Out	/	/	/
Shellside Nozzle Sizes; In/Out	/	/	/
Design Press; psig Shell/Tubes	/	/	/
Test Press; psig Shell/Tubes	/	/	/
MATERIALS			
Tubes			
Tube Sheet			
Channel			
Shell			
Shell Baffles			
Impingement Plate			

REMARKS

1 Unit to have complete stability at zero load and to be designed for 25 % overcapacity. ______
2 Tail pipes furnished by ______
3 Interconnecting piping by ______

REMARKS | APP'R'V | DATE | SUPRV. | CHKD | BY | REV

DES SUPRV. | CHECKED BY | DRAWN BY | INITIATOR

JOB NO. ______ PLANT LOCATION ______
PROJECT NO. ______ REF. DWG. NO. ______
TITLE ______
POSITION NO ______ SERIES ______ STEP NO ______
ISSUE DATE
SCALE
BLDG. NO.
DRAWING NO.

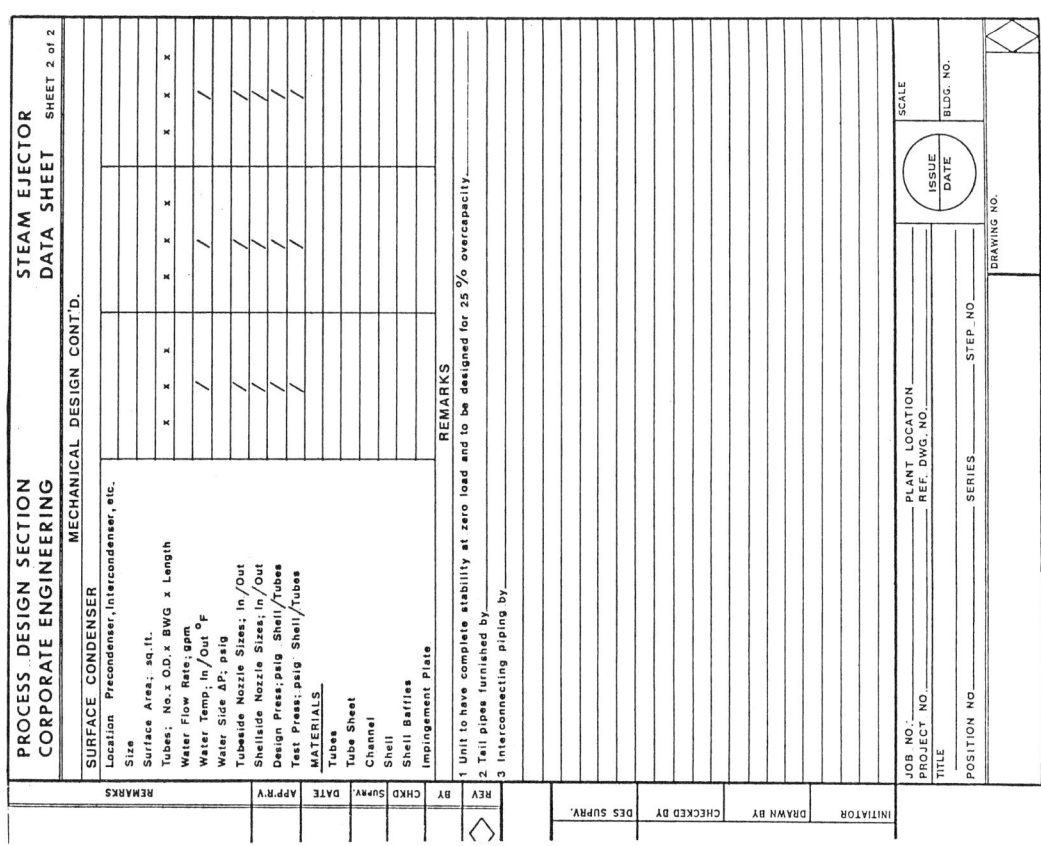

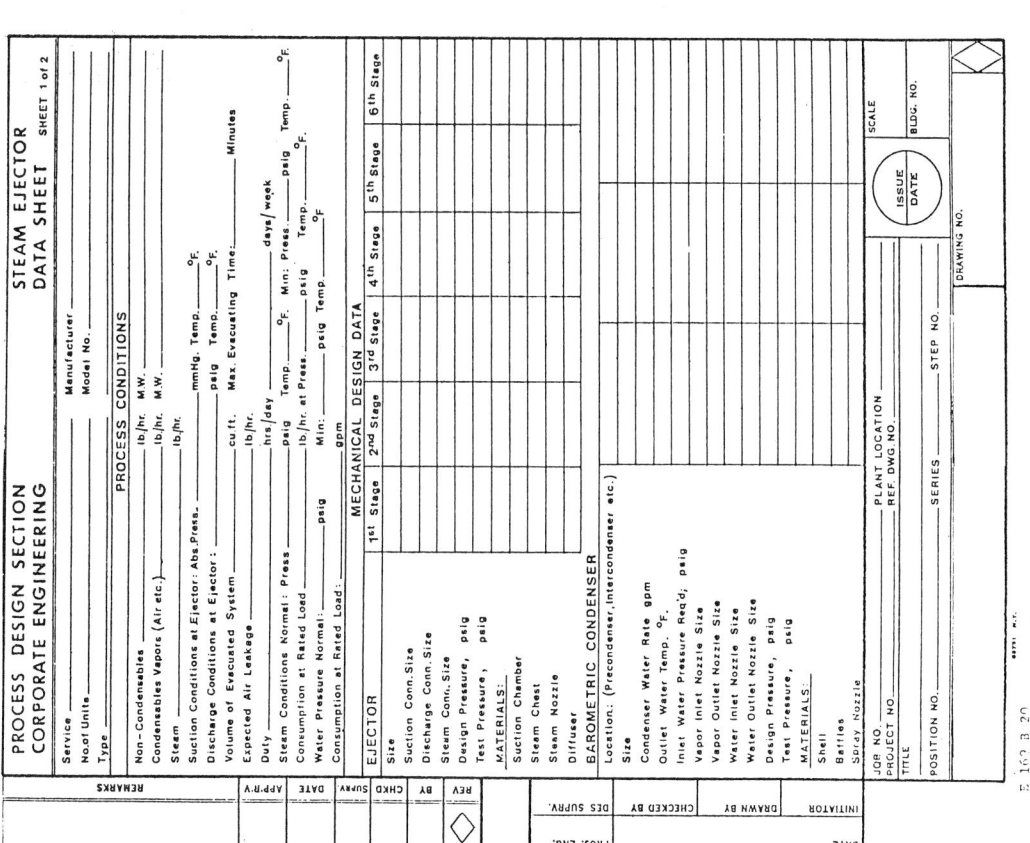

B11

FORM NO. 2829
1M 4-56 MP&S

SOUTHWESTERN ENGINEERING COMPANY

4800 SANTA FE AVENUE — LOS ANGELES 58, CALIFORNIA

SWECO

SHEET ___ OF ___

ELECTRIC MOTOR DATA SHEET

REV

1	CUSTOMER		EQUIP. NO.	
2	PLANT LOCATION		FILE NO.	
3	SERVICE			
4	MANUFACTURER		TYPE	FRAME NO.
5	INDUCTION MOTOR		SYNCHRONOUS MOTOR	
6	HORSEPOWER		HORSEPOWER	
7	RPM		RPM	
8	VOLTS		VOLTS	
9	PHASES		PHASES	
10	CYCLES		CYCLES	
11	FULL LOAD AMPERES		REQUIRED TORQUE	
12	LOCKED ROTOR AMPERES		LOAD WR^2	
13	STARTING TORQUE - PERCENT		FULL LOAD AMPERES	
14	BREAKDOWN TORQUE - PERCENT		LOCKED ROTOR AMPERES	
15	FULL LOAD RPM		STARTING TORQUE - PERCENT	
16	EFFICIENCY - PERCENT		PULL-IN TORQUE - PERCENT	
17	FULL LOAD		PULL-OUT TORQUE - PERCENT	
18	¾ LOAD		POWER FACTOR RATING	
19	½ LOAD		FIELD AMPS @ FULL LOAD	
20	POWER FACTOR		EFFICIENCY - PERCENT	
21	FULL LOAD		FULL LOAD	
22	¾ LOAD		¾ LOAD	
23	½ LOAD		½ LOAD	
24	ELEV. ABOVE SEA LEVEL - FT.		POWER FACTOR	
25	SERVICE FACTOR		FULL LOAD	
26	TEMPERATURE RISE		¾ LOAD	
27	ROTATION FACING END OPP. COUP.		½ LOAD	
28	CLASSIFICATION		ELEV. ABOVE SEA LEVEL - FT.	
29	GENERAL PURPOSE		SERVICE FACTOR	
30	SPLASH PROOF		ROTATION FACING END OPP. COUP.	
31	TEFC		ENCLOSURE	
32	TEFC EXPLOSION PROOF		RADIAL BEARINGS	
33	MOUNTING		LUBRICATION	
34	HORIZONTAL		DRIVE CONNECTION	
35	VERTICAL		BASE	
36	SHAFT - SOLID, HOLLOW		WEIGHT	
37	SHAFT DIAMETER		OUTLINE DRAWING	
38	LOAD THRUST - LBS.		SERIAL NUMBER	
39	MAX. ALLOWABLE THRUST		EXCITER - TYPE	
40	TYPE INSULATION		DRIVE	
41	RADIAL BEARINGS - BALL OR SLEEVE		(IF MOTOR, DESCRIBE AT LEFT)	
42	THRUST BEARINGS - BALL OR SLEEVE		RPM	
43	LUBRICATION - OIL OR GREASE		VOLTS, D.C.	
44	DRIVE CONNECTION		AMPS @ FULL LOAD	
45	BASE		SHUNT OR COMPOUND WOUND	
46	WEIGHT		SERVICE FACTOR	
47	OUTLINE DRAWING NUMBER		TEMPERATURE RISE	
48	SERIAL NUMBER		ENCLOSURES	
49	NEMA CODE LETTER		TYPE INSULATION	
50	MISCELLANEOUS		TYPE BEARINGS	
51	OIL COOLER - TYPE		LUBRICATION	
52	COOLING WATER @ °F & GPM		BASE	
53	FORCED FEED LUBRICATION		WEIGHT	
54	PUMP BUILT-IN OR MOTOR DRIVEN		OUTLINE DRAWING	
55	MOTOR HP - VOLTS - PHASE - CYCLES		SERIAL NUMBER	
56	REMARKS:			
57				

REVISION 1 — APPROVED — BY — DATE

DATE — BY — CHECKED — ENGINEER — PROJ. ENG.

HIGH VOLTAGE INDUCTION MOTORS (LESS THAN 1500 HP/1119 KW)

○ Motor Design Data ○ English ○ Metric Units
Mfr. ___ Model No. ___
Serial No. ___ Qty ___
Driven Equip. ___
Site Data:
Altitude ___ Amb.Temp.Max. ___ °C Min. ___ °C
Dust/Fumes/Other ___
Area Class ___ Gr. ___ Div. ___ /Non-hazardous
Basic Data:
Volts ___ Phase ___ Hertz ___
Nameplate H.P./KW ___ S.F. ___
RPM ___
Insulation: Class ___ Type ___
Temp. Rise ___ °C above ___ °C By ___
NEMA Design ___ Type: ○ Hor. ○ Vert.
Starting:
○ Full Voltage ○ Reduced Voltage ___ %
○ Loaded ○ Unloaded
○ Voltage Dip ___ %
Drive Systems:
○ Direct Connected ○ Gear
○ Other ___
Enclosure:
○ Open Drip Proof ○ Weather Protected ___
○ Totally Enclosed Fan Cooled
○ Other ___
Driven Equipment Inertia ___
Bearing Type ___ Lub. ___
No. Starts: ___ /hr, Motor @ Amb.Temp. ___ °C
No. Starts: ___ /hr, Motor @ Rated Temp. ___ °C
Rotation Facing Coupling End: ___
Vibration: ○ NEMA Std. ○ ___
Noise: ○ NEMA Std. ○ ___
Space Heaters: ○ Yes ○ No
Volts ___ Phase ___ Hertz ___
Maximum Sheath Temp. ___ °C
Accessories:
○ Baseplates ○ Soleplates
○ Mfr. Std. Fans ○ Non-Sparking Fans
Winding Temp. Detector: No./Phase ___
Type ___
Selector Switch & Ind. By: ○ Purch. ○ Mfr.
Winding Temp. Detector & Space Htr. Leads:
○ Separate Conduit Box ○ Same Conduit Box
Specifications:

○ Shop Inspection and Test

	Required	Witness
Shop Inspection	○	○
Testing Per NEMA(MG-1)	○	○
Mfr. Std. Shop Test	○	○
Special Tests Listed Below:		
___	○	○
___	○	○
___	○	○

Couplings Supplied By: ___
Mfr. ___ Model No. ___
○ Motor Mfr. ○ Driven Equip.Mfr. ○ Purch.
to Mount Motor Half.
Painting: Mfr. Std. ___ Spec. No. ___
Shipment: ○ Domestic ○ Export
○ Export Boxing Required
○ Outdoor Storage for Over Three (3) Months
□ Manufacturer's Data
Frame No.: ___ Full Load RPM ___
Efficiency: F.L. ___ 3/4L ___ 1/2L ___
Power Factor: F.L. ___ 3/4L ___ 1/2L ___
Current (Rated Volt). F.L. ___ L.R. ___
Locked Rotor Power Factor ___
Lock.Rtr. W/Stnd.Time (Cold Start) ___ Sec.
Lock.Rtr. W/Stnd.Time (Hot Start) ___ Sec.
Torques: Full Load ___
Locked Rotor ___
Pull-Up ___
Breakdown ___
Acceleration Time (Motor & Load @ 100%/80% Rated Voltage) ___ Sec./ ___ Sec.
Rotor WK^2 ___
Allow. WK^2 @ Motor Shaft ___
Min. Allowable Acceleration Volt. ___
Total Shaft End Float ___
Limit End Float To ___
Space Heater KW ___
Max. Stator Winding Temp.: ___ °C Alarm
___ °C for Shutdown.
Thrust Brg.Cap.(Vert.): Up ___ Down ___
Dimensions:
Length ___ Width ___ Height ___
Weight:
Net ___ Rotor ___ Shipping ___
Note: ___

Note: ○ Indicates information to be completed by Purchaser; □ By Manufacturer

REVISIONS NO	DATE	ENGINEER	CHK/REVIEW	APPROVE
6				
5				
4				
3				
2				
1				
0				

KELLOGG — MANAGING OFFICE — ENGINEERING OFFICE

EQUIPMENT DATA SHEET

CLIENT ___
LOCATION ___ JOB NO. ___
ITEM ___
ITEM NO. ___ SPEC. NO. ___
REQ'N. NO. ___ PAGE ___ OF ___

ENG EQUIP 38 (3-78) KELLOGG The M.W. Kellogg Company

JOB NO.______ ITEM NO.______
REQ. NO.______
PAGE NO.____ OF ____ BY:______
DATE:______ REV. NO.______

GAS ENGINE DATA

RATED HP ______ RATED RPM ______
MODEL ______ BMEP ______ PSI
☐ INTEGRAL ☐ SEPARATE
BORE ______ IN. STROKE ______ IN.
NO. POWER CYLINDERS ______ ENGINE CYCLES ______
FUEL RATE ______ BTU / BHP HR. @ RATED LOAD
RPM MAXIMUM ______ MINIMUM ______
POWER CYLINDERS ☐ WET ☐ DRY
☐ LINED ☐ UNLINED
☐ TURBOCHARGED ☐ SUPERCHARGED
☐ LOW FIRE HAZARD IGNATION ☐ STANDARD IGNITION

GOVERNOR

☐ CONSTANT SPEED ☐ VARIABLE SPEED
RESET BY: ☐ PNEUMATIC SIGNAL
☐ ELECTRONIC SIGNAL
☐ MANUAL
SPEED RANGE: RPM ______ MAX. ______ MIN.
SIGNAL RANGE: RPM ______ MAX. ______ MIN.
ON SIGNAL FAILURE VALVE TO: ☐ OPEN ☐ CLOSE
GOVERNOR MFR. ______ TYPE: ______
REGULATION: NEMA CLASS ______

WEIGHTS AND DIMENSIONS

NET WEIGHT: ______
MAX. ERECTION WEIGHT ______ MAX. MAINT. WT. ______
APPROX. FLOOR SPACE LENGTH ______
WIDTH ______
HEIGHT ______
ADDITIONAL DISTANCE TO REMOVE POWER RODS ______

ACCESSORIES

AIR INLET FILTER ☐ DRY ☐ OIL BATH ☐ MFR. STD.
MANUFACTURER ______ MODEL ______
EXHAUST SILENCER
MANUFACTURER ______ MODEL ______
STYLE ______
EXHAUST MANIFOLD ☐ WATERCOOLED ☐ INSULATED
TACHOMETER ______
☐ ELECT. ____ ENCL. ☐ MECH. ☐ VIBR. REED
TACHOMETER RANGE ______
FLYWHEEL TURNING BAR AND STAND
☐ MANUAL ☐ AIR JACK AIR ______ PSIG
LUBE OIL FILTER ☐ FULL FLOW ☐ SLIPSTREAM

FUEL GAS SURGE DRUM BY: ☐ PURCHASER ☐ ENG MFR
RECOMMENDED VOLUME ______ FT 3
STARTING AIR COMPR. BY: ☐ PURCHASER ☐ ENG MFR
DRIVER: ☐ MOTOR ☐ GAS ☐ GASOLINE
NO. REQUIRED ______ CAPACITY ______ CFM
MANUFACTURER ______ MODEL ______
☐ AUTOMATIC START / STOP CONTROL
STARTING AIR RECEIVER BY: ☐ PURCHASER ☐ ENG MFR.
NO. RECEIVER ______ ☐ HORIZ. ☐ VERT.
RECEIVER CAPACITY ____ FT PRESSURE ____ PSIG
NO. CONSECUTIVE STARTS ______ STARTS / HR. ______
CU. FT. / START ______

AUXILIARY SYSTEMS

☐ ENGINE MANUFACTURER SHALL FURNISH POWER CYLINDER COOLING WATER PIPING FROM A SINGLE INLET FLANGE TO A SINGLE DISCHARGE FLANGE.
SURGE TANK, ENGINE JACKET COOLER, CIRCULATING PUMP BY ☐ PURCHASER ☐ ENGINE MANUFACTURER
☐ CIRCULATING WATER PUMP DRIVEN BY ENGINE SHAFT.
☐ SEPARATELY MOUNTED PUMP AND DRIVER BY ☐ PURCHASER ☐ ENGINE MANUFACTURER
LUBE OIL INTERCONNECTING PIPING AND FITTINGS BY ☐ PURCHASER ☐ ENGINE MANUFACTURER

REMARKS: ______

REVISION	0	1	2	3	4	5	6	7	8	9	10	11	12
NO.	0	1	2	3	4	5	6	7	8	9	10	11	12
DATE													
BY													
CHK.													

GENERAL PURPOSE STEAM TURBINE SPECIFICATION

1	CONSTRUCTION FEATURES			hp
2	☐ Potential Max. Horsepower at Normal Steam Conditions			
3	☐ Outline Dwg. No.	☐ Sectional Dwg. No.		
4	Turbine Mounting	○ Vertical	○ Horizontal	
5	Support	☐ Centerline	☐ Foot	
6	Case Split	○ Radial	○ Axial	
7	☐ No. of Stages	☐ Wheel Diam.		
8	☐ Single Stage	☐ 2 Row ☐ 3 Row	☐ Reentry	
9	Interstage Gland Seal	○ Carbon	○ Labyrinth	
10	End Gland Seal	○ Carbon	○ Labyrinth	
11	☐ No. of Rings/Box			
12	Bearing Type	☐ Radial	☐ Thrust	
13	☐ rpm Max. Cont.	Trip		
14	☐ rpm Max. Allowable		1st Critical	
15	☐ Case Max. Allow. Working Press.	Inlet	psig Exhaust	psig
16	Max. Inlet Temp. °F Max. Exhaust Temp.		°F	
17	Min. Allowable Exhaust Pressure		psig	
18	☐ R.V. Set Press. Exh.		psig Capacity	lb/hr
19	☐ Hydrotest Press.	Inlet	psig Exhaust	psig
20	Lube System ○ Ring Oil	○ Purchasers	○ Other	
21	○ Corrosion Resistant Drain Lines			
22	☐ Oil Required		gpm	psig
23	MATERIALS			
24	☐ High Press. Casing	☐ Exhaust Casing		
25	☐ Nozzles	☐ Blades		
26	☐ Shaft	☐ Wheels		
27	☐ Under Packing			
28	☐ Trip Valve Body	☐ Gov. Valve Trim		
29	CONNECTIONS			

		SIZE	ANSI RATING	FACING	POSITION
30		SIZE	ANSI RATING	FACING	POSITION
31	Inlet				
32	Exhaust				
33	Drain				
34	Cooling Water				

35	☐ Manufacturer to give allowable piping forces & moments on nozzles on separate sheet.					
36	TESTS					
37		REQ'D.	WITN.		REQ'D	WITN.
38	Hydrostatic	⊗	○	Aux. Equipment	○	○
39	No Load Run	⊗	○	Disassembly After Test	○	○
40	Performance	○	○	Test Data Sheets	○	
41	Gov. Response to Contr. Signal	○	○	Inspection ○ Surface ○ Parts		
42						
43	WEIGHTS					
44	☐ Turbine lb ☐ Cooler lb ☐ Mounted lb.					
45	APPLICABLE SPECIFICATIONS					
46	○ API 611 Gen. Purpose Steam Turbines			○ -R-1 Painting		
47	○ -GN-1 Noise			○		
48						
49	REMARKS:					
50						
51						
52						
53						
54						
55						
56						

BADGER ENGINEERS, INC. — ISSUE — DATE — SHEET OF

BE-UE-P-5-2 (Rev. 5/83)

FORM NO. 2868
1M 4-56 MP&S

SOUTHWESTERN ENGINEERING COMPANY

SWECO

4800 SANTA FE AVENUE LOS ANGELES 58, CALIFORNIA

SHEET____OF____

GEAR DRIVE DATA SHEET

REV

No.				
1	CUSTOMER		EQUIP. NO.	
2	PLANT LOCATION		FILE NO.	
3	SERVICE			
4	MANUFACTURER	TYPE	SIZE	
5	OPERATION		BEARINGS	
6	CONTINUOUS OR INTERMITTENT		H.S. SHAFT - OUTBOARD	
7	GEAR RATING - BHP		H.S. SHAFT - INBOARD	
8	THERMAL RATING - HP		INTERMEDIATE SHAFT - OUTBOARD	
9	SERVICE RATING - HP		INTERMEDIATE SHAFT - INBOARD	
10	MECHANICAL HP RATING		L.S. SHAFT - OUTBOARD	
11	AGMA SERVICE FACTOR		L.S. SHAFT - INBOARD	
12	AGMA CLASS GEARS			
13	GEAR RATIO		COUPLINGS	
14	EFFICIENCY @ BHP RATING		H.S. SHAFT COUPLING - TYPE	
15	NO. OF REDUCTIONS OR INCREASES		SIZE	
16	HIGH SPEED SHAFT RPM		MFG.	
17	LOW SPEED SHAFT RPM		L.S. SHAFT COUPLING - TYPE	
18	ACTUAL DRIVEN MACHINE RPM		SIZE	
19	ROTATION H.S. SHAFT FACING DRIVER		MFG.	
20	ROTATION L.S. SHAFT FACING DRIVER		COUPLING GUARDS	
21				
22	CONSTRUCTION & MATERIAL		TESTING	
23	CASE TYPE		WITNESS PERFORMANCE TEST	
24	CASE MATERIAL		INSPECTION	
25	H. S. PINION OR WORM			
26	H.S. GEAR		MISCELLANEOUS	
27	H.S. SHAFT		WEIGHT	
28	INTERMEDIATE PINION OR WORM		OUTLINE DWG. NO.	
29	INTERMEDIATE GEAR		CROSS SECTION DWG. NO.	
30	INTERMEDIATE SHAFT		SERIAL NO.	
31	L.S. PINION OR WORM			
32	L. S. GEAR		PRIME MOVER	
33	L.S. SHAFT		TYPE - MOTOR, TURBINE, OTHER	
34	H.S. SHAFT - SINGLE OR DB'L. EXT.		RATED HP	
35	L.S. SHAFT - SINGLE OR DB'L. EXT.		DRIVER RPM /FULL LOAD RPM	/
36	H.S. SHAFT O.D. AT CLPG.		ROTATION FACING END OPP. CLPG.	
37	L.S. SHAFT O.D. AT CLPG.		MANUFACTURER	
38	CASE GASKETS		TYPE /SIZE	/
39	CASE BOLTS		WEIGHT	
40	BEDPLATE		NOM. SHAFT O.D. AT CLPG.	
41			DRAWING NO.	
42				
43	LUBRICATION		DRIVEN MACHINERY	
44	TYPE LUBRICATION - GEARS		MACHINE DRIVEN	
45	TYPE LUBRICATION - BEARINGS		BHP REQUIRED	
46	OIL PUMP TYPE		REQUIRED DRIVEN RPM	
47	OIL PUMP DRIVE		ROTATION FACING SHAFT END	
48	VISIBLE LUBRICATOR SIGHT GLASS		MANUFACTURER	
49	OIL FILTER OR STRAINER		TYPE /SIZE	/
50	OIL COOLER REQUIRED		WEIGHT	
51	TYE OIL COOLER		NOM. SHAFT O.D. AT CLPG.	
52	COOLING WATER @ °F & GPM		DRAWING NO.	
53	SIZE COOLING WATER CONN.			
54				
55				
56	REMARKS:			
57				

REVISION | 1 | DATE | BY | APPROVED

DATE | BY | CHECKED | ENGINEER | PROJ. ENG.

STANDARDS OF TUBULAR EXCHANGER MANUFACTURERS ASSOCIATION

EXCHANGER SPECIFICATION SHEET

No.			
1			JOB NO.
2	CUSTOMER		REFERENCE NO.
3	ADDRESS		INQUIRY NO.
4	PLANT LOCATION		DATE
5	SERVICE OF UNIT		ITEM NO.
6	SIZE	TYPE	CONNECTED IN
7	GROSS SURFACE PER UNIT	SHELLS PER UNIT	SURFACE PER SHELL
8	PERFORMANCE OF ONE UNIT		
9		SHELL SIDE	TUBE SIDE
10	FLUID CIRCULATED		
11	TOTAL FLUID ENTERING		
12	VAPOR		
13	LIQUID		
14	STEAM		
15	NON-CONDENSABLES		
16	FLUID VAPORIZED OR CONDENSED		
17	STEAM CONDENSED		
18	GRAVITY—LIQUID		
19	VISCOSITY—LIQUID		
20	MOLECULAR WEIGHT—VAPORS		
21	SPECIFIC HEAT—LIQUIDS	BTU/LB/°F	BTU/LB/°F
22	LATENT HEAT—VAPORS	BTU/LB/°F	BTU/LB/°F
23	TEMPERATURE IN	°F	°F
24	TEMPERATURE OUT	°F	°F
25	OPERATING PRESSURE	PSI	PSI
26	NUMBER OF PASSES PER SHELL		
27	VELOCITY	FT./SEC.	FT./SEC.
28	PRESSURE DROP	PSI	PSI
29	FOULING RESISTANCE		
30			
31	HEAT EXCHANGED—BTU/HR.	M.T.D. (Corrected)	
32	TRANSFER RATE—SERVICE	EFF. SURFACE PER UNIT	

CONSTRUCTION

34	DESIGN PRESSURE			PSI	PSI
35	TEST PRESSURE			PSI	PSI
36	DESIGN TEMPERATURE			°F	°F
37	TUBES	NO.	O.D. BWG.	LENGTH	PITCH
38	SHELL		I.D. O.D.		
39	SHELL COVER			FLOATING HEAD COVER	
40	CHANNEL			CHANNEL COVER	
41	TUBE SHEETS—STATIONARY			FLOATING	
42	BAFFLES—CROSS		TYPE		
43	BAFFLE—LONG		TYPE		
44	TUBE SUPPORTS				
45	GASKETS				
46	CONNECTIONS—SHELL—IN		OUT	RATING	PSI
47	CHANNEL—IN		OUT	RATING	PSI
48	CORROSION ALLOWANCE—SHELL SIDE			TUBE SIDE	
49	CODE REQUIREMENTS			TEMA CLASS	
50	WEIGHTS—EACH SHELL		BUNDLE	FULL OF WATER	
51	NOTE: INDICATE AFTER EACH PART WHETHER STRESS RELIEVED (S. R.) AND WHETHER RADIOGRAPHED (X-R)				
52	REMARKS:				

EDD FORM 2000 11-63

AIR COOLED EXCHANGER SPECIFICATION SHEET

Northern Natural Gas Company

LOCATION	FILE NO.
ITEM NO. JOB NO.	W. O. NO.
PREPARED BY	DATE
CHECKED BY	DATE
APPROVED BY	DATE
△ INDICATES REVISION	

SERVICE OF UNIT		NO. OF BAYS
SIZE TYPE		
SURFACE ITEM--EXTERNAL	BARE TUBE	SQ FT
HEAT EXCHANGED--BTU/HR	EFFECTIVE MTD	°F
TRANSFER RATE--EXT SURF.	BARE TUBE SURF.	BTU/HR SQ FT °F

PERFORMANCE DATA

TUBE SIDE

FLUID CIRCULATED		TEMPERATURE IN °F
TOTAL FLUID ENTERING	LBS/HR	TEMPERATURE OUT °F
VAPOR		INLET PRESSURE PSIG
LIQUID		GRAVITY–LIQUID
STEAM		VISCOSITY CPS@°F
NON–CONDENSABLES		VISCOSITY CPS@°F
VAPOR CONDENSED		MOLECULAR WEIGHT
STEAM CONDENSED		SPECIFIC HEAT BTU/LB°F
DENSITY: VAPOR–LIQUID MIXTURE	LBS/CU FT	LATENT HEAT BTU/LB
CONDUCTIVITY	BTU FT/HR SQ FT °F	ALLOW. PRESS. DROP PSI
FOULING RESISTANCE I. S.	HR SQ FT °F/BTU	DESIGN PRESS. DROP PSI

AIR SIDE

AIR QTY LBS/HR SCFM	TEMP IN °F	ALTITUDE FT
AIR QTY/FAN ACFM	TEMP OUT °F	

CONSTRUCTION

DESIGN PRESSURE PSI	TEST PRESSURE PSI	DESIGN TEMP °F
SECTION	HEADER	TUBE
SIZE ROWS	TYPE	MATERIAL ☐ SEAMLESS
NO./BAY	MATERIAL	☐ WELDED
ARRANGEMENT:	NO. PASSES	OD IN. BWG AVG MIN WALL
SECTIONS IN PARALLEL	PLUG–DESIGN	NO./SECTION
SECTIONS IN SERIES	MATERIAL	LENGTH FT
BAYS IN PARALLEL	GASKET MATERIAL	PITCH IN. △☐◇
BAYS IN SERIES	CORROSION ALLOWANCE IN.	FIN
	SIZE INLET NOZZLE IN.	MATERIAL
MISC.	SIZE OUTLET NOZZLE IN.	OD
STRUCTURE	RATING	NO./IN.
	CODE	TYPE
	VENT DRAIN	

MECHANICAL EQUIPMENT

FAN	DRIVER	SPEED REDUCER
TYPE	MFR	MFR
NO./BAY	TYPE	TYPE
TIP SPEED FT/MIN	NO./BAY	NO./BAY
DIAM FT RPM	HP/DRIVER RPM	MODEL
HP/FAN PITCH ☐ ADJUST.	ENCLOSURE	AGMA HP RATING
NO. BLADES ☐ AUTO.VAR		RATIO /1
BLADE MATL		
HUB MATL		

NOTES:

DIMENSIONS: LENGTH WIDTH HEIGHT SHIP. WT LBS

REPRODUCTION OF STANDARD NO. 53.01–3

SPECIFICATIONS FOR FIN FAN HEAT EXCHANGERS

FLUOR-GRISCOM RUSSELL INTEROFFICE

No.			
1	Customer		Proposal No.
2	Address		Date
3	Plant		
4	Fin Fan Model		
5			
6			
7	Service		
8		**INSIDE TUBES**	**AIR DATA**
9	Class of Fluid		Altitude Feet
10	Specific Gravity @ 60 F		Inlet Air Temp. °F.
11	Specific Heat @ Ave. Temp.		Air #/Hr.
12	Latent Heat @ Ave. Temp.	B. T. U./#	Outlet Air Temp. °F.
13	Density @ Ave. Conditions	#/Cu. Ft.	Static Pressure "H_2O
14	Molecular Weight		**MECHANICAL EQUIPMENT**
15	Conductivity		Fans: Model
16	Viscosity	Centipoises at °F.	Number
17		Centipoises at °F.	Diameter
18	Quantity—Liquid	#/Hr.	No. of Blades
19	Vapor or Gas	Cu. Ft./24 Hr. @ 14.7# Abs. & 60°F.	R.P.M.
20	Vapor	#/Hr.	Horsepower per Fan
21	Steam	#/Hr.	
22	N. C. Gas	#/Hr.	Gears: Model
23	Fluid Vap. or Condsd.	#/Hr.	Make
24	Steam Condensed	#/Hr.	
25	Operating Temperature	In °F. : Out °F.	Drive Shaft:
26	Operating Pressure	#/☐"	
27	Pressure Drop	#/☐"	Motors: Type
28	Design Pressure	#/☐"	Make
29	Test Pressure	#/☐"	Number
30	Design Temperature	°F.	H.P.
31			R.P.M.
32			Volts
33	Heat Exchanged	B. T. U./Hr.	Phase
34	Surface per Section	Sq. Ft.	Cycle
35	Total Surface	Sq. Ft.	
36		**MATERIALS AND CONSTRUCTION**	Turbines: Model
37	Headers		Make
38	Thrust Members		Number
39	Tubes		H.P.
40	Fins		R.P.M.
41	Tube Plugs		Steam In. #/☐"
42	Retarders		Steam Ex. #/☐"
43	Connections	In	Water Rate #/H.P. Hr.
44		Out	
45	Number of Passes		Gas Engines: Model
46			Make
47			Number
48			H.P.
49			R.P.M.

Price: F.O.B. Shipping Point, not erected . . . . $

Freight . . . . $

TOTAL PRICE $

Estimated Erection Labor . . . . $

Shipping Weight . . . . lbs.

FORM NO. 724

THE GRISCOM-RUSSELL CO.

G-FIN TANK HEATER SPECIFICATIONS

1	Customer		Item No.
2	Address		Inq. No.
3	Plant		Date
4	Unit Required		
5			
6	Service		Type Internal External
7		SHELLS	TUBES
8	Class of Fluid		
9	Specific Gravity @ 60°F.		
10	Specific Heat @ Ave. Temp.		
11	Latent Heat @ Ave. Temp.	B.T.U./#	B.T.U./#
12	Viscosity	S.S.U. at °F.	
13	Viscosity	Centipoises at °F.	Centipoises at °F.
14	Viscosity	Centipoises at °F.	Centipoises at °F.
15	Quantity	GPM #/Hr.	#/Hr.
16	Operating Temperature	In °F. : Out °F.	In °F. : Out °F.
17	Operating Pressure	#/□"	#/□"
18	No. Passes per Shell	1	2
19	Design Pressure	#/□"	#/□"
20	Test Pressure	#/□"	#/□"
21	Design Temperature (Max. Metal Temp.)	°F.	°F.
22	Heat Exchanged	BTU/Hr. Surface Per Unit	Sq. Ft.
23	Corrected M.T.D. Design Rate	Surface Per Shell	Sq. Ft.
24	MATERIALS AND CONSTRUCTION		
25	SHELL: Steel	CHANNEL: Steel	Baffles: Steel Standard Spacing
26	INLET CONN. SIZE	INLET CONN. SIZE:	Tie Rods and Spacers: Steel
27	Drilling: 150#	Drilling: 150# P.T.	Gaskets: Durabla
28	OUTLET CONN. SIZE	OUTLET CONN. SIZE	Saddles: Steel (Type HSU-CSU only)
29	Drilling: 150#	Drilling: 150# P.T.	Structural Codes
30	Tank Manhole Flange Size : O.D.	Tubes—¾" IPS Seamless Steel	Inspection by:
31	Bolt Circle No. Dia.	Fins—Steel. Nom. Length:	OVERALL LENGTH
32	Tube Sheet—Steel with Cone Seat Tube Joints and Lock Nuts		OUTSIDE SHELL DIAMETER
33	Support Plate on Open Shell Unit Steel 1" Thick		Reference Print
34	GUARANTEE: This unit will give full capacity with tank liquid level 2 Ft. or more above unit.		

SUGGESTIONS:

1. Steam trace all oil lines.
2. Use steam trap large enough to handle indicated steam condensate.
3. Vent steam side of unit.
4. Provide drains for steam side to take care of shut down during cold weather.
5. This is an instantaneous heater; steam required only when pumping oil.

KELLOGG

COMPRESSOR COOLER DATA SHEET

JOB NO. ________ ITEM NO. ________
REQ. NO. ________
PAGE NO. ______ OF ______

ENG EQUIP 37 (6-77)

INTERCOOLERS AND AFTERCOOLERS

ITEM NUMBER OR LOCATION
BETWEEN STAGE NUMBER
COOLER FURNISHED BY:
□ COMPRESSOR MANUFACTURER □ PURCHASER

COOLER MANUFACTURER
COOLER MODEL OR TYPE

SPECIFICATIONS — MANUFACTURERS STANDARDS
OTHER

TYPE
SHELL AND TUBE
CONCENTRIC PIPE
PIPE LINE
FINTUBE

MATERIALS
SHELL OR OUTER PIPE
TUBES OR INNER PIPE
TUBE SHEETS
TUBE SIZE. O.D., INCHES AND BWG.

COOLING WATER— GPM
TEMPERATURE IN °F
TEMPERATURE OUT °F
PRESSURE DROP PSI
MAXIMUM WORKING PRESSURE, PSIG

GAS SIDE
FLOW RATE AT INTAKE, LBS./HR
TEMPERATURE AT INTAKE °F
TEMPERATURE AT OUTLET °F
INLET PRESSURE, PSIG
PRESSURE DROP, PSIG

DESIGN
DESIGN PRESSURE, PSIG
DESIGN TEMPERATURE °F
EFFECTURE SURFACE, SQUARE FEET
RECOMMENDED RELIEF VALVE SETING, PSIG
HYDROSTATIC TEST PRESSURE PSIG

MOISTURE SEPARATORS

FURNISHED BY: □ COMPRESSOR MANUFACTURER □ PURCHASER □ SEPARATE □ INTEGRAL WITH COOLERS
MANUFACTURER
TYPE AND MODEL
SHELL MATERIAL
TRAPS OR LIQUID COLLECTION DRUMS, GAUGE GLASS, DRAIN OUT VALUE, HIGH LIQUID LEVEL ALARM CONTACTS
TO BE FURNISHED BY: □ COMPRESSOR MANUFACTURER □ PURCHASER

AIR RECEIVER

FURNISHED BY: □ COMPRESSOR MANUFACTURER □ PURCHASER
CAPACITY ________ CUBIC FEET
RATED PRESSURE ________ PSIG
□ RECEIVER TO COMPLY WITH ________ PRESSURE VESSEL CODE
□ RECEIVER TO FURNISHED WITH SUPPORTING STAND FOR FLOOR MOUNTING
□ RECEIVER COMPLETE WITH SAFETY VALVE, PRESSURE GAUGE, AND DRAIN VALVE.

REVISIONS											
NO.	0	1	2	3	4	5	6	7	8	9	10
DATE											
BY											
CHECK											
APPRVD.											

	ENG'R	DATE
REP'D BY		
CHKD BY		
APPR'D BY		

PLATE & FRAME EXCHANGER SPECIFICATION SHEET

REV.	BY	CHKD	DATE

PROJECT	POS. NO.
PROJ. NO.	F. S. REF.
BLDG. NO.	NO. REQ'D.
SERVICE OF UNIT	MODEL
SIZE	NO. OF PLATES
SURFACE PER UNIT	SURFACE PER PLATE

PERFORMANCE OF ONE UNIT

		HOT SIDE		COLD SIDE	
		IN	OUT	IN	OUT
FLUID CIRCULATED					
TOTAL FLUID ENTERING	LB/HR				
LIQUID					
STEAM					
STEAM CONDENSED					
GRAVITY-LIQUID					
VISCOSITY-LIQUID	CENTIPOISE				
SPECIFIC HEAT	BTU/LB. °F				
THERMAL CONDUCTIVITY	BTU/HR., FT², °F/FT				
LATENT HEAT-VAPORS	BTU/LB.				
TEMPERATURE	°F				
OPERATING PRESSURE	PSIG				
NUMBER OF PASSES					
VELOCITY	FT/SEC.				
PRESSURE DROP - ALLOWABLE/CALC.	PSI				
FOULING RESISTANCE (MIN.)					
HEAT EXCHANGED-BTU/HR.			M.T.D. (CORRECTED). °F - THEORETICAL		
TRANSFER RATE-SERVICE		CLEAN			

CONSTRUCTION

DESIGN PRESSURE	PSI		
TEST PRESSURE	PSI		
DESIGN TEMPERATURE	°F		

MATERIALS OF CONSTRUCTION	PRESSURE PLATE	CONNECTING PLATE
ENDS PLATES	THERMAL PLATES	FRAME
EXTERNAL BOLTING	INTERNAL BOLTING	HOT SIDE GASKET
PROTECTION SHROUD	PAINT	COLD SIDE GASKET

CONNECTIONS - HOT SIDE	IN	OUT	SERIES
COLD SIDE	IN	OUT	SERIES
FABRICATION SPEC			
INSPECTION & TEST		TEST PRESSURE, PSIG	
WEIGHT: LBS.-EMPTY	FULL OF WATER	OPERATING	

REMARKS:

F 2397

ENG EQUIP 705-1 (2-76)

KELLOGG

STEAM SURFACE CONDENSER DATA SHEET

CUSTOMER ______ EST. NO. ______
LOCATION ______ JOB NO. ______
PAGE NO. ______
SERVICE OF UNIT ______ ITEM NO. ______
DATE ______

A. Condenser

1. **Manufacturer**
   a. Name: ______
   b. Location: ______
2. Condenser surface (total effective): ______ sq ft
3. Number of water passes: ______
4. Design basis — operating conditions and performance
   a. Nature and source of circulating water: ______
   b. Duty (net heat rejected to circulating water): ______ Btu per hr
   c. Circulating water quantity: ______ gpm
   d. Circulating water inlet temperature: ______ F
   e. Cleanliness factor: ______ %
   f. Average circulating water velocity in tubes: ______ fps
   g. Absolute pressure (to be measured at point specified in the Heat Exchange Institute Standards): ______ in. Hg
   h. Deaeration — maximum free oxygen in condensate (42) (14) (7) — (Indicate): ______ ppb
   i. Circulating water friction loss through clean tubes and waterboxes: ______ ft
   j. Supplementary (expected) performance on curve sheets: ______
5. Tubes
   a. Main Condensing Sections:
      (1) No.: ______ Size: ______ in. OD Gauge: ______ BWG
      (2) Length, effective: ______ Overall: ______ ft
      (3) Material: ______
   b. Gas Removal Sections:
      (1) No.: ______ Size: ______ in. OD Gauge: ______ BWG
      (2) Length, effective: ______ Overall: ______ ft
      (3) Material: ______
   c. Impingement Sections:
      (1) No.: ______ Size: ______ in. OD Gauge: ______ BWG
      (2) Length, effective: ______ Overall: ______ ft
      (3) Material: ______
   d. Method of connection — Inlet end: ______ Outlet end: ______
   e. Furnished by: ______
   f. Installed by: ______
6. Waterboxes — Covers (if required)
   a. Material (Waterbox): ______
   b. Material (Covers): ______
   c. Divided or non-divided: ______
   d. Design pressure: ______ psig Test pressure: ______ psig
   e. Number and size — Inlet nozzles: ______ Outlet nozzles: ______

ENG EQUIP 705-2 (2-76)

STEAM SURFACE CONDENSER DATA SHEET (con't.)

CUSTOMER ____________ EST. NO. ____________
LOCATION ____________ JOB NO. ____________
SERVICE OF UNIT ____________ PAGE NO. ____________
ITEM NO. ____________
DATE ____________

7. Shell
   a. Material: ____________
   b. Thickness: ____________ Number of sections: ____________
   c. Method of connecting sections: ____________
   d. Provision for tube expansion: ____________
   e. Number and size of venting connections: ____________

8. Hotwell
   a. Type: ____________ Capacity: ____________ gals
   b. Material: ____________
   c. Thickness: ____________ Number of sections: ____________

9. Tube sheets
   a. Number and size: ____________ Thickness: ____________
   b. Material: ____________

10. Tube support plates
   a. Number: ____________ Thickness: ____________
   b. Material: ____________

11. Steam inlet neck and exhaust connection
   a. Material: ____________
   b. Thickness: ____________ Number of sections: ____________
   c. Connection to shell — (flanged) or (welded): ____________ To turbine — (flanged) or (welded): ____________
   d. Height — top of waterbox flange to turbine exhaust including expansion joint (if any): ____________
   e. Provision for vertical expansion (condenser hung) (spring supports) (expansion joint).
   f. Spring supports (if used). Type: ____________
   g. Expansion joint (if used). Type: ____________
      Material: ____________
      Height: ____________ Ends — (flanged) or (welded): ____________

12. Approximate weights, condenser with tubes
   a. Dry: ____________ lbs In service: ____________ lbs
   b. Steam space only filled with water: ____________ lbs

13. Preliminary drawing(s): ____________

14. Descriptive bulletin(s) (or specifications): ____________

ENG EQUIP 705-3 (2-76)

STEAM SURFACE CONDENSER DATA SHEET (con't.)

CUSTOMER ____________ EST. NO. ____________
LOCATION ____________ JOB NO. ____________
SERVICE OF UNIT ____________ PAGE NO. ____________
ITEM NO. ____________
DATE ____________

B. Steam Jet Air Ejector(s)

1. Manufacturer
   a. Name: ____________
   b. Location: ____________
2. Total number of Ejector Units: ____________
3. Size and type: ____________ No. of elements: ____________ No. of stages: ____________
   mounted on: ____________ surface inter- and after-condensers
4. Operating Conditions and Performance
   a. Design capacity — 100 per cent —
      Suction pressure: ____________ in. Hg abs
      Suction temperature: ____________ F
      Dry air leakage: ____________ SCFM corresponding to: ____________ lbs per hr
      Non-condensible gases other than air leakage: ____________ (________) ____________ lbs per hr
      ____________ (________) ____________ lbs per hr
      ____________ (________) ____________ lbs per hr
      Associated saturated water vapor: ____________ lbs per hr
      Total gas-vapor mixture: ____________ lbs per hr
      Design capacity each element: ____________ per cent
      Number of elements operating for 100% capacity: ____________
      Curve(s) of approximate capacity vs. suction pressure: ____________
   b. Steam conditions —
      Maximum initial pressure: ____________ psig Total temp: ____________ F
      Minimum operating pressure (at nozzles): ____________ psig Total temp: ____________ F
   c. Steam Consumption (100 per cent design capacity): ____________ lbs per hr
5. Condenser Data
   a. Surface sq ft: ____________ Inter-condenser ____________ After-condenser
   b. Surface: ____________ per cent design capacity
   c. Tube size: ____________ in. OD: ____________ BWG: ____________ Length (effective-overall)
   d. Method of fastening tubes: ____________
   e. Cooling Water (condensate) (raw water): ____________
      Minimum flow required at: ____________ per cent design capacity: ____________ gpm
      Maximum design flow: ____________ gpm
      Friction loss at maximum design flow: ____________ feet of water
   f. Waterboxes — Maximum design pressure: ____________ psig
6. Materials
   a. Suction chambers: ____________
   b. Diffusers: ____________
   c. Steam nozzles: ____________
   d. Steam chests (nozzle heads): ____________
   e. Condenser shells: ____________

Steam Surface Condenser—*(continued)*

ENG EQUIP 705-4 (2-76)

KELLOGG

STEAM SURFACE CONDENSER DATA SHEET (con't.)

EST. NO. ______
CUSTOMER ______ JOB NO. ______
LOCATION ______ PAGE NO. ______
ITEM NO. ______
SERVICE OF UNIT ______ DATE ______

f. Water chambers: ______

g. Tube sheets: ______

h. Tubes: ______

7. Fittings and Accessories included

a. Air leakage meter: ______

b. Other: ______

______

8. Approximate weight, each ejector unit: ______ lbs

9. Preliminary drawing(s): ______

10. Descriptive bulletin(s) (or specifications): ______

C. Hogging Ejectors

1. Manufacturer

a. Name: ______

b. Location: ______

2. (Steam) (Air) (Water) Operated. Number per condenser: ______ Size: ______

3. Operating Conditions and Performance

a. Capacity: ______ lb per hr dry air at ______ in. Hg abs Suction Pressure

b. Steam consumption: ______ lb per hr at ______ psig ______ F

c. Maximum design steam conditions: ______ psig ______ F

d. Air consumption: ______ SCFM at ______ psig

e. Water consumption: ______ gpm at ______ psig

4. Approximate weight, each: ______ lbs

5. Preliminary drawing(s): ______

6. Fittings and accessories included: ______

7. Descriptive bulletin(s) (or specifications): ______

______

D. Remarks: ______

______

______

______

______

______

______

______

______

______

______

______

______

SWECO

SOUTHWESTERN ENGINEERING COMPANY

4800 SANTA FE AVENUE LOS ANGELES 58, CALIFORNIA SHEET ___ OF ___

FIRED HEATER SPECIFICATION SHEET

REV

REVISION NO.	1			
DATE				
BY				
APPROVED				

DATE ___
BY ___
CHECKED ___
ENGINEER ___
PROJ. ENG. ___

CUSTOMER		EQUIP. NO.	
PLANT LOCATION		FILE NO.	
SERVICE			
MANUFACTURER	TYPE	SIZE	
	THERMAL DESIGN		
FLUID CIRCULATED			
TOTAL FLUID ENTERING #/HR			
	INLET	OUTLET	
TOTAL VAPOR #/HR			
TOTAL LIQUID			
GRAVITY LIQUID °API			
DENSITY VAPOR #/CU. FT.			
VISCOSITY LIQUID CP			
VISCOSITY VAPOR CP			
SPECIFIC HEAT BTU/#			
LATENT HEAT BTU/#			
TEMPERATURE °F			
OPERATING PRESSURE PSIG			
MOLECULAR WEIGHT OF VAPOR			
ALLOWABLE PRESSURE DROP PSI	CALCULATED PRESSURE DROP PSI		
TYPE OF FLOW	SERIES	PARALLEL	
EFFECTIVE SURFACE SQ. FT.	RADIANT	CONVECTION	
HEAT ABSORPTION BTU/HR.	RADIANT	CONVECTION	
HEAT ABSORPTION BTU/HR.	TOTAL		
LIQUID TEMPERATURE °F			
RADIANT TRANS. RATE BTU/SQ. FT./HR.	AVERAGE	MAXIMUM	
CONVECTION TRANS. RATE BTU/SQ. FT./HR.	AVERAGE	MAXIMUM	
EFFICIENCY % - LHV/HHV			
FLUE GAS TEMP. (APPROX.) °F			
TYPE OF FUEL			
LHV OF FUEL BTU/			
PRESSURE AVAILABLE AT BURNER - PSIG	FOR GAS	NORMAL	MAXIMUM
	FOR FUEL OIL	NORMAL	MAXIMUM
	FOR STEAM	NORMAL	MAXIMUM
	MECHANICAL DESIGN		
NO. OF TUBES	RADIANT	CONVECTION	
EFFECTIVE TUBE LENGTH	RADIANT	CONVECTION	
RADIANT TUBES	INCHES O.D. TUBEWALL	INCHES MATERIAL	
CONVECTION TUBES	INCHES O.D. TUBEWALL	INCHES MATERIAL	
RETURN BENDS TYPE		MATERIAL	
NOZZLES INLET		OUTLET	
DESIGN RADIANT	PRESSURE PSIG	TEMPERATURE °F	
CONVECTION	PRESSURE PSIG	TEMPERATURE °F	
CODE	CUSTOMERS SPECIFICATION		
BURNERS - NO. TYPE	MFG.	PILOTS REQ'D.	
HEATER TYPE	HEATER WEIGHT		
TYPE SHELL CONSTRUCTION	TYPE ASSEMBLY		
REMARKS:			

Steam Surface Condenser—(*continued*)

PROCESS DESIGN SECTION
CORPORATE ENGINEERING

WIPED FILM EVAPORATOR DATA SHEET

SHEET 1 OF 4

SCOPE: ____________________

PERFORMANCE REQUIREMENTS

	FEED	DISTILLATE	CONCENTRATE
IDENTIFICATION			
QUANTITY: LBS/HR.			
NORMAL OPERATING			
DESIGN			
COMPOSITION:			
PHYSICAL PROPERTIES			
MOLECULAR WEIGHT			
SPECIFIC GRAVITY			
VISCOSITY; CP			
SPECIFIC HEAT: BTU/#°F			
LATENT HEAT: BTU/#			
THERMAL CONDUCTIVITY – BTU/#°F FT.			
VAPOR PRESS: MMH_G			
MELTING POINT – °C			

REMARKS | APP'R'V | DATE | SUPRV | CHKD | BY | REV

INITIATOR | DRAWN BY | CHECKED BY | DES SUPRV.

JOB NO. | PLANT LOCATION
PROJECT NO. | REF. DWG. NO.
TITLE
POSITION NO. | SERIES | STEP NO.
ISSUE DATE | SCALE | BLDG. NO.
DRAWING NO.

E230B46*
SHEET 1

PROCESS DESIGN SECTION
CORPORATE ENGINEERING

WIPED FILM EVAPORATOR DATA SHEET

SHEET 2 OF 4

OPERATING CONDITIONS

	PROCESS	JACKET	CONDENSER
FLUID DESCRIPTION			
TEMPERATURE	FEED ____ °C DISTILLATE ____ °C CONCENT. ____ °C	INLET ____ °C OUTLET ____ °C	
VAPORIZATION TEMP.	____ °C	____ °C	____ °C
DESIGN TEMP.	____ °C	____ °C	____ °C
MAX. ALLOW. TEMP.	____ °C	____ °C	____ °C
OPERATING PRESS.	____ PSIG	____ PSIG	____ PSIG
DESIGN PRESS.	____ PSIG	____ PSIG	____ PSIG
NORMAL FLOW RATE			
MAX. FLOW RATE			
PRESSURE DROP			
MAX. ALLOW. PRESS DROP			
MAX. AIR LEAKAGE RATE			
HEAT DUTIES:			
PRE-HEAT			
VAPORIZATION			
CONDENSING			
SURFACE REQUIRED			
VAPORIZATION			
CONDENSATION			

MATERIALS

MAIN VESSEL ____ MECH. SEAL
JACKET ____ ROTATING FACE ____
SHAFT ____ STATIONARY FACE ____
WIPER BLADES ____ SEAL RING ____
BEARINGS ____ "O" RINGS ____
INTERNAL COND. TUBES ____ MAIN GASKETS ____
INTERNAL COND. TUBESHEET ____

MOTOR DRIVE DATA

ITEM NO. ____ MTD BY. ____ VOLTS/PHASE/HZ: ____ / ____ / ____
HP ____ RPM ____ FRAME ____ BEARINGS ____ LUBE ____
MFR. ____ FULL LOAD AMPS: ____ S.F. ____
TYPE ____ INSUL. ____ L.R. AMPS ____
ENCLOSURE: CLASS 1 GROUP D-DIV.

REMARKS | APP'R'V | DATE | SUPRV | CHKD | BY | REV

INITIATOR | DRAWN BY | CHECKED BY | DES SUPRV.

JOB NO. | PLANT LOCATION
PROJECT NO. | REF. DWG. NO.
TITLE
POSITION NO. | SERIES | STEP NO.
ISSUE DATE | SCALE | BLDG. NO.
DRAWING NO.

E230B46
SHEET 2

PROCESS DESIGN SECTION
CORPORATE ENGINEERING

WIPED FILM EVAPORATOR
DATA SHEET

SHEET 3 OF 4

VENDOR DATA (RETURNED BY VENDOR)

UNIT SIZE & TYPE: ______

NOZZLES	SIZE	RATING	FACING	POSITION
FEED				
VAPOR(VAC)				
RESIDUE				
DISTILLATE				
JACKET-IN				
JACKET-OUT				

HEATING SURFACE ______ SQ. FT.
CONDENSING SURFACE: ______ SQ. FT.
PACKING: TYPE ______
SIZE ______ NO. RINGS ______
PRESS. ON PACKING BOX. ______ PSIG
MECH. SEAL: TYPE ______
MFR. ______
SEAL FLUID ______

MOTOR: SIZE & TYPE ______ MFG: ______
RPM ______ PERIPHERAL SPEED ______ FT./SEC.
DRIVE TYPE ______
VENDOR TO SUPPLY DIMENSIONAL PRINTS OF ALL UNITS.
WEIGHT OF UNIT: EMPTY ______ FULL OF WATER ______

UTILITIES AVAILABLE

STEAM	______ #/HR @ ______	PSIG @ ______ °F
COOLING WATER	______ GPM @ ______	PSIG @ ______ °F
HOT OIL	______ GPM @ ______	PSIG @ ______ °F
ELECTRICITY	______ VOLTS ______	PHASE ______ CYCLES

MISCELLANEOUS REQUIREMENTS

QUOTATIONS:
VENDOR QUOTATION SHOULD INCLUDE THE FOLLOWING:
1 - EQUIPMENT DESCRIPTION WITH DEFINITIVE SPECIFICATIONS INCLUDING SIZE, OPERATING AND DESIGN PRESSURES, AND MATERIALS OF CONSTRUCTION ON ALL PARTS.
2 - VENDOR SHALL STATE THE REQUIRED FLOWS, PRESSURE DROPS AND DESIGN DUTIES FOR EACH JACKET, STILL CONDENSER, AND FREEZEOUT TRAP.
3 - SUMMARY OF UTILITY REQUIREMENTS.
4 - RECOMMENDED INSTRUMENTATION.
5.- RECOMMENDED LAYOUTS AND REQUIRED STRUCTURES.
6 - RECOMMENDED SPARE PARTS LIST AND COST OF EACH ITEM

TESTS
ALL EQUIPMENT AND MANIFOLDS DESCRIBED IN THIS SPECIFICATION SHALL BE ASSEMBLED AND TESTED UNDER DESIGN VACUUMS AT THE VENDORS SHOPS TO ASSURE PROPER FIT, VACUUM TIGHTNESS AND THE GENERAL OPERABILITY OF THE DESIGN. ROCHE INSPECTION AND APPROVAL OF THE RUN-IN TESTS IS REQUIRED PRIOR TO SHIPMENT.

GUARANTEES:
THE VENDOR SHALL GUARANTEE THE MECHANICAL OPERATION OF THE EQUIPMENT AND SYSTEM PROVIDED FOR A MINIMUM OF ONE YEAR AFTER START-UP. HLR EXPECTS PROCESS GUARANTEES ON THE SPECIFIED DISTILLATION RATES, CONDENSER PERFORMANCE, AND MAXIMUM PRESSURE DROPS. ENTRAINMENT, AIR LEAKAGE RATES AND UTILITY CONSUMPTIONS SHALL BE SPECIFIED. ALL PRODUCT AND PROCESS SPECIFICATIONS WHICH THE VENDOR CANNOT GUARANTEE SHALL BE ENUMERATED.

REMARKS | APP'R'V | DATE | CHKD SUPRV | BY | REV

DES SUPRV. | CHECKED BY | DRAWN BY | INITIATOR

JOB NO. | PLANT LOCATION
PROJECT NO. | REF. DWG. NO.
TITLE
POSITION NO. | SERIES | STEP NO.
ISSUE DATE | SCALE | BLDG. NO. | DRAWING NO.

E230B46
SHEET 3

PROCESS DESIGN SECTION
CORPORATE ENGINEERING

WIPED FILM EVAPORATOR
DATA SHEET

SHEET 4 OF 4

GENERAL NOTES

1 - VESSEL FABRICATION TO BE IN ACCORDANCE WITH THE FOLLOWING:
☐ ______ ASME CODE, SECTION VIII. STAMP REQ'D. ☐ NOT REQ'D. ☐
☐ SPEC. UNFIRED PRESSURE VESSELS.

2 - SPECIFICATIONS TO BE COMPLETED BY VENDOR AND RETURNED WITH PROPOSAL.

REMARKS | APP'R'V | DATE | CHKD SUPRV | BY | REV

DES SUPRV. | CHECKED BY | DRAWN BY | INITIATOR

JOB NO. | PLANT LOCATION
PROJECT NO. | REF. DWG. NO.
TITLE
POSITION NO. | SERIES | STEP NO.
ISSUE DATE | SCALE | BLDG. NO. | DRAWING NO.

E230B46
SHEET 4

Wiped Film Evaporator—*(continued)*

PROCESS DESIGN SECTION
CORPORATE ENGINEERING

ROTARY VACUUM DRYER
DATA SHEET

PAGE 1 OF 2

OPERATING CONDITIONS

MAT'L. TO BE DRIED: ______
CHARACTERISTICS: ______
BULK DENSITY: AVG. ______ LB./CU. FT.
MIN. ______ LB./CU. FT.
MOISTURE: ______
INITIAL MOISTURE CONTENT: ______ WT. %
FINAL MOISTURE CONTENT: ______ WT. %
AMT. OF MAT'L. PER BATCH ______ LBS.
WORKING CAPACITY REQ'D. ______ CU. FT.
MAX. TEMP: WET ______ °F DRY ______ °F
DRYING RATE: ______ LB./HR.
EST. DRYING TIME/BATCH ______ HRS.
MOISTURE RECOVERY: REQ'D. ☐ NOT REQ'D. ☐

OPERATING PRESS. ______ PSIA
DESIGN PRESS. ______ PSIG
JACKET DESIGN PRESS. ______ PSIG
HEAT TRANSFER RATE: ______ BTU/HR.
UTILITIES AVAILABLE:
ELECTRICITY: VOLTS/PHASE/CYCLE/- ______
WATER: TEMP. MIN. °F MAX. ______ °F
PRESS. ______ PSIG
STEAM: PRESS. ______ PSIG
TEMP. ______ °F
BRINE: PRESS. ______ PSIG
TEMP. ______ °F
OTHER: ______

MECHANICAL DATA

MFR. ______
SIZE & TYPE ______
TOP OPENING: SIZE ______ TYPE ______
BOT. OUTLET: SIZE ______ TYPE ______
DRYER RPM: ______
DRIVE: MFR. ______
TYPE ______

VACUUM SEAL: TYPE ______
PACKING: TYPE ______
SIZE ______ NO. RINGS ______
MECH. SEAL: MFR. ______
TYPE ______
COUPLING: MFR. ______
TYPE ______

MATERIALS

SHELL: ______
INTERNAL FINISH #: ______
JACKET: ______
VACUUM TUBE: ______
FILTER: ______
PACKING: ______

MECH. SEAL: ROTATING FACE: ______
STATIONARY FACE: ______
BOT. OUTLET VALVE: ______
VACUUM PIPING: ______
JACKET PIPING: ______
GASKETS: ______

MOTOR DRIVE DATA

	DRYER	VAC. PUMP	
ITEM NO./MTD. BY	/	/	/
HP/RPM/FRAME	/ /	/ /	/ /
MFR.			
TYPE/INSUL.	/	/	/
ENCLOSURE			
VOLTS/PHASE/CYCLE	/ /	/ /	/ /
BEARINGS/LUBE.	/	/	/
FULL LOAD AMPS/SF	/	/	/
L.R. AMPS			

REMARKS | APP'R'V | DATE | SUPRV | CHKD | BY | REV

DES SUPRV. | CHECKED BY | DRAWN BY | INITIATOR

JOB NO. | PLANT LOCATION
PROJECT NO. | REF. DWG. NO.
TITLE
POSITION NO. | SERIES | STEP NO.
ISSUE DATE | SCALE | BLDG. NO.
DRAWING NO.

E219B35

PROCESS DESIGN SECTION
CORPORATE ENGINEERING

ROTARY VACUUM DRYER
DATA SHEET

PAGE 2 OF 2

MISCELLANEOUS DATA

LIST OF ACCESSORIES TO BE FURNISHED BY VENDOR:

- ☐ COMPLETE VACUUM SYSTEM INCL. PUMP, CONDENSER, RECEIVER AND PIPING
- ☐ COMPLETE HEATING SYSTEM INCL. PUMP, EXCHANGER AND PIPING
- ☐ VACUUM CONTROL SYSTEM
- ☐ HEATING CONTROL SYSTEM
- ☐ AUTOMATIC JOGGING AND POSITIONING SYSTEM FOR DRYER
- ☐ INSTRUMENT SPECIFICATIONS FOR PURCHASE BY H.L.R.
- ☐ BOTTOM OUTLET VALVE
- ☐ THERMOWELL
- ☐ VIBRATOR
- ☐ UNMOUNTED EXPANDABLE SLEEVES

REMARKS

REMARKS | APP'R'V | DATE | SUPRV | CHKD | BY | REV

DES SUPRV. | CHECKED BY | DRAWN BY | INITIATOR

JOB NO. | PLANT LOCATION
PROJECT NO. | REF. DWG. NO.
TITLE
POSITION NO. | SERIES | STEP NO.
ISSUE DATE | SCALE | BLDG. NO.
DRAWING NO.

E219B35

KG-1135 (6-81)

BOILER DATA SHEET

BOILER SHALL BE FURNISHED IN ACCORDANCE WITH THE M. W. KELLOGG COMPANY SPEC.

1. OPERATING CONDITIONS

RATED CAPACITY: ________ PPH CONTINUOUS PER UNIT
STEAM PRESSURE: ________ PSIG AT S.H. OUTLET
STEAM TEMP.: ________ F AT S.H. OUTLET
ON ________ FIRED OPERATION
FEED WATER TEMP. ________ F. BLOWDOWN ________ %

FUEL			
A.P.I GRAVITY			
% BY			
ASH			
HYDROGEN			
CARBON			
SULPHUR			
H.H. V. BTU/			
L.H.V. BTU/			
PSIG @ BURNER			
TEMP. @ BURNER F.			

TYPE OF INSTALLATION: ________
CASING: ________

AMBIENT TEMP: ________ F
ALTITUDE: ________ FT. ABOVE SEA LEVEL
DESIGN WIND LOADING: ________ LB/SQ. FT.
EARTH QUAKE LOADING: ________ ZONE.
INSTRUMENT AIR: ________ PSIG

START-UP

11. EXPECTED PERFORMANCE

	STEAM ACTUAL M. LB/HR				
	BLOWDOWN %				
	FUEL				
	LOAD DURATION				
	FURNACE LIBERATION				
	M.B.T.U./CU. FT./HR.				
	FURNACE HEAT RATE				
	M.B.T.U./SQ. FT./HR.				
	EXCESS AIR LEAVING BLR. %				
QUANT. M LB/HR	FUEL (GAS CFM)				
	FLUE GAS ENT. A.H.				
	FLUE GAS LEAVING A.H.				
	AIR LEAVING A.H.				
	STEAM AT S.H. OUTLET – PSIG				
TEMP. F	SUPER HEATED STEAM				
	FLUE GAS LEAVING BLR				
	FLUE GAS LEAVING				
	WATER ENT. ECON.				
	WATER ENT. BLR.				
	AIR ENT. A.H.				
	AIR LEAVING A.H.				
DRAFT LOSS IN. OF WATER	FURNACE				
	BOILER AND S.H.				
	ECONOMIZER				
	AIR HEATER				
	FLUES				
	NET DRAFT LOSS				
	ACROSS INST. CONNECT				
AIR RESISTANCE IN. OF WATER	BURNERS WINDBOX				
	DUCTS				
	AIR HEATER				
	NET RESISTANCE				
	EFFICIENCY %				

		LB/HR	IN. H_2O	F
F.D. FAN	NET RATING			
	TEST RATING			
F.D. FAN	NET RATING			
	TEST RATING			

REVISIONS				
6				
5				
4				
3				
2				
1				
0				
NO.	DATE	ENGINEER	CHK/REVIEW	APPROVE

KELLOGG | MANAGING OFFICE | ENGINEERING OFFICE

EQUIPMENT DATA SHEET

CLIENT ________
LOCATION ________ JOB NO. ________
ITEM ________
ITEM NO. ________ SPEC. NO. ________
REQ'N. NO. PAGE OF

KG-1124 (7-81)

PACKAGE BOILER DATA SHEET

III EQUIPMENT

BOILER
TYPE ________
DESIGN PRESSURE, PSIG ________
FURNACE TYPE ________
FURNACE VOLUME, CU. FT. ________
HEATING SURFACE
(RADIANT SURFACE: FLAT PROJECTED BASIS.
CONVENTION SURFACE: CIRCUMFERNTIAL BASIS)
FURNACE, SQ. FT. ________
BOILER, SQ. FT. ________
TOTAL, SQ. FT. ________
DRUMS: ________
NO. REQUIRED ________
UPPER DIMENSION ________
LOWER DIMENSION ________
VERT. HT. (C.L. TO C.L.) ________

SUPER-HEATER
TYPE ________
TEMP. CONTROL ________
SURFACE AREA, SQ. FT. ________

ECONO MISER
TYPE ________
SURFACE AREA, SQ. FT. ________

AIR HTR.
TYPE ________
SURFACE AREA, SQ. FT. ________

BURNERS

	NO. REQ.	TYPE MAKE	LOAD RANGE	PRESS RANGE
OIL				
GAS				

COMBUSTION CONTROLS
FURNISHED BY ________

STACKS
IN ACCORDANCE WITH KELLOGG SPEC. ________
TYPE ________
HEIGHT, FT: ________ DIAMETER ________
ARRANGEMENT ________

DUCT FLUES
FURNISHED BY ________

PLAT-FORM LOADERS
FURNISHED BY ________
IN ACCORDANCE WITH KELLOGG SPEC. ________

SOOT BLOWERS
BOILER ________
SUPERHEATER ________
AIR HEATER ________

DIMEN-SIONS
SETTING (OVERALL) ________
UNIT (OVERALL) ________

CONNEC-TIONS
STEAM ________
WATER FEED ________
CHEMICAL FEED ________
BLOW DOWN ________

WEIGHT
SHIPPING ________ LBS
OPERATING ________ LBS
MAX. ERECTION ________ LBS

IV GUARANTEE CONDITIONS

FUEL ________
CAPACITY, P.P.H. ________
STEAM TEMP. DEG. F. ________
(@ S.H. OUTLET) ________
THERMAL EFFIC. % ________
MAX. NET DRAFT LOSS ________
MAX. NET AIR RESISTANCE ________
SOLIDS CARRY OVER P.P.M. ________
BLOW-DOWN ________
FEED WATER TEMP. DEG. F. ________
MAX ALLOWABLE BOILER ________
CONCENTRATION P.P.M. ________

REVISIONS				
6				
5				
4				
3				
2				
1				
0				
NO	DATE	ENGINEER	CHK/REVIEW	APPROVE

KELLOGG | MANAGING OFFICE | ENGINEERING OFFICE

EQUIPMENT DATA SHEET

CLIENT ________
LOCATION ________ JOB NO. ________
ITEM ________
ITEM NO. ________ SPEC. NO. ________
REQ'N. NO. PAGE OF

INDUCED DRAFT COOLING TOWER

SERVICE: ______ ENGLISH/METRIC UNIT
MANUFACTURER: ______ MODEL NO. ______
SITE ELEVATION: ______ PREVAILING WIND DIRECTION ______

TOWER DESIGN

DIMENSION: LENGTH ______ WIDTH ______ HEIGHT ______
HEIGHT: BASIN CURB TO FAN DECK ______ FAN STACK ______
H. W. INLET ABOVE BASIN CURB ______
COLUMN EXTENSIONS (FT. BELOW BASIN CURB): PERIMETER ______ INTERNAL ______
ACCESS TO TOP OF TOWER: ______
AIR TRAVEL THROUGH FILL: ______ AVG. FILL HEIGHT ______
TOTAL VERTICAL FILL HEIGHT: ______ AIR VELOCITY THROUGH FILL ______
FILL SLAT SPACING: HORIZONTAL ______ VERTICAL ______
LOUVER ANGLE ______ GROSS LOUVERED AREA ______ FAN DECK LIVE LOAD ______

MATERIALS OF CONSTRUCTION

FAN STACK: ______ PARTITIONS ______ SHEATHING ______
ITEMS RECEIVING PRESERVATIVE TREATMENT: ______

FAN BLADES: ______ STAIRWAYS ______ FILL SUPPORTS ______
DRIFT ELIMINATOR SUPPORTS ______ STRUCTURAL CONN. ______
RING JOINT CONN: ______ MECHANICAL EQUIPMENT SUPPORT ______
FAN HUB ______ DRIVE SHAFT ______

EQUIPMENT

FANS: MANUFACTURER ______ MODEL NO. ______
HUB DIAMETER ______ AIR DELIVERED A. C. F. M. ______
SPEED REDUCER: MANUFACTURER ______ MODEL NO. ______
DRIVE SHAFT: MANUFACTURER ______ MODEL NO. ______
RATED H. P. ______ COUPLING ______
DRIVER MANUFACTURER ______ MODEL NO. ______ MOTOR HEATERS: YES/NO
Elect. Area Class: Class ___ Gr. ___ Div. ___/Non-Hazardous; Volts ___ Ph. ___ Hz
Enclosure ______ Insulation ______ Temp. Rise ___ °C above ___ °C By ___ @ ___ S.F. Temp Code ___

Item No.	HP KW	Current: Full Load	Current: Locked Rotor	Eff. 4/4	Eff. 3/4	P.F. 4/4	P.F. 3/4	Full Load RPM	Frame No.	Brg. Lub. (1)	Weight LB/KG	Allow-Thrust Up/Down LB/KG	Space Heater Watts	S.F.

FANSTACK ______ VELOCITY RECOVERY YES/NO. HEIGHT ______
DISCHARGE AREA ______ DISCHARGE VELOCITY ______
DRIFT ELIMINATORS ______ TYPE ______ MODEL NO. ______
PRESSURE DROP ______ DRIFT LOSS ______
VALVES: ______ STOP ______
FLOW CONTROL ______ FLOAT CONTROL ______

ERECTION

LABOR BY: ______ UNLOADING & HAULING BY ______ SUPERVISION BY ______
EST. MANHOURS REQ. FOR ERECTION ______ FREIGHT ______

MISCELLANEOUS

MW KELLOGG SPECIFICATION: U41-F2. SNOW LOAD SPECIFICATION ______
WIND & EARTHQUAKE SPECIFICATION: ______ PAINT SPECIFICATION: ______
NOTES: ______

REVISIONS NO.	DATE	ENGINEER	CHK/REVIEW	APPROVE
6				
5				
4				
3				
2				
1				
0				

KELLOGG — MANAGING OFFICE — ENGINEERING OFFICE

DATA SHEET

CLIENT ______
LOCATION ______ JOB NO. ______
ITEM INDUCED DRAFT COOLING TOWER
ITEM NO. ______ SPEC. NO. ______
REQ'N. NO. ______ PAGE ___ OF ___

S & W FORM C-1407

Item ______
Page ______
Preliminary ______
Final ______

STONE & WEBSTER ENGINEERING CORPORATION

INSTRUMENT SPECIFICATION

Client ______ J. O. No. ______
Apparatus ______ Project No. ______
Service ______ Date ______ By ______
Based on ______ Process Page ______ Dated ______

INSTRUMENT CASE		BULBS OR THERMOCOUPLES		
Location		Location		
Type		Bulb Type		
Case Finish		Thermocouple Matl		
Connections	Back — Bottom	Socket Material		
Mounting	Flush — Projecting	Socket Length		
Chart or Dial	No Size in. width/diameter	Ext Neck Length		
Range	Flow Static Press.	Standard Pipe Thread		
	Temp. Diff. Press.	Bulb Distance From Instr	Above/Below	Above/Below
Control Point	No of pens	Tubing or Lead		
Chart Volume	Measured on °F lb. Abs/Ga Basis	RESPONSES		
Clock	Electric — Spring Wound, 7 Day/24 Hour	Throttling Band		
Current	Volts AC/DC Cycles	Reset	Automatic/Manual	
Accessories	Ink, 100 Charts etc to be included	Rate Response	Yes — No	
		Pneumatic Set		

INSTRUMENT		CONTROLLED VALVE — Item No.	
Location		Service	
Type			
Float Body	Cage/Flanged Matl Trim	Valve	Size in. Drilling
Float Flange Conn.	Size Drilling	Plug Type Valve	Balanced Parabolic V-Port or Equal
Float	Size in. Matl	Type	Diaphragm Spring/Weight Loaded
Stuffing Box	With Grease Seal	Body	Matl Trim
Orifice Flanges	Matl Drilling	Stem	Sliding — Rotary
Orifice Plate	Matl	Stuffing Box	With Grease Seal
Instr Body Diff Range	0 — in. Water Column	Operating Medium	Gas/Comp Air @ lb ga Corrosive Yes/No
Instr Body Located	ft above/below liquid level/orifice plate	Valve Action	Open/Close with Fall of
Instr Body Distance	ft from liquid level/orifice plate		Open/Close with Failure of
Mercury Included		Valve Positioner	Yes - No with Bypass
Sealing Medium		Radiating Fins	Yes - No

CONDITIONS AND REQUIREMENTS

In		Control Valve	In		Control Valve
Fluid			Nor Quantity @ °F P		
SpGr @ 60° F Air/Water = 1			Max Quantity @ °F P		
SpGr @ Nor T & P Air/Water = 1			Min Quantity @ °F P		
Viscosity	cp @ °F	cp @ °F	Nor Pressure (Up Stream)	lb ga	lb ga
Corrosion Present	Yes — No	Yes — No	Max Pressure (Up Stream)	lb ga	lb ga
Line Size	in.	in.	Down Stream Pressure	lb ga	lb ga
Normal Temperature	°F	°F	Pressure Drop	lb	lb
Max Temperature	°F	°F	Amount of Superheat	°F	°F

Supply Metal Tag, securely wired, marked with Item Number

REMARKS

For Typical Installation Details See ______
General Specifications for Instrument, Page ______ ; for Valve, Page ______

Copy to	Date	Checked	Date	Approved	Date
Revised:					

DIFFERENTIAL PRESSURE TYPE FLOW INSTRUMENT

SERVICE DATA	PRESENT	NEW
ITEM NO.		
LOCATION		
SERVICE		
LINE SIZE		
ORIFICE SIZE (Bevel)		
MANOMETER (Range & Mfg.)		
MANOMETER TYPE (Mercury, Aneroid)		
CONNECTIONS (Flange & Series)		
FLOWING FLUID		
SP. GR. AT 60F		
FLOWING TEMP.		
FLOWING PRESSURE		
SEAL MATERIAL		
SEAL TEMP.		
SEAL GRAVITY AT 60F		
MAX. FLOW		
CHART NO.		

INSTRUMENT

	REMARKS
PRIMARY METER	
TYPE - Recording, Indicating, Blind, Transmitting	
MFR. MODEL	
MANOMETER - 1500 psi - Carbon Steel	
RECEIVER	
TYPE - Conventional, Miniature, None	
MFR. MODEL	
MOUNTING - Control Board, Local	
CHART TYPE - Circular 24 Hour, Strip	
CHART DRIVE - Spring, Pneumatic	
CONTROLLER	
TYPE - Conventional, Force Balance, Pneumatic	
MFR. MODEL	
MOUNTING - Control Board, Field	
MANUAL CONTROL BYPASS - Yes, No	
PNEUMATIC SET - Adjustable, Fixed, None	
CONTROL FEATURES - Prop. Band, Reset, Rate	
INCREASE IN FLOW, OUTPUT TO - Increase, Decrease	
REQUISITION OR REQUEST -	
REF. SPEC. SHEET-	
REF. DRAWING -	
INST. TAG DATA	

By | App'd. | Date | By | App'd. | REV. 2 | Date | REV. —

By ____ App'd. ____ Date ____

HUMBLE OIL & REFINING CO.
BAYTOWN ENGINEERING DIVISION
INSTRUMENT SPECIFICATION

Appr'n. ____ Sub ____ Zone ____ NO. ____

E-172

N. Y. PURCHASING DEPT.

MATERIAL REQUISITION
FOSTER WHEELER CORPORATION
165 BROADWAY
NEW YORK, U. S. A.

TEMP.-PRESS. INSTRUMENTS

CUSTOMER'S NAME ____
CONTRACT No. ____
ITEM ____
SERVICE ____

REQUISITION No.
DATE ____
PAGE No. ____
CHANGE No. ____

INSTRUMENT

No. Req'd ____ Type ____
Location ____ Case ____
Mounting ____
Range ____ Scale ____
Chart ____
Thermal System ____
Press. Element ____
Control Point ____ No. Pens ____
Control ____
Clock ____
Current ____ Volts ____ Cycle ____ Phase
Connections ____
Accessories: ____ type air reducing valve,
air filter ____ charts, ink set and air gauges
Special ____

— TEMPERATURE BULBS —
Location ____
Type ____
No. Req'd ____ Bulb Mat'l ____
Bulb. Elev. ____ Instrument
Type Socket ____ Mat'l ____
Length Below Threads ____
Ext. Above Threads ____
S. P. T. Size ____
Tubing Mat'l ____
Tubing Length ____
Special ____

CONTROLLED VALVE

Type ____
Location ____
No. Req'd ____ Size ____ Diaphragm Control Valve
Body Material ____ Ends ____
Faced and Drilled ____
Trim ____
Valve Plug ____
Stuffing Box ____ Sealed, Isolating Valve ____

Operating Medium ____ @ ____ Lbs. Ga.
Valve to ____ With Increase in ____
Valve to ____ With Failure of Air
Air-Cooled Bonnet Required ____
Valve Positioner Required ____
Extra Reducing Valve and Filter Required ____
Special ____

SERVICE CONDITIONS AND REQUIREMENTS

		VALVE			VALVE
Fluid			Nor. Quan. @ ____ ° F.		
Sp. Gr. @ 60° F. Air = 1 Water = 1			Max. Quan. @ ____ ° F.		
Sp. Gr. @ Nor. Con. (T & P)			Nor. Press. (Upstream) Lb. Ga.		
Viscosity @ ____ ° F.			Max. Press. (Upstream) Lb. Ga.		
Line Size			Downstream Press.		
Normal Temp., ° F			Pressure Drop		
Max. Temp., ° F					

REMARKS: ____

1000 Sets 5-42 Printed in U. S. A. Form D41-A

FORM NO. 2788

Northern Natural Gas Company

LEVEL INSTRUMENTS SPECIFICATION SHEET

LOCATION ____ FILE NO. ____
ITEM NO. ____ JOB NO. ____ W.O. NO. ____
PREPARED BY ____ DATE ____
CHECKED BY ____ DATE ____
APPROVED BY ____ DATE ____
△ INDICATES REVISION

	GENERAL					
1	TYPE					
2						
3	TAG NO.					
4	VESSEL OR EQUIPMENT NO.					
	BODY					
5	MATERIAL					
6	TOP CONN LOCATION					
7	BTM CONN LOCATION					
8	CONN-SIZE					
9	CONN SCREWED OR FLANGED					
10	CASE MOUNTING					
11	FLANGE ORIENTATION					
12	ROTATABLE HEAD					
13						
	FLOAT OR DISPLACER					
14	DIAMETER OR LENGTH					
15	EXTENSION					
16	MATERIAL					
17	TORQUE TUBE MATERIAL					
18	AIR FIN					
19						
	TRANSMITTER					
20	TYPE					
21	OUTPUT					
22	RECEIVERS ON SHEET NO.					
	CONTROL					
23	TYPE					
24	PROPORTIONAL-% RESET					
25	OUTPUT					
26	ON LEVEL INCREASE; OUTPUT					
27						
	ACCESSORIES					
28	FILTER AND REGULATOR					
29	GAGE GLASS CONNECTIONS					
30	GAGE GLASS					
31	PURGE CONNECTION					
32	ELECTRIC SWITCH					
33						
34						
	SERVICE CONDITIONS					
35	UPPER LIQUID					
36	LOWER LIQUID					
37	SP GR UPPER LOWER					
38	PRESS MAX. NORM					
39	TEMP MAX. NORM					
40						
41						

REMARKS:

REPRODUCTION OF STANDARD NO. 53.02-1

EDD FORM 2792 1-64

Northern Natural Gas Company

GAUGE GLASSES AND COCKS SPECIFICATION SHEET

LOCATION ____ FILE NO. ____
ITEM NO. ____ JOB NO. ____ W.O. NO. ____
PREPARED BY ____ DATE ____
CHECKED BY ____ DATE ____
APPROVED BY ____ DATE ____
△ INDICATES REVISION

GAUGE GLASSES

1 SUPPLY ☐ GAUGES ONLY ☐ GAUGES & COCKS
NO NIPPLES REQUIRED
2 TYPE ☐ TRANSPARENT ☐ TUBULAR ☐ REFLEX
3 CONNECTIONS ☐ 1/2" ☐ 3/4" ☐ ____"
☐ TOP & BOTTOM ☐ SIDE ☐ BACK
☐ WELDING PAD OTHER ____
4 MATERIAL ☐ STEEL OTHER ____
5 MINIMUM RATING ____ PSIG @ ____ °F

ACCESSORIES

6 ILLUMINATORS: ____
7 SHIELDS: ____
8 HEATING-COOLING } INTERNAL ____
9 CHAMBERS } EXTERNAL ____
10 NON-FROSTING TYPE: ____
11 CALIBRATED SCALE ____
12 SUPPORT PLATES: ____
13 GUARD RODS: ____
14 OTHER ____

NOTES:

GAUGE COCKS

15 SUPPLY ☐ COCKS ONLY
16 TYPE ☐ OFFSET ☐ ANGLE
17 CONNECTIONS-NPT

VESSEL MALE	GAUGE FEMALE	DRAIN FEMALE
☐ 1/2"	☐ 1/2"	☐ 1/2"
☐ 3/4"	☐ 3/4"	☐ 3/4"
☐ OTHER	☐	☐

		BODY	TRIM
18	MAT'L		

19 MINIMUM RATING ____ PSIG @ ____ °F

20 CONST. ☐ PLAIN CLOSING ☐ QUICK CLOSING
☐ HANDWHEEL ☐ LEVER HANDLE
21 VESSEL ☐ PLAIN UNION ☐ SOLID SHANK
CONN. ☐ SPHERICAL UNION ____ OTHER
22 GAUGE ☐ PLAIN UNION ☐ PLAIN
CONN. ☐ SPHERICAL UNION ____ OTHER
23 BONNET ☐ SCREWED ☐ UNION ☐ BOLTED
24 SCREW ☐ INSIDE ☐ OUTSIDE
25 RENEWABLE SEAT ☐ YES ☐ NO
26 BALL CHECKS ☐ YES ☐ NO
27 PACKING ☐ MFR STD ____ OTHER
28 MFR MODEL NO. ____

NOTES:

REV.	QUAN.	TAG NO.	NO. OF SECT.	VISIBLE LENGTH	℄ CPLGS	MFR. MOD. NO.	OPER. PRESS	OPER. TEMP.	SERVICE	ACCES-SORIES	NOTES

REPRODUCTION OF STANDARD NO. 53.02-2

CONTROL VALVE SPECIFICATION SHEET

SPECIFICATION No.________

B/M NUMBER			
ITEM NUMBER			
TAG NUMBER			
INSTRUMENT OR CONTROL SERVICE			
PROCESS CONDITIONS			
FLUID THROUGH VALVE			
CORROSIVE DUE TO			
INLET OP. TEMP., °F NORMAL			
MAXIMUM			
INLET OP. PRESS., PSIG, NORMAL			
MAXIMUM			
PRESSURE DROP, PSI, NORMAL			
MAXIMUM			
NORM. FLOW AT NORM. OP. TEMP., VAPOR			
LIQUID			
TOTAL			
MAX. FLOW AT NORM. OP. TEMP., VAPOR			
LIQUID			
TOTAL			
M.W. OF VAPOR			
SP. GR. OF LIQ. REL. TO WATER AT 60°F			
AT NORM. OP. TEMP.			
VIS. OF LIQUID, CPS @ °F			
CPS @ °F			
CONSTRUCTION			
CATALOG OR TYPE NO.			
NOMINAL SIZE, INCHES			
REDUCED AREA PORTS			
BODY CONNECTIONS & RATING			
FLANGE FACING			
LUBRICATOR REQUIRED			
MATERIALS: BODY			
PLUG			
SEAT			
TYPE OF PLUG			
TYPE OF STUFFING BOX			
ACCESSORIES: VALVE POSITIONER			
PILOT RANGE			
FINS REQUIRED			
ACTUATING FLUID			
VALVE TO: OPEN WITH INCREASE OF			
CLOSE WITH INCREASE OF			
ON ACTUATING FLUID FAILURE, VALVE TO:			

REMARKS:

REVISION	1	2	3	4	5	5	7	8	9	10	11
BY											
DATE											

MADE BY:	CHECKED BY:	W.N. NO.	PAGE
DATE:	DATE:	CONT. NO.	OF

ARTHUR G. McKEE & COMPANY — CLEVELAND, OHIO

FORM 459

DD FORM 2834 5-64 — PAGE 1 OF ___

Northern Natural Gas Company

VALVE OPERATOR SPECIFICATION SHEET

LOCATION ________ FILE NO. ________
ITEM NO. ________ JOB NO. ________ W.O. NO. ________
PREPARED BY ________ DATE ________
CHECKED BY ________ DATE ________
APPROVED BY ________ DATE ________
△ INDICATES REVISION ________

GENERAL INFORMATION

1 NUMBER OF UNITS REQUIRED ________

OPERATOR

2 ☐ ELECTRIC ☐ PNEUMATIC ☐ HYDRAULIC
☐ PNEUMATIC HYDRAULIC

3 MODEL NO. ________

4 TIME: FULL OPEN MIN ______ MAX ______
FULL CLOSE MIN ______ MAX ______

5 MINIMUM TORQUE OUTPUT W/FULL POWER GAS PRESSURE ________

VALVE

6 MANUFACTURE: ________

7 FIGURE NO. ______ SIZE ______ LOT ______

8 DESIGN PRESSURE ________ PSI

9 MAX DIFFERENTIAL ________ PSI

10 VALVE STEM POSITION:
☐ HORIZONTAL ☐ VERTICAL

11 GEAR RATIO ________

12 TORQUE REQUIREMENT ________

OPERATOR

ELECTRIC

13 POWER SUPPLY:
VOLTAGE ______ PHASE ______ FREQ. ______

14 MOTOR ENCLOSURE:
☐ CLASS I GP. D ☐ GENERAL PURPOSE
☐ OTHER ________
☐ BREATHER AND DRAIN

15 DUTY RATING:
☐ 5 MIN. ☐ 15 MIN. ☐ CONTINUOUS

16 BEARINGS TYPE ________

PNEUMATIC OR HYDRAULIC

17 POWER GAS:
☐ AIR ☐ GAS

18 PRESSURE ________ PSI

19 DRIVER MOTOR:
☐ TURBINE ☐ VANE ☐ PISTON
☐ VANE MOTOR W/GEARING

CONTROL

PILOT VALVE

20 ELECTRIC:
☐ CLASS I GP. D ☐ GENERAL PURPOSE
☐ OTHER ________

21 VALVE ACTUATION:
PILOT VALVE
☐ ACTUATES ☐ DEACTUATES

22 VOLTAGE ________

23 PNEUMATIC:
PILOT PRESSURE FOR VALVE ACTUATION
☐ INCREASES ☐ DECREASES

24 CONTROL SYSTEM:
☐ ONE VALVE ☐ TWO VALVES

25 ☐ EMERGENCY SHUT DOWN SYSTEM

26 ESD OPERATION WHEN PILOT:
☐ PRESSURIZED ☐ DE-PRESSURIZED

27 SYSTEM PRESSURE ________

28 VALVE ON SHUTDOWN:
☐ TO OPEN ☐ TO CLOSE

29 ESD TO OVER-RIDE: ☐ REMOTE OPERATION
☐ LOCAL

MANUAL OPERATION

30 OVER-RIDE:
☐ MANUAL ☐ AUTOMATIC

PUSHBUTTON

31 CONTACT: ☐ MOMENTARY ☐ MAINTAINED

32 MOUNTED: ☐ REMOTE ☐ INTEGRALLY

ELECTRIC CONTACTER

33 MOUNTING: ☐ INTEGRAL

34 ENCLOSURE: ☐ CLASS 1 GROUP D
☐ GENERAL PURPOSE
☐ OTHER ________

35 HEATERS: ☐ OVERLOAD HEATERS
☐ SPACE HEATERS
☐ WITH THERMOSTAT

ACCESSORIES

36 LIMIT LIGHTS: ☐ LOCAL ☐ INTG. MTD.

37 ☐ REMOTE SLIDE WIRE TRANSMITTER

38 ☐ REMOTE POSITIVE INDICATOR W/PWR. SUP.

39 ☐ INTEGRALLY MOUNTED REVERSING STARTER

40 ☐ MANUAL HAND VALVE:

41 ___ NO. OF RELAYS ☐ TIME DELAY ☐ AUXILIARY

42 LIMIT SWITCHES: ☐ ELECTRIC ☐ PNEUMATIC
☐ ADJUSTABLE POSITION
☐ GEAR DRIVE

________ TYPE OF ELECT. SWITCH
______ NORMALLY OPEN IN EA. DIR. OF TRAVEL
______ NORMALLY CLOSED IN EA. DIR. OF TRAVEL

43 ADJ. TORQUE SW: ☐ OPENED END ☐ CLOSED END

44 ENCLOSURE: ☐ EXPLOSION PROOF
☐ GENERAL PURPOSE

45 ☐ MANUAL HAND PUMP

46 ☐ GAS HYDRAULIC TANKS

47 ☐ ALL NECESSARY ADAPTERS TO FIT UNIT TO VALVE

48 ☐ POWER STORAGE TANKS

MISCELLANEOUS

49 ☐ ALL WIRING BROUGHT TO EXPLOSION PROOF JUNCTION BOX

50 ☐ ALL TUBING BROUGHT TO BULKHEAD FITTINGS

REPRODUCTION OF STANDARD NO 53.02-3

ARTHUR G. McKEE & COMPANY

RELIEF VALVES

	REVISION DATE			
	ITEM No.			
	TAG No.			
	VALVE SERVICE			
	MANUFACTURER & TYPE No.			
	NUMBER VALVES REQ'D.			
	NORMAL SYSTEM PRESS., PSIG.			
	NORMAL SYSTEM TEMP., °F.			
	GOVERNING UPSET CONDITION			
	ACCUMULATION, PERCENT			
VALVE SIZING CONDITIONS	FLOWING FLUID			
	FLOW QUANTITY, GPM, SCFM, #/HR.			
	FLOW SP. GR. OR MW.			
	FLOW TEMP. °F.			
	FLOW VISCOSITY			
	SET PRESSURE, PSIG.			
	ACCUM. INLET PRESSURE, PSI. ABS.			
	BACK PRESS., PSIG. (STATIC-VALVE CL.)			
	REQ'D. ORIFICE AREA, SQ. IN.			
CONSTRUCTION — SIZE/RATING	NOMINAL SIZE, INS.			
	ORIFICE AREA, SQ. IN./VALVE			
	TOTAL ACTUAL AREA, SQ. IN.			
	BODY CONN. & RATING—INLET			
	—OUTLET			
MAT'L	BODY & BONNET			
	TRIM			
	SPRING			
ACCESSORIES	RADIATING BONNET			
	STYLE TOP			
	LIFTING GEAR—REG./PACKED			
	TEST ROD			

MADE BY___ DATE___ APP. BY___ DATE___

MADE BY___ DATE___ APP. BY___ DATE___

REMARKS:

WN No. PAGE OF CONT. No. B/M No.

FORM 495 500 3-56

CENTRIFUGAL PUMP DATA SHEET

FOR	JOB NO.
PLANT	PUMP NO.
LOCATION	NO. UNITS
SERVICE	MOTOR DRIVE
VENDOR - SIZE & MODEL	TURBINE DRIVE
TYPE	SERIAL NO.

OPERATING CONDITIONS

FLUID	NORMAL GPM @ PT	NPSH AVAIL FT. FLUID
PUMP TEMP. °F	DESIGN GPM @ PT	NPSH REQ'D. FT. WATER
SP. GR. @ P.T.	DISCHARGE PRESS.-PSIG	NPSH REQ'D. FT. FLUID
VAP. PR @ PT - PSIA	SUCTION PRESS - PSIG	
VISC @ PT - SSU	Δ P PSI FT.	

DESIGN

CASE SPLIT	RADIAL BRG TYPE	SHAFT - MAX. DIAM. " AT CPLG "
SUPPORT	THRUST BRG TYPE	WEAR RING CLEARANCE
IMPELLER TYPE	BRG LUBE	STUFFING BOX ID " DEPTH "
CORROSSION ALLOW.	VISIBLE LUBRICATOR	BASE PLATE
CW PIPING FURNISH BY		CPLG GUARD

COOLING WATER °F. @	PSIG		NOZZLES	POSITION	SIZE	RATING			
	REQ'D	GPM					TESTS	REQ'D.	WITNESSED
C.W. JACKETS			SUCT				SHOP INSPECTION		
RAD BRG			DISCH.				HYDROSTATIC		
THRUST BRG			VENT				PERFORMANCE		
PACKING			DRAIN				NPSH		
PEDESTAL			SEAL						

MATERIALS

CASE	PACKING	THROAT BUSHING
IMPELLER	No & SIZE RING	CASING GASKET
CASE WRG RING	MECH. SEAL	GLAND
IMP WRG RING	ROTATING FACE	BASE PLATE
SHAFT	STATIONARY FACE	COUPLING
SHAFT SLEEVE	SEAL RING	
LANTERN RING	AUX. GLAND	

PERFORMANCE

NO. STAGES	ROTATION FACING PUMP CPLG			WEIGHTS:
MAX. IMPELLER - INCHES	SHUTOFF HD W/DESIGN IMP PSI			PUMP & CPLG. #
BID IMPELLER - INCHES	MAX. W.P. PSIG @ °F			BASE #
MIN. IMPELLER - INCHES	HYDRO TEST PRESSURE PSIG			TOTAL #
EYE AREA - SQ. INCHES	FURNISHED BY	VENDOR	R.M.PARSONS	SPACE REQUIREMENTS : WITH DRIVER
SPEED RPM	PUMP			OVERALL LENGTH INCHES
DESIGN EFF %	BASE			OVERALL WIDTH INCHES
DESIGN BHP	COUPLING			OUTLINE DRG #
MAX. BHP, BID IMP.	CPLG GUARD			CROSS-SECTION #
MAX. BHP, MAX IMP.	MOTOR			PERFORMANCE CURVE #
	TURBINE			GFP #

DRIVER

MOTOR HP RPM	TURBINE HP RPM	
VOLTS PHASE CYCLE	STEAM PSIG °F SUPER HT.	
ENCL.	EXHAUST PSIG WR#/HP-HR	
FRAME No WT. #	TOTAL STEAM #/HR. WT. #	
SPEC. No	SPEC. No	

THE RALPH M. PARSONS COMPANY
LOS ANGELES

No	DATE	REVISIONS	BY	CH'K.	APP'D.	APP'D.

ED-325-1 PRINTED IN U.S.A.

PAGES PAGE

C F BRAUN & CO

SPECIFICATION

CENTRIFUGAL PUMP

SERVICE		DRIVE	QTY	SERIAL	ITEM
	OPER				
	SPARE				

GENERAL SPEC	MFR
PUMP CLASS	SIZE AND TYPE

PROCESS-DATA

LIQUID	GPM HOT	SUBMERGENCE AVAIL FT
PUMP-TEMP F	DISCH PRESS PSIG	SUBMERGENCE REQD FT
SP GR @ PT	SUCT PRESS PSIG	TOTAL SUCT LIFT FT
VAP PR @ PT PSIA	DIFF HEAD PSI	NPSH AVAIL FT
VIS @ PT SSU	DIFF HEAD FT	NPSH REQD FT WATER
CORR DUE TO	MAX SUCT-PRESS PSIG	NPSH REQD FT OIL
SOLIDS	IF CRITICAL, MIN GPM	

REQUIREMENTS

AXIS			MIN CASE CORR-ALLOW				CPLG GUARD		
SPLIT			SUPPORT				BASEPLATE		
IMPELLER			BRG LUBE				DRIVERS BY		
COOLING WATER F@ PSIG			FLUSH	REQD	FLUID	GPM	DRIVERS MTD BY		
C-W JACKETS	REQD	GPM	CAGE RING				TESTS	REQD	WITNESS
RADIAL BEARING			WEAR RING				SHOP INSPECTION		
THRUST BEARING			GLAND				HYDROSTATIC		
PACKING BOX			THROAT BUSH				RUNNING		
PEDESTAL			SEAL				NPSH		

MATERIALS

CASE	PACKING BOX	DIFFUSERS
INNER CASE	THROAT-BUSH OR -RING	BAL DRUM
IMPELLER	PACKING	BAL SLEEVE
CASE WRG-RING	CAGE RING	BASEPLATE
IMP WRG-RING	GLAND AND BUSH	CASE STUDS
SHAFT	MECHANICAL SEAL	GLAND STUDS
SHAFT SLEEVE	ROTATING FACE	CASE GASKET
SPACER SLEEVE	STATIONARY FACE	DISCH COLUMN
CASE BUSH	O-RING	SURFACE HEAD
DIAPHRAGMS	AUX GLAND	SUCT BELL

DESIGN

NO STAGES	ROTATION FACING CPLG				CPLG
MAX IMPELLER	SEAL				CPLG SIZE
BID IMPELLER	PACKING SIZE				RADIAL BRG
MIN IMPELLER	NO OF PACKING RINGS				THRUST BRG
EYE-AREA SQ IN	SHUTOFF HD. IMP. PSI				COLUMN BRG
SPEED RPM	MAX WP PSIG @ F				DESIGN THRUST, LB
DESIGN EFF	HYDRO TEST-PRESS PSIG				MAX THRUST, LB
DESIGN BHP	NOZZLE	ORIENT	SIZE	RATING	THRUST DIRECTION
MAX BHP, BID IMPELLER	SUCT				WT W/O DRIVER, LB
MAX BHP, MAX IMPELLER	DISCH				OUTLINE
MOTOR HP	VENT				CROSS SECTION
TURBINE BHP	DRAIN				PERFORMANCE

REMARKS SUPPLIER FILLS IN ALL DATA MARKED X

THE C.W. NOFSINGER COMPANY
KANSAS CITY, MISSOURI

RECIPROCATING PUMP SPECIFICATION SHEET

SHEET NO. ____
ITEM NO. ____
CWN JOB NO. ____
CUST. JOB NO. ____

CUSTOMER ____ DATE ____
LOCATION ____ BY ____
SERVICE ____ NO. REQUIRED ____ REVISED ____
MFR. ____ MODEL NO. ____ SERIAL NO. ____

OPERATING CONDITIONS			PERFORMANCE	
1	LIQUID	GPM AT PT: NORM	41	CURVE NO.
2	PT. °F	RATED	42	OUTLINE DWG. NO.
3	SP GR AT PT.	DISCH PRESS, PSIG	43	PUMP SPEED, RPM
4	VAP PRESS AT PT	SUCT PRESS, PSIG	44	PISTON: STROKES/MIN FPS
5	VIS AT PT: SSU - CP	DIFF PRESS, PSI	45	EFF AT DESIGN
6	CORR/EROS PROPERTIES	DIFF HEAD, FT	46	BHP: DES MAX
7		NPSHA, FT	47	SHUT-OFF HEAD, FT

CONSTRUCTION

8	ARRANGEMENT: HORIZ - VERT				
9	DIRECT - SINGLE - DOUBLE ACTING				
10	SIMPLEX - DUPLEX - TRIPLEX POW. FRAME: PISTON - PLUNGER				
11	CASE DES PRESS, PSIG SWP, PSIG				
12	TAPPED CONNECTIONS: VENT - DRAIN - GAUGE				
13	NOZZLES	SIZE	RATING	FACE	LOCATION
14	SUCTION				
15	DISCHARGE				
16					
17	PUMP SIZE, IN: BORE STROKE				
18	CYLINDER LINER				
19	VALVES(LIQ) TYPE NUMBER AREA, SQ. IN.				
20	SUCTION				
21	DISCHARGE				
22	BEARINGS: RADIAL THRUST				
23	LUBE: OIL - GREASE OILER: YES - NO				
24	CPLG: MFR MODEL GUARD: YES - NO				
25	PACKING BASE: MFR. STD. - OTHER				

48	RELIEF VA: INT - EXT
49	SETTING, PSIG
50	CLG WATER, GPM
51	BEARINGS JACKET
52	PEDESTAL
53	PRESS, PSIG T, °F
54	FLUSH
55	JACKETED: YES - NO
56	STM PRESS, PSIG T, °F
57	

SHOP TESTS

58	SHOP INSPECTION
59	PERFORMANCE
60	HYDROSTATIC, PSIG/°F
61	WITNESS
62	
63	WEIGHTS:
64	PUMP DRIVER
65	REDUCER SHPG

LIQUID CYLINDER MATERIALS			DRIVER CYLINDER MATERIALS	
26	CYLINDER	VALVES	CYLINDER	VALVES
27	CYL LINER	PISTON ROD	PISTON	GASKETS
28	LANTERN RINGS	GASKETS	PISTON RINGS	
29	PISTON - PLUNGER		PISTON ROD	

MOTOR DRIVER			TURBINE/PISTON DRIVER	
30	SUPPLIED BY	MOUNTED BY	SUPPLIED BY	MOUNTED BY
31	MFR	TYPE	MFR	TYPE
32	ENCLOSURE	RPM	HP RPM	WATER RATE, LB/HP-HR
33	HP	SF	INLET STM PRESS, PSIG: NORM	MAX
34	FRAME	INSULATION	INLET STM TEMP, °F: NORM	MAX
35	VOLTS/PH/HZ	TEMP RISE, °F	EXHAUST STM PRESS, PSIG: NORM	OTHER
36	BEARINGS	LUBE	BEARINGS	LUBE
37			NOZZLES SIZE RATING FACE	LOCATION
38	SPEED REDUCER: INTEGRAL - SEPARATE		INLET	
39	MFR	MODEL	EXHAUST	
40	RATIO	CLASS	API 611: YES - NO	SEPARATE DRIVER SPEC: YES - NO

VENDOR TO SUPPLY INFORMATION MARKED ____

FORM 21-A3. 4-76

ENG EQUIP 46 (5-77) KELLOGG

ROTARY PUMP DATA SHEET

FOR
LOCATION
ITEM NO
SERVICE
NO. REQUIRED ____ MOTOR DRIVE ____ TURBINE DRIVE ____

JOB NO
SHEETNO.
PAGE ____ OF ____

OPERATING CONDITIONS

PRODUCT HANDLED	DISCHARGE PRESSURE ____ PSIG
CORROSIVE DUE TO	SUCTION PRESSURE ____ PSIG
PUMPING TEMPERATURE ____ DEG F VAPOR PRESS ____ PSIA	DIFFERENTIAL PRESSURE ____ PSI
GRAVITY AT 60 DEG F SP GR ____ API GR	DESIGN HEAD
SP GR AT SUC COND ____ SP GR AT DIS COND	NPSH* AVAIL ____ FT NPSH REQUIRED ____ FT
VISCOSITY ____ AT ____ DEG F	NPSH INCLUDES ____ FT FT STATIC HEAD — LIFT
CAPACITY — G P D AT 60 DEG F NORMAL	STEAM PRESSURE ____ PSIG
DESIGN	% SATUR OR TOTAL TEMP
G P M AT PUMP TEMP NORMAL	EXHAUST STEAM PRESSURE
DESIGN	MOTOR CHARACTERISTICS ____ VOLTS ____ PHASE ____ CYCLE

SPECIFICATIONS

PUMP SHALL BE FURNISHED IN ACCORDANCE WITH

MANUFACTURER	HYD H P —NORMAL ____ DESIGN
TYPE	PUMP EFF %—NORMAL ____ DESIGN
SIZE	BRAKE H P —NORMAL ____ DESIGN
RATED CAPACITY ____ G P M AT ____ RPM	B H P AT RATED CAPACITY
ROTATION—COUNTER—CLOCKWISE FACING COUPLING END	MOTOR H P ____ TURBINE H P
SUC FLANGE SIZE	DRIVER (S) TO BE FURNISHED BY
DIS FLANGE SIZE	PUMP BASE
CASE DESIGN PRESS	DRIVER BASE
RELIEF VALVE TO BE FURN BY	COUPLING ____ WITH GUARD
MAX PEL VALVE SET TO PREVENT MOTOR OVERLOAD ____ PSIG	

MATERIALS AND DETAILS

CASE ____ MIN THICK	LUBRICATION
END COVERS	PACKING BOX—TYPE ____ NO
FLANGES	SIZE—I D ____ O D ____ LENGTH
SHAFT	PACKING ARRANGEMENT
SHAFT SLEEVES	
ROTORS	SPARE PACKING
DRIVE GEARS	PACKING GLAND—TYPE
RADIAL BEARING	GLAND STUDS
THRUST BEARING	LANTERN RING

GENERAL INFORMATION

COOLING WATER ____ DEG F ____ G P M	INSULATION BY
PACKING BOXES ____ G P M	INSPECTION—HYDRO TEST
COOLING OIL LANTERN RINGS ____ DEG F ____ G P M	PERFORM TEST
FLUSHING OIL TO THROAT BUSHING ____ G P M	SPECIAL TESTS
COOLING WATER PIPING ____ SYSTEM BY	DIMEN PRINT NO ____ SERIAL NO
GLAND OIL PIPING BY	NET WEIGHT—PUMP AND BASE ____ LB
FLUSHING OIL PIPING BY	FOUNDATION BOLTS—TO BE FURNISHED BY KELLOGG
BYPASS CONTROL—TO BE FURNISHED BY KELLOGG	FLANGE STUDS—TO BE FURNISHED BY PUMP MFR

* NPSH AVAILABLE MEASURED TO TOP OF PUMP FOUNDATION

REMARKS

REVISION	NO	0	1	2	3	4	5	6	7	8	9	10	11	12
	DATE													
	BY													
	CHECK													
	APPRVD													

CHEMICAL PROCESS PLANTS DEPARTMENT
CORPORATE ENGINEERING

ROTARY PUMP DATA SHEET Sheet 1 of 2

REMARKS / APP'R'V / DATE / SUPRV / CHKD / BY / REV.

OPERATING CONDITIONS

FLUID PUMPED	CAPACITY NORMAL ____ GPM
FLUID CHARACTERISTICS	CAPACITY DESIGN ____ GPM
TEMP. MAX ____ °F MIN. ____ °F NORMAL ____ °F	TDH DESIGN ____ FT. FLUID
VISCOSITY AT MIN. P.T.	SUCTION PRESS. ____ DISCH. PRESS.
SPECIFIC GRAVITY AT MIN. P.T.	NPSH AVAILABLE ____ FT. FLUID
VAPOR PRESS. AT MAX. P.T.	HYDRAULIC HP

TYPE OF OPERATION: ☐ CONTINUOUS ☐ INTERMITTENT HRS./DAY
LOCATION: ☐ INDOORS ☐ OUTDOORS

MATERIALS

BODY	MECH. SEAL:
HEADS	ROTATING FACE
ROTORS	STATIONARY FACE
SHAFT	METAL PARTS
PACKING GLAND	RELIEF VALVE
LANTERN RING	CASING GASKET

CONSTRUCTION DATA

PUMP TYPE:
PACKING: TYPE ____ MFR.
MECH. SEAL: TYPE ____ MFR. ____ CODE
NOTES:
SEAL FLUID:
RELIEF VALVE: BUILT-IN ____ SEPARATE

MOTOR DRIVE DATA

ITEM NO. ____ MTD. BY
H.P. ____ RPM ____ FRAME
MFR. ____ CPLG. GUARD
TYPE ____ INSUL.
ENC'L: CLASS I-GROUP ____ DIVISION 1
OTHER
VOLTS/PHASE/Hz:
BEARINGS ____ LUBE
FULL LOAD AMPS ____ SF
LR AMPS

TESTS

SHOP TESTS	NOT WITNESSED	WITNESSED	CERT. DATA
Running Perf.			
NPSH			

HYDROSTATIC ____ PSIG

PROJ. ENG. / DES SUPRV. / CHECKED BY / DRAWN BY / DATE / INITIATOR

MANUFACTURER'S DATA

MFR.
MODEL NO. ____ RPM

NOZZLES	SIZE	ASA RATING	FACING	POSITION
SUCTION				
DISCH				

MAX. ALLOW. W.P. ____ PSIG TEMP. ____ °F
NPSH REQ'D. (WATER) FT.
DESIGN EFF. ____ DESIGN BHP
SHUT-OFF HEAD FT.
ROTATION FACING CP'LG. END
CP'LG. MFR.
TYPE

PUMP BEARINGS
PUMP LUBRICATION
PRESS. IN SEALING CHAMBER ____ PSIG
SHAFT DIA. IN SEALING CHAMBER ____ INCHES
WATER COOLING AVAIL. AT ____ PSIG ____ °F
BEARINGS ____ GPM
STUFF. BOX ____ GPM
PEDESTAL ____ GPM
TOTAL WATER REQ'D. ____ GPM
PACKING LUBRICATION
SEAL LUBRICATION

JOB NO. ____ PLANT LOCATION
PROJECT NO. ____ REF. DWG. NO.
TITLE
POSITION NO. ____ SERIES ____ STEP NO.

ISSUE DATE

SCALE
BLDG. NO.
DRAWING NO.

E 142810* (REV. 4/83)

CHEMICAL PROCESS PLANTS DEPARTMENT
CORPORATE ENGINEERING

ROTARY PUMP DATA SHEET — Sheet 2 of 2

MANUFACTURER'S DATA CONT'D.

WEIGHTS: PUMP ______ BASE ______ DRIVE ______
SEAL FLUSH ______
SEAL QUENCH & DRAIN ______
SPEED REDUCER: MFR. ______
TYPE ______
BASEPLATE ______

SPECIAL REQUIREMENTS & NOTES

1. VENDOR IS TO QUOTE MOST ENERGY-EFFICIENT PUMP-DRIVE COMBINATION AS ALTERNATE IF DIFFERENT THAN UNIT(S) OFFERED.
2. GROUNDING LUG TO BE PROVIDED IN MOTOR CONDUIT BOX.
3. MOTOR SHALL BE IN CONFORMANCE WITH FOLLOWING STANDARD PROCUREMENT SPECIFICATION:
   - ☐ E16 - 150P, PROCESS ELECTRICAL EQUIPMENT - LOW VOLTAGE MOTORS, ALTERNATING CURRENT.
   - ☐ E16 - 151P, PROCESS ELECTRICAL EQUIPMENT - MEDIUM VOLTAGE MOTORS, ALTERNATING CURRENT.
   - ☐ E16 - 160P, PROCESS ELECTRICAL EQUIPMENT - VARIABLE FREQUENCY AC MOTORS
4. MOTOR SHALL BE PROVIDED IN ACCORDANCE WITH DATA GIVEN ON STANDARD DATA SHEET FORM E242P1 ELECTRICAL ALTERNATING CURRENT MOTORS.

REV.	BY	CHKD	SUPRV.	DATE	APP'R'V	REMARKS

DATE	INITIATOR	DRAWN BY	CHECKED BY	DES SUPRV.	PROJ. ENG.

JOB NO. ______ PLANT LOCATION ______
PROJECT NO. ______ REF. DWG. NO. ______
TITLE ______
POSITION NO. ______ SERIES ______ STEP NO. ______
ISSUE DATE
SCALE
BLDG. NO.
DRAWING NO.

E 142B10* (REV. 4/83)

Rotary Pump—*(continued)*

MOTOR DRIVEN PROPORTIONING & RECIPROCATING PUMP DATA SHEET

KELLOGG

KG 0977 (4-81)

FOR: ______ JOB NO. ______
LOCATION: ______ ITEM NO. ______
SERVICE: ______ PAGE: ______ OF ______
NO. REQUIRED: ______
______ SPARES: ______

SERVICE CONDITIONS

FLUID PUMPED ______	DISCHARGE PRESSURE ______ PSIG
CORROSIVE DUE TO ______	INLET PRESSURE ______ PSIG
INLET TEMPERATURE ______	DIFFERENTIAL PRESSURE ______ PSI
SPECIFIC GRAVITY AT INLET TEMP. ______	VAPOR PRESSURE AT INLET TEMP. ______ PSIA
CAPACITY AT INLET TEMP NORMAL ______	NPSH* AVAILABLE AT TOP OF PUMP FOUNDATION ______ FT
CAPACITY AT INLET TEMP. DESIGN ______	NPSH REQUIRED AT TOP OF PUMP FOUNDATION ______ FT

MECHANICAL DATA (PUMPS TO BE FURNISHED IN ACCORDANCE WITH KELLOGG SPEC ______ DATED ______

MANUFACTURER ______	BRAKE KW NORMAL ______ DESIGN ______
SIZE AND TYPE ______	INLET SIZE AND RATING ______
NO. CYLINDERS ______ SPM ______	DISCH—SIZE AND RATING ______
LIQUID END DESIGN PRESSURE ______ PSIG	STEAM IN SIZE ______
STALLING PRESSURE ______ PSIG	EXHAUST SIZE ______

PROPORTIONING PUMP DATA	RECIPROCATING PUMP DATA
CAPACITY CYL. A MAX. ______ MIN ______	STEAM PRESSURE ______
CAPACITY CYL. B MAX. ______ MIN ______	% SATURATION OR TOTAL TEMP ______
TYPE CAPACITY ADJ ______	EXHAUST STEAM PRESSURE ______
RANGE OF ADJ ______	TOTAL STEAM LB/HR NORMAL ______ DESIGN ______
BORE CYL. A ______ CYL. B ______	VALVE TYPE ______
STROKE CYL. A ______ CYL. B ______	STEAM END PACKING ______

MATERIALS

CYLINDER ______	CASE ______
PLUNGER ______	LINER ______
VALVE ______	PISTON ______
VALVE SEAT ______	PISTON ROD ______
VALVE SPRINGS ______	PISTON RING ______

MECHANICAL DETAILS

COMMON BASE FOR PUMP AND MOTOR BY ______	PACKING BOX TYPE ______ NO. OF BOXES ______
COUPLING WITH GUARD BY ______	PACKING BOX SIZE ______ O.D. ______ LENGTH
LUBRICATOR ______	PACKING TYPE ______
REVOLUTION COUNTER ______	☐ 100% ☐ 200% SPARE PACKING (BY MFGR.)
INSULATION: STM END BY ______ LIQ END BY ______	☐ LANTERN RING IN CENTER WITH CONN. PLUGGED
☐ HYDROSTATIC TEST WITNESSED ______	☐ GLAND OIL ______ GPM
☐ PERFORMANCE TEST WITNESSED ______	☐ FLUSHING OIL ______ GPM
☐	☐ COOLING WATER ______ GPM

MOTOR DATA (MOTOR TO BE FURNISHED IN ACCORDANCE WITH KELLOGG SPEC ______ DATED ______)

MANUFACTURER ______	FRAME ______
KILOWATTS ______ RPM ______	RATED LOAD CURRENT ______
VOLTS ______ PHASE ______ CYCLE ______	LOCKED ROTOR CURRENT ______
ENCLOSURE ______	EFFICIENCY 4/4 ______ 3/4 ______ 1/2 ______
SERVICE FACTOR ______	POWER FACTOR 4/4 ______ 3/4 ______ 1/2 ______
TEMPERATURE RISE ______ C	SPEED REDUCER TYPE ______ RATIO ______

GENERAL

APPROX. BASE SIZE ______ TOTAL WEIGHT ______ KG ______
EQUIPMENT TO BE SUITABLE FOR OUTDOOR INSTALLATION
SERIAL NO. ______
* NPSH — NET POSITIVE SUCTION HEAD TO TOP OF PUMP FOUNDATION
MANUFACTURER SHALL SUBMIT THIS SHEET FULLY COMPLETED WITH HIS QUOTATION
REMARKS: ______

REVISION	NO	0	1	2	3	4	5	6	7	8	9	10	11	12
	DATE													
	BY													
	CHECK													
	APPRVD													

PACKAGED REFRIGERATION SYSTEM

Refrigerant: ______ Indoor/Outdoor; English/Metric Units
Mfr: ______ Model No. ______ Qty. ____
Design: ASME Code Section VIII, Division 1, ______ Edition and all Mandatory Addenda.
Stamped: Yes/No, TEMA ______ Edition ______
Dry Bulb Temp. _____ Wet Bulb Temp. _____ Altitude _____ Barometric Press. ______

Item No.	Service	Refrig. Temp. (In/Out)	Refrig. Press.(In/Out)	Duty
	Condenser	/	/	
	Evaporator	/	/	
	Oil Cooler	/	/	
		/	/	

Cooling Water: Supply Press. ______ Maximum Δ Pressure ______
Temp. In/Out ______ Total Flowrate ______
Compressor: Item No. ______ Type ______ Qty. ______
Maximum Capacity ______ @ ______ RPM ______ BHP
Condenser (TEMA ___): Item No. ______ Type ______ Size ______ Qty. ______
Design Press. Shell ______ Tubes ______ Mat'l.: Shell ______ Tubes ______
Cooling Medium: Flow Inlet ______ Outlet ______ Δ Press. ______
Receiver: Item No. ______ Type ______ Size ______ Mat'l. ______ Qty. ____
Design Press./Temp. ______ / ______ Oper. Press./Temp. ______ / ______
Knock-Out Drum: Item No. ______ Type ______ Size ______ Mat'l. ______ Qty. ____
Design Press./Temp. ______ / ______ Oper. Press./Temp. ______ / ______
Sub-Cooler: Yes/No Item No. ______ Type ______ Size ______ Mat'l. ______ Qty. ____
Design Press/Temp. ______ / ______ Oper. Press./Temp. ______ / ______
Oil Separator: Item No. ______ Type ______ Size ______ Mat'l. ______ Qty. ____
Design Press./Temp. ______ / ______ Oper. Press./Temp. ______ / ______
Oil Heater: Qty. ______ Volts ______ Phase ______ Hertz ______ Watts ______
Oil Cooler (TEMA ___): Item No. ______ Type ______ Size ______ Qty. ______
Design Press.: Shell ______ Tube ______ Mat'l.: Shell ______ Tube ______
Cooling Medium ______ Flow ______ Maximum Δ Press. ______
Oil Filters: Item No. ______ Type ______ Size ______ Micron Size ______ Qty. ____
Design Press. ______ Oper. Press./Temp. ______ / ______ Mat'l. ______
Flow Rate: Normal ______ Maximum Δ Press. ______
Oil Pump: Mfr. ______ Model No. ______ Type ______ Qty. ______
Press.: Discharge ______ Flow Max. Cap. ____ @ ____ BHP; Mat'l: ______
Purger: Mfr. ______ Model No. ______ Type ______ Qty. ______
Size ______ Mat'l. ______
Motor: Compressor (less than 200 HP) ______ Oil Pump
Mfr./Mod. No./Qty. ______ / ______ / ______ ______ / ______ / ______
Volts/Phase/Hertz ______ / ______ / ______ ______ / ______ / ______
Enclosure/Insulation ______ / ______ ______ / ______
Area Class. Class ______ Group ______ Division ______ /Non-Hazardous.
Temp. Rise ______ °C above ______ °C by ______ @ ____ S.F.; Temp. Code ______
Space Heater Required: Yes/No: Heater; Volts ______ Phase ____ Ø ____ Hertz ______

Item No.	HP/KW	Current: Full Load	Current: Locked Rotor	Effy. 4/4	Effy. 3/4	P.F. 4/4	P.F. 3/4	Full Load RPM	Frame No.	Brg. Lub.	Weight LB/KG	Mtg. H/V	Allow-Thrust Up/Down LB/KG	Space Htr. Watts	S.F.

Remarks: ______

REVISIONS NO.	DATE	ENGINEER	CHK/REVIEW	APPROVE
6				
5				
4				
3				
2				
1				
0				

MW KELLOGG | MANAGING OFFICE | ENGINEERING OFFICE

EQUIPMENT DATA SHEET

CLIENT ______
LOCATION ______ JOB NO. ______
ITEM ______
ITEM NO. ______ SPEC. NO. ______
REQ'N. NO. ______ PAGE ____ OF ____

PACKAGED REFRIGERATION SYSTEM

Oper. Parameters	LOW STAGE				HIGH STAGE				Units
Capacity %									
Inlet Flow									
Inlet Temp.									
Inlet Press.									
Outlet Temp.									
Outlet Press.									
BHP									
Pockets/Valve Open									

RECIPROCATING COMPRESSOR

Mfr. ______ Model No. ______ Qty. ______

Cylinder Data:	Low Stg.	High Stg.		Low Stg.	High Stg.
Cylinder Per Stg.			Type of Valves		
Cylinder Type			Piston Speed Norm/Max.		
Cyl. Liner Yes/No			Rod Diameter		
Cyl. Liner Wet/Dry			Rod Load. Tension Norm/Max.		
Bore/Stroke (In.)	/	/	Rod Load. Comp. Norm/Max.		
Piston Displ.			Cyl. Press/Temp. Max.		
Vol. Eff. %			Recommended R.V. Press.		
No. In/Out	/	/	Hydro. Test Pressure		

Mat'l of Const:	Low Stg.	High Stg.		Low Stg.	High Stg.
Cylinders			Valve Seats		
Cylinder Liners			Valve Springs		
Pistons			Valve Stops		
Piston Rings			Valve Plates		
Piston Rods					

Cooling Water:	Comp.	Cyl.		Comp.	Cyl.
Flow Rate			Press. Drop		
Temp. In/Out:	/		Max. Press.		

Packing: Cooling Water; Yes/No. Vented to ______
Capacity Control: Variable Speed ____ % to ____ %. Mfr. Std. Control: Yes/No;
Start/Stop; Yes/No, 2/3/5 Step Control. Suction Valve Lifters: Manual/Automatic
Air/Power Failure, Compressor shall Load/Unload.

Accessories:	Gear		Low Speed CPLG	Belt Drive
Mfr.		Mfr.		
Model No. Size		Model No.		
H.P. Rating/AGMA S.F.	/	Type		
Ratio/Efficiency	/	No. Required		

Separate Moisture Separators with Traps: By Compressor Mfr. ______
Frame Lubrication: Splash System/Pressure System w/Comp. Shaft Driven Oil Pump/Elect.
Driven Oil Pump/Hand Operated Pump for Starting.
System Oil Capacity ______
Electric Heater w/Thermostat: Setting ______ No. Heaters ______ Watts ______
Cylinder Lubrication: Lubricator to be driven by Comp. Shaft/Electric Motor
System Oil Capacity ______ No. of Compartments ______
Electric Heater w/Thermostat: Setting ______ No. Heaters ______ Watts ______
Remarks: ______

REVISIONS NO.	DATE	ENGINEER	CHK/REVIEW	APPROVE
6				
5				
4				
3				
2				
1				
0				

MW KELLOGG | MANAGING OFFICE | ENGINEERING OFFICE

EQUIPMENT DATA SHEET

CLIENT ______
LOCATION ______ JOB NO. ______
ITEM ______
ITEM NO. ______ SPEC. NO. ______
REQ'N. NO. ______ PAGE ____ OF ____

PACKAGED REFRIGERATION SYSTEM (CONT'D)

POSITIVE DISPLACEMENT ROTARY COMPRESSOR

Mfr.: ______ Model No. ______
Speed: Max. Allow. RPM ______ Critical RPM 1st ______ 2nd ______
Casing: Casing split: ______ Thickness ______ Corr. Allow. ______
Max. Working Press. ______. Max. Rated Press. ______ Temp.: Max. Work ___ Min. Oper. ___
Test Pressure Hydro. ______
Rotor: Diameter ______ Type ______ Material ______
Type Fabrication: ______ No. of Lobes: Male ______ Female ______
Rotor Length to Dia. Ratio (L/D) ______ Rotor Clearance ______ Max. Deflection ______
Shaft: Dia. @ Rotor ______ Dia. @ Coupling ______ Shaft End: Tapered/Cylindrical
Sleeves at Shaft Seals ______ Shaft Seal Type ______ Seal Sys. Type ______
Inter Oil Leakage Guarantee (Vol./Day/Seal) ______ Type Buffer Gas ______
Buffer Gas Flow (Per Seal) ______
Normal ______ #/Mininum ______ Press. ______
Maximum ______ #/Minimum ______ Press. ______
Bearings: Housing Type (Separate/Integral) ______ Split ______
Radial Bearing Type ______ Span ______ Area ______ Loading: Act. ______ Allow. ______
Thrust Bearing Type ______ Area ______ Loading: Act. ______ Allow. ______
Gas Loading ______ CPLG. Slip Load ______ CPLG. Coeff. Frict. ______
CPLG. Gear Pitch Dia. ______ Balance Piston Compensating Load ______
Timing Gears: Size ______ Type ______
Couplings: Driver-Comp./Gear Gear/Comp. ______ Driver-Comp./Gear Gear/Comp.
Make ______ ______ Limited End Float Req'd. ______ ______
Model ______ ______ Idling Adaptor Req'd. ______ ______
Lubrication ______ ______ CPLG. Rating (HP/100 RPM) ______ ______
Mt. CPLG. Halves ______ ______ Keyed(1) or (2) or Hydr. CPLG. ______ ______
Spacers Required ______ ______
Material of Construction: Casing ______ Rotor ______ Shaft ______
Shaft Sleeves ______ Timing Gear ______ Bearings ______

INSPECTION AND TEST

Mfr. Standard Shop Test/Running Test with Shop Driver/Lube and Seal System/Hydrostatic/
Mechanical Run/Vibration-Axial Displacement/Noise Level/Consoles/Other: ______

Purchaser will Witness: ______

ACCESSORIES

Relief Valves/Pulsation Dampeners (Volume Bottles)/Inter-Connecting Piping/Initial
Charges (Oil, Refrigerant, Etc.) ______
Remarks: ______

REVISIONS NO.	DATE	ENGINEER	CHK/REVIEW	APPROVE
6				
5				
4				
3				
2				
1				
0				

KELLOGG — MANAGING OFFICE — ENGINEERING OFFICE

EQUIPMENT DATA SHEET

CLIENT ______
LOCATION ______ JOB NO. ______
ITEM ______
ITEM NO. ______ SPEC. NO. ______
REQ'N. NO. ______ PAGE ___ OF ___

PACKAGED REFRIGERATION SYSTEM (CONT'D)

The vendor shall furnish as a minimum the equipment and components shown in the sketch and as called for on the data sheets.

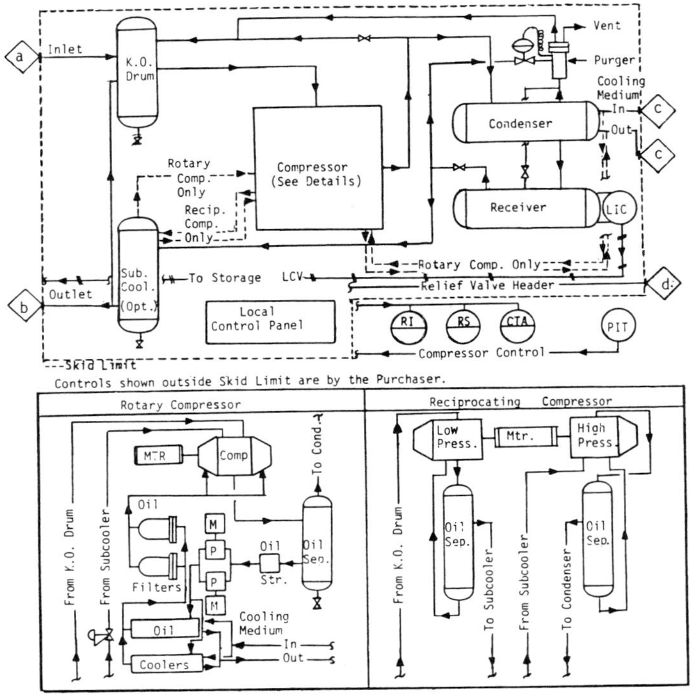

Sym ◇	Size	Rating	Facing
a			
b			
c			
d			

Dimensions	Length	Width	Height
Skid No. 1			

Weights	LBS/KG
Erection	
Maintenance	
Operating	
Shipping	

REVISIONS NO.	DATE	ENGINEER	CHK/REVIEW	APPROVE
6				
5				
4				
3				
2				
1				
0				

KELLOGG — MANAGING OFFICE — ENGINEERING OFFICE

EQUIPMENT DATA SHEET

CLIENT ______
LOCATION ______ JOB NO. ______
ITEM ______
ITEM NO. ______ SPEC. NO. ______
REQ'N. NO. ______ PAGE ___ OF ___

Packaged Refrigeration System—*(continued)*

PACKAGED REFRIGERATION SYSTEM (CONT'D)

INSTRUMENTATION AND CONTROLS

Panel: Freestanding/Other ______ Covered/Uncovered

Electrical Instruments: NEMA ______ Enclosures

Panel: Vibration Isolators/Strip Heaters/Purge/Other ______

Panels Required: Local/Lube Oil/Other______

MINIMUM DISPLAY ON LOCAL PANEL	MINIMUM NO. ALARMS AND SHUTDOWN	ALARM	SHUTDOWN
Lube Oil Pressure Gauge	Lube Oil Low Pressure	*	*
Lube Oil Temperature Gauge	Lube Oil High Pressure		
Compressor Suction Pressure Gauge	Compressor Low Suction Press.		
Compressor Suction Temperature Gauge	Compressor High Suction Press.	*	
Compressor Discharge Pressure Gauge	Compressor High Disch. Press.	*	
Compressor Discharge Temperature Gauge	System High Discharge Temp.	*	
	K.O. Drum High Level	*	
	K.O. Drum High-High Level		*
	Sub-Cooler High Level	*	
	Sub-Cooler High-High Level		*
	High Water Temperature	*	
	Receiver High Level	*	

Remarks:

1) Shutdown switches shall incorporate time delay relays, where required, to permit starting-up the system.
2) Packaged refrigeration system shall operate completely automatic after start-up.
3) Manual/automatic selector switch shall be provided, by the vendor, for manual start/stop of each refrigeration unit.
4) Condenser cooling water valve shall "fail open" and provide no less than minimum rated flow.
5) Liquid sub-cooler level control valve shall "fail open" on instrument air failure.
6) All control valves, furnished by the vendor, shall have blocks and by-passes.
7) A pair of "N.O." contacts shall be provided, by vendor, for purchasers shutdown alarm to be activated only when the system malfunctions.
8) A horn shall be mounted on the unit and wired to the shutdown circuit.
9) Copper, brass, bronze, aluminum and aluminum alloys shall not be used where exposed to ammonia, ammonia vapors, ammonia solutions or the general plant atmosphere, without the specific approval of M.W. Kellogg Engineering.
10) Vendor shall supply a completely piped and wired skid mounted packaged refrigeration system, which shall include but not be limited to the following: all components, piping, valves, instruments, alarms, shutdowns, control, control panels, special tools and equipment necessary to operate the unit at its rated capacity.
11) All relief valves shall be routed to the common manifold header, (if required).
12) All required utilities shall be piped to the skid limit for purchaser connections.

REVISIONS NO	DATE	ENGINEER	CHK/REVIEW	APPROVE
6				
5				
4				
3				
2				
1				
0				

KELLOGG | MANAGING OFFICE | ENGINEERING OFFICE

EQUIPMENT DATA SHEET

CLIENT ______

LOCATION ______ JOB NO. ______

ITEM ______

ITEM NO. ______ SPEC. NO. ______

REQ'N. NO. ______ PAGE ___ OF ___

Packaged Refrigeration System—*(continued)*

PROCESS DESIGN SECTION CORPORATE ENGINEERING — VIBRATING FEEDER DATA SHEET

REMARKS | REV | BY | CHKD | SUPRV. | DATE | APP'R'V

FEED MATERIAL CHARACTERISTICS

FEED MATERIAL: ______ WEIGHT % MOISTURE: ______

BULK DENSITY: ______ LBS./CU. FT. TEMPERATURE ______ °F

PARTICLE SIZE: AVERAGE: ______ ABRASIVENESS: ______

MAXIMUM: ______ TENDENCY TO PACK: ______

PRODUCT DESCRIPTION: ______ DESCRIPTION: ______

OPERATING CONDITIONS

FEED RATE – LBS/HR. NORMAL ______ LOCATION: INDOOR ☐ OUTDOOR ☐

MAX. ______ MAX. ROOM TEMP: ______ °F

REQ'D. WIDTH OF FEED ______ INLET TO OUTLET LENGTH: ______

REQ'D. DEPTH OF FEED ______ ATMOSPHERIC CONDITIONS: DUSTY ☐

MODE OF OPERATION ______ CORROSIVE ☐ EXPLOSIVE ☐

OTHER ______

DESIGN DATA

TROUGH TYPE: ______ COVER CLAMPS: ______

LENGTH/WIDTH/DEPTH ______ FREQUENCY – CPM: ______

THICKNESS: ______ SPEED OF MAT'L. CONVEYED–FPM: ______

COVER: REQ'D. ______ NATURAL FREQUENCY: REQ'D: ______

TYPE ______ SPRINGS: ______

MOUNTING: FLOOR ____ SUSPENSION ______ AMPLIFICATION ______

INCLINE ____ DECLINE ______ ISOLATION ______

CONTROL ______ HANGERS– TYPE: ______

BASE ______ NO. ______

MATERIALS OF CONSTRUCTION

TROUGH ______ SPRINGS – AMPLIFICATION ______

COVER ______ ISOLATION ______

BASE ______ DRIVE COVER ______

GASKETS ______

MOTOR DRIVE DATA

ITEM NO. ______ MTD. BY ______ BEARINGS ______ LUBE ______

MFG. ______ FULL LOAD AMPS ______ SF ______

TYPE ______ FRAME ______ L.R. AMPS ______

HP ______ RPM ______

VOLTS/PHASE/CYCLE ___/___/___ MOUNTING: BOTTOM ☐ OVERHEAD ☐

MISCELLANEOUS

TOTAL WEIGHT ______ VENDOR DWG. NO. ______

PAINTING ______ FEEDER SERIAL NO. ______

GALVANIZING ______

FINISH REQ'D. ______

DES SUPRV. | CHECKED BY | DRAWN BY | INITIATOR

JOB NO. ______ PLANT LOCATION ______

PROJECT NO. ______ REF. DWG. NO. ______

TITLE ______

POSITION NO. ______ SERIES ______ STEP NO. ______

ISSUE DATE | SCALE | BLDG. NO. | DRAWING NO.

E227B43

PROCESS DESIGN SECTION
CORPORATE ENGINEERING

SCREW CONVEYOR
DATA SHEET

SHEET 1 OF 2

REMARKS | APP'R'V | DATE | SUPRV | CHKD | BY | REV

FEED MATERIAL CHARACTERISTICS

FEED MATERIAL ______ ABRASIVENESS ______
BULK DENSITY: ______ ______ LBS./CU. FT. STICKINESS ______
PARTICLE SIZE ______ DUSTINESS ______
MAXIMUM SIZE ______ CHEMICAL ACTION ______
ANGLE OF REPOSE ______ TEMPERATURE ______ °F
PRODUCT DESCRIPTION ______

OPERATING CONDITIONS

CAPACITY: LBS./HR. NORMAL ______ LOADING CLASSIFICATION ______
MAX. ______ LOCATION: INDOOR ☐ OUTDOOR ☐
RPM OF SCREW ______

CONSTRUCTION AND MATERIALS

TROUGH: TYPE ______ SHAFT DIAM. ______
MAT'L. ______ TROUGH END ______
THICKNESS ______ TYPE HANGERS ______
COVER: TYPE ______ SADDLES ______
MAT'L. ______ DUST SEALS ______
THICKNESS ______ BEARINGS: TYPE – HARD IRON ______
FLIGHT: PITCH ______ BABBITED ______
MAT'L. ______ BRONZE ______
THICKNESS ______ ANTI-FRICTION ______
FLAT VALVE PLATE GATE ______ JACKETED TROUGH ______
RACK & PINION GATE – FLAT VALVE PLATE ______
CURVED VALVE PLATE ______

DRIVE DATA

CHAIN-ASA NO. ______
SPROKETS – DRIVER NO. TEETH ______
DRIVEN NO. TEETH ______
V-BELT NO. & SIZE ______
SHEAVES – DRIVER-PITCH DIA. ______
DRIVEN-PITCH DIA. ______
REDUCER – MFGR. ______
MODEL NO. ______
SERVICE FACTOR ______
INPUT RPM ______
OUTPUT RPM ______
COUPLING – MFG'R. ______
TYPE ______

MOTOR DATA

ITEM NO. ______ MTD. BY ______
MFG. ______
TYPE ______ FRAME ______
HP ______ RPM ______
VOLTS/PHASE/CYCLE ______ / ______ / ______
BEARINGS ______ LUBE ______
FULL LOAD AMPS ______ SF ______
LR AMPS ______
RANGE ______

DES SUPRV. | CHECKED BY | DRAWN BY | INITIATOR

JOB NO. PLANT LOCATION
PROJECT NO. REF. DWG. NO.
TITLE
POSITION NO. SERIES STEP NO.
ISSUE DATE
SCALE
BLDG. NO.
DRAWING NO.

E228B44

PROCESS DESIGN SECTION
CORPORATE ENGINEERING

SCREW CONVEYOR
DATA SHEET

SHEET 2 OF 2

REMARKS | APP'R'V | DATE | SUPRV | CHKD | BY | REV

MISCELLANEOUS

PAINTING: ______
GALVANIZING ______
TOTAL WEIGHT ______
MFG. ______
TYPE & MODEL NO. ______
VENDOR DWG. NO. ______
CONVEYOR SERIAL NO. ______
FINISH OF INTERNAL PARTS: ______

REMARKS

DES SUPRV. | CHECKED BY | DRAWN BY | INITIATOR

JOB NO. PLANT LOCATION
PROJECT NO. REF. DWG. NO.
TITLE
POSITION NO. SERIES STEP NO.
ISSUE DATE
SCALE
BLDG. NO.
DRAWING NO.

E228B44
SHEET 2

PROCESS DESIGN SECTION
CORPORATE ENGINEERING

BUCKET ELEVATOR DATA SHEET

REMARKS | APP'R'V | DATE | SUPRV | CHKD | BY | REV

OPERATING CONDITIONS

MATERIAL CONVEYED: ______
PRODUCT DESCRIPTION: ______
BULK DENSITY: ______ LBS./CU. FT.
SCREEN ANALYSIS: ______ %
PRODUCT TEMP: ______ °F
MOISTURE CONTENT: ______ WGT. %
CAPACITY: TONS/HR. NORMAL ______ MAX. ______
MODE OF OPERATION: ______
BUCKET SPEED: ______ FPM
ANGLE OF INCLINE: ______
INSTALLATION: INDOOR ☐ OUTDOOR ☐

MECHANICAL DATA

TYPE: CENTRIFUGAL ☐ CONTINUOUS ☐
DISTANCE BETWEEN HEAD & TAIL SHAFTS: ______
CASING – CROSS SECTION DIMENSION: ______
CASING GAUGE ______
CASING HD. SECT. – GAUGE DISCH. OPENING: ______
BOOT – TYPE: ______
GAUGE: ______
LOADING OPENINGS ______
CHAIN CARRIER FOR BUCKETS
TYPE: ______
HEAD SHAFT SPROCKET P.D: ______
TAIL SHAFT P.D: ______
SIZE OF BUCKETS: ______ INCHES
BUCKET SPACING: ______ INCHES
BUCKET BOLTS–TYPE: ______
NO. PER BUCKET: ______
BELT CARRIER FOR BUCKETS:
WIDTH: ______ NO. OF PLIES: ______
WT. OF DUCK: ______ COVER THICKNESS: ______
TYPE OF BELT SPLICE: ______
HEAD PULLEY DIA: ______ FACE WIDTH: ______
TAIL PULLEY DIA: ______ FACE WIDTH: ______
HEAD SHAFT DIA: ______ TAIL SHAFT DIA: ______
TYPE OF BEARINGS: ______
HEAD SHAFT SPEED: ______ RPM
TYPE OF TAKE-UP: ______
LOCATIONS: ______
INLET CHUTE ANGLE: ______
DISCH. CHUTE ANGLE: ______
INLET OPENING W/LOADING LEG: ______
FEED HOPPER ______

MATERIALS OF CONSTRUCTION

CASING: ______
HEAD SECTION: ______
BOOT: ______
BUCKETS: ______
HEAD & TAIL SHAFTS: ______
HEAD & TAIL SHAFT BEARINGS: ______
BELT & COVERS: ______
HEAD & TAIL PULLEYS: ______
CHAIN: ______
HEAD & TAIL SPROCKETS: ______
TAKE-UP DEVICE: ______
BUCKET BOLTS: ______
BUCKET FILLER PIECES: ______
GASKETS: ______
JOINTS: ______
DOORS & MANWAYS: ______

MOTOR DRIVE DATA

ITEM NO. ______ MTD. BY. ______
HP ______ RPM ______ FRAME ______
MFG. ______
TYPE ______ INSUL. ______
ENCLOSURE: CLASS I GROUP D
VOLTS/PHASE/CYCLE ____/____/____
BEARINGS ______ LUBE ______
FULL LOAD AMPS. ______ S.F. ______
LR AMPS ______
REDUCTION UNIT:
TYPE ______ RATIO ______
MFG. ______
TYPE DRIVE: CHAIN ☐ BELT ☐
MOTOR COUPLING: TYPE ______
MFG. ______

DES SUPRV. | CHECKED BY | DRAWN BY | INITIATOR

JOB NO. | PLANT LOCATION
PROJECT NO. | REF. DWG. NO.
TITLE
POSITION NO. | SERIES | STEP NO.
ISSUE DATE | SCALE | BLDG. NO.
DRAWING NO.

E225B41

PROCESS DESIGN SECTION
CORPORATE ENGINEERING

PRESSURE LEAF FILTER DATA SHEET

REMARKS | APP'R'V | DATE | SUPRV | CHKD | BY | REV

OPERATING CONDITIONS

FLUID: ______
CHARACTERISTICS: ______
SPECIFIC GRAVITY: ______
pH: ______
TEMPERATURE: ______ °F
VISCOSITY: ______ CP
SOLIDS CONTENT: ______ WT. %
FLOW RATE: ______ GPM
WORKING PRESS. ______ PSIG
ALLOWABLE ΔP-MAX. ______ PSIG
FILTER MEDIA: ______
FILTER AID: ______
WASH MEDIA: ______
TYPE OPERATION: ______
TIME ON STREAM: ______
CLEANING FREQUENCY: ______

MECHANICAL DATA

MFR. ______
SIZE & TYPE ______
FILTRATION AREA: ______ SQ. FT.
NO. OF LEAVES ______ SPACING ______
BUBBLE RING ______
LAKE CAPACITY ______
LAKE THICKNESS ______
SLUICER ______ DRIVE ______
JACKET ______ FOR ______ PSIG
TANK: VERTICAL ☐ HORIZONTAL ☐
DIAM. ______ VOLUME ______
COVER LIFT ______
CARRIAGE DRIVE ______
COVER CLOSURE ______
LEAF SPACERS ______ VIBRATOR ______
FILTRATE MANIFOLD ______
CONNECTIONS-TYPE ______
CODE STAMP: REQ'D. ☐ NOT REQ'D. ☐

MATERIALS

TANK: ______
LINING: ______
LEAVES: ______
SURFACE MEMBER ______
INTERMED. MEMBER ______
O-RING LEAF GASKETS ______
COVER GASKETS ______
MANIFOLD PIPING ______
JACKET ______
FINISH ______ INTERIOR ______ EXTERIOR

ACCESSORIES

☐ PRE-COAT OR SLURRY TANK:
VERTICAL ☐ HORIZONTAL ☐
DIAM. ______ VOLUME ______
MATERIAL ______
FINISH ______
LINING ______
COVER TYPE ______
LEVEL INDICATOR: YES ☐ NO ☐
☐ AGITATOR
RPM ______
MFR ______
SIZE & TYPE ______
MOTOR: H.P. ______ RPM ______
VOLTS/PHASE/CYCLE ____/____/____
☐ PRECOAT OR FILTER FEED PUMP:
CAPACITY: ______ GPM @ ______ TDH
MATERIAL: ______
MFR: ______
SIZE & TYPE: ______
MOTOR: H.P. ______ RPM ______
VOLTS/PHASE/CYCLE ______
☐ FILTERAID INJECTION PUMP:
CAPACITY: ______ GPM @ ______ TDH
MATERIAL: ______
MFR: ______
SIZE & TYPE ______
MOTOR H.P. ______ RPM ______
VOLTS/PHASE/CYCLE ______

NOTE: ALL MOTORS TO BE CLASS-1-GROUP-D EXP. PROOF

DES SUPRV | CHECKED BY | DRAWN BY | INITIATOR

JOB NO. | PLANT LOCATION
PROJECT NO. | REF. DWG. NO.
TITLE
POSITION NO. | SERIES | STEP NO.
ISSUE DATE | SCALE | BLDG. NO.
DRAWING NO.

E221B37

INTAKE AIR FILTER

Service ______ Indoor/Outdoor; English/Metric Units
Mfr. ______ Model No. ______ Qty. ______
Site: Alt. ______ Amb. Temp.: Max. ______ Min. ______; Wind Velocity: Max. ______
Unusual Conditions: Dust/Fumes/Sand/Snow/Other ______
Air Flow: Normal ______ Rated ______ @ Temp. ______ Press. ______
AFI Dust Spot Efficiency (Atmospheric Dust): ______%.
Unit Shall Remove ______/______ of all Particles ______/______ Microns & Larger.

Rated Data:	Stage No. 1	Stage No. 2
Filter Type		
Model No.		
Max. Possible Air Capacity		
Pressure Drop (Clean Filter)		
Max. Press. Drop (Dirty Filter)		
No. of Elements		
Size of each Element		
Filtration Area		
Filtration Material		
Face Velocity		
Dust Holding Capacity		
Weight of Element		
Element Retention Method in Frame		

Louvers @ Filter Inlet: Size ______ Type ______ Material ______
Inlet Trash Screen: Size ______ Wire Gauge ______ Material ______
Filter Housing: Material ______ Internal/External Paint
Access Doors: Qty. ___. Blow in Doors: Qty. ___. Filter Outlet: Side/Bottom/Back Conn.
Motor: Mfr. ______ Model No. ______ Qty. ______
Area Class ___ Gr. ___ Div. ___ /Non-Hazardous; Volts ______ Phase ______ Hertz ______
Enclosure ______ Insulation ______; Temp. Rise ______ °C above ______ by ______ @ ______ S.F.

Service	PWR	Current		Effy.	P.F.	Full		Brg.			Allow. Thrust	Space	
	HP	Full	Locked	4/4 3/4	4/4 3/4	Load	Frame	Lub.	Weight	Mtg.	Up/Down	Heater	S.F.
	KW	Load	Rotor			RPM	No.		LB/KG	H/V	LB/KG	Watts	

Diff. Press. Switch: Range ______ Set Press. ______ Qty. ______
Diff. Press. Ind. Type ______ Range ______ Qty. ______
Specifications: Paint ______ Wind ______ M.W.K. ______
Earthquake ______

Notes: 1) The structure, consisting of the plenum chamber & filter, shall be of such design that it can be supported by means of four (4) steel columns, one (1) at each corner. 2) Plenum chamber shall be of sufficient strength & bracing to prevent wall, floor or ceiling from pulsating or oscillating. 3) For ammonia/urea plants or other installations with ammonia in the atmosphere, no copper, aluminum or their alloys shall be used for parts exposed to the air. 4) Seals & filter material shall be resistant to vapors present in the inlet air (ammonia, hydrocarbons, etc.). 5) Unit shall be completely assembled and shipped as one unit. 6) Instruments & controls to be completely piped and wired. 7) Outlet connection screen (0.25 in.) (6 mm) wire mesh, with back up bars to withstand seven (7) psi/0.5 kg/cm^2 ΔPress. across the screen.

REVISIONS NO	DATE	ENGINEER	CHK/REVIEW	APPROVE
6				
5				
4				
3				
2				
1				
0				

KELLOGG — MANAGING OFFICE — ENGINEERING OFFICE

EQUIPMENT DATA SHEET

CLIENT ______
LOCATION ______ JOB NO. ______
ITEM ______
ITEM NO. ______ SPEC. NO. ______
REQ'N. NO. ______ PAGE ___ OF ___

PROCESS DESIGN SECTION
CORPORATE ENGINEERING

LINE FILTER SPECIFICATION

SHEET 1 OF 2

REV	BY	CHKD	SUPRV.	DATE	APPR'V	REMARKS
◇						

DATE	INITIATOR	DRAWN BY	CHECKED BY	DES SUPRV.	PROJ. ENG.

1. FUNCTION: ______

2. SERVICE CONDITIONS

FLUID FLOWING: ______
FLOW RATE: NORMAL ______ GPM
MAX. ______ GPM
PRESSURE: ______ PSIG
TEMPERATURE: ______ °F
FLUID VISCOSITY: ______ CP
FLUID SPECIFIC GRAVITY: ______
PARTICLE SIZE ANALYSIS: ______
SOLIDS LOADING: ______
FILTER PRESSURE DROP: ______ PSIG
FILTER RATING: ______

3. TYPE AND DESCRIPTION: ______

JOB NO. ______ PLANT LOCATION ______
PROJECT NO. ______ REF. DWG. NO. ______
TITLE ______
POSITION NO. ______ SERIES ______ STEP NO. ______
ISSUE DATE — SCALE — BLDG. NO.
DRAWING NO. ______

E 121 85*
53060

PROCESS DESIGN SECTION
CORPORATE ENGINEERING

LINE FILTER SPECIFICATION

SHEET 2 OF 2

4. MECHANICAL

MATERIALS OF CONSTRUCTION

HOUSING: ____

FILTER ELEMENT: ____

GASKETS: ____

CONNECTIONS:

INLET: ____

OUTLET: ____

DESIGN PRESSURE: ____ PSIG

DESIGN TEMPERATURE: ____ °F

ACCESSORIES: ____

NO. OF UNITS REQUIRED: ____

A.S.ME CODE: REQUIRED WHEN FILTER IS LARGER THAN 6" IN DIAMETER OR HOLDS MORE THAN 5 CU. FT. IN VOLUME, OR IF DESIGN PRESSURE IS GREATER THAN 15 P.S.I.G.

REMARKS | APP'R'V | DATE | SUPRV. | CHKD | BY | REV

DES SUPRV. | CHECKED BY | DRAWN BY | INITIATOR

JOB NO. ____ PLANT LOCATION ____
PROJECT NO. ____ REF. DWG. NO. ____
TITLE
POSITION NO. ____ SERIES ____ STEP NO. ____
ISSUE DATE
SCALE
BLDG. NO.
DRAWING NO.

E 121 B5*

PROCESS DESIGN SECTION
CORPORATE ENGINEERING

CENTRIFUGE DATA SHEET

SHEET 1 OF 2

OPERATING CONDITIONS

FEED MATERIAL: ____
CHARACTERISTICS: ____
DENSITY OF SLURRY: ____ LB./CU. FT.
PERCENT SOLIDS: ____
VISCOSITY OF SLURRY: ____ CP
TEMPERATURE OF SLURRY: ____ °F
FEED RATE: NORM. ____ MAX. ____
SG. SOLIDS ____ SG. LIQUID ____

CAKE CAPACITY: ____ CU. FT.
CAKE DENSITY: ____ LB./CU. FT.
CAKE MOISTURE: ____ %
SOLIDS PARTICLE SIZE: ____ MESH
WASH REQUIRED: ____
MAT'L. ____
RATE ____ GAL./LB. CAKE
DESIGN PRESS. ____ PSIG
REQ'D. CAPACITY: ____ LB/HR

MECHANICAL DATA

MFR. ____
SIZE & TYPE ____
BASKET DIAM. ____
TYPE: PERFORATE ☐ INPERFORATE ☐
BASKET DEPTH: ____
BASKET RING OPENING DIAM. ____
CAKE THICKNESS ____
CAPACITY UNDER TOP RING ____ CU. FT.
BASKET RPM ____
DRIVE TYPE: ELECTRIC ☐
HYDRAULIC ☐
COVER TYPE: FULL ☐ HALF ☐
FUME TIGHT ☐ NON-FUME TIGHT ☐

CONNECTIONS	SIZE	TYPE
FEED PIPE		
SPRAY PIPE		
M.L. DRAIN		
CAKE DISCH.		
LIGHT GLASS		
SIGHT GLASS		
VENT		
PURGE		

UNLOADING PLOW: REQ'D. ☐ NOT REQ'D. ☐
SKIMMING TUBE: REQ'D. ☐ NOT REQ'D. ☐
INTERLOCK REQ'D. ☐

MATERIALS

CASE ____
LINING ____
CASE RING ____
SCREENS ____
RETAINING RING ____
BASKET ____
FEED CONE ____
COVER ____
SPINDLE ____
BEARINGS ____

FEED PIPE ____
SPRAY PIPE ____
DRAIN NOZZLE ____
DISCH. NOZZLE ____
VENT NOZZLE ____
PURGE NOZZLE ____
UNLOADING PLOW ____
SKIMMING TUBE ____
FILTER CLOTH ____
GASKETS ____

ACCESSORIES

☐ AUTOMATIC CENTRIFUGING CYCLE CONTROLS
(SEE REMARKS COLUMN FOR PROPOSED CYCLE)
☐ CONTROL PANEL

REMARKS | APP'R'V | DATE | SUPRV. | CHKD | BY | REV

DES SUPRV. | CHECKED BY | DRAWN BY | INITIATOR

JOB NO.
PROJECT NO.
TITLE
POSITION NO.
PLANT LOCATION
REF. DWG. NO.
SERIES
STEP NO.
ISSUE DATE
SCALE
BLDG. NO.
DRAWING NO.

E222B38

Line Filter Specification—*(continued)*

PROCESS DESIGN SECTION
CORPORATE ENGINEERING

CENTRIFUGE DATA SHEET

SHEET 2 OF 2

DRIVE DATA

MOTOR
ITEM NO. ______ MTD BY ______
MFR. ______
TYPE ______ FRAME ______
H.P. ______ RPM ______
VOLTS/PHASE/CYCLE ______ / ______ / ______

ENCLOSURE: CLASS 1 GROUP D
BEARINGS. ______ LUBE ______
FULL LOAD AMPS ______ SF ______
L.R. AMPS ______

HYDRAULIC
MFR. ______
SIZE & TYPE ______

OIL COOLER
MFR ______
SIZE & TYPE ______

REMARKS

1 - INTERIOR BASKET FINISH TO BE ______

REMARKS | APP'R'V | DATE | CHKD SUPRV | BY | REV

PROJ. ENG. | DES SUPRV. | CHECKED BY | DRAWN BY | INITIATOR | DATE

JOB NO. ______ PLANT LOCATION ______
PROJECT NO. ______ REF. DWG. NO. ______
TITLE ______
POSITION NO. ______ SERIES ______ STEP NO. ______
ISSUE DATE | SCALE | BLDG. NO. | DRAWING NO.

E222B38
SHEET 2

PROCESS DESIGN SECTION
CORPORATE ENGINEERING

VIBRATING SCREEN DATA SHEET

SHEET 1 OF 2

FEED MATERIAL CHARACTERISTICS

FEED MATERIAL: ______
BULK DENSITY ______ LBS./CU. FT.
ANGLE OF REPOSE ______
WEIGHT % MOISTURE ______
PRODUCT DESCRIPTION ______

ABRASIVENESS ______
STICKINESS ______
CHEMICAL ACTION ______
DUSTINESS ______
TEMPERATURE ______ °F

MATERIAL TO BE SCREENED OUT ______
PART. SIZE ON 1ST SCREEN/% RETAINED ______
PART. SIZE ON 2ND SCREEN/% RETAINED ______
PART. SIZE ON 3RD SCREEN/% RETAINED ______

OPERATING CONDITIONS

FEED RATE – TONS/HR: ______ NORMAL/MAX. ______
% OF SOLIDS ______
FEED LINE SIZE ______
METHOD OF FEED ______

SUCTION IN DISCH. LINE ______
PRESS. IN DISCH. LINE ______
LOCATION: INDOOR ☐ OUTDOOR ☐
MODE OF OPERATION: ______

DESIGN DATA

NO. OF SCREENS REQ'D. ______
SCREEN TYPE/SIZE ______
1ST SCREEN MESH ______
2ND SCREEN MESH ______
3RD SCREEN MESH ______
1ST SCREEN DISCH. SIZE ______
2ND SCREEN DISCH. SIZE ______
3RD SCREEN DISCH. SIZE ______

DISCH. SIZE TABLE ______
FEED DISTRIBUTOR REQ'D. ______
SCREEN ANGLE: DESIGN ______
MAX. ______
COVER REQ'D. ______
MOUNTING: FLOOR ☐ SUSPENSION ☐
TYPE OF SCREEN CLEANER ______
INLET DIAM. ______

SCREEN AREA FREE/GROSS 1ST SCREEN SQ. FT. ______
SCREEN AREA FREE/GROSS 2ND SCREEN SQ. FT. ______
SCREEN AREA FREE/GROSS 3RD SCREEN SQ. FT. ______
SCREEN BACKING REQ'D./TYPE 1ST SCREEN ______
SCREEN BACKING REQ'D./TYPE 2ND SCREEN ______
SCREEN BACKING REQ'D./TYPE 3RD SCREEN ______

MATERIALS OF CONSTRUCTION

SCREEN DECKS ______
SCREENS ______
SCREEN SUPPORT RINGS ______
BASE ______

HOUSING ______
FEED DISTRIBUTOR ______
SPRINGS ______
COLLECTING PANS ______

TABLE ASSY. (IN CONTACT WITH PRODUCT) ______
TABLE ASSY. (NOT IN CONTACT WITH PRODUCT) ______
MISC: ______

REMARKS | APP'R'V | DATE | CHKD SUPRV | BY | REV

DES SUPRV. | CHECKED BY | DRAWN BY | INITIATOR

JOB NO. ______ PLANT LOCATION ______
PROJECT NO. ______ REF. DWG. NO. ______
TITLE ______
POSITION NO. ______ SERIES ______ STEP NO. ______
ISSUE DATE | SCALE | BLDG. NO. | DRAWING NO.

E226B42

Centrifuge Data Sheet—(*continued*)

PROCESS DESIGN SECTION
CORPORATE ENGINEERING

VIBRATING SCREEN DATA SHEET — SHEET 2 OF 2

MOTOR DRIVE DATA

ITEM NO. ______ MTD. BY ______ BEARINGS ______ LUBE ______
MFG. ______ FULL LOAD AMPS ______ SF. ______
TYPE ______ FRAME ______ L.R. AMPS ______
HP ______ RPM ______
VOLTS/PHASE/CYCLES ____/____/____ V BELT DRIVE ______
GEAR REDUCER ______

MISCELLANEOUS

PAINTING ______
GALVANIZING ______
TOTAL WEIGHT ______
VENDORS DWG. NO. ______
SIFTER SERIAL NO. ______
INLET & OUTLET FLEX. CONNECTIONS ______
VENDORS MODEL NO. ______

REMARKS

REMARKS | APP'R'V | DATE | SUPRV | CHKD | BY | REV

PROJ. ENG. | DES SUPRV. | CHECKED BY | DRAWN BY | INITIATOR | DATE

E226B42 SHEET 2

JOB NO. | PLANT LOCATION
PROJECT NO. | REF. DWG. NO.
TITLE
POSITION NO. | SERIES | STEP NO.
ISSUE DATE | SCALE | BLDG. NO.
DRAWING NO.

Vibrating Screen—*(continued)*

E-1280-A

COLUMN SPECIFICATIONS

ITEM NO.	NO. REQ'D	DESCRIPTION

PERFORMANCE DATA

FLUIDS PROCESSED: ______

PRESSURE:
OPERATING: ______
DESIGN: ______
TEST: ______

TEMPERATURE:
OPERATING: ______
DESIGN: ______

CONSTRUCTION DATA

☐ CODE: ______

☐ MANHOLES:
SIZE ______ SHAPE ______ NO. ______
LOCATION ______

☐ HANDHOLES:
SIZE ______ SHAPE ______ NO. ______
LOCATION: ______

INSULATION:
REMARKS: FOR INTERIOR CONSTRUCTION AND SUPPORT DATA SEE ATTACHED SHEETS.

DESCRIPTION OF FUNCTION OF VESSEL AND REMARKS

______

NOTE:
DATA FOR ITEMS MARKED THUS ☐ ARE NORMALLY SUPPLIED BY ENG'R'G DEPT. DATA ENTRIES FOR THESE ITEMS MADE BY OTHER THAN ENG'R'G DEPT. INDICATE MANDATORY REQUIREMENTS DICTATED BY PROCESS, PROPOSAL, CONTRACT, OR CUSTOMER.

SHELL DATA

☐ CONICAL HEAD

☐ DISHED HEAD
TYPE ______

☐ BLIND FLANGE COVER

☐ FLANGED TOP CLOSURE

☐ STRAIGHT SHELL
DIAMETER ______ I. D.
______ O. D.

LENGTH ______
TO TANG. LINE OR FACE OF FLANGE OR FLANGES.

NOTE:
WHERE INTERMEDIATE FLANGES OR OTHER SHELL DETAILS ARE REQUIRED, INDICATE BY SKETCH ON ADJACENT DRAWING AND LOCATE BY DIMENSIONS.

☐ MATERIAL OF CONSTRUCTION: ______

REMARKS: ______

APPR. | APPR. | APPR. | APPR. | CHKD. | MADE

REFERENCES

CHEMICAL PLANTS DIVISION
BLAW-KNOX CONSTRUCTION COMPANY
PITTSBURGH, PENNA.

CHKD.	MADE	DATE	NO.

REVISIONS

PAGE OF

S & W FORM B-1347

STONE & WEBSTER ENGINEERING CORPORATION

BUBBLE TOWER SPECIFICATION

Item ______ Page ______ Rev. 1 ______ Rev. 2 ______

Client	Location	J. O. No.
Apparatus		Project No.
		Date / By
Based on		Rating Page / Dated

DESIGN CONDITIONS

Oper Pr	lb/sq in ga	Oper Temp	°F	Hammer Test	Yes No
Des Pr	lb/sq in ga	Des Temp	°F	Field Hyd Test	Yes No
Wind - Proj Area	lb/sq ft	Self Supporting	Yes No	Other Tests	

Max. Horizontal Deflection at Top Tray in

Earthquake 0.2 Operating Weight applied at c.g. Yes No

MECHANICAL DATA

Code	Stress Relieve Yes No	Shell & Heads ASTM Spec
	Radiograph Yes No	Type of Heads

THICKNESS:	Calculated	Corrosion	Total Thick	Liner Mat	Liner Thick
Shell	in	in	in		in
	in	in	in		in
	in	in	in		in
Top Head	in	in	in		in
Intermediate Head	in	in	in		in
	in	in	in		in
Bottom Head	in	in	in		in
Cone Section	in	in	in		in
Skirt	in	in	in		in
Insulation			in		

Stiffener Rings Number Req'd Size in × in

NOZZLES:	Number	Type	Series	Size	Mark No.
Feed				in	
O H Vapor				in	
Reflux In				in	
Reflux Out				in	
Bottoms				in	
Drains				in	
Reboiler Vapor				in	
Reboiler Liquid				in	
				in	
				in	
Safety Valve				in	
Manholes				in	
Thermocouples				in	
				in	
				in	
Gage Glass				in	
Pr Gage				in	
Level Control				in	
				in	
Vent				in	
Steam out				in	

TRAYS:					CAPS:		
Tray No.	Diameter	Spacing	Type	Materials	Size	Type	Materials
	′ ″	″					
	′ ″	″					
	′ ″	″					

Estimated Weight lb

Baffles, etc.

Insulation Supports: Yes No: Number Req'd Size in × in: Shop Field

Platform Clips: Yes No: Number Req'd Shop Field | Ladder Clips: Yes No: Number Req'd Shop Field

Manhole Davits Required For All Covers Weighing Over 75 Lb

For Further Details Refer To:

Remarks:

Copy To	Date	Checked	Date	Approved	Date

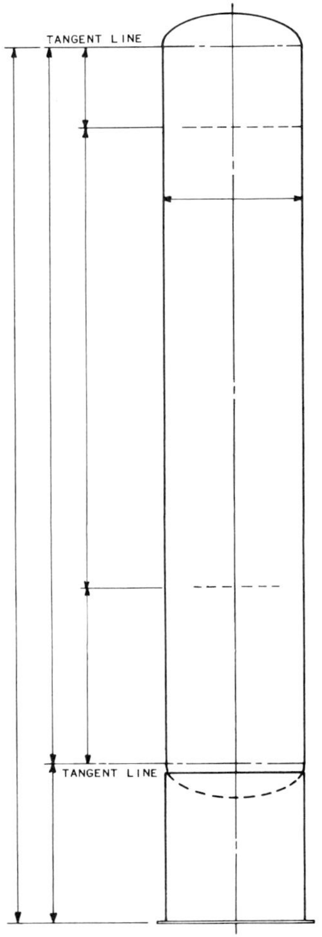

BY ______

DATE ______

CHK'D ______

APP'D ______

ST-L1-094

DESIGN DATA

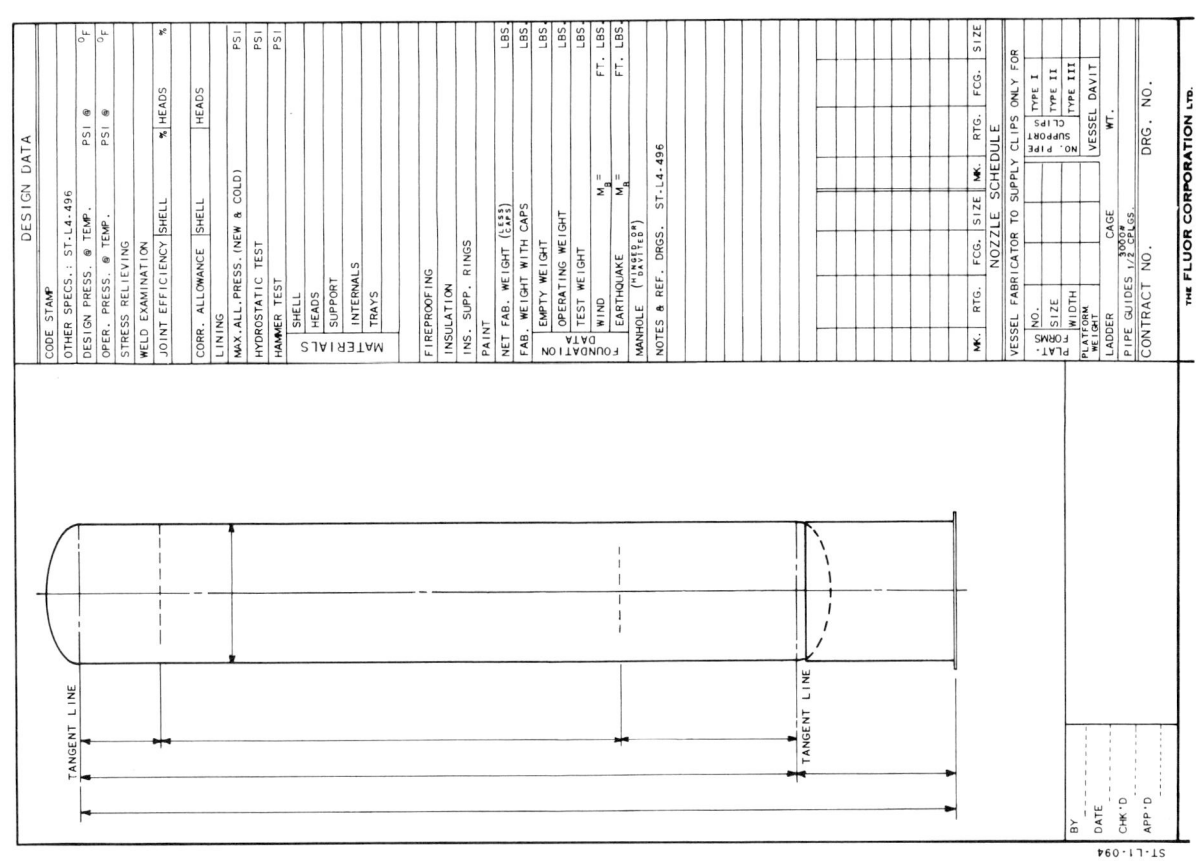

CODE STAMP			
OTHER SPECS.: ST-L4-496			
DESIGN PRESS. @ TEMP.	PSI @		°F
OPER. PRESS. @ TEMP.	PSI @		°F
STRESS RELIEVING			
WELD EXAMINATION			
JOINT EFFICIENCY	SHELL %	HEADS %	
CORR. ALLOWANCE	SHELL	HEADS	
LINING			
MAX.ALL.PRESS.(NEW & COLD)			PSI
HYDROSTATIC TEST			PSI
HAMMER TEST			PSI
MATERIALS	SHELL		
	HEADS		
	SUPPORT		
	INTERNALS		
	TRAYS		
FIREPROOFING			
INSULATION			
INS. SUPP. RINGS			
PAINT			
NET FAB. WEIGHT (LESS CAPS)			LBS.
FAB. WEIGHT WITH CAPS			LBS.
FOUNDATION DATA	EMPTY WEIGHT		LBS.
	OPERATING WEIGHT		LBS.
	TEST WEIGHT		LBS.
	WIND	M_B =	FT. LBS.
	EARTHQUAKE	M_B =	FT. LBS.
MANHOLE (HINGED OR DAVITED)			
NOTES & REF. DRGS. ST-L4-496			

NOZZLE SCHEDULE

MK.	RTG.	FCG.	SIZE	MK.	RTG.	FCG.	SIZE

VESSEL FABRICATOR TO SUPPLY CLIPS ONLY FOR

PLAT-FORMS	NO.	NO. PIPE SUPPORT CLIPS	TYPE I
	SIZE		TYPE II
	WIDTH		TYPE III
PLATFORM WEIGHT		VESSEL DAVIT	
LADDER	CAGE	WT.	
PIPE GUIDES 1/2 3000# CPLGS.			
CONTRACT NO.		DRG. NO.	

THE FLUOR CORPORATION LTD.

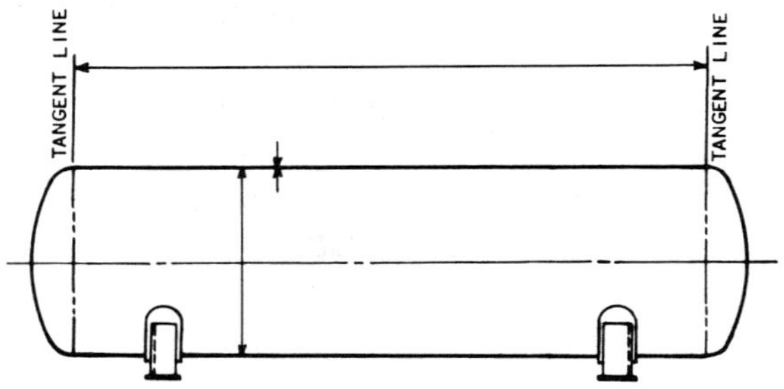

DESIGN DATA			
CODE STAMP			
OTHER SPECS.: ST-L4-496			
DESIGN PRESS. @ TEMP.		PSI @	°F
OPER. PRESS. @ TEMP.		PSI @	°F
STRESS RELIEVING			
WELD EXAMINATION			
JOINT EFFICIENCY	SHELL	% HEADS	%
CORR. ALLOWANCE	SHELL	HEADS	
LINING			
MAX. ALL. PRESS. (NEW & COLD)			PSI
HYDROSTATIC TEST			PSI
HAMMER TEST			PSI
MATERIALS	SHELL		
	HEADS		
	SUPPORT		
	INTERNALS		
FIREPROOFING			
INSULATION			
PAINT			
NET FAB WEIGHT			LBS.
EMPTY WEIGHT			LBS.
OPERATING WEIGHT			LBS.
TEST WEIGHT			LBS.
MANHOLE (HINGED OR DAVITED)			

BY ________ DATE ________ CHK'D ________ APP'D ________

NOTES & REF. DRGS. ST-L4-496

NOZZLE SCHEDULE

MK.	RTG.	FCG.	SIZE	MK.	RTG.	FCG.	SIZE

VESSEL FABRICATOR TO SUPPLY CLIPS ONLY FOR

PLATFORM WT.

LADDER CAGE WT.

CONTRACT NO. DRG. NO.

THE FLUOR CORPORATION LTD.

ST-L1-097

PROCESS DESIGN SECTION
CORPORATE ENGINEERING

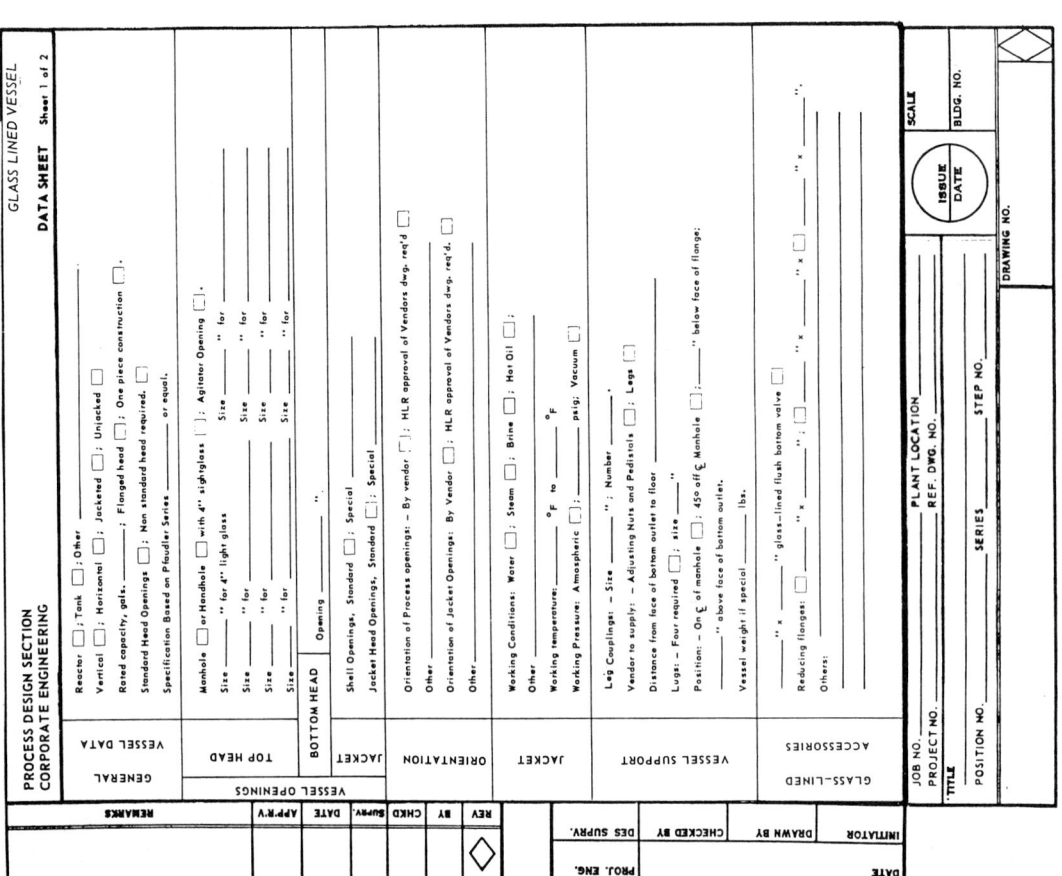

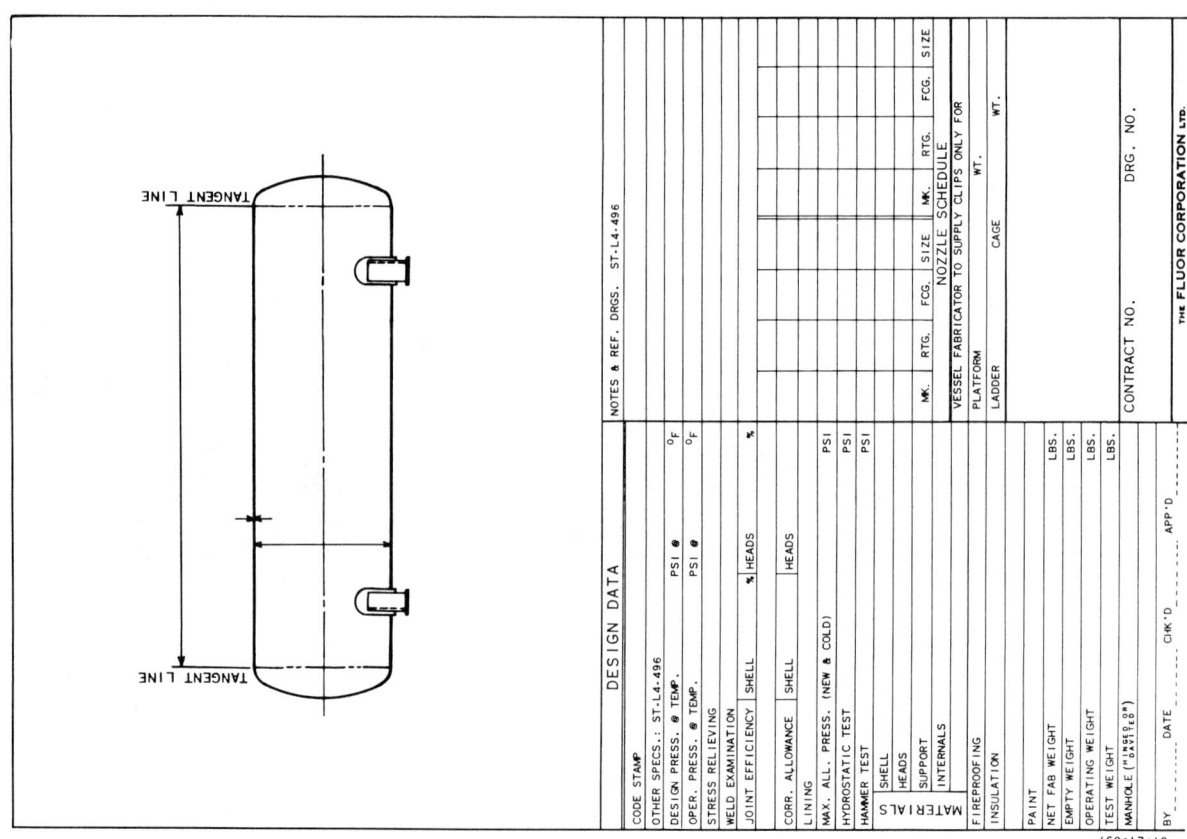

GLASS LINED VESSEL

DATA SHEET Sheet 1 of 2

GENERAL — VESSEL DATA

Reactor ☐; Tank ☐; Other ________.

Vertical ☐; Horizontal ☐; Jacketed ☐; Unjacked ☐

Rated capacity, gals. ________; Flanged head ☐; One piece construction ☐.

Standard Head Openings ☐; Non standard head required. ☐

Specification Based on Pfaudler Series ________ or equal.

VESSEL OPENINGS — TOP HEAD

Manhole ☐ or Handhole ☐ with 4" sightglass ☐; Agitator Opening ☐.

Size ____ " for 4" light glass — Size ____ " for ________

Size ____ " for ________ — Size ____ " for ________

Size ____ " for ________ — Size ____ " for ________

Size ____ " for ________ — Size ____ " for ________

BOTTOM HEAD

Opening ________ ".

JACKET

Shell Openings, Standard ☐; Special ________

Jacket Head Openings, Standard ☐; Special ________

ORIENTATION

Orientation of Process openings: – By vendor ☐; HLR approval of Vendors dwg. req'd ☐

Other ________

Orientation of Jacket Openings: By Vendor ☐; HLR approval of Vendors dwg. req'd. ☐

Other ________

JACKET

Working Conditions: Water ☐; Steam ☐; Brine ☐; Hot Oil ☐;

Other ________

Working temperature: ________ °F to ________ °F

Working Pressure: Atmospheric ☐; ________ psig; Vacuum ☐

VESSEL SUPPORT

Leg Couplings: – Size ________ "; Number ________.

Vendor to supply: – Adjusting Nuts and Pedistals ☐; Legs ☐

Distance from face of bottom outlet to floor ________

Lugs: – Four required ☐; size ____ "

Position: – On ℄ of manhole ☐; 45° off ℄ Manhole ☐; ________ " below face of flange;

________ " above face of bottom outlet.

Vessel weight if special ________ lbs.

GLASS-LINED ACCESSORIES

________ " x ________ " glass-lined flush bottom valve ☐

Reducing flanges: ☐ ____ " x ____ "; ☐ ____ " x ____ " x ☐ ____ " x ____ ".

Others: ________

REV	BY	CHKD	SUPRV.	DATE	APP'R'V	REMARKS
◇						

DATE	INITIATOR	DRAWN BY	CHECKED BY	DES SUPRV.	PROJ. ENG.

JOB NO. ________ PLANT LOCATION ________

PROJECT NO. ________ REF. DWG. NO. ________

TITLE ________

POSITION NO. ________ SERIES ________ STEP NO. ________

ISSUE DATE

SCALE

BLDG. NO.

DRAWING NO.

PROCESS DESIGN SECTION
CORPORATE ENGINEERING

GLASS LINED VESSEL

DATA SHEET Sheet 2 of 2

REMARKS | APP'R'V | DATE | SUPRV. | CHKD | BY | REV

PROJ. ENG. | DES SUPRV. | CHECKED BY | DRAWN BY | INITIATOR | DATE

GENERAL NOTES

Insulation Thickness ________ ". By Hoffmann-La Roche.

Matching flanges to be supplied by HLR ☐; Piping contractor ☐.

Material: -s/s ☐; Other ________

Vendor to supply: -Approval drawing ☐; Certified drawing ☐.

Vendor to fill in all missing information on: -Approval drawing ☐; Certified drawing ☐

AGITATION

DUTY

For: -Simple mixing ☐; Extraction ☐; Reaction ☐; Heat transfer ☐.

Gas incorporation ☐; Solids incorporation ☐.

1 Phase liquid ☐; d_4^- ____; Viscosity Cp/____ °C ____

Bottom layer, d_4^- ____; Viscosity Cp/____ °C ____

Liquid-gas ☐; Liquid d_4^- ____; Viscosity Cp/____ C ____; Gas ____

Liquid-solid ☐; Liquid d_4^- ____; Viscosity Cp/____ °C ____; Solid ____.

Change during process: ________

TYPE

Anchor ☐; Impeller ☐; Other____; rpm ____.

Agitation quality: -Mild ☐; Good ☐; Vigorous ☐; Violent ☐.

BAFFLES

None ☐; Thermowell ☐; Flat ☐; Std. up ☐.

Std. down ☐; Long ☐; To be used for thermowell ☐;

G.L. restricted tip ☐; Tantalum tip ☐; Hastelloy tip ☐.

Dimensions by instrument eng. I.D. ____ "; Length ____".

MOTOR

HP____; 3 phase, 60 cycle, 220/440 volts, rpm ____.

Enclosure: Class 1 Group D ☐ Other ________.

WORKING CONDITIONS

Level at start: -gals, ____; Straight shell ____ ".

Level at end: -gals, ____; Straight shell ____ ".

Pressure: -Atmospheric ☐; ____ psig; Vacuum ☐

Temperature range____ °F to ____ °F

Chemicals: ________

Plugs permitted: - Tantalum ☐; Hastelloy ☐; Other________.

PROCESS DESCRIPTION

____

JOB NO. ________ PLANT LOCATION ________
PROJECT NO. ________ REF. DWG. NO. ________
TITLE ________
POSITION NO. ________ SERIES ________ STEP NO. ________
ISSUE DATE | SCALE | BLDG. NO. | DRAWING NO.

Glass Lined Vessel—*(continued)*

PROCESS DESIGN SECTION
CORPORATE ENGINEERING

FRACTIONATOR TRAY DESIGN
DATA SHEET

REMARKS | APP'R'V | DATE | SUPRV. | CHKD | BY | REV

PROJ. ENG. | DES SUPRV. | CHECKED BY | DRAWN BY | INITIATOR | DATE

REBOILER TYPE (KETTLE OR THERMOSYPHON) ________ SERVICE ________

TRAYS NUMBERED FROM TOP OR BOTTOM

TRAY NUMBER					
TOWER INSIDE DIAM. (SEE NOTE 1)					
TOTAL TRAYS IN SECTION					
TRAY SPACING (INCHES) (SEE NOTE 1)					
NO. OF LIQUID PASSES (SEE NOTE 1)					
MAX. ΔP ( ) /TRAY					
TRAY EFFICIENCY (SEE NOTE 1)					
ENTRAINMENT (SEE NOTE 1)					
INTERNAL CONDITIONS					
VAPOR TO TRAY	MAX.				
RATE (#/HR.)	MIN.				
DENSITY (# FT.³)					
PRESSURE (PSIA)					
TEMPERATURE (°F)					
LIQUID FROM TRAY	MAX.				
RATE (# /HR)	MIN.				
DENSITY (# FT³)					
TEMPERATURE (°F)					
VISCOSITY, CP					
SURFACE TENSION (DYNES/CM)					
FOAMING TENDENCY					

MECHANICAL DATA

TOWER MANHOLE ________ I.D. (INCHES) (NOTE 2)

MATERIAL
- DECK ________
- CAP ________
- HOLDOWN ________
- NUTS & BOLTS ________
- SUPPORT RING ________

DECK THICKNESS (GAGE). (NOTE 1) ________

SUPPORT RING WIDTH & THK. (INCHES) (NOTE 1) ________

DOWNCOMER BOLT BAR THK. (INCHES) (NOTE 1) ________

CORROSION ALLOWANCE ________
- TRAYS (INCHES) ________
- TOWER ATTACHMENTS (INCHES) ________

TRAYS INSTALLED FROM TOP OR BOTTOM ________

TYPE OF TRAY INSTALLATION:
REMOVABLE ☐, WELDED ☐,
BOLTED ☐
HATCHWAY ________

NOTES:
1. VENDOR SHALL FILL IN ALL BLANKS AND RETURN WITH QUOTATION.
2. SMALLEST I.D. THROUGH WHICH TRAY PARTS MUST PASS.
3. WEIR ADJUSTABILITY OF ± 1/2" REQUIRED.
4. VENDORS GUARANTEE: THE VENDOR SHALL BE EXPECTED TO GUARANTEE HIS DESIGN: IF HE CANNOT DO THIS WITHIN THE LIMITS SPECIFIED FOR DIAMETER, TRAY SPACING OR DOWNCOMER AREA, HE SHOULD SUBMIT A DESIGN WHICH HE CAN GUARANTEE FOR THE SPECIFIED SERVICE.

JOB NO. ________ PLANT LOCATION ________
PROJECT NO. ________ REF. DWG. NO. ________
TITLE ________
POSITION NO. ________ SERIES ________ STEP NO. ________
ISSUE DATE | SCALE | BLDG. NO. | DRAWING NO.

E 144812

Catalytic Construction Company
PHILADELPHIA 2, PA.

TRAY SPECIFICATION

1 I.D. VESSEL	
2 ITEM NO.	
3 DESIGN TEMP. °F *	
4 I.D. VESSEL MANHOLE	
5 NO. & TYPE TRAYS	
6 TRAY MATERIAL	
7 TRAY CORROSION ALLOW. (Total)	
8 D'CH'R AREA-TOP % Tower Area Sq. Inches	
9 D'CH'R AREA-BOT. % Tower Area Sq. Inches	
10 OUTLET WEIR PERIPHERY	
11 OUTLET WEIR HEIGHT	
12 OUTLET WEIR ADJUSTMENT	
13 OUTLET WEIR NOTCH	
14 INLET WEIR HEIGHT	
15 TRAY MANWAY **	
16 TRAY SPACING	
17 D'CH'R CLEARANCE	
18 SUMP & DRAWOFF REQUIREMENTS	
19 NO. CAPS PER TRAY	
20 CAP & RISER MATERIAL	
21 CAP & RISER ASSY. TYPE	
22 SIZE OF CAPS	
23 CAP SPACING	
24 CAP SKIRT CLEARANCE	
25 CAP SLOT SPEC.	
26 TOP OF SLOT ABOVE TRAY	
27 TYPE OF SLOT IN CAP	
28 I.D. & HEIGHT OF RISERS	
29 SIZE OF TRAY SUPPORT BAR	
30 TRAY SUPPORT - MATERIAL	

REMARKS:

REVISIONS 1 2 3 4 5 6

* - Trays to be designed for 1/8" Max. Deflection at Design Temp.
** - Tray Manways to be removable from top and bottom

FORM CC-220

APPROVAL: DATE: DWG. NO. REV.

PROCESS DESIGN SECTION
CORPORATE ENGINEERING

TOWER INTERNALS DATA SHEET

REMARKS | APPR'V | DATE | SUPRV. | CHKD | BY | REV

DISTRIBUTOR
TOWER INSIDE DIA. ___
MATERIAL ___
CORROSION ALLOWANCE ___
LIQUID FLOW–NORMAL ___
MAX. ___ MIN. ___
LIQUID DENSITY ___
DESIGN TEMP. ___°F or ___°C
MIN. DIA. OF ACCESS ___
SPECIAL REQUESTS ___

SUPPORT PLATE
TOWER INSIDE DIA. ___
MATERIAL ___
CORROSION ALLOWANCE ___
DESIGN LOAD (LIQUID) ___ LB/SQ. FT/HR
DESIGN LOAD (VAPOR) ___ LB/SQ. FT/HR
DESIGN TEMP. ___°F or ___°C
MIN. DIA. OF ACCESS ___
PACKING SIZE & TYPE ___
PACKING DEPTH ___
SPECIAL REQUESTS ___

HOLD–DOWN PLATE
TOWER INSIDE DIA. ___
MATERIAL ___
CORROSION ALLOWANCE ___
DESIGN TEMP. ___°F or ___°C
MIN. DIA. OF ACCESS ___
PACKING SIZE & TYPE ___
SPECIAL REQUESTS ___

BED LIMITER
TOWER INSIDE DIA. ___
MATERIAL ___
CORROSION ALLOWANCE ___
DESIGN TEMP. ___°F or ___°C
MIN. DIA. OF ACCESS ___
PACKING SIZE & TYPE ___
SPECIAL REQUESTS ___

PROJ. ENG. | DES SUPRV. | CHECKED BY | DRAWN BY | INITIATOR | DATE

COLLECTOR PLATE
TOWER INSIDE DIA. ___
MATERIAL ___
CORROSION ALLOWANCE ___
DESIGN TEMP. ___°F or ___°C
LIQUID FLOW–NORMAL ___
MAX. ___ MIN. ___
MIN. DIA. OF ACCESS ___
SUPPORT LEDGE ___ WIDE x ___ THK.
MAX. % LEAKAGE ___
NO. OF SUMPS ___
GAS FLOW ___
SPECIAL REQUESTS ___

REDISTRIBUTOR
TOWER INSIDE DIA. ___
MATERIAL ___
CORROSION ALLOWANCE ___
LIQUID FLOW–NORMAL ___
MAX. ___ MIN. ___
LIQUID DENSITY ___
SUPPORT LEDGE ___ WIDE x ___ THK.
DESIGN TEMP. ___°F or ___°C
GAS FLOW ___
MIN. DIA. OF ACCESS ___
SPECIAL REQUESTS ___

JOB NO. ___ PLANT LOCATION ___
PROJECT NO. ___ REF. DWG. NO. ___
TITLE
POSITION NO. ___ SERIES ___ STEP NO. ___
ISSUE DATE
SCALE
BLDG. NO.
DRAWING NO.

E 162818

CHLORINATION SYSTEM

Service: ______ Indoor/Outdoor; English/Metric Unit
Type of Water to be Chlorinated: ______
Chlorine: Liquid/Gas; Dosage: Rate/Schedule ______ / ______
Reg. Chlorine In. Pressure ______ Chlorinator/Ejector Distance ______
Chlorinator: Item No. ______ Qty. ______ Capacity ______
Mfr. ______ Model No. ______
Min./Max./Shock Flow ______ / ______ / ______ Material ______
Evaporator: Item No. ______ Qty. ______ Water In. Press./Temp. ______ / ______
Mfr. ______ Model No. ______
Min./Max./Shock Flow ______ / ______ / ______ Material ______
Ejector: Item No. ______ Qty. ______ Material ______
Mfr. ______ Model No. ______
Press.: Water In. ______ Chlor./Inj. ______ Chlor./Water Out ______
Chlorine Residual Analyzer: Item No. ______ Qty. ______ Mat'l. ______
Mfr. ______ Model No. ______
Electrical: Area Class ___ Gr. ___ Div. ___ Volts ___ PH ___ HZ ___ Power ______ KW
Chlorinator House: Item No. ______ Qty. ______ Mat'l. of Const. ______
Ventilation System, Type: ______
Accessories: Control ______

Water Test Kit: Mfr. ______ Model No. ______ Range ______
Specifications: Painting ______
The Vendor shall furnish as a minimum the equipment and components shown on this sheet.

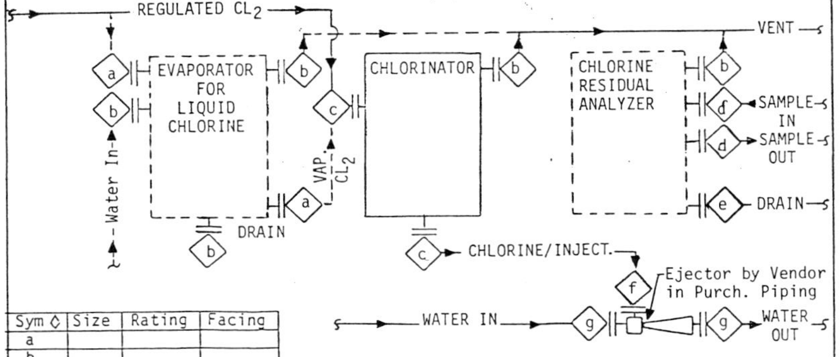

Sym ◊	Size	Rating	Facing
a			
b			
c			
d			
e			
f			
g			

Dimensions	Length	Width	Height	Weight	LBS/KG
Analyzer				Anal. (Inst)	
Chlorinator				Chlor. (Inst)	
Ejector				Evap. (Inst)	
Evaporator				Shipping	

REVISIONS NO.	DATE	ENGINEER	CHK/REVIEW	APPROVE
6				
5				
4				
3				
2				
1				
0				

KELLOGG — MANAGING OFFICE — ENGINEERING OFFICE

EQUIPMENT DATA SHEET

CLIENT ______
LOCATION ______ JOB NO. ______
ITEM ______
ITEM NO. ______ SPEC. NO. ______
REQ'N. NO. ______ PAGE ___ OF ___

COALESCERS

Service: ______ Indoor/Outdoor; English/Metric Units
Mfr. ______ Model No. ______ Qty. ______
Unit: Type ______ Size ______ Support (Type) ______
Design: ASME Code Section VIII, Div. 1, ______ Edition and all Mandatory Addendas.
Mech. Design Press. ______ Temperature ______ Corr. Allow. ______ Stamped Yes/No
Des. Vel.: Acid Serv. -15 FPM/Caustic or Water Serv.-10 FPM/Hydrocarbon-Vendor Stc.
Material Handled: ______ Corr. Mat'l. ______ Mol. % ______
Rated Capacity: ______ Oper. Press. ______ Temp. ______
Vapor Capacity: ______ Mol. Wt. ______ Mol. % ______
Liquid Capacity: ______ Mol. Wt. ______ Mol. % ______
Liquid Removal %: ______ Vendors Guaranteed Removal % ______
Sp. Gr. @ Oper. Temp: ______ Visc. @ Oper. Temp. ______ Allow Press.Drop ______
Contaminants: Liq.% ______; Solids % ______; Part. Size (Range) ______ Micron
Inhibitors: Type ______ Concentration ______ pH of Liq. ______
Gas Coalescing Media: ______
Liquid Coalescing Media: Excelsior/Comb. Pack/Other ______
Materials of Construction: Heads ______ Shell ______
Nozzles ______ Headgasket ______ Support (Int.) ______
Sump ______ Media ______
Coalescer: Head.Flanged/Quick Opening/Hinged/Davit; Supports: Saddles/Legs.
Instruments: By Purchaser/Vendors; Diff. Press. Gauge, Level Gauge, Press. Gauge, Relief Valve, Temp. Indicator with Thermowell.
Specifications: Wind ______ Earthquake ______ Paint ______

Notes: 1) Elastomer O-rings are not permitted in hydrocarbon service. 2) Slip-on flgs. Require Purch. Approval. 3) Combination pack shall consist of four (4) demister pads separated from each other by glass wool mats (three required) and held together by end support grids to form a removable cartridge. Demister pads shall be four (4) inch thick herringbone knitted stainless steel (AISI Type 316) wire mesh. Glass wool mats shall consist of a four (4) inch thickness, compressed to two (2) inch of 15 lb per cu. ft., standard glass wool batting combined with an inert binder suitable for the service conditions.

The vendor shall furnish as a minimum the equipment and components shown on this sheet.

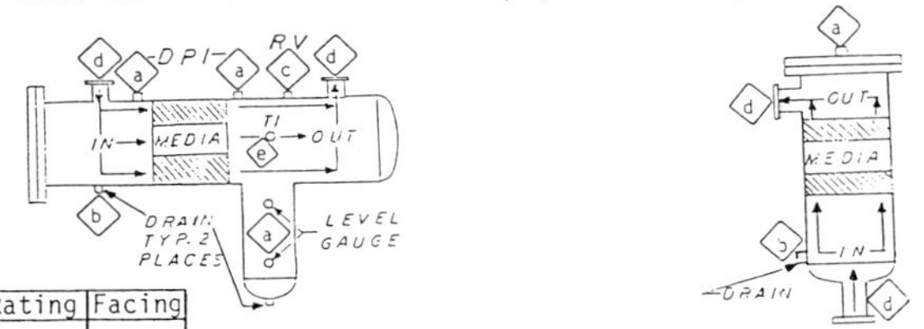

Sym	Size	Rating	Facing
a	3/4"		
b			
c			
d			
e	1"		

Dimensions	Length	Width	Height	Weight	LBS/KG
				Weight	
				Shipping	
				Operating	

REVISIONS NO.	DATE	ENGINEER	CHK/REVIEW	APPROVE
6				
5				
4				
3				
2				
1				
0				

KELLOGG — MANAGING OFFICE — ENGINEERING OFFICE

EQUIPMENT DATA SHEET

CLIENT ______
LOCATION ______ JOB NO. ______
ITEM ______
ITEM NO. ______ SPEC. NO. ______
REQ'N. NO. ______ PAGE ___ OF ___

DEAERATOR (SPRAY-TRAY)

MANUFACTURER ______ MODEL ______ QTY. ____ ENGLISH/METRIC UNITS
WATER STORAGE CAPACITY ______ HEATER: +HORIZONTAL/VERTICAL
DESIGN PRESSURE: ASME BPV CODE SECT. VIII DIV. I PRESS. ______ AND FULL VACUUM

OPERATING DATA

[1]	RATED WATER INLET	QTY/PRESS/TEMP	/ /
[2]	RATED 2 PH. Cond. Ret.	QTY/PRESS/TEMP	/ /
[3]	RATED STEAM INLET	QTY/PRESS/TEMP	/ /
[4]	RATED VENT	QTY/PRESS/TEMP	/ /
[5]	RATED WATER OUTLET	QTY/PRESS/TEMP	/ /

LIMITING INLET COND. AT NOZZLES: (COND. SHOWN WILL NOT OCCUR SIMULTANEOUSLY)

	CONDENSATE AND MAKE-UP	2 PH. CONDENSATE RETURN
MAX. QTY/PRESS/TEMP:	/ /	/ /
MIN. QTY/PRESS/TEMP:	/ /	/ /

	STEAM
MAX. QTY/PRESS/TEMP:	/ /
MIN. QTY/PRESS/TEMP:	/ /

MECHANICAL DATA

DEAERATOR UNIT:

	DEAERATOR AND STORAGE MATERIAL: ASME	SA
	HEATER AND STORAGE/CORROSION ALLOWANCE:	/
	HEATER INTERNAL MATERIALS:	
	TRAY MATERIAL:	
	SPRAY VALVE MATERIAL:	
	MANWAYS SIZE HEATER/STORAGE:	/
	VORTEX BREAKERS:	
[1]	WATER AND I PH. CONDENSATE NOZZLE: SIZE	
[2]	2PH. Condensate Nozzle: Size	
[3]	LOW PRESSURE STEAM NOZZLE: SIZE	
[5]	DEAERATOR WATER Outlet Nozzle: Size	
	PUMP BALANCE LINE NOZZLE: SIZE/QTY.	/
	DEAERATOR DRAIN NOZZLE: SIZE	
	PUMP KICKBACK/Sparger Nozzle: Size	/

ACCESSORIES:

	RELIEF VALVE/VAC. BRKR: SIZE/SET. PRESS.	/ / /
[4]	VENT VALVE: SIZE	

INLET CONTROL:	BY: VENDOR/KELLOGG
LEVEL CONTROL MANUF./MODEL NO.	/
OVERFLOW CONTROL:	BY: VENDOR/KELLOGG
CONTROL VALVE: SIZE/MFR/MOD. NO.	/ /
CONTROL VALVE MATERIALS:	
LEVEL CONTROL MANUF./MODEL NO.	
INSTRUMENT AIR PRESSURE/AREA CLASSIFICATION	/
INSTRUMENTS:	BY: VENDOR/KELLOGG
LEVEL SWITCHES/LEVEL GAUGE MFR/MOD. NO.	/ / /
PRESS. GAUGE/THERMOMETERS MFR/MOD. NO.	/ / /

NOTES: 1. LADDER/PLATFORM SPEC. ______
2. PAINTING SPEC. ______
3. INSULATION CLIPS SPEC. ______
4. WIND/EARTHQUAKE SPEC. ______
+VENDORS OPTION

REMARKS:

REVISIONS NO	DATE	ENGINEER	CHK/REVIEW	APPROVE
6				
5				
4				
3				
2				
1				
0				

KELLOGG — MANAGING OFFICE — ENGINEERING OFFICE

EQUIPMENT DATA SHEET

CLIENT ______
LOCATION ______ JOB NO. ______
ITEM ______
ITEM NO. ______ SPEC. NO. ______
REQ'N. NO. ______ PAGE ___ OF ___

DEAERATOR (SPRAY-TRAY)

The vendor shall furnish as a minimum the equipment and components shown on the sketch and as called for on the data sheets.

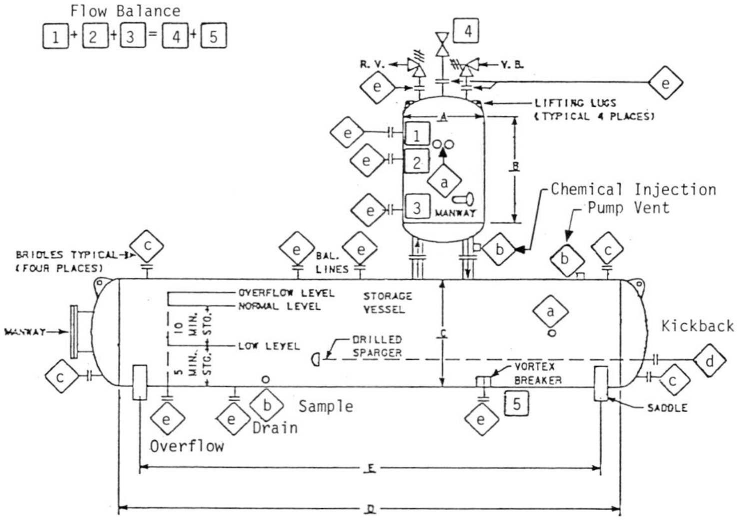

Sym ◇	Size	Rating	Facing
a	3/4"	6000#	Threaded
b	1"	6000#	Threaded
c	2"	150#	R.F.
d		150#	R.F.
e		150#	R.F.
Size by Vendor			

Dimensions	
A	
B	
C	
D	
E	

Weight	LBS/KG
Shipping	
Deaerator	
Storage	
Operating	
Flooded	

REVISIONS NO	DATE	ENGINEER	CHK/REVIEW	APPROVE
6				
5				
4				
3				
2				
1				
0				

KELLOGG — MANAGING OFFICE — ENGINEERING OFFICE

EQUIPMENT DATA SHEET

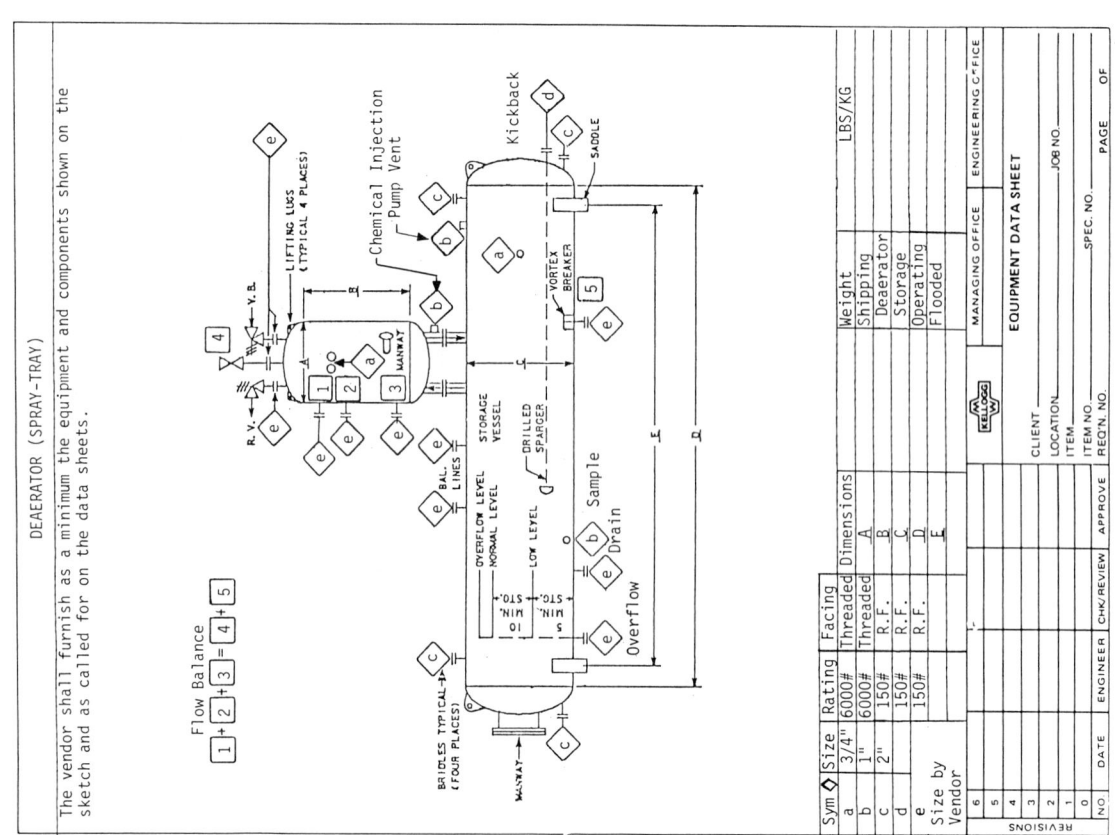

CLIENT ______
LOCATION ______ JOB NO. ______
ITEM ______
ITEM NO. ______ SPEC. NO. ______
REQ'N. NO. ______ PAGE ___ OF ___

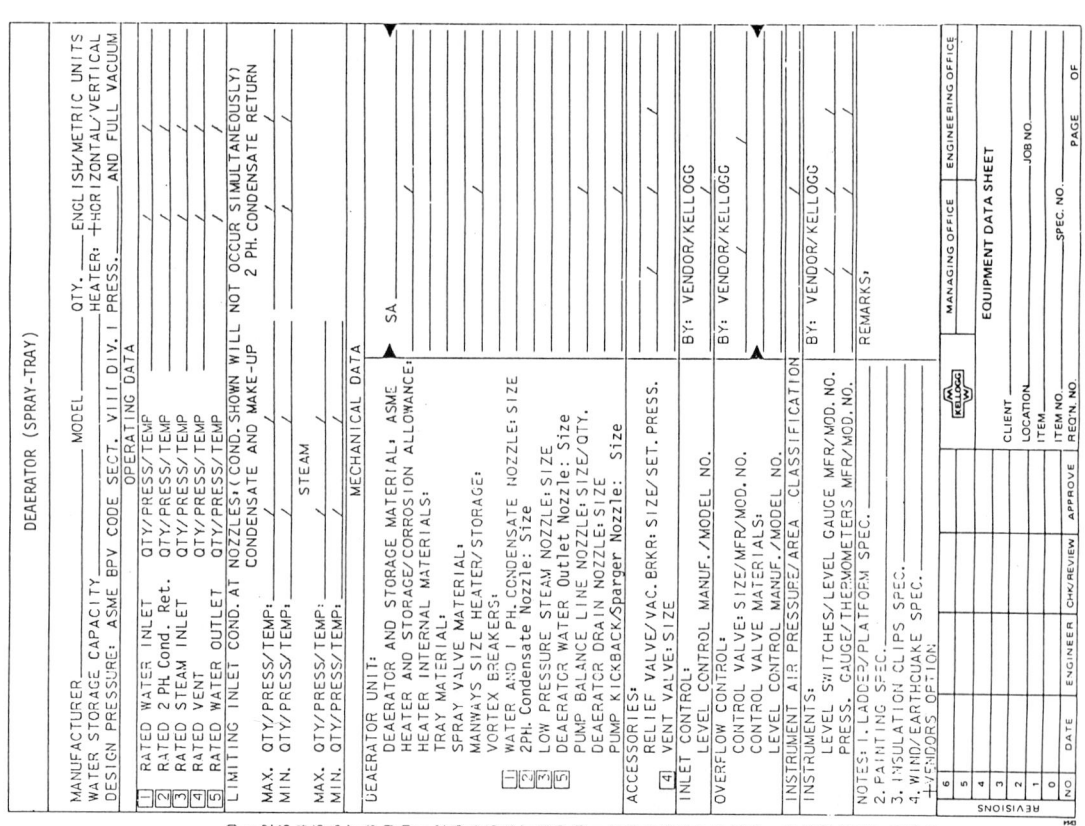

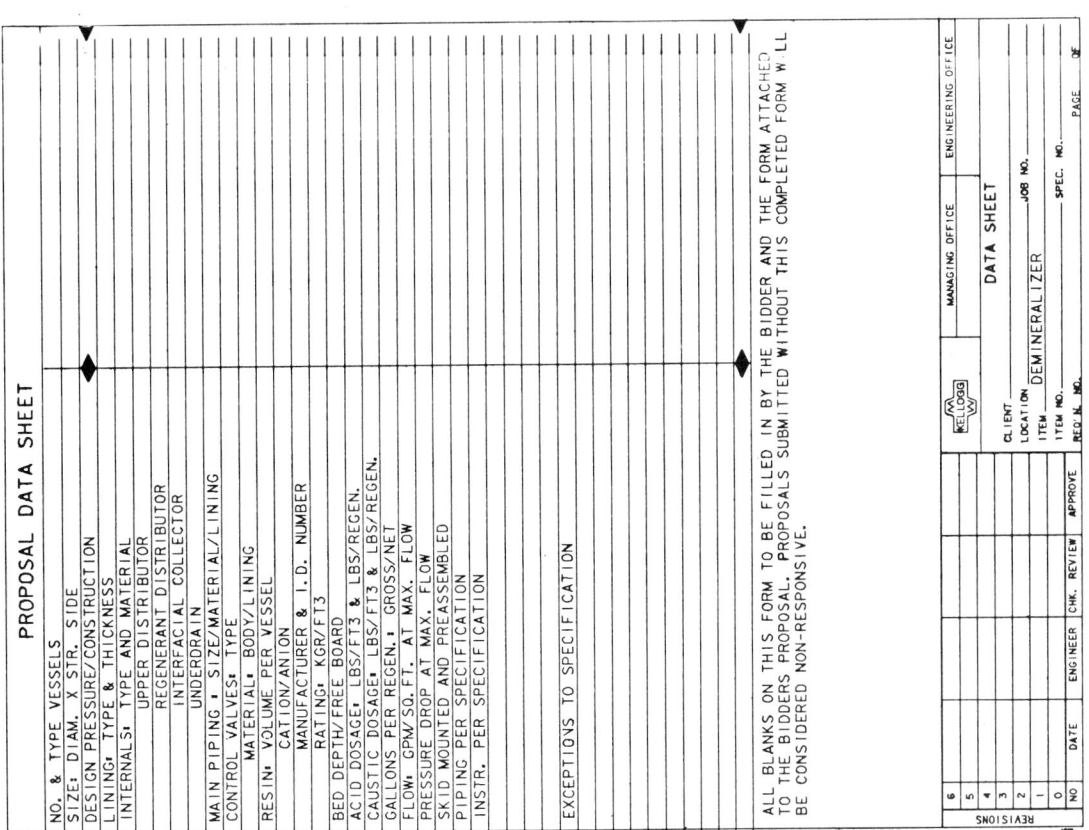

PROPOSAL DATA SHEET

1	NO. & TYPE VESSELS	
2	SIZE: DIAM. X STR. SIDE	
3	DESIGN PRESSURE/CONSTRUCTION	
4	LINING: TYPE & THICKNESS	
5	INTERNALS: TYPE AND MATERIAL	
6	UPPER DISTRIBUTOR	
7	REGENERANT DISTRIBUTOR	
8	INTERFACIAL COLLECTOR	
9	UNDERDRAIN	
10	MAIN PIPING : SIZE/MATERIAL/LINING	
11	CONTROL VALVES: TYPE	
12	MATERIAL: BODY/LINING	
13	RESIN: VOLUME PER VESSEL	
14	CATION/ANION	
15	MANUFACTURER & I.D. NUMBER	
16	RATING: KGR/FT3	
17	BED DEPTH/FREE BOARD	
18	ACID DOSAGE: LBS/FT3 & LBS/REGEN.	
19	CAUSTIC DOSAGE: LBS/FT3 & LBS/REGEN.	
20	GALLONS PER REGEN.: GROSS/NET	
21	FLOW: GPM/SQ. FT. AT MAX. FLOW	
22	PRESSURE DROP AT MAX. FLOW	
23	SKID MOUNTED AND PREASSEMBLED	
24	PIPING PER SPECIFICATION	
25	INSTR. PER SPECIFICATION	
26		
27		
28		
29		
30	EXCEPTIONS TO SPECIFICATION	
31		
32		
33		
34		
35		
36		
37		
38		
39		
40		

ALL BLANKS ON THIS FORM TO BE FILLED IN BY THE BIDDER AND THE FORM ATTACHED TO THE BIDDERS PROPOSAL. PROPOSALS SUBMITTED WITHOUT THIS COMPLETED FORM WILL BE CONSIDERED NON-RESPONSIVE.

REVISIONS NO	DATE	ENGINEER	CHK. REVIEW	APPROVE
6				
5				
4				
3				
2				
1				
0				

KELLOGG — MANAGING OFFICE — ENGINEERING OFFICE

DATA SHEET

CLIENT
LOCATION JOB NO.
ITEM DEMINERALIZER
ITEM NO. SPEC. NO.
REQ'N. NO. PAGE OF

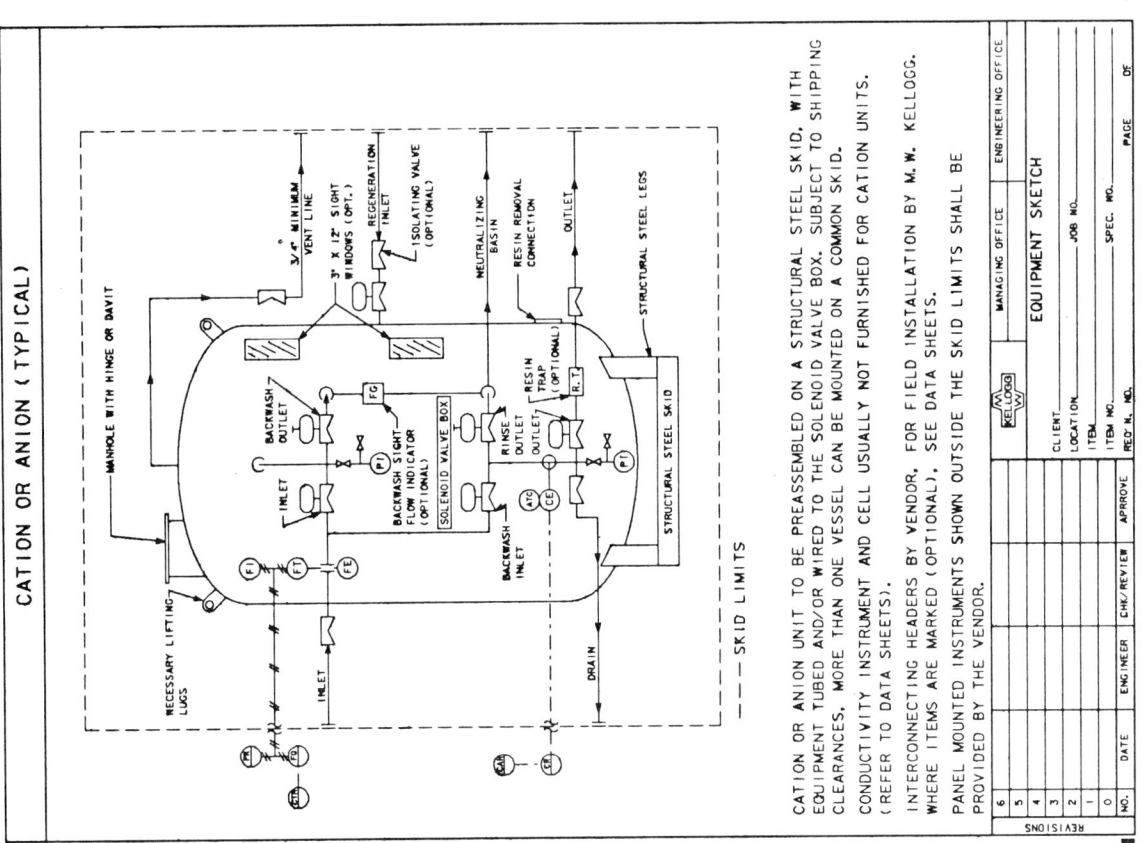

CATION OR ANION (TYPICAL)

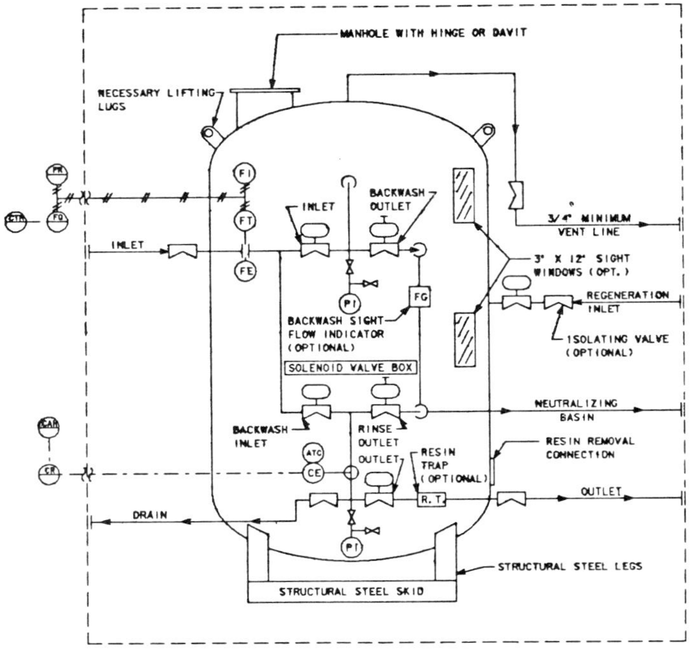

— — — SKID LIMITS

CATION OR ANION UNIT TO BE PREASSEMBLED ON A STRUCTURAL STEEL SKID, WITH EQUIPMENT TUBED AND/OR WIRED TO THE SOLENOID VALVE BOX. SUBJECT TO SHIPPING CLEARANCES, MORE THAN ONE VESSEL CAN BE MOUNTED ON A COMMON SKID.

CONDUCTIVITY INSTRUMENT AND CELL USUALLY NOT FURNISHED FOR CATION UNITS. (REFER TO DATA SHEETS).

INTERCONNECTING HEADERS BY VENDOR, FOR FIELD INSTALLATION BY M.W. KELLOGG. WHERE ITEMS ARE MARKED (OPTIONAL), SEE DATA SHEETS.

PANEL MOUNTED INSTRUMENTS SHOWN OUTSIDE THE SKID LIMITS SHALL BE PROVIDED BY THE VENDOR.

REVISIONS NO.	DATE	ENGINEER	CHK/REVIEW	APPROVE
6				
5				
4				
3				
2				
1				
0				

KELLOGG — MANAGING OFFICE — ENGINEERING OFFICE

EQUIPMENT SKETCH

CLIENT
LOCATION JOB NO.
ITEM
ITEM NO. SPEC. NO.
REQ'N. NO. PAGE OF

KG 1093 2 (5 81)

AIR DRIERS (HEATLESS)

Service: ______ Indoor/Outdoor; English/Metric Units
Mfr.: ______ Model No.: ______ Qty.: ______
Design: ASME Code Section VIII, Div. 1 Latest Edition and all Mandatory Addendas.
Pressure ______ Temp. ______ Stamped Yes/No
Drier Inlet: Flow ______ Press. ______ Temp. ______ Moisture Content ______
DrierOutlet Dewpoint ______; @ Line Press.
Drier: Desiccant Type ______ Wt. per TWR. ______
Reg. Time Cycle ____ Min.; Purge Fl. ____ Guar. Fl. ____; Drying Time Cycle ____ Min.
Press. Drop Across Drier ______ Mat'l. of Const. ______ Corr. Allow ______
Oil Separator: Yes/No; Mfr. ______ Model No. ______ Qty. ______
Material of Construction ______
Prefilter: Mfr. ______ Model No. ______ Qty. ______
Mat'l. of Const. ______ ΔP Clean ______ Micron Size ______
After Filter: Mfr. ______ Model No. ______ Qty. ______
Mat'l. of Const. ______ ΔP Clean ______ Micron Size ______
ΔP Across Entire Skid @ Max. Cap. (Allowed Max./Calc.) ______
Area Class: Class ____ Group ____ Div. ____; Volts ____ Phase ____ Hertz ____ Watts ____
Inst. & Controls: Pneu./Elect./Automatic/Other ______
Specifications ______
Accessories: Alarms, By-passes (for filters only), Control Panel, Diff. Press. Ind., Flow Ind., Ind. Lights, Moisture Ind., Piping, Pressure Gauges, Relief Valves, Switching Valves, Temp. Ind., Timer, Transformer and Valves.
Notes: 1) Provide Dry Contacts (Isolated) to Transmit C.T.A. Signal to Purch. Ann.

The Vendor shall furnish as a minimum the equipment & components shown on this sheet.

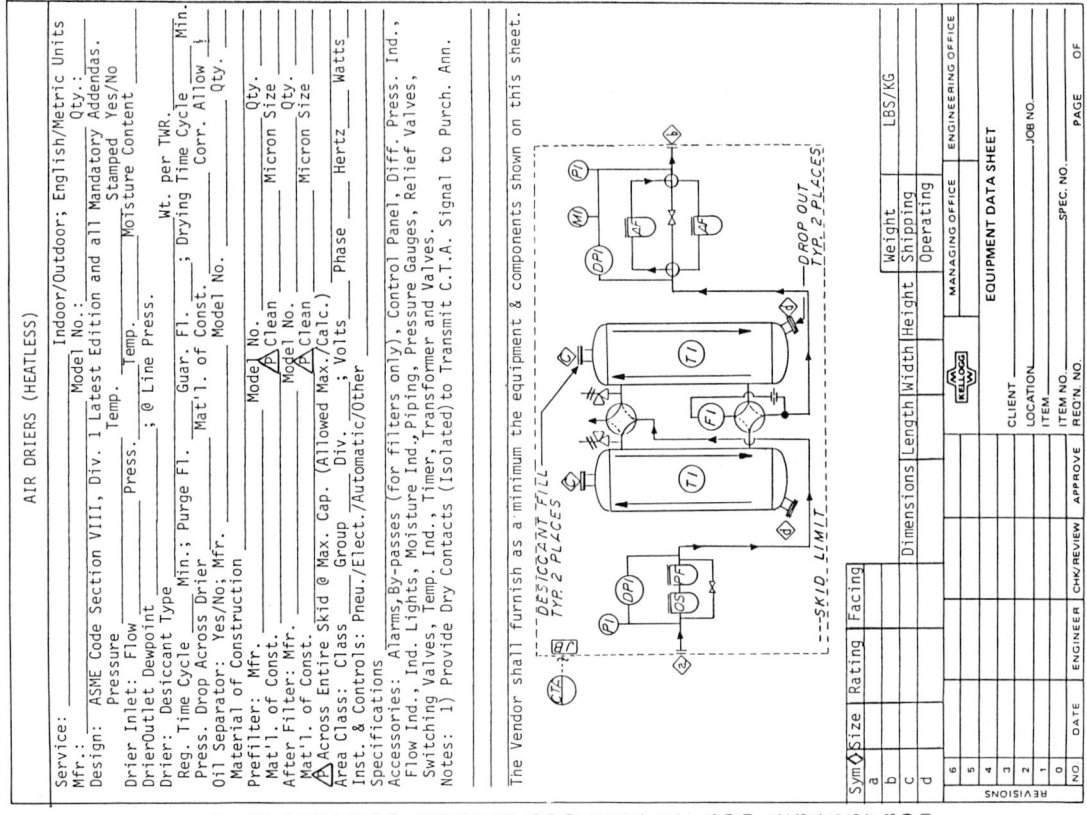

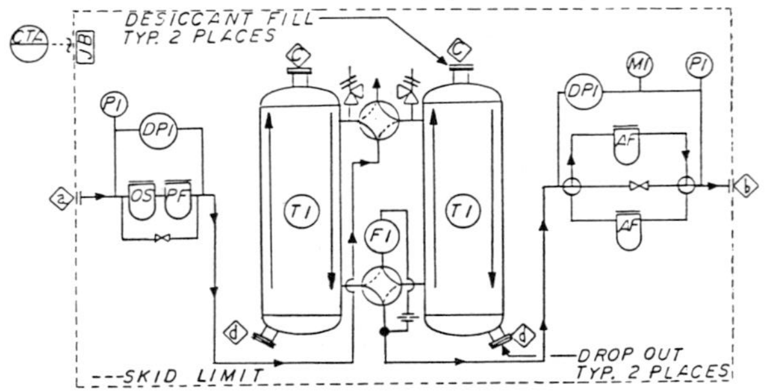

Sym◇	Size	Rating	Facing
a			
b			
c			
d			

Dimensions	Length	Width	Height

Weight	LBS/KG
Shipping	
Operating	

REVISIONS NO	DATE	ENGINEER	CHK/REVIEW	APPROVE
6				
5				
4				
3				
2				
1				
0				

KELLOGG

MANAGING OFFICE | ENGINEERING OFFICE

EQUIPMENT DATA SHEET

CLIENT ______
LOCATION ______ JOB NO. ______
ITEM ______
ITEM NO. ______ SPEC. NO. ______
REQ'N. NO. ______ PAGE ____ OF ____

PROCESS DESIGN SECTION, CORPORATE ENGINEERING

MECHANICAL SCALE DATA SHEET

REV	BY	CHKD	SUPRV	DATE	APP'R'V	REMARKS

SERVICE	LOCATION
NO. REQUIRED	MANUFACTURER
DUTY	MODEL NO.
TYPE	

PROCESS CONDITIONS

CHARACTERISTICS OF MATERIAL

MATERIAL				
STATE OF MATERIAL				
ANGLE OF REPOSE				
TEMP. °F				
BULK DENSITY-LB./FT.^{3}				
AMOUNT/BATCH				

TOTAL VOLUME OF MATERIAL	
AVERAGE DENSITY	
TOTAL WEIGHT OF MATERIAL	
ACCURACY	SENSITIVITY
ATMOSPHERE	

MECHANICAL DESIGN DATA

DESIGN CAPACITY NET WEIGHT	
WEIGHT OF CONTAINER EMPTY	
SIZE OF CONTAINER	
LOCATION OF DIAL WITH RESPECT TO SCALE (HEIGHT)	
LOCATION OF SUPPORTING LEVEL WITH RESPECT TO SCALE (HEIGHT)	
TOTAL SCALE CAPACITY	
MAXIMUM MOVEMENT WHEN FULLY LOADED	
DIAL GRADUATIONS (MINIMUM)	
ADJUSTABLE MARKERS REQUIRED	
TARE BEAM	CAPACITY BEAM
DUNNAGE	
DIAL DIAMETER	SCALE PLATFORM SIZE

REMARKS

INITIATOR | DRAWN BY | CHECKED BY | DES SUPRV.

JOB NO. ______ PLANT LOCATION ______
PROJECT NO. ______ REF. DWG. NO. ______
TITLE ______
POSITION NO. ______ SERIES ______ STEP NO. ______
ISSUE DATE
SCALE
BLDG. NO.
DRAWING NO.

E218B34

PROCESS DESIGN SECTION
CORPORATE ENGINEERING

ROTARY VALVE
DATA SHEET

OPERATING DATA

THROUGHPUT: ______ LBS/HR.
PARTICLE SIZE: ______
BULK DENSITY: ______ LBS./HR.
SOLIDS ARE: CRYSTALLINE ☐ STICKY OR GUMMY ☐ LIGHT OR FLUFFY ☐
SOLIDS: WET ☐ DRY ☐ ______ % MOISTURE
OPERATING PRESS: UPSTREAM ______ PSIG – DOWNSTREAM ______ PSIG
OPERATING TEMP: ______ °F
PRODUCT DESCRIPTION: ______

MECHANICAL DATA

MFGR: ______
SIZE & TYPE: ______
ROTOR TYPE: ______
VALVE TO HAVE OUTBOARD BEARING AND STUFFING BOX.

MATERIALS OF CONSTRUCTION

CASING: ______
ROTOR: ______
SHAFT: ______
PACKING: TYPE ______ SIZE ______ NO. OF RINGS ______
PACKING GLAND ______
GASKETS: ______

MOTOR DRIVE DATA

ITEM NO: ______ MTD. BY: ______
HP: ______ RPM: ______ FRAME: ______
MFG'R: ______
TYPE: ______ INSUL. ______
ENCLOSURE: CLASS I GROUP D
VOLTS/PHASE/CYCLE ______ / ______ / ______
BEARINGS: ______ LUBE: ______
FULL LOAD AMPS: ______ S.F. ______
L.R. AMPS: ______

SPEED REDUCER: MFG'R. ______
TYPE: ______ RATIO ______

MISCELLANEOUS

TOTAL WEIGHT: ______
VENDOR DWG. NO: ______
VALVE SERIAL NO: ______

REMARKS	APP'R'V	DATE	SUPRV
CHKD	BY	REV	
DES SUPRV.	CHECKED BY	DRAWN BY	INITIATOR

JOB NO. | PLANT LOCATION
PROJECT NO. | REF. DWG. NO.
TITLE
POSITION NO. | SERIES | STEP NO.
ISSUE DATE | SCALE | BLDG. NO.
DRAWING NO.

E224B40

Appendix C

QUESTIONNAIRES OF EQUIPMENT SUPPLIERS

Equipment is supplied on the basis of information about the needs provided by the ultimate purchaser and user. Although the more information, within limits, the better, suppliers do require a certain minimum amount before they can make recommendations of equipment and price. Some kinds of equipment are sufficiently standardized to allow their specification by standard forms like those of Appendix B. The questionnaires of individual Suppliers sampled in Appendix C are mostly of specialized equipment that requires custom designing to a greater extent. This is a random selection and other suppliers of each type are available to provide service. Following is a listing of questionnaires contained in this Appendix C. The number appearing in italic is the page number on which the form appears.

STURTEVANT
MILL COMPANY

103 Clayton Street
Boston, MA 02122
617·825·6500
Cable EMERYSTONE
Telex 94·0677

Request for Information

To allow us to better assist you in solving dry processing problems, please use this form when requesting information on Sturtevant equipment.

1. Material to be Processed: ____

2. Processing Method(s) Required (Number in order needed):
   - a. Crushing ____
   - b. Grinding ____
   - c. Air Separating ____
   - d. Micronizing ____
   - e. Blending ____
   - f. Mixing ____
   - g. Screening ____
   - h. Conveying ____
   - i. Elevating ____
   - j. Other (specify) ____

3. Characteristics of Material being Processed:
   - a. Specific Gravity ____
   - b. Hardness ____
   - c. Feed Size ____
   - d. Moisture Content ____ %
   - e. Handling Hazards ____
   - f. Temperature Limits ____ °F. or ____ °C.

4. Production Requirements:
   - a. Product Size Required ____
   - b. Capacity Required ____
   - c. Contamination Tolerance ____

5. If Air Separating is required:
   - a. Amount of product entrained in feed ____ %
   - b. Grinding media preceding separation ____
   - c. Preferred load if in closed circuit ____
   - d. Is heating or cooling required in separation ____
   - e. Product size tolerance, if any ____

6. Any comments which would help our engineers evaluate your requirements:

Name ____ Title ____

Company ____ Address ____

City ____ State ____ Zip Code ____

WILLIAMS

Williams Patent Crusher & Pulverizer Company

2701 N. Broadway • St. Louis, Missouri 63102 • 314/621-3348

Thank you for your interest in Williams size reduction equipment and/or systems. As a first step toward giving you a highly personalized answer to your need, we would appreciate your supplying the following information.

Company Name ____

Address ____ Zip ____

Your Name ____ Title ____

Type of Feed ____ Nominal Size ____

Compressive Strength or Work Index ____

Abrasiveness (contains quartz, silica, etc.) ____

Moisture Content ____ % Oil, Fat, etc. Content ____ %

Temperature Limitations: Softens ____ °F. Melts ____ °F. Decomposes ____ °F.

Chilling Permissible ____ Heating Permissible ____

Special Characteristics (volatile, toxic, hygroscopic, corrosive, etc.) ____

____

Volume of Feed in Tons Per Hour, U.S. (Metric) ____ (____)

Product Size Desired ____ Moisture Desired ____

Fines Desired (Min. - Max.) ____ % Bulk Density Desired ____

Air Classification of Product Desired? Yes ____ No ____

Additional Comments on Product ____

____

Thank you for your help. We will be back to you soon.

Form No. 943

DUCON CONVEYING TECHNOLOGY
A DIVISION OF THE DUCON COMPANY INCORPORATED

SYSTEMS DATA

DATE: ______

CUSTOMER: ______ REPRESENTATIVE:

______

______

______

PRODUCT CHARACTERISTICS

Name ______ ______

Particle size ______ % minus ______ U.S. mesh ______ % minus ______ U.S. mesh

______ % minus ______ U.S. mesh ______ % minus ______ U.S. mesh

	GIVEN	ASSUMED		GIVEN	ASSUMED
Bulk density lbs/cu. ft.			Hygroscopic		
Angle of repose			Corrosive		
% Silica			Moh's hardness number		
Particle shape			Temp of product °F		
Surface texture			Specific heat		
Change of state temp °F			Is breakage objectionable?		
Hazardous (explos., toxic, etc)			If so, what % permissible?		
Tendency to bridge					
Moisture content % free					

SYSTEM REQUIREMENTS

	1	2	3	4
Capacity - lbs/min (instantaneous/avg)				
Horizontal Distance - feet				
Vertical lift - feet				
45° Elbows				
90° Elbows				
Ambient temp at blower inlet °F				
Material of construction in contact with product				

DESCRIPTION (Use back for simple system sketch)

______

______

______

______

______

GENERAL

Starters: furnished ______; not furnished ______

Available current ______ v ______ phase ______ Hz

Available water ______ gpm ______ psig ______ °F

Available air ______ scfm ______ psig

Available steam ______ #/hr @ ______ psig

Motor enclosure

Open drip proof ______

Totally enclosed ______

Explosion proof ______

840 FIRST AVENUE • KING OF PRUSSIA, PA. 19406 • PHONE (215) 337-3770 • TELEX 84 - 6325

HAPMAN

Hapman Conveyor Co.
6002 E. Kilgore Road
Kalamazoo, Michigan 49003
(616) 343-1675
Telex: 224468

data sheet
(Fill out completely)

DATE ______

CUSTOMER ______

ADDRESS ______ ZIP ______

ATTN: ______ TITLE ______ PHONE (___) ______

______

LOCATION OF INSTALLATION ______

( ) TUBULAR CONVEYOR

( ) BUCKET CONVEYOR

( ) OSCILLATING CONVEYOR

( ) OTHER ______

DESCRIPTION OF MATERIAL TO BE HANDLED

1. Material ______

______

2. Particle Size (Mesh or Fraction of Inch) ______

______ Max. ______ Min. ______

Proportions ______

3. Actual Weight per Cubic Foot ______

4. Percentage and Type of Liquid Present ______

______

5. Temperature of Material at Conveyor Receiving

Point ______

6. Temperature Present at Conveyor Discharge

Point ______

7. Flowability: Very Free ______ Free ______

Sluggish ______

8. Abrasiveness: No ______ Mild ______ Very

9. Special Characteristics: Corrosive ______

Explosive ______ Light and Fluffy ______

Aerates and becomes Fluid ______ Packs under

Pressure ______

Describe any possible contamination problems

______

______

Pressure Zones ______

______

IMPORTANT: FURNISH SKETCH ON BACK OF THIS SHEET OR ON SEPARATE SHEET. Show plan, end elevation and/or side elevation properly indicated as such with all pertinent dimensions. Show your idea of preferred conveyor path together with possible alternates. Also indicate all clearances and obstructions to allow us *to select the most practical circuit should this differ from your sketches.*

SERVICE REQUIREMENTS

Required Capacity (in cubic feet/hour or min.)

______

Operation: Continuous ______

Intermittent ______

Operating Period: Hours ______

Frequency ______

______

MOTOR CHARACTERISTICS

Voltage ______ AC/DC ______

Phase: ______ Cycles ______

( ) Totally Enclosed — Fan Cooled

( ) Explosion Proof: Class ______ Group ______

CONSTRUCTION SPECIFICATIONS

1. Casing — Mild Steel ______

#304 S.S. ______ #316 S.S. ______

2. Chain — Sealed Pins ______

Mild St'l. ______ #304 S.S. ______ #316 S.S. ______

Open Pin Construction ______

3. Flight Material — Factory Spec. ______

Other ______

______

Miscellaneous ______

______

SALES REPRESENTATIVE:

______

Note: When sending in Sample, attach label showing Kind of Material, Company Name, Representative, and Date. Three to four cubic feet of material are required for complete testing.

OVER

Form H-100

MEYER MACHINE COMPANY

P.O. BOX 5460, SAN ANTONIO, TEXAS 78201
512/736-1811

DESIGNERS AND MANUFACTURERS OF CONVEYING AND PROCESSING EQUIPMENT

Gentlemen:

Please quote us on a SIMPLEX CONVEYING ELEVATOR to meet the specifications and approximate dimensions as indicated below and on the reverse side of this sheet.

PRODUCT TO BE HANDLED ____

APPROXIMATE WEIGHT OF PRODUCT PER CUBIC FOOT ____

DESIRED MAXIMUM CAPACITY, pounds per hour ____

IS PRODUCT CORROSIVE? ____ DUSTY? ____ FINE POWDER? ____ GRANULES? ____

STICKY? ____ ABRASIVE? ____ PIECES? ____ AVERAGE SIZE OF PIECES ____

TYPE OF BUCKETS PREFERRED: Fabricated Stainless Steel ____ Mild Steel ____ Plastic ____

Cast Aluminum Alloy ____ Other ____

IS DUST-TIGHT/WEATHER-TIGHT ENCLOSURE REQ'D? ____

END DISCHARGE ONLY ____ WITH SIDE DISCHARGES ____ NUMBER OF SIDE DISCHARGES ____

(Indicate on drawing (Fig. 3) on reverse side if discharges are to be Left Hand or Right Hand when viewed from intake end of SIMPLEX. Also indicate approximate location of discharges).

SIDE DISCHARGES TO BE MANUALLY CONTROLLED ____ or SOLENOID CONTROLLED for remote operation ____.

TYPE OF FEEDER/HOPPER PREFERRED: STANDARD INTAKE HOPPER only ____ BUILT-IN FEEDER, less hopper (for feeding from bin) ____ VIBRA-FLEX FEEDER with LARGE HOPPER built separately from SIMPLEX, (for manual feeding) ____ with ECCENTRIC DRIVE ____ with ELECTRO-MAGNETIC DRIVE ____ Feed Trough and Hopper to be Stainless Steel ____ Mild Steel ____ or Galvanized ____.

If head room is restricted, state ceiling height ____

Per ____

Title ____

Company ____

Address ____

____

Date ____ Phone No. ____

QUESTIONNAIRE DRYING SYSTEMS

BOWEN

The information below is required to enable us to prepare a design and quotation. All information given will be treated in strict confidence. We appreciate that some data may not be available or relevant. Please circle chosen alternates. Please return to:

BOWEN ENGINEERING INC., P.O. Box 898, Somerville, New Jersey 08876, U.S.A.
Telephone: (201) 725 3232 Telex: 4754087 (ITT)

COMPANY

Company Name: .... Ref./Project No: ....
Address: .... Personnel: ....
....
....
Telephone: ....
Telex: .... Plant Location: ....

MATERIAL/PROCESS

Name/Composition: ....
Chemical formulae: ....
Process: dry/cool/heat/agglomerate/react/absorb/congeal/calcine/ ....
Type of dryer envisaged: spray ....
Process gas flow: once through/partial recycle/closed cycle/self-inertized/inert/ ....
Heat source: direct/indirect: gas/LPG/oil/steam/electricity/HT oil. ....
Has material been dried before: yes/no. If so, how .... Temperatures ....

CAPACITY DATA

	Feed rate lb/h	at Feed conc. % solids	Evap. rate lb/h
For minimum evaporation rate :	....	....	....
For design evaporation rate :	....	....	....
For maximum evaporation rate:	....	....	....

Product rate: lb/h : min ...., normal ...., max ...., design ....
Feed rate : lb/h : min ...., max ....

FEED DATA

Type: solution/suspension/emulsion .... Corrosive: yes/no
From: tank/filter/ .... Toxic : yes/no
Feed rheology (liquid): Newtonian/pseudoplastic/dilatant/Bingham plastic/thixotropic/rheopectic/viscoelastic ....
Feed rheology (solid): free flowing/friable/cohesive/sticky/paste/ ....
Does rheology change markedly with feed concentration? yes/no. If so, add notes overleaf.
Feed concentration: % solids: min ...., normal ...., max ....
Moisture: water/other (see solvent data overleaf). Impurities in feed: ....
If feed is suspension, what concentration of solubles in liquid ....
Does feed form stable foam? yes/no. Material of construction in contact with feed: ....
Feed temperature: .... °F Viscosity: .... cp at .... °F pH range: ....
Density/bulk density: .... lb/ft³ at .... °F

Please turn over.

PRODUCT DATA

Moisture content: ... % normal, ... % design, ... % maximum.
Is product heat sensitive: yes/no. If yes, what maximum temperature for dry solids ... °F
Material of construction in contact with dry solid: ...
Required particle size: min... micron, av ... micron, max ... micron.
Required bulk density range: ... to ... lb/ft³. Required final discharge temperature: ... °F
Melting point: ... °F Specific heat: ... BTU/lb. °F Heat of crystallisation: ... BTU/lb.
Other special properties: ...
Toxic: yes/no. Flammable: yes/no. Explosive: yes/no. Abrasive: yes/no. Hygroscopic: yes/no.
Thermoplastic: yes/no. If yes, at what temperature: ... °F

SOLVENT DATA

Name: ... Toxic: yes/no. Corrosive: yes/no.
Impurities: ...
Flammable: yes/no. If yes, explosive limits: ... Flash point ...
Boiling point: ... °F Latent heat of vaporisation: ... BTU/lb.
Specific heat liquid ... BTU/lb. °F Specific heat vapour: ... BTU/lb. °F.
Other data: ...

ANALYTICAL METHODS

Moisture content: ...
Other: ...

ENVIRONMENT/SAFETY

Permissable emmission in exhaust air: ... grn/ft³ Effluent water limits ...
Maximum noise level: In plant ... dBA at 3 ft. Community ... dBA at ... ft.
If area classified hazardous, give division, group, etc. ...
Is dust flammable: yes/no. Ignition temperature, cloud ... °F layer ... °F at ... inch thick
Is dust explosive: yes/no. Dust class ... Max. explosive pressure ... Max. rate press. rise ...

LOCAL CONDITIONS

Plant location: indoors/outdoors. Existing building: yes/no. If yes, submit drawings.
Altitude: ... ft. above sea level. Max. wind velocity ... Earthquake: Richter ...
Ambient temperature: Minimum ... °F Average ... °F Maximum ... °F
Range ambient conditions for design evaporation rate: Winter °F at %RH, Summer °F at %RH max.
Special conditions: high dust loading/freezing fog/ ...

UTILITIES

Steam pressure at plant: ... psig. Towns water: ... °F at ... psig. at grade.
Compressed air: ... psig. Instrument air: ... psig. at ... °F dew point.
Electricity: Power ... V, ... ph, ... Hz; Control ... V AC/DC.
Fuel gas: Name ..., Gross calorific value ... BTU/scf. Pressure ... psig./inch WG.
Fuel oil: Grade ... viscosity ..., Gross calorific value ... BTU/lb.
Cooling water: Supply ... °F Return ... °F max, Available pressure differential ...
Other: ...

REMARKS

Signed ... *Date* ...

GOSLIN-BIRMINGHAM, INC.

EVAPORATOR INFORMATION SHEET

COMPANY ____

NAME OF INDIVIDUAL ____

POSITION ____

MAILING ADDRESS ____

PLANT ADDRESS OR JOBSITE ____

FEED SOLUTION ____ PPH ____ % Dissolved Solids ____

Temperature ____ Specific Gravity ____ Specific Heat ____

Analysis ____

% Suspended Solids ____ Viscosity ____ pH ____

Temperature Limit ____ Thermal Conductivity ____

PRODUCT ____ % Dissolved ____ % Suspended Solids ____

Viscosity ____ at Temperature ____ Specific Gravity ____

Scaling? ____ Foamy? ____ Specific Heat ____

Thermal Conductivity ____

OTHER DATA Available Steam ____

Low Pressure ____ Psig ____ Temp. ____ Cost ____

High Pressure ____ Psig ____ Temp. ____ Cost ____

Cooling Water ____ High Temp. ____ Low Temp. ____

Power ____ AC ____ DC ____ Cost/KWH

____ Volts ____ Hertz ____ Phase

Material of Construction

Liquid ____

Vapors ____

REMARKS ____

Spray Drying—(*continued*)

DRYER QUESTIONNAIRE

KRAUSSMAFFEI CORPORATION

COMPANY ____ KMC PROJECT NO. ____

CONTACT PERSON ____ KMC CONTACT PERSON ____

1. PRODUCT: formula, trade name

2. DRYER CAPACITY:
- 2.1 Wet product feed ____ lb/h @ ____% moisture
- 2.2 Dry prod.discharge ____ lb/h @ ____% moisture
- 2.3 Evaporated liquid ____ lb/h
- 2.4 Bone dry product ____ lb/h

3. WET PRODUCT FEED
- 3.1 Originates from: filter centrifuge other ____
- 3.2 Feed: continuous batch other ____
- 3.3 Volume/batch ____ gal.
- 3.4 Batch cycle ____ min.
- 3.5 Consistency: liquid sticky pasty friable thixotropic other ____
- 3.6 What conveyors worked satisfactorily?
- 3.7 Particle size analysis
- 3.8 Wet product temperature ____ °F
- 3.9 Composition of product moisture (vapor pressure curves?) ____ % ____ / ____ % ____ / ____ % ____
- 3.10 Moisture determination:
  - Drying chamber ____ °F
  - ____ min.
  - constant weight ____
  - under vacuum ____ in Hg
  - other ____
- 3.11 Wet bulk density (loose) ____ lbs/ft³
- 3.12 Dry bulk density (loose) ____ lbs/ft³
- 3.13 True product bulk density ____ lbs/ft³ way of measuring the bulk density?
- 3.14 Melting Point ____ °F
- 3.15 Spec. heat wet product ____ BTU/lb
- 3.16 Spec. heat solids ____ BTU/lb
- 3.17 Spec. heat moisture ____ BTU/lb
- 3.18 ph value ____
- 3.19 Max. admissible wet product temp. ____ °F
- 3.20 Dryer atmosphere: air, inert, vacuum

4. DRY PRODUCT:
- 4.1 Requested form of dry product
- 4.2 Admissible dry product temp. ____ °F
- 4.3 Dry product to be cooled ____ yes ____ no to ____ °F
- 4.4 Is there some isothermal absorption known of the dry product? (Please enclose) ____ yes ____ no
- 4.5 Electrostatic charging ____ yes ____ no
- 4.6 Risk of oxidation ____ yes ____ no
- 4.7 Is the product hygroscopic? ____ yes ____ no
- 4.8 Melting point ____ °F
- 4.9 Conveyability?

5. Do you dry the product in your plant now? What dryer type? With what results?

6. If moisture is a solvent, is recovery required?

7. Which materials of construction should be used or must not be used?
   - For contact w/wet product ____
   - For contact w/dry product ____
   - For contact w/fresh gas ____
   - For contact w/exhaust gas ____

8. SPECIAL REFERENCES, REQUESTS, SUGGESTIONS
   PLEASE ENCLOSE SAFETY DATA SHEETS RELATING TO HAZARDOUS PROPERTIES (TOXIC, IGNITABLE, INFLAMMABLE, ETC.)

9. AMBIENT AIR CONDITIONS
- 9.1 Temperature ____ °F
- 9.2 Humidity ____% or ____ Lbs. H_2O/lb. Dry Air

10. UTILITIES
- 10.1 Electrical characteristics ____ V, ____ PH, ____ Cycles
- 10.2 Heating Media:
  - Steam ____ @ ____ PSIG
  - OIL ____ @ ____ BTU/lb
  - GAS ____ @ ____ BTU/ft³

WOOTEN-WICHITA

SELAS
DEHYDRATOR DATA SHEET

Date ____

Selas Corporation of America
Dresher, Pa.
Attn: General Industry Division

Gentlemen:

I am interested in a Selas Dehydrator to meet the following specifications. Please send proposal, with the understanding that I am under no obligation:

GAS AS SUPPLIED TO DEHYDRATOR FOR DRYING
- Type of gas: ....
- Volume at standard atmospheric conditions .... cfm or at working pressure: .... cfm
- Gas temperature entering Dehydrator: .... °F
- Working pressure: .... psig
- Moisture content: saturated at entering temperature ☐ or what % of saturation ....
- Specific gravity: ....
- Specific heat: ....

GAS LEAVING DEHYDRATOR
- Final moisture content: dewpoint .... °F
- or .... grains per cubic foot at .... psig
- Final temperature .... °F

DEHYDRATOR OPERATING SCHEDULE
- Continuous ☐, or, if intermittent, ____ hours on, ____ hours off.

TYPE OF OPERATION
- ☐ Manual ☐ Semi-automatic ☐ Fully Automatic

TYPE OF ELECTRICAL CONSTRUCTION
- ☐ Standard open ☐ Weatherproof ☐ Explosion proof
- ☐ Other ....

UTILITIES AVAILABLE AT OPERATING POINT
- Electricity: Control ☐ AC ☐ DC ____ Volts ____ Phase ____ Cycle
- Power ☐ AC ☐ DC ____ Volts ____ Phase ____ Cycle
- Steam Pressure: .... psig
- Cooling water temperature: Summer ____ °F, Winter ____ °F.
- Cooling water pressure: ____ psig, minimum.
- Preferred heat source for desiccant reactivation ....

MECHANICAL CONSTRUCTION
- ☐ ASME Code ☐ Non-Code ☐ Other ....

REMARKS: ____

FIRM NAME ____
PLANT ADDRESS: ____
SUBMITTED BY: ____
TITLE: ____

M 318

ROTARY EQUIPMENT INQUIRY

Date . . . .

TO: VULCAN IRON WORKS, INC.
Miners National Bank Bldg., Rm. 1050
Wilkes-Barre, Pennsylvania 18701

We are interested in equipment for the following process:
drying . . . . . . , cooling . . . . . . , calcining . . . . . . (other) . . . .

Our interest is related to our:
preliminary design . . . . . . economic evaluation . . . . . .
budget estimate . . . . . . final analysis & purchasing . . . .

Process data available includes:
- type of material . . . .
- material wt./cu. ft.; wet . . . . , dry . . . .
- material specific heat . . . .
- material particle size . . . . (screen analysis desirable)
- capacity–feed rate . . . . or discharge rate . . . .
- moisture content, initial . . . . % and final . . . . %.
- material temperature, initial . . . . and final . . . .
- maximum temperature of material without injury . . . .
- duty cycle . . . . hours per day and . . . . days per week.
- final product temperature desired . . . .
- type of fuel to be used:
  - Oil @ . . . . . . btu/gal., ... psig and . . . . . . SSU.
  - Gas @ . . . . . . btu/cu. ft., and . . . . . . psig.
  - Other . . . .

Burner is to be automatically . . . . or manually . . . . controlled.
Material can . . . . . . , cannot . . . . . . be processed in contact with gases of combustion.
The . . . . sticky, ... abrasive, ... corrosive nature of the material requires special consideration.

Motors: . . . . TEFC . . . . Open . . . . Phase . . . . Cycle . . . . Volts
Installation is to be indoor, ... outdoor . . . .
Elevation of installation, ... feet above sea level.
Dust collection equipment to be included . . . . (yes or no).
Exhaust fan to be included . . . . (yes or no).

Your response on or before . . . . is desired.
(Date)

Please reply to: (Company) . . . .
(Address) . . . .
(Attention) . . . .
(Telephone) . . . .

GLITSCH, INC.

PLANT & HOME OFFICE LOCATION: 4900 SINGLETON BLVD. • DALLAS, TEXAS • 214/631-3841 • **TELEX 73-329**

MIST ELIMINATION AND LIQUID DROPLET COALESCING EQUIPMENT APPLICATION DATA

General Information:

Operating: Pressure __________ Temperature __________

Allowable Pressure drop __________

Vapor (Gas) Phase:

Composition __________ wt. % or mol. %
__________ (state which)
____
____

Rate __________

Liquid Phase:

Composition __________ wt. % or mol. %
__________ (state which)
____
____

Rate __________

Viscosity __________ @ operating temp.

Droplet size range __________

Source of droplets (condensation, spray, etc.) __________

Does liquid present a fouling problem when collected?
Yes ☐ No ☐

If yes, what type __________

Please describe any problems encountered __________

Does the vapor or liquid phase have any insoluble solids in it? Yes ☐ No ☐ If yes, describe their nature.

Does the liquid phase have any dissolved solids in it?
Yes ☐ No ☐ If yes, describe their nature. __________

If the potential application involves a number of units, please state the anticipated annual requirement. __________

MIST ELIMINATORS

Is the mist eliminator to go into an existing vessel?
Yes ☐ No ☐

If yes, please provide a detailed sketch or blueprint of the vessel to be used. __________

Do you wish to have Glitsch quote on a suitable vessel?
Yes ☐ No ☐

LIQUID DROPLET COALESCERS

Note: The coalescers are generally sold only as a complete unit with vessel.

SPACE LIMITATION

Are there any physical limitations due to vessel size?
Yes ☐ No ☐ If yes, please describe __________

SUITABLE MATERIALS OF CONSTRUCTION
(See list below.)

Shell: __________ *

Mist eliminator and/or coalescer __________ *

Support grids __________ *

304 Stainless	Inconel	Copper
316 Stainless	Carpenter 20	Carbon Steel
317 Stainless	Incoloy	Bethanized Steel
430 Stainless	Hastelloy C	Kynar ®
434A Stainless	Titanium	Polypropylene
Monel	Tantalum	Teflon ®
Nickel	Aluminum	

Other (specify) __________

*A corrosion allowance is acceptable for the shell and support grids. Other wetted parts should have the ability to withstand corrosion.

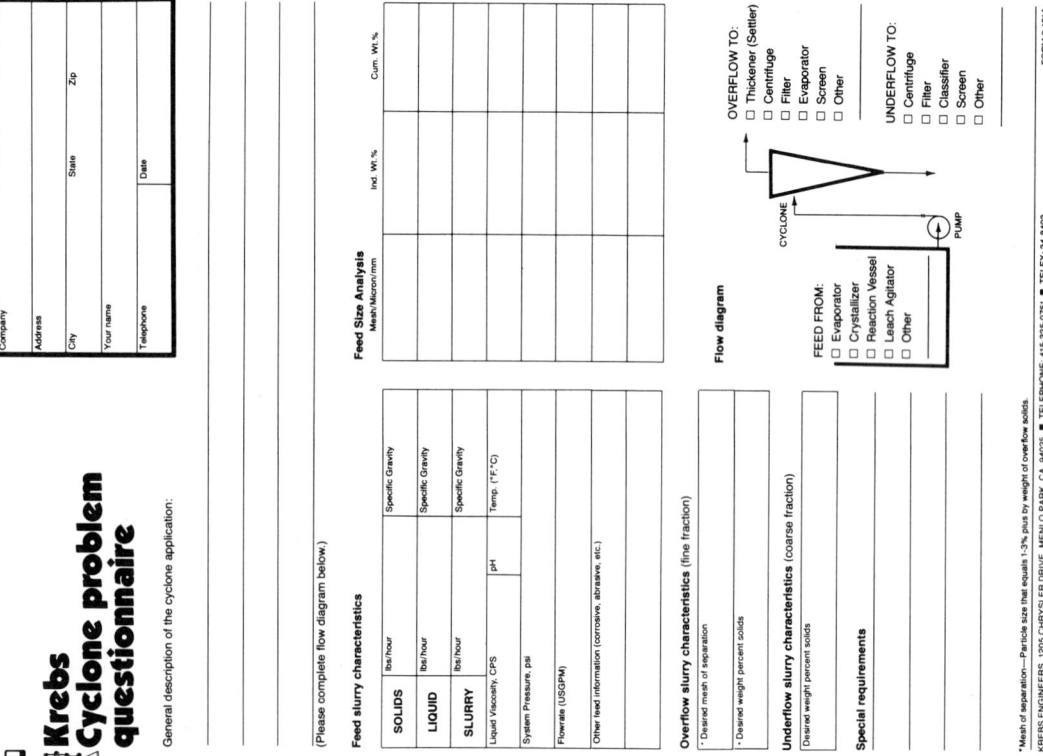

Inquiry data sheet

Raymond® Jet-Stream™ Classifiers

C-E Raymond's experience in the classification of powdered materials and the success of the *Jet-Stream™* classifier in providing consistently sharp cut-points at full production capacities can be put to work for you to help provide the kind of quality product you demand. For a no-cost evaluation of your application, simply remove this sheet, answer the questions below, fold and mail. If you need assistance, call us at [312] 236-4044.

(please print)

Your name ______________ Company ______________

Address ______________ City ______________ State ______ Zip ______

Telephone (____) ______________ Best time to call: ______________ am/pm

Material to be classified: ______________ Bulk density (lbs/cu ft) ______________

Specific gravity ______________ Feed temperature (°F) ______________

Moisture in feed (% by weight) ______________ Feed capacity (lbs/hr) ______________

Elevation above sea level (ft) ______________

Fineness Requirements

Micron Size	Cumulative percent of product smaller than micron size shown		
	Feed to Classifier	Fine Product	Coarse Product

Method of analysis (*Ro-Tap®* sieve shaker, *SediGraph®* particle size analyzer, or other equipment)* ______________

Electric power available: Phase ______ Hz ______ Volts ______

Type of final dust collection equipment ______________

Special materials of construction required ______________

Remarks ______________

* *Ro-Tap®* is a registered trademark of C-E Tyler, Combustion Engineering, Inc.
SediGraph® is a licensed trademark of Micromeritics Corp.

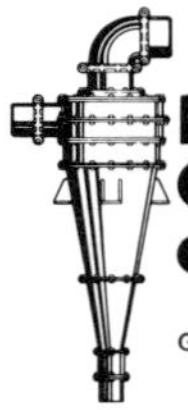

Krebs Cyclone problem questionnaire

Company		
Address		
City	State	Zip
Your name		
Telephone	Date	

General description of the cyclone application:

____

(Please complete flow diagram below.)

Feed slurry characteristics

SOLIDS	lbs/hour	Specific Gravity
LIQUID	lbs/hour	Specific Gravity
SLURRY	lbs/hour	Specific Gravity
Liquid Viscosity, CPS	pH	Temp. (°F,°C)
System Pressure, psi		
Flowrate (USGPM)		
Other feed information (corrosive, abrasive, etc.)		

Feed Size Analysis

Mesh/Micron/mm	Ind. Wt.%	Cum. Wt.%

Overflow slurry characteristics (fine fraction)

* Desired mesh of separation
* Desired weight percent solids

Underflow slurry characteristics (coarse fraction)

Desired weight percent solids

Special requirements

____

Flow diagram

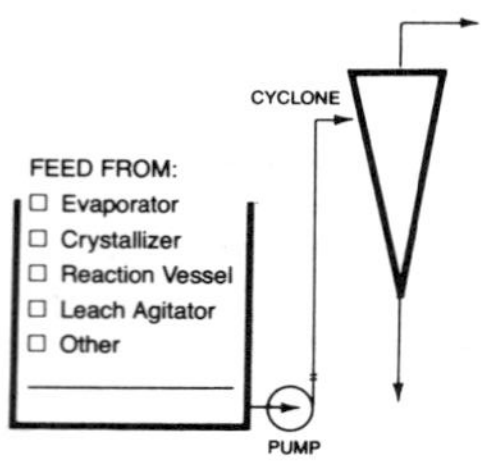

FEED FROM:
- ☐ Evaporator
- ☐ Crystallizer
- ☐ Reaction Vessel
- ☐ Leach Agitator
- ☐ Other

OVERFLOW TO:
- ☐ Thickener (Settler)
- ☐ Centrifuge
- ☐ Filter
- ☐ Evaporator
- ☐ Screen
- ☐ Other

UNDERFLOW TO:
- ☐ Centrifuge
- ☐ Filter
- ☐ Classifier
- ☐ Screen
- ☐ Other

*Mesh of separation—Particle size that equals 1-3% plus by weight of overflow solids.

KREBS ENGINEERS, 1205 CHRYSLER DRIVE, MENLO PARK, CA 94025 ■ TELEPHONE: 415-325-0751 ■ TELEX: 34-8403

FORM 2-131A

FILTER QUESTIONNAIRE

D. R. SPERRY & CO.

112 N. Grant Street
North Aurora, Illinois 60542

Date ______

Company ______

Address ______

Telephone No. ______

Name of Individual Reporting ______

Title ______

Type of Operation (Chemical, Tannery, Brewery, etc. ______

______ Location ______

______

(Please answer as many questions as possible)

Description of Material to be Filtered ______

Liquid ______

Solids ______

Specific Gravity of Slurry ______ @ ______ °F

Emulsion ______ ? Solution ______ ?

Viscosity ______ ______ (Units) pH ______

Avg. Particle Size of Solids ______

Concentration of Solids ______ (ppm, lbs., %, etc.)

Physical Nature of Solids ______ (Sticky, granular, flocculent, etc.)

Specific Gravity of Solids ______ gms/cc

Chemical Composition of Solids ______

______

Nature of Filtrate ______ (Water, alcohol, varnish, mixture, etc.)

Specific Gravity of Filtrate ______ gms/cc @ ______ °F

Vapor Pressure of Filtrate ______ mmHg

Dissolved Salts ______

Solubility: @ 32°F ______ Units ______

@ 68°F ______ Units ______

@ °F ______ Units

Conditions and Recommendations

General Comments ______

Filtration Temperature ______ °F Max. Permissible ______ °F.

Min. Permissible ______ °F.

Size of Batch ______ gals. ______

Closed or Open Discharge Desired ______

Filtration Pressure ______ psig

Filter Cycle Desired ______ hrs.

Is Product ______ Cake ______ Filtrate ______ Both ______

Cake Washing Necessary (Yes, No) ______ Name of Wash Liquid ______

Temperature of Wash Liquid ______ °F.

Desirable Dryness of Cake: % Solids ______

Physical Condition of Cake ______ (pastry, sticky, dry, cracks)

Thickness of Cake ______ in.

Quality of Filtrate - Desire or Allowed ______ Brilliant ______ Clear ______

Turbid ______ Decolorized ______

If Turbid - Max. Allowable particle size of solids ______ microns

If Filtrate is the Product

Filter aid to be used Yes ______ No ______

If Yes: Precoat ______ Quantity ______

Body Feed ______ % ______

Both ______

Type, Brand, Grade ______ % ______

Activated Carbon
or Bleaching Earth used Yes ______ No ______ % ______
or
Flocculation Agent ______ Type ______ Brand ______ % ______

Recommended Materials of Construction ______

______

Materials of Construction not Permitted ______

______

Filter Media: Cloth ______ Paper ______

Type ______ Grade ______

Filter Area ______ ft.^{2}

Filter Cake Volume ______ ft.^{3}

Laboratory Filter Tests

Filter Type ______

Area of Filter ______

Cake Volume of Filter ______

Media ______

Test Conditions ______

RESULTS ______

______

CF 101

BIRD Application Data Sheet
SOLIDS/LIQUIDS SEPARATION

PRODUCT DATA

Company ____
Address ____
City ____ State ____
Plant address ____
City ____ State ____
Project References ____
Bid date ____
Completion date ____

Project Status:
- ☐ R&D Laboratory
- ☐ Pilot Plant
- ☐ Semi-works
- ☐ In production

Customer Contacts

R&D ____
Engineering ____
Process ____
Purchasing ____

PROCESS DATA

Present problems ____
____
Present results ____
____

Feed

% Susp. solids (ss) ____
Sp. gr.____ Viscosity____(cp) at____(°F)

Process

Temp. ____(°F) Pressure____(PSIG)
Rate____(GPM) ____(lbs/hr ss)
Valuable Component(s) ____
BOD or COD (for waste) ____
Type (origin) of waste ____

Solids

Composition ____
____
☐ Slimy ☐ Soft ☐ Crystalline
☐ Fibrous ☐ Abrasive ☐ Amorphous

Size Distribution

____Sieve ____% On ____ Cumul. %
+ ____ M ____ ____
+ ____ M ____ ____
+ ____ M ____ ____
− ____ M ____ ____
Sp. gr. ____ Shape ____
Ash content ____
Grease content ____
Permissable Flocculants ____
Bulk density ____

Mother Liquor

Composition ____
____
Dissolved: Solids____% Impurities____%
Sp. gr. ____ at____(°F) pH ____

[continued next page]

Bird Machine Company, Inc., South Walpole, Massachusetts 02071 • (617) 668-0400

PROCESSING REQUIREMENTS

Max. liquid content in solids ____
Max. impurities in solids ____
Max. ss in clarified ML ____
Material of const. ____
Vaportight/pressure const. ____
Wash separation from ML ____

Wash Requirements

Temp.____°F ☐ Spray ☐ Drip pipe
☐ Countercurrent ☐ Other ____
____
Composition ____
Ratio ____
Rate ____lbs/hr. wet cake
Disposition of wash ____
____

Disposition of cake product ____
____
Disposition of Mother Liquor (ML) ____
____

Testing Information

☐ Large scale ☐ Preliminary ☐ Field
Equipment preference ____
____
Size of sample available ____
Precautions necessary ____
____
To be witnessed: ☐ Yes ☐ No
By: ____
Disposal: of testing residue ____
____
of test samples ____

FLOW SHEET

Comments ____
____
____

Bird Machine Company, Inc., South Walpole, Massachusetts 02071 • (617) 668-0400

ROBATEL INC.

Centrifuge Application Information

1. Company__________
2. Address__________
3. Submitted by__________Telephone________Date________

. . . .

A. PROCESS REQUIREMENTS

4. Name of Product__________
5. Production rate desired______lbs/hr., bone dry basis, or other______
6. Purpose (circle those which apply): recover solids, dewater solids, wash solids, classify solids, recover liquor, clarify liquor, separate 2 liquids, other__________

B. MATERIAL TO BE CENTRIFUGED

7. Name of solids in the feed__________Amount of solids_____% by wt.
   Amount of solids_____% by vol.
8. Feed. Temp____°F min,____°F max,____°F normal. Pressure________psig.
   Viscosity______Cps at______°F. Bulk density______lbs/cu ft. pH______
   Solids variation by wt______% max,______% min.
9. Is feed produced continuously or by batch. For______hrs/24 hrs.
10. Solids characteristics (circle those which apply)

explosive	amorphous	fragile	mineral	granular
inflammable	colloidal	free draining	natural	hygroscopic
toxic	crystalline	slow draining	synthetic	slimy
abrasive	fibrous	gelatinous	organic	

   Other__________

11. Solid particle size distribution. Apparatus used: Tyler, other____Tested wet,dry.

retained on___mesh____% by wt.			retained on___mesh____% by wt.		
" ____	" ____	"	" ____	" ____	"
" ____	" ____	"	" ____	" ____	"
" ____	" ____	"	passing ____	" ____	"

12. Solids specific gravity. Real__________Apparent__________
13. Liquor in feed

Name	% by vol.	Vic. (Cps)	Sp. gr.	pH
______	______	______	______	______
______	______	______	______	______

14. Liquor characteristics (circle those which apply): explosive, inflammable, toxic, foamy, heat sensitive, dermal sensitive, corrosive, other__________

C. OPERATING CONDITIONS

15. Feed Rate______gpm min,______gpm max______gpm normal.
16. Solids discharge requirements___lbs/hr., bone dry basis, Residual moisture by wt______% min,______% max,______% normal, how determined__________Bulk density______lbs/cu ft.
   Allowable crystal breakage________%
17. Liquor discharge requirements______gpm. Allowable solids______% by wt.
18. Is washing of solids required________Name of wash liquor______
   Allowable wash temp____to____°F. Amount____lbs wash liquor/lb solids.
19. Must wash liquor be kept separate from mother liquor__________
20. Are additives permitted________type and amount__________
21. For liquid-liquid separation______% of liquid (name)________ is allowable in liquid (name)_____Density______% solids in either phase________
22. Type of operation (circle one): Batch manual, batch automatic, continuous.
23. Operating requirements (circle those which apply): vapor tight, venting, inert atmosphere, pressurized. Other__________

D. CENTRIFUGE MACHINE SPECIFICATIONS & MISC.

24. Materials of construction for parts in contact with process material__________
25. Motor enclosure (circle one): drip proof, TEFC, explosion proof, other________
26. Available power: 1 phase,_____volts_____hertz_____
   3 phase,_____volts_____hertz_____
27. What handling precautions are necessary__________
28. Is product now being separated______Machine type__________
   Diameter & length of bowl or basket____x____" Cake Thickness________"
   Bowl Capacity per cycle in lbs_____or Feed Rate in lbs/hr or GPM______
   RPM Feed______; RPM Spin______; RPM Discharge______;
   Feed Time______; Spin Time______; Wash Time______; Spin Time______
   Discharge Time_____; Automatic, Semi-Automatic, Manual (circle one)
   What problems exist (circle those which apply): erosion, corrosion, capacity, efficiency, blinding, crystal breakage, cake moisture, other__________
29. Attach Flow or Process Diagram.

PAUL O. ABBE INC

QUESTIONNAIRE FOR MIXING OR BLENDING EQUIPMENT

In order to help our Engineering staff recommend the correct type and size mixer to satisy your individual needs, it is essential that you supply us with as much information as possible. The answers to the questions listed here will greatly assist us in making a prompt, intelligent reply to your inquiry. All information received will be held in strict confidence.

Company__________ Department__________

Address__________ Location of Plant__________

Inquiry initiated by__________ Title__________ Date__________

1. Name (if possible) material to be mixed or give names of similar well-know materials - also bulk density__________
2. Is this material currently being mixed by you?__________ If so, what type and size mixer is being used and what HP is required?__________
3. Will the material be mixed in dry, liquid or paste form?__________
4. Should this operation be batch or continuous?__________
5. What size batch do you wish to mix at one time, or how much production is required per hour, or per day of__________ hours?__________
6. What is the approximate weight per cu. ft. or gallon of the combined ingredients?__________
7. Describe the consistency or viscosity of the material__________
8. Is the material abrasive?__________
9. Will it be necessary to heat or cool the material during the mixing process?__________
   __________ If so, what temperature is required?__________
10. If steam heat is to be used, what pressure will be used in the Jacket?__________
11. Is the mixing to be done under pressure or vacuum? If so, to what extent?__________
12. Is special dust tight construction required?__________
13. What clearance is desired under the discharge?__________
14. Should parts coming in contact with your material be made of special materials such as stainless steel, bronze, aluminum, Monel, etc?__________ If so, what material is preferred?__________
15. Should motor be standard open, totally enclosed or explosion proof?__________
    Give voltage, phase, and cycles of your electric current__________
16. Outline any specific problems which may be involved in mixing this material and let us have any other information which you feel would be helpful to us in providing the best recommendations__________

PAUL O. ABBE INC. 148 CENTER AVENUE LITTLE FALLS, NEW JERSEY

bēpex corporation

MIXING TEST INQUIRY DATA

Name:__________ Title:__________

Company:__________ Division:__________

Street:__________

City:__________ State:__________ Zip:__________ Telephone: (____) ____ - ________

Name of Material:__________ Amount:__________ Value:__________

Shipped Prepaid from:__________ (Name of Company) Via:__________ Date:__________

Sample of Finished Product being sent: Yes:__________ No:__________ Via:__________ Date:__________

Return Tested Material to:__________ (Name of Company) Attn:__________

Address:__________ Amount:__________

Other Disposition Instructions:__________

TECHNICAL DATA

Statement of Mixing Application (Include Proportions of Ingredients):__________

Blending or Dispersion Required:__________

Capacity Per Hour Wanted — Min. and Max.:__________

Caution If Explosive, Inflammable, Poisonous, Noxious:__________

Characteristics Unusual as Hygroscopic, Volatile, Corrosive:__________

Temperature Limitations: Softens:__________ Melts:__________ Decomposes:__________

Chilling Permissible:__________ Heating Permissible:__________

Normal Moisture Content:__________ %. Oil, Fat, Etc., Content:__________ %

Drying Permissible to:__________ %. Addition of Water Permissible to:__________ %

Abrasive As Containing Silica, Quartz, etc.:__________

Bulk Density:__________ lbs./cu. ft.; Product:__________ ; Ingredients:__________

Method of Packing and Preserving Perishable Materials:__________

Solubility, and Fluids Recommended for Cleaning Equipment:__________

Power Available: Volts:__________ Phase:__________ Cycles:__________

Information Request Form

For Internal use:
P.P.________
date________

MAIL TO ONE OF THE ADDRESSES SHOWN BELOW

I would like a preliminary estimate of the cost and performance of Prism® separators for the application described below.

Note: If more than one stream is to be considered, please machine copy this form and complete the information requested for each stream.

1. Feed stream data
   a. Source________
   b. Temperature________ c. Pressure________
   d. Flowrate Maximum________ Expected________ Minimum________
   e. Feed gas composition

Component	Vol. %	Component	Vol. %

2. Desired Results:

	Recovery	Purity	Pressure
permeate			
non-permeate			

3. If a tradeoff between purity and recovery is necessary, which is more important?
   ☐ Purity
   ☐ Recovery
4. Use/Special Requirements for the non-permeate gas
____
____

NAME:________
TITLE:________
COMPANY:________
ADDRESS:________
CITY:________ STATE:________ ZIP:________
COUNTRY:________
TELEPHONE: (   )________
TELEX:________
PLANT LOCATION:________

Monsanto Company
Separations Business Group
800 No. Lindbergh Blvd.
St. Louis, Missouri 63167 USA
Telephone: (314) 694-8000
Telex: 44-7282

Monsanto Europe S.A.
Separations Business Group
270-272 Av. de Tervuren
1150 Brussels, Belgium
Telephone: 2/762-11-12
Telex: 62927

Monsanto Japan Limited
520 Kokusai Building
1-1, Marunouchi 3-chome
Chiyoda-ku
Tokyo 100, Japan
Telephone: (03) 287-1251
Telex: J22614

Monsanto Far East Limited
Separations Business Group
1304 Great Eagle Centre
23 Harbour Road
Hong Kong
Telephone: 5-740738
Telex: HX 73440

Monsanto Australia Limited
Separations Business Group
East Tower Princes Gate
151 Flinders Street
Melbourne, Australia
Telephone: (03) 658-6666
Telex: AA 30288

FRITZ W. GLITSCH & SONS, INC.
P. O. Box 6227 • Dallas, Texas 75222

D I S T I L L A T I O N
Process Design Data Sheet

Item No. or Service				
Tower diameter, I.D.				
Tray spacing, inches				
Total trays in section				
Max. Δ P, mm Hg				
Conditions at Tray No.				
Vapor to tray, °F				
Pressure,				
Compressibility				
•Density, lb./cu. ft.				
•Rate, lb./hr.				
cu. ft./sec. (cfs)				
cfs $\sqrt{D_V/(D_L-D_V)}$				
Liquid from tray, °F				
Surface tension				
Viscosity, cp				
•Density, lb./cu. ft.				
•Rate, lb./hr.				
GPM hot liquid				
Foaming tendency	None____	Moderate____	High____	Severe____

•These values are required in this form for direct computer input.

NOTES:

1. This form may be used for several sections of trays in one tower, for several towers, or for various loading cases. Use additional sheets if necessary.
2. Is maximum capacity at constant vapor-liquid ratio desired?________
3. Minimum rate as % of design rate:________%
4. Allowable downcomer velocity (if specified):________ft/sec
5. Number of flow paths or passes:________ FWG Choice;________
   Bottom tray downcomer: Total draw________; Other________
6. Trays numbered: top to bottom________; bottom to top________
7. Enclose tray and tower drawings for existing columns.
8. Manhole size, I.D.,________ inches.
9. Manways removable: top________; bottom________; top & bottom________
10. Corrosion allowance: c.s.________; other________
11. Adjustable weirs required: yes________ no________
12. Packing material if required________; not required________
13. Tray material and thickness________
14. Valve material________
15. Ultimate user________
16. Plant location________
17. Other________

Form No. PE-8

10-69-2.5M-BENNETT'S-DALLAS

ROBATEL INC.
SOLVENT EXTRACTION QUESTIONNAIRE
Page 2

1. NATURE OF THE OPERATION

---

2. PHASE TO EXTRACT

. Nature of diluent
. Nature of solution to extract
. Concentration in solution
. Density
. Output

---

3. SOLVENT

. Nature of solvent
. Density
. Output

---

4. REQUIRED CHARACTERISTICS OF EXTRACT IMPOVERISHED IN SOLUTION

. Concentration in solution
. Approximate density

---

5. REQUIRED CHARACTERISTICS OF EXTRACT ENRICHED IN SOLUTION

. Concentration in solution
. Approximate density

---

6. NUMBER OF THEORETIC STAGES CORRESPONDING TO THE REQUIRED EXTRACTION AMOUNT

(if possible, enclose a diagram of balance or the maximum information concerning the miscibility of the solvent and of the diluent).

---

7. OPTIMUM TEMPERATURE OF EXTRACTION

---

8. SOLIDS IN SUSPENSION

Does one of the 2 phases comprise any solid in suspension before extraction?

Which one?

ROBATEL INC.
SOLVENT EXTRACTION QUESTIONNAIRE
Page 3

Are these particles soluble in the other phase?

Can these particles be eliminated before introduction into the extractor?

. by filtration
. by centrifugal decantation

If particles can be eliminated by centrifugal decantation state:

. time of decantation
. Decanter bowl diameter
. Speed of rotation

With the two initial phases having no particles in suspension, do any precipitates appear during the process of extraction? In this case, after centrifugation are these precipitates situated:

. within one of the phases
. on the surface of the light phase
. at the bottom of the heavy phase
. in the zone of separation between the 2 phases

---

9. SPECIAL CHARACTERISTICS

Does an emulsion appear when the two phases are vigorously stirred?

In case of emulsion:

. If the phases separate themselves by static decantation state:
  - time required

. If the phases separate themselves by centrifugal decantation state:
  - time required
  - decanter bowl diameter
  - speed of rotation

After decantation is the surface of separation between the 2 phases very distinct?

Does one of the 2 phases stay cloudy?

Which one?

---

10. DIVERSE INFORMATION

Materials of construction
Electricity supply available
Explosion and flameproofing requirements
Type of supporting base or structure

SWENSON® CRYSTALLIZER CHECKLIST

CHECKLIST OF DATA REQUESTED ON CRYSTALLIZER INQUIRIES

Fill in or attach as much information as is available so the most economic design can be prepared.

1. Product being crystallized ________ Production Rate ________
2. Liquid Properties

	Feed	Mother Liquor	Purge
Rate	____	____	____
Temperature	____	____	
Composition ____	____	____	
____	____	____	
____	____	____	
____	____	____	
____	____	____	
Density	____	____	
Viscosity	____	____	
K (thermal conductivity)	____	____	
Cp (specific heat)	____	____	
Boiling point elevation	____	____	

3. Evaporation Rate ____
   For organic vapors only give latent heat of evaporation/specific volume/vapor viscosity ________
4. Solubility of Solute in solvent at various temperatures (solubility curve) or phase diagram if several solutes are present.
5. Product

Heat of Crystallization	Specific Gravity	Heat of Reaction if Reaction in Crystallizer Produces Crystals	Specific Heat
____	____	____	____

6. Crystal shape if known ________
7. Product size required ________
8. Suitable materials of construction for operating temperature range:
   Bodies ________ Tubes ________ Pumps ________ Vacuum Equipment ________
9. Type of seals required in agitators or pumps and their materials of construction ________
10. Gasket material required if known ________
11. Utilities Available

Steam Pressure	Cooling Water Temperature	Current Characteristics
____	____	____

12. Describe flowsheet required or used and whether one or several bodies are required (or the economic factors which are to be used in determining the number of stages or effects. These factors include steam cost, water cost, electricity cost, amortization period, etc.)
13. Any other data which the customer feels would be of help such as heat transfer data; present conditions under which the product is produced and type of equipment used; present crystal size; and effects of agitation, additives and/or temperature on crystal size, shape, recovery, or purity.
14. If the solution is organic and test work is desired, give the flash point of the solution ________

VULCAN IRON WORKS, INC.

1050 UNITED PENN BANK BUILDING
WILKES-BARRE, PENNSYLVANIA 18701
(717) 822-2161 TELEX: 831-831

Questionnaire

INQUIRY DATA

A. PROCESS REQUIREMENTS (PLEASE FURNISH ALL RELATED INFORMATION THAT IS READILY AVAILABLE)

B. INCINERATION (THERMAL OXIDATION) WITH OR WITHOUT HEAT RECOVERY. (PLEASE FURNISH ALL INFORMATION THAT IS READILY AVAILABLE.)
TYPE OF PLANT AND PROCESS ________

WASTES

SOLIDS ____ QUANTITY LBS./HR.
SOLIDS ____ SOLIDS WASTE TYPE
SOLIDS ____ HEAT VALUE BTU/LB
SOLIDS ____ ASH CONTENT %
SOLIDS ____ MOISTURE CONTENT %
SOLIDS ____ QUANTITY LBS./HR.
SLUDGES ____ TYPE OF SLUDGE
SLUDGES ____ ASH CONTENT %
SLUDGES ____ WATER CONTENT %
SLUDGES ____ COMBUSTIBLES CONTENT %
SLUDGES ____ HEAT CONTENT OF COMBUSTIBLES CONTENT %
SLUDGES ____ HEAT CONTENT OF COMBUSTIBLES BTU/HR.
LIQUIDS ____ QUANTITY GAL./HR.
LIQUIDS ____ WASTE LIQUID TYPE
LIQUIDS ____ HEAT VALUE BTU/GAL
LIQUIDS ____ VISCOSITY
LIQUIDS ____ SPECIFIC GRAVITY
FUMES ____ QUANTITY SCFM
FUMES ____ TEMP.
FUMES ____ EFFLUENT TYPE
FUMES ____ EFFLUENT QUANTITY

TYPE OF HEAT RECOVERY THAT CAN BE UTILIZED

STEAM GENERATION:
____ LBS./HR. REQUIRED
____ PSI OPERATING PRESSURE
____ °F. OPERATING TEMP.

THERMAL LIQUID
____ HEAT LOAD BTU/HR.
____ HEATING STATIONS
____ °F. TEMP. AT STATIONS

FUME PRE-HEATER
(GAS TO GAS HEAT EXCHANGER):
____ HEAT RECOVERY EFFICIENCY
____ °F. GAS INLET TEMP.
____ °F. HOT GAS OUTLET TEMP.

HOT WATER HEATING
____ GAL./MIN.
____ TEMP. RISE

ASPHALT HEATING
____ GAL./MIN.
____ °F. TEMP. RISE

INDICATE ADDITIONAL INFORMATION AND COMMENTS ON REVERSE SIDE

COMPANY ________ TELEPHONE ________
ADDRESS ________ ZIP ________
LOCATION OF PROJECT ________
INDIVIDUAL ________ TITLE ________ DATE ________

MAIL TO: VULCAN IRON WORKS, INC.
1050 UNITED PENN BANK BLDG. • WILKES-BARRE, PA. 18701 U.S.A.
TELEX: 831-831

bēpex corporation

CENTRAL LABORATORY TEST INQUIRY DATA
THERMAL PROCESSING

Project No.: ___
Date: ___

Company: ___
Address: ___
Contact: ___

MATERIAL SHIPMENT AND RETURN

Test Material: ___ Lbs Of Dry Material, ___ Lbs Wet Feed Is Being Shipped Prepaid
From: ___ Via: ___ Date: ___ Container: ___
Samples Of Feed ☐ Desired Product ☐ Shipped Via: ___ Date: ___
Test Material Return Instructions: **MATERIALS NOT DISPOSABLE IN SANITARY LANDFILL MUST BE RETURNED**

PURPOSE OF TEST

___

MATERIAL DATA

Dry Material Name: ___
Liquid Name: ___
Spec. Gravity, Solids: ___
Loose Bulk Density, Lb/Ft^3: ___ Wet ___ Dry
Angle Of Repose: ___ Wet ___ Dry
Spec. Heat: ___ Solids: ___ Liquid
Heat, Fusion: ___ Hydration: ___ Crystalliz. ___
Latent Heat, Liquid: ___ Boiling Point ___
Molecular Weight Of Liquid: ___
Initial Moisture (Wet Basis): ___
Chemically Free: ___ % Combined: ___

Final Moisture (Wet Basis): ___
Chemically Free: ___ % Combined: ___ %
Feed Consistency: ___
Melting: ___ °F, Softening: ___ °F, Sticking: ___ °F
Abrasive: ___ Hygroscopic: ___
Soluble: ___ (If Soluble In Liquid, Attach Solubility Curve)

Particle Size	Feed	Prod. Req'd.	Prod. Allowed
%–On ___			
%–On ___			
%–On ___			
%–On ___			

ANALYTICAL PROCEDURES

Partical Size: ___
Volatiles: ___
Other: ___

PROCESS DATA

Production Rate: ___ Lbs/Year At: ___ Hrs/Year = ___ Lbs/Hr
Temperature In: ___ °F
Inert Gas Required: ___ Contact With Combustion Products OK ? ___
Quality Criteria: ___
From What Equipment Is Feed Delivered ? ___
Ultimate Product Use: ___ Construction Materials: ___
Present Process And Equipment: ___
Equipment Cleaning Method: ___

UTILITIES

Electricity: ___ Steam At ___ Psig ___ °F, Dowtherm At ___ Psig ___ °F
Hot Oil At ___ Psig ___ °F, Molten Salt At ___ Psig ___ °F
Fuel Gas At ___ Btu/Cu Ft, ___ Fuel Oil At ___ Btu/Lbs, Cooling Water At ___ Psig ___ °F

ENVIRONMENT

Plant At ___ Ft Above Sea Level, ___ To ___ °F. Design For ___ % Relative Humidity

---

3-11
AGGLOMERATION

FEECO INTERNATIONAL AGGLOMERATION DATA SHEET

FEECO INTERNATIONAL, INC.
3913 ALGOMA RD.
GREEN BAY, WISCONSIN 54301
U. S. A.
PHONE 414—468-1000
TELEX 263456

Company ___
Address ___
Commercial Name of Material to be pelletized ___
Chemical Formula ___ Specific Heat ___

CAPACITY
Capacity Required ___ lbs./hr. Feed Rate ☐ or Discharge rate ☐
Agglomerator will operate ___ hrs. per day; ___ days per week; ___ weeks per year
is feed uniform in quantity ___ moisture ___ temperature ___

MATERIAL
Moisture Content ___ % before pelletizing ___ % product
LBS./CU. FT. ___
Particle Size Average ___ Smallest ___ Largest ___ Hygroscopic ___

Furnish screen analysis if possible

Mesh					
Passing %					

Material is sticky ___ Abrasive ___ Corrosive ___ Plastic ___ Brittle ___ Dusty ___ Hazardous ___
Special additive (binder) required?
Describe the process before and after pelletizing.

POWER 110 volt phase ___ cycle ___ AC ___ DC ___
220 volt phase ___ cycle ___ AC ___ DC ___
440 volt phase ___ cycle ___ AC ___ DC ___

AGGLOMERATION EXPERIENCE
We have been Agglomeration ___ material. Size of disc ___. Operating conditions ___
Capacity ___
Agglomerator to be made of carbon steel ☐ stainless steel ☐ other ☐ ___

DELIVERY
Drawings required ___ Proposal required ___
Quotation Date ___ Equipment delivery date ___
Information furnished by ___
Return to ___

3913 ALGOMA ROAD GREEN BAY, WISCONSIN 54301

bēpex corporation
10225 Higgins Road · Rosemont, Illinois 60018 · (312) 825-8010 · Telex: 25-3279

K–G
Rietz
Strong-Scott

A P P L I C A T I O N   D A T A

BRIQUETTING/COMPACTING – GRANULATING MACHINERY & SYSTEMS

Please provide as much information as is presently available, so that we may determine your application requirements. This information, which we will consider confidential, is essential to determine equipment that would be necessary for laboratory tests, budget estimates, and occasionally, to allow us to supply a firm quotation.

1 - We are interested in:

( ) Briquetting for a definite shape and size

( ) Compaction/granulation for a granular material

( ) Agglomerating systems in general

2 - Our immediate need is for:

( ) Costs for laboratory evaluation

( ) Budget estimate for economic evaluation

( ) A detailed proposal and quotation, if obtainable from this information

3 - Material to be processed ____________________

Bulk density (lbs./cu. ft.) ____________________

Specific gravity ____________________

Screen analysis ____% + ____mesh   ____% + ____mesh

____% + ____mesh   ____% + ____mesh

Moisture ____% Free ____% Combined

( ) Water ( ) Other ____________________

Temperature required for removal ____________

Maximum temperature the material can withstand without damage ____________

Estimated feed temperature to the machine ____________

Estimated maximum temperature of the feed material from the previous processing step that could be considered for machine feed ____________

Physical nature of the material:

( ) Sticky   ( ) Abrasive   ( ) Corrosive

( ) Hygroscopic   ( ) Thixotropic   ( ) Toxic

( ) Aerated   ( ) Irritant

If corrosive or abrasive, state some typical materials of construction you consider acceptable with this application.

____

If toxic or irritating, give safety precautions and the LD50 value, if known ____________________

____

4 - The material ( ) can ( ) cannot be processed in contact with the gases of combustion

( ) A special atmosphere is required for processing

Specify ____________________

( ) The material contains combustible matter which could be considered as a source of heat if the material is heated to processing temperature by direct flame.

Combustible matter ____________________

Percentage by weight ____________________

Estimate BTU/lb. ____________________

5 - What operation(s) will be performed immediately preceeding the agglomeration step ____________________

6 - Estimated production rate required ________ ( ) tons ( ) pounds per hour. Operation will be ( ) continuous ( ) intermittent

Estimated ______hrs./day ______days/week ______days/yr.

7 - Final product desired:

( ) Briquets or pellets ______" x ______" x ______"

( ) Granular material – ______mesh + ______mesh

( ) Other ________________

Moisture (solvent) ____________ that may remain ____ %

8 - Most materials are briquetted or compacted by pressure alone. However, in some cases, a binder must be utilized to supplement the mechanical forces required for agglomeration. The binder has to be inert and non-contaminating to the material. Please state whether this approach would be permissible with your material.

( ) Yes ( ) No Why? ________________

____

9 - Electrical power available:

________ Volts __________ Cycles __________ Hz.

All motors over ____ H.P. are to be for ____ volt service.

All motors are to be ( ) TEFC ( ) Explosion proof,

Class ________ ( ) Open ( ) Industrial standard

( ) Other ________________

10 - How is the agglomeration equipment to be used.

( ) Completely new product proposed for production

( ) Installation in a current process to improve or change the material's properties

( ) Expansion of a current process using agglomeration equipment

( ) Improvement of a current agglomeration process

11 - Why is agglomeration being considered at this time?

( ) Improve product properties

( ) Reduce dusting

( ) Improve efficiency in a current process

( ) To produce a specific size and shape

( ) Meet a competitive condition

( ) Reduce process costs

U.S. DEPARTMENT OF LABOR
Occupational Safety and Health Administration

Form Approved
OMB No. 44-R1387

MATERIAL SAFETY DATA SHEET

Required under USDL Safety and Health Regulations for Ship Repairing, Shipbuilding, and Shipbreaking (29 CFR 1915, 1916, 1917)

SECTION I

MANUFACTURER'S NAME	EMERGENCY TELEPHONE NO.
ADDRESS *(Number, Street, City, State, and ZIP Code)*	
CHEMICAL NAME AND SYNONYMS	TRADE NAME AND SYNONYMS
CHEMICAL FAMILY	FORMULA

SECTION II - HAZARDOUS INGREDIENTS

PAINTS, PRESERVATIVES, & SOLVENTS	%	TLV (Units)	ALLOYS AND METALLIC COATINGS	%	TLV (Units)
PIGMENTS			BASE METAL		
CATALYST			ALLOYS		
VEHICLE			METALLIC COATINGS		
SOLVENTS			FILLER METAL PLUS COATING OR CORE FLUX		
ADDITIVES			OTHERS		
OTHERS					

HAZARDOUS MIXTURES OF OTHER LIQUIDS, SOLIDS, OR GASES	%	TLV (Units)

SECTION III - PHYSICAL DATA

BOILING POINT (°F.)		SPECIFIC GRAVITY (H_2O=1)	
VAPOR PRESSURE (mm Hg.)		PERCENT, VOLATILE BY VOLUME (%)	
VAPOR DENSITY (AIR=1)		EVAPORATION RATE (________ =1)	
SOLUBILITY IN WATER			
APPEARANCE AND ODOR			

SECTION IV - FIRE AND EXPLOSION HAZARD DATA

FLASH POINT (Method used)	FLAMMABLE LIMITS	Lel	Uel
EXTINGUISHING MEDIA			
SPECIAL FIRE FIGHTING PROCEDURES			
UNUSUAL FIRE AND EXPLOSION HAZARDS			

PAGE (1) (Continued on reverse side) Form OSHA-20 Rev. May 72

SECTION V - HEALTH HAZARD DATA

THRESHOLD LIMIT VALUE	
EFFECTS OF OVEREXPOSURE	
EMERGENCY AND FIRST AID PROCEDURES	

SECTION VI - REACTIVITY DATA

STABILITY	UNSTABLE		CONDITIONS TO AVOID
	STABLE		
INCOMPATABILITY *(Materials to avoid)*			
HAZARDOUS DECOMPOSITION PRODUCTS			
HAZARDOUS POLYMERIZATION	MAY OCCUR		CONDITIONS TO AVOID
	WILL NOT OCCUR		

SECTION VII - SPILL OR LEAK PROCEDURES

STEPS TO BE TAKEN IN CASE MATERIAL IS RELEASED OR SPILLED	
WASTE DISPOSAL METHOD	

SECTION VIII - SPECIAL PROTECTION INFORMATION

RESPIRATORY PROTECTION *(Specify type)*		
VENTILATION	LOCAL EXHAUST	SPECIAL
	MECHANICAL *(General)*	OTHER
PROTECTIVE GLOVES		EYE PROTECTION
OTHER PROTECTIVE EQUIPMENT		

SECTION IX - SPECIAL PRECAUTIONS

PRECAUTIONS TO BE TAKEN IN HANDLING AND STORING	
OTHER PRECAUTIONS	

PAGE (2) GPO 934-110 Form OSHA-20 Rev. May 72

Material Safety Data Sheet—*(continued)*

Index